WETZEL'S LIMNOLOGY

WETZEL'S LIMNOLOGY

LAKE AND RIVER ECOSYSTEMS

FOURTH EDITION

Edited by

Ian D. Jones

John P. Smol

Academic Press is an imprint of Elsevier
125 London Wall, London EC2Y 5AS, United Kingdom
525 B Street, Suite 1650, San Diego, CA 92101, United States
50 Hampshire Street, 5th Floor, Cambridge, MA 02139, United States
The Boulevard, Langford Lane, Kidlington, Oxford OX5 1GB, United Kingdom

Notices

Knowledge and best practice in this field are constantly changing. As new research and experience broaden our understanding, changes in research methods, professional practices, or medical treatment may become necessary.

Practitioners and researchers must always rely on their own experience and knowledge in evaluating and using any information, methods, compounds, or experiments described herein. In using such information or methods they should be mindful of their own safety and the safety of others, including parties for whom they have a professional responsibility.

To the fullest extent of the law, neither the Publisher nor the authors, contributors, or editors, assume any liability for any injury and/or damage to persons or property as a matter of products liability, negligence or otherwise, or from any use or operation of any methods, products, instructions, or ideas contained in the material herein.

ISBN: 978-0-12-822701-5

For information on all Academic Press publications visit our website at https://www.elsevier.com/books-and-journals

Cover image is of Lough Furnace, Ireland. Photo credit: Mikkel Andersen
Publisher: Peter B. Linsley
Acquisition Editor: Rakhshan Rizwan
Editorial Project Manager: Sara Valentino
Publishing Services Manager: Shereen Jameel
Project Manager: Gayathri S
Designer: Christian J. Bilbow

Printed in Canada

Last digit is the print number: 9 8 7 6 5 4 3 2 1

Tributes to Robert Wetzel

A Family Affair As remembered by Paul R. Wetzel[1]

[1]Center for the Environment, Ecological Design & Sustainability, Smith College, Northampton, MA, United States

FIGURE 1 Robert Wetzel in the mid-1970s sitting in his office in the J.W. Stack Research Building at the Kellogg Biological Station (Michigan State University, Hickory Corners, Michigan, United States). *(Photo credit: Family of Robert Wetzel.)*

In the fall of 1973 my dad began organizing his home office for the task of writing a textbook. His home office was located on the lowest level of our modest trilevel house and opened to the family room, where it was common for my siblings and me to play games, wrestle, and watch TV. The office was large enough for a desk, a typewriter, a comfortable chair for visitors, a filing cabinet, and some low bookshelves with small mementos from around the world displayed on the top. There was no telephone in the room and, of course, no computer. A single window looked into our suburban backyard. With a 6-month sabbatical from his normal duties at the Michigan State University Kellogg Biological Station and a book contract from W.B. Saunders Company in hand, Dad, who was in his 30s at the time, started moving his work office to his home office. At one end of the family room were floor-to-ceiling built-in bookshelves and cabinets. At first, only a few shelves were commandeered for the book project. But eventually, there were hundreds of books and thousands of carefully organized reprints occupying the entire end of the family room. At this time in our family, there were four kids, ages 3, 6, 9, and 11, and my mother at home. Ironically, Dad's home office next to the family room was much quieter and less distracting than his office at work.

All of us kids knew that Dad was writing a book. A book about his work—limnology—a word that each of us would learn to pronounce and define at an early age. He was a scientist, after all. Beyond that, we had no idea what he was doing in that little home office. We were under strict orders not to bother him. If you got hurt or needed a dispute settled, you went to Mom—Dad was not there. Each night as dinner was going on the table, Mom would tell one of us, "Go get your dad." You would quietly open the door and tell him in almost a whisper that dinner was ready. He would look up, smile, and say, "I'll be right there."

When I look back on that time with the distance of over 48 years, I'm not surprised that my dad started writing the first edition of his textbook next to the family room. Science was his passion. The line between work and family affairs for my dad was quite fluid, even nonexistent, completely entwining our childhood experiences with his scientific career. Not one of us can talk about family lore for very long without mentioning science. For years, our mother took us to Kellogg Biological Station nearly every day during the summer to swim in Gull Lake. Dad hosted many international visitors, and it was his custom to invite them to the house for dinner. All the kids were included and everyone sat down to dinner at the same time. Only later did I learn that his graduate students wished that they could have joined those Wetzel kids that regularly dined with famous limnologists. Dad also held regular seminars with his graduate students in our home. We were introduced and then expected to disappear. But sometimes, we would watch the group from the top of the stairs to see who was falling asleep.

As we got older, each of us started working in Dad's laboratory washing glassware. I worked in his lab after school and during the summer for 6 years. Usually, I assisted graduate students by recording measurements, collecting samples, and running laboratory analyses. Through all these interactions, the graduate students knew the Wetzel kids and we knew them. Many of those graduate students became a part of our family and remained family friends after they moved on in their careers.

The three editions of *Limnology* represent a physical, intellectual, and spiritual manifestation of Dad's passion for limnology. It was his gift to students and colleagues who shared his love of freshwater lakes. No one creates such a gift alone. So Dad brought the people that he loved, his family, students, and peers, along with him on his scientific life journey. It never occurred to him to do otherwise.

Bob Wetzel: A Student's Perspective Reminiscences by James Cotner[2]

[2]Dept. Ecology, Evolution and Behavior, University of Minnesota, St. Paul, MN, United States

FIGURE 2 Robert Wetzel talks about periphyton to a class in Innsbruck, Austria, 1993. *(Photo Credit: Family of Robert Wetzel.)*

You could easily say that I grew up as a limnologist on "Wetzel." Of course, that's what this book has always been called. As an undergraduate at Wittenberg University back in the early 1980s, it was the book from which I first learned about stratification patterns and dissolved oxygen solubility. It had everything in it from diel migrations of zooplankton to how lakes are formed and, unfortunately, Professor Hobbs expected us to know most of it.

The second edition came out as I was finishing my master's degree at Kent State University and planning to begin my graduate work with Bob. I remember interviewing with him in the Stack Building at the Kellogg Biological Station and being completely intimidated to be in his presence given his status in the field. He had sent me some literature ahead of the visit and one of the documents outlined his relationship with his graduate students, referring to it at one point as a "marriage" to emphasize the level of commitment that he both provided and expected. Being in my early 20s and single at the time, this was a little scary and overwhelming, but not so much so that I did not jump at the opportunity to work with him when he offered.

Looking back on these early encounters, a few things stand out about Professor Wetzel. First, his focus and commitment were like those of nobody I had ever encountered before or since. It was palpable in the lab and in meetings with him. It most certainly had a huge impact on me and my desire to accomplish great things in the field. But the commitment was not a one-way endeavor. He truly loved working with students and postdocs, of which there were many. When he was in his office, the door was always open. Never did I ask to meet with him when he did not make the time for me. He also welcomed his lab group and other students that were interested in aquatic sciences into his home once a month to discuss current topics in the field.

As an illustration of the two-way commitment, I found myself in a dilemma early on in our relationship when I was developing my PhD research program. Bob clearly wanted me to work on characterizing dissolved organic matter (DOM), which he recognized as a key regulator of ecosystem function in freshwaters (Rich and Wetzel 1978; Wetzel 1984, 1995). At the time, a great deal of work by the likes of George Aiken, Henk de Haan, and others was focused on isolating different fractions of DOM and characterizing it using wet chemistry, nuclear magnetic resonance, and mass spectrometry. I tried my best to develop a project that might fit in with this focus area, but I never felt like I was enough of a chemist to do it justice and, besides, I was more interested in microbial processes. But I knew I could not just walk into his office and tell him what I *didn't* want to do, without presenting an alternative. So I worked for weeks on a proposal that focused on microbial P-cycling dynamics until finally I had the courage to confront him with it, knowing full well how disappointed he would be. After presenting my proposed new project, he looked at me and said, "That sounds great, Jim," with nary a trace of disappointment.

Bob's approach to limnology and his textbook was to read as much as possible and incorporate the relevant parts into the book. This encyclopedic nature of *Limnology* made it a perfect resource for teaching students, but it didn't necessarily make for a true "page-turner." Nonetheless, my approach for studying for my doctoral comprehensive exam was to read the second edition cover to cover...twice, making me among the few, other than his editor, to have done so, I'm sure.

Bob never stopped working and he was always writing the "next" edition of his book. Having completed the second edition shortly before I started working with him, he was already well along to the third edition during my tenure with him, first at Kellogg and subsequently at the University of Michigan. The third edition was not just an updated second edition, but he added chapters focused on paleolimnology, humans, and water quality issues, as well as new text focused on riverine systems. He certainly didn't know it at the time, but it was his swan song, and I have felt great pride over the years in using it to teach my own students.

However, Bob certainly was more than his textbook, even though it was a huge part of his life. He was also passionate about the *International Society of Limnology* (SIL) and making sure that information about freshwaters was available in the developing world. He edited several books focused on the limnology of many of these places. He also loved classical music and very often had it playing quietly in his office while he worked. For me, being less of a classical and more of a rock fan, I often gave him a hard time about how quiet the music was playing in his office and how he needed to "crank it up." He also loved art and expressed himself via oil paintings throughout his life and more so as he got older. In fact, the cover of the third edition was one of his own creations.

Aside from being remembered in freshwater science for this book, Bob also helped change the direction of the field. He published 30 books and over 400 research articles (https://limnology.org/notable-limnologists/robert-g-wetzel/). He recognized the important role of aquatic plants and littoral processes in influencing and even controlling aquatic functions. One of his strongest "soap boxes" (of which there were many!) was the importance of small lakes and their role in landscapes (Wetzel 2000, 2001). He was way ahead of his time on this issue, as we now recognize that small lakes are incredibly dynamic and important, especially with respect to carbon cycling and losses of greenhouse gases to the atmosphere. Lastly, he was also ahead of his time in recognizing the significant role of microbes in the water column, attached to plants, and in the sediments to the biogeochemistry of freshwaters (Stewart and Wetzel 1982; Coveney and Wetzel 1989).

This edited revision of his masterpiece certainly would make him happy, given that he always wanted it to be as up-to-date as possible, although I'm sure he would be quick to point out that it took over 50 limnologists and two editors to do what he managed to do all on his own!

References

Coveney, M.F., Wetzel, R.G., 1989. Bacterial metabolism of algal extracellular carbon. Hydrobiologia 173, 141–149.

Rich, P.H., Wetzel, R.G., 1978. Detritus in the lake ecosystem. Am. Nat. 112, 57.

Stewart, A.J., Wetzel, R.G., 1982. Influence of dissolved humic materials on carbon assimilation and alkaline phosphatase activity in natural algal-bacterial assemblages. Freshw. Biol. 12, 369–380.

Wetzel, R.G., 1984. Detrital dissolved and particulate organic carbon functions in aquatic ecosystems. Bull. Mar. Sci. 35, 503–509.

Wetzel, R.G., 1995. Death, detritus, and energy flow in aquatic ecosystems. Freshw. Biol. 33, 83–89.

Wetzel, R.G., 2000. Freshwater ecology: Changes, requirements, and future demands. Limnology 1, 3–9. https://doi.org/10.1007/s102010070023.

Wetzel, R.G., 2001. Limnology: Lake and river ecosystems, 3rd ed. Academic Press.

The Editors

IAN D. JONES, PhD is a physical limnologist in the Biological and Environmental Sciences Division at the University of Stirling (UK), having previously worked for many years at the UK's Centre for Ecology & Hydrology in Lancaster (UK). As well as physical limnology studies, his research has focused on the interaction between lake physics and the biology and chemistry in lakes, studying interactions with phosphorus, carbon, oxygen, bacteria, macrophytes, phytoplankton, zooplankton, and fish.

JOHN P. SMOL, OC, PhD, FRSC, FRS is a Distinguished University Professor in the Department of Biology at Queen's University (Canada). Smol founded and codirects the Paleoecological Environmental Assessment and Research Lab (PEARL), a group of about 35 students and other scientists dedicated to the study of long-term global environmental change, and especially as it relates to lake ecosystems. John has authored more than 700 journal publications and chapters, as well as completed 24 books. Smol was the founding Editor of the *Journal of Paleolimnology* (1987–2007) and is current Editor of *Environmental Reviews* (2004–present). Since 1990 he has received six honorary doctorates and has been awarded more than 70 research and teaching awards and fellowships, including the International Society of Limnology *Naumann-Thienemann Medal*, both the *Hutchinson Award* and the *Margalef Award* from the Association for the Sciences of Limnology and Oceanography, the *International Ecology Institute Prize*, and the *NSERC Herzberg Gold Medal* as Canada's top scientist or engineer. He was named an Officer of the *Order of Canada* for his environmental work and a *Fellow of the Royal Society (London)* and was elected President of the Academy of Science, Royal Society of Canada (2019–2022).

Contributors

Shuqing An School of Life Science and Institute of Wetland Ecology, Nanjing University, Nanjing, China; Nanjing University Ecological Research Institute of Changshu, Suzhou, China

Beatrix E. Beisner Department of Biological Sciences, University of Quebec at Montreal, Montreal, Quebec, Canada

Meryem Beklioğlu Limnology Laboratory, Department of Biological Sciences and Centre for Ecosystem Research and Implementation (EKOSAM), Middle East Technical University, Ankara, Turkey

Michael T. Bogan School of Natural Resources and the Environment, University of Arizona, Tucson, Arizona, United States

Núria Bonada Departament de Biologia Evolutiva, Ecologia i Ciències Ambientals, Facultat de Biologia, Institut de Recerca de la Biodiversitat (IRBio), Universitat de Barcelona (UB), Barcelona, Catalonia/Spain

Michele Astrid Burford Australian Rivers Institute, Griffith University, Nathan, Queensland, Australia; School of Environment and Sciences, Griffith University, Nathan, Queensland, Australia

Cayelan C. Carey Virginia Tech, Blacksburg, Virginia, United States

Patricia A. Chambers Burlington, Ontario, Canada

Jonathan J. Cole[†] Cary Institute of Ecosystem Studies, Millbrook, New York, United States

Raoul-Marie Couture Department of Chemistry and Centre for Northern Studies (CEN), Laval University, Quebec City, Quebec, Canada

Tonya DelSontro Department of Earth and Environmental Sciences, University of Waterloo, Waterloo, Ontario, Canada

Lisette N. de Senerpont Domis Netherlands Institute of Ecology, Wageningen, the Netherlands

Paula de Tezanos Pinto Instituto de Botánica Darwinion, Consejo Nacional de Investigaciones Científicas y Técnicas, Argentina

Peter J. Dillon School of the Environment, Trent University, Peterborough, Ontario, Canada

Hilary A. Dugan Center for Limnology, University of Wisconsin–Madison, Madison, Wisconsin, United States

Lluís Gómez-Gener Centre for Research on Ecology and Forestry Applications (CREAF), Edifici C, Campus de Bellaterra (UAB), Cerdanyola del Vallès, Barcelona, Spain

Takashi Gomi Graduate School of Bioagricultural Science, Nagoya University, Nagoya, Japan

Irene Gregory-Eaves Department of Biology, McGill University, Montreal, Quebec, Canada

Bruna Grizzetti European Commission Joint Research Centre (JRC), via E. Fermi 2749, 21027 Ispra (VA), Italy

David P. Hamilton Australian Rivers Institute, Griffith University, Nathan, Queensland, Australia

Stephanie E. Hampton Biosphere Sciences and Engineering, Carnegie Institution for Science, Pasadena, California, United States

Erin R. Hotchkiss Department of Biological Sciences, Virginia Polytechnic Institute and State University, Blacksburg, Virginia, United States

Sanet Janse van Vuuren Unit for Environmental Sciences and Management, North-West University, Potchefstroom, South Africa

Nasreen Jeelani Department of Geography and Regional Research, University of Vienna, Vienna, Austria

Erik Jeppesen Department of Ecoscience, Aarhus University, Aarhus, Denmark; Sino-Danish College, University of Chinese Academy of Sciences, Beijing, P.R. China; Limnology Laboratory, Department of Biological Sciences and Centre for Ecosystem Research and Implementation, Middle East Technical University, Ankara, Turkey; Institute of Marine Sciences, Middle East Technical University, Erdemli-Mersin, Turkey; Institute for Ecological Research and Pollution Control of Plateau Lakes, School of Ecology and Environmental Sciences, Yunnan University, Kunming, China

Ian Jones Biological and Environmental Sciences, University of Stirling, Stirling, UK

Ismael Kimirei Tanzania Fisheries Research Institute, Dar es Salaam, Tanzania

Emma S. Kritzberg Department of Biology, Lund University, Lund, Sweden

Michio Kumagai Research Center for Lake Biwa and Environmental Innovation, Ritsumeikan University, Kusatsu, Shiga, Japan

Bernhard Lehner Department of Geography, McGill University, Montreal, Quebec, Canada

[†] Deceased.

Elena Litchman Kellogg Biological Station and Department of Integrative Biology, Michigan State University, East Lansing, Michigan, United States; Department of Global Ecology, Carnegie Institution for Science, Stanford, California, United States

Rex Lowe Department of Biological Sciences, Bowling Green State University, Bowling Green, Ohio, United States; Center for Limnology, University of Wisconsin, Madison, Wisconsin, United States

Jing Lu Australian Rivers Institute, Griffith University, Nathan, Queensland, Australia

Stephen C. Maberly Lake Ecosystems Group, UK Centre for Ecology & Hydrology, Lancaster, UK

Sally MacIntyre Department of Ecology, Evolution, and Marine Biology, University of California, Santa Barbara, California, United States

Zhigang Mao State Key Laboratory of Lake Science and Environment, Nanjing Institute of Geography and Limnology, Chinese Academy of Sciences, Nanjing, China

Rafael Marcé Catalan Institute for Water Research (ICRA), Girona, Spain; Universitat de Girona, Girona, Spain

Katherine D. McMahon Department of Civil and Environmental Engineering, University of Wisconsin Madison, Madison, Wisconsin, United States; Department of Bacteriology, University of Wisconsin Madison, Madison, Wisconsin, United States

Mariana Meerhoff Department of Ecology and Environmental Management, Centro Universitario Regional del Este, Universidad de la República, Maldonado, Uruguay; Department of Ecoscience, Aarhus University, Aarhus, Denmark

Lewis A. Molot Faculty of Environmental & Urban Change, York University, Toronto, Ontario, Canada

Ryan J. Newton School of Freshwater Sciences, University of Wisconsin Milwaukee, Milwaukee, Wisconsin, United States

Rich Pawlowicz Department of Earth, Ocean and Atmospheric Sciences, University of British Columbia, Vancouver, British Columbia, Canada

Sandra Poikane European Commission Joint Research Centre (JRC), via E. Fermi 2749, 21027 Ispra (VA), Italy

Yves T. Prairie Department of Biological Sciences, University of Quebec at Montreal, Montreal, Quebec, Canada

Belinda J. Robson Harry Butler Institute & Environmental & Conservation Sciences, Murdoch University, Murdoch, Western Australia, Australia

Kevin C. Rose Department of Biological Sciences, Rensselaer Polytechnic Institute, Troy, New York, United States

Sapna Sharma Biology Department, York University, Toronto, Ontario, Canada

Roy C. Sidle Mountain Societies Research Institute, University of Central Asia, Khorog, Tajikistan

John P. Smol Paleoecological Environmental Assessment and Research Lab (PEARL), Department of Biology, Queen's University, Kingston, Ontario, Canada

Stephen J. Thackeray Lake Ecosystems Group, UK Centre for Ecology & Hydrology, Lancaster, UK

Yvonne Vadeboncoeur Department of Biological Sciences, Wright State University, Dayton, Ohio, United States

Jos T.A. Verhoeven Ecology and Biodiversity, Department of Biology, Utrecht University, CH Utrecht, the Netherlands

Warwick F. Vincent Department of Biology and Centre for Northern Studies (CEN), Laval University, Quebec City, Quebec, Canada

Pietro Volta CNR-IRSA Water Research Institute, Verbania Pallanza, Italy

John Wehr Louis Calder Center and Department of Biological Sciences, Fordham University, Armonk, New York, United States

Ram Yerubandi Environment and Climate Change Canada, Canada Center for Inland Waters, Burlington, Canada

Shenglai Yin Nanjing University Ecological Research Institute of Changshu, Suzhou, China; College of Life Science, Nanjing Normal University, Nanjing, China

Contents

Video chapter summaries are available at: https://limnology.org/resources/wetzel-videos/

CHAPTER

1

Prologue

Ian Jones[1] *and John P. Smol*[2]

[1]Biological and Environmental Sciences, University of Stirling, Stirling, UK

[2]Paleoecological Environmental Assessment and Research Lab (PEARL), Department of Biology, Queen's University, Kingston, Ontario, Canada

OUTLINE

I. Limnology in the 21st century

Water is essential to life on Earth. For human beings, the surface waters on land are especially vital because these are the waters that we access for sustaining our daily lives, providing a plethora of ecosystem services (see Chapter 2). Inland waters encompass a remarkably broad array of forms, as shown in Figs. 1-1 to 1-4. Nevertheless, inland surface waters only represent a small fraction of the water on the planet, dwarfed in volume by that in the salty ocean or trapped in ice. The study of these inland waters, in their various guises, and the crucial roles they play in our global environment, is imperative for our continuing existence. This study is limnology.

Not only are inland waters fundamental to our current lives, but they are also under unprecedented threat from human activities. Burgeoning population growth and expanding economic development place demands on available water never previously experienced, while the various pressures from climatic change, nutrient enrichment, invasive species, and more put these finite resources under an unparalleled level of stress. The vulnerability of inland waters to these twin burdens of increased demand and increased stress presents a singular challenge that must be overcome to ensure a sustainable future. Our ability to meet such a challenge is predicated on our ability to understand the processes at play. The study of these inland waters, though, encompasses not one but multiple disciplines: the limnologist must be a physicist, a chemist, and a biologist; an engineer, a hydrologist, and a statistician; an ecologist, a geologist, and a social scientist. The spatial scales of interest range from nano to global. The relevant timescales span milliseconds to millennia. We have much work to do.

II. Robert G. Wetzel (1936–2005)

There was an increasing understanding of the urgency to study inland waters back in the early 1970s, even when global demands on water were a fraction of those today, and some of the stressors that occupy us now were unknown or little discussed. Robert Wetzel recognized this need and set about producing a comprehensive textbook on lakes that resulted in the first edition of *Limnology* in 1975. The book provided an inclusive base of information on all aspects of lakes. It was aimed at upper-year undergraduate courses and was a first reference for seasoned limnologists and other specialists, quickly becoming the standard textbook in the field. The need for such a text did not diminish over the years, with an enlarged second edition published in 1983, and then a much revised and expanded third edition in 2001, that included river ecosystems

Wetzel's Limnology, Fourth Edition
https://doi.org/10.1016/B978-0-12-822701-5.00001-X

FIGURE 1-1 **A variety of inland waters around the world.** Photo credits in brackets. (a) Lough Ouler, Ireland (Mikkel Andersen); (b) Adams Stream, McMurdo Dry Valleys, Antarctica (Warwick Vincent); (c) Amazonian wetland, Peru (John Smol); (d) Lake Bogoria, Kenya (Emma Tebbs); (e) Niagara Falls, Canada/USA boundary (Ian Jones); (f) headwaters west of Tokyo, Japan (Roy Sidle).

FIGURE 1-2 **A variety of inland waters around the world.** Photo credits in brackets. (a) Headwater stream in Daintree Rainforest, Queensland, Australia (Roy Sidle); (b) Lake Erken, Sweden (Mikkel Andersen); (c) Amazonian flood plain lake, Brazil (Warwick Vincent); (d) thermokarst lakes and ponds in subarctic Quebec, Canada (Warwick Vincent); (e) pond in Ontario, Canada (Ian Jones); (f) shallow lake in Wuhan, China (Ian Jones).

FIGURE 1-3 **A variety of inland waters around the world.** Photo credits in brackets. (a) High Arctic Thores Glacier and proglacial Thores Lake (Warwick Vincent); (b) plunge pool in New Zealand (Ian Jones); (c) headwater stream in Ala Archa National Park, Kyrgyzstan (Roy Sidle); (d) Windermere from Ferry House, UK (Glenn Rhodes); (e) Lago Chirripo, Costa Rica (Neal Michelutti); (f) River Nile, Egypt (John Smol).

FIGURE 1-4 **A variety of inland waters around the world.** Photo credits in brackets. (a) Small lake on Ellesmere Island, High Arctic, Canada (John Smol); (b) Lough Furnace and inflow, Ireland (Mikkel Andersen); (c) Panj River, Tajikistan/Afghanistan (Roy Sidle); (d) small mountain stream, Morocco (Ian Jones); (e) Chaca Cocha, Peru, with alpaca in foreground (Neal Michelutti); (f) Caspian Sea from Iran (Mahtab Yaghouti).

and stretched to over 1000 large-format pages. For many limnologists, it has been the principal text they used from their student years to their professional careers.

The third edition has endured as one of the definitive limnology texts but cannot continue to do so perpetually. Science moves forward relentlessly. Some core principles may remain indefinitely, at least until a revolutionary understanding replaces them, but much of science is a continuous process with layer after layer of new information enhancing our understanding of the world. Though the third edition came out in 2001, much of it was surely written in the preceding decade. In the world of science, 20 to 30 years is a long time. Much has changed in our understanding of limnology during that time, and new threats have emerged to be tackled. Tragically, Robert Wetzel passed away in 2005, and for many years subsequently, it seemed that the third edition would be the last edition, not the latest. The idea that one person could put together a new volume on the scale of Wetzel's earlier editions appeared unthinkable. Nevertheless, with the knowledge that the third edition would eventually become obsolete, with the sense that a text as comprehensive as the previous editions was still required, and with the full cooperation of the Wetzel family, it was ultimately proposed that it was time to produce a fourth edition.

III. The fourth edition

A recurring theme in putting this volume together has been "How on Earth did one person do this?!" Indeed, it has taken nearly 60 authors and two editors to produce this volume—60 times as many people to produce a book of roughly the same length as one person had done before. Our respect for Robert Wetzel's knowledge and industry is overwhelming.

Wetzel's name is still on the cover of the book, and with good reason. He is still an author on this edition, albeit a "silent" one, unnamed in the individual chapters. The overall scope of the book, the narrative thread, and some of his original text, figures, and tables are still evident throughout the book. On rare occasions, some material and data that were unique to the earlier book have a citation to the third edition, to indicate that origin; in general, though, text and figures kept in this edition from the last edition are uncited and intermingled with new text, as if Robert Wetzel were named as one of the chapter authors. The passing of the years, of course, means that much also needed to be added or to be removed from the previous edition, following the developments in our understanding of science and reflected in the large number of post-2001 references present. This work remains a new edition of the original textbook, not a new book: an evolution, rather than a revolution, with Robert Wetzel still its central figure.

Putting a new edition together does, though, necessitate a dizzying array of choices and decisions, the easiest of which was to rule out the notion that just one or two people would produce the new text. Instead, a new model was adopted in which each individual chapter is now written by one or a small number of experts—specialists who are aware of the most important themes and latest developments in their own field. Contributors include authorities on biological, chemical, and physical aspects of aquatic science. These authors could decide how much of the original text they wished to keep, how much to update, and which new topics needed incorporating.

As editors, it was necessary to determine what to include and what to exclude, where to grow and where to prune, and how to order the material to maintain a functioning narrative. Each chapter needed simultaneously to stand alone and to fit into a greater whole. Chapters in previous editions had been notoriously variable in size; while there are still large discrepancies, chapter lengths are now more standardized than before. Some topics that were dealt with relatively quickly in old editions now have substantially more space devoted to them. Some particularly large chapters have been split into two, and a few new chapters have spawned. Given the explosion in limnological studies in the last 20 years, a challenge for all authors was how to keep the overall size manageable and comparable to the previous edition.

It is perhaps inevitable with science, given our finite understanding of its complexity, that some phenomena are considered slightly differently by different people, engendering debate on those topics. Evidence can be contradictory, confusing, and incomplete; multiple hypotheses can plausibly explain the same findings; awareness of different processes can influence interpretation. Often the topics stimulating the most vibrant discussions evolve through time. Some topics that were perceived from contrasting angles 20 or 30 years ago have subsequently transmuted into themes enjoying a broad consensus, while other subjects now stimulate a spirited dialogue instead. Having a book with multiple authors allows diverse and interesting perspectives to be communicated and enables, through time, new consensuses to emerge and scrutiny to fall on other issues.

The style of the book has been modernized. Color images and photographs abound, options that would not have been readily available to scientific texts over two decades ago. A similar narrative to earlier editions has been followed, though, with the physical elements of limnology the focus of earlier chapters, moving through to the chemical and then on to the biological aspects. To an extent, that follows a traditional view that the physics impacts the chemistry and they both influence the biology, but, of course, there are multiple feedbacks between all the subjects. Similarly, there are multiple feedbacks among chapters. Concepts can be central to more than one chapter, and linkages move back and forth throughout the text. For a book of this size, such overlaps are inevitable and help to reinforce important limnological concepts. As in previous editions, short summaries are provided at the end of each chapter to offer a synopsis of the key points within that field. The book remains aimed toward upper-year undergraduate students and for use as a first source for the professional limnologist when seeking information outside of their own limnological experience.

It is fair to say that, though lakes and rivers get equal billing in the title of the book, the 2001 edition focused more on the lake aspect than the river aspect. This bias persists somewhat in this edition, although there is an increase in the consideration given to rivers, not least because of the impact that rivers themselves have on lakes. Another theme to gain more weight in this edition is that of the impacts of environmental change and multiple stressors, particularly climate change, on inland aquatic ecosystems. These threats are too important to omit from a comprehensive textbook on limnology.

Robert Wetzel worked in the USA throughout his career but was an early advocate of the need for a global perspective. While the North American roots of this textbook are still evident, with many authors based in that continent, this edition is international. Authors stem from multiple countries, encompassing six continents. We anticipate that future editions will be even more international in scope. Limnology is a global subject and a collective undertaking. Inland waters are distributed worldwide and need to be understood in a worldwide context, both in terms of the geographical influence on ecosystem function and their relevance to society in different parts of the planet. As argued in Chapter 31, working collaboratively across the globe is a necessity for better understanding our lakes, rivers, and wetlands and for tackling the threats they currently face. This edition represents only a stepping stone in that journey toward understanding the world of inland waters.

Acknowledgments

The timing of this edition has not been fortuitous. During the conception and initial stages, no one had heard of COVID-19; the idea of a disease overturning the lives of billions worldwide was not on people's minds. The entire book, though, has been written and edited under the shadow of the pandemic. The direct or indirect impacts have affected all authors, causing untold impositions, obstacles, and hardships, yet still, they have managed to produce these chapters. As well as the 50+ authors officially listed in the chapters, there was much aiding and assisting by colleagues, students, and collaborators. Brian Cumming, Irene Gregory-Eaves, Erik Jeppesen, Neal Michelutti, Mariana Meerhof, Andrew Paterson, Kathleen Rühland, Sapna Sharma, Warwick Vincent, and Paul Wetzel kindly read drafts of this prologue and, indeed, each chapter underwent "friendly" reviews by other scientists: ultimately the book represents a community endeavor. We are thankful for the contributions of all. This edition originated at Elsevier with Louisa Munroe and would not have happened without her vision. Sara Valentino has subsequently been at the helm for Elsevier, guiding everyone forward at each step. Both have been indispensable. We are particularly grateful to the Wetzel family, whose approval for proceeding with the book was a prerequisite for us to take it forward. Special thanks go to Jim Cotner and Paul Wetzel for generously agreeing to share their memories of Robert Wetzel. It remains his book: the greatest thanks go to him.

C H A P T E R

2

The Importance of Inland Waters

Bruna Grizzetti and Sandra Poikane

European Commission Joint Research Centre (JRC), via E. Fermi 2749, 21027 Ispra (VA), Italy

O U T L I N E

A basic feature of the Earth is an abundance of water, which extends over 71% of its surface (Eakins and Sharman, 2010). Almost 99% of this immense *hydrosphere* lies in ocean depressions and polar ice deposits (Shiklomanov, 1993). In the hydrosphere most of the water is saline and only 2.5% is freshwater, with lakes, reservoirs, and rivers representing only 0.26% of fresh water and the rest being mainly groundwater (29.9%) and ice and permanent snowcover (68.7%; Shiklomanov, 2000; see Chapter 4 for rivers and lakes distribution on land and water retention time). The relatively small amounts of water that occur in freshwater lakes and rivers belie their fundamental importance in the maintenance and survival of terrestrial life.

I. Our freshwater resources

Any analysis of inland water resources must address the preeminence of human growth and utilization of freshwater. Humans must be recognized for what they are: animals whose population is growing. In spite of its absurdity, a belief prevails that the Earth's supply of finite water resources can be increased constantly to meet exponential demands. Freshwaters are a finite resource that can be increased only slightly. For example, the desalinization of ocean water requires tremendous energy expenditures for the treatment process and distribution of the product once obtained. In the past society has tended to underappreciate the role that humans have played as influential factors in the maintenance of lake and river ecosystems. Our understanding of anthropogenic impacts has been increasing steadily over the last several decades. Freshwater utilization is governed by the spiraling relationships in which supply is constantly expanded in response to growing demands. The unfortunate effect of growth is that consumption increases in response to rising supply.

At the initial stages of economic growth, consumption has grown uncontrollably, but then actions have been taken to reduce water use and pollution and decouple economic growth from water consumption. Since the Earth Summit in Rio de Janeiro, Brazil, in 1992, several international action plans have been proposed to build global partnerships for sustainable development and were adopted under the auspices of the United Nations. In 2015 all United Nations Member States adopted the 2030 Agenda for Sustainable Development that provides a shared plan for peace and prosperity for people and the planet, setting 17 *Sustainable Development Goals* (SDGs) (United Nations, 2016; 2021). Among the SDGs, SDG6 aims to ensure the availability and sustainable management of water and sanitation for all; SDG15 calls for the conservation, restoration, and sustainable use of terrestrial and inland freshwater ecosystems and their services;

Wetzel's Limnology, Fourth Edition
https://doi.org/10.1016/B978-0-12-822701-5.00002-1

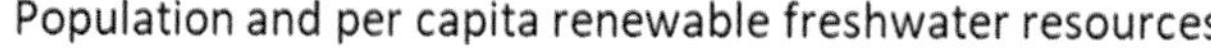

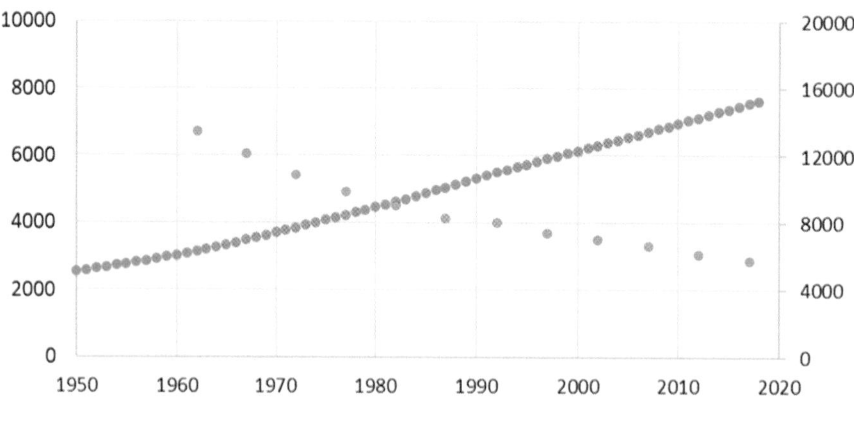

FIGURE 2-1 Changes in population and renewable water resources since the 1950s. *(Data sources: FAO. Annual population. License: CC BY-NC-SA 3.0 IGO. Extracted from: http://www.fao.org/faostat/en/#data [Accessed in November 24 2022]. World Bank. Renewable internal freshwater resources per capita (cubic meters). World Development Indicators. The World Bank Group. Extracted from: https://data.worldbank.org/indicator/ER.H2O.INTR.PC [Accessed in November 24 2022].)*

and SDG14 focuses on the conservation and sustainable use of coastal and marine resources.

II. Drivers of global change

Continued rapid population growth poses challenges for sustainable development, putting additional pressure on already strained water resources. The world's population reached 7.8 billion as of January 2021, having added 1 billion people since 2007 and 2 billion since 1994. It is expected to reach about 8.5 billion by 2030 and further increase to 9.7 billion by 2050 (United Nations, 2019). Not only population growth but also economic development have far-reaching implications for water supplies and their use. Humanity's water demands are rapidly increasing as developing countries undergo economic transformations and their living standards rise (Fig. 2-1). The degradation of the biosphere and water resources occurs as a result of the utilization of the environment for production and consumption.

Although about 106,000 km^3 of precipitation, the ultimate source of the freshwater supply, falls on the land surface per year, about two-thirds of that amount is returned to the atmosphere by evaporation and plant transpiration, collectively referred to as evapotranspiration (see Chapter 5). Only about 37% of terrestrial precipitation (c. 39,600 km^3 year^{-1}) forms a renewable freshwater supply recharging groundwater, rivers, and lakes. A substantial portion of this water supply is inaccessible to humans due to its remoteness or inability to store seasonal flows. From the total renewable water supply, 75% or 29,700 km^3 year^{-1} is accessible (Vörösmarty et al., 2005). The part of this water being extracted from groundwater aquifers, rivers, and lakes for human use is called *water withdrawal*.

Between 1900 and 2020, global water withdrawal increased by nearly eight times, from 500 to 4000 km^3 per year (FAO, 2021). Agriculture, which includes irrigation, animal husbandry, and aquaculture, is the world's main water user, accounting for 69% of annual water withdrawals. Industry, including power generation, accounts for 19% of the total, while households account for 12%; however, these proportions vary strongly among countries and continents (FAO, 2021).

If the global water withdrawal, 4000 km^3 year^{-1}, were divided evenly among the 7.8 billion humans now on Earth (2021), each person could potentially use 512 m^3 of water annually, or 1400 L per day. These amounts can seem high in relation to the physiological requirements of humans: the World Health Organization (Howard et al., 2020) estimates that humans need between 50 and 100 L of water per day. However, this is insufficient in view of modern technological demands and the water resources are far from being distributed evenly among Earth's population.

The average per capita water withdrawal in the United States is 1367 m^3 per year, or 3745 L per person per day (FAO, 2021), making it one of the most water-intensive countries. In the United States the two most important uses of water are thermoelectric power (41% of total withdrawals for all uses) and irrigation (42%; Dieter et al., 2018). Another example is Greece, with a total water withdrawal per capita of 962 m^3 per year, or 2636 L per person per day. Agriculture uses the majority of this amount (80.1%), while municipal water withdrawal accounts for 16.7% and industries use only 3.2% of this amount (FAO, 2021).

Water stress is defined as the ratio of annual water withdrawal to annual renewable water resources minus environmental flow requirements. The global indicator of water stress provides an efficient estimate of pressures that human activities exert over natural freshwater resources. A country is water scarce if this ratio lies in the range of 25%–70% and severely water scarce if this ratio exceeds 70%. Among the 53 countries worldwide currently experiencing water scarcity, in 22 countries the scarcity is severe, including 11 countries (such as Libya, Saudi Arabia, the United Arab Emirates, and Kuwait) where the water stress index is above 100% and freshwater demand is largely met by desalination (United Nations, 2018).

Global water demand is expected to continue rising due to increasing population and socioeconomic development. At the same time, climate change is projected to alter water availability, with a decrease in water resources in many midlatitude and dry subtropical regions and an increase in high-latitude and many humid midlatitude regions (Jiménez Cisneros et al., 2014). "Continued global warming is projected to further intensify the global water cycle, including its variability, [...] and the severity of wet and dry events" (IPCC, 2021). Climate change has already caused widespread impacts on human and natural systems. In the future, climate change risks can be mitigated only if substantial reductions of greenhouse gas emissions would take place, together with adaptation options (IPCC, 2014). In order to cope effectively with natural and induced climatic changes, we must maintain sufficient flexibility in our use of freshwater resources and an adequate margin below the maximum carrying capacity.

III. Human impact on freshwater ecosystems

The real freshwater supply is much smaller than the potential total because of many factors. First, rainfall is not evenly distributed over land surfaces, and humans themselves are not distributed over land in proportion to water availability. This disparity results in a great expense of resources and energy for distribution systems to move water from places of water abundance to places where it is inadequate to support human activities. Second, total consumption has increased exponentially with population and economic growth. Expansion of distribution systems to areas of low precipitation, such as for irrigation of semiarid regions, results in disproportionately high use of water because of very high losses by evapotranspiration. Third, the degradation of water quality from contaminants can severely reduce the water supply available for other purposes.

Fresh waters of the world are collectively experiencing markedly accelerating rates of qualitative and quantitative degradation (Wetzel, 1992). Population growth and economic development are the most important drivers of pressures on freshwater quantity and quality (Fig. 2-2). Human activities affect water resources through direct water consumption for drinking, sanitation, and household use, as well as indirect consumption through water-intensive goods and services, including food and energy. The concept of the *water footprint* has been coined as an indicator of water use that includes both direct and indirect water use of a consumer or producer (Hoekstra and Hung, 2002). Traditional water-use indicators like withdrawals report (gross) volumes removed from a water body, whereas the water footprint reports (net) water consumption. The present average water footprint is 3800 L per person per day, most of which is used indirectly for food production (Hogeboom, 2020).

In addition, human activities exert multiple pressures on freshwater ecosystems through water, air, and soil pollution, changes in the natural landscape (e.g., catchment drainage and river channelization, disconnection of flood plains, river continuity interruption), overfishing, and the introduction of invasive species. As described in subsequent chapters in this book, the pressures lead to an altered ecosystem state, involving changes in physicochemical variables (e.g., dissolved oxygen, pH, nutrients) and in the composition and abundance of biological communities and food webs (changes in biodiversity). These effects have major impacts on the provision of goods and benefits for people, which in turn have an impact on societal welfare. Therefore measures are needed to prevent/mitigate adverse changes, which can include legislative actions on pressures and activities, technological solutions, societal changes, and restoration techniques such as meandering and buffer strips for rivers and sediment removal, phosphorus precipitation, and food-web biomanipulation for lakes.

IV. Ecosystem condition and ecosystem services

Human life depends on nature. The concept of *ecosystem services* has been introduced to highlight how human well-being is connected to nature. Ecosystem services are defined as the benefits that people obtain from ecosystems (MEA, 2005), and the direct and indirect contributions of ecosystems to human well-being (TEEB, 2010). Understanding the linkages between the natural and socioeconomic systems can lead to an improved and more sustainable management of ecosystems (Guerry et al., 2015).

Ecosystem functioning and biodiversity provide multiple benefits to humans. Examples of water ecosystem services are drinking water; water provision for

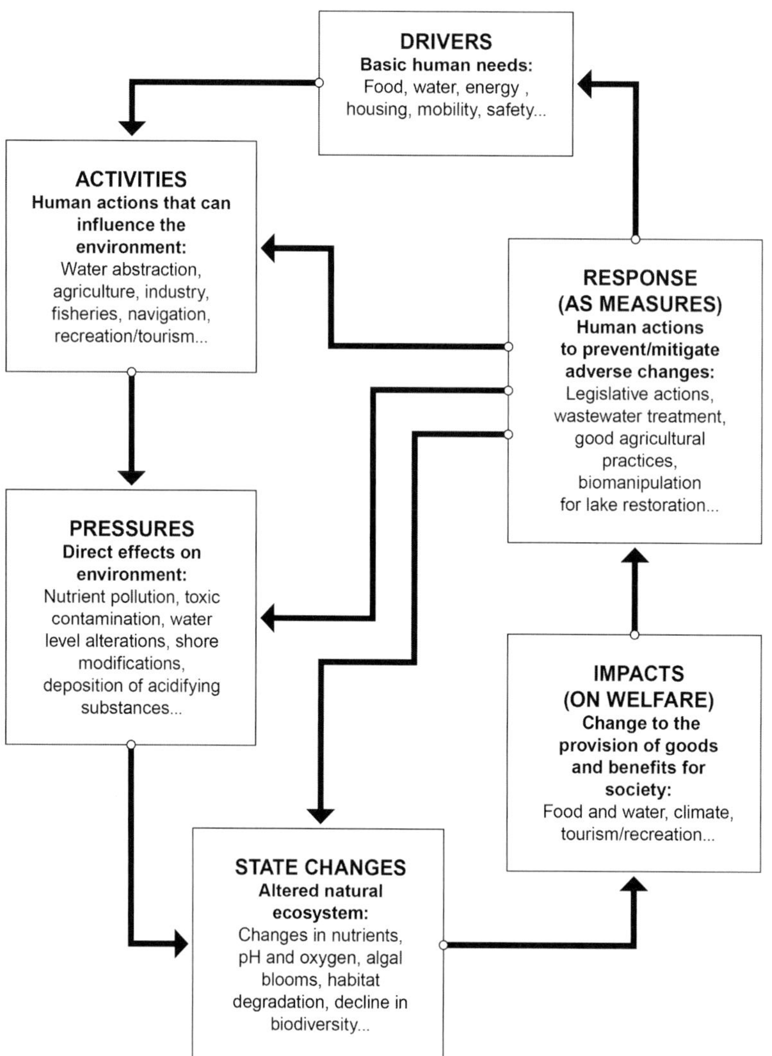

FIGURE 2-2 Conceptual framework for water management. *(Modified from Elliott et al., 2017; Poikane et al., 2020.)*

industry, irrigation, and transportation systems; fishing (food provision); water purification (e.g., from bacterial and contaminant pollution); lifecycle maintenance; flood protection; local climate regulation; and recreation (e.g., swimming, boating, nature viewing, and recreational fishing) (Grizzetti et al., 2016).

An ecosystem's condition influences the delivery of ecosystem services (Grizzetti et al., 2019). Good ecosystem status supports high delivery of services. Anthropogenic pressures affect the condition of aquatic ecosystems and in turn the delivery of their ecosystem services (Fig. 2-3). We need to protect and restore water ecosystems to conserve and enhance biodiversity and water resources, and to ensure the capacity of ecosystems to continue delivering a whole range of ecosystem services essential to human well-being (Maes, 2020). Looking back at the repetitious history of responses to impending environmental disasters, we can be optimistic about the future only until such time as our

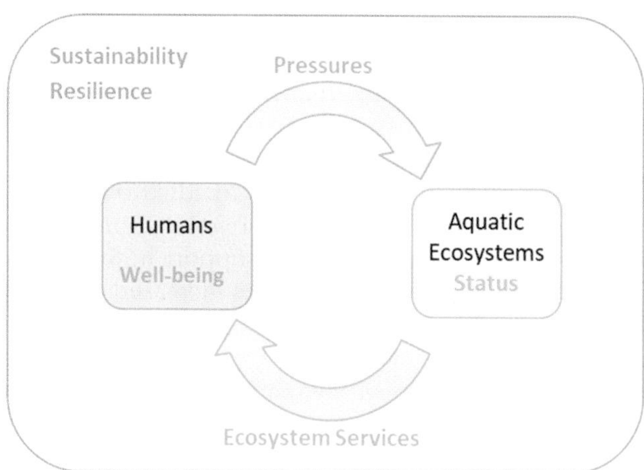

FIGURE 2-3 Relationships between humans and aquatic ecosystems. Human activities generate multiple pressures on aquatic ecosystems. Pressures affect the biodiversity and the condition of the aquatic ecosystem, which can result in a change in the ecosystem services. *(From Grizzetti et al., 2016.)*

understanding of the operation of the biosphere, and our knowledge of freshwater ecosystems in particular, is adequate to allow us to recognize the point of irreversibility (Francko and Wetzel, 1983; Wetzel, 1992).

The biodiversity and resilience of ecosystems are interconnected. Biodiversity increases the stability and resilience of ecosystem functioning (Cardinale et al., 2012; IPBES, 2018; Isbell et al., 2015). Global changes are threatening the stability and resilience of ecosystems (IPBES, 2018; Zampieri et al., 2021). Safe planetary boundaries have been overstepped for biosphere integrity (biodiversity loss and extinctions), nitrogen and phosphorus biogeochemical flows, and climate and land-system changes (Rockström et al., 2009; Steffen et al., 2015). Crossing planetary boundaries increases the risk of irreversible environmental changes.

It is of the utmost importance therefore that we understand the structure and function of freshwater ecosystems, which is the overall aim of limnology, including humans as a key component of these ecosystems.

Understanding the metabolic responses of aquatic ecosystems is essential in order to confront and offset the effects of these alterations and in order to achieve maximum, effective management of freshwater resources. A well-documented effect of human impacts on aquatic ecosystems is *eutrophication*, a multifaceted term associated with increased productivity, simplification of biotic communities, and a reduction in the ability of the metabolism of the organisms to adapt to the imposed loading of nutrients. These conditions lead to reduced stability of the ecosystem. In this condition of eutrophication excessive inputs often exceed the capacity of the ecosystem to be balanced. In reality, however, the ecosystems are out of equilibrium only with respect to the freshwater chemical and biotic characteristics that are desired by humans for specific purposes. In order to have any hope of effectively integrating humans as a component of aquatic ecosystems, and of monitoring their utilization of these resources, it is mandatory that we comprehend in some detail the functional properties of fresh waters. Only then can we evaluate and predict, with reasonable certainty, the influence human activities will have on the metabolic characteristics of these ecosystems.

Knowledge about the structure and functioning of water ecosystems is fundamental to informing management and policy decisions for a sustainable use of aquatic ecosystems and water resources. The analysis should embrace a holistic view that encompasses the dynamics of the water system. First of all, water follows river basin boundaries, rather than national borders, and within the river basin all waters are connected; that is, there is a link between upstream and downstream waters, groundwater and surface waters, inland and coastal waters, and of course, via the atmosphere. Second, the drivers of changes and anthropogenic pressures, such as pollution, hydromorphological alterations, and changes in species abundance and composition, can have both a local and a global dimension. Third, in aquatic ecosystems a lag time can occur between anthropogenic pressures and the detection of impacts, and between restoration interventions and effective recovery, as ecological processes and water retention time vary greatly across different freshwater ecosystems. Actions to protect and restore inland waters, such as rivers, lakes, and wetlands, should consider all these connections in the land-sea continuum.

V. Summary

1. Almost 99% of the immense amount of water in the biosphere occurs in the oceans and polar ice deposits. In the hydrosphere most of the water is saline and only 2.5% is freshwater, with lakes, reservoirs, and rivers representing only 0.26% of fresh water and the rest being mainly groundwater and ice and permanent snowcover.
2. Finite water resources are being exploited and degraded at an accelerating rate by the activities of humankind. Demands upon surface and groundwater supplies result from both population growth and expanding utilization and consumption. Biospheric degradation results from both biological population expansion and the growth of technological production and consumption. In addition, climate change is projected to alter water resources.
3. Human activities affect water resources through direct water consumption, pollution, changes in the natural landscape (e.g., catchment drainage and river channelization, disconnection of flood plains, river continuity interruption), overfishing, and the introduction of invasive species.
4. Human life and well-being depend on nature. Pressures affect the condition and biodiversity of the aquatic ecosystems, and this can change their ecosystem services, that is, the benefits to people.
5. We need to understand the structure and function of freshwater ecosystems, which is the overall aim of limnology, adopting a holistic view of the human–environment system to protect and enhance the aquatic ecosystems and to secure sustainable use of freshwater resources.

Acknowledgments

The authors would like to thank Dr. Peeter Nõges, Dr. Tiina Nõges, and Eliza Barbara Poikane for kindly reviewing the chapter.

References

Cardinale, B.J., Duffy, J.E., Gonzalez, A., Hooper, D.U., Perrings, C., Venail, P., Narwani, A., Mace, M.M., Tilman, D., Wardle, D.A., Kinzig, A.P., Daily, G.C., Loreau, M., Grace, J.B., Larigauderie, A., Srivastava, D.S., Naeem, S., 2012. Biodiversity loss and its impact on humanity. Nature 486, 59–67.

Dieter, C.A., Maupin, M.A., Caldwell, R.R., Harris, M.A., Ivahnenko, T.I., Lovelace, J.K., Barber, N.L., Linsey, K.S., 2018. Estimated use of water in the United States in 2015. US Geol. Surv. Circular 1441, 65. https://doi.org/10.3133/cir1441. (Accessed in March 2021).

Eakins, B.W., Sharman, G.F., 2010. Volumes of the world's oceans from ETOPO1. NOAA National Geophysical Data Center, Boulder. https://ngdc.noaa.gov/mgg/global/etopo1_ocean_volumes.html. (Accessed in October 2021).

Elliott, M., Burdon, D., Atkins, J.P., Borja, A., Cormier, R., De Jonge, V.N., Turner, R.K., 2017. And DPSIR begat DAPSI (W) R (M)!"—a unifying framework for marine environmental management. Mar. Pollut. Bull. 118 (1–2), 27–40.

FAO, 2021. AQUASTAT database. AQUASTAT Website. http://www.fao.org/nr/water/aquastat/main/index.stm. (Accessed in March 2021).

Francko, D.A., Wetzel, R.G., 1983. To Quench Our Thirst. The Present and Future Status of Freshwater Resources of the United States. University of Michigan Press, Ann Arbor, p. 153.

Grizzetti, B., Lanzanova, D., Liquete, C., Reynaud, A., Cardoso, A.C., 2016. Assessing water ecosystem services for water resource management. Environ. Sci. Pol. 61, 194–203.

Grizzetti, B., Liquete, C., Pistocchi, A., Vigiak, O., Zulian, G., Bouraoui, F., De Roo, A., Cardoso, A.C., 2019. Relationship between ecological condition and ecosystem services in European rivers, lakes and coastal waters. Sci. Total Environ. 671, 452–465.

Guerry, A.D., Polasky, S., Lubchenco, J., Chaplin-Kramer, R., Daily, G.C., Griffin, R., Ruckelshaus, M., Bateman, I.J., Duraiappah, A., Elmqvist, T., Feldman, M.W., Folke, C., Hoekstra, J., Kareiva, P.M., Kareiva, B.L., Li, S., McKenzie, E., Ouyang, Z., Reyers, B., Ricketts, T.H., Rockström, J., Tallis, H., Vira, B., 2015. Natural capital and ecosystem services informing decisions: from promise to practice. Proc. Natl. Acad. Sci. USA 112, 7348–7355.

Hoekstra, A.Y., Hung, P.Q., 2002. Virtual water trade: a quantification of virtual water flows between nations in relation to international crop trade. UNESCO-IHE Institute for Water Education, Delft, The Netherlands. Value of Water Research Report Series No. 11. http://www.waterfootprint.org/Reports/Report11.pdf. (Accessed 24 November 2022).

Hogeboom, R.J., 2020. The water footprint concept and water's grand environmental challenges. One Earth 2 (3), 218–222. https://doi.org/10.1016/j.oneear.2020.02.010.

Howard, G., Bartam, J., Williams, A., Overbo, A., Fuente, D., Geere, J.A., 2020. Domestic Water Quantity, Service Level and Health, second ed. World Health Organization, Geneva.

IPBES, 2018. The IPBES regional assessment report on biodiversity and ecosystem services for Europe and Central Asia. In: Rounsevell, M., Fischer, M., Torre-Marin Rando, A., Mader, A. (Eds.), Secretariat of the Intergovernmental Science-Policy Platform on Biodiversity and Ecosystem Services. Bonn, Germany, p. 892.

IPCC, 2014. Climate change 2014: synthesis report. contribution of working groups i, ii and iii to the fifth assessment report of the intergovernmental panel on climate change. In: [Core Writing Team, Pachauri, R.K., Meyer, L.A. (Eds.)]. IPCC, Switzerland, Geneva, p. 151.

IPCC, 2021. Summary for policymakers. climate change 2021: the physical science basis. contribution of working group i to the sixth assessment report of the intergovernmental panel on climate change. In: Masson-Delmotte, V., Zhai, P., Pirani, A., Connors, S.L., Péan, C., Berger, S., Caud, N., Chen, Y., Goldfarb, L., Gomis, M.I., Huang, M., Leitzell, K., Lonnoy, E., Matthews, J.B.R., Maycock, T.K., Waterfield, T., Yelekçi, O., Yu, R., Zhou, B. (Eds.). Cambridge University Press, Cambridge, United Kingdom and New York, NY, USA, pp. 3–32. https://doi.org/10.1017/9781009157896.001.

Isbell, F., Craven, D., Connolly, J., Loreau, M., Schmid, B., Beierkuhnlein, C., Bezemer, T.M., Bonin, C., Bruelheide, H., De Luca, E., Ebeling, A., Griffin, J.N., Guo, Q., Hautier, Y., Hector, A., Jentsch, A., Kreyling, J., Lanta, V., Manning, P., Meyer, S.T., Mori, A.S., Naeem, S., Niklaus, P.A., Polley, H.W., Reich, P.B., Roscher, C., Seabloom, E.W., Smith, M.D., Thakur, M.P., Tilman, D., Tracy, B.F., Van Der Putten, W.H., Van Ruijven, J., Weigelt, A., Weisser, W.W., Wilsey, B., Eisenhauer, N., 2015. Biodiversity increases the resistance of ecosystem productivity to climate extremes. Nature 526, 574–577. https://doi.org/10.1038/nature15374 (Accessed on 28 November 2022).

Jiménez Cisneros, B.E., Oki, T., Arnell, N.W., Benito, G., Cogley, J.G., Döll, P., Jiang, T., Mwakalila, S.S., 2014. Freshwater resources. In: C.B., Barros, V.R., Dokken, D.J., Mach, K.J., Mastrandrea, M.D., Bilir, T.E., Chatterjee, M., Ebi, K.L., Estrada, Y.O., Genova, R.C., Girma, B., Kissel, E.S., Levy, A.N., MacCracken, S., Mastrandrea, P.R., White, L.L. (Eds.), Climate Change 2014: Impacts, Adaptation, and Vulnerability. Part A: Global and Sectoral Aspects. Contribution of Working Group II to the Fifth Assessment Report of the Intergovernmental Panel on Climate Change [Field. Cambridge University Press, Cambridge and New York, pp. 229–269.

Maes, J., Teller, A., Erhard, M., Condé, S., Vallecillo, S., Barredo, J.I., Paracchini, M.L., Abdul Malak, D., Trombetti, M., Vigiak, O., Zulian, G., Addamo, A.M., Grizzetti, B., Somma, F., Hagyo, A., Vogt, P., Polce, C., Jones, A., Marin, A.I., Ivits, E., F., Mauri, A., Rega, C., Czúcz, B., Ceccherini, G., Pisoni, E., Ceglar, A., De Palma, P., Cerrani, I., Meroni, M., Caudullo, G., Lugato, E., Vogt, J.V., Spinoni, J., Cammalleri, C., Bastrup-Birk, A., San Miguel, J., San Román, S., Kristensen, P., Christiansen, T., Zal, N., de Roo, A., Cardoso, A.C., Pistocchi, A., Del Barrio Alvarellos, I., Tsiamis, K., Gervasini, E., Deriu, I., La Notte, A., Abad Viñas, R., Vizzarri, M., Camia, A., Robert, N., Kakoulaki, G., Garcia Bendito, E., Panagos, P., Ballabio, C., Scarpa, S., Montanarella, L., Orgiazzi, A., Fernandez Ugalde, O., Santos-Martín, F., 2020. Mapping and assessment of ecosystems and their services: an EU ecosystem assessment. EUR 30161 EN. Publications Office of the European Union, Ispra (Accessed on 28 November 2022).

MEA, 2005. Millennium Ecosystem Assessment. Ecosystems and Human Well-Being: Synthesis. Island Press, Washington, DC.

Poikane, S., Salas Herrero, F., Kelly, M.G., Borja, A., Birk, S., van de Bund, W., 2020. European aquatic ecological assessment methods: a critical review of their sensitivity to key pressures. Sci. Total Environ. 740, 140075.

Rockström, J., Steffen, W., Noone, K., Persson, Å., Chapin, F.S., Lambin, E.F., Lenton, T.L., Scheffer, M., Folke, C., Schellnhuber, H.J., Nykvist, B., de Wit, C.A., Hughes, T., van der Leeuw, S., Rodhe, H., Sörlin, S., Snyder, P.K., Costanza, C., Svedin, U., Falkenmark, M., Karlberg, L., Corell1, R.W., Fabry, V.J., Fabry, J., Walker1, W., Liverman, D., Richardson, K., Crutzen, P., Foley, J.A., 2009. A safe operating space for humanity. Nature 461, 472–475. https://doi.org/10.1038/461472a (Accessed on 28 November 2022).

Shiklomanov, I., 1993. World fresh water resources. In: Gleick, P.H. (Ed.), Water in crisis: a guide to the world's fresh water resources. Oxford University Press, New York.

Shiklomanov, I., 2000. Appraisal and assessment of world water resources. Water Int. 25, 11–32. https://doi.org/10.1080/02508060008686794.

Steffen, W., Richardson, K., Rockström, J., Cornell, S.E., Fetzer, C., Bennett, E.M., Biggs, R., Carpenter, S.R., de Vries, W., de Wit, C.A., Folke, C., Gerten, D., Heinke, J., Mace, G.M., Persson, L.M., Ramanathan, V., Reyers, B., Sörlin, S., 2015. Planetary boundaries: guiding human development on a changing planet. Science 347, 736.

TEEB, 2010. The Economics of Ecosystems and Biodiversity Ecological and Economic Foundations. Earthscan, London and Washington.

United Nations, 2016. Report of the inter-agency and expert group on sustainable development goal indicators. [E/CN.3/2016/2/Rev.1]. United Nations Economic and Social Council, New York, p. 49.

United Nations, 2018. Sustainable development goal 6 synthesis report 2018 on water and sanitation. https://www.unwater.org/publication_categories/sdg-6-synthesis-report-2018-on-water-and-sanitation/. (Accessed in March 2021).

United Nations, 2019. World population prospects 2019. https://population.un.org/wpp. (Accessed in March 2021).

United Nations, 2021. https://sdgs.un.org/. (Accessed in March 2021).

Vörösmarty, C.J., Leveque, C., Revenga, C., 2005. Fresh Water. Millennium Ecosystem Assessment, Volume 1: Conditions and Trends Working Group Report. Island Press, Washington, DC, pp. 165–207.

Wetzel, R.G., 1992. Clean water: a fading resource. Hydrobiologia 243/244, 21–30.

Zampieri, M., Grizzetti, B., Toreti, A., de Palma, P., Collalti, A., 2021. Rise and fall of vegetation annual primary production resilience to climate variability projected by a large ensemble of Earth System Models' simulations. Environ. Res. Lett. 16, 105001. https://doi.org/10.1088/1748-9326/ac2407 (Accessed on 28 November 2022).

CHAPTER

3

Water as a Substance

Rich Pawlowicz[1] *and Ram Yerubandi*[2]

[1]Department of Earth, Ocean and Atmospheric Sciences, University of British Columbia, Vancouver, Canada
[2]Environment and Climate Change Canada, Canada Center for Inland Waters, Burlington, Canada

OUTLINE

I. The characteristics of water

Water dominates the composition of liquids on Earth, including its lakes and oceans, as well as the composition of all living organisms. It is therefore an important fact that the physical and chemical properties of water are rather different from those of many liquids of similar molecular structure, with, for example, significantly higher melting and boiling points, surface tension, and viscosity. In addition, the properties of water in the liquid state often change in complex ways with temperature and pressure to a degree not seen in other liquids. Many of these subtle variations are crucial to regulating the physical, chemical, and biological processes that govern natural waters. In turn, aquatic biota have developed many adaptations that take advantage of these features to improve sustained productivity.

For practical purposes, when studying lakes, it is therefore important to understand and quantify the properties of water. Macroscopic physical properties of a liquid can be classified into several categories. *Thermodynamic properties*, like heat capacity, density, and the speed of pressure waves (i.e., sound), describe the state of the liquid, related to the type and arrangement of molecules within it. *Transport properties*, like viscosity, as well as thermal and electrical conductivity, describe the rate at which quantities like momentum, heat, and charge are transferred. Finally, *interface properties*, like surface tension and refractive index, govern changes that are associated with the boundaries between different fluids.

Although water in nature is never found in a pure state, the mass fraction of other dissolved matter in freshwaters (the "salinity," see Chapter 12) is usually small enough that its presence, with individual molecules separated by tens of water molecules, causes only small changes in most properties relative to those of the pure substance. The properties of pure water can therefore be a useful first approximation to the properties of natural waters in many, although not all, cases. Exceptions include electrical conductivity, which is mostly controlled by the small number of dissolved ions present in all natural waters, and the properties of brines (including seawater) and hypersaline lakes, where as few as two to four water molecules can separate those of the dissolved salts.

Wetzel's Limnology, Fourth Edition
https://doi.org/10.1016/B978-0-12-822701-5.00003-3

A. Molecular structure

Many of the anomalous properties of water can be explained by its unique chemical structure (Millero, 2001). Pure water is made up of hydrogen and oxygen atoms, which combine to form the water molecule H_2O. In equilibrium, the nuclei of the H_2O water molecule form an isosceles triangle, with a slightly obtuse bond angle of just under 105° separating the hydrogen atoms at the oxygen nucleus. As a result of this angled shape, charge distributions around the molecule are not symmetric, even though the total charge is balanced. There is an apparent negative charge located near the central oxygen atom and positive charges near the hydrogen atoms. Thus the molecule has an electrical dipole moment.

The polar nature of the water molecule allows for a special type of bonding, called *hydrogen bonding*, between water molecules and between water molecules and those of other dissolved substances. Hydrogen bonding is mostly electrostatic, but due to the sharing of unpaired electrons, it is also partly covalent in nature. Even in a liquid, this leads to the formation of lattices, clathrate cages, and polymers. Water is then a highly structured substance, although these structures form and reform on very short time scales. In ice the time scales are around 10^{-5} s and in liquid water about 10^{-11} s.

There are many consequences of this tendency toward a structured form. In pure water these structures are difficult to disrupt, leading to anomalously high melting and boiling points, as well as a large heat capacity, compared to other liquids. However, these structures are also much less space efficient than would otherwise be possible, so water contains "voids." The normally occurring form of water ice on the Earth's surface contains a tetrahedral structure with many voids, so its density is significantly less than that of liquid water. As water melts and warms, molecules reorient and rearrange themselves into a denser packing with fewer voids, helped by a slight bending of the hydrogen bonds. Liquid water then sinks below lake ice during surface cooling, and this ice in turn insulates the waters below from further heat loss. However, the greater thermal agitation that arises from warming also causes intermolecular distances to rise, acting to decrease density. In liquid water the changing balance between the tendency toward closer packing by reorientation and the tendency toward greater spacing by thermal agitation, as temperatures increase, results in the density reaching a maximum at 3.98°C instead of at the freezing point.

When dissolved matter is present in the water, the tendency of polar water molecules to form structures around ions, also called *hydration*, itself has many ramifications. Hydration greatly increases the solubility of many substances in water relative to their solubility in other liquids by acting to separate them into positive and negative ions. It leads to unexpectedly low electrical conductivities associated with some ions, as applied electrical fields must accelerate not only the ions but also water molecules attached to them. The formation of polymers also complicates the reactive behavior of elements like silica in water.

Although the most abundant forms of hydrogen and oxygen that make up water are ^{1}H and ^{16}O, elements in the natural world are found in a variety of isotopes (Meija et al., 2016a, 2016b). ^{2}H or D (deuterium) is a stable isotope of hydrogen, typically making up between 60 and 180 of every million (i.e., parts per million or ppm) hydrogen atoms in nature. However, the unstable ^{3}H (tritium) isotope, with a half-life of 12.5 years, was noticeably present only during the era of atmospheric nuclear bomb testing. Stable isotopes of oxygen found in nature are ^{17}O and ^{18}O (in amounts ranging from 370 to 400 and 1880 to 2220 ppm, respectively). A reference state for the isotopic composition of so-called "ordinary water" is *Vienna Standard Mean Ocean Water* (VSMOW), produced via the distillation of seawater (by far the largest reservoir of water on the planet's surface), in which ^{2}H makes up 156 ppm of hydrogen atoms and ^{17}O and ^{18}O, respectively, make up 379 and 2000 ppm of oxygen atoms.

Although isotopes do not behave differently in chemical reactions, their different masses cause small changes in physical properties. For example, density and freezing points are higher for water containing the heavier isotopes of hydrogen. More importantly, isotopically different water molecules evaporate and condense at different rates, and these differences depend on temperature (Rozanski et al., 2001). Thus their ratios in a water sample from one stage of the hydrologic cycle can be different from their ratios in a water sample from another stage, and these differences may vary with temperature. If water molecules are then incorporated into solid compounds or into organisms formed within a lake, which then die and are preserved over geological time, measurements of isotope ratios within them can be used to estimate temperature conditions at the time of their formation.

B. Thermodynamic properties

Properties of water have traditionally been determined through purely empirical measurements at different temperatures and pressures, to which analytic functions (e.g., polynomials or ratios of polynomials with specified coefficients) are fitted. Property values can then be calculated at any temperature or pressure using those functions, even if no measurements have ever been made at exactly those conditions.

These so-called correlation equations are often determined independently for different properties. However, different thermodynamic properties are linked through theoretical relationships, and so inconsistencies can arise in their estimated numerical values, depending on how the property was calculated. In recent years it has been recognized that a better approach is to define *thermodynamic state functions* for liquids like pure water or seawater. These state functions are not directly measurable, but all thermodynamic properties can be uniquely derived from specific mathematical derivatives and/or sets of derivatives of these state functions, codifying the thermodynamic interrelationships. Thus it is possible, by integration and fitting, to consistently summarize measurements of all thermodynamic properties of a particular liquid in a single thermodynamic state function. The availability of a state function also allows useful but otherwise unmeasurable thermodynamic properties like enthalpy and entropy to be quantified for these liquids, as they too can be derived from the state functions.

A thermodynamic state function for pure water is specified by the IAPWS-95 standard (IAPWS, R6-95, 2018). In some situations the effects of salinity are important, and an extension of IAPWS-95, which includes salinity, called TEOS-10 (IOC et al., 2010), may be useful as a description of saline surface waters even though it officially[1] describes seawater. The mathematical functions defining these state equations are quite complex and it is recommended that existing software, carefully written to compute thermodynamic properties according to these standards, is used (different packages are available at www.teos-10.org and https://web.mit.edu/seawater). However, the composition of dissolved matter in continental surface waters is rather different from that found in seawater and the accuracy of the resulting property estimates using seawater equations with a nonzero salinity, except for density and electrical conductivity, has not been evaluated for these saline surface waters (Pawlowicz and Feistel, 2012).

i. Density (water and ice)

The density of water is probably its most important property for understanding the physical dynamics of natural waters. Although the IAPWS-95 standard specifies the recommended mathematical function for numerically computing any of the thermodynamic properties of water at any temperature or pressure, for pure liquid water density ρ in the restricted range of temperatures t from 0°C to 40°C at sea level pressure (with temperature measured according to the ITS-90 standard; Preston-Thomas, 1990) a simpler equation may be useful (Tanaka et al., 2001):

$$\rho = a_5\left[1 - \frac{(t+a_1)^2(t+a_2)}{a_3(t+a_4)}\right] \tag{3-1}$$

where

$$a_1/°\text{C} = -3.983035$$
$$a_2/°\text{C} = 301.797$$
$$a_3/°\text{C}^2 = 522528.9$$
$$a_4/°\text{C} = 69.34881$$
$$a_5/(\text{kg m}^{-3}) = 999.974950$$

The density of pure liquid water is within 0.5% of 1000 kg m^{-3} over all typical conditions on the Earth's continents and is about 8% greater than the density of ice (around 920 kg m^{-3}; Fig. 3-1[a]) but 2.5% less than the density of seawater (typically around 1024 kg m^{-3}). However, the small variations in liquid water density that occur with changes in temperature, pressure, and sometimes salinity are critical in aquatic dynamics. This is because even these small changes in density can lead to stratification of the water column, and, in turn, motions within a stratified water column are strongly restricted to lie in the horizontal (see Chapters 7 and 8).

At atmospheric pressure, the density of pure water increases from 999.843 kg m^{-3} at the freezing point to a maximum of 999.975 kg m^{-3} at 3.98°C (Fig. 3-1[b]) and then decreases at higher temperatures. Water is also nearly, but not quite, incompressible. For every meter of depth increase, the ambient pressure in a water column increases by about 1 dbar from its sea level value of approximately 10 dbar. As this water column pressure increase reaches 100 dbar (typically near depths of 100 m), densities increase by about 0.5 kg m^{-3}. More subtly, the freezing point and temperature of maximum density also decrease to −0.07°C and 3.78°C, respectively, at that pressure.

Usually, water is in contact with the atmosphere and hence contains dissolved air. Air-saturated water has a density that is about 0.005 kg m^{-3} less than that of pure water at 0°C but only 0.002 kg m^{-3} less at 25°C (Harvey et al., 2005). The isotopic composition of water

1 Official descriptions or standards are important for scientific intercomparisons and for summarizing the best available knowledge on a subject and may be necessary for legal reasons when laws are signed or contracts entered into. Standards are developed and/or endorsed by intergovernmental treaty organizations like the International Bureau of Weights and Measures under the guidance of the International Committee of Weights and Measures (known by their French acronyms as BIPM and CIPM, respectively) and nongovernmental professional organizations like the International Association for the Properties of Water and Steam (IAPWS). However, in many cases simpler functions can be used to approximate specific properties with less computational effort than is often needed with the official standards.

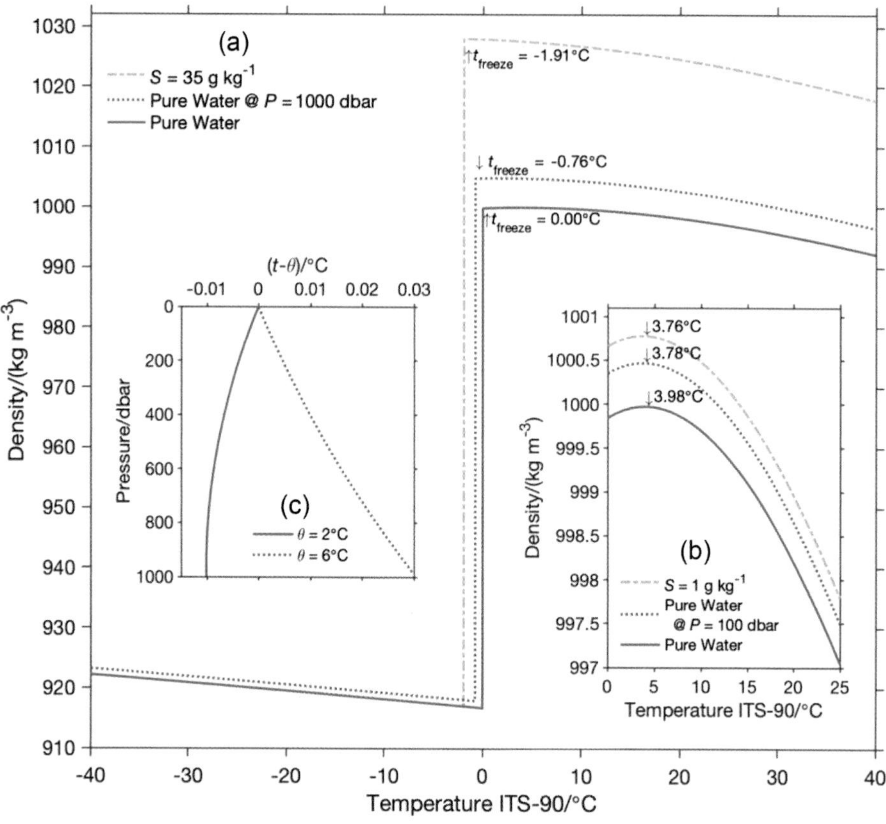

FIGURE 3-1 Density of water. (a) Densities of pure water and ice at atmospheric pressure and at 1000 dbar and of seawater (salinity S of 35 g kg^{-1}) at atmospheric pressure. Freezing temperatures are indicated. (b) Densities of pure water at atmospheric pressure and at 100 dbar, and of very slightly saline water, over warm temperatures only. Temperatures of maximum density are indicated. (c) The difference between *in situ* temperature t and the corresponding potential temperature θ, for water with potential temperatures of 2°C and 6°C. Data for this figure was generated using the GSW software toolbox available at www.teos-10.org.

can also change its density. The pure water values given here are for an isotopic composition found for water in the ocean (i.e., for VSMOW), but this tends to be a little different from the isotopic composition of water in rivers and lakes, which usually has fewer heavy isotopes. Heavy water, containing deuterium (^2H) atoms, has a density of about 1100 kg m^{-3} but the range of density variation in typical water, due to changes in isotopic composition from different natural conditions, is no more than about 0.008 kg m^{-3}. In most situations it is the density differences between different but adjacent water masses that are of interest and so correction factors arising from the presence of gases or differing isotope composition, which do not change much with temperature or location within a drainage basin, are usually ignored.

More importantly, the effect of adding dissolved matter to pure water (i.e., increasing its salinity) usually results in an increase in density. This increase is approximately but not completely linear with the salinity; the density increase for a given increase in the dissolved matter is less at higher salinities. However, the amount of increase may also depend on the particular substances added. The density of seawater increases by about 0.8 kg m^{-3} as concentrations of sea salt increase by a mass fraction of 1 g kg^{-1} (Fig. 3-1[b]), and this is also a reasonable estimate for the variations of density with mass fraction salinity in typical lakes in which anion content is dominated by HCO_3^-. However, if the dominant dissolved anion is CO_3^{2-} or SO_4^{2-}, then the density change can be up to 50% larger for the same salinity change (Pawlowicz and Feistel, 2012). In addition, the temperature of maximum density and the freezing point decrease with increasing salinity. For seawater with a salinity of 35 g kg^{-1}, densities are greatest at the freezing point of −1.91°C.

Finally, due to the nonlinearity of the equation of state, the density of a mixture of waters with different initial temperatures and salinities is greater than the average of the initial densities. There is a slight contraction with mixing, most pronounced in colder waters, which can lead to downward convection (*cabbeling*) in areas where different water masses of the same density (e.g., with temperatures from above and below the temperature of maximum density, or with different salinities and temperatures) come together. Also, as the compressibility of water also depends on temperature and salinity, two water masses with the same density at one depth may have different densities at another depth (the *thermobaric effect*). However, although the thermobaric effect has important consequences in the ocean, it is usually insignificant in its effect in lakes.

ii. Potential temperature and potential density

Not only is water slightly compressible, but its temperature will also change slightly as pressure is increased *isentropically* (i.e., without a loss or addition of heat energy or mass). The same phenomenon is responsible for the cooling temperatures with increasing altitude in the atmosphere, with drops of 4–10°C per 1000 m altitude change (depending on the moisture content of the air), described by the *adiabatic lapse rate*, as pressure drops with altitude. The effect is much smaller in water, however, with changes of only a few hundredths of a degree in temperature for a pressure change of 1000 dbar (i.e., depth change of about 1000 m). Another difference with the behavior of air is that whereas above the temperature of maximum density, the temperature in water increases with pressure as it does in air, below this point, the temperature in water decreases with pressure (Fig. 3-1[c]).

A common practice when the above changes are significant is to use a derived quantity called the *potential temperature*, often denoted by the symbol θ, in order to decide whether one fluid mass has more heat energy than another. The potential temperature of a fluid parcel is the temperature that would occur if the fluid was brought to a standard pressure (usually atmospheric pressure at sea level) isentropically, and thus it removes pressure dependence. *Potential density* is the fluid density at standard pressure, with a temperature given by the potential temperature; it similarly accounts for pressure-related changes when deciding if one water mass is denser than another. Potential temperature and potential density are widely used in studies of the ocean and atmosphere and may be useful in studies of deep lakes where temperatures lie in a narrow range. For water, they are best calculated using TEOS-10 software (IOC et al., 2010).

iii. Specific heat capacity and latent heat

The isobaric heat capacity of water, or *specific heat capacity* (the energy required to raise a unit mass by 1°C or 1 K at constant pressure), for water is about 4200 J kg^{-1} K^{-1} (IOC et al., 2010; see also Fig. 3-2[c]) and is very high

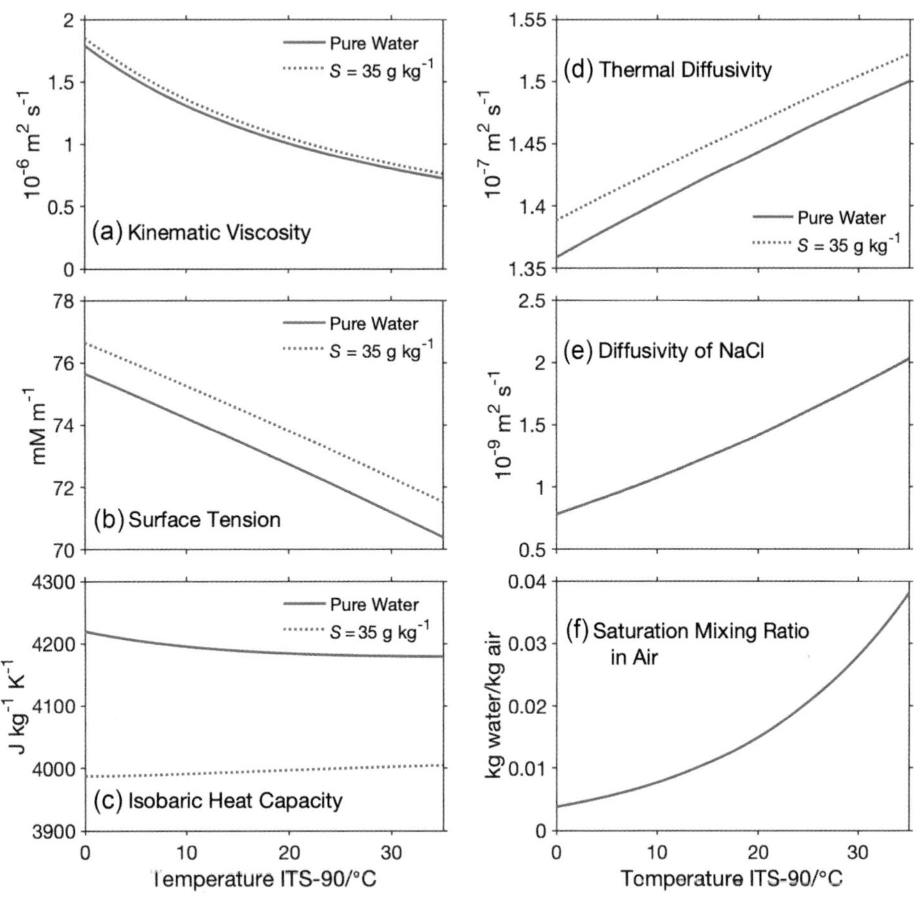

FIGURE 3-2 (a) Kinematic viscosity of pure water and seawater (salinity S of 35 g kg^{-1}). (b) Surface tension of pure water and seawater. (c) Isobaric heat capacity of pure water and seawater. (d) Thermal diffusivity of pure water and seawater. (e) Molecular diffusivity of NaCl in water. (f) The saturation mixing ratio in air at sea level. Much of the data in these figures was generated using the GSW toolbox available at www.teos-10.org (IOC et al., 2010) and the seawater toolbox available at https://web.mit.edu/seawater (Nayar et al., 2016, Sharqawi et al., 2010).

relative to values for most substances. In particular, it is much higher than the heat capacity of rock and soil (typically around 400–1200 J kg^{-1} K^{-1}), air (around 1000 J kg^{-1} K^{-1}), and ice (less than 2100 J kg^{-1} K^{-1}). Thus the daily and seasonal cycle of temperature in natural waters is usually much smaller than that in the air above the water, and in and over the surrounding land surface. In turn, these differences give rise to a wide range of weather conditions where open water exists: daily lake breeze/land breeze effects, milder winters and cooler summers near the water, fogs that appear in the fall, and the apparent "steaming" of the water surface on cold calm days.

The heat capacity of water itself varies slightly with temperature and salinity, so temperature is not conserved under mixing. That is, the temperature of a mixture of waters of different initial temperatures is slightly different from the average of the initial temperatures. However, the effect is small (e.g., the final temperature after mixing equal masses of pure water at 0°C and 20°C is 9.98°C), and for most limnological purposes, temperature can be considered an approximately conservative measure of heat. In the oceans, however, best practice for heat content calculations is to use a thermodynamic property called *enthalpy*, since it is by definition conserved under mixing at constant pressure, or a derived thermodynamic variable defined in TEOS-10 called *Conservative Temperature*, based on potential enthalpy, instead of the measured temperature.

Energy must be absorbed from the environment to convert ice into water (the *latent heat of melting* is about 330 kJ kg^{-1}), and considerably more must be absorbed to convert water into water vapor (the *latent heat of evaporation* is about 2500 kJ kg^{-1}); the same amount of energy is released for phase changes in the opposite direction (IOC et al., 2010). Evaporation occurs when the air above water is undersaturated, with warm air being able to hold considerably more water vapor than cold air (Fig. 3-2[f]). Sudden cooling of moist air can then lead to the formation of fog, clouds, or precipitation in the form of rain or snow. Freezing of surface waters is a complex process in which water molecules form a crystalline solid structure while rejecting the dissolved salts; the resulting brines have a much lower freezing point and can therefore remain liquid even when surrounded by ice. The presence of these *brine pockets* makes the properties of sea ice (and saline lake ice) very different than those of freshwater ice (Weeks and Ackley, 1986).

C. Transport properties

The transport properties of a fluid govern the rate at which quantities within a fluid are transferred from one location to another. The kinematic viscosity, thermal diffusivity, and molecular diffusivities govern the rate of transfer of momentum, temperature, and dissolved substances, respectively. However, the transfer of forces and heat energy are better described by the dynamic viscosity, which is the product of the kinematic viscosity and density, and the thermal conductivity, which is equal to the product of thermal diffusivity, density, and *specific heat capacity*.

i. Viscosity

The *viscosity* of water is the resistance it gives to moving objects. The kinematic viscosity of water decreases as temperature increases, and for pure water, the viscosity at 25°C is about half of its value at the freezing point (Sharqawy et al., 2010; see also Fig. 3-2[a]). Salinity increases the viscosity very slightly.

Although the kinematic viscosity of water is somewhat less than the viscosity of air, its density is very much greater, so viscous drag forces are about 100 times larger in water than in air. Resistance to movement caused by the viscosity of water is generally most important for the dynamics of microscopic particles. This is because moving objects are usually surrounded by a viscous boundary layer less than 1 mm thick; the motions of larger objects are thus more typically controlled by turbulent rather than viscous drag. Viscous drag is the major determinant of the sinking rate of small particles and the rise rate of small bubbles. Viscous drag also affects planktonic life at small scales, and these organisms have developed a variety of ingenious propulsion mechanisms to overcome its effects (Purcell, 1977). At even smaller scales, temperature-dependent changes in viscosity are the largest (although not the only) source of the temperature dependence of electrical conductivity.

ii. Diffusivity

The *diffusivity* of a substance in a liquid measures the degree to which that substance will disperse. In water the *thermal diffusivity* (Sharqawy et al., 2010; see also Fig. 3-2[d]), characterizing the spread of temperature from warm to cold areas (and vice versa), is approximately 100 times larger than the molecular diffusivity of most dissolved salts in water, which are typically within a factor of two or so of that for NaCl (Fig. 3-2 [e]). Unfortunately, it is not possible to precisely define the diffusivity of a mixture of salts with a single number, as the different ions will not diffuse at the same rate. Salt content in lakes and oceans may then fractionate when diffusion is important. Also, it is important to note that molecular diffusivities, which are a property of a particular substance in a fluid, should not be confused with eddy diffusivities, which are a property of turbulent flow with a particular geometry. Eddy diffusivities in turbulent flows are many orders of magnitude greater than molecular diffusivities and identically affect heat and salt.

The large numerical difference between thermal and molecular diffusivities gives rise to a variety of subtle effects in situations where shear-driven turbulence is weak. In water columns where density increases with depth, but the large-scale temperature profile is destabilizing while the salinity profile is stabilizing (or vice versa), the difference in diffusivities can give rise to so-called double-diffusive instabilities (Radko, 2013). These are spontaneously generated instabilities that change the water column into a highly layered state in which property profiles have a characteristic "staircase" appearance with steps in both temperature and salinity. These steps are composed of turbulent, well-mixed layers, from tens of centimeters to meters in height, separated by thinner stratified diffusive layers. Such staircases can sometimes be seen near the bottom of lakes with reasonably high salinities and geothermal heat sources (e.g., Schmid et al., 2010).

iii. Electrical conductivity

If ions are dissolved in a fluid and move under the influence of an applied electric field, the fluid is said to be electrically conductive. The electric field accelerates the positive and negative ions in opposite directions, resulting in a current flow, but the speed of these ions is also limited by viscosity. *Electrical conductivity* is the rate of this flow relative to the strength of the applied field and is typically expressed in units of micro- or millisiemens per centimeter ($\mu S\ cm^{-1}$ or $mS\ cm^{-1}$, see Chapter 12). As an aside, conductivity is a property of the fluid, but instruments that estimate this property typically measure the *conductance* (inverse of electrical resistance) in a measurement cell. Conductance is related to conductivity by a cell constant with units of inverse length, whose magnitude can be determined from the geometry of the cell but is usually quantified by standardization with a solution of known conductivity.

Pure water is very slightly conductive due to small concentrations of dissociated H^+ and OH^- ions. More typically, when in contact with air, the dissolution of CO_2 gas and its reaction with water to produce H^+, HCO_3^-, and CO_32^- ions gives rise to a conductivity of about 1 $\mu S\ cm^{-1}$ in otherwise pure water (Wu et al., 1991; see also Fig. 3-3[a]). A number of small lakes in western North America with very pure natural water have been found to have conductivities of less than 3 $\mu S\ cm^{-1}$, approaching this air-saturated limit (Eilers et al., 1990).

The electrical conductivity of natural waters, however, is usually much higher (10–2000 $\mu S\ cm^{-1}$ for lakes, and even higher for seawater; Fig. 3-3[a]), due entirely to the presence of dissolved salts (see Chapter 12 for more information). The magnitude of the conductivity is roughly proportional to the concentration of dissolved salts in all but the most saline natural waters, but the actual scaling factor that relates the two depends on the relative composition of the salts in question. Extremely precise measurements have been made of this relationship for seawater, whose relative composition is constant to a large degree, so seawater salinity estimates are primarily made through conductance measurements using the *Practical Salinity Scale 1978* (PSS-78; UNESCO, 1981). In lake and river waters, however, the scaling factor can be quite different from that for seawater. The mixture of salts in lakes whose dominant anion is HCO_3^- is only about 60% as conductive as seawater of the same mass fraction salinity. Other approximate relations of this sort are available for different water types (Pawlowicz and Feistel, 2012), and a number of attempts have been made to develop numerical models that relate arbitrary chemical compositions to conductivity (McCleskey et al., 2012), but for highest precision in a specific lake, laboratory measurements are still required to develop a salinity scale for that lake.

Electrical conductivity is also highly dependent on temperature, increasing by about 1.9% for every degree increase in temperature for most salts (see also Chapter 12). This change largely arises from changes in the viscosity (Robinson and Stokes, 1970). However, this temperature dependence is much lower for the H^+ ion, which migrates differently than other ions using a "proton jump" mechanism not really affected by viscosity, so very acidic lakes (with high concentrations of H^+ ions) can have a different temperature sensitivity for their electrical conductivity.

In any case, conductance measurements are only a useful proxy for salinity when combined with temperature measurements. *Specific conductance* is defined by limnologists as a conductance measurement that is both calibrated and corrected for the effects of temperature (usually to a temperature of 25°C) and is then a widely used measure of salinity; as specific quantities usually imply only a normalization to unit mass or volume, a better (but less used) term for this temperature-corrected measurement is *reference conductivity.* Conductance measurements do not, however, account for the presence of nonconductive elements such as silica, or organic matter, which may be a significant component of the dissolved matter in some lakes.

D. Interface properties

i. Surface tension

The intermolecular hydrogen bonding in water, balanced in all directions, is interrupted at the water's surface. Water molecules at the surface are then subject

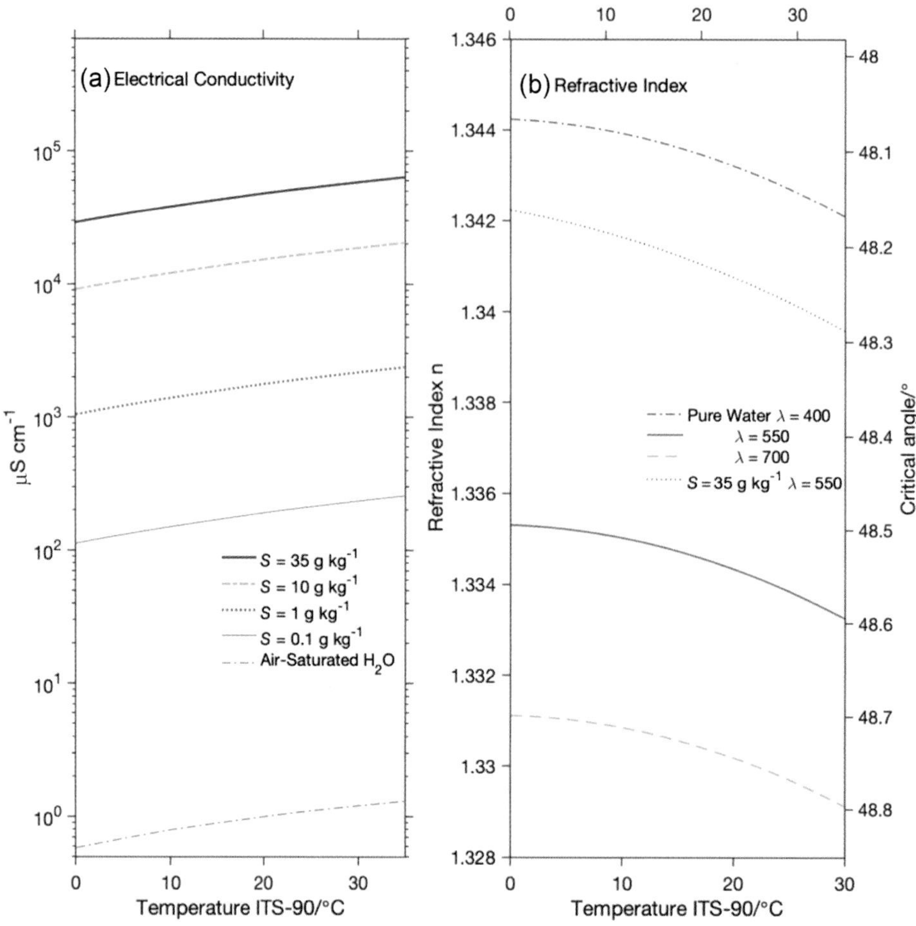

FIGURE 3-3 (a) Electrical conductivity of air-saturated pure water and seawater of different salinities *S*. (b) The refractive index of pure water and seawater, at different wavelengths and salinities (Quan and Fry, 1995). The corresponding critical angle is given on the scale to the right.

to an unbalanced force directed inward toward other water molecules, which increases the internal pressure, and also directed sideways as *surface tension*, acting to reduce the surface area. The surface tension at the air–water interface of pure water is higher than that for any other liquid except mercury. It decreases with increasing temperature and increases slightly with salinity (Nayar et al., 2016; see also Fig. 3-2[b]).

The restoring force from this strong surface tension is the cause of centimeter-scale capillary waves on the water surface; it also quickly damps these waves when the forces that cause them die down. Generation of capillary waves by localized wind gusts roughens the water surface, making the locations of these gusts visible as darker *cat's paws*. It also provides a mechanism for estimating surface wind speed from spaceborne radar systems, which are sensitive to the small wavelengths of capillary waves.

Surface tension is also strong enough that air bubbles of less than about 1 mm in diameter in water are essentially rigid spheres, with pressures higher than ambient inside them. In contrast, larger bubbles have more complex and even time-varying shapes that affect their rise rates in complicated ways (Tomiyama et al., 1998). Since the pressure inside bubbles generated by breaking surface waves is larger than ambient pressure, air–water gas transfer is enhanced and can lead to supersaturation of surface waters. Finally, laboratory studies show that the extremely high pressures within collapsing gas bubbles <1 µm in size in water can cause temperatures within the bubbles to exceed 10,000 K, resulting in the generation of light, a process called sonoluminescence.

In relatively calm areas a variety of organisms (collectively called *neuston*), which would otherwise sink, live on the surface supported by the surface tension (see Chapter 17 for details). However, the surface tension of water can be greatly reduced by the addition of natural or anthropogenic organic compounds. Surface tensions in eutrophic lakes and bogs can be as little as 75% of the pure water values. Decreases in surface tension also decrease the capillary wave roughness, making areas with higher concentrations of these substances visible as slicks when viewed at low angles.

ii. Refractive index

The *refractive index* is a measure of the speed of light in a substance, and it governs the change in the direction of light rays as they propagate through a surface.

Refractive effects in water are responsible for many fascinating and beautiful features visible in, on, and above lakes (Lynch and Livingston, 2001). The most well-known of these include rainbows, the twinkling of stars, and the sun-dappling of lake bottoms through the formation of caustic networks.

When slanted light rays strike the water's surface, some of the energy is reflected, and some enters the water. On entering the water, these slanted light rays refract toward a perpendicular to the surface with angles related to the refractive index using *Snell's Law*; conversely, they refract away from the perpendicular as they leave. Light within water will not be transmitted back to the air if it strikes the surface at angles larger than a critical angle of about 48° (Quan and Fry, 1995; see also Fig. 3-3[b]); instead, it will be reflected back into the water. Refraction in water is also slightly greater at shorter wavelengths, so white light will be split up into different colors if it passes through a water interface at an angle.

The refractive index of pure water is about 1.336 at 25°C, and its variation with temperature and salinity is very roughly proportional to the associated density changes. A change of 1 g kg^{-1} in salinity will have about the same effect on the refractive index as a 10°C change in temperature. Refractometers are therefore a useful method to quickly obtain the salinity of very small quantities of water (a few drops), if salinities are in the seawater range and temperatures are controlled.

Refraction at the water surface makes it difficult to view lake bottoms through a surface wave field; conversely, strong density gradients near the surface, containing many small ripples and turbulent eddies, are sometimes visible to the naked eye by the blurry nature of light upwelling through the associated refractive index variations. Refractive index changes at water surfaces, and/or at changes in density within water, also allow for a variety of flow visualization techniques in laboratory studies of fluid processes.

II. Summary

1. Water is a unique substance, whose unusual properties extensively influence physical, chemical, and biological processes in freshwater ecosystems.
2. Hydrogen bonding and the structured nature of water molecules is responsible for many of the unusual properties of water.
3. The density of freshwater is a maximum at around 3.98°C and is much greater than the density of ice.
4. Additions of dissolved matter increase density, as does increased pressure. Density also increases when water masses mix.
5. The heat capacity of water is far larger than that of land and air, so lake temperatures are more stable than those around them.
6. Viscous drag governs the dynamics of small bubbles and organisms.
7. Electrical conductivity in natural waters is largely governed by salinity, although exact relationships are somewhat dependent on the chemical composition of the dissolved salts.
8. Surface tension at the water's surface is high, capable of supporting an entirely separate ecosystem, but this tension can be greatly reduced by the addition of dissolved organic compounds.
9. The refractive index of water is responsible for many of the beautiful features seen in, on, and above lakes.

References

Eilers, J.M., Sullivan, T.J., Hurley, K.C., 1990. The most dilute lake in the world? Hydrobiologia 199 (1), 1–6.

Harvey, A.H., Kaplan, S.G., Burnett, J.H., 2005. Effect of dissolved air on the density and refractive index of water. Int. J. Thermophys. 26 (5), 1495–1514.

IAPWSR6-95, 2018. Revised Release on the IAPWS Formulation 1995 for the Thermodynamic Properties of Ordinary Water Substance for General and Scientific Use. Available at http://www.iapws.org.

IOC, SCOR, and IAPSO, 2010, The International Thermodynamic Equation of Seawater—2010: Calculation and Use of Thermodynamic Properties. Intergovernmental Oceanographic Commission, Manuals and Guides No. 56, UNESCO (English), p. 196, available at www.teos-10.org (GSW software available at www.teos-10.org).

Lynch, D.K., Livingston, W., 2001. Color and Light in Nature, second ed. Cambridge University Press, p. 292.

McCleskey, R.B., Nordstrom, D.K., Ryan, J.N., 2012. Comparison of electrical conductivity calculation methods for natural waters. Limnol. Oceanogr. Methods 10 (11), 952–967.

Meija, J., Coplen, T.B., Berglund, M., Brand, W.A., De Bièvre, P., Gröning, M., Prohaska, T., 2016a. Atomic weights of the elements 2013 (IUPAC Technical Report). Pure Appl. Chem. 88 (3), 265–291.

Meija, J., Coplen, T.B., Berglund, M., Brand, W.A., De Bièvre, P., Gröning, M., Prohaska, T., 2016b. Isotopic compositions of the elements 2013 (IUPAC Technical Report). Pure Appl. Chem. 88 (3), 293–306.

Millero, F.J., 2001. Physical Chemistry of Natural Waters. Wiley-Interscience, New York, p. 654.

Nayar, K.G., Sharqawy, M.H., Banchik, L.D., 2016. Thermophysical properties of seawater: a review and new correlations that include pressure dependence. Desalination 390, 1–24.

Pawlowicz, R., Feistel, R., 2012. Limnological applications of the Thermodynamic Equation of Seawater 2010 (TEOS-10). Limnol Oceanogr. Methods 10 (11), 853–867.

Preston-Thomas, H., 1990. The International temperature scale of 1990 (ITS-90). Metrologia 27 (1), 3–10.

Purcell, E.M., 1977. Life at low Reynolds number. Am. J. Phys. 45 (3), 3–11.

Quan, X., Fry, E.S., 1995. Empirical equation for the index of refraction of seawater. Appl. Opt. 34 (18), 3477–3480.

Radko, T., 2013. Double-Diffusive Convection. Cambridge University Press, New York, p. 342.

Robinson, R.A., Stokes, R.H., 1970. Electrolyte Solutions, second ed. Butterworth and Co., London, p. 571.

Rozanski, K., Froelich, K., Mook, W.G., 2001. Vol 3: Surface Water. In: Mook, W.G. (Ed.), Environmental Isotopes in the Hydrological Cycle, Principles and Applications, IHP-V Technical Documents in Hydrology, vol. 39. UNESCO, Paris, p. 117.

Schmid, M., Busbridge, M., Wuest, A., 2010. Double-diffusive convection in Lake Kivu. Limnol. Oceanogr. 55 (1), 225–238.

Sharqawy, M.H., Lienhard, V.J.H., Zubair, S.M., 2010. Thermophysical properties of seawater: a review of existing correlations and data. Desalination Water Treat. 16, 354–380.

Tanaka, M., Girard, G., Davis, R., Peuto, A., Bignell, N., 2001. Recommended table for the density of water between 0°C and 40°C based on recent experimental reports. Metrologia 30, 301–309.

Tomiyama, A., Kataoka, I., Zun, I., Sakaguchi, T., 1998. Drag coefficients of single bubbles under normal and micro gravity conditions. JSME International Journal Series B Fluids and Thermal Engineering 41 (2), 472–479.

UNESCO 1981, Background Papers and Supporting Data on the Practical Salinity Scale 1978, UNESCO Technical Papers in Marine Science vol. 37, Paris.

Weeks, W.F., Ackley, S.F., 1986. The Growth, structure, and properties of sea ice. In: Untersteiner, N. (Ed.), The Geophysics of Sea Ice. NATO ASI Series (Series B: Physics). Springer, Boston, p. 1195.

Wu, Y.C., Koch, W.F., Pratt, K.W., 1991. Proposed new electrolytic conductivity primary standards for KCl solutions. Journal of Research of the National Institute of Standards and Technology 96 (2), 191.

CHAPTER

4

Rivers and Lakes—Their Distribution, Origins, and Forms

Bernhard Lehner[1]

[1]Department of Geography, McGill University, Montreal, Quebec, Canada

OUTLINE

I. Distribution of inland surface waters

The amount of terrestrial surface water on Earth is very small in comparison to the water of the oceans, but inland waters have much more rapid renewal times. On a volumetric basis, terrestrial surface water is concentrated in the deep waterbodies of several very large lakes. The saline Caspian Sea forms the largest lake on Earth with a volume of 75,600 km^3, representing 41% of global lake volume, followed by Lake Baikal of Siberian Russia, the largest freshwater lake, which contains another 13%. The number of individual depressions of smaller lakes and reservoirs, however, is extremely large and most of these lakes are located in temperate and subarctic regions of the Northern Hemisphere. Most lakes are shallow, and their form is quite transient; i.e., they are continuously changing their shape over short or long timescales.

Large-scale events in past geologic epochs, driven largely by glacial, tectonic, and volcanic activities, have aggregated many lakes into lake districts. The morphometry and geological substrata of the drainage basins are very important because these properties influence the formation of interconnected river and lake

Wetzel's Limnology, Fourth Edition
https://doi.org/10.1016/B978-0-12-822701-5.00004-5

ecosystems, their sediment–water interactions, and the resultant productivity. Shallow lakes (see Chapter 26), which have more sediment area per unit of water volume, are generally more productive than deep lakes, and a greater proportion of their total productivity is attributable to communities of the littoral and wetland areas at the land–water interface.

Water presence on the Earth's terrestrial surface is highly variable in space and time, both seasonally and in the long term. The consistent acquisition of satellite imagery since the early 1980s has revealed how lakes and river networks expand, contract, and shift with seasons, climatic changes, and human activities. Current estimates indicate that rivers and lakes cover less than 3% of the Earth's terrestrial surface area, with a combined total of about 4.0×10^6 km^2 (Allen and Pavelsky, 2018; Messager et al., 2016). Their extent is fluctuating due to natural and anthropogenic causes; for example, from 1984 to 2015, about 14% of mapped permanent waterbodies were spatially or temporally variant (Pekel et al., 2016). Further adding wetlands, floodplains, and transitional systems, the total extent of terrestrial surface water may exceed 12.1×10^6 km^2 (Davidson et al., 2018), representing more than 8% of the Earth's land surface.

Rivers, covering an estimated area of 0.8×10^6 km^2 (Allen and Pavelsky, 2018), form large networks that dissect continents with a great geomorphic and hydrologic diversity of characteristics from mountain headwaters (see Chapter 5) to ocean outlets. Their width and depth are far more dynamic than those of lakes as they are more responsive to the hydrological cycle and its seasonality, and it has been estimated that more than half of all rivers and streams cease to flow at some point during the year (Messager et al., 2021). Given their importance for water supply, most human settlements are located along the meandering path of a river.

Lakes comprise the majority of the world's permanent inland surface waters, with an estimated area of 3.2×10^6 km^2 (lakes $\geq$1 ha; Messager et al., 2016). Most are located above 40°N in the Northern Hemisphere and can be categorized by various types of origin. As part of this coverage, humans have created about 0.3×10^6 km^2 of artificial reservoir surface area, mostly over the past century (Lehner et al., 2011).

A. Rivers and streams

Rivers and streams are estimated to cover an area of 773,000 ± 79,000 km^2 (Allen and Pavelsky, 2018), which represents about 0.6% of the Earth's land surface (excluding Greenland and Antarctica). All river channels combined store a total of about 2000 km^3 of water, which represents less than 0.005% of all freshwater stocks on Earth (Oki and Kanae, 2006). Despite these low quantities, running waters are of enormous significance for aquatic ecosystems and play a key role in biogeochemical cycles. Erosion and sediment transport in rivers move large amounts of dissolved and particulate matter from the land to the sea, estimated at about 19 billion tonnes annually (Milliman and Farnsworth, 2011). Rivers also provide a range of critical services that directly benefit humans, such as energy for hydropower generation, transport pathways, and water provision, as most water that is withdrawn for human use originates from rivers.

The global network of all rivers and streams with a long-term mean annual flow of at least 100 L s^{-1} (0.1 m^3 s^{-1}) has been mapped to exceed 24 million km of total stream length (Fig. 4-1) (Linke et al., 2019). Small streams and rivers often flow from high-elevation mountainous areas, feed into lakes, and eventually converge to form large river systems. Seven major river basins exceed 1 million km^2 in drainage basin area and reach or exceed 5000 km in mainstem length (Amazon, Congo, Mississippi–Missouri, Nile, Ob–Irtysh, Yangtze, and Yenisei–Angara) (Table 4-1). Among them, the Amazon has by far the highest river discharge, with a mean annual flow of more than 200,000 m^3 s^{-1}; even its main tributaries exceed the size of most other global rivers. The ecosystem processes, biodiversity, and services, however, that the world's largest rivers support are threatened by anthropogenic impacts on their connectivity, mostly caused by dam and reservoir construction (see section 4-VI below). Only 37% of rivers longer than 1000 km remain free-flowing over their entire length and only 23% flow uninterrupted to the ocean (Grill et al., 2019).

B. Lakes and reservoirs

The total number of lakes on Earth is not exactly known but has been estimated to be around 21 million, considering lakes with an area of at least 1 ha (0.01 km^2); however, only <1% of them are larger than 1 km^2 (Messager et al., 2016). When including even smaller lakes and ponds, estimates reach between 117 million (Verpoorter et al., 2014) and 304 million (Downing et al., 2006). More recently, humans have built a substantial number of reservoirs and artificial ponds (see section 4-VI below).

Most of the world's lakes were formed by glacial, tectonic, or volcanic processes, and thus many lakes are organized in lake districts that comprise waterbodies of similar age and origin. Glacial activity during the most recent Pleistocene period of major ice advance and retreat, ending ~10,000 years ago, created millions of small depressions that subsequently filled with water. Consequently, large numbers of lakes are found in regions of the Northern Hemisphere, where the land masses of northern North America, Europe, and Asia

FIGURE 4-1 Global river network drawn from the RiverATLAS v1.0 database (Linke et al., 2019). The distribution of river surface area by latitude (*right panel*) is derived from data provided by Allen and Pavelsky (2018) using 1-degree bins.

interfaced with glacial movements. In particular, countless small, shallow lakes are found throughout the Arctic, subarctic, and northern temperate zones (Fig. 4-2 and Table 4-2).

While lakes can be grouped according to their origin and formation (see section 4-III below), they can also be distinguished based on their salinity (see Chapter 12) into fresh or salt water. Most inland (i.e., athalassic) salt or saline lakes occur in landlocked depressions without a surface outlet, also called terminal or *endorheic* lakes, which can mainly be found in semiarid and arid regions. In endorheic lakes, dissolved salts and minerals enter the lake through inflowing river or groundwater fluxes and accumulate as the water evaporates from the lake's surface. Over long timescales, this process increases the salt concentration in the lake. Many endorheic lakes that occur in arid climates are ephemeral, forming large salt flats that only episodically fill with a shallow layer of water after heavy rainfall events (e.g., Kati Thanda—Lake Eyre, Australia). Other, *exorheic* saline lakes exist due to geologic reasons (e.g., highly saline groundwater influx) or along the marine coastline. Saline lakes can be classified according to their salinity (in grams salt per liter), ranging from slightly saline or brackish to saline and hypersaline (see, e.g., Williams and Sherwood, 1994), although there are different conventions on what values are used to separate out different classes (see Chapter 12).

The total water volume stored in global natural lakes that exceed 1 ha in surface area has been estimated to be 182,900 km^3 (Messager et al., 2016), which constitutes about 94% of all nonfrozen surface inland water stocks on Earth (~194,000 km^3; Oki and Kanae, 2006). The majority of this water, however, is contained in a small number of very large lakes (Figs. 4-3 and 4-4). The Caspian Sea, a saline lake, is the largest inland waterbody, both in terms of surface area (377,000 km^2) and volume (75,600 km^3). While Lake Superior (81,800 km^2) and Lake Victoria (67,200 km^2) are the largest freshwater lakes in terms of surface area, the most freshwater volume is contained in Lake Baikal (23,600 km^3) and Lake Tanganyika (18,900 km^3) (Table 4-2 and Fig. 4-4). The Laurentian Great Lakes of North America, namely, Lakes Superior, Michigan, Huron, Erie, and Ontario, constitute a collective volume of 22,600 km^3 and an area of 244,100 km^2, making them the largest contiguous area of nonfrozen freshwater in the world. Lake Vostok, the largest known subglacial lake on Earth, lies approximately 4 km under the surface of the East Antarctic Ice Sheet and is estimated to be the sixth largest lake in the world by volume (~5000 km^3). About 350 lakes exceed 500 km^2 in surface area and contain approximately 172,800 km^3 of water, i.e., 95% of all lake volume. Of these, ~20% are saline lakes (including lagoons), and ~25% are human-made reservoirs. About 90% of all lakes occur north of 40°N, and more than two-thirds of the world's lakes are located in North America (Table 4-2), while other areas of high lake densities can be found in Scandinavia and Siberia (Fig. 4-2).

TABLE 4-1 Characteristics for Selected Major River Drainage Basins, by Continent. All data from the HydroATLAS v1.0 database (Linke et al., 2019) except where stated otherwise. Additional hydro-environmental variables are available for all catchments and rivers of the world in HydroATLAS

River basin (by continent)	Main channel length[a] (km)	Drainage area[b] (10^3 km^2)	Mean annual discharge (m^3 s^{-1})	Limnicity[e] (%)	Forest extent (%)	Cropland extent (%)	Population as of 2010 (10^6)
Africa		**29,950**	**130,100[c]**	**0.9**	**22**	**7**	**1041.9**
Congo	4960	3710	39,600[c]	1.4	71	4	86.1
Niger	4250	2120	6900[c]	0.2	9	19	100.2
Nile	6850	3060	2800[d]	3.1	9	11	192.9
Orange	2540	980	310[c]	0.2	0	8	15.4
Zambezi	3150	1380	3900[c]	2.8	48	6	39.0
Asia and Middle East		**28,500**	**256,800[c]**	**1.0**	**17**	**17**	**3821.2**
Ganges–Brahmaputra	3350	1580	40,400[c]	0.4	17	39	636.6
Indus	3370	1170	5600[d]	0.4	9	35	244.5
Mekong	4850	780	15,300[c]	0.8	22	18	53.1
Tigris–Euphrates	3050	940	1800[d]	1.0	1	19	57.0
Yangtze	5920	1920	31,100[c]	1.1	36	24	481.0
Europe and Russia		**23,260**	**192,100[c]**	**2.1**	**41**	**14**	**741.2**
Danube	2950	800	6700[c]	0.6	39	37	79.1
Lena	5080	2450	15,800[c]	0.5	72	0	1.1
Ob–Irtysh	5490	3090	13,800[c]	1.9	47	11	27.7
Rhine	1270	160	2500[c]	1.0	43	23	49.6
Volga	3830	1400	8100[c]	1.8	47	30	60.0
North America		**22,110**	**223,800[c]**	**5.8**	**35**	**13**	**542.3**
Colorado	2600	660	580[d]	0.3	33	2	11.3
Mackenzie	4300	1800	9200[c]	9.6	36	2	0.5
Mississippi–Missouri	6020	3240	16,800[d]	0.9	27	38	78.6
Rio Grande	2750	850	100[d]	0.2	18	6	17.7
St. Lawrence	2990	1050	12,300[c]	27.5	51	17	46.5
South America		**17,950**	**382,000[c]**	**0.8**	**47**	**6**	**400.0**
Amazon (excl. Tocantins)	5990	5910	205,600[c]	0.4	80	2	29.7
Magdalena	1580	260	7200[c]	1.6	18	8	30.6
Orinoco	2940	940	37,100[c]	0.6	48	3	14.3
Parana–Paraguay	3860	2650	18,300[c]	0.8	31	13	85.1
Sao Francisco	2720	640	2700[c]	0.7	16	7	14.8
Oceania		**11,190**	**171,200[c]**	**0.6**	**26**	**9**	**377.6**
Murray–Darling	2830	1060	770[d]	0.3	17	13	2.1
Global		**135,010**	**1,368,900[c]**	**1.9**	**30**	**11**	**6924.2**

[a]*Rounded. Calculated from ocean outlet to most distant location on watershed divide, independent of river naming conventions.*

[b]*Rounded. Drainage area may differ from other sources due to inclusion/exclusion of dryland regions and/or tributaries near or at the ocean delta.*

[c]*Rounded. Calculated using results from the global hydrological model WaterGAP for the years 1971–2000, assuming naturalized flow conditions (for details, see Linke et al., 2019); confirmed to be within similar ranges in other sources.*

[d]*Compiled from a variety of sources, scholarly articles, or independent reports (including online).*

[e]*Calculated as percentage of total lake surface area within total basin area.*

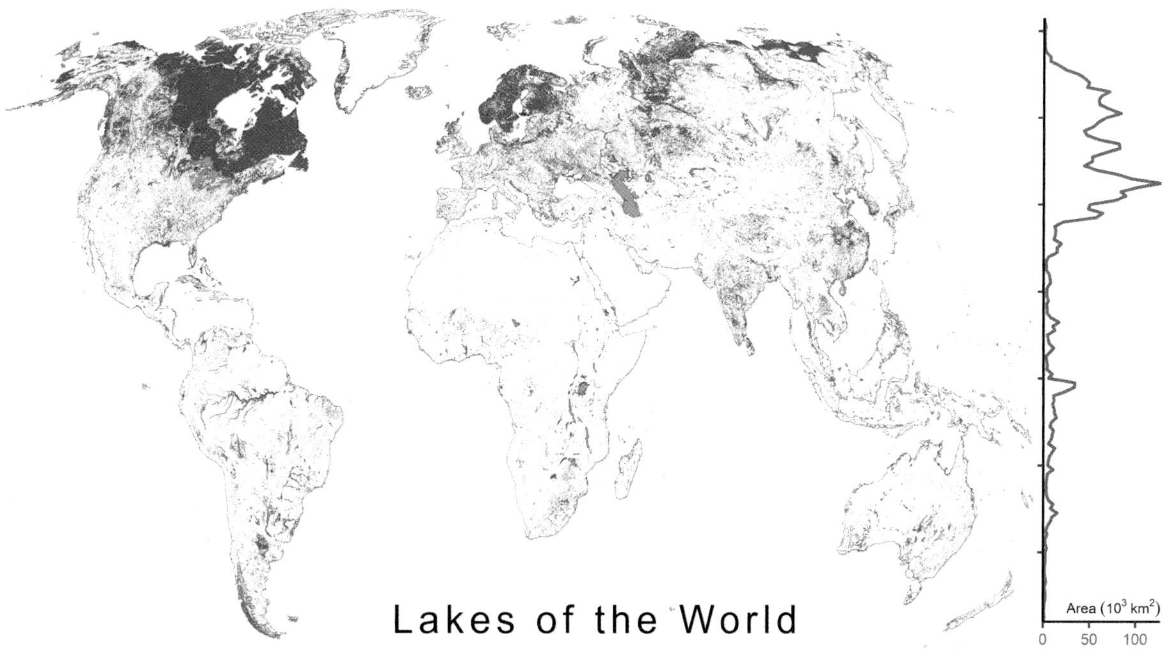

FIGURE 4-2 Global distribution of lakes ≥10 hectares drawn from the HydroLAKES v1.0 database (Messager et al., 2016). The distribution of lake surface area by latitude (*right panel*, including reservoirs) is derived using 1-degree bins. Note that regions of the highest lake density (in parts of Canada, Scandinavia, and northern Siberia) visually appear as contiguous lake areas due to the very high number of mostly small lakes that are individually drawn on the map.

About 35 lakes are known to be extremely deep, in excess of 400 m. Lake Baikal is the deepest lake on Earth, with a maximum depth of 1620 m. Nearly all extremely deep lakes are of tectonic or volcanic origin or have formed from fjords that have subsequently become fresh. Although a few lakes of glacial origin, such as the Laurentian Great Lakes and Great Slave Lake of Canada, reach great depths, few exceed 300 m. Most of the deep lakes are found in mountainous regions along the western portions of North and South America, of Europe, and of central Africa and Asia. Lakes Baikal and Tanganyika are the only lakes known to have maximum depths in excess of 1000 m and mean depths of over 500 m (Table 4-2).

Numbers and size estimates of very small lakes and ponds are typically extrapolated using statistical techniques and scaling laws (Fig. 4-5). Most of both natural and artificial lakes are small and relatively shallow, usually <20 m in depth (Messager et al., 2016). A vast number of lakes in the subarctic regions are particularly shallow, with mean depths of about a meter. In addition, over half of the total global lake shoreline length of 7.2×10^6 km for lakes ≥10 ha is found around the smallest lakes in the range of 10−100 ha (Table 4-2). Because of these morphometric characteristics of small and shallow lakes, a large proportion of their water volume is exposed to and interacts with the chemical and metabolic processes of soils, sediments, and coastlines; and their ability to develop sessile littoral flora generally results in markedly increased overall productivity.

II. River and stream characteristics

Rivers are watercourses that flow over the land surface, eventually entering lakes or other rivers at confluences. River systems are formed by a network of connected tributaries that drain toward a common mainstem and typically discharge into the ocean. Some rivers end at an inland depression, often represented by an endorheic lake. In arid or karst regions rivers can terminate without reaching a confluence or endorheic lake if their waters evaporate, infiltrate into the ground, or enter sinkholes or caves to continue underground.

The water flowing through rivers is originally provided by precipitation that falls over the drainage basin, which then, depending on environmental conditions, may transition through melting, soil infiltration, and groundwater recharge processes before entering the river channel as land surface runoff or groundwater baseflow (see Chapter 5). Affected by diverse hydrologic and geomorphologic influences, the world's rivers display varied physical, chemical, and biological characteristics.

TABLE 4-2 Lake Characteristics by Surface Area Size Class and by Continent. All data from the HydroLAKES v1.0 database for lakes ≥10 ha (Messager et al., 2016)

Spatial unit	Number of lakes (10^3)	Total surface area (10^3 km^2)	Total volume (10^3 km^3)	Total shoreline length (10^3 km)	Mean depth (m)	Mean residence time (years)
By surface area size class (km^2)						
0.1–1	1241.2	348.4	1.3	3637.6	3.5	4.4
1–10	165.1	411.0	2.5	2022.3	5.4	8.2
10–100	13.4	331.6	3.8	862.4	10.4	17.5
100–1000	1.22	313.5	6.5	412.3	19.6	66.1
1000–10,000	0.115	313.3	10.0	158.5	32.3	81.4
>10,000	0.018	959.1	157.8	79.0	139.6	102.4
By continent						
North America[a]	991.9	1229.5	36.6	4990.9	3.7	4.5
Europe[b]	280.7	781.2	103.8	1264.5	4.6	7.4
Asia[c]	66.2	274.8	7.3	391.7	2.9	5.9
South America	53.8	103.7	3.1	296.7	3.3	1.6
Africa	15.2	232.0	30.6	120.1	2.5	3.5
Oceania[d]	13.2	55.7	0.4	108.3	2.8	7.7
Global	**1421.0**	**2676.9**	**181.9**	**7172.2**	**3.8**	**4.9**
Largest lakes by volume						
Caspian Sea		377.0	75.6	15.8	201	295.4
Lake Baikal		32.0	23.6	2.7	739	374.6
Lake Tanganyika		32.8	18.9	2.1	577	402.6
Lake Superior		81.8	12.0	5.2	147	132.5
Lake Malawi		29.5	7.7	1.7	261	218.5
Lake Michigan		57.7	4.9	2.9	84	82.0
Lake Huron		59.4	3.6	8.9	60	12.3
Lake Victoria		67.2	2.6	7.4	41	50.4
Great Bear Lake		30.5	2.2	5.3	72	130.3
Kara–Bogaz–Gol		18.7	1.9	1.0	101	n.a.

[a]*Includes Mexico, the Caribbean, and Central America.*
[b]*Includes all of Russia.*
[c]*Includes Middle East and Turkey.*
[d]*Includes Australia, New Zealand, Micronesia, and Polynesia.*

Many expressions exist for smaller rivers, including streams, brooks, and creeks. Following natural scaling laws, the number and combined length of small streams exceed those of large rivers (Table 4-3). Their location within the upstream (headwater) parts of the drainage basin is reflected in their limited flow quantity, drainage basin area, and high average elevation. While more streams are generally found in regions of higher humidity (Fig. 4-1), stream length and density dynamically adjust when drainage basins become wetter or drier over seasons and precipitation events; i.e., river networks extend and retract as stream channels alternate between flowing and drying phases (Prancevic and Kirchner, 2019).

A. Running waters: lotic ecosystems

Rivers and streams within their drainage basins are central to surface water ecosystems. A number of characteristics differentiate running waters from lake

FIGURE 4-3 Comparison of the surface areas of selected lakes of the world, all drawn to the same scale; their location is marked on the center global inset map. *Red colors* indicate loss of surface area; *blue colors* indicate gain in surface area over the time period 1984–2019 (data from Pekel et al., 2016). The Aral Sea has experienced catastrophic reductions in area because of diversions of water for agriculture. The Amazon River is drawn for size comparison. Lake names in *italics* indicate human-made reservoirs. Lake Vostok (Antarctica) is a subglacial lake. (Design after Ruttner, 1963.)

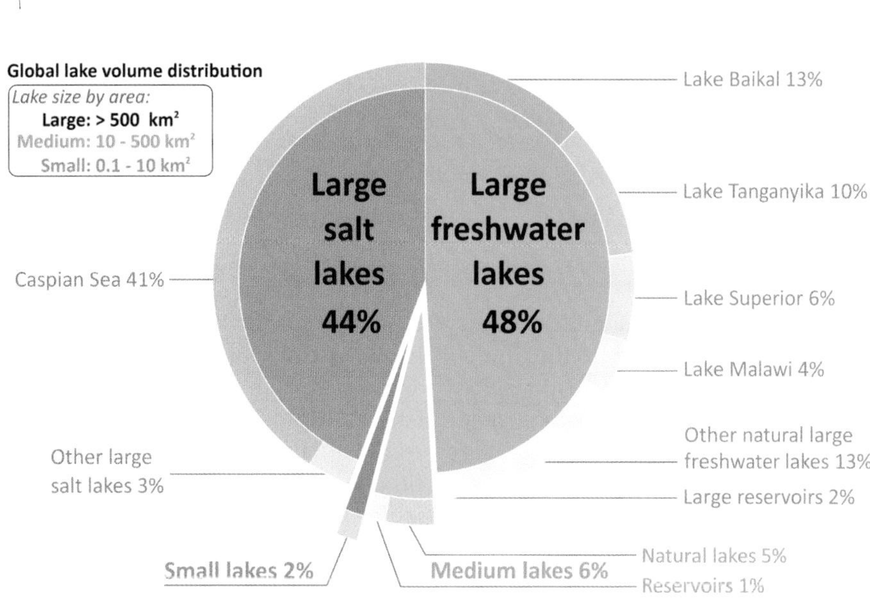

FIGURE 4-4 Global distribution of water volume stored in lakes and reservoirs with a surface area of at least 10 ha. Total volume is 187,900 km^3. *(From Messager et al., 2016.)*

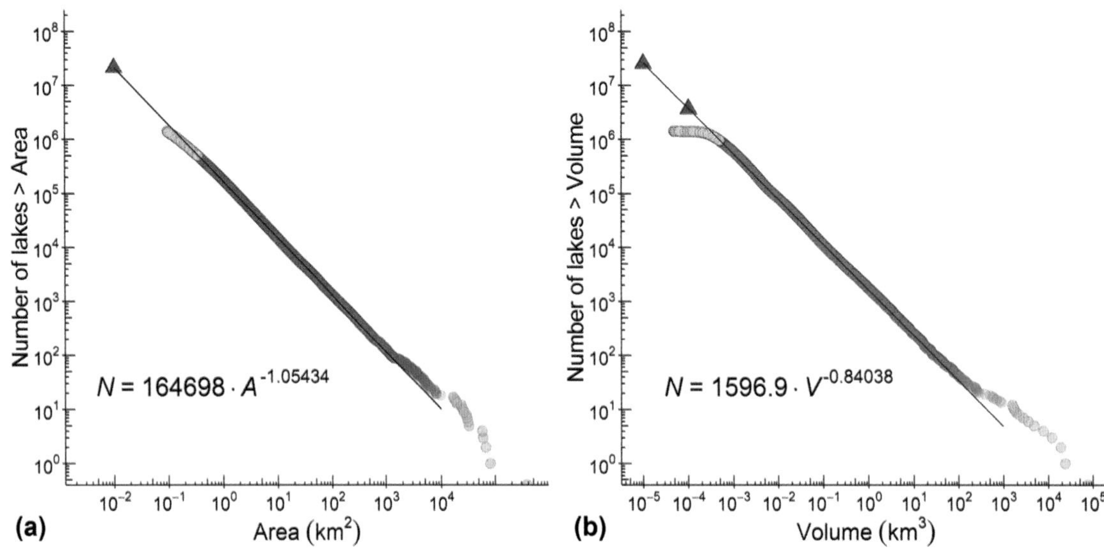

FIGURE 4-5 Global size and volume distributions of natural lakes using a Pareto model. Distributions are plotted as the total number of lakes larger than a given surface area (a) or volume (b) derived from data contained in the HydroLAKES v1.0 database (Messager et al., 2016). *Yellow points* represent data that were not included for fitting the log-log regression. *Red triangles* represent extrapolated values based on the log-log regression. *(From Messager et al., 2016.)*

TABLE 4-3 Drainage Network Characteristics for River Reaches From the RiverATLAS v1.0 Database (Linke et al., 2019). River reaches are here defined as sections between two confluences, or between the initiation of a stream and the first confluence. RiverATLAS contains a total of 6.3 million river reaches with a mean annual discharge ≥ 0.1 m^3 s^{-1} (100 L s^{-1}), representing 24.2 million km of total global river length

Mean annual discharge (m^3 s^{-1})	Number of river reaches	Total river length of all reaches (km)	Average upstream drainage area of all reaches (km^2)	Average elevation of all reaches (m.a.s.l.)	Percentage of nonperennial reaches[a] (% length)
0.1–1	3,883,222	15,151,005	90	604	47
1–10	1,526,711	6,226,902	629	519	35
10–100	584,019	1,917,905	4858	445	26
100–1000	211,459	665,082	36,929	345	9
1000–10,000	53,294	169,470	317,851	223	1
10,000–100,000	8078	24,802	1,239,182	68	0
Exceeding 100,000	360	1319	4,589,679	7	0

[a]*Nonperennial reaches are defined as those that cease to flow (i.e., discharge equal to zero) at least 1 day per year on average (model estimates; Messager et al., 2021).*

ecosystems. Flowing freshwater environments are called *lotic* (*lotus*, from *lavo*, to wash), which refers to the unidirectional water movement along a slope in response to gravity. Lotic ecosystems are contrasted to *lentic* (*lenis*, to make calm) or lake ecosystems.

The directional movement of water is a fundamental property of lotic ecosystems. In fact, the importance of variable and typically rapid throughput of water and materials contained in it is evident in the biology of most organisms inhabiting running waters. Dissipation of energy from moving masses of water affects the morphology of streams, sedimentation patterns, water chemistry, and the biology of organisms inhabiting them (Fig. 4-6). When the energy of flowing water is abruptly dissipated, as in the transitional zone between a river and lake, the change from lotic to lentic characteristics is rapid.

Most lakes are open and have distinct flows (or fluxes) into, through, and out of them, which can be variable and are often low in comparison to the lake's water volume. Thus a main distinction between running waters and lakes is focused on the relative *residence times* of their waters (also called turnover time, flushing time, renewal time, or water age, i.e., the mean time it takes to replace all water in the waterbody) or the related *water renewal rates* (percent water volume that is replaced

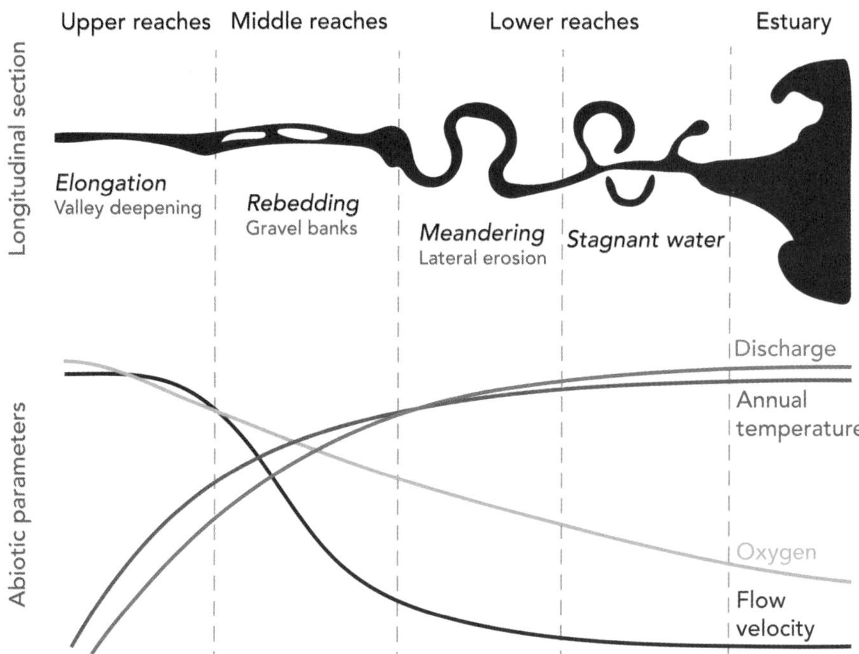

FIGURE 4-6 Sections of different fluvial geomorphologic characteristics and corresponding abiotic parameters along a river course from upper reaches to coastal outflow. *(Modified from Niemeyer-Lüllwitz and Zucchi, 1985.)*

per unit time). Residence time is defined as the ratio between storage volume and mean throughflow and is generally much shorter in rivers than in lakes, where it can exceed 100 years in large lakes (Table 4-2).

Not all rivers and streams show continuous surface flow throughout the year. Water ceases to flow for at least 1 day per year along 51%–60% of the world's rivers by length (Messager et al., 2021; see Table 4-3 for rivers $\geq$0.1 m^3 s^{-1}). The frequency, duration, and timing of flow cessation and resumption are the main determinants of the ecology and biogeochemistry of those nonperennial lotic ecosystems. In some cases surface water persists in pools of standing water. But without flow resumption, water eventually evaporates or infiltrates and river channels can fall dry for extended periods of time. Most rivers and streams in semiarid and arid landscapes are nonperennial, but flow cessation is also common in small streams across more humid climates.

B. Drainage basin characteristics

The *drainage basin* (or drainage area) is the area of land drained by tributary streams that coalesce into a main channel. Note that whereas the general term drainage basin is used here to describe any size of drainage area (i.e., from small headwater basins to large river basins), hydrologists commonly use the equivalent expressions *catchment* (see Chapter 5) or *watershed* (mostly in North America) when referring to small- or medium-sized drainage basins. Usually, surface features (topographic divides) are used to define the extent of drainage basins, although subsurface flows may have different boundaries (phreatic divides), particularly in areas underlain with relatively soluble or permeable rocks.

The world's largest drainage basin is formed by the Amazon River and its tributaries, reaching an area of about 6 million km^2 (Table 4-1). Drainage basins can be subdivided hierarchically into subbasins, where each tributary forms its own nested drainage basin. The continual downgradient movement of water, dissolved substances, and suspended particles in streams and rivers is derived primarily from the land area draining into the stream channel. The hydrological, chemical, and biological characteristics of a stream or river thus reflect the climate, geology, and vegetational cover of its drainage basin. These specific characteristics can be used to highlight similarities and differences between river systems and classify them into river types (e.g., Ouellet Dallaire et al., 2019).

The spatial and temporal patterns of streamflow (or *discharge*) are governed by the hydrological cycle (see Chapter 5). Discharge (Q) is defined as the volume of water passing through the cross-sectional area of a stream channel per unit time and can be calculated as:

$$Q = Av$$

where Q = discharge in m^3 s^{-1}, A = cross-sectional area of channel in m^2, and v = mean velocity in m s^{-1}. Discharge can vary greatly from season to season and is directly related to the extent of the drainage basin and its characteristics. Given this relationship, landscape characteristics within the drainage basin are often used as initial descriptors of possible hydrological behavior, including variables of physiography (e.g., slope), climate, land cover, soils, and geology, as well as geometric features of the drainage basin, such as elongated versus compact shapes (Tables 4-1 and 4-4). The length of the main channel (L_c, Table 4-4) increases in a relatively constant way with drainage basin area (A) as:

$$L_c = jA^m$$

where j and m are derived from many measurements and average about 1.4 and 0.6, respectively. Thus, as drainage basins increase in size, they tend to elongate (Leopold et al., 1964).

C. Drainage network patterns

Several methods have been used for ordering the tributary streams in a drainage network. In the *Classic stream order system* (also called *Hack's* or *Gravelius' stream orders*) order 1 refers to the mainstem (or "backbone") river that drains into the ocean, and every tributary is numbered one order greater than the respective river or stream into which they discharge. Other systems have been developed that rank streams from smallest (first order) to largest (highest order), including those by Horton (1945), Strahler (1957), and Shreve (1966) (see also Chapter 5). In the most widely used *Strahler method* (Strahler, 1957) the smallest permanent stream is designated as the first order, and the confluence of two first-order streams creates a second order (Fig. 4-7). The order then sequentially increases by one when two tributaries of equal order meet. Using this metric, the number and total length of streams within a drainage basin can be statistically estimated (Wetzel and Likens, 2000). Deficiencies, such as scale dependencies, in this popular method of stream ordering, as well as alternatives, are reviewed by Gordon et al. (1992). More recently, scale-independent ordering systems have been proposed that assign stream orders based on mean annual discharge (see Linke et al., 2019) (Table 4-3).

Many patterns of channel networks are observed (Fig. 4-8). These patterns depend on geologic structure, rock types, tectonic activity (presence of faults), landform shapes, and exposure to climate and vegetation. Dendritic channel networks typically develop in relatively homogeneous geology or where geology exerts less control, while nondendritic drainage patterns (e.g., trellised) occur in areas underlain by different lithologies and tectonic activity.

D. Channel morphology

The trough in the landscape, usually within a valley, that contains the flowing water is the stream or river *channel*. The channel is described physically in terms of length, width, depth, cross-sectional area, slope, aspect, and other parameters (Table 4-4; see Leopold et al., 1964; Rosgen, 1996; Wetzel and Likens, 2000). The channel is typically bordered on one or both sides by a flat area called the *floodplain*. Much of the soil of the floodplain is connected hydrologically to the water of the channel.

River channels can contain low to moderately high discharges without overflowing the banks. A *bankfull*

TABLE 4-4 Drainage Basin and River Channel Characteristics[a]

	Definitions of topographic characteristics
Basin length (L_B)	Straight-line distance from basin outlet to the point on the drainage divide used to determine the main channel length, L_c.
Basin width (W_B)	Mean width of the basin determined by dividing the area, A, by the basin length, L_B: $W_B = A/L_B$
Basin perimeter (P_B)	The length of the line that defines the (surface) drainage divide of the basin.
Basin land slope (S_B)	Average land slope (hillslope) calculated at points uniformly distributed throughout the basin. Typically calculated from high-resolution digital elevation models.
Basin diameter (B_D)	The diameter of the smallest circle that will encompass the entire drainage basin.
Basin shape (SH_B)	A measure of the shape of the basin computed as the ratio of the length of the basin, L_B, to its mean width, W_B: $SH_B = \frac{(L_B)^2}{A}$
Compactness ratio (CR_B)	The ratio of the perimeter of the basin to the circumference of a circle of equal area, computed from A and P_b as follows: $CR_B = \frac{P_B}{2(\pi A)^{1/2}}$
Main channel length (L_C)	The length of the main channel from the mouth to the drainage divide.
Main channel slope (S_C)	An index of the slope of the main channel, computed from the difference in stream-bed altitude between the point at 10% (E_{10}) and the point at 85% (E_{85}) of the distance along the main channel from the mouth to the drainage divide: $S_C = \frac{(E_{85} - E_{10})}{0.75L_C}$
Sinuosity ratio (P)	The ratio of the main channel length, L_C, to the basin length, L_B: $P = \frac{L_C}{L_B}$

[a]*Modified from Winter (1985).*

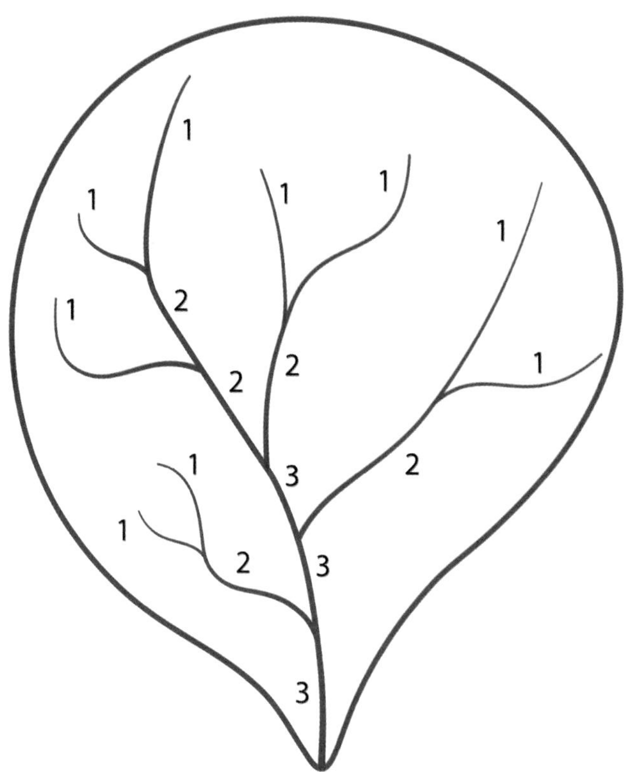

FIGURE 4-7 Stream ordering according to the Strahler method. *(Modified from Wetzel and Likens, 2000.)*

discharge is a flow that fills entirely the cross-sectional stream channel without overflowing onto the floodplain. The bankfull stage, namely, the water level at which the river begins to spill into the floodplain, is equaled or exceeded on average once every 1.5 years (Leopold et al., 1964). Therefore discharges are usually much smaller than at the bankfull stage.

River channels rarely flow over an even downward gradient and are never straight over an appreciable distance. As the energy of flowing water is expended against irregularities in channel morphology, with time, there is a tendency to erode the channel to produce a more-even gradient and a more-uniform dissipation of potential energy per unit length of stream. The bending and *meandering* of a mature channel (see Fig. 4-6) approaches the theoretical equilibrium of an even energy gradient much more closely than do straight channels. Meandering channels also increase the total volume of the channel and thus its capacity to receive and transport water from a river's source to its mouth.

Rivers are erosional where moderate to rapid flow prevails and are depositional of silt and organic particles in sections of slow current. River channel characteristics are constantly changing in relation to variables of discharge, width, depth, substrata resistance, and transport of sediments. The introduction of a dam and its

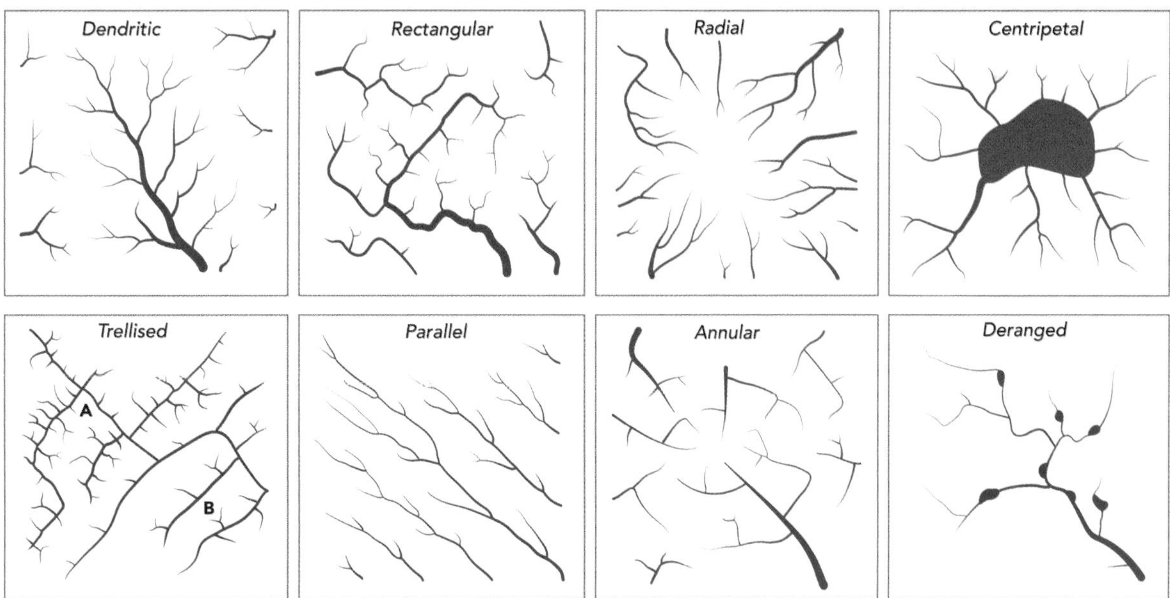

FIGURE 4-8 Common types of drainage network patterns. *(Modified from Gregory and Walling, 1973.)*

associated reservoir management practices can strongly affect these dynamics.

Stream and river channels gradually move laterally within the floodplain because of the erosion and deposition of sediments. Old, abandoned portions of river channels are common among most floodplains. As a particular feature, channel bars are accumulations of bed materials below the average water surface level that often move or are deformed by changing water velocities and that may emerge during low water levels. The morphology of rivers has been classified and organized on the basis of many criteria of fluvial morphology (e.g., Kellerhals and Church, 1989), such as drainage basin area, discharge, width–depth ratio, bed material size, or channel slope.

Most rivers and streams flow in valleys that exert some lateral and vertical control over the river. Reaches of lotic ecosystems are considered *constrained* when the valley floor is narrower than twice the active stream channel width (Gregory et al., 1991). Constrained river systems occur where natural geological or human-made features constrict the valley floor and limit lateral mobility and the development of adjacent plant communities. Rivers in valleys without contemporary floodplains are also called *incised* or *entrenched*. Canyons are an extreme case of river incision. *Unconstrained* river reaches have valley floors that are wider than twice the active channel width and lack major lateral geological or artificial constraints (Gregory et al., 1991). Most streams and rivers are of this type and are characterized by active migrations of the channel that expand into extensive floodplains and often form braided channels (Kellerhals and Church, 1989). The extent of the lateral constraint depends on river size and is quite variable from stream to stream as well as within the same river ecosystem.

The forms of river channels, as seen above, also termed river or channel *planform*, are very diverse and deviate from four general classes: *straight*, *meandering*, *anastomosing*, and *braided* (Fig. 4-9). These four planform types are often used to indicate general trends of stream

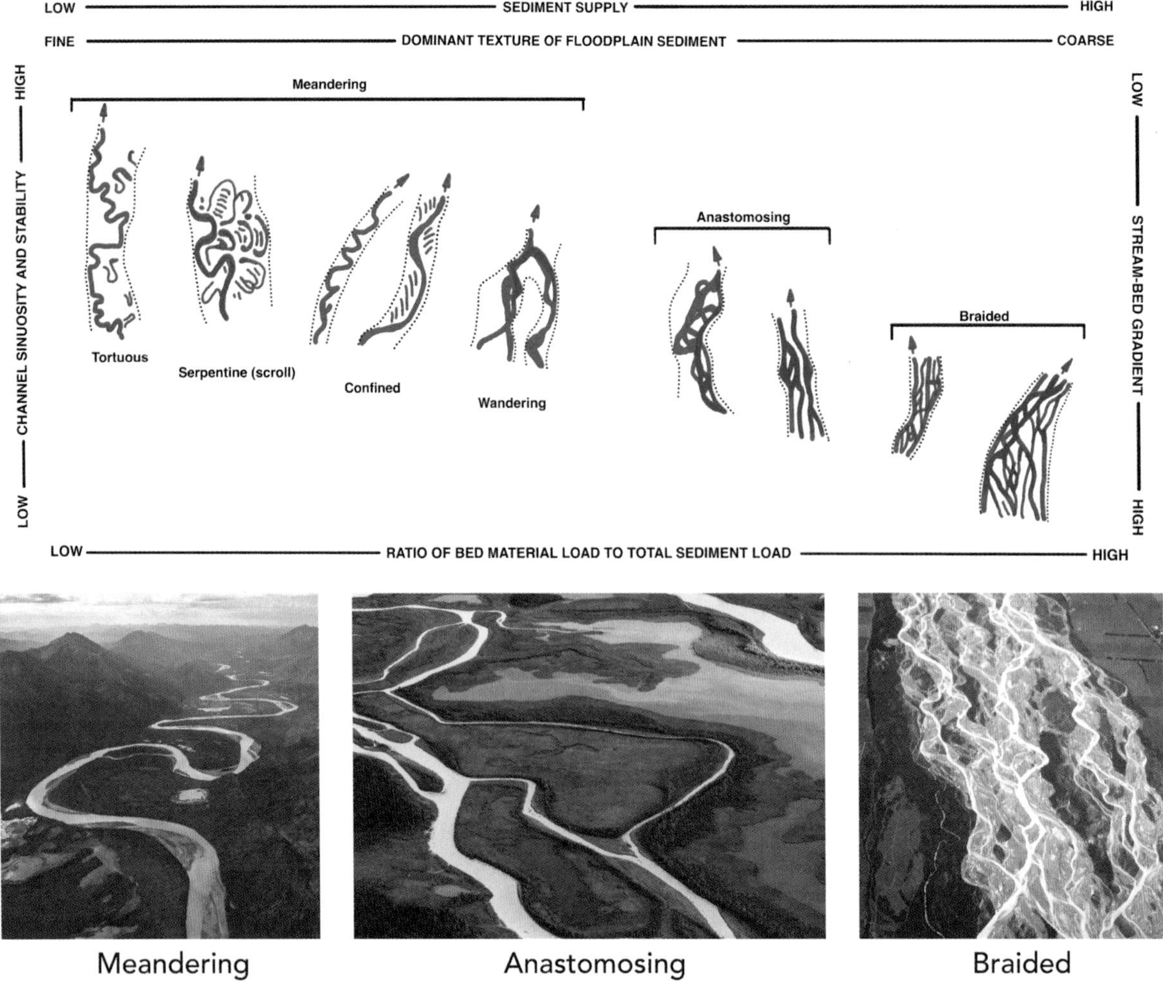

FIGURE 4-9 *Top panel*: form and gradient of alluvial stream channels and the type, supply, and dominant particle sizes of sediments. Great variations in channel patterns exist and the figure depicts examples within the range of channel types and the extent of anastomosing around channel islands. *Bottom panel*: three examples of meandering, anastomosing, and braided river types. *(Modified from Selby, 1985.)*

parameters such as erosion potential, sediment supply, effects of vegetation, and sensitivity to geomorphic disturbances. It should be emphasized, however, that such classification systems, while informative, cannot fully represent the great variations within channel patterns and should not be relied on to predict specific local situations, as different climatic conditions and hydrological regimes can strongly affect these parameters.

Although most rivers flow in valleys, exceptions are common, such as rivers on deltas, fans, or broad plains (Kellerhals and Church, 1989). *Deltas* are river-deposited continuations of land built by sedimentation out into a body of water. *Fans* are alluvial, fan-shaped surfaces deposited where a narrow valley emerges onto a broad surface. *Broad plains* are usually alluvial, lacustrine beds of drained lakes, or emergent coastal plains.

III. Origin of lakes

The origins of lakes and their morphometries are of much more than casual interest. The geomorphology of lakes is intimately tied to physical, chemical, and biological processes occurring in the waterbody and plays a major role in the control of a lake's metabolism, within the climatological constraints of its location. The geomorphology of a lake controls the nature of its surface and underground influxes and outfluxes, including the inputs of nutrients to the lake, and determines the relation of influx to storage volume, i.e., the lake's water renewal rate. These patterns, in turn, can affect the distribution of dissolved gases, nutrients, and organisms, so the entire metabolism of freshwater systems is influenced to varying degrees by the geomorphology of the waterbody and how it has been modified throughout its subsequent history.

The shape of a lake's waterbody, namely, the depression occupied by the lake water (also called the *lake basin*), can influence its productivity. Steep-sided U- or V-shaped lake depressions, often formed by tectonic forces, are usually deep and relatively unproductive. In such lakes a proportionally smaller volume of water is in contact with sediments. Shallow depressions with a greater percentage of contact of water with the sediments generally exhibit higher productivity.

The following brief resumé on the origin of lakes is based on Hutchinson's (1957) summary of the subject, which was drawn from an array of sources. Hutchinson differentiated 76 types of lakes on the basis of geomorphological inception. This detailed classification has been variously modified or amended subsequently but essentially stands intact. The discussion here is limited to the 11 main groups of lake types (slightly rearranged from Hutchinson, 1957), each formed by different processes. Some lakes have gone through stages of mixed origins and formations.

A. Lakes formed by glacial activity

By far, the most important agents in the formation of lakes are the gradual but nonetheless catastrophic erosional and depositional effects of glacial ice movements. Land surfaces that are now glaciated include Greenland and several smaller Arctic islands, Antarctica, and numerous small areas in high mountains throughout the world. These contemporary glacial activities are small, however, in comparison to the massive Pleistocene glaciation that advanced and receded in four major episodes of activity in the Northern Hemisphere. With the retreat of the last Pleistocene glaciers (~10,000 years ago), an immense number of small lakes were created. Lakes of glacial origin are far more numerous than lakes formed by all other processes combined. The action of glaciers in mountainous regions of high relief usually produces lake types that are quite different from those that result from the movements of large, continental ice sheets on regions of more mature and gentle relief.

A number of lakes, often temporary in nature, occur on the surfaces of, within, or beneath existing glacial ice masses in areas of transitory thaw. For example, numerous meltwater *sub-ice lakes* several kilometers in diameter occur 3000–4000 m below the Antarctic ice sheet (Oswald and Robin, 1973). In high mountain regions the fronts and lateral arms of glaciers, as well as terminal morainal deposits supported by the ice, often function as effective dams in river valleys that impound lakes which fill almost entirely with glacial meltwater (Fig. 4-10).

Glacial ice-scour lakes include a vast number of small lakes formed by ice moving over relatively flat, mature rock surfaces that are jointed and contain fractures. The ice-scour lakes are particularly common in mountainous regions where glacial movements have removed loosened rock material along fractures. When the glaciers retreat, the scoured depressions in the rock formations fill with meltwater. Such glacial-scour lakes can be found in large numbers in the vast Canadian Shield region, on the upland peneplains of Scandinavia, or in the United Kingdom.

A frequently occurring type of ice-scour lake forms in the upper portion of glaciated valleys in mountainous areas, where the valleys are shaped into structures resembling amphitheaters by freezing and thawing ice action. The amphitheater-like formation is referred to as a cirque, and lakes within such depressions are *cirque lakes* (also called *tarns*) (Figs. 4-11 and 4-12). Water is held in the cirque depression either by a rock lip above the depression or by morainal deposits. Cirque lakes, which are generally small and relatively shallow (<50 m), are often found in tandem arrangement within a glaciated trough of a mountain valley, with the higher lake "hanging" above the lower succeeding lake in a

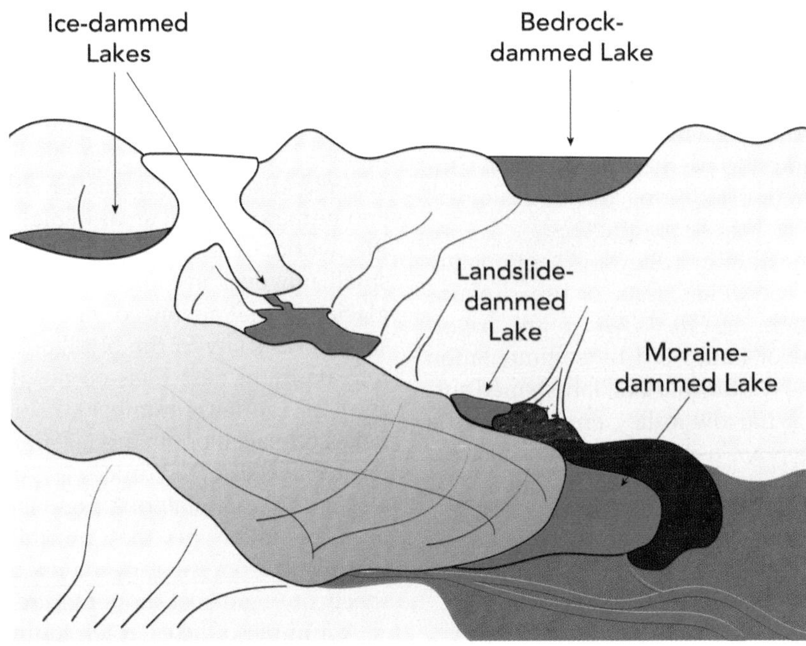

FIGURE 4-10 Examples of glacial lake types. *(Modified from Tweed and Carrivick, 2015.)*

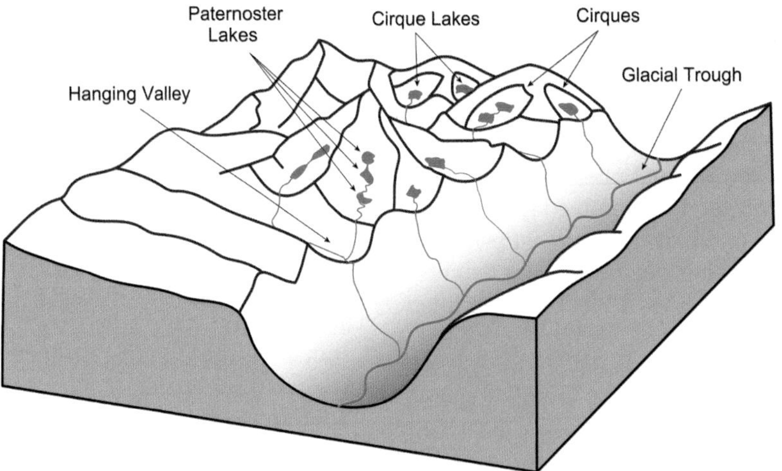

FIGURE 4-11 Several small cirque lakes (or tarns) and paternoster lakes within a mountain group from which several converging, cirque-headed branch troughs all join the same trunk trough. *(Modified from Davis, 1909.)*

stairway-like fashion. True cirque lakes, which are common to all major mountainous regions, are formed at approximately the firn line, i.e., the line across a glacier that indicates the transition from constant freezing to seasonal thawing. When the glaciers extend well below the firn line, the corrosive action of the ice can form rock depressions within the glacial valley. When such depressions form a chain of small, connected lakes at successively lower elevations in a glaciated valley trough, they are referred to as *paternoster lakes* (Fig. 4-11). Where there are mountains adjacent to the sea, as in many areas of Norway and western Canada, *fjord lakes* can be formed within narrow, deep depressions in glacially scoured valleys.

Where continental ice sheets encountered weak areas in primary basal rock formations, glacial scouring of lake depressions occurred on a massive scale in non-mountainous piedmont areas. Great Slave Lake and Great Bear Lake in the central Canadian subarctic region are examples of scouring of preexisting valleys to very great depths by the ice sheet movements. The most impressive examples, however, of large lakes produced by glacial continental ice erosion are the Laurentian Great Lakes of the St. Lawrence drainage.

As continental glacial ice sheets retreated in the late Pleistocene, vast amounts of rock debris, moved and incorporated into the ice during former advances, were deposited in terminal moraines and laterally to

FIGURE 4-12 Examples of different lake types and origins. The tectonic lake is Lake Baikal in Russia.

the lobes of the retreating glacier. These deposits dammed up valleys and depressions in an irregular way and filled up with water either as they were below later groundwater levels or by meltwater and drainage from the surrounding topography. *Morainal damming* of preglacial valleys (Fig. 4-10), usually at one end but occasionally at both ends, created many of the numerous lakes of the glaciated northern United States, including the distinct Finger Lakes of New York State (USA).

In regions glaciated by continental ice sheets a very common process of the formation of lakes was associated with the deposition of meltwater outwash left at the border of the retreating ice mass or with blocks of ice buried in this debris. The major features of the so-called *kettle lakes* that formed in this manner, which accounts for the origin of thousands of small lakes of northern North America, are shown in Fig. 4-13. Glacial drift material was deposited terminally and washed out into plains, often containing large segments of ice broken from the decaying glacier. As the ice blocks melted, the morphometry of the resulting depressions was modified variously by the overlying rock debris. The ensuing lake depressions are highly irregular in shape, size, and slope, corresponding to the irregularities of the original ice blocks. Kettle lakes often exhibit variable underwater relief, with multiple depressions separated by irregular ridges and mounds. These lakes rarely exceed 50 m in depth; their shallow waterbody is related to the limiting depth of crevasse formation in the fracturing of the terminal glacier.

Most Arctic lakes result from glacial activity, as discussed earlier, or are *cryogenic lakes* formed from the effects of permafrost (Pielou, 1998). The most abundant type of waterbody found in the Arctic is the shallow pond formed inside of an ice-wedge polygon of raised soil banks that grow above the permafrost from water seepage through cracks in the surface of the ground. Eventually, polygonal networks of ridges are formed that often contain ponds from 10 to 50 m in diameter. Millions of these ponds exist in coastal northern Alaska, Canada, Greenland, the flat regions of Siberia, and other high-latitude areas. The shallow ponds may coalesce into larger ponds, or large amounts of ice deeper in the permafrost may melt, especially if the plant cover is disturbed or destroyed, thereby forming *thermokarst lakes* (Fig. 4-14).

Some lakes form in the thawed craters of *pingos*, small conical hills scattered over tundra plains, especially in

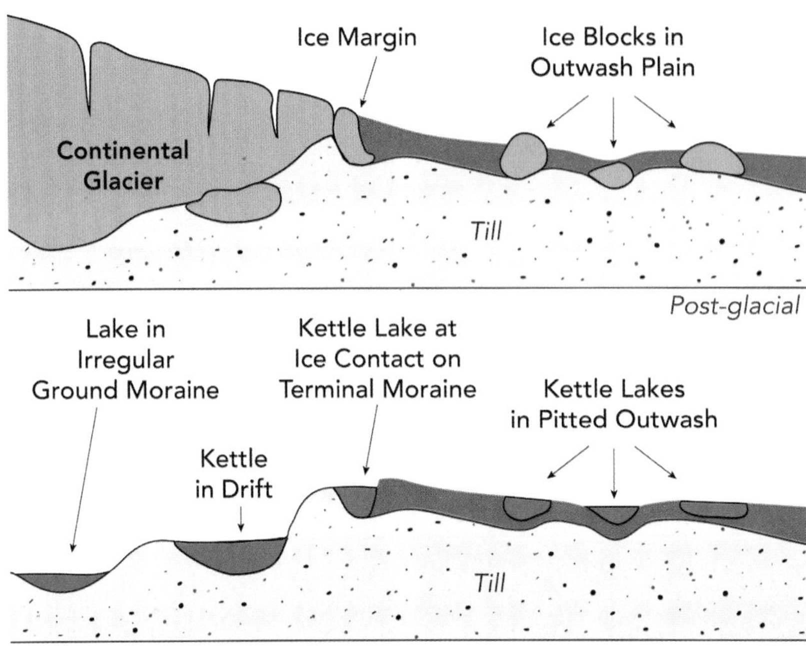

FIGURE 4-13 The formation of various types of kettle lakes. *Top panel*: an outwash plain of retreating continental ice containing ice blocks. *Bottom panel*: lakes formed in the outwash plain and morainal till. *(Modified from Hutchinson, 1957, after Zumberge, 1952.)*

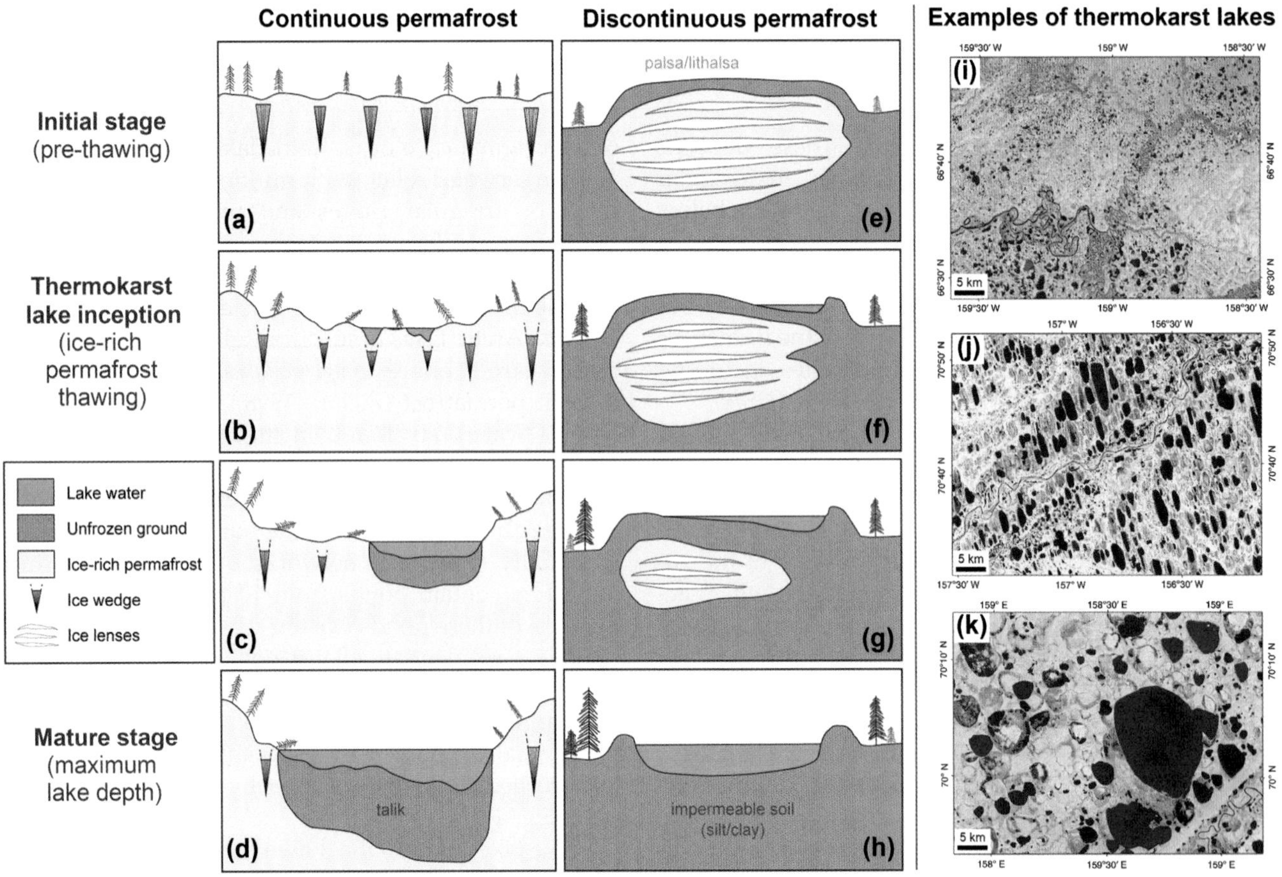

FIGURE 4-14 Thermokarst lake formation and evolution in ice-rich permafrost in the (a–d) continuous and (e–h) discontinuous zones. *Right panel*: Landsat-5 TM satellite image subsets showing thermokarst lake-rich regions: (i) Selawik Wildlife Refuge, Interior Alaska; (j) North Slope Coastal Plain, Alaska; (k) Chukochya River region, NE Siberia. *(Modified from Bouchard et al., 2016, and Grosse et al., 2013.)*

the North American Arctic regions. Water from previously insulated soils can rise by hydraulic pressure through cracks in the permafrost but is overlain with alluvium and tundra vegetation (Pielou, 1998). The water freezes and forms a mound, or pingo. The overlying soil can rupture, and the ice of the depression can thaw to form small *pingo lakes*.

B. Lakes formed by tectonic activity

Tectonic lakes occur in depressions formed by movements of deeper portions of the Earth's crust and are differentiated from lakes that resulted from volcanic activity. The main type of tectonic lakes forms as a result of *faulting*, in which depressions occur between the bases of a single fault displacement or in downfaulted troughs (Fig. 4-15). The latter type of depression is referred to as a *graben*, and multiple graben can form a rift valley. Lakes formed by this mode of origin are called *graben* (or *rift*) *lakes* and include most of the world's *relict lakes*, referring to their nature as being remnants of preexisting formations. Arguably the most spectacular among these is Lake Baikal (Fig. 4-12), the deepest and oldest lake in the world, which has a continuous lacustrine history from at least the early Tertiary period (25–30 million years ago). Lake Baikal, and many other relict lakes, are of particular interest because they contain a large number of relict endemic species. For example, of 2595 aquatic animal species and subspecies found in Lake Baikal, over 56% are endemic to this lake (Timoshkin et al., 2004). Lake Tanganyika of equatorial Africa is a similar deep graben lake (maximum depth = 1435 m), formed in rift valley displacements along crustal fractures, and contains many relict endemic species of plants and animals (Coulter, 1991). The Dead Sea of the Middle East represents a unique hypersaline graben lake with its surface at ~430 m below sea level.

Tectonic movements causing moderate *uplifting* of the marine seabed have isolated several very large lakes. The relict marine depressions of eastern Europe, which include the Caspian Sea and the Aral Sea, were separated by the formation of uplifted mountain ranges in the Miocene period. Similarly, the somewhat more localized *upwarping* of the Earth's crust in lesser degrees also has resulted in the formation of many large lake systems. Lake Okeechobee in Florida (USA) resulted from minor depressions (maximum depth ~4 m) in the sea floor as it uplifted in the Pliocene epoch to form the Floridian peninsula, and it forms one of the largest lakes (approximately 1880 km^2) completely within the United States. Similarly, Lake Victoria in central Africa resulted from the upwarping of the margins of the plateau in which it lies. Upwarping can also occur in formerly

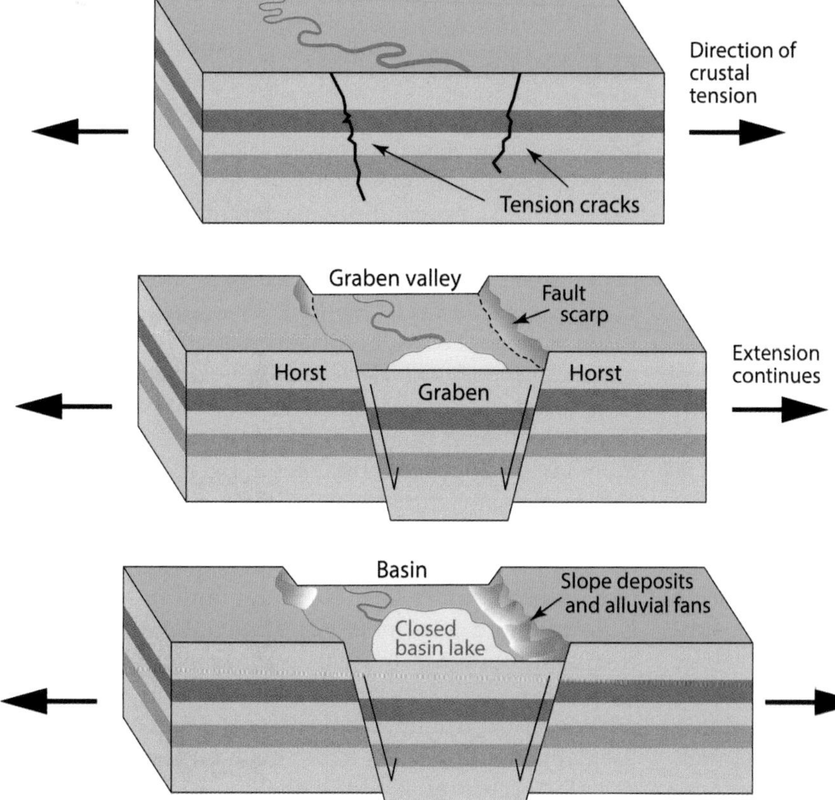

FIGURE 4-15 Origin of a tectonic lake. Due to crustal tension and extension, a depressed fault block (graben) occurs between two higher fault blocks (horst). While erosion and deposition can fill in some of the graben valley, large and deep tectonic lakes may eventually fill the depression. *(Modified from US National Park Service, 2020; original illustration by Trista L. Thornberry-Ehrlich.)*

glaciated regions where the release of pressure due to receding ice sheets may result in crustal rebound. This can modify—or be modified by—glacial scouring activities, such as the case, for example, in the formation of the Laurentian Great Lakes.

Tectonic lakes are occasionally formed in areas of localized *subsidence* that result from earthquake activity. Many of these depressions are dry or only temporarily contain water, depending on the porosity of the underlying material, while others become permanent lakes, usually open with outlet drainage.

C. Lakes formed by volcanic activity

Catastrophic events associated with *volcanic activity* can generate lakes in several different ways. As volcanic materials are ejected upward and create a void, or as released magma cools and is distorted in various ways, depressions and cavities are created. If these depressions are undrained, they may contain a lake. Because of the basaltic nature of the underlying rocks and the often very small drainage basin areas, many lakes associated with volcanic activity contain low concentrations of nutrients and are relatively unproductive.

Small *crater lakes* are occasionally found occupying cinder cones of quiescent volcanic peaks. Alternatively, crater lakes can occur in depressions formed by the violent ejection of magma (Fig. 4-12), or by the collapse of surface materials where underlying magma has been ejected, termed *maars*, and are generally small with diameters less than 2 km. Maars are usually nearly circular in shape and can be extremely deep (>100 m) in relation to their small surface area. Lakes formed by the subsidence of the roof of a partially emptied magmatic chamber are termed *caldera lakes* and can be somewhat larger than maars (minimum diameter about 5 km). Among the most spectacular of lakes formed by the collapse of the center of a volcanic cone is Crater Lake, Oregon, with an area of 64 km^2 and a depth of 594 m. Caldera lakes can be very large and be modified in various ways. For example, Lake Toba in Indonesia, the largest volcanic lake in the world with an area of 1130 km^2, occupies the caldera of a supervolcano and contains a resurgent dome that partially fills the original caldera.

Some volcanic lakes originated from a combination of large-scale volcanic and tectonic processes. In some situations caldera collapse occurred on such a large scale that extensive portions of the surrounding land subsided in addition to the central portion of the volcano. Such subsidence usually takes place along preexisting fault fractures. Some of the largest lakes associated with volcanic activity were formed in this manner, with many examples in equatorial Asia and New Zealand (see, e.g., Larson, 1989). Also, lava streams from volcanic activity can flow into a preexisting river valley and form a dam, behind which a lake can collect.

D. Lakes formed by landslides

Sudden movements of large quantities of unconsolidated material in the form of landslides into the floors of stream valleys can cause dams and create lakes, often of very large size (Fig. 4-12). Such landslide dams may result from rockfalls, mudflows, ice-slides, and even flows of large amounts of peat, but they are mostly found in glaciated mountains (Fig. 4-10). The landslides are usually caused by abnormal meteorological events, such as excessive rains acting on unstable slopes. More spectacular slides are occasionally initiated by earthquake activity. Lakes formed behind landslide dams can be large and persistent, such as Sarez Lake in Tajikistan, which was formed by an earthquake-induced landslide in 1911 and represents one of the deepest lakes in the world (maximum depth = 505 m). More often, however, landslide lakes are transitory and exist only for a few weeks to several months. This short period is due to the fact that, unless the slide is very massive, the unconsolidated dam is susceptible to rapid erosion by the effluent of the newly formed lake. Many disastrous floods have resulted from the rapid erosion of the dam material by the effluent, which, once started, can quickly empty the accumulated lake.

E. Lakes formed by river activity

The running waters of rivers possess considerable erosive power that may create lake depressions along their courses from elevated land to large lakes or the sea (Fig. 4-16). In the upper reaches of rivers where gradients are steep, excavation by water can produce rock depressions that may persist as lakes after the course of the river has been diverted. *Plunge-pool lakes*, excavated at the foot of waterfalls, provide a rare but spectacular example of such destructive fluvial action.

A combination of destructive erosional and obstructive depositional processes occurs as rivers flow down more gentle slopes to form lakes in the river floodplains. Many lakes are formed along large rivers when sediments of the main channel are deposited as levees across the mouths of tributary streams. In this way the obstruction of the tributary flow continues until the side valley is flooded and a *lateral lake* is formed. Lateral lakes are frequently found in tributary valleys along major river systems on all continents, especially in the upper portion of the drainage. The reverse situation of fluviatile dams holding lakes in the main river channel occasionally occurs as a result of deposition by a lateral tributary. If more sediments than the main channel can

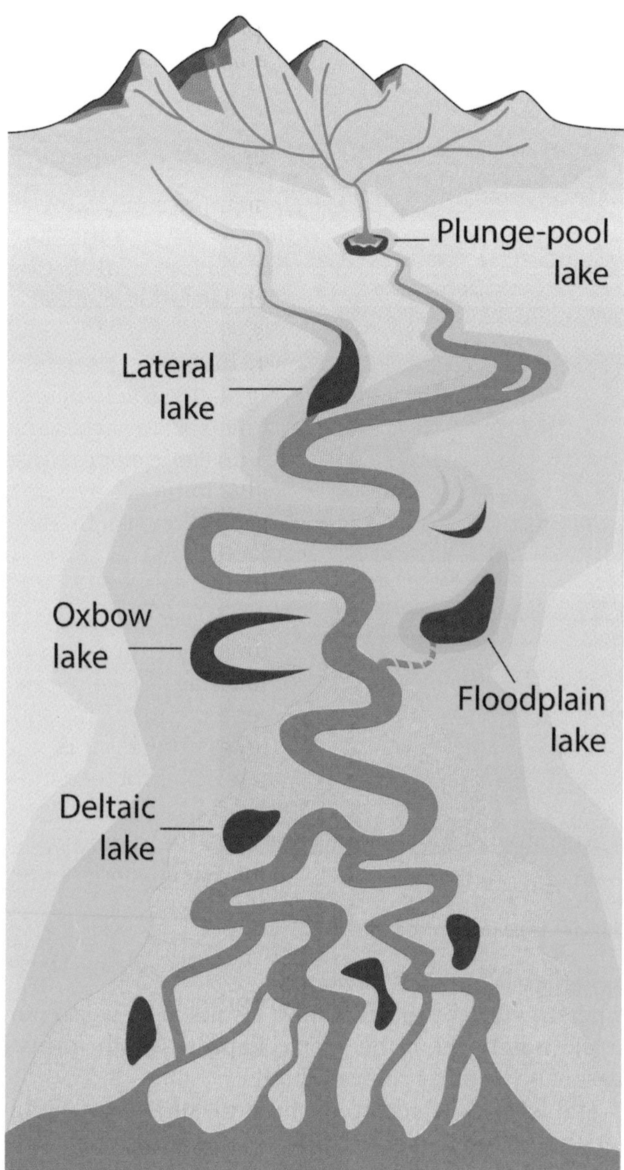

FIGURE 4-16 Different lake types created by river activity. (Illustration by Sarah Bridges.)

remove are deposited, the resulting lake may be permanent.

As a river meanders within the irregularities of the topography, greater turbulence and erosion occur on the outer, concave side of the river bend, while deposition occurs on the inner, convex side, where currents and turbulence are reduced. With time, continued erosion and concavity take place until the U-shaped meander of the river closes in upon itself. The main course of the river cuts a channel through the initial portion of the meander, and levee deposits eventually isolate the loop, referred to as an *oxbow lake*, from the river channel (see also Fig. 4-12). The outer, erosional side of oxbow lakes is usually deeper than the inner, convex side.

Floodplain lakes (also known under various regional names such as *billabongs* in Australia or *várzea* and *ria* lakes in Brazil) contain water throughout the year. When flooded, the floodplain of a river is usually covered continuously with water (Hamilton et al., 1992). As the river level recedes, water drains from the floodplain. Lakes persist in depressions perched above the river or in deep depressions that remain connected with the river.

Where rivers enter the relatively quiescent waters of a lake or the sea, sedimentation results in the formation of deltas, often of very large size. *Deltaic lakes* are formed in the deltas of all major rivers of the world. As the river velocity becomes reduced and sediments are deposited, the water tends to flow around the sediments in a U-shaped pattern where the open end extends seaward. As alluvial deposition extends further seaward, it eventually encloses and isolates inward depressions to form shallow lakes, often of large size. Subject to the influence of tides and other water movements of the sea, these deltaic lakes often receive saltwater, making them brackish.

F. Lakes formed by shoreline activity

When the coastline of a large body of water, such as the sea or a large lake, possesses some irregularity or indentation, the potential exists for the formation of a bar across the depression to form a *lagoon* or *coastal lake* (Fig. 4-12). A longshore current flowing along the shoreline and carrying sediments will, on encountering a bay, deposit the material in the form of a bar or spit across the mouth of the indentation. Often the spit can eventually separate the bay from the sea (or from the large lake) to form a coastal lake.

Marine coastal lakes commonly result from sand bar formation across the mouths of old estuaries that have been inundated by rising water levels or slight subsidence of the land. Often, river discharge and tidal currents are sufficient to prevent complete separation of the lake from the sea. The result is an alternation between fresh and brackish water in the lake, and the salinity depends upon the ratio of freshwater inputs and saltwater intrusions. Other coastal lakes are completely separated from the sea.

Numerous coastal lakes are also found inland, adjacent to large lakes. The formation of these waterbodies occurs in an analogous fashion, by the deposition of bars and spits across bays and river valleys.

G. Lakes formed by solution

Lake depressions can be created in any area where deposits of soluble rock are slowly dissolved by

FIGURE 4-17 Examples of different lake types and origins. The meteorite lake is Lake Manicouagan in Canada.

percolating water. Although many rock formations are readily soluble (salts such as sodium chloride [NaCl], calcium sulfate [$CaSO_4$], and ferric and aluminum hydroxides), most *solution lakes* are formed in depressions resulting from the solution of limestone (calcium carbonate [$CaCO_3$]) by slightly acidic water containing carbon dioxide (CO_2). Solution lakes in the form of *karst outcropping lakes* (Fig. 4-17) are thus common in many limestone regions of the world, such as the karst regions around the Adriatic Sea, especially in the former Yugoslavia, Southeast Europe, the Alps of Central Europe, central and southern China, and various regions in North and Central America, including Florida and the Mexican Yucatán peninsula.

Solution lakes often occur in very circular and conically shaped sinks, termed *dolines*, which develop from the solution and gradual erosion of the soluble rock stratum (Fig. 4-17). Percolating surface or groundwater dissolves limestone most readily at the joints and points of fault fractures through which it drains. Adjacent dolines may eventually fuse to form compound depressions that are more irregular in conformity. Alternatively, the dissolution of limestone frequently occurs in large subterranean caves. Continued erosion of the superstructure by groundwater results in its weakening to the point where the roof collapses, which forms a reasonably regular conical doline.

The level of water in a solution lake is often highly variable. Usually, the depressions are sufficiently deep to extend well into the groundwater table and permanently contain water. Other depressions that just reach the water table can undergo fluctuations in water level in response to seasonal and long-term variations in groundwater supply.

H. Lakes formed by wind activity

Wind action operates in arid regions to create lakes by *deflation* (i.e., wind erosion of loose material, which over long timescales can scoop out hollows or blow-outs), or by redistribution of sand to block existing river valleys, thus forming *aeolian* or *dune lakes* (Fig. 4-17). Such lake depressions may be solely or partially the result of wind action, and the water they contain is often temporary and dependent upon fluctuations in climate. Different types of dune lakes can be found in inland desert regions of the world, as well as in coastal areas. For example, organic additions from vegetation can assist in creating impervious, organically bonded sand-rock in dune depressions which may fill with surface water. In contrast, deflation depressions that are permeable to water can form lakes when they extend below an extensive groundwater table. All of these lakes are extremely transitory.

Deflation lakes are also common where material is moved and eroded from horizontal strata of rock or clay by wind activity, permitting the formation of large pans on nearly level areas, often called *playas*. The center of the pan then fills with water during rainy seasons or wet periods, creating an aeolian lake that becomes increasingly saline with evaporation, followed by desiccation during opposing seasons or dry periods. These ephemeral lakes are common in large portions of Australia, southern Africa, endorheic regions of Asia and South America, and the plains and arid regions of the United States.

I. Lakes formed by meteorites

As a rare form of exogenic origin, *meteorite lakes* (or *impact crater lakes*) are water accumulations inside a depression caused by the drastic impact of a meteorite. Meteorite lakes occur on all continents and can be very deep and old. As a special form, *annular lakes* are impact crater lakes shaped like a ring. One of the largest examples is Lake Manicouagan (Fig. 4-17) in Canada (surface area 1942 km^2, albeit increased by damming; maximum depth = 350 m; estimated age of crater >200 million years), which constitutes an annular lake with a diameter of about 70 km across, surrounding an inner island plateau that was formed by postimpact uplift.

J. Lakes of organic origin

The array of lakes created by the damming action of plant growth and associated detritus is not fully understood. It is known, however, that the growth of emergent and floating aquatic plants can be sufficiently profuse to dam the outlet of shallow depressions and alter the drainage patterns of lakes, particularly in floodplains of tropical river systems (Beadle, 1974). Such alterations of drainage by plant growth are common in *dystrophic lakes*, which contain large amounts of humic undecomposed organic matter. Also, the photosynthetic activities of attached Cyanobacteria and other algae (Chapters 17 and 25), as well as aquatic macrophytes (Chapter 24), can induce massive precipitation of $CaCO_3$ from calcareous waters. In time, these deposits can form barriers and isolate small bodies of water from stream systems (Golubić, 1973).

As an iconic "lake forming" animal species native to the temperate Northern Hemisphere, the beaver is particularly effective in constructing dams across river valleys to impound water into lakes (Fig. 4-17). While most of these structures are temporary, some become permanent by means of sediments deposited against the dams. For example, the widespread North American beaver (*Castor canadensis*) has created numerous large, long-lived lakes throughout the boreal and temperate ecoregions of Canada.

K. Lakes of anthropogenic origin

Humans have created artificial lakes by damming streams for at least 4000 years. The pace and scale of damming, however, increased dramatically in the mid-20th century. Thousands of large dams have been built to create anthropogenic *reservoirs* to store water for irrigation, hydroelectricity, and other purposes (Fig. 4-17). Some of them, such as Lake Volta (Ghana), Lake Nasser (Egypt), Lake Kariba (Zimbabwe/Zambia), Smallwood Reservoir (Canada), Sobradinho Reservoir (Brazil), or Bratsk Reservoir (Russia), are comparable in size to the world's largest natural lakes. Small, shallow reservoirs, in the form of farm ponds and moderately sized inundations, outnumber the large ones by the millions (Lehner et al., 2011). The shallow morphometry of these reservoirs can create extensive areas where macrophytic vegetation thrives, and such plant life can radically alter the productivity of the lake system. Moreover, as these ponds are most commonly associated with agricultural landscapes, small reservoirs generally receive high nutrient inputs in relation to their volume, which further increases their productive capacity. For more information on reservoirs, see section 4-VI below.

IV. Succession of lakes

Lakes are geologically ephemeral, and even if some exist for millions of years, they gradually disappear. Succession generally relates to the process of filling up the lake with sediments and organic matter and can be interpreted as the aging of the lake. Along with the filling come changes in the physical, chemical, and biological properties of the lake, such as nutrient content, oxygen demand, light penetration, biota, and bathymetry. Succession typically starts with sediment influx, which depends on many factors such as the underlying geology of the drainage basin, its vegetation cover, and regional climate. The rate of succession can accelerate due to biological activity within the lake, foremost plant growth.

The succession process is typically accompanied by different stages of nutrient loading (oligotrophic, mesotrophic, eutrophic, and hypereutrophic). These nutrient stages may occur in sequence, and they can repeat

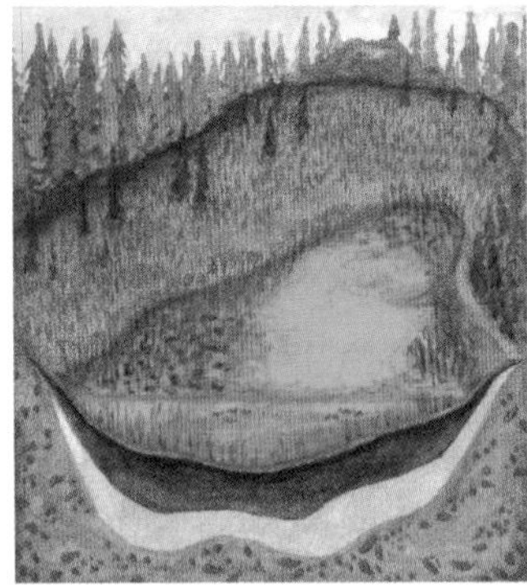

FIGURE 4-18 Lake succession. From left to right: over time, an open lake fills with sediments and organic materials and increased nutrient fluxes accelerate plant growth. The lake ecosystem eventually transitions into a wetland and ultimately land ecosystem. (Artwork by Olivia del Giorgio.)

themselves several times during a lake's existence depending on changes in climate or land cover (Harper, 1992). If there is a marked increase in nutrient loading, due either to natural or anthropogenic causes, the process is more specifically called *eutrophication*, which can lead to excessive algal and plant growth, and a fast transition from a lake to a wetland (Chapter 29) ecosystem (Fig. 4-18).

V. Morphology of lakes

The shape and size of a lake's waterbody affect nearly all physical, chemical, and biological parameters of the lake. The physical forms of lakes are extremely varied and reflect their modes of origin, how water movements have subsequently modified the lake's shape and morphometry, and the extent of loading of materials from the surrounding drainage basin.

The morphometry of a lake is best described by a *bathymetric map*, which is required for the determination of all major morphometric parameters. Such a map can be prepared by a survey of the shoreline by standard surveying methods, often in combination with aerial photography. From the map of the shoreline, a detailed bathymetric map of depth contours must be constructed from a series of accurate depth soundings along intersecting transects. Methods commonly employed for both lake and stream mapping are discussed, for example, by Håkanson (1981) and Wetzel and Likens (2000). Accurate sonar transects, which permit the acquisition of much greater detail than was previously possible by manual sounding methods, are now used almost exclusively.

A. Morphometric parameters

The many morphometric parameters that can be determined from a detailed bathymetric map are discussed at length by Hutchinson (1957). The most commonly used parameters are defined here. Today, most of these parameters can be routinely calculated within geographic information system (GIS) software using 2D or 3D functionality.

Area (A). The area of the lake surface (A_0) and each contour area at depth z. It can be determined by planimetry (see, e.g., Wetzel and Likens, 2000), or by digital analysis using a GIS.

Maximum length (l). The straight-line distance on the lake surface between the two most distant points on the lake shore. This length is the maximum effective length or fetch for wind to interact on the lake without land interruption.

Maximum width or breadth (b). The maximum distance on the lake surface at a right angle to the line of maximum length between the shores.

Mean width ($\bar{b}$). The surface area divided by the maximum length: $\bar{b} = A_0/l$.

Volume (V). The volume of the lake is the integral of the areas of each stratum at successive depths from the surface to the point of maximum depth. If not digitally computed in a GIS, the volume can be approximated by plotting the areas of contours, as closely spaced as possible, against depth. The integrated area of this curve corresponds to the lake volume. Alternatively, the volume can be estimated by summation of the frusta of a series of truncated cones of the strata (compare Fig. 4-19a):

$$V = \frac{h}{3(A_1 + A_2 + \sqrt{A_1 A_2})}$$

where h is the vertical depth of the stratum, A_1 the area of the upper surface, and A_2 the area of the lower surface of the stratum whose volume is to be determined.

Maximum depth (z_m). The greatest depth of the lake.

Mean depth ($\bar{z}$). The volume divided by its surface area: $\bar{z} = V/A_0$.

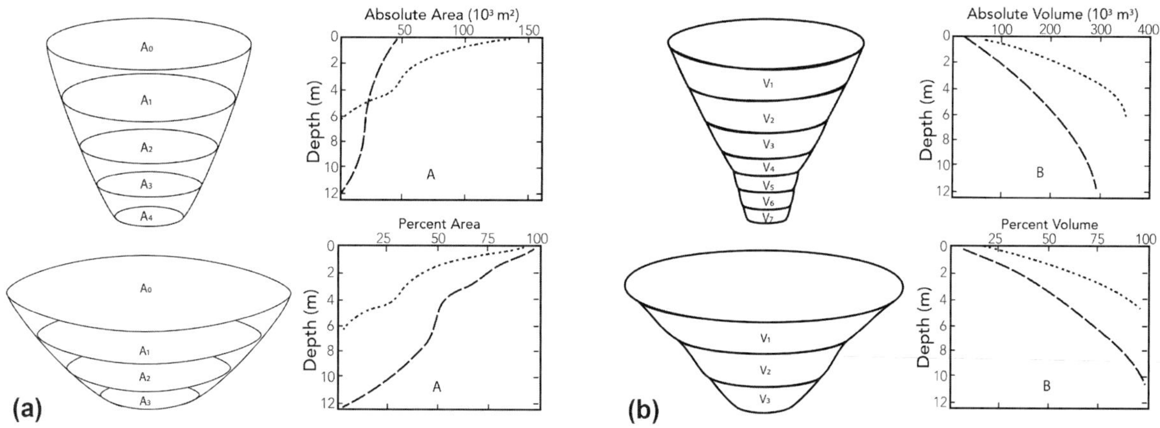

FIGURE 4-19 Hypsographic (depth–area) curves (a) and depth–volume curves (b) of oligotrophic Lawrence Lake (*upper cone shapes* and *dashed lines*) and eutrophic Wintergreen Lake (*lower cone shapes* and *dotted lines*), southwestern Michigan. (Graphs are based on unpublished data from Wetzel, 2001.)

Relative depth (z_r). The ratio of the maximum depth as a percentage of the mean diameter of the lake at the surface, expressed as a percentage.

$$z_r = \frac{50 z_m \sqrt{\pi}}{\sqrt{A_0}}$$

Most lakes have a relative depth of less than 2%, whereas deep lakes with a small surface area usually have $z_r > 4\%$.

Shoreline length (L). The length of the shoreline is measured along the intersection of the land with the lake's water surface. The shoreline length is nearly constant in most natural lakes but can fluctuate widely in ephemeral lakes and especially in reservoirs in response to variations in precipitation and discharge. The length of the shoreline can be determined directly in the field, from analog maps with measurement tools (see, e.g., Wetzel and Likens, 2000), or by use of a GIS.

Shoreline development (D_L). The ratio of the length of the shoreline (L) to the circumference of a circle with an area equal to that of the lake:

$$D_L = \frac{L}{2\sqrt{\pi A_0}}$$

Very circular lakes, such as crater lakes and some kettle lakes, approach the minimum shoreline development value of unity. The conformation of most lakes, however, deviates strongly from the circular. Many are subcircular or elliptical in form, with D_L values of about 2. A more elongated morphometry increases the value of D_L markedly, as, for example, is found in the dendritic outlines of lakes occupying flooded river valleys. Shoreline development is of considerable interest because it reflects the potential for greater development of littoral communities in proportion to the volume of the lake.

B. Hypsographic and volume curves

The shape of a lake's waterbody can influence its biological productivity. In comparative analyses of productivity among lakes, examining the relationships of lake surface area to the area of sediments exposed to specific volumes of water at different depths permits insights into potential parameters that influence productivity.

The *hypsographic curve* (depth–area curve) is a graphical representation of the relationship of the depth-dependent contour area A_z of a lake to its depth (Fig. 4-19a). The hypsographic curve represents the proportion of the bottom area of the lake that is below the depth under consideration. The hypsographic curve is, however, only an approximation of the area of the exposed lake bottom because areal measurements are related to the horizontal plane of the lake surface, whereas the actual sediments are slightly sloped and greater in area. For example, if the bottom slope were 35° from the horizontal, the benthic area under 1 m^2 at the water surface would be 1.22 m^2. The hypsographic curve may be expressed in absolute terms (m^2, ha, or km^2) or as a percentage of the total lake surface area that is below a specific depth.

The *depth–volume curve* is closely related to the hypsographic curve and represents the relationship of lake volume to depth (Fig. 4-19b). Similarly, the depth–volume curve can be expressed in actual volume units (m^3 or km^3) or as the percentage of total lake volume that is above a specific depth. The depth–volume curve can be used to estimate readily the approximate mean depth of a body of water, i.e., the depth above which 50% of the lake volume occurs.

Among lakes with otherwise comparable conditions, biological productivity is generally greater in those with more sediments in contact with water that receive sufficient light to support photosynthesis (Thienemann, 1927; Rawson, 1955). The extent of shallow water in a lake is a determining factor in the interrelationship of the zones of photosynthetic production and of decomposition, and it determines the area available for the growth of rooted aquatic plants and associated littoral communities.

The *ratio of mean to maximum depth* ($\bar{z}/z_m$) is an expression similar to the ratio of the volume of the lake to that of a cone of basal area A and height z_m $\left(A\bar{z}/\frac{1}{3}z_m A = 3\bar{z}/z_m\right)$. The ratio $\bar{z}/z_m$ thus gives a comparative value of the form of the lake in terms of volume development. For most lakes, the value $\bar{z}/z_m$ is >0.33, which is the value that would be given by a perfect conical depression. The ratio exceeds 0.5 in many caldera, graben, and fjord lakes, whereas most lakes in easily eroded rock usually have ratios between 0.33 and 0.5. Very low values of $\bar{z}/z_m$ occur only in generally shallow lakes with small deep holes, such as solution or kettle lakes. Examination of the morphometry of a large number of lakes has shown that the average shape of lake depressions approximates an elliptic sinusoid (Neumann, 1959) and that the mean depth of an average lake is slightly less than one-half (0.46) of its maximum depth.

Since the pioneering work of Thienemann (1927), much attention has been focused on the importance of lake morphometry, especially mean depth, to lake productivity in relation to the effects of climatic and edaphic factors (e.g., Rawson, 1955; Hayes and Anthony, 1964). Mean depth was regarded as the best single index of morphometric conditions, and mean depth exhibits a general inverse correlation to productivity at many trophic levels among large lakes.

VI. Reservoirs

While much of our limnological understanding originates from natural lake ecosystems, it is also important to consider the special role of the millions of reservoirs that have been built to serve human purposes. These purposes include domestic water supply, irrigation, flood control, hydroelectricity, navigation, and recreation. Although humans have been building reservoirs for thousands of years, the pace and magnitude at which they were built increased in the last century. In 1950 there were approximately 5000 large reservoirs in operation worldwide, and by 2018, that number had increased to nearly 60,000 (ICOLD, 2020). While the bulk of the construction boom occurred in five countries (United States, China, Spain, Japan, and India), large dams were established as a global phenomenon by 2000 (WCD, 2000) and are now rapidly proliferating in previously undeveloped drainage basins including in the Balkans and across the Amazon and Mekong river basins (Grill et al., 2019). Today's reservoirs are estimated to cover a new water surface area of over 300,000 km^2 and store more than 7000 km^3 of water (Table 4-5), which represents about one-sixth of the world's total annual river flow (Table 4-1). Smaller reservoirs and farm ponds are estimated to number in the millions (Lehner et al., 2011).

A. Reservoir characteristics

Reservoirs can be built in-stream or, less frequently, off-stream. *In-stream reservoirs* are constructed by damming a river valley; water then accumulates behind the dam. Water released from the dam is typically regulated to manage incoming flows, such as for flood control, or to support downstream water uses. *Off-stream reservoirs* can constitute small artificial ponds built in valley depressions, where inflows occur by surface flows and/or seepage. Off-stream excavations of soil for construction or mining can also result in reservoirs or artificial lakes as these excavations can cut into the water table of groundwater aquifers or be filled with mining effluents (e.g., tailings ponds, strip mining lakes, "barrow pit" lakes). Outflows from these excavations generally only occur by subaqueous seepage, evaporation, or breeched levees.

Large reservoirs bring about physical and chemical changes that influence aquatic biota in many ways and can be deleterious to human welfare and the environment (e.g., WCD, 2000; Richter et al., 2010). Most fundamentally, dams and their reservoirs alter the diurnal, seasonal, or interannual flow patterns; reduce downstream flooding; and are the main cause of the loss of fluvial connectivity found in natural river systems. As a result, the majority of the world's large rivers are no longer free-flowing, with unfragmented rivers primarily left in remote regions such as the Arctic or the Amazon and Congo river basins (Grill et al., 2019). Movements of diadromous and nondiadromous migratory fish species (see Chapter 22) are impeded or inhibited (Barbarossa et al., 2020). As they replace flowing with standing waters, reservoirs alter the chemical composition of the water they store and release. This includes changes in water temperature, oxygen content, and nutrient concentrations, which in turn affect biogeochemical cycling processes (Maavara et al., 2020). Reservoirs also impede sediment transport, thus affecting downstream fluvial geomorphology and aquatic habitat characteristics (Syvitski and Kettner, 2011). Overall, the effects of dams and reservoirs can be felt far downstream, sometimes for hundreds of kilometers or all the way to the river delta (Richter et al., 2010).

Due to their various effects, reservoirs can cause serious dislocations of traditional human societies, either by displacement or by indirect economic and livelihood changes (Duflo and Pande, 2007). The newly created stagnant waterbody of reservoirs can contribute to the spread of certain diseases that rely on waterborne vectors while they may reduce the incidence of others (Sleigh, 2006). Although reservoirs are often built to alleviate water scarcity, they can render cities ill-equipped to handle drought conditions or push water scarcity hotspots downstream, as upstream regions benefit from stored water to the detriment of those downstream (Veldkamp et al., 2017).

While large reservoirs number in the tens of thousands, millions of smaller reservoirs dot the landscape (Lehner et al., 2011), particularly in agricultural regions. An estimated 1.8 million small farm ponds in Australia store nearly 11 km^3 of water (Malerba et al., 2021) and more than 1 million dams fragment Europe's rivers, though not all of these create a reservoir (Belletti et al., 2020). While large reservoirs have customarily garnered more scientific attention, the individual and cumulative effects of small reservoirs are now being investigated and found to be considerable (Habets et al., 2018).

B. Zonation in reservoirs

As a result of the common linear morphology in a river valley, three distinct zones tend to develop along the longitudinal gradient of a typical medium-to-large reservoir: a *riverine zone*, a *zone of transition*, and a *lacustrine zone* (Fig. 4-20). Each zone possesses unique and dynamic physical, chemical, and biological characteristics. The width of the riverine zone is often relatively narrow and depends on the river's geomorphology. Although water velocities decrease as the water enters the reservoir, advective transport by currents is still sufficient to move significant quantities of fine suspended particulates (silts, clays, and organic matter) (Thornton, 1990). High particulate turbidity commonly reduces light penetration and limits primary production. Loading of organic matter from allochthonous sources is high in proportion to water volume in the riverine zone. High decomposition rates often result in high

TABLE 4-5 Reservoir Characteristics by Surface Area Size Class and Globally. All data from the GRanD v1.3 database (Lehner et al., 2011). Not including regulated natural lakes

Area size class (km^2)	Number of reservoirs	Total surface area (10^3 km^2)	Total volume (10^3 km^3)	Mean depth (m)
0.1–1[a]	63,464	17.8	575.9	*n.a.*
1–10[a]	10,515	29.5	865.9	*n.a.*
10–100	1843	55.1	1402.3	24.5
100–1000	353	94.9	2216.5	24.2
1000–10,000	46	105.7	2072.0	18.7
Global	**76,221**	**303.0**	**7,132.6**	***n.a.***
Largest reservoirs by volume (not including regulated natural lakes)				
Lake Kariba		5.3	185	35.1
Bratskoye Reservoir		4.8	169	35.1
Lake Nasser		5.4	162	30.1
Lake Volta		6.0	148	24.5
Caroni Reservoir		3.7	135	36.9
Williston Reservoir		1.7	74	43.2
Krasnoyarskoye Reservoir		1.6	73	45.0
Zeyskoye Reservoir		2.2	68	30.6
Cahora Bassa Reservoir		2.0	63	30.8
Robert Bourassa Reservoir		2.8	62	22.4

[a]*Derived using a statistical Pareto distribution model (Lehner et al., 2011).*

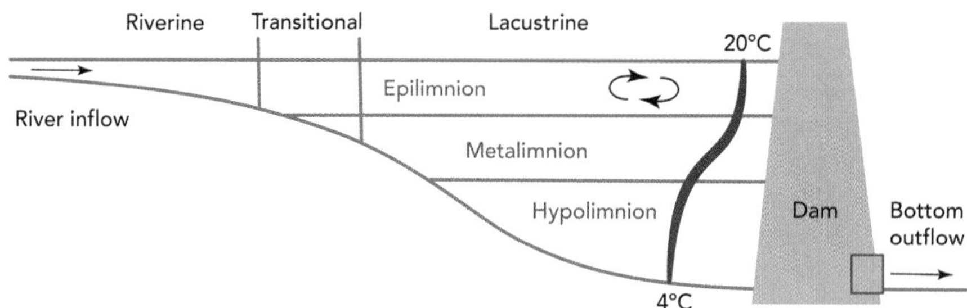

FIGURE 4-20 Generalized zones along longitudinal gradients in reservoirs. The curve ending in 4°C represents a stereotypical summer thermocline in a stratified deep reservoir. *(Modified from Ruhl et al., 2014.)*

consumption of dissolved oxygen, but aerobic conditions generally prevail in the shallow, well-mixed riverine zone.

Riverine water velocities decrease as energy is dispersed over wider areas in the transitional zone (Fig. 4-20). An appreciable portion of the turbidity load settles out of the upper water strata as a result. Decreased turbidity results in an enhanced depth of light penetration and increased rates of photosynthetic productivity by phytoplankton. Gradually, a shift occurs to an increasing percentage of total organic matter loading from phytoplankton and, in some shallow reservoirs, from rooted vascular plants.

Within the lacustrine zone, characteristics become more like those of lake ecosystems (Fig. 4-20). This portion of the reservoir often stratifies thermally and assumes many of the properties of natural lakes when it comes to planktonic production, nutrient limitations, sedimentation of organic matter, and decomposition in the hypolimnion. Stratification and water movements can be modified or complicated, both spatially and temporally, by hypolimnetic or bottom withdrawal of water from the reservoir at the dam (for a review of beneficial and detrimental effects of hypolimnetic withdrawal, see Nürnberg, 2007).

C. Geomorphological differences between reservoirs and lakes

Reservoirs differ in significant ways from natural lake ecosystems. Yet studies of reservoir ecosystems also indicate many functional similarities between artificial and natural lakes, such as in metabolic functioning and community interrelationships of the biota. Comparative analyses of the geomorphological characteristics and properties between natural lakes and managed reservoirs, including their integration within the river system, can thus be instructive to point out both structural differences and functional similarities in these ecosystems. A summary of comparable hydrological and morphological attributes is presented in Fig. 4-21 and Table 4-6.

Because reservoirs are typically formed in river valleys, the morphometry of the reservoir's waterbody reflects the nature of these valleys by being dendritic, narrow, and elongated (Table 4-6). These physical characteristics affect biological processes in many complex ways, the most important of which are light and nutrient availability.

Reservoirs for irrigation or drinking water are created predominantly in regions where natural lakes are sparse or unsuitable (e.g., too saline) for human exploitation. In these regions the climate tends to be warm and dry, resulting in relatively high evaporative losses that can outweigh precipitation inputs (Wetzel, 1990b). Reservoirs for hydropower production include some of the largest artificial lakes in the world. Their demand-driven management for electricity production can result in highly fluctuating water releases to serve short-duration peak requirements (so-called *hydropeaking*) creating downstream flows that are in stark contrast to natural conditions.

Given their purpose to collect and store water, many reservoirs are strategically built on larger rivers. Therefore the ratios between inflowing discharge and storage volume (i.e., the water renewal rates) of reservoirs tend to exceed those of natural lakes. The larger rivers also exhibit higher energy for erosion and thus carry larger sediment loads. Since reservoir inflows are primarily channelized and often not intercepted by energy-dispersive and biologically active wetlands and littoral interface regions, water influxes tend to be larger, are more directly coupled to precipitation events, and carry dissolved and particulate loads much farther into the reservoir than would be the case in most natural lakes (Table 4-6). These properties result in high, irregularly pulsed nutrient and sediment loading to the reservoir. As the reduced flow velocities inside the impoundment then cause sediment deposition, reservoirs generally serve as both recipients and collectors of riverine sediment loads. In consequence, the longevity of reservoirs can be very short, often less than 100 years, before their entire storage volume is filled with sediments.

Water-level fluctuations tend to be small and stable in natural lakes, and shoreline erosion is localized to wind-generated currents. In contrast, extreme and irregular water-level fluctuations are common in reservoirs as a

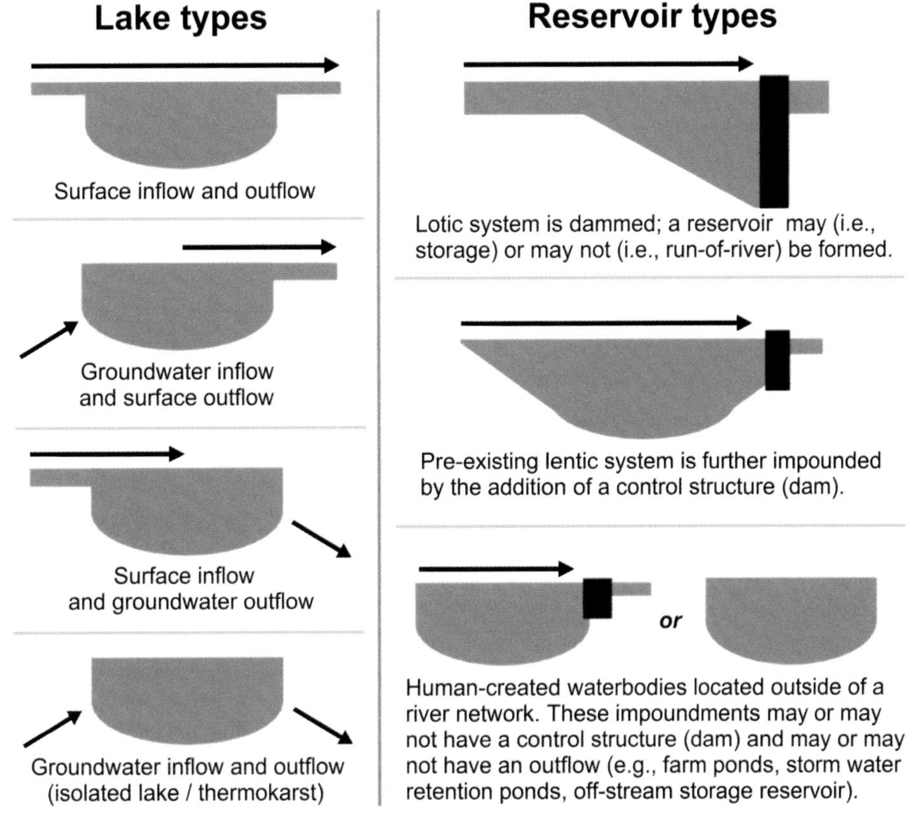
Lake types
Surface inflow and outflow
Groundwater inflow
and surface outflow
Surface inflow
and groundwater outflow
Groundwater inflow and outflow
(isolated lake / thermokarst)
Reservoir types
Lotic system is dammed; a reservoir may (i.e., storage) or may not (i.e., run-of-river) be formed.
Pre-existing lentic system is further impounded by the addition of a control structure (dam).
or
Human-created waterbodies located outside of a river network. These impoundments may or may not have a control structure (dam) and may or may not have an outflow (e.g., farm ponds, storm water retention ponds, off-stream storage reservoir).

FIGURE 4-21 *Left panel*: Natural lakes are water-filled depressions in the landscape that can vary in their inflow and outflow conditions. *Right panel*: Reservoirs, in contrast, are typically formed by damming (*black bars*) a river and can turn the lotic into a lentic system (although run-of-river dams create little storage); dams can also regulate or expand an original natural lake or be built off the river network. *(Modified from Hayes et al., 2017.)*

result of their specific flood inflow characteristics (in particular if built for flood prevention) and their often short-term water storage and withdrawal dynamics (Table 4-6). This has multiple effects on sediment loadings and causes a more unstable shoreline morphology. For example, reservoir drawdown events associated with irrigation requirements and hydroelectricity generation can lead to large areas of sediments along shorelines being alternately inundated and exposed. These shifts between aerobic and anaerobic conditions enhance greenhouse gas and nutrient release. Ultimately, the water-level manipulations can prohibit the establishment of productive, stabilizing wetlands and littoral plant communities around the reservoir and interfere with macrobiotic lifecycles (Richter and Thomas, 2007). Overall, this causes much-reduced nutrient and physical filtering capacities compared to those of most natural lake ecosystems (Wetzel, 1979, 1990a, 1990b).

VII. Climatic and anthropogenic effects

Climate change is altering the global hydrological cycle through changed precipitation inputs, increased snow and glacier melt, and overall increased evaporation rates. In addition, humans are modifying surface water flows either directly by extracting water for irrigation, industrial, and domestic uses, or indirectly by changing land uses (which in turn affects runoff processes), or both. These alterations lead to complex changes in global water availability and streamflow, which can ultimately cause the expansion, shrinking, formation, or disappearance of rivers and lakes, both in space and time. Furthermore, entirely new, artificial waterbodies are created by humans through dam construction (see section 4-VI above).

Overall, there has been a global net increase in permanent surface waters between 1984 and 2019 (Pekel et al., 2016; Pickens et al., 2020), with reservoir filling being the main cause. A similar net increasing trend has been found to persist for large natural lakes for the period 1992–2019, which is mostly attributed to climate variation as well as long-term climate change (Fig. 4-22; Kraemer et al., 2020). Examples of spatial expansion include some lakes of the Tibetan Plateau, which have grown since the late 20th century due to increasing precipitation, glacier and snow melt, and permafrost degradation following rising temperatures at high altitudes

TABLE 4-6 Comparative Geomorphological and Hydrodynamic Characteristics and Properties Among Natural Lakes and Anthropogenic Reservoirs[a]

Property	Natural lakes	Anthropogenic reservoirs
Shape	Predominantly circular to elliptical	Variable, often narrow, elongated, ovoid, dendritic, or triangular
Mean depth	Moderate to high; average <10 m	Mostly moderate, rarely exceeding 30 m
Depth gradient	Deepest usually remote from shoreline	Shallow in riverine portions, increasing toward transitional and lacustrine zones
Drainage basin	Lake usually central or in higher portion of drainage basin; drainage area often small in comparison to lake area (~10:1)	Reservoir often at base of drainage basin; drainage area large in comparison to reservoir area (~100:1 to 300:1)
Shoreline development	Relatively low; stable	Relatively high; astatic
Shoreline erosion and substrata distribution	Localized, induced by wind-generated waves and currents	Extensive in riverine areas by water currents; less from wind-induced currents in lacustrine zones
Sediment loading	Low to very low; deltas small, broad, gradation slow	Large with large drainage basin area; deltas large, channelized, gradation rapid
Deposition of sediments	Low, limited dispersal; relatively constant rates across seasons	High in riverine zone, decreasing exponentially down reservoir; greatest in old riverbed valley; highly variable across seasons
Sediments suspended in water (turbidity)	Low to very low; turbidity low	High, variable; high percentage of clay and silt particles; turbidity high
Water-level fluctuations	Small; stable	Large; irregular
Inflow	Runoff to lake via tributaries (often low stream orders) and diffuse sources; penetration into stratified waters small and dispersive	Most runoff to reservoir via river tributaries (high stream orders); penetration into stratified waters complex (over-, inter-, and underflows); flow is often directed along old riverbed valley
Outflow (withdrawal)	Relatively stable; usually via surface outflow or shallow groundwater	Highly irregular with water use; withdrawals from surface layers or from hypolimnion
Residence time	Often long, relatively constant (one to many years)	Often short, variable (days to several weeks); disruption of stratification with hypolimnetic withdrawal

[a]*Assembled from many sources, particularly the synthesis on lakes and reservoirs by Wetzel (1990b) and references cited therein.*

(Zhu et al., 2010; Liao et al., 2013). However, not all regions of the world are experiencing surface water expansions; severe depletion of surface waters occurs, for example, in the Middle East and the southwest United States (Fig. 4-22). A striking example is the Aral Sea (Central Asia), which lost 79% of its water surface area between 1960 and 2007, due to upstream irrigation withdrawals (Fig. 4-3; Aladin et al., 2009).

Air temperatures are rising at twice the average global rate in Arctic regions, and the thawing of permafrost is a main driver in both the disappearance and expansion of lakes in high-latitude areas. Overall, lake areas and numbers are increasing on continuous permafrost soils, which are permanently frozen soils that are thawing at the top only, whereas many lakes are permanently drained into the ground in regions with

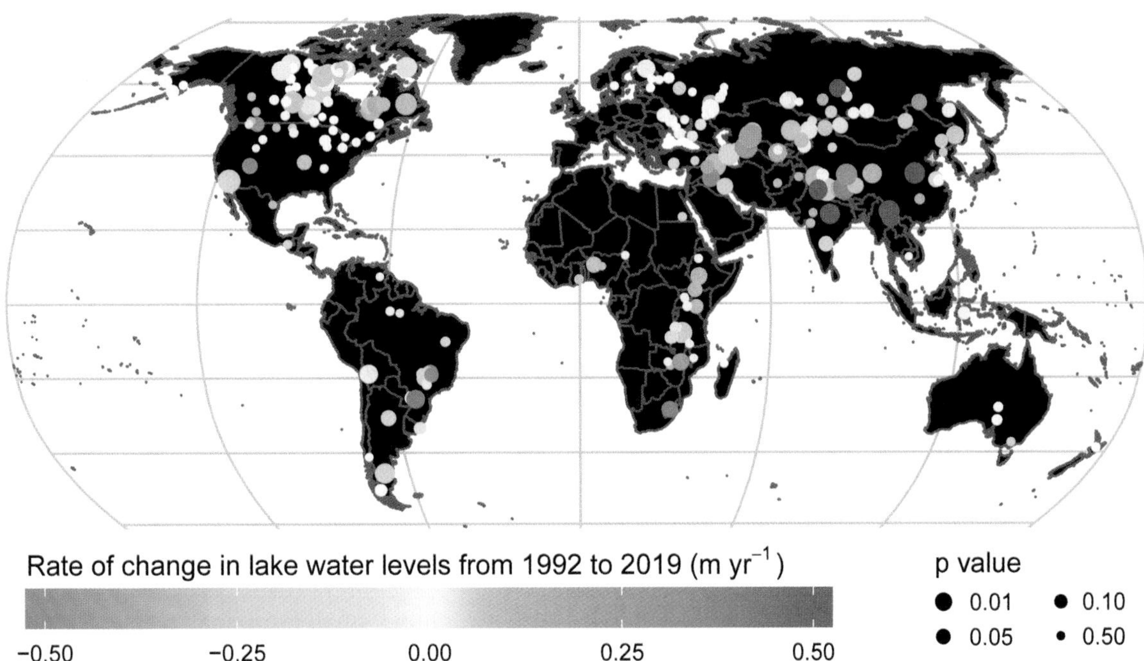

FIGURE 4-22 Long-term trends (1992–2019) in water levels for 200 large lakes. Colors indicate the rate of change and the p-value indicates the significance of the trend. *(Modified from Kraemer et al., 2020.)*

discontinuous permafrost soils (Smith et al., 2005). In addition, some high Arctic ponds, which have been permanent waterbodies for millennia, are now starting to dry during polar summers due to increased evaporation (Smol and Douglas, 2007).

VIII. Summary

1. The present distribution and prevalence of global rivers and lakes is governed by multiple factors related to both the origin and progression of these waterbodies over time. Past (and ongoing) geologic, climatologic, physiographic, and geomorphologic characteristics and processes are largely responsible for the initial shape of lake depressions and river channels. After their formation, hydroclimatic and landscape factors within the contributing drainage basins determine the hydrological cycle and thus the fluxes of water and sediments into and out of rivers and lakes, which cause continuous and dynamic transitions in their form and properties.
2. Estimates show that there may be more than 24 million km of interconnected rivers (with mean annual flow $\geq$100 L s^{-1}) draining the Earth's land surface. River density is highest in wet tropical climates, including the Amazon and Congo basins, as well as in colder landscapes in the north.
3. Rivers account for only an extremely small amount (<0.005%) of all freshwater stocks on Earth but are major transporters of dissolved and particulate matter from the land to the sea.
   a. Streamflow (i.e., discharge) is directly related to climatologic factors (e.g., precipitation, temperature), physiographic characteristics of the drainage basin (e.g., area, soils), and their interaction with vegetation (e.g., evapotranspiration) (see also Chapter 5).
   b. Stream and river channels move laterally within the floodplain because of erosion and redeposition of sediments.
   c. Forms of river channels deviate from four general classes: straight, meandering, braided, and anastomosing. River morphology varies widely with geology, drainage basin area, discharge, width–depth ratios, riverbed material size, and channel slope.
4. The global land surface is covered by an estimated 21 million lakes ($\geq$1 ha) with a total storage volume of about 183,000 km^3. Lakes are ubiquitous in formerly glaciated regions where countless depressions were created that filled with water, foremost in the subarctic and temperate regions of the Northern Hemisphere (Canada, Scandinavia, Siberia). In contrast, particularly large and old lakes can be found inside tectonic rifts such as Lake Baikal in Russia and the African Rift Valley lakes.
5. About two-thirds of the total volume of freshwater lakes is contained in only the four largest lakes (Lakes Baikal, Tanganyika, Superior, and Malawi).

Most lakes are much smaller, however, and relatively shallow, usually <10 m in depth.

6. Natural lakes are formed by either relatively fast (catastrophic) geological events or by more gradual processes. They can be grouped into several main types of origin:
   a. *Glacial lakes* are formed by the erosional and depositional activity of glaciers and represent the highest abundance of lakes globally. Main processes include scouring in glaciated valleys (e.g., cirque lakes, tarns, paternoster lakes), damming by morainal deposits, depressions created by ice blocks in the outwash of morainal debris (kettle lakes), or melting and freezing processes (cryogenic lakes).
   b. *Tectonic lakes* are depressions formed by displacements of the Earth's crust, creating many of the deepest lakes of the world (e.g., graben lakes, rift lakes). Uplifting of the Earth's crust can also form lake depressions.
   c. *Volcanic lakes* form in volcanic cones or in depressions resulting from the collapse of partially emptied magmatic chambers within a volcano (e.g., maar lakes, caldera lakes).
   d. *Landslides* can dam stream valleys and thus create temporary or permanent lakes.
   e. *River activity* in the form of erosional and depositional processes can isolate depressions that fill with water and turn into lakes (e.g., plunge-pool lakes, oxbow lakes, deltaic lakes).
   f. *Coastal lakes* often form along irregularities in the shoreline of the sea (or of a large lake), where currents deposit sediments in bars or spits that eventually isolate a fresh or brackish-water lake.
   g. *Solution lakes* result from sinks, termed dolines, formed by the gradual dissolution of rock, such as limestone.
   h. *Wind erosion* can form shallow depressions that often contain water temporarily or seasonally (e.g., aeolian lakes, dune lakes, playa lakes).
   i. More rarely, a lake can be formed in the depression caused by the impact of a *meteorite.*
   j. Natural lakes can also have *organic* origins, such as lakes created by the damming action of plant growth and associated detritus, or the activity of animals (in particular beavers).
7. After lake formation, succession processes start to fill the lake with the influx of river sediments and the accumulation of organic matter that grows in the lake. The succession process may be accompanied by different stages of nutrient loading (oligotrophic, mesotrophic, eutrophic, and hypereutrophic).
8. There is a great variation in lake morphology. Most waterbodies of lakes approximate an elliptic sinusoid shape. The mean depth of lakes, on average, is about half (0.46) of the maximum depth.
9. The morphology of a lake's waterbody has profound effects on nearly all physical, chemical, and biological properties of the lake. The morphometry of the lake and the geological substrates of the drainage basin influence sediment–water interactions and lake productivity, especially that of extremely productive littoral communities. The greater productivity of small, shallow lakes is usually correlated with the higher water–sediment interface area per water volume (i.e., lower mean depth).
10. Reservoirs are impoundments created by humans through the damming of river valleys. These dams act as barriers and fragment the connected network of naturally free-flowing rivers. The retention of fluvial sediments in reservoirs can have far-reaching consequences on downstream river morphology. If one includes small farm ponds, reservoirs have been estimated to be in the millions, with a total water storage capacity of over 7000 km^3. Given their purpose to store and release water for specific purposes, reservoirs are typically more dynamically fluctuating systems than lakes. Where high rates of sedimentation exist, reservoirs or small farm ponds can be very short-lived.
11. Besides the anthropogenic effects of reservoir construction, climate change is expected to both create new lakes, for example, through permafrost melting, and to remove lakes, for example, through desiccation caused by drier or hotter climates.

Acknowledgments

The revision of this chapter was supported by contributions, data, reviews, and visual art provided by colleagues, students, and emerging scholars including Penny Beames, Heloisa Ehalt Macedo, Olivia del Giorgio, Maartje C. Korver, Mathis L. Messager, Florence Tan, and Tianqi Xing.

References

Aladin, N.V., Plotnikov, I.S., Micklin, P., Ballatore, T., 2009. Aral Sea: water level, salinity and long-term changes in biological communities of an endangered ecosystem—past, present and future. Nat. Resour. Environ. Issues 15, 36.

Allen, G.H., Pavelsky, T.M., 2018. Global extent of rivers and streams. Science 361, 585–588.

Barbarossa, V., Schmitt, R.J.P., Huijbregts, M.A.J., Zarfl, C., King, H., Schipper, A.M., 2020. Impacts of current and future large dams on the geographic range connectivity of freshwater fish worldwide. Proc. Natl. Acad. Sci. 117.

Beadle, L.C., 1974. The Inland Waters of Tropical Africa: An Introduction to Tropical Limnology. Longman Group Ltd., London, p. 365.

Belletti, B., Garcia de Leaniz, C., Jones, J., Bizzi, S., Börger, L., Segura, G., et al., 2020. More than one million barriers fragment Europe's rivers. Nature 588, 436–441.

Bouchard, F., MacDonald, L.A., Turner, K.W., Thienpont, J.R., Medeiros, A.S., Biskaborn, B.K., et al., 2016. Paleolimnology of thermokarst lakes: a window into permafrost landscape evolution. Arct. Sci. 3 (2), 91–117.

Coulter, G.W. (Ed.), 1991. Lake Tanganyika and Its Life. Oxford University Press, Oxford, p. 354.

Davidson, N., Fluet-Chouinard, E., Finlayson, C., 2018. Global extent and distribution of wetlands: trends and issues. Mar. Freshw. Res. 69, 620–627.

Davis, W.M., 1909. Geographical Essays. Ginn & Company, Boston, p. 777.

Downing, J.A., Prairie, Y.T., Cole, J.J., Duarte, C.M., Tranvik, L.J., Striegl, R.G., et al., 2006. The global abundance and size distribution of lakes, ponds, and impoundments. Limnol. Oceanogr. 51, 2388–2397.

Duflo, E., Pande, R., 2007. Dams. Q. J. Econ. 122, 601–646.

Golubić, S., 1973. The relationship between blue-green algae and carbonate deposits. In: Carr, N.G., Whitton, B.A. (Eds.), The Biology of Blue-Green Algae. University of California Press, Berkeley, pp. 434–472.

Gordon, N.D., McMahon, T.A., Finlayson, B.L., 1992. Stream hydrology: an introduction for ecologists. John Wiley & Sons, Chichester, p. 526.

Gregory, K.J., Walling, D.E., 1973. Drainage Basin Form and Process. Edward Arnold Ltd., London.

Gregory, S.V., Swanson, F.J., McKee, W.A., Cummins, K.W., 1991. An ecosystem perspective of riparian zones. Bioscience 41 (8), 540–551.

Grill, G., Lehner, B., Thieme, M., Geenen, B., Tickner, D., Antonelli, F., et al., 2019. Mapping the world's free-flowing rivers. Nature 569, 215–221.

Grosse, G., Jones, B., Arp, C., 2013. Thermokarst lakes, drainage, and drained basins. In: Shroder, J., Giardino, R., Harbor, J. (Eds.), Treatise on Geomorphology, Volume 8: Glacial and Periglacial Geomorphology. Academic Press, San Diego, pp. 325–353.

Habets, F., Molénat, J., Carluer, N., Douez, O., Leenhardt, D., 2018. The cumulative impacts of small reservoirs on hydrology: a review. Sci. Total Environ. 643, 850–867.

Håkanson, L., 1981. A Manual of Lake Morphometry. Springer-Verlag, New York, p. 78.

Hamilton, S.K., Melack, J.M., Goodchild, M.F., Lewis Jr., W.M., 1992. Estimation of the fractal dimension of terrain from lake size dimensions. In: Carling, P.A., Petts, G.E. (Eds.), Lowland Floodplain Rivers: Geomorphological Perspectives. John Wiley & Sons Ltd., London, pp. 145–163.

Harper, D.M., 1992. Eutrophication of Freshwaters: Principles, Problems and Restoration. Chapman and Hall, London, p. 333.

Hayes, F.R., Anthony, E.H., 1964. Productive capacity of North American lakes as related to the quantity and the trophic level of fish, the lake dimensions, and the water chemistry. Trans. Amer. Fish. Soc. 93, 53–57.

Hayes, N.M., Deemer, B.R., Corman, J.R., Razavi, N.R., Strock, K.E., 2017. Key differences between lakes and reservoirs modify climate signals: a case for a new conceptual model. Limnol. Oceanogr. 2, 47–62.

Horton, R.E., 1945. Erosional development of streams and their drainage basins: hydrophysical approach to quantitative morphology. Geol. Soc. Am. Bull. 56 (3), 275–370.

Hutchinson, G.E., 1957. A Treatise on Limnology. I. Geography, Physics, and Chemistry. John Wiley & Sons, New York, p. 1015.

International Commission on Large Dams (ICOLD), 2020. General synthesis. Available at. http://www.icold-cigb.net/GB/world_register/general_synthesis.asp.

Kellerhals, R., Church, M., 1989. The morphology of large rivers: characterization and management. In: Dodge, D.P. (Ed.), Proceedings of the International Large River Symposium, Can. Spec. Publ. Fish. Aquat. Sci., vol. 106. Department of Fisheries and Oceans, Ottawa, pp. 31–48.

Kraemer, B.M., Seimon, A., Adrian, R., McIntyre, P.B., 2020. Worldwide lake level trends and responses to background climate variation. Hydrol. Earth Syst. Sci. 24 (5), 2593–2608.

Larson, G.L., 1989. Geographical distribution, morphology and water quality of caldera lakes: a review. Hydrobiologia 171, 23–32.

Lehner, B., Reidy Liermann, C., Revenga, C., Vörömsmarty, C., Fekete, B., Crouzet, P., et al., 2011. High-resolution mapping of the world's reservoirs and dams for sustainable river-flow management. Front. Ecol. Environ. 9, 494–502.

Leopold, L.B., Wolman, M.G., Miller, J.P., 1964. Fluvial Processes in Geomorphology. W. H. Freeman and Co., San Francisco, p. 522.

Liao, J., Shen, G., Li, Y., 2013. Lake variations in response to climate change in the Tibetan Plateau in the past 40 years. Int. J. Digit. Earth 6 (6), 534–549.

Linke, S., Lehner, B., Ouellet Dallaire, C., Ariwi, J., Grill, G., Anand, M., et al., 2019. Global hydro-environmental sub-basin and river reach characteristics at high spatial resolution. Sci. Data 6, 283.

Maavara, T., Chen, Q., van Meter, K., Brown, L.E., Zhang, J., Ni, J., Zarfl, C., 2020. River dam impacts on biogeochemical cycling. Nat. Rev. Earth Environ. 1, 103–116.

Malerba, M.E., Wright, N., Macreadie, P.I., 2021. A continental-scale assessment of density, size, distribution and historical trends of farm dams using deep learning convolutional neural networks. Remote Sens. 13, 1–17.

Messager, M.L., Lehner, B., Cockburn, C., Lamouroux, N., Pella, H., Snelder, T., et al., 2021. Global prevalence of non-perennial rivers and streams. Nature 594, 391–397.

Messager, M.L., Lehner, B., Grill, G., Nedeva, I., Schmitt, O., 2016. Estimating the volume and age of water stored in global lakes using a geo-statistical approach. Nat. Commun. 7, 13603.

Milliman, J.D., Farnsworth, K.L., 2011. River Discharge to the Coastal Ocean: A Global Synthesis. Cambridge University Press, Cambridge, p. 384.

Neumann, J., 1959. Maximum depth and average depth of lakes. J. Fish. Res. Board Can. 16, 923–927.

Niemeyer-Lüllwitz, A.Z., Zucchi, H. (Eds.), 1985. Fließgewässerkunde: Ökologie fließender Gewässer unter besonderer Berücksichtigung wasserbaulicher Eingriffe. Sauerländer Verlag, Berlin, Frankfurt/M.

Nürnberg, G.K., 2007. Lake responses to long-term hypolimnetic withdrawal treatments. Lake Reserv. Manag. 23, 388–409.

Oki, T., Kanae, S., 2006. Global hydrological cycles and world water resources. Science 313 (5790), 1068–1072.

Oswald, G.K.A., Robin, G.Q., 1973. Lakes beneath the Antarctic ice sheet. Nature 245, 251–254.

Ouellet Dallaire, C., Lehner, B., Sayre, R., Thieme, M., 2019. A multidisciplinary framework to derive global river reach classifications at high spatial resolution. Environ. Res. Lett. 14 (2), 024003.

Pekel, J.F., Cottam, A., Gorelick, N., Belward, A.S., 2016. High-resolution mapping of global surface water and its long-term changes. Nature 540, 418–422.

Pickens, A.H., Hansen, M.C., Hancher, M., Stehman, S.V., Tyukavina, A., Potapov, P., et al., 2020. Mapping and sampling to characterize global inland water dynamics from 1999 to 2018 with full Landsat time-series. Remote Sens. Environ. 243, 111792.

Pielou, E.C., 1998. Fresh Water. University of Chicago Press, Chicago, p. 275.

Prancevic, J.P., Kirchner, J.W., 2019. Topographic controls on the extension and retraction of flowing streams. Geophys. Res. Lett. 46, 2084–2092.

Rawson, D.S., 1955. Morphometry as a dominant factor in the productivity of large lakes. Verh. Internat. Verein. Limnol. 12, 164–175.

Richter, B.D., Postel, S., Revenga, C., Scudder, T., Lehner, B., Churchill, A., Chow, M., 2010. Lost in development's shadow: the downstream human consequences of dams. Water Altern. 3, 14–42.

Richter, B.D., Thomas, G.A., 2007. Restoring environmental flows by modifying dam operations. Ecol. Soc. 12, 1–26.

Rosgen, D.L., 1996. Applied River Morphology. Wildlife Hydrology Books, Pagosa Springs, p. 365.

Ruhl, N., DeAngelis, H., Crosby, A.M., Roosenburg, W.M., 2014. Applying a reservoir functional-zone paradigm to littoral bluegills: differences in length and catch frequency? PeerJ 2, e528.

Ruttner, F., 1963. Fundamentals of Limnology. University of Toronto Press, Toronto, p. 295.

Selby, M.J., 1985. Earth's Changing Surface: An Introduction to Geomorphology. Oxford University Press, Oxford, p. 607.

Shreve, R.L., 1966. Statistical law of stream numbers. J. Geol. 74 (1), 17–37.

Sleigh, A.C., 2006. Water, dams and infection: Asian challenges. In: Population Dynamics and Infectious Diseases in Asia. World Scientific, Singapore, pp. 57–71.

Smith, L.C., Sheng, Y., MacDonald, G., Hinzman, L., 2005. Disappearing Arctic lakes. Science 308, 1429.

Smol, J.P., Douglas, M.S.V., 2007. Crossing the final ecological threshold in high Arctic ponds. Proc. Natl. Acad. Sci. U. S. A. 104 (30), 12395–12397.

Strahler, A.N., 1957. Quantitative analysis of watershed geomorphology. Trans. Am. Geophys. Union 38 (6), 913–920.

Syvitski, J.P.M., Kettner, A., 2011. Sediment flux and the anthropocene. Philos. Trans. R Soc. 369, 957–975.

Thienemann, A., 1927. Der Bau des Seebeckens in seiner Bedeutung für den Ablauf des Lebens im See. Verh. Zool. Bot. Ges. Wien 77, 87–91.

Thornton, K.W., 1990. Perspectives on reservoir limnology. In: Thornton, K.W., Kimmel, B.L., Payne, F.E. (Eds.), Reservoir Limnology: Ecological Perspectives. John Wiley & Sons, New York, pp. 1–13.

Timoshkin, O.A., Sitnikova, T.Y., Rusinek, O.T., Pronin, N.M., Proviz, V.I., 2004. Index of animal species inhabiting Lake Baikal and its catchment area 1. Nauka, Novosibirsk, pp. 74–113.

Tweed, F.S., Carrivick, J.L., 2015. Deglaciation and proglacial lakes. Geol. Today 31, 96–102.

US National Park Service, 2020. Horst and graben. Available at. https://www.nps.gov/articles/horst-and-graben.htm.

Veldkamp, T.I.E., Wada, Y., Aerts, J.C.J.H., Döll, P., Gosling, S.N., Liu, J., et al., 2017. Water scarcity hotspots travel downstream due to human interventions in the 20th and 21st century. Nat. Commun. 8, 15697.

Verpoorter, C., Kutser, T., Seekell, D.A., Tranvik, L.J., 2014. A global inventory of lakes based on high-resolution satellite imagery. Geophys. Res. Lett. 41, 6396–6402.

WCD (World Commission on Dams), 2000. Dams and Development: A New Framework for Decision-Making. Earthscan, London.

Wetzel, R.G., 1979. The role of the littoral zone and detritus in lake metabolism. Arch. Hydrobiol. Beih. Ergebn. Limnol. 13, 145–161.

Wetzel, R.G., 1990a. Land-water interfaces: metabolic and limnological regulators. Verh. Internat. Verein. Limnol. 24, 6–24.

Wetzel, R.G., 1990b. Reservoir ecosystems: conclusions and speculations. In: Thornton, K.W., Kimmel, B.L., Payne, F.E. (Eds.), Reservoir Limnology: Ecological Perspectives. Wiley-Interscience, New York, pp. 227–238.

Wetzel, R.G., Likens, G.E., 2000. Limnological Analyses, third ed. Springer-Verlag, New York, p. 429.

Wetzel, R.G., 2001. Limnology: Lake and River Ecosystems, Third ed. Academic Press, San Diego, p. 1006.

Williams, W.D., Sherwood, J.E., 1994. Definition and measurement of salinity in salt lakes. Int. J. Salt Lake Res. 3, 53–63.

Winter, T.C., 1985. Physiographic setting and geologic origin of Mirror Lake. In: Likens, G.E. (Ed.), An Ecosystem Approach to Aquatic Ecology: Mirror Lake and Its Environment. Springer-Verlag, New York, pp. 40–53.

Zhu, L., Xie, M., Wu, Y., 2010. Quantitative analysis of lake area variations and the influence factors from 1971 to 2004 in the Nam Co basin of the Tibetan Plateau. Chin. Sci. Bull. 55, 1294–1303.

Zumberge, J.H., 1952. The Lakes of Minnesota. Their Origin and Classification. University of Minnesota Press, Minneapolis, p. 99.

C H A P T E R

5

Hydrological Systems

Roy C. Sidle[1] *and Takashi Gomi*[2]

[1]Mountain Societies Research Institute, University of Central Asia, Khorog, Tajikistan [2]Graduate School of Bioagricultural Science, Nagoya University, Nagoya, Japan

OUTLINE

I. The hydrological cycle and relevant processes

A. General description of the hydrological cycle

Freshwater bodies of the Earth, consisting of lakes, rivers, and streams, are regulated by inputs, outputs, and storage changes. The timing of these processes is critical to the fluxes of materials within these systems, as well as their interactions with proximate ecosystems. Precipitation in all forms is the main input to the hydrological cycle, but long-term reservoirs of glacial and permafrost water are also significant inputs in high-elevation and high-latitude environments during warm periods. All these inputs supply water to soils, groundwater, and surface waters, each of which has different storage and transit times depending on their inherent characteristics and seasonal effects. Thus water reaching lake systems can arrive as direct precipitation onto the lake surface, inflow from streams and rivers, overland flow from surrounding areas, groundwater and soil water seepage, glacial melt, and permafrost thaw.

The global hydrological cycle involves complex interactions and exchanges of liquid, gaseous, and solid phases of water among the geosphere and biosphere components, including important evaporative exchange from oceans with subsequent precipitation on landscapes. This exchange of water among reservoirs in the atmosphere, hydrosphere (including lakes), cryosphere, and terrestrial biosphere is expected to increase in warming climates. Land-use and infrastructure changes and expansion, such as agricultural development, urbanization, forest conversion, and road and reservoir construction, exert increasing impacts on the hydrological cycle, including effects on lake systems (e.g., Mango et al., 2011; Soranno et al., 2015). Adverse effects of these activities are most prevalent in developing countries, particularly dryland and mountainous regions, and many lakes in these regions have already been impacted. Climate change is also affecting the water sources and distribution, but in complex ways that are not well understood and often intertangled with land-use impacts. It is generally recognized that water

Wetzel's Limnology, Fourth Edition
https://doi.org/10.1016/B978-0-12-822701-5.00005-7

supplies in high-elevation and high-latitude regions will be disproportionally affected by climate change due to direct connections between warming and melt processes from snowpacks, glaciers, and permafrost (Immerzeel et al., 2010).

B. Water sources

Precipitation constitutes by far the largest input of water to the Earth, with about 95% of global precipitation occurring as rainfall. Rainfall generates an estimated 97.5% of global runoff to rivers, streams, and lakes. Snowfall is a significant water supply in higher latitudes and many high-elevation areas. Other forms of solid precipitation (hail and sleet) constitute minor sources of water. Snow and ice (e.g., glaciers, permafrost) on and within the land retain water and slowly release it to surface waters during melt seasons, constituting important water sources in high-latitude and high-elevation regions. Snow and ice constitute about 76.4% of the freshwater stores.

(a) Precipitation. Most precipitation in the mid to low latitudes falls as rain; however, the distribution and types of rain events vary widely. The highest global rainfall occurs in a band that seasonally moves north and south of the equator, causing the wet tropics and subtropics to experience severe, high-intensity rainstorms during typhoon, cyclone, monsoon, and hurricane seasons. Typically, land areas north of the equator receive much of their rainfall between November and April, whereas areas south of the equator experience heavy rainfall from May to October. Temperate environments often receive long-term, low-intensity rainfall events during frontal storm systems that are seasonally driven. In these midlatitudes the highest rainfall tends to occur along the east coasts of continents (Adler et al., 2017). Dry or seasonally dry regions typically receive short-duration, high-intensity convective storms. Depending on rainfall intensity, soil moisture, ground cover, and other environmental factors, a portion of the rain is rapidly routed to surface water bodies, and a portion is stored in soil and recharges groundwater reservoirs (Fig. 5-1).

At latitudes above 45° and in high elevations, snow contributions become increasingly important and are the dominant water sources of high elevation areas in East and Central Asia, known as the Water Towers of the world. The *snowline* decreases with increasing latitude across most continents, and snow cover tends to exhibit the greatest interannual variability near the snowline (Hammond et al., 2018). Intermittent, seasonal, and permanent snow cover affects northerly latitudes to a much greater extent than in the Southern Hemisphere due to much less land mass at high latitudes; however, where snow falls at higher latitudes (40–65°), it tends to be a bit greater on Southern Hemisphere lands (135 mm year^{-1}) compared to the Northern Hemisphere (125 mm year^{-1}) (Adhikari et al., 2018). In the Northern Hemisphere snowfall is more geographically fragmented than in the Southern Hemisphere.

Mountain ranges play an important role in the distribution of precipitation. The most common phenomenon is that windward flanks of mountain ranges receive more precipitation than leeward flanks, the latter residing in a precipitation shadow; however, many other

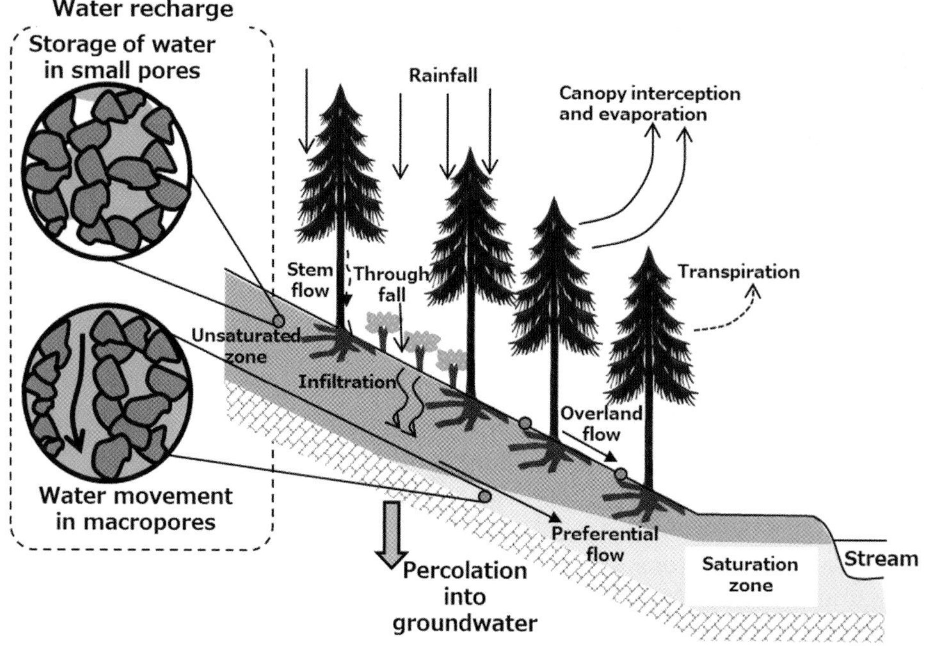

FIGURE 5-1 Schematic illustration of important hydrological processes and pathways on vegetated hillslopes.

orographic (induced by mountains) interactions with precipitation occur. Orographic effects on precipitation at smaller scales tend to be more complex. For example, on small hills, the maximum precipitation tends to occur near the crest. Typhoons and hurricanes are also modified by mountain topography (Roe, 2005). Although precipitation generally increases with elevation, the patterns and correlations between these attributes are highly variable from basin to basin. For example, in mountainous Central Japan a relatively low overall positive correlation ($r \approx 0.3-0.4$) was observed between elevation and monthly precipitation; however, when individual river basins were assessed separately, most basins had higher correlations ($r > 0.4-0.8$), suggesting that complex topography exerts a major control on orographic precipitation (Sasaki and Kurihara, 2008). In the Pamirs of Central Asia, large differences in annual precipitation occur at high elevations due to combinations of orographic blocking of westerly air masses, high-elevation plateaus, and complex topography, and much of the high-elevation East Pamir is an alpine desert.

A type of precipitation not captured in standard gages (i.e., *occult precipitation*) involves the condensation of water vapor onto vegetation foliage and the subsequent release of this water by dripping or melting. These inputs are generally minor, but in areas with dense forest canopies and heavy cloud cover, fog drip can constitute up to 10% of the total atmospheric hydrological input (Chang et al., 2006). Dew represents the condensation of water vapor into droplets on surfaces usually during nighttime cooling. Although this is a minor occult precipitation input, it helps sustain desert ecosystems. In colder regions fog can freeze on foliage and other surfaces, creating *rime*, which later melts.

(b) Glacial melt and permafrost thaw. Glaciers and permafrost represent largely ancient stores of precipitation that are seasonally augmented by snowfall and ablated during warmer periods. Although glaciers and ice caps outside of Greenland and the Antarctic cover only about 0.1% of the global land area, they contain the largest source of freshwater. Most glaciers worldwide have been losing mass in the past century, although at different rates and with some exceptions. With warming temperatures, more water is released from glaciers creating glacial lakes. These lakes typically form at the foot of the glacier (see Chapter 4) but also can form behind unstable terminal moraines or within depressions in the glacier where, if breached, catastrophic *glacial lake outburst floods* (GLOFs) can result as experienced in the eastern Himalayas (Veh et al., 2020). The risk of GLOFs also occurs in transitional areas where the glacial front migrates, such as those experienced in South America and Central Asia. However, such disasters are very sporadic and localized, and the contribution of melt from glaciers to receiving waters in most high latitudes is less than from snowmelt.

Although poorly studied, permafrost is estimated to occupy 24% of Northern Hemisphere lands and, as such, is a much more pervasive source of water release in a warming climate compared to glacial melt. *Permafrost* is defined as the subsurface soil–rock–water matrix that is continuously below 0°C for more than 2 years, but much of the permafrost formed many centuries ago during glacial periods. In addition to CO_2 and CH_4 emissions, contaminant releases, and soil stability problems, permafrost thaw associated with climate warming will cause thaw lakes and wetlands to expand, and resultant ground collapses will create new lakes (see Chapter 4). However, in some regions permafrost thaw can also result in lake shrinkage and lake drainage (e.g., Smith et al., 2005).

C. Evaporation and transpiration

Although *evaporation* and *transpiration* are often estimated as a lumped parameter, *evapotranspiration*, it is important to examine these processes separately related to water sources and budgets (Fig. 5-1). Evaporation represents a direct loss of water from open water bodies, soil surfaces, or as water evaporated from vegetation and litter cover that has not reached the soil. Transpiration is the physiological process whereby plants extract water from the soil and subsequently release a portion of this to the atmosphere to support metabolism and growth. Both processes affect water availability on the land surface as well as delivery to water bodies, but in different ways.

(a) Evaporation. Evaporation is driven by the availability of energy at water and land surfaces and occurs as the result of a vapor pressure gradient between the evaporating surface and overlying air mass, as well as the important influence of wind on water vapor transport. Evaporation from free water surfaces, including lakes, constitutes an important loss of water to the atmosphere. Average annual evaporation losses (mm d^{-1}) from lakes vary widely, with lakes in dry regions typically having the greatest losses (Table 5-1). In deep lakes, controls on evaporation are more complex than just the effects of surface humidity gradients and wind; in these deepwaters the thermodynamic exchange and storage within the entire water mass must be considered. In cold winter months, when the water surface temperature of lakes is higher than the overlying air, deep lakes, which retain heat, have the highest evaporation rates. In contrast, shallow lakes lose heat rapidly in cooler weather and may freeze over, blocking evaporation. Shallow lakes experience the highest evaporation during warm periods (Table 5-1).

TABLE 5-1 Evaporation Estimates and Limnological Features From Selected Lakes in Different Climatic and Geographic Regions

Lake name	*Region/coordinates*	*Lake surface area (km^2)*	*Mean lake depth (m)*	*Climate*	*Average evaporation loss (mm y^{-1})*
Lake Superior	47.72°N, 86.94°W North Central USA/ Canada	82,800	147	Cold, humid continental	584
Lake Ontario	43.25°N, 77.83°W Northeast USA/ Canada	18,960	86	Cool, humid continental	744
White Bear Lake	45.09°N, 93.01°W Twin Cities, Minnesota, USA	9.7	25	Humid continental	701
Lake Tahoe	39.10°N, 120.03°W Nevada/California, USA	490	305	Dry summer subtropical	1095
Lake Mead	Southwest USA 36.14°N, 114.41°W	640	69	Low desert, reservoir	1971
Great Salt Lake	Utah, USA 41.12°N, 112.48°W	4400	6	Semiarid, saline lake	1011
Last Mountain Lake	Saskatchewan, Canada 51.09°N, 105.26°W	215	7.6	Continental, cold winters, hot summers	702
Issyk-Kul	Kyrgyzstan 42.39°N, 77.29°E	6236	280	Dry steppe	934
Tucurui Reservoir	Pará, Brazil 3.50°S, 49.30°W	2875	20	Tropical wet	1784
Lake Victoria	Tanzania, Uganda, Kenya 1.00°S, 33.00°E	68,800	40	Tropical wet	1593
Doosti Reservoir	Iran/Turkmenistan 35.57°N, 61.10°E	35	15	Semiarid	1997
Lake Vegoritis	Northern Greece 40.47°N, 21.80°E	44	29	Semidry Mediterranean	945
Lake Ikeda	Kyushu, Japan 31.15°N, 130.56°E	11	233	Subtropical, caldera lake	828
Lake Biwa	Shiga, Japan 35.14°N, 130.56°E	674	40	Humid subtropical	754
Lake Titicaca	Peru/Bolivia Andes 15.45°S, 69.25°W	8372	105	Subtropical highland/ alpine	1700
Lake Tuusulanjärvi	Southern Finland 60.44°N, 25.06°E	6.0	3.1	Humid continental, borderline subarctic	482 (ice-free months)
Serling Lake	Qiangtang area, Tibet 31.50°N, 89.00°E	2391	124	Plateau continental high-elevation salt lake	985
Sea of Galilee Lake Kinneret	Israel 32.82°N, 35.59°E	1699	26	Hot, dry Mediterranean	1490

Data Compiled from Various Sources that Used Different Methods to Estimate Lake Evaporation.

Evaporation from land surfaces typically constitutes only about 10% of total evapotranspiration during warm summers (Tezza et al., 2019). During wet periods, soil evaporation is largely controlled by solar radiation, whereas during dry periods, evaporation is more regulated by soil hydrological properties. Heavily vegetated areas experience less soil evaporation due to shading. Evaporation can also occur from litter on the ground surface. Because organic litter can store water and can be hydrophobic when dry, as much as 20% of throughfall from tree canopies can evaporate from litter surfaces before reaching the soil (Gerrits et al., 2010). Litter and decomposed organic matter on the soil surface also serve to reduce evaporation losses from soils.

Vegetation, particularly dense tree canopies, intercept significant amounts of rainfall and snow, allowing this water to evaporate or sublimate back to the atmosphere without reaching the soil. Interception losses in forests typically represent 10% to more than 40% of gross precipitation depending on tree species, canopy density, and climate. Tree and shrub canopies can store considerable rainwater when dry, but once saturated, most of the rain falls to the forest floor (Fig. 5-1). High winds during intense rains also reduce canopy storage. Multitiered forests, such as in the tropics, have a high capacity to intercept rainfall, as water can be stored at various levels within the forest structure. In temperate climates conifers intercept more rainfall in winter than deciduous trees due to leaf fall from the latter. In regions with winter snowpacks, conifers intercept considerably more snow than deciduous trees. Evaporation of snow from tree canopies is promoted by moderate velocity, dry winds (strong winds may blow snow from trees), persistence of snow in trees, and strong radiation.

A related, albeit minor, water loss is *sublimation*, whereby snow or ice converts to water vapor without melting. Sublimation more readily occurs when warm, dry winds pass over snowpacks or ice at relatively high velocities. High-altitude conditions are also subject to sublimation due to low air pressure.

(b) Transpiration. Transpiration is a complex physiological process in which water loss from vegetation is controlled by stomatal openings on the surface of foliage, rooting depth, type of vegetation and stage of growth, cover density, vegetation height, and dormancy. Climatic controls on transpiration include wind speed, humidity, radiation balance, and shading. Availability of soil moisture is a major limitation on transpiration; if soil water is freely available, transpiration does not vary greatly for different vegetation types. Transpiration declines when soils become dry and when solar radiation decreases. Rates of transpiration from grasslands are considerably lower than from forests, largely due to the ability of forest species to extract water from greater depths in the soil. Various methods have been applied to estimate transpiration, including measuring sap flow, canopy fluxes, and catchment water balances; however, accurate estimation of transpiration remains challenging, particularly across wide areas. Phreatophytes, deep-rooted plants that extract water from saturated zones, are known to deplete baseflow in streams and potentially water levels in lakes via high transpiration rates when they grow near water bodies. Helophytes transpire water directly from lake margins, with water losses exceeding evaporation from open water surfaces in some cases during the growing season (Grabowska et al., 2016).

D. Surface and shallow subsurface runoff processes

Both surface and subsurface flow pathways route precipitation and stored water from landscapes to streams, rivers, and lakes (Fig. 5-1). The relative importance of these pathways depends on climate, ambient precipitation patterns, vegetation, topography, soils, and geology. These pathways determine the timing of water delivery to streams and lakes during storms or melting, as well as water quality due to the interaction and residence time of water within different "reservoirs" (e.g., surface soil, subsoil, bedrock).

(a) Surface runoff processes. Surface runoff in catchments occurs via two mechanisms: (1) *infiltration-excess overland flow* (i.e., *Hortonian overland flow*); and (2) *saturation overland flow.* Hortonian overland flow (HOF) occurs when precipitation intensity (or snowmelt intensity) at any time exceeds the infiltration capacity of the soil; thus the excess water that cannot infiltrate runs off on the land surface. This does not mean that this runoff cannot reinfiltrate into the soil as it flows downslope to water bodies (Gomi et al., 2008). Dry landscapes with sparse vegetation cover support more HOF compared with temperate areas where vegetation is denser and soils have higher infiltration capacities due to the accumulation of organic matter in surface horizons. Because of high decomposition rates in the warm tropics, HOF can occur, especially if the thin surface litter cover is displaced. The condition of exposed soil surfaces affects HOF. If crusts form due to raindrop impact, infiltration capacity is reduced, causing more precipitation to run off (van de Giesen et al., 2000). Stones, organic debris, and other roughness elements on the surface of otherwise bare soils reduce the velocity of overland flow and may promote some reinfiltration. HOF is the most responsive and rapid pathway of all water delivery mechanisms to streams and lakes, assuming it is continuously connected along the flow path. Arid and semiarid climates also promote HOF because the small amount of annual rainfall that occurs often falls as high-intensity storms, favoring HOF on sparsely vegetated slopes. Certain dry surface organic materials are hydrophobic, causing HOF during the first portion of rain events, but these effects typically disappear as the surface wets. Soil surfaces exposed to hot fires generate more and longer lasting HOF due to the combustion of organic material and coating of mineral soils with hydrophobic residues.

Saturation overland flow (SOF) emerges when rain or snowmelt occurs on fully saturated soils with no opportunity for infiltration. In areas where soils are nearly saturated, a small amount of precipitation or melt is required to generate SOF. As such, areas of SOF often expand during long-duration storms or as wet seasons progress, particularly where gently sloping topography allows such expansion. Stream or lake-side riparian corridors and wetlands are ecosystem components most prone to SOF and facilitate the rapid transport of overland flow to receiving waters. The amount of SOF entering lakes depends on the spatial extent of the riparian zone surrounding the water body as well as the degree of saturation in these proximate soils. Other landscape features where SOF may occur include geomorphic hollows (i.e., swales or zero-order basins), slope breaks, and areas of poorly drained soils where perched water tables form. A special type of SOF, known as *return flow*, occurs at the base of hillslopes or hillslope discontinuities where soils are shallow and subsurface water is forced to the surface or where hillslopes intersect riparian corridors. Return flow can also occur when subsurface flow from upslope encounters a saturated zone, which exceeds the transmission capacity of water within the soil, forcing some of the flow onto the surface as return flow. As with all overland flow, return flow is a rapid transmission vector of water to streams and lakes, especially when it is generated near surface water margins; however, return flow typically contributes to the later portion of the *storm hydrograph* (Fig. 5-2).

(b) Subsurface runoff processes. Subsurface flow occurs in all environments but is a dominant runoff process in relatively undisturbed and vegetated landscapes, particularly in temperate, subtropical, and tropical regions.

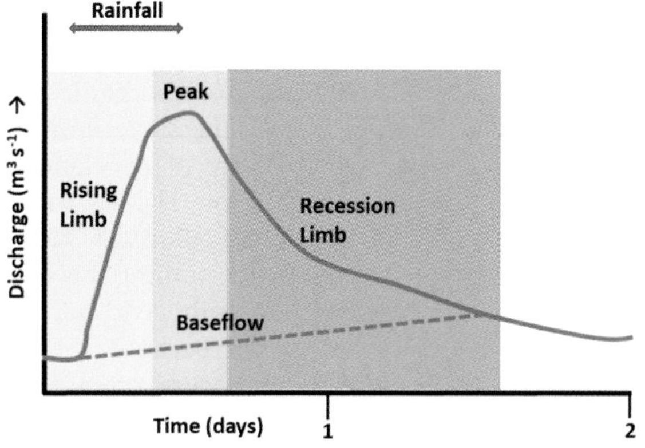

FIGURE 5-2 Simplified example of a storm hydrograph showing the various hydrograph components during a rainstorm. The *yellow-shaded portion* of the rising limb represents rapid runoff sources that are near or highly connected to the stream (e.g., SOF, HOF); the *blue-shaded area* surrounding the peak is influenced by these processes, but also other subsurface pathways that take time to develop (e.g., preferential flow, saturated subsurface matrix flow, return flow); and the *red area* on the recession limb is mostly influenced by subsurface flow (both preferential and matrix flow) and drainage from remote parts of the catchment.

These areas are characterized by soils with relatively high infiltration capacities that exceed most rainfall intensities or snowmelt rates that occur; thus little overland flow occurs. Once infiltrated, this subsurface flow either recharges deep groundwater or travels downslope to augment streamflow or lakes. These recharge and transport processes can take different paths with various travel times depending on soil depth, hydraulic conductivity, and the presence or absence of preferential flow pathways (Fig. 5-1). Subsurface flow can either be saturated or unsaturated, the former being much more rapid than the latter. Unsaturated flow can be important over the long term to redistribute water in landscapes or cause evaporation from soil surfaces, but these are slow processes. Saturated subsurface runoff can directly contribute to stormflow, especially in small mountain catchments with steep slopes and narrow riparian corridors. Saturated flow is characterized as either *matrix flow*—that is, flow through the fine pores of the soil matrix—or *preferential flow*—that is, more rapid subsurface flow that occurs in a small portion of the soil associated with macropores, wetting front instabilities, soil pipes, and funneled or deflected flow (Fig. 5-1).

Saturated matrix flow can be described by *Darcy's law*, which states that flow velocity in porous media is a function of the permeability of the substrate and the hydraulic gradient. In a soil without significant preferential flow pathways, this can be described as:

$$V = K_S(\Delta H/L)$$

where V is the flow velocity through the porous media; K_S is the saturated hydraulic conductivity; and $\Delta H/L$ is the hydraulic gradient, in which ΔH is the change in total hydraulic head (pressure head + elevation head) from one position to another and L is the linear flow path over which the hydraulic head change is measured. Thus for lateral subsurface flow in relatively steep hillslopes, clearly the elevation head dominates over the pressure head in determining flow velocity. Darcy's law is only valid for *laminar* (nonturbulent flow), and this subsurface velocity is much slower than overland flow. However, steep vegetated hillsides can contribute substantial amounts of subsurface matrix flow to streams during storm events, contributing to both the peak and recession limb of storm hydrographs (e.g., Hewlett and Hibbert, 1967; Sidle et al., 1995) (Fig. 5-2). K_S of the soil matrix is strongly affected by soil texture, with sandy soils having much higher values than clayey soils. Because K_S is usually assessed in small core samples, the derived values do not capture effects of "secondary porosity" of water transmission—for example, effects of soil structure, macropores, and other preferential flow pathways.

Saturated preferential flow is common in vegetated soils where live and decayed root channels reside, as well as in soils that experience shrinking and swelling, freezing-thaw, subsurface erosion, and dissolution processes (Fig. 5-1). Preferential flow is much more rapid than matrix flow in soils but slower than overland flow. In many cases preferential flow initiates in wet soils and involves the connectivity of various macropores across portions of the hillslope (Sidle et al., 2000). In these cases it takes time for these pathways to self-organize and preferential flow typically contributes to the peak and falling limb of storm hydrographs (Fig. 5-2). In peat soils, loess deposits, and other highly erodible substrates, larger, more continuous soil pipes may develop that rapidly convey water to streams and lakes during storms (Wilson et al., 2016). In larger macropores and soil pipes, flow is often turbulent and cannot be described by Darcy's law.

E. Groundwater flow

Groundwater exists in saturated substrates such as soil, unconsolidated sediment, or bedrock, where the uppermost saturated surface is the water table. Groundwater comprises more than 40 times the amount of water compared to surface freshwater (rivers, streams, and lakes), but the residence time of groundwater is much longer. An *aquifer* is the saturated substrate that supports the movement of groundwater from a region of higher to lower hydraulic head, as described by Darcy's law. Rates of groundwater movement commonly vary from 7 to 60 cm d^{-1} depending on the aquifer characteristics; in comparison, velocities of mountain rivers and streams are at least 10^5 times higher. When the water table is high, groundwater can feed streams and lakes (influent) and when it is low, water seeps from these water bodies (effluent) into the groundwater reservoir. Aquifers are recharged by infiltration into and through the soil mantle, a process that is spatially variable based on topography and substrate characteristics. Compared to surface water dynamics, the effects of short-term climate fluctuations on groundwater are temporally damped; however, long-term climate change may exert substantial effects on groundwater systems, including how these affect the supply of water to streams, lakes, and wetlands (Alley et al., 2002). Feedbacks among dynamic atmospheric inputs, hydraulic properties and pathways in the porous media, and groundwater storage and release are becoming more challenging in an era of climate change.

As in soils, geological substrate has both *primary* (fine pores within the subsurface matrix) and *secondary porosity*. In many cases the secondary porosity in aquifers (e.g., fractures, dissolution cavities) provides the most effective transport of groundwater. Karst aquifers are notorious for their large solution cavities that can transmit groundwater at rapid rates, thus providing a high potential, albeit somewhat elusive, supply of water. Mapping surface fracture traces has proven to be an effective method of locating productive wells in carbonate aquifers (Lattman and Parizek, 1964). Fractures and joints along bedding planes in sandstone aquifers provide important flow paths. In crystalline and volcanic rocks, almost all flow occurs along fractures and joints. Fractures can also develop in otherwise low-permeability tills (glacial or marine), originating from desiccation, glaciotectonic, and unloading processes, greatly influencing the rate of recharge to deeper aquifers (Nilsson et al., 2001).

Aquifers are classified as unconfined and confined. In an *unconfined aquifer* water within the porous media is not under pressure and the water table freely rises and falls during changing conditions of recharge and water withdrawals. In *confined aquifers* the saturated zone is under pressure and typically recharged from a distance or through seepage. Both the upper and lower boundaries of a confined aquifer have very low permeability; thus if a well is drilled into this aquifer, the water table will rise above the saturated zone (i.e., an *artesian well*).

The *residence time of groundwater* is quantified as the volume of the groundwater reservoir divided by the rate of input or output. Residence time strongly affects the chemistry of groundwater as the water interacts with rock materials with implications for point and nonpoint source pollution. For example, pollutants migrating into aquifers with long residence times pose severe problems for remediation. Conversely, pollutants that move rapidly through solution channels or large fractures will not be strongly adsorbed on the substrate and will progress directly toward wells or receiving surface water bodies. Groundwater seepage into lakes is difficult to quantify but has been shown to be significant in some systems (LaBaugh et al., 1997; Alifujiang et al. 2017).

II. Catchment structure and runoff generation

A. Catchment concept and delineation

A *catchment* (also termed watershed, drainage basin, or drainage area, see Chapter 4) is defined as a topographical delimited area that gathers and transmits water from a higher elevation site and transports this water to a designated downstream location. Catchments are separated from each other by topographic divides such as ridges, peaks, and saddles. Mountainous catchments typically comprise steep upland slopes, which supply water, sediments, and nutrients to lower-lying areas and floodplains as runouts, accumulations, and depositions. The catchment structure is typically used to assess material (e.g., sediments, nutrients) budgets in the continuum from hillslopes to channels. The analysis of such budgets allows us to understand the sources, transport, storage, and runoff of materials (Dietrich and Dunne, 1978). Routing and storage of water and sediment from uplands to lowlands depends on the characteristics of channel networks in a catchment.

Water and sediment from steep hillslopes first flow into small streams called *headwater channels*. This uppermost area of the channel network within a catchment is the headwater system (Gomi et al., 2002). A headwater system consists of multiple units, including hillslopes, zero-order basins, and first- and second-order stream channels. *Zero-order basins* or *swales* have concave topography (i.e., hollows) without defined channels and no perennial surface flow. Other hillslope segments are typically divergent or planar surfaces with no channelized flow. Following the stream ordering system of Strahler (1957), *first- and second-order channels* have either perennial and/or ephemeral flow depending on runoff generation from hillslopes and bedrock. Other channel ordering systems have been developed (see Chapter 4) and ordering also depends on the resolution of maps and whether only perennial channels are included in the assessment (Gomi et al., 2002). Some later hydrological and geomorphic studies suggest that the relative upper size limit of first-order streams is likely 1 km^2 (Woods et al., 1995; Montgomery and Foufoula-Georgiou, 1993).

Water and sediment are transported from headwater systems to larger water bodies, such as lakes and rivers, depending on the connectivity of hydrological processes within catchments. Related to the transport of such materials within catchments, connectivity from upstream to downstream is important, as explained by the River Continuum Concept (Vannote et al., 1980). Furthermore, connectivity from numerous tributaries to main channels, including the presence of floodplains and alluvial fans within channel networks, is a key geomorphic component that alters water and sediment dynamics (Sidle et al., 2017). Hence hydrological processes from hillslopes to channels that define the headwater area vary depending on vegetation, geology, and soil formation (Gomi et al., 2002). Such differences alter the order of magnitude of connectivity within catchments.

Drainage patterns represent key structures of catchments. Drainage patterns depend on geologic structure, rock types, tectonic activity (presence of faults), landform shapes, and anthropogenic alternation. Dendritic pattern is a typical channel network in relatively homogeneous geology or where geology exerts less control, whereas nondendritic drainage patterns (e.g., trellis) occur in areas underlain by different lithologies and tectonic activity. Such network structures alter the timing and accumulation of runoff, sediment, and nutrient transport from headwaters to lakes. For instance, the mobility of mass movements (landslides and debris flows) is affected by tributary junction angles. Thus the locations of lakes and ponds within catchments with respect to the channel network are important for understanding routing and storage of water and sediment, as well as for examining human impacts on lakes and ponds. Furthermore, locations of dams and reservoirs that control water and sediment movement need to be considered as key components within catchments.

B. Flow evolution from hillslopes to catchments

Hydrologic and geomorphic processes are closely coupled from hillslopes to stream channels within confined and steep valleys of headwater systems. Stormflow generation in headwater channels is affected by the degree of soil saturation in hillslopes and zero-order basins and by changing antecedent moisture conditions (Sidle et al., 2000). Such dynamic changes in soil moisture and runoff associated with rainfall is called the variable source area concept. This conceptual framework describes the expansion of source areas for rapid runoff occurring from the onset of rainfall to later periods when soils are saturated.

During dry conditions before rainfall, streamflow is usually generated by subsurface flow or bedrock flow; this is termed *baseflow* (Fig. 5-2). In dry conditions rainfall typically infiltrates into the soil matrix; occasionally, HOF occurs during intense precipitation. Pore spaces in the soil, both large and small, play different roles in relation to water supplied by rainfall (Fig. 5-1). In the early stages of rainfall, water is stored in small pores between soil grains. In this state, water is present in some parts of the soil, termed the "unsaturated" state. Unsaturated water is retained in the soil or flows slowly within the soil matrix under an energy gradient. When precipitation continues and sufficient water accumulates in the soil, water flows into and out of large pores causing drainage. The rate of water movement in the saturated state is 100 to 1000 times faster than in the unsaturated state (Tani et al., 2020). As rainfall proceeds, subsurface flow within the soil matrix emerges at the foot of hillslopes and in riparian areas, gradually increasing with increasing wetness of the basins. When subsurface water reaches the stream, streamflow begins to increase. Concurrently, surface runoff also initiates in some ephemeral channels.

With increasing wetness in catchments, zero-order basins accumulate water and initially those with relatively shallow soils start contributing to surface runoff; as more water accumulates, zero-order basins with deeper soils contribute to surface runoff (Sidle et al., 2000; Fig. 5-3). Preferential flow from hillslopes also augments

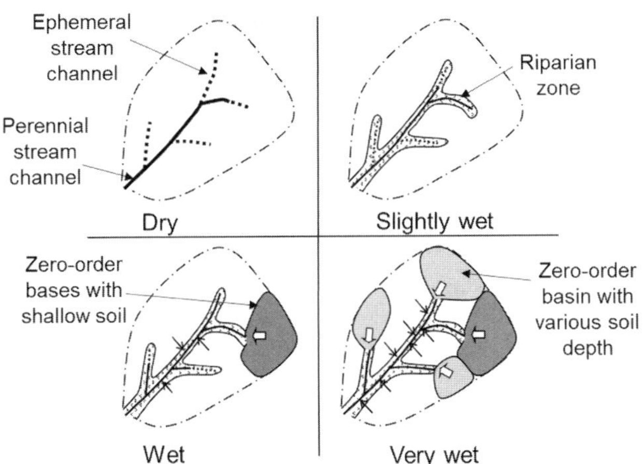

FIGURE 5-3 Conceptual view of flow process evolution from hillslopes to catchments. (Modified from Sidle et al., 2000.) During dry conditions, riparian zones and direct precipitation on channels are the only active sites of flow generation. Throughflow from the soil matrix in the foot of hillslopes and riparian areas gradually activates with increasing wetness. Zero-order basins with relatively shallow soils (*dark-shaded areas*) begin to contribute surface runoff (*large arrows*) during wet conditions, whereas preferential flow (*small arrows*) from hillslopes contributes less to stream flow. Flow begins to occur in transitional channels emerging from zero-order basins during wet conditions. Zero-order basins, including those with deeper soils (*lighter-shaded areas*) and preferential flow, actively contribute to storm flow during very wet conditions.

stormflow once a threshold of soil water is reached (Sidle et al., 2001). Ephemeral channels emerging from zero-order basins typically have high flows during major storms preceded by very wet conditions. Some of the soil water percolates into bedrock and is stored for longer periods. In such cases the presence of a soil layer allows water to flow slowly through the soil and promotes infiltration into the bedrock (Kosugi et al., 2006). Runoff generation from areas of exposed soil or bedrock caused by anthropogenic or natural disturbances (e.g., soil displacement during earthquake-induced landslides) also should be considered as "variable sources" for a more comprehensive understanding of runoff generation in catchments, because significant changes in flow pathways may occur (Sidle et al., 2018). During rain and rain-on-snow events, nearly saturated conditions in hydrologically responsive areas (e.g., zero-order basins or swales) may induce landslides and subsequent debris flows (Sidle et al., 1985).

III. Water balance in catchments with lakes

A. General water balance concept

Globally, more water evaporates from oceans than is returned by precipitation, whereas on land, overall, more water falls as precipitation than is lost by

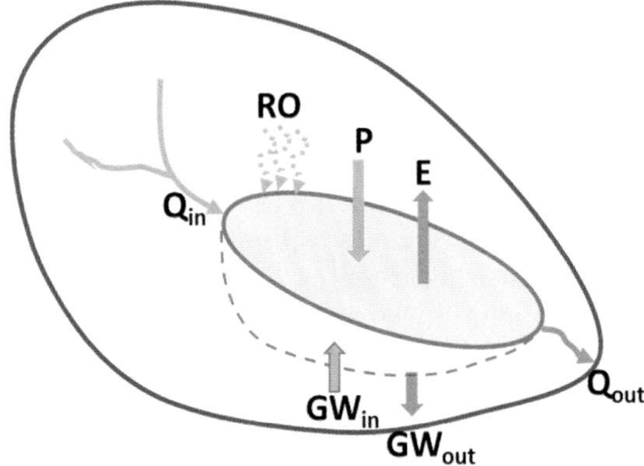

FIGURE 5-4 Example of water balance of a lake. Inputs are *blue arrows*: P is precipitation; GW_{in} is groundwater seepage; Q_{in} is streamflow inputs; and RO is catchment runoff. Outputs are *red arrows*: E is evaporation; GW_{out} is losses to groundwater; and Q_{out} is streamflow output. Catchment is not to scale.

evaporation. A large part of the landfall precipitation originates from evaporation of oceanic water. Massive water demands and diversions for agriculture, urban and industrial development, and land-clearing effects and climate change are creating changes in what would otherwise be a relatively stable global water balance. Furthermore, construction of large reservoirs and other impounded river waters can modify local climates, particularly in dry and seasonally dry regions (Degu et al., 2011).

A *water balance* for a catchment or a surface water body like a lake can be simply described as water inputs (inflow) equals outputs (outflow) plus changes in storage. For lake systems, the water balance is largely regulated by inputs of stream and river flow; runoff from the surrounding catchment; direct precipitation; groundwater and soil water seepage; and glacier and permafrost melt, whereas outputs are associated with outflow of surface water, evaporation, seepage losses, and sublimation (Fig. 5-4). In hydrologically modified systems, water diversions can also significantly contribute to lake inputs and outputs. Changes in water storage require monitoring of lake-level changes, as well as bathymetry, over a designated period. This process can be simplified by conducting the water balance over a period where lake levels are the same at the beginning and end. Projecting how precipitation timing and magnitude, evaporation, and runoff will change over several decades can provide insights into how lake water levels might change in a warming climate (Hunt et al., 2013).

(a) Inputs. Precipitation that falls directly onto lakes generally contributes only a small input to lake systems. Exceptions include when lake catchment areas are very small and lake surfaces are relatively large. For example, precipitation over the Great Lakes, where the lake areas

constitute about one-third of the total contributing basin, is the largest input of the water balance (Neff and Nicholas, 2005). The reliability of precipitation inputs depends on the number and position of gages around the lake and across the lake surface. Uncertainties in direct precipitation inputs to lakes arise from inadequate gage coverage and underestimation of rainfall and particularly snowfall catch by gages, especially during windy conditions. In larger lake systems, advances in remotely sensed data collection (e.g., active and passive data from satellites, radar-based precipitation estimates) provide an alternative method for acquiring data over these vast water bodies.

Stream and small river flows into lakes often constitute major inputs of water, particularly in small to medium-sized lakes. Oftentimes multiple small streams enter these lakes, presenting a major effort for monitoring via stream gaging. The largest uncertainty in measuring streamflow arises during large flood events when streams overtop their banks introducing errors in peak flow estimation. An alternative approach to estimating surface water inputs is to use catchment hydrology models, which require considerable catchment data for calibration. Another input is direct overland runoff into lakes; this will typically not be captured by stream gages but can be estimated by process-based catchment models. Assuming that all incoming streams are gaged and/or a robust catchment model is used, these surface water inputs into lakes likely have the least error of all water inputs.

Groundwater seepage contributions into lakes (*exfiltration*) are difficult to quantify and have largely been either ignored or estimated as a residual in most lake water balances. Seepage into lakes occurs when the lake bottom lies below the local groundwater table. Seepage meters can be used to directly measure groundwater inputs, but only in shallow waters near the perimeter of lakes. Indirect methods for estimating seepage have employed isotope balance approaches using oxygen-18 composition of lake water, precipitation, and groundwater, as well as using vertical profiles of radon as a groundwater tracer. Both indirect approaches require rather extensive sampling, employ expensive analytical methods, and have limitations related to climate and geological settings. As such, groundwater seepage estimates typically have the highest uncertainty of all inputs. Models of three-dimensional (3-D) groundwater transport can help quantify seepage; however, hydraulic properties and relevant boundary conditions must be specified. Groundwater springs that enter lakes from surrounding catchments can be quantified as contributions to surface water runoff, either by stream discharge or direct overland flow.

(b) Outputs. Evaporation from lakes constitutes a substantial portion of outputs, particularly in shallow lakes with large surface area to volume ratios (Table 5-1). Lakes in dry regions with no outlets (closed lakes) lose their water primarily by evaporation. Quantification of evaporation from lakes is typically accomplished using an energy budget approach requiring meteorological monitoring on and proximate to the lake. Lacking such resources, evaporative losses are estimated as residuals in the water balance. Aquatic plants also transpire water from lakes. This process is highly seasonal in temperate lake systems but can persist year-round in the tropics and subtropics, at times surpassing water losses due to evaporation depending on the level of biomass. Sublimination (the phase transformation of water from a solid to vapor form) only affects water loss from lakes that are frozen most of the year (e.g., in Antarctica), where high winds persist.

Similar to surface water inflows, outflows from lakes often constitute the major portion of the output budget. Outflows in single or multiple channels can be monitored at gaging stations. Unlike inflows, it is not possible to apply catchment hydrology models to estimate lake outflows. Such monitored flows have a relatively high degree of certainty. Outflows from lakes have direct effects on the water balance given their high velocities in contrast to losses (or gains) by groundwater seepage.

Seepage losses from lakes to groundwater reservoirs (*infiltration*) typically occur in dry sites where water tables are well below the lake bottom. In addition to the location of the water table, permeable lake bottoms (e.g., sandy or gravelly materials) are more likely to lose water by seepage compared to lakes with rock or clay substrate. Groundwater models can potentially be used to estimate seepage losses with the same constraints as for seepage gains into the lake.

(c) Changes in storage. Quantifying the change in lake storage over time allows us to close the lake water balance. This is basically accomplished by monitoring lake-level changes combined with assumptions or measurements of lake bathymetry. A recent global-scale study assessed seasonal changes in water storage of large lakes using remotely sensed data from multiple sources together with assumptions related to lake bathymetry (Chen et al., 2021). Although these lake volume dynamics provide useful data related to climate change and lake ecology, they cannot address the specific inputs and outputs that cause these changes.

B. Lake water balance examples

Constructing a complete water balance for lakes has proven to be a difficult task, especially for larger lakes. As noted, quantifying groundwater seepage both into and out of large lake systems is only possible by estimating this as a residual in the overall water balance, thus subjecting these estimates to large errors. As a result, groundwater fluxes are often ignored in water balances of large lakes (see Table 5-2). Of the eight lake systems we summarize where water balances were estimated, only one of the six large lakes (lake area >2000 km^2) considered groundwater inflows and none estimated any seepage losses (Table 5-2). Other than the two very small lakes, only Lake Seminole (a reservoir in southwest Georgia, USA) has estimates of both groundwater inputs (16% of total water inputs) and seepage losses (2.9% of total water outputs) (Healy et al., 2007). Groundwater inputs in the two small lakes, Williams Lake and Mirror Lake, constituted 74% and 16%, respectively, of total water inputs. The two midsize lakes with closed outlets (Urmia Lake and Issyk-Kul) and the small, closed lake (Williams Lake) have high ratios of precipitation to total water inputs (0.30–0.77), although the very large Lake Victoria with a large lake to catchment area ratio had the highest precipitation to water input ratio (0.82).

IV. Management implications

Human activities have played a significant role in modifying water sources, the timing and pathways of water transport, and the ultimate delivery of water to rivers, streams, and lakes. These modifications have resulted from extensive land-use practices and land-cover change, as well as more intensive anthropogenic practices (e.g., road and trail construction, residential development, and urbanization). Furthermore, climate change and variability have impacted water sources and water delivery to freshwater bodies in ways that have been difficult to assess and decouple from anthropogenic activities (Li et al., 2020).

A. Effects of land-cover change

Forest conversion to other land-cover types is one of the most widespread anthropogenic activities affecting catchment hydrology. Much higher rates of evapotranspiration occur from forest stands compared to areas converted to grasslands, brushlands, or agricultural croplands. As such, soil moisture is usually higher in these converted sites than in forested lands, thus generating more annual runoff to receiving streams. However, the pathways that water takes on its way to streams during storms or snowmelt may be affected by individual conversion practices. These hydrological pathways are also influenced by the generally lower amount of organic matter in surface soils of converted sites compared to forests. Surface organic matter serves to promote infiltration and thus subsurface runoff, especially true in temperate regions. When substantial organic matter has been removed, hydrological pathways can change from being subsurface flow dominated to surface runoff dominated, thus accelerating water delivery to streams and enhancing peak flows (Ziegler et al., 2004).

Fire is widely used to convert forests to other land uses, particularly in the developing world. Hot fires impart *hydrophobicity* in surface soils that may last for several years after the initial burn. This water repellency is most pronounced during the first portion of rainstorms when the surface soils are dry; once the soils become wetter, the effect of hydrophobicity is reduced (Robichaud, 2000). Nevertheless, fire-affected landscapes can increase overland flow, peak flows in streams, and surface erosion.

Forest harvesting followed by reforestation (i.e., sustainable timber harvesting) exerts little influence on catchment hydrology compared with forest conversion practices when surface soils are only moderately disturbed. The organic horizon in most forest soils provides a buffer against surface runoff by enhancing the infiltration capacity of surface soils. Surface runoff from forested areas is mostly associated with roads and trails that are used in the logging, thinning, and site maintenance practices. During extended periods after forest harvesting and extensive thinning, both total annual streamflow and baseflow typically increase due to lower evapotranspiration (Bent, 2001; Dung et al., 2012). In areas that develop significant winter snowpacks, forest harvesting increases the snow accumulation on the ground, resulting in more spring runoff and potentially higher peak snowmelt discharges. Tropical forests and dryland forests are more sensitive to site disturbances because of sparse organic matter in surface soils, the former due to high rates of decomposition and the latter due to fewer trees and less biomass.

Mechanized agricultural practices can also enhance runoff. Excessive cultivation of surface soils destroys natural soil aggregates and in the long term promotes more overland flow compared to no-till or limited-till agriculture. Also, poor irrigation management—for example, overirrigation or irrigating already wet soils using saline water—causes excess surface runoff that can deliver sediments and agrochemicals to receiving waters. Poor cultivation techniques, like plowing up and down hillsides, also exacerbate overland flow and sediment transport. Plow pans, compacted zones in the subsoil, can cause SOF to develop in sites where water would otherwise naturally percolate deeper into the soil profile (Sidle et al., 2006). These practices can all increase peak flows in streams during storms and periods of peak snowmelt.

Heavily grazed pastures also produce increased amounts of storm and snowmelt runoff due to

TABLE 5-2 Water Budgets for Selected Lakes of Different Sizes and Geographic Areas

Lake and location	Lake and catchment areas; other attributes	Precipitation inputs	Surface water inputs	Groundwater inputs (gains)	Evaporation losses	Surface water outputs	Groundwater leakage (losses)	Change in storage	Time frame of estimates	Reference
					($km^3\ y^{-1}$)					
Lake Victoria, east-central Africa	LA: 68,800 km^2 CA: 193,000 km^2 2nd largest freshwater lake	117	26	N/M	105	33	N/M	N/M	1950–2004 data	Sewagudde, 2009
Lake Seminole, Georgia, USA	LA: 152 km^2 CA: 44,548 km^2 reservoir	0.18	10.5	2.05	0.27	11.6	0.36	0.089	April 2000–September 2001	Healy et al., 2007
Lake Erie, NE Canada–USA border	LA: 25,700 km^2 CA: 78,000 km^2	22.8	187	N/M	23.2	184	N/M	N/M	1913–1960	Neff and Nicholas, 2005
Lake Taihu, Yangtze Delta, eastern China	LA: 2338 km^2 CA: 36,895 km^2	3.24	10.5	N/M	2.2	10.2	N/M	0.67	1998–2000	Kelderman et al., 2005
Urmia Lake, northwest Iran	LA: 6000 km^2 CA: 51,875 km^2 hypersaline closed lake	1.65	3.78	N/M	6.11	0.0	N/M	N/M	1985–2010	(Ghale et al., 2018)
Issyk-Kul, northeast Kyrgyzstan	LA: 6236 km^2 CA: 15,844 km^2 deep, endorheic	1.95	2.42	1.87	5.58	0.0	Assume zero	Land use effects	1980–2012	Alifujiang et al., 2017
Williams Lake, central Minnesota, USA	LA: 0.364 km^2 CA: 2.26 km^2 completely closed basin	0.135×10^{-3}	0 (runoff negligible)	0.39×10^{-3}	0.216×10^{-3}	0 (no outflow)	0.49×10^{-3}	-0.183×10^{-3}	July 1991–June 1992	LaBaugh et al., 1997
Mirror Lake, New Hampshire, USA	LA: 0.15 km^2 CA: 1.03 km^2 oligotrophic	0.182×10^{-3}	0.417×10^{-3}	0.113×10^{-3}	0.077×10^{-3}	0.257×10^{-3}	0.347×10^{-3}	0.016×10^{-3}	1981–2000	Healy et al., 2007

compaction caused by ungulate trampling. Such effects are most pronounced in hydrologically active areas of catchments where animals tend to congregate because of more lush forage. Such riparian corridors, swales, and other seasonally moist areas are closely coupled to stream channels, and these changes in runoff from subsurface to overland flow can enhance peak flows in streams and potentially reduce baseflow due to restricting infiltration and water storage in soils. Additionally, in heavily grazed sites extensive cattle trails have been shown to transport and concentrate overland flow, causing initiation or expansion of gully systems (Koci et al., 2020).

B. Effect of intense anthropogenic activities

Anthropogenic practices that compact soils, seal soil surfaces, and concentrate runoff onto limited areas can significantly increase HOF and thus deliver water more rapidly to streams during rainfall and snowmelt events. The compacted or sealed surfaces of roads and trails facilitate the generation of HOF and this runoff concentrates and discharges at culvert outlets or low-lying discharge nodes along the road where it can be rapidly connected to streams (Sidle et al., 2004). Furthermore, when secondary roads are cut into steep hillslopes, they have the potential to intercept subsurface flow, where it is converted to rapid overland runoff on the road surface. If not intercepted, this water continues to travel to streams via slower subsurface flow. Even though roads and trails typically make up only a small portion of the catchment (often <1% of the area), they contribute a disproportional amount of rapid runoff (Ziegler et al., 2004). Ultimately, it is the degree of connectivity of these roads and trails with stream systems that determines the extent to which they enhance storm runoff.

Urban and residential areas contain a complex array of highly connected sealed surfaces (e.g., paved areas, buildings, roads) together with drainage infrastructures, all of which facilitate the rapid transport of HOF and contaminants to receiving waters (Miller et al., 2014). Because of the extensive nature of sealed surfaces and drains, the connectivity of these rapid runoff paths to streams is high. The resulting storm hydrographs in urbanized catchments are flashier—that is, peak flows are higher and appear earlier compared to a vegetated catchment, whereas duration of storm runoff is shorter. Additionally, when residential construction takes place on hillsides, excavations into the hillslope as well as placement of fill materials alter natural hydrological pathways, typically redirecting subsurface flow to rapid overland flow pathways. Some urban areas are incorporating green infrastructure, such as infiltration gullies and basins, vegetated rooftops, roadside green areas, and permeable asphalt and parking lots, into their stormwater management plans to reduce direct runoff into drainage systems that directly link to streams. Benefits of green infrastructure include reducing peak flows, mitigating pollutant runoff, and decreasing stream temperatures.

Development and agricultural activities around lakes and near streams that drain into lakes affect the transport of not only water but also sediments, nutrients, and contaminants into lake systems (Karakoç et al., 2003). Sealed surfaces and compacted areas facilitate rapid transport of materials into lakes and runoff from agricultural fields and home yards may transport agrochemicals and pesticides. As such, better planning is required to establish vegetated buffers to break the hydrological connectivity from source areas into the lake. A further problem exists when septic systems are used in residences around lakes, causing subsurface transport of contaminants, which may also contribute to algal blooms and pathogen contamination.

C. Climate change effects

Climate change and climate variability have important implications for small catchment and river basin hydrology, but the effects are often quite locally dependent, and the long-term trajectories are uncertain. Even the well-recognized temperature increases are not consistent globally and have been periodically affected by volcanic and solar activities. Evidence of long-term trends in precipitation is less convincing, with both annual increases and decreases in annual precipitation found in different regions (Blöschl and Montanari, 2010). Although climate-change effects on increases in large storms and flooding have been implicated in many venues, these remain much less predictable compared to temperature changes. This is partly due to the complicating aspects of land-use change, local orographic and geomorphic effects, and other dynamic environmental variables. In some temperate regions that receive substantial snowfall, peak flows due to rainfall and baseflows have increased over the past few decades, but the larger snowmelt peak flows have decreased (Novotny and Stefan, 2007). Uncertainties associated with using Global Circulation Models (GCMs) to estimate large precipitation events increase substantially at more local scales and for larger events (Blöschl et al., 2007). Additionally, in regions where runoff is generated by multiple sources (e.g., rainfall, snowmelt, glacial melt), the estimation of climate change effects on runoff becomes very complicated. For example, in high-elevation regions like Central Asia and the Himalayas, snowpacks may become more variable and eventually decline, but glaciers will release more water in a warming climate until they become disconnected from streams or no longer exist. Predicting these changes together with changing rainfall patterns becomes a daunting task and will depend substantially on local conditions. Thus large-scale climate change generalizations derived from GCMs are not very useful for predicting local hydrological changes.

Temperature-driven hydrological changes (e.g., droughts, low flows) are more predictable than extreme precipitation inputs. Climate models project that in many regions not significantly affected by snow accumulation and melt, discharge during periods of low flow will decline (Wanders et al., 2015). Patterns of droughts are generally expected to increase in the long-term, particularly in dry climates where, if coupled with poor land management practices (e.g., overgrazing, unsustainable cultivation), they can induce desertification.

V. Summary

1. The global hydrological cycle involves interactions and exchanges of liquid, gaseous, and solid phases of water within components of Earth and the biosphere.
   a. Freshwater inputs, outputs, and storage changes determine fluxes of materials in ecosystems. Typically, freshwater bodies with rapid turnover rates (e.g., rivers, streams) have the smallest storage of water at any given time, whereas freshwater bodies with very slow turnover (e.g., glaciers, groundwater) contain much more water.
   b. Rainfall is the dominant input of water to the Earth, but at latitudes above 45° and in high elevations, snow contributions become increasingly important. Mountains affect the distribution and type of precipitation, but often in complex ways.
   c. Dense vegetation, particularly forest cover, intercepts large amounts of incoming precipitation, which may evaporate back into the atmosphere. Snow interception by conifer forests is the greatest water loss. In contrast, evaporation from land surfaces is relatively minor. Transpiration of soil water by various plants depends on rooting depth, age, and other factors and can be modified by climate controls and available soil moisture.
   d. Surface runoff occurs when rainfall (or snowmelt) intensity exceeds the infiltration capacity of the soil or when the soil is already saturated, thus not permitting precipitation to infiltrate. The former is a very rapid and responsive flow pathway during storms that contributes substantial water to streams and lakes in dry, sparsely vegetated regions.
   e. Subsurface flow contributes to storm runoff in streams but is more important in temperate and tropical regions where vegetated soils have high infiltration rates.
   f. Groundwater supports baseflow in streams and rivers and recharges influent lakes.
   g. A hydrograph is a graph of streamflow (discharge) versus time. Storm hydrographs depict discharge changes during a precipitation event, enabling the examination of stream response to water inputs and a qualitative assessment of the contributions of various flow paths.
2. Catchments are topographically delimited areas that collect and transmit water from higher to lower elevations feeding streams, rivers, and lakes.
   a. The geomorphic structure of catchments affects the movement of water, sediments, and nutrients in the continuum from hillslopes to freshwater bodies.
   b. Drainage patterns that develop over time reflect catchment geology, tectonic activity, landform shapes, and anthropogenic alterations.
   c. Locations of lakes within the catchment drainage structure are important for understanding the routing and storage of sediment and nutrients.
   d. In sloping terrain, headwater streams are often fed by concave landforms (zero-order basins) that collect subsurface flow and release it to streams once a threshold of water has accumulated. Subsurface flow in hillslopes (including preferential flow) also responds to increasing levels of soil moisture. Both of these processes increase the response of storm runoff in headwaters.
   e. During a continuum from dry to wet catchment conditions (see Fig. 5-3), runoff is therefore associated with variable source areas.
   f. The rate of subsurface water movement in saturated soils is several orders of magnitude greater than in unsaturated soils.
3. The water balance of a catchment or a lake describes the water inputs, outputs, and changes in water storage.
   a. Water budgets for lake systems are often dominated by inflows and outflows from streams or rivers, but evaporation can be a significant output, particularly in dry areas. Additionally, precipitation is a major input in closed lakes and lakes with very large surface areas.
   b. Assessing groundwater seepage into and out of the lake poses challenges and is often ignored.

c. Estimating changes in lake water storage requires monitoring of lake levels over time and lake bathymetry.
d. Surface runoff into lakes is usually not estimated but may be significant for shallow lakes.

4. Land management practices and other human activities have modified water sources; timing and pathways of water transport; and the delivery of water to rivers, streams, and lakes.
   a. Land-cover changes affect water sources and pathways as follows:
      i. Vegetation conversion from forests to agriculture or grasslands reduces evapotranspiration and modifies soil surfaces, often inducing more overland flow, especially when fire is used in the conversion process.
      ii. Hot fires used in land clearing induce water repellency in surface soils, causing more surface runoff during the first part of storms until soil becomes wetter.
      iii. Excessive cultivation of surface soils and plowing up and down hillslopes promotes overland flow and can increase peak flows in streams.
      iv. Poor irrigation management causes excessive overland flow.
      v. Heavy animal grazing, especially near streams, converts storm runoff from subsurface to overland flow, enhances peak flows in streams, and potentially reduces baseflow.
   b. Intense anthropogenic activities affect water sources and pathways as follows:
      i. Roads and other compacted or sealed surfaces do not allow infiltration into the soil, increasing rapid overland flow and contributing to higher peak flows.
      ii. Urban and residential development contains a complex array of highly connected sealed surfaces and drainage infrastructures, which facilitate rapid overland runoff and transport of contaminants to receiving waters.
   c. Climate change is affecting the radiation balance of the Earth, which in turn has implications for hydrological processes, but in very complex ways, such as:
      i. Climate change appears to be most pronounced in higher latitude regions where snowfall constitutes a substantial portion of the overall precipitation inputs.
      ii. However, multiple sources of precipitation and water release in high-latitude regions pose prediction challenges of climate change impacts on hydrology.
      iii. In most areas it is extremely difficult to disentangle transparently the effects of land management changes from those due to climate change.
      iv. Thus current broad-scale climate change predictions are not useful for local-scale water resource planning or for forecasting hydrological changes in lake dynamics.

References

Adhikari, A., Liu, C., 2018. Global distribution of snow precipitation features and their properties from 3 years of GPM observations. J. Climate 31, 3731–3754.

Adler, R.F., Gu, G., Sapiano, M., Wang, J.-J., Huffman, G.J., 2017. Global precipitation: means, variations and trends during the satellite era (1979–2014). Surv. Geophys. 38, 679–699.

Alifujiang, Y., Abuduwaili, J., Ma, L., Samat, A., Groll, M., 2017. System dynamics modeling of water level variations of Lake Issyk-Kul. Kyrgyzstan. Water 9, 989.

Alley, W.M., Healy, R.W., LaBaugh, J.W., Reilly, T.E., 2002. Flow and storage in groundwater systems. Science 296, 1985–1990.

Bent, G.C., 2001. Effects of forest-management activities on runoff components and ground-water recharge to Quabbin Reservoir, central Massachusetts. Forest. Ecol. Manage. 143, 115–129.

Blöschl, G., 2010. Climate change impacts—throwing the dice? Hydrol. Process. 24, 374–381.

Blöschl, G., et al., 2007. At what scales do climate variability and land cover change impact on flooding and low flows? Hydrol. Process. 21, 1241–1247.

Chang, S.-C., Yeh, C.-F., Wu, M.-J., Hsia, Y.-J., Wu, J.-T., 2006. Quantifying fog water deposition by in situ exposure experiments in a mountainous coniferous forest in Taiwan. Forest. Ecol. Manage 224, 11–18.

Chen, T., Song, C., Ke, L., Wang, J., Liu, K., Wu, Q., 2021. Estimating seasonal water budgets in global lakes by using multi-source remote sensing measurements. J. Hydrol. 593, 125781.

Degu, A.M., Hossain, F., Niyogi, D., Pielke, R., Shepherd, J.M., Voisin, N., et al., 2011. The influence of large dams on surrounding climate and precipitation patterns. Geophys. Res. Lett. 38, L04405.

Dietrich, W.E., Dunne, T., 1978. Sediment budget for small catchment in mountainous terrain. Z. Geomorph. N.F. Suppl. Bd. 29, 191–206.

Dung, B.X., Gomi, T., Miyata, S., Sidle, R.C., Kosugi, K., Onda, Y., 2012. Runoff responses to forest thinning at plot and catchment scales in a headwater catchment draining Japanese cypress forest. J. Hydrol 444-445, 51–62.

Gerrits, A.M.J., Pfister, L., Savenije, H.H.G., 2010. Spatial and temporal variability of canopy and forest floor interception in a beech forest. Hydrol. Processes 24, 3011–3025. https://doi.org/10.1002/hyp.7712.

Ghale, Y.A.G., Altunkaynak, A., Unal, A., 2018. Investigation anthropogenic impacts and climate factors on drying up of Urmia Lake using water budget and drought analysis. Water. Resour. Manage. 32, 325–337.

Gomi, T., Sidle, R.C., Richardson, J.S., 2002. Understanding processes and downstream linkages of headwater systems. Bioscience 52, 905–916.

Gomi, T., Sidle, R.C., Miyata, S., Kosugi, K., Onda, Y., 2008. Dynamic runoff connectivity of overland flow on steep forested hillslopes: scale effects and runoff transfer. Water Resour. Res. 44 (8), W08411.

Grabowska, K., Borowiak, D., Nowińsk, K., 2016. The impact of helophyte transpiration on the vertical water exchange in water bodies. Limnol. Rev. 16 (3), 129–140.

Hammond, J.C., Saavedra, F.A., Kampf, S.K., 2018. Global snow maps and trends in snow persistence 2001–2016. Int. J. Climatol. 38, 4369–4383.

Healy, R.W., Winter, T.C., LaBaugh, J.W., Franke, O.L., 2007. Water budgets: foundations for effective water resources and environmental management. US Geol. Surv. Circular 1308, 90.

Hewlett, J.D., Hibbert, A.R., 1967. Factors affecting the response of small watersheds to precipitation in humid areas. In: Sopper, W.E., Lull, H.W. (Eds.), Proceedings of the International Symposium on Forest Hydrology. Pergamon, New York, pp. 275–290.

Hunt, R.J., Walker, J.F., Selbig, W.R., Westenbroek, S.M., Regan, R.S., 2013. Simulation of climate-change effects on streamflow, lake water budgets, and stream temperature using GSFLOW and SNTEMP, Trout Lake Watershed, Wisconsin. U.S. Geol. Surv. Sci. Investigations Rep. 2013– 5159, 118.

Immerzeel, W.W., van Beek, L.P.H., Bierkens, M.F.P., 2010. Climate change will affect the Asian water towers. Science 338, 1382–1385.

Karakoç, G., Erkoç, F.Ü., Katırcıoğlu, H., 2003. Water quality and impacts of pollution sources for Eymir and Mogan Lakes (Turkey). Environ. Int. 29, 21–27.

Kelderman, P., Wei, Z., Maessen, M., 2005. Water and mass budgets for estimating phosphorus sediment-water exchange in Lake Taihu (China P.R.). Hydrobiologia 544, 167–175.

Koci, J., Sidle, R.C., Jarihani, B., Cashman, M.J., 2020. Linking hydrological connectivity to gully erosion in savanna rangelands tributary to the Great Barrier Reef using Structure-from-Motion photogrammetry. Land. Degrad. Dev. 31, 20–36.

Kosugi, K., Katsura, S., Katsuyama, M., Mizuyama, T., 2006. Water flow processes in weathered granitic bedrock and their effects on runoff generation in a small headwater catchment. Water Resour. Res. 42, W02414.

LaBaugh, J.W., Winter, T.C., Rosenberry, D.O., Schuster, P.F., Reddy, M.M., Aiken, G.R., 1997. Hydrological and chemical estimates of the water balance of a closed-basin lake in north central Minnesota. Water Resour. Res. 33, 2799–2812.

Lattman, L.H., Parizek, R.R., 1964. Relationship between fracture traces and the occurrence of ground water in carbonate rocks. J. Hydrol. 2, 73–91.

Li, B., Shi, X., Lian, L., Chen, Y., Chen, Z., Sun, X., 2020. Quantifying the effects of climate variability, direct and indirect land use change, and human activities on runoff. J. Hydrol. 584, 124684.

Mango, L.M., Melesse, A.M., McClain, M.E., Gann, D., Setegn, S.G., 2011. Land use and climate change impacts on the hydrology of the upper Mara River Basin, Kenya: results of a modeling study to support better resource management. Hydrol. Earth. Syst. Sci. 15, 2245–2258.

Miller, J.D., Kim, H., Kjeldsen, T.R., Packman, J., Grebby, S., Dearden, R., 2014. Assessing the impact of urbanization on storm runoff in a peri-urban catchment using historical change in impervious cover. J. Hydrol. 515, 59–70.

Montgomery, D.R., Foufoula-Georgiou, E., 1993. Channel network source representation using Digital Elevation Model. Water Resour. Res. 29, 3925–3934.

Neff, B.P., Nicholas, J.R., 2005. Uncertainty in the Great Lakes Water Balance. U.S. Geol. Surv. Sci. Invest. Rep. 42, 2004–5100.

Nilsson, B., Sidle, R.C., Klint, K.E., Bøggild, C.E., Broholm, K., 2001. Mass transport and scale-dependent hydraulic tests in a heterogeneous glacial till–sandy aquifer system. J. Hydrol. 243, 162–179.

Novotny, E.V., Stefan, H.G., 2007. Stream flow in Minnesota: indicator of climate change. J. Hydrol. 334, 319–333.

Robichaud, P.R., 2000. Fire effects on infiltration rates after prescribed fire in Northern Rocky Mountain forests, USA. J. Hydrol. 231–232, 220–229.

Roe, G.H., 2005. Orographic precipitation. Annu. Rev. Earth Planet Sci. 33, 645–671.

Sasaki, H., Kurihara, K., 2008. Relationship between precipitation and elevation in the present climate reproduced by the non-hydrostatic regional climate model. Sci. Online Lett. Atm. (SOLA) 4, 109–112.

Sewagudde, S.M., 2009. Lake Victoria's water budget and the potential effects of climate change in the 21st century. African J. Trop. Hydrobiol. Fish. 12, 22–30.

Shreve, R.L., 1966. Statistical law of stream numbers. J. Geol. 74, 17–38.

Sidle, R.C., Pearce, A.J., O'Loughlin, C.L., 1985. Hillslope Stability and Land Use. Am. Geophys. Union, Water Resour. Mono. No. 11, Washington, D.C., p. 140

Sidle, R.C., Tsuboyama, Y., Noguchi, S., Hosoda, I., Fujieda, M., Shimizu, T., 1995. Seasonal hydrologic response at various spatial scales in a small forested catchment, Hitachi Ohta, Japan. J. Hydrol. 168, 227–250.

Sidle, R.C., Tsuboyama, Y., Noguchi, S., Hosoda, I., Fujieda, M., Shimizu, T., 2000. Stormflow generation in steep forested headwaters: a linked hydrogeomorphic paradigm. Hydrol. Process. 14, 369–385.

Sidle, R.C., Noguchi, S., Tsuboyama, Y., Laursen, K., 2001. A conceptual model of preferential flow systems in forested hillslopes: evidence of self-organization. Hydrol. Process. 15, 1675–1692.

Sidle, R.C., Sasaki, S., Otsuki, M., Noguchi, S., Abdul Rahim, N., 2004. Sediment pathways in a tropical forest: effects of logging roads and skid trails. Hydrol. Process. 18, 703–720.

Sidle, R.C., Ziegler, A.D., Negishi, J.N., Abdul Rahim, N., Siew, A., Turkelboom, F., 2006. Erosion processes in steep terrain—truths, myths, and uncertainties related to forest management in Southeast Asia. Forest. Ecol. Manage. 224 (1–2), 199–225.

Sidle, R.C., Gomi, T., Loaiza Usuga, J.C., Jarihani, B., 2017. Hydrogeomorphic processes and scaling issues in the continuum from soil pedons to catchments. Earth Sci. Rev. 175, 75–96.

Sidle, R.C., Gomi, T., Akasaka, M., Koyanagi, K., 2018. Ecosystem changes following the 2016 Kumamoto earthquakes in Japan: future perspectives. Ambio 47 (6), 721–734.

Smith, L.C., Sheng, Y., MacDonald, G.M., Hinzman, L.D., 2005. Disappearing arctic lakes. Science 308, 1429.

Soranno, P.A., Cheruvelil, K.S., Wagner, T., Webster, K.E., Bremigan, M.T., 2015. Effects of land use on lake nutrients: the importance of scale, hydrologic connectivity, and region. PLoS One 10 (8), e0135454.

Strahler, A.N., 1957. Quantitative analysis of watershed geomorphology. Am. Geophys. Union, Trans. 38, 913–920.

Tani, M., Matsushi, Y., Sayama, T., Sidle, R.C., Kojima, N., 2020. Characterization of vertical unsaturated flow reveals why storm runoff responses can be simulated by simple runoff-storage relationship models. J. Hydrol. 588, 124982.

Tezza, L., Häusler, M., Conceição, N., Ferreira, M.I., 2019. Measuring and modelling soil evaporation in an irrigated olive orchard to improve water management. Water 11, 2529. https://doi.org/10.3390/w11122529.

van de Giesen, N.C., Stomph, T.J., de Ridder, N., 2000. Scale effects of Hortonian overland flow and rainfall-runoff dynamics in a West African catena landscape. Hydrol. Process. 14, 165–175.

Vannote, R.L., Minshall, W.G., Cummins, K.W., Sedell, J.R., Cushing, C.E., 1980. The river continuum concept. Can. J. Fish. Aquat. Sci. 37, 130–137.

Veh, G., Korup, O., Walz, A., 2020. Hazard from Himalayan glacier lake outburst floods. Proc. Nat. Acad. Sci. 117, 907–912.

Wanders, N., Wada, Y., Van Lanen, H.A.J., 2015. Global hydrological droughts in the 21st century under a changing hydrological regime. Earth. Syst. Dynam. 6, 1–15.

Wilson, G., Rigby, J., Ursic, M., Dabney, S., 2016. Soil pipe flow tracer experiments: 1. Connectivity and transport characteristics. Hydrol. Process. 30, 1265–1279.

Woods, R., Sivapalan, M., Duncan, M., 1995. Investigating the representative elementary area concept: an approach based on field data. Hydrol. Process. 9, 291–312.

Ziegler, A.D., et al., 2004. Hydrological consequences of landscape fragmentation in mountainous northern Vietnam: evidence of accelerated overland flow generation. J. Hydrol. 284, 124–146.

CHAPTER

6

Light in Inland Waters

Kevin C. Rose

Department of Biological Sciences, Rensselaer Polytechnic Institute, Troy, NY, United States

OUTLINE

Solar radiation is a fundamental regulator of aquatic ecosystem dynamics. The absorption of solar energy and its dissipation as heat have profound effects on the thermal structure, stratification of water masses, and circulation patterns of lakes and other water bodies. Light influences numerous aspects of nutrient cycling and the distribution and behavior of dissolved gases and biota. Nearly all energy that controls the metabolism of lakes and streams is derived from solar energy via photosynthesis.

I. Light as an entity

Ecologically, the term *light* often does not have a consistent use due to the different ways in which it is measured and the term applied. First and foremost, light is energy, something that is capable of doing work and of being transformed from one form into another but can neither be created nor destroyed. Radiant energy can be transformed into heat that modifies the vertical structure of aquatic ecosystems as well as the metabolic rates of organisms, or into potential energy via biochemical reactions, such as photosynthesis. Energy transformations are far less than 100% efficient in a system such as a lake, and most radiant energy is dissipated as heat.

A. Units and definitions

Here, the term light refers to the portion of the *electromagnetic spectrum* that reaches Earth's surface and provides energy for biological, biogeochemical, and other ecological processes (Fig. 6-1). All electromagnetic radiation has both particle (i.e., photon) and wave properties. *Wavelength* (λ) is an important quantitative parameter of any periodic wave motion, including light and water movements, and is defined as the linear distance between adjacent wave crests. A *photon* is a pulse of electromagnetic energy and has an associated

Wetzel's Limnology, Fourth Edition
https://doi.org/10.1016/B978-0-12-822701-5.00006-9

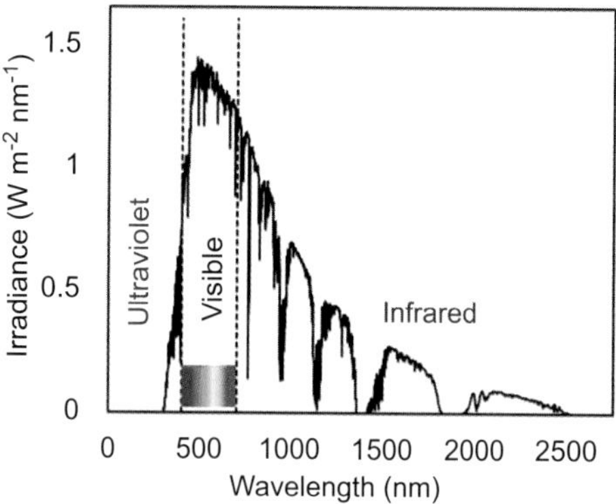

FIGURE 6-1 Solar irradiance at the surface of the Earth at sea level, showing major absorption bands from atmospheric O_2, O_3, and water vapor. Ultraviolet, visible, and infrared represent ~5%, 45%, and 50%, respectively, of the solar radiation energy reaching Earth's surface. Created using the SMARTS model (Myers & Gueymard, 2004), version 2.9.5 to generate a solar spectrum similar to ASTM Standard G-173-03.

wavelength and frequency (v), which are inversely related by the equation:

$$\lambda = c/v \tag{6-1}$$

where c is the speed of light (2.998×10^8 m s^{-1} in a vacuum). The electromagnetic spectrum is expressed as units of frequency (cycles per time) and wavelength (typically nanometers: 10^{-9} m). Wavelength can also be expressed as a wave number (v^1, or k), which is the number of wavelengths per centimeter.

For most practical purposes, and with few exceptions such as hot springs, solar radiation constitutes all of the energy input to aquatic systems. Colors in the portion of the electromagnetic spectrum that can be perceived by the human eye are all associated with wavelengths ranging from violet (~400–450 nm) to red (~650–700 nm). However, it should be noted that the limits of the visible spectrum are not sharply defined and vary among individuals, and many aquatic organisms can perceive wavelengths completely outside the range of the human eye. Visible wavelengths are similar to the portion of the solar spectrum used by photosynthesizing organisms, otherwise known as *photosynthetically active radiation* (PAR, 400–700 nm). This relatively narrow wavelength range accounts for about 45% of the energy reaching Earth's surface. At wavelengths longer than visible wavelengths (>700 nm) is *infrared radiation*, which accounts for about half of all incoming solar energy. At wavelengths shorter than visible wavelengths is *ultraviolet radiation*, including UV-A (typically defined as 320–400 nm), UV-B (typically defined as 280–320 nm), and UV-C (typically defined as 200–280 nm). UV-C and some UV-B are not found naturally on the modern Earth's surface. The shortest wavelength penetrating through the atmosphere and reaching Earth's surface is typically around 290 nm (Fig. 6-1). UV accounts for about 5% of the energy reaching Earth's surface.

The amount of light per unit area, otherwise known as *radiant flux density* or *irradiance*, is quantified by counting quanta or measuring energy. *Visible light* or PAR can be expressed either as moles of photons per unit area per unit time (μmol m^{-2} s^{-1}; otherwise known as *photosynthetic photon flux density*), or *PAR irradiance* (W m^{-2}; otherwise known as *photosynthetic radiant flux density*). A related term is the *photosynthetic photon flux fluence rate*, which is the total number of photons incident on a point from all directions. PAR is typically measured in quantum units (e.g., μmol m^{-2} s^{-1}), while higher-energy (shorter λ) UV radiation is typically measured in energetic units (e.g., W m^{-2}), but conventions vary among subdisciplines and it is possible to convert between the two.

At noon on a summer day without clouds at temperate latitudes, PAR is approximately 2000 μmol m^{-2} s^{-1}. This value can be converted to energetic units using *Planck's law*, which describes how the energy content (measured in Joules) of photons of a known wavelength is inversely related to wavelength via:

$$E = hv = hc/\lambda \tag{6-2}$$

where h = Planck's constant, 6.63×10^{-34} J s. Expressed energetically, a PAR of about 2000 μmol m^{-2} s^{-1} is about 430 W m^{-2}. Converting PAR from energetic to quantum units requires information on the spectral composition of the light field. In the absence of those data an approximate conversion factor of 2.77×10^{18} quanta J^{-1} (equivalent to 4.7 μmol quanta J^{-1}) can be used (Morel & Smith, 1974).

B. Dynamics of incident solar radiation

Atmospheric characteristics regulate the amount of solar radiation impinging on the surface of aquatic ecosystems, and dynamics in the atmosphere introduce variability in surface irradiance over time scales from seconds to millennia. As solar radiation penetrates and diffuses in the atmosphere, the energy of wavelengths is differentially attenuated by absorption and scattering processes. According to *Rayleigh's law*, light scattering by gas molecules in the atmosphere is proportional to $1/\lambda^4$, meaning that shorter wavelengths are scattered substantially more than longer wavelengths. This effect gives the sky a blue color to the human eye on cloudless days, and the reflection of the scattered skylight contributes to the blue color of many water bodies. Atmospheric light scattering also regulates the amount of indirect solar radiation reaching Earth's surface. In contrast to indirect solar radiation, direct solar radiation is largely regulated by the sun angle and the atmospheric distance through which the light must pass.

The amount of light received at a water body's surface, known as *incident solar radiation*, varies with sun angle, elevation, cloud cover, atmospheric aerosols, *ozone* (O_3), and other gas concentrations. Ozone strongly absorbs short-wavelength UV-B rays, and this is largely responsible for wavelengths below about 290 nm not reaching Earth's surface. However, ozone does not substantially affect UV wavelengths above about 330 nm. Water vapor and carbon dioxide (CO_2) strongly absorb infrared radiation and hence are important gases contributing to the greenhouse effect. Anthropogenic air pollution also contributes to the *attenuation* (i.e., reduction of light with depth or distance; see also Section III) of incoming solar radiation and is variable through time and space. Often, high concentrations of anthropogenic air pollution in the form of aerosols give the sky a brownish hue, which can be visible over some urban areas.

Sun angle is the most important regulator of variability in irradiance over the timescale of days to seasons. The sun angle is measured relative to an angle directly overhead and is known as the *solar zenith angle*. This angle regulates the distance through the atmosphere that light travels before reaching an object (called the *air mass effect*). Solar zenith angles vary substantially with latitude, day of year, and time of day (Fig. 6-2). In equatorial regions sunlight impinges closer to vertically throughout the year, leading to muted seasonal variation in energy inputs. In temperate and polar areas, however, the sun's angle changes substantially seasonally and thereby regulates many phenological processes. In polar extremes direct solar energy decreases to zero for over one-third of the year, and polar waters then receive thermal radiation only from indirect sources. Time of day is another factor that strongly influences the solar flux reaching the surface of water bodies, as the sun angle impacts the distance that the path of light must travel through the absorbing atmosphere. In addition to the above-described factors, *scattering* and *absorption* also increase in moist air, which is more common on the downwind side of large water bodies. The elevation of a lake also influences the quantity of atmosphere through which solar radiation must travel and therefore the amount and spectral composition of incident light reaching the water body.

The amount of solar irradiance on the Earth's surface varies over time. Decadal-scale increases and decreases in anthropogenic air pollution have caused corresponding decreases and increases in incident solar radiation. A widespread decline in incident solar radiation was observed for decades up to around 1990 across Europe, North America, and East and Southern Asia; this phenomenon has been referred to as *global dimming*. Since then, observations generally indicate a partial recovery and increases in solar radiation in some regions across Europe and North America, a phenomenon referred to as *brightening* (Wild, 2009; Wild et al., 2005). These trends are believed to be driven predominantly by long-term increases and then decreases in anthropogenic aerosols (Wild et al., 2021).

Long-term changes in ultraviolet radiation have also occurred due to changes in the ozone layer. Chemical pollutants (primarily chlorine and bromine in *chlorofluorocarbons* or CFCs) led to a thinning of the ozone layer and consequent long-term increases in ultraviolet radiation, particularly in Antarctica and adjacent regions in the latter half of the 20th century (Madronich & Gruijl, 1994). Total ozone column declines of about 3% per decade were observed until the mid-1990s (Chipperfield et al., 2017). International agreements such as the 1987 Montreal Protocol have since been successful in reducing emissions of ozone-depleting substances. Consequent ozone stabilization and recovery has begun, with a return to preimpacted ozone layer thickness

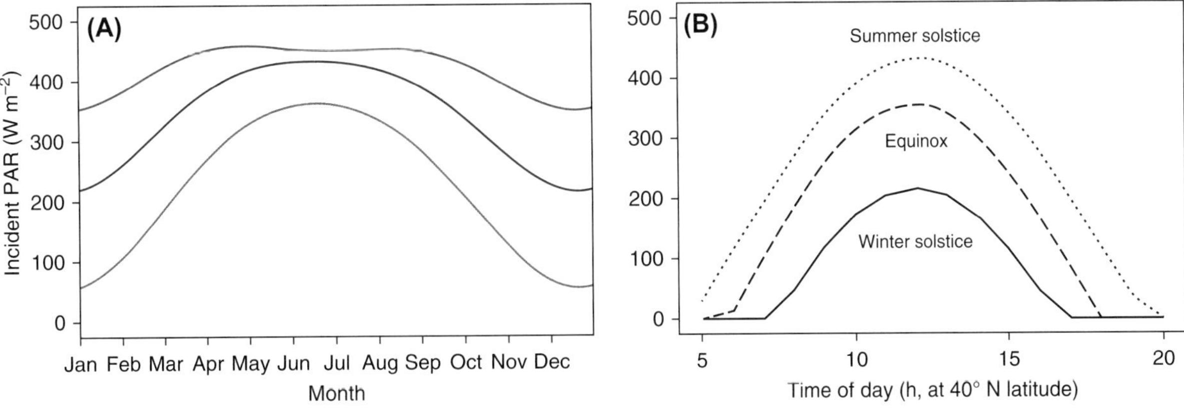

FIGURE 6-2 Variations in incident photosynthetically active radiation (PAR) associated with variations in solar zenith angle. (a) Noontime PAR across seasons at different latitudes (*red* = 20° N, *blue* = 40° N, *purple* = 60° N) (b) over the course of a day during different times of the year at 40° latitude (b). *(Data generated using the SolaR R package version 0.45 (Perpiñán Lamigueiro, 2012).)*

predicted by the mid-21st century (Newman et al., 2006; Solomon et al., 2016). However, recovery may be delayed due to climate change (Chipperfield et al., 2017).

II. At the water surface

Not all of the solar radiation that impinges upon the surface of a water body penetrates the water (Fig. 6-3). Surface reflections alter the direction of light, and snow and ice covering the surface can reflect and absorb solar radiation. These losses mean that the amount of light underwater can be substantially lower than that incident at the surface.

A. Reflection and albedo

A significant portion of incident light is *reflected* from water body surfaces and is lost from a system, unless it returns to the water after being backscattered from the atmosphere or surrounding topography. The extent of reflectivity of solar radiation varies greatly with the angle of incidence, the surface characteristics of the water, the surrounding topography, and the meteorological conditions. The reflection (R) of unpolarized direct sunlight as a fraction of the incident light is a function of *Fresnel's law*:

$$R = \frac{1}{2}\left[\frac{\sin^2(i-r)}{\sin^2(i+r)} + \frac{\tan^2(i-r)}{\tan^2(i+r)}\right] \tag{6-3}$$

where i = angle of incidence (the zenith angle of the sun) and r = angle of refraction (the angle of the light transmitted into the water). The angle of refraction can be calculated using *Snell's law* and the *refractive index* of water (n_w = 1.33) relative to air (n_a = 1.0). The zenith angle can be calculated as a function of latitude, solar declination (which is affected by time of year), and hour (e.g., Woolway et al., 2015). The *albedo* describes the ratio of reflected to incident irradiance. An albedo of 0 describes a surface that absorbs all incoming energy, and an albedo of 1 describes a surface that reflects all incoming energy. The reflectance off a flat water surface with the sun directly overhead is about 2%, which equates to an albedo of 0.02. Reflectance increases rapidly as zenith angles approach 90° (Fig. 6-4). An average reflectance loss value of 6.5% is common, but there is substantial variability through the year and with latitude (Cogley, 1979; Hummel & Reck, 1979). For example, albedo can exceed 30% in January at high northern latitudes, whereas it remains under 7% all year near the equator (Cogley, 1979). Reflective losses can also be lower if surrounding vegetation or topography—mountains, for example—moderate low-angle radiation. Surface waves can increase reflectance. The increase can be as much as 20% at zenith angles over 60°, but the effect is small when the sun is close to directly overhead. Qualitatively, light in the red portion of the spectrum is reflected to a slightly greater extent than light of higher frequencies, particularly at low angles of incidence.

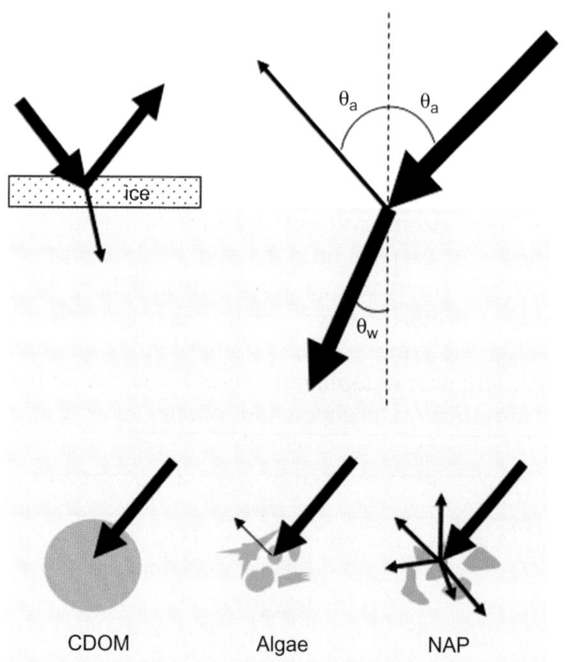

FIGURE 6-3 Conceptual diagram of light impinging on water and interacting with optically active substances underwater. θ_a and θ_w are the zenith angle of incidence and refraction, respectively, and are related to the refractive indices of air and water by Snell's law. Ice is usually highly scattering due to air bubbles and the presence of snow. In addition to pure water, chromophoric dissolved organic matter (CDOM), algae, and nonalgal particulates (NAP) are the primary optically active substances in most inland water bodies. CDOM strongly absorbs light. Algae absorb light and are weakly scattering. NAP is weakly absorbing and strongly scattering.

B. Ice and snow

Light transmission through clear, colorless, so-called *black ice* is similar to that through liquid water from the ultraviolet to mid-infrared wavelengths (Warren, 2019). Indeed, the refractive index of ice (1.31) is very similar to that of water (1.33). However, most ice includes air bubbles that scatter solar radiation, and ice is often covered in snow. Molecules in air are not incorporated into the ice crystal lattice upon freezing and therefore come out of solution as bubbles, which are frequently trapped between crystals as ice grows. Dense concentrations of air bubbles in ice (so-called *white ice*) as well as cracks scatter light. In turn, this reduces light transmission and increases reflectivity (Mullen & Warren, 1988).

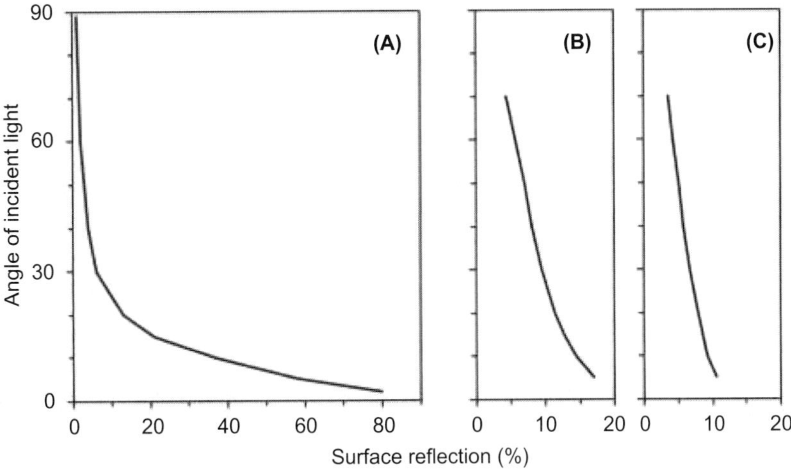

FIGURE 6-4 Surface reflection and backscattering as a percentage of total solar radiation at angles of incidence varying from the horizontal. (a) Clear, cloudless conditions; (b) reflection of diffuse light under moderate cloud cover; (c) heavily overcast conditions. *(Plots generated from data extracted from Wetzel (2001).)*

During ice-covered periods, the overall amount of reflectance (albedo), absorbance, and transmittance through a lake surface is often regulated more so by the presence versus absence of snow cover on top of ice and snow texture than by ice itself. For example, in a study of a perennially ice-covered Arctic lake only 0.45% of incident PAR was transmitted through 2 m thick ice and 0.5 m thick snow cover (Belzile et al., 2001). Removal of snow increased underwater PAR by 13-fold and UV by 16-fold (Belzile et al., 2001). Coarse-grained snow typically has a lower albedo than fine-grained snow because incident light travels a longer distance between crystals and is therefore less likely to be reflected (Warren, 2019). The presence of white ice and snow can delay ice-off due to its high reflectivity (Leppäranta, 2015). However, even slight snow melt causes a rapid increase in grain size, which decreases albedo. While freshly fallen snow generally has a high albedo (>0.8), once snow begins to melt, the albedo can fall to 0.2–0.5 (Henneman & Stefan, 1999; Perovich et al., 1998). Melt pools on top of ice also substantially increase overall transmittance (Matthes et al., 2020). Overall, the thickness of snow and ice can vary by as much as a factor of two within short distances (<25 m) and lead to considerable patchiness in both reflectance of light and the amount of light underwater (e.g., Perovich et al., 1998).

III. Light attenuation in the water column

Light that penetrates through the atmosphere and water's surface is available to do physical, chemical, and biological work in aquatic ecosystems. Different wavelengths of sunlight participate in or regulate a range of ecological and biogeochemical processes. The spectrum of light underwater differs from the spectrum of light reaching the water surface and is highly dynamic across space and time. The reduction of light with depth as it is absorbed and scattered by constituents in the water column is referred to as *light attenuation*. Spectral and temporal variability in light attenuation underwater can have as large or larger effects on the total amount of light than variability in atmospheric or surface processes.

A. Inherent optical properties

The absorption coefficient, scattering coefficient, and volume scattering function are *inherent optical properties* because they depend on the concentrations and characteristics of the substances in water and not the geometry of the light field. The fraction of light absorbed or scattered, divided by the layer thickness (also referred to as pathlength), is the *absorption coefficient* (a) or *scattering coefficient* (b), respectively. The volume scattering function describes the angular distribution of light that is dispersed after being scattered. Another inherent optical property is the *beam attenuation coefficient* (c), which is the sum of a and b and represents the fraction of incident light that is absorbed and scattered, divided by the layer thickness. These coefficients have units of 1/length and are typically expressed as per meter (m^{-1}).

Light scattering is the result of the deflection of light by the molecular components of the water and its solutes and, often to a larger extent, by particulate materials suspended in the water. Scattering increases the pathlength of a photon, increasing the likelihood of it being absorbed. *Backscattering* toward the surface can result in photons leaving a water body. Scattering of light is a spectrally dependent process, meaning that wavelengths have different associated *scattering coefficients* (c_λ). Scattering coefficients generally increase exponentially with decreasing wavelength. However, the exact spectral shape of scattering is greatly influenced by the geometry and composition of the particulate material suspended in water, and total scattering is influenced by the concentration of particulate material. As the quantity and size of suspensoids increase, longer wavelengths are scattered preferentially. Glacially influenced lakes, rivers, and coastal zones often contain high concentrations of glacial "flour" that is highly scattering, giving these water bodies a milky or opalescent color (e.g., Gallegos et al., 2008). A significant amount of light can also be reflected and scattered from sunlit sediments. Sediments rich in calcium carbonate (marl; $CaCO_3$) reflect considerably more light than dark-colored sediments of high organic content. Hardwater lakes with high concentrations of suspended $CaCO_3$ particles characteristically backscatter light that is predominantly blue-green; water bodies rich in suspended organic materials appear more green or yellow. In oligotrophic water bodies water appears blue in part because blue light is the shortest visible wavelength backscattered that is detected by the eye; however, in reality, UV, which is not detectable by the human eye, scatters more than visible light.

Light scattering can vary substantially with depth, season, and location in a water body and in response to variations in the concentration or sources of particulate matter. For example, particulate matter can become concentrated in the middle zone of great density change (e.g., metalimnion) of a thermally stratified lake, due to reduced sinking as the particles encounter increased water densities or as a result of the development of large populations of plankton. While scattering from algae and other organic particles is usually low, it can increase in these layers. Scattering can also increase markedly in areas of a lake where wind-induced currents and wave action agitate and temporarily suspend littoral and shore deposits of particulate matter (e.g., Ostrovsky et al., 1996). When dimictic or amictic lakes (Chapter 7) undergo complete circulation, a significant portion of the recent sediments of the lake basin are brought into resuspension (Davis, 1973; White & Wetzel, 1975) and can affect the scattering properties of the lake for weeks. Similarly, the variable influxes of suspended inorganic and organic matter from stream inflows to reservoirs can radically increase the scattering of light nonuniformly within the lake basin. For example, cold or saline, silt-laden river inflows can move as subsurface density currents along the metalimnion or in the hypolimnion far out into a lake or reservoir. These alterations in optical properties can be short- or long-lived depending upon the composition and density of the material and the density characteristics of the water column in the recipient water body.

Absorption is an inherent optical property defined as diminution of light energy with depth by the transformation of light to chemical energy or heat energy. Absorption of light energy by atoms and molecules occurs when the electrons of the atoms and molecules resonate at frequencies that correspond to an energy state of a photon. In the collision of an electron and a photon the electron gains the quantum of energy lost by the photon. The energy transferred via absorption breaks chemical bonds and supports photosynthesis via generated electrons.

Light may be absorbed by water, dissolved substances, or particulate substances. While scattering is predominantly a function of the concentration and characteristics of particulate matter, absorption is primarily a function of dissolved substances, although organic particles such as algal cells can be important in some circumstances (see optically active substances section below). Absorption typically increases rapidly with decreasing wavelength.

Light absorption is typically measured in a *spectrophotometer* by measuring the *absorbance* or *optical density* (D), which is defined as the base 10 logarithm of the ratio of the incident light intensity (I_0) to the light intensity (I) transmitted through a water sample. Beer's law describes how absorbance is proportional to the concentration of a solution. Lambert's law describes how absorbance is proportional to the pathlength of light through a medium. These laws are often combined, and the resulting *Beer–Lambert law* is used to calculate a wavelength-specific absorption coefficient (a_λ, m^{-1}) as 2.303 (coefficient to convert $\log_{10}$ to natural logs) multiplied by the measured absorbance (D), divided by the pathlength (L) through the medium (Kirk, 1994). In some literature, the term *absorption coefficient* takes a decadic form (i.e., absorbance normalized to pathlength, where α [m^{-1}] is equivalent to D/L), whereas here, and in other literature, it takes a Napierian form (a is equivalent to $\alpha \times \ln 10$) (Hu et al., 2002).

B. Apparent optical properties

Apparent optical properties represent those characteristics of the light field that depend on both inherent optical properties (described above) and the geometry of the light field (e.g., solar zenith angle). One of the most commonly measured apparent optical properties is the *diffuse attenuation coefficient*, also known as the *light extinction coefficient*. The coefficient is typically estimated empirically by measuring light at multiple depths in the water column using the following equation:

$$E_{dz} = E_{d0}\, e^{-k_d z} \quad (6\text{-}4)$$

where E_{dz} is the downwelling irradiance at depth z, E_{d0} is the downwelling irradiance just below the water's surface (0 depth), and K_d is the diffuse attenuation coefficient. Depths (z) are positive values. K_d has units of reciprocal depth (typically m^{-1}) and describes the slope of the exponential decay of light in the water column. Accurate estimation of K_d is often reduced in very near-surface measurements due to wave action, and shallow depths are often excluded. As K_d is an apparent optical property, it depends on the geometry of the light field. Thus standard practice is to obtain light profiles to calculate a K_d within a few hours of when the sun is highest in the sky.

Coefficients for PAR K_d for natural waters vary from less than 0.1 m^{-1} in very clear lakes, such as Crater Lake (Oregon, USA), to above 4.0 m^{-1} in highly stained lake waters or lakes with very high algal biomass or suspended sediments (Fig. 6-5). PAR K_d is the diffuse attenuation coefficient for PAR, 400–700 nm. Values for PAR K_d in excess of 18 m^{-1} have been recorded (e.g., Rudorff et al., 2011). However, K_d depends strongly on wavelength and can vary substantially across the solar spectrum. In almost all cases K_d for UV wavelengths is greater than for PAR (Laurion et al., 2000; Morris et al., 1995; Scully et al., 1995; Williamson et al., 1996). However, the ratio of attenuation for UV versus PAR depends on the quality and quantity of optically active substances and can vary substantially (Fig. 6-6).

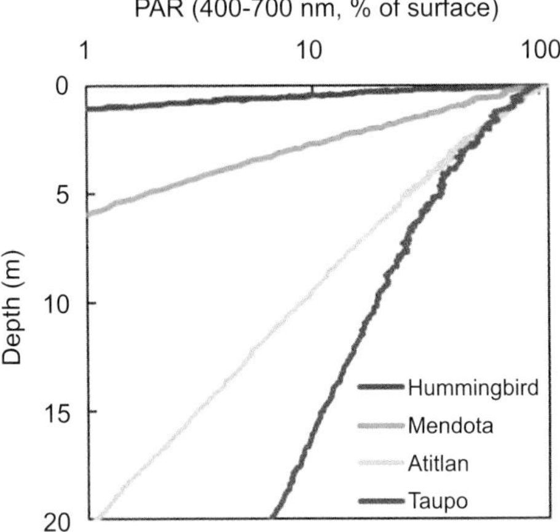

FIGURE 6-5 Variation in photosynthetically active radiation (PAR) with depth across four contrasting lakes, namely, Hummingbird Bog, Michigan, USA ($K_d = 4.4$); Lake Mendota, Wisconsin, USA ($K_d = 0.72$); Lake Atitlan, Guatemala ($K_d = 0.22$); and Lake Taupo, New Zealand ($K_d = 0.11$). Note that the horizontal x-axis is logarithmic.

It is common practice to convert a K_d to a predicted 1% depth ($Z_{1\%}$); that is, the depth at which 99% of light is attenuated:

$$Z_{1\%} = \ln(100)/k_d \quad (6\text{-}5)$$

The PAR $Z_{1\%}$ depth approximates the *compensation depth*, which is defined as the depth at which daily photosynthesis is equal to autotrophic respiration. Consistent with reporting PAR $Z_{1\%}$ depth, the equivalent $Z_{1\%}$ depth for other wavelengths is also often reported (e.g., 320 nm UV $Z_{1\%}$). In reality, the amount of light at the $Z_{1\%}$ depth can vary substantially between water bodies at different latitudes, elevations, or seasons due to differences in the amount of incident solar radiation. Thus the compensation depth may be substantially deeper than the PAR $Z_{1\%}$ depth in some water bodies, such as tropical high-elevation lakes, due to high incident irradiances at the water's surface.

In addition to the properties of the water column, the characteristics and amount of light penetrating the littoral zone of lakes are greatly altered by the type and extent of development of floating leaves and emergent higher aquatic plants, as well as by reflection and scattering from lake sediments. Dense vegetative stands can reduce underwater irradiance by 50% to over 90%, but effects on attenuation can vary substantially among species (Andersen et al., 2017; Carpenter & Lodge, 1986; Ondok, 1978).

C. Optically active substances

The diffuse attenuation coefficient is influenced by the absorption and scattering of water and substances in water. Thus K_d can be partitioned into partial diffuse attenuation coefficients, such as for dissolved and particulate substances. The K_d at a given wavelength or waveband (e.g., PAR) is a function of the concentration of *optically active substances*, and variations in K_d across the solar spectrum correspond to variations in the absorption and scattering spectra of these substances. The most commonly described categories of optically active substances include chromophoric dissolved organic matter (CDOM), algal biomass, nonalgal particulates, and water itself (Fig. 6-3). The spectral shape and magnitude of absorption and scattering depend on the composition and concentration of each optically active substance. In most lakes of moderate clarity it is common for the green portion of the spectrum to penetrate most deeply. The waveband of maximum penetration is shifted to shorter wavelengths in oligotrophic water bodies, whereas in lakes that are darkly stained with CDOM (see below) practically no light of wavelengths shorter than 600 nm penetrates below 1 m.

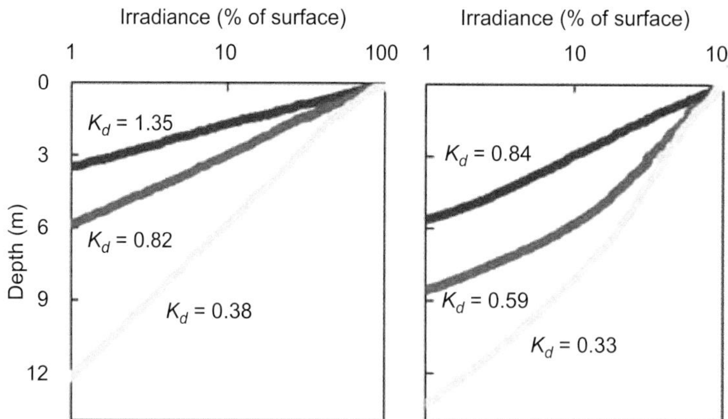

FIGURE 6-6 Diffuse attenuation coefficients (K_d) measured in spring (*left*) and summer (*right*) for Crystal Lake, Wisconsin, USA, for 320 nm UV (*purple*), 380 nm UV (*blue*), and photosynthetically active radiation (PAR; *yellow*). All wavelengths penetrated deeper in summer than in spring, with more pronounced seasonality in short-wavelength UV than in PAR. Additionally, light attenuation exhibited heterogeneity through the water column (i.e., it is not log-linear) during summer, associated with higher chromophoric dissolved organic matter (CDOM) and algal biomass below the thermocline. Note that the horizontal x-axes are logarithmic.

i. Chromophoric dissolved organic matter

Chromophoric dissolved organic matter (CDOM) represents the fraction of *dissolved organic matter* (DOM) that absorbs light. It is typically quantified by measuring dissolved absorbance (*a*; see Section III). However, other substances, such as iron can also be dissolved and contribute to absorption (Poulin et al., 2014). DOM is a polydisperse, heterogeneous mixture of organic compounds originating from the incomplete decomposition of organic matter, and the DOM pool in most inland waters is primarily derived from *allochthonous* organic matter, such as terrestrial plants and soil organic matter. DOM is often quantified by measuring the dissolved organic carbon (DOC) concentration (Chapter 28). Wetlands extent within a catchment area is a top predictor of DOC and CDOM in inland waters (Gergel et al., 1999; Winn et al., 2009). *Autochthonous* organic matter from algae and other microbes in inland water bodies also comprise the DOM pool. However, absorption of light by CDOM is often more associated with allochthonous sources of organic matter than autochthonous sources. Throughout the solar UV-B and UV-A range, K_d and $Z_{1\%}$ are well estimated with a univariate power model based on DOC concentration, particularly in waters of low to moderate productivity, and differences in CDOM explain >90% of the among-lake variation in UV K_d coefficients (Fig. 6-7). The penetration of UV into the water column increases rapidly when DOC concentrations decline below ~2 mg L^{-1}.

The wavelength-specific dissolved absorption coefficient relative to the DOC concentration is referred to as

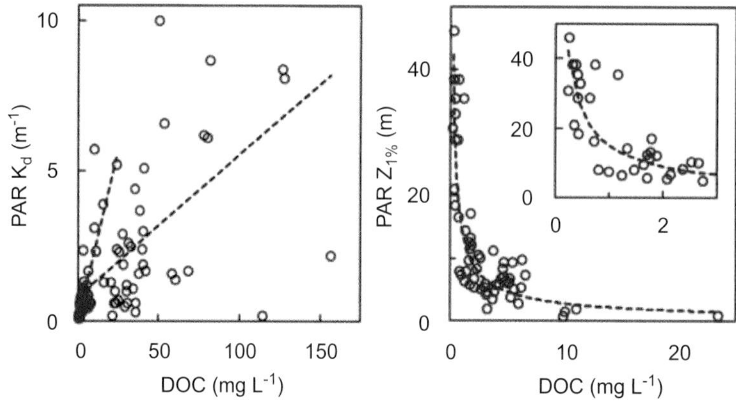

FIGURE 6-7 *Left:* Relationship between photosynthetically active radiation (PAR) K_d and dissolved organic carbon (DOC) concentration from 65 freshwater lakes in North and South America (*blue*; data from Morris et al. (1995)) and 44 mixed freshwater and saline lakes (*black*; data from Arts et al. (2000)). *Right*: Data from Morris et al. (1995) replotted to show the relationship between PAR $Z_{1\%}$ and DOC concentration. Light penetration into the water column is a power function of DOC concentration, and light penetrates especially deep when DOC concentrations are below about 2 mg L^{-1} (inset figure).

a DOC-specific absorption coefficient and is an indicator of DOM quality. A PAR DOC-specific absorption coefficient of about 0.22 $m^2\ g^{-1}$ is common (Morris et al., 1995; Read et al., 2015). The DOC-specific absorption coefficient at 254 nm, otherwise known as the *specific UV absorbance* (SUVA), is related to the aromatic carbon content of the DOM pool. The process of *photobleaching*, or the loss of dissolved absorbance when DOM is exposed to sunlight, can reduce dissolved absorbance and DOC-specific absorption coefficients. In water bodies with long residence times, such as many saline lakes, the DOC-specific absorption coefficient can be substantially lower than in stained freshwater lakes (Arts et al., 2000).

CDOM exhibits an approximately exponential increase in absorbance with decreasing wavelength (Fig. 6-8). The shape of this absorbance spectrum is characterized by a *spectral slope* (S) over a given wavelength range (e.g., $S_{275-295}$ nm). The shape of the spectral slope is related to the DOM source and extent of *photobleaching* (John et al., 2008). In the UV portion of the solar spectrum (290–400 nm) CDOM is usually the most important light-attenuating substance. For example, even during a severe harmful algal bloom, CDOM accounted for over 50% of UV attenuation in Lake Erie, USA/Canada (Cory et al., 2016). Only when DOC is very low (e.g., <0.3 mg L^{-1}) do other optically active substances become important regulators of UV K_d (Laurion et al., 2000). High CDOM often gives water bodies a brownish hue and absorbs very little above ~600 nm (Fig. 6-9).

The dissolved absorption coefficient at 440 nm (a_{440}) is commonly used to describe the *true color* of a water body, which is measured by filtering the water and thus represents color associated with dissolved substances. A related measurement called *apparent color* represents the color of a whole water sample, including both dissolved and particulate substances. Several other color scales have been devised to empirically compare the true color of lake water, after filtration to remove particulates, to various combinations of inorganic compounds in serial dilutions. The *Platinum units* (Pt units) scale is a much widely used comparative scale in the United States based on comparisons to a mixture of potassium hexachloroplatinate (K_2PtCl_6), cobaltic chloride hexahydrate ($CoCl_2 \cdot 6H_2O$), hydrochloric acid (HCl),

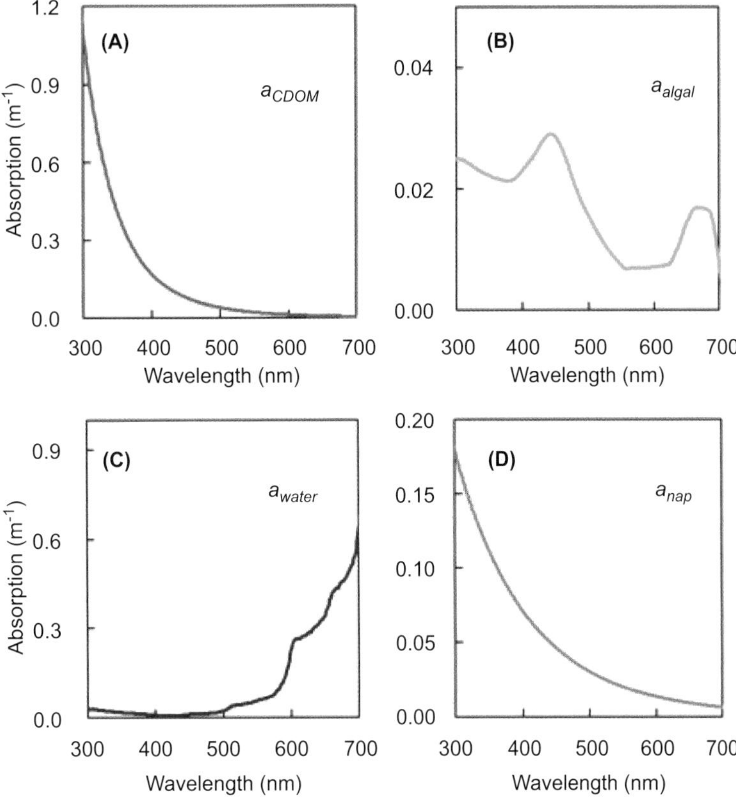

FIGURE 6-8 Example absorption spectra for the Rhode River, Maryland, USA, for (a) chromophoric dissolved organic matter (CDOM) (a_{CDOM}, per unit DOC, mg L^{-1}); (b) algae (a_{algal}, per unit chlorophyll *a*, μg L^{-1}); (c) water (a_{water}); and (d) nonalgal particulates (a_{nap} per unit total suspended solids, mg L^{-1}). Note that y-axis scales differ among plots to highlight spectral variations. *(Absorption data for CDOM, chlorophyll, and nonalgal particulates from Rose et al. (2019). Pure water absorption data from Quickenden and Irvin (1980) in the UV-B region and Pope and Fry (1997) for the UV-A and visible regions.)*

FIGURE 6-9 High concentrations of chromophoric dissolved organic matter (CDOM) give water bodies a brown color. *Left*: Bear Brook, New York, USA (credit: Barry Baldigo); *right*: John Brown Pond, New York, USA (credit: Kevin Rose).

and water (Hongve & Akesson, 1996). It is possible to estimate a_{440} from Pt units (Cuthbert & del Giorgio, 1992). Very clear lake water has a color of 0 Pt units, while darkly stained bog water is about 300. In Europe the *Forel-Ule color scale*, involving comparisons to alkaline solutions of cupric sulfate ($CuSO_4$), potassium chromate (K_2CrO_4), and cobalt sulfate ($CoSO_4$), is more commonly used. The scale classifies lake color into 22 categories in the range of blue, green, yellow, and brown.

ii. Algal biomass

Algal biomass contributes to light attenuation primarily through absorption. Organic particulate material is usually only weakly scattering (Fig. 6-3). In general, algal biomass exhibits a peak in absorption that corresponds to the chlorophyll *a* absorption spectra, with peaks in the blue and red ends of the spectrum, thus primarily reflecting (and appearing) green (Figs. 6-8 and 6-10). While the absorbance of chlorophyll varies with wavelength, empirical evidence shows that algal biomass has a PAR attenuation coefficient of about 0.07 per unit chlorophyll (Morris et al., 1995). However, algal cells can also contain accessory pigments that absorb light, such as carotenoids and mycosporine-like amino acids, which are believed to protect against UV-induced damage and absorb strongly in the UV portion of the spectrum. At high concentrations, these compounds can alter the perceived color of water (Fig. 6-11). The perceived color of a water body is frequently associated with the concentration of algae or, less frequently, with pigmented bacteria or microcrustaceans. Where Cyanobacteria or diatom blooms occur, they may produce blue-green or yellowish-brown colors, respectively. Blood-red color occurs commonly in lakes where conditions are temporarily ideal for the massive development of populations of certain algae, such as the dinoflagellate *Glenodinium* (Fig. 6-11).

iii. Nonalgal particulates

Nonalgal particulates can be diverse in source and geometry and are often low in standing waters due to settling. However, they can be high in rivers and water bodies with substantial upstream erosion, such as in glacially fed lakes and agricultural reservoirs. In these systems K_d can be approximated by the concentration of nonalgal particulates (Rose et al., 2014). Suspended particles absorb as well as scatter photons, and smaller particles tend to be more effective scatterers than large ones. The absorption spectra for nonalgal particulates often increase exponentially with decreasing wavelength but at a slower rate of increase than CDOM (Fig. 6-8).

Turbidity is a visual property of water and implies a reduction or lack of clarity that results from the presence of suspended particles or suspensoids. A related term, *seston*, refers to all particulate material present in the free water. Seston includes both *bioseston* (plankton

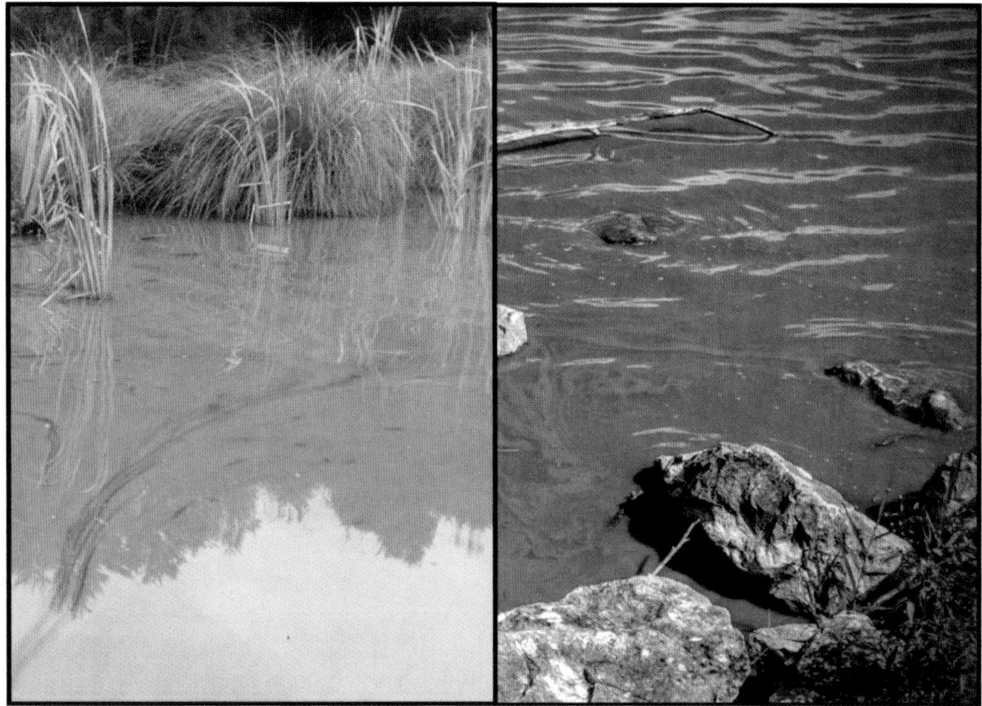

FIGURE 6-10 High algal biomass gives water bodies a green color. *Left*: Lake Rotorua, South Island, New Zealand (credit: David Hamilton); *right*: Lake Pamvotida, Greece (credit: Kevin Rose).

FIGURE 6-11 Bright red-colored water can be seen when algae and bacteria produce high concentrations of photoprotective pigments such as carotenoids. *Left*: Laguna Simba, near San Pedro de Atacama, Chile (elevation 5913 m asl; credit: Kevin Rose) and *right*: Lake Hillier, Australia (credit: Ken McGrath).

FIGURE 6-12 High concentrations of fine particulates can give water bodies a milky or turquoise color. *Left*: Crater Lake, Mt. Ruapehu, New Zealand, where fine silt comes from loose volcanic ash (credit: Kevin Rose); *right*: Zigadenus Lake, Alberta, Canada, where fine silt comes from nearby glacial action (credit: Mark Olson).

and nekton) and *abioseston* or *tripton* (nonliving particulate material). When the density of particulate matter suspended in the water becomes great, a *seston color* can be imparted to the water (Fig. 6-12).

Turbidity usually consists of inorganic particles and originates from the erosion of soil of the catchment basin and from resuspension of the bottom sediments but in many eutrophic water bodies also includes algal cells. Inorganic turbidity tends to be higher in reservoirs than in natural lakes, in part often associated with larger fetch distances and shallower depths for much of the basin, and in some regions by extensive agricultural soil disturbance within the catchment. These inorganic particles often range in size from 0.2 to 2 μm and consist of silicate minerals (mica, illite, montmorillonite, vermiculite, kaolinite) and aluminum and iron oxides (Kirk, 1994).

A particularly common suspensoid in hardwater lakes is associated with colloidal and particulate $CaCO_3$, almost entirely as calcite crystals (see detailed discussion in Chapter 13). These suspensions of carbonate become particularly abundant and visually conspicuous as *whitings* in temperate lakes during spring and early summer. During this period, supersaturated conditions are often exceeded because of increasing temperatures and photosynthetic utilization of CO_2 and production of hydroxyl ions, which result in the precipitation of particles in the range of 1–20 μm. As these particles settle, turbidity increases markedly, with large increases in the backscattering of light, particularly in the blue-green portion of the spectrum (e.g., Effler & Johnson, 1987; Weidemann et al., 1985). In certain lakes, such as Blind Lake in southeastern Michigan, USA (Schelske et al., 1962), the suspension is so fine and production so intense from supersaturated groundwater sources that the colloidal suspension renders the water nearly opaque. Light is backscattered with a brilliant blue-green opalescence. Where turbidity is extremely high, such as in some reservoirs near major river inflows under flood conditions or in lakes receiving fine materials, such as volcanic ash or glacial runoff, extinction coefficients in excess of 10 m^{-1} are not unusual.

iv. Pure water

Pure water can make a substantial contribution to the attenuation of UV and PAR in oligotrophic lakes. In distilled water absorption decreases as wavelength increases from UV into PAR, with a minimum in the short wavelength blue light. Absorption then increases rapidly from about 430 nm into the infrared range (Fig. 6-8). The absorption of infrared radiation is over 100 times greater than that of UV and PAR, and therefore usually over half of the total light energy is transformed into heat in the first meter.

D. Seasonal and long-term patterns in light attenuation

Light attenuation can vary substantially over days to seasons. Many lakes exhibit clear seasonal patterns in light attenuation that correspond with variations in the concentration of dominant optically active substances (Fig. 6-13). In lakes where algal biomass is high and CDOM is low the seasonal dynamics of algal biomass regulate the seasonality of light attenuation. Many lakes exhibit a springtime clear-water phase, which corresponds to a decrease in UV and PAR K_d (Williamson et al., 2007). The clear-water phase typically follows spring algal blooms when there is a rapid decline in algal biomass associated with increased zooplankton grazing

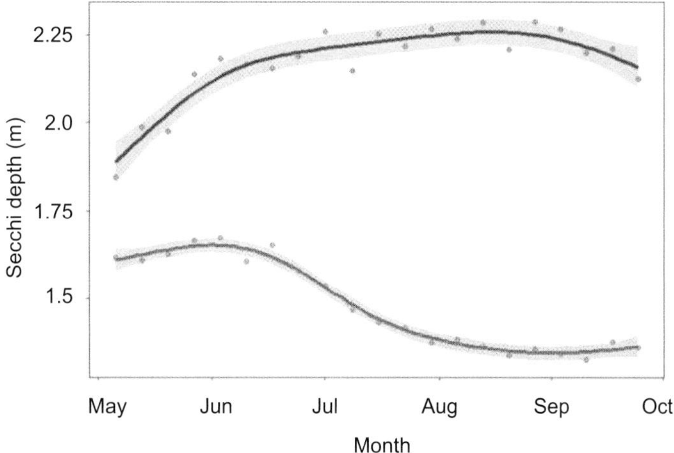

FIGURE 6-13 Typical seasonality in water clarity in temperate lakes where light attenuation is frequently dominated by algal biomass (*green*) or chromophoric dissolved organic matter (CDOM; *brown*). Figure is based on unpublished remotely sensed median estimates of water clarity from 50,000 randomly selected US lakes following methods described in Topp et al. (2021).

(Lampert et al., 1986; Sommer et al., 1986). In more eutrophic water bodies light attenuation is typically high (i.e., water clarity is low) during warm summer months following the clear water phase due to conditions that favor algal blooms (Soranno, 1997). In contrast to eutrophic lakes, lakes where CDOM is the dominant optically active substance regulating underwater UV and PAR tend to exhibit a parabolic seasonal pattern in light attenuation, with peak water clarity (lowest K_d) often corresponding to peak incident light exposure and stratification strength (Morris & Hargreaves, 1997). Initially, light attenuation is highest in early spring due to limited photobleaching, deep mixing, and often substantial allochthonous DOM inputs with snowmelt where that occurs. Seasonal decreases in light attenuation are associated with seasonality in photobleaching and density stratification, and seasonality is greater in shorter-wavelength UV relative to PAR (Morris & Hargreaves, 1997). When stratification begins to breakdown in the fall, deeper waters with higher absorbance are entrained into sunlit surface waters, and K_d increases as a result. Separately, disturbance events such as severe storms can severely reduce water clarity, but these effects are often temporary.

Long-term observations of light attenuation and water clarity reveal complex and diverse trends that vary among lakes and regions. In lakes of the upper Midwest of the United States about 10% of lakes have exhibited significant decadal-scale trends in light attenuation, with about equal numbers of increases and decreases (Lottig et al., 2014). In other regions, such as throughout part of the Northeast United States and Northwest Europe, light attenuation has mostly increased due to CDOM increases associated with recovery from acidification (e.g., Leach et al., 2019). Similarly, many studied lakes in China have increased in light attenuation associated with an increase in eutrophication (Zhang et al., 2020). Overall, trends in light attenuation and water clarity reflect corresponding changes in the concentrations of optically active substances, which respond to both local and regional drivers, including land use/land cover and climate (Rose et al., 2017).

E. Measuring light attenuation and water clarity

As discussed above, empirical estimation of K_d often relies on measurements of light at multiple depths in the water column. A *submersible radiometer* is commonly used to measure light underwater. Radiometers often vary in their spectral sensitivity. When measuring light attenuation, a second above-surface radiometer (deck cell) is often used to correct for changes in incident (above-surface) light that could otherwise impact the estimation of K_d.

Often, several other terms are used to describe light attenuation, including *water clarity* and *transparency*. Water clarity and transparency are often used interchangeably and inversely with light attenuation. An approximate evaluation of the transparency of water to light was devised by Pietro Angelo Secchi (pronounced "sekki") in the mid-1800s, now termed the *Secchi disk depth*. This method continues to be widely employed due to its simplicity and long historical record (Lottig et al., 2014). The Secchi disk depth is the mean depth of the point where a weighted disk, typically white or divided white and black and 20 cm in diameter, disappears when viewed from the shaded side of a vessel and that point where it reappears upon raising it after

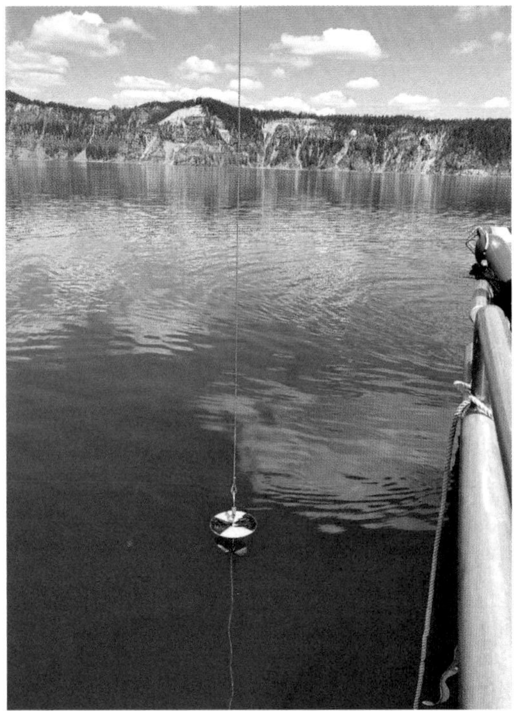

FIGURE 6-14 A Secchi disk being lowered into Crater Lake, Oregon, USA, next to a US National Park Service research vessel. Crater Lake is one of the clearest lakes in the world, with Secchi disk depths regularly exceeding 30 m (credit: Kevin Rose).

it has been lowered beyond visibility. Secchi disk depths range from a few centimeters in turbid reservoirs to over 30 m in ultraoligotrophic lakes, such as in Crater Lake (Oregon, USA) (Fig. 6-14).

The Secchi disk depth is essentially a function of the contrast related to the reflection of light from its surface versus the background and is therefore influenced by the absorption and scattering characteristics of the water and substances in it. Both theoretical analyses and a large number of empirical observations have shown that Secchi disk depth measurements are more sensitive to variations in the suspended particles than dissolved substances. This correlation is particularly true in very productive lakes and in a very generalized way has been used to estimate the approximate density of phytoplankton populations. However, where even moderate amounts of dissolved substances or nonalgal particles are present, attempts to estimate algal biomass for trophic state classification from Secchi depth data are often inappropriate (Lind, 1986).

An approximate relationship exists (Idso & Gilbert, 1974; Poole & Atkins, 1929) between Secchi disk depth, Z_{SD} (m), and PAR K_d (m^{-1}):

$$K_d = 1.7/Z_{SD}$$

However, variations in the relative concentrations of dissolved and particulate substances can influence this relationship, with coefficients up to 3.8 reported for stained water bodies where CDOM regulates light attenuation and as low as 0.5 for highly turbid systems (Koenings & Edmundson, 1991). Thus, while the Secchi disk depth typically corresponds to about 10% of surface PAR, this percentage can be as low as 3.6% in stained lakes or as high as about 22% in turbid water bodies. Although the Secchi disk depth is independent of surface light intensity to a significant extent, results become erratic near dawn and dusk, and the Secchi disk should be used preferably at midday.

IV. Ecological role of solar radiation

The amount and quality of underwater light is a master variable, controlling many aspects of aquatic ecosystems. In most lakes underwater light is controlled by both incident solar radiation and the concentration and spectral properties of optically active substances in water. In many smaller water bodies adjacent canopy cover regulates light exposure (Savoy et al., 2021). In turn, light regulates numerous aspects of water temperature and stratification, ecosystem productivity and organism behavior, carbon cycling, contaminant toxicity, and overall economic valuations of water quality.

A. Light and thermal structure

The amount of incident light and underwater light attenuation together regulate numerous physical processes in aquatic ecosystems (see Chapter 7) such as heat transfer (e.g., Schmid & Koster, 2016), and the depth and strength of thermal stratification, especially in smaller lakes (Fee et al., 1996). In surface waters, heat absorbed is redistributed by wind-induced mixing, and this process interacts with K_d to determine the depth of stratification. Small lakes that are wind sheltered tend to have very shallow mixed layers and thermal profiles that mirror the exponential decrease in solar energy input with depth (Read & Rose, 2013).

Lakes with a high PAR K_d tend to have warmer surface waters relative to lakes with higher water clarity. For example, light-absorbing algal blooms can amplify surface temperatures by up to 1.5°C relative to nearby water bodies without blooms (Ibelings et al., 2003; Jones et al., 2005; Kahru et al., 1993). The effects of changing water clarity are most pronounced when PAR K_d is below 0.5 (Rinke et al., 2010). However, in some circumstances surface water temperatures in low-clarity lakes may not be amplified over higher-clarity lakes because greater heat absorption in darker waters can be offset by greater outward heat fluxes. Thus increases in K_d (i.e., clarity losses) may increase surface temperatures during warming periods (e.g., spring to early summer) but also facilitate more rapid heat losses during cooling periods (e.g., late summer to fall; Persson & Jones, 2008).

In general, the effects of K_d on deepwater temperature are often far greater than on surface water temperatures. Lakes with a high K_d (low clarity) typically have colder deep waters, volumetrically averaged lower water temperatures, and stronger and shallower stratification relative to lakes with a low K_d (Houser, 2006; Rinke et al., 2010; Rose et al., 2016). Overall, water clarity losses, for example, due to increases in algal biomass or DOM, can suppress climate warming impacts on lake temperatures, leading to net cooling, despite rising air temperatures (Bartosiewicz et al., 2019; Rose et al., 2016; Tanentzap et al., 2008).

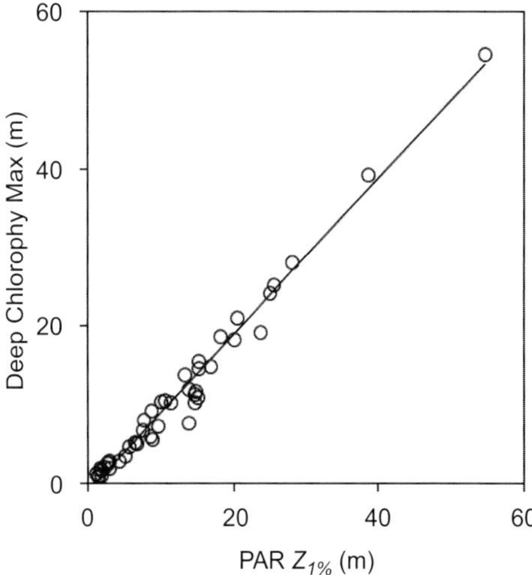

FIGURE 6-15 The relationship between the depth of the deep chlorophyll maximum (DCM) and photosynthetically active radiation (PAR) $Z_{1\%}$ depths across 47 lakes. *(Data from Leach et al. (2018) in lakes where light attenuation was measured using a radiometer [R^2 = 0.98].)*

B. Light and aquatic food webs

Solar radiation is the major energy source driving the productivity of aquatic ecosystems, and light often limits their productivity. Algae and macrophytes use between 4 and 9 quanta of light energy per molecule of CO_2 reduced. Even in nutrient-poor aquatic ecosystems, total productivity is frequently light limited. Light frequently limits benthic primary production, with corresponding reductions in invertebrates and fish (Karlsson et al., 2009). However, higher light exposure does not necessarily equate with greater productivity, as *photoinhibition* can occur when photosynthesizers are exposed to excess PAR and UV light (Neale, 2001). In an analysis of >900 days of lake metabolism data from surface mixed layers, 43% had moderate to strongly light-saturated primary production during midday summer conditions, and photoinhibition occurred on 77% of these days (Staehr et al., 2016).

Habitats within aquatic ecosystems are determined in part by the amount of light they receive. The *euphotic zone* is determined by the vertical range over which daily photosynthesis exceeds autotrophic respiration (see also *compensation depth*, discussed previously). The extent of the littoral zone is also commonly defined as the depth over which attached algae and plants can photosynthesize. Many lakes exhibit a subsurface peak in algal biomass, termed a *deep chlorophyll maximum* (DCM), which often occurs when the PAR $Z_{1\%}$ exceeds the depth of the surface mixed layer (Leach et al., 2018). The depth of the DCM often corresponds closely with the PAR $Z_{1\%}$ depth (Fig. 6-15). A common perception is that a DCM is primarily a feature of oligotrophic lakes. However, "deep" is a term relative to the thermocline, and thermoclines can be very shallow in small water bodies with high CDOM. For example, Hummingbird Bog is a highly stained and wind-sheltered small pond in Northern Michigan (USA), with a DOC concentration often over 20 mg L^{-1}. It had a thermocline at 0.6 m deep, a PAR K_d of 4.4 ($Z_{1\%}$ = 1.1 m), and a DCM peak at 1.3 m deep (Leach et al., 2018). Exclusion of productivity measurements from sunlit deep sources, whether from a DCM or benthic habitats, may substantially underestimate whole ecosystem productivity.

In addition to supporting photosynthesis, light impacts aquatic organisms in a number of other ways. For example, solar UV radiation is mutagenic and carcinogenic and can kill sensitive organisms in under a single summer day of exposure in clear surface waters. The two primary mechanisms of UV-induced damage are direct absorption and the production of reactive oxygen species. DNA strongly absorbs UV wavelengths, and exposure can create pyrimidine dimers (DNA lesions formed from thymine or cytosine bases). UV damage typically increases exponentially with decreasing wavelength, but since there are few photons underwater at very short wavelengths (e.g., 290–310 nm), peak UV-induced damage usually occurs at longer wavelength UV-B (e.g., ~320 nm). However, solar UV exposure can also be beneficial to organisms (Williamson et al., 2001). For example, vitamin D synthesis requires exposure to UV-B radiation, and longer-wavelength UV-A or visible light can be used by some organisms to repair DNA damage using the photolyase enzyme.

Organisms have evolved a number of mechanisms to deal with damaging UV exposure, including the synthesis or sequestration of photoproductive compounds such as carotenoids, mycosporine-like amino acids, and melanin. Other mechanisms include behavioral adaptations to avoid high UV environments and mechanisms to repair UV-induced damage. For example, *Daphnia* use light as a cue in body orientation while swimming. Zooplankton species, including many in the genus *Daphnia*, exhibit daily migrations through the water column, downward during the day and upward at night in response to changing light cues. In low-water-clarity water bodies these daily migrations are often associated with avoidance of visually predating fish, whereas in high-water-clarity water bodies these migrations are associated with avoidance of UV exposure (Williamson et al., 2011). *Daphnia* in particular can detect, and are *negatively phototactic* (swim away from) to, UV wavelengths (Storz & Paul, 1998). In contrast, some zooplankton, including some calanoid copepod species, are *positively phototactic*, or attracted to, UV radiation (Overholt et al., 2015). Some larval fish also have the ability to detect and avoid UV radiation and may use UV to increase prey capture (Leech et al., 2009). More generally, water clarity impacts fisheries productivity and species interactions. High UV can exclude some invasive fish species from otherwise suitable habitats (Tucker et al., 2010), and some species exhibit preferences for specific thermal-optical habitats (Hansen et al., 2019).

Many lakes experience seasonal ice cover. Snow on ice severely reduces underwater light and can have profound effects on the whole ecosystem's metabolism. In more productive lakes the loss of light under ice, together with the consumption of dissolved oxygen by the catabolic process, can exceed production by photosynthesis, leading to severe seasonal reductions in dissolved oxygen or even total anoxia. Seasonal oxygen reduction or anoxia often leads to the death of many organisms. When fish die under these conditions, this phenomenon is termed *winterkill*. Additionally, the patchiness in light beneath the ice–snow cover can, in turn, influence the horizontal distribution of algal photosynthesis and the behavior of planktonic animals.

C. Light and carbon cycling

The exposure of organic matter to light, and UV in particular, represents an important process in the global carbon cycle. The absorption of light by DOM can lead to photobleaching and photochemical oxidation of DOM to CO_2, facilitating the release of substantial quantities of CO_2 from aquatic ecosystems. Photochemical oxidation has been estimated to account for about 10–25% of all boreal lake CO_2 emissions (Koehler et al., 2014; Vachon et al., 2016). However, these are conservative estimates because they only consider direct photomineralization. In the Arctic photomineralization accounts for 70–95% of all DOC processing in lakes and rivers and about one-third of total CO_2 emissions from surface waters (Cory et al., 2014). Even when mineralization does not occur following light exposure, photobleaching can increase the biolability of DOM by breaking larger molecules into smaller ones and by partial oxidation of DOM (Ward et al., 2017). The process by which sunlight partially oxidizes DOM and changes its chemical composition can stimulate microbial biomass production by 1.5–6 times, increasing ecosystem respiration (Bowen et al., 2020; Moran & Zepp, 1997). Partial photooxidation of DOM is underestimated by at least three times in inland waters (Ward & Cory, 2020). Long-term exposure (e.g., weeks to months) of allochthonous DOM to sunlight commonly results in a substantial (>50–75%) decline in DOC concentration and an even larger loss of dissolved absorbance. However, some UV-insensitive nonchromophoric compounds tend to be more stable and can accumulate, resulting in the accumulation of nonabsorbing or poorly absorbing DOM in water bodies with long residence times (e.g., Arts et al., 2000).

D. Light and aquatic contaminants

Light exposure can amplify the toxicity of some contaminants in aquatic ecosystems, a process known as *phototoxicity* (Roberts et al., 2017). For example, exposure of polycyclic aromatic hydrocarbons (PAHs) to UV radiation enhances their toxicity to a wide range of aquatic organisms through the formation of reactive oxygen species. PAHs are formed during incomplete combustion, including during forest fires and the burning of fossil fuels. Separately, light exposure, and UV in particular, plays an important role in mercury demethylation in aquatic ecosystems and represents a sink for methylmercury (Lehnherr and St. Louis, 2009). Light may also play an important role in the degradation of other contaminants, including microplastics. Finally, DOM can be chemically associated (complexed) with metals such as iron, which are released into the water following exposure to sunlight and may impact metal bioavailability to aquatic organisms (Nalven et al., 2020).

E. Light and economic valuations of aquatic ecosystems

Water clarity is often used by the public as an indicator or proxy for overall water quality (Votruba & Corman, 2020). Water clarity is also a central component of the *Trophic State Index* (Carlson, 1977), which is a common classification system used to evaluate water bodies and assist in the determination of appropriate management actions. Water clarity is also a sentinel of broader ecological health and integrity (Adrian et al., 2009), and so a common management goal is maintaining or improving water clarity. High water clarity is often viewed as an indicator of unimpacted aquatic ecosystems, and property values are higher around lakes of higher water clarity (e.g., Gibbs et al., 2002).

Additionally, people are willing to drive farther for recreation at higher-water-clarity lakes to avoid nearby lower-water-clarity ones (Keeler et al., 2015).

V. Summary

1. Solar radiation is the major energy source for aquatic ecosystems and can be measured in quantum or energetic units. The shortest, highest-energy wavelengths penetrating the atmosphere are ~290 nm UV-B radiation. UV extends up to 400 nm. Photosynthetically active radiation (PAR) includes wavelengths from 400 to 700 nm, and infrared radiation is longer than 700 nm. Infrared represents about 50% of all light reaching Earth's surface; PAR represents about 45%, and UV about 5%.
2. The amount and spectral composition incident on a water body surface is influenced by factors such as sun angle, elevation, cloud cover, atmospheric aerosols, ozone, and other gas concentrations. The greater the departure of the angle of the sun from the perpendicular, the greater the reflection. Short-wavelength radiation of higher frequency is scattered more than longer-wavelength light (hence the blue appearance of clear sky at midday). Indirect solar radiation from the sky results from atmospheric molecular scattering. Decadal-scale changes in incident solar radiation have occurred due to changes in emissions of aerosols and ozone-depleting substances. At the water's surface, reflection reduces the amount of light entering water. Reflectivity increases with ice and snow cover and solar zenith angle.
3. Of the total light energy entering the water, a portion is scattered and the remainder is absorbed by the water or substances in water. Scattering and absorption are inherent optical properties in that they do not rely on the geometry of the incident light field. The total diminution of radiant energy by both scattering and absorption is called light attenuation, commonly quantified by the diffuse attenuation coefficient (K_d), an apparent optical property. K_d often exhibits substantial spectral variability. Chromophoric dissolved organic matter is typically by far the most important optically active substance regulating light attenuation at short wavelengths, while CDOM and algal biomass both are important regulators of PAR, depending on their relative concentrations. Nonalgal particulates can be important in regulating light attenuation in some water bodies, such as glacially fed lakes or reservoirs. The absorption of pure water is an important regulator of the attenuation of long-wavelength PAR and infrared radiation.
4. Light attenuation can be a more important factor regulating the amount of light underwater than variability in atmospheric properties. Light attenuation often exhibits seasonal patterns that depend on the phenology of the optical properties and concentration of optically active substances, including CDOM and algal biomass. Many lakes are exhibiting long-term trends in light attenuation. This is because light attenuation is responsive to drivers including land use and climate changes.
5. Light attenuation is commonly measured using a submersible radiometer. The Secchi disk depth provides an approximation of light attenuation and provides a simple way to estimate light attenuation. There are many long-term Secchi disk depth records. However, the relationship between Secchi disk depth and light attenuation depends on the relative concentrations of optically active substances. In lakes where CDOM regulates light attenuation a Secchi disk depth is often at a depth where there is a lower percent surface PAR relative to water bodies where algal biomass regulates light attenuation.
6. Light is a regulator of many ecological features, including thermal structure, food web dynamics and organismal behavior, carbon cycling, contaminant toxicity, and economic or social valuations of water quality. Thus light has marked attendant effects on nearly all chemical cycles, metabolic rates, and population dynamics. The productivity, metabolism, and carbon cycling of aquatic ecosystems are driven and controlled by energy derived directly from the solar energy utilized in photosynthesis, and many aquatic habitats are defined by the amount of light in them. Photosynthetic biochemical conversions utilizing solar energy occur within fresh waters both directly by aquatic flora as well as indirectly by terrestrial and wetland plants within the drainage basin and are imported to the streams and lakes as organic matter. UV radiation often has detrimental effects on organisms, but many have evolved mechanisms to deal with UV, and UV also has some beneficial effects.

Acknowledgments

The author acknowledges support from the US National Science Foundation (grants numbers 1754265 and 2048031) and constructive comments from Rose Cory, Sally MacIntyre, Patrick Neale, and Craig Williamson. Research at Laguna Simba was supported by a SETI Institute NASA Astrobiology Team. Fig. 6-12 (*right*) was supported by US National Science Foundation grant DEB 1754181. Fig. 6-11 (*left*) was supported by the SETI Institute NASA Astrobiology Team; Fig. 6-11 (*right*) was supported by the Extreme Microbiome Project. The author also acknowledges Jonathan Stetler, who contributed Fig. 6-2, and Max Glines, who contributed Fig. 6-13.

References

Adrian, R., O'Reilly, C.M., Zagarese, H., Baines, S.B., Hessen, D.O., Keller, W., et al., 2009. Lakes as sentinels of climate change. Limnology and Oceanography 54 (6 Part 2), 2283–2297. https://doi.org/10.4319/lo.2009.54.6_part_2.2283.

Andersen, M.R., Sand-Jensen, K., Iestyn Woolway, R., Jones, I.D., 2017. Profound daily vertical stratification and mixing in a small, shallow, wind-exposed lake with submerged macrophytes. Aquatic Sciences 79 (2), 395–406. https://doi.org/10.1007/s00027-016-0505-0.

Arts, M.T., Robarts, R.D., Kasai, F., Waiser, M.J., Tumber, V.P., Plante, A.J., et al., 2000. The attenuation of ultraviolet radiation in high dissolved organic carbon waters of wetlands and lakes on the northern Great Plains. Limnology and Oceanography 45 (2), 292–299. https://doi.org/10.4319/lo.2000.45.2.0292.

Bartosiewicz, M., Przytulska, A., Lapierre, J.F., Laurion, I., Lehmann, M.F., Maranger, R., 2019. Hot tops, cold bottoms: Synergistic climate warming and shielding effects increase carbon burial in lakes. Limnology and Oceanography Letters 4 (5), 132–144. https://doi.org/10.1002/lol2.10117.

Belzile, C., Vincent, W.F., Gibson, J.A.E., Hove, P.V., 2001. Bio-optical characteristics of the snow, ice, and water column of a perennially ice-covered lake in the High Arctic. Canadian Journal of Fisheries and Aquatic Sciences 58 (12), 2405–2418. https://doi.org/10.1139/cjfas-58-12-2405.

Bowen, J.C., Kaplan, L.A., Cory, R.M., 2020. Photodegradation disproportionately impacts biodegradation of semi-labile DOM in streams. Limnology and Oceanography 65 (1), 13–26. https://doi.org/10.1002/lno.11244.

Carlson, R.E., 1977. A trophic state index for lakes. Limnology and Oceanography 22 (2), 361–369. https://doi.org/10.4319/lo.1977.22.2.0361.

Carpenter, S.R., Lodge, D.M., 1986. Effects of submersed macrophytes on ecosystem processes. Aquatic Botany 26 (C), 341–370. https://doi.org/10.1016/0304-3770(86)90031-8.

Chipperfield, M.P., Bekki, S., Dhomse, S., Harris, N.R.P., Hassler, B., Hossaini, R., et al., 2017. Detecting recovery of the stratospheric ozone layer. Nature 549 (7671), 211–218. https://doi.org/10.1038/nature23681.

Cogley, J.G., 1979. The albedo of water as a function of latitude. Monthly Weather Review 107 (6), 775–781.

Cory, R.M., Davis, T.W., Dick, G.J., Johengen, T., Denef, V.J., Berry, M.A., et al., 2016. Seasonal dynamics in dissolved organic matter, hydrogen peroxide, and cyanobacterial blooms in Lake Erie. Frontiers in Marine Science 3 (Apr), 1–17. https://doi.org/10.3389/fmars.2016.00054.

Cory, R.M., Ward, C.P., Crump, B.C., Kling, G.W., 2014. Sunlight controls water column processing of carbon in arctic fresh waters. Science 345 (6199), 925–928.

Cuthbert, I.D., del Giorgio, P., 1992. Toward a standard method of measuring color in freshwater. Limnology and Oceanography 37 (6), 1319–1326. https://doi.org/10.4319/lo.1992.37.6.1319.

Davis, M.B., 1973. Redeposition of pollen grains in lake sediment. Limnology and Oceanography 18 (1), 44–52. https://doi.org/10.4319/lo.1973.18.1.0044.

Effler, S.W., Johnson, D.L., 1987. Calcium carbonate precipitation and turbidity measurements in Otisco Lake, New York. Water Resources Bulletin 23 (1), 73–79.

Fee, E.J., Hecky, R.E., Kasian, S.E.M., Cruikshank, D.R., 1996. Effects of lake size, water clarity, and climatic variability on mixing depths in Canadian Shield lakes. Limnology and Oceanography 41 (5), 912–920.

Gallegos, C.L., Davies-Colley, R.J., Gall, M., 2008. Optical closure in lakes with contrasting extremes of reflectance. Limnology and Oceanography 53 (5), 2021–2034. https://doi.org/10.4319/lo.2008.53.5.2021.

Gergel, S.E., Turner, M.G., Kratz, T.K., 1999. Dissolved organic carbon as an indicator of the scale of watershed influence on lakes and rivers. Ecological Applications 9 (4), 1377–1390. https://doi.org/10.1890/1051-0761(1999)009[1377:DOCAAI]2.0.CO;2.

Gibbs, J.P., Halstead, J.M., Boyle, K.J., Huang, J.-C., 2002. An hedonic analysis of the effects of lake water clarity on New Hampshire lakefront properties. Agricultural and Resource Economics Review 31 (1), 39–46. https://doi.org/10.1017/s1068280500003464.

Hansen, G.J.A., Winslow, L.A., Read, J.S., Treml, M., Schmalz, P.J., Carpenter, S.R., 2019. Water clarity and temperature effects on walleye safe harvest: An empirical test of the safe operating space concept. Ecosphere 10 (5). https://doi.org/10.1002/ecs2.2737.

Henneman, H.E., Stefan, H.G., 1999. Albedo models for snow and ice on a freshwater lake. Cold Regions Science and Technology 29 (1), 31–48. https://doi.org/10.1016/S0165-232X(99)00002-6.

Hongve, D., Akesson, G., 1996. Spectrophotometric determination of water colour in Hazen units. Water Research 30 (11), 2771–2775.

Houser, J.N., 2006. Water color affects the stratification, surface temperature, heat content, and mean epilimnetic irradiance of small lakes. Canadian Journal of Fisheries and Aquatic Sciences 63 (11), 2447–2455. https://doi.org/10.1139/F06-131.

Hu, C., Muller-Karger, F.E., Zepp, R.G., 2002. Absorbance, absorption coefficient, and apparent quantum yield: A comment on common ambiguity in the use of these optical concepts. Limnology and Oceanography 47 (4), 1261–1267. https://doi.org/10.4319/lo.2002.47.4.1261.

Hummel, J.R., Reck, R.A., 1979. A global surface albedo model. Journal of Applied Meteorology 18 (3), 239–253.

Ibelings, B.W., Vonk, M., Los, H.F.J., Van Der Molen, D.T., Mooij, W.M., 2003. Fuzzy modeling of cyanobacterial surface waterblooms: Validation with NOAA-AVHRR satellite images. Ecological Applications 13 (5), 1456–1472. https://doi.org/10.1890/01-5345.

Idso, S.B., Gilbert, R.G., 1974. On the universality of the Poole and Atkins Secchi disk-light extinction equation. The Journal of Applied Ecology 11 (1), 399. https://doi.org/10.2307/2402029.

John, R.H., Stubbins, A., Ritchie, J.D., Minor, E.C., Kieber, D.J., Mopper, K., 2008. Absorption spectral slopes and slope ratios as indicators of molecular weight, source, and photobleaching of chromophoric dissolved organic matter. Limnology and Oceanography 53 (3), 955–969. https://doi.org/10.4319/lo.2009.54.3.1023.

Jones, I., George, G., Reynolds, C., 2005. Quantifying effects of phytoplankton on the heat budgets of two large limnetic enclosures. Freshwater Biology 50 (7), 1239–1247. https://doi.org/10.1111/j.1365-2427.2005.01397.x.

Kahru, M., Leppanen, J.M., Rud, O., 1993. Cyanobacterial blooms cause heating of the sea surface. Marine Ecology Progress Series 101 (1–2), 1–8. https://doi.org/10.3354/meps101001.

Karlsson, J., Byström, P., Ask, J., Ask, P., Persson, L., Jansson, M., 2009. Light limitation of nutrient-poor lake ecosystems. Nature 460 (7254), 506–509. https://doi.org/10.1038/nature08179.

Keeler, B.L., Wood, S.A., Polasky, S., Kling, C., Filstrup, C.T., Downing, J.A., 2015. Recreational demand for clean water: Evidence from geotagged photographs by visitors to lakes. Frontiers in Ecology and the Environment 13 (2), 76–81. https://doi.org/10.1890/140124.

Kirk, J.T.O., 1994. Light and Photosynthesis in Aquatic Ecosystems. Cambridge University Press.

Koehler, B., Landelius, T., Weyhenmeyer, G.A., Machida, N., Tranvik, L.J., 2014. Sunlight-induced carbon dioxide emissions from inland waters. Global Biogeochemical Cycles 1199–1214. https://doi.org/10.1002/2014GB004832.Received.

Koenings, J.P., Edmundson, J.A., 1991. Secchi disk and photometer estimates of light regimes in Alaskan lakes: Effects of yellow color and turbidity. Limnology and Oceanography 36 (1), 91–105. https://doi.org/10.4319/lo.1991.36.1.0091.

Lampert, W., Fleckner, W., Rai, H., Taylor, B.E., 1986. Phytoplankton control by grazing zooplankton: A study on the spring clear-water phase. Limnology and Oceanography 31 (3), 478–490. https://doi.org/10.4319/lo.1986.31.3.0478.

Laurion, I., Ventura, M., Catalan, J., Psenner, R., Sommaruga, R., 2000. Attenuation of ultraviolet radiation in mountain lakes: Factors controlling the among- and within-lake variability. Limnology and Oceanography 45 (6), 1274–1288. https://doi.org/10.4319/lo.2000.45.6.1274.

Leach, T.H., Beisner, B.E., Carey, C.C., Pernica, P., Rose, K.C., Huot, Y., et al., 2018. Patterns and drivers of deep chlorophyll maxima structure in 100 lakes: The relative importance of light and thermal stratification. Limnology and Oceanography 63 (2), 628–646. https://doi.org/10.1002/lno.10656.

Leach, T.H., Winslow, L.A., Hayes, N.M., Rose, K.C., 2019. Decoupled trophic responses to long-term recovery from acidification and associated browning in lakes. Global Change Biology 25 (5), 1779–1792. https://doi.org/10.1111/gcb.14580.

Leech, D.M., Boeing, W.J., Cooke, S.L., Williamson, C.E., Torres, L., 2009. UV-enhanced fish predation and the differential migration of zooplankton in response to UV radiation and fish. Limnology and Oceanography 54 (4), 1152–1161. https://doi.org/10.4319/lo.2009.54.4.1152.

Lehnherr, I., St. Louis, V.L., 2009. Importance of ultraviolet radiation in the photodemethylation of methylmercury in freshwater ecosystems. Environmental Science & Technology 43 (15), 5692–5698.

Leppäranta, M., 2015. Freezing of Lakes and the Evolution of Their Ice Cover. Springer Science & Business Media.

Lind, O.T., 1986. The effect of non-algal turbidity on the relationship of Secchi depth to chlorophyll a. Hydrobiologia 140, 27–35.

Lottig, N.R., Wagner, T., Henry, E.N., Cheruvelil, K.S., Webster, K.E., Downing, J.A., et al., 2014. Long-term citizen-collected data reveal geographical patterns and temporal trends in lake water clarity. PLoS ONE 9 (4). https://doi.org/10.1371/journal.pone.0095769.

Madronich, S., Gruijl, F.R.d., 1994. Stratospheric ozone depletion between 1979 and 1992: Implications for biologically active ultraviolet-B radiation and non-melanoma skin cancer incidence. Photochemistry and Photobiology 59 (5), 541–546. https://doi.org/10.1111/j.1751-1097.1994.tb02980.x.

Matthes, L.C., Mundy, C.J., L.-Girard, S., Babin, M., Verin, G., Ehn, J.K., 2020. Spatial heterogeneity as a key variable influencing spring-summer progression in UVR and PAR transmission through Arctic sea ice. Frontiers in Marine Science 7 (March), 1–15. https://doi.org/10.3389/fmars.2020.00183.

Moran, M.A., Zepp, R.G., 1997. Role of photoreactions in the formation of biologically labile compounds from dissolved organic matter. Limnology and Oceanography 42 (6), 1307–1316. https://doi.org/10.4319/lo.1997.42.6.1307.

Morel, A., Smith, R.C., 1974. Relation between total quanta and total energy for aquatic photosynthesis. Limnology and Oceanography 19 (4), 591–600. https://doi.org/10.4319/lo.1974.19.4.0591.

Morris, D.P., Hargreaves, B.R., 1997. The role of photochemical degradation of dissolved organic carbon in regulating the UV transparency of three lakes on the Pocono Plateau. Limnology and Oceanography 42 (2), 239–249. https://doi.org/10.4319/lo.1997.42.2.0239.

Morris, D.P., Zagarese, H., Williamson, C.E., Balseiro, E.G., Hargreaves, B.R., Modenutti, B., et al., 1995. The attenuation of solar UV radiation in lakes and the role of dissolved organic carbon. Limnology and Oceanography 40 (8), 1381–1391. https://doi.org/10.4319/lo.1995.40.8.1381.

Mullen, P.C., Warren, S.G., 1988. Theory of the optical properties of lake ice. Journal of Geophysical Research 93 (D7), 8403–8414. https://doi.org/10.1016/0375-9601(70)90860-1.

Myers, D.R., Gueymard, C.A., 2004. Description and availability of the SMARTS spectral model for photovoltaic applications. Organic Photovoltaics V 5520, 56–67. https://doi.org/10.1117/12.555943.

Nalven, S.G., Ward, C.P., Payet, J.P., Cory, R.M., Kling, G.W., Sharpton, T.J., et al., 2020. Experimental metatranscriptomics reveals the costs and benefits of dissolved organic matter photo-alteration for freshwater microbes. Environmental Microbiology 22. https://doi.org/10.1111/1462-2920.15121.

Neale, P.J., 2001. Modeling the effects of ultraviolet radiation on estuarine phytoplankton production: impact of variations in exposure and sensitivity to inhibition. Journal of Photochemistry and Photobiology B: Biology 62 (1–2), 1–8.

Newman, P.A., Nash, E.R., Kawa, S.R., Montzka, S.A., Schauffler, S.M., 2006. When will the Antarctic ozone hole recover? Geophysical Research Letters 33 (12), 1–5. https://doi.org/10.1029/2005GL025232.

Ondok, J.P., 1978. Radiation Climate in Fishpond Littoral Plant Communities. In: Dykyjová, D., Květ, J. (Eds.), Pond Littoral Ecosystems, vol 28. Ecological studies, Springer.

Ostrovsky, I., Yacobi, Y.Z., Walline, P., Kalikhman, I., 1996. Seiche-induced mixing: Its impact on lake productivity. Limnology and Oceanography 41 (2), 323–332. https://doi.org/10.4319/lo.1996.41.2.0323.

Overholt, E.P., Rose, K.C., Williamson, C.E., Fischer, J.M., Cabrol, N.A., 2015. Behavioral responses of freshwater calanoid copepods to the presence of ultraviolet radiation: Avoidance and attraction. Journal of Plankton Research 38 (1), 16–26. https://doi.org/10.1093/plankt/fbv113.

Perovich, D.K., Roesler, C.S., Pegau, W.S., 1998. Variability in Arctic sea ice optical properties. Journal of Geophysical Research 103 (C1), 1193–1208.

Perpiñán Lamigueiro, O., 2012. solaR: Solar radiation and photovoltaic systems with R. Journal of Statistical Software 50 (9), 1–32. https://doi.org/10.18637/jss.v050.i09.

Persson, I., Jones, I.D., 2008. The effect of water colour on lake hydrodynamics: A modelling study. Freshwater Biology 53 (12), 2345–2355. https://doi.org/10.1111/j.1365-2427.2008.02049.x.

Poole, H.H., Atkins, W.R.G., 1929. Photo-electric measurements of submarine illumination throughout the year. Journal of the Marine Biological Association of the United Kingdom 16 (1), 297–324. https://doi.org/10.1017/S0025315400029829.

Pope, R.M., Fry, E.S., 1997. Absorption spectrum (380–700 nm) of pure water. II. Integrating cavity measurements. Applied Optics 36 (33), 8710–8723.

Poulin, B.A., Ryan, J.N., Aiken, G.R., 2014. Effects of iron on optical properties of dissolved organic matter. Environmental Science and Technology 48 (17), 10098–10106. https://doi.org/10.1021/es502670r.

Quickenden, T.I., Irvin, J.A., 1980. The ultraviolet absorption spectrum of liquid water. The Journal of Chemical Physics 72 (8), 4416–4428.

Read, J.S., Rose, K.C., 2013. Physical responses of small temperate lakes to variation in dissolved organic carbon concentrations. Limnology and Oceanography 58 (3), 921–931. https://doi.org/10.4319/lo.2013.58.3.0921.

Read, J.S., Rose, K.C., Winslow, L.A., Read, E.K., 2015. A method for estimating the diffuse attenuation coefficient (KdPAR) from paired temperature sensors. Limnology and Oceanography: Methods 13 (2), 53–61. https://doi.org/10.1002/lom3.10006.

Rinke, K., Yeates, P., Rothhaupt, K.O., 2010. A simulation study of the feedback of phytoplankton on thermal structure via light extinction. Freshwater Biology 55 (8), 1674–1693. https://doi.org/10.1111/j.1365-2427.2010.02401.x.

Roberts, A.P., Alloy, M.M., Oris, J.T., 2017. Review of the photo-induced toxicity of environmental contaminants. Comparative Biochemistry and Physiology Part C: Toxicology & Pharmacology 191, 160–167.

Rose, K.C., Greb, S.R., Diebel, M., Turner, M.G., 2017. Annual precipitation regulates spatial and temporal drivers of lake water clarity. Ecological Applications 27 (2), 632–643. https://doi.org/10.1002/eap.1471.

Rose, K.C., Hamilton, D.P., Williamson, C.E., Mcbride, C.G., Fischer, J.M., Olson, M.H., et al., 2014. Light attenuation characteristics of glacially-fed lakes. Journal of Geophysical Research: Biogeosciences 119, 1446–1457. https://doi.org/10.1002/2014JG002674.Received.

Rose, K.C., Neale, P.J., Tzortziou, M., Gallegos, C.L., Jordan, T.E., 2019. Patterns of spectral, spatial, and long-term variability in light attenuation in an optically complex sub-estuary. Limnology and Oceanography 64 (S1), S257–S272.

Rose, K.C., Winslow, L.A., Read, J.S., Hansen, G.J.A., 2016. Climate-induced warming of lakes can be either amplified or suppressed by trends in water clarity. Limnology and Oceanography Letters 1, 44–53.

Rudorff, C.M., Melack, J.M., MacIntyre, S., Barbosa, C.C.F., Novo, E.M.L.M., 2011. Seasonal and spatial variability of CO_2 emission from a large floodplain lake in the lower Amazon. Journal of Geophysical Research: Biogeosciences 116 (4), 1–12. https://doi.org/10.1029/2011JG001699.

Savoy, P., Bernhardt, E., Kirk, L., Cohen, M.J., Heffernan, J.B., 2021. A seasonally dynamic model of light at the stream surface. Freshwater Science 40 (2), 286–301. https://doi.org/10.1086/714270.

Schelske, C.L., Hooper, F.F., Haertl, E.J., 1962. Responses of a marl lake to chelated iron and fertilizer. Ecology 43 (4), 646–653.

Schmid, M., Koster, O., 2016. Excess warming of a Central European lake driven by solar brightening. Water Resouces Research 52, 8103–8116. https://doi.org/10.1111/j.1752-1688.1969.tb04897.x.

Scully, N.M., Lean, D.R.S., McQueen, D.J., Cooper, W.J., 1995. Photochemical formation of hydrogen peroxide in lakes: Effects of dissolved organic carbon and ultraviolet radiation. Canadian Journal of Fisheries and Aquatic Sciences 52 (12), 2675–2681. https://doi.org/10.1139/f95-856.

Solomon, S., Ivy, D.J., Kinnison, D., Mills, M.J., Neely, R.R., Schmidt, A., 2016. Emergence of healing in the Antarctic ozone layer. Science 353 (6296), 269–274. https://doi.org/10.1126/science.aae0061.

Sommer, U., Gliwicz, Z.M., Lampert, W.I., Duncan, A., 1986. The PEG-model of seasonal succession of planktonic events in fresh waters. Archiv für Hydrobiologie 106 (January), 433–471. http://doi.wiley.com/10.1111/j.1469-185X.1969.tb01218.x.

Soranno, P.A., 1997. Factors affecting the timing of surface scums and epilimnetic blooms of blue-green algae in a eutrophic lake. Canadian Journal of Fisheries and Aquatic Sciences 54 (9), 1965–1975. https://doi.org/10.1139/f97-104.

Staehr, P.A., Brighenti, L.S., Honti, M., Christensen, J., Rose, K.C., 2016. Global patterns of light saturation and photoinhibition of lake primary production. Inland Waters 6 (4), 593–607.

Storz, U.C., Paul, R.J., 1998. Phototaxis in water fleas (*Daphnia magna*) is differently influenced by visible and UV light. Journal of Comparative Physiology—A Sensory, Neural, and Behavioral Physiology 183 (6), 709–717. https://doi.org/10.1007/s003590050293.

Tanentzap, A.J., Yan, N.D., Keller, B., Girard, R., Heneberry, J., Gunn, J.M., et al., 2008. Cooling lakes while the world warms: Effects of forest regrowth and increased dissolved organic matter on the thermal regime of a temperate, urban lake. Limnology and Oceanography 53 (1), 404–410. https://doi.org/10.4319/lo.2008.53.1.0404.

Topp, S.N., Pavelsky, T.M., Stanley, E.H., Yang, X., Griffin, C.G., Ross, M.R.V., 2021. Multi-decadal improvement in US lake water clarity. Environmental Research Letters 16 (5). https://doi.org/10.1088/1748-9326/abf002.

Tucker, A.J., Williamson, C.E., Rose, K.C., Oris, J.T., Connelly, S.J., Olson, M.H., et al., 2010. Ultraviolet radiation affects invasibility of lake ecosystems by warm-water fish. Ecology 91 (3), 882–890. https://doi.org/10.1890/09-0554.1.

Vachon, D., Lapierre, J.F., Del Giorgio, P.A., 2016. Seasonality of photochemical dissolved organic carbon mineralization and its relative contribution to pelagic CO_2 production in northern lakes. Journal of Geophysical Research: Biogeosciences 121 (3), 864–878. https://doi.org/10.1002/2015JG003244.

Votruba, A.M., Corman, J.R., 2020. Definitions of water quality: A survey of lake-users of water quality-compromised lakes. Water (Switzerland) 12 (8). https://doi.org/10.3390/W12082114.

Ward, C.P., Cory, R.M., 2020. Assessing the prevalence, products, and pathways of dissolved organic matter partial photo-oxidation in Arctic surface waters. Environmental Science: Processes and Impacts 22 (5), 1214–1223. https://doi.org/10.1039/c9em00504h.

Ward, C.P., Nalven, S.G., Crump, B.C., Kling, G.W., Cory, R.M., 2017. Photochemical alteration of organic carbon draining permafrost soils shifts microbial metabolic pathways and stimulates respiration. Nature Communications 8 (1), 1–7. https://doi.org/10.1038/s41467-017-00759-2.

Warren, S.G., 2019. Optical properties of ice and snow. Philosophical Transactions of the Royal Society A: Mathematical, Physical and Engineering Sciences 377 (2146). https://doi.org/10.1098/rsta.2018.0161.

Weidemann, A.D., Bannister, T.T., Effler, S.W., Johnson, D.L., 1985. Particulate and optical properties during $CaCO_3$ precipitation in Otisco Lake. Limnology and Oceanography 30 (5), 1078–1083. https://doi.org/10.4319/lo.1985.30.5.1078.

Wetzel, R.G., 2001. Limnology: Lake and River Ecosystems. Gulf Professional Publishing.

White, W.S., Wetzel, R.G., 1975. Nitrogen, phosphorus, particulate and colloidal carbon content of sedimenting seston of a hard-water lake. SIL Proceedings, 1922–2010. 19 (1), 330–339. https://doi.org/10.1080/03680770.1974.11896072.

Wild, M., 2009. Global dimming and brightening: A review. Journal of Geophysical Research Atmospheres 114 (12), 1–31. https://doi.org/10.1029/2008JD011470.

Wild, M., Gilgen, H., Roesch, A., Ohmura, A., Long, C.N., Dutton, E.C., et al., 2005. From dimming to brightening: Decadal changes in solar radiation at earth's surface. Science 308 (5723), 847–850. https://doi.org/10.1126/science.1103215.

Wild, M., Wacker, S., Yang, S., Sanchez-Lorenzo, A., 2021. Evidence for clear-sky dimming and brightening in Central Europe. Geophysical Research Letters 48 (6), 1–9. https://doi.org/10.1029/2020GL092216.

Williamson, C.E., De Lange, H.J., Leech, D.M., 2007. Do zooplankton contribute to an ultraviolet clear-water phase in lakes? Limnology and Oceanography 52 (2), 662–667. https://doi.org/10.4319/lo.2007.52.2.0662.

Williamson, C.E., Fischer, J.M., Bollens, S.M., Overholt, E.P., Breckenridgec, J.K., 2011. Toward a more comprehensive theory of zooplankton diel vertical migration: Integrating ultraviolet radiation and water transparency into the biotic paradigm. Limnology and Oceanography 56 (5), 1603–1623. https://doi.org/10.4319/lo.2011.56.5.1603.

Williamson, C.E., Neale, P.J., Grad, G., De Lange, H.J., Hargreaves, B.R., 2001. Beneficial and detrimental effects of UV on aquatic organisms: Implications of spectral variation. Ecological Applications 11 (6), 1843–1857. https://doi.org/10.1890/1051-0761(2001)011[1843:BADEOU]2.0.CO;2.

Williamson, C.E., Stemberger, R.S., Morris, D.P., Frost, T.M., Paulsen, S.G., 1996. Ultraviolet radiation in North American lakes: Attenuation estimates from DOC measurements and implications for plankton communities. Limnology and Oceanography 41 (5), 1024–1034. https://doi.org/10.4319/lo.1996.41.5.1024.

Winn, N., Williamson, C.E., Abbitt, R., Rose, K., Renwick, W., Henry, M., et al., 2009. Modeling dissolved organic carbon in subalpine and alpine lakes with GIS and remote sensing. Landscape Ecology 24 (6), 807–816. https://doi.org/10.1007/s10980-009-9359-3.

Woolway, R.I., Jones, I.D., Hamilton, D.P., Maberly, S.C., Muraoka, K., Read, J.S., et al., 2015. Automated calculation of surface energy fluxes with high-frequency lake buoy data. Environmental Modelling and Software 70, 191–198. https://doi.org/10.1016/j.envsoft.2015.04.013.

Zhang, Y., Qin, B., Shi, K., Zhang, Y., Deng, J., Wild, M., et al., 2020. Radiation dimming and decreasing water clarity fuel underwater darkening in lakes. Science Bulletin 65 (19), 1675–1684. https://doi.org/10.1016/j.scib.2020.06.016.

CHAPTER

7

Fate of Heat

Sally MacIntyre[1] *and David P. Hamilton*[2]

[1]Department of Ecology, Evolution, and Marine Biology, University of California, Santa Barbara, California, United States

[2]Australian Rivers Institute, Griffith University, Nathan, Queensland, Australia

OUTLINE

The heating of ponds, lakes, streams, rivers, and wetlands is largely due to the absorption of solar radiation. Heat is distributed via physical processes acting on the water surface, at boundaries, and within the waterbodies, creating density structure in the water column in which organisms live and interact and in which chemical reactions occur. Thermal structure in individual lakes and flowing waters is dynamic and changes hourly, daily, seasonally, and over longer time scales depending on rates of heating and cooling and the extent of mixing in the water column. Whether lakes and streams heat and stratify or cool and mix is determined by surface energy budgets, which quantify rates of heating and cooling in the upper water, or a heat budget which also accounts for heat gains or losses from inflowing and outgoing streams and rivers, groundwater, and geothermal heating. Thermal structure varies with climate, elevation, morphometry, and optical characteristics. Typical stratification in lakes and ponds and, at times, rivers consists of an upper layer directly influenced by atmospheric forcing, a strongly stratified layer, and a weakly stratified layer below. Depth-dependent stratification is quantified by the *buoyancy frequency*, and stratification at a whole lake scale is characterized by *Schmidt stability*. The sensitivity of lakes to wind-induced mixing across diurnal and seasonal thermoclines is quantified by two dimensionless indices, the *Wedderburn* and *Lake numbers*, and the downward flux of heat indicative of turbulent mixing can be used to quantify the *coefficient of eddy diffusivity*, an index of mixing used for computing fluxes of heat and scalars such as nutrients and phytoplankton. Annual patterns of stratification have been classified and depend on latitude, elevation, and lake morphometry. These patterns depend on the surface energy budget and the redistribution of heat by physical processes.

I. Distribution of heat in lakes

A. Brief introduction to thermal structure

Heating and cooling of water in aquatic ecosystems combined with the action of wind on the water surface

Wetzel's Limnology, Fourth Edition
https://doi.org/10.1016/B978-0-12-822701-5.00007-0

leads to thermal stratification and mixing over diel, seasonal, and annual periods. As a result, diverse and dynamic habitats are created for organisms and the biogeochemical and ecological effects are wide-ranging. The descriptions that follow apply to lakes, ponds, and reservoirs. In lakes deep enough to stratify during warming periods, thermal structure is typically characterized by an upper warmer, weakly stratified layer, called the upper mixed layer or *epilimnion* (Figs. 7-1 to 7-4). It overlays a more strongly stratified *thermocline* or *metalimnion*, with the term thermocline used interchangeably with metalimnion or distinguished as the most strongly stratified region within the metalimnion. If the water column is sufficiently deep, a weakly stratified region of colder and denser water called the *hypolimnion* occurs below. Annual cycles of stratification are characterized by the frequency of periods in which the water column is stratified and by the periods in which temperatures are nearly *isothermal* (the same temperature throughout the water column) and characterized as mixing. These distinctions can be seen for temperate Lawrence Lake (Michigan, USA), where the density stratification is stable during summer and winter and temperatures are near isothermal in spring and autumn (Fig. 7-4). In summer the warmest temperatures are near the surface. In winter when ice forms and water temperatures are less than ~4°C, the temperature of maximum density in freshwater (see Chapter 3), the water is coolest near the surface and warmer below. The temperature profile is then characterized as *inversely stratified*. These distinctions of periods with stratification and mixing belie the dynamics that occur within lakes. As can be seen in Fig. 7-4, the upper mixed layer is increasing and decreasing in depth during the summer, indicating mixing within this layer, and while sometimes a water body is mixing when temperatures are near isothermal, at other times, it is not. Lakes such as Lawrence Lake (Michigan, USA), with two periods annually in which stratification occurs over the full water column and two periods when temperatures are nearly uniform throughout, are classified as *dimictic*, a common classification for moderately deep and deep lakes in temperate climates. Annual patterns of thermal stratification for lakes at different latitudes and of different sizes are described in Section IIC.

High-resolution instrumentation and more frequent sampling indicate more complex and dynamic structures in the upper mixed layer and hypolimnion than implied in earlier descriptions (Imberger 1985; Brainerd and Gregg 1993; Wüest and Lorke 2003). Over diel cycles, solar heating tends to exceed heat losses such that warmer, less dense water is found near the surface (Figs. 7-1 and 7-2). At night, heat losses dominate, eroding stratification which develops in the day. For example, a temperature profile taken midday during light winds in a warm season would have a near-surface, weakly stratified upper layer called the *surface layer* or *actively mixing layer*, a layer with a larger temperature change below called a *diurnal thermocline*, a weakly stratified region below called the *subsurface layer*, a persistent region with a larger temperature gradient called the *seasonal thermocline*, and, depending on the depth of the lake and the penetration of light, a weakly stratified deeper layer called the *hypolimnion* (Fig. 7-1).

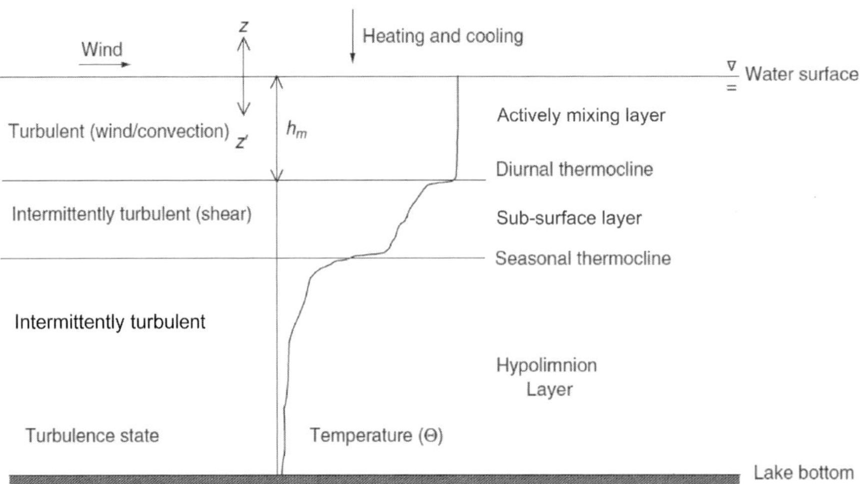

FIGURE 7-1 Sketch of a temperature profile in a lake during the day showing the various layers which form and whether they are likely to be *turbulent* and mixing or intermittently turbulent, as opposed to *laminar* (not mixing) (Chapter 8 Section IA). The vertical axis is depth (z) with z' the distance downward from the surface. The actively mixing layer, diurnal thermocline, and subsurface layer comprise the upper mixed layer, also known as the epilimnion. The seasonal thermocline and hypolimnion are below it. The actively mixing layer is also called the surface layer and its depth here is identified as h_m. Its depth and that of the diurnal thermocline varies over the course of a day (Figs. 7-2 and 7-3). The horizontal lines labeled diurnal and seasonal thermocline, respectively, bisect the layers where temperatures rapidly decrease with depth. The diurnal and seasonal thermoclines are also intermittently turbulent. *(Redrawn from Monismith and MacIntyre (2009).)*

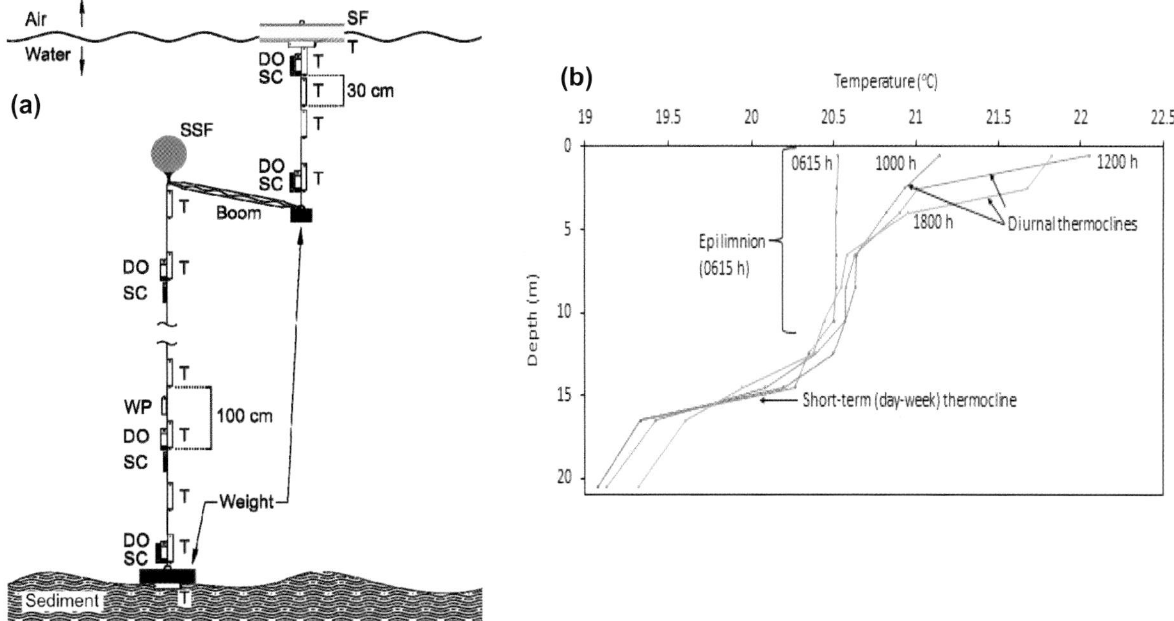

FIGURE 7-2 (a) Example of a mooring design to obtain time series temperature (*T*), specific conductance (*SC*), dissolved oxygen (*DO*), and water pressure (*WP*) measurements in a lake. The array of sensors has a deeper component called a taut-line mooring with a subsurface float (*SF*) with sufficient buoyancy to support the sensors and the line to which they are attached and an anchor heavy enough to counter the buoyancy of the float and any drag on the line. Sensors are suspended along the line with depth intervals at the user's discretion. Attached to the lower arrays is an upper one with a flat float at the surface below which is a temperature sensor deployed horizontally so that temperature can be measured within 0.05 m of the air–water interface and shielded from direct solar insolation. Suspended from the surface float are additional sensors. The upper array is connected to the lower via a line in a short plastic tube to help prevent twisting. The upper float will move up and down with surface waves, whereas the sensors on the taut-line mooring will be at fixed depths. Sensors point down or the near-surface ones are wrapped in reflective tape to avoid direct heating from solar radiation. Such arrays were used to collect the data in the figures in Section IIA. Depths between sensors are selected in order to quantify dynamics in the actively mixing layer and below and also depend on the depth of the lake. For accuracy in depths, the lines should have minimal stretch. Hardware and lines used will depend on the depth and size of a lake. Jansen et al. (2021) provide a sketch for designs for winter studies in lakes that become ice covered. (b) Temperature profiles obtained from an array as described in (a) with temperature sensors positioned at depths of 0.5, 2.5, 4.5, 6.5, 8.5, 10.5, 12.5, 14.5, 16.5, and 20.5 m in Lake Rotorua (North Island, New Zealand) at four times of the day (*colored lines*) on 5 February 2017. The thermocline at 15 m persisted for several days, and the depth of the shallow diurnal thermocline varied over the course of the day. *(Unpublished data provided by Chris McBride.)*

The extent to which these layers are turbulent depends on the magnitudes of wind, heating, and cooling. If winds are light during a period of heating, the diurnal thermocline may reach the surface (Fig. 7-2). With stronger winds or the onset of cooling, mixing can lead to the development of a surface layer that is near isothermal, and with continued cooling, this upper layer may deepen, potentially reaching the seasonal thermocline at night or when winds are strong (Fig. 7-3). Such changes over a diel cycle are illustrated for a lake in New Zealand in Fig. 7-2, and the changes in the depth of the layers over diel cycles are illustrated in Fig. 7-3. The density structure that develops in the day moderates ecosystem function. For instance, it influences the extent to which phytoplankton are mixed and the light they receive as well as the efficacy of grazing by zooplankton.

The depths of the upper mixed layer, diurnal and seasonal thermoclines, and hypolimnion vary with heating and cooling, and with changes in wind velocity and direction, such that the depth of these layers can vary along the length of a lake (Section ID: *Wind-driven mixing*). High-resolution temperature profiles often show temperature steps within the thermocline and hypolimnion, with these steps related to *internal wave* motions (Section ID, Chapter 8 III, IV). No strict definition of a thermocline based on temperature change is appropriate given the nonlinearity of the curve relating temperature to density (Chapter 3) and the variable heating, cooling, and magnitudes of wind over lakes. Hence, in some water bodies, the temperature change across the thermocline may be a few degrees Celsius per meter, whereas in tropical lakes it may be only a few tenths of a degree per meter. The hypolimnion is sometimes divided into two layers, a near-bottom one, called the *bottom (or benthic) boundary layer*, where stratification is weaker and more mixing occurs, and a quiescent layer above, where stratification is slightly stronger (Wüest and Lorke 2003) (Chapter 27). These authors, in fact, propose a

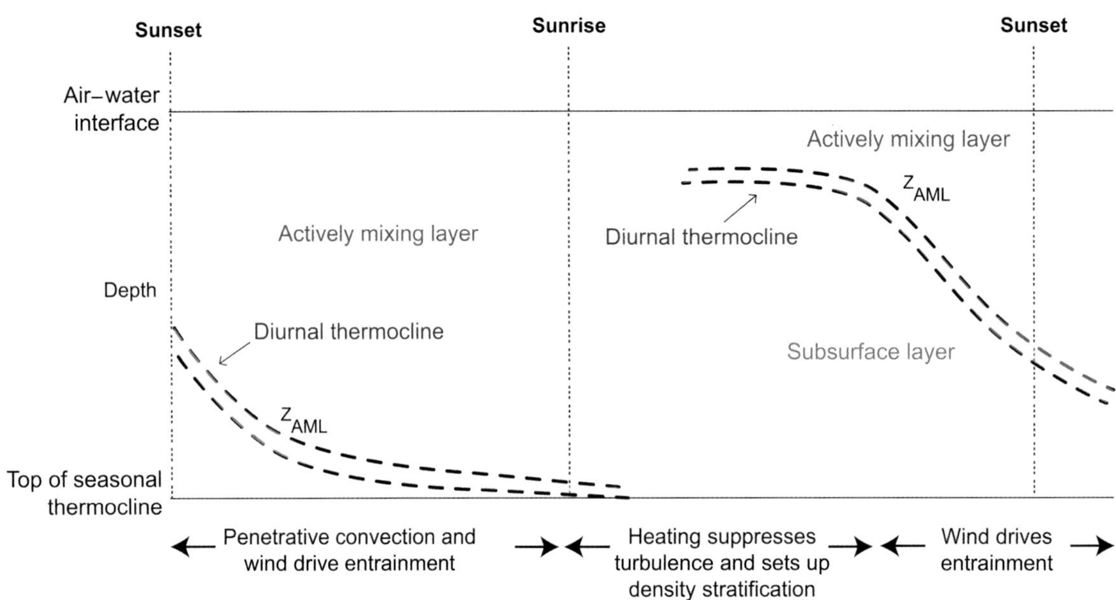

FIGURE 7-3 Layered structures within the upper mixed layer and their changes over diel cycles. The actively mixing layer is above the diurnal thermocline and the subsurface layer is below it. The diurnal thermocline forms in the day during heating. If heating is ongoing, it may deepen in the afternoon from wind-induced mixing. As the lake begins to cool, both penetrative convection and wind-induced deepening continue at night. The diurnal thermocline reforms after sunrise with increased heating. z_{AML} indicates the depth of the actively mixing layer. During stormy periods, the entire upper mixed layer may be an actively mixing layer (not shown). Concepts are derived from Imberger (1985) and Brainerd and Gregg (1993). *(Redrawn from Tedford et al. (2014).)*

three-layer classification scheme with surface and bottom boundary layers that frequently mix and a quiescent interior (Fig. 8-6) (Chapter 8 III). In their scheme the thermocline and upper hypolimnion would comprise the interior layer. These distinctions point to the dynamic nature of thermal structure over diel to annual periods.

Annual cycles of heating and cooling in lakes demonstrate patterns dependent on latitude, local climate, optical conditions, and lake depth. For example, Lawrence Lake, a small, shallow lake in Michigan, was ice covered from December through March, with temperatures increasing from the surface downward (Fig. 7-4). With the coldest temperatures immediately under the ice, the lake had *inverse temperature stratification*. Following ice-off, temperatures were nearly isothermal at 4°C, the temperature of the maximum density of freshwater. Temperatures rose with increased solar radiation entering the water, and thermal stratification developed by May. Several periods with heating and cooling occurred in summer such that the upper mixed layer had the structure evident in Fig. 7-1 and the changes over diel cycles as in Figs. 7-2 and 7-3. By August, as incoming solar radiation declined, the mixed layer began to deepen and the lake was isothermal by November. This period is called *fall turnover*, as with near-surface temperatures cooling, water near the surface becomes denser, causing mixing. The term turnover is applied when the water column becomes isothermal, although high-resolution instrumentation indicates isothermy is rarely sustained. Ice-on occurred when temperatures in much of the water column had declined to ~3°C and were near zero at the surface. The patterns in Lawrence Lake are representative of lakes in regions with cold winters and warm summers.

Given the variability in climate as a function of latitude and altitude, methods to quantify heat budgets in lakes are described in Section IB and in streams and rivers in Section IC. Indices derived from heat budgets that indicate the susceptibility of lakes to mixing are described in Section ID. The changes in the variables described provide an understanding of linkages among heating, cooling, wind-induced mixing, and thermal structure. Examples of the upper mixed layer are provided in Sections IE. A brief introduction to other aspects of hydrodynamics that further shape thermal structure is included, a topic more fully addressed in Chapter 8. Metrics used to classify the stability of lakes and their susceptibility to wind forcing are applied to patterns of stratification as a function of latitude and lake depth in Section IIA. Thermal structure and processes influencing water movement under the ice are described in Section IIB, and classification schemes are presented in Sections IIC, IID, and IIE. Changes in temperature and thermal structure with ongoing climate and land use change are briefly described in Section III.

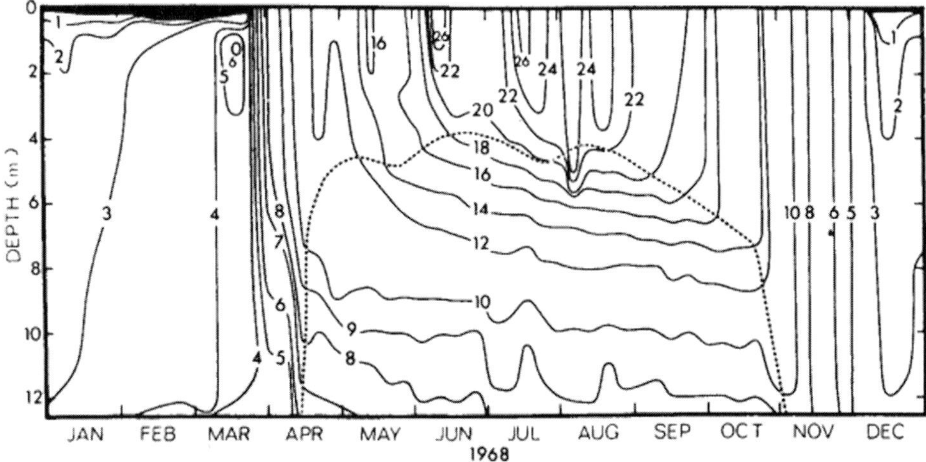

FIGURE 7-4 Time depth diagram illustrating the annual cycle of heating and cooling for Lawrence Lake, a temperate lake in Michigan, USA, in 1968. Each *line* is a line of constant temperature. The *dotted line* shows the depth of the upper mixed layer. Immediately below the upper mixed layer, the thermocline can be discriminated by abrupt temperature changes, but the full region below the upper mixed layer is characterized as a metalimnion due to the relatively strong stratification. Ice cover (*blackened region* near the surface from January to March and in December) drawn to scale. Lawrence Lake is characterized as dimictic, with two periods of near isothermy in spring and autumn and stable density stratification in summer and under the ice.

B. Thermal energy content: heat budgets of lakes

The thermal structure of lakes is determined by the fluxes of energy across the air–water interface, sediment–water interface, and incoming water as it mixes within the lake, and the action of wind (Fig. 7-5). The fluxes across the air–water interface are quantified via a surface energy budget computed from time series measurements of meteorology and water temperature. A full heat budget additionally includes discharge into and out of the system, rainfall, other heat sources or sinks such as groundwater and sediments, and any geothermal heating. The extent of thermal stratification varies with heating and cooling and wind-induced mixing.

A surface energy budget includes the processes that introduce heat or cause cooling at the water surface (Fig. 7-5). The budget includes *incoming shortwave radiation* (SW_i) and *outgoing shortwave radiation* (SW_o), that is, the fraction of SW_i that is lost by reflection, with their sum called *net shortwave radiation* (SW_{net}); *outgoing and incoming longwave radiation* (LW_o, LW_i), with the sum called *net longwave radiation* (LW_{net}); *latent heat flux* (*LE*), which results from *evaporation* (*E*) and leads to heat loss or from *condensation*, which results in heat being gained; and *sensible heat exchange* (*SE*) also known as *conduction* (C). These terms are the predominant sources of heating and cooling for many lakes and can be obtained with radiation measurements plus wind speed, air and surface water temperatures, and relative humidity. The terms which induce heating have a positive sign convention in this chapter. SW_i, LW_i, and condensation always are positive; SW_o, LW_o, and evaporation are always negative; and the sign changes for *SE* and *LE*. While LW_i is a large positive term, LW_{net} is negative and contributes to cooling. Equations for the calculations have been presented by Fischer et al. (1979), Imberger and Patterson (1990), MacIntyre et al. (2002), Verburg and Antenucci (2010), Fink et al. (2014), MacIntyre et al. (2014), and Woolway et al. (2015).

The fluxes at the air–water interface are generally the largest sources of heating and cooling within a lake. The surface energy budget is calculated as follows with H the net heat flux (W m^{-2}):

$$H = SW_i + SW_o + LW_i + LW_o + LE + SE \quad (7\text{-}1)$$

To determine how much heat is being added, or lost, over a period of time, one multiplies by surface area and time, noting that W m^{-2} is equivalent to Joules m^{-2} s^{-1}. In studies of the upper mixed layer, the equation is modified to only consider the heat stored in the actively mixing layer (Imberger 1985), and the heat flux into it is called the effective heat flux ($H*$). Buoyancy flux, with units m^2 s^{-3}, a measure of kinetic energy per unit time, is computed as:

$$\beta = \frac{g\alpha H}{C_{pw}\rho_o} \quad (7\text{-}2)$$

where g is gravity, α is the thermal coefficient of expansion which varies with temperature (Bouffard and Wüest (2019) (2.1×10^{-4} °C^{-1} at 20°C), c_{pw} is the specific heat capacity of water (4184 J kg^{-1} °C^{-1}), and ρ_o is the density of water at the surface. Buoyancy flux into the actively mixing layer, effective buoyancy flux, $\beta*$, is similarly calculated with $H*$ instead of H. β quantifies the extent to which the incoming and outgoing energy

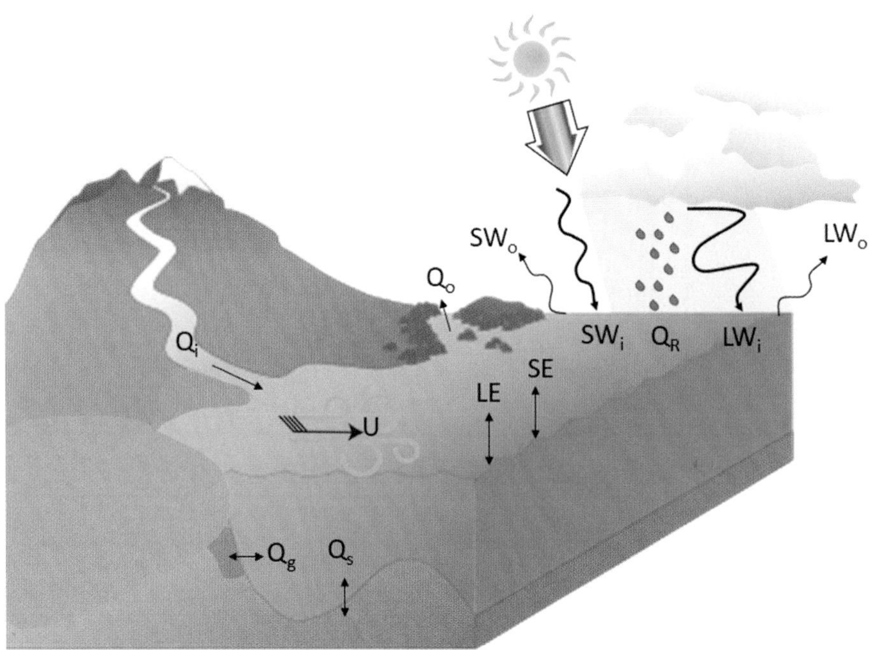

FIGURE 7-5 Illustration of the terms for a heat budget. SW_i is incoming shortwave radiation, SW_o is outgoing (reflected) shortwave radiation, LW_i is incoming longwave radiation, LW_o is outgoing longwave radiation, LE is latent heat flux, and SE is sensible heat flux. Q_i is heat flux from inflows, Q_o is heat flux from outflows, Q_s is heat flux between the bottom sediments and water, Q_R is heat flux from rainfall, and Q_g is heat flux from inflows and outflows of groundwater. U is wind speed, which acts to redistribute heat by mixing and advection associated with the energy input at the surface. The *colored arrow* representing the visible heating component of penetrative shortwave radiation is embedded within a *larger arrow* representing the nonvisible heating component (ultraviolet and infrared components (Table 7-1)).

will cause stratification (β positive) and mixing (β negative) in the surface layer (Figs. 7-1 and 8-6).

Other sources of heating or cooling include incoming and outgoing stream water, groundwater, rainwater, and heat fluxes from sediments (Fig. 7-5). Typically, their contributions are minor. Sediment heating of the overlying water column can be important under ice cover. The contribution from sediment heating occurs at the sediment—water interface, so its contribution to the total heat flux is calculated by multiplying the flux, Q_s, by the surface area of sediments involved and the time period. Fluxes from groundwater, Q_g, usually occur in nearshore or shallow water. Cold groundwater intrusions can cool lakes and, in some shallow lakes, Q_g can exceed latent heat fluxes for parts of the summer (Kettle et al. 2012). Fluxes from rainfall, Q_R, are usually variable over the surface of a lake. In the case of inflows, if the volume of incoming water is large relative to the lake volume, their contribution to heat flux can be significant. The heat flux from inflows (Q_i) is computed as:

$$Q_i = c_{pw}\,\rho\,Q\,(T_{\text{in}} - T_w)\,A_o^{-1} \tag{7-3}$$

where c_{pw} is specific heat capacity, ρ is density (kg m^{-3}), Q is discharge, that is, the volume of incoming water per second (m^3 s^{-1}) (Chapter 8 IIB), A_o is the surface area of the lake, and T_{in} and T_w are temperature (°C) of incoming and lake water, respectively (Fink et al. 2014). By convention in the literature, heat flux terms are identified using the letter Q, as is discharge; hence we use the same letter and include the definitions. The temperature of outgoing water is often assumed to be similar to that of surface water, but in reservoirs outflows could be deeper and have a temperature significantly lower than the surface water. In tectonically active regions geothermal heating can occur either from hot springs or from the Earth's heat warming the sediments.

Heat fluxes (HF) can be budgeted as:

$$HF = SW_i + SW_o + LW_i + LW_o + LE + SE + Q_i + Q_o + Q_s + Q_g + Q_R \tag{7-4}$$

Terms are assigned positive or negative values according to whether the heat flux is incoming to the lake or outgoing from the lake, respectively. Here we use the convention that the loss terms are negative. Examples using this approach are included in Imboden and Wüest (1995) and Fink et al. (2014).

Surface energy budgets—Net shortwave radiation is the largest source of heat. Despite LW_i being a large term, LW_o is larger; hence LW_{net} is negative. The depth to which shortwave radiation penetrates, enabling the water to warm and providing energy for photosynthesis, depends on the transparency of the water (Fig. 7-6).

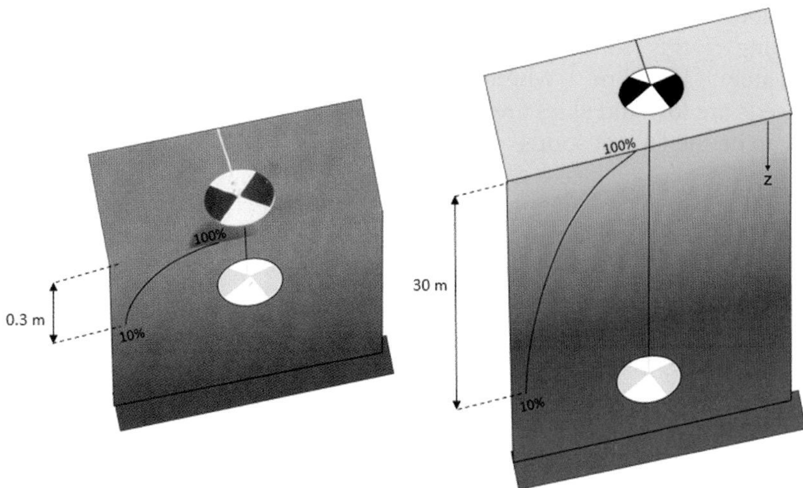

FIGURE 7-6 Comparison of shortwave attenuation (*curved line* representing exponential decay with depth) in a lake with low (*left*) and high optical clarity (*right*). The Secchi depth is indicative of water clarity, and it disappears from view by an observer just above the lake surface at a depth where visible light (approximately 45% of incoming shortwave radiation) is about 10% of the surface value. Eq. 7-5 shows a mathematical description of the relation of shortwave radiation to depth (z).

Solar radiation in the visible spectrum penetrates most deeply in the water column. Calculation of penetration of shortwave radiation using *Beer's Law* was introduced in Chapter 6. Here we expand upon the formulation in Chapter 6 to include penetration for the full spectra using the symbol I for *irradiance*.

$$I(z) = Io \int_{\lambda 1}^{\lambda n} e^{-k_{i,n} z} \, d\lambda \tag{7-5}$$

where $\lambda_{i:n}$ are the seven wavelength bands, $k_{i:n}$ are the corresponding attenuation coefficients (Table 7-1), and Io and $I(z)$ are irradiance at the surface and at depth z, respectively.

Based on the fraction of solar radiation in each wavelength band and the attenuation coefficients, solar radiation in the visible range and that in the near-infrared range are responsible for most of the heating. As illustrated in Fig. 7-6, the penetration of light declines exponentially with depth and the decline is more rapid when attenuation is greater (Chapter 6). Heating will be greater near the surface in more turbid or highly stained lakes. The contribution of light in the near infrared to surface heating is greater for optically clear lakes than for turbid ones. Net shortwave radiation is zero at night and has maxima, when cloud cover is low, of about 800–1000 W m^{-2}, with maxima depending on latitude (Section IIA).

Heat is lost at a lake's surface via several processes. Heat loss from outgoing *longwave radiation* (LW_o) can be high, up to around −400 W m^{-2}. Downwelling longwave radiation (LW_i) from the sky is also large, and larger during cloudy conditions, ~350 W m^{-2}. Thus the net is negative. A small fraction of LWi is reflected, 3% (LW_{ir}), and is included in direct measurements of

TABLE 7-1 Wavelength bands, fraction of shortwave radiation in each wavelength band, and attenuation coefficient in each band ($k_{i:n}$) based on transmission in pure water (Hale and Querry 1973) except for ultraviolet (280–400 nm) and visible (400–700 nm) light, which vary within each lake. Attenuation of ultraviolet radiation and visible radiation depends on chromophoric dissolved organic matter (CDOM) (Chapter 28), which stains the water, and particulates in the water (Chapter 6). The attenuation in the ultraviolet, $k_{280-400}$, in Mono Lake, California (USA) was 26 m^{-1} (Jellison and Melack 1993). The attenuation coefficient of ultraviolet radiation from 280 to 315 nm (UVB) in boreal lakes in the Canadian Shield is ~3 m^{-1} (Xenopoulos et al. 2001) and in Crystal Lake, Wisconsin (USA), is ~ 1 m^{-1} (Chapter 6). The diffuse attenuation coefficient for visible radiation (400–700 nm), known as k_d, ranges from ~0.1 m^{-1} or less for optically clear lakes such as Lake Tahoe (California, Nevada; USA), is commonly in the range from 0.2 to 0.8 m^{-1} for lakes with low algal biomass that are either clear or colored with dissolved organic carbon (DOC) concentrations up to ~5 mg L^{-1} (examples being Emerald Lake, California, in the Sierra Nevada [USA] and Toolik Lake, Alaska [USA], respectively; see Chapter 6) and of order 10 m^{-1} for lakes that are highly turbid due to sediments or abundant phytoplankton. (Fractions sum to 1.02 due to rounding errors; fraction in the visible varies slightly.)

Range of λ (nm)	280–400	400–700	700–800	800–900	900–1000	1200–1800	1800–2800
Fraction	0.04	0.45	0.13	0.09	0.04	0.25	0.02
Attenuation m^{-1}	Variable	Variable	1.1	3.4	26	870	7800

outgoing longwave radiation. Net longwave radiation is around −50 W m^{-2} during cloudy conditions and is more negative, approximately −125 W m^{-2}, when skies are clear. When wind speeds are low and skies are clear, net longwave radiation can be the largest source of heat loss.

Sensible heat exchange depends on the temperature difference between air and the water surface and wind speed. It is generally the smallest term in a surface energy budget. In smaller lakes it may be positive during the day, indicating heating, when air temperatures are higher than surface water temperatures, but it is typically negative at night, indicating heat loss, when air temperatures are cooler than surface water. On a seasonal basis, sensible heat exchange is more often positive during spring in temperate or Arctic regions when lake temperatures are colder than air temperatures and can reach 100 W m^{-2} (MacIntyre et al. 2009a). In other seasons, and often year-round in subtropical and tropical lakes, sensible heat loss is negative and of order −20 W m^{-2} (MacIntyre et al. 2009a, 2014).

Heat is lost through evaporation, a component of *latent heat flux*. The rate of evaporation increases with higher temperatures, reduced vapor pressure, lower barometric pressure, and increased wind speed but decreases with increasing salinity. Evaporation is usually the largest source of heat loss from lakes, with values over tropical lakes during light winds around −100 W m^{-2} and reaching −600 to −700 W m^{-2} during squalls. Evaporation is greater from warm than cold water bodies for the same wind speed (MacIntyre and Melack 2009). Latent heat fluxes are also higher in lakes in more arid regions with their low vapor pressure. When cold air flows over a slightly warmer lake, the water vapor condenses over the water (Fig. 7-7), and heat is released, contributing a slight heat input to a lake.

i. The surface energy budget and stratification

The diel patterns of heating and cooling recorded in the upper water column of lakes result, in part, because solar radiation penetrates into the water, whereas the heat losses from evaporation, conduction, and net longwave radiation are restricted to the upper few centimeters of water. When incoming solar radiation exceeds heat losses and when wind speeds are low, the water column is warmed, creating stable stratification in which less dense water overlays cooler water (Figs. 7-1 to 7-3). In contrast, when solar radiation is less than the sum of the heat loss terms, the near surface becomes cooler, leading to unstable stratification, in which denser water overlays less dense water. This condition is conducive to mixing by *penetrative convection*, a process whereby *thermal plumes*, or *convection cells*, form when surface waters cool, become denser, and sink (Chapter 8 IIIC). The associated mixing occurs at night and when surface waters cool during shifts in meteorological conditions, such as when cloud cover increases, winds increase, leading to increased evaporation, cold rain falls, or seasonal decreases in air temperature. At night, this mixing can erode the diurnal thermocline (Figs. 7-2 and 7-3), and in cooling seasons it contributes to the erosion of the seasonal thermocline (Fig. 7-4).

The thermal structure of a lake over an annual cycle is related to the surface energy budget (Fig. 7-8). As illustrated for north temperate Lake Mendota, Wisconsin, net radiative heating begins in January and peaks in July (Fig. 7-8a). Since the lake is cold following ice-off, conduction is initially positive as air temperature is warmer than surface water temperature, and conduction becomes negative in summer as the lake warms. Evaporation becomes a progressively larger heat loss term as the lake's surface water temperature warms. The sum of these terms, the *net heat flux*, indicates

FIGURE 7-7 Conditions conducive to heat input to a lake from *condensation*. As air above the lake becomes cooler, saturation levels of water vapor decrease and water condenses at the surface. Condensation releases heat. *(Photograph credit D.P. Hamilton.)*

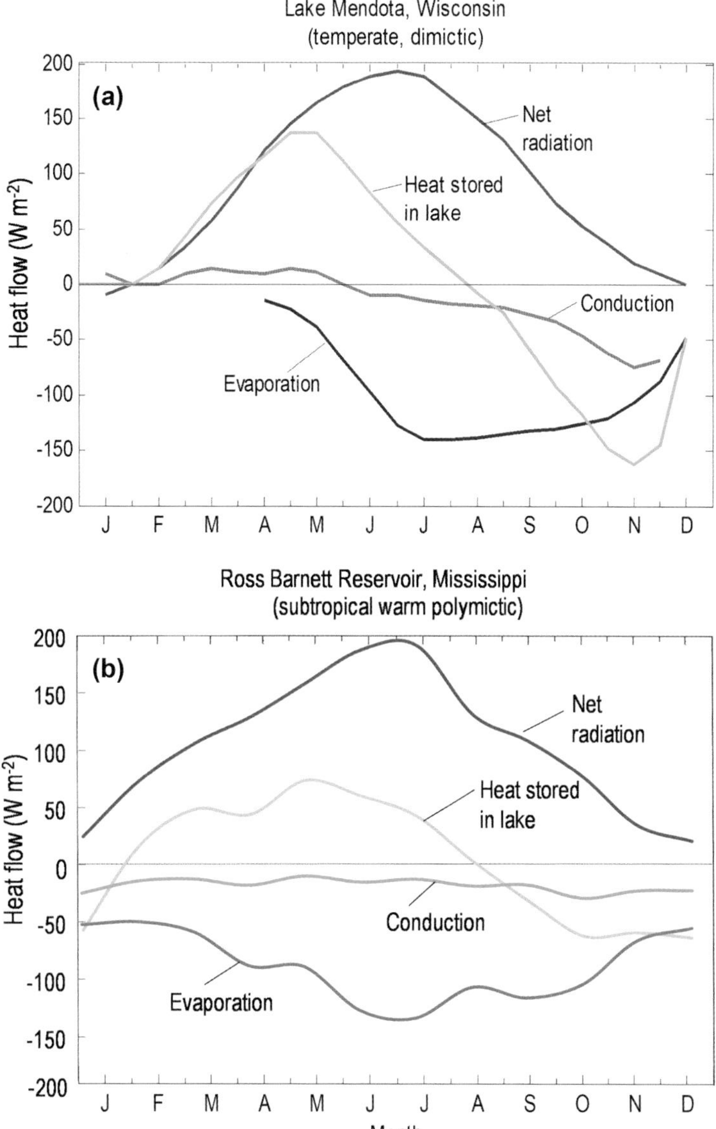

FIGURE 7-8 Annual surface energy budget for (a) dimictic Lake Mendota, Wisconsin (USA), and (b) Ross Barnett Reservoir, Mississippi (USA). Heat flow is equivalent to heat flux (H), and net radiation is the sum of SW_{net} and LW_{net}. Note the synchrony of heating and cooling and the rate of heat storage in the surface energy budget of Lake Mendota with the annual cycle of stratification in Lawrence Lake (Fig. 7-4). The monthly means of evaporation (latent heat flux) and conduction (sensible heat flux) in the subtropical lake have less variability than those for the temperate dimictic lake (a). As the water depth varies from 4 to 8 m and winds over the lake are frequently high, even though water temperatures increase in the summer, the lake frequently mixes and has a mixing regime classified as warm polymictic. *(The upper panel is redrawn from Horne and Goldman (1994) based on Ragotzki (1978), and the* lower panel *was created using data from Liu et al. (2012).)*

whether the lake is gaining or losing heat. Net heat flux becomes negative by late summer. While Lawrence Lake is smaller than Lake Mendota, its annual cycle of heating and cooling follows the pattern established by the surface energy budget (Figs. 7-4 and 7-7). Similar correspondence between thermal structure and the surface energy budget is found in lakes at other latitudes. For Ross Barnett Reservoir in subtropical Mississippi, the monthly surface energy budget is similar to that for Lake Mendota but with reduced seasonality, evaporation and conduction are important components of the heat budget throughout the year. Patterns over diel and seasonal cycles are included for Arctic and tropical lakes in Section IID, with the additional inclusion of *Lake numbers* (Section ID) to illustrate the influence of wind on mixing across the thermocline.

ii. Additional sources of heating and cooling

(a) *Heat fluxes associated with ice*—Fluxes of heat occur with the formation and loss of lake ice. When water at 0°C freezes, 3.34×10^5 J kg^{-1} of heat are released, and conversely, when ice at 0°C melts to water at 0°C, an equal amount of heat is absorbed. In addition, as ice sublimes to water vapor, 2.62×10^4 J kg^{-1} is absorbed. Once ice has formed, its *albedo*, that is, its reflectance, can be high at the ice surface if the ice is white, as occurs when there has been an interaction of ice and snow, or the surface is snow covered, and is lower when the ice is clear (Chapter 6). The transmissivity of light through ice also depends on the clarity of the ice. When the ice is clear, k_d can be similar to that of lake water (Belzile et al. 2001), whereas transmission can be considerably reduced with white ice (i.e., high k_d) (Chapter 6). When the ice is clear, water below the ice can be warmed by solar radiation, algal growth can occur, and, depending on the stratification and incoming solar radiation, mixing can be induced by penetrative convection (Chapter 8).

(b) *Heat losses and gains associated with the sediments*—Sediments of lakes absorb heat from the water during the warmer periods of the year (Fig. 7-9). However, this component of the heat budget (see Fig. 7-5) is generally small compared to the direct absorption of solar radiation by the water. That said, for shallower lakes, and depending on the magnitude of k_d, the heat budget of the sediments can be significant. Sediments in shallow regions of lakes can absorb quantities of solar radiation, and some of this heat may be transferred to the water on time scales of a day (Fang and Stefan 1996), and heat losses from the sediments at night can create convective motions that induce mixing (Bednarz et al. 2009). Under ice, heat accumulated over the summer is released by the sediments and heats the overlying water, with implications for circulation and thermal structure (Section IIB).

Few detailed studies of sediment heat fluxes over annual periods are available, making those from Lake Mendota particularly notable (Birge et al. 1927). Temperatures of surface sediments at a location where the water depth was 8 m closely followed the annual cycle of the epilimnetic water temperatures (Fig. 7-9). This thermal variation is attenuated with depth in the sediments so that at a depth of 5 m into the sediments, seasonal oscillations were muted. A similar relation was observed in hypolimnetic sediments at a depth of 23.5 m, but the amplitude of the temperature changes was reduced (Fig. 7-9). Changes in temperature of the sediments lagged those in the hypolimnion, and the lag increased by about 1 month with each 1 m increase in depth of measurement. The heat flux between the bottom sediments and water (Q_s, Fig. 7-5) can be approximated by the temperature gradient over a sediment–water boundary layer of distance z_{sed}:

$$Q_s = K_w \frac{T_s - T_w}{z_{sed}} \qquad (7\text{-}6)$$

where K_w is thermal conductivity (W m^{-1} °C^{-1}), T_s is the sediment temperature at the lower edge of the boundary, and T_w is the water temperature at the upper edge of the boundary (Hamilton et al. 2018). As noted in Fig. 7-9, z_{sed} can be 5 m. For purposes of modeling, the temporal variations in the sediment temperature profiles are included and the seasonal temperature variations are assumed to be damped at a depth of 10 m below the sediment–water interface (Fang and Stefan 1996).

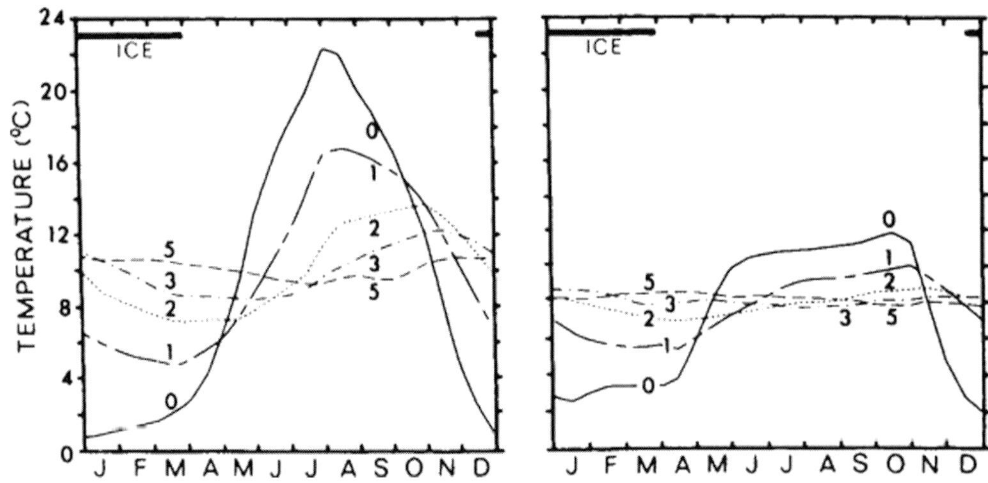

FIGURE 7-9 Annual temperature curves at 0, 1, 2, 3, and 5 m within the sediments of Lake Mendota, Wisconsin (means for 1918, 1919, and 1920). *Left:* sediments at a water depth of 8 m. *Right:* sediments at 23.5 m. *(Modified from data of Birge et al. 1927.)*

(c) *Geothermal heating*—In regions with tectonic activity sediment temperatures can be higher than water temperatures at depths below the surface and water from hot springs may also flow into the lakes. Examples include Lake Kivu, Rwanda Degens et al. 1973; Bärenbold et al. 2022), and Lake Rotowhero, New Zealand (Brookes et al. 2013). Geothermal heating in volcanic lakes can create strong convection currents that disrupt the seasonal stratification cycle (Lake Rotowhero) or mix the hypolimnion of deep lakes (Priscu et al. 1986). Indirect heating also can be significant in deep lakes (Finckh 1981). Figures illustrating the influence of geothermal heating are included in Sections IIC and IID. Geothermal heating also occurs within the subglacial lakes of Antarctica and contributes to horizontal and vertical circulation in these water bodies (Couston and Siegert 2021).

(d) *Heat fluxes from incoming and outgoing streams*—Incoming water from streams, rivers, or groundwater can either warm or cool the receiving water. The inflow term, Q_i, is usually only important in lakes that receive significant percentages of their volume from surface runoff, either during one storm event, throughout a rainy season, or when there is considerable snowmelt flowing into a lake (Sadro et al. 2018; Smits et al. 2020). Small or shallow lakes, reservoirs that have short retention times, and canyon reservoirs are likely to be affected (Jones et al. 2008; Xie et al. 2017; Posada Bedoya et al. 2021). Heat can also be lost through outflows (Q_o), for example, when warmer surface water is transported out of the lake. Some hydroelectric and drinking water reservoirs release deep, cool water in summer months, which leads to warmer mean annual temperatures than if the deep water was retained. The effect of incoming water on the temperature within a lake will also depend on the extent to which the water spreads and mixes laterally, as quantified by the term *horizontal dispersion* (Chapter 8 IB), relative to the speed with which it exits the lake. Time scales for these processes are included in Chapter 8 IB.

(e) *Consequences of spatial variability of heating and cooling*—Advection can be caused by greater heating, cooling, or wind-driven mixing in different regions of a lake. For example, shallow areas may heat or cool more rapidly, and exposed regions may be more affected by wind and cooler due to the combination of deeper mixing and higher latent heat fluxes (Imberger and Patterson 1990). In large lakes spatial variations in the surface energy budget may occur (Verburg et al. 2011; MacIntyre et al. 2014). These spatial differences in heating and cooling cause density differences, resulting in *density-driven flows* (Chapter 8 IIIC; Figs. 8-31, 8-32). *Overflows* are expected from locations where density decreases, as when heating is greater at temperatures above 4°C, and *underflows* originate from locations where density increases, as when cooling is greater at temperatures above 4°C (Monismith et al. 1990). The likelihood of such flows can be determined if meteorological and time series temperature measurements are taken at more than one location on a lake, and the resultant flows computed from the differences in the surface energy budget or the heat budget.

The differences in the heating of sediments at different depths or locations can also induce density-driven flows. Spatial variations of temperature within the sediments at similar water depths can be due to the type of adjacent land forms in the basin (Matveev 1964) and also vary with proximity to shore, sometimes being warmer nearshore and sometimes offshore (Likens and Johnson 1969; Reif 1969). Shallower lakes may have greater overall sediment heating (Ding and Mao 2021). Sediments heating overlying water can also cause vertical mixing by convection. Density-driven flows occurring under the ice in winter are described in Section IIB and Chapter 8 IIIC (MacIntyre et al. 2018).

C. Heat budgets in streams and rivers

The heat content of rivers and streams reflects the balance among inputs, storage, and outputs, as in Fig. 7-5 for lakes. As for lakes, all the terms of the surface energy budget apply, as do the advective terms, which often make the largest contribution. The heat influx from upstream dominates in larger rivers and the energy budget also includes the fluxes from tributaries. The incoming heat per unit time (H, in joules per second) from a stream is given by:

$$H = \rho c_{pw} T Q \qquad (7\text{-}7)$$

where ρ is water density (kg m^{-3}), c_{pw} is specific heat capacity, T is the temperature (°C) of the stream water, and Q is discharge (m^3 s^{-1}).

As in lakes, heat can be absorbed by the sediments, particularly in shallow streams, and can warm overlying water by conduction. Similarly, exchanges with the *hyporheic zone* (Chapter 8 IIC, Chapter 27), the wetted area around and below a stream, can modify within-stream temperatures (King and Neilson 2019). For modeling stream temperatures, rather than assessing all the terms in the surface energy budget, a simplified approach is sometimes used which assumes the variation of stream temperature depends on air temperature and advection from upstream sources (Toffolon and Piccolroaz 2015).

Such simplifications can be justified by long adjustment times, that is, transport over long distances for appreciable changes in heat content (Råman Vinnå et al. 2017) and observations of linear relations between air and river water temperatures. They will not apply when air temperatures drop below freezing (Smith 1981; Crisp and Howson 1982) and when inflows from melting snow keep temperatures well below that of air for weeks or longer. Precipitation on land can result in inflows of surface or ground waters with temperatures quite different from air temperature for considerable periods (days).

Rivers and streams have diel variations in water temperatures. These are more pronounced when they are shallow and exposed to direct solar irradiance, whereas streams sheltered from direct solar insolation by trees or other vegetation have muted diel changes. The speed of water flow and associated turbulence within the water column influence whether there is temperature stratification during the day (Guseva et al. 2021). The amplitude of annual oscillations of temperature varies with latitude and altitude (Walling and Webb 1992). For river systems that flow along elevational gradients, maximum temperatures may increase progressively from the headwaters to the mouths of rivers. These downstream changes in water temperature formed the foundation for a river zonation system in relation to changes in benthic invertebrate and fish communities (Müller 1951; Ilies 1953). Within river reaches, temperatures can vary in response to vegetation cover along the margins, groundwater seepage, channel depth and shape, orientation, substratum conditions, and silt content of the water (Walling and Webb 1992). Submersed aquatic vegetation (Chapter 24), if it absorbs considerable solar radiation, may increase annual mean temperatures and the amplitude of daily fluctuations (Crisp et al. 1982). Within the hyporheic zone of stream sediments, stream water infiltration occurs at the head of riffles and can alter temperatures of the interstitial water to depths of 50 cm (White et al. 1987; Crisp 1990).

Ice formation on rivers reduces heat losses from the water. Complete ice cover is rare outside the Arctic, Antarctica, or at high elevations; often, small areas remain open to the atmosphere during winter, particularly over riffles. Ice can also form on submerged substrata. Crystals of *frazil ice*, small discs of ice (1–4 mm in diameter and 1–100 μm in thickness), can form in turbulent, slightly supercooled water (Martin 1981). When moved with the water, frazil ice can scour the substratum and associated organisms. *Anchor ice* (Hobbs 1974) can form on riffle substrata in shallow water and impede water flow, increase water levels in pool areas, and cause flooding (Prowse 1994). Increased snowmelt and ice break-up in rivers at high altitudes and latitudes may lead to large increases in current velocity, stage (water elevation), concentrations of suspended materials, and scouring of the substrata (Scrimgeour et al. 1994). Major disturbances to biota associated with the substrata result.

D. Heat budgets and measures of stratification and mixing

Heat budgets can be computed within lakes to assess whether the heating and cooling expected from the surface energy budget are in balance with changes within the lake. When they are out of balance, other processes may be contributing, such as advection. *Heat content* (H, in joules) is a function of the mass (M, in kg) of the substance, temperature (T, in °C), and c_{pw}, the specific heat capacity of water (Chapter 3). Then,

$$H = M\,T\,c_{pw} \tag{7-8}$$

Heat flux (Q_T) in a layer of thickness Δz, that is, the change in heat content over time Δt in a layer, can be computed from time series measurements of temperature using the following equation:

$$Q_T(\Delta t, z_c) = \rho c_{pw}\left[\frac{T(z_c,t_2) - T(z_c,t_1)}{\Delta t}\right]\Delta z \tag{7-9}$$

where T is temperature, t_1 and t_2 are times 1 and 2, respectively, such that Δt is the time difference between them, z_c is the measurement depth, Δz is a depth interval above and below z_c, and ρ is density (kg m^{-3}). To obtain the change in heat content for the full lake, heat flux in each layer is summed, accounting for the volume of each layer. The relations between depth, area, and volume are obtained from a hypsographic curve (Chapter 4). Change in temperature is calculated as the inverse of Eq. 7-9:

$$\Delta T = \frac{Q_T}{\rho c_{pw} l}\,\Delta t \tag{7-10}$$

where l is a vertical dimension.

The historical work on this problem, known as *Birgean heat budgets*, addressed changes in heat content over the summer beginning with temperatures at 4°C, over the winter beginning with minimum temperatures up to 4°C, and over an annual cycle from the time when temperatures were minimal to the time when heat content was maximal (Birge and Juday 1914; Birge 1915, 1916; Hutchinson 1957, Stewart 1973; Ragotzkie 1978).

In contemporary analyses heat budgets are used for further understanding of a lake's hydrodynamics. They can be used to infer the extent of turbulence and mixing in the thermocline and below (Jassby and Powell 1975). By comparing changes in heat content within a lake to the surface energy budget, advection can be assessed, that is, changes in heat from incoming snowmelt, rain, streams, or rivers (Smits et al. 2020) or from transport elsewhere in a lake (Augusto-Silva et al. 2019). Additionally, heat budgets can be used to determine if changes in heat content at specific depths in ice-covered lakes are due to either direct solar heating or mixing (Cortés and MacIntyre 2020) and to validate fluxes with eddy covariance systems (Vesala et al. 2006). Comparison of one-dimensional heat budgets from multiple sites within a lake can be used to quantify the flow of heat from one region to another due to changes in wind stress or to differential heating and cooling (Monismith et al. 1990; Augusto-Silva et al. 2019). Thus heat budgets are valuable tools for assessing physical processes occurring in lakes (Chapter 8).

Efforts to quantify the work required to fully mix a lake, that is, a lake's *stability* (S), began with Birge (1916) with the calculations related to heat budgets. *Stability per unit area* of a lake is the quantity of work or mechanical energy required to mix the entire volume of water to a uniform temperature without addition or subtraction of heat (Birge 1915; Schmidt 1915, 1928; Idso 1973). The term is called *Schmidt stability* (S_t), and seasonal changes are generally determined based on changes in S_t from the time of minimum heat content to the time of maximum heat content. Following Idso (1973):

$$S_t = \frac{g}{A_0}\int_0^{z_m} (z - z_g)(\rho_z - \overline{\rho})A_z \partial z \qquad (7\text{-}11)$$

where A_o and A_z are surface area and area at depth z, g is gravity, z_g is the depth of the center of gravity when the lake is completely mixed (i.e., depth where the mean value of density would be located if the lake was unstratified; also known as the center of volume), z_m is the maximum depth of the lake, and $\overline{\rho}$ and ρ_z are the mean density and the density at depth z, respectively. The calculation determines the amount of work to mix the lake completely (see Fig. 8-6). Work equals force times distance. The force, $g(\rho_z - \overline{\rho})A_z dz$ in each volume element is multiplied by the distance away from the center of volume, $(z - z_g)$. These products are summed from the surface to the maximum depth of the lake. The unit is Joules per square meter (J m^{-2}). Schmidt stability is scale dependent, and larger, deeper lakes will have larger S_t after controlling for climate and altitude (Hutchinson 1957).

Density stratification is quantified by the square of the *buoyancy frequency* (N), also known as the *Brunt–Väisälä frequency*, where

$$N^2 = \frac{-g\left(\frac{\partial \rho}{\partial z}\right)}{\rho} \qquad (7\text{-}12)$$

and g is gravity, ρ is density, and z is depth. In this metric stability varies with depth and will be highest where density gradients are largest, as at diurnal or seasonal thermoclines or at *chemoclines* where density increases due to increases in solutes (Chapter 12). When density is only a function of temperature and temperatures are above 4°C, the density profile is approximately the mirror image of the temperature profile. As shown in Fig. 7-10, increased values of N occur where the density gradients are largest. The units of N are radians per second or cycles per hour (cph), recalling there are 2π

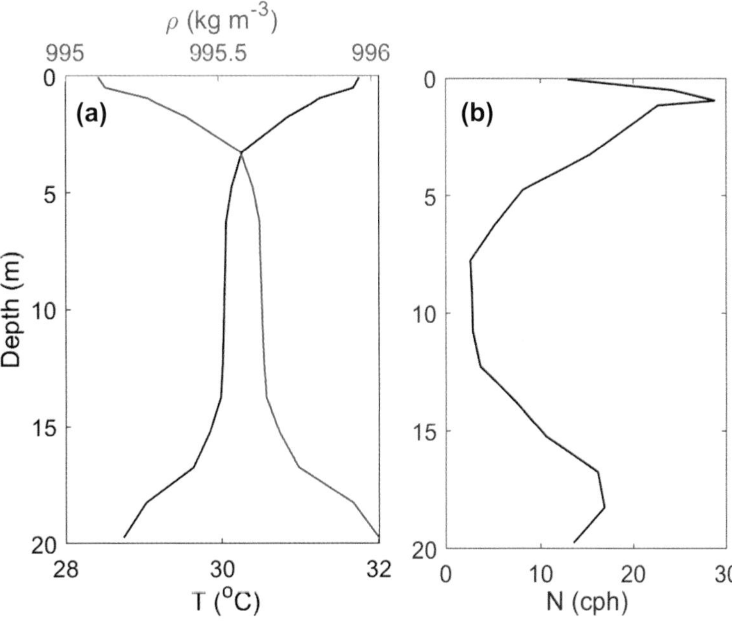

FIGURE 7-10 Profiles of (a) temperature (T) (*black line*) and of density (ρ) (*red*) showing the diurnal thermocline from near-surface warming, the subsurface layer, and the stratification in the upper part of the seasonal thermocline; (b) buoyancy frequency (N). *(Data are from the upper 20 m of Balbina Reservoir, Brazil, at noon on 19 July 2013.)*

radians per cycle. The time series of N in lakes at different latitudes are presented in Section IIA. Numerous oceanographic studies have shown stratification is weak where N is less than 5 cph, moderate between 5 and 25 cph, and strongly stratified when greater than 25 cph. Hence, with N approaching 30 cph, the diurnal thermocline in Balbina Reservoir (Brazil) at noon in July was strongly stratified (Fig. 7-10). However, values of N within diurnal and seasonal thermoclines in lakes frequently exceed 25 cph. The extent of stratification determines how effective wind or cooling will be in deepening the upper mixed layer or causing mixing in and across the thermocline (Section II, Chapter 8 IIIB, C). The greater the stratification, that is, the larger the value of N, the greater the cooling or wind speed required for mixing. Additionally, the amplitude and other attributes of internal waves depend on N.

Coefficient of eddy diffusivity—Vertical mixing redistributes the heat introduced at a lake's surface as well as any solutes or particles that are not uniformly distributed. Were it not for this mixing, temperature profiles would follow those of irradiance as in Fig. 7-6. Instead, lakes have more complex temperature profiles, as demonstrated by Figs. 7-1 and 7-2. The downward mixing of heat, the flux (F), is conceptualized as the product of a *coefficient of eddy diffusivity* (K_z) and the temperature gradient (dT/dz) and including specific heat capacity, c_{pw} (4184 J kg^{-1} °C^{-1}), and density of water, ρ_w, to convert temperature to heat. That is,

$$F = c_{pw} \rho K_z \, dT/dz \qquad (7\text{-}13)$$

(similar to Eq. 8-10), where K_z has the unit m^2 s^{-1}, the product of the units of velocity and length (see Chapter 8 IB). Thus K_z represents the product of the average size of turbulent eddies multiplied by their average velocity, a concept derived from the kinetic theory of gases. The process is analogous to molecular diffusion but often several orders of magnitude higher (Fig. 8-9e).

Mixing ($b_{z,t}$) below each depth (z) and at each time (t) can be assessed from changes in heat content over time intervals by removing increases in heat due to solar radiation (Jassby and Powell 1975). Dividing $b_{z,t}$ by dT/dz and surface area at each depth and time yields K_z for each depth and time. The heat budget approach provides K_z within the thermocline and hypolimnion as long as there is minimal advection (Jassby and Powell 1975; MacIntyre et al. 2009a,b) or heating by sediments (Benoit and Hemond 1996). The advective effect of internal wave motions is removed by filtering time series temperature data or by averaging temperature data obtained by profiling at multiple stations. Typical values of K_z determined by the heat budget method, tracer studies, and microstructure profilers are illustrated in Fig. 7-11.

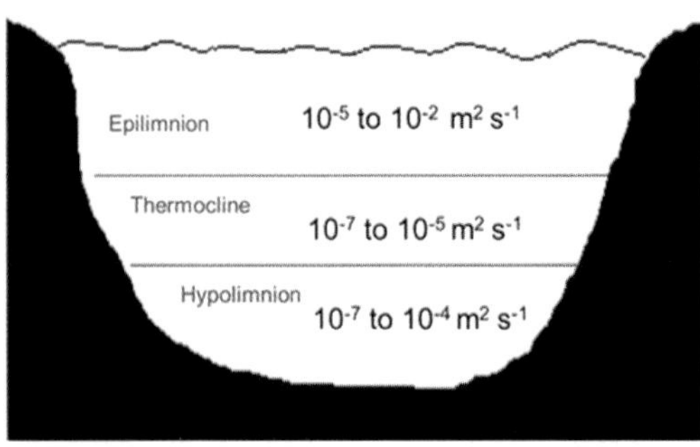

FIGURE 7-11 The range of values of K_z typically measured in the epilimnion, thermocline, and hypolimnion of lakes. K_z is highest in the epilimnion where heat loss and wind combine to cause mixing and least in the more strongly stratified thermocline. K_z in the hypolimnion is also low but possibly higher than in the thermocline (see Figs. 7-20 to 7-23 and 7-28; Jassby and Powell 1975; MacIntyre 1993; MacIntyre and Melack 1995). Molecular diffusivity of heat (see Chapter 3) is ~1.4 × 10^{-7} m^2 s^{-1} at 20°C and sets a lower boundary for calculating K_z using the heat budget method (Jassby and Powell 1975). K_z can also be obtained from tracer studies in which solutes are injected at some depth in the water column; K_z is calculated from their rate of spread (Chapter 8). Values of K_z from such studies may include lower values as the molecular diffusivities of many solutes are approximately two orders of magnitude less than that of heat.

Wind-driven mixing in the upper mixed layer—Wind induces near-surface currents with velocity, u, that decrease with distance, z, below the air–water interface (Fig. 8-21). That is, wind causes *shear* in the water column, du/dz (Fig. 8-5). The currents and the associated shear cause near-surface mixing. The shear stress (τ) that generates these currents can be quantified by measurements of wind velocity at several heights above a lake or of flow speeds at several depths within a lake. However, due to considerable experimentation, formulas have been developed enabling calculations from wind speed measured at one height:

$$\tau = \rho_a \, C_D \, U^2 \qquad (7\text{-}14)$$

where ρ_a is the density of air, C_D is the *drag coefficient* at instrument height, and U is the wind speed at instrument height. The shear stress is also expressed as an air friction velocity, u_{*a}, which is representative of the rate of increase of winds above the water surface within the *atmospheric boundary layer*. The boundary layer is the layer within the atmosphere where winds are reduced due to friction. Shear stress is assumed to be equal on both sides of the air–water interface. The equivalent expressions are $\tau = \rho_a C_D U^2 = \rho_a u_{*a}{}^2 = \rho_w u_{*w}^2$ (Eq. 8-25), where ρ_w is the density of water and u^*_w is the water friction velocity. Similar to u_{*a}, u_{*w} depends on the change in current speed with the logarithm of distance from the air–water interface (Chapter 8 IA, IIA). Empirically derived drag coefficients enable the momentum transfer to the water to be quantified from wind speed (Smith 1988; Imberger and Patterson 1990; (Fairall et al. 1996, 2003); Woolway et al.

2015). Similarly, these studies provide the transfer coefficients that enable the calculation of latent and sensible heat fluxes. For accurate work on lakes, anemometers need to be within the lower boundary layer where wind speeds increase logarithmically with height above the lake's surface. For small lakes, the anemometers are usually 2–3 m above the water surface. By convention, measured wind speeds are converted to those at 10 m, a calculation, along with that of the drag coefficient at instrument height, which requires taking into account atmospheric stability. For a neutral atmosphere and light to moderate winds, C_D at 10 m is approximately 0.0013. Values increase at higher winds (Fairall et al. 2003). The atmosphere is unstable when denser air overlays less dense air, as occurs when air temperatures are colder than surface water temperature, stable when the density of air progressively decreases above the water surface, or neutral when the density of air does not change with height (at least immediately above the lake's surface). Transfer coefficients can be calculated as a function of atmospheric stability using the Monin–Obukhov similarity theory (MOST) (Monin and Obukhov 1954; Csanady 2001).

The stability of the atmosphere influences drag coefficients appreciably when winds are light to moderate (Fig. 7-12). Atmospheric stability changes over diel cycles and with the passage of cold fronts, particularly over small lakes away from the tropics. The atmosphere is more frequently unstable over tropical lakes (Verburg and Antenucci 2010). Air temperatures are often warmer than surface water temperatures during the day and cooler at night and during stormy periods over small Arctic lakes (MacIntyre et al. 2009a). Consequently, atmospheric stability often varies over diel cycles and with the prevalence of cold fronts resulting in considerable variability in drag coefficients as a function of wind speed (Fig. 7-12). For the tropical lake, the air temperatures were nearly always cooler than the surface water temperatures, indicating the atmosphere was unstable. For the Arctic lake, for winds of 2 m s^{-1}, C_D was 3.6×10^{-4} under stable conditions and 2.6×10^{-3} under unstable conditions. The respective shear stresses are 0.0014 N m^{-2} and 0.01 N m^{-2}, approximately a sevenfold difference. Thus, when the atmosphere is unstable at low wind speeds, more shear is generated to mix incoming heat downward and to generate surface waves. For the same wind speeds and temperature differences, C_D was similar for the tropical lake and the Arctic lake.

The transfer coefficients used to compute latent and sensible heat fluxes also depend on atmospheric stability and have similar variability, although somewhat larger

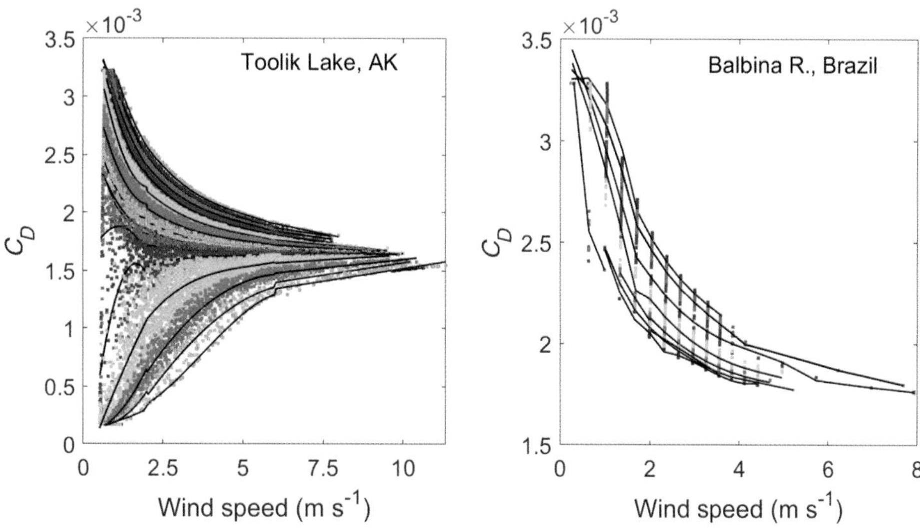

FIGURE 7-12 Drag coefficient (C_D) at instrument height as a function of wind speed, showing variation with temperature difference between that in the air and surface water for Toolik Lake, Alaska, in summer 2010 (*left panel*) and Balbina Reservoir, Brazil, in July 2013 (*right panel*). Wind speed has been corrected to that for 10 m height and a neutral atmosphere. Here the temperature difference is a proxy for atmospheric stability, with the higher values of C_D when the atmosphere is unstable. The temperature differences (°C) illustrated, beginning with the most negative, ≤ -11 (*green*), $-11 > \Delta T \leq -9$ (*red*), $-9 > \Delta T \leq -7$ (*orange*), $-7 > \Delta T \leq -5$ (*purple*), $-5 > \Delta T \leq -3$ (*pale blue*), $-3 > \Delta T \leq -1$ (*blue*), $-1 > \Delta T \leq -0.5$ (*yellow*), $-0.5 > \Delta T \leq 0.01$ (*green*), $-0.01 > \Delta T \geq 0$ (*orange*), and, for positive temperature differences, $0 \leq \Delta T \leq 0.5$ (*purple*), $0.5 < \Delta T \leq 1$ (*pale blue*), $1 < \Delta T \leq 3$ (*blue*), $3 < \Delta T \leq 5$ (*yellow*), $5 < \Delta T \leq 7$ (*green*), $7 < \Delta T \leq 9$ (*purple*). Data acquisition for Toolik Lake is described by MacIntyre et al. (2009a) and for Balbina Reservoir by MacIntyre et al. (2021a). Equations for C_D are from Imberger and Patterson (1990, Section III) with calculations as in MacIntyre et al. (2002). Black curves for each range are computed by nonlinear curve fitting. Under unstable conditions, C_D is appreciably increased above values for a neutral atmosphere, which, for the heights of the anemometers in these studies, was ~0.0018. The atmosphere over the tropical lake was always unstable, whereas over the Arctic lake, the atmosphere shifted between stable and unstable. For the stable atmosphere, momentum transfer to the water can be very small.

magnitude than C_D at low wind speeds. An implication of this difference is that for the same solar heating when winds are light and temperatures are similar, the duration of the period of heating can be longer when the atmosphere is stable due to lower sensible and latent heat fluxes.

The question often arises as to whether wind or heat loss causes the most mixing in the upper mixed layer. This problem is addressed by computing the *Monin–Obukhov length scale* (L_{MO}) in the water and comparing it with the depth of the actively mixing layer.

$$L_{MO} = \frac{u{*}_w^3}{\kappa\beta{*}} \quad (7\text{-}15)$$

where κ (= 0.41) is the von Kármán constant, $u{*}_w$ is the water friction velocity, and β_* is the buoyancy flux defined previously. β_*^+ applies to heating and β_*^- to cooling. Within the actively mixing layer at depths above L_{MO}, turbulence production by shear exceeds that from β_*^-; below it, turbulence production by β_*^- dominates. Even for the lightest winds, shear production of turbulence dominates that from buoyancy flux near the air–water interface. The sign and magnitude of L_{MO} change over diel cycles and during storms. The ratio L_{MO}/z_{AML} indicates the fraction of the mixing layer in which shear predominates over turbulence production (Imberger 1985). During heating, buoyancy flux contributes to stabilizing the upper water column such that a larger fraction of the energy input from the wind is dissipated rather than used for mixing (MacIntyre et al. 2021a). As is illustrated in Section IIC, when the ratio is small and positive, turbulent eddies are small and more of the incoming heat is stored in a shallow diurnal thermocline. As the ratio exceeds 1, more of the heat is mixed downward (Section IIA). When the ratio is negative, turbulent eddies often extend throughout the actively mixing layer. If winds or buoyancy flux are high enough, as will be discussed below, the actively mixing layer may deepen.

Wind-driven mixing across diurnal or seasonal thermoclines—Indices that assess whether wind will cause mixing in a stratified lake incorporate stability either as the density difference across the thermocline or as *Schmidt stability* (Eq. 7-11) (Imberger and Patterson 1990). The relevant indices are the *Wedderburn (W)* and *Lake numbers (L_N)*. Both indices include the effects of stratification across the thermocline, and hence the difficulty in causing mixing across it, relative to the wind stress which induces mixing. Additionally, they incorporate basin morphometry (Chapter 4). Thus the interpretation of the magnitude of *W* and L_N applies to lakes of all sizes; that is, these indices are scale independent. When wind passes over a lake, currents are generated, which are proportional to the wind speed (Chapter 8 IIIA). The accumulation of water downwind causes a pressure gradient that depresses the thermocline. As the wind decreases or ceases, the thermocline tilts in the opposite direction, creating a *seiche*, a basin-scale standing wave along the thermocline and other types of internal waves (Fig. 7-13, Chapter 8 IIIB). Tilting occurs where stratification is greatest, as at diurnal and seasonal thermoclines (Figs. 7-1 and 7-10). Shear (du/dz) (Figs. 8-2 and 8-5), occurs across the thermocline. When shear is sufficiently large relative to stratification,

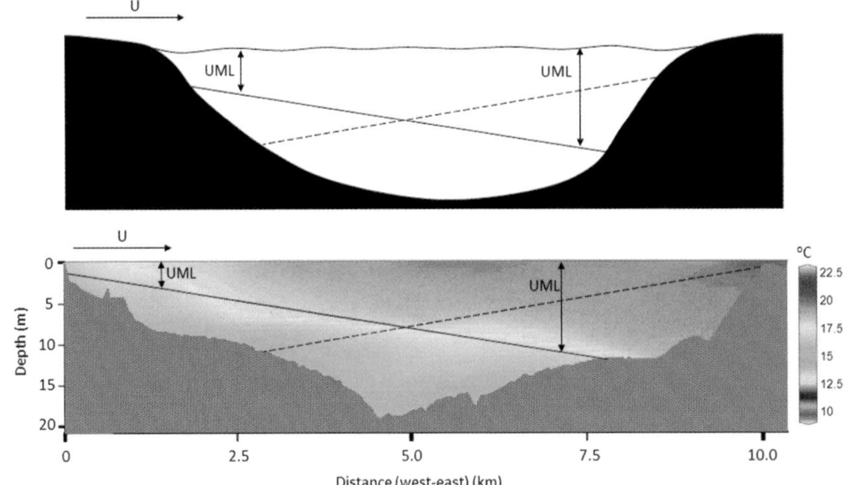

FIGURE 7-13 (Top) Wind (*U*) induces the thermocline to tilt; that is, it upwells at the upwind end of the lake and downwells downwind (—). This movement results from wind-driven currents transporting surface water downwind. On relaxation of the wind, the thermocline will reverse its direction of tilt (—), initiating an internal seiche, that is, a standing wave along the thermocline, as well as other types of internal waves (Chapter 8). Because of the horizontal movement of water and related tilting of the thermocline, the depth of the upper mixed layer (UML) can vary along the length of the lake at any one time, independently of changes in the surface energy budget and of mixing. (Bottom) Conceptualization in (top) is demonstrated in the response of Lake Rotorua (North Island, New Zealand) to a strong westerly wind. When the wind relaxes, the thermocline will reverse its tilt (—). Data were obtained by synoptic sampling of the lake.

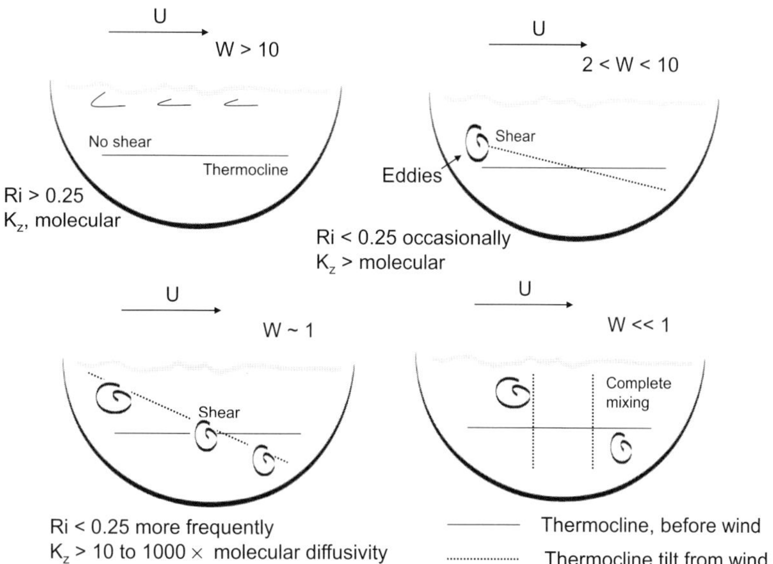

FIGURE 7-14 Wedderburn number and resultant tilting of the thermocline. For $W > \sim 10$, the thermocline does not tilt; for $2 < W < 10$, partial tilting occurs; for $W = 1$, full upwelling occurs; and for $W << 1$, complete mixing occurs. See Chapter 8 for further development of the development of the concept of the Wedderburn number (Fig. 8-23) and description of the types of internal waves expected in the various regimes (Figs. 8-24 to 8-28). Changes in thermal structure when $W < 1$ are described in Section II, *Comparative stratification in temperate, Arctic, and tropical lakes* (Fig. 7-25 and Chapter 8, Figs. 8-27 and 8-28). Interpretations have been developed empirically from laboratory and field experiments (Monismith 1986; Stevens and Lawrence 1997; MacIntyre et al. 2009a,b). As tilting increases, the *Richardson number* (*Ri*, Chapter 8 IA) decreases to values less than 0.25, the threshold for shear-induced mixing, and the coefficient of eddy diffusivity (K_z) increases above molecular values, which for heat is $\sim 1.4 \times 10^{-7}\ m^2\ s^{-1}$. *Ri* and K_z are indicative of the extent of mixing due to turbulence (Chapter 8). Interpretations with Lake number are similar when applied to the seasonal thermocline. *(From MacIntyre (2008).)*

mixing is induced (Chapter 8 IIIB). The extent of tilting can be predicted from W and L_N (Figs. 7-14 and 8-2; Chapter 8 IIIB). Similarly, these indices can indicate whether the resultant shear across the thermocline is sufficiently large to overcome stratification and induce mixing.

The Wedderburn number is defined as:

$$W = \frac{g\Delta\rho h^2}{\rho_0 u*_w{}^2 L} \tag{7-16}$$

where g is gravity, ρ_0 is mean density, $\Delta\rho$ is the density difference across the metalimnion, h is the depth of the upper mixed layer, u_{*w} is water friction velocity computed from shear stress on the water (Eqs. 7-14 and 8-25) and, when the atmosphere is neutrally stable, approximately $0.001U$ where U is wind speed, and L is the length of the lake, or more accurately fetch, the maximum length of open water in the direction of the wind.

Schmidt stability (S_t) (Eq. 7-11), indicative of the stratification stabilizing the lake at the time of measurement, and shear stress on the water surface (a function of $\rho_w u_{*w}^2$) are implicit in the definition of L_N. With S_t in the numerator and shear stress in the denominator, and other terms taking into account basin morphometry, higher values of L_N indicate less tilting of the thermocline and as a result less shear across it and less mixing.

Technically, the definition of the Lake number is derived from moments around the center of volume computed when the wind stress acting to overturn the density structure is in equilibrium with the restoring force (Imberger and Patterson 1990):

$$L_N = \frac{S_t\left(1 - \frac{z_T}{z_m}\right)}{\rho_w u*_w{}^2 A_0^{0.5}\left(1 - \frac{z_g}{z_m}\right)} \tag{7-17}$$

$$z_g = \frac{\int_0^{z_m} zA(z)\partial z}{\int_0^{z_m} A(z)\partial z} \tag{7-18}$$

where z is depth, z_g is height to the center of volume of the lake, z_T is the height to the center of the metalimnion, z_m is the maximum depth of the lake, $A(z)$ is area at depth z, A_o is surface area, and $\rho_w u_{*w}^2$ is shear stress on the water surface. L_N can be readily calculated using LakeAnalyzer (Read et al. 2011). Wedderburn numbers can be obtained similarly but it is valuable practice to estimate them by examining observational data and using algorithms for density in LakeAnalyzer or other sources.

Wedderburn and Lake numbers indicate the extent of tilting of the thermocline, that is, the amount of up- and downwelling induced by wind (Fig. 7-14). The degree of tilting of the thermocline is indicative of the amount of mixing that will occur within and across it (Chapter 8 IIIB). Besides the pronounced upwelling when W or L_N approach or drop below 1 and the resultant direct exchanges between the layers in a lake, the tilting induces internal waves that can break near boundaries or in the central portion of a lake (Chapter 8 IIIB). Regression equations have been developed to predict K_z from W and L_N (Romero et al. 1998; Yeates and Imberger 2004).

Because the Wedderburn and Lake numbers are scale independent, they can be used to predict the extent of upwelling and downwelling of the thermocline for lakes of various sizes and shapes. Thus, if W or L_N equals 1 in a 0.01 km^2 lake or a 10 km^2 or a 100 km^2 lake, full upwelling would be expected in each. Similarly, the mixing associated with the increased shear would presumably be independent of lake size. When diurnal thermoclines are present, their behavior can be decoupled from that of the seasonal one, and it is important to calculate W for the diurnal thermocline based on the density gradient across it and the thickness of the actively mixing layer above it (Fig. 7-15) (Robertson and Imberger 1994). W would also be computed for the seasonal thermocline and L_N for the lake as a whole. Similarly, for lakes with persistent stratification (e.g., *meromictic* and *oligomictic* lakes, Section II D and E), the Wedderburn number can be calculated for the various layers and the Lake number for the lake as a whole.

Wedderburn and Lake numbers, along with surface energy budgets, are the appropriate metrics to use when addressing whether lakes are mixing in a manner that will redistribute solutes and particulates. Additionally, changes in the magnitude of these indices can indicate whether a lake's dynamics are changing with ongoing changes in land use or climate change. They can be used to quantify the frequency of events in which dissolved oxygen depletion is alleviated in the lower water column and, along with it, the potential for the release of reduced solutes from lake sediments (Robertson and Imberger 1994; Cortés et al. 2021). When combined with *residence time*, that is, lake volume divided by the volume of incoming water, they can be used to evaluate the persistence of phytoplankton including harmful algal species in the euphotic zone (Tundisi et al. 2002). Time series of these values are included in Section IIA.

E. Processes causing stratification and mixing in the upper mixed layer

Wind influences the depth of the diurnal thermocline via four processes. Wind generates currents that transport surface water downwind (Fig. 8-21). This flow can cause the diurnal thermocline to tilt, as well as the seasonal thermocline, and moderate the depth of each horizontally across a lake (Fig. 7-15). The extent of tilting of the diurnal thermocline is predicted from the Wedderburn number (Imberger 1985). Current speeds are highest at the surface and decrease with depth, creating *shear* (τ), which is calculated as the change in velocity (u) with depth, that is, $\tau = du/dz$. Mixing is induced near the

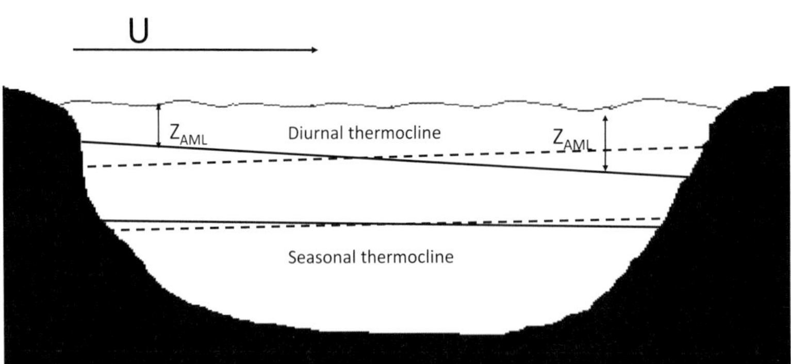

FIGURE 7-15 Variable responses of the diurnal and seasonal thermocline to wind forcing. In the example here the Wedderburn number of the diurnal thermocline is low enough that considerable tilting occurs. Its value is likely between 2 and 7 (Fig. 7-14). In contrast, the seasonal thermocline is barely deflected. The value of the Wedderburn number across it would be closer to 10. As the wind induces the diurnal thermocline to tilt, water near the surface flows in the direction of the wind, leading to the deepening of the actively mixing layer downwind. When the wind relaxes, the diurnal thermocline upwells downwind and water near the surface will flow in the opposite direction. These flows lead to cross-basin exchange (Chapter 8). Thermoclines during the initial period of wind forcing (—) and after the wind relaxes or changes direction (–). Depth of the actively mixing layer (z_{AML}).

surface by the shear, where it is often called *stirring* to differentiate it from the shear-driven mixing across the metalimnion. Additionally, as wind speeds increase, evaporation increases. Thus, during periods of cooling, the diurnal thermocline can deepen from both instabilities from heat loss and wind shear. Mixing across the diurnal and seasonal thermoclines is induced by shear and, as discussed above, the extent of mixing can be obtained by calculating Wedderburn and Lake numbers (Imberger and Patterson 1990; Chapter 8 IIIB). Additionally, if the Lake number decreases below 10 for a lake as a whole, upwelling of the seasonal thermocline can occur upwind, and cooler water can be mixed into the upper mixed layer, creating horizontal density gradients (Fig. 8-21). With continued winds, cooler water can be advected horizontally and further contribute to mixing due to the resultant density instabilities. When W and L_N are below critical values (Fig. 7-14; Chapter 8 IIIC; Horn et al. 2001), and with the transition from heating to cooling, rapid changes can occur in the thermal structure of the upper mixed layer (Figs. 7-2, 7-3, and 7-15 to 7-17).

The combination and interactions of processes such as those described above cause variations in temperature, thermal structure, and mixing between days and over diel cycles (Fig. 7-16). The observations made in Mono Lake (CA, USA) in September 2007 illustrate these processes. Incoming solar radiation was nearly identical on all days of the study, with maxima of ~900 W m^{-2}, but winds were light on day of year (DOY) 254, and the timing of the onset of high winds varied on the other two days (DOYs 255, 256). Lake numbers exceeded 10 on the first day of the study, were above 10 during the morning of the second, and then dropped to 1 when winds rose to 10 m s^{-1}, and values remained below 3 through the remainder of the second day and throughout the third day. Consequently, cooler water upwelled nearshore and was mixed into the upper mixed layer nearshore. The difference in wind speed led to appreciable differences in evaporation such that net heat fluxes (Eq. 7-1) varied. On the first day, they reached 700 W m^{-2} and had maxima 200 W m^{-2} lower on the last day. On all nights, winds and heat losses were sufficient to cause mixing to the seasonal thermocline at 12 m. With the differences in wind speed, near-surface temperatures were 22°C on the first day and heat was stored in a diurnal thermocline. Within the upper mixed layer, $W > 1$ when the diurnal thermocline was established. On the second and third days, maximal surface temperatures were 21°C and 20°C, respectively.

A diurnal thermocline formed on the second day, and mixing, as indicated by elevated values of the coefficient of eddy diffusivity, K_z, occurred within it but not immediately below (Fig. 7-17). It descended when wind speeds increased in the afternoon and W dropped below 1. Values of K_z were elevated within it but not immediately below. A diurnal thermocline did not form on the third day when winds had remained high. While the upper mixed layer heated on the second day, little heating occurred on the third day even when surface heat fluxes were positive. The cooler temperatures on the third day

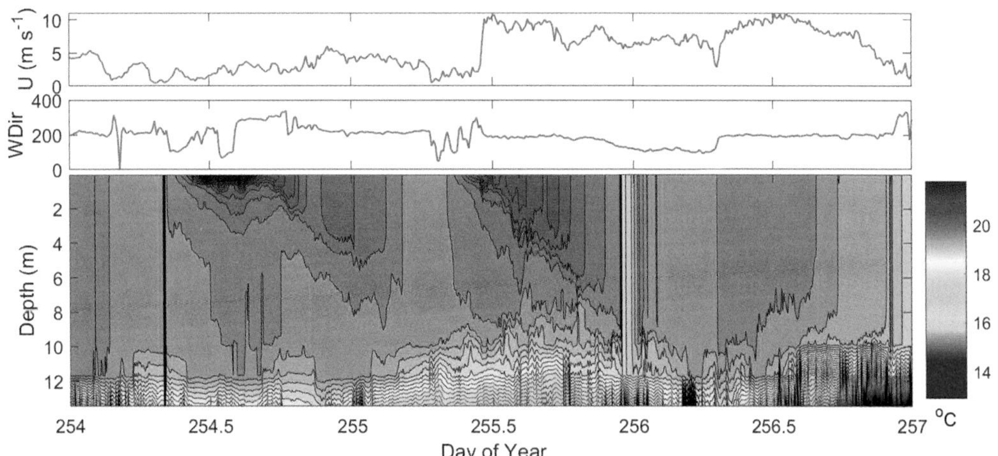

FIGURE 7-16 Time series of wind speed (*U*, *upper panel*), wind direction (WDir, degrees, *middle panel*), and temperature contours on days of year (DOYs) 254 through 256 (11–13 September 2007) in Mono Lake, California (USA). Each *black contour line* indicates one temperature (isotherm). As a result of cooling at night and some wind-driven mixing, stratification was minimal at night above the seasonal thermocline located below ~12 m. Wind speeds during the three days differed. A shallow diurnal thermocline built up and persisted on DOY 254 when winds were light throughout the day, and a diurnal thermocline developed during light winds on the morning of DOY 255 and progressively downwelled through the afternoon with southerly winds of 10 m s^{-1}. With persistent high winds on DOY 256, heat was mixed downward during the day and a diurnal thermocline did not form. Variations in the depth of mixing and stratification, such as documented for the three days here, mean that phytoplankton may not be mixed or may be mixed to different depths with consequences for the ambient light to which they are exposed and the extent of photoinhibition (Brookes et al. 2003; Chapter 18).

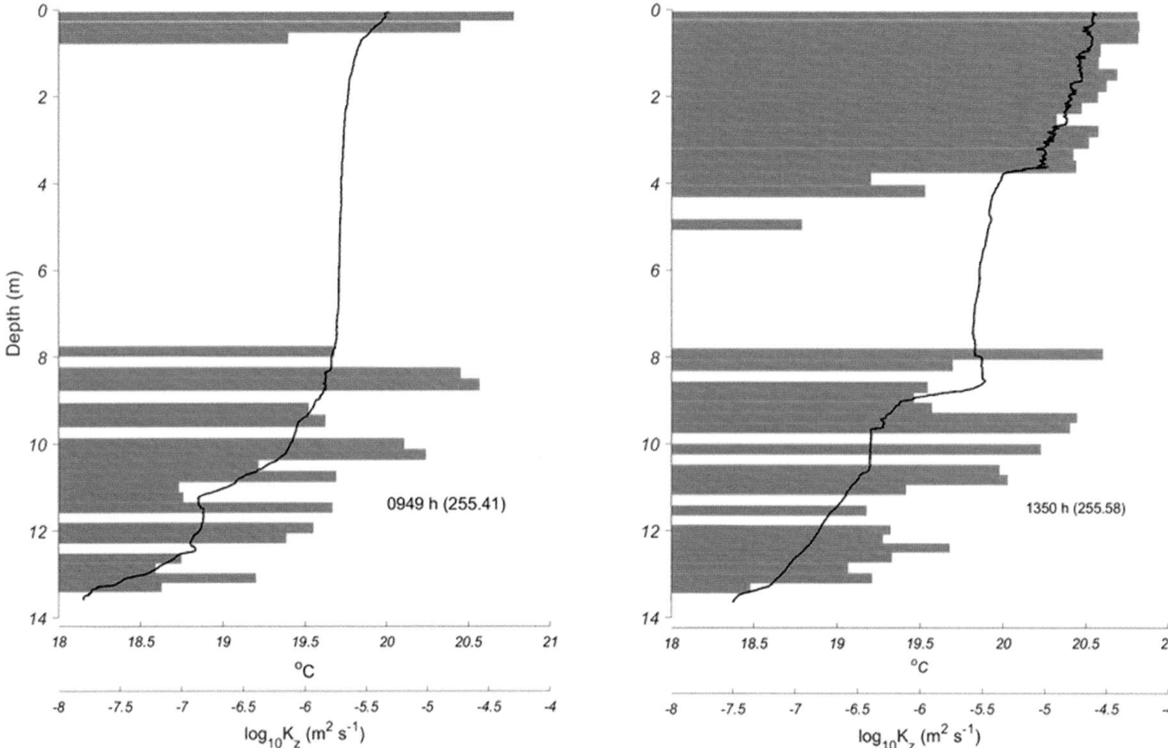

FIGURE 7-17 High-resolution profiles of temperature (*black lines*) and calculated values of the coefficient of eddy diffusivity (K_z, *purple*) above 14 m in the 30 m deep western basin of Mono Lake, California (USA), on DOY 255 (12 September 2007). The diurnal thermocline was forming at 0949 h (*left panel*) and had deepened in the afternoon with high winds (1350 h, *right panel*). High-resolution instruments, such as the *Self-Contained Microstructure Profiler* (SCAMP) used here, enabled limnologists to recognize the complex thermal structure found in the epilimnion and to calculate K_z (Chapter 8 I). During the light winds in the morning, near-surface mixing was constrained to within the diurnal thermocline. When winds increased, horizontal advection led to tilting of the diurnal thermocline and increased shear. As a result, the temperature inversions associated with Kelvin–Helmholtz billows (Chapter 8 I) are evident throughout the 4 m deep diurnal thermocline. Values of K_z approached 10^{-4} $m^2 s^{-1}$ near the surface and were $\sim 10^{-5}$ $m^2 s^{-1}$ at the base of the diurnal thermocline. Here, the actively mixing layer extended through the diurnal thermocline. As the afternoon progressed, and the surface heat budget became negative, temperature structure became more similar to that in Fig. 7-1 with a near-isothermal layer above the diurnal thermocline. With the high winds, turbulence was maintained in the near-isothermal layer and in the diurnal thermocline below. The elevated values of K_z in the more stratified water below 8 m were due to shear from internal wave motions. Mono Lake is stratified by both solutes and temperature; hence the anomalous warm layer at 8 m at 1350 h has higher specific conductance and is stable relative to the cooler water above.

resulted from the greater downwelling at the measurement site such that incoming heat was distributed in a deeper layer, horizontal transport of cooler water from the nearshore where upwelling had occurred, and *entrainment* of cooler water from the seasonal thermocline (Fig. 7-18). Entrainment implies the upward mixing of water from deeper in the water column.

Comparison of the time scales of mixing, τ_{mix}, and of horizontal advection, τ_H, further enables dominant processes to be assessed (Section 8 IB). The time scale of mixing is:

$$\tau_{mix} = l^2/K_z \tag{7-19}$$

where l is a depth interval and K_z is the coefficient of eddy diffusivity, and the time scale for horizontal advection is:

$$\tau_H = L/u \tag{7-20}$$

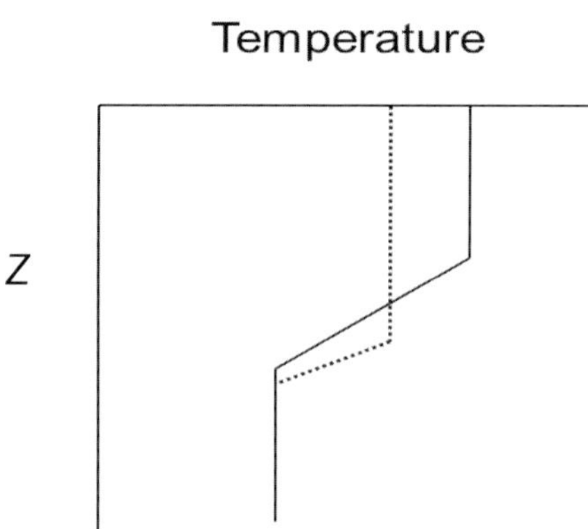

FIGURE 7-18 Deepening of the upper mixed layer due to cooling and possibly wind-induced mixing. As a result, water from the thermocline is *entrained* (mixed upward) into the mixed layer. The y-axis is depth, here denoted as z, with an initial temperature profile (—) and subsequent profile (…).

where L is the horizontal distance traveled and u is mean velocity. The diurnal thermocline at 1.8 m downwelled ~1 m ($l = 1$ m) in just over an hour beginning on DOY 255.565 (Fig. 7-16). During that period, K_z varied from 3×10^{-5} m^2 s^{-1} (Fig. 7-17) to 10^{-4} m^2 s^{-1}, giving τ_{mix} between 2 and 6 h. τ_H was ~1 hour given the measurement site was ~1 km offshore and assuming u was 3% of the wind speed. The downwelling occurred in an hour, pointing to the contribution of horizontal advection as in Fig. 7-15. Hence changes in the depth of the diurnal thermocline result from the combination of horizontal advection and vertical mixing.

As the diurnal thermocline deepened on DOY 255, water in the subsurface layer below it would have been entrained (Fig. 7-18). Water from the seasonal thermocline was entrained into the upper mixed layer on DOY 256 (Fig. 7-16) when Lake numbers were less than 3. Entrainment is evident by the isotherms rising from the seasonal thermocline. Entrainment at the measurement site contributed to cooler temperatures in the upper mixed layer on DOY 256, as did horizontal transport of cooler water entrained as upwelling occurred near shore. Any solutes (e.g., dissolved nutrients) or phytoplankton within the seasonal thermocline would be mixed upward as well. As a result of these interacting processes, temperature, and thus density, is rarely uniform in the upper mixed layer during the day, although isothermy may occur at night (Figs. 7-1 to 7-3 and 7-15 to 7-17; Imberger and Hamblin 1982; Imberger and Patterson 1990). These analyses point to the inherently two-dimensional nature of lakes when Wedderburn and Lake numbers are below critical values (Figs. 7-14 and 7-15). That is, vertical fluxes vary spatially, and horizontal advection contributes to transport.

II. Stratification

A. Thermal stratification

The processes that cause stratification and mixing are similar for all lakes, yet variations occur as a function of latitude and altitude as they influence solar radiation, near-surface temperatures, and the presence or absence of ice, resulting in several types of annual stratification (see Section IID). First, using time series of temperature from an Arctic lake, typical changes in stratification over the ice-free period are illustrated. Then, comparisons are provided for Arctic, temperate, and tropical lakes to illustrate changes in thermal structure, buoyancy frequency (N), and K_z and address how the differences among lakes result from differing surface energy budgets, optical properties, wind speeds, Lake numbers, and morphometry.

The setup and persistence of summer stratification in an Arctic lake—Stratification is induced when incoming solar radiation exceeds heat losses and the energy from wind is insufficient to erode the stratification. For example, following ice-off in 1.5 km^2 Toolik Lake (Alaska, USA) (Fig. 7-19), stratification was weak. Incoming solar radiation had daily maxima approaching 800 W m^{-2}. With a diffuse attenuation coefficient (k_d) of 0.5 m^{-1} (wavelength band 400 to 700 nm), and other attenuation coefficients as in Table 7-1, more than 80% of the incoming shortwave radiation was absorbed in the upper 2 m. With cold surface water temperatures relative to air temperature, sensible heat was initially positive and latent heat fluxes were small compared to later in the season despite wind speed maxima of ~5 m s^{-1}. The heat loss terms of the surface energy budget were appreciably less than the incoming shortwave radiation during the day, enabling heating (MacIntyre et al. 2009a). As heat was introduced, the temperature and therefore density gradient between the upper mixed layer and water immediately below increased due to the cold temperatures following ice-off. As maximal wind speeds occurred during heating, any mixing was due to shear across the developing thermocline. However, the low values of turbulence, as rates of dissipation of turbulent kinetic energy (Chapter 8 I), indicate that shear was insufficient relative to the stratification to cause much mixing across the thermocline (Figs. 7-20 and 8-2). Net heat fluxes at night were maximally −100 to −200 W m^{-2}. The instabilities due to heat loss at night were insufficient to cause appreciable deepening given the strong stratification which developed. Hence seasonal stratification developed.

Horizontal differences in the rate of heating may also have contributed to the onset of stratification. In lakes like Toolik Lake, ice cover often recedes earlier near the margins and persists for a longer period over the deeper parts, in part due to the direction of prevailing winds. Thus surface temperatures may be warmer in the shallow regions. While temperatures at Midlake stations, as in Fig. 7-19a, may not be as warm as those near shore, temperatures there were warmer than over the deeper site following ice-off (Fig. 7-19b). Thus the wind-induced advection of water from locations where warming occurs earlier to those where ice cover persists longer may also contribute to increases in stratification following ice-off.

The changes in temperature and thermal structure over the rest of the summer varied with periods of heating, cooling, intervals with higher winds, and discharge from an incoming stream after a large rain event (Fig. 7-21). Diurnal stratification and nocturnal mixing occurred in the upper mixed layer above the seasonal thermocline. Periods of heating and cooling led to changes in temperature within the upper mixed layer with minimal change in the depth of the thermocline until a cold front occurred, with higher winds and

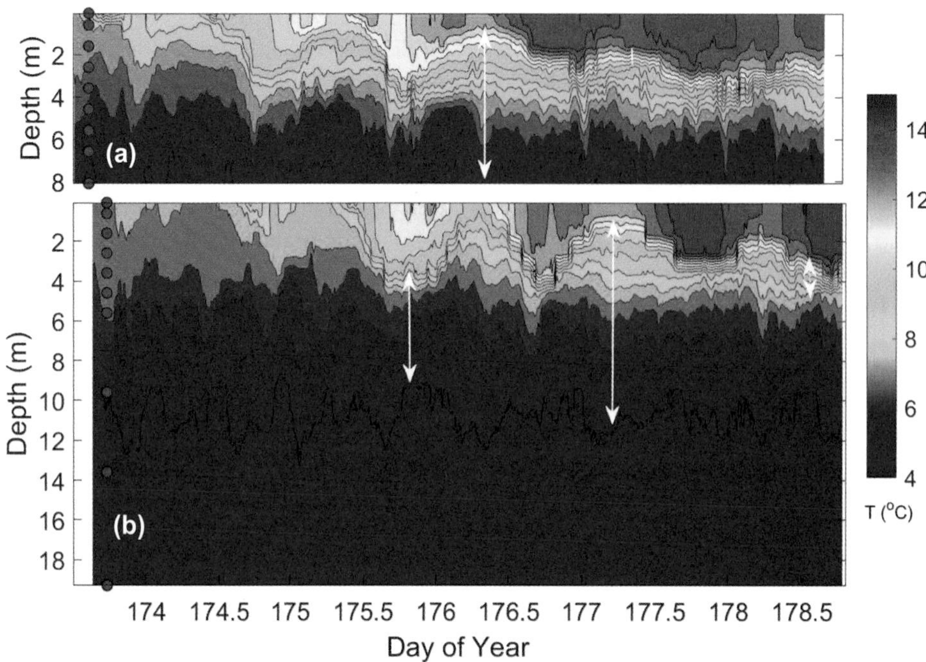

FIGURE 7-19 Time series of temperatures in Toolik Lake (Alaska, USA), shortly after ice-off in 1999 at (a) a midlake station about 10 m deep and (b) in a deeper basin in the lake illustrating the rapid development of the seasonal thermocline. *Colors* indicate temperature, and each *contour line* (*black*) is indicative of one temperature and called an *isotherm*. *Red dots* indicate the locations of the temperature sensors. Surface waters heat rapidly from 7°C to 14°C. Diurnal heating is made evident by increases and nocturnal cooling by decreases in temperature. The up and down motions of the isotherms are due to wind-induced internal waves (Chapter 8 IIIB). The deepening of the upper mixed layer in the day is due to afternoon winds that depress the thermocline at the measurement sites (e.g., Fig. 7-13). As winds were northerly during the day, the mixed layer was deeper at such times in the more southerly basin (Fig. 7-19b). The shoaling of the upper mixed layer at night is due to the thermocline upwelling at the measurement site when winds declined. As winds were often southerly at night, the upwelling was greater to the south such that initially metalimnetic water reached the surface (Fig. 7-19b, DOY 176.2). *White lines* and *arrows* extending less than 2 m in (a) and (b) indicate the depth of the thermocline using the strict definition that it is the region with the sharpest temperature gradient. The *white arrows* extending over a larger range of depth indicate the location and thickness of the metalimnion, or thermocline, using the term thermocline synonymously with metalimnion. The thickness of the metalimnion changes due to expansions and contractions from internal wave motions (Fig. 8-23).

rainfall leading to increased discharge into the lake on DOY 199. The incoming stream flowed into the main basin of the lake at the top of the metalimnion, injecting water with a temperature of 10–11°C (DOY 199–DOY 202). Under cloudy conditions and with cold air temperatures, the upper mixed layer subsequently remained cool. Strong winds, 8–10 m s^{-1}, such that the Lake number dropped below 1, occurred on DOY 212. The rapid descent of the thermocline is evidence of the strong mixing that occurred such that rapid exchange of water occurred between the epilimnion and hypolimnion (Fig. 7-14; Chapter 8 IIIB). Subsequently, with increased solar heating and lighter winds, the epilimnion warmed again and, by DOY 218, a secondary thermocline had formed above the metalimnion.

Comparative stratification in temperate, Arctic, and tropical lakes—Comparison of similarly sized lakes in the temperate zone and Arctic illustrate how differences in temperature and stratification result from the higher solar radiation, lower wind speed, and longer ice-free period in a temperate lake (Lawrence Lake) relative to an Arctic lake (N2) (Figs. 7-22 and 7-23). Both lakes had similar attenuation coefficients. In both lakes stratification sets up similarly after ice-off, often after a period of wind-induced mixing. Initially, sensible heat fluxes are positive and latent heat fluxes are often low, less than −50 W m^{-2}. Net longwave radiation is of the order of −100 W m^{-2}. As a result, the contribution of cooling to mixing is small compared to conditions a few weeks later. Thus wind contributes to the downward mixing of incoming heat. Typically, incoming solar radiation exceeds 500 W m^{-2}. When the winds are stronger, Lake numbers can be less than 1, meaning the water column is comprised of up- and downwelling internal waves that serve to mix the incoming heat downward. In Lawrence Lake the combination of heating and mixing led to a seasonal thermocline developing around 6 m. In contrast, it initially was at 3 m in the Arctic lake. After a period with lighter winds, such that $L_N > 10$, stratification set up in the upper few meters. Diurnal heating and cooling occurred, but the depth of nocturnal mixing was shallow, leading to stratification and buoyancy frequencies in the thermocline that were high, 20 to 30 cph in Lawrence Lake and Lake N2. Afternoon winds

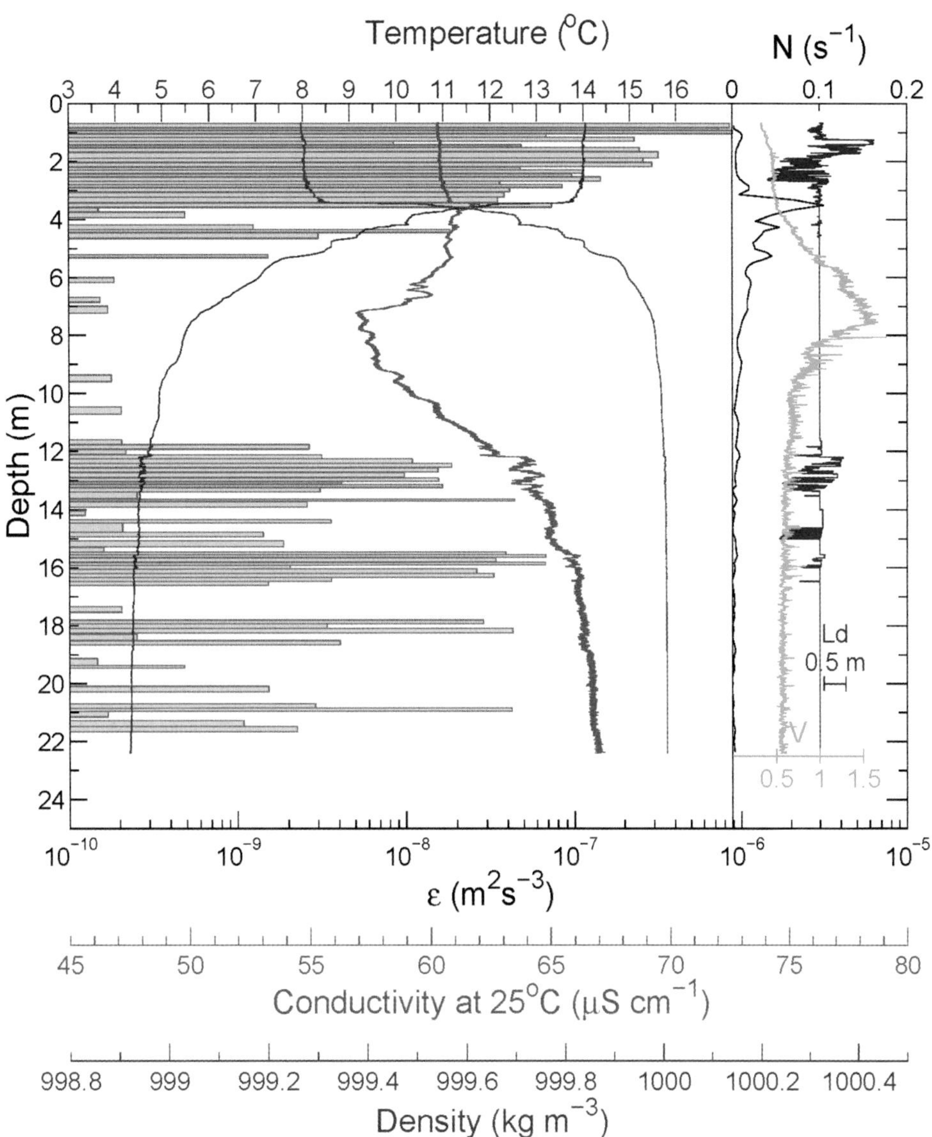

FIGURE 7-20 Pronounced near-surface thermal stratification as the water column warmed in Toolik Lake (Alaska, USA), following ice-off in 1999. *Left panel*: profiles of temperature (*blue*), specific conductance (*red*), density (*purple*), and turbulence as dissipation rate of turbulent kinetic energy (ε, $m^2\ s^{-3}$, *gray bars*) (Chapter 8 I). Dissipation rates are low when less than $10^{-8}\ m^2\ s^{-3}$, moderate from $10^{-8}\ m^2\ s^{-3}$ to $10^{-6}\ m^2\ s^{-3}$, and high when greater than $10^{-6}\ m^2\ s^{-3}$. *Right panel*: buoyancy frequency (*N*, *black*, radians per second), overturning length scales (*Ld*, *blue*, Chapter 8 I, also written as L_T), and fluorescence signal (*green*, *V*, indicative of algal biomass) in the deep basin of Toolik Lake on DOY 179 (28 June 1999) at 1638 h Alaska Standard Time. With the strong thermal stratification ($N = 0.1\ s^{-1}$), which is 57 cycles per hour (cph) across the thermocline, turbulence was constrained to the upper mixed layer, as indicated by $\varepsilon > 10^{-7}\ m^2\ s^{-3}$ and overturning motions (*Ld*) (Chapter 8 I) similar in size to that of the mixed layer. Turbulence was suppressed in the thermocline. Turbulence occurred intermittently in the lower water column due to shear from internal wave motions (Fig. 7-19).

typically occurred and induced thermocline tilting and internal waves, but these did not cause appreciable mixing unless $L_N < 10$.

The deepening of the upper mixed layer by convection depends on the extent of heat loss and the underlying stratification. This relation is given by:

$$\frac{dh}{dt} = \frac{\beta}{hN^2} \tag{7-21}$$

where β is buoyancy flux, that is, the magnitude of cooling computed from the heat flux (Section IB, Eq. 7-4), h is the depth of the mixed layer, and N^2 is the buoyancy frequency squared. In spring values of β

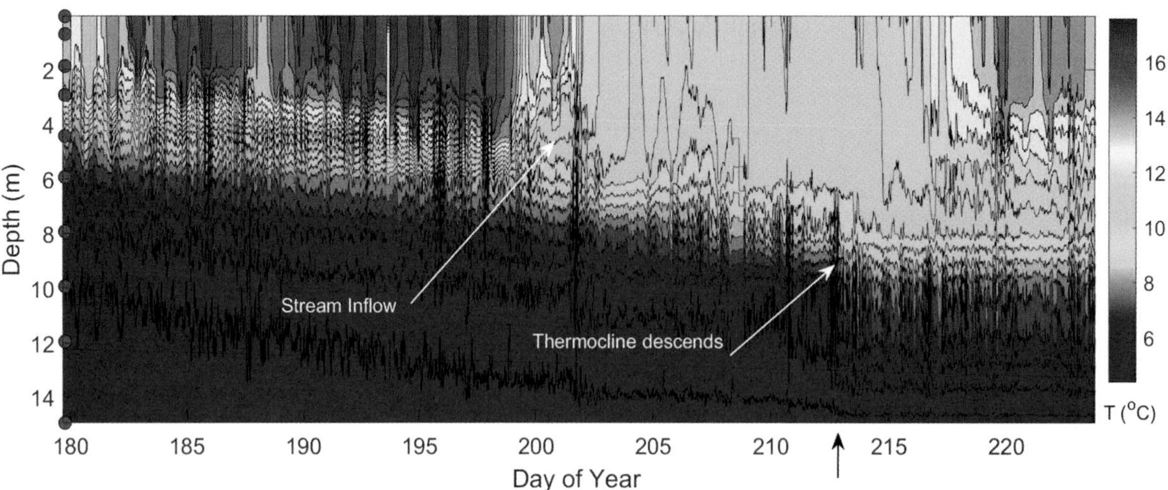

FIGURE 7-21 Temperature contours over the summer season in Toolik Lake, Alaska (USA), 1999. Data begin after that in Fig. 7-19 and illustrate alternate warming and cooling periods, the formation of diurnal thermoclines, the seasonal thermocline (metalimnion) from 3 to 8 m in the early season with the most strongly stratified region of the thermocline between 3 and 5 m, the weakening of the thermocline by incoming storm water on DOYs 199-202 between 3 and ~6 m (marked with *white arrow*), and a major wind-driven deepening of the upper mixed layer and abrupt depression of the seasonal thermocline on DOYs 212-213 (*black arrow* on x-axis and *white arrow* in figure). This depression was induced by the intense mixing associated with nonlinear internal waves and Lake number of 1 (Chapter 8 IIIB). While not shown here, such mixing also warms water in the hypolimnion. Diffuse attenuation coefficients varied from 0.5 m^{-1} after ice-off, increased to 0.9 m^{-1} with the incoming stream water, decreased slightly subsequently, and increased again by late summer to 1 m^{-1}. *(Redrawn from MacIntyre et al. 2006.)*

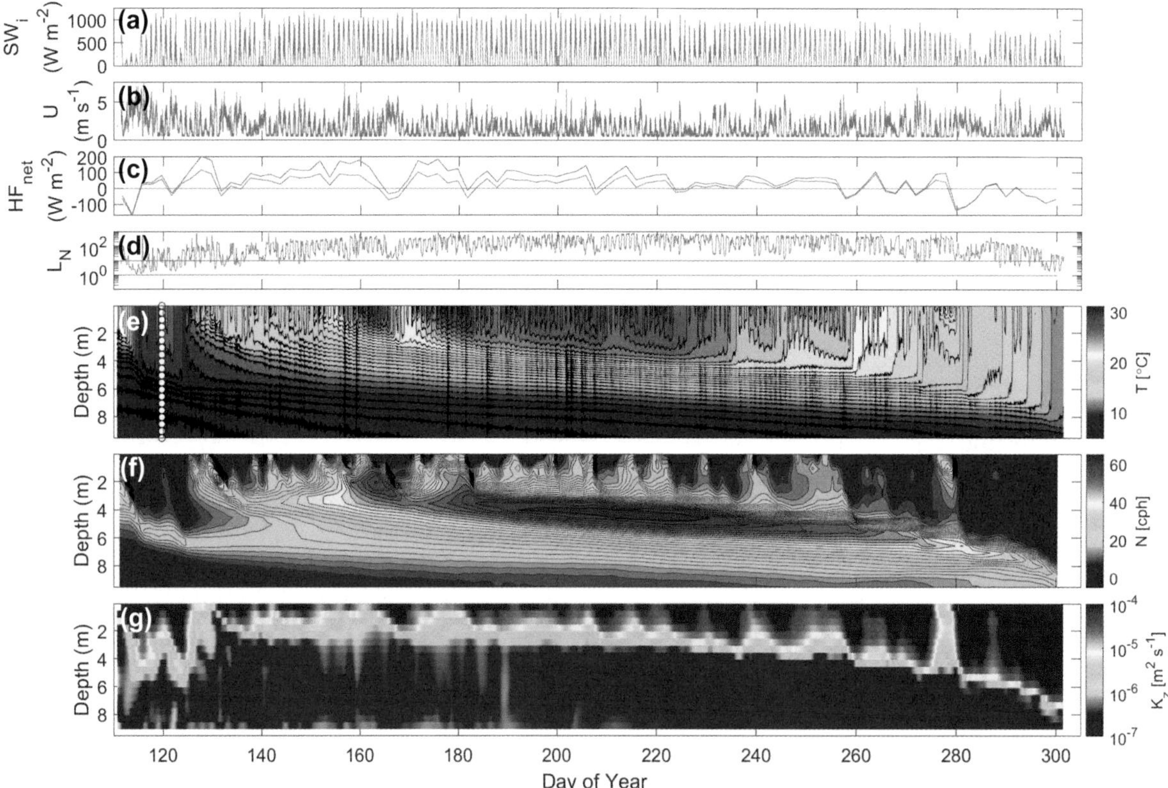

FIGURE 7-22 Time series data for Lawrence Lake, Michigan (USA), in 2005, of (a) incoming shortwave radiation (SW_i, W m^{-2}), (b) wind speed at instrument height (U, m s^{-1}), (c) 3 day averaged net heat flux (*blue*) and 3-day averaged net heat flux into the actively mixing layer (*orange*) (HF_{net}, W m^{-2}), (d) hourly averaged Lake number (L_N) for the full lake (*blue*), (e) temperature contours, (f) buoyancy frequency (N, cycles per hour, cph), and (g) 3-day averaged coefficient of eddy diffusivity (K_z, m^2 s^{-1}) below the actively mixing layer computed following Jassby and Powell (1975). This approach is most accurate within the thermocline and below and when there is no advection. K_z in the actively mixing layer was estimated as the product of turbulent velocity and length scales (see Chapter 8 I). The higher values of K_z in the upper water column provide an indication of the greater mixing there but have been scaled to have a maximum of 10^{-4} m^2 s^{-1} and thus do not illustrate the changes associated with diel cycles. Data are from a meteorological station on the lake and temperatures within the lake were measured with thermistors with 0.002°C accuracy sampling at 20-second intervals. Depths of the thermistors are indicated with *white dots*. Depth of the actively mixing layer defined by a change of 0.02°C relative to surface temperature measured within 0.05 m of the surface. The diffuse attenuation coefficient was 0.4 m^{-1}.

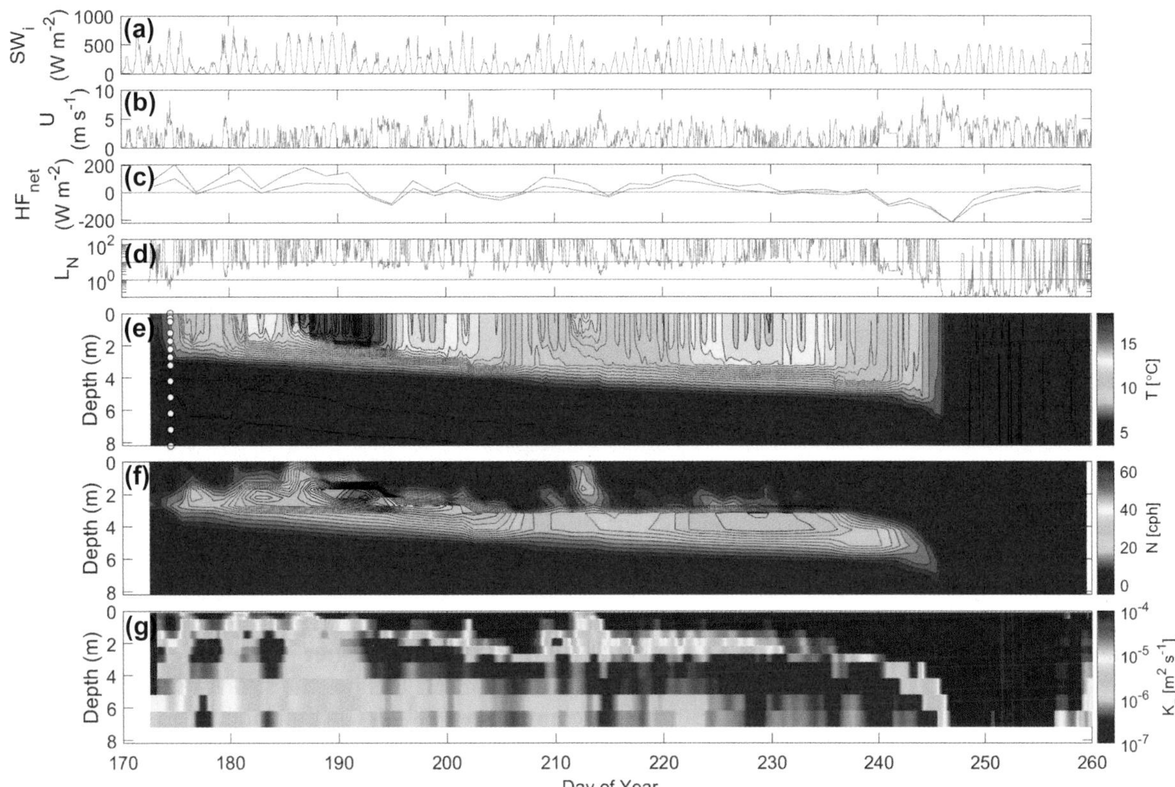

FIGURE 7-23 As described in Fig. 7-22, except that the heat budget is averaged over 2 days. Data are from 0.016 km^2 Lake N2 in Arctic Alaska in 2014. Data are from a meteorological station on the lake shore and from a thermistor array using sensors with 0.2°C accuracy but intercalibrated to 0.01°C accuracy. Sampling was hourly. The diffuse attenuation coefficient was 0.4 m^{-1}. Range for N is 0 to 65 cph, as for Fig. 7-22.

are $\sim -1 \times 10^{-8}$ m^2 s^{-3} in the temperate lake. For h of 1 m and N^2 of 0.0012 s^{-2}, only 0.1 m would be entrained over a 12-hour night, and the upper water column would restratify in the day, enabling a continued buildup of near-surface stratification. However, with cloudy days and increased winds such that L_N again dropped to values near 1, the increased mixing transports the heat downward, contributing to a thickening of the seasonal thermocline (Fig. 7-22). Thus the stratification in Lawrence Lake consisted of an upper mixed layer and a metalimnion. The patterns were similar in the Arctic lake, where buoyancy frequencies reached 20–30 cph (N^2 = 0.0012–0.0027 s^{-2}) in the thermocline within about 2 weeks of ice-off (Fig. 7-23).

Summer stratification was stronger and surface temperatures were higher in the temperate than in the Arctic lake because of a longer period with high solar radiation. Solar radiation often exceeded 1000 W m^{-2} and near-surface temperatures reached 30°C in Lawrence Lake. With a k_d of 0.4 m^{-1}, solar radiation reaches the bottom of the lake. As the summer progressed, stratification intensified in the seasonal thermocline, with buoyancy frequency reaching 60 cph. The stratification is so strong that Lake numbers are above 10 for much of the summer, indicating minimal, if any, tilting of the seasonal thermocline and minimal shear-induced mixing across it. That the coefficient of eddy diffusivity was around molecular levels, 10^{-7} m^2 s^{-1}, below the mixed layer attests to the lack of mixing. Wedderburn numbers for the diurnal thermocline began to reach values near 1 with afternoon winds after midsummer. This result implies that the diurnal thermocline up- and downwelled, which would have caused shear from currents above and below it. The combination of the resulting shear at the base of the upper mixed layer, shorter day lengths, and potentially some flow of cooler water from shallow regions (Doda et al. 2022) contributed to a slow erosion of the thermocline until the end of October. Then, with higher winds such that Lake numbers dropped to values of 1, indicating full upwelling, and with increased cloud cover such that the heat budget became negative, the thermocline was finally eroded.

Summer stratification was weaker in the Arctic lake than in similarly sized Lawrence Lake due to lower solar insolation and more variable cloud cover combined with intermittent periods with higher winds. Maximum values of incoming shortwave were $\sim$800 W m^{-2}, periods of cloud cover lasted several days such that HF_{net}

was more often negative, and winds were such that L_N dropped below 10 for much of the summer and occasionally, as in early summer, below 2. While the buoyancy frequency in early summer did reach 50 cph in the thermocline, with the subsequent increased mixing from heat loss and shear, stratification within the seasonal thermocline decreased to values of 20−30 cph, lower than in Lawrence Lake. Values of K_z within and below the thermocline were higher than in Lawrence Lake and occurred with decreases in L_N. Following ice-off, values of K_z within the hypolimnion were often between 10^{-6} and 10^{-5} m^2 s^{-1}. As temperatures in the lower water column of Lake N2 are often below 4°C at that time and the anoxia produced over the winter persists, along with reduced chemical species from anaerobic respiration (MacIntyre et al. 2018; Cortés and MacIntyre 2020), this mixing can alleviate anoxia. Values of K_z decreased as temperatures gradually warmed in the thermocline and hypolimnion, and after DOY 190, values were near molecular in the thermocline. Mixing in late summer occurred with a cold front and high winds such that $L_N < 0.1$ and shear was sufficient across the seasonal thermocline to fully erode it. The stratification within Lake N2 varied between summers. In some summers the lake had a distinct upper mixed layer, thermocline, and hypolimnion, and was mixed by late August, as in 2014. In others the thermocline was thicker, as in Lawrence Lake, and stratification persisted a month longer.

The influence of lake size on mixing dynamics in lakes at the same latitude is evident from comparisons between 0.016 km^2 Lake N2 and 1.50 km^2 Toolik Lake (Fig. 7-24). The patterns during the ice-free period in Toolik Lake in 2014 were similar to those in Lake N2 with respect to the onset of heating and cooling periods and the time of full mixing. However, with Toolik Lake's somewhat higher wind speeds due to its more exposed location, combined with its larger surface area, Lake numbers were consistently lower and often reached 1. In Toolik Lake the greater wind-induced mixing across the thermocline resulted in a maximum buoyancy frequency in early summer of 35 cph (as opposed to 52 in Lake N2). It decreased to 20 cph (as opposed to 32 in Lake N2) in late summer. Values of K_z were up to an order of magnitude higher in the hypolimnion and consistently higher in the thermocline, indicating greater connectivity across the thermocline for much of the summer. Midsummer, ~DOY 205, the temperature decrease

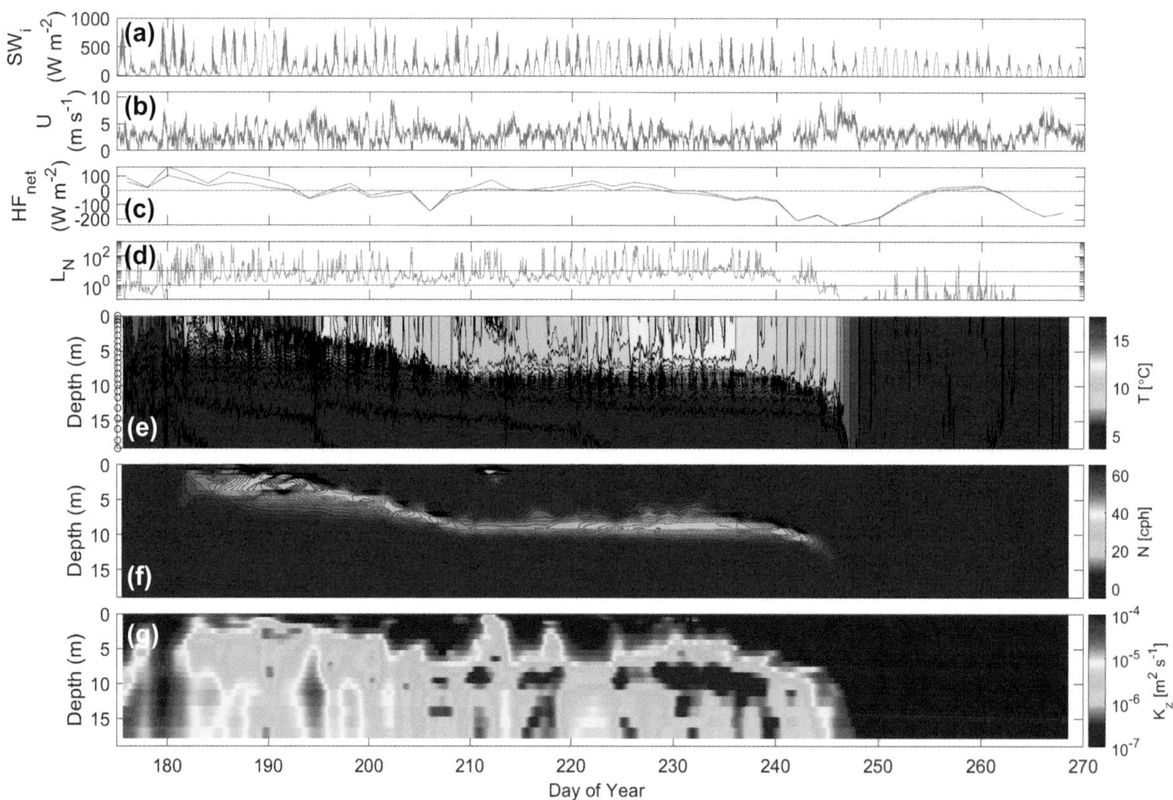

FIGURE 7-24 As described in Fig. 7-22, except that the heat budget is averaged over 2 days. Data are from 1.50 km^2 Toolik Lake in Arctic Alaska in 2014. Data are from a meteorological station on the lake and from thermistor arrays comprised of sensors with 0.002°C accuracy and sampling every 10 s. Lower panel is ranged so that the maximum value of K_z is 10^{-4} m^2 s^{-1}. Actively mixing layer defined by temperatures cooler than the surface temperature by 0.02°C. The diffuse attenuation coefficient averaged 0.7 m^{-1} in that summer.

in the mixed layer was induced by the passage of a cold front. With $L_N \leq 1$ in Toolik Lake, a multibasin kettle lake, wind-induced mixing often leads to between-basin exchanges such that water from the inlet basin flows into the main basin, either in the epilimnion or metalimnion. These exchanges moderate temperature and buoyancy frequency (Rueda and MacIntyre 2009; 2010). The final mixing event, with L_N dropping to 10^{-2}, was so intense that it caused the thermocline to descend from 11 to 19 m in 3 days. Over the final day, it descended 5 m. The contribution from buoyancy flux (heat loss) is computed as $dh/dt = \beta/(hN^2)$ (Eq. 7-21), where h is the depth at the onset of cooling. Based on the buoyancy fluxes and N on DOY 246, the thermocline would have descended 1.5 m. Hence much of the mixing can be attributed to the shear from the internal wave motions. As an approach to quantify the mixing, we calculated K_z based on the inverse of the time scale of mixing ($\tau_{mix} = l^2/K_z$) (Eq. 7-19). That is, $K_z = l^2/\tau_{mix}$. Here l is the change in depth of the thermocline. For the 3-day period, K_z was 2.4×10^{-4} m^2 s^{-1}, and for the final day, K_z was 2.9×10^{-4} m^2 s^{-1}, high values in a stratified region.

The above comparisons illustrate how stratification dynamics can vary in two lakes of similar size at different latitudes and in two lakes of different sizes at the same latitude. The weaker stratification and greater vertical transports in the small Arctic lake than in the temperate lake result from the lower irradiance, longer period of ice cover, and higher winds in the Arctic. At the same latitude, longer fetch and typically higher afternoon winds contribute to consistently lower Lake numbers and therefore more frequent mixing across the metalimnion in the larger Arctic lake (Fig. 7-14).

The magnitude of the Lake number varied in these three lakes over the stratified period. In assessing the importance of Lake number to mixing dynamics, quantifying the frequency of events when values drop below critical values of 5 and 1, and the duration of periods with values in excess of 10 or 100, enables meaningful comparisons. Long-term (e.g., seasonal) averages are generally not helpful. It is essential to quantify the frequency of events that enable the vertical transports, which redistribute solutes and particulate material and therefore influence primary and bacterial production. Low values of Wedderburn and Lake numbers are also indicative of the extent of horizontal advection and internal wave motions that modify the transport and distribution of zooplankton, with implications for grazing (Sprules et al. 2022).

The dynamics within the upper mixed layer were similar in all the lakes described above. That is, diel cycles of heating and cooling occurred, leading to the formation of diurnal thermoclines and their erosion at night (Fig. 7-3). Diurnal thermoclines were not evident during periods of cooling lasting over a day and typically when $L_N \leq 1$. The values of N presented in Figs. 7-21 to 7-23 are averaged as indicated in the figure captions. The unaveraged data indicate that values of $N < 1$ in the upper mixed layer only occur during periods of cooling when penetrative convection has caused mixing. Similar to the observations in Mono Lake (Figs. 7-16 and 7-17), the high-frequency internal waves that often lead to mixing (Chapter 8 IIIB) were prevalent in the diurnal thermocline of Lawrence Lake at both inshore and offshore locations on days when the Wedderburn number was near 1, indicating the potential for shear-induced mixing from such waves. The figures below for tropical lakes further illustrate dynamics in the surface layer.

Seasonal variability in mixing dynamics in response to low values of the Lake number—The consequences of low values of the Lake number vary over the stratified period (Fig. 7.25). During spring or early summer, when heat has been accumulating near the surface, wind events strong enough to cause L_N to drop below 1 mix the heat downward, often thickening the

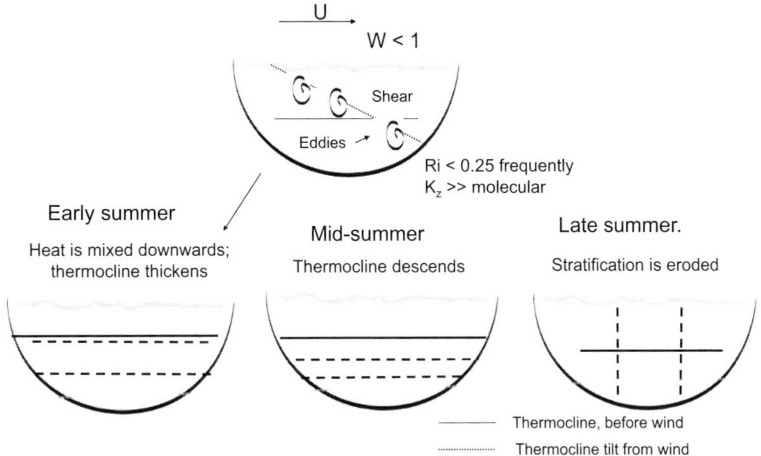

FIGURE 7-25 Consequences of low values of the Wedderburn or Lake numbers during the ice-free stratified period. See Chapter 8 IIIB for the mechanisms that enhance mixing on the basin scale. (*Solid lines*, prewind; *dashed lines*—after wind mixing event).

thermocline (Fig. 7-24). Observations are similar at ice-off in Lake N2 (Cortés and MacIntyre 2020) and in Mono Lake, California (USA), in spring (MacIntyre et al. 2009b) (Figs. 8-28 and 8-29). The coefficient of eddy diffusivity in the hypolimnion of Toolik Lake exceeded 10^{-4} m^2 s^{-1}, attesting to the considerable transfer of heat (Fig. 7-24). During midsummer, the thermocline abruptly descends (Fig. 7-21); and in late summer, it is eroded (Figs. 7-23 and 7-24). K_z was $\sim 10^{-4}$ m^2 s^{-1} across the thermocline and in the hypolimnion at that time, similar to predictions in Yeates and Imberger (2004) and attesting to the considerable mixing (Fig. 7-11). The considerable up- and downwelling of the thermocline in events with W or $L_N < 1$ may allow direct exchange between the epilimnion and hypolimnion near the lake margins, similar to Fig. 8-21, a process known as *edge leakage*. Such exchanges are difficult to quantify with K_z, which inherently implies exchanges are occurring in one dimension, that is, the vertical. Although the examples here are for the seasonal thermocline of dimictic lakes, the movement of the diurnal thermocline at low W is similar, and the associated mixing, especially if it occurs near the lake margins where diffusive fluxes from the sediments or inflowing stream waters may have led to concentration differences between inshore and offshore waters, could lead to offshore transport of these solutes (MacIntyre et al. 2021b). Additionally, as will be seen for polymictic lakes, those that are assumed to mix frequently, Wedderburn and Lake numbers can fall below 1, but the subsequent return of stratification as the wind relaxes implies that the wind had induced thermocline tilting and that the lakes were not well mixed.

Surface energy budgets, stratification, and mixing in tropical lakes—Variations in solar radiation and air temperature over tropical water bodies are muted in comparison with those at higher latitudes (Talling and Lemoalle 1998). Consequently, patterns of stratification vary with the monsoonal winds that cause movement of the Intertropical Convergence Zone and associated cloud cover and rainfall. Thus stratification is least during periods when the winds increase with the monsoons, and these periods are often times with decreased cloud cover, cooler air temperatures, and lower relative humidity. The converse can be true during rainy seasons. That said, there is considerable variability in meteorological patterns in tropical regions. However, diel stratification and nocturnal mixing are prevalent in tropical lakes. Their relation to the surface energy budget and wind over tropical lakes is illustrated with examples from Balbina Reservoir, Brazil, and Lake Janauacá, Brazil (Figs. 7-26 and 7-27). Balbina Reservoir, a 170 km long reservoir, is located a few degrees above the Equator and Lake Janauacá is a component of the Amazon floodplain. Data illustrated were obtained during the dry season. For both lakes, incoming shortwave radiation maxima were ~ 1000 W m^{-2}, with fluctuations due to cloud cover. Wind speeds were low, < 4 or 5 m s^{-1}, but punctuated by squalls due to convective storms. While the net heat flux leading to near-surface stratification was similar to that in the northern lakes described above, with the low winds, temperatures increased rapidly in the morning and pronounced diurnal thermoclines developed. In Balbina Reservoir, with $k_d \approx 0.5$ m^{-1}, the buoyancy frequency in the diurnal thermocline was maximally 25 cph, whereas in Lake Janauacá, with $k_d \approx 1.7$ m^{-1}, values of N in the diurnal thermocline during morning heating were between 30 and 80 cph depending on cloud cover and solar radiation. In both lakes the net heat fluxes at night, -100 to -400 W m^{-2}, were sufficient to erode a considerable portion of the stratification that formed in the day such that values of N dropped below 1, though stratification was maintained at deeper depths. The patterns in Balbina Reservoir are similar to those in Fig. 7-2. Internal wave activity occurred in the subsurface layer, which caused the diurnal thermocline to upwell during cooling (MacIntyre et al. 2021b). Nocturnal mixing often reached the seasonal thermocline at night in Balbina Reservoir and values of N dropped below 1 above it.

In Lake Janauacá, despite nocturnal mixing, some residual stratification was maintained, as evidenced by the upwelling of colder water in the morning, even following the squall on DOY 237 (Fig. 7-27). The spatial variability in the temperature field is evident in modeled cross-sections of the thermal structure of Lake Janauacá for a $\sim$24-h period on DOYs 238 to 239 (25–26 August 2016) (Fig. 7-28). The cross-sections illustrate the greater deepening downwind with increased winds followed by flow reversals, as in Fig. 7-13. Horizontal variability in temperature is further illustrated via simulations in Fig. 8-40. Thus, depending on the time of sampling, the depth of the diurnal thermocline and the temperature of the water column varies in the horizontal. Lakes like Janauacá are classified as *polymictic*, meaning mixing frequently occurs to the bottom. Here, we can see that one side of a lake can appear mixed, due to the downwelling of the upper mixed layer or upwelling of cooler water, while the other remains stratified. Hence inferring polymixis on the basis of temperature measurements at one location can be misleading. The results from Lake Janauacá illustrate the three-dimensionality of the physical structure and extent of mixing. That is, temperatures and mixing vary not only in the vertical, and not only in one horizontal direction, but also along other cross-sections (Fig. 8-41).

The persistent stratification in Lake Janauacá and in other polymictic lakes (Section IIC) is surprising given the often-appreciable heat losses at night ($\beta = -1 \times 10^{-7}$ m^2 s^{-3}) and expected vertical mixing. Several

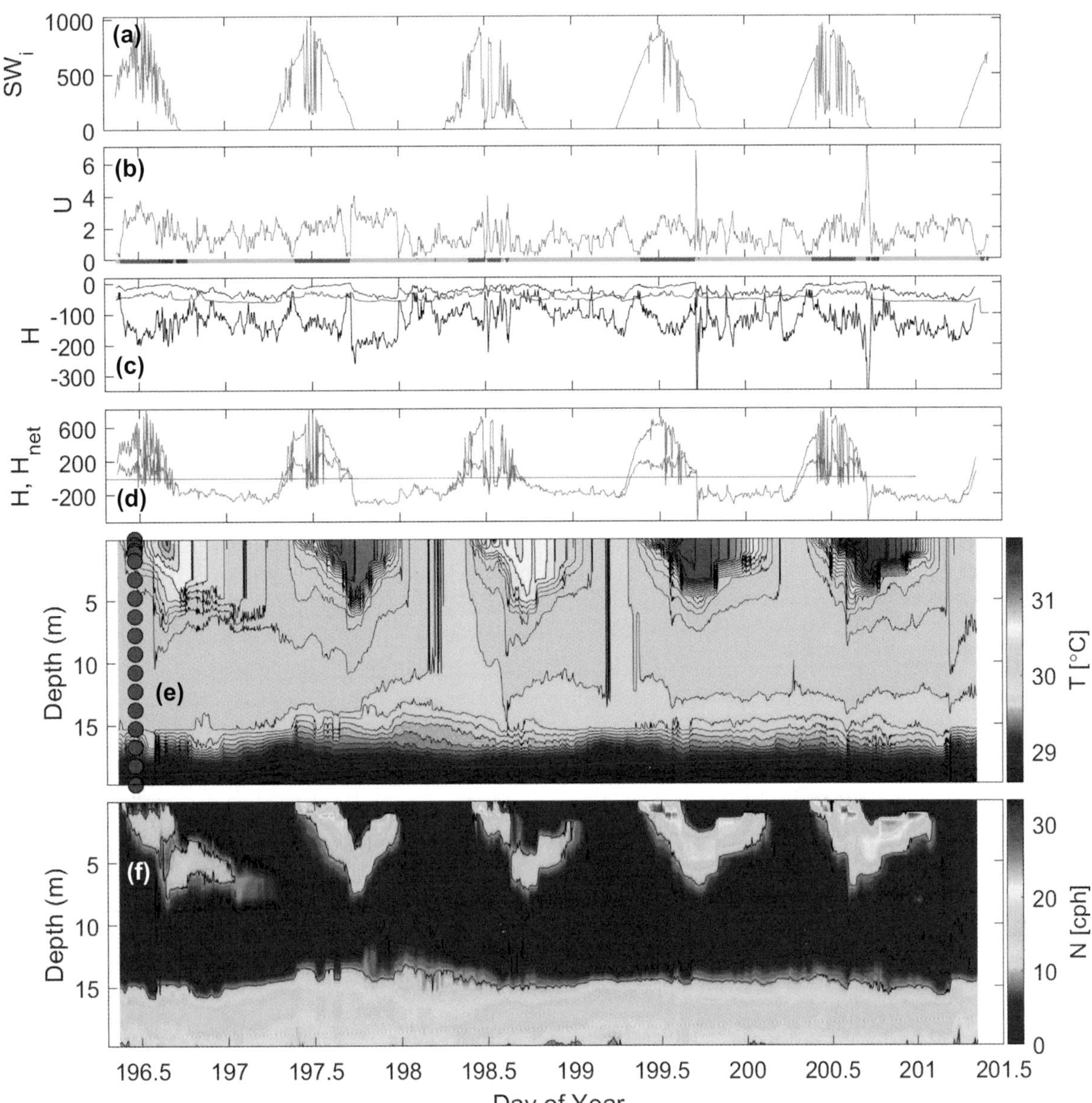

FIGURE 7-26 Five-minute averaged time series for Balbina Reservoir, Brazil (01°54′38.5″ S; 59°28′08.5″ W) during the dry season in July 2013. (a) Incoming shortwave radiation (SW_i); (b) wind speed corrected to 10 m height (U) with wind directions as underbars with west (*red*) generally coinciding with heating, east (*green*) typical during cooling, and north (*blue*); (c) heat loss terms of the surface energy budget: latent (*black*) and sensible heat fluxes (*blue*) and net longwave radiation (*red*); (d) heat fluxes with total heat flux (*blue*) and heat flux into the actively mixing layer (*orange*); (e) temperature contours; and (f) buoyancy frequencies in cycles per hour (cph). Data are from a meteorological station on the reservoir, and temperatures were measured with sensors with 0.002°C accuracy sampling at 10-sec intervals (MacIntyre et al. 2021a). k_d was 0.5 m^{-1}. Temperature data were obtained in the upper 20 m of the 34 m deep reservoir. With the strong solar heating and light winds in the morning, near-surface waters stratify, creating a diurnal thermocline that downwells with slight changes in wind speed during the heating period. Once cooling begins, unlike in most situations, the diurnal thermocline begins to upwell while losing heat. The upwelling is in response to internal wave motions in the subsurface layer. With continued heat loss at night, the diurnal thermocline was eroded on all but one night, and occasionally the depth of convective mixing reached the seasonal thermocline at 15 m.

factors can lead to the persistent stratification. Under convection, the mixing of stratified water occurs partly by shear and thus occurs with a lower mixing efficiency than by convection alone (Chapter 8 IB) (Kirkpatrick et al. 2019). The rate of entrainment depends on β and the buoyancy frequency, N. That is, $dh/dt = \beta/(hN^2)$ (Eq. 7-21). Upwelled water in the lower water column of Lake Janauacá sometimes has N of 10 cph, and sometimes as much as 20 cph. For those two values of N, using typical buoyancy flux at night of 10^{-7} m^2 s^{-3} and a mixing layer depth of 4 m, dh/dt over 12 h is 3.6 m and 0.8 m, respectively. Thus the unexpected persistence

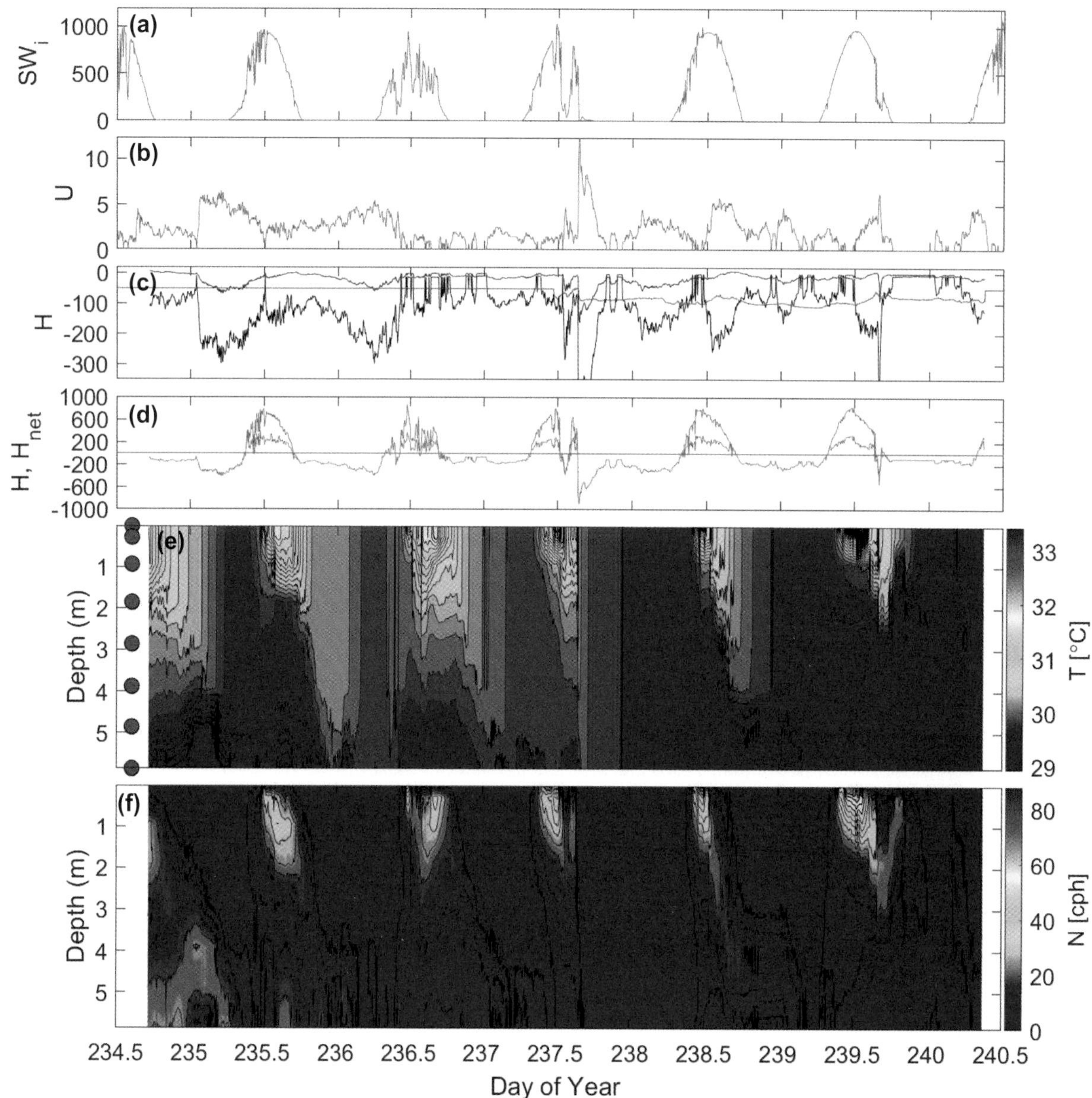

FIGURE 7-27 Similar to Fig. 7-26 for Lake Janauacá, Brazil (3°23′ S; 60°18′ W), during the dry season in 2016. The surface area of the lake was ~40 km^2 and maximum depth ~6 m. Thermistors with 0.002°C accuracy were deployed at the depths indicated by the red dots in panel (e) with 5-sec sampling intervals. A meteorological station was located on the lake. k_d was 1.7 m^{-1}.

of stratification in the lower water column of shallow polymictic lakes like Lake Janauacá may result from horizontal advection, inefficient mixing where the water column is stratified, and slow rates of entrainment. As will be discussed in Section IIC and Chapter 8 Section IIIC, inflows of cooler water from shallow regions may also contribute to the stratification.

Near-surface turbulence and the coefficient of eddy diffusivity vary with diurnal heating and cooling and changes in wind speed in the surface layer of lakes (Imberger 1985; MacIntyre 1993; MacIntyre et al. 1999, 2018, 2021a,b; Pernica et al. 2014; Tedford et al. 2014). Quantifying turbulence in the upper mixed layer requires the use of profiling instruments or acoustic Doppler current profilers and can be estimated from meteorological and time series temperature data using algorithms in the references cited immediately above (Chapter 8 IB). Turbulence can be identified by instabilities in the water column known as turbulent eddies or Thorpe scales (L_T) and by the rate of dissipation of turbulent kinetic energy (ε) (Chapter 8 I). These terms and the buoyancy frequency N enable the calculation of K_z and the gas transfer velocity using the surface renewal model (Chapter 8 IIIE). These are used to compute fluxes in the water column and at the air–water interface, respectively.

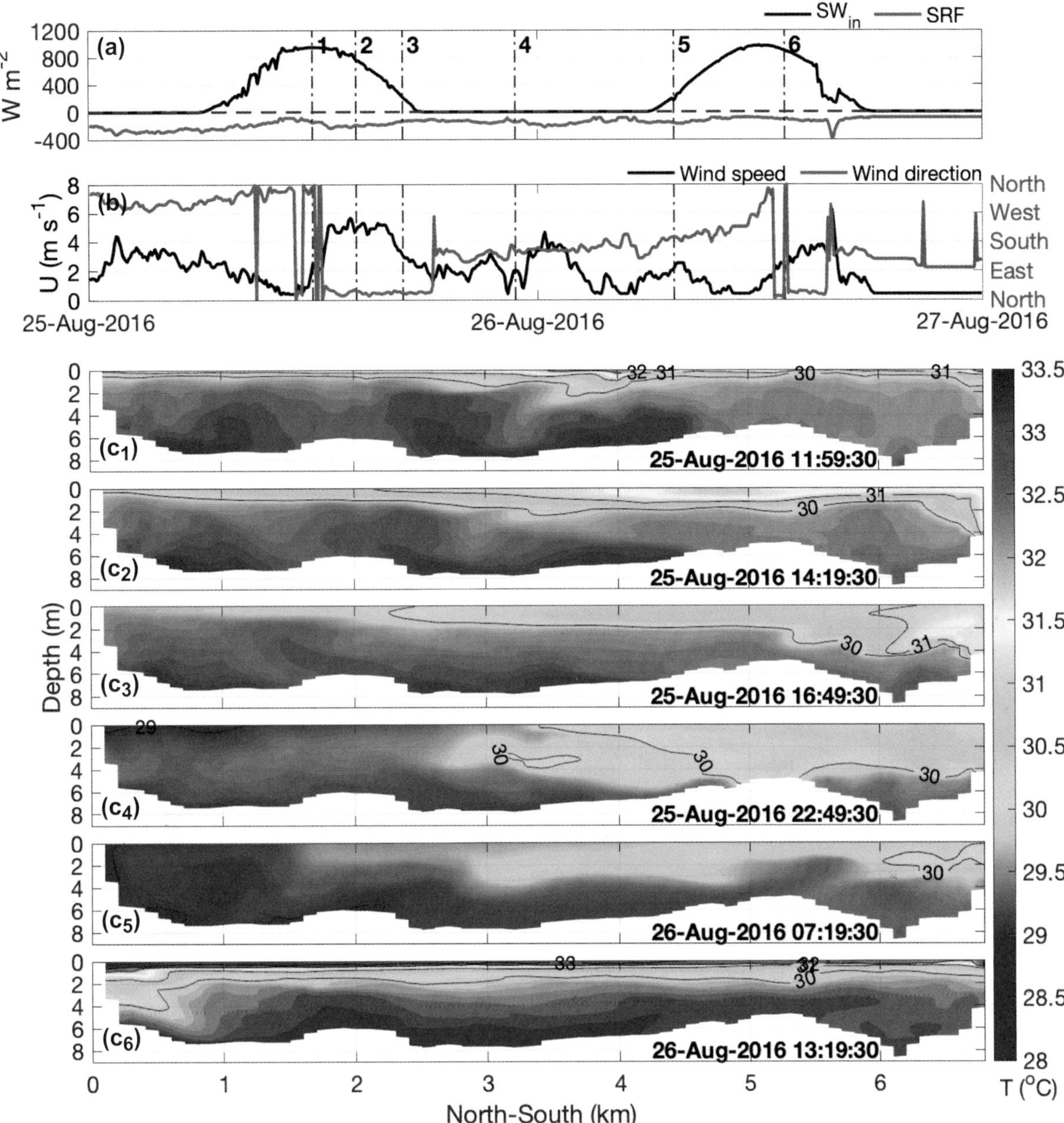

FIGURE 7-28 Cross-sections of temperature along the north to south axis of Lake Janauacá on 25–26 August 2016 (DOYs 238–239 in Fig. 7-27) obtained via three-dimensional hydrodynamic modeling (see Chapter 8 IV). (a) Incoming shortwave (*black*) and net heat loss terms (*red*); (b) wind speed (*black*) and wind direction (*red*). (c1) With morning heating and minimal wind, near-surface waters stratified although, with northerly winds, the diurnal thermocline was slightly deeper to the south. (c2) With increased northerly winds, heat was mixed downward and advected across the lakes such that the diurnal thermocline downwelled to the south. (c3) With the continued winds, further upwelling occurred to the north with concomitant increased downwelling of warmer water to the south. (c4) With southerly winds, the warmer water to the south flowed northward. With nocturnal cooling, temperatures decreased. (c5) With light southerly winds through much of the night, the warmer water to the south continued to flow north reducing cross-basin temperature differences. In c5 heating has just begun. With light winds before noon, surface waters warmed across the lake, and temperatures at noon were uniformly distributed, as in c1. (c6) By 1319 h, the winds had shifted direction and were northerly. However, the warmer surface temperatures and the apparent downwelling to the north were a result of greater heating to the north due to shallower regions there, a process called differential heating (e.g., Fig. 8-31). With the continued northerly winds, the warm water flowed to the south (not shown). The data from Lake Janauacá illustrate that spatial variability can develop in the temperature field. With these spatial gradients and their up and downwelling with changes in wind velocity, the lake did not fully mix in the vertical or horizontal despite considerable heat loss at night.

Patterns in Balbina Reservoir (Brazil) are similar to those elsewhere under calm conditions (Fig. 7-3). Changes depend on the relative proportions of wind shear (as u_{*w}^3) and buoyancy flux (β_*) quantified via the Monin–Obukhov length scale ($L_{MO} = u_{*w}^3/\kappa\beta_*$, where κ is the von Kármán constant of 0.41, Eq. 7-15). During nocturnal convection, when L_{MO} is negative, indicating heat loss, overturns typically fill the upper mixed layer. As heating begins and when winds are light, L_{MO}/z_{AML} is positive and approaches 0 (Fig. 7-29a), stratification develops (Fig. 7-29b), and turbulent eddies diminish in size near the surface but persist below (Fig. 7-29c). Due to wind shear, ε increases near the surface (Fig. 7-29d). As heating continues, stratification intensifies and, with light winds and some near-surface mixing, the diurnal thermocline descends (Fig. 7-29b). Dissipation rates remain highest near the surface and can increase in the diurnal thermocline when shear increases from internal wave activity. With the onset of cooling in the afternoon, the ratio L_{MO}/z_{AML} becomes negative. Overturns become larger, and when L_{MO}/z_{AML} approaches zero, as in this data

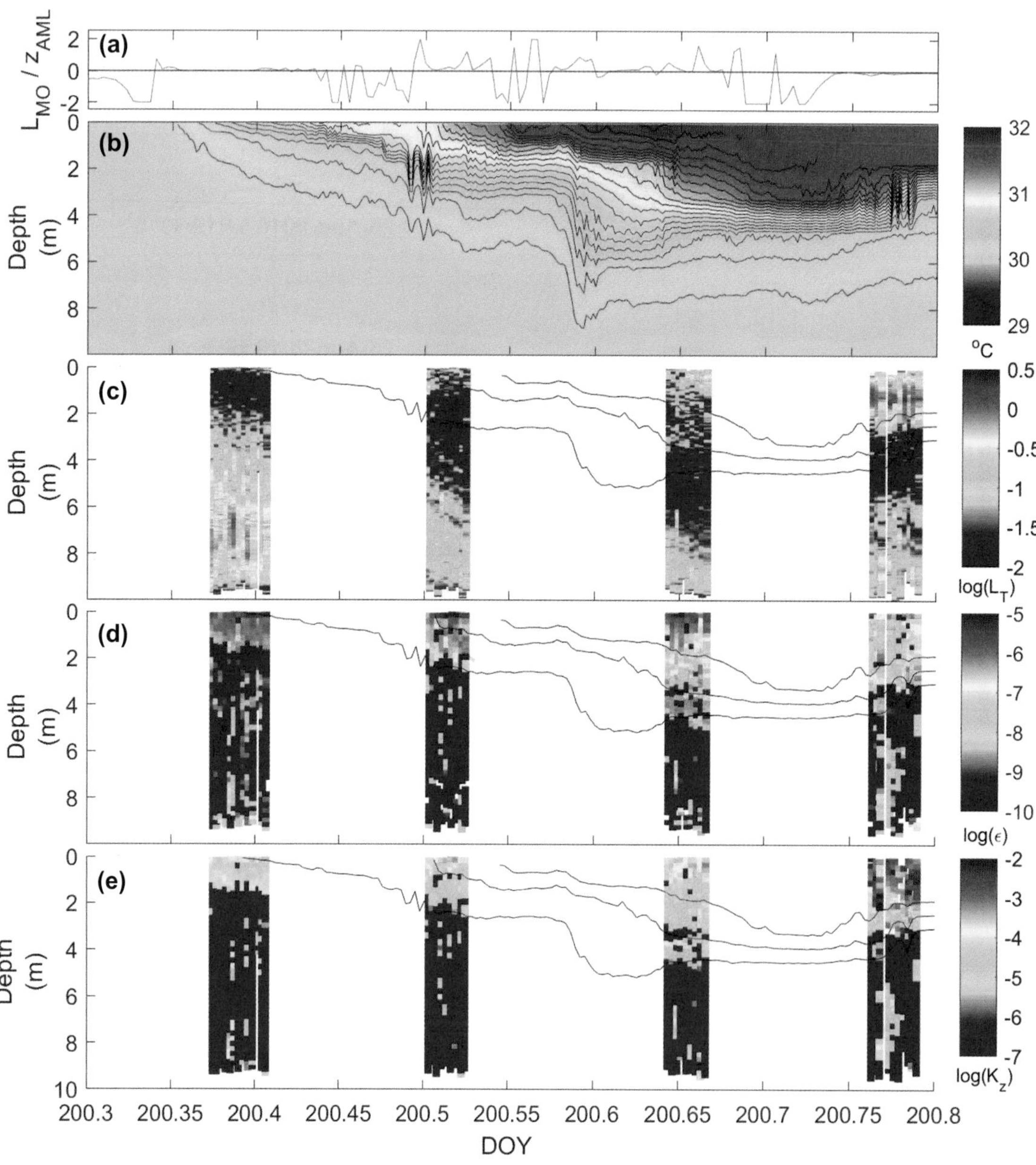

FIGURE 7-29 Time series of (a) Monin–Obukhov length scale divided by depth of the actively mixing layer (L_{MO}/z_{AML}), (b) temperature contours, (c) logarithm of *overturning length scales* (m) with maximal value centered at the midpoint of the overturn (L_T), (d) *rate of dissipation of turbulent kinetic energy* (ε, m^2 s^{-3}), and (e) K_z (m^2 s^{-1}) from early morning to evening on 19 July 2013 (DOY 200) in Balbina Reservoir, Brazil. Data were obtained with a temperature-gradient microstructure profiler with ca. 0.001 m resolution in the vertical. See Chapter 8 I for descriptions of variables used to describe turbulence. K_z was calculated as described in Chapter 8 IB, *Turbulence and its measurement*. Calculated values of K_z during cooling are approximately an order of magnitude higher than during heating due to higher mixing efficiency under cooling. *(Revised from MacIntyre et al. (2021a).)*

set, dissipation rates decrease except for a thin region near the surface energized by wind shear. These changes all moderate K_z in the upper water column (Fig. 7-29e). Due to differences in mixing efficiency when turbulence is induced by shear versus convection from heat loss (Chapter 8 IB), values of K_z are often lower in the day and can be higher as cooling begins in late afternoon and at night even when dissipation rates are lower. Typical values near the surface are 10^{-4} m^2 s^{-1} during heating and 10^{-2} and 10^{-3} m^2 s^{-1} during cooling. Values in the diurnal thermocline are about an order of magnitude lower, and negligible in the subsurface layer in the day. The gas transfer velocities under light to moderate winds depend on ε near the air–water interface (Fig. 8-36). Dissipation rates were appreciably higher under heating than under cooling despite similar wind speeds (Figs. 7-26b and 7-29d). With respect to fluxes of dissolved gases across the air–water interface, fluxes will depend on wind speed and the relative magnitude of heating and cooling; when winds are light to moderate, they may be higher in the day (MacIntyre et al. 2021b, 2021a)

Time scales of mixing, $\tau_{mix} = l^2/K_z$, from the base of the diurnal thermocline to the air–water interface (~2 m) during the day are long, $\tau_{mix} = 2 \times 2/10^{-5} \approx$ 4 days, indicating that the mixing processes during the day are not sufficient to disrupt the stratification. In contrast, by late afternoon when convection occurred, $\tau_{mix} = 2 \times 2/10^{-3} \approx 0.5$ days. At night when there were light winds as well as cooling, $\tau_{mix} = 6 \times 6/10^{-2} \approx 1$ h. Thus, while dissolved gases, nutrients, or populations of phytoplankton may be stratified in the day, mixing under convection can redistribute them, and, when convective mixing is deep at night, entrain solutes from the seasonal thermocline and redistribute them throughout the upper mixed layer (Figs. 7-26 and 8-37).

B. Under-ice stratification

Inverse temperature stratification is commonly found below the ice in freshwater lakes, but patterns differ when solutes are present and modify the density structure. Inverse stratification results because the temperature of maximum density of freshwaters is close to 4°C (see Chapter 3), and temperatures immediately below the ice are 0°C and increase below. Thus the temperature profile is inversely stratified. The time required for surface temperatures to reach 0°C and ice to form depends on the heat content of a lake, with larger, deeper lakes having greater heat content and therefore later ice-on relative to smaller lakes in the same region. Prior to ice-on, lakes often continue to circulate and cool to well below the temperature of maximum density. Water temperatures of between 3 and 4°C are common in small lakes before ice cover forms, and even colder temperatures may be found before ice-on in small Arctic lakes (Figs. 7-30 and 7-31). Isothermy to less than 1°C has

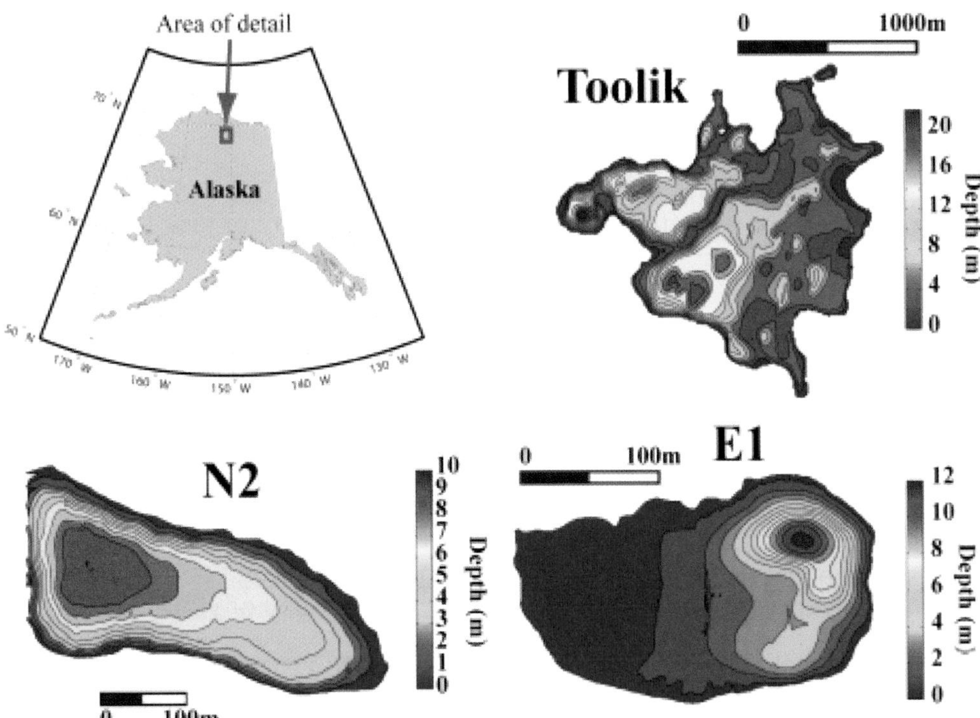

FIGURE 7-30 Location and bathymetry of Toolik Lake (SA 1.5 km^2, 24 m deep) and Lakes N2 (0.016 km^2, 10.4 m) and E1 (0.029 km^2, 12 m with a side basin to the west) in the Alaskan Arctic. Their differing bathymetry leads to differences in thermal and density structure under ice.

been observed before the formation of ice cover, particularly in large lakes that are subject to considerable wind action (Yang et al. 2020). The variability in water temperature, driven by the prevailing meteorological conditions prior to ice-on, thus affects the initial temperature and density stratification. The layer below the ice where temperature changes rapidly is called the *surface boundary layer* (Forrest et al. 2008). Typically, temperature changes in deeper water are more gradual until reaching the sediments, where they may again increase.

Variability in thermal structure under the ice depends on lake morphometry, the heat budget, and sediment respiration (Hutchinson 1957; Mortimer and Mackereth 1958). For three lakes in the Alaskan Arctic, temperatures in fall continued to decrease even after they fell below ~4°C, the temperature of maximum density of freshwater (Figs. 7-30 and 7-31). Temperatures appeared to be isothermal, although examination of high-resolution data indicated that up- and downwelling from internal wave motions persisted even as temperatures decreased. When water temperatures are less

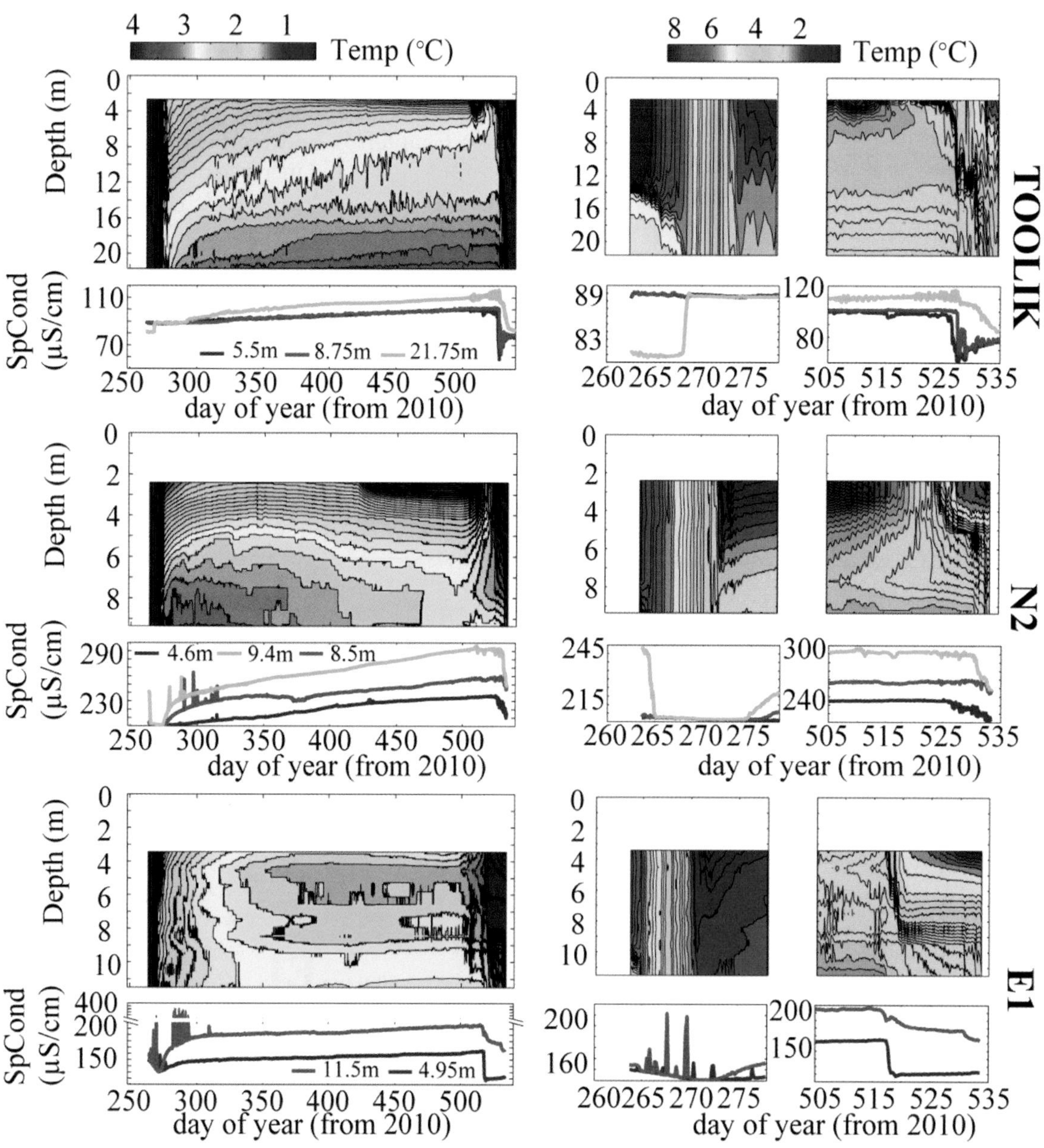

FIGURE 7-31 Thermal structure and time series of changes in specific conductance (SpCond) during winter (*left-most panels*), fall (*middle panels*), and shortly before and after ice-off (*right panels*) in the three lakes beginning in fall 2010 (DOY 250, 7 September) and ending in spring 2011 (indicated as DOY 535, 19 June 2011). Ice-on in fall can be identified by temperatures becoming inversely stratified. Ice-off can be identified in spring as the time when temperatures rapidly increased in the upper water column. Because of the increased density from increased solutes in the lower water column of all three lakes and additional density gradients as snowmelt freshened near-surface waters, mixing from penetrative convection prior to and at ice-off either did not occur or did not penetrate through the full water column (Cortés et al. 2017).

than 4°C, incoming solar radiation, in warming the surface waters, increases density near the surface, which can induce convective mixing (*penetrative convection*), whereas heat loss can contribute to stratification. Solar heating of shallow regions can create denser water, which can flow offshore as gravity currents at deeper depths, stabilizing the water column. These processes lead to variations in thermal structure prior to and after ice-on.

Despite weak stratification prior to ice-on, the water column can quickly stratify under the ice (Fig. 7-31). Several cooccurring processes contribute. Heat fluxes from the sediments warm the overlying water and create downslope flows such that warm water accumulates in the deepest locations (Figs. 7-4 and 8-33; Mortimer and Mackereth 1958; Terzhevik et al. 2009). Sediment respiration, and potentially intrusions of warmer groundwater, may lead to an increase in solutes, which also increase near-bottom density and contribute to the flow of gravity currents downslope (Mortimer and Mackereth 1958; MacIntyre et al. 2018; Jansen et al. 2021). At temperatures under the ice in freshwater lakes, solutes contribute to density even in lakes as fresh as Toolik Lake (Alaska), where specific conductance (Chapter 12) in summer is low, 60–80 μS cm^{-1}. Calculated velocities of the gravity currents are on the order of a few millimeters per second (Malm 1998). Thus flow to the bottom of a small lake, assuming distances of 50–200 m, would only take a few hours. Additionally, if the water column mixes immediately prior to ice-on, the temperature of the mixed water may vary across the basin, being warmer over the deeper region. The cooler, less dense water will flow over the warmer water on relaxation of the wind, leading to an overall increase in stratification.

In both Toolik Lake and nearby Lake N2 in Alaska, near-bottom heating is apparent in the days following ice-on. The flow from gravity currents occurs beginning at ice-on and may extend for several months or longer, depending on the amount of heat accumulated in the sediments over the summer and the organic matter available for respiration. As the gravity currents flow to the bottom, the water immediately above must rise such that the warmer water rises through the water column (Fig. 7-31). The thickness of the surface boundary layer, with its colder water, increases over time. Ultimately, in shallower N2 temperatures begin to decline throughout the water column due to heat flux to the ice. In contrast, in deeper Toolik Lake temperatures continue to increase to at least the mid region of the water column. Increases in specific conductance, indicative of sediment respiration, continue over the winter, with the largest increases near the bottom. In some lakes in warmer regions where surrounding soils have not frozen, groundwater intrusions may contribute to the flow from gravity currents. The combined influence of increased specific conductance and increased temperature near the bottom leads to an appreciable increase in near-bottom density. Consequently, as winter progresses, gravity currents from continued heat loss from the sediments or sediment respiration would flow over the denser bottom waters at progressively shallower depths (Fig. 8-33) (MacIntyre et al. 2018).

Other processes may also contribute to flow under the ice. For example, temperature fluctuations occur in the middle of the water column in Toolik Lake (Fig. 7-31). These are coherent with changes in wind velocity, implying they are the signature of internal waves. These motions have been shown to induce slow, oscillatory flow in other ice-covered lakes (Malm et al. 1998; Malm 1999; Bengtsson et al. 1996). Malm's (1999) analysis implies that they do not cause mixing in the interior of a lake. The extent to which they induce mixing near the boundaries is not known (Kirillin et al. 2012). *Cryoconcentration* (the exclusion of solutes when ice forms) may lead to solute concentrations immediately below the ice that are higher than those below. With temperatures increasing below the ice, these combined conditions can lead to convective motions known as *double diffusion* (Chapter 8 IIIC). Modeling indicates that instabilities can develop under the ice that ultimately modify stratification (Olsthoorn et al. 2020).

The thermal structure of Lake E1 (Alaska, USA) differs from that of the other two lakes discussed above in that the layer of warmest water is in the midwater column. The layering of warmer water has been observed in this lake for all the years for which data are available. The warm water is attributed to the offshore flow of water from the shallow western basin of the lake, which had been warmed by heat flux from the sediments (Fig. 7-30). Rather than sink to the bottom, the water intrudes offshore at depths in the main basin where the density is similar.

Calculations of complete heat budgets under the ice are rare. One of the earliest detailed studies on winter heating showed that most (>75%) of the temperature increase is from solar radiation (Birge et al. 1927). This large fraction may result from the temperate latitude of the lake studied and is only possible if there is little snow on the ice. The contribution from the sediments was computed based on the profiling of temperatures into the sediments over an annual cycle (Fig. 7-9). The relative contribution of heat stored in the sediments during the summer, which enables heat flux to the overlying water under the ice, will vary from lake to lake in relation to morphometry, summer heating conditions, and other factors (Hutchinson 1957). Measurements of sediment heat fluxes indicate they are of the order of 5 W m^{-2}, with higher values in shallower sediments (Bengtsson and Svensson 1996). Heat budgets for Toolik Lake indicate solar heating warms the water only for a few weeks after ice cover with maximal values of ~15 W m^{-2} when the ice is snow-free. Mortimer and Mackereth's (1958) analysis indicated that heat flux from the sediments and production of solutes by sediment respiration would produce gravity currents that transport the heat and slightly more solute-rich water to the deeper parts of the lakes. The increased density induced by the downward flow of solutes released by sediment respiration enables near-bottom temperatures above 4°C, as have been observed in some shallow lakes.

A transition in the mixing dynamics of ice-covered lakes can occur in spring when temperatures are below 4°C, the temperature of maximum density. Solar radiation penetrating through the ice can create instabilities as it warms water closer to the surface to temperatures exceeding those in the water below. This process is known as *radiatively driven convection* (RDC) or *penetrative convection*. The turbulent eddies that form are initially small as the instabilities begin but progressively deepen. In Toolik Lake they can reach the bottom (24 m) in years with minimal chemical stratification from solutes. These eddies help maintain negatively buoyant phytoplankton (e.g., diatoms) in suspension enabling blooms under the ice. When water heats more rapidly inshore than offshore, nearshore water may flow offshore, penetrating at depths with similar density. This process, *horizontal convection*, is another form of RDC and can lead to offshore transport of water with higher concentrations of oxygen or phytoplankton which have been growing nearshore (Vehmaa and Salonen 2009). Density stratification under the ice is often accentuated near the bottom by dissolved solutes, and at times in a layer below the ice due to fresher water from snowmelt or ice-melt. In such cases mixing under the ice by RDC may not occur and full mixing is delayed until after ice-off, as in Fig. 7-31. Different scenarios for the success of phytoplankton under the ice with or without RDC are presented in Jansen et al. (2021). The convective mixing beginning under the ice and in some cases continuing or beginning afterward can be highly energetic, and in large lakes the convection cells may reach 185 m and be 50 m wide (Austin 2019; Cannon et al. 2021). The thermal bar seen in large lakes is described in Chapter 8 IIIC and see Carmack (1978) for warming processes in deep lakes with large river inflows.

The extent to which incoming snowmelt or river water is mixed under the ice also depends on whether there is sufficient solar radiation to induce penetrative convection and whether density gradients restrict the ensuing mixing. For example, in Fig. 7-4 water whose temperature exceeded 6°C was centered around 1 m in late March in Lawrence Lake. The ice at this time was weak and porous from warm rains. Water low in dissolved solutes, which had entered through the ice or possibly from the margins of the lake, was less dense than that of the water below due to higher concentrations of dissolved ions in the lake water. By this means, solar heating through the ice increased the temperatures in this dilute layer without producing instability. Temperature anomalies such as this one may be transitory (Kózminski and Wisniewski 1935) and are likely related in part to air temperatures rising above freezing and advective inflows of warm water from the periphery of the lake (Ellis et al. 1991). Cortés et al. (2017) illustrate the initially cold snowmelt intrusions that flowed directly under the ice. Over time, as stream temperatures rose above 4°C, temperatures in the inflows were warmer than the lake water but inflows remained near the surface due to their lower solute concentrations and correspondingly lower densities. The fresher water (lower specific conductivity) in these inflows relative to lake water precluded or reduced the depth of penetrative convection (Fig. 7-31). Pasche et al. (2019) illustrate the persistence of a CO_2-enriched river inflow under the ice at a depth of ~10 m in a year when penetrative convection extended to 10 m and the velocity of the thermals was low. The plume persisted when there was some snow cover on the ice and the magnitude of the diffuse attenuation coefficient (k_d) in the water was 3.2 m^{-1}. In a year when convective mixing reached 24 m, snow cover on the ice was minimal and k_d was 1.2 m^{-1}. Incoming river plumes are expected to be dispersed throughout the water column in such years. The extent of such mixing can then affect the distribution of incoming nutrients and the rate at which climate forcing trace gases such as CO_2 are emitted into the atmosphere.

A detailed résumé of the types and characteristics of lake ice is given in Hutchinson (1957). Two books by Pivovarov (1973) and Ficke and Ficke (1977) summarize the diverse literature on the properties of ice and the thermal characteristics of lakes and rivers under differing conditions of ice cover. More recent work includes Leppäranta (2015) and summaries by Kirillin et al. (2012). Model studies involving thermodynamic considerations have often separated three layers with different attenuation coefficients, for example, blue or black ice (which is clear ice), white ice (which is ice with snow interleaved), and snow-covered ice (Jansen et al. 2021), with these distinctions important for calculating the heating of the water below the ice (Hamilton et al. 2018).

C. Annual patterns of stratification

Classification schemes describing stratification in lakes are based on the frequency of circulation occurring throughout the entire water column on an annual basis

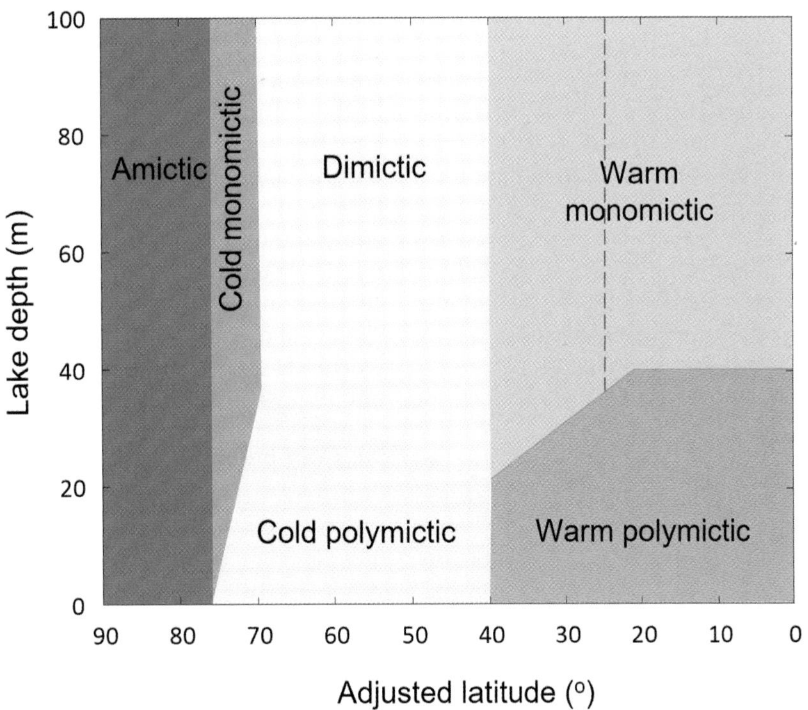

FIGURE 7-32 Classification schemes on an annual basis for lakes based on latitude (degrees north or south of the equator), adjusted for altitude, and depth. Monomictic lakes, which circulate in only one season each year, are further subdivided into *cold monomictic lakes* in which water temperatures are not above 4°C or are close to 4°C and circulate at or below 4°C and *warm monomictic lakes*, which circulate in the winter or coolest season in the tropics at or above 4°C and are not ice covered. Both are stably stratified for the remainder of the year. Ice cover forms on *dimictic lakes*, which circulate intermittently in spring and fall and have more pronounced density stratification in summer and winter. *Polymictic lakes*, which mix more than twice each year, are also subdivided as *cold polymictic* or *warm polymictic*. Ice cover forms on cold polymictic lakes, and they may circulate at temperatures near 4°C or at warmer temperatures. The depths discriminating polymictic lakes from dimictic ones are not exact as drawn here. Instead, they will vary with surface area, optical properties, and degree of sheltering from wind. The *vertical dashed line* delineates the approximate boundary of tropical and temperate lakes. Adjusted latitude includes a weighting for altitude. *(Redrawn from Lewis (1983).)*

(Fig. 7-32). Patterns of stratification and mixing vary with local or regional differences in climate combined with lake-specific factors such as differences in morphometry, exposure to wind and radiation, and chemistry. Classification schemes were developed by F. A. Forel, who is often referred to as the "father of limnology" (Vincent and Bertola 2014; Vincent 2018). His extensive monographs on Lac Léman (Lake Geneva), Switzerland, in 1892, 1895, and 1904, served as a foundation for limnology. Subsequent classifications were developed by Hutchinson and Löffler (1956) and Hutchinson (1957), extended by Ruttner (1963) to shallow lakes, and further refined by Lewis (1983). Here we distinguish annual stratification patterns following Lewis (1983), in which the patterns are generalized by latitude, altitude, and depth (Fig. 7-32). Lakes at far northern or southern latitudes that are ice covered year-round and do not fully circulate are classified as *amictic*. Lakes in which complete circulation occurs at least once a year, or at least in which stratification throughout the water column becomes quite weak at some time during the year, are called *holomictic*. Holomictic lakes are further discriminated by the number of periods per year in which the lakes are classified as circulating or mixing on the basis of near isothermy. High-resolution instrumentation indicates that stratification is often present in such periods and that mixing throughout the water column is intermittent. That said, on the basis of near isothermy, *monomictic lakes* have one period in which water temperatures are near isothermal, *dimictic lakes* are those with two such periods, and *polymictic lakes* have more than two such periods. Cold polymictic lakes have ice cover in winter, whereas warm polymictic lakes do not. We remove the distinction between *continuous* and *discontinuous polymixis*, which had been distinguished on the basis of *fetch*, the length of the lake along the axis of prevailing wind. Recent analysis indicates that even small lakes may stratify for periods of several days and that the apparent mixing at night may be a result of thermocline tilting (Figs. 7-14 and 7-28). The depth dependence illustrated in Fig. 7-32 is not precise. For example, both Lawrence Lake and Lake N2 described above are 10 m deep, yet both are dimictic. Kirillin and Shatwell (2016) provide approaches to predict whether lakes will be polymictic or stratify seasonally based on depth, optical properties, and mean Monin–Obukhov length scale, and Shatwell et al. (2016) link within and between year changes in classification type to seasonal changes in algal biomass due to zooplankton grazing. There are two other stratification types. *Oligomictic lakes* undergo full circulation of the water column in intermittent years (Section IIE). *Meromictic lakes* (Section IID) do not undergo full mixing due to their great depth and/or to the presence of a *chemocline*, which further increases the density stratification and the difficulty inducing mixing. In the following further details and examples are provided for lakes in each classification scheme.

i. Amictic lakes

Lakes defined as *amictic* are sealed perennially by ice from most of the annual variations in temperature. Amictic, perennially ice-covered lakes are rare, and largely limited to Antarctica and the High Arctic or, more rarely, to high mountains. That said, with ongoing climate warming, some of the lakes in the High Arctic are beginning to open with only a small amount of residual ice in summer (Lehnherr et al. 2018; Michelutti et al. 2020; Bégin et al. 2021), while mixing is expected to be enhanced in lakes that continue to be amictic (Spigel et al. 2018). With ongoing climate change, these lakes, which have been amictic, may even shift to cold monomictic (Bégin et al. 2021). Few amictic lakes have been recorded in the Arctic, mostly in Greenland and Ellesmere Island (Canada), as elsewhere, conditions necessary for the formation of a permanent ice cover are rare. Heating of these lakes depends on light penetrating through the ice and absorption by pigmented microbial communities (Vincent et al. 2008a). The balance of heat inputs and losses tends to be relatively constant over annual cycles. While these lakes do not fully mix, they can have complex thermal structure due to the combination of thermal and chemical stratification and horizontal convective circulation due to lateral heat inputs or losses (Fig. 7-31; Chapter 8 IIIC) (Spigel and Priscu 1998). In summer, with the melting of ice on the margins and stream inflows, additional mechanisms for advective exchange can occur; microstructure data indicated low levels of turbulence in exchange flows between basins (Spigel and Priscu 1998).

ii. Cold monomictic lakes

Lakes with water temperatures equal to or below 4°C, ice-covered most of the year, and with only one period of circulation in the summer are described as *cold monomictic* (Fig. 7-33a). The category comprises, for the most part, Arctic and mountain lakes, which are ice-free for brief periods in the summer and may be in contact with glaciers or permafrost. The thermal cycles of Arctic lakes have large variations in relation to their location, depth, and summer climate, such that not all are cold monomictic and the stratification type can vary between years. Lake Schrader in the Brooks Range, Alaska (USA), was dimictic and stratified in 1958, with a temperature in the epilimnion of 10°C and the hypolimnion of 4°C. The following year, summer ice breakup occurred a month later, and the lake did not warm above 4°C; that is, it was cold monomictic (Hobbie 1961). Broadening the temperature range for classifying cold monomictic lakes may be appropriate given the small changes in density above and below 4°C. For example, Char Lake, Cornwallis Island, Canada (latitude 76°), warms to 4–5°C and is technically dimictic but does not always stratify in summer.

iii. Cold polymictic lakes

Mixing occurs more than twice per year in polymictic lakes. *Cold polymictic lakes* differ from warm polymictic lakes in that they are ice-covered for part of the year. When ice-free during the warm season, temperatures may be close to 4°C, as in nearby cold monomictic lakes, but even at the most northern latitudes, due to their shallow depth, they may warm to temperatures above 4°C such that they stratify. Mixing can occur due to heat losses and wind at night or after cloudy periods. Besides some Arctic lakes, many shallow north-temperate lakes have this circulation pattern (e.g., Harvey and Coombs 1971). Cold polymictic lakes are also found at higher altitudes in equatorial regions with high wind and low humidity, and where little seasonal change in air temperature occurs. At high altitudes in equatorial regions, cold polymictic lakes gain a significant amount of heat during the day, but until recently, nocturnal losses were sufficient to permit mixing during the night (see Section III).

Although shallow, polymictic lakes are assumed to mix on a near daily basis or at least frequently, interpretation of high-resolution time series data using metrics such as the Lake number indicates dynamics are more complex (Fig. 7-34). Strong near-surface stratification built up in Lake St. Augustin (Quebec, Canada) during morning heating, and the mixed layer deepened with afternoon winds as L_N dropped to 1 or below. The return of cold water in the lower water column at night when winds declined indicates that the apparent mixing was the result of the upper mixed layer downwelling at the measurement site in the day (Fig. 7-13). Thus, on many days, the lake had not fully mixed despite near isothermy in the afternoon.

Lake E6, in subarctic Alaska (USA), provides an example of a shallow lake with brief periods of stratification (Fig. 7-35). Lake E6 has a surface area of 0.019 km^2 and is 3 m deep. Depending on the year, periods of stratification persisted for several days up to 10 days. Thermoclines develop in these periods, with buoyancy frequencies higher than in the larger, nearby lakes

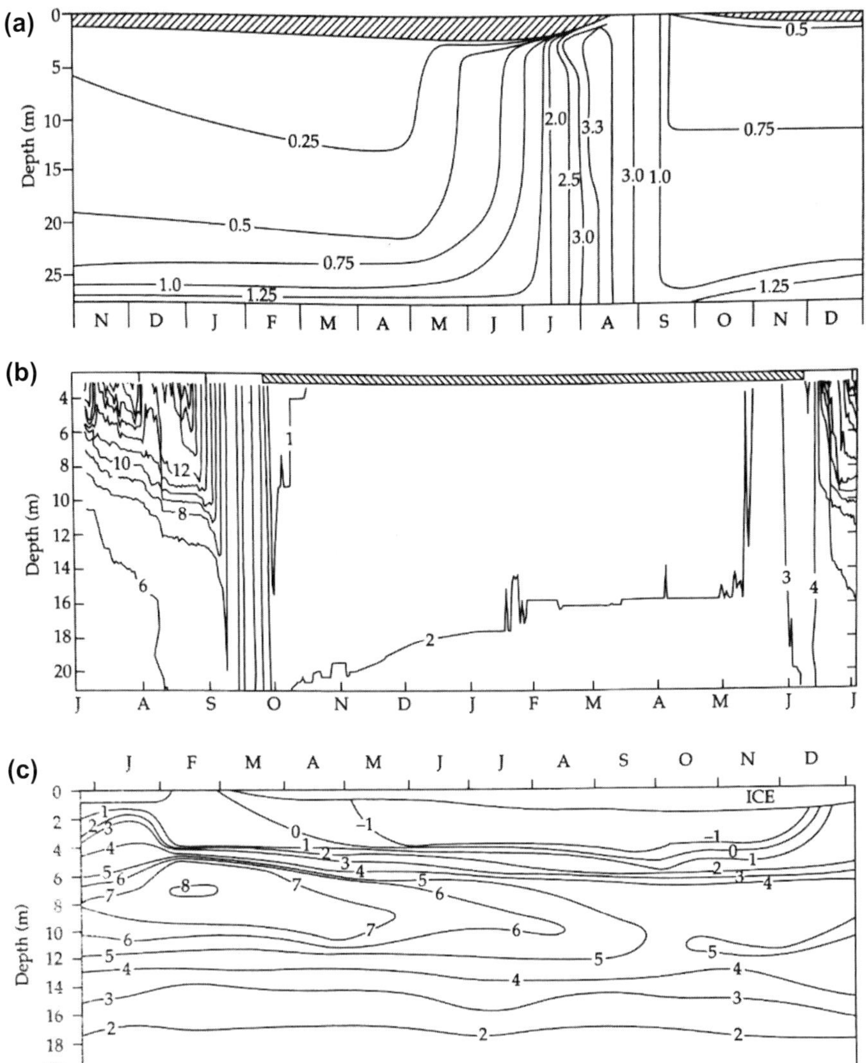

FIGURE 7-33 Isotherms over an annual cycle for (a) cold monomictic Char Lake (Nunavut, Canadian High Arctic), (b) dimictic Toolik Lake (Alaska, USA), and (c) meromictic Ace Lake (Antarctica). *(Figure from Vincent et al. (2008b).)*

(Figs. 7-23 and 7-24). N reached 55 cph around DOY 190 in 2014. During these periods, L_N decreased to ~1 over diel cycles. When cooling lasted several days, along with L_N dropping below 1, stratification was alleviated (e.g., as by DOY 194.5). During the periods with weak stratification, diurnal thermoclines formed, and Lake numbers cycled between 0.1 and 10, indicating the up- and downwelling on the basin scale (e.g., DOYs 195–200, 220–235), as in Lake St. Augustin and Lake Janauacá. Even when the lake was least stratified, as from DOY 230–240, cooler water flowed to the lower water column at night and sustained stratification. This influx of water was evident when winds dropped or changed direction and so may be indicative of the upwelling of cooler water on the relaxation of the wind. However, the relatively extensive shallow regions could cool faster at night than the deeper regions. Heat losses at night during this period were -200 W m^{-2}, which is considerable, and 2-day averaged heat fluxes were negative. The downslope flow of cooler water from shallow regions, a process called *differential cooling* (Chapter 8 IIIC), may have contributed to the persistent weak stratification at night. With cool water flowing from the shallows to the bottom, and warmer surface water flowing to replace it, a *thermal siphon* is set up (Monismith et al. 1990; Doda et al. 2022; Ulloa et al. 2022). Thus the lake circulates by horizontal convection. Overall, diverse processes can contribute to reestablishing stratification in polymictic lakes over diel cycles even during periods of net cooling.

iv. Dimictic lakes

Dimictic lakes are prevalent in sub-Arctic, boreal, and temperate regions of the world where ice forms in winter. Their characteristics were described above and illustrated for spring through autumn in Figs. 7-4 and

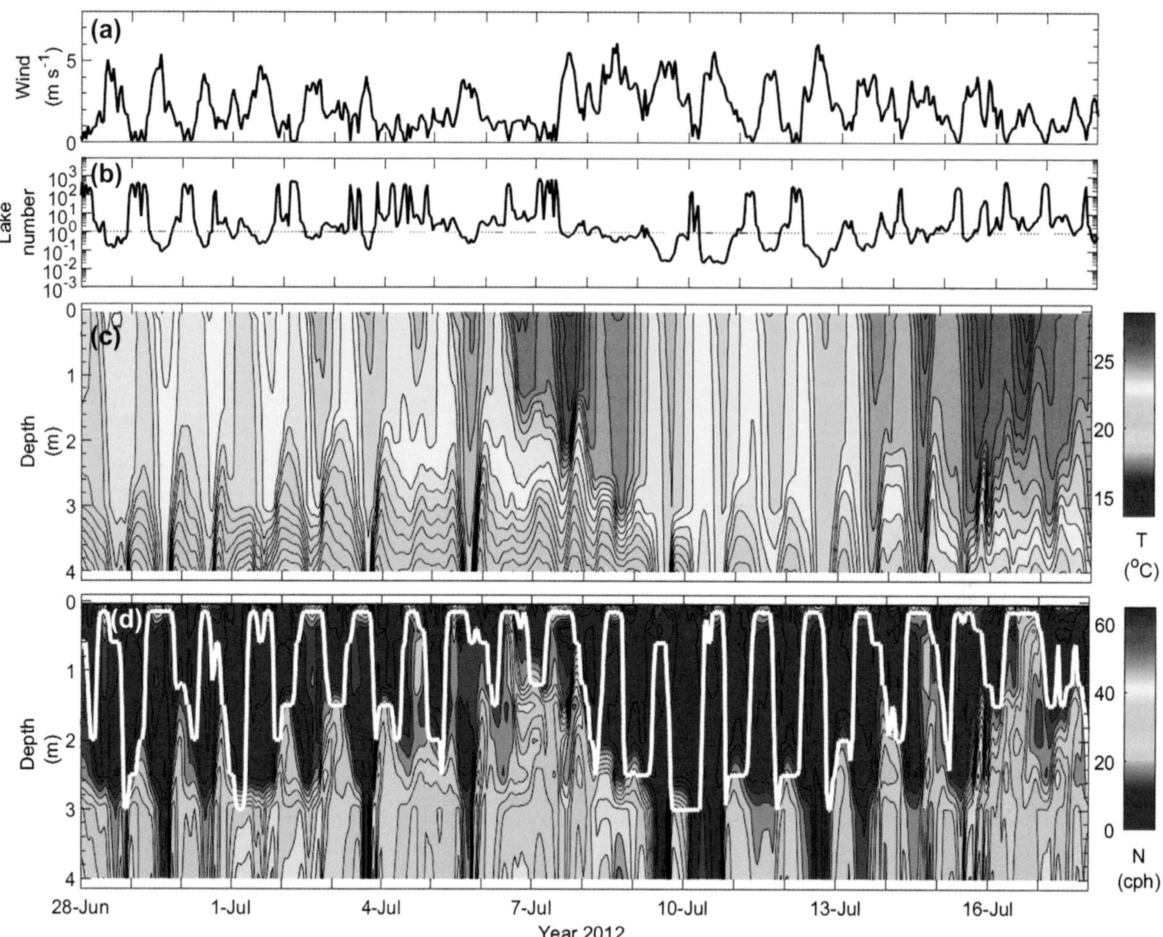

FIGURE 7-34 Time series of (a) wind speed, (b) Lake number, (c) temperature contours, and (d) buoyancy frequency N with white overlay indicating the layer with temperature stratification less than 0.02°C, identified as the actively mixing layer in polymictic Lake St. Augustin (Quebec, Canada) in 2012. Data illustrate the high surface temperatures (T = 28°C) and strong near-surface stratification ($N = 60$ cph) that can result when wind speeds are low. The Lake number decreased below 0.1 on many days, with warm water reaching nearly to the bottom and $N < 5$ cph. This downwelling of the upper mixed layer was followed by upwelling of cool water as wind speeds decreased to less than 1 m s^{-1} (i.e., the wind forcing relaxed) and N increased above 30 cph in the lower water column. This pattern implies that the thermocline tilted so much that the water at the measurement site comprised the upper mixed layer in the day. As the winds decreased and L_N increased above 1, cooler water from the thermocline upwelled and the upper mixed layer shoaled (Fig. 7-14). Thus, complete mixing of the lake occurred less frequently than typically expected for polymictic lakes. Estimated k_d ranged from 1.2 to 2.2 m^{-1} over this period. *(From Bartosiewicz et al. (2019).)*

7-22 to 7-24, and for winter in Fig. 7-26. A full annual cycle for a subarctic lake is illustrated in Fig. 7-33b. Dimictic lakes are also found at high elevations in subtropical latitudes.

Our understanding of dimictic lakes has been altered with the use of high-resolution instrumentation. The term dimictic resulted from the assumption that lakes were freely circulating in autumn and spring, and assumed to be mixing, in contrast to being more stably stratified in summer and winter. In fact, some stratification is also present in autumn and spring, albeit weak, and interrupted by periods of mixing. Mixing periods are often ones in which the stratification supports large internal waves where water both upwells and downwells. With respect to temperature stratification, dimictic lakes are stably stratified when temperatures are above the temperature of maximum density, which, when solute concentrations are low, is precisely 3.98°C (ρ = 999.972 kg m^{-3}), and inversely stratified when temperatures are below this value. Thus density stratification can be stable in autumn prior to ice-on, yet temperatures can be inversely stratified. Under the ice in winter, temperatures are often inversely stratified, but as ice melts and snowmelt flows into the lakes in spring, freshening near-surface waters, near-surface temperatures can reach 8°C. Stable temperature stratification can occur in the dilute near-surface boundary layer, and inverse temperature stratification can persist below where solute concentrations are higher.

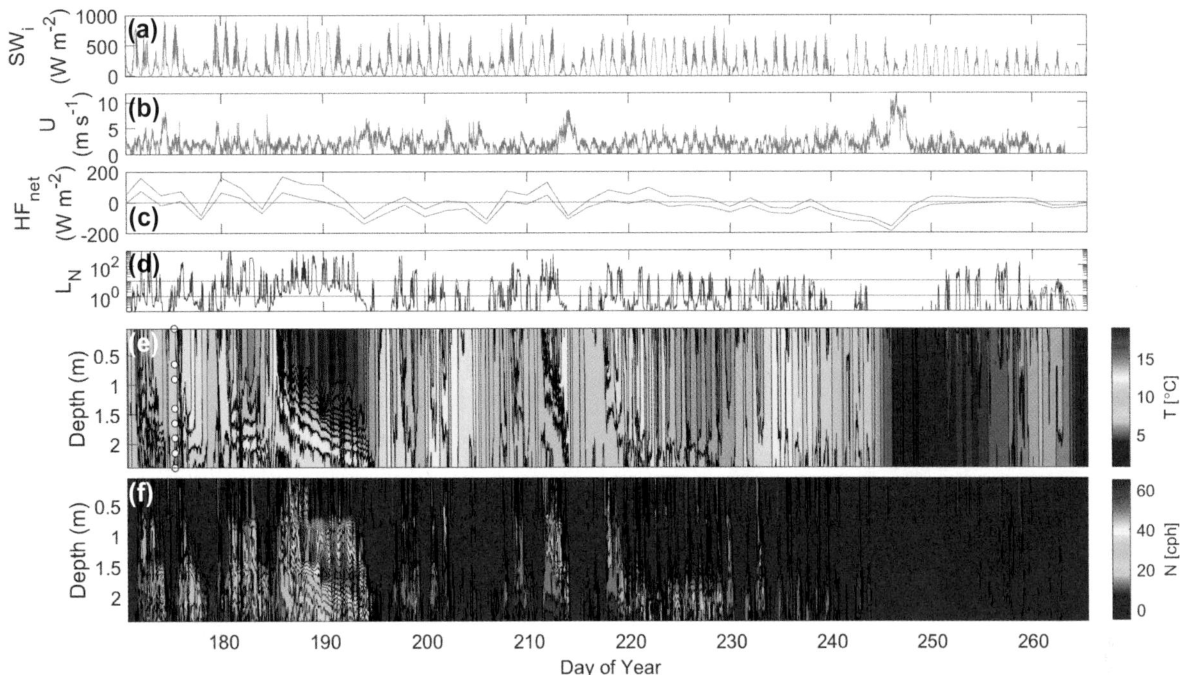

FIGURE 7-35 As described in Figs. 7-22 to 7-24, except that N is 1-h-averaged for 0.019 km^2 polymictic Lake E6 in Arctic Alaska in 2014. The diffuse attenuation coefficient was 1.2 m^{-1}. Instrumentation and sampling as for Lake N2 (Fig. 7-23). Despite being only 3 m deep, stratification persists for periods of several days such that thermoclines develop with diurnal thermoclines above (e.g., DOYs 186–194). These thermoclines can become strongly stratified, with N often being 20 cph and reaching 57 cph. For much of the summer, diel stratification alternates with nocturnal mixing such that N in the diurnal thermoclines is 5–20 cph. At night, N approaches zero in much of the water column; however, intrusions of cooler water occur when winds drop or change direction at night. The timing of these intrusions corresponds with L_N increasing, implying that the inflows may be a result of the thermocline tilting in the day and reversing direction as in Lake St. Augustin. Alternatively, they may be inflows of cooler water from shallow regions. Such gravity currents were especially likely from DOY 230–240 as the daily heat budget was negative or briefly positive and heat losses were ~ −200 W m^{-2}.

The duration of autumn cooling and the onset of ice formation in dimictic lakes varies between lakes and depends upon basin morphometry, the volume of water cooled, and prevailing meteorological conditions. For example, Lawrence Lake, with a surface area of 0.05 km^2 and a volume of 0.29×10^6 m^3, cooled to and below the temperature of maximum density by early December (Fig. 7-4). Waters of nearby Gull Lake (Michigan, USA), with an area of 8.27 km^2 and a volume of 102×10^6 m^3, usually circulate well into January before reaching comparable temperatures. Large lakes, such as the Laurentian Great Lakes, which are of similar latitudes to Gull and Lawrence lakes, circulate all winter and, depending on latitude, may only have ice cover in nearshore regions. In large, deep lakes, for example, 450 m deep Great Bear Lake of the Canadian Northwest Territories, late summer circulation often does not extend to the lower depths (e.g., below 200 m), because stability is sufficiently strong to prevent mixing (Johnson 1966). The temperature of the maximum density of the water below 200 m is 3.53°C because of hydrostatic pressure. In cold years, with slow spring heating, when the upper 200 m of water are cooled to 3.53°C in the autumn, the upper waters reach the same temperature of maximum density as the water strata below 200 m. Circulation then extends to the bottom. Changes in mixing patterns in Lake Michigan over the last 75 years are described by Anderson et al. (2021), and Cannon et al. (2021) quantify the turbulence both as the rate of dissipation of turbulent kinetic energy and the coefficient of eddy diffusivity in the upper 55 m during summer stratification, the convective period in winter and spring when both are much higher, and the transitional periods in between.

Dimictic lakes are expected to mix near ice-off as near-surface temperatures reach the temperature of maximum density and exceed those below, leading to mixing throughout the water column. While the density stratification under the ice can be weak, solute production by respiration and the flow of gravity currents can lead to near-bottom chemoclines and *pycnoclines* (regions where density increases rapidly from temperature, solutes, or both). Buoyancy frequencies in these regions can reach 20 cph, less than the highest values recorded in summer, yet still substantial. Hence, despite the warming of surface waters in the three Arctic lakes described in Section IIB, none of them circulated fully at ice-off and the subsequent development of stratification was rapid

(Figs. 7-23 and 7-24). Thus the lakes did not fully mix following ice-off, although the larger lake does in some years. This pattern differed from that in Lawrence Lake in 1968 (Fig. 7-4), although observations of incomplete mixing are reported for other lakes nearby at that time. Incomplete mixing at ice-off has been noted in thermokarst thaw lakes where specific conductance increases near the bottom over the winter from cryoconcentration and respiration (Deshpande et al. 2015), in thermokarst thaw lakes in the Mackenzie Delta, Northwest Territories, Canada (McIntosh Marcek et al. 2021), whereas more complete mixing is expected in lakes with optically clear ice and minimal stratification from solutes produced over the winter (Bouffard et al. 2019). Human activities, that is, the introduction of salts due to deicing of roads in winter, are leading to a greater incidence of incomplete mixing in dimictic lakes in spring (Ladwig et al. 2021).

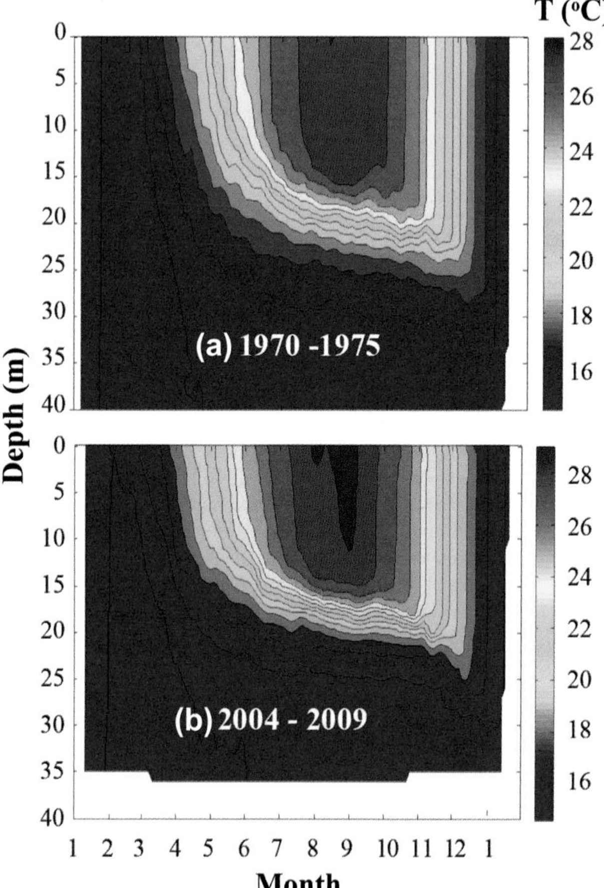

FIGURE 7-36 Thermal structure of warm monomictic Lake Kinneret, Israel, based on averages of profile data (a) from 1970 to 1975 and (b) from 2004 to 2009 showing the earlier and greater warming and the intensified thermocline in the later period (Ostrovsky et al. 2013). The changes are largely a result of increased pumping of water and changes in lake level, with a smaller contribution from increased air temperature in spring and summer.

v. Warm monomictic lakes

In *warm monomictic* lakes temperatures do not drop below 4°C. These lakes intermittently circulate in the winter or during a cooling season, and they stratify stably in the summer or in a warming season. Warm monomictic lakes are common in warm regions of the temperate zones of the southern and northern hemispheres, particularly those influenced by oceanic climates, in mountainous areas of subtropical latitudes, and in the lowland tropics, or in northern temperate regions where deep lakes accumulate considerable heat over the summer and despite cooling in winter, their surface temperatures do not decrease sufficiently to form ice. Warm monomixis is prevalent in many coastal regions of North America and northern Europe. Examples include Lake Tahoe and, in some years, Mono Lake in California (USA), Lake Constance on the northern edge of the European Alps (Straile et al. 2003), Lake Biwa (Japan), and Lake Kinneret, Israel (Fig. 7-36). They are found in tropical regions with mixing during a dry windy season or during a season with clear nights (e.g., Lake Victoria, East Africa (Fig. 7-37) (Talling 1966; MacIntyre et al. 2014).

vi. Warm polymictic lakes

Lakes with frequent mixing to the near bottom and in which ice cover does not form are called *warm polymictic*. Diel stratification and mixing is prevalent in shallow tropical lakes (Talling 1966; Melack and Kilham 1974; Augusto-Silva et al. 2019). As observed for cold polymictic lakes, high-resolution instrumentation indicates that

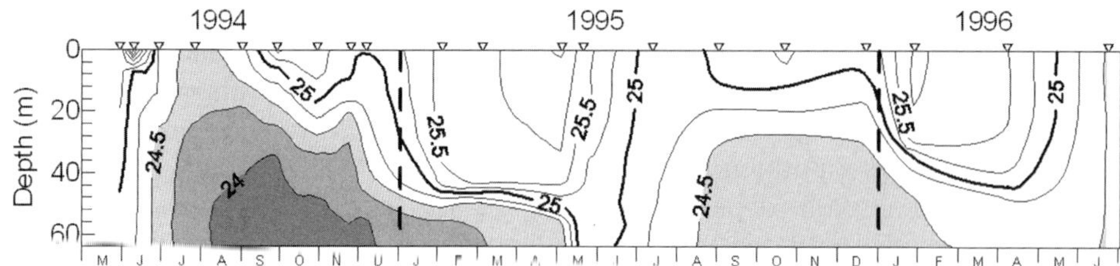

FIGURE 7-37 Thermal structure of warm monomictic Lake Victoria, East Africa. Full water column mixing occurs during the southeast monsoon when winds are higher over East Africa. Winds increase beginning in May and, depending on the phase of the El Niño–Southern Oscillation, may remain elevated into August. *(From MacIntyre et al. (2014).)*

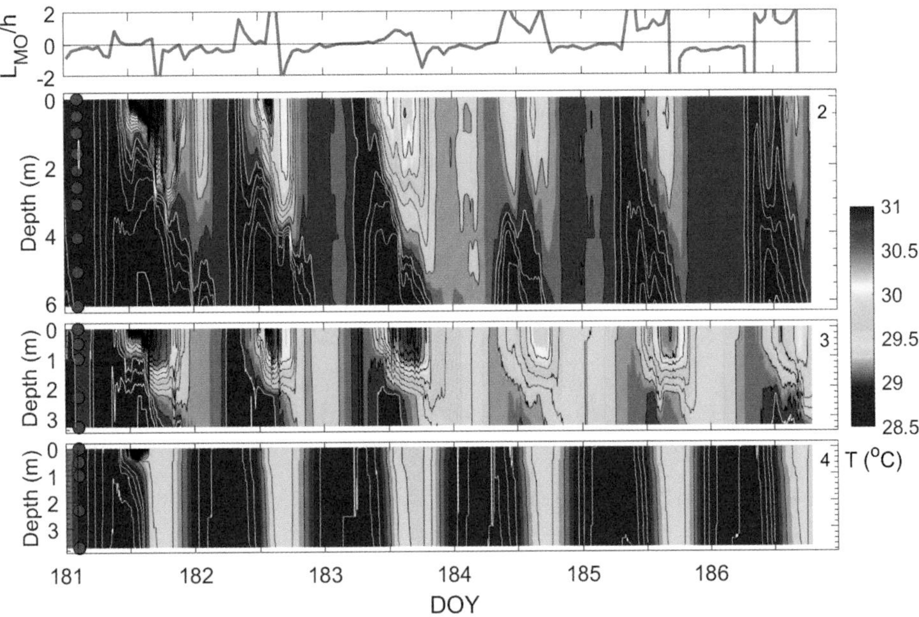

FIGURE 7-38 Spatial variability of thermal structure at three stations within warm polymictic Lake Paranaptinga (Brazil). (Stations identified with numbers to the right) (2) A midlake station, (3) a more inshore station, (4) and a more northerly station closer to the location where water from the Amazon River flows into the lake. This lake is part of the extensive Lake Grande do Curai at high water. Sites are several kilometers apart. The upper panel illustrates the Monin–Obukhov length scale divided by the depth of the actively mixing layer (L_{MO}/h) for station 2. During heating and light winds (positive $L_{MO}/h \rightarrow 0$), strong diurnal thermoclines form in the day as on DOY 181 with buoyancy frequency reaching 60 cph. In contrast, when winds are high during heating ($L_{MO}/h > 1$), incoming heat is mixed deeper in the water column (DOYs 184–186). In both cases the stratification that builds up in the day is eroded by cooling. At the onset of cooling, conditions were windy ($L_{MO}/h < -1$) and the diurnal thermocline deepened as it cooled. For much of the night, $-1 < L_{MO}/h < 0$, which indicates wind mixing near the surface and mixing by convection throughout the actively mixing layer as in Balbina Reservoir (Fig. 7-29). The upwelling of cold water in the morning at stations 2 and 3 indicated the thermocline had tilted at night, as in Lake Janauacá, and the lake had not fully mixed despite the near isothermy at the measurement site at night. The persistent near isothermy at Station 4 resulted from the greater mixing associated with the incoming river water. *(Revised from Augusto-Silva et al. (2019).)*

the near isothermy at a measurement site over diel cycles may be due to wind-induced thermocline tilting with stratification evident on the relaxation of the wind (Fig. 7-38). The use of multiple thermistor arrays aligned in the prevailing wind direction and three-dimensional hydrodynamic modeling have confirmed this perspective (Augusto-Silva et al. 2019; Fig. 7-25). More persistent stratification is found in deeper Clear Lake, California (USA) (Cortés et al. 2021), and the mixing regimes of tropical floodplain lakes vary as a function of water depth (MacIntyre and Melack 1986). Polymixis can also be induced in shallow lakes that stratify during warm periods but in which intermittent high discharge events replace a considerable portion of the lake water (Fig. 7-39) and by geothermal heating in shallow lakes such as Lake Rotowhero (New Zealand) (Fig. 7-40) and in caldera lakes such as Ruapehu Crater Lake (New Zealand).

D. Meromixis

Lakes that do not undergo complete circulation are termed *meromictic* (Findenegg 1935; Hutchinson 1937; Boehrer and Schultze 2008). In meromictic lakes the upper layer which periodically circulates is called the *mixolimnion*. The deeper *monimolimnion* does not mix with the overlying water. The two layers may be separated by a thermocline or by a *chemocline*, where solute concentrations increase rapidly. The density gradients, be they due to temperature alone or the combination of temperature and solutes, are called *pycnoclines*. Meromixis develops as a result of conditions leading to chemical stratification or when the depth of mixing possible from wind and cooling is shallower than the depth of the lake. Nomenclature describing meromixis was developed by Hutchinson (1937) and Walker and Likens (1975), as noted in Dickman and Hartman (1979), and Zadereev et al. (2017).

Ectogenic meromixis results when an external event brings salt water into a freshwater lake or freshwater into a saline lake. A superficial layer of less dense, less saline water overlies a monimolimnion of saline, denser water. Such situations are mainly found along marine coastal regions where intrusions of salt water occur. A stricter definition of ectogenic meromictic lakes includes only those that are isolated from marine influxes of water. Such isolation may be recent or long-standing. An example of the latter is the southern Norwegian

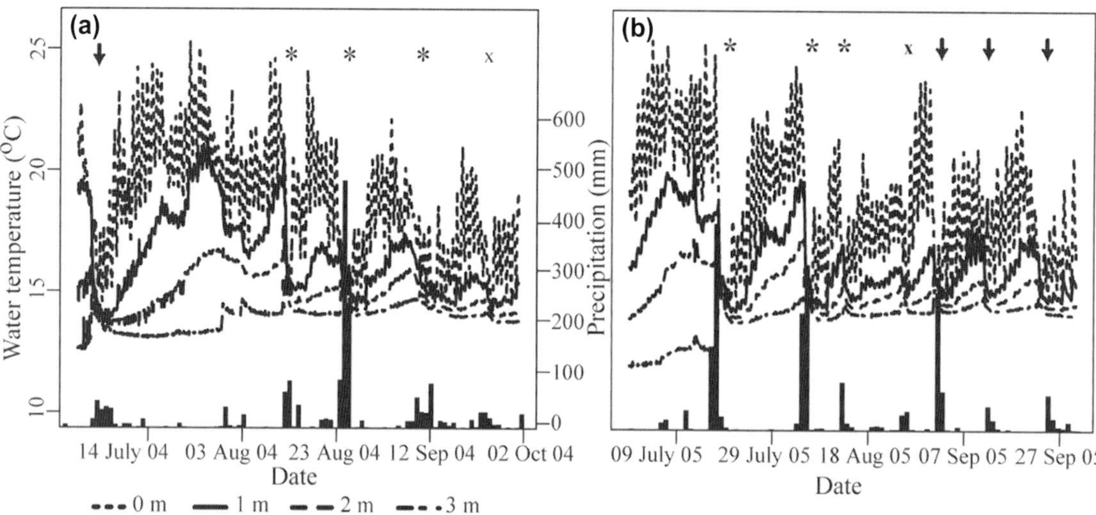

FIGURE 7-39 Water temperature at depths of 0, 1, 2, and 3 m in Yuan Yang Lake, Taiwan, for periods between 2004 and 2005. Temperatures cool appreciably and stratification is usually diminished or lost when large inflow events occur and displace lake water. These result from increased precipitation as indicated by *vertical black bars* (Jones et al. 2008). *Downward arrows* indicate typhoons that caused mixing and the *x* indicates a typhoon that did not cause mixing. Data in panel (a) are from 2004 and in panel (b) from 2005.

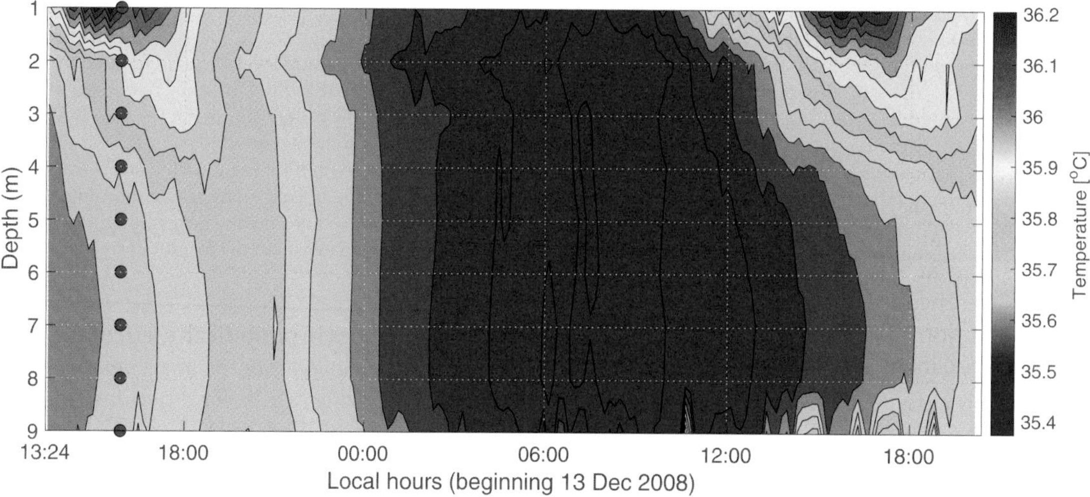

FIGURE 7-40 Thirty-hour record of water column temperature in geothermal Lake Rotowhero, North Island, New Zealand. The water column mixes at night due to the combination of heat loss at the surface, differential cooling, and convective mixing induced by the upward rise of warmer, less dense water (note warmer water at 9 m depth). It stratifies during the day from incoming solar radiation. *(Data replotted from Brookes et al. (2013).)*

Lake Tokke, which most likely has been permanently meromictic for about 6000 years; the isolation of the lake probably took place during a period when the former fjord depression and the surrounding land were elevated some 60 m above sea level (Strøm 1955). The paleohistory of meromictic, ice-covered Ace Lake, Antarctica, is similar (Fig. 7-33c). Lake Sonachi, a volcanic crater lake in Kenya, has periodic changes in specific conductance due to alternating multiyear periods of drought followed by periods with increased rainfall. Ectogenic meromixis is maintained by the freshening of the upper waters by rainfall and increased specific conductance of the deeper waters, presumably by the increased flow of groundwater with dissolved evaporates (MacIntyre and Melack 1982).

An interesting variant of ectogenic meromixis is found in saline lakes of arid regions (Chapter 12) or in small depressions that occasionally receive saltwater intrusions (Hudec and Sonnenfeld 1974). Often, these saline lakes overlie deposits of soluble salts, such as magnesium sulfate, and are only a few meters in depth. The input of freshwater, either naturally during wet periods or artificially as a result of increased irrigation of nearby land, creates meromictic stratification.

Whatever the causes of meromixis, some lakes have marked differences in temperature with depth, in which the chemocline is functioning as a heat trap of solar energy. Such *heliothermic lakes* contain a solar-heated stratum of warm, saline water beneath a surface layer of cooler, less saline water (reviewed by Cohen et al. 1977; Kirkland et al. 1983, and see Fig. 7-33c). Much of the sunlight that penetrates the chemocline is transformed to heat and retained due to increases in solute concentrations at the depths of heating. As a result, temperatures rise within the chemocline. The corresponding changes in the distribution of oxygen and nutrients, as in Ace Lake, Antarctica (Burch 1988; Rankin et al. 1999; Laybourn-Parry and Bell 2014), create habitats favorable for different species of microbes.

A striking example of heliothermy is Hot Lake, a shallow (3.5 m) saline body of water occupying a former Epsom salt excavation in north central Washington, USA (Anderson 1958; Kirkland et al. 1983). Solar radiation energy passing through the overlying mixolimnion of freshwater accumulated as heat in the *chemolimnion* and monimolimnion, where circulation was absent. The resulting temperatures can be in excess of 50°C, and even when the surface of the lake is frozen, a temperature of nearly 30°C can be found at a depth of 2 m. These *dichothermic* conditions, in which the monimolimnion is considerably warmer than the overlying water of the chemolimnion, are observed in shallow meromictic lakes.

Crenogenic meromixis is a type of ectogenic meromixis that results when submerged saline springs deliver dense water to deep portions of lake basins. The saline water displaces the water of the mixolimnion. The chemolimnion stabilizes at a depth related to the rate of influx, density differences, and the degree of wind mixing of the mixolimnion. Meromictic lakes of this type occur in, for example, interior Alaska, where subsurface springs introduce saline water into small, deep lakes originating in thawed craters of *pingos* (Likens and Johnson 1966). Pingos are formed when water, rising by hydraulic pressure through gaps in the permafrost, freezes and uplifts a mound of ice covered by a layer of alluvium. The overburden usually ruptures and the ice of the depression can thaw to form small lakes. In some cases geothermal and saline waters form complex stratification patterns, with warm, saline waters underlying cooler waters (see Drago 1989).

Biogenic or *endogenic meromixis* results from an accumulation of solutes in the monimolimnion, often liberated by the decomposition of organic matter in the sediments or water column. Often biogenic meromixis is initiated when abnormal meteorological conditions prevent the circulation of a normally dimictic or monomictic lake. *Morphological meromixis* results when mixing from wind and cooling cannot penetrate the full depth of a lake. Over time, solutes may accumulate in the monimolimnion, but the resulting chemoclines would not necessarily be sufficient to preclude mixing in a shallower lake. Temperature inversions can occur in the monimolimnion. Elevated temperatures observed in the chemolimnion may result from the absorption of solar radiation by dense bacterial plates that occur at that level, residual warmer water from the last period of mixing, or geothermal heating.

Nearly all deep lakes in the equatorial tropics are meromictic. Some of these are small crater lakes in which sheltering from wind precludes deep mixing (Melack 1978; Kling 1988). Notorious among these are Lakes Nyos and Monoun in Cameroon (Kling 1987). With the long period of meromixis and the interaction of ground water with magmatically produced CO_2, concentrations of CO_2 became high in the monimolimnion. At Lake Nyos a landslide disrupted the stratification and led to the venting of the dissolved gases, and, as CO_2 has a higher density than ambient air, the formation of a plume of CO_2-enriched air that flowed downslope and suffocated villagers in the towns below the lake resulted in loss of 1746 lives on 21 August 1986. The hazard has since been reduced somewhat with siphon systems that vent the CO_2 to the atmosphere (Kling et al. 2005).

Tectonically formed Lakes Malawi, Tanganyika, and Kivu in East Africa are meromictic. With depths exceeding 500 m, and reaching 1460 m in Lake Tanganyika, the enhanced mixing during the southeast monsoon causes deepening to only ~150 m in Malawi and Tanganyika (Fig. 7-41) and, with lower winds, to 50 or 60 m in Lake Kivu (Fig. 7-42). Sediment cores from these lakes indicate that the lakes have undergone periods of mono- and meromixis over geological time, with monomixis prevalent during dry, windy periods when the lakes were shallower than today and meromixis during wetter periods as now (Haberyan and Hecky 1987; Scholz et al. 2011).

Stratification in Lake Kivu (the Democratic Republic of the Congo/Rwanda) results from the interaction of freshwater inputs in the upper layer combined with incoming solutes from geothermal springs and decomposition in the lower water column (Pasche et al. 2009). The first chemocline is set by the depth of mixing during the southeast monsoon. Above it, the water column is oxygenated during the monsoon. The other chemoclines occur at depths where subterranean springs introduce water, some of which is CO_2 saturated and solute rich. Thermoclines often occur with the chemoclines, as do large gradients in CO_2 and CH_4. The methane from the lake is currently being extracted and used for the production of electricity for Rwanda and neighboring countries. The extraction is designed to preclude the partial pressure of the gases from reaching saturation, which should avoid a catastrophic release as in the two lakes in Cameroon. As natural processes can induce such events, the lake is being carefully monitored.

Another, quite different cause of endogenic meromixis apparently occurs as a result of the precipitation of salts in the upper water strata by "freezing out"

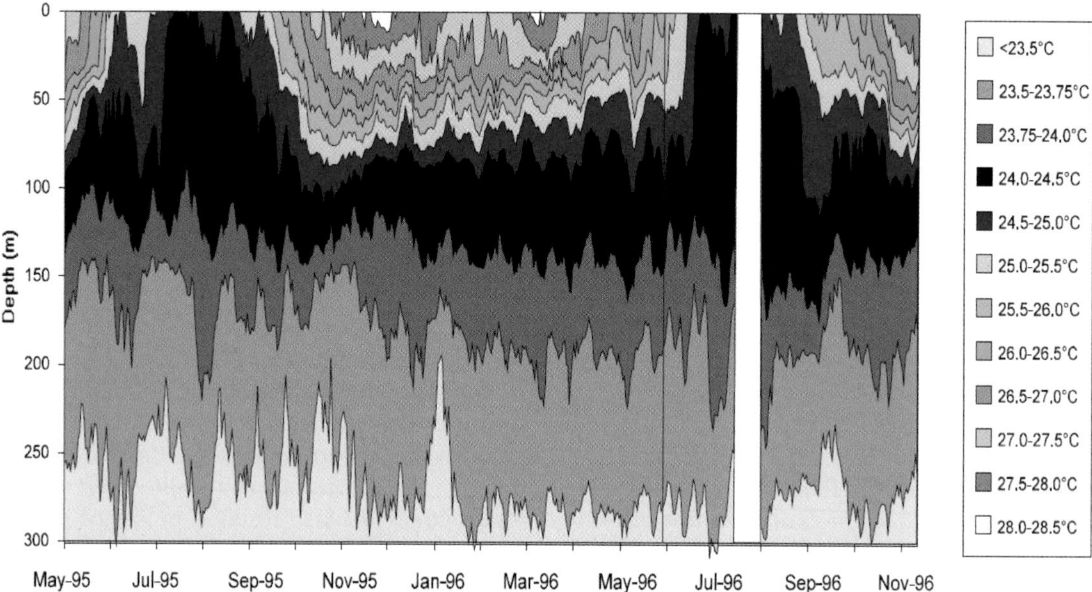

FIGURE 7-41 Time series temperatures from the southern basin in Lake Tanganyika to 300 m. Rather than mixing to the bottom, the upwelling and mixing are constrained to the surface ~100 or 175 m during the dry, windy conditions of the southeast monsoon (May—August). The lake subsequently restratifies in the upper 100 m. Large amplitude internal waves are evident below 100 m. *(From Huttula (1997).)*

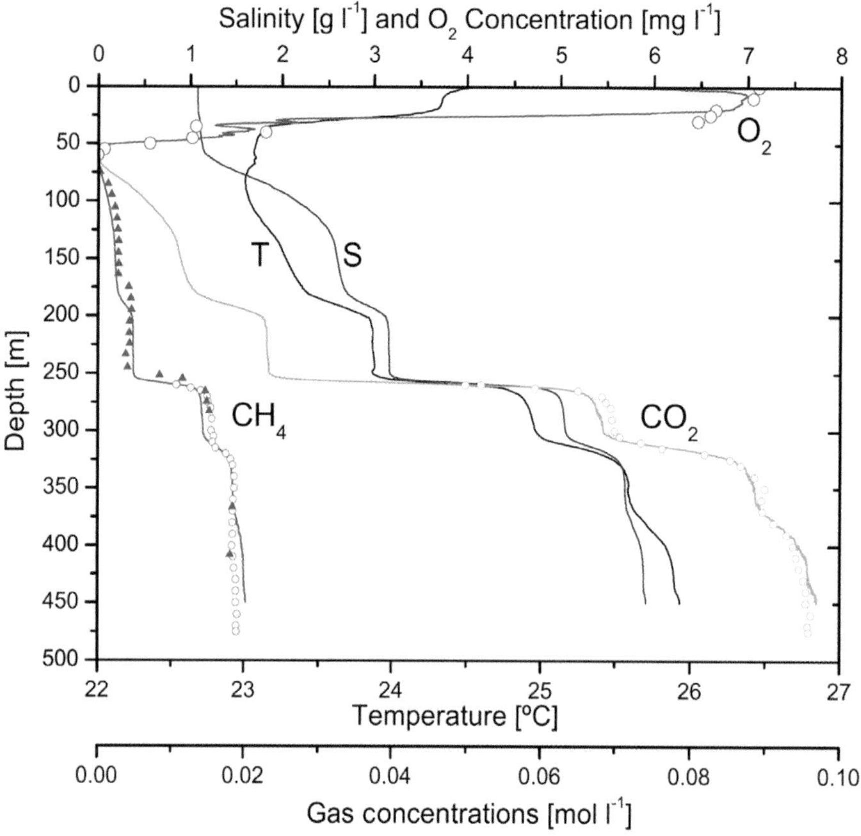

FIGURE 7-42 Stratification of temperature (T), oxygen (O_2), salinity (S), CH_4, and CO_2 in Lake Kivu. *(From Schmid et al. (2005).)*

from the surface ice layer. The precipitating salts may accumulate in deeper waters in sufficient concentrations to cause meromixis. In Algal Lake, Antarctica, this condition has been termed *cryogenic meromixis* (Goldman et al. 1972; cf. also Priddle and Heywood 1980).

Human activities can lead to meromixis. For example, creating a connecting channel between a freshwater lake and the sea for purposes of navigation can result in an intrusion of saline water into the lake and meromixis. The runoff from street deicing has increased salt concentrations (Chapter 12) in the deeper water of small lakes resulting in meromixis (Judd 1970; Sibert et al. 2015; Chapter 12). The increased solutes have led to delays in spring mixing in larger lakes with the potential to lead to meromixis (Ladwig et al. 2021). Mining voids, that is, pits where mining occurred, are common in several areas of the world (Boehrer et al. 2014). These initially fill with saline groundwater but gradually become overlain by a freshwater layer, either from runoff or artificial diversions, such that meromixis results.

E. Oligomixis

Oligomixis occurs among lakes of small surface area and moderate depth that are sheltered, especially in continental regions that experience long winters. In such locations a dimictic lake may skip a circulation period (usually in spring), and increased solutes from decomposition in the lower strata of the lake can persist under unusual weather conditions (Findenegg 1937). For example, in lakes in which solutes accumulate near the bottom during winter, and near-bottom temperatures are slightly above or below 4°C, mixing in spring may be incomplete, and stratification can persist for much of the summer (Einsele 1941; Wetzel 1972, 1973). Persistent stratification at ice-off and in the week following is illustrated for three Arctic lakes in Fig. 7-31. The rapid downward mixing of heat at ice-off that precluded full mixing in the smaller of these lakes is described in Cortés and MacIntyre (2020). Near-bottom chemical stratification persists in the smaller lakes for much of the summer, whereas when the fetch is longer, as in Toolik Lake, near-bottom pycnoclines can be eroded more rapidly.

Human and natural causes led to oligomixis in endorheic Mono Lake, California (Melack and Jellison 1998; Melack et al. 2017). Meromixis results from increased inputs of freshwater in years with large snowpack in the adjacent Sierra Nevada and from reduced diversions of incoming stream water (Fig. 7-43). The freshening of the upper water column creates chemical stratification. Precipitation patterns in California vary over the years, sometimes with a snowpack up to 300% of average followed by several years with below-average precipitation. Evaporation during the summer is high. The resulting increases in solutes in the mixolimnion during drought years reduce the density gradient between the mixolimnion and monimolimion such that monomixis is again established. Stratification in Mono Lake has features in common with Lake Kivu in that there is an upper oxygenated layer with chemoclines below. These gradients and accompanying variations in solutes create habitats favorable for a wide range of microbes that occupy discrete layers according to redox conditions and either oxidize or reduce the substrates.

III. Changes in seasonal and annual stratification with climate change

Shifts in stratification regimes have been observed over long time periods (Haberyan and Hecky 1987; Scholz et al. 2011). However, current concerns are changes in temperature and mixing regimes (Hampton et al. 2008; O'Reilly et al. 2015; Sahoo et al. 2016; Yankova et al. 2017; Pilla et al. 2020; Mesman et al. 2021; Råman Vinnå et al. 2021) and modifications of the thermal habitat of organisms in lakes as air temperatures warm with recent climate change (Kraemer et al. 2021). Changes within lakes may also vary over time from cyclic changes in meteorological forcing due to atmospheric cycles such as the El Niño–Southern Oscillation (ENSO), North Atlantic Oscillation (NAO), Pacific Decadal Oscillation (PDO), and the Arctic Oscillation (AO) (Dokulil et al. 2006; Wolff et al. 2011). Thus assessing change due to climate warming requires data collected over time periods that exceed the higher frequency changes from these cycles. Changes in heat content may be independent of changes in lake surface water temperature (LSWT), indicating that changes in LSWT may not always be the correct metric to assess responses to climate change (Ye et al. 2019). In addition to increasing temperature, some regions are experiencing increased precipitation, others less. The timing of precipitation is also changing, with in some cases less as snow and more as rain. Changes in precipitation and changes in land use can together modify the loading of solutes and particulates, which influence the optical properties of water. Increases in attenuation coefficients lead to more heat being absorbed in the upper water column (Fig. 7-6) and an intensification of stratification (Houser 2006; Pilla et al. 2020), whereas decreases lead to deeper mixed layers (Schindler et al. 1996). In other cases, decreases in water level combined with slight increases in air temperature in some seasons lead to near-surface warming (Rimmer et al. 2011). The integrated effects of these and other changing variables on the future states of lakes can be predicted using hydrodynamic models, a topic discussed in Chapter 8 IV. The following is a brief summary of changes or expected

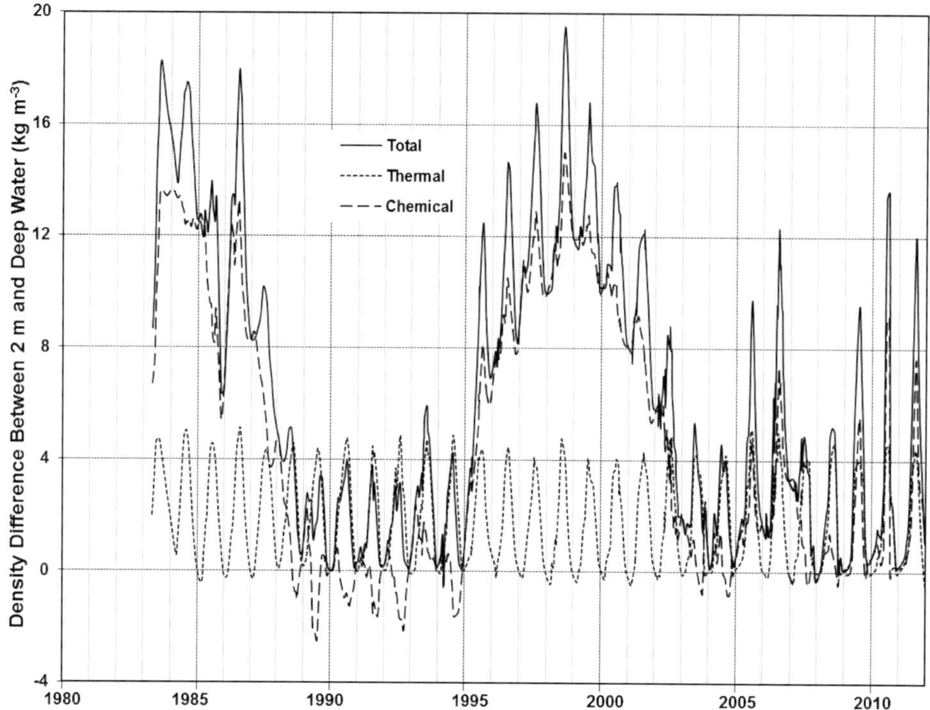

FIGURE 7-43 Density differences between 2 and 28 m depths in Mono Lake due to thermal stratification from seasonal variations in temperature, chemical stratification due to freshening of the upper water column, and total stratification, that is, the sum of the two contributions. Between year variations in thermal stratification are minor. Persistent stratification was induced in 1983 due to increased chemical stratification due to runoff from a snowpack in the Sierra Nevada that was 300% above average. Seasonal mixing was restored in 1988 as increased solutes from evaporation during summer reduced the chemical stratification during years with below average snowpack. Persistent stratification was induced again in 1995 with reductions in water diversions to the City of Los Angeles (California, USA), with chemical stratification further accentuated by increased runoff from elevated snowpack in 1997–1998 which persisted until 2004. *(Figure from Melack et al. (2017).)*

changes in mixing dynamics on a regional basis. For a comprehensive assessment of climate change effects on diverse lakes and rivers, see Goldman et al. (2013).

In the High Arctic, partial loss of ice cover in previously amictic lakes has led to cold monomixis in some years with consequences for the stability of near-bottom sediments with their extensive cyanobacterial mats (Bégin et al. 2021). In High Arctic Lake Hazen (Ellesmere Island, Nunavut, Canada), the mean daily ice-free area from early May to early September ranged from 45% to 65% from 2000 to 2012, with an overall increase in open water, which led to concomitant increases in pelagic diatoms (Lehnherr et al. 2018; St. Pierre et al. 2019). For smaller Arctic water bodies, increased temperature led to increased evaporation and the loss of ponds (Smol and Douglas 2007). Many small lakes in the subarctic are currently dimictic and, based on their Lake numbers, are in a regime in which wind-driven mixing will persist (Figs. 7-23 and 7-24) (MacIntyre et al. 2009a). However, in years in which synoptic activity is reduced the enhanced winds associated with frontal systems do not occur. Hence Lake numbers remain above 10 through much of the ice-free period such that heat is retained in the upper mixed layer and the metalimnion thickens. Thus shifts in meteorological patterns can induce shifts in the mixing dynamics in subarctic lakes such that they are more like those in similarly sized temperate lakes. Water temperatures under the ice depend on summer air temperatures and heat losses in fall as they modify the heating of the sediments, as well as direct solar heating. With increased heat in the sediments, near-bottom stratification with anoxia may occur over a larger volume with concomitant effects on concentrations of climate forcing trace gases and loss of habitat for fish. These changes can occur in dimictic lakes.

In many monomictic and deep lakes, temperatures have warmed over time and the period of stratification has lengthened (Livingstone 2003; Anderson et al. 2021). In some cases the changes may be due to anthropomorphic influences rather than from changing air temperatures (Fig. 7-36) (Ostrovsky et al. 2013). The combination of stronger and more persistent stratification in temperate lakes has led to greater oxygen drawdown in the lower water column with diverse ecosystem implications (Foley et al. 2012; Sakamoto 2013; Tsugeki and Urabe 2013; Adrian et al. 2009; Jane et al. 2021). In northern lakes with increased heat content in summer, a longer period is required in autumn to cool lakes. Such delays, combined with warmer autumn and winter air temperatures, could then lead to warmer minimum lake temperatures and delays in the onset of inverse temperature stratification and ice-on (Woolway et al. 2019, 2021). Combined with overall warming such that ice-off is earlier, the duration of ice cover has decreased and will decrease with projected climate warming (Magnuson et al. 2000; Sharma et al. 2019, 2020, 2021). In many lakes ice may no longer form and dimictic lakes will become monomictic. For deeper lakes in which ice does not form, for example, Lake Tahoe (California/Nevada, USA), Lake Taupō (New Zealand), and Lake Biwa (Japan), the concern is that they will not fully mix with accompanying loss of oxygen in the lower water column (Straile et al. 2003; Coats et al. 2006; Adrian et al. 2009; Hamilton et al. 2013; Sakamoto 2013; Sahoo et al. 2016). A number of deep lakes in central Europe and Japan either are shifting or have shifted from holomictic to oligomictic (Kitazawa 2013; Yankova et al. 2017). For example, complete mixing in Lake Geneva (Switzerland, France) is expected to shift from once every 4.5 years to once every 8 years (Schwefel et al., 2016). Anoxia is building up in the lower water column in winter, and with previous attempts to reduce eutrophication in these lakes, shifts in nutrient stoichiometry have occurred with the combined effect of a loss of centric diatoms in the spring bloom and increased concentrations of cyanobacteria (*Planktothrix rubescens*), particularly in autumn, with consequences for higher trophic levels. In high-altitude tropical lakes deeper than 17 m stratification has shifted from polymictic to stratified for periods of several months over the last few decades (Giles et al. 2018; Labaj et al. 2018).

Results from deep lakes illustrate the importance of evaluating heat content to identify change and the variability associated with ENSO cycles. The changing meteorological conditions over Lake Michigan over the past 75 years have led to increased heat being mixed downward, a shortened period of overturning, warmer temperatures at the onset of stratification in spring, and a longer period of stratification with warmer temperatures (Anderson et al. 2021). In a global study with 26 lakes of differing sizes, increases in heat content and Schmidt stability were larger in deeper lakes such as the East African Great Lakes, even though changes in surface temperature of such lakes were more muted (Kraemer et al. 2015). Metrics such as the Wedderburn (W, Eq. 7-16) and Lake number (L_N, Eq. 7-17) may contribute toward determining if the ecological functioning of such lakes has changed. For Lake Tanganyika, calculated values of W during the southeast monsoon are <1. This result is obtained by calculating density using temperatures near the surface and at 150 m and an upper mixed layer depth of 50 m just before the monsoon (Fig. 7-41) and mean wind speeds of 5 m s^{-1} at the start of and during the monsoon (and see Coulter and Spigel 1991). The result implies that considerable tilting of the thermocline can occur. Heat can be mixed downward and recycled nutrients can be mixed upward to support productivity. Of additional importance for assessing change, the depth and/or duration of mixing in tropical East African lakes depend on the duration of winds in the southeast monsoon, which varies with ENSO cycles (Wolff et al. 2011; MacIntyre et al. 2014; Tuyisenge 2018). For example, the depth of mixing during the southeast monsoon in Lake Tanganyika was 200 m in some years and 100 m in others (Fig. 7-41) (Huttula 1997). Such variability adds further challenges to quantifying long-term changes to mixing dynamics. Continued efforts are required to evaluate changes in the mixing dynamics of these large, deep lakes as well as lakes of other sizes in order to evaluate the influence of climate on their ecological functioning. The efforts in the studies referenced above point to the value of, and need for, long-term monitoring data to assess the effects of climate and land-use change on the mixing dynamics of lakes and resulting ecosystem changes (Plisnier et al. 2018, 2022).

IV. Summary

1. The absorption of incoming solar radiation leads to the heating of inland waters. In lakes heat is distributed and altered by a number of coupled physical processes. They include further heating, cooling, near-surface currents, wind-driven motion of the thermocline, density-driven flows, and incoming and outgoing flows, with many of these processes modified by basin morphometry. The mechanisms are more fully described in Chapter 8. Resulting patterns of density-induced stratification influence the physical and chemical properties of lakes and, in turn, govern the organisms that live within them and their productivity and other metabolic processes. Similar processes influence the temperature of rivers and streams, but advection has a larger effect.
2. Temperature depends on heat stored in a volume of water. Heat is measured in Joules and is the product of the mass of the substance (in kilograms), temperature (°C), and the specific heat (c_{pw}, J kg^{-1} °C^{-1}). Changes in temperature can be computed from $\Delta T = \frac{Q}{\rho c_{pw} l} \Delta t$, where Q is heat flux (J m^{-2} s^{-1} or W m^{-2}), ρ is density (kg m^{-3}), l is a vertical dimension, and Δt is change in time. To compute changes in temperature or heat content over a whole lake, the volume of each layer l must be summed.
3. Temperatures of streams, rivers, and lakes change over diel, seasonal, and annual cycles. Annual

variability in surface and mean temperatures tends to increase with greater latitude and altitude. Temperatures of streams usually increase downstream.

4. Heat inputs result from several processes:
   a. Direct absorption of solar radiation; usually the dominant source
   b. Incoming longwave radiation. As outgoing longwave is generally larger, their sum, net longwave radiation, is negative.
   c. Transfer of heat from the air (conduction or sensible heat exchange)
   d. Condensation of water vapor at the water surface (a component of latent heat flux)
   e. Transfer of heat from sediments to the water
   f. Heat transfer from terrestrial sources via precipitation, surface runoff, and groundwater and occasionally geothermal inputs. While inflows always add heat, they may or may not increase temperature.
5. Heat losses occur by:
   a. Conduction (sensible heat flux) of heat to the air and, to a lesser extent, to the sediments
   b. Evaporation (a component of latent heat flux; usually the largest of the heat loss terms)
   c. Outgoing longwave radiation
   d. Outflows. While outflows are a heat loss term, whether or not they lead to increases or decreases in temperature depends on whether the outflow is at the surface, as in lakes, or deeper, as in reservoirs, or when water enters or leaves as groundwater.
6. Surface energy budgets and heat budgets are computed to determine the contributions of ambient meteorology and advective flows to the heating and cooling of lakes and other water bodies. The net heat flux computed from a surface energy budget is the sum of incoming and outgoing shortwave and longwave radiation and latent and sensible heat fluxes. The unit is in watts per square meter ($W\ m^{-2}$). The net heat flux within the actively mixing layer takes into account the differing absorption of shortwave radiation at different wavelengths within the actively mixing layer. The inclusion of advective terms must be done with similar units. The magnitudes of each of these variables fall into typical ranges, which enable comparisons of the efficacy of each in determining heating and cooling in lakes and flowing waters over time and between latitudes.
7. Heat budgets within water bodies are computed from temperature profiles or time series temperature data with sensors at multiple depths. These calculations provide heat content at any time or change in heat content from one time to the next. When such budgeting is done, and compared with the surface energy budget, the effect of advective flows on the overall heat budget can be ascertained. Incoming heat per unit time from a stream can be calculated from discharge, water temperature, and specific heat capacity of water. Storms can potentially induce changes in the heat budget. Birgean heat budgets, or annual heat budgets, are the total amount of heat accumulated between the period of a lake's lowest and highest heat content. The annual income and loss of heat in tropical lakes is small in comparison to that of lakes at higher latitudes and altitudes. Annual heat budgets of lakes of the temperate zone generally increase with lake volume.
8. Mixing redistributes heat in a lake as surface waters cool, causing convection, and as wind induces currents and shear. The magnitude of these processes varies seasonally and by latitude, particularly as latent heat fluxes are larger for the same wind speeds in warmer water bodies. The efficacy of these processes depends on stratification across the metalimnion, and this difference depends on the degree of heating in summer and cooling in winter. Near-surface temperatures vary by latitude depending on the magnitude of solar radiation, length of the warming period, mechanisms that redistribute heat, and the optical properties of the water. The efficacy of wind in redistributing heat across diurnal and seasonal thermoclines can be computed using the Wedderburn and Lake numbers and depends on wind shear, stratification across the metalimnion, and basin morphometry. The efficacy of cooling depends on heat loss relative to stratification across the metalimnion.
9. Seasonal thermal stratification occurs in most lakes of moderate depth ($> \sim 10$ m).
   a. During warmer periods of the year, the surface waters are heated, largely by solar radiation, more rapidly than the heat is distributed by mixing. As the surface waters are warmed and become less dense, stratification between the upper and lower layers increases. Stratification, as it varies with depth, is quantified by the buoyancy frequency, with the highest values occurring at diurnal and seasonal thermoclines.
   b. Lakes become stratified into multiple zones:
      i. Epilimnion: An upper stratum of less dense water. This region is also called the upper mixed layer. It may or may not be mixing at the time of measurement, and even with weak temperature stratification, stratification of solutes and particulates may be appreciable.
      ii. Hypolimnion: The lower stratum of denser, cooler water in the lower water column.
      iii. Metalimnion: The transitional region with the largest changes in temperature with depth. Interchangeably called the thermocline.

c. The seasonal thermocline can be defined as the plane or surface of the maximum rate of decrease of temperature in the metalimnion and can be identified as the depths where the buoyancy frequency is highest within the metalimnion. The term can also be used interchangeably with metalimnion.

d. The resistance of a lake's density stratification to mixing depends on stability, which varies with the distribution of heat and solutes in a lake. Heat budgets enable stability, the likelihood of mixing, and the depths where mixing is expected to occur to be quantified. A number of metrics are used.

i. **a.** For a lake as a whole, Schmidt stability (S_t) is the quantity of work or mechanical energy per unit area required to mix the entire volume of water to a uniform temperature by the wind without the addition or loss of heat.

b. The buoyancy frequency (N, radians per second or cycles per hour, cph) quantifies the extent of stratification as a function of depth. Typically, it will be least in the upper mixed layer, with values close to zero at night, and also low in the bottom boundary layer. However, over diel cycles of stratification and mixing, values of N are highly variable in the upper mixed layer. N in diurnal thermoclines often exceeds 20 cph and values can exceed 80 cph. N in the subsurface layer often exceeds 10 cph. Thus the upper mixed layer is typically stratified in the day unless there have been several cloudy, windy days such that it mixes. As cooling begins and the diurnal thermocline is eroded, values of N above the diurnal thermocline decrease, sometimes dropping below 1. As a result of considerable mixing during cold fronts, N may also approach or drop below 0 in the upper mixed layer. At such times, the upper mixed layer has mixed or will mix. The depth of the epilimnion, or upper mixed layer, thus depends on the depth of mixing at night unless stormy conditions have prevailed. If so, its depth will depend upon the extent of deepening during the storm. Comparisons of N in the diurnal and seasonal thermoclines in lakes of different latitudes provide a means to determine the strength of stratification. For example, maximal values of N in the thermocline of small temperate lakes may be 50 cph and persist for over a month, whereas in Arctic lakes maximal values may be 25–50 cph and persist for shorter periods. In large tropical lakes N may be 20 cph in the seasonal thermocline and higher in diurnal thermoclines. To evaluate whether the stratification is strong enough to resist mixing by wind or cooling, further calculations are required.

ii. Wedderburn and Lake numbers provide a dynamic assessment of the stability of a lake. These numbers take into account the stratification across the thermocline, the depth of the center of gravity relative to the surface or depth of the lake, wind shear, and the fetch over which wind acts. Wedderburn numbers can also be computed for diurnal thermocline. High values, that is, greater than 10, indicate strong resistance to wind mixing and a stable system. In contrast, values approaching one and/or dropping below one indicate the thermocline will tilt, shear will increase, leading to instability, and wind-induced exchange can occur between different layers of a lake. Both the Wedderburn and Lake numbers predict the extent of tilting of the thermocline. Thus the indices quantify whether wind is enabling fluxes between different depths, and they indicate whether vertical exchanges occur primarily in one dimension or whether exchanges occur in two or three dimensions. For the latter, vertical fluxes may occur near lateral boundaries and horizontal advection will transport the water offshore. In such cases, to quantify fluxes, the two-dimensionality of the system must be included.

iii. The coefficient of eddy diffusivity (K_z) is an indicator of the efficacy of mixing and is used to compute fluxes of heat and of scalars such as nutrients and dissolved gases. Hence estimates of K_z are valuable in biogeochemical studies and when computing lake metabolism. Using the heat budget method, K_z can be computed for depths below the upper mixed layer by accounting for incoming solar radiation and changes in heat content. Other approaches are used for the upper mixed layer. Highest values of K_z are typically found in the upper mixed layer and lowest values occur in the thermocline. Variability occurs in response to the intensity and frequency of events with higher winds and cooling. Values of K_z in the stratified regions can be predicted from Lake and Wedderburn numbers.

10. Loss of stratification towards the end of the stratified period occurs when heat losses consistently exceed heat inputs over diel cycles, surface waters of the epilimnion cool, become denser, and, when denser than the water below, mix with water below. The process is called penetrative convection. Strong wind events can also lead to loss of stratification. Given that heat loss often accompanies increases in winds, the mixing can be by both wind and penetrative convection. Which process dominates can be determined from the Monin–Obukhov length scale. In autumn, or in a dry windy season in tropical regions, the metalimnion is progressively eroded as the density across it is reduced. Eventually, the entire volume of water is included in the circulation, and turnover is initiated. Turnover continues with continued wind and cooling. In cold regions temperatures often decrease to and below the temperature of maximum density, ~4°C.

11. Ice cover can form on the surface under calm, cold conditions in cold regions. Water temperatures are usually inversely stratified by that time, and ice cover may form when mean temperatures are 3°C or lower. The mean temperature when ice forms is often lower for deep lakes.

12. Inverse temperature stratification results when water temperatures decrease below 4°C, the temperature of maximum density. The density profile will be stable. Inverse temperature stratification may occur in regions where water temperatures become cold in winter but not cold enough for ice to form. It also occurs under ice, with exceptions when there are gradients in solutes, as in lakes in the McMurdo Dry Valleys, Antarctica, and the Arctic. The accumulated solutes contribute to changes in density with depth and can modify the expected influence of temperature alone. Water motions under the ice include gravity currents produced by heat and solute fluxes from the sediments and internal wave motions. Penetrative convection can occur in fall and winter when sufficient light penetrates through the ice to warm near-surface water such that its density exceeds that of the water immediately and to some depths below. Mixing will be induced.

13. As solar radiation increases in spring, mixing may occur under the ice or may be induced after ice-off as temperatures increase and cause instabilities, often leading to full or partial turnover of the water column. This process is known as radiatively driven convection (RDC) or penetrative convection. When water heats more rapidly inshore than offshore, nearshore water may flow offshore, penetrating at depths with similar density. This process, horizontal convection, is another form of RDC. Density stratification under the ice is often accentuated near the bottom by dissolved solutes and at times in a layer below the ice due to fresher water from snowmelt or ice-melt. In such cases, mixing under the ice by RDC may not occur and full mixing is delayed until after ice-off. Loss of ice cover may be patchy in the horizontal; earlier recession of ice cover around the margins can lead to the warming of surface water by solar radiation in the nearshore environment. When temperatures exceed 4°C only in inshore waters, advection of water that warms more rapidly in inshore than offshore locations may lead to formation of a thermal bar. If the warming over 4°C is lakewide, the advection of warmer water from inshore may contribute to the rapid onset of stratification after ice off.

14. Lakes undergoing complete circulation twice in a 12-month period, separated by periods of thermal stratification, are classified as dimictic. Moderately sized temperate, boreal, and Arctic lakes are often dimictic. Many other types of density stratification patterns occur due to the interacting effects of climate, morphometry, and optical and chemical conditions. These include amictic lakes that are permanently ice covered and do not fully circulate in the vertical each year and lakes classified as holomictic, meaning they fully circulate in the vertical at least once each year. Monomictic lakes have one period of circulation, and polymictic lakes mix multiple times a year. Monomictic and polymictic lakes are further classified based on latitude and associated differences in water temperature or, in the case of polymictic lakes, the presence of ice cover in winter. Lakes that are not ice covered yet either do not mix or only mix to the bottom in some years are classified as meromictic or oligomictic, respectively.

15. The layers in meromictic lakes are called:

a. Mixolimnion: The upper layer that stratifies and mixes over diel and seasonal cycles

b. Monimolimnion: The deeper stratum of water that is isolated

c. Pycnocline: The density gradient between the mixolimnion and the monimolimnion. In deeper lakes it may primarily be due to changes in temperature. In lakes in which the accumulation of solutes has led to meromixis, the density gradient may primarily be due to chemical stratification and called a chemocline. The density profile within such lakes is calculated taking into account the contributions from temperature and solutes. In some lakes, the pycnocline results from changes in both temperature and solutes.

16. Climate change is having multiple direct and indirect effects on lakes associated with the warming of surface waters; later onset and earlier loss of seasonal ice cover (with many dimictic lake examples), longer periods of stratification, and reduced mixing of surface with bottom waters. The loss of ice cover may alter mixing regimes, for example, amictic to cold monomictic and dimictic to monomictic. Some shallow lakes (~5–10 m deep) may transition from polymictic to monomictic. Changes in the timing and magnitude of precipitation can also lead to changes in mixing dynamics. These changes trigger an array of biogeochemical and biological processes, most notably expressed by a greater loss of dissolved oxygen from bottom waters due to reduced mixing with surface waters, and cascading impacts on other biogeochemical cycles with implications for nutrient supply and fluxes of climate forcing trace gases.

Acknowledgments

We thank John Melack, Robert Schwefel, and Warwick Vincent for their thoughtful comments on the chapter. Wencai Zhou modeled the changes in thermal structure in Lake Janauacá in August 2016 using the three-dimensional hydrodynamic model AEM3D. He additionally contributed to graphics. Megan Melack and Javier Vidal also contributed to graphics.

References

Adrian, R., O'Reilly, C.M., Zagarese, H., Baines, S.B., Hessen, D.O., Keller, W., et al., 2009. Lakes as sentinels of climate change. Limnol. Oceanogr. 54 (6), 2283–2297. https://doi.org/10.4319/lo.2009.54.6_part_2.2283.

Anderson, E.J., Stow, C.A., Gronewold, A.D., Mason, L.A., McCormick, M.J., Qian, S.S., Ruberg, S.A., Beadle, K., Constant, S.A., Hawley, N., 2021. Seasonal overturn and stratification changes drive deep-water warming in one of Earth's largest lakes. Nat. Commun. 12, 1688. https://doi.org/10.1038/s41467-021-21971-1.

Anderson, G.C., 1958. Some limnological features of a shallow saline meromictic lake. Limnol. Oceanogr. 3 (3), 259–270. https://doi.org/10.4319/lo.1958.3.3.0259.

Augusto-Silva, P.B., MacIntyre, S., de Moraes Rudorff, C., Cortés, A., Melack, J.M., 2019. Stratification and mixing in large floodplain lakes along the lower Amazon River. J. Great Lakes Res. 45 (1), 61–72. https://doi.org/10.1016/j.jglr.2018.11.001.

Austin, J., 2019. Observations of radiatively driven convection in a deep lake. Limnol. Oceanogr. 64 (5), 2152–2160. https://doi.org/10.1002/lno.11175.

Bärenbold, F., Kipfer, R., Schmid, M., 2022. Dynamic modelling provides new insights into development and maintenance of Lake Kivu's density stratification. Environ. Model. Softw. 147, 105251. https://doi.org/10.1016/j.envsoft.2021.105251.

Bartosiewicz, M., Przytulska, A., Deshpande, B.N., Antoniades, D., Cortes, A., MacIntyre, S., Lehmann, M.F., Laurion, I., 2019. Effects of climate change and episodic heat events on cyanobacteria in a eutrophic polymictic lake. Sci. Total Environ. 693, 13341. https://doi.org/10.1016/j.scitotenv.2019.07.220.

Bednarz, T.P., Lei, C.W., Patterson, J.C., 2009. An experimental study of unsteady natural convection in a reservoir model cooled from the water surface. Exp. Therm. Fluids. 32, 841856. https://doi.org/10.1016/j.expthermflusci.2007.10.007.

Bégin, P.N., Tanabe, Y., Rautio, M., Wauthy, M., Laurion, I., Uchida, M., Culley, A.I., Vincent, W.F., 2021. Water column gradients beneath the summer ice of a High Arctic freshwater lake as indicators of sensitivity to climate change. Sci. Rep. 11, 2868. https://doi.org/10.1038/s41598-021-82234-z.

Belzile, C., Vincent, W.F., Gibson, J.A., Van Hove, P., 2001. Bio-optical characteristics of the snow, ice, and water column of a perennially ice-covered lake in the High Arctic. Can. J. Fish. Aquat. Sci. 58 (12), 2405–2418. https://doi.org/10.1139/cjfas-58-12-2405.

Bengtsson, L., Malm, J., Terzhevik, A., Petrov, M., Boyarinov, P., Glinsky, A., Palshin, N., 1996. Field investigation of winter thermo- and hydrodynamics in a small Karelian lake. Limnol. Oceanogr. 41, 1502–1513. https://doi.org/10.4319/lo.1996.41.7.1502.

Bengtsson, L., Svensson, T., 1996. Thermal regime of ice covered Swedish lakes. Hydrol. Res. 37, 39–56. https://doi.org/10.2166/nh.1996.0018.

Benoit, G., Hemond, H.F., 1996. Vertical eddy diffusion calculated by the flux gradient method: significance of sediment-water heat exchange. Limnol. Oceanogr. 41 (1), 157–168.

Birge, E.A., 1915. The heat budgets of American and European lakes. Trans. Wis. Acad. Sci. Arts Lett. 18, 166–213.

Birge, E.A., 1916. The work of the wind in warming a lake. Trans. Wis. Acad. Sci. Arts Lett. 18, 341–391.

Birge, E.A., Juday, C.A., 1914. A limnological study of the Finger Lakes. Bull. Bur. Fish. 33, 525–609.

Birge, E.A., Juday, C., March, H.W., 1927. The temperature of the bottom deposits of Lake Mendota; a chapter in the heat exchanges of the lake. Trans. Wis. Acad. Sci. Arts Lett. 23, 187–231.

Boehrer, B., Kiwel, U., Rahn, K., Schultze, M., 2014. Chemocline erosion and its conservation by freshwater introduction to meromictic salt lakes. Limnologica 44, 81–89. https://doi.org/10.1016/j.limno.2013.08.003.

Boehrer, B., Schultze, M., 2008. Stratification of lakes. Rev. Geophys. 46, RG2005. https://doi.org/10.1029/2006RG000210.

Bouffard, D., Wüest, A., 2019. Convection in lakes. Ann. Rev. Fluid Mech. 51, 189–215. https://doi.org/10.1146/annurev-fluid-010518-040506.

Bouffard, D., Zdorovennova, G., Bogdanov, S., Efremova, T., Lavanchy, S., Palshin, N., et al., 2019. Under-ice convection dynamics in a boreal lake. Inland Waters 9 (2), 142–161. https://doi.org/10.1080/20442041.2018.1533356.

Brainerd, K.E., Gregg, M.C., 1993. Diurnal restratification and turbulence in the oceanic surface mixed layer: 1. Observations. J. Geophys. Res. 98 (C12), 22645–22656. https://doi.org/10.1029/93JC02297.

Brookes, J.D., O'Brien, K.R., Burford, M.A., Bruesewitz, D.A., Hodges, B.R., McBride, C., Hamilton, D.P., 2013. Effects of diurnal vertical mixing and stratification on phytoplankton productivity in geothermal Lake Rotowhero, New Zealand. Inland Waters 3, 369–376. https://doi.org/10.5268/IW-3.3.625.

Brookes, J.D., Regel, R.H., Ganf, G.G., 2003. Changes in the photochemistry of *Microcystis aeruginosa* in response to light and mixing. New Phytol. 158 (1), 151–164. https://doi.org/10.2307/1514088.

Burch, M.D., 1988. Annual cycle of phytoplankton in Ace Lake, an ice covered, saline meromictic lake. Hydrobiologia 165, 59–75. https://doi.org/10.1007/BF00025574.

Cannon, D.J., Troy, C., Bootsma, H., Liao, Q., MacLellan-Hurd, R., 2021. Characterizing the seasonal variability of hypolimnetic mixing in a large, deep lake. J. Geophys. Res. Oceans 126. https://doi.org/10.1029/2021JC017533, e2021JC017533.

Carmack, E.C., 1978. Combined influence of inflows and lake temperatures on spring circulation in a riverine lake. J. Phys. Oceanogr. 9, 422–434.

Coats, R., Perez-Losada, J., Schladow, G., Richards, R., Goldman, C., 2006. The warming of Lake Tahoe. Clim. Change 76, 121–148. https://doi.org/10.1007/s10584-005-9006-1.

Cohen, Y., Krumbein, W., Goldbert, M., Shilo, M., 1977. Solar Lake (Sinai). 1. Physical and chemical limnology 1. Limnol. Oceanogr. 22 (4), 597–608.

Cortés, A., Forrest, A.L., Sadro, S., Stang, A.J., Swann, M., Framsted, N.T., Thirkill, R., Sharp, S.L., Schladow, S.G., 2021. Prediction of hypoxia in eutrophic polymictic lakes. Water Resour. Res. 57 (6). https://doi.org/10.1029/2020wr028693. e2020WR028693.

Cortés, A., MacIntyre, S., 2020. Mixing processes in small arctic lakes during spring. Limnol. Oceanogr. 65 (2), 260–288. https://doi.org/10.1002/lno.11296.

Cortés, A., MacIntyre, S., Sadro, S., 2017. Flowpath and retention of snowmelt in an ice-covered arctic lake. Limnol. Oceanogr. 62, 2023–2044. https://doi.org/10.1002/lno.10549.

Coulter, G.W., Spigel, R.H, 1991. Hydrodynamics. In: Coulter, G.W. (Ed.), Lake Tanganyika and Its Life. Oxford University Press, London, pp. 49–75.

Couston, L.-A., Siegert, M., 2021. Dynamic flows create potentially habitable conditions in Antarctic subglacial lakes. Sci. Adv. 7, eabc3972.

Crisp, D.T., 1990. Water temperature in a stream gravel bed and implications for salmonid incubation. Freshw. Biol. 23 (3), 601–612. https://doi.org/10.1111/j.1365-2427.1990.tb00298.x.

Crisp, D.T., Howson, G., 1982. Effect of air temperature upon mean water temperature in streams in the north Pennines and english lake district. Freshw. Biol. 12 (4), 359–367. https://doi.org/10.1111/j.1365-2427.1982.tb00629.

Crisp, D.T., Matthews, A.M., Westlake, D.F., 1982. The temperatures of nine flowing waters in southern England. Hydrobiologia 89 (3), 193–204. https://doi.org/10.1007/BF00005705.

Csanady, G.T., 2001. Air-Sea Interaction: Laws and Mechanisms. Cambridge University Press.

Degens, E.T., von Herzen, R.P., Wong, H.K., Deuser, W.G., Jannasch, H.W., 1973. Lake Kivu: structure, chemistry and biology of an East African rift lake. Geol. Rundsch. 62, 245–277. https://doi.org/10.1007/BF01826830.

Deshpande, B.N., MacIntyre, S., Matveev, A., Vincent, W.F., 2015. Oxygen dynamics in permafrost thaw lakes: anaerobic bioreactors in the Canadian subarctic. Limnol. Oceanogr. 60 (5), 1656–1670. https://doi.org/10.1002/lno.10126.

Dickman, M.D., Hartman, J.S., 1979. A rationale for the subclassification of biogenic meromictic lakes. Int Rev. Ges. Hydrobiol. 64 (2), 189–192. https://doi.org/10.1002/iroh.19790640204.

Ding, F., Mao, Z., 2021. Observation and analysis of water temperature in Ice-covered shallow lake: case study in Qinghuahu Lake. Water 13 (21), 3139. https://doi.org/10.3390/w13213139.

Doda, T., Ramón, C.L., Ulloa, H.N., Wüest, A., Bouffard, D., 2022. Seasonality of density currents induced by differential cooling. Hydrol. Earth Syst. Sci. 26 (2), 331–353. https://doi.org/10.5194/hess-26-331-2022.

Dokulil, M.T., Jagsch, A., George D., G., Anneville, O., Jankowski, T., Wahl, B., Lenhart, B., Blenckner, T., Teubner, K., 2006. Twenty years of spatially coherent deepwater warming in lakes across Europe related to the North Atlantic Oscillation. Limnol. Oceanogr. 51 (6), 2787–2793. https://doi.org/10.4319/lo.2006.51.6.2787.

Drago, E.C., 1989. Morphological and hydrological characteristics of the floodplain ponds of the Middle Parana river (Argentina). Rev. Hydrobiol. Trop. 22 (3), 183–190.

Einsele, W., 1941. Die Umsetzung von zugeführtem, anorganischen Phosphat im eutrophen See und ihre Rüchwirkungen auf seinen Gesamthaushalt. Z. Fisch. 39, 407–488.

Ellis, C.R., Stefan, H.G., Gu, R., 1991. Water temperature dynamics and heat transfer beneath the ice cover of a lake. Limnol. Oceanogr. 36 (2), 324–334. https://doi.org/10.4319/lo.1991.36.2.0324.

Fairall, C.W., Bradley, E.F., Hare, J.E., Grachev, A.A., Edson, J.B., 2003. Bulk parameterization of air-sea fluxes updates and verification for the COARE algorithm. J. Clim. 16, 571–591.

Fairall, C.W., Bradley, E.F., Rogers, D.P., Edson, J.P., Young, G.S., 1996. Bulk parameterization of air-sea fluxes for tropical ocean-global atmosphere coupled-ocean atmosphere response experiment. J. Geophys. Res. 101, 3747–3764.

Fang, X., Stefan, H.G., 1996. Dynamics of heat exchange between sediment and water in a lake. Water Resour. Res. 32 (6), 1719–1727. https://doi.org/10.1029/96WR00274.

Ficke, E.R., Ficke, J.F., 1977. Ice on rivers and lakes: A bibliographic essay. U.S. Geological Survey. Water-Resources Investigations Report. https://doi.org/10.3133/wri779. Series number 77-9.

Finckh, P., 1981. Heat-flow measurements in 17 perialpine lakes. Geol. Soc. Am. Bull. 92 (3_Part_II), 452–514. https://doi.org/10.1130/GSAB-P2-92-452.

Findenegg, I., 1935. Limnologische untersuchungen in Karntner seengebiete. Int. Rev. Ges. Hydrobiol. 32, 369–423.

Findenegg, I., 1937. Holomiktische und meromiktische Seen. Int. Rev. Ges. Hydrobiol. 35 (1–6), 586–610. https://doi.org/10.1002/iroh.19370350130.

Fink, G., Schmid, M., Wahl, B., Wolf, T., Wuest, A., 2014. Heat flux modifications related to climate-induced warming of large European lakes. Water Resour. Res. 50 (3), 2072–2085. https://doi.org/10.1002/2013WR014448.

Fischer, H.B., List, J.E., Koh, C.R., Imberger, J., Brooks, N.H., 1979. Mixing in inland and coastal waters. Academic Press, New York, p. 484. https://doi.org/10.1016/C2009-0-22051-4.

Foley, B., Jones, I.D., Maberly, S.C., Rippey, B., 2012. Long-term changes in oxygen depletion in a small temperate lake: effects of climate change and eutrophication. Freshw. Biol. 57 (2), 278–289. https://doi.org/10.1111/j.1365-2427.2011.02662.x.

Forrest, A.L., Laval, B., Doble, M.J., Yeo, R., Magnusson, E., 2008. AUV measurements of under-ice thermal structure. Oceans 2008, 1–10. https://doi.org/10.1109/Oceans., 2008.5152046.

Giles, M.P., Michelutti, N., Grooms, C., Smol, J.P., 2018. Long-term limnological changes in the Ecuadorian páramo: comparing the ecological responses to climate warming of shallow waterbodies versus deep lakes. Freshw. Biol. 63 (10), 1316–1325. https://doi.org/10.1111/fwb.13159.

Goldman, C.R., Kumagai, M., Robarts, R.D., 2013. Climatic change and global warming of inland waters. Wiley-Blackwell, p. 472.

Goldman, C.R., Mason, D.T., Wood, B.J., 1972. Comparative study of the limnology of two small lakes on Ross Island, Antarctica. In: Llano, G.A. (Ed.), Antarctic Terrestrial Biology. American Geophysical Union, Washington, DC, pp. 1–50. https://doi.org/10.1002/9781118664667.ch1.

Guseva, S., Aurela, M., Cortes, A., Kivi, R., Lotsari, E., MacIntyre, S., 2021. Variable physical drivers of near-surface turbulence in a regulated river. Water Resour. Res. 57 (11). https://doi.org/10.1029/2020WR027939. e2020WR027939.

Haberyan, K.A., Hecky, R.E., 1987. The late Pleistocene and Holocene stratigraphy and paleolimnology of lakes Kivu and Tanganyika. Palaeogeogr. Palaeoclimatol. Palaeoecol. 61, 169–197. https://doi.org/10.1016/0031-0182(87)90048-4.

Hale, G.M., Querry, M.R., 1973. Optical constants of water in the 200-nm to 200-μm wavelength region. Appl. Opt. 12 (3), 555–563. https://doi.org/10.1364/AO.12.000555.

Hamilton, D.P., Magee, M.R., Wu, C.H., Kratz, T.K., 2018. Ice cover and thermal regime in a dimictic seepage lake under climate change. Inland Waters 8 (3), 381–398. https://doi.org/10.1080/20442041.2018.1505372.

Hamilton, D.P., McBride, C.G., Özkundakci, D., Shallenberg, M., Verburg, P., de Winton, M., et al., 2013. Effects of climate change on New Zealand Lakes. In: Goldman, C.R., Kumagai, M., Robarts, R.D. (Eds.), Climatic Change and Global Warming of Inland Waters. Wiley-Blackwell, pp. 337–366.

Hampton, S.E., Izmest'eva, L.R., Moore, M.V., Katz, S.L., Dennis, B., Silow, E.A., 2008. Sixty years of environmental change in the world's largest freshwater lake—Lake Baikal, Siberia. Glob. Chang.

Biol. 14 (8), 1947–1958. https://doi.org/10.1111/j.1365-2486.2008.01616.x.

Harvey, H.H., Coombs, J.F., 1971. Physical and chemical limnology of the lakes of Manitoulin Island. J. Fish. Res. 28 (12), 1883–1897. https://doi.org/10.1139/f71-284.

Hobbie, J.E., 1961. Summer temperatures in Lake Schrader, Alaska. Limnol. Oceanogr. 6 (3), 326–329. https://doi.org/10.4319/lo.1961.6.3.0326.

Hobbs, P.V., 1974. Ice Physics. Clarendon Press, Oxford, p. 352.

Horn, D.A., Imberger, J., Ivey, G.N., 2001. The degeneration of large-scale interfacial gravity waves in lakes. J. Fluid Mech. 434, 181–207. https://doi.org/10.1017/S0022112001003536.

Horne, A.J., Goldman, C.R., 1994. Limnology. McGraw-Hill, New York, pp. 1–576.

Houser, J.N., 2006. Water color affects the stratification, surface temperature, heat content, and mean epilimnetic irradiance of small lake. Can. J. Fish. Aquat. Sci. 63 (11), 2447–2455. https://doi.org/10.1139/f06-131.

Hudec, P.P., Sonnenfeld, P., 1974. Hot brines on Los Roques, Venezuela. Science 185, 440–442. https://doi.org/10.1126/science.185.4149.440.

Hutchinson, G.E., 1937. A contribution to the limnology of arid regions: primarily founded on observations made in the Lahontan basin. Trans. Conn. Acad. Arts Sci. 33, 47–132.

Hutchinson, G.E., 1957. A treatise on limnology. In: Geography, Physics and Chemistry, 1. Chapman and Hall, London, p. 1015. https://doi.org/10.1017/S0016756800062634.

Hutchinson, G.E., Löffler, H., 1956. The thermal classification of lakes. Proc. Natl. Acad. Sci., U. S. A. 42 (2), 84. https://doi.org/10.1073/pnas.42.2.84.

Huttula, T. (Ed.), 1997. Flow, thermal Regime and Sediment Transport Studies in Lake Tanganyika, vol. 73. Kuopio University Publications C. Natural and Environmental Sciences, p. 173.

Idso, S.B., 1973. On the concept of lake stability. Limnol. Oceanogr. 18 (4), 681–683. https://doi.org/10.4319/lo.1973.18.4.0681.

Illies, J., 1953. Die Besiedlung der Fulda (insbes. das Benthos der Salmonidenregion) nach dem jetzigen Stand der Untersuchung. Berl. Limnol. Flußstat. Freudenthal 5, 1–28.

Imberger, J., 1985. The diurnal mixed layer. Limnol. Oceanogr. 30 (4), 737–770. https://doi.org/10.4319/lo.1985.30.4.0737.

Imberger, J., Hamblin, P.F., 1982. Dynamics of lakes, reservoirs and cooling ponds. Annu. Rev. Fluid Mech. 14, 153–187. https://doi.org/10.1146/annurev.fl.14.010182.001101.

Imberger, J., Patterson, J., 1990. Physical limnology. Adv. Appl. Mech. 27, 303–475. https://doi.org/10.1016/S0065-2156(08)70199-6.

Imboden, D.M., Wüest, A., 1995. Mixing mechanisms in lakes. In: Lerman, A., et al. (Eds.), Physics and Chemistry of Lakes. Springer-Verlag Berlin Heidelberg, New York.

Jane, S.F., Hansen, G.J.A., Kraemer, B.M., Leavitt,, P.R., Mincer, J.L., North, R.L., et al., 2021. Widespread deoxygenation of temperate lakes. Nature 594, 66–70. https://doi.org/10.1038/s41586-021-03550-y.

Jansen, J., MacIntyre, S., Barrett, D.C., Chin, Y.P., Cortes, A., Forrest, A.L., et al., 2021. Winter limnology: how do hydrodynamics and biogeochemistry shape ecosystems under ice? J. Geophys. Res. Biogeosci. 126. https://doi.org/10.1029/2020JG006237 e2020JG00637.

Jassby, A., Powell, T., 1975. Vertical patterns of eddy diffusion during stratification in Castle Lake, California. Limnol. Oceanogr. 20 (4), 530–543. https://doi.org/10.4319/lo.1975.20.4.0530.

Jellison, R., Melack, J.M., 1993. Meromixis in hypersaline Mono Lake, California: vertical mixing and density stratification during the onset, persistence, and breakdown of meromixis. Limnol. Oceanogr. 38, 1008–1019. https://doi.org/10.4319/lo.1993.38.5.1008.

Johnson, L., 1966. Temperature of maximum density of fresh water and its effect on circulation in Great Bear Lake. J. Fish. Res. Board Can. 23, 963–973. https://doi.org/10.1139/f66-089.

Jones, S.E., Chiu, C.Y., Kratz, T.K., Wu, J.T., Shade, A., McMahon, K.D., 2008. Typhoons initiate predictable change in aquatic bacterial communities. Limnol. Oceanogr. 53 (4), 1319–1326. https://doi.org/10.4319/lo.2008.53.4.1319.

Judd, J.H., 1970. Lake stratification caused by runoff from street deicing. Water Res. 4, 521–532. https://doi.org/10.1016/0043-1354(70)90002-3.

Kettle, A.J., Hughes, C., Unazi, G.A., Birch, L., Mohie-El-Din, H., Jones, M.R., 2012. Role of groundwater exchange on the energy budget and seasonal stratification of a shallow temperate lake. J. Hydrol. 470–471, 12–27. https://doi.org/10.1016/j.jhydrol.2012.07.004.

King, T.V., Neilson, B.T., 2019. Quantifying reach-average effects of hyporheic exchange on Arctic river temperatures in an area of continuous permafrost. Water Resour. Res. 55 (3), 1951–1971. https://doi.org/10.1029/2018WR023463.

Kirillin, G., Leppäranta, M., Terzhevik, A., Granin, N., Bernhardt, J., Engelhardt, C., et al., 2012. Physics of seasonally ice-covered lakes: a review. Aquat. Sci. 74 (4), 659–682. https://doi.org/10.1007/s00027-012-0279-y.

Kirillin, G., Shatwell, T., 2016. Generalized scaling of seasonal thermal stratification in lakes. Earth-Sci. Rev. https://doi.org/10.1016/j.earscirev.2016.08.008.

Kirkland, D.W., Bradbury, J.P., Dean, W.E., 1983. The heliothermic lake—a direct method of collecting and storing solar energy. Arch. Hydrobiol. Suppl. 65, 1–60.

Kirkpatrick, M.P., Williamson, N., Armfield, S.W., Zecevic, V., 2019. Evolution of thermally stratified open channel flow after removal of the heat source. J. Fluid Mech. 876, 356–412. https://doi.org/10.1017/jfm.2019.543.

Kitazawa, D., 2013. Numerical simulation of future overturn and ecosystem impacts for deep lakes in Japan. In: Goldman, C.R., Kumagai, M., Robarts, R.D. (Eds.), Climatic Change and Global Warming of Inland Waters. Wiley-Blackwell, Hoboken, NJ, p. 472.

Kling, G.W., 1987. Seasonal mixing and catastrophic degassing in tropical lakes, Cameroon, West Africa. Science 237 (4818), 1022–1024. https://doi.org/10.1126/science.237.4818.1022.

Kling, G.W., 1988. Comparative transparency, depth of mixing, and stability of stratification in lakes of Cameroon, West Africa. Limnol. Oceanogr. 33 (1), 27–40. https://doi.org/10.4319/lo.1988.33.1.0027.

Kling, G.W., Evans, W.C., Tanyileke, G., Kusakabe, M., Ohba, T., Yoshida, Y., et al., 2005. Degassing lakes Nyos and Monoun: defusing certain disaster. Proc. Natl. Acad. Sci. 102 (40), 14185–14190. https://doi.org/10.1073/pnas.0502274102.

Kózminski, Z., Wisniewski, J., 1935. Über die Forfrühlingthermik der Wigry-Seen. Arch. Hydrobiol. 28, 198–235. https://doi.org/10.1073/pnas.0502274102.

Kraemer, B.M., Anneville, O., Chandra, S., Dix, M., Kuusisto, E., Livingstone, D.M., et al., 2015. Morphometry and average temperature affect lake stratification responses to climate change. Geophys. Res. Lett. 42 (12), 4981–4988. https://doi.org/10.1002/2015GL064097.

Kraemer, B.M., Pilla, R.M., Woolway, R.I., Anneville, O., Ban, S., Colom-Montero, W., et al., 2021. Climate change drives widespread shifts in lake thermal habitat. Nat. Clim. Change 11 (6), 521–529. https://doi.org/10.1038/s41558-021-01060-3.

Labaj, A.L., Michelutti, N., Smol, J.P., 2018. Annual stratification patterns in tropical mountain lakes reflect altered thermal regimes in response to climate change. Fund. Appl. Limnol. 191, 267–275. https://doi.org/10.1127/fal/2018/1151.

Ladwig, R., Rock, L.A., Dugan, H.A., 2021. Impact of salinization on lake stratification and spring mixing. Limnol. Oceanogr. Lett. https://doi.org/10.1002/lol2.10215.

Laybourn-Parry, J., Bell, E.M., 2014. Ace Lake: three decades of research on a meromictic, Antarctic lake. Polar Biol. 37, 1685–1699. https://doi.org/10.1007/s00300-014-1553-3.

Lehnherr, I., St, Louis, V.L., Sharp, M., Gardner, A.S., Smol, J.P., Schiff, S.L., et al., 2018. The world's largest High Arctic lake responds rapidly to climate warming. Nat. Commun. 9 (1), 1–9. https://doi.org/10.1038/s41467-018-03685-z.

Leppäranta, M., 2015. Freezing of Lakes and the Evolution of their Ice Cover. Springer, Berlin. https://doi.org/10.1007/978-3-642-29081-7.

Lewis Jr., W.M., 1983. A revised classification of lakes based on mixing. Can. J. Fish. Aquat. Sci. 40 (10), 1779–1787. https://doi.org/10.1139/f83-207.

Likens, G.E., Johnson, N.M., 1969. Measurement and analysis of the annual heat budget for the sediments in two Wisconsin lakes. Limnol. Oceanogr. 14 (1), 115–135. https://doi.org/10.4319/lo.1969.14.1.0115.

Likens, G.E., Johnson, P.L., 1966. A chemically stratified lake in Alaska. Science 153 (3738), 875–877. https://doi.org/10.1126/science.153.3738.875.

Liu, H., Zhang, Q., Dowle, G., 2012. Environmental controls on the surface energy budget over a large southern inland water in the United States: an analysis of one-year eddy covariance flux data. J. Hydrometeorol. 13 (6), 1893–1910. https://doi.org/10.1175/JHM-D-12-020.1.

Livingstone, D.M., 2003. Impact of secular climate change on the thermal structure of a large temperate central European lake. Clim. Change 57, 205–225. https://doi.org/10.1023/A:1022119503144.

MacIntyre, S., 1993. Vertical mixing in a shallow, eutrophic lake: possible consequences for the light climate of phytoplankton. Limnol. Oceanogr. 38, 798–817. https://doi.org/10.4319/lo.1993.38.4.0798.

MacIntyre, S., 2008. Describing fluxes within lakes using temperature arrays and surface meteorology. Verh. Internat. Verein. Limnol. 30, 339–344. https://doi.org/10.1080/03680770.2008.11902139.

MacIntyre, S., Amaral, J.H.F., Melack, J.M., 2021. Enhanced turbulence in the upper mixed layer under light winds and heating: implications for gas fluxes. J. Geophys. Res. Oceans 126. https://doi.org/10.1029/2020JC017026, e2020JC017026.

MacIntyre, S., Bastviken, D., Arneborg, L., Crowe, A.T., Karlsson, J., Andersson, A., 2021b. Turbulence in a small boreal lake: consequences for air–water gas exchange. Limnol. Oceanogr. 66 (3), 827–854. https://doi.org/10.1002/lno.11645.

MacIntyre, S., Clark, J.F., Jellison, R.J., Fram, J.P., 2009b. Turbulent mixing induced by non-linear internal waves in Mono Lake, CA. Limnol. Oceanogr. 54 (6), 2255–2272. https://doi.org/10.4319/lo.2009.54.6.2255.

MacIntyre, S., Crowe, A.T., Cortés, A., Arneborg, L., 2018. Turbulence in a small arctic pond. Limnol. Oceanogr. 63 (6), 2337–2358. https://doi.org/10.1002/lno.10941.

MacIntyre, S., Flynn, K.M., Jellison, R.M., Romero, J.R., 1999. Boundary mixing and nutrient flux in Mono Lake, CA. Limnol. Oceanogr. 44, 512–529.

MacIntyre, S., Fram, J.P., Kushner, P.J., Bettez, N.D., O'Brien, W.J., Hobbie, J.E., et al., 2009a. Climate-related variations in mixing dynamics in an Alaskan arctic lake. Limnol. Oceanogr. 54 (6), 2401–2417. doi: 10.4319lo.2009.54.6_part_2.2401.

MacIntyre, S., Melack, J.M., 1982. Meromixis in an equatorial African soda lake. Limnol. Oceanogr. 27, 595–609.

MacIntyre, S., Melack, J.M., 1986. Vertical mixing in Amazon floodplain lakes Verh. Internat. Verein. Limnol. 22, 1283–1287.

MacIntyre, S., Melack, J.M., 1988. Frequency and depth of vertical mixing in an Amazon floodplain lake (L. Calado, Brazil). Verh. Internat. Verein. Limnol. 23, 80–85. https://doi.org/10.1080/03680770.1987.11897906.

MacIntyre, S., Melack, J.M., 1995. Vertical and horizontal transport in lakes: linking littoral, benthic, and pelagic habitats. J. North Am. Benthol. Soc. 14 (4), 599–615. https://doi.org/10.2307/1467544.

Macintyre, S., Melack, J.M., 2009. Mixing dynamics in lakes across climatic zones. In: Likens, G.E. (Ed.), Encyclopedia of Inland Waters. Elsevier, Amsterdam, pp. 603–612. https://doi.org/10.1016/B978-012370626-3.00040-5.

MacIntyre, S., Romero, J.R., Kling, G.W., 2002. Spatial-temporal variability in surface layer deepening and lateral advection in an embayment of Lake Victoria, East Africa. Limnol. Oceanogr. 47 (3), 656–671. https://doi.org/10.4319/lo.2002.47.3.0656.

MacIntyre, S., Romero, J.R., Silsbe, G.M., Emery, B.M., 2014. Stratification and horizontal exchange in Lake Victoria, East Africa. Limnol. Oceanogr. 59 (6), 1805–1838. https://doi.org/10.4319/lo.2014.59.6.1805.

MacIntyre, S., Sickman, J.O., Goldthwait, S.A., Kling, G.W., 2006. Physical pathways of nutrient supply in a small, ultraoligotrophic arctic lake during summer stratification. Limnol. Oceanogr. 51 (2), 1107–1124. https://doi.org/10.4319/lo.2006.51.2.1107.

Magnuson, J.J., Robertson, D.M., Benson, B.J., Wynne, R.H., Livingstone, D.M., Arai, T., et al., 2000. Historical trends in lake and river ice cover in the Northern Hemisphere. Science 289 (5485), 1743–1746. https://doi.org/10.1126/science.289.5485.1743.

Malm, J., 1998. Bottom buoyancy layer in an ice-covered lake. Water Resour. Res. 34, 1981–2993. https://doi.org/10.1029/98WR01904.

Malm, J., 1999. Some properties of currents and mixing in a shallow ice-covered lake. Water Resour. Res. 35 (1), 221–232.

Malm, J., Bengtsson, L., Terzhevik, A., Boyarinov, P., Glinsky, A., Palshin, N., et al., 1998. Field study on currents in a shallow, ice-covered lake. Limnol. Oceanogr. 43 (7), 1669–1679.

Martin, S., 1981. Frazil ice in rivers and oceans. Annu. Rev. Fluid Mech. 3 (1), 379–397. https://doi.org/10.1146/annurev.fl.13.010181.002115.

Matveev, V.P., 1964. O vertikal'nom raspredelenii temperatury v donnykh otlozheniyakh Ozer Dolgogo (Pitkayarvi) i Volochaevskogo (Vuotyarvi). (On the vertical distribution of temperature in the bottom deposits of Lake Dolgom [Pitkayarvi] and Volochaevskom [Vuotyarvi]). Ozera Karel'skogo leresheika. Izdatel'stvo Nauka, Moscow, pp. 45–50.

McIntosh Marcek, H.A., Lesack, L.F., Orcutt, B.N., Wheat, C.G., Dallimore, S.R., Geeves, K., et al., 2021. Continuous dynamics of dissolved methane over 2 years and its carbon isotopes ($\delta^{13}C$, $\Delta^{14}C$) in a small Arctic lake in the Mackenzie Delta. J. Geophys. Res., Biogeosci. 126 (3). https://doi.org/10.1029/2020JG006038, e2020JG006038.

Melack, J.M., 1978. Morphometric, physical and chemical features of the volcanic crater lakes of western Uganda. Arch. Hydrobiol. 84, 430–453.

Melack, J.M., 1988. Primary producer dynamics associated with evaporative concentration in a shallow, equatorial soda lake (Lake Elmenteita, Kenya). Hydrobiologia 158, 1–14. https://doi.org/10.1007/978-94-009-3095-7_1.

Melack, J.M., Jellison, R., 1998. Limnological conditions in Mono Lake: contrasting monomixis and meromixis in the 1990s. Hydrobiologia 384, 21–39.

Melack, J.M., Jellison, R., MacIntyre, S., Hollibaugh, J.T., 2017. Mono Lake: plankton dynamics over three decades of meromixis or monomixis. In: Gulati, R.D., Zadereev, E., Degermendzhi, A.G. (Eds.), Ecology of Meromictic Lakes. Springer, New York, pp. 325–351. https://doi.org/10.1007/978-3-319-49143-1_11.

Melack, J.M., Kilham, P., 1974. Photosynthetic rates of phytoplankton in East African alkaline, saline lakes. Limnol. Oceanogr. 19, 743–755. https://doi.org/10.4319/lo.1974.19.5.0743.

Mesman, J.P., Stelzer, J.A., Dakos, V., Goyette, S., Jones, I.D., Kasparian, J., et al., 2021. The role of internal feedbacks in shifting deep lake mixing regimes under a warming climate. Freshw. Biol. 66 (6), 1021–1035. https://doi.org/10.1111/fwb.13704.

Michelutti, N., Douglas, M.S., Antoniades, D., Lehnherr, I., St, Louis, V.L., St. Pierre, K., et al., 2020. Contrasting the ecological effects of decreasing ice cover versus accelerated glacial melt on the High Arctic's largest lake. Proc. Royal Soc. B 287, 1929. https://doi.org/10.1098/rspb.2020.1185.

Monin, A.S., Obukhov, A.M., 1954. Basic laws of turbulent mixing in the ground layer of the atmosphere. Trans. Geophys. Inst. Akad. Nauk. USSR 24 (151), 163–187.

Monismith, S., 1986. An experimental study of the upwelling response of stratified reservoirs to surface shear stress. J. Fluid Mech. 171, 407–439. https://doi.org/10.1017/S0022112086001507.

Monismith, S.G., Imberger, J., Morison, M.L., 1990. Convective motions in the sidearm of a small reservoir. Limnol. Oceanogr. 35 (8), 1676–1702. https://doi.org/10.4319/lo.1990.35.8.1676.

Monismith, S.G., MacIntyre, S., 2009. The surface mixed layer in lakes and reservoirs. In: Likens, G.E. (Ed.), Encyclopedia of Inland Waters. Academic Press, Oxford, pp. 636–665.

Mortimer, C.H., Mackereth, F.J.H., 1958. Convection and its consequences in ice-covered lakes. Verh. - Int. Ver. Theor. Angew. Limnol. 13 (2), 923–932. https://doi.org/10.1080/03680770.1956.11895490.

Müller, K., 1951. Fisch und Fischregionen der Fulda Berl. Limnol. Flußstat. Freudenthal. 2, 18–23.

Olsthoorn, J., Bluteau, C.E., Lawrence, G.A., 2020. Under-ice salinity transport in low-salinity waterbodies. Limnol. Oceanogr. 65 (2), 247–259. https://doi.org/10.1002/lno.11295.

O'Reilly, C.M., Sharma, S., Gray, D.K., Hampton, S.E., Read, J.S., Rowley, R.J., et al., 2015. Rapid and highly variable warming of lake surface waters around the globe. Geophys. Res. Lett. 42, 10773–10781. https://doi.org/10.1002/2015GL066235.

Ostrovsky, I., Rimmer, A., Yacobi, Y.Z., Nishri, A., Sukenik, A., Hadas, O., Zohary, T., 2013. Long-term changes in the Lake Kinneret ecosystem: the effects of climate change and anthropogenic factors. In: Goldman, C.R., Kumagai, M., Robarts, R.D. (Eds.), Climatic Change and Global Warming of Inland Waters. Wiley-Blackwell, Hoboken, NY, p. 472.

Pasche, N., Dinkel, C., Müller, B., Martin Schmid, M., Alfred Wüest, A., Wehrli, B., 2009. Physical and biogeochemical limits to internal nutrient loading of meromictic Lake Kivu. Limnol. Oceanogr. 54, 1863–1873.

Pasche, N., Hofmann, H., Bouffard, D., Schubert, C.J., Lozovik, P.A., Sobek, S., 2019. Implications of river intrusion and convective mixing on the spatial and temporal variability of under-ice CO_2. Inland Waters 9 (2), 162–176. https://doi.org/10.1080/20442041.2019.1568073.

Pernica, P., Wells, M.G., MacIntyre, S., 2014. Persistent weak thermal stratification inhibits mixing in the epilimnion of north-temperate Lake Opeongo, Canada. Aquat. Sci. 76 (2), 187–201. https://doi.org/10.1007/s00027-013-0328-1.

Pilla, R.M., Williamson, C.E., Adamovich, B.V., Adrian, R., Anneville, O., Chandra, S., et al., 2020. Deeper waters are changing less consistently than surface waters in a global analysis of 102 lakes. Sci. Rep. 10 (1), 1–15. https://doi.org/10.1038/s41598-020-76873-x.

Pivovarov, A.A., 1973. Thermal conditions in freezing lake and rivers. John Wiley & Sons, New York, p. 136.

Plisnier, P.D., Kayanda, R., MacIntyre, S., Obiero, K., Okello, W., Vodacek, A., et al., 2022. Need for harmonized long-term multi-lake monitoring of African Great Lakes. J. Great Lakes Res. 44 (6), 1194–1208. https://doi.org/10.1016/j.jglr.2018.05.019.

Plisnier, P.D., Nshombo, M., Mgana, H., Ntakimazi, G., 2018. Monitoring climate change and anthropogenic pressure at Lake Tanganyika. J. Great Lakes Res. 44 (6), 1194–1208. https://doi.org/10.1016/j.jglr.2018.05.019.

Posada-Bedoya, A., Gómez-Giraldo, A., Román-Botero, R., 2021. Effects of riverine inflows on the climatology of a tropical Andean reservoir. Limnol. Oceanogr. 66 (9), 3535–3551. https://doi.org/10.1002/lno.11897.

Priddle, J., Heywood, R.B., 1980. Evolution of Antarctic lake ecosystems. Biol. J. Linn. Soc. 14 (1), 51–66. https://doi.org/10.1111/j.1095-8312.1980.tb00097.x.

Priscu, J.C., Spigel, R.H., Gibbs, M.M., Downes, M.T., 1986. A numerical analysis of hypolimnetic nitrogen and phosphorus transformations in Lake Rotoiti, New Zealand: a geothermally influenced lake. Limnol. Oceanogr. 31 (4), 812–831. https://doi.org/10.4319/lo.1986.31.4.0812.

Prowse, T.D., 1994. Environmental significance of ice to streamflow in cold regions. Freshw. Biol. 32 (2), 241–259. https://doi.org/10.1111/j.1365-2427.1994.tb01124.x.

Ragotzkie, R.A., 1978. Heat budgets of lakes. In: Lerman, A. (Ed.), Lakes: Chemistry, Geology, Physics. Springer-Verlag, New York, pp. 1–19. https://doi.org/10.1007/978-1-4757-1152-3_1.

Råman Vinnå, L., Medhaug, I., Schmid, M., Bouffard, D., 2021. The vulnerability of lakes to climate change along an altitudinal gradient. Commun. Earth Environ. 2, 35. https://doi.org/10.1038/s43247-021-00106-w.

Råman Vinnå, L., Wüest, A., Bouffard, D., 2017. Physical effects of thermal pollution in lakes. Water Resour. Res. 53 (5), 3968–3987. https://doi.org/10.1002/2016WR019686.

Rankin, L.M., Gibson, A.E., Franzrnann, P.D., Burton, H.R., 1999. The chemical stratification and microbial communities of Ace Lake, Antarctica: a review of the characteristics of a marine-derived meromictic lake. Polarforschung 66 (1/2), 33–52.

Read, J.S., Hamilton, D.P., Jones, I.D., Muraoaka, K., Winslow, L.A., Kroiss, R., et al., 2011. Derivation of lake mixing and stratification indices from high-resolution lake buoy data. Environ. Model. Softw. 26, 1325–1336. https://doi.org/10.1016/j.envsoft.2011.05.006.

Reif, C.B., 1969. Temperature profiles and heat flow in sediments of Nuangola. Proc. Pennsylvania Acad. Sci. 43, 98–100.

Rimmer, A., Gal, G., Opher, T., Lechinsky, Y., Yacobi, Y.Z., 2011. Mechanisms of long-term variations in the thermal structure of a warm lake. Limnol. Oceanogr. 56 (3), 974–988. https://doi.org/10.4319/lo.2011.56.3.0974.

Robertson, D.M., Imberger, J., 1994. Lake Number, a quantitative indicator of mixing used to estimate changes in dissolved oxygen. Int. Revue. Ges. Hydrobiol. 79 (2), 159–176. https://doi.org/10.1002/iroh.19940790202.

Romero, J.R., Jellison, R., Melack, J.M., 1998. Stratification, vertical mixing, and upward ammonium flux in hypersaline Mono Lake, California. Arch. Hydrobiol. 142 (3), 283–315. https://doi.org/10.1127/archiv-hydrobiol/142/1998/283.

Rueda, F.J., MacIntyre, S., 2009. Flow paths and spatial heterogeneity of stream inflows in a small multibasin lake. Limnol. Oceanogr. 54 (6), 2041–2057. https://doi.org/10.4319/Lo.2009.54.2041.

Rueda, F.J., MacIntyre, S., 2010. Modelling the fate and transport of negatively buoyant storm–river water in small multi-basin lakes. Environ. Model. Softw. 25 (1), 146–157. https://doi.org/10.1016/j.envsoft.2009.07.002.

Ruttner, F., 1963. Fundamentals of limnology. In: English—3rd Edition. Translated [from the German Version by Frey, D. G., Fry, F. E. J. University of Toronto Press, Toronto, p. 295.

Sahoo, G.B., Forrest, A.L., Schladow, S.G., Reuter, J.E., Coats, R., Dettinger, M.D., et al., 2016. Climate change impacts on lake thermal dynamics and ecosystem vulnerabilities. Limnol. Oceanogr. 61 (2), 496–507. https://doi.org/10.1002/lno.10228.

Sakamoto, M., 2013. Biogeochemical ecosystem dynamics in Lake Biwa under anthropogenic impacts and global warming. In: Goldman, C.R., Kumagai, M., Robarts, R.D. (Eds.), Climatic Change and Global Warming of Inland Waters. Wiley-Blackwell, Hoboken, NY, p. 472.

Schindler, D.W., Bayley, S.E., Parker, B.R., Beaty, K.G., Cruikshank, D.R., Fee, E.J., et al., 1996. The effects of climatic warming on the properties of boreal lakes and streams at the Experimental Lakes Area, northwestern Ontario. Limnol. Oceanogr. 41 (5), 1004–1017. https://doi.org/10.4319/lo.1996.41.5.1004.

Schmidt, W., 1915. Über den energie-gehalt der seen. Mit beispielen vom Lunzer Untersee nach messungen mit einen enfachen temperaturlot. Int. Rev. Hydrobiol. 6. Suppl. (Not seen in original).

Schmidt, W., 1928. Über temperatur und stabilitäts verhältnisse von seen. Geogr. Ann. 145–177. https://doi.org/10.1080/20014422.1928.11880475.

Schmid, M., Halbwachs, M., Wehrli, B., Wüest, A., 2005. Weak mixing in Lake Kivu: new insights indicate increasing risk of uncontrolled gas eruption. Geochem. Geophys. Geosyst. 6, Q07009. https://doi.org/10.1029/2004GC000892.

Scholz, C.A., Cohen, A.S., Johnson, T.C., King, J., Talbot, M.R., Brown, E.T., 2011. The Lake Malawi scientific drilling project: an overview. Palaeogeogr. Paleoclimatol. Palaeoecol. 303, 3–19.

Schwefel, R., Gaudard, A., Wüest, A., Bouffard, D., 2016. Effects of climate change on deepwater oxygen and winter mixing in a deep lake (Lake Geneva): comparing observational findings and modeling. Water Resour. Res. 52 (11), 8811–8826.

Scrimgeour, G.J., Prowse, T.D., Culp, J.M., Chambers, P.A., 1994. Ecological effects of river ice break-up: a review and perspective. Freshw. Biol. 32 (2), 261–275. https://doi.org/10.1111/j.1365-2427.1994.tb01125.x.

Sharma, S., Blagrave, K., Magnuson, J.J., O'Reilly, C.M., Oliver, S., Batt, R.D., et al., 2019. Widespread loss of lake ice around the Northern Hemisphere in a warming world. Nat. Clim. Change 9 (3), 227–231. https://doi.org/10.1038/s41558-018-0393-5.

Sharma, S., Meyer, M.F., Culpepper, J., Yang, X., Hampton, S., Berger, S.A., et al., 2020. Integrating perspectives to understand lake ice dynamics in a changing world. J. Geophys. Res. Biogeosci. 125 (8). https://doi.org/10.1029/2020JG005799. e2020JG005799.

Sharma, S., Richardson, D.C., Woolway, R.I., Imrit, M.A., Bouffard, D., Blagrave, K., et al., 2021. Loss of ice cover, shifting phenology, and more extreme events in Northern Hemisphere lakes. J. Geophys. Res. Biogeosci. 126 (10). https://doi.org/10.1029/2021JG006348 e2021JG006348.

Shatwell, T., Adrian, R., Kirillin, G., 2016. Planktonic events may cause polymictic-dimictic regime shifts in temperate lakes. Sci. Report. 6, 24361. https://doi.org/10.1038/srep24361.

Sibert, R.J., Koretsky, C.M., Wyman, D.A., 2015. Cultural meromixis: effects of road salt on the chemical stratification of an urban kettle lake. Chem. Geol. 395, 126–137. https://doi.org/10.1016/j.chemgeo.2014.12.010.

Smith, K., 1981. The prediction of river water temperatures. Hydrol. Sci. Bull. 26, 19–32. https://doi.org/10.1080/02626668109490859.

Smith, S.D., 1988. Coefficients for surface wind stress, heat flux, and wind profiles as a function of wind speed and temperature. J. Geophys. Res. 93, 15467–15472.

Smits, A.P., MacIntyre, S., Sadro, S., 2020. Snowpack determines relative importance of climate factors driving summer lake warming. Limnol. Oceanogr. Lett. 5 (3), 271–279. https://doi.org/10.1002/lo12.10147.

Smol, J.P., Douglas, M.S., 2007. Crossing the final ecological threshold in High Arctic ponds. Proc. Natl. Acad. Sci. USA 104 (30), 12395–12397. https://doi.org/10.1073/pnas.0702777104.

Spigel, R.H., Priscu, J.C., 1998. Physical limnology of the mcmurdo dry Valleys lakes. In: Priscu, J.C. (Ed.), Ecosystem Dynamics in a Polar Desert: The McMurdo Dry Valleys, Antarctica, vol. 72. Antarctic Research Series, pp. 153–187. https://doi.org/10.1029/AR072p0153.

Spigel, R.H., Priscu, J.C., Obryk, M.K., Stone, W., Doran, P.T., 2018. The physical limnology of a permanently ice-covered and chemically stratified Antarctic lake using high resolution spatial data from an autonomous underwater vehicle. Limnol. Oceanogr. 63, 1234–1252. https://doi.org/10.1002/lno.10768.

Sprules, W.G., Cyr, H., Menza, C.W., 2022. Multiscale effects of wind-induced hydrodynamics on lake plankton distribution. Limnol. Oceanogr. 67, 1631–1646. https://doi.org/10.1002/lno.12158.

Stevens, C.L., Lawrence, G.A., 1997. Estimation of wind-forced internal seiche amplitudes in lakes and reservoirs, with data from British Columbia, Canada. Aquat. Sci. 59 (2), 115–134. https://doi.org/10.1007/BF02523176.

Stewart, K.M., 1973. Detailed time variations in mean temperature and heat content of some Madison lakes. Limnol. Oceanogr. 18, 218–226. https://doi.org/10.4319/lo.1973.18.2.0218.

Straile, D., Jöhnk, K., Henno, R., 2003. Complex effects of winter warming on the physicochemical characteristics of a deep lake. Limnol. Oceanogr. 48 (4), 1432–1438. https://doi.org/10.4319/lo.2003.48.4.1432.

Strøm, K.M., 1955. Waters and sediments in the deep of lakes. Mem. Ist. Ital. Idrobiol. 8 (Suppl.), 345–356.

St. Pierre, K.A., Louis, S., V. L, Lehnherr, I., Schiff, S.L., Muir, D.C.G., Poulain, A.J., et al., 2019. Contemporary limnology of the rapidly changing glacierized watershed of the world's largest High Arctic lake. Sci. Rep. 9 (1), 1–15. https://doi.org/10.1038/s41598-019-39918-4.

Talling, J.F., 1966. The annual cycle of stratification and phytoplankton growth in Lake Victoria (East Africa). Int. Rev. ges. Hydrobiol. 51, 545–562. https://doi.org/10.1002/iroh.19660510402.

Talling, J.F., Lemoalle, J., 1998. Ecological Dynamics of Tropical Inland Waters. Cambridge University Press.

Tedford, E.W., MacIntyre, S., Miller, S.D., Czikowsky, M.J., 2014. Similarity scaling of turbulence in a small temperate lake during fall cooling. J. Geophys. Res. Oceans. 119, 1–25. https://doi.org/10.1002/2014JC010135.

Terzhevik, A., Golosov, S., Palshin, N., Mitrokhov, A., Zdorovennov, R., Zdorovennova, G., et al., 2009. Some features of the thermal and dissolved oxygen structure in boreal, shallow ice-covered Lake Vendyurskoe, Russia. Aquat. Ecol. 43 (3), 617–627. https://doi.org/10.1007/s10452-009-9288-x.

Toffolon, M., Piccolroaz, S., 2015. A hybrid model for river water temperature as a function of air temperature and discharge. Environ. Res. Lett. 10 (11), 114011. https://doi.org/10.1088/1748-9326/10/11/114011.

Tsugeki, N.K., Urabe, J., 2013. Eutrophication, warming and historical changes in the plankton community in Lake Biwa during the twentieth century. In: Goldman, C.R., Kumagai, M., Robarts, R.D. (Eds.), Climatic Change and Global Warming of Inland Waters. Wiley-Blackwell, Oxford, p. 472.

Tundisi, J., Arantes, J.D., Matsuura-Tundisi, T., 2002. The Wedderburn and Richardson numbers applied to shallow reservoirs in Brazil. Verh. Int. Ver. Theoret. Angew. Limnol. 28, 663–666. https://doi.org/10.1080/03680770.2001.11901796.

Tuyisenge, J., 2018. Assessing the temporal mixing and stratification in Lake Kivu. In: thesis, M.S. (Ed.). UNESCO-IHE Institute for Water Education, Delft. https://doi.org/10.25831/9b10-z912.

Ulloa, H.N., Ramon, C.L.C., Doda, T., Wuest, A., Bouffard, D., 2022. Development of overturning circulation in sloping waterbodies due to surface cooling. J. Fluid Mech. 930, A18. https://doi.org/10.1017/fjm2021.883.

Vehmaa, A., Salonen, K., 2009. Development of phytoplankton in Lake Pääjärvi (Finland) during under-ice convective mixing period. Aquat. Ecol. 43, 693–705. https://doi.org/10.1007/s10452-009-9273-4.

Verburg, P., Antenucci, J.P., 2010. Persistent unstable atmospheric boundary layer enhances sensible and latent heat loss in a tropical great lake: lake Tanganyika. J. Geophys. Res. 115, D11109. https://doi.org/10.1029/2009JD012839.

Verburg, P., Antenucci, J.P., Hecky, R.E., 2011. Differential cooling drives large-scale convective circulation in Lake Tanganyika. Limnol. Oceanogr. 56 (3), 910–926. https://doi.org/10.4319/lo.2011.56.3.0910.

Vesala, T., Huotari, J., Rannik, Ü., Suni, T., Smolander, S., Sogachev, A., et al., 2006. Eddy covariance measurements of carbon exchange and latent and sensible heat fluxes over a boreal lake for a full open-water period. J. Geophys. Res. Atmos. 111 (D11). https://doi.org/10.1029/2005JD006365.

Vincent, A.C., Mueller, D.R., Vincent, W.F., 2008a. Simulated heat storage in a perennially ice-covered High Arctic lake: sensitivity to climate change. J. Geophys. Res. 113, C04036. https://doi.org/10.1029/2007JC0004360.

Vincent, W.F., 2018. Lakes: a very short introduction. Oxford University Press, Oxford.

Vincent, W.F., Bertola, C., 2014. Lake physics to ecosystem services: forel and the origins of limnology. Limnol. Oceanogr., e-Lectures. https://doi.org/10.4319/lol.2014.wvincent.cbertola.8.

Vincent, W.F., MacIntyre, S., Spigel, R.H., Laurion, I., 2008b. The physical limnology of high-latitude lakes. In: Vincent, W.F. (Ed.), Polar Lakes and Rivers: Limnology of Arctic and Antarctic Aquatic Systems, J. Laybourn-Parry. Oxford University Press, Oxford, p. 377.

Walker, K.F., Likens, G.E., 1975. Meromixis and a reconsidered typology of lake circulation patterns. Verh. Int. Ver. Theor. Angew. Limnol. 19, 442–458. https://doi.org/10.1080/03680770.1974.11896084.

Walling, D.E., Webb, B.W., 1992. Water quality. I. Physical characteristics. In: Calow, P., Petts, G.E. (Eds.), The Rivers Handbook. I. Hydrological and Ecological Principles. Blackwell Scientific Pubs, pp. 48–72.

Wetzel, R.G., 1972. The role of carbon in hard-water marl lakes. In: Likens, G.E. (Ed.), Nutrients and Eutrophication: The Limiting-Nutrient Controversy. Special Symposium. Amer. Soc. Limnol. Oceanogr., vol. 1, pp. 84–97.

Wetzel, R.G., 1973. Productivity investigations of interconnected lakes. I. The eight lakes of the Oliver and Walters chains, northeastern Indiana. Hydrobiol. Stud. 3, 91–143.

White, D.S., Elzinga, C.H., Hendricks, S.P., 1987. Temperature patterns within the hyporheic zone of a northern Michigan river. J. N. Am. Benthol. Soc. 6, 85–91. https://doi.org/10.2307/1467218.

Wolff, C., Haug, G.H., Timmermann, A., Sinninghe Damsté, J.S., Brauer, A., Sigman, D.M., Cane, M.A., Verschuren, D., 2011. Reduced interannual rainfall variability in East Africa during the last ice age. Science 333, 743–747. https://doi.org/10.1126/science.1203724.

Woolway, R.I., Jones, I., Hamilton, D.P., Maberly, S.C., Muraoka, K., Read, J.S., Smyth, R.L., Winslow, L.A., 2015. Automated calculation of surface energy fluxes with high-frequency lake buoy data. Environ. Model. Softw. 70, 191–198. https://doi.org/10.1016/j.envsoft.2015.04.013.

Woolway, R.I., Sharma, S., Weyhenmeyer, G.A., Debolskiy, A., Golub, M., Mercado-Bettín, D., et al., 2021. Phenological shifts in lake stratification under climate change. Nat. Commun. 12 (1), 1–11. https://doi.org/10.1038/s41467-021-22657-4.

Woolway, R., Weyhenmeyer, G.A., Schmid, M., Dokulil, M.T., de Eyto, E., 2019. Substantial increase in minimum lake surface temperatures under climate change. Clim. Change 155, 81–94. https://doi.org/10.1007/s10584-019-02465-y.

Wüest, A., Lorke, A., 2003. Small-scale hydrodynamics in lakes. Ann. Rev. Fluid Mech. 35, 373–412. https://doi.org/10.1146/annurev.fluid.35.101101.161220.

Xenopoulos, M., Schindler, D., 2001. The environmental control of near-surface thermoclines in boreal lakes. Ecosystems 4, 699–707. https://doi.org/10.1007/s10021-001-0038-8.

Xie, Q., Liu, Z., Fang, X., Chen, Y., Li, C., MacIntyre, S., 2017. Understanding the temperature variations and thermal structure of a subtropical deep river-run reservoir before and after impoundment. Water 9, 603. https://doi.org/10.3390/w9080603.

Yang, B., Wells, M.G., Li, J., Young, J., 2020. Mixing, stratification, and plankton under lake-ice during winter in a large lake: implications for spring dissolved oxygen levels. Limnol. Oceanogr. 65 (11), 2713–2729. https://doi.org/10.1002/lno.11543.

Yankova, Y., Neuenschwander, S., Köster, O., Posch, T., 2017. Abrupt stop of deep water turnover with lake warming: drastic consequences for algal primary producers. Sci. Rep. 7 (1), 1–9. https://doi.org/10.1038/s41598-017-13159-9.

Ye, X., Anderson, E.J., Chu, P.Y., Huang, C., Xue, P., 2019. Impact of water mixing and ice formation on the warming of Lake Superior: a model-guided mechanism study. Limnol. Oceanogr. 64, 558–574. https://doi.org/10.1002/lno.11059.

Yeates, P.S., Imberger, J., 2004. Pseudo two-dimensional simulations internal and boundary fluxes in a stratified lakes and reservoirs. Int. J. River Basin Manag. 1 (4), 1–23. https://doi.org/10.1007/s10750-006-0111-6.

Zadereev, E.S., Boehrer, B., Gulati, R.D., 2017. Introduction: meromictic lakes, their terminology and geographic distribution. In: Gulati, R.D., Zadereev, E.S., Degermendzhi, A.G. (Eds.), Ecology of Meromictic Lakes. Springer, Cham, pp. 1–11. https://doi.org/10.1007/978-3-319-49143-1_1.

CHAPTER

8

Water Movements

David P. Hamilton[1] *and Sally MacIntyre*[2]

[1]Australian Rivers Institute, Griffith University, Brisbane, Australia [2]Department of Ecology, Evolution and Marine Biology, University of California, Santa Barbara, California, United States

OUTLINE

Lakes, ponds, reservoirs, and flowing waters are dynamic systems in which organisms live, grow, and actively move or are passively transported and where a diverse set of biogeochemical reactions occur. The movement of water within these aquatic ecosystems influences how organisms are distributed and where, when, and how fast reactions occur. Through theory and experimentation, both in the laboratory and in the field, the mechanisms influencing the flow of water have been identified and evaluated. Hydrodynamic modeling has extended our understanding of the individual processes at scales up to the whole lake or whole river. In this chapter we cover pertinent physical processes, in particular turbulence, and its influence on dispersion and mixing over large and small scales. Turbulence is enhanced in boundary layers, such as the surface of lakes and rivers directly influenced by the atmosphere, and the bottoms and sides of lakes and rivers where currents interact with the substrate. Wind, cooling of surface waters, and the flow of water generate large-scale flows, such as currents, organized spiral motions called Langmuir cells, surface waves, flows driven by differences in density, and internal waves. Curvature in streams and rivers and variability in lake morphometry and the wind field induce spiraling motions in streams and gyres in lakes. We describe these processes and provide equations pertinent to quantifying the flows and mixing. On the system scale, flows interact with boundaries and density stratification, and the magnitudes of currents and turbulent mixing change. We provide examples of the resulting spatial and temporal variability of physical perturbations and their importance for ecosystem processes. Similarly, we review one- and three-dimensional hydrodynamic models and provide examples of how they can be used for predictions of changes in the hydrodynamics of lakes over time and in ecosystem studies.

Wetzel's Limnology, Fourth Edition
https://doi.org/10.1016/B978-0-12-822701-5.00008-2

I. Hydrodynamics and physical limnology

The flow of water within streams, rivers, lakes, ponds, and other water bodies is governed by principles from the field of hydrodynamics. The ensuing motions determine the habitats for aquatic organisms and create conditions that influence biogeochemical reactions. *Hydrodynamics* deals with the movement of parcels of water, each containing a large number of molecules, not the movement of individual molecules. Understanding hydrodynamics requires considerations of the properties of water (and ice), including density, specific latent heat and specific heat capacity, viscosity, and molecular diffusivity (see Chapter 3; Imboden, 2003), the forces acting on the water, and the ensuing response. In aquatic systems water motion is induced by gravity, as in streams flowing downhill, by wind that induces currents, and by uneven heating or cooling, which causes density differences that drive flows. Water motion is also influenced by shortwave and longwave radiation, evaporation, and conduction, which influence density stratification (Chapter 7). The presence of boundaries modifies flows. For example, in streams and rivers friction slows currents near the bottom, the curves in streams and rivers create eddies and meanders, and the friction from wind acting on the surface of the water creates currents. The regions in which the flow is modified are called *boundary layers*.

As the flow of water modifies the habitat of organisms and transports solutes critical for the growth of organisms or pollutants that are detrimental, a key goal in physical limnology is to quantify *advection* (how fast the water flows), *dispersion* by turbulent motions (how fast it spreads), and *mixing* (how fast it homogenizes by spreading and diffusion). Quantifying advection, dispersion, and mixing relies on conservation laws for mass, momentum, and energy that are used in mathematical formulations to determine the velocities, pressures, and temperatures of water. These formulations are the basis of hydrodynamic models used to simulate the dispersion of heat and solutes (Hodges, 2009; see Section IV), as well as the foundation for water quality models that describe the way in which planktonic organisms, solutes, and particulate material are redistributed by water movements.

Theoretical, laboratory, and observational studies have been essential for identifying critical hydrodynamic processes. Both measurements in the field, enabled by recent advances in sensor technology, and hydrodynamic models have greatly extended theoretical and laboratory studies by providing assessments of mixing dynamics in inland water bodies, by identifying flow patterns, and by quantifying the resulting advection, dispersion, and mixing. The fluxes due to *turbulence*, that is, spreading and mixing, are quantified using *coefficients of dispersion* and *eddy diffusivity* and, for wind speeds less than ~10 m s^{-1}, gas transfer velocities. Understanding these topics is essential for describing how physical processes moderate ecosystem functions, such as metabolism, and for calculating the fluxes of climate forcing trace gases.

A. Laminar, transitional, and turbulent flows

Water in lakes and rivers is not still, with its motion categorized as laminar, turbulent, or transitional between the two. *Laminar flow* is characterized by linear motions, and *transitional flow* is characterized by distinctive curving motions, while *turbulent flow* is denoted by irregular, chaotic fluctuations in velocity and pressure that generate mixing (Fig. 8-1). The types of flows that occur in a system can be predicted by *dimensionless indices*. These indices are developed by considering the forces acting on a parcel of water or a water body, evaluating the dimensions, and collapsing the terms.

Dimensionless indices commonly used in studies of inland waters to understand turbulent flows are the *Reynolds, Richardson,* and *Froude numbers* (Chapter 8 II.A) and the *Wedderburn* and related *Lake numbers* (Chapter 7 I.D and II.A; Chapter 8 Section IIIB). Other metrics describe the dominant forcing mechanisms. For instance, the *Monin–Obukhov length scale* (Chapter 7) indicates the depth to which turbulence production from wind-induced shear exceeds turbulence production by near-surface cooling, and the *Damkohler number* is used, for example, to infer whether there is adequate time for a chemical reaction to occur prior to flow out of a system.

The *Reynolds number* (Re) is the ratio of the inertial force of the water to its viscous force (Reynolds, 1883):

$$\mathrm{Re} = -\frac{\mathrm{u}l}{\upsilon} \tag{8-1}$$

where u is the mean velocity of the fluid (m s^{-1}), l is a characteristic length scale (m), and υ is the kinematic viscosity of water (~10^{-6} m^2 s^{-1}). The *inertial forces* (F_I) are derived from Newton's second law of motion, $F_I = \rho A u^2$, where ρ is density, A is surface area, and the *viscous forces* (F_V) are given by $F_V = \mu A u/l$, where μ is dynamic viscosity, a measure of the stickiness of the fluid, or strictly, the resistance to deformation. For this case, l is a length scale with $S = l^2$ (Vogel, 2020). The ratio of the terms

$$\frac{F_I}{F_V} = \frac{\rho A u^2}{\left(\frac{\mu A u}{l}\right)} = \frac{\rho l u}{\mu} \tag{8-2}$$

is the Reynolds number. μ/ρ is kinematic viscosity as used in Eq. 8-1 and conveniently is nearly constant

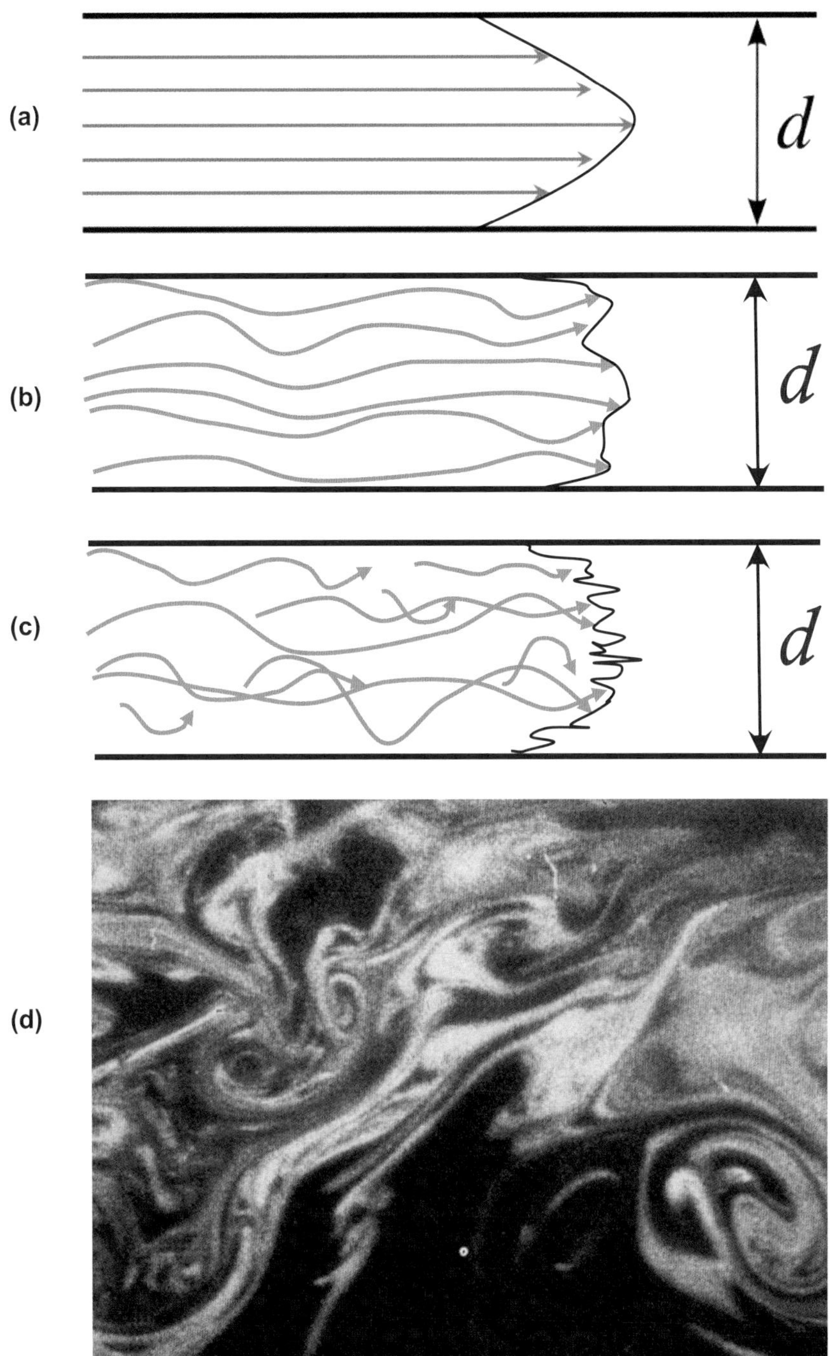

FIGURE 8-1 (a) Laminar, (b) transitional, and (c) turbulent flow in an along-stream direction of lateral dimension *d*. *Blue lines* in (a) to (c) illustrate pathways of water in the flow. In (a) the velocities along each flow path do not vary; in (b) and (c) there is a mean velocity with fluctuations. In (b) the fluctuations are periodic, whereas those in (c) are chaotic. *(From SimScale: https://www.simscale.com/docs/simwiki/cfd-computational-fluid-dynamics/what-is-laminar-flow/.)* (d) Turbulent eddies in a flow (Van Dyke, 1982). The stretching of the fluid by the larger eddies creates smaller ones, and these in turn become smaller and gradients become progressively thinner, such that final mixing (not shown) is by molecular diffusion.

in water. When inertial forces dominate, the fluid will tend to stay in motion once set in motion. When viscous forces dominate, the motion will tend to cease due to friction. The transition between laminar and turbulent flow varies depending on scale and application, but within aquatic ecosystems, turbulent flows are prevalent (see Eckhardt, 2008). For calculating Re in a stream, l is equivalent to the hydraulic radius (R_h), defined by the area of the cross-section of the stream channel divided by its wetted perimeter. In lakes *l* can be the depth of the upper mixed layer. When calculated using conservative velocities in the upper mixed layer of a lake (e.g., 0.01 m s^{-1}) and a mixed layer of, say, 4 m, Re equals 40,000, indicating turbulent flow. However, the upper mixed layer is often stratified for part of the day (Chapter 7), so for addressing whether the flow will become turbulent in the presence of vertical density gradients, dimensionless indices such as the Richardson number are used.

Considerable experimentation has been done to find the transitional values of Reynolds numbers for different types of flows. These include experiments with flow over flat plates, analogous to the flow over the surface of a lake, flow around cylinders, which is analogous to flow around stems in beds of aquatic macrophytes (Nepf, 2012; Wingenroth et al., 2021), and flow due to plumes descending from the surface of a lake as it cools. A basic use is for flow around organisms, in which case the transitional value is ~1. Thus, for larger organisms that swim quickly, such as fish and large zooplankton, inertial forces govern their world. In contrast, for small planktonic organisms, Re < 1 and viscosity dominates, and adaptations are required for locomotion (Vogel, 2020). Hence, in evaluating whether flow is laminar or turbulent, it is critical to consider the application.

The Richardson number (Ri; see Chapter 7) is used to determine whether a stratified flow is turbulent. Stratification is considered *stable* when less dense water overlays denser fluid (Fig. 8-2). This condition occurs in freshwater when warm water overlays cooler water at temperatures above 4°C and the converse occurs at temperatures below 4°C. Ri is the ratio of the buoyancy forces due to the stratification that resist mixing to the inertial ones that induce it:

$$\mathrm{Ri} = -\frac{g}{\rho}\frac{d\rho/dz}{(du/dz)^2} \tag{8-3}$$

where g is gravity, ρ is density, z is depth, $d\rho/dz$ is the change in density with depth, and du/dz is the change in horizontal velocity with depth (Fig. 8-2). The equation can be rewritten as:

$$\mathrm{Ri} = N^2/S^2 \tag{8-4}$$

where

$$N = \left(-\frac{g}{\rho}\frac{d\rho}{dz}\right)^{\frac{1}{2}} \tag{8-5}$$

and N is the *buoyancy frequency*, also known as the *Brunt–Väisälä frequency* (s^{-1}) (see Chapter 7), and S (s^{-1}) is the vertical shear of the horizontal velocity (du/dz). In this form Ri is called the gradient Richardson number. For Ri ≤ 0.25, flow can become *unstable* such that denser water overlays less dense water, resulting in mixing. *Kelvin–Helmholtz billows* form as the flow becomes unstable (Figs. 8-2 and 8-3) (Thorpe, 2007). In lakes, the vertical dimension often depends on the degree of stratification, with overturns of 0.25 m vertical scale at times in the metalimnion.

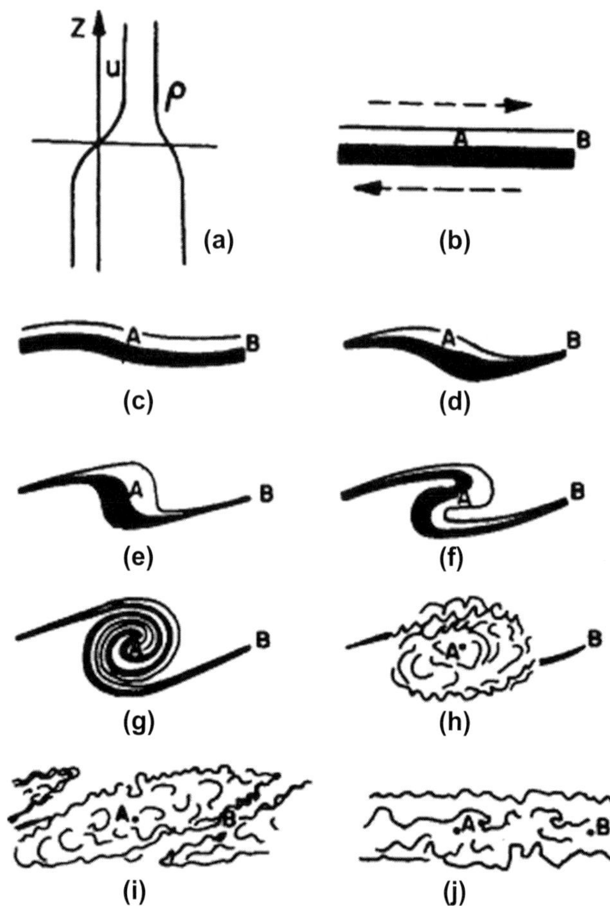

FIGURE 8-2 Growth of shear instability leading to turbulent mixing in a stratified fluid with velocity (u) and density (ρ) profiles with respect to depth (z) shown in (a) and initial stable stratification with less dense water (*white*) over denser water (*black*) and flow directions indicated by *arrows* in (b). *A* and *B* are fixed points, and the *lines* represent surfaces of equal density. As flow begins, the motions become wavelike in (c) and (d). The wavelike features become unstable in (e) and (f) and further roll into Kelvin–Helmholtz instabilities (g) before collapsing into a turbulent patch (h), with (i) and (j) illustrating the final stages of mixing in which the fluid is nearly homogeneous. *(From Mortimer [1974], redrawn from Thorpe [1969].)*

Overturns such as Kelvin–Helmholtz billows become *turbulent eddies*. Thorpe (1977) illustrated that overturns can be identified with high-resolution temperature sensors and that they occur more frequently in the upper mixed layer than in the thermocline. These observations provided a framework for many subsequent studies of turbulence. The overturning regions are called *Thorpe scales* (L_T), and turbulent length scales are often assessed by determining the size and frequency of the Thorpe scales (Section IIIE).

Turbulence and the Energy Cascade – Turbulent eddies cover a range of sizes (Fig. 8-4). For instance, they can be as large as the depth of the upper mixed layer, which can range from 50 to 100 m or a meter or less, depending on lake size, magnitude of prevailing winds, heat loss at night, and clarity, as well as the width and depth of a channel. In stratified flows, as in the thermocline of lakes, the largest size may be that of a Kelvin–Helmholtz billow and considerably smaller than in the mixed layer. Due to stretching, folding, and *vorticity*, that is, spinning induced by velocity gradients, the largest eddies spawn progressively smaller ones down to the scale of viscosity. This process was penned as a poem by L. F. Richardson (1922):

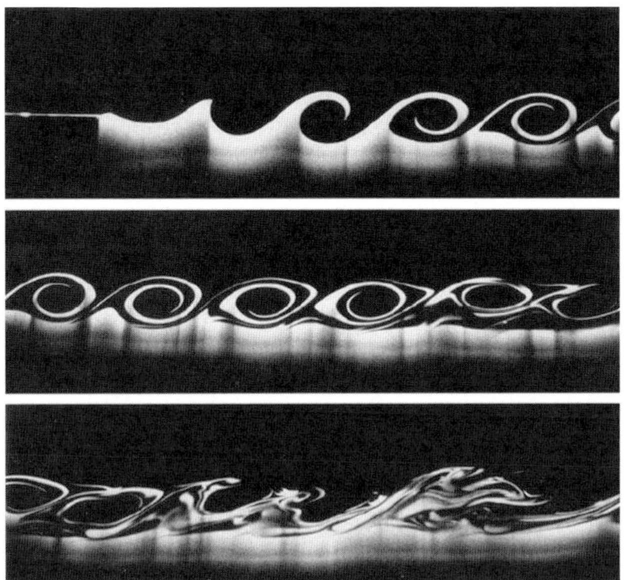

FIGURE 8-3 Kelvin–Helmholtz billows with an upper layer that is slightly less dense and flowing faster than the fluid below. The fluid movement progresses from waves (*upper panel*) to billows (*middle panel*) and is beginning to collapse into turbulent motions (*lower panel*). The photographs are from a laboratory experiment in which fluorescent dye dissolved in the lower layer is illuminated by a thin sheet of light. *(Photo credit Gregory A. Lawrence, University of British Columbia.)*

> Big whirls have little whirls,
>
> That feed on their velocity,
>
> Little whirls have lesser whirls,
>
> And so on to viscosity
>
> – in the molecular sense.

Eddies can be described as a rotating region of a fluid. They can also be characterized by a turbulent velocity fluctuation (u) with a certain length scale (l) (Mortimer, 1974). This approach enables a turbulent flow field to be envisioned as covering a spectrum of wavelengths, each of which applies to the diameter of an eddy. The larger eddies obtain their energy from the forces generating turbulence, that is, wind, cooling, or gravity, also known as the stirring agent. At the largest scale (Fig. 8-4b), the turbulent velocity

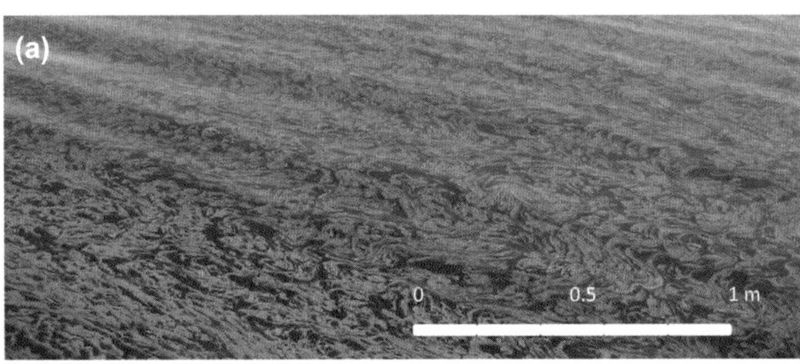

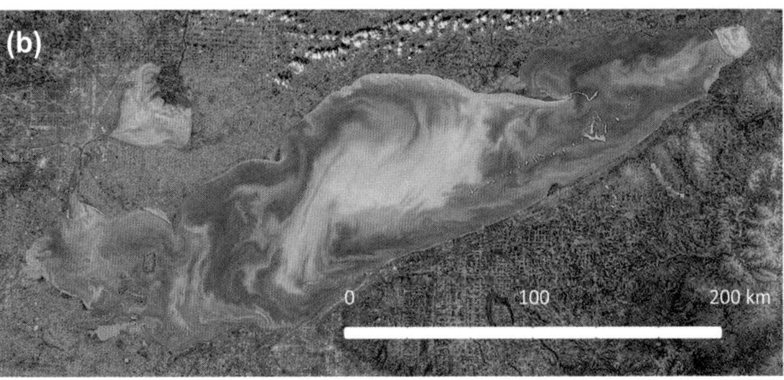

FIGURE 8-4 Examples of turbulent eddies and *gyres* (a rotating transitional flow) acting at two different scales to disperse and concentrate Cyanobacteria at the water surface of (a) Lake Ohinewai, Waikato, New Zealand *(Photograph by Adam Daniel)*, and (b) Lake Erie, USA/Canada *(NASA Earth Science Program)*.

fluctuations may not be the same in the horizontal and vertical, indicating that the turbulence is *anisotropic*. The energy cascades to eddies of moderate scale in which the turbulence is *isotropic*; that is, the magnitude of velocity fluctuations is the same in all directions. Here, the energy cascades as a function of the *wavenumber*, the inverse of the wavelength, in a predictable manner, and eddies in this size range are classified as being in the inertial subrange. The smallest eddies are classified as being in the inertial/viscous subrange, where some dissipation of energy by viscosity occurs, and finally in the molecular subrange the energy of turbulence is transferred to heat.

The concepts related to the size and velocity of turbulent eddies led to what is called the *statistical theory of turbulence* (Tennekes and Lumley, 1972; Mortimer, 1974). Spreading and mixing are conceptualized as the product of the turbulent velocity (u) and length scales (l). That is, coefficients (K) describing turbulent diffusion (units of $m^2\ s^{-1}$) can be written as:

$$K = c_1 ul \tag{8-6}$$

where c_1 is an empirically determined coefficient. The large-scale eddies are related to the smallest scales through the rate of dissipation of turbulent kinetic energy (ε), that is, the rate that energy is lost from the turbulent motions by viscosity (units of $m^2\ s^{-3}$):

$$\varepsilon = u^3/l \tag{8-7}$$

These relations, experimentally verified as will be seen below, have been critical for understanding hydrodynamic processes and their biogeochemical implications in aquatic systems.

Measurements of turbulence require instruments capable of sampling fast enough to capture the turbulent velocity fluctuations or the fluctuations of scalars in the flow, such as temperature (see Section IB: Measurement of turbulence). These data enable the quantification of the turbulent kinetic energy of the flow, its production, transport, and decay. This latter quantity, the rate of dissipation of turbulent kinetic energy, is the most commonly measured of these terms. With it, the turbulence within the mixed layer, thermocline, and hypolimnia of lakes have been described, as well as flow in rivers. Temperature sensors with sufficient resolution can be used to identify turbulent length scales, that is, instabilities such as those from Kelvin–Helmholtz billows (Figs. 8-2 and 8-3) or near-surface cooling, which lead to penetrative convection (Thorpe, 1977). Additionally, ε enables a description of the smallest scales of turbulent flow with application to the environment of phytoplankton and other microbes, as well as the fluxes of dissolved gases.

Boundary layers — The concept of a *boundary layer*, the region where velocity gradients form due to friction, is critical for understanding flows in lakes and rivers. Boundary layers form wherever the flow is in contact with either the atmosphere or a substrate. Due to friction, a velocity gradient results, du/dz, and parcels of water are stretched (Fig. 8-5). The eddies typical of turbulence result because of the spin imparted due to these spatial differences in velocity experienced by the parcels of water. Hence boundary layers tend to be turbulent except in thin layers adjacent to the atmosphere and near a substrate. A boundary layer forms near the surface of water bodies where currents induced by wind decay as a function of depth. Similarly, boundary layers form in streams, rivers, and lakes (e.g., the benthic boundary layer; Fig. 8-6) as currents interact with the margins. Velocity (u) decreases from the surface downward or increases from a substrate upward in a region called the

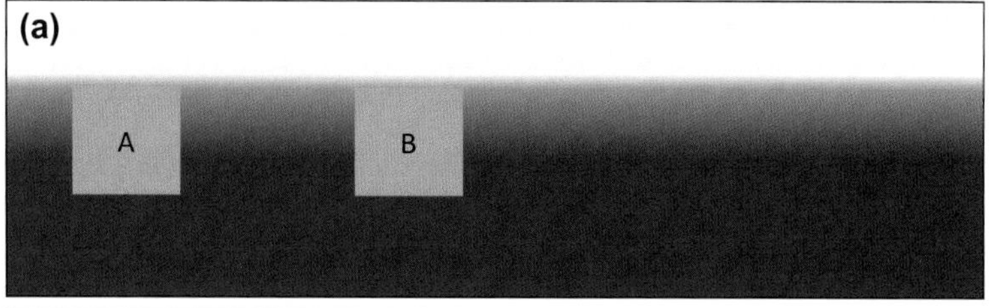

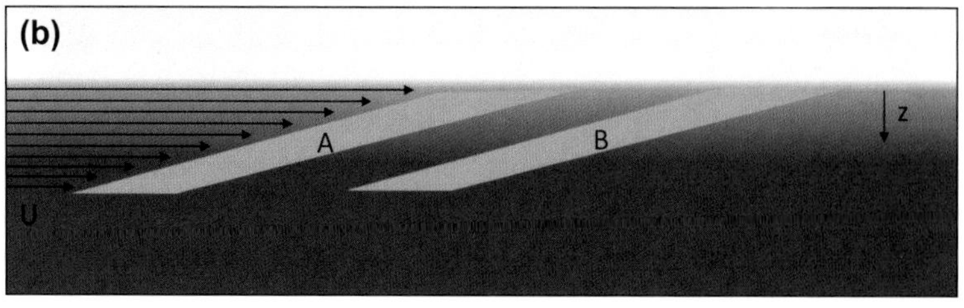

FIGURE 8-5 Parcels of water A and B (a) are initially stationary and (b) are stretched due to faster flow near their surface and slower flow near their bottom in a boundary layer. The change in velocity with depth (z), du/dz, is known as *shear*. The example here is for laminar flow. While this mechanism has been shown over large scales to create thin layers of phytoplankton in stratified flows, within the boundary layer, the difference in flow speed means the parcels experience torque, which causes rotation, such that the parcels become turbulent eddies (Fig. 8-1d). *(Adapted from Eckart [1948].)*

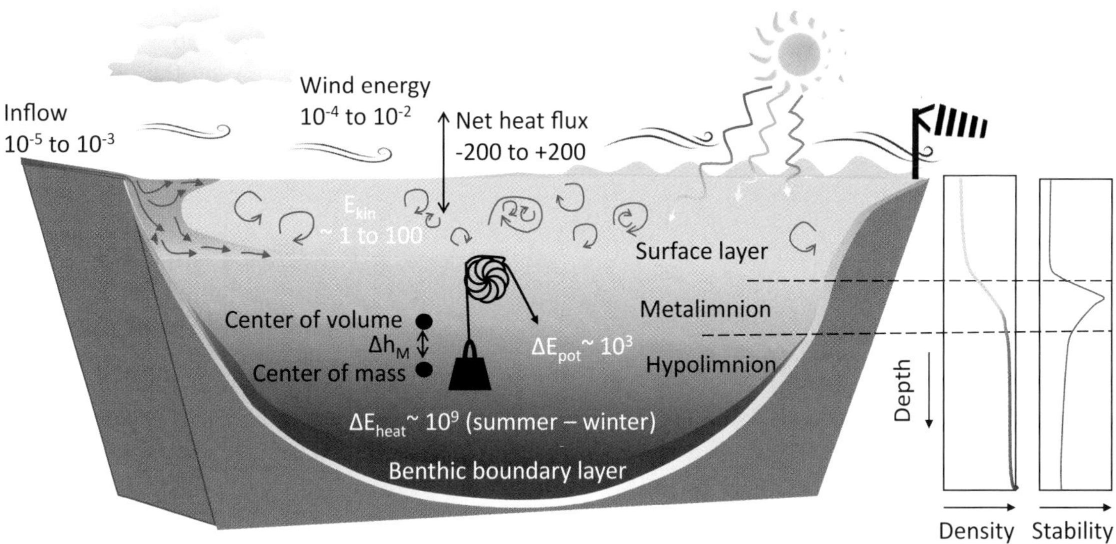

FIGURE 8-6 Schematic of a stratified lake using boundary layer classification. The upper layer is the surface layer in contact with the atmosphere, also described as the actively mixing layer in Chapter 7, and the lower water column adjacent to the sediment–water interface is classified as the bottom, or benthic, boundary layer. The interior layer, which comprises the metalimnion and hypolimnion away from the bottom, is not considered a boundary layer. Also included are the energy fluxes from incoming streams, the wind, and heat flux (and following the convention in Chapter 7, negative ones causing instability and mixing and positive ones causing stratification). Values (*black font*) are in W m^{-2}. The kinetic energy in the surface layer is included (E_{kin}, J m^{-2}). The figure also illustrates that, when a lake is stratified, the center of mass is below the center of volume, with the distance between them Δh_M, and work to overcome the resulting difference in potential energy, ΔE_{pot}, is required to mix the lake and for Δh_M to therefore be zero. This concept underlies the calculation of Schmidt stability and Lake number (Chapter 7). Changes in heat content over the summer, ΔE_{heat}, are shown in *white font* with units in J m^{-2}. Inserts to the right illustrate the density (*line*) and temperature structure (indicated by the color of the *line*; *yellow* is warmer than *blue*) and those to the far right illustrate stability as obtained by differentiating the density profile, the basis of the calculation of buoyancy frequency (Chapter 7). Dissipation rates, indicative of the extent of turbulence, can be orders of magnitude higher in the surface layer and benthic boundary layer than in the interior. *(Redrawn from Wüest and Lorke [2009].)*

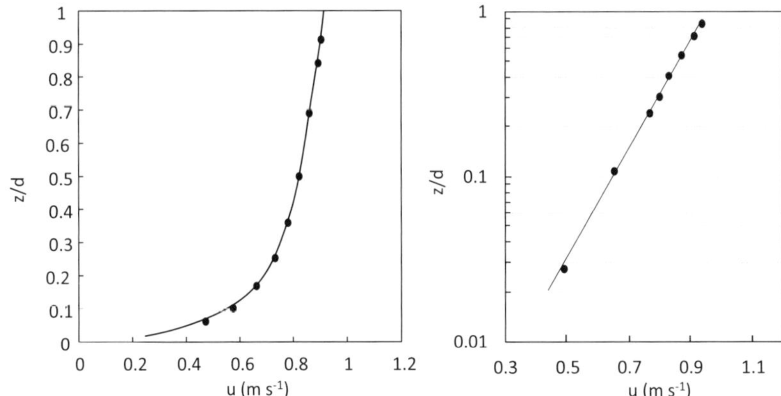

FIGURE 8-7 Boundary layers are identified by changes in velocity with depth due to friction. Velocity, u, decreases due to friction on the bottom of a stream are apparent when plotted against normalized depth (z/d) on a linear scale (*left panel*). Here, depth z was normalized by dividing by the full depth (d) of the channel. Except in a thin layer near the bottom (not shown), changes in velocity occur as a logarithmic function of depth, yielding a *straight line* (*right panel*). The slope of the line is called u*, where u* is known as the *friction velocity.* Flow within the logarithmic portion of the boundary layer is generally turbulent. The channel was 0.09 m deep and 0.851 m wide. In this study, except very close to the substrate, the full depth of the stream was a boundary layer. *(Adapted from Vanoni and Brooks [1957].)*

logarithmic layer where *law of the wall* (LOW) scaling applies (Fig. 8-7):

$$u = \left(\frac{u_*}{\kappa}\right)\ln\left(\frac{z}{z_0}\right) \quad (8\text{-}8)$$

where κ is the Von Kármán constant (≈0.41), u* is the *friction velocity,* z is the depth, and z_0 is the roughness parameter, which depends on, but is much smaller than, any distortions of the substrate, such as rocks or submerged plants on the stream bottom, or surface waves. Friction velocities are thus equal to the slope of the line relating the logarithm of normalized depth (height) to flow speed (Fig. 8-7). Measurements of u at two heights allow estimation of u* without consideration of z_o. Friction velocities are often distinguished with respect to the atmosphere, u^*_a, and to water, u^*_w.

The thin layer adjacent to the interface, only micrometers thick near the air–water interface and millimeters thick on the bottom of lakes, is comprised of the *viscous sublayer* with the *diffusive sublayer* within it (see section below, Small scales in turbulence, Section I; and Chapter 27 on the benthic boundary layer). Here the magnitude of turbulent diffusivity is less than kinematic viscosity, ν (Wüest and Lorke, 2009), and molecular diffusivity. The thickness of the viscous sublayer, δ, is computed as:

$$\delta = 11\nu/u^*_W \quad (8\text{-}9)$$

Variations from law of the wall scaling occur when the water column is stratified or when vegetation is present (Csanady, 2001; Nepf, 2009) and at the air–water interface when surface waves are present (Terray et al., 1996). As will be discussed later, turbulence in boundary layers has many biological and biogeochemical implications, from circulating phytoplankton throughout the upper mixed layer and modifying their light climate (MacIntyre, 1993) to causing variations in annual succession in species composition (Berman and Shteinman, 1998) and influencing redox profiles in the sediments (Frindte et al., 2015; Chapter 27).

B. Mixing and dispersion

Flow in lakes and streams is primarily turbulent (Tennekes and Lumley, 1972). These chaotic flows lead to the spreading or diffusion of the solutes and particles within them. Velocities (u) of a turbulent flow can be averaged to have a mean component, $\overline{u}$, and fluctuations, u'. The latter, by definition, would average to zero, so the product of the square of the fluctuations is computed, averaged over time, and then the square root taken, yielding the *root mean square velocity* (rms) ($u_{rms} = \langle u'^2 \rangle^{0.5}$) where the angled bars indicate averaging. This calculation is one approach to obtain the turbulent velocity scale (u) defined earlier. Here the velocity u implicitly applies to all three directions of motion (see *Turbulence and its measurement* for the full expansion). Transport by the mean flow is referred to as *advection*. The turbulent eddies, which have the velocity scale described above and an average eddy size, l, cause turbulent diffusion (K), quantified as proportional to the product of the turbulent velocity and length scales, that is, $K \approx u_{rms} l$ (Richardson, 1920; Tennekes and Lumley, 1972). Spreading in the horizontal is more often quantified with a *horizontal dispersion coefficient* (K_H) and mixing in the vertical with a vertical *eddy diffusivity* (K_z), although the usage is not always consistent. Both have units $m^2\ s^{-1}$.

Due to the challenges in quantifying turbulence, key concepts have been extended from the understandings of molecular flow. One of these is the concept of *Fickian diffusion*, in which the flux (F) of mass of solute per unit area per unit time is proportional to the gradient of concentration. In the vertical direction and for molecular diffusion (D) the flux is:

$$F = -D(dC/dz) \quad (8\text{-}10)$$

where C is concentration. The turbulent flux of momentum in the vertical, τ, is the density of water, ρ_w, times the time-averaged product of the horizontal, u', and vertical, w', velocity fluctuations, $\tau = \rho_W \langle u'\ w' \rangle$, with this term known as a Reynolds stress. Conceptually, if the fluctuations in both directions are correlated and the time average is positive, flux occurs. Similar reasoning is true for the vertical flux of solutes, which can be calculated as $F = \langle w'c' \rangle$, where c' is the fluctuation in concentration due to turbulence. Applying the reasoning of Fickian diffusion to the solutes in a turbulent flow gives:

$$\langle w'c' \rangle = K_Z(dC/dz) \quad (8\text{-}11)$$

Thus if the eddy diffusivity coefficient and the concentration gradients of solutes are measured, the flux of the solutes due to turbulence can be quantified. The reasoning behind Eq. 8-11 led to the heat budget approach for computing the coefficient of eddy diffusivity in lakes (Jassby and Powell, 1975), described in Chapter 7 I.D with examples included in Chapter 7 I.E and II.A and Chapter 8 Section III.E.

In the next four sections, approaches for quantifying turbulence are provided as well as brief summaries of the magnitude of dispersion and eddy diffusivity coefficients in streams, rivers, and lakes. Additional perspectives on turbulence production and the use of K_H and K_z will be provided in the context of the physical processes occurring in these water bodies, with descriptions in Sections II and III of this chapter.

Tracer approaches to quantify dispersion and eddy diffusivity coefficients – Injection of a tracer, such as a colored dye or a gas that will not react with lake or stream water, is frequently used to quantify dispersion. Figs. 8-8

FIGURE 8-8 Confluence of the turbid, light-colored Queen River (right) and clear, dark-colored King River (Tasmania, Australia). Dispersion and spreading of each tributary inflow are characterized by a dispersion coefficient (K_H, $m^2\ s^{-1}$) and the mean velocity u. *(Photo: Stuart Bunn.)*

and 8-9a–d illustrate the spreading of tracers in the horizontal in a river and lake, respectively, to obtain horizontal dispersion/diffusivity coefficients. Fig. 8-9e illustrates the expected spreading of a tracer injected at 22 m in a 44 m deep lake. Values of K are computed from the rate of change over time of the spreading (Fischer et al., 1979). Horizontal dispersion coefficients have been found to scale with the size of the dye patch, thus increasing as patch size increases and increasingly larger eddies contribute to the spread. The experiment illustrated in Fig. 8-9 confirmed that the relations obtained in ocean waters and in the Laurentian Great Lakes also applied to small lakes (Fig. 8-10). The scaling $K \approx c_1 ul$, where c_1 is a coefficient to relate the contribution of turbulent eddies of different length scales and velocities (u) causing dispersion and ultimately mixing. In the example here the relation has been simplified to only consider the size of the turbulent eddies.

Turbulence and its measurement — Turbulence is quantified as turbulent kinetic energy (TKE, per unit mass in J kg^{-1}; Csanady, 2001), given by:

$$\mathrm{TKE} = 1/2(<u'^2> + <v'^2> + <w'^2>) \quad (8\text{-}12)$$

where u' and v' are velocity fluctuations at 90° angles to each other in the horizontal, also referred to as the x and y directions, and w' are vertical velocity fluctuations, in what is called the z direction. The turbulent velocity scale described previously is computed from the velocity fluctuations in all directions as $u_{rms} = (<u'^2> + <v'^2> + <w'^2>)^{1/2}$. Measurements of turbulence take into consideration the budget of turbulent kinetic energy and simplifications that make the measurements feasible.

The turbulent kinetic energy budget is

$$\frac{d\mathrm{TKE}}{dt} = \mathrm{P} + \mathrm{J_b} - \varepsilon + \frac{d\mathrm{TKE}}{dz} \quad (8\text{-}13)$$

where the left-hand term is the change of turbulent kinetic energy with time, P is the rate of production, J_b is buoyancy flux, ε is the rate of dissipation of turbulent kinetic energy, and $d\mathrm{TKE}/dz$ is the vertical transport of TKE including pressure fluctuations. Typically, the vertical transport term is small, and, assuming the turbulence is at steady state, the equation reduces to $P = -J_b + \varepsilon$. Each variable has units of $m^2\ s^{-3}$. The production term is quantified as $P = -<u'w'>du/dz$, which, following Eq. 8-11, is equivalent to $K_z(du/dz)^2$. The buoyancy flux term is used to evaluate whether the vertical velocity fluctuations (w') are lifting heavier parcels of water upward, as when mixing a stratified flow, or whether heavy particles are falling downward and generating turbulence. The buoyancy flux is calculated as:

$$J_b = -g < w'\rho' > /\rho = -g\alpha < w'T' > \quad (8\text{-}14)$$

where g is gravity, ρ is density, and ρ′ are the density fluctuations. The density fluctuations can also be calculated as α, the coefficient of thermal expansion, times the temperature fluctuations T′, given that the fluctuations

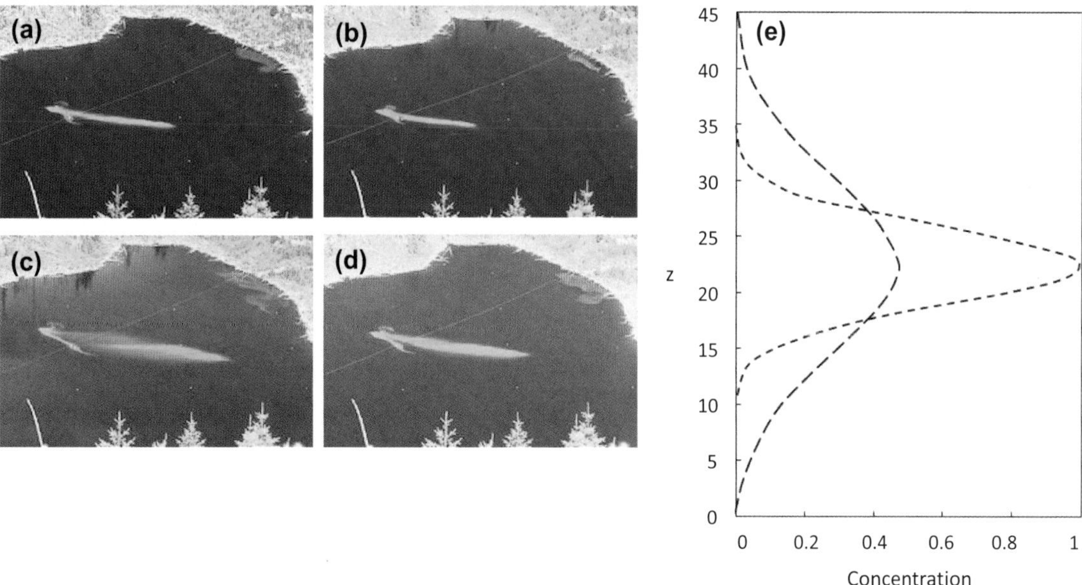

FIGURE 8-9 Dispersion of a central release of sodium fluorescein (*green*) dye and a shore release of Rhodamine WT (*red*) dye into Twin West Lake (British Columbia, Canada) on 24 August 1988. The central release was at 12:48 h and the shore release at 12:59 h. The photographs were taken at (a) 13:10, (b) 13:18, (c) 13:25, and (d) 13:36 h. *(From Lawrence et al., 1995.)* (e) Spreading of a tracer injected 22 m above the bottom in a 45 m deep lake. Here depth z is positive upward from the lake bottom. The coefficient of eddy diffusivity (K_z, $m^2\ s^{-1}$) determines the vertical rate of spreading as indicated by the *dashed lines*, with first sampling having higher maximum concentration. With continued sampling, the tracer would spread even farther from the injection point.

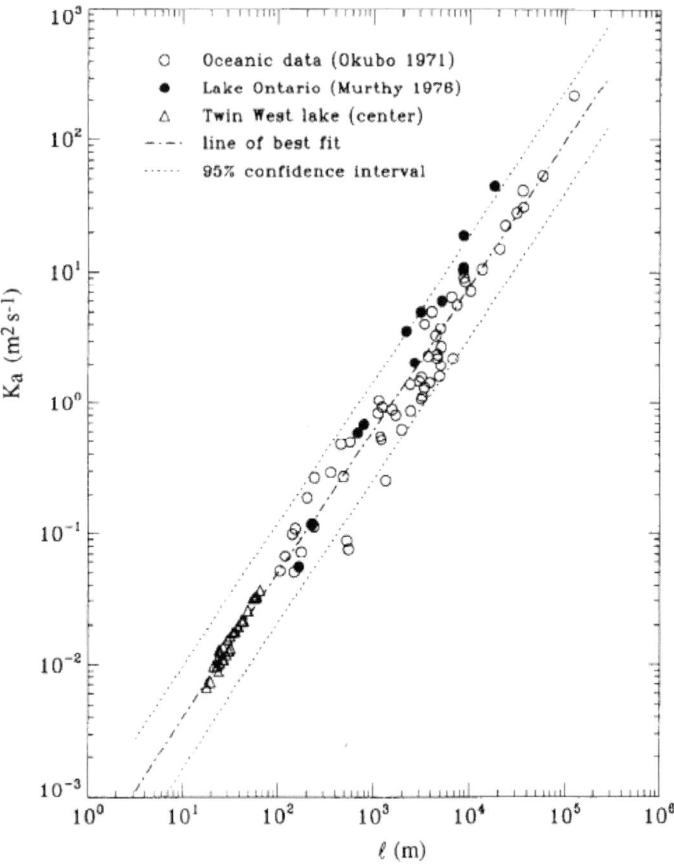

FIGURE 8-10 K_a, apparent horizontal diffusivity, as it increases with patch size l based on data from the ocean and Laurentian Great Lakes and extended to smaller lakes (*triangles*). The line of best fit is $K_a = 3.2 \times 10^{-4}\, l^{1.1}$. An apparent diffusivity is calculated from the rate of spread of the dye patch and the word "apparent" takes into account inherent undersampling and other complexities when doing field work. *(From Lawrence et al. [1995].)*

are small enough for α to apply to the mean temperature and solutes are not contributing to density. J_b is negative when mixing a stratified flow and positive when contributing to turbulence production.

Accurately quantifying the coefficient of eddy diffusivity requires calculating the *flux Richardson number*, R_f, and the *mixing efficiency*, Γ. Under steady state, R_f is defined as:

$$R_f = \frac{J_b}{P} = \frac{J_b}{J_b + \varepsilon} \tag{8-15}$$

For turbulence in stratified flows, the *mixing efficiency*, $\Gamma = R_f/(1 - R_f)$, has an upper bound of ~ 0.2 and can decrease to values of ~ 0.05 when close to boundaries or when buoyancy frequencies are high relative to dissipation rates (Ivey and Imberger, 1991; Gregg et al., 2018; Monismith et al., 2018; MacIntyre et al., 2021a). Γ is larger in *convective flows* in which cooling induces instabilities (Ulloa et al., 2018). For stratified flows, where mixing efficiency is $\leq \sim 0.2$, turbulence production (P) is approximately balanced by dissipation, that is, $P \approx \varepsilon$.

Returning to the concept of the energy cascade, the kinetic energy per unit mass is proportional to the turbulent velocity scale squared, u^2 (also u_{rms}^2) and, near boundaries, is also approximated by u^{*2}, where u^* is the friction velocity, and the transfer of energy is proportional to u^3/l, where l approximates the size of the largest eddies in the flow. The analysis above then links the larger turbulent scales to the smallest scales such that the dissipation rate can be estimated as $\varepsilon \approx u^3/l$ (Eq. 8-7), as verified experimentally in field studies by Luketina and Imberger (2001). This relation indicates that the viscous decay of energy can be approximated by the large-scale dynamics that are independent of viscosity (Tennekes and Lumley, 1972). In the following, we see how this approximation enables methods to compute coefficients of eddy diffusivity.

Measurements of turbulence are usually based on quantifying its rate of production or dissipation (ε) (Thorpe, 2007; Monismith and MacIntyre, 2009). Both approaches require instruments that can sample at frequencies high enough to enable measurements of the fluctuations in velocity or temperature, which are the signatures of turbulence (Thorpe, 2005). Quantifying turbulence production requires instruments that measure both velocity gradients and velocity fluctuations. *Acoustic Doppler velocimeters* (ADVs) and *acoustic Doppler current profilers* (ADCPs) have been used. Measurements of ε have been obtained with multiple approaches. One approach takes advantage of the rate of transfer of turbulent kinetic energy to progressively smaller scales, that is, the Kolmogorov -5/3 power law within a subrange of wavenumbers (Tennekes and Lumley, 1972; Thorpe, 2007). Applications include measurements at a fixed point as with an ADV and along the beams of high-resolution ADCPs (Guseva et al., 2021). Calculations can also be based on the distance over which turbulent motions are coherent (Wiles et al., 2006; Lorke, 2007) or with large eddy approaches (Greene et al., 2015). Fluxes of scalars, such as dissolved oxygen, can be obtained from the product of oxygen fluctuations and vertical velocity fluctuations (Berg et al., 2009), a procedure known as *eddy covariance*. Such fluxes can also be obtained with high-resolution profiles of oxygen near the sediment–water interface combined with high-resolution velocity measurements (Bryant et al., 2010).

Velocity and temperature-gradient microstructure profilers enable the rate of dissipation of turbulent kinetic energy (ε) to be measured as a function of depth (Osborn and Cox, 1972; Osborn, 1980; Imberger, 1985; MacIntyre, 1993). The use of temperature-gradient microstructure profilers was enabled by critical theoretical work that established the relation of small-scale temperature fluctuations to dissipation rates (Batchelor, 1959).

The assumption that turbulence production is balanced by dissipation enables K_z to be calculated from profiling data. The derivation is based on $<w'\rho'> = K_z\, d\rho/dz$ (similar to Eq. 8-10), the buoyancy frequency $N^2 = (g/\rho)(d\rho/dz)$ (Eq. 8-3), and the definitions of R_f and Γ above. That is,

$$K_z = \Gamma\varepsilon/N^2 \tag{8-16}$$

(Osborn, 1980; Thorpe, 2007).

Estimating K_z accurately requires accounting for the forcing mechanisms and proximity to interfaces. Examples of the variation in values of K_z are provided in Chapter 7 and below under the subsection *Turbulence in lakes and ponds*.

Turbulence in streams and rivers – Flow in streams is mostly turbulent, driven by the channel bed slope and modified by the stream morphology. Water is advected downstream, while shear at interfaces (e.g., at the stream bed and banks) generates turbulence that rapidly removes any patchiness in concentrations and, if strong enough, may also scour the channel and modify its shape (see Section IID). Velocity is near zero at the sediment–water interface and, as illustrated in Fig. 8-7, increases logarithmically within the bottom boundary layer, reaches a maximum in the central channel, and then decreases just below the water surface due to shear at the air–water interface. As stream velocity decreases (e.g., in downstream regions where slope decreases, and stream depth and width usually increase), turbulence may decrease. For bulk flow in streams and rivers, the horizontal dispersion, K_H, is of the order of 10^{-2} to 10 $m^2\ s^{-1}$ (Fischer et al., 1979; Deng et al., 2002). Velocity and turbulence are, however, not evenly distributed vertically or horizontally in streams. Benthic boundary layers (Chapter 27) provide shelter for macroinvertebrates in fast-flowing streams or during spates when boundary layers form adjacent to cobbles and boulders. Flow in the nearshore may be somewhat disconnected from the mainstem flow and associated turbulence. The quieter waters support the development of complex lateral habitats (e.g., *sidearms*) where aquatic plants may proliferate, and organisms in the associated food webs are not advected downstream or scoured from the streambed.

Turbulence in lakes and ponds – Eddy diffusivities in lakes vary over predictable ranges in the upper mixed layer, the thermocline, and the hypolimnion (Chapter 7, Figs. 7-11, 7-17, 7-22 to 7-24, and 7-29). The highest eddy diffusivities may be of order $10^{-2}\ m^2\ s^{-1}$ in the upper mixed layer on a windy day. The values of K_z may approach molecular values, $\sim 10^{-9}\ m^2\ s^{-1}$ for solutes and $1.4 \times 10^{-7}\ m^2\ s^{-1}$ for heat, in a quiescent thermocline or deep in an ice-covered lake. K_z may vary horizontally across the thermocline, from approximately $2 \times 10^{-8}\ m^2\ s^{-1}$ at central locations to $1 \times 10^{-4}\ m^2\ s^{-1}$ where there is strong boundary mixing from the interaction of internal waves with the bottom boundary (MacIntyre et al., 1999; see Chapter 7; Section IIIB). Values of K_z also vary seasonally with stratification (Chapter 7).

Spreading in the horizontal is much faster than in the vertical in lakes, with initial scaling indicating spreading in the horizontal would be proportional to the patch size, l, raised to the 4/3 power, that is, $l^{4/3}$ (Okubo, 1971). Lawrence et al. (1995) extended the scaling to smaller lakes (Fig. 8-10). Simulations in 1.5 km^2 Toolik Lake (Alaska, USA) were of order $10^{-1}\ m^2\ s^{-1}$ for small patches (100 m) during light winds and increased to the order of 1 $m^2\ s^{-1}$ during higher winds as spreading increased to the basin scale, in agreement with the predictions in Fig. 8-10 (Rueda and MacIntyre, 2010). Under light winds, it would take 10 days for a patch to disperse, whereas, under higher winds, it takes only a day. Imboden and Wüest (1995) present models that take into account velocity shear; Monismith (1986) provides analytical procedures to calculate K_H.

Ecological importance of K_H and K_z – Quantifying K_H and K_z is of considerable ecological and biogeochemical importance. When concentration gradients of solutes or particles, such as algae or sediments, occur, *fluxes* (F) can be calculated as described in Eq. 8-10, that is, $F = K(dC/dx)$, where dC/dx is the concentration gradient over a defined distance x, where x can be a horizontal or vertical dimension, and K is either a dispersion or eddy diffusivity coefficient. Units are moles or mass per unit area per unit time (mol m^{-2} s^{-1} or g m^{-2} s^{-1}). Expressed in three dimensions (x, y, and z), the rate of change of concentration due to turbulent diffusion is given by:

$$\partial C / \partial t = \frac{\partial}{\partial x}\left(K_x \frac{\partial C}{\partial x}\right) + \frac{\partial}{\partial y}\left(K_y \frac{\partial C}{\partial y}\right) + \frac{\partial}{\partial z}\left(K_z \frac{\partial C}{\partial z}\right) \quad (8\text{-}17)$$

K_x and K_y are usually horizontal dispersion coefficients in the x and y directions, and K_z is the coefficient of eddy diffusivity in the vertical direction. In the vertical this equation implies calculating the flux of solute in units of moles or mass across a plane into a series of volumes over a defined time and then, by accounting for the changes in moles or mass in each volume, determining the change in concentration in each volume per unit time.

To determine how the concentration of a solute C (g m^{-3}) changes with time due to both advection and diffusion, an *advection–diffusion equation* is useful and is the basis of many numerical models of hydrodynamics. The term representing advection is velocity times the concentration gradient in the direction of motion. Eq. 8-17 is the advection–diffusion equation in one dimension assuming any solutes are unreactive or conserved. The equation would be applied for determining tracer concentrations with distance along a river or with depth in a lake:

$$\partial C/\partial t = K_x \frac{\partial^2 C}{\partial x^2} - u\frac{\partial C}{\partial x}. \quad 0 \le x \le L,\ 0 < t \le T \quad (8\text{-}18)$$

where t is time (s), x is distance (m) representing either the longitudinal coordinate in a stream or the vertical coordinate in a stratified lake, u is mean advective velocity (m s^{-1}), K_x is the local dispersion coefficient (or coefficient of eddy diffusivity, m^2 s^{-1}), L is the length of the stream channel or the lake water depth (m), and T is the time frame considered (s). If applied in the vertical for a lake, u would likely be negligible but the sinking or rising speeds of particles such as phytoplankton would be used instead. The equation can be extended to two or three dimensions and reaction rates can be included such that the interactions of flow with biogeochemical (non-conservative) processes are included.

Time scale approaches — While expressions such as Eq. 8-18 are complex and made more so by adding the reaction terms, scaling arguments are often used to simplify analyses. They can determine, for instance, how fast the transport would be and then can be evaluated against the time scale of biological or chemical processes (see Section A: Laminar, transitional, and turbulent flows, specifically the Damkohler number). To first order, the *time scale of advection* is L/U where L is the distance of travel and U is the mean velocity. The *time scale of mixing* is $\tau_{mix} = l/u$ or l^2/K, where l is an appropriate length scale, u is the turbulent velocity scale defined above, and K is K_H or K_z, depending on the problem. For example, to determine whether a solute would flow downstream in a channel faster than it would spread laterally, one first divides the downstream distance L by U to obtain the advective time scale. Spreading would occur over the width of the channel and is initially concentrated at the center. Thus l would be one-half the width of the channel and K_H would be obtained by experimentation or from the analyses summarized in Fig. 8-10. Approaches such as these have been applied to evaluate whether pollutants flow out of lakes before dispersing, and whether there are biogeochemical transformations of a chemical prior to it leaving the lake (Knauer et al., 2000). They can also be used to determine whether phytoplankton cells would grow and persist rather than sinking or being mixed below the euphotic zone (Huisman et al., 2002; O'Brien et al., 2003; see Small scales in turbulence below and Section IIIE).

Small scales in turbulence — The attributes of turbulent eddies on the smallest scale, the *Kolmogorov microscales* of length, time, and velocity, can be computed from the rate of dissipation of turbulent kinetic energy (ϵ) and kinematic viscosity (ν). The Kolmogorov microscales are used in models to quantify fluxes at the benthic and atmospheric boundary layers (Lorke and Peeters, 2006; Zappa et al., 2007). The size of the smallest turbulent eddies, the Kolmogorov microscale (η), is:

$$\eta = \left(\nu^3/\epsilon\right)^{0.25} \quad (8\text{-}19)$$

The Kolmogorov timescale (τ_K), indicative of the lifetime of a turbulent eddy, is

$$\tau_K = \left(\nu^3/\epsilon\right)^{0.5} \quad (8\text{-}20)$$

The Kolmogorov velocity scale (v_K) is:

$$v_K = (\epsilon\nu)^{0.25} \quad (8\text{-}21)$$

The Batchelor length scale, l_B, the scale over which the smallest gradients are removed by molecular diffusion (Batchelor, 1959), is:

$$l_B = \eta Sc^{-n} \quad (8\text{-}22)$$

where the *Schmidt number*, Sc, is the ratio of molecular diffusion of the solute of interest to kinematic viscosity, ν. The exponent n is $-1/2$ near fluid boundaries and $-2/3$ near solid boundaries.

Typical dissipation rates in the upper mixed layer vary from 10^{-9} to 10^{-6} m^2 s^{-3}, implying Kolmogorov length scales of 6×10^{-3} m to 1×10^{-3} m, Kolmogorov time scales of 31 s to 1 s, and Kolmogorov velocities of 2×10^{-4} m s^{-1} to 1×10^{-3} m s^{-1}. The Kolmogorov length scale has been proposed as the size of the viscous sublayer adjacent to the sediment–water interface and at the air–water interface (Lorke and Peeters, 2006), assuming minimal influence by surface waves (Terray et al., 1996). For the dissipation rates above and for heat mixing in the water column, l_B would range from 2×10^{-3} m to 3×10^{-5} m. The *Batchelor scale* is also considered the scale of the diffusive sublayer, the thin layer over which solute flux occurs by molecular diffusion at interfaces (Section B: Mixing and diffusion and Section IIIE) (Lorke et al., 2003; Schwefel et al., 2017). For CO_2 diffusing across the air—water interface at 20°C and for the dissipation rates above, l_B ranges from 230 to 40 μm. Dissipation rates tend to be lower near the benthic boundary layer. However, for the dissipation rates above and for dissolved oxygen at 25°C, l_B would range from 1×10^{-4} m (100 μm) to 2×10^{-5} m (20 μm) at the benthic boundary layer. These calculations illustrate that, as turbulence increases, the viscous sublayer and diffusive sublayer progressively thin, with the potential for enhanced fluxes across interfaces (see Section IIIE).

Phytoplankton, bacteria, and other microorganisms tend to be smaller than the Kolmogorov microscale and hence are embedded within these eddies. The diffusive boundary layer around a phytoplankton cell is about 10 times larger than its radius (Lazier and Mann, 1989). Thus, for phytoplankton cells that range in size from ~1 μm to 100 μm, the diffusive boundary layer would range from 1×10^{-5} m to 1×10^{-3} m, such that in most cases the layer is smaller than or equal to η, and thus itself will not become turbulent. Instead, the boundary layer can be deformed by the turbulence in the flow. For appreciably larger phytoplankton or those in colonial form (i.e., ~400 μm diameter), the boundary layer thickness would exceed η; however, very little energy is contained in turbulent eddies even five times larger than η; hence their diffusive boundary layer is not expected to become turbulent. For larger algal cells, however, including colonial and filamentous morphotypes, there is increased capacity to take up nutrients relative to being in stationary flow (Mann and Lazier, 2006). The increased uptake points toward thinning of the boundary layer by the shear and pressure fluctuations of the small eddies, enabling the more rapid uptake.

A challenge for phytoplankton is to remain in the euphotic zone where there is adequate light to support growth. Diverse strategies have developed, including being positively buoyant (e.g., gas-vacuolate Cyanobacteria), neutrally buoyant, or having flagella (e.g., Dinoflagellata) which enable swimming (Chapters 17 and 18). Some algae, like diatoms, are negatively buoyant, that is, denser than ambient water. The *sinking/floating velocity* (w_s) of negatively/positively buoyant phytoplankton is derived from *Stokes' law* (Stokes, 1851) under idealized laminar flows:

$$w_s = \frac{\left(\rho_p - \rho_w\right) g D_p^2}{18\mu} \tag{8-23}$$

where ρ_p is the particle density (kg m^{-3}), ρ_w is the water density (kg m^{-3}), g is gravity (m s^{-2}), D_p is the particle diameter (m), and μ is the dynamic viscosity of water (kg m^{-1} s^{-1}) (Mann and Lazier, 2006). In laboratory experiments the sinking speeds of growing phytoplankton are generally less than those of senescent cells, and both growing and senescent cells have sinking rates lower than those predicted by Stokes' law. Various morphological adaptations lead to sinking or floating speeds differing from those predicted by Stokes' law. Similarly, accumulation of lipids may reduce the density of, for example, diatoms and slow their sinking speeds. Even positively buoyant phytoplankton may have evolved strategies to improve their competitive success. Accumulation of lipids can make colonies of the chlorophyte *Botryococcus braunii* even more buoyant (Wake and Hillen, 1981), resulting in blooms similar to those of the gas-vacuolate Cyanobacteria shown in Fig. 8-4.

Understanding many aspects of aquatic life, as well as biogeochemical transformations, requires knowledge of turbulence. The small-scale velocity fluctuations associated with the eddies in which cells are embedded will cause subtle changes in their position, somewhat larger eddies will potentially sweep them away from grazers, and even larger eddies will act to redistribute them, modify their residence time in the water column, and expose them to fluctuating light. Thus turbulent eddies at a range of scales, with the smallest motions having lifetimes of 10^{-3} to 10^3 s (Imboden, 2003), modify the growth of phytoplankton and influence organisms of different sizes in lakes and flowing waters.

II. Water movement in rivers and streams

A. Principles of open channel flow

In rivers and streams, the stream banks and bottom constrain flow to a channel that is in contact with the atmosphere. The flow in the channel is called *open channel flow*. Changes in water pressure in the stream, from incoming or outgoing water, lead to increases or decreases in water depth because the relative change in atmospheric pressure is negligible. Gravitational forces that induce water movement depend on the bottom slope and oppose the frictional forces acting at the streambed and bank that impede the flow. The flow in streams is mostly turbulent (e.g., Figs. 8-11 and 8-12), driven by the *channel bed slope* (S) and modified by the stream morphology (denoted as the characteristic length scale, l, or hydraulic radius, R_h). As stream velocity decreases (e.g., in downstream regions where slope decreases, and stream depth and width usually increase), the flow becomes less turbulent. Stream velocity u (m s^{-1}) can be described by the

FIGURE 8-11 Riffles in an Austrian stream create high levels of turbulence that ensure rapid dispersion and homogeneity of the bulk water within the stream. Pools are downstream, denoted by smoother areas of the water surface. *(Photo credit David P. Hamilton.)*

FIGURE 8-12 An example of turbulent flow at the Rhine Falls (Switzerland), where the water depth is highly variable. *(Photo credit David P. Hamilton.)*

Gauckler–Manning formula, usually referred to as *Manning's equation* (Manning, 1891):

$$u = \frac{R_h^{2/3} S^{1/2}}{n} \tag{8-24}$$

where n is "Manning's n" (s $m^{-1/3}$), which is dependent on the depth and characteristics of the streambed that alter its friction, including the nature of the substrate (e.g., clay, cobbles, boulders), streambank morphology, and presence of vegetation. A guide for selecting Manning's n values for natural stream channels and floodplains is given by Arcement and Schneider (1984), but there will be some variations among stream reaches, with flow conditions, and with stream sinuosity not accounted for in the guide.

Flow regimes in streams and rivers can be categorized by a dimensionless *Froude number* (Fr) that considers the ratio of the inertial disturbing force given by water velocity to the opposing gravitational weight of water based on a hydraulic mean depth (h_m) (Jowett, 1993):

$$Fr = \frac{u}{\sqrt{gh_m}} \tag{8-25}$$

The Froude number is important for defining flow regimes. When u varies little at a point (e.g., under baseflow conditions), the flow is categorized as steady, as opposed to unsteady flow with large fluctuations in velocity (e.g., Fig. 8-12). Here baseflow refers to residual stream flow mostly associated with groundwater input to the stream and stable in the short term, with gradual changes at longer time scales corresponding to fluctuations in groundwater levels (Allan et al., 2021) and more abrupt changes that lead to velocities above baseflow associated with rainfall (see below). Values of $Fr < 1$ correspond to subcritical flow when *gravity waves* (see Section IIIA) disturb the uniformity of downstream flow, denoted, for example, by surging in backflows at the edges of a stream channel. The speed that an individual wave propagates is wave *celerity*, represented by $\sqrt{gh_m}$ in the denominator of the Froude number (Eq. 8-25). *Supercritical flow* ($Fr > 1$) means that these waves do not disturb the continuity and uniformity of downstream flow. Various other classifications are in use, including turbulent and laminar streamflow, which are differentiated by the critical Reynolds number (see Section IA above). Between the two is a range of Re values corresponding to transitional flows (see Fig. 8-1b). The variability of flow along a stream provides a diversity of habitats, often associated longitudinally with runs, riffles (Fig. 8-11), and pools, and laterally with a *thalweg* (deepest part of the main channel) and a shear zone separating it from the quiescent nearshore region (Jowett, 1993).

B. Hydrographs

The volume of water per unit time passing a point is known as *discharge* or flow rate, Q (m^3 s^{-1}). It can be visualized as the product of flow speed and cross-sectional area. Discharge varies by location and time, with notable increases during storms. A *hydrograph* (see also Chapter 5) is a plot of Q as a function of time t for a fixed location (e.g., Fig. 8-13). *Baseflow* occurs when flow is steady, for example, ~1 m^3 s^{-1} in Fig. 8-13, whereas *stormflow*, also known as *quickflow*, is characterized by changes (Fig. 8-13). For example, the rising limb of the hydrograph occurs when overland surface runoff and shallow groundwater flow increase with rainfall. The duration of stormflow depends on the intensity of rainfall and the relative contributions of surface runoff and shallow and deep groundwater, with the offset between the peak of rainfall and the peak of stream discharge reflecting storm intensity and the

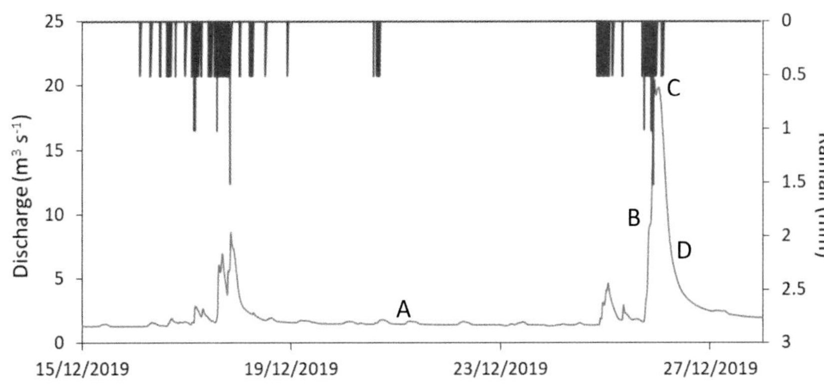

FIGURE 8-13 Hydrograph showing "baseflow" and "stormflow" discharge (*blue line*, 15-min resolution) for Utuhina Stream, Rotorua, New Zealand, and rainfall at Whakarewarewa station within the catchment (*red bars*, 1-h resolution), 15 −27 December 2019. Letters show periods of A: baseflow, B: rising limb of hydrograph corresponding to peak of rainfall, C: peak of discharge, D: descending limb of hydrograph. *(Data from Bay of Plenty Regional Council.)*

relative contributions of surface and groundwater sources. A recession limb follows the peak discharge, when rainfall in temporary storages, mostly shallow groundwater, is gradually depleted and the discharge decreases toward baseflow.

The shapes of hydrographs vary spatially, as well as temporally, based on factors such as land use and land cover, which affect the balance of rainfall and evapotranspiration; soil infiltration capacity (affected by soil composition and extent of impervious surface); the size of the unsaturated (*vadose*) zone and the saturated (*phreatic*) groundwater zone; slope, which affects the rate of water delivery (Allan et al., 2021); and the rate of snowmelt in regions with seasonal snow cover. Hydrographs differ within a stream depending on distance from the headwaters. High in the catchment, streams are classified as *lower order*, and away from the headwaters, streams are usually classified as *higher order* (see also Chapters 4 and 5). For example, in the *Strahler system* (Strahler, 1952) headwater streams are denoted as first order and as streams join, the resulting stream reach has the same order as the higher value of the two joining stream reaches (see also Chapters 4 and 5). Hydrographs of lower-order streams have relatively steep gradients in the rising and recession limbs compared with the higher-order part of the stream network. Where groundwater is strongly depleted, for example, seasonally in Mediterranean climates or with high levels of extraction, baseflow declines to zero, leading to an intermittent stream or a perennial stream that dries out. These processes indicate the importance of baseflow in maintaining key stream ecosystem processes and habitat for biota, as well as for providing water supply for humans.

The frequency of flow of a given magnitude generally involves the application of a *flow–duration curve*, which relates flow (on a logarithmic scale) to a percentage or frequency of exceedance (Fig. 8-14). Flow duration curves are useful for considerations of the median daily flow (50% probability) and derivation of the probability of extreme events (e.g., 1% = 1 in 100 years and 10% = 1

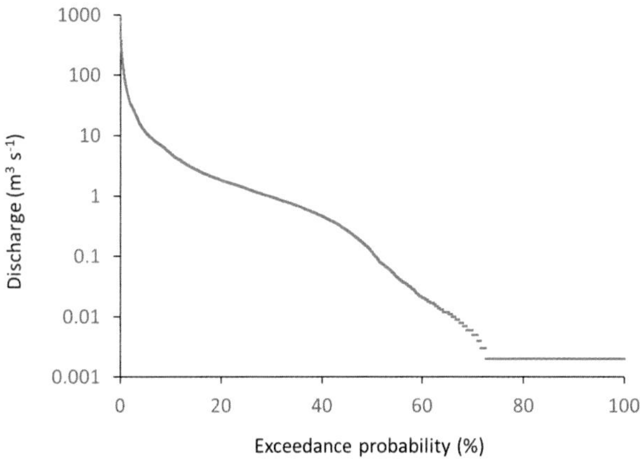

FIGURE 8-14 Flow duration curve (mean daily flow) for the subtropical ephemeral Condamine River at Yarramalong Weir, Queensland, Australia, from 1989 to 2016. The river is dry for more than one-quarter of the time (i.e., discharge not detectable, denoted by the horizontal baseline), median daily flow is 0.019 $m^3 s^{-1}$, and stormflows in the summer wet season can exceed 100 $m^3 s^{-1}$. *(Data from the Australian Bureau of Meteorology/Sunwater [www.bom.gov.au/waterdata].)*

in 10 years) that inform the design of flood control measures. These probabilities are strictly applicable only to the instrumental period and their direct application to a future period assumes stationarity of rainfall–runoff, including for periods under the influence of climate change (Vogel et al., 2011).

C. Exchanges at boundaries

The *hyporheic zone* (see also Chapter 27) underlies the streambed where surface water overlies permeable sediments (Fig. 8-15). This saturated area has interstitial spaces that enable exchange between the relatively slow-moving groundwater and the stream itself, and additionally hosts communities of microbes and benthic invertebrates (see Chapter 27). A two-way exchange occurs between the stream and hyporheic zone over small scales (centimeters to tens of meters), which distinguishes it from groundwater, which has unidirectional

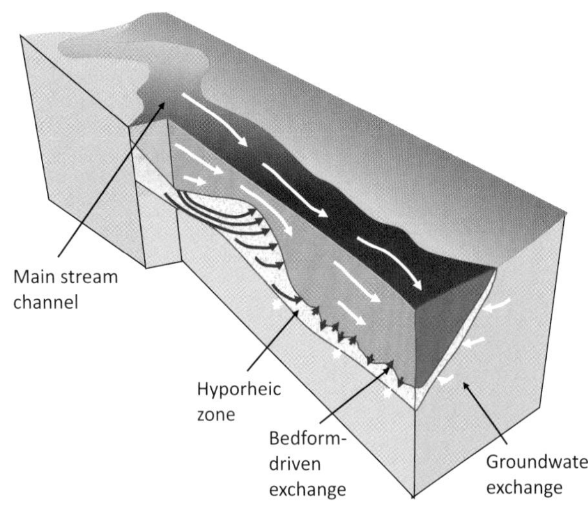

FIGURE 8-15 Schematic of the hyporheic zone and the downwellings and upwellings that enable exchange between a stream and its adjacent hyporheic zone. Groundwater flows into the hyporheic zone and then into the stream. *Dark blue* denotes surface view and *pale blue* indicates a cross-section. *(Redrawn from Stonedahl et al. [2010].)*

travel (discharge or recharge) over extended areas (Boano et al., 2014). Exchange between the stream and the hyporheic zone is mediated by the morphometry of the streambed, which creates water-level fluctuations and drives vertical hydraulic gradients across and along the streambed (Fig. 8-15), resulting in localized areas of the stream that may gain or lose water from hyporheic exchanges. The gains (upwellings) to the stream are from small decreases in the hydraulic head and the losses (downwellings) from increases. Most of the water in the hyporheic zone is derived from the stream channel, ~98% at the top of the zone and 10–98% in the interactive zone that intersects with groundwater at its lower and lateral boundaries (Triska et al., 1989).

Changes in the topography of the streambed contribute to variations in exchange with the hyporheic zone. Stream water typically enters the hyporheic zone in areas where water pools above boulders or bed constrictions, as well as where there are small-scale hydraulic gradients like riffles (see Fig. 8-11). Water returns from the hyporheic zone to the stream mostly in downstream areas as the slope of the stream bed decreases and pools predominate in the stream reaches. Exchanges are also mediated by the permeability of the sediments, stream sinuosity, and the length scale of roughness features, that is, the presence of rocks and boulders. Approaches to modeling the exchanges are described by Marzadri et al. (2016).

D. River and stream geomorphology, currents, and sediment loads

Flow velocity and distribution in rivers and streams depend on the extent of curvature (Fig. 8-16). Flow in straight sections of the main channel is parabolic, with the highest velocities near the center and lower velocities near the margins due to friction. As a channel curves, flow accelerates at the outer margin of the curve and decelerates on the inner margin. The changes in stream flow additionally lead to changes in cross-stream flow, which lead to surface waters exchanging with bottom waters (Sukhodolov et al., 2009).

High discharge affects stream morphology and sediment load. During high flows, strong shear forces are generated, which affect the form of a stream channel (i.e., *channel-forming flows*) and the suspended sediment load entrained in the bulk water of the channel (Costa and O'Connor, 1995). The increased load of suspended material is maintained in the water column by turbulent eddies. Suspended load concentrations are relatively uniform in the bulk stream channel (above the sediment–water boundary layer), making measurement straightforward using grab samples (although an integrated water column sample is preferable), but with a caveat that large variations occur on the rising and falling limbs of a hydrograph, when high-frequency sampling should be used to estimate sediment loads. The rising limb of the hydrograph is usually characterized by higher sediment loads than the falling limb because fine particles are initially washed into the stream and entrained within the channel, but their supply is gradually depleted over the duration of a major storm event (Allan et al., 2021). Suspended loads can be quantified using *turbidity* as a surrogate measure, usually reported in *Nephelometric Turbidity Units* (NTU). The relation of turbidity to suspended sediment concentrations varies with the size, shape, and refractive index of the suspended material (Kirk, 1984). Suspended material includes organic and colloidal particles, as well as inorganic suspended sediments. The total stream suspended load also includes *bed load*, that is, sediments that move along the streambed in close proximity to it. Sediment traps and tracer studies can be used to approximate bed loads. These are usually smaller than suspended loads but increase substantially during flood periods (Allan et al., 2021).

Rivers generally migrate across floodplains gradually but sometimes changes occur abruptly. Such changes are called *avulsions* and nourish the floodplain with water, sediments, and nutrients but may also cause drastic flooding (Brooke et al., 2022). Avulsions can occur, for example, when after several years of drought, sediment particles that have deposited where river waters slow are abruptly scoured during flood events, leading to changes in the river's flow path.

Widespread alteration of streamflow by humans (e.g., through abstraction, damming, wetland drainage, and other changes) has led to the recognition of the importance of *environmental flows* (eFlows) that seek to return

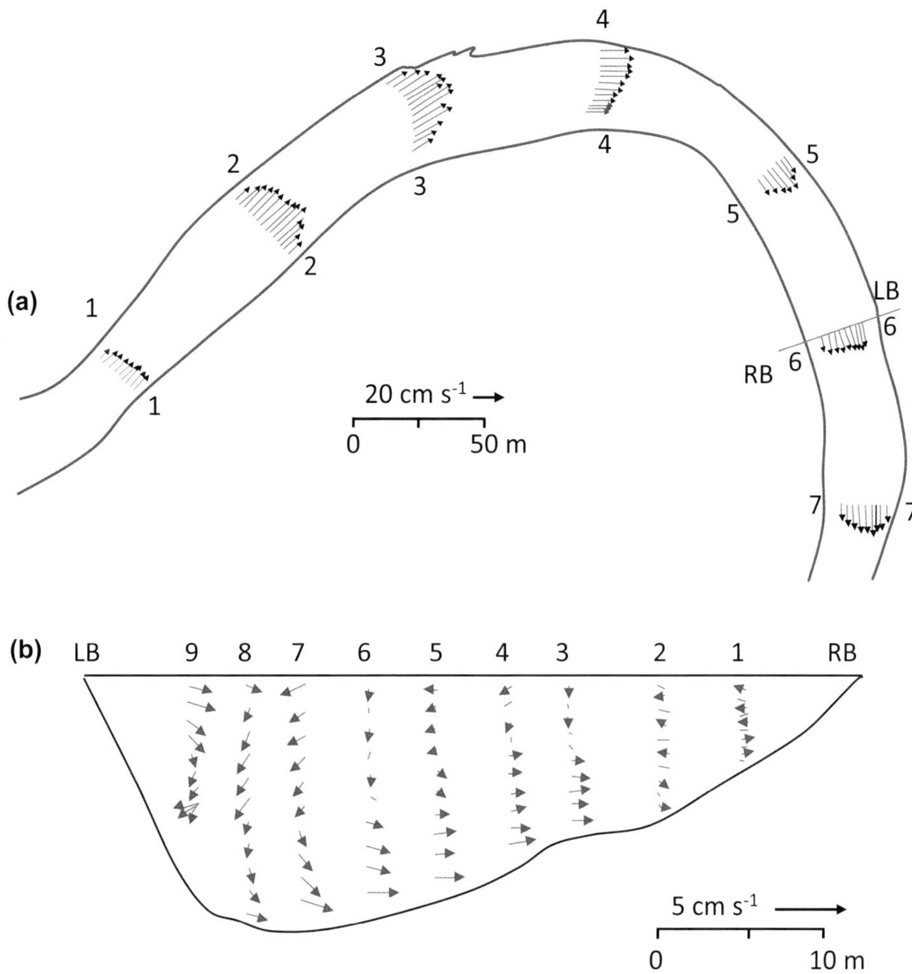

FIGURE 8-16 (a) Variations in flow velocity along the channel of the Spree River, Germany, due to curvature, and (b) secondary currents in the river's cross-section at location 6-6 with depth enlarged five times relative to breadth. LB and RB refer to left bank and right bank, respectively. *(Redrawn from Sukhodolov et al. [2009].)*

the hydrological regime of the stream toward its natural state and provide benefits for nature and humans (Arthington et al., 2006; Chen and Olden, 2017).

III. Water movement in lakes

Lakes are in motion from wind on the water surface that transfers energy and momentum, heat loss that induces instabilities and convectively induced motion, *rotational effects* of the Earth's rotation, and intrusions from streams and rivers (Fig. 8-17). The energetics of the various processes are illustrated in Fig. 8-6. Wind induces motions at the surface, including currents, surface waves, surface seiches, Langmuir cells, and gyres. An approximation of energy transfer from wind is given by Imboden (2003), who indicates that ~20% of the wind energy input at 10 m above the water surface is transferred into surface waves and an even smaller fraction into currents, seiches, and turbulence. Wind causes the thermocline to tilt, which induces internal waves such as seiches or nonlinear waves that cause mixing (Lemmin, 2012). Flows from differential heating and cooling are common, as discussed in Chapter 7, reinforced by the complexity of basin morphometry and the presence of plants that can shade the water. Basin size and shape, as well as the prevailing energetic forcing (e.g., by wind), critically influence the period, intensity, and location of basin-scale water movements. Imboden and Wüest (1995) summarize the major types of waves, from capillary motions to topographic in scale. Many of these motions cause the transport and mixing that are essential elements of ecosystem structure and function.

A. Wind-driven motions near the surface

The movement of surface water is strongly influenced by the strength, duration, and direction of wind.

FIGURE 8-17 Lake Dunstan (foreground), with wind-induced surface waves, some of which are white-capping, and incoming Clutha River (Central Otago, New Zealand) in the background. Mixing zones where the river enters the lake are apparent as changes in color, as is scouring along the river margins, indicated by increased turbidity, which will transport sediments into the lake. *(Photo credit David P. Hamilton.)*

FIGURE 8-18 "Windrows" on Lake Mendota (Wisconsin, USA) due to Langmuir circulation in the surface mixed layer. The windrows align roughly parallel to wind direction and the direction of wave propagation. Foam and light organic matter accumulate at the convergence zones, making the windrows, also known as streaks, visible. *(Photo credit David P. Hamilton.)*

Wind-driven motions in the surface layer include currents, surface waves, gyres, and Langmuir circulations (Figs. 8-18 and 8-19). Equations and theory for these processes are summarized in Monismith and MacIntyre (2009). When wind blows over a calm ("glassy") lake, it generates shear, which results in surface currents, moving in the direction of the wind at approximately 3% of the wind speed. Due to friction, currents decrease with depth from the surface downward (Fig. 8-5). When the upper mixed layer is not heating or cooling, a velocity profile follows the *"law of the wall"* scaling presented in Eq. 8-8 (see Section I: Boundary layer flows). A similar expression describes velocity with height above the water surface with u^*_a, the air friction velocity. Shear stress on the water surface, τ, is the product of the density and the friction velocity squared and is assumed equivalent on both sides of the air−water interface:

$$\tau = \rho_a u^{*2} = \rho_w u^* w^2 = \rho a C_D U^2 \qquad (8\text{-}26)$$

where u^*_w is the water friction velocity, ρ_w is the density of water, ρ_a is the density of air, C_D is the drag coefficient at the height of the anemometer above water, and U is wind speed. Note that this shear stress is equivalent to the Reynolds stress described in Section IB. Based on experimentation, *drag coefficients*, C_D, have been developed such that shear stress can be quantified from wind speed and atmospheric stability (see also Chapter 7 I: *Wind-driven mixing in the upper mixed layer*).

Many lakes are sheltered by trees or hills, which affects the wind field, while there is also a lag time for lakes to respond to the prevailing wind forcing (Smith, 1975). Markfort et al. (2010) have approaches to correct

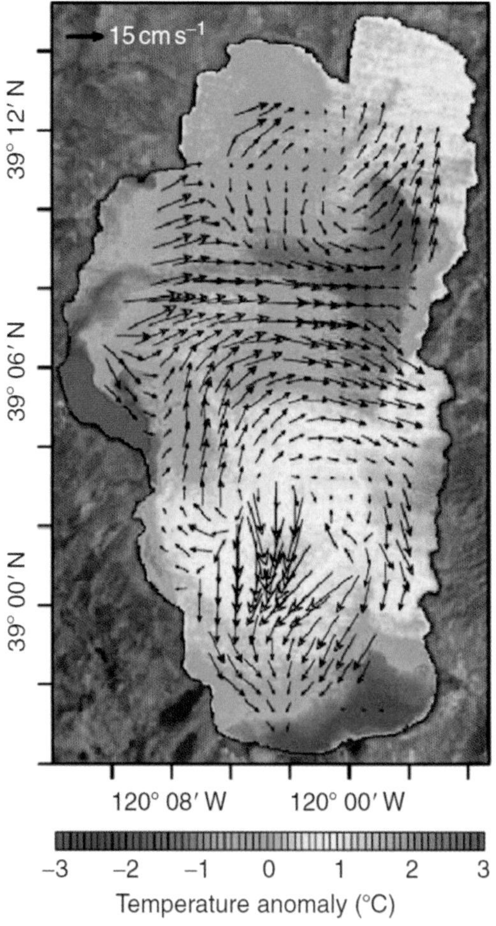

FIGURE 8-19 Gyres in Lake Tahoe (California, USA) detected using remote sensing (Landsat Enhanced Thematic Mapper Plus). Current vectors are *black arrows*; scale is in the *upper left* and *colors* indicate temperature anomalies, that is, the difference in temperature relative to the mean value. *(From Rueda and Vidal [2009] and adapted from Steissberg et al. [2005].)*

for the sheltering and Vesala et al. (2006) illustrate the flow of air over such lakes. In fact, the flow consists of turbulent eddies. In larger lakes wind accelerates as it flows over water distances of about a kilometer. Wind can also vary spatially over larger lakes.

The upper mixed layer undergoes cycles of heating and cooling during the day. During the day, unless conditions are exceptionally windy, it is divided into an actively mixing layer, a diurnal thermocline, and a subsurface layer (Chapter 7, Fig. 7-1). The diurnal thermocline, which sometimes reaches the surface, and the subsurface layer are stratified. At night due to cooling, stratification is eroded, often to the depth of the seasonal thermocline. Shear in the upper mixed layer varies with the extent of cooling and heating and can be quantified with *Monin–Obukhov similarity theory* (Csanady, 2001). Law of the wall scaling applies when wind forcing is appreciable relative to heating and cooling. However, shear, and the related dissipation of turbulent kinetic energy, can be enhanced over predictions from law of the wall scaling when winds are moderate to light. This enhancement applies during both cooling and heating but with a greater augmentation under heating (MacIntyre et al., 2021a,b) and an even larger augmentation due to the breaking of surface waves (Terray et al., 1996). The energy of the wind not only induces currents but also causes surface waves and Langmuir cells (see below). Because the upper mixed layer is often stratified, it can support internal wave motions with velocities in the upper mixed layer not only dependent on shear but also on the characteristics of these waves (Imberger, 2012). Thus studies quantifying flows and the turbulence in the upper mixed layer need to include the effects of these processes.

As wind speed increases above 3 m s^{-1}, small erratic surface deformations, known as *capillary waves*, form that exceed the resisting force of surface water tension. As wind speeds continue to increase, *surface gravity waves* form (Chapter 27, Fig. 27-7). Water is displaced upward and returned to equilibrium by gravity along a circular path; therefore the waves are called *gravity waves*. They are coherent (though not uniform) and characterized by their wave period (T), wavelength (L), and significant wave height (H_S). H_S is commonly denoted by the average height measured from trough to crest of the highest one-third of the waves. Due to the various controls on H_S, including fetch limitation, empirical equations have been developed to characterize it and other characteristics of surface waves (Hasselmann et al., 1976, USACE, 1984, Mao et al., 2016).

As surface waves travel, they are called *progressive waves*. As they enter shallower water, their velocity decreases proportionally to the square root of depth. A reduction in wavelength occurs with a marked increase in wave height, such that the waves become asymmetrical and are unstable and break. In a *plunging breaker* the forward face of the wave becomes convex and the crest curls over and collapses, whereas the crest of a *spilling breaker* collapses forward, spilling water downward over the front of the wave.

Wind speed, fetch, and depth affect the evolution of surface waves across a lake, with waves usually limited by fetch at the windward end and depth at the leeward end. White-capping and breaking occur offshore as winds exceed 6–8 m s^{-1}, and wave breaking inshore can be accompanied by near-bottom shear (Chapter 27, Fig. 27-7). This transition from "deep" to "shallow" water waves occurs where the water depth (z) is less than approximately one-half of the wavelength ($z < 0.5L$). At this point, the idealized exponentially decreasing orbital motions of deepwater waves become elliptical and flattened, generating shear at the boundary from the oscillatory motion of the waves (Chapter 27, Fig. 27-7). The shear stress can resuspend sediments and mobilize nutrients and metals (Geng et al., 2022) and resuspended fine sediments can be transported by currents to deeper areas (Håkanson and Jansson, 1983). High sustained winds from one direction can lead to *storm surges*, exposing bottom sediments at the windward end of the lake and piling up water at the leeward end. For example, Hamblin (1979) recorded a 4.5 m difference in water level between two ends of Lake Erie during a storm, when wind direction was aligned with the longest axis of the lake.

Surface seiches are examples of *long-standing waves* produced in response to winds. Gravitational adjustment to water that accumulates at one end of the lake leads to flow toward the other; that is, an oscillatory motion develops with a node in the middle of the lake. These waves affect the entire water mass of a lake, regardless of whether it is stratified, and have their largest amplitude (displacement) at the surface. In most cases the amplitude of surface seiches is small, in the order of millimeters. Occasionally on large lakes such as Lake Geneva (Switzerland/France) or Lake Erie (USA/Canada), amplitudes can be one or two meters and serve to flush river delta and harbor areas. Surface seiches have been known since at least the 19th century; for example, Forel (1895) noted them in Lake Geneva. Flows from surface seiches were one of the first wind-induced motions recorded under ice in winter (Malm et al., 1998). The *seiche period* in a well-mixed lake can be calculated as (Bengtsson, 2012):

$$T = 2\frac{L}{\sqrt{g\bar{z}}} \tag{8-27}$$

where T is the wave period, L is the length of the lake along its long axis, and $\bar{z}$ is the mean depth. Reported periods are on the order of 26 min in Lake Mendota

(Wisconsin, USA), 73 min in Lake Geneva, and 14 h in Lake Erie.

Langmuir circulations or *Langmuir cells* are often evident on the water surface as windrows once winds exceed ~3 m s^{-1} (Langmuir, 1938). Vertical counter-rotating helical currents form perpendicular to the wind direction. Areas of convergence and divergence are visible at the water surface, with light organic material and foam accumulating and forming coherent streaks in convergence zones. The streaks are aligned parallel to the wind at the water surface (Fig. 8-18) with the helical motions below. The spacing between streaks is usually about twice the mixed layer depth in stratified systems (Dethleff and Kempema, 2007). Langmuir cells result from the interaction of surface waves with wind-driven surface currents (Craik and Leibovich, 1976; Craik, 1977; Leibovich, 1977). They are sometimes considered an organized form of turbulence capable of transporting appreciable momentum downward. However, many questions remain as to their contribution to mixing in the surface layer.

Influence of the Earth's rotation on surface currents and effect of stratification – The *Coriolis force* due to the Earth's rotation causes currents to be deflected relative to the direction of the wind. With respect to surface currents and the formation of gyres, the influence of the Earth's rotation can be determined from the Rossby number, Ro = u/(Lf), where u is a characteristic velocity, L is basin dimension, and f is the Coriolis parameter, which varies with latitude. Rotation is important for small values of the Rossby number (less than ~0.2; Mortimer, 1974). For example, for f = 10^{-4} s^{-1}, as at latitude 45°N, a medium size lake with L = 5 km, and assuming surface currents are 0.1 m s^{-1}, Ro = 0.2, indicating an influence of rotation on surface currents. *Inertial currents* are those influenced by rotation. The combined effects of wind and geostrophic deflection cause surface water to move downwind and to the right in the Northern Hemisphere and to the left in the Southern Hemisphere. The resulting wind-drift current is deflected 45° from the direction of the wind in a spiral manner (the *Ekman spiral*) in open waters of large, deep lakes. As the size, and especially the depth, of the lakes decreases, the magnitude of the deflection angle decreases until, in a lake with a depth < ~20 m, the angle of declination becomes insignificant.

When considering rotational effects within stratified lakes, the internal *Rossby radius of deformation* (R_I) must be smaller than the basin dimension. The equation for the Rossby radius accounts for the Coriolis force and the phase speed of the internal waves (c), and the rate at which a wave propagates, which depends on stratification. The internal Rossby radius is calculated as L_I = c/f. In a "two-layer" lake, that is, one with an epilimnion and a hypolimnion and a sharp discontinuity between them,

$$c = \frac{g(\rho_1 - \rho_2)z_1 z_2}{\sqrt{\rho_0 z}} \tag{8-28}$$

where z is the depth of the lake. The hypolimnion would have depth z_1 and density ρ_1, and the epilimnion would have depth z_2 and density ρ_2, giving a positive value for c, and the mean density is ρ_o. For example, in 150 km^2 Mono Lake (California, USA), L_I was ~8 km based on the stratification in 1995, just less than the length of the lake at the depth of the thermocline. Hence internal waves were expected to be affected by rotation, which three-dimensional (3-D) hydrodynamic modeling subsequently proved to be the case (Vidal et al., 2013).

Gyres – The surface currents of moderate and large lakes often have large swirls known as *gyres* (Figs. 8-4b and 8-19). Lake Michigan (USA), for example, is typically exposed to predominantly westerly winds during summer, and its large, conspicuous gyres are clearly influenced by geostrophic rotational forces (Noble, 1967). Gyres in lakes the size of Lake Michigan are not simply the result of geostrophic forces; their direction is strongly modified by large longwaves and especially by shifts in the duration of strong prevailing winds. Similarly, Lake Biwa, the largest lake in Japan, has three counter-rotating gyres that have been observed by ADCP measurements (Kumagai et al., 1998). In lakes of all sizes, gyres may be induced due to promontories and other irregularities along their coast or spatial differences in wind speed and direction that lead to the swirls (see Fig. 8-4b). Excellent examples of gyres in lakes with and without the influence of rotation are provided in Rueda and Vidal (2009). The movement of gyres can be complicated by other water movements such as internal waves. The geostrophic right-hand acceleration of currents (Northern Hemisphere) also occurs under ice cover, even in small lakes, due to weak stratification under the ice. Consequently, gyres can form under the ice in midsized lakes (Forrest et al., 2013) and large lakes such as Lake Baikal (Russia; Kouraev et al., 2016, 2019).

B. Wind-driven motions in the interior

As wind induces surface currents, it displaces water downwind. The resultant pressure gradient causes the thermocline to tilt (Figs. 8-20 and 8-21) and the shear induces Kelvin–Helmholtz billows (Fig. 8-3) and mixing between the upper and deeper layers of a lake (Fig. 8-20).

Internal wave motions – The onset of wind causes the thermocline to tilt. On relaxation, as surface water flows

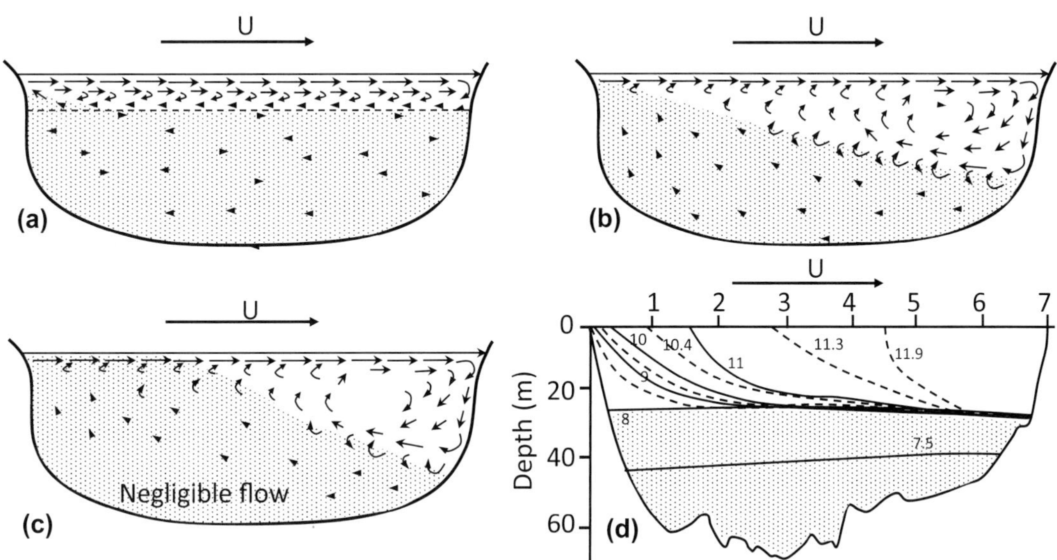

FIGURE 8-20 Wind-driven surface currents and associated tilting of the thermocline as surface water accumulates downwind in the northern basin of Windermere (UK). In this conceptualization, Windermere is divided into an upper mixed layer with the hypolimnion below (panels [a] to [c]). The thermocline is indicated by a *dashed line* in (a). The magnitude of the surface currents is indicated by the length of the *arrows*, with currents of highest speed at the surface. They induce downwelling and a return flow above the thermocline. They also induce flow in the hypolimnion. Indications of speed and direction of flow are indicated by arrows. In the example the wind (U) is strong enough and was persistent long enough to induce considerable upwelling of hypolimnetic water. As can be seen in (d), upwelling results in cross-basin differences in temperature (indicated by labeled isotherms) and deepening of the thermocline. The *curved arrows* in (b) are indicative of shear-induced mixing, mostly Kelvin–Helmholtz billows. These combined processes occur when Wedderburn numbers are low (see Chapters 7 and 8 IIIB). *(Redrawn from Mortimer [1961].)*

in the opposite direction, the thermocline begins to tilt in the other direction. Due to inertia, the thermocline continues to upwell and downwell at successive ends of a lake basin (Fig. 8-21). This continuous rocking back and forth of the thermocline is known as *seiching*. The up-and-down motion of the *seiche* has its maximum amplitude at the lake margins, the *antinode*, and minimal vertical movement at the *node*. Seiches are *standing waves*; that is, they do not travel and are generally classified as a type of linear internal gravity wave. Internal waves can also be classified as progressive, meaning they can travel across a lake, transporting energy. Such waves can also steepen abruptly, similar to the steepening of wind-induced surface waves prior to their breaking in the nearshore (Section IIIA; Chapter 27, Fig. 27-7). Such waves are classified as nonlinear and may contribute to mixing within a lake.

The thickness of the thermocline and the stratification across it moderate whether internal wave modes are *first (1st) vertical mode*, as in Fig. 8-21, or of higher (e.g., 2nd, 3rd) mode. When the thermocline is relatively thin, as it first forms in spring or after storms, first vertical mode waves typically occur. Here the upper and lower boundaries of the thermocline remain approximately parallel (Figs. 8-22a and b). As the thermocline thickens, it can support higher-order internal waves. For example, in a *second (2nd) vertical mode internal wave*, the thermocline spreads at one end of the lake and compresses at the other (Fig. 8-22d). As the thermocline oscillates, the compression and spreading alternate between the sides of the lake. Second vertical mode waves are common within the metalimnion of Toolik Lake (Alaska, USA) in summer (Chapter 7, Fig. 7-19). Frequency analysis shows that the internal wave field is comprised of a composite of these wave types.

Seiche periods (T) can be calculated from the length of a lake (L) and the phase speed (c) as T = 2L/c, where c is given by $c = g(\rho_1 - p_2)\, z_1 z_2/\sqrt{\rho_0 \bar{z}}$ (Eq. 8-28) for the first vertical mode internal seiche. The periods of higher mode internal seiches can be computed considering the density stratification throughout the water column.

As the thermocline upwells and downwells across a lake, shear develops across it, which leads to mixing (see mixing examples in Fig. 8-2). Spigel and Imberger (1980) developed a quantitative scheme based on the Richardson number and basin morphometry to illustrate how the extent of tilting of the thermocline, the density difference across it, the depth of the upper mixed layer, and the length of the lake would moderate shear within the thermocline and the extent of mixing within a lake.

This conceptualization, formalized to the *Wedderburn number* (Thompson and Imberger, 1980; Chapter 7), W, and later extended to the *Lake number* (Imberger and

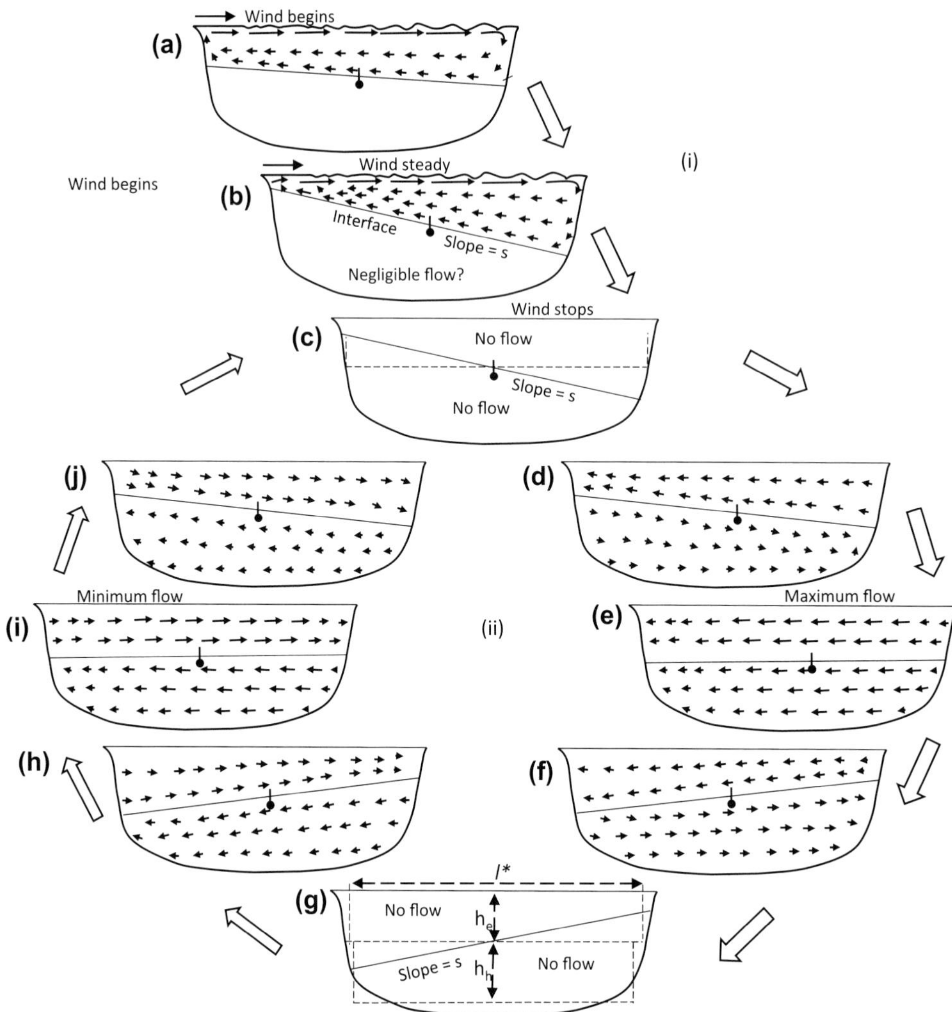

FIGURE 8-21 Movement caused by (*i*) wind stress and (*ii*) a subsequent internal seiche in a hypothetical two-layered lake, neglecting friction. Direction and velocity of flow are approximately indicated by *arrows*. h_e and h_h are depth of the epilimnion and hypolimnion, respectively, and l^* is approximate basin length. *(Redrawn from Mortimer [1952].)*

Patterson, 1990; Chapter 7), L_N, has led to a quantitative understanding of controls on mixed layer deepening, the contribution of internal waves to mixing within the thermocline, and to exchange between the epilimnion and hypolimnion. Laboratory experiments illustrated the extent of mixed layer deepening for different values of W (Monismith, 1986). For $W < 1$, results indicated compression of the thermocline downwind and its expansion upwind as in Windermere in Fig. 8-20d. This motion is not the result of a second vertical mode internal wave (Fig. 8-22d) but is instead from shear at the density interface, resulting in the direct exchange of water from the hypolimnion into the epilimnion and considerable mixing (Monismith, 1986), similar to the observations in Toolik Lake, Alaska (Figs. 7-21 and 7-24). Field experiments illustrated that when W dropped to 2, ϵ and K_z were calculated to be 3 to 4 orders of magnitude higher in the thermocline nearshore than offshore (MacIntyre et al., 1999). The mixing was enhanced by internal wave breaking near boundaries. Tracer experiments similarly indicated the enhanced mixing near boundaries (Goudsmit et al., 1997).

The Wedderburn number also predicts the type of internal waves that result as they steepen and become nonlinear (Fig. 8-23) (Horn et al., 2001). Surges and *solitons* (i.e., *solitary waves*) occur for Wedderburn numbers in the range from ~1 to 5 (Figs. 8-24 and 8-25). The steepening of the thermocline is called a *surge* and a packet of solitons forms in its rear. Laboratory experiments and modeling indicate that the solitons will break at the lake margins and cause mixing (Boegman et al., 2003, 2005a,b).

For larger wind forcing relative to stratification, internal waves become nonlinear (Horn et al., 2001). In the Kelvin–Helmholtz regime, the progression begins with a tilted thermocline and is followed by *internal bores*

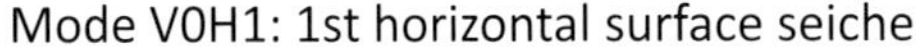

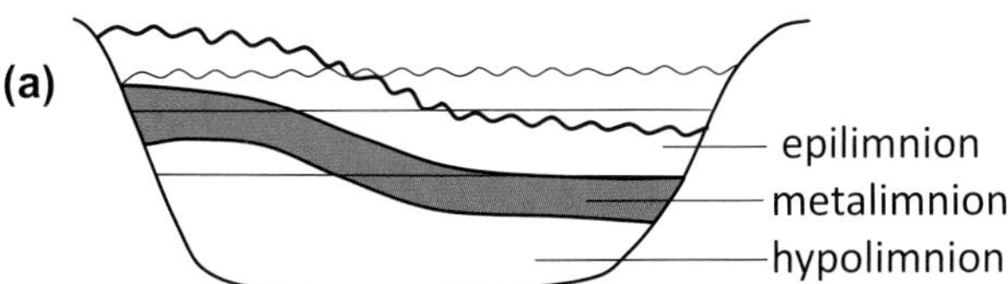

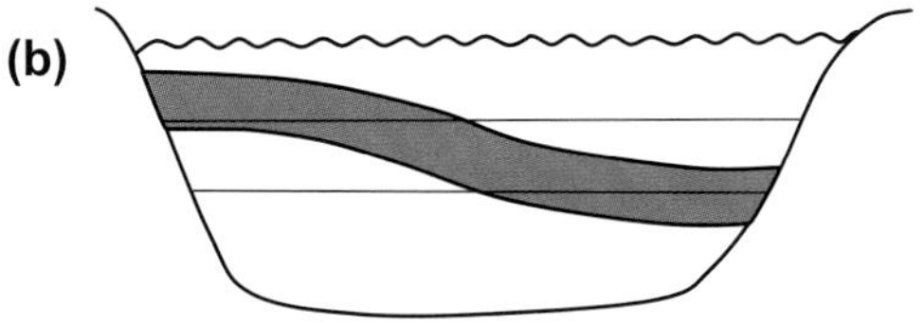

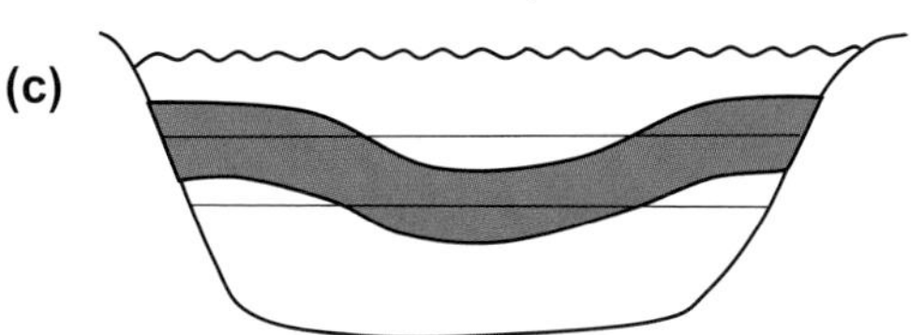

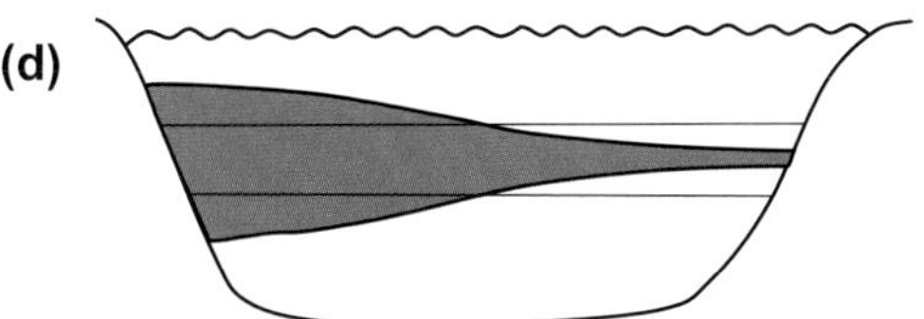

FIGURE 8-22 (a) Horizontal surface seiche identified by the change in surface elevation; (b) first vertical, first horizontal (V1H1) seiche in which the thermocline thickness is the same as it upwells and downwells; the upwelling and downwelling occur relative at a stationary node in the center of the lake; the antinodes, where vertical motions are largest, are at the ends of the lake; (c) first vertical, second horizontal (V1H2) seiche in which again the thermocline thickness does not vary across the lake and the up- and downwelling is relative to two nodes; and (d) second vertical, first horizontal (V2H1) seiche in which the thermocline alternately compresses and expands at each end of the lake as the thermocline upwells and downwells around one node. *(Redrawn from Munnich et al. [1992].)*

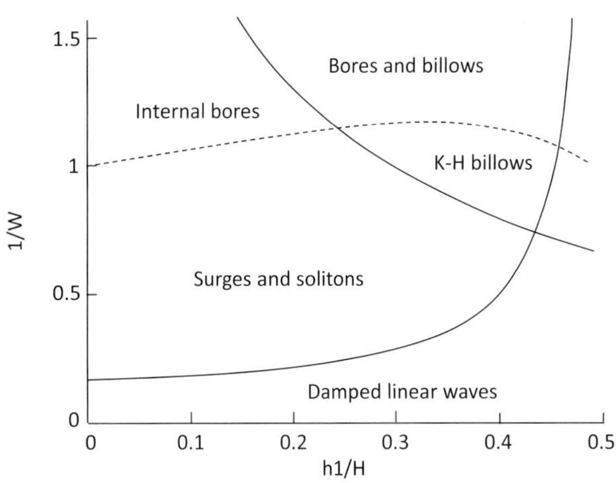

FIGURE 8-23 Phase space illustrating whether linear or nonlinear waves will form in a lake as a function of the inverse of the Wedderburn number (1/W) and the depth of the mixed layer (h1) over the depth of the lake basin (H). Linear waves such as seiches are expected to be damped when $1.4 < W < 5$ depending on h1/H. Nonlinear waves such as surges and solitons are predicted for $1 < W < 5$. When $W < 1$, the predicted waves are highly nonlinear and intense mixing is expected across the thermocline and at the lateral boundaries. *(Redrawn from Horn et al. [2001].)*

(a highly nonlinear internal wave) forming on both ends of the thermocline, horizontal movement of the bores inward from the margins, and, with increased shear, the formation and instability of Kelvin–Helmholtz billows (Fig. 8-25). The intense mixing from a wind event that induced these features in Lake Cayuga (New York, USA) was illustrated using 3-D hydrodynamic modeling (Dorostkar and Boegman, 2013). Fluxes of solutes and particles from the sediments may additionally result from the interaction of the nonlinear internal waves with the boundary (Boegman and Ivey, 2009; Valipour et al., 2017). Mixing near the boundary of lakes is also caused by near-bottom currents induced by seiches, not necessarily due to the breaking of internal waves (Lorke et al., 2005; Nielson and Henderson, 2022).

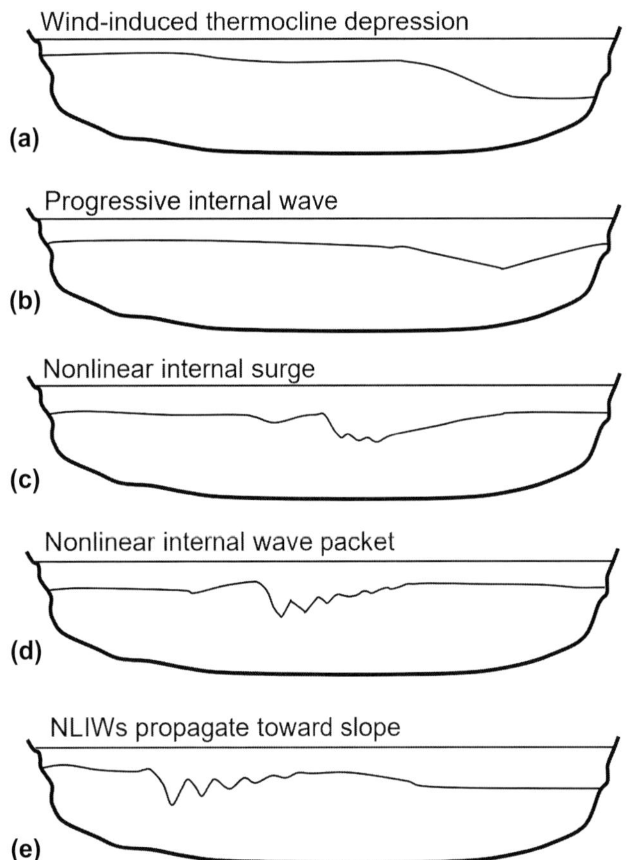

FIGURE 8-24 Evolution of a nonlinear internal wave (NLIW) packet with (a) initial wind-induced depression of the thermocline, (b) and (c) development of a surge, (d) and (e) further evolution with the surge and solitons. Depending on the slope of the bottom boundary, these may be reflected or become unstable (and break), causing mixing at the boundary. *(From Boegman [2009].)*

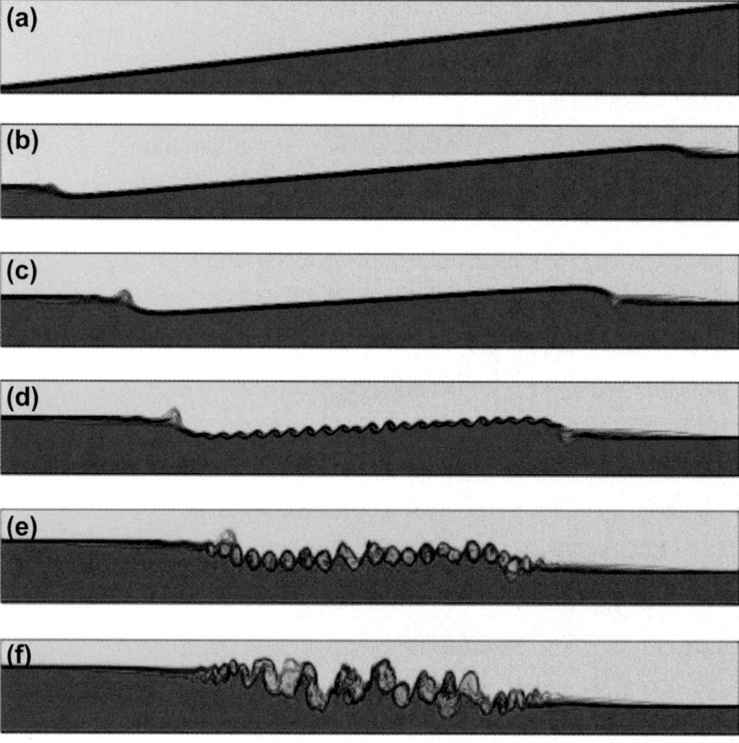

FIGURE 8-25 Model results illustrating (a) thermocline tilt; (b) and (c) the formation of internal bores at the lake margins; (d) and (e) subsequent formation of Kelvin–Helmholtz billows in the thermocline due to the strong shear that develops; and (f) their collapse and mixing. *(From Stashchuk et al. [2005], with figures provided by V. Vlasenko.)*

The influence of the Earth's rotation on internal waves – Coriolis force leads to rotational waves in lakes large enough for the water motion to be influenced by the Earth's rotation. This influence depends on the extent of stratification, the diameter of the lake at the depth of the thermocline, and the latitude. *Kelvin waves* form in the nearshore, with upwelling regions rotating around the lake (Fig. 8-26). In moderately sized lakes just large enough for rotation to influence the internal wave dynamics, the Kelvin wave may extend a considerable distance offshore. *Poincaré waves*, another type of rotational wave, form in the central regions (Hodges et al., 2000; Bouffard and Boegman, 2012). Rotational waves can also become nonlinear, as discussed above, and contribute to mixing.

Seasonal variation of basin-scale response to low values of W and L_N – The previous descriptions of internal waves and associated mixing for given values of W and L_N were largely based on theory and laboratory investigations. Within lakes, other processes may modify the expected response to increased winds. In particular, changes to the internal wave dynamics may occur in response to heating and cooling over the stratified period (Fig. 7-25). Field data indicate that the expected nonlinear waves form and intense mixing occurs as W and L_N decrease to critical values. However, in springtime incoming heat in the upper water column can be mixed downward, such that the thermocline thickens and the internal waves shift from mode 1 to mode 2 (Figs. 7-24, 8-27, and 8-28). The nonlinearity of the *mode 2 waves* is indicated by their asymmetry; that is, they are steep sided and not sinusoidal. An additional response occurred in the two lakes of Figs. 8-27 and 8-28. The second and third vertical mode internal waves, which encompassed the thermocline, had waves of higher frequency at their upper and lower extents. The result is an increase in the shear at the top and bottom of the metalimnion which contributes to local mixing and horizontal advection (Gomez-Giraldo et al., 2008). Thus events with W decreasing to 0.3 contributed to the downward mixing of heat and thickening of the thermocline, which modified the stratification and subsequent responses to events with low W.

Mixing across the thermocline was in fact incomplete during an event with W of 0.3 in Mono Lake (California, USA) (MacIntyre et al., 2009). The differences in mixing inshore and offshore were illustrated by a tracer experiment that began prior to and extended for another week after the period shown in Fig. 8-27. By the time of the high winds, the tracer had formed a layer 3–4 m thick across much of the lake. The intense mixing implied by the thermocline compression on DOY 113 (April 23) did lead to upward and downward mixing of the tracer near the lake margins. However, the tracer remained concentrated in the 3–4 m thick layer of the metalimnion offshore. Two reasons are likely. The time scale for mixing, τ_{mix}, was long. On a lake-wide basis, the coefficient of eddy diffusivity increased to 10^{-5} m^2 s^{-1} for the 2-day

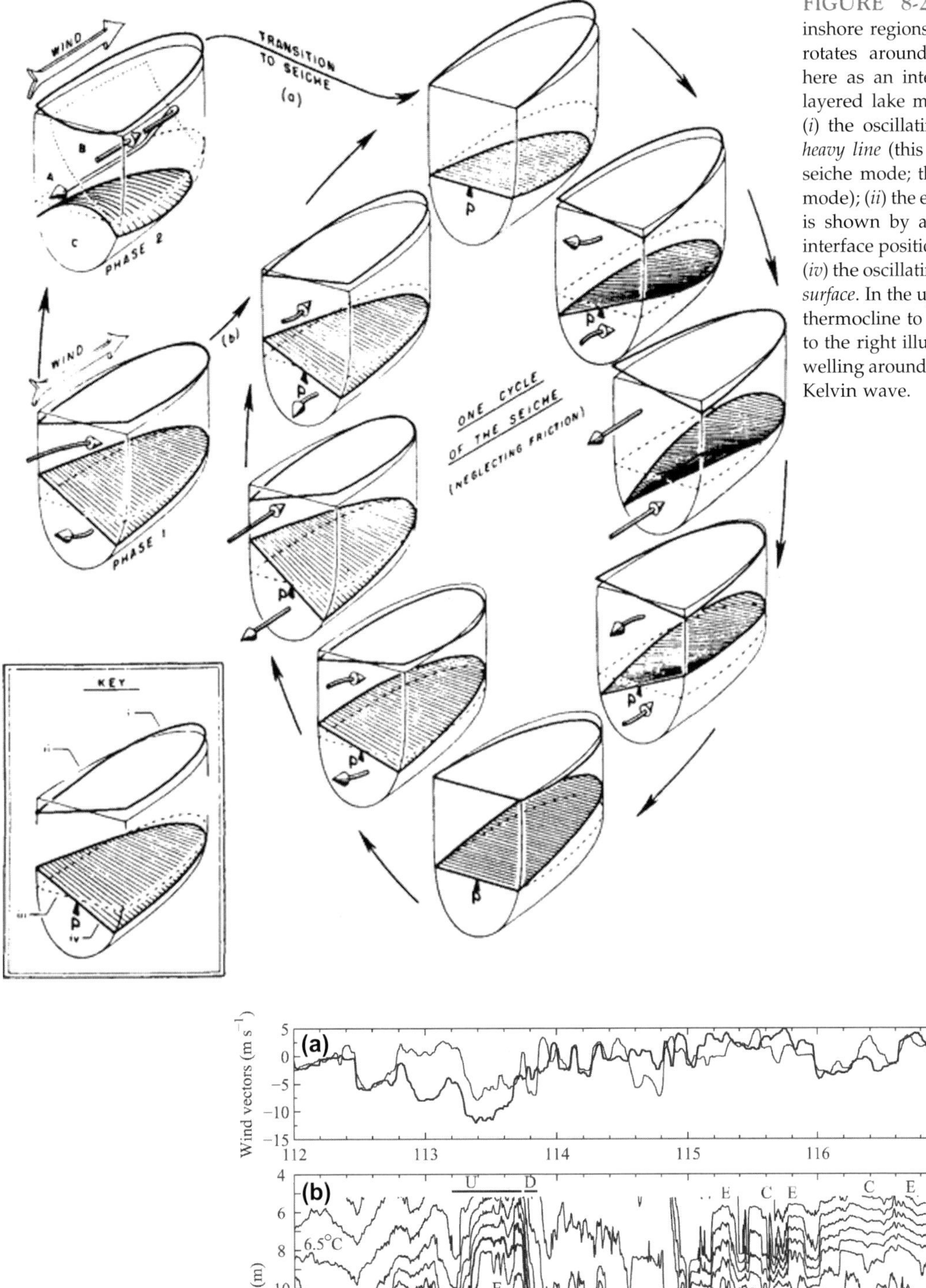

FIGURE 8-26 Rotational waves in the inshore regions of a lake cause upwelling that rotates around the lake margins, illustrated here as an internal seiche in a rotating two-layered lake model. In the inset key diagram (*i*) the oscillating lake surface is shown by a *heavy line* (this is the signature of the internal seiche mode; that is, it is not a surface seiche mode); (*ii*) the equilibrium lake surface position is shown by a *thin line*; (*iii*) the equilibrium interface position is shown by a *broken line*; and (*iv*) the oscillating interface is shown by a *shaded surface*. In the upper right the wind induces the thermocline to deflect, and the eight scenarios to the right illustrate the movement of the upwelling around the lake. Such a wave is called a Kelvin wave.

FIGURE 8-27 (a) Wind vectors (north to south [N–S], with N positive, *thin line*; east to west [E–W], with E positive, *darker line*), and (b) temperature time series data illustrating changes in stratification when W and L_N decreased to 0.3 and 2, respectively, on day of year (DOY) 113 (see Figs. 7-14 and 7-25). Internal waves shifted from first vertical mode when the thermocline was approximately 4 m thick prior to DOY 113 to second vertical mode and higher after a strong wind event on DOY 113 caused heat to be mixed into the thermocline such that this layer thickened (DOY 115 and following). The internal waves illustrated are nonlinear (see Fig. 8-24). The 4°C contour is in bold. U indicates upwelling and D downwelling; markings of E and C indicate expansion and contraction of the internal wave field as in Fig. 8-22d. First arrow indicates a second vertical mode internal wave, which looks like a shock wave; second arrow indicates a third vertical mode internal wave. *(Temperature data are from a nearshore station in 150 km² Mono Lake [California, USA], 22–27 April 1998 [DOYs 112–117]; From MacIntyre et al. [2009].)*

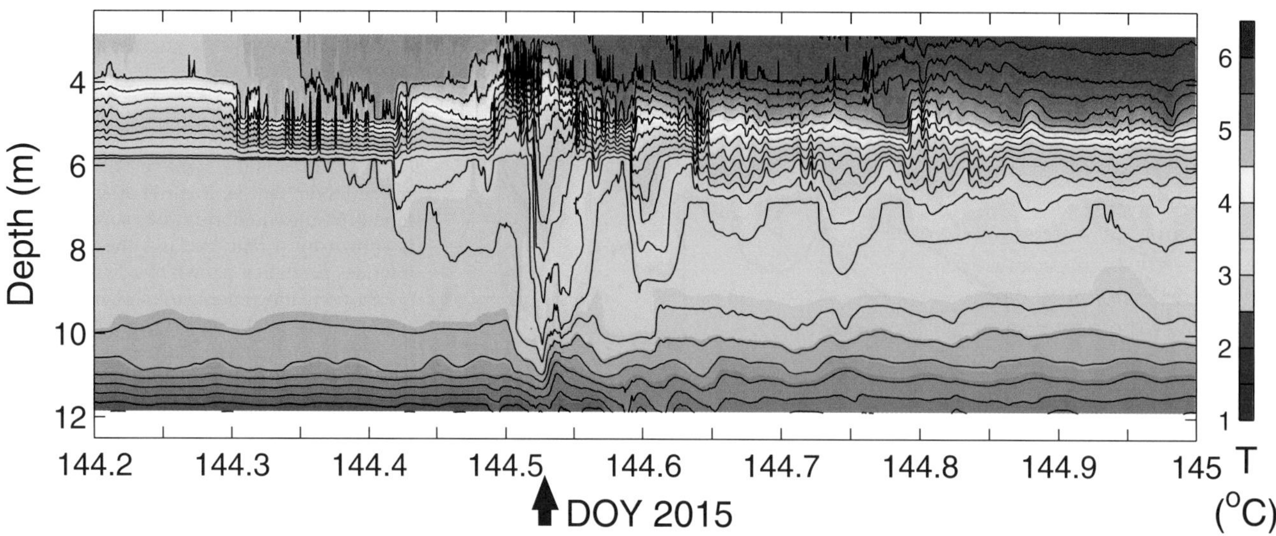

FIGURE 8-28 Temperature time series for 0.029 km^2 Lake E1 (Alaska) illustrating the thickening of the thermocline and formation of nonlinear second vertical mode internal waves when the wind increased at ice-off (indicated by *black arrow*) such that the Wedderburn number decreased to 0.6 across the thermocline and the calculated Lake number, which included the near bottom chemical stratification, was 0.5 for the main basin of the two-basin kettle lake. The downward mixing of heat intensified the stratification such that near-bottom stratification persisted. Thus, even when upwelling is appreciable, as when $W < 1$, full mixing of a lake may not result. *(From Cortés and MacIntyre [2020].)*

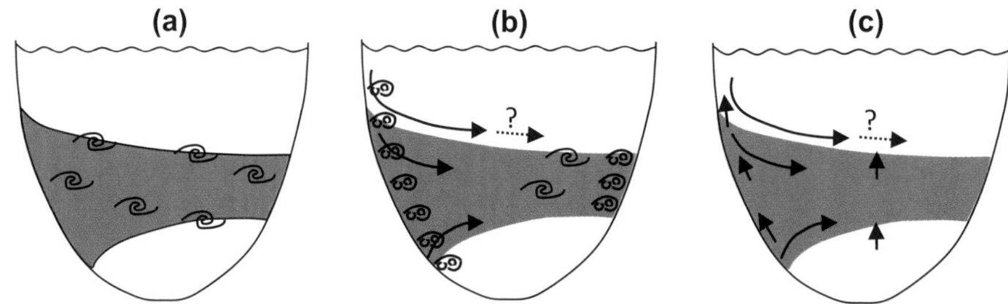

FIGURE 8-29 Development of perspectives as to where mixing occurs within the metalimnion. In earlier models (a) it was assumed to occur via Kelvin–Helmholtz billows throughout. When $1 < W < 10$, models and field data indicated it occurs near the margins (b). However, a tracer experiment in springtime indicates that even when $W < 1$, mixing is concentrated near the margins, with intermittent Kelvin–Helmholtz billowing offshore. The mixed parcels of water are assumed to flow offshore as intrusions (c). The mechanisms causing vertical fluxes and offshore flow will determine the efficacy of supply of nutrients and other subsidies to organisms living in the pelagic zone. *(From MacIntyre et al. [1999].)*

period, during which W dropped to 0.3. For a thermocline that is 3 m thick, $\tau_{mix} \approx l^2/K_z$, that is, 9 $m^2/10^{-5}$ $m^2\,s^{-1}$, which is 10 days. Mixing across the thermocline would not be expected horizontally across the lake on the time scale from the strong wind event. Additionally, Mono Lake is just large enough for the Earth's rotation to influence the movement of internal waves after a strong wind event (Vidal et al., 2013). That is, the wind-induced upwelling rotates around the lake as expected with the rotational waves known as Kelvin waves (Fig. 8-26). With larger amplitude and greater nonlinearity near the lake margins, internal wave-induced mixing was greater in this location. Thus, despite the intense mixing expected for low values of W, vertical mixing may be largest near the boundaries and supply of nutrients or other subsidies to the interior may require offshore flow of intrusions of mixed water (Fig. 8-29). The alternate expansions and contractions of the metalimnion would enable horizontal flow but not necessarily vertical mixing. A number of processes related to internal wave motions have been demonstrated to cause the required intrusions (Wain and Rehman, 2010; Wain et al., 2013).

The effects of strong winds generating low values of W and L_N vary over the remainder of the stratified period (Fig. 7-25). During midsummer, the thermocline can deepen abruptly, as expected when strong up- and downwelling occur across the lake, and generate appreciable shear (Figs. 8-27 and 7-21). Later in the stratified period, events with W and L_N near 1 can cause rapid mixing to the bottom in lakes 10–20 m deep (Figs. 7-23 and 7-24)

When W and L_N approach 1 or drop below 1, the tilting of the thermocline and resultant exchange of water from the different strata of lakes are appreciable, and we refer to the lakes as being *two-dimensional* (2-D). In such cases considerable exchange of nutrients, dissolved gases, and plankton can occur between the upper and lower strata of a lake, particularly near the lake margins (Fig. 8-20). A *one-dimensional* (1-D) perspective assumes that vertical fluxes occur uniformly across the thermocline. Considerable effort has been expended to include appropriate parameterizations in 1-D models for the influence of processes near the boundaries (Fig. 8-29; Section III).

Two-dimensionality can also be expected within the upper mixed layer when a diurnal thermocline is present. That is, the flows associated with up- and downwelling of the thermocline will cause inshore-offshore exchange. Monismith (1986) predicted that the small-scale instabilities from enhanced shear by nonlinear internal waves would be expected in the diurnal thermocline. High-frequency internal waves with abrupt up- and downwellings indicative of nonlinear waves occurred within a diurnal thermocline in Toolik Lake (Alaska, USA) (Fig. 7-21), in shallow regions of Lawrence Lake (Michigan, USA) (data not shown) when the Lake number for the whole lake was ~30, in Balbina Reservoir (Brazil) (Figs. 7-26 and 7-29), and in Lake Janauacá (Brazil) (Fig. 7-27). The data were filtered, that is, smoothed, in several figures so that the shapes of the internal waves can only be well resolved in Balbina Reservoir (Fig. 7-29). Forces associated with these wave motions (Olsthoorn et al., 2012) and from surface waves (Hofmann et al., 2011; Fig. 8-17) could entrain solutes or particles near the bottom into the water column and, along with currents associated with the up- and downwelling of the diurnal thermocline, lead to exchange from inshore to offshore waters. Such transport is another example of the 2-D structure in lakes driven by internal wave motions.

C. Motions induced by density differences

Convective motions result when water in one region becomes denser than that in another. The resulting flows can be in the vertical, as in descending plumes of cooler, near-surface water into warmer water below when heat is lost from the surface of a lake, or in the horizontal, when density differences result from horizontal temperature gradients. The diverse suite of processes driving convective motions have been reviewed in MacIntyre and Melack (1995), in Peeters and Kipfer (2009), and comprehensively in Bouffard and Wüest (2019). The most well-known of these motions is *penetrative convection* induced by cooling of the surface of a lake when temperatures are above 4°C. The cooling creates denser water, which then sinks and mixes with the water below (Fig. 8-30). This process also occurs under the ice in spring when water temperatures are below 4°C (Farmer, 1975; Bouffard et al., 2019), with biological implications reviewed in Jansen et al. (2021). The velocity of the downwelling thermals, w*, can be estimated as $w^* = (\beta z)^{1/3}$, where β is the surface buoyancy flux (Chapter 7), dependent on the rate of heat loss at temperatures above 4°C and the length of the thermals, which is estimated as the depth of the upper mixed layer, z.

Convection also occurs in the horizontal. If inshore regions become warmer than offshore regions as water warms over the course of a day, the warmer, less dense water will flow offshore and cooler, denser water will flow onshore to replace it. This process is called *differential heating* (Fig. 8-31). Warm water may also be associated with dense vegetation canopies in the nearshore and shade the underlying water (Coates and Ferris, 1994; Coates and Folkard, 2009). Gravity currents from *differential cooling* result when inshore waters cool more

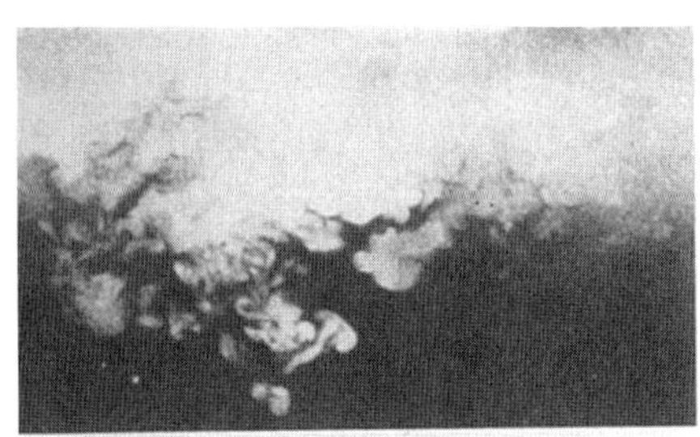
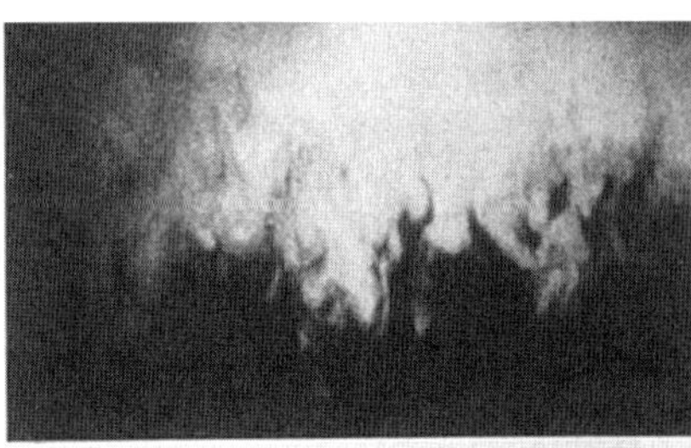
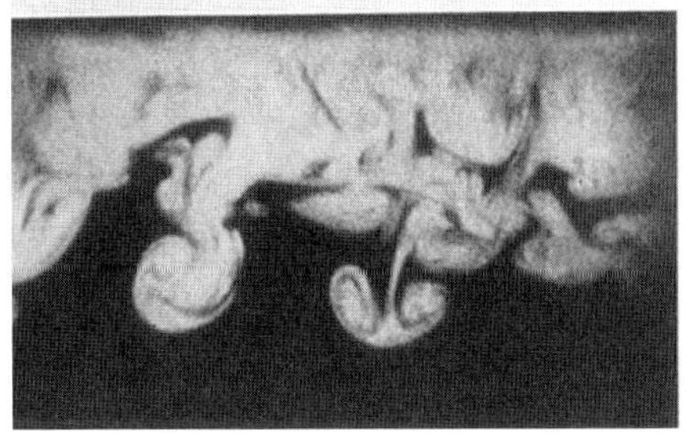
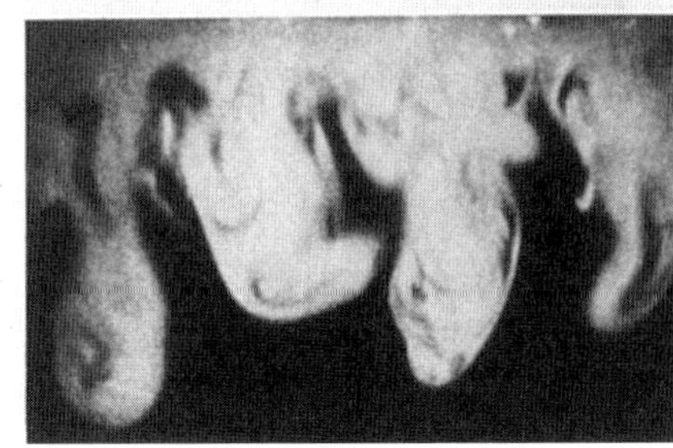

FIGURE 8-30 Turbulent eddies created due to cooling of lake surface water at temperatures above 4°C. The process is known as penetrative convection, and the resultant motions can be called thermals or eddies. Whether the eddies remain distinct depends on the Reynolds number of the flow, with Re in the *upper panels* considerably higher than in the *panels below*. For the lower Re, the descending eddies remain distinct. Upward flow will occur between the downward, more energetic eddies. This figure has been inverted from that in Van Dyke (1982), which illustrates thermals produced by near-bottom heating.

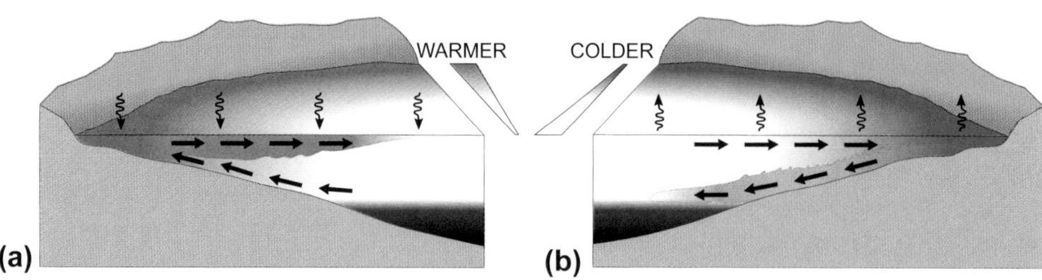

FIGURE 8-31 Cross sections of a lake indicating (a) differential heating and (b) differential cooling. The *downward* or *upward arrows* indicating uniform heating or cooling across the lake. The greater heating/cooling inshore result from the shallower water depth. In some lakes, there may be spatial differences in the extent of heating or cooling (e.g., Fig. 8-32). The *horizontal arrows* in the lake indicate the direction of water flow. *Burnt red shading* indicates warmer water; *blue shading* indicates cooler water. In (a) warmer water occurs in the shallows and will flow offshore at the surface with a return flow below; in (b) cooler water forms in the shallows and will flow offshore deeper in the water column and intrude at a depth of similar density. When water flows out of the shallows, water flows in from offshore to replace it. *(Adapted from Monismith et al. [1990]. Drawn by Tomy Doda.)*

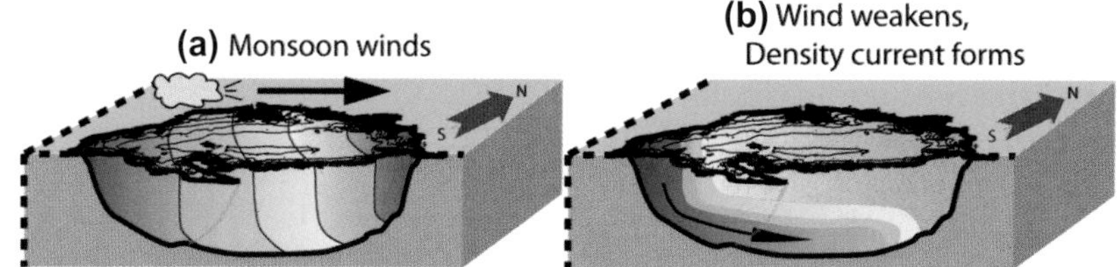

FIGURE 8-32 (a) Example of conditions leading to horizontal convection, in this case, differential cooling, and (b) the resultant gravity current. This scenario was developed for Lake Victoria (East Africa) due to the considerably larger latent heat fluxes to the south during the Southeast Monsoon and the near-bottom stratification observed to the north in August on cessation of the windy period (Talling, 1966; MacIntyre et al., 2014). Evaluating processes occurring in lakes can be challenging when data are scarce. The winds may also have caused downwelling of a persistent thermocline; on cessation of the wind, the cooler, hypolimnetic water would flow north. *(From MacIntyre [2013].)*

rapidly than those offshore, leading to offshore flow at depth (Fig. 8-31).

The flow of cool water out of shallow regions as downslope currents and its replacement by warmer water has been called a *thermal siphon* (Monismith et al., 1990). Recent observational and modeling studies indicate that the flushing of the shallow regions can occur from approximately midsummer onward toward winter as lakes begin to cool in temperate and subtropical regions (Doda et al., 2021), as was evident in polymictic Lake E6 (Alaska, USA) (Fig. 7-35). Recent modeling illustrates how wind and heat loss combine to moderate the flows (Ramón et al., 2022). Horizontal flows also occur if there is more wind-induced mixing offshore such that the water there is cooler (Parker and Imberger, 1986). In that case warmer water from inshore will flow offshore. These flows can occur within small lakes and large ones and can contribute to the development of stratification in offshore waters (Wells and Sherman, 2001). Such flows are attributed to the restratification of 5600 km^2 Lake Albert, East Africa (Talling, 1963), and the northern waters of 85,000 km^2 Lake Victoria (East Africa) on the cessation of the southeast monsoon, resulting in greater cooling to the south in this East African Great Lake (Fig. 8-32) (Talling, 1966; MacIntyre et al., 2014).

Horizontal convection can be a continuous process. In ice-covered lakes, as flow moves toward the deeper locations, an upward flow occurs to balance it, and at times not only one but two circulating layers may develop (Phillips, 1970). When one lateral boundary of a lake is warmer or cooler than the other, horizontal convective circulation can also develop. Such has been modeled for Lake Tanganyika (East Africa), given the much greater heating of this lake to the north and the considerable cooling to the south (Verburg et al., 2011), and is hypothesized based on field observations in permanently ice-covered Lake Bonney in the Dry Valleys of Antarctica (Spigel and Priscu, 1998). There, due to the cold glacial sidewalls of the lake, and also taking into account the salinity of the lake, which contributes to its density, a near-continuous flow is expected with warmer water rising on one side and cooler water descending on the other with horizontal flows resulting from continuity. The observational data includes evidence for the expected intrusions.

Double diffusive convection, motions that result from the different diffusivities of heat and solutes, occurs in lakes in which both temperature and salinity contribute to density. When the stratification consists of cold, fresher water over warmer, somewhat saltier water, *salt fingering* may result and lead to a characteristic staircase structure in a temperature profile (Thorpe, 2007). When the stratification is reversed, with warm salty water over cooler, fresher water, double diffusion results and convective motions result again. Conditions that enable these two types of motion are reviewed in Thorpe (2005, 2007). The staircase temperature profiles that can result from double diffusive convection in a lake were first noted in 540 m deep, meromictic Lake Kivu (Democratic Republic of the Congo and Rwanda) (Newman, 1976). Subsequent efforts quantified the increased fluxes of heat and solutes resulting from this process (Schmid et al., 2010; Carpenter et al., 2012). The stepped temperature profiles have also been noted in Lake Vanda and Lake Miers in the Dry Valleys of Antarctica (Spigel and Priscu, 1998).

Convective flows provide a mechanism whereby solutes, dissolved gases, or particulates produced at boundaries can flow offshore. Laboratory studies have provided equations that allow the calculation of expected flow velocities and flushing of nearshore regions (Nepf and Oldham, 1997; Sturman et al., 1999). The velocities (u) of the flows can be computed at steady state as:

$$u \approx (g\,\alpha\,\Delta T\,H)^{1/2} \qquad (8\text{-}29)$$

where g is gravity, α is the thermal coefficient of expansion, ΔT is the temperature difference between inshore and offshore waters, and H is the thickness of the intrusion. Tracer studies have identified the offshore flows of solutes from differential heating and cooling (Fig. 8-32) (James and Barko, 1991). Due to the rapidity of the flows from horizontal convection, it may be difficult to discern whether increased solutes at depth in the water column are produced locally or result from the convective motions. The use of 3-D hydrodynamic models with tracers embedded illustrates, for example, when increases in dissolved gases such as CO_2 result from respiration in nearshore regions as opposed to respiration in offshore sediments (Amaral et al., 2021).

Convective motions in ice-covered lakes – Horizontal convective motions also occur under the ice and drive circulation when the action of wind is limited. In that case, since the temperature of maximum density of freshwater is ~4°C, sediments heating the overlying water can increase density and cause a gravity current that then flows downslope (Fig. 8-33) (Mortimer and Mackereth, 1958; Malm, 1998; MacIntyre et al., 2018). These flows may be enhanced by further increases in density resulting from mineralization in littoral sediments and freeze-out and any settling of solutes during ice formation above shallow waters. The result is the well-known inverse temperature stratification found in ice-covered lakes; the products of respiration also accumulate in the deeper locations and anoxia tends to develop (Terzhevik et al., 2009, 2010). Tracer studies illustrated the horizontal convective circulation in ice-covered lakes (Likens and Hasler, 1962; Likens and Ragotzki 1965; Welch and Bergmann, 1985). In springtime heating below the ice when temperatures are below 4°C leads to penetrative convection (Farmer, 1975). Greater heating of inshore water at temperatures below 4°C increases the density such that offshore flow results, causing heating offshore; another example of the thermal siphon described above (Ulloa et al., 2019; Cortés and MacIntyre, 2020).

Considerable effort has also been conducted to quantify the convective flows under the ice. Such efforts include the methodology to calculate the velocity of gravity currents (Malm, 1998; Rizk et al., 2014), the velocity and pathway of the full convective circulation (Phillips, 1970; Malm, 1998), and the velocity and nature of the plumes from penetrative convection (Bouffard et al., 2019). The velocity of the gravity currents is of the order of millimeters per second and the resultant upward flows are much lower, such that it could take months for a parcel of water starting at the lake bottom to reach the surface (MacIntyre et al., 2018). Downwelling plumes from penetrative convection also have velocities of millimeters per second. Penetrative and horizontal convection have been modeled by Ulloa et al. (2018, 2019), and Ramón et al. (2021) have quantified the influence of rotation on the extent of horizontal convection. Modeling has shown that the combination of the rapid decrease in temperature immediately below the ice combined with the *cryoconcentration* of solutes

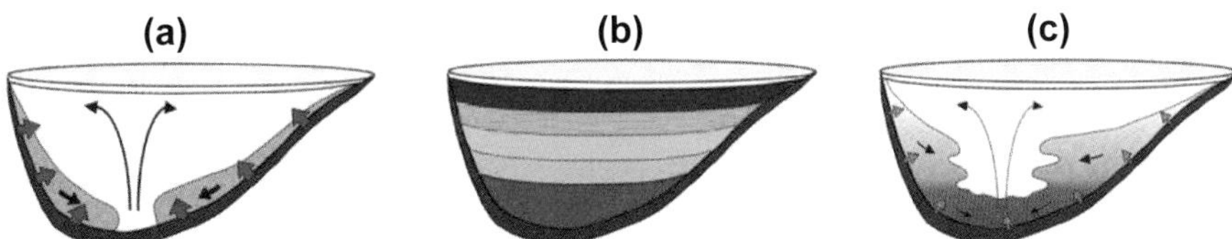

FIGURE 8-33 Horizontal convective circulation and stratification in lakes under the ice. (a) At temperatures below 4°C, sediment heating of the overlying water creates density currents that flow downslope; respiration at the sediment–water interface produces solutes that contribute to increased density and the near-bottom flow (Mortimer and Mackereth, 1958). Upward flows result in an overturning circulation (Welch and Bergmann, 1985; Malm, 1998; Rizk et al., 2014). (b) Temperature and solutes stratify as warmer, solute-enriched water flows to the deepest points of lakes. (c) Later intrusions flow at progressively shallower depths due to the increased density in the lower water column. With reduced near-bottom flow, anoxia develops and methane may accumulate. *(From MacIntyre et al. [2018].)*

can create the temperature and salinity stratification conducive to double diffusion in the near-surface boundary layer (Olsthoorn et al., 2020). The extent of mixing deeper in the water column is expected to depend on the strength of the respective temperature and salinity gradients.

Thermal bar – An additional convective circulation is set in motion as a result of density gradients arising when shallow, nearshore waters heat more rapidly than the open water mass following ice off. In Lake Ontario (USA and Canada), for example, inshore shallow waters stratify, while offshore water temperatures are below 4°C and may be inversely stratified (Fig. 8-34). The narrow transition zone between inshore and offshore waters is called a *thermal bar*. It consists of a nearly vertical 4°C isotherm. Water flowing from inshore and offshore locations converges at the thermal bar and flows downward. The density-driven circulation is dependent on the bottom topography, with a more pronounced circulation and considerable descending motions inside the thermal bar zone of convergence in lakes with steep slopes (Malm, 1995). The downward density-mediated currents outside the thermal bar can penetrate to deep strata or bottom layers

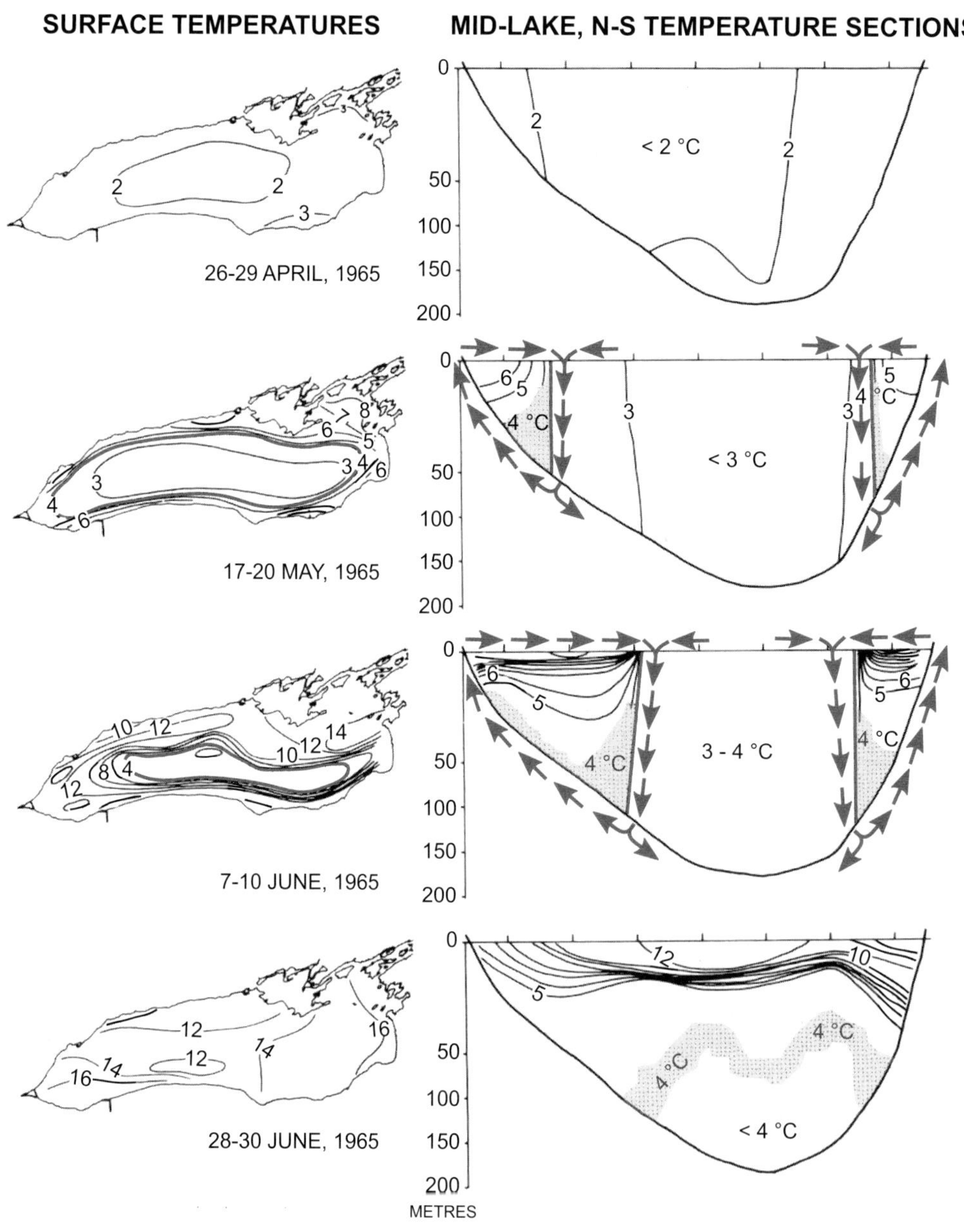

FIGURE 8-34 Formation and progression of a thermal bar in Lake Ontario in spring (26–29 April and 17–20 May 1965), through early summer (7–10 June 1965), and loss of the thermal bar with the progression of summer (28–30 June 1965). Figures to the *left*, surface view with the 4°C isotherm in *red*. Figures to the *right*, cross-sections, with arrows indicating the direction of water flow. 4°C water is shaded *red*. Note the vertical orientation of isotherms indicating the thermal bar. *(Redrawn from Rodgers [1966], with further revisions by Tomy Doda.)*

and cause appreciable convective mixing (e.g., Lake Baikal; Shimaraev et al., 1993). The Earth's rotation combines with the density gradient to induce a counterclockwise coastal current inside the bar.

The thermal bar moves progressively further from shore as the heat influx continues to warm the larger open water mass (Fig. 8-34). Little mixing occurs between inshore and offshore waters, and in some locations a significant portion of inshore water may originate from runoff (Spain et al., 1976). Finally, thermal differences between the inshore and offshore regions are reduced to the point where stratification of the whole basin occurs (Rodgers, 1966).

Thermal bars commonly occur to some extent in lakes with a period of ice cover. In small lakes the phenomenon is transitory, often lasting only a few days. In large lakes, however, the transition to stratification of the whole basin may take weeks, as seen in the example from Lake Ontario (Fig. 8-34). As a result of the thermal bar, inshore waters are isolated and can warm more rapidly than those offshore. With the warmer temperatures and potential increased nutrient supply from surface runoff and river discharge, productivity can increase earlier in the spring in the nearshore waters compared to the open water.

D. Pathways of stream or river inflows

The pathway of a stream or river as it flows into a lake or reservoir depends on its density and discharge, with the initial density of the incoming water determined by its temperature, solutes, and the quantity of suspended material. At the river mouth, streams and rivers displace and mix with the lake water, and density differences between the inflow and the surface lake water cause pressure gradients (Imboden, 2003). Thus, the density of the inflow changes depending on its momentum. As the incoming water continues to flow into a lake, mixing may continue, further changing its density. If the density of the incoming water (ρ_0) remains less than that of the ambient lake water (ρ_1), as is the case when stream or river water is warmer than lake water in summer, the inflow will travel over the lake water such that the incoming water is an *overflow* (Fig. 8-35a). Otherwise, as for cold (and therefore denser) inflows entering a warmer lake, the incoming water will plunge and then

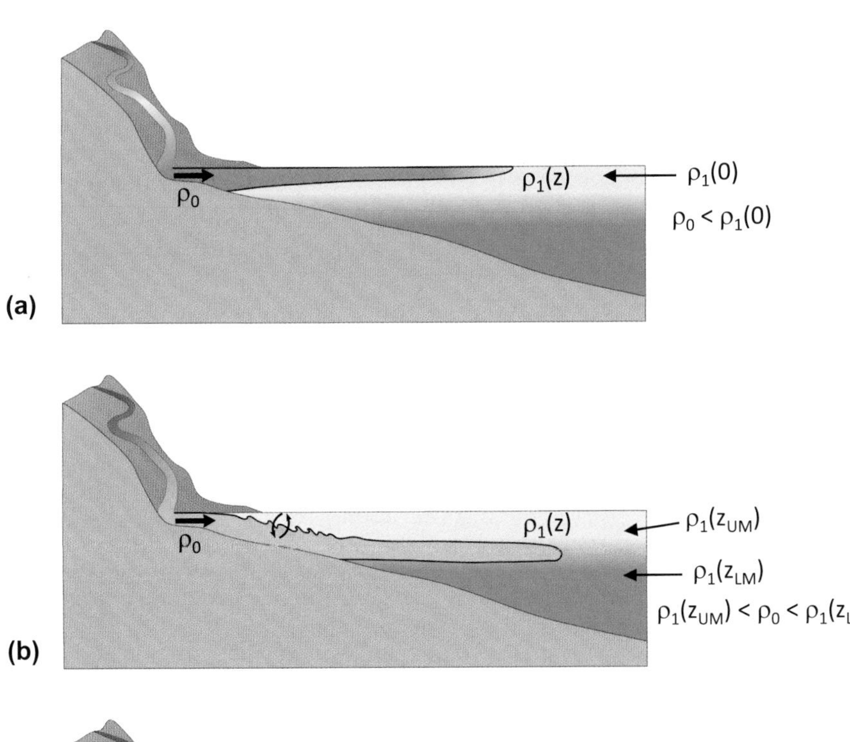

FIGURE 8-35 Cross sections of a lake showing entry of an inflow from a stream or river in the *upper left*. (a) Overflow (*burnt red* = warm) when incoming stream water is warmer and less dense than lake water and flows over the surface of a lake; (b) interflow where the density of the incoming water, after initially mixing with lake water, is intermediate (*gray*) between that at the surface (*pink* = warm) and deeper (*blue* = cold) in the water column; (c) underflow (*blue* = cold), when incoming water is denser than receiving water and flows to the bottom and then offshore. In all panels incoming stream or river water initially displaces some of the lake water. ρ_o is the density of the incoming water after its initial mixture, if any, with lake water, ρ_1 is the density of ambient lake water, and z refers to depth. $\rho_1(z)$ refers to the density of the lake at each depth, $\rho_1(0)$ to the density at the lake surface, $\rho_1(z_{UM})$ to the density of the upper metalimnion, and $\rho_2(z_{LM})$ to the density of the lower metalimnion. Curves associated with the plume indicate mixing with lake water. *(Adapted from communication with Alicia Cortés and drawn by Tomy Doda.)*

travel along the lake bottom for some distance. The resultant shear contributes to further mixing between the plume and the lake water and a further modification of its density, typically decreasing the difference between it and the ambient lake water. The plume will either flow offshore into the lake at a depth equivalent to this new density as an *interflow* (Fig. 8-35b), or, if its density remains greater than that of the lake water, it will flow along the bottom as an *underflow* (Fig. 8-35c). In fact, incoming inflows often split such that they intrude at multiple depths. Whether they do so depends in part on the momentum of the plume relative to any density differences within it, on the stratification within the lake, and on the suspended sediment load. Detailed analyses of these mixing characteristics and the dimensional indices used to predict the behavior of the plume are given in Imberger and Hamblin (1982), Imberger and Patterson (1990), Rueda et al. (2007), Marti et al. (2011), Wells and Wettlaufer (2007), Rueda and MacIntyre (2010), Cortés et al. (2014), Ramón et al. (2020), and Wells and Dorrell (2021).

Discharge varies widely seasonally, not only in volume but in the accompanying load of dissolved and suspended materials. For example, in alpine situations the density of river water is typically greater (cold, high dissolved and particulate load) than the recipient water and underflow is common, particularly if glacial erosion at the head of the river contributes to the suspended load (Robb et al., 2021). During the summer, although alpine river inflow water is still cold, its sediment load and therefore density can be reduced by dilution with runoff from snow and icemelt. In the summer, then, alpine rivers may flow into the metalimnion. This situation has been documented frequently (e.g., Chikita et al., 1985). A particularly striking case is seen in Lake Constance (Central Europe), where light penetration is abruptly attenuated in the metalimnion over large areas of the lake by the intrusion of river water with a high suspended sediment load (Lehn, 1965).

The pathway and extent of mixing of inflows with ambient lake water can modify nutrient cycling and biological productivity (Vincent et al., 1991; Gibbs, 1992). The extent of mixing with lake water depends in part on discharge rates. At high discharge rates relative to lake volumes, overflows and interflows may mix with and displace a considerable portion of the ambient lake water (Fig. 7-39). When discharge is lower, inflows may mix more slowly with lake water. For example, incoming storm water first mixed near the inlet and then intruded into the top of the metalimnion and base of the upper mixed layer in Toolik Lake (Alaska, USA), where it was subsequently mixed by processes such as penetrative convection (Fig. 7-21). In other years inflows penetrated into the metalimnion and were not mixed into the epilimnion for several weeks (Rueda and MacIntyre, 2010). Similar pathways have been noted in warm water reservoirs with thickened metalimnia (Cortés et al., 2014). In regions with persistent inflows, whether stratification is due to solar radiation or the inflow, or whether stratification breaks down due to wind mixing, can be predicted from dimensionless indices similar to the comparisons inherent in the Monin–Obukhov length scale (Chapter 7). These indices include buoyancy flux from heating and from the inflow, and wind shear (Xing et al., 2014). In larger lakes the pathway of incoming water is modified by the Coriolis force.

The extent that incoming snowmelt flows into lakes under the ice and mixes can be important for their light climate, nutrient supply, and the fate of climate-forcing trace gases. Early experiments indicated that while lakes were still ice covered, much of the incoming snowmelt was an overflow and flowed out of lakes before mixing (Welch and Bergmann, 1985). In 1.5 km^2 Toolik Lake the incoming water is initially an overflow due to its cold temperature. Even as the stream water warmed, it continued to flow near the surface under the ice as it was more dilute than lake water. The extent to which the incoming water mixed downward depended on time scales of advection relative to vertical mixing from penetrative convection. Based on a mass balance, only 10–15% of the dissolved organic carbon (DOC; see Chapter 28) was retained within the lake (Cortés et al., 2017). In smaller Arctic lakes incoming snowmelt tends to flow through the lakes; however, when discharge increases rapidly, it can displace water in shallow embayments, such that denser, near-bottom water in these locations flows offshore as a gravity current. In small lakes in the Sierra Nevada (California, USA) discharge is sufficiently large that it displaces appreciable volumes of lake water, including solutes released by sediments over the winter, and supplies terrestrial organic matter that fuels subsequent microbial activity (Nelson et al., 2009; Sadro et al., 2011; Melack et al., 2021). Incoming glacier melt water can carry large concentrations of suspended material whose influence on lake ecology depends upon its depth of insertion (Robb et al., 2021). When these cold inflows occur after lakes have warmed and stratified, much of the plume flows deeper into the water column. Some of the suspended load enters the upper mixed layer as the water initially mixes with lake water before plunging and some may be upwelled and mixed into the upper water column by strong wind events. When settling rates of particulates are rapid enough relative to rates of horizontal dispersion along the main axis of the lake, downwelling irradiance may be sufficient for photosynthesis downstream of the incoming plume.

E. Combined processes, mixing, and ecological implications

Wind forcing and buoyancy flux, that is, heating and cooling, and internal wave motions act continuously to modify the stratification within lakes and the fluxes of biogeochemically and biologically important scalars. Examples of the forcing, thermal stratification, buoyancy frequency, and eddy diffusivities have been provided over seasonal and diel time scales in Chapter 7 (Figs. 7-22 to 7-24 and 7-26 to 7-29) and consequences on the small scale in the near surface are shown in Fig. 8-36.

Influence of turbulence on solute fluxes – In the following, we illustrate the changes in turbulence as the rate of dissipation of turbulent kinetic energy (ϵ) and overturning length scales (eddy sizes) in Lake Pleasant (New York, USA). Changes are illustrated with respect to the turbulent velocity scale for wind, u^*_w, and the turbulent velocity scale for cooling, w^*, computed from heat loss and the depth of the mixing layer, $w^* = (\beta z)^{1/3}$ (Section IIC). The data illustrated in Fig. 8-37 were taken after several windy days that caused the upper mixed layer to downwell to the bottom of the northern basin and then upwell. The Lake number was less than 1 and stratification weakened with the ensuing mixing. Cloudy days and cool nights further contributed to the weakening of the stratification.

Upwelling, combined with cooling such that turbulent eddies entrained CO_2 from the seasonal thermocline, led to increased concentrations of dissolved gases in the upper mixed layer of Lake Pleasant (Fig. 8-37). With a shift in wind direction, the seasonal thermocline, with elevated concentrations of CO_2, upwelled at the measurement site (afternoon of DOY 275). The turbulent eddies (Lc) induced by the cooling entrained CO_2 into the upper water column that night and into the following morning. Heating commenced at nearly the same time as an increase in wind speed, as made evident by u^*_w increasing from 0.001 to 0.005 m s^{-1}. The increased wind led to further upwelling of the thermocline and to increases in ϵ in the upper 2 m; however, concentrations of dissolved gas did not increase (DOY 276.5). With the increased stratification, turbulent eddies were small, about 0.1 m, and even smaller immediately below the diurnal thermocline and in the upwelled thermocline (e.g., Chapter 7, Fig. 7-1). However, with the onset of cooling early in the afternoon of DOY 276 and continued moderate winds, turbulent eddies once again increased in size, as did concentrations of CO_2 in the upper water column. The rate of deepening of the upper mixed layer, that is, the entrainment (Chapter 7, Fig. 7-18) possible due to the surface buoyancy flux, is computed as $dh/dt = \beta/(zN^2)$ (Chapter 7 II.A). The surface buoyancy flux β was 1×10^{-7} m^2 s^{-3}, buoyancy frequency N in the upwelled thermocline was 10 cph, equivalent to 0.02 s^{-1}, and the overturning scales shortly after the onset of cooling were 6 m. The rate of deepening would have been 5.5×10^{-5} m s^{-1}. Over a 5-h period, the upper mixed layer could deepen by nearly a meter, similar to the change shown in Fig. 8-37. This study shows the importance of wind-induced upwelling of the thermocline combined with convective mixing for increasing concentrations of solutes in the upper mixed layer. However, care must be taken to evaluate whether the increases are not due to horizontal motions, and the analyses in Czikowsky et al. (2018) for two different time periods illustrate when local convection caused entrainment and when the elevated CO_2 was advected.

These combined processes will likely be important for entrainment in many lakes (Chapter 7, Fig. 7-22). That

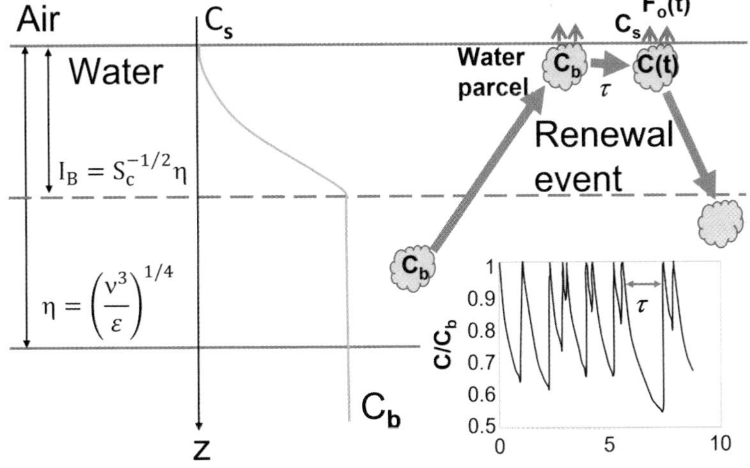

FIGURE 8-36 Conceptualization based on Kolmogorov microscales (η) and Batchelor length scale (l_B) of fluxes of dissolved gases across the air–water interface (see Eqs. 8-19 and 8-22); S_c is Schmidt number, ϵ is turbulent kinetic energy, and ν is kinematic viscosity. Concentrations in the water near the interface are illustrated with the *green line*. The thickness of the viscous sublayer at the air–water interface scales with the Kolmogorov microscale for length. The concentration gradient occurs over the diffusive sublayer scaled as the Batchelor length scale. Dissolved gas with concentration C_b is transported by the small-scale eddies to the interface where concentrations are in equilibrium with those in the atmosphere (C_s). Flux at the interface over time, $F_o(t)$, occurs when there is a difference in concentration. The upward transport is conceptualized as a renewal event whose duration is the Kolmogorov microscale for time (τ in figure, τ_K in text). The remaining gas after a time t is C(t). Concentrations, C, at the air–water interface over several renewal events are scaled as C/C_b. z is depth. *(Redrawn from Katul and Liu [2017].)*

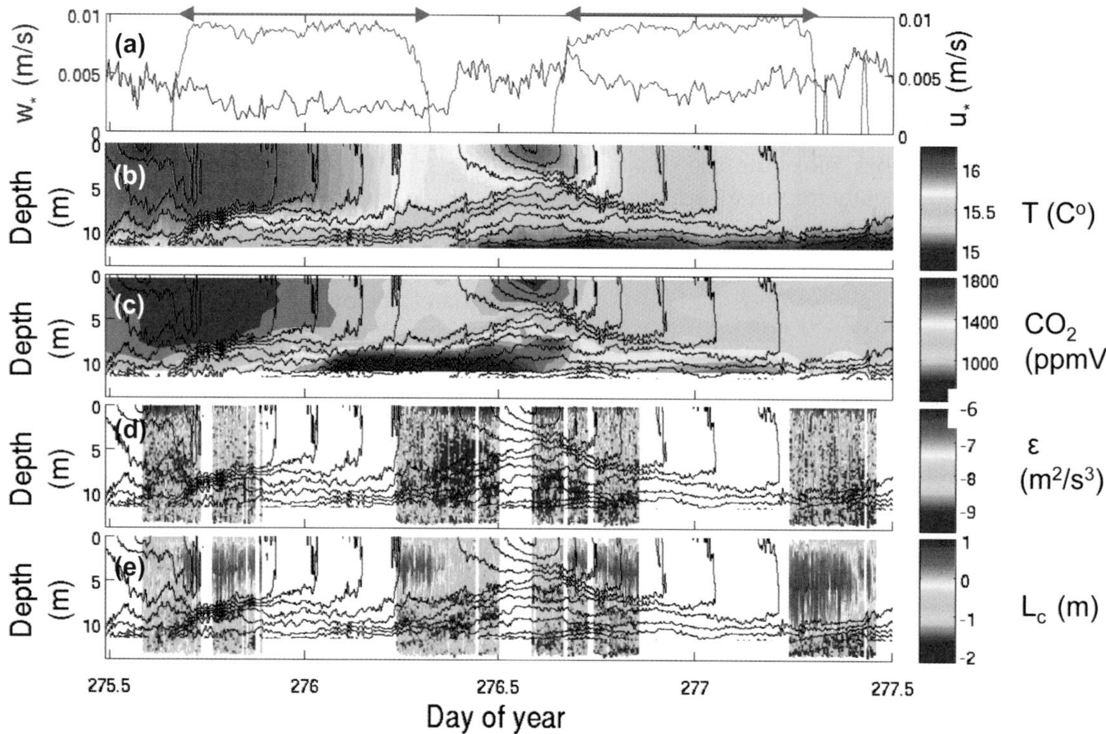

FIGURE 8-37 Time series of turbulent velocity scales for (a) cooling (w*) and wind (u^*_w), (b) temperature contours, (c) CO_2 concentrations (parts per million volume [ppmv]), (d) logarithm of ε, and (e) logarithm of centered overturning scales (L_c, eddy sizes). Maximum overturn size is in the middle of the eddy (e.g., the 10 m overturn at DOY 276.8 extends throughout the actively mixing layer). 0.1°C isotherms overlaid in panels (b) through (e). *Blue overbars* indicate cooling. *(Data are from Lake Pleasant [New York, USA] on days of year [DOYs] 275–277,* (2–4 October 2010.) *The lake is 4 km long with a deeper southern basin and a shallower, 10 m deep northern basin where measurements were taken. See Tedford et al. [2014] and Czikowski et al. [2018] for additional details.)*

said, the intense mixing associated with Lake numbers dropping to values below 1 (e.g., Chapter 7, Fig. 7-24) has led to large increases in CO_2 concentrations in the surface waters of Toolik Lake in the Alaskan Arctic. Hence both wind-driven and convective motions can lead to the entrainment of solutes.

Relation of small to large scales of turbulence (example from Lake Pleasant, New York, USA) — A key concept in turbulence is that the small scales are related to the large, with $\epsilon = u^3/l$ representing the relation (Eq. 8-7, Section I). Taking the near-surface dissipation rates (1×10^{-6} $m^2\ s^{-3}$) around noon on DOY 276 and overturn scales of 0.1 m in Lake Pleasant, and solving for u, gives 0.005 m s^{-1}, similar to u^*_W (Fig. 8-37a). Doing the same calculation under cooling and using the measured value of ϵ (1×10^{-7} $m^2\ s^{-3}$) at the surface and a length scale of 6 m, $u = 0.008$ m s^{-1}, which rounds to 0.01 m s^{-1}, similar to w*. Thus the large and small scales of turbulence are related. Additionally, the turbulent velocity scales computed from dissipation rates and overturning scales are similar to those computed from meteorological measurements, that is, wind speed and buoyancy flux (Chapter 7). The calculation $K_Z \approx ul$ is reasonable, where l is the depth of the overturns. These are usually much smaller under heating than cooling (Chapter 7, Fig. 7-29; Fig. 8-37) such that within the upper mixed layer, K_Z can be one to two orders of magnitude higher under cooling than heating (Chapter 7, Fig. 7-29).

Time scales for mixing can be l/u or l^2/K_Z. Here, great care must be taken. The mixing efficiency of shear-driven flow, as occurs during stratification, is lower than that driven by cooling (Section IB: Turbulence and its measurement). Under cooling, l/u is approximately 10 min based on the length scale of 6 m and velocity scale of 0.008 m s^{-1}, as calculated above. That is, a parcel of water would circulate throughout the 6 m deep upper mixed layer in around 10 min. Under heating, it would be more appropriate to do the calculation as l^2/K_Z with K_Z carefully calculated to take into account the lower mixing efficiency.

Influence of turbulence on fluxes of dissolved gases — Dissipation rates as in Fig. 8-37 can be used in calculating fluxes of dissolved gases for studies of metabolism and for carbon budgets in streams and in lakes when wind speeds are light to moderate. Gas fluxes (F; see also Chapters 10 and 13) are calculated as

$$F = k(C_b - Cs) \tag{8-30}$$

where k is a gas transfer velocity and C_b is gas concentration in the water (often notated C_w) and C_s is the gas concentration at the top of the interface in equilibrium with the atmosphere (often notated C_{eq}). Gas transfer velocities depend on turbulence (ϵ) when wind speeds are below those that will induce breaking surface waves, with the incidence of wave breaking increasing as winds increase above 8 m s^{-1}. Similarly, the dependency would decrease with appreciable white-capping in streams. Gas transfer velocities are modeled using the *surface renewal model*,

$$k = c_1(\epsilon v)^{1/4} Sc^{-n} \tag{8-31}$$

where c_1 is a coefficient of ~0.4, $(\epsilon v)^{1/4}$ is the *Kolmogorov microscale* for velocity (v_K, Eq. 8-21), Sc is the *Schmidt number*, which is the ratio of kinematic viscosity of the water (v) to the molecular diffusivity (D) for the gas of interest, and n depends on the fluidity of the water surface (Lamont and Scott, 1970; Lorke and Peeters, 2006; Zappa et al., 2007; MacIntyre et al., 2010; Katul and Liu, 2017). Dissipation rates can be measured, as in Fig. 8-37, or calculated from meteorological data. When the latter approach is taken, they depend on u_w^{*3} divided by the Von Kármán constant (≈0.41) and depth, with a slight augmentation depending on the rate of heating or cooling of the water column (MacIntyre et al., 2021a,b).

The concepts underpinning the exchange are illustrated in Fig. 8-36. The thickness of the viscous sublayer scales with the *Kolmogorov microscale* for length (η, Eq. 8-19), the thickness of the diffusive sublayer scales with the *Batchelor length scale*, $l_B = Sc^{-1/2}\eta$ (Eq. 8-22), where l_B is the length scale at which the smallest eddies diffuse, and the frequency of renewal events scales with the Kolmogorov timescale (τ_K, Eq. 8-20) (Lorke et al., 2003; Lorke and Peeters, 2006).

The comparison of Kolmogorov microscales and the Batchelor length scale computed for typical dissipation rates in the Lake Pleasant study Fig. 8-37) shows that the velocities are twice as fast, the viscous sublayer and l_B are half as thick, and the time scale at which eddies replenish the air–water interface is five times faster during a windy afternoon than at night (Table 8-1). The computed gas transfer velocities for CO_2 at 20°C were 7.6 and 3.2 cm h^{-1}, respectively. These calculations are in agreement with the results of the study, which showed that the large eddies at night entrained

TABLE 8-1 Typical dissipation rates at night and in the day for Lake Pleasant (New York, USA) in Fig. 8-37 and calculated Kolmogorov microscales for velocity, v_K, length, η, time, τ_K, the Batchelor Length Scale, l_B, and gas transfer velocity (k_{600}) computed with the surface renewal model. By convention, units for gas transfer velocities are cm h^{-1} and for purposes of comparison are normalized using the Schmidt number to values for CO_2 at 20°C (i.e., k_{600}).

	ϵ $m^2\,s^{-3}$	v_K $m\,s^{-1}$	η m	l_B m	τ_K s	k_{600} cm h^{-1}
Night	1×10^{-7}	5.6×10^{-4}	1.8×10^{-3}	7×10^{-5}	3.2	3.2
Day	3×10^{-6}	1.3×10^{-3}	8×10^{-4}	3×10^{-5}	0.6	7.6

CO_2 from the upwelled thermocline but that fluxes were appreciably higher during afternoon winds (Czikowsky et al., 2018).

When particle image velocimetry has been used to quantify turbulence, ϵ can be measured even closer to the water surface and values are appreciably higher, indicating the influence of turbulence in supporting fluxes by surface renewal may be higher than in these calculations (Wang et al., 2015).

Advection, mixing, and inflows – Evaluating the combined influence of advection, mixing, and inflows on biological and biogeochemical processes can be approached via experimentation or modeling. For example, assessing whether stream inflows moderate primary production requires a number of measurements. These include measurements of discharge and the concentrations of nutrients in the stream. This information allows the calculation of the nutrient loading to the lake. Meteorological and time series temperature measurements would be obtained so that graphs and calculations could be done as in this chapter and Chapter 7. Sampling would also be done on transects. The profile data can include conductivity, temperature, depth (CTD) profiles, as well as microstructure profiling, and samples for nutrients, phytoplankton, and primary production. Instrumentation can be placed on arrays along and adjacent to the expected flow path of the incoming water. Similar approaches can be applied to quantify the effects of storm events with their increased winds and mixing. Studies designed to quantify lake metabolism would have a similar design with the instrument arrays including oxygen sensors. Examples of studies using these approaches are Frenette et al. (1996a,b), Robarts et al. (1998), MacIntyre et al. (1999), and MacIntyre et al. (2006). Antenucci et al. (2013) describe how to quantify lake metabolism accounting for transport and dispersion.

Time scale and simulation approaches to assess biogeochemical and biological outcomes – Predictive approaches include computing time scales for advection, diffusion, and reaction rates (Knauer et al., 2000). By comparing the different time scales, the dominant fluxes can be determined. Time scales for advection and mixing were presented in Section I. The ratio of transport and transformation time scales is known as the *Damkohler number.* Huisman et al. (2002) and O'Brien et al. (2003) use this approach to address whether sinking phytoplankton can grow when the euphotic zone is turbulent. O'Brien et al. (2003) determined that only a small fraction of the sinking cells had their residence time increased by turbulence and that growth and blooms would only occur if growth rates were sufficiently rapid relative to sinking rates. These approaches can be applied to diverse lakes if typical advective flow speeds are known, if values of the coefficient of eddy diffusivity are known or can be computed (see Chapter 8 Section IB *Turbulence and its measurement* and Fig. 7-29), and if growth rates as a function of irradiance are known for the phytoplankton community. Similar approaches can be applied to predict community structure and energy flows under the ice in winter (Jansen et al., 2021).

Simulations can be conducted on the basis of the 1-D equation for change of solute concentration (C) with depth given in Section II (ecological importance of K_H and K_z):

$$dC/dt = K_z \frac{d^2C}{dz^2} - u\frac{dC}{dz} \qquad (8\text{-}32)$$

where t is time (s), z is depth (m), u is mean advective velocity ($m\ s^{-1}$), and K_z is coefficient of eddy diffusivity ($m^2\ s^{-1}$). This approach has been used by solving for K_z using the heat budget method (Chapter 7 I.D) and applied to understand whether respiration rates or fluxes from the anoxic benthic boundary layer determined the rate of oxygen change in the hypolimnion of Lake Kinneret (Israel) (Nishri et al., 2011).

Coupled horizontal and vertical movements influence the functioning of aquatic ecosystems. Upper mixed layers are not always fully mixed and when winds are strong enough, horizontal advection may modify concentrations of nutrients and microbes. For example, during the high winds over Mono Lake (California, USA) in September 2007 (Chapter 7, Fig. 7-17), the time scale of vertical mixing was 11 h and the actual more rapid deepening of the mixed layer resulted, in part, from horizontal advection, as in Fig. 8-20. High-resolution profiles of fluorescence, indicative of algal biomass, showed increases and decreases with subtle (<0.05°C) changes in temperature, implying advection of phytoplankton of different concentrations from near-shore regions. Similarly, when Wedderburn and Lake numbers decrease below 1, exchange can occur between the epilimnion and hypolimnion. In productive lakes with nutrient-rich hypolimnia, nutrient fluxes may be larger than predicted from eddy diffusivities calculated from 1-D approaches. Similarly, in oligotrophic lakes plankton communities in the euphotic zone may be diluted. Consideration of these coupled processes will lead to improved understanding of controls on ecosystem structure and function.

IV. Modeling water movements

Water movements are modeled to improve understanding of discrete physical processes, such as the internal bores in Fig. 8-27 or the temporal variation of the full range of physical processes occurring within a lake or a river system. To meet these goals, models have different degrees of complexity. In some cases all the terms that affect water flow, both advective and turbulent, are included, as when modeling discrete processes. For full lake and river models, simplifying assumptions are essential to reduce computational costs. Modeling is based on using conservation equations for momentum and continuity (e.g., upward motions are balanced by downward ones). A critical simplifying assumption for full lake models is that *hydrostatic pressure*, that is the pressure on the lake from the weight of water and overlying atmosphere exceeds the changes in pressure from motions such as acceleration, the *nonhydrostatic pressure component*. Neglect of this term in models of lakes and rivers means that they can be modeled as parcels of water and gridded with a much larger horizontal length scale than in the vertical, albeit with care required in selecting these scales (Chen, 2005). As with all models, outcomes require validation with observational data or theory. Equations in the models include changes in velocity with respect to time, advective and diffusive transport, changes in density, force per unit mass, and shear stress on the water surface and within the water column. The terms included depend upon model complexity. For studies of full lakes and river systems, 1-D or 3-D *hydrodynamic models* are usually applied.

A. Hydrodynamic models

Hydrodynamic models are used to describe the physical processes within lakes and rivers and their changes over time. They are used to better understand physical-biological coupling in lakes and are increasingly being used to predict changes in thermal structure and mixing under climate change, and, for decades, have been an essential component in predicting changes in water quality required for water management. Hydrodynamic models applied to lakes and reservoirs can be one-dimensional (1-D, horizontally averaged and assuming minimal horizontal differences in thermal structure across a lake), two-dimensional (2-D horizontally or vertically averaged), or three-dimensional (3-D). Examples of 1-D models include DYRESM (Imberger and Patterson, 1981; Yeates and Imberger, 2004); GLM, which is derived from DYRESM (Bruce et al., 2018; Hipsey et al., 2019); LAKE 2.0 (Stepanenko et al., 2016); Simstrat (Goudsmit et al., 2002; Gaudard et al., 2019); and Flake (Kirillin et al., 2011). Two-dimensional models are often used in water quality modeling (2-D, horizontally averaged), including CE-QUAL-W2 (Cole and Buchak, 1995). Three-dimensional (3-D) formulations used in research include the Aquatic Ecosystem Model 3D (AEM3D) (Hodges and Dallimore, 2018); ELCOM, from which AEM3D was derived (Hodges et al., 2000); and Si3D (Rueda and Schladow, 2003; Chen et al., 2017, 2018). For rivers, 1-D (laterally and vertically averaged) models include HEC-RAS (USACE, 2010) and MIKE 11 (DHI, 2011), while CE-QUAL-W2 may also include a vertical component.

Models of reduced dimension (i.e., 1-D or 2-D), using vertical (depth) or horizontal (width or whole lake) averages, reduce computational times. For 1-D lake models, the assumption is that horizontal density gradients are small relative to vertical ones. Lake number can be used to check the validity of the 1-D assumption under strong winds when the upwelling of the lower layer indicates that modeling with higher dimensionality is advisable, including accounting for boundary mixing at the thermocline. In contrast, in 1-D models of rivers, an assumption is that vertical density gradients are smaller than those in the horizontal. Processes such as up- and downwelling of the thermocline or internal wave breaking are not explicitly identified in 1-D models due to horizontal averaging. To compensate, approaches to capture the resultant shear and mixing are included in some 1-D models (e.g., DYRESM; Yeates and Imberger, 2004).

B. Model forcing and validation data

1-D and 3-D hydrodynamic models require data on basin morphometry, meteorology, and the magnitude and density of inflows and outflows. For 1-D models of lakes, area at depth intervals within the lake (hypsographic table; Chapter 4) is required, whereas 3-D models require spatial (horizontal) maps of depths across the whole basin, including morphometric features. Requirements are similar for rivers. For example, depth and cross-section area along the mainstem are required for 1-D (longitudinally resolved) models.

Meteorological data required for the models include shortwave and longwave radiation, air temperature, relative humidity, and wind speed and direction (Chapter 7). Temperature profiles are needed to begin model simulations, and time series data are needed for validation of model output. Meteorological data are sometimes collected from locations distant from the lake of interest, and these data may need to be adjusted to improve model comparisons with *in situ* data. This approach is often relatively successful if there is a network of meteorological stations in a region. The use of on-lake meteorological stations with 5-min averaged data, and in some cases the use of high-frequency sensors with digital data loggers, enables direct calculation of fluxes of momentum and of sensible and latent heat exchange, which can be used to validate calculations for these variables based on the algorithms in the model. Small lakes and rivers can be sheltered by riparian vegetation or hills, which reduce radiative fluxes, surface water temperature, and wind, requiring modification of the radiation input and correction of wind in hydrodynamic models. Heat, momentum, and mass transfer coefficients provide the link between meteorological input data and heat content and movement of water simulated by hydrodynamic models. Early reports (e.g., Tennessee Valley Authority, 1972) form the basis of many of the transfer coefficients used in current hydrodynamic models. Considerable research since that time has quantified how the coefficients vary with atmospheric stability (Imberger and Patterson, 1990; Fairall et al., 1996, 2003; Woolway et al., 2014). Varying accuracy in quantifying surface heat fluxes led to appreciable discrepancies in thermal structure and mixing when 1-D models were contrasted with measured data on Lake Kivu (Central Africa). Such comparisons and subsequent tuning led to model improvement (Thiery et al., 2014). Recent texts provide valuable insights into how to interpret and quantify meteorological data (e.g., Guffie and Henderson-Sellers, 2014).

Attributes of inflows and outflows are additional inputs to river and lake models. For rivers and lakes or reservoirs with large hydrologic flows, accurate specification of the volume of discharge and density of the inflows is critical as the conditions within the model domain depend on those at the boundary. For all lakes, the density of the inflow and discharge is important for resolving the depth of insertion of the inflow (i.e., overflow, interflow, or underflow; Fig. 8-35) and the entrainment of the inflow with the adjacent lake water. Model input data related to outflows usually requires information on the volume of the discharge, which will be known for reservoirs and some lakes, and the depth of the outflow relative to the surface of the lake. Many times, the discharge at the outlet is a model result and can be compared with available discharge data for validation. In reservoirs the density of the outgoing water from deep outlets depends on density stratification in the lower water column and the speed of withdrawal (e.g., selective withdrawal of water in reservoirs; see Farrow and Hocking, 2006). Recent model improvements have included modules for the prediction of snow and ice cover and sediment heating (Hamilton et al., 2018), with additional appropriate inputs such as temperature sensors in the ice and sediments.

Evaluation of 1-, 2- and 3-D hydrodynamic models includes comparisons of temperature and thermal stratification over periods of time ranging from days to several years. For example, the evaluation of Simstrat, a 1-D model that explicitly parameterizes the wind energy transferred to internal waves and the resultant mixing in the benthic boundary layer, was conducted in two steps (Goudsmit et al., 2002). Results from the upper 16 m of 5.2 km^2 Lake Baldegg (Switzerland) illustrated the model correctly simulated temperatures, the depth of the upper mixed layer and seasonal thermocline in the stratified period, the onset of fall cooling, and the onset of the stratified period (Fig. 8-38). Simulations in 34 m deep Lake Alpnach (Switzerland, 4.2 km^2) extended to the bottom. While the modeling could not resolve the internal wave motions, it correctly simulated temperatures and thermal structure throughout the lake. Calculated eddy diffusivities using the heat budget method (Chapter 7) and simulations were both $\sim 3.5 \times 10^{-5}$ m^2 s^{-1} in the benthic boundary layer, an order of magnitude higher than in the interior layer, attesting to the accuracy in incorporating enhanced internal wave-induced shear and mixing in this region. Yeates and Imberger (2004) extended the 1-D Dynamic Reservoir Simulation Model (DYRESM) to simulate internal wave-induced mixing and resultant increased fluxes in the benthic boundary layer by parameterizations using the Lake number, whose interpretation is similar to that of the Wedderburn number (Chapter 7, Figs. 7-14 and 7-25). Their efforts were successful in modeling the thermal structure in Mono Lake (California, USA) and Lake Kinneret (Israel), both of which have surface areas of $\sim$150 km^2 and frequent events with Lake numbers approaching 1 during the stratified period. Considerable additional development has occurred in these models and others including incorporating other hydrodynamic processes, as described in this chapter (Gaudard et al., 2019; Hipsey et al., 2019). Comparative studies have quantified the accuracy of the same 1-D model in lakes at other latitudes (Bruce et al., 2018) and a suite of models for one lake (Perroud et al., 2009; Thiery et al., 2014), with such studies indicating which models need improvement in calculating heat fluxes and any tuning parameters.

3-D hydrodynamic models are validated against observational data with thermistor chains deployed at multiple sites in a lake. Examples of such careful

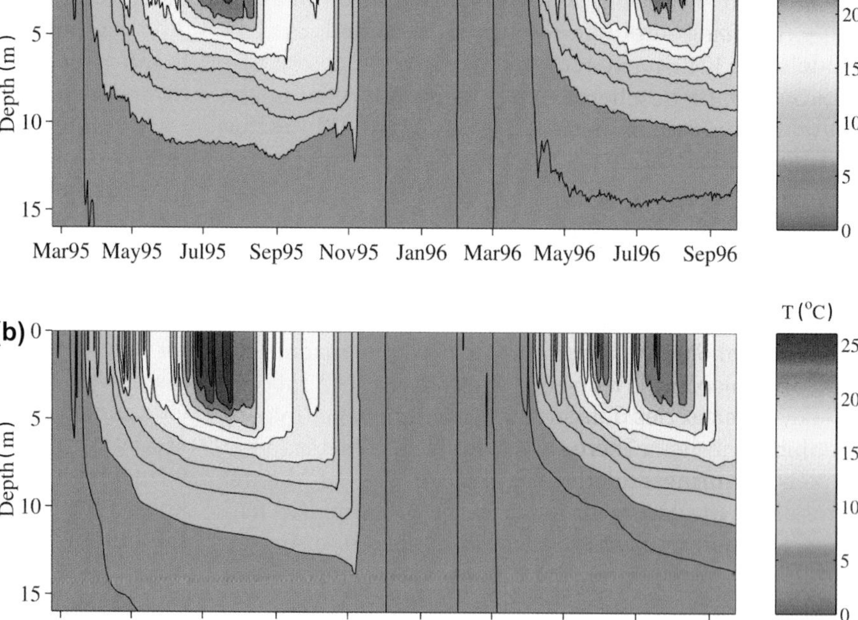

FIGURE 8-38 (a) Measured and (b) simulated temperature and thermal structure in Lake Baldegg (Switzerland) in 1995–1996 with Simstrat. *(From Goudsmit et al. [2002].)*

validation for the Estuary, Lake, and Coastal Model (ELCOM) include studies by Hodges et al. (2000), Laval et al. (2003b), and Vidal et al. (2013). The rotation of the modeled Kelvin wave in Hodges et al. (2000) is quite similar to that in Mortimer (1974) (Fig. 8-26).

C. Choosing 1-D versus 3-D models

Hydrodynamic models are used to predict thermal structure and water quality over various time scales and to further ecosystem understanding. They can be used to predict the future states of lakes under the influence of climate and land use change. These predictions can be valuable for water management and weather prediction. The degree of stratification and ventilation of the lower water column can be simulated to identify whether anoxia will become persistent in the hypolimnion or the fate of greenhouse gases.

There are trade-offs in selecting 1-, 2-, or 3-D models. Selection depends considerably on the goals of a project. Computational time is one consideration. For instance, the computational time involved in running 3-D hydrodynamic models is high; hence 1-D models are often used to predict the consequences of climate change. An alternate approach is to run 3-D models at less frequent intervals over, say, 5-year intervals over a 100-year period (Birt et al., 2021). Given the simplifying assumptions made in 1-D models, comparisons with respect to future predictions are often made with several models developed with somewhat different conceptual underpinnings. Thus, if predictions are similar from the diverse models, there is more confidence in the results (Golub et al., 2022).

1-D models can provide time series of the coefficient of vertical diffusivity, which enables a better understanding of biogeochemical and biological processes. For example, the estimates of K_z made using Simstrat for Lake Baikal (Russia) show relatively low values of K_z under the ice in early winter and increases to values of $\sim 10^{-2}$ m^2 s^{-1} due to penetrative convection (Fig. 8-39) (Schmid et al., 2008). After ice-off, high values of K_z are found to depths of nearly 200 m. Once summer stratification sets up, they are only high in the surface mixed layer. These changes are mirrored in measurements of chlorophyll *a* and Secchi depth (Chapter 6) obtained as time series data in other years (Hampton et al., 2008). Both indicate increases in biomass when light increases in spring and diatoms in offshore waters are maintained in suspension by the turbulence from penetrative convection below the ice (Hampton et al., 2008). They also indicate decreases in biomass with deep mixing that transports cells to depths with insufficient light for growth. Biomass increased again in summer with the

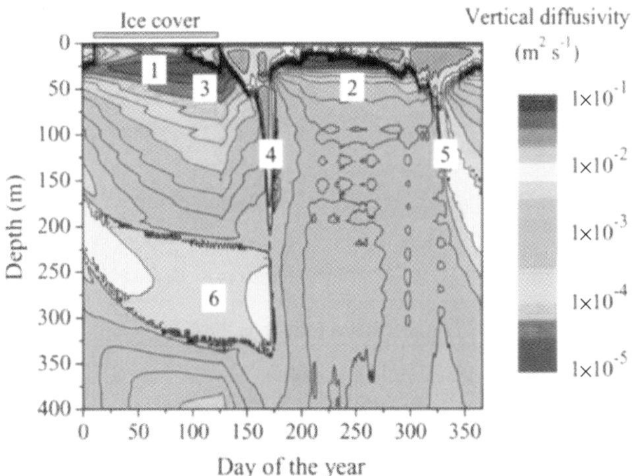

FIGURE 8-39 Time series of the coefficient of eddy diffusivity (K_z) in Lake Baikal (Russia) simulated with the one-dimensional model Simstrat. *Numbers* indicate features of the stratification: (*1*) formation of a thermocline with weak mixing under the ice ($K_z = 10^{-5}$ m^2 s^{-1}); (*2*) formation of a thermocline in summer; (*3*) increased mixing and depth of mixing under the ice due to penetrative convection in spring ($K_z >$ 10^{-2} m^2 s^{-1}); (*4*) and (*5*) deep mixing from convection after ice-off in June (*4*) and November (*5*); (*6*) increased mixing deep beneath ice under low-density gradients. *(From Schmid et al. [2007].)*

decreased depth of mixing, although species composition would differ.

3-D models also lead to improved understanding and visualization of hydrodynamic processes and can be an invaluable tool in ecosystem studies. The following examples from Lake Janauacá (Brazil), a lake within the Amazon floodplain, illustrate the importance of wind-driven motions in the 40 km^2 main basin and of differential heating and cooling within an embayment. The example from 24 m deep Lake Pleasant (New York, USA, 4 km^2) illustrates the importance of upwelling from internal wave motions for the movement of dissolved gases to the air–water interface.

Wind-induced currents transported water warmed by solar radiation across Lake Janauacá such that the upper mixed layer downwelled to a depth of 2 m in the ~4–6 m deep lake and led to a temperature difference across the lake of 2°C in the afternoon (Fig. 8-40). On relaxation of the winds, or with a change in wind direction, the thermocline downwells in another direction (Fig. 7-26). The sustained up- and downwelling means that there are cross-basin flows, as in Figs. 8-20 and 8-21. The upwelling of water from deeper depths enables exchange across the air–water interface. That is, if dissolved oxygen has become undersaturated in the lower water column, it can be replenished by the flux of oxygen from the atmosphere during upwelling. Similarly, fluxes of dissolved gases such as CO_2 or CH_4 can evade during upwelling. Thus the 3-D model provides

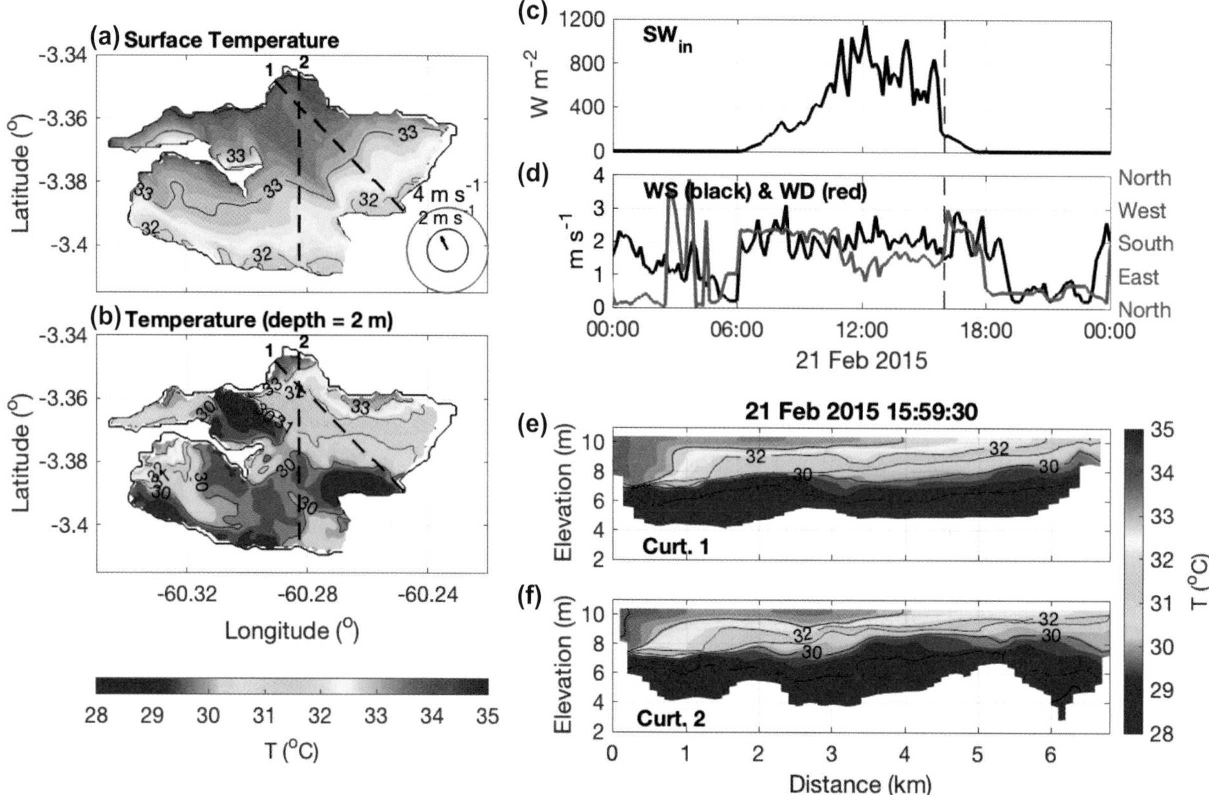

FIGURE 8–40 Plan view of (a) simulated surface temperatures and wind velocity (inset) and (b) simulated temperature at 2 m, (c) incoming shortwave (SW_{in}) with snapshot at time of *blue vertical dashed line*, (d) wind speed (WS, *black line*) and wind direction (WD in quadrats, *red line*), (e) and (f) cross-sections of temperature in the vertical lake on axes indicated by *dashed lines* labeled *1* and *2* in panel (a). Simulations are for Lake Janauacá (Brazil), 21 February 2015 at 1600 h using AEM3D. *(Modeling by Wencai Zhou.)*

perspectives important for ecosystem-level questions, and in this case, with near-surface shear and dissipation rates as additional outputs of the model, more accurate predictions of the fluxes of climate forcing trace gases can be made than are possible from one station alone.

Adding numerical tracers to 3-D models can provide additional perspective on the physical processes, which moderate ecosystem function (Figs. 8-41 and 8-42). Tracers are quantitative tools that illustrate the flow path of solutes or particulates (Rueda and MacIntyre, 2010). For example, dissolved gases can be produced by respiration associated with the roots of floating vegetation along the shores of Amazonian floodplain lakes (Amaral et al., 2021). The water under such meadows is cooler than offshore water and can sink to a depth of neutral buoyancy and then flow offshore (Fig. 8-41). During the day, water in the beds warms more than the adjacent offshore water and flows offshore. At night, the dissolved gases can be mixed vertically in the open water and evade to the atmosphere. Thus the combination of differential cooling and differential heating leads to the offshore flow of dissolved gases and mixing by penetrative convection brings them to the air–water interface where they can evade. The addition of a tracer for CO_2 when modeling thermally stratified Lake Pleasant (New York, USA) showed the upwelling of cold, CO_2-enriched water into the upper mixed layer of the southern basin when winds were from the southwest and the cross-basin differences in temperature and CO_2 (Fig. 8-42). The presence of an island precluded the northerly flow of appreciable quantities of the dissolved gas. Accurate estimates of CO_2 emissions need to account for spatial variability in concentrations. Applying 3-D models can improve ecosystem understanding and in cases such as these, the modeling will lead to improved estimates of emissions of climate forcing trace gases.

D. Applications of 1-D and 3-D models

1-D models have been applied in water quality applications for several decades. With ongoing land-use change and climate change, their value in predicting future states of lakes, the likelihood of ecosystem change, and for improving weather forecasting has led to models with several computational schemes and ongoing workshops and user groups for comparisons and further improvements (Thiery et al., 2014; Golub et al., 2022). Efforts for inclusion of lakes in improved weather forecasting and climate prediction include Mironov (2005), Kirillin et al. (2011), MacKay et al. (2009),

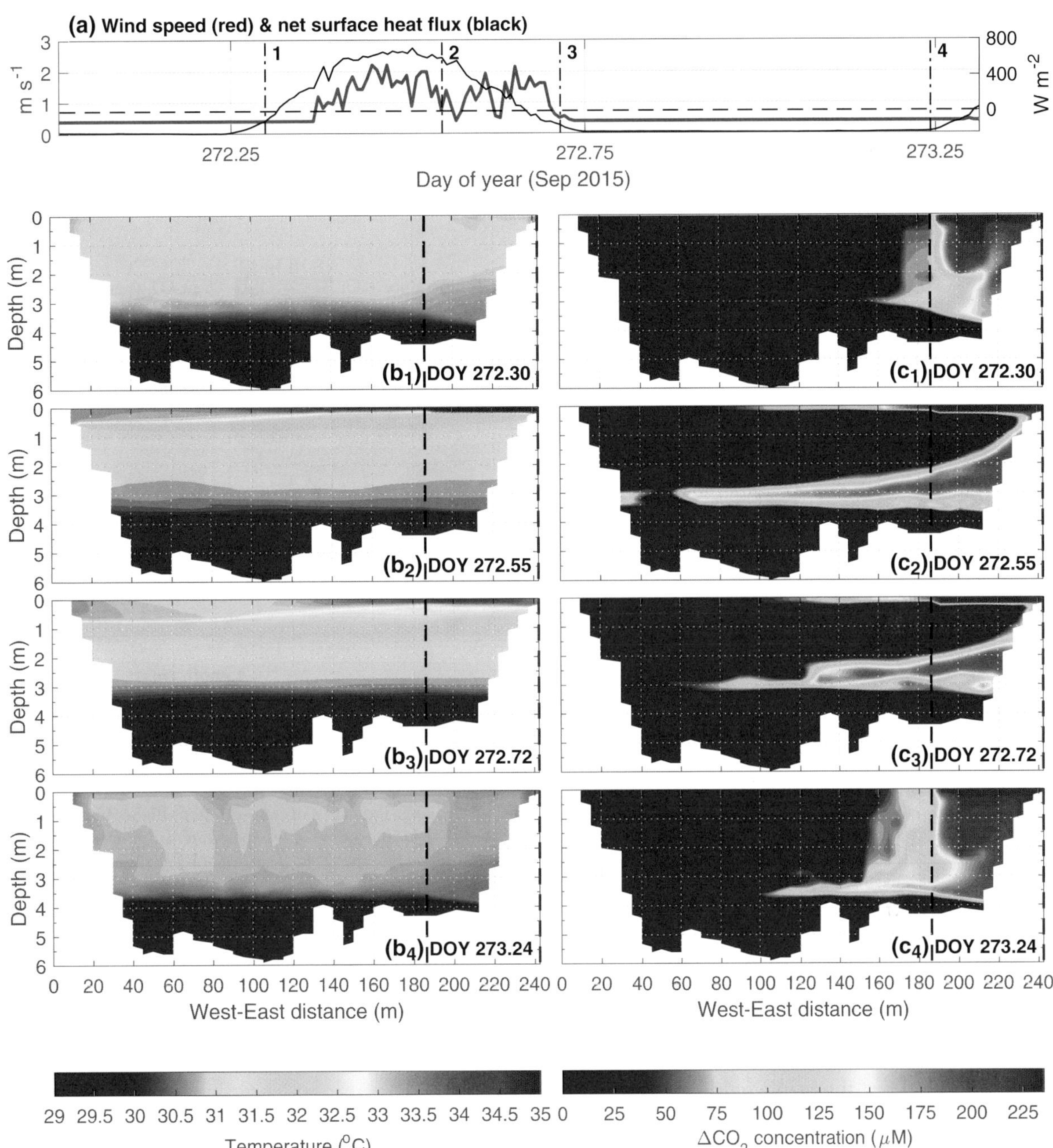

FIGURE 8-41 Cross sections illustrating the results of hydrodynamic simulations over a diel cycle for an embayment of Lake Janauacá (Brazil). (a) Wind speed and net surface heat flux, (b_1)–(b_4) thermal structure from early morning DOY 272 to DOY 273 (29–30 September 2015) at the times marked with the *vertical dashed lines* in (a), and (c_1)–(c_4) tracer indicating CO_2 concentrations at the four times with an estimated concentration of 235 μM in the macrophyte beds at the site of the tracer injection. The *black dashed lines* in (b) and (c) indicate the horizontal extent of the mat of floating vegetation. At this site, differential heating and cooling led to offshore flows enriched in CO_2 with convective mixing at night bringing the dissolved gas to the air–water interface where it can evade. *(From Amaral et al. [2021].)*

and MacKay (2012). Examples of using 1-D models include forecasting water quality (Carey et al., 2022) and evaluation of changes in thermal structure, with implications for oxygen dynamics (Sahoo et al., 2013; Schwefel et al., 2016) and for climate change (Woolway and Merchant, 2019; Woolway and Maberly, 2020; Woolway et al., 2021, 2022). Ongoing challenges include addressing errors in 1-D models associated with the two and three dimensionality (e.g., Figs. 8-40 to 8-42) of lakes and uncertainties in the reanalysis data and climate projections used as input data to model lakes over large spatial areas under climate change.

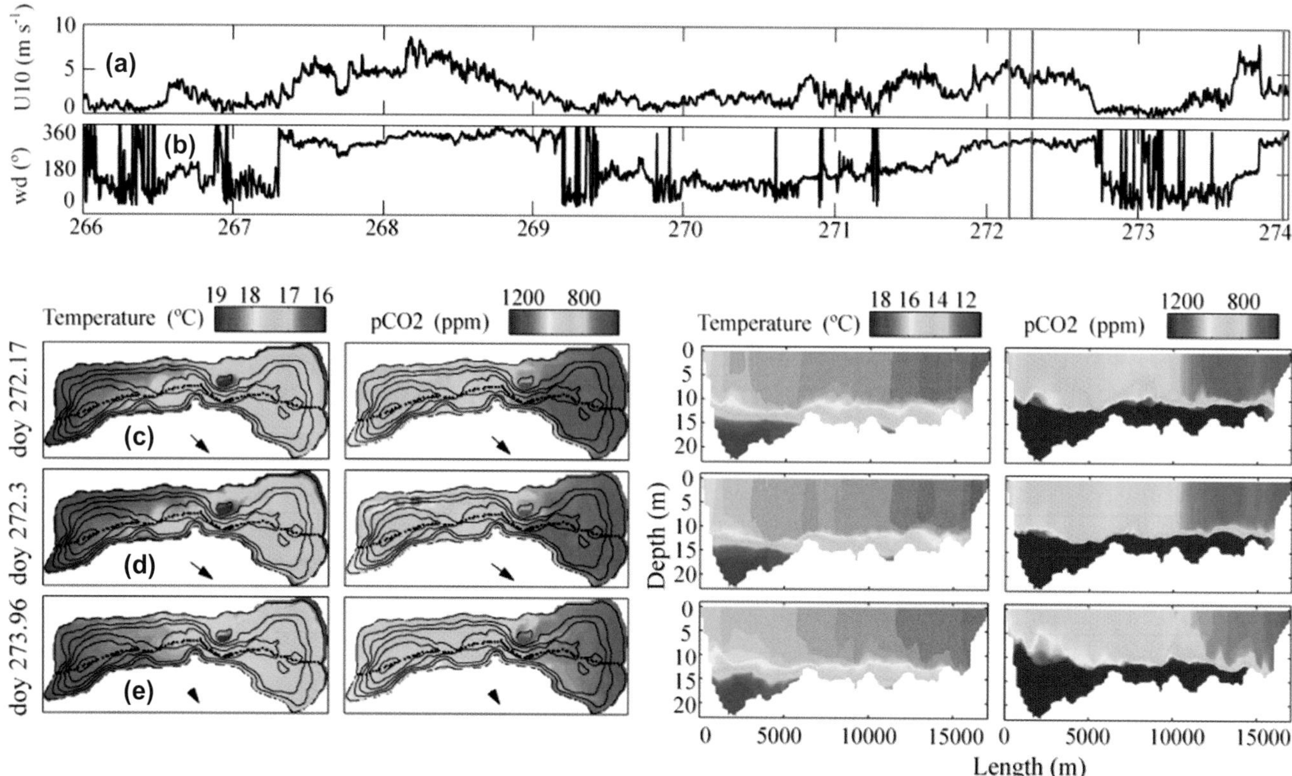

FIGURE 8-42 Wind-induced upwelling and cross-basin transport in Lake Pleasant (New York, USA). (a) Wind speed (U10, wind speed at 10 m height), (b) wind direction (wd) for days of year 266 to 274 (23 September to 1 October 2010). To the *left*, plan views, and to the *right*, cross-sections along the *dashed black line* in the plan view of simulated temperature and CO_2 for the three times indicated by the *horizontal red lines* in panels (a) and (b). The CO_2 concentrations in the numerical tracer were designed to be similar to those in the lake at the time. In order to see concentrations in the upper mixed layer, the range has been compressed. That is, the *brown color* is indicative of concentrations higher than 1200 ppm. The lake is oriented from the south to the north along its long axis. Results show the wind-induced flow of warmer water from the south to the north and the upwelling of colder, CO_2-rich water from the thermocline to the south. *(Details of the experiment are in Tedford et al. [2014] and Czikowski et al. [2018]; modeling by Javier Vidal using ELCOM.)*

3-D modeling has provided considerable understanding of lake hydrodynamics. These models can represent processes such as the *Kelvin and Poincare waves* in basins in which rotation is important and gyre formation due to complex basin morphometry or spatial variability in the wind field (Hodges et al., 2000; Laval et al., 2003b; Rueda et al., 2005). Laval et al. (2003a) provided an approach to reduce numerical diffusion in the vertical, that is, the apparent mixing due to computational error. Subsequent challenges were to capture the nonlinearity in the internal wave field. Dorostkar and Boegman (2013) provided the first demonstration of the highly nonlinear internal waves predicted in Horn et al. (2001) and quantified the considerable increase in eddy diffusivity from these internal waves. Vidal et al. (2013) captured the nonlinearity of the internal wave field in Mono Lake (California, USA). Further, they showed upwelling regions rotating around the lake due to Kelvin waves when wind directions remained stationary but also due to changes in wind direction. This latter observation was anticipated by Thorpe (1995) and implies that upwelling regions can rotate around basins too small to be influenced by the Earth's rotation. By changing the morphometry of the lake from its horseshoe shape to progressively less complex shapes, modeling results illustrated that complex basin shape reduced the potential for mixing, an additional result allowing extensions and predictions to other lakes. Other examples include the modeling in Lake Janauacá and Pleasant Lake with implications for predicting evasion of climate forcing trace gases (Figs. 8-40 to 8-42); establishing the links between storm inflows, variable weather patterns, and primary production in Arctic lakes (Rueda and MacIntyre, 2009); illustrating the movement of sediment-laden river water with high nutrient concentrations from an embayment to the main basin in Lake Victoria (East Africa) with implications for the development of harmful algal blooms (Romero et al., 1995); and the links between the spatial distribution of large- and small-bodied zooplankton to internal wave dynamics (Sprules et al., 2022). 3-D modeling also contributes to improved accuracy in computing lake metabolism (Antenucci et al., 2013).

Earlier and ongoing 1-D and 3-D modeling efforts, combined with observational data, have provided considerable understanding of hydrodynamic processes and the functioning of aquatic ecosystems.

V. Summary

1. Water movements in lakes and rivers occur in response to wind, cooling, and gravity. Surface waves and currents form at the surface of the water body, and within stratified water bodies, internal waves develop, which induce periodic currents. The energy from many of these movements induces turbulence, which spreads and mixes the water and the chemicals and organisms within it.
2. Below a certain speed of flow, water movement is smooth and undisturbed along interfaces. This ordered movement is called laminar flow. As water moves along an interface, such as the bottom of a stream or water of a different density, shear stresses result due to friction. As the velocity increases, vortices occur as a result of the shear, that is, the change in velocity away from the boundary or at the interface. The flow is considered transitional. With further increases in velocity or with disturbances at an interface, such as rocks on a stream bed, the flow becomes chaotic and dispersive, that is, turbulent. In a turbulent flow, the largest eddies generated by the stirring agent break down into progressively smaller ones until, at the scale of the smallest eddies, gradients in heat or solutes are removed by diffusion and the fluid is "mixed." Since the velocities needed to induce turbulent flow are low, laminar flow is only found in a few locations in aquatic systems, as in a thin layer adjacent to sediments or the atmosphere. Dimensionless indices can be used to predict whether the flow will become turbulent. These include the Reynolds number and the Richardson number, with the latter used in stratified flows.
3. The velocity of a turbulent flow is the sum of the mean velocity (u) and the velocity fluctuations (u'). Here u and u' imply velocities in all three directions. However, when the velocities are fully specified, u and v are horizontal velocities perpendicular to each other and w is vertical velocity. As the velocity fluctuations average to zero, the root mean square velocity (rms) is computed, that is, $\overline{u'^2}$, with this term called the turbulent velocity scale (u). It is approximately equal to the water friction velocity (u^*_w), which is the slope relating the logarithm of distance from a boundary to current speed. Turbulence is characterized by chaotic motions in three dimensions such that the turbulent kinetic energy (TKE) of the flow per unit mass is: TKE = ½ ($<u'^2> + <v'^2> + <w'^2>$) with u', v', and w' velocity fluctuations in the three directions described above. The angle brackets denote averaging. Changes in TKE are the sum of production, any transport of turbulence from elsewhere, mixing, pressure fluctuations, and dissipation, that is, the loss of turbulent kinetic energy at the smallest scales. This latter term is known as the rate of dissipation of turbulent kinetic energy (ϵ). Measurements of turbulence are made with instruments that can measure the rapid fluctuations in velocity and temperature as well as ones that can characterize the size of the turbulent eddies. These include acoustic Doppler velocimeters, acoustic Doppler current profilers, and microstructure profilers that measure the small-scale fluctuations in either velocity or temperature and the sizes of turbulent eddies.
4. The amount of spreading induced by turbulence is quantified by dispersion coefficients and the amount of spreading and mixing is quantified by the coefficient of eddy diffusivity. Both are identified by the symbol K and are proportional to the product of the turbulent velocity, u, and length scale (l), that is, $K = cul$, where c is an empirically determined coefficient. Subscripts on the turbulence coefficients denote whether the coefficients refer to the horizontal or vertical. Turbulence on the largest and smallest scales are related due to the flow of turbulent kinetic energy from the largest to the smallest scales, such that $\epsilon = u^3/l$. This relation enables calculations describing turbulence when only some of the key variables can be measured. The magnitude of horizontal dispersion (K_H) is usually orders of magnitude higher than that in the vertical, with empirical work showing how K_H varies with the length scale of the diffusing patch.
5. Fluxes of momentum in the vertical are the product of turbulent velocity fluctuation in the horizontal and vertical, that is, $<u'w'>$, and fluxes of heat or other scalars are similar, that is, $<T'w'>$, where T' are the fluctuations of temperature or another scalar. These fluxes can be calculated as the product of the coefficient of eddy diffusivity (K_z) and the velocity gradient or temperature gradient. Thus the magnitude of K_z provides an estimate of the capacity of a turbulent fluid to induce vertical mixing. The capacity, related to mixing efficiency, varies depending on whether the turbulence is produced by shear or by heat loss, that is, penetrative convection. Mixing efficiency is lower for turbulence induced by shear than by convection. Values of K_z range over several orders of magnitude within the different layers of lakes depending on the forcing from wind, heat loss, or incoming streams and the stratification. They are highest in the surface layer,

that is, the upper water column directly energized by wind or heat loss. They are smallest in the more stably stratified metalimnion and may increase again in the less stably stratified hypolimnion, particularly in the bottom boundary layer. The magnitude of K_z varies with the Wedderburn and Lake numbers, which take into account the density stratification, morphometry, and wind forcing. For two lakes of differing sizes but similar stratification and wind forcing, the smaller one will have lower values of K_Z.

6. Water in streams and rivers flows directionally along decreasing elevational gradients. Turbulence results from the shear from the friction of the water against the sides and bottom of the channel. Current velocities are low near the substrate and increase toward the center of the channel. The change in velocity, except in a thin layer near the substrate, is proportional to the logarithm of distance from the bottom, with this pattern called law of the wall scaling. The flow is turbulent where this relation holds. Resistance along the channel and banks results in a spiraling flow pattern that causes scouring and deposition along the stream channel such that the morphometry of the stream channel meanders. Obstructions (e.g., fallen trees, aquatic plants) of any type can modify flow patterns and greatly increase habitat and flow complexity.
7. The hyporheic zone is the wetted region below a stream bed into which groundwater flows. As the morphometry of the stream channel changes, stream water downwells into the hyporheic zone and subsequently upwells. This enables nutrient fluxes to occur and communities to live in the hyporheic zone. It also enables the flow of groundwater into the stream. Flowing water can be forced by advection into the interstitial spaces of stream sediments, particularly sands and gravel. The hyporheic zone may extend for several centimeters to well over a meter in depth.
8. Movement of air over the water induces currents and, depending on wind speed, oscillating motions called capillary waves or surface gravity waves. While water entrained in surface waves oscillates considerably up and down, horizontal movement is small. For deepwater gravity waves, the wavelength is much less than the water depth. In shallow water the circular motions of surface gravity waves become elliptical and extend to the bottom, causing shear that may induce entrainment. As surface waves enter shallower water, their velocity slows. A reduction in wavelength occurs with a marked increase in wave height. With increased height, the waves become asymmetrical and unstable and break.
9. Currents are nonperiodic water movements. Although many external forces contribute to the generation of currents, wind is the dominant driver of surface currents in open water. The velocity of wind-driven currents near the water surface is about 3% of wind speed. Depending on the size of a lake basin and its latitude, the Coriolis force from the Earth's rotation may cause geostrophic effects by deflecting the direction of surface currents from that of the wind. Deflection is to the right of wind direction in the Northern Hemisphere and to the left in the Southern Hemisphere. The currents in lakes may take the form of gyres, that is, circular or elliptical flows also characterized as swirls. These motions may be induced by the Coriolis force or by promontories or islands in lakes.
10. Langmuir circulations are helical motions that result from the interaction of wind and surface waves. They form at wind velocities above $\sim 3 \text{ m s}^{-1}$. They move water and entrained particles in a spiral manner approximately parallel to the wind direction and that of surface waves. The streaks, which are visible to the eye, are the result of foam and other light materials collecting where the water converges.
11. Surface seiches result when wind-driven currents cause water to accumulate downwind. They affect the motion of the entire water mass of the lake, whether stratified or not, and have maximum amplitude at the surface. The amplitude of surface seiches is generally small but occasionally reaches several meters in large lakes and can cause significant flooding and erosion of the shoreline.
12. Wind causes surface currents that move water downwind. A return current occurs above the metalimnion. The increased volume of water downwind causes the thermocline to downwell downwind. The degree of deflection depends on the Wedderburn and Lake numbers, and it determines the shear across the thermocline and any mixing between the different layers of lakes. Depending on the strength and persistence of the winds relative to stratification, as well as basin morphometry, metalimnetic and, in some cases, even hypolimnetic water may reach the surface at the upwind end of the basin. There it will mix with epilimnetic water. Downwind, epilimnetic and metalimnetic water may exchange with hypolimnetic water. The degree of exchange can be predicted based on the Wedderburn and Lake numbers. After the wind ceases, surface water that has accumulated downwind will flow to the other end of the lake.
13. The thermocline tilts downwind in response to the increased pressure when wind-induced currents transport surface water downwind. When the wind stops, the thermocline tilts in the opposing direction. The tilting results in long standing waves called internal seiches with wavelengths approximating the dimensions of the basin. The surface of the

metalimnion oscillates up and down at the lake ends like a seesaw about a pivot, the node, which has no vertical motion. Maximum vertical motion occurs at the ends, the antinodes, often at the ends of the lake basin. Horizontal movement is maximum at the nodal line and nil at the antinodes. The horizontal movements reverse direction with each long oscillation. Seiches eventually return to equilibrium as a result of gravity and friction.

14. Internal waves can be described based on their mode and the number of nodes. That is, a first vertical mode, first horizontal mode (V1H1) internal wave has uniform thickness at its upper and lower extents across a lake basin and oscillates about a node at the center of the lake. A second vertical mode, first horizontal node (V2H1) internal wave alternately separates across the thermocline at one end and compresses at the other while rocking around the node at the lake center. Higher vertical modes are more prevalent as heat has been mixed downward and the metalimnion has thickened. That said, a strong wind event will compress the thermocline downwind and cause it to spread upwind. The modal structures above result subsequently.

15. The extent of wind-induced tilting of the thermocline varies with the stratification across the metalimnion, the thickness of the upper mixed layer, wind shear, and the length of the lake in the direction of wind. These variables are incorporated in the Wedderburn (W) and related Lake numbers (L_N). Partial upwelling occurs when these dimensionless indices drop below 10 and full upwelling occurs when they decrease below 1. The shear across the thermocline varies accordingly, leading to instability, vortices, and internal waves that travel across the lake and are steeper than seiches and known as nonlinear internal waves. Depending on the value of W or L_N, increased mixing will result near the boundaries or across the full thermocline. The increased mixing leads to energy loss from the internal waves and the up- and downwelling of the thermocline will cease until the next strong wind event. With spatial differences in mixing across a basin, the full two and three dimensionality of lakes must be considered when computing fluxes.

16. Internal waves of various types occur in stratified lakes. Internal waves can be induced by wind, as described above, and can also result from disturbances, such as those from a river inflow flowing as a gravity current within or at the bottom of a lake. Shear across the metalimnion can be large enough to generate a range of nonlinear waves, which are examples of internal progressive waves. These waves have amplitudes considerably larger than those of surface waves, but they propagate and break internally much as surface waves do. The types of waves and whether they break at a lake's boundaries or lake-wide can be predicted based on the inverse of the Wedderburn number and the ratio of mixed layer depth to mean depth. The geostrophic effects of Coriolis force increase toward either pole north and south of the Equator. When geostrophic effects cause upwelling regions to rotate around a lake, the resulting waves are known as Kelvin waves. Shear from currents and internal waves can cause Kelvin–Helmholtz billows and other instabilities that cause the mixing of water, solutes, and organisms within and at the interfaces between the upper mixed layer, metalimnion, and hypolimnion.

17. Exchanges between inshore and offshore regions are caused by density differences in the horizontal. These flows are a form of horizontal convection. When water inshore is warmer, it flows offshore near the surface in a process called differential heating. When water inshore is cooler, it flows offshore as a density or gravity current and intrudes offshore at a depth of similar density. This process is called differential cooling. A related process is differential wind-mixing, in which wind mixing offshore creates cooler water such that surface water flows from inshore to offshore and offshore water flows inshore at depth. These flows can stop and start over the course of a day due to the persistence of temperature differences. In water bodies in which one region is persistently warmer, the flows can be near continuous and called horizontal convective circulation.

18. In winter under the ice, slow currents occur in lakes even despite ice cover reducing the influence of wind and the lack of appreciable inflows and outflows. As water temperatures are below 4°C, the temperature of maximum density of freshwater, heat flowing from the sediments and solutes produced by respiration increase the density of water immediately above the sediments. The increased density causes gravity currents that flow downslope. The resultant temperature profile is warmer near the bottom in most cases. Most of the heat released from the sediments during winter is from accumulation during the previous summer. To balance the downward flow of water to the bottom, upward flows occur, causing horizontal convective circulation. Downslope current velocities tend to be greater than vertical velocities. The rise times of isopycnals originating near the bottom can be many months, attesting to the slow vertical flows. In spring, and possibly in fall, when sunlight penetrates through the ice, warming of water closer to the surface can cause its density to increase and exceed that below creating density instabilities and mixing, a process known as radiatively driven convection or penetrative convection. Similarly, if heating under the ice in spring is greater nearshore, the warmer, denser water can flow offshore to deeper depths and contribute to warming offshore.

19. In spring a narrow zone of warm water can develop between the nearshore and offshore. The shallow nearshore waters can continue to heat more rapidly than the open waters such that their temperature exceeds 4°C, whereas that offshore is below 4°C. The resultant density-driven circulation perpendicular to the shore consists of two circulation cells. A convergence zone, the *thermal bar*, occurs between them, with downwelling water with temperature ~4°C. Little mixing occurs between inshore and offshore waters until surface waters over the whole lake exceed 4°C.

20. River water flowing into a lake or reservoir initially mixes somewhat with lake water and, dependent on the resultant density, either flows into the lake at the surface as an overflow, intrudes at depths where its density is similar to that of the lake water, or, if its density exceeds that of lake water, flows towards and along the bottom. The initial density of an inflow depends on its temperature, the concentration of dissolved solutes, and suspensions of particulates and can vary seasonally and between different lakes. It also changes as it initially mixes with lake water. It is this density that determines the depth to which it will flow in the lake. The extent to which the intrusion subsequently mixes with ambient lake water depends on its temperature and momentum and on the extent of mixing within the lake due to wind, cooling, internal wave motions, and the stratification within the lake.

21. The motions within lakes consist of many interacting processes acting over a range of temporal and spatial scales. Wind-driven upwelling can occur when the surface layer is heating in spring, and the accompanying mixing from shear as the thermocline tilts may thicken the metalimnion. However, if wind-driven upwelling and downwelling occurs at other times or during cooling, fluxes from the stratified portions of lakes into the upper mixed layer may be appreciable. These fluxes may support the growth of phytoplankton or contribute to fluxes of climate forcing trace gases across the air–water interface. Dissipation rates can be used to calculate the smallest scales of turbulent flow, the Kolmogorov microscales. The fluxes of solutes across interfaces occur at these scales. Discriminating the different flows and their contribution to transport is a challenge for limnologists and enables improved understanding of the coupling between hydrodynamics and ecosystem processes.

22. By use of time scales, such as those for advection (mean velocity divided by the length of a lake), mixing (l^2/K_z, where l is a length scale for the region of interest), and reaction rates such as chemical conversions or growth rates, outcomes of the interaction of hydrodynamic processes and chemical or biological ones can be predicted. Many such interactions are of fundamental importance, including determining whether phytoplankton within the euphotic zone can grow fast enough to support a bloom rather than sink or be mixed out of the euphotic zone.

23. Hydrodynamic models have been developed whose complexity varies from describing an individual process to the many interacting processes as they change over time. Hydrodynamic models including the processes described in this chapter may include one, two, or three dimensions and model selection depends upon water quality or research goals. 1-D models are often used to illustrate the changing temperatures and stratification in lakes due to land use and climate change, and 3-D models enable a greater understanding of the physical processes that moderate ecosystem function.

Acknowledgments

We thank Wencai Zhou and Javier Vidal for three-dimensional modeling and Wencai Zhou, Javier Vidal, and Alicia Cortés for contributing to graphics production; Andreas Lorke, Greg Lawrence, Cayelan Carey, Ryan Shojinaga and Robert Schwefel provided helpful comments; and John Melack, Warwick Vincent, and Michio Kumagai critically read the chapter.

References

Allan, J.D., Castillo, M.M., Capps, K.A., 2021. Stream ecology: structure and function of running waters, third ed. Springer Nature, Switzerland, p. 485. https://doi.org/10.1007/978-3-030-61286-3.

Amaral, J.H.F., Melack, J.M., Barbosa, P.M., Borges, A.V., Kasper, D., Cortés, A.C., Zhou, W., MacIntyre, S., Forsberg, B.R., 2021. Inundation, hydrodynamics and vegetation influence carbon dioxide concentrations in Amazon floodplain lakes. Ecosystems 25, 1–20. https://doi.org/10.1007/s10021-021-00692-y.

Antenucci, J.P., Tan, K.M., Eikaas, H.S., Imberger, J., 2013. The importance of transport processes and spatial gradients on in situ estimates of lake metabolism. Hydrobiol. (Sofia) 700, 9–21. https://doi.org/10.1007/s10750-012-1212-z.

Arcement Jr., G.J., Schneider, V.R., 1984. Guide for selecting Manning's roughness coefficients for natural channels and flood plains. Water-Resource Paper 2339. U.S. Geological Survey, Denver, CO, USA. https://doi.org/10.3133/wsp2339.

Arthington, A.H., Bunn, S.E., Poff, N.L., Naiman, R.J., 2006. The challenge of providing environmental flow rules to sustain river ecosystems. Ecol. Appl. 16 (4), 1311–1318. https://doi.org/10.1890/1051-0761(2006)016[1311:TCOPEF]2.0.CO;2.

Batchelor, G.K., 1959. Small-scale variation of convected quantities like temperature in turbulent fluid. Part 1. General discussion and the case of small conductivity. J. Fluid Mech. 5, 113–133. https://doi.org/10.1017/S002211205900009X.

Bengtsson, L., 2012. Circulation processes in lakes. In: Bengtsson, L., Herschy, R.W., Fairbridge, R.W. (Eds.), Encyclopedia of Lakes and Reservoirs. Encyclopedia of Earth Sciences Series. Springer, Dordrecht. https://doi.org/10.1007/978-1-4020-4410-6_8.

Berg, P., Glud, R.N., Hume, A., Stahl, H., Oguri, K., Meyer, V., Kitazato, H., 2009. Eddy correlation measurements of oxygen uptake in deep ocean sediments. Limnol. Oceanogr.-Meth. 7, 576–584. https://doi.org/10.4319/lom.2009.7.576.

Berman, T., Shteinman, B., 1998. Phytoplankton development and turbulent mixing in Lake Kinneret (1992–1996). J. Plankton Res. 20 (4), 709–726. https://doi.org/10.1093/plankt/20.4.709.

Birt, D., Wain, D., Slavin, E., Zang, J., Luckwell, R., Bryant, L.D., 2021. Stratification in a reservoir mixed by bubble plumes under future climate scenarios. Water 13, 2467. https://doi.org/10.3390/w13182467.

Boano, J.W., Harvey, Marion, A., Packman, A.I., Revelli, R., Ridolfi, L., Worman, A., 2014. Hyporheic flow and transport processes: mechanisms, models, and biogeochemical implications. Rev. Geophys. 52, 603–679. https://doi.org/10.1002/2012RG000417.

Boegman, L. (2009). Currents in stratified water bodies 2: internal waves. Encyclopedia of Inland Waters, 1: 539–558, Elsevier, Amsterdam. https://doi.org/10.1016/B978-012370626-3.00081-8.

Boegman, L., Imberger, J., Ivey, G.N., 2005a. The degeneration of internal waves in lakes with sloping topography. Limnol. Oceanogr. 50 (5), 1620–1637. https://doi.org/0.4319/lo.2003.48.2.0895.

Boegman, L., Imberger, J., Ivey, G.N., 2005b. The energetics of large-scale internal wave degeneration in lakes. J. Fluid Mech. 531, 159–180. https://doi.org/10.1017/S0022112005003915.

Boegman, L., Imberger, J., Ivey, G.N., Antenucci, J.P., 2003. High-frequency internal waves in large stratified lakes. Limnol. Oceanogr. 48 (2), 895–919. https://doi.org/0.4319/lo.2003.48.2.0895.

Boegman, L., Ivey, G.N., 2009. Flow separation and resuspension beneath shoaling nonlinear internal waves. J. Geophys. Res. Oceans. 114 (C2). https://doi.org/10.1029/2007JC004411.

Bouffard, D.D., Boegman, P.L., 2012. Basin-scale internal waves. In: Bengtsson, L., Herschy, R.W., Fairbridge, R.W. (Eds.), Encyclopedia of Lakes and Reservoirs. Springer, The Netherlands, pp. 102–107. https://doi.org/10.1007/978-1-4020-4410-6.

Bouffard, D., Wüest, A., 2019. Convection in lakes. Ann. Rev. Fluid Mech. 51, 189–215. https://doi.org/10.1146/annurev-fluid-010518-040506.

Bouffard, D., Zdorovennova, G., Bogdanov, S., Efremova, T., Lavanchy, S., Palshin, N., Terzhevik, A., Vinnå, L.R., Volkov, S., Wüest, A., Zdorovennov, R., 2019. Under-ice convection dynamics in a boreal lake. Inland Waters 9 (2), 142–161. https://doi.org/10.1080/20442041.2018.1533356.

Brooke, S., Chadwick, A., Silvestre, J., Lamb, M.P., Edmonds, D.A., Ganti, V., 2022. Where rivers jump course. Science 376 (6596), 987–990. https://doi.org/10.1126/science.abq1166.

Bruce, L.C., Frassl, M.A., Arhonditsis, G.B., Gal, G., Hamilton, D.P., Hanson, P.C., et al., 2018. A multi-lake comparative analysis of the General Lake Model (GLM): stress-testing across a global observatory network. Environ. Model. Softw. 102, 274–291. https://doi.org/10.1016/j.envsoft.2017.11.016.

Bryant, L.D., Claudia, L., McGinnis, D.F., Brand, A., Wüest, A., Little, J.C., 2010. Variable sediment oxygen uptake in response to dynamic forcing. Limnol. Oceanogr. 55, 950–964. https://doi.org/10.4319/lo.2010.55.2.0950.

Carey, C.C., Woelmer, W.M., Lofton, M.E., Figueiredo, R.J., Bookout, B.J., et al., 2022. Advancing lake and reservoir water quality management with near-term, iterative ecological forecasting. Inland Waters 12 (1), 107–120. https://doi.org/10.1080/20442041.2020.1816421.

Carpenter, J., Sommer, T., Wüest, A., 2012. Simulations of a double-diffusive interface in the diffusive convection regime. J. Fluid Mech. 711, 411–436. https://doi.org/10.1017/jfm.2012.399.

Chikita, et al., 2005. A comparison of hydrostatic and nonhydrostatic pressure components in seiche oscillations. Math. Comput. Model. 41 (8–9), 887–902. https://doi.org/10.1016/j.mcm.2004.08.005.

Chen, S., Lei, C., Carey, C.C., Gantzer, P.A., Little, J.C., 2017. A coupled three-dimensional hydrodynamic model for predicting hypolimnetic oxygenation and epilimnetic mixing in a shallow eutrophic reservoir. Water Resour. Res. 53, 470–484. https://doi.org/10.1002/2016WR019279.

Chen, S., Little, J.C., Carey, C.C., McClure, R.P., Lofton, M.E., Lei, C., 2018. Three-dimensional effects of artificial mixing in a shallow drinking-water reservoir. Water Resour. Res. 54, 425–441. https://doi.org/10.1002/2017WR021127.

Chen, W., Olden, J.D., 2017. Evaluating transferability of flow–ecology relationships across space, time and taxonomy. Freshwater Biol 63, 817–830. https://doi.org/10.1111/fwb.13041.

Chikita, K., Hattori, M., Hagiwara, E., 1985. A field study on river-induced currents: an intermountain lake, Lake Okotanpe, Hokkaido. Jap. J. Limnol. 46 (4), 256–267. https://doi.org/10.3739/rikusui.46.256.

Coates, M., Ferris, J., 1994. The radiatively driven natural convection beneath a floating plant layer. Limnol. Oceanogr. 39 (5), 1186–1194. https://doi.org/10.4319/lo.1994.39.5.1186.

Coates, M., Folkard, A., 2009. The effects of littoral zone vegetation on turbulent mixing in lakes. Ecol. Model. 220 (20), 2714–2726. https://doi.org/10.1016/j.ecolmodel.2009.06.042.

Cole, T., Buchak, E.M., 1995. CE-QUAL-W2: A two-dimensional, laterally averaged, hydrodynamic and water quality model, version 2.0, user manual. U.S. Army Corps of Engineers, Waterways Experiment Station, Vicksburg, Mississippi.

Cortés, A., MacIntyre, S., 2020. Mixing processes in small arctic lakes during spring. Limnol. Oceanogr. 65, 260–288. https://doi.org/10.1002/lno.11296.

Cortés, A., MacIntyre, S., Sadro, S., 2017. Flowpath and retention of snowmelt in an ice-covered arctic lake. Limnol. Oceanogr. 62 (5), 2023–2044. https://doi.org/10.1002/lno.10549.

Cortés, A., Rueda, F.J., Wells, M.J., 2014. Experimental observations of the splitting of a gravity current at a density step in a stratified water body. J. Geophys. Res. Oceans 119, 1038–1053. https://doi.org/10.1002/2013JC009304.

Costa, J.E., O'Connor, J.E., 1995. Geomorphically effective floods. In: Costa, J., Miller, A., Potter, K., Wilcock, P. (Eds.), Natural and Anthropogenic Influences in Fluvial Geomorphology, vol. 89. Geophysical Monograph Series, pp. 45–56. https://doi.org/10.1029/GM089p0045.

Craik, A.D.D., 1977. The generation of Langmuir circulations by an instability mechanism. J. Fluid Mech. 81, 206–223. https://doi.org/10.1017/S0022112077001980.

Craik, A.D.D., Leibovich, S., 1976. A rational model for Langmuir circulations. J. Fluid Mech. 73, 401–426. https://doi.org/10.1017/S0022112076001420.

Csanady, G.T., 2001. Air-sea interaction: laws and mechanisms. Cambridge University Press, Cambridge, UK, p. 239. https://doi.org/10.1017/CBO9781139164672.

Czikowsky, M.J., MacIntyre, S., Tedford, E.W., Vidal, J., Miller, S.D., 2018. Effects of wind and buoyancy on carbon dioxide distribution and air-water flux of a stratified temperate lake. J. Geophys. Res. Biogeosci. 123, 2305–2322. https://doi.org/10.1029/2017JG004209.

Danish Hydraulics Institute [DHI], 2011. Mike 11, reference manual. Technical report, DHI, Denmark.

Deng, Z.Q., Bengtsson, L., Singh, V.P., Adrian, D.D., 2002. Longitudinal dispersion coefficient in single-channel streams. J. Hydraul. Eng. 128 (10), 901–916. https://doi.org/10.1061/(ASCE)0733-9429(2002)128:10(901).

Dethleff, D., Kempema, E.W., 2007. Langmuir circulation driving sediment entrainment into newly formed ice: tank experiment results with application to nature (Lake Hattie, United States; Kara Sea,

Siberia). J. Geophys. Res. 112, C02004. https://doi.org/10.1029/2005JC003259.

Doda, T., Ramón, C.L., Ulloa, H.N., Wüest, A., Bouffard, D., 2021. Seasonality of density currents induced by differential cooling. Hydrol. Earth Syst. Sci. 26, 331–353. https://doi.org/10.5194/hess-26-331-2022.

Dorostkar, A., Boegman, L., 2013. Internal hydraulic jumps in a long narrow lake. Limnol. Oceanogr. 58 (1), 153–172. https://doi.org/10.4319/lo.201358.1.0153.

Eckart, C., 1948. An analysis of stirring and mixing processes in incompressible fluids. J. Mar. Res. 7, 265–275.

Eckhardt, B., 2008. Introduction. Turbulence transition in pipe flow: 125th anniversary of the publication of Reynolds' paper. Philos. Trans. R. Soc. A 367, 449–455. https://doi.org/10.1098/rsta.2008.0217.

Fairall, C.W., Bradley, E.F., Hare, J.E., Grachev, A.A., Edson, J.B., 2003. Bulk parameterization of air-sea fluxes: updates and verification for the COARE algorithm. J. Clim. 16, 571–591. https://doi.org/10.1175/1520-0442(2003)016<0571:BPOASF>2.0. CO.

Fairall, C.W., Bradley, E.F., Rogers, D.P., Edson, J.B., Young, G.S., 1996. Bulk parameterization of air-sea fluxes for tropical ocean-global atmosphere coupled-ocean atmosphere response experiment. J. Geophys. Res. 101, 3747–3764. https://doi.org/10.1029/95JC03205.

Farmer, D.M., 1975. Penetrative convection in the absence of mean shear. Q. J. R. Meteorol. Soc. 101 (430), 869–891. https://doi.org/10.1002/qj.49710143011.

Farrow, D.E., Hocking, G.C., 2006. A numerical model for withdrawal from a two-layer fluid. J. Fluid Mech. 549, 141–157. https://doi.org/10.1017/S0022112005007561.

Fischer, H.B., List, E.J., Koh, R.C.Y., Imberger, J., Brooks, N.H., 1979. Mixing in inland and coastal waters. Academic, New York, p. 483. https://doi.org/10.1016/C2009-0-22051-4.

Forel, F.A., 1895. Le léman: monographie limnologique. tome 2: méchanique, chimie, thermique, optique, acoustique. F. Rouge, Lausanne.

Forrest, A.L., Laval, B.E., Pieters, R., Lim, D.S.S., 2013. A cyclonic gyre in an ice-covered lake. Limnol. Oceanogr. 58 (1), 363–375. https://doi.org/10.4319/lo.2013.58.1.0363.

Frenette, J.J., Vincent, W.F., Legendre, L., Nagata, T., 1996a. Size-dependent phytoplankton responses to atmospheric forcing in Lake Biwa. J. Plankton Res. 18, 371–391. https://doi.org/10.1093/plankt/18.3.371.

Frenette, J.J., Vincent, W.F., Legendre, L., Nagata, T., 1996b. Size-dependent changes in phytoplankton C and N uptake in the dynamic mixed layer of Lake Biwa. Freshwater Biol 36, 221–236. https://doi.org/10.1046/j.1365-2427.1996.00083.x.

Frindte, K., Allgaier, M., Grossart, H.-P., Eckert, W., 2015. Microbial response to experimentally controlled redox transitions at the sediment water interface. PLoS One 10 (11), e0143428. https://doi.org/10.1371/journal.pone.0143428.

Gaudard, A., Råman Vinnå, L., Bärenbold, F., Schmid, M., Bouffard, D., 2019. Toward an open access to high-frequency lake modeling and statistics data for scientists and practitioners—the case of Swiss lakes using Simstrat v2.1. Geosci. Model Dev. (GMD) 12, 3955–3974. https://doi.org/10.5194/gmd-12-3955-2019.

Geng, N., Bai, Y., Pan, S., 2022. Research on heavy metal release with suspended sediment in Taihu Lake under hydrodynamic condition. Environ. Sci. Pollut. Res. 29, 28588–28597. https://doi.org/10.1007/s11356-021-17666-1.

Gibbs, M.M., 1992. Influence of hypolimnetic stirring and underflow on the limnology of Lake Rotoiti, New Zealand. N. Z. J. Mar. Freshwater Res. 26 (3–4), 453–463. https://doi.org/10.1080/00288330.1992.9516538.

Golub, M., Thiery, W., Marcé, R., Pierson, D., Vanderkelen, I., Mercado-Bettin, D., Woolway, R.I., Grant, L., Jennings, E., Kraemer, B.M., Schewe, J., 2022. A framework for ensemble modelling of climate change impacts on lakes worldwide: the ISIMIP Lake Sector. Geosci. Model Dev. (GMD) 15 (11), 4597–4623. https://doi.org/10.5194/gmd-15-4597-2022.

Gomez-Giraldo, A., Imberger, J., Antenucci, J.P., Yeates, P.S., 2008. Wind-shear—generated high-frequency internal waves as precursors to mixing in a stratified lake. Limnol. Oceanogr. 53 (1), 354–367. https://doi.org/10.4319/lo.2008.53.1.0354.

Goudsmit, G.-H., Burchard, H., Peeters, F., Wüest, A., 2002. Application of k-ϵ turbulence models to enclosed basins: the role of internal seiches. J. Geophys. Res. Oceans 107 (C12), 3230. https://doi.org/10.1029/2001JC000954.

Goudsmit, G.-H., Peeters, F., Gloor, M., Wüest, A., 1997. Boundary versus internal diapycnal mixing in stratified natural waters. J. Geophys. Res. 102 (C13), 27903–27914. https://doi.org/10.1029/97JC01861.

Greene, A.D., Hendricks, P.J., Gregg, M.C., 2015. Using an ADCP to estimate turbulent kinetic energy dissipation rate in sheltered coastal waters. J. Atmos. Ocean. Technol. 32 (2), 318–333. https://doi.org/10.1175/JTECH-D-13-00207.1.

Gregg, M.C., D'Asaro, E.A., Riley, J.J., Kunze, E., 2018. Mixing efficiency in the ocean. Ann. Rev. Mar. Sci 10 (1), 443–473. https://doi.org/10.1146/annurev-marine-121916-063643.

Guffie, K.M., Henderson-Sellers, A., 2014. The climate modelling primer, fourth ed. Wiley-Blackwell, West Sussex, p. 439.

Guseva, S., Aurela, M., Cortés, A., Kivi, R., Lotsari, E., et al., 2021. Variable physical drivers of near-surface turbulence in a regulated river. Water Resour. Res. 57, e2020WR027939. https://doi.org/10.1029/2020WR027939.

Håkanson, I., Jansson, M., 1983. Principles of lake sedimentology. Springer-Verlag, New York, p. 320. https://doi.org/10.1002/iroh.19850700318.

Hamblin, P.F., 1979. Great lakes storm surge of April 6, 1979. J. Great Lakes Res. 5 (3–4), 312–315. https://doi.org/10.1016/S0380-1330(79)72157-5.

Hamilton, D.P., Magee, M.R., Wu, C.H., Kratz, T.K., 2018. Ice cover and thermal regime in a dimictic seepage lake under climate change. Inland Waters 8 (3), 381–398. https://doi.org/10.1080/20442041.2018.1505372.

Hampton, S.E., Izmest'eva, L.R., Moore, M.V., Katz, S.L., Dennis, B., Silow, E.A., 2008. Sixty years of environmental change in the world's largest freshwater lake—Lake Baikal, Siberia. Glob. Chang. Biol. 14 (8), 1947–1958. https://doi.org/10.1111/j.1365-2486.2008.01616.x.

Hasselmann, K., Ross, D., Muller, P., Sell, W., 1976. A parametric wave prediction model. J. Phys. Oceanogr. 6, 200–228. https://doi.org/10.1175/1520-0485(1976)006<0200:APWPM>2.0.CO;2.

Hipsey, M.R., Bruce, L.C., Boon, C., Busch, B., Carey, C.C., Hamilton, D.P., Hanson, P.C., Read, J.S., de Sousa, E., Weber, M., Winslow, L.A., 2019. A general lake model (GLM 3.0) for linking with high-frequency sensor data from the global lake ecological observatory network (GLEON). Geosci. Model Dev. (GMD) 12, 473–523. https://doi.org/10.5194/gmd-12-473-2019.

Hodges, B.R., 2009. Hydrodynamical modeling. In: Likens, G.E. (Ed.), Encyclopedia of Inland Waters, vol. 1. Elsevier, Oxford, pp. 613–627. https://doi.org/10.1016/B978-012370626-3.00088-0.

Hodges, B., Dallimore, C., 2018. Aquatic ecosystem model: AEM3D. v1.0 user manual. HydroNumerics, Melbourne.

Hodges, B.R., Imberger, J., Saggio, A., Winters, K.B., 2000. Modeling basin-scale internal waves in a stratified lake. Limnol. Oceanogr. 7, 1603–1620. https://doi.org/10.4319/lo.2000.45.7.1603.

Hofmann, H., Locke, A., Peeters, F., 2011. Wind and ship wave-induced resuspension in the littoral zone of a large lake. Water Resour. Res. 47, W09505. https://doi.org/10.1029/2010WR010012.

Horn, D.A., Imberger, J., Ivey, G.N., 2001. The degeneration of large-scale interfacial gravity waves in lakes. J. Fluid Mech. 434, 181–207. https://doi.org/10.1017/S0022112001003536.

Huisman, J., Arrayás, M., Ebert, U., Sommeijer, B., 2002. How do sinking phytoplankton species manage to persist? Am. Nat. 159 (3), 245–254. https://doi.org/10.1086/338511.

Imberger, J., 1985. The diurnal mixed layer 1. Limnol. Oceanogr. 30 (4), 737–770. https://doi.org/10.4319/lo.1985.30.4.0737.

Imberger, J., 2012. Environmental fluid dynamics: flow processes, scaling, equations of motion, and solutions to environmental flows. Academic Press, Cambridge, MA, p. 460.

Imberger, J., Hamblin, P.F., 1982. Dynamics of lakes, reservoirs and cooling ponds. Ann. Rev. Fluid Mech. 14, 153–187. https://doi.org/10.1146/annurev.fl.14.010182.001101.

Imberger, J., Patterson, J.C., 1981. A dynamic reservoir simulation model—DYRESM: 5. In: Fischer, H.B. (Ed.), Transport Models for Inland and Coastal Waters. Academic Press, New York, pp. 310–361. https://doi.org/10.1016/B978-0-12-258152-6.50014-2.

Imberger, J., Patterson, J.C., 1990. Physical limnology. Adv. Appl. Mech. 27, 303–475. https://doi.org/10.1016/S0065-2156(08)70199-6.

Imboden, D.M., 2003. The motion of lake waters. In: O'Sullivan, P., Reynolds, C. (Eds.), The Lakes Handbook. Blackwell Publishing Ltd., Oxford, UK. https://doi.org/10.1002/9780470999271.ch6.

Imboden, D.M., Wüest, A., 1995. Mixing mechanisms in lakes. In: Lerman, A., Imboden, D.M., Gat, J.R. (Eds.), Physics and Chemistry of Lakes. Springer-Verlag, Berlin, pp. 83–138. https://doi.org/10.1007/978-3-642-85132-2.

Ivey, G.N., Imberger, J., 1991. On the nature of turbulence in a stratified fluid. Part I: the energetics of mixing. J. Phys. Oceanogr. 21, 650–659. https://doi.org/10.1175/1520-0485(1991)021<0650:OTNOTI>2.0.CO;2.

James, W.F., Barko, J.W., 1991. Littoral-pelagic phosphorus dynamics during nighttime convective circulation. Limnol. Oceanogr. 36 (5), 949–960. https://doi.org/10.4319/lo.1991.36.5.0949.

Jansen, J., MacIntyre, S., Barrett, D.C., Chin, Y.P., Cortés, A., Forrest, A.L., Hrycik, A.R., Martin, R., McMeans, B.C., Rautio, M., Schwefel, R., 2021. Winter limnology: how do hydrodynamics and biogeochemistry shape ecosystems under ice? J. Geophys. Res. Biogeosci. 126. https://doi.org/10.1029/2020JG006237. e2020JG006237.

Jassby, A., Powell, T., 1975. Vertical patterns of eddy diffusion during stratification in Castle Lake, California. Limnol. Oceanogr. 20 (4), 530–543. https://doi.org/10.4319/lo.1975.20.4.0530.

Jowett, I.G., 1993. A method for objectively identifying pool, run, and riffle habitats from physical measurements. N. Z. J. Mar. Freshwater Res. 27 (2), 241–248. https://doi.org/10.1080/00288330.1993.9516563.

Katul, G., Liu, H., 2017. Multiple mechanisms generate a universal scaling with dissipation for the air-water gas transfer velocity. Geophys. Res. Lett. 44 (4), 1892–1898. https://doi.org/10.1002/2016GL072256.

Kirillin, G., Hochschild, J., Mironov, D., Terzhevik, A., Golosov, S., Nützmann, G., 2011. FLake-Global: online lake model with worldwide coverage. Environ. Model Softw. 26, 683–684. https://doi.org/10.1016/j.envsoft.2010.12.004.

Kirk, J.T.O., 1984. Light and photosynthesis in aquatic ecosystems, second ed. Cambridge University Press, Cambridge, p. 509.

Knauer, K., Nepf, H.M., Hemond, H.F., 2000. The production of chemical heterogeneity in Upper Mystic Lake. Limnol. Oceangr. 45 (7), 1647–1654. https://doi.org/10.4319/lo.2000.45.7.1647.

Kouraev, A.V., Zakharova, E.A., Rémy, F., Kostianoy, A.G., Shimaraev, M.N., Hall, N.M., Suknev, A.Y., 2016. Giant ice rings on Lakes Baikal and Hovsgol: Inventory, associated water structure and potential formation mechanism. Limnol. Oceanogr. 61 (3), 1001–1014. https://doi.org/10.1002/lno.10268.

Kouraev, A.V., Zakharova, E.A., Rémy, F., Kostianoy, A.G., Shimaraev, M.N., Hall, N.M., Zdorovennov, R.E., Suknev, A.Y., 2019. Giant ice rings on lakes and field observations of lens-like eddies in the Middle Baikal (2016–2017). Limnol. Oceanogr. 64 (6), 2738–2754. https://doi.org/10.1002/lno.11338.

Kumagai, M., Asada, Y., Nakano, S., 1998. Gyres measured by ADCP in Lake Biwa. In: Imberger, J. (Ed.), Physical Processes in Lakes and Oceans, vol. 54. Coastal and Estuarine Studies, American Geophysical Union. https://doi.org/10.1029/CE054p0199.

Lamont, J.C., Scott, D.S., 1970. An eddy cell model of mass transfer into the surface of a turbulent liquid. AIChE J. 16 (4), 513–519. https://doi.org/10.1002/aic.690160403.

Langmuir, I., 1938. Surface motion of water induced by wind. Science 87, 1119–1123. https://doi.org/10.1126/science.87.2250.119.

Laval, B., Hodges, B.R., Imberger, J., 2003a. Reducing numerical diffusion effects with pycnocline filter. J. Hydraul. Eng., ASCE 129 (3), 215–224. https://doi.org/10.1061/(ASCE)0733-9429(2003)129:3(215).

Laval, B., Imberger, J., Hodges, B.R., Stocker, R., 2003b. Modeling circulation in lakes: spatial and temporal variations. Limnol. Oceanogr. 48 (3), 983–994. https://doi.org/10.4319/lo.2003.48.3.0983.

Lawrence, G.A., Ashley, K.I., Yonemitsu, N., Ellis, J.R., 1995. Natural dispersion in a small lake. Limnol. Oceanogr. 40 (8), 1519–1526. https://doi.org/10.4319/lo.1995.40.8.1519.

Lazier, J.R.N., Mann, K.H., 1989. Turbulence and the diffusive layers around small organisms. Deep Sea Research Part A. Oceanogr. Res. Pap. 36 (11), 1721–1733. https://doi.org/10.1016/0198-0149(89)90068-X.

Lehn, H., 1965. Zur Durchsichtigkeitsmessung im Bodensee. Schrift. Ver. Geschichte Bodensees Umgebung 83, 32–44.

Leibovich, S., 1977. On the evolution of the system of wind drift currents and Langmuir circulations in the ocean. Part 1. Theory and the averaged current. J. Fluid Mech. 79, 716–743. https://doi.org/10.1017/S0022112077001803.

Lemmin, U., 2012. Surface seiches. In: Bengtsson, L., Herschy, R.W., Fairbridge, R.W. (Eds.), Encyclopedia of Lakes and Reservoirs. Encyclopedia of Earth Sciences Series. Springer, Dordrecht. https://doi.org/10.1007/978-1-4020-4410-6_226.

Likens, G.E., Hasler, A.D., 1962. Movements of radiosodium (Na^{24}) within an ice-covered lake. Limnol. Oceanogr. 7, 48–56. https://doi.org/10.4319/lo.1962.7.1.0048.

Likens, G.E., Ragotzkie, R.A., 1965. Vertical water motions in a small ice-covered lake. J. Geophys. Res. 70, 2333–2344. https://doi.org/10.1029/JZ070i010p02333.

Lorke, A., 2007. Boundary mixing in the thermocline of a large lake. J. Geophys. Res. 112, C09019. https://doi.org/10.1029/2006JC004008.

Lorke, A., Müller, B., Maerki, M., Wüest, A., 2003. Breathing sediments: the control of diffusive transport across the sediment—water interface by periodic boundary-layer turbulence. Limnol. Oceanogr. 48 (6), 2077–2085. https://doi.org/10.4319/lo.2003.48.6.2077.

Lorke, A., Peeters, F., 2006. Toward a unified scaling relation for interfacial fluxes. J. Phys. Oceanogr. 36 (5), 955–961. https://doi.org/10.1175/JPO2903.1.

Lorke, A., Peeters, F., Wüest, A., 2005. Shear-induced convective mixing in bottom boundary layers on slopes. Limnol. Oceanogr. 50 (5), 1612–1619. https://doi.org/10.4319/lo.2005.50.5.1612.

Luketina, D.A., Imberger, J., 2001. Determining turbulent kinetic energy dissipation from Batchelor curve fitting. J. Atmos. Ocean. Technol. 18, 100–113. https://doi.org/10.1175/1520-0426(2001)018<0100:DTKEDF>2.0.CO;2.

MacIntyre, S., 1993. Vertical mixing in a shallow, eutrophic lake: possible consequences for the light climate of phytoplankton. Limnol. Oceanogr. 38 (4), 798–817. https://doi.org/10.4319/lo.1993.38.4.0798.

MacIntyre, S., Amaral, J.H., Melack, J.M., 2021a. Turbulence in the upper mixed layer under light winds and heating: implications for gas fluxes. J. Geophys. Res.: Oceans 126. https://doi.org/10.1029/2020JC017026 e2020JC017026.

MacIntyre, S., Bastviken, D., Arneborg, L., Crowe, A.T., Karlsson, J., Andersson, A., 2021b. Turbulence in a small boreal lake: consequences for air—water gas exchange. Limnol. Oceanogr. 66 (3), 827–854. https://doi.org/10.1002/lno.11645.

MacIntyre, S., Clark, J.F., Jellison, R., Fram, J.P., 2009. Turbulent mixing induced by nonlinear internal waves in Mono Lake, California. Limnol. Oceanogr. 54 (6), 2255–2272. https://doi.org/10.4319/lo.1999.44.3.0512.

MacIntyre, S., Cortés, A., Sadro, S., 2018. Sediment respiration drives circulation and production of CO_2 in ice-covered Alaskan arctic lakes. Limnol. Oceanogr. Lett. 3, 302–310. https://doi.org/10.1002/lol2.10083.

MacIntyre, S., Flynn, K.M., Jellison, R., Romero, J.R., 1999. Boundary mixing and nutrient fluxes in Mono Lake, California. Limnol. Oceanogr. 44 (3), 512–529. https://doi.org/10.4319/lo.1999.44.3.0512.

MacIntyre, S., Jonsson, A., Jansson, M., Aberg, J., Turney, D.E., Miller, S.D., 2010. Buoyancy flux, turbulence, and the gas transfer coefficient in a stratified lake. Geophys. Res. Lett. 37 (24), L24604. https://doi.org/10.1029/2010GL044164.

MacIntyre, S., Melack, J.M., 1995. Vertical and horizontal transport in lakes: linking littoral, benthic, and pelagic habitats. J. North Am. Benthol. Soc. 14 (4), 599–615. https://doi.org/10.2307/1467544.

MacIntyre, S., Romero, J.R., Silsbe, G.M., Emery, B.M., 2014. Stratification and horizontal exchange in Lake Victoria, East Africa. Limnol. Oceanogr. 59 (6), 1805–1838. https://doi.org/10.4319/lo.2014.59.6.1805.

MacIntyre, S., Sickman, J.O., Goldthwait, S.A., Kling, G.W., 2006. Physical pathways of nutrient supply in a small, ultraoligotrophic arctic lake during summer stratification. Limnol. Oceanogr. 51 (2), 1107–1124. https://doi.org/10.4319/lo.2006.51.2.1107.

MacKay, M.D., 2012. A process-oriented small lake scheme for coupled climate modeling applications. J. Hydrometeorol. 13, 1911–1924. https://doi.org/10.1175/JHM-D-11-0116.1.

MacKay, M.D., Neale, P.J., Arp, C.D., De Senerpont Domis, L.N., Fang, X., Gal, G., Jöhnk, K.D., Kirillin, G., Lenters, J.D., Litchman, E., MacIntyre, S., 2009. Modeling lakes and reservoirs in the climate system. Limnol. Oceanogr. 54 (6), 2315–2329. https://doi.org/10.431lo.2009.54.6_part_2.2315.

Malm, J., 1995. Spring circulation associated with the thermal bar in large temperate lakes. Nordic Hydrol 26, 331–358. https://doi.org/10.2166/nh.1995.0019.

Malm, J., 1998. Bottom buoyancy layer in an ice-covered lake. Water. Resour. Res. 34 (11), 2981–2993. https://doi.org/10.1029/98WR01904.

Malm, J., Bengtsson, L., Terzhevik, A., Boyarinov, P., Glinsky, A., Palshin, N., Petrov, M., 1998. Field study on currents in a shallow, ice-covered lake. Limnol. Oceanogr. 43 (7), 1669–1679. https://doi.org/10.4319/lo.1998.43.7.1669.

Mann, K.H., Lazier, J.R.N., 2006. Dynamics of marine ecosystems. biological-physical interactions in the oceans. Wiley-Blackwell, Hoboken, NJ, p. 496.

Manning, R., 1891. On the flow of water in open channels and pipes. Inst. Civil Engin. Ireland 20, 161–207.

Mao, M.A., van der Westhuysen, J., Xia, M., Schwab, D.J., Chawla, A., 2016. Modeling wind waves from deep to shallow waters in Lake Michigan using unstructured SWAN. J. Geophys. Res. Oceans 121, 3836–3865. https://doi.org/10.1002/2015JC011340.

Markfort, C.D., Perez, A.L.S., Thill, J.W., Jaster, D.A., Porté-Agel, F., Stefan, H.G., 2010. Wind sheltering of a lake by a tree canopy or bluff topography. Water Resour. Res. 46, W03530. https://doi.org/10.1029/2009WR007759.

Marti, C.L., Mills, R., Imberger, J., 2011. Pathways of multiple inflows into a stratified reservoir: Thomson Reservoir, Australia. Adv. Water Resour. 34, 551–561. https://doi.org/10.1016/j.advwatres.2011.01.003.

Marzadri, A., Tonina, D., Bellin, A., Valli, A., 2016. Interfaces, fluxes, residence times and redox conditions of the hyporheic zones induced by dune-like bedforms and ambient groundwater flow. Adv. Water Resour. 88, 139–151. https://doi.org/10.1016/j.advwatres.2015.12.014.

Melack, J.M., Sadro, S., Sickman, J.O., Dozier, J., 2021. Lakes and watersheds in the Sierra Nevada of California: responses to environmental change. University of California Press, Oakland, California. https://doi.org/10.2307/j.ctv17hm9sr.

Mironov, D.V., 2005. Parameterization of lakes in numerical weather prediction. Part 1: description of a lake model. German Weather Service, Offenbach am Main, Germany.

Monismith, S.G., 1986. An experimental study of the upwelling response of stratified reservoirs to surface shear stress. J. Fluid Mech. 171, 407–439. https://doi.org/10.1017/S0022112086001507.

Monismith, S.G., Imberger, J., Morison, M.L., 1990. Convective motions in the sidearm of a small reservoir. Limnol. Oceanogr. 35 (8), 1676–1702. https://doi.org/10.4319/lo.1990.35.8.1676.

Monismith, S.G., Koseff, J.R., White, B.L., 2018. Mixing efficiency in the presence of stratification: when is it constant? Geophys. Res. Lett. 45, 5627–5634. https://doi.org/10.1029/2018GL077229.

Monismith, S.G., Macintyre, S., 2009. The surface mixed layer in lakes and reservoirs. In: Likens, G.E. (Ed.), Encyclopedia of Inland Waters, vol. 1. Elsevier, Oxford, pp. 636–650. https://doi.org/10.1016/B978-012370626-3.00088-0.

Mortimer, C.H., 1952. Water movements in lakes during summer stratification: evidence from the distribution of temperature in Windermere. Proc. Royal Soc. London, Series B 236, 355–404. https://doi.org/10.1098/rstb.1952.0005.

Mortimer, C.H., 1961. Motion in thermoclines. Verh. Internat. Verein. Limnol. 14, 79–83. https://doi.org/10.1080/03680770.1959.11899249.

Mortimer, C.H., 1974. Lake hydrodynamics. Mitteilungen Int. Ver. Limnol. 20, 124–197. https://doi.org/10.1080/05384680.1974.11923886.

Mortimer, C.H., Mackereth, F.J.H., 1958. Convection and its consequences in ice-covered lakes. Verh. Internat. Ver. Limnol. 13 (2), 923–932. https://doi.org/10.1080/03680770.1956.11895490.

Munnich, M., Wuest, A., Imboden, D., D. M, 1992. Observations of the second vertical mode of the internal seiche in an alpine lake. Limnol. Oceanogr. 37, 1705–1719. https://doi.org/10.4319/lo.1992.37.8.1705.

Nelson, C.E., Sadro, S., Melack, J.M., 2009. Contrasting the influences of stream inputs and landscape position on bacterioplankton community structure and dissolved organic matter composition in high-elevation lake chains. Limnol. Oceanogr. 54 (4), 1292–1305. https://doi.org/10.4319/lo.2009.54.4.1292.

Nepf, H.M., 2009. Flow modification by submerged vegetation. In: Likens, G.E. (Ed.), Encyclopedia of Inland Waters, vol. 1. Elsevier, Oxford, pp. 606–612. https://doi.org/10.1016/B978-012370626-3.00088-0.

Nepf, H.M., 2012. Flow and transport in regions with aquatic vegetation. Annu. Rev. Fluid Mech. 44 (1), 123–142. https://doi.org/10.1146/annurev-fluid-120710-101048.

Nepf, H.M., Oldham, C.E., 1997. Exchange dynamics of a shallow contaminated wetland. Aquat. Sci. 59 (3), 193–213. https://doi.org/10.1007/BF02523273.

Newman, F.C., 1976. Temperature steps in Lake Kivu: a bottom heated saline lake. J. Phys. Oceanogr. 6, 157–163. https://doi.org/10.1175/1520-0485(1976)006<0157:TSILKA>2.0.CO;2.

Nielson, J.R., Henderson, S.M., 2022. Bottom boundary layer mixing processes across internal seiche cycles: dominance of downslope flows. Limnol. Oceanogr. 67 (5), 1111–1125. https://doi.org/10.1002/lno.12060.

Nishri, A., Rimmer, A., Wagner, U., Rosentraub, Z., Yeates, P., 2011. Physical controls on spatial variability in decomposition of organic matter in Lake Kinneret, Israel. Aquat. Geochem. 17 (3), 195–207. https://doi.org/10.1007/s10498-011-9119-2.

Noble, V.E., 1967. Evidences of geostrophically defined circulation in Lake Michigan. Proc. Conf. Great Lakes Res., Int. Assoc. Great Lakes Res. 20, 289–298.

O'Brien, K.R., Ivey, G.N., Hamilton, D.P., Waite, A.M., Visser, P.M., 2003. Simple mixing criteria for the growth of negatively buoyant phytoplankton. Limnol. Oceanogr. 48 (3), 1326–1337. https://doi.org/10.4319/lo.2003.48.3.1326.

Okubo, A., 1971. Oceanic diffusion diagrams. Deep-Sea Res. I: Oceanogr. Res. Pap. 18 (8), 789–802. https://doi.org/10.1016/0011-7471(71)90046-5.

Olsthoorn, J., Bluteau, C.E., Lawrence, G.A., 2020. Under-ice salinity transport in low-salinity waterbodies. Limnol. Oceanogr. 65 (2), 247–259. https://doi.org/10.1002/lno.11295.

Olsthoorn, J., Stastna, M., Soontiens, N., 2012. Fluid circulation and seepage in lake sediment due to propagating and trapped internal waves. Water Resour. Res. 48, W11520. https://doi.org/10.1029/2012WR012552.

Osborn, T.R., 1980. Estimates of the local rate of vertical diffusion from dissipation measurements. J. Phys. Oceanogr. 10 (1), 83–89. https://doi.org/10.1175/1520-0485(1980)010<0083:EOTLRO>2.0.CO;2.

Osborn, T.R., Cox, C.S., 1972. Oceanic fine structure. Geophys. Fluid Dyn 3 (4), 321–345. https://doi.org/10.1080/03091927208236085.

Parker, G.J., Imberger, J., 1986. Differential mixed-layer deepening in lakes and reservoirs. In: De Deckker, P., Williams, W.D. (Eds.), Limnology in Australia. Monographiae Biologicae, Volume 61. Springer, Dordrecht. https://doi.org/10.1007/978-94-009-4820-4_3.

Peeters, F., Kipfer, R., 2009. Currents in stratified water bodies 1: density-driven flows. In: Likens, G.E. (Ed.), Encyclopedia of Inland Waters, vol. 1. Elsevier, Oxford, pp. 530–538. https://doi.org/10.1016/B978-012370626-3.00088-0.

Perroud, M., Goyette, S., Martynov, A., Beniston, M., Anneville, O., 2009. Simulation of multiannual thermal profiles in deep Lake Geneva: a comparison of one-dimensional lake models. Limnol. Oceanogr. 54 (5), 1574–1594. https://doi.org/10.4319/lo.2009.54.5.1574.

Phillips, O.M., 1970. On flows induced by diffusion in a stably stratified fluid. Deep Sea Res. Oceanogr. Abs. 17, 435–443. https://doi.org/10.1016/0011-7471(70)90058-6.

Ramón, C.L., Priet-Mahéo, M.C., Rueda, F.J., Andradóttir, H., 2020. Inflow dynamics in weakly stratified lakes subject to large isopycnal displacements. Water Resour. Res. 56 (8), e2019WR026578. https://doi.org/10.1029/2019WR026578.

Ramón, C.L., Ulloa, H.N., Doda, T., Bouffard, D., 2022. Flushing the lake littoral region: the interaction of differential cooling and mild winds. Water Resour. Res. 58 (3), e2021WR030943. https://doi.org/10.1029/2021WR030943.

Ramón, C.L., Ulloa, H.N., Doda, T., Winters, K.B., Bouffard, D., 2021. Bathymetry and latitude modify lake warming under ice. Hydrol. Earth Syst. Sci. 25, 1813–1825. https://doi.org/10.519-hess25-1813-2021.

Reynolds, O., 1883. An experimental investigation of the circumstances which determine whether the motion of water shall be direct or sinuous, and of the law of resistance in parallel channels. Philos. Trans. R. Soc. 174, 935–982. https://doi.org/10.1098/rstl.1883.0029.

Richardson, L.F., 1920. The supply of energy from and to atmospheric eddies. Proc. Royal Soc. Lond. Series A 97 (686), 354–373. https://doi.org/10.1098/rspa.1920.0039.

Richardson, L.F., 1922. Weather prediction by numerical process. Cambridge University Press, Cambridge, UK, p. 236. https://doi.org/10.1002/qj.49704820311.

Rizk, W., Kirillin, G., Leppäranta, M., 2014. Basin-scale circulation and heat fluxes in ice-covered lakes. Limnol. Oceanogr. 59 (2), 445–464. https://doi.org/10.4319/lo.2014.59.2.0445.

Robarts, R.D., Waiser, M.J., Hadas, O., Zohary, T., MacIntyre, S., 1998. Relaxation of phosphorus limitation due to typhoon-induced mixing in two morphologically distinct basins of Lake Biwa, Japan. Limnol. Oceanogr. 43 (6), 1023–1036. https://doi.org/10.4319/lo.1998.43.6.1023.

Robb, D.M., Pieters, R., Lawrence, G.A., 2021. Fate of turbid glacial inflows in a hydroelectric reservoir. Env. Fluid Mech. 21, 1201–1225. https://doi.org/10.1007/s10652-021-09815-4

Rodgers, G.K., 1966. The thermal bar in Lake Ontario, spring 1965 and winter 1965–66. Publ. Great Lakes Res. Div, 15. University of Michigan, pp. 369–374.

Romero, J.R., Alexander, R., Antenucci, J.P., Attwater, G., et al., 1995. Management implications of the physical limnological studies of Rusinga Channel and Winam Gulf in Lake Victoria. In: Odada, E.O., Olago, D.O., Ocholoa, W., Ntiba, M., Wandiga, S., Gichuki, N., Oyieke, H. (Eds.), Proceedings of the 11th World Lakes Conference, 31 October–4 November 2005, Nairobi, Kenya/Volume II, pp. 63–68.

Rueda, F.J., Fleenor, W.E., de Vicente, I., 2007. Pathways of river nutrients towards the euphotic zone in a deep reservoir of small size: Uncertainty analysis. Ecol. Model. 202, 345–361. https://doi.org/10.1016/j.ecolmodel.2006.11.006.

Rueda, F.J., MacIntyre, S., 2009. Flow paths and spatial heterogeneity of stream inflows in a small multibasin lake. Limnol. Oceanogr. 54, 2041–2057, 0.4319/lo.2009.54.6.2041.

Rueda, F.J., MacIntyre, S., 2010. Modelling the fate and transport of negatively buoyant storm–river water in small multi-basin lakes. Environ. Model. Softw. 25 (1), 146–157. https://doi.org/10.1016/j.envsoft.2009.07.002.

Rueda, F.J., Schladow, S.G., 2003. Dynamics of a large polymictic lake. II: numerical simulations. J. Hydraul. Eng. 129, 92–101. https://doi.org/10.1061/(ASCE)0733-9429(2003)129:2(92).

Rueda, F.J., Schladow, S.G., Monismith, S.G., Stacey, M.T., 2005. On the effects of topography on wind and the generation of currents in a large multi-basin lake. Hydrobiologia 532 (1), 139–151. https://doi.org/10.1007/s10750-004-9522-4.

Rueda, F.J., Vidal, J., 2009. Currents in the upper mixed layer and in unstratified water bodies. In: Likens, G.E. (Ed.), Encyclopedia of Inland Waters, vol. 1. Elsevier, Oxford, pp. 568–582. https://doi.org/10.1016/B978-012370626-3.00088-0.

Sadro, S., Nelson, C.E., Melack, J.M., 2011. Linking diel patterns in community respiration to bacterioplankton in an oligotrophic high-elevation lake. Limnol. Oceanogr. 56, 540–550. https://doi.org/10.4319/lo.2011.56.2.0540.

Sahoo, G.B., Schladow, S.G., Reuter, J.E., Coasts, R., Dettinger, M., Riverson, J., Wolfe, B., Coasta-Cabral, M., 2013. The response of Lake Tahoe to climate change. Clim. Change 116, 71–95. https://doi.org/10.1007/s10584-012-0600-8.

Schmid, M., Batist, M.D., Granin, N.G., Kapitanov, V.A., McGinnis, D.F., Mizandrontsev, I.B., Obzhirov, A.I., Wüest, A., 2007. Sources and sinks of methane in Lake Baikal: a synthesis of measurements and modeling. Limnol. Oceanogr. 52, 1824–1837. https://doi.org/10.4319/lo.2007.52.5.1824.

Schmid, M., Budnev, N.M., Granin, N.G., Sturm, M., Schurter, M., Wüest, A., 2008. Lake Baikal deepwater renewal mystery solved. Geophys. Res. Lett. 35, L09605. https://doi.org/10.1029/2008GL033223.

Schmid, M., Busbridge, M., Wüest, A., 2010. Double-diffusive convection in Lake Kivu. Limnol. Oceanogr. 55, 225–238. https://doi.org/10.4319/lo.2010.55.1.0225.

Schwefel, R., Gaudard, A., Wüest, A., Bouffard, D., 2016. Effects of climate change on deepwater oxygen and winter mixing in a deep lake (Lake Geneva): comparing observational findings and modeling. Water Resour. Res. 52, 8811–8826. https://doi.org/10.1002/2016WR019194.

Schwefel, R., Hondzo, M., Wüest, A., Bouffard, D., 2017. Scaling oxygen microprofiles at the sediment interface of deep stratified waters. Geophys. Res. Lett. 44 (3), 1340–1349. https://doi.org/10.1002/2016GL072079.

Shimaraev, M.N., Granin, N.G., Zhdanov, A.A., 1993. Deep ventilation of Lake Baikal waters due to spring thermal bars. Limnol. Oceanogr. 38 (5), 1068–1072. https://doi.org/10.4319/lo.1993.38.5.1068.

Smith, I.R., 1975. Turbulence in lakes and rivers. Freshwater Biological Association, Ambleside, Cumbria, UK, p. 79.

Spain, J.D., Wernert, G.M., Hubbard, D.W., 1976. The structure of the spring thermal bar in Lake Superior, II. J. Great Lakes Res. 2 (2), 296–306. https://doi.org/10.1016/S0380-1330(76)72294-9.

Spigel, R.H., Imberger, J., 1980. The classification of mixed-layer dynamics of lakes of small to medium size. J. Phys. Oceanogr. 7, 1104–1121. https://doi.org/10.1175/1520-0485(1980)010<1104:TCOMLD>2.0.CO;2.

Spigel, R.H., Priscu, J.C., 1998. Physical limnology of the McMurdo dry Valleys lakes. In: Ecosystem Dynamics in a Polar Desert: The McMurdo Dry Valleys, vol. 72, pp. 153–187, Antarctic Research Series. https://doi.org/10.1029/AR072p0153.

Sprules, W.G., Cyr, H., Menza, C.W., 2022. Multiscale effects of wind-induced hydrodynamics on lake plankton distribution. Limnol. Oceanogr. 67, 1631–1646. https://doi.org/10.1002/lno.12158.

Stashchuk, N., Vlasenko, V., Hutter, K., 2005. Numerical modelling of disintegration of basin-scale internal waves in a tank filled with stratified water. Nonlinear Process Geophys. 12, 955–964. https://doi.org/10.5194/npg-12-955-2005.

Steissberg, T.E., Hook, S.J., Schladow, S.G., 2005. Measuring surface currents in lakes with high spatial resolution thermal infrared imagery. Geophys. Res. Lett. 32 (11), L11402. https://doi.org/10.1029/2005GL022912.

Stepanenko, V., Mammarella, I., Ojala, A., Miettinen, H., Lykosov, V., Vesala, T., 2016. Lake 2.0: a model for temperature, methane, carbon dioxide and oxygen dynamics in lakes. Geosci. Model Dev. (GMD) 9, 1977–2006. https://doi.org/10.5194/gmd-9-1977-2016.

Stokes, G. G. (1851). On the effect of internal friction of fluids on the motion of pendulums. Biol. Rev. Camb. Philos. Soc., 9: 10 pp. Reprinted 2009: Cambridge University Press, Cambridge. https://doi.org/10.1017/CBO9780511702266.002.

Stonedahl, S.H., Harvey, J.W., Wörman, A., Salehin, M., Packman, A.I., 2010. A multiscale model for integrating hyporheic exchange from ripples to meanders. Water Resour. Res. 46, W12539. https://doi.org/10.1029/2009WR008865.

Strahler, A.N., 1952. Dynamic basis of geomorphology. Geol. Soc. Am. Bull. 63, 923–938. https://doi.org/10.1130/0016-7606(1952)63[923:DBOG]2.0.CO;2.

Sturman, J.J., Oldham, C.E., Ivey, G.N., 1999. Steady convective exchange flows down slopes. Aquat. Sci. 61 (3), 260–278. https://doi.org/10.1007/s000270050065.

Sukhodolov, A., Bertoldi, W., Wolter, C., Surian, N., Tubino, M., 2009. Implications of channel processes for juvenile fish habitats in Alpine rivers. Aquat. Sci. 71 (3), 338–349. https://doi.org/10.1007/s00027-009-9199-x.

Talling, J.F., 1963. Origin of stratification in an African rift lake. Limnol. Oceanogr. 8, 68–78. https://doi.org/10.4319/lo.1963.8.1.0068.

Talling, J., 1966. The annual cycle of stratification and phytoplankton growth in Lake Victoria (East Africa). Int. Rev. ges. Hydrobiol. 51, 545–562. https://doi.org/10.1002/iroh.19660510402.

Tedford, E.W., MacIntyre, S., Miller, S.D., Czikowsky, M.J., 2014. Similarity scaling of turbulence in a temperate lake during fall cooling. J. Geophys. Res. Oceans 119, 4689–4713. https://doi.org/10.1002/2014JC010135.

Tennekes, H., Lumley, J.L., 1972. A first course in turbulence. MIT Press Academic, Cambridge, MA, p. 300.

Tennessee Valley Authority, 1972. Heat and mass transfer between a water surface and the atmosphere. In: Water Resources Research Laboratory Report No. 14. Tennessee Valley Authority Report No. 0-6803.

Terray, E.A., Donelan, M.A., Agrawal, Y.C., Drennan, W.L., Kahma, K.K., et al., 1996. Estimates of kinetic energy dissipation under breaking waves. J. Phys. Oceanogr. 26 (5), 792–807. https://doi.org/10.1175/1520-0485(1996)026<0792:EOKEDU>2.0.CO;2.

Terzhevik, A., Golosov, S., Palshin, N., Mitrokhov, A., Zdorovennov, R., Zdorovennova, G., Kirillin, G., Shipunova, E., Zverev, I., 2009. Some features of the thermal and dissolved oxygen structure in boreal, shallow ice-covered Lake Vendyurskoe, Russia. Aquatic Ecol 43 (3), 617–627. https://doi.org/10.1007/s10452-009-9288-x.

Terzhevik, A.Y., Pal'shina, N.I., Golosov, S.D., Zdorovennov, R.E., Zdorovennova, G.E., et al., 2010. Hydrophysical aspects of oxygen regime formation in a shallow ice-covered lake. Water Resour. 37 (5), 662–673. https://doi.org/10.1134/S0097807810050064.

Thiery, W.I.M., Stepanenko, V.M., Fang, X., Jöhnk, K.D., Li, Z., Martynov, A., Perroud, M., Subin, Z.M., Darchambeau, F., Mironov, D., Van Lipzig, N.P., 2014. LakeMIP Kivu: evaluating the representation of a large, deep tropical lake by a set of one-dimensional lake models. Tellus A: Dyn. Meteorol. Oceanogr. 66 (1), 21390. https://doi.org/10.3402/tellusa.v66.21390.

Thompson, R., Imberger, J., 1980. Response of a numerical model of a stratified lake to wind stress. Proc. 2nd Int. Symp. Stratified Flows 1, 562–570.

Thorpe, S.A., 1969. Experiments on the stability of stratified shear flows. Radio Sci. 4, 1327–1331.

Thorpe, S.A., 1977. Turbulence and mixing in a Scottish loch. Philos. Trans. R. Soc. Lond. A 286, 125–181. https://doi.org/10.1098/rsta.1977.0112.

Thorpe, S.A., 1995. Dynamical processes of transfer at the sea surface. Prog. Oceanogr. 35 (4), 315–352. https://doi.org/10.1016/0079-6611(95)80002-B.

Thorpe, S.A., 2005. The turbulent ocean. Cambridge University Press, Cambridge, UK, p. 439. https://doi.org/10.1017/CBO9780511819933.

Thorpe, S.A., 2007. An introduction to ocean turbulence. Cambridge University Press, Cambridge, UK, p. 240. https://doi.org/10.1017/CBO9780511801198.

Triska, F.J., Kennedy, V.C., Avanzino, R.J., Zellweger, G.W., Bencala, K.E., 1989. Retention and transport of nutrients in a third-order stream in northwestern California: hyporheic processes. Ecology 70, 1893–1905. https://doi.org/10.2307/1938120.

Ulloa, H.N., Wüest, A., Bouffard, D., 2018. Mechanical energy budget and mixing efficiency for a radiatively heated ice-covered waterbody. J. Fluid Mech. 852, R1. https://doi.org/10.1017/jfm.2018.587.

Ulloa, H.N., Winters, K.B., Wüest, A., Bouffard, D., 2019. Differential heating drives downslope flows that accelerate mixed-layer warming in ice-covered waters. Geophys. Res. Lett. 46, 13872–13882. https://doi.org/10.1029/2019GL085258.

United States Army Corps of Engineers [USACE], 1984. Shore protection manual. In: Volumes I and II, fourth ed. Coastal Engineering Research Center, Vicksburg, Mississippi.

United States Army Corps of Engineers [USACE] 2010. HEC-RAS: river analysis system. User's Manual v. 4.1. U.S. Army Corps of Engineers, Hydrologic Engineering Center, Davis, California, 407 pp.

Valipour, M., Sefidkouhi, M.A.G., Raeini, M., 2017. Selecting the best model to estimate potential evapotranspiration with respect to climate change and magnitudes of extreme events. Agric. Water Manag. 180, 50–60. https://doi.org/10.1016/j.agwat.2016.08.025.

Van Dyke, M., 1982. An album of fluid motion. Parabolic Press, Stanford, California, p. 177.

Vanoni, V.A., Brooks, N.H., 1957. Laboratory studies of the roughness and suspended load of alluvial streams. report no. E-68. Sedimentation Laboratory, California Institute of Technology, Pasadena, California.

Verburg, P., Antenucci, J.P., Hecky, R.E., 2011. Differential cooling drives large-scale convective circulation in Lake Tanganyika. Limnol. Oceanogr. 56 (3), 910–926. https://doi.org/10.4319/lo.2011.56.3.0910.

Vesala, T., Huotari, J., Rannik, Ü., Suni, T., Smolander, S., Sogachev, A., Launiainen, S., Ojala, A., 2006. Eddy covariance measurements of carbon exchange and latent and sensible heat fluxes over a boreal lake for a full open-water period. J. Geophys. Res. 111, D11101. https://doi.org/10.1029/2005JD006365.

Vidal, J., MacIntyre, S., McPhee-Shaw, E.E., Shaw, W.J., Monismith, S.G., 2013. Temporal and spatial variability of the internal wave field in a lake with complex morphometry. Limnol. Oceanogr. 58 (5), 1557–1580. https://doi.org/10.4319/lo.2013.58.5.1557.

Vincent, W.F., Gibbs, M.M., Spigel, R.H., 1991. Eutrophication processes regulated by a plunging river inflow. Hydrobiologia 226, 51–63. https://doi.org/10.1007/BF00007779.

Vogel, R.M., Yaindl, C., Walter, M., 2011. Nonstationarity: flood magnification and recurrence reduction factors in the United States. J. Am. Water Resour. Assoc. 47, 464–474. https://doi.org/10.1111/j.1752-1688.2011.00541.x.

Vogel, S., 2020. Life in moving fluids: the physical biology of flow—revised and expanded, second ed. Princeton University Press, Princeton, p. 400. https://doi.org/10.1515/9780691212975.

Wain, D.J., Kohn, M.S., Scanlon, J.A., Rehmann, C.R., 2013. Internal wave-driven transport of fluid away from the boundary of a lake. Limnol. Oceanogr. 58 (2), 429–442. https://doi.org/10.4319/lo.2013.58.2.0429.

Wain, D.J., Rehmann, C.R., 2010. Transport by an intrusion generated by boundary mixing in a lake. Water Resour. Res. 46, W08517. https://doi.org/10.1029/2009WR008391.

Wake, L.V., Hillen, L.W., 1981. Nature and hydrocarbon content of blooms of the alga *Botryococcus braunii* occurring in Australian freshwater lakes. Mar. Freshw. Res. 32, 353–367. https://doi.org/10.1071/MF9810353.

Wang, B., Liao, Q., Fillingham, J.H., Bootsma, H.A., 2015. On the coefficients of small eddy and surface divergence models for the air-water gas transfer velocity. J. Geophys. Res. Oceans. 120, 2129–2146. https://doi.org/10.1002/2014JC010253.

Welch, H.E., Bergmann, M.A., 1985. Water circulation in small arctic lakes in winter. Can. J. Fish. Aquat. Sci. 42 (3), 506–520. https://doi.org/10.1139/F85-068.

Wells, M.G., Dorrell, R.M., 2021. Turbulence processes within turbidity currents. Ann. Rev. Fluid Mech. 53 (1), 59–83. https://doi.org/10.1146/annurev-fluid-010719-060309.

Wells, M.G., Sherman, B., 2001. Stratification produced by surface cooling in lakes with significant shallow regions. Limnol. Oceanogr. 46 (7), 1747–1759. https://doi.org/10.4319/lo.2001.46.7.1747.

Wells, M.G., Wettlaufer, J.S., 2007. The long-term circulation driven by density currents in a two-layer stratified basin. J. Fluid Mech. 572, 37–58. https://doi.org/10.1017/S0022112006003478.

Wiles, P.J., Rippeth, T.P., Simpson, J.H., Hendricks, P.J., 2006. A novel technique for measuring the rate of turbulent dissipation in the marine environment. Geophys. Res. Lett. 33, 1–5. https://doi.org/10.1029/2006GL027050.

Wingenroth, J., Yee, C., Nghiem, J., Larsen, L., 2021. Effects of stem density and Reynolds number on fine sediment interception by emergent vegetation. Geosciences 11, 136. https://doi.org/10.3390/geosciences11030136.

Woolway, R.I., Albergel, C., Frölicher, T.L., Perroud, M., 2022. Severe lake heatwaves attributable to human-induced global warming. Geophys. Res. Lett. 49, e2021GL097031. https://doi.org/10.1029/2021GL097031.

Woolway, R.I., Maberly, S.C., 2020. Climate velocity in inland standing waters. Nat. Clim. Change 10 (12), 1124–1129. https://doi.org/10.1038/s41558-020-0889-7.

Woolway, R.I., Maberly, S.C., Jones, I.D., Feuchtmayr, H., 2014. A novel method for estimating the onset of thermal stratification in lakes from surface water measurements. Water Resour. Res. 50 (6), 5131–5140. https://doi.org/10.1002/2013WR014975.

Woolway, R.I., Merchant, C.J., 2019. Worldwide alteration of lake mixing regimes in response to climate change. Nat. Geosci. 12 (4), 271–276. https://doi.org/10.1038/s41561-019-0322-x.

Woolway, R.I., Sharma, S., Weyhenmeyer, G.A., Debolskiy, A., Golub, M., et al., 2021. Phenological shifts in lake stratification under climate change. Nat. Commun. 12 (1), 1–11. https://doi.org/10.1038/s41467-021-22657-4.

Wüest, A., Lorke, A., 2009. Small-scale hydrodynamics in lakes. Annu. Rev. Fluid Mech. 35 (1), 373–412. https://doi.org/10.1146/annurev.fluid.35.101101.161220.

Xing, Z., Fong, D.A., Yat-Man Lo, E., Monismith, S.G., 2014. Thermal structure and variability of a shallow tropical reservoir. Limnol. Oceanogr. 59 (1), 115–128. https://doi.org/10.4319/lo.2014.59.01.0115.

Yeates, P.S., Imberger, J., 2004. Pseudo two-dimensional simulations of internal and boundary fluxes in stratified lakes and reservoirs. Int. J. River Basin Manag. 1, 297–319. https://doi.org/10.1016/0304-3800(94)90111-2.

Zappa, C.J., McGillis, W.R., Raymond, P.A., Edson, J.B., Hintsa, E.J., Zemmelink, H.J., Dacey, J.W., Ho, D.T., 2007. Environmental turbulent mixing controls on air-water gas exchange in marine and aquatic systems Geophys. Res. Lett. 34, L10601. https://doi.org/10.1029/2006GL028790.

CHAPTER

9

Structure and Productivity of Aquatic Ecosystems

Lisette N. de Senerpont Domis[1] *and Belinda J. Robson*[2]

[1]Netherlands Institute of Ecology, Wageningen, the Netherlands [2]Harry Butler Institute & Environmental & Conservation Sciences, Murdoch University, Murdoch, Western Australia, Australia

OUTLINE

I. The ecosystem concept

An *ecosystem* comprises all the life forms (microbes, plants, animals) and the inorganic context (sediment types, water regimes, water quality, etc.) that they live in. Together, the life forms interact with each other. They influence, and are influenced by, their environment, carrying out ecosystem processes such as the transport of nitrogen and carbon. Ecosystems are highly complex systems and have features such as emergent properties and thresholds of transformability. Indeed, ecological studies of freshwater ecosystems have played a vital part in understanding the dynamics of complex systems, such as the identification of thresholds for regime shifts in shallow lakes (e.g., Folke et al., 2004; Chapter 26). Scientists usually name ecosystems based on their physical features (e.g., shallow lake ecosystems, headwater stream ecosystems), but these places are grouped because they show the same dynamics and controlling factors. Indeed, the International Union for Conservation of Nature (IUCN) has classified and described the features and dynamics of every ecosystem type on Earth in their Global Ecosystem Typology 2.0 (Keith et al., 2020). In this chapter we introduce freshwater ecosystems, how they are structured, and how they function.

Ecotone is the term used to describe habitats that lie at the boundary between two ecosystems. The ecotone between rivers and the land is called the *riparian zone,*

Wetzel's Limnology, Fourth Edition
https://doi.org/10.1016/B978-0-12-822701-5.00009-4

between lakes and the land is called the *fringing zone*, and between rivers and the ocean lies another ecosystem type, the *estuary*. There is also a fourth dimension to inland water ecosystems, which is the *groundwater* (Chapter 5), that underlies many waterbodies. The ecotone between surface and groundwater is called the *hyporheic zone*. Understanding ecosystem function in rivers and lakes requires an understanding of riparian, fringing, and hyporheic zones, and often of estuaries too.

So far, discussion in this book has been directed to the basic physical aspects of water itself (Chapter 3), how water is distributed to river and lake basins, the origins of rivers and of lake basins and how the catchment basin morphology is modified with time (Chapter 4), and the hydrology of inland waters (Chapter 5). We then considered both the penetration of solar radiation and the distribution of solar heat energy in various fresh waters (Chapter 6) and explained how absorbed heat interacts with wind energy to influence water movements, particularly in standing waters (Chapters 7 and 8). Here, the flows of nutrients and energy throughout the catchment will be examined. The biota is inseparably coupled with the dynamics of many chemical elements, in particular nutrients and the biogeochemical regulation of nutrient fluxes and recycling. Before discussing the major chemical constituents in detail, however, a brief overview of the ecosystem components and how they interrelate is useful.

II. Catchment concept

As introduced in Chapters 4 and 5, a catchment is the area of land surrounding a waterbody, through and over which water flows into that waterbody. The size and geological composition of catchments thus influences freshwater ecosystems and catchment geomorphology influences the rates and volumes of water (and materials carried by that flow) entering waterways. Catchment boundaries are delineated by the high points of land at their edge (Chapter 4). Catchment size may vary from very small (such as the tiny catchments of volcanic crater lakes, delineated by the upper rim of their crater) to enormous (such as the Amazon River basin). River catchments may be divided into *subcatchments*, which comprise tributaries of the mainstem river. River catchments may also include lakes, such as glacial lakes in river headwaters or floodplain lakes on lowland rivers.

A. Geomorphology

Terrestrial and aquatic ecosystems are linked by the movements of water and materials through the drainage basin to recipient rivers and lakes, and by the movement of materials back into terrestrial and estuarine ecosystems from inland waters (Fig. 9-1). Chemical and biological processes modify the composition of the materials dissolved within and carried by the water. Geological

FIGURE 9-1 Terrestrial and aquatic ecosystems are linked by movements of water and materials through the drainage basin to recipient rivers and lakes, and by the movement of materials back into terrestrial and estuarine ecosystems from inland waters. *(Source illustrations adapted from Tracey Saxby & Jane Hawkey, Integration and Application Network (ian.umces.edu/media-library).)*

features of the landscape govern the directions of movement and particularly the residence time of water during movement in surface and ground waters toward rivers and lakes. The duration of contact with soil and associated microbiota influences the content of dissolved salts and organic products conveyed in the water. Terrestrial and wetland vegetation in the drainage basin influence the chemical content of both water falling through their leaf canopy as well as water in the soil by selective assimilation and storage within their tissues. The movements and fluxes of a number of these compounds will be treated separately in subsequent chapters (Chapters 13–15).

From vegetated catchments, such as in forests, usually less than half of the incoming precipitation leaves as runoff and groundwater seepage to streams and rivers. Removal of vegetation greatly accelerates the flow of runoff from the landscape and increases the amount, particularly when vegetation is replaced by impervious surfaces altering *evapotransporation* (e.g., concrete surfaces [Walsh et al., 2005], see also Chapters 4 and 5). During periods of high precipitation, water in streams may exceed the capacity of the river channel and may inundate the floodplain (where present), often for long periods (e.g., months). During this inundation, organic matter (OM) and nutrients are deposited on floodplains and in floodplain wetlands resulting in highly productive aquatic habitats. As floodwaters recede, organic carbon fixed on the floodplain returns to river channels in the form of detritus, fish, and other animals (Warfe et al., 2011). Thus the geomorphology of the land partly determines the soil and ionic composition, slope, and (in combination with climate) the vegetation cover. Within the context of climate, vegetation and soil composition influence not only the amount of runoff but the composition and quantity of OM entering streams and lakes.

Human effects on global climate and element cycling are now so profound that we live in what many believe will soon be defined as a new era, the *Anthropocene*. These human impacts, especially changes in climate and land use, have reduced the influence of catchment geology on aquatic ecosystems (Rockström et al., 2014). Instead, water regime, as defined below, is now recognized as an important factor affecting aquatic ecosystems.

B. Water regimes in the Anthropocene

Water regime is the term used to describe the annual or multiannual pattern of flow in rivers or inundation in lakes and wetlands. In some regions (e.g., boreal) the vast majority of rivers and lakes are perennially flowing/inundated, despite freezing over each winter. In other regions (e.g., arid, semiarid, Mediterranean climates) the majority of rivers and lakes dry out. The frequency and intensity of drying varies between climates (Mediterranean climates tend to have greater annual predictability of flows than arid climates) but also between regions with the same climate. To make sense of all this variability, water regimes have been divided into spatial and temporal components, each of which describes an aspect of the water regime (Boulton et al., 2014).

The spatial components of the water regime comprise extent (area inundated) and depth, volume, discharge (volume of water flowing through a known cross-sectional area in a specified time), and variability (amount of change at a range of spatial scales). The temporal components of the flow regime are timing (timing when water is present, both within and among years), frequency (how often inundation and drying occur), duration (period of inundation), and variability (amount of change at a range of timescales) (Boulton et al., 2014). These components are often combined into classifications. For example, the term *episodic* is often used to denote waterbodies that fill unpredictably, such as those that result from cyclonic rainfall in arid zones. The term *seasonal* is used for predictable patterns of wetting and drying in climates where there are distinct wet and dry seasons, such as Mediterranean climates, temperate climates, and some tropical climates. These components of water regime, separately and in combination, are important for life in inland waters. For example, biota may be adapted for episodic filling of arid zone lakes by being able to lie dormant in sediments for periods of years. However, when such lakes are the terminus of episodic rivers, water may be present for long periods and so biota may be adapted to breed when inundation occurs, because even though it occurs rarely, the duration will be sufficient for successful population recruitment. In most types of freshwater ecosystems many species will be adapted to the water regime that has occurred historically in that system. This may be in relation to the timing of reproduction or dispersal or to the presence of habitats necessary for the reproduction or juvenile or adult survival. Because of the strong relationship between water regime and populations of microbes, plants, and animals, changing this most fundamental of variables is fraught with risk and known to alter ecosystem function.

Global warming is changing global climate in many ways, most of which will affect freshwater ecosystems. Relative to increases in mean climatic conditions, there is growing consensus that extreme events (increasing in frequency and intensity) may more strongly affect aquatic ecosystem functioning (Srivastava et al., 2020; Stockwell et al., 2020). Furthermore, although rising temperatures affect freshwater species and ecosystems, changes to hydrological variables have even more substantial impacts (e.g., changing rainfall volume and intensity, glacial melt, replacement of snowfall with rainfall) because they alter water regimes (Dudgeon,

2019). In some regions rainfall is increasing, and snow and ice cover is decreasing, so rivers and lakes may receive higher volumes of warmer water as well as more light. In other regions rainfall is decreasing and has caused threshold changes in lakes and streams (perennial to seasonal, seasonal to ephemeral), causing some to dry out for the first time, altering biotic communities (Carey et al., 2021; Crabot et al., 2021). On top of this, human water demands for sanitation, irrigation, and hydropower have led to increased water abstraction and river regulation, both of which may cause the drying of wetlands or downstream river channels (e.g., Piano et al., 2020; Rouse et al., 1997).

III. Streams and rivers

In flowing (*lotic*) waters we refer to the water regime as the *flow regime*. This is because the unidirectional nature of flow in rivers and streams distinguishes them from other ecosystems (Giller and Malmqvist, 1998; Hynes, 2001; Vannote et al., 1980). Furthermore, because lotic ecosystems are linear in form, they have an enormous surface (riverbank) area-to-volume ratio, which heavily exposes them to the influence of their catchment, and human impacts that occur there. In flowing channels the substratum may be unstable and rearranged by flowing water, causing interactions between flow and geology structure river habitats (Frissell et al., 1986). Because of the primacy of flow, especially in upland rivers and streams, many lotic plants and animals are adapted to persist in, and in some cases even use, fast flowing water.

When studying river ecosystems, ecologists distinguish between upland and lowland rivers, because there are basic physical and functional differences between them. *Upland rivers* are less likely to have floodplains than lowland rivers, and they are generally erosive (and so often have a hard, stony substratum), whereas *lowland rivers* are generally depositional (and so have soft sediments). These differences in substratum and flow influence ecosystem function, the balance between *autotrophic* and *heterotrophic* production, and the importance of benthic versus pelagic food webs.

IV. Lakes

In *lentic* or standing waters we often refer to the water regime as the *water balance*. The water balance of a lake is determined by the surface and subsurface inflows and outflows, precipitation, and evaporation. As such, a lake ecosystem consists of the lake itself and its entire catchment. Lakes that drain into a river or ocean are called *exorheic* or *open lakes*, whereas lakes that only drain through the subsurface or through evaporation are called *closed* or *endorheic* lakes (Chapter 4).

Lakes have been recognized as sentinels of global change, as they integrate information from the entire catchment (Adrian et al., 2009; Schindler, 2006), reflected in their loading of inorganic and organic nutrients, but also synthetic pollutants such as pharmaceuticals and pesticides. Through an increased understanding of both terrestrial and aquatic biomarkers, our understanding of how terrestrial biota regulate loading to streams and recipient lakes and vice versa has vastly improved in recent decades (Kayler et al., 2019; Marín-Spiotta et al., 2014). For instance, inland waters are no longer perceived as passive pipes of organic carbon to the ocean, but rather their role in actively processing organic carbon has been recognized (Butman and Raymond, 2011; Chapter 28).

Catchment areas input both particulate organic matter (POM) and especially dissolved organic matter (DOM) to lakes, and this dissolved and colloidal OM, once it reaches the lake, regulates many ecosystem processes (as it does for rivers, discussed above). Extremely heterogeneous and productive wetland–fringing areas often lie at the interface between the terrestrial drainage basin and the open-water zone of the lake (Fig. 9-2) and also occur on floodplains of lowland rivers. These complex wetland–littoral areas are exceedingly important in regulating metabolism in the receiving waterbody (Wetzel, 1979, 1990, 1995). Since the majority of lakes globally are small and relatively shallow (Downing, 2008), the metabolically active wetland and littoral components dominate the productivity of most lakes of the world.

The effects that terrestrial, wetland, and littoral biota have on the quality and quantity of inorganic and organic loading to a lake can be profound. Water laden with inorganic and organic substances flows from higher elevations to the recipient lake basin or river channel in both ground and surface flows. Chemical and biological reactions occur en route that selectively modify the quality (balance between different macro- and micronutrients) and quantity of nutrients and organic substances entering the waterbody. Surface flows often pass through the wetland–littoral complex and selectively lose or gain inorganic and organic compounds before reaching the open water. Finally, appreciable loading of inorganic and organic substances to the atmosphere is common from the industrial and agricultural activities of humans. These compounds may reach the drainage basin and lake itself via dryfall and with wet and frozen precipitation. This source of loading is often a significant portion of the total nutrient and contaminant loading to a lake and was a substantial problem prior to the 1980s when acid rain acidified many lakes in the Northern Hemisphere (Hildrew, 2018).

V. Spatial structure and terminology in lakes and rivers

Apart from thermal structuring, due to the distribution of heat through water (Chapter 7), lakes and rivers also display other forms of spatial structures. The bottom of a lake basin is separable from the free open water, the

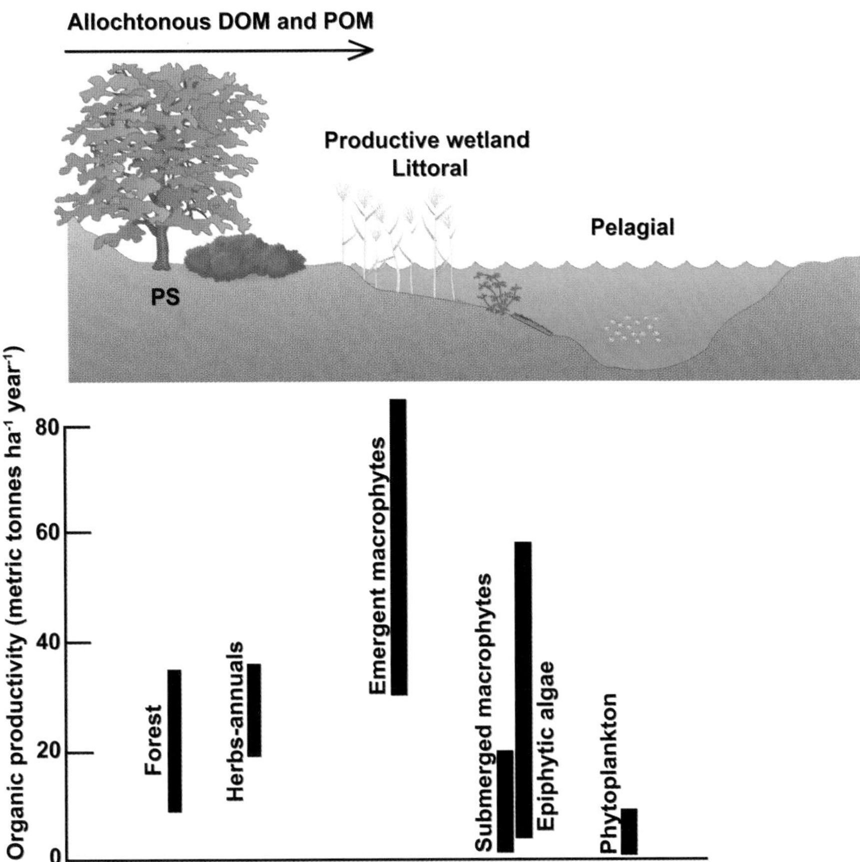

FIGURE 9-2 The lake ecosystem, showing the drainage basin with terrestrial photosynthesis (PS) of organic matter; movement of nutrients and dissolved organic matter (DOM), and particulate organic matter (POM) in surface and groundwater flows toward the lake basin. Shown are the chemical and biotic alterations of these materials en route, especially as they pass through the highly productive and metabolically active wetland–littoral zone of the lake (net organic productivity in metric tons per hectare per year). *(Source illustrations adapted from Tracey Saxby, Integration and Application Network (ian.umces.edu/media-library).)*

pelagic zone, and is further divisible into a number of rather distinct transitional zones from the shoreline to the deepest point (Fig. 9-3). The *supralittoral* zone also lies entirely above the water level but is subject to spray from waves. The *littoral* zone encompasses the shoreline region between the highest and lowest seasonal water levels together with the rest of the shallow perimeter where aquatic plants can grow because light reaches the lake-bed. *Emergent aquatic plants* (whose stems and leaves protrude above the water surface), *submerged aquatic plants* (the majority of stems and leaves are below the water surface), and *floating aquatic plants* (either free-floating or with roots in the sediment) are common in undisturbed littoral zones (see Chapter 24). Thus, in shallow lakes (see Chapter 26), primary production is possible across the entire lake-bed and volume, so the littoral can be viewed as existing across the entire lake-bed (and may be completely filled by aquatic plants). Below the littoral in deeper lakes lies a transitional zone, the *littoriprofundal*, which is occupied by scattered photosynthetic algae and bacteria (or a microbial mat) and is often adjacent to the metalimnion of stratified lakes. The remainder of the sediments, which consist of exposed fine sediment free of vegetation (due to the lack of light), is referred to as the *profundal* zone. Upshore from the littoral is the *fringing zone,* which is generally vegetated with plants capable of tolerating inundation during the wet season. The fringing zone is analogous to the riparian zone of rivers.

In rivers a range of different terms are used. The *riparian* zone refers to the riverbanks and their vegetation, which can tolerate periods of inundation during floods. Rivers may be divided into *headwaters* (small, erosive streams at the upland end of a catchment), *middle reaches,* and *lowland rivers* (usually wide and slowly flowing, may have a broad floodplain filled with lakes). However, a variety of different geological forms (e.g., waterfalls, gorges) may exist at different points along rivers. Therefore rivers are often divided into a nested hierarchy of structural forms (Frissell et al., 1986, see Fig. 9-4). These include river *segments,* divided by tributary junctions. Nested within these are *reaches* divided by structures such as waterfalls, debris dams, or large pools. Nested within reaches are habitats such as *pools, runs,* and *riffles,* each of which is defined by relative depth and flow speed (Davis

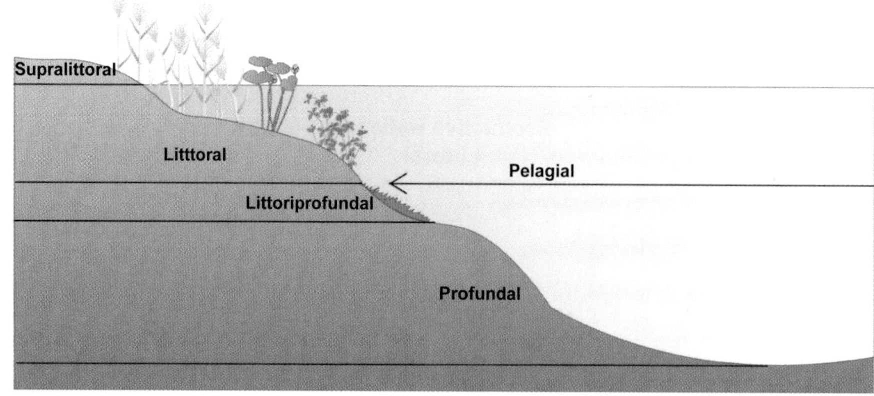

FIGURE 9-3 Spatial structure of lakes (see text for explanation). *(Source illustrations adapted from Tracey Saxby, Integration and Application Network (ian.umces.edu/media-library).)*

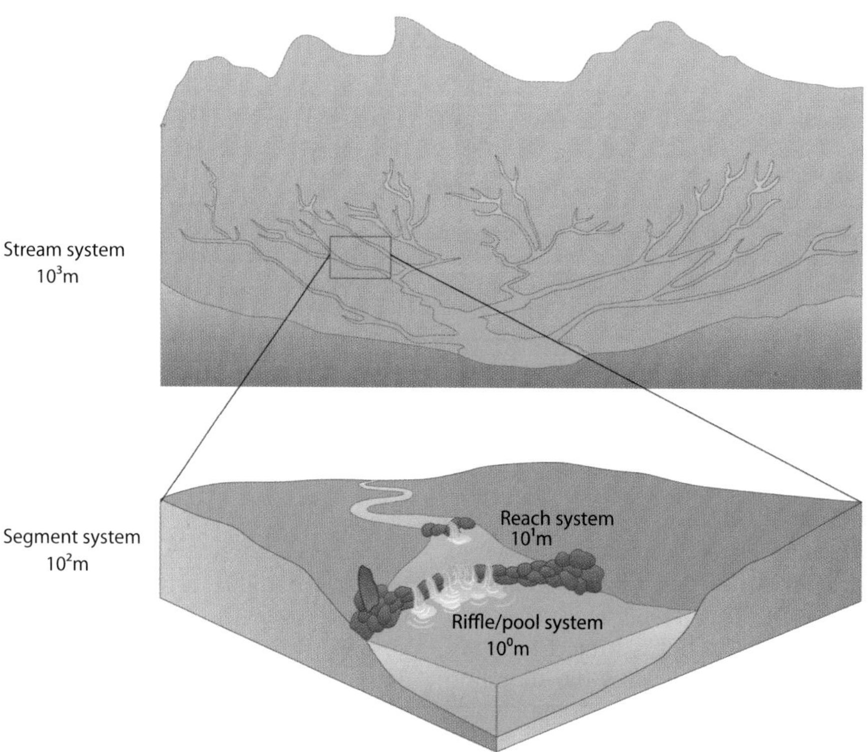

FIGURE 9-4 Hierarchical structure of streams and its habitat systems. Linear scale of each of the systems indicated in meters (m).

and Barmuta, 1989): pools have very slow flow and are deeper than runs, which have faster, smoother linear flow and are also deeper than riffles, which are shallow with fast turbulent flow created by stones.

Some terms are used in both rivers and lakes. The term *benthos*, originating from the Greek word for "bottom," is defined broadly to include the assemblage of organisms associated with the bottom or, better, any solid–liquid interface in aquatic systems (Chapter 21). *Periphyton*, although variously used, usually refers to microbial and/or algal growth upon substrata (Chapter 25). It is a broad term that applies to microbiota living on any substratum, living or dead, plant, animal, or nonliving. *Biofilm* is another term more frequently used to describe much the same thing: it refers to attached communities of algae, fungi, and bacteria growing on any submerged surface.

VI. Subsidies and productivity in lakes and rivers

The living biota of inland waters constitute only a very small portion of the total OM of ecosystems, even in the most productive of freshwater habitats. Most of the OM in aquatic ecosystems is nonliving and is collectively referred to as detritus. *Detritus* consists of all nonliving OM, in both dissolved and particulate forms. In aquatic ecosystems, in general, nearly all of the OM consists of *dissolved organic carbon* (DOC) compounds and *particulate organic carbon compounds* (POC). The ratio of DOC to POC is usually between 6:1 and 10:1 in both lake and running water ecosystems (Wetzel, 1984). The functioning of aquatic ecosystems centers on the cycling of organic carbon between living and nonliving components. Because of the pivotal relationships between the productivity of the living components and the massive amounts of nonliving OM, the subject will be discussed in detail later in Chapters 23 and 27.

Autochthonous productivity means carbon fixed within a water body, by microbes, algae, and aquatic plants. *Allochthonous productivity* arrives in the water body through terrestrial subsidy. In most rivers and lakes both forms of subsidy occur, with the relative importance depending on the morphology of the water body, the geology of the catchment, and the abundance of riparian (rivers) or fringing (lakes) vegetation.

Autochthonous production may be carried out by microbes (photosynthetic bacteria), algae, or aquatic plants. These photosynthetic organisms all require light to fix carbon. Therefore freshwaters that have low light availability through shading by vegetation or through the presence of DOM (e.g., humic substances) or inorganic particles (e.g., clay) decreasing water transparency will have little subaqueous autochthonous production. However, this does not mean that autochthonous production is not important in those waterbodies. In turbid intermittent rivers during the dry season a thin ring of algae and Cyanobacteria grows in the shallow water at the edge of perennial pools. Although tiny, this highly productive "bathtub ring" (Fig. 9-5) supports the entire ecosystem (Bunn et al., 2003). Where there is sufficient light, the form of the substratum and flow speeds determine which photosynthetic organisms occur. For example, many aquatic plants have roots in the substratum and cannot grow on bedrock or among boulders. If they also have foliose leaves (having numerous broad leaves), they will not be able to persist in fast currents. For these reasons, autochthonous production in stony fast-flowing rivers is done by microbes and algae that live in the biofilm on stones, and by bryophytes (Chapter 24), low-growing, tough plants that can attach to stones and withstand fast flows. In lentic water bodies such as lakes (no flow) or slower flowing rivers (mid-lowland rivers and streams), substrata are often finer (i.e., sand, clay, or gravel) and more suitable for the roots of aquatic (including foliose) plants. If there is sufficient light, dense stands of macroalgae and vascular plants will grow. Depth is also a factor here, because some aquatic plants require shallow water to obtain sufficient light

FIGURE 9-5 A thin ring ("bathtub ring," as indicated by the *white arrow*) of algae and Cyanobacteria growing in the shallow water at the edge of perennial pools. This highly productive "bathtub ring" may support an entire ecosystem. *(Photo credit: Belinda J. Robson.)*

for growth. In deeper waters most autochthonous production will be done by phytoplankton (Chapters 17 and 18). These are algae and microbes capable of floating or swimming in the water column to obtain sufficient light but can be washed out during fast flows.

Allochthonous productivity arrives in the waterbody through terrestrial subsidy, from terrestrial or fringing plants. *Allochthonous* organic material is divided into size classes: dissolved organic carbon (DOC <0.045 mm), *fine particulate organic matter* (*FPOM* 0.045–1 mm in longest dimension), *coarse particulate organic matter* (*CPOM* >1–10 mm diameter), and *large particulate organic matter* (*LPOM*) or woody debris (>10 mm diameter). FPOM and CPOM enter rivers and lakes through leaves, bark, or nuts falling (or being blown or washed) into water bodies. LPOM enters water bodies through branches or entire trees falling in. DOC is released by the leaching of all forms of OM when they enter the water. DOC may also enter a waterbody when soil that is rich in OM becomes inundated and OM dissolves (leaches) out of it. By darkening the color of water, DOC can limit photosynthesis and therefore autochthonous production. Furthermore, it is itself a food source for aquatic microbes and in high concentration can stimulate heterotrophic carbon pathways in aquatic food webs.

FPOM is consumed by a wide diversity of freshwater invertebrates; it may be nutritionally enriched by the presence of microbes on it and it also includes fecal pellets from invertebrates and pollen from riparian plants (which is very nutritious, Masclaux et al., 2013). CPOM is colonized by a biofilm of microbes, fungi, and algae which increase its nutritional value to the few invertebrates (shredders) who can consume it.

CPOM also provides a habitat for invertebrates, whereas LPOM or woody debris also provides a habitat for larger animals (e.g., fish) in addition to invertebrates.

Rivers vary in the predominance of autochthonous or allochthonous production. The first model to recognize this was the *River Continuum Concept* (RCC) (Chapter 5), which predicted that as headwater streams were erosive and shaded by forest, their ecosystems depended on allochthonous carbon (Vannote et al., 1980). Further downstream, rivers widened, so there was assumed to be more light, thus increasing autochthonous production. In their lower reaches rivers are generally much wider and deeper, and so phytoplankton were thought to be the dominant producers, but considerable amounts of FPOM were also deposited in those reaches from sources upstream. Later research has shown that there is considerable variation in the balance between autochthonous and allochthonous production along and between rivers (Bowes et al., 2020) and that algal carbon is usually the best quality carbon source for consumers, even in upland streams (Brett et al., 2017).

In lakes allochthonous subsidy may be less important than in rivers due to their larger surface area to perimeter ratio. They still receive CPOM and LPOM, but these sources of carbon are considered less influential than in rivers. The emergence of fatty acids and stable isotope research shows, however, that terrestrial POM can enrich phytoplankton-based food webs (Masclaux et al. 2013; Stoler and Relyea, 2020), with the importance of terrestrial subsidy increasing with decreasing surface area to perimeter ratio in lakes (Polis et al., 1997).

VII. Evaluation of biomass and production

In order to evaluate the *productivity* of a system, it is important to realize that *production* is a rate and is often expressed as a unit of mass or energy production per volume (or surface area) per time. Confusingly, *biomass* or *standing stock* is often used as a proxy for the productivity of a given system, but this is inherently wrong. Biomass is the mass or weight of living material in a unit area at a given instant in time and is often calculated as the product of mean individual weight (W) and density (N). Biomass has the dimensions of mass per unit area or volume (e.g., ash-free dry mass m^{-2}). Biomass can be calculated for certain functional groups, such as algal biomass, zooplankton biomass, macrophyte biomass, and fish biomass, and as such be used for calculating the productivity of these groups. Another useful term in the context of productivity is *carrying capacity*, the maximum biomass a specific group or even an ecosystem can achieve given the prevailing abiotic and biotic factors. Similar to biomass, production can be evaluated for a single species or a specific group (e.g., primary productivity or bacterial productivity) but also on higher organizational levels, up to the ecosystem level.

A. Primary productivity

The primary production of phytoplankton has received an extraordinary amount of attention in limnology and has been measured in great detail in a number of aquatic ecosystems. In a process called *photosynthesis* autotrophic species such as phytoplankton produce biomass by using light as a source of energy to convert CO_2 and H_2O into glucose and oxygen (see Chapter 10, Eqs. 10-1, 10-2, 10-3, and 10-4). *Respiration*, on the other hand, is a process where glucose and oxygen are used to produce CO_2 and energy. Whereas *gross primary production* (GPP) is defined as the total rate at which autotrophs produce biomass or energy, *net primary production* is defined as the rate at which autotrophs produce biomass or energy considering the energy or biomass used for respiration.

The *light- and dark-bottle techniques* for estimating primary production have received broad applications (Fig. 9-6). In the oxygen method samples of phytoplankton populations are incubated at the depths from which they were collected in clear and opaque bottles. The initial concentration of dissolved oxygen (c1) can be expected to decline to a lower value (c2) by respiration in the darkened bottles and increase to a higher concentration (c3) in the clear bottles, according to the difference between photosynthetic production and respiratory consumption. The difference (c1 − c2)

FIGURE 9-6 The light- and dark-bottle technique for estimating primary production. The bottles wrapped in tin foil are the "dark bottles" excluding light and allowing for the assessment of respiration in the absence of primary productivity. The clear bottles are the "light bottles" that allow for primary production. *(Photo credit Miquel Lürling.)*

represents the respiratory activity per unit volume over the time interval of incubation. The difference (c3 − c1) is equal to the net photosynthetic activity, and the sum (c3 − c1) + (c1 − c2) = (c3 − c2) corresponds to the gross photosynthetic activity. Numerous assumptions are made in the method that can alter the photosynthetic measurements appreciably; for example, respiration rates are not necessarily the same in light and dark, since photorespiration clearly occurs in algae, other processes such as photooxidative consumption use oxygen separately from apparent respiratory uptake, nonphotolysis of water by bacterial photosynthesis, and so forth. Under many circumstances, these errors are small, but the technique can be considered only as a reasonable estimate. It is probable, however, that these and analytical errors in the determination of oxygen concentrations are appreciably less than sampling errors in the analyses of heterogeneous plankton populations in the lake. Large portions of methodological works are devoted to detailed discussions of this and the following ^{14}C techniques (see especially Strickland, 1960; Strickland and Parson, 1972; Vollenweider, 1969; Wetzel and Likens, 2000).

The incorporation of ^{14}C tracers into the OM of phytoplankton during photosynthesis has been used as a sensitive measure of the rate of primary production. If the total CO_2 content of the experimental water is known (which can be estimated from a lake water alkalinity analysis), and if a known amount of $^{14}CO_2$ is added to the water, then the determination of the content of labeled carbon in the phytoplankton after incubation permits the estimation of the total amount of carbon assimilated. Numerous methodological and physiological problems confront the application of the ^{14}C light-and-dark technique. With care, however, most technical problems can be overcome, and errors evaluated; for example, respiratory losses of CO_2 and secretion rates of soluble organic products of photosynthesis. Rates of respiration are difficult to evaluate directly with this technique. In many situations the ^{14}C method yields a measure close to net photosynthetic rates. Comparison of the oxygen and ^{14}C methods under optimal conditions shows close agreement, with a photosynthetic quotient (i.e., the volume of oxygen released in photosynthesis as a proportion of the volume of CO_2 used in that process; $PQ = \Delta O_2 / -\Delta CO_2$, by volume) somewhat greater than unity (Fogg, 1963). The photosynthetic quotient varies from near unity when carbohydrates are the principal photosynthetic products, to as high as 3.0 during the synthesis of lipids. Assuming a photosynthetic quotient of 1.2 and a statistical significance level of $p < 0.05$, the smallest amount of photosynthesis that the oxygen light-and-dark technique can measure under ideal conditions is about 10 mg cm^{-3} h^{-1} (Strickland, 1960). The limit of sensitivity of the ^{14}C method is some three orders of magnitude lower, on the order of 0.01 mg cm^{-3} h^{-1} (Wetzel and Likens, 2000).

B. Secondary production

Secondary production in fresh waters by invertebrates and vertebrates is much more difficult to estimate accurately than is primary production. Trophic relationships are complex and often change during the life cycle of a species or from one ecosystem to another. The size of animals varies greatly from immature to adult stages, and the diet can also vary greatly throughout their life history. Because most animals are mobile, they actively distribute themselves in response to environmental stimuli, which results in heterogeneous distributions and makes accurate sampling more difficult.

Production measurements of animals are based on the estimation of numbers, biomass, and growth rates. The sampling accuracy of changes in population size decreases from zooplankton to benthic organisms to fish. In contrast, growth rates are easier to evaluate in many temperate fishes and in long-lived invertebrates than they are in small, rapidly reproducing zooplankton, other invertebrates, and protists.

When evaluating the population dynamics and production of any animal, estimating food incorporation or ingestion and the utilization of ingested food is imperative. *Assimilation* means the absorption of food from the digestive system, and the efficiency of assimilation refers to the percentage fraction of ingested food that is digested and absorbed into the body. Assimilation efficiency is not constant and varies greatly with the food quality and rates of food ingestion. Measurements of assimilation are based on the simple relation:

$$\text{Assimilation} = \text{ingestion} - \text{egestion}$$

or

$$\text{Assimilation} = \text{growth} + \text{respiration}$$

Although these relationships are simple, their accurate measurement among natural populations of zooplankton and larger organisms is problematic. The methods employed are discussed critically by Downing and Rigler (1984), Edmondson and Winberg (1971), and Wetzel and Likens (2000). Although ingestion can be measured *in situ* with reasonable sophistication, accurate *in situ* measurement of egestion rates is complicated by interindividual and temporal variation (Halvorson and Atkinson, 2019).

Around the turn of the millennium, these conventional *in situ* measurements were replaced by stable isotope analyses used to estimate time-integrated trophic positions and biomass flows to consumers (Post, 2002; Vander Zanden and Rasmussen, 1999). Elements, such as C, N, or O, can occur in different masses or isotopes in the environment due to chemical and physical processes. To exemplify, ^{12}C, ^{13}C, and ^{14}C are three isotopes of the element carbon. Stable isotopes, as opposed to radioactive isotopes such as ^{14}C, do not decay in other elements and are therefore used as tracers in food web studies. In aquatic systems frequently used tracers are ^{13}C and ^{15}N, which occur in nominal amounts in the aquatic environment. These stable isotope tracers are compared to the more abundant and lighter stable isotopes ^{12}C and ^{14}N, in stable isotope ratios, e.g., $\delta^{15}N$ and $\delta^{13}C$.

As consumers typically become enriched relative to their food sources in $\delta^{15}N$ by 3–4‰ (Deniro and Epstein, 1981; Minagawa and Wada, 1984; Peterson and Fry, 1987; Vander Zanden and Rasmussen, 1999), this stable nitrogen isotope ratio is often used as a time-integrated indicator of biomass flows in a food web. Stable isotope ratios of carbon ($\delta^{13}C$), on the other hand, are more conserved throughout the food web but do show distinct variation between the carbon/energy sources of the food web base of primary producers and are often used as an indicator of terrestrial vs. aquatic subsidies of aquatic food webs (Middelburg, 2014; Peterson and Fry, 1987; Post, 2002). This is possible because aquatic plants (algae, macrophytes) usually have different $\delta^{13}C$ signatures than terrestrial plants. Using $\delta^{15}N$ and $\delta^{13}C$ as stable isotope tracers also allows for direct measurements of consumer assimilation and ingestion in experimental settings (Verschoor et al., 2005).

C. *Ecosystem metabolism*

Ecosystem metabolism describes the turnover of biomass and energy. Measuring all the inputs and outputs of energy and biomass can be a very time-consuming way to calculate *net ecosystem production* (NEP). NEP is the difference between GPP (see above) and respiration. With the increased availability of high-frequency dissolved oxygen probes, the diel O_2 method has become widely used to determine the ecosystem metabolism of inland waters (Peeters et al., 2016; Tengberg et al., 2006). To facilitate comparison and use of different metabolism methods, Winslow et al. (2016) developed a software package that implements different models that estimate lake metabolism based on *in situ* high-frequency time series of dissolved oxygen and temperature. As with other methods that quantify aquatic metabolism, this method comes with its own set of limitations and assumptions (Staehr et al., 2010). For instance, an underlying assumption of the diel O_2 method is that respiration measured during nighttime resembles respiration during daylight, which may lead to a significant underestimation of respiration during daylight (Markager and Sand-Jensen, 1989). In addition, the uncertainties associated with this method decrease considerably when precise measurements of wind speed at the water surface are used as input, which is rarely done in practice (Staehr et al., 2010).

VIII. Aquatic food webs

In self-sustaining biological communities functionally similar organisms can be grouped into a series of operational levels of producers and consumers, together constituting a food web. The trophic structure of a community refers to the pathways by which energy is transferred and nutrients are cycled through the community trophic levels. The basis of the aquatic food web is formed by primary producers that are able to fix carbon by photosynthesis. Primary producers in aquatic food webs include phytoplanktonic algae and photosynthetic bacteria (Chapters 17, 18, and 23), benthic algae and photosynthetic bacteria (Chapter 25), and aquatic plants (Chapter 24).

Primary producers are eaten by primary consumers (herbivores or omnivores), which are in turn eaten by secondary consumers such as carnivores and omnivores. Consumers in aquatic food webs include zooplankton (Chapters 19 and 20), benthic animals (Chapter 21), fish (Chapter 22), terrestrial animals (amphibians, reptiles, birds, mammals), and estuarine/marine animals (fish, reptiles, mammals). Often neglected in studying aquatic food webs, parasitism is arguably the most commonly occurring consumer lifestyle (Frenken et al., 2017; Ibelings et al., 2004; Wommack and Colwell., 2000). Autochthonous plant material not consumed until after it dies joins the pool of detritus (which also includes allochthonous dead plant material, dead animals, and feces) consumed by omnivores and detritivores.

Omnivory is prevalent in ecosystems; few species feed entirely on either live plants, live animals, or detritus. In most ecological communities individuals of a species feed from several trophic levels (cf. review by Persson, 1999). Furthermore, many animals that feed on live or dead plant material are actually gaining much of their nutrition from insects or microbes living on the plant material. The use of stable isotopes (rare isotopes of carbon, nitrogen, hydrogen, sulfur, and oxygen), together with feeding observations and gut contents, has enabled ecologists to separate what an organism consumes from what it assimilates into its tissues. This has confirmed that omnivory is prevalent and that species differ in their capacity to change their diet in response to food availability or competition (Thompson et al., 2012). Indeed, it is now perhaps better to divide species into flexible or inflexible consumers than into herbivores, omnivores, detritivores, carnivores, or scavengers. Omnivory also occurs across life histories as a result of individuals using different resources during different parts of their life cycle. Many of these differences in feeding, particularly during the different life stages and sizes of a species, will be detailed in the discussions that follow in this work.

The amount of energy available for metabolism decreases with each increase in trophic level because energy is lost with each transformation. Since organisms expend a considerable amount of energy for maintenance, only a portion of the energy of one trophic level is available for transfer and use by higher trophic levels. Total biomass and numbers of individuals at each successively higher trophic level also usually decrease, although some exceptions will be discussed later.

In addition to flows of energy, elements such as phosphorus and nitrogen are also transferred through the food web. Ecological stoichiometry provides a compelling framework to understand the effect of the (im)balance of such chemical elements on ecological interactions (Sterner and Elser, 2002). Stoichiometric constraints may arise when, through human-induced increase in atmospheric nitrogen deposition (Elser et al., 2009) or climate warming (De Senerpont Domis et al., 2014), the changes in the chemical balance of the primary food-web base results in lower food quality for higher trophic levels.

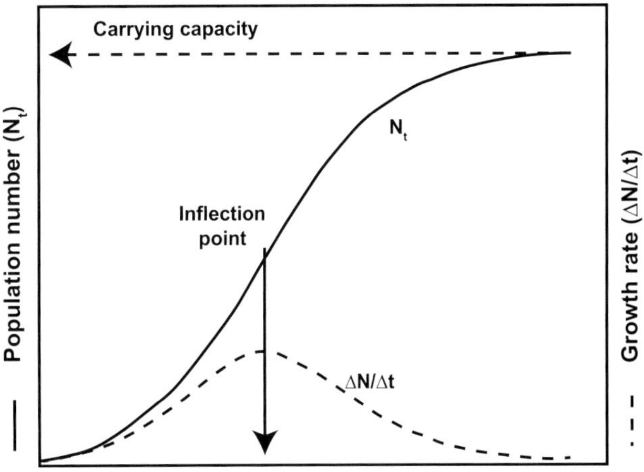

FIGURE 9-7 Common sigmoid growth curve of a population with time (N_t) as it approaches the environmental carrying capacity (K). The growth rate ($\Delta N/\Delta t$) denotes the difference in the number of organisms over each time interval.

IX. Population structure, growth, and regulation

A *population* comprises a subgroup of a single species that are interbreeding; ecologists use molecular genetics to define the spatial and temporal extent of aquatic populations. The number of individuals per unit area or volume of water, the *population density*, is an index of population size at any given time. Ecologists sometimes assume that the representatives of a species at a particular location are in a different population from those at other locations. This assumption may appear reasonable when comparing a number of lakes but is less acceptable when comparing sites in the same river catchment, which may be connected by flow. However, molecular genetics has shown clearly that any sample of invertebrate species from a single site will contain some species that do make up a single population at a site, and others where a single population covers a much wider area. For vertebrates (e.g., fish, water birds), a single population may occupy an area of hundreds of square kilometers.

The area that a population occupies is determined by three variables: the dispersal ability of the species (determines its capacity to reach new sites and interbreed with individuals there); the distribution of suitable habitats (including biotic and abiotic variations) within the surrounding area (determines where the species can establish populations); and the connectivity permitted by the landscape. For freshwater species, landscape *connectivity* may mean connections via surface flow or underground water movement, ribbons of fringing vegetation along rivers, or overland via flight (insects, birds; Figuerola and Green, 2002; Minchin, 2006). Unfortunately, human activity in landscapes (including climate change) has often altered both connectivity and the distribution of suitable habitats and, in doing so, has altered the distribution and population size of microbes, plants, and animals in freshwaters.

The growth of a population can be very rapid among smaller organisms, but restrictions imposed by the environment, predation, and (intra- and interspecific) competition quickly limit the size of populations, either by increasing the death rate or by decreasing the birth rate (or both). Populations have an *intrinsic growth rate*, that is, how quickly they can reproduce under ideal conditions, but mostly there is a range of limitations on population growth. *Density-dependent limitations* (i.e., limitations caused by population density) are imposed by competition with others in the population for available environmental resources (e.g., space, nutrients, light), by the production of waste products, and by behavioral factors (e.g., territories). As a population increases toward the *carrying capacity* (maximum number that can be supported by available resources), fewer resources are available for reproduction, so reproductive rates decline. This produces the *logistic growth curve*, which has a sigmoid shape (Fig. 9-7) and has been commonly observed among natural and experimental populations. However, it is important to note that species may "cheat" on these equations, which are a simplification of reality. They cheat by use of "trade-offs"; that is, they trade off body condition to achieve reproductive success. For example, a population of a species in a location where resources are scarce may reproduce at a smaller body size or younger age than under good conditions, and they may produce fewer or smaller offspring. Those offspring may start life at a disadvantage in size, numbers, or body condition (i.e., fat content, gonad size) but may be able to make up for that disadvantage if they experience better conditions for growth and development than their parents.

When sampling a population, it may be possible to identify a large group of individuals of the same, or nearly the same, age or size, referred to as a *cohort*. Cohorts may be followed through time, to estimate survivorship and understand a species' phenology. *Phenology* is the time at which important life cycle events (e.g., pollination, metamorphosis, breeding) occur. However, it can be very difficult to identify a cohort in a species that reproduces continuously (asynchronous reproduction), as do many populations of freshwater plants and animals. Also, many freshwater animals have complex life cycles where the juveniles rely on quite different resources and may even live in quite a different habitat to the adults. For example, many juvenile amphibians (i.e., tadpoles) rely on the presence and duration of water and aquatic food, but as adults, they obtain most of their resources from the terrestrial environment, only returning to water to breed. Modeling such complexity is challenging as the different life stages are exposed to different predators, environmental risks, and intra- and interspecific competition. For this reason, the use of structured population analyses such as life tables (De Roos et al., 2003) and individual-based modeling approaches (Mooij et al., 2019) have waned in popularity among freshwater scientists. Time series analyses have shown that the phenology of lake processes may be impacted by climate change (Winder and Schindler, 2004), which could interrupt the trophic transfer of energy when producers and consumers are affected differently (De Senerpont Domis et al., 2013; Winder and Schindler, 2004).

A. *Metapopulations*

Metapopulation theory acknowledges that individuals within species are not necessarily sedentary but may disperse among populations to interbreed. This interbreeding is essential to maintain genetic diversity within species. Essentially, a *metapopulation* is a group of patches occupied by subpopulations and connected by the dispersal of individuals among the patches, where interbreeding occurs. Most populations of freshwater species are likely to be part of a metapopulation. Freshwater ecosystems are dynamic and disturbances (such as flooding or drying) may extirpate a species from a water body. In a metapopulation individuals will arrive back at that patch and reestablish the species there, so metapopulations allow freshwater species to survive in spatially and temporally dynamic ecosystems. In a metapopulation patches of habitat may differ in their function: some patches may be sources, some may be sinks, and others may be stepping stones. That is, a species may be present in some patches (and recorded there by ecologists) even though it cannot complete its life cycle or breed in that patch. In *source patches* the birth rate exceeds the death rate and the emigration rate is therefore high. *Stepping-stone patches* enable individuals to move longer distances across landscapes by providing sufficient resources (perhaps rest or food) to enable species to survive there briefly. Stepping-stones patches have no birth rate, a low death rate, and high emigration and immigration rates. *Sink patches* also have no birth rate and they have a death rate equal to the number of immigrants. Sink patches (or ecological traps) may or may not provide resources to species, but they are insufficient to support successful reproduction or recruitment.

Understanding metapopulation dynamics is important for species conservation. For example, conservation reserves should aim to include source patches and be linked by stepping-stones, if needed. A conservation reserve that contains only sink patches for a species will not support that species for the future, even though the species has been recorded there. As climate change continues to alter water regimes, and therefore the availability and connectivity of freshwater habitat in landscapes, it will alter metapopulation dynamics and the relative frequency of source, stepping-stone, and sink patches in landscapes.

X. Communities

Although variously defined, a *community* usually refers to a group of interacting populations of different species (microbes, plants, and animals) within a defined spatial area. Some practical spatial boundaries are most often applied to studies of communities (e.g., planktonic communities) because habitats are diverse and often organisms move from one habitat to another. Communities are generally defined by a researcher interested in a particular group of species, usually occupying a defined environment. Sometimes, the term is used erroneously, such as when it is used to describe a single taxonomic group (e.g., trichopteran community). The term community should only be used where there are a variety of different taxa (e.g., rock-pool community). Where only one taxon is described, or one group of taxa, the term *assemblage* should be used (e.g., dragonfly assemblage, invertebrate assemblage).

Community ecologists focus on the patterns and processes in the living parts of ecosystems. They are interested in biodiversity and the processes that sustain it, the effects of disturbances on communities, and pathways for recovery from disturbances. They are interested in interactions between species, such as predation, competition, mutualisms, and parasitism. They also seek explanations for the patterns of multispecies distributions in different places and at different times. These questions are important for managing ecosystems to sustain biodiversity and the natural processes that support it, and to help ecosystems be resilient to environmental change. The rapid progress of global warming and other global changes (habitat destruction, urbanization, pollution, etc.) has increased efforts to understand community function, respond to negative impacts, and use restoration to try to sustain the Earth's remarkable biodiversity. Freshwaters are under exceptional pressure because of their importance to humans for water supply, power generation, transport and waste disposal, and their vulnerability to climate change while also occupying a disproportionally small area of the Earth. Ensuring healthy, functioning freshwater ecosystems for the future is the main aim of community ecology in freshwaters.

A. *Community assembly*

There is an enormous amount of literature covering the debate about how ecological communities are structured. This debate has changed over the decades, but there is still vigorous discussion. Here we provide only a beginner's guide to some of these ideas, noting that there will be further developments over the lifetime of this book.

Why do species occur where they do? How are communities assembled? These are two of the most enduring questions in ecology. A range of factors are known to determine assemblage composition and their relative importance varies from place to place and between times in a single location. These factors include environmental conditions (i.e., the inorganic environment, including disturbances), species traits (inherited characteristics of species), interactions between species (predation, competition, mutualisms, parasitism), dispersal (the intrinsic ability for individuals to move across landscapes), and connectivity (the permeability of landscapes for species with particular traits).

Density dependence, as mentioned above, was one of the first concepts regarding limits to populations. It includes both *threshold limitations*, such as the availability of territories or nest sites, and *continuous limitations* such as food supply, the effect of which intensifies as population size increases. One aspect of density-dependent controls on populations is that they may only be observable at particular times. Ecologists may observe only the after-effects of a density-dependent variable without being aware that it has occurred.

In another theory Southwood (1988) defined the *habitat template* as "the habitat provides the template on which evolution forges characteristic life history strategies." This theory is based on the premise that habitat conditions select for certain traits that reflect adaptions for survival and reproduction. This idea was most often applied by ecologists studying disturbances, which are natural events such as floods and droughts or anthropogenic events such as pesticide influx into freshwaters. However, unless an ecosystem is being disturbed very frequently, this theory does not represent all the dynamics we observe in freshwater ecosystems.

More recently, *complex systems theory* has been applied to freshwater ecosystems. Indeed, many of the studies that led to the development of this theory occurred in lakes (e.g., Carpenter and Brock, 2006). Complex systems display fluctuating nonlinear dynamics, so short-term phenomena may be hard to predict, but they are more predictable in the long term because they operate within particular boundaries. Indeed, because freshwater ecosystems provide so many ecosystem services (Chapter 2), they are understood to be embedded in complex socioecological systems (Schlüter and Pahl-Wostl, 2007). *Resilience* is the ability of a system to recover following disturbance (Folke et al., 2004). It is a key element of complex systems. Therefore managing resilience requires a thorough understanding of the functioning of both the social and the aquatic system, plus the interdependencies between them (Pelletier et al., 2020). The alternative stable ecosystem states observed in shallow lakes (i.e., the macrophyte-dominated state vs. the phytoplankton-dominated state) are an extreme form of nonlinear dynamics and will be elaborated on in Chapter 26.

Interactions between species, especially *predation* and *competition*, have long been studied by community ecologists. Predators are known to be able to control the densities of prey in some ecosystems, establishing a *trophic cascade* (see also Chapter 22). Trophic cascades occur when predation controls the density of prey who themselves are controlling the density of their prey, which is then released from predation pressure. Trophic cascades are known to occur in both rivers and lakes (Brett and Goldman, 1996). Often, the top predator is a fish that keeps the density of an algal-grazing prey species low. As a result, algal growth is not consumed and standing stocks of algae increase. Trophic cascades may have more than three levels and may occur only at particular times or places. For example, they may only occur once the population size of algal grazers is large enough to control algal biomass, or only on certain substratum types (e.g., Power et al., 2008). Interspecific competition occurs in freshwater ecosystems (e.g., between different plant types), but it is often highly temporally variable (e.g., among different phytoplankton species). Competition, predation, and other interactions between species (e.g., symbioses, commensalisms) are vulnerable to decoupling as species' phenologies are altered by climate change. There remains considerable uncertainty around what this will mean for ecosystems in the future, but the development of novel ecosystems (combinations of species not previously seen on Earth) is one possibility (Craig et al., 2017).

Many community ecologists focus on whether environmental filtering or niche partitioning control assemblage composition. The concept of *environmental filtering* is that species present at a site will be those that have successfully passed through the filter of environmental conditions at that site, as determined by their species traits (including their tolerances for environmental conditions and disturbance) (Lake et al., 2007). Only once species have arrived at the site can local factors influence community composition. This theory has been criticized for ignoring the role of species dispersal capacity and landscape connectivity in determining which species can arrive at a site. Freshwater species face greater dispersal challenges than do marine or terrestrial species because freshwater bodies are often isolated from each other by terrestrial landscapes. Human actions often increase this isolation by removing habitat (draining or drying out lakes and rivers) or creating dispersal barriers (e.g., weirs). Also, the restoration of high-quality habitat has shown that dispersal limitation influences community composition. Both dispersal and environmental filtering may influence community composition, but their relative roles at any particular waterbody or ecosystem type will vary.

Where environmental filtering exerts strong pressure on communities, the convergence of species traits (i.e., multiple species with similar traits) may occur. For example, microbes in hot springs are limited to *extremophiles* that tolerate (or need) very high water temperatures. Those types of communities tend to be relatively low in diversity. Where trait convergence occurs, species may be regarded as occupying similar "niches." The concept of the *niche* has been criticized and defined in various ways and there is insufficient space to encompass that debate here (but see McInerny and Etienne, 2012). For our purposes, the niche can be defined as the suite of environmental requirements and tolerances of a species. *Niche partitioning* describes how species with similar niches divide up habitat in space and time to coexist. Over evolutionary timescales, competition should create selection pressure for species to differentiate, termed *niche differentiation*. This is a fundamental way in which evolution creates biodiversity. However, competition between species must affect sexual selection, or at least select individuals prior to reproduction, for evolution to cause differentiation. Studies in freshwater show that both niche partitioning and environmental filtering occur. The pressure exerted by environmental filters varies in space and time, and this means that the pressure on similar species to diverge will also vary, potentially enabling similar species to coexist much of the time, or to coexist in different habitat patches within an ecosystem.

How the freshwater biota form communities with particular structure and composition is still the subject of research. However, the variables important to different parts of the biota are now well understood for many groups and are described in subsequent chapters. Overall, it is important to remember the main factors that structure communities: environmental conditions, species traits, biotic interactions, and landscape connectivity. Of these, the effects of global warming may alter environmental conditions and landscape connectivity, influencing biotic interactions. Importantly, how species respond to these changes will be determined by whether their species traits enable persistence, either *in situ* or through dispersal to cooler locations (Parmesan, 2006).

B. *Metacommunity dynamics*

Metacommunities (Leibold et al., 2004) are communities connected by the dispersal of individuals among them, analogous to *metapopulations*. They are often conceptualized as a cluster of lakes, each of which provides a range of environmental conditions and community composition, a subset of which supports metapopulations of multiple species. The dynamics of metacommunities are therefore reliant on dispersal and connectivity among the subcommunities in the metapopulations. Most lakes, rivers, and river reaches are now regarded as being part of a metacommunity, connected by dispersal. Restoration of one lake within a metacommunity will see recolonization of the restored lake by species capable of dispersal among lakes, provided there are no barriers to that dispersal. Similarly, a river reach containing a series of riffles, pools, and runs will be connected by dispersal to other reaches. Species that prefer minimal flow will form metapopulations among the pools within their dispersal range and together these metapopulations form a metacommunity. *Rheophilic species* that live in riffles will also form metapopulations and a metacommunity. This understanding of community dynamics across larger spatial scales has changed the way freshwater ecosystems are managed.

It is now regarded as essential that lakes are managed in groups rather than as individual waterbodies and that connectivity along rivers (often disrupted by dams and other human impacts) is important to maintaining ecosystem function and river biodiversity.

C. The metaecosystem concept

The *metaecosystem concept* recognizes that ecosystems are connected by flows of organisms, energy, and materials across ecosystem boundaries (Loreau et al., 2003). It progresses beyond the RCC, because it views riverscapes as networks of rivers connected by flows of genes, species, materials, and energy in multiple directions. It thus moves away from the predominantly longitudinal (downstream) flows encapsulated by the RCC. Lentic systems such as lakes and ponds are also part of this metaecosystem, because they are connected through rivers and groundwater flows but also through animal movement or passive, wind-driven dispersal (Gounand et al., 2018).

XI. Diversity and diversity metrics

Diversity may be simply described as the number of species living in a defined area, which can be anything from a single site or water body or scaled upward to the size of a continent. When diversity is determined at the habitat level, it is referred to as *alpha diversity*, whereas diversity at the landscape or regional level is referred to as *gamma diversity* (Whittaker, 1972). *Species turnover* between different habitat patches within the landscape is defined by Whittaker (1972) as *beta diversity*. This is an intuitively easy concept, but in practice, measuring diversity can be difficult. Many species of microbes, algae, and invertebrates have not been described or named and thus may be difficult to identify and quantify. Measures of diversity rely on the species concept, but this itself can be difficult to apply to microbes.

There are many ways to quantify diversity (see Magurran, 2013). *Species richness* is the count of the number of species in a defined area (or volume or amount of sampling effort) and is often used to quantify diversity. If similar sampling methods are used, species richness can be easily compared between locations. As species richness estimates are heavily influenced by sampling effort, species saturation or accumulation curves are often used to assess whether a sufficient volume or area has been sampled to provide a representative estimate of species richness. Only when the curve reaches an asymptote can the sample be considered sufficient for reliable quantification of richness. These species accumulation curves also allow researchers to compare species richness across different populations.

However, anyone who has studied an ecological community will soon see that some species are common while others are rare, and others lie somewhere in between. Species therefore differ in their *abundance* (total number in a defined area or volume) or *density* (number per unit area or volume). *Relative abundance* is often used to compare species; it is defined as the abundance of each species collected using the same sampling method. Relative abundance can be very useful because it facilitates comparisons, whereas absolute abundances are difficult to estimate for most species and are often less useful. Species evenness considers the number of species and the relative abundance in a certain area and is often expressed as the Shannon(−Wiener) index or the Simpson index (Borda-de-Agua et al., 2002). Therefore, although we might measure species richness, the species counted will occur at a wide range of different abundances. What does this mean and how does it affect how we quantify diversity?

There are two well-known relationships that affect species richness and therefore affect how diversity is quantified: the species−area relationship and the species−abundance relationship. As the area (or volume) of habitat increases, the number of species encountered (species richness) will increase; this is the *species−area relationship*. For this reason, when comparing different places, researchers endeavor to use the same size and number of samples so that the probability of encountering species is the same in each place and thus species richness can be compared. Similarly, as the total number of individuals (of plants or animals) collected increases, the probability of encountering species will increase; this is the *species−abundance* relationship. Where two places are compared and one has higher abundance, it will likely also have more species. Thus, when comparing species richness between places, researchers often also measure abundance. For both these relationships, *rarefaction* can be used to estimate species richness either at equal numbers of samples or equal abundances of individuals. Rarefaction is a statistical technique that randomly resamples species from a dataset to show the relationship between the average number of species found in each of these random samples and the number of samples (or individuals). Rarefaction curves usually show a steep increase at first, when every new sample results in a large number of new species found, and then reaches an asymptote when additional samples (or individuals) only result in the identification of a few additional rare species. Rarefied estimates therefore provide more comparable measures of species richness as they take sampling effort (or abundance) into account.

Differing abundances of individuals in different places may be a simple product of the amount of energy and nutrients available in that ecosystem. For example, lakes that receive excess nitrogen and phosphorus from agricultural activities in their catchments often have high algal productivity and density. This provides a large food resource for animals, which can also reach very high densities. However, in such situations diversity often declines because environmental conditions are only suitable for a narrow range of species. Where the ecosystems to be compared have apparently similar levels of productivity (e.g., comparing a group of headwater streams), and the same sampling effort is used to collect data, simple species richness (i.e., presence/absence data) is often readily comparable.

Species richness provides information on the number of taxa present at a time and place. Where those taxa are based on reliable phylogenies (*phylogeny* describes the evolutionary relationships among taxa), it can also be used to measure *phylogenetic richness*. With modern molecular genetics methods, phylogenies can be constructed through genetic analyses of the species collected. This may explain why some locations have a greater number of species within some taxa than others. For example, comparisons of taxonomic richness made at large spatial scales (e.g., continents) will in part have been determined by the evolutionary history of those lands (e.g., their history of glaciation or aridity). Phylogenetic diversity can distinguish between historical contingency and contemporary differences in richness that can tell us about factors currently affecting diversity.

Molecular genetics may also provide a better description of species richness than taxonomic studies if there are many cryptic species present. *Cryptic species* are "hidden" within populations, as they belong to two or more species that appear morphologically identical but are genetically distinct. Differences between species' physical appearance can be very subtle, yet animals and plants may perceive these differences such that they only breed with genetically compatible individuals. Cryptic species are often identified among aquatic plants and animals, sometimes in surprising numbers, and morphological, physical, or behavioral differences between those species are often subsequently observed.

One recent way to measure richness is to use *environmental DNA (eDNA)* (Valentini et al., 2016). Each animal, plant, or microbe living in freshwaters will shed DNA into that water and eDNA methods then sequence those genetic materials. A list of gene sequences is then compared with libraries such as GenBank so that names can be put against each sequence. In much of the world where species are poorly known, eDNA provides long lists of species against which it is impossible to put species names. In better-studied parts of the world eDNA can provide an accurate list of species present in the waterbody at the time of sampling, including cryptic species and those that are secretive and hard to sample. eDNA is therefore an increasingly popular way to measure diversity in freshwaters.

Another way in which ecologists often measure diversity is through *functional* or *trait diversity. Functional groups* are groupings of species that fulfill similar functions in ecosystems. *Trophic groups*, such as predators and grazers or carnivores, herbivores, and omnivores, are examples of functional groups. Food webs and ecosystems may differ in *functional diversity* (i.e., the number of functional groups that they contain). *Species traits* are heritable characteristics that determine how species interact with their environment and each other. They include variables such as body size, method of respiration, and dispersal method. Species may be assigned to different combinations of traits within a matrix and then differences observed when ecosystems are compared. For example, shallow, highly vegetated lakes may show marked diel fluctuations in dissolved oxygen due to high levels of photosynthesis during the day and equally high respiration at night. Aquatic animals present in those lakes may be predominantly air-breathers (e.g., turtles, frogs, some aquatic snails, and insects), for whom low dissolved oxygen is irrelevant, and gill-breathing species (e.g., crustaceans) may be quite rare. Thus the composition of species traits present within an ecosystem can often be related to environmental conditions or the impacts of disturbances to those conditions.

XII. Summary

1. An *ecosystem* comprises all the life forms (microbes, plants, animals) and the inorganic context (sediment types, water regimes, water quality, etc.) that they live in.
2. A *catchment* is the area of land surrounding a waterbody, through and over which water flows into that waterbody. The size and geological composition of catchments thus influences freshwater ecosystems and catchment geomorphology influences the rates and volumes of water (and materials carried by that flow) entering waterways.
3. Human effects on global climate and element cycling are now so profound that we live in a potentially new era, the Anthropocene. These human impacts, especially changes in climate and land use, have reduced the influence of catchment geology on aquatic ecosystems. Instead, water regime is now understood to be one of the most important factors affecting aquatic ecosystems. Water regimes are being altered all around the world by the impacts of climate change.
4. In flowing (*lotic*) waters we refer to the water regime as the flow regime. This is because the unidirectional

nature of flow in rivers and streams distinguishes them from other ecosystems.

5. In *lentic* or standing waters we often refer to the water regime as the water balance. The water balance of a lake is determined by the surface and subsurface inflows and outflows, precipitation, and evaporation.
6. A lake is separated into the open-water *pelagic* zone and the *littoral* zone, the latter consisting of the bottom of the lake basin colonized by rooted macrophytes (aquatic plants). The sediments free of vegetation that lie below the pelagic zone are referred to as the *profundal* zone (Fig. 9-3). The *littoriprofundal* zone is the transitional area of the sediments between the littoral and profundal zones occupied by scattered benthic algae or a microbial mat.
7. In rivers the *riparian zone* refers to the riverbanks and their vegetation, which can tolerate periods of inundation during floods. Rivers are divided into a nested hierarchy of structural forms. River *segments* are lengths divided by tributary junctions, and nested within these are *reaches* that are divided by structures such as waterfalls, debris dams, or large pools. Nested within reaches are habitats such as *pools, runs*, and *riffles*, each of which is defined by relative depth and flow speed.
8. As water moves through the catchment and fringing zone and into a waterbody, the types of nutrients and OM in the water, and their concentrations, are modified.
9. *Autochthonous production* means carbon fixed within a waterbody, by microbes, algae, and aquatic plants. *Allochthonous productivity* arrives in the water body through terrestrial subsidy from terrestrial or fringing plants. In most rivers and lakes both forms of subsidy occur, with the relative importance depending on the morphology of the water body, the geology of the catchment, and the abundance of riparian (rivers) or fringing (lakes) vegetation.
10. In order to evaluate the *productivity* of a system, it is important to realize that production is a rate, and often expressed as a unit of mass or energy production per volume (or surface area) per time. Whereas *gross primary production* (*GPP*) is defined as the total rate at which autotrophs produce biomass or energy, *net primary production* is defined as the rate at which autotrophs produce biomass or energy, taking into account the energy or biomass used for respiration.
11. In self-sustaining biological communities functionally similar organisms can be grouped into a series of operational levels of producers and consumers, together constituting a food web. Primary producers (algae, photosynthetic bacteria, aquatic plants) are able to fix carbon by photosynthesis and form the basis of the food web. These are consumed by herbivores and omnivores (and by detritivores, when plant material dies). Carnivores, scavengers, and parasites feed on other consumers, and mutualisms between species (e.g., symbiosis, commensalism) also occur. Omnivory is the most common diet in freshwaters, especially if changes in diet are viewed across the life cycle of a species.
12. Since organisms expend a considerable amount of energy for maintenance, only a portion of the energy at one trophic level is available for transfer and use by higher trophic levels. Thus the amount of energy available for metabolism decreases with each increase in trophic level because energy is lost with each transformation.
13. A *population* comprises a subgroup of a single species that are interbreeding; ecologists use molecular genetics to define the spatial and temporal extent of aquatic populations. The number of individuals per unit area or volume of water, the *population density*, is an index of population size at any given time.
14. The area that a population occupies is determined by three variables: the dispersal ability of the species (determines its capacity to reach new sites and interbreed with individuals there), the distribution of suitable habitats (including biotic and abiotic variations) within the surrounding area (determines where the species can establish populations), and the connectivity permitted by the landscape.
15. *Phenology* is the time at which important life cycle events (e.g., pollination, metamorphosis, breeding) occur for a particular species.
16. A *metapopulation* is a group of patches occupied by subpopulations and connected by the dispersal of individuals among the patches, where interbreeding occurs. This interbreeding is essential to maintain genetic diversity within species.
17. A community refers to a group of interacting populations of different species (microbes, plants, and animals) within a defined spatial area.
18. How the freshwater biota form communities with particular structure and composition is still the subject of research. Overall, it is important to remember that the main factors that structure communities are environmental conditions, species traits, biotic interactions, and landscape connectivity.
19. *Metacommunities* are communities connected by the dispersal of individuals among them, analogous to metapopulations. The dynamics of metacommunities are therefore reliant on dispersal and connectivity among the subcommunities in the metapopulations.

20. The *metaecosystem concept* recognizes that ecosystems are connected by flows of organisms, energy, and materials across ecosystem boundaries. It views riverscapes as networks of rivers connected by flows of genes, species, materials, and energy in multiple directions. Lentic systems are part of this metaecosystem, because they are connected through rivers, groundwater flows, and animal movement.
21. *Diversity* may be simply described as the number of species living in a defined area, which can be anything from a single site or waterbody up to a continent. Another way in which ecologists often measure diversity is through functional, phylogenetic, or trait diversity. Functional groups are groupings of species that fulfill similar functions in ecosystems.

Acknowledgments

We want to thank Ellen van Donk and Sven Teurlincx (NIOO-KNAW) for their constructive comments on earlier drafts of this chapter. In addition, we are grateful to Miquel Lürling for providing the picture displayed in Fig. 9-6.

References

Adrian, R., O'Reilly, C.M., Zagarese, H., Baines, S.B., Hessen, D.O., Keller, W., Livingstone, D.M., Sommaruga, R., Straile, D., Van Donk, E., Weyhenmeyer, G.A., Winder, M., 2009. Lakes as sentinels of climate change. Limnol. Oceanogr. 54, 2283–2297. https://doi.org/10.4319/lo.2009.54.6_part_2.2283.

Borda-de-Agua, L., Hubbell, S.P., McAllister, M., 2002. Species-area curves, diversity indices, and species abundance distributions: a multifractal analysis. Am. Nat. 159, 138–155.

Boulton, A.J., Brock, M.A., Robson, B.J., Ryder, D.S., Chambers, J.M., Davis, J.A., 2014. Australian Freshwater Ecology: Processes and Management, second ed. Wiley-Blackwell, Melbourne.

Bowes, R.E., Thorp, J.H., Delong, M.D., 2020. Reweaving river food webs through time. Freshw. Biol. 65, 390–402. https://doi.org/10.1111/fwb.13432.

Brett, M.T., Bunn, S.E., Chandra, S., Galloway, A.W.E., Guo, F., Kainz, M.J., Kankaala, P., Lau, D.C.P., Moulton, T.P., Power, M.E., Rasmussen, J.B., Taipale, S.J., Thorp, J.H., Wehr, J.D., 2017. How important are terrestrial organic carbon inputs for secondary production in freshwater ecosystems? Freshw. Biol. 62, 833–853. https://doi.org/10.1111/fwb.12909.

Brett, M.T., Goldman, C.R., 1996. A meta-analysis of the freshwater trophic cascade. Proc. Natl. Acad. Sci. USA 93, 7723–7726.

Bunn, S.E., Davies, P.M., Winning, M., 2003. Sources of organic carbon supporting the food web of an arid zone floodplain river. Freshw. Biol. 48, 619–635. https://doi.org/10.1046/j.1365-2427.2003.01031.x.

Butman, D., Raymond, P., 2011. Significant efflux of carbon dioxide from streams and rivers in the United States. Nat. Geosci. 4, 839–842. https://doi.org/10.1038/ngeo1294.

Carey, N., Chester, E.T., Robson, B.J., 2021. Life-history traits are poor predictors of species responses to flow regime change in headwater streams. Global Change Biol. 27, 3547–3564. https://doi.org/10.1111/gcb.15673.

Carpenter, S.R., Brock, W.A., 2006. Rising variance: a leading indicator of ecological transition. Ecol. Lett. 9, 308–315.

Crabot, J., Polášek, M., Launay, B., Pařil, P., Datry, T., 2021. Drying in newly intermittent rivers leads to higher variability of invertebrate communities. Freshw. Biol. 66, 730–744. https://doi.org/10.1111/fwb.13673.

Craig, L.S., Olden, J.D., Arthington, A.H., Entrekin, S., Hawkins, C.P., Kelly, J.J., Kennedy, T.A., Maitland, B.M., Rosi, E.J., Roy, A.H., Strayer, D.L., Tank, J.L., West, A.O., Wooten, M.S., 2017. Meeting the challenge of interacting threats in freshwater ecosystems: a call to scientists and managers. Elementa: Science of the Anthropocene 5. https://doi.org/10.1525/elementa.256.

Davis, J.A., Barmuta, L.A., 1989. An ecologically useful classification of mean and near-bed flows in streams and rivers. Freshw. Biol. 21, 271–282. https://doi.org/10.1111/j.1365-2427.1989.tb01365.x.

De Roos, A.M., Persson, L., McCauley, E., 2003. The influence of size-dependent life-history traits on the structure and dynamics of populations and communities. Ecol. Lett. 6, 473–487. https://doi.org/10.1046/j.1461-0248.2003.00458.x.

De Senerpont Domis, L.N., Elser, J.J., Gsell, A.S., Huszar, V.L.M., Ibelings, B.W., Jeppesen, E., Kosten, S., Mooij, W.M., Roland, F., Sommer, U., van Donk, E., Winder, M., Lürling, M., 2013. Plankton dynamics under different climate conditions in tropical freshwater systems (a reply to the comment by Sarmento, Amado & Descy, 2013). Freshw. Biol. 58, 2211–2213. https://doi.org/10.1111/fwb.12203.

De Senerpont Domis, L.N., Van de Waal, D.B., Helmsing, N.R., Van Donk, E., Mooij, W.M., 2014. Community stoichiometry in a changing world: combined effects of warming and eutrophication on phytoplankton dynamics. Ecology 95, 1485–1495.

Deniro, M.J., Epstein, S., 1981. Influence of diet on the distribution of nitrogen isotopes in animals. Geochem. Cosmochim. Acta 45, 341–351. https://doi.org/10.1016/0016-7037(81)90244-1.

Downing, J., 2008. Emerging global role of small lakes and ponds: little things mean a lot. Limnética 29, 9–24.

Downing, J.A., Rigler, F.H., 1984. A Manual on Methods for the Assessment of Secondary Productivity in Fresh Waters. IBP Handbook No. 17, second ed. Blackwell Scientific Publications, p. 500.

Dudgeon, D., 2019. Multiple threats imperil freshwater biodiversity in the Anthropocene. Curr. Biol. 29, R960–R967. https://doi.org/10.1016/j.cub.2019.08.002.

Edmondson, W.T., Winberg, G.G., 1971. A Manual on Methods for the Assessment of Secondary Productivity in Fresh Waters. IBP Handbook No. 17. F.A. Davis, Philadelphia. Blackwell Scientific Publications, Oxford and Edinburgh xxiv + 358 p.

Elser, J.J., Andersen, T., Baron, J.S., Bergström, A.-K., Jansson, M., Kyle, M., Nydick, K.R., Steger, L., Hessen, D.O., 2009. Shifts in lake N:P stoichiometry and nutrient limitation driven by atmospheric nitrogen deposition. Science 326, 835–837. https://doi.org/10.1126/science.1176199.

Figuerola, J., Green, A.J., 2002. Dispersal of aquatic organisms by waterbirds: a review of past research and priorities for future studies. Freshw. Biol. 47, 483–494. https://doi.org/10.1046/j.1365-2427.2002.00829.x.

Fogg, G.E., 1963. The role of algae in organic production in aquatic environments. Br. Phycol. Bull. 2, 195–205. https://doi.org/10.1080/00071616300650021.

Folke, C., Carpenter, S., Walker, B., Scheffer, M., Elmqvist, T., Gunderson, L., Holling, C.S., 2004. Regime shifts, resilience, and biodiversity in ecosystem management. Annual Review of Ecology, Evolution, and Systematics 35, 557–581. https://doi.org/10.1146/annurev.ecolsys.35.021103.105711.

Frenken, T., Alacid, E., Berger, S.A., Bourne, E.C., Gerphagnon, M., Grossart, H.-P., Gsell, A.S., Ibelings, B.W., Kagami, M., Küpper, F.C., Letcher, P.M., Loyau, A., Miki, T., Nejstgaard, J.C., Rasconi, S., Reñé, A., Rohrlack, T., Rojas-Jimenez, K.,

Schmeller, D.S., Scholz, B., Seto, K., Sime-Ngando, T., Sukenik, A., Waal, D.B.V. de, Wyngaert, S.V. den, Donk, E.V., Wolinska, J., Wurzbacher, C., Agha, R., 2017. Integrating chytrid fungal parasites into plankton ecology: research gaps and needs. Environ. Microbiol. 19, 3802–3822. https://doi.org/10.1111/1462-2920.13827.

Frissell, C.A., Liss, W.J., Warren, C.E., Hurley, M.D., 1986. A hierarchical framework for stream habitat classification: viewing streams in a watershed context. Environ. Manag. 10, 199–214. https://doi.org/10.1007/BF01867358.

Giller, P.S., Malmqvist, B., 1998. The Biology of Streams and Rivers. Oxford University Press, Oxford.

Gounand, I., Harvey, E., Little, C.J., Altermatt, F., 2018. Meta-ecosystems 2.0: rooting the theory into the field. Trends Ecol. Evol. 33, 36–46. https://doi.org/10.1016/j.tree.2017.10.006.

Halvorson, H.M., Atkinson, C.L., 2019. Egestion versus excretion: a meta-analysis examining nutrient release rates and ratios across freshwater fauna. Diversity 11, 189. https://doi.org/10.3390/d11100189.

Hildrew, A.G., 2018. Freshwater Acidification: Natural History, Ecology and Environmental Policy. Volume 27 of Excellence in Ecology. International Ecology Institute, Oldendorf/Luhe, Germany, p. 194.

Hynes, H.B.N., 2001. The Ecology of Running Water. The Blackburn Press, p. 555.

Ibelings, B.W., De Bruin, A., Kagami, M., Rijkeboer, M., Brehm, M., van Donk, E., 2004. Host parasite interactions between freshwater phytoplankton and chytrid fungi (Chytridiomycota). J. Phycol. 40, 437–453. https://doi.org/10.1111/j.1529-8817.2004.03117.x.

Kayler, Z.E., Premke, K., Gessler, A., Gessner, M.O., Griebler, C., Hilt, S., Klemedtsson, L., Kuzyakov, Y., Reichstein, M., Siemens, J., Totsche, K.-U., Tranvik, L., Wagner, A., Weitere, M., Grossart, H.-P., 2019. Integrating aquatic and terrestrial perspectives to improve insights into organic matter cycling at the landscape scale. Front. Earth Sci. 7. https://doi.org/10.3389/feart.2019.00127.

Keith, D.A., Ferrer-Paris, J.R., Nicholson, E., Kingsford, R.T. (Eds.), 2020. IUCN Global Ecosystem Typology 2.0: Descriptive profiles for biomes and ecosystem functional groups. IUCN, Gland, Switzerland, p. 192.

Lake, P.S., Bond, N., Reich, P., 2007. Linking ecological theory with stream restoration. Freshw. Biol. 52, 597–615. https://doi.org/10.1111/j.1365-2427.2006.01709.x.

Leibold, M.A., Holyoak, M., Mouquet, N., Amarasekare, P., Chase, J.M., Hoopes, M.F., Holt, R.D., Shurin, J.B., Law, R., Tilman, D., Loreau, M., Gonzalez, A., 2004. The metacommunity concept: a framework for multi-scale community ecology. Ecol. Lett. 7, 601–613.

Loreau, M., Mouquet, N., Holt, R.D., 2003. Meta-ecosystems: a theoretical framework for a spatial ecosystem ecology. Ecol. Lett. 6, 673–679. https://doi.org/10.1046/j.1461-0248.2003.00483.x.

Magurran, A.E., 2013. Measuring Biological Diversity. Wiley-Blackwell, p. 272.

Marín-Spiotta, E., Gruley, K.E., Crawford, J., Atkinson, E.E., Miesel, J.R., Greene, S., Cardona-Correa, C., Spencer, R.G.M., 2014. Paradigm shifts in soil organic matter research affect interpretations of aquatic carbon cycling: transcending disciplinary and ecosystem boundaries. Biogeochemistry 117, 279–297. https://doi.org/10.1007/s10533-013-9949-7.

Markager, S., Sand-Jensen, K., 1989. Patterns of night-time respiration in a dense phytoplankton community under a natural light regime. J. Ecol. 77, 49–61. https://doi.org/10.2307/2260915.

Masclaux, H., Perga, M.-E., Kagami, M., Desvilettes, C., Bourdier, G., Bec, A., 2013. How pollen organic matter enters freshwater food webs. Limnol. Oceanogr. 58, 1185–1195. https://doi.org/10.4319/lo.2013.58.4.1185.

McInerny, G.J., Etienne, R.S., 2012. Ditch the niche—is the niche a useful concept in ecology or species distribution modelling? J. Biogeogr. 39, 2096–2102. https://doi.org/10.1111/jbi.12033.

Middelburg, J.J., 2014. Stable isotopes dissect aquatic food webs from the top to the bottom. Biogeosciences 11, 2357–2371. https://doi.org/10.5194/bg-11-2357-2014.

Minagawa, M., Wada, E., 1984. Stepwise enrichment of 15N along food chains: further evidence and the relation between δ15N and animal age. Geochem. Cosmochim. Acta 48, 1135–1140. https://doi.org/10.1016/0016-7037(84)90204-7.

Minchin, D., 2006. The transport and the spread of living aquatic species. In: Davenport, J., Davenport, J.L. (Eds.), The Ecology of Transportation: Managing Mobility for the Environment, Environmental Pollution. Springer Netherlands, Dordrecht, pp. 77–97. https://doi.org/10.1007/1-4020-4504-2_5.

Mooij, W.M., van Wijk, D., Beusen, A.H., Brederveld, R.J., Chang, M., Cobben, M.M., DeAngelis, D.L., Downing, A.S., Green, P., Gsell, A.S., Huttunen, I., Janse, J.H., Janssen, A.B., Hengeveld, G.M., Kong, X., Kramer, L., Kuiper, J.J., Langan, S.J., Nolet, B.A., Nuijten, R.J., Strokal, M., Troost, T.A., van Dam, A.A., Teurlincx, S., 2019. Modeling water quality in the Anthropocene: directions for the next-generation aquatic ecosystem models. Current Opinion in Environmental Sustainability, Environmental Change Assessment 36, 85–95. https://doi.org/10.1016/j.cosust.2018.10.012.

Parmesan, C., 2006. Ecological and evolutionary responses to recent climate change. Annu. Rev. Ecol. Evol. Systemat. 37, 637–669.

Peeters, F., Atamanchuk, D., Tengberg, A., Encinas-Fernández, J., Hofmann, H., 2016. Lake metabolism: comparison of lake metabolic rates estimated from a Diel CO_2- and the Common Diel O_2-Technique. PLoS One 11, e0168393. https://doi.org/10.1371/journal.pone.0168393.

Pelletier, M.C., Ebersole, J., Mulvaney, K., Rashleigh, B., Gutierrez, M.N., Chintala, M., Kuhn, A., Molina, M., Bagley, M., Lane, C., 2020. Resilience of aquatic systems: review and management implications. Aquatic Science 82, 1–44. https://doi.org/10.1007/s00027-020-00717-z.

Persson, L., 1999. Trophic cascades: abiding heterogeneity and the trophic level concept at the end of the road. Oikos 85, 385–397. https://doi.org/10.2307/3546688.

Peterson, B.J., Fry, B., 1987. Stable isotopes in ecosystem studies. Annu. Rev. Ecol. Systemat. 18, 293–320. https://doi.org/10.1146/annurev.es.18.110187.001453.

Piano, E., Doretto, A., Mammola, S., Falasco, E., Fenoglio, S., Bona, F., 2020. Taxonomic and functional homogenisation of macroinvertebrate communities in recently intermittent Alpine watercourses. Freshw. Biol. 65, 2096–2107. https://doi.org/10.1111/fwb.13605.

Polis, G.A., Anderson, W.B., Holt, R.D., 1997. Toward an integration of landscape and food web ecology: the dynamics of spatially subsidized food webs. Annu. Rev. Ecol. Systemat. 28, 289–316. https://doi.org/10.1146/annurev.ecolsys.28.1.289.

Post, D.M., 2002. Using stable isotopes to estimate trophic position: models, methods, and assumptions. Ecology 83, 703–718. https://doi.org/10.1890/0012-9658(2002)083[0703:USITET]2.0.CO.

Power, M.E., Parker, M.S., Dietrich, W.E., 2008. Seasonal reassembly of a river food web: floods, droughts, and impacts of fish. Ecol. Monogr. 78, 263–282. https://doi.org/10.1890/06-0902.1.

Rockström, J., Falkenmark, M., Allan, T., Folke, C., Gordon, L., Jägerskog, A., Kummu, M., Lannerstad, M., Meybeck, M., Molden, D., Postel, S., Savenije, H.H.G., Svedin, U., Turton, A., Varis, O., 2014. The unfolding water drama in the Anthropocene: towards a resilience-based perspective on water for global sustainability. Ecohydrology 7, 1249–1261. https://doi.org/10.1002/eco.1562.

Rouse, W.R., Douglas, M.S.V., Hecky, R.E., Hershey, A.E., Kling, G.W., Lesack, L., Marsh, P., Mcdonald, M., Nicholson, B.J., Roulet, N.T.,

Smol, J.P., 1997. Effects of climate change on the freshwaters of arctic and subarctic North America. Hydrol. Process. 11, 873–902. https://doi.org/10.1002/(SICI)1099-1085(19970630)11:8<873::AID-HYP510>3.0.CO;2-6.

Schindler, D.W., 2006. Recent advances in the understanding and management of eutrophication. Limnol. Oceanogr. 51, 356–363. https://doi.org/10.4319/lo.2006.51.1_part_2.0356.

Schlüter, M., Pahl-Wostl, C., 2007. Mechanisms of resilience in common-pool resource management systems: an agent-based model of water use in a river basin. Ecol. Soc. 12. https://doi.org/10.5751/ES-02069-120204.

Southwood, T.R.E., 1988. Tactics, strategies and templets. Oikos 52, 3–18. https://doi.org/10.2307/3565974.

Srivastava, D.S., Céréghino, R., Trzcinski, M.K., MacDonald, A.A.M., Marino, N.A.C., Mercado, D.A., Leroy, C., Corbara, B., Romero, G.Q., Farjalla, V.F., Barberis, I.M., Dézerald, O., Hammill, E., Atwood, T.B., Piccoli, G.C.O., Ospina-Bautista, F., Carrias, J.-F., Leal, J.S., Montero, G., Antiqueira, P.A.P., Freire, R., Realpe, E., Amundrud, S.L., de Omena, P.M., Campos, A.B.A., 2020. Ecological response to altered rainfall differs across the Neotropics. Ecology 101. https://doi.org/10.1002/ecy.2984.

Staehr, P.A., Bade, D., Bogert, M.C.V. de, Koch, G.R., Williamson, C., Hanson, P., Cole, J.J., Kratz, T., 2010. Lake metabolism and the diel oxygen technique: state of the science. Limnol Oceanogr. Methods 8, 628–644. https://doi.org/10.4319/lom.2010.8.0628.

Sterner, R.W., Elser, J.J., 2002. Ecological Stoichiometry: The Biology of Elements from Molecules to the Biosphere. Princeton University Press, p. 464.

Stockwell, J.D., Doubek, J.P., Adrian, R., Anneville, O., Carey, C.C., Carvalho, L., Senerpont, L.N.D., Gaël, D., Grossart, M.A.F.H., Ibelings, B.W., Lajeunesse, M.J., Rinke, K., Rudstam, L.G., Rusak, J.A., Salmaso, N., 2020. Storm Impacts on Phytoplankton Community Dynamics in Lakes. Glob. Chang. Biol 26, 2756–2784.

Stoler, A.B., Relyea, R.A., 2020. Reviewing the role of plant litter inputs to forested wetland ecosystems: leafing through the literature. Ecol. Monogr. 90, e01400. https://doi.org/10.1002/ecm.1400.

Strickland, J.R., Parson, T.R., 1972. *A Practical Handbook of Seawater Analysis*. second ed. Bulletins of the Fisheries Research Board of Canada.

Strickland, J.R., 1960. Measuring the Production of Marine Phytoplankton (No. 167). Fisheries Research Board of Canada, Ottawa, Canada, p. 172.

Tengberg, A., Hovdenes, J., Andersson, H.J., Brocandel, O., Diaz, R., Hebert, D., Arnerich, T., Huber, C., Körtzinger, A., Khripounoff, A., Rey, F., Rönning, C., Schimanski, J., Sommer, S., Stangelmayer, A., 2006. Evaluation of a lifetime-based optode to measure oxygen in aquatic systems. Limnol Oceanogr. Methods 4, 7–17. https://doi.org/10.4319/lom.2006.4.7.

Thompson, R.M., Dunne, J.A., Woodward, G., 2012. Freshwater food webs: towards a more fundamental understanding of biodiversity and community dynamics. Freshw. Biol. 57, 1329–1341. https://doi.org/10.1111/j.1365-2427.2012.02808.x.

Valentini, A., Taberlet, P., Miaud, C., Civade, R., Herder, J., Thomsen, P.F., Bellemain, E., Besnard, A., Coissac, E., Boyer, F., Gaboriaud, C., Jean, P., Poulet, N., Roset, N., Copp, G.H., Geniez, P., Pont, D., Argillier, C., Baudoin, J.-M., Peroux, T., Crivelli, A.J., Olivier, A., Acqueberge, M., Brun, M.L., Møller, P.R., Willerslev, E., Dejean, T., 2016. Next-generation monitoring of aquatic biodiversity using environmental DNA metabarcoding. Mol. Ecol. 25, 929–942. https://doi.org/10.1111/mec.13428.

Vander Zanden, M.J., Rasmussen, J.B., 1999. Primary consumer δ13C and δ15N and the trophic position of aquatic consumers. Ecology 80, 1395–1404. https://doi.org/10.1890/0012-9658(1999)080[1395:pccana]2.0.co;2.

Vannote, R.L., Minshall, G.W., Cummins, K.W., Sedell, J.R., Cushing, C.E., 1980. The River Continuum Concept. Can. J. Fish. Aquat. Sci. 37, 130–137. https://doi.org/10.1139/f80-017.

Verschoor, A.M., Boonstra, H., Meijer, T., 2005. Application of stable isotope tracers to studies of zooplankton feeding, using the rotifer *Brachionus calyciflorus* as an example. Hydrobiologia 546, 535–549. https://doi.org/10.1007/s10750-005-4296-x.

Vollenweider, R.A., 1969. A Manual on Methods for Measuring Primary Production in Aquatic Environments. IBP Handbook. Blackwell Scientific Publications, Oxford.

Walsh, C.J., Roy, A.H., Feminella, J.W., Cottingham, P.D., Groffman, P.M., Morgan, R.P., 2005. The urban stream syndrome: current knowledge and the search for a cure. J. North Am. Benthol. Soc. 24, 706–723. https://doi.org/10.1899/04-028.1.

Warfe, D.M., Pettit, N.E., Davies, P.M., Pusey, B.J., Hamilton, S.K., Kennard, M.J., Townsend, S.A., Bayliss, P., Ward, D.P., Douglas, M.M., Burford, M.A., Finn, M., Bunn, S.E., Halliday, I.A., 2011. The "wet–dry" in the wet–dry tropics drives river ecosystem structure and processes in northern Australia. Freshw. Biol. 56, 2169–2195. https://doi.org/10.1111/j.1365-2427.2011.02660.x.

Wetzel, R.G., 1979. The role of the littoral zone and detritus in lake metabolism. pp. 145–161. In: Likens, G.E., Rodhe, W., Serruya, C. (Eds.), Symposium on Lake Metabolism and Lake Management. Ergebnisse Limnol., Arch. Hydrobiol. 13, pp. 145–161.

Wetzel, R.G., 1984. Detrital dissolved and particulate organic carbon functions in aquatic ecosystems. Bull. Mar. Sci. 35, 503–509.

Wetzel, R.G., 1990. Land-water interfaces: metabolic and limnological regulators. SIL Proceedings, 1922–2010 24, 6–24. https://doi.org/10.1080/03680770.1989.11898687.

Wetzel, R.G., 1995. Death, detritus, and energy flow in aquatic ecosystems. Freshw. Biol. 33, 83–89. https://doi.org/10.1111/j.1365-2427.1995.tb00388.x.

Wetzel, R.G., Likens, G., 2000. Limnological Analyses. Springer, New York, NY, USA.

Whittaker, R.H., 1972. Evolution and measurement of species diversity. Taxon 21, 213–251. https://doi.org/10.2307/1218190.

Winder, M., Schindler, D.E., 2004. Climatic effects on the phenology of lake processes. Global Change Biol. 10, 1844–1856.

Winslow, L.A., Zwart, J.A., Batt, R.D., Dugan, H.A., Woolway, R.I., Corman, J.R., Hanson, P.C., Read, J.S., 2016. LakeMetabolizer: an R package for estimating lake metabolism from free-water oxygen using diverse statistical models. Inland Waters 6, 622–636. https://doi.org/10.1080/iw-6.4.883.

Wommack, K.E., Colwell, R.R., 2000. Virioplankton: Viruses in aquatic ecosystems. Microbiol. Mol. Biol. Rev. 64, 69–114. https://doi.org/10.1128/MMBR.64.1.69-114.2000.

CHAPTER

10

Water as a Chemical Environment

Lewis A. Molot[1] *and Peter J. Dillon*[2]

[1]Faculty of Environmental & Urban Change, York University, Toronto, Ontario, Canada

[2]School of the Environment, Trent University, Peterborough, Ontario, Canada

OUTLINE

I. Water as a solvent

An aquatic ecosystem is as much a chemical environment as it is a physical one. Because of its very high polarity, water is a solvent capable of dissolving an extraordinarily large number of substances (i.e., solutes, many of which enter the food web, especially at the microbial level). These include metabolically essential substances (nutrients, trace metals, and gases) as well as harmful substances.

Solubility (defined as the maximum solubility at equilibrium under a given set of conditions) is influenced by numerous factors, such as a solute's chemical structure as well as environmental conditions including temperature, pH, and oxidation–redox (redox) potential. Nonpolar compounds are characterized by very low (maximum) solubility. To illustrate the importance of chemical structure to solubility, compare the nonpolar aromatic ring, benzene (C_6H_6), with a solubility in water of 1.8 g L^{-1} (23 mmol L^{-1}) at room temperature to a slightly different C-6 compound, phenol (C_6H_5OH), with a solubility of 8.3 g L^{-1} (88 mmol L^{-1}). Substitution of an H^+ for an OH^- group increases the ring's polarity slightly and its molar solubility over threefold. Another C-6 molecule, glucose ($C_6H_{12}O_6$), is much more polar and has a solubility of 909 g L^{-1} (5050 mmol L^{-1}).

Water as a solvent not only provides a medium from which aquatic organisms can acquire substances but also brings into contact solutes that may chemically react to produce new substances (products) while lowering reactant concentrations. This means that ecosystem productivity and ecological structure are profoundly dependent on a substance's capacity for dissolving in water and factors which remove and add it to solution.

Solubility is to be distinguished from the actual concentration of a solute, which is often less than maximum, in which case it is said to be undersaturated. Solute concentration is affected by rates of input from sources external to lakes, discharge downstream, adsorption to particulate matter, chemical reactions that use or produce the solute, precipitation after reacting with other solutes, degassing and microbial assimilation, excretion, and decomposition. These rates can vary significantly over time and space.

This chapter begins with a brief summary of the basic principles governing chemical reactions and then examines factors that affect solute concentrations. For the purposes of this chapter, two categories of solutes are

Wetzel's Limnology, Fourth Edition
https://doi.org/10.1016/B978-0-12-822701-5.00010-0

recognized: (1) substances with a solid phase within temperature and pressure ranges found in lakes such as major nutrients (phosphate, ammonia, and nitrate), trace metals, major ions (chloride, sulfate, calcium, magnesium, potassium, sodium), and organic compounds; and (2) gases such as CO_2, O_2, N_2, H_2S, and N_2O. There is some overlap between these categories; for example, metals are often bound to organic carbon, and ammonia (especially above pH 9) and elemental mercury are *volatile* (vaporizes readily).

II. Chemical reactions in freshwater

The five major classes of reactions that occur in freshwaters are *acid–base reactions*, most of which involve protons (H^+), *redox reactions* that involve the transfer of electrons, *photochemical reactions* involving free radical production via the breaking of covalent bonds by photons to produce chemical species with single free electrons (rather than pairs of electrons as in ions), *complexation reactions* involving weak bonding between dissolved molecules or atoms, and *adsorption–desorption reactions*, which involve the interaction of solutes with solids.

A. Acid–base reactions

An acid–base reaction (*Brønsted–Lowry definition*) is a chemical reaction that involves an exchange of one or more hydrogen ions (H^+) between molecules that may be neutral (e.g., water, H_2O) or electrically charged (e.g., ammonium, NH_4^+; hydroxide, OH^-; or carbonate, CO_3^{2-}). In the *Lewis theory of acid–base reactions* an acid is any substance, such as H^+, that can accept a pair of nonbonding electrons. Meanwhile, a base is any substance, such as OH^-, that can donate a pair of nonbonding electrons.

Carbonic acid (H_2CO_3) produced through the dissolution of carbon dioxide in water is the predominant acid in most unpolluted freshwaters. It is also the source of many of the inorganic solutes in water through its interaction with the lake or river catchment's soils and rock (i.e., weathering which is the dissolution of rocks and minerals on the Earth's surface). Dissolution of sedimentary substrate (limestone, dolomite) by carbonic acid provides high levels of base cations (Ca, Mg, Na, K), as well as bicarbonate ions, leading to waters with high *ionic strength* (a measure of the concentration of ions in a solution), while the interaction of carbonic acid with crystalline siliceous substrates provides the same cations but in much lesser amounts; low ionic strength waters with much lower bicarbonate concentrations result. The carbonic acid–bicarbonate–carbonate system is the dominant *buffering* system in natural waters (a buffer is a solution that resists changes in pH when an acid or alkali is added to it), leading to pH values generally in the circumneutral range (6.0–8.0). The presence of high levels of organic acids (humic and fulvic) resulting from the decay of vegetative material can in some cases lower the pH significantly (as low as 4 in some cases), while high primary productivity can alter the carbonic acid–bicarbonate–carbonate equilibrium to favor carbonate, raising the pH to well above 9. Excessive inputs of anthropogenic acids (H_2SO_4, HNO_3 (acid rain)) can also drastically lower the pH by overwhelming the buffering capacity of the water.

The carbonate system is discussed in more detail in Chapters 12 and 13.

B. Redox reactions

A *redox reaction* consists of two parts or half-reactions: an oxidation reaction in which a substance loses or donates electrons, which is coupled to a reduction reaction in which a substance gains or accepts electrons. An oxidation reaction and a reduction reaction must always be coupled because "free" electrons cannot exist in solution and electrons must be conserved. Although these reactions may involve the transfer of oxygen, other substances may replace oxygen as electron acceptor.

One of the most important oxidation reactions in natural waters is photosynthesis:

$$CO_2 + H_2O = CH_2O + O_2 \qquad (10\text{-}1)$$

where CH_2O is a simplified generic formula for organic matter. In this example CO_2 is reduced while water is oxidized. The converse reaction, respiration of organic matter, can be represented as:

$$CH_2O + O_2 = CO_2 + H_2O \qquad (10\text{-}2)$$

which is the sum of two half-reactions:

$$CH_2O + H_2O = CO_2 + 4H^+ + 4e^- \text{ (oxidation half-reaction)} \qquad (10\text{-}3)$$

$$O_2 + 4H^+ + 4e^- = 2H_2O \text{ (reduction half-reaction)} \qquad (10\text{-}4)$$

When free O_2 gas is not present in the water, a situation typically found in the hypolimnia of productive waters (Chapter 11), other electron acceptors including nitrate, ferric iron, and sulfate may replace the role of oxygen.

$$2NO_3^- + 12H^+ + 10e^- = N_2 + 6H_2O \text{ (denitrification)} \qquad (10\text{-}5)$$

$$FeOOH + 3H^+ + e^- = Fe^{2+} + 2H_2O \text{ (ferric iron reduction)} \quad (10\text{-}6)$$

$$SO_4^{2-} + 10H^+ + 8e^- = H_2S + 4H_2O \text{ (sulfate reduction)} \quad (10\text{-}7)$$

These and other processes can serve as the half-reactions that support the oxidation of organic carbon (respiration).

The sequence of redox reactions that can occur as oxidants are depleted (the *redox ladder* or *cascade*) is shown in Fig. 16-1 and Fig. 27-24. In this figure *pE* or *redox potential*, a measure of electron potential or activity (in some ways analogous to pH, the measure of proton activity in waters), determines when each half-reaction can occur (see also Chapter 16). Thus only when O_2 is fully utilized will NO_3^- serve as an oxidizing agent, with denitrification (forming N_2 gas) occurring followed by ammonification (conversion of nitrate to ammonium). Subsequently, reduction of manganese dioxide (MnO_2), then ferric iron reduction to produce ferrous iron, and then sulfate (SO_4^{2-}) reduction to produce sulfide (S^-) can occur. Finally, carbon dioxide (CO_2) can be reduced to methane (CH_4) at extremely low pE values (Grundl et al., 2011). The reduction half-reactions are particularly important in the *anaerobic* hypolimnia of lakes (anaerobic means O_2 and nitrate are depleted), within lake sediments, and in wetlands that typically have very high organic matter content. Trace (transition) metals (Chapter 16) are especially important to cell metabolism because of their capacity (i.e., their low ionization energy) for electron transfer during redox reactions and their affinity for protein complexation (see section 10-D, Complexation Reactions below) (Sunda, 1989).

C. Photochemical reactions

Unlike acid–base reactions and redox reactions that involve breaking covalent bonds to produce species with paired electrons, photochemical reactions typically result in the production of uncharged chemical species, each with a single free electron, that is, *free radicals*. Instead of both electrons in a covalent bond migrating to one of the two bonded atoms, a photochemical reaction results in a single electron on each atom. Photochemical reactions are initiated by the absorption of photons with sufficient energy to break a covalent bond, that is, at wavelengths mostly in the ultraviolet region, a process also referred to as *photolysis*. A *chromophore* is that part of the molecule where absorption occurs.

Free radicals such as the hydroxyl radical (depicted as OH•, not to be confused with the hydroxyl ion, OH^-) are very reactive and rapidly undergo further reactions with other substances, which has implications for the chemical composition of natural waters close enough to the surface to be irradiated with wavelengths energetic enough to break bonds. Hydroxyl radicals can be produced via several mechanisms, including photolysis of hydrogen peroxide (H_2O_2), ozone (O_3), nitrate (NO_3^-), nitrite (NO_2^-), dissolved organic matter (DOM), and the reaction of Fe^{2+} with H_2O_2 (called the photo-Fenton reaction when Fe^{2+} and H_2O_2 are photochemically produced) (Brezonik and Fulkerson-Brekken, 1998; Vaughn and Blough, 1998).

A significant portion of chromophoric or colored DOM degradation in surface waters occurs as a result of direct photochemical decay and reaction with photochemically produced hydroxyl radicals. This results in the gradual loss of *chromophores* (*photobleaching*) and progressive degradation to smaller organic molecules and CO_2 (photodecay or photomineralization) (Molot et al., 2005). In boreal lakes chromophoric DOM is dominated by humic and fulvic acids exported from wetlands. These compounds are relatively resistant to microbial respiration (i.e., *refractory*) until irradiated. Photodegradation rates of chromophoric DOM are much higher than those of nonchromophoric DOM (Vähätalo and Wetzel, 2004). Photochemical reactions are an important part of an abiotic pathway that transfers dissolved humic acids to particulate organic matter (POM) in sediments (Porcal et al., 2013).

The pH of the water has a very strong effect on photodecay rates of DOM, with low pH favoring rapid loss of DOM via enhanced production of hydroxyl radicals (Molot et al., 2005). The apparent mechanism for this is the iron-mediated photo-Fenton pathway (Zepp et al., 1992; Voelker et al., 1997). Reoxidation of photochemically produced ferrous iron (Fe^{2+}) to ferric iron (Fe^{3+}) is rapid in circumneutral and alkaline waters but acidic pH inhibits reoxidation, thus ensuring greater production of hydroxyl radicals.

D. Complexation reactions

Complexes are chemical species that are composed of two or more entities that are capable of independent existence. There are two general types of complexes:

a) ion pairs that are held together by the attraction of oppositely charged ions, for example, CdI^+, $HgCl_3^-$

b) coordination complexes where the main binding force is like a covalent bond with direct electron sharing between a ligand and an ion, usually a metal.

Potential *ligands* (a ligand is an ion or molecule that binds to a metal atom to make a complex) only require an electron pair that can be donated to the metal. Some common examples of ligands are H_2O, OH^-, NH_3, and CH_3COO^- (the latter molecule is the acetate anion, where COO^- is shorthand for O=C—O^-). When water is the ligand, the process is called *hydration*. Organic ligands that have more than one site at which complexation to a metal ion can occur are called *multidentate ligands* and the complexes formed are called *chelates*.

In natural waters there are many dissolved substances, both inorganic and organic, that can act as ligands to form complexes. The DOM (Chapter 28) in natural waters that is composed of humic and fulvic substances is a very important group of ligands that has a critical role in the aquatic environment affecting the bioavailability and toxicity of metals and other substances.

E. Adsorption–desorption reactions

Adsorption–desorption reactions are a two-phase partition process typically involving adsorption, that is, attachment, of a solute (adsorbate) to the surface of a solid phase (adsorbent). Desorption refers to the release of an adsorbed substance from an adsorbent. The solid phase can be settled sediments or suspended particulate matter, either live or dead biological material, or inorganic suspended material. This process can be approximately described using a *Freundlich isotherm equation*:

$$R = K_f \left(C_{aq}\right)^{1/n} \tag{10-8}$$

where R is the ratio of adsorbed solute to adsorbent by weight, C_{aq} is the solute concentration, K_f is the *Freundlich constant* (an empirical partitioning constant), and n is an empirical value. K_f and n are coefficients specific for a given adsorbate and adsorbent pair at a given temperature. As in the case of complexation reactions, adsorption–desorption reactions can affect the bioavailability of a substance and, for some substances, its toxicity.

III. Factors regulating concentrations of nongaseous solutes

The concentration of a solute in the water column of a surface water at any point in time is the net result of dynamic processes, defined as the difference between inputs from all sources and all outputs including transport processes and chemical reactions. Inputs include external inputs (tributaries, groundwater, precipitation, and the atmosphere in the case of gases) and internal inputs (release from sediments, biota excretion, release from decomposition, and chemical reactions in the water column that produce it). Outputs include discharge through the outflow, sedimentation, degassing or volatilization to the atmosphere, microbial assimilation, and chemical reactions in the water column that consume it. Not all inputs and outputs are applicable to every substance and every instance. For example, metals, other than mercury, do not have gaseous forms and metals cannot be photodegraded. Furthermore, rates for a specific process can vary with location in a lake and time.

This *mass balance* concept describing concentration as a function of inputs and outputs can be expressed with a simple differential equation:

$$dC_x/dt = (\text{d Inputs}/dt - \text{d Outputs}/dt)/V \tag{10-9a}$$

where C_x is the mean concentration of substance X at time t; inputs and outputs are the instantaneous mass fluxes of X into and out of a lake at time t; and V is the lake volume. Over the time interval Δt (i.e., $t_2 - t_1$), Eq. 10-9a becomes

$$\Delta C_x = (\Delta\text{Inputs} - \Delta\text{Outputs})/V \tag{10-9b}$$

where Δ means the difference over the time interval. In Eqs. 10-9a and 10-9b individual inputs and losses are implicit, but the equations form the conceptual basis of more complex deterministic models where separate inputs and outputs are explicitly described with process-specific formulae.

Dynamic processes are the norm in lakes, but steady-state conditions can be assumed under some conditions, simplifying the model. Steady-state conditions assume that a lake is well mixed and has constant input and output rates, conditions which can be approximated in small lakes using mean concentrations and fluxes averaged over periods that span several years. At steady state, Eq. 10-9a becomes

$$C_x = \text{Inputs}/(v + q) \tag{10-10}$$

where v is a generalized loss coefficient that represents all potential net losses from the water column and q is a water discharge coefficient representing loss via the outflow (this is discussed in more detail in Chapter 15). While Eq. 10-10 works well for predicting long-term whole-lake average concentrations, it does not work well for predicting a whole-lake concentration on a given day, week, or even a given year because fluxes and biological process rates can be very dynamic (see Chapters 14 and 15 for discussion of N and P cycles). However, long-term averages smooth the short-term peaks and valleys, approximating a steady state value (Molot and Dillon, 1993; Dillon and Molot, 1996).

Eq. 10-10 works well for predicting the mean whole-lake concentrations of total dissolved and particulate phosphorus (P) and total nitrogen (N) over many years in small oligotrophic lakes in central Ontario (Canada), even though total P and total N inputs to lakes vary seasonally and annually, as do concentrations within a lake. Eq. 10-10 works reasonably well for predicting mean, long-term ammonium (NH_4^+) and nitrate (NO_3^-) concentrations in central Ontario lakes with a variation of about threefold in net losses, v (Molot and Dillon, 1993). The N cycle consists of numerous microbially mediated transformations producing and consuming NH_4^+ to NO_3^- that are not explicitly described in Eq. 10-10.

Eq. 10-10 has also been successfully used to predict long-term mean concentrations of total dissolved organic carbon (DOC) and the humic or colored fraction of DOC in central Ontario lakes (Dillon and Molot, 1997; Molot and Dillon, 1997). The major external source of DOC to these lakes is peatlands (Dillon and Molot, 1997) and the major losses are from photochemical oxidation and microbial decomposition, which transfer CO_2 to the atmosphere and DOC to sediments, perhaps via adsorption to settling Fe hydroxide precipitates. The humic fraction of DOC is especially sensitive to solar radiation in the UVB and UVA regions, which cause photobleaching (loss of color), loss of metal binding sites, and eventually, oxidation of the carbon fraction to CO_2.

Steady-state Eq. 10-10 is useful for predicting long-term, multiyear averages, but there is too much uncertainty when used to predict outcomes over shorter time periods. For shorter time periods, dynamic process-oriented models should be used (e.g., Markelov et al., 2019).

Dissolution of inorganic Fe is affected by the *oxidation state* of Fe (oxidation state is the number of electrons lost if the number is positive [or gained, if the number is negative] by an atom). There are two common oxidation states of Fe in aquatic systems with relative abundances or distributions between these two states controlled by pH and dissolved oxygen (O_2) (see Chapter 16). The oxidized form, ferric iron (Fe^{3+}), is sparingly soluble in the presence of O_2 at circumneutral pH and tends to precipitate as an iron hydroxide in the absence of a ligand such as DOC:

$$4Fe^{3+}_{(aq)} + 12H_2O \rightarrow 4Fe(OH)_{3(s)} + 12H^+_{(aq)} \quad (10\text{-}11)$$

where subscripts denote the phases: aq—aqueous, g—gas, and s—solid. The concentration of uncomplexed or free Fe^{3+} in natural waters is too small to be measured directly and is instead estimated from chemical equilibrium models. High total dissolved Fe^{3+} concentrations in circumneutral, oxygenated natural waters are only possible when bound to DOC with binding sites, for example, humic DOC. Complexation to DOC maintains Fe in solution, thereby lowering the total Fe settling rate, v (Molot and Dillon, 2003).

The reduced form, ferrous iron (Fe^{2+}), is rapidly and spontaneously oxidized to Fe^{3+} (oxidation is defined as losing one or more electrons; reduction is the gaining of electrons) in the presence of O_2 above pH 6; consequently, free Fe^{2+} concentrations are much lower than free Fe^{3+} in circumneutral, oxygenated waters and are said to be vanishingly small:

$$4Fe^{2+}_{(aq)} + O_{2(g)} + 4H^+_{(aq)} \rightarrow 4Fe^{3+}_{(aq)} + 2H_2O \quad (10\text{-}12)$$

However, Fe^{2+} concentrations can be high in waters that are anoxic (complete absence of O_2) or acidic (pH <5). Low pH inhibits oxidation to Fe^{3+}; hence Fe^{2+} produced by the photochemical reduction of Fe^{3+} in surface waters (initiated by the absorption of photons in the UV region by photoreactive organic molecules to which Fe^{3+} is bound) can accumulate. Fe^{2+} concentrations can also accumulate over Fe-rich sediments in anoxic waters where the source is the microbial reductive dissolution of Fe^{3+} hydroxides because rapid oxidation of Fe^{2+} does not occur in the absence of O_2. Thus Fe solubility is a function of several factors including oxidation state, pH, O_2, and the chelating capacity of DOC.

Since free Fe^{2+} and Fe^{3+}, PO_4^{3-}, NH_4^+, NO_3^-, and solutes such as other trace metals and "major ions" are metabolically essential nutrients, concentrations of these nutrients in the water column can be affected by phytoplankton demand, which removes them from solution. Cellular uptake transfers them from a dissolved phase across the cell membrane, which is then operationally defined as particulate because it no longer passes through a small-pore filter. In some inland lakes in Canada along Lake Superior, relatively abundant populations of Cyanobacteria, which have an especially high metabolic demand for Fe, draw down free Fe^{3+} to concentrations ranging from 10^{-15} to 10^{-10} nmol L^{-1} (Sorichetti et al., 2014). Subsequent sedimentation of senescent phytoplankton cells transfers particulate Fe^{3+} to sediments where Fe is diagenetically processed into ferric hydroxides. Should surficial sediments become anoxic, microbial reduction of Fe hydroxides will release Fe^{2+} into overlying water, a process called *internal loading*. Phytoplankton productivity, especially pico-Cyanobacteria, can sometimes be limited by low dissolved Fe concentrations (Twiss et al., 2000; De Wever et al., 2008; Downs et al., 2008).

Metalloproteins (metal–protein complexes) are important cell constituents. The metal enzyme cofactors Fe, manganese (Mn), and zinc (Zn) are used by approximately 30% of all enzymes, while copper (Cu), molybdenum (Mo), nickel (Ni), and cobalt (Co) are used in smaller amounts (Foster et al., 2014). Fe is in greatest demand, constituting more than 50% of the intracellular molar trace metal content in marine phytoplankton and the bacterium *Desulfovibrio* (Ho et al., 2003; Barton et al., 2007). Thus trace metals are involved in thousands of cellular redox reactions including those in photosynthesis, respiration, nitrate reduction, urea decomposition, neutralization of reactive oxygen species, and nitrogen fixation. Hence we can expect aquatic dissolved trace metal concentrations to be low where biological demand is high and supply rates are low. These are most likely to be productive waters in agricultural watersheds with low trace metal geology, little urbanization, and negligible industrialization (Neal and Robson, 2000; Downs et al., 2008; Yuan, 2017).

IV. Factors regulating concentrations of dissolved gases in water

Several gases are important to aquatic systems because of their metabolic roles (e.g., O_2, CO_2, and N_2), their effect on pH (CO_2), or their role in climate change (e.g., the greenhouse gases CO_2, CH_4, N_2O). Similar to nongaseous solutes, dissolved gas concentrations in lakes are affected by a host of internal processes that produce and consume gases (photosynthetic fixation, respiration, N fixation, anaerobic respiration, denitrification, etc.), which are discussed in Chapters 9, 11, 13, 14, and 23. Dissolved gas concentrations are also influenced by atmospheric partial pressure, pH, temperature, and turbulence. This section discusses atmospheric exchange as a starting point.

Movement, or *flux*, of a gas across the air–water boundary occurs when the dissolved gas concentration is not at equilibrium with the atmosphere, with the magnitude and direction of the flux depending on the difference between the actual and equilibrium concentrations. At equilibrium, the concentration of a dissolved gas in water is a function of the partial pressure of the gas in the atmosphere. The equilibrium relationship is given by

$$K_H = C_x/pX \tag{10-13}$$

where K_H is *Henry's law* constant for a substance X, C_x is the dissolved concentration of the substance at the air-water boundary, and pX is the partial pressure of X in the atmosphere. K_H is negatively affected by increasing temperature, with cold water dissolving more gas than warm water. If X is oversaturated relative to the equilibrium concentration, then X "degasses" from the water surface into the atmosphere until equilibrium is achieved. For example, when high rates of photosynthetically produced O_2 concentrations exceed equilibrium values, O_2 will degas to the atmosphere. If a surface water is undersaturated, that is, if the gas concentration is less than the equilibrium value set by the atmospheric concentration, then atmospheric gas will invade and dissolve in the lake. Highly productive lakes draw down CO_2 below saturation levels during the growing season, allowing temporary atmospheric invasion (see Chapter 13). In contrast, oligotrophic lakes in central Ontario (Canada) are typically supersaturated with CO_2 because of high inputs from streams (Koprivnjak et al., 2010) and the photomineralization of DOC. Hence many streams and lakes in that region are CO_2 sources for the atmosphere.

Under nonequilibrium conditions, it may be useful to estimate the magnitude of the flux or diffusive transfer of a gas across the air–water boundary, which is driven by the difference between the actual dissolved concentration at the air–water boundary and the equilibrium concentration. The flux can be given by

$$F = k\left(X_{surf} - X_{equ}\right) \tag{10-14}$$

where F is the flux (mass of X transferred across the boundary per unit area and per unit time), k is an apparent constant (the piston velocity), and X_{surf} and X_{equ} are the measured and equilibrium concentrations of the gas at the water surface, respectively (Cole et al., 2010) (see Chapter 13 for the case with CO_2). The constant k varies with gas, temperature, and turbulence at the surface of the water body. Gases transferred to the atmosphere can be transported long distances, which is an especially important consideration for volatile pollutants in surface waters and soils such as mercury (Gaffney and Marley, 2014) and hexachlorocyclohexane (Shen et al., 2004).

V. Summary

1. Water is a solvent capable of dissolving an extraordinarily large number of substances, that is, solutes, providing a medium from which aquatic organisms can acquire substances.
   a. Water also brings into contact solutes which may chemically react to produce new substances (products) while lowering reactant concentrations.
   b. Ecosystem productivity and ecological structure are profoundly dependent on a substance's capacity for dissolving in water and factors that remove and add it to solution.
2. The five major classes of reactions that occur in freshwaters are:
   a. acid–base reactions, most of which involve protons (H^+);
   b. redox reactions that involve the transfer of electrons;

c. photochemical reactions involving free radical production via the breaking of covalent bonds by photons to produce chemical species with single free electrons (rather than pairs of electrons as in ions);
d. complexation reactions involving weak bonding between dissolved molecules or atoms;
e. adsorption–desorption reactions, which involve the interaction of solutes with solids.

3. The concentration of a solute in the water column of a surface water at any point in time is the net result of dynamic processes, defined as the difference between inputs from all sources, and all outputs, including transport processes and chemical reactions.
4. Several gases are important in aquatic systems because of their metabolic roles (e.g., O_2, CO_2, and N_2), their effect on pH (CO_2), or their role in climate change (e.g., the greenhouse gases CO_2, CH_4, N_2O).
   a. Similar to nongaseous solutes, dissolved gas concentrations in lakes are affected by a host of internal processes that produce and consume gases, such as photosynthetic fixation, respiration, N fixation, anaerobic respiration, and denitrification.

References

Barton, L.L., Goulhen, F., Bruschi, M., Woodards, N.A., Plunkett, R.M., Rietmeijer, F.J., 2007. The bacterial metallome: composition and stability with specific reference to the anaerobic bacterium *Desulfovibrio desulfuricans*. Biometals 20, 291–302. https://doi.org/10.1007/s10534-006-9059-2.

Brezonik, P.L., Fulkerson-Brekken, J., 1998. Nitrate-induced photolysis in natural waters: controls on concentrations of hydroxyl radical photo-intermediates by natural scavenging agents. Environ. Sci. Technol. 32, 3004–3010.

Cole, J.J., Bade, D.L., Bastviken, D., Pace, M.L., Van de Bogert, M., 2010. Multiple approaches to estimating air-water gas exchange in small lakes. Limnol. Oceanogr. Methods 8, 285–293.

De Wever, A., Muylaert, K., Langlet, D., Alleman, L., Descy, J.-P., André, L., Cocquyt, C., Vyverman, W., 2008. Differential response of phytoplankton to additions of nitrogen, phosphorus and iron in Lake Tanganyika. Fresh Biol. 53, 264–277.

Dillon, P.J., Molot, L.A., 1996. Long-term phosphorus budgets and an examination of the steady state mass balance model for central Ontario lakes. Wat. Res. 30, 2273–2280.

Dillon, P.J., Molot, L.A., 1997. Dissolved organic and inorganic carbon mass balances in central Ontario lakes. Biogeochem. 36, 29–42.

Downs, T.M., Schallenberg, M., Burns, C.W., 2008. Responses of lake phytoplankton to micronutrient enrichment: a study in two New Zealand lakes and an analysis of published data. Aquat. Sci. 70, 347–360.

Foster, A.W., Osman, D., Robinson, N.J., 2014. Metal preferences and metallation. J. Biol. Chem. 289, 28095–28103. https://doi.org/10.1074/jbc.R114.588145.

Gaffney, J., Marley, N., 2014. In-depth review of atmospheric mercury: sources, transformations, and potential sinks. Energy Emiss. Control Technol. 2, 1–21.

Grundl, T.J., Haderlein, S., Nurmi, J.T., Tratnyek, P.G., 2011. Introduction to aquatic redox chemistry. Chapter 1. In: Tratnyek, P.G., Grundl, T.J., Haderlein, S.B. (Eds.), Aquatic Redox Chemistry. American Chemical Society. Symposium Series 1071. https://doi.org/10.1021/bk-2011-1071.ch001.

Ho, T.-Y., Quigg, A., Finkel, Z.V., Milligan, A.J., Wyman, K., Falkowski, P.G., et al., 2003. The elemental composition of some marine phytoplankton. J. Phycol. 39, 1145–1159. https://doi.org/10.1111/j.0022-3646.2003.03-090.x.

Koprivnjak, J.-F., Dillon, P.J., Molot, L.A., 2010. Importance of CO_2 evasion from small boreal streams. Global Biogeochem. Cycles 24, GB4003. https://doi.org/10.1029/2009GB003723.

Markelov, I., Couture, R.-M., Fischer, R., Haande, S., Van Cappellen, P., 2019. Coupling water column and sediment biogeochemical dynamics: modeling internal phosphorus loading, climate change responses, and mitigation measures in Lake Vansjø, Norway. J. Geophys. Res.-Biogeo. 124, 3847–3866.

Molot, L.A., Dillon, P.J., 1993. Nitrogen mass balances and denitrification rates in central Ontario lakes. Biogeochem 20, 195–212.

Molot, L.A., Dillon, P.J., 1997. Colour—mass balances and colour—dissolved organic carbon relationships in lakes and streams in central Ontario. Can. J. Fish. Aquat. Sci. 54, 2789–2795.

Molot, L.A., Dillon, P.J., 2003. Variation in iron, aluminum and dissolved organic carbon mass transfer coefficients in lakes. Wat. Res. 37, 1759–1768.

Molot, L.A., Hudson, J.J., Dillon, P.J., Miller, S.A., 2005. Effect of pH on photo-oxidation of dissolved organic carbon by hydroxyl radicals in a coloured, softwater stream. Aquat. Sci. 67, 189–195.

Neal, C., Robson, A.J., 2000. A summary of river water quality data collected within the Land Ocean Interaction Study: core data for Eastern UK rivers draining to the North Sea. Sci. Total Environ. 251/252, 585–665. https://doi.org/10.1016/S0048-9697(00)00397-1.

Porcal, P., Dillon, P.J., Molot, L.A., 2013. Photochemical production and decomposition of particulate organic carbon in a freshwater stream. Aquat. Sci. 75, 469–482.

Shen, L., Wania, F., Lei, Y.D., Teixeira, C., Muir, D.C.G., Bidleman, T.F., 2004. Hexachlorocyclohexanes in the North American atmosphere. Environ. Sci. Technol. 38, 965–975.

Sorichetti, R.J., Creed, I.F., Trick, C.G., 2014. Evidence for iron-regulated cyanobacterial predominance in oligotrophic lakes. Fresh. Biol. 59, 679–691.

Sunda, W.G., 1989. Trace metal interactions with marine phytoplankton. Biol. Oceanogr. 6, 411–442. https://doi.org/10.1080/01965581.1988.10749543.

Twiss, M.R., Auclair, J.-C., Charlton, M.N., 2000. An investigation into iron-stimulated phytoplankton productivity in epipelagic Lake Erie during thermal stratification using trace metal clean techniques. Can. J. Fish. Aquat. Sci. 57, 86–95.

Vähätalo, A.V., Wetzel, R.G., 2004. Photochemical and microbial decomposition of chromophoric dissolved organic matter during long (months—years) exposures. Mar. Chem. 89, 313–326.

Vaughn, P.P., Blough, N.V., 1998. Photochemical formation of hydroxyl radical by constituents of natural waters. Environ. Sci. Technol. 32, 2947–2953.

Voelker, B.M., Morel, F.M.M., Sulzberger, B., 1997. Iron redox cycling in surface waters: effects of humic substances and light. Environ. Sci. Technol. 31, 1004–1011.

Yuan, F., 2017. A multi-element sediment record of hydrological and environmental changes from Lake Erie since 1800. J. Paleolimnol. 58, 23–42. https://doi.org/10.1007/s10933-017-9953-3.

Zepp, R.G., Faust, B.C., Hoigne, J., 1992. Hydroxyl radical formation in aqueous reaction (pH 3-8) of Iron(II) with hydrogen peroxide: the photo-Fenton. Environ. Sci. Technol. 26, 313.

CHAPTER

11

Oxygen

Rafael Marcé[1,2], *Lluís Gómez-Gener*[3] *and Cayelan C. Carey*[4]

[1]Catalan Institute for Water Research (ICRA), Girona, Spain [2]Universitat de Girona, Girona, Spain [3]Centre for Research on Ecology and Forestry Applications (CREAF), Edifici C, Campus de Bellaterra (UAB), Cerdanyola del Vallès, Barcelona, Spain [4]Virginia Tech, Blacksburg, Virginia, United States

OUTLINE

I. The oxygen content of inland waters

Oxygen is the most fundamental parameter of lakes and streams, aside from water itself. The rates of supply of oxygen from the atmosphere, photosynthesis, and water mass transport are counterbalanced by physical processes and consumption by biotic and abiotic chemical reactions (Fig. 11-1). The interaction of all these biological, chemical, and physical processes within and among freshwater ecosystems results in a rich diversity of dissolved oxygen distributions in space and time. Their generating mechanisms, description, and ecological implications are the main focus of this chapter.

Oxygen content is related to a myriad of biogeochemical processes that control the concentration of substances (e.g.,

Wetzel's Limnology, Fourth Edition
https://doi.org/10.1016/B978-0-12-822701-5.00011-2

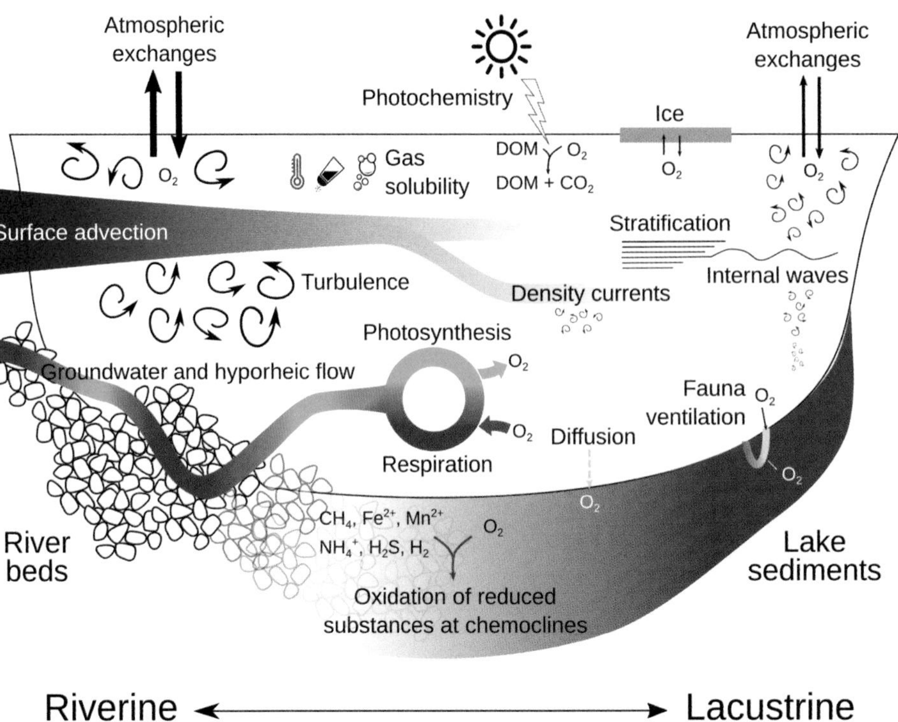

FIGURE 11-1 **Conceptual depiction of the processes affecting oxygen concentration in lotic and lentic ecosystems.** The plot summarizes the most relevant transport and reaction processes affecting oxygen dynamics in lotic (*left side*) and lentic (*right side*) ecosystems, while processes toward the center of the plot are relevant in both types of ecosystems. The size of the *arrows* is proportional to the relevance of the process. In streams and rivers advective transport and exchange with the atmosphere dominates. In lakes advective transport is less important and stratification limits turbulent diffusion. The oxidation of reduced substances happens in lake and river sediments but can also occur in the water column if oxyclines are established. *DOM*, Dissolved organic matter.

dissolved metals), ecosystem functioning, and the usability of water for human purposes. Dissolved oxygen is essential to the respiratory metabolism of most aquatic organisms. Consequently, environmental and water-use regulations around the world almost invariably consider oxygen content as one of the master parameters to define thresholds for ecological status. Low oxygen levels are generally associated with poor water quality and degraded habitat for organisms (e.g., as a driver of fish kills).

Oxygen deficits (see Section V) are common in inland waters and have increased due to climate change and other anthropogenic pressures during the last century (e.g., Jenny et al., 2016b; Jane et al., 2021), necessitating oxygen monitoring at multiple temporal and spatial scales. Oxygen deficits are associated with events of *hypoxia* (when oxygen concentrations decrease below an acceptable threshold, such as the minimum oxygen level needed for the survival of deepwater fish) and *anoxia* (when oxygen concentrations are close to zero). For example, eutrophic streams exhibit large fluctuations of oxygen between day and night, and eutrophic lakes can exhibit high oxygen concentrations in surface layers but anoxic deep waters at the same time. Metrics of oxygen content and deficit are thus a fundamental component of limnological studies.

Following Hutchinson (1957), "A skillful limnologist can probably learn more about the nature of a lake from a series of oxygen determinations than from any other kind of chemical data" (p. 575): such is oxygen's informative power for illuminating both the ecological functioning of the system and the suitability of water for human use.

II. Processes determining dissolved oxygen concentration

Dissolved oxygen concentrations in inland waters are the net result of a complex interplay of physical, chemical, and biological processes, operating at different spatial and temporal scales (Fig. 11-1). This makes the interpretation of oxygen spatial and temporal patterns a challenge for limnologists. Oxygen concentrations are highly variable, ranging from zero to more than

20 mg O_2 L^{-1} in very productive waters. Under certain circumstances, such as in water under ice in permanently ice-covered Antarctic lakes, concentrations can reach as high as ~50 mg O_2 L^{-1} (Wharton et al., 1986).

A. Measuring dissolved oxygen

Here, we provide an overview of dissolved oxygen analysis, as methods for measuring dissolved oxygen are reviewed in detail elsewhere (e.g., Baird and Bridgewater, 2017; Carignan et al., 1998; Wetzel and Likens, 2000; Hauer and Lamberti, 2017). The foundation of historical oxygen analysis is the *Winkler titration method*, and its numerous modifications (Wilkin et al., 2002), which has served as a benchmark for laboratory comparisons and calibration of electrochemical (e.g., polarographic and galvanic electrodes), optical (e.g., luminescent sensors), and colorimetric techniques (e.g., indigo carmine and rhodazine methods). However, advances in sensor technology during the last decades have now enabled reasonably accurate, instantaneous measures of dissolved oxygen in the field (e.g., handheld and field-deployed sensors) and the lab (e.g., sensor spots for transparent vessels), with an ever-growing diversity of sensor designs. For example, microelectrodes allow micrometer scale profiling in sediments and the water column (Bryant et al., 2010), and some sensor designs allow measurement of oxygen concentrations as low as 10 nmol O_2 L^{-1} (Kirf et al., 2014). These new sensor technologies have greatly enabled high-frequency automatic monitoring of oxygen in inland waters, which has changed the way researchers and practitioners use oxygen measurements to solve fundamental and applied problems (sensu Marcé et al., 2016).

B. Solubility of oxygen in water

Oxygen dissolves in water as a gas; therefore understanding the solubility of oxygen in water is of paramount importance to understanding oxygen dynamics in inland waters. Oxygen solubility is controlled by *Henry's law* (see Chapter 10), which states that the amount of gas that dissolves in water (solubility or saturation concentration) is proportional to its partial pressure:

$$C = K_h P_{gas} \quad (11\text{-}1)$$

where C is the solubility concentration at a fixed temperature (mg O_2 L^{-1}), K_h is the temperature-dependent Henry's law coefficient (mg O_2 L^{-1} atm^{-1}), and P_{gas} is the oxygen partial pressure (atm). For precise results, the effects of nonideal gas behavior, density of water, salinity, and water vapor pressure on K_h and P_{gas} must be considered (Benson and Krause, 1980). For oxygen, these corrections are accounted for by nonlinear fits between the solubility concentration at 1 atm and the combination of temperature and salinity (Fig. 11-2). The solubility concentration of oxygen decreases nonlinearly with increasing temperature and salt content (Fig. 11-2).

Variations in atmospheric pressure (and thus elevation) also influence oxygen solubility concentration (Eq. 11-1). At atmospheric pressures other than 1 atm, solubility concentrations calculated with equations in Fig. 11-2 should be corrected for atmospheric pressure and water vapor content using:

$$C^{sol}_{P\neq 1} = C^{sol}P\left[\frac{\left(1-\frac{P_{wv}}{P}\right)(1-\theta_0 P)}{(1-P_{wv})(1-\theta_o)}\right] \quad (11\text{-}2)$$

where C^{sol} is the solubility concentration at 1 atm, and P is the actual atmospheric pressure in atm. For the water vapor pressure (P_{wv}, in atm) and the coefficient for accounting for nonideal gas behavior (θ_0), we must solve (Benson and Krause, 1980):

$$\ln P_{wv} = 11.8571 - \left(3840.70\,T^{-1}\right) - \left(216961\,T^{-2}\right) \quad (11\text{-}3)$$

$$\theta_0 = 0.000975 - \left(1.426\cdot 10^{-5}\,t\right) + \left(6.436\cdot 10^{-8}\,t^2\right) \quad (11\text{-}4)$$

where T and t denote temperature in K and °C, respectively.

The amount of dissolved gas is also influenced by the hydrostatic pressure exerted by the overlying stratum of water. If dissolved oxygen partial pressure in water is higher than the combined atmospheric (P) and hydrostatic pressures (which was called absolute saturation by Ricker, 1934), bubbles will form and escape to the atmosphere, removing gases from the waterbody. If the hydrostatic pressure is high, oxygen supersaturation relative to the atmosphere can occur without the formation of bubbles. The combination of atmospheric and hydrostatic pressure (Pz, atm) at a given *depth* (m) is:

$$P_Z = P + 0.0967\,depth \quad (11\text{-}5)$$

The importance of the above relationship lies in its impact on absolute saturation concentration and its control of bubble formation. The oxygen concentration necessary for bubble growth increases with depth and hydrostatic pressure, and it is facilitated by turbulence and the presence of surfaces (e.g., plant tissues) (McGinnis and Little, 2002). As a consequence, bubble formation is more common in shallow waters of less than 4 m (Ramsey, 1962; Wik et al., 2013). Oxygen bubbles can also form in the blood of freshwater animals if the solubility concentration suddenly falls due to a water temperature increase or pressure fluctuations promoted by extreme turbulence, resulting in gas bubble trauma in fish and mortality events (Pleizier et al., 2020).

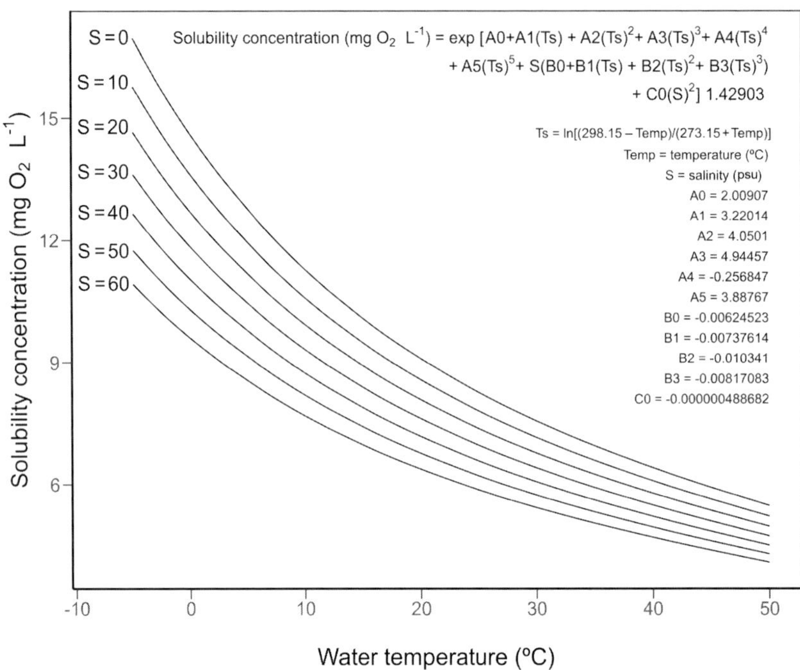

FIGURE 11.2 **Solubility concentration of dissolved oxygen in water at a pressure of 1 atm as a function of water temperature (°C) and salinity (S, psu).** The figure includes the equations and parameters necessary for calculating the saturation concentration from temperature and salinity, derived by Garcia and Gordon (1992). In waters showing specific conductivity below 2000 μS cm^{-1}, S = 0 can be used if very small errors are acceptable (errors always below 0.76%). In other situations measured S values must be considered. Salinity can be approximated from conductivity using $S = 5.572 \cdot 10^{-4}\ SC + 2.02 \cdot 10^{-9}\ SC^2$, where SC is specific conductance at 25°C (USGS, 2011). However, in saline environments with ionic composition very different from the sea, ad hoc equations for oxygen solubility may be preferable (Millero et al., 2002a, 2002b). If atmospheric pressure at the site is not 1 atm, the final solubility concentration must be corrected for the change in atmospheric pressure and water vapor pressure using Eqs. 11-2 to 11-4.

A special case of the effects of solubility on oxygen levels is in permanently ice-covered lakes, where the exclusion of oxygen from the ice lattice as water freezes can result in very high oxygen supersaturation in the water below the ice (Hood et al., 1998). Oxygen saturation levels >300% have been measured in these amictic lakes due to this mechanism (Wharton et al., 1986).

Solubility concentration is thus a convenient and important reference needed to interpret oxygen content, because it provides a baseline that accounts for the effects of temperature, elevation, atmospheric pressure, and salinity on oxygen concentrations. This is why oxygen content is often reported as *percent saturation* with respect to the solubility concentration at atmospheric pressure at the surface of a waterbody (i.e., the hydrostatic pressure is not considered for estimating percent saturation). Because of differences in temperature, solutes, and elevation among waterbodies, percent saturation can provide a better indicator of oxygen status in some cases than oxygen concentration. Water with oxygen concentration equal to the solubility concentration at the same temperature and dissolved solids content will thus be at 100% oxygen saturation, whereas water with an oxygen content half the solubility concentration will be at 50% oxygen saturation (undersaturation), and so on. Large deviations from 100% saturation indicate imbalances between processes producing and consuming oxygen.

The dependence of oxygen solubility on temperature and salinity has important implications for the biogeographical distribution of organisms inhabiting inland waters, particularly for those requiring high oxygen concentrations (e.g., early life stages in salmonid fishes). The dependence of oxygen solubility on temperature also results in less oxygen available for organisms during the warm months, when metabolic activity peaks, which may lead to oxygen depletion and fauna kills.

C. Oxygen exchange with the atmosphere

The bidirectional exchange of oxygen across the water–atmosphere boundary is frequently the most relevant process affecting dissolved oxygen concentration in surface waters (Fig. 11-1). The transfer of oxygen between the atmosphere and freshwaters is governed by the interplay of the partial pressure gradient with *turbulent and molecular diffusion processes* (see Chapter 10). Therefore, processes adding turbulence to the water, such as water currents, wind, convective cooling, and surface and internal waves (see Chapter 8), are important drivers of water-atmosphere exchange. Other processes affecting the stability of the boundary layer (e.g., presence of surfactants on the water surface) usually play a minor role.

Water–atmosphere exchange by diffusion pushes oxygen content toward 100% saturation, either by increasing or decreasing its concentration. Air–water oxygen exchange rates are tremendously variable, but generally, rivers and streams show higher transfer rates than lentic ecosystems because of greater turbulence. Considering gas transfer velocities in rivers and streams, atmospheric exchanges could range between 1 and 10,000 mmol O_2 m^{-2} h^{-1} (Hall and Ulseth, 2020), while in lakes the maximum atmospheric exchanges rarely exceed 300 mmol O_2 m^{-2} h^{-1}.

Bubble-mediated gas exchange also contributes to the air–water oxygen transfer. Breaking waves in lakes and high turbulence in streams help the formation of air bubbles that may exchange gases with the surrounding water before being released back into the atmosphere. This mechanism is particularly relevant in saline lakes (Chapter 12), where the solubility of oxygen is low (Gat and Shatkay, 1991). In steep, turbulent streams, bubbles likely cause most of the gas exchange (Hall et al., 2012). An extreme case is waterfalls, where air bubbles can promote oxygen *supersaturation*. A similar process may also happen downstream of hydropower plants or spillways, which is a major cause of bubble trauma in fish (Marking, 1987). Bubbles rich in oxygen will typically form on the surface of photosynthetic tissues of freshwater plants and may eventually escape to the atmosphere. Very small bubbles can be stabilized by an adsorbed organic film and accumulate on the water surface as foam (Ramsey, 1962). Oxygen can also be emitted to the atmosphere via methane bubbles generated in the sediments that entrain oxygen during their ascent to the surface (i.e., *oxygen ebullition*). Those bubbles can contain up to 34% oxygen (Koschorreck et al., 2017) and can substantially contribute to oxygen exchange with the atmosphere in some cases.

D. Advection, dispersion, and biota-mediated transport

Advection (the transport of a substance by the flow of water) and *dispersion* (spreading of a substance from concentrated to less concentrated areas) due to turbulent and molecular diffusion are the main transport mechanisms of dissolved oxygen in inland waters. Molecular diffusion of oxygen (~ 0.2 m^2 s^{-1} at 20°C) is only relevant at small spatial distances (e.g., at the air–water boundary layer) or in situations where turbulence and advection are absent (see Chapter 8). The relative importance of the different transport mechanisms on oxygen concentration varies among waterbodies, and across space and time. For example, advection is usually more relevant than dispersion for the fate of oxygen in rapidly flowing rivers (Benedini and Tsakiris, 2013), while dispersion may account for most transport at the surface of a lake during very low wind conditions. In lake sediments, oxygen transport may be entirely controlled by molecular diffusion, restricting oxygen penetration to a few top millimeters of the sediments (Fig. 11-3a).

In rivers, transport is influenced by local geomorphological characteristics such as slope, roughness, and meandering, which influence water velocity and turbulence, and thus favor advection over dispersion. Other relatively small-scale features in rivers, such as the presence of logs, sand bars, or backwaters, influence dissolved oxygen concentrations by lowering water velocity and enhancing subsurface *hyporheic flow*. This increases the residence time of water and solutes in the hyporheic compartment, which can be responsible for a substantial share of oxygen consumption in river networks (Boano et al., 2014).

In lakes, advection and dispersion transport is shaped by the overall water mixing and stratification dynamics of the water body. *Pycnoclines* derived from heat distribution or dissolved and suspended solids may constitute formidable barriers for dispersion and even advective transport, constituting one of the fundamental causes of spatial and temporal oxygen patterns and resulting in the distinct oxygen regimes described below (see Section IV). Subsurface flows are also frequent in lakes, particularly in human-made reservoirs (Thornton et al., 1990), leading to metalimnetic maxima or minima (see Section IV).

The transport of oxygen is also mediated by organisms. For example, burrowing fauna facilitate the diffusion of oxygen into sediments (Kristensen et al., 2012). This greatly enhances oxygen penetration (by up to several centimeters; Fig. 11-3a, b), which increases aerobic respiration in surface sediments (Baranov et al., 2016) and favors phosphorus burial (Hupfer et al., 2019). In vegetated sediments plants transport oxygen through the *aerenchyma* (a spongy tissue that forms spaces/air channels in leaves, stems, and roots of halophytic plants), which is eventually lost in roots (Fig. 11-3c), generating small oxic zones that exert strong control over sediment dissolved oxygen dynamics (Koop-Jakobsen et al., 2018).

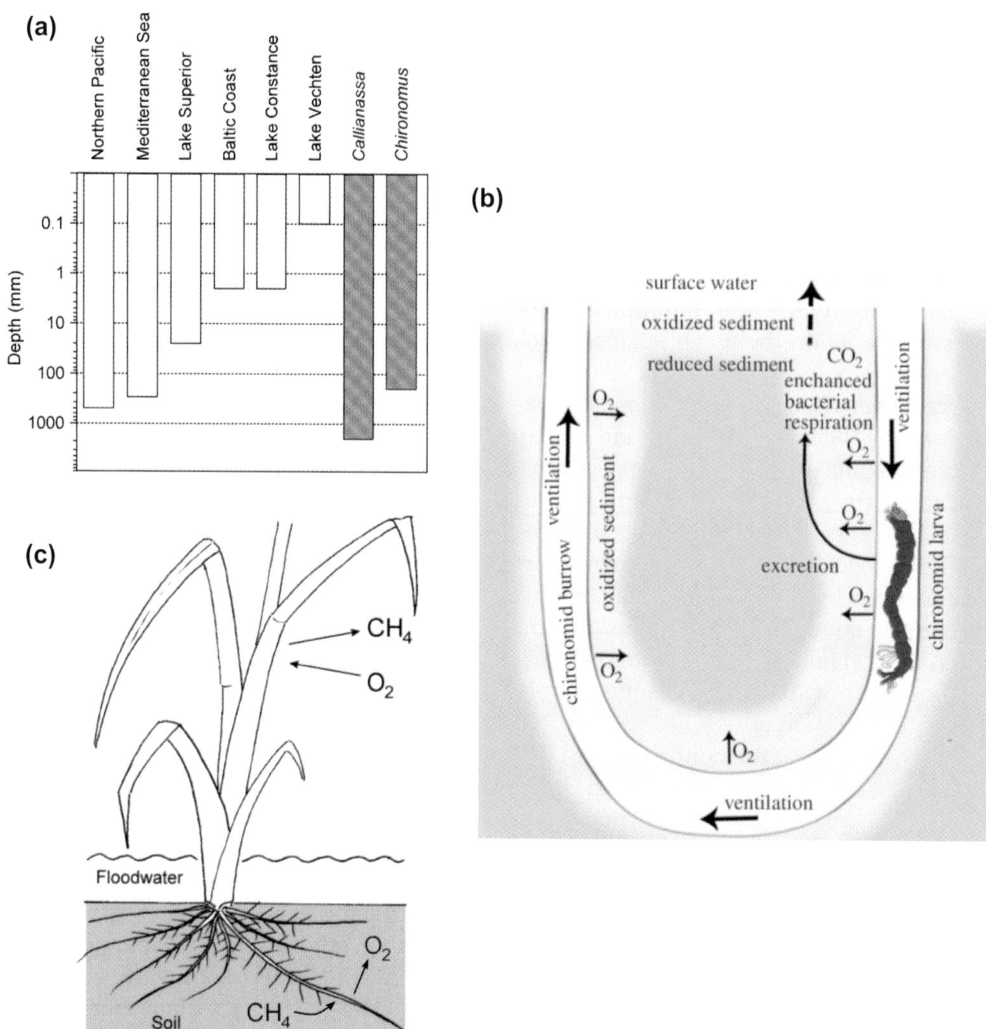

FIGURE 11-3 **Effect of organisms on oxygen transport in sediments.** (a) Oxygen penetration depth in several marine and freshwater sediments where transport is either almost entirely controlled by diffusion (*white bars*) or greatly enhanced by advective transport through the burrows of fauna (*shaded bars*). *(From Brune et al., 2000.)* The *bar* for *Chironomus* corresponds to a freshwater example. (b) Scheme of the impacts of chironomid larvae's (Diptera, Chironomidae) bioturbation on sediment biogeochemistry. *(From Baranov et al., 2016.)* (c) Transport of oxygen and methane mediated by plants. O_2 (oxygen) is transported from the atmosphere to the roots, where it may be released into the sediments. At the same time, CH_4 (methane) produced in the anoxic sediments is transported through the aerenchyma and emitted into the atmosphere. *(Modified from Brune et al., 2000.)*

E. Photosynthesis and photorespiration

Oxygenic photosynthesis is the fundamental source of dissolved oxygen supersaturation in productive inland waters. The availability of light and nutrients for photosynthesis is the net result of interacting biogeochemical processes acting at differing spatial (from microns to watersheds) and temporal (from seconds to decades) scales (see Chapter 9). In response to these drivers, the photosynthetic production of oxygen in inland waters varies across multiple orders of magnitude, from 10^{-3} to 10^{5} mmol O_2 m^{-3} h^{-1} (Krausen-Jensen and Sand-Jensen, 1998), showing higher values for benthic primary producers and macrophytes. As a reference, the highest rates of production ($\sim 10^{5}$ mmol O_2 m^{-3} h^{-1}) would bring a volume of anoxic water to saturation (~ 0.37 mmol O_2 L^{-1}) in ~ 10 s. Expressed as an areal rate (i.e., per unit area), photosynthesis in phytoplankton and macrophytes reaches maximum values of 80 mmol O_2 m^{-2} h^{-1} (Krausen-Jensen and Sand-Jensen, 1998).

Photorespiration is the light-dependent oxygen consumption by the enzyme RuBisCO in photosynthetic organisms (Maberly and Spence, 1989; Raven et al., 2020). Photorespiration rates in freshwaters have been considered low in comparison to rates in terrestrial plants due to the low relative oxygen solubility in water (Birmingham et al., 1982) and the presence of carbon concentration mechanisms (Beer et al., 1991). However, high photorespiration rates accounting for 25% of net photosynthesis have been observed in macrophytes (Maberly and Spence, 1989), suggesting that this mechanism could be important in some situations.

F. Aerobic respiration

Aerobic respiration of organic matter is frequently the most important oxygen-consuming process in freshwaters. Aerobic respiration and photosynthesis together are the dominant biological mechanisms shaping oxygen concentrations in freshwaters. The use of oxygen as the electron acceptor for oxidizing organic matter always yields the highest amounts of energy compared to any other acceptor, and the presence of oxygen suppresses most anaerobic metabolic pathways (Petsch, 2003; Stumm and Morgan, 2012). The dramatic changes in aerobic and anaerobic respiration metabolism across oxyclines in water and sediments thus become fundamental for matter and energy cycling in aquatic ecosystems (McClure et al., 2021).

Similar to photosynthesis, respiration rates in freshwaters respond to a complex combination of biogeochemical processes acting at differing spatial and temporal scales (del Giorgio and Williams, 2005). In lakes, respiration usually does not show the large variability shown by photosynthesis (Pace and Prairie, 2005), but it may be more variable than photosynthesis in rivers (Song et al., 2018). However, different methods reveal different respiration rates. For example, bottle incubations yield volumetric respiration rates in the water column from 10 to 4000 mmol O_2 m^{-3} h^{-1}, which corresponds to areal rates between 240 and 4500 mmol O_2 m^{-2} h^{-1}. Sediment core incubations show a range from 40 to 800 mmol O_2 m^{-2} h^{-1}, and whole-ecosystem respiration estimates derived from open-water diel oxygen curves reach ~24,000 mmol O_2 m^{-2} h^{-1} (Hoellein et al., 2013).

Methane oxidation is a type of aerobic methylotrophic metabolism. Methane diffusing from anoxic sediments, from rising methane bubbles, or produced in the water column is oxidized in the presence of oxygen by methane-oxidizing bacteria (Hanson and Hanson, 1996). Methane oxidation usually occurs in *oxyclines* in the water column or sediments, as well as the hyporheic zone of rivers (Trimmer et al., 2010). Freshwater methane oxidation rates range from 0 to 833 mol CH_4 m^{-3} h^{-1} (Bastviken, 2009; Trimmer et al., 2010) and can result in the consumption of up to 98% of stored methane in a waterbody during seasonal mixing (Zimmermann et al., 2021).

G. Oxidation of reduced inorganic compounds

The oxidation of reduced substances can be an important contribution to the oxygen budget of waterbodies if there are contrasting redox conditions across a spatial gradient (e.g., across the sediment-water interface). The sediment porewater and hyporheic zone can be enriched with reduced ions (described below) that consume large quantities of oxygen, especially when released into the oxic water column (Fig. 11-4; Matzinger et al., 2010; Müller et al., 2012; Carignan and Lean, 1991). A rich mosaic of biogeochemical processes is involved in the oxidation of reduced compounds with oxygen. Almost invariably, sediment–water interactions largely control the amount of reduced substances available for oxidation in the water column of lakes and rivers.

Oxidation of iron and manganese (see also Chapter 16)—Ferrous iron (Fe^{2+}) oxidation involves both biotic and abiotic reactions (Melton et al., 2014), but as a simplification, biotic processing of Fe^{2+} outcompetes abiotic oxidation at low oxygen and high Fe^{2+} concentrations. These conditions are typical at chemoclines with opposing gradients of oxygen and Fe^{2+}, for example, in river biofilms (Chiellini et al., 2018) and in the water column of meromictic lakes (Bravidor et al., 2015). In contrast, Mn^{2+} oxidation is mainly mediated by biological activity (Munger et al., 2016; Tipping et al., 1984), as the abiotic oxidation of reduced manganese (Mn^{2+}) is very slow at pH values usually found in inland waters (Hansel and Learman, 2015; Godwin et al., 2020).

Oxidation of sulfur compounds (see also Chapter 16)—Reduced sulfur compounds (mainly sulfide and elemental sulfur) are oxidized mostly by biologically mediated reactions (Luther et al., 2011). Bacteria and Archaea (Chapter 23) comprise most of the microorganisms that oxidize reduced forms of sulfur in large quantities (~50,000 μmol L^{-1} day^{-1}; Luther et al., 2011). The oxidation of sulfide can revert the sediments from a sink to a source of sulfate to the overlying water seasonally, particularly in oligotrophic systems where oxygen penetrates deep in the sediments (Holmer and Storkholm, 2001). In highly productive, sulfate-rich lakes, sulfide can be released into the anoxic water column and be oxidized upon mixing, accounting for a substantial

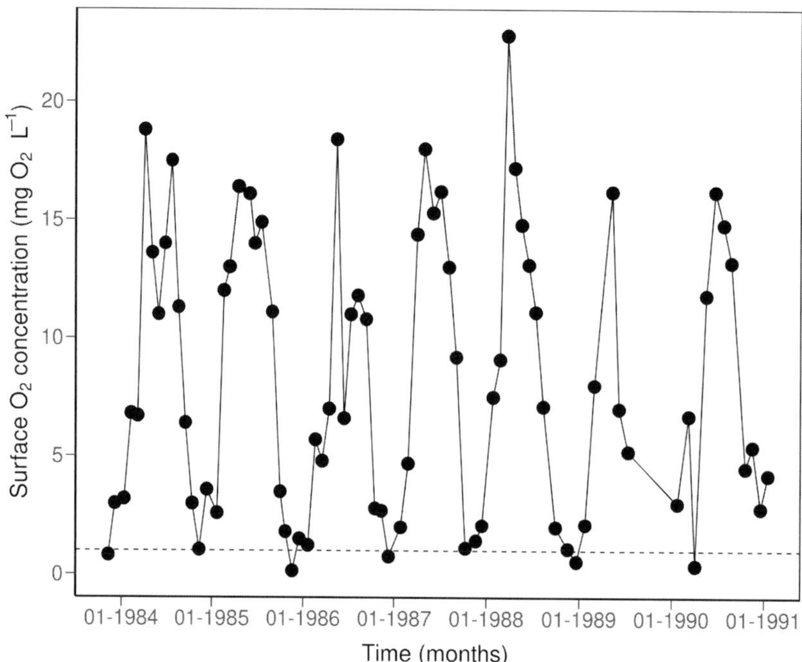

FIGURE 11-4 **Monthly dissolved oxygen concentration at the surface waters (<0.5 m) of Sau Reservoir (Spain) during 7 years of eutrophication.** Note the extremely high values during summer due to primary production (which resulted in >200% saturation) and the extremely low values during fall turnover, when the oxidation of reduced compounds from the hypolimnion (mainly ammonium) consumed almost all oxygen. This is an example of how intense oxidation of reduced substances can create anoxic conditions even in well-mixed surface waters. A common threshold for defining anoxia in freshwaters (<1 mg O_2 L^{-1}) is indicated as a horizontal dashed line for reference. *(Data from Marcé et al., 2010.)*

share of oxygen consumption in the hypolimnion (Gelda et al., 1995), even causing fish kills.

Oxidation of ammonium—Although abiotic ammonium oxidation has been measured in soils, nitrification is considered a biologically mediated process in inland waters. If large amounts of ammonium are present, nitrification (Chapter 14) can consume large amounts of oxygen (Fig. 11-4). Nitrification is partially responsible for oxygen consumption in the hypolimnia of lakes during summer stratification (Clevinger et al., 2014) and under ice (Powers et al., 2017), and it is responsible for anoxic conditions in rivers downstream from point sources rich in ammonium (Gowda, 1983; Brion and Billen, 2000).

Oxidation of hydrogen—Aerobic hydrogen-oxidizing bacteria usually thrive in the epilimnia of lakes, where hydrogen that has escaped from the sediments and hydrogen produced by nitrogen fixation in the surface layers can accumulate (Aragno and Schlegel, 1981; Conrad et al., 1983). However, the concentration of hydrogen in the oxic layers of freshwaters is usually orders of magnitude lower than that of CH_4, and therefore hydrogen oxidation rates likely do not consume relevant quantities of oxygen (Schütz et al., 1988).

H. Photochemical oxidation of organic matter

Reactive oxygen species (singlet oxygen, superoxide, and hydrogen peroxide) formed by the photoexcitation of organic molecules can also react with organic matter, consuming oxygen (Vähätalo, 2009). Although photochemical oxidation processes are likely masked in very productive waters by intense respiration, photochemical reactions drive oxygen dynamics and affect metabolism estimates (Brothers and Vadeboncoeur 2021) in freshwaters with high concentrations of chromophoric (colored) organic matter (Chapter 28). For instance, photochemical oxidation in highly colored lakes can be as high as 0.12 mg O_2 L^{-1} h^{-1} (Brezonik, 2002) and can result in substantial CO_2 production (and oxygen consumption) in the surface layers of many freshwaters (Amon and Benner, 1996; Vähätalo, 2009; Cory et al., 2014).

I. Chemical weathering

Rocks containing chemically reduced minerals (e.g., olivine, pyroxenes, amphiboles, metal sulfides) undergo oxidation when in contact with oxygenated waters. However, the rates of mineral oxidation from bedrock or regolith are deemed too low to have a discernible effect on oxygen concentration in most surface waters (Huang et al., 2018). Chemical weathering likely only has a measurable impact in waterbodies with very long water residence times and where other oxygen consumption mechanisms are very low (e.g., amictic lakes; Siegert et al., 2001).

III. Distribution of dissolved oxygen in running waters

Flow is the major structuring force in rivers (in this chapter we refer to rivers as running waters of any size, from small headwater streams to large lowland rivers). The flow of water in rivers maintains a constant supply of nutrients, produces physical turbulence that constantly mixes the water column, promotes reoxygenation via atmospheric exchange, and results in frequent scouring that limits permanent sedimentation (Poff et al., 1997). Water flows along the surface of river channels, and through sediment and rock interstices within channel boundaries (i.e., from *hyporheic*, *parafluvial*, and *riparian zones* to the stream, and vice versa; Boano et al., 2014; Harvey and Gooseff, 2015; Jones and Holmes, 1996). These lateral and vertical pathways are also involved in both the advection of dissolved oxygen as well as other substances indirectly influencing riverine dissolved oxygen dynamics (e.g., organic matter, nutrients, and electron acceptors). In addition to this spatial variability, oxygen concentrations in rivers may vary temporally on diel, seasonal, annual, and decadal scales (Garvey et al., 2007). This rich spectrum of temporal signals on multiple scales combines to shape the oxygen dynamics of a river, which integrate the biogeochemical processes occurring within the river and in its catchment.

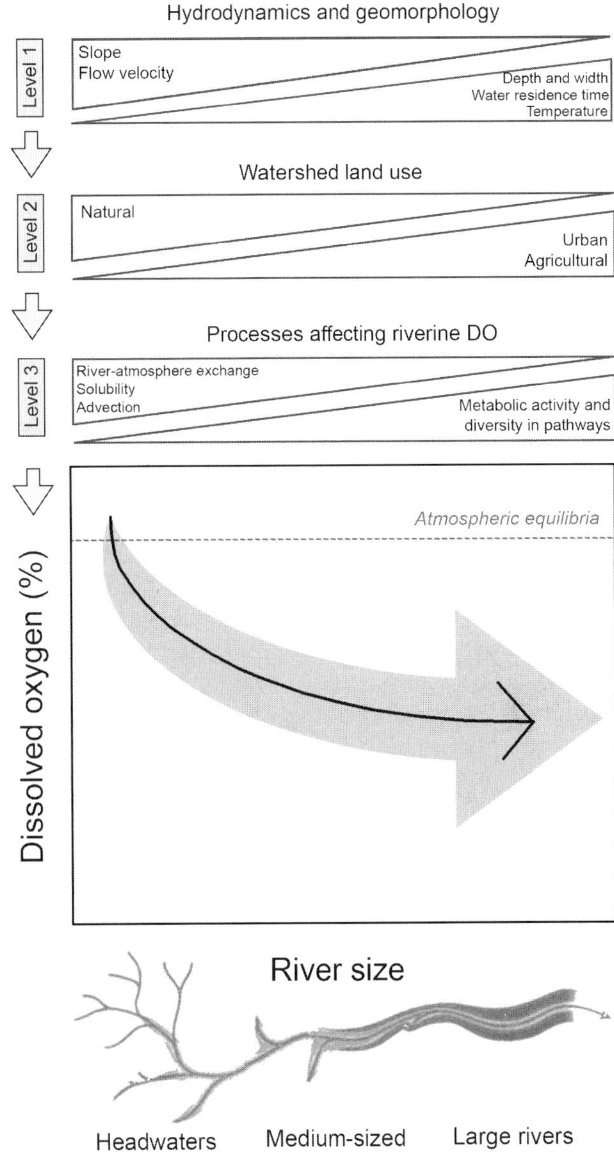

FIGURE 11-5 Conceptual model of patterns and drivers of dissolved oxygen (DO) concentrations along the river continuum (from headwaters to larger rivers) in an idealized river network. The interactions between factors associated with channel hydro-geomorphological dynamics (*Level 1*), watershed land uses (*Level 2*), and riverine physical and biogeochemical processes (*Level 3*) generates expectations for patterns and variability of dissolved oxygen saturation along river networks.

A. Spatial variation

i. River network scale

River hydrodynamics, geomorphic characteristics, and watershed land uses generally change in a predictable way from small headwaters to larger rivers. As these attributes largely determine the physical dynamics (e.g., rates of gas exchange, temperature, light availability) and biogeochemical processes (e.g., the balance between photosynthesis and respiration) driving the content of oxygen in water (see Section II), the *River Continuum Concept* (sensu Vannote et al., 1980; see also Chapter 5) can be used as a simple hierarchical framework to provide general expectations for dissolved oxygen concentrations along river networks (Fig. 11-5).

Slope and flow velocity correlates with factors such as river depth, surface area, water residence time, and temperature to influence atmospheric exchange (see Section II). Because high reaeration can override signals of riverine metabolic activity and thus limit low oxygen conditions, the oxygen content of small headwater streams is expected to be stable and approximately at, or somewhat above, saturation (Fig. 11-5; Boix-Canadell et al., 2021b; Garvey et al., 2007; Owens et al., 1964). However, biotic consumption of oxygen and reduced solubility as a response to increasing water temperature is more pronounced in smaller streams (with a smaller ratio of water volume to streambed active area) than in larger rivers, leading to more variable oxygen concentrations in low-gradient headwater streams that are less influenced by gas exchange (Blaszczak et al., 2022; Garvey et al., 2007). Conditions of increased temperature in smaller systems are usually associated with anthropogenic water extraction (Pardo and García, 2016), river impoundment (Zaidel et al., 2021), or riparian tree cover loss (Dodds et al., 2015) but can also occur during flow stagnation or extended low flow periods during droughts (Gómez-Gener et al., 2020; Vazquez et al., 2011). Droughts and flow intermittency are common to many rivers worldwide, a phenomenon that may become even more widespread with global change (Messager et al., 2021). As flow recedes and river networks contract, water residence times and temperature increase, and, as a consequence, oxygen concentration decreases.

In contrast, larger rivers draining lowlands show lower and more spatially variable oxygen concentrations than smaller streams (Fig. 11-5). Factors that control oxygen concentrations, such as organic matter supply, temperature, and vertical stratification, are more spatially variable in low-gradient systems draining urbanized or agricultural catchments (Fig. 11-5). In addition, increased geomorphological and habitat complexity in unmodified rivers (e.g., nonchannelized rivers) is linked with enhanced activity and diversity of metabolic pathways, and thus larger oxygen variability (Peterson, 2001). For example, in backwater areas of large rivers water residence times are longer, limiting the advection of oxygen and promoting marked variations in oxygen content due to photosynthesis and respiration.

Reduced turbulence (and advection) in low-gradient rivers limits the replenishment of oxygen-depleted waters during downstream transport, leading to a progressive reduction in dissolved oxygen. Oxygen consumption from chemical reactions with dissolved organic and inorganic reduced compounds can be large (see Section II). In very productive wetland floodplains, as in the Pantanal wetland of the Paraguay River (Brazil), high rates of organic matter decomposition can lead to high bacterial production of methane (Hamilton et al., 1995). Subsequent bacterial methane oxidation can deplete dissolved oxygen in backwater areas (see Section II).

Moreover, lowland rivers are more prone to be in urban and agricultural watersheds. Rivers in human-modified watersheds are often subsidized with high organic matter and nutrient loading (e.g., from untreated sewage inputs or non-point source pollution). Riverine eutrophication can result in biochemical oxygen demand that exceeds physical reoxygenation and autotrophic production (Arroita et al., 2019; de Carvalho Aguiar et al., 2011).

River oxygen depletion can also be the result of hydrologic alterations associated with damming or water extraction. Hydroelectric dams that release deep water from reservoirs often create *hypoxic* conditions downstream due to both low oxygen and high nutrient concentrations (Miranda and Krogman, 2014). In addition, the presence of large dams and/or the withdrawal of water for irrigation can cause the *lentification* (when water stops flowing, increasing water residence time to values comparable to stagnant waters) of running waters, leading to reduced atmospheric exchange rates and lower oxygen supply in downstream reaches (Pardo and García, 2016). This lotic-lentic interplay can occur naturally (e.g., wetland—river) or in human-modified (e.g., reservoir—river) river networks and determines the extent of well-mixed versus more spatially and temporally heterogeneous river flows. Consequently, these dynamics are also a key factor in predicting dissolved oxygen dynamics in river networks.

ii. Reach and subreach scale

The hydraulic connections between surface water and groundwater are a major driver of spatial variability in oxygen concentrations, both when groundwater flows into a river reach (hydrologically gaining conditions) or when surface water flows downward to the groundwater (hydrologically losing conditions). Groundwater becomes naturally depleted in oxygen due to lack of recharge, aerobic respiration, and chemical weathering (Bloomer et al., 2016; Bu et al., 2020). For example, in alluvial aquifer-river systems flow zones (at the hundreds of meters scale) often determine the location of seasonally variable low oxygen sites within river reaches (Fig. 11-6a). However, as oxygen-depleted water reaches the river channel surface, reaeration occurs. Conversely, losing conditions do not normally imply a net gain or loss of oxygen with respect to surface water (Piatka et al., 2021).

At a smaller spatial scale (tens of meters), the transfer of oxygen can also occur between surface water and the riverbed due to the presence of logs and sand bars, and the sequence of alternating topographic highs (riffles) and lows (pools) (Boulton et al., 1998; Brunke and Gonser, 1997). In such cases, water infiltrates the subsurface and returns to the surface after spending some time in the hyporheic zone (i.e., hyporheic exchange; Boano et al., 2014; Harvey and Gooseff, 2015; Zarnetske et al., 2011). With surface waters often near oxygen saturation, and the subsurface often oxygen limited, the flux of oxygen from the surface to the hyporheic zone dictates the magnitude of oxidative processes. Thus, oxygen consumption rates are determined by the hyporheic exchange rate and the biological activity of the microbial communities in the hyporheic zone (Beaulieu et al., 2011; Mulholland et al., 1997; Jones and Holmes, 1996). Overall, the interplay between physical transport processes and microbially mediated biogeochemical reactions generate high spatial variability in oxygen at the subreach scale (Fig. 11-6b).

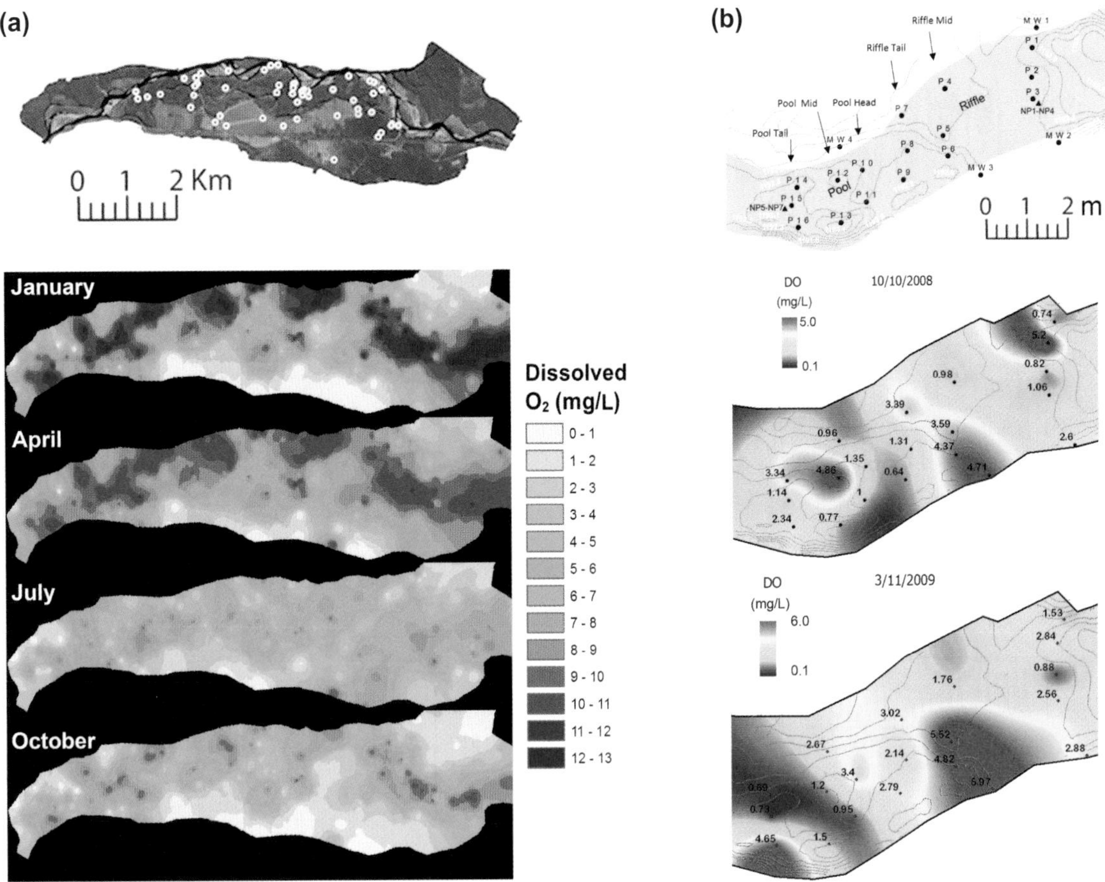

FIGURE 11-6 The spatial distribution of dissolved oxygen concentration varies over time within (a) a shallow alluvial aquifer-floodplain-river system, and (b) a hyporheic zone of a riffle-pool sequence. River flows from right to left in both cases. *Upper panels* show the location of sampling sites including piezometers and shallow riparian monitoring wells. (a) The Nyack Floodplain of the Middle Fork Flathead River located in northwest Montana, USA. *(Modified from Helton et al., 2012.)* (b) The Truckee River riffle-pool study area located in northwest Nevada, USA. *(Modified from Naranjo et al., 2015).*

The length of hyporheic flow paths determines the residence time of water, influencing the biogeochemical processes operating within a river reach (Malard et al., 2002). For example, a suite of redox reactions may occur as surface water moves through the sediment of a bar or riffle so that the concentrations of electron donors and acceptors, particularly of dissolved oxygen, would vary following a predictable thermodynamic sequence (Champ et al., 1979; Hedin et al., 1998). First, dissolved oxygen is consumed by aerobic respiration, and next, nitrification of ammonium produced by the mineralization of organic nitrogen increases nitrate concentration along the hyporheic flow path (Fig. 11-7). As dissolved oxygen is depleted, denitrification (Chapter 14) may remove nitrate produced at the upstream end of the flow path. Consequently, hyporheic flow systems within short sand bars are normally well oxygenated and are a source of nitrate to the river (Jones et al., 1995), whereas longer bars would be a sink of nitrate caused by the increased residence time and low oxygen subsurface water.

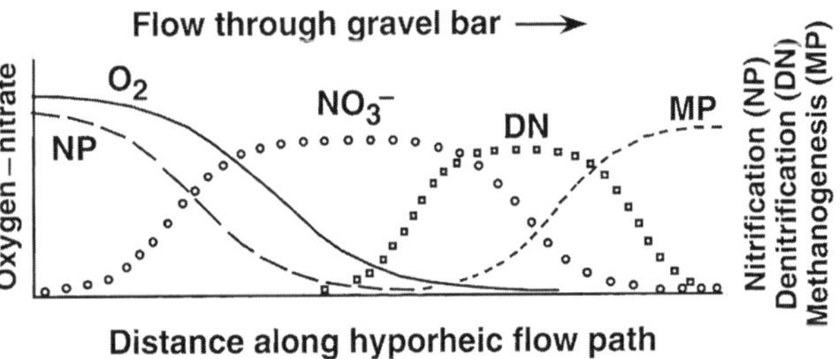

FIGURE 11-7 Hypothetical trends of dissolved oxygen (O_2) and nitrate (NO_3^-) concentration, and nitrification (NP), denitrification (DN), and methanogenesis (MP) along a hyporheic flow path within a gravel bar. *(From Malard et al., 2002.)*

B. Temporal variation

Rivers exhibit a broad range of dissolved oxygen fluctuations associated with processes operating on temporal scales of seconds to decades (Fig. 11-8). Each temporal scale of oxygen fluctuation provides different insights into the structure and function of rivers. Below,

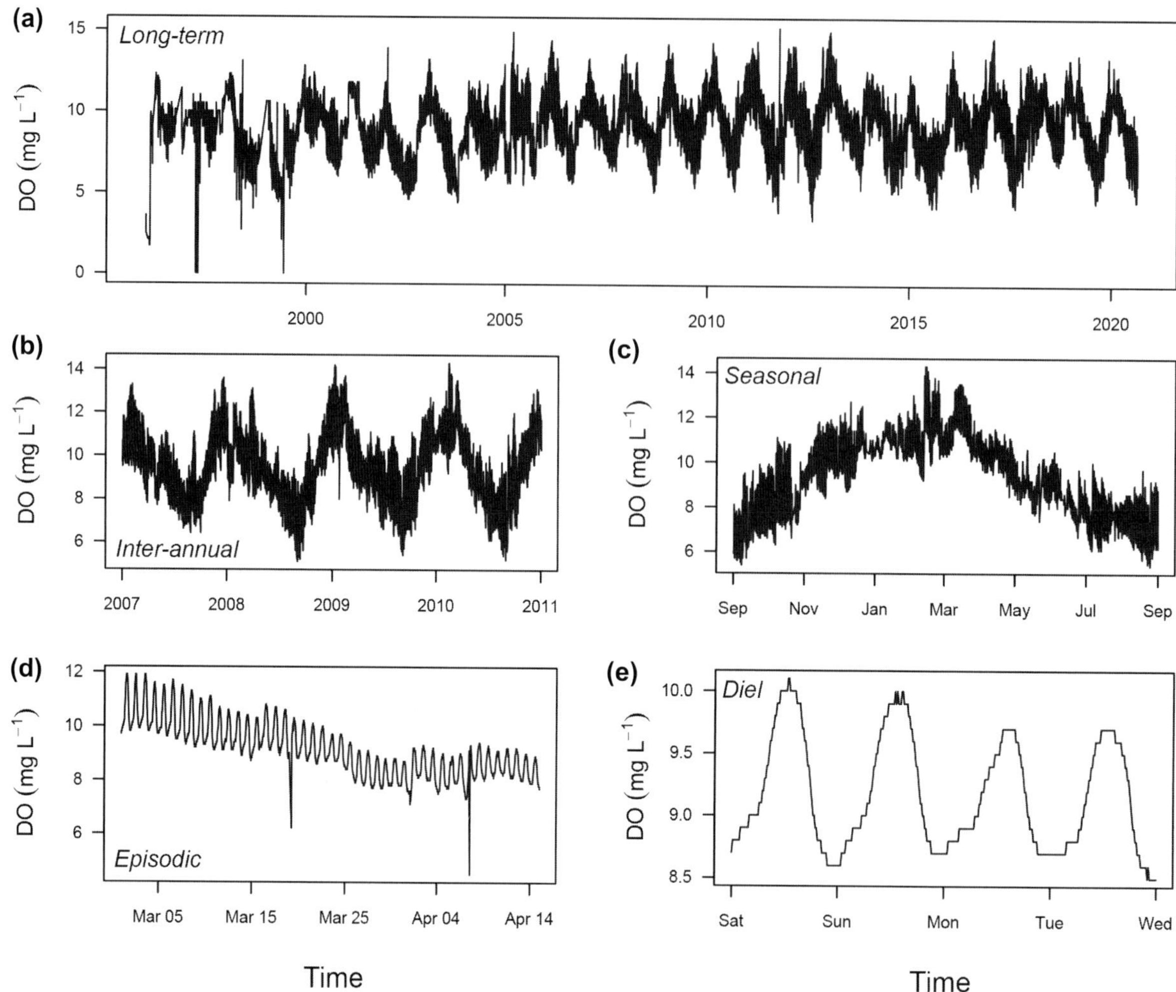

FIGURE 11-8 Time series of dissolved oxygen at the surface of the Cinca River (Ebro River Basin, Spain) illustrating how observed patterns can vary for (a) decadal, (b) multiannual, (c) seasonal, (d) episodic, and (e) diel timescales. This variability results from the rich spectrum of short- and long-term dynamics of biotic and abiotic origin and leads to different insights into the structure and function of rivers. *(Data from "Confederación Hidrográfica del Ebro" accessed through the SAICA [Automated Water Quality Information System] Portal [https://www.saicaebro.com/].)*

we discuss temporal variability in oxygen dynamics at different scales of ecological and biogeochemical relevance for river ecosystems. This includes the signals of processes operating over fractions of seconds (e.g., biochemical reactions), from seconds to days (e.g., photosynthesis), from days to weeks or even several months (e.g., solar cycles or snowmelt events), and longer-term patterns, including annual to decadal variability (e.g., severe droughts).

i. Annual patterns

The relative importance of the physical, chemical, and biological drivers of dissolved oxygen content in rivers changes throughout the year. These interacting processes determine a river's annual oxygen pattern. Annual oxygen patterns are mostly driven by the seasonal fluctuations of photosynthesis and respiration dependent on flow, light, and temperature conditions (Bernhardt et al., 2018; Savoy et al., 2019). Below, we describe different examples of river annual oxygen dynamics, classified according to their degree of seasonality (seasonal versus aseasonal rivers), and address potential controls.

Annual patterns in seasonal rivers.

The highly productive Menominee River, a large river in northern Wisconsin (USA), shows a regular and predictable annual pattern, with large summer peaks and constant low oxygen under ice during winter (Fig. 11-9a). This river is an example of how both low hydrological disturbance and unlimited light availability drive summer productivity peaks, as depicted by large daily oxygen swings. As fall turns into winter, rivers in snow-dominated regions normally enter a "dormant period" in which biophysical processes have a minimal effect on oxygen saturation (Boix-Canadell et al., 2021b). However, oxidative

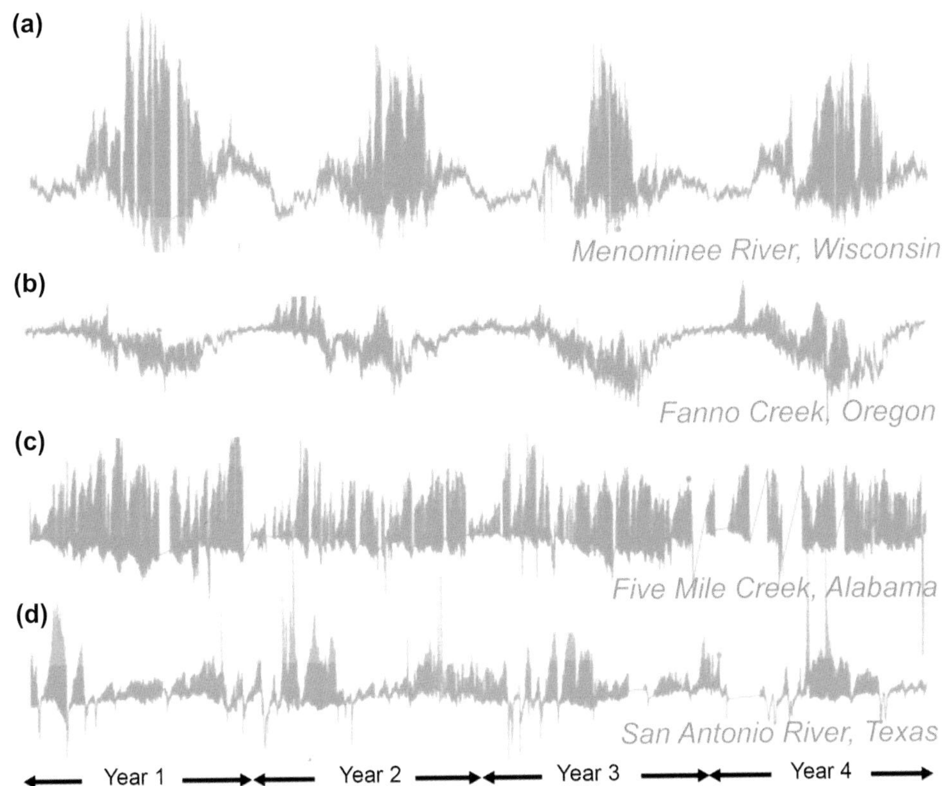

FIGURE 11-9 Temporal trends in dissolved oxygen (DO, % saturation) for four contrasting rivers over a 4-year period. (a) Menominee River in northern Wisconsin, USA. (b) Fanno Creek in western Oregon, USA. (c) Five Mile Creek in Alabama, USA. (d) San Antonio River in Texas, USA. The scale of both axes is the same for all four of the time series; note that the y-axes have been removed to focus on the differences in annual patterns across streams. *(Modified from Bernhardt et al., 2018.)*

processes at the sediments in large rivers can promote some depletion of oxygen throughout the winter, although at a lower rate than in the summer because of colder temperatures (Akomeah and Lindenschmidt, 2017). In small rivers, ice formation during winter interrupts the exchange of oxygen with the atmosphere and groundwater chemistry can impose a constant baseline of oxygen saturation (Boix-Canadell et al., 2021b). As winter turns into spring, elevated discharge sustained by snowmelt increases light attenuation and physical disturbance, which are both unfavorable for oxygen production by benthic algae (Malard et al., 2006; Uehlinger et al., 2003). At the same time, increasing discharge also increases turbulence and hence atmospheric exchange, keeping oxygen near saturation (Boix-Canadell et al., 2021b). The Fanno Creek (Fig. 11-9b), a heavily shaded and frequently flooded small river in western Oregon (USA), shows how the phenology of terrestrial vegetation leads to transitions in the light regime during the leaf-out and litterfall periods that can be particularly pronounced in temperate streams (Hill and Dimick, 2002). Although it is well-known that light regime primarily drives riverine productivity variation over time (Roberts et al., 2007), regular and predictable litterfall patterns in the fall fuel high in-stream oxygen demand, which results in persistent low oxygen conditions (Acuña et al., 2004; Bernhardt et al., 2018; Carter et al., 2021). Dissolved organic matter leached from soils following precipitation events or loaded from organic sewage sources can similarly increase microbially mediated consumption of oxygen at shorter time scales (Demars, 2019).

In nutrient-limited oligotrophic rivers, such as those draining high-latitude and high-elevation catchments, the temporal variation in nutrient availability can also drive seasonal productivity (Myrstener et al., 2021). In such cases early spring (initial stages of snowmelt) and fall represent critical "windows of opportunity" when discharge, turbidity, riverbed stability, temperature, and light conditions align to promote high rates of river productivity and thus result in the highest oxygen variability exhibited within a year (e.g., Battin et al., 2004; Boix-Canadell et al., 2021a, 2021b; Malard et al., 2006). In streams influenced by glaciers, high levels of turbulence coupled with reduced light availability additionally limit benthic oxygen production and push oxygen saturation back to equilibrium during summer, the period of maximum ice ablation (Boix-Canadell et al., 2021b). Therefore physical drivers of oxygen dynamics prevail over biological drivers in the period between "shoulder" (i.e., fall or spring) seasons in these high-elevation rivers.

Annual patterns in aseasonal rivers.

Five Mile Creek in Alabama and San Antonio River in Texas (both USA) show little seasonality (Fig. 11-9c and d, respectively). Five Mile Creek exhibits large diel variations in oxygen, while the more urbanized San Antonio River shows more erratic behavior. This stochastic oxygen pattern is characteristic of river ecosystems draining highly urbanized areas subject to more "flashy" hydrographs (Blaszczak et al., 2019; Walsh et al., 2005). Short-term episodic hydrological events (e.g., from one to few days), such as storms, dampen regular oxygen oscillations through a combination of increased water volume, shading due to turbidity, and removal or burial of photosynthetically active biomass (Savoy et al., 2019; Bernhardt et al., 2018; Reisinger et al., 2017). The intensity of hydrological disturbance thus determines the duration for oxygen dynamics to recover back to baseline conditions (Bernhardt et al., 2018; Carter et al., 2021). For example, recovery times for both primary production and respiration (and consequently oxygen) are expected to be longer after large floods that disturb riverbeds compared with small floods that only generate turbidity without disturbing the riverbed (Hall et al., 2012; O'Connor et al., 2012). In addition, storms can also lead to episodic hypoxia (Blaszczak et al., 2019; Carter et al., 2021), coinciding with pulses of low-oxygen water or increasing organic matter from anthropogenic waste or natural storage (Dutton et al., 2018; Whitworth et al., 2012).

ii. Diel patterns

Diurnal cycles in solar radiation impose a well-known periodicity on river biogeochemical processes, creating diel (i.e., 24-h periods) patterns for many solutes and gases, including nutrients, dissolved organic matter, metals (Nimick et al., 2003), and dissolved oxygen (Hensley and Cohen, 2016). Photosynthesis increases oxygen concentrations on a diurnal basis, particularly in relatively slow-moving rivers (Müller and Weise, 1987). In productive rivers, diurnal oscillations in dissolved oxygen concentrations are even more magnified (Fig. 11-10). For example, in a wide, unshaded reach of an urban tributary draining the New Hope Creek watershed (North Carolina, USA), hypoxia occurs daily at night, with diel oxygen concentrations oscillating as much as 10 mg O_2 L^{-1} day^{-1} (130% saturation) in response to very high primary productivity (Carter et al., 2021). Heterotrophic and autotrophic respiration at night results in repeated oxic-hypoxic cycles (e.g., downstream of sewage inputs).

At the watershed scale, land use is a major determinant of a river's daily oxygen dynamics. Land use controls light availability within channels and alters the intensity of hydrologic disturbances as well as the delivery of limiting nutrients and organic matter (Stets et al., 2020). For example, forested catchments are typically characterized by dense canopies surrounding stream channels. Light availability for photosynthesis in those rivers is limited (Koenig et al., 2019; Savoy and Harvey, 2021), resulting in damped diel cycles of dissolved oxygen (dos Reis Oliveira et al., 2019; Fig. 11-11). For other types

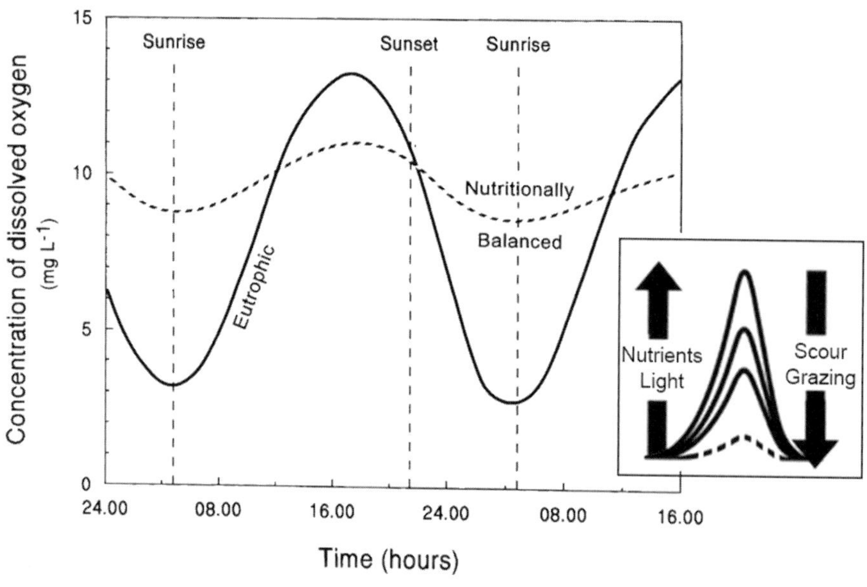

FIGURE 11-10 Diurnal cycles of dissolved oxygen concentrations in a nutritionally balanced (i.e., oligotrophic) and a eutrophic river. *Inset* represents major biophysical factors driving changes in the amplitude of diel oxygen cycles (e.g., higher nutrient inputs in rivers increase the amplitude of the diel cycles by subsidizing autotrophic primary production). *(Main figure from Walling and Webb, 1994; Inset modified from Bernhardt et al., 2018.)*

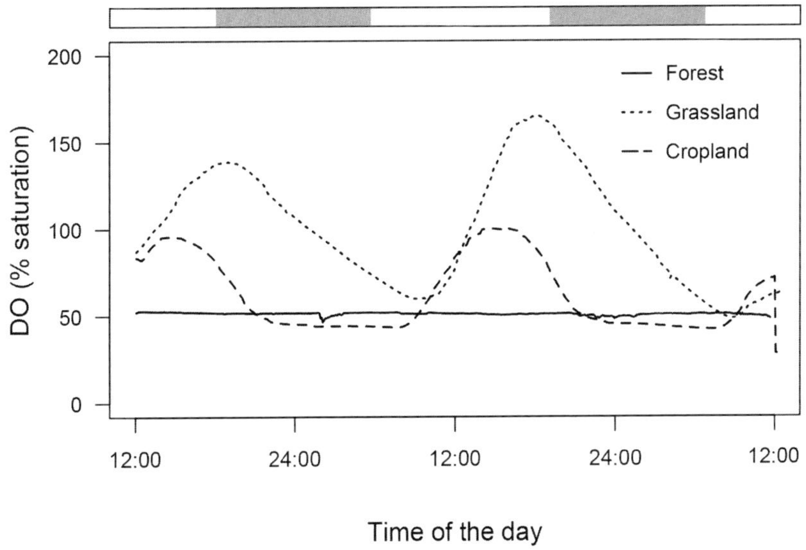

FIGURE 11-11 Examples of daily trends in dissolved oxygen content (DO, % saturation) measured over 48 h (two daily cycles) in three rivers draining contrasted land-use types. Light and dark periods are represented by *white* and *gray rectangles*, respectively. *(From dos Reis Oliveira et al., 2019.)*

of land uses (e.g., open-canopy grasslands and croplands), high nutrient inputs and unlimited light availability make aquatic primary production and diel oxygen cycles more likely (dos Reis Oliveira et al., 2019; Fig. 11-11). Other processes also vary at subdaily timescales and drive diel changes in dissolved oxygen, including light-induced photochemical oxidation of organic matter (see Section II), diel changes in discharge and associated scouring (Figure 11-10), and lateral export of water with various amounts of oxygen from adjacent terrestrial environments (Bloomer et al., 2016; Bu et al., 2020). Regardless of the driving forces, the magnitude and extent of diel oxygen changes in running waters remain poorly studied except for a few well-studied cases (e.g., Carter et al., 2021; Venkiteswaran et al., 2015).

C. Changes in river oxygen concentrations at decadal and continental scales

Long-term oxygen data series are key to revealing both improvements and declines in river oxygen content. For example, long-term monitoring data have enabled the detection of a decline in biochemical oxygen demand trends in Europe since 1992, following a general improvement in wastewater treatment induced by the EU Water Framework Directive (WFD) (European Community, 2000). These long-term data (e.g., Fig. 11-8a) have also detected ecological recovery patterns of highly altered rivers following sewage abatement (Arroita et al., 2019). However, data from the Global Riverine Dissolved Oxygen database (GRDO, https://doi.org/10.5066/P99X6SIR), which is composed of 125,158 unique river locations spanning more than 90 countries, show that riverine hypoxia is more widely distributed than had been previously assumed, and more likely to occur in warmer, smaller, and lower-gradient rivers draining urban land cover (Blaszczak et al., 2022).

Despite these advances, larger-scale spatial and temporal patterns in dissolved oxygen remain largely unknown. There is still a clear bias toward river monitoring in temperate regions and several major regions on the planet (central Africa, eastern South America, and almost all of Asia) remain woefully understudied (Piatka et al., 2021). Consequently, the rivers described in this section are isolated examples, although promising recent studies are rapidly expanding our understanding of river oxygen dynamics (e.g., Segatto et al., 2021; Zhi et al., 2021).

IV. Distribution of dissolved oxygen in lakes

Oxygen concentrations in lakes can vary both temporally on diel, seasonal, and annual scales and spatially on horizontal gradients between littoral to pelagic sites and on vertical gradients with depth. Below, we first discuss temporal variability in oxygen dynamics in dimictic, north temperate lakes as an illustrative conceptual model, and then address a broader suite of observed dynamics in lakes of different mixing regimes.

On time scales of minutes to hours, oxygen concentrations can respond rapidly to meteorological forcing, river inflows, biogeochemical cycling (e.g., chemical oxidation of reduced ions), and plankton populations (Langman et al., 2010). There is generally greater variability in oxygen within the epilimnion than the

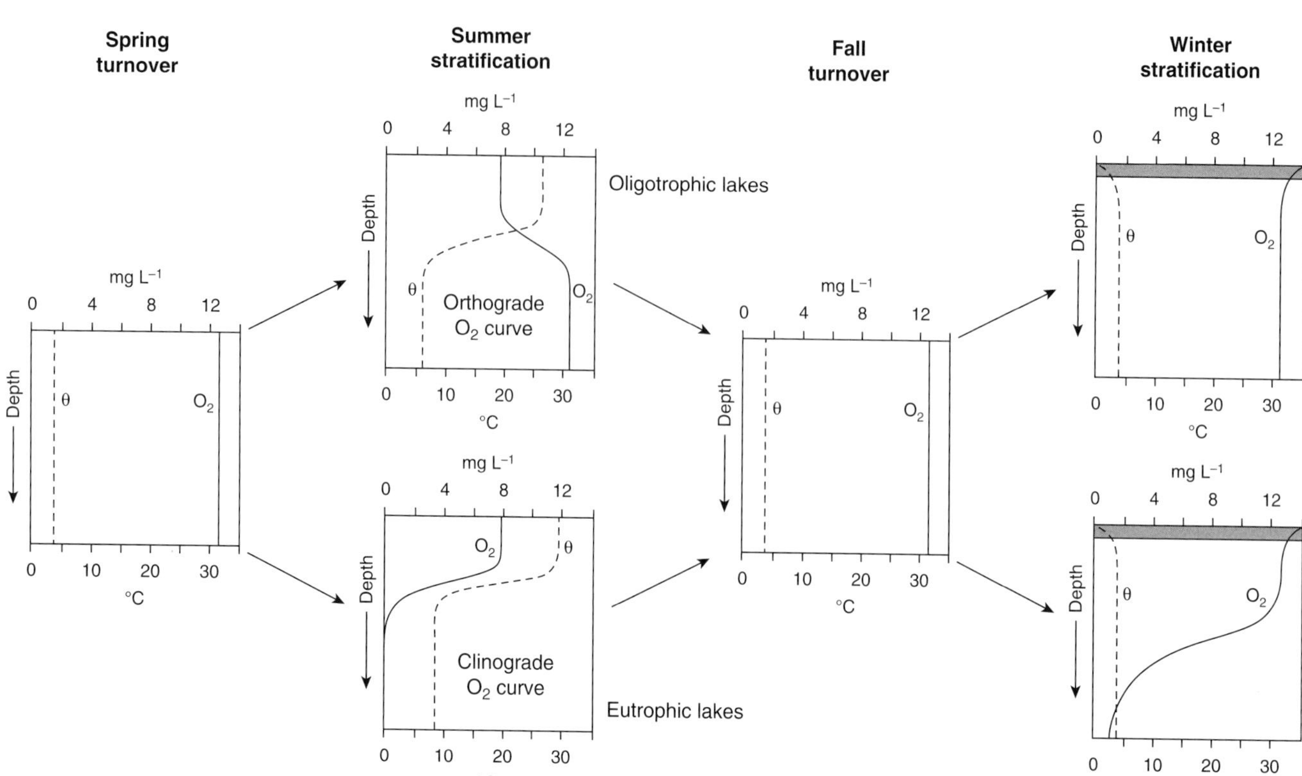

FIGURE 11-12 Idealized vertical distribution of oxygen concentrations and water temperature (θ) during the four seasons of an oligotrophic and a eutrophic dimictic lake.

hypolimnion on daily and subdaily scales during thermally stratified summer periods. During the day, oxygen in the epilimnion increases due to primary production and decreases during the night due to respiration. In productive lakes, photosynthesis during the day can result in epilimnetic *supersaturation* until there is an evasion of the excess oxygen to the atmosphere. Mixing events due to storms or inflows can also result in short-term increases in oxygen if surface oxic water is entrained into deeper layers that are undersaturated. Internal seiches (Chapter 8) may also leave a measurable imprint on oxygen dynamics at daily time scales by promoting vertical and horizontal advection in surface and deeper layers (Fernández Castro et al., 2021).

On seasonal time scales, lakes can exhibit distinct vertical oxygen profiles, enabling the differentiation of lakes based on their oxygen dynamics (Fig. 11-12). These profiles emerge from physical, chemical, and biological interactions within the waterbody, and thus limnologists can infer valuable information about a lake's ecosystem functioning from its oxygen profile. It is important to note that the classification of lake oxygen profiles is based on profiles collected during thermally stratified conditions. This is because oxygen concentrations can be at or near 100% saturation throughout a vertical profile when the lake is mixing, making it thus impossible to differentiate among profile types during mixing periods.

Mixing periods are an important time for increasing oxygen availability in the bottom waters of lakes. Because molecular diffusion is very slow (see Section II), generally, the only way for the entire water column to reach 100% dissolved oxygen is for the water column to mix, as occurs during spring and fall turnover. After the onset of mixing, 100% saturation is established relatively quickly, usually in a few days in most lakes. Very deep lakes, however, require longer periods for complete equilibrium saturation, and saturation may not be achieved before the onset of stratification. In an idealized deep lake the oxygen concentration throughout the water column during circulation is at or near 100% saturation (Fig. 11-12).

Once thermal stratification sets up in the summer in a temperate lake, four distinct oxygen profiles can emerge: *orthograde*, *clinograde*, *positive heterograde*, and *negative heterograde* (Fig. 11-13). Below, we address each of these oxygen profiles, focusing on the example of an idealized dimictic, deep lake.

A. Orthograde oxygen profile

For lakes that are very *oligotrophic* (low in nutrient inputs and productivity), primary production and respiration can be so low that oxygen concentrations are

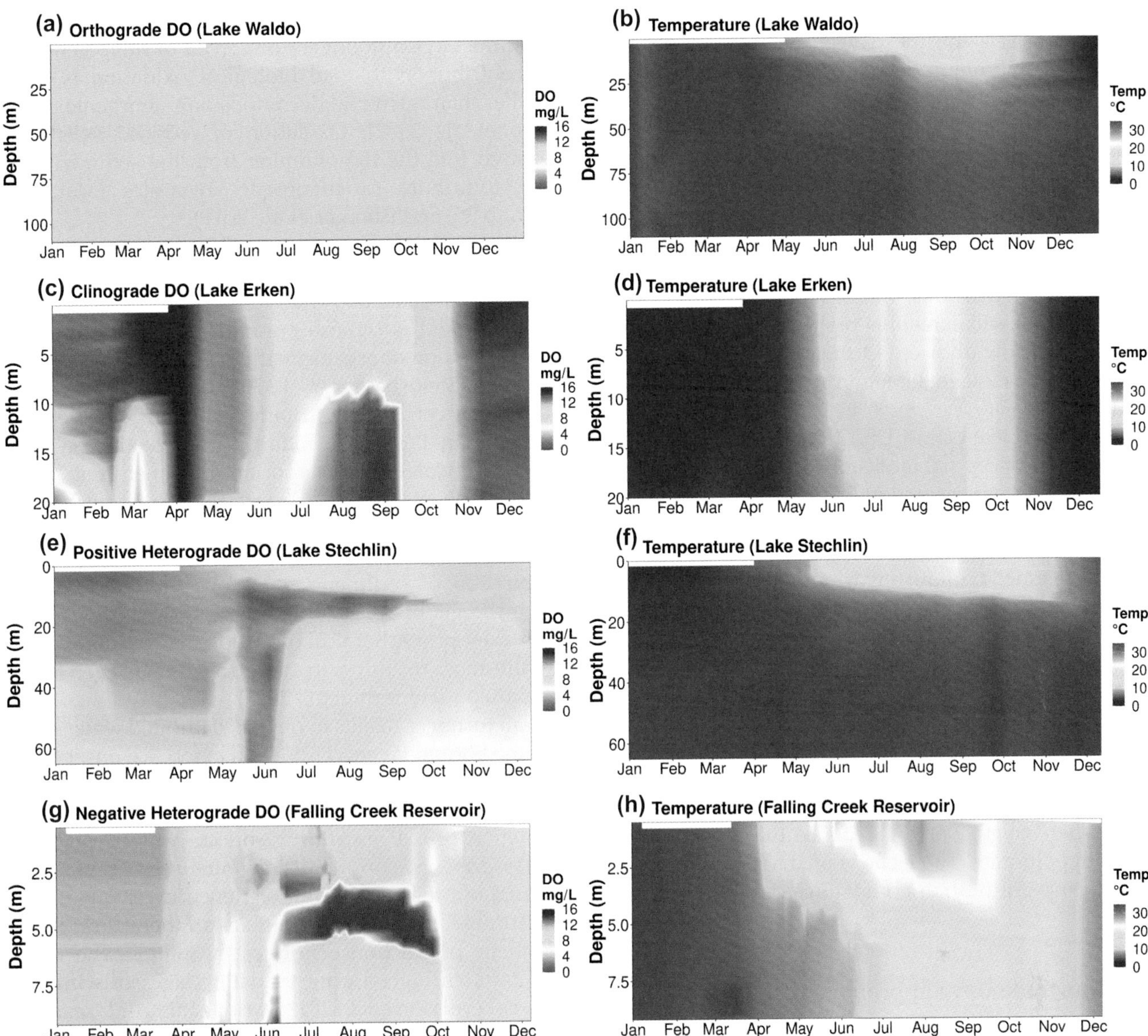

FIGURE 11-13 Time-depth dissolved oxygen (DO, *left column*) and temperature (*right column*) observations from January to December illustrate (a and b) orthograde dynamics in Lake Waldo (USA); (c and d) clinograde dynamics in Lake Erken (Sweden); (e and f) metalimnetic oxygen maximum (positive heterograde) dynamics in Lake Stechlin (Germany); and (g and h) metalimnetic oxygen minimum (negative heterograde) dynamics in Falling Creek Reservoir (USA). The oxygen and temperature scales are the same for all four dimictic lakes for comparison. *White bars* on the surface of lakes represent ice-covered periods. *(Oxygen and temperature data for Lake Waldo are derived from Salinas, 2000 and Sytsma et al., 2004. Due to the few profiles available, most of these data are interpolated from observations collected from May to October. Oxygen and temperature data for Lake Erken are from Stetler et al., 2021, with ice data from Benson et al., 2000. Lake Stechlin data were provided by the Department of Experimental Limnology, Leibniz Institute of Freshwater Ecology and Inland Fisheries (IGB). Falling Creek Reservoir oxygen and temperature data are from Carey et al., 2021, with ice data from Carey, 2021.)*

regulated largely by physical processes. Throughout the year, the lake exhibits ~100% saturation at each depth, regardless of whether it is mixed or thermally stratified. Due to the colder waters in the hypolimnion, oxygen concentrations actually increase in the deeper waters. During stratification, this oxygen profile has been termed *orthograde* (Åberg and Rodhe, 1942). Because most lakes have at least some biological activity, orthograde conditions are rare and found only consistently in a few extremely unproductive lakes or in moderately oligotrophic lakes during the early stages of summer stratification. Because of increased primary productivity in many waterbodies due to climate and land-use change, the number of lakes exhibiting orthograde profiles may be decreasing (Stoddard et al., 2016).

Due to their very low biological activity, differences in oxygen concentrations in orthograde depth profiles are driven primarily by the varying solubility of oxygen at different temperatures (Fig. 11-12). As summer thermal stratification sets up, the oxygen concentration in the circulating epilimnion decreases as its temperature increases and solubility decreases, promoting evasion of the excess oxygen to the atmosphere. In contrast, the hypolimnion remains at the same oxygen concentration as it exhibited during spring turnover. As a result, oxygen concentrations in the hypolimnion are higher than in the epilimnion during summer stratification, though the depth profile remains at 100% saturation at all depths. In winter, due to inverse stratification, a lake with orthograde profiles will exhibit its highest oxygen concentrations just below the ice. Orthograde profiles exhibit minimal to no oxygen depletion at the sediments during stratification.

Data from Lake Waldo, USA (Fig. 11-13a, b), generally support the conceptual model of a lake exhibiting orthograde dynamics presented in Fig. 11-12. Lake Waldo is a glacially formed alpine lake with summer Secchi depths >30 m (Larson, 2000; Sytsma et al., 2004). The lake has a small watershed without any tributaries, contributing to its low productivity (Salinas, 2000). In the summer, hypolimnetic oxygen depletion is not observed, and the oxygen profile is marked by relatively low variability throughout the year (Fig. 11-13a). Unlike the conceptual model in Fig. 11-12, however, Lake Waldo exhibits hypolimnetic supersaturation in the summer. Its high transparency enables solar radiation to reach deep layers and slightly warms the hypolimnion (Fig. 11-13b), decreasing oxygen solubility. Since the excess oxygen in the hypolimnion cannot be vented to the atmosphere, oxygen supersaturation occurs (Fig. 11-13a).

B. Clinograde oxygen profile

In contrast to orthograde profiles, *clinograde* profiles are very common in lakes. Oxygen in the hypolimnion of *eutrophic* (high in nutrient loading and productivity) lakes is depleted rapidly during stratified periods, often after a few days to weeks of stratification onset. The resulting profile, in which the hypolimnion exhibits oxygen depletion, potentially reaching hypoxia or even anoxia (see Section V), is termed *clinograde* (Fig 11-12). In the absence of turbulence, the horizontal influx of oxic water, density currents along basin sediments, or human intervention, the hypolimnion will remain anaerobic throughout the stratified period until oxygen is replenished during fall turnover.

The primary mechanism driving hypolimnetic oxygen depletion in most eutrophic lakes is aerobic respiration of organic matter by microbes, both in the water column and especially at the sediment–water interface, with oxidation of reduced substances secondary (Livingstone and Imboden, 1996; Schwefel et al., 2018). Although plant and animal respiration can occasionally consume large amounts of dissolved oxygen, generally, its contribution to total biological oxidation is much smaller than microbial decomposition of organic matter (Seto et al., 1982). Oxidation of reduced substances released into the hypolimnion from the sediments can also contribute to clinograde dynamics (Matzinger et al., 2010; Steinsberger et al., 2017).

The relative importance of different mechanisms of hypolimnetic oxygen depletion varies among lakes (Müller et al., 2012). In lakes less than ~100 m, deep oxygen demand is typically dominated by the microbial decomposition of organic matter at the sediment–water interface (Steinsberger et al., 2020). The flux of reduced substances can also contribute to hypolimnetic oxygen depletion in these waterbodies, with hypolimnion depth inversely related to its importance (Müller et al., 2012). In very deep lakes (>100 m) the decomposition of autochthonous organic matter is mostly complete by the time it reaches the sediments, and thus water column respiration of sedimenting organic matter dominates (Steinsberger et al., 2020). Conversely, in highly stained bog lakes, photochemical oxidation may assume greater significance, resulting in undersaturation, even in the epilimnion (Hanson et al., 2006).

Fall turnover results in oxygen-saturated water reaching the bottom layers (Fig. 11-12) before ice onset ceases the exchange of oxygen with the atmosphere. Without circulation, most eutrophic lakes will exhibit some depletion of oxygen during the ice-covered period at the sediments. Oxygen-consuming processes occur throughout the winter, although at a lower rate than in the summer because of the colder temperatures. Photosynthetic production continues throughout the winter and is often high during the later stages of winter ice cover (Bertilsson et al., 2013; Twiss et al., 2012), matched in some lakes by high respiration (Brentrup et al., 2021). Light penetration is variable with changing conditions of ice and snow cover, but the photic zone is generally confined to the upper layers (Obertegger et al., 2017).

The annual stages of a lake exhibiting clinograde dynamics shown in Fig. 11-12 are for the most part exhibited by Lake Erken, Sweden (Fig. 11-13c, d), but these data reveal overall much greater within-season variability. Noteworthy is the variability in oxygen under the ice from January to late March, likely due to physical processes (e.g., horizontal currents), followed by high oxygen during spring circulation in early April (Fig. 11-13c). Oxygen depletion at the sediments is evident despite mixing occurring in the upper portion of the water column from mid-April to the end of May. Depletion accelerates after the onset of thermal stratification in early June, resulting in anoxia throughout the hypolimnion by late summer, which favors the accumulation of reduced substances (Wendt-Potthoff et al.,

2014). Summer variability in epilimnetic oxygen is likely due to phytoplankton, which is halted by fall turnover in mid-September. Uniform oxygen concentrations throughout the depth profile at 100% saturation persist through the remainder of fall as the water temperature gets colder (Fig. 11-13c).

C. Metalimnetic oxygen maxima

Two variations in the vertical distributions of dissolved oxygen from orthograde and clinograde conditions are often observed. The first is an increase in oxygen in the metalimnion during stratification, which is referred to as either a *metalimnetic oxygen maximum* or a *positive heterograde curve* (Åberg and Rodhe, 1942). As the solubility of the epilimnetic water decreases with increasing summer temperatures, thereby decreasing oxygen in the surface waters, and hypolimnetic depletion decreases oxygen in the bottom waters, an oxygen maximum in the metalimnion can emerge relative to the rest of the profile (Fig. 11-14). Metalimnetic oxygen maxima can sometimes be extremely pronounced, with supersaturated values recorded above 300% (Eberly, 1964) and even above 400% (Birge and Juday, 1911).

The development of metalimnetic oxygen maxima can be attributed to both physical and biological processes (Wilkinson et al., 2015). The dominant physical process is the trapping of a layer of cold water with high oxygen concentrations under the thermocline as thermal stratification sets up after spring mixing. Solar radiation warms this layer, decreasing its oxygen solubility and thereby increasing its percent saturation above 100%. This oxic layer can persist until mixing occurs, or the oxygen is depleted by biological and chemical oxidation. Interflow of oxic water from inflowing streams into the metalimnion may also contribute to the formation of an oxygen maximum.

Biological processes can also form metalimnetic oxygen maxima. Deep chlorophyll maxima are widely observed in many lakes (Leach et al., 2018), driven by subsurface aggregations of multiple phytoplankton taxa (Lofton et al., 2020). In some lakes it is estimated that up to 60% of aerial primary production is attributable to deep chlorophyll maxima (Moll et al., 1984; Pannard et al., 2020). The phytoplankton optimize their vertical distribution to ensure sufficient temperature and light for growth as well as access to higher nutrients at depth (Cullen, 2015), which are ideal conditions for mixotrophs (Tittel et al., 2003). High densities of submerged macrophytes are also associated with metalimnetic oxygen maxima (Dubay and Simmons, 1979; Ruttner, 1963).

The relative contribution of physical versus biological processes in driving metalimnetic oxygen maxima is debated. Wilkinson et al. (2015) observed that physical processes dominated the formation of metalimnetic oxygen maxima in 17 small temperate lakes. In contrast, Moll et al. (1984) found that physical processes were less important than biological processes in driving Lake Michigan's chlorophyll maxima. If a metalimnetic layer is extremely supersaturated, it is likely that biological processes dominate because only primary production can increase oxygen saturation much above 100%.

Data from Lake Stechlin, Germany (Fig. 11-13e, f), illustrate a persistent metalimnetic oxygen maximum throughout summer stratification. Both physical and biological processes likely contributed to its development. First, the maximum developed as stratification set up in May, trapping cold oxic water below the thermocline. Solar radiation warmed this layer, decreasing its solubility and gradually increasing its percent saturation to 123% by late August (Fig. 11-13e). Second, the lake has a deep chlorophyll maximum of picoplankton that repeatedly sets up at ~10 m (Padisák et al., 1997), coinciding with the oxygen maximum.

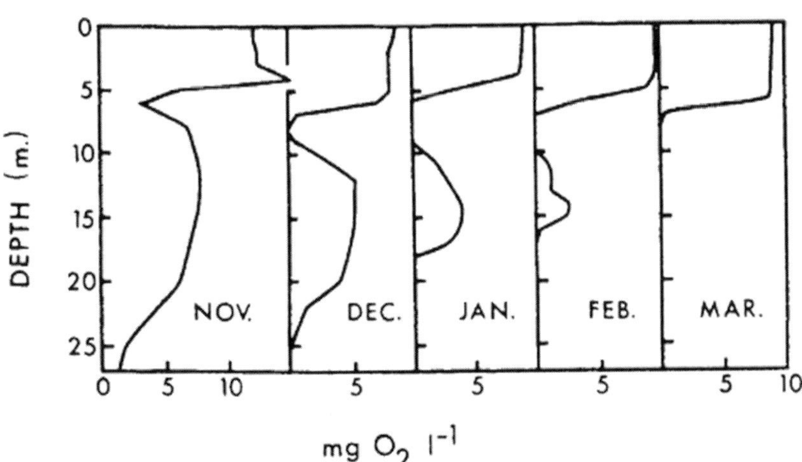

FIGURE 11-14 Depth profiles of dissolved oxygen concentrations during the summer (southern hemisphere) in hypereutrophic Lake Johnson, New Zealand. In November (Nov.) the lake exhibited a metalimnetic oxygen maximum just above 5 m due to high primary production, with a minimum located just below due to high respiration of sedimenting organic matter. The phytoplankton population in the chlorophyll maximum had collapsed by December (Dec.), thereby increasing the metalimnetic oxygen minimum, which persisted until March (Mar.). *(Modified from Mitchell and Burns, 1979.)*

D. Metalimnetic oxygen minima

The opposite condition, a *metalimnetic oxygen minimum* or a *negative heterograde curve*, is also commonly observed in many waterbodies (reviewed by Cole and Hannan, 1990; McClure et al., 2018). Numerous physical and biological mechanisms have been attributed to causing metalimnetic oxygen minima.

First, the basin morphometry of a waterbody can contribute to the formation of metalimnetic oxygen minima. For lakes with a gradual basin slope and substantial contact with the metalimnetic layer, a larger area of the sediments will be in contact with metalimnetic water than hypolimnetic water. The oxygen reduction in the metalimnion due to high sediment microbial decomposition may be sufficient to extend laterally over the entire lake, or, as in the case of Skärshultsjön, Sweden (Alsterberg, 1927), it may occur more strongly in the metalimnion closer to the lake perimeter, near the sediment–metalimnion interface.

Second, interflow from incoming streams can introduce low-oxygen water or water with high sediment or nutrient loads in the middle of the water column, thereby resulting in greater decomposition in the metalimnion than the layers above and below (Nix, 1981).

Third, a major biological driver of metalimnetic oxygen minima is the sinking and subsequent respiration of autochthonous organic matter in the water column (Kreling et al., 2017). During stratified periods, sinking rates slow when the organic matter encounters denser metalimnetic water, allowing more time for decomposition (Birge and Juday, 1911; Thienemann, 1928). Moreover, decomposition rates are usually higher in the metalimnion, where temperatures are greater, than in the hypolimnion. It is also possible that a phytoplankton population responsible for forming a metalimnetic oxygen maximum could collapse, driving a subsequent metalimnetic oxygen minimum (Wentzky et al., 2019), as shown in Fig. 11-14.

Fourth, in certain situations large populations of zooplankton in the metalimnion can contribute to a severe reduction of oxygen. Such respiratory consumption was most likely a major cause of the metalimnetic minima observed in Zürichsee, Switzerland (Minder, 1923), and Lake Washington, USA, where copepods can reach high populations (Shapiro, 1960).

Reservoirs are prone to exhibit metalimnetic oxygen minima due to their morphometry and the prevalence of *interflows* and *outflow extraction*, which favor water currents in the middle of the water column (Cole and Hannan, 1990). Reservoir water moving horizontally over the sediments downstream can become depleted of oxygen. As the reservoir's channel transitions from the riverine zone to the lacustrine zone, this layer of water can become a metalimnetic oxygen minimum if the water column is stratified as the basin deepens at the dam. If water is being extracted from metalimnetic depths, the minimum can be further intensified by currents that entrain this low-oxygen water from upstream into the deep basin. During storms or periods of high inflow, interflow can further accelerate oxygen consumption in a reservoir's metalimnion (LaBounty and Horn, 1997; Nix, 1981).

Falling Creek Reservoir, USA, is a eutrophic reservoir with a strong metalimnetic minimum in the summer stratified period (Fig. 11-13e, f). The reservoir's ice-off occurs in March, with well-mixed conditions until March. Similar to Lake Erken (Fig. 11-13c), Falling Creek Reservoir exhibited early oxygen depletion at its sediments while spring circulation occurred throughout the rest of the water column (Fig. 11-13e). The reservoir exhibited clinograde dynamics until the onset of a metalimnetic oxygen minimum in May, which became more pronounced in June and persisted until turnover in October. The high oxygen in the epilimnion during May to mid-July was due to high phytoplankton production. This epilimnetic production likely contributed to the metalimnetic oxygen minimum throughout the rest of the summer. The oxygen minimum was also favored by mid-water column extraction and the gradual upstream slope that implies high exposure of the metalimnion to sediments (McClure et al., 2018). One of the major takeaways from the Falling Creek Reservoir data is the substantial variability in its oxygen profiles over the year, driven by a combination of human management, variable inflow volumes, high productivity, and the reservoir's morphometry. Falling Creek Reservoir exemplifies the dynamism in oxygen conditions that many eutrophic reservoirs exhibit.

E. Oxygen dynamics in lakes with different mixing regimes

While the examples provided in Figs. 11-12 and 11-13 are useful for illustrating oxygen dynamics in north temperate dimictic lakes, these waterbodies represent only a subset of lakes. Below, we analyze oxygen profiles from lakes with amictic, warm monomictic, cold monomictic, polymictic, meromictic, and oligomictic mixing regimes (Chapter 7) located around the globe.

i. Amictic

The oxygen profiles in amictic Lake Bonney (Fig. 11-15a, b) are marked by extremely high oxygen concentrations and consistently stratified conditions. Lake Bonney is located in Antarctica and has permanent ice cover that is 3–5 m deep and very high salinities at depth (Spigel et al., 2018; Spigel and Priscu, 1998). Like in other amictic lakes in this region (e.g., Craig et al., 1992; Wharton et al.,

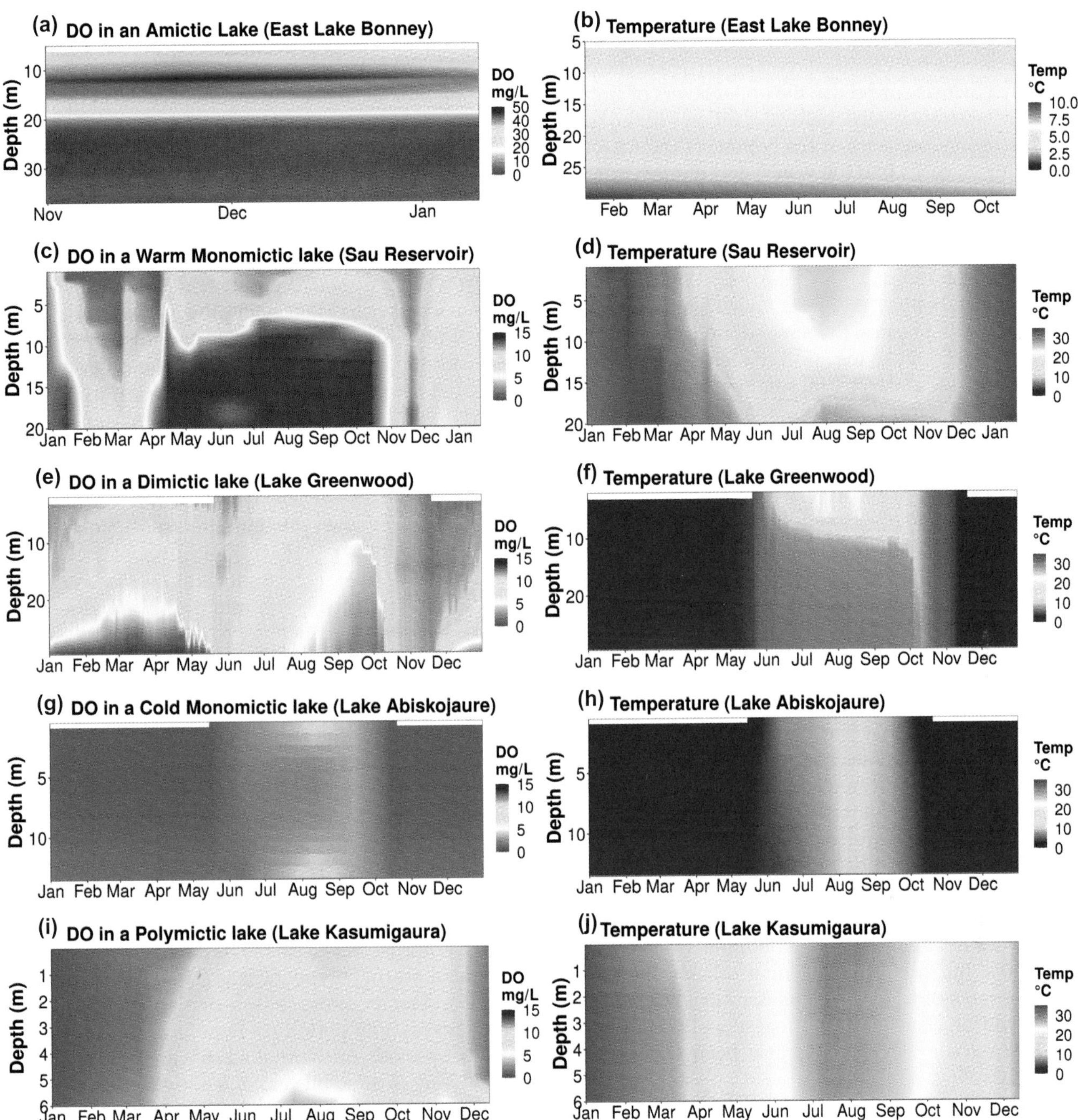

FIGURE 11-15 Time-depth dissolved oxygen (DO, *left column*) and water temperature (*right column*) observations illustrate dynamics in: (a and b) amictic East Lobe of Lake Bonney (Antarctica); (c and d) warm monomictic Sau Reservoir (Spain); (e and f) dimictic Lake Greenwood (USA); (g and h) cold monomictic Lake Abiskojaure (Sweden); and (i and j) continuous warm polymictic Lake Kasumigaura (Japan). All lakes show January to December data except for Bonney, which shows data from the austral summer of November to January, and Sau, which shows data from January to January to highlight episodic surface anoxia during some winter mixing periods. Note that the oxygen scale for East Lake Bonney extends from 0 to 50 mg O_2 L^{-1} and the temperature scale extends from 0 to 10°C, whereas the other lakes all share the same oxygen scale from 0 to 15 mg O_2 L^{-1} and temperature scale of 0 to 35°C. *White bars* on the surface of lakes represent ice-covered periods. *(Lake Bonney oxygen data are available from Lyons, 2016 and temperature data from Priscu, 2019. Sau data were provided by R. Marcé. Greenwood data were provided by the Sentinel Lakes Program, Minnesota Department of Natural Resources. Abiskojaure data were obtained from Stetler et al., 2021. Kasumigaura data were obtained from the Lake Kasumigaura Long-term Environmental Monitoring Program of the National Institute for Environmental Studies (NIES), Japan, and available at NIES, 2016.)*

1986), gases in Lake Bonney are supersaturated, resulting in oxygen concentrations exceeding 200% saturation above an anoxic hypersaline layer (Fig. 11-15a). The processes that increase oxygen in the upper layers of amictic Antarctic lakes are glacial meltwater inflows in the summer, the formation of ice at the bottom of the ice sheet (which forces gases into the water) and photosynthesis, while the processes decreasing oxygen in the lower layers are ablation through the ice as bubbles, respiration, and diffusion from the meltwater perimeter of the lake in the summer (Craig et al., 1992; Wharton et al., 1986). In Lake Bonney and other amictic lakes with meters of ice cover, physical processes dominate the annual oxygen budget. Due to logistical challenges, no data are available outside of the austral summer, but it is unlikely that oxygen concentrations substantively change during the year.

ii. Warm monomictic

In contrast to amictic lakes, warm monomictic lakes, such as Sau Reservoir, Spain, never have any ice cover and exhibit mixing and concomitant constant oxygen concentrations throughout the vertical profile during at least part of the year (Fig. 11-15c, d). As lakes exhibiting orthograde profiles are generally found in alpine or higher latitude locations, where ice cover is more likely (thus precluding a warm monomictic mixing regime), it would be expected that warm monomictic lakes would exhibit clinograde profiles during their stratified period. Clinograde conditions are especially likely if there is a long stratified period for the warm monomictic lake, decreasing the exposure of sediments to oxygen for most of the year, as in Sau Reservoir. The brief mixing period for Sau Reservoir lasts for 1 to 2 months per year prior to the onset of stratification, with subsequent hypoxic and then anoxic conditions setting up throughout the entire hypolimnion quickly thereafter. Thus, for most of the year, Sau Reservoir's oxygen profiles would be classified as clinograde, although negative heterograde curves have also been observed in some years.

iii. Dimictic

Oxygen dynamics in dimictic, north temperate lakes are discussed above at length, but we include Lake Greenwood, USA, for comparison with the other mixing regimes (Fig. 11-15e, f). Lake Greenwood exhibits classic clinograde profiles during both its winter and summer stratified periods. The lake is ice covered from November until May and exhibits anoxia near the sediments throughout most of the ice period. Interestingly, there are lower oxygen concentrations at the sediments in the winter than in the summer, potentially because the winter stratified period has a longer duration (6 months) than the summer stratified period (~3 months). Greenwood exhibits substantial daily to weekly swings in oxygen in the upper layers of the lake below the ice from March until ice-off in May, reflecting biological activity during this period.

iv. Cold monomictic

Oxygen profiles from Lake Abiskojaure, Sweden, reflect typical cold monomictic dynamics: low variability through the year, and the presence of orthograde profiles (Fig. 11-15g, h). Abiskojaure is a clear-water, oligotrophic arctic lake (latitude 68°N) that experiences ice from October to May. Under the ice, oxygen profiles are at or near 100% saturation. During the mixed period from July to October, some minor oxygen depletion is evident at the sediments, but the profile stays overall oxic, $\geq$10 mg O_2 L^{-1}. The lack of variability in oxygen dynamics is primarily due to the low productivity in the lake, similar to Lake Waldo (Fig. 11-13a). Although water temperatures reach above 4°C in the summer, wind-driven mixing prevents thermal stratification from establishing.

v. Polymictic

Lake Kasumigaura, Japan, provides an example of a continuous warm polymictic lake (Fig. 11-15i, j). The lake's water column exhibits an isothermal temperature profile throughout the year, with circulation occurring near-continuously due to its shallow depth and strong winds. Kasumigaura is hypereutrophic, with chlorophyll *a* concentrations exceeding 50 μg L^{-1} most of the year (Fukushima and Matsushita, 2021), which explains the oxygen depletion at the sediments despite isothermal conditions. The oxygen depletion is particularly evident in the warm summer months, when water temperatures >20°C occur throughout the water column.

Lake Rotorua, New Zealand, is a second example of a continuous warm polymictic lake (Fig. 11-16). Dissolved oxygen and temperature sensors deployed at 1 and 20 m depths reveal brief periods of hypolimnetic anoxia during short periods of thermal stratification until continuous mixing is reestablished.

vi. Meromictic

Extreme clinograde oxygen profiles can occur in permanently stratified *meromictic* lakes, such as Lake Kivu, located on the border of Rwanda and the Democratic Republic of the Congo (Fig. 11-17a, b). The monimolimnion receives inputs of sedimented organic matter and has high concentrations of reduced substances, which rapidly consume any oxygen intrusions due to mixing. In Lake Kivu the monimolimnion has very high ammonium, sulfide, and iron concentrations (Hategekimana et al., 2020), resulting in persistent anoxia (Fig. 11-17a). Brief partial mixing periods in the winter (July to October) resulted in hypoxia at the

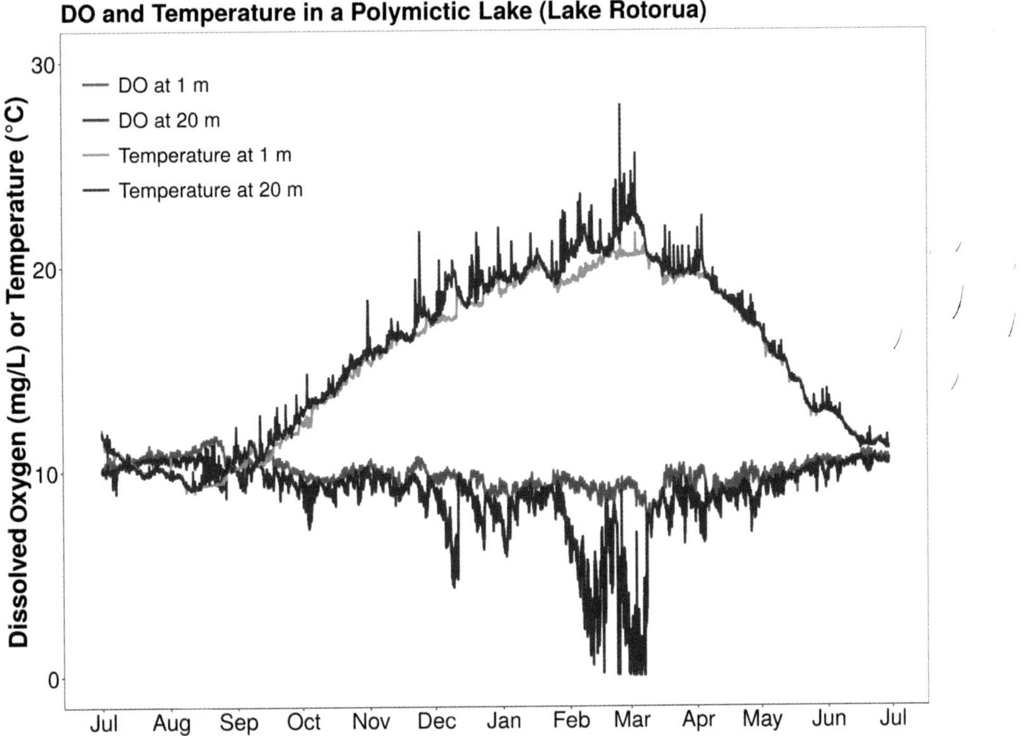

FIGURE 11-16 High-frequency dissolved oxygen (DO, *red*) and water temperature (*blue*) measurements at two depths of the polymictic Lake Rotorua (New Zealand) during 1 year. The temperature time series shows the flashy nature of stratification and mixing events at the lake, mirrored by oxygen chemocline dynamics. *(Data provided by C. McBride.)*

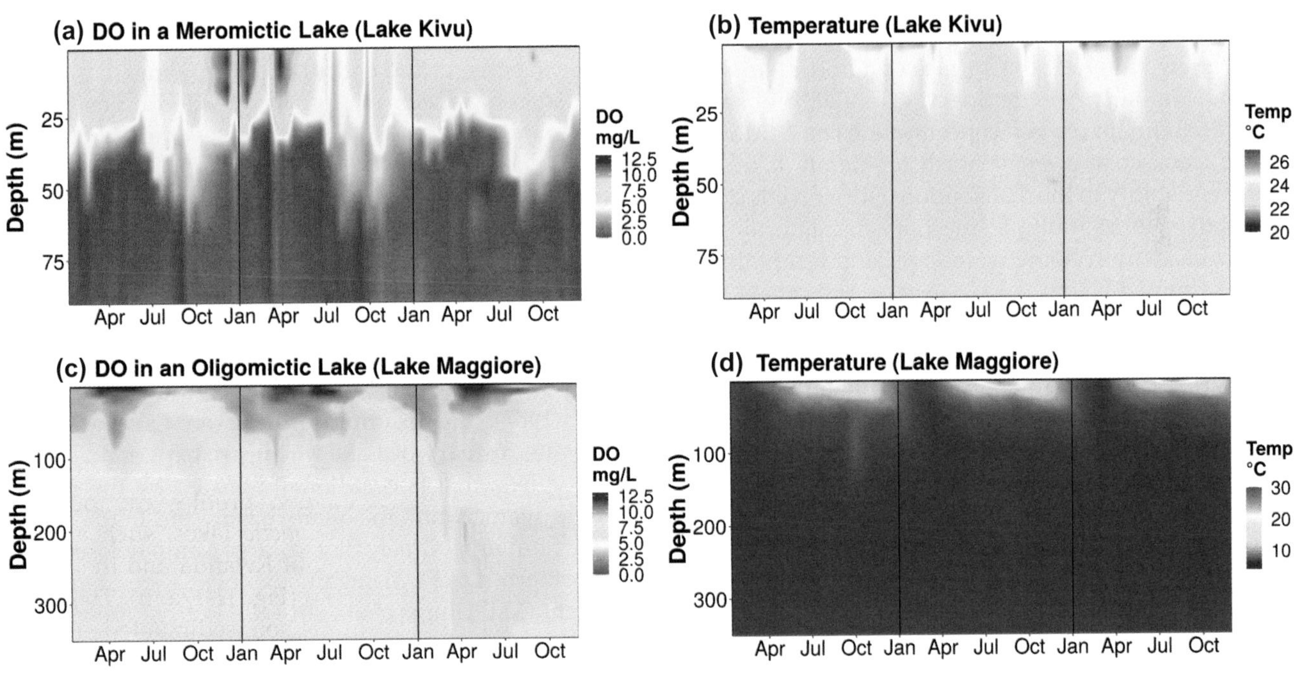

FIGURE 11-17 Three years of time-depth dissolved oxygen (DO, *left column*) and water temperature (*right column*) observations illustrate dynamics in (a and b) meromictic Lake Kivu (Democratic Republic of the Congo/Rwanda) and (c and d) oligomictic Lake Maggiore (Italy). The *vertical lines* separate years; note the differences in oxygen and temperature scales between lakes. *(Data are from Stetler et al., 2021.)*

surface, which was particularly evident in the second year of data. In comparison, increases in oxygen in the surface waters in the summer (November to April) were likely due to increases in phytoplankton productivity.

vii. Oligomictic

Oligomictic Lake Maggiore, Italy, experiences complete mixing for a few weeks in the winter only every few years (Tonolli, 1969) (Fig. 11-17c, d). During the prolonged stratified period, the lower layer is unable to be replenished with oxygen, reinforcing a clinograde profile. Because of Lake Maggiore's relatively low productivity and subalpine location (Arfè et al., 2019), the lake does not exhibit the extreme clinograde curves exhibited by Kivu. In the third year of data shown in Fig. 11-17, full mixing to the sediments occurs in March, increasing the oxygen in the lower strata until depletion becomes evident again at the sediments in August. It is notable, however, that the mixing is extremely brief and the establishment of an oxygen minimum occurs at ~150 m, and an oxygen maximum occurs at the surface, likely driven by an increase in phytoplankton productivity.

F. Horizontal variability in oxygen distributions

Monitoring with high-frequency sensors has revealed that there can be substantial horizontal variability in oxygen concentrations within a lake. Crawford et al. (2015) observed a range of 75% to 94% oxygen saturation within a few meters at the surface of a north temperate bog lake. Similarly, Van de Bogert et al. (2007) observed large variability in oxygen concentrations on a littoral to pelagic transect in a small north temperate lake. This spatial variability in the distribution of dissolved oxygen can occur due to multiple mechanisms: littoral plant growth, basin morphometry, inflows, and reservoir continua, examined below.

i. Littoral plants

Aquatic plants (Chapter 24) often inhabit large areas of shallow lakes and the littoral perimeter of productive deep lakes. These macrophytes, attached algae, and benthic mats can produce large amounts of oxygen during the day but with high respiration at night, resulting in much larger diel swings in littoral oxygen concentrations than in the pelagic zone (Fig. 11-18). Sadro et al. (2011) observed that the magnitude of diel oxygen concentrations in Emerald Lake, USA, was approximately two times higher in the littoral zone than in the pelagic zone. Within the littoral zone, the vertical distribution of oxygen in macrophyte beds can be highly stratified, where upper macrophytes are photosynthesizing and lower macrophytes are predominately respiring (Fig. 11-18).

The oxygen content of the littoral zone can periodically become severely reduced (e.g., Pokorný et al., 1987). Such is the case when large populations of macrophytes senesce at the end of their growing season. Severe oxygen deficits caused by the decomposition of this organic matter can persist for several months and may extend into the open water (Thomas, 1960).

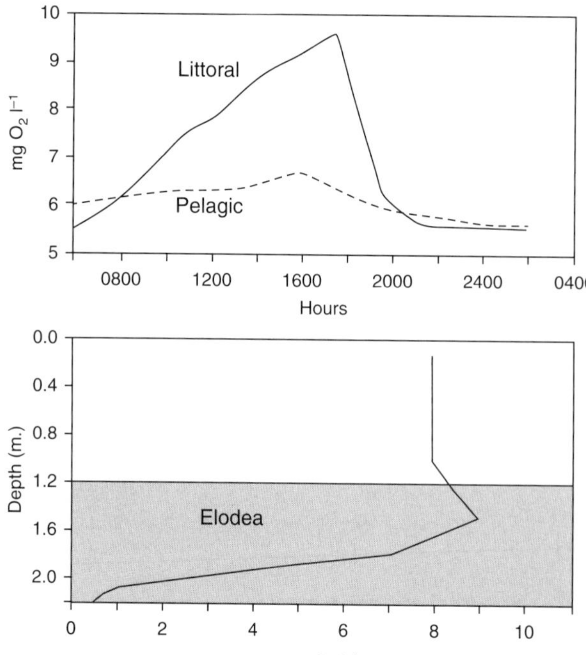

FIGURE 11-18 *Upper*: changes in dissolved oxygen in the littoral and open-water areas over a diurnal period during summer in eutrophic Winona Lake, USA. *(From data of Scott, 1924.) Lower*: vertical stratification of oxygen within the littoral zone of Parvin Lake, USA, in a luxuriant stand of the submersed macrophyte Elodea. *(Generated from data of Buscemi, 1958.)*

ii. Basin morphometry and inflows

Large horizontal variations in oxygen are found in lakes of complex basin morphometry. When a lake consists of many bays or multidepressions, oxygen concentrations can vary substantially among embayments, which may effectively function as individual lakes (Lind, 1987). For example, Muskegon Lake (USA), a freshwater estuary of Lake Michigan, experiences summer hypoxia that is occasionally abated by the mixing of oxic water from the main basin (Biddanda et al., 2018). Similarly, Green Bay, a large semienclosed gulf in northern Lake Michigan, exhibits a large oxygen gradient (Val Klump et al., 2018). In both of these cases, high-nutrient inflows are a major driver of their horizontal variability in oxygen, which can have a large effect on littoral dynamics (e.g., MacIntyre et al., 2006).

iii. Reservoirs

In reservoirs, the distribution of oxygen is highly variable both horizontally and vertically (Cole and Hannan, 1990; Fig. 11-19). The complex hydrodynamics of reservoirs are unique in relation to variations in inflows, morphometry, draw-down and outflow discharge, and other factors. As illustrated by Falling Creek Reservoir, USA (Fig. 11-20), a hypolimnetic anoxic zone often develops in the summer in reservoirs, first in the thalweg of the transition zone during summer stratification, and then expands horizontally and vertically until fall

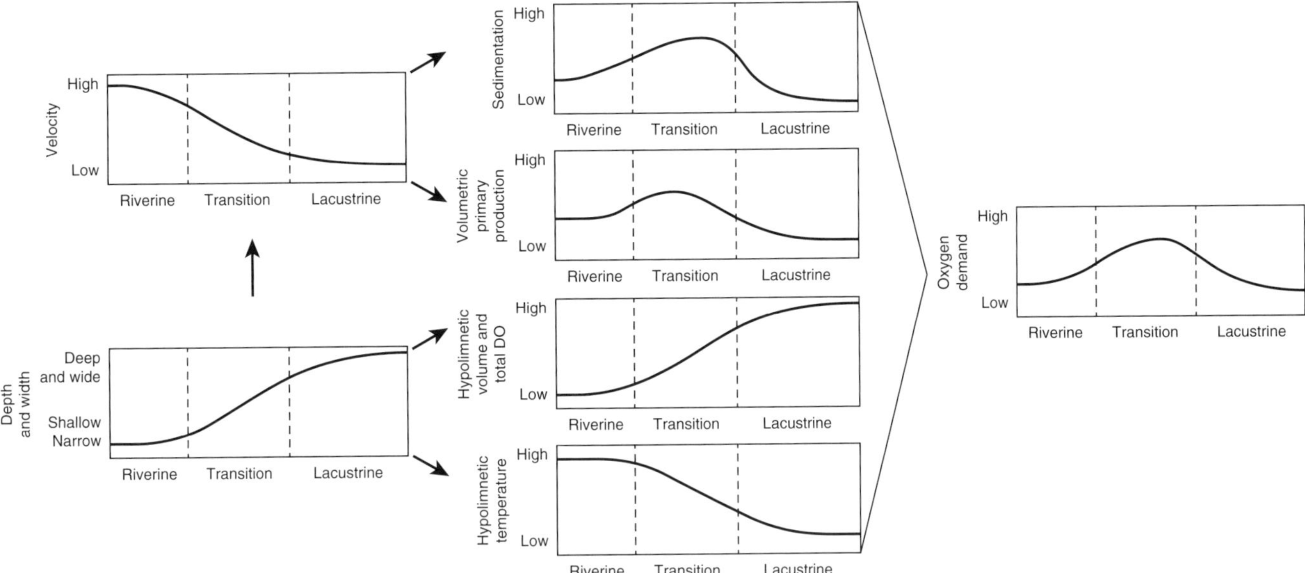

FIGURE 11-19 Relationships of velocity and depth to sedimentation, areal primary production of phytoplankton, hypolimnetic volume and total dissolved oxygen (DO), and hypolimnetic temperature in determining oxygen demand along the length of a reservoir. *(Modified from Cole and Hannan, 1990.)*

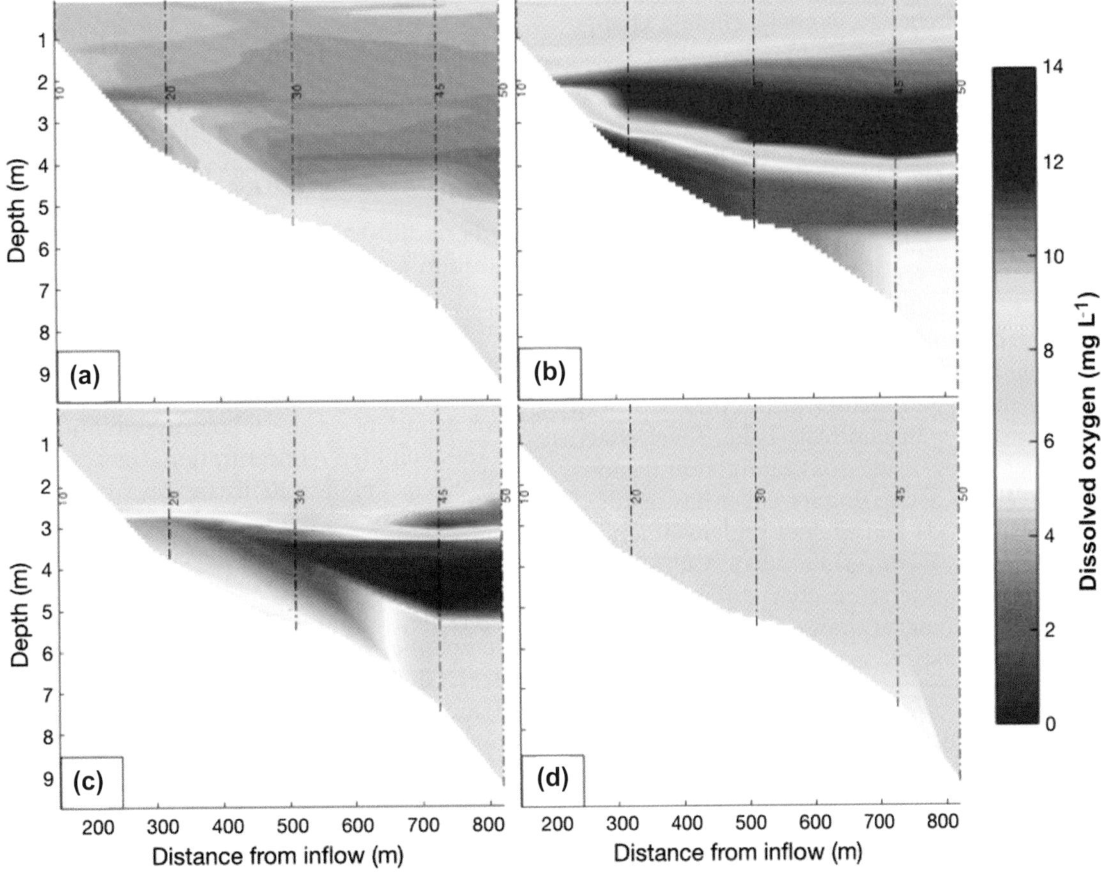

FIGURE 11-20 Dissolved oxygen concentrations measured at five sites on a longitudinal transect in Falling Creek Reservoir, USA, from the inflow tributary to the dam on four select dates: (a) in April, during spring mixing; (b) in June, during the onset of a metalimnetic oxygen minimum; (c) in September, when the metalimnetic oxygen minimum was established throughout the reservoir; and (d) in October, after the onset of fall mixing. *(Figure modified from McClure et al., 2018.)* The dashed vertical lines *represent the sampling locations; the intervening data were interpolated.*

turnover. As in naturally formed lakes, the distribution of oxygen and rates of loss in reservoirs are governed by a balance among mixing, inflow intrusions, photosynthesis, and losses by oxidative consumption. The processes are similar among all freshwater ecosystems, but the rates of change in reservoirs are generally much more variable and dynamic than in naturally formed lakes.

G. Changes in lake oxygen concentrations at decadal and continental scales

Human activities are altering lake oxygen concentrations at large temporal and spatial scales. In North America and Europe, monitoring data from 393 lakes show a median decrease in bottom-water oxygen concentrations of 0.12 mg O_2 L^{-1} decade^{-1}, attributed to strengthening thermal stratification and decreasing water clarity (Jane et al., 2021). Surface-water oxygen concentrations were more variable: while some lakes exhibited decreases due to warmer temperatures (and thus lower solubility), others increased due to greater productivity (Jane et al., 2021). Paleolimnological records (Chapter 30) from 365 lakes support an increasing incidence of hypolimnetic anoxia over the past century due to nutrient inputs (Jenny et al., 2016a); importantly, this study also revealed the lack of recovery to well-oxygenated conditions in lakes that had multiple decades of nutrient abatement.

In synthesis, global change, namely, climate change and eutrophication, may decrease hypolimnetic oxygen concentrations in some lakes at the decadal scale, while at shorter time scales, lake oxygen may potentially be more variable. For example, 41 years of monitoring data of Blelham Tarn, UK, showed that the duration of hypolimnetic anoxia significantly increased over time, but year-to-year hypolimnetic oxygen depletion was associated with spring wind and chlorophyll *a* levels (Foley et al., 2012). Similarly, a 44-year analysis of Sau Reservoir, Spain, by Marcé et al. (2010) observed greater anoxia over time due to declining inflow, with year-to-year variability in oxygen related to El Niño Southern Oscillation periodicity. In contrast, Lake Tovel, Italy, exhibited higher deep-water oxygen concentrations over eight decades, as decreasing ice cover has resulted in longer mixing periods and greater replenishment of hypolimnetic oxygen (Flaim et al., 2020). Collectively, these three studies underscore that different global change drivers will interact at a range of spatial and temporal scales to affect lake oxygen.

V. Metrics for assessing anoxia and hypoxia in inland waters

When processes consuming oxygen exceed processes supplying it, oxygen concentrations may decrease below an acceptable threshold (e.g., the minimum oxygen level needed for fish survival), a condition called *hypoxia*. The threshold for hypoxia is variable but usually set between 2 and 6 mg O_2 L^{-1} (Nürnberg, 2002; Saari et al., 2018). If oxygen concentrations decrease even further, *anoxia* develops, with a threshold usually set between 0.5 and 2 mg O_2 L^{-1}. Anoxia is loosely associated with the redox conditions at which anaerobic metabolic pathways occur (Chapra and Dobson, 1981), because most anaerobic metabolisms are inhibited by oxygen concentrations above 2 mg O_2 L^{-1}. Anoxia and hypoxia are among the most acute environmental problems facing inland waters (Friedrich et al., 2014), as their prevalence has rapidly increased (Jenny et al., 2016a, 2016b; Saari et al., 2018) and likely will become even more frequent in the future (Jane et al., 2021). Appropriate metrics to assess anoxic and hypoxic conditions are thus fundamental for managing inland waters.

Beyond metrics based on oxygen concentrations (e.g., duration curves, Carter et al., 2021; volume-weighted concentrations, Quinlan et al., 2005), there are two major approaches for assessing oxygen deficit in freshwaters: *rate-based metrics* focus on the velocity of oxygen consumption in a water parcel, while *extent-based metrics* focus on aggregated estimates of the spatial and temporal extent of anoxia or hypoxia.

A. Rate-based metrics

A.i. The *Volumetric Hypolimnetic Oxygen Depletion* (VHOD) rate (mg O_2 m^{-3} d^{-1}) is defined as:

$$\text{VHOD} = \text{d}[O_2]\,\text{d}t^{-1} \tag{11-6}$$

where d[O_2] and dt are differentials for oxygen concentration and time. VHOD results from the slope of a linear regression between a time series of decreasing oxygen concentration (mg O_2 m^{-3}) and time (days), from which values below 2 mg O_2 L^{-1} should be discarded (Fig. 11-21a). It can be calculated for individual strata, or for a part of or the whole hypolimnion by using a volume-weighted concentration time series (Quinlan et al., 2005). VHOD usually ranges from 5 to 250 mg O_2 m^{-3} d^{-1}.

A.ii. The *Areal Hypolimnetic Oxygen Depletion* (AHOD) rate (mg O_2 m^{-2} d^{-1}) was proposed by Hutchinson (1938) to account for the effect of varying hypolimnetic volumes in the VHOD rate. However, both VHOD and AHOD are the result of an interaction of volumetric and areal processes (Livingston and Imboden, 1996; Rhodes et al., 2017). AHOD is calculated as:

$$\text{AHOD} = \text{d}mO_2\,A^{-1}\,\text{d}t^{-1} \tag{11-7}$$

where dmO_2 is the differential of the mass of oxygen and A is the upper horizontal boundary area of the

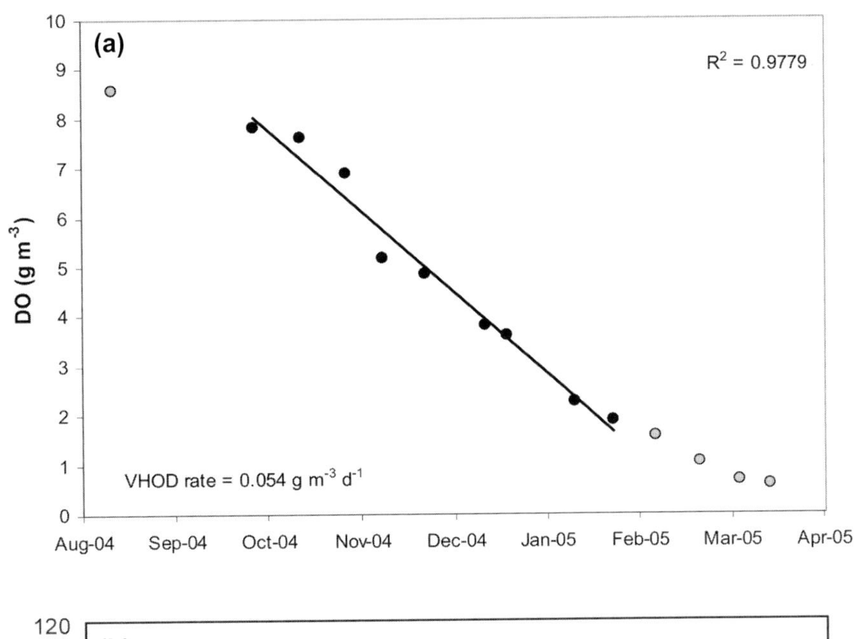

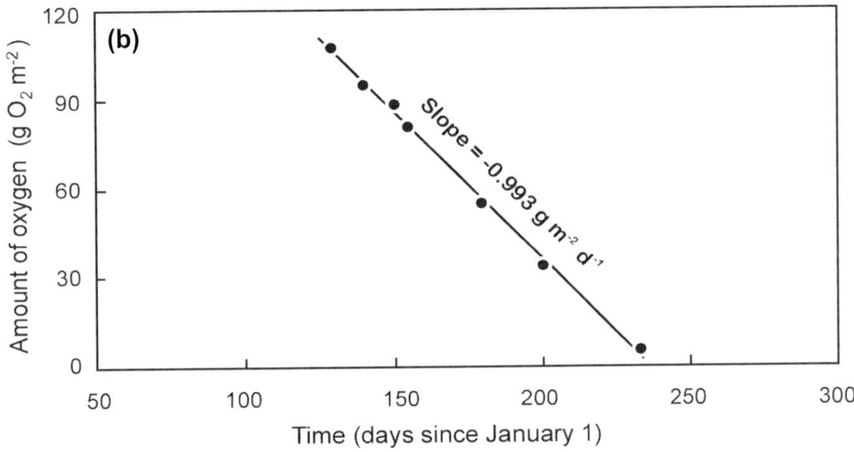

FIGURE 11-21 Calculation of (a) volumetric hypolimnetic oxygen deficit rate (VHOD) in Lake Pupuke, New Zealand, in 2004–2005 (Gibbs, 2005); and (b) areal hypolimnetic oxygen deficit rate (AHOD) in Lake Raduńskie Górne, Poland, in 2009 (Borowiak et al., 2012). Both examples use linear regression between oxygen concentration or areal mass against time (days). Note the gray points not used in (a).

hypolimnion. Linear regression of a time series of mO_2 A^{-1} versus *time* is used to obtain AHOD (Fig. 11-21b). If oxygen concentration profiles are available, mO_2 is calculated for each layer and summed to obtain an overall mass before dividing by A. Values usually range from 50 to 1000 mg O_2 m^{-2} d^{-1}.

A.iii. The *Areal Hypolimnetic Mineralization* (AHM) rate is an extension of the AHOD rate, which takes into account the oxygen equivalents that reduced substances would consume when oxidized (Matzinger et al., 2010). Therefore AHM can be used when hypolimnia are anoxic. AHM is calculated in the same way as AHOD but using *Equivalent Oxygen Concentration* $[EqO_2]$ instead of oxygen concentration:

$$[EqO_2] = [O_2] - 0.5[NO_2^-] - 2[NH_4^+] - 2[S^{-2}] - 2[CH_4] - 0.25[Fe^{+2}] - 0.5[Mn^{+2}] \quad (11\text{-}8)$$

where the concentration of oxygen and reduced substances are expressed in molar concentrations (mol m^{-3}) and subsequently transformed into oxygen mass units as in Eq. 11-7. The rate can be calculated using a linear regression of $[EqO_2]$ against time (Fig. 11-22) and then normalizing per area or by directly regressing $mEqO_2$ A^{-1} (mg m^{-2}) against time. Values usually range from 100 to 1500 mg O_2 m^{-2} d^{-1}.

A.iv. The *Ecosystem Oxygen Demand* (EOD) metric has been proposed to assess the rate at which a river reach

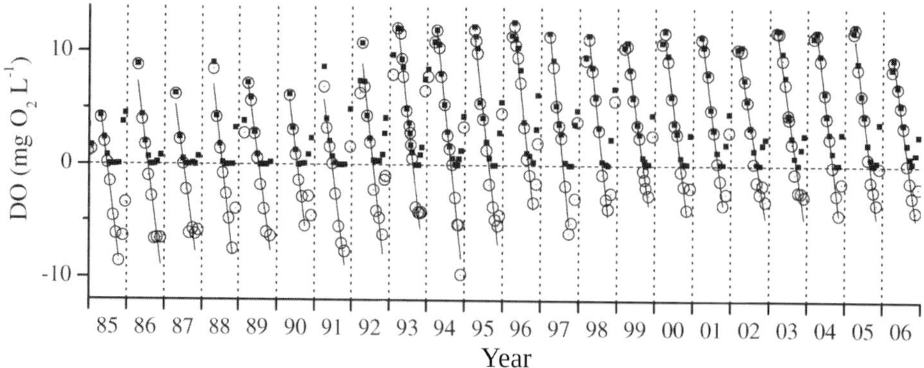

FIGURE 11-22 Measured DO concentrations (*squares*) and calculated [EqO_2] (*circles*) in the hypolimnion of Pfäffikersee, Switzerland, between 1985 (85) and 2006 (06). [EqO_2] was calculated following Eq. 11-8. *Fine lines* are linear regressions, used to estimate the areal hypolimnetic mineralization (AHM). *(Modified from Matzinger et al., 2010.)*

goes from 100% oxygen saturation to hypoxia after a storm (Carter et al., 2021):

$$EOD = -d\%O_{2min}\, dt^{-1} \qquad (11\text{-}9)$$

where $d\%O_{2min}$ is the differential of the daily minimum oxygen percent saturation. It is calculated by regressing the daily minimum oxygen percent saturation (obtained from a high-frequency time series of oxygen concentrations) against time. Values $\geq$20% saturation reduction per day are exhibited by systems that develop anoxia very fast after aeration.

All rate-based metrics in the hypolimnia of lakes are affected by downward fluxes of oxygen across the thermocline, which can be corrected by deriving vertical heat exchange coefficients from temperature data (Matthews and Effler, 2006). Another complication is the subjectivity when defining the period for calculating the linear regressions (Fig. 11-21a) and the fact that for some rates, calculations are no longer possible when anoxia is established.

B. Extent-based metrics

B.i. *Anoxic Factor* (AF) is a metric of the number of days per year or season that an area of sediment is overlain by anoxic water (Nürnberg, 1995). To compute the AF (in days $year^{-1}$ or days $season^{-1}$), the depth of the 1 or 2 mg O_2 L^{-1} isopleth must be determined in a series of oxygen profiles measured during a year or season, and the anoxic area at that isopleth (a, m^2) calculated with a hypsographic curve:

$$AF = \sum_{i=1}^{n} \frac{t_i a_i}{A_{0i}} \frac{1}{period} \qquad (11\text{-}10)$$

where n is the number of profiles available during the year (e.g., 12 in case of monthly sampling), t is the time (days) for which anoxic area at sampling i (a_i) applies, A_{0i} is the surface area of the lake at sampling i (m^2), and *period* is the time span of the calculation (i.e., 1 year or season). Other approaches to integrate over the calculation period have been used (e.g., interpolation to obtain daily a and A_0 values). If the threshold for the oxygen isopleth is >2 mg O_2 L^{-1}, we refer to the calculation usually as the *Hypoxic Factor,* indicating the specific threshold used. The AF upper bound is the length of the period considered. For instance, for AF calculated for a year, the upper bound is 365 days $year^{-1}$, which would mean that the whole lake volume and area has been anoxic during the whole year. This intuitive upper bound makes AF a useful metric for assessing and comparing anoxia extent across and within lakes (Fig. 11-23) and for investigating anoxia drivers (e.g., Nürnberg, 1995; Ladwig et al., 2021; Marcé et al., 2010; Schwefel et al., 2016).

B.ii. For streams, the *Probability of Hypoxia or Anoxia* (Carter et al., 2021) is calculated from a time series of oxygen measurements as the probability of having measurements below a hypoxic or anoxic threshold (h):

$$P_{anoxia} = P([O_2] \leq h) = \frac{n([O_2] \leq h)}{n([O_2])} \qquad (11\text{-}11)$$

where $n([O_2])$ is the number of available oxygen content measurements (in percent saturation or concentration), and $n([O_2] \leq h)$ is the number of cases for which oxygen content is equal or below the threshold h. There is not much information on the probability of anoxia across streams, but anoxia may be substantial in lowland, slow-moving rivers (Carter et al., 2021). For example, Pampean streams (Argentina) are characterized by very low water velocity, limiting atmospheric exchanges. Combined with anthropogenic pressures (e.g., cattle ranching), this may result in extreme diel oxygen concentration oscillations including anoxia during the night time (Fig. 11-24).

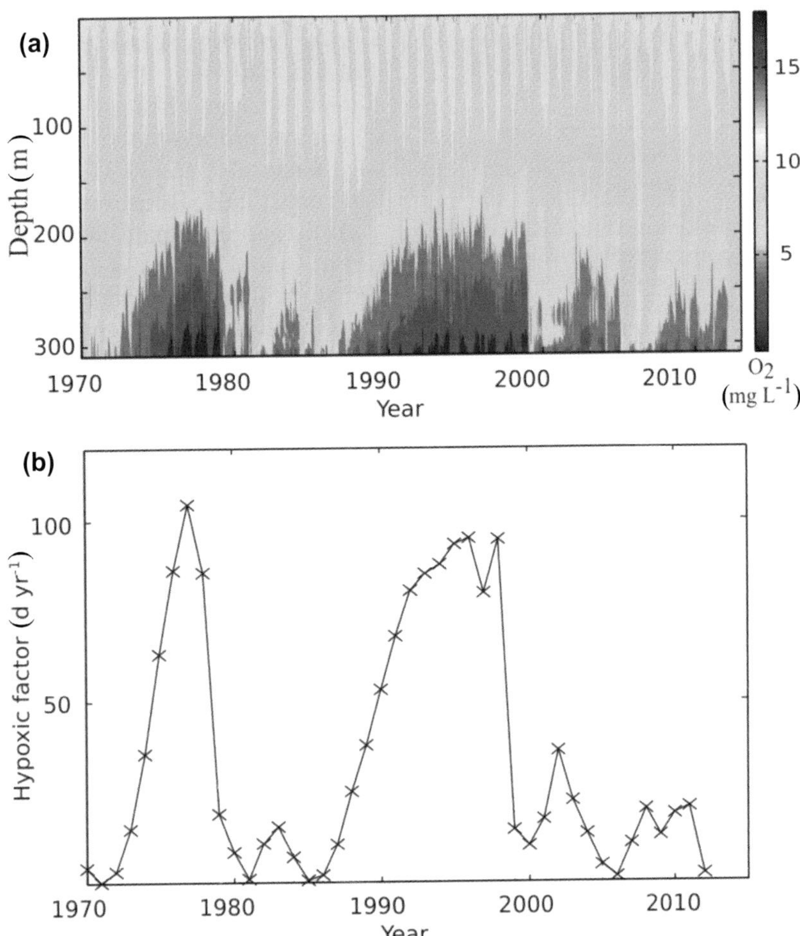

FIGURE 11-23 (a) Contour plot of oxygen concentration in Lake Geneva, Switzerland, for the period 1970–2012, compared to (b) the Hypoxic Factor calculated with the same data using a threshold oxygen concentration of 4 mg O_2 L^{-1}. *(Modified from Schwefel et al., 2016.)*

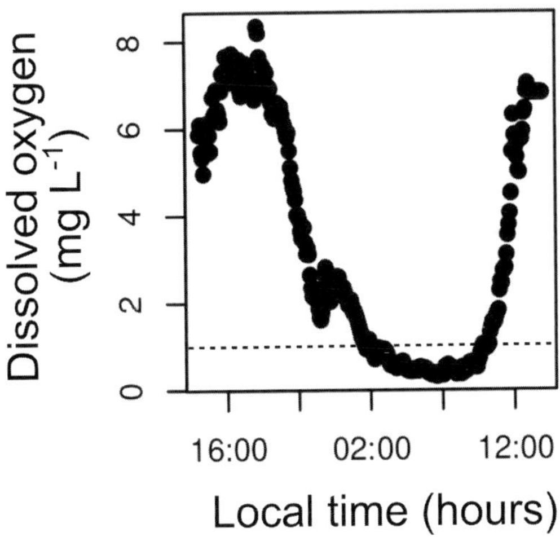

FIGURE 11-24 Diel cycle of dissolved oxygen concentrations in a tributary of the Río Salado, in the Argentinean Pampa. The *dashed line* shows the threshold for anoxia (1 mg O_2 L^{-1}). In this example the probability of anoxia (P_{anoxia}) is 0.34. *(Data courtesy of C. Feijoó and M. Arroita.)*

C. Predicting anoxia: a look at the past and the future

Anoxia has generally been associated with eutrophication, and the development of metrics to assess oxygen deficit in water bodies ran parallel to the discussion of oxygen deficit as a proxy for productivity and trophic status in lakes. Many attempts at predicting metrics of anoxia at annual time scales have used regression-based approaches using mean phosphorus concentration or areal load, water temperature, chlorophyll *a* content, and even trophic state indices as independent variables (Matzinger et al., 2010; Müller et al., 2019; Nürnberg, 2019; Steinsberger et al., 2020). Additionally, almost all statistical models include at least one term to account for lake morphometry (e.g., area, mean depth, volume), because lake morphometry interacts with the relationship between oxygen deficit and trophic status (Cornett and Rigler, 1979; Deeds et al., 2021). Lakes with smaller hypolimnia are expected to have a higher

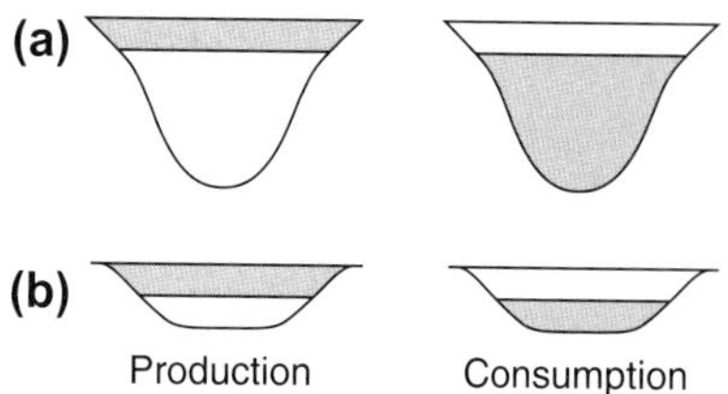

FIGURE 11-25 Diagrammatic representation of a (a) deep lake and a (b) shallow lake of equal production in the trophogenic zone (where photosynthesis occurs) but of differing volumes of their tropholytic zones (where respiration predominates). A small hypolimnion volume relative to the epilimnion implies a larger probability of developing bottom anoxia. *(Modified from Thienemann, 1926, 1928.)*

probability of anoxia (Fig. 11-25). Thus AHOD was considered by many to be a better metric than VHOD, because it was suggested to account for the effect of varying hypolimnetic volumes (Hutchinson, 1938). However, in some cases, such as in reservoirs with large hypolimnetic volume fluctuations due to deep withdrawals, VHOD may be a better alternative. In any case, both VHOD and AHOD are the result of a combination of processes at the water column (volumetric) and the sediments (areal); thus they cannot be independent of lake morphometry (Livingston and Imboden, 1996; Rhodes et al., 2017) and interpretation of metrics of anoxia as proxies of productivity across lakes must be cautious. In reservoirs, variables associated with primary production (nutrient loads or concentrations, chlorophyll *a* content) are not good predictors of annual anoxia, but variables associated with organic matter load from tributaries (river dissolved organic carbon and chloride concentrations) show strong prediction power (Marcé et al., 2008, 2010), suggesting that in some reservoirs the degradation of organic carbon from tributaries is a major driver of anoxia.

The cumulative understanding of the processes affecting oxygen concentrations in inland waters (see Section II) and the ever-increasing computing power available during the last decades have prompted the development of numerous process-based, dynamical models that can predict oxygen concentrations and anoxia in rivers and lakes (Fu et al., 2019; Bai et al., 2022; Vinçon-Leite and Casenave, 2019). Dynamical process-based models attempt to mimic the essential features of inland waters, describing how system properties change over time, usually using a time step of days or even shorter. The number of physical and biogeochemical processes included in different models is very diverse, but the main processes shaping oxygen concentrations are frequently present (solubility, transport processes, exchange with the atmosphere, primary production, aerobic respiration). Although using these models can be more difficult than applying regression-based approaches, they are better suited to simulate oxygen deficit in inland waters subject to multifaceted changes (e.g., hydrological variability, land-use changes, nutrient and organic matter point sources, shifts in stratification patterns). This is why most projections of climate change impacts on inland water anoxia use dynamical process-based models of varying complexity (e.g., Chapra et al., 2021; Golosov et al., 2012; Danladi Bello et al., 2017; Piccolroaz and Toffolon, 2018).

Anticipating climate change impacts on oxygen deficits in inland waters is challenging, because shifts in climate will be accompanied by changes in many other aspects influencing oxygen concentrations (e.g., hydrology and human uses of water, land uses, wastewater inputs, fertilizers application in agriculture, and damming). Rising temperatures are expected to decrease solubility and thus oxygen concentrations, a phenomenon already observed in lakes worldwide (Jane et al., 2021). Modeling exercises have also suggested that the decrease in oxygen concentrations due to reduced solubility in a warmer climate will increase the probability of anoxia in the world's rivers (Chapra et al., 2021). Lakes and reservoirs are expected to stratify for longer in many regions (Woolway et al., 2021), and a longer hypolimnetic isolation during stratification favors anoxia development (Ladwig et al., 2021). Warmer temperatures may also favor anoxia due to the dependence of aerobic respiration rates on temperature. Although temperature is often included in statistical models of anoxia development (Cornett and Rigler, 1979), the relevance of this mechanism in a climate change context is still unclear. Anoxia has been attributed to decreasing streamflows promoted by recent climate change in reservoirs (Marcé et al., 2010), which has potential implications for the myriad of reservoirs located in arid and semiarid regions that will suffer severe reductions of streamflow during the upcoming decades. Nutrient point and diffusive sources have been the main driver of oxygen deficit in lakes during the 20th century (Jenny et al., 2016b); therefore future increases in wastewater inputs to inland waters and fertilizers application in agriculture may also result in more anoxia in lakes due to concomitant increases in primary production and subsequent respiration in deep layers. More frequent extreme weather events in the future may also increase anoxia by promoting overflows of sewer systems during storms that may lead to river anoxia (Chapra et al., 2021). A complete understanding of the dynamics of anoxia across inland waters in a changing climate requires resolving the interactions between all of these co-occurring processes on oxygen concentrations, and we are just starting to develop the necessary scientific infrastructures to deal with such a complex undertaking (Golub et al., 2022a,b). Consequently, refining the predictions for anoxia in inland waters is a fundamental challenge that limnologists will face in the upcoming decade.

VI. Summary

1. Dissolved oxygen is essential to the respiratory metabolism of most aquatic organisms, the ecosystem functioning of freshwaters, and the ecological quality of lakes and rivers. In addition to altering organism habitat, oxygen distributions are important for the metabolism of many organisms and affect many biogeochemical cycles, as well as the productivity of aquatic ecosystems. Low oxygen levels are generally associated with poor water quality and degraded habitats for organisms.
2. The solubility of oxygen in water decreases as temperature increases. The solubility of oxygen decreases with lower atmospheric pressures at higher elevations and increases with greater hydrostatic pressures at depth within lakes. Oxygen solubility decreases exponentially with increases in salt content. Oxygen content is often reported as percent saturation with respect to the solubility concentration at atmospheric pressure at the surface of the water body.
3. The variability of oxygen dynamics in inland waters results from the combination of exchanges with the atmosphere; advection, dispersion, and biota-mediated transport; photosynthesis and photorespiration; aerobic respiration; oxidation of reduced inorganic compounds; photochemical oxidation of organic matter; and chemical weathering. Most of these processes are directly or indirectly mediated by organisms.
4. Surface and subsurface water flow are of paramount importance to understanding oxygen dynamics in rivers. Lateral and vertical flow pathways are involved in the advection of dissolved oxygen as well as other substances indirectly influencing riverine dissolved oxygen dynamics (e.g., organic matter, nutrients, and electron acceptors).
5. While the oxygen content within small, high-gradient streams is usually near or above saturation due to the high turbulent mixing of water, marked variations occur spatially and temporally in low-gradient streams and rivers and are often coupled to variations in the discharge and loading of nutrients and organic matter.
6. Lateral (between surface water and groundwater) and vertical (between surface water and the riverbed) hydrologic exchanges represent a major source of dissolved oxygen variability in rivers and also affect other key substances, indirectly influencing riverine dissolved oxygen dynamics (e.g., organic matter, nutrients, and electron acceptors).
7. Oxygen concentrations in rivers can vary temporally on diel, seasonal, annual, and decadal scales. This spectrum of temporal signals combines to shape the oxygen dynamics of a river, which depends on the hydrological and biogeochemical conditions of the river's catchment.
8. Similar to rivers, oxygen concentrations in lakes vary both temporally on diel, seasonal, and annual scales and spatially on horizontal gradients between littoral to pelagic sites and vertical gradients with depth. On an annual scale, dimictic, north temperate lakes exhibit four distinct oxygen vertical profiles: orthograde, clinograde, positive heterograde, and negative heterograde. These oxygen profiles are determined by a lake's productivity, mixing regimes, and climate, as well as shorter-term factors influencing the development of metalimnetic oxygen maxima and minima.
9. Examining dissolved oxygen profiles in lakes with amictic, warm monomictic, dimictic, cold monomictic, polymictic, meromictic, and oligomictic mixing regimes provides insights into the substantial variability in dissolved oxygen dynamics that lakes can exhibit. In general, the many lakes with clinograde oxygen profiles exhibit greater variability in dissolved oxygen dynamics during thermally stratified periods than the few lakes showing orthograde oxygen profiles.
10. Lakes globally are exhibiting decreased oxygen concentrations in both their surface and bottom waters due to climate and land-use change. Ultimately, clinograde oxygen profiles in lakes will likely become more frequent due to global change, resulting in long-term changes to lake ecosystem functioning.
11. Anoxia and hypoxia are among the most acute environmental problems facing inland waters, as their prevalence has rapidly increased and likely will become even more frequent in the future. Metrics to assess anoxic and hypoxic conditions are based on oxygen concentrations, on the calculation of rates of oxygen consumption in the hypolimnion, and on the estimation of the temporal and spatial extent of anoxia in lakes and running waters.
12. Annual anoxia extent in lakes can be predicted using regression-based models with variables related to lake morphometry, hydrology, and trophic state as predictors. In reservoirs variables related to hydrology and organic carbon load from tributaries work better. Dynamical process-based models are used to predict oxygen deficits at smaller time steps (e.g., daily), and the inclusion of the multifaceted processes affecting oxygen concentration makes them useful tools to assess the impacts of climate change on oxygen deficit in inland waters.
13. Anticipating climate change impacts on oxygen deficit in inland waters is challenging, because shifts in climate will be accompanied by changes in many other aspects influencing oxygen concentrations.

Impacts of climate change on oxygen solubility, stratification, respiration rates, streamflow, nutrient point and diffusive sources, and extreme weather events will shape the future prevalence of oxygen deficits in inland waters.

Acknowledgments

RM led the development and writing of this chapter. RM and CCC worked closely together to craft the chapter's structure and initiated the major revisions in this edition to the best of their ability following Wetzel's vision. RM led Sections I, II, and V; LG led Section III; and CCC led Section IV, with all authors working in tandem to integrate the revised chapter. We thank B. Obrador for insights into the structure and contents of this chapter and for initiating this work. We thank R.P. McClure for generating figures and H.-P. Grossart, J. Priscu, T. Martin, C. McBride, C. Feijoó, M. Arroita, and D.P. Hamilton for sharing data. We thank J. Priscu, J.-P. Jenny, J. Catalán, L. Camarero, A.S. Lewis, W.M. Woelmer, N.G. Hairston Jr., and K.C. Weathers for helpful feedback and support. We are deeply indebted to colleagues who reviewed this chapter: J.R. Blaszczak, C. Gutiérrez, A. Helton, M. Koschorreck, G. Nürnberg, and G. Rocher. Any errors, omissions, or conceptual limitations are the authors' exclusive responsibility.

References

Åberg, B., Rodhe, W., 1942. Über die Milieufaktoren in einigen südschwedischen Seen. Symbol. Bot. Upsalien. 5, 256.

Acuña, V., Giorgi, A., Muñoz, I., Uehlinger, U., Uehlinger, S., 2004. Flow extremes and benthic organic matter shape the metabolism of a headwater Mediterranean stream. Freshw. Biol. 49, 960–971. https://doi.org/10.1111/j.1365-2427.2004.01239.x.

Akomeah, E., Lindenschmidt, K.-E., 2017. Seasonal variation in sediment oxygen demand in a northern chained river-lake system. Water 9, 254. https://doi.org/10.3390/w9040254.

Alsterberg, G., 1927. Die sauerstoffschichtung der Seen. Bot. Not. 255–274.

Amon, R.M.W., Benner, R., 1996. Photochemical and microbial consumption of dissolved organic carbon and dissolved oxygen in the Amazon River system. Geochem. Cosmochim. Acta 60, 1783–1792. https://doi.org/10.1016/0016-7037(96)00055-5.

Aragno, M., Schlegel, H.G., 1981. The hydrogen-oxidizing bacteria. In: Starr, M.P., Stolp, H., Trüper, H.G., Balows, A., Schlegel, H.G. (Eds.), The Prokaryotes: A Handbook on Habitats, Isolation, and Identification of Bacteria. Springer, Berlin, Heidelberg, pp. 865–893.

Arfè, A., Quatto, P., Zambon, A., MacIsaac, H.J., Manca, M., 2019. Long-term changes in the zooplankton community of Lake Maggiore in response to multiple stressors: a functional principal components analysis. Water 11, 962.

Arroita, M., Elosegi, A., Hall, R.O., 2019. Twenty years of daily metabolism show riverine recovery following sewage abatement. Limnol. Oceanogr. 64, S77–S92. https://doi.org/10.1002/lno.11053.

Bai, J., Zhao, J., Zhang, Z., Tian, Z., 2022. Assessment and a review of research on surface water quality modeling. Ecol. Model. 466, 109888.

Baird, R., Bridgewater, L., 2017. Standard Methods for the Examination of Water and Wastewater, 23rd ed. American Public Health Association, Washington, DC.

Baranov, V., Lewandowski, J., Krause, S., 2016. Bioturbation enhances the aerobic respiration of lake sediments in warming lakes. Biol. Lett. 12, 20160448. https://doi.org/10.1098/rsbl.2016.0448.

Bastviken, D., 2009. Methane. In: Likens, G.E. (Ed.), Encyclopedia of Inland Waters. Academic Press, pp. 783–805. https://doi.org/10.1016/B978-012370626-3.00117-4.

Battin, T.J., Wille, A., Psenner, R., Richter, A., 2004. Large-scale environmental controls on microbial biofilms in high-alpine streams. Biogeosciences 1, 159–171.

Beaulieu, J.J., Tank, J.L., Hamilton, S.K., Wollheim, W.M., Hall, R.O., Mulholland, P.J., et al., 2011. Nitrous oxide emission from denitrification in stream and river networks. Proc. Natl. Acad. Sci. 108, 214–219.

Beer, S., Sand-Jensen, K., Vindbeak Madsen, T., Nielsen, S.L., 1991. The carboxylase activity of Rubisco and the photosynthetic performance in aquatic plants. Oecologia 87, 429–434. http://www.jstor.org/stable/4219716.

Benedini, M., Tsakiris, G., 2013. Dispersion in rivers and streams. In: Water Quality Modelling for Rivers and Streams. Water Science and Technology Library, vol. 70. Springer, Dordrecht.

Benson, B.B., Krause, D., 1980. The concentration and isotopic fractionation of gases dissolved in freshwater in equilibrium with the atmosphere. 1. Oxygen. Limnol. Oceanogr. 25, 662–671. https://doi.org/10.4319/lo.1980.25.4.0662.

Benson, B., Magnuson, J., Sharma, S., 2000, updated 2020. Global lake and river ice phenology database, version 1. Boulder, Colorado, USA. NSIDC: National Snow and Ice Data Center. https://doi.org/10.7265/N5W66HP8. (Accessed 2021-08-31).

Bernhardt, E.S., Heffernan, J.B., Grimm, N.B., Stanley, E.H., Harvey, J.W., Arroita, M., et al., 2018. The metabolic regimes of flowing waters. Limnol. Oceanogr. 63, S99–S118.

Bertilsson, S., Burgin, A., Carey, C.C., Fey, S.B., Grossart, H.P., Grubisic, L.M., et al., 2013. The under-ice microbiome of seasonally frozen lakes. Limnol. Oceanogr. 58 (6), 1998–2012.

Biddanda, B.A., Weinke, A.D., Kendall, S.T., Gereaux, L.C., Holcomb, T.M., Snider, M.J., et al., 2018. Chronicles of hypoxia: time-series buoy observations reveal annually recurring seasonal basin-wide hypoxia in Muskegon Lake—a Great Lakes estuary. J. Great Lakes Res. 44, 219–229.

Birge, E.A., Juday, C., 1911. The inland lakes of Wisconsin. The dissolved gases of the water and their biological significance. Bull. Wis. Geol. Nat. Hist. Survey 22, Sci. Ser. 7, 259.

Birmingham, B.C., Coleman, J.R., Colman, B., 1982. Measurement of photorespiration in algae. Plant Physiol. 69, 259–262.

Blaszczak, J.R., Delesantro, J.M., Urban, D.L., Doyle, M.W., Bernhardt, E.S., 2019. Scoured or suffocated: urban stream ecosystems oscillate between hydrologic and dissolved oxygen extremes. Limnol. Oceanogr. 64, 877–894.

Blaszczak, J.R., Koenig, L.E., Mejia, F.H., Gómez-Gener, L., Duton, C.L., Carter, A.M., Grimm, N.B., Harvey, J.W., Helton, A.M., Cohen, M.J., 2022. Extent, patterns, and drivers of hypoxia in the world's streams and rivers. Limnol. Oceanogr. Letters. https://doi.org/10.1002/lol2.10297.

Bloomer, J., Sear, D., Dutey-Magni, P., Kemp, P., 2016. The effects of oxygen depletion due to upwelling groundwater on the posthatch fitness of Atlantic salmon (*Salmo salar*). Can. J. Fish. Aquat. Sci. 73, 1830–1840.

Boano, F., Harvey, J.W., Marion, A., Packman, A.I., Revelli, R., Ridolfi, L., et al., 2014. Hyporheic flow and transport processes: mechanisms, models, and biogeochemical implications. Rev. Geophys. 52, 603–679. https://doi.org/10.1002/2012RG000417.

Boix-Canadell, M., Gómez-Gener, L., Clémençon, M., Lane, S.N., Battin, T.J., 2021a. Daily entropy of dissolved oxygen reveals different energetic regimes and drivers among high-mountain stream types. Limnol. Oceanogr. 66, 1594–1610.

Boix-Canadell, M., Gómez-Gener, L., Ulseth, A.J., Clémençon, M., Lane, S.N., Battin, T.J., 2021b. Regimes of primary production and their drivers in Alpine streams. Freshw. Biol. 66, 1449–1463.

Borowiak, D., Nowiński, K., Barańczuk, J., Marszelewski, W., Skowron, R., Solarczyk, A., 2012. Relationship between areal hypolimnetic oxygen depletion rate and the trophic state of five lakes in northern Poland. Limnol. Rev. 11, 135–142.

Boulton, A.J., Findlay, S., Marmonier, P., Stanley, E.H., Valett, H.M., 1998. The functional significance of the hyporheic zone in streams and rivers. Annu. Rev. Ecol. Syst. 29, 59–81.

Bravidor, J., Kreling, J., Lorke, A., Koschorreck, M., 2015. Effect of fluctuating oxygen concentration on iron oxidation at the pelagic ferrocline of a meromictic lake. Environ. Chem. 12, 723.

Brentrup, J.A., Richardson, D.C., Carey, C.C., Ward, N.K., Bruesewitz, D.A., Weathers, K.C., 2021. Under-ice respiration rates shift the annual carbon cycle in the mixed layer of an oligotrophic lake from autotrophy to heterotrophy. Inland Waters 11, 114–123.

Brezonik, P.L., 2002. Chemical Kinetics and Process Dynamics in Aquatic Systems, 2nd ed. Routledge.

Brion, N., Billen, G., 2000. Wastewater as a source of nitrifying bacteria in river systems: the case of the River Seine downstream from Paris. Water Res. 34, 3213–3221. https://doi.org/10.1016/S0043-1354(00)00075-0.

Brothers, S., Vadeboncoeur, Y., 2021. Shoring up the foundations of production to respiration ratios in lakes. Limnol. Oceanogr. 66, 2762–2778. https://doi.org/10.1002/lno.11787.

Brune, A., Frenzel, P., Cypionka, H., 2000. Life at the oxic–anoxic interface: microbial activities and adaptations. FEMS Microbiol. Rev. 24, 691–710. https://doi.org/10.1111/j.1574-6976.2000.tb00567.x.

Brunke, M., Gonser, T., 1997. The ecological significance of exchange processes between rivers and groundwater. Freshw. Biol. 37, 1–33.

Bryant, L.D., McGinnis, D.F., Lorrai, C., Brand, A., Little, J.C., Wüest, A., 2010. Evaluating oxygen fluxes using microprofiles from both sides of the sediment-water interface. Limnol. Oceanogr-Meth. 8, 610–627. https://doi.org/10.4319/lom.2010.8.0610.

Bu, J., Sun, Z., Ma, R., Liu, Y., Gong, X., Pan, Z., et al., 2020. Shallow groundwater quality and its controlling factors in the Su-Xi-Chang region, Eastern China. Int. J. Environ. Res. Public Health 17, 1267.

Buscemi, P.A., 1958. Littoral oxygen depletion produced by a cover of *Elodea canadensis*. Oikos 9, 239–245.

Carey, C.C., 2021. Ice cover data for Falling Creek Reservoir, Vinton, Virginia, USA for 2013–2021 ver 3. Environmental Data Initiative. https://doi.org/10.6073/pasta/a23233527aa90638b2cd3075627c91e6. (Accessed 2021-08-31).

Carey, C.C., Lewis, A.S., McClure, R.P., Gerling, A.B., Chen, S., Das, A., et al., 2021. Time series of high-frequency profiles of depth, temperature, dissolved oxygen, conductivity, specific conductivity, chlorophyll *a*, turbidity, pH, oxidation-reduction potential, photosynthetic active radiation, and descent rate for Beaverdam Reservoir, Carvins Cove Reservoir, Falling Creek Reservoir, Gatewood Reservoir, and Spring Hollow Reservoir in Southwestern Virginia, USA 2013–2020 ver 11. Environmental Data Initiative (Accessed 2021-08-31).

Carignan, R., Blais, A.M., Vis, C., 1998. Measurement of primary production and community respiration in oligotrophic lakes using the Winkler method. Can. J. Fish. Aquat. Sci. 55, 1078–1084. https://doi.org/10.1139/f97-319.

Carignan, R., Lean, D., 1991. Regeneration of dissolved substances in a seasonally anoxic lake: the relative importance of processes occurring in the water column and in the sediments. Limnol. Oceanogr. 36, 683–707. https://doi.org/10.4319/lo.1991.36.4.0683.

Carter, A.M., Blaszczak, J.R., Heffernan, J.B., Bernhardt, E.S., 2021. Hypoxia dynamics and spatial distribution in a low gradient river. Limnol. Oceanogr. 66, 2251–2265. https://doi.org/10.1002/lno.11751.

Champ, D.R., Gulens, J., Jackson, R.E., 1979. Oxidation–reduction sequences in ground water flow systems. Can. J. Earth Sci. 16, 12–23.

Chapra, S.C., Camacho, L.A., McBride, G.B., 2021. Impact of global warming on dissolved oxygen and BOD assimilative capacity of the world's rivers: modeling analysis. Water 13, 2408.

Chapra, S.C., Dobson, H.F.H., 1981. Quantification of the lake trophic typologies of Naumann (surface quality) and Thienemann (oxygen) with special reference to the Great Lakes. J. Great Lakes Res. 7, 182–193. https://doi.org/10.1016/S0380-1330(81)72044-6.

Chiellini, C., Miceli, E., Bacci, G., Fagorzi, C., Coppini, E., Fibbi, D., et al., 2018. Spatial structuring of bacterial communities in epilithic biofilms in the Acquarossa River (Italy). FEMS Microbiol. Ecol. 94, fiy181. https://doi.org/10.1093/femsec/fiy181.

Clevinger, C.C., Heath, R.T., Bade, D.L., 2014. Oxygen use by nitrification in the hypolimnion and sediments of Lake Erie. J. Great Lakes Res. 40, 202–207. https://doi.org/10.1016/j.jglr.2013.09.015.

Cole, T.M., Hannan, H.H., 1990. Dissolved oxygen dynamics. In: Thornton, K.W., Kimmel, B.L., Payne, F.E. (Eds.), Reservoir Limnology: Ecological Perspectives. J. Wiley & Sons, New York, pp. 71–107.

Conrad, R., Aragno, M., Seiler, W., 1983. Production and consumption of hydrogen in a eutrophic lake. Appl. Environ. Microbiol. 45, 502–510. https://doi.org/10.1128/aem.45.2.502-510.1983.

Cornett, R.J., Rigler, F.H., 1979. Hypolinimetic oxygen deficits: their prediction and interpretation. Science 205, 580–581. https://doi.org/10.1126/science.205.4406.580, 17729679.

Cory, R.M., Ward, C.P., Crump, B.C., Kling, G.W., 2014. Sunlight controls water column processing of carbon in arctic fresh waters. Science 345, 925–928. https://doi.org/10.1126/science.1253119.

Craig, H., Wharton, R.A., McKay, C.P., 1992. Oxygen supersaturation in ice-covered Antarctic lakes: biological versus physical contributions. Science 255, 318–321.

Crawford, J.T., Loken, L.C., Casson, N.J., Smith, C., Stone, A.G., Winslow, L.A., 2015. High-speed limnology: using advanced sensors to investigate spatial variability in biogeochemistry and hydrology. Environ. Sci. Technol. 49, 442–450.

Cullen, J.J., 2015. Subsurface chlorophyll maximum layers: enduring enigma or mystery solved? Ann. Rev. Mar. Sci. 7, 207–239.

Danladi Bello, A.A., Hashim, N.B., Mohd Haniffah, M.R., 2017. Predicting impact of climate change on water temperature and dissolved oxygen in tropical rivers. Climate 5, 58.

de Carvalho Aguiar, V.M., Neto, J.A.B., Rangel, C.M., 2011. Eutrophication and hypoxia in four streams discharging in Guanabara Bay, RJ, Brazil, a case study. Mar. Pollut. Bull. 62, 1915–1919.

Deeds, J., Amirbahman, A., Norton, S.A., Suitor, D.G., Bacon, L.C., 2021. Predicting anoxia in low-nutrient temperate lakes. Ecol. Appl. 31, e02361.

del Giorgio, P., Williams, P., 2005. Respiration in Aquatic Ecosystems. Oxford University Press, Oxford. https://doi.org/10.1093/acprof:oso/9780198527084.001.0001.

Demars, B.O.L., 2019. Hydrological pulses and burning of dissolved organic carbon by stream respiration. Limnol. Oceanogr. 64, 406–421.

Dodds, W.K., Gido, K., Whiles, M.R., Daniels, M.D., Grudzinski, B.P., 2015. The Stream Biome Gradient Concept: factors controlling lotic systems across broad biogeographic scales. Freshw. Sci. 34, 1–19. https://doi.org/10.1086/679756.

dos Reis Oliveira, P.C., van der Geest, H.G., Kraak, M.H.S., Verdonschot, P.F.M., 2019. Land use affects lowland stream ecosystems through dissolved oxygen regimes. Sci. Rep. 9, 19685.

Dubay, C.I., Simmons Jr., G.M., 1979. The contribution of macrophytes to the metalimnetic oxygen maximum in a montane, oligotrophic lake. Amer. Midland Nat. 101, 108–117.

Dutton, C.L., Subalusky, A.L., Hamilton, S.K., et al., 2018. Organic matter loading by hippopotami causes subsidy overload resulting in downstream hypoxia and fish kills. Nat. Commun. 9, 1951. https://doi.org/10.1038/s41467-018-04391-6.

Eberly, W.R., 1964. Further studies on the metalimnetic oxygen maximum, with special reference to its occurrence throughout the world. Invest. Indiana Lakes Streams 5, 103–139.

European Community, 2000. Directive 2000/60/EC of October 23 2000 of the European parliament and of the council establishing a framework for community action in the field of water policy. Official Journal of the European Community L327, 1–72.

Fernández Castro, B., Chmiel, H.E., Minaudo, C., Krishna, S., Perolo, P., Rasconi, S., et al., 2021. Primary and net ecosystem production in a large lake diagnosed from high-resolution oxygen measurements. Water Resour. Res. 57, e2020WR029283. https://doi.org/10.1029/2020WR029283.

Flaim, G., Andreis, D., Piccolroaz, S., Obertegger, U., 2020. Ice cover and extreme events determine dissolved oxygen in a placid mountain lake. Water Resour. Res. 56, e2020WR027321.

Foley, B., Jones, I.D., Maberly, S.C., Rippey, B., 2012. Long-term changes in oxygen depletion in a small temperate lake: effects of climate change and eutrophication. Freshw. Biol. 57, 278–289.

Friedrich, J., Janssen, F., Aleynik, D., Bange, H.W., Boltacheva, N., Çagatay, M.N., et al., 2014. Investigating hypoxia in aquatic environments: diverse approaches to addressing a complex phenomenon. Biogeosciences 11, 1215–1259. https://doi.org/10.5194/bg-11-1215-2014.

Fu, B., Merritt, W.S., Croke, B.F., Weber, T.R., Jakeman, A.J., 2019. A review of catchment-scale water quality and erosion models and a synthesis of future prospects. Environ. Model. Softw. 114, 75–97.

Fukushima, T., Matsushita, B., 2021. Limiting nutrient and its use efficiency of phytoplankton in a shallow eutrophic lake, Lake Kasumigaura. Hydrobiologia 848, 3469–3487.

Garcia, H.E., Gordon, L.I., 1992. Oxygen solubility in seawater: better fitting equations. Limnol. Oceanogr. 37, 1307–1312. https://doi.org/10.4319/lo.1992.37.6.1307.

Garvey, J.E., Whiles, M.R., Streicher, D., 2007. A hierarchical model for oxygen dynamics in streams. Can. J. Fish. Aquat. Sci. 64, 1816–1827.

Gat, J.R., Shatkay, M., 1991. Gas exchange with saline waters. Limnol. Oceanogr. 36, 988–997.

Gelda, R.K., Auer, M.T., Effler, S.W., 1995. Determination of sediment oxygen demand by direct measurement and by inference from reduced species accumulation. Mar. Freshw. Res. 46, 81–88. https://doi.org/10.1071/MF9950081.

Gibbs, M., 2005. Summer Oxygen Depletion in Lake Pupuke during 2004–05 with summary of historic data. Technical Publication No. 276, Auckland Regional Council, Hamilton, New Zealand. ISBN 1-877353-96-5.

Godwin, C.M., Zehnpfennig, J.R., Learman, D.R., 2020. Biotic and abiotic mechanisms of manganese (II) oxidation in Lake Erie. Front. Environ. Sci. 8, 57. https://doi.org/10.3389/fenvs.2020.00057.

Golosov, S., Terzhevik, A., Zverev, I., Kirillin, G., Engelhardt, C., 2012. Climate change impact on thermal and oxygen regime of shallow lakes. Tellus A: Dyn. Meteorol. Oceanogr. 64, 1. https://doi.org/10.3402/tellusa.v64i0.17264.

Golub, M., Thiery, W., Marcé, R., Pierson, D., Vanderkelen, I., et al., 2022a. A framework for ensemble modelling of climate change impacts on lakes worldwide: the isimip lake sector. Geoscientific Model Development Discussions [preprint]. https://doi.org/10.5194/gmd-2021-433.

Golub, M., Thiery, W., Marcé, R., Pierson, D., Vanderkelen, I., Mercado-Bettin, D., et al., 2022b. A framework for ensemble modelling of climate change impacts on lakes worldwide: the ISIMIP Lake Sector. Geosci. Model Dev. 15, 4597–4623. https://doi.org/10.5194/gmd-15-4597-2022.

Gómez-Gener, L., Lupon, A., Laudon, H., Sponseller, R.A., 2020. Drought alters the biogeochemistry of boreal stream networks. Nat. Commun. 11, 1795. https://doi.org/10.1038/s41467-020-15496-2.

Gowda, T.H., 1983. Modelling nitrification effects on the dissolved oxygen regime of the Speed River. Water Res. 17, 1917–1927.

Gower, A.M., 1980. Ecological effects of changes in water quality. In: Water Quality in Catchment Ecosystems. Wiley & Sons, Chichester, pp. 145–171.

Hall, R.O., Kennedy, T.A., Rosi-Marshall, E., 2012. Air-water oxygen exchange in a large whitewater river. Limnol. Oceanogr. Fluids Environ. 2, 1–11. https://doi.org/10.1215/21573689-1572535.

Hall, R.O., Ulseth, A.J., 2020. Gas exchange in streams and rivers. WIREs Water 7, e1391. https://doi.org/10.1002/wat2.1391.

Hamilton, S.K., Sippel, S.J., Melack, J.M., 1995. Oxygen depletion and carbon dioxide and methane production in waters of the Pantanal wetland of Brazil. Biogeochemistry 30, 115–141.

Hansel, C.M., Learman, D.R., 2015. Geomicrobiology of manganese. In: Ehrlich, H.L, Newman, D.K., Kappler, A. (Eds.), Ehrlich's Geomicrobiology, 6th ed. CRC Press, New York.

Hanson, P.C., Carpenter, S.R., Armstrong, D.E., Stanley, E.H., Kratz, T.K., 2006. Lake dissolved inorganic carbon and dissolved oxygen: changing drivers from days to decades. Ecol. Monogr. 76, 343–363. http://www.jstor.org/stable/27646047.

Hanson, R.S., Hanson, T.E., 1996. Methanotrophic bacteria. Microbiol. Rev. 60, 439–471. https://doi.org/10.1128/mr.60.2.439-471.1996.

Harvey, J., Gooseff, M., 2015. River corridor science: hydrologic exchange and ecological consequences from bedforms to basins. Water Resour. Res. 51, 6893–6922.

Hategekimana, F., Ndikuryayo, J.D., Habimana, E., Mugerwa, T., Christian, K., Rwabuhungu, R.D.E., 2020. Lake Kivu water chemistry variation with depth over time, Northwestern Rwanda. Rwanda Journal of Engineering, Science, Technology and Environment 3, 1–20. https://doi.org/10.4314/rjeste.v3i1.5.

Hauer, F.R., Lamberti, G.A., 2017. Methods in Stream Ecology, 3rd ed. vol. 1. Academic Press. ISBN 9780124165588. https://doi.org/10.1016/B978-0-12-416558-8.05001-0.

Hedin, L.O., von Fischer, J.C., Ostrom, N.E., Kennedy, B.P., Brown, M.G., Robertson, G.P., 1998. Thermodynamic constraints on nitrogen transformations and other biogeochemical processes at soil-stream interfaces. Ecology 79, 684.

Helton, A.M., Poole, G.C., Payn, R.A., Izurieta, C., Stanford, J.A., 2012. Scaling flow path processes to fluvial landscapes: an integrated field and model assessment of temperature and dissolved oxygen dynamics in a river-floodplain-aquifer system. J. Geophys. Res. 117, G00N14. https://doi.org/10.1029/2012JG002025.

Hensley, R.T., Cohen, M.J., 2016. On the emergence of diel solute signals in flowing waters. Water Resour. Res. 52, 759–772.

Hill, W.R., Dimick, S.M., 2002. Effects of riparian leaf dynamics on periphyton photosynthesis and light utilisation efficiency. Freshw. Biol. 47, 1245–1256.

Hoellein, T.J., Bruesewitz, D.A., Richardson, D.C., 2013. Revisiting Odum (1956): a synthesis of aquatic ecosystem metabolism. Limnol. Oceanogr. 58, 2089–2100. https://doi.org/10.4319/lo.2013.58.6.2089.

Holmer, M., Storkholm, P., 2001. Sulphate reduction and sulphur cycling in lake sediments: a review. Freshw. Biol. 46, 431–451.

Hood, E.M., Howes, B.L., Jenkins, W.J., 1998. Dissolved gas dynamics in perennially ice-covered Lake Fryxell, Antarctica. Limnol. Oceanogr. 43, 265–272. https://doi.org/10.4319/lo.1998.43.2.0265.

Huang, J., Huang, J., Liu, X., Li, C., Ding, L., Yu, H., 2018. The global oxygen budget and its future projection. Sci. Bull. 63, 1180–1186. https://doi.org/10.1016/j.scib.2018.07.023.

Hupfer, M., Jordan, S., Herzog, C., Ebeling, C., Ladwig, R., Rothe, M., et al., 2019. Chironomid larvae enhance phosphorus burial in lake sediments: insights from long-term and short-term experiments. Sci. Total Environ. 663, 254–264. https://doi.org/10.1016/j.scitotenv.2019.01.274.

Hutchinson, G.E., 1938. On the relation between the oxygen deficit and the productivity and typology of lakes. Int. Rev. Gesamten Hydrobiol. 36, 336–355. https://doi.org/10.1002/iroh.19380360205.

Hutchinson, G.E., 1957. A Treatise on Limnology. vol. 1. Geography, physics and chemistry. Wiley, New York, p. 1015.

Jane, S.F., Hansen, G.J.A., Kraemer, B.M., et al., 2021. Widespread deoxygenation of temperate lakes. Nature 594, 66–70. https://doi.org/10.1038/s41586-021-03550-y.

Jenny, J.-P., Francus, P., Normandeau, A., Lapointe, F., Perga, M.-E., Ojala, A., et al., 2016a. Global spread of hypoxia in freshwater ecosystems during the last three centuries is caused by rising local human pressure. Glob. Change Biol. 22, 1481–1489. https://doi.org/10.1111/gcb.13193.

Jenny, J.-P., Normandeau, A., Francus, P., Taranu, Z.E., Gregory-Eaves, I., Lapointe, F., et al., 2016b. Urban point sources of nutrients were the leading cause for the historical spread of hypoxia across European lakes. Proc. Natl. Acad. Sci. 113, 12655–12660.

Jones, J.B., Fisher, S.G., Grimm, N.B., 1995. Nitrification in the hyporheic zone of a desert stream ecosystem. J. N. Am. Benthol. Soc. 14, 249–258.

Jones, J.B., Holmes, R.M., 1996. Surface-subsurface interactions in stream ecosystems. Trends Ecol. Evol. 11, 239–242.

Kirf, M.K., Dinkel, C., Schubert, C.J., Wehrli, B., 2014. Submicromolar oxygen profiles at the oxic–anoxic boundary of temperate lakes. Aquat. Geochem. 20, 39–57. https://doi.org/10.1007/s10498-013-9206-7.

Koenig, L.E., Helton, A.M., Savoy, P., Bertuzzo, E., Heffernan, J.B., Hall, R.O., et al., 2019. Emergent productivity regimes of river networks. Limnol. Oceanogr. Lett. 4, 173–181.

Koop-Jakobsen, K., Mueller, P., Meier, R.J., Liebsch, G., Jensen, K., 2018. Plant-sediment interactions in salt marshes—an optode imaging study of O_2, pH, and CO_2 gradients in the rhizosphere. Front. Plant Sci. 9, 541. https://doi.org/10.3389/fpls.2018.00541.

Koschorreck, M., Hentschel, I., Boehrer, B., 2017. Oxygen ebullition from lakes. Geophys. Res. Lett. 44, 9372–9378. https://doi.org/10.1002/2017GL074591.

Krausen-Jensen, D., Sand-Jensen, K., 1998. Light attenuation and photosynthesis of aquatic plant communities. Limnol. Oceanogr. 43, 396–407. https://doi.org/10.4319/lo.1998.43.3.0396.

Kreling, J., Bravidor, J., Engelhardt, C., Hupfer, M., Koschorreck, M., Lorke, A., 2017. The importance of physical transport and oxygen consumption for the development of a metalimnetic oxygen minimum in a lake. Limnol. Oceanogr. 62, 348–363.

Kristensen, E., Penha-Lopes, G., Delefosse, M., Valdemarsen, T., Quintana, C.O., Banta, G.T., 2012. What is bioturbation? The need for a precise definition for fauna in aquatic sciences. Mar. Ecol. Prog. Ser. 446, 285–302.

LaBounty, J.F., Horn, M.J., 1997. The influence of drainage from the Las Vegas Valley on the limnology of Boulder Basin, Lake Mead, Arizona-Nevada. Lake Reserv. Manag. 13, 95–108.

Ladwig, R., Hanson, P.C., Dugan, H.A., Carey, C.C., Zhang, Y., Shu, L., et al., 2021. Lake thermal structure drives interannual variability in summer anoxia dynamics in a eutrophic lake over 37 years. Hydrol. Earth Syst. Sci. 25, 1009–1032. https://doi.org/10.5194/hess-25-1009-2021.

Langman, O.C., Hanson, P.C., Carpenter, S.R., Hu, Y.H., 2010. Control of dissolved oxygen in northern temperate lakes over scales ranging from minutes to days. Aquat. Biol. 9, 193–202.

Larson, D.W., 2000. Waldo Lake, Oregon: eutrophication of a rare, ultraoligotrophic, high-mountain lake. Lake Reserv. Manag. 16, 2–16.

Leach, T.H., Beisner, B.E., Carey, C.C., Pernica, P., Rose, K.C., Huot, Y., et al., 2018. Patterns and drivers of deep chlorophyll maxima structure in 100 lakes: the relative importance of light and thermal stratification. Limnol. Oceanogr. 63, 628–646.

Lind, O.T., 1987. Spatial and temporal variation in hypolimnetic oxygen deficits of a multidepression lake. Limnol. Oceanogr. 32, 740–744.

Livingstone, D.M., Imboden, D.M., 1996. The prediction of hypolimnetic oxygen profiles: a plea for a deductive approach. Can. J. Fish. Aquat. Sci. 53, 924–932. https://doi.org/10.1139/f95-230.

Lofton, M.E., Leach, T.H., Beisner, B.E., Carey, C.C., 2020. Relative importance of top-down vs. bottom-up control of lake phytoplankton vertical distributions varies among fluorescence-based spectral groups. Limnol. Oceanogr. 65, 2485–2501.

Luther, G.W., Findlay, A., MacDonald, D., Owings, S., Hanson, T., Beinart, R., et al., 2011. Thermodynamics and kinetics of sulfide oxidation by oxygen: a look at inorganically controlled reactions and biologically mediated processes in the environment. Front. Microbiol. 2, 62. https://doi.org/10.3389/fmicb.2011.00062.

Lyons, B., 2016. McMurdo Dry Valleys lakes dissolved oxygen (Winkler values), 1994/95 season ver 5. environmental data initiative (Accessed 2021-08-31).

Maberly, S.C., Spence, D.H.N., 1989. Photosynthesis and photorespiration in freshwater organisms: amphibious plants. Aquat. Bot. 34, 267–286. https://doi.org/10.1016/0304-3770(89)90059-4.

MacIntyre, S., Sickman, J.O., Goldthwait, S.A., Kling, G.W., 2006. Physical pathways of nutrient supply in a small, ultraoligotrophic arctic lake during summer stratification. Limnol. Oceanogr. 51, 1107–1124. https://doi.org/10.4319/lo.2006.51.2.1107.

Malard, F., Tockner, K., Dole-Oliver, M.J., Ward, J.V., 2002. A landscape perspective of surface–subsurface hydrological exchanges in river corridors. Freshw. Biol. 47, 621–640.

Malard, F., Uehlinger, U., Zha, R., Tockner, K., 2006. Flood-pulse and riverscape dynamics in a braided glacial river. Ecology 87, 704–716.

Marcé, R., George, G., Buscarinu, P., Deidda, M., Dunalska, J., de Eyto, E., et al., 2016. Automatic high frequency monitoring for improved lake and reservoir management. Environ. Sci. Technol. 50, 10780–10794. https://doi.org/10.1021/acs.est.6b01604.

Marcé, R., Moreno-Ostos, E., López, P., Armengol, J., 2008. The role of allochthonous inputs of dissolved organic carbon on the hypolimnetic oxygen content of reservoirs. Ecosystems 11, 1035–1053.

Marcé, R., Rodríguez-arias, M.A., García, J.C., Armengol, J., 2010. El Niño Southern Oscillation and climate trends impact reservoir water quality. Glob. Chang. Biol. 16, 2857–2865. https://doi.org/10.1111/j.1365-2486.2010.02163.x.

Marking, L.L., 1987. Gas supersaturation in fisheries: causes, concerns, and cures. Fish and Wildlife Leaflet 9. United States Department of the Interior, Fish and Wildlife Service, Washington, DC. Accessed in August 2021 at. https://apps.dtic.mil/sti/pdfs/ADA322709.pdf.

Matthews, D.A., Effler, S.W., 2006. Long-term changes in the areal hypolimnetic oxygen deficit (AHOD) of Onondaga Lake: evidence of sediment feedback. Limnol. Oceanogr. 51, 702–714. https://doi.org/10.4319/lo.2006.51.1_part_2.0702.

Matzinger, A., Müller, B., Niederhauser, P., Schmid, M., Wüest, A., 2010. Hypolimnetic oxygen consumption by sediment-based reduced substances in former eutrophic lakes. Limnol. Oceanogr. 55, 2073–2084. https://doi.org/10.4319/lo.2010.55.5.2073.

McClure, R.P., Hamre, K.D., Niederlehner, B.R., Munger, Z.W., Chen, S., Lofton, M.E., et al., 2018. Metalimnetic oxygen minima alter the vertical profiles of carbon dioxide and methane in a managed freshwater reservoir. Sci. Total Environ. 636, 610–620. https://doi.org/10.1016/j.scitotenv.2018.04.255.

McClure, R.P., Schreiber, M.E., Lofton, M.E., Chen, S., Krueger, K.M., Carey, C.C., 2021. Ecosystem-scale oxygen manipulations alter terminal electron acceptor pathways in a eutrophic reservoir. Ecosystems 24, 1281–1298. https://doi.org/10.1007/s10021-020-00582-9.

McGinnis, D.F., Little, J.C., 2002. Predicting diffused-bubble oxygen transfer rate using the discrete-bubble model. Water Res. 36, 4627–4635.

Melton, E.D., Swanner, E.D., Behrens, S., Schmidt, C., Kappler, A., 2014. The interplay of microbially mediated and abiotic reactions in the biogeochemical Fe cycle. Nat. Rev. Microbiol. 12, 797–808.

Messager, M.L., Lehner, B., Cockburn, C., Lamouroux, N., Pella, H., Snelder, T., et al., 2021. Global prevalence of non-perennial rivers and streams. Nature 594, 391–397.

Millero, F.J., Huang, F., Laferiere, A.L., 2002a. Solubility of oxygen in the major sea salts as a function of concentration and temperature. Mar. Chem. 78, 217–230.

Millero, F.J., Huang, F., Laferiere, A.L., 2002b. The solubility of oxygen in the major sea salts and their mixtures at 25°C. Geochem. Cosmochim. Acta 66, 2349–2359.

Minder, L., 1923. Studien über den Sauerstoffgehalt des Zürichsees. Arch. Hydrobiol. Suppl. 3, 197, 155.

Miranda, L.E., Krogman, R.M., 2014. Environmental stresses afflicting tailwater stream reaches across the United States. River Res. Appl. 30, 1184–1194.

Mitchell, S.F., Burns, C.W., 1979. Oxygen consumption in the epilimnia and hypolimnia of two eutrophic, warm-monomictic lakes. N. Z. J. Mar. Freshw. Res. 13, 427–441.

Moll, R.A., Brache, M.Z., Peterson, T.P., 1984. Phytoplankton dynamics within the subsurface chlorophyll maximum of Lake Michigan. J. Plankton Res. 6, 751–766. https://doi.org/10.1093/plankt/6.5.751.

Mulholland, P.J., Marzolf, E.R., Webster, J.R., et al., 1997. Evidence that hyporheic zones increase heterotrophic metabolism and phosphorus uptake in forest streams. Limnol. Oceanogr. 42, 443–451.

Müller, B., Bryant, L.D., Matzinger, A., Wüest, A., 2012. Hypolimnetic oxygen depletion in eutrophic lakes. Environ. Sci. Technol. 46, 9964–9971. https://doi.org/10.1021/es301422r.

Müller, B., Steinsberger, T., Schwefel, R., Gächter, R., Sturm, M., Wüest, A., 2019. Oxygen consumption in seasonally stratified lakes decreases only below a marginal phosphorus threshold. Sci. Rep. 9, 18054.

Müller, J., Weise, G., 1987. Oxygen budget of a river rich in submerged macrophytes (River Zschopau in the south of the GDR). Int. Rev. Gesamten Hydrobiol. Hydrogr. 72, 653–667.

Munger, Z.W., Carey, C.C., Gerling, A.B., Hamre, K.D., Doubek, J.P., Klepatzki, S.D., et al., 2016. Effectiveness of hypolimnetic oxygenation for preventing accumulation of Fe and Mn in a drinking water reservoir. Water Res. 106, 1–14.

Myrstener, M., Gómez-Gener, L., Rocher-Ros, G., Giesler, R., Sponseller, R.A., 2021. Nutrients influence seasonal metabolic patterns and total productivity of Arctic streams. Limnol. Oceanogr. 66, S182–S196. https://doi.org/10.1002/lno.11614.

Naranjo, R.C., Niswonger, R.G., Davis, C.J., 2015. Mixing effects on nitrogen and oxygen concentrations and the relationship to mean residence time in a hyporheic zone of a riffle-pool sequence. Water Resour. Res. 51, 7202–7217. https://doi.org/10.1002/2014WR016593.

National Institute for Environmental Studies, 2016. Lake Kasumigaura Database. National Institute for Environmental Studies, Japan. Accessed via. https://db.cger.nies.go.jp/gem/moni-e/inter/GEMS/database/kasumi/index.html on 2021-04-23.

Nimick, D.A., Gammons, C.H., Cleasby, T.E., Madison, J.P., Skaar, D., Brick, C.M., 2003. Diel cycles in dissolved metal concentrations in streams: occurrence and possible causes. Water Resour. Res. 39 (9), 1247. https://doi.org/10.1029/2002WR001571.

Nix, J., 1981. Contribution of hypolimnetic water on metalimnetic dissolved oxygen minima in a reservoir. Wat. Resour. Res. 27, 329–332.

Nürnberg, G.K., 1995. Quantifying anoxia in lakes. Limnol. Oceanogr. 40, 1100–1111. https://doi.org/10.4319/lo.1995.40.6.1100.

Nürnberg, G.K., 2002. Quantification of oxygen depletion in lakes and reservoirs with the hypoxic factor. Lake Reserv. Manag. 18, 299–306.

Nürnberg, G.K., 2019. Quantification of anoxia and hypoxia in water bodies. In: Maurice, P.A. (Ed.), Encyclopedia of Water: Science, Technology, and Society, 2nd ed. John Wiley & Sons, Inc., Hoboken.

Obertegger, U., Obrador, B., Flaim, G., 2017. Dissolved oxygen dynamics under ice: three winters of high-frequency data from Lake Tovel. Italy. Water Resour. Res. 53, 7234–7246.

O'Connor, B.L., Harvey, J.W., McPhillips, L.E., 2012. Thresholds of flow-induced bed disturbances and their effects on stream metabolism in an agricultural river. Water Resour. Res. 48, W08504.

Owens, M., Edwards, R., Gibbs, J., 1964. Some reaeration studies in streams. Int. J. Air Water Pollut. 8, 469–486.

Pace, M.L., Prairie, Y.T., 2005. Respiration in lakes. In: del Giorgio, P., Williams, P. (Eds.), Respiration in Aquatic Ecosystems. Oxford University Press, Oxford. https://doi.org/10.1093/acprof:oso/9780198527084.003.0007.

Padisák, J., Krienitz, L., Koschel, R., Nedoma, J., 1997. Deep-layer autotrophic picoplankton maximum in the oligotrophic Lake Stechlin, Germany: origin, activity, development and erosion. Eur. J. Phycol. 32, 403–416.

Pannard, A., Planas, D., Le Noac'h, P., Bormans, M., Jourdain, M., Beisner, B.E., 2020. Contribution of the deep chlorophyll maximum to primary production, phytoplankton assemblages and diversity in a small stratified lake. J. Plankton Res. 42, 630–649. https://doi.org/10.1093/plankt/fbaa043.

Pardo, I., García, L., 2016. Water abstraction in small lowland streams: unforeseen hypoxia and anoxia effects. Sci. Total Environ. 568, 226–235.

Peterson, B.J., 2001. Control of nitrogen export from watersheds by headwater streams. Science 292, 86–90.

Petsch, S.T., 2003. The global oxygen cycle. In: Holland, H.D., Turekian, K.K., Schlesinger, W.E. (Eds.), Treatise on Geochemistry, vol. 8. Elsevier, Biogeochemistry, pp. 515–555. https://doi.org/10.1016/B0-08-043751-6/08159-7.

Piatka, D.R., Wild, R., Hartmann, J., Kaule, R., Kaule, L., Gilfedder, B., et al., 2021. Transfer and transformations of oxygen in rivers as catchment reflectors of continental landscapes: a review. Earth Sci. Rev. 220, 103729.

Piccolroaz, S., Toffolon, M., 2018. The fate of Lake Baikal: how climate change may alter deep ventilation in the largest lake on Earth. Clim. Change 150, 181–194.

Pleizier, N.K., Algera, D., Cooke, S.J., Brauner, C.J., 2020. A meta-analysis of gas bubble trauma in fish. Fish Fish. 21, 1175–1194. https://doi.org/10.1111/faf.12496.

Poff, N.L., Allan, J.D., Bain, M.B., Karr, J.R., Prestegaard, K.L., Richter, B.D., et al., 1997. The natural flow regime. Bioscience 47, 769–784.

Pokorný, J., Hammer, L., Ondok, J.P., 1987. Oxygen budget in the reed belt and open water of a shallow lake. Arch. Hydrobiol. Beih. Ergebn. Limnol. 27, 185–201.

Powers, S.M., Baulch, H.M., Hampton, S.E., Labou, S.G., Lottig, N.R., Stanley, E.H., 2017. Nitrification contributes to winter oxygen depletion in seasonally frozen forested lakes. Biogeochemistry 136, 119–129. https://doi.org/10.1007/s10533-017-0382-1.

Priscu, J., 2019. McMurdo dry valleys ctd profiles in lakes ver 15. environmental data initiative. https://doi.org/10.6073/pasta/fcde3058c2d599572dc02e528f99db1e. (Accessed 2021-08-31).

Quinlan, R., Paterson, A.M., Smol, J.P., Douglas, M.S.V., Clark, B.J., 2005. Comparing different methods of calculating volume-weighted hypolimnetic oxygen (VWHO) in lakes. Aquat. Sci. 67, 97–103.

Ramsey, W.L., 1962. Bubble growth from dissolved oxygen near the sea surface. Limnol. Oceanogr. 7, 1–7.

Raven, J.A., Beardall, J., Quigg, A., 2020. Light-driven oxygen consumption in the water-water cycles and photorespiration, and light stimulated mitochondrial respiration. In: Larkum, A., Grossman, A., Raven, J. (Eds.), Photosynthesis in Algae: Biochemical and Physiological Mechanisms. Advances in Photosynthesis and Respiration (Including Bioenergy and Related Processes, vol. 45. Springer, Cham. https://doi.org/10.1007/978-3-030-33397-3_8.

Reisinger, A.J., Rosi, E.J., Bechtold, H.A., Doody, T.R., Kaushal, S.S., Groffman, P.M., 2017. Recovery and resilience of urban stream metabolism following Superstorm Sandy and other floods. Ecosphere 8, e01776.

Rhodes, J., Hetzenauer, H., Frassl, M.A., Rothhaupt, K.-O., Rinke, K., 2017. Long-term development of hypolimnetic oxygen depletion rates in the large Lake Constance. Ambio. 46, 554–565.

Ricker, W.E., 1934. A critical discussion of various measures of oxygen saturation in lakes. Ecology 15, 348–363.

Roberts, B.J., Mulholland, P.J., Hill, W.R., 2007. Multiple scales of temporal variability in ecosystem metabolism rates: results from 2 years of continuous monitoring in a forested headwater stream. Ecosystems 10, 588–606.

Ruttner, F., 1963. Fundamentals of limnology. In: Translat, Frey, D.G., Fry, F.E.J. (Eds.). University of Toronto Press, Toronto, p. 295.

Saari, G.N., Wang, Z., Brooks, B.W., 2018. Revisiting inland hypoxia: diverse exceedances of dissolved oxygen thresholds for freshwater aquatic life. Environ. Sci. Pollut. Res. 25, 3139–3150. https://doi.org/10.1007/s11356-017-8908-6.

Sadro, S., Melack, J.M., MacIntyre, S., 2011. Depth-integrated estimates of ecosystem metabolism in a high-elevation lake (Emerald Lake, Sierra Nevada, California). Limnol. Oceanogr. 56, 1764–1780.

Salinas, J., 2000. Thermal and chemical properties of Waldo Lake, Oregon. Lake Reserv. Manag. 16, 40–51.

Savoy, P., Appling, A.P., Heffernan, J.B., Stets, E.G., Read, J.S., Harvey, J.W., et al., 2019. Metabolic rhythms in flowing waters: an approach for classifying river productivity regimes. Limnol. Oceanogr. 64, 1835–1851. https://doi.org/10.1002/lno.11154.

Savoy, P., Harvey, J.W., 2021. Predicting light regime controls on primary productivity across CONUS river networks. Geophys. Res. Lett. 48, e2020GL092149.

Schütz, H., Conrad, R., Goodwin, S., Seiler, W., 1988. Emission of hydrogen from deep and shallow freshwater environments. Biogeochemistry 5, 295–311. https://doi.org/10.1007/BF02180069.

Schwefel, R., Gaudard, A., Wuest, A., Bouffard, D., 2016. Effects of climate change on deepwater oxygen and winter mixing in a deep lake (Lake Geneva): comparing observational findings and modeling. Water Resour. Res. 52, 8811–8826. https://doi.org/10.1002/2016WR019194.

Schwefel, R., Steinsberger, T., Bouffard, D., Bryant, L.D., Müller, B., Wüest, A., 2018. Using small-scale measurements to estimate hypolimnetic oxygen depletion in a deep lake. Limnol. Oceanogr. 63, S54–S67. https://doi.org/10.1002/lno.10723.

Scott, W., 1924. The diurnal oxygen pulse in Eagle (Winona). Lake. Proc. Indiana Acad. Sci. 33, 311–314.

Segatto, P.L., Battin, T.J., Bertuzzo, E., 2021. The metabolic regimes at the scale of an entire stream network unveiled through sensor data and machine learning. Ecosystems 24, 1792–1809.

Seto, M., Nishida, S., Yamamoto, M., 1982. Dissolved organic carbon as a controlling factor in oxygen consumption in natural and man-made waters. Jpn. J. Limnol. 43, 96–101.

Shapiro, J., 1960. The cause of a metalimnetic minimum of dissolved oxygen. Limnol. Oceanogr. 5, 216–227.

Siegert, M.J., Cynan Ellis-Evans, J., Tranter, M., Mayer, C., Petit, J.-R., Salamatin, A., et al., 2001. Physical, chemical and biological processes in Lake Vostok and other Antarctic subglacial lakes. Nature 414, 603–609. https://doi.org/10.1038/414603a.

Song, C., Dodds, W.K., Rüegg, J., Argerich, A., Baker, C.L., Bowden, W.B., et al., 2018. Continental-scale decrease in net primary productivity in streams due to climate warming. Nat. Geosci. 11, 415–420. https://doi.org/10.1038/s41561-018-0125-5.

Spigel, R.H., Priscu, J.C., 1998. Physical limnology of the McMurdo dry valleys lakes. In: Priscu, J.C. (Ed.), Ecosystem Dynamics in a Polar Desert: The McMurdo Dry Valleys, Antarctica. https://doi.org/10.1029/AR072p0153.

Spigel, R.H., Priscu, J.C., Obryk, M.K., Stone, W., Doran, P.T., 2018. The physical limnology of a permanently ice-covered and chemically stratified Antarctic lake using high resolution spatial data from an autonomous underwater vehicle. Limnol. Oceanogr. 63, 1234–1252. https://doi.org/10.1002/lno.10768.

Steinsberger, T., Schmid, M., Wüest, A., Schwefel, R., Wehrli, B., Müller, B., 2017. Organic carbon mass accumulation rate regulates the flux of reduced substances from the sediments of deep lakes. Biogeosciences 14, 3275–3285. https://doi.org/10.5194/bg-14-3275-2017.

Steinsberger, T., Schwefel, R., Wüest, A., Müller, B., 2020. Hypolimnetic oxygen depletion rates in deep lakes: effects of trophic state and organic matter accumulation. Limnol. Oceanogr. 65, 3128–3138. https://doi.org/10.1002/lno.11578.

Stetler, J.T., Jane, S.F., Mincer, J.L., Sanders, M.N., Rose, K.C., 2021. Long-term lake dissolved oxygen and temperature data, 1941–2018 Ver 2. environmental data initiative. https://doi.org/10.6073/pasta/841f0472e19853b0676729221aedfb56. (Accessed 2021-08-31).

Stets, E.G., Sprague, L.A., Oelsner, G.P., Johnson, H.M., Murphy, J.C., Ryberg, K., et al., 2020. Landscape drivers of dynamic change in water quality of U.S. Rivers. Environ. Sci. Technol. 54, 4336–4343.

Stoddard, J.L., Van Sickle, J., Herlihy, A.T., Brahney, J., Paulsen, S., Peck, D.V., et al., 2016. Continental-scale increase in lake and stream phosphorus: are oligotrophic systems disappearing in the United States? Environ. Sci. Technol. 50, 3409–3415. https://doi.org/10.1021/acs.est.5b05950.

Stumm, W., Morgan, J.J., 2012. Aquatic Chemistry: Chemical Equilibria and Rates in Natural Waters, vol. 126. John Wiley & Sons, New York.

Sytsma, M., Rueter, J., Petersen, R., Koch, R., Wells, S., Miller, R., et al., 2004. Waldo Lake research in 2003. Report 97201-0751. Center for Lakes and Reservoirs, Department of Environmental Sciences and Resources, Department of Civil and Environmental Engineering, Portland State University, Portland, Oregon.

Thienemann, A., 1926. Der Nahrungskreislauf im Wasser. Verh. Deutsch. Zool. Gesell. 31, 29–79.

Thienemann, A., 1928. Der Sauerstoff im eutrophen und oligotrophen See. Ein Beitrag zur Seetypenlehre. Die Binnengewässer 4, 175.

Thomas, E.A., 1960. Sauerstoffminima und Stoffkreislaüfe im ufernahen Oberflächenwasser des Zürichsees (Cladophora- und Phragmites-Gürtel). Monatsbull. Schweiz. Ver. Gas- Wasserfachmännern 1960 (6), 1–8.

Thornton, K.W., Kimmel, B.L., Payne, F.E., 1990. Reservoir limnology: ecological perspectives. John Wiley and Sons, New York.

Tipping, E., Thompson, D.W., Davison, W., 1984. Oxidation products of Mn(II) in lake waters. Chem. Geol. 44, 359–383. https://doi.org/10.1016/0009-2541(84)90149-9.

Tittel, J., Bissinger, V., Zippel, B., Gaedke, U., Bell, E., Lorke, A., et al., 2003. Mixotrophs combine resource use to outcompete specialists: implications for aquatic food webs. Proc. Natl. Acad. Sci. 100, 12776–12781.

Tonolli, L., 1969. Holomixy and oligomixy in Lake Maggiore: inference on the vertical distribution of zooplankton. SIL Proc., 1922-2010 17, 231–236.

Trimmer, M., Maanoja, S., Hildrew, A.G., Pretty, J.L., Grey, J., 2010. Potential carbon fixation via methane oxidation in well-oxygenated river bed gravels. Limnol. Oceanogr. 55, 560–568. https://doi.org/10.4319/lo.2010.55.2.0560.

Twiss, M.R., McKay, R.M.L., Bourbonniere, R.A., Bullerjahn, G.S., Carrick, H.J., Smith, R.E.H., et al., 2012. Diatoms abound in ice-covered Lake Erie: an investigation of offshore winter limnology in Lake Erie over the period 2007 to 2010. J. Great Lakes Res. 38, 18–30.

Uehlinger, U., Kawecka, B., Robinson, C.T., 2003. Effects of experimental floods on periphyton and stream metabolism below a high dam in the Swiss Alps (River Spöl). Aquat. Sci. 65, 199–209.

U.S. Geological Survey, 2011. U.S. geological survey office of water quality technical memorandum 2011.03, 2 P. plus attachment accessed 6 January 2020, at http://water.usgs.gov/admin/memo/QW/qw11.03.pdf.

Vähätalo, A., 2009. Light, photolytic reactivity and chemical products. In: Likens, G.E. (Ed.), Encyclopedia of Inland Waters, vol. 2. Academic Press, pp. 761–773.

Val Klump, J., Brunner, S.L., Grunert, B.K., et al., 2018. Evidence of persistent, recurring summertime hypoxia in Green Bay, Lake Michigan. J. Great Lakes Res. 44, 841–850. https://doi.org/10.1016/j.jglr.2018.07.012.

Van de Bogert, M.C., Carpenter, S.R., Cole, J.J., Pace, M.L., 2007. Assessing pelagic and benthic metabolism using free water measurements. Limnol. Oceanogr. Methods 5, 145–155.

Vannote, R.L., Minshall, G.W., Cummins, K.W., Sedell, J.R., Cushing, C.E., 1980. The River Continuum Concept. Can. J. Fish. Aquat. Sci. 37, 130–137. https://doi.org/10.1139/f80-017.

Vazquez, E., Amalfitano, S., Fazi, S., Butturini, A., 2011. Dissolved organic matter composition in a fragmented Mediterranean fluvial system under severe drought conditions. Biogeochemistry 102, 59–72. https://doi.org/10.1007/s10533-010-9421-x.

Venkiteswaran, J.J., Schiff, S.L., Taylor, W.D., 2015. Linking aquatic metabolism, gas exchange, and hypoxia to impacts along the 300-km Grand River, Canada. Freshw. Sci. 34, 1216–1232.

Vinçon-Leite, B., Casenave, C., 2019. Modelling eutrophication in lake ecosystems: a review. Sci. Total Environ. 651, 2985–3001.

Walling, D.E., Webb, B.W. (Eds.), 1994. Erosion and sediment yield: global and regional perspectives. International Association of Hydrological Sciences, Wallingford.

Walsh, C.J., Roy, A.H., Feminella, J.W., Cottingham, P.D., Groffman, P.M., Morgan, R.P., 2005. The urban stream syndrome: current knowledge and the search for a cure. J. North Am. Benthol. Soc. 24, 706–723.

Wendt-Potthoff, K., Kloß, C., Schultze, M., Koschorreck, M., 2014. Anaerobic metabolism of two hydro-morphological similar pre-dams under contrasting nutrient loading (Rappbode Reservoir System, Germany). Int. Rev. Hydrobiol. 99, 350–362.

Wentzky, V.C., Frassl, M.A., Rinke, K., Boehrer, B., 2019. Metalimnetic oxygen minimum and the presence of *Planktothrix rubescens* in a low-nutrient drinking water reservoir. Water Res. 148, 208–218.

Wetzel, R.G., Likens, G.E., 2000. Limnological analyses. Springer, New York. https://doi.org/10.1007/978-1-4757-3250-4_6.

Wharton, R.A., McKay, C.P., Simmons, G.M., Parker, B.C., 1986. Oxygen budget of a perennially ice-covered Antarctic lake. Limnol. Oceanogr. 31, 437–443. https://doi.org/10.4319/lo.1986.31.2.0437.

Whitworth, K.L., Baldwin, D.S., Kerr, J.L., 2012. Drought, floods and water quality: drivers of a severe hypoxic blackwater event in a major river system (the southern Murray–Darling Basin, Australia). J. Hydrol. 450–451, 190–198.

Wik, M., Crill, P.M., Varner, R.K., Bastviken, D., 2013. Multiyear measurements of ebullitive methane flux from three subarctic lakes. J. Geophys. Res. Biogeosci. 118, 1307–1321. https://doi.org/10.1002/jgrg.20103.

Wilkin, R.T., Ludwig, R.D., Ford, R.G. (Eds.), 2002. Workshop on Monitoring Oxidation-Reduction Processes for Ground-Water Restoration. National Risk Management Research Laboratory, Office of Research and Development, U.S. Environmental Protection Agency, Cincinnati. Report PA/600/R-02/002. Subsurface Protection and Remediation Division.

Wilkinson, G.M., Cole, J.J., Pace, M.L., Johnson, R.A., Kleinhans, M.J., 2015. Physical and biological contributions to metalimnetic oxygen maxima in lakes. Limnol. Oceanogr. 60, 242–251.

Woolway, R.I., Sharma, S., Weyhenmeyer, G.A., Debolskiy, A., Golub, M., et al., 2021. Phenological shifts in lake stratification under climate change. Nat. Commun. 12, 2318. https://doi.org/10.1038/s41467-021-22657-4.

Zaidel, P.A., Roy, A.H., Houle, K.M., Lambert, B., Letcher, B.H., Nislow, K.H., et al., 2021. Impacts of small dams on stream temperature. Ecol. Indic. 120, 106878.

Zarnetske, J.P., Haggerty, R., Wondzell, S.M., Baker, M.A., 2011. Dynamics of nitrate production and removal as a function of residence time in the hyporheic zone. J. Geophys. Res. 116, G01025.

Zhi, W., Feng, D., Tsai, W.-P., Sterle, G., Harpold, A., Shen, C., et al., 2021. From hydrometeorology to river water quality: can a deep learning model predict dissolved oxygen at the continental scale? Environ. Sci. Technol. 55, 2357–2368.

Zimmermann, M., Mayr, M.J., Bürgmann, H., Eugster, W., Steinsberger, T., Wehrli, B., Brand, A., Bouffard, D., 2021. Microbial methane oxidation efficiency and robustness during lake overturn. Limnol. Oceanogr. Lett. 6, 320–328. https://doi.org/10.1002/lol2.10209.

C H A P T E R

12

Salinity and Ionic Composition of Inland Waters

Hilary A. Dugan

Center for Limnology, University of Wisconsin–Madison, Madison, WI, United States

OUTLINE

Salinity refers to the mass of dissolved inorganic solids found in water (Table 12-1) and is important in limnology both for its influence on water density (Chapter 3), and therefore water movements (Chapter 8), and for the individual constituents the term encompasses (Day, 1990). Owing to the power of water as a solvent (Chapter 3), salinity comprises numerous ions and molecules, but for most limnological purposes, salinity is considered the mass fraction (i.e., mg kg^{-1}) or mass concentration (i.e., mg L^{-1}) of the major cations (Na^+, Mg^{2+}, Ca^{2+}, K^+), anions (Cl^-, SO_4^{2-}), and carbonate species (HCO_3^-, CO_3^{2-}) dissolved in water. When high precision measurement is required, the small fraction of mass contributed by silica ($Si(OH)_4$, notably inert), minor ions (e.g., Fe, Mn, Al, F), and nutrients (e.g., NO_3^-, PO_4^{3-}) may be important. Salinity is often considered interchangeable with *total dissolved solids* (TDS), but TDS includes the mass of dissolved organic matter (DOM, Table 12-1). This subtle difference is often insignificant but becomes important where DOM concentrations are high, as DOM can affect density.

In physical oceanography the definition of salinity has a long history (Pawlowicz et al., 2012), driven by its important influence on water density and ocean mixing. Approximating salinity became necessary because comprehensively measuring the complex mixture of constituents in water was (and still is) a massive undertaking. The term salinity was originally proposed after quantitative analyses revealed that of the numerous elements dissolved in ocean water, only water and salts were of consequence to the overall mass (Forchhammer, 1865). Because salinity is dominated by ionic constituents that are positively and negatively charged, salinity can be approximated by the *electrical conductivity* (EC), a measure

https://doi.org/10.1016/B978-0-12-822701-5.00012-4

TABLE 12-1 Properties Related to the Ionic Composition of Inland Waters (Day, 1990; USGS, 2019)

Property	Common SI unit	Definition
Salinity	mg kg^{-1} or mg L^{-1}	Mass of dissolved inorganic solids in water
Total dissolved solids (TDS)	mg kg^{-1} or mg L^{-1}	Mass of dissolved solids in water (includes dissolved organic matter)
Electrical conductance	μS	Ability of water to conduct an electric current
Electrical conductivity (EC)	μS cm^{-1}	Electrical conductance normalized to a unit length and cross-sectional area
Specific conductance (SpC)	μS cm^{-1} @ 25°C	Electrical conductivity of 1 cubic centimeter (cm^3) of solution at 25°C

Note: Dissolved gases also influence water density but are not typically accounted for in measurements of TDS or salinity.

TABLE 12-2 Salinity and ionic concentrations (mg L^{-1}) of Swedish Bicarbonate Freshwaters in Relation to Specific Conductance (SpC at 20°C, μS cm^{-1})[a]

Salinity	Ca^{2+}	Mg^{2+}	Na^+	K^+	CO_3	SO_4^{2-}	Cl^-	SpC (20°C)
10.5	2.5	0.4	0.7	0.3	4.4	1.5	0.7	20
32.9	7.9	1.3	2.2	0.8	13.8	4.7	2.2	60
56.5	13.5	2.3	3.8	1.4	23.7	8.0	3.8	100
92.1	22.0	3.7	6.2	2.3	38.6	13.1	6.2	160
117.3	28.1	4.7	7.9	2.9	49.2	16.6	7.9	200
155.1	37.1	6.2	10.4	3.8	65.2	22.0	10.4	260
180.8	43.3	7.3	12.2	4.4	75.8	25.6	12.2	300
219.7	52.6	8.8	14.8	5.4	92.1	31.2	14.8	360
246.2	59.0	9.9	16.6	6.0	103.2	34.9	16.6	400

[a]*From data of Hutchinson (1957) and Rodhe (1948).*

of a solution's ability to conduct electrical flow, expressed in micro- or millisiemens per centimeter (μS cm^{-1} or mS cm^{-1}). Older literature often used the unit μmhos cm^{-1}, which is equivalent to μS cm^{-1}, but the former term is now considered obsolete. By definition, EC is the reciprocal of the resistance of a solution measured between two electrodes 1 cm^2 in area and 1 cm apart. In fresh to saline waters the resistance of water to electron flow declines with increasing ion content. Hence the purer that water is, that is, the lower its salinity, the greater its resistance and the lower the EC. EC is also highly temperature dependent. Temperature increases from 0°C to 30°C both decrease water viscosity and increase the solubility of most salts, increasing EC about 1.9% per degree Celsius. Thus *specific conductance* is often reported as EC corrected to 25°C (USGS, 2019) (see also Chapter 3).

The relationship between EC and salinity (and density) is dependent on the relative proportion of dissolved ions, as ions are not equal in their influence on conductance, and therefore the relationship of EC to salinity is site specific and only holds for a constant relative ratio of dissolved ions (Pawlowicz and Feistel, 2012; Talling, 2009; Williams and Sherwood, 1994). The homogeneity of the composition of sea salt led to the use of the *practical salinity scale* in oceanography in the 1980s, a unitless scale defined as a conductivity ratio. Similar EC:salinity and EC:density relationships can be devised for inland waters where the proportions of the individual ions are relatively constant (Table 12-2) (Moreira et al., 2016). However, as noted, EC only reflects ionic species and overlooks neutral species such as dissolved silica, which can be present in high concentrations in some inland waters (McManus et al., 1992). Due to this limitation and other thermodynamic considerations, oceanographers have since adopted the term *absolute salinity* (IOC et al., 2010), which is a true measure of total dissolved inorganic solids.

Limnologists rarely have the luxury of working in a homogenous setting and must be cognizant that relative proportions of dissolved ions are typically in flux. The specific conductance of the common bicarbonate-type of lake and river water is closely proportional (see Table 12-2) to concentrations of the major ions (Rodhe, 1948). Once the specific conductance and concentrations of the major ions are known, changes in specific conductance reflect proportional changes in ionic concentrations. This relationship is not true, however, for minor constituents of lake water (e.g., N, Fe, Mn, Sr, or especially P), as a large change in relative concentration may have an indistinguishable effect on specific conductance (Rodhe, 1950). As would be anticipated, a positive correlation exists between specific conductance and pH in the intermediate pH range of bicarbonate freshwaters (Talling, 2009), but this relationship deteriorates among lakes of low salinity and high DOM content (Strøm, 1947). Whether or not to report EC, specific conductance, salinity, or TDS will depend on the application,[1] and a

[1] In this chapter data are presented in their original measurement units. Since the density of pure water is 1 kg L^{-1}, concentration units presented as mass/mass or mass/volume are often considered equal for freshwaters. This is not true for saline waters.

detailed description of the methods used to calculate these values should be provided.

I. Salinities and ionic composition of inland waters

The beauty of inland waters is in their diversity, and salinity is no exception. Lakes range in salinity from almost pure water (glacially fed lakes) to the record-setting Gaet'ale Pond in Ethiopia with a TDS of 433 g kg^{-1}, 12 times the concentration of seawater (Pérez and Chebude, 2017). Given the natural gradient of salinities in inland waters, it is useful to differentiate freshwater from saline water, even though there is no ecological basis for such an exact threshold (Beadle, 1969), and opinions on terminology have varied through time (see Hammer, 1986). Nevertheless, for practical purposes, the boundary between fresh and saline water has been generally accepted as 3 g kg^{-1}, as originally put forth by Williams (1981). For some applications, more regionally meaningful thresholds may be appropriate, and suggestions have included 0.5, 1, and 5 g kg^{-1}. For instance, 99% of Canadian lakes are <3 g kg^{-1}, as compared to only 50% of Australian lakes (Hammer, 1986). Freshwater for drinking purposes is often restricted to TDS < 0.5 g L^{-1} for aesthetic effects (US EPA, 2019).

Lakes with salinities >3 g kg^{-1} are referred to as saline lakes, and further differentiated into those of marine origin and may be *brackish* (salinities ranging from the freshwater cut-off up to ocean salinity of 30–35 g kg^{-1}) and those of nonmarine origin, termed *athalassic lakes*. The water in saline lakes is referred to as *brine*, meaning a salty solution. Notably, many of the largest lakes on Earth are saline, including the Caspian Sea, Issyk-Kul, Lake Van, and Lake Turkana, but the majority of the hundreds of millions of lakes on Earth are freshwater lakes. A more exhaustive classification system of athalassic saline waters by Hammer (1986) includes freshwater (<0.5 g L^{-1}), subsaline (0.5–3 g L^{-1}), hyposaline (3–20 g L^{-1}), mesohaline (20–50 g L^{-1}), and hypersaline (>50 g L^{-1}). Well-known hypersaline lakes include the Great Salt Lake (USA), the Dead Sea, and Don Juan Pond (Antarctica) (Fig. 12-1b). Lastly, *poikilohaline* is a term used to refer to water of variable salinity. Detailed summaries of inland saline lakes have been made in several studies (e.g., Williams 1981; Hammer 1986; Deocampo and Jones 2014; Wurtsbaugh et al. 2017) and are discussed in Section III.

The ionic composition of freshwaters is dominated by dilute solutions of alkalis (Na^+, K^+) and alkaline earth compounds (Ca^{2+}, Mg^{2+}), particularly bicarbonates, carbonates, sulfates, and chlorides. The concentrations of four major cations, Ca^{2+}, Mg^{2+}, Na^+, and K^+, and four major anions, HCO_3^-, CO_3^{2-}, SO_4^{2-}, and Cl^-, usually

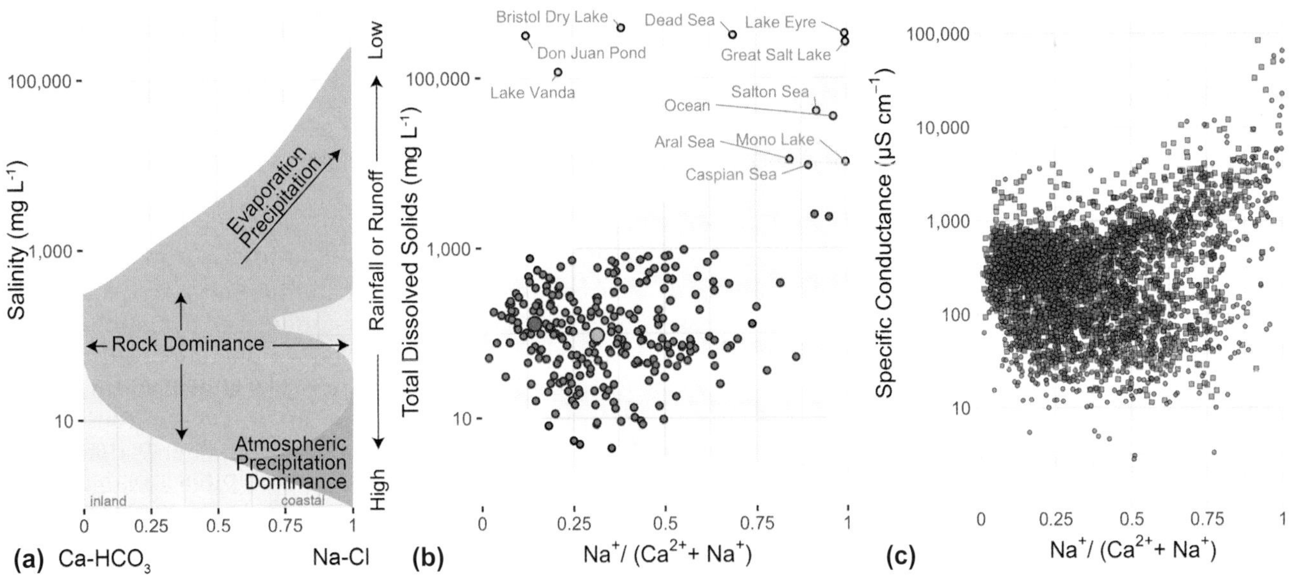

FIGURE 12-1 (a) Gibbs diagram. Conceptual representation of the processes controlling the ionic chemistry of inland surface waters (Gibbs, 1970, 1992). (b) Prevalence of sodium in relationship to calcium versus TDS of the GEMS-GLORI database of major world rivers as well as 10 saline lakes (Deocampo and Jones, 2014). Large circles represent the world weighted average (orange) and world spatial median (red) of global rivers (Meybeck, 2003). (c) Prevalence of sodium in relationship to calcium versus specific conductance of 1230 lakes (circles) and 2261 rivers (squares) from the United States EPA 2013–2014 National Rivers and Streams Assessment and 2012 National Lakes Assessment (US EPA, 2020, 2016).

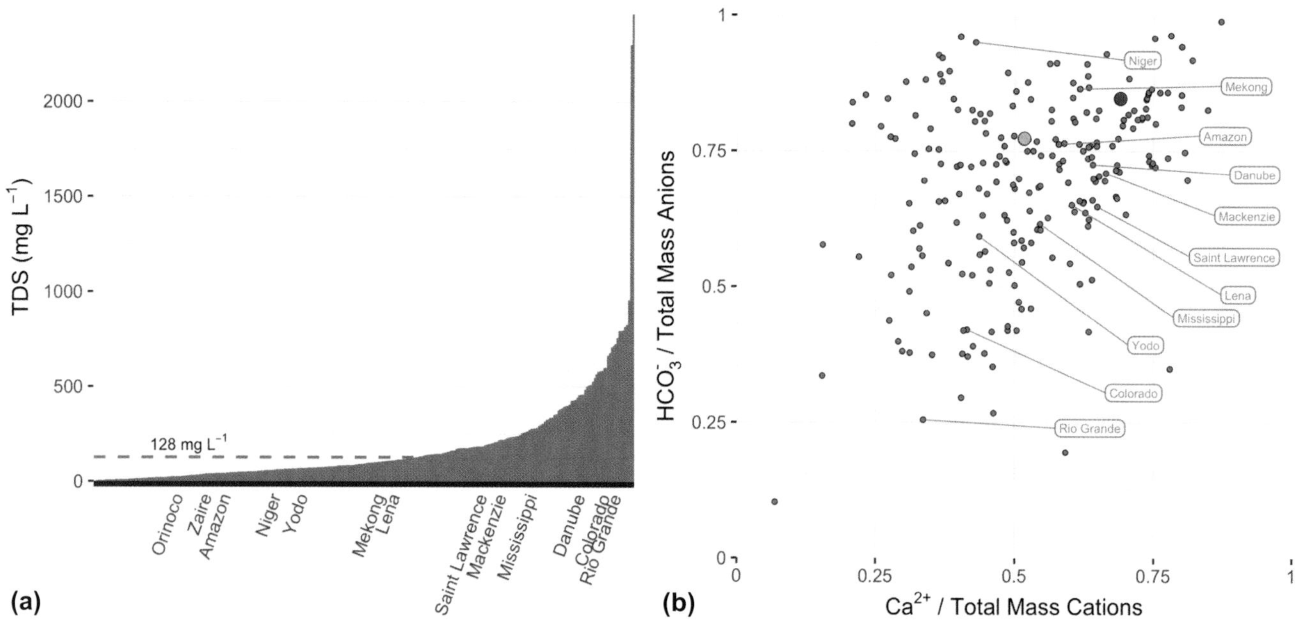

FIGURE 12-2 (a) Total dissolved solids (TDS) of the world's major rivers. The dashed line represents the mean global river TDS of 128 mg L^{-1}. (b) Ratios of bicarbonate:total anions and calcium:total cations for the same rivers. Data from the GEMS-GLORI world river discharge database (Meybeck and Ragu, 2012). Large circles represent the world weighted average (*orange*) and world spatial median (*red*) of global rivers (Meybeck, 2003).

constitute the *total salinity* of the water for all practical purposes. Major ion composition of inland waters is commonly expressed in mass concentrations (mg L^{-1} or μg L^{-1}) or molar equivalents (meq L^{-1} or μeq L^{-1}), the latter allowing a check of ionic balances, where $\sum$ cations (meq L^{-1}) = $\sum$ anions (meq L^{-1}). Because the conversion between mass and molar equivalents changes ionic ratios (Fig. 12-2), the choice of units must be clearly stated.

It is commonplace for freshwaters to be characterized as *soft water*, if waters are dilute and ions are derived from the drainage of acidic igneous rocks. In contrast, *hardwaters* contain high concentrations of Ca^{2+} and Mg^{2+}, usually derived from the drainage of calcareous deposits (see also Chapter 13). The amounts of silicic acid, which occur largely in undissociated form, are usually small but occasionally significant. Concentrations of ionized components of other elements such as nitrogen (N), phosphorus (P), and iron (Fe) and numerous minor elements are of immense biological importance but are minor contributors to total salinity.

A. Control mechanisms on salinity

The variation in ionic composition in the world's inland waters has led limnologists to try to find commonality among the diversity of systems. The *Gibbs model* is perhaps the most well-known and delineates the roles of rock weathering, atmospheric precipitation, and the evaporation–precipitation process on the mass ratios of Na^+: (Na^+ + Ca^{2+}) and Cl^-: (Cl^- + HCO_3^-) versus salinity or TDS (Gibbs, 1970) (Fig. 12-1a). Revisions of Gibbs' original boomerang diagram have expanded the envelope to include a wider range of possible ionic compositions, including dilute lakes that Gibbs eloquently described as "crystalline bowls that simply catch rain water" (Gibbs, 1992). While there have been many documented departures from the boomerang (Eilers et al., 1992; Kilham, 1990), the Gibbs diagram has not been displaced in limnology textbooks as its characterizations hold true for many surface waters (Fig. 12-1). Waters at the rock-dominated end of the spectrum are rich in Ca^{2+} and HCO_3^- and are more or less in equilibrium with materials of their drainage basins. Positions within this grouping depend on the climate, basin relief, and particularly the composition of the rock material in the basin (Table 12-3). The chemical compositions of low-salinity waters, typically dominated by Na^+ and Cl^-, are influenced by dissolved salts derived from atmospheric precipitation acquired from the ocean. These freshwaters are limited to immediate maritime coastal regions. In very chemically dilute inland lakes with a strong precipitation dominance and little marine influence, appreciable sodium can be derived from weathering (Eilers et al., 1992; Wu and Gibson, 1996). The major ionic composition of these lakes can fall outside (lower left, salinity below 10 mg L^{-1}) the primary envelope of Fig. 12-1a. Inland tropical humid areas of South America and Africa generally have high rainfall with an ionic composition that is either influenced primarily by terrestrial biological

TABLE 12-3 Global Average Chemical Composition (μeq L^{-1} Except as Noted) of Unpolluted Rivers and Variations in Composition According to Dominant Rock Type[a]

	Global average[b]							
	μeq L^{-1}	*mg L^{-1}*	*Granite*	*Gneiss*	*Volcanic rocks*	*Sandstone*	*Shale*	*Carbonate rock*
Ca^{2+}	670	13.43	39	60	154	88	404	2560
Mg^{2+}	259	3.15	31	57	161	63	240	640
Na^{+}	159	3.66	88	80	105	51	105	34
K^{+}	32	1.25	8	10	14	21	20	13
HCO_3^-	835	50.94	128	136	425	125	580	3195
SO_4^{2-}	163	7.83	31	56	10	95	143	85
Cl^-	86	3.05	0	0	0	0	20	0
SiO_2 (μmol L^{-1})	173	10.40	150	130	200	150	150	100

[a]*Extracted from data of Meybeck and Helmer (1989).*
[b]*Global discharge-weighted natural concentrations, corrected for oceanic cyclic salts.*

emissions and particulate dust or, rapid, intense rock weathering (Kilham, 1990; Stallard and Edmond, 1981). Kilham (1990) showed that plotting data from African lakes resulted in "an alchemist's retort than a boomerang." In the upper right of the boomerang, surface water salinity is further influenced by the evaporation and sedimentation of mineral salts. Surface waters typically extend in a series from calcium- and bicarbonate-rich low-salinity waters of rock dominance to sodium-dominated, high-salinity waters (Fig. 12-1). Lakes at the extreme saline end of this series are generally located in arid regions and display a wide range of ionic compositions (see Section III).

The examples discussed above represent average concentrations and general distributions on a global basis. Where the Gibbs diagram falls short is recognizing the inherent variability in river and lake water chemistry as concentrations continually fluctuate in time due to the interplay of solute inputs, precipitation/evaporation, and human influences. Within regions, the distribution of salinity of inland waters is highly variable and localized. Hence some areas or lake districts have, because of uniform rock formations, ionic proportions that are relatively consistent (Table 12-2). In contrast, in some areas, such as in southwestern Michigan, USA, extremely softwater seepage lakes may be found immediately adjacent to very hardwater calcareous lakes because of complex, varying patterns in the deposition of glacial till.

Any concept of "average" river or lake chemistry is useful for scaling global elemental cycling, although not particularly applicable for studies of individual systems. The most comprehensive estimate of the global mean salinity of river water is 128 mg L^{-1}, based on 1200 pristine and subpristine basins (Table 12-4, Fig. 12-3; Meybeck, 2003), a concentration remarkably close to Livingston's (1963) estimate of 120 mg L^{-1}, with rivers spanning a range of 4.7 (1% quantile; Q1) to 2314 mg L^{-1} (Q99). In a randomly stratified sampling of United States rivers and lakes mean specific conductance in 2261 rivers was 629 μS cm^{-1} (Q1: 20 μS cm^{-1}, Q50: 292 μS cm^{-1}, Q99: 4609 μS cm^{-1}) and 688 μS cm^{-1} (Q1: 11 μS cm^{-1}, Q50: 214 μS cm^{-1}, Q99: 7516 μS cm^{-1}) in 1230 lakes (US EPA, 2016, 2020).

The dissolved composition of rivers is dominated by HCO_3^- and Ca^{2+}, followed by SO_4^{2-} and Mg^{2+} and Na^+ (Table 12-4). Over large regions of the temperate zone, the ionic composition of lakes tends to mirror rivers, with dominance by calcium and bicarbonate ions (Table 12-5).

TABLE 12-4 Median Composition of the River Waters of the World (mg L^{-1}). (Adapted from Meybeck, 2003; the world spatial median values are derived from the PRISRI dataset ($n = 1200$) of 3200–200,000 km^2 endorheic and exorheic global basins.)

	SiO_2	*Ca^{2+}*	*Mg^{2+}*	*Na^+*	*K^+*	*Cl^-*	*SO_4^{2-}*	*HCO_3^-*	*Sum*
World spatial median	8.1	20.0	4.6	3.4	1.0	3.4	10.5	76.6	128
99th quantile	40.9	186.4	71.7	333.4	19.7	602.7	696.4	363.0	2314
1st quantile	0.2	0.6	0.1	0.4	0.2	0.1	0.2	2.9	4.7

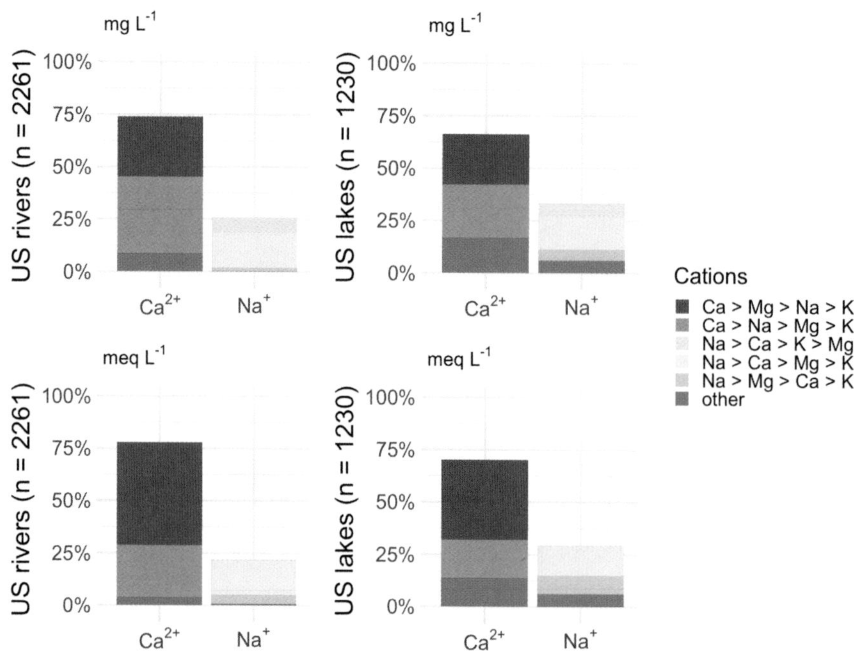

FIGURE 12-3 The prevalence of cations in a randomly stratified sampling of United States rivers and lakes in 2012 and 2013 shown in both mg L^{-1} and meq L^{-1}. *(Data from the United States EPA 2013–2014 National Rivers and Streams Assessment [US EPA, 2020)] and 2012 National Lakes Assessment [US EPA, 2016].)*

TABLE 12-5 Mean Ionic Salinity in Equivalent Percentages of Several Inland Waters

Natural waters	Ca^{2+}	Mg^{2+}	Na^{+}	K^{+}	CO_3^{2-}	SO_4^{2-}	Cl^{-}
Wisconsin softwaters (Juday, *et al.*, 1938; Lohuis *et al.*, 1938)	46.9	37.7	10.9	4.8	69.9	20.5	9.9
N. German softwaters (Ohle, 1955a)	36.0	14.3	43.0	6.7	42.4	14.1	43.5
Water from igneous rock (Conway, 1942)	48.3	14.2	30.6	6.9	73.3	14.1	12.6
Hubbard Brook, New Hampshire (Likens *et al.*, 1970)	25.2	5.6	67.6	1.6	<5.3	66.3	28.4
Mean sedimentary source material (Hutchinson, 1957, after Clarke, 1924)	53.2	34.0	8.0	4.8	93.8	6.2	—
Swedish rivers, average entire Sweden (Ahl, 1980)	54.9	18.4	22.6	4.1	42.7	39.1	16.5
Ugandan river and lake waters (Viner, 1975)	32.5	26.9	32.1	8.7	62.9	28.0	9.1
Upland Swedish sources (Rodhe, 1948, after Lohammar)	67.3	16.9	13.6	2.2	74.3	16.2	9.5
Hardwater lakes of N.E. Indiana (Wetzel, 1966b)	79.3	14.4	3.3	3.0	60.0	37.9	2.1
Hypereutrophic Sylvan Lake, Indiana[a] (Wetzel, 1966a)	57.3	11.8	26.9	3.4	48.7	34.7	16.6
Average river content of world (Livingstone, 1963)	52.6	24.0	19.3	4.1	67.1	16.3	15.4

[a] *A hardwater lake that received large amounts of domestic effluents that had high sodium content, most likely from detergents. Nutrient loading to this lake has subsequently been reduced.*

Cations: $Ca^{2+} > Mg^{2+}$ and $Na^{+} > K^{+}$

Anions: $HCO_3^- / CO_3^{2-} > SO_4^{2-} > C1^-$

These relationships generally hold regardless of if ions are compared by mass or molar equivalents (Fig. 12-3). Drainage from igneous rock commonly has a salinity of less than 50 mg L^{-1} and cationic proportions of $Ca^{2+} > Na^{+} > Mg^{2+} > K^{+}$. The proportions of anions in softwater systems shift toward a decrease in carbonates and an increase in halides: $Cl^- > SO_4^{2-} > HCO_3^- / CO_3^{2-}$.

II. Sources of ions

The salinity of inland waters of the world is highly variable and depends upon ionic influences of drainage and exchange from the surrounding land and groundwater, atmospheric sources, and human activity. In lakes, more so than in rivers, salinity is further influenced by evaporation, biotic and abiotic losses from the water column, and equilibrium and exchange with sediments. In addition, there are many coastal lakes and ponds of marine origin, where ionic composition has evolved from trapped seawater.

A. Weathering of soil and rock

The composition of soil and rock and their ion exchange capacities influence both rates of weathering and ion supply to runoff and groundwater. Four general processes of weathering control ion supply (Carroll, 1970):

(a) *Chemical weathering from the dissociation of carbonic acid* (Chapter 10) is of major importance in the weathering of soils and rocks. Water and carbon dioxide combine to form carbonic acid, which dissociates to H^+ and HCO_3^-. This process is heightened in soils, where microbial decomposition of organic matter increases the concentration of CO_2. Chemical weathering reactions include the alteration of minerals (Example 1) and the complete dissolution of minerals (Example 2):
Example 1: Hydrolysis reaction of orthoclase, a common mineral in igneous rock, which yields silicic acid and potassium ions

$$2KAlSi_3O_8 + 2H^+ + 9H_2O \rightarrow H_4Al_2Si_2O_9 + 4H_4SiO_4 + 2K^+$$

Example 2: Carbonic acid dissolves calcite, a common mineral in sedimentary rock, to calcium and bicarbonate ions

$$CaCO_3 + H_2CO_3 \rightarrow Ca^{2+} + 2HCO_3^-$$

The importance of carbonic acid in weathering is evidenced by the high proportion of bicarbonate ions in most river waters. The proportion is high even in waters of low salinity (<50 mg L^{-1}) draining igneous rocks (Table 12-3). Strong acids (H_2SO_4, HNO_3) in rainfall originating from air pollution accelerate the rate of weathering (Likens and Bormann, 1974).

(b) *Ordinary solution* not involving hydrolysis or acid weathering is primarily important in sedimentary deposits rich in soluble salts. Leaching of marine salt deposits results in an enrichment of sodium, potassium, and chloride relative to other ions in recipient lake and river waters. Relative leaching rates of different ions are time dependent, and thus marked fluctuations in ionic proportions of runoff water occur during periods of high (flooding) or low rates of runoff because of differences in solubility.

(c) *Oxidation and reduction* processes primarily affect iron, manganese, sulfur, nitrogen, phosphorus, and carbon compounds in soils. Iron sulfides are common constituents of rocks and water-saturated soils. Oxidation of these sulfides can be a major source of sulfate for natural waters, usually in the form of dilute sulfuric acid, which solubilizes other rock and soil constituents as well. Microbial decomposition of sulfur-containing organic compounds, particularly in woodland soils and peaty bog waters, adds sulfate to natural waters, usually as sulfuric acid. Within lakes, oxidation–reduction processes play dominant roles in the sulfur cycle (see Chapter 16).

(d) *Certain soluble organic molecules in soils can bind ions,* especially metals such as iron and manganese, thereby inhibiting these ions from reacting with others.

Overall, the adsorption and release of ions is dependent upon: (a) the cationic availability; (b) ionic concentrations and proportions in a soil solution or leaching water; (c) the nature and number of exchange sites on the exchange complex of soil particles; and (d) the volume of water in contact with the exchange complex.

Soil particles are generally negatively charged, with the amount of charge dependent on soil characteristics. The *cation exchange capacity* (CEC) of soils is a measure of the magnitude of negative charge. Soils with high CEC can more readily bind positively charged cations, thereby slowing their movement through soils. Anions are more mobile, apart from phosphate that is readily sorbed to soil particles. The exchange of ions between soils and the soil solution is highly variable and is governed by exchange equilibria between hydrogen ions and ions attached to the soil particles.

Along with ion supply, the depth and mode of water percolation through soils will influence the loading of ions to natural waters. Lakes receiving relatively large amounts of surface runoff are usually very dilute in comparison to lakes that drain deeper soil horizons, because the salinity of groundwater is generally much higher than that of surface runoff.

B. Atmospheric precipitation and fallout

Atmospheric deposition can be a significant source of salinity for many dilute freshwaters and for some saline lakes. The delivery of ions from the atmosphere to catchments is largely through *wet deposition*, the washing out of atmospheric gases or aerosols by rain, snow, or fog. In arid regions *dry deposition* of salts (i.e., direct fallout in the absence of precipitation) can occur in significant quantities and often exceeds wet deposition. All of the major anions in natural waters are cycled in part through the atmosphere as gases, as well as in dissolved and particulate form. The traditional suite of ions measured in wet deposition is slightly different from those measured in surface waters. Because of the low salinity and low pH of rain, H^+ and nitrogen species (NH_4^+, NO_3^-) can account for an appreciable percent of total salinity in rainwater.

Variation in atmospheric ion loading to surface waters is significant. Globally, most precipitation has a salinity < 3 mg L^{-1} (Fig. 12-4a). Higher salinities, up to 30 mg L^{-1}, are found near the coast, and downwind of industrial pollution. The ocean is a major source of atmospheric sodium, chloride, magnesium, and sulfate (Vet et al., 2014). Sea spray carries large amounts of seawater into the atmosphere that, on evaporation, forms salt particles that can act as nuclei for cloud and raindrop condensation (Fig 12-5a). Atmospheric salinity can be carried for great distances. Although most atmospheric salinity is precipitated with rainfall in the coastal regions, decreasing amounts are carried inland before deposition occurs. Ignoring H^+, which can be the dominant cation in many locations, ionic composition of coastal wet deposition tends to follow: $Na^+ > NH_4^+ \sim Mg^{2+} > Ca^{2+} > K^+$ and $Cl^- > SO_4^{2-} > NO_3^- > HCO_3^-$. The effects of atmospheric transport of ions can be seen in lakes enriched with Na^+ and Cl^- in coastal maritime regions (Fig. 12-4d). Continental rain generally contains less sea salt–derived ions and tends to follow: $NH_4^+ > Ca^{2+} > Na^+ \sim Mg^{2+} > K^+$ and $SO_4^{2-} > NO_3^- > Cl^- > HCO_3^-$ (Fig. 12-4b, c).

Windblown dust from the soil also contributes salts, especially calcium and potassium, to surface waters (Ballantyne et al., 2011; Lawrence and Neff, 2009; Psenner, 1999). Wind-transported salts from salt pans in arid regions, such as in Russia, western Australia, or the United States, can be moved large distances to drainage basins of other river and lake systems.

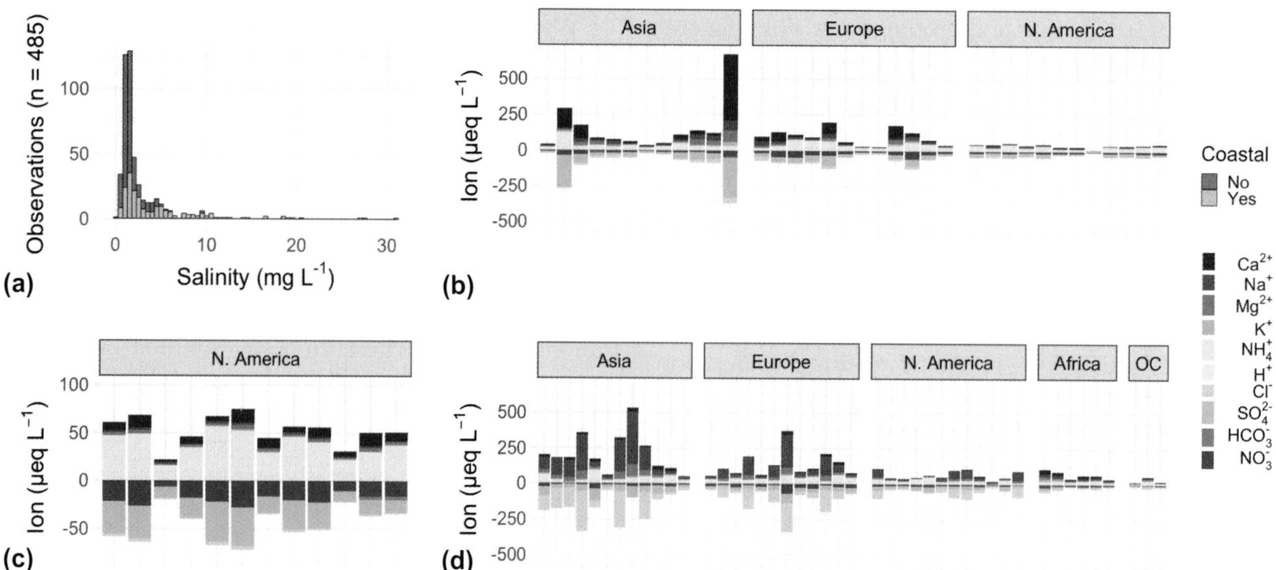

FIGURE 12-4 (a) Distribution of salinity of atmospheric precipitation in coastal and noncoastal sites calculated from wet deposition observations from 2005 to 2007. Breakdown of ionic composition at (b) continental sites, including (c) a zoom-in on North America and (d) coastal sites. *(Data from Vet et al. 2014.)*

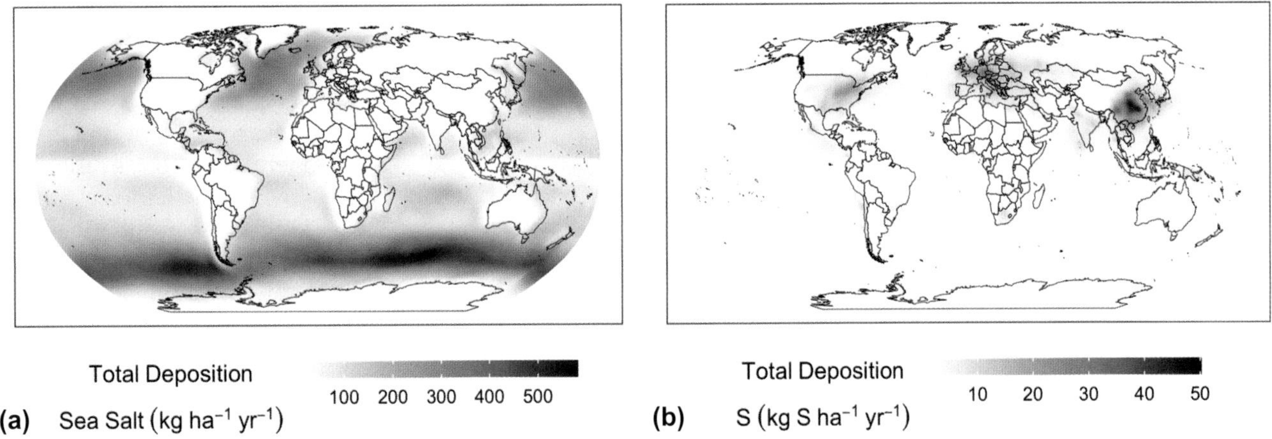

FIGURE 12-5 Atmospheric deposition rates of (a) sea salt and (b) sulfur from 2001 ensemble-mean models by the Task Force on Hemispheric Transport of Atmospheric Pollutants (TF HTAP). *(Data from Vet et al. 2014.)*

C. Anthropogenic sources

The exploitation of geological resources for human use, agricultural intensification, industrialization, and urbanization, all contribute to changing ion contributions to surface waters. Point source pollutants include, but are not limited to, industrial outflows, wastewater effluents, and mining waste (Schulz and Cañedo-Argüelles, 2019). A notable point source pollutant that quickly became globally pervasive was the industrial emission of sulfur and nitrogen compounds, namely, from the burning of fossil fuels by power plants. Sulfuric and nitric acids in the atmosphere led to the widespread acidification of aquatic ecosystems, particularly in northeastern North America and Scandinavia (Schindler, 1988). The devastation of "acid rain" on the chemistry of surface waters in the 1960–1980s led to global action to curtail its effects. In extreme cases lakes were "limed" with $CaCO_3$ to buffer the acidification (Bengtsson et al., 1980). Coordinated policy actions by multiple countries have drastically curtailed acid rain, and 50 years later, it is looked upon as an environmental policy success (Grennfelt et al., 2020), and many lakes have recovered or are well on their way to recovery (Garmo et al., 2014; Strock et al., 2014). However, aquatic ecosystems are still at risk of atmospheric pollutants, especially in industrializing countries such as China and India (Fig. 12-5b, Aas et al., 2019; Li et al., 2017).

Non-point source pollutants of major ions are equally prevalent. Common agricultural fertilizers include potash (commonly KCl) and sulfur additions (Hinckley et al., 2020). The addition of ammonium fertilizer has also been found to mobilize base cations via soil acidification (Aquilina et al., 2012). Livestock manure applied to cultivated lands can similarly contribute ions. In many cold regions road deicing salts (NaCl, Ca_2Cl, Mg_2Cl) applied to impervious surfaces have resulted in increasing chloride concentrations in numerous lakes (Dugan et al., 2017a), and they are a leading cause of salinization in many regions of the United States and Canada (Fig. 12-6). In the United States over 20 million metric tonnes are applied to US roadways annually (as of 2015, Bock et al., 2018). All of this salt ends up in terrestrial and aquatic environments. In addition, urbanization has increased anion, cation, and silica loading to surface waters through both direct sources such as concrete weathering and sewage leakage (Kaushal et al., 2017), and indirect feedback from land-use change (Carey and Fulweiler, 2012) and elevated urban temperatures (Bhatt et al., 2014).

Humans are altering natural salt concentrations in inland waters not only through changing ion contributions but through the diversion and withdrawals of freshwater flows (Albert et al., 2021; Rodell et al., 2018). Where lakes inflow is reduced, the shift in water balance leads to ion concentration via evaporation. This salinization threat is compounded by human-induced climate change (see Section IID).

D. Environmental influences on salinity

Climate affects the balance between precipitation and evaporation and thus the salinity of surface waters. In closed drainage basins the salinity of lake waters is governed not only by inputs of dissolved ions from runoff but by the fate of these materials upon evaporation. As water evaporates from a lake basin, ions become concentrated and salinity increases. Most closed-basin lakes, termed *endorheic lakes*, occur in regions with long-term (several years) fluctuating climates and are often exposed to periods of severe aridity. If shallow, these lakes may evaporate completely or sufficiently to expose large expanses of sediments. Diversions of river inputs

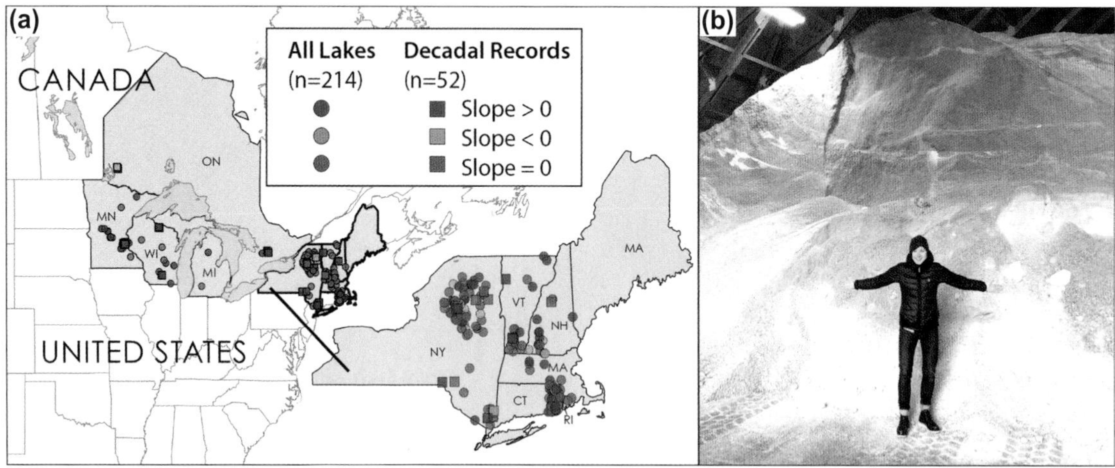

FIGURE 12-6 (a) Chloride trends for North American freshwater lakes in areas of heavy road salt application (Dugan et al., 2017a). Points are colored by the slope value of linear regression models (*red*, positive slope; *yellow*, negative slope; *purple*, zero or nonsignificant slope). Squares denote lakes with at least biennial chloride concentrations recorded from 1985 to 2010 ($n = 52$). (b). Author in a salt storage shed. In many cold regions salt (mostly NaCl) is used to deice roads. In the United States over 20 million metric tonnes of road salt is applied in a year (Bock et al., 2018). All deicing salts end up in the terrestrial and aquatic environment.

by human activities, as in the tragedy of the Aral Sea (Williams and Aladin, 1991), can also lead to the exposure of large areas of sediments. Loss of dried salts then occurs via wind *deflation*, the removal of loose/small surface material by wind.

Other significant climatic factors influencing salinity are temperature, wind, and ice formation. Temperature influences the rate of rock weathering (Kopáček et al., 2017). Tropical waters, for example, which drain strongly weathered soils, are usually poor in electrolytes, and a large part of their total composition consists of silica. The type of vegetation growing on the drainage basin and its requirements for major ions are also influenced by climate. Mineral cycling in tropical perennial forests of dense vegetative cover, for example, differs greatly in higher rates of utilization and leaching of soil nutrients as compared to nutrient cycling in deciduous vegetation of temperate regions. Wind direction and speed may affect the chemical composition of atmospheric precipitation by altering the amount of incorporated sea salinity and sites of deposition inland. Lastly, for lakes and rivers that experience winter ice cover, ion rejection during ice formation will increase ion concentrations in the water column. This can temporally, and significantly, increase ion concentrations in shallow water bodies (Dugan et al., 2017b).

The magnitude and variability of environmental influences on inland water salinity are predicted to increase in the future due to climate change (Cañedo-Argüelles et al., 2016). Regionally, the effects of climate change will differ (IPCC, 2021). In coastal areas sea-level rise threatens freshwater habitats through seawater intrusion and the salinization of groundwater (Schuerch et al., 2018). As the Earth's surface warms, evaporation of surface water will increase. The concentrating effect of evaporation may be offset by increasing precipitation or exacerbated by increased aridity and the occurrence of drought. Overall, it is expected that increased evaporation and a decrease in precipitation driven by climate change pose a salinization risk for many inland waterbodies (Iglesias, 2020; Jeppesen et al., 2020). Even saline lakes in historically arid regions are at risk of desiccation in the wake of climate change (Hassani et al., 2020; Wurtsbaugh et al., 2017)

III. Saline lakes

Saline lakes are found on every continent on Earth (Fig. 12-7). *Athalassic*, meaning "not the sea," was a term originally intended to differentiate inland waters with ions derived from terrestrial and atmospheric sources from those of marine origin (such as the Caspian Sea). While the definition does not provide a salinity threshold, and therefore all freshwater lakes are by nature athalassic (Williams, 1981), the term athalassic is typically reserved for saline lakes.

Non-marine-influenced saline lakes form and persist when: (a) outflow of water is restricted, as in hydrologically closed basins; (b) evaporation exceeds inflows; and (c) the inflow is sufficient to sustain a standing body of water (Eugster and Hardie, 1978). Saline lakes display a range of brine compositions that are a result of the interplay of hydrology, evaporation, and geology (Deocampo and Jones, 2014). Ionic composition often evolves toward sodium-rich brines due to the precipitation of alkaline-earth minerals. Such lakes include Great Salt Lake and Mono Lake, USA, and Lake Eyre,

FIGURE 12-7 Global diversity of saline lakes, including (a) Mono Lake, California, USA *(Photo: Celia Symons)*; (b) Lake Corangamite, Australia *(Photo: Wikimedia)*; (c) Lake Urmia, Iran *(Photo: NASA ISS)*; (d) Lake Bogoria, Kenya *(Photo: Jonathan Grey)*; (e) Lake Bonney and Blood Falls, Antarctica *(Photo: Hilary Dugan)*; (f) A road to Great Salt Lake, Utah, USA *(Photo: Urvish Prajapati)*; (g) Chaka Salt Lake, Qinghai, China *(Photo: James Elser)*; (h) Don Juan Pond, Wright Valley, Antarctica *(Photo: Hilary Dugan)*.

Australia (Fig. 12-1b). Calcium-rich brines do exist where groundwater/catchment inputs dominate over evaporative processes (Fig. 12-1b, Don Juan Pond, Antarctica). Chloride very commonly predominates, due to the precipitation of carbonate and sulfate minerals. Certain elements, normally present in only trace amounts in surface waters, can occur in very high concentrations in saline lakes (e.g., bromine, strontium, phosphate, or boron) (Eugster and Hardie, 1978; Hammer, 1986), which has led to the economic use of salt lakes as a source of minerals. The salinities of the world's saline lakes are often in flux due to the tight coupling of ion concentration with climate (both inflows and evaporation rates) and anthropogenic activities (such as water diversions). In addition to being hydrologically and chemically fascinating, saline lakes are also biologically peculiar and host important biodiversity among their highly specialized biota (Saccò et al., 2021).

IV. Distribution of major ions in freshwaters

All major ions are essential to sustaining life in inland waters. The spatial and temporal distribution of the major cations and anions are separable into: (1) *conservative ions,* whose concentrations within a lake and many streams undergo relatively minor changes from biological and geochemical processes, and (2) nonconservative or *dynamic ions,* whose concentrations can be influenced strongly by biological and geochemical processes. Of the major cations, magnesium, sodium, and potassium ions are relatively conservative both in their chemical reactivity under typical freshwater conditions and their small biotic requirements. Calcium is more reactive and can exhibit marked seasonal and spatial dynamics. Of the major anions, inorganic carbon is so fundamental to the metabolism of freshwaters that it is treated in a separate chapter (Chapter 13) and later coupled with organic carbon cycling (Chapter 28). Similarly, sulfate is greatly influenced by microbial cycling and the chemical milieu and is treated separately (Chapter 16), along with iron and silica. Chloride is highly conservative.

The total ionic salinity, composed almost entirely of the eight major ions (Ca^{2+}, Mg^{2+}, Na^{+}, K^{+}, HCO_3^-, CO_3^{2-}, SO_4^{2-}, and Cl^-), is of major importance in the osmotic regulation of metabolism and in the distribution of biota. The biological importance of these ions is briefly discussed below, and in Chapter 13 (HCO_3^-, CO_3^{2-}) and Chapter 16 (SO_4^{2-}).

A. Calcium

Calcium influences the growth and population dynamics of freshwater flora and fauna both directly and indirectly. Plants require calcium for development, regulatory processes, and signal transduction (Kudla et al., 2010), and it is also a fundamental component of the oxygen evolving complex in photosystem II in phytoplankton, including cyanobacteria. In animals calcium is an essential mineral. It is needed for bone and exoskeleton formation and molecular signaling and can limit growth and development at low concentrations.

Some organisms require high calcium concentrations and low concentrations can limit their distributions. For example, certain species of the zooplankton Cladocera *Daphnia*, which is a dominant keystone grazer in many temperate freshwater lakes, has a Ca-rich chitinous exoskeleton. When calcium concentrations are <2 mg L^{-1}, the fitness of certain large *Daphnia* species decreases (Jeziorski et al., 2008; Jeziorski and Smol, 2017). In some Canadian lakes a decline in calcium has resulted in *Daphnia* being replaced by *Holopedium glacialis*, a Cladocera with one-tenth the calcium requirement of *Daphnia* (Jeziorski et al., 2015). Similarly, crayfish require calcium for their exoskeletons (Capelli and Magnuson, 1983) and mussels require calcium for their shells (Whittier et al., 2008). Low calcium concentrations may protect lakes from the spread of invasive species such as rusty crayfish ($Ca^{2+} < 2.5$ mg L^{-1}) and zebra and quagga mussels ($Ca^{2+} < 12$ mg L^{-1}). Many inland waters have calcium concentrations close to these thresholds. A global analysis of rivers and lakes found median calcium concentrations of 3.76 and 5.50 mg L^{-1}, respectively (Weyhenmeyer et al., 2019). The lowest calcium concentrations were found in boreal regions of Fennoscandia and Canada and some tropical regions of South America (Fig. 12-8).

Dissolved calcium is removed from the water column, both biogenically from the formation of calcareous skeletons and other organismal structures and through biogenic and inorganic precipitation of carbonate minerals (Khan et al., 2022). Whiting events are striking phenomena where calcium carbonate precipitates in the water column turning lakes a chalky white color (see Chapter 13). Whiting events are known to be triggered by large phytoplankton blooms that rapidly raise the pH of surface waters, thereby decreasing calcite solubility.

As discussed previously, the major source of calcium in inland waters is rock weathering of calcium carbonate and calcium silicate, and input rates should be relatively constant. However, when acidified from

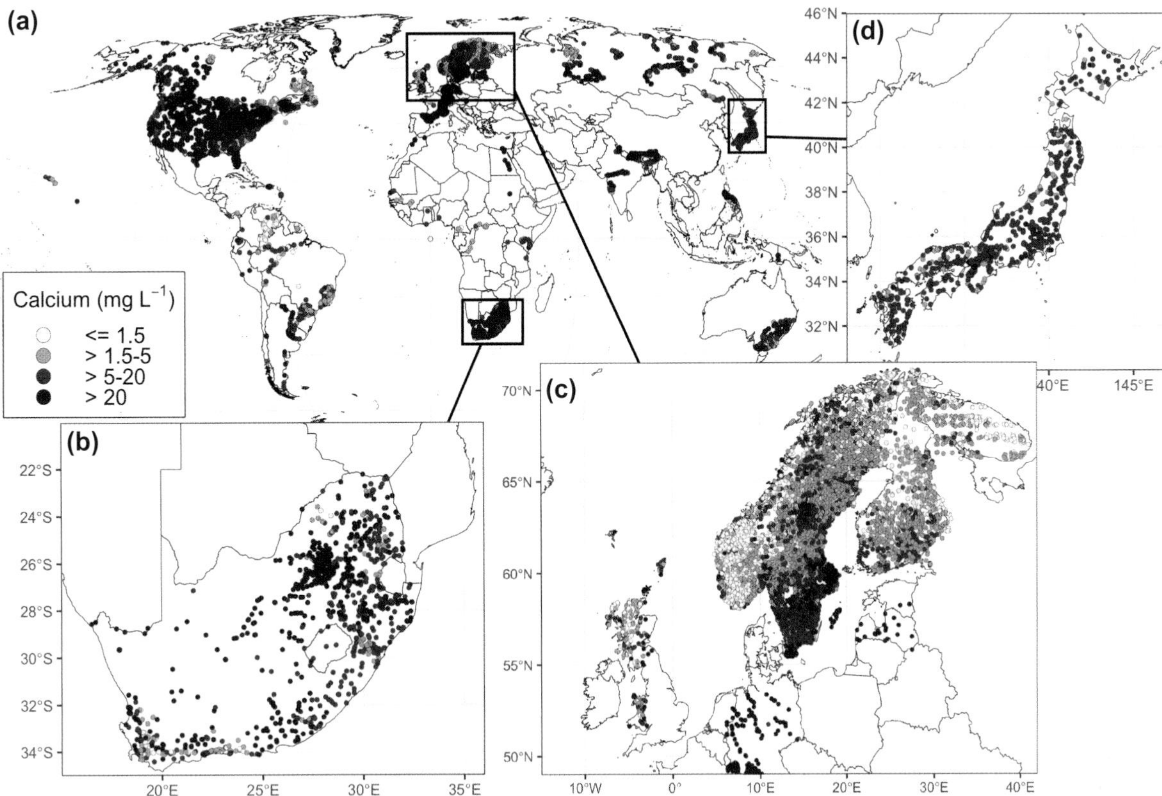

FIGURE 12-8 (a) Global distribution of long-term median calcium concentrations (mg L^{-1}) in lakes and running waters. *(Adapted from Weyhenmeyer et al. 2019.)* Insets show (b) South Africa and Lesotho, (c) Northern Europe, and (d) Japan.

acid rain, terrestrial landscapes leach cations, through cation exchange displacement by hydrogen and aluminum ions in soils (Kirchner and Lydersen, 1995). Through this process, calcium (and magnesium) concentrations are depleted in soils and increase in surface runoff. As rivers and lakes recover from acidification, one consequence is a long-term decline in calcium concentrations (Jeziorski and Smol, 2017), even to concentrations below preacidification concentrations due to the depletion of cations in catchment soils. Globally, many lakes affected by acid rain are currently in the recovery phase.

The calcium content of softwater lakes and streams remains well below saturation levels, and these concentrations exhibit minor seasonal variations with depth (Fig. 12-9a). The amount of calcium used by the biota is usually so small in comparison to existing concentrations that reduction or depletion by biota cannot be seen in normal analyses. Decomposition processes can lead to some calcium accumulation in the hypolimnion of productive softwater lakes during stratification.

The calcium content of hardwater lakes, however, undergoes marked seasonal dynamics. The changes depicted in Fig. 12-9b for a hardwater lake in southern Wisconsin are quite typical. Between rather uniform concentrations during spring and fall periods of mixing, a conspicuous stratification occurs that is repeated annually with only minor variations. The concentrations of calcium decreased markedly in the epilimnion (0–5 m in Figs. 12-9b) as a result of precipitation of $CaCO_3$ during the summer months from May through September. Similar losses are known to take place during the period of winter stratification. In winter, decreases beneath lake ice are associated in part with dilution by rains permeating the ice and melting ice (Canfield et al., 1983), as well as with increases in photosynthesis just before ice loss. Decreases in the concentrations of calcium and of inorganic carbon in the epilimnion and metalimnion are related largely to rapid increases in the rates of photosynthesis by phytoplankton and littoral flora (Otsuki and Wetzel, 1974) and indicate the major role of photosynthetic alteration of carbonate equilibria and induction of epilimnetic decalcification (cf. review of Küchler-Krischun and Kleiner 1990). The loss of $CaCO_3$ influences the metabolism of hardwater lakes via the coprecipitation of inorganic nutrients, such as phosphorus, and selective removal of humic organic acids and other organic compounds by adsorption (Stewart and Wetzel, 1981).

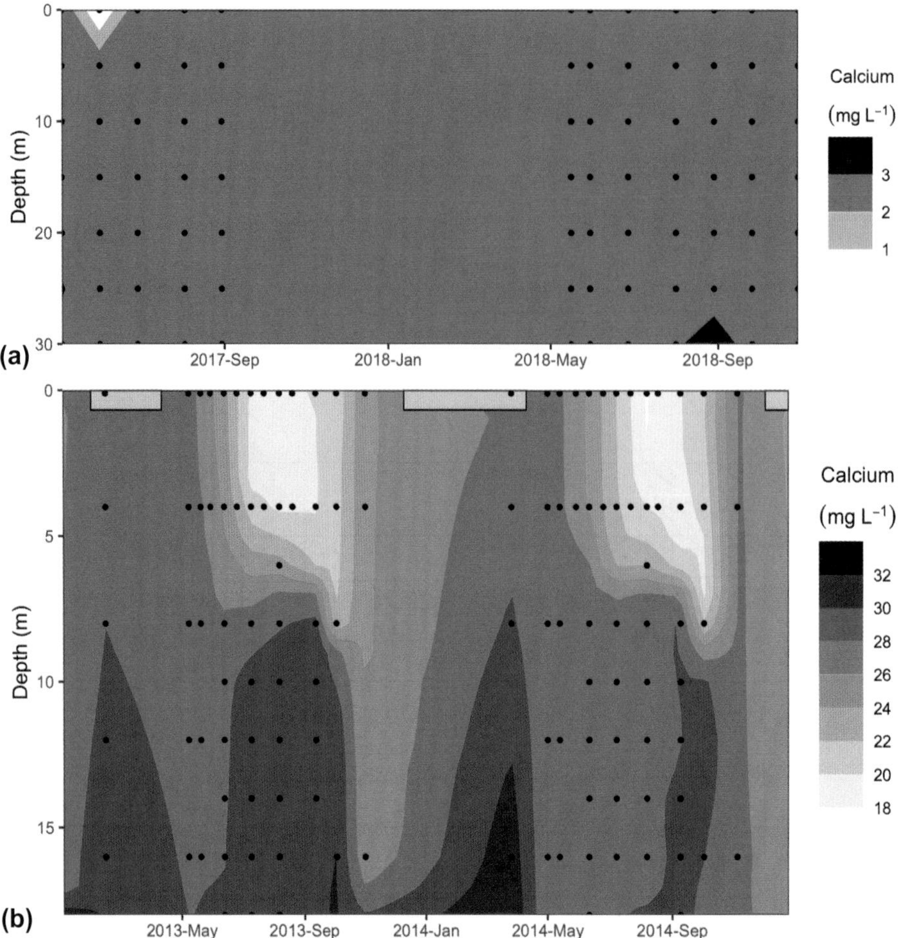

FIGURE 12-9 (a) Depth–time distribution over 2 years of calcium concentrations (mg L^{-1}) in a softwater lake, Lake 239, Canada, and (b) in a hardwater lake, Fish Lake, USA. The *grey rectangles* in panel (b) show the duration of ice cover on Fish Lake. Contour lines interpolate between manual samples (*black dots*). *(Data courtesy of IISD-ELA and the North Temperate Lakes Long-Term Ecological Research program [NTL-LTER], 2020.)*

Epilimnetic decalcification is reflected simultaneously in the distribution of HCO_3^- of hardwater lakes (Fig. 12-10). Because concentrations of Mg^{2+}, Na^+, K^+, and Cl^- are relatively conservative, as discussed further on, the specific conductance follows changes in Ca^{+2} and HCO_3^- concentrations in nearly a 1:1 relationship ($r = 0.997$; Otsuki and Wetzel, 1974). A portion of the precipitating $CaCO_3$ is resolubilized in the hypolimnion, which is reflected in both the increased concentrations of Ca^{2+} and the specific conductance in hypolimnetic waters. Some of the $CaCO_3$ is entrained permanently in the sediments and commonly constitutes >30% of the sediments by weight in moderately hardwater lakes (Wetzel, 1970). Adsorption of organic compounds to $CaCO_3$ lowers the rates of dissolution in hypolimnetic strata of reduced pH. Further, calcium complexes with humic acids, especially in the sediments, alter exchange equilibria in the hypolimnion (Hering and Morel, 1988; Stewart and Wetzel, 1981)

Biogenically induced decalcification of the epilimnion reaches an extreme in very productive hardwaters. For example, the calcium concentrations of the epilimnion of hypereutrophic Wintergreen Lake, Michigan, USA, were reduced from nearly 50 mg L^{-1} to below analytical detectability in a few weeks (Wetzel, unpublished data). The magnitude of this loss from the trophogenic zone and the resulting increase in monovalent:divalent cation ratios should be recalled in the subsequent discussion of the effects of the calcium content and of cationic ratios on species distribution.

B. Magnesium

Magnesium is required universally by photosynthetic plants for the magnesium porphyrin component of chlorophyll molecules and as a micronutrient in enzymatic transformations, especially in transphosphorylations by algae, fungi, and bacteria. The demands for magnesium in metabolism are minor in comparison to

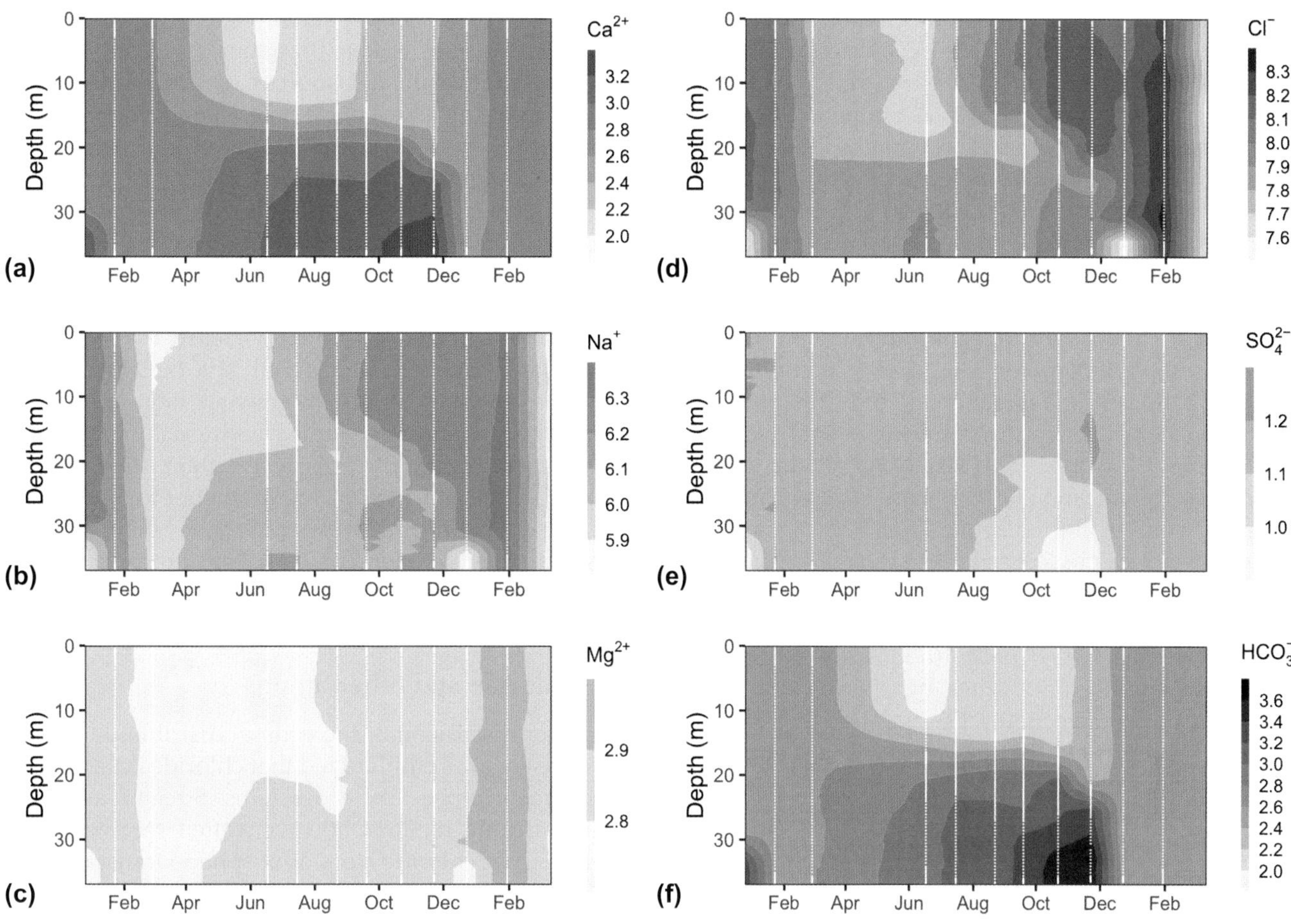

FIGURE 12-10 Depth–time distribution of major ion concentrations (a) calcium, (b) sodium, (c) magnesium, (d) chloride, (e) sulfate, and (f) bicarbonate (meq L^{-1}) in Lake Kinneret, Israel, from 2002 to early 2003. Contour lines interpolate between manual samples (*white dots*). *(Adapted from Katz and Nishri (2013). Potassium is not shown, as concentrations were constant between 0.179 and 0.198 meq L^{-1} over this time period.)*

quantities generally available in freshwaters. For example, a large phytoplankton population of 1000 μg chlorophyll *a* L^{-1} would be sustained by concentrations of 30 μg Mg L^{-1} (Reynolds, 2006). The low availability of magnesium has been implicated as one of several factors influencing phytoplanktonic productivity in an extremely oligotrophic Alaskan lake (Goldman, 1960). However, such conditions are exceedingly rare in comparison to limitations imposed by the restricted availability of other nutrients.

Magnesium compounds are much more soluble than their calcium counterparts. As a result, significant amounts of magnesium rarely precipitate. The monocarbonates of hardwaters are usually >95% $CaCO_3$ under ordinary CO_2 pressures (e.g., Murphy and Wilkinson 1980). $MgCO_3$ and magnesium hydroxide precipitate significantly only at very high pH values (>10) under most natural conditions. The concentrations of magnesium are extremely high in certain closed-basin saline lakes (Deocampo and Jones, 2014).

Because of magnesium's solubility characteristics and its minor biotic demand, concentrations of magnesium are relatively conservative and fluctuate little both in softwater streams and lakes (e.g., Likens, 1985) and in hardwater streams (e.g., Wetzel and Otsuki, 1974) and lakes (Fig. 12-10c). This attribute has been used to advantage by employing the magnesium budget of inputs and outflows to determine groundwater influxes to a lake by magnesium mass balance (Wetzel and Otsuki, 1974).

C. Sodium, potassium, and minor cations

The monovalent cations sodium and potassium are involved primarily in cellular ion transport and exchange. An absolute sodium requirement has been demonstrated in only a few plants. The sodium requirements are particularly high in some species of Cyanobacteria and sodium is required for photosynthesis, bicarbonate transport, intracellular pH regulation, nitrogen fixation and nitrate reduction, and uptake of phosphate (Maeso et al., 1987; Oleson and Makarewicz, 1990; Rees, 1984; Valiente and Avendaño, 1993; Ward and Wetzel, 1975). The highest reported sodium requirement for phytoplankton is 4-5 mg L^{-1} for a cultivated *Dolichospermum* (*Anabaena*) (Kratz and Myers, 1955), but many other algae and Cyanobacteria have much lower thresholds (Reynolds, 2006). Potassium and other monovalent cations, namely, lithium (Li), rubidium (Rb), and cesium (Cs), cannot substitute for sodium. The potassium limitation of phytoplankton is unlikely to occur in nature (Talling, 2010). In an experimental study the growth of the freshwater diatom *Asterionella* was unaffected above 27.5 μg K L^{-1} (Jaworski et al., 2003), and another diatom, *Diatoma elongatum*, and cryptomonad, *Plagioselmis nannoplanctica*, were unaffected above 31 μg K L^{-1}.

Cyanobacteria possess separate systems for the active transport of both CO_2 and HCO_3^-. The active transport system for molecular CO_2 assimilation is so efficient that extracellular CO_2 concentrations are reduced far below the equilibrium values by a transport system that recognizes molecular CO_2 and transports it into the cells (Miller et al., 1991). Cyanobacteria also possess active transport systems for HCO_3^-. The bicarbonate transport system occurs in cells growing at low concentrations of extracellular CO_2, although bicarbonate transport appears to have no effect on CO_2 transport and CO_2 transport appears to have no effect on bicarbonate transport. One of the HCO_3^- transport systems is dependent on the presence of millimolar concentrations of sodium (Espie et al., 1991; Smith and Ferry, 2000). The active transport of CO_2, however, requires only micromolar levels of Na^+ (Miller et al., 1990). Therefore sodium, along with many other factors, can influence the development of Cyanobacteria (de Lima et al., 2020; Seale et al., 1987).

Sodium and potassium occur in relative abundance as highly soluble cations of numerous salts, and their spatial and temporal distribution in lakes is relatively uniform. Alteration of their concentrations in natural waters is not common except under conditions of pollution, for example, salinity from industry, road deicers, water softening, or agricultural fertilizers (Kaushal et al., 2018).

Only small seasonal variations are observed, particularly for sodium, in accordance with the conservative nature of these ions (Fig. 12-10). Moderate epilimnetic reductions in potassium concentrations have been observed in productive lakes and in fertilized, productive farm ponds (Barrett, 1957). This reduction is related to potassium utilization by the massive algal populations and by submersed macrophytes and their epiphytes (Barko, 1982; Mickle and Wetzel, 1978). Potassium is actively assimilated into submersed plant tissues with a light-dependent exchange process of reciprocal sodium efflux (Brammer and Wetzel, 1984). Lake sediments are net sources of potassium during summer stratification, and some hypolimnetic enrichment and export occurs in stratified, eutrophic lakes (see also Stauffer and Armstrong 1986).

The concentrations of rarer alkaline earth and alkali cations of natural waters vary considerably in relation to the lithology of the drainage basins. Their distribution is discussed by Durum and Haffty (1961) and Livingstone (1963) in general, and in closed-basin lakes in particular by Whitehead and Feth (1961). General ratios of major to minor cations are given in Table 12-6. Strontium cycling is closely coupled with calcium, and Sr^{2+}:Ca^{2+} ratios maintain a nearly constant stoichiometry (Stabel, 1989). Nutritional requirements for these relatively rare elements have not been demonstrated, although they can sometimes substitute for the major cations in metabolic pathways.

D. Chloride and other anions

Earlier discussion of the general distribution of chloride ions in lakes emphasized that chloride is usually not dominant in open lake ecosystems. Streams and lakes near maritime regions, however, often receive significant input of chloride from atmospheric transport from the sea. Anthropogenic sources of chloride can greatly modify natural concentrations, so much so that salinization is an emerging global threat to freshwater resources (Cañedo-Argüelles et al., 2016; Herbert et al., 2015; Kaushal et al., 2018; Schuler et al., 2019). In North America many north-temperate lakes in urban areas have seen a rise in chloride concentrations from road deicers. Fig 12-11a shows the rise in chloride concentrations in Lake Mendota, USA, primarily as a result of road deicer inputs (Dugan et al., 2017a). An analysis of Lake Constance, a large 254 m deep European lake, revealed that the threefold increase in chloride concentrations was a result of road salts (52%), wastewater (23%), farming (11%), soil weathering (9%), and precipitation and solid waste incineration (3%) (Fig. 12-11b; Müller and Gächter 2011). Chloride is influential in general osmotic salinity balance and ion exchange, but variations

TABLE 12-6 Approximate Ratios of Minor Alkaline Earths and Alkalis to Sodium and Calcium[a]

Na/Li	1500	Ca/Ba	400
Na/Rb	3600	Ca/Sr	3000–4500
Na/Cs	31,900	Ca/Be	c. 40,000
		Ca/Ra	5×10^{10}

[a]*From data of Livingstone (1963, p. 19).*

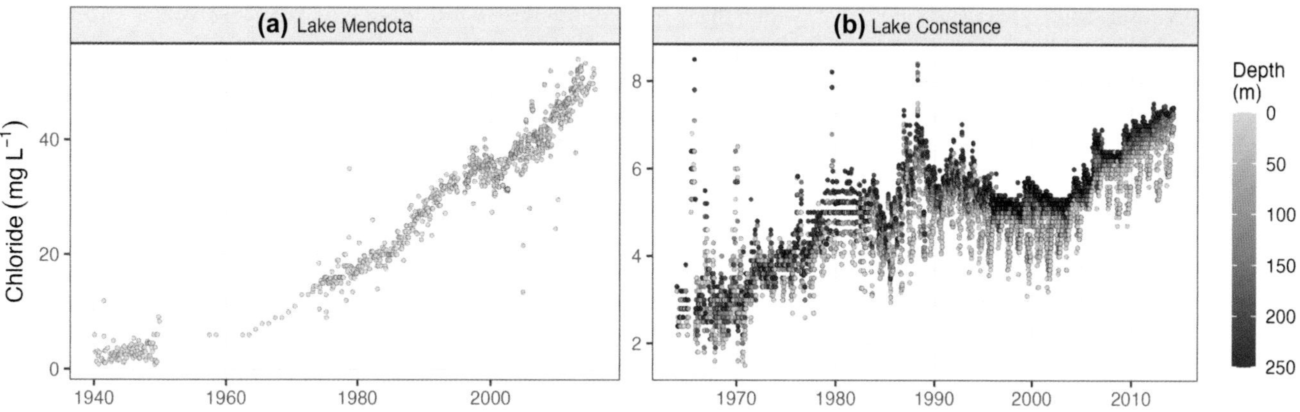

FIGURE 12-11 Chloride concentrations (mg L^{-1}) in (a) Lake Mendota, USA, and (b) Lake Constance, Europe. *(Data from Dugan et al. [2017c].)*

observed in many lakes are associated with the hydrology of the basin and seasonal fluctuations, and not metabolic utilization (Fig. 12-10d).

Concentrations of minor halides (Table 12-7) vary somewhat with the lithology within the drainage basin, and higher concentrations often occur in lakes and streams in proximity to marine regions or in those that possess marine rock formations within their drainage basins. Boron is of limnological interest because it is a micronutrient required by many algae and other organisms. Concentrations of boron in natural waters are relatively high in comparison to other minor elements. The average concentration in the ocean is 4.6 mg B L^{-1}, but typically less than 0.5 mg B L^{-1} in freshwaters, with major sources being sea salt aerosol production, volcanic production, and anthropogenic activities (Howe, 1998; Parks and Edwards, 2005). Concentrations reach exceedingly high levels (1 g B L^{-1}) in certain closed, saline lakes. As is the case with most micronutrients, high concentrations are generally toxic to most organisms (Schoderboeck et al., 2011; Soucek et al., 2011), as was exemplified by the well-known excessive use of the micronutrient copper as a biocide.

TABLE 12-7 Approximate Average Concentrations of Halides and Boron in Natural Freshwaters[a]

Element	Average concentration (mg L^{-1})	Chloride ratio	
Chloride	8.3		
Fluoride	0.26	Cl/F	32
Bromine	0.006	Cl/Br	1400
Iodine	0.0018	Cl/I	4600
Boron	c. 0.01		

[a]*From data of Livingstone (1963, p. 19).*

E. Cation ratios

The ratio of monovalent to divalent cations (M:D) has been studied in relation to the distribution and dynamics of algae and higher aquatic plants in freshwaters. Early studies, particularly by Provasoli and coworkers (Provasoli, 1958), indicated that the M:D ratio was significant to the observed growth of some algae. For instance, diatoms dominate the algal flora of very hardwater lakes with M:D ratios much less than 1.5; however, this relationship was more strongly correlated with the concentrations of calcium (Shoesmtth and Brook, 1983). There is little conclusive research to indicate that the stoichiometry of M:D impacts lake ecology (Moss, 1972), but M:D ratios and divalent cation ratios (Bogart et al., 2019) may be of importance under certain conditions, or seasonally, due to biogenically induced decalcification in early summer in calcareous waters, or sequestering of divalent cations by chelation with dissolved organic compounds.

F. Ionic budgets within a drainage basin

The temporal and spatial loading, distribution, and fate of ions vary greatly with the lithology, climatic conditions, drainage and limnological characteristics, and biotic activities of soils, streams, and lakes. Examination of a representative example of an ionic budget from Mirror Lake, New Hampshire, USA, is instructive for evaluating relative proportions of inputs and fates of ions.

The ionic budget of the Mirror Lake ecosystem within the Hubbard Brook Valley of New Hampshire, USA, was studied for many years (Likens, 1985). In this drainage basin nearly all of the precipitation runoff not lost to evapotranspiration flowed through the soil and was collected in streamflow. At the time of the study, precipitation was acidic (pH 4.1) and strongly influenced by air pollution. Sulfate was found to be the principal ion in precipitation (Fisher et al., 1968) and

TABLE 12-8 Allocation of Elements (in Percent) within the Hubbard Brook Experimental Forest Drainage Basin[a]. Contribution of Minor Cations to Total Percent Is Considered Negligible.

	Ca^{2+}	Mg^{2+}	Na^+	K^+	N	S
Source						
Precipitation input	9	15	22	11	31	65
Net gas or aerosol input	—	—	—	—	69	31
Weathering release	91	85	78	89	—	4
Storage or loss Biomass accumulation						
Vegetation	35	17	2	68	43	6
Forest floor	6	5	<1	4	37	4
Streamflow						
Dissolved substances	59	74	95	22	19	90
Particulate matter	<1	5	3	6	<1	<1

[a]*After Likens and Bormann (1995).*

supplied most of the sulfate that was discharged by the streams (Table 12-8). The input of ammonium and nitrate exceeded the discharge of these constituents. The annual deposition of hydrogen ions exceeded that of sulfate and was a major determinant of the pH of the waters of that region. Cationic loadings in stream water greatly exceeded inputs in bulk precipitation, as gains were obtained by weathering, evapotranspiration, biomass, and exchange processes within the soils (Table 12-9). Dissolved cations in stream water draining aggrading forests remained relatively constant despite extremes in hydrologic output and climatic variations. Particulate loadings and outputs were very small and were dependent on storm peaks (Table 12-9). Inputs to the lake and outputs from the lake were largely via cations dissolved in the stream water. Regulation of fluxes to the lake were affected by disturbances within the drainage basin, such as deforestation and loadings from atmospheric, sewage, road salt leaching, and other sources of pollution. For example, clearcutting forest disturbance resulted in large losses of potassium, in particular to stream water (Likens et al., 1994). Of the major cations, potassium was the slowest to recover from clearcutting disturbances.

V. Salinity, osmoregulation, and distribution of biota

A. Origins and distribution of freshwater biota

The salinity of inland waters is generally very low in comparison to the global oceans, although in semiarid regions the salinity of closed-basin lakes can exceed that of seawater by several times. The distribution of biota in freshwaters has been influenced by a long evolutionary history of physiological mechanisms and adaptation of osmotic regulation to a wide range of salinities. These mechanisms developed against a background of large differences between the salinity of the environment and that of the cytoplasm or body fluids. The general distribution of the freshwater biota and their tentative origins in terrestrial, freshwater, or marine sources are summarized in the detailed review by Hutchinson (1967). The distribution of biota in estuaries that interface marine and freshwater environments provides insight into the salinity ranges occupied by various species (Whitfield et al., 2012) (Fig. 12-12). In these transitional waters, as in the spectrum of salinities of inland waters, the salinity range occupied by a species depends on the efficiency of the physiological mechanisms by which it is adapted to changes in salinity in the environment.

Few freshwater species are truly *euryhaline* (i.e., able to tolerate a range of salinities). In ecosystems threatened by freshwater salinization the impact of increasing ion concentrations on freshwater organisms is complex and is dependent on the magnitude and ratios of individual ions and their temporal variability, the cooccurrence of other abiotic and biotic stressors (such as temperature), and the evolutionary context (Cañedo-Argüelles et al., 2019).

B. Osmotic adaptations of aquatic plants and animals

The function of osmotic regulation is to maintain cellular ion concentrations at appropriate operational physiological levels. Most freshwater bacteria and Cyanobacteria are relatively *homoiosmotic* and tolerate only a narrow range of salinity but can adapt to increasing salinity relatively rapidly by means of genetic change. Extensive adaptive radiation is seen among these groups. Members of the Protista (algae, fungi, and protozoa with mitochondria and, if they are photosynthetic, with chromoplastids), which are largely single celled, retain considerable evolutionary *euryhalinity,* that is, broad salinity tolerance, and are widely distributed with respect to salinity, although some groups, especially among the green algae, are restricted to freshwater. The *contractile vacuole* is the primary osmoregulatory organelle among the Protista. Among the higher plants, only 2% of angiosperm species are aquatic (Cook, 1996), and very few groups extend into saline waters of brackish or marine areas or into hypersaline closed-basin lakes.

TABLE 12-9 Average Annual Ionic Budgets (kg Year^{-1}) for Mirror Lake, New Hampshire, USA 1970–1976[a]

	Ca^{2+}	Mg^{2+}	Na^{+}	K^{+}	Cl^{-}	H^{+}
Inputs						
Precipitation (bulk)	18.3	4.5	19	6.1	88	14.1
Litter	10	1.0	<1	4.6	<1	—
Fluvial						
Dissolved	1943	406	946.0	339.0	882	0.90
Particulate	18	16	21.0	44.0	<1	—
Domestic sewage/road salt seepage	?	?	>4[b]	?	>6[e]	—
Total inputs	1980	427	990	394.0	976	15.0
Outputs						
Fluvial						
Dissolved	1881	393	970	380.0	735	0.67
Particulate	0	0	0	—	<1	—
Insect emergence	0.35	0.33	0.35	2.0	0.2	—
Permanent sedimentation	42[c]–117[d]	16[c]–68[d]	32[c]–221[d]	39[c]–205[d]	207[b]	—
Total outputs	1923–1998	409–461	1002–1191	421–587	942	0.67
Change in lake storage	0	0	+ 4	0	+ 34	0

[a]*Extracted from Likens (1985).*
[b]*By difference.*
[c]*Spatially integrated sedimentation since c. AD 1840.*
[d]*Extrapolation of precultural deposition rate and chemical content from a sediment depth of 25 cm to 50% of the lake area.*
[e]*Based on Na value and assumption of NaCl.*

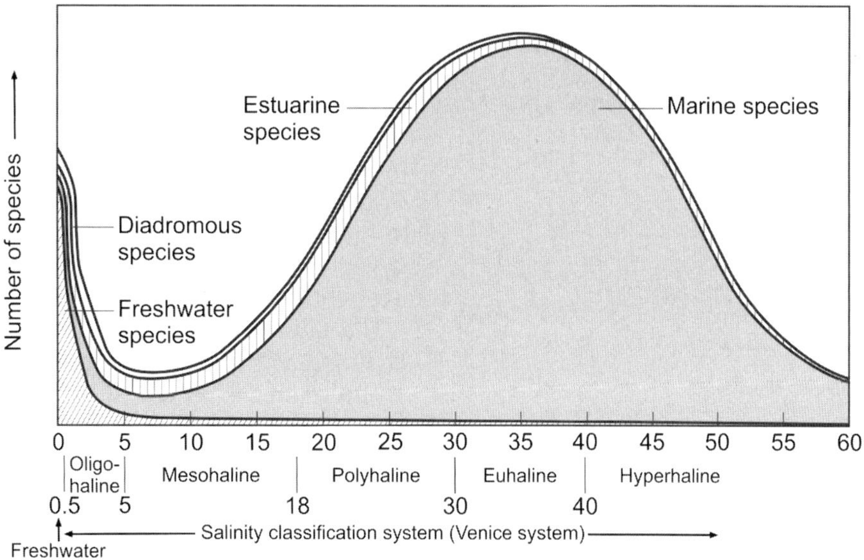

FIGURE 12-12 Proposed conceptual model for estuarine biodiversity (species) changes covering the salinity continuum from freshwater to hyperhaline conditions. *(Reproduced with permission from Whitfield et al. 2012.)*

Nearly all of the higher freshwater animals originated from the sea, although most aquatic insects are of terrestrial origin. Both in terms of evolutionary and contemporary life in freshwaters, osmoregulation is a major problem for which diverse mechanisms have developed to regulate salt and water content. Adaptation to low salinities by some marine animals has been achieved without osmoregulation. Such *poikilosmotic* animals adjust the osmotic pressure of their body fluids to become more or less isotonic with the salinity of the medium. In contrast, a *homoiosmotic* animal will tend to retain its initial internal osmotic concentration upon being exposed to modest changes in the salinity of the medium. The general relations between the osmotic pressure among different types of animals, expressed as the salinity of the blood versus the salinity of the external water, can be visualized in Fig. 12-13. The range depicted by area A extends over a wide variation in the osmotic pressure of body fluids that is found in brackish-water animals, which tend to be more poikilosmotic at high concentrations and more homoiosmotic at lower salinities. The range of the osmotic pressure curve extends from the most homoiosmotic (a_1) to the most poikilosmotic species (a_2, the isosmotic line). The lower salinity tolerances are more variable, represented by the undefined left-hand edge of area A, but all of these species have failed to colonize freshwater.

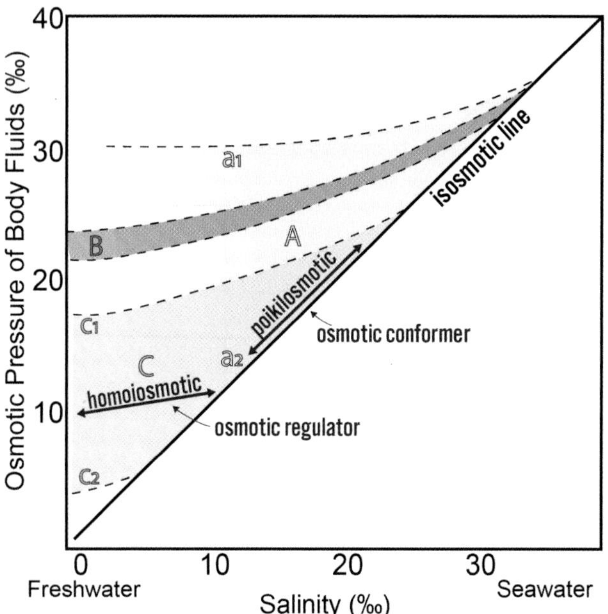

FIGURE 12-13 Relationship between the osmotic pressure, expressed as salinity, of body fluids of brackish-water organisms (*A, yellow area*) and freshwater animals (*C, pink area*) and the salinity of external water. Relationships of the curves are detailed in the text. *(Modified from Beadle, 1959.)*

A few species have succeeded in penetrating freshwaters without a renal osmoregulatory mechanism. These brackish-water animals are partially homoiosmotic (area B, Fig. 12-13) and maintain a very high osmotic pressure in hypertonic body fluids by the active uptake of ions from the water. Excretory organs are not involved since the urine produced is isotonic with the blood.

In most freshwater animals, however, osmotic pressures of the body fluids have decreased to levels equivalent to 5–15‰ salinity to reduce osmotic gradients. These organisms have developed excretory organs that effectively recover ions and produce urine hypotonic to the body fluids. Most freshwater animals therefore effect osmoregulation by the active uptake of ions and by a renal mechanism of ion retention. Curves c_1 and c_2 bound the extremes in osmotic pressures of body fluids in freshwater animals (area C, Fig. 12-13). The isotonic line along the right-hand edge of area C indicates the upper salinity tolerance limit of most freshwater animals and that they are incapable of hypotonic regulation. Although most freshwater animals can live in water of low salinity, the adaptation of body fluids of low osmotic pressure is apparently irreversible, and with few exceptions, freshwater organisms are restricted to salinities of $<10 \text{ g L}^{-1}$.

Because of the slow diffusion of oxygen in water relative to that in air, the movement of water over permeable membranes or tissue surfaces for respiratory needs is almost universal among aquatic animals. The pumping process places high energetic demands on the animals and additionally exposes cellular surfaces to osmotic gradients. Mechanisms for taking up salts against a concentration gradient vary greatly among freshwater animals. In a few organisms the incorporation of salts with food may be adequate, but more often, organs have developed for this purpose. Aside from resorption mechanisms of the excretory organs, which are advantageous energetically, active uptake mechanisms for ions, especially sodium and chloride, are often associated with the respiratory organs (commonly gills) of many invertebrates and vertebrates.

The fauna of extremely saline inland lakes is relatively insensitive to the high salinity of these lakes and to large interseasonal fluctuations in the chemical composition of the water (Beadle, 1969). Most of the animals of saline lakes are of freshwater origin and include representatives of the aquatic insects, phyllopods, copepod and cladoceran Crustacea, and rotifers, all of which belong to predominantly freshwater groups (see Chapters 19 and 20). The blood of these saline-inhabiting animals maintains osmotic pressures at levels characteristic of those living in freshwaters. The body surfaces of these animals exhibit very low permeability, and they possess effective excretory mechanisms for maintaining the body fluids strongly hypotonic to the external medium. Furthermore, the water balance of saline inland waters frequently fluctuates widely; freshwater animals with resting stages capable of withstanding desiccation have a distinct advantage over marine animals, which lack strong development of this characteristic.

VI. Summary

1. In inland waters total salinity, the mass of dissolved inorganic solids, is usually dominated by four major cations [calcium (Ca^{2+}), magnesium (Mg^{2+}), sodium (Na^{+}), and potassium (K^{+})] and the major anions [bicarbonate (HCO_3^-), carbonate (CO_3^{2-}), sulfate (SO_4^{2-}), and chloride (Cl^-)].
2. Inland waters range in salinity from near 0 to over 400 g kg^{-1}.
   - **a.** The delineation of fresh versus saline waters is generally accepted as 3 g kg^{-1}.
   - **b.** Saline lakes (salinity > 3 g kg^{-1}) are present on every continent on Earth, and their ion composition is a result of the interplay of hydrology, evaporation, and geology.
3. The salinity of inland waters is a composite of ionic contributions from rock and soil weathering, atmospheric precipitation and deposition, and the balance between evaporation and precipitation.
   - **a.** The salinity of global river waters has an average concentration of about 128 mg L^{-1} but varies among continents and with the lithology of the land masses and drainage basins (Table 12-4).
   - **b.** Concentrations of major ions of many surface waters of the world exist in the proportions of $Ca^{2+} > Mg^{2+} \geq Na^{+} > K^{+}$ and $HCO_3^-/CO_3^{2-} > SO_4^{2-} > Cl^-$. In soft waters and in surface waters of coastal regions, Na^+ and Cl^- often occur in greater proportions (Table 12-5).
4. The supply of ions from rocks and soils is controlled by the processes of chemical weathering, solution and oxidation–reduction, and the formation of organic complexes. The depth and mode of water percolation through soils will influence the loading of ions to natural waters.
5. Concentrations of the cations, magnesium, sodium, and potassium, as well as the major anion chloride, are relatively conservative and undergo only minor spatial and temporal fluctuations within a lake from biotic utilization or biotically mediated environmental changes. Calcium, inorganic carbon, and sulfate are dynamic, and concentrations of these ions are influenced strongly by microbial metabolism.
   - **a.** Proportional concentrations of major cations and the ratios of monovalent:divalent cations can influence the metabolism of many organisms, particularly certain algae and submersed macrophytes.
   - **b.** Therefore processes that influence the availability of some cations disproportionately to others (e.g., organic complexing of calcium or biologically induced decalcification of the epilimnion) can indirectly affect seasonal population succession and productivity.
6. The relatively low salinity of freshwaters has influenced the distribution of biota and their long evolutionary history of physiological adaptations for osmotic and ionic regulation in an extremely hypotonic environment.
   - **a.** Although some groups of microbes and algae are relatively homoiosmotic and can tolerate only a narrow range of salinity, many lower flora and fauna are euryhaline, that is, adaptable to a wide range of salinity.
   - **b.** Most higher-order freshwater animals originated from the sea or from land and adapted to freshwater secondarily. In comparison to marine forms, nearly all of these organisms have reduced osmotic pressures of body fluids and have developed efficient mechanisms for the active uptake of ions and renal mechanisms for ion retention.
7. Inland water salinity is predicted to increase in the future due to climate change, human activities, and heightened demands for freshwater resources.
   - **a.** Salt loading to freshwaters is a byproduct of resource extraction, agricultural intensification, industrialization, and urbanization.
   - **b.** Salinity will rise due to increased evaporation, drought, sea-level rise, and freshwater diversions.

Acknowledgments

Examples provided in this chapter would not have been possible without open data sharing from limnologists worldwide. Further thanks to colleagues and students who reviewed this chapter.

References

Aas, W., Mortier, A., Bowersox, V., Cherian, R., Faluvegi, G., Fagerli, H., et al., 2019. Global and regional trends of atmospheric sulfur. Sci. Rep. 9, 953. https://doi.org/10.1038/s41598-018-37304-0.

Albert, J.S., Destouni, G., Duke-Sylvester, S.M., Magurran, A.E., Oberdorff, T., Reis, R.E., et al., 2021. Scientists' warning to humanity on the freshwater biodiversity crisis. Ambio 50, 85–94. https://doi.org/10.1007/s13280-020-01318-8.

Aquilina, L., Poszwa, A., Walter, C., Vergnaud, V., Pierson-Wickmann, A.-C., Ruiz, L., 2012. Long-term effects of high nitrogen loads on cation and carbon riverine export in agricultural catchments. Environ. Sci. Technol. 46, 9447–9455. https://doi.org/10.1021/es301715t.

Ballantyne, A.P., Brahney, J., Fernandez, D., Lawrence, C.L., Saros, J., Neff, J.C., 2011. Biogeochemical response of alpine lakes to a recent increase in dust deposition in the southwestern US. Biogeosciences 8, 2689–2706. https://doi.org/10.5194/bg-8-2689-2011.

Barko, J.W., 1982. Influence of potassium source (sediment vs. open water) and sediment composition on the growth and nutrition of a submersed freshwater macrophyte (*Hydrilla verticillata*) (L.f.) Royle). Aquat. Bot. 12, 157–172. https://doi.org/10.1016/0304-3770(82)90011-0.

Barrett, P.H., 1957. Potassium concentrations in fertilized trout lakes. Limnol. Oceanogr. 2, 287–294. https://doi.org/10.1002/lno.1957.2.3.0287.

Beadle, L.C., 1959. Osmotic and ionic regulation in relation to the classification of brackish and inland saline waters. Arch Ocean. Limnol 11, 143–151.

Beadle, L.C., 1969. Osmotic regulation and the adaptation of freshwater animals to inland saline waters. Int. Ver. Für Theor. Angew. Limnol. Verhandlungen 17, 421–429. https://doi.org/10.1080/03680770.1968.11895868.

Bengtsson, B., Dickson, W., Nyberg, P., 1980. Liming acid lakes in Sweden. Ambio 9, 34–36.

Bhatt, M.P., McDowell, W.H., Gardner, K.H., Hartmann, J., 2014. Chemistry of the heavily urbanized Bagmati River system in Kathmandu Valley, Nepal: export of organic matter, nutrients, major ions, silica, and metals. Environ. Earth Sci. 71, 911–922. https://doi.org/10.1007/s12665-013-2494-9.

Bock, A.R., Falcone, J.A., Oelsner, G.P., 2018. Estimates of road salt application across the conterminous United States, 1992–2015. US Geol. Surv. Data Release. https://doi.org/10.5066/P96IX385.

Bogart, S.J., Azizishirazi, A., Pyle, G.G., 2019. Challenges and future prospects for developing Ca and Mg water quality guidelines: a meta-analysis. Philos. Trans. R. Soc. B Biol. Sci. 374, 20180364. https://doi.org/10.1098/rstb.2018.0364.

Brammer, E., Wetzel, R., 1984. Uptake and release of K^+, Na^+ and Ca^{2+} by the water soldier. *Stratiotes aloides* L. Aquat. Bot. 19, 119–130. https://doi.org/10.1016/0304-3770(84)90012-3.

Cañedo-Argüelles, M., Hawkins, C.P., Kefford, B.J., Schäfer, R.B., Dyack, B.J., Brucet, S., et al., 2016. Saving freshwater from salts. Science 351, 914–916. https://doi.org/10.1126/science.aad3488.

Cañedo-Argüelles, M., Kefford, B., Schäfer, R., 2019. Salt in freshwaters: causes, effects and prospects—introduction to the theme issue. Philos. Trans. R. Soc. B Biol. Sci. 374, 20180002. https://doi.org/10.1098/rstb.2018.0002.

Canfield, D.E., Bachmann, R.W., Hoyer, M.V., 1983. Freeze-out of salts in hard-water lakes. Limnol. Oceanogr. 28, 970–977. https://doi.org/10.4319/lo.1983.28.5.0970.

Capelli, G.M., Magnuson, J.J., 1983. Morphoedaphic and biogeographic analysis of crayfish distribution in northern Wisconsin. J. Crustac Biol. 3, 548–564. https://doi.org/10.1163/193724083X00210.

Carey, J.C., Fulweiler, R.W., 2012. Human activities directly alter watershed dissolved silica fluxes. Biogeochemistry 111, 125–138. https://doi.org/10.1007/s10533-011-9671-2.

Carroll, D., 1970. Rock Weathering. Plenum Press, New York.

Cook, C.D.K., 1996. Aquatic Plant Book. SPB Academic Pub, Amsterdam/New York.

Day, J.A., 1990. Pitfalls in the presentation of chemical data. South. Afr. J. Aquat. Sci. 16, 2–15. https://doi.org/10.1080/10183469.1990.10557364.

de Lima, D.V.N., Pacheco, A.B.F., Goulart, C.L., e Azevedo, S.M.F. de O., 2020. Physiological responses of *Raphidiopsis raciborskii* (Cyanobacteria) strains to water conductivity: effect of sodium and magnesium ions. Hydrobiologia 847, 2449–2464. https://doi.org/10.1007/s10750-020-04265-3.

Deocampo, D.M., Jones, B.F., 2014. Geochemistry of saline lakes. Treatise Geochem 5, 437–469. https://doi.org/10.1016/B0-08-043751-6/05083-0.

Dugan, H.A., Bartlett, S.L., Burke, S.M., Doubek, J., Krivak-Tetley, F., Skaff, N.K., et al., 2017a. Salting our freshwater lakes. Proc. Natl. Acad. Sci. 114, 4453–4458. https://doi.org/10.1073/pnas.1620211114.

Dugan, H.A., Helmueller, G., Magnuson, J.J., 2017b. Ice formation and the risk of chloride toxicity in shallow wetlands and lakes. Limnol. Oceanogr. Lett. 2, 150–158. https://doi.org/10.1002/lol2.10045.

Dugan, H.A., Summers, J.C., Skaff, N.K., Krivak-Tetley, F.E., Doubek, J.P., Burke, S.M., et al., 2017c. Long-term chloride concentrations in North American and European freshwater lakes. Sci. Data 4, 170101. https://doi.org/10.1038/sdata.2017.101.

Durum, W.H., Haffty, J., 1961. Occurrence of Minor Elements in Water. U.S. Geological Survey Circular 445, 11.

Eilers, J., Brakke, D., Henriksen, A., 1992. The inapplicability of the Gibbs model of world water chemistry for dilute lakes. Limnol. Oceanogr. 37, 1335–1337. https://doi.org/10.4319/lo.1992.37.6.1335.

Espie, G.S., Miller, A.G., Kandasamy, R.A., Canvin, D.T., 1991. Active HCO_3^- transport in cyanobacteria. Can. J. Bot. 69, 936–944. https://doi.org/10.1139/b91-120.

Eugster, H.P., Hardie, L.A., 1978. Saline lakes. In: Lerman, A. (Ed.), Lakes: Chemistry, Geology, Physics. Springer, New York, NY, pp. 237–293. https://doi.org/10.1007/978-1-4757-1152-3_8.

Fisher, D.W., Gambell, A.W., Likens, G.E., Bormann, F.H., 1968. Atmospheric contributions to water quality of streams in the Hubbard Brook experimental forest, New Hampshire. Water Resour. Res. 4, 1115–1126. https://doi.org/10.1029/WR004i005p01115.

Forchhammer, G., 1865. IV. On the composition of sea-water in the different parts of the ocean. Philos. Trans. R. Soc. Lond. 155, 203–262. https://doi.org/10.1098/rstl.1865.0004.

Garmo, Ø.A., Skjelkvåle, B.L., de Wit, H.A., Colombo, L., Curtis, C., Fölster, J., et al., 2014. Trends in surface water chemistry in acidified areas in Europe and North America from 1990 to 2008. Water. Air. Soil Pollut 225, 1880. https://doi.org/10.1007/s11270-014-1880-6.

Gibbs, R.J., 1970. Mechanisms controlling world water chemistry. Science 170, 1088–1090. https://doi.org/10.1126/science.170.3962.1088.

Gibbs, R.J., 1992. A reply to the comment of Eilers et al. Limnol. Oceanogr. 37, 1338–1339. https://doi.org/10.4319/lo.1992.37.6.1338.

Goldman, C.R., 1960. Primary productivity and limiting factors in three lakes of the Alaska Peninsula. Ecol. Monogr. 30, 207–230. https://doi.org/10.2307/1948552.

Grennfelt, P., Engleryd, A., Forsius, M., Hov, Ø., Rodhe, H., Cowling, E., 2020. Acid rain and air pollution: 50 years of progress in environmental science and policy. Ambio 49, 849–864. https://doi.org/10.1007/s13280-019-01244-4.

Hammer, U.T., 1986. Saline Lake Ecosystems of the World. Dr W. Junk Publishers, Dordrecht.

Hassani, A., Azapagic, A., D'Odorico, P., Keshmiri, A., Shokri, N., 2020. Desiccation crisis of saline lakes: a new decision-support framework for building resilience to climate change. Sci. Total Environ. 703, 134718. https://doi.org/10.1016/j.scitotenv.2019.134718.

Herbert, E., Boon, P., Burgin, A.J., Neubauer, S.C., Franklin, R.B., Ardon, M., et al., 2015. A global perspective on wetland salinization: ecological consequences of a growing threat to freshwater wetlands. Ecosphere 6, 1–43. https://doi.org/10.1890/ES14-00534.1.

Hering, J.G., Morel, F.M., 1988. Humic acid complexation of calcium and copper. Environ. Sci. Technol. 22, 1234–1237. https://doi.org/10.1021/es00175a018.

Hinckley, E.-L.S., Crawford, J.T., Fakhraei, H., Driscoll, C.T., 2020. A shift in sulfur-cycle manipulation from atmospheric emissions to agricultural additions. Nat. Geosci. 13, 597–604. https://doi.org/10.1038/s41561-020-0620-3.

Howe, P.D., 1998. A review of boron effects in the environment. Biol. Trace Elem. Res. 66, 153–166. https://doi.org/10.1007/BF02783135.

Hutchinson, G.E., 1957. A Treatise on Limnology. I. Geography, Physics, and Chemistry. John Wiley & Sons, New York.

Hutchinson, G.E., 1967. A Treatise on Limnology. II. Introduction to Lake Biology and the Limnoplankton. John Wiley & Sons, New York.

Iglesias, M.C.-A., 2020. A review of recent advances and future challenges in freshwater salinization. Limnética 39, 185–211. https://doi.org/10.23818/limn.39.13.

IOC, SCOR, IAPSO, 2010. The International Thermodynamic Equation of Seawater—2010: Calculation and Use of Thermodynamic Properties. Intergovernmental Oceanographic Commission. Manuals and Guides No. 56, UNESCO (English).

Jaworski, G.H.M., Talling, J.F., Heaney, S.I., 2003. Potassium dependence and phytoplankton ecology: an experimental study. Freshw. Biol. 48, 833–840. https://doi.org/10.1046/j.1365-2427.2003.01051.x.

Jeppesen, E., Beklioğlu, M., Özkan, K., Akyürek, Z., 2020. Salinization increase due to climate change will have substantial negative effects on inland waters: a call for multifaceted research at the local and global scale. Innovation 1, 100030. https://doi.org/10.1016/j.xinn.2020.100030.

Jeziorski, A., Smol, J.P., 2017. The ecological impacts of lakewater calcium decline on softwater boreal ecosystems. Environ. Rev. 25, 245–253. 10.1139/er-2016-0054.

Jeziorski, A., Tanentzap, A.J., Yan, N.D., Paterson, A.M., Palmer, M.E., Korosi, J.B., et al., 2015. The jellification of north temperate lakes. Proc. R. Soc. B Biol. Sci. 282, 20142449. https://doi.org/10.1098/rspb.2014.2449.

Jeziorski, A., Yan, N.D., Paterson, A.M., DeSellas, A.M., Turner, M.A., Jeffries, D.S., et al., 2008. The widespread threat of calcium decline in fresh waters. Science 322, 1374–1377. https://doi.org/10.1126/science.1164949.

Katz, A., Nishri, A., 2013. Calcium, magnesium and strontium cycling in stratified, hardwater lakes: Lake Kinneret (Sea of Galilee), Israel. Geochim. Cosmochim. Acta 105, 372–394. https://doi.org/10.1016/j.gca.2012.11.045.

Kaushal, S.S., Duan, S., Doody, T.R., Haq, S., Smith, R.M., Newcomer Johnson, T.A., et al., 2017. Human-accelerated weathering increases salinization, major ions, and alkalinization in fresh water across land use. Appl. Geochem., Urban Geochemistry 83, 121–135. https://doi.org/10.1016/j.apgeochem.2017.02.006.

Kaushal, S.S., Likens, G.E., Pace, M.L., Utz, R.M., Haq, S., Gorman, J., et al., 2018. Freshwater salinization syndrome on a continental scale. Proc. Natl. Acad. Sci. 115, E574–E583. https://doi.org/10.1073/PNAS.1711234115.

Khan, H., Marcé, R., Laas, A., Obrador, B., 2022. The relevance of pelagic calcification in the global carbon budget of lakes and reservoirs. Limnética 41, 17–25. https://doi.org/10.23818/limn.41.02.

Kilham, P., 1990. Mechanisms controlling the chemical composition of lakes and rivers: data from Africa. Limnol. Oceanogr. 35, 80–83. https://doi.org/10.4319/lo.1990.35.1.0080.

Kirchner, J.W., Lydersen, E., 1995. Base cation depletion and potential long-term acidification of Norwegian catchments. Environ. Sci. Technol. 29, 1953–1960. https://doi.org/10.1021/es00008a012.

Kopáček, J., Kaňa, J., Bičárová, S., Fernandez, I.J., Hejzlar, J., Kahounová, M., et al., 2017. Climate change increasing calcium and magnesium leaching from granitic alpine catchments. Environ. Sci. Technol. 51, 159–166. https://doi.org/10.1021/acs.est.6b03575.

Kratz, W.A., Myers, J., 1955. Nutrition and growth of several blue-green algae. Am. J. Bot. 42, 282–287. https://doi.org/10.1002/j.1537-2197.1955.tb11120.x.

Küchler-Krischun, J., Kleiner, J., 1990. Heterogeneously nucleated calcite precipitation in Lake Constance. A short time resolution study. Aquat. Sci. 52, 176–197. https://doi.org/10.1007/BF00902379.

Kudla, J., Batistič, O., Hashimoto, K., 2010. Calcium signals: the lead currency of plant information processing. Plant Cell 22, 541–563. https://doi.org/10.1105/tpc.109.072686.

Lawrence, C.R., Neff, J.C., 2009. The contemporary physical and chemical flux of aeolian dust: a synthesis of direct measurements of dust deposition. Chem. Geol., Combined Ecological and Geologic Perspectives in Ecosystem Studies 267, 46–63. https://doi.org/10.1016/j.chemgeo.2009.02.005.

Li, C., McLinden, C., Fioletov, V., Krotkov, N., Carn, S., Joiner, J., et al., 2017. India is overtaking China as the world's largest emitter of anthropogenic sulfur dioxide. Sci. Rep. 7, 14304. https://doi.org/10.1038/s41598-017-14639-8.

Likens, G.E., 1985. An Ecosystem Approach to Aquatic Ecology: Mirror Lake and Its Environment. Springer-Verlag, New York.

Likens, G.E., Bormann, F.H., 1974. Acid rain: a serious regional environmental problem. Science 184, 1176–1179. https://doi.org/10.1126/science.184.4142.1176.

Likens, G., Bormann, F., 1995. Biogeochemistry of a Forested Ecosystem, second ed. Springer-Verlag, New York.

Likens, G.E., Driscoll, C.T., Buso, D.C., Siccama, T.G., Johnson, C.E., Lovett, G.M., et al., 1994. The biogeochemistry of potassium at Hubbard Brook. Biogeochemistry 25, 61–125. https://doi.org/10.1007/BF00000881.

Livingstone, D.A., 1963. Chemical composition of rivers and lakes. USGS Rep 440. https://doi.org/10.3133/pp440G.

Maeso, E.S., Piñas, F.F., Gonzalez, M.G., Valiente, E.F., 1987. Sodium requirement for photosynthesis and its relationship with dinitrogen fixation and the external CO_2 concentration in cyanobacteria. Plant Physiol 85, 585–587. https://doi.org/10.1104/pp.85.2.585.

IPCC, 2021. Climate Change 2021: The Physical Science Basis. Contribution of Working Group I to the Sixth Assessment Report of the Intergovernmental Panel on Climate Change. In: Masson-Delmotte, V., Zhai, P., Pirani, A., Connors, S.L., Péan, C., Berger, S., et al. (Eds.). Cambridge University Press, Cambridge, United Kingdom and New York, NY, USA.

McManus, J., Collier, R.W., Chen, C.-T.A., Dymond, J., 1992. Physical properties of Crater Lake, Oregon: a method for the determination of a conductivity- and temperature-dependent expression for salinity. Limnol. Oceanogr. 37, 41–53. https://doi.org/10.4319/lo.1992.37.1.0041.

Meybeck, M., 2003. Global occurrence of major elements in rivers. In: Treatise on Geochemistry, Volume 5: Surface and Ground Water, Weathering, and Soils. Elsevier Science, Amsterdam, pp. 207–223.

Meybeck, M., Helmer, R., 1989. The quality of rivers: from pristine stage to global pollution. Glob. Planet. Change 1, 283–309. https://doi.org/10.1016/0921-8181(89)90007-6.

Meybeck, M., Ragu, A., 2012. GEMS-GLORI world river discharge database. Lab. Géologie Appliquée Univ. Pierre Marie Curie Paris Fr. https://doi.org/10.1594/PANGAEA.804574.

Mickle, A.M., Wetzel, R.G., 1978. Effectiveness of submersed angiosperm-epiphyte complexes on exchange of nutrients and organic carbon in littoral systems. I. Inorganic nutrients. Aquat. Bot. 4, 303–316. https://doi.org/10.1016/0304-3770(78)90028-1.

Miller, A.G., Espie, G.S., Canvin, D.T., 1990. Physiological aspects of CO_2 and HCO_3^- transport by cyanobacteria: a review. Can. J. Bot. 66, 1291–1302. https://doi.org/10.1139/b90-165.

Miller, A.G., Espie, G.S., Canvin, D.T., 1991. Active CO_2 transport in cyanobacteria. Can. J. Bot. 69, 925–935. https://doi.org/10.1139/b91-119.

Moreira, S., Schultze, M., Rahn, K., Boehrer, B., 2016. A practical approach to lake water density from electrical conductivity and temperature. Hydrol. Earth Syst. Sci. 20, 2975–2986. https://doi.org/10.5194/hess-20-2975-2016.

Moss, B., 1972. The influence of environmental factors on the distribution of freshwater algae: an experimental study: I. Introduction and the influence of calcium concentration. J. Ecol. 60, 917–932. https://doi.org/10.2307/2258575.

Müller, B., Gächter, R., 2011. Increasing chloride concentrations in Lake Constance: characterization of sources and estimation of loads. Aquat. Sci. 74, 101–112. https://doi.org/10.1007/s00027-011-0200-0.

Murphy, D.H., Wilkinson, B.H., 1980. Carbonate deposition and facies distribution in a central Michigan marl lake. Sedimentology 27, 123–135. https://doi.org/10.1111/j.1365-3091.1980.tb01164.x.

NTL-LTER, 2020. North Temperate Lakes LTER: chemical limnology of primary study lakes 1981—current. Environ. Data Initiat. https://doi.org/10.6073/pasta/a457e305538a0d8e669b58bb6f35721f.

Oleson, D.J., Makarewicz, J.C., 1990. Effect of sodium and nitrate on growth of *Anabaena Flos-aquae* (cyanophyta). J. Phycol. 26, 593–595. https://doi.org/10.1111/j.0022-3646.1990.00593.x.

Otsuki, A., Wetzel, R.G., 1974. Calcium and total alkalinity budgets and calcium carbonate precipitation of a small hard-water lake. Arch. Hydrobiol. 14–30. https://doi.org/10.1127/archiv-hydrobiol/73/1974/14.

Parks, J., Edwards, M., 2005. Boron in the environment. Crit. Rev. Environ. Sci. Technol. 35, 81–114. https://doi.org/10.1080/10643380590900200.

Pawlowicz, R., Feistel, R., 2012. Limnological applications of the thermodynamic equation of seawater 2010 (TEOS-10). Limnol Oceanogr. Methods 10, 853–867. https://doi.org/10.4319/lom.2012.10.853.

Pawlowicz, R., McDougall, T.J., Feistel, R., Tailleux, R., 2012. An historical perspective on the development of the thermodynamic equation of seawater—2010. Ocean Sci. 8, 161–174.

Pérez, E., Chebude, Y., 2017. Chemical analysis of Gaet'ale, a hypersaline pond in Danakil depression (Ethiopia): new record for the most saline water body on earth. Aquat. Geochem. 23, 109–117. https://doi.org/10.1007/s10498-017-9312-z.

Provasoli, L., 1958. Nutrition and ecology of protozoa and algae. Annu. Rev. Microbiol. 12, 279–308. https://doi.org/10.1146/annurev.mi.12.100158.001431.

Psenner, R., 1999. Living in a dusty world: airborne dust as a key factor for alpine lakes. Water. Air. Soil Pollut 112, 217–227. https://doi.org/10.1023/A:1005082832499.

Rees, T.A.V., 1984. Sodium dependent photosynthetic oxygen evolution in a marine diatom. J. Exp. Bot. 35, 332–337. https://doi.org/10.1093/jxb/35.3.332.

Reynolds, C.S., 2006. The Ecology of Phytoplankton, Ecology, Biodiversity and Conservation. Cambridge University Press, Cambridge. https://doi.org/10.1017/CBO9780511542145.

Rodell, M., Famiglietti, J.S., Wiese, D.N., Reager, J.T., Beaudoing, H.K., Landerer, F.W., et al., 2018. Emerging trends in global freshwater availability. Nature 557, 651–659. https://doi.org/10.1038/s41586-018-0123-1.

Rodhe, W., 1948. The ionic composition of lake waters. Int. Ver. Für Theor. Angew. Limnol. Verhandlungen 10, 377–386. https://doi.org/10.1080/03680770.1948.11895170.

Rodhe, W., 1950. Minor constituents in lake waters. Int. Ver. Für Theor. Angew. Limnol. Verhandlungen 11, 317–323. https://doi.org/10.1080/03680770.1950.11895243.

Saccò, M., White, N.E., Harrod, C., Salazar, G., Aguilar, P., Cubillos, C.F., et al., 2021. Salt to conserve: a review on the ecology and preservation of hypersaline ecosystems. Biol. Rev. 96, 2828–2850. https://doi.org/10.1111/brv.12780.

Schindler, D.W., 1988. Effects of acid rain on freshwater ecosystems. Science 239, 149–157. https://doi.org/10.1126/science.239.4836.149.

Schoderboeck, L., Mühlegger, S., Losert, A., Gausterer, C., Hornek, R., 2011. Effects assessment: boron compounds in the aquatic environment. Chemosphere 82, 483–487. https://doi.org/10.1016/j.chemosphere.2010.10.031.

Schuerch, M., Spencer, T., Temmerman, S., Kirwan, M.L., Wolff, C., Lincke, D., et al., 2018. Future response of global coastal wetlands to sea-level rise. Nature 561, 231–234. https://doi.org/10.1038/s41586-018-0476-5.

Schuler, M.S., Cañedo-Argüelles, M., Hintz, W.D., Dyack, B., Birk, S., Relyea, R.A., 2019. Regulations are needed to protect freshwater ecosystems from salinization. Philos. Trans. R. Soc. B Biol. Sci. 374, 20180019. https://doi.org/10.1098/rstb.2018.0019.

Schulz, C.-J., Cañedo-Argüelles, M., 2019. Lost in translation: the German literature on freshwater salinization. Philos. Trans. R. Soc. B Biol. Sci. 374, 20180007. https://doi.org/10.1098/rstb.2018.0007.

Seale, D., Boraas, M., Warren, G., 1987. Effects of sodium and phosphate on growth of cyanobacteria. Water Res. 21, 625–631. https://doi.org/10.1016/0043-1354(87)90072-8.

Shoesmtth, E.A., Brook, A.J., 1983. Monovalent–divalent cation ratios and the occurrence of phytoplankton, with special reference to the desmids. Freshw. Biol. 13, 151–155. https://doi.org/10.1111/j.1365-2427.1983.tb00667.x.

Smith, K.S., Ferry, J.G., 2000. Prokaryotic carbonic anhydrases. FEMS Microbiol. Rev. 24, 335–366. https://doi.org/10.1111/j.1574-6976.2000.tb00546.x.

Soucek, D.J., Linton, T.K., Tarr, C.D., Dickinson, A., Wickramanayake, N., Delos, C.G., et al., 2011. Influence of water hardness and sulfate on the acute toxicity of chloride to sensitive freshwater invertebrates. Environ. Toxicol. Chem. 30, 930–938. https://doi.org/10.1002/etc.454.

Stabel, H.-H., 1989. Coupling of strontium and calcium cycles in Lake Constance. In: Sly, P.G., Hart, B.T. (Eds.), Sediment/Water Interactions, Developments in Hydrobiology. Springer Netherlands, Dordrecht, pp. 323–329. https://doi.org/10.1007/978-94-009-2376-8_30.

Stallard, R.F., Edmond, J.M., 1981. Geochemistry of the Amazon: 1. Precipitation chemistry and the marine contribution to the dissolved load at the time of peak discharge. J. Geophys. Res. Oceans 86, 9844–9858. https://doi.org/10.1029/JC086iC10p09844.

Stauffer, R., Armstrong, D.E., 1986. Cycling of iron, manganese, silica, phosphorus, calcium and potassium in two stratified basins of Shagawa Lake, Minnesota. Geochim. Cosmochim. Acta 50, 215–229. https://doi.org/10.1016/0016-7037(86)90171-7.

Stewart, A.J., Wetzel, R.G., 1981. Dissolved humic materials: photodegradation, sediment effects, and reactivity with phosphate and calcium carbonate precipitation. Arch. Hydrobiol. 92, 265–286.

Strock, K.E., Nelson, S.J., Kahl, J.S., Saros, J.E., McDowell, W.H., 2014. Decadal trends reveal recent acceleration in the rate of recovery from acidification in the northeastern U.S. Environ. Sci. Technol. 48, 4681–4689. https://doi.org/10.1021/es404772n.

Strøm, K.M., 1947. Correlation between x_{18} and pH in lakes. Nature 159, 782–783. https://doi.org/10.1038/159782b0.

Talling, J.F., 2009. Electrical conductance—a versatile guide in freshwater science. Freshw. Rev. 2, 65–78. https://doi.org/10.1608/FRJ-2.1.4.

Talling, J.F., 2010. Potassium—a non-limiting nutrient in fresh waters? Freshw. Rev. 3, 97–104. https://doi.org/10.1608/FRJ-3.2.1.

US EPA, 2016. National Lakes Assessment 2012: A Collaborative Survey of Lakes in the United States.

US EPA, 2019. Secondary Drinking Water Standards: Guidance for Nuisance Chemicals 40 CFR Ch. I (7–1–19 Edition).

US EPA, 2020. National Aquatic Resource Surveys: National Rivers and Streams Assessment 2013–2014.

USGS, 2019. Specific conductance. In: U.S. Geological Survey Techniques and Methods, Book 9. U.S. Department of the Interior, U.S. Geological Survey, Reston, VA, p. 15. Chap. A6.3.

Valiente, E.F., Avendaño, M. del C., 1993. Sodium-stimulation of phosphate uptake in the cyanobacterium *Anabaena* PCC 7119. Plant Cell Physiol. 34, 201–207. https://doi.org/10.1093/oxfordjournals.pcp.a078407.

Vet, R., Artz, R.S., Carou, S., Shaw, M., Ro, C.U., Aas, W., et al., 2014. A global assessment of precipitation chemistry and deposition of sulfur, nitrogen, sea salt, base cations, organic acids, acidity and pH, and phosphorus. Atmos. Environ. 93, 3–100. https://doi.org/10.1016/j.atmosenv.2013.10.060.

Ward, A.K., Wetzel, R.G., 1975. Sodium: some effects on bluegreen algal growth. J. Phycol. 11, 357–363. https://doi.org/10.1111/j.1529-8817.1975.tb02796.x.

Wetzel, R.G., 1970. Recent and postglacial production rates of a marl lake. Limnol. Oceanogr. 15, 491–503. https://doi.org/10.4319/lo.1970.15.4.0491.

Wetzel, R.G., Otsuki, A., 1974. Allochthonous organic carbon of a marl lake. Arch. Hydrobiol. 31–56. https://doi.org/10.1127/archiv-hydrobiol/73/1974/31.

Weyhenmeyer, G.A., Hartmann, J., Hessen, D.O., Kopáček, J., Hejzlar, J., Jacquet, S., et al., 2019. Widespread diminishing anthropogenic effects on calcium in freshwaters. Sci. Rep. 9, 10450. https://doi.org/10.1038/s41598-019-46838-w.

Whitehead, H.C., Feth, J.H., 1961. Recent chemical analyses of waters from several closed-basin lakes and their tributaries in the western United States. GSA Bull 72, 1421–1425. https://doi.org/10.1130/0016-7606(1961)72[1421.RCAOWF]2.0.CO;2. RCAOWF]2.0.CO;2.

Whitfield, A.K., Elliott, M., Basset, A., Blaber, S.J.M., West, R.J., 2012. Paradigms in estuarine ecology—a review of the Remane diagram with a suggested revised model for estuaries. Estuar. Coast Shelf Sci. 97, 78–90. https://doi.org/10.1016/j.ecss.2011.11.026.

Whittier, T.R., Ringold, P.L., Herlihy, A.T., Pierson, S.M., 2008. A calcium-based invasion risk assessment for zebra and quagga mussels (*Dreissena* spp). Front. Ecol. Environ. 6, 180–184. https://doi.org/10.1890/070073.

Williams, W.D., 1981. Inland salt lakes: an introduction. Hydrobiologia 81, 1–14. https://doi.org/10.1007/BF00048701.

Williams, W.D., Aladin, N.V., 1991. The Aral Sea: recent limnological changes and their conservation significance. Aquat. Conserv. Mar. Freshw. Ecosyst. 1, 3–23. https://doi.org/10.1002/aqc.3270010103.

Williams, W.D., Sherwood, J.E., 1994. Definition and measurement of salinity in salt lakes. Int. J. Salt Lake Res. 3, 53–63.

Wu, Y., Gibson, 1996. Mechanisms controlling the water chemistry of small lakes in Northern Ireland. Water Res. 30, 178–182. https://doi.org/10.1016/0043-1354(95)00140-G.

Wurtsbaugh, W.A., Miller, C., Null, S.E., DeRose, R.J., Wilcock, P., Hahnenberger, M., et al., 2017. Decline of the world's saline lakes. Nat. Geosci. 10, 816–821. https://doi.org/10.1038/ngeo3052.

CHAPTER

13

The Inorganic Carbon Complex

Jonathan J. Cole[1,†] *and Yves T. Prairie*[2]

[1]Cary Institute of Ecosystem Studies, Millbrook, NY, United States

[2]Department of Biological Sciences, University of Quebec at Montreal, Montreal, QC, Canada

OUTLINE

I. The occurrence of inorganic carbon in freshwater systems

A. Carbon dioxide and its solution in water

1. The basic equations governing CO_2 gas in water

The carbon dioxide (CO_2) content of the atmosphere varies with locality and with enrichment from industrial pollution. The global average is approximately 0.042% by volume, as of 2021, and has been increasing recently at a rate of about 2 ppm per year due largely to the emissions of fossil fuel C (Gates 1993; Schlesinger and Bernhardt 2020). Atmospheric CO_2 in the preindustrial age is estimated to be about 270 ppm (or 0.027%). As a gas, carbon dioxide is very soluble in water, some 200 times more so than oxygen, and obeys normal solubility laws within the conditions of temperature and pressure encountered in lakes. At equilibrium, the partial pressure of a dissolved gas is equal to the partial pressure of that gas in the overlying atmosphere.

$$pCO_{2(water)} = pCO_{2(atmosphere)} \quad (13\text{-}1)$$

Partial pressure is usually expressed as μatm. So, if the atmosphere has a pCO_2 of 400 μatm, surface water at equilibrium with that atmosphere will also have a pCO_2 of 400 μatm. However, even when the water and atmospheric CO_2 partial pressures are identical, their concentrations per unit volume in the two media will be very different, as it depends on the solubility of the particular gas in water. The concentration of CO_2 ($[CO_2]$) in the water is controlled by $pCO_{2(water)}$ and a temperature-dependent solubility coefficient known as Henry's constant (K_H).

$$[CO_2] = K_H * pCO_{2(water)} \quad (13\text{-}2)$$

[†]Deceased author.

Wetzel's Limnology, Fourth Edition
https://doi.org/10.1016/B978-0-12-822701-5.00013-6

K_H has units of μmol L^{-1} μatm^{-1} (which is the same as mol L^{-1} atm^{-1}) and is about 0.04 at 25°C (see also Chapter 10).

CO_2 behaves like most dissolved gases. That is, like O_2, CO_2 is more soluble as the water gets colder. Another way to say this is that K_H increases with decreasing temperature. At an atmospheric partial pressure of 400 μatm (e.g., 0.04%), the amount of CO_2 dissolved in water at equilibrium with the atmosphere is about 30 μmol L^{-1} at 0°C, 19 μmol L^{-1} at 15°C, and 12 μmol L^{-1} at 30°C. As we will see, dissolved CO_2 in the surface waters of lakes, rivers, and streams is often not at equilibrium with the atmosphere.

2. *The basic equations governing the distribution of the forms of dissolved inorganic carbon (DIC)*

As CO_2 dissolves in water, the solution contains unhydrated CO_2 at about the same concentration by volume (approximately 10 μmol L^{-1}) as in the atmosphere (reviewed extensively by Butler, 1992; Stumm and Morgan, 1995):

$$CO_2(\text{air}) \leftrightarrow CO_2(\text{dissolved}) + H_2O \quad (13\text{-}3)$$

Dissolved CO_2 hydrates by a slow reaction (a half-time of approximately 18 s at 25°C):

$$CO_2 + H_2O \leftrightarrow H_2CO_3 \quad (13\text{-}4)$$

This reaction predominates at a pH of less than 8, with the equilibrium concentration of H_2CO_3 only about 1/400 that of the unhydrated CO_2. Above a pH of 10, $CO_2 + OH^- \leftrightarrow HCO_3^-$ is the dominant reaction. Thus, for most freshwaters, H_2CO_3 can be ignored compared to CO_2.

H_2CO_3 is a fairly weak acid that dissociates rapidly relative to the hydration reaction:

$$H_2CO_3 \leftrightarrow H^+ + HCO_3^- \quad (13\text{-}5)$$

$$HCO_3^- \leftrightarrow H^+ + CO_3^{2-} \quad (13\text{-}6)$$

And:

$$K_1 = \frac{[H^+][HCO_3^-]}{[CO_{2(aq)}]} \quad (13\text{-}7)$$

K_1 in Eq. 13-7 is the equilibrium constant for the combined reactions of Eqs. 13-4 and 13-5. The pK_1 (e.g., $-\log(K_1)$) *dissociation value* of the first reaction, including both hydrated and unhydrated CO_2 as the undissociated molecule, is 6.43 at 15°C in dilute freshwater. A useful way to think about pK_1 is that at pH = 6.43, the concentration of CO_2 and HCO_3^- will be equal (Fig. 13-1). Note that K_1 is not constant but varies with temperature, ionic strength, and some other ions, notably calcium. K_2 is given in Eq. 13-11 and it is the equilibrium constant for Eq. 13-6. Look up the values of pK_1 and pK_2 for seawater and you will see the difference, especially for pK_2 (below). To get the correct values in saline lakes, you need to know the ionic strength and the ionic composition of the salts.

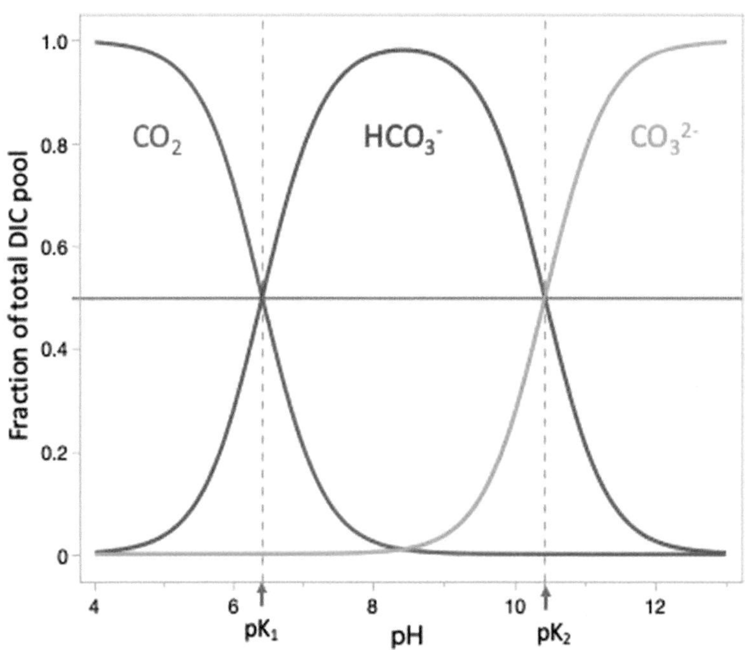

FIGURE 13-1 **Relation between pH and the relative proportions of inorganic carbon species of $CO_2(+H_2CO_3)$, HCO_3^-, and CO_3^{2-} in solution.** This diagram is common in limnology and water chemistry textbooks but often causes some confusion. First, it is actually the proportions of the forms of DIC that regulate the pH in natural waters rather than vice versa, as the diagram might suggest. Second, the y-axis is the fraction, not the absolute amount. It is useful to remember that at pH 6.3, the amount of CO_2 and HCO_3^- are equal. Note that the values for pK_1 and pK_2 are appropriate for dilute freshwaters. *(Modified from Stumm and Morgan, (1995).)*

Computer programs such as PHREECQ can do the complex calculations (Parkhurst and Appelo, 2013).

The bicarbonate and carbonate ions also dissociate to establish an equilibrium:

$$HCO_3^- + H_2O \leftrightarrow H_2CO_3 + OH^- \quad (13\text{-}8)$$

$$CO_3^{2-} + H_2O \leftrightarrow HCO_3^- + OH^- \quad (13\text{-}9)$$

$$H_2CO_3 \leftrightarrow H_2O + CO_2 \quad (13\text{-}10)$$

And:

$$K_2 = \frac{[H^+][CO_3^{2-}]}{[HCO_3^-]} \quad (13\text{-}11)$$

The pK_2 of the second reaction in dilute freshwater is 10.43 at 15°C. So at pH = 10.43, the concentration of HCO_3^- and CO_3^{2-} will also be equal (Fig. 13-1). K_2 is also affected by the factors that affect K_1, but even more strongly. For example, in seawater, pK_2 at 15°C is about 9. This means that at the same concentrations of Ca and CO_3^{2-}, carbonate will precipitate more in the sea than in freshwater.

B. Range and origin of DIC in freshwaters

1. *Range of DIC concentrations*

DIC varies in natural freshwaters in the range of less than 20 μmol L^{-1} in dilute, acidic streams and lakes to more than 5000 μmol L^{-1} in hardwater systems (greater than twice the level in seawater), and even higher in some saline lakes (Finlay et al., 2009). At the pH of most freshwaters (>6.8 and <10), HCO_3^- is the dominant form of DIC. pCO_2 is typically within 0.5- to 10-fold that of equilibrium with the atmosphere, so from about 200 to 4000 μatm. Thus the surface water concentration of carbon dioxide $[CO_2]$ ranges from about 6 to 120 μM in typical systems. CO_2 is lowest during intense photosynthetic blooms and may be undetectable, as has been observed during some experimental fertilizations of lakes (Cole et al., 2000; Schindler, 1977). CO_2 is highest when heterotrophic respiration is much higher than photosynthesis or when external inputs in groundwater are high. By contrast, $[HCO_3^-]$ is typically between about 10 μM in mildly acidic water and up to more than 5000 μmol L^{-1} in typical hardwaters and even higher in some *soda lakes*. In hardwater lakes 90% or more of the DIC in hardwaters is as HCO_3^-, which is often a major, or one of the major, anions in freshwater. The average concentration of HCO_3^- in the world's major rivers is about 1000 μmol L^{-1} (Livingstone, 1963; Schlesinger and Bernhardt, 2020), and values near this are common in many lakes, rivers, and streams but depend on the geologic settings. Because pK_2 is so high in freshwater, above 10, CO_3^{2-} can often (but not always) be effectively ignored for many calculations. For example, at pH = 8 and 15°C, the CO_3^{2-} is only 0.3% as abundant as HCO_3^-; at pH = 9, it is more significant (3%) but still small in comparison. So, if HCO_3^- is 1000 μmol L^{-1}, CO_3^{2-} is only 3 μmol L^{-1} at pH 8 and still only 30 μmol L^{-1} at pH 9. On the other hand, while CO_3^{2-} is usually small in comparison to HCO_3^-, it can still represent a lot of C, especially if the total DIC is quite high.

2. *Origins of DIC*

DIC enters aquatic ecosystems through several pathways. All three components (dissolved CO_2, HCO_3^-, and CO_3^{2-}) can be imported via groundwater or from upstream systems, and all three can be exported in outflowing water. The CO_2 gas dissolved in groundwater is typically quite high due to the respiration of soil and plant roots. Concentrations of CO_2 near 1000 μM are not uncommon in streams when they first exit the ground and are even higher in actual groundwater (Crawford et al., 2014). This is equivalent to a pCO_2 above 28,000 μatm or about 70 times greater than the atmosphere. CO_2 additionally is captured and transformed into organic matter by photosynthesis and chemosynthesis, and organic matter is converted to CO_2 by both aerobic and anaerobic respiration (Table 13-1).

CO_2 is also consumed in the weathering of carbonate or silicate rocks, carbonate shells, or the biogenic silica of diatom frustules. CO_2 is generated by the precipitation of solid carbonate or shells and siliceous diatom frustules. Further, CO_2 exchanges with the atmosphere and surface waters can be either a source or sink for atmospheric CO_2. If the pCO_2 in the water is greater than that in the overlying atmosphere, the system acts as a source, and vice versa.

Bicarbonate (HCO_3^-) is usually the largest component of DIC. HCO_3^- is a product of weathering of both carbonate (limestone and dolomite) rock and silicates as well. The weathering is caused by CO_2 in the air and in soil water due to the respiration of microbes and plants. For *calcium carbonate* ($CaCO_3$), a generic weathering equation is:

$$CO_2 + H_2O + CaCO_3 \rightarrow Ca^{++} + 2HCO_3^- \quad (13\text{-}12)$$

And similarly for silicate:

$$2CO_2 + 3H_2O + CaSiO_3 \rightarrow Ca^{++} + 2HCO_3^- + H_4SiO_4 \quad (13\text{-}13)$$

You will see many versions of these equations that may include magnesium, water, or other factors, but the overall idea is the same. Note that the atmosphere is the source of the CO_2 in weathering either directly (rock exposed to air) or more commonly rock exposed to CO_2 in soil water. That is, the CO_2 in either case started out in the atmosphere. It was taken up during photosynthesis and was then respired either by the

TABLE 13-1 Processes that either remove or add CO_2 to the water column of the aquatic system.

Removes CO_2	Process type	Adds CO_2	Comments
Atmospheric exchange	Physical process	Atmospheric exchange	Direction depends on pCO_2 gradient
Photosynthesis	Algae, macrophytes, periphyton	Aerobic respiration, see Table 13-2	All respiration including heterotrophs and autotrophs, aerobic bacteria
Chemosynthesis	Oxidation of methane, sulfur, ammonia, or other electron acceptors coupled with CO_2 uptake (e.g., *Thiobacillus*)	Anaerobic respiration, see Table 13-2	Using SO_4, NO_3, Fe as electron acceptor (e.g., *Desulfovibrio*)
Carbonate dissolution by CO_2	Carbonate weathering	Groundwater inputs or surface water inputs	
		Anoxygenic photosynthesis	Purple or green sulfur bacteria – often in stratified lakes
	May be either chemical or biological process	Formation of calcite precipitation in water column or on plants and macroalgae (e.g., *Chara*)	Also shells of snails, bivalves, etc. Rare for pelagic organisms in freshwater but common in benthos

plants themselves or during the decomposition of organic matter. Globally, slightly more than half of the DIC transported by rivers to the ocean is derived from the weathering of carbonate rock and slightly less than half from silicate rock (Meybeck 1993; Stallard 1998; Gaillardet et al., 1999). These values are obtained by tracking the cation and anion concentrations and discharge from the world's major rivers (Gaillardet et al., 1999). There are large regional differences in these values. For North America, about 80% of the DIC alkalinity discharge is attributable to carbonate rock dissolution and only 20% to the weathering of aluminosilicate rocks (Morel and Hering, 1993). Gaillardet et al., (1999) present a table with 50 of the world's major rivers (by discharge). The Seine, Rhone, and St. Lawrence rivers are totally dominated by the carbonate weathering pathway (>90%). The Orinoco, Zaire, and Irrawaddy rivers are dominated by the silicate pathway (>80%), while the Amazon, the largest river by discharge in the world, is about 60% silicate dominated. At the local scale, both DIC and *acid neutralizing capacity* (ANC) concentrations are usually much higher in carbonate geology than in silicate geology. DIC and ANC are lowest in granitic, schist, and gneiss regions of North American and northern Scandinavia. Olivine and basalt, particularly in the climate of the humid tropics, weather more easily than does granite or gneiss in the colder climates at high latitudes. So these tropical areas can have high to moderately high DIC and ANC. Because river water discharge is also high in the humid tropics, these large rivers make the silicate pathway of ANC delivery to the ocean as large as it is.

The carbonate ion can also be in solution and enter from upstream and from groundwater if the DIC and/or pH are high enough. Further, CO_3^{2-} can be produced by *"reverse weathering"* (e.g., Eq. 13-12 running to the left). This occurs during the drawdown of CO_2 during intense periods of primary production. If the resulting concentration of $CaCO_3$ is high enough, it will precipitate, as the solubility of $CaCO_3$ is not large. Fine particles of $CaCO_3$ can form *whiting events* (see also Chapter 12), which are common in productive hardwater lakes (Hodell et al., 1998; Walsh et al., 2019). These whiting events can even be tracked by remote sensing (Nouchi et al., 2019) and even by satellite in very large lakes (Binding et al., 2007). Somewhat counterintuitively, the precipitation of $CaCO_3$ represents a source of CO_2 into the water column just as the dissolution of $CaCO_3$ is a sink for CO_2. Some freshwater organisms make shells of carbonate, notably snails and freshwater mussels (Table 13-1). It is intriguing that freshwater pelagic organisms with carbonate shells are very rare in lakes, whereas in marine systems they are common. For example, coccolithophorids are very common in the ocean and absent in freshwaters. Many macrophytes (Chapter 24) will precipitate $CaCO_3$ (*marl*) on their exterior during photosynthesis. The macroalga *Chara* (Chapter 17) is often seen with calcium encrustations (Herbst and Schubert, 2018). It is interesting that even in very soft freshwaters, there can be perfectly healthy populations of benthic mussels and snails. For example, Strayer et al., (1981) found that the unionid clam, *Elliptio complanata*, had normal growth rates in Mirror Lake, which has an ANC of less than 70 μEq L^{-1} and a pH between 6.5 and 7.

$CaCO_3$ can also be found in *colloidal form*. The importance of these is poorly understood; Wetzel, (2001) reviewed the literature on this topic, which was mainly old and has not been seriously followed up. The major importance of this is that the *organic coatings* reduce the rate of dissolution of sedimenting $CaCO_3$ in lakes and can form a major sink for inorganic and organic detrital carbon in hardwater ecosystems (Kleiner and Stabel, 1989; Otsuki and Wetzel, 1972); Wetzel, 1970).

C. Alkalinity, acidity, and pH of natural waters

1. Alkalinity

Natural waters exhibit wide variations in relative acidity and alkalinity, not only in actual pH values but also in the amount of dissolved material producing the acidity or alkalinity. The concentration of these compounds and the ratio of one to another determine the actual pH and the buffering capacity of a given water body. Since the lethal effects of most acids begin to appear near pH 4.5 and of most alkalis near pH 9.5, that buffering can be of major importance in the maintenance of life.

Alkalinity is historically a term that referred to the *buffering capacity* of the carbonate system in water. The terms *alkalinity, carbonate alkalinity, alkaline reserve, titratable base,* or *acid-binding capacity* are frequently used to express the total quantity of base (usually in equilibrium with carbonate or bicarbonate) that can be determined by titration with a strong acid (Hutchinson, 1957). The milliequivalents of acid necessary to neutralize the hydroxyl, carbonate, and bicarbonate ions in a liter of water are known as the *total alkalinity.* Alkalinity is numerically the equivalent concentration of titratable base and is determined by titration with a standard solution of a strong acid to equivalency points dictated by pH values at which the alkaline contributions of hydroxide, carbonate, and bicarbonate are neutralized.

The term *hardness* is frequently used as an assessment of the quality of water supplies. The hardness of a water is governed by the content of calcium and magnesium salts, largely combined with bicarbonate and carbonate (*temporary hardness*) and with sulfates, chlorides, and other anions of mineral acids (*permanent hardness*). The *carbonate hardness* can be removed by boiling, which causes the precipitation of $CaCO_3$. The fraction of calcium and magnesium remaining in solution as sulfates, chlorides, and nitrates after boiling constitutes the *residual noncarbonate hardness.* The extent of hardness has been expressed numerically in a remarkably heterogeneous system of scales among different countries. When water is "hard," it causes problems for plumbing pipes, hot water heaters, and coffee makers, among other issues. This is why many homes need to use a "water softener" to lengthen the life of plumbing and appliances. Limnologists still sometimes categorize lakes as hard or softwater systems.

Alkalinity is now used interchangeably with *acid neutralizing capacity (ANC),* which is the capacity to neutralize strong inorganic acids. ANC is the amount of acid it would take to move a water sample from its current pH to reach a pH corresponding to a solution of pure CO_2 (Butler, 1992), which is near pH 4. Another way to say this is that ANC is the amount of acid required to convert all the DIC to CO_2. One can write a long chemical equation to describe alkalinity that includes all of the acids and bases that might be in a water sample. In most natural freshwaters the simplified equation below is sufficient where the carbonate system totally dominates ANC:

$$\mathrm{ANC} = \left[\mathrm{HCO_3^-}\right] + 2\left[\mathrm{CO_3^{2-}}\right] + \left[\mathrm{OH^-}\right] - \mathrm{H^+} \tag{13-14}$$

ANC is expressed in units of $\mu Eq\ L^{-1}$. A low ANC system would have ANC less than approximately 50 $\mu Eq\ L^{-1}$, and a high ANC system would have greater than 2000 $\mu Eq\ L^{-1}$.

Measurement and calculation of alkalinity and DIC

In the older literature alkalinity was usually expressed in units of $CaCO_3\ L^{-1}$. Thus limnologists measured alkalinity by titration with acid and expressed it as if all the C was in the CO_3^{2-} anion and calcium was the cation. Since the latter half of the 20th century, DIC is typically measured by either *gas chromatography* (GC) or by a dedicated *infrared gas analyzer* (IRGA) after all forms of inorganic carbon are converted to CO_2 by acidification. Correspondingly, in modern literature, ANC is calculated from direct measurements of pH and measurements of DIC. Dissolved organic compounds are also a weak source of alkalinity (typically 1–3 μEq per mg DOC). When carbonate alkalinity is very low (<50 $\mu Eq\ L^{-1}$), however, these organic acids can constitute a significant fraction of the total alkalinity.

Properties of alkalinity

ANC has some useful properties. For one, unlike total DIC, it is unaffected by changes in CO_2. This makes ANC relatively conservative especially in oxic surface waters. Further, pH is affected by changes in CO_2. Thus in many systems one can take continuous measurements of pH and use these to calculate the changes in CO_2 if ANC holds relatively constant. These changes in pH alone can be used to calculate ecosystem metabolism under the right conditions. There are some reactions that generate or consume ANC. Obviously, the addition of acids lowers it. Acid can also be generated by the oxidation of iron or sulfur compounds. Several processes can generate ANC. These include: the uptake of nitrate by phytoplankton and nitrate and sulfate reduction in anoxic waters, among others (Kelly, 1988; Giblin et al., 1990). These reactions that change ANC tend to be relatively small compared to ANC in most systems, especially in surface waters. On the other hand, ANC can increase over depth, sometimes greatly,

TABLE 13-2 Some of the Key Aerobic and Anaerobic Respiratory Pathways in Inland Waters and Their Effects on ANC.

Process	Equation	Change in ANC	Types of organisms
Aerobic respiration	$CH_2O + O_2 \rightarrow CO_2 + H_2O$	No change	Aerobic bacteria, algae, animals, etc.
Aerobic methane oxidation	$CH_4 + 2O_2 \rightarrow CO_2 + 2H_2O$	No change	Methanotrophic bacteria, *Methylocystis*
Anaerobic respiration			
Nitrate reduction	$CH_2O + 0.5NO_3^- + 0.5H_2O \rightarrow 0.5NH_4^+ + HCO_3$	Increase	Denitrifying bacteria, *Pseudomonas*
Sulfate reduction	$CH_3COOH + SO_4^{2-} + 2\ H^+ \rightarrow HS- + 2\ HCO_3^- + 3\ H^+$	Increase	Sulfate reducing bacteria, *Desulfovibirio*
Iron reduction	$HCOO- + 2\ Fe(III) + H_20 \rightarrow HCO_3^- + 2\ Fe(II) + 2H^+$	Increase	Iron reducing bacteria, *Geobacter*
Methanogenesis - acetoclastic (fermentation)	$CH_3COOH \rightarrow CH_4 + CO_2$	No change	Methanogenic bacteria, *Methanosarcina, Methanobacterium*

Modified from Stumm and Morgan, 1995 and Lovley et al., 2004. Note that some of the equations are not balanced stoichiometrically in order to make them easier to read. Aerobic and anaerobic respiration both have CO_2 as the oxidized C end product. However, in some of the anaerobic respiratory processes OH^- is also generated. This OH^- combines with CO_2 to produce HCO_3^-, as shown in the equations see Stumm and Morgan, 1995.

if there is sufficient anaerobic respiration (Table 13-2), as several types of anaerobic respiration generate ANC.

Secondly, there is a great deal of historical data with measurements of alkalinity, some also with pH, and some going back nearly 100 years. Raymond et al., (2008) used archived USGS data to show that the export of alkalinity by the Mississippi River (USA) has been increasing over the past century. Further, there is a great deal of alkalinity and pH data from regions of the world where direct DIC measurement has not been routine and from field stations that do not have the most modern equipment. These data (along with temperature) have been instrumental in revealing patterns about both DIC and pCO_2 in surface waters of lakes, rivers, and streams (Rebsdorf et al., 1991; Cole et al., 1994; Regnier et al., 2013).

2. *Acidity*

Acidity is used infrequently as a parameter in limnological investigations. In practice, acidity is a measure of the quantity of strong base per liter required to attain a pH equal to that of a solution of sodium carbonate (Na_2CO_3) equivalent to the total inorganic carbon. Like alkalinity, it is measured by titration. As such, acidity is not particularly useful. On the other hand, $[H^+]$ is both a useful and necessary parameter.

3. *Hydrogen ion activity and pH*

In practice, it is common to use the term acidity rather interchangeably with pH, even though this is not strictly correct. When limnologists talk about acidic water bodies, they mean water bodies with high levels of hydrogen ion activity. We cannot directly measure $[H^+]$. Instead, we can measure *hydrogen ion activity*. This is what is actually measured with a pH meter. $pH = -\log_{10}$ (hydrogen ion activity). In the above equations the H^+ term comes from pH measurements and we write it as the concentration of H^+. All of the constants in the carbonate system equations were determined in reference to pH measurements (i.e., hydrogen ion activity), so the system is completely consistent.

Pure water dissociates weakly to H^+ and OH^- ions. The *dissociation constant* is very small ($\approx 10^{-14}$), however, and the amounts of H^+ and OH^- present are 10^{-7} mol L^{-1}. Natural waters are, of course, not pure, and salts, acids, and bases contribute to the H^+ and OH^- ions in varying ways, depending on the individual circumstances. Since the dissociation constant of water is fixed, the addition of one ion will result in a decrease in the other. The pH is defined as the logarithm of the reciprocal of the concentration of free hydrogen ions. The "p" of pH refers to the power (from the French term *puissance*) of the hydrogen ion activity. Therefore more H^+ activity in an acid reaction increases the power from neutrality (10^{-7} or pH 7) to, say, 10^{-4} (pH 4). In more alkaline reactions H^+ ion activity is decreased from neutrality to, for example, 10^{-10} (pH 10).

4. *Range of pH in natural waters*

The pH of most natural waters ranges between the extremes of pH less than 2 and 12. Nearly all natural, unpolluted waters with pH values less than 4 occur in volcanic regions that receive strong mineral acids, particularly sulfuric acid. Some of these lakes are extremely acidic. For example, a crater lake in Indonesia (Kawah Ijen) has a remarkably low pH of about 0.3, which is much more acidic than the most acidic soft drinks (Lohr et al., 2006). The oxidation of pyrite of rocks and clays in drainage basins can result in sulfuric acid drainage to lakes. Several acid-mine drainage lakes in China have a pH in the range of 2 to 3 (Xin et al., 2021).

Moderately low pH values (e.g., acidic) are often found in natural waters rich in dissolved organic matter (Chapter 28) with low amounts of DIC. An appreciable portion (often >50%) of dissolved organic matter occurs as organic acids derived from humic compounds, which are comprised of a large number of complex ligands that have numerous p*K* and *charge density* sites. Charge density means the number of sites (expressed as moles) per mole of carbon (Lydersen, 1998). The concentration of organic acids varies from 5 to 22 μEq mg C^{-1}, often with an average of carboxylic acids of approximately 10 μEq mg C^{-1}. *Organic acids* can modify both the acidity of surface waters and changes in strong acid inputs in waters with low or little bicarbonate alkalinity. Organic acids may depress pH by 0.5 to 2.5 pH units in waters with an ANC range of 0–50 μeq L^{-1} (Lydersen, 1998). Despite these direct additions to acidity, organic acids have a substantial buffering capacity for strong acids and can reduce pH fluctuations from strong acids in precipitation.

Low pH values are particularly common in bogs and bog lakes (Chapter 29) that are dominated in the littoral mat by the moss *Sphagnum*. The pH of *Sphagnum* bogs is usually in the range of 3.3–4.5 (0.5–0.03 meq H^+ L^{-1}). The sources of the H^+ ion activity are several (Clymo, 1964, 1965). The metabolism of proteins and the reduction of $SO_4^=$ by sulfur-metabolizing bacteria may contribute some H^+ ions to these waters, but their contributions are likely to be small. Although live *Sphagnum* plants secrete organic acids, concentrations of these acids are usually insufficient to account for the observed acidity. Most of the H^+ ion concentrations appear to result from the active *cation exchange* by the cell walls of *Sphagnum*, during which H^+ is released.

Another cause of acidified lakes and streams is *acid rain* (Likens et al., 2021), or more correctly referred to as *acid precipitation* as it includes snow, sleet, and other forms of precipitation. The emissions of S and N from the burning of fossil fuel and high-temperature combustion produce sulfuric and nitric acids, which the precipitation returns to Earth. We think of pure water that is neither acidic nor basic as having a pH of 7. Rain water is naturally acidic because it is in equilibrium with atmospheric CO_2. The pH of rain in a pristine atmosphere at current CO_2 levels is about 5.6 but can be below 4 in polluted areas, so quite acid. In poorly buffered areas that receive this rain the lakes and streams reflect that acidity. For example, in the Adirondack Region of New York State (USA) (Asbury et al., 1990) and in areas of Quebec (Canada) and Sweden, many lakes and streams have had pH values well below 5 (Baker et al., 1991). Much controversy has arisen over the natural versus anthropogenic components of the acidification of natural waters. In remote areas natural processes can dominate. These include complex humic and other organic acids, for example, from bog drainage (Schindler et al., 1986), as well as H^+ ions generated from the nitrification of NH_4^+, manganese oxidation, ferrous iron oxidation, and sulfide oxidation (Table 13-2). However, the overwhelming loading of acidity to sensitive surface softwaters has resulted from anthropogenic pollution from fossil fuel combustion and the production of H^+ ionized from several strong acids (see, e.g., Galloway et al., 1984; Oppenheimer et al., 1985; Stumm and Morgan, 1995). With stronger regulation on the use of high-sulfur coal, as well as other acidic emissions, pH values of lakes and rivers are recovering slowly from peak acidity (Lawrence et al., 2008; Waller et al., 2012). Another major source of anthropogenic acidity is *acid mine drainage*, which is caused by the oxidation of reduced iron and sulfur compounds that are brought up to the surface during the mining of coal and other minerals. When these oxidize, they can cause extremely low pH in the waters that drain them, and if this water finds its way into lakes and streams, it can result in strong acidification (Blodau, 2006).

The highest pH values in lakes are usually found in *endorheic regions*, where lake water contains exceedingly high concentrations of soda (e.g., Na_2CO_3), and are therefore often referred to as *soda lakes*. For example, highly saline soda lakes include Mono Lake in California (USA), which has a pH of about 9.5 (Carini and Joye, 2008), and lakes Nakuru and Bogoria in Kenya, which have pH values above 10.

The range of pH of most open (drainage) lakes is between 6 and 9. Most of these lakes are the "bicarbonate type"; that is, they contain varying amounts of carbonate and are regulated by the proportions of carbon dioxide, bicarbonate and carbonate in the water. Calcareous hardwater lakes commonly are buffered strongly at pH values greater than 8. *Seepage lakes* and lakes within igneous-rock catchment areas are less well buffered and more acidic, with pH values usually somewhat less than 7.

5. CO_2 dynamics

Carbon dioxide gas is a master variable that connects the inorganic and organic C cycles through productivity and respiration. Further, dissolved CO_2 gas is the major regulator of pH in most inland waters. Finally, pCO_2 connects surface waters and controls exchange with flux from the atmosphere.

Rewriting Eq. 13-7 reveals an important truth: it is the ratio of CO_2 to HCO_3^- that regulates the pH of most natural waters for pH values between about 5 and 9.

$$K_{1}* \left[CO_{2aq}\right]/\left[HCO_3^-\right] = \left[H^+\right] \qquad (13\text{-}15)$$

Fig. 13-1 gives some the impression that pH controls the proportions of the species of inorganic C. This is not incorrect in that changing $[H^+]$ will increase $[CO_{2aq}]$ at the expense of $[HCO_3^-]$. However, in most natural waters it is, in fact, the change in $[CO_{2aq}]$ at relatively constant $[HCO_3^-]$ or constant ANC that actually controls the pH. Increasing $[CO_{2aq}]$ lowers the pH; decreasing $[CO_{2aq}]$ increases the pH and causes CO_3^{2-} to form and/or precipitate.

Since $[CO_{2aq}]$ is a gas, one might expect that in surface waters that are in contact with the overlying air, pCO_2 in the air and water would simply be equal. This is not the case. Kling et al., (1991) showed that lakes and rivers in the Arctic were supersaturated with CO_2 and a significant regional source of CO_2 to the atmosphere. The first extensive global survey of the partial pressure of CO_2 in the surface waters from a large number (1835) of lakes showed that only a small proportion were in or near atmospheric equilibrium (Fig. 13-2a; Cole et al., 1994). Instead, the average partial pressure of dissolved CO_2 was about three times the value in the overlying atmosphere. Sobek et al., (2005) extended this analysis to more than 4900 lakes and Raymond et al., (2013) to more than 7900 lakes and reservoirs and came to very similar conclusions. For rivers, the case is even more extreme (Fig. 13-2b). Most rivers and streams are quite oversaturated in CO_2 (Cole and Caraco, 2001; Mann et al., 2014; Duarte and Prairie, 2005; Raymond et al., 2013). Raymond et al., (2013) assessed a global database of streams and rivers, which included 6708 sampling locations. They found the average pCO_2 value was about 3000 μatm, which is about 7.5-fold greater than equilibrium with the atmosphere. It is not uncommon for many rivers to be supersaturated during the entire year (Fig. 13-3a,b).

If pCO_2 water > pCO_2 air, the water body is a net source of CO_2 to the atmosphere. Thus, on the global scale, lakes, rivers, and streams are on average net sources of CO_2 to the atmosphere and these sources are significant to the C balance of the continents (Cole et al., 2007; Duarte and Prairie, 2005; Raymond et al., 2013). However, pCO_2 is quite dynamic in time in some systems and in many systems it is near equilibrium or consistently undersaturated with respect to the atmosphere, especially in places where primary production is high (Tank et al., 2008). It is quite common for a lake to be near equilibrium, or slightly undersaturated, during the summer and then oversaturated for the rest of the year. On shorter time scales, pCO_2 in surface waters tends to go up at night and down during the day in response to changes in the balance between primary production and respiration (Fig. 13-4a,b). Over depth

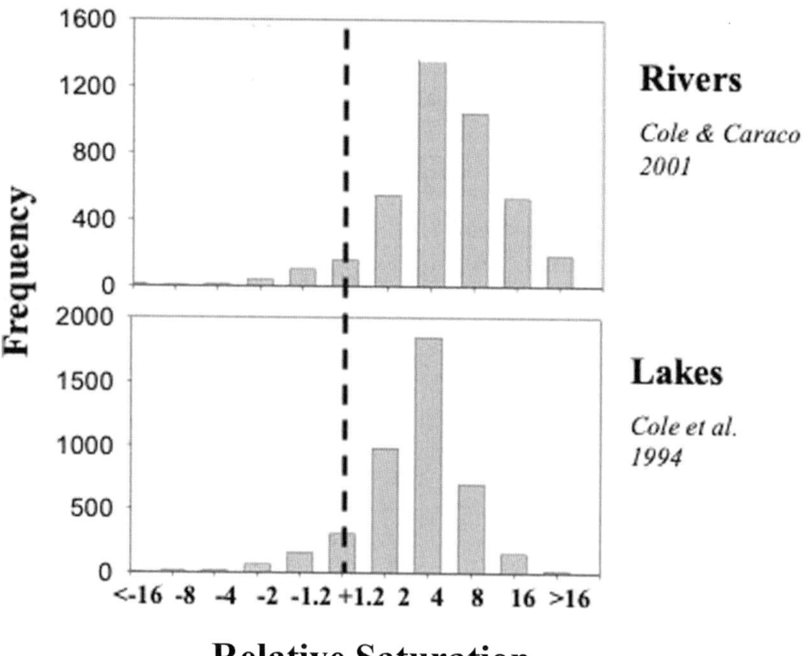

FIGURE 13-2 **Dissolved CO_2 in the surface waters of lakes (*upper*) and rivers (*lower*) with a global distribution.** Relative saturation is the ratio of actual CO_2 to the CO_2 the water would have if it were in equilibrium with the atmosphere. Lakes with surface waters near atmospheric equilibrium would have relative saturation values of 1.0 on this scale and be on the *dotted line*. Positive values denote supersaturation; negative values denote undersaturation. A value of 4 means the water body has four times as much dissolved CO_2 as could be explained by atmospheric equilibrium; −4 means the water body has four times less (or one-fourth as much dissolved CO_2 as could be explained by atmospheric equilibrium). *(Lake data are for 1835 lakes from Cole et al., 1994; river data are for 80 of the world's major rivers from Cole and Caraco, 2001, both with a global distribution. See also Duarte and Prairie, 2005.)*

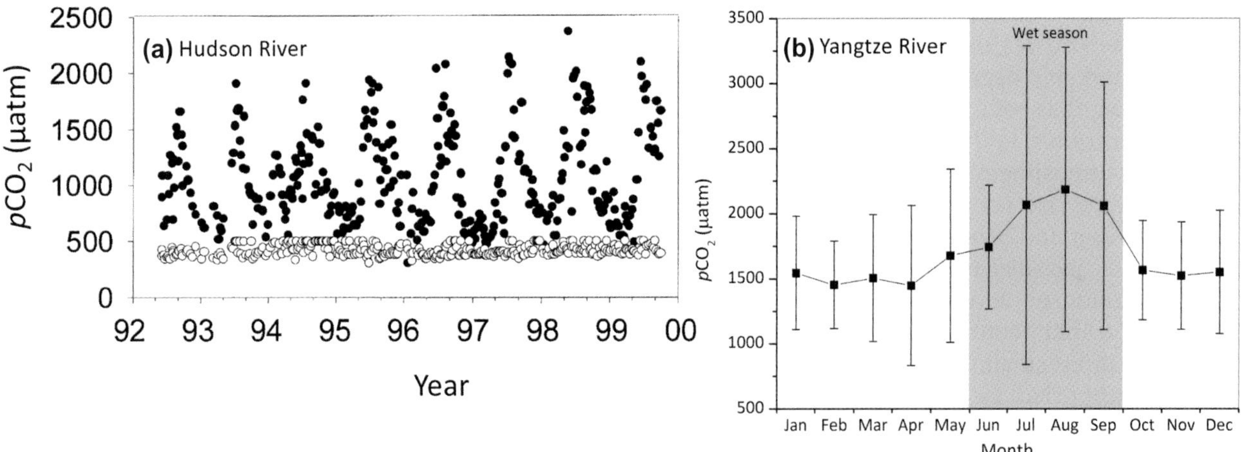

FIGURE 13-3 Time series of pCO_2 in the surface waters of two rivers: the Hudson River near Rhinebeck, New York, USA (Cole and Caraco, 2001) and the Yangtze in China (Ran et al., 2017). For the Hudson River (panel a), we show the partial pressures of CO_2 both in the water (*filled circles*) and in the overlying atmosphere. For the Yangtze River (panel b), the wet season is marked with *green shading*.

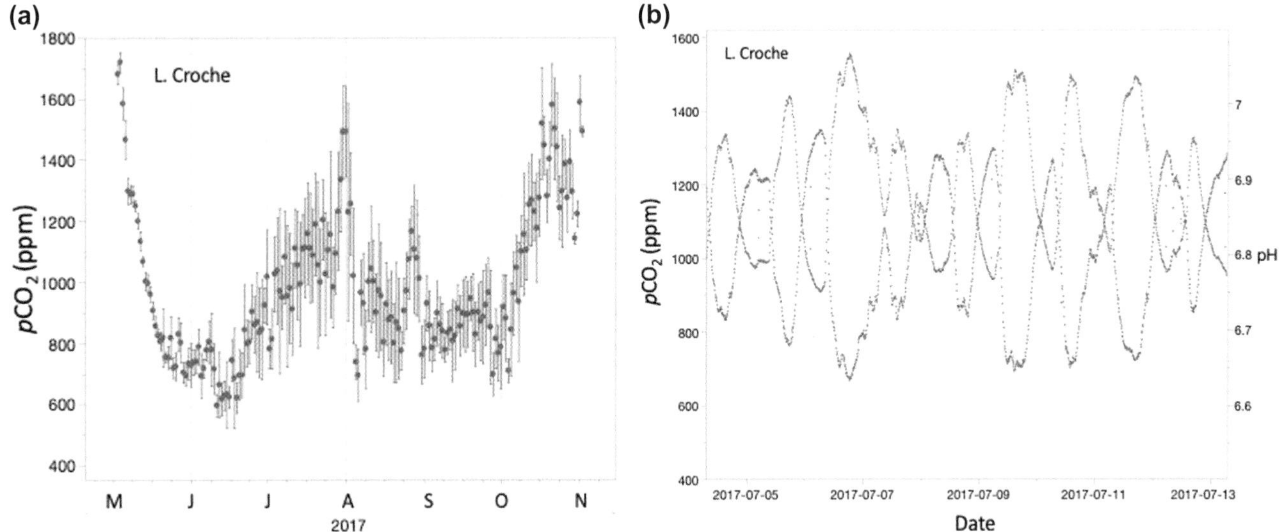

FIGURE 13-4 **Continuous measurements of pCO_2 in the surface water of Lac Croche, Quebec, Canada.** (a) Data for a full ice-free season. *Points* show the mean for the day; the *bars* show the range on that day. (b) Detail showing several days of pCO_2 in the same lake along with pH. Note how pH (*red dots*) mirrors the changes in pCO_2 (*blue dots*). *(Data are from del Giorgio and Prairie, 2017.)*

in a stratified lake, as light becomes too low for photosynthesis, pCO_2 increases as O_2 decreases and accumulates over the course of the stratified period (discussed later).

Of the diverse processes that can lead to CO_2 supersaturation, two are the most quantitatively important in the majority of inland waters: (1) importation of CO_2 in inflowing surface and groundwater; and (2) respiration in excess of gross primary production (often called "*net heterotrophy*"; see del Giorgio et al., 1997). In systems that are supersaturated in CO_2 and have very long residence times, or high metabolism compared to residence time, such as Lake Superior, Lake Biwa, the lower Amazon, or the Hudson River, the excess CO_2 is largely the result of net heterotrophy (Cole and Caraco, 2001; Richey et al., 2002; Roehm et al., 2009; Urban et al., 2004; Urabe et al., 2005). In systems with short residence times, such as headwater streams, the excess CO_2 comes from groundwater (Butman et al., 2016; Denfeld et al., 2013).

In recent years a number of attempts have been made to quantify the sources of CO_2 efflux in small lakes. There is a remarkable degree of consensus between the studies showing that both net heterotrophy and hydrologic input contribute to the efflux of CO_2 from inland waters to the atmosphere. During the ice-free season, many small lakes are net heterotrophic (Prairie et al., 2002; Sand-Jensen and Staehr, 2009). However, this net heterotrophy is insufficient to account for all of the CO_2 efflux. Wilkinson and collaborators (Wilkinson et al., 2016) found that about half the summertime efflux of CO_2 from three small, seepage lakes in Wisconsin (USA) was supported by net heterotrophy, with the remaining 50% from groundwater input. Vachon et al., (2017) showed that for three lakes in Quebec (Canada), the contribution to lake CO_2 emissions varied with season. In the spring, after ice-out, CO_2 from hypolimnetic build-up (e.g., net heterotrophy) was the major contributor (50%–100%). In summer 34%–80% of the flux was the result of respiration in excess of CO_2 in the benthos and pelagic parts of the lakes. In fall hydrologic inputs accounted for 39%–55% of the flux.

Measurement and calculation of pCO_2. Like DIC, pCO_2 can be measured directly or calculated from temperature, pH, and DIC or ANC. The calculation is highly sensitive to the pH. In dilute waters with circumneutral pH a great deal of care is required to obtain an accurate pH value (Stauffer, 1990). In fact, one might do better by measuring both DIC and pCO_2 and calculating the pH (Herczeg et al., 1985). Nevertheless, if an accurate pH value can be measured, calculating pCO_2 works well. Discrete measurement options include *headspace equilibration* in a sealed container and then measurement of CO_2 via gas chromatography (GC) or infrared gas analyzer (IRGA). The idea in a headspace equilibration is to add a small volume of gas to a sealed container and force, by shaking, the dissolved gas to equilibrate with that headspace (Chapter 28). If the water volume is extremely large compared to the headspace gas volume, the gas concentration in the headspace is a direct measure of the partial pressure. Of course, care is required to maintain environmental temperature (or correct for changes). Further, the larger the water volume is compared to the headspace volume, the closer this approach comes to a direct measurement. Calculations are always required to correct for the displaced inorganic carbon equilibrium during the headspace equilibration process (Koschorreck et al., 2021). Over the past few decades, it has become possible to measure pCO_2 directly and continuously using *in situ* equilibrators and an IRGA mounted on a raft (Sellers et al., 1995) or an infrared CO_2 probe (Vachon and del Giorgio, 2014). Continuous records of pCO_2 are shown for a full ice-free season in Lac Croche in Fig. 13-4a. In this lake pCO_2 is consistently elevated above the value of atmospheric equilibrium and the lake is a net source of CO_2 to the atmosphere. A several-day period on the same lake is shown in Fig. 13-4b. Note how CO_2 goes up at night and down during daylight due to the change in the balance between primary production and respiration. Similarly, pCO_2 increases over depth, as does DIC (Fig. 13-5).

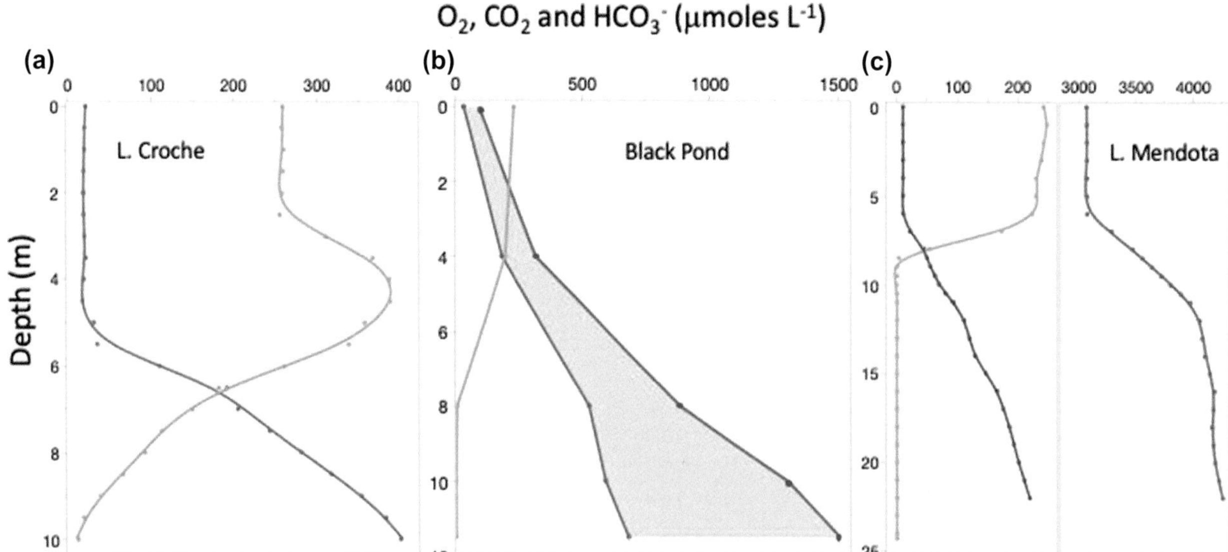

FIGURE 13-5 Depth profiles from three stratified lakes: (a) Lac Croche is a softwater oligotrophic lake (Mercier-Blais and Prairie, 2014); (b) Black Pond (New Hampshire, USA) is a softwater oligomictic (possibly meromictic) lake (Caraco et al., 1989, 1990, 1992); in this plot concentrations of CO_2 and HCO_3^- are stacked; and (c) Lake Mendota (Wisconsin, USA), a hardwater, eutrophic lake (data from the NTL-LTER website courtesy of E. Stanley). All three lakes show dissolved CO_2 (*blue dots*) and dissolved O_2 (*red dots*). We also show the concentration of HCO_3^- (*orange dots*) for Black Pond (*shaded area*) and Lake Mendota (*separate subpanels*). Note the scale break for Lake Mendota HCO_3^- concentrations.

D. CO_2 exchange between the atmosphere and water

When the pCO_{2water} does not equal $pCO_{2atmosphere}$, there will be a net flux of CO_2 from air to water or vice versa (see Chapter 10). We can estimate the magnitude of this flux and then use that to obtain important insights as to how C is cycled in inland waters. To do this, we need to consider the physics of gas diffusion at the air–water interface.

A *diffusive flux* is proportional to the concentration (in this case the *partial pressure*) gradient existing at the interface. It can be described by a simple equation:

$$\text{Flux} = k \times K_H \left(pCO_{2(water)} - pCO_{2(atmosphere)}\right) \quad (13\text{-}16)$$

K_H is the temperature-dependent *Henry's constant* in Eq. 13-2. $K_H \times pCO_{2(water)}$ gives the concentration of CO_2 in water in μmol L^{-1}. Similarly, $K_H \times pCO_{2(atmosphere)}$ gives the concentration the water would have had it been in equilibrium with the atmosphere. This equation can be rewritten as:

$$\text{Flux} = k \times \left(\left[CO_{2(water)}\right] - \left[CO_{2(equilibrium)}\right]\right) \quad (13\text{-}17)$$

where k is the gas exchange velocity, often termed *piston velocity*. In this equation $[CO_{2(equilibrium)}]$ means the concentration the water would have had it been in equilibrium with the atmosphere. The piston velocity is a way to represent the physical rate of gas exchange with the atmosphere. The variable k can be thought of as the height of water that exchanges gas per unit time with the atmosphere, as if a piston pushed the excess CO_2 (e.g., in the case of oversaturation) through that height of water. The value of k is the result of turbulent energy exchange with the atmosphere and increases with factors such as wind and current speed. Embedded in k is a *diffusion constant* for a given gas. Thus k is a specific value for a given gas (e.g., CO_2, O_2, etc.) at a given value of k for one gas and temperature, it can be calculated for any other gas and temperature by using the *Schmidt number* (*Sc*) equation from Jahne et al., (1987):

$$k_{gas1}/k_{gas2} = \left(Sc_{gas1}/Sc_{gas2}\right)^n \quad (13\text{-}18)$$

where n can vary from unity to −0.67; −0.5 is a commonly assumed value for n (Cole and Caraco, 1998; Holtgrieve et al., 2010).

Wanninkhof, (1992) provides a useful table that gives equations to calculate the *Sc* number for several gasses at any temperature. For CO_2, the equation is:

$$1911.1 - 118.11 * T + 3.4527 * T^2 - 0.04132 * T^3 \text{ (where } T \text{ is temperature in °C)} \quad (13\text{-}19)$$

The *Sc* number for CO_2 at 20°C is 600. When k is measured, it is usually expressed as k_{600}, meaning the k value for a gas and temperature combination that equals 600 in Schmidt number. While not straightforward to do, k can be measured directly. The approaches include adding an inert gas tracer such as SF_6; using floating chambers (but correcting for chamber effects on turbulence); modeling short-term changes in the temperature profile; and employing more complex statistical methods using either multiple isotopes of a gas (Holtgrieve et al., 2010) or multiple different gases (Pennington et al., 2018). Where they have been compared, these approaches agree reasonably well (Cole et al., 2010; Vachon et al., 2010; Read et al., 2012). From the measurement of k along with other factors, we know that k is greatly affected by factors affecting the turbulent regime of surface waters, such as wind, current speeds, and convection. While several other factors are known to influence gas exchange velocities, the most widely used models predict k from wind speed in lakes, from wind and current in large rivers, and current and water depth and slope in shallow rivers. In a lake with low winds k would be near 0.4 m day^{-1}; about 1 to 3 m day^{-1} in a windy lake or a turbulent but deep river; higher (>10 m day^{-1}) in shallow streams running over rocks; and extremely high in alpine streams (>>100 m day^{-1}, Ulseth et al., 2019). In small streams researchers often model k as a linear distance or reaeration length or time (e.g., gas exchanges per day) rather than as a piston velocity (Genereux and Hemond, 1992). There are a number of useful empirical relationships by which one can predict k from wind, current speed, and other easy-to-measure features (e.g., lake size) of a given system (Wanninkhof, 1992; Clark et al., 1995; Cole and Caraco, 1998; Read et al., 2012; Vachon and Prairie, 2013). If we express k in m day^{-1} and realize that CO_2 in μmol L^{-1} is the same as mmol m^{-3}, flux, in Eq. 13-17, has the useful units of mmol CO_2 m^{-2} day^{-1}. When $CO_{2(water)} > CO_{2(atmosphere)}$, the flux is from the water to the air and the water body is a source of CO_2 to the atmosphere. The opposite is true when $CO_{2(water)} < CO_{2(atmosphere)}$.

Chemically Enhanced Diffusion

At very low CO_2 concentrations and concomitantly very high pH, CO_2 exchange with the atmosphere is controlled in part by factors other than regular diffusion. What we know is that under these conditions, CO_2 enters the water much faster (7–21 times faster) than can be explained by the diffusion equations above. This is called *chemically enhanced diffusion* (Emerson 1975; Wanninkhof and Knox, 1996). The kinetic and isotopic evidence suggests that CO_2 from the air reacts directly with the hydroxyl ion in the surface water more quickly than it can diffuse across the air–water interface, forming bicarbonate.

$$[CO_2] + [OH^-] = [HCO_3^-] \quad (13\text{-}20)$$

We express this chemical enhancement as β, the *enhancement factor*, which is simply the ratio of measured k under enhanced conditions (k_{enh}) to what k would be without the enhancement. Bade and Cole, (2006) found reasonable agreement between measurements of CO_2 flux using floating chambers and models of k_{enh} using Wanninkhof and Knox, (1996) and (Hoover and Berkshire, 1969) to predict β. In the studied low-wind lake β ranged from three-fold to six-fold, increased with increasing pH, and had strong effects on the isotopic fractionation of the CO_2 that was exchanged. Using a time-series modeling approach on the concentration of ^{13}C-DIC in the surface water, the use of β caused closer agreement with measured ^{13}C-DIC than without chemical enhancement (Bade and Cole, 2006).

As discussed later in this chapter, CO_2 from the respiratory processes of decomposition can also accumulate to large quantities in the hypolimnia of lakes.

II. Spatial and temporal distribution of total inorganic carbon and pH in rivers and lakes

Rivers and streams

In many streams and rivers decomposition dominates over in-channel photosynthetic production, leading to the production of CO_2 being higher than its consumption by photosynthesis. In addition, inflowing water from runoff of soils of both surface and groundwater sources would typically have large concentrations of CO_2 from bacterial and plant root respiration.

Carbon dioxide in high concentrations above equilibrium values is relatively rapidly lost to the atmosphere, particularly in the water turbulence of streams. Respiration within the stream and river water is high, however, and CO_2 production can exceed evasive losses to the atmosphere and photosynthetic utilization. For example, several Danish streams contained about eight times more free CO_2 ($pCO_2 = 10^{-2.6}$ atm) than water in equilibrium with air ($pCO_2 = 10^{-3.5}$ atm) (Rebsdorf et al., 1991). As a result, variations in the CO_2 concentrations are common; higher concentrations and reduced pH can be found in quiescent regions or times of reduced flows, particularly near wetland areas of high loadings of dissolved and particulate organic matter. Changes in alkalinity and pH can also be induced by rapid changes in discharge. For example, alkalinity depressions in stream water can result from dilution by snowmelt water (Molot et al., 1989). Although discharge was positively correlated with alkalinity in circumneutral streams, in acidic streams with little buffering capacity the snowmelt can induce marked short-term reductions in pH and create conditions that can be lethal to certain biota.

In hardwater bicarbonate rivers, such as the Rhine and the Rhone rivers, total CO_2, largely as bicarbonate, increased downstream from the source, mostly as a result of increased organic loadings and decomposition (Golterman and Meyer, 1985). Seasonality was evident in both the bicarbonate concentrations and pH, largely as a result of temperature on the solubility of CO_2. Upon entering reservoirs, many of the features discussed later for lakes apply. However, the complexities of reservoir morphometry, inflow volumes, and density-mediated distributions, as well as retention times, make generalizations difficult (cf. Thornton et al., 1990).

A. DIC in stratified lakes

DIC in stratified lakes shows a number of interesting features. During periods of mixing, DIC is distributed nearly uniformly with depth in typical dimictic and monomictic lakes and in shallow lakes of insufficient depth for thermal stratification. The DIC content of the water is dependent upon the equilibria established between atmospheric CO_2, the bicarbonate–carbonate system, external loadings, contributions from metabolic respiration, and utilization in photosynthesis. Metabolism is markedly influenced by numerous parameters; in the spring, increasing light and temperature exert major controls on the photosynthetic uptake of CO_2, and generation rates of CO_2 from microbial decomposition of organic matter are temperature and oxygen dependent.

1. Vertical distribution of DIC

During the period of thermal stratification, several conspicuous changes occur in the vertical distribution of DIC, largely due to the changing balance between primary production, which requires light, and respiration, which does not. Below the upper mixed layer, DIC usually (largely in the form of CO_2) increases and mirrors the decrease in dissolved oxygen (Fig. 13-5a,b,c). In deep oligotrophic lakes these increases in DIC (and declines in O_2) are usually not very large and are limited by the low supply of organic matter from the trophogenic zone. In shallow or more eutrophic lakes, where this supply of organic C is larger, or if the mixing is less frequent, O_2 can be completely consumed with a concomitant increase in dissolved CO_2 and DIC

(Fig. 13-5a). In systems where there is a sufficient supply of alternative electron acceptors, such as nitrate, sulfate, or iron, anaerobic respiration becomes dominant in the hypolimnion and the DIC increase (as CO_2, but also HCO_3^-) can be very large indeed (Fig. 13-5b,c) (Caraco et al., 1989). For example, in Sider's Pond, a coastal pond near Wood's Hole, Massachusetts, USA, DIC increased from about 841 μmol L^{-1} at the surface to 12,236 μmol L^{-1} at 13 m depth (Caraco et al., 1990). This system, which is probably meromictic and usually anoxic below about 5 m, has very high sulfate concentrations (3500 μmol L^{-1}) in surface water because it is brackish. That amount of sulfate could generate 14,000 μmol L^{-1} DIC if all the sulfate were used in sulfate reduction. So the observed 11,393 μmol L^{-1} increase in DIC can potentially be accounted for by sulfate reduction alone in this system.

Even in lakes with relatively low sulfate, DIC can build up to high levels through multiple processes. Black Pond in the White Mountains of New Hampshire, USA, was also in the Caraco et al., (1990) data set. This lake has about 50 μmol L^{-1} sulfate in surface waters. The lake is mesotrophic, well stratified, and anoxic in the hypolimnion. Black Pond is oligomictic and builds up about 1400 μmol L^{-1} DIC above the surface values (Fig. 13-5b). In Black Pond both CO_2 and DIC concentrations are much higher than can be accounted for by aerobic decomposition alone. Also, much of the DIC accumulation is as HCO_3^- (Fig. 13-5b), unlike the case in Lac Croche, where it was all in the form of CO_2 (Fig. 13-6). In the bottom waters of Black Pond, then, one or more of the alkalinity-generating oxidation pathways must dominate (Table 13-2). Some of the CO_2 in Black Pond came from *acetoclastic methanogenesis* as bottom water methane (CH_4) reached over 600 μmol L^{-1}. Sulfate reduction could add, at most, 80 μmol L^{-1} DIC (as ANC) and Fe reduction perhaps another 100 μM (as ANC). In every case in which water is oxic, we can completely account for the increase in CO_2 from oxic decomposition. However, in the anoxic waters we are missing major processes that generated more than half of the observed increase in HCO_3^-. That is, more inorganic C is generated than we can account for by the measured electron acceptors in the water column. The

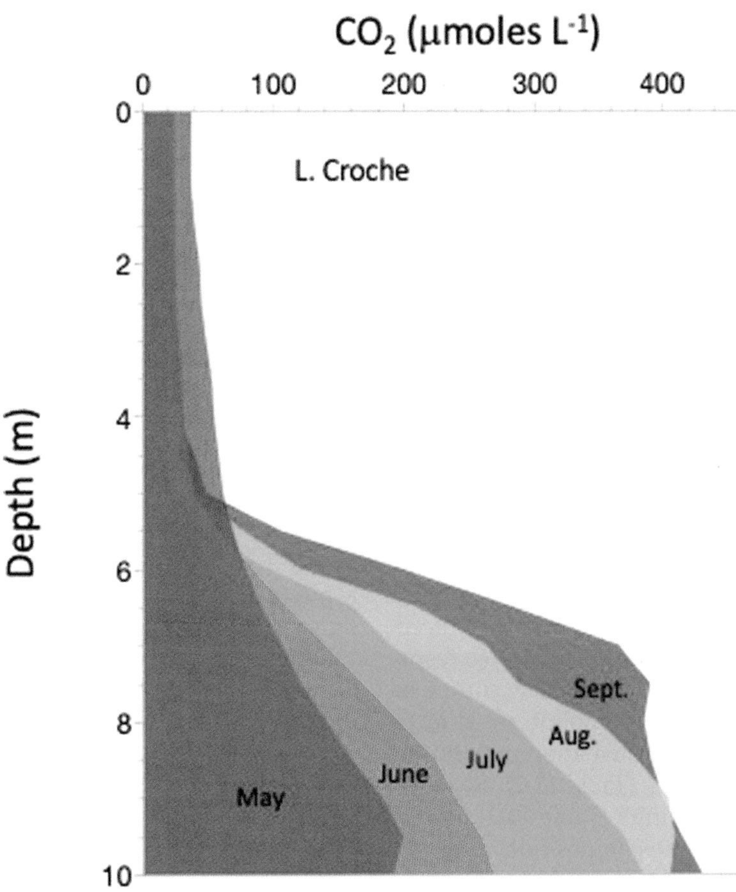

FIGURE 13-6 Depth profile of pCO_2 in Lac Croche (Mercier-Blais and Prairie, 2014). The time series of the increase of hypolimnetic CO_2 can be used to calculate hypolimnetic CO_2 accumulation. For lakes with significant amounts of anaerobic respiration, one would need to use DIC instead of CO_2 because of the generation of ANC during those processes.

remainder might come from manganese reduction or some ionic exchange with the sediments.

Lake Mendota (USA) is a hardwater eutrophic lake that is typically dimictic. In this lake we again see that the increase in CO_2 is balanced by the decrease in O_2 in the oxic zone, and below this, both DIC and ANC increase (Fig. 13-5c). While sulfate is moderately high (about 180 μmol L^{-1}), it is insufficient to explain the large increase in DIC in the anoxic water even if all the sulfate were reduced. We need to assume again that other unmeasured electron acceptors were involved. In studies that have tried to explain all of the electron acceptors in anaerobic waters, it is often the case that we come up with an *electron acceptor deficit*. That is, there is more DIC accumulation than we can account for by the measured electron acceptors (Mattson and Likens, 1993; Matthews et al., 2008).

The vertical distributions of total DIC and pH are strongly influenced by various biologically mediated redox reactions. Most conspicuous is the photosynthetic utilization of DIC in the trophogenic zone, which tends to reduce CO_2 content and to increase pH, and the respiratory generation of CO_2 throughout the water column and sediments, which tends to decrease pH. In addition to the heterotrophic degradation of organic matter, the generation of CO_2 and reduction of pH are augmented by microbial methane fermentation, nitrification of ammonia, and sulfide oxidation (Table 13-2). Further, denitrification of nitrate to molecular nitrogen, reduction of sulfate to sulfide, and iron and manganese reduction can result in a net increase in pH and alkalinity (e.g., Kling et al., 1991; Dillon and Molot, 1997) (Table 13-2). The combination of decompositional processes results in an increase in the DIC of the hypolimnetic waters and a decrease in pH compared to that in surface waters.

As the intensity of decomposition increases in the tropholytic zone, the amount of CO_2 and especially of HCO_3^- increases markedly. The accumulation of the sum of CO_2 plus HCO_3^- exceeds the supply of oxygen, and decomposition shifts from aerobic to anaerobic as the hypolimnion becomes anoxic (Fig. 13-5b). At equilibrium during spring mixis, cold water has about 300 to 400 μM O_2 depending on temperature and altitude. Thus only about 300 to 400 μM DIC can be produced from the complete utilization of this O_2 in aerobic respiration in the hypolimnion, which has no new source of O_2. Increases in DIC beyond this amount are due to either anaerobic respiration or fermentation in both the water column and the sediments (Table 13-2). Some of these processes also affect alkalinity (Table 13-2). If aerobic respiration is the key decomposition process, measuring CO_2 over depth will give an accurate picture of the balance between photosynthesis and respiration in the hypolimnion. However, once the system is anoxic, total DIC is needed because of the ANC generated during many of the key anaerobic respiratory processes (Table 13-2; Fig. 13-5).

In extremely hardwater lakes that are also eutrophic, there are additional processes that increase HCO_3^-. For example, part of the $CaCO_3$ sedimenting from the epilimnion (e.g., whiting events) undergoes dissolution in the colder, more acidic hypolimnion, resulting in increases in HCO_3^- concentrations that are larger than the respiratory increase in CO_2. These same events can also strip the euphotic zone of some of its DIC, causing declines in the DIC curve, and have been called *epilimnetic decalcification* (Minder, 1923; Pia, 1933). Further, ion exchange with the sediments can also influence the alkalinity.

The DIC profile in a lake can show some bumps or wiggles rather than a simple monotonic increase over depth. Where photosynthetic activity in a layer is particularly intense, as might occur in the metalimnion, you might observe a local O_2 maximum (Chapter 11) associated with a DIC minimum. On the other hand, many lakes have a local O_2 maximum that is caused largely by physical (e.g., heating and cooling or mixing) rather than biological processes (Wilkinson et al., 2015), so no CO_2 minimum would be expected. The opposite situation, that is, a DIC maximum, or a negative heterograde oxygen curve, with concomitant DIC and pH curves, is occasionally found in strata of intensive respiration.

In regions where the soils and rock deposits contain high carbonates (such as limestone bedrock), acidified precipitation can increase rates of carbonate weathering and result in increased alkalinity and equilibrium pH in softwater seepage lakes (Kilham, 1982). In drainage basins and waters with little buffering capacity, however, acid deposition has resulted in the loss of acid-neutralizing capacity (alkalinity), decreased pH, increases in sulfate and in dissolved aluminum concentrations, and marked alterations in ionic speciation and ratios. These acidification changes can persist on a long-term basis or, in less affected areas, chemical changes occur reversibly for short-term periods. In many softwater streams and lakes, for example, melting of acidic snow accumulations can result in rapid (a few days), precipitous declines in pH and alkalinity (Catalan and Camarero, 1992), referred to as "*spring shock*." Lower pH values occurring during late winter and early spring in softwater seepage lakes, on the other hand, were attributed to increases in pCO_2 under the ice because total alkalinity did not change at that time (Kratz et al., 1987).

B. Extremely high CO_2 and exploding lakes

Under certain conditions of lake basin morphometry, underlying volcanism, and lack of frequent deep mixing, as in certain deep African rift lakes, CO_2 and CH_4 can accumulate to extraordinarily high concentrations. Decompression of CO_2-saturated water can, with changes in seasonal stabilities of stratification in the upper waters, erupt to the surface at explosive rates (range, 50–90 m s^{-1}) (Zhang, 1996). So much CO_2 is released by this exsolution and evasion to the atmosphere that the dense CO_2 over the lake surface can temporarily cascade over the surrounding terrain. Such massive evasions occurred in Lake Nyos and Lake Monoun in Cameroon, and over 1700 humans and many terrestrial animals died due to asphyxiation (Kling et al., 1987, Kling et al., 1991). There were no measurements of hypolimnetic gas concentrations in these lakes prior to these eruptions. Following the eruptions, gas concentrations built up quickly so that by 2004, the deepwater CO_2 concentration was greater than 150,000 μmol L^{-1} in Lake Monoun and more than 350,000 μmol L^{-1} in Lake Nyos (Kling et al., 2005). These concentrations are the equivalent of more than 6 atm in Monoun and more than 12 atm of CO_2 in Nyos and were probably even higher prior to the eruptions. For scale, the atmosphere has a partial pressure of CO_2 of about 400 μatm or 0.04 atm. So the partial pressure of CO_2 at depth in these lakes exceeded the total gas pressure on land by more than 150- to 300-fold. Based on the ^{14}C age of the CO_2, most of it has to be of volcanic origin (Kling et al., 2005, 2015). Limnologists and engineers came up with a practical and relatively inexpensive method to relieve lakes of extreme hypolimnetic CO_2 gas. They put open pipes from the surface of the lake into the hypolimnion. The gas pressure in the hypolimnion caused these pipes to act as gas fountains and thereby prevent further large build-up of CO_2 (Kling et al., 1994, 2015; Fig. 13.7).

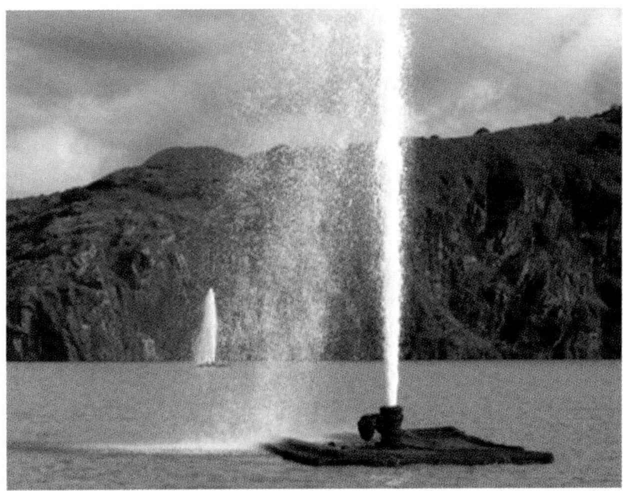

FIGURE 13-7 **Gas release vents in Lake Nyos, Cameroon.** Lake Nyos built up enormous concentrations of CO_2 gas in its deep hypolimnion. Most of this gas is from the tectonic activity below the lake's sediments. After the catastrophic outgassing of CO_2 from this lake in 1986, limnologists and engineers worked toward a simple and inexpensive solution. The gas fountains in the picture are the tops of long pipes that extend into the hypolimnion. The huge pCO_2 in the hypolimnion drives the release of gas and water from the pipes (Kling et al., 2015). *(Photo courtesy of G.W. Kling.)*

III. Hypolimnetic CO_2 accumulation in relation to lake metabolism

While hypolimnetic metabolism is often estimated by the rate of oxygen depletion (see Chapter 11; Muller et al., 2012), it can also be derived from the accumulation of CO_2 or DIC (Fig. 13-6). Being based on carbon, hypolimnetic DIC accumulation is a more fundamental metric of organic matter metabolism than the removal of the main respiratory terminal electron acceptor, namely, oxygen. In principle, the two approaches should yield similar estimates in systems where the degradation of organic material is largely aerobic and the accumulated end-product will stay in the form of CO_2. Although often treated as a single ecosystem attribute, *hypolimnetic metabolism* is really the sum of the two decomposition processes, one occurring in the water column and the other in the sediments. Both processes are positively linked to the overall productivity of the system (Pace and Prairie, 2005), and the relative contribution of the water column and benthic processes depends on the main sources of organic matter fueling the catalytic metabolism, as well as on the bathymetric shape of lakes (Livingstone and Imboden, 1996). For example, while autochthonously produced organic matter (algal and macrophyte photosynthesis) sustains most of the hypolimnetic respiration in eutrophic conditions, dissolved organic matter of terrestrial origin can also be an abundant and important substrate (Houser et al., 2003) contributing to water column respiration. In deep, stratified lakes, most of the respiration occurs in the water column. In shallower systems with thinner hypolimnia, or in systems that were formerly eutrophic, a significant fraction of the hypolimnetic metabolism is benthic (Matzinger et al., 2010). In the benthic environment the transport of dissolved substances within the sediment matrix is largely constrained by molecular diffusion with the consequence that conditions quickly become anoxic, usually within a few millimeters of the sediment–water interface (Sobek et al., 2009), even if the overlying water is fully oxygenated. Thus, depending on the rate at which new particulate organic matter settles at the sediment surface, sedimentary organic carbon can be exposed to oxygen for as little as a few days or weeks, after which decomposition proceeds largely anaerobically, that is, without immediately consuming O_2. Therein lies the significant advantage of using the rate of DIC accumulation over oxygen depletion as a measure of hypolimnetic metabolism: it encompasses the end-products of several respiratory processes, both aerobic and anaerobic (see Table 13-2). Nevertheless, if the reduced substances that are released to the water column are later exposed to oxygen within the hypolimnetic volume, they can be oxidized, ultimately consuming hypolimnetic oxygen.

First introduced by (Ruttner, 1926) and further developed by (Ohle, 1952, 1956), the rate of DIC accumulation was originally conceived as an indirect way of measuring the intensity of whole-lake metabolism. Today, it is more appropriately viewed as a component of the lake ecosystem that is dominated by respiratory processes. Estimated using sequential profiles of DIC over the stratified period, together with the hypsography of the lake, DIC accumulation rates offer an integrative metric of whole hypolimnion respiration and how it varies through time. Even the exact shape of a DIC (or CO_2) profile is informative. The observation that the rate of increase in CO_2 concentration is faster at greater depths is a reflection of how the sediment area per unit volume of a hypolimnetic layer increases with depth, thereby increasing the relative contribution of sediment respiration at that depth. Livingstone and Imboden, (1996) have used this property to introduce a simple method to estimate the relative contribution of the water column and sediment respiration processes to whole hypolimnetic metabolism. Applied to the CO_2 profiles illustrated in Fig. 13-6, the method yields an average sediment respiration rate (J_A) of 13 mmol C m^{-2} day^{-1} and a water column respiration rate (J_V) of 0.2 mmol m^{-3} day^{-1}. Note the different units for the area respiration and the volume respiration. Total hypolimnion respiration can then be estimated as J_A*$Area_{(hypo.sed.)}$ + J_V*$Vol_{(hypo.)}$.

In temperate regions the CO_2 accumulated in the hypolimnion during the stratified period will quickly mix to the surface during autumn turnover, creating a large emission pulse of CO_2 to the atmosphere. During the ice-covered period of the winter, a similar process operates with CO_2 accumulating under the ice and suddenly released after ice break-up. In many lakes these two short-lived emission pulses (a few weeks at most) account for a significant fraction of the total annual CO_2 flux to the atmosphere (Striegl et al., 2001; Ducharme-Riel et al., 2015) and are therefore important to consider when quantifying the annual carbon budget of a lake. In deep lakes these pulses can account for the majority of the annual CO_2 efflux.

IV. Utilization of carbon by photoautotrophs and chemolithotrophs

A. Photoautotrophy and chemolithotrophy

A truly *autotrophic organism*, in theory, gets all of its carbon from inorganic C and none of its energy from the oxidation of organic C. If the sole source of energy is sunlight, we call the organism a *photoautotroph*. If the sole source of energy is from the oxidation of chemical compounds, like hydrogen sulfide, we call the organism a *chemoautotroph* or *chemolithotroph* (Table 13-3). These are useful distinctions, but in the world of algae and bacteria there are many cases of organisms that are mixtures of these classifications. For example, both *Euglena* and *Utricularia* (and many other algae) are both photosynthetic and predaceous. That is, they are capable of eating small organisms such as bacteria (Bird and Kalff, 1986) and in the case of *Utricularia*, small zooplankton. Numerous species of phytoplankton are both *photosynthetic* and *osmotrophic* (Rengefors et al., 2008; Sanders, 2011). That is, these photosynthetic algae can utilize some dissolved organic matter.

We can divide photoautotrophs into two main categories: those that produce oxygen during photosynthesis (*oxygenic photosynthesis*) and those that do not

TABLE 13-3 Some of the Key Pathways of the Photoautotrophic and Chemoautotrophic Uptake of CO_2 in Inland Waters

Process	Equation	Kinds of organisms	Change in ANC
Oxygenic photosynthesis	$CO_2 + H_2O \rightarrow CH_2O + O_2$	Algae (most), macrophytes, emergent plants etc.	No change
Anoxigenic photosynthesis	$2CO_2 + H_2S + 2H_2O \rightarrow 2(CH_2O) + H_2SO_4$	Purple and green sulfur bacteria - *Chromatium, Chlorobium*	Decrease
Chemolithotrophy			
sulfide oxidizers	$CO_2 + H_2S + O_2 \rightarrow CH_2O + H_2SO_4$	*Thiobacillus*	Decrease
nitrification	$CO_2 + NH_4^+ + O_2 \rightarrow (CH_2O) + HNO_3$(shown unbalanced for simplicity)	*Nitrobacter*, Nitrosomonas in sequence. Many other genera	Decrease
hydrogenotrophic methanogenesis	$CO_2 + 4\,H_2 \rightarrow CH_4 + 2\,H_2O$	*Methanobacterium* (does both types of methanogenesis)	No change

(*anoxygenic photosynthesis*; Table 13-3). Oxygenic photosynthesis is the type observed in most aerobic phytoplankton, periphyton, and macrophytes.

Anoxygenic photosynthesis. Anoxygenic photosynthesizers include the purple and green photosynthetic bacteria. These organisms are photosynthetic, but instead of oxidizing water to hydrogen and oxygen, they oxidize other reduced species, including sulfur compounds (*purple and green sulfur bacteria*) and other reduced compounds such as hydrogen (*purple and green nonsulfur bacteria*) (Table 13-3). Because these anoxygenic bacteria require both light and a source of reduced compounds like sulfide, they tend to exist in narrow layers in lakes where both light and sulfide are available and where oxygen is low or absent, so near or in the anoxic zone. They can also occur in hot springs where there is a geothermal source of sulfide and abundant light. There are also both *purple and green nonsulfur bacteria*. These groups are not taxonomic units. The term *nonsulfur* was used because it was thought that these organisms could not use sulfide. Many of these bacteria, especially the green and purple nonsulfur bacteria, are *photoheterotrophs*, subsidizing anoxygenic photosynthesis with the heterotrophic uptake of some carbon compounds.

Chemolithotrophs. These organisms, all bacteria, use the energy from the oxidation of reduced compounds such as sulfide, ammonium, iron, and hydrogen to fix inorganic C into organic materials. Similar to the situation for the anoxygenic photosynthetic bacteria, many of these organisms also derive some of their carbon from organic sources (Rojas et al., 2021). The majority are aerobic using oxygen as the electron acceptor (Table 13-3). The common pathways include the oxidation of reduced sulfur compounds or iron-sulfur compounds such as pyrite (*Thiobacillus*) or the oxidation of ammonium and nitrate (*Nitrosomonas* and *Nitrobacter*). These oxidation pathways generate acid and reduce alkalinity (Table 13-3). As these organisms need a source of the reduced compound and a source of oxygen, their range is limited. In stratified lakes they are most likely to be found just above the *oxycline* (Chapter 11) and in streams with high sources of reduced iron and/or reduced sulfur. For example, thick mats of *Beggiatoa* are found in the hot springs in Yellowstone Park (USA). In the case of streams, these organisms produce organic matter at the expense of energy imported from outside the system's boundaries. In stratified lakes the situation is different because the energy that reduced the sulfate to sulfide came either from photosynthesis within the lake or in the lake's watershed.

There are also some examples of *anaerobic chemolithotrophs. Thiobacillus ferrooxidans* has the remarkable ability to oxidize reduced iron compounds using sulfate as the electron acceptor. In hydrogenotrophic methanogenesis CO_2 is reduced to CH_4 using H_2 as the electron donor (Table 13-3).

Oxygenic photosynthesis. Oxygenic photosynthesis is quantitatively the most important biological assimilation of inorganic C in freshwaters. This is the type of photosynthesis performed by all green plants including eukaryotic algae, most cyanobacteria, and submerged and emergent macrophytes. From the point of view of C assimilation alone, oxygenic photosynthesis increases the pH but has no direct effect on ANC (Table 13-2). However, ANC is affected by the transformation of N and P during cell growth. For example, if NO_3^- is the sole N source, ANC is increased; if NH_4^+ is the sole N source, ANC is decreased (Stumm and Morgan, 1995). This effect on ANC is usually quite small.

Abundant physiological evidence indicates that free CO_2 is most readily utilized by nearly all algae and larger aquatic plants. A number of algae and submersed macrophytes, particularly the mosses, can utilize only free CO_2. Many algae and aquatic vascular plants are capable of utilizing CO_2 from bicarbonate ions when free CO_2 is in very low supply and HCO_3^- is abundant; a few species of algae require HCO_3^- and cannot grow with free CO_2 alone (Maberly and Spence, 1983). There is no clear evidence that algae or higher aquatic plants assimilate CO_3^{2-} directly as a carbon source.

1. The CO_2 supply problem

The biochemistry of inorganic C assimilation by algae and higher plants is interesting and complex. Despite the solubility of CO_2 and relatively turbulent water strata of the pelagic trophogenic zone, the photosynthetic carbon supply is not very favorable. Rates of CO_2 diffusion in water are 10,000 times lower than in air, and diffusive transport can be a major limitation. Much of the total carbon is in the form of bicarbonate and carbonate, which is not readily available to all aquatic plants. Rapid photosynthesis can rapidly reduce the total DIC and increase the pH of the water and shift dissociation equilibria so that dissolved $[CO_2]$ is nearly zero. This can happen during intense algal blooms and also in dense stands of macrophytes or periphyton. Experimental evidence on freshwater photosynthetic utilization of inorganic carbon indicates a strong relationship between physiological availability and the forms of inorganic carbon, for example, CO_2 versus HCO_3^- (Hough and Wetzel, 1977; Maberly and Spence, 1983; Raven, 1970; Raven et al., 1985; Smith and Walker, 1980). These experimental results have been supported by a global survey on the occurrence of taxa that can photosynthesize using HCO_3^- (Iversen et al., 2019).

The primary enzyme of inorganic carbon fixation in all photosynthetic oxygen-evolving organisms is ribulose-1,5-bisphosphate carboxylase/oxygenase (RuBisCO). This carboxylase is involved in fixing some 95% of the carbon assimilated in plants with C_3 metabolism, in which 3-phosphoglycerate is the first major product. Most algae and aquatic macrophytes have C_3 biochemistry, and thus at least 95% of the carbon fixed must be supplied as CO_2 at the carboxylase site (Raven et al., 1985, 1995). The enzyme carboxydismutase is involved in the carboxylation reaction, which produces 3-phosphoglycerate as the first product of carbon fixation, and uses free, unhydrated CO_2 as its immediate substrate.

Carbonic anhydrase, an enzyme that catalyzes the reversible hydration of carbon dioxide, is found universally in photosynthetic cells of plants. Where CO_2 is the carbon species entering the cell, there is no obvious biochemical role for carbonic anhydrase found in plants, although some evidence indicates that this enzyme accelerates the diffusion of CO_2 and transport to sites of biochemical fixation by carboxylases (Raven 1995). Bicarbonate ions have been implicated as a critical factor in the evolution of oxygen during photosynthesis (Raven et al., 1985; Stemler and Govindjee, 1973). At high pH values and HCO_3^- concentrations, bicarbonate moves less effectively from binding sites to the chloroplast. When HCO_3^- enters cells, carbonic anhydrase is needed to supply CO_2 to RuBisCO. Carbonic anhydrase activity is found in plants that cannot use bicarbonate as well as those that can, although such activity generally increases among algae and certain submersed macrophytes that can utilize bicarbonate (Weaver and Wetzel, 1980; Raven, 1995).

The ability to assimilate bicarbonate ions is variable among planktonic algae, macroalgae, and submersed angiosperms (Raven 1970, 1984; Maberly and Spence, 1983). When this ability does occur, an additional reaction is needed for HCO_3^- assimilation that is not required for CO_2 assimilation. Active bicarbonate transport followed by HCO_3^- dehydration within the cytoplasm is apparently required and is coupled with a similarly active stoichiometric excretion of hydroxyl ions from the cells. Assimilated bicarbonate ions are used to produce high internal CO_2 pools by action of carbonic anhydrase.

When aquatic plants have similar affinities for CO_2 and the HCO_3^- ion, utilization of bicarbonate generally occurs when the bicarbonate concentration exceeds that of CO_2 by more than 10 times. Free CO_2 concentrations are about 10 to 20 μM at equilibrium with the atmosphere and somewhat higher (perhaps 300 μM) in many lakes and rivers. Most freshwaters contain bicarbonate concentrations far in excess of five times that quantity. Equilibrium concentrations of dissolved CO_2, particularly in alkaline hardwater lakes with a pH greater than 8.5, are inadequate to saturate photosynthesis in plants adapted to utilize bicarbonate. As these waters become more productive, and in densely populated littoral zones of less productive lakes, pH is rapidly altered by metabolism on a diurnal basis (pH ranges from a low of 6 to a high of 10 or more per 24 h) and can be associated with reduced carbon fixation and bicarbonate assimilation (Spencer et al., 1994). Under stagnant conditions, common to heavily colonized littoral zones of lakes, the shift to bicarbonate metabolism, as well as the increased pH, is often associated with a severe reduction in CO_2.

Stagnant layers around plant cells (algae) and tissues (macrophytes) clearly have major effects on limiting the assimilation of both CO_2 and HCO_3^- by actively photosynthesizing plants (Smith and Walker, 1980). Even when the plants are exposed to water movements by turbulence or sinking, a *residual layer* (approximately 10–100 μm thick) parallel to the surface remains. The rate of solute movement across this residual layer approximates that of molecular diffusion. Diffusion of CO_2 and HCO_3^- through these stagnant layers is an important rate-limiting process to both availability of CO_2 and membrane transport of HCO_3^- ions. Mucilage sheaths surrounding algal cells, especially among Cyanobacteria, can further reduce the uptake rates of inorganic carbon.

Bicarbonate assimilation in media of high pH assumes greater significance in large aquatic macrophytes that have morphologically long diffusion paths. Active transport of HCO_3^- originating from the dissociation of calcium bicarbonate and the secretion of OH^- ions results in the precipitation of $CaCO_3$. Many submersed angiosperms and algae can utilize bicarbonate. Aquatic mosses are able to utilize only free CO_2 and are restricted to waters of relatively low pH and abundant CO_2 (e.g., springs, mountain streams, bogs, or shallow waters) or the lower trophogenic zone where free CO_2 is higher than elsewhere in the lake. CO_2 of the rhizosphere can diffuse into rooting tissues, move to the leaves via internal gas lacunae, and be fixed photosynthetically (Loczy et al., 1983; Steinberg and Melzer, 1983; Wetzel et al., 1985). Although this benthic source of inorganic carbon is widely utilized in submersed macrophytes, the percentage of the total carbon fixed photosynthetically from the rooting tissues is greater among small plants growing in poorly mineralized waters. Moreover, many angiosperms with large intercellular gas lacunae refix the CO_2 of respiration and photorespiration rather efficiently (Hough and Wetzel, 1972; Søndergaard and Wetzel, 1980). The efficiency of refixation and the photosynthetic efficiency of carbon fixation must be highly plastic, related in part to induced shifts to bicarbonate assimilation, and can affect rates of net primary production significantly.

B. The effects of inorganic C concentrations on primary production

It is reasonable to ask if primary production or biomass in aquatic ecosystems is ever limited by the supply of inorganic C. For the past 40 years or so, the answer has been a resounding no (Schindler et al., 1972). There are several reasons for this. First, there are strong correlations between the concentrations of total P and algal biomass in lakes. Second, in nearly every experiment performed in bottles, mesocosms, or whole lakes and rivers, the addition of N and P results in a great increase in primary production and plant biomass. Third, as we have seen above, many algae and some higher plants can grow on bicarbonate when CO_2 is in low supply. Low CO_2 and its associated high pH may select for different algal species (often Cyanobacteria) if the low CO_2 condition is sustained. Fourth, CO_2 is resupplied by diffusion from the atmosphere and the rate of diffusive resupply increases as CO_2 decreases. Once the pH has increased due to CO_2 drawdown, the rate of resupply is increased even more by chemically enhanced diffusion. So, while there may be short-term limitations of photosynthesis due to extremely low CO_2, given time, we do not expect to see inorganic C limit the biomass attained by primary producers.

In some contrast to the long-held idea that phytoplankton biomass is not limited by the supply of inorganic C, several recent studies show that phytoplankton productivity and biomass does sometimes respond to increases in either pCO_2 or ANC (Hammer et al., 2019). Jansson et al. (2012) found that the supersaturation of CO_2 often observed in lakes of the boreal region enhanced both the rate of photosynthesis and the resulting algal biomass. Similarly, (Kragh and Sand-Jensen, 2018) found the usual correlation between TP and chlorophyll *a* in a series of 204 Danish lakes. Splitting the data into those lakes with ANC greater than 800 μEq and ANC less than 800 μEq revealed an intriguing pattern. For the same level of total phosphorus (TP), chlorophyll *a* was consistently higher in the higher ANC group compared to the lower ANC group. The difference was most striking at high TP. Kragh and Sand-Jensen, 2018 followed up with a series of mesocosm experiments in which they varied both ANC and TP. The experimental results were similar to the correlation work. Primary production was about five times higher under the higher ANC conditions when TP was sufficient. Based on an analysis of how fast CO_2 could be resupplied by the atmosphere, including chemically enhanced diffusion, Kragh and Sand-Jensen, 2018 found that the stimulation in photosynthesis was due to the uptake of HCO_3^- directly. So, perhaps under conditions of nutrient sufficiency, there is a case for C limitation.

Rising atmospheric CO_2 caused marine scientists to ask what effect this might have on phytoplankton in the ocean. Experiments with marine phytoplankton subjected to different CO_2 levels have shown a general increase in algal growth and an increase in the C:N and C:P ratios with increasing CO_2 (Burkhardt et al., 1999; Pardew et al., 2018). The ocean is in near equilibrium with the atmosphere, so rising atmospheric CO_2 causes rising CO_2 in the ocean. In freshwater systems the concentration of CO_2 in surface water ranges hugely among systems, much more so than the small effect of atmospheric increases in CO_2. Urabe et al., (2003) grew *Scenedesmus acutus* at three levels of CO_2 to span the range recorded in most freshwaters: pCO_2 of 360 μatm, 1500 μatm, and 3000 μatm. Algal biomass (as mg C L^{-1}) was 7 and 10 times higher in the elevated CO_2 treatments compared to the ambient atmospheric treatment. Further, the C:P ratio increased across the CO_2 treatment from about 285 in the ambient treatment to about 1000 in both of the elevated CO_2 treatments. Urabe et al., (2003) concluded that CO_2 limited the growth of this alga in culture. Interestingly, the high-CO_2 algae were a poorer food source for *Daphnia* than those grown at ambient CO_2. Vogt et al., (2017) surveyed 69 boreal lakes that varied in pCO_2 from undersaturation to about 2000 μatm and found that both chlorophyll *a* and gross primary production (GPP) increased along the pCO_2 gradient. For GPP, there was a nearly linear increase across the pCO_2 range. For chlorophyll, the relationship only held true for those lakes (most of the ones in this data set) that were above atmospheric level. That is, in the few very low CO_2 lakes chlorophyll *a*, but not GPP, was high.

V. Summary

1. In typical freshwaters, DIC is dominated by the HCO_3^- anion and this carbon pool is usually as large or larger than that of dissolved organic C, particulate organic C, and planktonic organisms. The major exception is soft, acidic, boggy waters, where DOC is high and DIC can be quite low. DIC concentrations range from as low as 10 μmol L^{-1} to more than 5000 μmol L^{-1} in typical lakes and rivers, although there are places where it can be much higher such as lakes in karst regions or at the bottom of deep lakes with underlying vulcanism.
2. The complex equilibrium reactions of inorganic carbon and the distribution of the chemical species of total dissolved inorganic have been well understood in considerable detail for some time. The DIC system

is the major regulator of both the pH and buffering capacity of typical freshwater ecosystems. DIC = $CO_2 + HCO_3^- + CO_3^{2-}$. The pH of the water is largely controlled by the ratios of the species of DIC.

3. As carbonic acid percolates over rock and through soils, carbonates and silicates are solubilized; ionized Ca^{++} and HCO_3^- are released into the water. The dilute solution of bicarbonate of many freshwaters is weakly alkaline because slightly greater concentrations of OH^- ions than H+ ions result from the dissociation of HCO_3^-, CO_3^{2-}, and H_2CO_3. The concentration of DIC tends to be highest in carbonate geology, but at a global scale, the delivery of DIC to the ocean is nearly equal from carbonate and silicate geological regions. This is due to the very large hydrologic flux of moderately high DIC from the humid tropics, where silicate or basalt geology often dominates.
4. Losses of CO_2 by photosynthetic utilization or additions of CO_2 from biotic respiration are the major factors that change the pH (H^+ ion activity) of a given water body over time. CO_2 and pH vary inversely over diel and seasonal scales. The buffering capacity (or ANC) of the water, however, tends to resist changes in pH as long as the equilibria of the DIC complex are operational.
5. The distribution of DIC and pH in surface waters varies both seasonally and vertically in lakes in relation to loading from allochthonous sources, physical conditions, and biotic inputs and consumption. Alkalinity often increases in the anaerobic portion of stratified systems due to anaerobic respiratory processes such as sulfate reduction.
6. Most lakes, rivers, and streams are supersaturated in terms of CO_2. Thus there is a constant net efflux of excess CO_2 from most inland waters. This net CO_2 evasion to the atmosphere derives from both inputs of CO_2 in groundwater from the watershed and an excess of respiration over gross primary production (e.g., net heterotrophy) within the water body. Net heterotrophy is supported by the input and respiration of terrestrial organic matter.
7. In thermally stratified lakes of low to moderate productivity with minimal inputs of organic matter from outside the basin, the accumulation of CO_2 in the hypolimnion is positively correlated to rates of organic production in the trophogenic zone.
8. Inorganic carbon is a major nutrient of photosynthetic metabolism. However, phosphorus and nitrogen limit photosynthesis more frequently than does inorganic carbon, which occurs in much greater abundance. Because assimilation of CO_2 and HCO_3^- occurs more rapidly than resupply, and conditions of reduced availability result from diffusion resistance at the uptake sites, potential photosynthetic productivity by both algae and submersed macrophytes is not always fully realized. Many algae and vascular aquatic plants have developed compensatory mechanisms to enhance the utilization and recycling of respired CO_2.

References

Allen, E.D., Spence, D.H.N., 1981. The differential ability of aquatic plants to utilize the inorganic supply in fresh waters. New Phytol. 87, 269–283.

Asbury, C.E., Mattson, M.D., Vertucci, F.A., Likens, G.E., 1990. Acidification of Adirondack lakes. Environmental Science & Technology 24, 387–390.

Bade, D.L., Cole, J.J., 2006. Impact of chemically enhanced diffusion on dissolved inorganic carbon stable isotopes in a fertilized lake. J. Geophys. Res.: Oceans 111, C01014. https://doi.org/10.1029/2004JC002684.

Baker, L.A., Herlihy, A.T., Kaufmann, P.R., Eilers, J.M., 1991. Acidic lakes and streams in the United States—the role of acidic deposition. Science 252, 1151–1154.

Binding, C.E., Jerome, J.H., Bukata, R.P., Booty, W.G., 2007. Trends in water clarity of the lower Great Lakes from remotely sensed aquatic color. J. Great Lake. Res. 33, 828–841.

Bird, D., Kalff, J., 1986. Bacterial grazing by planktonic algae. Science 231, 493–495.

Blodau, C., 2006. A review of acidity generation and consumption in acidic coal mine lakes and their watersheds. Sci. Total Environ. 369, 307–332.

Burkhardt, S., Riebesell, U., Zondervan, I., 1999. Effects of growth rate, CO_2 concentration, and cell size on the stable carbon isotope fractionation in marine phytoplankton. Geochem. Cosmochim. Acta 63, 3729–3741.

Butler, J.N., 1992. Carbon dioxide equilibria and their applications. Lewis Publishers, Chelsea, Michigan.

Butman, D., Stackpoole, S., Stets, E., McDonald, C.P., Clow, D.W., Striegl, R.G., 2016. Aquatic carbon cycling in the conterminous United States and implications for terrestrial carbon accounting. Proceedings of the National Academy of Sciences of the United States of America 113, 58–63.

Caraco, N.F., Cole, J.J., Likens, G.E., 1989. Evidence for sulphate-controlled phosphorus release from sediments of aquatic systems. Nature 341, 316–318.

Caraco, N., Cole, J.J, Likens, G.E., 1990. A comparison of phosphorus immobilization in sediments of freshwater and coastal marine sediments. Biogeochemistry 9, 277–290.

Caraco, N.F., Cole, J.J., Likens, G.E., 1992. Sulfate control of phosphorus availability in lakes: a test and re-evaluation of Hasler and Einsele's model. Hydrobiologia 253, 275–280.

Carini, S.A., Joye, S.B., 2008. Nitrification in Mono Lake, California: activity and community composition during contrasting hydrological regimes. Limnol. Oceanogr. 53, 2546–2557.

Catalan, J., Camarero, L., 1992. Seasonal Changes in Alkalinity and pH in 2 Pyrenean Lakes of Very different Water Redisence Time. Congress of the International Association of Theoretical and Applied Limnology Barcelona, Spain, 1, 749–753.

Clark, J.F., Schlosser, P., Wanninkhof, R., Simpson, H.J., Schuster, W.S.F., Ho, D.T., 1995. Gas transfer velocities for SF_6 and ^3He in a small pond at low wind speeds. Geophys. Res. Lett. 22, 93–96.

Clymo, R.S., 1964. The origin of acidity in *Sphagnum* bogs. Bryologist 67, 427–431.

Clymo, R.S., 1965. Experiments on breakdown of *Sphagnum* in two bogs. J. Ecol. 53, 747–758.

Cole, J.J., Bade, D.L., Bastviken, D., Pace, M.L., Van de Bogert, M., 2010. Multiple approaches to estimating air-water gas exchange in small lakes. Limnol Oceanogr Methods 8, 285–293.

Cole, J.J., Caraco, N.F., 1998. Atmospheric exchange of carbon dioxide in a low-wind oligotrophic lake measured by the addition of SF_6. Limnol. Oceanogr. 43, 647–656.

Cole, J.J., Caraco, N.F., 2001. Carbon in catchments: connecting terrestrial carbon losses with aquatic metabolism. Mar. Freshw. Res. 52, 101–110.

Cole, J.J., Caraco, N.F., Kling, G.W., Kratz, T.K., 1994. Carbon dioxide supersaturation in the surface waters of lakes. Science 265, 1568–1570.

Cole, J.J., Pace, M.L., Carpenter, S.R., Kitchell, J.F., 2000. Persistence of net heterotrophy in lakes during nutrient addition and food web manipulations. Limnol. Oceanogr. 45, 1718–1730.

Cole, J.J., Prairie, Y.T., Caraco, N, McDowell, W.H., Tranvik, L.J., Streigl, R.G., Duarte, C.M., Kortelainen, P., Downing, J., Middelburg, J., Melack, J., 2007. Plumbing the global carbon cycle: integrating inland waters into the terrestrial carbon budget. Ecosystems 10, 171–184.

Crawford, J.T., Lottig, N.R., Stanley, E.H., Walker, J.F., Hanson, P.C., Finlay, J.C., et al., 2014. CO_2 and CH_4 emissions from streams in a lake-rich landscape: patterns, controls, and regional significance. Global Biogeochem. Cycles 28, 197–210.

del Giorgio, P.A., Cole, J.J., Cimbleris, A., 1997. Respiration rates in bacteria exceed phytoplankton production in unproductive aquatic systems. Nature 385, 148–151.

del Giorgio, P.A., Prairie, Y.T. 2017. Lac Croche monitoring data, unpublished.

Denfeld, B.A., Frey, K.E., Sobczak, W.V., Mann, P.J., Holmes, R.M., 2013. Summer CO_2 evasion from streams and rivers in the Kolyma River basin, north-east Siberia. Polar Res. 32 (19704). https://doi.org/10.3402/polar.v32i0.19704.

Dillon, P.J., Molot, L.A., 1997. Dissolved organic and inorganic carbon mass balances in central Ontario lakes. Biogeochemistry 36, 29–42.

Duarte, C.M., Prairie, Y.T., 2005. Prevalence of heterotrophy and atmospheric CO_2 emissions from aquatic ecosystems. Ecosystems 8, 862–870.

Ducharme-Riel, V., Vachon, D., del Giorgio, P.A., Prairie, Y.T., 2015. The relative contribution of winter under-ice and summer hypolimnetic CO_2 accumulation to the annual CO_2 emissions from northern lakes. Ecosystems 18, 547–559.

Emerson, S., 1975. Chemically enhanced CO_2 exchange in a eutrophic lake: a general model. Limnol. Oceanogr. 20, 743–753.

Finlay, K., Leavitt, P.R., Wissel, B., Prairie, Y.T., 2009. Regulation of spatial and temporal variability of carbon flux in six hard-water lakes of the northern Great Plains. Limnol. Oceanogr. 54, 2553–2564.

Gaillardet, J., Dupre, B., Louvat, P., Allegre, C.J., 1999. Global silicate weathering and CO_2 consumption rates deduced from the chemistry of large rivers. Chem. Geol. 159, 3–30.

Galloway, J.N., Likens, G.E., Hawley, M.E., 1984. Acid precipitation: natural versus anthropogenic components. Science 226, 829–831.

Genereux, D.P., Hemond, D.P., 1992. Determination of gas exchange rate constants for a small stream on Walker Branch Watershed, Tennessee. Water Resources Research 28, 2365–2374.

Giblin, A.E., Likens, G.E., White, D., Howarth, R.W., 1990. Sulfur storage and alkalinity generation in New England lake sediments. Limnol. Oceanogr. 35, 852–869.

Golterman, H.L., Meyer, M.L., 1985. The geochemistry of hard water rivers, the Rhine and the Rhone. Hydrobiologia 126, 11–19.

Hammer, K.J., Kragh, T., Sand-Jensen, K., 2019. Inorganic carbon promotes photosynthesis, growth, and maximum biomass of phyto plankton in eutrophic water bodies. Freshw. Biol. 64, 1956–1970.

Herbst, A., Schubert, H., 2018. Age and site-specific pattern on encrustation of charophytes. Botanical Studies 59, 31.

Herczeg, A.L., Broecker, W.S., Anderson, R.F., Schiff, S.L., Schindler, D.W., 1985. A new method for monitoring temporal trends in the acidity of freshwaters. Nature 315, 133–135.

Hodell, D.A., Schelske, C.L., Fahnenstiel, G.L., Robbins, L.L., 1998. Biologically induced calcite and its isotopic composition in Lake Ontario. Limnol. Oceanogr. 43, 187–199.

Holtgrieve, G.W., Schindler, D.E., Branch, T.A., A'Mar, Z.T., 2010. Simultaneous quantification of aquatic ecosystem metabolism and reaeration using a Bayesian statistical model of oxygen dynamics. Limnol. Oceanogr. 55, 1047–1063.

Hoover, T.E., Berkshire, D.C., 1969. Effects of hydration on carbon dioxide exchange across an air-water interface. J. Geophys. Res. 74, 456–464.

Hough, R.A., Wetzel, R.G., 1977. Photosynthetic pathways of some aquatic plants. Aquat. Bot. 3, 297–313.

Houser, J.N., Bade, D.L., Cole, J.J., Pace, M.L., 2003. The dual influences of dissolved organic carbon on hypolimnetic metabolism: organic substrate and photosynthetic reduction. Biogeochemistry 64, 247–269.

Hutchinson, G.E., 1957. A Treatise on Limnology, Geography, Physics and Chemistry. John Wiley and Sons, New York, pp. 1–1015.

Iversen, L.L., Winkel, A., Baastrup-Spohr, A., Hinke, L., Alahuhta, A.B., Baattrup-Pedersen, A., et al., 2019. Catchment properties and the photosynthetic trait composition of freshwater plant communities. Science 366, 878–881.

Jahne, B., Heinz, G., Dietrich, W., 1987. Measurement of diffusion coefficients of sparingly soluble gases in water. J. Geophys. Res. 92, 10767–10776.

Jansson, M., Karlsson, J., Jonsson, A., 2012. Carbon dioxide supersaturation promotes primary production in lakes. Ecol. Lett. 15, 527–532.

Kelly, C.A., 1988. Toward improving comparisons of alkalinity generation in lake basins. Limnol. Oceanogr. 33, 1635–1637.

Kilham, P., 1982. Acid precipitation—its role in the alkalization of a lake in Michigan. Limnol. Oceanogr. 27, 856–867.

Klaus, M., Vachon, D., 2020. Challenges of predicting gas transfer velocity from wind measurements over global lakes. Aquat. Sci. 82 (3), 83.

Kleiner, J., Stabel, H.-H., 1989. Phosphorus transport to the bottom of Lake Constance. Aquat. Sci. 51 (3), 181–191.

Kling, G.W., Evans, W.C., Tanyileke, G.Z., 2015. The Comparative limnology of lakes Nyos and Monoun, Cameroon. In: Rouwet, D., Christenson, B., Tassi, F., Vandemeulebrouck, J. (Eds.), Volcanic Lakes. Springer, Berlin, pp. 401–425.

Kling, G.W., Evans, W.C., Tanyileke, G., Kusakabe, M., Ohba, T., Yoshida, Y., Hell, J.V., 2005. Degassing lakes Nyos and Monoun: defusing certain disaster. Proceedings of the National Academy of Sciences of the United States of America 102, 14185–14190.

Kling, G.W., Evans, W.C., Tuttle, M.L., Tanyileke, G., 1994. Degassing of lake Nyos. Nature 368, 405–406.

Kling, G.W., Kipphut, G.W., Miller, M.C., 1991. Arctic lakes and streams as gas conduits to the atmosphere: implications for tundra carbon budgets. Science 251, 298–301.

Koschorreck, M., Prairie, Y.T., Kim, J., Marce, R., 2021. Technical note: CO_2 is not like CH_4—Limits of and corrections to the headspace method to analyse pCO_2 in fresh water. Biogeosciences 18, 1619–1627.

Kragh, T., Sand-Jensen, K., 2018. Carbon limitation of lake productivity. Proceedings of the Royal Society B: Biological Sciences 285 (1891), 20181415.

Kratz, T.K., Cook, R.B., Bowser, C.J., Brezonik, P.L., 1987. Winter and spring pH depressions in northern Wisconsin lakes caused by increases in pCO_2. Can. J. Fish. Aquat. Sci. 44, 1082–1088.

Lawrence, G.B., Roy, K.M., Baldigo, B.P., Simonin, H.A., Capone, S.B., Sutherland, J.W., Nierzwicki-Bauer, S.A., Boylen, C.W., 2008.

Chronic and episodic acidification of Adirondack streams from acid rain in 2003–2005. J. Environ. Qual. 37, 2264–2274.

Likens, G.E., Butler, T.J., Claybrooke, R., Vermeylen, F., Larson, R., 2021. Long-term monitoring of precipitation chemistry in the US: insights into changes and condition. Atmos. Environ. 245, 118031.

Livingstone, D.A., 1963. Chemical composition of lakes and rivers. Data of Geochemistry. USGS.

Livingstone, D.M., Imboden, D.M., 1996. The prediction of hypolimnetic oxygen profiles: a plea for a deductive approach. Can. J. Fish. Aquat. Sci. 53, 924–932.

Loczy, S., Carignan, R., Planas, D., 1983. The role of roots in carbon uptake by the submersed macrophytes *Myriophyllum spicatum, Vallisneria americana*, and *Heteranthera dubia*. Hydrobiologia 98, 3–7.

Lohr, A.J., Sluik, R., Olaveson, M.M., Ivorra, N., Van Gestel, C.A.M., Van Straalen, N.M., 2006. Macroinvertebrate and algal communities in an extremely acidic river and the Kawah Ijen crater lake (pH <0.3), Indonesia. Arch. Hydrobiol. 165, 1–21.

Lovley, D.R., Holmes, D.E., Nevin, K.P., 2004. Dissimilatory Fe(III) and Mn(IV) reduction. In: Poole, R.K. (Ed.), Advances in Microbial Physiology, vol. 49, pp. 219–286.

Lydersen, E., 1998. Humus and acidification. In: Hessen, D.O., Tranvik, L.J. (Eds.), Aquatic Humic Substances: Ecology and Biogeochemistry. Springer-Verlag, Berlin, pp. 63–92.

Maberly, S.C., Spence, D.H.N., 1983. Photosynthetic inorganic carbon use by fresh-water plants. J. Ecol. 71, 705–724.

Mann, P.J., Spencer, R.G.M., Dinga, B.J., Poulsen, J.R., Hernes, P.J., Fiske, G., Salter, M.E., Wang, Z.A., Hoering, K.A., Six, J., Holmes, R.M., 2014. The biogeochemistry of carbon across a gradient of streams and rivers within the Congo Basin. J. Geophys. Res.: Biogeosciences 119, 687–702.

Matthews, D.A., Effler, S.W., Driscoll, C.T., O'Donnell, S.M., Matthews, C.M., 2008. Electron budgets for the hypolimnion of a recovering urban lake, 1989–2004: response to changes in organic carbon deposition and availability of electron acceptors. Limnol. Oceanogr. 53, 743–759.

Mattson, M.D., Likens, G.E., 1993. Redox reactions of organic matter decomposition in a soft water lake. Biogeochemistry 19, 149–172.

Matzinger, A., Muller, B., Niederhauser, P., Schmid, M., Wuest, A., 2010. Hypolimnetic oxygen consumption by sediment-based reduced substances in former eutrophic lakes. Limnol. Oceanogr. 55, 2073–2084.

Mercier-Blais, S., Prairie, Y.T. 2014. Lac Croche monitoring data, unpublished.

Meybeck, M., 1993. Riverine transport of atmospheric carbon: sources, global typology and budget. Water Air Soil Pollut. 70, 443–463.

Minder, L., 1923. Studien über den Sauerstoffgehalt des Zürichsees. Arch. Hydrobiol./Suppl. 3, 107–155.

Molot, L.A., Dillon, P.J., Lazerte, B.D., 1989. Factors affecting alkalinity concentrations of streamwater druing snowmelt in central Ontario. Can. J. Fish. Aquat. 46, 1658–1666.

Morel, F.M.M., Hering, J.G., 1993. Principles and Applications of Aquatic Chemistry. John Wiley & Sons, New York, p. 374.

Muller, B., Bryant, L.D., Matzinger, A., Wuest, A., 2012. Hypolimnetic oxygen depletion in eutrophic lakes. Environmental Science & Technology 46, 9964–9971.

Nouchi, V., Kutser, T., Wuest, A., Muller, B., Odermatt, D., Baracchini, T., Bouffard, D., 2019. Resolving biogeochemical processes in lakes using remote sensing. Aquat. Sci. 81 (2), 27.

Ohle, W., 1952. Die Kohlendioxyd-Akkumulation als produktionsbiologischer Indikator. Archiv für Hydrobiologie 46, 153–285.

Ohle, W., 1956. Bioactivity, Production, and Energy Utilization of Lakes. Limnol. Oceanogr 1 (3), 139–149.

Oppenheimer, M., Galloway, J.N., Likens, G.E., Norton, S.A., 1985. Acid deposition. Science 227 (4691), 1154–1156. &.

Otsuki, A., Wetzel, R.G., 1972. Co-precipitation of phosphate with carbonates in a marl lake. Limnol. Oceanogr 17 (5), 763–767.

Pace, M.L., Prairie, Y.T., 2005. Respiration in lakes. In: del Giorgio, P.A., Williams, P.J.l.B. (Eds.), Respiration in Aquatic Systems. Oxford University Press, Oxford, UK, pp. 103–121.

Pardew, J., Pimentel, M.B., Low-Decarie, E., 2018. Predictable ecological response to rising CO_2 of a community of marine phytoplankton. Ecol. Evol. 8, 4292–4302.

Parkhurst, D.L., Appelo, C.A.J., 2013. Description of Input and Examples for PHREEQC Version 3—A Computer Program for Speciation, Batch-Reaction, One-Dimensional Transport, and Inverse Geochemical Calculations. US Geological Survey Techniques and Methods, Book 6. USGS.

Pennington, R., Argerich, A., Haggerty, R., 2018. Measurement of gas-exchange rate in streams by the oxygen-carbon method. Freshw. Sci. 37, 222–237.

Pia, J., 1933. Kohlensäure und Kalk. Die Binnengewässer 13, 1–183.

Prairie, Y.T., Bird, D.F., Cole, J.J., 2002. The summer metabolic balance in the epilimnion of southeastern Quebec lakes. Limnol. Oceanogr. 47, 316–321.

Ran, L.S., Lu, X.X., Liu, S.D., 2017. Dynamics of riverine CO_2 in the Yangtze River fluvial network and their implications for carbon evasion. Biogeosciences 14, 2183–2198.

Raven, J.A., 1970. Exogenous inorganic carbon sources in plant photosynthesis. Biol. Rev. 45, 167–221.

Raven, J.A., Lucas, W.J., Berry, J.A., 1985. Energy costs of DIC accumulation. Inorganic carbon uptake by aquatic photosynthetic organisms. American Society of Plant Physiologists, Rockville, MD, pp. 305–324.

Raven, J. A. 1995. Photosynthetic and non-photosynthetic roles of carbonic anhydrase in algae and cyanobacteria. Phycologia 34: 93–101. https://doi.org/10.2216/i0031-8884-34-2-93.1.

Raymond, P.A., Hartmann, J., Lauerwald, R., Sobek, S., McDonald, C., Hoover, M., Butman, D., Striegl, R., Mayorga, E., Humborg, C., Kortelainen, P., Durr, H., Meybeck, M., Ciais, P., Guth, P., 2013. Global carbon dioxide emissions from inland waters. Nature 503, 355–359.

Raymond, P.A., Oh, N.H., Turner, R.E., Broussard, W., 2008. Anthropogenically enhanced fluxes of water and carbon from the Mississippi River. Nature 451, 449–452.

Read, J.S., Hamilton, D.P., Desai, A.R., Rose, K.C., MacIntyre, S., Lenters, J.D., Smyth, R.L., Hanson, P.C., Cole, J.J., Staehr, P.A., Rusak, J.A., Pierson, D.C., Brookes, J.D., Laas, A., Wu, C.H., 2012. Lake-size dependency of wind shear and convection as controls on gas exchange. Geophys. Res. Lett. 39 (9), L09405.

Rebsdorf, A., Thyssen, N., Erlandsen, M., 1991. Regional and temporal variation in pH, alkalinity and carbon dioxide in Danish streams, related to soil type and land use. Freshw. Biol. 25, 419–435.

Regnier, P., Friedlingstein, P., Ciais, P., Mackenzie, F.T., Gruber, N., Janssens, I.A., Laruelle, G.G., Lauerwald, R., Luyssaert, S., Andersson, A.J., Arndt, S., Arnosti, C., Borges, A.V., Dale, A.W., Gallego-Sala, A., Godderis, Y., Goossens, N., Hartmann, J., Heinze, C., Ilyina, T., Joos, F., LaRowe, D.E., Leifeld, J., Meysman, F.J.R., Munhoven, G., Raymond, P.A., Spahni, R., Suntharalingam, P., Thullner, M., 2013. Anthropogenic perturbation of the carbon fluxes from land to ocean. Nat. Geosci. 6, 597–607.

Rengefors, K., Palsson, C., Hansson, L.A., Heiberg, A., 2008. Cell lysis of competitors and osmotrophy enhance growth of the bloom-forming alga Gonyostomum semen. Aquat. Microb. Ecol. 51, 87–96.

Richey, J.E., Melack, J.M., Aufdenkampe, A.K., Ballester, V.M., Hess, L.L., 2002. Outgassing from Amazonian rivers and wetlands as a large tropical source of atmospheric CO_2. Nature 416, 617–620.

Roehm, C.L., Prairie, Y.T., del Giorgio, P.A., 2009. The pCO_2 dynamics in lakes in the boreal region of northern Quebec, Canada. Global Biogeochem. Cycles 23 (3), GB3013.

Rojas, C.A., Torio, A.D., Park, S., Bosak, T., Klepac-Ceraj, V., 2021. Organic electron donors and terminal electron acceptors structure anaerobic microbial communities and interactions in a permanently stratified sulfidic lake. Front. Microbiol. 12, 620424.

Ruttner, F., 1926. Über die Kohlensäureassimilation einiger Wasserpflanzen in verschiedenen Tiefen des Lunzer Untersees. International Review of Hydrology 15 (1–2), 1–30.

Ruttner, F, 1967. Fundamentals of Limnology-3rd Edition translated [from the German] by D.G. Frey and F.E.J. Fry. University of Toronto Press, Toronto.

Sanders, R.W., 2011. Alternative nutritional strategies in protists: symposium introduction and a review of freshwater protists that combine photosynthesis and heterotrophy. Eukaryotic Microbiology 58, 181–184.

Sand-Jensen, K., Staehr, P.A., 2009. Net heterotrophy in small Danish lakes: a widespread feature over gradients in trophic status and land cover. Ecosystems 12, 336–348.

Schindler, D.W., 1977. Evolution of phosphorus limitation in lakes. Science 195, 260–262.

Schindler, D.W., Brunskill, G.J., Emerson, S., Broecker, W.S., Peng, T.H., 1972. Atmospheric carbon dioxide: its role in maintaining phytoplankton standing stocks. Science 177, 1192–1194.

Schindler, D.W., Turner, M.A., Stainton, M.P., Linsey, G.A., 1986. Natural sources of acid neutralizing capacity in low alkalinity lakes of the Precambrian Shield. Science 232, 844–847.

Schlesinger, W.H., Bernhardt, E.S., 2020. Biogeochemistry. An Analysis of Global Change. 4th Edition. Elsevier Press, pp. 1–762.

Sellers, P., Hesslein, R.H., Kelly, C.A., 1995. Continuous measurement of CO_2 for estimation of air-water fluxes in lakes: an in situ technique. Limnol. Oceanogr. 40, 575–581.

Smith, F.A., Walker, N.A., 1980. Photosynthesis by aquatic plants—effects of unstirred layers in relation to assimilation of CO_2 and HCO_3^- and to carbon isotopic discrimination. New Phytol. 86, 245–259.

Sobek, S., Durisch-Kaiser, E., Zurbrugg, R., Wongfun, N., Wessels, M., Pasche, N., Wehrli, B., 2009. Organic carbon burial efficiency in lake sediments controlled by oxygen exposure time and sediment source. Limnol. Oceanogr. 54, 2243–2254.

Sobek, S., Tranvik, L.J., Cole, J.J., 2005. Temperature independence of carbon dioxide supersaturation in global lakes. Global Biogeochem. Cycles 19 (2), 1–10. GB2003.

Søndergaard, M., Wetzel, R.G., 1980. Photo-respiration and internal recycling of CO_2 in the submersed angiosperm *Scirpus subterminalis*. Canadian Journal of Botany-Revue Canadienne de Botanique 58, 591–598.

Spencer, W.E, Teeri, J., Wetzel, R.G., 1994. Acclimation of photosynthetic phenotypes to environmental heterogeneity. Ecology 75, 301–314.

Stallard, R.F., 1998. Terrestrial sedimentation and the carbon cycle: coupling weathering and erosion to carbon burial. Global Biogeochemistry Cycles 12, 231–257.

Stauffer, R.E., 1990. Electrode pH error, seasonal epilimnetic pCO_2, and the recent acidification of the Maine lakes. Water Air Soil Pollut. 50, 123–148.

Steinberg, C., Melzer, A., 1983. Uptake, translocation and release of carbon by submersed macrophytes from running water habitats. Swiss J. Hydrol. 45, 333–344.

Stemler, A., Govindjee, A, 1973. Bicarbonate ion as a critical factor in photosynthetic oxygen evolution. Plant Physiol. 51, 119–123.

Strayer, D.L., Cole, J.J., Likens, G.E., Buso, D.C., 1981. Biomass and annual production of the fresh-water mussel *Elliptio complanata* in an oligotrophic softwater lake. Freshw. Biol. 11, 435–440.

Striegl, R.G., Kortelainen, P., Chanton, J.P., Wickland, K.P., Bugna, G.C., Rantakari, M., 2001. Carbon dioxide partial pressure and 13C content of north temperate and boreal lakes at spring ice melt. Limnol. Oceanogr. 46, 911–945.

Stumm, W., Morgan, J.J., 1995. Aquatic Chemistry: Chemical Equilibria and Rates in Natural Waters, 3rd Edition. Wiley, New Jersey.

Tank, J.L., Rosi-Marshal, E.J., Baker, M.A., Hall, R.O, 2008. Are rivers just big streams? A pulse method to quantify nitrogen demand in a large river. Ecology 89, 2935–2945.

Thornton, K.W., Kimmel, B.L., Payne, F.E., 1990. Reservoir limnology: ecological perspective. Wiley Interscience, New York, NY.

Ulseth, A.J., Hall, R.O., Canadell, M.B., Madinger, H.L., Niayifar, A., Battin, T.J., 2019. Distinct air-water gas exchange regimes in low- and high-energy streams. Nat. Geosci. 12 (4), 259–263.

Urabe, J., Togari, J., Elser, J.J., 2003. Stoichiometric impacts of increased carbon dioxide on a planktonic herbivore. Global Change Biol. 9, 818–825.

Urabe, J., Yoshida, T., Gurung, T.B., Sekino, T., Tsugeki, N, Nozaki, K., 2005. The production-to-respiration ratio and its implication in Lake Biwa, Japan. Ecol. Res. 20, 367–375.

Urban, N.R., Apul, D.S., Auer, M.T., 2004. Community respiration rates in Lake Superior. J. Great Lakes Res. 30, 230–244.

Vachon, D., del Giorgio, P.A., 2014. Whole-lake CO_2 dynamics in response to storm events in two morphologically different lakes. Ecosystems 17, 1338–1353.

Vachon, D., Prairie, Y.T., 2013. The ecosystem size and shape dependence of gas transfer velocity versus wind speed relationships in lakes. Can. J. Fish. Aquat. Sci. 70, 1757–1764.

Vachon, D., Prairie, Y.T., Cole, J.J., 2010. The relationship between near-surface turbulence and gas transfer velocity in freshwater systems and its implications for floating chamber measurements of gas exchange. Limnol. Oceanogr. 55, 1723–1732.

Vachon, D., Solomon, C.T., del Giorgio, P.A., 2017. Reconstructing the seasonal dynamics and relative contribution of the major processes sustaining CO_2 emissions in northern lakes. Limnol. Oceanogr. 62, 706–722.

Vogt, R.J., St-Gelais, N.F., Bogard, M.J., Beisner, B.E., del Giorgio, P.A., 2017. Surface water CO_2 concentration influences phytoplankton production but not community composition across boreal lakes. Ecol. Lett. 20, 1395–1404.

Waller, K., Driscoll, C., Lynch, J., Newcomb, D., Roy, K., 2012. Long-term recovery of lakes in the Adirondack region of New York to decreases in acidic deposition. Atmos. Environ. 46, 56–64.

Walsh, J.R., Corman, J.R., Munoz, S.E., 2019. Coupled long-term limnological data and sedimentary records reveal new control on water quality in a eutrophic lake. Limnol. Oceanogr. 64, S34–S48.

Wanninkhof, R., 1992. Relationship between gas exchange and wind speed over the ocean. J. Geophys. Res. 97, 7373–7381.

Wanninkhof, R., Knox, M., 1996. Chemical enhancement of CO_2 exchange in natural waters. Limnol. Oceanogr. 41, 689–697.

Weaver, C.I., Wetzel, R.G., 1980. Carbonic-anhydrase levels and internal lacunar CO_2 concentrations in aquatic macrophytes. Aquatic Botany 8, 173–186.

Wetzel, R.G., 1970. Recent and postglacial production rates of a marl lake. Limnol. Oceanogr 15, 491–503.

Wetzel, R.G., 2001. Limnology: Lake and river ecosystems. Academic Press, San Diego, CA.

Wetzel, R.G., Brammer, E.S., Lindström, K., Forsberg, C., 1985. Photosynthesis of submersed macrophytes in acidified lakes II. Carbon limitation and utilization of benthic CO_2 sources. Aquatic Botany 22 (2), 107–120.

Wilkinson, G.M., Buelo, C.D., Cole, J.J., Pace, M.L., 2016. Exogenously produced CO_2 doubles the CO_2 efflux from three north temperate lakes. Geophys. Res. Lett. 43, 1996–2003.

Wilkinson, G.M., Cole, J.J., Pace, M.L., Johnson, R.A., Kleinhan, M.J., 2015. Physical and biological contributions to metalimnetic oxygen maxima in lakes. Limnol. Oceanogr 60, 242–251.

Xin, R., J. F. Banda, C. Hao, and others. 2021. Contrasting seasonal variations of geochemistry and microbial community in two adjacent acid mine drainage lakes in Anhui Province, China. Environ Pollut 268: 115826. https://doi.org/10.1016/j.envpol.2020.115826

Zhang, Y.X., 1996. Dynamics of CO_2-driven lake eruptions. Nature 379, 57–59.

C H A P T E R

14

The Nitrogen Cycle

Michele A. Burford[1,2] *and Jing Lu*[1]

[1]Australian Rivers Institute, Griffith University, Nathan, QLD, Australia [2]School of Environment and Sciences, Griffith University, Nathan, QLD, Australia

OUTLINE

I. Introduction

Nitrogen (N) is abundant on the surface of the Earth but less than 2% is available to organisms (Galloway, 1998). A major source of nitrogen input to streams and lakes is from surface land drainage and groundwater sources. Atmospheric sources and nitrogen fixation are other mechanisms that contribute nitrogen. An obvious loss from freshwater systems is via inorganic and organic (both dissolved and particulate) compounds in effluents flowing out of catchments/watersheds. Further losses occur as a result of the permanent burial of partially decomposed biota and inorganic and organic nitrogen compounds adsorbed to particulate matter in the sediments. Nitrogen can also be lost by the volatilization of compounds from the water surface, for example, ammonia (NH_3) at high pH and as gaseous nitrogen forms mediated by microbial processes.

Nitrogen undergoes many complex reactions in aquatic systems (Fig. 14-1, Table 14-1), and our

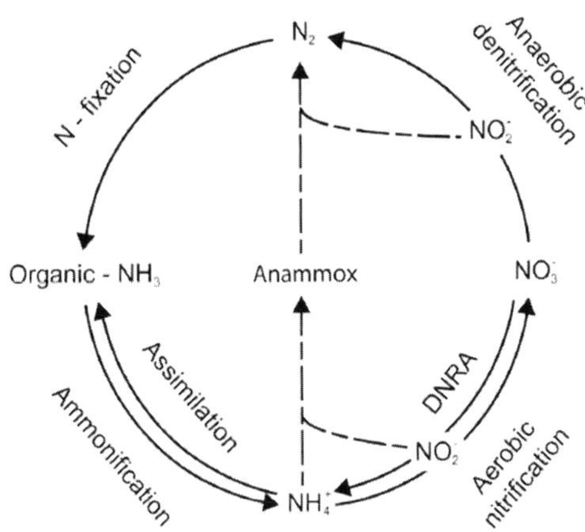

FIGURE 14-1 Simplified N cycle showing the complexity of nitrogen transformations in freshwater. NO_3^- is nitrate, NO_2^- is nitrite, NH_4^+ is ammonium, NH_3 is ammonia, N_2 is dinitrogen, and DNRA is dissimilatory NO_3^- reduction to NH_4^+. Anammox is anaerobic NH_4^+ oxidation. *(From Trimmer et al., 2003.)*

Wetzel's Limnology, Fourth Edition
https://doi.org/10.1016/B978-0-12-822701-5.00014-8

TABLE 14-1 Definition of Major Nitrogen Transformations and Its Chapter Section

Nitrogen transformations	Process definition	Sections in the chapter
N-fixation	Atmospheric nitrogen is converted by bacteria or Cyanobacteria to NH_3	IA; IIB
Ammonification (same as mineralization)	Bacteria convert organic nitrogen to NH_4^+	IA
Assimilation	Organisms assimilate NH_4^+ and convert it to organic nitrogen	IA
Aerobic nitrification	Biological oxidation of NH_4^+ to NO_2^- followed by the oxidation of NO_2^- to NO_3^-	IB
Anaerobic denitrification	Microbes reduce NO_3^- ultimately to dinitrogen (N_2) through a series of intermediate gaseous nitrogen oxide products	IIC
Dissimilatory NO_3^- reduction to NH_4^+ (DNRA; same as NO_3^-/NO_2^- ammonification)	Microbes oxidize organic matter and reduce NO_3^- to NO_2^-, then NH_4^+ in anaerobic conditions	IIC
Anaerobic NH_4^+ oxidation (anammox)	Microbes convert 1:1 molecule ratio of $NH_4^+$and NO_3^-/NO_2^- to dinitrogen in the absence of oxygen and without carbon supply	IID

knowledge of the importance of a wide range of physical, chemical, and biological drivers of these reactions continues to grow over time. The following sections of this chapter focus on sources, transformations, and sinks/outputs of nitrogen from a range of freshwater systems.

Combined nitrogen occurs as ammonium (NH_4^+), hydroxylamine (NH_2OH), nitrite (NO_2^-), nitrate (NO_3^-), and dissolved and particulate organic nitrogen (Table 14-2). NH_4^+ can range from undetectable levels to 5 mg L^{-1} in surface waters, with concentrations in unpolluted waters typically on the low end of this scale. Concentrations of NO_3^- range from undetectable levels in unpolluted waters to 10 mg L^{-1} or above in polluted waters. Groundwater may be an important source. NO_2^- levels of natural oxygenated waters are generally very low, in the range of 0 to 0.01 mg L^{-1}. Concentrations of NO_2^- increase in the anaerobic hypolimnion of lakes under reducing conditions (e.g., Brezonik and Lee, 1968; Overbeck, 1968) and in streams and lakes receiving

TABLE 14-2 Nitrogen Species and Their Chemical Formulas

Nitrogen species	Chemical formula
Volatilized ammonia	NH_3
Ammonium	NH_4^+
Hydroxylamine	NH_2OH
Nitrite	NO_2^-
Nitrate	NO_3^-
Dissolved organic nitrogen (DON)	not available[a]
Particulate organic nitrogen (PON)	not available[a]
Particulate inorganic nitrogen (PIN)	not available[a]
Dinitrogen	N_2
Nitrous oxide	N_2O

[a]*These forms of nitrogen are a group of compounds rather than individual chemicals.*

heavy organic matter pollution. Dissolved organic nitrogen (DON), much of which occurs in forms resistant to rapid bacterial degradation, commonly accounts for more than one-half of the total dissolved nitrogen (TDN). Much of the particulate organic nitrogen (PON) is bound up in algae, bacteria, and detritus, and the relative contribution to the total nitrogen (TN) load of a waterway can vary substantially. Particulate inorganic nitrogen, where nitrogen is sorbed to colloidal and particulate matter, can also vary substantially in its contribution to the nitrogen load between water bodies.

Lake productivity (Chapter 9) is linked to nitrogen concentrations, and this productivity can be categorized to provide useful definitions of scales of productivity, with inferences about ecosystem health (Table 14-3).

A. Ammonium

Much of the NH_4^+ in aquatic systems arises as a primary end product of the decomposition of organic matter by heterotrophic bacteria from the deamination of proteins, amino acids, urea, and other nitrogenous organic compounds. Intermediate nitrogen compounds are formed in the progressive degradation of organic material but rarely accumulate, because deamination by bacteria proceeds rapidly. Phagotrophic protists, particularly bacterivorous flagellates, can appreciably accelerate the regeneration of NH_4^+ by the consumption of bacteria and the release of NH_4^+ ions (Suzuki et al., 1996). NH_4^+ can also be generated by *dissimilatory NO_3^- reduction to NH_4^+* (DNRA) (Fig. 14-1), as will be discussed later in this chapter. Although NH_4^+ is an excretory product of higher aquatic animals, this nitrogen source is quantitatively minor in comparison to that generated by bacterial decomposition.

TABLE 14-3 General Relationship of Lake Productivity to Mean Concentrations of Epilimnetic Nitrogen. nd = no data.

General level of lake productivity	Change in alkalinity in epilimnion in summer (meq L^{-1})	Inorganic N (mg m^{-3})	Approximate mean organic N (mg m^{-3})
Ultraoligotrophic	<0.2	<200	<200
Oligomesotrophic	0.6	200–400	200–400
Mesoeutrophic	0.6–1.0	300–650	400–700
Eutrophic	nd	500–1500	700–1200
Hypereutrophic	>1.0	>1500	>1200

Wetzel, 2001.

In addition to NH_4^+, undissociated NH_4OH can also be present, which is highly toxic to many organisms, especially fish (Trussell, 1972). However, the proportion of NH_4OH is typically low at intermediate pH and temperatures. The approximate ratios of NH_4^+ to NH_4OH are as follows (Hutchinson, 1957):

pH 6	3000:1
pH 7	300:1
pH 8	30:1
pH 9.5	1:1

Detailed dissociation relationships with pH and temperature are given by Trussell (1972) and Emerson et al. (1975). NH_4^+ can be strongly sorbed to particulate and colloidal matter, especially in alkaline lakes containing high concentrations of humic dissolved organic matter.

At high pH values, NH_3 gas can be formed by the deprotonization of NH_4^+. The rate of NH_3 volatilization from wetlands has been shown to increase with increasing concentrations of NH_4^+ and pH of the water overlying the sediments, as well as with greater wind velocities and temperatures (Bouwmeester and Vlek, 1981; Jones et al., 1982).

The distribution of NH_4^+ in freshwaters is highly variable regionally, seasonally, and spatially within streams and lakes and depends upon the level of productivity and the extent of pollution from organic matter. Although generalizations are difficult to make, concentrations of NH_4^+ in well-oxygenated waters are usually low. In the trophogenic (photic) zone NH_4^+ is rapidly assimilated by algae and represents the most significant source of nitrogen for the plankton in many lakes (Liao and Lean, 1978). The energy necessary to assimilate nitrogen by plants is lowest for NH_4^+ and increases for other nitrogen sources, such as NO_3^- (and N_2–N for N_2-fixing Cyanobacteria). Some experiments show that Cyanobacteria can grow faster on NH_4^+, while others have comparable growth rates using NH_4^+ and NO_3^-, as well as urea (Chaffin and Bridgeman, 2014; Erratt et al., 2018). Other studies show that NO_3^- uptake can be suppressed by NH_4^+ concentrations as low as 2 μg NH_4–N L^{-1} (Priscu et al., 1989). A few algae, such as a species of *Chrysochromulina* (a haptophyte), are apparently only able to utilize NH_4^+ as an inorganic nitrogen source (Wehr et al., 1987).

Turnover rates of NH_4^+ can be in the order of a few hours (Burford et al., 2006). Thus concentrations of NH_4^+ are commonly low in unproductive oligotrophic waters, in the trophogenic zones of most lakes, and in most lakes after periods of circulation (Fig. 14-2).

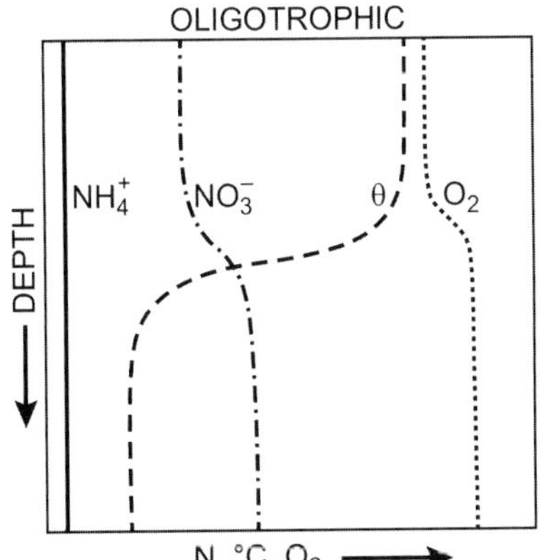

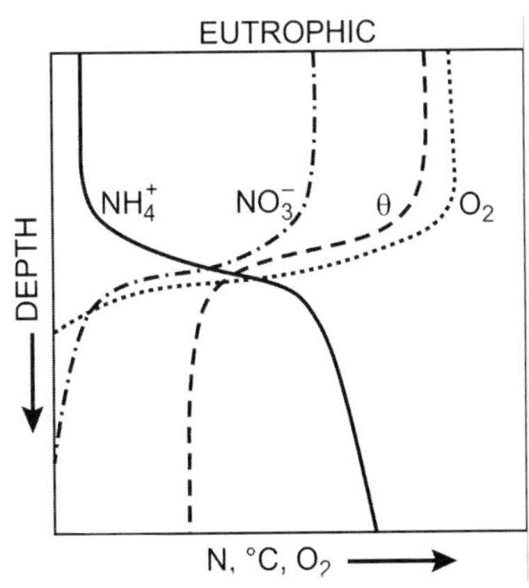

FIGURE 14-2 Generalized vertical distribution of NH_4^+ and NO_3^- in stratified lakes of very low (oligotrophic) and high productivity (eutrophic). θ is temperature.

When appreciable amounts of organic matter sink into the hypolimnion of stratified lakes, NH_4^+ can accumulate due to microbial activity. The accumulation of NH_4^+ greatly accelerates as the hypolimnion becomes anoxic. Under anaerobic conditions, bacterial nitrification of NH_4^+ to NO_2^- and NO_3^- ceases, as the redox potential is reduced to below about +0.4 V.

In the sediments a large percentage of the NH_4^+ is adsorbed on sediment particles. With the loss of the oxidized microzone at the sediment–water interface under anoxic hypolimnetic conditions (Chapter 11), the adsorptive capacity of the sediments is greatly reduced (Kamiyama et al., 1977; Verdouw et al., 1985). A marked release of NH_4^+ from the sediments then occurs. Diffusion transport into the overlying water approximately equals production and release into the interstitial waters of the sediments. These diffusion rates of NH_4^+ into the overlying water can be increased severalfold by the activities of benthic invertebrates such as chironomid larvae, tubificid worms, and bivalve mollusks (e.g., Fukuhara and Yasuda, 1989; Svensson, 1997). The NH_4^+ enhancement apparently results more from water movements associated with respiratory behavior rather than from direct excretion by fauna. Excretion rates of NH_4^+ of four species of chironomids and tubificids ranged from 0.33 to 2.87 μg N mg DW (dry weight animal)$^{-1}$ day^{-1} at 15°C.

If light reaches the sediments in amounts sufficient to support benthic algae, these organisms can assimilate NH_4^+ from the interstitial water and prevent the flux of NH_4^+ from the sediment to the water (Jansson, 1980; Reuter et al., 1986; Risgaard-Petersen et al., 1994). Because the uptake rates are in part dependent upon photosynthesis, a marked diurnal variation in release rates can occur, with reduced uptake during periods of darkness. These sources of interstitial nitrogen often serve as the dominant inorganic nitrogen source, or supplement N_2 fixation, specifically for the benthic algae and microbes attached at the sediment–water interface, particularly in nitrogen-deficient lakes and streams. These attached microbial communities can release appreciable amounts of dissolved organic nitrogen into the water above the sediments.

If aquatic macrophytes, i.e., freshwater plants (Chapter 24) are rooted in the sediments, these plants are capable of absorbing large quantities of nitrogen from the sediments. Much of the nitrogen can be immobilized in the belowground rooting tissues and in the above-sediment foliage tissues. Much of the nitrogen is assimilated as NH_4^+ (e.g., Dean and Biesboer, 1985) and can remain for long periods in living tissues and particulate detritus after senescence. Thus, in littoral and wetland areas of freshwater ecosystems nitrogen storage in living and senescent macrophytes can be an important component (e.g., Sarvala et al., 1982; Reddy and Patrick, 1984; Bowden, 1987) However, the nutrients stored in macrophytes can be released back to the water column again when environmental factors change, such as when plants die back at low temperatures during winter, when there is low light availability due to eutrophication, and when experiencing hydrological stress. For example, both decayed macrophytes and sediments have been shown to release a substantial amount of nutrients after drying, followed by rewetting (Bostic and White, 2007; Wilson and Baldwin, 2008; Kerr et al., 2011; Steinman et al., 2014; Lu et al., 2017), which might be further exacerbated by climate change (see Section IVB).

B. Nitrate and nitrification

Nitrification may be broadly defined as the biological conversion of organic and inorganic nitrogenous compounds from a reduced state to a more oxidized state (Alexander, 1965). Of the numerous oxidation and reduction stages outlined in Fig. 14-1, initial nitrification by bacteria, fungi, and autotrophic organisms involves (Kuznetsov, 1970):

$$NH_4^+ + 1\frac{1}{2}O_2 \leftrightarrow 2H^+ + NO_2^- + H_2O$$

$$\left(\Delta G_0' = -276.1 \text{ KJ mol}^{-1}\right)$$

$\Delta G_0'$: Gibbs energy of formation, kilojoules per mole (KJ mol^{-1}), which is the thermodynamic potential that is minimized when a system reaches chemical equilibrium at constant pressure and temperature.

This proceeds by a series of oxidation stages through hydroxylamine and pyruvic oxime to nitrous acid:

$$NH_4^+ \rightarrow NH_2OH \rightarrow H_2N_2O_2 \rightarrow HNO_2$$

The intermediate products are labile to physical and heterotrophic oxidation and are only found rarely in significant quantities relative to other forms of combined nitrogen (cf. Baxter et al., 1973). Much of the energy (total exothermic energy = −351.5 KJ mol^{-1}) released by the oxidation series is used to reduce carbon dioxide (CO_2) in the formation of organic matter; detailed reactions are discussed by Alexander (1965), Kuznetsov (1970), and Fenchel and Blackburn (1979).

The nitrifying bacteria capable of oxidation of $NH_4^+ \rightarrow NO_2^-$ are largely confined to NH_4^+ oxidizing archaea and NH_4^+ oxidizing bacteria (Shiozaki et al., 2016). These bacteria are mesophilic (i.e., grow best at a moderate temperature), but with a wide temperature tolerance range (1–37°C), and grow optimally at a pH near neutrality.

Oxidation of NO_2^- proceeds further to NO_3^- by:

$$NO_2^- + \frac{1}{2}O_2 \leftrightarrow NO_3^-$$

$$\left(\Delta G_0' = -75.3 \text{ KJ mol}^{-1}\right)$$

NO_3^- oxidizing bacteria are involved in this oxidation (Könneke et al., 2005). *Nitrobacter* is the primary bacterial genus. *Nitrobacter* is somewhat less tolerant of low temperatures and high pH, conditions that can lead to some accumulation of NO_2^-. The release of energy for the synthesis of organic matter by the oxidation of NO_2^- is much lower, at -75.3 KJ mol^{-1}, than that of NH_4^+ to NO_2^-.

The overall nitrification reaction is therefore:

$$NH_4^+ + 2O_2 \rightarrow NO_3 - + H_2O + 2H^+$$

This reaction requires 2 moles of oxygen for the oxidation of each mole of NH_4^+.

Nitrification is a key component of the nitrogen cycle in lakes. However, the recent discovery of complete NH_4^+ oxidizers (*comammox*) has shown that the oxidization of NH_4^+ to NO_3^- can be achieved by a single microorganism (Daims et al., 2015; Kessel et al., 2015). The genome of this microbe contains genes encoding NH_4^+ monooxygenase, hydroxylamine dehydrogenase, and NO_2^- oxidoreductase, indicating that they have the potential for full-scale NH_4^+ oxidation (Daims et al., 2015; Kessel et al., 2015).

Although conditions must be aerobic for nitrification to occur, these processes will continue until concentrations of dissolved oxygen decline to about 0.3 mg L^{-1}. Below this concentration, diffusion rates of oxygen to the bacteria become critical. Nitrification is greatly reduced in undisturbed sediments because oxygen is very low or absent (Chen et al., 1972). NO_3^- may diffuse to the water following nitrification in the well-oxygenated surficial sediments of the littoral zone or during periods of circulation of the water and disturbance of sediments (cf. Landner and Larsson, 1973; Laurent and Badia, 1973).

Nitrification is severely inhibited by certain dissolved organic compounds (Chapter 28), especially by tannins and their decompositional derivatives (Rice and Pancholy, 1972, 1973). Therefore it is likely that rates of nitrification are lower in neutral or alkaline waters containing high concentrations of dissolved humic organic matter. Moreover, nitrification proceeds more slowly in acidic waters, such as in acid bogs and acidic bog lakes, where the pH is 5 or less. NO_3^- produced in such lakes is probably utilized as rapidly as it is formed, so most of the time, only very low or undetectable quantities are found.

C. Dissolved and particulate organic nitrogen

Concentrations of dissolved organic nitrogen (DON) are frequently higher than dissolved inorganic nitrogen ($DIN = NH_4^+ + NO_3^-$) (Bronk et al., 2007), with the DON fraction often exceeding 50% of the total dissolved nitrogen (TDN) pool in freshwaters (Berman and Bronk, 2003), especially in highly developed catchments (Petrone et al., 2009). In Lake Taihu, China, average DON concentrations accounted for up to 50% of TDN (Zhang et al., 2015). In a study of lakes and streams in Wales, DON export from catchments was related to soil type and vegetation cover, with more vegetation (conifers) leading to less DON loss (Willett et al., 2004). There was no seasonality in DON concentrations reflecting the *refractory* nature (difficult to degrade) of many of these compounds.

Many of the DON compounds are relatively refractory. Less than one-third occurs as free amino nitrogen. The simple amino acids are substrates that are readily utilized by microbes; rates of decomposition are high and result in low instantaneous concentrations of free amino acids in freshwaters.

As with organic carbon, the DON of lakes and streams can be from 5 to 10 times greater than particulate organic nitrogen (PON) contained in the plankton and seston (e.g., Serruya et al., 1975; Takahashi and Saijo, 1981; Zehr et al., 1988). The ratios of DON to PON decrease as the lakes become more eutrophic, are closer to 1:1 in the *trophogenic* (photic) zone, and increase in the *tropholytic* (aphotic) zone. More organic nitrogen is typically synthesized by small phytoplanktonic algae (<10 μm) per unit cell volume than by larger forms (Manny, 1972). A significant portion of algal intracellular nitrogen (10–20% in the cyanobacterium *Oscillatoria*) is released extracellularly, mainly as protein and NH_4^+ nitrogen with smaller amounts of NO_2^- and amino acid nitrogen (Meffert and Zimmermann-Telschow, 1979). Bacterial utilization of these organic compounds, particularly the amino acids, is extremely rapid (Gardner et al., 1987, 1989) and linked with organic carbon availability (Horňák and Pernthaler, 2020). Different types of bacteria have been shown to have preferences for different DON forms; for example, one study found bacterioplankton groups had distinct preferences for a range of low-molecular-weight organic compounds (Salcher et al., 2012).

Algae, and especially Cyanobacteria, also excrete polypeptides and other organic compounds that are capable of forming complexes with metals, such as iron and copper, as well as phosphates. These complexed organic compounds differ in their solubility and physiological availability their solubility and physiological availability (Fogg and Westlake, 1955; Murphy et al., 1976; Tuschall and Brezonik, 1980). Similar nitrogenous compounds have been found to be secreted by larger aquatic plants (Wetzel and Manny, 1972); in some situations where the littoral zone is extensively vegetated, the release of DON by macrophytes can form a major source of organic nitrogen for the lake. Furthermore, as aquatic vascular vegetation decomposes, large quantities of organic nitrogen are released (Nichols and Keeney, 1973). Much of this organic nitrogen is absorbed by the sediments and is utilized by attached microbiota, where decomposition can rapidly become limited by inorganic nitrogen, especially under anaerobic conditions.

There is limited information on which forms of DON are bioavailable to phytoplankton or bacteria (Bronk et al., 2007; Su et al., 2016). However, it has been established that many species of algae from diverse groups are able to utilize urea, simple amino acids, and a range of other simple nitrogen compounds as a nitrogen source. Among the algae of Lake Biwa, Japan, for example, relative assimilation rates were highest for NH_4^+ (average 74% of total), followed by urea (average 20%), then NO_3^- (average 6% of total) (Mitamura and Saijo, 1986). A study of the green alga, *Raphidocelis subcapitata*, found that a wide range of bioavailable compounds, such as amino acids, urea, DNA, and RNA, as well as some humic acids, could be used for growth (Fan et al., 2018). The filamentous Cyanobacteria i.e., *Planktothrix rubescens*, the *Microcystis* complex, and *Spirulina platensis* are known to assimilate several low-molecular-weight organic compounds (Dai et al., 2019; Salcher et al., 2012; Shanthi et al., 2018). Conversely, at higher concentrations, the amino acids proline and arginine have been found to inhibit Cyanobacterial growth (Dai et al., 2019; Neilen et al., 2020).

D. Catchment scale nitrogen budgets

One way to quantify inputs, outputs, and retention of nitrogen by freshwater systems is to use nitrogen budgets where all sources are accounted for. The quantification of nitrogen inputs and outputs for rivers and streams is challenging at the watershed scale, so it is normally estimated using hydrological and water quality models, including simple export coefficient models (Johnes, 1996); statistical models (Grizzetti et al., 2005); mechanistic models, such as SWAT (Witte, 2009); and hybrid mechanistic-statistical models, such as SPARROW (Elliot et al., 2005).

Terrestrial sources are the main nitrogen sources of rivers and streams during storm events. Groundwater nitrogen deliveries can be significant especially when groundwater is enriched with nitrogen via leaching and infiltration from surface soils (Durand et al., 2011), especially when fertilizers are applied to soils (Van Drecht et al., 2003). Studies also showed that atmospheric deposition can be a significant nitrogen source in some rivers (Lawrence et al., 2000; Boyer et al., 2002), with nitrogen fixation (use of nitrogen gas by organisms for growth; Section IIB) being important in some lentic systems. The output of nitrogen is mainly through denitrification (Section IIC), *anammox*, that is, *anaerobic* NH_4^+ *oxidation* (Section IID), and export to oceans. According to a review of nitrogen removal in streams of agricultural catchments, in-stream processing can remove 10–70% of the total nitrogen load to drainage networks at the watershed scale on an annual basis (Birgand et al., 2007).

Maranger et al. (2018) made global estimates of nitrogen inputs and outputs to and from freshwaters and determined that the major nitrogen input to freshwaters was from catchment (= watershed) inputs, with atmospheric inputs being approximately 10% of this (Table 14-4). Approximately 45% of inputs from runoff and the atmosphere were lost to the atmosphere, with only 15% buried and the remainder exchanged at the

TABLE 14-4 Estimates of C, N, and P Terrestrial Inputs, Atmospheric Inputs and Losses, Reservoir Processing, and Losses to Burial Through the Freshwater Pipe as well as Export to the Coastal Ocean.

		Tg year^{-1}					Molar ratio			
		DOC	POC	DIC	N	P	C:N	C:P	N:P	C:N:P
Catchment inputs		1189	116	1595	98	9	15.5	374.6	24.1	375:24:01
Atmospheric flux[a]	Input	—	2288	—	10	0.04	266.9	—	—	—
	Output	1018	2056	—	49	0.04	73	—	—	—
	Net	−786[a]	−1064[b]	−39	—	—	—	—	—	—
Total burial		—	150	—	16	5	11	77.5	7	78:7:1
Coastal exports		171	198	531	43	4	10	238.3	23.8	238:24:1
Reservoir processing[c]		—	60	38.6	6.5	1.3	10.8	119.2	11.1	119:11:1

[a]*Calculated as the difference between gross primary production (atmospheric inputs to POC) minus respiration of DOC and POC (DOC and POC atmospheric flux output).*
[b]*Calculated as the difference between watershed inputs of DIC and DIC coastal exports.*
[c]*POC and P is burial; DIC is CO_2 respiratory losses; N is both denitrification and burial; C:N ratios are sum of losses to atmosphere and burial.*
Molar ratios from different terms also provided; only organic carbon estimates are used in ratios.
From Maranger et al. (2018). C = Carbon, DIC = dissolved inorganic carbon, DOC = dissolved organic carbon, N = nitrogen, P = phosphorus, POC = particulate organic carbon.

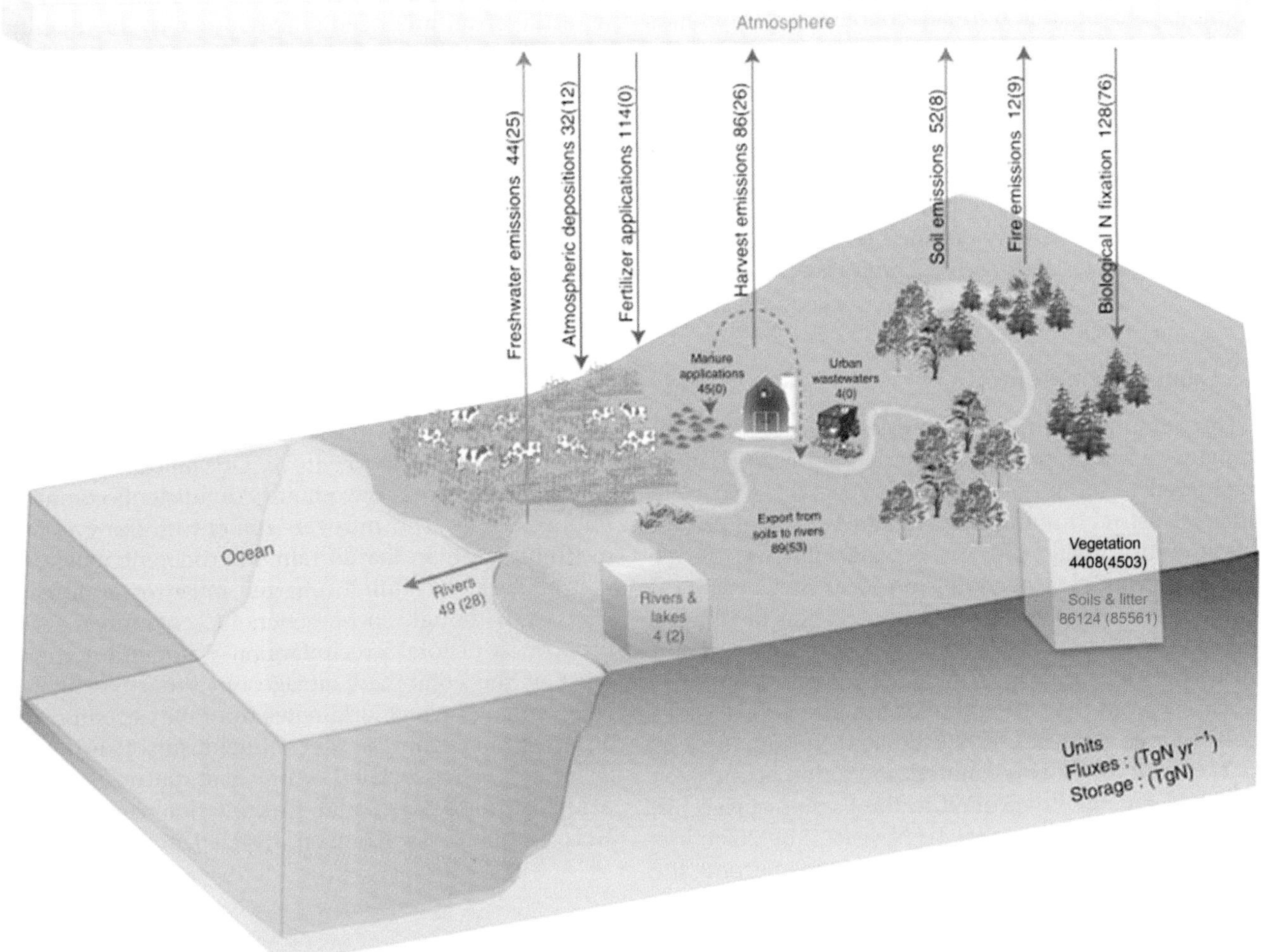

FIGURE 14-3 Changes in the global N budget from contemporary to preindustrial periods. *Numbers* represent global land N storage in Tg N or annual N exchange fluxes in Tg N year^{-1} for contemporary (1991–2005 average) and preindustrial (1831–1860 average in parentheses) times. *(These results are summarized, discussed, and compared with reported estimates from various scientific studies in Supplementary Table 1 and Supplementary Note 1 from Lee et al., 2019.)*

mouth. Residence time was shown to play a critical role in the processing of nitrogen, phosphorus, and carbon through the riverine pipe, where longer water residence times from streams to lakes result in substantial increases in carbon:nitrogen (C:N), carbon:phosphorus (C:P), and nitrogen:phosphorus (N:P) ratios. Nitrogen removal through *anammox* (to be discussed in this chapter in Section IID) has also been shown to be important in turbid river systems (Xia et al., 2018).

Nitrogen budgets at a global scale can also be used to highlight the impact of anthropogenic activities. A comparison of preindustrial nitrogen inputs and outputs from catchments showed that annual fluxes have increased substantially (Fig. 14-3).

II. Nitrogen sources, transformations, and fate in lakes and reservoirs

Nitrogen occurs in freshwaters in numerous forms: NH_4^+, NO_2^-, and NO_3^-, dissolved molecular N_2, organic compounds ranging from amino acids and amines to proteins and recalcitrant humic compounds of low nitrogen content. Sources of nitrogen include: (1) precipitation falling directly onto the lake surface, (2) nitrogen fixation both in the water and the sediments, and (3) inputs from surface and groundwater drainage. Losses of nitrogen occur via: (1) effluent outflow from the basin, (2) reduction of NO_3^- to N_2 by bacterial denitrification and anammox with subsequent return of N_2 gas to the atmosphere, and (3) permanent sedimentation loss of inorganic and organic nitrogen-containing compounds to the sediments.

Key processes and inputs outlined below are: atmospheric inputs to lakes and reservoirs, including nitrogen fixation; nitrogen transformation processes within these systems; and atmospheric removal processes. Hydrological outputs and inputs of nitrogen are also important and are briefly outlined in the section on nitrogen budgets in lakes and reservoirs (Section IIE).

A. Atmospheric inputs

The amount of influent nitrogen to lakes and their drainage areas from atmospheric sources is often considered to be minor in comparison with that from direct terrestrial runoff. However, nitrogen from precipitation and direct bulk (particulate) fallout can be extremely variable depending on local meteorological conditions, wind patterns, and the location of streams and lakes with respect to industrial and agricultural outputs. Globally, atmospheric nitrogen deposition has increased more than 10-fold over the past few decades (Galloway et al., 2008). It is estimated that in 1860, 34 Tg N $year^{-1}$ of N was emitted as oxides of nitrogen and NH_3 and then deposited to the Earth's surface; in 1995 it had increased to 100 Tg N $year^{-1}$; by 2050, it is projected to be 200 Tg N $year^{-1}$. China has become the largest global emitter of reactive nitrogen in the last two decades (Liu et al., 2013). The combination of higher atmospheric loads and the building of more reservoirs has doubled nitrogen inputs to lakes, rivers, and reservoirs in China from the 1990s to the 2010s (Gao et al., 2020). In Lake Taihu, China, atmospheric deposition accounts for 12% of riverine nitrogen inputs to the lake (Chen et al., 2018).

Nitrogen deposition into lakes, reservoirs, and rivers is having significant effects on nitrogen concentrations and productivity. Nitrogen mass-balance techniques applied to a large number of drainage basins and lakes in central Ontario (Canada), for example, indicated that, on a regional area-weighted basis, 67% of bulk atmospheric TN was stored or denitrified terrestrially, 12% was denitrified in lakes, 4% was stored in lake sediments, and 17% was exported from the lakes (Molot and Dillon, 1993). For example, in relatively oligotrophic mountainous regions of granitic bedrock, precipitation is a major source of nitrogen (Likens et al., 1977). Nitrogen deposition in mountain lakes can also increase the rates of denitrification (Palacin-Lizarbe et al., 2020). Inorganic nitrogen input by precipitation similarly was found to be a major source of loading to the drainage basin of Lake Tahoe, California–Nevada (Coats et al., 1976; Jassby et al., 1994). A global analysis has shown that increased atmospheric nitrogen deposition in lakes as a result of anthropogenic activities may have shifted some lakes from nitrogen to phosphorus limitation. This impacts nitrogen cycling and higher trophic levels (Elser et al., 2009). Critical thresholds in atmospheric nitrogen for nutrient enrichment and acidification of lakes across the United States have been identified to guide water quality management (Baron et al., 2011).

Atmospheric nitrogen may enter a drainage basin in many forms: as dissolved N_2, nitric acid, NH_4^+, NO_3^-, as NH_4^+ adsorbed to inorganic particulate matter, and as organic compounds, which can occur in either dissolved or particulate phases. No direct relationship exists between the volume of rainfall or snowfall and the quantity of nitrogen influx per area of land or water (Chapin and Uttormark, 1973). Dry fallout can contain as much as 10 times the quantity of nutrients commonly found in rain. The nitrogen content of snow is often much higher than that of rain and can contribute up to half of the total annual nitrogen influx to a stream or lake, even though snow generally constitutes a small proportion of total precipitation. In nonpolluted areas most of the combined nitrogen in the atmosphere is NH_3, much of which originates from the decomposition of terrestrial organic matter (Hutchinson, 1944). Atmospheric NH_3 associated with dust particles can be oxidized to NO_3^- so that precipitation contains both NH_4^+ and NO_3^- (Hutchinson, 1944, 1975).

B. N_2 fixation

Nitrogen fixation, in which atmospheric nitrogen is utilized by microbes or Cyanobacteria for growth, plays an important role in the productivity of streams and lakes, particularly where dissolved inorganic nitrogen is depleted. In the open waters of lakes Cyanobacteria that possess *heterocysts* (Chapters 17 and 18) can be an important source of nitrogen (Fogg, 1971, 1974; Riddolls, 1985). Heterocysts (also called heterocytes) are specialized cells that occur singly in some filamentous Cyanobacteria and are the site of nitrogen fixation (Wolk, 1973; Haselkorn and Buikema, 1992). Nitrogen fixation has also been found to occur in some unicellular forms that do not produce heterocysts (Fogg, 1974). However, among many species of Cyanobacteria, such as *Dolicospermum* spp. (previously *Anabaena*), numbers of heterocysts correspond approximately to the observed nitrogen-fixing capacity (e.g., Horne et al., 1972, 1979; Riddolls, 1985; Willis et al., 2016).

Nitrogen fixation is primarily light dependent in that the process requires reducing power and adenosine triphosphate (ATP), both of which are generated in photosynthesis. Electrons for the reduction of N_2 are supplied by dinitrogenase reductase in a highly *endergonic reaction* (a chemical reaction where energy is absorbed) that requires about 12–15 moles of ATP per mole of N_2 reduced (Fay, 1992). Nitrogen-fixing photosynthetic bacteria can fix limited quantities of N_2 in the dark, but at rates usually <10% of maximum daytime rates (Horne, 1979; Levine and Lewis, 1984; Livingstone et al., 1984; Storch et al., 1990). In full sunlight N_2 fixation is often inhibited at the surface (as is primary production). Below this, there is a nearly exponential decrease as light attenuates (e.g., Horne, 1979; Lewis and Levine, 1984). The pattern of nitrogen fixation rates through the water column is similar to that of photosynthesis (Ward and Wetzel, 1980).

The use of nitrogen fixation for growth by Cyanobacteria is energetically expensive compared to the use of dissolved inorganic sources such as NH_4^+ or NO_3^-. Therefore species will preferentially use the inorganic sources and growth is typically faster when these sources are available (Burford et al., 2006). Additionally, alternating between nitrogen fixation and the use of dissolved inorganic sources may not happen immediately, particularly for species that lose their heterocysts when nitrogen fixation is not occurring (Willis et al., 2016).

Nitrogen fixation has also been correlated positively with concentrations of DON occurring in the water (Horne and Fogg, 1970; Horne et al., 1972; Paerl, 1985). Algae and Cyanobacteria release extracellularly many simple and complex organic carbon and nitrogen compounds. Heterotrophic bacteria colonize common nitrogen-fixing Cyanobacteria, such as *Anabaena*, particularly during the maxima of N_2-fixing blooms (Paerl, 1978, 1985; Paerl and Prufert, 1987). These Cyanobacterial-heterotrophic bacterial aggregates form internal microenvironments with microzones of redox conditions distinct from those of the surrounding pelagic or benthic habitats. The oxygen-depleted microzones adjacent to the heterocysts protect the highly oxygen-sensitive nitrogenase. In temperate winters N_2 fixation is nonexistent or greatly reduced (Billaud, 1968; Horne and Fogg, 1970; Toetz, 1973). However, in productive tropical lakes where the periodicity of physicochemical factors and Cyanobacteria is less marked, nitrogen fixation rates can extend throughout the year (e.g., Moyo, 1991; Burford et al., 2006).

Evaluations of nitrogen cycles of lakes or streams require estimates of the total nitrogen fixed per annum. Planktonic nitrogen fixation tends to be low in oligotrophic and mesotrophic lakes (generally <<0.1 g N m^{-2} $year^{-1}$) and may be higher in eutrophic lakes (0.2–9.2 g N m^{-2} $year^{-1}$) (Howarth et al., 1988a, 1988b). Benthic fixation in oligotrophic ecosystems is often dominated by Cyanobacteria (Howarth et al., 1988b) and tends to increase in cyanobacterial mats of very productive, shallow environments (1–76 g N m^{-2} $year^{-1}$). Despite the relatively small areas of such mats, their nitrogen contributions to the entire ecosystem can be disproportionally large.

Nitrogen fixation may be insufficient at balancing system nitrogen deficits in lakes and reservoirs, and therefore nitrogen will not accumulate relative to phosphorus (Grantz et al., 2014). A study using a mass balance model suggested that approximately half of oligotrophic lakes in the United States had a stoichiometric nitrogen deficit relative to phosphorus, while a higher percentage of eutrophic and hypereutrophic lakes (72–89%, respectively) had a similar nitrogen deficit (Scott et al., 2019). The process of nitrogen fixation cannot be viewed in isolation. Phosphorus is a critical element needed to support nitrogen fixation rates. Rates of nitrogen fixation are greatly enhanced when the productivity of a lake is increased by phosphorus fertilization (e.g., Lean et al., 1978; Lundgren, 1978).

Nitrogen fixation capabilities are widespread throughout other photosynthetic bacteria. Heterotrophic nitrogen fixation is commonly disregarded, on the premise that these nitrogen-fixing bacteria are limited by the low availability of exogenous carbohydrate. About 1 to 25 mg of nitrogen can be fixed per gram of carbohydrate utilized (Stewart, 1969; Hill, 1992). Sufficiently large quantities of soluble carbohydrate are rarely available in natural waters, and there is an intense competition for these substrates by heterotrophic bacteria incapable of N_2 fixation. The most common heterotrophic N_2-fixing bacteria comprise several species of *Azotobacter* and *Clostridium pasteurianum*, while several methane-oxidizing bacteria are also capable of N_2 fixation (reviewed by Zeikus, 1977; Rudd and Taylor, 1980).

In addition to the inputs of nitrogen from drainage runoff from terrestrial sources, lakes and streams often are bordered by dense stands of shrublike trees, some of which can fix nitrogen from the atmosphere. Common species of the genera *Alnus* (alder) and *Myrica* (bayberry) are nonleguminous angiosperms that form large nodules containing an actinomycetal fungal endophyte at or just below the soil surface. *Alnus* trees have been implicated as a significant nitrogen source for streams and lakes in glaciated regions of Alaska that are particularly nitrogen deficient (Goldman, 1960; Dugdale and Dugdale, 1961). A New Zealand study found that invasive nitrogen-fixing species, such as European Gorse (*Ulex europaeus*), can be a significant NO_3^- contributor to rivers and creeks compared with native forest, thus impacting water quality (Stewart et al., 2019).

Microbes and Cyanobacteria attached to macrophytes can also be a significant contributor of nitrogen to waterways. A study of the wetland sedge, *Schoenoplectus californicus*, found that microbes associated with the plant contributed fixed nitrogen to increase the nitrogen content by 13.8 to 32.5% (Rejmánková et al., 2018). The aquatic fern, *Azolla*, contains the symbiotic heterocyst-forming, N_2-fixing Cyanobacterium *Anabaena azollae* (Lumpkin and Plucknett, 1980; Peters and Meeks, 1989; Wagner, 1997). *Azolla* species are actively cultivated in rice agriculture to supplement alternative sources of nitrogen fertilization.

C. Nitrate reduction and denitrification

Microbes can reduce NO_3^- to N_2/N_2O through *denitrification* and anaerobic *NH_4^+ oxidation (anammox)* pathways, or to NH_4^+ through the dissimilatory NO_3^- reduction to NH_4^+ (DNRA) pathway (Fig. 14-4). Microbes involved in denitrification, anammox, and DNRA therefore compete for the same substrate (van den Berg et al., 2017; Keren et al., 2019) and can coexist in natural environments (Dong et al., 2011; Song et al., 2013; Luvizotto et al., 2019), including both sediments and the water column (Hamersley et al., 2009; Wenk et al., 2013).

The assimilation of NO_3^- and its reduction by aquatic plants can be a dominant process of NO_3^- removal in the trophogenic zone of lakes. As much as 60% of the photo-assimilated NO_3^- can be excreted as DON compounds, some of which are simple amino acids readily utilized by microbes (Chan and Campbell 1978).

However, NO_3^- and NO_2^- can also be denitrified by bacteria, producing nitrogen gas, with concomitant oxidation of organic matter. The general sequence of this process is:

$$NO_3- \rightarrow NO_2- \rightarrow N_2O \rightarrow N_2$$

which results in a significant reduction of combined nitrogen that can, in part, be lost from the ecosystem.

Many facultative anaerobic bacteria, particularly of the genera *Pseudomonas*, *Achromobacter*, *Escherichia*, *Bacillus*, and *Micrococcus*, can utilize NO_3^- (Alexander, 1961). The denitrification reactions are associated with the enzyme nitrogen reductase and require the cofactors iron and molybdenum. Denitrification operates similarly under both aerobic and anaerobic conditions (Bandurski, 1965).

An exemplary reaction of the oxidation of glucose and the concomitant reduction of NO_3^- is (Hutchinson, 1957):

$$C_6H_{12}O_6 + 12NO_3- \leftrightarrow 12NO_2- + 6CO_2 + 6H_2O$$

$$\left(\Delta G_0' = -1924.3 \text{ KJ mol}^{-1}\right)$$

and for the reduction of NO_2^- to molecular nitrogen:

$$C_6H_{12}O_6 + 8NO_2- \leftrightarrow 4NO_2 + 2CO_2 + 4CO_3^{2-} + 6H_2O$$

$$\left(\Delta G_0' = -3012.5 \text{ KJ mol}^{-1}\right)$$

Approximately as much free energy results as in the aerobic oxidation of glucose by dissolved O_2 ($\Delta G_0' = -2924.6$ KJ mol^{-1}). The denitrification reactions occur intensely in anaerobic environments, such as in the hypolimnia of eutrophic lakes (Fig. 14-2) and in anoxic sediments, where oxidizable organic substrates are relatively abundant. However, aerobic denitrifiers have also been found in natural environments, and some bacteria can simultaneously nitrify and denitrify under aerobic conditions in the water column (Guo et al., 2013; Ji et al., 2015; Zhao et al., 2018). In eutrophic lakes, littoral areas, and wetlands, denitrification is a major process influencing the distribution of NO_3^-. A range of organic carbon compounds, commonly found in agricultural and domestic pollutants, provide the energy sources in the NO_3^- reduction. Glucose, acetate, and particularly formate, among others, have been shown to induce high NO_2^- accumulation, caused, in part, by partial inhibition of denitrification (Kelso et al., 1997, 1999).

NO_3^- can also be denitrified concurrently with the oxidation of sulfur (Fig. 14-4). The process is accomplished by denitrifying sulfur bacteria, particularly *Thiobacillus denitrificans*, that utilize sulfur, or reduced sulfur compounds such as thiosulfate (Kuznetsov, 1970):

$$5S + 6KNO_3 + 2H_2O \rightarrow 3N_2 + K_2SO_4 + 4KHSO_4$$

$$5Na_2S_2O_3 + 8KNO_3 + 2NaHCO_3 \rightarrow 6Na_2SO_4 + 4K_2SO_4 + 2CO_2 + H_2O + 4N_2$$

Both processes occur chemosynthetically under dark, anaerobic conditions and yield relatively small changes in free energy. In a similar manner the *chemolithotrophic* (organisms capable of using inorganic, rather than organic, molecules for fueling reactions) iron bacterium *Gallionella ferruginea* reduces NO_3^- to N_2 via the

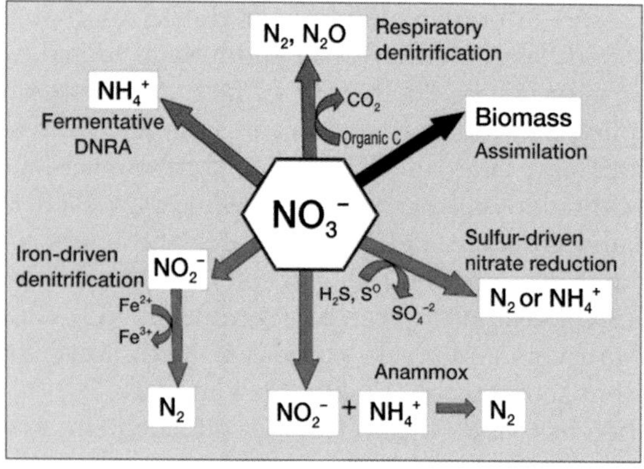

FIGURE 14-4 A conceptual diagram of NO_3^- removal pathways. *Blue arrows* denote autotrophic (microbes use inorganic material for energy and nutrients) pathways, while *purple arrows* denote heterotrophic (microbes use organic material for energy and nutrients) pathways. Fe^{2+} = ferrous, Fe^{3+} = ferric, H_2S = hydrogen sulfide, SO_4^{2-} = sulfate. *(From Burgin and Hamilton, 2007.)*

oxidation of ferrous ions under anaerobic conditions (Gouy et al., 1984).

The reduction of NO_3^- coupled to iron (Fe) cycling also exists, which can take place through both biotic pathways mediated by microbes (Weber et al., 2006) and abiotic pathways (Davidson et al., 2003) (Fig. 14-4). The abiotic pathway was found in an aquifer with a lower anoxic zone characterized by ferrous-rich water (Postma et al., 1991) and in forest soils (Dail et al., 2001). The biotic pathway might be more likely to occur in surface waters than the equivalent abiotic reaction (Burgin and Hamilton, 2007).

Global rates of denitrification have been estimated and compared across soils, groundwater, lakes and rivers, and estuaries (Seitzinger et al., 2006). Denitrification in groundwater can be significant, particularly in some areas of Europe, while denitrification in rivers appears to be more important in warmer areas of the world (Fig. 14-5). A caveat in this global data is that more studies have been done in Europe and North America than many other areas of the world, so this may bias interpretation.

The rate of denitrification, as of nitrification, decreases in acidic waters (Keeney, 1973). Denitrification rates also decrease with lower temperatures (Breitenbeck and Bremner, 1987; Brin et al., 2017). When substrates (NO_2^-/NO_3^-) are limited in the system, the winter denitrification rate can be higher than the

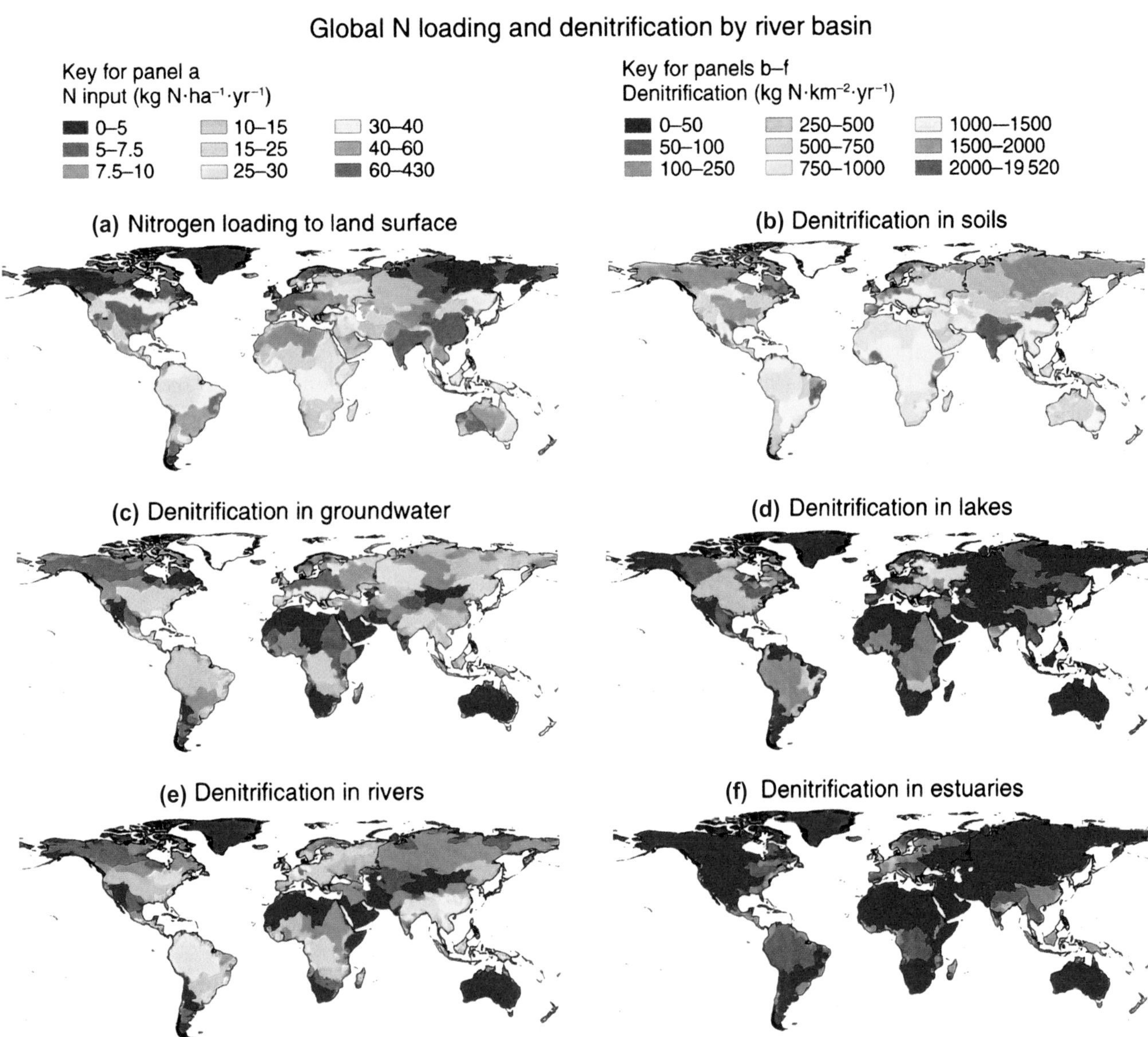

FIGURE 14-5 Model predictions of (a) N loading (kg N ha^{-1} $year^{-1}$) and denitrification of land-based N sources in terrestrial and aquatic systems (kg N km^{-2} $year^{-1}$), (b) Denitrification in soils, (c) Denitrification in groundwater, (d) Denitrification in lakes, (e) Denitrification in rivers, (f) Denitrification in estuaries. All rates are mapped as average rates for catchments. Note the difference in scale for N loading compared to denitrification. *(From Seitzinger et al., 2006.)*

summer denitrification rate, as has been found in multiples lakes in Canada and Japan (Hasegawa and Okino, 2004; Cavaliere and Baulch, 2018). Optimum rates of denitrification occur well above the temperatures of most natural freshwaters. Rates of denitrification by NO_3^--reducing bacteria in a eutrophic lake, Plußsee, Germany, showed marked seasonal and depth variations (Tan and Overbeck, 1973). Rates of NO_3^- reduction were particularly high and were correlated with denitrifier cell numbers during the early portion of summer stratification, before hypolimnetic NO_3^- concentrations were greatly reduced. High concentrations of oxygen and low levels of NO_3^- depressed rates of NO_3^- reduction but had relatively little effect on cell numbers, especially in winter. In the Baltic Sea of northern Europe the denitrification was more strongly driven by the chemolithotrophic process and limited to the water layers with NO_3^-/NO_2^- coexisting with sulfide (Hietanen et al., 2012; Dalsgaard et al., 2013).

Denitrification can be coupled with nitrification, resulting in low concentrations of NO_3^- (Kuenen and Robertson, 1994; Zou et al., 2014; Racchetti et al., 2017). In lake sediments denitrification of added $^{15}NO_3^-$ was rapid; within 2 h, up to 90% of added NO_3^- was reduced, much to $^{15}N_2$ gas (Chen et al., 1972). Similar results were found in the sediments of several oligotrophic lakes, where denitrification in the sediments was functionally coupled to the supply of NO_3^- by nitrification (Klingensmith and Alexander, 1983; Dodds and Jones, 1987). Much of the NO_3^- from lake sediments is incorporated into bacterial organic matter. Keeney et al. (1971) found that up to 37% of experimentally added NO_3^- (at the level of 2 mg $NO_3{-}N\ L^{-1}$) became incorporated into the organic fraction. The remainder was denitrified.

Denitrification rates of sediments can be three to four orders of magnitude greater than those of the overlying water within the same system (Table 14-5). Denitrification activity within sediments of both lakes and streams is related to the microgradients of reducing conditions (Jones, 1979a, 1979b; Nielsen et al., 1990). In sediments of the littoral zone or those in contact with oxygenated water denitrification rates can be lower at the sediment surface and increase at sediment depths (10–15 mm) where reducing conditions increase (electrode potential, E_h, of 210 mV or less). However, in the riparian zone in Baiyangdian Lake, China, with a groundwater level of around 50 cm, the denitrification rate increased with soil depth up to 100 cm (Wang et al., 2019). Denitrification rates have also been found to increase with a higher macrophyte root surface, regulated by organic substrates excreted from the roots (Christensen and Sorensen, 1986). Macrophytes can also provide suitable attachment surfaces for denitrifying bacteria (Zhang et al., 2016) and the macrophyte litter decomposition is likely to favor denitrification, via oxygen consumption and carbon source supply (Bastviken et al., 2005). Denitrification rates show diel and seasonal variations (Christensen and Sorensen, 1986; Piña-Ochoa and Álvarez-Cobelas, 2006). Oxygen production by algae attached to the sediments in the light can increase the thickness of the aerobic layer and reduce denitrification in the sediments by as much as 70%. After dark, oxygen is rapidly consumed by bacterial respiration at rates faster than it diffuses from overlying water, and denitrification again increases as anoxic and reducing conditions return.

Nitrous oxide (N_2O) is an intermediate product of the denitrification pathway and is also a greenhouse gas. N_2O is usually rapidly reduced to N_2, but its distribution is quite variable. Although N_2O has not been found in many lakes in appreciable quantities (e.g., Goering and Dugdale, 1966; Kuznetsov, 1970; Macgregor and Keeney, 1973; Kaplan and Wofsy, 1985; Mengis et al., 1996), large N_2O accumulations were repeatedly observed in the strata of greatly reduced oxygen of eutrophic lakes, concomitant with large concentrations of NO_2^- (Yoh et al., 1988, 1990; Yoh, 1992; Mengis et al., 1997). Although nitrification contributed to the accumulations of both compounds to some extent, denitrification was the dominant process of N_2O formation in these lakes. Rates of N_2O and NO_2^- formation were maximal with very low concentrations of dissolved oxygen (~0.1 mg L^{-1}) and inhibited by reducing conditions in the lower hypolimnion sufficient to reduce iron and sulfide ions.

In anaerobic littoral and wetland sediments denitrification dominates as the process producing N_2O, but N_2O-to-N_2 ratios are greatest at low rates of denitrification (van Cleemput, 1994). Nitrous oxide has been found to be released from aerobic riverine sediments into the overlying water (Wissmar et al., 1987) or from wetlands into the atmosphere during summer months, for example, in Sanjiang Plain, China (Song et al., 2009). In a series of seven amictic, permanently ice-covered lakes of Antarctica, a detailed study indicates that N_2O maxima were largely a product of nitrification and that denitrification was a sink for this gas in anoxic water (Priscu, 1997). Maxima of N_2O were common in-depth gradients where oxygen concentrations and redox potentials were decreasing, and N_2O was nearly absent in anoxic zones of low redox potential. The ice barrier results in an appreciable supersaturation of N_2O (Cavaliere and Baulch, 2018; Soued et al., 2016), some of which evades to the atmosphere through the brief period of the summer moat of open water at the edge of the ice.

The process of DNRA also contributes to NO_3^- removal in aquatic ecosystems, but the relative importance of DNRA and denitrification on NO_3^- removal is quite variable between different aquatic systems (Burgin and Hamilton, 2007). A study of a eutrophic lake in China showed that denitrification was the dominant pathway to remove NO_3^- in this eutrophic lake, with

TABLE 14-5 Approximate Rates of Bacterial Denitrification in Lake and River Water and Sediments by Direct Measurements[a]

	Rate of denitrification		
Lake	Water (μg N L^{-1} day^{-1})	Sediment (mg N m^{-2} day^{-1})	Source
Lake 227, Ontario	0–30	15	Chan and Campbell (1980)
Norrviken, Sweden	0.53	100	Tirén et al. (1976)
Smith Lake, Alaska	15	(90)	Goering and Dugdale (1966)
Lake Mendota, Wisconsin	8–26	—	Brezonik and Lee (1968); Keeney et al. (1971)
Ramsjön, Sweden	—	10.1	Svensson et al. (2001)
Enriched drainage ditch, Netherlands	—	160	Van Kessel (1978)
Lake Kasumiga-ura, Japan	—	3–74[b]	Yoshida et al. (1979)
Boyrup Langsø, Denmark	—	57.5	Andersen (1977)
Kvindsø, Denmark	—	34.2	
Eight lakes of Ontario and New York, average summer		26.4	Rudd et al. (1986)
Fukami-Ike, Japan	0–27	—	Terai et al. (1987)
Lake Kizaki, Japan	0–22	—	Terai (1987)
Stream sediments, Ontario			
Without worms	—	50	Chatarpaul et al. (1980)
With tubificid worms	—	90	
Littoral sediments			Chan and Knowles (1979)
Lake St. George, Ontario			
Without submerged macrophytes	—	2.6	
With submerged macrophytes	—	2.2	
Rivers	0	0–116	Seitzinger (1988)
Sugar Creek, USA	—	40.3	Böhlke et al. (2004)
Walker Branch, USA	—	4	Mulholland et al. (2004)
Green Creek, Antarctica	—	5	Gooseff et al. (2004)
Lakes			
Oligo mesotrophic	3–27	1.5–19	Seitzinger (1988)
Eutrophic	3–500	1–57	
Wetlands	—	0–1000	Bowden (1987); Groffman (1994)

[a] *Note: Measurements often use $^{15}NO_3^-$; other approaches use mass balance techniques, but one can only calculate total lake denitrification via the mass balance approach.*
[b] *Approximate extrapolations from values expressed as rates per weight of sediment and the area sampled by the coring device.*

the denitrification rate being 7.4–8.5 times the DNRA rate (Shen et al., 2020).

D. Anammox

The process of *anaerobic NH_4^+ oxidation*, abbreviated to *anammox*, has only been described relatively recently (Kuenen, 2008). This process converts 1:1 molecule ratio of NH_4^+ and NO_3^-/NO_2^- to dinitrogen (Fig. 14-4) in the absence of oxygen and without the need for a carbon supply (Kuenen, 2008; Kartal et al., 2011):

$$NH_4^+ + NO_2- \rightarrow N_2 + 2H_2O$$

Anammox is known to be mediated by five genera of chemoautotrophic bacteria, that is, *Brocadia*, *Kuenenia*, *Anammoxoglobus*, *Jettenia*, and *Scalindua* (Jetten et al., 2009).

Anammox, in the natural environment, was first discovered in marine sediments (Thamdrup and Thamdrup, 2002) and in the marine water column (Dalsgaard et al., 2003). Recent studies have shown that anammox also occurs in freshwater ecosystems, for example, freshwater lakes (Schubert et al., 2006; Hamersley et al., 2009; Wenk et al., 2013; Crowe et al., 2017), rivers (Zhang et al., 2017), reservoirs (Shen et al., 2017), wetlands (Erler et al., 2008; He et al., 2012; Hou et al., 2015), and rice paddies (Nie et al., 2015).

A comparison of the percentage N_2 production from anammox versus denitrification in a range of studies has shown highly variable results within systems (Fig. 14-6; Burgin and Hamilton 2007). However, at times, this process can be significant in terms of nitrogen removal. Anammox, as with other nitrogen transformation processes, varies spatially and temporally driven by chemical and physical factors, including substrate availability (Risgaard-Petersen et al., 2004; Trimmer et al., 2005; Nicholls and Trimmer, 2009), temperature (Rysgaard et al., 2004), salinity (Rich et al., 2008; Lisa et al., 2014), and organic carbon content (Trimmer et al., 2003).

E. Nitrogen budgets in lakes and reservoirs

The discussion of nitrogen sources, transformations, and fate in lakes and reservoirs in the sections above highlights the complexity of the nitrogen cycle. The use of nitrogen budgets for lakes and reservoirs provides a way of simplifying this. Typically, nitrogen loading has been shown to be an excellent predictor of nitrogen retention in lakes and wetlands in a study of 23 wetlands and lakes in north America and Europe (Saunders and Kalff, 2001). Wetlands can retain twice as much nitrogen as lakes, and denitrification was the primary mechanism for nitrogen retention.

Groundwater inputs of nitrogen are not well studied, in part because the methodology is challenging and it is difficult to get accurate estimations, particularly for whole-system nitrogen budgets. Groundwater often has high concentrations of dissolved inorganic nitrogen, particularly NO_3^- (Lewandowski et al., 2015). NO_3^- that has seeped into the shallower depths of the sediment of lakes and reservoirs, where there may be a redox potential created by settling organic matter, may be rapidly denitrified.

A global study of lakes and reservoirs found that theses water bodies played a major role as sinks for nitrogen, retaining 19.7 Tg N year^{-1} globally (Fig. 14-7) (Harrison et al., 2009). Small lakes were responsible for almost half of this global retention. The degree of retention was shown to be a function of climate and the density of lakes and reservoirs across the landscape. Reservoirs play a greater role in nitrogen retention than lakes due to a combination of higher drainage ratios (catchment surface area:lake or reservoir surface area), higher apparent settling velocities for nitrogen, and greater average nitrogen loading rates in reservoirs compared to lakes. Maranger et al. (2018) showed that 6% of global inputs of nitrogen to freshwater was processed by reservoirs as either denitrification or burial (Table 14-4).

Most nitrogen budgets have been constructed for temperate areas of Europe and North America, where seasonal effects may be more pronounced than in tropical and subtropical areas. One study of semiarid lakes

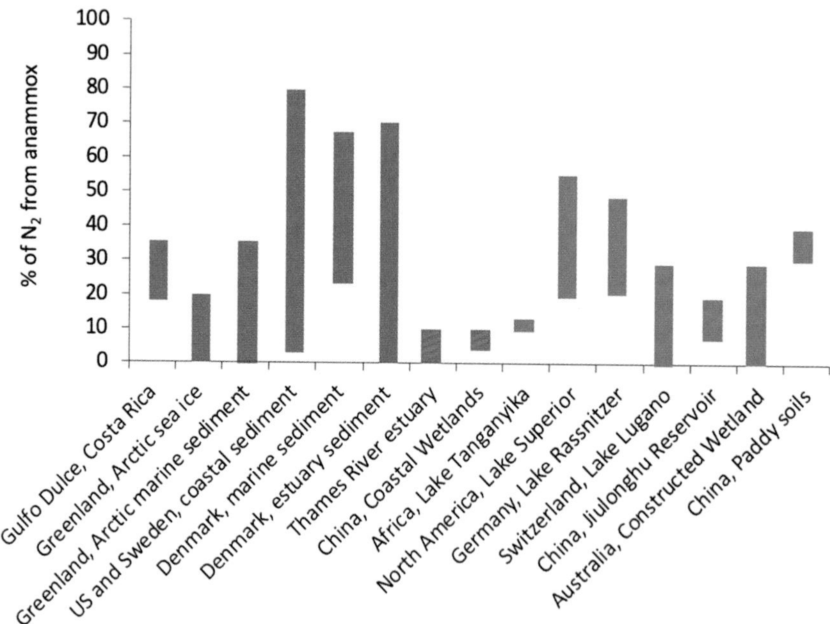

FIGURE 14-6 Anammox estimates across a variety of aquatic ecosystems globally. The *bars* represent the ranges of total N_2 production that can be attributed to anammox in a given study site. *Purple bars* designate marine and brackish ecosystems; *blue bars* designate freshwaters. The Thames River estuary and the coastal wetlands, China, are hatched because the study spanned a range of freshwater and marine-influenced sites. *(Updated from Burgin and Hamilton, 2007.)*

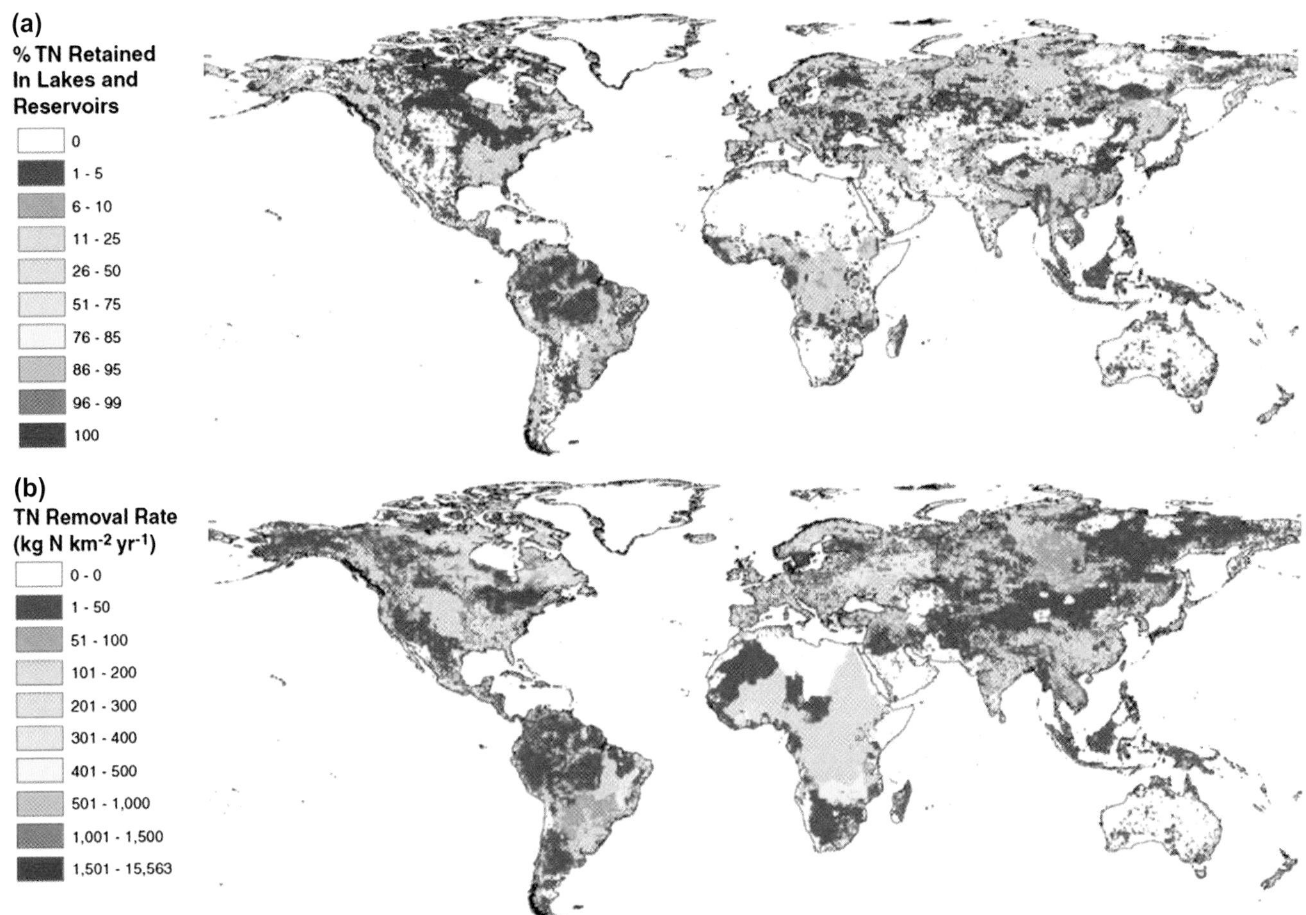

FIGURE 14-7 (a) NiRReLa-modeled global distribution of percent N removal by lakes and reservoirs, and (b) N removal by lakes and reservoirs, kg N km^{-2} year^{-1}. *(From Harrison et al., 2009.)*

in Austalia found that only 7% of nitrogen in lakes was retained, a percentage that is lower than in lakes in the United States (Fig. 14-8; Cook et al., 2010). Despite nitrogen fixation inputs in this study, it was counteracted by similar rates of nitrogen burial in the sediment. The contribution of denitrification was highly uncertain, as estimates were based on measured relationships between nitrogen loading and denitrification rates in temperate regions (Seitzinger et al., 2006).

III. Nitrogen sources, transformations, and fate in streams and rivers

Nutrients move unidirectionally within lotic systems (streams and rivers) and as such have some different characteristics in terms of nitrogen cycling compared with lakes and reservoirs. Dissolved substances move downstream, where they may be bound or assimilated for a period of time and later released for further movement downstream.

Hydrological processes physically move water containing dissolved and particulate components to reactive sites. Exchanges at reactive sites include chemical ionic transformations, sorption and desorption, and metabolically mediated uptake and assimilation by biota (Fig. 14-9). Materials can be transferred from the water column to the stationary streambed. Some of these materials will be retained, permanently removed from the system via gaseous forms, utilized by incorporation into living organisms, potentially transferred to other organisms, and subsequently released by animal excretion or decomposition to the water column and further transported downstream.

As nitrogen cycles among biota and abiotic components of the stream ecosystem, they are transported downstream in processes that resemble spirals and have been termed *nutrient spiraling* (Webster and Patten, 1979; Newbold et al., 1981; Stream Solute Workshop, 1990). Although upstream movements of nutrients can occur in backflows from eddies, fish migration, and flight of adult aquatic insects, net fluxes are downstream.

FIGURE 14-8 An approximate budget for N and P in the Lower Lakes at the terminus of the Murray River, Australia, spanning the years 1979–1996. Quantities shown are the average annual rate normalized to the surface area of the lake over the 17-year study period with units in mmol m^{-2} $year^{-1}$. *TN* denotes total nitrogen, *TP* denotes total phosphorus, *FRP* denotes filterable reactive phosphorus, *TKN* denotes total Kjeldahl nitrogen (the total concentration of organic nitrogen and NH_4^+), *Den* denotes denitrification, and *Nfix* denotes nitrogen fixation. *(From Cook et al., 2010.)*

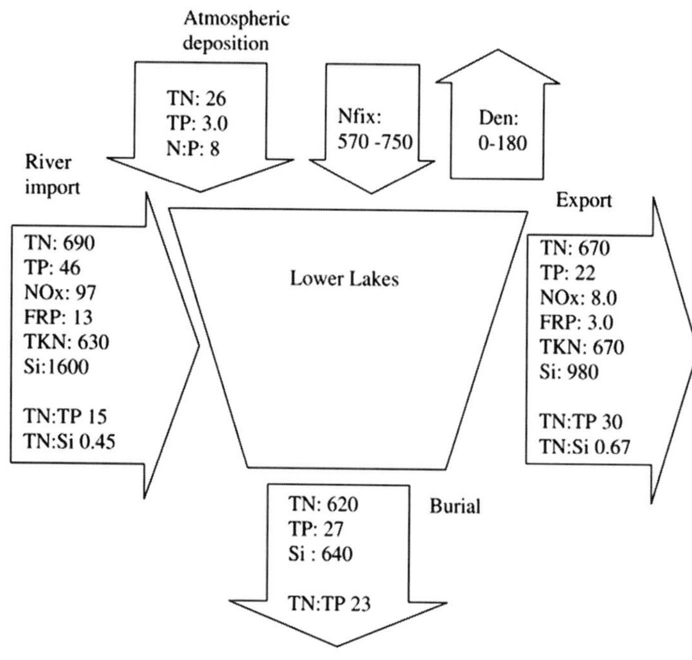

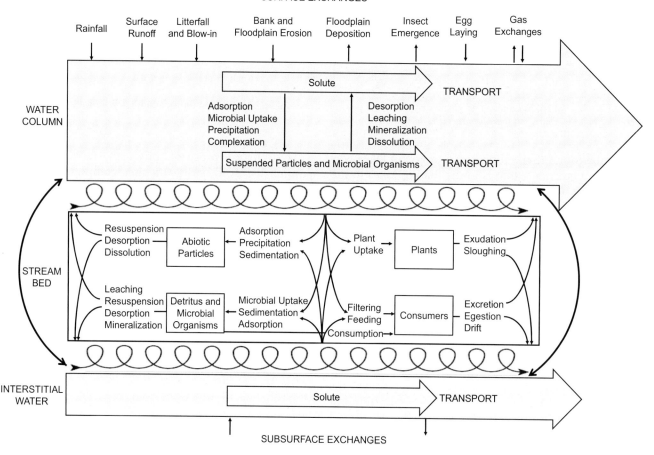

FIGURE 14-9 Solute processes in streams. The *two spirals* represent the continuous exchange of solutes and particle-bound chemical substances between the water column and the streambed and between the streambed and interstitial water. Materials in the water column and interstitial water are moving downstream, while the streambed materials are stationary. *(Revised from Stream Solute Workshop, 1990.)*

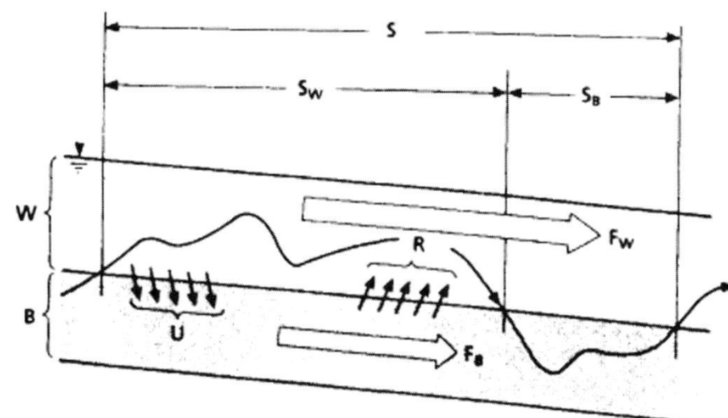

FIGURE 14-10 Spiraling in a river ecosystem of two compartments: water (W) and biota (B). The spiraling length (S) is the average distance a nutrient atom travels downstream during one cycle through the water and biotic compartments. S = the sum of the uptake length (S_W) and the turnover length (S_B) estimated from the downstream nutrient fluxes (F_W and F_B) and the exchange fluxes of uptake (U) and retention (R). *(From Newbold, 1992.* The Rivers Handbook*: Vol. 1. Blackwell Science Ltd. Reproduced with permission.)*

Spiraling length (S) is the average distance a nutrient atom travels downstream during one cycle through the water and biotic compartment. The S equals the sum of distance traveled until uptake ("uptake length," S_W) and the downstream distance traveled within the biota until regeneration ("turnover length," S_B) (Fig. 14-10). The S can be calculated from the downstream nutrient fluxes (mass per unit width of river per unit time in the water [W] component, F_W, and in the biota [B] component, F_B) and the exchange fluxes of biotic utilization (U) of nutrients from the water compartment or regeneration (R) from the biota in mass per unit area per unit time (Fig. 14-10).

The average downstream velocity of the nutrient may be near that of the water in large rivers but is very much slower in streams and rivers where nutrients reside in the sediments and microbiota for a high proportion of the time. The spiraling length of a nutrient suggests the extent of availability and utilization rates. A shorter spiraling length of one nutrient versus another could imply that it is in greater demand and is possibly limiting the potential growth and productivity of the community. The retention of nutrients for at least some time is necessary in order to maintain ecosystem processes in streams (Merill and Tonjes, 2014), so the spiraling concept also illuminates stream ecological dynamics (Triska et al., 1989; Ensign and Doyle, 2006). Perhaps a more functional application is to determine the spiraling rates of nutrients among different dissolved, particulate, and animal consumer groups (e.g., Newbold et al., 1983). Such analyses demonstrate the rapid recycling rates and very short spiraling lengths of the attached microbiota in comparison to abiotic particulate materials and consumer metazoans.

Water flow in streams and rivers can continuously bring dissolved nutrients to primary producers (especially attached ones); however, nutrient limitation in streams and rivers has been found in many studies (e.g., Keck and Lepori, 2012; Reisinger et al., 2016; Tank et al., 2017) and shows regional and seasonal variation (Reisinger et al., 2016). From a global review of nitrogen and phosphorus limitation on primary producers in freshwater, marine, and terrestrial ecosystems (Elser et al., 2007), the frequency of phosphorus limitation in freshwater systems (including streams and rivers) was as common as in marine and terrestrial systems. This global review also demonstrated that compared to other systems, marine systems have a higher response to nitrogen enrichment, but freshwater systems have a higher response to simultaneous nitrogen and phosphorus addition (Fig. 14-11).

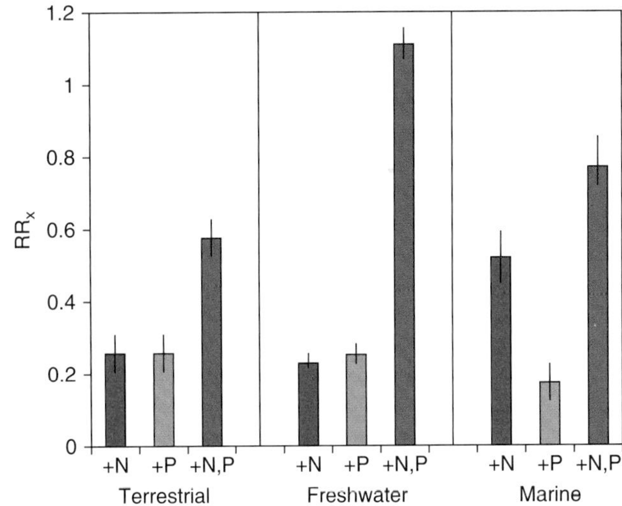

FIGURE 14-11 Responses of autotrophs to single enrichment of N (*red*) or P (*blue*) or to combined N + P enrichment (*purple*) in terrestrial, freshwater, and marine ecosystems. Data are given as natural-log transformed response ratios (RRx) in which autotroph biomass or production in the enriched treatment is divided by its value in the control treatment and then ln-transformed. Thus a value of 0.5 indicates a value in the manipulated treatment that is ~1.6 times its value in the control, while a value of 1.0 indicates a 2.7-fold increase. Sample sizes +N, +P, and +N&P treatments were 112, 107, and 126 for terrestrial studies; 509, 506, and 618 for freshwater studies; and 149, 141, and 197 for marine systems, respectively. *Error bars* indicate plus or minus one standard error. *(From Elser et al., 2007.)*

A global-scale riverine nitrogen flux calculation estimated that approximately 20% of nitrogen from catchments (watersheds) was retained in rivers/streams prior to reaching the coast (Boyer et al., 2006). Headwater streams have been considered key sites to process and remove nutrients, thus reducing their downstream export (Alexander et al., 2000; Peterson et al., 2001; Tank et al., 2017). A large-scale study including 123 tributaries in the Mississippi River Basin (USA) showed that the first-order nitrogen loss rate decreased from 0.45 to 0.005 d^{-1} with increased channel size (Alexander et al., 2000). This study also concluded that the proximity of sources to large streams and rivers, rather than the distance of sources to the coastal waters, was an important determinant of nitrogen delivery to the estuary in the Mississippi basin, and possibly also in other large river basins (Alexander et al., 2000). A ^{15}N stable isotope tracer study of nitrogen dynamics in headstreams of all biomes throughout North America showed that nitrogen uptake length increased significantly with the volume of river discharges (Peterson et al., 2001). Seasonal variation in nitrogen removal rates has been found in a range of studies (Simon et al., 2005; Mulholland et al., 2006; Hoellein et al., 2007; Alexander et al., 2009). From this, it can be inferred that there are a range of biogeochemical and hydrological controlling factors, such as vegetation, nutrient and carbon concentrations, temperature, and discharge volume, on nutrient removal that are significant and may confound the stream size impact (Alexander et al., 2009). Ensign and Doyle (2006) also argued that the role of headstreams on nutrient retention might be biased by the fact that more sampling and experimental work may have been done in lower- rather than higher-order streams. More research on nutrient spiraling controlling factors in larger rivers and entire river networks would be valuable to estimate nutrient removal from streams and rivers to coastal waters (Ensign and Doyle, 2006; Alexander et al., 2009).

Dissolved nutrient retention mechanisms in streams and rivers include biological uptake by primary producers, microbial activity (such as coupled nitrification-denitrification, and anammox), volatilization, physical adsorption, and sedimentation (Peterson et al., 2001; Bernot and Dodds, 2005) (Fig. 14-12). Evidence suggests that in lower-order streams, the greater dominance of biota attached to substrata results in higher retention of nutrients and intensive nutrient recycling within the attached communities (cf. Paul and Duthie, 1989; Wetzel, 1993; Burns, 1998). Removal of the attached microbial community, as was the case

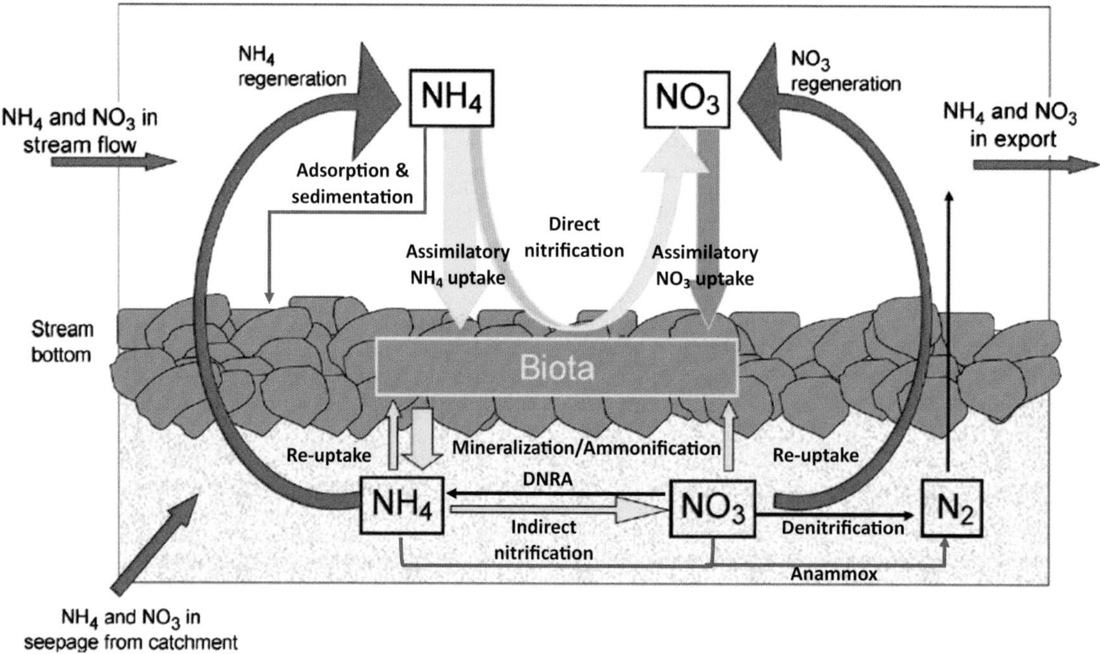

FIGURE 14-12 Conceptual model of DIN dynamics in headwater stream ecosystems. NH_4^+ and NO_3^- enter the stream reach via stream flow and lateral seepage. NH_4^+ removal is due to uptake by primary producers, bacteria, and fungi, plus direct nitrification and physical adsorption to sediment particles followed by sedimentation. Indirect nitrification is the conversion of NH_4^+ mineralized from organic matter to NO_3^-. NO_3^- removal from the water is primarily via assimilation by biota and denitrification on the channel bottom. Regeneration is the release of NH_4^+ and NO_3^- from the stream bottom back to the water column and is the net result of several interacting processes, including mineralization, indirect nitrification, denitrification, DNRA, anammox, and reuptake by organisms. NO_3^- and NH_4^+ remaining in the water are exported downstream. *(Updated from Peterson et al., 2001.)*

in a desert stream following a flood that eliminated most of the biota, reduced the retention of nitrogen to very low levels, slowed recycling rates, and increased spiraling length (Grimm, 1987). Nitrogen retention increased rapidly as the attached algal and bacterial community reestablished. Riparian zones were found to mitigate in-stream nutrient retention either by changing in-stream ecological features or providing energy inputs (heat, and organic matter source) into streams (Sabater et al., 2000; McMillan et al., 2014). A range of studies showed that natural/ecological restoration of streams and rivers can increase nutrient retention (Kronvang et al., 1999; Craig et al., 2008; Faulkner, 2008). Clear evidence of riparian vegetation assimilating ^{15}N-isotopes of inorganic nitrogen from a Sonoran Desert (USA) stream has been reported, suggesting a strong hydrological linkage between the stream and riparian zone in arid areas (Schade et al., 2005). The enhanced interaction between sediment and water in hyporheic zones can reduce nutrient uptake length and therefore increase nutrient retention (Mulholland and Deangelis, 2000).

Denitrification and anammox are important permanent nitrogen removal pathways. A series of ^{15}N tracer additions to 72 streams in the United States and Puerto Rico showed that denitrification accounted for a median of 16% of total NO_3^- uptake across all streams and exceeded 43% of total uptake in a quarter of the streams. These data also suggested that the total uptake of NO_3^- was related to ecosystem photosynthesis, while denitrification was related to ecosystem respiration (Mulholland et al., 2008). Nitrogen retention via anammox has also been determined in coastal rivers, for example, New River Estuary, USA (Lisa et al., 2014), and two eutrophic coastal rivers in Northeast China (Li et al., 2020).

A review study of 79 hydrologically reconnected streams and rivers showed that restoration of the hydrology of streams and rivers in urban and agricultural watersheds significantly increased nutrient retention (Newcomer Johnson et al., 2016). Welter and Fisher (2016) showed that storm characteristics (intensity, time, and size) affect hydrological connectivity in intermittent rivers and thus affect denitrification and nitrogen retention. The changes in river flow regime, as a result of dam construction, were also found to change nitrogen retention, based on comparisons between sites upstream and downstream from dams (von Schiller et al. 2016). One study found equally high variation in nutrient retention when intermittent and constant flow rivers were compared, suggesting high resilience of biological communities responsible for nutrient uptake and/or significant control from the local catchment and climate conditions (von Schiller et al., 2008).

A. Dissolved and particulate nitrogen

The concentrations and proportions of dissolved inorganic (NO_3^-, NO_2^-, and NH_4^+) and organic nitrogen compounds in rivers can vary substantially over space and time (e.g., Spalding and Exner, 1993) because of many competing reactions occurring in the nitrogen cycle. For example, high concentrations of NO_2^- can occur in large rivers in summer under warm, slow-flowing conditions as a result of dissimilatory NO_3^- reduction in anaerobic sediments (Kelso et al., 1997, 1999). Catchment characteristics, such as land use, as well as hydrology, can play an important role. Higher NO_3^- and NH_4^+ concentrations are found among rivers influenced by anthropogenic activities, such as fertilizer use in agricultural land and industrial effluent emissions (Donner et al., 2004; Du et al., 2017). Maximum NO_3^- concentrations during storm flows were directly related to the magnitude of the storms and resulting high discharge and were inversely related to the frequency of storm events (Tate, 1990; Triska et al., 1990). A study from 57 European rivers showed that both NO_3^- concentrations and its proportion of TN increased along a gradient from ultraoligotrophic to hypertrophic rivers, and that DON was the important secondary constituent of the total nitrogen, even in the most eutrophic rivers (Durand et al., 2011). DON constitutes a major part (world average ~40%) of the TDN concentrations (Wetzel and Manny, 1977; Meybeck, 1982). In subarctic and humic tropical rivers DON can constitute over 90% of the dissolved nitrogen.

Terrestrial vegetation in riparian zones influences NO_3^- and NH_4^+ loadings from catchments to streams, with nitrogen concentrations from runoff generally being higher during periods of vegetation dormancy or following harvesting or fire (e.g., Likens, 1985; Spencer and Hauer, 1991; McClain et al., 1994). The width of the fully vegetated riparian zone on both sides of the river can be an important indicator of its inorganic nitrogen removal efficiency (Mayer et al., 2007; Sweeney and Newbold, 2014).

Although dissolved inorganic nitrogen concentrations and discharge can vary widely on a diurnal basis and seasonally, DON concentrations remain relatively constant (Manny and Wetzel, 1973). Concentrations of particulate nitrogen in streams and rivers can increase dramatically during flooding events (Mitchell et al., 1997) and in catchments with higher rates of soil erosion and deforestation (Ittekkot and Zhang, 1989).

B. Nitrogen cycling

Channel and riparian sediments can sorb a significant amount of NH_4^+ from the water column. The physical sorption of NH_4^+ to sediment particles is dynamically coupled to sources from groundwater, ammonification, and transformations of dissolved inorganic nitrogen by nitrification, denitrification, and NO_3^- reduction (Fig. 14-12). The duration of the storage of sorbed NH_4^+ can be highly variable. Some studies indicate that appreciable retention can occur in summer months and contribute from 12 to 25% of NO_3^- released subsequently by nitrification in winter (e.g., Richey et al., 1985).

The riparian zone, when water saturated, and *parafluvial* (the part of the active stream channel without surface water that is connected hydrologically with the surface stream water) and *hyporheic* zones (areas of the streambed and near-stream aquifers through which stream water flows) are often major internal sources of NH_4^+ and DON to streams during baseflow conditions. Once within a stream, nitrogen uptake by attached microbes (bacteria, fungi, and algae) may be the primary mechanism controlling spatial and seasonal variation in the water. This in turn is regulated by local environments, including light, temperature, discharges, and water velocity (Butturini and Sabater, 1998). However, an NH_4^+ enrichment study in a mountainous tropical river network (seven streams) in Puerto Rico showed that nitrification can rival microbial and algal assimilation of NH_4^+ in the streamwater, thus resulting in additional NO_3^- exports from those streams (Koenig et al., 2017). Reductions in stream NO_3^- concentrations at baseflow during the day compared with night suggest that uptake by photoautotrophs can be an important retentive process (Burns, 1998). This has been shown in cooler seasons when forest vegetation is dormant or deciduous and light levels are relatively high (Mulholland et al., 2006).

Sediment denitrification rates in streams may be controlled by hydrologic residence time (e.g., Flewelling et al., 2012; Kaushal et al., 2008; Seitzinger et al., 2006), NO_3^- concentrations (e.g., Fischer et al., 2005; Mulholland et al., 2009), and organic carbon quality and quantity (Bradley et al., 1995; Fork and Heffernan, 2014). A metaanalysis study showed that the relative availability of carbon and nitrogen (i.e., organic carbon: NO_3^- ratio) can be a more important factor limiting denitrification rates (Taylor and Townsend, 2010).

The nitrification rate is likely to be limited by substrate (NH_4^+) availability (Strauss et al., 2002). However, in a study of 18 agricultural and urban streams from the Kalamazoo River catchment (southwestern Michigan, USA), nitrification was found positively related to sediment carbon content, but not to NH_4^+ concentrations, probably due to benthic decomposition providing NH_4^+ via mineralization (Arango and Tank, 2008). The processes of nitrification and denitrification often function simultaneously and reciprocally in running water sediments, where many microzones of steep redox gradients occur in the hyporheic zone of the streambed (Gross, 2003; Bastviken et al., 2005; Munroe et al., 2018). The effectiveness of the denitrification and nitrification processes in the hyporheic zone of stream sediments depends greatly upon the redox conditions within the interstitial water. Flow conditions within the sediment interstices are quite variable spatially because of localized differences in sediment composition and texture. In organic-rich sediments the redox profiles can be quite similar to those in lake sediments, in which the zone of nitrification can be restricted to the upper 2–3 mm and intensive denitrification occurs below the stratum of nitrifier activity (e.g., Cooke and White, 1987a, 1987b; Birmingham et al., 1994). Denitrification also occurs in shallow riparian sediments connected to the stream by the hyporheic zone (Duff and Triska, 1990). The presence of suspended sediments in two large rivers from China, the Yellow and Yangtze Rivers, was found to enhance nitrification, coupled nitrification-denitrification, and anammox in rivers by providing both oxic and anoxic microenvironments, and larger surface areas for contact between bacteria and nitrogen substrates (Xia et al., 2004, 2009, 2017; Zhang et al., 2017).

Compared to lakes and reservoirs, hydrological regimes can have strong impacts on nutrient dynamics, especially for streams and rivers in arid, semiarid, and Mediterranean areas (von Schiller et al., 2017). Intermittent rivers were conservatively estimated to account for 30% of the total length and discharge of the global river network (Tooth, 2000) and are common around the world (Larned et al., 2010; Datry et al., 2014), especially in Australia (Sheldon et al., 2010). Intensive water abstraction and regulation, as well as climate change, can also shift flow regimes and result in disrupted flows in streams and rivers (Sabater et al., 2018). Nutrient sources, in-stream nutrient processes, and hydrological interactions between the surface stream, hyporheic, parafluvial, and riparian zones can differ significantly between dry, wet, and transitional phases in intermittent river systems (von Schiller et al., 2017).

Cessation of surface water flow causes rivers to retract and disconnected waterholes (kettle holes) might emerge. Persistent waterholes can have high primary productivity during dry periods and become hotspots for nutrient processing when flow reconnects (Stanley et al., 1997; Fisher et al., 2001; Faggotter et al., 2013; Datry et al., 2014). NO_3^- concentrations in waterholes can decrease due to denitrification, which is enhanced by hypoxic conditions (Kemp and Dodds, 2002; von Schiller et al., 2011). In contrast, increased NH_4^+ concentrations mainly resulted from the rapid mineralization of accumulated organic detritus (Acuña et al., 2005) or enhanced DNRA and inhibited nitrification processes in hypoxic conditions (Baldwin and Mitchell, 2000; Arce et al., 2015). The impact of desiccation on nitrogen

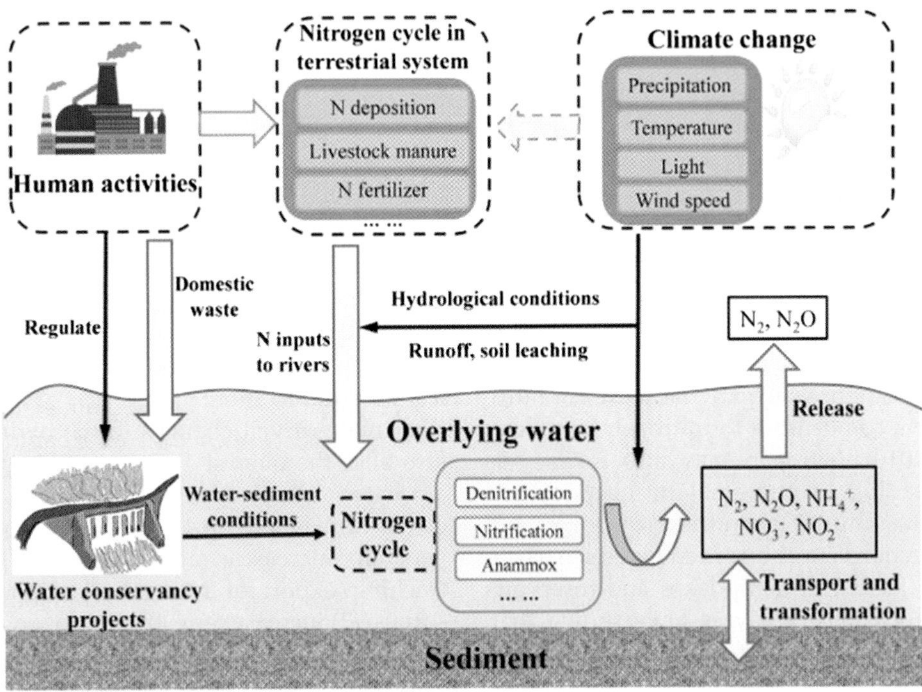

FIGURE 14-13 The cycle of nitrogen in river systems: sources, transformation, and flux. *(From Xia et al., 2018.)*

processes in the dry sediment has been shown to vary substantially between studies. For example, nitrification and denitrification have been shown to increase, decrease, or remain unchanged in dry sediments of intermittent rivers, depending on the study (Mitchell and Baldwin, 1999; Austin and Strauss, 2011; Gómez et al., 2012; Merbt et al., 2016). This is probably due to differences in desiccation duration and subsurface flows (Reverey et al., 2016). Large amounts of nutrients (both particulate and dissolved forms) can be transported from catchments into rivers during these first flush events (Tzoraki et al., 2007; Obermann et al., 2009; von Schiller et al., 2011). The in-stream nitrogen processes, such as denitrification and nitrification, can be activated rapidly upon rewetting (McIntyre et al., 2009; Austin and Strauss, 2011), together with the recovery of nitrogen retention in streams and rivers after the first flow event (Martí et al., 1997).

IV. Effect of human activities

A. Eutrophication

Enrichment of freshwaters with nutrients commonly occurs as a result of catchment losses from agricultural fertilization, loading from sewage and industrial wastes, and enrichment via atmospheric pollutants (especially NO_3^- and NH_4^+). Excessive loading of these nutrients permits increased plant growth, including algal growth, until other nutrients or light availability ultimately limit further plant biomass accumulation. The pathways of anthropogenic nitrogen inputs to freshwaters can be seen in Fig. 14-13.

At the catchment scale, nitrogen fertilizer application for agriculture is a major contributor to eutrophication. It is estimated to double the amount of bioavailable nitrogen to the terrestrial biosphere annually (Galloway et al., 2008; Fowler et al., 2013). The tropics are a major contributor, producing 56 ± 6% of global land nitrogen pollution despite covering only 34% of the global land area (Lee et al., 2019). Over the last few decades, ammonium nitrate synthesis has been replaced by the chemical synthesis of urea and it is now the most widely used nitrogenous fertilizer (Glibert et al., 2014). Studies have shown that urea concentrations can be more than two-fold greater in lakes receiving nitrogen from cities than in agriculturally affected basins (Bogard et al., 2012). Furthermore, the nitrogen:phosphorus ratio of global fertilizer use has increased over time. This is likely to have altered the ratios in inflows to freshwater systems, with resultant effects on biogeochemical processes (Glibert et al., 2014).

Increased nitrogen loads to waterways can stimulate excessive algal productivity, negatively affecting the health of aquatic ecosystems (Xu et al., 2015, 2021). Increased nitrogen loading also reduces microalgal and aquatic plant biodiversity, changes the population structure, and is associated with the eventual loss of macrophyte communities (Moss et al., 2013; Phillips et al., 2016). The subsequent decomposition of lost macrophytes could be an important internal nutrient source for the water column (Lu et al., 2017, 2018).

Historically, phosphorus was considered the key nutrient controlling algal growth in freshwater systems (Schindler, 1977). This information has been used to determine the effect of phosphorus loading on productivity in lakes (Vollenweider, 1968; chapter 15). However, in recent years the P-limitation paradigm has been challenged (e.g., Lewis and Wurtsbaugh, 2008), with arguments made that controlling only phosphorus may not be sufficient for reducing productivity in some lakes or reservoirs (Xu et al., 2010; Ma et al., 2015). This may be because nitrogen fixation, as an external source of nitrogen, cannot always compensate for nitrogen loss (Scott and McCarthy, 2010). Nitrogen may also escape via denitrification, leading to perpetual nitrogen deficits (Paerl et al., 2016). Thus dual nutrient (nitrogen and phosphorus) reductions may be needed to reduce harmful algal blooms (HABs) in many lakes and reservoirs (Paerl et al., 2016). Global analyses have shown that, despite differences in marine, terrestrial, and freshwater habitats, significant synergistic effects of combined nitrogen and phosphorus enrichment are seen across many ecosystems (Elser et al., 2007; Fig. 14-11).

An important step in minimizing eutrophication is to reduce external nitrogen loads to waterways. However, nitrogen release and cycling across the sediment–water interface within freshwater systems is also an important mechanism leading to eutrophication (Sondergaard et al., 2003). In lakes or reservoirs in which external loading has been reduced, internal nutrient loading may prevent the reduction of eutrophication (Xu et al., 2021). Increased atmospheric nitrogen deposition in lakes, as a result of anthropogenic activities, has also contributed to eutrophication (Elser et al., 2009). A global analysis has shown that increased atmospheric nitrogen deposition may have shifted some lakes from nitrogen to phosphorus limitation (Elser et al., 2009). This impacts nitrogen cycling and higher trophic levels.

The impact of long-term increases in anthropogenic nitrogen has been shown in sediment and fish in Lake Ontario (Guiry et al., 2020). The nitrogen stable isotope was used to reflect transformation processes with an increase in the ratio of the heavier ^{15}N isotope around the 1830s, at a time when industrial-scale forestry increased. Since that time, the ratio of the ^{15}N isotope has continued to increase. There are a range of likely sources of increased nitrogen runoff including industrial-scale forestry and broadscale clearing for agriculture.

B. Climate change

Climate change can also impact nitrogen budgets and transformations in freshwater systems (Fig. 14-13). Warmer temperatures increase enzyme activity affecting rates of microbial nitrogen transformation, until a temperature optimum is reached (Greaver et al., 2016). Climate change is also expected to intensify the hydrologic cycle (e.g., Gleick, 1989; Mamuye and Kebebewu, 2018; Chapter 5), including changes such as more frequent and intense heavy rainfall events, potential deepening and lengthening of dry periods, altered snow accumulation and melt, and changes in evapotranspiration. This will affect the rates of microbial nitrogen transformations, such as decomposition, mineralization, nitrification, denitrification, and biological nitrogen fixation.

Climate-driven changes to the hydrologic cycle will also alter the quantity of nitrogen transported through a system via waterborne transport and soil water-content-mediated nitrogen cycling. Greater precipitation generally increases water flow, which may: (1) increase leaching/export of nitrogen through terrestrial landscapes; (2) increase terrestrial nitrogen inputs to streams and rivers; and (3) increase nitrogen transport rates through streams and rivers, though adaptation by microbes and plants may increase their ability to retain nitrogen as flushing increases, thereby potentially limiting some of the overall impact of increased precipitation and flow.

Beyond simply the total volume of precipitation, precipitation intensity influences the rate of nitrogen flow through ecosystems and affects nitrogen spiraling rates. Increased precipitation intensity of storm systems, as is expected to continue with global warming, would likely increase the frequency of high nitrogen loading events to aquatic systems. Due to the limited capacity for instream removal of nitrogen during high flow pulse events, most nitrogen is transported downstream. A study showed that North European temperate lakes, for example, have projected increases in winter rainfall and higher agricultural cropping practices, which will increase runoff and nitrogen transport annually (Jeppesen et al., 2011). This, combined with higher temperatures, may increase the risk of cyanobacterial blooms, including harmful blooms, in lakes and reservoirs. In contrast, in warm, arid lakes, increased temperatures will increase evapotranspiration, concentrating nutrients in the shallow systems and exacerbating water quality issues.

Such hydrologic cycle changes are also expected to affect the timing of nitrogen transport. Changes in the seasonality of precipitation, and in particular snowmelt, will tend to alter the timing of nitrogen flushing through the ecosystem. There have already been widespread instances of earlier snowmelt and increased winter thaws associated with warming over the past few decades. Timing changes can ultimately alter the magnitude of N-export to downstream water bodies, particularly if the timing of flushing changes relative to the timing of biologically mediated uptake in either terrestrial or aquatic ecosystems.

V. Summary

1. Nitrogen, along with carbon, hydrogen, and phosphorus, are major constituents of the cellular protoplasm of organisms. Nitrogen is a major nutrient that affects the productivity of freshwaters.
2. The nitrogen cycle is a complex biochemical process in which nitrogen in various forms is altered (Fig. 14-1). This includes nitrogen fixation, assimilation, remineralization, anammox, and denitrification, to name a few. For all practical purposes, the nitrogen cycle of lakes and rivers is microbial in nature: bacterial oxidation and reduction of nitrogen compounds are coupled with photosynthetic assimilation and utilization by algae, Cyanobacteria, and larger aquatic plants. The direct role of animals in the nitrogen cycle is typically very small; under certain conditions, however, their grazing activities can influence microbial populations and nitrogen transformation rates as well as nitrogen utilization rates by photosynthetic organisms.
3. Dominant forms of nitrogen in freshwaters include: (1) dissolved molecular N_2; (2) ammonium (NH_4^+); (3) nitrite (NO_2^-); (4) nitrate (NO_3^-); and (5) a suite of organic compounds (e.g., amino acids, amines, nucleotides, proteins, refractory humic compounds of low nitrogen content) and (6) particulate nitrogen.
4. The nitrogen budget consists of a balance between nitrogen inputs to and nitrogen losses from an aquatic ecosystem.
   - **a.** Sources of nitrogen include: (1) atmospheric; (2) nitrogen fixation both in the water and the sediments; and (3) inputs of nitrogen from surface and groundwater drainage.
   - **b.** Losses of nitrogen occur by: (1) outflow from the basin; (2) denitrification and anammox; and (3) sedimentation of inorganic and organic nitrogen-containing compounds to the sediments.
5. Microbial fixation of molecular N_2 in soils by bacteria is a major source of nitrogen. In lakes and streams N_2 fixation by heterotrophic bacteria and certain Cyanobacteria is quantitatively less significant, except under certain conditions of severe depletion of combined inorganic nitrogen compounds. N_2 fixation by Cyanobacteria is usually much greater than fixation by heterotrophic bacteria. In Cyanobacteria N_2 fixation is light dependent and usually coincides with the spatial and temporal distribution of these microbes. NH_4^+ assimilation requires less energy expenditure than N_2 fixation by Cyanobacteria, and hence N_2 fixation increases when NH_4^+ and NO_3^- concentrations decrease in the trophogenic zone.
6. NH_4^+ is generated by heterotrophic bacteria as the primary nitrogenous end product of the decomposition of proteins and other nitrogenous organic compounds. NH_4^+ can also be generated in anaerobic conditions via dissimilatory NO_3^- reduction to NH_4^+ (DNRA). NH_4^+ is readily assimilated by plants in the trophogenic zone.
   - **a.** NH_4^+ concentrations are usually close to detection limits in aerobic waters because of utilization by plants in the photic zone. Additionally, bacterial nitrification occurs, resulting in NH_4^+ being oxidized through several intermediate compounds to NO_2^- and NO_3^-.
   - **b.** When the hypolimnion of a eutrophic lake becomes anaerobic, bacterial nitrification of NH_4^+ ceases. The oxidized microzone at the sediment–water interface is also lost, which reduces the adsorptive capacity of the sediments for NH_4^+. A marked increase in the release of NH_4^+ from the sediments then occurs. As a result, the NH_4^+ concentrations of the hypolimnion increase (Fig. 14-2).
   - **c.** Bacterial nitrification normally proceeds in two stages: (1) the oxidation of $NH_4^+ \rightarrow NO_2^-$, largely by NH_4^+ oxidizing archaea and NH_4^+ oxidizing bacteria; and (2) the oxidation of $NO_2^- \rightarrow NO_3^-$, in which NO_3^- oxidizing bacteria are involved. However, the recent discovery of complete NH_4^+ oxidizers (comammox) showed that the oxidization of NH_4^+ to NO_3^- can also be done by a single microorganism.
   - **d.** NO_2^- is readily oxidized and rarely accumulates except in the metalimnion, upper hypolimnion, or interstitial water of sediments of eutrophic lakes. Concentrations are usually close to detection limits unless organic pollution is high.
7. NO_3^- is assimilated and aminated into organic nitrogenous compounds within organisms. This organic nitrogen is bound and cycled in photosynthetic and microbial organisms. During the normal metabolism of these organisms, and at death, much of their nitrogen is liberated as NH_4^+. Additionally, organisms release a variety of organic nitrogenous compounds that are resistant to proteolytic deamination and ammonification by heterotrophic bacteria to varying degrees.
   - **a.** NO_3^- is the common form of inorganic nitrogen entering freshwaters from the drainage basin in surface waters, groundwater, and precipitation. In certain oligotrophic waters in basaltic rock formations, NO_3^- loading from atmospheric sources, especially if contaminated by

human-produced combustion emission products, can dominate nitrogen loading.

b. Bacterial denitrification is the biochemical reduction of oxidized nitrogen anions (NO_3^- and NO_2^-) concomitant with the oxidation of organic matter: $NO_3^- \rightarrow NO_2^- \rightarrow N_2O \rightarrow N_2$. Nitrous oxide ($N_2O$) is rapidly reduced to N_2 and has rarely been found in lakes in appreciable quantities. Denitrification is accomplished by many genera of facultative anaerobic bacteria, which utilize NO_3^- as an exogenous terminal hydrogen acceptor in the oxidation of organic substrates. Denitrification occurs in anaerobic environments, such as in the hypolimnia of eutrophic lakes (Fig. 14-2) or in anoxic sediments, where oxidizable substrates are relatively abundant.

c. The process of anammox has only been described relatively recently. This process converts 1:1 molecule ratio of NH_4^+ and NO_3^-/NO_2^- to N_2 in the absence of oxygen and without the need for a carbon supply.

d. The process of *dissimilatory NO_3^- reduction to NH_4^+* (DNRA) also contributes to NO_3^- removal in aquatic ecosystems, but the relative importance of DNRA and denitrification on NO_3^- removal is quite variable between different aquatic systems.

8. Dissolved organic nitrogen (DON) often constitutes over 50% of the total soluble nitrogen in freshwaters. Over half of the DON occurs as amino nitrogen compounds, mostly as polypeptides and complex nitrogen compounds. Simpler forms, such as urea and dissolved free amino acids, are rapidly utilized by bacteria and algae.

9. Much of the total nitrogen occurs in the sediments in forms that are relatively unavailable for biotic utilization. Of the readily available nitrogen, a majority occurs in soluble form in the water and in the interstitial water of surficial sediments (and in littoral vegetation in shallow, productive lakes). Turnover rates of NH_4^+ are rapid in water but slower in the sediments. In contrast, NO_3^- turnover is slower in the water than in sediments, where, under anoxic conditions in eutrophic lakes, NO_3^- may be denitrified.

10. In running waters nitrogen is used repeatedly as it passes downstream. The rate of utilization and release depends upon physical and biological retentiveness, largely by the microbiota attached to the streambed. The average distance a nutrient atom travels downstream during one cycle through the water, biotic, and substrata compartments is referred to as the spiraling length (Figs. 14-9 and 14-10).

11. Water flow in streams and rivers can continuously bring dissolved nutrients to primary producers (especially attached algae and other plants); however, nutrient limitation often occurs in streams and rivers, in the absence of excessive nutrient inputs.

12. The processes of nitrogen cycling in the water of streams and rivers are similar to those of lakes and are influenced to a large extent by bacterial, fungal, and other microbial metabolism. Attached bacteria, fungi, and algae are the primary organisms controlling the seasonal spatial and temporal variations within the water. The processes of nitrification and denitrification often function simultaneously and reciprocally in running water sediments, where many microzones of steep redox gradients occur in the hyporheic zone of the streambed.

13. Increased loading of inorganic nitrogen to rivers and lakes results from agricultural activities, sewage, and anthropogenic atmospheric pollution. Increased nitrogen loads to waterways can stimulate excessive algal productivity, reduce plant biodiversity, are associated with loss of macrophyte communities, and have effects on the quality and quantity of food available for aquatic fauna. Fertilizer use for crops is a major source of nitrogen to freshwaters globally and is changing the ratio of nitrogen to phosphorus, with impacts on the functioning of aquatic systems.

14. Climate change can also impact nitrogen budgets and transformations in freshwater systems. Warmer temperatures increase enzyme activity affecting rates of microbial nitrogen transformation. Climate change is also expected to intensify the hydrologic cycle, with more frequent and intense heavy rainfall events, potential deepening and lengthening of dry periods, altered snow accumulation and melt, and changes in evapotranspiration. This will affect the rates of microbial nitrogen transformations.

Acknowledgments

We thank Sian Taylor for the inclusion of references and Hannah Franklin for their constructive review of the chapter.

References

Acuña, V., Muñoz, I., Giorgi, A., Omella, M., Sabater, F., Sabater, S., 2005. Drought and postdrought recovery cycles in an intermittent Mediterranean stream: structural and functional aspects. J. North Am. Benthol. Soc. 24, 919–933.

Alexander, M., 1961. Introduction to Soil Microbiology. John Wiley & Sons, New York, p. 472.

Alexander, M., 1965. Nitrification. Agronomy 10, 307–343.

Alexander, R.B., Böhlke, J.K., Boyer, E.W., David, M.B., Harvey, J.W., Mulholland, P.J., Seitzinger, S.P., Tobias, C.R., Tonitto, C., Wollheim, W.M., 2009. Dynamic modeling of nitrogen losses in river networks unravels the coupled effects of hydrological and biogeochemical processes. Biogeochemistry 93, 91–116.

Alexander, R.B., Smith, R.A., Schwarz, G.E., 2000. Effect of stream channel size on the delivery of nitrogen to the Gulf of Mexico. Nature 403, 758–761.

Andersen, J.M., 1977. Rates of denitrification of undisturbed sediment from six lakes as a function of nitrate concentration, oxygen and temperature. Arch. Hydrobiol. 80, 147–159.

Arango, C.P., Tank, J.L., 2008. Land use influences the spatiotemporal controls on nitrification and denitrification in headwater streams. J. North Am. Benthol. Soc. 27, 90–107.

Arce, M.I., Sánchez-Montoya, M., del, M., Gómez, R., 2015. Nitrogen processing following experimental sediment rewetting in isolated pools in an agricultural stream of a semiarid region. Ecol. Eng. 77, 233–241.

Austin, B.J., Strauss, E.A., 2011. Nitrification and denitrification response to varying periods of desiccation and inundation in a western Kansas stream. Hydrobiologia 658, 183–195.

Baldwin, D.S., Mitchell, A.M., 2000. The effects of drying and reflooding on the sediment and soil nutrient dynamics of lowland river-floodplain systems: a synthesis. Regul. Rivers Res. Manag. 16, 457–467.

Bandurski, R.S., 1965. Biological reduction of sulfate and nitrate. In: Bonner, J., Varner, J.E. (Eds.), Plant Chemistry. Academic Press, New York, pp. 467–490.

Baron, J.S., Driscoll, C.T., Stoddard, J.L., Richer, E.E., 2011. Empirical critical loads of atmospheric nitrogen deposition for nutrient enrichment and acidification of sensitive US lakes. Bioscience 61, 602–613.

Bastviken, S.K., Eriksson, P.G., Premrov, A., Tonderski, K., 2005. Potential denitrification in wetland sediments with different plant species detritus. Ecol. Eng. 25, 183–190.

Baxter, R.M., Wood, R.B., Prosser, M.V., 1973. The probable occurrence of hydroxylarnine in the water of an Ethiopian lake. Limnol. Oceanogr 18, 470–472.

Berman, T., Bronk, D.A., 2003. Dissolved organic nitrogen: A dynamic participant in aquatic ecosystems. Aquat. Microb. Ecol. 31, 279–305.

Bernot, M.J., Dodds, W.K., 2005. Nitrogen retention, removal, and saturation in lotic ecosystems. Ecosystems 8, 442–453.

Billaud, V.A., 1968. Nitrogen fixation and the utilization of other inorganic nitrogen sources in a subarctic lake. J. Fish. Res. Board Can. 25, 2101–2110.

Birgand, F., Skaggs, R.W., Chescheir, G.M., Gilliam, J.W., 2007. Nitrogen removal in streams of agricultural catchments—a literature review. Crit. Rev. Environ. Sci. Technol. 37, 381–487.

Birmingham, M.W., Bachmann, R.W., Crumpton, W.G., 1994. Nitrate uptake by stream sediments: the influence of sediment character. Verh. Internat. Verein. Limnol. 25, 1467–1470.

Bogard, M.J., Donald, D.B., Finlay, K., Leavitt, P.R., 2012. Distribution and regulation of urea in lakes of central North America. Freshw. Biol. 57, 1277–1292.

Böhlke, J.K., Harvey, J.W., Voytek, M.A., 2004. Reach-scale isotope tracer experiment to quantify denitrification and related processes in a nitrate-rich stream, midcontinent United States. Limnol. Oceanogr 49, 821–838.

Bostic, E.M., White, J.R., 2007. Soil phosphorus and vegetation influence on wetland phosphorus release after simulated drought. Soil Sci. Soc. Am. J. 71, 238–244.

Bouwmeester, R.J.B., Vlek, P.L.G., 1981. Rate control of ammonia volatilization from rice paddies. Atmos. Environ. 15, 131–140.

Bowden, W.B., 1987. The biogeochemistry of nitrogen in freshwater wetlands. Biogeochemistry 4, 313–348.

Boyer, E.W., Goodale, C.L., Jaworski, N.A., Howarth, R.W., 2002. Anthropogenic nitrogen sources and relationships to riverine nitrogen export in the northeastern U.S.A. Biogeochemistry 57–58, 137–169.

Boyer, E.W., Howarth, R.W., Galloway, J.N., Dentener, F.J., Green, P.A., Vörösmarty, C.J., 2006. Riverine nitrogen export from the continents to the coasts. Global Biogeochem. Cycles 20, 1–9.

Bradley, P.M., McMahon, P.B., Chapelle, F.H., 1995. Effects of carbon and nitrate on denitrification sediments of an effluent-dominated river. Water Resour. Res. 31, 1063–1068.

Breitenbeck, G.A., Bremner, J.M., 1987. Effects of storing soils at various temperatures on their capacity for denitrification. Soil Biol. Biochem. 19, 377–380.

Brezonik, P.L., Lee, G.F., 1968. Denitrification as a nitrogen sink in Lake Mendota, Wisconsin. Environ. Sci. Technol. 2, 120–125.

Brin, L.D., Giblin, A.E., Rich, J.J., 2017. Similar temperature responses suggest future climate warming will not alter partitioning between denitrification and anammox in temperate marine sediments. Glob. Chang. Biol. 23, 331–340.

Bronk, D.A., See, J.H., Bradley, P., Killberg, L., 2007. DON as a source of bioavailable nitrogen for phytoplankton. Biogeosciences 4, 283–296.

Burford, M.A., Mcneale, K.L., Mckenzie-Smith, F.J., 2006. The role of nitrogen in promoting the toxic cyanophyte *Cylindrospermopsis raciborskii* in a subtropical water reservoir. Freshw. Biol. 51, 2143–2153.

Burgin, A.J., Hamilton, S.K., 2007. Have we overemphasized the role of denitrification in aquatic ecosystems? A review of nitrate removal pathways. Front. Ecol. Environ. Times 5, 89–96.

Burns, D.A., 1998. Retention of NO_3^- in an upland stream environment: A mass balance approach. Biogeochemistry 40, 73–96.

Butturini, A., Sabater, F., 1998. Ammonium and phosphate retention in a Mediterranean stream: Hydrological versus temperature control. Can. J. Fish. Aquat. Sci. 55, 1938–1945.

Cavaliere, E., Baulch, H.M., 2018. Denitrification under lake ice. Biogeochemistry 137, 285–295.

Chaffin, J.D., Bridgeman, T.B., 2014. Organic and inorganic nitrogen utilization by nitrogen-stressed cyanobacteria during bloom conditions. J. Appl. Phycol. 26, 299–309.

Chan, Y.K., Campbell, N.E.R., 1978. Phytoplankton uptake and excretion of assimilated nitrate in a small Canadian shield lake. Appl. Env. Microbiol. 35, 1052–1060.

Chan, Y.K., Campbell, N.E.R., 1980. Denitrification in Lake 227 during summer stratification. Can. J. Fish. Aquat. Sci. 37, 506–512.

Chan, Y.K., Knowles, R., 1979. Measurement of denitrification in two freshwater sediments by an in situ acetylene inhibition method. Appl. Environ. Microbiol. 37, 1067–1072.

Chapin, J.D., Uttormark, P.D., 1973. Atmospheric contributions of nitrogen and phosphorus. Tech. Rep. Wat. resources Ctr, vols. 73–2. Univ. Wis, p. 35.

Chatarpaul, L., Robinson, J.B., Kaushik, N.K., 1980. Effects of tubificid worms on denitrification and nitrification in stream sediment. Can. J. Fish. Aquat. Sci. 37, 656–663.

Chen, R.L., Keeney, D.R., Graetz, D.A., Holding, A.J., 1972. Denitrification and nitrate reduction in Wisconsin lake sediments. J. Environ. Qual. 1, 158–162.

Chen, X., Wang, Y.H., Ye, C., Zhou, W., Cai, Z.C., Yang, H., Han, X., 2018. Atmospheric nitrogen deposition associated with the eutrophication of Taihu Lake. J. Chem. 4017107. https://doi.org/10.1155/2018/4017107.

Christensen, P.B., Sorensen, J., 1986. Temporal variation of denitrification activity in plant-covered, littoral sediment from Lake Hampen. Denmark. Appl. Environ. Microbiol. 51, 1174–1179.

Coats, R.N., Leonard, R.L., Goldman, C.R., 1976. Nitrogen uptake and release in a forested watershed, Lake Tahoe Basin, California. Ecology 57, 995–1004.

Cook, P.L.M.M., Aldridge, K.T., Lamontagne, S., Brookes, J.D., 2010. Retention of nitrogen, phosphorus and silicon in a large semi-arid riverine lake system. Biogeochemistry 99, 49–63.

Cooke, J.G., White, R.E., 1987a. The effect of nitrate in stream water on the relationship between gentrification and nitrification in a stream-sediment microcosm. Freshw. Biol. 18, 213–226.

Cooke, J.G., White, R.E., 1987b. Spatial distribution of denitrifying activity in a stream draining an agricultural catchment. Freshw. Biol. 18, 509–519.

Craig, L.S., Palmer, M.A., Richardson, D.C., Filoso, S., Bernhardt, E.S., Bledsoe, B.P., Doyle, M.W., Groffman, P.M., Hassett, B.A., Kaushal, S.S., Mayer, P.M., Smith, S.M., Wilcock, P.R., 2008. Stream restoration strategies for reducing river nitrogen loads. Front. Ecol. Environ. 6, 529–538.

Crowe, S.A., Treusch, A.H., Forth, M., Li, J., Magen, C., Canfield, D.E., Thamdrup, B., Katsev, S., 2017. Novel anammox bacteria and nitrogen loss from. Lake Superior. Sci. Rep. 7, 1–7.

Dai, R., Zhou, Y., Chen, Y., Zhang, X., Yan, Y., An, D., 2019. Effects of arginine on the growth and microcystin-LR production of *Microcystis aeruginosa* in culture. Sci. Total Environ. 651, 706–712.

Dail, D.B., Davidson, E.A., Chorover, J.O.N., 2001. Rapid abiotic transformation of nitrate in an acid forest soil. Biogeochemistry 54, 131–146.

Daims, H., Lebedeva, E.V., Pjevac, P., Han, P., Herbold, C., Jehmlich, N., Palatinszky, M., Vierheilig, J., Bulaev, A., Kirkegaard, R.H., Bergen, M., Von, Rattei, T., Bendinger, B., Nielsen, P.H., Wagner, M., 2015. Complete nitrification by *Nitrospira* bacteria. Nature 528, 504–509.

Dalsgaard, T., Canfield, D.E., Petersen, J., Thamdrup, B., Acuna-Gonzalez, J., 2003. N_2 production by the anammox reaction in the anoxic water column of Golfo Dulce, Costa Rica. Nature 422, 606–608.

Dalsgaard, T., De Brabandere, L., Hall, P.O.J., 2013. Denitrification in the water column of the central Baltic Sea. Geochim. Cosmochim. Acta. 106, 247–260.

Datry, T., Larned, S.T., Tockner, K., 2014. Intermittent rivers: A challenge for freshwater ecology. Bioscience 64, 229–235.

Davidson, E., Chorover, J., Dail, D., 2003. A mechanism of abiotic immobilization of nitrate in forest ecosystems: The ferrous wheel hypothesis. Glob. Chang. Biol. 9, 228–236.

Dean, J.V., Biesboer, D.D., 1985. Loss and uptake of ^{15}N-Ammonium in submerged soils of a cattail marsh. Am. J. Bot. 72, 1197–1203.

Dodds, W.K., Jones, R.D., 1987. Potential rates of nitrification and denitrification in an oligotrophic freshwater sediment system. Microb. Ecol. 14, 91–100.

Dong, L.F., Sobey, M.N., Smith, C.J., Rusmana, I., Phillips, W., Stott, A., Osborn, A.M., Nedwell, D.B., 2011. Dissimilatory reduction of nitrate to ammonium, not denitrification or anammox, dominates benthic nitrate reduction in tropical estuaries. Limnol. Oceanogr. 56, 279–291.

Donner, S.D., Kucharik, C.J., Foley, J.A., 2004. Impact of changing land use practices on nitrate export by the Mississippi River. Global Biogeochem. Cycles 18, GB1028.

Du, Y., Ma, T., Deng, Y., Shen, S., Lu, Z., 2017. Sources and fate of high levels of ammonium in surface water and shallow groundwater of the Jianghan Plain, Central China. Environ. Sci. Process. Impacts 19, 161–172.

Duff, J.H., Triska, F.J., 1990. Denitrification in sediments from the hyporheic zone adjacent to a small forested stream. Can. J. Fish. Aquat. Sci. 47, 1140–1147.

Dugdale, R.C., Dugdale, V.A., 1961. Sources of phosphorus and nitrogen for lakes on Afognak Island. Limnol. Oceanogr 6, 13–23.

Durand, P., Breuer, L., Johnes, P.J., Billen, G., Butturini, A., Pinay, G., van Grinsven, H., Garnier, J., Rivett, M., Reay, D.S., Curtis, C., Siemens, J., Maberly, S., Kaste, Ø., Humborg, C., Loeb, R., de Klein, J., Hejzlar, J., Skoulikidis, N., Kortelainen, P., Lepistö, A., Wright, R., 2011. Nitrogen processes in aquatic ecosystems. Eur. Nitrogen Assess 126–146.

Elliot, A.H., Alexander, R.B., Schwarz, G.E., Shankar, U., Sukias, J.P.S., McBride, G.B., 2005. Estimation of nutrient sources and transport for New Zealand using the hybrid mechanistic-statistical model SPARROW. J. Hydrol. N. Z 44, 1–27.

Elser, J.J., Andersen, T., Baron, J.S., Bergström, A.K., Jansson, M., Kyle, M., Nydick, K.R., Steger, L., Hessen, D.O., 2009. Shifts in lake N: P stoichiometry and nutrient limitation driven by atmospheric nitrogen deposition. Science 326, 835–837.

Elser, J.J., Bracken, M.E.S., Cleland, E.E., Gruner, D.S., Harpole, W.S., Hillebrand, H., Ngai, J.T., Seabloom, E.W., Shurin, J.B., Smith, J.E., 2007. Global analysis of nitrogen and phosphorus limitation of primary producers in freshwater, marine and terrestrial ecosystems. Ecol. Lett. 10, 1135–1142.

Emerson, K., Russo, R.C., Lund, R.E., Thurston, R.V., 1975. Aqueous ammonia equilibrium calculations: Effect of pH and temperature. J. Fish. Res. Board Can. 32, 2379–2383.

Ensign, S.H., Doyle, M.W., 2006. Nutrient spiraling in streams and river networks. J. Geophys. Res. Biogeosci. 111, 1–13.

Erler, D.V., Eyre, B.D., Davison, L., 2008. The contribution of anammox and denitrification to sediment N_2 production in a surface flow constructed wetland. Environ. Sci. Technol. 42, 9144–9150.

Erratt, K.J., Creed, I.F., Trick, C.G., 2018. Comparative effects of ammonium, nitrate and urea on growth and photosynthetic efficiency of three bloom-forming cyanobacteria. Freshw. Biol. 63, 626–638.

Faggotter, S.J., Webster, I.T., Burford, M.A., 2013. Factors controlling primary productivity in a wet-dry tropical river. Mar. Freshw. Res. 64, 585–598.

Fan, L., Brett, M.T., Li, B., Song, M., 2018. The bioavailability of different dissolved organic nitrogen compounds for the freshwater algae *Raphidocelis subcapitata*. Sci. Total Environ. 618, 479–486.

Faulkner, B.R., 2008. Bayesian modeling of the assimilative capacity component of nutrient total maximum daily loads. Water Resour. Res. 44, 1–10.

Fay, P., 1992. Oxygen relations of nitrogen fixation in cyanobacteria. Microbiol. Rev. 56, 340–373.

Fenchel, T., Blackburn, T.H., 1979. Bacteria and Mineral Cycling. Academic Press, London.

Fischer, H., Kloep, F., Wilzcek, S., Pusch, M.T., 2005. A river's liver—Microbial processes within the hyporheic zone of a large lowland river. Biogeochemistry 76, 349–371.

Fisher, S.G., Welter, J., Schade, J., Henry, J., 2001. Landscape challenges to ecosystem thinking: Creative flood and drought in the American Southwest. Sci. Mar. 65, 181–192.

Flewelling, S.A., Herman, J.S., Hornberger, G.M., Mills, A.L., 2012. Travel time controls the magnitude of nitrate discharge in groundwater bypassing the riparian zone to a stream on Virginia's coastal plain. Hydrol. Process. 26, 1242–1253.

Fogg, G.E., 1971. Nitrogen fixation in lakes. Plant Soil 1971, 393–401.

Fogg, G.E., 1974. Nitrogen fixation. In: Stewart, W.D.P. (Ed.), Algal Physiology and Biochemistry. University of California Press, Berkeley, pp. 560–582.

Fogg, G.E., Westlake, D.F., 1955. The importance of extracellular products of algae in freshwater. Verh. Internat. Verein. Limnol 12, 219–232.

Fork, M.L., Heffernan, J.B., 2014. Direct and indirect effects of dissolved organic matter source and concentration on denitrification in northern florida rivers. Ecosystems 17, 14–28.

Fowler, D., Coyle, M., Skiba, U., Sutton, M.A., Cape, J.N., Reis, S., Sheppard, L.J., Jenkins, A., Grizzetti, B., Galloway, J.N., Vitousek, P., Leach, A., Bouwman, A.F., Butterbach-Bahl, K., Dentener, F., Stevenson, D., Amann, M., Voss, M., 2013. The global nitrogen cycle in the twenty-first century. Philos. Trans. R. Soc. B Biol. Sci. 368, 0130164.

Fukuhara, H., Yasuda, K., 1989. Ammonium excretion by some freshwater zoobenthos from a eutrophic lake. Hydrobiologia 173, 1–8.

Galloway, J.N., 1998. The global nitrogen cycle: Changes and consequences. Environ. Pollut. 102, 15–24.

Galloway, J.N., Townsend, A.R., Erisman, J.W., Bekunda, M., Cai, Z., Freney, J.R., Martinelli, L.A., Seitzinger, S.P., Sutton, M.A., 2008. Transformation of the nitrogen cycle: Recent trends, questions, and potential solutions. Science 320, 889–892.

Gao, Y., Zhou, F., Ciais, P., Miao, C., Yang, T., Jia, Y., Zhou, X., Klaus, B.B., Yang, T., Yu, G., 2020. Human activities aggravate nitrogen-deposition pollution to inland water over China. Natl. Sci. Rev. 7, 430–440.

Gardner, W.S., Chandler, J.F., Laird, G.A., 1989. Organic nitrogen mineralization and substrate limitation of bacteria in Lake Michigan. Limnol. Oceanogr. 34, 478–485.

Gardner, W.S., Chandler, J.F., Laird, G.A., Carrick, H.J., 1987. Sources and fate of dissolved free amino acids in epilimnetic Lake Michigan water. Limnol. Oceanogr. 32, 1353–1362.

Gleick, P.H., 1989. Climate change, hydrology, and water resources. Rev. Geophys. 27, 329–344.

Glibert, P.M., Maranger, R., Sobota, D.J., Bouwman, L., 2014. The Haber Bosch–harmful algal bloom (HB-HAB) link. Environ. Res. Lett. 9, 105001.

Goering, J.J., Dugdale, V.A., 1966. Estimates of the rates of denitrification in a subarctic lake. Limnol. Oceanogr. 11, 113–117.

Goldman, C.R., 1960. Primary productivity and limiting factors in three lakes of the Alaska Peninsula. Ecol. Monogr. 30, 207–230.

Gómez, R., Arce, M.I., Sánchez, J.J., Sánchez-Montoya, M., del, M., 2012. The effects of drying on sediment nitrogen content in a Mediterranean intermittent stream: A microcosms study. Hydrobiologia 679, 43–59.

Gooseff, M.N., Mcknight, D.M., Runkel, R.L., Duff, J.H., 2004. Denitrification and hydrologic transient storage in a glacial meltwater stream, McMurdo Dry Valleys, Antarctica. Limnol. Oceanogr. 49, 1884–1895.

Gouy, J.-L., Bergé, P., Labroue, L., 1984. *Gallionella ferruginea*, facteur de dénitrification dans les eaux pauvres en matière organique. C R Seances Acad Sci Série 3, Sciences de la vie 298, 153–156.

Grantz, E.M., Haggard, B.E., Thad Scott, J., 2014. Stoichiometric imbalance in rates of nitrogen and phosphorus retention, storage, and recycling can perpetuate nitrogen deficiency in highly-productive reservoirs. Limnol. Oceanogr. 59, 2203–2216.

Greaver, T.L., Clark, C.M., Compton, J.E., Vallano, D., Talhelm, A.F., Weaver, C.P., Band, L.E., Baron, J.S., Davidson, E.A., Tague, C.L., Felker-Quinn, E., Lynch, J.A., Herrick, J.D., Liu, L., Goodale, C.L., Novak, K.J., Haeuber, R.A., 2016. Key ecological responses to nitrogen are altered by climate change. Nat. Clim. Chang 6, 836–843.

Grimm, N.B., 1987. Nitrogen dynamics during succession in a desert stream. Ecology 68, 1157–1170.

Grizzetti, B., Bouraoui, F., De Marsily, G., Bidoglio, G., 2005. A statistical method for source apportionment of riverine nitrogen loads. J. Hydrol. 304, 302–315.

Groffman, P.M., 1994. Denitrification in freshwater wetlands. Curr. Top. Wetl. Biogeochem. 1, 15–35.

Gross, E.M., 2003. Differential response of tellimagrandin II and total bioactive hydrolysable tannins in an aquatic angiosperm to changes in light and nitrogen. Oikos 103, 497–504.

Guiry, E.J., Buckley, M., Orchard, T.J., Hawkins, A.L., Needs-Howarth, S., Holm, E., Szpak, P., 2020. Deforestation caused abrupt shift in Great Lakes nitrogen cycle. Limnol. Oceanogr. 65, 1921–1935.

Guo, L., Chen, Q., Fang, F., Hu, Z., Wu, J., Miao, A., Xiao, L., Chen, X., Yang, L., 2013. Application potential of a newly isolated indigenous aerobic denitrifier for nitrate and ammonium removal of eutrophic lake water. Bioresour. Technol. 142, 45–51.

Hamersley, R.M., Woebken, D., Boehrer, B., Schultze, M., Lavik, G., Kuypers, M.M.M., 2009. Water column anammox and denitrification in a temperate permanently stratified lake (Lake Rassnitzer, Germany). Syst. Appl. Microbiol. 32, 571–582.

Harrison, J.A., Maranger, R.J., Alexander, R.B., Giblin, A.E., Jacinthe, P.A., Mayorga, E., Seitzinger, S.P., Sobota, D.J., Wollheim, W.M., 2009. The regional and global significance of nitrogen removal in lakes and reservoirs. Biogeochemistry 93, 143–157.

Hasegawa, T., Okino, T., 2004. Seasonal variation of denitrification rate in Lake Suwa sediment. Limnology 5, 33–39.

Haselkorn, R., Buikema, W.J., 1992. Nitrogen fixation in cyanobacteria. In: Stacey, G., Burris, R.H., Evans, H.J. (Eds.), Biological Nitrogen Fixation. Chapman & Hall, New York, pp. 166–190.

He, Y., Tao, W., Wang, Z., Shayya, W., 2012. Effects of pH and seasonal temperature variation on simultaneous partial nitrification and anammox in free-water surface wetlands. J. Environ. Manage 110, 103–109.

Hietanen, S., Jäntti, H., Buizert, C., Jürgens, K., Labrenz, M., Voss, M., Kuparinen, J., 2012. Hypoxia and nitrogen processing in the Baltic Sea water column. Limnol. Oceanogr. 57, 325–337.

Hill, S., 1992. Physiology of nitrogen fixation in free-living heterotrophs. In: Stacey, G., Burris, R.H., Evans, H.J. (Eds.), Biological Nitrogen Fixation. Chapman & Hall, New York, pp. 87–134.

Hoellein, T.J., Tank, J.L., Rosi-Marshall, E.J., Entrekin, S.A., Lamberti, G.A., 2007. Controls on spatial and temporal variation of nutrient uptake in three Michigan headwater streams. Limnol. Oceanogr. 52, 1964–1977.

Horňák, K., Pernthaler, J., 2020. Homeostatic regulation of dissolved labile organic substrates by consumption and release processes in a freshwater lake. Limnol. Oceanogr. 65, 939–950.

Horne, A.J., 1979. Nitrogen fixation in Clear Lake, California. 4. Diel studies on *Aphanizomenon* and *Anabaena* blooms. Limnol. Oceanogr. 24, 329–341.

Horne, A.J., Dillard, J.E., Fujita, D.K., Goldman, C.R., 1972. Nitrogen fixation in Clear Lake, California. II. Synoptic studies on the autumn *Anabaena* bloom. Limnol. Oceanogr. 17, 693–703.

Horne, A.J., Fogg, G.E., 1970. Nitrogen fixation in some English lakes. Proc. R. Soc. London. Ser. B Biol. Sci. 175, 351–366.

Horne, A.J., Sandusky, J.C., Carmiggelt, C.J.W., 1979. Nitrogen fixation in Clear Lake, California. 3. Repetitive synoptic sampling of the spring *Aphanizomenon* blooms. Limnol. Oceanogr. 24, 316–328.

Hou, L., Zheng, Y., Liu, M., Li, X., Lin, X., Yin, G., Gao, J., Deng, F., Chen, F., Jiang, X., 2015. Anaerobic ammonium oxidation and its contribution to nitrogen removal in China's coastal wetlands. Sci. Rep. 5, 1–11.

Howarth, R.W., Marino, R., Cole, J.J., 1988a. Nitrogen fixation in freshwater, estuarine, and marine ecosystems. 2. Biogeochemical controls. Limnol. Oceanogr. 33, 688–701.

Howarth, R.W., Marino, R., Lane, J., Cole, J.J., 1988b. Nitrogen fixation in freshwater, estuarine, and marine ecosystems. 1. Rates and importance. Limnol. Oceanogr. 33, 669–687.

Hutchinson, G.E., 1944. Nitrogen in the biogeochemistry of the atmosphere. Am. Sci. 32, 178–195.

Hutchinson, G.E., 1957. A Treatise on Limnology. I. Geography, Physics, and Chemistry. John Wiley & Sons, New York.

Hutchinson, G.E., 1975. A Treatise on Limnology. III. Limnological Botany. John Wiley & Sons, New York, p. 660.

Ittekkot, V., Zhang, S., 1989. Pattern of particulate nitrogen transport in world rivers. Global Biogeochem. Cycles 3, 383–391.

Jansson, M., 1980. Role of benthic algae in transport of nitrogen from sediment to lake water in a shallow clearwater lake. Arch. Hydrobiol. 89, 101–109.

Jassby, A.D., Reuter, J.E., Axler, R.P., Goldman, C.R., Hackley, S.H., 1994. Atmospheric deposition of nitrogen and phosphorus in the annual nutrient load of Lake Tahoe (California-Nevada). Water Resour. Res. 30, 2207–2216.

Jeppesen, E., Kronvang, B., Olesen, J.E., Audet, J., Søndergaard, M., Hoffmann, C.C., Andersen, H.E., Lauridsen, T.L., Liboriussen, L., Larsen, S.E., Beklioglu, M., Meerhoff, M., Özen, A., Özkan, K., 2011. Climate change effects on nitrogen loading from cultivated catchments in Europe: Implications for nitrogen retention, ecological state of lakes and adaptation. Hydrobiologia 663, 1–21.

Jetten, M.S.M., Niftrik, L., Strous, van, Kartal, M., Keltjens, B., Op Den Camp, J.T., H.J.M, 2009. Biochemistry and molecular biology of anammox bacteria. Crit. Rev. Biochem. Mol. Biol. 44, 65–84.

Ji, B., Yang, K., Zhu, L., Jiang, Y., Wang, H., Zhou, J., Zhang, H., 2015. Aerobic denitrification: A review of important advances of the last 30 years. Biotechnol. Bioprocess Eng. 20, 643–651.

Johnes, P.J., 1996. Evaluation and management of the impact of land use change on the nitrogen and phosphorus load delivered to surface waters: The export coefficient modelling approach. J. Hydrol 183, 323–349.

Jones, J.G., 1979a. Microbial nitrate reduction in freshwater sediments. J. Gen. Microbiol. 115, 27–35.

Jones, J.G., 1979b. Microbial activity in lake sediments with particular reference to electrode potential gradients. J. Gen. Microbiol. 115, 19–26.

Jones, J.G., Simon, B.M., Horsley, R.W., 1982. Microbiological sources of ammonia in freshwater lake sediments. J. Gen. Microbiol. 128, 2823–2831.

Kamiyama, K., Okuda, S., Kawai, A., 1977. Studies on the release of ammonium nitrogen from the bottom sediments in freshwater regions. II. Ammonium nitrogen in dissolved and absorbed form in the sediments. Jpn. J. Limnol. 38, 100–106.

Kaplan, W.A., Wofsy, S.C., 1985. The biogeochemistry of nitrous oxide: A review. Adv. Aquat. Microbiol. 3, 181–206.

Kartal, B., Maalcke, W.J., De Almeida, N.M., Cirpus, I., Gloerich, J., Geerts, W., Op Den Camp, H.J.M., Harhangi, H.R., Janssen-Megens, E.M., Francoijs, K.J., Stunnenberg, H.G., Keltjens, J.T., Jetten, M.S.M., Strous, M., 2011. Molecular mechanism of anaerobic ammonium oxidation. Nature 479, 127–130.

Kaushal, S.S., Groffman, P.M., Mayer, P.M., Striz, E., Gold, A.J., 2008. Effects of stream restoration on denitrification in an urbanizing watershed. Ecol. Appl. 18, 789–804.

Keck, F., Lepori, F., 2012. Can we predict nutrient limitation in streams and rivers? Freshw. Biol. 57, 1410–1421.

Keeney, D.R., 1973. The nitrogen cycle in sediment-water systems. J. Environ. Qual. 2, 15–29.

Keeney, D.R., Chen, R.L., Graetz, D.A., 1971. Importance of denitrification and nitrate reduction in sediments to the nitrogen budgets of lakes. Nature 233, 66–67.

Kelso, B.H.L., Smith, R.V., Laughlin, R.J., 1999. Effects of carbon substrates on nitrite accumulation in freshwater sediments. Appl. Environ. Microbiol. 65, 61–66.

Kelso, B.H.L., Smith, R.V., Laughlin, R.J., Lennox, S.D., 1997. Dissimilatory nitrate reduction in anaerobic sediments leading to river nitrite accumulation. Appl. Environ. Microbiol. 63, 4679–4685.

Kemp, M., Dodds, W., 2002. The influence of ammonium, nitrate, and dissolved oxygen concentrations on uptake, nitrification, and denitrification rates associated with prairie stream substrata. Limnol. Oceanogr. 47, 1380–1393.

Keren, R., Lawrence, J.E., Zhuang, W., Jenkins, D., Banfield, J.F., Alvarez-Cohen, L., Zhou, L., Yu, K., 2019. Increased replication of dissimilatory nitrate-reducing bacteria leads to decreased anammox bioreactor performance. Microbiome 8, 1–21.

Kerr, J.G., Burford, M.A., Olley, J.M., Bunn, S.E., Udy, J., 2011. Examining the link between terrestrial and aquatic phosphorus speciation in a subtropical catchment: The role of selective erosion and transport of fine sediments during storm events. Water Res. 45, 3331–3340.

Kessel, M.A.H., van, J., Speth, Albertsen, D.R., Nielsen, M., Huub, P.H., Camp, J.M., den, Kartal, O., Jetten, B., Lücker, S, M.S.M., 2015. Complete nitrification by a single microorganism. Nature 528, 555–559.

Klingensmith, K.M., Alexander, V., 1983. Sediment nitrification, denitrification, and nitrous oxide production in a deep arctic lake. Appl. Environ. Microbiol. 46, 1084–1092.

Koenig, L.E., Song, C., Wollheim, W.M., Rüegg, J., McDowell, W.H., 2017. Nitrification increases nitrogen export from a tropical river network. Freshw. Sci. 36, 698–712.

Könneke, M., Bernhard, A.E., De La Torre, J.R., Walker, C.B., Waterbury, J.B., Stahl, D.A., 2005. Isolation of an autotrophic ammonia-oxidizing marine archaeon. Nature 437, 543–546.

Kronvang, B., Hoffmann, C.C., Svendsen, L.M., Windolf, J., Jensen, J.P., Dørge, J., 1999. Retention of nutrients in river basins. Aquat. Ecol. 33, 29–40.

Kuenen, J.G., 2008. Anammox bacteria: From discovery to application. Nat. Rev. Microbiol. 6, 320–326.

Kuenen, J.G., Robertson, L.A., 1994. Combined nitrification-denitrification processes. FEMS Microbiol. Rev. 15, 109–117.

Kuznetsov, S.I., 1970. Mikroflora ozer i ee geokhimicheskaya deyatel'-nost'. (Microflora of lakes and their geochemical activities.). In: Russian. Izdatel'stvo Nauka, Leningrad, p. 440.

Landner, L., Larsson, T., 1973. Indications of disturbances in the nitrification process in a heavily nitrogen-polluted water body. Ambio 2, 154–157.

Larned, S.T., Datry, T., Arscott, D.B., Tockner, K., 2010. Emerging concepts in temporary-river ecology. Freshw. Biol. 55, 717–738.

Laurent, M., Badia, J., 1973. Étude comparative du cycle biologique de l'azote dans deux étangs. Ann. d'Hydrobiologie 4, 77–102.

Lawrence, G.B., Goolsby, D.A., Battaglin, W.A., Stensland, G.J., 2000. Atmospheric nitrogen in the Mississippi River Basin—emissions, deposition and transport. Sci. Total Environ. 248, 87–100.

Lean, D.R.S., Liao, C.F., Murphy, T.P., Painter, D.S., 1978. The importance of nitrogen fixation in lakes. Ecol. Bull. 26 (In Environmental role of nitrogen-fixing blue-green algae and asymbiotic bacteria), 41–51.

Lee, M., Shevliakova, E., Stock, C.A., Malyshev, S., Milly, P.C.D., 2019. Prominence of the tropics in the recent rise of global nitrogen pollution. Nat. Commun. 10, 1437.

Levine, S.N., Lewis, W.M., 1984. Diel variation of nitrogen fixation in Lake Valencia, Venezuela. Limnol. Oceanogr. 29, 887–893.

Lewandowski, J., Meinikmann, K., Nützmann, G., Rosenberry, D.O., 2015. Groundwater—The disregarded component in lake water and nutrient budgets. Part 2: Effects of groundwater on nutrients. Hydrol. Process. 29, 2922–2955.

Lewis, W.M., Levine, S.N., 1984. The light response of nitrogen fixation in Lake Valencia, Venezuela. Limnol. Oceanogr. 29, 894–900.

Lewis, W.M., Wurtsbaugh, W.A., 2008. Control of lacustrine phytoplankton by nutrients: Erosion of the phosphorus paradigm. Int. Rev. Hydrobiol. 93, 446–465.

Li, J., Yu, S., Qin, S., 2020. Removal capacities and environmental constrains of denitrification and anammox processes in eutrophic riverine sediments. Water. Air. Soil Pollut 231, 274.

Liao, C.F.-H., Lean, D.R.S., 1978. Nitrogen transformations within the trophogenic zone of lakes. J. Fish. Res. Board Can. 35, 1102–1108.

Likens, G.E. (Ed.), 1985. An Ecosystem Approach to Aquatic Ecology: Mirror Lake and Its Environment. Springer-Verlag, New York, p. 516.

Likens, G.E., Bormann, F.H., Pierce, R.S., Eaton, J.S., Johnson, N.M., 1977. Biogeochemistry of a forested ecosystem. Springer-Verlag, New York.

Lisa, J.A., Song, B., Tobias, C.R., Duernberger, K.A., 2014. Impacts of freshwater flushing on anammox community structure and activities in the New River Estuary, USA. Aquat. Microb. Ecol. 72, 17–31.

Liu, X., Zhang, Y., Han, W., Tang, A., Shen, J., Cui, Z., Vitousek, P., Erisman, J.W., 2013. Enhanced nitrogen deposition over China. Nature 494, 459–463.

Livingstone, D., Pentecost, A., Whitton, B.A., 1984. Diel variations in nitrogen and carbon dioxide fixation by the blue-green alga *Rivularia* in an upland stream. Phycologia 23, 125–133.

Lu, J., Bunn, S.E., Burford, M.A., 2018. Effects of water level fluctuations on nitrogen dynamics in littoral macrophytes. Limnol. Oceanogr. 63, 833–845.

Lu, J., Faggotter, S.J., Bunn, S.E., Burford, M.A., 2017. Macrophyte beds in a subtropical reservoir shifted from a nutrient sink to a source after drying then rewetting. Freshw. Biol. 62, 854–867.

Lumpkin, T.A., Plucknett, D.L., 1980. *Azolla*: Botany, physiology, and use as a green manure. Econ. Bot. 34, 111–153.

Lundgren, A., 1978. Nitrogen fixation induced by phosphorus fertilization of a subarctic lake. Ecol. Bull. (In Environmental role of nitrogen-fixing blue-green algae and asymbiotic bacteria) 26, 52–59.

Luvizotto, D.M., Araujo, J.E., Silva, M.D.C.P., Dias, A.C.F., Kraft, B., Tegetmeye, H., Strous, M., Andreote, F.D., 2019. The rates and players of denitrification, dissimilatory nitrate reduction to ammonia (DNRA) and anaerobic ammonia oxidation (anammox) in mangrove soils. An. Acad. Bras. Cienc 91, 1–14.

Ma, J., Qin, B., Wu, P., Zhou, J., Niu, C., Deng, J., Niu, H., 2015. Controlling cyanobacterial blooms by managing nutrient ratio and limitation in a large hyper-eutrophic lake: Lake Taihu, China. J. Environ. Sci. (China) 27, 80–86.

Macgregor, A.N., Keeney, D.R., 1973. Denitrification in lake sediments. Environ. Lett. 5, 175–181.

Mamuye, M., Kebebewu, Z., 2018. Review on impacts of climate change on on watershed hydrology. J. Environ. Earth Sci. 8, 91–99.

Manny, B.A., 1972. Seasonal changes in organic nitrogen content of net- and nannophytoplankton in two hardwater lakes. Arch. Hydrobiol. 71, 103–123.

Manny, B.A., Wetzel, R.G., 1973. Diurnal changes in dissolved organic and inorganic carbon and nitrogen in a hardwater stream. Freshw. Biol. 3, 31–43.

Maranger, R., Jones, S.E., Cotner, J.B., 2018. Stoichiometry of carbon, nitrogen, and phosphorus through the freshwater pipe. Limnol. Oceanogr. Lett. 3, 89–101.

Martí, E., Grimm, N.B., Fisher, S.G., 1997. Pre- and post-flood retention efficiency of nitrogen in a Sonoran Desert stream. J. North Am. Benthol. Soc. 16, 805–819.

Mayer, P.M., Reynolds, S.K., McCutchen, M.D., Canfield, T.J., 2007. Meta-analysis of nitrogen removal in riparian buffers. J. Environ. Qual. 36, 1172.

McClain, M.E., Richey, J.E., Pimentel, T.P., 1994. Groundwater nitrogen dynamics at the terrestrial-lotic interface of a small catchment in the central amazon basin. Biogeochemistry 27, 113–127.

McIntyre, R.E.S., Adams, M.A., Ford, D.J., Grierson, P.F., 2009. Rewetting and litter addition influence mineralisation and microbial communities in soils from a semi-arid intermittent stream. Soil Biol. Biochem. 41, 92–101.

McMillan, S.K., Tuttle, A.K., Jennings, G.D., Gardner, A., 2014. Influence of restoration age and riparian vegetation on reach-scale nutrient retention in restored urban streams. J. Am. Water Resour. Assoc. 50, 626–638.

Meffert, M.-E., Zimmermann-Telschow, H., 1979. Net release of nitrogenous compounds by axenic and bacteria-containing cultures of *Oscillatoria raedekei* (Cyanophyta). Arch. Hydrobiol. 87, 125–138.

Mengis, M., Gachter, R., Wehrli, B., 1996. Nitrous oxide emissions to the atmosphere from an artificially oxygenated lake. Limnol. Oceanogr. 41, 548–553.

Mengis, M., Gächter, R., Wehrli, B., 1997. Sources and sinks of nitrous oxide (N_2O) in deep lakes. Biogeochemistry 38, 281–301.

Merbt, S.N., Proia, L., Prosser, J.I., Marti, E., Casamayor, E.O., von Schiller, D., 2016. Stream drying drives microbial ammonia oxidation and first-flush nitrate export. Ecology 97, 2192–2198.

Merill, L., Tonjes, D.J., 2014. A review of the hyporheic zone, stream restoration, and means to enhance denitrification. Crit. Rev. Environ. Sci. Technol. 44, 2337–2379.

Meybeck, M., 1982. Carbon, nitrogen, and phosphorus transport by world rivers. Am. J. Sci. 282, 401–450.

Mitamura, O., Saijo, Y., 1986. Urea metabolism and its significance in the nitrogen cycle in the euphotic layer of Lake Biwa. I. In situ measurement of nitrogen assimilation and urea decomposition. Arch. Hydrobiol. 107, 23–51.

Mitchell, A.M., Baldwin, D.S., 1999. The effects of sediment desiccation on the potential for nitrification, denitrification and methanogenesis in an Australian reservoir. Hydrobiologia 392, 3–11.

Mitchell, A.W., Bramley, R.G.V., Johnson, A.K.L., 1997. Export of nutrients and suspended sediment during a cyclone-mediated flood event in the Herbert River catchment, Australia. Mar. Freshw. Res. 48, 79–88.

Molot, L.A., Dillon, P.J., 1993. Nitrogen mass balances and denitrification rates in central Ontario Lakes. Biogeochemistry 20, 195–212.

Moss, B., Jeppesen, E., Søndergaard, M., Lauridsen, T.L., Liu, Z., 2013. Nitrogen, macrophytes, shallow lakes and nutrient limitation: Resolution of a current controversy? Hydrobiologia 710, 3–21.

Moyo, S.M., 1991. Cyanobacterial nitrogen fixation in Lake Kariba, Zimbabwe. Verh. Internat. Verein. Limnol. 24, 1123–1127.

Mulholland, P.J., Deangelis, D.L., 2000. Surface-subsurface exchange and nutrient spiraling. In: Jones, J.B., Mulholland, P.J. (Eds.), Streams and Ground Waters. Academic Press, San Diego, pp. 149–166.

Mulholland, P.J., Hall, R.O., Sobota, D.J., Dodds, W.K., Findlay, S.E.G., Grimm, N.B., Hamilton, S.K., McDowell, W.H., O'Brien, J.M., Tank, J.L., Ashkenas, L.R., Cooper, L.W., Dahm, C.N., Gregory, S.V., Johnson, S.L., Meyer, J.L., Peterson, B.J., Poole, G.C., Valett, H.M., Webster, J.R., Arango, C.P., Beaulieu, J.J., Bernot, M.J., Burgin, A.J., Crenshaw, C.L., Helton, A.M., Johnson, L.T., Niederlehner, B.R., Potter, J.D., Sheibley, R.W., Thomasn, S.M., 2009. Nitrate removal in stream ecosystems measured by 15N addition experiments: Denitrification. Limnol. Oceanogr. 54, 666–680.

Mulholland, P.J., Helton, A.M., Poole Jr., G.C., Hamilton, R.O.H., Peterson, S.K., Tank, B.J., Ashkenas, J.L., Cooper, L.R., Dahm, L.W., Dodds, C.N., Findlay, W.K., Gregory, S.E.G., Grimm, S.V., Johnson, N.B., Mcdowell, S.L., Meyer, W.H., Valett, J.L., Webster, H.M., Arango, J.R., Beaulieu, C.P., Bernot, J.J., Burgin, M.J., Crenshaw, A.J., Johnson, C.L., Niederlehner, L.T., Brien, B.R., Potter, J.M.O., Sheibley, J.D., Sobota, R.W., Thomas, D.J., S.M, 2008. Stream denitrification across biomes and its response to anthropogenic nitrate loading. Nature 452, 1–5.

Mulholland, P.J., Thomas, S.A., Valett, H.M., Webster, J.R., Beaulieu, J., 2006. Effects of light on NO_3^- uptake in small forested streams: Diurnal and day-to-day variations. J. North Am. Benthol. Soc. 25, 583–595.

Mulholland, P.J., Valett, H.M., Webster, J.R., Thomas, S.A., Cooper, L.W., Hamilton, S.K., Peterson, B.J., Kellogg, W.K., Peterson, B.J., 2004. Stream denitrification and total nitrate uptake rates measured using a field ^{15}N tracer addition approach. Limnol. Oceanogr. 49, 809–820.

Munroe, S.E.M., Coates-Marnane, J., Burford, M.A., Fry, B., 2018. A benthic bioindicator reveals distinct land and ocean–based influences in an urbanized coastal embayment. PLoS One 13, 1–28.

Murphy, T.P., Lean, D.R.S., Nalewajko, C., 1976. Blue-green algae: Their excretion of iron-selective chelators enables them to dominate other algae. Science 192, 900–902.

Neilen, A.D., Carroll, A.R., Hawker, D.W., O'Brien, K.R., Burford, M.A., 2020. Identification of compounds from terrestrial dissolved organic matter toxic to cyanobacteria. Sci. Total Environ. 749, 141482.

Newbold, J.D., 1992. Cycles and spiral of nutrients. In: Calow, P., Petts, G.E. (Eds.), The Rivers Handbook. Vol. I, Hydrological and Ecological Principles. Blackwell Scientific Publications, Oxford, pp. 379–408.

Newbold, J.D., Elwood, J.W., O'Neill, R.V., Winkle, W., Van, 1981. Measuring nutrient spiralling in streams. Can. J. Fish. Aquat. Sci. 38, 860–863.

Newbold, J.D., Elwood, J.W., Schulze, M.S., Stark, R.W., Barmeier, J.C., 1983. Continuous ammonium enrichment of a woodland stream: Uptake kinetics, leaf decomposition, and nitrification. Freshw. Biol. 13, 193–204.

Newcomer Johnson, T., Kaushal, S., Mayer, P., Smith, R., Sivirichi, G., 2016. Nutrient retention in restored streams and rivers: A global review and synthesis. Water 8, 116.

Nicholls, J.C., Trimmer, M., 2009. Widespread occurrence of the anammox reaction in estuarine sediments. Aquat. Microb. Ecol. 55, 105–113.

Nichols, D.S., Keeney, D.R., 1973. Nitrogen and phosphorus release from decaying water milfoil. Hydrobiologia 42, 509–525.

Nie, S., Li, H., Yang, X., Zhang, Z., Weng, B., Huang, F., Zhu, G.B., Zhu, Y.G., 2015. Nitrogen loss by anaerobic oxidation of ammonium in rice rhizosphere. ISME J. 9, 2059–2067.

Nielsen, L.P., Bondo Christensen, P., Revsbech, N.P., Sørensen, J., 1990. Denitrification and photosynthesis in stream sediment studied with microsensor and wholecore techniques. Limnol. Oceanogr. 35, 1135–1144.

Obermann, M., Rosenwinkel, K.H., Tournoud, M.G., 2009. Investigation of first flushes in a medium-sized mediterranean catchment. J. Hydrol. 373, 405–415.

Overbeck, H.-J., 1968. Bakterien im Gewässer—ein Beispiel für die gegenwärtig Entwicklung der Limnologie. Mitt. Max-Planck-Ges 3, 165–182.

Paerl, H.W., 1978. Role of heterotrophic bacteria in promoting N_2 fixation by *Anabaena* in aquatic habitats. Microb. Ecol. 4, 215–231.

Paerl, H.W., 1985. Microzone formation: Its role in the enhancement of aquatic N_2 fixation. Limnol. Oceanogr. 30, 1246–1252.

Paerl, H.W., Prufert, L.E., 1987. Oxygen-poor microzones as potential sites of microbial N_2 fixation in nitrogen-depleted aerobic marine waters. Appl. Environ. Microbiol. 53, 1078–1087.

Paerl, H.W., Scott, J.T., McCarthy, M.J., Newell, S.E., Gardner, W.S., Havens, K.E., Hoffman, D.K., Wilhelm, S.W., Wurtsbaugh, W.A., 2016. It takes two to tango: When and where dual nutrient (N & P) reductions are needed to protect lakes and downstream ecosystems. Environ. Sci. Technol. 50, 10805–10813.

Palacin-Lizarbe, C., Camarero, L., Hallin, S., Jones, C.M., Catalan, J., 2020. Denitrification rates in lake sediments of mountains affected by high atmospheric nitrogen deposition. Sci. Rep. 10, 1–10.

Paul, B.J., Duthie, H.C., 1989. Nutrient cycling in the epilithon of running waters. Can. J. Bot. 67, 2302–2309.

Peters, G.A., Meeks, J.C., 1989. The *Azolla-Anabaena* symbiosis: Basic biology. Annu. Rev. Plant Physiol. Plant Mol. Biol. 40, 193–210.

Peterson, B.J., Wollheim, W.M., Mulholland, P.J., Webster, J.R., Meyer, J.L., Tank, J.L., Marti, E., Bowden, W.B., Valett, H.M., Hershey, A.E., McDowell, W.H., Dodds, W.K., Hamilton, S.K., Gregory, S., Morrall, D.D., 2001. Control nitrogen export from watersheds by headwater streams. Science 292, 86–90.

Petrone, K.C., Richards, J.S., Grierson, P.F., 2009. Bioavailability and composition of dissolved organic carbon and nitrogen in a near coastal catchment of South-Western Australia. Biogeochemistry 92, 27–40.

Phillips, G., Willby, N., Moss, B., 2016. Submerged macrophyte decline in shallow lakes: What have we learnt in the last forty years? Aquat. Bot. 135, 37–45.

Piña-Ochoa, E., Álvarez-Cobelas, M., 2006. Denitrification in aquatic environments: A cross-system analysis. Biogeochemistry 81, 111–130.

Postma, D., Boesen, C., Kristiansen, H., Larsen, F., 1991. Nitrate reduction in an unconfined sandy aquifer—Water chemistry, reduction processes, and geochemical modeling. Water Resour. Res. 27, 2027–2045.

Priscu, J.C., 1997. The biogeochemistry of nitrous oxide in permanently ice-covered lakes of the McMurdo Dry Valleys, Antarctica. Glob. Chang. Biol. 3, 301–315.

Priscu, J.C., Vincent, W.F., Howard-Williams, C., 1989. Inorganic nitrogen uptake and regeneration in perennially ice-covered Lakes Fryxell and Vanda, Antarctica. J. Plankton Res. 11, 335–351.

Racchetti, E., Longhi, D., Ribaudo, C., Soana, E., Bartoli, M., 2017. Nitrogen uptake and coupled nitrification–denitrification in riverine sediments with benthic microalgae and rooted macrophytes. Aquat. Sci. 79, 487–505.

Reddy, K.R., Patrick, W.H., 1984. Nitrogen transformations and loss in flooded soils and sediments. CRC Crit. Rev. Environ. Control 13, 273–309.

Reisinger, A.J., Tank, J.L., Dee, M.M., 2016. Regional and seasonal variation in nutrient limitation of river biofilms. Freshw. Sci. 35, 474–489.

Rejmánková, E., Sirová, D., Castle, S.T., Bárta, J., Carpenter, H., 2018. Heterotrophic N_2-fixation contributes to nitrogen economy of a common wetland sedge, *Schoenoplectus californicus*. PLoS One 13, 1–22.

Reuter, J.E., Loeb, S.L., Goldman, C.R., 1986. Inorganic nitrogen uptake by epilithic periphyton in a N-deficient lake. Limnol. Oceanogr. 31, 149–160.

Reverey, F., Grossart, H.P., Premke, K., Lischeid, G., 2016. Carbon and nutrient cycling in kettle hole sediments depending on hydrological dynamics: A review. Hydrobiologia 775, 1–20.

Rice, E.L., Pancholy, S.K., 1972. Inhibition of nitrification by climax ecosystems. Am. J. Bot. 59, 1033–1040.

Rice, E.L., Pancholy, S.K., 1973. Inhibition of nitrification by climax ecosystems. II. Additional evidence and possible role of tannins. Am. J. Bot. 60, 691–702.

Rich, J.J., Dale, O.R., Song, B., Ward, B.B., 2008. Anaerobic ammonium oxidation (anammox) in Chesapeake Bay sediments. Microb. Ecol. 55, 311–320.

Richey, J.S., McDowell, W.H., Likens, G.E., 1985. Nitrogen transformations in a small mountain stream. Hydrobiologia 124, 129–139.

Riddolls, A., 1985. Aspects of nitrogen fixation in Lough Neagh. I. Acetylene reduction and the frequency of *Aphanizomenon flos-aquae* heterocysts. Freshw. Biol. 15, 289–297.

Risgaard-Petersen, N., Meyer, R.L., Schmid, M., Jetten, M.S.M., Enrich-Prast, A., Rysgaard, S., Revsbech, N.P., 2004. Anaerobic ammonium oxidation in an estuarine sediment. Aquat. Microb. Ecol. 36, 293–304.

Risgaard-Petersen, N., Rysgaard, S., Nielsen, L.P., Revsbech, N.P., 1994. Diurnal variation of denitrification and nitrification in sediments colonized by benthic microphytes. Limnol. Oceanogr. 39, 573–579.

Rudd, J.W.M., Kelly, C.A., Furutani, A., 1986. The role of sulfate reduction in long term accumulation of organic and inorganic sulfur in lake sediments. Limnol. Oceanogr. 31, 1281–1291.

Rudd, J.W.M., Taylor, C.D., 1980. Methane cycling in aquatic environments. Adv. Aquat. Microbiol. 2, 77–150.

Rysgaard, S., Glud, R.N., Risgaard-Petersen, N., Dalsgaard, T., 2004. Denitrification and anammox activity in Arctic marine sediments. Limnol. Oceanogr. 49, 1493–1502.

Sabater, S., Bregoli, F., Acuña, V., Barceló, D., Elosegi, A., Ginebreda, A., Marcé, R., Muñoz, I., Sabater-Liesa, L., Ferreira, V., 2018. Effects of human-driven water stress on river ecosystems: A meta-analysis. Sci. Rep. 8, 1–11.

Sabater, F., Butturini, A., Martí, E., Muñoz, I., Romaní, A., Wray, J., Sabater, S., 2000. Effects of riparian vegetation removal on nutrient retention in a Mediterranean stream. J. North Am. Benthol. Soc. 19, 609–620.

Salcher, M.M., Posch, T., Pernthaler, J., 2012. In situ substrate preferences of abundant bacterioplankton populations in a prealpine freshwater lake. ISME J 7, 896–907.

Sarvala, J., Kairesalo, T., Koskimies, I., Lehtovaara, A., Ruuhijärvi, J., Vähä-Piikkiö, I., 1982. Carbon, phosphorus and nitrogen budgets of the littoral *Equisetum* belt in an oligotrophic lake. Hydrobiologia 86, 41–53.

Saunders, D.L., Kalff, J., 2001. Nitrogen retention in wetlands, lakes and rivers. Hydrobiologia 443, 205–212.

Schade, J.D., Welter, J.R., Martí, E., Grimm, N.B., 2005. Hydrologic exchange and N uptake by riparian vegetation in an arid-land stream. J. North Am. Benthol. Soc. 24, 19–28.

Schindler, D.W., 1977. Evolution of phosphorus limitation in lakes. Science 195, 260–262.

Schubert, C.J., Durisch-Kaiser, E., Wehrli, B., Thamdrup, B., Lam, P., Kuypers, M.M.M., 2006. Anaerobic ammonium oxidation in a tropical freshwater system (Lake Tanganyika). Environ. Microbiol. 8, 1857–1863.

Scott, J.T., McCarthy, M.J., 2010. Nitrogen fixation may not balance the nitrogen pool in lakes over timescales relevant to eutrophication management. Limnol. Oceanogr. 55, 1265–1270.

Scott, J.T., McCarthy, M.J., Paerl, H.W., 2019. Nitrogen transformations differentially affect nutrient-limited primary production in lakes of varying trophic state. Limnol. Oceanogr. Letts 4, 96–104.

Seitzinger, S.P., 1988. Denitrification in freshwater and coastal marine ecosystems: Ecological and geochemical significance. Limnol. Oceanogr. 33, 702–724.

Seitzinger, S.P., Harrison, J.A., Böhlke, J.K., Bouwman, A.F., Lowrance, R., Tobias, C., Drecht, G., van, 2006. Denitrification across landscapes and waterscapes: A synthesis. Ecol. Appl. 16, 2064–2090.

Serruya, A.C., Pollingher, U., Gophen, M., 1975. N and P distribution in Lake Kinneret (Israel) with emphasis on dissolved organic nitrogen. Oikos 26, 1–8.

Shanthi, G., Premalatha, M., Anantharaman, N., 2018. Effects of L-amino acids as organic nitrogen source on the growth rate, biochemical composition and polyphenol content of *Spirulina platensis*. Algal Res. 35, 471–478.

Sheldon, F., Bunn, S.E., Hughes, J.M., Arthington, A.H., Balcombe, S.R., Fellows, C.S., 2010. Ecological roles and threats to aquatic refugia in arid landscapes: dryland river waterholes. Mar. Freshw. Res. 61, 885–895.

Shen, L.-d, Cheng, H.-x, Liu, X., Li, J.-h, Liu, Y., 2017. Potential role of anammox in nitrogen removal in a freshwater reservoir, Jiulonghu Reservoir (China). Environ. Sci. Pollut. Res. 24, 3890–3899.

Shen, Y., Huang, Y., Hu, J., Li, P., Zhang, C., Li, L., Xu, P., Zhang, J., Chen, X., 2020. The nitrogen reduction in eutrophic water column driven by *Microcystis* blooms. J. Hazard Mater. 385, 121578.

Shiozaki, T., Ijichi, M., Isobe, K., Hashihama, F., Nakamura, K.I., Ehama, M., Hayashizaki, K.I., Takahashi, K., Hamasaki, K., Furuya, K., 2016. Nitrification and its influence on biogeochemical cycles from the equatorial Pacific to the Arctic Ocean. ISME J. 10, 2184–2197.

Simon, K.S., Townsend, C.R., Biggs, B.J.F., Bowden, W.B., 2005. Temporal variation of N and P uptake in 2 New Zealand streams. J. North Am. Benthol. Soc. 24, 1–18.

Sondergaard, M., Jensen, J.P., Jeppersen, E., 2003. Role of sediment and internal loading of phosphorus in shallow lakes. Hydrobiologia 506–509, 135–145.

Song, G.D., Liu, S.M., Marchant, H., Kuypers, M.M.M., Lavik, G., 2013. Anammox, denitrification and dissimilatory nitrate reduction to ammonium in the East China Sea sediment. Biogeosciences 10, 6851–6864.

Song, C., Xu, X., Tian, H., Wang, Y., 2009. Ecosystem-atmosphere exchange of CH_4 and N_2O and ecosystem respiration in wetlands in the Sanjiang Plain. Northeastern China. Glob. Chang. Biol. 15, 692–705.

Soued, C., Del Giorgio, P.A., Maranger, R., 2016. Nitrous oxide sinks and emissions in boreal aquatic networks in Quebec. Nat. Geosci. 9, 116–120.

Spalding, R.F., Exner, M.E., 1993. Occurrence of nitrate in groundwater—A review. J. Environ. Qual. 22, 392–402.

Spencer, C.N., Hauer, F.R., 1991. Phosphorus and nitrogen dynamics in streams during a wildfire. J. North Am. Benthol. Soc. 10, 24–30.

Stanley, E.H., Fisher, S.G., Grimm, N.B., 1997. Ecosystem expansion and contraction in streams. Bioscience 47, 427–435.

Steinman, A.D., Ogdahl, M.E., Weinert, M., Uzarski, D.G., 2014. Influence of water-level fluctuation duration and magnitude on sediment-water nutrient exchange in coastal wetlands. Aquat. Ecol. 48, 143–159.

Stewart, W.D., 1969. Biological and ecological aspects of nitrogen fixation by free-living micro-organisms. Proc. R. Soc. London. Ser. B Biol. Sci. 172, 367–388.

Stewart, S.D., Young, M.B., Harding, J.S., Horton, T.W., 2019. Invasive nitrogen-fixing plant amplifies terrestrial-aquatic nutrient flow and alters ecosystem function. Ecosystems 22, 587–601.

Storch, T.A., Saunders, G.W., Ostrofsky, M.L., 1990. Diel nitrogen fixation by cyanobacterial surface blooms in Sanctuary Lake, Pennsylvania. Appl. Environ. Microbiol. 56, 466–471.

Strauss, E.A., Mitchell, N.L., Lamberti, G.A., 2002. Factors regulating nitrification in aquatic sediments: Effects of organic carbon, nitrogen availability, and pH. Can. J. Fish. Aquat. Sci. 59, 554–563.

Stream Solute Workshop, 1990. Concepts and methods for assessing solute dynamics in stream ecosystems. J. North Am. Benthol. Soc. 9, 95–119.

Su, M., Zhang, J., Huo, S., Xi, B., Hua, F., Zan, F., Qian, G., Liu, J., 2016. Microbial bioavailability of dissolved organic nitrogen (DON) in the sediments of Lake Shankou, Northeastern China. J. Environ. Sci. (China) 42, 79–88.

Suzuki, M.T., Sherr, E.B., Sherr, B.F., 1996. Estimation of ammonium regeneration efficiencies associated with bacterivory in pelagic food webs via a ^{15}N tracer method. J. Plankton Res. 18, 411–428.

Svensson, J.M., 1997. Influence of *Chironomus plumosus* larvae on ammonium flux and denitrification (measured by the acetylene blockage- and the isotope pairing-technique) in eutrophic lake sediment. Hydrobiologia 346, 157–168.

Svensson, J.M., Enrich-Prast, A., Leonardson, L., 2001. Nitrification and denitrification in a eutrophic lake sediment bioturbated by oligochaetes. Aquat. Microb. Ecol. 23, 177–186.

Sweeney, B.W., Newbold, J.D., 2014. Streamside forest buffer width needed to protect stream water quality, habitat, and organisms: A literature review. J. Am. Water Resour. Assoc. 50, 560–584.

Takahashi, M., Saijo, Y., 1981. Nitrogen metabolism in Lake Kizaki, Japan. II. Distribution and decomposition of organic nitrogen. Arch. Hydrobiol. 92, 359–376.

Tan, T.L., Overbeck, J., 1973. Ökologische Untersuchungen über nitratreduzierende Bakterien im Wasser des Pluss-Sees (Schleswig-Holstein). Z. Allg. Mikrobiol. 13, 71–82.

Tank, J.L., Reisinger, A.J., Rosi, E.J., 2017. Nutrient limitation and uptake, Methods in stream ecology. Methods in Stream Ecology, , 3rd ed.Volume 2: Ecosystem Function. Elsevier, pp. 147–171.

Tate, C.M., 1990. Patterns and controls of nitrogen in tallgrass prairie streams. Ecology 71, 2007–2018.

Taylor, P.G., Townsend, A.R., 2010. Stoichiometric control of organic carbon-nitrate relationships from soils to the sea. Nature 464, 1178–1181.

Terai, H., 1987. Studies on denitrification in the water column of Lake Kizaki and Lake Fukami-Ike. Jpn. J. Limnol. 48, 257–264.

Terai, H., Yoh, M., Saijo, Y., 1987. Denitrifying activity and population growth of denitrifying bacteria in Lake Fukami-Ike. Jpn. J. Limnol. 48, 211–218.

Thamdrup, B., Dalsgaard, T., 2002. Production of N_2 through anaerobic ammonium oxidation coupled to nitrate reduction in marine sediments. Appl. Environ. Microbiol. 68, 1312–1318.

Tirén, T., Thorin, J., Nommik, H., 1976. Denitrification measurements in lakes. Acta Agric. Scand 26, 175–184.

Toetz, D.W., 1973. The limnology of nitrogen in an Oklahoma reservoir: Nitrogenase activity and related limnological factors. Am. Midl. Nat. 89, 369–380.

Tooth, S., 2000. Process, form and change in dryland rivers: A review of recent research. Earth Sci. Rev. 51, 67–107.

Trimmer, M., Nicholls, J.C., Deflandre, B., 2003. Anaerobic ammonium oxidation measured in sediments along the Thames estuary, United Kingdom. Appl. Environ. Microbiol. 69, 6447–6454.

Trimmer, M., Nicholls, J.C., Morley, N., Davies, C.A., Aldridge, J., 2005. Biphasic behavior of anammox regulated by nitrite and nitrate in an estuarine sediment. Appl. Environ. Microbiol. 71, 1923–1930.

Triska, F.J., Kennedy, V.C., Avanzino, R.J., Zellweger, G.W., 1989. Retention and transport of nutrients in a third-order stream in northwestern California: Hyporheic processes. Ecology 70, 1893–1905.

Triska, F.J., Kennedy, V.C., Avanzino, R.J., Zellweger, G.W., Bencala, K.E., 1990. In situ retention-transport response to nitrate loading and storm discharge in a third-order stream. J. North Am. Benthol. Soc. 9, 229–239.

Trussell, R.P., 1972. The percent un-ionized ammonia in aqueous ammonia solutions at different pH levels and temperatures. J. Fish. Res. Board Can. 29, 1505–1507.

Tuschall, J.R., Brezonik, P.L., 1980. Characterization of organic nitrogen in natural waters: Its molecular size, protein content, and interactions with heavy metals. Limnol. Oceanogr. 25, 495–504.

Tzoraki, O., Nikolaidis, N.P., Amaxidis, Y., Skoulikidis, N.T., 2007. Instream biogeochemical processes of a temporary river. Environ. Sci. Technol. 41, 1225–1231.

van Cleemput, O., 1994. Biogeochemistry of nitrous oxide in wetlands. Curr. Top. Wetl. Biogeochem. 1, 3–14.

van den Berg, E.M., Rombouts, J.L., Kuenen, J.G., Kleerebezem, R., van Loosdrecht, M.C.M., 2017. Role of nitrite in the competition between denitrification and DNRA in a chemostat enrichment culture. AMB Express 7, 91.

Van Drecht, G., Bouwman, A.F., Knoop, J.M., Beusen, A.H.W., Meinardi, C.R., 2003. Global modeling of the fate of nitrogen from point and nonpoint sources in soils, groundwater, and surface water. Global Biogeochem Cycles 17, 1115.

Van Kessel, J.F., 1978. The relation between redox potential and denitrification in a water-sediment system. Water Res. 12, 285–290.

Verdouw, H., Boers, P.C.M., Dekkers, E.M.J., 1985. The dynamics of ammonia in sediments and hypolimnion of Lake Vechten (the Netherlands). Arch. Hydrobiol. 105, 79–92.

Vollenweider, R.A., 1968. Scientific fundamentals of the eutrophication of lakes and flowing waters, with particular reference to nitrogen and phosphorus as factors in eutrophication. Rep. Organisation for Economic Co-operation and Development, Paris, p. 192. DAS/CSI/68.27.

von Schiller, D., Acuña, V., Graeber, D., Martí, E., Ribot, M., Sabater, S., Timoner, X., Tockner, K., 2011. Contraction, fragmentation and expansion dynamics determine nutrient availability in a Mediterranean forest stream. Aquat. Sci. 73, 485–497.

von Schiller, D., Aristi, I., Ponsati, l., Arroita, M., Acuna, V., Arturo, E., Sabater, S., 2016. Regulation causes nitrogen cycling discontinuities in Mediterranean rivers. Sci. Total Environ. 540, 168–177.

von Schiller, D., Bernal, S., Dahm, C.N., Martí, E., 2017. Chapter 3.2: Nutrient and organic matter dynamics in intermittent rivers and ephemeral streams. In: Datry, T, Bonada, N, Boulton, A (Eds.), Intermittent Rivers and Ephemeral Streams. Academic Press, New York, pp. 135–160.

von Schiller, D., Martí, E., Riera, J.L., Ribot, M., Argerich, A., Fonollà, P., Sabater, F., 2008. Inter-annual, annual, and seasonal variation of P and N retention in a perennial and an intermittent stream. Ecosystems 11, 670–687.

Wagner, G.M., 1997. *Azolla*: A review of its biology and utilization. Bot. Rev. 63, 1–26.

Wang, S., Wang, W., Zhao, S., Wang, X., Hefting, M.M., Schwark, L., Zhu, G., 2019. Anammox and denitrification separately dominate microbial N-loss in water saturated and unsaturated soils horizons of riparian zones. Water Res. 162, 139–150.

Ward, A.K., Wetzel, R.G., 1980. Interactions of light and nitrogen source among planktonic blue-green algae. Arch. Hydrobiol. 90, 1–25.

Weber, K.A., Urrutia, M.M., Churchill, P.F., Kukkadapu, R.K., Roden, E.E., 2006. Anaerobic redox cycling of iron by freshwater sediment microorganisms. Environ. Microbiol. 8, 100–113.

Webster, J.R., Patten, B.C., 1979. Effects of watershed perturbation on stream potassium and calcium dynamics. Ecol. Monogr. 49, 51–72.

Wehr, J.D., Brown, L.M., O'Grady, K., 1987. Highly specialized nitrogen metabolism in a freshwater phytoplankter, *Chrysochromulina breviturrita*. Can. J. Fish. Aquat. Sci. 44, 736–742.

Welter, J.R., Fisher, S.G., 2016. The influence of storm characteristics on hydrological connectivity in intermittent channel networks: Implications for nitrogen transport and denitrification. Freshwater Biol. 61, 1214–1227.

Wenk, C.B., Blees, J., Zopfi, J., Veronesi, M., Bourbonnais, A., Schubert, C.J., Niemann, H., Lehmann, M.F., 2013. Anaerobic ammonium oxidation (anammox) bacteria and sulfide-dependent denitrifiers coexist in the water column of a meromictic southalpine lake. Limnol. Oceanogr. 58, 1–12.

Wetzel, R.G., 1993. Microcommunities and microgradients: Linking nutrient regeneration, microbial mutualism, and high sustained aquatic primary production. Netherl. J. Aquat. Ecol. 27, 3–9.

Wetzel, R.G., 2001. Limnology, Lake and River Ecosystems. Academic Press, San Diego, p. 1006.

Wetzel, R.G., Manny, B.A., 1972. Secretion of dissolved organic carbon and nitrogen by aquatic macrophytes. Verh. Internat. Verein. Limnol 18, 162–170.

Wetzel, R.G., Manny, B.A., 1977. Seasonal changes in particulate and dissolved organic carbon and nitrogen in a hardwater stream. Arch. Hydrobiol. 80, 20–39.

Willett, V.B., Reynolds, B.A., Stevens, P.A., Ormerod, S.J., Jones, D.L., 2004. Dissolved organic nitrogen regulation in freshwaters. J. Environ. Qual. 33, 201–209.

Willis, A., Chuang, A.W., Burford, M.A., 2016. Nitrogen fixation by the diazotroph *Cylindrospermopsis raciborskii* (Cyanophyceae). J. Phycol. 52, 854–862.

Wilson, J., Baldwin, D.S., 2008. Exploring the "Birch effect" in reservoir sediments: Influence of inundation history on aerobic nutrient release. Chem. Ecol. 24, 379–386.

Wissmar, R.C., Lilley, M.D., DeAngelis, M., 1987. Nitrous oxide release from aerobic riverine deposits. J. Freshw. Ecol. 4, 209–218.

Witte, F., 2009. SWAT theory. Via Medici 13, 52–55.

Wolk, C.P., 1973. Physiology and cytological chemistry blue-green algae. Bacteriol. Rev. 37, 32–101.

Xia, X., Liu, T., Yang, Z., Michalski, G., Liu, S., Jia, Z., Zhang, S., 2017. Enhanced nitrogen loss from rivers through coupled nitrification-denitrification caused by suspended sediment. Sci. Total Environ. 579, 47–59.

Xia, X.H., Yang, Z.F., Huang, G.H., Zhang, X.Q., Yu, H., Rong, X., 2004. Nitrification in natural waters with high suspended-solid content—A study for the Yellow River. Chemosphere 57, 1017–1029.

Xia, X., Yang, Z., Zhang, X., 2009. Effect of suspended-sediment concentration on nitrification in river water: Importance of suspended sediment—Water interface. Environ. Sci. Technol. 43, 3681–3687.

Xia, X., Zhang, S., Li, S., Zhang, L., Wang, G., Zhang, L., Wang, J., Li, Z., 2018. The cycle of nitrogen in river systems: Sources, transformation, and flux. Environ. Sci. Process. Impacts 20, 863–891.

Xu, H., McCarthy, M.J., Paerl, H.W., Brookes, J.D., Zhu, G., Hall, N.S., Qin, B., Zhang, Y., Zhu, M., Hampel, J.J., Newell, S.E., Gardner, W.S., 2021. Contributions of external nutrient loading and internal cycling to cyanobacterial bloom dynamics in Lake Taihu, China: implications for nutrient management. Limnol. Oceanogr. 66, 1–18.

Xu, H., Paerl, H.W., Qin, B., Zhu, G., Gao, G., 2010. Nitrogen and phosphorus inputs control phytoplankton growth in eutrophic Lake Taihu, China. Limnol. Oceanogr. 55, 420–432.

Xu, H., Paerl, H.W., Qin, B., Zhu, G., Hall, N.S., Wu, Y., 2015. Determining critical nutrient thresholds needed to control harmful cyanobacterial blooms in eutrophic Lake Taihu, China. Environ. Sci. Technol. 49, 1051–1059.

Yoh, M., 1992. Marked variation in lacustrine N_2O accumulation level and its mechanism. Jpn. J. Limnol. 53, 75–81.

Yoh, M., Terai, H., Saijo, Y., 1988. Nitrous oxide in freshwater lakes. Arch. Hydrobiol. 113, 273–294.

Yoh, M., Yagi, A., Terai, H., 1990. Significance of low-oxygen zone for nitrogen cycling in a freshwater lake: Production of N_2O by simultaneous denitrification and nitrification. Jpn. J. Limnol. 51, 163–171.

Yoshida, T., Aizaki, M., Asami, T., Makishima, N., 1979. Biological nitrogen fixation and denitrification in Lake Kasumiga-ura. Jpn. J. Limnol. 40, 1–9. In Japanese.

Zehr, J.P., Paulsen, S.G., Axler, R.P., Goldman, C.R., 1988. Dynamics of dissolved organic nitrogen in subalpine Castle Lake, California. Hydrobiologia 157, 33–45.

Zeikus, J.G., 1977. The biology of methanogenic bacteria. Bacteriol. Rev. 41, 514–541.

Zhang, Y., Huo, S., Zan, F., Xi, B., Zhang, J., 2015. Dissolved organic nitrogen (DON) in seventeen shallow lakes of Eastern China. Environ. Earth Sci. 74, 4011–4021.

Zhang, S., Pang, S., Wang, P., Wang, C., Guo, C., Addo, F.G., Li, Y., 2016. Responses of bacterial community structure and denitrifying bacteria in biofilm to submerged macrophytes and nitrate. Sci. Rep. 6, 1–10.

Zhang, S., Xia, X., Liu, T., Xia, L., Zhang, L., Jia, Z., Li, Y., 2017. Potential roles of anaerobic ammonium oxidation (anammox) in overlying water of rivers with suspended sediments. Biogeochemistry 132, 237–249.

Zhao, J., Wang, X., Li, X., Jia, S., Peng, Y., 2018. Advanced nutrient removal from ammonia and domestic wastewaters by a novel process based on simultaneous partial nitrification-anammox and modified denitrifying phosphorus removal. Chem. Eng. J 354, 589–598.

Zou, S., Yao, S., Ni, J., 2014. High-efficient nitrogen removal by coupling enriched autotrophic-nitrification and aerobic-denitrification consortiums at cold temperature. Bioresour. Technol. 161, 288–296.

CHAPTER

15

The Phosphorus Cycle

Peter J. Dillon[1] and Lewis A. Molot[2]

[1]School of the Environment, Trent University, Peterborough, Ontario, Canada [2]Faculty of Environmental & Urban Change, York University, Toronto, Ontario, Canada

OUTLINE

Wetzel's Limnology, Fourth Edition
https://doi.org/10.1016/B978-0-12-822701-5.00015-X

I. Phosphorus in fresh waters

No other element in freshwaters has been studied as intensively and extensively as phosphorus. Ecological interest in phosphorus stems from its major role in biological metabolism combined with the relatively small amounts of phosphorus available in the hydrosphere. In comparison to the rich natural supply of other major nutritional and structural components of the biota (carbon, hydrogen, nitrogen, oxygen, and sulfur), phosphorus is the least abundant and most commonly limits biological productivity in freshwaters.

A great body of quantitative data exists on the sources and sinks of phosphorus in lakes, the seasonal distribution of phosphorus in lakes, its quantitative effects on the trophic status of lakes and flowing waters, and the role of lakes' catchments in the phosphorus cycle. Five decades of modeling of the phosphorus cycle and its role in freshwaters have contributed greatly to our ability to manage water quality and reduce the impact of detrimental anthropogenic activities on lakes and rivers.

II. The distribution of organic and inorganic phosphorus in lakes and streams

It is instructive to discuss first the forms and distribution of phosphorus that occur in freshwaters before analyzing the dynamics of exchange between the compartments.

A very large proportion, usually greater than 90%, of the phosphorus in freshwater occurs as *organic phosphates* and cellular constituents either within the biota (both living and dead) or adsorbed to inorganic and dead particulate organic materials. A much smaller but most significant component of the total phosphorus pool is the *inorganic phosphorus*, most of which is *orthophosphate* (PO_4^{3-}), the most biologically important component.

Phosphine (PH_3) is a volatile, highly toxic constituent of the global biogeochemical phosphorus cycle. Generation of phosphine occurs by anaerobic enzymatic reduction of phosphate and has been found in anoxic sediments of freshwaters, wetlands of flood plains and rice fields, and sewage treatment facilities (Dévai et al., 1988; Gassmann and Schorn, 1993; Gassmann, 1994; Glindemann et al., 1996). In very shallow areas of organic-rich sediments or saturated hydrosoils, phosphine can evade directly into the atmosphere; this could result in transport to other distant drainage basins and subsequent deposition in precipitation as phosphate. Because phosphine is highly reactive under oxic conditions, it has not been detected in most lake or river waters. Within sediments, however, concentrations of phosphine have been found in the nanomolar ranges (0–25 nmol L^{-1}).

Total inorganic and organic phosphorus have been separated in various ways for chemical analyses; in general, these fractions are operationally defined, meaning that their value depends to some extent on the methodology employed. Similarly, the distinction between dissolved and particulate fractions is dependent on methodology; typically a filter pore size of 0.45 μm has been used but other sizes are also common. Additionally, very large (e.g., >75 μm) particulate matter is sometimes excluded from the particulate fraction measurement. Often, these fractions relate poorly to the way in which phosphorus is metabolized, that is, to its bioavailability. Because of this, the most important, and most often employed, measure is the total phosphorus content of unfiltered water, which consists of the phosphorus in the particulate and in "dissolved" compartments (Juday et al., 1927; Ohle, 1938). The rationale for this is that although some forms of phosphorus such as the particulate fraction may not be readily available to biota, they may in time be converted to form(s) that are. *Particulate phosphorus* includes (1) phosphorus in organisms as (a) the relatively stable nucleic acids DNA, RNA, and phosphoproteins and phospholipids, which are not involved in rapid cycling of phosphorus; (b) low-molecular-weight esters of enzymes, vitamins, etc.; and (c) nucleotide phosphates, such as adenosine diphosphate (ADP), adenosine 5-triphosphate (ATP), and nicotinamide adenine dinucleotide phosphate (NADPH) used in biochemical pathways of respiration and CO_2 assimilation; (2) mineral phases of rock and soil, such as hydroxyapatite, in which phosphorus is adsorbed onto inorganic complexes such as clays, carbonates, and ferric hydroxides; and (3) phosphorus adsorbed onto dead particulate organic matter or in macroorganic aggregations. In contrast to the phosphorus of particulate matter, dissolved phosphorus is composed of (1) orthophosphates (PO_4^{3-}); (2) *polyphosphates*, often originating from synthetic detergents; (3) organic colloids or phosphorus combined with adsorptive colloids; and (4) low-molecular-weight *phosphate esters*.

Because of the fundamental importance of phosphorus as a nutrient and major cellular constituent, much emphasis has been placed on its analytical evaluation. Chemical analyses of phosphorus center on the reactivity of orthophosphate with molybdate to form a blue complex that can be measured colorimetrically. Complex forms of phosphorus compounds, which are unreactive with molybdate, can be converted to "reactive" orthophosphate by enzymatic and acidic hydrolysis. Four operational categories result: (a) *soluble reactive phosphorus* (SRP); (b) *soluble unreactive phosphorus* (SUP); (c) *particulate reactive phosphorus* (PRP); and (d) *particulate unreactive phosphorus* (PUP). However, these operational methods do not necessarily correspond to either the chemical species of phosphorus or to their role in biotic cycling of phosphorus. Thus SRP cannot be equated to orthophosphate as the acidic hydrolysis of some phosphorus species may occur, leading to an SRP value greater than the true orthophosphate concentration.

Most of the phosphorus data for freshwaters refer to total phosphorus and inorganic soluble phosphorus (orthophosphate), although in more detailed studies four general fractions have been identified (Hutchinson, 1957). These four fractions are similar to the four operational groups already cited: (a) *soluble phosphate phosphorus*; (b) *acid-soluble suspended (sestonic) phosphorus*, mainly ferric phosphate and calcium phosphate; (c) *organic soluble and colloidal phosphorus*; and (d) *organic suspended (sestonic) phosphorus*.

Total phosphorus concentrations in nonpolluted natural waters extend over a very wide range from <1 μg L^{-1} to more than 200 mg L^{-1} in some closed saline lakes, with phosphorus concentrations of most uncontaminated surface waters falling between 5 and 50 μg P L^{-1}. Variation is high, however, and can be related to characteristics of regional geology. Phosphorus levels of freshwaters are generally lowest in regions of crystalline bedrock geomorphology and increase in lowland waters derived from sedimentary or metamorphic rock deposits. Lakes in catchments with volcanic soils are often particularly enriched in phosphorus. Lakes rich in organic matter, such as bogs and bog lakes, tend to exhibit high total phosphorus concentrations because of the high content of dissolved organic matter (Chapter 28). A few sedimentary coastal areas, such as in the southeastern United States, are rich in phosphatic rock. Lakes with drainage from these deposits have abnormally high phosphorus levels.

In a detailed treatment relating phosphorus and nitrogen to lake productivity, Vollenweider (1968) demonstrated using several criteria that lake productivity generally increases with the lake's total phosphorus concentration (Table 15-1). Although there are a number of exceptions to this relationship, it demonstrates a general principle that is useful when dealing with applied questions of eutrophication. The quantitative relationships of total phosphorus content to loading rates of phosphorus from the drainage basin, the atmosphere, and lake sediments are discussed in the concluding sections of this chapter.

Separation of the total phosphorus into inorganic and organic fractions in an appreciable number of lakes indicates that most of the total phosphorus is in an organic phase (Table 15-2). Of the total organic phosphorus, about 70% or more is within the particulate (sestonic) organic material, and the remainder is present as dissolved or colloidal organic phosphorus. Rigler (1964a) and Lean (1973b) demonstrated that centrifugation and paper filtration methods of separation underestimated the importance of the sestonic organic phosphorus since soluble organic phosphorus includes a significant quantity of phosphorus in a colloidal state. Inorganic soluble phosphorus is consistently very low, constitutes only a

TABLE 15-1 General Relationship of Lake Productivity to Average Concentrations of Epilimnetic Total Phosphorus[a]

General level of lake productivity	Change (reduction) in alkalinity in epilimnion during summer (meq L^{-1})	Total phosphorus (μg L^{-1})
Ultraoligotrophic	<0.2	<5
Oligomesotrophic	0.6	5–10
Mesoeutrophic	0–6–1.0	10–30
Eutrophic	??	30–100
Hypereutrophic	>1.0	>100

[a]*Modified from Vollenweider (1968).*

TABLE 15-2 Fractionation of Total Phosphorus in Lakes Analyzed Using Different Techniques of Separation

	Soluble inorganic P		Soluble organic P		Sestonic P		Total organic P		Total P
Lakes	($\mu g\ L^{-1}$)	(%)	($\mu g\ L^{-1}$)	(%)	($\mu g\ L^{-1}$)	(%)	($\mu g\ L^{-1}$)	(%)	($\mu g\ L^{-1}$)
Northern Wisconsin[a] (Juday and Birge, 1931)	3	13.0	14	60.9	6	26.1	20	87.0	23
Michigan Lakes[b] (Tucker, 1957)	1.5	12.0	5.7	46.9	5.0	41.1	10.7	88.0	12.2
Linsley Pond, Connecticut[c] (Hutchinson, 1957)	2	9.5	6	28.6	13	61.9	19	90.5	21
Ontario Lakes[c] (Rigler, 1964a)	–	5.9	–	28.7	–	65.4	–	94.1	–

[a] *Centrifugation techniques.*
[b] *Paper (No. 44 Whatman) filtration.*
[c] *Membrane filtration (0.5 μm).*

few percent of total phosphorus in unpolluted waters, and, as will be seen further on, is cycled very rapidly in the zones of utilization. The ratio of inorganic soluble phosphorus to other forms of phosphorus is approximately 1:20, or <5%, as inorganic phosphate phosphorus is remarkably constant in a large variety of lakes and streams within the temperate zone. The percentage of total phosphorus occurring as truly ionic orthophosphate is probably considerably <5% in most unproductive natural waters (e.g., Prepas and Rigler, 1982; Tarapchak et al., 1982; Bradford and Peters, 1987).

The phosphorus distribution within the fractions just discussed is the picture generally observed in the trophogenic zones of lakes. Phosphate, pyrophosphate, triphosphate, and higher polyphosphate anions additionally form complexes, chelates, and insoluble salts with a number of metal ions (Stumm and Morgan, 1996). The extent of complexing and chelation between various phosphates and metal ions in natural waters depends upon the relative concentrations of the phosphates and the metal ions, the pH, and the presence of other ligands (sulfate, carbonate, fluoride, and organic species).

Because phosphate concentrations are generally low in unpolluted waters, complex formations involving these major cations and various phosphate anions will have little effect on the distribution of metal ions but may have marked effects on the phosphate distribution (cf. Golachowska, 1971). On the other hand, metal ions such as ferric iron, manganous manganese, zinc, and copper are present in concentrations comparable to or lower than those of phosphates. For these ions, complex formation can significantly affect the distribution of the metal ion, the phosphates, or both. For example, the solubility of aluminum phosphate ($AlPO_4$) is minimal at pH 6 and increases at both higher and lower pH values. Ferric phosphate ($FePO_4$) behaves similarly, although it is more soluble than $AlPO_4$. Calcium concentration influences the formation of hydroxylapatite, [$Ca_5(OH)(PO_4)_3$]. In an aqueous solution lacking other compounds, a calcium concentration of 40 mg L^{-1} at a pH of 7 limits the solubility of phosphate to approximately 10 μg P L^{-1}. A calcium level of 100 mg L^{-1} lowers the maximum equilibrium of phosphate to 1 μg P L^{-1}. Elevation of the pH of waters containing typical concentrations of calcium leads to apatite formation (Kümmel, 1981). Moreover, increasing pH leads to the formation of calcium carbonate, which coprecipitates phosphate with carbonates (Otsuki and Wetzel, 1972). Sorption of phosphates and polyphosphates on surfaces is well known, particularly onto clay minerals (cf. Stumm and Morgan, 1996), by chemical bonding of the anions to positively charged edges of the clays and by substitution of phosphates for silicate in the clay structure. In general, high phosphate adsorption by clays is favored by low pH levels (approximately 5–6).

Following from these interactions and the distribution of phosphorus in inorganic and organic fractions, the general tendency is for unproductive lakes with orthograde oxygen curves (see Chapter 11) to show little variation in phosphorus content with depth (Fig. 15-1). Similarly, during periods of fall and spring circulation, the vertical distribution of phosphorus is more or less uniform. Oxidized metals, such as iron, and major cations, particularly calcium, can react with and precipitate phosphorus.

Lakes exhibiting clinograde oxygen curves (see Chapter 11) during the periods of stratification,

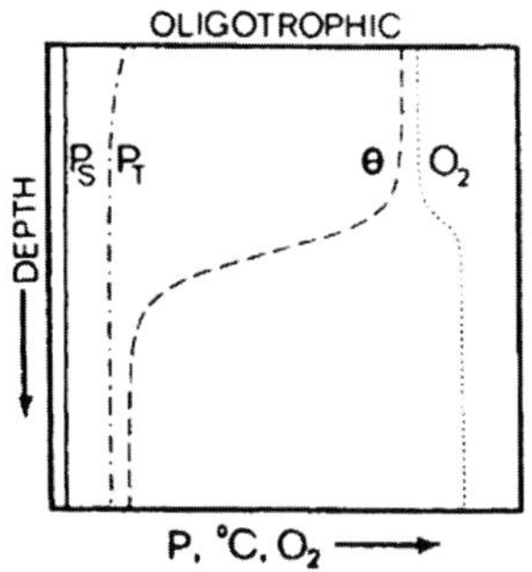

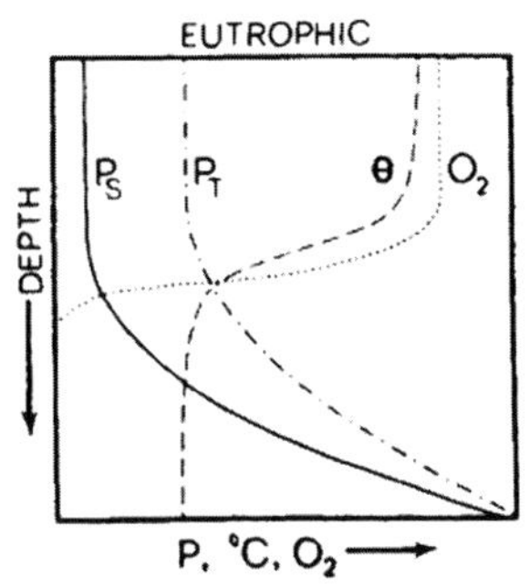

FIGURE 15-1 Generalized vertical distribution of soluble (P_S) and total (P_T) phosphorus in stratified lakes of very low (oligotrophic) and of high (eutrophic) productivity. θ refers to the vertical distribution of temperature.

however, possess much more variable vertical distributions of phosphorus. Commonly, there is a marked increase in phosphorus content in the lower hypolimnion, particularly during the later phases of thermal stratification (Fig. 15-1). Much of the hypolimnetic increase is in soluble phosphorus near the sediments. The sestonic phosphorus fraction is highly variable with depth. Sestonic phosphorus in the epilimnion fluctuates widely with oscillations in plankton populations and loadings from littoral areas. Sestonic phosphorus in the metalimnion and hypolimnion varies with the sedimentation of plankton, depth-dependent rates of decomposition, and the development of deep-living populations of bacterial and other plankton (e.g., euglenophyceans).

III. Phosphorus cycling in running waters

A. Dissolved and particulate phosphorus in rivers

Dissolved inorganic phosphorus (DIP), more correctly referred to as *soluble reactive phosphorus* (SRP) because the analytical method can include some organic phosphorus, averages about 10 μg P L^{-1} worldwide among unpolluted rivers (Meybeck, 1982, 1993). Total dissolved phosphorus (SUP and SRP) averages about 25 μg L^{-1}.

SRP in river water often increases by a factor of two-to four-fold during and following increases in discharge from heavy rainfall events or in the early stages of snowmelt. SRP in rivers often increases to levels of 50–100 μg P L^{-1} or more from agricultural runoff and to over 1000 μg P L^{-1} from municipal sewage sources (Meybeck, 1982, 1993). Uptake occurs both by physical adsorption to benthic substrata and to particulate seston, as well as through assimilation by attached biota. Although uptake of SRP by planktonic seston may be small (c. 1% of total uptake) in small streams (Newbold et al., 1983; Mulholland et al., 1990), sestonic uptake increases markedly in larger streams, particularly when dominated by phytoplankton (Newbold, 1992). Some of the seston and adsorbed phosphorus may settle out during base flows and then later be transported downstream during storm-induced high discharge. Exchange from these particles to the water may then occur in these displaced locations.

B. Biological utilization

Biological uptake of phosphorus by algae, Cyanobacteria, bacteria, and larger aquatic plants generally follows Michaelis–Menten kinetics (see Section VII). Higher trophic levels cannot transport phosphate from surrounding water and are dependent on ingesting food to meet their phosphorus needs. Growth rate at steady state generally follows Monod growth rate kinetics, which is mathematically similar to Michaelis–Menten kinetics (Rhee 1973; Jiang et al., 2019). Many algae and bacteria assimilate phosphorus at rates more rapidly than used for growth and store the excess. As a result, cells accumulate phosphorus and steady-state growth (e.g., in chemostats) occurs at concentrations much lower than the half-saturation constant (the concentration at which uptake is half its maximum rate). Uptake rates of phosphorus by coarse particulate organic matter and microbiota associated with it commonly reach a maximum in 15–20 days and then decline precipitously (e.g., Mulholland et al., 1984).

Average sestonic algal abundance has been positively correlated with total phosphorus and average phosphorus concentrations of many streams at moderate levels of nutrient loadings (Jones et al., 1984; Søballe and Kimmel, 1987; Nieuwenhuyse and Jones, 1996) but becomes increasingly less responsive, as in lakes, to nutrient loadings at higher phosphorus concentrations. Summer mean chlorophyll *a* concentration of seston among temperate streams exhibits a curvilinear relationship with summer mean total phosphorus concentrations (Fig. 15-2). The amount of nutrient loading to a river increases with greater catchment area. The analyses indicated that the phosphorus–sestonic chlorophyll *a* relationship increased some 2.3-fold as the area increased from 100 to 100,000 km^2 (Fig. 15-2). It is important to note that the primary productivity of rivers is generally much less responsive to increased nutrient loadings than is the productivity of lakes. As will be discussed later in the sections on algal productivity, nutrient availability is generally higher in rivers than in lakes. Nutrient losses in lakes are greater and large quantities are stored in parts of the ecosystem that are not readily available to the producers.

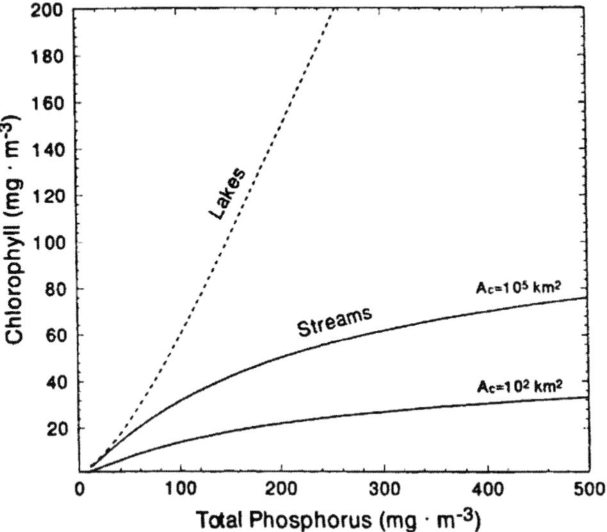

FIGURE 15-2 Phosphorus–sestonic chlorophyll *a* relationships in temperate streams and lakes. *Solid curves* indicate predicted chlorophyll *a* concentrations in rivers of differing catchment areas (A_c) of 100 and 100,000 km^2. The *dashed curve* depicts the trajectory of chlorophyll *a* concentration in *P*-limited lakes in which total N:P > 25 (Forsberg and Ryding, 1980). *(After Nieuwenhuyse and Jones, 1996.)*

C. Adsorption, release, and recycling

The kinetics of abiotic adsorption and desorption of phosphorus onto and from organic and inorganic particles, such as natural sediment and suspended particles, generally comply with *Langmuir isotherms* (see Chapters 8 and 28). Fine particles (<0.1 mm) account for nearly all of the sorption capacity for phosphorus (0.1–1.0 μg P g^{-1} sediment per μg P L^{-1} water; Meyer, 1979; Logan, 1982; Stabel and Geiger, 1985). The adsorption capacities of particles saturate and reach quasi–steady state in which uptake and release are about balanced.

The abundance of phosphate is also regulated by solubility reactions, particularly in relation to solid–solution associations with colloids or particles. Phosphorus concentrations in turbid rivers with low calcium concentrations are influenced by a solid ferric hydroxide-phosphate present in colloidal suspensions or on suspended particulates (Fox, 1993). In hardwater streams the solubility of inorganic phosphorus can decrease as pH increases above 8.5, as is common in areas of intensive photosynthesis (Diaz et al., 1994). Precipitation of inorganic phosphorus as calcium phosphate, highly stable as hydroxyapatite, increased to >60% at Ca concentrations approaching 100 mg L^{-1} and pH values of 9.0 or greater. If calcium carbonate precipitates under these conditions (see Chapter 13), phosphorus can be coprecipitated into the crystals and also adsorbed to the surfaces of the carbonates (Otsuki and Wetzel, 1972). These carbonate deposits are common in travertine encrustations in rivers that receive groundwaters supersaturated with calcium and bicarbonates.

Thus the major divalent metal ions in water, calcium and magnesium, tend to destabilize particles and enhance aggregation and sedimentation rates. In contrast, dissolved organic matter tends to stabilize colloidal particles in stream and lake water, slowing aggregation rates and thereby lowering sedimentation rates (Tipping and Higgins, 1982; Ali et al., 1985). Separation distances and efficacy of van der Waals forces for attraction and aggregation, however, vary greatly with different types and concentrations of natural dissolved organic compounds and with the ionic strengths of charged colloids or particles (O'Melia, 1998).

As particulate organic matter increases, as commonly results from leaf fall in late fall and winter in temperate woodland streams, phosphorus uptake can increase as the particulate organic matter increases (Mulholland et al., 1985a, 1985b; Klotz, 1985). It is during this time as well that fungal activity is maximal on particulate organic matter (e.g., Suberkropp, 1995) and bacterial activities are maximal on dissolved organic matter leaching from the particulate organic matter (Wetzel and Manny, 1972). *Spiraling* (see Chapter 14) describes the cycling of nutrients in streams as they are assimilated from the water column into benthic biomass, temporarily retained, and mineralized back into the water column, where they are transported downstream. Phosphorus uptake length, and hence total spiraling length, decreases in winter. These rates are counteracted at other times when detrital particulate organic matter is reduced by high discharge associated with storms (Ensign and Doyle, 2006).

D. Recycling of phosphorus in streams

As discussed earlier in terms of the biota of lakes, phosphorus can be released by excretion in inorganic and organic forms from living microbiota or as the organisms senesce, die, and lyse. Phosphorus can also be released during the egestion of consuming animals. As in lakes, the dissolved organic phosphorus compounds in streams are utilized enzymatically appreciably more slowly than DIP (reviewed by Newbold, 1992). Because of their slower rates of utilization, an accumulation of phosphorus compounds of high molecular weight (>5000 Da) occurs, and these compounds are exported downstream for subsequent utilization (Mulholland et al., 1988).

The spiraling of phosphorus has been measured in only a few streams (Ensign and Doyle, 2006). In a first-order woodland stream in Tennessee (USA) the spiraling length was 190 m, most of it (165 m) associated with the water movement, and the remainder (25 m) in the particulate components, mostly coarse and fine particulate matter (Newbold et al., 1983). Less than 3% of the total passed through the invertebrate and vertebrate consumers. Phosphorus uptake lengths ranged from 21 to 165 m and varied inversely with the quantity of detritus on the stream sediments (Mulholland et al., 1985a, 1985b). In a larger river in Michigan (USA) phosphorus uptake lengths varied between 1100 and 1700 m (Ball and Hooper, 1963). Most streams, however, have considerable uptake capacity for phosphorus, with uptake lengths in the range of 5–200 m (e.g., Hart et al., 1991).

Long-term evaluations of phosphorus dynamics in streams are rare. Examination of an average annual phosphorus budget for a forested second-order stream ecosystem of New Hampshire (USA) illustrates some of the potential transport and transformation processes (Table 15-3). Phosphorus inputs were dominated by dissolved and fine particulate phosphorus (63%) and falling and windblown litter (23%). Subsurface inflows (10%) and precipitation (5%) were relatively small sources of phosphorus in this ecosystem. The geologic export of phosphorus in the stream water was the only removal vector of consequence. On an annual basis, no annual net retention of phosphorus occurred in the stream. However, over short periods of time, inputs exceeded exports, with phosphorus accumulation. The accumulated phosphorus was exported in large pulses during precipitation-mediated episodes of high-stream discharge. Similar exports have been observed in many other stream ecosystems (e.g., Long and Cooke, 1978; Verhoff et al., 1982; Munn and Prepas, 1986; Wetzel, 1989). For example, in a detailed analysis of daily phosphorus dynamics in a small second-order stream in Michigan passing through a wetland to a lake, between 60 and 80% of the annual PO_4–P from this stream to the receiving lake occurred during three major precipitation events (Wetzel, 1989). This is also true of the many catchments in the Boreal ecozone that contain wetlands, home to a large portion of the world's lakes and streams. The wetlands retain some of their phosphorus input for short periods (Devito and Dillon, 1993) but release it downstream during periods of high flow or during periods when the wetland vegetation senesces.

TABLE 15-3 Phosphorus Budget for a Forested Second-Order Stream, Bear Brook, New Hampshire[a]

	mg P m^{-2} yr^{-1}		*% of Total*	
Inputs				
Dissolved		346		28
Precipitation	63		5	
Tributaries	152		12	
Subsurface	131		10	
Particulate		900		72
Fine: tributaries	459		37	
Coarse:	441		35	
Litter	283		23	
Tributaries	158		13	
Total inputs		1246		100
Exports (fluvial):				
Dissolved		242		19
Particulate		1059		81
Fine	807		62	
Coarse	252		19	
Total exports		1301		100

[a]*Rounded to whole numbers; modified from Meyer and Likens (1979) and several papers cited therein.*

IV. External natural and anthropogenic sources of phosphorus

A. Terrestrial runoff via tributaries

Surface drainage is usually a major contributor of phosphorus to streams and lakes. In general, the regional chemical characteristics of surface waters are closely related to the soil characteristics of their drainage basins and land use (Keup, 1968; Vollenweider, 1968; Dillon and Kirchner, 1975; Dillon and Molot, 1997; Lal, 1998). Soils, in turn, reflect the regional geological and climatic characteristics. The quantities of phosphorus transported to surface drainage vary with the amount of phosphorus in soils, topography, vegetative cover, quantity and duration of runoff flow, land use, and pollution. For example, wetlands are a major source of phosphorus to headwater lakes on the Precambrian Shield in North America (Dillon and Molot, 1997).

The parent rock material from which soil evolves by weathering is extremely variable in phosphorus content, and this variability increases with the thickness and heterogeneity of the stratified soil layers. Basic igneous rock contains relatively little phosphorus usually as apatite; phosphorus percentages of other substrates are typically as follows: sandstone is approximately 0.02, gneiss 0.04, unweathered loess 0.07, andesite 0.16, and diabase 0.03%. Limestone containing approximately 1.3% P and rare deposits of rock phosphate (10–15% P), in which biotic accumulations either from the organisms themselves or from guano were concentrated, are largely of sedimentary origin (cf. Hutchinson, 1950). Surface layers of soil are relatively rich in organic phosphorus from plant detritus in various stages of

TABLE 15-4 Concentrations of Nitrogen and Phosphorus in Runoff from Miami Silt Loam of Differing Gradients[a]

	$g\ m^{-2}\ yr^{-1}$	
Gradient (%)	*N*	P
8	2.0	0.06
20	4.25	0.2

[a]*After data cited in Vollenweider (1968).*

TABLE 15-5 Relative Erosion from Soil in Relation to Vegetative Cover and Land Use, Pacific Northwest of United States[a]

Crop or practice	Relative erosion
Forest duff	0.001–1
Pastures, humid region, or irrigated	0.001–1
Range or poor pasture	5–10
Grass/legume hayland	5
Lucerne	10
Orchards, vineyards with cover crops	20
Wheat, fallow, stubble not burned	60
Wheat, fallow, stubble burned	75
Orchards, vineyards, without cover crops	90
Row crops and fallow	100

[a]*From data cited in Biggar and Corey (1969), after Musgrave.*

decomposition by the action of soil fungi and bacteria. The exchange capacity of soils for phosphorus depends on the composition of the soil and increases with greater quantities of organic and inorganic colloids. Phosphorus is most available and easily leached from soils having a pH of 6–7. At lower pH values, phosphorus combines readily with aluminum, iron, and manganese. Therefore, where drainage basins are acidified from atmospheric pollution, inputs of phosphorus in runoff from acidified soils are reduced (Jansson et al., 1986). At pH 6 and above, progressively greater amounts of phosphate are associated with calcium as apatites and calcium phosphates.

The topography of the catchment basin influences the flow path of runoff and extent of erosion and the subsequent export of nutrients. Flat lands with little runoff and relatively high infiltration rates contribute less nutrient load to runoff than similar lands with steeper gradients (Table 15-4). Further, the relative erosion is influenced markedly by the type of vegetation and use to which the land is put (Table 15-5). Disturbances to the land cover can also have marked effects on nutrient releases to runoff water. For example, phosphorus and nitrogen concentrations of stream waters increased from 5- to 60-fold over background levels within the first 2 days after forest wildfires impacted the stream catchments (Schindler et al., 1980; Spencer and Hauer, 1991). Much of the nutrient loading was leached from ash. Returns to background concentrations of nutrients ranged from several weeks to years. The literature on this subject is large, and the reader is referred to the reviews examining phosphorus inputs from runoff by Vollenweider (1968), Biggar and Corey (1969), Cooper (1969), Keup (1968), Griffith et al. (1973), Dillon and Kirchner (1975), Meyer and Likens (1979), Loehr et al. (1980), Beaulac and Reckhow (1982), Grobler and Silberbauer (1985), Likens (1985), Lewis (1986), Stibe and Fleischer (1991), Likens and Bormann (1995), and Lal (1998).

An attempt has been made to classify natural and agricultural areas on the basis of the quantity of nitrogen and phosphorus in runoff (Table 15-6). Applications of fertilizers and land management practices, both in agriculture and forestry, will modify and generally increase these values considerably (see, e.g., Clesceri et al., 1986; Skaggs et al., 1994; Jordan and Weller, 1996). Urbanization results in increases in the amount of phosphorus discharged to surface waters in approximate proportion to population densities (Weibel, 1969). Phosphorus originating from heavy residential fertilization and storm sewer drainage (Cowen and Lee, 1973) can increase phosphorus inputs to surface drainage. For example, in the largest lake of southwestern Michigan (USA), Gull Lake, approximately 24% of the total phosphorus entering this eutrophic lake was from fertilization of lakeside lawns. The soils of 75% of the lawns, however, were saturated with phosphorus, making additions unnecessary (Moss, 1972; Moss et al., 1980). In contrast, loadings of phosphorus to a large, undisturbed Arctic Lake were nearly entirely from stream inputs, largely as dissolved P (Table 15-7).

In many cases lakes receive drainage and thus phosphorus from upstream lakes; that is, they are not headwater lakes. The portion of phosphorus input to a lake that leaves via outflowing waters is largely dependent on the lake's hydrologic budget, particularly its flushing rate. This, in turn, determines how much of the input is lost to the lake's sediments (see Section VIII on modeling). Since all but those lakes with extremely fast flushing rates act as nutrient sinks, downstream lakes receive less phosphorus as a result.

Some surface runoff receives very large quantities of phosphorus from domestic sewage. The importance of this source varies greatly with population size and the efficiency of the treatment of the sewage for nutrient

TABLE 15-6 Export of Nitrogen and Phosphorus from Soils of Natural and Agricultural Areas

Drainage basin type	Total dissolved inorganic nitrogen losses, kg N km^{-2} yr^{-1}	Total phosphorus losses, kg P km^{-2} yr^{-1}	Source
Undisturbed temperate forest	c. 200	c. 2	Hobbie and Likens (1973)
Undisturbed boreal forest	90–160	3–9	Ahl (1975)
Cleared forest, igneous watershed	–	c. 5	Dillon and Kirchner (1975)
Cleared forest, sedimentary watershed	–	c. 11	
Cleared forest, volcanic watershed	–	72	
Pasture, low intensity	100–1000	8–20	Ahl (1975); Johnston et al. (1965)
Pasture, high intensity	2000–25,000	–	Harper (1992)
Arable, cereals	4000–6000	–	Harper (1992)
Arable, cash crops	4000–<10,000	–	Letey et al. (1977)
Arable, intensive (UK, USA, Netherlands)	–	7–190	Cooke and Williams (1973);
			Kohlenbrander (1972); Johnston et al. (1965); Schuman et al. (1973)
Mixed upland (Northern UK)	530–630	c. 34	Atkinson et al. (1986)
Groundwater leachates	427–638	35–93	Reynolds (1979)
Urban runoff	c. 1000	c. 100	Harper (1992)
General soil productivity			Vollenweider (1968)
Low	<500	<20	
Medium	500–2500	20–50	
High	>2500	>50	

removal. Improved treatment in the form of tertiary phosphorus removal has reversed eutrophication in a number of well-documented instances in North America (Lake Erie, Lake Simcoe, Gravenhurst Bay, Lake Washington) (Schindler et al., 2016). Average values for the contributions via domestic sewage are given in Table 15-8. Industrial inputs, especially those associated with food processing, can be exceedingly high.

Ironically, cleaning detergents have been major sources of phosphorus that contribute marked fertilization effects to many freshwaters. Where legally permitted, synthetic detergents include phosphate builders as a major constituent, mainly sodium pyrophosphate and polyphosphates, to complex and inactivate cations of water supplies and permit more effective cleaning action. Before detergent phosphorus content was regulated, 7–12% of the weight of detergents was phosphorus, which contributed greatly to the load to waste treatment plants. In the United States, Canada, and the European Union, phosphorus has been banned in laundry detergents, in some jurisdictions as long ago as the 1970s. The situation with dishwasher detergents is less uniform, with some jurisdictions allowing low levels of phosphorus. Where phosphates are banned as a constituent of detergents, the phosphorus loadings to treatment facilities and surface waters declined by 50% to >80% (e.g., Maki et al., 1984; Pallesen et al., 1985; Hoffman and Bishop, 1994).

B. Groundwater

The phosphorus content of ground water is generally low; concentrations are about 20 μg P L^{-1} or less even in areas where soils contain relatively large quantities of phosphorus. The low phosphorus content is a result of the relatively insoluble nature of phosphate-containing minerals and the scavenging of surface phosphate by biota and soil particles.

TABLE 15-7 Phosphorus Budget for a Deep, Arctic Tundra Lake, Toolik Lake, Alaska[a]

	mmol m^{-2} yr^{-1}		
	Fractional P	*Total P*	*% of Total*
Inputs:			
Stream inflows			
Dissolved P	3.28		70.7
Particulate P	1.12	4.40	24.1
			94.8
Direct precipitation			
Dissolved P	0.09		1.9
Particulate P	0.15	0.24	3.3
			5.2
Total inputs			
Dissolved P	3.37		72.6
Particulate P	1.27	4.64	27.4
			100.0
Exports:			
Sedimentation from water column	1.4–1.7	1.4–1.7	30.1–36.6
Stream outflow			
Dissolved P	2.02		42.1
Particulate P	1.23	3.25	25.6
			67.7
Total outputs			
Dissolved P	2.02		42.1
Particulate P	2.63–2.93	4.65–4.95	56.6–59.2
			100.00

[a]*Numbers rounded; from data cited in Whalen and Cornwell (1985).*

C. Atmospheric deposition

Phosphorus in wet atmospheric precipitation and dry particulate material fallout originates from fine particles of soil and rocks, from living and dead organisms, as volatile compounds released from plants as well as from natural fires and the burning of fossil fuels (Newman, 1995). In inland regions the major source of phosphorus in precipitation is dust generated over the land from soil erosion and from urban, agricultural, and industrial contamination of the atmosphere. In heavily fertilized agricultural regions the phosphorus content of precipitation is much higher during the active growing season than in winter even in mountainous regions (Chapin and Uttormark, 1973; Lewis et al., 1985; Cole et al., 1990). However, nutrients accumulate in snowpacks and on the ice during winter and can be released rapidly in large amounts during the spring thaw (e.g., English, 1978; Adams et al., 1979).

Although the total phosphorus content of precipitation is generally low (about <30 μg P L^{-1}) in unpopulated regions, it can increase to well over 100 μg P L^{-1} around urban-industrial-agricultural environments. Even the lower concentrations are often substantially greater than that of the lakes receiving the deposition; thus atmospheric deposition of phosphorus can be a significant contributor to the overall P budget of a lake. Atmospheric contributions of phosphorus in wet and dry deposition range from approximately 0.01 to 0.65 g m^{-2} yr^{-1} (0.1–6.5 kg ha^{-1} yr^{-1}), with most values in the lower portion (<0.2) of that range (Kortmann, 1980; Gibson et al., 1995; Newman, 1995; Winter et al., 2007). These inputs can represent highly significant loadings when compared, for example, with the 0.07 g m^{-2} yr^{-1} value that Vollenweider (1968) estimated as a general permissible phosphorus loading rate for lakes. In some cases atmospheric loadings of phosphorus can

TABLE 15-8 Surface Water Loadings of Nitrogen and Phosphorus per Unit Area from Human Excrement and Other Sources Based on an Average of 12 g N $Person^{-1}$ Day^{-1} and 2.25 g P $Person^{-1}$ Day^{-1}[a]

Population density (persons km^{-2}*)*	*Nitrogen (g* m^{-2} yr^{-1}*)*	*Phosphorus (g* m^{-2} yr^{-1}*)*
50	0.22	0.04
100	0.44	0.08
150	0.66	0.12
200	0.88	0.16
300	1.32	0.24
500	2.20	0.40
1000	4.40	0.80
2500	11.00	2.00
5000	22.00	4.05

[a]*After Vollenweider (1968).*

constitute 40% or more of the annual loadings (e.g., Kowalczewski and Rybak, 1981; Psenner, 1984; Winter et al., 2007). The relative importance of atmospheric inputs of phosphorus to the total input to a lake will be strongly influenced by the ratio of the lake's catchment area to the lake surface area, with smaller drainage-to-lake area ratios typically resulting in the increased significance of the atmospheric inputs.

V. Phosphorus and the sediments: internal loading

The exchange of phosphorus between sediments and the overlying water is a major component of the phosphorus cycle in natural waters. There is an apparent net movement of phosphorus into the sediments in most lakes. The effectiveness of the net phosphorus sinks to the sediments and the rapidity of processes regenerating the phosphorus back to the water depend upon an array of physical, chemical, and biological factors. The correlation between the amount of phosphorus in the sediments and the productivity of the overlying water is modest to weak, and the phosphorus content of the sediments can be several orders of magnitude greater than that of the water. The important factors are (1) the ability of the sediments to retain phosphorus, (2) the conditions of the overlying water, and (3) the biota within the sediments that alter exchange equilibria and effect phosphorus transport back to the water.

A. Exchanges across the sediment—water interface: overview

The deposition of phosphorus into lake sediments occurs via five mechanisms (Williams and Mayer, 1972; Boström et al., 1988; Wetzel, 1990): (1) sedimentation of particulate phosphorus minerals imported from the drainage basin. Most of this material settles rapidly and is deposited largely in nearshore areas, (2) adsorption or precipitation of dissolved phosphorus with inorganic compounds via three mechanisms: (a) phosphorus coprecipitation with iron and manganese; (b) adsorption to clays, amorphous oxyhydroxides, and similar materials; and (c) phosphorus associated with carbonates. It is difficult to distinguish the direct adsorption of dissolved phosphorus of lake water onto particles in the sediments from diagenetic and transfer processes within surface sediments, (3) sedimentation of phosphorus with allochthonous organic matter, (4) sedimentation of phosphorus with autochthonous organic matter, (5) uptake of phosphorus from the water column by attached algal and other attached microbial communities and to a lesser extent by submersed macrophytes, with eventual transport back to the sediments by translocation and deposition with detritus.

Once the phosphorus is within the sediment in various forms, exchanges across the sediment—water interface are regulated by: mechanisms associated with mineral—water equilibria; sorption processes, particularly ion exchange, oxygen and other electron acceptor-dependent redox interactions; and the physiological and behavioral activities of many biota (bacteria, algae, fungi, macrophytes, invertebrates, and fish) associated with the sediments. The sediment—water interface separates into two distinct domains. In all but the upper few millimeters of sediment, exchange is controlled by motions on molecular scales with correspondingly low diffusion rates (Duursma, 1967). In the water exchange is regulated by much higher and more variable rates of turbulent diffusion (Mortimer, 1971). The exchange rates between various deposits of phosphorus and the interstitial water of the sediments depend on local adsorption and diffusion coefficients and their alteration by enzyme-mediated reactions of the microbiota.

Numerous processes operate, often simultaneously, to mobilize phosphorus from particulate stores to dissolved interstitial phosphorus and then to transport that dissolved phosphorus across the sediment—water interface into the overlying water (Fig. 15-3). Orthophosphate is bound to particles by physical adsorption as well as chemical binding of different strengths (complex, covalent, and ionic bonds). Physical and chemical mobilization includes desorption, dissolution of phosphorus-containing compounds, particularly assisted by microbially mediated acidity, and ligand exchange mechanisms between phosphate and hydroxide ions or organic chelating agents (Boström et al., 1982; Stumm and Morgan, 1996; Hupfer and Lewandowski, 2008; Orihel et al., 2017). Microbial biochemical mobilization processes include mineralization by hydrolysis of phosphate—ester bonds, release of phosphorus from living cells as a result of changing environmental conditions, particularly redox, and autolysis of cells.

The *internal loading* of phosphorus to a lake from the sediments depends on hydrodynamic and biotic mechanisms that transport dissolved phosphorus from the sediments to the lake water. Because of the steep concentration gradients of phosphorus between interstitial water and the overlying water, molecular diffusion is a primary transport into the overlying anaerobic water (Fig. 15-3). Currents from wind-induced water turbulence can disrupt gradients and resuspend sediment particles. Disturbance of sediments by benthic invertebrates living on or in the sediments and by bottom-feeding fishes, when occurring in large densities, can cause appreciable bioturbation of sediments (Boudreau, 1998). Microbial generation of gases, particularly CO_2,

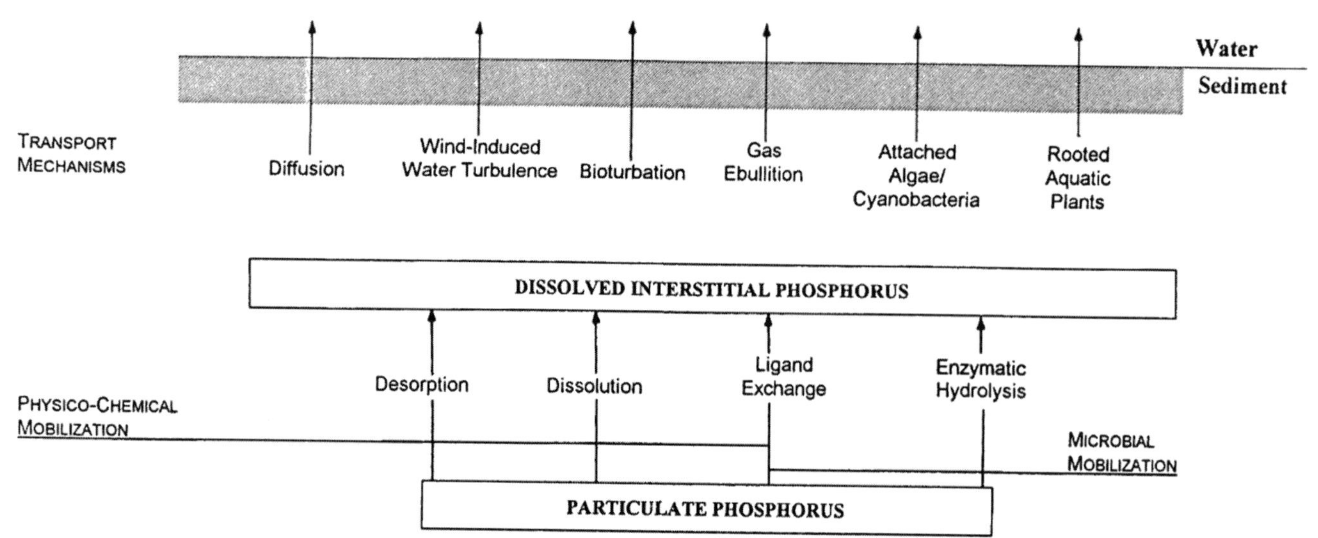

FIGURE 15-3 Processes involved in the mobilization of phosphorus from particulate stores into dissolved states of interstitial water of the sediments and transport across the sediment–water interface into the overlying water. *(Modified extensively from Boström et al., 1982, 1988; Wetzel, 1999.)*

CH_4, and N_2, can accumulate, form bubbles, and rise to the surface (Bastviken et al., 2008). Such ebullition can disrupt gradients near the sediment/water interface and may accelerate the diffusion of phosphorus upward. The metabolism and growth of aquatic macrophytes (Chapter 24) living on and within the sediments can both suppress and enhance the transport of phosphorus across the sediment–water interface. Because of the major importance of the movement of phosphorus from large accumulations in the sediment to the overlying water, each of these processes is discussed in greater detail in subsequent sections.

B. Oxygen content of the microzone

The most conspicuous regulatory features of the sediment boundary are the mud–water interface and the oxygen content. The oxygen content at this *microzone* is influenced primarily by the metabolism of bacteria, algae, fungi, planktonic invertebrates that migrate to and live within the interface, and sessile benthic invertebrates. Microbial degradation of dead particulate organic matter that settles into the hypolimnion and onto the sediments is the primary consumptive process of oxygen, which is depleted within a few centimeters of the surface in oligotrophic lakes but autotrophic nitrification also plays a role. The rate of heterotrophic oxygen depletion at the sediment–water boundary is governed by the rate of organic loading to the sediments and the rate of oxygen supply from turbulent mixing from above (Chapter 11). For example, it has been estimated that 88% of the hypolimnetic oxygen consumption in the central basin of Lake Erie resulted from bacterial degradation of algae sedimenting from the trophogenic zone (Burns and Ross, 1971), while Clevinger et al. (2014) estimated that by 2008–2010 heterotrophic respiration consumed 70% of the depleted oxygen and autotrophic nitrification 30%. Nitrification rates depend on supplies of ammonium and other metabolically essential nutrients including phosphorus and oxygen (see Chapter 14). Decomposition of more labile organic fractions, largely of plant origin, occurs while it is settling into the sediments, and, depending on water column depth and rates of input and sedimentation, sediments in deeper water can receive organic residues that are relatively resistant to rapid decomposition. Beneath the sediment surface, oxygen concentrations rapidly disappear with sediment depth because oxygen diffusion rates are too low to meet the biological and chemical demands.

Sediment demand for oxygen is high and is governed by the intensity of microbial and respiratory metabolism and by the fact that in the anaerobic zone, oxidizable (i.e., reduced) inorganic elements, such as Fe^{2+}, accumulate from the decomposition of organic matter. Diffusion regulates transport within the sediments, unless the superficial sediments are disturbed by overlying water turbulence, which increases the supply rate of oxygen. Oxygen from well-aerated overlying water, as in oligotrophic lakes or in more productive lakes at periods of complete circulation, will penetrate only a few centimeters into the sediments by diffusion. However, the limited diffusion of oxygen beyond surficial sediments can be overcome to some extent by centimeters-long microbial cables that connect redox couples in the aerobic and anaerobic zones (Teske, 2019).

The superb early experimental and observational work of Mortimer (1941,1942, 1971) demonstrated the importance of an *oxidized microzone* to chemical exchanges, especially of phosphorus, from the sediments. At the sediment surface, a difference of a few millimeters in oxygen penetration is the critical factor regulating the exchange between sediment and water. These relationships are exemplified by two lakes, as described below, with organic sediments: the first lake maintained oxygen concentrations at the sediment interface >8 mg L^{-1} throughout summer stratification, while in the second lake oxygen levels at the interface decreased to <1 mg L^{-1}.

In the first situation illustrated by Mortimer's (1941, 1942) studies of Windermere, England, oxygen concentration at the sediment surface did not fall below 1 or 2 mg L^{-1}. Electrode potentials, which approximate composite redox potentials (Chapter 16), of the oxygenated overlying water and surficial sediments to a depth of approximately 5 mm were uniformly high (+200 to 300 mV). Below 40–50 mm in the sediments, the potentials were uniformly low (approximately −200 mV), indicative of extreme reducing conditions and total anoxia (Fig. 15-4). The sediment remained oxidized to a depth of about 5 mm throughout the period of summer stratification. Seasonal differences were observed in the sediment depth at which the transition from high to low potential occurred, but the region of low potential never extended into the water. After 5 months of thermal stratification, the point of 0 mV moved toward the surface of the sediments, to −5 mm from approximately −12 mm at the time of spring turnover and moved downward to −10 mm during fall circulation. The integrity of the oxidized microzone was maintained in a thin but operationally very significant layer during stratification periods. The oxidized microzone was further maintained by diffusion and by turbulent displacement of the uppermost sediments to the overlying water during turnover periods (cf. Gorham, 1955). The effectiveness of the oxidized microzone in preventing a significant release of soluble components from the interstitial waters of the sediments to the overlying water was demonstrated in experimental chambers for over 5 months (Fig. 15-5, *left*). Phosphorus, in particular, was prevented from migrating upward.

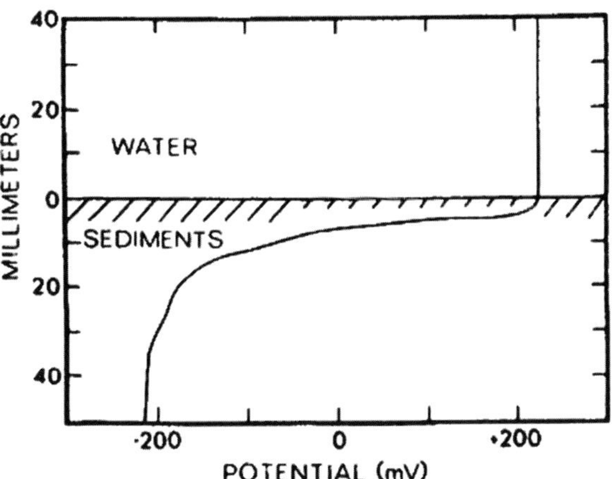

FIGURE 15-4 Diagrammatic profile of composite electrode potentials, not corrected for pH variations, across the sediment–water interface in undisturbed cores from the deepest portion of Windermere before, during, and after stratification. *(Based on data from Mortimer, 1971.)*

The ability of sediments to retain phosphorus beneath an oxidized microzone at the interface is related to several interacting factors. Much of the organic phosphorus reaching the sediments by sedimentation is decomposed and hydrolyzed (Sommers et al., 1970). Most of the sediment phosphorus is inorganic, for example, apatite, derived from the watershed, and phosphate adsorbed onto clays and aluminum and ferric hydroxides (Frevert, 1979a, 1979b; Detenbeck and Brezonik, 1991; Andersen and Jensen, 1992; Danen-Louwerse et al., 1993). Additionally, phosphate coprecipitates with iron, manganese, and carbonates (Mackereth, 1966; Harter, 1968; Wentz and Lee, 1969; Otsuki and Wetzel, 1972; Boström et al., 1988) with proportions differing from watershed geology (Orihel et al., 2017). Work on Wisconsin lake sediments and the Great Lakes indicated that phosphorus was present in the sediments predominantly as apatites, organic phosphorus, and orthophosphate ions covalently bonded to hydrated iron oxides (Shukla et al., 1971; Williams et al., 1970, 1971a–c; Williams and Mayer, 1972; Golterman, 1982, 1995). In calcareous sediments of hardwater lakes containing 30–60% $CaCO_3$ by weight, $CaCO_3$ levels were not directly related to inorganic and total phosphorus. These sediments had a lower capacity to adsorb inorganic phosphorus than noncalcareous sediments (cf. Stauffer, 1985; Olila and Reddy, 1993). The observations imply that $CaCO_3$ sorption is less important than iron–phosphate complexes in controlling the concentrations of phosphorus in sediments (Andersen, 1975; Frevert, 1980). Major P-bound fractions responsible for release can vary within a large lake. For example, in Lake Simcoe in central Ontario (Canada), P release from the sediments from deep glacially formed Kempenfelt Bay with a highly urbanized catchment is dominated by the redox-sensitive P fraction, whereas in shallow and agriculturally impacted Cook's Bay, the main P binding form that can be mobilized through diagenesis is carbonate-bound P (Dittrich et al., 2013).

Although phosphorus exchange by adsorption and desorption within the sediments between sediment particles and interstitial water can be as rapid as a few minutes (Hayes and Phillips, 1958; Li et al., 1972), the rate of transfer across the sediment–water interface depends on the state of the microzone. The oxidized layer forms an efficient trap for iron and manganese

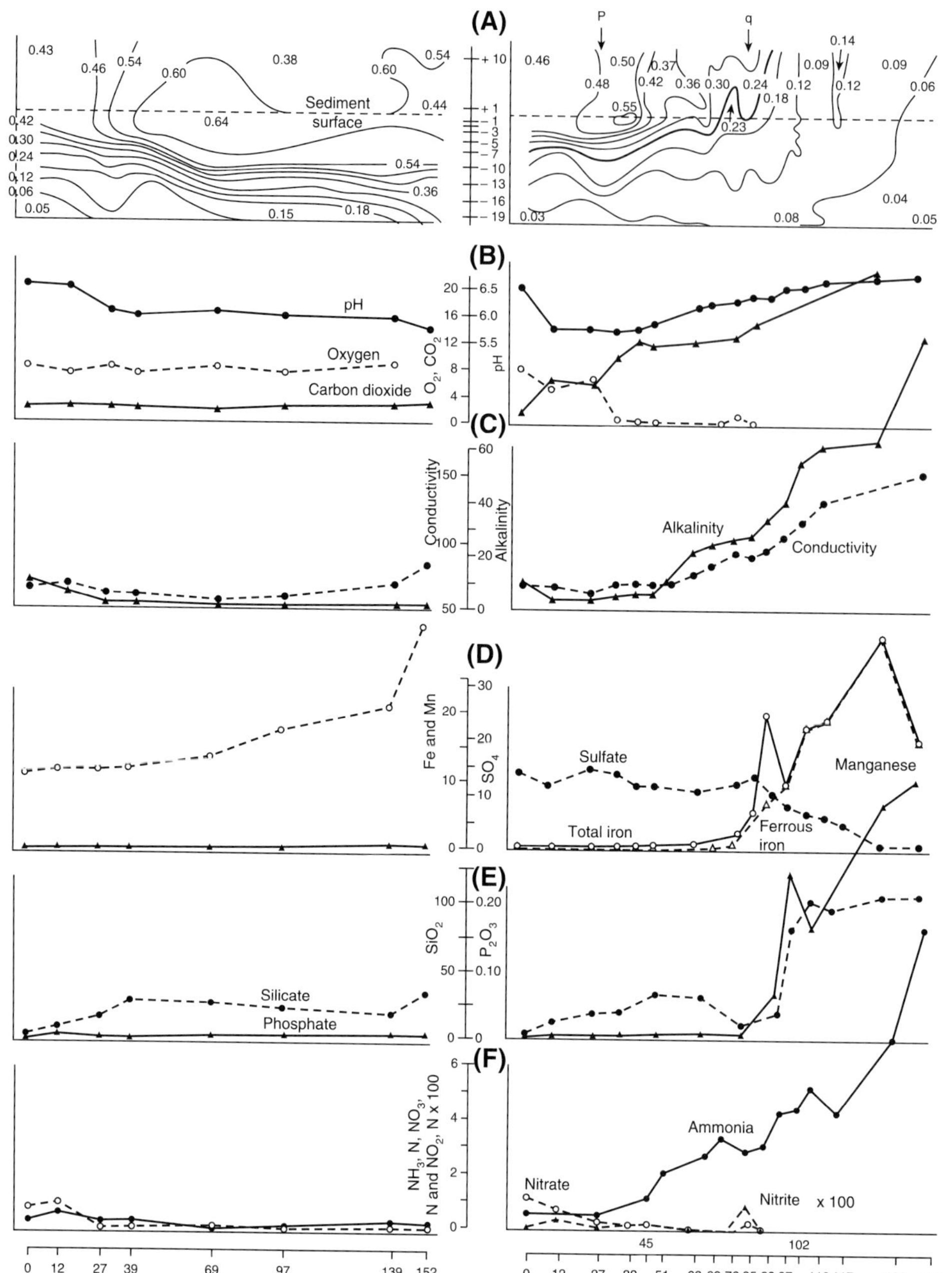

FIGURE 15-5 Variation in the chemical composition of water overlying deepwater Windermere sediments over 152 days in experimental sediment–water tanks. *Left-hand series*: aerated chamber; *right-hand series*: anoxic chamber. (a) Distribution of redox potential (E_7; Eh adjusted to pH 7) across the sediment–water interface in mm; (b) pH, concentrations of O_2, and CO_2 in mg L^{-1}; (c) alkalinity expressed as mg $CaCO_3$ L^{-1} and specific conductance, in μS cm^{-1} at 18°C; (d) iron (total and ferrous as Fe) and SO_4 in mg L^{-1}; (e) phosphate as P_2O_5 and SiO_2 in mg L^{-1}; (f) nitrate, nitrite ×100, and ammonia, all as N, in mg L^{-1}. *(From Mortimer, 1941, 1971.)*

(see Chapter 16) as well as for phosphate, thereby greatly reducing the transport of materials into the water and scavenging materials such as phosphate from the water.

As the oxygen content of water near the sediment interface declines, the oxidized microzone barrier weakens. As seen from Mortimer's experiments in Windermere (Fig 15-5, *right-hand series*), the release of phosphorus, iron, and manganese increased markedly as the redox potential decreased. With the reduction of ferric hydroxides and complexes, ferrous iron and adsorbed phosphate were mobilized and diffused into the water. The same general reactions were observed in the hypolimnetic water just overlying the sediments in eutrophic Esthwaite Water, England (Fig. 15-6), a pattern that has been observed repeatedly in productive

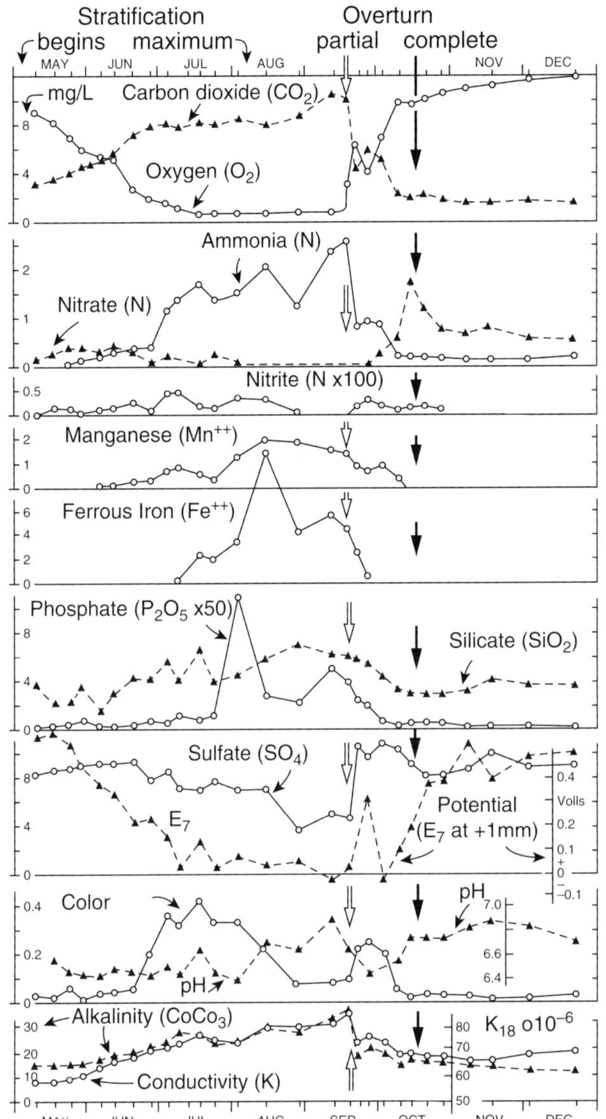

FIGURE 15-6 Seasonal distribution in composition (mg L^{-1}) and properties of water within 30 cm of the sediments at 14 m in Esthwaite Water, England. Components as in Fig. 15-5; color in arbitrary units. *(From Mortimer, 1971.)*

dimictic lakes since its initial detailed description by Einsele (1936). A sudden release of ferrous iron and phosphate into the water occurs at the time when the +0.20 isovolt (E_7 = +200 mV) emerged above the interface surface. This event is preceded by nitrate reduction and the slow release of bases (alkalinity), CO_2, and ammonia. As long as an abundance of nitrate remains and nitrate reduction continues in the overlying water (c. >0.1–1 mg NO_3–N L^{-1}), no release of iron-bound phosphate occurred from the sediments to the anoxic hypolimnion (Andersen, 1982; Tirén and Pettersson, 1985; Foy, 1986; Beutel et al., 2016). Manganese is reduced and mobilized at a higher redox potential than iron.

The introduction of oxygen during, for example, autumnal circulation to anaerobic waters causes ferrous iron to be oxidized and produces a simultaneous reduction of phosphate, part as ferric phosphate, which is less soluble than ferric hydroxide, and part by adsorption onto ferric hydroxide and $CaCO_3$. Although manganese is oxidized more slowly than iron, it nonetheless is effectively precipitated at the time of overturn. Ferrous iron released from the sediments is often in excess of phosphate, and when oxidized, it precipitates much of the phosphate. Some of the ferric phosphate in particulate form may slowly hydrolyze and restore some phosphate to the upper waters and littoral areas (Hutchinson, 1957). Although most phosphate is returned eventually to the sediments, much (>50%) of the hypolimnetic phosphorus upwelled during autumnal circulation is available biologically (Nürnberg, 1985). Often, however, other growth conditions (temperature, light) are not optimal at that time of year.

In very productive lakes, where hypolimnetic decomposition of sedimenting organic matter produces anoxic conditions and hydrogen sulfide, some ferrous sulfide (FeS) is precipitated. Ferrous sulfide, like many other metal sulfides, is exceedingly insoluble and forms at a redox potential of about −100 mV. If large quantities of FeS precipitate, sufficient iron can be removed to permit some of the phosphate accumulated in the hypolimnion to remain in solution during autumnal circulation, although seasonal oxidation of FeS precipitates can affect long-term accumulation rates (Howard and Evans, 1993; Leonard et al., 1993). The addition of sulfate to a lake in order to increase the bacterial production of hydrogen sulfide (H_2S) and to accelerate the loss of iron has been suggested as a method of fertilizing lakes by regenerating phosphate from the sediments (Hasler and Einsele, 1948).

Iron interacts with dissolved humic substances (Chapter 16) to bind phosphorus at acidic to near-neutral pH values, irrespective of redox conditions, in the surficial sediments of certain softwater lakes (Jackson and Schindler, 1975; Francko, 1986). This process should enhance sediment retention of phosphorus in lakes with high loading of iron and humic substances.

C. Phosphorus release from the sediments

Because of the obvious importance of phosphorus as a nutrient that often accelerates the productivity of freshwaters, much interest has been devoted to the phosphorus content of sediments and its movement into the overlying water (see review by Orihel et al., 2017). Lake sediments contain much higher concentrations of phosphorus than the overlying water and represent a large potential pool for phytoplankton under certain conditions (Olsen, 1958, 1964; Holden, 1961; Hepher, 1966; Holdren et al., 1977; Boström et al., 1982; and many others). Under aerobic conditions, the exchange equilibria are largely unidirectional toward the sediments, although low release rates from aerobic sediments into overlying waters can occur. Under anaerobic conditions, however, inorganic exchange at the sediment–water interface is strongly influenced by redox conditions (Frevert, 1979b) unless aluminum/iron ratios are high (Kopacek et al., 2005, 2007). When these ratios are high, phosphorus is sorbed primarily to aluminum hydroxides, which are redox insensitive and do not release phosphorus under low redox potential.

The depth of the sediment involved in the active migration of phosphorus to the water is considerable (e.g., Carignan and Flett, 1981). In undisturbed anoxic sediments, given sufficient time (2–3 months), phosphorus moved upward readily from a depth of at least 10 cm below the sediment surface to the overlying water, regardless of whether the sediments were calcareous eutrophic muds or acidic and peaty in nature (Hynes and Greib, 1970). Comparison of movement in sterile sediments and sediments with anaerobic bacteria showed no significant difference; thus diffusion predominated.

Phosphorus-mobilizing bacteria, especially of the genera *Pseudomonas*, *Bacterium*, and *Chromobacterium*, were abundant to at least 15 cm depth in reservoir sediments (Gak, 1959, 1963). Their abundance and vertical distribution varied with the type of sediments. Low numbers occurred in sandy sediments with small amounts of silt, and the bacteria were concentrated near the interface. Their numbers increased more uniformly with depth in sandy sediments with moderate amounts of organic matter and silts. The greatest numbers of bacteria were found in silts high in organic matter. Hayes and Anthony (1959) and Anthony and Hayes (1964) examined the relationship of bacterial densities and the organic content of sediments in detail and found only a weak correlation in a large number of lakes. Bacterial numbers in the sediments increased proportionally, especially at the interface, with several indices of increasing lake productivity in lakes of neutral or alkaline pH and with low concentrations of humic compounds. In general, however, bacterial production rates are correlated with phosphorus and carbon in the sediments (Eckerrot and Pettersson, 1993). Bacterial biomass in sediments of organically stained acidic bog lakes was also high.

Although bacteria are of major importance in the dynamics of phosphorus cycling in the water, as will be discussed in the following section, their role in expediting phosphorus exchange across the sediment interface is relatively minor in comparison to chemical equilibria processes (Hayes, 1964). Bacterial decomposition is proportional to bacterial densities at the interface and directly related to the general productivity of the lake because their metabolic activity is driven largely by the supply rate of organic matter (Hayes and MacAuley, 1959; Hargrave, 1972). The sediment microflora is important in increasing the concentrations of phosphorus dissolved in the interstitial water of the sediments (Fleischer, 1978). Bacterial metabolism at the interface, however, has relatively little effect on biogenic fixation and removal of phosphate from the overlying water. In sterilized and natural sediments microbial fixation and transport of phosphorus from the water to the sediments amounted to <5% of the total movement under anaerobic reducing conditions (Hayes, 1955; Macpherson et al., 1958; Olsen, 1958; Pomeroy et al., 1965). Under aerobic conditions, bacteria of the interface did significantly increase the microbial transport of phosphorus to the sediments (Hayes, 1955; Hayes and Phillips, 1958), and this loss was related to the amount of microbial phosphorus sedimenting to the interface (cf. also Frevert, 1979a).

Mobilization of sediment-associated P to the overlying water has been ascribed largely to PO_4^{3-} release from Fe(III) oxide, as these compounds are reduced by anaerobic conditions in surface sediments and the overlying water, a process referred to as reductive dissolution (Mortimer, 1941; Boström et al., 1982). Mobilization can, however, occur in a variety of other ways that do not involve Fe(III) oxide, such as desorption of organic and inorganic P bound to mineral surfaces, dissolution of minerals including PO_4^{3-} within their crystal structure, mineralization of organic matter, excretion of PO_4^{3-} by macrophyte roots and bacteria, and dissociation of ternary P complexes with humics (Orihel et al., 2017). While PO_4^{3-} release rates are undoubtedly higher in many lakes under anaerobic conditions, such conditions alone may not lead to higher internal P loading. For example, high amounts of aluminum hydroxide to iron hydroxide in sediments may limit PO_4^{3-} release under anaerobic conditions (Kopacek et al., 2005). When the release of PO_4^{3-} is compared on a molar ratio to PO_4^{3-} liberated per unit of organic carbon mineralized by bacterial metabolism, the availability of alternate electron-acceptor compounds, such as sulfate (SO_4^{2-}), becomes as important within the interstitial waters of the sediments as the oxidative conditions of the overlying water (Hasler and Einsele, 1948; Sugawara et al., 1957; Caraco et al., 1989, 1990; Roden and Edmonds, 1997; Gächter and Muller, 2003; Orihel et al., 2015). PO_4^{3-} release from sediments can increase as sulfate concentrations increase from natural or anthropogenic sources. However, the potential for higher internal loading in acidified oligotrophic, softwater lakes is more than offset by lower external P inputs, possibly from increased P retention by acidified soils (Jansson et al., 1986). Conversely, declining sulfate deposition could result in lower internal loading.

Iron sulfide formation coupled with sulfate reduction can reduce the abundance of Fe compounds that can complex PO_4^{3-} and thereby promote release of PO_4^{3-} into sediment porewater. When sulfate content is low or absent in anaerobic sediments, microbial reduction generates Fe(II) compounds from the microbial reduction of Fe(III) oxide. PO_4^{3-} can be retained effectively with Fe(II) compounds (vivianite), but Fe-associated PO_4^{3-} is quantitatively released when amorphous Fe(III) oxide of the sediment is converted to iron sulfides (Roden and Edmonds, 1997). Conversion of reactive sediment Fe compounds to iron sulfides by sulfate-reducing bacteria leads to a more efficient release of Fe-associated PO_4^{3-} than does direct microbial Fe(III) oxide reduction (Fig. 15-7).

Other processes can contribute to the composite release of Fe-associated PO_4^{3-} from sediments. For example, the uptake of excessive amounts of P by aerobic bacteria (e.g., Fleischer, 1986) and storage as *polyphosphates* could be rapidly degraded and released with the onset of anaerobic conditions (DeMontigny and Prairie, 1993; Gachter and Meyer, 1993). Because organic sediments are aerobic for only up to several millimeters, this contribution is likely small in comparison to other mechanisms.

The sorption capacities of Fe(III) oxide decrease as pH levels increase above 6.5 (Stumm and Morgan, 1996). Large increases in pH values can occur within the sediments for several millimeters as a result of diurnal fluctuations in photosynthesis by epipelic algae (Fig. 15-8) and by submersed aquatic macrophytes and associated epiphytic algae. Simultaneous increases in oxygen concentrations (Fig. 15-8) and adsorption to, or with, photosynthetically induced precipitation of $CaCO_3$ (e.g., Mickle and Wetzel, 1978) can counteract the availability of desorbed PO_4^{3-}.

The rate of phosphorus release from lake sediments increases (about doubles) markedly if the sediments are disturbed by agitation from turbulence (Zicker et al., 1956). Covering anaerobic sediments with sand or polyethylene sheeting greatly impedes the loss of oxygen from the overlying water and decreases sediment release of phosphorus, iron, and ammonium (Hynes and Greib, 1970).

In thermally stratified lakes the internal phosphorus loading rate, expressed as mass P $lake^{-1}$ day^{-1}, increases with time as the hypolimnetic anoxic area increases, which generally happens during the thermally

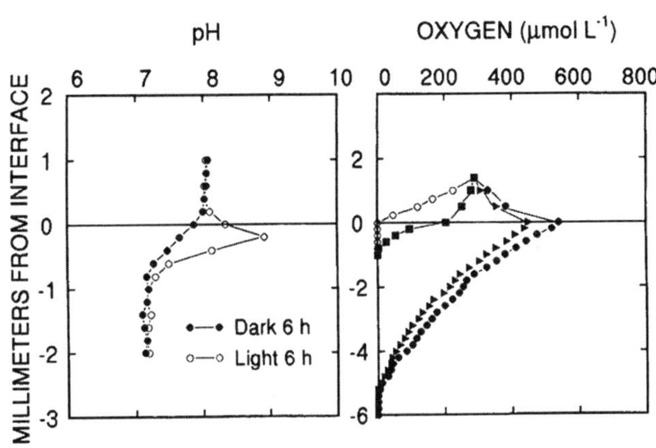

FIGURE 15-8 Distribution of pH (*left*; illumination 30 μmol quanta m^{-2} s^{-1}) and dissolved oxygen (*right*) immediately above and within sediments colonized by epipelic microalgae and bacterial communities. Oxygen microprofiles (*right*) during darkness (○—○), and after 1 h (■—■), 8 h (▲—▲), and 10 h (●—●) illumination with 10 μmol quanta m^{-2} s^{-1}. *(Extracted from data of Carlton and Wetzel, 1987, 1988.)*

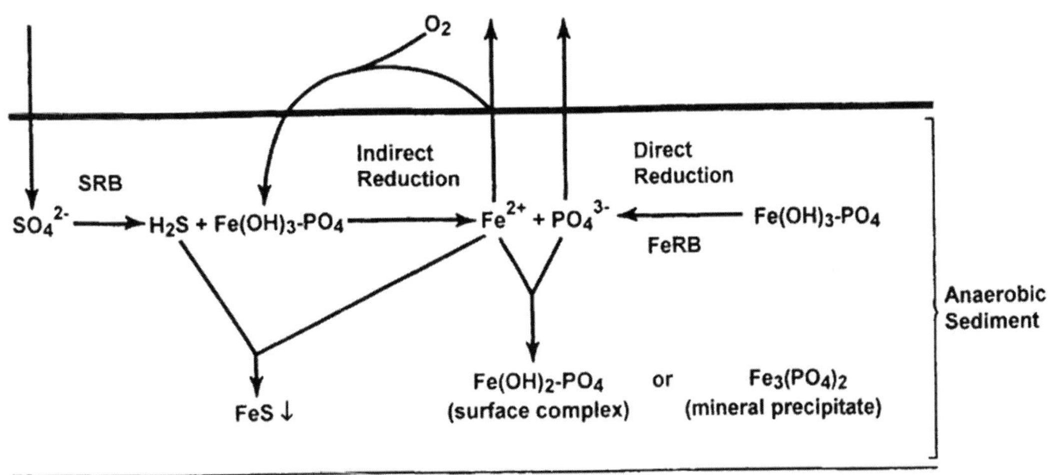

FIGURE 15-7 Interactions of sulfate on the reduction of Fe(III) and release of Fe(II) and phosphate. SRB = sulfate-reducing bacteria; FeRB = iron-reducing bacteria. *(Modified from Roden and Edmonds, 1997.)*

stratified season. It is thus a function of the areal and temporal extent of sediment anoxia. The latter can be summarized in terms of an annual parameter, the *anoxic factor* (AF), defined as the duration of anoxia multiplied by anoxic sediment area/lake area (Nürnberg 1995; see Chapter 11). Empirical relationships between the AF and annual internal loading rates have been developed (Tammeorg et al., 2020). Most studies of internal loading have focused on hypolimnia in thermally stratified lakes where oxygen depletion is easily detected with relatively large oxygen sensors. However, internal loading has also been detected in shallow waters of eutrophic thermally stratified and polymictic lakes where anoxic layers are thinner and episodic (they form overnight or during periods of low wind speed), thus requiring different sensors (Jabbari et al., 2019, 2021; Tammeorg et al., 2020). In thermally stratified lakes internally loaded phosphorus becomes photosynthetically available as anoxic hypolimnetic water is entrained into metalimnetic waters over the summer and after turnover, while internally loaded phosphorus in shallow waters is entrained into the mixed layer when wind speed increases. Hypoxic bottom water can also be translocated during episodic coastal upwelling events from hypoxic hypolimnia to shallower waters in large lakes, for example, in Lake Erie (Jabbari et al., 2019; Valipour et al., 2021). It follows that translocated hypoxic waters can increase phosphorus concentrations in shallower waters if they are phosphorus enriched.

Algae and Cyanobacteria growing on sediments are able to effectively utilize phosphorus from the sediments (Golterman et al., 1969; Björk-Ramberg, 1985). Moreover, phytoplankton suspended in water with various particulate inorganic compounds of extremely low solubility were capable of extracting sufficient phosphorus for growth; without sediment phosphorus sources, the phosphorus content of the water-limited phytoplankton growth under experimental conditions. The presence or absence of bacteria had little effect on the algal utilization of phosphorus. These results stress the importance of extractable phosphates in the sediments if they are agitated into the water column, as in shallow lakes (Chapter 26), even though their solubilities may be extremely low.

Some phosphorus release occurs from littoral sediments under aerobic conditions, particularly as temperatures warm above 10–15°C; however, fluxes are limited compared to anaerobic release rates. Release rates under aerobic conditions increase as phosphate concentrations in the overlying water decline below equilibrium levels and as pH increases from the intense photosynthetic activity of submersed macrophytes and attached algae (Twinch and Peters, 1984; Drake and Heaney, 1987; Boers and van Hese, 1988; Carlton and Wetzel, 1988; Boers, 1991; James and Barko, 1990). Because the volume of the littoral water is often small in comparison to that of the total lake, the littoral water often cools more rapidly at night than the pelagic zone. Convective flows from the littoral as interflow to the base of the epilimnion can move appreciable quantities of phosphorus from the littoral areas to the pelagic zone, especially if the littoral zone is large relative to the pelagic zone.

D. Benthic algae and phosphorus release from sediments

Oxygen (Chapter 11) is a primary factor influencing the release of phosphorus from sediments. If the sediments receive light, even at very low intensities (<50 μmol quanta m^{-2} s^{-1}), photosynthesis of *epipelic* algal communities growing on the sediments (Chapter 25) can quickly (minutes) produce high, often markedly supersaturated concentrations of oxygen within the community, usually <2 mm in thickness (Fig. 15-8) (Revsbech et al., 1983; Carlton and Wetzel, 1987). This oxygen can diffuse several millimeters into the interstitial water of the supporting sediments at rates greater than it is consumed by bacterial respiration and chemical oxidations and thus suppress internal phosphorus loading. If respiration rates are high enough, surficial sediments can, however, become anoxic overnight. By this mechanism, diurnal changes occur in the oxidized microzone of sediments shallow enough to receive light from fully oxidized in the daylight hours to fully reducing at nighttime. It was shown experimentally that the rate of phosphorus efflux from sediments is inversely related to the extent of sediment oxygenation and the magnitude of epipelic algal photosynthesis (Carlton and Wetzel, 1988). Although much of the reduced phosphorus efflux is caused by direct chemical redox changes, the microbial community also assimilates and complexes nutrients in organic compounds (cf. Kelderman et al., 1988; Hansson, 1989).

E. Phosphorus translocation by migrating phytoplankton

Several algae and Cyanobacteria of lakes exhibit vertical migrations from nutrient-rich sediments or lower water strata to the euphotic trophogenic zone on a seasonal or daily basis. For example, a motile cryptomonad alga and a dinoflagellate may migrate vertically from phosphorus-rich strata below the metalimnion at night to surface waters in early morning in sufficient quantities to enrich the epilimnion (Salonen et al., 1984; Taylor et al., 1988). Migrations of a large part of certain benthic-inhabiting Cyanobacteria to the epilimnion have been implicated in the translocation of benthic phosphorus sufficient to increase phytoplanktonic productivity

(Barbiero and Welch, 1992). Vacuolate Cyanobacteria are able to regulate their position in the water column when turbulence is low, probably to optimize access to nutrients and light (Oliver, 1994; Camacho et al., 1996, 2000; Head et al., 1999; Gervais et al., 2003; Molot et al., 2014). They have been observed migrating upwards at 11 m depth in a metalimnion and have been recorded migrating into anoxic waters, a distance that can be covered under low turbulence within several hours depending on the species, cell size, and starting position in the water column. For example, maximum migration rates for *Microcystis aeruginosa* ranged from 0.4 to 10.8 m h^{-1} downward and 1.3 m hr^{-1} upward, while for *Anabaena* (recently renamed *Dolichospermum*) and *Aphanizomenon*, downward and upward rates were 0.04 and 0.14 m hr^{-1}, respectively (Reynolds et al., 1987).

F. Phosphorus cycling mediated by aquatic plants and epiphytic microflora

Lakes, reservoirs, and rivers are coupled with their drainage basins by intervening wetlands and littoral zones, the land–water interface zones, through which much runoff and seepage water flows before entering the main water bodies. The aquatic plants of the interface zones are major biotic components in many, if not most, small lakes and in the flood plains of river ecosystems (Chapter 24). These plants often occupy much of the epilimnion and trophogenic zone, both of which extend virtually to the same depth in many small lakes. The littoral macrophytes and their attached microflora function mutualistically as conservative retainers of phosphorus and can modify greatly phosphorus loadings and budgets among aquatic ecosystems (Wetzel, 1990). The importance of rooted plants to lake-wide phosphorus cycling will depend on the proportion of sediments receiving enough light to support plants.

1. *Aquatic plants*

Emergent, rooted, floating-leaved, and submersed higher aquatic plants take up most of their phosphorus from the interstitial waters of the sediments, where the concentrations are several orders of magnitude greater than in the overlying water (Denny, 1972; Barko and Smart, 1980, 1981; Brock et al., 1983). Even though the structural and vascular systems of submersed aquatic angiosperms are simplified and reduced, nutrient uptake is largely by the root-rhizome system from the sediments (Chapter 24). Early studies provided conflicting results. For example, the presence of the macrophytes *Eriocaulon* or *Utricularia* in experimental sediment–water systems increased the movement of radiophosphorus from the water to the sediments (Hayes, 1955). Absorption and translocation of phosphorus, studied in bacteria-free cultures of the macroalga *Chara* common to hardwater lakes, showed all parts of this alga could absorb ^{32}P about equally (Littlefield and Forsberg, 1965). About the same proportion of adsorbed phosphorus was translocated from the apices or from rhizoids to other parts of the alga. Similar results were obtained for other freshwater plants (e.g., Schwoerbel and Tillmanns, 1964; DeMarte and Hartman, 1974).

Rates of phosphorus uptake and excretion by the roots and leaves of submersed macrophytes were found to be dependent on the phosphorus concentrations of the medium (McRoy et al., 1972; Kussatz et al., 1984; Brix and Lyngby, 1985). Because the interstitial sediment phosphorus concentrations are very much greater than those of the water, uptake from the sediments predominates (e.g., Gabrielson et al., 1984; Smith and Adams, 1986). Critical experimental studies of phosphorus uptake, both from the sediments and from the water by submersed macrophytes that possessed their normal epiphytic microflora showed, however, that nearly all phosphorus was obtained from the sediments (e.g., Moeller et al., 1988). Nearly all of the nutrient uptake from the water was by the attached microbiota and very little was transferred to the supporting macrophyte.

As foliage matures, a gradual, partial senescence of leaves often occurs, particularly among submersed macrophytes, from which nutrients leach. Some senescing leaves slough off of the living plants and collect among detrital accumulations of the sediments. For example, the leaching of phosphorus from dead macrophytes under sterile conditions was rapid and resulted in a loss of from 20 to 50% of total phosphorus content in a few hours and 65–85% over longer periods (Solski, 1962; Nichols and Keeney, 1973; Landers, 1982). Rates of leaching were greater from roots than from leaves.

Macrophytes can also limit the release of phosphorus from sediments. Most aquatic angiosperms are herbaceous perennials with multiple cohorts (Chapter 24). Upon completion of a growth cycle, a major portion (25–75%) of phosphorus of the foliage is translocated to the rooting tissues (cf. review of Granéli and Solander, 1988). Some of this phosphorus store can be permanently interred in senescing root tissues that decompose very slowly in anaerobic sediments. Oxygen release from roots of submerged macrophytes in anaerobic sediments can lead to iron crust formation and immobilization of phosphorus by iron-coated roots (Hupfer and Dollan, 2003), impeding phosphorus release from sediments. Sediment resuspension is lower in macrophyte beds because of lower turbulence (Horppila and Nurminen, 2003).

2. *Epiphytic microbiota*

The surfaces of submersed vegetation are colonized with epiphytic microflora (Chapter 25). The physiology and growth of attached microflora are coupled to the physical and physiological dynamics of the living substrata upon which they grow. The productivity of epiphytic bacteria and algae is very high. This community acquires nutrients both from the water within and passing through the littoral zone and from the supporting "host" macrophyte tissues. Although relatively little of the total phosphorus pool within actively growing macrophytes is released, this release can be important to the epiphytes. For example, certain algal species that grow adnate to the macrophyte can obtain over 60% of their phosphorus from the macrophyte (Moeller et al., 1988). Even when phosphorus concentrations are very high in the water, some nutrients are obtained from the macrophyte simply because diffusion within the complex epiphytic community is too slow to meet demands.

The phosphorus of the littoral water is very actively assimilated by the loosely attached epiphytic periphyton, incorporated into the periphyton, and intensively recycled (e.g., Riber and Wetzel, 1987; Wetzel, 1993). The periphyton, rather than the submersed macrophytes, function as the primary scavenger for limiting nutrients such as phosphorus from the water.

During dormancy phases and senescence of aboveground macrophyte tissues, releases of phosphorus are readily utilized by the periphytic community, which tends to develop profusely during autumn and winter periods. As the senescing macrophytes with their epiphyte communities collapse to the detrital mass near the sediments, much of the released phosphorus is rapidly retained and recycled by epiphytic bacteria and algae (Fig. 15-9) (Wetzel, 1990, 1993). Therefore the attached microflora, particularly epiphytes on submersed macrophytes, can function as an effective phosphorus scavenger from the water (e.g., Howard-Williams and Allanson, 1981). The retention capacity is regulated by many environmental parameters and can be exceeded if the loading of phosphorus is very high or the rate of water flushing through this biological sieve is too rapid to allow time for uptake and retention. Rapid water movements through littoral zones, such as by natural storm events (Kairesalo and Matilainen, 1988) or from artificial human-induced runoff (e.g., Adams and Prentki, 1982), can result in flushing movements of nutrients into the receiving open waters at rates too fast to be retained biologically. Once phosphorus is cycling within the submersed macrophyte–periphyton community, however, it is improbable that much of it will be exported to the open waters.

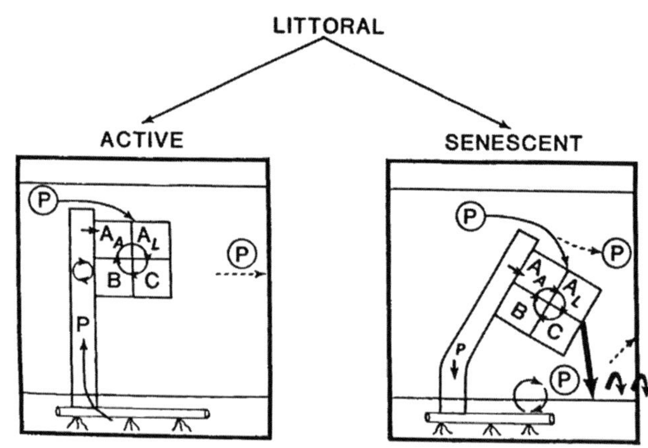

FIGURE 15-9 Fluxes of phosphorus (P) from the sediments to submersed littoral macrophytes and among the epiphytic microflora of the periphyton. A_A = adnate algae; A_L = loosely attached algae; B = bacteria; and C = inorganic or organic particulate detritus, such as calcium carbonate. *(From Wetzel, 1990.)*

G. Littoral phosphorus fluxes and loadings

Major sites of phosphorus fluxes in lakes have been examined by dividing them into three general compartments: (a) the pelagic open water and organisms of the epilimnion; (b) the littoral macrophytes and attached microorganisms; and (c) the hypolimnion and sediments. Early tracer experiments employing such compartmental analyses indicated that phosphorus in the epilimnion is extremely mobile (Rigler, 1964a). The *turnover time* of phosphorus in the epilimnion (i.e., the time in which an amount of phosphorus equivalent to the total amount in a compartment leaves that compartment and a similar amount enters it) of a small lake was found to be 3.6 days (Rigler, 1956). Within 20 min of the time it entered, over 95% of the added phosphorus was taken up by the plankton; it had a turnover time of <20 min. In Toussaint Lake (Quebec, Canada), a small acidic lake with a well-developed littoral zone of rooted aquatic plants, the littoral region was the most important contributor to the turnover and retention of phosphorus in the epilimnion; phosphorus was lost to this compartment 10 times more rapidly than to the hypolimnion and sediments, and 50 times more rapidly than its loss through the outlet (Rigler, 1956). In comparable experiments in an acidic bog lake in Nova Scotia (Canada) with extensive developments of littoral *Sphagnum* moss, nearly all of the tracer phosphorus was taken up by the *Sphagnum* and plankton (Coffin et al., 1949; Hayes and Coffin, 1951). Essentially no phosphorus reached the hypolimnion in the short term.

Rate constants of phosphorus loss from the epilimnion cannot be derived from these data. However, reanalyses of the results of several studies indicated that the

TABLE 15-9 Calculated Rate Constants of Phosphorus Transport in Three Lakes of Differing Littoral Development[a]

Lake	Area (ha)	Littoral vegetation (rank)	P turnover time (days)	Rate constants	
				k (out of epilimnion)	k (sedimentation from epilimnion)
Toussaint	4.7	1	20	0.05	0.01
Upper Bass	5.8	2	27	0.04	–
Linsley Pond	9.4	3	45	0.02	0.02

[a]*Modified from Rigler (1973); lakes are ranked in order of decreasing amount of littoral vegetation, subjectively estimated.*

turnover time of phosphorus in the epilimnion ranged from 20 to 45 days and was inversely correlated with areas of the lakes and estimates of development of the littoral vegetation (Table 15-9). In 200 L aquaria with a steady-state flow of dechlorinated water and regulated steady-state inputs of phosphorus to the aquaria, attached filamentous algae in simulated "littoral zones" removed the majority of phosphorus before it reached pelagic algae in the "epilimnion" or open water (Confer, 1972). The equilibrium exchange between phosphorus inputs to or releases from the littoral flora and from or to the epilimnetic water varies with the amounts of phosphorus input from the drainage basin (Chamberlain, 1968).

Phosphorus uptake by and release from senescing littoral macrophytes and attached algae will vary with the physical constraints of littoral development determined, in part, by basin morphometry and water exchange patterns. This relationship can shift with seasonal changes in active growth of the littoral macrophytes (e.g., Sarvala et al., 1982). The physiology and growth of attached microflora are coupled to the physical and physiological dynamics of the living substrata upon which they grow. By means of intensive recycling of limiting dissolved nutrients and gases to maintain high sustained growth and biomass, losses are minimized and imported external nutrients can be directed primarily to net growth (Wetzel, 1990, 1993). Most aquatic plants are perennial with numerous cohorts and continuous growth and turnover, particularly in regions of moderate climate where littoral growth is more or less continuous (Chapter 24). Thus the littoral complex of macrophytes and attached microbiota is often a major component of phosphorus retention and exchange with epilimnetic water, but it is constantly changing in the effectiveness of nutrient retention and loading.

H. Benthic invertebrates and the transport of phosphorus

Evidence for how effectively benthic invertebrates can influence the rates and directions of phosphorus exchange between sediments and the water has been controversial. Clearly, these organisms can enhance rates of transport of phosphorus from the interstitial waters of sediments under certain environmental conditions. In other cases their influences are minor. Examination of the transport mechanisms related to benthic invertebrate activities on phosphorus fluxes need to address (a) feeding activities of the invertebrates, intestinal decomposition, and excretion; (b) feeding and respiratory behavioral activities that alter redox gradients in sediments and phosphorus solubilities; (c) movement of organisms from the sediment habitat to other habitats with displaced phosphorus release; and (d) their size and population densities. This discussion only pertains to areas where there is sufficient oxygen to sustain benthic invertebrates, but benthic invertebrate behavior is not viewed as a major contributor of phosphorus to the water column because phosphorus transfer rates under aerobic conditions are relatively low and benthic invertebrates are absent under anoxic conditions.

Benthic macroinvertebrates are extremely diverse in their feeding, respiratory, growth, and reproductive behaviors (Chapter 21). This diversity contributes to the conflicting, inconclusive, and occasionally uncritical conclusions drawn on the effects of organisms on phosphorus and nitrogen fluxes (cf. reviews of Kamp-Nielsen et al., 1982; Andersson et al., 1988). Simply because phosphorus may be transported from interstitial waters to water immediately above the sediments does not necessarily mean that these compounds will be transported in available chemical forms for use in the trophogenic zones of lakes or rivers. One must retain an ecosystem-scale perspective in evaluating these fluxes.

Dominant macroinvertebrate groups can include the oligochaete worms, amphipods, bivalve mollusks, mayflies, and immature chironomid midges. All inhabit and feed in the sediments differently (Chapter 21). Oligochaete worms feed shallowly on surface sediments and do not burrow extensively. Amphipods inhabit and feed on the rich microbiota and detrital particles at the sediment–water interface. Bivalve mollusks move through and disturb surficial sediments and filter particles from the water. Chironomid larvae feed on surficial detrital particles or filter particles from water as they inhabit tubes within the sediments. Their undulating body movements transport water in and out of these tubes.

As populations of benthic invertebrates develop, phosphorus is incorporated into the fauna from the organic material fed upon in the sediments. Adsorption or direct assimilation of inorganic phosphorus is quantitatively insignificant, at least among the microcrustaceans (Rigler, 1964b). Ingested materials, however, are partially digested and released feces can accelerate the microbial regeneration of nutrients. Direct excretion of dissolved phosphorus, largely as inorganic phosphorus, occurs in small quantities; rates decrease markedly at temperatures below 15°C (e.g., Nalepa et al., 1983; Fukuhara and Yasuda, 1985).

The movement of water by the respiratory and feeding activities of macroinvertebrates is often invoked as a primary means of transporting interstitial water of higher nutrient content from microbial remineralization activities in the sediments to the overlying water. Considerable experimental evidence supports such advective movements of water and particles mediated by benthic animals (*bioturbation* [Aller, 1982]). For example, chironomid midge larvae and mayfly nymphs can cause increases in the phosphorus content of the overlying water (Gallepp et al., 1978; Gallepp, 1979; Fukuhara and Sakamoto, 1987; Chaffin and Kane, 2010). Larger species cause a greater effect. The release of total phosphorus increased approximately linearly (0.3–9.4 mg P m^{-2} d^{-1}) over a range of 0–6585 larvae m^{-2}. In contrast, Davis et al. (1975) showed that tubificid worms only slightly influenced phosphorus release and may actually enhance phosphorus deposition at the sediment–water interface. Most of the phosphorus released to the overlying water was as SRP, which is readily inactivated chemically under oxidizing conditions or is readily assimilated by bacteria (Johannes, 1964a, 1964b, 1964c) or algae if light is available. Under oxygenated conditions of the sediment–water interface, only when larval invertebrate densities reach extreme levels (e.g., >100,000 m^{-2}; Lindegaard and Jónasson, 1979; Wisniewski, 1991) do their activities influence the transport of phosphorus across the sediment–water interface. Densities of large mayfly nymph of 417 m^{-2} resulted in significant release of phosphorus (Chaffin and Kane, 2010).

The role of macroinvertebrate activity at the sediment–water interface is unclear in relation to the transport of phosphorus to the water. Ciliates associated with the sediments are capable of hydrolyzing dissolved organic acids and releasing inorganic phosphate into the water (Hooper and Elliot, 1953). Low oxygen concentrations, however, not only produce an unfavorable environment for most ciliates but also inhibit the release of phosphate by the cells.

Negatively phototactic cladoceran and mysid zooplankton (Chapters 19 and 20), which migrate to the sediment interface region during daylight, presumably feed actively on the rich microbiota of that region (e.g., Kasuga and Otsuki, 1984). The extent of phosphorus transport to the epilimnion during their subsequent nighttime migration is unknown but worthy of further investigation. Although the experimental addition of snails and ostracod microcrustaceans to the sediments altered plankton abundance in ponds, these benthic organisms produced no corresponding changes in uptake rates of phosphorus in open water (Confer, 1972). When benthic invertebrates emerge as adults, they may emigrate from the sediments, thereby transporting phosphorus to other compartments of the ecosystem. For example, in streams a significant upstream migration of phosphorus by fish and invertebrates has been found (Ball and Hooper, 1963). However, this displacement of phosphorus by invertebrates plays only a small part in the overall quantitative cycling of phosphorus in aquatic ecosystems.

I. Contributions of birds and fishes to nutrient loadings and cycling

Vertebrate excreta, particularly of birds, are well known to import large quantities of nutrients, especially nitrogen and phosphorus, to inland waters (Hutchinson, 1950). Nutrient loadings from waterfowl can be large as well as highly variable seasonally in relation to behavioral and migratory patterns of the use of rivers, wetlands, and lakes. A few examples illustrate the potential magnitude of nutrient loadings. In Wintergreen Lake, southwestern Michigan (USA), Canada, geese and ducks contributed 69% of all carbon, 27% of all nitrogen, and 70% of all phosphorus that entered the lake from external sources (Manny et al., 1975, 1994). In a small kettle seepage lake of Massachusetts (USA) nearly half the annual loading of phosphorus resulted from two species of gulls (Portnoy and Soukup, 1990). Up to 37% of the annual inputs of nitrogen and 95% of phosphorus of the largest natural lake in France, Grand-Lieu, emanate from large concentrations of resident and migratory birds (Marion et al., 1994). Many other examples exist, but eutrophic conditions uniformly result from such high external loadings. Waterfowl loadings must be incorporated into nutrient load–response models, as proposed by Manny et al. (1994), in order to accurately evaluate levels of eutrophication under high bird-use environments. The relative impact of waterfowl on the nutrient dynamics of a lake will be heavily influenced by the size and morphology of the lake.

Fish (Chapter 22) can impact the distribution of phosphorus and nitrogen in aquatic ecosystems in several ways. As we will see in subsequent discussions, the total amount of phosphorus in fish communities is small in comparison to other inorganic and organic pools in the lake ecosystem (water, littoral zone, sediments, and dissolved and particulate organic matter). However, fish activities can promote and accelerate the transport of phosphorus and nitrogen from sediments to overlying water. In particular, the feeding activities by benthic fishes, such as carp (*Cyprinus*) and bullheads (*Ictalurus*), can disturb sediments and increase diffusion to overlying waters (e.g., Lamarra, 1975; Keen and Gagliardi, 1981; Schaus et al., 1997). Some phosphorus and nitrogen, obtained from food and detrital sources in the sediments, can be transported into the upper strata of lakes and a portion released as soluble phosphorus and ammonia and urea nitrogen with urine and feces. When external loadings to lakes are small, as in dry summer periods, these internal redistributions can be important for planktonic algae and other microbiota (Brabrand et al., 1990). Similarly in streams, fish-derived nitrogen and phosphorus can be moved upstream for large distances by migrating salmon and subsequently released and recycled for use by microbiota, algae, and macroinvertebrates (Schuldt and Hershey, 1995; Naiman et al., 2002). Feeding history affects the N:P ratios of fish excrements. For example, rates of phosphorus excretion decreased more rapidly after feeding than those of nitrogen, and hence the N:P ratio of excrement increased with time from feeding (Mather et al., 1995).

As discussed earlier, fish predation–induced shifts in the size structure of zooplankton communities can alter rates of sedimentation of seston from the epilimnetic trophogenic zone. Thus loss rates of phosphorus via sedimenting seston can decline somewhat with a shift by predation to smaller-sized plankton (Mazumder et al., 1989, 1992; cf. also Kairesalo and Seppälä, 1987; Pérez-Fuentetaja et al., 1996). These effects appear to be significant only in very oligotrophic lakes.

Other large vertebrates can influence phosphorus and other nutrient loadings to small waters. For example, hippopotamuses graze on land at night and defecate considerable amounts of leachable phosphorus into lakes and rivers during the day (Kilham, 1982). Moose (*Alces alces*) forage in wetlands and drier upland areas during the summer, which can translocate phosphorus in feces from one area to the other (Laforge et al., 2016).

VI. Phosphorus cycling within the epilimnion

The classical studies of Einsele (1941) and many subsequent theoretical and applied analyses of the circulation and fate of phosphorus in the open water of oligotrophic epilimnia have shown that phosphorus is very rapidly incorporated into planktonic algae and bacteria. Open-water recycling helps limit daily phosphorus losses to sediments. A recent study has focused on rates of movement of phosphorus among biologically important forms in lake water. Two broad classes of investigations have resulted: (1) biotically mediated phosphorus-transfer mechanisms; and (2) abiotic complexation reactions. The former includes studies on the transfer of phosphorus among seston and various forms of dissolved phosphorus, including the biotically mediated formation of colloidal phosphorus and the enzyme-mediated utilization of dissolved phosphate esters. The latter category includes studies on the sorption and desorption of phosphorus to dissolved humic compounds, colloidal calcium carbonate, and other particles.

The early analyses of Frank Rigler and his coworkers have generated a framework that has stimulated considerable further study of phosphorus dynamics (Lean, 1973a, 1973b; Rigler, 1973; Lean and Rigler, 1974). They proposed a quantitative, steady-state model of phosphorus exchanges between different phosphorus pools in epilimnetic water in the summer, with the following composition (Fig. 15-10): (a) *particulate phosphorus*, the fraction removed when the water is filtered through a 0.5-μm pore-size filter, which contains the bulk of the phosphorus; (b) *reactive inorganic soluble orthophosphate* (PO_4^{3-}), which has an extremely short turnover time especially during the summer (Rigler, 1964a); (c) a low-molecular-weight (approximately 250 Da) *organic phosphorus compound (XP)*; and (d) a soluble macromolecular *colloidal phosphorus* of a molecular weight >5,000,000 Da. An exchange mechanism predominates between the inorganic phosphate and the particulate fraction, but some phosphorus is excreted by the microorganisms in the form of the low-molecular-weight compound (XP). Polycondensation of the low-molecular-weight compound (XP) produces the high-molecular-weight colloidal compound. Both fractions, but primarily the latter, release phosphate in the soluble inorganic fraction, which then becomes available to the plankton. Fig. 15-10 illustrates phosphorus dynamics in an epilimnetic zone in midsummer.

Sedimentation of particulate phosphorus (as cellular detritus, aggregated colloids or adsorbed to organic and inorganic sestonic particles) represents a slow but continuous loss from the epilimnion. Thus, during the active growth of phytoplankton and heterotrophic bacteria, phosphorus must be continuously replaced through input from influents or from the littoral zone or recycled before settling. Inorganic phosphate can be released by actively growing as well as senescent cells, enzymatic hydrolysis of polyphosphates, and zooplankton ingestion and subsequent excretion (see

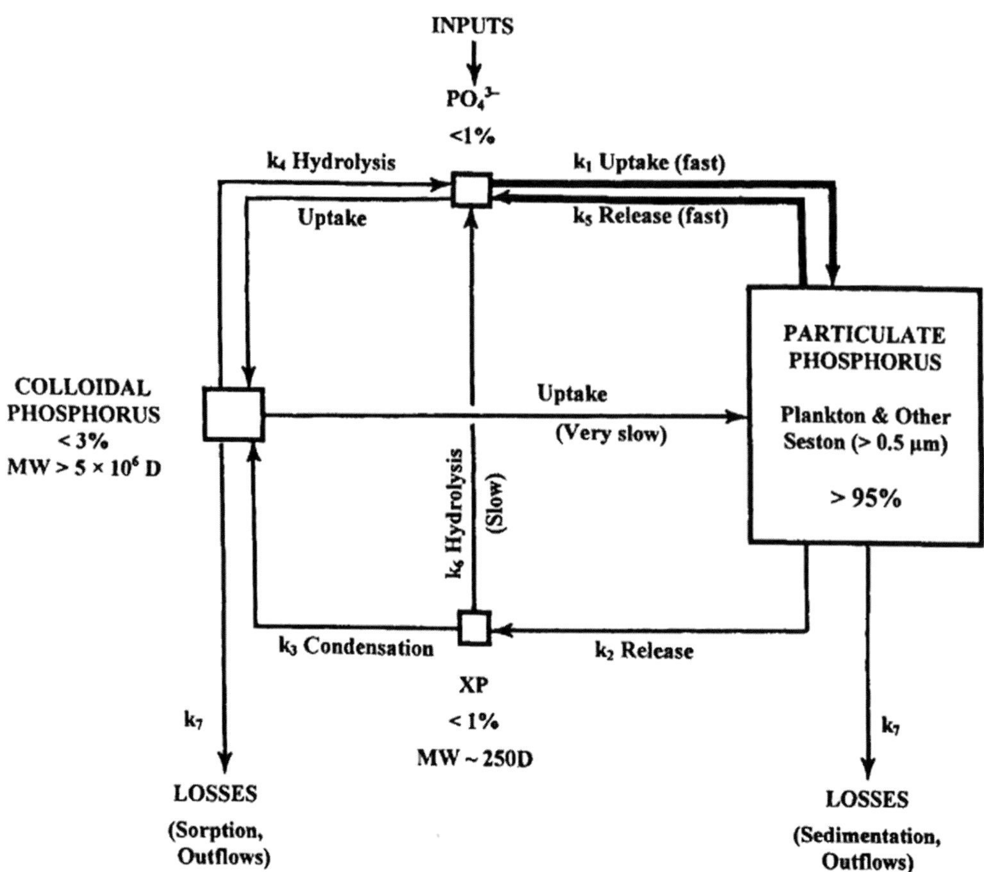

FIGURE 15-10 Phosphorus movement within the epilimnetic zone of lakes, elucidating the exchange mechanisms between the phosphate and the particulate fractions. Rate constants describing movement between pools are indicated by k_i. MW is molecular weight; D is Dalton; XP is a low molecular weight compound (see text for details). *(Modified and expanded from Lean, 1973a, 1973b.)*

the following discussion) (Kuenzler, 1970; Fogg, 1971; Lean, 1973a; Tarapchak and Herche, 1985; Rijkeboer et al., 1991). In experimental systems much of the colloidal fraction settles out and becomes biologically unavailable after 1–5 days. Other workers, however, have shown that high-molecular-weight colloidal phosphorus compounds can be used directly as a phosphorus source by algae possessing alkaline phosphatase enzymatic activity (Paerl and Downes, 1978). These rates of particulate loss are congruous with those found in natural waters to which phosphorus additions have been made.

From these initial insights into and subsequent studies on the cycling of phosphorus in open water, it is apparent that sestonic organic phosphorus must be separated into at least two fractions: (a) a rapidly cycled fraction, which is exchanged with soluble forms. In this fraction phosphate is transferred rapidly through the particulate phase to low-molecular-weight compounds; and (b) a fraction of the sestonic phosphorus that is released more slowly. The transfer of colloidal phosphorus material from the phytoplankton to the water appears within minutes after the uptake of soluble phosphate by bacteria and algae (Lean and Nalewajko, 1976; Paerl and Lean, 1976; Cembella et al., 1984). In addition to the formation of high-molecular-weight fibrillar and amorphous particles of colloidal size (0.05–0.1 μm), bacteria and algae excrete significant amounts of dissolved organic phosphorus compounds. As discussed later, the uptake of phosphate is greater by bacteria than algae, but since algal biomass is greater than bacterial biomass, algal phosphate incorporation usually dominates uptake and release pathways in the epilimnion.

The uptake and turnover kinetics of phosphorus have been studied in a number of lakes of low to high productivity and at different seasons of the year (Table 15-10). Phosphorus turnover rates are extremely rapid in the summer periods of high demand, warm temperatures, and relatively low loading inputs but become as much as two orders of magnitude slower during the winter periods. Phosphorus turnover rates are faster under more oligotrophic conditions of greater phosphorus deficiency (e.g., Peters, 1979; Cembella et al., 1983, 1984).

Phytoplankton contain many complex phosphorus esters (e.g., sugar phosphates, nucleotide phosphates,

TABLE 15-10 Turnover Times of Phosphate in Freshwaters[a]

Lake	*Turnover times (min)*	*Source*
Lago Maggiore, Italy		
Epilimnion, summer	10–200	Peters (1975)
Epilimnion, winter	200–10,000	
Hypolimnion	1000–100,000	
15 European lakes (epilimnia)	4–74,400	Peters (1975)
Southern Ontario lakes (epilimnia)		Rigler (1964); Planas and Hecky (1984)
Summer	2–14	
Winter	10–10,000	
Lake Kinneret, Israel		Halmann and Stiller (1974)
Summer	11–280	Halmann and Elgavish (1975)
Winter and spring	40–545	
East African lakes	1–1000	Peters and MacIntyre (1976)
Lake 227, Ontario		Levine (1975)
Summer	0.4–15	
Winter	120–11,700	
Lake 302, Ontario (summer)	5–26	Levine (1975)
Lake Memphremagog, Quebec		Peters (1979)
Summer	9–26	
Winter	83–770	
Jordan River, Israel (winter)	630	Halmann and Stiller (1974)
Southern Indian Reservoir, Manitoba	150–91,000	Planas and Hecky (1984)
Albertan lakes		Prepas (1983)
Shallow, mixed		
Spring	17–1020	
Summer	2–2160	
Deep, stratified		
Spring	3–42	
Summer	3–16	
Humic lakes, Finland		Jones (1990)
Summer	60–4800	
Winter	100–72,000	

[a]*The turnover time is the reciprocal of the turnover rate, which is the fraction of phosphate incorporated by the seston per unit time.*

and polyphosphates). These low-molecular-weight phosphorylated compounds can be released into freshwater upon cell death or by the active release by microorganisms (Berman, 1970; Kuenzler, 1970). A very large portion (>50%) of the phosphorus content of algal cells can be lost during sedimentation over a period of several days (Otto and Benndorf, 1971), but daily losses are typically less than 2%.

Enzyme-mediated hydrolysis of naturally occurring phosphate esters is one of the most plausible mechanisms for phosphorus regeneration in the epilimnion. Direct enzymatic breakdown of DIP compounds and polyphosphates to DIP by relatively nonspecific *phosphatase (phosphomonoesterase)* activity has been the subject of intensive study. Numerous investigations demonstrate that membrane-bound algal phosphatase activity increases as phosphorus deficiency becomes acute. Increases in *alkaline phosphatase activity* (APA) under conditions of phosphate limitation, either from increased rates of enzyme synthesis or from derepression of preexisting enzyme, can enhance algal competitive ability by permitting algae to utilize organophosphate or inorganic polyphosphate substrates as alternative phosphorus sources (e.g., Berman, 1970; Jones, 1972; Heath and Cooke, 1975; Jansson, 1976; Stevens and Parr, 1977; Pettersson, 1980; Wetzel, 1981; Jansson et al., 1988; Wynne and Rhee, 1988; Boon, 1994; Newman et al., 1994). Although free, dissolved, alkaline phosphatase is short-lived (Reichardt et al., 1967; Pettersson, 1980), substantial quantities of APA can be found in the dissolved phase (Pettersson, 1980; Wetzel, 1981; Stewart and Wetzel, 1982a; Cotner and Heath, 1988; Rai and Jacobson, 1993). The importance of phosphatase activity to phosphorus cycling in lakes is further confounded because bacterial production of alkaline phosphatase occurs in pelagic environments (Jones, 1972) but need not be induced strictly by phosphate limitation (e.g., Wilkins, 1972; Chróst and Overbeck, 1987; Jansson et al., 1988).

Soluble APA and APA associated with algae and nonalgal particulate matter are highly variable seasonally and spatially (Wetzel, 1981; Stewart and Wetzel, 1982a; Cotner and Wetzel, 1991a, 1991b). Epilimnetic APA is commonly high during the spring and summer periods of maximal phosphorus demand. During years when spring circulation is incomplete or a lake experiences temporary meromixis, extremely high APA and phosphorus limitations were found in the phytoplankton during spring and summer months (Wetzel, 1981).

Phosphomonoesters (PME) were found in lake water in substantial quantities (up to 55 μg PME-P L^{-1} in eutrophic Twin Lakes, Ohio; Heath and Cooke, 1975). An inverse relationship was found between PME concentrations and APA during summer stratification. Potential hydrolysis rates of PME of >0.05 mmol h^{-1} by APA were found. These values are similar to those found in marine systems using algal alkaline phosphatase as the hydrolytic enzyme (Rivkin and Swift, 1980). These results, in addition to widely differing chemical analyses of the complex phosphorus pool in lake water (e.g., Lean, 1973a, 1973b; Lean and Nalewajko, 1976; Paerl and Downes, 1978; Downes and Paerl, 1978; Francko and Heath, 1979, 1981, 1982; Boavida and Marques, 1995), suggest that the composition of the soluble phosphorus pool varies greatly among different lake types. Although some lakes may approach the model of phosphorus cycling presented in Fig. 15-10, other lakes may possess a phosphorus cycle dominated by the production and biotically mediated hydrolysis of PME.

Physicochemical mechanisms can also influence the cycling of phosphorus and, as a result, affect primary productivity. Certain complex humic materials of both low and high molecular weights contain phosphorus, such as phosphate esters and inositol hexaphosphate (e.g., Anderson and Hance, 1963; Koenings and Hooper, 1976). High-molecular-weight humic compounds can be photoreduced by low-intensity ultraviolet light, as occurs in sunlight (see Photochemical Reactions in Chapter 10; Francko and Heath, 1979; Stewart and Wetzel, 1981). Orthophosphate adsorbed to ferric-humic compounds can be released by UV-induced photoreduction of ferric iron (Francko and Heath, 1982; Jones et al., 1988). Low-molecular-weight humic compounds can release orthophosphate upon exposure to alkaline phosphatase.

Phosphorus-humic complexes can interact with phytoplankton and bacterioplankton in complex ways. For example, rates of carbon assimilation by phytoplankton were reduced when phytoplankton were exposed to low concentrations of dissolved humic materials of littoral and wetland origin (Stewart and Wetzel, 1982b). The effects were species selective and greatest among smaller (1–5 μm) phytoplankton growing under low light conditions. The production of alkaline phosphatase activity was greatly stimulated by low concentrations of dissolved humic compounds of low molecular weight. Evidence indicates that these humic materials can act as sequestering agents for phosphate ions, organophosphorus compounds, and iron, thereby reducing phosphorus and possibly iron availability to phytoplankton (Jackson and Hecky, 1980; Chow-Fraser and Duthie, 1983; Jones et al., 1988, 1993; de Haan et al., 1990; Münster, 1994; Shaw, 1994). These relationships can be important to both phosphorus cycling and phytoplankton productivity, for major rain events can simultaneously introduce substantial quantities of phosphorus and dissolved humic materials into the pelagial areas of small and moderate-sized lakes.

Additionally noteworthy is that in acidified lakes where aluminum (Al(III)) and iron (Fe(III)) concentrations can be markedly increased, the production of acid phosphatases is enhanced (Jansson, 1981). Although the Al(III) and Fe(III) at low pH do not affect the enzymes directly, the metal ions combine with the phosphate group on the substrate and inhibit enzymatic hydrolysis. The organisms may compensate, at considerable energetic cost, for these losses by increased production of phosphatases.

A. Phosphorus losses by outflow, inactivation, and sedimentation

Accurate nutrient budgeting in a lake or river ecosystem requires quantification of routes of reduced availability for utilization by biota as well as direct losses. Direct losses by movement with water in outflows from lakes and rivers occur in inorganic and organic dissolved and particulate forms. Knowledge of concentrations and discharge volumes allows reasonably accurate evaluations of losses by outflow. Losses by seepage through the sediments, particularly sediments near the shore line, are quite variable (Chapter 5). However, in most lakes the basin seal is relatively complete and seepage outflows are small in relation to surface outflow. Most of the losses of phosphorus by seepage are via dissolved inorganic and organic compounds.

Sedimentation of nutrients occurs in settling (a) living and dead biota and (b) inorganic particles that are imported to the water from the catchment or are formed within the water and sorb nutrients. As particles settle to the bottom, they can experience partial decomposition and microbial recycling of nutrients to the water or encounter altered chemical conditions that change nutrient sorptive properties of the particles. Once at and within the sediments, periodic resuspension occurs, especially from surficial sediments in shallow waters (Dillon et al., 1990). Major resuspension occurs at times of complete water circulation, during which several centimeters of the surficial sediments can be resuspended into the water column (e.g., Davis, 1973; White and Wetzel, 1975).

Most surface waters of the world are bicarbonate dominated and many, perhaps a quarter of them, lakes are very hard waters with sufficient bicarbonate and calcium to experience periodic precipitation of $CaCO_3$ (Chapters 12 and 13). Precipitation is induced and accelerated by high pH, and these conditions are often associated with microzones adjacent to rapidly photosynthesizing cells of the phytoplankton, attached algae, and submersed macrophytes (Otsuki and Wetzel, 1972, 1974; House and Donaldson, 1986). Much of the phosphorus associated with the $CaCO_3$ is adsorbed onto the crystals; a small portion is incorporated by coprecipitation into the growing crystals. Inhibitors of $CaCO_3$ growth such as magnesium and iron reduce the amount of phosphorus coprecipitated on calcite (House et al., 1986).

The morphology of calcite crystals can change markedly under different environmental conditions (Koschel, 1990, 1997; Raidt and Koschel, 1993). In particular, the complexity of the crystals increases with increasing supersaturation, and they shift from rhombohedral to complex dendritic and columnar crystals (Fig. 15-11). As the crystal complexity increases, the adsorptive surfaces also increase and are potentially more effective in the removal of nutrients and dissolved organic compounds.

Because phosphorus is coprecipitated with and adsorbed to nucleating $CaCO_3$, the process of formation and sedimentation of $CaCO_3$ can markedly alter phosphorus availability to biota. For example, about 35% of the annual total phosphorus removal from the epilimnion of Lake Constance in southern Germany was associated with autochthonous calcite precipitation (Kleiner, 1988), some 25% in eutrophic Wallersee of Austria (Jäger and Rohrs, 1990), and about 30% in hypereutrophic Onondaga Lake of central New York (Effler et al., 1996). In hypereutrophic prairie lakes, very high concentrations of phosphorus are stripped from the epilimnetic waters by periodic precipitation of $CaCO_3$ (Murphy et al., 1983).

B. Phosphorus and nitrogen uptake and recycling by protists and zooplankton

As we have seen, the phosphorus content of bacteria is about 10 times that of the algae. Bacteria are commonly responsible for 80% or more of the net uptake of phosphorus (Hessen and Andersen, 1990; Jürgens and Güde, 1990; Vadstein et al., 1993, 1995). Thus bacterioplankton (Chapter 23) are a potentially rich source of phosphorus.

Many studies have examined the ingestion of algal and bacterial particles by macrozooplankton and the potential regeneration of phosphorus by the feeding activities and from excreted feces, as discussed next. It should be emphasized that such uptake and excretion of phosphorus and other nutrients is a *recycling* process, not a *de novo* source of nutrients. The digestive and metabolic processes alter the storage time of the nutrients already in the ecosystem and potentially render them more available for reuse by other organisms.

Where the uptake and recycling by heterotrophic microflagellates were compared to the large zooplankton, the heterotrophic flagellates feeding on phosphorus-rich bacteria regenerated two-thirds or more of the phosphorus released by grazers (Jürgens and Güde, 1990; Vadstein et al., 1993, 1995). Phosphorus and nitrogen regeneration rates varied, however, among fast- and slow-growing species of heterotrophic flagellates (Eccleston-Parry and Leadbeater, 1995) and were much higher during exponential growth phases than during stationary growth phases (Andersen et al., 1986; Nakano, 1994a; Ferrier-Pagès et al., 1998). A portion of the bacteria senesce and lyse and thereby release nutrients. However, many of the bacteria function as consumers of phosphate, because of their high requirements for phosphorus, and thus serve as

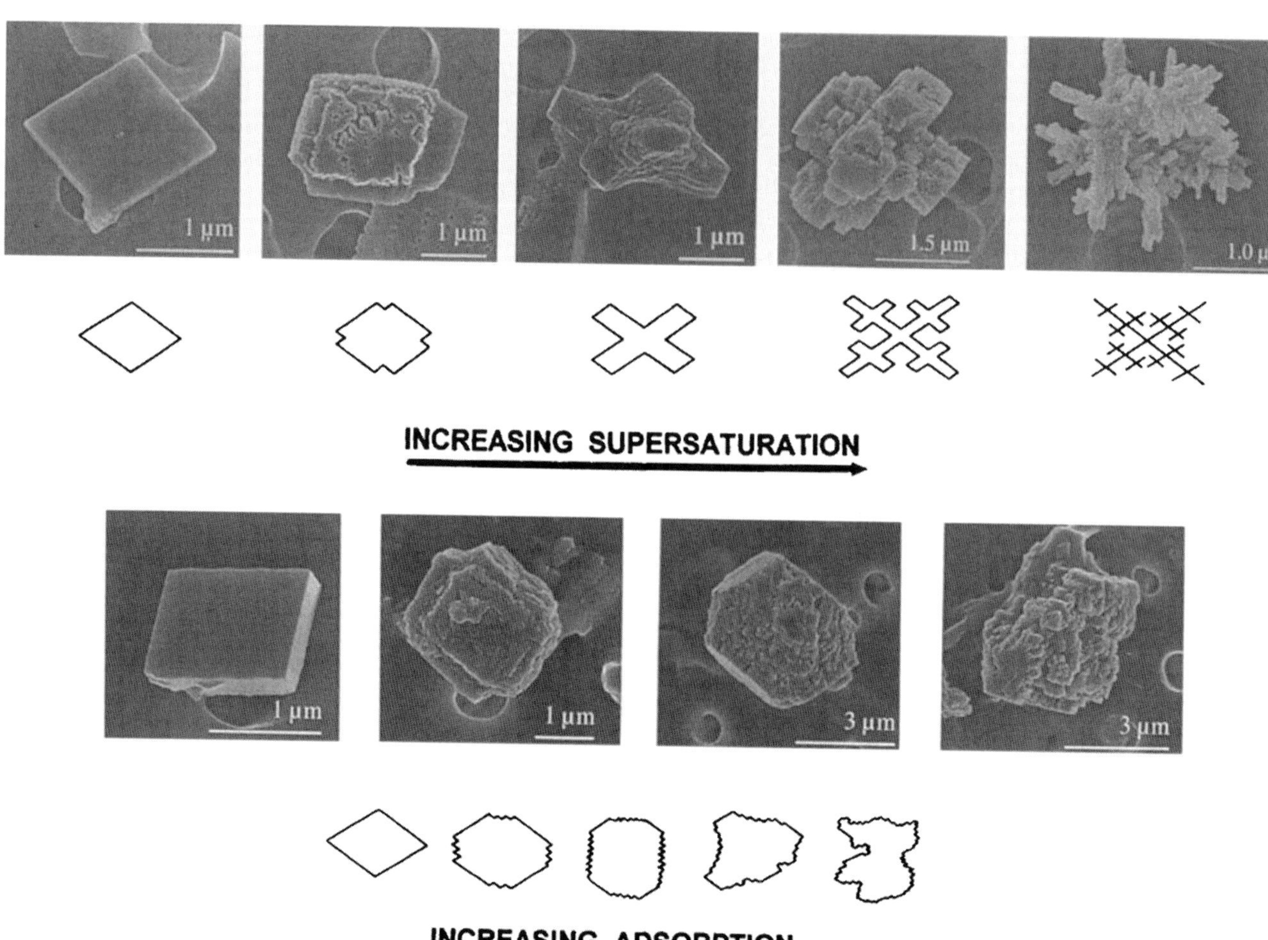

FIGURE 15-11 Calcite crystals from hardwater lakes under differing conditions of chemical saturation. Marked alterations of surface areas and adsorption also occur with changing morphology of the sedimenting crystals. *(Photomicrographs courtesy of Prof. R. Koschel, Neuglobsow, Germany.)*

phosphorus-rich particles for feeding by flagellates, other Protista, and larger zooplankton. The release rates of phosphorus from the bacterivorous feeding of flagellates can account for major recycling sources for phytoplanktonic primary production (e.g., Miyajima, 1992; Rothhaupt, 1992; Nakano, 1994b; Sterner et al., 1995). This recycling by planktonic consumers can be a significant source for the growth of bacteria and phytoplankton, particularly late in the summer period of stratification during periods of low inflow.

The uptake of nutrients, particularly phosphorus and nitrogen, by ingestion of food particles by herbivorous zooplankton and protists can be large at certain times of the year (Chapters 19 and 20). During egestion and subsequent leaching from feces, nutrient release occurs in the form of soluble phosphate ions (>70% of dissolved P), ammonium nitrogen (>80% of dissolved N), and some organic phosphorus and nitrogen compounds, particularly urea (usually <10% of dissolved N) (Vargo, 1979; Lehman, 1980a; Mitamura and Saijo, 1980; Madeira et al., 1982; Lenz et al., 1986; Urabe, 1993; Ferrier-Pagès and Rassoulzadegan, 1994; Nakano, 1994a, 1994b; Goma et al., 1996; Ferrier-Pagès et al. 1998). The release rates are highly variable, but the rates of N release from crustacean zooplankton (range, 0.2–2 μg N mg dry weight^{-1} h^{-1}) increased with increasing food abundance, whereas the rates of P release (0.05–0.3 μg mg dry weight^{-1} h^{-1}) and N:P release ratios were affected directly by the N:P ratio of the food. Release rates of ammonium N were similar per unit biomass for ciliates and phagotrophic flagellates.

When prey food organisms were saturated with phosphorus, release rates from grazing zooplankton were in the range of 1.1–1.5 μg P (mg DW)$^{-1}$ h^{-1} (Olsen and Østgaard, 1985; Lehman and Naumoski, 1985; Jürgens and Güde, 1990; van Donk et al., 1993). When planktonic bacteria and algae are phosphorus starved and contain less phosphorus, however, release rates from grazers are about 0.05 μg P (mg DW)$^{-1}$ h^{-1}, <5% of the rates under P-sufficient conditions. Detrital particles were also found to contain very low concentrations of phosphorus (Olsen et al. 1986).

Appreciable quantities of phosphorus, nitrogen, and carbon remain in the particulate fecal materials and dead zooplankton that settle out of the water column. Loss rates of nitrogen or phosphorus from molting or dead crustacean and rotifer zooplankton were the most rapid (>2% per h) during the first 24 h following death (Krause, 1961, 1962; Scavia and McFarland, 1982; Ravera and Gatti, 1993). Thereafter, rates of nutrient and carbon losses decreased but usually >75% was lost in the subsequent 6 days. Release rates were somewhat faster in the warmer water strata than in the hypolimnion.

The quantitative significance of these sources of nutrients from micro- and macrozooplankton to phytoplanktonic productivity is unclear. Where examined carefully, zooplanktonic excretion of nitrogen likely constitutes only a few percent (<5%) of that required for gross production on an annual basis. Recycling of phosphorus could be appreciably higher, particularly during certain periods of extensive zooplanktonic grazing. The difficulty is that very few quantitative measurements of rates of nutrient regeneration by bacteria, protists, and other processes (e.g., photolysis) have been examined simultaneously with the estimates of release from zooplankton. For example, in eutrophic Lake Loosdrecht (the Netherlands) in summer, <30% of the phosphorus required by phytoplankton emanated from zooplankton excretion (Den Oude and Gulati, 1988). Most regeneration of phosphorus resulted from excretion and release from bacteria and algae.

To ignore nutrient regeneration by rotifers and protists as insignificant in comparison to cladocerans is not warranted. Although the individual biomass of the smaller organisms is less by two-thirds to an order of magnitude or more, densities are often several orders of magnitude greater than those of cladocerans. Specific excretion rates are generally higher as biomass decreases (Taylor, 1984). Estimates indicate that microconsumer protists and rotifers can regenerate 50% or more of the total phosphorus and nitrogen that is regenerated by the total zooplankton community (Ejsmont-Karabin, 1983; Lenz et al., 1986; Gulati et al., 1989; Den Oude and Gulati, 1988; Taylor and Lean, 1991).

Nutrients released by zooplankton are rapidly reassimilated by phytoplankton. Under certain circumstances, when the concentrations of phosphorus in the trophogenic zone are low, phosphorus and nitrogen regenerated by the herbivorous zooplankton can constitute a significant portion of the phosphorus and nitrogen requirements of the algae (reviewed by Lehman, 1980a, 1980b; Gulati et al., 1995). For example, based on nutrient budgets and *in situ* measurements, epilimnetic zooplankton in Lake Washington (USA) recycle 10 times more phosphorus and three times more nitrogen to the surface-mixed layer during the summer months (June to September) than enters from all external sources combined. Examples of the rates of regeneration of phosphate and ammonia by zooplankton are given in Table 15-11.

The magnitude of nutrient uptake and recycling depends on the nutritional status of the algal cells (Lehman, 1980a; Blažka et al., 1982). Compared with nutrient-sufficient cells, nutrient-deficient algal cells assimilated phosphate and ammonia more rapidly and exhibited lower rates of remineralization.

Zooplankton grazing accelerates rates of phosphorus and nitrogen regeneration. It is also probable that planktonic bacteria, particularly those less than 1 μm in size, and very small algae immediately sequester released phosphorus (Dodds et al., 1991). Even though average phosphorus (and often combined nitrogen) concentrations in the epilimnion can be low, the release of nutrients from zooplankton can cause a microheterogeneous patchiness of nutrient concentrations, since the zooplankton distribution is far more clumped than the phytoplankton distribution.

Less clear is the importance of nutrient recycling to the overall rates of algal productivity on an annual basis. Removal of algae and bacteria by grazing zooplankton fluctuates widely on a seasonal and spatial basis (see also Chapters 19 and 20). Moreover, grazing of algae is often size selective. This selectivity may enhance the availability of nutrients to those algae that are not as effectively grazed. For example, assimilation rates of phosphorus by various dissolved and particulate fractions and different zooplankton species were examined in a mesotrophic lake (Lyche et al., 1996). The microconsumers (<20 μm) exhibited the largest specific phosphorus assimilation (0.28 day^{-1}) and regenerated about 45% of assimilated phosphorus. The larger microcrustaceans, such as *Daphnia*, acted largely as sinks for phosphorus, whereas the cyclopoids regenerated rapidly the small amounts of phosphorus assimilated because of their active feeding on the microconsumers. The zooplankton contributed significantly to the sedimentation of phosphorus at this time of the year (late summer) (c. 1% day^{-1}).

The interactions are furthermore complicated by seasonal patterns of selective predation on zooplankton by planktivorous fishes. As indicated in Chapter 22, planktivorous predation often results in reductions in zooplankton size and therefore grazing rates on phytoplankton. The smaller zooplankton can have higher biomass-specific excretion rates of phosphorus than larger zooplankton (Vanni and Findlay, 1990). Although fish excretion of phosphorus is large per unit biomass, the collective biomass is very small (see also Chapter 22). In other studies rates of phosphorus release from

TABLE 15-11 Regeneration Rates of NH_4–N and PO_4–P by Some Freshwater Zooplankton[a]

Species[b]	Temperature (°C)	μg dry wt animal^{-1}	μg (mg dry wt)$^{-1}$ day^{-1} N	P	Source
Daphnia magna (F)	20–22	250.0	–	0.8	Rigler (1961)
D. rosea (F)	20–22	13.4	–	2.0[c]	Peters and Lean (1973)
D. pulex (F)	15, 20, 25	26.0	5.1	–	Jacobsen and Comita (1976)
D. pulex (F)	20	20	27.9	5.6	Lehman (1980a)
D. longispina (U)	15–22	200–250	–	3.4	James and Salonen (1991)
Daphnia spp. (U)	15	10–48	–	31	Olsen and Østgaard (1985)
Mysis relicta (U)	4–7	–	4.5	3.3	Madeira et al. (1982)
Diacyclops thomasi	20	4.5	–	5.0	Bowers (1986)
Mostly Cladocera (U)	5	1.0	–	2.0	Barlow and Bishop (1965)
	20	0.9	–	4.6	
Mostly Copepoda (U)	5	3.6	–	1.9	
	20	2.9	–	6.0	
Rotifers (U)	15–25	–	–	0.10–83	Ejsmont-Karabin (1983)
Euchlanis dilatata	18–19	0.5–1.0	8.8	2.9	Gulati et al. (1989)
Mixed Cladocera,					
Copepoda, rotifers (U)					
Epilimnion	3–24				Ferrante (1976)
>0.3 mm			–	58.4[d]	
<0.3 mm			–	132.8	
Hypolimnion	4–9				
>0.3 mm			–	120.1	
<0.3 mm			–	170.1	
Mixed epilimnetic zooplankton (U)					
27–28 Sep. '77	Approx. 20		11.9	1.7	Lehman (1980b)
29–30 Sep. '77			12.4	2.0	
30 Sep.–1 Oct. '77			16.7	4.3	
15–16 Sep. '77	Approx. 20		19.2	2.0[e]	
				3.1[f]	
Mixed zooplankton (U)	20	10–70	–	13	Bartell (1981)
	18–20	1–5	–	4.1	Den Oude and Gulati (1988)
	20	–	6–25	1–7.2	Urabe (1993)
	<15	–	0–85	6–55	Carillo et al. (1996)

[a]*From the sources cited, especially Lehman (1980b); cf. also Taylor and Lean (1981).*
[b]*Water for the incubations was either filtered (F) or unfiltered (U) to remove phytoplankton.*
[c]*Total P, including dissolved organic P.*
[d]*Annual mean excretion rates; values fluctuated two orders of magnitude over an annual period.*
[e]*Soluble reactive P.*
[f]*Dissolved P, including organic P.*

pelagic zooplankton were reduced during June and July when fish predation reached its seasonal maximum and allochthonous supplies were at seasonal lows (Bartell and Kitchell, 1978; Schindler et al., 1993).

Finally, it must be recalled that most of the phosphorus is bound in particulate matter: algae, bacteria, and detritus. Rates of nutrient regeneration from these sources as the phytoplankton senesce and die are poorly understood. Certainly, these organisms with short generation times rapidly recycle nutrients in the trophogenic zone before they settle to lower depths. The relative magnitude of their recycling to that of the zooplankton is not clear but is variable spatially, seasonally, and diurnally and is probably affected by mean depth with greater recycling within the water column in deeper lakes (Schindler et al., 1993). Zooplankton grazing and concomitant nutrient recycling can enhance rates of primary productivity and algal growth during periods when heavy nutrient demands exceed loading inputs to the trophogenic zone from external sources.

C. Nearshore nutrient shunt

Even in large lakes, dense inshore populations of benthic filter feeders can redirect nutrients in particulate form and energy (i.e., organic carbon) away from offshore pelagic zones to inshore sediments. Invasive dreissenid mussels appear to be responsible for the higher content of nutrients and organic matter in nearshore sediments compared to offshore in Lake Erie (Hecky et al., 2004; Pennuto et al., 2014). Dreissenid filtering rates were high enough to lower mean ice-free chlorophyll *a* concentrations increase Secchi depths and increase minimum volumetric hypolimnetic dissolved oxygen concentrations in Lake Erie (Canada/USA) and Lake Simcoe (Ontario, Canada) (Conroy et al., 2005; Barbiero and Tuchman, 2004; Li et al., 2018). Hecky et al. (2004) conclude that dreissenids have not only substantially modified the inshore physical habitat, but their ecological activities may have also changed the availability of resources to other species at the ecosystem scale.

VII. Phytoplankton requirements for phosphorus

Compounds containing phosphorus influence nearly all phases of metabolism, particularly in the energy transformation of phosphorylation reactions during photosynthesis and respiration (Rao, 1997). Phosphorus is required in the synthesis of nucleotides, phospholipids, sugar phosphates, and other phosphorylated intermediate compounds. Further, phosphate is bonded, usually as an ester, in a number of low-molecular-weight enzymes and vitamins essential to algal metabolism.

The importance of phosphorus in phytoplankton physiology is of special interest to the limnologist because phosphorus is the least abundant element of the major nutrients required for algal growth in a large majority (c. 75%) of freshwaters. Furthermore, the most important form of phosphorus for plant nutrition is ionized inorganic PO_4 (orthophosphate). Although phytoplankton can utilize organic phosphate esters, such as glycerophosphates, this ability is variable among species, as is their ability to obtain phosphate groups enzymatically or by the release of exoenzymes to the water for catalytic dissociation (see reviews of Krauss, 1958, Provasoli, 1958; Overbeck, 1962, 1963; Cembella et al., 1984). Most phosphorus released to the water during the active growth of phytoplankton is inorganic soluble phosphate and organic esters, which are, in turn, very rapidly recycled. During cell lysis and decomposition, most of the phosphorus released by phytoplankton is organic and undergoes bacterial degradation (Krause, 1964).

A. Phosphorus concentrations required for growth and uptake

Phosphorus transport (uptake) across the outer cell membrane generally follows *Michaelis–Menten enzyme kinetics*, where the phosphorus uptake rate, V, versus extracellular concentration (S) curve is described by a rectangular hyperbola (Rhee 1973).

$$V = V_{\max} / (K_m + S) \quad (15\text{-}1)$$

where K_m is the half-saturation constant defined as the concentration where uptake is half its maximum uptake rate, $V_{\max}$. The initial slope of phosphorus uptake rate versus extracellular concentration, defined as $V_{\max}/K_m$, is an indicator of the relative competitive ability of a species to acquire phosphorus at very low concentrations (Molot and Brown, 1986). The half-saturation constant alone is not an indicator of competitive ability at low phosphorus, although it is sometimes used as such. At steady state, population growth rate versus substrate concentration also follows a rectangular hyperbola and kinetic equations similar to uptake equations can be used (the Monod growth equations) (Droop, 1973; Jiang et al. 2019). Cyanobacteria appear to have higher phosphate uptake affinities than eukaryotic algae and hence would have a competitive advantage at low phosphate concentrations unless limited by something other than phosphate (Molot and Brown, 1986; Molot et al., 2014).

B. Phosphate uptake and light

The uptake of phosphate from the water is influenced by numerous external factors. Among many phytoplankton studied in culture, the initial absorption of phosphate and subsequent uptake increase with light intensity and are also affected by spectral quality (Cloern, 1977; Esteves et al., 2020). Natural populations are more or less synchronized by daily photoperiods in their productive and growth cycles, and protein synthesis predominates in the initial portion of the daily light cycle (Soeder, 1965). In the green alga *Scenedesmus* the release of phosphate from the algae has been found to be greatest during the latter portion of the light period, both prior to and during cell division in the dark (Overbeck, 1962).

Absorption of phosphate per cell in the light is generally dependent on phosphorus concentration in the medium, within a range that is rather specific for a given algal species. Natural phytoplankton and bacterial populations, however, adapt to phosphorus-limited environments by synthesizing enzymatic transport systems capable of uptake at very low concentrations (Tarapchak and Herche, 1986). Uptake by phosphorus-deficient cells in the dark is independent of phosphate concentrations (Kuhl, 1962, 1974).

When phosphate is supplied in excess, phosphorus is often incorporated and stored as polyphosphates (Zaiss, 1985). Nitrogen and carbon limitation can lead to intensive polyphosphate production, and such phosphate accumulation is stimulated by calcium (Siderius et al., 1996). When phosphate concentrations become limiting, production of the polyphosphate-forming enzyme, polyphosphate kinase, decreases and the phosphate stores can be degraded in the light.

Phosphorus uptake rates are quite specific to groups of phytoplankton and often to species within a genus. Whereas nitrate uptake is independent of phosphate concentrations, optimal growth of many phytoplankton occurs at higher concentrations of phosphate when nitrate, rather than ammonia, is the nitrogen source. Phosphate uptake and transport through the cell membrane, the rate-limiting step, by Cyanobacteria is strongly regulated by cation concentrations, particularly Ca^{2+} and Mg^{2+} (Falkner et al., 1980; Budd and Kerson, 1987). Thus phosphate uptake by Cyanobacteria should be favored in hard waters with an abundance of divalent cations unless the Cyanobacteria are limited by another nutrient. However, Cyanobacteria may dominate softwater eutrophic lakes (Schindler et al., 2008), so low cation concentrations are not necessarily detrimental to their dominance.

Studies of growth of the natural phytoplankton communities under gradients of nutrients indicated that diatoms were superior competitors for phosphorus and dominated at high silica-to-phosphorus ratios, that is, when silica did not limit diatom growth than green algae (Kilham, 1986; Tilman et al., 1986), whereas green algae dominated at low Si:P ratios because they were superior acquirers of phosphorus under the experimental conditions. This is an example of resource competition theory, which states that the species with the significantly lower resource requirement is the superior competitor when they are limited by the same resource. Species will coexist only if each is limited by a different resource or if these species are limited by the same resource and do not differ significantly in their resource requirements (Tilman, 1981). This theory explains how competition for a limiting resource leads to a species becoming dominant within a phytoplankton community in a natural system under a given set of conditions (light, pH, hardness, etc.) in the absence of chemical warfare (allelopathy; Keating 1977).

C. Phosphate uptake and pH

Many species of phytoplankton exhibit optimal growth and uptake of phosphorus within a distinct range of pH of the medium (Gong et al., 2014). Changes in the medium pH can affect a variety of metabolic activities (Nalewajko and Paul, 1985), which may include altering rates of phosphate absorption by direct effects on the activity of enzymes and on the permeability of the cell membrane or by changing the degree of phosphate ionization. For example, the uptake of phosphorus by the diatom *Asterionella* is greatest at pH values between 6 and 7 (Mackereth, 1953). Uptake rates of phosphorus have been correlated directly with the presence of numerous ions and compounds in the water, such as potassium, availability of micronutrients, and organic compounds. The mechanisms involved are not clearly understood.

D. Phosphorus concentrations required for growth

Growth rates of algae and Cyanobacteria in both natural and laboratory cultures exhibit dependency on the amount of available phosphorus and the rate at which it is cycled between cells. Extensive investigations of minimal and maximal phosphorus concentrations, especially by Chu (1943) and Rodhe (1948), grouped freshwater algae into categories according to whether their tolerance ranges fell below, around, or above 20 μg PO_4–P L^{-1}:

a. Species whose optimum growth and upper tolerance limit is below 20 μg PO_4–P L^{-1} (e.g., *Uroglena*) and some species of the macroalga *Chara*

b. Species whose optimum growth is below 20 μg PO_4–P L^{-1} but whose tolerance limit is well above that level (e.g., *Asterionella* and other diatoms)
c. Species whose optimal growth and upper tolerance limit is above 20 μg PO_4–P L^{-1} (e.g., green algae such as *Scenedesmus*, *Ankistrodesmus*, and many others)

In these studies phosphorus concentrations of the culture media required for optimal growth were almost always higher than those of the water in natural habitats where the algae were growing. Many explanations were offered for this difference, such as the presence of unknown organic growth factors in the lake but not in the artificial media of the bacteria-free cultures. It is now apparent that the chemical mass of inorganic phosphate in the water is not the only factor affecting growth kinetics. Of overriding importance is the rapidity with which phosphorus is cycled and exchanged between the particulate phosphorus and soluble inorganic and organic phases, as discussed in detail earlier. Secondly, it is of great ecological importance that many algae and Cyanobacteria, when provided with a sufficient supply of phosphorus, can absorb phosphorus in quantities far in excess of their actual needs (Kuhl, 1974). A portion of this phosphorus is lost in normal active growth both as inorganic and organic compounds and is recycled rapidly, at least in part. A large majority (>95%) of the phosphorus is in the particulate phase of algae and dead organic seston in phosphorus-limited systems but is lower in nitrogen-limited eutrophic systems. Assimilation of surplus amounts of phosphorus by algae, often referred to as *luxury consumption*, can provide sufficient phosphorus to sustain algal growth in the epilimnion as the external concentration of phosphorus becomes very low.

Of course, this recycling and utilization of stored phosphorus reserves cannot persist for very long. Losses continually occur from the colloidal phosphorus component, as well as from sedimentation from the particulate phosphorus component and with outflows of both dissolved and particulate components. Inputs of phosphorus to the system are needed continually, either from the littoral zone and turbulent transfer from lower-depth strata and sediments or externally, in order to sustain growth for extended periods. The amount of phosphorus input or loading, then, is important to the sustenance of algal growth and is one of many factors influencing the types of phytoplankton that are growing in a particular lake at a particular time of year.

It is therefore more relevant to the question of increasing phytoplankton productivity to view phosphorus concentrations in terms of total phosphorus, since most of the phosphorus is bound in the particulate component at any given time. From the few studies available in which common phytoplankton species of lakes of differing productivity were studied in relation to phosphorus requirements, the minimum phosphorus required per cell volume can be evaluated (Table 15-12). The colonial diatom *Asterionella formosa* is often found in oligotrophic waters, has a low phosphorus requirement, and can reach maximum cell densities at very low phosphorus concentrations. The quantity of phytoplankton, expressed as cell volume, that may be produced by 1 μg P L^{-1}, is 2–5 mm^3 L^{-1} (Vollenweider, 1968). This quantity of algae is common in lakes of low productivity. *Tabellaria* and *Fragilaria* are diatoms that reach maximum population densities at phosphorus concentrations of approximately 45 μg PO_4–P L^{-1}, while *Scenedesmus* needs higher concentrations of around 500 μg L^{-1}. The Cyanobacterium *Oscillatoria rubescens* does not reach its maximum phosphorus content until the initial phosphate concentration of the media used is about 3000 μg PO_4–P L^{-1}. It is of ecological interest to note that cations stimulate phosphate uptake by the pico-Cyanobacterium *Synechococcus* (= *Anacystis*). Calcium ions in particular cause a pronounced reduction in the half-saturation concentration for orthophosphate uptake by increasing the active transport of phosphorus into the cells (Rigby et al., 1980; cf. also Lehman, 1976).

TABLE 15-12 The Minimal Phosphorus Requirements per Unit Cell Volume of Several Algae or Cyanobacteria Common to Lakes of Progressively Increasing Productivity[a]

Algae	*Minimum P requirement, in μg mm⁻³ cell volume*
Asterionella	<0.2
Fragilaria	0.2–0.35
Tabellaria	0.45–0.6
Scenedesmus	>0.5
Oscillatoria	>0.5
Microcystis	>0.5

[a] *From data of Vollenweider (1968).*

Studies on the kinetics of phosphate uptake and growth of *Scenedesmus* in continuous culture have shown that phosphorus uptake velocity is a function of both the concentrations of internal cellular phosphorus compounds and the concentrations of the external substrate (Rhee, 1973). The apparent *half-saturation constant* (K_m), the coefficient in the Michaelis–Menten formula, was 0.6 μmol L^{-1} (approximately 18 μg P L^{-1}) for phosphorus uptake, whereas the apparent half-saturation constant for growth rate (the Monod equation) was less than K_m for uptake by an order of magnitude. Growth was a function of cellular phosphorus concentrations. Internal polyphosphate content appeared to regulate growth rate, particularly in the initiation of cell division. The activity of alkaline phosphatase, the primary enzyme involved in the release of phosphates from polyphosphates, exhibited a relationship that was inversely related to growth rate and was correlated to both polyphosphates and internally stored surplus phosphorus. During phosphate limitation, polyphosphates are mobilized for the synthesis of cellular macromolecules (protein, RNA, and DNA). Similar results were found for species of diatoms (Fuhs et al., 1972) and for green algae and Cyanobacteria (Healey and Hendzel, 1975; Senft, 1978).

E. Phosphorus uptake and competition by phytoplankton versus bacteria

Early studies on the uptake kinetics and competition for phosphorus between phytoplankton and bacteria were inconclusive. For example, bacteria were found to have a lower affinity (V_m/K_m based on cell surface area) for phosphate than did two diatoms (Table 15-13). This implies that diatoms can transport more phosphate per unit of cell surface membrane than bacteria at low phosphorus under the experimental conditions. However, bacteria could potentially outgrow the diatoms *in situ* because of a more favorable surface-area-to-volume ratio and indeed, bacteria had higher growth rates in single species culture under the experimental conditions (Fuhs et al., 1972; Smith and Kalff, 1982). As cell size increases, volume increases faster than surface area, so the amount of phosphorus supplied from a unit of the outer membrane to a unit of cell volume diminishes. To better reflect competitive abilities, the affinities in Table 15-13 can be expressed per unit cell volume by multiplying values by the cell's surface area/volume ratio, which would be $3/r$ for a sphere and $2(l+r)/(lr)$ for a cylinder where r is radius and l is length. Assuming the bacteria in Table 15-13 are cylindrical with radius 0.5 and length 1 μm and the diatoms are spherical with radius 10 μm, their mean affinities of 1.23 and 3.8×10^{-6} L μm^{-2} day^{-1} would become 7.4 and 1.1×10^{-6} L μm^{-3} day^{-1} with bacteria now having higher transport affinities based on cell volume. Molot and Brown (1986) found that chlorophytes and a diatom had lower volume-based phosphate transport affinities than several small and large prokaryotic Cyanobacteria and a yeast, suggesting that the algae were not as competitive at low phosphorus concentrations. These studies suggest that if competition between pelagic microbial heterotrophs and eukaryotic algae were based only on phosphate, then the former would dominate and eukaryotic algae would be excluded (Rhee, 1972). However, microbial heterotroph abundance *in situ* depends on continual supplies of organic carbon, and thus their abundance may be limited to some extent by organic carbon supplied by phytoplankton (Brown et al., 1981; Grover, 2000). Intensive grazing of the smaller bacteria and pico-Cyanobacteria by microzooplankton may also allow larger eukaryotic algae to thrive in low-phosphorus environments (Cavender-Bares et al., 1999; Mann and Chisholm, 2000).

A plethora of studies have examined the phosphorus uptake kinetics and specific growth rates of bacterioplankton (which includes photosynthetic pico-Cyanobacteria and picoeukaryotic algae) in comparison to those of larger phytoplankton. Correlation analyses (the two groups were separated on the basis of size; Currie and Kalff, 1984a; Currie, 1990) and direct measurements of specific growth rates in relation to phosphorus availability and competition among bacteria and phytoplankton (Currie and Kalff, 1984b, 1984c; Vadstein et al., 1988, 1993; Toolan et al., 1991; Cotner and Wetzel, 1992; Coveney and Wetzel, 1992) indicated these primary points: (a) bacterioplankton have substantially higher phosphorus requirements than do larger phytoplankton; (b) bacterioplankton net consumption of phosphorus is larger (4–10× higher) than that of the phytoplankton; (c) bacterioplankton contain much greater amounts of phosphorus (c. 10×) than algae; (d) phosphorus as well as, or rather than, organic carbon commonly limits bacterioplankton; and (e) bacterioplankton can outcompete algae for phosphorus under a wide range of phosphorus supply rates. Nevertheless, since these two groups coexist, some factor(s) other than phosphorus constrain heterotrophic bacterioplankton populations *in situ*. Since the bacterioplankton community is operationally defined by size fractionation and will include pico-Cyanobacteria and picoeukaryotic algae, which are not heterotrophic, the extent to which organic carbon limits heterotrophic bacterioplankton populations remains unclear.

Phosphate is the preferred form of phosphorus for uptake into both phytoplankton and bacteria (Cotner and Wetzel, 1992). Phytoplankton use both phosphate and dissolved organic phosphorus compounds, particularly at high substrate concentrations. At most natural concentrations, however, the uptake of phosphate is completely dominated (>50 to nearly 100%) by bacteria (Currie and Kalff, 1984b, 1984c; Güde, 1991; Cotner and Wetzel, 1992; Vadstein et al., 1993). The phosphorus contained in the bacteria is then regenerated by death and autolysis of the cells or by feeding activities of heterotrophic flagellates. Because bacteria have high requirements for phosphorus, much of the phosphorus occurs in and is regenerated from these phosphorus-rich particles.

TABLE 15-13 Phosphorus-Dependent Growth Rate and Phosphorus Uptake Kinetics of Two Diatoms and Three Bacterial Species[a]

Phytoplankton and bacteria	*Mean cell Volume* (μm^3)	*Mean cell surface* (μm^2)	a_0 (10^{-15} *g-atom P*)	μ_m (*doublings per day*)	K_m (10^{-6} *g-atom P* L^{-1})	V_m (10^{-15} *g-atom P* μm^{-2} day^{-1})	V_m/K_m (10^{-6} *L* μm^{-2} day^{-1})
Diatoms							
Cyclotella nana	77.5	103	0.95	1.6	0.6	2.0	3.3
Thalassiosira fluviatilis	1570	826	12.5	1.6	1.7	7.3	4.3
Bacteria							
Corynebacterium bovis	0.71	6	0.19	4.8	6.7	7.7	1.1
Pseudomonas aeruginosa	0.41	4.2	0.10	48	12.2	17.9	1.5
Bacillus subtilis	0.39	3.9	0.15	12	11.3	12.5	1.1

[a]*Modified from Fuhs et al. (1972).*

a_0 = minimum P content per cell.

μ_m = growth rate during unrestricted growth at 22°C, other conditions at or near optimum.

K_m = Michaelis half-saturation constant of uptake mechanism.

V_m = maximum uptake rate for orthophosphate P per unit area of cell surface.

V_m/K_m = slope of Michaelis–Menten uptake curve near the origin, which represents relative competitive ability to sequester P at low concentration (Molot and Brown, 1986). These affinities are expressed per cell surface area but can be converted to cell volume if cell geometries are known.

VIII. Phosphorus and eutrophication

A. Eutrophication and trophic status

All phytoplankton species require phosphorus, nitrogen (Chapter 14), major ions (calcium, magnesium, sodium, potassium, sulfate, and chloride), several trace metals, and, in some species, silica (Chapter 16). The importance of phosphorus to lake ecology lies in the fact that, in most unpolluted lakes globally, phosphorus is in shortest supply. Therefore an increase in phosphorus inputs results in higher productivity, a phenomenon referred to as *Liebig's Law of the Minimum* or the *Principle of Limiting Factors*. Today it is sometimes referred to as the *phosphorus paradigm*.

The majority of monitored oligotrophic lakes globally are phosphorus limited, although some are not, for example, the nitrogen-limited lakes in low nitrogen deposition areas in northern Sweden (Bergström et al., 2005) and lakes in the arid American Great Basin (Reuter et al., 1993). Some oligotrophic lakes may be limited by iron or colimited by iron and phosphorus, or at least some of their communities with high metal requirements such as pico-Cyanobacteria may be growth limited for short periods of time (Sterner et al., 2004; Sterner, 2008; Twiss et al., 2005).

In naturally phosphorus-limited lakes that have become nutrient enriched or eutrophic, phosphorus is usually the only nutrient with natural or background levels low enough for it to be the basis of effective nutrient control programs that lower anthropogenic inputs to restore or maintain oligotrophic to mesotrophic conditions. For this reason, phosphorus has become the cornerstone of eutrophication management (Schindler et al., 2016). As whole-lake phosphorus fertilization experiments in the Experimental Lakes Area have repeatedly shown, even dilute (low solute) lakes have natural major ion and trace metal concentrations high enough to support large increases in productivity (Figs. 15-12 and 15-13). Nitrogen is a special case because, should a eutrophic lake become nitrogen-limited, *diazotrophic (nitrogen-fixing) Cyanobacteria* will convert dissolved N_2 to ammonia during warm periods in the summer, provided that trace metal cofactors are high enough to support *heterocyst* and *nitrogenase synthesis* (Schindler et al., 2008; Molot et al., 2021a).

As phosphorus inputs increase, lake characteristics progressively change with *oligotrophy*, signifying a state associated with low nutrient concentrations, and *eutrophy* signifying a state associated with excessive nutrient concentrations. *Mesotrophy* is in the middle of this continuum. Naumann (1929) introduced the general

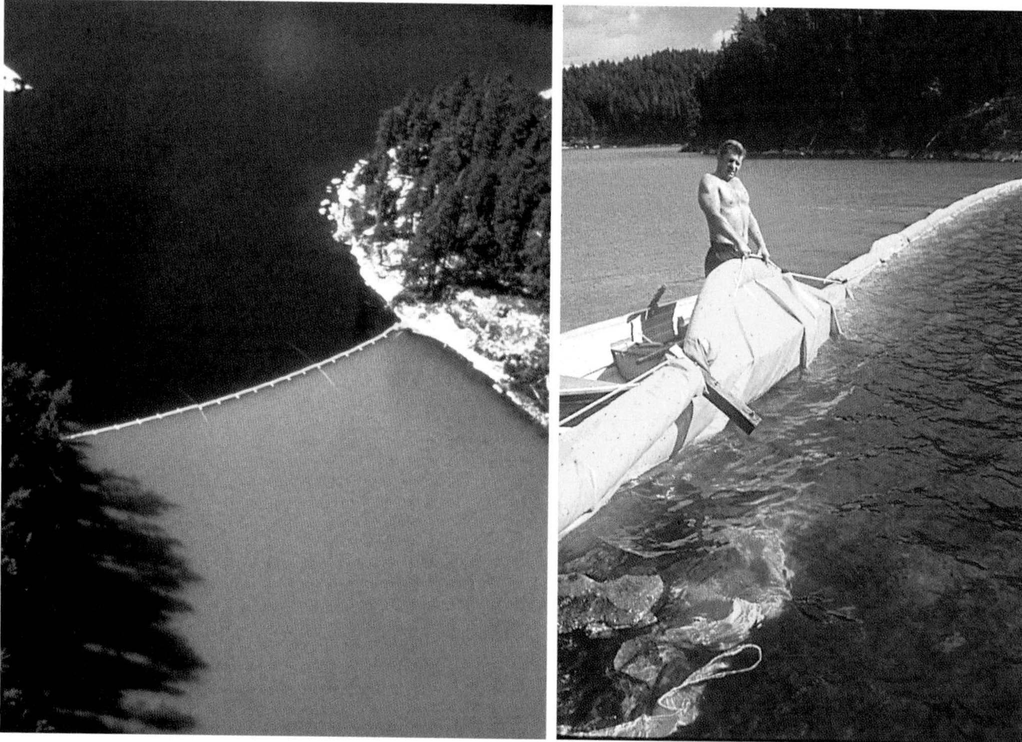

FIGURE 15-12 Photographs of Lake 226 of the Experimental Lakes Area (ELA) of northwestern Ontario, which was partitioned at the constriction of the basin, 4 September 1973 (*left*), and David Schindler, director of ELA from 1968–1989, installing the curtain (*right*). The far northeastern basin, fertilized with phosphorus, nitrogen, and carbon, was covered by a dense algal bloom within 2 months. No increases in algae or species changes were observed in the near basin, which received similar quantities of nitrogen and carbon but no phosphorus. *(From Schindler, 1974. Copyright 1974 by the American Association for the Advancement of Science. Photographs courtesy of D. W. Schindler and IISD-ELA.)*

FIGURE 15-13 Photograph of a nitrogen-fixing Cyanobacteria bloom in Lake 303 in the Experimental Lakes Area in northwestern Ontario on 1 September 2019, 12 weeks after the first weekly phosphorus addition. Although the lake was not fertilized with nitrogen during the experiment, nitrogen fixation doubled the total nitrogen concentration during the bloom (Molot et al., 2021a). *(Photograph courtesy of IISD-ELA.)*

concepts of oligotrophy and eutrophy and distinguished them on the basis of phytoplanktonic populations, a recognition that as lakes become nutrient enriched, they become more productive. He later elaborated on the concept (Naumann, 1929). Naumann emphasized that chemical factors, particularly phosphorus, combined nitrogen (ammonia, nitrite, and nitrate), and calcium, were primary determining factors of productivity rates, although today it is recognized that eutrophic softwater lakes with low calcium are similar in production and other characteristics to eutrophic, hardwater lakes (Schindler et al., 2008). Limnological evidence since Naumann's early work has shown a strong relationship between the biological dynamics and structure of a lake and its concentrations of nutrients. The consequences of *eutrophication*, that is, the development of characteristics that differ markedly from oligotrophic lakes such as loss of macrophyte beds, hypolimnetic oxygen loss, dense phytoplankton populations with the formation of harmful Cyanobacteria blooms, and changes in invertebrate and fish community structure, have been further elaborated on by many others including Thienemann (summarized in 1925), Likens (1972), Hutchinson (1973), Vallentyne (1974), and Paerl (1988). The terms *oligotrophy* for lakes of low production and *eutrophy* for lakes of high production evolved from similar concepts developed for bogs (Weber, 1907; Naumann, 1929, 1932; Hutchinson, 1967; Carlson, 1992; Carlson and Simpson, 1996).

Today, the *trophic state concept* is based on the notion that nutrient enrichment and production (biomass) are causally linked and does not try to incorporate all of the biological and abiotic consequences of that production. The concept has led to a useful system for classifying the extent of nutrient enrichment of a lake based on phosphorus concentration (and implicitly on biomass) with three and sometimes more categories used: *oligotrophic*, *mesotrophic*, *eutrophic*, and *hypereutrophic*. The phosphorus boundaries separating them are somewhat fluid, being based on lake characteristics that change with enrichment (Carlson, 1977, 1992; Carlson and Simpson, 1996; Gibson et al., 2000). In addition, different jurisdictions use different boundaries. While phosphorus and productivity run along a continuum, such classifications are very useful in public communication and environmental management (Table 15-1).

Combinations of many physical and biological variables as regulating trophic states have clouded true cause-and-effect relationships of rates of primary production. As Carlson (1977) and Carlson and Simpson (1996) emphasized, the trophic state concept is important as a critical organizing concept in limnology. The concept represents a continuum of trophic levels that is difficult to quantitatively delineate into trophic state subgroups but is done so nevertheless. Organization along a trophic continuum is but one facet of lake typology and uses biomass or rates of production as the simplest, and most useful, definition of the trophic state. Many variables can be evaluated in the context of algal productivity, for example, phosphorus loadings and concentrations, chlorophyll *a* concentrations as a proxy of phytoplankton biomass, phytoplankton productivity, phytoplankton biomass, hypolimnetic oxygen deficits, transparency (Secchi disk depth), etc. The *annual mean rate of production* is an excellent criterion but requires intensive and sophisticated methods for determination and thus is generally not included in routine monitoring programs. Biological food-web structure is highly susceptible to change along dynamic environmental gradients other than nutrients; hence it differs greatly globally. Therefore no portion of the biological structure other than primary production or its proxy is recommended as a criterion of the trophic concept. Nevertheless, certain ecological changes such as the formation of harmful Cyanobacteria blooms, loss of macrophyte beds in shallow systems, or loss of coldwater fish species in deep lakes due to loss of hypolimnetic oxygen, are often noted as hallmark characteristics of a eutrophic state where the risk of these occurring increases with increasing phosphorus concentration (Downing et al., 2001; Yamamichi et al., 2018; Müller et al., 2019).

B. The relationship between nutrient loading and productivity in aquatic ecosystems

Great emphasis has been placed on nutrient availability in regulating aquatic productivity, especially phosphorus, since the majority of oligotrophic lakes and many eutrophic lakes are phosphorus limited (excessive phosphorus inputs render some eutrophic lakes nitrogen limited). Strong linear and sigmoidal log–log correlations between chlorophyll *a* and total phosphorus (Dillon and Rigler, 1974; Nicholls and Dillon, 1978; Quinlan et al., 2021) along with numerous whole-lake fertilization and bioassay experiments are evidence of the causal nature of this relationship. As will be discussed later in treatments of biological productivity, many nonbiological factors other than phosphorus and nitrogen can regulate productivity such as temperature, light, and lake morphology (Quinlan et al., 2021).

The early analyses of differences in productivity among aquatic ecosystems focused on differences in nutrient concentrations. Increased information gradually demonstrated the rapid dynamics of the exchange of nutrients among algae as nutrients were assimilated, utilized, and released during and following growth, reproduction, and death. Still among these studies, however, a prevailing concept was to treat lakes as closed, self-sustained entities. The most rapid change in these viewpoints arose, largely in the late 1960s and 1970s, with full recognition that rivers and particularly lakes are open systems critically dependent upon couplings with their drainage basins.

Phosphorus and nitrogen had been recognized for several decades in agriculture as critical nutrients that often limit plant growth and productivity, and they were similarly recognized in phytoplankton productivity. The importance of phosphorus and nitrogen in the eutrophication of surface waters was also long recognized, but with the understanding that lakes are open systems, it became apparent that the *rates* of loading (i.e., importation) of these nutrients from the drainage basin were critical to trophic conditions. From a management viewpoint, one could also define loadings of nutrients in terms of whether they were acceptable or excessive in terms of the algal productivity generated under the lake conditions of morphometry and water retention times (Vollenweider and Kerekes, 1982). The trophic categories of oligo-, meso-, and eutrophic were gradually analyzed in increasingly quantitative ways, in terms of nutrient concentrations, phytoplankton biomass, chlorophyll *a* concentrations, and water transparency (e.g., Secchi disk readings), in addition to the classical characterization of hypolimnetic oxygen conditions. These data from many lakes allowed the development of dynamic quantitative modeling for the prediction of probable biotic conditions as a result of nutrient loadings from the drainage basin. A summary of some of these relationships, developed and extensively tested in the 1980s, is presented later.

In more recent times much attention has been directed to the internal recycling of nutrients within lakes and streams. In particular, internal loadings of nutrients from accumulations in sediments have been examined to determine how this recycling of nutrients supplements those allochthonous sources from the drainage basin and atmosphere. In addition, as is summarized in this chapter, much attention is directed to the recycling of nutrients through food ingestion and alteration by animals.

C. Effect of phosphorus concentration on lake productivity

The consensus among limnologists is that the term *eutrophication* is synonymous with increased production of the biota of lakes and that the rate of increasing productivity is accelerated over that rate that would have occurred in the absence of perturbations to the system. The most conspicuous, basic, and measurable criterion of accelerated productivity is an increased quantity of carbon assimilated by algae and larger plants per given area.

Under most lake conditions, the most important nutrients causing the shift from a lesser to a more productive state are phosphorus and nitrogen by virtue of their high cellular content, although other metabolically essential elements such as certain trace metals may also limit productivity (Downs et al., 2008). Typical plant organic matter of aquatic algae, Cyanobacteria, and macrophytes contains primarily phosphorus, nitrogen, and carbon approximately in ratios known as the *C:N:P Redfield ratios*, which are 41.6:7.3:1.0 (weight) or 106:16:1 (molar) (Redfield et al., 1963; Vallentyne, 1974;

Hecky et al., 1993; Ågren, 2004), with smaller amounts of sulfur and trace metals. These C:N:P ratios are relatively constant in marine plankton but are more variable in freshwater systems, depending upon luxury uptake and growth rate (see Section VIIID).

A comparison of the relative amounts of different elements required for algal growth with supplies available in freshwaters illustrates the general importance of phosphorus and nitrogen (Table 15-14). Similar proportional ratios were demonstrated by comparison of the demand among terrestrial plants with the accessible supply of elements from the lithosphere (Hutchinson, 1973). Even though variations in conditions that affect solubility or availability may at times make elements that are normally very abundant in Earth's crust (e.g., silicon, iron, and certain other micronutrients) almost unobtainable, phosphorus, and secondarily nitrogen, most frequently impose limitation on aquatic systems. This relationship is emphasized further by consideration of average demand–supply ratios in late winter prior to the spring algal maximum common in lakes of temperate regions and in midsummer during maximum sustained algal productivity (Table 15-15).

Oligotrophic lakes are usually limited by phosphorus and contain nitrogen in quantities in excess of demand from growth supported by available phosphorus. As

TABLE 15-15 Comparisons of the Ratios of Required Concentrations of Inorganic Nutrients to Average Supplies Available in Freshwaters[a]

	Ratio of demand to available supplies	
Element	*Late winter*	*Midsummer*
Phosphorus	80,000	Up to 800,000
Nitrogen	30,000	Up to 300,000
Carbon	5000	Up to 6000
Iron, silicon	Generally low, but variable	
All other elements	<1000	<1000

[a]*After Vallentyne (1974).*

TABLE 15-14 Proportions of Essential Elements for Growth in Living Tissues of Freshwater Plants (Requirements) in the Mean World River Water (Supply) and the Approximate Ratio of Concentrations Required to Those Available[a]

Element	*Average plant content or requirements (%)*	*Average supply in water (%)*	*Ratio of plant content to supply available*
Oxygen	80.5	89	1
Hydrogen	9.7	11	1
Carbon	6.5	0.0012	5000
Silicon	1.3	0.00065	2000
Nitrogen	0.7	0.000023	30,000
Calcium	0.4	0.0015	<1000
Potassium	0.3	0.00023	1300
Phosphorus	0.08	0.000001	80,000
Magnesium	0.07	0.0004	<1000
Sulfur	0.06	0.0004	<1000
Chlorine	0.06	0.0008	<1000
Sodium	0.04	0.0006	<1000
Iron	0.02	0.00007	<1000
Boron	0.001	0.00001	<1000
Manganese	0.0007	0.0000015	<1000
Zinc	0.0003	0.000001	<1000
Copper	0.0001	0.000001	<1000
Molybdenum	0.00005	0.0000003	<1000
Cobalt	0.000002	0.000000005	<1000

[a]*After Vallentyne (1974).*

the lakes become more productive, the primary effecting agent is increased loading of phosphorus. As discussed earlier, the instantaneous concentrations of phosphorus usually decrease and are quite variable, but the turnover rate increases markedly. Rates of loss increase, and high algal productivity requires sustained phosphorus loading of the system from allochthonous and littoral sources.

Some studies have indicated that nitrogen rather than phosphorus is limiting to nonfixing phytoplanktonic productivity of lakes, at least in certain water strata and at certain times of the year in phosphorus-limited lakes. Most oligotrophic lakes globally are phosphorus limited but nitrogen limitation is common in, for example, the arid American Great Basin and northern Sweden (e.g., Elser et al., 1990; Reuter et al., 1993; Bergström et al., 2005). Diazotrophic Cyanobacteria are only nitrogen limited when nitrogen fixation is resource limited, for example, by a trace metal enzyme cofactor such as molybdenum or by phosphorus (Molot et al., 2021b). When phosphorus is available in quantities in excess to support metabolism, nitrogen availability can become limiting to nonfixing biota. This limitation is often observed in the two extremes of the trophic spectrum. Under eutrophic conditions of high phosphorus loading, planktonic utilization of combined nitrogen can exceed inputs and literally deplete combined nitrogen supplies in the trophogenic zone in a few days or weeks (Chapter 14). When this occurs, nitrogen fixation by heterocystous Cyanobacteria can augment the nitrogen inputs to the lake, although it may be insufficient to alleviate nitrogen deficits in nonfixing biota. Although the metabolic cost is high, Cyanobacteria growth rates and biomass yields in batch cultures are relatively unaffected when trace metal supplies are adequate (Molot, 2017). In tropical lakes of low latitudes nitrogen limitation to phytoplanktonic growth can occur (cf. reviews of Salas and Martino, 1991, and of Lewis, 1996). Phytoplanktonic productivity of tropical waters is generally higher than at temperate latitudes, and higher temperatures can accelerate rates of denitrification and losses of nitrogen to the atmosphere.

Under oligotrophic conditions, phosphorus availability may be adequate to support modest levels of phytoplanktonic productivity, but sources and loadings of combined nitrogen are insufficient to support such productivity. Increasing the availability of nitrogen can result in increased algal photosynthesis until phosphorus again becomes limiting. Enrichments with both nitrogen and phosphorus can stimulate algal productivity markedly (e.g., Holmgren, 1984; White et al., 1986; Elser et al., 1990).

Phytoplanktonic productivity of reservoirs often fluctuates to a much greater extent than in natural lakes. Because of the frequent variations in inorganic turbidity in some reservoirs, the limiting effect of light penetration on photosynthetic activity can be more severe than that of nutrients (e.g., Henry et al., 1985). Nutrient loading to reservoirs from external sources can vary with long-term precipitation patterns and water inflows in comparison to the volumes of the productive zones. If flushing rates are high, turbidity can increase but also nutrient loading and availability can increase (e.g., Hoyer and Jones, 1983; Turner et al., 1983). Hypolimnetic withdrawal can also affect concentrations by discharging internally loaded phosphorus downstream.

The potential for limitation of eutrophication by inorganic carbon availability has been a subject of much discussion (Chapter 13). In a few exceedingly productive situations, such as sewage lagoons, in which phosphorus and nitrogen compounds are available in excess of any demands, carbon can become limiting to algal growth (Kerr et al., 1972). While there is some evidence for inorganic carbon limitation in exceedingly soft lakes (e.g., Allen, 1972), diffusion of atmospheric CO_2 is generally adequate to sustain the carbon requirements of phytoplanktonic populations in eutrophic lakes (Schindler and Fee, 1973). It is extremely unlikely that carbon limitation is of significance in the limitation of algal populations in a majority of harder waters, in which dense algae develop under eutrophic conditions.

The importance of phosphorus in comparison to nitrogen and carbon has been particularly well illustrated by large-scale fertilization experiments (Schindler, 1974). Lake 226, located in the Experimental Lakes Area (ELA) of northwestern Ontario in Precambrian Shield bedrock, was chemically and biologically similar to >50% of the waters draining to the northern Laurentian Great Lakes. Lake 226 was partitioned with a large plastic curtain into two basins at a constriction in the lake (Fig. 15-12). One basin was fertilized with phosphorus, nitrogen, and carbon, and the other with equivalent concentrations of nitrogen and carbon, made in 20 equal weekly increments. The N/P and C/P ratios were greater than in treated sewage, but the quantity of P added was not exceptionally high for lakes that commonly receive pollution from domestic sources. The phosphate-enriched basin quickly became highly eutrophic, while the basin receiving only nitrogen and carbon remained at prefertilization conditions (Fig. 15-12). In this phosphate-enriched basin, and in several other lakes receiving analogous treatments over a period of several years, algal biomass increased to two orders of magnitude over that of lakes receiving only nitrogen and carbon enrichment. Recovery to near prefertilization levels was very rapid when phosphate additions were discontinued. Lake 227 was the site of a 50-year experiment between 1970 and 2020 at a constant phosphorus fertilization rate, receiving nitrogen as nitrate for the first 20 years and only phosphorus for the last 30 years

(Schindler et al., 2008). The lake remained highly productive 30 years after nitrogen additions stopped, with increasing inputs from nitrogen fixation as legacy nitrogen was depleted (Higgins et al., 2018).

A distinct issue, but one related to increased productivity, is the displacement of eukaryotic algae by diazotrophic Cyanobacteria during blooms for short periods of time in nitrogen-limited eutrophic systems, that is, systems with excessive phosphorus inputs relative to nitrogen inputs. In these systems mean phytoplankton productivity during the ice-free season will be nitrogen limited but Cyanobacteria productivity will not, by virtue of their nitrogen-fixing ability as long as trace metal cofactor supplies, molybdenum, and iron, for the nitrogen-fixing enzyme, nitrogenase, are adequate (Molot, 2017). Note that some species are not fixers but will not dominate blooms in nitrogen-limited systems. It is important to distinguish between mean daily ice-free phytoplankton productivity, which includes eukaryotic algae and Cyanobacteria bloom productivity, and daily productivity during a bloom. The latter will be higher than the ice-free daily mean because of nitrogen fixation inputs during the bloom. For example, in nitrogen-limited Lake 227 a month-long nitrogen-fixing *Aphanizomenon* bloom occurred every summer during the 30-year period of phosphorus-only fertilization with a large biomass peak during the bloom (Schindler et al., 2008). In another experiment at the Experimental Lakes Area, oligotrophic Lakes 303 and 304 were also fertilized weekly with only phosphorus, producing dense, month-long blooms of nitrogen-fixing *Dolichospermum* species in both lakes within 9–12 weeks after fertilization began, turning the lakes visibly green without the addition of nitrogen (Fig. 15-13; Molot et al., 2021a). Nitrogen fixation doubled the nitrogen content during the blooms (Molot et al., 2021a). Hence reporting the mean daily ice-free average of, for example, chlorophyll *a*, without reporting chlorophyll *a* during a Cyanobacteria bloom will obscure the size of the bloom within the seasonal average, which may affect eutrophication management choices. These results suggest that many nitrogen control programs are unlikely to mitigate Cyanobacteria blooms, although they may lower Cyanobacteria toxicity and eukaryotic biomass. Phosphorus controls should remain a cornerstone for Cyanobacteria management.

The amount of evidence for the phosphorus–chlorophyll *a* relationships is so overwhelming, both from experimental and applied investigations, that it is difficult to appreciate how the bitter controversy over the importance of carbon rather than phosphorus as a limiting nutrient in freshwaters developed in Canada in the late 1960s. As a result of a combination of the magnification of a few findings from cyanobacterial cultures and a few rare cases of discovered low carbon availability in natural habitats in which excessive amounts of nitrogen and phosphorus occurred, phosphorus was implied to be of less importance in the acceleration of eutrophication than carbon. So convincing was the misinterpretation of results by a few scientists, and so effective was the exploitation of the situation by the detergent industry and irresponsible journalists, that legislation was urgently needed to reduce the effluent loading of surface waters with phosphorus was seriously impeded. The matter was finally put to rest by the rebuttal of Vallentyne (1970), numerous subsequent investigations (e.g., Schindler and Fee, 1973), and a national symposium on the subject (Likens, 1972).

D. Stoichiometry of carbon, nitrogen, and phosphorus in particulate organic matter

The elemental chemical composition of planktonic particulate organic matter reflects both the planktonic community structure and biochemical processes, as well as nutrient supply ratios and turnover rates that occurred in the formulation of the composition of the biomass (reviewed in Tilman et al., 1982; Kilham, 1990). For example, phosphorus or nitrogen limitation can lead to alterations in the ratios of protein, carbohydrate, and lipid content of cells, as well as the C:N ratio. Hence the C:N ratio indicates not only characteristics of the nutrient availability but also an approximate evaluation of the relative proportions of cellular proteins and nonprotein structural elements (cf. Vollenweider, 1985).

A relative constancy in the molar ratio of C:N:P of 106:16:1 (or 41:7.2:1 by weight) among marine plankton, termed the *Redfield ratio*, is generally supported by numerous studies (Redfield et al., 1963; Vallentyne, 1974; Hecky et al., 1993; Ågren, 2004). The variance in this ratio is small, usually $<20\%$. This constancy has been attributed to the relatively nutrient-sufficient growth conditions of marine plankton and the more homogeneous and stable nature of oceans (Goldman et al., 1979). In contrast, marked deviations in the sestonic C:N:P proportions occur in lakes. These particulate composition ratios have been coupled to physiological indicators, such as rates of growth and productivity, to estimate the nutrient conditions to which phytoplankton of lakes have been exposed. Such relationships of nutrient stoichiometry can also give insights into species composition, growth rates, and successional patterns of phytoplankton (Kilham, 1990; Sommer, 1990).

Elemental cellular stoichiometries of natural phytoplankton communities and the seston can reflect the type and extent of nutrient limitation and availability. Ratios of elements being loaded to a lake are reflected in the elemental composition of the phytoplankton

TABLE 15-16 Stoichiometric Ratios of Phytoplankton-Dominated Seston of Lakes as Approximate Indicators of Relative Nutrient Limitations[a]

Ratio	*Deficiency*	*Degree of nutrient limitation*[b]		
		None	*Moderate*	*Severe*
C:N	N	<8.3	8.3–14.6	>14.6
N:P	P	<23	–	>23
C:P	P	<133	133–258	>258
Si:P	Si	<20	–	>100
C:Chl *a*	General	<4.2	4.2–8.3	>8.3
APA:Chl *a*	P	<0.003	0.003–0.005	>0.005

[a]*After Healey and Hendzel (1980), Kilham (1990), and Hecky et al. (1993).*
[b]*Composition ratios of C:N, N:P, C:P in* $\mu mol\ \mu mol^{-1}$; *C:Chl* a *ratios as* $\mu mol\ \mu g^{-1}$; *and physiological ratio of alkaline phosphatase activity (APA): Chl* a *in* $(\mu mol\ \mu g^{-1})\ h^{-1}$.

community. For example, N:P ratios in phytoplankton (seston) are strongly correlated with N:P loading rates to lakes (Table 15-16). The C:P and N:P ratios of lake seston are generally higher than the Redfield ratio for marine waters. Phosphorus limitation of algal productivity in lakes tends to be much greater than nitrogen limitation, with nitrogen limitation more common in eutrophic lakes because of excessive phosphorus inputs (Healey and Hendzel, 1980; Hecky et al., 1993). Streams, shallow lakes, and reservoirs with short residence times have C:P ratios <350 and N:P ratios <26, whereas lakes with longer residence times (>6 months) differentiate from their inflows typically with C:P >400 and N:P >30. Tropical lakes tend to have relatively high C:N ratios, indicative of potential nitrogen limitations, although the number of lakes sampled was relatively small (Hecky et al., 1993).

Cellular elemental stoichiometry is affected by the availability of nutrients and therefore can be correlated with the extent of nutrient limitations (Table 15-16). Epilimnetic ratios, particularly of N:P, are often higher, indicative of high phosphorus demands in proportion to total inputs, than those of the hypolimnion, and C:P as well as N:P ratios decrease with depth (e.g., Jones, 1976; Gächter and Bloesch, 1985; Tezuka, 1985; Gálvez et al., 1991). These differences with depth are usually related to high phosphorus use in relation to supply, increased nutrient pool size with depth, and respiration of organic carbon as the seston settles.

The ratio of fluxes of nutrients into the dissolved nutrient resource pool (i.e., the supply rates of essential nutrients) can markedly influence the community structure of the phytoplankton (Tilman, 1982). As pointed out by Sterner et al. (1992), external sources of nutrient loading can be, and have been, evaluated with some degree of accuracy, but rates and controls of internal regeneration of nutrients by physical (e.g., resuspension and mixing) and biological (e.g., nitrogen fixation, denitrification, food web regeneration by all heterotrophs, including viruses) processes are much less quantified to levels that allow effective use of predictive models. For example, denitrification was the most important mechanism for reducing N:P ratios in the water column, whereas both nitrogen fixation and sediment resuspension raised N:P ratios (Levine and Schindler, 1992). Sediments tend to be a major nitrogen source, particularly in littoral areas and anoxic zones.

The N:P and Si:P supply ratios are fundamental axes along which structures of phytoplanktonic communities vary (Hecky and Kilham, 1988; Kilham, 1990). Diatoms and some chrysophyte algae have absolute silica requirements (Chapters 17 and 18). Although total phytoplankton biomass is not limited by silica availability, the phytoplankton community composition, interspecific competition, and succession can be altered markedly (Chapters 17 and 18). Species of diatoms with low Si:P requirements, for example, often develop maximally during seasonal periods of lowest Si:P ratios (Kilham, 1984). In another example many Cyanobacteria are capable of nitrogen fixation, which allows these species to maintain high growth rates in habitats low in available combined inorganic nitrogen where adequate supplies of trace metals needed for nitrogen fixation and heterocyst synthesis (e.g., iron, molybdenum, cobalt) are available. The relative competitive capacities and proportions of Cyanobacteria in epilimnetic phytoplankton communities are thus indicated by the N:P ratio; however, Cyanobacteria tend to be rare at N:P >29 by weight (Smith, 1983).

Zooplankton and larger animals usually constitute small portions of the total dissolved and seston pools of nitrogen and phosphorus. For example, in an alpine lake in California zooplankton always constituted <5% and <10% of the pelagic storage compartments for nitrogen and phosphorus, respectively (Elser and George, 1993). However, during brief periods of high zooplankton biomass, the zooplanktonic contributions to total sestonic elemental stoichiometry can be significant. Although the N:P and C:P ratios do not vary greatly intraspecifically among zooplankton, considerable differences occur interspecifically (Watanabe, 1990; Gulati et al., 1991; Sterner et al., 1992; Elser and George, 1993; Urabe, 1995). Zooplanktonic N:P ratios may be altered by the N:P ratio of seston as certain individual species populations of zooplankton, which grow and reproduce most efficiently at a given N:P supply ratio, come to dominate the zooplankton community. Because the biochemical requirements of crustacean zooplankton for an N:P ratio in their tissues is below the ratio of most sestonic assemblages, there is a tendency for crustacean zooplankton to retain P preferentially to N. As a result of this process, nutrient regeneration via feces of zooplankton can be skewed toward nitrogen and thereby enhance phosphorus limitations to phytoplankton during short periods of intense cladoceran grazing (e.g., "clear water phase").

In examining the chemical composition of particulate organic matter, large zooplankton can be effectively separated from smaller particles. However, detrital particles nearly always dominate (>50–80%) over algal and bacterial biomass in the remaining seston (Saunders, 1972; Uehlinger and Bloesch, 1987; Gálvez et al. 1989). If the planktonic microbiotas are reproducing rapidly, much of the detritus present can have chemical composition relatively similar to that of the living plankton. Appreciable loadings of particulate organic matter from the structural tissues of terrestrial and littoral plants, however, can result in seston relatively rich in carbon and lesser amounts of nitrogen and lead to high C:N, C:P, and N:P ratios. Although much of the particulate organic matter from external sources tends to be deposited in nearshore regions in natural lakes, importation from the drainage basin is much greater in reservoir ecosystems. In addition, because of the high ratio of drainage basin area to reservoir area, large quantities of inorganic particles (e.g., clay) can be imported to the pelagic zone. These inorganic particles of the seston can contain high and variable quantities of nutrients that are undifferentiated from particulate organic matter of biota and can lead to skewed results, requiring care in interpretation. However, in most lakes with water residence times of several months or longer, sestonic composition ratios reflect the general status of the availability of nitrogen and phosphorus for planktonic growth in these waters.

Furthermore, the predictive value of nutrients present in the seston as a reflection of nutrient availability has been extended by the obvious relationship of energy available as light for photosynthesis to nutrient availability and incorporated into the seston (Sterner et al., 1997). When phosphorus availability is low, C:P ratios of particulate organic matter will tend to be low because photosynthetic productivity is low. Differences in nutrient use efficiency may occur at different light availability, and these ratios will be translated to higher trophic levels feeding upon them. Low light:phosphorus ratios of seston likely indicate simultaneous carbon and organic matter (energy) limitation in other dependent trophic levels, whereas high light:phosphorus ratios suggest phosphorus limitation among several dependent trophic levels.

E. Effects of decreasing phosphorus on productivity

The reduction of the productivity of lakes by decreasing phosphorus loading can be very effective. Many examples are given in reviews by Vollenweider (1968) and Schindler et al. (2016). Among the most frequently discussed examples is Lake Washington, Washington, USA (Edmondson, 1972, 1991). Lake Washington was enriched with increasing volumes of effluent from secondary sewage treatment facilities during the period from 1941 to 1953. Production increased markedly, and the phytoplankton became more abundant. Phosphate concentrations also increased proportionally much more than those of nitrate or carbon dioxide. Effluent was diverted away from the lake in 1963, and by 1969, phosphate had decreased to 28% of its 1963 value. Summer chlorophyll *a* concentrations had decreased about as much, but nitrate and total CO_2 fluctuated from year to year at relatively high values. Reduction in phytoplankton and phosphorus continued to a relatively stable level in the 1990s (Edmondson, 1969, 1991; Edmondson and Lehman, 1981). A very similar situation occurred in Gravenhurst Bay (Ontario, Canada) with tertiary treatment of wastewaters resulting in the same improvements as wastewater diversion in the Lake Washington case (Dillon et al., 1978).

The rate at which the productivity of a lake will revert toward conditions existing prior to increased phosphorus loading is variable and depends upon basin morphometry, water chemistry, the nature of the phosphorus sources (diffuse or concentrated at point sources), and the rate of internal phosphorus loading. The loading of lakes with nitrogen and phosphorus (expressed per unit area of lake area) is further influenced by the ratio of the surface area of the drainage basin to that of the lake (Ohle, 1965). Depending on the percentage losses from the land, some of which may be under cultivation, loading increases roughly in proportion to this ratio.

Upon reduction of loading rates to or below those of the lake prior to accelerated inputs, recovery will require 2–10 years for lakes of average size and average hydrological replenishment time (Schindler et al., 2016; Chorus et al., 2020). The relative residence time, or, inversely, the flow-through rate of phosphorus, is directly coupled to the efficiency of sediment removal and retention capacities in the regulation of phytoplanktonic productivity (e.g., Janus and Vollenweider, 1984; Vollenweider, 1985, 1990). Whole-lake experiments at the Experimental Lakes Area in northwestern Ontario show that oligotrophic lakes will respond rapidly to large increases in loads (Schindler, 1974; Molot et al., 2021a). In contrast, eutrophic lakes with a long eutrophication history and large accumulations of nutrients in the sediments will respond slowly to decreasing loads because of continued phosphorus release from anaerobic sediments (internal loading). The rate of recovery is much slower in shallower lakes where the internal areal loading of phosphorus from sediment sources is high relative to lake volume. In eutrophic Shagawa Lake, Minnesota (USA), for example, where external phosphorus loading was reduced by >70%, predicted reduced equilibrium levels were not achieved (observed levels of phosphorus concentrations after diversion were >100% in excess of those predicted; Larsen et al., 1975). Shagawa Lake is shallow, has an anaerobic hypolimnion, and has an extensively developed littoral zone of submersed macrophytes, all of which contribute to the enhanced release of sediment phosphorus. Sediment P is the main source of P for macrophytes (Barko and Smart, 1980, 1981) which can release P to the water column when they senesce. Internal loading from sediment phosphorus sources has been observed in many other shallow lakes, reservoirs, or lakes with anoxic hypolimnia (e.g., Cooke et al., 1977, 1993; Grobler and Davies, 1981; Hillbricht-Ilkowska and Lawacz, 1983; Nürnberg, 1984). The magnitude of internal phosphorus loading can be predicted from the product of an average rate of phosphorus release from anoxic sediments, the surface area of the anoxic sediment, and the period of anoxia (Nürnberg, 2009). Such internal loading slows the rate of lake recovery implemented by the reduction of phosphorus loading from external allochthonous sources. In most lakes, however, the "internal loading" contribution is a relatively small proportion of the external sources (e.g., Bannerman et al., 1974; Edmondson and Lehman, 1981; Smith and Shapiro, 1981; Rast et al., 1983) so that regulation of loading rates can often be effective in controlling phytoplankton productivity.

The retention capacities for phosphorus can differ considerably between stratified natural lakes, polymictic lakes, and reservoirs. Much of the difference in phosphorus retention focuses on marked differences in *water retention times*. Water retention time can be estimated by the number of days needed in a given year to fill the lake or reservoir if it were empty. Water retention can be correlated to the phosphorus retention capacity estimated as the percentage of the total phosphorus loadings remaining and not lost from the water body via its outflow (Straškraba, 1996). Although the phosphorus retention capacities are similar, >50% of phosphorus loading, for water retention times of a year or longer (Fig. 15-14), the water retention times of lakes are much longer (1–7 years on average) than those of reservoirs (often <100 days and highly variable). The differences are most pronounced at water retention times <100 days (Fig. 15-14). These differences are related, in part, to more rapid sedimentation of particulate phosphorus and removal of outflow water deep in the stratified waters in reservoirs versus the slower

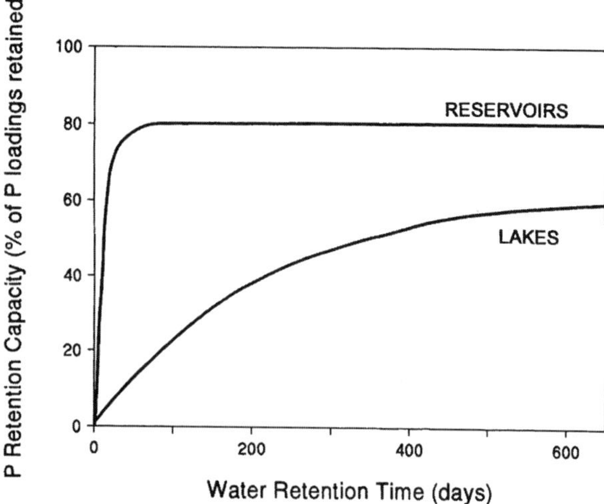

FIGURE 15-14 Phosphorus retention capacity of lakes and reservoirs as a percentage of total phosphorus loadings retained in relation to water retention times. *(Modified from Straškraba, 1996.)*

sedimentation of biotic particles and water removal by surface outflows that predominate in lakes.

F. Effects of decreasing phosphorus on other lake characteristics

Changes in ecological structure and physical-chemical characteristics can also be expected with reductions in phosphorus concentration (Downing et al., 2001; Müller et al., 2019). Water clarity will increase as phytoplankton densities decline and oxygen concentrations in surficial sediments will improve as the supply of new organic matter to sediments declines. While this is most evident in the hypolimnia of thermally stratified lakes, the same holds true for the sediment/water boundaries in polymictic lakes. Hypolimnetic benthic and cold-water fish communities will be less impaired as oxygen levels improve.

While phosphorus directly regulates productivity in most lakes (until loading becomes excessive relative to nitrogen loading and a system becomes nitrogen limited), it indirectly regulates Cyanobacteria bloom formation via the impact of productivity on sediment redox because blooms are promoted by highly reduced surficial sediments (Molot et al., 2021b). Therefore the risk of Cyanobacteria blooms decreases as phosphorus loading declines in response to declines in the spatial and temporal extent of highly reduced surficial sediments. In turn, this will lead to the dominance of phytoplankton communities by eukaryotic algae. Collectively, these changes have ramifications for the entire food web.

IX. Modeling relationships between nutrient loading and phytoplankton productivity

The cornerstone of eutrophication management since the 1970s has been control of phosphorus inputs to aquatic systems. Successful restoration of eutrophic systems and preventing systems from becoming eutrophic requires decisions as to what constitutes a maximum acceptable phosphorus concentration. This is typically based on maintaining desirable lake characteristics such as nonhypoxic hypolimnetic waters, preventing Cyanobacteria blooms, protecting macrophyte beds, and maintaining phytoplankton densities below levels that would turn a lake green. Maximum acceptable phosphorus concentrations can then be used to estimate maximum acceptable phosphorus loads. In contemporary language loads and concentrations above the maximum acceptable criteria are considered unsustainable. Open-water chlorophyll *a* and total phosphorus concentrations are often used as indicators of sustainable conditions.

The pressing need for decision-making tools to guide eutrophication management efforts in the 1970s led to an early emphasis on developing empirical and simplified steady-state models (Reckhow and Chapra, 1983; Peters, 1986). These models do not explicitly describe all key processes and therefore have fewer data requirements than process-oriented dynamic models, yet they can still be reasonably accurate and therefore useful. Early empirical models predicted chlorophyll *a* as a function of total phosphorus loading (discussed below) bypassing the phosphorus concentration–chlorophyll *a* relationship, which was implicit in these models. More complex empirical and steady-state models have been developed (Håkanson and Bryhn, 2008; Paterson et al., 2009; Young et al., 2011) as well as non-steady-state process-oriented dynamic models (e.g., Reckhow and Chapra, 1999; Robson, 2014).

When phosphorus is added as a short-term pulse to unproductive lakes or ponds, either experimentally, for purposes of intentional fertilization, or in effluents resulting from human activities, the usual response is a very rapid increase in phytoplankton productivity (e.g., Einsele, 1941; Maciolek, 1954; Mortimer and Hickling, 1954; Vinberg and Liakhnovich, 1965; Vollenweider, 1968; Correll, 1998). Increased productivity is not sustained as nutrient loadings are reduced, however, but decreases rather rapidly in a few weeks or months to levels near those prior to the addition. Losses in the colloidal fraction and from sedimentation of particulate phosphorus result in continuous reductions of phosphorus from the trophogenic zone. Inputs to the system must be maintained in a continuous or pulsed manner in order to sustain the increased productivity. In other words, steady phosphorus loading of the system is critical to sustaining increased productivity in most lakes of low or medium biological productivity.

Conversely, in order to reduce the productivity of a lake that is receiving continuous loading of nutrients, phytoplankton growth is generally decreased most effectively by reduction of phosphorus input rates to below existing internal loss rates within the lake. Reduction of total phosphorus loading to the water column from external and internal sources is the objective, since phosphorus is the major nutrient in greatest demand in relation to supply. Since suppression of internal phosphorus loading requires oxidized sediments, external loading targets should be set at levels that prevent excessive production of organic matter.

It has been argued that reductions in nitrogen loading should also be pursued in conjunction with phosphorus reduction to lower productivity and decrease Cyanobacteria blooms, especially in systems where responses to phosphorus reductions have not achieved adequate goals. This requires removing enough nitrogen to shift a lake to nitrogen limitation since it is unlikely that nitrogen removal will lower phytoplankton biomass while it remains phosphorus limited. Although eukaryotic productivity might decrease if enough nitrogen is removed, Molot et al. (2021b) argue that, as long as sediments remain anaerobic, nitrogen controls will not prevent or mitigate (except perhaps to a limited extent) Cyanobacteria blooms unless nitrogen fixation is limited by supplies of trace metal cofactors such as molybdenum and iron. Indeed, decreasing nitrate loading might exacerbate freshwater blooms by promoting lower sediment redox, which seems to have happened in experimentally fertilized Lake 227 (Schindler et al., 2008). However, nitrogen reductions may decrease *cyanotoxin* production and might be warranted in some watersheds to deal with inorganic nitrogen toxicity and eutrophication of marine coastal waters. It should be noted that only phosphorus reductions achieve all of the major freshwater eutrophication management objectives: lower ecosystem productivity, prevention of Cyanobacteria blooms and associated cyanotoxicity, and improved hypolimnetic oxygen levels. Moreover, phosphorus is chemically reactive and therefore technologically easier to remove from water than nitrogen and does not have major reservoirs in the atmosphere that replenish losses.

A. Modeling nutrient loading and concentration

The *nutrient loading concept* implies that a relationship exists between the quantity of nutrients entering a water body, the lake's concentration of that nutrient, and the lake's biological response to that nutrient input. The effects of this relationship can be expressed by some quantifiable index of productivity or water-quality parameter (e.g., chlorophyll *a* concentrations and water transparency).

The loading concept was recognized some time ago as applicable to lake changes in response to phosphorus and nitrogen enrichment (e.g., Rawson, 1939, 1955; Ohle, 1956; Edmondson, 1961). Sawyer (1947) suggested that, if critical levels of dissolved inorganic nitrogen (300 μg N L^{-1}) and phosphorus (10 μg P L^{-1}) measured at the time of spring turnover in Wisconsin lakes were exceeded, nuisance populations of phytoplankton would likely develop during the growing season. Vollenweider (1966, 1968) first formulated definitive quantitative loading criteria for phosphorus and nitrogen and expected trophic conditions in water bodies. He defined boundaries between oligotrophic and eutrophic lakes by relating nutrient loadings to mean depth (as a measure of lake volume) and later refined these relationships (Vollenweider, 1969, 1975). Because phosphorus is commonly the initial limiting nutrient to freshwater phytoplankton, as discussed earlier, and because the many processes of nitrogen cycling (nitrification, denitrification, nitrogen fixation; Chapter 14) complicate accurate measurements of nitrogen loading of lakes, the loading criteria and models emphasized phosphorus relationships. Attempts to combine phosphorus and nitrogen loadings into lake enrichment models have also been empirically successful (e.g., Bachmann, 1984).

The concentration-loading relationships are all based on the mass balance equation of a substance *M* between its sources and losses (sinks) in an open system (Vollenweider et al., 1980):

$$\Delta M / \Delta t = I - O - (S - R) \tag{15-2}$$

where

$\Delta M/\Delta t$ = storage gain or loss of nutrient *M* over time Δt
I = external nutrient load
O = nutrient loss by outflow
S = nutrient loss to sediments
R = nutrient regeneration from sediments (internal loading)

While *I* and *O* can often be measured directly with accuracy, $(S - R)$ is more difficult to measure and is often obtained by difference.

Under steady-state conditions, $\Delta M/\Delta t = 0$. Although, in reality, steady state does not exist in a lake lakes that oscillate around relatively constant nutrient loads over periods of several years exhibit a set of trophic characteristics and may be viewed as a tendency toward a steady state. Often this steady state is defined as the mean nutrient storage content as measured at spring turnover or as the annual storage content over several years. The models based on mass balance further assume that the nutrient load is instantaneously and completely mixed throughout the lake water (continuously stirred mixed reactor), an assumption that is only partially true for most lakes. The models, consequently, are particularly applicable to a large population of lakes rather than to describing the specific behavior of an individual lake. The predictive ability of these simplistic steady-state models is quite good for small lakes.

A simple mass balance equation for predicting lake concentration that takes into account volumetric loading of phosphorus as well as losses by flushing and sedimentation is given by (Vollenweider and Kerekes, 1980):

$$\frac{\bar{d}[P]_\lambda}{dt} = \left(\frac{1}{\bar{\tau}_w}\right)[P]_i - \left(\frac{1}{\bar{\tau}_p}\right)[P]_\lambda \tag{15-3}$$

where

$[P]_\lambda$ = average (total) lake concentration (both dissolved and particulate phosphorus components)
$[P]_i$ = average inflow concentration of total phosphorus
$\bar{\tau}_p$ = average residence time of phosphorus
$\bar{\tau}_w$ = average residence time of water

in which the right-hand terms represent the average rate of supply to and the average rate of loss of total phosphorus from the lake, respectively, and the left-hand term represents the corresponding temporal variations of the average lake concentration. Then, assuming steady state and simplifying Eq. 15-3:

$$[P]_\lambda = (\bar{\tau}_p / \bar{\tau}_w)[P]_i \tag{15-4}$$

The premise of Eq. 15-4 can also be expressed as:

$$[P]_\lambda = (1 - R)[P]_i \tag{15-5}$$

where R = the phosphorus-retention coefficient:

$$R = \frac{[P]_i - [P]_\lambda}{[P]_i} = 1 - \bar{\tau}_p / \bar{\tau}_w \tag{15-6}$$

or statistically approximated by

$$\bar{\tau}_p/\bar{\tau}_w = \frac{1}{1 + \sqrt{\bar{\tau}_w}} \tag{15-7}$$

(Larsen and Mercier, 1976; Vollenweider, 1976). An approximate retention coefficient as a function of hydraulic load (q) was found to be (Kirchner and Dillon, 1975)

$$R = 0.426e^{-0.27q} + 0.574e^{-0.00949q} \tag{15-8}$$

where q = areal water load (hydraulic load) to the lake (m yr^{-1}), calculated as the lake outflow volume divided by the lake surface area. These two retention values, while not identical, yield similar results for lakes of mean depths between 10 and 30 m and hydraulic loads between 3 and 90 m yr^{-1} (Vollenweider and Kerekes, 1980). When either of these models was used and data plotted to show the extent of prediction of spring turnover phosphorus concentration from loading, they described the average statistical behavior of lakes in terms of the relationship between phosphorus load and phosphorus accumulation within the lake (Vollenweider, 1976, 1979; Rast and Lee, 1978). Deviant lakes could be shown to have mechanisms operating to overestimate or underestimate the loading (e.g., high sedimentation rates, high regeneration of phosphorus from the sediments, particularly in lakes that developed anoxic deepwaters; Nürnberg, 1984).

The steady-state relationship between phosphorus loading and concentration can also be expressed as (Dillon and Rigler 1975a, 1975b; Dillon and Molot, 1996; cf. Chapter 10, Eq. 10-10):

$$[\mathrm{TP}] = \mathrm{Load}/(q + v) \qquad (15\text{-}9)$$

where Load is the load of total phosphorus from all external and internal sources (mg m^{-2} yr^{-1}), [TP] is the mean annual whole-lake total phosphorus concentration (mg m^{-3}, equivalent to the commonly used μg L^{-1}), q is the water discharge rate (m yr^{-1}), and v is the generalized loss coefficient representing settling loss of phosphorus to the sediments (m yr^{-1}) because there is no gas phase and thus no loss to the atmosphere. In this formulation the fraction of the load retained by the lake is given by $R = v/(q + v)$. This version of the mass balance relationship forms the foundation of the Lakeshore Capacity Model, which was developed to predict the impact of shoreline development on water quality in forested areas of central Ontario (Paterson et al., 2009). Dillon and Molot (1996) viewed multiyear runoff, loading, and concentration means to be an approximation of steady state with differences between model predictions and observations for a given year often being large compared to differences between predictions and multiyear means.

Empirical models should be calibrated with high-quality regional data for predicting relationships within that region, but when data are not available, models from other regions or models calibrated with combined data sets from different studies could be used with the understanding that lake responses may deviate substantially from model predictions.

B. Modeling biological responses to nutrient loading

Based on these nutrient loading relationships, the loadings to a lake should be related to biological responses such as phytoplankton biomass (e.g., chlorophyll *a* concentrations) and productivity, which is a rate, measured as, for example, g C m^{-2} yr^{-1}. Early empirical predictions of chlorophyll *a* levels from phosphorus concentrations in lakes at spring turnover or during the growing season (Sakamoto, 1966; Dillon and Rigler, 1974; Jones and Bachmann, 1976; Schindler, 1978; Nicholls and Dillon, 1978) were later extended further with larger worldwide data sets (Vollenweider, 1976; Oglesby and Schaffner, 1978; Rast and Lee, 1978; Canfield and Bachmann, 1981; Vollenweider and Kerekes, 1982; Watson et al., 1992). Within certain boundaries, the regression between phosphorus inflow concentration, $[P_i]$, as a proxy for loading and average annual chlorophyll *a* concentration, [chl.a], is approximated by:

$$\overline{[\mathrm{chl}.a]} = 0.55\{[\mathrm{P}]_i/(1 + \sqrt{\tau_w})\}^{0.76} \qquad (15\text{-}10)$$

where τ_w is the water residence time. Examples of this relationship are given in Fig. 15-15; similar relationships have been found for many other lakes with only minor deviations.

The annual rate of phytoplankton primary productivity, C, has also been related to the predicted phosphorus concentration (Fig. 15-16). The nonlinearity of the log–log plot results from the light-reducing, self-shading effects of dense phytoplankton populations (high biomass) at high levels of productivity and nitrogen limitation, which can replace phosphorus limitation in highly eutrophic lakes.

Multicompartment dynamic models that predict biological responses as a function of nutrient loading have also been developed (Reckhow and Chapra, 1999; Robson, 2014).

C. Application of predictive models and data

The evaluation of the trophic status of a lake has great practical importance. The extent of eutrophication must be known before remedial corrective measures can be implemented in relation to the desired uses for any lake. Based upon a large data set obtained during an international program on eutrophication of the Organization for Economic Cooperation and Development (OECD), nutrient load–eutrophication response relationships for lakes and reservoirs were quantified and evaluated (Vollenweider, 1968; Rast and Lee, 1978; Canfield and Bachmann, 1981; Ortiz and Martinez, 1984). The analyses provide a potential basis by which predictions can be made of the changes in water quality that will occur from changes in the phosphorus loadings to phosphorus-limited water bodies. Analyses of over 200 water bodies permitted a general classification of lakes based on water transparency and concentrations of

FIGURE 15-15 Regression relationships between predicted total phosphorus concentration and annual mean chlorophyll *a* concentration for selected alpine lakes. *(Modified slightly from Vollenweider, 1979.)*

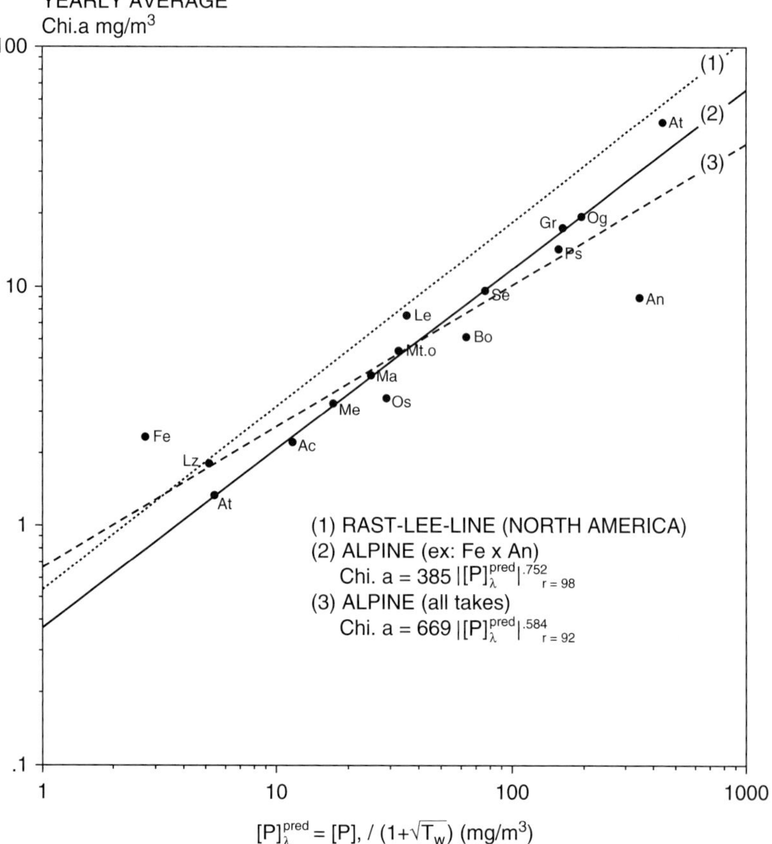

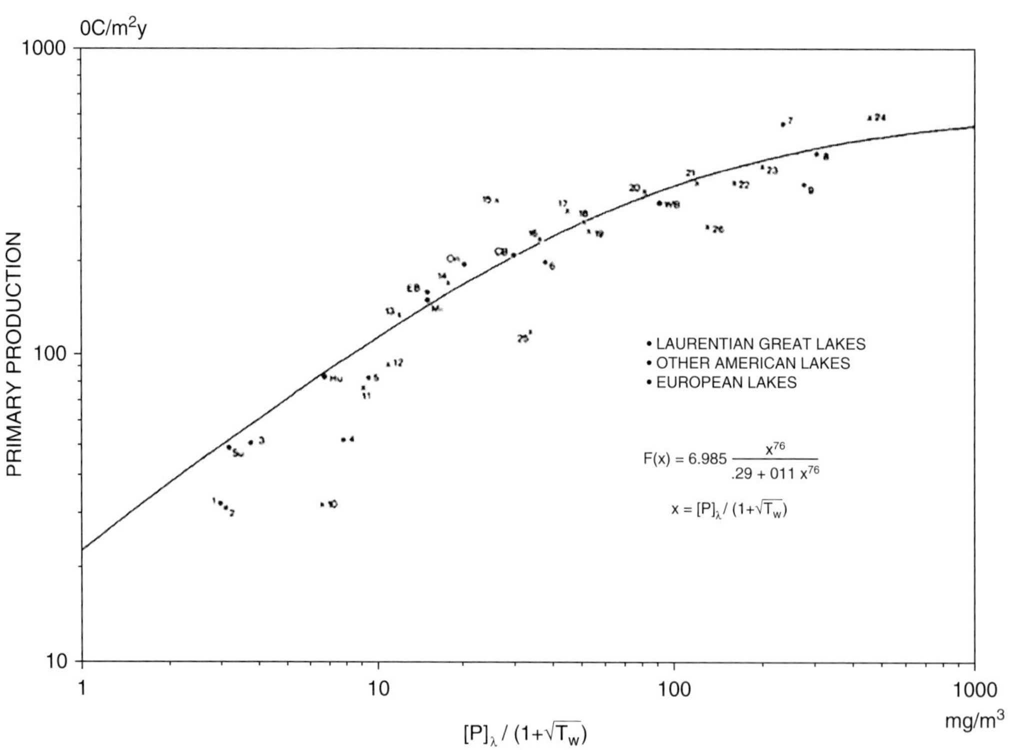

FIGURE 15-16 Annual primary productivity (g C m^{-2} yr^{-1}) as a function of predicted total phosphorus concentration. *(From Vollenweider, 1979.)*

TABLE 15-17 General Trophic Classification of Lakes and Reservoirs in Relation to Phosphorus and Nitrogen.[a]

Parameter (annual mean values)	*Oligotrophic*	*Mesotrophic*	*Eutrophic*	*Hypereutrophic*
Total phosphorus (mg m^{-3})				
Mean	8.0	26.7	84.4	–
Range	3.0–17.7	10.9–95.6	16–386	750–1200
N	21	19	71	2
Total nitrogen (mg m^{-3})				
Mean	661	753	1875	–
Range	307–1630	361–1387	393–6100	–
N	11	8	37	–
Chlorophyll *a* (mg m^{-3}) of phytoplankton				
Mean	1.7	4.7	14.3	–
Range	0.3–4.5	3–11	3–78	100–150
N	22	16	70	2
Chlorophyll *a* maxima (mg m^{-3}) ("worst case")				
Mean	4.2	16.1	42.6	–
Range	1.3–10.6	4.9–49.5	9.5–275	–
N	16	12	46	–
Secchi transparency depth (m)				
Mean	9.9	4.2	2.45	–
Range	5.4–28.3	1.5–8.1	0.8–7.0	0.4–0.5
N	13	20	70	2

[a]*Based on data from an international eutrophication program. Trophic status based on the opinions of the experienced investigators of each lake. (Modified from Vollenweider, 1979.)*
N is number of observations.

phosphorus, nitrogen, and phytoplankton pigment (Table 15-17). Although any attempt at rigid classification of water bodies is subjective and fraught with exceptions, the general classification is useful.

The models permit predictions of the probability of oligotrophic, intermediate, or eutrophic conditions developing in phosphorus-limited lakes in response to various loading regimes (Fig. 15-17). The literature on this subject is very large, for the models have been continuously tested, refined, and improved (e.g., Reckhow, 1979; Chapra and Reckhow, 1979, 1983). As with all modeling efforts of natural and disturbed ecosystems, one must not permit the mathematical manipulations to become an end in themselves (e.g., Pielou, 1981).

Models based on many data points permit estimates of permissible loading rates of phosphorus and nitrogen while still allowing tolerable conditions of productivity (Paterson et al., 2009). Considering a combination of influencing conditions, Vollenweider (1968) estimated very approximate provisional loading rates of nitrogen and phosphorus required to maintain lakes in a steady state (Table 15-18). Rapid eutrophication is likely to result as loading is increased. Although the predictive models have improved considerably since this 1968 analysis appeared, the tolerable loading level of 10 mg P m^{-3} remains unaltered, while the excessive loading level has been increased slightly to 25 mg P m^{-3} (Vollenweider and Kerekes, 1980).

A simple production-based *trophic state index* (TSI) uses phytoplankton biomass as a basis for a continuum of trophic states of lakes and reservoirs under both nutrient-limiting and non-nutrient-limited conditions (Carlson, 1977, 1980, 1992; Carlson and Simpson, 1996). TSI is a unitless empirical parameter with values mostly between 0 and 100: TSI values increase as a lake becomes more productive. Log-linear regressions for TSI as functions of Secchi depth, chlorophyll *a*, and total phosphorus have been developed using pooled data sets:

$$\mathrm{TSI(SD)} = 60 - 14.41\ \ln(\mathrm{SD}) \qquad (15\text{-}11)$$

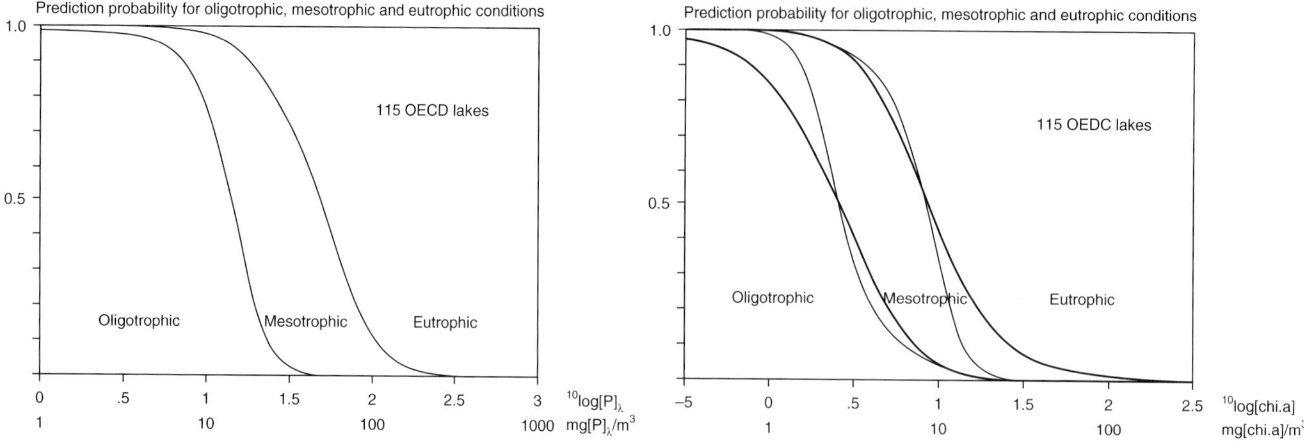

FIGURE 15-17 Probable boundaries of the degrees of trophy of water bodies with differing annual mean values of total phosphorus concentrations (*left*) and chlorophyll *a* concentrations (*right*). *(Modified from Vollenweider, 1979.)*

TABLE 15-18 Provisional Permissible Loading Levels for Total Nitrogen and Total Phosphorus (Biochemically Active) in g m^{-2} yr^{-1}[a]

	Permissible loading		*Dangerous loading*	
Mean depth (m)	*N*	*P*	*N*	*P*
5	1.0	0.07	2.0	0.13
10	1.5	0.10	3.0	0.20
50	4.0	0.25	8.0	0.50
100	6.0	0.40	12.0	0.80
150	7.5	0.50	15.0	1.00
200	9.0	0.60	18.0	1.20

[a]*After Vollenweider (1968).*

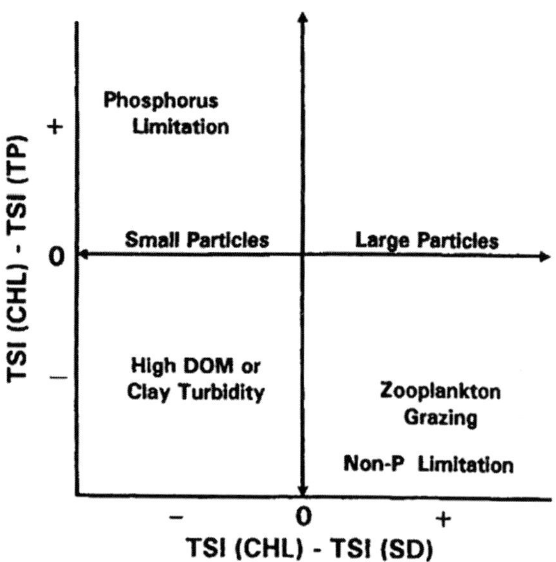

FIGURE 15-18 Potential nutrient-limited and non-nutrient-limited causes for deviation of biomass-based trophic state index (TSI). *(See text for details; based upon developments by Carlson, 1992.)*

$$TSI(CHL) = 9.81 \ln(CHL) + 30.6 \quad (15\text{-}12)$$

$$TSI(TP) = 14.42 \ln(TP) + 4.15 \quad (15\text{-}13)$$

where SD = Secchi disk transparency (m), CHL = chlorophyll *a* pigment concentrations (mg m^{-3}), and TP = total phosphorus concentrations (mg m^{-3}). Since the parameters exhibit seasonal trends, ice-free means are probably best for calculating TSI. TSI values of <30 are common among lakes and reservoirs of classical oligotrophy, and from 50 to 70 correspond to classical eutrophy. Hypereutrophic conditions are common at TSI values of >70.

Deviations in the TSI relationships can be clarified by graphical expression (Fig. 15-18). Values below the zero x-axis indicate the likelihood of something other than phosphorus limitation; points above this axis suggest increasing possibility of phosphorus limitation. Points to the right of the y-axis indicate transparency is greater than predicted from the chlorophyll *a* index. Points along the lower left diagonal to the left of the y-axis indicate reduced transparency from nonalgal factors, such as high concentrations of dissolved organic matter or small particles of nonalgal turbidity.

The general trophic classification system for lakes has been extended to streams. Rather than use the chlorophyll *a* of seston, the chlorophyll *a* content of benthic algae has been correlated to total nitrogen and phosphorus content in temperate streams (Dodds et al., 1998). The correlations between total phosphorus and chlorophyll *a* for streams are distinctly weaker than is the case for lakes, but a set of proximate trophic categories has been proposed (Table 15-19)

TABLE 15-19 Suggested Boundaries for Trophic Classification of Streams Based on Distributions of Benthic Chlorophyll Content and Nutrient Concentrations[a]

Variable	*Oligotrophic–mesotrophic boundary*	*Mesotrophic-eutrophic boundary*
Mean benthic chlorophyll *a* (mg m^{-2})	20	70
Maximum benthic chlorophyll *a* (mg m^{-2})	60	200
Sestonic chlorophyll *a* (mg L^{-1})	10	30
Total phosphorus (μg L^{-1})	25	75
Total nitrogen (μg L^{-1})	700	1500

[a]*After Dodds et al. (1998).*

X. Climate change and the phosphorus cycle

Climate change is an emerging topic in limnology as it is in many areas of environmental science. While numerous studies of climate impacts on aquatic systems have been conducted to date, the impacts on lake phosphorus cycles and what the future may hold remain unclear. A starting place to fill this knowledge gap is to identify factors that affect phosphorus export and internal lake dynamics based on our current understanding of these processes and combine them with our knowledge of how climate change affects those factors. While speculative, informed speculation is the first step to designing research studies. Using this approach, we can identify climate-related factors such as precipitation patterns, storm intensity, frequency of droughts and floods, changes in soil moisture and runoff, atmospheric stilling of winds, lake flushing rates, and decreased ice-cover for their potential to affect phosphorus export from catchments and lake phosphorus dynamics (Klink, 1999; McVicar et al., 2008; Hou et al., 2017; Sharma et al., 2019).

The majority of external phosphorus loading to many lakes enters via surface runoff; hence, changes in discharge volume, storm intensity, and timing will affect export rates. Large storm events can flush dissolved and particulate soil phosphorus resulting in large nutrient pulses to lakes, but prolonged storms will gradually dilute runoff concentrations (Heathwaite and Dils, 2000; Jordan et al., 2005; Jarvie et al., 2006). Large storms may also decrease lake residence time depending on the volume of runoff relative to lake volume. The frequency of large storms has increased in some regions in the continental USA (Karl and Knight, 1998), which suggests that long-term and short-term phosphorus export and timing of pulses may have changed. Storms or large snow melts that coincide with crop fertilization or buildup of soil phosphorus may export relatively more during storm runoff events. Droughts deprive lakes of new external loads of P while temporarily increasing residence times (Schindler and Donahue, 2006). However, baseflow phosphorus concentrations are relatively high during prolonged periods between storms and continuous baseflow can export significant amounts of phosphorus downstream (Jordan et al., 2005). Clarification of the impacts of changing storm and runoff patterns on phosphorus export will require long-term high-frequency monitoring of runoff and its phosphorus fractions in numerous catchments with varying land uses in different climatic regimes.

Atmospheric stilling of winds over the past several decades in some regions likely reduced average turbulence in lakes, but changing frequency of high-speed wind storms will affect the frequency of episodes of wind-induced turbulence, which are important to resuspension and elimination of thermal gradients, especially in polymictic systems. Shorter periods of ice-cover result in earlier onset and longer periods of thermal stratification, which means hypolimnetic waters may be deprived of fresh oxygen supplies for longer periods, increasing the risk of anoxia and internal phosphorus loading (Woolway et al., 2021). Less ice cover also leads to prolonged exposure to wind-induced mixing events. Warmer temperatures will increase microbial metabolic rates but temperature increases will vary with season and region. Productivity and therefore phosphorus cycling indirectly is also affected by cloud cover, which affects temperature and solar radiation (Shuvo et al., 2021).

A note of caution: changes in climate-related factors, such as those described above, may not always be driven by climate change (Crossman et al., 2013, 2014; Oni et al., 2014). In addition, they are not the only factors that can influence export and lake dynamics. Changes in land use, fertilization practices, agricultural and urban best management practices, upgrades to wastewater treatment plants, and expanding human populations will also affect export (Raney and Eimers, 2014; Jin et al., 2015) but their impacts may be difficult to distinguish from those caused by climate change when they cooccur. Coupled dynamic watershed-lake models may be helpful in evaluating their relative importance (Jin et al., 2013).

XI. Summary

1. Phosphorus plays a major role in biological metabolism. In comparison to other macronutrients required by biota, phosphorus is the least abundant and commonly is the first element to limit biological productivity.
2. Many quantitative studies exist on the seasonal and spatial distribution of phosphorus in streams and lakes, as well as the loading rates to recipient waters from drainage basins.
   - **a.** Orthophosphate (PO_4^{-3}) is the most directly utilizable form of soluble inorganic phosphorus. Phosphate is extremely reactive and interacts with many cations (e.g., Fe and Ca) to form, especially under oxidizing conditions, relatively insoluble compounds that precipitate out of the water. Availability of phosphate is also reduced by adsorption to inorganic colloids and particulate compounds (e.g., clays, carbonates, and hydroxides).
   - **b.** A large proportion of phosphorus in freshwaters is bound in organic phosphates and cellular constituents of organisms, both living and dead, and within or adsorbed to organic colloids.
   - **c.** The range of total phosphorus in freshwaters is large, from <5 μg L^{-1} in very unproductive waters to >100 μg L^{-1} in highly eutrophic waters. Most uncontaminated freshwaters contain between 5 and 50 μg total P L^{-1}.
   - **d.** Although concentrations of total and soluble phosphorus of oligotrophic lakes exhibit little variation with increasing depth, eutrophic lakes with strongly clinograde oxygen profiles commonly show a marked increase in phosphorus content in the lower hypolimnion (Fig. 15-1). Much of the hypolimnetic increase is from soluble phosphorus near the sediment–water interface.
3. Dissolved phosphorus in rivers is generally higher than in lakes and often increases markedly following rainfall and snowmelt events on the drainage basin.
   - **a.** At low concentrations in streams, phosphorus can be limiting and microbial productivity often increases with increased loading of phosphorus.
   - **b.** In addition to biological uptake, phosphorus abundance is regulated by abiotic adsorption and desorption reactions with organic and inorganic particles. Dissolved organic phosphorus compounds are utilized more slowly than inorganic forms and tend to be exported downstream for considerable distances.
   - **c.** Most of the recycling of phosphorus is associated with microbiota attached to particles. Less than 5% is associated with macroinvertebrates and higher organisms.
4. Exchanges of phosphorus across the sediment–water interface are regulated by oxidation–reduction (redox) interactions dependent on oxygen supply, mineral solubility, and sorptive mechanisms; the metabolic activities of bacteria and fungi; and turbulence from physical and biotic activities.
   - **a.** In all but the upper few millimeters of sediment, exchange is slow and controlled by low diffusion rates.
   - **b.** If water above the sediments is oxygenated (approximately >1 mg O_2 L^{-1}), an oxidized microzone is formed below the sediment–water interface (0 to −5 mm), below which the sediments usually become extremely reducing (Fig. 15-4). The oxidized microzone effectively prevents phosphorus (which is solubilized under reducing conditions in the sediments) from migrating by diffusion upward into the water column (Fig. 15-5, *left*).
   - **c.** As the hypolimnion becomes anoxic in productive lakes, the oxidized microzone is lost. The release of phosphate and ferrous iron into the water occurs readily when reducing conditions reach a redox potential <−100 mV uncorrected for pH (or about +200 mV when corrected for pH 7, E_7) (Figs. 15-4 and 15-5a, *right*).
   - **d.** Soluble phosphorus can accumulate in large quantities in anaerobic hypolimnia. With the advent of autumnal circulation, ferrous iron is rapidly oxidized and precipitates much of the phosphate as ferric phosphate.
   - **e.** Bacterial metabolism of organic matter is the primary mechanism by which organic phosphorus is converted to phosphate in the sediments and for creating the reducing conditions required for the release of phosphate to the water.
   - **f.** Movement of phosphorus from sediment interstitial water can be accelerated by physical turbulence and by biota. If light reaches sediments, the photosynthetic activity of epipelic algae can create a highly oxidized microzone in the surficial sediments and regulate phosphorus release to overlying water on a diurnal basis.
     - **i.** Rooted aquatic macrophytes often obtain phosphorus from the sediments and can release some back into the water both during active growth and particularly upon senescence and death. Most of the released phosphorus is

retained and recycled by attached algae, bacteria, and other microbiota.

ii. High population densities of sediment-dwelling invertebrates, such as midge larvae, can increase the exchange of phosphorus across the sediment–water interface.

5. Recent studies of phosphorus cycling in the trophogenic zone have demonstrated that the exchange of phosphorus between its various forms is often rapid and involves numerous complex pathways.
   a. A large portion, often >95%, of the phosphorus is bound in the particulate phase of living biota, especially bacteria, algae, and other microbes.
   b. Organic phosphorus of the seston of the open water consists of at least two major fractions:
      i. A rapidly cycled fraction that is exchanged with soluble forms. In this fraction phosphate is transferred rapidly through the particulate phase to low-molecular-weight compounds.
      ii. A fraction of dissolved organic and colloidal phosphorus material that is released and cycled more slowly.
   c. Phosphorus uptake and turnover kinetics in numerous lakes show that turnover rates are extremely rapid (5–100 min) during summer periods of high demand and low loading inputs. During winter periods, turnover rates become as much as two orders of magnitude slower. Phosphorus turnover rates are generally faster under more oligotrophic conditions of lower phosphorus availability.
   d. Flagellates, other Protista, and zooplankton that feed on bacteria and other seston excrete soluble phosphorus and ammonia. These nutrients are rapidly utilized by algae and bacteria. When supplies of phosphorus are low, this source of recycled phosphorus can be critical to the growth and succession of phytoplankton.
   e. Phytoplankton requirements for phosphorus are variable among species and can lead to selective advantages of certain algae as phosphorus supplies change seasonally or as phosphorus loadings to freshwater change over a longer period of time. Phytoplankton must compete with bacteria for phosphorus sources; if organic substrates for bacterial growth are high, algal growth may be seriously impeded.
6. Sedimentation of particles results in constant losses of phosphorus from the trophogenic zone. As a result, new phosphorus supplies must enter the ecosystem in order to maintain or increase productivity.
   a. Phosphorus enters freshwaters from atmospheric precipitation and from groundwater and surface runoff. The loading rates of phosphorus vary greatly with patterns of land use, geology, and morphology of the drainage basin, soil productivity, human activities, pollution, and other factors.
   b. When phosphorus is added to unproductive freshwaters, either experimentally or as a result of human activities, the usual response is a rapid increase in phytoplankton productivity. Inputs must be more or less continuous, however, to maintain a higher level of productivity. The risk of harmful Cyanobacteria blooms increases with increasing phosphorus concentration. Whole lake experiments have shown that anthropogenic nitrogen is not needed for blooms to form.
   c. Numerous steady-state and dynamic mass-balance models have been developed to predict, on the basis of phosphorus loading and retention times, the anticipated responses of algal biomass and productivity. The models predict a reasonably accurate estimation of permissible phosphorus loading needed to achieve a certain level of productivity if loading is lowered.
   d. In certain shallow lakes with greater-than-average turbulence, large littoral areas, and small, anaerobic hypolimnia, reduced productivity does not always occur as rapidly as reductions in external loading, as predicted from some models. In these lakes phosphorus release from sediment sources ("internal loading") is much greater than values (10–30% of total loading) for deeper, more stratified lakes.
7. The impacts of climate change on phosphorus dynamics are not well understood yet, but climate-related changes in precipitation patterns, storm intensity, frequency of droughts and floods, changes in soil moisture and runoff, atmospheric stilling of winds, lake flushing rates, and decreased ice-cover have potential to affect phosphorus export rates from catchments and lake phosphorus dynamics.

Acknowledgments

The authors would like to thank the editors for their constructive help in preparing this chapter.

References

Adams, M.S., Prentki, R.T., 1982. Biology, metabolism and function of littoral submersed weedbeds of Lake Wingra, Wisconsin, USA: a summary and review. Arch. Hydrobiol. Suppl. 62, 333–409.

Adams, W.P., English, M.C., Lasenby, D.C., 1979. Snow and ice in the phosphorus budget of a lake in South Central Ontario. Water Res. 13, 213–215.

Ågren, G.I., 2004. The C:N:P stoichiometry of autotrophs—theory and observations. Ecol. Lett. 7, 185–191.

Ahl, T., 1975. Effects of man-induced and natural loadings of phosphorus and nitrogen on the large Swedish lakes. Verh. Int. Verein. Limnol. 19, 1125–1132.

Ali, W., O'Melia, C.R., Edzwald, J.K., 1985. Colloidal stability of particles in lakes: measurement and significance. Water Sci. Tech. 17, 701–712.

Allen, H.L., 1972. Phytoplankton photosynthesis, micronutrient interactions, and inorganic carbon availability in a soft-water Vermont lake. In: Likens, G.E. (Ed.), Nutrients and Eutrophication: The Limiting-Nutrient Controversy. Special Symposium, vol. 1. Amer. Soc. Limnol. Oceanogr, pp. 63–83.

Aller, R.C., 1982. The effects of macrobenthos on chemical properties of marine sediment and overlying water. In: McCall, P.L., Tevesz, M.J.S. (Eds.), Animal-sediment Relations: The Biogenic Alteration of Sediments. Plenum Press, New York, pp. 53–102.

Andersen, J.M., 1975. Influence of pH on release of phosphorus from lake sediments. Arch. Hydrobiol. 76, 411–419.

Andersen, J.M., 1982. Effect of nitrate concentrations in lake water on phosphate release from the sediment. Water Res. 16, 1119–1126.

Andersen, F.Ø., Jensen, J.S., 1992. Regeneration of inorganic phosphorus and nitrogen from decomposition of seston in a freshwater sediment. Hydrobiologia 228, 71–81.

Anderson, G., Hance, R.J., 1963. Investigation of an organic phosphorus component of fulvic acid. Plant Soil 19, 296–303.

Andersen, O.K., Goldman, J.C., Caron, D.A., Dennett, M.R., 1986. Nutrient cycling in a microflagellate food chain. III. Phosphorus dynamics. Mar. Ecol. Progr. Ser. 31, 47–55.

Andersson, G., Granéli, W., Stenson, J., 1988. The influence of animals on phosphorus cycling in lake ecosystems. Hydrobiologia 170, 267–284.

Anthony, E.H., Hayes, F.R., 1964. Lake water and sediment. VII. Chemical and optical properties of water in relation to the bacterial counts in the sediments of twenty-five North American lakes. Limnol. Oceanogr. 9, 35–41.

Atkinson, K.M., Elliott, J.M., George, D.G., Jones, J.G., Haworth, E.Y., Heaney, S.I., Mills, C.A., Reynolds, C.S., Tailing, J.F., 1986. A General Assessment of Environmental and Biological Features of Windermere and Their Susceptibility to Change. Rept. Freshwater Biological Association, Ambleside, UK (Cited in Reynolds, 1997.).

Bachmann, R.W., 1984. Calculation of phosphorus and nitrogen loadings to natural and artificial lakes. Verh. Int. Verein. Limnol. 22, 239–243.

Ball, R.C., Hooper, E.F., 1963. Translocation of phosphorus in a trout stream ecosystem. In: Schultz, V., Klement Jr., A.W. (Eds.), Radioecology, First National Symposium on Radioecology. Reinhold Publishing Corp., New York, pp. 217–228.

Bannerman, R.T., Armstrong, D.E., Holdren, G.C., Harris, R.F., 1974. Phosphorus mobility in lake ontario sediments. Proc. Conf. Great Lakes Res. 17, 158–178.

Barbiero, R.P., Tuchman, M.L., 2004. Long-term dreissenid impacts on water clarity in Lake Erie. J. Great Lakes Res. 30, 557–565.

Barbiero, R.P., Welch, E.B., 1992. Contribution of benthic blue-green algal recruitment to lake populations and phosphorus translocation. Freshwat. Biol. 27, 249–260.

Barko, J.W., Smart, R.M., 1980. Mobilization of sediment phosphorus by submersed freshwater macrophytes. Freshw. Biol. 10, 229–238.

Barko, J.W., Smart, R.M., 1981. Sediment-based nutrient of submersed macrophytes. Aquat. Bot. 10, 339–352.

Barlow, J.P., Bishop, J.W., 1965. Phosphate regeneration by zooplankton in Cayuga Lake. Limnol. Oceanogr. 10 (Suppl. l), R15–R25.

Bartell, S.M., 1981. Potential impact of size-selective planktivory on phosphorus release by zooplankton. Hydrobiolgia 80, 139–145.

Bartell, S.M., Kitchell, J.F., 1978. Seasonal impact of planktivory on phosphorus release by Lake Wingra zooplankton. Verh. Int. Verein. Limnol. 20, 466–474.

Bastviken, D., Cole, J.J., Pace, M.L., Van de Bogert, M.C., 2008. Fates of methane from different lake habitats: connecting whole-lake budgets and CH_4 emissions. J. Geophys. Res. 113, G02024. https://doi.org/10.1029/2007JG000608.

Beaulac, M.N., Reckhow, K.H., 1982. An examination of land use-nutrient export relationships. Water Resour. Bull. 18, 1013–1024.

Bergström, A.K., Blomqvist, P., Jannson, M., 2005. Effects of atmospheric nitrogen deposition on nutrient limitation and phytoplankton biomass in unproductive Swedish lakes. Limnol. Oceanogr. 50, 987–994.

Berman, T., 1970. Alkaline phosphatases and phosphorus availability in Lake Kinneret. Limnol. Oceanogr. 15, 663–674.

Beutel, M.W., Duvil, R., Cubas, F.J., Matthews, D.A., Wilhelm, F.M., Grizzard, T.J., Austin, D., Horne, A.J., Gebremariam, S., 2016. A review of managed nitrate addition to enhance surface water quality. Crit. Rev. Environ. Sci. Tec. 46, 673–700.

Biggar, J.W., Corey, R.B., 1969. Agricultural drainage and eutrophication. In: Eutrophication: Causes, Consequences, Correctives. National Academy of Sciences, Washington, DC, pp. 404–445.

Björk-Ramberg, S., 1985. Uptake of phosphate and inorganic nitrogen by a sediment-algal system in a subarctic lake. Freshwat. Biol. 15, 175–183.

Blažka, P., Brandi, Z., Procházková, L., 1982. Oxygen consumption and ammonia and phosphate excretion in pond zooplankton. Limnol. Oceanogr. 27, 294–303.

Boavida, M.J., Marques, R.T., 1995. Low activity of alkaline phosphatase in two eutrophic reservoirs. Hydrobiologia 297, 11–16.

Boers, P.C.M., 1991. The influence of pH on phosphate release from lake sediments. Water Res. 25, 309–311.

Boers, P.C.M., van Hese, O., 1988. Phosphorus release from the peaty sediments of the Loosdrecht lakes (The Netherlands). Water Res. 22, 355–363.

Boon, P.I., 1994. Interactions Between Suspended Solids and Phosphatase Activity in Turbid Rivers.

Boström, B., Andersen, J.M., Fleischer, S., Jansson, J., 1988. Exchange of phosphorus across the sediment-water interface. Hydrobiologia 170, 229–244.

Boström, B., Jansson, M., Forsberg, C., 1982. Phosphorus release from lake sediments. Arch. Hydrobiol. Beih. Ergebn. Limnol. 18, 5–59.

Boudreau, B.P., 1998. Mean mixed depth of sediments: the wherefore and the why. Limnol. Oceanogr. 43, 524–526.

Bowers, J.A., 1986. Phosphorus regeneration by the predatory copepod *Diacyclops thomasi*. Can. J. Fish. Aquat. Sci. 43, 361–365.

Brabrand, Å., Faafeng, B.A., Nilssen, J.P.M., 1990. Relative importance of phosphorus supply to phytoplankton production: fish excretion versus external loading. Can. J. Fish. Aquat. Sci. 47, 364–372.

Bradford, M.E., Peters, R.H., 1987. The relationship between chemically analyzed phosphorus fractions and bioavailable phosphorus. Limnol. Oceanogr. 32, 1124–1137.

Brix, H., Lyngby, J.E., 1985. Uptake and translocation of phosphorus in eelgrass (*Zostera marina*). Mar. Biol. 90, 111–116.

Brock, T.C.M., Bongaerts, M.C.M., Heijnen, G.J.M.A., Heijthuijsen, J.H.F.G., 1983. Nitrogen and phosphorus accumulation and cycling by *Nymphoides peltata* (Gmel.) O. Kuntze (Menyanthaceae). Aquat. Bot. 17, 189–214.

Brown, E.J., Button, D.K., Lang, D.S., 1981. Competition between heterotrophic and autotrophic microplankton for dissolved nutrients. Microb. Ecol. 7, 199–206.

Budd, K., Kerson, G.W., 1987. Uptake of phosphate by two cyanophytes: cation effects and energetics. Can. J. Bot. 65, 1901–1907.

Burns, N.M., Ross, C., 1971. Nutrient relationships in a stratified eutrophic lake. Proc. Conf. Great Lakes Res., Int. Assoc. Great Lakes Res. 14, 749–760.

Camacho, A., Garcia-Pichel, F., Vicente, E., Castenholz, R.W., 1996. Adaptation to sulfide and to the underwater light field in three cyanobacterial isolates from Lake Arcas (Spain). FEMS Microbiol. Ecol. 21, 293–301.

Camacho, A., Vicente, E., Miracle, M., 2000. Ecology of a deep-living *Oscillatoria* (=*Planktothrix*) population in the sulphide-rich waters of a Spanish karstic lake. Archiv für Hydrobiol 148, 333–355.

Canfield Jr., D.E., Bachmann, R.W., 1981. Prediction of total phosphorus concentrations, chlorophyll a, and Secchi depths in natural and artificial lakes. Can. J. Fish. Aquat. Sci. 38, 414–423.

Caraco, N., Cole, J., Likens, G.E., 1990. A comparison of phosphorus immobilization in sediments of freshwater and coastal marine systems. Biogeochemistry 9, 277–290.

Caraco, N., Cole, J.J., Likens, G.E., 1989. Evidence for sulphate-controlled phosphorous release from sediments of aquatic systems. Nature 341, 316–318.

Carignan, R., Flett, R.J., 1981. Postdepositional mobility of phosphorus in lake sediments. Limnol. Oceanogr. 26, 361–366.

Carlson, R.E., 1977. A trophic state index for lakes. Limnol. Oceanogr. 22, 361–369.

Carlson, R.E., 1980. More complications in the chlorophyll-Secchi disk relationship. Limnol. Oceanogr. 25, 378–382.

Carlson, R.E., 1992. Expanding the trophic state concept to identify non-nutrient limited lakes and reservoirs. In: Proceedings of a National Conference on Enhancing the States' Lake Management Programs. Monitoring and Lake Impact Assessment, Chicago, pp. 59–71.

Carlson, R.E., Simpson, J., 1996. Trophic state. In: A Coordinator's Guide to Volunteer Lake Monitoring Methods. North American Lake Management Society, 7–1–7–20.

Carlton, R.G., Wetzel, R.G., 1987. Distributions and fates of oxygen in periphyton communities. Can. J. Bot. 65, 1031–1037.

Carlton, R.G., Wetzel, R.G., 1988. Phosphorus flux from lake sediments: effects of epipelic algal oxygen production. Limnol. Oceanogr. 33, 562–570.

Carrillo, P., I. Reche and L. Cruz-Pizarro. 1996. Quantification of the phosphorus released by zooplankton in an oligotrophic lake (La Caldera, Spain): regulating factors and adjustment to theoretical models. J. Plankton Res. 18:1567–1586.

Cavender-Bares, K.K., Mann, E.L., Chisholm, S.W., Ondrusek, M.E., Bidigare, R.R., 1999. Differential response of equatorial Pacific phytoplankton to iron fertilization. Limnol. Oceanogr. 44, 237–246.

Cembella, A.D., Antia, N.J., Harrison, P.J., 1983. The utilization of inorganic and organic phosphorous compounds as nutrients by eukaryotic microalgae: a multidisciplinary perspective. Part 1. CRC Crit. Rev. Microbiol. 10, 317–391.

Cembella, A.D., Antia, N.J., Harrison, P.J., 1984. The utilization of inorganic and organic phosphorous compounds as nutrients by eukaryotic microalgae: a multidisciplinary perspective. Part 2. CRC Crit. Rev. Microbiol. 11, 13–81.

Chaffin, J.D., Kane, D.D., 2010. Burrowing mayfly (Ephemeroptera: ephemeridae: Hexagenia spp.) bioturbation and bioirrigation: a source of internal phosphorus loading in Lake Erie. J. Great Lakes Res. 36, 57–63.

Chamberlain, W.M., 1968. A Preliminary Investigation of the Nature and Importance of Soluble Organic Phosphorus in the Phosphorus Cycle of Lakes. Ph.D. Diss., University of Toronto, Ontario, p. 232.

Chapin, J.D., Uttormark, P.D., 1973. Atmospheric contributions of nitrogen and phosphorus. Tech. Rep. Wat. Resources Ctr. Univ. Wis. 73–2, 35.

Chapra, S.C., Reckhow, K.H., 1979. Expressing the phosphorus loading concept in probabilistic terms. J. Fish. Res. Board Can. 36, 225–229.

Chapra, S.C., Reckhow, K.H., 1983. Engineering Approaches for lake Management. II. Mechanistic Modelling. Ann Arbor Science-Butterworth Publishers, Boston.

Chorus, I., Köhler, A., Beulker, C., Fastner, J., van de Weyer, K., Hegewald, T., Hupfer, M., 2020. Decades needed for ecosystem components to respond to a sharp and drastic phosphorus load reduction. Hydrobiologia 847, 4621–4651.

Chow-Fraser, P., Duthie, H.C., 1983. An interpretation of phosphorus loadings in dystrophic lakes. Arch. Hydrobiol. 97, 109–121.

Chróst, R.J., Overbeck, J., 1987. Kinetics of alkaline phosphatase activity and phosphorus availability for phytoplankton and bacterioplankton in Lake Plußsee (North German eutrophic lake). Microb. Ecol. 13, 229–248.

Chu, S.P., 1943. The influence of the mineral composition of the medium on the growth of planktonic algae. II. The influence of the concentration of inorganic nitrogen and phosphate phosphorus. J. Ecol. 31, 109–148.

Clesceri, N.L., Curran, S.J., Sedlak, R.I., 1986. Nutrient loads to Wisconsin lakes. Part 1. Nitrogen and phosphorus export coefficients. Water Resour. Bull. 22, 983–990.

Clevinger, C.C., Heath, R.T., Bade, D.L., 2014. Oxygen use by nitrification in the hypolimnion and sediments of Lake Erie. J. Great Lakes Res. 40, 202–207.

Cloern, J.E., 1977. Effects of light intensity and temperature on *Cryptomonas ovata* (Cryptophyceae) growth and nutrient uptake rates. J. Phycol. 13, 389–395.

Coffin, C.C., Hayes, F.R., Jodrey, L.H., Whiteway, S.G., 1949. Exchange of materials in a lake as studied by the addition of radioactive phosphorus. Can. J. Res. 27, 207–222.

Cole, J.J., Caraco, N.F., Likens, G.E., 1990. Short-range atmospheric transport: a significant source of phosphorus to an oligotrophic lake. Limnol. Oceanogr. 35, 1230–1237.

Confer, J.L., 1972. Interrelations among plankton, attached algae, and the phosphorus cycle in artificial open systems. Ecol. Monogr. 42, 1–23.

Conroy, J.D., Kane, D.D., Dolan, D.M., Edwards, W.J., Charlton, M.N., Culver, D.A., 2005. Temporal trends in Lake Erie plankton biomass: roles of external phosphorus loading and dreissenid mussels. J. Great lakes. Res. 31, 89–110.

Cooke, G.D., Welch, E.B., Peterson, S.A., Newroth, P.R., 1993. Restoration and Management of Lakes and Reservoirs, second ed. Lewis Publishers, Chelsea, MI, p. 548.

Cooke, G.D., McComas, M.R., Waller, D.W., Kennedy, R.H., 1977. The occurrence of internal phosphorus loading in two small, eutrophic glacial lakes in northeastern Ohio. Hydrobiologia 56, 129–135.

Cooke, G.W., Williams, R.J.B., 1973. Significance of man-made sources of phosphorus: fertilizers and farming. Water Res. 7, 19–33.

Cooper, C.F., 1969. Nutrient output from managed forests. In: Eutrophication: Causes, Consequences, Correctives. National Academy of Sciences, Washington, DC, pp. 446–463.

Correll, D.L., 1998. The role of phosphorus in the eutrophication of receiving waters: a review. J. Environ. Qual. 27, 261–266.

Cotner, J.B., Wetzel, R.G., 1991a. Bacterial phosphatase from different habitats in a small, hardwater lake. In: Chróst, R.J. (Ed.), Microbial Enzymes in Aquatic Environments. Springer-Verlag, New York, pp. 187–205.

Cotner, J.B., Wetzel, R.G., 1991b. 5′-nucleotidase activity in a eutrophic lake and an oligotrophic lake. Appl. Environ. Microbiol. 57, 1306–1312.

Cotner, J.B., Wetzel, R.G., 1992. Uptake of dissolved inorganic and organic phosphorus compounds by phytoplankton and bacterioplankton. Limnol. Oceanogr. 37, 232–243.

Cotner Jr., J.B., Heath, R.T., 1988. Potential phosphate release from phosphomonoesters by acid phosphatase in a bog lake. Arch. Hydrobiol. 111, 329–338.

Coveney, M.F., Wetzel, R.G., 1992. Effects of nutrients on specific growth rate of bacterioplankton in oligotrophic lake water cultures. Appl. Environ. Microbiol. 58, 150–156.

Cowen, W.F., Lee, G.F., 1973. Leaves as source of phosphorus. Environ. Sci. Technol. 7, 853–854.

Crossman, J., Futter, M., Oni, S., Whitehead, P.G., Jin, L., Baulch, H.M., Dillon, P.J., 2013. Impacts of climate change on hydrology and water quality: the need to future proof management strategies in the Lake Simcoe watershed. Canada. J. Great Lakes Res. 39, 19–32.

Crossman, J., Futter, M.N., Whitehead, P.G., Stainsby, E., Baulch, H.M., Jin, L., Oni, S.K., Dillon, P.J., 2014. Flow pathways and nutrient transport mechanisms drive hydrochemical sensitivity to climate change across catchments with different geology and topography. Hydrol. Earth Sys. Sci. 18, 5125–5148.

Currie, D.J., 1990. Large-scale variability and interactions among phytoplankton, bacterioplankton, and phosphorus. Limnol. Oceanogr. 35, 1437–1455.

Currie, D.J., Kalff, J., 1984a. The relative importance of bacterioplankton and phytoplankton in phosphorus uptake in freshwater (*sic*). Limnol. Oceanogr. 29, 311–324.

Currie, D.J., Kalff, J., 1984b. Can bacteria outcompete phytoplankton for phosphorus: a chemostat test. Microb. Ecol. 10, 205–216.

Currie, D.J., Kalff, J., 1984c. A comparison of the abilities of freshwater algae and bacteria to acquire and retain phosphorus. Limnol. Oceanogr. 29, 298–310.

Danen-Louwerse, H., Lijklema, L., Coenraats, M., 1993. Iron content of sediment and phosphate adsorption properties. Hydrobiologia 253, 311–317.

Davis, M.B., 1973. Redeposition of pollen grains in lake sediment. Limnol. Oceanogr. 18, 44–52.

Davis, R.B., Thurlow, D.L., Brewster, F.E., 1975. Effects of burrowing tubificid worms on the exchange of phosphorus between lake sediment and overlying water. Verh. Int. Verein. Limnol. 19, 382–394.

de Haan, H., Jones, R.I., Salonen, K., 1990. Abiotic transformations of iron and phosphate in humic lake water revealed by double-isotope labeling and gel filtration. Limnol. Oceanogr. 35, 491–497.

DeMarte, J.A., Hartman, R.T., 1974. Studies on absorption of ^{32}P, ^{59}Fe, and ^{45}Ca by water-milfoil (*Myriophyllum exalbescens* Fernald). Ecology 55, 188–194.

DeMontigny, C., Prairie, Y.T., 1993. The relative importance of biological and chemical processes in the release of phosphorous from a highly organic sediment. Hydrobiologia 253, 141–150.

Den Oude, P.J., Gulati, R.D., 1988. Phosphorus and nitrogen excretion rates of zooplankton from the eutrophic Loosdrecht lakes, with notes on other P sources for phytoplankton requirements. Hydrobiologia 169, 379–390.

Denny, P., 1972. Sites of nutrient absorption in aquatic macrophytes. J. Ecol. 60, 819–829.

Detenbeck, N.E., Brezonik, P.L., 1991. Phosphorus sorption by sediments from a soft-water seepage lake. 2. Effects of pH and sediment composition. Environ. Sci. Technol. 25, 403–409.

Dévai, I., Felföldy, L., Wittner, I., Plósz, S., 1988. Detection of phosphine: new aspects of the phosphorus cycle in the hydrosphere. Nature 333, 343–345.

Devito, K.J., Dillon, P.J., 1993. The influence of hydrologic conditions and peat oxia on the phosphorus and nitrogen dynamics of a conifer swamp. Water Resour. Res. 29, 2675–2685.

Diaz, O.A., Reddy, K.R., Moore Jr., P.A., 1994. Solubility of inorganic phosphorus in stream water as influenced by pH and calcium concentration. Water Res. 28, 1755–1763.

Dillon, P.J., Rigler, F.H., 1974. The phosphorus-chlorophyll relationship in lakes. Limnol. Oceanogr. 19, 767–773.

Dillon, P.J., Kirchner, W.B., 1975. The effects of geology and land use on the export of phosphorus from watersheds. Water Res. 9, 135–148.

Dillon, P.J., Rigler, F.H., 1975a. A simple method for predicting the capacity of a lake for development based on lake trophic status. J. Fish. Res. Board Can. 32, 1519–1531.

Dillon, P.J., Rigler, F.H., 1975b. A test of a simple nutrient budget model predicting the phosphorus concentration in lake water. J. Fish. Res. Board Can. 31, 1771–1778.

Dillon, P.J., Molot, L.A., 1996. Long-term phosphorus budgets and an examination of the steady state mass balance model for central Ontario lakes. Water Res. 30, 2273–2280.

Dillon, P.J., Molot, L.A., 1997. Effect of landscape form on export of dissolved organic carbon, iron, and phosphorus from forested stream catchments. Water Resour. Res. 33, 2591–2600.

Dillon, P.J., Nicholls, K.H., Robinson, G., 1978. Phosphorus removal at Gravenhurst Bay, Ontario: an 8-year study on water quality changes. Verh. Int. Verein. Limnol. 20, 263–271.

Dillon, P.J., Evans, R.D., Molot, L.A., 1990. Retention and resuspension of phosphorus, nitrogen, and iron in a central Ontario lake. Can. J. Fish. Aquat. Sci. 47, 1269–1274.

Dittrich, M., Chesnyuk, A., Gudimov, A., McCulloch, J., Quazi, S., Young, J., Winter, J., Stainsby, E., Arhonditsis, G., 2013. Phosphorus retention in a mesotrophic lake under transient loading conditions: Insights from a sediment phosphorus binding form study. Water Res. 47, 1433–1447.

Dodds, W.K., Ellis, B.K., Priscu, J.C., 1991. Zooplankton induced decrease in inorganic phosphorus uptake by plankton in an oligotrophic lake. Hydrobiologia 211, 253–259.

Dodds, W.K., Jones, J.R., Welch, E.B., 1998. Suggested classification of stream trophic state: distributions of temperate stream types by chlorophyll, total nitrogen, and phosphorus. Water Res. 32, 1455–1462.

Downes, M.T., Paerl, H.W., 1978. Separation of two dissolved reactive phosphorus fractions in lakewater. J. Fish. Res. Board Can. 35, 1636–1639.

Downing, J.A., Watson, S.B., McCauley, J.E., 2001. Predicting Cyanobacteria dominance in lakes. Can. J. Fish. Aquat. Sci. 58, 1905–1908.

Downs, T.M., Schallenberg, M., Burns, C.W., 2008. Responses of lake phytoplankton to micronutrient enrichment: a study in two New Zealand lakes and an analysis of published data. Aquat. Sci. 70, 347–360.

Drake, J.C., Heaney, S.I., 1987. Occurrence of phosphorus and its potential remobilization in the littoral sediments of a productive English lake. Freshwat. Biol. 17, 513–523.

Droop, M.R., 1973. Some thoughts on nutrient limitation in algae. J. Phycol. 9, 264–272.

Duursma, E.K., 1967. The mobility of compounds in sediments in relation to exchange between bottom and supernatant water. In: Golterman, H.L., Clymo, R.S. (Eds.), Chemical Environment in the Aquatic Habitat. N.V. Noord-Hollandsche Uitgevers Maatsahappij, Amsterdam, pp. 288–296.

Eccleston-Parry, J.D., Leadbeater, B.S.C., 1995. Regeneration of phosphorus and nitrogen by four species of heterotrophic nanoflagellates feeding on three nutritional states of a single bacterial strain. Appl. Environ. Microbiol. 61, 1033–1038.

Eckerrot, Å., Pettersson, K., 1993. Pore water phosphorus and iron concentrations in a shallow, eutrophic lake—indications of bacterial regulation. Hydrobiologia 253, 165–177.

Edmondson, W.T., 1961. Changes in Lake Washington following an increase in the nutrient income. Verh. Int. Verein. Limnol. 14, 167–175.

Edmondson, W.T., 1969. Eutrophication in north America. In: Eutrophication: Causes, Consequences, Correctives. National Academy of Sciences, Washington, DC, pp. 124–149.

Edmondson, W.T., 1972. Nutrients and phytoplankton in lake Washington. In: Likens, G.E. (Ed.), Nutrients and Eutrophication: The Limiting-Nutrient Controversy. Special Symposium, vol. 1. Amer. Soc. Limnol. Oceanogr, pp. 172–193.

Edmondson, W.T., 1991. The Uses of Ecology: Lake Washington and beyond. Univ. Washington Press, Seattle, p. 329.

Edmondson, W.T., Lehman, J.T., 1981. The effect of changes in the nutrient income on the condition of Lake Washington. Limnol. Oceanogr. 26, 1–29.

Effler, S.W., Auer, M.T., Johnson, N., Penn, M., Rowell, H.C., 1996. Sediments. In: Effler, S.W. (Ed.), Limnological and Engineering

Analysis of a Polluted Urban Lake. Springer-Verlag, New York, pp. 600–666.

Einsele, W., 1936. Über die Beziehungen des Eisenkreislaufs zum Phosphatkreislauf im eutrophen See. Arch. Hydrobiol. 29, 664–686.

Einsele, W., 1941. Die Umsetzung von zugeführtem, anorganischen Phosphat im eutrophen See und ihre Rüchwirkungen auf seinen Gesamthaushalt. Zeitsch. f. Fischerei 39, 407–488.

Ejsmont-Karabin, J., 1983. Ammonia nitrogen and inorganic phosphorus excretion by the planktonic rotifers. In: Pejler, B., Starkweather, R., Nogrady, T. (Eds.), Biology of Rotifers. Developments in Hydrobiology, 14. Springer, Dordrecht. https://doi.org/10.1007/978-94-009-7287-2_29.

Elser, J.J., George, N.B., 1993. The stoichiometry of N and P in the pelagic zone of Castle Lake, California. J. Plankton Res. 15, 977–992.

Elser, J.J., Marzolf, E.R., Goldman, C.R., 1990. Phosphorus and nitrogen limitation of phytoplankton growth in the freshwaters (*sic*) of North America: a review and critique of experimental enrichments. Can. J. Fish. Aquat. Sci. 47, 1468–1477.

English, M.C., 1978. The magnitude and significance of the terrestrial snowpack and white ice contribution to the phosphorus budget of a lake in the Canadian Shield Region of the Kawartha lakes. Proc. Eastern Snow Conf. 35, 173–189.

Ensign, S.H., Doyle, M.W., 2006. Nutrient spiraling in streams and river networks. J. Geophys. Res. 111, G04009. https://doi.org/10.1029/2005JG000114.

Esteves, A.F., Soares, O.S.G.P., Vilar 2, V.J.P., Pires, J.C.M., Gonçalves, A.L., 2020. The effect of light wavelength on CO_2 capture, biomass production and nutrient uptake by green microalgae: a step forward on process integration and optimisation. Energies 13, 333. https://doi.org/10.3390/en13020333.

Falkner, G., Horner, F., Simonis, W., 1980. The regulation of the energy-dependent phosphate uptake by blue-green alga *Anacystis nidulans*. Planta 149, 138–143.

Ferrante, J.G., 1976. The role of zooplankton in the intrabiocoenotic phosphorus cycle and factors affecting phosphorus excretion in a lake. Hydrobiologia 49, 203–214.

Ferrier-Pagès, C., Rassoulzadegan, F., 1994. N remineralization in planktonic protozoa. Limnol. Oceanogr. 39, 411–419.

Ferrier-Pagès, C., Karner, M., Rassoulzadegan, F., 1998. Release of dissolved amino acids by flagellates and ciliates grazing on bacteria. Oceanol. Acta 21, 485–494.

Fleischer, S., 1978. Evidence for the anaerobic release of phosphorus from lake sediments as a biological process. Naturwissenschaften 65, 109.

Fleischer, S., 1986. Aerobic uptake of Fe(III)-precipitated phosphorus by microorganisms. Arch. Hydrobiol. 107, 269–277.

Fogg, G.E., 1971. Extracellular products of algae in freshwater. Arch. Hydrobiol. Beih. Ergebn. Limnol. 5, 25.

Forsberg, C., Ryding, S.O., 1980. Eutrophication parameters and trophic state indices in 30 Swedish waste-receiving lakes. Arch. Hydrobiol. 89, 189–207.

Fox, L.E., 1993. The chemistry of aquatic phosphate: inorganic processes in rivers. Hydrobiologia 253, 1–16.

Foy, R.H., 1986. Suppression of phosphorus release from lake sediments by the addition of nitrate. Water Res. 20, 1345–1351.

Francko, D.A., 1986. Epilimnetic phosphorus cycling: influence of humic materials and iron on coexisting major mechanisms. Can. J. Fish. Aquat. Sci. 43, 302–310.

Francko, D.A., Heath, R.T., 1979. Functionally distinct classes of complex phosphorus compounds in lake water. Limnol. Oceanogr. 24, 463–473.

Francko, D.A., Heath, R.T., 1981. Aluminum sulfate treatment: short-term effect on complex phosphorus compounds in a eutrophic lake. Hydrobiologia 78, 125–128.

Francko, D.A., Heath, R.T., 1982. UV-sensitive complex phosphorus: association with dissolved humic material and iron in a bog lake. Limnol. Oceanogr. 27, 564–569.

Frevert, T., 1979a. Phosphorus and iron concentrations in the interstitial water and dry substance of sediments of Lake Constance (Obersee). I. General discussion. Arch. Hydrobiol. Suppl. 55, 298–323.

Frevert, T., 1979b. The pE redox concept in natural sediment-water systems; its role in controlling phosphorus release from lake sediments. Arch. Hydrobiol. Suppl. 55, 278–297.

Frevert, T., 1980. Dissolved oxygen dependent phosphorus release from profundal sediments of Lake Constance (Obersee). Hydrobiologia 74, 17–28.

Fuhs, G.W., Demmerle, S.D., Canelli, E., Chen, M., 1972. Characterization of phosphorus-limited plankton algae (with reflections on the limiting-nutrient concept). In: Likens, G.E. (Ed.), Nutrients and Eutrophication: The Limiting-Nutrient Controversy. Special Symposium, vol. 1. Amer. Soc. Limnol. Oceanogr., pp. 113–133

Fukuhara, H., Yasuda, K., 1985. Phosphorus excretion by some zoobenthos in a eutrophic freshwater lake and its temperature dependency. Jpn. J. Limnol. 46, 287–296.

Fukuhara, H., Sakamoto, M., 1987. Enhancement of inorganic nitrogen and phosphate release from lake sediment by tubificid worms and chironomid larvae. Oikos 48, 312–320.

Gabrielson, J.O., Perkins, M.A., Welch, E.B., 1984. The uptake, translocation and release of phosphorus by *Elodea densa*. Hydrobiologia 111, 43–48.

Gächter, R., Bloesch, J., 1985. Seasonal and vertical variation in the C:P ratio of suspended and settling seston of lakes. Hydrobiologia 128, 193–200.

Gachter, R., Meyer, J.S., 1993. The role of microorganisms in mobilization and fixation of phosphorous in sediments. Hydrobiologia 253, 103–121.

Gächter, R., Muller, B., 2003. Why the phosphorus retention of lakes does not necessarily depend on the oxygen supply to their sediment surface. Limnol. Oceanogr. 48, 929–933.

Gak, D.Z., 1959. Fiziologicheskaia aktivnost' i sistematicheskoe polozhenie mobilizuiushchikh fosfor mikroorganizmov, vydelennykh iz vodoemov pribaltiki. Mikrobiologiya 28, 551–556.

Gak, D.Z., 1963. Vertikal'noe raspredelenie mobilizuiushchikh fosfor bakterii v gruntakh Latviiskikh vodoemov. Mikrobiologiya 32, 838–842.

Gallepp, G.W., 1979. Chironomid influence on phosphorus release in sediment-water microcosms. Ecology 60, 547–556.

Gallepp, G.W., Kitchell, J.F., Bartell, S.M., 1978. Phosphorus release from lake sediments as affected by chironomids. Verh. Int. Verein. Limnol. 20, 458–465.

Gálvez, J.A., Niell, F.X., Lucena, J., 1989. Seston vertical flux model for a eutrophic reservoir. Arch. Hydrobiol. Beih. Ergebn. Limnol. 33, 9–18.

Gálvez, J.A., Niell, F.X., Lucena, J., 1991. C:N:P ratio of settling seston in a eutrophic reservoir. Verh. Int. Verein. Limnol. 24, 1390–1395.

Gassmann, G., 1994. Phosphine in the fluvial and marine hydrosphere. Mar. Chem. 45, 197–205.

Gassmann, G., Schorn, F., 1993. Phosphine from harbor surface sediments. Naturwissenschaften 80, 78–80.

Gervais, F., Siedel, U., Heilmann, B., Weithoff, G., Heisig-Gunkel, G., Nicklisch, A., 2003. Small-scale vertical distribution of phytoplankton, nutrients and sulphide below the oxycline of a mesotrophic lake. J. Plank. Res. 25, 273–278.

Gibson, C.E., Wu, Y., Pinkerton, D., 1995. Substance budgets of an upland catchment: the significance of atmospheric phosphorus inputs. Freshwat. Biol. 33, 385–392.

Gibson, G., Carlson, R., Simpson, J., Smeltzer, E., Gerritson, J., Chapra, S., Heiskary, S., Jones, J., Kennedy, R., 2000. Nutrient

Criteria Technical Guidance Manual. Lakes and Reservoirs. United States Environmental Protection Agency. EPA-822-B00-001.

Glindemann, D., Stottmeister, U., Bergmann, A., 1996. Free phosphine from the anaerobic biosphere. Environ. Sci. Pollut. Res. 3, 17–19.

Golachowska, J.B., 1971. The pathways of phosphorus in lake water. Pol. Arch. Hydrobiol. 18, 325–245.

Goldman, J.C., McCarthy, J.J., Peavey, D.G., 1979. Growth rate influence on the chemical composition of phytoplankton in oceanic waters. Nature 279, 210–214.

Golterman, H.L., 1982. Loading concentration models for phosphate in shallow lakes. Hydrobiologia 91, 169–174.

Golterman, H.L., 1995. The role of the iron hydroxide-phosphate-sulphide system in the phosphate exchange between sediments and overlying water. Hydrobiologia 297, 43–54.

Golterman, H.L., Bakels, C.C., Jakobs-Mögelin, J., 1969. Availability of mud phosphates for the growth of algae. Verh. Int. Verein. Limnol. 17, 467–479.

Goma, R.H., Aizaki, M., Fukushima, T., Otsuki, A., 1996. Significance of zooplankton grazing activity as a sources of dissolved organic nitrogen, urea and dissolved free amino acids in a eutrophic shallow lake: experiments using outdoor continuous flow pond systems. Jpn. J. Limnol. 57, 1–13.

Gong, Q., Feng, Y., Kang, L., Luo, M., Yang, J., 2014. Effects of light and pH on cell density of *Chlorella vulgaris*. Energy Proc. 61, 2012–2015.

Gorham, E., 1955. On some factors affecting the chemical composition of Swedish fresh waters. Geochim. Cosmochim. Acta 7, 129–150.

Granéli, W., Solander, D., 1988. Influence of aquatic macrophytes on phosphorus cycling in lakes. Hydrobiologia 170, 245–266.

Griffith, E.J., Beeton, A., Spencer, J.M., Mitchell, D.T., 1973. Environmental Phosphorus Handbook. John Wiley & Sons, New York, p. 718.

Grobler, D.C., Daves, E., 1981. Sediments as a source of phosphate: a study of 38 impoundments. Water S.A. 7, 54–60.

Grobler, D.C., Silberbauer, M.J., 1985. The combined effect of geology, phosphate sources and runoff on phosphate export from drainage basins. Water Res. 19, 975–981.

Grover, J.P., 2000. Resource competition and community structure in aquatic microorganisms: experimental studies of algae and bacteria along a gradient of organic carbon to inorganic phosphorus supply. J. Plank. Res. 22, 1591–1610.

Güde, H., 1991. Participation of bacterioplankton in epilimnetic phosphorus cycles of Lake Constance. Verh. Int. Verein. Limnol. 24, 816–820.

Gulati, R.D., Ejsmont-Karabin, J., Rooth, J., Siewertsen, K., 1989. A laboratory study of phosphorus and nitrogen excretion of Euchlanis dilatata lucksiana. In: Ricci, C., Snell, T.W., King, C.E. (Eds.), Rotifer Symposium V. Developments in Hydrobiology, vol. 52. Springer, Dordrecht, pp. 347–354. https://doi.org/10.1007/978-94-009-0465-1_42.

Gulati, R.D., Siewertsen, K., Van Liere, L., 1991. Carbon and phosphorus relationships of zooplankton and its seston food in Loosdrecht lakes. Mem. Ist. Ital. Idrobiol. 48, 279–298.

Gulati, R.D., Perez Martinez, C., Siewertsen, K., 1995. Zooplankton as a compound mineralising and synthesizing system: phosphorus excretion. Hydrobiologia 315, 25–37. https://doi.org/10.1007/BF00028628.

Håkanson, L., Bryhn, A.C., 2008. Eutrophication in the Baltic Sea. Environmental Science and Engineering. Springer-Verlag, Berlin Heidelberg.

Halmann, M., Elgavish, A., 1975. The role of phosphate in eutrophication. Stimulation of plankton growth and residence times of inorganic phosphate in Lake Kinneret water. Verh. Int. Verein. Limnol. 19, 1351–1356.

Halmann, M., Stiller, M., 1974. Turnover and uptake of dissolved phosphate in freshwater. A study in Lake Kinneret. Limnol. Oceanogr. 19, 774–783.

Hansson, L.-A., 1989. The influence of a periphytic biolayer on phosphorus exchange between substrate and water. Arch. Hydrobiol. 115, 21–26.

Hargrave, B.T., 1972. Aerobic decomposition of sediment and detritus as a function of particle surface area and organic content. Limnol. Oceanogr. 17, 583–596.

Harper, D.M., 1992. Eutrophication of Freshwaters (*Sic*): Principles, Problems and Restoration. Chapman and Hall, London.

Hart, B.T., Freeman, P., McKelvie, I.D., Pearse, S., Ross, D.G., 1991. Phosphorus spiralling in myrtle creek, victoria, Australia. Verh. Int. Verein. Limnol. 24, 2065–2070.

Harter, R.D., 1968. Adsorption of phosphorus by lake sediment. Proc. Soil Sci. Soc. Amer. 32, 514–518.

Hasler, A.D., Einsele, W.G., 1948. Fertilization for increasing productivity of natural inland waters. Trans. N. Amer. Wildl. Conf. 13, 527–555.

Hayes, F.R., 1955. The effect of bacteria on the exchange of radiophosphorus at the mud-water interface. Verh. Int. Verein. Limnol. 12, 111–116.

Hayes, F.R., 1964. The mud-water interface. Oceanogr. Mar. Biol. Rev. 2, 121–145.

Hayes, F.R., Coffin, C.C., 1951. Radioactive phosphorus and the exchange of lake nutrients. Endeavor 10, 78–81.

Hayes, F.R., Phillips, J.E., 1958. Lake water and sediment. IV. Radiophosphorus equilibrium with mud, plants, and bacteria under oxidized and reduced conditions. Limnol. Oceanogr. 3, 459–475.

Hayes, F.R., Anthony, E.H., 1959. Lake water and sediment. VI. The standing crop of bacteria in lake sediments and its place in the classification of lakes. Limnol. Oceanogr. 4, 299–315.

Hayes, F.R., MacAulay, M.A., 1959. Lake water and sediment. V. Oxygen consumed in water over sediment cores. Limnol. Oceanogr. 4, 291–298.

Head, R.M., Jones, R.I., Bailey-Watts, A.E., 1999. Vertical movements by planktonic cyanobacteria and the translocation of phosphorus: implications for lake restoration. Aquat. Conserv. 9, 111–120.

Healey, F.P., Hendzel, L.L., 1975. Effect of phosphorus deficiency on two algae growing in chemostats. J. Phycol. 11, 303–309.

Healey, F.P., Hendzel, L.L., 1980. Physiological indicators of nutrient deficiency in lake phytoplankton. Can. J. Fish. Aquat. Sci. 37, 442–453.

Heath, R.T., Cooke, G.D., 1975. The significance of alkaline phosphatase in a eutrophic lake. Verh. Int. Verein. Limnol. 19, 959–965.

Heathwaite, A.L., Dils, R.M., 2000. Characterising phosphorus loss in surface and subsurface hydrological pathways. Sci. Tot. Environ. 251/252, 523–538.

Hecky, R.E., Kilham, P., 1988. Nutrient limitation of phytoplankton in freshwater and marine environments: a review of recent evidence on the effects of enrichment. Limnol. Oceanogr. 33, 796–882.

Hecky, R.E., Campbell, P., Hendzel, L.L., 1993. The stoichiometry of carbon, nitrogen, and phosphorus in particulate matter of lakes and oceans. Limnol. Oceanogr. 38, 709–724.

Hecky, R.E., Smith, R.E.H., Barton, D.R., Guildford, S.J., Taylor, W.D., Charlton, M.N., Howell, T., 2004. The nearshore phosphorus shunt: a consequence of ecosystem engineering by dreissenids in the Laurentian Great Lakes. Can. J. Fish. Aquat. Sci. 61, 1285–1293.

Henry, R., Hino, K., Gentil, J.G., Tundisi, J.G., 1985. Primary production and effects of enrichment with nitrate and phosphate on phytoplankton in the Barra Bonita reservoir (state of sao paulo, Brazil). Int. Revue ges. Hydrobiol. 70, 561–573.

Hepher, B., 1966. Some aspects of the phosphorus cycle in fish ponds. Verh. Int. Verein. Limnol. 16, 1293–1297.

Hessen, D.O., Andersen, T., 1990. Bacteria as a source of phosphorus for zooplankton. Hydrobiologia 206, 217–223.

Higgins, S.N., Paterson, M.J., Hecky, R.E., Schindler, D.W., Venkiteswaran, J.J., Findlay, D.L., 2018. Biological nitrogen fixation prevents the response of a eutrophic lake to reduced loading of nitrogen: evidence from a 46-year whole-lake experiment. Ecosystems 21, 1088–1100.

Hillbricht-Ilkowska, A., Lawacz, W., 1983. Biotic structure and processes in the lake system of R. Jorka watershed (Masurian Lakeland, Poland). Ekol. Pol. 31, 539–585.

Hobbie, J. E. and G. E. Likens. 1973. Output of phosphorus, dissolved organic carbon, and fine particulate carbon from Hubbard Brook watersheds. Limnol. Oceanogr. 18:734–742.

Hoffman, F.A., Bishop, J.W., 1994. Impacts of a phosphate detergent ban on concentrations of phosphorus in the James River, Virginia. Water Res. 28, 1239–1240.

Holden, A.V., 1961. The removal of dissolved phosphate from lake waters by bottom deposits. Verh. Int. Verein. Limnol. 14, 247–251.

Holdren Jr., G.C., Armstrong, D.E., Harris, R.F., 1977. Interstitial inorganic phosphorus concentrations in lakes Mendota and Wingra. Water Res. 12, 1041–1047.

Holmgren, S.K., 1984. Experimental lake fertilization in the Kuokkel Area, northern Sweden. Phytoplankton biomass and algal composition in natural and fertilized subarctic lakes. Int. Revue ges. Hydrobiol. 69, 781–817.

Hooper, F.F., Elliott, A.M., 1953. Release of inorganic phosphorus from extracts of lake mud by protozoa. Trans. Amer. Microsc. Soc. 72, 276–281.

Horppila, J., Nurminen, L., 2003. Effects of submerged macrophytes on sediment resuspension and internal phosphorus loading in Lake Hiidenvesi (southern Finland). Wat. Res. 37, 4468–4474.

Hou, E., Chen, C., Luo, Y., Zhou, G., Kuang, Y., Zhang, Y., Heenan, M., Lu, X., Wen, D., 2017. Effects of climate on soil phosphorus cycle and availability in natural terrestrial ecosystems. Global Change Biol. 24, 3344–3356.

House, W.A., Donaldson, L., 1986. Adsorption and coprecipitation of phosphate on calcite. J. Colloid Interface Sci. 112, 309–324.

House, W.A., Casey, H., Donaldson, L., Smith, S., 1986. Factors affecting the coprecipitation of inorganic phosphate with calcite in hardwaters. I. Laboratory studies. Water Res. 20, 917–922.

Howard, D.E., Evans, R.D., 1993. Acid-volatile sulfide (AVS) in a seasonally anoxic mesotrophic lake: seasonal and spatial changes in sediment AVS. Environ. Toxicol. Chem. 12, 1051–1057.

Howard-Williams, C., Allanson, B.R., 1981. Phosphorus cycling in a dense *Potamogeton pectinatus* L. Bed. Oecologia 49, 56–66.

Hoyer, M.V., Jones, J.R., 1983. Factors affecting the relation between phosphorus and chlorophyll *a* in midwestern reservoirs. Can. J. Fish. Aquat. Sci. 40, 192–199.

Hupfer, M., Dollan, A., 2003. Immobilisation of phosphorus by iron-coated roots of submerged macrophytes. Hydrobiologia 506 (509), 635–640.

Hupfer, M., Lewandowski, J., 2008. Oxygen controls the phosphorus release from lake sediments—a long-lasting paradigm in limnology. Internat. Rev. Hydrobiol. 93, 415–432.

Hutchinson, G.E., 1950. The biogeochemistry of vertebrate excretion. Bull. Amer. Mus. Nat. Hist. 96, 554.

Hutchinson, G.E., 1957. A Treatise on Limnology. I. Geography, Physics, and Chemistry. John Wiley & Sons, New York, p. 1015.

Hutchinson, G.E., 1967. A Treatise on Limnology. II. Introduction to Lake Biology and the Limnoplankton. John Wiley & Sons, New York, p. 1115.

Hutchinson, G.E., 1973. Eutrophication. The scientific background of a contemporary practical problem. Am. Sci. 61, 269–279.

Hynes, H.B.N., Greib, B.J., 1970. Movement of phosphate and other ions from and through lake muds. J. Fish. Res. Board Can. 27, 653–668.

Jabbari, A., Ackerman, J.D., Boegman, L., Zhao, Y., 2019. Episodic hypoxia in the western basin of Lake Erie. Limnol. Oceanogr. 64, 2220–2236.

Jabbari, A., Ackerman, J.D., Boegman, L., Zhao, Y., 2021. Increases in Great Lake winds and extreme events facilitate interbasin coupling and reduce water quality in Lake Erie. Sci. Reports 11, 5733. https://doi.org/10.1038/s41598-021-84961-9.

Jackson, T.A., Schindler, D.W., 1975. The bio-geochemistry of phosphorus in an experimental lake environment: evidence for the formation of humic-metal-phosphate complexes. Verh. Int. Verein. Limnol. 19, 211–221.

Jackson, T.A., Hecky, R.E., 1980. Depression of primary productivity by humic matter in lake and reservoir waters of the boreal forest zone. Can. J. Fish. Aquat. Sci. 37, 2300–2317.

Jacobsen, T.R., Comita, G.W., 1976. Ammonia-nitrogen excretion in *Daphnia pulex*. Hydrobiologia 51, 195–200.

Jäger, P., Röhrs, J., 1990. Phosphorfällung über Calciumcarbonat im eutrophen Wallersee (Salzburger Alpenvorland, Österreich). Int. Revue ges. Hydrobiol. 75, 153–173.

James, W.F., Barko, J.W., 1990. Macrophyte influences on the zonation of sediment accretion and composition in a north-temperate reservoir. Arch. Hydrobiol. 120, 129–142.

James, W.F., Salonen, K., 1991. Zooplankton-phytoplankton interactions and their importance in the phosphorus cycle of a polyhumic lake. Archive Hydrobiol 123, 37–51.

Jansson, M., 1976. Phosphatases in lakewater: characterization of enzymes from phytoplankton and zooplankton by gel filtration. Science 194, 320–321.

Jansson, M., 1981. Induction of high phosphatase activity by aluminum in acid lakes. Arch. Hydrobiol. 93, 32–44.

Jansson, M., Persson, G., Broberg, O., 1986. Phosphorus in acidified lakes: the example of Lake Gardsjon, Sweden. Hydrobiologia 139, 81–96.

Jansson, M., Olsson, H., Pettersson, K., 1988. Phosphatases: origin, characteristics and function in lakes. Hydrobiologia 170, 157–175.

Janus, L.L., Vollenweider, R.A., 1984. Phosphorus residence time in relation to trophic conditions in lakes. Verh. Int. Verein. Limnol. 22, 179–184.

Jarvie, H.P., Neal, C., Withers, P.J.A., 2006. Sewage-effluent phosphorus: a greater risk to river eutrophication than agricultural phosphorus? Sci. Tot. Environ. 360, 246–253.

Jiang, M., Zhou, Y., Cao, X., Ji, X., Zhang, W., Huang, W., Zhang, J., Zheng, Z., 2019. The concentration thresholds establishment of nitrogen and phosphorus considering the effects of extracellular substrate-to-biomass ratio on cyanobacterial growth kinetics. Sci. Tot. Environ. 662, 307–312.

Jin, L., Whitehead, P.G., Baulch, H.M., Dillon, P.J., Butterfield, D., Oni, S.K., Futter, M.N., Crossman, J., O'Connor, E.M., 2013. Modelling phosphorus in Lake Simcoe and its subcatchments: scenario analysis to assess alternative management strategies. Inland Waters 3, 207–220.

Jin, L., Whitehead, P.G., Sarkar, S., Sinha, R., Futter, M.N., Butterfield, D., Caesar, J., Crossman, J., 2015. Assessing the impacts of climate changes and socioeconomic changes on flow and phosphorus flux in the Ganga River system. Environ. Sci.: Process. Impacts 17, 1098–1110.

Johannes, R.E., 1964a. Uptake and release of phosphorus by a benthic marine amphipod. Limnol. Oceanogr. 9, 235–242.

Johannes, R.E., 1964b. Uptake and release of dissolved organic phosphorus by representatives of a coastal marine ecosystem. Limnol. Oceanogr. 9, 224–234.

Johannes, R.E., 1964c. Phosphorus excretion and body size in marine animals: microzooplankton and nutrient regeneration. Science 146, 923–924.

Johnston, W.R., Ittihadieh, F., Daum, R.M., Pillsbury, A.F., 1965. Nitrogen and phosphorus in tile drainage effluent. Proc. Soil Sci. Soc. Amer. 29, 287–289.

Jones, J.G., 1972. Studies on freshwater bacteria: association with algae and alkaline phosphatase activity. J. Ecol. 60, 59–75.

Jones, J.G., 1976. The microbiology and decomposition of seston in open water and experimental enclosures in a productive lake. J. Ecol. 64, 241–278.

Jones, J.R., Bachmann, R.W., 1976. Prediction of phosphorus and chlorophyll levels in lakes. J. Water Pollut. Control Fed. 48, 2176–2182.

Jones, J.R., Smart, M.M., Sebaugh, J.N., 1984. Factors related to algal biomass in Missouri Ozark streams. Verh. Int. Verein. Limnol. 22, 1867–1875.

Jones, R.I., 1990. Phosphorus transformations in the epilimnion of humic lakes: biological uptake of phosphate. Freshwat. Biol. 23, 323–337.

Jones, R.I., Salonen, K., de Haan, H., 1988. Phosphorus transformations in the epilimnion of humic lakes: abiotic interactions between dissolved humic materials and phosphate. Freshwat. Biol. 19, 357–369.

Jones, R.I., Shaw, P.J., de Haan, H., 1993. Effects of dissolved humic substances on the speciation of iron and phosphate at different pH and ionic strength. Environ. Sci. Technol. 27, 1052–1059.

Jordan, P., Arnscheidt, J., McGrogan, H., McCormick, S., 2005. High resolution phosphorus transfers at the catchment scale: the hidden importance of no-storm transfers. Hydrol. Earth Syst. Sci. 9, 685–691.

Jordan, T.E., Weller, D.E., 1996. Human contributions to terrestrial nitrogen flux. Bioscience 46, 655–664.

Juday, C., Birge, E.A., 1931. A second report on the phosphorus content of Wisconsin lake waters. Trans. Wis. Acad. Sci. Arts Lett. 26, 353–382.

Juday, C., Birge, E.A., Kemmerer, G.I., Robinson, R.J., 1927. Phosphorus content of lake waters in northeastern Wisconsin. Trans. Wis. Acad. Sci. Arts Lett. 23, 233–248.

Jürgens, K., Güde, H., 1990. Incorporation and release of phosphorus by planktonic bacteria and phagotrophic flagellates. Mar. Ecol. Prog. Ser. 59, 271–284.

Kairesalo, T., Matilainen, T., 1988. The importance of low flow rates to the phosphorus flux between littoral and pelagial zones. Verh. Int. Verein. Limnol. 23, 2210–2215.

Kairesalo, T., Seppälä, T., 1987. Phosphorus flux through a littoral ecosystem: the importance of cladoceran zooplankton and young fish. Int. Rev. ges. Hydrobiol. 72, 385–403.

Kamp-Nielsen, L., Mejer, H., Jørgensen, S.E., 1982. Modelling the influence of bioturbation on the vertical distribution of sedimentary phosphorus in L. Esrom. Hydrobiologia 91, 197–206.

Karl, T.R., Knight, R.W., 1998. Secular trends of precipitation amount, frequency, and intensity in the United States. Bull. Amer. Meteorol. Soc. 79, 231–242. https://doi.org/10.1175/1520-0477(1998)079<0231:STOPAF>2.0.CO;2.

Kasuga, S., Otsuki, A., 1984. Phosphorus release by stirring up sediments and mysids feeding activities. Res. Rep. Natl. Inst. Environ. Stud. Jpn. 51, 141–155.

Keating, K.I., 1977. Allelopathic influence on blue-green bloom sequence in a eutrophic lake. Science 196, 885–887.

Keen, W.H., Gagliardi, J., 1981. Effect of brown bullheads on release of phosphorus in sediment and water systems. Prog. Fish-Cult. 43, 183–185.

Kelderman, P., Lindeboom, H.J., Klein, J., 1988. Light dependent sediment-water exchange of dissolved reactive phosphorus and silicon in a producing microflora mat. Hydrobiologia 159, 137–147.

Kerr, P.C., Brockway, D.L., Paris, D.F., Barnett Jr., J.T., 1972. The interrelation of carbon and phosphorus in regulating heterotrophic and autotrophic populations in an aquatic ecosystem, Shriner's Pond. In: Likens, G.E. (Ed.), Nutrients and Eutrophication: The Limiting-Nutrient Controversy. Special Symposium, vol. 1. Amer. Soc. Limnol. Oceanogr., pp. 41–62

Keup, L.E., 1968. Phosphorus in flowing waters. Water Res. 2, 373–386.

Kilham, P., 1982. Acid precipitation: its role in the alkalization of a lake in Michigan. Limnol. Oceanogr. 27, 856–867.

Kilham, P., 1984. Sulfate in African inland waters: sulfate to chloride ratios. Verh. Int. Verein. Limnol. 22, 296–302.

Kilham, S.S., 1986. Dynamics of Lake Michigan natural phytoplankton communities in continuous cultures along a Si:P loading gradient. Can. J. Fish. Aquat. Sci. 43, 351–360.

Kilham, S.S., 1990. Relationship of phytoplankton and nutrients to stoichiometric measures. In: Tilzer, M.M., Serruya, C. (Eds.), Large Lakes: Ecological Structure and Function. Springer-Verlag, New York, pp. 403–414.

Kirchner, W.B., Dillon, P.J., 1975. An empirical method of estimating the retention of phosphorus in lakes. Water Resour. Res. 11, 182–183.

Kleiner, J., 1988. Coprecipitation of phosphate with calcite in lake water: a laboratory experiment modelling phosphorus removal with calcite in Lake Constance. Water Res. 22, 1259–1265.

Klink, K., 1999. Trends in mean monthly maximum and minimum surface wind speeds in the coterminous United States, 1961 to 1990. Clim. Res. 13, 193–205.

Klotz, R.L., 1985. Factors controlling phosphorus limitation in stream sediments. Limnol. Oceanogr. 30, 543–553.

Koenings, J.P., Hooper, F.F., 1976. The influence of colloidal organic matter on iron and iron-phosphorus cycling in an acid bog lake. Limnol. Oceanogr. 21, 684–696.

Kohlenbrander, G.J., 1972. The eutrophication of surface water by agriculture and the urban population. Stickstoff 13, 56–67.

Kopacek, J., Borovec, J., Hejzlar, J., Ulrich, K.-U., Norton, S.A., Amirbahman, A., 2005. Aluminum control of phosphorus sorption by lake sediments. Environ. Sci. Technol. 39, 8784–8789.

Kopacek, J., Marešová, M., Hejzlar, J., 2007. Natural inactivation of phosphorus by aluminum in preindustrial lake sediments. Natural inactivation of phosphorus by aluminum in preindustrial lake sediments. Limnol. Oceanogr. 52, 1147–1155.

Kortmann, R.W., 1980. Benthic and atmospheric contributions to the nutrient budgets of a soft-water lake. Limnol. Oceanogr. 25, 229–239.

Koschel, R., 1990. Pelagic calcite precipitation and trophic state of hardwater lakes. Arch. Hydrobiol. Beih. Ergebn. Limnol. 33, 713–722.

Koschel, R., 1997. Structure and function of pelagic calcite precipitation in lake ecosystems. Verh. Int. Verein. Limnol. 26, 343–349.

Kowalczewski, A., Rybak, J.I., 1981. Atmospheric fallout as a source of phosphorus for Lake Warniak. Ekol. Polska 29, 63–71.

Krause, H.R., 1961. Einige Bemerkungen über den postmortalen Abbau von Süsswasser-Zooplankton unter Laboratoriums- und Freilandbedingungen. Arch. Hydrobiol. 57, 539–543.

Krause, H.R., 1962. Investigation of the decomposition of organic matter in natural waters. FAO Fish. Biol. Report 34, 19 (FB/R34).

Krause, H.R., 1964. Zur Chemie und Biochemie der Zersetzung von Süsswasserorganismen, unter besonderer Berücksichtigung des Abbaues der organischen Phosphorkomponenten. Verh. Int. Verein. Limnol. 15, 549–561.

Krauss, R.W., 1958. Physiology of the fresh-water algae. Ann. Rev. Plant Physiol. 9, 207–244.

Kuenzler, E.J., 1970. Dissolved organic phosphorus excretion by marine phytoplankton. J. Phycol. 6, 7–13.

Kuhl, A., 1962. Inorganic phosphorus uptake and metabolism. In: Lewin, R.A. (Ed.), Physiology and Biochemistry of Algae. Academic Press, New York, pp. 211–229.

Kuhl, A., 1974. Phosphorus. In: Stewart, W.D.P. (Ed.), Algal Physiology and Biochemistry. Univ. California Press, Berkeley, pp. 636–654.

Kümmel, R., 1981. Zur Phosphateliminierung durch Fällung mit Calciumionen. Acta Hydrochim. Hydrobiol. 9, 585–588.

Kussatz, C., Gnauck, A., Jorga, W., Mayer, H.-G., Schürmann, L., Weise, G., 1984. Untersuchungen zur Phosphataufnahme durch Unterwasserpflanzen. Acta Hydrochim. Hydrobiol. 12, 659–677.

Laforge, M.P., Michel, N.L., Wheeler, A.L., Brook, R.K., 2016. Habitat selection by female moose in the Canadian prairie ecozone. J. Wildlife Manage. 80, 1059–1068. https://doi.org/10.1002/jwmg.21095.

Lal, R., 1998. Soil erosion impact on agronomic productivity and environment quality. Crit. Rev. Pl. Sci. 17, 319–464.

Lamarra Jr., V.J., 1975. Digestive activities of carp as a major contributor to the nutrient loading of lakes. Verh. Int. Verein. Limnol. 19, 2461–2468.

Landers, D.H., 1982. Effects of naturally senescing aquatic macrophytes on nutrient chemistry and chlorophyll *a* of surrounding waters. Limnol. Oceanogr. 27, 428–439.

Larsen, D.P., Mercier, H.T., 1976. Phosphorus retention capacity of lakes. J. Fish. Res. Board Can. 33, 1742–1750.

Larsen, D.P., Malueg, K.W., Schults, D.W., Brice, R.M., 1975. Response of eutrophic Shagawa Lake, Minnesota, U.S.A., to point-source phosphorus reduction. Verh. Int. Verein. Limnol. 19, 884–892.

Lean, D.R.S., 1973a. Phosphorus dynamics in lake water. Science 179, 678–680.

Lean, D.R.S., 1973b. Movements of phosphorus between its biologically important forms in lake water. J. Fish. Res. Board Can. 30, 1525–1536.

Lean, D.R.S., Nalewajko, C., 1976. Phosphate exchange and organic phosphorus excretion by freshwater algae. J. Fish. Res. Board Can. 33, 1312–1323.

Lean, D.R.S., Rigler, F.H., 1974. A test of the hypothesis that abiotic phosphate complexing influences phosphorus kinetics in epilimnetic lake water. Limnol. Oceanogr. 19, 784–788.

Lehman, J.T., 1976. Ecological and nutritional studies on *Dinobryon* Ehrenb.: Seasonal periodicity and the phosphate toxicity problem. Limnol. Oceanogr. 21, 646–658.

Lehman, J.T., 1980a. Release and cycling of nutrients between planktonic algae and herbivores. Limnol. Oceanogr. 25, 620–632.

Lehman, J.T., 1980b. Nutrient recycling as an interface between algae and grazers in freshwater communities. In: Kerfoot, W.C. (Ed.), Evolution and Ecology of Zooplankton Communities. Univ. Press New England, Hanover, NH, pp. 251–263.

Lehman, J.T., Naumoski, T., 1985. Content and turnover rates of phosphorus in *Daphnia pulex:* Effect of food quality. Hydrobiologia 128, 119–125.

Lenz, P.H., Melack, J.M., Robertson, B., Hardy, E.A., 1986. Ammonium and phosphate regeneration by the zooplankton of an Amazon floodplain lake. Freshwat. Biol. 16, 821–830.

Leonard, E.N., Mattson, V.R., Benoit, D.A., Hoke, R.A., Ankley, G.T., 1993. Seasonal variation of acid volatile sulfide concentration in sediment cores from three northeastern Minnesota lakes. Hydrobiologia 271, 87–95.

Letey, J., Blair, J.W., Devitt, D., Lund, L.J., Nash, P., 1977. Nitrate-nitrogen in effluent from agricultural tile drains in California. Hilgardia 45, 289–319.

Levine, S., 1975. Orthophosphate concentration and flux within the epilimnia of two Canadian Shield lakes. Verh. Int. Verein. Limnol. 19, 624–629.

Levine, S.N., Schindler, D.W., 1992. Modification of the N:P ratio in lakes by in situ processes. Limnol. Oceanogr. 37, 917–935.

Lewis Jr., W.M., 1986. Nitrogen and phosphorus runoff losses from a nutrient-poor tropical moist forest. Ecology 67, 1275–1282.

Lewis Jr., W.M., 1996. Tropical lakes: How latitude makes a difference. In: Schiemer, F., Boland, K.T. (Eds.), Perspectives in Tropical Limnology. SPB Academic Publishing BV, Amsterdam, pp. 43–64.

Lewis Jr., W.M., Grant, M.C., Hamilton, S.K., 1985. Evidence that filterable phosphorus is a significant atmospheric link in the phosphorus cycle. Oikos 45, 428–432.

Li, W.C., Armstrong, D.E., Williams, J.D.H., Harris, R.F., Syers, J.K., 1972. Rate and extent of inorganic phosphate exchange in lake sediments. Proc. Soil Sci. Soc. Amer. 36, 279–285.

Li, J., Molot, L.A., Palmer, M.E., Winter, J.G., Young, J.D., Stainsby, E.A., 2018. Long-term changes in hypolimnetic dissolved oxygen in a large lake: Effects of invasive mussels, eutrophication and climate change on Lake Simcoe, 1980–2012. J. Great Lakes Res. 44, 779–787.

Nutrient and Eutrophication: The Limiting-Nutrient Controversy. In: Likens, G.E. (Ed.), Special Publ. Amer. Soc. Limnol. Oceanogr. 1, 328.

Likens, G.E. (Ed.), 1985. An Ecosystem Approach to Aquatic Ecology: Mirror Lake and its Environment. Springer-Verlag, New York, p. 516.

Likens, G.E., Bormann, F.H., 1995. Biogeochemistry of a Forested Ecosystem, second ed. Springer-Verlag, New York, p. 159.

Lindegaard, C., Jónasson, P.M., 1979. Abundance, population dynamics and production of zoobenthos in Lake Myvatn, Iceland. Oikos 32, 202–227.

Littlefield, L., Forsberg, C., 1965. Absorption and translocation of phosphorus–32 by *Chara globularis* Thuill. Physiol. Plant. 18, 291–296.

Loehr, R.C., Martin, C.S., Rast, W. (Eds.), 1980. Phosphorus Management Strategies for Lakes. Ann Arbor Science Publ., Inc., Ann Arbor, MI, p. 490.

Logan, T.J., 1982. Mechanisms for release of sediment-bound phosphate to water and the effects of agricultural land management on fluvial transport of particulate and dissolved phosphate. Hydrobiologia 92, 519–530.

Long, E.T., Cooke, G.D., 1978. Phosphorus variability in three streams during storm events: Chemical analysis vs. algal assay. Mitt. Internat. Verein. Limnol 21, 441–452.

Lyche, A., Andersen, T., Christoffersen, K., Hessen, D.O., Hansen, P.H.B., Klysner, A., 1996. Mesocosm tracer studies. 1. Zooplankton as sources and sinks in the pelagic phosphorus cycle of a mesotrophic lake. Limnol. Oceanogr. 41, 460–474.

Maciolek, J.A., 1954. Artificial fertilization of lakes and ponds. A review of the literature. USFWS, Spec. Sci. Rep. Fish. 113, 41.

Mackereth, F.J.H., 1953. Phosphorus utilization by *Asterionella formosa* Hass. J. Exp. Bot. 4, 296–313.

Mackereth, F.J.H., 1966. Some chemical observations on postglacial lake sediments. Phil. Trans. Roy. Soc. London (Ser. B) 250, 165–213.

Macpherson, L.B., Sinclair, N.R., Hayes, F.R., 1958. Lake water and sediment. III. The effect of pH on the partition of inorganic phosphate between water and oxidized mud or its ash. Limnol. Oceanogr. 3, 318–326.

Madeira, P.T., Brooks, A.S., Seale, D.B., 1982. Excretion of total phosphorus, dissolved reactive phosphorus, ammonia, and urea by Lake Michigan *Mysis relicta*. Hydrobiologia 93, 145–154.

Maki, A.W., Porcella, D.B., Wendt, R.H., 1984. The impact of detergent phosphorus bans on receiving water quality. Water Res. 18, 893–903.

Mann, E.L., Chisholm, S.W., 2000. Iron limits the cell division rate of *Prochlorococcus* in the eastern equatorial Pacific. Limnol. Oceanogr. 45, 1067–1076.

Manny, B.A., Wetzel, R.G., Johnson, W.C., 1975. Annual contribution of carbon, nitrogen, and phosphorus to a hard-water lake by migrant Canada geese. Verh. Int. Verein. Limnol. 19, 949–951.

Manny, B.A., Johnson, W.C., Wetzel, R.G., 1994. Nutrient additions by waterfowl to lakes and reservoirs: Predicting their effects on productivity and water quality. Hydrobiologia 279/280, 121–132.

Marion, L., Clergeau, P., Brient, L., Bertru, G., 1994. The importance of avian-contributed nitrogen (N) and phosphorus (P) to Lake Grand-Lieu, France. Hydrobiologia 279/280, 133–147.

Mather, M.E., Vanni, M.J., Wissing, T.E., Davis, S.A., Schaus, M.H., 1995. Regeneration of nitrogen and phosphorus by bluegill and gizzard shad: Effect of feeding history. Can. J. Fish. Aquat. Sci. 52, 2327–2338.

Mazumder, A., Taylor, W.D., McQueen, D.J., Lean, D.R.S., 1989. Effects of fertilization and planktivorous fish on epilimnetic phosphorus and phosphorus sedimentation in large enclosures. Can. J. Fish. Aquat. Sci. 46, 1735–1742.

Mazumder, A., Taylor, W.D., Lean, D.R.S., McQueen, D.J., 1992. Partitioning and fluxes of phosphorus: Mechanisms regulating the size-distribution and biomass of plankton. Arch. Hydrobiol. Beih. Ergebn. Limnol. 35, 121–143.

McRoy, C.P., Barsdate, R.J., Nebert, M., 1972. Phosphorus cycling in an eelgrass (*Zostera marina* L.) ecosystem. Limnol. Oceanogr. 17, 58–67.

McVicar, T.R., Van Niel, T.G., Li, L.T., Roderick, M.L., Rayner, D.P., Ricciardulli, L., Donohue, R.J., 2008. Wind speed climatology and trends for Australia, 1975–2006: Capturing the stilling phenomenon and comparison with near-surface reanalysis output. Geophys. Res. Lett. 35, L20403. https://doi.org/10.1029/2008GL035627.

Meybeck, M., 1982. Carbon, nitrogen, and phosphorus transport by world rivers. Am. J. Sci. 282, 401–450.

Meybeck, M., 1993. Natural sources of C, N, P and S. In: Wollast, R., Mackenzie, F.T., Chou, L. (Eds.), Interactions of C, N, P and S Biogeochemical Cycles and Global Change. Springer-Verlag, Berlin, pp. 163–193.

Meyer, J.L., 1979. The role of sediments and bryophytes in phosphorus dynamics in a headwater stream ecosystem. Limnol. Oceanogr. 24, 365–375.

Meyer, J.L., Likens, G.E., 1979. Transport and transformation of phosphorus in a forest stream ecosystem. Ecology 60, 1255–1269.

Mickle, A.M., Wetzel, R.G., 1978. Effectiveness of submersed angiosperm-epiphyte complexes on exchange of nutrients and organic carbon in littoral systems. I. Inorganic nutrients. Aquat. Bot. 4, 303–316.

Mitamura, O., Saijo, Y., 1980. Urea supply from decomposition and excretion of zooplankton. J. Oceanogr. Soc. Jpn. 36, 121–125.

Miyajima, T., 1992. Recycling of nitrogen and phosphorus from the particulate organic matter associated with the proliferation of bacteria and microflagelates. Jpn. J. Limnol. 53, 133–138.

Moeller, R.G., Burkholder, J.M., Wetzel, R.G., 1988. Significance of sedimentary phosphorus to a submersed freshwater macrophyte (*Najas flexilis*) and its algal epiphytes. Aquat. Bot. 32, 261–281.

Molot, L.A., 2017. The effectiveness of cyanobacteria nitrogen fixation: Review of bench top and pilot scale nitrogen removal studies and implications for nitrogen removal programs. Environ. Rev 25, 292–295.

Molot, L.A., Brown, E.J., 1986. A method for determining the temporal response of phosphate transport affinity in phosphorus-limited, perturbed continuous culture. Appl. Environ. Microbiol. 51, 524–531.

Molot, L.A., Watson, S.B., Creed, I.F., Trick, C.G., McCabe, S.K., Verschoor, M.J., Sorichetti, R.J., Powe, C., Venkiteswaran, J.J., Schiff, S.L., 2014. A novel model for cyanobacteria bloom formation: The critical role of anoxia and ferrous iron. Freshw. Biol. 59, 1323–1340.

Molot, L.A., Higgins, S.N., Schiff, S.L., Venkiteswaran, J.J., Paterson, M.J., Baulch, H.M., 2021a. Phosphorus-only fertilization rapidly initiates large nitrogen-fixing cyanobacteria blooms in two oligotrophic lakes. Environ. Res. Lett. 16. https://doi.org/10.1088/1748-9326/ac0564.

Molot, L.A., Schiff, S.L., Venkiteswaran, J.J., Baulch, H.M., Higgins, S.N., Zastepa, A., Verschoor, M.J., Walters, D., 2021b. Low sediment redox promotes cyanobacteria across a trophic range: Implications for bloom management. Lake Reserv. Manage. 37. https://doi.org/10.1080/10402381.2020.1854400.

Mortimer, C.H., 1941. The exchange of dissolved substances between mud and water in. lakes (Parts I and II). J. Ecol. 29, 280–329.

Mortimer, C.H., 1942. The exchange of dissolved substances between mud and water in lakes (Parts III, IV, summary, and references). J. Ecol. 30, 147–201.

Mortimer, C.H., 1971. Chemical exchanges between sediments and water in the Great Lakes—speculations on probable regulatory mechanisms. Limnol. Oceanogr. 16, 387–404.

Mortimer, C.H., Hickling, C.F., 1954. Fertilizers in Fishponds, vol. 5. Fish. Publ. U.K. Colonial Office, London, p. 155.

Moss, B., 1972. Studies on Gull Lake, Michigan. II. Eutrophication—evidence and prognosis. Freshwat. Biol. 2, 309–320.

Moss, B., Wetzel, R.G., Lauff, G.H., 1980. Studies on Gull Lake, Michigan. III. Annual productivity and phytoplankton changes between 1979 and 1974. Freshwat. Biol. 10, 113–121.

Mulholland, P.J., Steinman, A.D., Elwood, J.W., 1990. Measurement of phosphorus uptake length in streams: Comparison of radiotracer and stable PO_4 releases. Can. J. Fish. Aquat. Sci. 47, 2351–2357.

Mulholland, P.J., Newbold, J.D., Elwood, J.W., Ferren, L.A., 1985a. Phosphorus spiralling in a woodland stream: Seasonal variations. Ecology 66, 1012–1023.

Mulholland, P.J., Elwood, J.W., Newbold, J.D., Ferren, L.A., 1985b. Effect of a leaf-shredding invertebrate on organic matter dynamics and phosphorus spiralling in heterotrophic laboratory streams. Oecologia 66, 199–206.

Mulholland, P.J., Elwood, J.W., Newbold, J.D., Webster, J.R., Ferren, L.A., Perkins, R.E., 1984. Phosphorus uptake by decomposing leaf detritus: Effect of microbial biomass and activity. Verh. Int. Verein. Limnol. 22, 1899–1905.

Mulholland, P.J., Minear, R.A., Elwood, J.W., 1988. Production of soluble, high molecular weight phosphorus and its subsequent uptake by stream detritus. Verh. Int. Verein. Limnol. 23, 1190–1197.

Müller, B., Steinsberger, T., Schwefel, R., Gächter, R., Sturm, M., Wüest, A., 2019. Oxygen consumption in seasonally stratified lakes decreases only below a marginal phosphorus threshold. Sci. Rep 9, 18054. https://doi.org/10.1038/s41598-019-54486-3.

Munn, N., Prepas, E., 1986. Seasonal dynamics of phosphorus partitioning and export in two streams in Alberta, Canada. Can. J. Fish. Aquat. Sci. 43, 2464–2471.

Münster, U., 1994. Studies on phosphatase activities in humic lakes. Environ. Int. 20, 49–59.

Murphy, T.P., Hall, K.J., Yesaki, I., 1983. Coprecipitation of phosphate with calcite in a naturally eutrophic lake. Limnol. Oceanogr. 28, 58–69.

Naiman, R.J., Bilby, R.E., Schindler, D.E., Helfield, J.M., 2002. Pacific salmon, nutrients, and the dynamics of freshwater and riparian ecosystems. Ecosystems 5, 399–417.

Nakano, S., 1994a. Rates and ratios of nitrogen and phosphorus released by a bacterivorous flagellate. Jpn. J. Limnol. 55, 115–123.

Nakano, S., 1994b. Estimation of phosphorus release rate by bacterivorous flagellates in Lake Biwa. Jpn. J. Limnol. 55, 201–211.

Nalepa, T.F., Gardner, W.S., Malczyk, J.M., 1983. Phosphorus release by three kinds of benthic invertebrates: Effect of substrate and water medium. Can. J. Fish. Aquat. Sci. 40, 810–813.

Nalewajko, C., Paul, B., 1985. Effects of manipulations of aluminum concentrations and pH on phosphate uptake and photosynthesis of planktonic communities in two Precambrian Shield lakes. Can. J. Fish. Aquat. Sci. 42, 1946–1953.

Naumann, E., 1929. The scope and chief problems of regional limnology. Int. Rev. ges. Hydrobiol. 22, 423–444.

Naumann, E., 1932. Grundzüge der regionalen Limnologie. Die Binnengewässer 11, 176.

Newbold, J.D., 1992. Cycles and spirals of nutrients. In: Calow, P., Petts, G.E. (Eds.), The Rivers Handbook. I. Hydrological and Ecological Principles. Blackwell Sci. Publs., Oxford, pp. 379–408.

Newbold, J.D., Elwood, J.W., O'Neill, R.V., Sheldon, A.L., 1983. Phosphorus dynamics in a woodland stream ecosystem: A study of nutrient spiralling. Ecology 64, 1249–1265.

Newman, E.I., 1995. Phosphorus inputs to terrestrial ecosystems. J. Ecol. 83, 713–726.

Newman, S., Aldridge, F.J., Phlips, E.J., Reddy, K.R., 1994. Assessment of phosphorus availability for natural phytoplankton populations from a hypereutrophic lake. Arch. Hydrobiol. 130, 409–427.

Nicholls, K.H., Dillon, P.J., 1978. An evaluation of phosphorus-chlorophyll-phytoplankton relationships for lakes. Int. Rev. Gesamt. Hydrobiol. 63, 141–154.

Nichols, D.S., Keeney, D.R., 1973. Nitrogen and phosphorus release from decaying water milfoil. Hydrobiologia 42, 509–525.

Nieuwenhuyse, E.E.V., Jones, J.R., 1996. Phosphorus-chlorophyll relationship in temperate streams and its variation with stream catchment area. Can. J. Fish. Aquat. Sci. 53, 99–105.

Nürnberg, G.K., 1984. The prediction of internal phosphorus load in lakes with anoxic hypolimnia. Limnol. Oceanogr. 29, 111–124.

Nürnberg, G.K., 1985. Availability of phosphorus upwelling from iron-rich anoxic hypolimnia. Arch. Hydrobiol. 104, 459–476.

Nürnberg, G.K., 1995. Quantifying anoxia in lakes. Limnol. Oceanogr. 40, 1100–1111.

Nürnberg, G.K., 2009. Assessing internal phosphorus load—problems to be solved. Lake Reserv. Manage. 25, 419–432.

Oglesby, R.T., Schaffner, W.R., 1978. Phosphorus loadings to lakes and some of their responses. Part 2. Regression models of summer phytoplankton standing crops, winter total P, and transparency of New York lakes with phosphorus loadings. Limnol. Oceanogr. 23, 135–145.

Ohle, W., 1938. Zur Vervolkommnung der hydrochemischen Analyse. III. Die Phosphorbestimmung. Angew. Chem. 51, 906–911.

Ohle, W., 1956. Bioactivity, production, and energy utilization of lakes. Limnol. Oceanogr. 1, 139–149.

Ohle, W., 1965. Nährstoffanreicherung der Gewässer durch Düngemittel und Meliorationen. Münchner Beiträge 12, 54–83.

Olila, O.G., Reddy, K.R., 1993. Phosphorus sorption characteristics of sediments in shallow eutrophic lakes of Florida. Arch. Hydrobiol. 129, 45–65.

Oliver, R.L., 1994. Floating and sinking in gas-vacuolate cyanobacteria. J. Phycol. 30, 161–173.

Olsen, S., 1958. Phosphate adsorption and isotopic exchange in lake muds. Experiments with P32; Preliminary report. Verh. Int. Verein. Limnol. 13, 915–922.

Olsen, S., 1964. Phosphate equilibrium between reduced sediments and water. Laboratory experiments with radioactive phosphorus. Verh. Int. Verein. Limnol. 15, 333–341.

Olsen, Y., Østgaard, K., 1985. Estimating release rates of phosphorus from zooplankton: Model and experimental verification. Limnol. Oceanogr. 30, 844–852.

Olsen, Y., Jensen, A., Reinertsen, H., Børsheim, K.Y., Heldal, M., Langeland, A., 1986. Dependence of the rate of release of phosphorus by zooplankton on the P:C ratio in the food supply, as calculated by a recycling model. Limnol. Oceanogr. 31, 34–44.

O'Melia, C.R., 1998. Coagulation and sedimentation in lakes, reservoirs and water treatment plants. Wat. Sci. Tech. 37, 129–135.

Oni, S.,K., Futter, M.N., Molot, L.A., Dillon, P.J., Crossman, J., 2014. Uncertainty assessments and hydrological implications of climate change in two adjacent agricultural catchments of a rapidly urbanizing watershed. Sci. Tot. Environ. 473–474, 326–337.

Orihel, D.M., Schindler, D.W., Ballard, N.C., Graham, M.D., O'Connell, D.W., Wilson, L.R., Vinebrooke, R.D., 2015. The "nutrient pump:" Iron-poor sediments fuel low nitrogen-to-phosphorus ratios and cyanobacterial blooms in polymictic lakes. Limnol. Oceanogr. 60, 856–871.

Orihel, D.M., Baulch, H.M., Casson, N.J., North, R.L., Parsons, C.T., Seckar, D.C.M., Venkiteswaran, J.J., 2017. Internal phosphorus loading in Canadian fresh waters: A critical review and data analysis. Can. J. Fish. Aquat. Sci. 74, 2005–2029.

Ortiz, J.L., Martinez, R.P., 1984. Applicability of the OECD eutrophication models to Spanish reservoirs. Verh. Int. Verein. Limnol. 22, 1521–1535.

Otsuki, A., Wetzel, R.G., 1972. Coprecipitation of phosphate with carbonates in a marl lake. Limnol. Oceanogr. 17, 763–767.

Otsuki, A., Wetzel, R.G., 1974. Calcium and total alkalinity budgets and calcium carbonate precipitation of a small hard-water lake. Arch. Hydrobiol. 73, 14–30.

Otto, G., Benndorf, J., 1971. Über den Einfluss des physiologischen Zustandes sedimentierender Phytoplankter auf die Abbauvorgänge während der Sedimentation. Limnologica 8, 365–370.

Overbeck, J., 1962. Untersuchungen zum Phosphathaushalt von Grünalgen. III. Das Verhalten der Zellfraktionen von *Scenedesmus quadricauda* (Turp.) Bréb. im Tagescyclus unter verschiedenen Belichtungsbedingungen und bei verschidenen Phosphatverbindungen. Arch. Mikrobiol. 41, 11–26.

Overbeck, J., 1963. Untersuchungen zum Phosphathaushalt von Grünalgen. VI. Ein Beitrag zum Polyphosphatstoffwechsel des Phytoplanktons. Ber. Deut. Bot. Gesellsch. 76, 276–286.

Paerl, H.W., 1988. Nuisance phytoplankton blooms in coastal, estuarine, and inland waters. Limnol. Oceanogr. 33, 823–847.

Paerl, H.W., Lean, D.R.S., 1976. Visual observations of phosphorus movement between algae, bacteria, and abiotic particles in lake water. J. Fish. Res. Board Can. 33, 2805–2813.

Paerl, H.W., Downes, M.T., 1978. Biological availability of low versus high molecular weight reactive phosphorus. J. Fish. Res. Board Can. 35, 1639–1643.

Pallesen, L., Berthouex, P.M., Booman, K., 1985. Environmental intervention analysis: Wisconsin's ban on phosphate detergents. Water Res. 19, 353–362.

Paterson, A.M., Dillon, P.J., Hutchinson, N.J., Futter, M.N., Clark, B.J., Mills, R.B., Reid, R.A., Scheider, W.A., 2009. A review of the components, coefficients and technical assumptions of Ontario's Lakeshore Capacity Model. Lake Reserv. Manage. 22, 7–18.

Pennuto, C.M., Burlakova, L.E., Karatayev, A.Y., Kramer, J., Fischer, A., Mayer, C., 2014. Spatiotemporal characteristics of nitrogen and phosphorus in the benthos of nearshore Lake Erie. J. Great Lakes Res. 40, 541–549.

Pérez-Fuentetaja, A., McQueen, D.J., Ramcharan, C.W., 1996. Predator-induced Bottom-Up.

Peters, R.H., 1975. Orthophosphate turnover in central European lakes. Mem. Ist. Ital. Idrobiol. 32, 297–311.

Peters, R.H., 1979. Concentrations and kinetics of phosphorus fractions along the trophic gradient of Lake Memphremagog. J. Fish. Res. Board Can. 36, 970–979.

Peters, R.H., 1986. The role of prediction in limnology. Limnol. Oceanogr. 31, 1143–1159.

Peters, R., Lean, D., 1973. The characterization of soluble phosphorus released by limnetic zooplankton. Limnol. Oceanogr. 18, 270–279.

Peters, R.H., MacIntyre, S., 1976. Orthophosphate turnover in East African lakes. Oecologia 25, 313–319.

Pettersson, K., 1980. Alkaline phosphatase activity and algal surplus phosphorus as phosphorus-deficiency indicators in Lake Erken. Arch. Hydrobiol. 89, 54–87.

Pielou, E.C., 1981. The usefulness of ecological models: A stock-taking. Quart. Rev. Biol. 56, 17–31.

Planas, D., Hecky, R.E., 1984. Comparison of phosphorus turnover times in northern Manitoba reservoirs with lakes of the Experimental Lakes Area. Can. J. Fish. Aquat. Sci. 41, 605–612.

Pomeroy, L.R., Smith, E.E., Grant, C.M., 1965. The exchange of phosphate between estuarine water and sediments. Limnol. Oceanogr. 10, 167–172.

Portnoy, J.W., Soukup, M.A., 1990. Gull contributions of phosphorus and nitrogen to a Cape Cod kettle pond. Hydrobiologia 202, 61–69.

Prepas, E.E., 1983. Orthophosphate turnover time in shallow productive lakes. Can. J. Fish. Aquat. Sci. 40, 1412–1418.

Prepas, E.E., Rigler, F.H., 1982. Improvements in quantifying the phosphorus concentration in lake water. Can. J. Fish. Aquat. Sci. 39, 822–829.

Provasoli, L., 1958. Nutrition and ecology of protozoa and algae. Ann. Rev. Microbiol. 12, 279–308.

Psenner, R., 1984. The proportion of empneuston and total atmospheric inputs of carbon, nitrogen and phosphorus in the nutrient budget of a small mesotrophic lake (Piburger See, Austria). Int. Rev. ges. Hydrobiol. 69, 23–39.

Quinlan, R., Filazzola, A., Mahdiyan, O., Shuvo, A., Blagrave, K., Ewins, C., Moslenko, L., Gray, D.K., O'Reilly, C.M., Sharma, S., 2021. Relationships of total phosphorus and chlorophyll in lakes worldwide. Limnol. Oceanogr. 66, 392–404.

Rai, H., Jacobsen, T.R., 1993. Dissolved alkaline phosphatase activity (APA) and the contribution of APA by size fractionated plankton in Lake Schöhsee. Verh. Int. Verein. Limnol. 25, 164–169.

Raidt, H., Koschel, R., 1993. Variable morphology of calcite crystals in hardwater lakes. Limnologica 23, 85–89.

Rao, I.M., 1997. The role of phosphorus in photosynthesis. In: Pessarakli, M. (Ed.), Handbook of Photosynthesis. M. Dekker, Inc., New York, pp. 173–194.

Raney, S., Eimers, M.C., 2014. Unexpected declines in stream phosphorus concentrations across southern Ontario. Can. J. Fish. Aquat. Sci. 71, 337–342.

Rast, W., Lee, G.F., 1978. Summary Analysis of the North American (U.S. Portion) OECD Eutrophication Project: Nutrient Loading-lake Response Relationships and Trophic State Indices. U.S. Environmental Protection Agency Rept, p. 455. EPA-600/3–78–008.

Rast, W., Jones, R.A., Lee, G.F., 1983. Predictive capability of U.S. OECD phosphorus loading-eutrophication response models. J. Water Poll. Contr. Fed. 55, 990–1003.

Ravera, O., Gatti, M.C., 1993. Release of carbon, nitrogen and hydrogen from dead zooplankton. Verh. Int. Verein. Limnol. 25, 766–769.

Rawson, D. S. 1939. Some physical and chemical factors in the metabolism of lakes. *In* Problems of Lake Biology. Publ. Amer. Assoc. Adv. Sci. vol. *10*:9–26.

Rawson, D.S., 1955. Morphometry as a dominant factor in the productivity of large lakes. Verh. Int. Verein. Limnol. 12, 164–175.

Reckhow, K.H., 1979. Empirical lake models for phosphorus: Development, applications, limitations and uncertainty. In: Scavia, D., Robertson, A. (Eds.), Perspectives on Lake Ecosystem Modeling. Ann Arbor Sci. Publs., Ann Arbor, MI, pp. 193–221.

Reckhow, K.H., Chapra, S.C., 1983. Engineering Approaches for Lake Management. I. Data Analysis and Empirical Modeling. Butterworths, Boston, p. 340.

Reckhow, K.H., Chapra, S.C., 1999. Modeling excessive nutrient loading in the environment. Environ. Pollut. 100, 197–207.

Redfield, A.C., Ketchum, B.H., Richards, F.A., 1963. The Influence of organisms on the composition of the sea water. In: Hill, M.N. (Ed.), The Sea, Vol. 2. Interscience Publishers, New York, pp. 26–77.

Reichardt, W., Overbeck, J., Steubing, L., 1967. Free dissolved enzymes in lake waters. Nature 216, 1345–1347.

Reuter, J.E., Rhodes, C.L., Lebo, M.E., Kotzman, M., Goldman, C.R., 1993. The importance of nitrogen in Pyramid Lake (Nevada, USA), a saline, desert lake. Hydrobiologia 267, 179–189.

Revsbech, N.P., Jørgensen, B.B., Blackburn, T.H., Cohen, Y., 1983. Microelectrode studies of the photosynthesis and O_2, H_2S, and pH profiles of a microbial mat. Limnol. Oceanogr. 28, 1062–1074.

Reynolds, C.S., 1979. The limnology of the eutrophic meres of the Shropshire-Cheshire Plain: A review. Field Stud. 5, 93–173.

Reynolds, C.S., Oliver, R.L., Walsby, A.E., 1987. Cyanobacterial dominance: The role of buoyancy regulation in dynamic lake environments. N. Z. J. Mar. Freshwat. Res. 21, 379–390.

Rhee, G.-Y., 1972. Competition between an alga and an aquatic bacterium for phosphate. Limnol. Oceanogr. 17, 505–514.

Rhee, G.-Y., 1973. A continuous culture study of phosphate uptake, growth rate and polyphosphate in *Scenedesmus sp.* J. Phycol. 9, 495–506.

Riber, H.H., Wetzel, R.G., 1987. Boundary-layer and internal diffusion effects on phosphorus fluxes in lake periphyton. Limnol. Oceanogr. 32, 1181–1194.

Rigby, C.H., Craig, S.R., Budd, K., 1980. Phosphate uptake by *Synechococcus leopoliensis* (Cyanophyceae): Enhancement by calcium ions. J. Phycol. 16, 389–393.

Rigler, F.H., 1956. A tracer study of the phosphorus cycle in lake water. Ecology 37, 550–562.

Rigler, F.H., 1961. The uptake and release of inorganic phosphorus by *Daphnia magna* Straus. Limnol. Oceanogr. 6, 165–174.

Rigler, F.H., 1964a. The phosphorus fractions and the turnover time of inorganic phosphorus in different types of lakes. Limnol. Oceanogr. 9, 511–518.

Rigler, F.H., 1964b. The contribution of zooplankton to the turnover of phosphorus in the epilimnion of lakes. Can. Fish Culturist 32, 3–9.

Rigler, F.H., 1973. A dynamic view of the phosphorus cycle in lakes. In: Griffith, E.J., Beeton, A., Spencer, J.M., Mitchell, D.T. (Eds.), Environmental Phosphorus Handbook. John Wiley & Sons, New York, pp. 539–572.

Rijkeboer, M., de Bles, F., Gons, H.J., 1991. Role of sestonic detritus as a P-buffer. Mem. Ist. Ital. Idrobiol. 48, 251–260.

Rivkin, R.B., Swift, E., 1980. Characterization of alkaline phosphatase and organic phosphorus utilization in the oceanic dinoflagellate *Pyrocystis noctiluca*. Mar. Biol. 61, 1–8.

Robson, B.J., 2014. State of the art in modelling of phosphorus in aquatic systems: Review, criticisms and commentary. Environ. Model. Softw. 61, 339–359.

Roden, E.E., Edmonds, J.W., 1997. Phosphate mobilization in anaerobic sediments: Microbial Fe(III) oxide reduction versus iron-sulfide formation. Arch. Hydrobiol. 139, 347–378.

Rodhe, W., 1948. Environmental requirements of freshwater plankton algae. Experimental studies in the ecology of phytoplankton. Symbol. Bot. Upsalien. 10 (1), 149.

Rothhaupt, K.O., 1992. Stimulation of phosphorus-limited phytoplankton by bacterivorous flagellates in laboratory experiments. Limnol. Oceanogr. 37, 750–759.

Sakamoto, M., 1966. Primary production of phytoplankton community in some Japanese lakes and its dependence on lake depth. Arch. Hydrobiol. 62, 1–28.

Salas, H.J., Martino, P., 1991. A simplified phosphorus trophic state model for warm-water tropical lakes. Water Res. 25, 341–350.

Salonen, K., Jones, R.I., Arvola, L., 1984. Hypolimnetic phosphorus retrieval by diel vertical migrations of lake phytoplankton. Freshwat. Biol. 14, 431–438.

Sarvala, J., Kairesalo, T., Koskimies, I., Lehtovaara, A., Ruuhijärvi, J., Vähä-Piikkiö, I., 1982. Carbon, phosphorus and nitrogen budgets of the littoral *Equisetum* belt in an oligotrophic lake. Hydrobiologia 86, 41–53.

Saunders, G.W., 1972. The transformation of artificial detritus in lake water. Mem. Ist. Ital. Idrobiol. 29 (Suppl. l.), 261–288.

Sawyer, C.N., 1947. Fertilization of lakes by agricultural and urban drainage. J. New England Water Works Assoc. 61, 109–127.

Scavia, D., McFarland, M.J., 1982. Phosphorus release patterns and the effects of reproductive stage and ecdysis in *Daphnia magna*. Can. J. Fish. Aquat. Sci. 39, 1310–1314.

Schaus, M.H., Vanni, M.J., Wissing, T.E., Bremigan, M.T., Garvey, J.E., Stein, R.A., 1997. Nitrogen and phosphorus excretion by detritivorous gizzard shad in a reservoir ecosystem. Limnol. Oceanogr. 42, 1386–1397.

Schindler, D.W., 1974. Eutrophication and recovery in experimental lakes: Implications for lake management. Science 184, 897–899.

Schindler, D.W., 1978. Predictive eutrophication models. Limnol. Oceanogr. 23, 1080–1081.

Schindler, D.W., Fee, E.J., 1973. Diurnal variation of dissolved inorganic carbon and its use in estimating primary production and CO_2 invasion in Lake 227. J. Fish. Res. Board Can. 30, 1501–151.

Schindler, D.W., Donahue, W.F., 2006. An impending water crisis in Canada's western prairie provinces. Proc. Nat. Acad. Sci. 103, 7210–7216. https://doi.org/10.1073/pnas.0601568103.

Schindler, D.W., Newbury, R.W., Beatty, K.G., Prokopowich, J., Ruszczynski, T., Dalton, J.A., 1980. Effects of a windstorm and forest fire on chemical losses from forested watersheds and on the quality of receiving streams. Can. J. Fish. Aquat. Sci. 37, 328–334.

Schindler, D.W., Hecky, R.E., Mills, K.H., 1993. Two decades of whole lake eutrophication and acidification experiments. In: Rasmussen, L., Brydges, T., Mathy, P. (Eds.), Experimental Manipulations of Biota and Biogeochemical Cycling in Ecosystems. Commission of the European Communities, Brussels, pp. 294–304.

Schindler, D.W., Hecky, R.E., Findlay, D.L., Stainton, M.P., Parker, B.R., Paterson, M.J., Beaty, K.G., Lyng, M., Kasian, S.E.M., 2008. Eutrophication of lakes cannot be controlled by reducing nitrogen input: Results of a 37-year whole-ecosystem experiment. Proc. Natl. Acad. Sci. 105, 11254–11258.

Schindler, D.W., Carpenter, S.R., Chapra, S.C., Hecky, R.E., Orihel, D.M., 2016. Reducing phosphorus to curb lake eutrophication is a success. Environ. Sci. Technol. 50, 8923–8929.

Schuldt, J.A., Hershey, A.E., 1995. Effect of salmon carcass decomposition on Lake Superior tributary streams. J. N. Am. Benthol. Soc. 14, 259–268.

Schuman, G.E., Burwell, R.E., Priest, R.F., Spomer, R.C., 1973. Nitrogen losses in surface runoff from agricultural watersheds on Missouri Valley loess. J. Environ. Qual. 2, 299–302.

Schwoerbel, J., Tillmanns, G.C., 1964. Konzentrationsabhängige Aufnahme von wasserlöslichem PO_4-P bei submersen Wasserpflanzen. Naturwissenschaften 51, 319–320.

Senft, W.H., 1978. Dependence of light-saturated rates of algal photosynthesis on intracellular concentrations of phosphorus. Limnol. Oceanogr. 23, 709–718.

Sharma, S., Blagrave, K., Magnuson, J.J., O'Reilly, C., Oliver, S., Batt, R.D., Magee, M.R., Straile, D., Weyhenmeyer, G.A., Winslow, L., Woolway, R.I., 2019. Widespread loss of lake ice around the Northern Hemisphere in a warming world. Nat. Clim. Change 9, 227–231.

Shaw, P.J., 1994. The effect of pH, dissolved humic substances, and ionic composition on the transfer of iron and phosphate to particulate size fractions in epilimnetic lake water. Limnol. Oceanogr. 39, 1734–1743.

Shukla, S.S., Syers, J.K., Williams, J.D.H., Armstrong, D.E., Harris, R.F., 1971. Sorption of inorganic phosphate by lake sediments. Proc. Soil Sci. Soc. Amer. 35, 244–249.

Shuvo, A., O'Reilly, C.M., Blagrave, K., Ewins, C., Filazzola, A., Gray, D., Mahdiyan, O., Moslenko, L., Quinlan, R., Sharma, S., 2021. Total phosphorus and climate are equally important predictors of water quality in lakes. Aquat. Sci. 83, 16. https://doi.org/10.1007/s00027-021-00776-w.

Siderius, M., Musgrave, A., van den Ende, H., Koerten, H., Cambier, P., van der Meer, P., 1996. *Chlamydomonas eugametos* (Chlorophyta) stores phosphate in polyphosphate bodies together with calcium. J. Phycol. 32, 402–409.

Skaggs, R.W., Brevé, M.A., Gilliam, J.W., 1994. Hydrologic and water quality impacts of agricultural drainage. Crit. Rev. Environ. Sci. Technol. 24, 1–32.

Smith, C.S., Adams, M.S., 1986. Phosphorus transfer from sediments by *Myriophyllum spicatum*. Limnol. Oceanogr. 31, 1312–1321.

Smith, R.E.H., Kalff, J., 1982. Size-dependent phosphorus uptake kinetics and cell quota in phytoplankton. J. Phycol. 18, 275–284.

Smith, V.H., 1983. Low nitrogen to phosphorus ratios favor dominance by blue-green algae in lake phytoplankton. Science 221, 669–671.

Smith, V.H., Shapiro, J., 1981. Chlorophyll-phosphorus relations in individual lakes. Their importance to lake restoration strategies. Environ. Sci. Technol. 15, 444–451.

Søballe, D.M., Kimmel, B.L., 1987. A large-scale comparison of factors influencing phytoplankton abundance in rivers, lakes, and impoundments. Ecology 68, 1943–1954.

Soeder, C.J., 1965. Some aspects of phytoplankton growth and activity. Mem. Ist. Ital. Idrobiol. 18 (Suppl. l.), 47–59.

Solski, A., 1962. Mineralizacja roslin wodnych. I. Uwalnianie fosforu i potasu przez wymywanie. Pol. Arch. Hydrobiol. 10, 167–196.

Sommer, U., 1990. The role of competition for resources in phytoplankton succession. In: Sommer, U. (Ed.), Plankton Ecology: Succession in Plankton Communities. Springer-Verlag, New York, pp. 57–106.

Sommers, L.E., Harris, R.F., Williams, J.D.H., Armstrong, D.E., Syers, J.K., 1970. Determination of total organic phosphorus in lake sediments. Limnol. Oceanogr. 15, 301–304.

Spencer, C.N., Hauer, F.R., 1991. Phosphorus and nitrogen dynamics in streams during a wildfire. J. N. Amer. Benthol. Soc. 10, 24–30.

Stabel, H.-H., Geiger, M., 1985. Phosphorus adsorption to riverine suspended matter: Implications for the P-budget of Lake Constance. Water Res. 19, 1347–1352.

Stauffer, R.E., 1985. Relationships between phosphorus loading and trophic state in calcareous lakes of southeast Wisconsin. Limnol. Oceanogr. 30, 123–145.

Sterner, R.W., 2008. On the phosphorus limitation paradigm for lakes. Internat. Rev. Hydrobiol. 93, 433–445.

Sterner, R.W., Elser, J.J., Hessen, D.O., 1992. Stoichiometric relationships among producers, consumers and nutrient cycling in pelagic ecosystems. Biogeochemistry 17, 49–67.

Sterner, R.W., Elser, J.J., Fee, E.J., Guildford, S.J., Chrzanowski, T.H., 1997. The light:nutrient ratio in lakes: The balance of energy and materials affects ecosystem structure and process. Am. Nat. 150, 663–684.

Sterner, R.W., Chrzanowski, T.H., Elser, J.J., George, N.B., 1995. Sources of nitrogen and phosphorus supporting the growth of bacterio- and phytoplankton in an oligotrophic Canadian shield lake. Limnol. Oceanogr. 40, 242–249.

Sterner, R.W., Smutka, T.M., McKay, R.M.L., Xiaoming, Q., Brown, E.T., Sherrell, R.M., 2004. Phosphorus and trace metal limitation of algae and bacteria in Lake Superior. Limnol. Oceanogr. 49, 495–507.

Stevens, R.J., Parr, M.P., 1977. The significance of alkaline phosphatase activity in Lough Neagh. Freshwat. Biol. 7, 351–355.

Stewart, A.J., Wetzel, R.G., 1981. Dissolved humic materials: Photodegradation, sediment effects, and reactivity with phosphate and calcium carbonate precipitation. Arch. Hydrobiol. 92, 265–286.

Stewart, A.J., Wetzel, R.G., 1982a. Phytoplankton contribution to alkaline phosphatase activity. Arch. Hydrobiol. 93, 265–271.

Stewart, A.J., Wetzel, R.G., 1982b. Influence of dissolved humic materials on carbon assimilation and alkaline phosphatase activity in natural algal-bacterial assemblages. Freshwat. Biol. 12, 369–380.

Stibe, L., Fleischer, S., 1991. Agricultural production methods—impact on drainage water nitrogen. Verh. Int. Verein. Limnol. 24, 1749–1752.

Straškraba, M., 1996. Lake and reservoir management. Verh. Int. Verein. Limnol. 26, 193–209.

Stumm, W., Morgan, J.J., 1996. Aquatic Chemistry: Chemical Equilibria and Rates in Natural Waters, third ed. J. Wiley & Sons, New York, p. 1040.

Suberkropp, K., 1995. The influence of nutrients on fungal growth, productivity, and sporulation during leaf breakdown in streams. Can. J. Bot. 73 (Suppl. 1), S1361–S1369.

Sugawara, K., Koyama, T., Kamata, E., 1957. Recovery of precipitated phosphate from lake muds related to sulfate reduction. Chem. Inst. Fac. Sci. Nagoya Univ. 5, 60–67.

Tammeorg, O., Nürnberg, G., Niemistö, J., Haldna, M., Horppila1, J., 2020. Internal phosphorus loading due to sediment anoxia in shallow areas: implications for lake aeration treatments. Aquat. Sci. 82, 54. https://doi.org/10.1007/s00027-020-00724-0.

Tarapchak, S.J., Herche, L.R., 1985. Perspectives in epilimnetic phosphorus cycling. In: Dubinsky, Z., Steinberger, Y. (Eds.), Environmental Quality and Ecosystem Stability. Vol. 3A/B. Bar-Ilan Univ. Press, Ramat-Gan, Israel, pp. 245–255.

Tarapchak, S.J., Herche, L.R., 1986. Phosphate uptake by microorganisms in lake water: Deviations from simple Michaelis-Menten kinetics. Can. J. Fish. Aquat. Sci. 43, 319–328.

Tarapchak, S.J., Bigelow, S.M., Rubitschun, C., 1982. Overestimation of orthophosphorus concentrations in surface waters of southern Lake Michigan: Effects of acid and ammonium molybdate. Can. J. Fish. Aquat. Sci. 39, 296–304.

Taylor, W.D., 1984. Phosphorus flux through epilimnetic zooplankton from Lake Ontario: Relationship with body size and significance to phytoplankton. Can. J. Fish. Aquat. Sci. 41, 1702–1712.

Taylor, W.D., Lean, D.R.S., 1981. Radiotracer experiments on phosphorus uptake and release by limnetic microzooplankton. Can. J. Fish. Aquat. Sci. 38, 1316–1321.

Taylor, W.D., Lean, D.R.S., 1991. Phosphorus pool sizes and fluxes in the epilimnion of a mesotrophic lake. Can. J. Fish. Aquat. Sci. 48, 1293–1301.

Taylor, W.D., Barko, J.W., James, W.F., 1988. Contrasting diel patterns of vertical migration in the dinoflagellate *Ceratium hirundinella* in relation to phosphorus supply in a north temperate reservoir. Can. J. Fish. Aquat. Sci. 45, 1093–1098.

Teske, A., 2019. Cable bacteria, living electrical conduits in the microbial world. Proc. Nat. Acad. Sci. 116, 18759–18761.

Tezuka, Y., 1985. C:N:P ratios of seston in Lake Biwa as indicators of nutrient deficiency in phytoplankton and decomposition process of hypolimnetic particulate matter. Jpn. J. Limno1. 46, 239–246.

Thienemann, A., 1925. Die Binnengewässer Mitteleuropas. Eine limnologische Einführung. Die Binnengewässer. 1, 255.

Tilman, D., 1981. Tests of resource competition theory using four species of Lake Michigan algae. Ecology 62, 802–815.

Tilman, D., 1982. Resource Competition and Community Structure. Princeton Univ. Press, Princeton, NJ, p. 296.

Tilman, D., Kiesling, R., Sterner, R., Kilham, S.S., Johnson, F.A., 1986. Green, bluegreen and diatom algae: Taxonomic differences in competitive ability for phosphorus, silicon and nitrogen. Arch. Hydrobiol. 106, 473–485.

Tilman, D., Kilham, S.S., Kilham, P., 1982. Phytoplankton community ecology: The role of limiting nutrients. Ann. Rev. Ecol. Syst. 13, 349–372.

Tipping, E., Higgins, D.C., 1982. The effect of adsorbed humic substances on the colloid stability of hematite particles. Colloids Surfaces 5, 85–92.

Tirén, T., Pettersson, K., 1985. The influence of nitrate on the phosphorus flux to and from oxygen depleted lake sediments. Hydrobiologia 120, 207–223.

Toolan, T., Wehr, J.D., Findlay, S., 1991. Inorganic phosphorus stimulation of bacterioplankton production in a meso-eutrophic lake. Appl. Environ. Microbiol. 57, 2074–2078.

Tucker, A., 1957. The relation of phytoplankton periodicity to the nature of the physico-chemical environment with special reference to phosphorus. I. Morphometrical, physical and chemical conditions. Am. Midl. Nat. 57, 300–333.

Turner, R.R., Laws, E.A., Harriss, R.C., 1983. Nutrient retention and transformation in relation to hydraulic flushing rate in a small impoundment. Freshwat. Biol. 13, 113–127.

Twinch, A.J., Peters, R.H., 1984. Phosphate exchange between littoral sediments and overlying water in an oligotrophic north-temperate lake. Can. J. Fish. Aquat. Sci. 41, 1609–1617.

Twiss, M.R., Gouvêa, S.P., Bourbonniere, R.A., McKay, R.M.L., Wilhelm, S.W., 2005. Field investigations of trace metal effects on Lake Erie phytoplankton productivity. J. Great Lakes Res. 31, 168–179.

Uehlinger, U., Bloesch, J., 1987. Variation in the C:P ratio of suspended and settling seston and its significance for P uptake calculations. Freshwat. Biol. 17, 99–108.

Urabe, J., 1993. N and P cycling coupled by grazers' activities: Food quality and nutrient release by zooplankton. Ecology 74, 2337–2350.

Urabe, J., 1995. Direct and indirect effects of zooplankton on seston stoichiometry. Ecoscience 2, 286–296.

Vadstein, O., Jensen, A., Olsen, Y., Reinertsen, H., 1988. Growth and phosphorus status of limnetic phytoplankton and bacteria. Limnol. Oceanogr. 33, 489–503.

Vadstein, O., Brekke, O., Andersen, T., Olsen, Y., 1995. Estimation of phosphorus release rates from natural zooplankton communities feeding on planktonic algae and bacteria. Limnol. Oceanogr. 40, 250–262.

Vadstein, O., Olsen, Y., Reinertsen, H., Jensen, A., 1993. The role of planktonic bacteria in phosphorus cycling in lakes: Sink and link. Limnol. Oceanogr. 38, 1539–1544.

Valipour, R., León, L.F., Howell, T., Dove, A., Yerubandi, Y.R., 2021. Episodic nearshore-offshore exchanges of hypoxic waters along the north shore of Lake Erie. J. Great Lakes Res. 47, 419–436. https://doi.org/10.1016/j.jglr.2021.01.014.

Vallentyne, J.R., 1970. Phosphorus and the control of eutrophication. Can. Res. Development (May–June 1970) 36–43, 49.

Vallentyne, J.R., 1974. The Algal Bowl—Lakes and Man. Misc. Special Publ. 22. Dept. of the Environment, Ottawa, p. 185.

van Donk, E., Faafeng, B.A., Hessen, D.O., Kllqvist, T.S., 1993. Use of immobilized algae for estimating bioavailable phosphorus released by zooplankton. J. Plank. Res. 15, 761–769.

Vanni, M.J., Findlay, D.L., 1990. Trophic cascades and phytoplankton community structure. Ecology 71, 921–937.

Vargo, G.A., 1979. The contribution of ammonia excreted by zooplankton to phytoplankton production in Narragansett Bay. J. Plankton Res. 1, 75–84.

Verhoff, F.H., Melfi, D.A., Yaksich, S.M., 1982. An analysis of total phosphorus transport in river systems. Hydrobiologia 91, 241–252.

Vinberg, G.G., Liakhnovich, V.P., 1965. Udobrenie Prudov. (Fertilization of Fish Ponds. Izdatel'stvo "Pishchevaia Promyshlennost," Moscow, p. 271 (English translat. 1969. Fish. Res. Board Can. Translation Ser. No. 1339, 482 pp.).

Vollenweider, R.A., 1966. Advances in defining critical loading levels for phosphorus in lake eutrophication. Mem. Ist. Ital. Idrobiol. 33, 53–83.

Vollenweider, R.A., 1968. Scientific Fundamentals of the Eutrophication of Lakes and Flowing Waters, with Particular Reference to Nitrogen and Phosphorus as Factors in Eutrophication. Rep. Organisation for Economic Cooperation and Development, Paris. DAS/CSI/68.27, 192 pp.; Annex, 21 pp.; Bibliography, 61 pp.

Vollenweider, R.A., 1969. Möglichkeiten und Grenzen elementarer Modelle der Stoffbilanz von Seen. Arch. Hydrobiol. 66, 1–36.

Vollenweider, R.A., 1975. Input-output models, with special reference to the phosphorus loading concept in limnology. Schweiz. Z. Hydrol. 37, 53–84.

Vollenweider, R.A., 1976. Advances in defining critical loading levels for phosphorus in lake eutrophication. Mem. Ist. Ital. Idrobiol. 33, 53–83.

Vollenweider, R.A., 1979. Das Nährstoffbelastungskonzept als Grundlage für den externen Eingriff in den Eutrophierungsprozess stehender Gewässer und Talsperren. Z. Wasser-u. Abwasser-Forschung 12, 46–56.

Vollenweider, R.A., 1985. Elemental and biochemical composition of plankton biomass; some comments and explorations. Arch. Hydrobiol. 105, 11–29.

Vollenweider, R.A., Kerekes, J., 1980. The loading concept as a basis for controlling eutrophication philosophy and preliminary results of the OECD Programme on eutrophication. Prog. Water Technol. 12, 5–18.

Vollenweider, R.A., Kerekes, J., 1982. Eutrophication of Waters, Monitoring, Assessment and Control. Organization for Economic Co-Operation and Development, Paris, p. 154.

Vollenweider, R.A., Rast, W., Kerekes, J., 1980. The phosphorus loading concept and Great Lakes eutrophication. In: Loehr, R.C., Martin, C.S., Rast, W. (Eds.), Phosphorus Management Strategies for Lakes. Ann Arbor Science Publs., Ann Arbor, MI, pp. 207–234.

Watanabe, Y., 1990. C:N:P ratios of size-fractionated seston and planktonic organisms in various trophic levels. Verh. Int. Verein. Limnol. 24, 195–199.

Watson, S., McCauley, E., Downing, J.A., 1992. Sigmoid relationships between phosphorus, algal biomass, and algal community structure. Can. J. Fish. Aquat. Sci. 49, 2605–2610.

Weber, C.A., 1907. Aufbau und Vegetation der Moore Norddeutschlands. Beibl. Bot. Jahrb. 90, 19–34.

Weibel, S.R., 1969. Urban drainage as a factor in eutrophication. In: Eutrophication: Causes, Consequences, Correctives. Nat. Acad. Sciences, Washington, DC, pp. 383–403.

Wentz, D.A., Lee, G.F., 1969. Sedimentary phosphorus in lake cores—analytical procedure. Environ. Sci. Technol. 3, 750–754.

Wetzel, R.G., 1981. Longterm dissolved and particulate alkaline phosphatase activity in a hardwater lake in relation to lake stability and phosphorus enrichments. Verh. Int. Verein. Limnol. 21, 337–349.

Wetzel, R.G., 1989. Wetland and littoral interfaces of lakes: Productivity and nutrient regulation in the Lawrence Lake ecosystem. In: Sharitz, R.R., Gibbons, J.W. (Eds.), Freshwater Wetlands and Wildlife. U.S. Dept. Energy, Office Sci. Technical Information, Oak Ridge, TN, pp. 283–302.

Wetzel, R.G., 1990. Land-water interfaces: Metabolic and limnological regulators. Verh. Int. Verein. Limnol. 24, 6–24.

Wetzel, R.G., 1993. Microcommunities and microgradients: Linking nutrient regeneration, microbial mutualism, and high sustained aquatic primary production. Netherlands J. Aquat. Ecol. 27, 3–9.

Wetzel, R.G., 1999. Organic phosphorus mineralization in soils and sediments. In: Reddy, K.R., O'Connor, G.A., Schelske, C.L. (Eds.), Phosphorus Biogeochemistry of Subtropical Ecosystems. CRC Press, Inc., Boca Raton, FL, pp. 225–245.

Wetzel, R.G., Manny, B.A., 1972. Decomposition of dissolved organic carbon and nitrogen compounds from leaves in an experimental hard-water stream. Limnol. Oceanogr. 17, 927–931.

Whalen, S.C., Cornwell, J.C., 1985. Nitrogen, phosphorus, and organic carbon cycling in an arctic lake. Can. J. Fish. Aquat. Sci. 42, 797–808.

White, E., Payne, G., Pickmere, S., Woods, P., 1986. Nutrient demand and availability related to growth among natural assemblages of phytoplankton. N. Z. J. Mar. Freshwat. Res. 20, 199–208.

White, W.S., Wetzel, R.G., 1975. Nitrogen, phosphorus, particulate and colloidal carbon content of sedimenting seston of a hard-water lake. Verh. Int. Verein. Limnol. 19, 330–339.

Wilkins, A.S., 1972. Physiological factors in the regulation of alkaline phosphatase synthesis in *Escherichia coli*. J. Bacteriol. 110, 616–623.

Williams, J.D.H., Mayer, T., 1972. Effects of sediment diagenesis and regeneration of phosphorus with special reference to lakes Eire and Ontario. In: Allen, H.E., Kramer, J.R. (Eds.), Nutrients in Natural Waters. John Wiley & Sons, New York, pp. 281–315.

Williams, J.D.H., Syers, J.K., Harris, R.F., Armstrong, D.E., 1970. Adsorption and desorption of inorganic phosphorus by lake sediments in a 0.1M NaCl system. Environ. Sci. Technol. 4, 517–519.

Williams, J.D.H., Syers, J.K., Shukla, S.S., Harris, R.F., Armstrong, D.E., 1971a. Levels of inorganic and total phosphorus in lake sediments as related to other sediment parameters. Environ. Sci. Technol. 5, 1113–1120.

Williams, J.D.H., Syers, J.K., Harris, R.F., Armstrong, D.E., 1971b. Fractionation of inorganic phosphate in calcareous lake sediments. Proc. Soil Sci. Soc. Amer. 35, 250–255.

Williams, J.D.H., Syers, J.K., Armstrong, D.E., Harris, R.F., 1971c. Characterization of inorganic phosphate in noncalcareous lake sediments. Proc. Soil Sci. Soc. Amer. 35, 556–561.

Winter, J.G., Eimers, M.C., Dillon, P.J., Scott, L.D., Scheider, W.A., Willox, C.C., 2007. Phosphorus inputs to Lake Simcoe from 1990 to 2003: Declines in tributary loads and observations on lake water quality. J. Great Lakes. Res. 33, 381–396.

Wisniewski, R.J., 1991. The role of benthic biota in the phosphorus flux through the sediment-water interface. Verh. Int. Verein. Limnol. 24, 913–916.

Woolway, R.I., Sharma, S., Weyhenmeyer, G.A., Debolskiy, A., Golub, M., Mercado-Bettín, D., Perroud, M., Stepanenko, V., Tan, Z., Grant, L., Ladwig, R., Mesman, J., Moore, T.N., Shatwell, T., Vanderkelen, I., Austin, J.A., DeGasperi, C.L., Dokulil, M., La Fuente, S., Mackay, E.B., Schladow, S.G., Watanabe, S., Marcé, R., Pierson, D.C., Thiery, W., Jennings, E., 2021. Phenological shifts in lake stratification under climate change. Nature 12, 2318. https://doi.org/10.1038/s41467-021-22657-4.

Wynne, D., Rhee, G.-Y., 1988. Changes in alkaline phosphatase activity and phosphate uptake in P-limited phytoplankton, induced by light intensity and spectral quality. Hydrobiologia 160, 173–178.

Yamamichi, M., Kazama, T., Tokita, K., Katano, I., Doi, H., Yoshida, T., Hairston Jr., N.G., Urabe, J., 2018. A shady plankton paradox: When phytoplankton increases under low light. Proc. R. Soc. A B 285, 20181067. https://doi.org/10.1098/rspb.2018.1067.

Young, J.D., Winter, J.G., Molot, L.A., 2011. A re-evaluation of the empirical relationships connecting dissolved oxygen and phosphorus loading after zebra mussel invasion in Lake Simcoe. J. Great Lakes Res. 37, 7–14.

Zaiss, U., 1985. Physiologische und ökologische Untersuchungen zur Regulation der Phosphatspeicherung bei *Oscillatoria redekei*. II. Der Einfluss ökologischer Parameter auf die Regulation des Polyphosphatstoffwechsels. Arch. Hydrobiol. Suppl. 72, 166–219.

Zicker, E.L., Berger, K.C., Hasler, A.D., 1956. Phosphorus release from bog lake muds. Limnol. Oceanogr. 1, 296–303.

CHAPTER

16

Other Important Elements

Emma S. Kritzberg

Department of Biology, Lund University, Lund, Sweden

OUTLINE

I. Biogeochemical cycling of micronutrients and minor elements

In addition to the major nutrients carbon, nitrogen, and phosphorus, many minor elements are essential to the survival and function of the freshwater biota and are referred to as *micronutrients*. Other minor elements affect the biota by being toxic. But elements can also play a major role in the limnic system indirectly, by interacting with other elements and influencing their mobility and availability. Often elements influence the limnic system in multiple ways; for instance, iron, which is a *cofactor* in essential enzymes, is toxic in high concentrations and exerts a major control on the transport and mobility of, for example, phosphorus (Chapter 15). This chapter also includes the biogeochemical cycle of sulfur, which is in fact an abundant macronutrient but is included here because of its close interaction with iron and manganese. The chapter also includes silicon, which is an element of key importance to specific organisms in freshwaters.

The biogeochemical cycling of these elements is regulated to a large extent by variations in oxidation–reduction states, which in turn are mediated by photosynthetic and heterotrophic metabolism, as well as by photochemical reactions. The biogeochemical cycling of many elements is also influenced by pH and complexation by organic matter. Many of the reactions among different elements are coupled, and the state of one can influence the availability of another. Although the

Wetzel's Limnology, Fourth Edition
https://doi.org/10.1016/B978-0-12-822701-5.00016-1

coupled reactions must be viewed simultaneously, the following discussion separates the components as much as possible.

II. Oxidation–reduction potentials in freshwater systems

Life in freshwater systems depends ultimately on solar energy (Chapter 6). Solar light drives photosynthesis, whereby the reduction of CO_2 and oxidation of H_2O produces organic compounds of reduced state, and the solar energy is converted into chemical bonds. The products of photosynthesis are thermodynamically unstable, and by catalytical decomposition through energy-yielding reactions, such as respiration or fermentation, organisms tend to restore equilibrium. It is through such reactions that nonphotosynthetic organisms obtain energy for their metabolic demands (Fig. 16-1). The decomposition of the photosynthetic products also provides the reducing power to drive the cycling of oxygen, nitrogen, manganese, iron, sulfur, and a number of other elements in the aquatic environment. During respiration and fermentation, the organisms act as *redox catalysts* by mediating the reactions and transfer of electrons; the organisms themselves do not oxidize substrates or reduce compounds. There are a few elements (C, O, N, S, Fe, Mn) that are predominant reactants in redox processes in natural waters.

The *oxidation–reduction potential* (see also Chapter 10), also known as the *redox potential*, is a composite measure of the oxidizing or reducing intensity of a solution (Stumm and Morgan, 1996). It reflects the tendency of a solution to receive or provide electrons. Redox potentials (E_h) of aqueous solutions are estimated by measuring the electron flow between a platinum electrode and a stable reference electrode (reference point of zero volts) and are expressed in volts (V) or millivolts (mV). The E_h measured corresponds to the voltage that is required to stop electrons flowing between the reference electrode and the solution. In complex freshwater environments many redox reactions take place simultaneously and conditions are not in equilibrium, and therefore absolute redox potentials are impossible to measure accurately. Moreover, while, for example, reduced and oxidized iron (Fe(II) and Fe(III)) are among the most electroactive redox reactants in natural water systems, the redox components of organic carbon, nitrogen, and sulfur are not electronegative and yield reversible potentials only following enzyme-mediated changes. As a result, redox measurements in natural waters do not lend themselves to quantitative interpretation and comparison. For instance, in pH neutral and fully oxygenated water redox potentials slightly greater than 500 mV are obtained, which is considerably less than the theoretical E_h of 800 mV (Fig. 16-2). However, qualitative and relative comparisons of E_h, representing mixed composite potentials, can be extremely instructive to predict which redox reactions are favored as well as what concentrations and forms of iron, manganese, sulfur, and several trace metals prevail in environments with specific conditions.

The oxidation–reduction state of an environment can also be expressed as *electron activity* [$pE = -\log(e^-)$], just

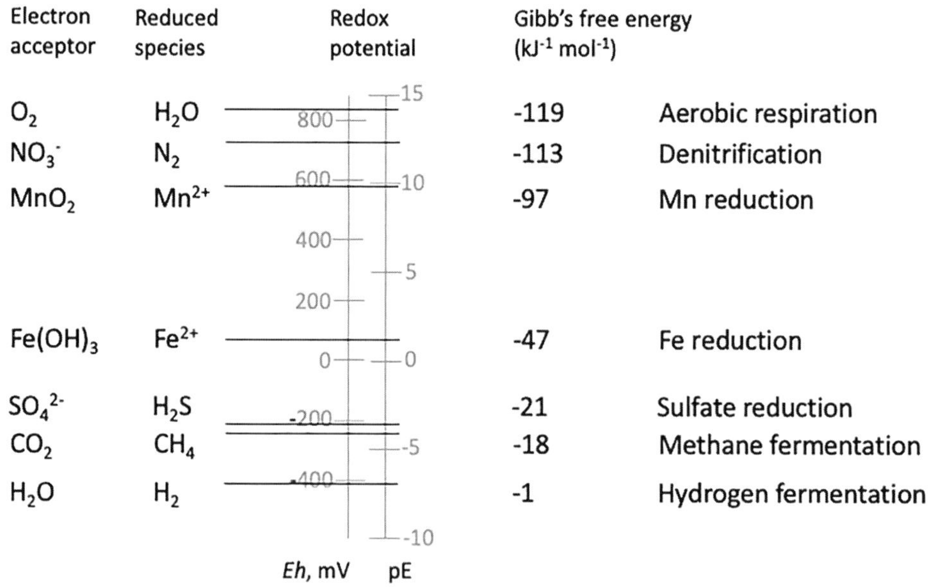

FIGURE 16-1 Redox ladder showing some of the major reactions in freshwaters under varying redox potential at pH 7 and the free energy associated with these processes. *(Data from Morel and Hering, 1993)*

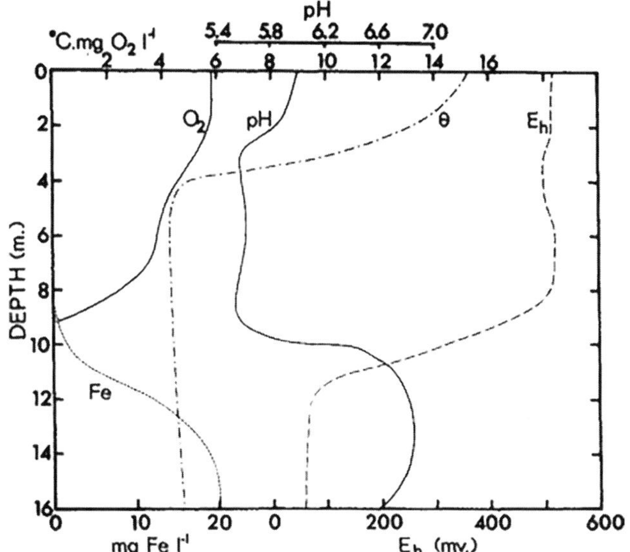

FIGURE 16-2 Vertical distribution of temperature (theta, θ), oxygen, pH, total iron, and redox potential Eh in permanently meromictic Lake Skiennungen, Norway, June 1967. *(After Kjensmo, 1970.)*

like pH defines proton activity [pH = $-\log(H^+)$]. pE is large and positive in strongly oxidizing solutions (low electron activity), just as pH is high in strongly alkaline solutions (low proton activity). Negative pE indicates relatively reducing conditions. Thus both pH and pE are intensity factors of free energy levels.

The redox potential of water is relatively insensitive to a change in oxygen concentration and extent of saturation. If the oxygen content is decreased by 99%, the redox potential would be reduced by only about 30 mV. Redox reactions in aqueous solutions often involve H_2O, H^+, and OH^- (Eqs. 10-1 to 10-7), and therefore the redox potential is significantly altered by changes in the pH. It is customary to express the redox potential of redox reactions in natural waters at pH 7. Thus the redox potential is expressed as E_h or as E_7, in which a correction is made for the change in redox at the pH of the sample to a pH of 7. A change in pH of one unit is accompanied by a change in redox potential of 58 mV. Hence a common practice is to correct potentials to pH 7 by subtracting 58 mV for every pH unit on the acid side of neutrality and by adding 58 mV for every pH unit on the alkaline side of neutrality. E_h is influenced to only a small extent by temperature; for example, E_h of water = 860 mV at 0°C and 800 at 30°C, at pH 7.

In line with the preceding discussion, little change in redox potential is found with increasing depth as long as the water contains dissolved oxygen (Fig. 16-2). Provided that the water is not near anoxia, the E_h will remain positive and fairly high (300–500 mV). Where anoxic conditions prevail, as often in the lower hypolimnion and near the sediments, the E_h decreases sharply. Within the sediments, reducing conditions prevail, and the E_h declines to 0 mV or below within a few millimeters of the interface. Of the many reducing compounds that contribute to the reductions in E_h, ferrous iron (Fe(II)) of the sediments is the most important.

III. The iron and manganese cycle

The similarities in chemical reactivity between iron (Fe) and manganese (Mn) permit us to discuss them together. Although clear differences exist between the two metals, and the dynamics and cycling of iron is much more studied, they are both strongly affected by redox conditions and behave in a similar fashion in freshwater systems. A strong interaction exists between the cycling of these metals and sulfur (S). The biogeochemical fluxes of iron and manganese reflect the spatial and temporal variations in physical chemistry and bacterial metabolism and are also strongly influenced by interactions with organic matter.

A. Sources and forms of iron and manganese

Iron and manganese constitute ~5% and ~0.1% of the mass of Earth's crust, respectively, and following weathering and mobilization, they may enter into suspension. Under natural conditions, iron and manganese come primarily from catchment soils and enter via soilwaters that discharge into streams or via groundwater. Wetlands, soils with high organic matter content, and pyrite (FeS_2) can be significant sources, and air deposition is an important contributor in some regions (Dillon et al., 1988; Canfield et al., 2005a; Maranger et al., 2006; Björkvald et al., 2008). Drainage from acidic coniferous forests is often high in iron and manganese, indicating the importance of organic complexes in the effective transport of these easily oxidized metal ions. Sources of anthropogenic pollution to freshwaters include mine drainage, waste- and storm-water discharge, and atmospheric deposition resulting from the combustion of fossil fuels. Under certain conditions, internal loading from the sediment can be significant, fueled by reductive dissolution in the sediment and subsequent release into the water column (Nürnberg and Dillon, 1993). Nevertheless, most lakes are sinks of iron and manganese, as a result of particle sinking and sediment accumulation.

Iron exists in solution in either the *ferric*, Fe(III), or *ferrous*, Fe(II), form. Amounts of iron in solution in natural waters and the rate of oxidation and reduction are dependent primarily on pH, E_h, and temperature. The simultaneous influence of pH and E_h on the equilibria of aqueous iron is illustrated in Fig. 16-3, which shows

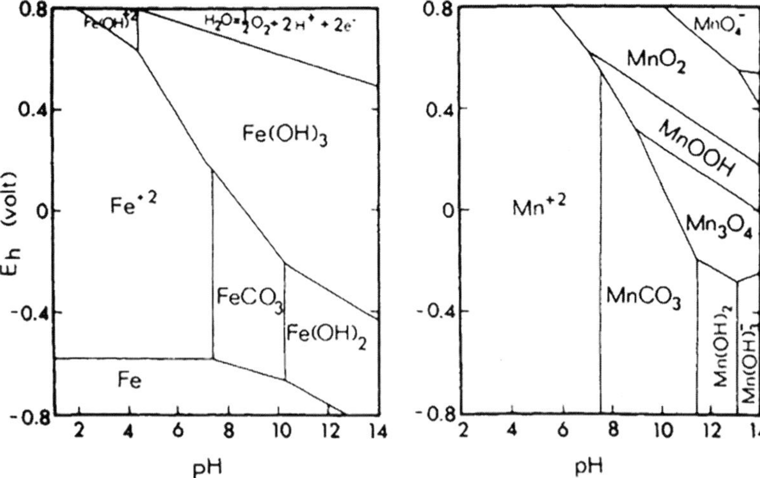

FIGURE 16-3 Approximate distribution of species of iron and manganese in relation to pH and redox potential E_h. Alkalinity is assumed to be equal to 2 meq L^{-1}. *Lines denote points at which the activities of soluble Fe and Mn are* 10^{-5} mol L^{-1}. *(Modified from Stumm and Lee, 1960; Stumm and Morgan, 1996.)*

that, in the absence of organic matter and based only on solubility criteria, a number of inorganic iron forms can exist in natural waters.

In the presence of oxygen the chemical oxidation of ferrous iron to ferric iron is greatly dependent on pH. At pH values below about 4, ferrous iron can dominate, but between pH 5 and 8, the oxidation rate is rapid and increases with pH (Morgan and Lahav, 2007). Oxygenation reactions of Fe(II) to Fe(III) are *exergonic,* and some of these reactions serve as an energy source for microorganisms. The most common species of ferric iron in natural waters are hydrated ferric hydroxides, Fe(oxy) hydroxides (Tipping et al., 1981; Sjöstedt et al., 2013). At equilibrium in the pH range of 5–8, $Fe(OH)_3$ precipitates, because its solubility is very low (equilibrium constant $\sim 10^{-36}$ at 25°C) and immediately converts to polymeric Fe(oxy)hydroxides. Other insoluble ferric salts are of less significance. For example, phosphate does not influence the solubility of Fe(III) when inorganic phosphorus concentration is $<10^{-4}$ mol L^{-1}, as is usually the case.

Ferric iron can be chemically, photochemically, or microbially reduced to ferrous iron. The latter is energetically favored only when dissolved oxygen, nitrate, and oxidized manganese have been depleted during microbially mediated oxidation of organic matter (Fig. 16-1). Chemical reduction can be mediated by reaction with sulfide or organic compounds (Bauer and Kappler, 2009). Photochemical reduction is described in Section IIIC.

Ferrous constituents tend to be more soluble than ferric ones. Soluble ferrous iron occurs mainly as hydrated Fe^{2+} and hydrated hydroxy ions, the solubility of which is determined largely by the solubility of ferrous hydroxide $Fe(OH)_2$, ferrous carbonate ($FeCO_3$), and ferrous sulfide (FeS). $Fe(OH)_2$ is exceedingly insoluble within the normal pH range of oxygenated natural waters. The solubility of ferrous iron is generally controlled by the solubility of $FeCO_3$. Even at low pH, carbonate concentrations are usually sufficient to limit solubilization. Soft waters with very low concentrations of bicarbonate usually contain somewhat higher concentrations of Fe(II), although most is oxidized to Fe(III). For example, a water free of bicarbonate at pH 7 would contain more than 1000 times as much Fe(II) as a water containing 2 milliequivalents (meq) L^{-1} alkalinity, an average value for many waters. Ferrous sulfides are also exceedingly insoluble and form both amorphous and stable, crystalline phases that darken the color of anaerobic sediments (Davison, 1991; Doyle, 1968). At low redox potentials in the anaerobic hypolimnia and sediment of productive or meromictic lakes, bacterial reduction of sulfate to sulfides is common. Excess of sulfide can decrease concentrations of Fe(II) through the formation of insoluble ferrous sulfides.

Interactions with organic matter are a key feature of iron dynamics in freshwater systems. Components of natural organic matter with strong complexing capacity can bind mononuclear Fe(III) and thereby protect it from hydrolyzation and precipitation as Fe(oxy)hydroxides (Tipping, 1981). Carboxylates appear to be the dominant functional group involved in these interactions (Karlsson and Persson, 2012). In acidic humic waters (pH 4–5) mononuclear organic complexes may dominate total iron and contribute to the intense yellow-brown color of bogwaters (Fig. 16-4). At higher pH, hydrolyzation is favored and Fe(oxy)hydroxide phases are often dominant (Neubauer et al., 2013). Organic matter interactions with Fe(oxy)hydroxides are also common (Tombácz et al., 2004; Vindedahl et al., 2016). Colloidal particles of Fe(oxy)hydroxides are commonly positively charged, and surface interactions with negatively charged organic matter can prevent polymerization and aggregation into larger

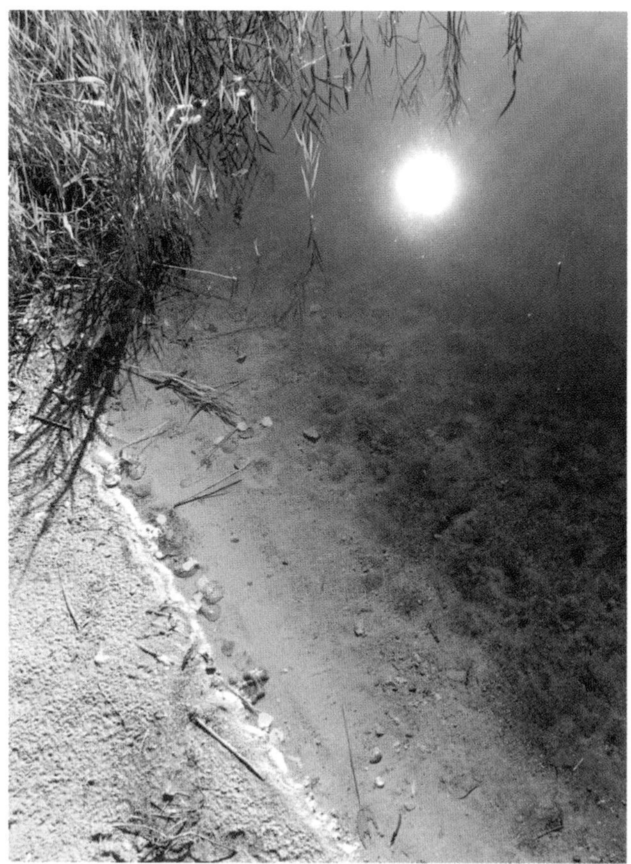

FIGURE 16-4 Brown water from a boreal lake (Lake Strålsjön, Sweden) with a high concentration of iron and dissolved organic matter originating from a peat bog in the catchment. *(Photo: Caroline Björnerås.)*

particles, which would settle more rapidly (Karlsson and Persson, 2012). These two types of organic matter interactions are why Fe concentrations in freshwaters are often much higher than would be predicted based on solubility criteria alone (Shapiro, 1966). Positively charged Fe(oxy)hydroxide colloids may also form complex with other negatively charged solids such as clays, ions (phosphate, trace metals), and other suspended solids, which can neutralize the charges on the colloidal particles. The uncharged aggregates can join to form a rapidly settling precipitate.

Interactions between iron and organic matter can influence the redox state of iron in intricate ways. For instance, humic acids can reduce Fe(III) (Bauer and Kappler, 2009) and preserve Fe(II) (Daugherty et al., 2017). The preservation is mainly due to complexes with carboxyl groups, similar to the complexation of Fe(III) but weaker (Rose and Waite, 2003). This function of organic matter as a complexant and redox buffer may explain why Fe(II) is sometimes present also in oxic riparian and stream waters, where equilibrium calculations and oxidation rates suggest that Fe(III) would be favored (Daugherty et al., 2017; Sundman et al., 2014). However, humic acids have also been demonstrated to promote oxidation of Fe(II), supposedly by semiquinone-like components, which are the result of irradiation (Garg et al., 2015). Thus the nature of interaction between iron and organic matter varies widely, depending on pH, functional composition of the organic matter, iron-to-organic matter ratios, and the redox state of the components.

As follows from the above, Fe is found in a wide range of molecular sizes in freshwaters (Pokrovsky and Schott, 2002). While mononuclear complexes may dominate in the small size range, and Fe(oxy)hydroxides and clays dominate the larger size fractions (Pokrovsky and Schott, 2002; Stolpe and Hassellöv, 2007), there is also a significant overlap in size between the Fe species (Herzog et al., 2020). Under oxic and circumneutral conditions Fe(oxy)hydroxides are often the dominating phase (Tipping et al., 1982; Sjöstedt et al., 2013), and a significant fraction of this is removed by 0.45 μm filtration (Herzog et al., 2020). However, a very finely divided precipitate of Fe(oxy)hydroxides with colloidal properties (0.001−0.5 μm) may also exist (e.g., (Herzog et al., 2020; Tipping, 1981).

A portion of total iron is contained in the living plankton, but this quantity is generally a small part of the whole. Iron in clays and other secondary minerals can also contribute to the total suspended iron as well as to sediments.

Manganese occurs in several valence states, but the most important species in freshwater environments are Mn(II), Mn(III), and Mn(IV). Similar to iron, the solubility of manganese decreases markedly with increases in E_h and pH. As with ferrous iron, divalent Mn occurs in the ionic form at low redox potentials and pH (Fig. 16-3). Mn(IV) is insoluble at most environmental pH values and forms various oxide minerals. Mn(III) is an intermediate and is thermodynamically unstable in aqueous solutions under normal conditions. It should be noted that oxides often contain a mix of Mn(II), Mn(III), and Mn(IV). Under oxidizing conditions of high pH and E_h, some form of oxidized Mn will be in equilibrium with Mn(II), and under reducing conditions, Mn(II) may be in equilibrium with manganese carbonate. Supersaturation of both ferrous and manganese carbonate has been observed in anoxic hypolimnia of eutrophic lakes (Verdouw and Dekkers, 1980). The redox equilibrium E_h values are higher and the rates of oxidation slower than for iron and, as a result, relatively high concentrations of manganese are commonly observed longer than comparable concentrations of iron under lake conditions. Manganese forms soluble complexes with bicarbonate and sulfate. When the pH exceeds 7, manganese adsorbs to iron oxides and coprecipitates with ferric hydroxide. In a manner analogous to iron organic molecules can form stable complexes with Mn(III) (Oldham et al., 2017), although their operation in aquatic systems is poorly understood.

B. Distribution of iron and manganese in freshwaters

Under oxidized conditions, as in the epilimnia of lakes and most streams, large amounts of iron are found only in waters that are acidic or with high concentrations of organic matter (Table 16-1). Acidic waters ($pH < 3-4$) include lakes of volcanic origin and influence (Yoshimura, 1936), as well as streams and lakes affected by runoff from mining or acidic sulfate soils (Boman et al., 2010). In these waters organic content is relatively low, and acidity usually results from sulfuric acid. Acid sulfate soils release high concentrations of iron following the oxidation of metal sulfides, and particularly under high flow conditions, iron and manganese are released into streams with detrimental effects on biota (Fältmarsch et al., 2008). When streams that receive acidic drainage are exposed to sunlight, photoreduction of ferric iron can result in a production of ferrous iron that by far outweighs the nighttime oxidation of ferrous iron (McKnight et al., 1988).

The quantity of total iron found in neutral or alkaline surface waters, however, which is in the range of 50–2000 $\mu g\ L^{-1}$, varies with lithology and is dominated by Fe(oxy)hydroxides, organically complexed iron, Fe-bearing clays, and primary silicates. The concentration is generally higher than would be expected from solubility criteria and is positively correlated with organic matter concentration (Kritzberg and Ekström, 2012), indicating the importance of organic matter interactions for keeping iron in suspension. Trends of increasing iron concentrations are found for wide regions in Northern Europe and North America, often concurrent with rising concentrations of organic matter and more frequently in catchments with high coniferous vegetation cover (Björnerås et al., 2017). Such rising concentrations of iron contribute to the frequently observed browning of waters, which is generally ascribed to increasing concentrations of terrestrially derived organic carbon (Kritzberg and Ekström, 2012).

The range of manganese concentrations ($\sim$10–850 $\mu g\ L^{-1}$) is also highly variable in relation to the lithology and drainage of the lake basins (Hutchinson, 1957; Livingstone, 1963; Hongve, 1980). The concentration of manganese is generally lower than that of iron. However, the ratio of Fe to Mn in water is generally considerably lower than that of the lithosphere (50:1) and indicates the relative enrichment of Mn with respect to Fe. This is in agreement with the reaction equilibria previously discussed.

The vertical distribution of iron and manganese is reflected in the distribution of redox potentials (Fig. 16-5). Concentrations of ionic iron of oxygenated waters of oligotrophic lakes, epilimnia of more productive lakes, and circulating waters are exceedingly low. Manganese is somewhat more soluble. Ferrous ions diffuse readily from the sediments when redox potentials decline to about 200 mV, while migration of Mn(II) from the sediments occurs at somewhat greater redox potentials and precedes that of Fe(II) (Robbins and Callender, 1975; Ostendorp and Frevert, 1979) (see Fig. 15-5 and 15-6 in Chapter 15). Reductive dissolution of settling oxide phases is also a source of Mn(II) and Fe(II) in the hypolimnion. As decomposition proceeds in the hypolimnion of very productive, thermally stratified lakes, the redox potential of hypolimnetic waters can decline to well below 100 mV. At an E_h below 100 mV, sulfate is reduced to hydrogen sulfide, which is toxic to all eukaryotes. Since ferrous iron is released in significant quantities from the sediments at a higher E_h of about 200 mV, Fe(II) is already present in the hypolimnion at the time of sulfide formation. The formation of iron sulfides can result in a significant reduction of iron concentration toward the end of summer stratification (Davison and Heaney, 1978). Manganous sulfide, on the other hand, is much more soluble and has little effect on the Mn(II) concentrations under normal lake conditions.

TABLE 16-1 Iron, Manganese, and Sulfate Concentrations in Some Waters of Varying Characteristics

Water body		*pH*	*DOC mg L^{-1}*	*Fe $\mu g\ L^{-1}$*	*Mn $\mu g\ L^{-1}$*	*SO_4^{2-} mg L^{-1}*
L. Fiolen[a]	Clear softwater	6.8	9.1	170	32	3.6
L. Skärshult[a]	Brown softwater	5.8	22.7	1400	120	2.2
L. Vomb[a]	Calcareous	8.4	8.7	59	170	24.9
St Kevin's Gulch[b]	Acidic stream	3.8	0.9	1600	4200	90.7

[a]*Data from Sweden's national surface water monitoring program (sampling August 2012) http://webstar.vatten.slu.se/db.html.*
[b]*McKnight et al. (1988).*

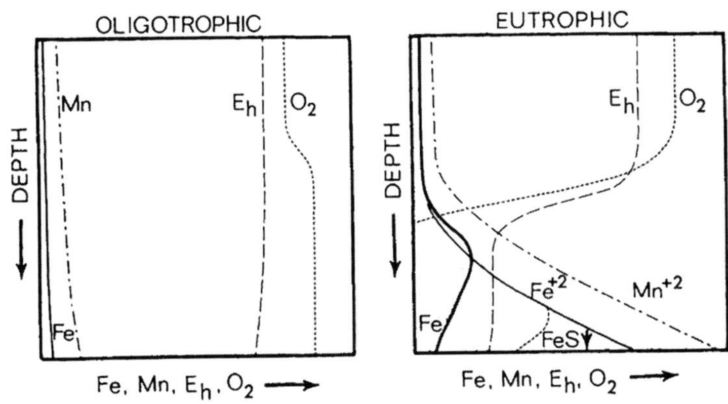

FIGURE 16-5 Generalized vertical distribution of iron, manganese, and redox potential (E_h) in stratified lakes of very low and high productivity.

Iron concentrations in the hypolimnia of softwater lakes can reach very high levels under conditions that prevail in small, deep basins, especially in bog waters receiving high concentrations of humic organic matter. Levels of sulfate are low in such waters, and sulfide concentrations seldom become sufficient to precipitate iron sulfides. Thus gradients of elevated iron concentrations are found in hypolimnia of humic lakes along with the development of anoxia, but not FeS precipitation. Stages along the continuum of declining redox potentials can be divided into four phases of hypolimnetic conditions in stratified lakes of increasing productivity or within a very productive lake during the period of summer stratification (Table 16-2). The hypolimnetic iron accumulations in wind-protected lakes can reach such levels (>250 mg L^{-1}) that the salinity gradient becomes adequate to render the lakes permanently meromictic (Kjensmo, 1968).

Although oxidation of Mn(II) by oxygen is thermodynamically favored under the pH of most lakes, it is kinetically limited. In the absence of biotic or abiotic catalysts the half-life of Mn(II) is in the order of months. In contrast, Fe(II) is rapidly oxidized by oxygen with a half-life of only minutes. Mn oxidation is accelerated by microbes, complexation with organic compounds, and oxide surfaces (Godwin et al., 2020). Nevertheless, the slower oxidation kinetics still results in larger mobility of manganese than of iron in oxic freshwaters.

A seasonal sequence is commonly observed in productive lakes in which the oxic/anoxic boundary migrates from the sediment–water interface well into the hypolimnion (Fig. 16-6). As the hypolimnion becomes more reducing, a progressive shift occurs from the amorphous particulate Fe(oxy)hydroxides to soluble ferrous iron. Where appreciable sulfide is formed by the biological reduction of sulfate, ferrous and total iron can be reduced by the formation of, and precipitation of, insoluble FeS (e.g., Hutchinson, 1957; Cook, 1984). Although the dissolved iron concentrations may be controlled in the lower strata during stratification by FeS precipitation, MnS phases are generally undersaturated and appreciably less than the solubility constants needed for precipitation (Balistrieri et al., 1992). The general pattern of seasonal distribution is given by examples of a mesotrophic lake and an interconnected eutrophic lake that undergo hypolimnetic oxygen reduction (Figs. 16-7 and 16-8).

TABLE 16-2 Changes in Iron During the Continuum of Declining Redox Potentials in the Hypolimnia of Stratified Lakes of Increasing Productivity[a]

Lake status	$[O_2]$	E_h	Fe^{2+}	H_2S	PO_4^{3-}
Oligotrophic	High (orthograde)	400–500 mV	Absent	Absent	Very low
	↓	↓	↓	↓	↓
↓	Much reduced (clinograde)	400–500 mV	Absent	Absent	Very low
	↓	↓	↓	↓	↓
Eutrophic	Much reduced (clinograde)	Approx. 250 mV	High	Absent	High
↓	↓	↓	↓	↓	↓
Hypereutrophic	Much reduced (or absent)	<100 mV	Decreasing	High	Very high

[a]*After discussion of Hutchinson (1957).*

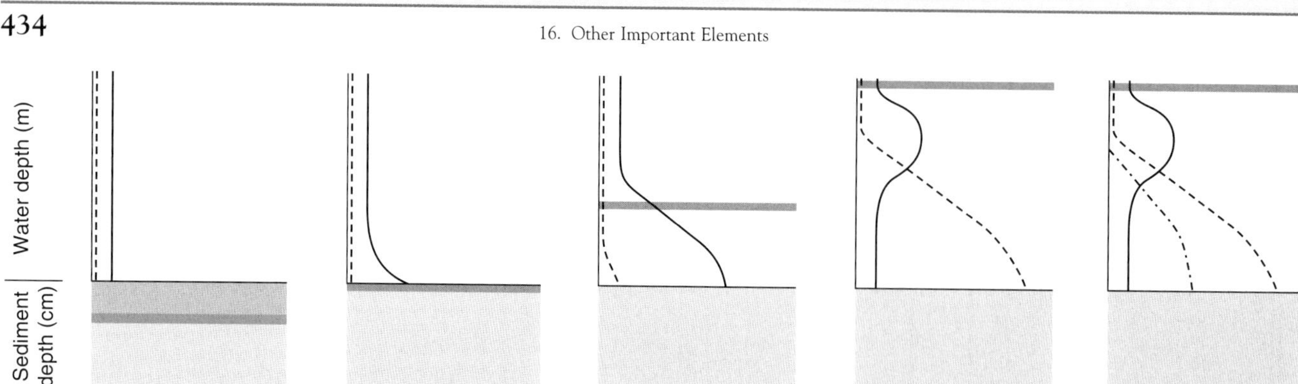

FIGURE 16-6 Changes in the concentration of particulate iron (—) and Fe(II) (- - - -) in the water column of a lake as the oxic/anoxic boundary (thick gray line) undergoes seasonal migration. The sediment is represented in gray, and elevated concentrations of particulate iron in the sediment are represented by darker gray. The alternative late summer conditions (b) represent marked reducing conditions where sulfide (-·-·-) and FeS precipitation can contribute to the alterations of vertical distributions. *(Modified from Davison and De Vitre, 1992.)*

C. Cycling of iron and manganese

The spatial and temporal distribution of iron and manganese depends upon the balance of many physical, chemical, and biological parameters. Temperature and wind affect the hydrodynamic stability of stratification (Chapter 7). Solar radiation influences speciation, the nature of interactions with organic matter, and thereby the stability of iron and manganese in the water column. Biologically mediated chemical processes regulate the redox conditions and the availability of electron acceptors, and the extent of reduction to allow the accumulation of sulfide and subsequent precipitation of metal sulfides.

A schematic model of iron transport processes in a lake under stratified (left) and mixed conditions (right) is depicted in Fig. 16-9. Iron loading (Fig. 16-9(A)) is significantly larger than export (B) from most lakes, due to processes that result in sinking and loss to the sediment (C–F). An example of an iron budget is given in Table 16-3. Internal loading by resuspension or reductive dissolution from sediments (G) can vary from insignificant to many times larger than external loading (Nürnberg and Dillon, 1993). While lakes in general are efficient sinks of iron, the fraction of total loading that is retained declines with the importance of internal loading. Moreover, loss to sediments tends to increase with the increasing residence time of the water body, leaving more time for loss processes (Nürnberg and Dillon, 1993; Weyhenmeyer et al., 2014). Fractions of high density are lost rapidly by particle sinking as a result of reduced water velocity (C). Other losses are linked to in-lake transformations, which promote the formation of colloids/particles and aggregation (Björnerås et al., 2021). Such transformations may be triggered by increasing pH, which favors the oxidation of Fe(II), hydrolyzation of Fe(III), and precipitation of Fe(oxy) hydroxides (Neubauer et al., 2013). Transformations can also be light induced.

High energy UV radiation and lower wavelengths of the visual region have the capacity to chemically reduce Fe(III) to Fe(II). This can be either direct, by photon absorption and ligand to metal charge transfer (LMCT), where an electron is transferred from the ligand to the Fe(III), or indirect, when photochemically produced reactive oxygen species (ROS) reduce Fe(III). LMCT reactions can happen when carboxylate groups that complex mononuclear Fe(III) transfer the charge, but also when ligands interact with surfaces of minerals or colloids, and can then lead to mineral dissolution (Lueder et al., 2020a). Photoreductive dissolution can also take place in the absence of organic complexes by charge transfer from the surface hydroxide ion to Fe(III). However, reductive dissolution is generally decreasing with increasing pH and is efficient only in acidic waters (Sulzberger et al., 1994; Borer et al., 2009).

Photolysis of organic matter can produce ROS, such as superoxide ($O_2^{-}\bullet$), hydroxyl radicals ($OH\bullet$), and hydrogen peroxide (H_2O_2). Superoxide is most likely to mediate the reduction of Fe(III) in circumneutral freshwaters and can reduce both Fe(III) associated and disassociated from ligands but also colloidal fractions (Xing et al., 2019). In addition to the photolytic reactions of organic matter, superoxide can be formed by microbes (Vähätalo et al., 2021) and from the oxidation of Fe(II) by oxygen.

Photochemical redox reactions of Fe(III) compounds are an important source of Fe(II) in surface waters. Fe(II) can exist as Fe_{aq}^{2+}, or in organic complex with ligands that can serve as a redox buffer. Fe_{aq}^{2+} and organic Fe(II) complexes are generally more bioavailable than Fe(III) and photochemically induced reduction thereby provides a source of nutrients to aquatic organisms. Photoreduction results in diel cycles with enhanced

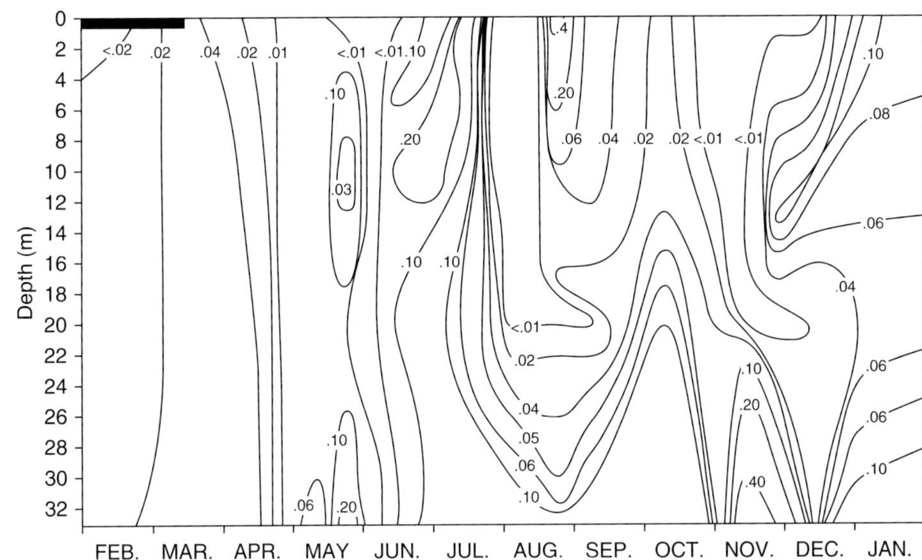

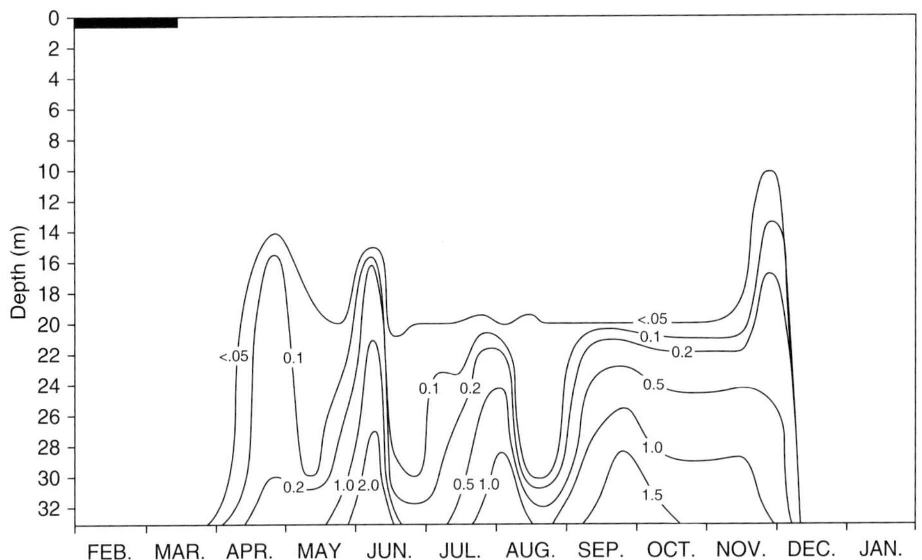

FIGURE 16-7 Depth–time diagrams of isopleths of total iron (*upper*) and manganese (*lower*) in mg L^{-1} of mesotrophic hardwater Crooked Lake, northeastern Indiana, 1963. *(Graphs are based on unpublished data from R.G. Wetzel.)*

concentrations of Fe(II) during daytime and close to the surface (Emmenegger et al., 2001). However, in oxic circumneutral waters Fe(II) is rapidly oxidized so that Fe(III) dominates, either in the form of mononuclear organic complexes or Fe(oxy)hydroxides. Freshly formed colloidal Fe(oxy)hydroxides may remain in suspension when stabilized by surface interactions with organic matter or may be a source of iron to the sediment (D). Only in strongly acidic waters can Fe(III) reduction outweigh Fe(II) oxidation so that considerable amounts of Fe(II) prevail (McKnight et al., 1988). In such waters, and in some surface sediments, photoinduced Fe(III) reduction can be sufficient to provide substrate for Fe(II) oxidizing bacteria (Lueder et al., 2020b).

At the oxic/anoxic interface of the water column, there is also redox cycling, where Fe(II) from the anoxic water is diffusing into the oxic epilimnetic water. This Fe(II) originates from the reductive dissolution of sinking ferric particles or from the sediment and is transported laterally from the slopes of the basin and vertically by currents and eddy diffusion (G). When it reaches the epilimnion, it is rapidly oxidized to Fe(III), forming Fe(oxy)hydroxides which at least partly sink back into anoxic water (E). The lake-internal replenishment of Fe(oxy)hydroxides and subsequent sinking is held to be a significant vector for the vertical transport of phosphate and organic carbon by coprecipitation and surface interactions (Mortimer, 1941; von Wachenfeldt et al., 2008; Sundman et al., 2016).

In highly productive waters, which maintain strongly reducing conditions in the hypolimnion and sediment, FeS precipitation is a potential loss process (F). Under

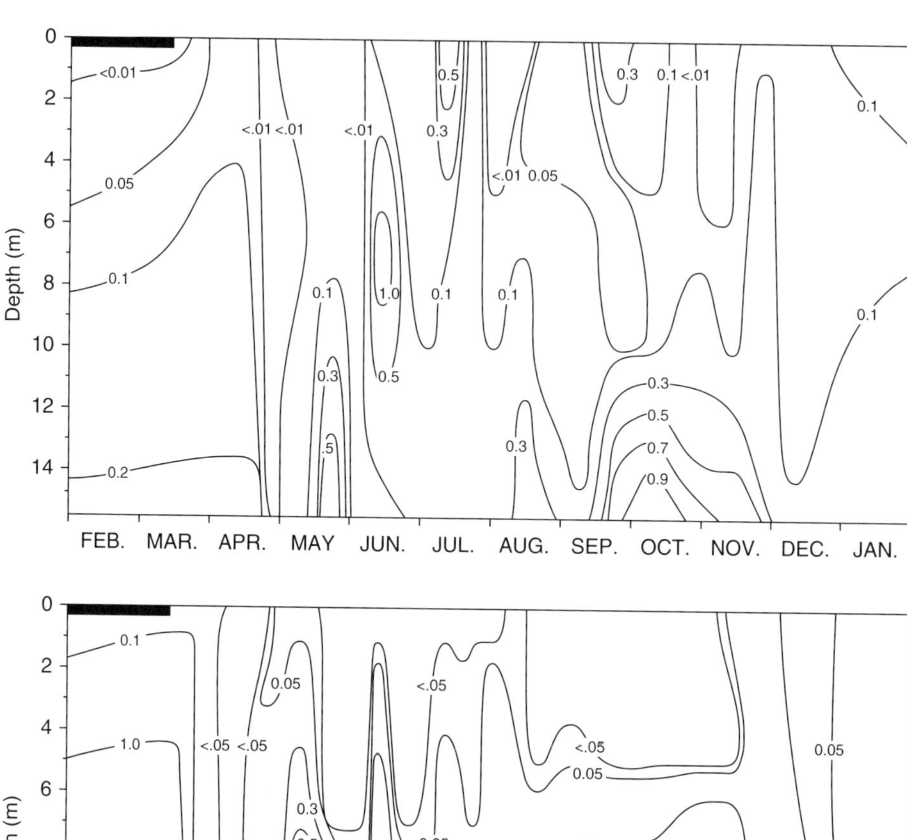

FIGURE 16-8 Depth–time diagrams of isopleths of total iron (*upper*) and manganese (*lower*) in mg L^{-1} of eutrophic hardwater Little Crooked Lake, northeastern Indiana, USA, 1963. *(Graphs are based on unpublished data from R.G. Wetzel.)*

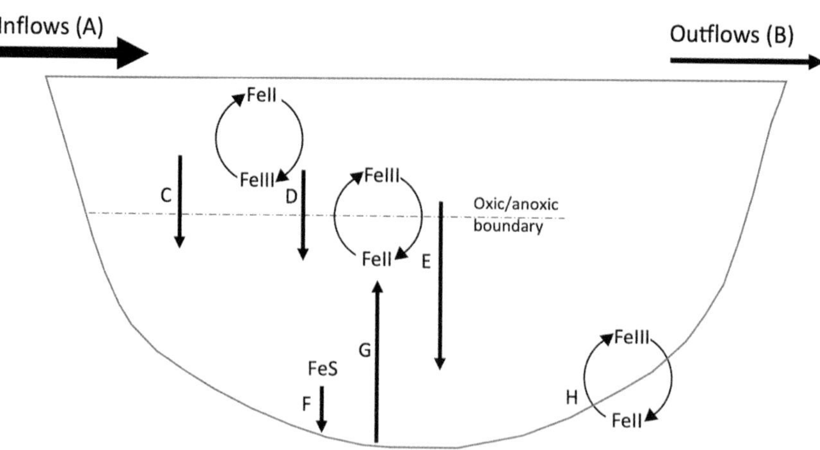

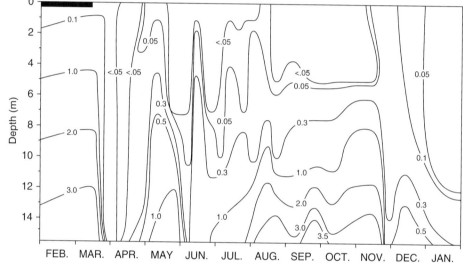

FIGURE 16-9 Transport processes of iron in a lake with an anoxic hypolimnion (*left part*) and during mixed conditions (*right part*). Each of the processes, A–H, is described in the text.

TABLE 16-3 Budget of Iron Sources and Sinks, Esthwaite Water, England[a]

	tons $year^{-1}$
Inputs:	
Minimum annual inflows	20
Annual dissolution from sediments	1.8–3.6
Outputs:	
Annual outflow from lake	6.4
Annual accumulation in sediments	15–60
Minimum net loss from water column	14

[a]*Extracted from data of Davison et al. (1980).*

conditions when the bottom water and surface sediment is oxic, Fe(II) that is formed by reduction in anoxic layers of the sediment and diffuses upward is oxidized and precipitated as Fe(III) in the surface sediment (H). While all the processes above have been demonstrated individually, their relative contribution to the transformations and fluxes of iron are poorly known.

Cycling and fluxes of manganese in limnic systems are much less studied, and much of what is known about the transformations and behavior of different Mn phases is derived from marine work. In the wider sense the distribution, speciation, and transformations of manganese in lakes resemble that of iron, i.e., it is strongly influenced by reductive processes, which mostly take place in the hypolimnion and promote soluble Mn(II), and oxidation that results in insoluble Mn(oxy)hydroxides which tend to largely sediment (Davison, 1993; Godwin et al., 2020). A major difference is the slow oxidation of Mn(II) by molecular oxygen. This means that biotic and oxide mineral-catalyzed oxidation is much more important for Mn(II) than for Fe(II), and also that Mn(II) is much more stable and abundant in oxic water than Fe(II). This, and the reduction of Mn oxides at higher levels of E_h, contribute to the preferential loss of Fe and declining Fe:Mn ratios when comparing soils to waters along the aquatic conduit (Munger et al., 2016). While organic matter complexation may play a major role in stabilizing and transforming Mn, for example, stability constants for Mn(III) binding to many ligands are similar to or greater than for Fe(III), our understanding of this in freshwaters is sparse (Oldham et al., 2017).

Fluxes to, and mobilization of, Fe and Mn from the sediments have already been treated in some detail here and in Chapter 15 in relation to the release of phosphorus from sediments. The cycling and import of Fe and Mn to the sediments must be evaluated within redox conditions in the lake strata or river system, the settling and focusing of particles, particularly Fe and Mn (oxy)hydroxides, and the resuspension of surficial sediments. Iron and manganese remobilization from the sediments is regulated by oxygen-mediated redox and rates of sedimenting organic matter. Under the prevailing anoxic conditions of nearly all sediments of lakes, Fe and Mn oxides are subject to dissimilatory microbial reduction in which Fe(III) or Mn(IV) are used as an electron acceptor in metabolism-consuming organic matter. However, the abundance of reactive Fe(oxy)hydroxides also in anoxic lake sediments (Lalonde et al., 2012; Björnerås et al., 2021) suggests that either Fe reduction is impaired by interactions with organic matter or clays, or limited by organic substrates that can act as reductants. Fe(III) oxide concentrations can be regenerated by rapid oxidation of Fe(II) compounds; for example, reductively dissolved Fe(II) from anoxic sediments that migrates upward can be trapped by oxidation and results in the accumulation of Fe(oxy)hydroxides in the surface sediment (Engstrom and Wright, 1984). In littoral and wetland areas radial oxygen release from the roots of aquatic plants (Chapter 24) induces the oxidation of Fe(II) and the formation of appreciable deposits of iron plaque on the roots (Roden and Wetzel, 1996), which can have both positive and negative effects on the physiology of the plants. In oligotrophic lakes and bogs of low pH and with high iron concentrations, oxidation of Fe(II) can generate deposits known as *lake or bog ore*, which are high in iron oxides, but also manganese and other trace metals, and were historically used for iron smelting. Where oxidation is not available to the sediments, and sulfide is not produced, Fe(II) accumulates and migrates into the overlying anoxic water. When lakes are acidic or become acidic from anthropogenic causes, Mn and Fe tend to be more soluble and behave more conservatively, with lower rates of sediment retention (White and Driscoll, 1987; Wällstedt and Borg, 2005).

As is apparent from the above, a diverse suite of reactions cycle iron in freshwater systems. When these overlap, oxidation and reduction cycles of iron can be cyclical and result in cryptic iron cycling, which is not reflected in the speciation of the iron pool. For example, Fe(II) oxidation by phototrophic bacteria was masked by the rapid microbial reduction of Fe(III) in meromictic Lake Cadagno (Berg et al., 2016), and by light-induced reduction of Fe(III) in a laboratory setting (Peng et al., 2019). The cryptic cycling of iron may also be linked to the cycling of nitrogen and sulfur (Kappler et al., 2021).

D. Utilization and transformations of iron and manganese

Both iron and manganese are essential micronutrients for most living organisms, acting as catalysts of numerous enzyme systems (Ferreira and Straus, 1994). Iron is required in the enzymatic pathways of chlorophyll and protein synthesis, the protein integrity of cell membranes, and in the respiratory enzymes. The function of iron in cytochromes and as the basic component of hemoglobin in higher animals is well-known and Fe-S proteins are ubiquitous and include nitrogenase, which is key to nitrogen fixation. Manganese plays a critical role in the photosystem of the chloroplast (Barber, 2008), is involved in fatty acid synthesis, in the neutralization of oxygen radicals, and is a functional component of nitrate assimilation through the reduction of nitrogen to ammonia. Iron is the trace element required in the highest amount in almost all organisms. Although the iron requirement varies between species and as a function of iron availability, an approximation of elemental stoichiometry of plankton is (Canfield et al., 2005a):

$$\begin{aligned} &C:N:P:S:\ \mathbf{Fe}:Zn:\{Cu,\ \mathbf{Mn},\ Ni,\ Cd\} \\ &= 106:16:1:1:\mathbf{0.005}:0.002:\mathbf{0.0004} \end{aligned}$$

Since manganese is required in lower amounts and the slower oxidation kinetics makes it more accessible, the requirement of iron is of particular biogeochemical interest.

The mechanisms for the assimilation of iron from the forms available in oxygenated natural waters are highly complex and emerging. While Fe(II) is taken up across the cell membrane by active transport systems, concentrations in oxic waters are generally exceedingly low. Concentrations of Fe(II) may be replenished by photolytic reduction and biological reduction of inorganic or organic complexes at the cell surface (Kranzler et al., 2011; Lis et al., 2015). Rapid reoxidation hinders uptake but may to some extent be impeded by ligand complexation. Fe(II) complexes are weaker and kinetically more labile than those of Fe(III), which facilitates cell uptake. Low solubility and slow reaction of Fe(III) complexes impedes the uptake of Fe(III). To enhance and mediate iron uptake under iron-limiting conditions, some fungi and bacteria, including Cyanobacteria, produce *siderophores*, a wide variety of low-molecular-weight organic chelators with high specificity and affinity for Fe(III) (Wilhelm and Trick, 1994; Hider and Kong, 2010). Major groups of siderophores include catecholates, hydroxamates, and carboxylates. The Fe-siderophore complex is either anchored to the cell membrane or excreted into the medium. Either excreted siderophores are taken up, with iron released through reduction in the cell and the siderophore recycled into the medium, or the Fe-siderophore complex interacts with surface transporters leaving the chelator in the medium. Microbiota that do not synthesize and release siderophores may still have the capacity to transport siderophores or reduce siderophore-bound iron to trigger its release and subsequent uptake. Some acquire iron by phagotrophy (Maranger et al., 1998).

The metabolic demands for iron and manganese are usually sufficiently low so that biota do not materially reduce the concentrations of these metals in freshwater systems. However, since concentrations are low and availability restricted in the trophogenic zone of lakes under certain conditions, the availability of Fe can limit photosynthetic productivity and alter planktonic community composition. For instance, Fe limitation or co-limitation has been implied to constrain primary production in oligotrophic clearwater lakes (Vrede and Tranvik, 2006) and hardwater lakes where concentrations of Fe are exceedingly low (Wetzel et al., 1972; North et al., 2007). Since species vary in their iron requirement and uptake mechanisms, iron influences species composition and succession. Cyanobacteria, particularly nitrogen-fixing (diazotrophic) species, have a high iron requirement (Molot et al., 2014), and the importance of iron to Cyanobacteria has been shown in many ways. Iron additions enhanced the growth of diazotrophic Cyanobacteria, especially *Gloetrochia echinulata* in eutrophic Lake Erken (Sweden) (Hyenstrand et al., 2001), and also stimulated the growth of Cyanobacteria but not other phytoplankton groups in Lake Tanganyika (de Wever et al., 2008). Cyanobacteria have also been shown to benefit from the availability of Fe(II) in the hypolimnion, which they can reach by their capacity to migrate vertically on a diel basis (Molot et al., 2010). This, as well as the ability to efficiently compete for iron by siderophore-mediated uptake, and thereby to suppress access to eukaryotic algae (Sorichetti et al., 2014), provides Cyanobacteria with a competitive advantage. Recent studies propose a connection between the high abundance of the bloom-forming algae *Gonyostomum semen* and high iron concentrations. Batch culture experiments showed that the growth of *G. semen* required higher iron concentrations ($>200\ \mu g\ L^{-1}$) than other common phytoplankton species (Munzner et al., 2021), and a high abundance of *G. semen* was found primarily in lakes with iron concentrations $>200\ \mu g\ L^{-1}$ (Lebret et al., 2018).

Despite the biochemical importance of manganese, it is not considered to limit phytoplankton growth or primary productivity in freshwaters due to its high abundance. Nevertheless, it has been suggested that siderophores may play a role in manganese biochemistry, primarily by forming strong Mn(III)-siderophore complexes (Duckworth et al., 2009).

Iron and manganese can both be toxic at high concentrations. At elevated concentrations, iron can induce oxidative stress, disrupt cell membranes, proteins, pigments, and even damage DNA (Linton et al., 2007; Bakker et al., 2016). Detrimental effects of high iron concentrations can also be due to the mechanical disturbance of vital functions due to the precipitation of ferric hydroxide and Fe–humus precipitates on biological and inert surfaces (Vuori, 1995; Bakker et al., 2016). Adverse effects on human health have been associated with chronic exposure from drinking water (Wasserman et al., 2006), and the World Health Organization has recommended maximum contaminant limits for Fe and Mn concentrations in drinking water at 0.3 and 0.05 mg L^{-1}, respectively (World Health Organization, 2004).

E. Bacterial transformations of iron and manganese

Photosynthetic and bacterial metabolism greatly influence the oxidation–reduction conditions of lakes, which regulate the state and fluxes of iron and manganese. However, certain bacteria utilize iron and manganese directly for energetic transformations. These transformations are usually minor in comparison to the heterotrophic metabolism of organic substrates in most natural systems.

i. Iron and manganese oxidation

A number of bacteria are known to exploit the free energy that can be obtained from the oxidation of Fe(II) by oxygen or nitrate (Canfield et al., 2005a). Such bacteria may adopt a *chemolithoheterotrophic* strategy, where inorganic oxidation takes place during the process of metabolizing organic compounds, or a *chemolithoautotrophic/chemosynthetic* strategy, where all energy is obtained from chemical reactions and all organic compounds are synthesized from CO_2, or a combination of both strategies. Oxidation of Mn(II) yields less energy than that of Fe(II), and although manganese oxidation in many environments is largely microbially catalyzed, with Mn(III) or Mn(IV) as the oxidation product, this has not been coupled to microbial growth.

Several groups of bacteria catalyze the oxidation of Fe(II) with oxygen, and the ferric iron rapidly precipitates mainly as Fe(oxy)hydroxide:

$$4Fe^{2+} + O_2 + 4H^+ \rightarrow 4Fe^{3+} + 2H_2O$$

$$Fe^{3+} + 3H_2O \rightarrow Fe(OH)_3 + 3H^+$$

The ferrous iron utilized in the above reaction can be in ionic form but also mineral forms, including $FeCO_3$, FeS, and pyrite. Since Fe(II) is spontaneously oxidized at neutral pH and in the presence of oxygen, neutrophilic iron-oxidizing bacteria, with optimum growth at neutral pH, are restricted to zones of steep redox gradients, in which they can compete effectively with oxygen for reduced iron. Therefore iron bacteria develop mainly at the interface regions of iron-bearing rock seeps, swamps, and bogs and at the upper hypolimnetic areas and sediment layers, where the redox potential is sufficiently low for reduced iron to occur (Camacho et al., 2001; Santoro et al., 2013; Alfreider et al., 2017). These redox gradients can be as large as several meters of water strata in the metalimnion and upper hypolimnion of eutrophic lakes or only a few centimeters at oxic/anoxic interfaces, such as iron groundwater seeps into rivers. Acidophilic iron bacteria avoid competition with the chemical oxidation of Fe(II) by exploiting environments of extremely low pH and iron oxidation can be a major metabolic strategy in iron-rich acidic environments affected by mine drainage.

Chemosynthetic utilization of energy from inorganic oxidations is relatively inefficient, especially in the case of the oxidation of iron and manganese. Over 220 g of ferrous iron are required to produce 0.5 g of cellular carbon. As a result, oxidized iron will be precipitated on the sheaths of the bacteria and extruded materials (e.g., Emerson and Revsbech, 1994).

In addition to the bacteria that catalyze Fe(II) oxidation by O_2, namely the *microaerophiles*, nitrate-reducing Fe(II) oxidizing bacteria have been found in high numbers in lakes and sediments (Muehe et al., 2009) and support the following process (Straub et al., 1996):

$$10Fe^{2+} + 2NO_3^- + 24H_2O \rightarrow 10Fe(OH)_3 + N_2 + 18H^+$$

This metabolism is widespread among denitrifying bacteria and requires an organic cosubstrate.

Finally, microbial iron oxidation can also be completed by *photoferrotrophs*, which use light as an energy source and ferrous iron as an electron donor, in a process by which inorganic carbon is fixed into organic matter (Widdel et al., 1993). This process has been proposed to be one of the oldest forms of photoautotrophic metabolism on Earth. Photoferrotrophy is anaerobic and requires light and bicarbonate:

$$HCO_3^- + 4\,Fe^{2+} + 10\,H_2O \xrightarrow{h\nu} CH_2O + 4\,Fe(OH)_3 + 7\,H^+$$

Photoferrotrophs have been isolated from freshwater sediments and are probably restricted to depths where light reaches, and Fe(II) is diffusing upward, since only dissolved ferrous iron is available for phototrophic oxidation (Kappler and Newman, 2004). Spatially, nitrate-reducing Fe(II) oxidizing and photoferrotrophic organisms should overlap in shallow anoxic sediments, where conditions are optimal for both, and thereby compete for Fe(II). It has been proposed that their coexistence in these environments is possible due to diel fluctuations in light, which means that nitrate reducers oxidize Fe(II) during darkness and phototrophs dominate the oxidation of Fe(II) during daylight (Melton et al., 2012).

Following Fe(II) oxidation, Fe(III) is generally deposited on outer cell surfaces, and the bacterial cellular structures become encrusted with Fe(III). The presence of characteristic iron bacteria can be recognized from the typical iron crusts they deposit. *Leptothrix* is a genus of neutrophilic, filamentous, gram-negative rod bacteria that form chains, catalyze oxidation with oxygen, and produce iron and manganese encrusted sheaths (Fig. 16-10). *Gallionella* is a genus of neutrophilic, microaerophilic, bean-shaped bacteria, restricted to oxidizing only iron, producing helical stalk structures (Fig. 16-10). Some iron-oxidizing bacteria appear to avoid encrustation and precipitate amorphous oxide particles. Some other examples of important bacteria are *Acidithiobacillus ferrooxidans*, which are rod-shaped, acidophilic chemolithoautotrophs that use elementary sulfur, tetrathionate, and ferrous iron as electron donors, and *Metallogenium*, which oxidize manganese and form star-shaped manganese oxide minerals and manganese nodules in lakes (Miyajima, 1992).

ii. Iron and manganese reduction

Microorganisms can enzymatically catalyze the reduction of many metals. Fe(III) and Mn(IV) become reduced during microbially mediated oxidation of organic matter, when energetically more favorable electron acceptors (oxygen, nitrate) have been depleted. A variety of oxidized iron and manganese forms can be reduced, including free ions, soluble complexes, oxides, hydroxides, and clays. The metal reducers oxidize a variety of organic substrates, including fatty acids, alcohols, aromatic compounds, and sugars (Lovley, 2000), but are not known to oxidize complex organic matter. Examples with acetate as the electron donor are:

$$CH_3COO^- + 8FeOOH + 3H_2O \rightarrow 2HCO_3^- + 8Fe^{2+} + 15OH^-$$

$$CH_3COO^- + 4MnO_2 + 3H_2O \rightarrow 2HCO_3^- + 4Mn^{2+} + 7OH^-$$

The same organisms generally reduce both iron and manganese. Many groups fall into two distinct clusters, the *Geobacteraceae* family and the *Shewanella-Ferrimonas-Aeromonas* group. The former are anaerobes or microaerophiles and couple the metal reduction to complete oxidation of substrates, while the latter are facultative anaerobes that couple oxide reduction to hydrogen oxidation or to the incomplete oxidation of, for instance, lactate to acetate. Iron and manganese reducers are versatile and many can reduce elemental sulfur, nitrate, and humic substances, meaning that they can grow under conditions when iron and manganese reduction is not favorable. Some sulfate reducers reduce iron, and the presence of ferric iron may suppress their normal metabolism.

For the reduction of metal oxides, some groups require physical attachment on the mineral surface (Childers et al., 2002) and can move toward metal oxides by *chemotaxis* (Skerker and Berg, 2001). Other species do not need direct contact, since they rely on *electron-shuttling compounds*, which are reduced by the bacterium and reoxidized spontaneously at the mineral surface. Such electron shuttles can be excreted by bacteria, but naturally occurring compounds, including quinone-containing humic substances, can act as electron shuttles (Lovley et al., 1996; Kappler et al., 2004). The transfer of electrons from humic compounds to Fe(III) is an abiotic process, which can occur in the absence of microorganisms. Once oxidized by Fe(III), humic compounds may again accept electrons from humic-reducing microorganisms.

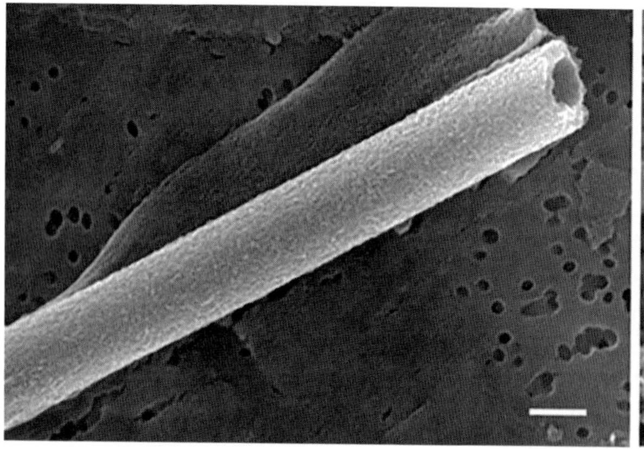

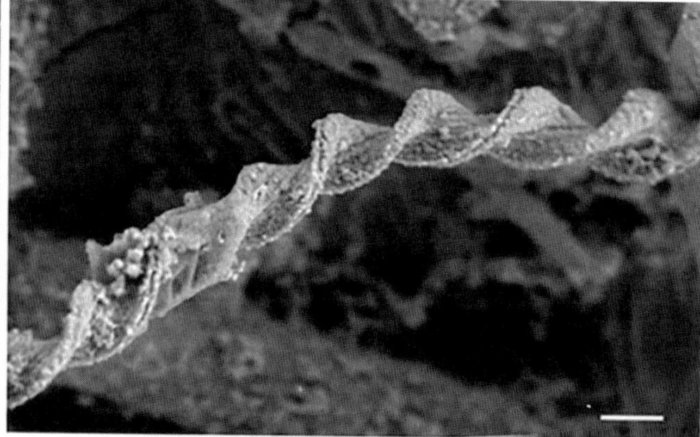

FIGURE 16-10 Scanning electron microscope image of iron-encrusted sheaths from *Leptothrix ochracea* stalk (*left*) and helical stalk from *Gallionella ferruginea* (*right*). All scale bars = 1 μm. (*Photo: courtesy of Chan et al., 2016.*)

Although Fe(III) reduction is a significant redox process in eutrophic lakes, it contributes to only a small portion of the total reducing potential. For example, it has been calculated that Fe(III) and Mn(IV) reduction accounted for <1% of the potential reducing equivalents that enter anoxic hypolimnia as sedimentary organic carbon from primary production (Verdouw and Dekkers, 1980; Davison et al., 1980).

IV. The sulfur cycle

Sulfur is a macronutrient that is utilized by all living organisms in both inorganic and organic forms and is required in similar amounts as phosphorus. Sulfate is reduced to sulfhydryl (—SH) groups in protein synthesis and is abundant in the common amino acids cysteine and methionine. Sulfur is further pervasive in various coenzymes, vitamins, and cofactors, which are essential to cellular metabolism. However, interest in the sulfur cycle of freshwaters extends beyond the nutritional demands of the biota, which are almost always met by the abundance and widespread distribution of sulfate, sulfide, and organic sulfur-containing compounds. Decomposition of organic matter containing sulfur, the anaerobic reduction of sulfate in stratified waters, and chemosynthetic and phototrophic sulfur oxidation, all contribute to altered conditions that markedly affect the cycling of other nutrients, ecosystem productivity, and distribution of the biota.

A. Sources and forms of sulfur

Sources of sulfur compounds to natural waters include solubilization from bedrock, fertilizers that have been applied in the catchment, and atmospheric precipitation and dry deposition. Sulfate (SO_4^{2-}) is released during the geochemical weathering of rocks and soils containing either sulfides or free sulfur, which are oxidized in the presence of water to form sulfuric acid (H_2SO_4) (ZoBell, 1973). Calcium sulfate ($CaSO_4$) is a common constituent of sedimentary rocks, and as a result, drainage from calcareous regions generally contains higher-than-average concentrations of sulfate (Nriagu and Hem, 1978; Kilham, 1984). Acid sulfate soils, which are found particularly in coastal but also inland areas of Asia, Australia, Scandinavia, and Africa (Ljung et al., 2009), contain substantial quantities of iron sulfide minerals or their oxidation reaction products and can release significant quantities of sulfuric acid upon exposure and oxidation (Karimian et al., 2018). Mine drainage and paper mills can also be significant point sources for sulfur pollution.

Large quantities of hydrogen sulfide (H_2S) are released into the atmosphere from volcanic gases and biogenic and industrial sources (Kuznetsov, 1964; Kellogg et al., 1972). Hydrogen sulfide undergoes a number of oxidative reactions to become sulfur dioxide (SO_2), sulfur trioxide (SO_3), and sulfuric acid (H_2SO_4). About 95% of the sulfur compounds from the burning of sulfur-containing fossil fuels consist of SO_2. The removal processes over land are sufficiently slow to result in markedly increased concentrations in areas hundreds to thousands of kilometers downwind. Although the oxidation of SO_2 to H_2SO_4 in air is slow (hours to days), SO_2 is rapidly oxidized to sulfuric acid as it dissolves in atmospheric water.

Acidic precipitation can increase weathering processes at rates equal to or greater than the carbonic acid system. Where soils are sensitive to acidic precipitation, aluminum silicate minerals weather (Dillon et al., 1983):

$$\text{cation} - \text{Al} - \text{silicate} + H_2SO_4 + H_2O \rightarrow H_2SiO_4 + \text{cation} + \text{Al} - \text{silicate} + SO_4^{2-}$$

Sulfur deposition thereby results in the export of cations (Ca^{2+}, Mg^{2+}) and metals (aluminum, iron, manganese) and a shift in anion content in the runoff water from bicarbonate to SO_4^{2-}.

Combustion of fossil fuels has released reactive sulfur that has more than doubled background concentrations (Lamarque et al., 2013) and resulted in acidification with far-reaching consequences for forested and aquatic ecosystems (Likens et al., 1996; Driscoll et al., 2001). Sulfur deposition rose drastically in regions of high industrial activity after World War II, especially in Europe and eastern North America. Acid rain dramatically lowered pH and mobilized trace metals into freshwaters, which resulted in, for example, toxicity from aluminum with detrimental effects on biota and drastically degraded water quality (Driscoll et al., 2001). After the introduction of techniques to reduce sulfur emissions from fossil fuel combustion and other industries (e.g., metal smelting) in the 1980s, sulfur deposition declined rapidly. However, while deposition in many regions has returned to near background levels, elevated concentrations of sulfur remain in the catchment soils (Akselsson et al., 2013). In nonindustrial areas the primary source of sulfate (SO_4^{2-}) in rain and snow is atmospherically oxidized H_2S that is produced by aquatic anaerobic bacteria along coastal regions (Jensen and Nakai, 1961). The SO_4^{2-} derived from sea spray is generally restricted to coastal lakes.

Another human modification of the sulfur cycle is the addition of sulfur to croplands as fertilizers and pesticides. It has been proposed that this has similar consequences for the soil and downstream aquatic ecosystems as those observed in regions historically impacted by acid rain, yet the mechanisms and impact of this environmental perturbation are not well studied (Hinckley et al., 2020).

The retention of sulfur compounds in soils of the drainage basin and their release to natural waters varies with regional lithology, agricultural application of sulfate-containing fertilizers and pesticides, and atmospheric sources. In calcareous areas of sedimentary rock atmospheric contributions can be a small portion of the total. In contrast, in crystalline rock areas wet and dry deposition supplies nearly all of the sulfate of natural waters. While much of the sulfur can accumulate adsorbed to soil particles and in snowfall, most of the total sulfur content is found in the organic matter of the mineral soils (e.g., ~90%; Houle and Carignan, 1995). Release of the sulfate and associated H^+ ions is often very high during the flushing process associated with commonly occurring rapid snowmelt periods. For example, in a small Canadian Shield basin, flushing of soluble SO_4^{2-} from organic and upper mineral soil horizons during the melt was four times the amount supplied in meltwater and precipitation (Steele and Buttle, 1994). Much of the loading of sulfur compounds to streams and lakes therefore is in the form of sulfate and soluble organic S constituents (e.g., David and Mitchell, 1985; Lelieveld et al., 1997).

B. Distribution of sulfur in freshwaters

Although sulfur content in organisms is significant, amounts of sulfur in the biota and detritus are generally small in comparison to the inorganic sulfur components of limnic systems. The predominant form of dissolved sulfur in water is sulfate (Fig. 16-11). Among the lowest concentrations of sulfate in oxic waters (often slightly <1 mg L^{-1}) are found in numerous African lakes situated in crystalline rock drainage basins (Talling and Talling, 1965). The other extreme is found in sulfate saline lakes (>50 g L^{-1}). Nearly all assimilation of sulfur is as sulfate, but during decomposition of organic matter, sulfur is released largely as hydrogen sulfide (H_2S). Since H_2S is rapidly oxidized under oxic conditions, little H_2S would be anticipated in aerated regions of aquatic systems. However, the strict application of redox and pH to the evaluation of reactions involving sulfur is difficult because some are chemically slow and mediated by bacterial metabolism. Although SO_4^{2-} and H_2S dominate, HS^- and very low concentrations of S^{2-} occur in strongly alkaline solutions (Hutchinson, 1957). Under certain conditions of low redox and pH, partial oxidation of sulfides occurs and free S^0 and numerous intermediate compounds (e.g., thiosulfate ($S_2O_3^{2-}$) and sulfite (SO_3^{2-})) may be formed.

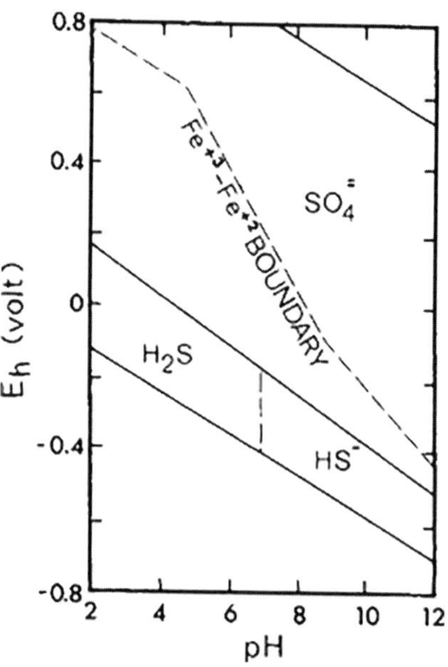

FIGURE 16-11 Approximate redox–pH fields of the stability of dissolved sulfur species likely to occur in natural water. *(Modified from Hem, 1960, with Chen and Morris, 1972.)*

Metal sulfides are exceedingly insoluble at neutral or alkaline pH. The equilibrium ion activity product of FeS of anaerobic lake sediments is $10^{-17.7}$ (Doyle, 1968). Therefore Fe^{2+} released from sediments reacts vigorously with H_2S to form FeS. In an anaerobic hypolimnion the water must be somewhat acidic in order for appreciable H_2S to accumulate. If the water is alkaline, H_2S will accumulate only after most of the Fe^{2+} has been precipitated as FeS. The reduction of SO_4^{2-} to sulfide, some of which is lost to the sediments as insoluble metal sulfides, and oxidation of H_2S to sulfate play a significant role in the modification of conditions for mobilization of phosphate (Chapter 15) and numerous other trace elements (Section V).

Volatile organic sulfur compounds (VOSCs) of biogenic origin [e.g., dimethyl sulfide (CH_3SH), dimethyl disulfide (CH_3SSCH_3), methanethiol (CH_3SH), carbonyl sulfide (COS), and carbon disulfide (CS_2)] are found in freshwaters. VOSC concentrations are low, due to the balance between production and consumption (Lomans et al., 2002). As a result, volatilization and emissions of VOSCs from stratified lakes are usually very small and not considered a major sulfur loss mechanism compared to processes such as sulfate reduction in the sediments. Concentrations of VOSCs are generally higher in shallow lakes and wetland/littoral habitats (Richards et al., 1991), as well as along gradients of increasing salinity (Richards et al., 1994).

A generalized vertical distribution of sulfate and hydrogen sulfide for stratified lakes is depicted in Fig. 16-12. Under oxic conditions, as is the case in many oligotrophic and mesotrophic lakes, and during

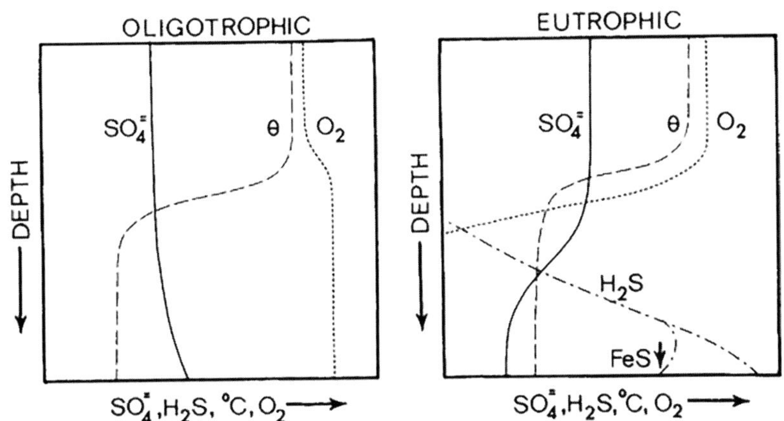

FIGURE 16-12 Generalized distribution of sulfate and hydrogen sulfide in relation to temperature (theta, sed on unpublished data from R.G. Wetzel.) (h these processesn

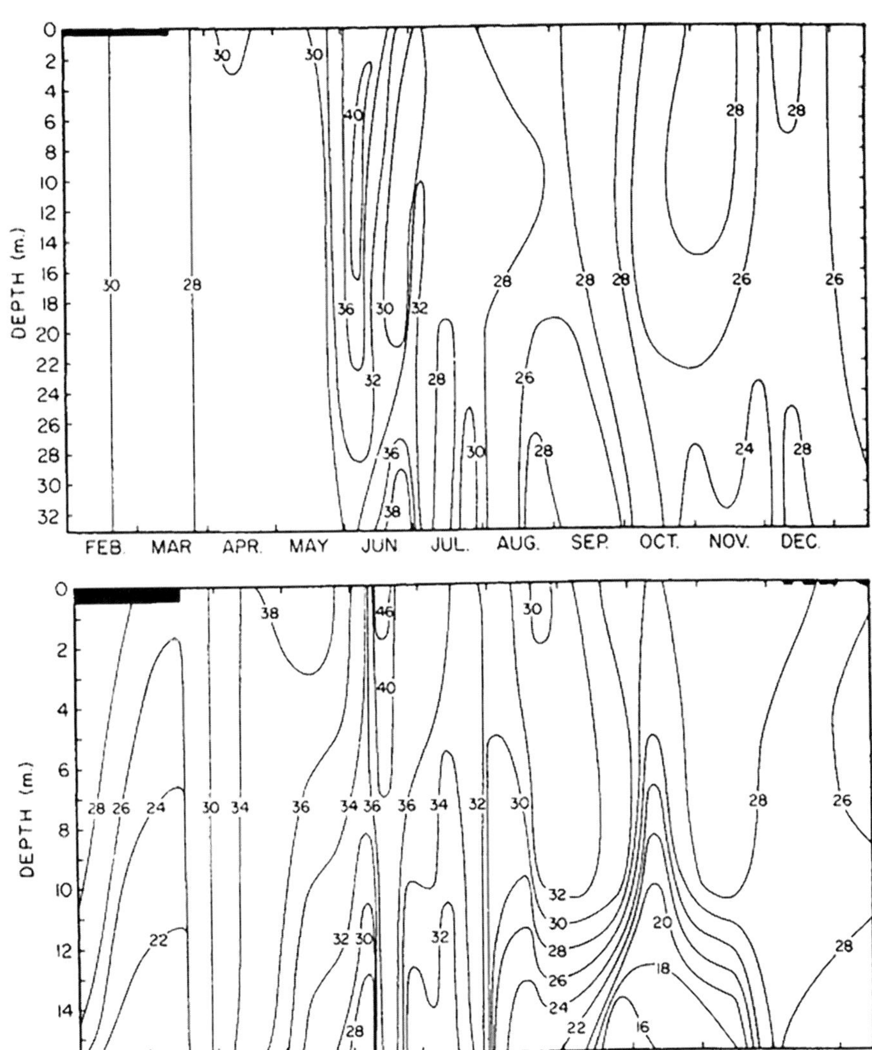

FIGURE 16-13 Depth–time diagrams of isopleths of sulfate concentrations (mg L^{-1}) of mesotrophic hardwater Crooked Lake (*upper*) and interconnected eutrophic Little Crooked Lake (*lower*), Noble–Whitley counties, northeastern Indiana, USA. *Opaque areas* = ice cover to scale. *(Graphs are based on unpublished data from R.G. Wetzel.)*

periods of circulation, H_2S is absent and SO_4^{2-} concentrations change little with depth. Some release of SO_4^{2-} occurs from the sediments, and this increase can become more pronounced in hypolimnia of mesotrophic or eutrophic lakes in the earlier phases of summer stratification (Fig. 16-13). Reduction of sulfate to H_2S occurs as the redox potential declines to less than about 100 mV as a result of bacterial decomposition in highly productive lakes. Particularly near the sediments, much of the H_2S reacts with Fe^{2+} ions to form insoluble FeS. In this way considerable quantities of sulfur can be lost to the sediments (Ingvorsen et al., 1981; Jones et al., 1982). Lakes receiving high concentrations of sulfate from inflowing water, such as meromictic lakes of *crenogenic* (salinity derived from saline springs) formation, or freshwaters fed by oxygen-depleted groundwater, often contain immense concentrations of H_2S in their anoxic monimolimnia. Analogous situations occur in certain anoxic stretches of rivers that are polluted with sulfate-rich organic wastes. Effluents from paper-producing industries are a common source of such pollution.

Most of the sulfur of the seston occurs as ester sulfates and protein sulfur. The bulk (up to 80%) of sulfur in the sediments of productive lakes consists of organic sulfur compounds, and the remainder is pyritic sulfur, acid-volatile sulfides, sulfides dissolved in interstitial water, elemental sulfur, and dissolved sulfates (King and Klug, 1980, 1982; Mitchell et al., 1981; Smith and Klug, 1981).

Sulfur-containing organic compounds are degraded more slowly than other organic compounds. In Lake Mendota, Wisconsin, about 45% of the sulfur precipitated as sulfide was estimated to be derived from the mineralization of organic matter, and the remainder (55%) originated from bacterial reduction of sulfates (Nriagu, 1968). Rates of sulfate reduction in the water column were 10^3 times lower than those of the surface sediment and, on an areal basis, accounted for less than 18% of the total sulfate reduction in the hypolimnion during summer stratification (Ingvorsen et al., 1981). Estimates of net mineralization of sestonic sulfur inputs to the sediments in hypereutrophic Wintergreen Lake, Michigan, USA, indicated that only about 45% to 50% of the total and ester sulfate sulfur inputs, and 75% of the protein sulfur inputs, were mineralized (King and Klug, 1982). Sulfur enrichment of organic matter in sediments can also be due to the incorporation of microbially reduced sulfur (Urban et al., 1999). About 3% of the total water column sulfur was permanently lost to the sediments each year.

C. Bacterial transformations of sulfur

Bacteria and many other organisms reduce sulfate for assimilation to form bioessential compounds at an energetic cost. Such *assimilatory reduction* generally involves the reduction of sulfate to thiol/sulfhydryl (—SH) but may also be based on the reduction of thiosulfate, and elemental sulfur, and is an energy requiring process (Canfield et al., 2005b). Further reduction to H_2S occurs upon the decomposition of organic material by heterotrophic bacterial metabolism.

In addition to assimilatory reduction, there are a number of bacteria that are capable of *dissimilatory reduction* in which they gain energy by the reduction of sulfate, sulfite, thiosulfate, hyposulfite, or elemental sulfur to hydrogen sulfide. Examples of common sulfur-reducing genera are *Desulfovibrio, Desulfobulbus, Desulfobacterium, Desulfomonile,* and *Desulfuromonas* (Diao et al., 2018). The sulfur-reducing bacteria are heterotrophic and anaerobic and use the sulfur compound as a terminal electron acceptor during oxidative metabolism. The reduction is linked to the oxidation of either organic matter or molecular hydrogen (Hamilton, 1985; Canfield 2005):

$$H_2SO_4 + 2(CH_2O) \rightarrow 2CO_2 + 2H_2O + H_2S$$

$$H_2SO_4 + 4H_2 \rightarrow 4H_2O + H_2S$$

Many sulfur reducers can employ oxidative metabolism in the presence of and even by oxygen but only grow in anoxic conditions. Although the above reduction reactions do not consume oxygen directly, the H_2S generated is readily oxidized and consumes oxygen upon intrusion into, or transport to, aerobic regions.

Oxidation of sulfide can be abiotic or microbial. Reaction with oxygen produces mostly sulfate but also sulfite and thiosulfate, although the kinetics are slow (Luther et al., 2011). The reactions between dissolved sulfide and poorly crystalline Fe and Mn oxides are considerably more rapid; for example, the half-lives of sulfide reacting with Fe(oxy)hydroxide and colloidal Mn oxide is 25 min and 50 sec, respectively, and produce elemental sulfur (Burdinge and Nealson, 1986; Millero, 1991). In this way Fe(oxy)hydroxides in sediments can buffer sulfide concentrations (Canfield, 1989). Nevertheless, several groups of bacteria oxidize sulfide to elemental sulfur and elemental sulfur to sulfate at much faster rates, suggesting that in most environments biotic oxidation will exceed abiotic oxidation (Luther et al., 2011).

The *sulfur-oxidizing bacteria* are commonly differentiated into two groups, the *chemosynthetic colorless sulfur bacteria* and the *photosynthetic colored sulfur* bacteria (Kuenen et al., 1985; Canfield et al., 2005b) (Fig. 16-14). The *chemosynthetic sulfur-oxidizing bacteria* are mostly aerobic or facultative anaerobic forms that couple the oxidation of H_2S, or intermediate forms of sulfur, to oxygen or nitrate reduction and are of two types. The first deposits sulfur *inside* the cell:

$$H_2S + \frac{1}{2}O_2 \rightarrow S^0 + H_2O$$

$$\Delta G_0' = -172 \text{ kJ mole}^{-1}$$

which accumulates as long as H_2S is available (see Chapter 14 for Gibbs energy of formation $\Delta G_0'$). As

FIGURE 16-14 Light microscope image of the photosynthetic purple sulfur bacteria *Chromatium okenii* (*large cells*) and *Thiodictyon synthrophicum* (*clusters of spherical cells*) sampled from the chemocline (*top*), and pure cultures of purple and green sulfur bacteria (*middle*), from meromictic Lake Cadagno in Switzerland (*bottom*). (*Photos: Mauro Tonolla.*)

sulfide sources are depleted, the internally stored sulfur is oxidized, and sulfate is released:

$$S^0 + 1\frac{1}{2}O_2 + H_2O \rightarrow H_2SO_4$$

$$\Delta G_0' = -494 \text{ kJ mole}^{-1}$$

Beggiatoa, a long, filamentous bacterium, and *Thiothrix* are common bacteria that oxidize H_2S with the deposition of sulfur intracellularly.

By similar reactions, a second type of chemosynthetic sulfur-oxidizing bacteria deposits sulfur *outside* of the cell. This assemblage is represented by the genus *Thiobacillus*, which oxidizes sulfide, S^0, and other reduced sulfur compounds such as thiosulfate:

$$2Na_2S_2O_3 + O_2 \rightarrow 2S^0 + 2Na_2SO_4$$

The anaerobe *Thiobacillus denitrificans* oxidizes thiosulfate in alkaline waters by reduction of nitrate to N_2:

$$5S_2O_3^{2-} + 8NO_3^- + 2HCO_3^- \rightarrow 10SO_4^{2-} + 2CO_3 + 2H_2 + 4N_2$$

$$5S^0 + 6NO_3^- + 2CO_3^{2-} \rightarrow 5SO_4^{2-} + 2CO_2 + 3N_2$$

$$\left(\Delta G_0'? = -749 \text{ kJ mole}^{-1}\right).$$

The elemental sulfur-oxidizing bacteria commonly adhere to sulfur granules, continuously utilizing a little at a time in the formation of sulfate. While most colorless sulfur bacteria are autotrophic, heterotrophy and mixotrophy also occur.

The anaerobic photosynthetic sulfur bacteria were long conveniently divided into the green sulfur bacteria and the purple sulfur bacteria only but have come to include also the purple nonsulfur bacteria, the green nonsulfur bacteria, and the heliobacteria. Historically, these groups were distinguished based on pigment composition, morphology, and physiology, and later the nonsulfur clades have been found to also have the capacity to oxidize sulfide. The photosynthetic sulfur bacteria require light as an energy source and use the sulfur of H_2S as an electron donor in the photosynthetic reduction of CO_2. Many strains can utilize low-molecular-weight organic substrates as their carbon source, singly or in combination with CO_2. Furthermore, many species can use elemental sulfur and thiosulfate, and several can utilize molecular hydrogen alone:

$$CO_2 + 2H_2S \xrightarrow{\text{light}} (CH_2O) + H_2O + 2S$$

$$2CO_2 + 2H_2O + H_2S \xrightarrow{\text{light}} 2(CH_2O) + H_2SO_4$$

$$2CO_2 + Na_2S_2O_3 + 3H_2O \xrightarrow{\text{light}} 2(CH_2O) + Na_2SO_4 + H_2SO_4$$

The green sulfur bacteria family, Chlorobiacaea, represented by *Chlorobium* and *Chlorobaculum*, are generally unicellular and nonmotile and produce sulfur granules outside of their cell membranes. At least four *bacteriochlorophylls* occur in these bacteria (BChl *a*, *c*, *d*, and *e*) that differ from chlorophyll *a* in a primary absorption maximum at higher wavelengths (770–780 nm versus 665 nm for chlorophyll *a*), and green sulfur bacteria are able to grow at particularly low light (Overmann, 2001; Gregersen et al., 2009; Llorens-Marès et al., 2017).

Members of the purple sulfur bacteria are generally large (5–10 μm), actively motile, and deposit free S^0 intracellularly. They are *γ-proteobacteria* divided into two families, and *Chromatium* (Fig. 16-14), *Thiodictyon*, *Thiocystis*, and *Thiorhodococcus* are some important genera. They synthesize both BChl *a* and *b*. In anaerobic environments where iron is abundant, phototrophic sulfur bacteria can use FeS as an electron donor as well as H_2S (Davison and Finlay, 1990; Garcia-Gil et al., 1990; Ehrenreich and Widdel, 1994). Thus, under conditions where light supports phototrophic sulfur bacteria in proximity to combined sources of reduced iron and sulfide, the sulfide is oxidized either directly or indirectly through FeS to sulfate at rates adequate to prevent iron sulfide from forming in the water column.

The purple nonsulfur bacteria are pigmented photosynthetic bacteria with the capacity to oxidize sulfide or hydrogen. They are physiological generalists, which can function in light or dark, in aerobic or anaerobic environments, but prefer *photoorganoheterotrophy* and are typically found in organic-rich environments with little or no oxygen. *Rhodopseudomonas*, *Rhodospirillum*, and *Rhodomicrobium* are some important genera. The green nonsulfur bacteria, also named the *filamentous anoxygenic phototrophic bacteria*, have *photoorganoheterotrophy* as their primary metabolism, as do the heliobacteria.

The occurrence and distribution of the various sulfur-oxidizing or sulfur-reducing bacteria are restricted by the redox and pH conditions in relation to oxygen and the state of sulfur compounds. The reducing conditions required by strictly anaerobic photosynthetic sulfur bacteria, for example, must coincide with adequate light of high wavelength before large populations can develop. Often, conditions required for optimal growth of sulfur bacteria occur in stratified lakes, as sharply defined layers with steep physical and chemical gradients, and result in thin layers or strata of bacterial populations. A distinct distribution of sulfur-transforming bacteria was seen in Lake Belovod in Russia (Fig. 16-15). In a shallow layer of the hypolimnion chemosynthesis was taking place and the major bacteria in this zone were several species of *Thiobacillus*, under which laid a dense population of photosynthetic purple sulfur bacteria. The most intensive rate of sulfate reduction was observed near sediments and in a shallower water layer in the hypolimnion, which was probably receiving organic material from the littoral slopes. Green sulfur bacteria are commonly found in profusion in a thin layer immediately below a dense population of purple sulfur bacteria at the interface of the oxic/anoxic layer of very productive lakes. The seasonal development of optimal conditions for the specific groups of sulfur bacteria can be transitory, so their contribution to the total annual productivity of the lake may be short-lived. Within meromictic lakes, *bacterial plates* can persist more or less continuously (Baker et al., 1985; Overmann et al., 1991).

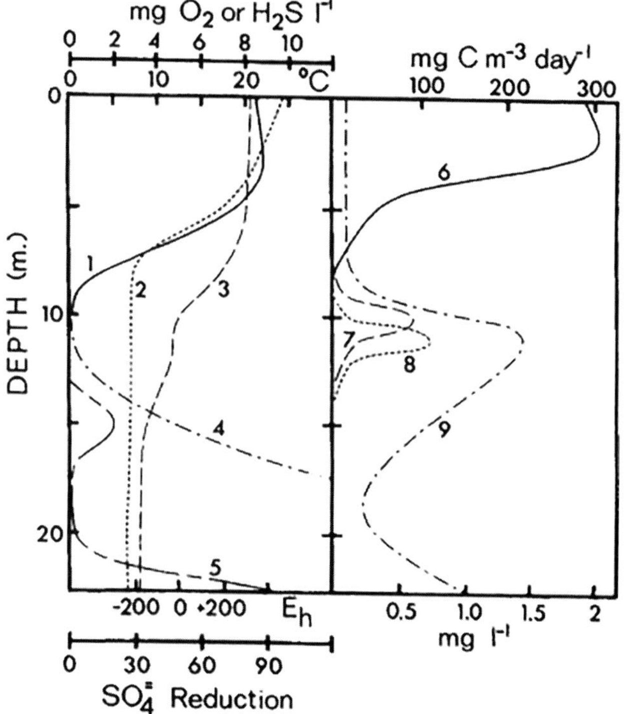

FIGURE 16-15 Distribution of midsummer characteristics and intensity of biological processes in the central depression of meromictic Lake Belovod, Russia; (*1*) oxygen, mg L^{-1}; (*2*) temperature, °C; (*3*) E_h, mV; (*4*) H_2S, mg L^{-1}; (*5*) rate of sulfate reduction, mg H_2S formed m^{-3} day^{-1}; (*6*) photosynthesis by algae, mg C m^{-3} day^{-1}; (*7*) chemosynthesis, mg C m^{-3} day^{-1}; (*8*) photosynthesis by purple sulfur bacteria, mg C m^{-3} day^{-1}; (*9*) biomass of bacteria, mg L^{-1}. *(Data from Sorokin, 1970.)*

It is clear that at certain periods in productive dimictic and in meromictic lakes, bacterial photosynthesis can easily exceed that of algae and macrophytes (Van Gemerden and Mas, 1995). In meromictic Lake Cadagno (Switzerland) photosynthetic sulfur bacteria were found to support up to 40% of inorganic photoassimilation

(Camacho et al., 2001). A detailed study was undertaken of the organic carbon budget and comparative productivity of phytoplankton and phototrophic sulfur bacteria in Mahoney Lake, a small, saline meromictic lake in British Columbia, Canada (Overmann, 1997). The purple sulfur bacterium *Amoebobacter purpureus* (Chromatiaceae) completely dominated (98%) the microorganisms and concentrated in a 20-cm plate in the monimolimnion between depths of 6 and 7 m. Fluxes of organic carbon were calculated from the rates of photosynthesis, sulfate reduction, and sedimentation (Fig. 16-16). Upwelling was a major loss process for *A. purpureus* (80% moved upward to the oxic strata). Because the production and sedimentation of phytoplankton were very small, most of the organic carbon that supported sulfate reduction and generated H_2S, which fueled the phototrophic bacterial production, emanated from allochthonous dissolved organic carbon (90 mg C L^{-1}). In this way allochthonous organic carbon is converted into readily degradable bacterial biomass, using light as the energy source and sulfide as an electron acceptor in the formation of biomass. In most situations, however, the contribution of bacterial photosynthesis to the entire system over an annual period is small, even though populations of photosynthetic sulfur bacteria may flourish locally and periodically.

Microbial processes involved in the sulfur cycle of lakes are represented in Fig. 16-17. The processes on the left-hand side of the figure are more characteristic of a lake with relatively high concentrations of sulfur in various forms. The gradient between the oxic upper strata and the lower H_2S-rich strata is steep but with a

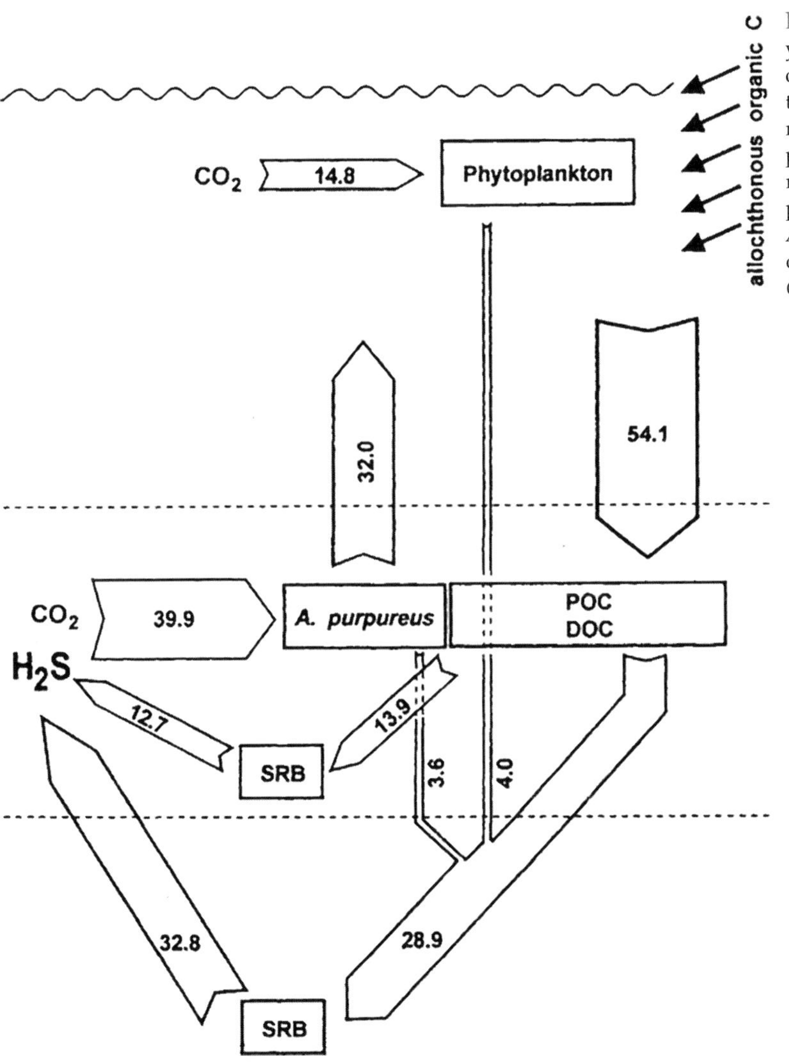

FIGURE 16-16 Carbon cycle (g C m^{-2} of lake surface yr^{-1}) in Mahoney Lake, British Columbia, Canada, based on rates of photosynthesis, sulfate reduction, sedimentation, and upwelling of the photosynthetic sulfur bacterium *Amoebobacter purpureus. Horizontal arrows*: photosynthesis; *oblique arrows*: carbon demand of sulfate-reducing bacteria (SRB); *vertical arrows*: sedimentation of particulate organic carbon or upwelling of the POC of *A. purpureus. Dotted lines*: upper and lower boundaries of the bacterial place between 6.7 and 6.9 m. *(Data from Overmann, 1997.)*

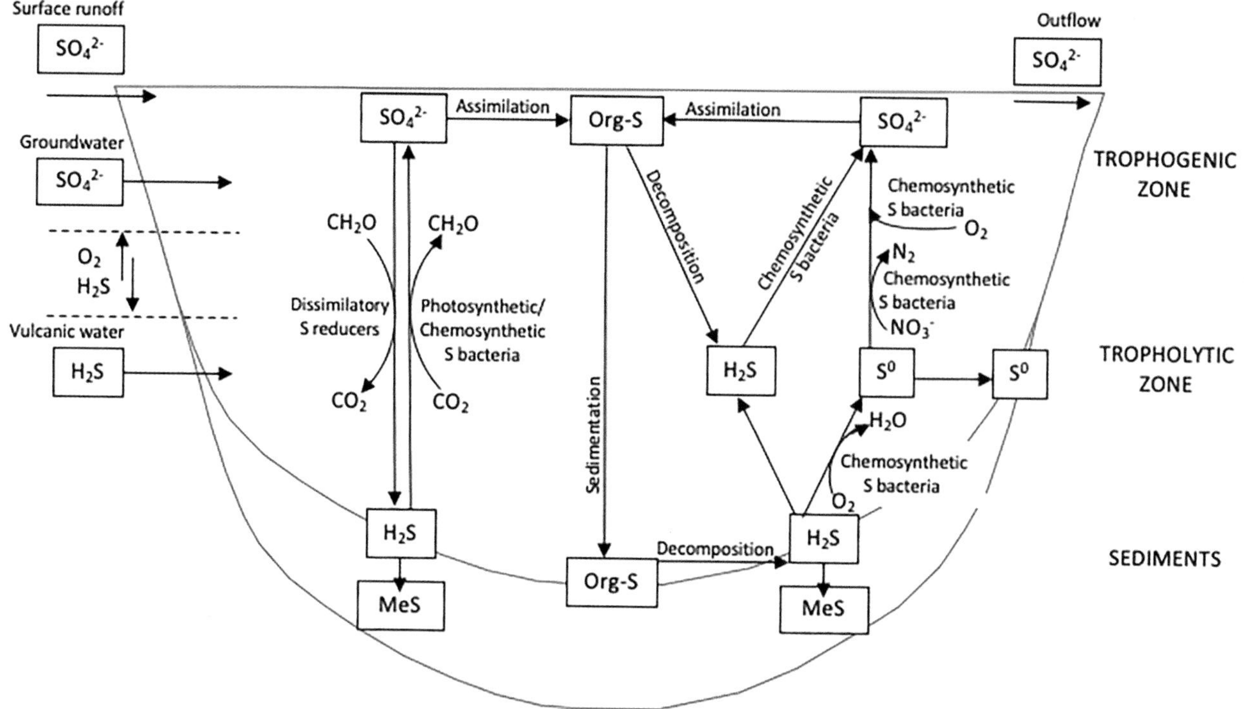

FIGURE 16-17 Composite representation of the sulfur cycle in a lake, with emphasis on the microbiological processes. *MeS* = metallic, primarily iron, sulfides. *(Modified from Kuznetsov, 1970.)*

diffusion interface zone where both oxygen and H_2S occur. Those processes on the right-hand side are more representative of lakes with lower sulfate content.

V. Minor elements

A. Micronutrients

Minor metallic elements, collectively referred to as *micronutrients*, include Fe, Mn, zinc (Zn), copper (Cu), boron (B), cobalt (Co), molybdenum (Mo), nickel (Ni), and vanadium (V), nearly all of which are required for the nutrition of plants and many animals. In some cases, as with iron and manganese, the essentiality of the element for biota is well established. In others it is known that one element can substitute another; vanadium, for example, can replace molybdenum in enzymes, and growth is enhanced by vanadium among certain algae and Cyanobacteria (e.g., Attridge and Rowell, 1997; Rowell et al., 1998) but not among others (Nalewajko et al., 1995). Common physiological functions of these metallic micronutrients and some others are summarized in Table 16-4.

Establishing information on micronutrient requirements is complicated by the risk of contamination but also by the fact that organisms respond differently to amendments depending on the composition and concentration of other minerals and an antagonism among these elements. Ultimately, the question of requirements and potential limitations concerns the *availability* of the micronutrient within the constraints of other ions, particle interactions, the extent of organic complexing of the micronutrient, and the metabolic demands of varying species. In general, concentrations and availability of micronutrients in most natural waters are adequate to sustain active populations of organisms within constraints of light, temperature, and macronutrient availability. There are, however, cases in which micronutrients can limit photosynthesis to a degree (see, e.g., Section IIID). Deficiency of micronutrients is found in some oligotrophic aquatic systems of granitic Arctic, alpine, and volcanic areas, in which a paucity of these elements is well-known (Goldman, 1972). In other situations, such as in hardwater calcareous lakes, the micronutrients are not deficient in the system but are present in forms unavailable for assimilation (Wetzel, 1972). In both situations the control of productivity can be quite transitory and effective on only certain microbial species, thereby indirectly influencing, for example, algal succession and productivity as well as higher trophic levels. The research concerning the effects of micronutrients and their availability in freshwaters has focused largely on phytoplankton. It is generally assumed that the requirements of animals are met by means of ingestion of and uptake from food.

TABLE 16-4 Primary Physiological Functions of Micronutrients Generally Accepted as Essential or Potentially Essential for Aquatic Organisms[a]

Element	Primary functions
Iron	Electron transport in redox systems of respiration and photosynthesis; enzyme activation, oxygen carrier in N_2 fixation
Manganese	Enzyme activation; electron transport reactions particularly in photosynthesis, detoxification of superoxide radicals; synthesis of secondary metabolites; ribosome structure
Zinc	Membrane integrity; enzyme activation, particularly carbonic anhydrase; gene structure, expression, and regulation; carbohydrate metabolism, anaerobic respiration, protein synthesis; ribosome structure; detoxification of superoxide radicals; phytohormone activity
Copper	Redox reactions of respiration and photosynthetic electron transport; detoxification of superoxide radicals; lignification; hemocyanin in aquatic invertebrates
Nickel	Iron absorption; nitrogen fixation; several constitutive enzymes, particularly urease; reproductive growth in plants
Boron	Cell wall formation and stabilization; membrane integrity; carbohydrate utilization; pentose phosphate metabolism; lignification; xylem differentiation; stomatal regulation; heterocyst structure and nitrogen fixation in Cyanobacteria
Molybdenum	Electron transfer reactions; nitrate reduction and nitrogen fixation; sulfate oxidation; protein synthesis
Chloride	Osmoregulation; cation uptake; photosynthesis; reactivity of enzymes
Selenium	Essential growth regulator among certain algae; antioxidant properties; enhancement of phosphorus metabolism; amino acid and protein synthesis
Cobalt	Essential for growth among many microbiota, particularly algae; photosynthesis; nitrogen fixation; essential component of vitamin B_{12}; potential substitution for zinc
Vanadium	Unclear functions and absolute requirements; enhancement of nitrogenase and nitrate reductase activities; substitution for molybdenum in some algae; phosphorylation
Cadmium	Unclear if a required nutrient in algae; implicated in functions in carbonic anhydrase activity; possible substitution for zinc

[a]*Summarized from Lee (1983), Vaishampayan (1983), Keating and Dagbusan (1984), Soeder and Engelmann (1984), Oliveira and Antia (1986), Willsky (1990), Läuchli (1993), Hamilton (1994), Hausinger (1994), Lee et al. (1995), Welch (1995), Attridge and Rowell (1997), Kisker et al. (1997), Blevins and Lukaszewski (1998), Taylor and Anstiss (1999), Lane et al. (2005), and other sources cited in the text.*

Most metallic micronutrients are very toxic when present in excess ionically, or when complexed organically to the point where their availability exceeds the physiological tolerance limits of biota. Copper, for instance, is essential in the aerobic respiration of all eukaryotes, but its toxicity when in high abundance has been used repeatedly as an herbicide (usually as $CuSO_4$) to control algal blooms and growth of larger aquatic plants (McKnight, 1981). While some metals are essential but potentially toxic, others lack physiological function and are solely toxic, such as mercury and lead.

B. Toxic elements

Among the minor elements that are toxic even at very low concentrations are the heavy metals. *Lead* (Pb) is a heavy metal that is widely known for its toxicity to aquatic organisms at relatively low levels. The cause of toxicity is that it binds to sulfhydryl groups of enzymes and displaces other metals as cofactors and thereby inhibits enzyme function. In freshwater systems most Pb is associated with dissolved organic substances and adsorbed to particle surfaces, such as oxide phases (Botelho et al., 1994; Hamilton-Taylor et al., 2002). Much of the Pb loadings to drainage basins has emanated from the combustion of fossil fuels. However, paleolimnological analyses of lake sediments (Chapter 30) showed that lead deposition increased from preindustrial uses of Pb in Europe, starting with the Greek and Roman cultures, more than 2600 years BP (Renberg et al., 1994). The Pb deposition accelerated in the 19th and particularly the 20th centuries, with a deposition maximum around 1970. Since the 1970s, legislation has restricted the use of lead additives in gasoline, and Pb in precipitation declined in many areas by 97% between 1976 and 1989 (Johnson et al., 1995; Nriagu, 1996). Accumulated Pb in forest soils has declined at much slower rates because of low loss rates to drainage waters. Pb content in lake sediments indicates a gradual decline.

At very low levels, *cadmium* (Cd) can function as a metal replacement for Zn in carbonic anhydrase when concentrations of Zn are low, and a Cd-dependent carbonic anhydrase has been found in some diatoms (Price and Morel, 1990; Lane et al., 2005). However, concentrations of a few μg Cd L^{-1} are highly toxic to many organisms, particularly plankton and fishes, primarily by enzyme inhibition (Wong et al., 1980). Much of the Cd is complexed with dissolved organic macromolecules or associated with particles (Sigg and Behra, 2005). Natural emissions (e.g., volcanism and soils) of Cd are completely dwarfed by anthropogenic emissions from industry and fuel combustion.

Mercury (Hg) in freshwaters has received intense interest because of its acute neurological toxicity and widespread findings of elevated Hg concentrations in fish in waters even far away from pollution sources. Natural sources include volcanoes, and anthropogenic emissions are derived mostly from coal combustion and production of other metals. The atmospheric deposition of Hg to drainage basins and surface waters occurs largely as inorganic Hg(0). In oxygenated waters reactive Hg(0) is converted to Hg(II), which forms complexes with inorganic ions (e.g., Cl^- and OH^-) and dissolved organic compounds or absorbs to particulate matter. In anoxic environments Hg(II) can be methylated by dissimilatory sulfate-reducing bacteria to form methylmercury (CH_3Hg, MeHg; Compeau and Bartha, 1985). MeHg is the more toxic form to organisms and is assimilated and concentrated in fish. The toxicity is due to MeHg binding to cysteine, forming complexes that impair function, but has also been linked to the inhibition of selenoenzymes (Ralston and Raymond, 2010; Manceau et al., 2021).

Interestingly, *selenium* (Se) has the potential to counteract Hg toxicity (Ralston and Raymond, 2010; Manceau et al., 2021). Hg has a high binding affinity for Se (10^{45}), surpassing that for S (10^{39}), and Se can demethylate Hg that has complexed to cysteine and form inert HgSe nanoparticles. Additionally, if Se is abundant, enough is available to replace that lost to HgSe complexes, and function of selenoenzymes is maintained.

Interactions with dissolved organic matter influence Hg(II) and MeHg in multiple ways, including enhancing transport, cycling, photolysis, and photoreduction, thereby affecting its fate and toxicity in aquatic systems (Sobczak and Raymond, 2015). Concentrations of total mercury and MeHg increase with increasing concentrations of dissolved organic matter and with the percentage of nearshore wetlands in the drainage basin. Bacteria can reduce Hg(II) to Hg(0), which can be volatilized to the atmosphere. Furthermore, within anoxic strata, Hg can form aqueous complexes with sulfide and precipitate as HgS.

Aluminum (Al), which is not a heavy metal, is highly abundant in natural environments and yet seems to have no biological function. The aqueous ionic Al is toxic to fish, principally by affecting the gills leading to dysfunction of ion regulation and osmoregulation and respiration (Driscoll et al., 1980). Within the range of pH and common limnological conditions, Al is a minor dissolved constituent. Much of the total dissolved Al is complexed with dissolved organic matter in acidic waters, with a binding capacity that increases from pH 3 to 5 but decreases above pH 5 via the formation of Al-hydroxy species (e.g., $Al(OH)_3$). Therefore, in acidic waters (pH < 5), concentrations of reactive Al can be 200–500 μg L^{-1} or more (e.g., Hongve, 1993). Acidic precipitation has caused high concentrations of aluminum to be leached from poorly buffered soils in many continental regions (Dillon et al., 1983), resulting in large increases of reactive Al to receiving streams and lakes, with severe effects on fish populations (Driscoll et al., 1980).

C. Distribution of minor elements in freshwaters

For most minor elements, fluvial inputs dominate loading to freshwaters, due to the strong association to particles and dissolved organic matter. The cycling and fate of elements within the aquatic system depends on the overall dynamics of the system, including particle settling, distribution of oxygen and sulfide, and the specifics of each element.

Although the solubilities of the metals vary somewhat, concentrations in ionic solution are usually very small in aerated surface waters. Most of each metal is adsorbed to crystalline solids or biologically produced particulate matter, and much of the remainder resides in organic complexes (Stumm and Morgan, 1996; Twiss and Campbell, 1998). In the circumneutral pH of many freshwaters both cations and anions interact strongly with oxide surfaces, which are therefore important for the distribution of many minor elements. Similar to what has been described for Fe and Mn in previous sections, many minor elements are transported by particle settling (Balistrieri et al., 1994; Achterberg et al., 1997). To the extent that Fe or Mn(oxy)hydroxides are the vector for vertical transport into anoxic waters, elements can be released upon reductive dissolution of the (oxy) hydroxide and therefore be found in higher concentrations in anoxic bottom waters than in surface waters. This is typically observed for lead (Pb) and arsenic (As). Similar vertical profiles for elements that are important nutrients, such as zinc (Zn), can also be the result of sinking biota and release upon decomposition. If sediments or bottom waters contain sulfide, metals may be removed directly by sulfide precipitation or by adsorption to FeS. For instance, Zn, Pb, Cu, and Co form stabile sulfide complexes. If the bottom waters and surface sediments are oxic, Fe and Mn(oxy)hydroxides can scavenge minor elements into the sediments.

In addition to particle adsorption, settling, and redox chemistry, the distribution of minor elements is strongly influenced by organic complexation. Many minor elements form stable complexes with organic compounds so that losses of free ions via the formation of insoluble hydroxides, sulfides, phosphates, and carbonates can be reduced appreciably (Groth, 1971; Hamilton-Taylor et al., 2002). This is also why concentrations and availability of many minor elements are often much higher in freshwaters than would be anticipated simply on the basis of inorganic solution chemistry and solubilities. Organic complexation seems particularly important for copper (Cu). The binding of trace elements to dissolved organic matter is generally attributed to association with carboxylic groups, which are abundant in humic substances. While thiols are much less abundant, for some metals, binding constants for these are much higher, making them more important than binding by carboxylic groups for elements such as Hg and Pb (Smith et al., 2002).

The distribution of minor elements and their major controls are illustrated by total amounts of Mn, Fe, Cu, Zn, Co, and Mo in the epi- and hypolimnion of Schöhsee in Germany (Table 16-5). The concentrations of Fe and Mn were found to be strongly related to redox conditions, while the primary source of the hypolimnetic accumulation of Co, Mo, and Zn was release from the dissolution of oxides and mineralization of sedimenting organic detritus. No significant differences in the vertical distribution of Cu were found, reflecting its tight association with dissolved organic matter.

VI. The silica cycle

Silica (SiO_2) is moderately abundant in freshwaters and it is of major significance to algae such as diatoms and chrysophytes (Chapter 17), as well as some higher aquatic plants, sponges, and a few other organisms. Diatoms assimilate large quantities of dissolved silica in the synthesis of their frustules (Fig. 16-18). Silicon is a major factor influencing algal production in many lakes, and diatom utilization of dissolved silica greatly modifies the flux rates of silicon in lakes and streams. The availability of dissolved silica can have a strong influence on the overall pattern of algal succession and productivity in lakes and streams.

TABLE 16-5 Average Concentrations (μg L^{-1}) of Minor Elements in the Epilimnion and Hypolimnion of Schöhsee, Northern Germany[a]

Stratum	Mn	Fe	Cu	Zn	Co	Mo
Epilimnion (E)	4.5	15	1.0	1.8	0.03	0.21
Hypolimnion (H)	590	425	0.9	1.9	0.07	0.30
Enrichment ratio of H/E	130	28	0.9	1.1	2.3	1.4

[a]*Data after Groth (1971).*

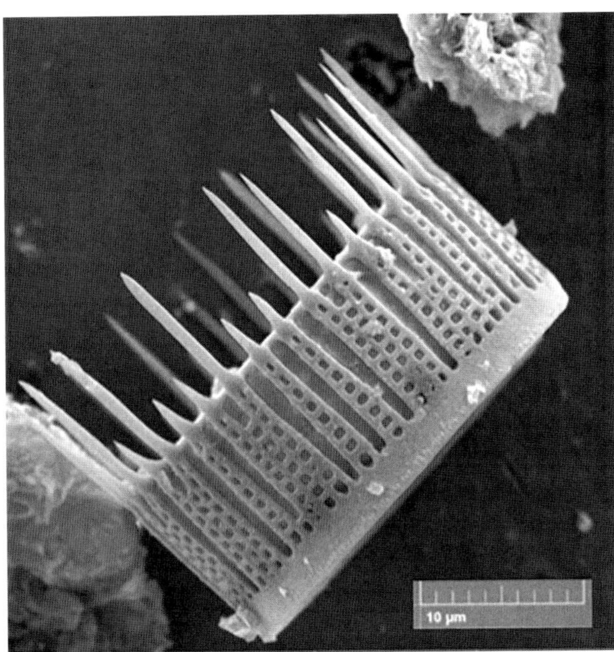

FIGURE 16-18 Scanning electron microscope image of *Aulacoseira agassizi* var. *malayensis* diatom frustule. *(Photo: Jeffrey Stone.)*

A. Sources and forms of silicon

Silicon is the second most abundant element on Earth, comprising 28% of the Earth's crust. It occurs in freshwaters in three major forms. (1) *Mineral silicates* are highly ordered Si- and O-containing structures that react slowly on biological time scales. Mineral silicates range from primary minerals to clays. They can be suspended in the water or deposited as sediments on the bottom. (2) *Dissolved silica* is *orthosilicic acid* with the chemical formula of H_4SiO_4. Dissolved silica originates primarily from the weathering of mineral silicates and is the chemical form used by organisms during biological uptake. (3) *Biogenic silica* is an amorphous form of Si biogenically precipitated by a variety of aquatic organisms including chrysophytes and sponges, but most importantly by the diatoms. In addition, biogenic silica is formed as *phytoliths* in plants.

The world average dissolved silica content of rivers is about 9.5 mg SiO_2 L^{-1} (Dürr et al., 2011), with the amount of dissolved silica transported by rivers to the oceans varying over more than one order of magnitude. The greatest concentrations of dissolved silica are found in waters in contact with volcanic rocks; intermediate amounts occur in association with plutonic rocks and sediments; and small amounts originate from marine sandstones. The lowest amounts of dissolved silica are found in water draining from carbonate rocks.

Dissolved silica loadings to lakes are largely surface waters from rivers or streams. In certain lakes, however, groundwaters can be an important source of water (Winter, 2002) and of dissolved silica. For example, even though the short-term, seasonal input of groundwater to a small, precipitation-dominated oligotrophic lake in northern Wisconsin, USA, was <10% of the annual water budget of the lake, groundwater influents accounted for nearly all of the external dissolved silica loading (Hurley et al., 1985). More recently, groundwater discharge was shown to be a main source of water and dissolved silica to a small, high-latitude subarctic lake (Lake 850, northern Sweden), with groundwater-derived dissolved silica inputs three times higher than those from ephemeral stream inlets (Zahajská et al., 2021).

B. Distribution of dissolved and biogenic silica in freshwaters

Concentrations of dissolved silica within lakes and reservoirs frequently exhibit marked seasonal and spatial variations. Even in oligotrophic waters, a conspicuous decrease in dissolved silica is often found in the epilimnetic strata during early winter, as well as in the spring during circulation and during thermal stratification (Fig. 16-19). In eutrophic lakes dissolved silica concentrations in the trophogenic zone are commonly reduced to near analytical undetectability. Reductions in dissolved silica concentrations within the epilimnetic and metalimnetic zones result in a *negative heterograde dissolved silica curve* against depth (Fig. 16-19). The heterograde silica distribution is clearly associated with the intensive assimilation of dissolved silica by diatoms in surface waters, which sediment their biogenic silica from the trophogenic zone more rapidly than dissolved silica is replaced by inputs from surface water and groundwater. Other sinks of biogenic silica are generally minor relative to the quantities transported by diatoms. Dilution by water low in dissolved silica can occur, as, for example, by rain percolating through decaying ice (March–April, Lawrence Lake, Fig. 16-20).

The seasonal cycle of dissolved silica, demonstrated in Fig. 16-20, has been observed frequently. In many lakes dissolved silica concentrations are usually highest in winter and lowest after the spring diatom bloom. The utilization of dissolved silica by diatoms occurs during photosynthesis and can continue in darkness to complete the division. The dissolved silica gradient becomes steeper in eutrophic lakes, and in most stratified lakes dissolved silica concentrations increase in the water immediately above the sediments from sediment regeneration of biogenic silica (Fig. 16-20). The annual cycle of dissolved silica in reservoirs is strongly influenced by residence time and the degree of hydrological alteration, which can reduce the land-sea flux of dissolved silica (Humborg et al., 2006).

Lakes and reservoirs provide low turbidity environments particularly conducive to biogenic silica production by diatoms and are important sinks for biogenic silica (Frings et al., 2014). The biogenic silica of diatoms can be recycled within the water column; however, often, a large fraction settles or is transported to the sediments.

Interstitial water is enriched in dissolved silica at concentrations far in excess of those of water entering the lake. Interstitial concentrations increase as the pH declines below 7, decrease between pH 7 and 9, and greatly increase above pH 9. Concentrations also increase at higher temperatures within the range found in freshwaters. The dissolved silica of interstitial water is not in equilibrium with amorphous silica but rather with chemically bound or adsorbed silica. Liberation of dissolved silica to the overlying water in relatively isolated lake sediments is governed by these equilibria. Sediment to water fluxes of dissolved silica can be related to the trophic status of the lake and are greatest in eutrophic lakes (Pearson et al., 2006).

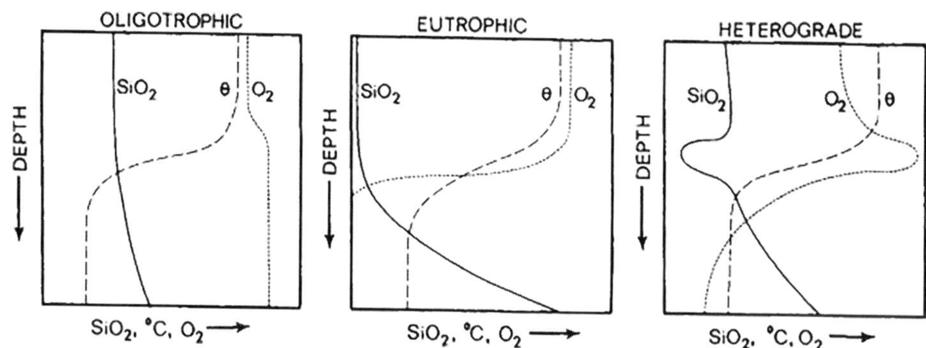

FIGURE 16-19 Generalized distribution of dissolved silica concentrations in relation to temperature (theta, θ) and oxygen in unproductive and very productive lakes, and in a lake exhibiting a metalimnetic development of diatom algae and a negative heterograde dissolved silica curve.

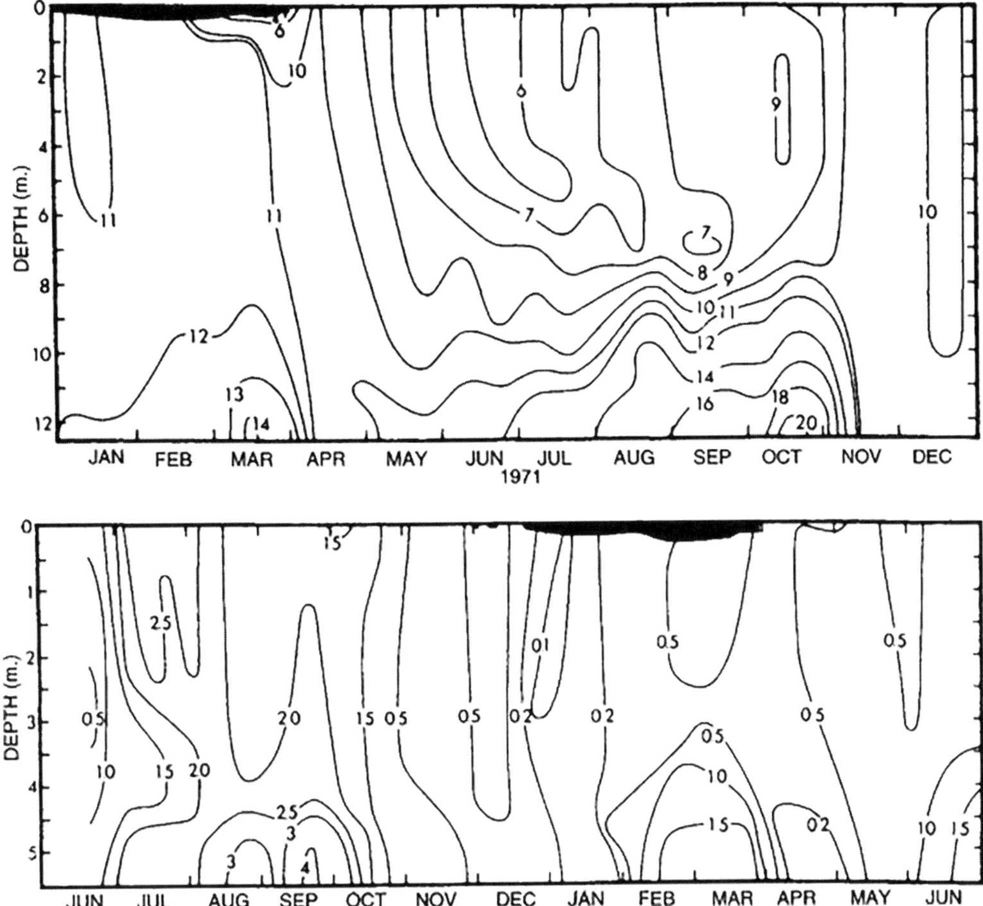

FIGURE 16-20 Depth–time diagram of isopleths of dissolved silica concentrations (mg SiO_2 L^{-1}) in oligotrophic hardwater Lawrence Lake, 1971 (*upper*) and hypereutrophic Wintergreen Lake, 1971–1972 (*lower*), southwestern Michigan, USA. *Opaque areas* = ice cover to scale. *(Graphs are based on unpublished data from R.G. Wetzel.)*

Exchange between sediment and water decreases concentrations of interstitial water and results in greater redissolution from the sediments. Sediment-dissolved silica release rates increase with rising temperatures, even though in stratified, moderately deep temperate lakes, most sediments are continually cold (<10°C). Sediment-dissolved silica fluxes range from c. 0.5 mmol m^{-2} day^{-1} in oligotrophic lakes to much higher values (1–14 mmol m^{-2} day^{-1}) in eutrophic lakes. Dissolved silica regeneration in nearshore regions of shallow sediments, subject to varying seasonal temperatures and high diatom production, can have much higher release rates than colder deep sediments (e.g., Quigley and Robbins, 1984). The rate of dissolved silica release also varies with differences in dissolved silica concentrations between the sediments and the overlying water. Equilibrium is not attained because of the slowness of diffusion (weeks). The difference between dissolved silica of interstitial water and in the overlying water is influenced by currents, movements produced by benthic organisms (e.g., larvae of chironomid insects; Granéli 1979), and gas bubbles escaping from the sediments.

In lakes dominated by diatom algae, large numbers of sedimenting diatom frustules can accumulate within the sediments (Chapter 30) and be lost permanently from the system (Frings et al., 2014). The extent of this permanent loss depends on rates of diatom productivity, on the morphometry of the lake basin, and on the percentage of the sediments located in the quiescent waters of the deep hypolimnion. In deep lakes many of the diatom frustules undergo partial dissolution before reaching the sediments. The biogeochemical silica cycle and economy of most lakes are regulated largely by autochthonous metabolism within the lake, but losses are balanced strongly by allochthonous inputs.

C. Utilization and role of dissolved silica

Silicified structures occur in many aquatic organisms, but none approaches the importance of the diatoms. All diatoms are enclosed in a silica wall or frustule in which silicic acid has been polymerized to form silica particles (Lewin, 1962). The vegetative cells of some species of yellow–brown algae (Synurophyceae) bear discrete siliceous scales and also form cysts with silicified walls (Chapter 17). These algae, however, have only an insignificant impact on silica cycling in comparison to the active utilization and more extensive distribution of the diatoms. Macrophytes can contain significant, but variable, amounts of biogenic silica (Schoelynck et al., 2010) and potentially influence the silica cycle, especially in shallow lakes. Siliceous sponges are rarely sufficient to alter the quantitative cycling of silica in lakes, although Conley and Schelske (1993) have shown the potential of the additional role of sponge spicules in influencing the silicon biogeochemistry of shallow lakes in Florida, USA.

Phytoplankton in temperate dimictic waters usually undergo a spring maximum (Fig. 16-21). This population development may begin beneath ice as light conditions improve but is most conspicuous during and following spring circulation when the water is relatively

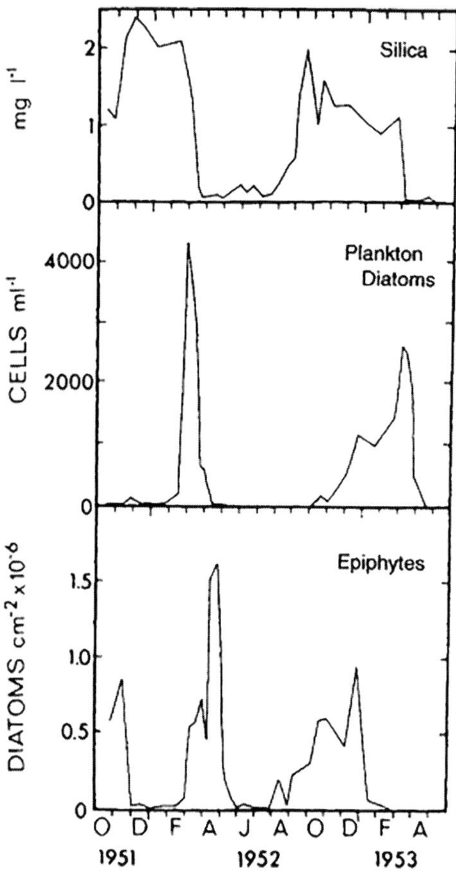

FIGURE 16-21 Variations in dissolved silica concentrations in the surface waters (*top*), numbers of the dominant phytoplanktonic diatom (*Stephanodiscus hantzschii*) (*middle*), and epiphytic diatoms per cm^{-2} of *Phragmites* stems at a depth of 20–30 cm, Furesø, Denmark (*bottom*). (*After data of Jørgensen, 1957.*)

rich in nutrients as the winter accumulations are mixed throughout the water column. In a large number of lakes diatoms constitute the predominant algae of the spring maximum. Increasing light and, to a lesser extent, a rise in water temperatures are major factors initiating the development of diatom populations from smaller residual winter planktonic populations. Circulation and turbulence are much higher in the spring than later in the season and assist in maintaining the relatively dense diatom cells in optimal intensities of light. A diatom succession within the spring bloom is commonly seen and results from interspecific differences in assimilation efficiencies and the effects of silicon concentration on growth. Other factors influencing the succession and development of the algal populations will be treated later (Chapter 18).

Collectively, the spring diatom maximum often declines abruptly as the dissolved silica concentrations fall below 0.5 mg L^{-1} and silica requirements (Conley et al., 1993) cannot be met (Fig. 16-21). In lakes where dissolved silica levels remain high, even though severely reduced during the summer productive period (e.g., Lawrence Lake, Fig. 16-20, *upper*), the spring diatom maximum persists longer into the summer and is overtaken gradually by a predominance of green algae. In very productive lakes a diatom peak can often be observed in the fall and early winter. During summer, other algae then quickly dominate and persist until the combined effects of increased inorganic nitrogen, reduced light and temperature, and renewed dissolved silica reoccur during circulation in late summer and early autumn.

The succession sequence of different species changes as lakes become more productive. In lakes in which dissolved silica concentrations are moderate to low, progressive long-term enrichment with phosphorus and nitrogen can lead to rapid biogenic reduction in dissolved silica levels so that diatoms cannot effectively compete and are replaced by nonsiliceous phytoplankton (Kilham, 1971). A common response to nutrient loading in northern temperate lakes, reservoirs, and even rivers under some circumstances is an increase in diatom growth and biomass. Nutrient enrichments commonly lead to increases in concentrations of total P and total N, but not dissolved silica. The greatly increased sedimentation of diatoms can result in rapid decline and even depletion of dissolved silica in trophogenic zones. As the reductions of dissolved silica from the photosynthetic zones continue over a number of years, more dissolved silica is removed to the sediments than is replenished from external sources and regeneration from the sediments (Schelske and Stoermer, 1971; Conley et al., 1993). Under such circumstances, diatoms are gradually outcompeted and excluded by green algae and Cyanobacteria during the summer period. If reductions in dissolved silica persist over long periods of time, other algae besides diatoms can permanently become dominant components of the phytoplankton.

VII. Summary

1. The biogeochemical cycling of minor elements and sulfur is regulated to a large extent by changes in oxidation–reduction (redox) states, which are governed largely by photosynthetic and bacterial metabolism.
   a. By conversion of light energy into chemical bonds, photosynthesis produces reduced states (negative E_h) of high free energy and nonequilibrium concentrations of carbon, nitrogen, and sulfur compounds. Respiratory and fermentative reactions tend to restore equilibrium by catalytically decomposing the products of photosynthesis. This is how nonphotosynthetic organisms obtain free energy for their metabolism.
   b. Complete redox equilibrium conditions do not occur in natural aquatic ecosystems because most redox transformations are slow and consist of several reactions of different rates. In addition, inputs of photosynthetic energy disrupt the tendency toward equilibrium.
   c. The predominant reactants in the redox processes in natural waters are carbon, oxygen, nitrogen, sulfur, iron, and manganese.
   d. Redox potentials remain positive (300–500 mV) as long as some dissolved oxygen is present. As oxygen concentrations approach zero, E_h decreases precipitously. Temperature and pH changes have minor effects on redox.
2. Concentrations of ionic iron are exceedingly low in aerated water. Most iron in oxygenated water occurs as Fe(oxy)hydroxides in colloidal and particulate form and as complexes with organic, especially humic, compounds. The solubility of manganese is somewhat higher than that of iron, but it is analogous to iron in the way it reacts.
   a. Under conditions of low pH and low redox potential (~250 mV), Fe and Mn(oxy)hydroxides are reduced to ferrous and manganous ions, which diffuse from the sediments and accumulate in anaerobic hypolimnetic water of productive lakes.
   b. Under strongly reducing conditions of very low redox potentials (<100 mV), sulfate is reduced to sulfide. Highly insoluble metallic sulfides, particularly ferrous sulfide (FeS), form under these conditions. Therefore, in hypereutrophic lakes, bacterial decomposition of sulfur-containing organic matter and reduction of sulfate result in high concentrations of hydrogen sulfide, which can precipitate significant amounts of dissolved iron (but not the more soluble manganese) as sulfides form.
   c. Iron complexes with numerous organic compounds, particularly dissolved humic substances, which can increase iron solubility and availability to organisms.
3. Iron and manganese are micronutrients that are essential to freshwater biota.
   a. Under certain conditions of restricted availability, photosynthetic productivity can be limited by these elements.
   b. Certain chemosynthetic bacteria can utilize the energy of inorganic oxidations of ferrous and manganous salts in relatively inefficient reactions involving CO_2 fixation. Microbial iron oxidation can also be completed by photoferrotrophs, which use light as an energy source and ferrous iron as an electron donor, in the process by which inorganic carbon is fixed into organic matter. These bacteria are restricted to zones of steep redox gradients between reduced metal ions and oxygenated water.
   c. There are also bacteria that enzymatically catalyze the reduction of Fe(III) and Mn(IV) during oxidation of organic matter, when energetically more favorable electron acceptors (oxygen, nitrate) have been depleted.
4. Sulfur is nearly always present in adequate concentrations to meet high requirements of biota for protein and sulfate ester synthesis. However, the dynamics of sulfate and hydrogen sulfide alter conditions that affect the cycling of other nutrients, ecosystem productivity, and distribution of biota.
   a. Atmospheric sulfur from the combustion of fossil fuels reaches land by precipitation and dry fallout of particles. This sulfur constitutes a major global source of sulfur to freshwaters, which in many natural waters exceeds inputs from rock and soil weathering.
   b. Sulfate is the primary dissolved form of sulfur in oxic waters. Hydrogen sulfide accumulates in anoxic zones of intensive decomposition in productive lakes where the redox potential is reduced below about 100 mV.
   c. Most of the sulfur in lake water is stored as dissolved sulfate and hydrogen sulfide. Sestonic sulfur-containing proteins and sulfate esters and dissolved sulfides are primary constituents of the sediments.
   d. The sulfur-reducing bacteria are heterotrophic and anaerobic and use various sulfur compounds as a terminal electron acceptor during oxidative metabolism. In doing so, energy is gained and H_2S is generated.

e. Sulfur-oxidizing bacteria consist of two general types: chemosynthetic aerobes that oxidize reduced sulfur compounds and elemental sulfur to sulfate; and photosynthetic sulfur bacteria that utilize light as an energy source and reduced sulfur compounds as electron donors in the photosynthetic reduction of CO_2. The redox requirements of the sulfur-oxidizing bacteria, especially those requiring light, are rather specific, and the distribution of species is often restricted to zones of steep gradients between anoxic and aerated water strata. When conditions are optimal, the photosynthetic bacteria often develop in extreme profusion and may contribute significantly to the annual productivity of lakes.

5. Quantitative information on the dynamics of other essential metallic micronutrients, namely, zinc, copper, cobalt, molybdenum, vanadium, nickel, and selenium, is limited. The cycling and availability of these elements are governed largely by biogeochemically mediated redox processes: production and degradation of organic matter, Fe and Mn redox cycling, and sulfide precipitation. These reactions can be altered by metal complexation with organic, particularly humic, substances.
   a. In most natural waters, micronutrient concentrations and availability are usually adequate to meet metabolic requirements within the constraints of light, temperature, and macronutrient availability. But limitations to photosynthetic productivity by micronutrient deficiencies or physiological unavailability have been demonstrated.
   b. Inputs of many trace elements and heavy metals to freshwaters are increasing as a result of pollution from industrial and combustion emissions to the atmosphere and subsequent deposition via precipitation. Metal enrichments to toxic levels are particularly widespread for cadmium, lead, mercury, and aluminum.
6. Silica occurs in relative abundance in natural waters as dissolved silica and particulate silica. Much of the particulate Si is biogenic Si associated with living or dead diatom frustules.
   a. Diatom algae assimilate large quantities of dissolved silica and markedly modify the flux rates of dissolved and biogenic silica in lakes and streams.
   b. Utilization of dissolved silica in the trophogenic zone of lakes by diatoms reduces the epilimnetic concentrations and induces, along with other factors, a seasonal succession of diatom species that have different assimilation efficiencies for dissolved silica and different growth rates.
   c. When the concentration of dissolved silica is reduced below about 0.5 mg L^{-1}, many diatom species cannot compete effectively with nonsiliceous algae, and their growth rates decline until dissolved silica supplies are renewed, usually during autumnal circulation.
   d. Nutrient enrichments of P and N often lead to increased diatom production and removal of dissolved silica by sedimenting diatoms at rates faster than renewed by inputs. Diatoms are gradually outcompeted and replaced by algae and Cyanobacteria not dependent on dissolved silica availability.

Acknowledgments

I gratefully thank Daniel J Conley for his invaluable effort and input on the section about silicon and Caroline Björnerås for reading and providing useful insights and feedback on the entire chapter. Thank you also to Caroline Björnerås, Clara Chan, Mauro Tonolla, and Jeffrey Stone for providing photographs.

References

Achterberg, E., vad der Berg, C.M.G., Boussemart, M., Davison, W., 1997. Speciation and cycling of trace metals in esthwaite water: a productive English lake with seasonal deep-water anoxia. Geochim. Cosmochim. Acta. 61, 5233–5253.

Akselsson, C., Hultberg, H., Karlsson, P.E., Karlsson, G.P., Hellsten, S., 2013. Acidification trends in south Swedish forest soils 1986–2008—slow recovery and high sensitivity to sea-salt episodes. Sci. Tot. Environ. 444, 271–287.

Alfreider, A., Baumer, A., Bogensperger, T., Posch, T., Salcher, M.M., Summerer, M., 2017. CO_2 assimilation strategies in stratified lakes: diversity and distribution patterns of chemolithoautotrophs. Environ. Microbiol. 19, 2754–2768.

Attridge, E.M., Rowell, P., 1997. Growth, heterocyst differentiation and nitrogenase activity in the cyanobacteria *Anabaena variabilis* and *Anabaena cylindrica* in response to molybdenum and vanadium. New Phytol. 135, 518–526.

Baker, A.L., Baker, K.K., Tyler, P.A., 1985. Fine-layer depth relationships of lakewater chemistry, planktonic algae and photosynthetic bacteria in meromictic Lake Fidler, Tasmania. Freshwat. Biol. 15, 735–747.

Bakker, E.S., Van Donk, E., Immers, A.K., 2016. Lake restoration by in-lake addition: a synopsis of iron impact on aquatic organisms and shallow lake ecosystems. Aquat. Ecol. 50, 121–135.

Balistrieri, L.S., Murray, J.W., Paul, B., 1992. The cycling of iron and manganese in the water column of Lake Sammamish, Washington. Limnol. Oceanogr. 37, 510–528.

Balistrieri, L.S., Murray, J.W., Paul, B., 1994. The geochemical cycling of trace elements in a biogenic meromictic lake. Geochim. Cosmochim. Acta. 58, 3993–4008.

Barber, J., 2008. Photosynthetic generation of oxygen. Philos. Trans. R. Soc. Lond. B Biol. Sci. 363, 2665–2674.

Bauer, I., Kappler, A., 2009. Rates and extent of reduction of Fe(III) compounds and O_2 by humic substances. Environ. Sci. Technol. 43, 4902–4908.

Berg, J.S., et al., 2016. Intensive cryptic microbial iron cycling in the low iron water column of the meromictic Lake Cadagno. Environ. Microbiol. 18, 5288–5302.

Björkvald, L., Buffan, I., Laudon, H., Mörth, C.M., 2008. Hydrogeochemistry of Fe and Mn in small boreal streams: the role of seasonality, landscape type and scale. Geochim. Cosmochim. Acta. 72, 2789–2804.

Björnerås, C., et al., 2017. Widespread increases in iron concentration in European and North American freshwaters. Global Biogeochem. Cycles 31, 1488–1500.

Björnerås, C., Persson, P., Weyhenmeyer, G.A., Hammarlund, D., Kritzberg, E.S., 2021. The lake as an iron sink—new insights on the role of iron speciation. Chem. Geol. 584, 120529.

Blevins, D.G., Lukaszewski, K.M., 1998. Boron in plant structure and function. Annu. Rev. Plant Physiol. Mol. Biol. 49, 481–500.

Boman, A., Frojodo, S., Backlund, K., Åström, M.E., 2010. Impact of isostatic land uplift and artificial drainage on oxidation of brackish-water sediments rich in metastable iron sulfide. Geochim. Cosmochim. Acta. 74, 1268–1281.

Borer, P., Sulzberger, B., Hug, S.J., Kraemer, S.M., Kretzschmar, R., 2009. Photoreductive dissolution of iron (III) (hydr)oxides in the absence and presence of organic ligands: experimental studies and kinetic modeling. Environ. Sci. Technol. 43, 1864–1870.

Botelho, C.M.S., Boaventura, R.A.R., Goncalves, M.L.S.S., Sigg, L., 1994. Interactions of lead(II) with natural river water. Part II. Particulate matter. Sci. Total Environ. 151, 101–112.

Burdinge, D.J., Nealson, K.H., 1986. Chemical and microbiological studies of sulfide-mediated manganese reduction. Geomicrobiol. J. 4, 361–387.

Camacho, A., Erez, J., Chicote, A., Florin, M., Squires, M.M., Lehman, C., Backofen, R., 2001. Microbial microstratification, inorganic carbon photoassimilation and dark carbon fixation at the chemocline of the meromictic Lake Cadagna (Switzerland) and its relevance to the food web. Aquat. Sci. 63, 91–106.

Canfield, D.E., 1989. Reactive iron in marine sediments. Geochim. Cosmochim. Acta. 53, 619–632.

The iron and manganese cycles. In: Canfield, D.E., Kristensen, E., Thamdrup, B. (Eds.), 2005a. Aquatic Geomicrobiology. Elsevier Academic Press, pp. 269–312.

The sulfur cycle. In: Canfield, D.E., Kristensen, E., Thamdrup, B. (Eds.), 2005b. Aquatic Geomicrobiology. Elsevier Academic Press, pp. 313–381.

Chan, C.S., McAllister, S.M., Leavitt, A.H., Glazer, B.T., Krepski, S.T., Emerson, D., 2016. The architecture of iron microbial mats reflects the adaptation of chemolithotrophic iron oxidation in freshwater and marine environments. Front. Microbiol. 7, 796.

Chen, K.Y., Morris, J.C., 1972. Kinetics of oxidation of aqueous sulfide by O_2. Environ. Sci. Technol. 6, 529–537.

Childers, S.E., Ciufo, S., Lovley, D.R., 2002. *Geobacter metallireducens* accesses insoluble Fe(III) oxide by chemotaxis. Nature 416, 767–769.

Compeau, G.C., Bartha, R., 1985. Sulfate-reducing bacteria—principal methylators of mercury in anoxic estuarine sediment. Appl. Environ. Microbiol. 50, 498–502.

Conley, D.J., Schelske, C.L., 1993. Potential role of sponge spicules in influencing the silicon biogeochemistry of Florida lakes. Can. J. Aquat. Sci. 50, 296–302.

Conley, D.J., Schelske, C.L., Stoermer, E.F., 1993. Modification of the biogeochemical cycle of silica with eutrophication. Mar. Ecol. Prog. Ser. 101, 179–192.

Cook, R.B., 1984. Distributions of ferrous iron and sulfide in an anoxic hypolimnion. Can. J. Fish. Aquat. Sci. 41, 286–293.

Daugherty, E.E., Gilbert, B., Nico, P.S., Borsch, T., 2017. Complexation and redox buffering of iron (II) by dissolved organic matter. Environ. Sci. Technol. 51, 11096–11104.

David, M.B., Mitchell, M.J., 1985. Sulfur constituents and cycling in waters, seston, and sediments of an oligotrophic lake. Limnol. Oceanogr. 30, 1196–1207.

Davison, W., 1991. The solubility of iron sulphides in synthetic and natural waters at ambient temperature. Aquat. Sci. 53, 309–329.

Davison, W., 1993. Iron and manganese in lakes. Earth Sci. Rev. 34, 119–163.

Davison, W., De Vitre, R., 1992. Iron particles in freshwater. In: Buffle, J., van Leeuwen, H.P. (Eds.), Environmental Particles, vol. 1. Lewis Publishers, Boca Raton, pp. 315–355.

Davison, W., Finlay, B.J., 1990. Ferrous iron and phototrophy as alternative sinks for sulphide in the anoxic hypolimnia of two adjacent lakes. J. Ecol. 74, 663–673.

Davison, W., Heaney, S.I., 1978. Ferrous iron-sulfide interactions in anoxic hypolimnetic waters. Limnol. Oceanogr. 23, 1194–1200.

Davison, W., Heaney, S.I., Tailing, J.F., Rigg, E., 1980. Seasonal transformations and movements of iron in a productive English lake with deep-water anoxia. Schweiz. Z. Hydrol. 42, 196–224.

De Wever, A., et al., 2008. Differential response of phytoplankton to additions of nitrogen, phosphorus and iron in Lake Tanganyika. Freshw. Biol. 53, 264–277.

Diao, M., Huisman, J., Muyzer, G., 2018. Spatio-temporal dynamics of sulfur bacteria during oxic-anoxic regime shifts in a seasonally stratified lake. FEMS Microbiol. Evol. 94, fiy040.

Dillon, P.J., Evans, H.E., Scholer, P.J., 1988. The effects of acidification on metal budgets of lakes and catchments. Biogeochemistry 5, 201–220.

Dillon, P.J., Yan, N.D., Harvey, H.H., 1983. Acidic deposition: effects on aquatic ecosystems. CRC Crit. Rev. Environ. Control 13, 167–194.

Doyle, R.W., 1968. Identification and solubility of iron sulfide in anaerobic lake sediment. Am. J. Sci. 266, 980–994.

Driscoll, C.T., Baker, J.P., Bisogni, J.J., Schofield, C.L., 1980. Effect of aluminum speciation on fish in dilute acidified waters. Nature 284, 161–164.

Driscoll, C.T., et al., 2001. Acidic deposition in the northeastern United States: sources and inputs, ecosystem effects, and management strategies. Bioscience 51, 180–198.

Duckworth, O.W., Bargar, J.R., Sposito, G., 2009. Coupled biogeochemical cycling of iron and manganese as mediated by microbial siderophores. Biometals 22, 605–613.

Dürr, H.H., Meybeck, M., Hartmann, J., Laruelle, G.G., Roubeix, V., 2011. Global spatial distribution of natural riverine silica inputs to the coastal zone. Biogeosciences 8, 597–620.

Ehrenreich, A., Widdel, F., 1994. Anaerobic oxidation of ferrous iron by purple bacteria, a new type of phototrophic metabolism. Appl. Environ. Microbiol. 60, 4517–4526.

Emerson, D., Revsbech, N.P., 1994. Investigation of an iron-oxidizing microbial mat community located near Aarhus, Denmark: field studies. Appl. Environ. Microbiol. 60, 4022–4031.

Emmenegger, L., Schönenberger, R., Sigg, L., Sulsberger, B., 2001. Light-induced redox cycling of iron in circumneutral lakes. Limnol. Oceanogr. 46, 49–61.

Engstrom, D.R., Wright Jr., H.E., 1984. Chemical stratigraphy of lake sediments as a record of environmental change. In: Haworth, E.Y., Lund, J.W.G. (Eds.), Lake Sediments and Environmental History. Univ. Minnesota Press, Minneapolis, pp. 11–67.

Fältmarsch, R.M., Åström, M.E., Vuori, K.M., 2008. Environmental risks of metals mobilized from acid sulphate soils in Finland: a literature review. Boreal Environ. Res. 13, 444–456.

Ferreira, F., Straus, N.A., 1994. Iron deprivation in cyanobacteria. J. Appl. Phycol. 6, 199–210.

Frings, P.J., Clymans, W., Jeppesen, E., Lauridsen, T.L., Struyf, E., Conley, D.J., 2014. Perspectives: lack of steady-state in the global biogeochemical Si cycle: emerging evidence from lake Si sequestration. Biogeochemistry 117, 255–277.

Garcia-Gil, L.J., Sala-Genoher, L., Esteva, J.V., Abella, C.A., 1990. Distribution of iron in Lake Banyoles in relation to the ecology of purple and green sulfur bacteria. Hydrobiologia 192, 259–270.

Garg, S., Jiang, C., Waite, T.D., 2015. Mechanistic insights into iron redox transformations in the presence of natural organic matter: impact of pH and light. Geochim. Cosmochim. Acta. 165, 14–34.

Gibson, C.E., Wang, G., Foy, B., 2000. Silica and diatom growth in Lough Neagh: the importance of internal recycling. Freshwat. Biol. 45, 285–293.

Goldman, C.R., 1972. The role of minor nutrients in limiting the productivity of aquatic ecosystems. In: Likens, G.E. (Ed.), Nutrients and Eutrophication: The Limiting-Nutrient Controversy. Special Symposium, vol. 1. Amer. Soc. Limnol. Oceanogr, pp. 21–38.

Godwin, C.M., Zehnpfenning, J.R., Learman, D.R., 2020. Biotic and abiotic mechanisms of manganese (II) oxidation in Lake Erie. Front. Environ. Sci. 8, 57.

Granéli, W., 1979. The influence of *Chironomus plumosus* larvae on the exchange of dissolved substances between sediment and water. Hydrobiologia 66, 149–159.

Gregersen, L.H., et al., 2009. Dominance of a clonal green sulfur bacterial population in a stratified lake. FEMS Microbiol. Biol. 70, 30–41.

Groth, P., 1971. Untersuchungen über einige Spurenelemente in Seen. Arch. Hydrobiol. 68, 305–375.

Hamilton, W.A., 1985. Sulphate-reducing bacteria and anaerobic corrosion. Ann. Rev. Microbiol. 39, 195–217.

Hamilton, E.I., 1994. The geobiochemistry of cobalt. Sci. Total Environ. 150, 7–39.

Hamilton-Taylor, J., Postill, A.S., Tipping, E., Harper, M.P., 2002. Laboratory measurements and modelling of metal-humic interactions under estuarine conditions. Geochim. Cosmochim. Acta. 66, 403–415.

Hausinger, R.P., 1994. Nickel enzymes in microbes. Sci. Total Environ. 148, 157–166.

Hem, J.D., 1960. Restraints on dissolved ferrous iron imposed by bicarbonate, redox potential, and pH. In: Chemistry of Iron in Natural Water, 1459-B. U.S. Geol. Surv. Water-Supply Pap., pp. 33–55

Herzog, S.D., Gentile, L., Olsson, U., Persson, P., Kritzberg, E.S., 2020. Characterization of iron and organic carbon colloids in boreal rivers and their fate at high salinity. J. Geophys. Res. Biogeosci. 125, e2019JG005517.

Hider, R.C., Kong, X., 2010. Chemistry and biology of siderophores. Nat. Prod. Rep. 27, 637–657.

Hinckley, E.L.S., Crawford, J.T., Fakhraei, H., Driscoll, C.T., 2020. A shift in sulfur-cycle manipulation from atmospheric emissions to agricultural conditions. Nat. Geosci. 13, 597–604.

Hongve, D., 1980. Chemical stratification and stability of meromictic lakes in the Upper Romerike district. Schweiz. Z. Hydrol. 42, 171–195.

Hongve, D., 1993. Total and reactive aluminum concentrations in non-turbid Norwegian surface waters. Verh. Internat. Verein. Limnol. 25, 133–136.

Houle, D., Carignan, R., 1995. Role of SO_4 adsorption and desorption in the long-term S budget of a coniferous catchment on the Canadian Shield. Biogeochemistry 28, 161–182.

Humborg, C., Pastuszak, M., Aigars, J., Siegmund, H., Mörth, C.-M., Ittekkot, V., 2006. Decreased silica land–sea fluxes through damming in the Baltic Sea catchment—significance of particle trapping and hydrological alterations. Biogeochemistry 77, 265–281.

Hurley, J.P., Armstrong, D.E., Kenoyer, G.J., Bowser, C.J., 1985. Ground water as a silica source for diatom production in a precipitation-dominated lake. Science 227, 1576–1578.

Hutchinson, G.E., 1957. A Treatise on Limnology. I. Geography, Physics, and Chemistry. John Wiley & Sons, New York, p. 1015.

Hyenstrand, P., Rydin, E., Gunnerhed, M., Linder, J., Blomqvist, P., 2001. Response of the cyanobacterium *Gloetrichia echinulate* to iron and boron additions—an experiment from Lake Erken. Freshw. Biol. 46, 735–741.

Ingvorsen, K., Zeikus, J.G., Brock, T.D., 1981. Dynamics of bacterial sulfate reduction in a eutrophic lake. Appl. Environ. Microbiol. 42, 1029–1036.

Jensen, M.L., Nakai, N., 1961. Sources and isotopic composition of atmospheric sulfur. Science 134, 2102–2104.

Johnson, C.E., Siccama, T.G., Driscoll, C.T., Likens, G.E., Moeller, R.E., 1995. Changes in lead biogeochemistry in response to decreasing atmospheric inputs. Ecol. Appl. 5, 813–822.

Jones, J.G., Simon, B.M., Roscoe, J.V., 1982. Microbiological sources of sulphide in freshwater lake sediments. J. Gen. Microbiol. 128, 2833–2839.

Jørgensen, E.G., 1957. Diatom periodicity and silicon assimilation. Dansk Bot. Arkiv. 18 (1), 54.

Kappler, A.M., Benz, B., Schink, Brune, A., 2004. Electron shuttling via humic acids in microbial iron(III) reduction in a freshwater sediment. FEMS Microbiol. Ecol. 1, 85–92.

Kappler, A., Bryce, C., Mansor, M., Lueder, U., Byrne, J.M., Swanner, E.D., 2021. An evolving view on biogeochemical cycling of iron. Nat. Rev. Microbiol. 19, 360–374.

Kappler, A., Newman, D.K., 2004. Formation of Fe(III)-minerals by Fe(II)-oxidizing photoautotrophic bacteria. Geochim. Cosmochim. Acta. 68, 1217–1226.

Karimian, N., Johnston, S.G., Burton, E.D., 2018. Iron and sulfur cycling in acid sulfate soil wetlands under dynamic redox conditions: a review. Chemosphere 197, 803–816.

Karlsson, T., Persson, P., 2012. Complexes with aquatic organic matter suppress hydrolysis and precipitation of Fe(III). Chem. Geol. 322–323, 19–27.

Keating, K.I., Dagbusan, B.C., 1984. Effect of selenium deficiency on cuticle integrity in the Cladocera (Crustacea). Proc. Natl. Acad. Sci. U.S.A. 81, 3433–3437.

Kellogg, W.W., Cadle, R.D., Allen, E.R., Lazrus, A.L., Martel, E.A., 1972. The sulfur cycle. Science 175, 587–596.

Kilham, P., 1971. A hypothesis concerning silica and the freshwater planktonic diatoms. Limnol. Oceanogr. 16, 10–18.

Kilham, P., 1984. Sulfate in African inland waters: sulfate to chloride ratios. Verh. Internat. Verein. Limnol. 22, 296–302.

King, G.M., Klug, M.J., 1980. Sulfhydrolase activity in sediments of Wintergreen Lake, Kalamazoo County, Michigan. Appl. Environ. Microbiol. 39, 950–956.

King, G.M., Klug, M.J., 1982. Comparative aspects of sulfur mineralization in sediments of a eutrophic lake basin. Appl. Environ. Microbiol. 43, 1406–1412.

Kisker, C., Schindelin, H., Rees, D.C., 1997. Molybdenum-cofactor-containing enzymes: structure and mechanism. Annu. Rev. Biochem. 66, 233–267.

Kjensmo, J., 1968. Iron as the primary factor rendering lakes meromictic, and related problems. Mitt. Internat. Verein. Limnol. 14, 83–93.

Kjensmo, J., 1970. The redox potentials in small oligo and meromictic lakes. Nordic. Hydrol. 1, 56–65.

Kranzler, C., Lis, H., Shaked, Y., Keren, N., 2011. The role of reduction in iron uptake processes in a unicellular, planktonic cyanobacterium. Environ. Microbiol. 13, 2990–2999.

Kritzberg, E.S., Ekström, S.E., 2012. Increasing iron concentrations in surface waters—a factor behind brownification? Biogeosciences 9, 1465–1478.

Kuenen, J.G., Robertson, L.A., Gemerden, H.V., 1985. Microbial interactions among aerobic and anaerobic sulfur-oxidizing bacteria. Adv. Microb. Ecol. 8, 1–58.

Kuznetsov, S.I., 1964. Biogeochemistry of sulphur. In: Lo Zolfo in Agricoltura, vol. 5. Simposio Int. Agrochimica, Palermo, Italy, pp. 312–330.

Kuznetsov, S.I., 1970. Mikroflora Ozer I Ee Geokhimicheskaya Deyatel'nost'. (Microflora of Lakes and Their Geochemical Activities). Izdatel'stvo Nauka, Leningrad, p. 440 (In Russian).

Lamarque, J.F., et al., 2013. Multi-model mean nitrogen and sulfur deposition from the atmospheric chemistry and climate model intercomparison project (ACCMIP): evaluation of historical and projected future changes. Atmos. Chem. Phys. 13, 7997–8018.

Lalonde, K., Mucci, A., Oullet, A., Gelinas, Y., 2012. Preservation of organic matter in sediments promoted by iron. Nature 483, 198–200.

Lane, T.W., Saito, M.A., George, G.N., Pickering, I.J., Prince, R.C., Morel, F.M.M., 2005. A cadmium enzyme from a marine diatom. Nature 435, 42.

Läuchli, A., 1993. Selenium in plants: uptake, functions, and environmental toxicity. Bot. Acta. 106, 455–468.

Lebret, K., Östman, Ö., Langenheder, S., Drakare, S., Guillemette, F., Lindström, E.S., 2018. High abundances of the nuisance raphidophyte *Gonyostomum semen* in brown water lakes are associated with high concentrations of iron. Sci. Rep. 8, 13463.

Lee, K., 1983. Vanadium in the aquatic ecosystem. In: Nriagu, J.O. (Ed.), Aquatic Toxicology. John Wiley & Sons, New York, pp. 155–187.

Lee, J.G., Roberts, S.B., Morel, F.M.M., 1995. Cadmium: a nutrient for the marine diatom *Thalassiosira weissflogii*. Limnol. Oceanogr. 40, 1056–1063.

Lelieveld, J., Roelofs, G.-J., Ganzeveld, L., Feichter, J., Rodhe, H., 1997. Terrestrial sources and distribution of atmospheric sulphur. Phil. Trans. R. Soc. Lond. B 352, 149–158.

Lewin, J.C., 1962. Silicification. In: Lewin, R.A. (Ed.), Physiology and Biochemistry of Algae. Academic Press, New York, pp. 445–455.

Likens, G.E., Driscoll, C.T., Buso, D.C., 1996. Long-term effects of acid rain: response and recovery of a forest ecosystem. Science 272, 244–246.

Linton, T.K., Pacheco, M.A.W., McIntyre, D.O., Clement, W.H., Goodrich-Mahoney, J., 2007. Development of bioassessment-based benchmarks for iron. Environ. Tox. Chem. 26, 1291–1298.

Lis, H., Kranzler, C., Keren, N., Shaked, Y., 2015. A comparative study of iron uptake rates and mechanisms amongst marine and fresh water cyanobacteria: prevalence of reductive iron uptake. Life 5, 841–860.

Livingstone, D.A. 1963. Chemical Composition of Rivers and Lakes. Chap. G. Data of Geochemistry. 6th Edition. Prof. Pap. U.S. Geol. Surv. 440-G, P. 64.

Ljung, K., Maley, F., Cook, A., Weinstein, P., 2009. Acid sulfate soils and human health—a millennium ecosystem assessment. Environ. Int. 35, 1234–1242.

Llorens-Marès, T., et al., 2017. Speciation and ecological success in dimly lit waters: horizontal gene transfer in a green sulfur bacteria bloom unveiled by metagenomic assembly. ISME J. 11, 201–211.

Lomans, B.P., van der Drift, C., Pol, A., Op den Camp, H.J.M., 2002. Microbial cycling of volatile organic sulfur compounds. Cell. Mol. Life Sci. 59, 575–588.

Lovley, D.R., 2000. Anaerobic benzene degradation. Biodegradation 11, 107–116.

Lovley, D.R., Coates, J.D., Blunt-Harris, E.L., Phillips, E.J.P., Woodward, J.C., 1996. Humic substances as electron acceptors for microbial respiration. Nature 382, 445–448.

Lueder, U., Jørgensen, B.B., Kappler, A., Schmidt, C., 2020a. Photochemistry of iron in aquatic environments. Environ. Sci. Processes Impacts 22, 12–24.

Lueder, U., Jørgensen, B.B., Kappler, A., Schmidt, C., 2020b. Fe(III) photoreduction producing Fe^{2+}_{aq} in oxic freshwater sediment. Environ. Sci. Technol. 54, 862–869.

Luther III, G.W., et al., 2011. Thermodynamics and kinetics of sulfide oxidation by oxygen: a look at inorganically controlled reactions and biologically mediated processes in the environment. Front. Microbiol. 2, 62.

Manceau, A., et al., 2021. Mercury isotope fractionation by internal demethylation and biomineralization reactions in seabirds: implications for environmental mercury science. Environ. Sci. Technol. 55, 13942–13952.

Maranger, R., Bird, D.F., Price, N.M., 1998. Iron acquisition by photosynthetic marine phytoplankton from ingested bacteria. Nature 396, 248–251.

Maranger, R., Canham, C.D., Pace, M.L., Papaik, M.J., 2006. A spatially explicit model of iron loading to lakes. Limnol. Oceanogr. 51, 247–256.

McKnight, D., 1981. Chemical and biological processes controlling the response of a freshwater ecosystem to copper stress: a field study of the $CuSO_4$ treatment of Mill Pond Reservoir, Burlington, Massachusetts. Limnol. Oceanogr. 26, 518–531.

McKnight, D.M., Kimball, B.A., Bencala, K.E., 1988. Iron photoreduction and oxidation in an acidic mountain stream. Science 240, 637–640.

Melton, E.D., Schmidt, C., Kappler, A., 2012. Microbial iron(II) oxidation in littoral freshwater lake sediment: the potential for competition between phototrophic vs. nitrate-reducing iron(II)-oxidizers. Front. Microbiol. 3, 197.

Millero, F.J., 1991. The oxidation of H_2S in Black Sea waters. Deep Sea Res. 38, S1139–S1150.

Mitchell, M.J., Landers, D.H., Brodowski, D.F., 1981. Sulfur constituents of sediments and their relationship to lake acidification. Water Air Soil Poll 16, 351–359.

Miyajima, T., 1992a. Biological manganese oxidation in a lake. I. Occurrence and distribution of *Metallogenium* sp. and its kinetic properties. Arch. Hydrobiol. 124, 317–335.

Molot, L.A., Li, G., Findlay, D.L., Watson, S.B., 2010. Iron-mediated suppression of bloom-forming cyanobacteria by oxine in a eutrophic lake. Freshw. Biol. 55, 1102–1117.

Molot, L.A., et al., 2014. A novel model for cyanobacteria bloom formation: the critical role of anoxia and ferrous iron. Freshw. Biol. 59, 1323–1340.

Morel, F.M.M., Hering, J.G., 1993. Principles and Applications of Aquatic Chemistry. John Wiley & Sons, New York, p. 374.

Morgan, B., Lahav, O., 2007. The effect of pH on the kinetics of spontaneous Fe(II) oxidation by O_2 in aqueous solution—basic principles and a simple heuristic description. Chemosphere 68, 2080–2084.

Mortimer, C.H., 1941. The exchange of dissolved substances between mud and water in lakes (Parts I and II). J. Ecol. 29, 280–329.

Muehe, E.M., Gerhardt, S., Schink, B., Kappler, A., 2009. Ecophysiology and the energetic benefit of mixotrophic Fe(II) oxidation by various strains of nitrate-reducing bacteria. FEMS Microbiol. Ecol. 70, 335–343.

Munger, Z.W., et al., 2016. Effectiveness of hypolimnetic oxygenation for preventing accumulation of Fe and Mn in a drinking water reservoir. Water Res. 106, 1–14.

Munzner, K., Gollnisch, R., Rengefors, K., Koreiviene, J., Lindström, E.S., 2021. High iron requirements for growth in the nuisance alga *Gonyostomum semen* (Raphidophyceae). J. Phycol. 57, 1309–1322.

Nalewajko, C., Lee, K., Jack, T.R., 1995. Effects of vanadium on freshwater phytoplankton photosynthesis. Water Air Soil Poll 51, 93–105.

Neubauer, E., Köhler, S.J., von der Kammer, F., Laudon, H., Hofman, T., 2013. Effect of pH and stream order on iron and arsenic speciation in boreal catchments. Environ. Sci. Technol. 47, 7120–7128.

North, R.L., Guilford, S.J., Smith, R.E.H., Havens, S.M., Twiss, M.R., 2007. Evidence for phosphorus, nitrogen, and iron colimitation of phytoplankton communities in Lake Erie. Limnol. Oceanogr. 52, 315–328.

Nriagu, J.O., 1968. Sulfur metabolism and sedimentary environment: Lake Mendota, Wisconsin. Limnol. Oceanogr. 23, 430–439.

Nriagu, J.O., 1996. A history of global metal pollution. Science 272, 223–224.

Nriagu, J.O., Hem, J.D., 1978. Chemistry of pollutant sulfur in natural waters. In: Nriagu, J.O. (Ed.), Sulfur in the Environment. II. Ecological Impacts. John Wiley & Sons, New York, pp. 211–270.

Nürnberg, G.K., Dillon, P.J., 1993. Iron budgets in temperate lakes. Can. J. Fish. Aquat. Sci. 50, 1728–1737.

Oldham, V.E., Mucci, A., Tebo, B.M., Luther, G.W., 2017. Soluble Mn(III)-L complexes are abundant in oxygenated waters and stabilized by humic ligands. Geochim. Cosmochim. Acta. 199, 238–246.

Oliveira, L., Antia, N.J., 1986. Nickel ion requirements for autotrophic growth of several marine microalgae with urea serving as nitrogen source. Can. J. Fish. Aquat. Sci. 43, 2427–2433.

Ostendorp, W., Frevert, T., 1979. Untersuchungen zur Manganfreisetzung und zum Mangangehalt der Sedimentoberschicht im Bodensee. Arch. Hydrobiol. Suppl. 55, 255–277.

Overmann, J., 1997. Mahoney Lake: a case study of the ecological significance of phototrophic sulfur bacteria. Adv. Microbiol. Ecol. 15, 251–288.

Overmann, J., 2001. Green sulfur bacteria. In: Boone, D.R., Castenholz, R.W. (Eds.), Bergey's Manual of Systematic Bacteriology, second ed., vol. 1. Williams & Wilkins, Baltimore, MD, USA, pp. 601–630.

Overmann, J., Beatty, J.T., Hall, K.J., Pfennig, N., Northcote, T.G., 1991. Characterization of a dense, purple sulfur bacterial layer in a meromictic salt lake. Limnol. Oceanogr. 36, 846–859.

Pearson, L.K., Hendy, C.H., Hamilton, D.P., 2006. Dynamics of silicon in lakes of the Taupo Volcanic Zone, New Zealand, and implications for diatom growth. Inland Waters 6, 185–198.

Peng, C., Bryce, C., Sundman, A., Kappler, A., 2019. Cryptic cycling of complexes containing Fe(III) and organic matter by phototrophic Fe(II)-oxidizing bacteria. Appl. Environ. Microbiol. 85, e02826–02818.

Pokrovsky, O.S., Schott, J., 2002. Iron colloids/organic matter associated transport of major and trace elements in small boreal rivers and their estuaries (NW Russia). Chem. Geol. 190, 141–179.

Price, N.M., Morel, F.M.M., 1990. Cadmium and cobalt substitution for zinc in a marine diatom. Nature 344, 658–660.

Quigley, M.A., Robbins, J.A., 1984. Silica regeneration processes in nearshore southern Lake Michigan. J. Great Lakes Res. 10, 383–392.

Ralston, N.V.C., Raymond, L.J., 2010. Dietary selenium's protective effects against methylmercury toxicity. Toxicology 278, 112–123.

Renberg, I., Persson, M.W., Emteryd, O., 1994. Pre-industrial atmospheric lead contamination detected in Swedish lake sediments. Nature 368, 323–326.

Richards, S.R., Kelly, C.A., Rudd, J.W.M., 1991. Organic volatile sulfur in lakes of the Canadian Shield and its loss to the atmosphere. Limnol. Oceanogr. 36, 468–482.

Richards, S.R., Rudd, J.W.M., Kelly, C.A., 1994. Organic volatile sulfur in lakes ranging in sulfate and dissolved salt concentrations over five orders of magnitude. Limnol. Oceanogr. 39, 562–572.

Robbins, J.A., Callender, E., 1975. Diagenesis of manganese in Lake Michigan sediments. Am. J. Sci. 275, 512–533.

Roden, E.E., Wetzel, R.G., 1996. Organic carbon oxidation and suppression of methane production by microbial Fe(III) oxide reduction in vegetated and unvegetated freshwater wetland sediments. Limnol. Oceanogr. 41, 1733–1748.

Rose, A.L., Waite, T.D., 2003. Effect of dissolved natural organic matter on the kinetics of ferrous iron oxygenation in seawater. Environ. Sci. Technol. 37, 4877–4886.

Rowell, P., James, W., Smith, W.L., Handley, L.L., Scrimgeour, C.M., 1998. ^{15}N discrimination in molybdenum- and vanadium-grown N_2-fixing *Anabaena variabilis* and *Azotobacter vinelandii*. Soil Biol. Biochem. 30, 2177–2180.

Santoro, A.L., Bastviken, D., Gudasz, C., Tranvik, L., Enrich-Prast, A., 2013. Dark carbon fixation: an important process in lake sediments. PLoS One 8, e65813.

Schelske, C.L., Stoermer, E.F., 1971. Eutrophication, silica depletion, and predicted changes in algal quality in Lake Michigan. Science 173, 423–424.

Schoelynck, J., Bal, K., Backx, H., Okruszko, T., Meire, P., Struyf, E., 2010. Silica uptake in aquatic and wetland macrophytes: a strategic choice between silica, lignin and cellulose? New Phytol. 186, 385–391.

Shapiro, J., 1966. The relation of humic color to iron in natural waters. Verh. Internat. Verein. Limnol. 16, 477–484.

Sigg, L., Behraa, R., 2005. Speciation and bioavailability of trace metals in freshwater environments. Met. Ions Biol. Syst. 44, 47–73.

Sjöstedt, C., Persson, I., Hesterberg, D., Kleja, D.B., Borg, H., Gustafsson, J.P., 2013. Iron speciation in soft-water lakes and soils as determined by EXAFS spectroscopy and geochemical modelling. Geochim. Cosmochim. Acta. 105, 172–186.

Skerker, J.M., Berg, H.C., 2001. Direct observation of extension and retraction of type IV pili. Proc. Natl. Acad. Sci. U.S.A. 98, 6901–6904.

Smith, R.L., Klug, M.J., 1981. Electron donors utilized by sulfate-reducing bacteria in eutrophic lake sediments. Appl. Environ. Microbiol. 42, 116–121.

Sobczak, W.V., Raymond, P.A., 2015. Watershed hydrology and dissolved organic matter export across time scales: minute to millennium. Freshw. Sci. 34, 392–398.

Soeder, C.J., Engelmann, G., 1984. Nickel requirement in *Chlorella emersonii*. Arch. Microbiol. 137, 85–87.

Sorichetti, R.J., Creed, I.F., Trick, C.G., 2014. Evidence for iron-regulated cyanobacterial predominance in oligotrophic lakes. Freshw. Biol. 59, 679–691.

Sorokin, J.I., 1970. Interrelations between sulphur and carbon turnover in meromictic lakes. Arch. Hydrobiol. 66, 391–446.

Steele, D.W., Buttle, J.M., 1994. Sulphate dynamics in a northern wetland catchment during snowmelt. Biogeochemistry 27, 187–211.

Stolpe, B., Hassellöv, M., 2007. Changes in size distribution of fresh water nanoscale colloidal matter and associated elements on mixing with seawater. Geochim. Cosmochim. Acta. 71, 3292–3301.

Straub, K.L., Bemz, M., Schink, B., Widdel, F., 1996. Anaerobic, nitrate-dependent microbial oxidation of ferrous iron. Appl. Environ. Microbiol. 62, 1458–1460.

Stumm, W., Lee, F.G., 1960. The chemistry of aqueous iron. Schweiz. Z. Hydrol. 22, 295–319.

Stumm, W., Morgan, J.J., 1996. Aquatic Chemistry: Chemical Equilibria and Rates in Natural Waters, third ed. J. Wiley & Sons, New York, p. 1040.

Sulzberger, B., Laubscher, H., Karametaxas, G., 1994. Photoredox reactions at the surface of iron(III) (hydr)oxides. In: Helz, G.R., Zepp, R.G., Crosby, D.G. (Eds.), Aquatic and Surface Photochemistry. Lewis Publishers, Boca Raton, pp. 53–73.

Sundman, A., Karlsson, T., Laudon, H., Persson, P., 2014. XAS study of iron speciation in soils and waters from a boreal catchment. Chem. Geol. 364, 93–102.

Sundman, A., Karlsson, T., Sjöberg, S., Persson, P., 2016. Impact of iron-organic matter complexes on phosphate concentrations. Chem. Geol. 426, 109–117.

Talling, J.F., Talling, I.B., 1965. The chemical composition of African lake waters. Int. Rev. ges. Hydrobiol. 50, 421–463.

Taylor, H.H., Anstiss, J.M., 1999. Copper and haemocyanin dynamics in aquatic invertebrates. Mar. Freshwat. Res. 50, 907–931.

Tipping, E., 1981. The adsorption of aquatic humic substances by iron oxides. Geochim. Cosmochim. Acta. 45, 191–199.

Tipping, E., Woof, C., Ohnstad, M., 1982. Forms of iron in the oxygenated waters of Esthwaite Water. U.K. Hydrobiologia 92, 383–393.

Tombácz, E., Libor, Z., Illes, E., Majzik, A., Klumpp, E., 2004. The role of reactive surface sites and complexation by humic acids on the interaction of clay mineral and iron oxide particles. Org. Geochem. 35, 257–267.

Twiss, M.R., Campbell, P.G.C., 1998. Trace metal cycling in the surface waters of Lake Erie: linking ecological and geochemical fates. J. Great Lakes Res. 24, 791–807.

Urban, N.R., Ernst, K., Bernascone, S., 1999. Addition of sulfur to organic matter during early diagenesis of lake sediments. Geochim. Cosmochim. Acta. 63, 837–853.

Vähätalo, A.V., Xiao, Y., Salonen, K., 2021. Biogenic Fenton process—a possible mechanism for the mineralization of organic carbon in fresh waters. Water Res. 188, 116483.

Vaishampayan, A., 1983. Vanadium as a trace element in the blue-green alga, *Nostoc muscorum:* influence on nitrogenase and nitrate reductase. New Phytol. 95, 55–60.

Van Gemerden, H., Mas, J., 1995. Ecology of phototrophic sulphur bacteria. In: Blankenship, R.E., Madigan, M.T., Bauer, C.E. (Eds.), Anoxygenic Photosynthetic Bacteria. Kluwer Academic Publishers, Amsterdam, pp. 49–85.

Verdouw, H., Dekkers, E.M.J., 1980. Iron and manganese in Lake Vechten (The Netherlands): dynamics and role in the cycle of reducing power. Arch. Hydrobiol. 89, 509–532.

Vindedahl, A.M., Strehlau, J.H., Arnold, W.A., Penn, R.L., 2016. Organic matter and iron oxide nano particles: aggregation, interactions, and reactivity. Environ. Sci. Nano. 3, 494–505.

Von Wachenfeldt, E., Sobek, S., Bastviken, D., Tranvik, L.J., 2008. Linking allochthonous dissolved organic matter and boreal lake sediment carbon sequestration: the role of light-mediated flocculation. Limnol. Oceanogr. 53, 2416–2426.

Vrede, T., Tranvik, L.J., 2006. Iron constraints on planktonic primary production in oligotrophic lakes. Ecosystems 9, 1094–1105.

Vuori, K.-M., 1995. Direct and indirect effects of iron on river ecosystems. Ann. Zool. Fennici 32, 317–329.

Wällstedt, T., Borg, H., 2005. Metal burdens in surface sediments of limed and nonlimed lakes. Sci. Total Environ. 336, 135–154.

Wasserman, G.A., et al., 2006. Water manganese exposure and children's intellectual function in Araihazar, Bangladesh. Environ. Health Perspect. 114, 124–129.

Welch, R.M., 1995. Micronutrient nutrition of plants. Crit. Rev. Plant Sci. 14, 49–82.

Wetzel, R.G., 1972. The role of carbon in hard-water marl lakes. In: Likens, G.E. (Ed.), Nutrients and Eutrophication: The Limiting-Nutrient Controversy. Special Symposium, vol. 1. Amer. Soc. Limnol. Oceanogr, pp. 84–97.

Wetzel, R.G., Rich, P.H., Miller, M.C., Allen, H.L., 1972. Metabolism of dissolved and particulate detrital carbon in a temperate hard-water lake. Mem. Ist. Ital. Idrobiol. 29 (Suppl.), 185–243.

Weyhenmeyer, G.A., Prairie, Y.T., Tranvik, L.J., 2014. Browning of boreal freshwaters coupled to carbon-iron interactions along the aquatic continuum. PLoS One 9, e88104.

White, J.R., Driscoll, C.T., 1987. Manganese cycling in an acidic Adirondack lake. Biogeochemistry 3, 87–103.

Widdel, F., Schnell, S., Heising, S., Ehrenreich, A., Assmus, B., Schink, B., 1993. Ferrous iron oxidation by anoxygenic phototrophic bacteria. Nature 362, 834–836.

Wilhelm, S.W., Trick, C.G., 1994. Iron-limited growth of cyanobacteria: multiple siderophore production is a common response. Limnol. Oceanogr. 39, 1979–1984.

Willsky, G.R., 1990. Vanadium in the biosphere. In: Chasteen, N.D. (Ed.), Vanadium in Biological Systems: Physiology and Biochemistry. Kluwer Acad. Publ., Dordrecht, pp. 1–24.

Winter, T.C., 2002. The concept of hydrologic landscapes. J. Am. Water Resour. Assoc. 37, 335–349.

Wong, P.T.S., Mayfield, C.I., Chau, Y.K., 1980. Cadmium toxicity to phytoplankton and microorganisms. In: Nriagu, J.O. (Ed.), Cadmium in the Environment. Part I. Ecological Cycling. John Wiley & Sons, New York, pp. 571–585.

World Health Organization, 2004. Guidelines for Drinking-Water Quality: Recommendations, third ed. World Health Organization, Geneva, Switzerland.

Xing, G., Garg, S., Waite, T.D., 2019. Is superoxide-mediated Fe(III) reduction important in sunlit surface waters? Environ. Sci. Techol. 53, 13179–13190.

Yoshimura, S., 1936. Contributions to the knowledge of iron dissolved in the lake waters of Japan. Second report. Jpn. J. Geol. Geogr. 13, 39–56.

Zahajská, P., Olid, C., Stadmark, J., Fritz, S.C., Opfergelt, S., Conley, D.J., 2021. Modern silicon dynamics of a small high-latitude subarctic lake. Biogeosciences 18, 2325–2345.

ZoBell, C.E., 1973. Microbial biogeochemistry of oxygen. In: Imshenetskii, A.A. (Ed.), Geokhimicheskaia Deiatel'nost' Mikroorganizmov V Vodoemakh I Mestorozhdeniiach Poleznykh Iskopaemykh. Tipografiia Izdatel'stva Sovetskoe Radio, Moscow, pp. 3–76.

CHAPTER

17

Algae and Cyanobacteria Communities

John Wehr[1] *and Sanet Janse van Vuuren*[2]

[1]Louis Calder Center and Department of Biological Sciences, Fordham University, Armonk, NY, United States

[2]Unit for Environmental Sciences and Management, North-West University, Potchefstroom, South Africa

OUTLINE

Algae are fundamental components of freshwater ecosystems. There have been countless studies on their composition, abundance, seasonal succession, and ecological function spanning more than a century (Tiffany, 1958; Round, 1981; Reynolds, 2006). The field of phytoplankton ecology has been characterized as one of the most popular and productive areas of research within all of aquatic ecology. Similarly, studies on benthic algal species (see Chapter 25) have revealed many complex and important roles of algae in lakes, rivers, springs, and wetlands (Stevenson, 1996; Necchi 2016; Steinman et al., 2017). Planktonic and benthic algae function as the main primary producers in lakes and rivers, and their diverse life histories, morphology, pigments, motility, physiology, size, reproduction, and ecological requirements enable algal species to occupy a vast range of ecological niches and roles in aquatic ecosystems. Algae are an essential, basal food source in the food webs of nearly all freshwater ecosystems (Thorp and Delong, 2002; Brett et al., 2017). Some species form symbiotic associations with plants or animals, serve as a structure and habitat for epiphytes, produce toxins, or shift their metabolism to feed on bacteria. In this chapter we aim to summarize this diversity, characterize some of the unique features of algal species, and

Wetzel's Limnology, Fourth Edition
https://doi.org/10.1016/B978-0-12-822701-5.00017-3

illustrate their many and important roles in freshwater ecosystems.

I. Diversity and composition of algae in inland waters

A. Classification of algae

What are Algae? The word *algae* (singular, *alga*) is the Latin word for seaweed. According to G.M. Smith, (1950), Linnaeus, (1754) used the term *Algae* as a formal order within the plant kingdom, but his concept of the group was vaguely described and included bryophytes and higher plants. Since then, many authors have endeavored to define the algae in more specific terms, often as a group of mainly aquatic organisms or protists capable of photosynthesis that use *chlorophyll a* as the main photosynthetic pigment and lack true roots, stems, leaves, and vascular tissue. In addition, unicellular forms can function as gametes, whereas in nearly all multicellular forms *gametangial (gamete-producing) cells* are not surrounded by *sterile (nonreproductive) cells* (Bold and Wynne, 1985; Van den Hoek et al., 1995). This definition, though more extensive, does not encompass the full diversity of organisms commonly referred to as algae, some of which are terrestrial, mixotrophic, or even parasitic. Algae today are simply recognized as a diverse but heterogeneous collection of organisms, which includes unrelated prokaryotic and eukaryotic taxa (Wehr et al., 2015).

Indeed, over 70 years ago, Prescott, (1951) observed that "although convenient, the term 'algae' has been applied to such a large variety of plant groups and has been given so many interpretations that it has no precise meaning." Modern molecular studies have since confirmed that "algae" is no longer a meaningful phylogenetic or taxonomic concept. That is, they cannot be classified into a single evolutionary group that has descended from a common ancestor. But despite this confusion, the term is still useful for limnologists who study phytoplankton, periphyton, primary production, or food webs (Sheath and Wehr, 2015), so long as one's results differentiate these from bryophytes, ferns, or seed plants. The organisms that are commonly described as algae span two domains in the tree of life: Bacteria (Cyanobacteria) and Eukaryota (multiple Eukaryote phyla; Guiry and Guiry, 2021). The eukaryotic algae alone are members of most of the recognized supergroups within the Eukaryota (Fig. 17-1). Research conducted on the Algal Tree of Life, through genomic sequence analyses, has revealed previously unknown

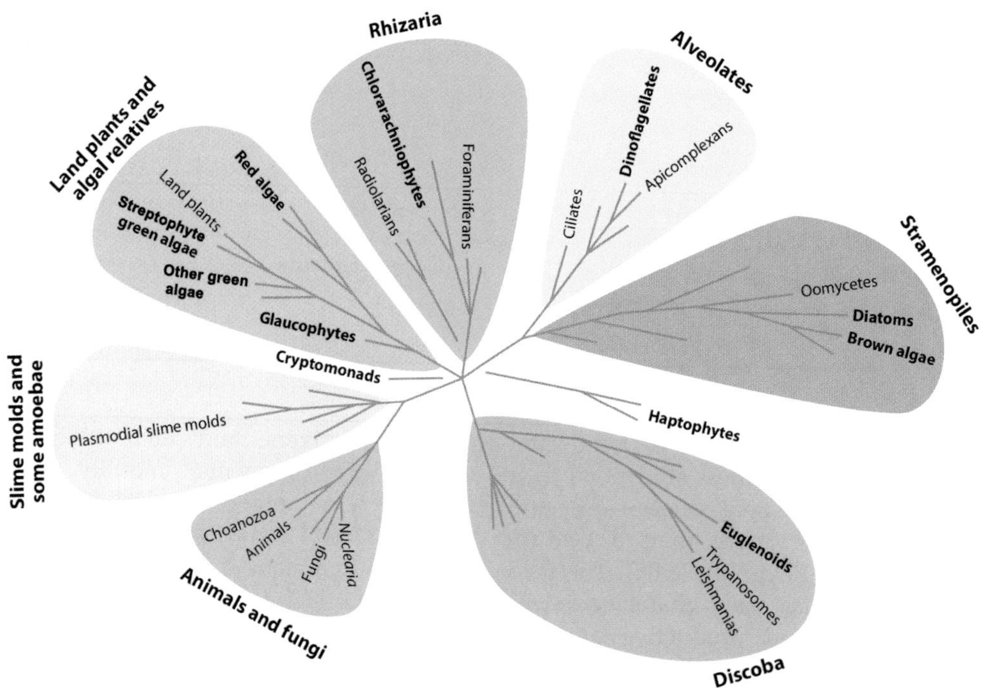

FIGURE 17-1 Simplified Eukaryote tree of life representing the phylogeny of eukaryotic lineages, with seven Supergroups (or Empires), with the majority of this diversity comprising microbial protists. Eukaryotic algal phyla or classes are shown in *bold (after Graham et al., 2016)*. Haptophytes have a close affinity to the Stramenopiles but have recently been assigned to a new supergroup, the Haptista, and cryptomonads were assigned to the Cryptista (Burki et al., 2020), making traditional "algae" classified within seven of the nine eukaryotic supergroups. Further genomic studies may identify additional diversity at the kingdom level. *(Modified from Graham et al., 2016.)*

diversity and many cryptic species and has created many new classifications. The present treatment of algal diversity and algal groups takes a practical rather than phylogenetic approach, with a focus on properties most relevant to freshwater ecology.

B. Pigment and cellular features

Cyanobacteria appeared on Earth 2.7 to 3.0 billion years ago through the development of photosynthetic *thylakoids* in early prokaryotic cells, which enabled chlorophyll-based photosynthesis. As they evolved, Cyanobacteria acquired several phycobiliprotein complexes, notably *phycocyanin* (blue) and *phycoerythrin* (red) as major photosynthetic pigments (Schopf, 2012; Komárek and Johansen, 2015). Eukaryotic algae later evolved over many separate evolutionary events involving the symbiotic acquisition of cyanobacterial cells into a eukaryotic host (primary endosymbiosis) or acquired plastids within a photosynthetic eukaryote that then became an endosymbiont within another eukaryotic host (secondary and tertiary endosymbioses; Graham et al., 2016). These processes, along with gains and losses of other pigments over time, have led to a huge diversity of pigments and other cellular properties, features that are used in classification schemes but also are indicative of the diversity of photosynthetic adaptations and ecological niches among the algae we see today (Table 17-1).

Differences in the suite of pigments in algal cells reflect the physiological differences and ecological complexity of species in a phytoplankton community. A metaanalysis of the photosynthetic properties of 56 species of freshwater phytoplankton suggests there are broad differences in the light utilization traits among the major taxonomic groups of algae, with many species of planktonic green algae adapted to higher-light environments (greater growth vs. irradiance) than many species of Cyanobacteria (Schwaderer et al., 2011). This general pattern, however, would seem to contrast with the common phenomenon of surface blooms of many species of Cyanobacteria. However, many planktonic cyanobacterial species undergo relatively rapid *photoadaptation* (via short-term changes in the photosynthetic apparatus) to minimize the effect of high light causing *photoinhibition* (Oliver et al., 2012). Further, adaptations in various algal species to different light fields differ among species and ecological factors such as nutrient and temperature conditions (Edwards et al., 2016) and can vary within an individual species. For example, the light-harvesting complexes of cryptomonads and golden algae, including diatoms, contain chlorophyll *c* (and *fucoxanthin* in some) in addition to chlorophyll *a* (Table 17-1). Variations in the relative amounts of these pigments in freshwater diatoms allow species to optimize for the capture of different quantities of light energy (photon flux rates) and different spectra and thereby to adapt to different light fields (Shi et al., 2016). The diversity of photosynthetic pathways results in a correspondingly wide range of photosynthetic storage products, many of which are polymers of glucose, such as starch in green algae, Floridean starch in red algae, paramylon in euglenoids, and cyanophycean granules in Cyanobacteria. Although their structural and cellular properties vary a great deal, the chief purpose of all these products is to store nitrogen and carbon. A few groups, in particular diatoms, also synthesize a variety of lipids, including ω3- and ω6-fatty acids, such as essential ω3-fatty acids, eicosapentaenoic acid (EPA) and docosahexaenoic acid (DHA), which are essential for the nutrition of aquatic consumers (Torres-Ruiz et al., 2007; Brett et al., 2017).

Many algal species have mechanisms to enable motility, which further shape their ecological requirements and characteristics. Among eukaryotes, those capable of movement may do so by means of one (e.g., *Euglena*), two equal (e.g., *Chlamydomonas*), two unequal (e.g., *Ochromonas*), or more (e.g., *Carteria*) *flagella* (thread-like appendages that function in motility; singular = *flagellum*). Motility in certain pennate diatoms is achieved by means of mucilage released from a slit known as a *raphe* in the *frustule* (silica walls composed of two *valves*; Section 17IIG). The mechanism for movement in Cyanobacteria is via external brush-like structures known as *pilli* (e.g., *Synechocystis*), allowing movement toward (= positive) or away from (= negative) a light source, known as *phototaxis* (Bhaya, 2004). Gliding movements are common among many filamentous Cyanobacteria (e.g., *Oscillatoria*), especially in benthic habitats where trichomes move within individual sheaths or microbial mats, such as species of *Microcoleus* and *Phormidium* (Fig. 17-2j).

C. Morphological and physiological diversity

The algae comprise a remarkable biodiversity, estimated to number somewhere between 10^5 and 10^7 species (Norton et al., 1996), with a corresponding diversity in morphology, size, and physiological attributes. In a single sample of lake water or a scraping of diatoms from a rock, one may easily observe more than 100 species based on their morphological features alone. The classic and intriguing question posed by G.E. Hutchinson in 1961 in the "Paradox of the Plankton" asked whether a biologically diverse assemblage of phytoplankton species can coexist in an apparently uniform environment, such as a lake, in the face of competition for limited resources. Some answers have

TABLE 17-1 Major Photosynthetic Pigments, Storage Reserves, External Covering, and Flagella in the Key Groups of Algae From Fresh and Other Inland Waters

Algal group (Phylum/Class)	Photosynthetic pigments[a]	Storage reserve[b]	External covering[c]	Flagella
Cyanobacteria (Cyanophyta)	Chl *a*, (*d*) β-carotene, PE, PC, APC, several xanthophylls	Cyanophycean	Peptidoglycan layer	0
Red algae (Rhodophyta)	Chl *a*, *e*, PE, PC, APC, several xanthophylls	Floridean starch	Walls with galactose polymer matrix	0
Cryptomonads (Cryptophyta)	Chl *a*, c_2, PC, PE, diadinoxanthin	True starch	Periplast	2 subequal
Euglenoid algae (Euglenophyta)	Chl *a*, *b*, β-carotene, several xanthophylls	Paramylon	Pellicle or lorica	1–2 emergent
Yellow-green algae Eustigmatophytes (Eustigmatophyceae)	Chl *a*, Violaxanthin, Vaucheriaxanthin	Chrysolaminarin	Cellulosic walls	1–2 unequal
Yellow-green algae (Raphidophyceae) (Xanthophyceae)	Chl *a*, c_1, c_2, several xanthophylls	Chrysolaminarin	Cellulosic walls, some naked	2 unequal
Chrysophyte algae (Chrysophyceae)	Chl *a*, c_1, c_2, c_3, Fucoxanthin, several xanthophylls, β-carotene	Chrysolaminarin, lipids	None, scales, or lorica	2 unequal
Haptophyte algae (Haptophyceae)	Chl *a*, c_1, c_2, c_3, fucoxanthin, several xanthophylls	Chrysolaminarin, fucoxanthin	Nonsiliceous scales	2 equal + haptonema
Synurophyte algae (Synurophyceae)	Chl *a*, c_1, c_2, c_3, fucoxanthin, several xanthophylls, β-carotene	Chrysolaminarin, fucoxanthin	Siliceous scales, bristles	2 unequal
Diatoms (Bacillariophyta)	Chl *a*, c_1, c_2, c_3, fucoxanthin, several xanthophylls	Chrysolaminarin, lipids	Siliceous frustule	1 (male gametes)
Dinoflagellates (Dinophyta)	Chl *a*, c_2, peridinin, several xanthophylls	True starch	Cellulosic theca	2 unequal: transverse + longitudinal
Brown algae (Phaeophyta)	Chl *a*, c_1, c_2, c_3, fucoxanthin	Laminarin	Cellulose walls with alginate matrices	2 unequal
Green algae (Chlorophyta) (Charophyta)	Chl *a*, *b*, β-carotene, xanthophylls	True starch	Cellulosic walls, scales, lorica	0 to many

[a] *APC = allophycocyanin (blue), fucoxanthin and peridinin (golden to brown), Chl = chlorophyll (green), PC = phycocyanin (blue), PE = phycoerythrin (red).*
[b] *All of the reserves are polymers of glucose; only true starch stains positively with iodine (purple or black).*
[c] *External coverings are external to the plasma membrane, except the pellicle and periplast, which are located within the membrane.*
Modified from Sheath and Wehr, (2015).

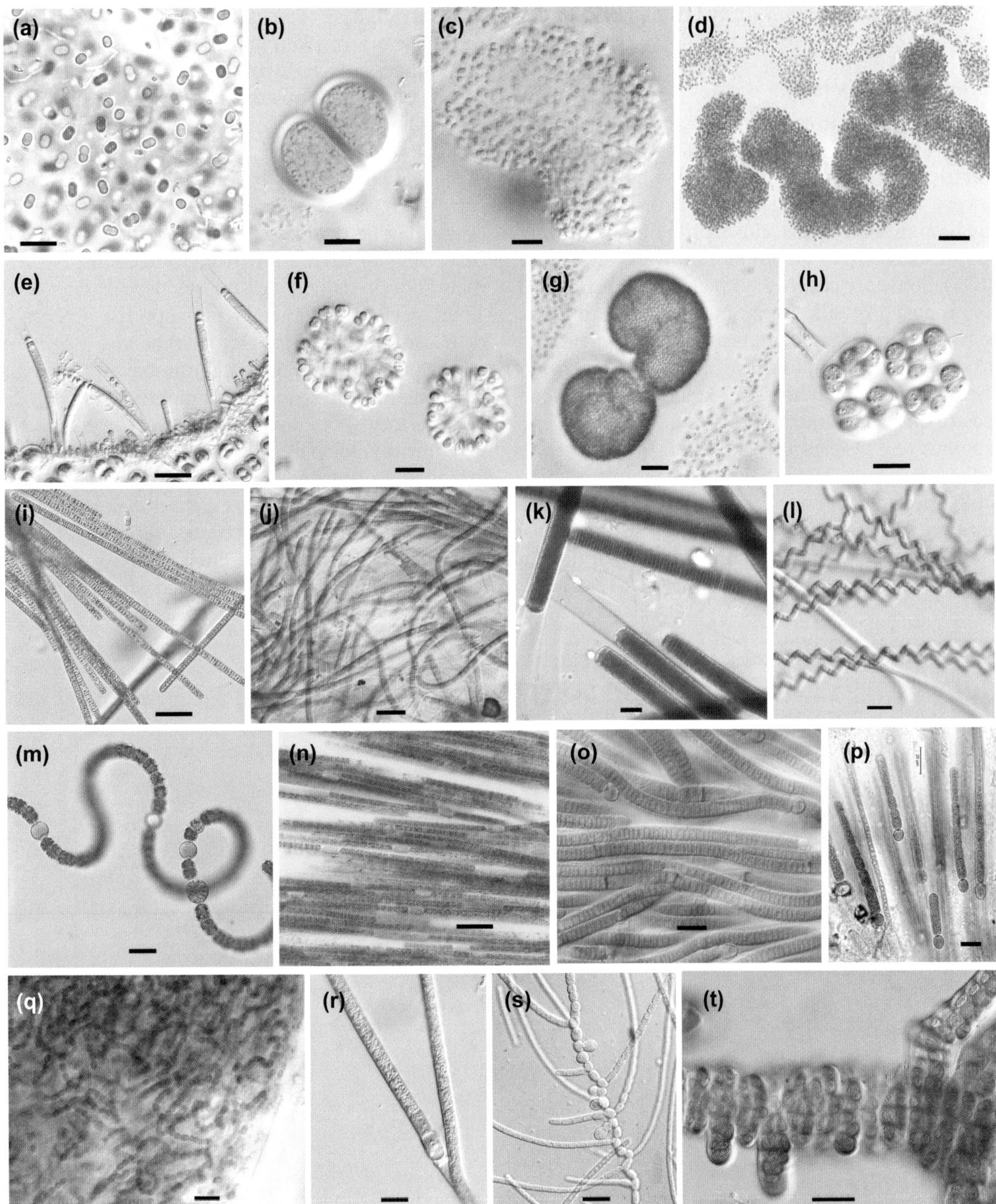

FIGURE 17-2 Representatives of Cyanobacteria in inland waters. Coccoid forms (a–h): (a) *Synechococcus* sp.; (b) *Chroococcus turgidus*; (c) *Aphanocapsa* cf. *conferta*; (d) *Microcystis aeruginosa*; (e) *Chamaesiphon confervicolus*; (f) *Snowella lacustris*; (g) *Woronichinia naegeliana*; (h) *Eucapsis* cf. *alpina*. Filamentous forms (i–t): (i) *Planktothrix* cf. *agardhii*; (j) *Phormidium* cf. *retzii*; (k) *Limnoraphis birgei*; (l) *Arthrospira* sp.; (m) *Dolichospermum* cf. *crassum*; (n) *Aphanizomenon flos-aquae*; (o) *Coleodesmium wrangelii*; (p) *Gloeotrichia* sp.; (q) *Nostoc verrucosum*; (r) *Tolypothrix distorta*; (s) *Nostochopsis lobate*; (t) *Stigonema mamillosum*. *Scale bars*: (a)–(c), (f), (h) = 10 μm; (d), (t) = 50 μm; (e), (g), (i)–(s) = 20 μm.

focused on the environment, explaining that a lake or river is not a simple or homogenous body of water but a complex ecosystem, with multiple microenvironments and periodic disturbances, within which many species can succeed (Sommer et al., 1993; Scheffer et al., 2003). But equally, it is the myriad morphological and physiological properties of algae that enable them to occupy highly specific niches, and which creates a complex and biologically diverse community (Tilman et al., 1982; Huisman et al., 2001).

i. Morphological diversity

Unicellular forms naturally exist as individual cells. They vary widely from simple spheres to an assortment of geometric and even complex shapes and occur in nearly every algal phylum. Their walls or other coverings may be composed of cellulose, protein, calcium, silica, or peptidoglycan, or they may lack walls altogether (Table 17-1). Some are covered with elaborate scales and bristles composed of silica (e.g., *Synura*), calcium (e.g., *Hymenomonas*), or organic materials, usually protein and carbohydrate, in haptophytes (e.g., *Chrysochromulina*). The walls of many others are extended into various processes (e.g., *Staurastrum*), horns (e.g., *Ceratium*; Fig. 17-7b), spines (e.g., *Golenkinia*), lobes (e.g., *Micrasterias*; Fig. 17-9s), or other appendage-like features, which may aid in buoyancy or have other functions. Some species produce gelatinous sheaths (e.g., *Chroococcus*; Fig. 17-2b) or are contained within a case or lorica, such as the euglenophyte *Trachelomonas* (Fig. 17-4f) and the chrysophyte *Epipyxis*. Many dinoflagellates have cellulosic armor-like plates (e.g., *Peridinium*; Fig. 17-3a). Still others attach to other algae or larger plants by means of a gelatinous disc or stalk (e.g., *Characium*).

Colonies form from multiple cells aggregated in variously loosely arranged to tightly organized structures, usually held together by a gelatinous envelope, a firm sheath, or various connection strands or processes. Colonies may be composed of an indeterminate number of cells, ranging from a few (e.g., *Chroococcus*; Fig. 17-2b) to many hundreds (e.g., *Microcystis*, *Volvox*; Figs. 17-2d and 17-9d). Cell number can be very specific, such as 4, 8, 16, or 32, as in the green alga *Scenedesmus* (Fig. 17-9e). In others, when a specific cell number or colony size is reached, they then produce "daughter cells" (e.g., *Sphaerocystis*, *Oocystis*) or "daughter colonies" (e.g., *Eudorina*; Fig. 17-9b) within the parent cell wall. Many colonial forms are arranged in organized patterns, whereas others appear haphazard and their size is indeterminate. Both flagellated and nonflagellated species occur in the form of colonies.

Filamentous forms have their cells arranged in linear series to form short or elongated strands. Some grow

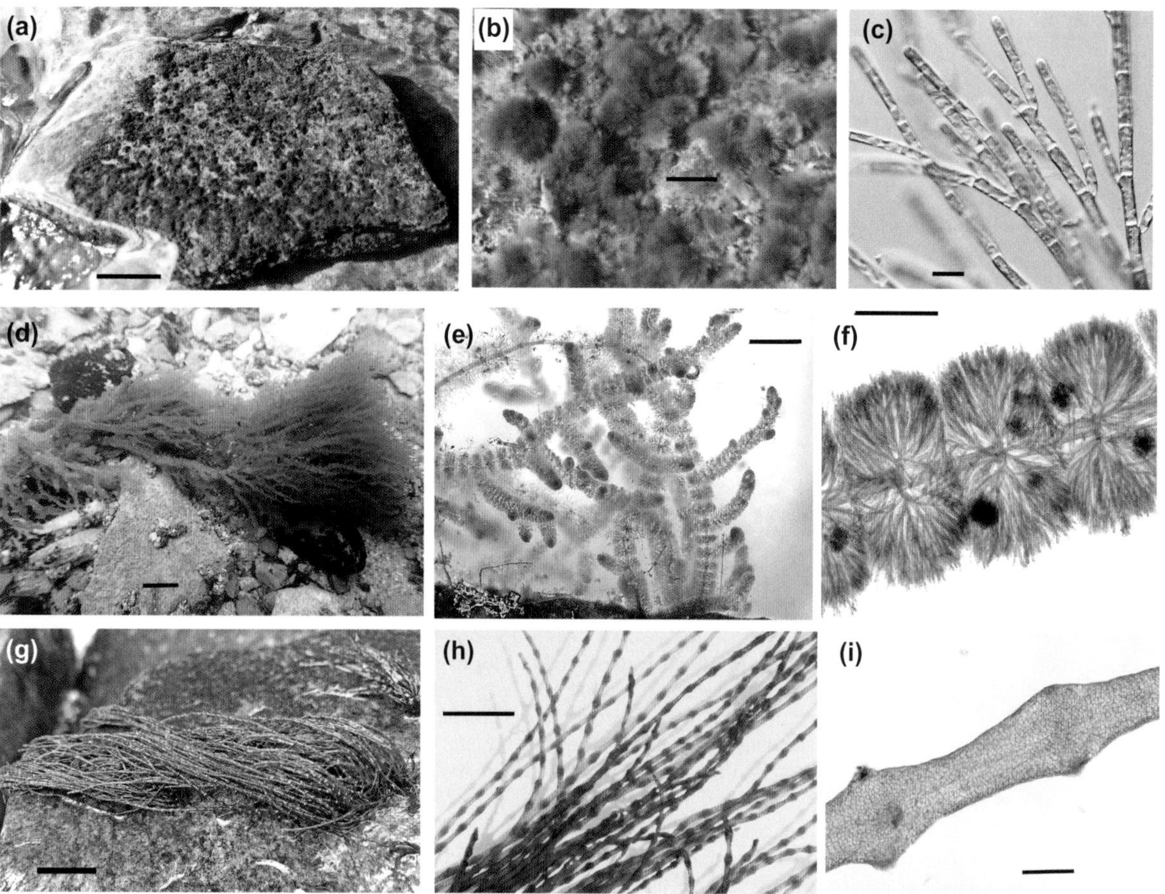

FIGURE 17-3 Representatives of freshwater red algae in inland waters. (a–c) *Audouinella hermannii*; (d–f): *Batrachospermum* spp.; (g–i): *Lemanea* spp. *Scale bars*: (a) = 10 cm; (b) = 5 mm; (c) = 25 μm; (d), (g), (h) = 2 cm; (e), (i) = 1 mm; (f) = 250 μm.

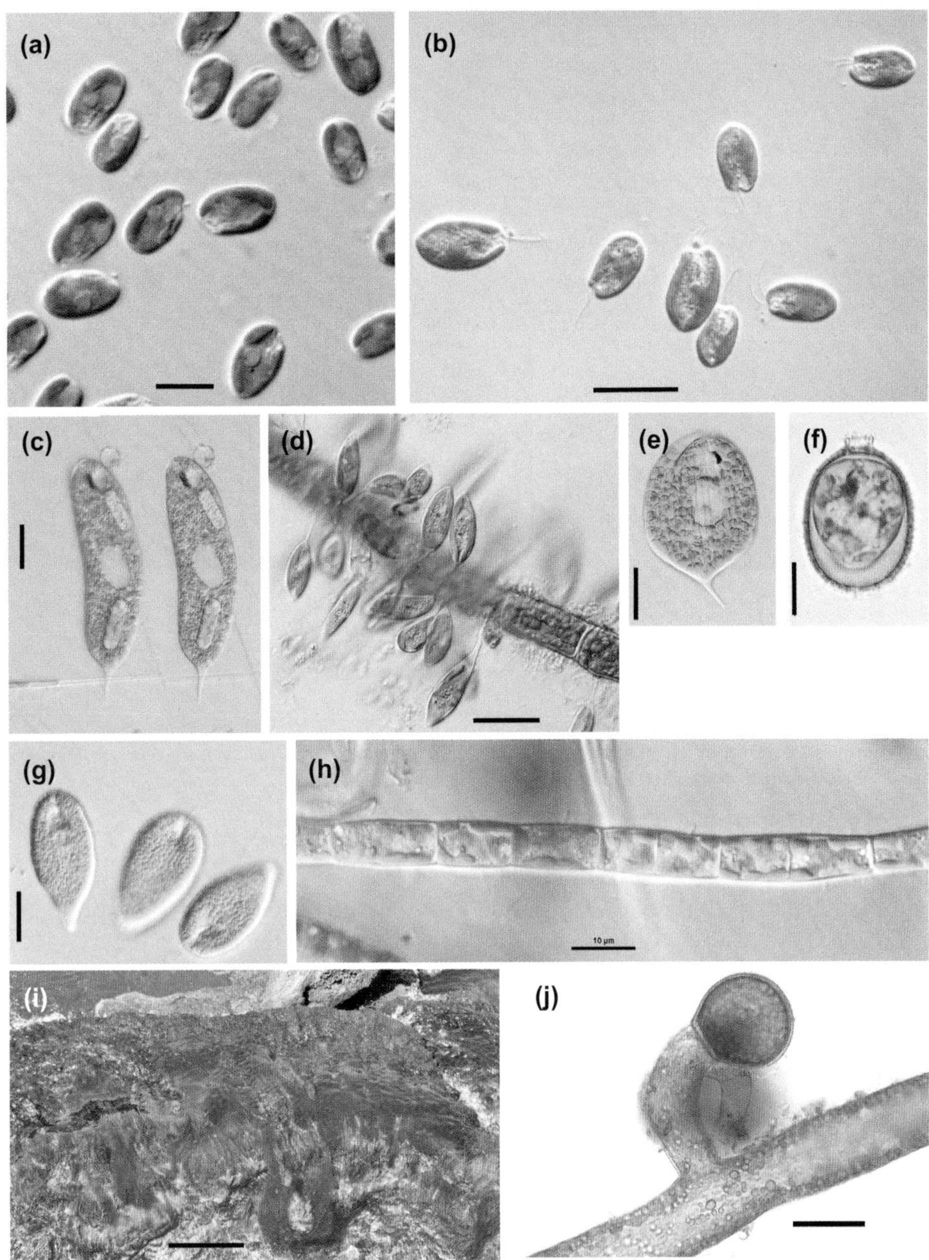

FIGURE 17-4 Representatives of cryptomonads (a and b), euglenoids (c–f), raphidophytes (g), and xanthophytes (h–j) in inland waters. (a) *Rhodomonas* sp.; (b) *Chroomonas* sp.; (c) *Euglena* sp. (surface/internal focus); (d) *Colacium* sp. attached to the green alga *Oedogonium*; (e) *Phacus orbicularis*; (f) *Trachelomonas* sp.; (g) *Gonyostomum semen*; (h) *Tribonema viride*; (i) mat of *Vaucheria* in a calcareous spring; (j) *Vaucheria* cf. *geminata*. *Scale bars*: (a), (b), (f), (h) = 10 μm; (c) = 25 μm; (d), (e), (g) = 20 μm; (i) = 10 cm; (j) = 50 μm.

to form large macroscopic growths, such as *Cladophora* (Fig. 17-9m) in rivers, or *Spirogyra* (Fig. 17-9q) on the surface of ponds. Strictly, filaments are those in which adjacent cells share a common cross wall, as in *Spirogyra*. However, the term is often used more broadly to refer to any morphology in a linear series of cells, because there is no clear-cut distinction among the different arrangements. For example, the cyanobacterium *Romeria*, the green alga *Geminella*, the red alga *Chroodactylon*, and the diatom *Aulacoseira* (Fig. 17-6g) form filamentous strands in which the cells have separate walls and appear as separate cells but are retained within a common sheath or are linked via spines or other structures. Some texts refer to these arrangements as *pseudofilaments*. Branching, or the lack of it, is generally regarded as a diagnostic feature in most algal groups, although ecological factors likely affect this property. Some Cyanobacteria exhibit false branching (e.g., *Scytonema*, *Tolypothrix*; Fig. 17-2r), in which the sheath splits and two or more separate trichomes share a common sheath. True

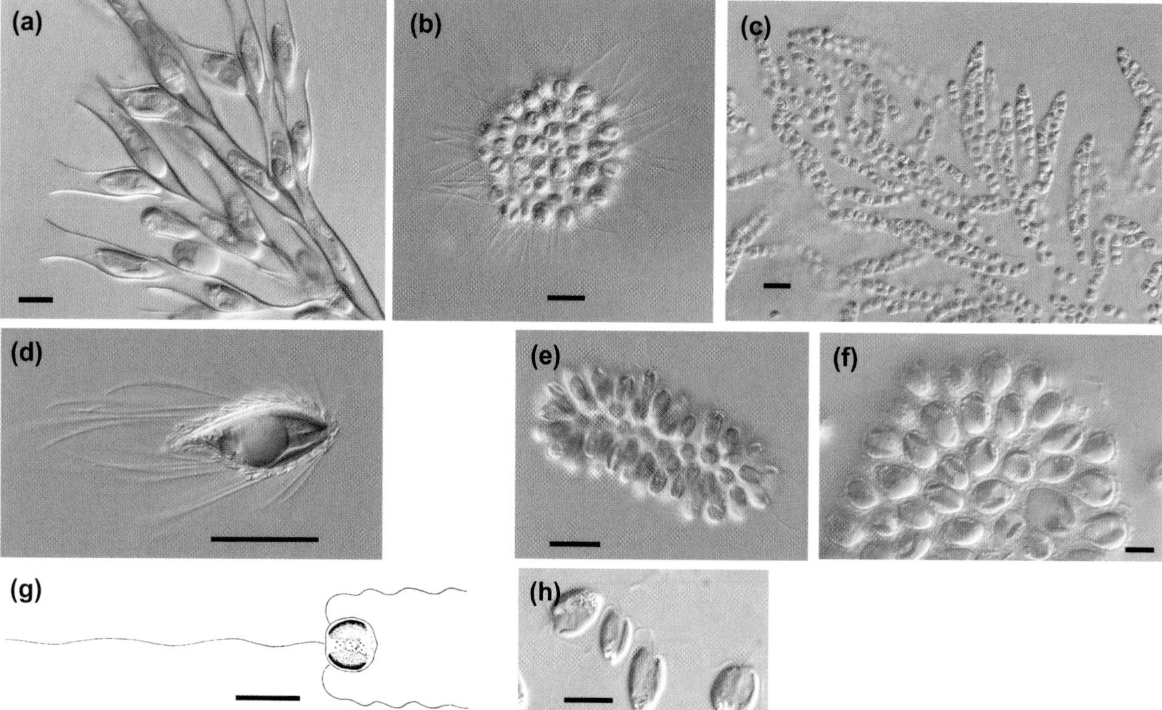

FIGURE 17-5 Representatives of chrysophyte and haptophyte algae in inland waters. (a) *Dinobryon divergens*; (b) *Chrysosphaerella longispina*; (c) *Hydrurus foetidus*; (d) *Mallomonas* cf. *caudata*; (e and f) *Synura* sp.; (g) *Chrysochromulina parva*; (h) *Prymnesium parvum*. *Scale bars*: (a), (f)–(h): 10 μm; (b)–(e): 20 μm.

branching, for example, *Audouinella* (Fig. 17-3a-c) and *Stigeoclonium* (Fig. 17-9k), is the result of direct cell divisions from the main axis, forming side branches. The main axis of filamentous forms can be *uniseriate* (single series of cells) or *multiseriate*. Some filamentous forms lack cross walls altogether, such as in the siphonous xanthophyte *Vaucheria* (Fig. 17-4j).

Pseudoparenchymatous and *Parenchymatous forms* are tissue-like thalli consisting of closely appressed and generally more complex arrangements of uniseriate or multiseriate cells or filaments. Some form broad expanses or sheets, such as the green alga *Prasiola*. Most tissue-like forms in freshwater habitats are simple tubes or blades, like the tubular green alga *Ulva flexuosa* (Fig. 17-9n), which consists of a single layer of cells. Crustose morphologies develop from a series of short, compacted filaments, as in the green alga *Gongrosira* and the brown alga *Heribaudiella* (Fig. 17-8d-f). True parenchyma cells are often differentiated into an outer photosynthetic layer (*cortex*) and an inner nonphotosynthetic region (*medulla*). The stream-dwelling red alga *Lemanea* (Fig. 17-3g-i) forms a cartilaginous tubular thallus that surrounds a layer of small, branched filamentous cells. *Chara* (Fig. 17-10b,c) is large and complex, up to 1 m in length, with a corticated multicellular primary axis bearing whorls of branches and prominent gametangia. The ovoid oogonium is surrounded by a ring of tube cells, and the spherical antheridia have shield cells on the surface (Fig. 17-10c).

ii. Physiological diversity

Countless studies have been conducted on the many physiological adaptations and biochemical pathways of algae, and the significance of these properties on their ecological roles in aquatic ecosystems can be profound (Stewart, 1974; Morris, 1980; Reynolds, 2006; Barsanti and Gualtieri, 2014). A review of this literature is far beyond the scope of this chapter. Aspects of primary productivity are discussed in Chapters 9 and 13, and the ecological requirements and limitations of freshwater phytoplankton are discussed in Chapter 18. Here a brief overview of two key aspects of algal physiological ecology serves to highlight the extraordinary functional diversity of algae in freshwater ecosystems.

Carbon Acquisition. The major photosynthetic pigment in all oxygen-evolving photosynthetic organisms is chlorophyll *a*, and thus the primary role of algae and Cyanobacteria in aquatic ecosystems may be seen as the synthesis and transfer of organic carbon. *Photoautotrophs* acquire inorganic carbon primarily in the forms of CO_2 and HCO_3^-, which is fixed through the Calvin cycle, as in higher plants. Many algal species rely solely or principally on photosynthesis for their carbon metabolism. Although autotrophic carbon fixation is

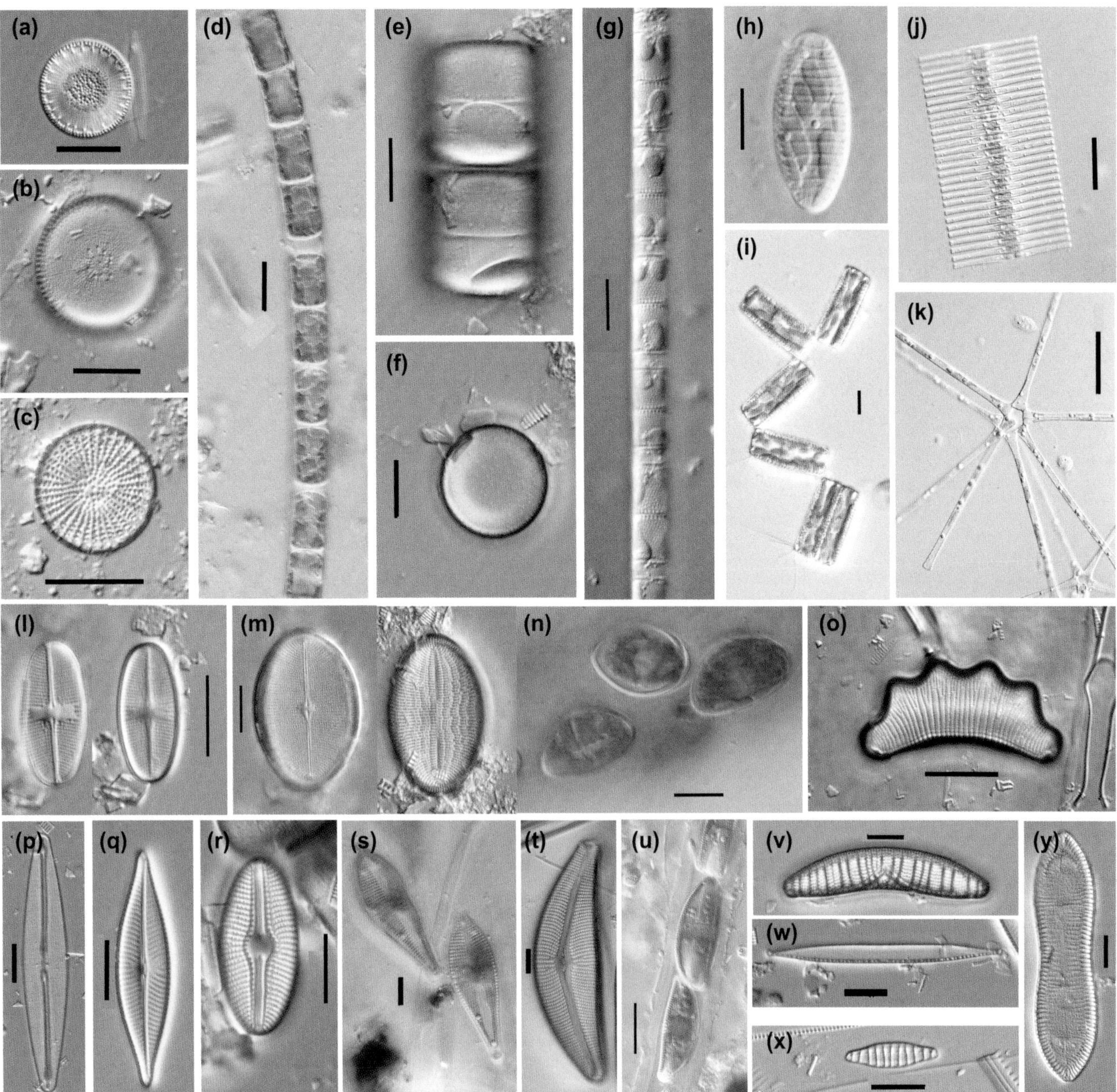

FIGURE 17-6 Representatives of diatoms in inland waters. (a) *Lindavia intermedia*; (b) *Thalassiosira weissflogii*; (c) *Stephanodiscus* cf. *hantzschii*; (d) *Melosira varians* (chain); (e and f) *Melosira varians* (girdle view, valve view); (g) *Aulacoseira* sp.; (h and i) *Diatoma vulgaris* (valve view, girdle view); (j) *Fragilaria crotonensis*; (k) *Asterionella formosa*; (l) *Psammothidium* cf. *bioretii* (raphe and rapheless valves); (m) *Cocconeis pediculus* (raphe and rapheless valves); (n) *Cocconeis pediculus* epiphytic on *Cladophora*; (o) *Eunotia diadema*; (p) *Frustulia amphipleuroides*; (q) *Navicula* cf. *rhynchocephala*; (r) *Diploneis* cf. *elliptica*; (s) *Gomphonema sphaerophorum*; (t) *Cymbella mexicana*; (u) *Encyonema leibleinii*; (v) *Epithemia* sp.; (w) *Nitzschia* sp. (x) *Denticula tenuis*; (y) *Cymatopleura solea*. *Scale bars*: (a)–(i), (l)–(t), (v)–(y) = 10 μm; (j), (k) = 25 μm; (u) = 20 μm.

widespread among all algal phyla (but not all species), many species acquire some or most of their organic carbon heterotrophically, as *mixotrophs* (Sanders, 2011). Within this broad category, *osmotrophs* absorb dissolved organic molecules such as polysaccharides, amino acids, nucleic acids, or polypeptides as an energy source. *Facultative* or *adaptive heterotrophy* has been documented in species of Cyanobacteria (e.g., *Microcystis*), diatoms (e.g., *Gomphonema*, *Encyonema*, *Nitzschia*), green algae (e.g., *Chlamydomonas*, *Keratococcus*, *Scenedesmus*), cryptomonads (e.g., *Cryptomonas*), and euglenoids (Gervais, 1997; Kamjunke and Tittel, 2008; Sanders, 2011; Beamud et al., 2014). The mechanisms and effects of the uptake of organic carbon sources are complex. Uptake may occur concurrently with photosynthesis, during low-light conditions (e.g., under ice), or during dark periods,

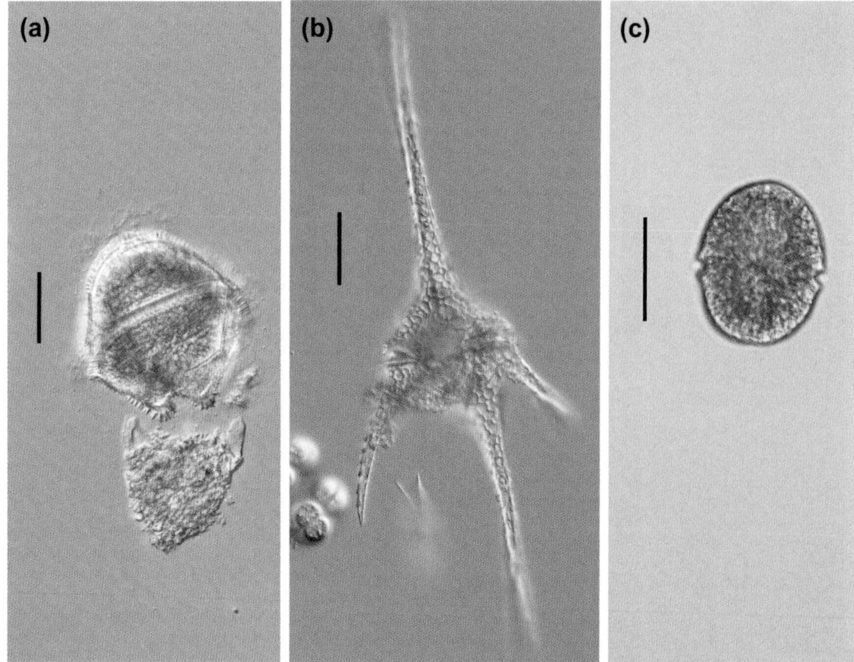

FIGURE 17-7 Representatives of dinoflagellates in inland waters. (a) *Peridinium willei*; (b) *Ceratium hirundinella*; (c) *Gymnodinium* sp. *Scale bars*: (a)–(c) = 25 μm.

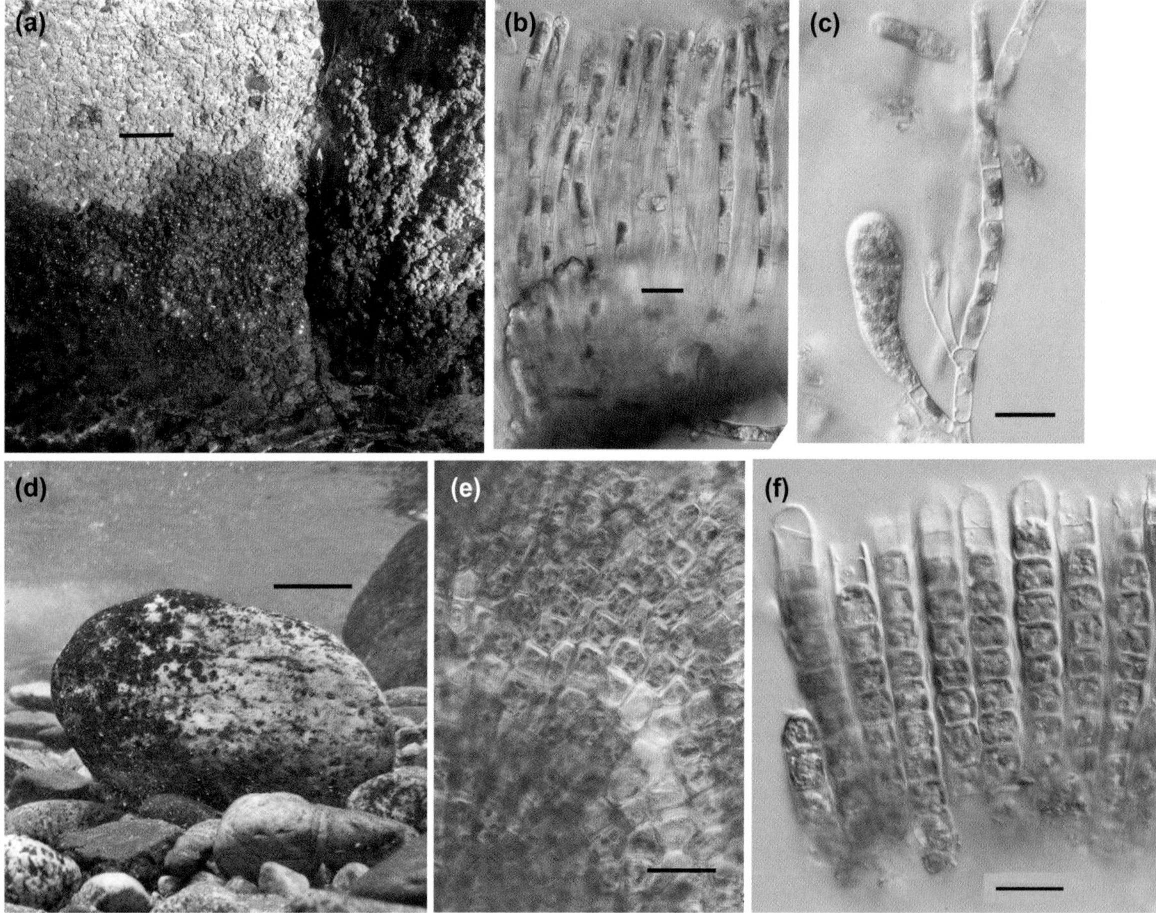

FIGURE 17-8 Representatives of brown algae in inland waters. (a) Calcified colonies of *Pleurocladia lacustris*; (b) filaments of *P. lacustris* showing basal $CaCO_3$ precipitate; (c) unilocular sporangium of *P. lacustris*, (d) reddish brown crust-like colonies of *Heribaudiella fluviatilis* on a rock; (e) prostrate morphology of *H. fluviatilis*; (f) upright morphology of *H. fluviatilis*. *Scale bars*: (a) = 25 mm; (b), (c), (e), (f) = 20 μm; (d) = 10 cm.

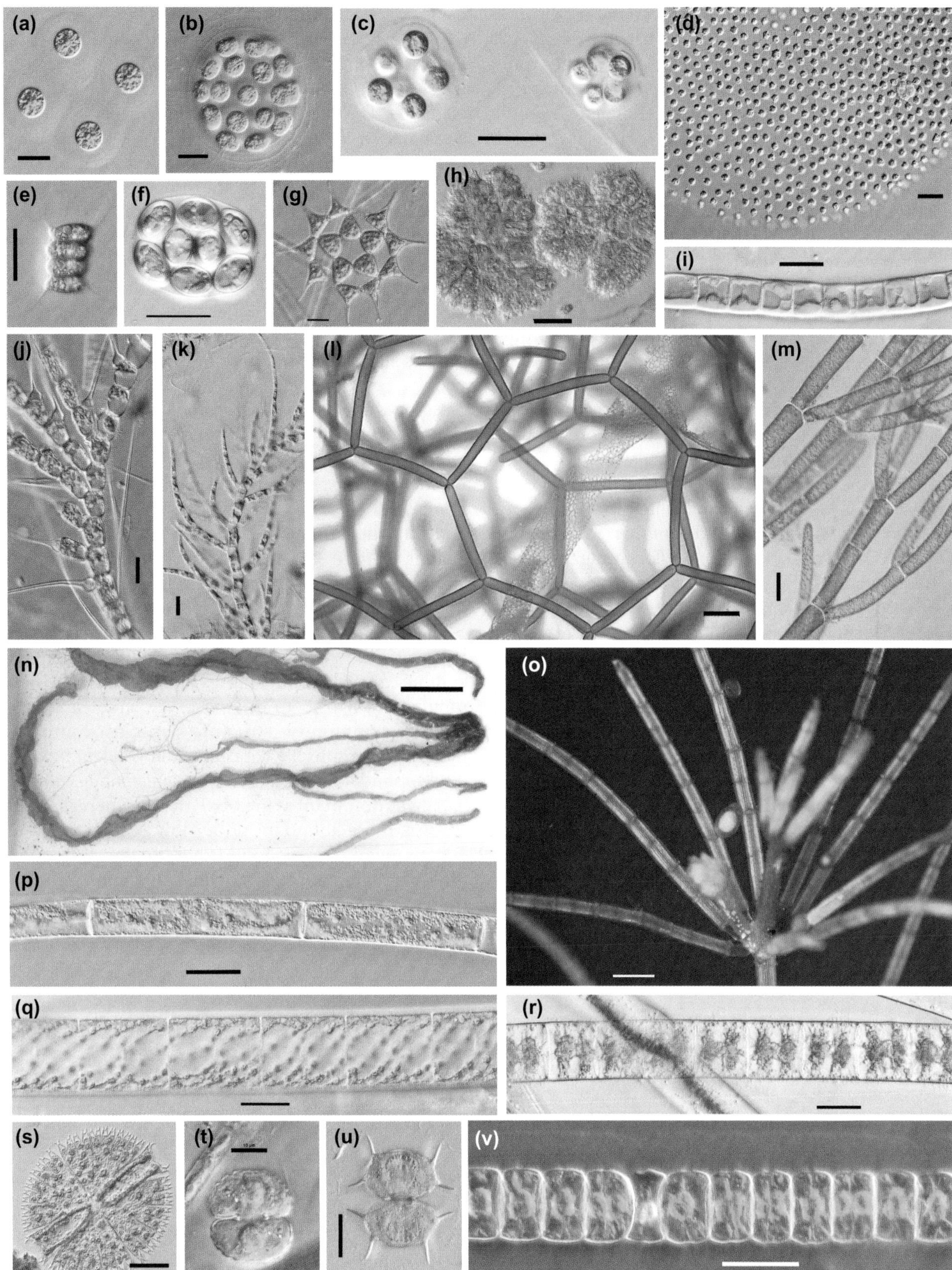

FIGURE 17-9 Representatives of green algae in inland waters. (a) *Asterococcus* cf. *limneticus*; (b) *Eudorina elegans*; (c) *Sphaerocystis schroeteri*; (d) *Volvox* sp.; (e) *Scenedesmus* cf. *quadricauda*; (f) *Oocystis* cf. *lacustris*; (g) *Pediastrum* cf. *simplex*; (h) *Botryococcus braunii*; (i) *Ulothrix* cf. *tenuissima*; (j) *Bulbochaete* sp.; (k) *Stigeoclonium tenue*; (l) *Hydrodictyon reticulatum*; (m) *Cladophora glomerata*; (n) *Ulva flexuosa*; (o) *Chara* cf. *foliolosa*; (p) *Mougeotia* sp.; (q) *Spirogyra* sp.; (r) *Zygnema* sp.; (s) *Micrasterias rotata*; (t) *Cosmarium* sp.; (u) *Xanthidium antilopaeum*; (v) *Hyalotheca mucosa*. *Scale bars*: (a)–(c), (e)–(k), (r) = 20 μm; (d), (u), (v) = 25 μm; (l) = 500 μm; (m) = 100 μm; (n) = 5 cm; (o) = 1 mm; (p), (q), (s) = 50 μm; (t) = 10 μm.

depending on the species and environmental conditions. Many freshwater diatoms, especially benthic forms, increase the heterotrophic utilization of organic compounds under dark conditions. Differential uptake of compounds, such as polysaccharides versus amino acids, may have positive or negative effects on the growth and fatty composition of an alga, an effect that is also species specific (Peltomaa and Taipale, 2020). The ability of the flagellated alga *Gonyostomum semen* (Fig. 17-4g) to form nuisance blooms in humic lakes has been attributed partly to its ability to take up dissolved organic matter, which is present in high concentrations in these systems (Rengefors et al., 2008). Viruses may participate in the transfer of organic solutes and nutrients to algae through the lysis of planktonic bacteria (Shelford and Suttle, 2018).

Freshwater algae may function as *phagotrophs*, ingesting particulate organic matter, either in the form of bacteria or detritus. *Bacterivory*, in particular, is frequently observed in the colonial chrysophyte *Dinobryon* (Fig. 17-5a), as well as in other algal flagellates, including species of *Cryptomonas*, *Uroglena*, and *Ceratium* (Sanders and Porter, 1988; Tranvik et al., 1989). Evidence suggests that this phenomenon is often light dependent, that is, necessary under low-light conditions, and is reduced when cells are functioning in autotrophic mode (Tranvik et al., 1989; Caron et al., 1993). Phagotrophy may also facilitate the uptake of particulate nitrogen and phosphorus when nutrients are limiting. It has been observed that rates of bacterivory were inversely proportional to concentrations of dissolved P or N (Urabe et al., 1999; Bergström et al., 2003; Terrado et al., 2017).

Nutritional Requirements. The role of nutrients in algal ecology is among the most widely studied subjects in all of limnology and has helped to develop major theories of competition and ecological stoichiometry. These works, initially based on phytoplankton studies, were later expanded to encompass a broader ecological theory (Tilman et al., 1982; Hessen et al., 2013). Many of these concepts and principles are discussed in Chapters 14–16 and 18. Briefly, most algal cells require a similar array of elements that are required by higher plants, that is, C, H, O, P, K, N, S, Ca, Fe, and Mg, plus some trace elements. The proportions of the elements in algal cells, combined with the demand for these elements, largely determine which nutrients most affect algal production in fresh waters. Although the proportions in algal cells vary with environmental conditions and among species, on average, the key elements for most species occur in similar proportions (Table 17-2; see also Table 15-14).

Although algal species in general have similar nutritional requirements to higher plants, with the important exception of the diatoms and most chrysophytes, which require silica, their demand for different elements varies widely. These differences significantly affect ecological patterns of growth, competition, and seasonal succession in lakes and rivers. For example, laboratory experiments have shown that the planktonic diatoms *Asterionella formosa* (Fig. 17-6k) and *Cyclotella meneghiniana* coexist because of differences in the demand for Si and P in the two species (Tilman and Kilham, 1976). A similar pattern of niche differentiation was demonstrated in field experiments explaining the potential dominance of green algae (Si:P <70) or diatoms (Si:P >300) in Lake Michigan (USA), and their stable coexistence at Si:P = 71 (Kilham, 1986). Experiments with benthic algae in streams with varying N:P ratios, ranging from 65:1 to 4:1, observed greater effects on diatom community structure (species composition) than on chlorophyll *a* concentration or algal biovolume (Stelzer and Lamberti, 2001). *Achnanthidium minutissimum* was dominant at high N:P ratios, whereas *Fragilaria* sp. was more abundant under low N:P. A study using experimental streams found that benthic algal communities with greater species richness had more efficient nitrogen uptake and nutrient processing, which was attributed to niche differences among the different species (Cardinale, 2011).

FIGURE 17-10 Marl lakes and charophyte flora. (a) Plitvice Lakes National Park, Croatia, with white $CaCO_3$ deposits on the lake bottom. (b–c) *Chara* cf. *vulgaris* from a calcium-rich pond, Potchefstroom, North West Province, South Africa. (b) Macroscopic view of the thallus, and (c) Detail, showing gametangia and $CaCO_3$ incrustation (*arrow*). *Scale bars*: (b) = 5 cm; (c) = 1 mm.

TABLE 17-2 Average Quantities ($\mu g\ mg^{-1}$ Dry Mass) and Proportions of Nutrients in Algae, Sorted in Order of Relative Quantity, in Atoms

Element	Average	Range	Relative amount (atoms)
H	65	29–100	8,140,000
C	430	175–650	4,460,000
O	275	205–330	2,120,000
N	55	10–140	487,000
Si[a]	54	0–230	237,000
K	17	1–75	55,000
P	11	0.5–33	43,800
Na	6.1	0.4–47	32,500
Mg	5.6	0.5–75	28,700
Ca	8.7	0–80	27,500
S	5.9	1.5–16	23,800
Fe	5.9	0.2–34	13,800
Zn	0.28	0.005–1.0	540
B	0.03	0.001–0.25	350
Cu	0.10	0.006–0.3	200
Mn	0.06	0.02–0.24	138
Co	0.06	0.0001–0.2	125
Se	0.04	0.02–0.08	100
Mo	0.0008	0.0002–0.001	1

[a]Si is required only in some species.
Adapted from Healey, (1973); Hecky and Kilham, (1988); Sterner and Elser, (2002); Baines et al., (2004).

Algae often experience limiting concentrations of nitrogen or phosphorus in freshwater systems (Chapters 14 and 15), and species have evolved a variety of traits to ensure their survival and growth under these conditions. Many algae employ phosphatase enzymes, such as phosphomonoesterase, to hydrolyze P from various organic forms when inorganic P is limiting (Jansson et al., 1988; Whitton et al., 1991). Before periods of nutrient depletion, some species of algae store excess, so-called luxury, phosphorus in the form of polyphosphate granules and later use this source when inorganic-P becomes limiting (Pettersson, 1985; Li et al., 2019). Algal cells can also store nitrogen in intracellular inorganic-N pools, proteins, or pigments (Dortch et al., 1984; Oliver et al., 2012). When concentrations of dissolved inorganic nitrogen (mainly NO_3^- and NH_4^+) in a water body decline relative to phosphorus (N: P <16), primary production does not necessarily decline, because many species of Cyanobacteria possess the nitrogenase enzyme, which enables them to fix atmospheric N_2 and reduce it to biologically useful forms such as NH_4^+ (Schindler, 1977; Smith, 1983). This includes many common bloom-forming planktonic forms (e.g., *Aphanizomenon*; Fig. 17-2n; *Dolichospermum*; Fig. 17-2m) and benthic (e.g., *Nostoc*, *Tolypothrix*; Fig. 17-2q,r) species that have one or more specialized cells known as *heterocysts* (also *heterocytes*; visible in Fig. 17-2m) that contain the nitrogenase enzyme and lack photosynthetic thylakoids (Oliver et al., 2012). No other group of algae has this ability. More recent studies indicate that many nonheterocytous Cyanobacteria have the *NIF* gene and can fix N_2 (Chapter 14), usually at night when photosynthetic activity is minimal, or in microaerobic conditions, such as multiple trichomes in a dense bundle (e.g., *Microcoleus*) or in benthic microbial mats (Stal, 1995). Few species have been shown to fix N_2 aerobically. Rates of nitrogen fixation may also be colimited by trace concentrations of iron or molybdenum (Howarth et al., 1998). Other trace elements may also be critical factors in algal nutrition and ecology, and requirements are often tightly coupled to other physiological processes. For example, trace amounts of nickel are required as a cofactor for urease activity, whereas alkaline phosphatase requires trace amounts of zinc for proper functioning (Price and Morel, 1991; Sterner et al., 2004). An absolute requirement for nanomolar quantities (0.01–1.0 nmol L^{-1}) of selenium has been demonstrated in several species of freshwater algae, including the dinoflagellate *Peridinium cinctum*, diatom *Stephanodiscus hantzschii*, and haptophyte *Chrysochromulina breviturrita* (Lindström, 1983; Wehr and Brown, 1985). At greater concentrations (>100 nmol L^{-1}), selenium becomes a pollutant in surface waters and can inhibit phytoplankton growth (Riedel et al., 1996). These multiple nutritional requirements, evolutionary strategies, and tolerances underscore the enormous ecological diversity of algae and their roles in freshwater ecosystems.

D. Importance of size

Algal cells, filaments, and thalli vary in size from about 0.2 to 1.0 μm, up to individual thalli of more than 1 m, a linear scale of at least six orders of magnitude. The definitions and methods used to categorize algal size have evolved over time. Plankton size was originally based on different mesh sizes of plankton nets and later based on the capture size of different membrane filters. Today, the scale most often used for planktonic organisms is divided into decadal categories based on direct (usually microscopical) measurements, a system originally proposed by Dussart, (1965) and later updated (Sieburth et al., 1978; Sicko-Goad and Stoermer, 1984) to be in accordance with SI units (Table 17-3). The smallest phytoplankton cells, the *picoplankton*, range from 0.2 to 2.0 μm (Table 17-3) and can be very

TABLE 17-3 Plankton Size Categories With Characteristics and Examples of Freshwater Algal Species in Each

Category	Size range	Morphologies	Generation times	Examples
Femtoplankton	0.02–0.2 μm	Unicells	minutes	*Cyanodictyon*
Picoplankton	0.2–2.0 μm	Unicells, microcolonies	minutes, hours	*Synechococcus, Nannochloris, Snowella*
Nanoplankton	2.0–20 μm	Unicells, colonies	hours	*Cryptomonas, Mallomonas, Peridinium*
Microplankton	20–200 μm	Large cells, colonies filaments	days	*Asterionella, Ceratium, Eudorina*
Mesoplankton	0.2–2.0 mm	Colonies, filaments	days	*Anabaena, Synura, Tribonema*
Macroplankton	2.0–20 mm	Filaments, aggregates	weeks	*Aphanizomenon, Spirogyra, Melosira*
Megaplankton	>20 mm	Filaments, macroalgae	weeks	*Cladophora, Hydrodictyon, Vaucheria*

Adapted from Dussart, (1965); Sieburth et al., 1978; Sicko-Goad and Stoermer, (1984).

abundant, as dense as 10^4 to 10^6 cells mL^{-1}, even in oligotrophic lakes (Callieri et al., 2012; Takasu et al., 2015). These tiny cells exist mainly as coccoid unicells or microcolonies. The intense interest in these tiny algal cells was spurred by the relatively recent recognition (detected using epifluorescence microscopy) of their ubiquitous abundance and rapid growth rates in oceans and lakes (Paerl, 1977; Johnson and Sieburth, 1979). Many members of the autotrophic picoplankton are Cyanobacteria in the order Chroococcales, with species of *Cyanobium*, *Synechococcus* (Fig. 17-2a), and *Synechocystis* most frequently reported (Callieri et al., 2012).

However, phylogenetic studies suggest that many cyanobacterial taxa are polyphyletic (Komárek and Johansen, 2015; Coutinho et al., 2016), so reports of "species" in different lakes based on very limited morphological traits are doubtful. In addition, picocyanobacterial cells exhibit marked plasticity in size, morphology, pigmentation, and tendency to form microcolonies. This plasticity is often in response to changing environmental conditions, including factors such as temperature, nutrients, and the presence or absence of predators (Callieri, 2017). Their apparent adaptability, as well as the many recent studies that have revalued the enormous diversity of picoplanktonic forms and/or ecotypes, and likely cryptic diversity, require further research using molecular techniques (Tromas et al., 2020). There are also *picoeukaryotes* in freshwater systems. A molecular survey (*rRNA* gene sequences) of eukaryotes 0.2 to 5.0 μm in size from oligotrophic Lake George (NY) revealed at least 77 phylotypes with clades including chrysophytes and cryptomonads, and several others with no clear affinity to any known higher-level taxonomic groups (Richards et al., 2005), although many of these cells were likely *heterotrophs* (organisms that obtain their organic matter from external sources). An enormous molecular diversity of photosynthetic picoeukaryotes was discovered from Lake Taihu and Lake Chaohu, China, using high-throughput sequencing combined with flow cytometry (Li et al., 2017). Among the pigmented forms, 210 unique green algal phylotypes (Chlorophyceae, 164; Trebouxiophyceae, 40; Ulvophyceae, 6) were identified, as well as 110 members of the Chrysophyta, with fewer cryptomonads (32), euglenophytes (22), synurophytes (18), and eustigmatophytes (10).

Photosynthetic picocyanobacteria and picoeukaryotes have been reported worldwide from oligotrophic, mesotrophic, and eutrophic waters, as well as in brackish, humic, and meromictic lakes. Picoplankton as a group are recognized as organisms with rapid growth rates of between 0.2 and 2.0 day^{-1} (Nagata et al., 1994; Lavallée and Pick, 2002; Callieri et al., 2012) and superior abilities for nutrient acquisition, in part due to their small size (Li et al., 2019). Variation in relative amounts of *phycocyanin* (blue) and *phycoerythrin* (red) pigments in picocyanobacteria further enables cells to adapt to different light regimes (chromatic adaptation) in surface and deep strata within lakes (Stomp et al., 2007; Somogyi et al., 2020). In many temperate lakes *autotrophic* (i.e., photosynthetic) picoplankton typically have two seasonal peaks, in late spring–early summer, and a second peak in summer–autumn (Callieri et al., 2012), and contribute between 1% and nearly 100% of the total primary production in lakes, with a negative correlation between total phytoplankton biomass (as chlorophyll *a*) and percent picophytoplankton (Bell and Kalff, 2001; Callieri, 2008).

Among the larger species and colonies in the planktonic environment, other adaptations are required. It is generally accepted that larger cells may be less efficient in acquiring nutrients than their smaller counterparts (the latter having higher surface area:volume ratios) but may be less susceptible to grazing by smaller herbivores such as rotifers and flagellates. Larger size also affects sinking rates and therefore the ability of photosynthetic cells to remain within the euphotic zone of lakes and large rivers. A key requirement is a mechanism to remain buoyant. Motility by means of one or more flagella is common, although not universal, across most eukaryotic algal groups. Regular diel vertical migration has been widely observed in many flagellate species from diverse algal groups, including *Cryptomonas*, *Gonyostomum*, *Mallomonas*, *Peridinium*, and many others (Happey-Wood, 1976; Salonen and Rosenberg, 2000). A population of the colonial green alga *Volvox* (Fig. 17-9d) in Lake Cahora Bassa (Mozambique) was observed migrating downward (rates of 1.8–3.6 m h^{-1}) below the euphotic zone at night, to a vertical distance of about 17 m, and returning to the epilimnion by morning (Sommer and Gliwicz, 1986). The dinoflagellate *Ceratium hirundinella* (Fig. 17-7b) is able to migrate vertically within the water column of stratified lakes (rates of $1.6–2.7 \times 10^{-4}$ m s^{-1}), in order to minimize photorespiration at the surface during periods of high light and can sequester dissolved nutrients from deep strata at night and then return to the epilimnion for photosynthesis (James et al., 1992; Whittington et al., 2000). Perhaps the most recognizable outcome of algal buoyancy and vertical migration is seen in the Cyanobacteria, which possess *gas vesicles* (clusters of vesicles termed *aerotopes*), which enable many cyanobacterial species, including large accumulations of bloom-forming species, notably *Aphanizomenon flos-aquae* (Fig. 17-2n), *Dolichospermum planctonicum*, *Microcystis aeruginosa* (Fig. 17-2d), and *Planktothrix agardhii* (Fig. 17-2i), to rise to the surface and thereby dominate eutrophic lakes (Reynolds, 1987; Oliver et al., 2012). The colonial cyanobacterium *Gloeotrichia echinulata* occupies nutrient-rich benthic habitats in eutrophic lakes but forms gas vesicles in late summer in order to migrate into surface waters to maximize photosynthetic activity (Barbiero and Welch, 1992). Planktonic diatoms, with silica frustules that increase their density in water, are especially susceptible to sinking out of the euphotic zone, particularly larger species in well-mixed or turbulent systems (Reynolds, 1994). Some may partly regulate their vertical position through changes in intracellular solutes such as lipids, ions, and inorganic nutrients (Pearre, 2003), in response to changes in nutrient and light conditions (Brookes and Ganf, 2001). Other algal taxa form spines or structures that increase their surface area, thereby slowing their sinking rates. Some long-term and paleolimnological data indicate that, due to climate change, there will be a selection for smaller species (change in community structure) and smaller size within species of diatoms, in response to a change in water density with warming of surface waters (Winder et al., 2009; Rühland et al., 2015). Presumably, these smaller cells would also be better competitors for dissolved nutrients, although current trends in eutrophication may have the opposite effect on the size structure of phytoplankton communities.

II. Major groups of algae

The present overview describes the main groups of algae that occur in freshwater ecosystems and summarizes the key characteristics of each with examples. Detailed information on the biology, evolution, and classification of algae can be found in recent phycology textbooks (e.g., Graham et al., 2016; Lee, 2018). A series of broad taxonomic works may be consulted for guidance with identification (John et al., 2011; Wehr et al., 2015).

A. Cyanobacteria

The Cyanobacteria are a unique and diverse group of photoautotrophic organisms within the domain Eubacteria, having prokaryotic cells, gram-negative cell walls composed of peptidoglycan, and, in addition to chlorophyll *a*, several phycobiliprotein pigments (Table 17-1). Chief among them are phycocyanin (blue) and phycoerythrin (red), which render characteristic blue-green pigmentation in many species. Rarely, chlorophylls *b* (e.g., *Prochlorothrix hollandica*) and *d* (e.g., *Acaryochloris marina*) occur (Zwart et al., 2007; Loughlin et al., 2013; Komarek and Johansen, 2015; Graham et al., 2016). Secondary pigments include various *xanthophylls* such as astaxanthin and β-carotene (Table 17-1). Although the common name "blue-green algae" is often used for these organisms, cells may appear almost any color, depending on the relative amounts of these pigments, from cyan to purple, red, brown, gray, or black. When viewed with a microscope or with the naked eye, the apparent color may be further altered by the presence of gelatinous sheaths, which can be variously colored due to pigments in the sheath, such as yellow-brown scytonemin (e.g., *Stigonema*, Fig. 17-2t) or red to blue gloeocapsin, which is thought to provide protection against UV radiation, particularly in shallow benthic habitats, ultraoligotrophic lakes, and terrestrial environments (Castenholz and Garcia-Pichel, 2012).

Morphologies of Cyanobacteria are also diverse, ranging from the simplest of unicells to complex filaments and macroscopic thalli (Fig. 17-2). Often cells are enclosed in a loose mucilage (e.g., *Aphanocapsa*, *Microcystis*, *Phormidium*), or a firm gelatinous sheath (e.g., *Chroococcus*, *Limnoraphis*, *Nostoc*, *Tolypothrix*; Fig. 17-2). As prokaryotes, Cyanobacteria lack membrane-bound organelles but have a huge diversity of cell types, as well as several specialized structures that serve specific functions. Many planktonic species have *aerotopes* (clusters of gas vesicles) in their cells to provide buoyancy, for example, *Microcystis* (Fig. 17-2d), *Dolichospermum* (Fig. 17-2m), *Aphanizomenon* (Fig. 17-2n), and *Limnoraphis* (Fig. 17-2k). With light microscopy, cells appear red or brownish due to the diffraction of light. Some Cyanobacteria produce reproductive structures such as *baeocytes* in *Chamaesiphon* (Fig. 17-2e), or thick-walled *akinetes*, which serve as resting stages under severe or otherwise limiting conditions, as in *Dolichospermum* (Fig. 17-2m), *Gloeotrichia* (Fig. 17-2p), and *Aphanizomenon* (Fig. 17-2n). *Heterocysts* are specialized cells lacking thylakoids that contain the nitrogenase enzyme to fix N_2 (e.g., *Dolichospermum, Aphanizomenon, Gloeotrichia, Nostoc, Tolypothrix*; Fig. 17-2m to r).

The above traits have contributed to making Cyanobacteria globally ubiquitous, colonizing perhaps the widest range of ecological conditions of all algae, from oligotrophic to hypertrophic lakes and rivers, to hot springs, deserts, alkaline springs, wastewater treatment works, hypersaline lakes, temporary pools, and as symbionts within higher plants or other algae. Some bloom-forming species produce potent toxins, including *M. aeruginosa* (e.g., *microcystin*, a hepatotoxin; Fig. 17-2d), *Dolichospermum* (*Anabaena*) *circinale* (*anatoxin*, a neurotoxin), *A. flos-aquae* (*saxitoxin*, a neurotoxin; Fig. 17-2n), and *Cylindrospermopsis raciborskii* (*cylindrospermopsin*, a hepatotoxin) (Li et al., 2017).

B. Red algae

The red algae are classified in the phylum Rhodophyta, whose members have photosynthetic pigments phycocyanin and phycoerythrin in addition to chlorophyll *a* and *d*, thylakoids that are not organized in stacks, and lack flagella in either vegetative or reproductive stages (Table 17-1). They are primarily a marine group; however, at least 27 genera of freshwater rhodophytes are recognized and observed in lakes and rivers from the Arctic to the tropics (Sheath and Vis, 2015). Nearly all species are macroscopic and benthic, and none are strictly planktonic. Most species of red algae, in which life histories are known, have heteromorphic gametophyte and sporophyte (termed the *chantransia* stage in many species) stages, with a diversity of reproductive strategies and modes. In some species the two stages are isomorphic. Gametophyte morphologies vary from simple filaments to complex branched forms and *pseudoparenchymatous thalli* (Fig. 17-3). They range from microscopic cells (e.g., *Porphyridium*) or pseudofilaments several μm in size (e.g., *Chroodactylon, Porphyridium*) to bright red crusts (e.g., *Hildenbrandia*), uniaxial branched filaments (e.g., *Audouinella* Fig. 17-3a-c), multiaxial filaments (e.g., *Bangia*), to macroscopic thalli ranging from 2 to 20 cm (e.g., *Batrachospermum* and *Lemanea* Fig. 17-3d-i), and others up to 2 m in length (e.g., *Thorea*).

Red algae can be major components of the algal flora in streams and rivers, and several larger forms are regularly recorded in macrophyte surveys of rivers (Holmes and Whitton, 1977; Schaumburg et al., 2004). Several species, such as *Batrachospermum gelatinosum* and *Paludicola* (formerly *Batrachospermum*) *turfosa*, are regarded as indicators of good water quality (low total nutrients), whereas *Audouinella hermannii* (Fig. 17-3a-c) occurred over much wider ranges of TP and NO_3^- in Austria and Norway (Schneider and Lindstrøm, 2011; Rott and Schneider, 2014). A species of *Lemanea* has been observed to tolerate moderate levels of heavy metal pollution (up to 1.216 mg Zn L^{-1}; Harding and Whitton, 1981). *Bangia atropurpurea* is an apparent invader of the North American Great Lakes, likely introduced via ballast water (Lin and Blum, 1977; Müller et al., 2003; Tittley and Neto, 2005; Shea et al., 2014). It may cooccur with, or potentially replace, filamentous green algae in some microhabitats (Garwood, 1982). Although most red algal species are inhabitants of streams and oligotrophic lakes (Sheath and Vis, 2015), others are able to tolerate extreme environments including acidic hot springs (Section 17IIID).

C. Cryptomonads

Cryptomonads are unicellular, biflagellate protists in the phylum Cryptophyta and are quite unlike other common freshwater algae. Cells are asymmetrical with oval, lunate, elongate, or in some, sigmoid shapes (Fig. 17-4a,b). Flagella are subapical and subequal, and most cells have a distinct depression called a *vestibulum*, with an apical longitudinal furrow or gullet (Clay, 2015). Photosynthetic pigments are chlorophyll *a* and c_2, phycocyanin, phycoerythrin (pigments from an ancient cyanobacterial symbiont), α- and β-carotene, and various xanthophylls (Table 17-1). When viewed with a light microscope, freshwater forms appear greenish, olive, reddish-brown, or golden-brown. Members of the genus *Chroomonas* (Fig. 17-4b) are typically blue-green in color, and lack a furrow, but have a shallow depression where the flagella are inserted (Kugrens et al., 1986; Hill, 1991). The external covering of cryptomonads is a *periplast*, in which the cell membrane is layered between two protein sheets. Cells of cryptomonads produce and release *ejectisomes*, organelles constructed as tightly coiled ribbons in the Golgi, and rapidly expelled in response to a stimulus, forcing the cell in the opposite direction, presumably as a means to escape capture by herbivores (Kugrens et al., 1994). The common mode of reproduction is asexual (mitotic division), but studies have documented sexual reproduction and a complex life cycle that results in two sizes of cells, known as a *diplomorph* (2n) and *haplomorph* (n), which are also distinct with respect to periplast structure, pigmentation, and flagellar apparatus (Clay, 2015; Van den Hoff et al., 2020).

Cryptomonads are common and occasionally abundant members of the phytoplankton in lakes, although many studies list very few genera or species, likely owing to few diagnostic features easily recognized with the light microscope. Community-level studies indicate cryptomonads may be more abundant in N- and P-sufficient (i.e., nonlimiting) conditions, but their importance in freshwater food webs has not been sufficiently studied (Klaveness, 1989; Munawar et al., 2017). However, cryptomonads are a high-quality food source for herbivores with a high content of polyunsaturated fatty acids (Brett and Müller-Navarra, 1997; Taipale et al., 2011). When supplied with a mixed phytoplankton assemblage, *Ceriodaphnia quadrangula* preferentially feed on *Cryptomonas erosa*, a subdominant species, rather than on more abundant species of diatoms of similar size (Gladyshev et al., 1999). Populations of cryptomonads exhibit seasonal patterns in small eutrophic lakes (Reynolds, 1988). In Austrian mountain lakes *Rhodomonas minuta* had a winter-spring surface maximum but persisted through the year at greater depths, whereas *Cryptomonas phaseolus* developed a deepwater population during summer (Rott, 1988). Peaks in concentration often follow declines of other phytoplankton species, perhaps due to relaxed competition or an increase in dissolved organic matter (Stewart and Wetzel, 1986). Smaller species, such as *Cryptomonas pusilla* and *Plagioselmis lacustris*, may persist in lakes over longer periods (many months), whereas larger species, such as *Cryptomonas ovata*, *C. rostratiformis*, and *C. tetrapyrenoidosa* exhibit distinct seasonal periodicity (Klaveness, 1989). Cryptomonads are often broadly referred to as mixotrophs, although a few photosynthetic cryptomonads rely on bacterivory as an alternative energy source, or do so on a very limited basis (Sanders and Porter, 1988; Tranvik et al., 1989). A metalimnetic population of *Cryptomonas* that ingested bacteria may have done so to obtain N and P when dissolved nutrients were limiting rather than as an energy source (Urabe et al., 2000). Heterotrophic (nonpigmented) species can, however, be major consumers of bacteria (Grujcic et al., 2018). Reports of *Cryptomonas* species at or below the thermocline (e.g., Gasol et al., 1993) suggest several possible ecological strategies such as osmotrophic nutrition, sequestering phosphorus supplies, or avoidance of grazers, although one study found that grazing on *Cryptomonas* cf. *erosa* could continue at or near the chemocline of one meromictic lake (Camacho et al., 2001).

D. Euglenoids

Euglenoids are a diverse, primarily unicellular flagellated group of protists with photosynthetic (phylum Euglenophyta) and nonphotosynthetic (Euglenozoa or Discoba) members (Fig. 17-4c-f). Cells are elongate, spindle- or disk-shaped, or twisted, and covered by a *pellicle*, a proteinaceous layer arranged in parallel and/or spiral strips, which can be seen with the light microscope (Fig. 17-4c,e; Ciugulea and Triemer, 2010). Among photosynthetic species, chloroplasts have chlorophyll *a*, *b*, β-carotene, and several xanthophylls and produce *paramylon*, an unbranched β-1,3-linked glucan as a storage product (rather than starch; Table 17-1, Fig. 17-4c,e), located in the cytoplasm (Triemer and Zakryś, 2015). Euglenoids are active swimmers, using one or more flagella that emerge from an apical invagination. There is also an adjacent red *eyespot* that protects the photoreceptor, which is involved in signals affecting locomotion. Reproduction is asexual through longitudinal division from the anterior to the posterior end (Graham et al., 2016). *Strombomonas* and *Trachelomonas* form an outer protective structure known as a *lorica* (Fig. 17-4f), with a diversity of shapes (species specific) and can be colored (yellow, reddish, brown) through the accumulation of iron, manganese, and other materials. Although there are about 13 common genera of photosynthetic euglenoids, there are hundreds of recognized species, as well as an unknown number of cryptic taxa (Triemer and Zakryś, 2015).

Photosynthetic euglenoids occur in many types of inland waters but are especially common in shallow ponds or pools with high organic content and in eutrophic lakes. They are less abundant in turbulent rivers or streams (Ciugulea and Triemer, 2010). Various species utilize autotrophy, osmotrophy, and/or phagotrophy for their nutrition. The loricate species in *Trachelomonas* (Fig. 17-4f) are common in eutrophic lakes, especially in littoral zones, but can become outcompeted or possibly inhibited by blooms of toxic Cyanobacteria, such as *Planktothrix agardhii* (Fig. 17-2i) and *Microcystis aeruginosa* (Fig. 17-2d; Crossetti and Bicudo, 2008; Grabowska and Wołowski, 2014). Shallow ponds can occasionally produce spectacular blood-red blooms of *Euglena sanguinea* (Fig. 17-11), and some strains produce the alkaloid toxin euglenophycin, which can cause fish kills (Zimba et al., 2017; Janse van Vuuren and Levanets, 2020; Section 17IIIA). *Colacium* (Fig. 17-4d) differs from other euglenoids, living attached via gelatinous stalks to various substrata, including nonliving surfaces, aquatic plants, algal filaments, and the carapace of copepods and cladocerans (Hoagland et al., 1982; Al-Dhaheri and Willey, 1996).

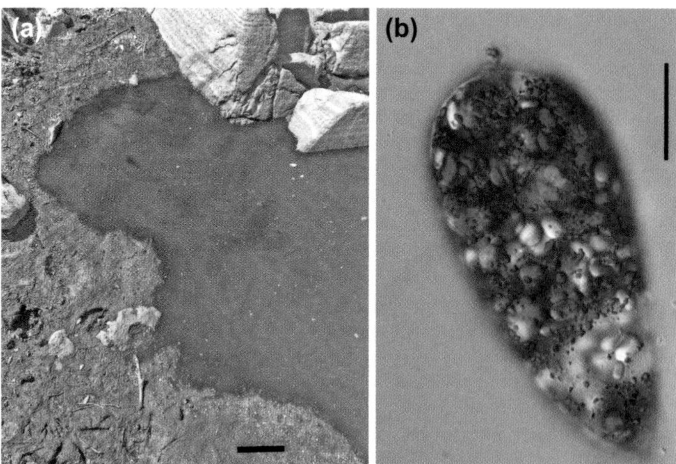

FIGURE 17-11 Bloom of *Euglena sanguinea* in a rainwater pond in Kruger National Park, South Africa. (a) Edge of pond with red surface scum and animal hoof prints. (b) Micrograph of *E. sanguinea*, showing dense accumulation of the red carotenoid pigment astaxanthin. *Scale bars*: (a) = 30 cm; (b) = 15 μm.

E. Yellow-green algae: eustigmatophytes, raphidophytes, xanthophytes

The collection of mostly yellow-green organisms, once regarded as a single group, is now recognized as three distinct, independent lineages, Eustigmatophyceae, Raphidophyceae, and Xanthophyceae, within the phylum Ochrophyta (Ott et al., 2015). Many of the taxa assigned to these groups have been rarely observed in nature or in culture, and their phylogenetic positions have not been established by molecular methods. Their life cycles are poorly known, except for a few common species such as *Tribonema* and *Vaucheria* (Fig. 17-4h-j). All have biflagellate *heterokont* stages (two different flagella) in their life cycle, chlorophyll *a* (not *b*), and several xanthophylls as photosynthetic pigments, giving cells a bright green or yellow-green color. Raphidophytes and xanthophytes also have chlorophyll *c*, but eustigmatophytes do not (Table 17-1). The many differences in the cellular organization of these diverse organisms are described in Ott et al., (2015).

Eustigmatophytes include several genera with very similar morphological features recorded in green algae (and may have been misidentified in past studies) with similar names such as *Characiopsis*, *Nannochloropsis*, and *Pseudostauratrum*, but careful observations reveal the characteristic and distinctive orange-red eyespot. For example, zooxanthellae are often reported in green sponges and attributed to species of *Chlorella*, a green alga (Wehr and Sheath, 2015), but some are populated by a eustigmatophyte, as observed in Little Rock Lake, Wisconsin, USA (Frost et al., 1997). Many eustigmatophyte species are unicellular or colonial forms and often occur in small ponds, shallow streams, bogs, and damp soils. *Characiopsis* is often reported from the littoral zones of lakes as an epiphyte on filamentous algae or among the plankton in small lakes, and *Chlorobotrys* and *Goniochloris* occur in lakes, as well as bogs and swamps (Prescott, 1951). *Nannochloropsis limnetica* is a picoplanktonic eustigmatophyte with rapid growth rates and has been reported from lakes in Europe and North America, as well as Lake Baikal, Siberia (Fietz et al., 2005; Fawley and Fawley, 2007).

Members of the Raphidophyceae are all flagellates with only a few freshwater species (in three genera). *Gonyostomum semen* (Fig. 17-4g) is a common and occasionally abundant phytoplankter forming nuisance blooms in humic and acidic lakes (pH 4.0–5.0) high in dissolved organic carbon (DOC). It has been labeled as an invasive species linked to climate warming in Finnish, Norwegian, and Swedish lakes (Hongve et al., 1988; Lepistö et al., 1994; Rengefors et al., 2012). The species' success may also be ascribed to its resistance to many grazers due to its large size (60–100 μm), although multiple factors such as warming, acidification, and browning of lakes (elevated DOC) have all been proposed (Hagman et al., 2020).

Morphological diversity within Xanthophyceae includes flagellated, coccoid, colonial, capsoid (cells embedded in mucilage), rhizoidal, filamentous (branched and unbranched), and siphonous forms (Ott et al., 2015). Many species are seldom reported, although several are common in freshwater plankton, growing attached to various substrata. Many of the smaller forms, such as *Botrydiopsis*, *Mischococcus*, *Ophiocytium*, and filamentous forms such as *Bumilleria*, are common metaphyton or epiphyton in lakes, pools, or boggy waters, associated with submersed vegetation or filamentous algae. *Tribonema* is a simple unbranched filament with golden or yellow-green chloroplasts (Fig. 17-4h), whose cell walls are formed from overlapping H-shaped pieces. Species are epilithic or metaphytic in cool lakes and streams mixed with other filamentous species or forming macroscopic masses in rapidly flowing streams. The most recognizable xanthophyte is *Vaucheria* (Fig. 17-4j), a macroscopic *coenocytic filament* (unbranched or sparsely branched) that forms conspicuous felt-like masses in springs, rivers, and lakes, as well as on damp soil (Fig. 17-4i). Ecological surveys indicate that most aquatic species have their optimal growth at lower temperatures, though current velocity and total dissolved ions strongly affect the presence and abundance of different species (Schagerl and Kerschbaumer, 2009). Some species, including *V. sessilis*, have a wide ecological range, whereas *V. geminata* (Fig. 17-4j) requires cool, unpolluted water (Ott et al., 2015). The use of *Vaucheria* species in bioassessments is, however, problematic because thalli are often sterile and cannot be identified at the species level (Stancheva et al., 2012).

F. Golden algae-1: chrysophytes, haptophytes, synurophytes

Members of the Ochrophyta, commonly referred to as chrysophytes or chrysomonads, comprise at least three (up to six) independent lineages but have the following features in common: vegetative or reproductive cells with heterokont flagella, chlorophylls a, c_1, c_2, c_3, the xanthophyll fucoxanthin as major photosynthetic pigments, and plastids appearing golden or golden-brown (Table 17-1; Fig. 17-5a-f). Food reserves include *chrysolaminarin* (β-1,3- and β-1,6-linked polysaccharide) and various lipids. Many are flagellates lacking a cell wall, both unicellular and colonial, some of which are contained within a lorica. There are also coccoid, palmelloid, filamentous forms, and a few have macroscopic thalli. The Chrysophyceae is the most diverse class in this group and have been collected on every continent, including Antarctica, occurring in a wide range of lakes, rivers, and streams (Nicholls and Wujek, 2015).

The Synurophyceae is a much smaller group with one unicellular (*Mallomonas* Fig. 17-5d) and three colonial (*Synura* [Fig. 17-5e,f], *Chrysodidymus*, *Tessellaria*) forms, all of which bear a variety of distinctively sculpted silica scales, with or without extensions called bristles (Siver, 2015). A great deal of interest has been placed on these organisms because of their broad ecological range among species and their use in paleolimnological studies based on their scales and cysts (Smol, 1995). Using molecular, ultrastructural, and pigment data, a few former chrysophyte taxa have been moved to the Phaeothamniophyceae (Bailey et al., 1998).

Most freshwater chrysophytes (in the broad sense) are photosynthetic, a few are unpigmented and heterotrophic, and perhaps many are mixotrophic. Reproduction is generally by simple mitotic division, but many species form silica-walled *stomatocysts* (or *statospores* in older literature) that serve as resting stages when a population declines and which remain in the sediments until specific ecological conditions occur and then serve to recover the population at a later time (Siver, 2019). The morphology and ornamentation of stomatocysts appear to be species specific (Duff et al., 1995; Wilkinson et al., 2001), and as such, they have been used in paleolimnological assessments (Chapter 30). Although many genera and species of freshwater chrysophytes are known, among those most commonly reported in freshwater plankton are *Chrysosphaerella*, *Dinobryon*, *Ochromonas*, *Uroglena*, and *Uroglenopsis*. Most lack external ornamentation, but a few, such as *Chrysosphaerella*, *Synura*, and *Mallomonas* (Fig. 17-5d), have siliceous scales and spines. Despite earlier generalizations, as a group, chrysophytes do not fall into a single trophic category and occur in oligotrophic to eutrophic lakes, with individual species having diverse temperature, nutrient, and pH optima (Nicholls and Wujek, 2015). Several species of *Dinobryon*, *Ochromonas*, *Spiniferomonas*, *Synura*, and *Uroglena* occasionally form surface blooms in lakes or ponds or form dense populations in the metalimnion, such as those of *Chrysosphaerella longispina* (Pick et al., 1984; Nicholls and Wujek, 2015). When abundant, chrysophyte species can cause taste and odor issues in drinking water supplies (Jüttner, 1981; Watson and Satchwill, 2003). Meanwhile, *Hydrurus foetidus* (Fig. 17-5c) is a gelatinous macroalga that occurs in cold mountain streams.

Haptophytes (phylum Haptophyta) are a distinct phylogenetic group with similar pigments and storage products as other golden algae but distinguished by having usually two equal flagella and a unique third hair-like appendage known as a *haptonema*. It has a simple structure and usually faces outward from the cell when swimming but does not participate in motility (Nicholls, 2015). The length of the haptonema may range from very short and stubby to very long (up to 12 times the length of the cell). The typically long haptonema functions in some species to capture prey or responds to other external stimuli. There are perhaps about a dozen species known from fresh or other inland waters, with *Chrysochromulina parva* (Fig. 17-5g) being the most common in large and small lakes, and large rivers. The species occurs worldwide and can become a major component of phytoplankton communities in a very broad range of ecological conditions, including eutrophic bays and rivers (Parke et al., 1962; Nicholls, 2015). In contrast, *C. breviturrita* is an inhabitant of moderately acidic (pH 5.0–6.0) softwater lakes and can form blooms associated with unpleasant odor episodes (Nicholls et al., 1982). *Prymnesium parvum* (Fig. 17-5h) is primarily a marine or brackish alga but has created blooms in inland saline and occasionally freshwater lakes and aquaculture facilities, where it produces the potent toxin prymensin, a cause of extensive fish kills in affected waters. Other members of the Haptophyta include the coccolithophorids, which have calcified or organic scales and are major contributors to the productivity of the world's oceans. There are just a few freshwater representatives, one of which is *Hymenomonas roseola*, which has been observed in scattered freshwater lakes, including Saginaw Bay of Lake Huron, USA (Stoermer and Sicko-Goad 1977).

G. Golden algae-2: diatoms

The diatoms (Bacillariophyta) are the most widely recognized and diverse group of protists on Earth. It is estimated that there are more than 7500 freshwater diatom taxa in North America alone (Kociolek et al., 2015a), and likely many more on other continents, especially in tropical locations. Their photosynthetic pigments, storage products, and motile (heterokont) reproductive cells are similar to that in other golden-brown groups within the Ochrophyta (Table 17-1). Diatoms are essentially unicellular organisms, sometimes arranged in colonies, chains, or filaments, whose cells are enclosed within two rigid silica (SiO_2) walls or *valves*, that together are called a *frustule*. Their surface ornamentation and symmetry (or lack thereof) of the valves are equally diverse, with complex patterns that are used for routine taxonomic identification. Diatoms occur in almost every type of inland water body and are major contributors to primary production in lakes, large rivers, and streams. They colonize virtually all ecological habitats in a water body, including free-floating within plankton, attached to various living and nonliving substrata (algae, plants, rocks), in and on sand or sediments, and attached to aquatic animals. Nearly all species are autotrophic, although many may utilize dissolved organic matter to supplement their carbon needs, especially those inhabiting sediments. Many reviews and major works have described the biology, diversity, and ecology of diatoms, a topic well beyond the scope of this brief summary but which can be consulted for detailed information (e.g., Round et al., 1990; Stevenson, 1996; Smol and Stoermer, 2010; Kociolek et al., 2015a, b; Letáková et al., 2018).

A general classification of diatoms usually considers two broad groups: those with radial symmetry, the commonly called "centric" forms (Coscinodiscophyceae; Fig. 17-6a-g), and those with bilateral symmetry, commonly called "pennate" forms (Fig. 17-6h-y), although these categories are not regarded as monophyletic (Kociolek et al., 2015a). Some of those with nonradial arrangements possess a *raphe* (Bacillariophyceae), which is a slit in one or both valves that allows cells to glide along surfaces (Fig. 17-6l-p). Some pennate forms lack a raphe and are not motile in the vegetative state (Fragilariophyceae; Fig. 17-6h-k). Raphe-bearing diatoms (e.g., *Frustulia*, *Navicula*, *Diploneis*; Fig. 17-6p-r) move along surfaces or within sediments by means of extruded mucilage composed of extracellular polymeric materials (Hoagland et al., 1993) and can glide at relatively rapid rates (1–7 $\mu m\ s^{-1}$; Lind et al., 1997) along a rock or plant surface. Other diatoms, such as *Encyonema* (Fig. 17-6u) and *Frustulia*, glide within a gelatinous tube secreted by multiple, separate cells. Diatoms colonize virtually all inland water habitats, with pennate taxa forming diverse assemblages attached to plants, rocks, and sediments (Stevenson et al., 1996). Diatoms are a high-quality primary food source for benthic herbivores in many rivers and lakes (e.g., Hardwick et al., 1992; Vadeboncoeur et al., 2002; Hill and Middleton, 2006) in part due to their high concentrations of essential fatty acids (Torres-Ruiz et al., 2007). Some species of diatoms can form dense mats in benthic habitats. One noteworthy species is the large diatom *Didymosphenia geminata*, which attaches to rocks in mostly oligotrophic rivers. Its massive production of mucilage creates nuisance problems and may hinder herbivore grazing (Whitton et al., 2009; Furey et al., 2014). The species significantly alters benthic communities in Chilean and New Zealand rivers, where it is regarded as invasive (Kilroy et al., 2009; Montecino et al., 2016), although fossil records indicate it existed worldwide in ancient times in various regions (Taylor and Bothwell, 2014).

H. Dinoflagellates

The dinoflagellates (kingdom Alveolata, phylum Miozoa, class Dinophyceae) are a monophyletic group of mostly unicellular protists that possess two uniquely arranged flagella (Fig. 17-7). The *transverse* and *longitudinal flagellum*, in combination, propel the cell through the water creating a whirling pattern (Carty and Parrow, 2015). They are also unique in having permanently condensed chromosomes in a nucleus known as a *dinokaryon*. Dinoflagellates have many photosynthetic and heterotrophic (phagotrophic and osmotrophic) members, and a few parasitic species. Autotrophic members use chlorophyll *a* and c_2 as primary photosynthetic pigments, as well as the carotenoid peridinin and several xanthophylls, and produce true starch, comparable to green algae and higher plants (Table 17-1). Many have a cell covering known as an *amphiesma*, which consists of cellulosic armor-like thecal plates (which may be thick, thin, or barely detectible with the light microscope), located within the plasma membrane. Athecate species lack walls.

Ceratium hirundinella (Fig. 17-7b) is a common, large species (200−400 μm long) whose horn-like extensions of the cellulose wall vary markedly over the wax and wane of a population, in response to differences in nutrient and light availability (Lindström, 1983; Van Ginkel et al., 2001). The species often occurs in middepth regions in mesotrophic and eutrophic lakes (Harris et al., 1979). Using its flagella, it can migrate downward within the water column in order to obtain nutrients and minimize photorespiration during periods of maximum sunlight and upward (as rapidly as 160−270 mm s^{-1}) toward the surface when sunlight is less intense (James et al., 1992; Whittington et al., 2000). Blooms of 10^3 to 10^4 cells mL^{-1} have been reported in temperate and tropical lakes, reservoirs, and larger rivers in North and South America, Europe, Africa, Australia, and Asia (Van Ginkel et al., 2001; Mac Donagh, 2005; Carty and Parrow, 2015). One massive bloom of *C. hirundinella*, with an associated fish kill, occurred in a lake previously managed to limit Cyanobacteria (Nicholls et al., 1980). Though less often reported, species of *Gymnodinium* (Fig. 17-7c), *Peridinium* (Fig. 17-7a), *Peridiniopsis*, and *Woloszynskia* also form blooms, including a remarkable instance of red-colored ice during winter, caused by *Borghiella* (formerly *Woloszynskia*) *pascheri* (Nicholls, 2017). Dinoflagellates persist over multiple years by producing resting cysts when populations decline and *excyst* (emerge from cyst) to initiate new increases when conditions match their environmental requirements (Mertens et al., 2012; Carty and Parrow, 2015). Facultative and obligate phagotrophy is common in many freshwater dinoflagellates, where they employ a feeding veil or *pallium*, which is an extension of cytoplasm used to capture bacteria, small algal cells, and other organic particles (Wilcox and Wedemayer, 1991; Calado and Moestrup, 1997).

I. Brown algae

The brown algae (Phaeophyceae) are almost exclusively a marine class within the Ochrophyta and are the least diverse group of algae in fresh waters. Only about seven species are currently recognized from lakes and rivers, and all are multicellular and benthic in habit (Wehr, 2015). Like other golden-brown groups, their main photosynthetic pigments are chlorophylls *a*, c_1, c_3, and the xanthophyll fucoxanthin (among others), but they produce laminarin as their main photosynthetic storage product (Table 17-1) and have cell walls composed of a 3:1:1 combination of alginates, sulfated fucans, and cellulose (Wehr, 2016). Some produce phlorotannins, polyphenolic compounds that may serve as an inducible chemical defense against herbivory (Lüder and Clayton, 2004; Wehr, 2016). Until recently, phaeophytes were regarded as rare or unimportant in freshwater systems, but regional and continent-wide studies indicate that at least one species, *Heribaudiella fluviatilis* (Fig. 17-8d-f), is a relatively common and widespread component of the benthic macroalgal flora of streams and rivers, and likely was overlooked in the past (Holmes and Whitton, 1977; Wehr and Stein, 1985; Wehr, 2015). Its evolutionary history in fresh waters is apparently very old. Several other species, including *Ectocarpus subulatus*, *P. lacustris* (Fig. 17-8a-c), and *Porterinema fluviatile*, occur in both fresh and brackish-water environments and apparently tolerate a broad range of total ions (Waern, 1952; Dop, 1979; West and Kraft, 1996; Wehr, 2016). These may be more recent introductions from marine environments.

J. Green algae and charophytes

The green algae represent a large, diverse, and ubiquitous group of organisms (Fig. 17-9), now recognized to be at least two major lineages, the Chlorophyta and the Streptophyta (Charophyta), which include at least 16 classes or orders of green algae (Leliaert et al., 2012). There is a huge diversity in morphology, including flagellates, coccoid cells, colonies, filaments, gelatinous masses, blade-like sheets, tubes, coenocytes, and complex multicellular forms.

Green algae range in size from tiny picoplanktonic unicells (<3 μm), such as *Mychonastes homosphaera* (Hanagata et al., 1999) and clusters of simple spherical cells such as *Asterococcus limneticus* (Fig. 17-9a), to large thalli (>50 cm in length) that form vast underwater meadows, such as *Chara foliolosa* (Fig. 17-9o). The biomass of *Chara hispida* may exceed 500 g dry mass m^{-2} and profoundly affect nutrient dynamics in shallow lakes (Andrews et al., 1984; Rodrigo et al., 2007). All green algae contain chlorophyll *a* and *b*, supplemented with β-carotene and xanthophylls, and produce starch as the major photosynthetic reserve (Table 17-1). Their plastids occur in many shapes and the number per cell may vary. Various types of sexual reproduction occur and in some cases are unique to a certain group. The majority have cell walls composed of cellulose and, if present, there are two equal flagella (Graham et al., 2016). Some representatives may, however, have more than two flagella (e.g., four in *Carteria*). Prasinophytes may also have a covering of minute, nonmineralized scales, composed primarily of polysaccharides and glycoproteins (Becker et al., 1994).

Green algae occur in nearly all types of inland waters and occupy planktonic, epiphytic, epipelic, metaphytic, and epilithic habitats. Others live in terrestrial environments and form part of symbiotic associations. A tremendous diversity of green algae may occur in a single body of water. This is especially true for desmids in oligotrophic or slightly acidic lakes and ponds, sometimes with dozens to more than a hundred species recognized from a single location (Oesel et al., 1978; Negro et al., 2003; Hall and Karol, 2016). The majority of green algal species are autotrophs, although a number of species are facultative heterotrophs (capable of growth in the dark), such as *Chlamydomonas reinhardtii*, *Haematococcus pluvialis*, and *Tetradesmus* (*Scenedesmus*) *obliquus*, which can grow on acetate as a sole carbon source (Abeliovich and Weisman, 1978; Hata et al., 2001; Boyle et al., 2017). Many species are motile or have motile reproductive stages, whereas many others are attached to various substrata. The globally distributed filamentous alga *Cladophora glomerata* (Fig. 17-9m) has been dubbed an "ecosystem engineer" for its effects on modifying the local environment by stabilizing sediments, altering water movements, and creating a habitat for epiphytic microalgae, bacteria, and associated consumers (Zulkifly et al., 2013). Few species of green algae are reported as components of phytoplankton blooms, although the colonial alga *Botryococcus braunii* (Fig. 17-9h), which forms golden or rusty orange surface blooms and whose cells produce large quantities of oils (and thereby gain buoyancy), may flourish in eutrophic waters (Aaronson et al., 1983; Janse van Vuuren and Levanets, 2019). Benthic environments can experience blooms of filamentous green algae. For example, large clouds of filamentous green algae (*Mougeotia*, *Spirogyra*, *Zygnema* spp.; Fig. 17-9p-r) form in softwater lakes undergoing early stages (pH ~5) of acidification (Turner et al., 1991), accompanying a diversity of desmid species (Fig. 17-9s-v). Dense mats of *Ulothrix zonata* and other filamentous greens have formed in nearshore areas displacing other species in Lake Baikal (Russia), apparently due to increasing eutrophication.

Charophytes (commonly known as "stoneworts") are members of the order Charales (phylum Charophyta, class Charophyceae). They are complex macroscopic algae that range in size from about 10 to 100 cm (Fig. 17-10b,c) and can produce extensive underwater meadows in the littoral zone of lakes. They possess a *corticated* central axis (central filament surrounded by a layer of covering filaments) with elongate *internodes*, along which whorls of branches form that may also be corticated (e.g., *Chara*) or uncorticated (e.g., *Nitella*), and may subdivide dichotomously several times. Species in this group have a complex, advanced *oogamous* mode of reproduction (sexual reproduction based on a large nonflagellated female gamete or oogonium, fertilized by a smaller flagellated male gamete or spermatium). Charophytes are attached to sediments in shallow lakes or slower-flowing rivers by means of rhizoids. Species that occur in hardwater systems typically become calcified.

III. Algal habitats in inland waters

Algae colonize an enormous diversity of habitats in inland waters, from small puddles, ponds, and pools to huge lakes, as well as springs, streams, and large rivers. Some inland habitats are extreme environments, which are also colonized by very specific groups of algae.

A. Ponds and lakes

The vast majority of studies on freshwater algae have been conducted on small- to moderate-sized lakes. Larger, more permanent water bodies typically may have a greater proportion of planktonic forms, although a substantial and highly diverse assemblage of periphyton often develops in shallower littoral zones of many lakes. Phytoplankton in the pelagic (limnetic) zones of lakes have adaptations to remain suspended in the water column, including flagella (e.g., *Ceratium*, *Synura*), spines or appendages that reduce the sinking rate (e.g., *Pediastrum*, *Staurastrum*), or oil droplets within their cells to gain buoyancy (common in many diatoms). Another notable adaptation is the presence of clusters of gas vesicles, or aerotopes, found in Cyanobacteria (Section 17IIA).

Algal species composition and diversity vary with lake trophic state, generally with lower species richness in more productive systems (Rawson, 1956). Characteristic algal assemblages over a trophic gradient of temperate lakes were summarized by Reynolds, (1980, 1984), Rott, (1984), and Willén, (1992, 2000). Algae characteristic of oligotrophic lakes include many diatoms (e.g., *Eunotia*, *Rhizosolenia*, *Tabellaria*) and chrysophytes (*Dinobryon*, *Mallomonas*, *Uroglena*), as well as a large number and variety of desmids (*Closterium*, *Desmidium*, *Spondylosium*, *Staurastrum*, *Staurodesmus*; Rawson, 1956; Reynolds, 1980; Willén, 2000). Although some species of *Cryptomonas* are common in oligotrophic lakes (where they often switch to bacterivory to obtain sufficient nutrients; Pålsson and Granéli, 2003), other species, such as *Cryptomonas reflexa* in Hamilton Harbor, Lake Ontario, Canada/USA (Munawar et al., 2017), may form blooms in eutrophic waters. The typical phytoplankton of eutrophic lakes include Cyanobacteria such as *Anabaena flos-aquae*, *Microcystis aeruginosa*, *Dolichospermum* spp., and *Planktothrix agardhii* (Fig. 17-2n,d,m,i), as well as many species of green algae. Several important species of planktonic diatoms are widely reported from more productive lakes worldwide, including *Aulacoseira granulata*, *Fragilaria crotonensis*, *Asterionella formosa* (Fig. 17-6g,j,k), and *Stephanodiscus* spp. (Fig. 17-6c; Rawson, 1956; Reynolds, 1980; Willén, 2000).

A rich algal flora occurs in the littoral zone and other benthic habitats of larger lakes, where they live between or on rooted vegetation. Many are mixtures of filamentous green algae, together with a variety of planktonic, tychoplanktonic, benthic, symbiotic, and even a few parasitic species (Saber et al., 2022). The littoral zones of softwater, slightly acidic lakes are often dominated by floating masses of filamentous algae in the family Zygnemataceae, such as *Zygogonium*, *Mougeotia*, *Spirogyra*, and *Zygnema* (Fig. 17-9p-r; Turner et al., 1991), whereas open-water habitats may have a variety of chrysophyte (e.g., *Chrysococcus*, *Dinobryon*; Wehr et al., 2015) and desmid (*Cosmarium*, *Netrium*, *Hyalotheca*, and *Desmidium*; Hall and Karol, 2016) species within the plankton (Wehr et al., 2015). Desmids may become particularly diverse in *dystrophic* lakes and ponds, waters with high amounts of humic substances (Chapter 28). Hardwater lakes, rich in calcium, have very different and characteristic benthic macroalgal assemblages, including lush *Chara* meadows (Rodrigo et al., 2007), as seen in Plitvice Lakes National Park in Croatia (Fig. 17-10a). The adnate green alga *Gongrosira* may cooccur with *Chara*, the latter of which becomes encrusted with $CaCO_3$ (Fig. 17-10b,c). *Hydrodictyon* (Fig. 17-9l) is also typical of hardwater conditions, occasionally blanketing the surface of ponds and small lakes.

Far less effort has been focused on small, ephemeral water bodies, such as temporary ponds, rainwater puddles, or roadside ditches. Studies indicate that unicellular and colonial planktonic forms are common pioneers in such systems (Hickman, 1974). These include unicellular and colonial green algae (e.g., *Carteria*, *Chlamydomonas*, *Chlorella*, *Desmodesmus*, *Oocystis*), a variety of diatoms (e.g., *Navicula*, *Nitzschia*, *Stauroneis*), euglenophytes (*Euglena*, *Trachelomonas*), and cryptophytes (particularly *Cryptomonas*), which may exhibit a seasonal or brief temporal wax and wane. If water persists for a few weeks, filamentous green algae (e.g., *Microspora*, *Mougeotia*, *Oedogonium*) may form dense benthic expanses, which later detach and float to the surface (Round, 1971; Cantonati, 2005). In ephemeral ponds algal populations can rapidly reach substantial densities during the brief growing season and have adaptations, such as cysts, akinetes, or zygospores, to withstand exposure and desiccation when pools dry up (Wehr et al., 2015). Euglenoids can become especially abundant in small water bodies, including farm ponds and shallow lakes with an abundance of decaying organic material (Bellinger and Sigee, 2015) or in organically enriched water bodies, such as sewage ponds (Palmer, 1969). Bright red blooms by some *Euglena* species, including *E. sanguinea* (Fig. 17-11), a potentially toxic species, can occur in farm ponds, waterholes on golf courses, and sewage lagoons (Oliveira and Calheiros, 2000). The red discoloration of the water is the result of a mixture of carotenoid pigments, with astaxanthin most abundant (Rosowski, 2003). *E. sanguinea* has recently formed extensive blooms in small ponds from which animals drink in the Kruger National Park, South Africa (Fig. 17-11; Janse van Vuuren and Levanets, 2020).

B. Streams and rivers

The characteristics of stream and river ecosystems are extremely diverse. Algal species occupying these systems largely vary with discharge and current velocity. In addition to the influence of current on the stability of planktonic and benthic habitats, turbulent conditions may also affect the underwater light availability. Benthic substrata range from silt and sand in slow-flowing systems to cobbles and boulders in more rapidly flowing environments and these differences influence the composition of algal assemblages.

An enormous diversity of epilithic and epiphytic diatoms occur in stony streams worldwide (Stevenson, 1996), including cosmopolitan species such as *Cocconeis placentula*, *Achnanthidium minutissimum*, and *Hannaea arcus* (Biggs, 1996; Riseng et al., 2004). Among macroalgal species, *Cladophora glomerata* (Fig. 17-9m) is likely the most common and widespread species in streams and rivers worldwide, including countless rivers across Europe and North America (Dodds and Gudder, 1992; Zulkifly et al., 2013), Egypt (Nile River) (Saber et al., 2022), Serbia (Simić et al., 2002), Australia (Entwisle, 1989), and New Zealand (Biggs and Price, 1987). Macroalgal species of Cyanobacteria, green, red, and brown algae may also dominate benthic assemblages in streams (Necchi, 2016). Rocky substrata in the upper reaches of fast-flowing streams serve as habitat for Cyanobacteria capable of adhering to surfaces, such as *Chamaesiphon* (Fig. 17-2e), *Homeothrix*, and *Phormidium* (Casamatta and Hašler, 2016; Rott and Wehr, 2016). *Phormidium retzii* (Fig. 17-2j) is widely reported in flowing water, where it can occur in small patches on rocks or form large macroscopic mats (Sheath and Cole, 1992; Casamatta and Hašler, 2016). Algal assemblages also develop on other substrata such as submerged wood, roots, and aquatic plants.

In larger rivers a variety of planktonic forms, the *potamoplankton*, develop in deeper sections, where the flow rate may be substantial, including the rivers Congo, Central Africa (Descy et al., 2017); Thames, UK (Bowes et al., 2012); Murray, Australia (Baker et al., 2000); Meuse, Western Europe (Gosselain et al., 1994); Mississippi, USA (Baker and Baker, 1981); and Ohio, USA (Wehr and Thorp, 1997). Potamoplankton assemblages are composed primarily of smaller forms with rapid growth rates, due to the negative effects of downstream transport (advection), turbidity, and mixing in and out of the photic zone (Reynolds, 1994; Reynolds and Descy, 1996). These typically include small centric diatoms such as *Cyclotella* spp., *Stephanodiscus hantzschii*, and *S. minutulus*, along with various colonial green algae, such as *Ankistrodesmus falcatus*, *Desmodesmus* spp., and *Selenastrum* spp. Although flagellates are less common in large rivers, representatives of green algae, golden algae, cryptomonads, dinoflagellates, and euglenophytes can be found. In slower-flowing large rivers and their backwaters, where downstream losses are less, blooms of planktonic Cyanobacteria can develop (Bowes et al., 2012; Giblin and Gerrish, 2020). Some Cyanobacteria, notably *Merismopedia* and *Pseudanabaena*, rest on the sediments of slow-flowing rivers and migrate into the water column to form planktonic populations (Casamatta and Hašler, 2016). Models suggest that climate change (global warming) may exacerbate bloom formation in the River Thames (UK; Bussi et al., 2016). The filamentous diatom *Melosira varians* (Fig. 17-6d-f), as well as unicellular diatom species that attach to substrates by means of mucilage stalks, for example, *Gomphoneis herculeana* or *Cymbella mexicana* (Fig. 17-6t), may predominate in slow-flowing rivers (Biggs, 1996; Stevenson, 1996). The huge diversity of diatom species and their wide array of water quality, pollution tolerance, and habitat requirements have made these organisms the primary group of algae used in bioassessments of water quality (Round, 1993; Stevenson and Smol, 2015). Benthic forms, such as the macroalgae *Ulva* (*Enteromorpha*) *flexuosa* (Fig. 17-9n), and species of *Chara*, are common in littoral zones and shallower reaches of larger rivers (Rybak and Gąbka, 2018) and may harbor many epiphytes (Azam et al., 2016).

C. Springs

Due to their isolation and hydrogeology, spring habitats often contain endemic species, particularly in arid and semiarid regions (Fensham et al., 2010). Although some springs harbor high biological diversity, anthropogenic stressors, such as agricultural nutrient inputs, climate change, groundwater depletion, and site-specific habitat destruction, have led to the extinction of endangered and rare spring-dependent species. These impacts have led to an urgent plea for the global protection of springs (Cantonati et al., 2020). An extensive summary of the most common and characteristic Cyanobacteria and algae in different major spring types is presented in Cantonati et al., (2012) and Saber et al., 2022. These studies indicate that the algal flora varies greatly according to spring type, geology, and various hydrological properties. Although many springs are sources of pristine, oligotrophic, fresh water, some (e.g., hot, geothermal springs, mineral springs, and acid springs) are inhospitable and can be colonized only by extremophilic algal species.

Diatoms are the most frequently encountered and most widely studied algal group in springs (Cantonati and Pipp, 2000; Żelazna-Wieczorek, 2011; Cantonati et al., 2012; Abdelahad et al., 2015; Taxböck et al., 2020). The majority of studies on spring diatoms have been conducted in mountainous regions of Europe. Several report extraordinary species diversity, especially among springs within a local region. For example, Taxböck et al., (2020) identified almost 540 diatom species from 71 springs in Switzerland. Studies also report endemic species or diatoms new to science, including the genus *Microfissurata* from seepages (*helocrenes*) and pool-springs (*limnocrenes*) (Cantonati et al., 2009), *Navicula fontana* from a limnocrenic spring in Poland (Żelazna-Wieczorek, 2020), *Cymbopleura margalefii* from alkaline and eutrophic springs in Majorca, Spain (Delgado et al., 2013), and *Geissleria gercekei* in shaded carbonate springs in the Italian Alps (Cantonati and Lange-Bertalot, 2009). Collected data suggest that pH, conductivity, inorganic N, substrate particle size, and shade are key variables affecting the composition of diatom assemblages in spring habitats (Cantonati et al., 2006).

Studies on Cyanobacteria suggest that, though they are less species rich than the diatom flora, they often form the dominant photoautotrophs in many springs (Cantonati et al., 2015). Their success in typically oligotrophic waters of pristine springs may be ascribed to their ability to fix atmospheric nitrogen, as well as the production of phosphatases and anti-UV compounds such as scytonemin in exposed habitats (Castenholz and Garcia-Pichel, 2012; Cantonati et al., 2015). Efforts using molecular methods have discovered hitherto unknown biodiversity in Florida springs, with 50 cyanobacterial taxa from Ichetucknee Springs alone, including many that do not currently fit any circumscribed species descriptions and are likely new to science (Perkerson et al., 2011). Knowledge of spring-dwelling or spring-specialist taxa among other algal groups is less complete. For example, Vis and Sheath, (1996) described a new species of the red alga *Batrachospermum* (*B. carpoinvolucrum*) from the outflow of the ancient Montezuma Well spring system in Arizona (USA). This spring habitat harbors four endemic species, two of which were infrageneric diatom taxa (Czarnecki and Blinn, 1979). Surveys of 23 freshwater springs in Northern Italy (Abdelahad et al., 2015) revealed a fairly high diversity of red algae, including *Batrachospermum atrum*, *B. gelatinosum*, *B. gelatinosum* f. *spermatoinvolucrum*,

Sheathia arcuata, and *S. boryana*. Collectively, these studies indicate that early views, which suggested that algal flora of springs are composed mainly of common or cosmopolitan species (Whitford, 1956; Whitford and Schumacher, 1963), may not be true; rather, they appear to be hotspots of biological diversity for many phyla of freshwater algae.

D. Extreme environments

Many studies have examined adaptation and evolution in algal species in response to a wide range of extreme physical and chemical conditions (Elster, 1999; Bell, 2012). Much of the interest among limnologists has centered on thermal environments (hot springs), cold environments, inland saline waters, and extremes of pH.

i. Thermal waters and hot springs

Cyanobacteria and other algae living in thermal springs have been studied extensively (Singh et al., 2018), occur all over the Earth, and are particularly abundant in volcanically active regions of Iceland, Italy, Japan, New Zealand, and western parts of North America (Emoto and Yoneda, 1942; Brock and Brock, 1971; Krienitz et al., 2003; Owen et al., 2008; Dadheech et al., 2013). Although thermal springs characteristically have water temperatures at or above 37°C (Pentecost et al., 2003), many emit water near the boiling point (90°C or higher; Stockner, 1967; Seckbach, 2007). These are typified by extensive, often thick microbial mats, which include Cyanobacteria (Pepe-Ranney et al., 2012; Ward et al., 2012; Kaštovský et al., 2014), some of which can tolerate water temperatures close to 75°C (Seckbach, 2007; Trivedi et al., 2010; Ward et al., 2012). Eukaryotic algae occur in slightly cooler waters at an upper temperature of ~55°C (Brock, 1967). Cyanobacteria in some thermal habitats produce secondary metabolites, including a variety of hepato- and neurotoxins (Tevena et al., 2005), such as those linked to the deaths of Lesser Flamingos adjacent to Lake Bogoria, Kenya (Krienitz et al., 2003).

Many thermophilic Cyanobacteria have restricted geographic distributions, likely due to their specific environmental requirements and tolerances (Ward et al., 1998; Miller and Castenholz, 2000), and exhibit a degree of endemism (Castenholz, 1996; Dadheech et al., 2013). Of the different Cyanobacteria dominating in geothermal waters, *Synechococcus* (Fig. 17-2a) is probably the most well-known and widely occurring genus (Seckbach, 2007), with the species *S. elongatus*, *S. lividus*, and *S. minuscula* commonly reported (Stal, 1995; McGregor and Rasmussen, 2008). Other common thermophilic Cyanobacteria include *Fischerella laminosa* (syn. *Mastigocladus laminosus*), *F. thermalis*, *Chroococcus thermalis*, *Johannesbaptistia schizodichotoma*, *Schizothrix calcicola*, and several species of *Cyanothece*, *Geitlerinema*, and *Phormidium* (Stockner, 1967; Sambamurty, 2005; McGregor and Rasmussen, 2008; Ward et al., 2012).

Diatoms can proliferate in hot springs with less extreme temperatures (35–50°C; pH 7.8–8.0; Pringle et al., 1993) and have been reported from thermal waters in Sardinia (Lai et al., 2019), northern Thailand (Pumas et al., 2018), Iceland, New Zealand (Hindák, 2001; Owen et al., 2008; Krienitz et al., 2012), and Russia (Nikulina and Kociolek, 2011), as well as several locations in Africa, including Burundi (Mpawenayo et al., 2005), Zambia (Compère and Delmotte, 1988), Kenya (Owen et al., 2008; Krienitz et al., 2012), Namibia (Schoeman and Archibald, 1988), and South Africa (Jonker et al., 2013). These studies revealed a huge diversity of species, some of which are novel, with marked differences among locations. For example, two new genera (*Naviculonema*, *Williamsella*) and six new species of diatoms were described from Blue Lake Warm Springs in Utah, USA (Graeff et al., 2013), and four new species were described from La Calera hot spring in Colca Canyon, Peru, including *Ulnaria colcae* and *Denticula thermaloides* (Van de Vijver and Cocquyt, 2009). Cosmopolitan taxa, such as *Anomoeoneis sphaerophora*, *Craticula cuspidata*, *Hantzschia amphioxys*, *Nitzschia palea*, *Rhoicosphenia abbreviata*, *Planothidium lanceolatum*, and *Synedra ulna*, are also reported.

Green algae cooccur with diatoms in the cooler range of thermal springs, although far less information exists on these associations. Fott, (1959) and Gruia, (1976) state a temperature limit of 50.5°C for the development of green algae. Commonly reported are members of the Charales, Chaetophorales, and Zygnematales (Stoyneva, 2003; Pentecost, 2005). Stoyneva, (2003) reported nearly 200 green algal species in thermal springs in Bulgaria. Thermal springs in South Africa, with temperatures ranging between 40°C and 67.5°C, also support a relatively diverse flora primarily of Cyanobacteria, as well as species of *Oocystis*, *Coelastrum*, *Chlorella*, and *Spirogyra*, in temperatures exceeding 60°C (Jonker et al., 2013).

Some thermal springs are highly acidic, and their species diversity appears more limited (Wehr and Sheath, 2015) and will be discussed later. The cyanobacterium *Geitlerinema sulphureum* is the dominant species in mineral hot springs with temperatures of ~70°C in Egypt (Hamed, 2008) and in the Himalayas (Singh et al., 2018).

ii. Cold environments

Algae also occupy extremely or continuously cold conditions, such as polar lakes in the Arctic and Antarctica, and permafrost regions of Alaska (USA), Russia, and Canada. *Cryophiles* that survive in or underneath ice have temperature optima for growth of ~1°C but range from −20°C to 10°C (Hoham, 1975). Microbes, many of which are Cyanobacteria, persist in ice and snow near the South Pole (Brambilla et al., 2001), 3.5 km deep in Vostok ice, Antarctica (Priscu et al., 1999a), meltwater ponds (Vincent, 1988), in the water column of permanently ice-covered lakes (Priscu et al., 1999b), and within permanent lake ice (Fritsen and Priscu, 1998; Priscu et al., 1998). Although there has been much focus on *snow algae* (fascinating algal species that color permanent or semipermanent snow fields and glaciers in alpine regions in various colors), these are not a major focus of limnological research. A comprehensive review of the ecology of snow algae can be found in Hoham and Duval, (2001).

iii. Saline environments

Algae colonize many saline or hypersaline inland systems, with salt concentrations that may exceed four or five times that of sea water. They comprise coastal lagoons, salt and soda lakes, lakes in desert regions, and salterns used for salt production. Many are well-known water bodies including the Dead Sea (Jordan/Israel), Great Salt Lake (USA), alkaline salt marshes and lakes in Wadi El Natrun (Egypt), Hutt Lagoon (Australia), and the inland saltern of La Mala (Spain) (Ali et al., 2016).

Halophilic species tolerate osmotic extremes through several mechanisms. Some avoid osmotic shock by increasing ion concentrations in their cytoplasm (Ortega et al., 2011), employ proteins modified to prevent denaturation, or use special UV-protecting pigments for protection from extreme sunlight (e.g., carotenoids), discoloring waters in various shades of red (Rich and Maier, 2015). Several authors (Fritz et al., 1993; Wilson et al., 1996; Balakrishnan et al., 2019) observed that algal species diversity in ponds and lakes decreases sharply with greater salinity. Mildly saline (see Chapter 12) lakes (TDS 500−2000 mg L^{-1}) are fairly rich in species and may contain several representatives of Cyanobacteria, diatoms, and green algae. As the salt concentration increases toward hypersalinity (>20−600 g L^{-1}), a progressive reduction in species number occurs and only those species with unique adaptations can survive.

Cyanobacteria adapted to high salinity levels often represent the main autotrophs in hypersaline environments (Komárek, 2003). *Aphanothece halophytica* is a widespread and well-known halophilic cyanobacterium (Mandal and Rath, 2015). Common colonial and filamentous Cyanobacteria found in mildly saline lakes include cosmopolitan species of *Aphanizomenon*, *Lyngbya*, *Microcystis*, and *Oscillatoria* (Wehr et al., 2015), whereas those in hypersaline waters are high-salinity specialists such as *Coleofasciculus chthonoplastes*, *Johannesbaptistia pellucida*, *Nodularia spumigena*, and species of *Arthrospira* (Fig. 17-2l) and *Leptolyngbya* (Komárek, 2003). Alkaline soda lakes (pH 9−12) also contain high sodium chloride content, together with other dissolved salts, and *Arthrospira fusiformis* (formerly *Spirulina platensis*) is perhaps the most common cyanobacterium. This species is adapted to a wide range of habitats, ranging from freshwater-alkaline to saline-alkaline and also hypersaline environments (Dadheech et al., 2010).

Among eukaryotes, the green alga *Dunaliella*, a unicellular biflagellate, is the most ubiquitous and best known for its high salinity tolerance and red color (due to carotenoid pigments). Blooms of this species in the Great Salt Lake can be seen from space (Javor, 1989). The salt optimum of *Dunaliella salina* and *D. terticola* is ~120 g L^{-1} (Rahman et al., 2020). *D. parva* is a major inhabitant of the Dead Sea, where very few organisms exist (Oren et al., 2008). A study of 41 saline lakes in Canada (Hammer et al., 1983) showed that green algae dominate lakes in which salinity exceeded 100 g L^{-1}, and filamentous species, such as *Ctenocladus circinatus*, were widely reported in western North America (Blinn and Stein, 1970; Hammer et al., 1983). A recently described green picoplanktonic species, *Picocystis salinarum*, was identified from a saline pond in California where it was associated with various *Dunaliella* species (Lewin et al., 2001). This species was later shown to be a successful competitor in East African Flamingo Lakes (Pálmai et al., 2020), where it replaced the cyanobacterium *A. fusiformis* as the dominant species (Krienitz et al., 2012). Rahman et al., (2020) indicated that *P. salinarum* is able to grow in a salinity range of 15−300 g L^{-1}. Large, filamentous green algal species, such as *Ulva* (*Enteromorpha*) *intestinalis*, traditionally found in marine environments, can also occur in inland saline waterbodies (Wehr et al., 2015).

Diatoms are also common in saline and hypersaline lakes. Species of *Amphora*, *Campylodiscus*, *Cyclotella*, *Fragilaria*, and *Rhopalodia* are common in mildly saline conditions, whereas hypersaline conditions usually sustain diatom species such as *Anomoeoneis sphaerophora*, *Navicula subinflatoides*, and *Nitzschia frustulum* (Wehr et al., 2015). *Cymbella pusilla* can be present in a wide range of salinities, ranging from 3 to 200 g L^{-1} (Rahman et al., 2020). Marine diatoms, such as *Chaetoceros muelleri* and *Thalassiosira pseudonana*, have been recorded from several saline inland waters (Baek et al., 2011).

Other algal groups have few representatives in hypersaline environments, with two notable exceptions. The toxic haptophyte *Prymnesium parvum* (Fig. 17-5h) causes fish kills in saline ponds and lakes and is a major and growing nuisance in the south-central United States (Roelke et al., 2011). The species has a global distribution and has been reported from inland saline waters in China, Europe, Australia, Morocco, Israel, and the United States (Brooks et al., 2011). This is particularly a concern in warmer regions of the world, with greater rates of evaporation and water extraction. Climate change will lead to increased salinity, making more lakes susceptible to invasion by this species (Roelke et al., 2011). A second, less troublesome example is the coccolithophorid, *Pleurochrysis carterae*, reported by Johansen et al., (1988) from an inland saline pool, New Mexico, USA.

iv. Acidic and alkaline environments

A large number of algal species are able to persist in water with a pH ~4.0, but a smaller number of species can tolerate extremely acidic environments with a pH less than 3.0 (Gimmler, 2001), and very few can tolerate a pH less than 2.0 (Gross, 2000). These acidophiles may also be resistant to a multitude of other environmental factors, including heavy metals, high temperatures, or oxidative stress. They occur in volcanic areas, hydrothermal sources, effluents from metal mining activities, acidic hot springs, or acid mine drainages (Gómez, 2011).

Studies on algae inhabiting highly acidic (nonthermal) environments indicated that they generally have similar species diversity and species composition (Lackey, 1938; Hargreaves et al., 1975). The most widespread and abundant algae in acidic and metal-polluted environments are *Euglena mutabilis* and *E. gracilis* (Hargreaves et al., 1975; Sheath et al., 1982; Pringle et al., 1993). With an optimum pH of 3.0 and tolerance of pH below 1.0, *E. mutabilis* is considered an indicator species of acid mine drainage (Lackey, 1968; Hargreaves et al., 1975). *Chlamydomonas acidophila* is also common in acidic waters with pH values of 2.0–3.2, where inorganic phosphorus may be limiting (Nishikawa and Tominaga, 2001; Gerloff-Elias et al., 2005; Spijkerman, 2005, 2007). A few other green algae in acidic environments include *Dunaliella acidophila* (Ñancucheo and Johnson, 2012) and the filamentous green algae *Klebsormidium rivulare* (Wehr and Sheath, 2015), *K. acidophilum*, and *Zygnema circumcarinatum* (Novis and Harding, 2007; Rowe et al., 2007). A few chrysophytes, such as *Ochromonas*, may also occur in acidic environments (Ñancucheo and Johnson, 2012). The enigmatic red alga *Cyanidium caldarium* is often the sole photosynthetic organism in acid (pH 2–4) hot springs up to about 55°C (Wehr and Sheath, 2015). *Cyanidioschyzon merolae*, *Galdieria sulphuraria*, and *G. maxima* are also frequently encountered in acid hot springs (Toplin et al., 2008) with pH 0.5–3.0 and upper temperatures of 50–56°C. They are found in widespread locations, such as Yellowstone National Park (Wyoming, USA; Doemel and Brock, 1971), Pisciarelli Fields (Italy; De Luca et al., 1981), and fumaroles in Mount Lawu (Java; Ciniglia et al., 2004). Cyanobacteria, on the other hand, are virtually absent from highly acidic waters, although cyanobacterial *16S rRNA* genes have been recovered from acid mine drainage sites (pH 2.8) in the Anhui Province, China (Hao et al., 2012), and from Los Rueldos mine in north-west Spain (pH 2.0 ± 0.95; Mesa et al., 2017).

Extremely alkaline habitats include naturally occurring soda lakes, marl lakes, and water polluted with soda lime, where pH is typically between 9 and 12 (Kempe and Kazmierczak, 2011). Their water chemistry is dominated by carbonate salts, including sodium carbonate and related salt complexes, and they may also be saline. Cyanobacteria are frequently abundant in highly alkaline systems. *Arthrospira fusiformis* often dominates in these systems, where it serves as the main food for Lesser Flamingos (Krienitz et al., 2013). Other Cyanobacteria in alkaline soda lakes in East Africa include *Anabaenopsis abijatae*, *A. elenkinii*, *Cyanospira capsulata*, *Haloleptolyngbya alcalis*, and *Picocystis salinarium* (Krienitz et al., 2013; Schagerl et al., 2015).

Eukaryotic microalgae occur in soda lakes in Kenya (Schagerl et al., 2015) and Russia (Afonina and Tashlykova, 2020). However, many of these habitats are poorly described in terms of their algal flora. Several new filamentous Cyanobacteria genera and species were described from Lake Dziani, a crater lake on the island of Pamanzi, north of Madagascar: *Desertifilum dzianense*, *Sodalinema komarekii*, *Sodaleptolyngbya stromatolitii*, and *Haloleptolyngbya elongata*, cooccurring with *Ankistrodesmus fusiformis* (Cellamare et al., 2018). Marl lakes, such as Plitvice Lake (Croatia; Fig. 17-9) and Fayetteville Green Lake in New York (USA), with pH 8–9, are extremely alkaline habitats, characterized by clear aquamarine water with deposits of $CaCO_3$ on algae, plants, and sediments, rendering the sediments white in color. Marl lakes have low or moderate phytoplankton productivity, due to the coprecipitation of phosphorus with carbonates (Murphy and Wilkinson, 1980). $CaCO_3$-encrusted *Chara* meadows (Fig. 17-10) are extremely common in these habitats (Pentecost et al., 2006).

IV. Types of algal associations in inland waters

Algae can be free-floating in the pelagic zone, associated with the surface layer at the air–water interface, grow attached, or can be associated with a variety of substrata, including on or within sediments, sand, attached to rocks, larger filamentous algae, aquatic plants, artificial surfaces, or to various aquatic animals (Chapter 25). Algae can also live in symbioses with other

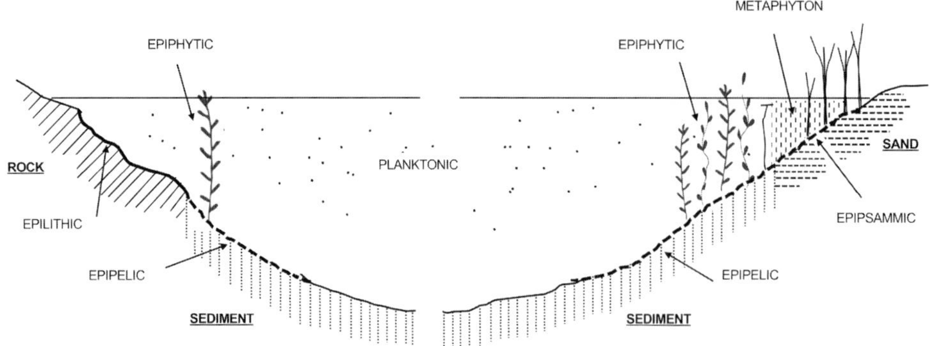

FIGURE 17-12 Diagram of the major algal habitats in a lake associated with different substrata.

organisms. The common algal associations in inland waters are shown in Fig. 17-12.

A. Plankton and neuston

Phytoplankton are algal species with limited means of locomotion, so that cells and colonies are freely floating, passively drifting in the water. Some can perform movements by means of flagella, mucous excretions, or other mechanisms that render them buoyant (Section 17IIIA). Most phytoplankters are also *holoplanktonic*, spending their entire life cycle in the plankton. Their size range spans at least six orders of magnitude (Section 17ID). Phytoplankton are most common in standing waters, although also common in large rivers with relatively low current velocities (Wetzel and Ward, 1992; Reynolds and Descy, 1996; Wehr and Thorp, 1997). Because phytoplankton are primarily photosynthetic, most spend their time within the euphotic zone, where there is sufficient light for photosynthesis. There are many planktonic species in inland waters from nearly all algal groups, with the exception of members of the Rhodophyceae and Phaeophyceae.

Organisms associated with the water surface are termed *neuston*, with those organisms living on the upper surface (less than 100 to a few hundred micrometers of thickness) of the air–water interface termed *epineuston*, and those attached to or living directly below the surface film known as *hyponeuston*. Both rely on the surface tension of the water to maintain their position at the surface and do so by means of suction pads, hydrophobic substances, or the formation of special groups of cells (Babenzien, 1966; Valkanov, 1968). Examples include specialized chrysophyte taxa, for example, *Chromulina neustophila*, *Epipyxis minuta*, and *Paraphysomonas vestita*, and the green alga *Nautococcus pyriformis* (Pentecost, 1984; Nicholls, 1992). Some species are apparently able to migrate in and out of this surface (Marshall and Gladyshev, 2009). This association does not include species whose position is due to buoyancy adaptations unrelated to surface tension, such as aerotopes in Cyanobacteria; hence examination of this microhabitat requires careful study. Burchardt and Marshall, (2003) and Nägeli and Schanz, (1990) determined that common members of the phytoplankton, or even benthic communities, may temporarily be associated with the surface film but should not be confused with true neuston.

B. Epilithic, epipelic, and epipsammic associations

Epilithic algae grow attached to submerged rocks, boulders, stones, and pebbles in benthic habitats of streams and rivers. Although epilithic algae vary from single cells to large thalli, their collective growths often create conspicuous, characteristic colors and macroscopic morphologies, which are recognizable, and in some cases identifiable, to genus (Sheath and Cole, 1992). Variously colored dash-like spots are formed on cobbles and boulders in alpine streams by members of the cyanobacterial genus *Chamaesiphon* (Fig. 17-2e), dark green to blue-green mucilaginous masses are created by species of *Phormidium* (Fig. 17-2j), and black cottony sheets are formed by *Scytonema* (Rott et al., 2006; Rott and Wehr, 2016). The red alga *Hildenbrandia* forms thin red crusts on rocks in hardwater streams. The cosmopolitan green algal genus *Gongrosira* is also common in hardwater streams, as well as in shallow lake margins or ponds, where it often becomes lime encrusted, rendering a pale-green color. Other common epilithic green algae include species of *Cladophora*, *Stigeoclonium*, and *Ulothrix* (Fig. 17-9m,k,i). Green cotton-like mats are created by networks of tangled filaments of the xanthophyte *Vaucheria* (Fig. 17-4i,j). Because of its profuse production of gelatinous stalks, the diatom

Didymosphenia geminata forms extensive tan-colored gelatinous blobs on rocks in oligotrophic rivers.

Epipelic Cyanobacteria and algae (termed *epipelon*) live freely on, or in association with, sediments consisting of mud, silt, and fine clay particles. Epipelic assemblages usually live within the first few millimeters of sediment and must be able to tolerate conditions of very low light or oxygen, making motility a distinct advantage. Epipelic algae include gliding diatoms such as *Navicula*, *Nitzschia*, and *Stauroneis* and flagellates such as *Euglena* (Round, 1981; Poulíčková et al., 2008). Vertical migration of epipelic algae has been linked to both positive and negative phototaxis (Round and Happey, 1965). The epipelic communities in ambient springs are dominated by diatoms (Poulíčková et al., 2005; Cantonati et al., 2006; Żelazna-Wieczorek and Maninska, 2006). The algal community composition of epipelon can vary across a trophic gradient (Lysáková et al., 2007) and in lakes undergoing acidification (Vinebrooke et al., 2002). In the latter case lowered pH suppresses filamentous Cyanobacteria and favors green algae, especially members of the Zygnematophyceae (*Cosmarium*, *Closterium*, *Spirogyra*, *Zygnema*), together with pennate diatoms such as *Eunotia* (Fig. 17-6o), *Frustulia* (Fig. 17-6p), and *Pinnularia* (Round, 1981; Graham and Vinebrooke, 1998). Epipelic Cyanobacteria in neutral to alkaline lakes are represented by species of *Komvophoron*, *Pseudanabaena*, *Phormidium*, and *Oscillatoria* (Hašler et al., 2008). In flowing waters species diversity is limited by sediment accumulation resulting in burial (Passy et al., 1999). A study of epipelic algal assemblage in the Al-Sadir River (Iraq) documented a diversity of diatoms (87%), Cyanobacteria (10%), and euglenophytes (3%; Al-Ghanimy 2020).

Epipsammic (also known as *episammic*) algae live among sand grains and occur in a variety of inland waterbodies, primarily in shallow habitats. In general, epipsammic algal communities are the least studied of the benthic associations (Wehr and Sheath, 2015), and published research mostly focuses on diatoms. An extensive description of the epipsammic algal flora from Shear Water (UK) revealed mainly minute raphid diatoms. More rarely, frustules of *Amphora pediculus*, *Staurosirella leptostauron*, and *S. martyi* may occur (Round, 1965; Hickman and Round, 1970). Due to wave action in streams and rivers, the epipsammic and epipelic communities are difficult to separate (Kingston et al., 1983). Sand as substratum is subject to abrasion as a result of movement of the grains and therefore strongly attached taxa are much better adapted for survival. In streams sand-dwelling diatoms occur mainly in slower-flowing, lower-altitude reaches (Ormerod et al., 1994). Diatom taxa from sandy substrates in the Congo River (Central Africa) include *Cavinula lilandae* and *Psammothidium marginulatum* (Cocquyt et al., 2013; Taylor and Cocquyt, 2016). Perhaps due to the limited research on this habitat, studies often describe new or apparently endemic species, such as *Cymbellonitzschia minima* in East African Great Lakes (Cocquyt and Jewson, 1994), *Plagiotropus arizonica* from Cholla Lake and Chevelon Creek in northern Arizona, USA (Czarnecki and Blinn, 1978), and *Encyonema* from West Lake Okoboji in Iowa, USA (Round et al., 1990). Although diatoms are by far the most common epipsammic algal group, Cyanobacteria can also contribute significantly to the algal biomass (Sundbäck et al., 1990; Asmus and Bauerfeind, 1994). These include several cosmopolitan species of *Oscillatoria*: *O. limosa*, *O. sancta*, *O. princeps*, and *O. proboscidea* (Stein and Borden, 1979; Sheath and Steinman, 1982). Less often, species of cryptomonads, desmids, colonial and filamentous green algae, and euglenoids have been reported (Wehr and Sheath, 2015).

C. Epiphytic, metaphytic, and tychoplanktonic associations

Epiphytic algae live on a variety of submersed and emergent plants, as well as on larger algal species, and are the most well-studied group of benthic algae across all aquatic habitats. Common macrophytes acting as hosts for epiphytic algal assemblages include *Ceratophyllum*, *Phragmites*, *Potamogeton*, *Stratiotes*, *Typha*, and *Vallisneria* (Toporowska et al., 2008; Bellinger and Sigee, 2015). These plants also create a nutrient-rich local environment for loosely associated algal forms, the *metaphyton*, which are not attached to the plants but live in close association. Attachment to other algae (mostly large filamentous forms or macroscopic algae) is particularly important in parts of lakes and rivers where macrophytes are absent. Many filamentous and larger algal species, such as *Cladophora* (Fig. 17-9m), *Mougeotia* (Fig. 17-9p), *Oedogonium*, *Rhizoclonium*, *Vaucheria* (Fig. 17-4i,j), and *Hydrodictyon* (Fig. 17-9l) are major hosts for epiphytic and metaphytic algal assemblages (Wehr and Sheath, 2015). Not all plants provide suitable substrates for epiphyte attachment; a study of submersed stems of *Iris pseudacorus* revealed very little attached algae (Bellinger and Sigee, 2015). Several studies report epiphytic Cyanobacteria, green algae, and diatoms as epiphytes on charophytes, such as *Chara*, *Nitella*, and *Lychnothamnus* (Azam et al., 2016). However, other studies have suggested that charophytes produce repellent (*allelopathic*) materials that exclude or inhibit certain species of algae (Anthoni et al., 1980; Gibbs, 1973; Wium-Andersen et al., 1982). Some reports are mainly correlative and based on the lack of epiphytes frequently associated with these species (Crawford, 1979; Wium-Andersen et al., 1982) and the direct evidence for this inhibition needs further research. *Tychoplankton* are those forms that spend part of their life cycle attached to plants or other substrata but are later released into the water column and become planktonic, either by wave action or a change in their life history, such as the production of flagella. Some authors regard this as similar to *meroplankton*, a term used more often for marine zooplankton or benthic animals that change their habitat between juvenile and adult life stages.

A number of Cyanobacteria, including *Calothrix*, *Chamaesiphon* (Fig. 17-2e), *Heteroleiblienia*, *Leptolyngbya*, and *Xenococcus*, are widely reported as epiphytes on many different species of algae, including the red alga *Paralemanea catenata*, the green alga *Cladophora glomerata* (Fig. 17-9m), and several larger Cyanobacteria including *Tolypothrix distorta* (Fig. 17-2r) and *Coleodesmium wrangelii* (Fig. 17-2o), as well as on aquatic mosses (Chen et al., 2012) and many submersed higher plants (Komárek and Anagnostidis, 1998; Kučera et al., 2006; Bellinger and Sigee, 2015). Other filamentous Cyanobacteria, such as *Hapalosiphon*, *Phormidium* (Fig. 17-2j), *Rivularia*, and *Tolypothrix* (Fig. 17-2r), are common members of metaphyton assemblages in littoral zones of lakes, often associated with reeds and other emergent aquatic plants.

Diatoms are also extremely common and often abundant in both epiphytic and metaphytic associations and have been used as ecological indicators in lakes (Kilroy et al., 2017). Typical planktonic diatom species may also occasionally attach to plants, acting as facultative epiphytes. Examples include many species of *Achnanthidium*, *Cocconeis*, *Cymbella*, *Encyonema*, *Epithemia*, *Eunotia*, *Gomphonema Navicula*, *Nitzschia*, *Rhoicosphenia*, *Synedra*, and *Tabellaria*. Epiphytic species, such as *Rhoicosphenia abbreviata*, form spectacular colonies with cells radiating outward from a single attachment point. Another striking example of a planktonic, and potentially epiphytic diatom, is *Tabellaria fenestrata*, having zig-zag chains that may attach to the host by means of a mucilaginous plug. *Encyonema*, *Frustulia*, and *Nitzschia* live within mucilage tubes attached to the host (Fig. 17-6u), whereas other taxa, such as *Cocconeis* and *Amphora*, are appressed to the host (*adnate forms*) with the concave raphe bearing valve facing the substrate (Fig. 17-6n; Taylor and Cocquyt, 2016). Studies by Poulíčková et al., (2004) and Hájková et al., (2011) indicated that some moss-dwelling diatoms are host specific to certain bryophytes. For example, the bryophyte *Sphagnum fallax* is often colonized by the diatoms *Adlafia bryophila*, *Pinnularia subcapitata*, *Chamaepinnularia mediocris*, *Tabellaria flocculosa*, and *Eunotia steineckii*. Macrophytes, such as *Potamogeton*, can also serve as hosts for a variety of diatoms, of which *Amphora*, *Cocconeis* (Fig. 17-6m,n), and *Diatoma* (Fig. 17-6h,i) are common (Sultana et al., 2004).

Apart from Cyanobacteria and diatoms, green algae are also common epiphytes. Genera with modifiers "chaete" or "chaeto" (referring to the bristles, setae, or hairs) in their names are particularly common green algal epiphytes. *Coleochaete* often colonizes the surfaces of aquatic macrophytes (e.g., *Vallisneria*, *Typha*, *Ipomoea*), and other algae, such as charophytes (*Chara* and *Nitella*; Mahakham and Theerakulpisut, 2010). *Chaetonema* grows associated with the mucilaginous sheaths of other freshwater macroalgae, including *Batrachospermum* (Fig. 17-3d-f), *Chaetophora*, and *Draparnaldia* (Sherwood, 2016). In nutrient-rich waters *Aphanochaete* and *Bulbochaete* (Fig. 17-9j) are common epiphytes on large filamentous green algae such as *Cladophora* and *Oedogonium*, and on submerged macrophytes (Bellinger and Sigee, 2015). Charophytes (stoneworts; Fig. 17-10b) are common hosts for a variety of green algae, such as *Bulbochaete* (Fig. 17-9j), *Characium*, *Cladophora* (Fig. 17-9m), *Gongrosira*, *Oedogonium*, and *Rhizoclonium* (Azam et al., 2016).

Macrophytic red algae are often hosts for algal epiphytes, but a few species are themselves epiphytic on mosses, higher plants, and macroalgae. Widespread red algal epiphytes include species of *Audouinella* (Fig. 17-3a-c) and *Chroodactylon*. *Audouinella meiospora* is a microscopic epiphyte on other red algae (*Nothocladus lindaueri*, *Compsopogon leptoclados*), Cyanobacteria (*Blennothrix ganeshii*), and *Vaucheria* (Necchi and Zucchi, 1995), whereas tufts of *Audouinella hermannii* (Fig. 17-3a-c) often grow on *Paralemanea* (Sheath and Vis, 2015). *Chroodactylon* is a common epiphyte in hardwater environments, often attached to filaments of *Cladophora* or *Rhizoclonium* in temperate and tropical freshwaters (Sheath and Morrison, 2004; Sheath and Vis, 2015).

There are a number of epiphytic chrysophytes, which are mainly small, loricated single cells, such as *Epipyxis*, which can be attached to filamentous species such as *Oedogonium* or on planktonic algae, including *Asterionella formosa* (Fig. 17-6k), *Dinobryon* spp. (Fig. 17-5a), and *Woronichinia naegeliana* (Fig. 17-2g; Hilliard, 1966; John et al., 2011). *Lagynion* is exclusively epiphytic and common, living on filamentous algae (Pentecost, 1984), especially in oligotrophic and slightly acidic freshwaters (Baranova et al., 2004), and has been reported from Europe, North America, Australia, New Zealand, Brazil, and South Africa (Day et al., 1995; O'Kelly and Wujek, 2001; Menezes, 2010). *Chrysopyxis* is surrounded by a vase-shaped lorica and occurs as epiphytes on green algal filaments, attached by means of basal projections (King, 1984). Others include *Chrysoamphipyxis*, *Chrysoamphitrema*, *Porostylon*, and *Stylococcus* (Nicholls, 1987) Epiphytic euglenophytes are not common, but

Euglena elastica and *Phacus* spp. are epiphytes on *Ceratophyllum* and *Myriophyllum* (Hassan et al., 2007), and *Colacium epiphyticum* (Fig. 17-4d) is an epiphyte on *Oedogonium* (John et al., 2011).

D. Epizoic and endozoic associations

In freshwater environments *epizoic* algae are found attached to a variety of living animals such as turtles, mollusks, alligators, crocodiles, tadpoles, toads, frogs, fish, and invertebrates. A large diversity of algae has been described as associated with freshwater turtles, including representatives of nearly every algal phylum (e.g., Edgren et al., 1953; Dixon, 1960; Belusz and Reed, 1969; Fayolle et al., 2016). For example, 77 species were cataloged by Fayolle et al., (2016) on the shells of European pond turtles (*Emys orbicularis*). Filamentous green algae of the order Cladophorales are especially common on turtle shells, and chief among these is *Arnoldiella* (formerly *Basicladia*) *chelonum*, which can form dense turfs on shells, heads, and tails (Belusz and Reed, 1969; Garbary et al., 2007). Several Cyanobacteria (e.g., *Komvophoron*, *Oscillatoria*) have also been recorded from turtle substrata (Edgren et al., 1953; Dixon, 1960; Sahoo and Baweja, 2005; Garbary et al., 2007). Freshwater snails, clams, and mussels are also common substrates for a variety of algae, especially filamentous chlorophytes (Campion, 1956; Sahoo and Baweja, 2005). *Pleurocladia* grows on the shells of zebra mussels in Lake Michigan, USA (Wehr, 2015). Epizoic diatoms are also frequently observed on crocodile skin (Cupul-Magaña and Cortés-Lara 2005). Much less literature is available on epizoic algae attached to tadpoles, toads, and frogs, but a study by Tumlison and Trauth, (2006) revealed the presence of the flagellated green alga *Chlorogonium* in the form of greenish blotches attached to the skin of tadpoles. Freshwater invertebrates are also common substrates for epizoic algae. The euglenophyte *Colacium* (Fig. 17-4d) is very commonly attached to the carapace of cladocerans and copepods (Al-Dhaheri and Willey, 1996). The unicellular green alga *Characium* was observed on the anterior leg, and *Characiopsis* on the posterior legs of *Branchipus*, a fairy shrimp living in inland waters (Awasthi, 2015), as well as on *Anopheles* larvae (Iyengar and Iyengar, 1932). Holland and Hergenrader, (1981) observed the green alga *Rhopalosolen saccatus* (syn. *Characium saccatum*) growing on at least two cladocerans, *Daphnia similis* and *Simocephalus vetulus*.

Endozoic Cyanobacteria and algae live inside the body of animals. *Zoochlorellae* are small, nonmotile (often *Chlorella*) or sometimes flagellated species (e.g., *Tetraselmis*, *Carteria*) living within the bodies of various freshwater protozoa and invertebrates. An example is the green hydra, *Hydra viridissima* (also known as *Hydra viridis* or *Chlorohydra viridissima*), in which *Chlorella vulgaris* lives (Massaro and Roscha, 2008). *Chlorella*, which supplies the host with the products of photosynthesis, can also live with ciliates such as *Ophrydium* (Queimaliños et al., 1999) and *Stentor* species (Woelfl et al., 2010).

E. Symbioses

Many algal species form symbiotic associations with other organisms, including with plants, fungi, and animals. Some of these are of importance in limnological studies. Besides *Chlorella* (discussed above), a mutualistic relationship exists between a flagellated green alga, *Oophila amblystomatis*, and amphibian larvae, often with salamander eggs and larvae. The alga lives within the egg jelly of the spotted salamander, *Ambystoma maculatum. Oophila amblystomatis* metabolizes the carbon dioxide produced by the embryo and provides oxygen produced during photosynthesis (Pinder and Friet, 1994). Tadpoles of the Dwarf American Toad (*Bufo americanus*) are colonized by cells of the green alga *Chlorogonium*, which are hypothesized to act as shields against intense solar radiation and supply oxygen to developing larvae (Marshall and Grigg, 1980; Ultsch et al., 1999). There are several aquatic lichens that occur in inland waters, with Cyanobacteria including *Calothrix* (in *Lichina*), *Nostoc* (in *Pyrenocollema*), and *Stigonema* (in *Ephebe*). In addition, several green algae, such as *Stichococcus* (in *Staurothele*) and *Coccobotrys* (in *Verrucaria*), are common in aquatic lichens; *Heterococcus* is a yellow-green algal photobiont in the lichen *Verrucaria* (Hawksworth, 2000).

V. Summary

1. Algae are the major primary producers in inland waters and fundamental members of freshwater food webs. From a phylogenetic perspective, the algae are a diverse and heterogeneous collection of prokaryotic and eukaryotic taxa spanning two domains in the tree of life: the Bacteria (Cyanobacteria) and the Eukaryota (many phyla). Limnologists retain the term "algae" to refer to those aquatic microorganisms or protists that are capable of photosynthesis, use chlorophyll *a* as the main photosynthetic pigment, and lack true stems, roots, leaves, and vascular tissue. Algal species exhibit an enormous range of photosynthetic pigments, storage products, sizes, morphologies, and life histories.
   a. Photosynthetic forms possess chlorophyll *a* as their primary light-harvesting pigment but differ in employing a diversity of additional pigments, including chlorophylls *b*, *c* (several forms),

β-carotene, various xanthophylls, phycoerythrin, and phycocyanin. Many of these pigments were acquired as plastids in their evolutionary history through primary and/or secondary endosymbioses of other algal cells. The combination of pigments among species enables them to occupy different light fields, colonize deeper regions of lakes, tolerate extremes of solar radiation, and adapt to changes in sunlight over time. The distinctive photosynthetic and biochemical pathways produce an array of carbon storage products including true starch, paramylon, chrysolaminarin, and a range of lipids.

b. The algal cells, colonies, filaments, and larger thalli in fresh waters span a size range of more than six orders of magnitude, from <1.0 μm to >1 m in length. The smallest cells, the femtoplankton (0.02–0.2 μm) and picoplankton (0.2–2.0 μm), can be very abundant in lakes, with rapid growth rates and efficient nutrient acquisition abilities. Algal morphologies range from simple unicells to colonies, filaments, mucilaginous sheets, tubes, blades, and complex, branched thalli. Many species are motile and employ hair-like flagella or release mucilage to move within the water column or among various substrata. Sessile species form complex mats and closely adherent crusts or develop macroscopic, shrub-like expanses that create physical habitats used by other organisms.

c. Algal species have diverse physiological properties and adaptations. Many species are strict photoautotrophs, whereas others are facultative or obligate osmotrophs (acquire dissolved organic matter) or mixotrophs (capture particulate carbon, bacteria). Some species have lost or have limited photosynthetic abilities. Common plant nutrients (N, P, K, Ca, Fe, etc.) are required by most algae, but the amounts and proportions differ greatly among species. Some groups of algae, such as diatoms and Si-scaled chrysophytes, require Si to form their cell walls or scales. Many species require trace amounts of specific elements, such as Mo, Ni, and Se. Other species have adapted to accumulate key nutrients in excess of needs ("luxury storage") or synthesize enzymes to make use of organic forms of nutrients (e.g., alkaline phosphatase), in order to maintain growth during periods of nutrient limitation. Many Cyanobacteria possess the nitrogenase enzyme, by which cells convert atmospheric N_2 to physiologically useful forms such as NH_4^+. This latter adaptation enables Cyanobacteria to dominate in eutrophic lakes and develop massive populations in P-rich but N-poor environments.

2. The major groups of algae have traditionally been classified according to their morphology and reproductive structures, photosynthetic pigments, storage reserves, cell coverings, means of cell division, and life histories. Newer molecular data partially support these classifications but have revealed new groupings and undiscovered biodiversity.

a. The Cyanobacteria are classified in a photoautotrophic phylum within the Eubacteria and possess the phycobiliprotein pigments phycocyanin (blue) and phycoerythrin (red), in addition to chlorophyll *a*. Morphologies range from simple unicells to complex filaments and macroscopic thalli, often enclosed in a loose mucilage or firm gelatinous sheath. Cyanobacteria occur worldwide and colonize an extraordinary range of ecological conditions, including oligotrophic to hypertrophic water bodies, hot springs, saline lakes, alkaline springs, wastewater treatment works, temporary pools, and as symbionts within plants or other algae. Some are important bloom-forming species under conditions of elevated nitrogen and phosphorus. Some species, including *Microcystis aeruginosa*, *Dolichospermum circinale*, *Aphanizomenon flos-aquae*, and *Cylindrospermopsis raciborskii*, may produce potent toxins.

b. Red algae (Rhodophyta) are primarily a marine group of algae that possess phycocyanin and phycoerythrin in addition to chlorophyll *a* and *d* and produce Floridean starch. Nearly all freshwater species are macroscopic and benthic in habit, many having heteromorphic gametophyte and sporophyte (chantransia) stages and different reproductive strategies. In some species the two stages are isomorphic. Gametophytes range from simple filaments (e.g., *Audouinella*, *Bangia*) to complex branched forms and pseudoparenchymatous thalli (e.g., *Batrachospermum*, *Lemanea*, *Thorea*). Many occur in softwater or nutrient-poor systems, although some may tolerate broad ranges of dissolved nutrients or metals, and a few are restricted to acidic hot springs.

c. Cryptomonads are unicellular, biflagellate protists (Cryptophyta) with oval, lunate, elongate, or sigmoid shapes and have chlorophyll *a* and c_2, phycocyanin, phycoerythrin, β-carotene, and various xanthophylls. Commonly reported species in phytoplankton communities include *Cryptomonas ovata*, *C. erosa*, and *Plagioselmis lacustris*. Cells produce *ejectisomes*, organelles rapidly expelled in response to various stimuli; the ecological importance of these structures within the plankton is not well understood. Ecological data on cryptomonads indicate that some may be more abundant in oligotrophic water, whereas others can flourish in eutrophic water. Some cryptomonads live as facultative osmotrophs and use dissolved organic matter as a source of nutrition.

d. Euglenoids are unicellular flagellates with photosynthetic (Euglenophyta) and nonphotosynthetic (Euglenozoa) phyla. They are characterized by a proteinaceous covering known as a pellicle, arranged in spiral or parallel strips visible in the light microscope (e.g., *Euglena, Phacus*). Some cells (e.g., *Strombomonas, Trachelomonas*) produce an additional bottle-like covering known as a lorica. Photosynthetic forms have multiple chloroplasts with chlorophyll *a*, *b*, β-carotene, and several xanthophylls and produce paramylon as a storage product rather than starch. Photosynthetic euglenoids occur in a range of inland waters; they are common in shallow ponds or pools with high organic content, eutrophic lakes, and systems rich in dissolved organic matter and are rare in turbulent rivers.

e. The yellow-green algae are a heterogeneous collection of organisms, currently classified into several separate classes within the phylum Ochrophyta. Three main groups are commonly reported in inland waters. The xanthophytes (Xanthophyceae) and raphidophytes (Raphidophyceae) possess two (heterokont) flagella either in vegetative or in reproductive cells, chlorophyll *a* (not *b*), chlorophyll *c*, and several xanthophylls as photosynthetic pigments, giving cells a bright green or yellow-green color. Common xanthophytes include filamentous (e.g., *Tribonema*) and siphonous (e.g., *Vaucheria*) forms, as well as flagellated, coccoid, colonial, and rhizoidal forms in planktonic and benthic aquatic habitats and on moist soils. Many species are seldom reported, although several are common in freshwater plankton, growing attached to various substrata. Raphidophytes are all flagellates, the most common of which is *Gonyostomum semen*, a large species that is occasionally abundant in the plankton of mildly acidic and humic lakes. Eustigmatophytes (Eustigmatophyceae) have chlorophyll *a* (lack chlorophylls *b* and *c*) and several xanthophylls. A diverse but poorly understood group, they include many genera with morphological features similar to (and mistaken for) members of the green algae (e.g., *Chlorobotrys, Goniochloris, Nannochloropsis*) but have a characteristic orange-red eyespot. Several are members of phytoplankton and epiphyton flora in softwater lakes and bogs.

f. The golden algae, commonly known as chrysophytes, include two classes within the Ochrophyta (recent molecular data may expand this number) with heterokont flagella, chlorophylls a, c_1, c_2, and c_3, and fucoxanthin as major photosynthetic pigments, golden-brown plastids, and produce chrysolaminarin and various lipids as storage products. The Chrysophyceae is a diverse group of unicellular (e.g., *Ochromonas*) and colonial (e.g., *Uroglena*) flagellates that are common in freshwater phytoplankton communities. Some produce a lorica around each cell (e.g., *Dinobryon, Epipyxis*), and others produce siliceous spines and scales (e.g., *Chrysosphaerella, Spiniferomonas*). Spring blooms of chrysophytes are common causes of taste and odor issues in water supplies. Some species are mixotrophic and significant consumers of bacterioplankton. The synurophytes are a small but important group characterized by having distinctive silica scales, with or without extensions called bristles (e.g., *Mallomonas, Synura*). A third group, the haptophytes (phylum Haptophyta), have similar pigments to other golden algae but are distinguished by having two equal flagella and a third hair-like appendage, the haptonema. Most haptophyte species occur in marine environments, but a few (e.g., *Chrysochromulina*) occur in fresh waters. *Prymnesium parvum* is a brackish species that forms toxic blooms in inland saline and a few freshwater lakes, causing fish kills.

g. Diatoms are the most diverse group of algae in fresh waters, characterized by cells that form rigid silica (SiO_2) walls, termed frustules, with an enormous diversity of surface ornamentation, shapes, and symmetry. Photosynthetic pigments and storage products are like those in chrysophytes. Diatoms are often the major primary producers in lake and river plankton (e.g., *Asterionella, Cyclotella, Fragilaria*) and in benthic habitats (e.g., *Cymbella, Encyonema, Frustulia*). Some are motile and glide by means of a raphe, along various surfaces or within secreted gelatinous tubes. Many are attached to plants and hard substrata by means of gelatinous stalks (e.g., *Gomphonema, Didymosphenia*). Diatoms are a key food source for herbivorous zooplankton and benthic macroinvertebrates and are important sources of essential fatty acids in aquatic food webs. Diatoms colonize aquatic and semiaquatic habitats on every continent, under a wide array of ecological conditions including pH, temperature, nutrients, salinity, and organic pollution. These diverse properties are used in bioassessments of water quality.

h. Dinoflagellates (phylum Miozoa, class Dinophyceae) are unicellular protists with two flagella inserted in a transverse and longitudinal arrangement. They possess chlorophyll *a* and c_2, the carotenoid peridinin, and several xanthophylls. True starch is the main photosynthetic storage product. There are many photosynthetic and heterotrophic (phagotrophic and osmotrophic) members and a few parasitic species. Many species have cellulose armor-like walls or thecae (e.g., *Ceratium, Peridinium, Woloszynskia*), whereas athecate forms (e.g., *Gymnodnium*) do not. Dinoflagellates can be important components of freshwater phytoplankton communities. Facultative and obligate phagotrophy is common, using a feeding veil or pallium, an extension of cytoplasm, to capture bacteria. *Ceratium hirundinella* actively migrates between deep strata and surface water and can form dense blooms in meso- and eutrophic lakes, often below the surface or at the metalimnion. This and other species have been associated with taste and odor problems.

i. Brown algae (class Phaeophyceae) is a primarily marine class that include kelps and many other seaweed species. Only a few species, all of which are benthic, occur in freshwater streams and lakes. These are filamentous forms that aggregate to form small but macroscopic thalli as dark brown crusts (e.g., *Heribaudiella*) or erect tufts (e.g., *Pleurocladia*), and occasionally are important members of the stream benthic flora.

3. Algae are morphologically and physiologically adapted to colonize a wide variety of aquatic habitats. Life in each of these habitats poses its own challenges, but the most excessive challenges are endured by algae living in extreme environments (extremely high or low temperatures, extremely high salinity levels, or extreme levels of acidity or alkalinity).

a. Ponds and lakes represent standing water bodies, presenting relatively stable environments in which a variety of algae can flourish. Small ephemeral water bodies are inhabited by pioneer algae (mostly unicellular and colonial forms), which are later succeeded by floating filamentous forms. Algae in these temporary habitats have specialized resistant cells enabling them to survive long periods of drought. Large, deep permanent lakes, on the other hand, sustain many planktonic algae, several of which have adaptations to slow their sinking rate and to remain suspended in the euphotic zone, where there is sufficient sunlight for photosynthesis. These adaptations include the production of oil droplets (less dense than water) in diatoms, the excretion of heavy (high molecular mass) divalent ions, while monovalent ions are retained, the production of mucilage that is less dense than water that enable cells to float, or aid with the motility of Cyanobacteria and pennate diatoms. Other cells can actively swim through the water by means of one to several flagella or Cyanobacteria can passively move up and down in the water column by means of aerotopes (gas-filled vesicles). Appendages (spines, horns, or other protuberances) increase the surface-to-volume ratio, thereby reducing the sinking rate. The abundance and species diversity are usually dependent on the trophic status of the water body. More productive systems usually sustain high algal concentrations and noxious blooms may occur, but these systems are usually characterized by low species diversity. Eutrophic lakes are frequently dominated by Cyanobacteria, whereas oligotrophic lakes are frequently inhabited by a variety of desmids and several chrysophytes. Euglenoids are common in organically enriched water bodies. Softwater acidic lakes also sustain many desmid species, as well as filamentous green algae such as *Spirogyra, Mougeotia,* and *Zygnema.* Hardwater lakes may have dense *Chara* populations encrusting $CaCO_3$ on their cell walls.

b. Streams and rivers are extremely complex habitats, and several factors influence algal assemblages in running waters. Water velocity and continuous mixing influence the turbidity of the water by keeping particles in suspension, and this affects the underwater light available for photosynthesis. Furthermore, different substrate types in flowing water (silt, mud, rock, boulders) may also influence the composition of algal assemblages. Common macroalgae in flowing waters include the filamentous green alga, *Cladophora glomerata,* and *Phormidium,* a cyanobacterium capable of forming large, macroscopic mats. Potamoplankton mainly include small centric diatoms as well as unicellular and colonial green algae. Planktonic species often form blooms in eutrophic, slow-flowing rivers. Benthic forms are more abundant in the littoral zones and shallow regions, where they attach to a variety of substrates.

c. Springs represent unique habitats that harbor high levels of algal diversity. Springs often contain endemic species as a result of their isolation and numerous species new to science have been described from spring habitats. Unfortunately, human activities have had a significant impact on the water quality of springs, endangering the existence of fragile species. The algal assemblages in springs are dependent on the geology, hydrology, and water quality of the spring and may vary widely between different springs. A variety of diatoms and Cyanobacteria are regularly found in springs. The presence of Cyanobacteria in pristine spring water is ascribed to the fact that those with heterocysts can fix atmospheric nitrogen if nitrogen concentrations in the water are limited. Some springs are inhospitable (e.g., thermal hot water springs with water up to boiling point) and only extremophilic algae can live here.

d. Many extreme environments serve as habitat for specially adapted algae. Thermal springs may emit water near the boiling point. Some Cyanobacteria (e.g., *Synechococcus*) can grow in water with temperatures up to 75°C, whereas eukaryotic algae such as diatoms and green algae generally have lower temperature limits (55°C). Cold environments, such as polar lakes and snow-covered soil, may be inhabited by cold-loving algae, called cryophiles. Snow algae may color the snow in a wide range of colors, depending on their pigments. Cyanobacteria and green algae are common in cold habitats. Hypersaline water bodies are inhabited by halophilic algal species that can tolerate osmotic exttremes by means of special adaptations. In general, the higher the salinity, the lower the species diversity. Some Cyanobacteria (e.g., *Arthrospira*), green algae (of which *Dunaliella* is probably the most well-known), and diatoms can tolerate hypersaline environments. *Dunaliella* color the water red due to carotenoid pigments and blooms can be seen from space. Blooms of the toxic haptophyte, *Prymmesium parvum*, are responsible for fish kills in saline ponds and lakes. Extremely acidic and alkaline habitats are also home to various algae. Acidophiles can grow in water having pH levels below 3, whereas alkaliphiles can live in water with pH levels exceeding 11. *Euglena mutabilis* is regarded as an indicator species of acid mine drainage and survives waters with pH levels between 1.0 and 3.0. A few members of several other algal groups (green algae, chrysophytes, red algae) can also tolerate highly acidic environments. Cyanobacteria rarely occur in highly acidic waters but may be found in alkaline habitats. *Arthrospira fusiformis* is a common cyanobacterium in alkaline soda lakes, whereas *Chara* (green alga) is often dominant in alkaline marl lakes.

4. Various groups of algae can be distinguished based on the type of habitat with which they are associated. Some are free-floating, and others are attached to the surface layer or water, or to other nonliving or living substrates. Many algae can also live in symbiosis with other organisms. Different terminologies are used for each of these groups of algae.

a. Free-floating, passively drifting algae are called phytoplankton. They have limited mechanisms of locomotion and will be swept away by strong currents. They are therefore most abundant in standing or slow-flowing water bodies. Neuston (epineuston and hyponeuston) lives in association with the water surface, where they rely on the surface tension to maintain their position.

b. The major difference between epilithic, epipelic, and epipsammic algae is the type of substrates with which they are associated. Epilithic algae grow attached to hard rocky substrates, such as stones, boulders, and pebbles, in streams or rivers, often rendering the stones various colors. Epipelic algae live in close association with sediments consisting of mud, silt, or clay particles, through which many can migrate vertically. Epipsammic algae live on and in sandy substrates and are mostly represented by diatoms and to a lesser extent Cyanobacteria, especially *Oscillatoria.*

c. Epiphytic algae, growing on other algae and plants, are extremely common and well studied. A variety of large filamentous algae as well as aquatic plants act as hosts to which epiphytic algae attach. Epiphytic algae can be found in most algal groups and representatives of Cyanobacteria, diatoms, green algae (often with "chaete" or "chaeto" in the name), red algae, and chrysophytes (especially small cells surrounded by loricas) are common as epiphytes. Metaphyton are also mostly loosely associated with these plants, but they do not physically attach to the plants.

d. Epizoic algae grow on the outer surface of animals, whereas endozoic algae live on the inside of animals (often as part of a symbiotic relationship). Algae, living as epiphytes on turtles, have been studied extremely well and members belonging to all algal groups are found. Other common animals to which epizoic algae attach include mollusks, alligators, crocodiles, tadpoles, toads, frogs, fish, and invertebrates. Examples of epizoic algae include several Cyanobacteria and green algae living within protozoa and invertebrates, to which they supply their products of photosynthesis.

e. Algae can live in symbiosis with a wide variety of plants, fungi, and animals. In most cases they supply food, produced during photosynthesis, to the host. Symbiotic relationships were also found between a green alga and salamander eggs and larvae, where the alga metabolizes CO_2 produced by the embryo and simultaneously supplies the embryo with O_2 produced during photosynthesis. Green algae can also grow on the skin of tadpoles, where they act as a shield against harmful UV rays and supply O_2, produced by photosynthesis, to the tadpoles. Lichens are also a symbiotic relationship between Cyanobacteria and/or green algae and fungi.

Acknowledgments

The authors thank the editors for many useful suggestions which have greatly improved this chapter.

References

Aaronson, S., Berner, T., Gold, K., Kushner, L., Patni, N.J., Repak, A., Rubin, D., 1983. Some observations on the green planktonic alga, *Botryococcus braunii* and its bloom form. J. Plankton Res. 5, 693–700. https://doi.org/10.1093/plankt/5.5.693.

Abdelahad, N., Bolpangi, R., Lasinio, G.J., Vis, M.L., Amadio, C., Laini, A., Keil, E.J., 2015. Distribution, morphology and ecological niche of *Batrachospermum* and *Sheathia* species (Batrachospermales, Rhodophyta) in the *fontanili* of the Po plain (northern Italy). Eur. J. Phycol. 50, 318–329. https://doi.org/10.1080/09670262.2015.1055592.

Abeliovich, A., Weisman, D., 1978. Role of heterotrophic nutrition in growth of the alga *Scenedesmus obliquus* in high-rate oxidation ponds. Appl. Environ. Microbiol. 35 (1), 32–37. https://doi.org/10.1128/AEM.35.1.32-37.1978.

Afonina, E.Y., Tashlykova, N.A., 2020. Fluctuations in plankton community structure of endorheic soda lakes of southeastern Transbaikalia (Russia). Hydrobiologia 847 (6), 1383–1398. https://doi.org/10.1007/s10750-020-04207-z.

Al-Dhaheri, R.S., Willey, R.L., 1996. Colonization and reproduction of the epibiotic flagellate *Colacium vesiculosum* (Euglenophyceae) on *Daphnia pulex*. J. Phycol. 32, 770–774. https://doi.org/10.1111/j.0022-3646.1996.00770.x.

Al-Ghanimy, G.D.B., 2020. Study of epipelic algae and epiphytic algae in Al-Sadir river, Al-Najaf, Iraq. EurAsia J. BioSci. 14, 763–772.

Ali, I., Prasongsuk, S., Akbar, A., Aslam, M., Lotrakul, P., Punnapayak, H., Rakshit, S.K., 2016. Hypersaline habitats and halophilic microorganisms. Maejo International Journal of Science and Technology 10 (3), 330–345.

Andrews, M., Box, R., McInroy, S., Raven, J.A., 1984. Growth of *Chara hispida*: II. Shade adaptation. J. Ecol. 72 (3), 885–895. https://doi.org/10.2307/2259538.

Anthoni, U., Christophersen, C., Madsen, J.O., Wium-Andersen, S., Jacobsen, N., 1980. Biological active sulfur compounds from the green alga *Chara globularis*. Phytochemistry 19, 1228–1229. https://doi.org/10.1016/0031-9422(80)83090-1.

Asmus, R.M., Bauerfeind, E., 1994. The microphytobenthos of Königshaven—spatial and seasonal distribution on a sandy tidal flat. Helgoländer Meeresunters 48, 257–276. https://doi.org/10.1007/BF02367040.

Awasthi, A.K., 2015. Textbook of Algae. Vikas Publishing House Pvt Ltd, New Delhi, p. 407.

Azam, S.M.G.G., Nahar, U., Diba, N.J., Moniruzzaman, M., Naz, S., 2016. A study on epiphytic algae growing on charophytes. International Journal of Environment, Science and Technology 2 (1), 1–12.

Babenzien, H.D., 1966. Untersuchungen zur Mikrobiologie des Neustons. Verhandlungen des Internationalen Verein Limnologie 16, 1503–1511.

Baek, S.H., Jung, S.W., Shin, K., 2011. Effects of temperature and salinity on growth of *Thalassiosira pseudonana* (Bacillariophyceae) isolated from ballast water. J. Freshw. Ecol. 26 (4), 547–552. https://doi.org/10.1080/02705060.2011.582696.

Bailey, J.C., Bidigare, R.R., Christensen, S.J., Andersen, R.A., 1998. Phaeothamniophyceae classis nova: a new lineage of chromophytes based upon photosynthetic pigments, *rbc*L sequence analysis and ultrastructure. Protist 149, 245–263. https://doi.org/10.1016/S1434-4610(98)70032-X.

Baines, S.B., Fisher, N.S., Doblin, M.A., Cutter, G.A., Cutter, L.S., Cole, B., 2004. Light dependence of selenium uptake by phytoplankton and implications for predicting selenium incorporation into food webs. Limnol. Oceanogr. 49, 566–578. https://doi.org/10.4319/lo.2004.49.2.0566.

Baker, K.K., Baker, A.L., 1981. Seasonal succession of the phytoplankton in the upper Mississippi River. Hydrobiologia 83, 295–301. https://doi.org/10.1007/BF00008280.

Baker, P.D., Brookes, J.D., Burch, M.D., Maier, H.R., Ganf, G.G., 2000. Advection, growth and nutrient status of phytoplankton populations in the lower River Murray, South Australia. Regul. Rivers Res. Manag. 16, 327–344. https://doi.org/10.1002/1099-1646(200007/08)16:4%3C327::AID-RRR576%3E3.0.CO;2-Q.

Balakrishnan, S., Santhanam, P., Jeyanthi, S., Divya, M., Srinivasan, M., 2019. Preliminary screening of halophilic microalgae collected from different salt pans of Tuticorin, southeast coast of India. Proc. Zool. Soc. 72, 90–96. https://doi.org/10.1007/s12595-018-0264-0.

Baranova, S.S., Anissimova, O.V., Nevo, E., Jarygin, M.M., Waser, S.P., 2004. Diversity and ecology of algae from Nahal Qishon river, northern Israel. Plant Biosyst. 138, 245–259. https://doi.org/10.1080/11263500400006985.

Barbiero, R.P., Welch, E.B., 1992. Contribution of benthic blue-green algal recruitment to lake populations and phosphorus translocation. Freshw. Biol. 27, 249–620. https://doi.org/10.1111/j.1365-2427.1992.tb00537.x.

Barsanti, L., Gualtieri, P., 2014. Algae: Anatomy, Biochemistry, and Biotechnology. CRC Press, Boca Raton, FL, p. 345.

Beamud, S.G., Karrasch, B., Pedrozo, F.L., Diaz, M.M., 2014. Utilisation of organic compounds by osmotrophic algae in an acidic lake of Patagonia (Argentina). Limnology 15, 163–172. https://doi.org/10.1007/s10201-014-0427-2.

Becker, B., Marin, B., Melkonian, M., 1994. Structure, composition, and biogenesis of prasinophyte cell coverings. Protoplasma 181, 233–244. https://doi.org/10.1007/BF01666398.

Bell, E.M. (Ed.), 2012. Life at Extremes: Environments, Organisms and Strategies for Survival. CABI Publishing, Wallingford, p. 576.

Bell, T., Kalff, J., 2001. The contribution of picophytoplankton in marine and freshwater systems of different trophic status and depth. Limnol. Oceanogr. 46, 1243–1248. https://doi.org/10.4319/lo.2001.46.5.1243.

Bellinger, E.G., Sigee, D.C., 2015. Freshwater Algae: Identification, Enumeration and Use as Bioindicators, second ed. John Wiley and Sons Ltd., The Atrium, Southern Gate, Chichester, West Sussex, p. 290.

Belusz, L.C., Reed, R.J., 1969. Some epizoophytes on six turtle species collected in Massachusetts and Michigan. Am. Midl. Nat. 81 (2), 598–601. https://doi.org/10.2307/2423999.

Bergström, A.K., Jansson, M., Drakare, S., Blomqvist, P., 2003. Occurrence of mixotrophic flagellates in relation to bacterioplankton production, light regime and availability of inorganic nutrients in unproductive lakes with differing humic contents. Freshw. Biol. 48, 868–877. https://doi.org/10.1046/j.1365-2427.2003.01061.x.

Bhaya, D., 2004. Light matters: phototaxis and signal transduction in unicellular cyanobacteria. Mol. Microbiol. 53, 745–754. https://doi.org/10.1111/j.1365-2958.2004.04160.x.

Biggs, B.J.F., 1996. Patterns of benthic algae in streams. In: Bothwell, M.L., Lowe, R.L. (Eds.), Algal Ecology: Freshwater Benthic Ecosystems. Academic Press, San Diego, CA, pp. 31–56.

Biggs, B.J.F., Price, G.M., 1987. A survey of filamentous algal proliferations in New Zealand rivers. N. Z. J. Mar. Freshw. Res. 21, 175–191. https://doi.org/10.1080/00288330.1987.9516214.

Blinn, D.W., Stein, J.R., 1970. Distribution and taxonomic reappraisal of *Ctenocladus* (Chlorophyceae: Chaetophorales). J. Phycol. 6, 101–105. https://doi.org/10.1111/j.1529-8817.1970.tb02365.x.

Bold, H.C., Wynne, M.J., 1985. Introduction to the Algae: Structure and Reproduction. Prentice-Hall, Englewood Cliffs, NJ, p. 720.

Bowes, M.J., Gozzard, E., Johnson, A.C., Scarlett, P.M., Roberts, C., Read, D.S., Armstrong, L.K., Harman, S.A., Wickham, H.D., 2012.

Spatial and temporal changes in chlorophyll-*a* concentrations in the River Thames basin, UK: are phosphorus concentrations beginning to limit phytoplankton biomass? Sci. Total Environ. 426, 45–55. https://doi.org/10.1016/j.scitotenv.2012.02.056.

Boyle, N.R., Sengupta, N., Morgan, J.A., 2017. Metabolic flux analysis of heterotrophic growth in *Chlamydomonas reinhardtii*. PLoS One 12 (5), e0177292. https://doi.org/10.1371/journal.pone.0177292.

Brambilla, E., Hippe, H., Hagelstein, A., Tindall, B.J., Stackbrandt, E., 2001. 16S rDNA diversity of cultured and uncultured prokaryotes of a mat sample from Lake Fryxell, McMurdo Dry Valleys, Antarctica. Extremophiles 5, 23–33. https://doi.org/10.1007/s007920000169.

Brett, M.T., Bunn, S.E., Chandra, S., Galloway, A.W., Guo, F., Kainz, M.J., Kankaala, P., Lau, D.C., Moulton, T.P., Power, M.E., Rasmussen, J.B., Taipale, S.J., Thorp, J.H., Wehr, J.D., 2017. How important are terrestrial organic carbon inputs for secondary production in freshwater ecosystems? Freshw. Biol. 62, 833–853. https://doi.org/10.1111/fwb.12909.

Brett, M.T., Müller-Navarra, D.C., 1997. The role of highly unsaturated fatty acids in aquatic foodweb processes. Freshw. Biol. 38, 483–499. https://doi.org/10.1046/j.1365-2427.1997.00220.x.

Brock, T.D., 1967. Life at high temperatures. Science 58 (3804), 1012–1019. https://doi.org/10.1126/science.158.3804.1012.

Brock, T.D., Brock, M.L., 1971. Microbiological studies of thermal habitats of the central volcanic region, North Island, New Zealand. N. Z. J. Mar. Freshw. Res. 5 (2), 233–258. https://doi.org/10.1080/00288330.1971.9515379.

Brookes, J.D., Ganf, G.G., 2001. Variations in the buoyancy response of *Microcystis aeruginosa* to nitrogen, phosphorus and light. J. Plankton Res. 23 (12), 1399–1411. https://doi.org/10.1093/plankt/23.12.1399.

Brooks, B.W., Grover, J.P., Roelke, D.L., 2011. *Prymnesium parvum*: an emerging threat to inland waters. Environ. Toxicol. Chem. 30, 1955–1964. https://doi.org/10.1002/etc.613.

Burchardt, L., Marshall, H.G., 2003. Algal composition and abundance in the neuston surface micro layer from a lake and pond in Virginia (U.S.A.). J. Limnol. 62 (2), 139–142. https://doi.org/10.4081/jlimnol.2003.139.

Burki, F., Roger, A.J., Brown, M.W., Simpson, A.G., 2020. The new tree of eukaryotes. Trends Ecol. Evol. 35, 43–55. https://doi.org/10.1016/j.tree.2019.08.008.

Bussi, G., Whitehead, P.G., Bowes, M.J., Read, D.S., Prudhomme, C., Dadson, S.J., 2016. Impacts of climate change, land-use change and phosphorus reduction on phytoplankton in the River Thames (UK). Sci. Total Environ. 572, 1507–1519. https://doi.org/10.1016/j.scitotenv.2016.02.109.

Calado, A.J., Moestrup, Ø., 1997. Feeding in *Peridiniopsis berolinensis* (Dinophyceae): new observations on tube feeding by an omnivorous, heterotrophic dinoflagellate. Phycologia 36, 47–59. https://doi.org/10.2216/i0031-8884-36-1-47.1.

Callieri, C., 2008. Picophytoplankton in freshwater ecosystems: the importance of small-sized phototrophs. Freshwater Reviews 1, 1–28. https://doi.org/10.1608/FRJ-1.1.1.

Callieri, C., 2017. *Synechococcus* plasticity under environmental changes. FEMS (Fed. Eur. Microbiol. Soc.) Microbiol. Lett. 364 (23), fnx229. https://doi.org/10.1093/femsle/fnx229.

Callieri, C., Cronberg, G., Stockner, J.G., 2012. Freshwater picocyanobacteria: single cells, microcolonies and colonial forms. In: Whitton, B.A. (Ed.), Ecology of Cyanobacteria II. Springer, Dordrecht, pp. 229–269.

Camacho, A., Vicente, E., Miracle, M.R., 2001. Ecology of Cryptomonas at the chemocline of a karstic sulfate-rich lake. Mar. Freshw. Res. 52, 805–815. https://doi.org/10.1071/MF00097.

Campion, M., 1956. A survey of the green algae epiphytic on the shells of some freshwater molluscs. Hydrobiologia 8 (1), 38–53. https://doi.org/10.1007/BF00047480.

Cantonati, M., 2005. Algae. In: Stoch, F. (Ed.), Pools, ponds and marshes. Quaderni habitat 11, pp. 28–37.

Cantonati, M., Angeli, N., Bertuzzi, E., Spitale, D., Lange-Bertalot, H., 2012. Diatoms in springs of the Alps: spring types, environmental determinants, and substratum. Freshw. Sci. 31, 499–524. https://doi.org/10.1899/11-065.1.

Cantonati, M., Fensham, R.J., Stevens, L.E., Gerecke, R., Glazier, D.S., Goldscheider, N., Knight, R.L., Richardson, J.S., Springer, A.E., Tockner, K., 2020. Urgent plea for global protection of springs. Conserv. Biol 35 (1), 378–382. https://doi.org/10.1111/cobi.13576.

Cantonati, M., Gerecke, R., Bertuzzi, E., 2006. Springs of the Alps—Sensitive ecosystems to environmental change: from biodiversity assessments to long-term studies. Hydrobiologia 562, 59–96. https://doi.org/10.1007/s10750–005–1806–9.

Cantonati, M., Komárek, J., Montejano, G., 2015. Cyanobacteria in ambient springs. Biodivers. Conserv. 24, 865–888. https://doi.org/10.1007/s10531–015–0884–x.

Cantonati, M., Lange-Bertalot, H., 2009. *Geissleria gereckei* sp. nov. (Bacillariophyta) from leaf-litter covered stones of very shaded carbonate mountain springs with extremely low discharge. Phycol. Res. 57, 171–177. https://doi.org/10.1111/j.1440-1835.2009.00536.x.

Cantonati, M., Pipp, E., 2000. Longitudinal and seasonal differentiation of epilithic diatom communities in the uppermost sections of two mountain spring-fed streams. Verhandlungen der Internationalen Vereinigung für Limnologie 27, 1591–1595. https://doi.org/10.1080/03680770.1998.11901507.

Cantonati, M., Van de Vijver, B., Lange–Bertalot, H., 2009. *Microfissurata* gen. nov. (Bacillariophyta), a new diatom genus from dystrophic and intermittently wet terrestrial habitats. J. Phycol. 45, 732–741. https://doi.org/10.1111/j.1529-8817.2009.00683.x.

Cardinale, B.J., 2011. Biodiversity improves water quality through niche partitioning. Nature 472, 86–89. https://doi.org/10.1038/nature09904.

Caron, D.A., Sanders, R.W., Lim, E.L., Marrasé, C., Amaral, L.A., Whitney, S., Aoki, R.B., Porter, K.G., 1993. Light–dependent phagotrophy in the freshwater mixotrophic chrysophyte *Dinobryon cylindricum*. Microb. Ecol. 25, 93–111. https://doi.org/10.1007/BF00182132.

Carty, S., Parrow, M.W., 2015. Dinoflagellates. In: Wehr, J.D., Sheath, R.G., Kociolek, J.P. (Eds.), Freshwater Algae of North America. Academic Press, San Diego, CA, pp. 773–807.

Casamatta, D.A., Hašler, P., 2016. Blue–green algae (cyanobacteria) in rivers. In: Necchi, O. (Ed.), River Algae. Springer International Publishing, Cham, Switzerland, pp. 5–34.

Castenholz, R.W., 1996. Endemism and biodiversity of thermophilic cyanobacteria. Nova Hedwigia 112, 33–47.

Castenholz, R.W., Garcia-Pichel, F., 2012. Cyanobacterial responses to UV radiation. In: Whitton, B.A. (Ed.), Ecology of Cyanobacteria II. Springer, Dordrecht, pp. 481–499.

Cellamare, M., Duval, C., Drelin, Y., Djediat, C., Touibi, N., Agogué, H., Leboulanger, C., Ader, M., Bernard, C., 2018. Characterization of phototrophic microorganisms and description of new cyanobacteria isolated from the saline-alkaline crater-lake Dziani Dzaha (Mayotte, Indian Ocean). FEMS (Fed. Eur. Microbiol. Soc.) Microbiol. Ecol. 94 (8), fiy108. https://doi.org/10.1093/femsec/fiy108.

Chen, X., Bu, Z., Yang, X., Wang, S., 2012. Epiphytic diatoms and their relation to moisture and moss composition in two montane mires, Northeast China. Fundamental and Applied Limnology 181 (3), 197–206. https://doi.org/10.1127/1863-9135/2012/0369.

Ciniglia, C., Yoon, H.S., Pollio, A., Pinto, G., Bhattacharya, D., 2004. Hidden biodiversity of the extremophilic Cyanidiales red algae. Mol. Ecol. 13, 1827–1838. https://doi.org/10.1111/j.1365-294X.2004.02180.x.

Ciugulea, I., Triemer, R.E., 2010. Color Atlas of Photosynthetic Euglenoids. Michigan State University Press, p. 204.

Clay, B.L., 2015. Cryptomonads. In: Wehr, J.D., Sheath, R.G., Kociolek, J.P. (Eds.), Freshwater Algae of North America. Academic Press, San Diego, CA, pp. 809–850.

Cocquyt, C., De Haan, M., Taylor, J.C., 2013. *Cavinula lilandae* (Bacillariophyta), a new diatom species from the Congo Basin. Diatom Res. 28 (2), 157–163. https://doi.org/10.1080/0269249X.2012.753952.

Cocquyt, C., Jewson, D.H., 1994. *Cymbellonitzschia minima* Hustedt (Bacillariophyceae), a light and electron microscopic study. Diatom Res. 9 (2), 239–247. https://doi.org/10.1080/0269249X.1994.9705304.

Compère, P., Delmotte, A., 1988. Diatoms in two hot springs in Zambia (Central Africa). In: Round, F.E. (Ed.), Proceedings of the Ninth International Diatom Symposium. Biopress Ltd, Bristol, pp. 29–39.

Coutinho, F., Tschoeke, D.A., Thompson, F., Thompson, C., 2016. Comparative genomics of *Synechococcus* and proposal of the new genus *Parasynechococcus*. PeerJ 4 (3), e1522. https://doi.org/10.7717/peerj.1522.

Crawford, S.A., 1979. Farm pond restoration using *Chara vulgaris* vegetation. Hydrobiologia 62, 17–31. https://doi.org/10.1007/BF00012559.

Crossetti, L.O., Bicudo, C.E.D.M., 2008. Adaptations in phytoplankton life strategies to imposed change in a shallow urban tropical eutrophic reservoir, Garças Reservoir, over 8 years. Hydrobiologia 614, 91–105. https://doi.org/10.1007/s10750-008-9539-1.

Cupul-Magaña, F.G., Cortés-Lara, M.C., 2005. Primer registro de epibiontes en ejemplares juveniles de *Crocodylus acutus* en el medio silvestre. Caldasia 27 (1), 147–149.

Czarnecki, D.B., Blinn, D.W., 1978. Observations on southwestern diatoms. I. *Plagiotropis arizonica* n. sp. (Bacillariophyta, Entomoneidaceae), a large mesohalobous diatom. Trans. Am. Microsc. Soc. 97 (3), 393–396. https://doi.org/10.2307/3225993.

Czarnecki, D.B., Blinn, D.W., 1979. Observations on southwestern diatoms, 2. *Caloneis latiuscula* var. *reimeri* n. var., *Cyclotella pseudostelligera* f. *parva* n. f. and *Gomphonema montezumense* n. sp., new taxa from Montezuma Well National Monument. Trans. Am. Microsc. Soc. 98, 110–114. https://doi.org/10.2307/3225945.

Dadheech, P., Ballot, A., Casper, P., Kotut, K., Novelo, E., Lemma, B., Pröschold, T., Krienitz, L., 2010. Phylogenetic relationship and divergence among planktonic strains of *Arthrospira* (Oscillatoriales, Cyanobacteria) of African, Asian and American origin, deduced by 16S–23S ITS and phycocyanin operon sequences. Phycologia 49, 361–372. https://doi.org/10.2216/09-71.1.

Dadheech, P.K., Glöckner, G., Casper, P., Kotut, K., Mazzoni, C.M., Mbedi, S., Krientiz, L., 2013. Cyanobacterial diversity in the hot spring, pelagic and benthic habitats of a tropical soda lake. FEMS (Fed. Eur. Microbiol. Soc.) Microbiol. Ecol. 85, 389–401. https://doi.org/10.1111/1574-6941.12128.

Day, S.A., Wickham, R.P., Entwisle, T.J., Tyler, P.A., 1995. Bibliographic checklist of non-marine algae in Australia. Flora of Australia Supplementary Series 4, 1–276.

De Luca, P., Musacchio, A., Taddei, R., 1981. Acidophilic algae from the fumaroles of Mount Lawu (Java), *locus classicus* of *Cyanidium caldarium* Geitler. G. Bot. Ital. 115, 1–9. https://doi.org/10.1080/11263508109427979.

Delgado, C., Ector, L., Novais, M.H., Blanco, S., Hoffmann, L., Pardo, I., 2013. Epilithic diatoms of springs and spring–fed streams in Majorca Island (Spain) with the description of a new diatom species *Cymbopleura margalefii* sp. nov. Fottea. Olomouc 13 (2), 87–104. https://doi.org/10.5507/fot.2013.009.

Descy, J.P., Darchambeau, F., Lambert, T., Stoyneva-Gaertner, M.P., Bouillon, S., Borges, A.V., 2017. Phytoplankton dynamics in the Congo River. Freshw. Biol. 62, 87–101. https://doi.org/10.1111/fwb.12851.

Dixon, J.R., 1960. Epizoophytic algae on some turtles of Texas and Mexico. Tex. J. Sci. 12, 36–38.

Dodds, W.K., Gudder, D.A., 1992. The ecology of *Cladophora*. J. Phycol. 28, 415–427. https://doi.org/10.1111/j.0022-3646.1992.00415.x.

Doemel, W.N., Brock, T.D., 1971. The physiological ecology of *Cyanidium caldarium*. Microbiology 67, 17–32. https://doi.org/10.1099/00221287-67-1-17.

Dop, A.J., 1979. *Porterinema fluviatile* (Porter) Waern (Phaeophyceae) in the Netherlands. Acta Bot. Neerl. 28, 449–458. https://doi.org/10.1111/j.1438-8677.1979.tb01169.x.

Dortch, Q., Clayton, J.R., Thoresen, S.S., Ahmed, S.I., 1984. Species differences in accumulation of nitrogen pools in phytoplankton. Marine Biology 81, 237–250. https://doi.org/10.1007/BF00393218.

Duff, K., Zeeb, B., Smol, J.P., 1995. Atlas of Chrysophycean Cysts. Kluwer Academic Publishers, Dordrecht, p. 189.

Dussart, B.H., 1965. Les different categories de plancton. Hydrobiologia 26, 72–74. https://doi.org/10.1007/BF00142255.

Edgren, R.A., Edgren, M.K., Tiffany, L.H., 1953. Some North American turtles and their epizoophytic algae. Ecology 34, 733–740. https://doi.org/10.2307/1931336.

Edwards, K.F., Thomas, M.K., Klausmeier, C.A., Litchman, E., 2016. Phytoplankton growth and the interaction of light and temperature: a synthesis at the species and community level. Limnol. Oceanogr. 61, 1232–1244. https://doi.org/10.1002/lno.10282.

Elster, J., 1999. Algal versatility in various extreme environments. In: Seckbach, J. (Ed.), Enigmatic Microorganisms and Life in Extreme Environments. In: Cellular Origin, Life in Extreme Habitats and Astrobiology, Vol 1. Springer, Dordrecht, pp. 215–227.

Emoto, Y., Yoneda, Y., 1942. Bacteria and algae of hot springs in Toyama Perfecture. Acta Phytotaxonomica Geobot. 11, 7–26 (In Japanese).

Entwisle, T.J., 1989. Macroalgae in the Yarra river basin: flora and distribution. Proc. Roy. Soc. Vic. 101, 1–76.

Fawley, K.P., Fawley, M.W., 2007. Observations on the diversity and ecology of freshwater *Nannochloropsis* (Eustigmatophyceae), with descriptions of new taxa. Protist 158, 325–336. https://doi.org/10.1016/j.protis.2007.03.003.

Fayolle, S., Moriconi, C., Oursel, B., Koenig, C., Suet, M., Ficheux, S., Logez, M., Olivier, A., 2016. Epizoic algae distribution on the carapace and plastron of the European pond turtle (*Emys orbicularis*, Linnaeus, 1758): a study from the Camargue, France. Cryptogam. Algol. 37 (4), 221–232. https://doi.org/10.7872/crya/v37.iss4.2016.221.

Fensham, R.J., Ponder, W.F., Fairfax, R.J., 2010. Recovery plan for the community of native species dependent on natural discharge of groundwater from the great artesian basin. Report to department of the environment, Water, Heritage and the Arts, Canberra. Queensland Department of Environment and Resource Management, Brisbane. http://www.environment.gov.au/system/files/resources/0cefc83a–3854–4cff–9128–abc719d9f9b3/files/great–artesian–basin–ec.pdf.

Fietz, S., Bleiß, W., Hepperle, D., Koppitz, H., Krienitz, L., Nicklisch, A., 2005. First record of *Nannochloropsis limnetica* (Eustigmatophyceae) in the autotrophic picoplankton from Lake Baikal. J. Phycol. 41, 780–790. https://doi.org/10.1111/j.0022-3646.2005.04198.x.

Fott, B., 1959. Algenkunde. Gustav Fischer Verlag, Jena, p. 482.

Fritsen, C.H., Priscu, J.C., 1998. Cyanobacterial assemblages in permanent ice covers of Antarctic lakes: distribution, growth rate, and temperature response of photosynthesis. J. Phycol. 34, 587–597. https://doi.org/10.1046/j.1529-8817.1998.340587.x.

Fritz, S.C., Juggins, S., Battarbee, R.W., 1993. Diatom assemblages and ionic characterization of freshwater and saline lakes of the Northern Great Plains, North America: a tool for reconstructing past salinity and climate fluctuations. Can. J. Fish. Aquat. Sci. 50, 1844–1856. https://doi.org/10.1139/f93-207.

Frost, T.M., Graham, L.E., Elias, J.E., Haase, M.J., Kretchmer, D.W., Kranzfelder, J.A., 1997. A yellow-green algal symbiont in the freshwater sponge, *Corvomeyenia everetti*: convergent evolution of symbiotic associations. Freshw. Biol. 38, 395–399. https://doi.org/10.1046/j.1365-2427.1997.00254.x.

Furey, P.C., Kupferberg, S.J., Lind, A.J., 2014. The perils of unpalatable periphyton: *Didymosphenia* and other mucilaginous stalked diatoms as food for tadpoles. Diatom Res. 29 (3), 267–280. https://doi.org/10.1080/0269249X.2014.924436.

Garbary, D.J., Bourque, G., Herman, T.B., McNeil, J.A., 2007. Epizoic algae from freshwater turtles in Nova Scotia. J. Freshw. Ecol. 22 (4), 677–685. https://doi.org/10.1080/02705060.2007.9664828.

Garwood, P.E., 1982. Ecological interactions among *Bangia, Cladophora*, and *Ulothrix* along the Lake Erie shoreline. J. Great Lake. Res. 8, 54–60. https://doi.org/10.1016/S0380-1330(82)71942-2.

Gasol, J.M., García-Cantizano, J., Massana, R., Guerrero, R., Pedrós-Alió, C., 1993. Physiological ecology of a metalimnetic *Cryptomonas* population: relationships to light, sulfide and nutrients. J. Plankton Res. 15, 255–275. https://doi.org/10.1093/plankt/15.3.255.

Gerloff-Elias, A., Spijkerman, E., Proschold, T., 2005. Effect of external pH on the growth, photosynthesis and photosynthetic electron transport of *Chlamydomonas acidophila* Negoro, isolated from an extremely acidic lake (pH 2.6). Plant Cell Environ. 28, 1218–1229. https://doi.org/10.1111/j.1365-3040.2005.01357.x.

Gervais, F., 1997. Light-dependent growth, dark survival, and glucose uptake by cryptophytes isolated from a freshwater chemocline. J. Phycol. 33, 18–25. https://doi.org/10.1111/j.0022-3646.1997.00018.x.

Gibbs, G.W., 1973. Cycles of macrophytes and phytoplankton in Pukepuke lagoon following a severe drought. Proc. N. Z. Ecol. Soc. 20, 13–20.

Giblin, S.M., Gerrish, G.A., 2020. Environmental factors controlling phytoplankton dynamics in a large floodplain river with emphasis on cyanobacteria. River Res. Appl. 36, 1137–1150. https://doi.org/10.1002/rra.3658.

Gimmler, H., 2001. Acidophilic and acidotolerant algae. In: Rai, L.C., Gaur, J.P. (Eds.), Algal Adaptation to Environmental Stresses. Springer, Berlin, Heidelberg, pp. 259–290.

Gladyshev, M.I., Temerova, T.A., Dubovskaya, O.P., Kolmakov, V.I., Ivanova, E.A., 1999. Selective grazing on *Cryptomonas* by *Ceriodaphnia quadrangula* fed a natural phytoplankton assemblage. Aquat. Ecol. 33, 347–353. https://doi.org/10.1023/A:1009916209394.

Gómez, F., 2011. Acidophile. In: Gargaud, M., Irvine, W.M. (Eds.), Encyclopedia of Astrobiology. Springer, Berlin, Heidelberg, pp. 10–11.

Gosselain, V., Descy, J.P., Everbecq, E., 1994. The phytoplankton community of the River Meuse, Belgium: seasonal dynamics (year 1992) and the possible incidence of zooplankton grazing. Hydrobiologia 289, 179–191. https://doi.org/10.1007/BF00007419.

Grabowska, M., Wołowski, K., 2014. Development of *Trachelomonas* species (Euglenophyta) during blooming of *Planktothrix agardhii* (Cyanoprokaryota). Int. J. Limnol. 50, 49–57. https://doi.org/10.1051/limn/2013070.

Graeff, C.L., Kociolek, J.P., Rushforth, S.R., 2013. New and interesting diatoms (Bacillariophyta) from Blue Lake Warm Springs, Tooele County, Utah. Phytotaxa 153, 1–38. https://doi.org/10.11646/phytotaxa.153.1.1.

Graham, L.E., Graham, J.M., Wilcox, L.W., Cook, M.E., 2016. Algae, third ed. LJLM Press, Madison, WI, p. 689.

Graham, M.D., Vinebrooke, R.D., 1998. Trade-offs between herbivore resistance and competitiveness in periphyton of acidified lakes. Can. J. Fish. Aquat. Sci. 55, 806–814. https://doi.org/10.1139/f97-309.

Gross, W., 2000. Ecophysiology of algae living in highly acidic environments. Hydrobiologia 433, 31–37. https://doi.org/10.1023/A:1004054317446.

Gruia, L., 1976. Alge din ecosuisteme tericole si biotopuri en temperaturi deosebite. In: Peterfi, S., Ionescu, A.L. (Eds.), Alge Termofile, Tratat de Algologie, Edidutura, Academiei Republicii Socialiste Romania, pp. 430–439. Bucuresti.

Grujcic, V., Nuy, J.K., Salcher, M.M., Shabarova, T., Kasalicky, V., Boenigk, J., Jensen, M., Simek, K., 2018. Cryptophyta as major bacterivores in freshwater summer plankton. ISME J. 12, 1668–1681. https://doi.org/10.1038/s41396-018-0057-5.

Guiry, M.D., Guiry, G.M., 2021. AlgaeBase. World-wide electronic publication. National University of Ireland, Galway. https://www.algaebase.org.

Hagman, C.H.C., Rohrlack, T., Riise, G., 2020. The success of *Gonyostomum semen* (Raphidophyceae) in a boreal lake is due to environmental changes rather than a recent invasion. Limnologica 84, 125818. https://doi.org/10.1016/j.limno.2020.125818.

Hájková, P., Bojková, J., Fránková, M., Opravilová, V., Hájek, M., Kintrová, K., Horsák, M., 2011. Disentangling the effects of water chemistry and substratum structure on moss-dwelling unicellular and multicellular micro-organisms in spring-fens. J. Limnol. 70, 54–64. https://doi.org/10.4081/jlimnol.2011.s1.54.

Hall, J.D., Karol, K.G., 2016. An inventory of the algae (excluding diatoms) of lakes and ponds of Harriman and Bear Mountain State Parks (Rockland and Orange Counties, New York, U.S.A). Brittonia 68, 148–169. https://doi.org/10.1007/s12228-016-9409-5.

Hamed, A.F., 2008. Biodiversity and distribution of blue-green algae/cyanobacteria and diatoms in some of the Egyptian water habitats in relation to conductivity. Aust. J. Basic Appl. Sci. 2 (1), 1–21.

Hammer, U.T., Shamess, J., Haynes, R.C., 1983. The distribution and abundance of algae in saline lakes of Saskatchwan, Canada. Hydrobiologia 105, 1–26. https://doi.org/10.1007/BF00028018.

Hanagata, N., Malinsky-Rushansky, N., Dubinsky, Z., 1999. Eukaryotic picoplankton, *Mychonastes homosphaera* (Chlorophyceae, Chlorophyta), in Lake Kinneret, Israel. Phycol. Res. 47, 263–269.

Hao, C., Zhang, L., Wang, L., Li, S., Dong, H., 2012. Microbial community composition in acid mine drainage lake of Xiang Mountain sulfide mine in Anhui Province, China. Geomicrobiol. J. 29 (10), 886–895. https://doi.org/10.1080/01490451.2011.635762.

Happey-Wood, C.M., 1976. Vertical migration patterns in phytoplankton of mixed species composition. Br. Phycol. J. 11, 355–369. https://doi.org/10.1080/00071617600650411.

Harding, J.P.C., Whitton, B.A., 1981. Accumulation of zinc, cadmium and lead by field populations of *Lemanea*. Water Res. 15, 301–319. https://doi.org/10.1016/0043-1354(81)90034-8.

Hardwick, G.G., Blinn, D.W., Usher, H.D., 1992. Epiphytic diatoms on *Cladophora glomerata* in the Colorado River, Arizona: longitudinal and vertical distribution in a regulated river. SW. Nat. 37, 148–156. https://doi.org/10.2307/3671663.

Hargreaves, J.W., Lloyd, E.J.H., Whitton, B.A., 1975. Chemistry and vegetation of highly acidic streams. Freshw. Biol. 5, 563–576. https://doi.org/10.1111/j.1365-2427.1975.tb00156.x.

Harris, G.P., Heaney, S.I., Talling, J.F., 1979. Physiological and environmental constraints in the ecology of the planktonic dinoflagellate *Ceratium hirundinella*. Freshw. Biol. 9, 413–428. https://doi.org/10.1111/j.1365-2427.1979.tb01526.x.

Hašler, P., Štěpánková, J., Špačková, J., Neustupa, J., Kitner, M., Hekera, P., Veselá, J., Burian, J., Poulíčková, A., 2008. Epipelic cyanobacteria and algae: a case study from Czech ponds. Fottea 8, 133–146. https://doi.org/10.5507/fot.2008.012.

Hassan, F.M., Salah, M.M., Salman, J.M., 2007. Quantitative and qualitative variability of epiphytic algae on three aquatic plants in Euphrates River, Iraq. Iraqi Journal of Aquaculture 1, 1–16. https://doi.org/10.21276/ijaq.2007.4.1.1.

Hata, N., Ogbonna, J.C., Hasegawa, Y., Taroda, H., Tanaka, H., 2001. Production of astaxanthin by *Haematococcus pluvialis* in a sequential heterotrophic-photoautotrophic culture. J. Appl. Phycol. 13, 395–402. https://doi.org/10.1023/A:1011921329568.

Hawksworth, D.L., 2000. Freshwater and marine lichen-forming fungi. Fungal Divers. 5, 1–7.

Healey, F.P., 1973. Inorganic nutrient uptake and deficiency in algae. CRC Crit. Rev. Microbiol. 3, 69–113. https://doi.org/10.3109/10408417309108746.

Hecky, R.E., Kilham, P., 1988. Nutrient limitation of phytoplankton in freshwater and marine environments: a review of recent evidence on the effects of enrichment. Limnol. Oceanogr. 33, 796–822. https://doi.org/10.4319/lo.1988.33.4part2.0796.

Hessen, D.O., Elser, J.J., Sterner, R.W., Urabe, J., 2013. Ecological stoichiometry: an elementary approach using basic principles. Limnol. Oceanogr. 58, 2219–2236. https://doi.org/10.4319/lo.2013.58.6.2219.

Hickman, M., 1974. The seasonal succession and vertical distribution of the phytoplankton in Abbot's Pond, North Somerset, UK. Hydrobiologia 44, 127–147. https://doi.org/10.1007/BF00036160.

Hickman, M., Round, F.E., 1970. Primary production and standing crops of epipsammic and epipelic algae. Br. Phycol. J. 5 (2), 247–255. https://doi.org/10.1080/00071617000650311.

Hill, D.R., 1991. *Chroomonas* and other blue-green cryptomonads. J. Phycol. 27, 133–145. https://doi.org/10.1111/j.0022-3646.1991.00133.x.

Hill, W.R., Middleton, R.G., 2006. Changes in carbon stable isotope ratios during periphyton development. Limnol. Oceanogr. 51, 2360–2369. https://doi.org/10.4319/lo.2006.51.5.2360.

Hilliard, D.K., 1966. New or rare chrysophytes from Lancashire County, England. Archive für Protistenkünde 109, 114–124.

Hindák, F., 2001. Thermal microorganisms from a hot spring on the coast of Lake Bogoria, vol. 123. Nova Hedwigia, Kenya, pp. 77–93.

Hoagland, K.D., Roemer, S.C., Rosowski, J.R., 1982. Colonization and community structure of two periphyton assemblages, with emphasis on the diatoms (Bacillariophyceae). Am. J. Bot. 69, 188–213. https://doi.org/10.1002/j.1537-2197.1982.tb13249.x.

Hoagland, K.D., Rosowski, J.R., Gretz, M.R., Roemer, S.C., 1993. Diatom extracellular polymeric substances: function, fine structure, chemistry, and physiology. J. Phycol. 29, 537–566. https://doi.org/10.1111/j.0022-3646.1993.00537.x.

Hoham, R.W., 1975. Optimum temperatures and temperature ranges for growth of snow algae. Arct. Alp. Res. 7 (1), 13–24. https://doi.org/10.2307/1550094.

Hoham, R.W., Duval, B., 2001. Microbial ecology of snow and freshwater ice with emphasis on snow algae. In: Jones, H.G., Pomeroy, J.W., Walker, D.A., Hoham, R.W. (Eds.), Snow Ecology: An Interdisciplinary Examination of Snow–Covered Ecosystems. Cambridge University Press, Cambridge, pp. 168–228.

Holland, R.S., Hergenrader, G.L., 1981. *Rhopalosolen saccatus* Fott: an epizoophyte from an alkaline lake in Nebraska. Am. Midl. Nat. 106 (2), 403–405. https://doi.org/10.2307/2425179.

Holmes, N.T.H., Whitton, B.A., 1977. The macrophytic vegetation of the river Tees in 1975: observed and predicted changes. Freshw. Biol. 7, 43–60. https://doi.org/10.1111/j.1365-2427.1977.tb01656.x.

Hongve, D., Løvstad, Ø., Bjørndalen, K., 1988. *Gonyostomum semen*—A new nuisance to bathers in Norwegian lakes. Internationale Vereinigung für Theoretische und Angewandte Limnologie: Verh Proc. Trav. SIL 23 (1), 430–434. https://doi.org/10.1080/03680770.1987.11897957.

Howarth, R.W., Marino, R., Cole, J.J., 1988. Nitrogen fixation in freshwater, estuarine, and marine ecosystems. 2. Biogeochemical controls. Limnol. Oceanogr. 33, 688–701. https://doi.org/10.4319/lo.1988.33.4part2.0688.

Huisman, J., Johansson, A.M., Folmer, E.O., Weissing, F.J., 2001. Towards a solution of the plankton paradox: the importance of physiology and life history. Ecol. Lett. 4, 408–411. https://doi.org/10.1046/j.1461-0248.2001.00256.x.

Hutchinson, G.E., 1961. The paradox of the plankton. Am. Nat. 95, 137–145.

Iyengar, M.O.P., Iyengar, M.O.T., 1932. On a *Characium* growing on *Anopheles* larvae. New Phytol. 31 (1), 66–69. https://doi.org/10.1111/j.1469-8137.1932.tb07434.x.

James, W.F., Taylor, W.D., Barko, J.W., 1992. Production and vertical migration of *Ceratium hirundinella* in relation to phosphorus availability in Eau Galle Reservoir, Wisconsin. Can. J. Fish. Aquat. Sci. 49, 694–700. https://doi.org/10.1139/f92-078.

Janse van Vuuren, S., Levanets, A., 2019. First record of *Botryococcus braunii* Kützing from Namibia. Bothalia, African Biodiversity & Conservation 49, 1–5. https://doi.org/10.4102/abc.v49i1.2382.

Janse van Vuuren, S., Levanets, A., 2020. Mass developments of *Euglena sanguinea* Ehrenberg in South Africa. Afr. J. Aquat. Sci. 46 (1), 1–13. https://doi.org/10.2989/16085914.2020.1799743.

Jansson, M., Olsson, H., Pettersson, K., 1988. Phosphatases; origin, characteristics and function in lakes. Hydrobiologia 170, 157–175. https://doi.org/10.1007/BF00024903.

Javor, B., 1989. *Dunaliella* and other halophilic, eucaryotic algae. In: Javor, B. (Ed.), Hypersaline Environments. In: Brock/Springer Series in Contemporary Bioscience. Springer, Berlin, Heidelberg, pp. 147–158.

Johansen, J.R., Doucette, G.J., Barclay, W.R., Bull, J.D., 1988. The morphology and ecology of *Pleurochrysis carterae* var. *dentata* var. nov. (Prymnesiophyceae), a new coccolithophorid from an inland saline pond in New Mexico, USA. Phycologia 27 (1), 78–88. https://doi.org/10.2216/i0031-8884-27-1-78.1.

John, D.M., Whitton, B.A., Brook, A.J. (Eds.), 2011. The Freshwater Algal flora of the British Isles—An Identification Guide to Freshwater and Terrestrial Algae, second ed. Cambridge University Press, Cambridge, p. 878.

Johnson, P.W., Sieburth, J.M., 1979. Chroococcoid cyanobacteria in the sea: a ubiquitous and diverse phototrophic biomass. Limnol. Oceanogr. 24, 928–935. https://doi.org/10.4319/lo.1979.24.5.0928.

Jonker, N., Van Ginkel, C., Olivier, J., 2013. Association between physical and geochemical characteristics of thermal springs and algal diversity in Limpopo Province, South Africa. WaterSA 39 (1), 95–103. https://doi.org/10.4314/wsa.v39i1.10.

Jüttner, F., 1981. Detection of lipid degradation products in the water of a reservoir during a bloom of *Synura uvella*. Appl. Environ. Microbiol. 41, 100–106. https://doi.org/10.1128/AEM.41.1.100-106.1981.

Kamjunke, N., Tittel, J., 2008. Utilisation of leucine by several phytoplankton species. Limnologica 38, 360–366. https://doi.org/10.1016/j.limno.2008.05.002.

Kaštovský, J., Gomez, E.B., Hladil, J., Johansen, J.R., 2014. *Cyanocohniella calida* gen. et sp. nov. (Cyanobacteria: Aphanizomenonaceae) a new cyanobacterium from the thermal springs from Karlovy Vary, Czech Republic. Phytotaxa 181 (5), 279–292. https://doi.org/10.11646/phytotaxa.181.5.3.

Kempe, S., Kazmierczak, J., 2011. Soda lakes. In: Reitner, J., Thiel, V. (Eds.), Encyclopedia of Geobiology, Encyclopedia of Earth Sciences Series. Springer, Dordrecht, pp. 824–829.

Kilham, S.S., 1986. Dynamics of Lake Michigan natural phytoplankton communities in continuous cultures along a Si:P loading gradient. Can. J. Fish. Aquat. Sci. 43, 351–360. https://doi.org/10.1139/f86-045.

Kilroy, C., Larned, S.T., Biggs, B.J.F., 2009. The non-indigenous diatom *Didymosphenia geminata* alters benthic communities in New Zealand rivers. Freshw. Biol. 54, 1990–2002. https://doi.org/10.1111/j.1365-2427.2009.02247.x.

Kilroy, C., Suren, A.M., Wech, J.A., Lambert, P., Sorrel, B.K., 2017. Epiphytic diatoms as indicators of ecological condition in New Zealand's lowland wetlands. N. Z. J. Mar. Freshw. Res. 51 (4), 1–23. https://doi.org/10.1080/00288330.2017.1281318.

King, J.M., 1984. The occurrence of *Chrysopyxis urna* (Chrysophyceae) in the United States. Trans. Am. Microsc. Soc. 103 (3), 317–319. https://doi.org/10.2307/3226193.

Kingston, J.C., Lowe, R.L., Stoermer, E.F., Ladewski, T.B., 1983. Spatial and temporal distribution of benthic diatoms in northern Lake Michigan. Ecology 64, 1566–1580. https://doi.org/10.2307/1937511.

Klaveness, D., 1989. Biology and ecology of the Cryptophyceae: status and challenges. Biol. Oceanogr. 6, 257–270. https://doi.org/10.1080/01965581.1988.10749530.

Kociolek, J.P., Theriot, E.C., Williams, D.M., Julius, M., Stoermer, E.F., Kingston, J.C., 2015a. Centric and araphid diatoms. In: Wehr, J.D., Sheath, R.G., Kociolek, J.P. (Eds.), Freshwater Algae of North America. Academic Press, San Diego, CA, pp. 653–708.

Kociolek, J.P., Spaulding, S.A., Lowe, R.L., 2015b. Bacillariophyceae: the raphid diatoms. In: Wehr, J.D., Sheath, R.G., Kociolek, J.P. (Eds.), Freshwater Algae of North America. Academic Press, San Diego, CA, pp. 709–772.

Komárek, J., 2003. Coccoid and colonial cyanobacteria. In: Wehr, J.D., Sheath, R.G. (Eds.), Freshwater Algae of North America. Academic Press, San Diego, CA, pp. 59–116.

Komárek, J., Anagnostidis, K., 1998. Cyanoprokaryota 1. Teil: Chroococcales. In: Ettl, H., Gärtner, G., Heynig, H., Mollenhauer, D. (Eds.), Süsswasser flora von Mitteleuropa 19/1. Gustav Fischer, Jena-Stuttgart-Lübeck-Ulm, pp. 1–548.

Komárek, J., Johansen, J.R., 2015. Coccoid cyanobacteria. In: Wehr, J.D., Sheath, R.G., Kociolek, J.P. (Eds.), Freshwater Algae of North America. Academic Press, San Diego, CA, pp. 75–133.

Krienitz, L., Ballot, A., Kotut, K., Wiegand, C., Putz, S., Metcalf, J.S., Codd, G.A., Pflugmacher, S., 2003. Contribution of hot spring cyanobacteria to the mysterious deaths of Lesser Flamingos at Lake Bogoria, Kenya. FEMS (Fed. Eur. Microbiol. Soc.) Microbiol. Ecol. 43, 141–148.

Krienitz, L., Bock, C., Kotut, K., Luo, W., 2012. *Picocystis salinarum* (Chlorophyta) in saline lakes and hot springs of East Africa. Phycologia 51 (1), 22–32. https://doi.org/10.1111/j.1574-6941.2003.tb01053.x.

Krienitz, L., Dadheech, P.K., Kotut, K., 2013. Mass developments of the cyanobacteria Anabaenopsis and Cyanospira (Nostocales) in the soda lakes of Kenya: ecological and systematic implications. Hydrobiologia 703, 79–93. https://doi.org/10.1007/s10750-012-1346-z.

Kučera, P., Uher, B., Komárek, O., 2006. Epiphytic cyanophytes *Xenococcus kerneri* and *Chamaeosiphon minutus* on the freshwater red alga *Paralemanea catenata* (Rhodophyta). Biologia, Bratislava 61 (1), 11–13. https://doi.org/10.2478/s11756-006-0002-3.

Kugrens, P., Lee, R.E., Andersen, R.A., 1986. Cell form and surface patterns in *Chroomonas* and *Cryptomonas* cells (Cryptophyta) as revealed by scanning electron microscopy. J. Phycol. 22, 512–522. https://doi.org/10.1111/j.1529-8817.1986.tb02495.x.

Kugrens, P., Lee, R.E., Corliss, J.O., 1994. Ultrastructure, biogenesis, and functions of extrusive organelles in selected non-ciliate protists. Protoplasma 181, 164–190. https://doi.org/10.1007/BF01666394.

Lackey, J.B., 1938. The fauna and flora of surface waters polluted by acid mine drainage. Publ. Health Rep. 53, 1499–1507.

Lackey, J.B., 1968. Ecology of *Euglena*. In: Buetow, D.E. (Ed.), The Biology of *Euglena*. I: General Biology and Ultrastructure. Academic Press, New York, NY, pp. 27–44.

Lai, G.G., Padedda, B.M., Wetzel, C.E., Cantonati, M., Sechi, N., Lugliè, A., Ector, L., 2019. Diatom assemblages from different substrates of the Casteldoria thermo-mineral spring (Northern Sardinia, Italy). Botany Letters 166 (1), 14–31. https://doi.org/10.1080/23818107.2018.1466726.

Lavallée, B.F., Pick, F.R., 2002. Picocyanobacteria abundance in relation to growth and loss rates in oligotrophic to mesotrophic lakes. Aquat. Microb. Ecol. 27, 7–46. https://doi.org/10.3354/ame027037.

Lee, R.E., 2018. Phycology, fifth ed. Cambridge University Press, Cambridge, p. 535.

Leliaert, F., Smith, D.R., Moreau, H., Herron, M.D., Verbruggen, H., Delwiche, C.F., De Clerck, O., 2012. Phylogeny and molecular evolution of the green algae. Crit. Rev. Plant Sci. 31, 1–46. https://doi.org/10.1080/07352689.2011.615705.

Lepistö, L., Antikainen, S., Kivinen, J., 1994. The occurrence of *Gonyostomum semen* (Ehr.) Diesing in Finnish lakes. Hydrobiologia 273, 1–8. https://doi.org/10.1007/BF00126764.

Letáková, M., Fránková, M., Poulíčková, A., 2018. Ecology and applications of freshwater epiphytic diatoms. Cryptogam. Algol. 39, 3–22. https://doi.org/10.7872/crya/v39.iss1.2018.3.

Lewin, R.A., Krienitz, L., Goericke, R., Takeda, H., Hepperle, D., 2001. *Picocystis salinarum* gen. et sp. nov. (Chlorophyta)—A new picoplanktonic green alga. Phycologia 39 (6), 560–565. https://doi.org/10.2216/i0031-8884-39-6-560.1.

Li, J., Plouchart, D., Zastepa, A., Dittrich, M., 2019. Picoplankton accumulate and recycle polyphosphate to support high primary productivity in coastal Lake Ontario. Sci. Rep. 9, 1–10. https://doi.org/10.1038/s41598-019-56042-5.

Li, S., Bronner, G., Lepère, C., Kong, F., Shi, X., 2017. Temporal and spatial variations in the composition of freshwater photosynthetic picoeukaryotes revealed by MiSeq sequencing from flow cytometry sorted samples. Environ. Microbiol. 19, 2286–2300. https://doi.org/10.1111/1462-2920.13724.

Lin, C., Blum, J., 1977. Recent invasion of a red alga (*Bangia atropurpurea*) in Lake Michigan. J. Fish. Res. Board Can. 24, 13–161. https://doi.org/10.1139/f77-326.

Lind, J.L., Heimann, K., Miller, E.A., Van Vliet, C., Hoogenraad, N.J., Wetherbee, R., 1997. Substratum adhesion and gliding in a diatom are mediated by extracellular proteoglycans. Planta 203, 213–221. https://doi.org/10.1007/s004250050184.

Lindström, K., 1983. Selenium as a growth factor for plankton algae in laboratory experiments and in some Swedish lakes. Hydrobiologia 101, 35–47. https://doi.org/10.1007/BF00008655.

Linnaeus, C., 1754. Genera plantarum. In: Engelmann, H.R. (Ed.), fifth ed. 1960 reprint, J. Cramer, p. 500.

Loughlin, P., Lin, Y., Chen, M., 2013. Chlorophyll *d* and *Acaryochloris marina*: current status. Photosynth. Res. 116, 277–293. https://doi.org/10.1007/s11120-013-9829-y.

Lüder, U.H., Clayton, M.N., 2004. Induction of phlorotannins in the brown macroalga *Ecklonia radiata* (Laminariales, Phaeophyta) in response to simulated herbivory—The first microscopic study. Planta 218, 928–937. https://doi.org/10.1007/s00425-003-1176-3.

Lysáková, M., Kitner, M., Poulíčková, A., 2007. The epipelic algae at fishponds of Central and Northern Moravia (the Czech Republic). Fottea 7, 69–75. https://doi.org/10.5507/fot.2007.006.

Mac Donagh, M.E., Casco, M.A., Claps, M.C., 2005. Colonization of a neotropical reservoir (Córdoba, Argentina) by *Ceratium hirundinella* (OF Müller) Bergh. J. Limnol. 41, 291–299. https://doi.org/10.1051/limn/2005020.

Mahakham, W., Theerakulpisut, P., 2010. Two new records of coleochaetalean algae (Coleochaetales, Chlorophyta) from Northeast Thailand. Int. J. Bot. 6, 144–150. https://doi.org/10.3923/ijb.2010.144.150.

Mandal, S., Rath, J., 2015. Extremophilic cyanobacteria for novel drug development. Springer, Dordrecht, p. 92.

Marshall, E., Grigg, G., 1980. Lack of metabolic acclimation to different thermal histories by tadpoles of *Limnodynastes peroni* (Anura: Leptodactylidae). Physiol. Zool. 53, 1–7. https://doi.org/10.1086/physzool.53.1.30155768.

Marshall, H., Gladyshev, M.I., 2009. Neuston in aquatic ecosystems. In: Likens, G.E. (Ed.), Encyclopedia of Inland Waters, first ed. Elsevier Academic Press, Amsterdam, Netherlands, pp. 97–102.

Massaro, F.C., Roscha, O., 2008. Development and population growth of *Hydra viridissima* Pallas. Braz. J. Biol. 68 (2), 379–383. https://doi.org/10.1590/S1519-69842008000200020.

McGregor, G.B., Rasmussen, J.P., 2008. Cyanobacterial composition of microbial mats from an Australian thermal spring: a polyphasic evaluation. FEMS (Fed. Eur. Microbiol. Soc.) Microbiol. Ecol. 63, 23–35. https://doi.org/10.1111/j.1574-6941.2007.00405.x.

Menezes, M., 2010. Chrysophyceae. In: Forzza, R.C. (Ed.), Catálogo de plantas e fungos do Brasil, Andrea jakobsson estúdio, 1. Instituto de Pesquisas Jardim Botânico do Rio de Janeiro, Rio de Janeiro, pp. 352–354.

Mertens, K.N., Rengefors, K., Moestrup, Ø., Ellegaard, M., 2012. A review of recent freshwater dinoflagellate cysts: taxonomy, phylogeny, ecology and palaeocology. Phycologia 51, 612–619. https://doi.org/10.2216/11-89.1.

Mesa, V., Gallego, J.L., González-Gil, R., Lauga, B., Sánchez, J., Méndez-García, C., Peláez, A.I., 2017. Bacterial, archaeal, and eukaryotic diversity across distinct microhabitats in an acid mine drainage. Front. Microbiol. 8, 1756. https://doi.org/10.3389/fmicb.2017.01756.

Miller, S.R., Castenholz, R.W., 2000. Evolution of thermotolerance in hot spring cyanobacteria of the genus *Synechococcus*. Appl. Environ. Microbiol. 66 (10), 4222–4229. https://doi.org/10.1128/AEM.66.10.4222-4229.2000.

Montecino, V., Molina, X., Bothwell, M., Muñoz, P., Carrevedo, M.L., Salinas, F., Kumar, S., Castillo, M.L., Bizama, G., Bustamante, R.O., 2016. Spatio temporal population dynamics of the invasive diatom *Didymosphenia geminata* in central-southern Chilean rivers. Sci. Total Environ. 568, 1135–1145. https://doi.org/10.1016/j.scitotenv.2016.03.080.

Morris, I., 1980. The Physiological Ecology of Phytoplankton. Blackwell Scientific Publishers, Oxford, p. 625.

Mpawenayo, B., Cocquyt, C., Nindorera, A., 2005. Diatoms (Bacillariophyta) and other algae from the hot springs of Burundi (Central Africa) in relation with the physical and chemical characteristics of the water. Belg. J. Bot. 138 (2), 152–164.

Müller, K.M., Cole, K.M., Sheath, R.G., 2003. Systematics of *Bangia* (Bangiales, Rhodophyta) in North America II. Biogeographic trends in karyology: chromosome numbers and linkage with gene sequence phylogenetic trees. Phycologia 42, 209–219. https://doi.org/10.2216/i0031-8884-42-3-209.1.

Munawar, M., Fitzpatrick, M., Niblock, H., Kling, H., Rozon, R., Lorimer, J., 2017. Phytoplankton ecology of a culturally eutrophic embayment: Hamilton Harbour, Lake Ontario. Aquat. Ecosys. Health Manag. 20, 201–213. https://doi.org/10.1080/14634988.2017.1307678.

Murphy, D.H., Wilkinson, B.H., 1980. Carbonate deposition and facies distribution in a central Michigan marl lake. Sedimentology 27, 123–135. https://doi.org/10.1111/j.1365-3091.1980.tb01164.x.

Nagata, T., Takai, K., Kawanobe, K., Kim, D.S., Nakazato, R., Guselnikova, N., Bondarenko, N., Mologawaya, O., Kostrnova, T., Drucker, V., Satoh, Y., 1994. Autotrophic picoplankton in southern Lake Baikal: abundance, growth and grazing mortality during summer. J. Plankton Res. 16, 945–959. https://doi.org/10.1093/plankt/16.8.945.

Nägeli, A., Schanz, F., 1990. Planktoneustonic algae in the surface films of Lake Zurich: occurrence and dependence on phytoplankton succession. Aquat. Sci. 52 (3), 269–280. https://doi.org/10.1007/BF00877284.

Ñancucheo, I., Johnson, D.B., 2012. Acidophilic algae isolated from mine–impacted environments and their roles is sustaining heterotrophic acidophiles. Front. Microbiol. 3, 1–8. https://doi.org/10.3389/fmicb.2012.00325.

Necchi Jr., O. (Ed.), 2016. River Algae. Springer International Publishing, Cham, Switzerland, p. 279.

Necchi Jr., O., Zucchi, M.R., 1995. Systematics and distribution of freshwater *Audouinella* (Acrochaetiaceae, Rhodophyta) in Brazil. Eur. J. Phycol. 30, 209–218. https://doi.org/10.1080/09670269500650991.

Negro, A.I., De Hoyos, C., Aldasoro, J.J., 2003. Diatom and desmid relationships with the environment in mountain lakes and mires of NW Spain. Hydrobiologia 505, 1–13. https://doi.org/10.1023/B:HYDR.0000007212.78065.c1.

Nicholls, K.H., 1987. *Chrysoamphipyxis* gen. nov.: a new genus in the Stylococcaceae (Chrysophyceae). J. Phycol. 23, 499–501. https://doi.org/10.1111/j.1529–8817.1987.tb02537.x.

Nicholls, K.H., 1992. Chrysophyte blooms in the plankton and neuston of marine and freshwater systems. Ontario Ministry of Environment (OME), Ontario, Toronto 66. https://doi.org/10.5962/bhl.title.30990.

Nicholls, K.H., 2015. Haptophyte algae. In: Wehr, J.D., Sheath, R.G., Kociolek, J.P. (Eds.), Freshwater Algae of North America. Academic Press, San Diego, CA, pp. 587–605.

Nicholls, K.H., 2017. Introduction to the biology and ecology of the freshwater cryophilic dinoflagellate *Woloszynskia pascheri* causing red ice. Hydrobiologia 784, 305–319. https://doi.org/10.1007/s10750-016-2885-5.

Nicholls, K.H., Beaver, J.L., Estabrook, R.H., 1982. Lakewide odours in Ontario and New Hampshire caused by *Chrysochromulina breviturrita* Nich. (Prymnesiophyceae). Hydrobiologia 96, 91–95. https://doi.org/10.1007/BF00006281.

Nicholls, K.H., Kennedy, W., Hammett, C., 1980. A fish-kill in Heart Lake, Ontario, associated with the collapse of a massive population of *Ceratium hirundinella* (Dinophyceae). Freshw. Biol. 10, 553–561. https://doi.org/10.1111/j.1365-2427.1980.tb01231.x.

Nicholls, K.H., Wujek, D.E., 2015. Chrysophyceae and Phaeothamniophyceae. In: Wehr, J.D., Sheath, R.G., Kociolek, J.P. (Eds.), Freshwater Algae of North America. Academic Press, San Diego, CA, pp. 537–586.

Nikulina, T.V., Kociolek, J.P., 2011. Diatoms from hot springs from Kuril and Sakhalin Islands (far east, Russia). In: Seckbach, J.E., Kociolek, J.P. (Eds.), The Diatom World, Cellular Origin, Life in Extreme Habitats and Astrobiology, 19. Springer, Dordrecht, Netherlands, pp. 335–363.

Nishikawa, K., Tominaga, N., 2001. Isolation, growth, ultrastructure, and metal tolerance of the green alga *Chlamydomonas acidophila* (Chlorophyta). Biosci., Biotechnol., Biochem. 65, 2650–2656. https://doi.org/10.1271/bbb.65.2650.

Norton, T.A., Melkonian, M., Andersen, R.A., 1996. Algal biodiversity. Phycologia 35, 308–326. https://doi.org/10.2216/i0031-8884-35-4-308.1.

Novis, P., Harding, J.S., 2007. Extreme acidophiles: freshwater algae associated with acid mine drainage. In: Seckbach, J. (Ed.), Algae and Cyanobacteria in Extreme Environments. Springer, Heidelberg, pp. 443–463.

O'Kelly, J.O., Wujek, D.E., 2001. Cell structure and asexual reproduction in *Lagynion delicatulum* (Stylococcacease, Chrysophycae). Eur. J. Phycol. 36, 51–59. https://doi.org/10.1080/09670260110001735198.

Oesel, P.F., Kwakkestein, R., Verschoor, A., 1978. Oligotrophication and eutrophication tendencies in some Dutch moorland pools, as reflected in their desmid flora. Hydrobiologia 61, 21–31. https://doi.org/10.1007/BF00019021.

Oliveira, M.D., Calheiros, D.F., 2000. Flood pulse influence on phytoplankton communities of the south Pantanal floodplain, Brazil. Hydrobiologia 427, 101–112. https://doi.org/10.1023/A:1003951930525.

Oliver, R.L., Hamilton, D.P., Brookes, J.D., Ganf, G.G., 2012. Physiology, blooms and prediction of planktonic cyanobacteria. In: Whitton, B.A. (Ed.), Ecology of Cyanobacteria II. Springer, Dordrecht, pp. 155–194.

Oren, A., Ionescu, D., Hindiyeh, M., Malkawi, H., 2008. Microalgae and cyanobacteria of the Dead Sea and its surrounding springs. Isr. J. Plant Sci. 56 (1), 1–13. https://doi.org/10.1560/IJPS.56.1-2.1.

Ormerod, S.D., Rundle, S.M., Wilkinson, G.P., Daly, K.M., Juttner, I., 1994. Altitudinal trends in the diatoms, bryophytes, invertebrates and fish of a Nepalese river system. Freshw. Biol. 32, 309–322. https://doi.org/10.1111/j.1365-2427.1994.tb01128.x.

Ortega, G., Laín, A., Tadeo, X., López–Méndez, B., Castano, D., Millet, O., 2011. Halophilic enzyme activation induced by salts. Sci. Rep. 1 (1), 1–6. https://doi.org/10.1038/srep00006.

Ott, D.W., Oldham-Ott, C.K., Rybalka, N., Friedl, T., 2015. Xanthophyte, eustigmatophyte, and raphidophyte algae. In: Wehr, J.D., Sheath, R.G., Kociolek, J.P. (Eds.), Freshwater Algae of North America. Academic Press, San Diego, CA, pp. 485–536.

Owen, R.B., Renaut, R.W., Jones, B., 2008. Geothermal diatoms: a comparative study of floras in hot spring systems of Iceland, New Zealand, and Kenya. Hydrobiologia 610, 175–192. https://doi.org/10.1007/s10750-008-9432-y.

Paerl, H.W., 1977. Ultraphytoplankton biomass and production in some New Zealand lakes. N. Z. J. Mar. Freshw. Res. 11, 297–305. https://doi.org/10.1080/00288330.1977.9515679.

Pálmai, T., Szabóm, B., Kotut, K., Krienitz, L., Padisák, J., 2020. Ecophysiology of a successful phytoplankton competitor in the African flamingo lakes: the green alga *Picocystis salinarum* (Picocystophyceae). J. Appl. Phycol. 32, 1813–1825. https://doi.org/10.1007/s10811-020-02092-6.

Palmer, G., 1969. A composite rating of algae tolerating organic pollution. J. Phycol. 5, 78–82. https://doi.org/10.1111/j.1529-8817.1969.tb02581.x.

Pålsson, C., Granéli, W., 2003. Diurnal and seasonal variations in grazing by bacterivorous mixotrophs in an oligotrophic clearwater lake. Arch. Hydrobiol. 157, 289–307. https://doi.org/10.1127/0003-9136/2003/0157-0289.

Parke, M., Lund, J.W.G., Manton, I., 1962. Observations on the biology and fine structure of the type species of *Chrysochromulina* (*C. parva* Lackey) in the English Lake District. Arch. Mikrobiol. 42, 333–352. https://doi.org/10.1007/BF00409070.

Passy, S.I., Pan, Y., Lowe, R.L., 1999. Ecology of the major periphytic diatom communities from the Mesta River, Bulgaria. Int. Rev. Gesamten Hydrobiol. 84, 129–174. https://doi.org/10.1002/iroh.199900017.

Pearre Jr., S., 2003. Eat and run? The hunger/satiation hypothesis in vertical migration: history, evidence and consequences. Biol. Rev. 78, 1–79. https://doi.org/10.1017/S146479310200595X.

Peltomaa, E.T., Taipale, S., 2020. Osmotrophic glucose and leucine assimilation and its impact on EPA and DHA content in algae. PeerJ 8, e8363. https://doi.org/10.7717/peerj.8363.

Pentecost, A., 1984. Introduction to Freshwater Algae. The Richmond Publishing Company, Slough, p. 247.

Pentecost, A., 2005. Travertine. Springer, Berlin, p. 445.

Pentecost, A., Andrews, J.E., Dennis, P.F., Marca–Bell, A., Dennis, S., 2006. Charophyte growth in small temperate water bodies: extreme isotopic disequilibrium and implications for the palaeoecology of shallow marl lakes. Palaeogeogr. Palaeoclimatol. Palaeoecol. 240, 389–404. https://doi.org/10.1016/j.palaeo.2006.02.008.

Pentecost, A., Jones, B., Renaut, R.W., 2003. What is a hot spring? Can. J. Earth Sci. 40, 1443–1446. https://doi.org/10.1139/e03-083.

Pepe-Ranney, C., Berelson, W.M., Corsetti, F.A., Treants, M., Spear, J.R., 2012. Cyanobacterial construction of hot spring siliceous stromatolites in Yellowstone National Park. Environ. Microbiol. 14 (5), 1182–1197. https://doi.org/10.1111/j.1462-2920.2012.02698.x.

Perkerson, R.B., Johansen, J.R., Kovácik, L., Brand, J., Kaštovský, J., Casamatta, D.A., 2011. A unique Pseudanabaenalean (Cyanobacteria) genus *Nodosilinea* gen. nov. based on morphological and molecular data. J. Phycol. 47, 1397–1412. https://doi.org/10.1111/j.1529-8817.2011.01077.x.

Pettersson, K., 1985. The availability of phosphorus and the species composition of the spring phytoplankton in Lake Erken. Internationale Revue der Gesamten Hydrobiologie und Hydrographie 70, 527–546. https://doi.org/10.1002/iroh.19850700407.

Pick, F.R., Nalewajko, C., Lean, D.R.S., 1984. The origin of a metalimnetic chrysophyte peak. Limnol. Oceanogr. 29, 125–134. https://doi.org/10.4319/lo.1984.29.1.0125.

Pinder, A.W., Friet, S.C., 1994. Oxygen transport in egg masses of the amphibians *Rana sylvatica* and *Ambystoma maculatum*: convection, diffusion and oxygen production by algae. J. Exp. Biol. 197, 17–30.

Poulíčková, A., Hájek, M., Rybníček, K., 2005. Ecology and Palaeoecology of Spring Fens of the West Carpathians. Palacký University Press, Olomouc, p. 209.

Poulíčková, A., Hájková, P., Křenková, P., Hájek, M., 2004. Distribution of diatoms and bryophytes on linear transects through spring fens. Nova Hedwigia 78, 411–424. https://doi.org/10.1127/0029-5035/2004/0078-0411.

Poulíčková, A., Hašler, P., Lysáková, M., Spears, B., 2008. The ecology of freshwater epipelic algae: an update. Phycologia 47, 437–450. https://doi.org/10.2216/07-59.1.

Prescott, G.W., 1951. Algae of the Western Great Lakes Area: exclusive of desmids and diatoms. Cranbrook Institute of Science, Bloomfield Hills, Michigan, 946.

Price, N.M., Morel, F.M., 1991. Colimitation of phytoplankton growth by nickel and nitrogen. Limnol. Oceanogr. 36, 1071–1077. https://doi.org/10.4319/lo.1991.36.6.1071.

Pringle, C.M., Rowe, G.L., Triska, F.J., Fernandez, J.F., West, J., 1993. Landscape linkages between geothermal activity and solute composition and ecological response in surface waters draining the Atlantic slope of Costa Rica. Limnol. Oceanogr. 38, 753–774. https://doi.org/10.4319/lo.1993.38.4.0753.

Priscu, J.C., Adams, E.E., Lyons, W.B., Voytek, M.A., Mogk, D.W., Brown, R.L., McKay, C.P., Takacs, C.D., Welch, K.A., Wolf, C.F., Kirstein, J.D., Avci, R., 1999a. Geomicrobiology of sub-glacial ice above Vostok Station. Science 286, 2141–2144. https://doi.org/10.1126/science.286.5447.2141.

Priscu, J.C., Wolf, C.F., Takacs, C.D., Fritsen, C.H., Laybourn-Parry, J., Roberts, E.C., Lyons, W.B., 1999b. Carbon transformations in the water column of a perennially ice-covered Antarctic Lake. Bioscience 49, 997–1008. https://doi.org/10.2307/1313733.

Priscu, J.C., Fritsen, C.H., Adams, E.E., Giovannoni, S.J., Paerl, H.W., McKay, C.P., Doran, P.T., Gordon, D.A., Lanoil, B.D., Pinckney, J.L., 1998. Perennial Antarctic lake ice: an oasis for life in a polar desert. Science 280, 2095–2098. https://doi.org/10.1126/science.280.5372.2095.

Pumas, C., Pruetiworanan, S., Peerapornpisal, Y., 2018. Diatom diversity in some hot springs of northern Thailand. Botanica (Delhi) 24, 69–86. https://doi.org/10.2478/botlit-2018-0007.

Queimaliños, C.P., Modenutti, B., Balseiro, E., 1999. Symbiotic association of the ciliate *Ophrydium naumanni* with *Chlorella* causing a deep chlorophyll a maximum in an oligotrophic South Andes lake. J. Plankton Res. 21 (1), 167–178. https://doi.org/10.1093/plankt/21.1.167.

Rahman, A., Agrawal, S., Nawaz, T., Pan, S., Selvaratnam, T., 2020. A review of algae-based produced water treatment for biomass and biofuel production. Water 12 (9), 2351. https://doi.org/10.3390/w12092351.

Rawson, D.S., 1956. Algal indicators of trophic lake types. Limnol. Oceanogr. 1, 18–25. https://doi.org/10.4319/lo.1956.1.1.0018.

Rengefors, K., Pålsson, C., Hansson, L.A., Heiberg, L., 2008. Cell lysis of competitors and osmotrophy enhance growth of the bloom-forming alga *Gonyostomum semen*. Aquat. Microb. Ecol. 51, 87–96. https://doi.org/10.3354/ame01176.

Rengefors, K., Weyhenmeyer, G.A., Bloch, I., 2012. Temperature as a driver for the expansion of the microalga *Gonyostomum semen* in Swedish lakes. Harmful Algae 18, 65–73. https://doi.org/10.1016/j.hal.2012.04.005.

Reynolds, C.S., 1980. Phytoplankton assemblages and their periodicity in stratifying lake systems. Holarctic Ecology 3, 141–159.

Reynolds, C.S., 1984. Phytoplankton periodicity: the interactions of form, function and environmental variability. Freshw. Biol. 14, 111–142. https://doi.org/10.1111/j.1365-2427.1984.tb00027.x.

Reynolds, C.S., 1987. Cyanobacterial water-blooms. Adv. Bot. Res. 13, 67–143. https://doi.org/10.1016/S0065-2296(08)60341-9.

Reynolds, C.S., 1988. The concept of ecological succession applied to seasonal periodicity of freshwater phytoplankton. Internationale Vereinigung für Theoretische und Angewandte Limnologie: Verh Proc. Trav. SIL 23, 683–691. https://doi.org/10.1080/03680770.1987.11899692.

Reynolds, C.S., 1994. The long, the short and the stalled: on the attributes of phytoplankton selected by physical mixing in lakes and rivers. Hydrobiologia 289, 9–21. https://doi.org/10.1007/BF00007405.

Reynolds, C.S., 2006. The ecology of phytoplankton. Cambridge University Press, Cambridge, p. 535. https://doi.org/10.1017/CBO9780511542145.

Reynolds, C.S., Descy, J.P., 1996. The production, biomass and structure of phytoplankton in large rivers. Large Rivers 10, 161–187. https://doi.org/10.1127/lr/10/1996/161.

Rich, V., Maier, R., 2015. Aquatic environments. In: Pepper, I.L., Gerba, C.P., Gentry, T.J. (Eds.), Environmental Microbiology, third ed. Academic Press, Amsterdam and Boston, MA, pp. 111–138. https://doi.org/10.1016/B978–0–12–394626–3.00006–5.

Richards, T.A., Vepritskiy, A.A., Gouliamova, D.E., Nierzwicki-Bauer, S.A., 2005. The molecular diversity of freshwater picoeukaryotes from an oligotrophic lake reveals diverse, distinctive and globally dispersed lineages. Environ. Microbiol. 7, 1413–1425. https://doi.org/10.1111/j.1462-2920.2005.00828.x.

Riedel, G.F., Sanders, J.G., Gilmour, C.C., 1996. Uptake, transformation, and impact of selenium in freshwater phytoplankton and bacterioplankton communities. Aquat. Microb. Ecol. 11, 43–51. https://doi.org/10.3354/ame011043.

Riseng, C.M., Wiley, M.J., Stevenson, R.J., 2004. Hydrologic disturbance and nutrient effects on benthic community structure in midwestern US streams: a covariance structure analysis. J. North Am. Benthol. Soc. 23, 309–326. https://doi.org/10.1899/0887-3593(2004)023%3C0309:HDANEO%3E2.0.CO;2. https://www.researchgate.net/journal/Journal-of-the-North-American-Benthological-Society-1937-237X.

Rodrigo, M.A., Rojo, C., Álvarez-Cobelas, M., Cirujano, S., 2007. *Chara hispida* beds as a sink of nitrogen: evidence from growth, nitrogen uptake and decomposition. Aquat. Bot. 87, 7–14. https://doi.org/10.1016/j.aquabot.2007.01.007.

Roelke, D.L., Grover, J.P., Brooks, B.W., Glass, J., Buzan, D., Southard, G.M., Fries, L., Gable, G.M., Schwierzke-Wade, L., Byrd, M., Nelson, J., 2011. A decade of fish-killing *Prymnesium parvum* blooms in Texas: roles of inflow and salinity. J. Plankton Res. 33, 243–253. https://doi.org/10.1093/plankt/fbq079.

Rosowski, J.R., 2003. Photosynthetic euglenoids. In: Wehr, J.D., Sheath, R.G., Kociolek, J.P. (Eds.), Freshwater Algae of North America. Academic Press, San Diego, CA, pp. 383–422.

Rott, E., 1984. Phytoplankton as biological parameter for the trophic characterization of lakes. Verhandlunger der Internationale Vereinigung für Limnologie 22, 1078–1085. https://doi.org/10.1080/03680770.1983.11897441.

Rott, E., 1988. Some aspects of the seasonal distribution of flagellates in mountain lakes. Hydrobiologia 161, 159–170. https://doi.org/10.1007/BF00044108.

Rott, E., Cantonati, M., Füreder, L., Pfister, P., 2006. Benthic algae in high altitude streams of the Alps—A neglected component of the aquatic biota. Hydrobiologia 562, 195–216. https://doi.org/10.1007/s10750-005-1811-z.

Rott, E., Schneider, S.C., 2014. A comparison of ecological optima of soft-bodied benthic algae in Norwegian and Austrian rivers and consequences for river monitoring in Europe. Sci. Total Environ. 475, 180–186. https://doi.org/10.1016/j.scitotenv.2013.08.050.

Rott, E., Wehr, J.D., 2016. The spatio-temporal development of macroalgae in rivers. In: Necchi Jr., O. (Ed.), River Algae. Springer-Verlag, Cham, Switzerland, pp. 159–195.

Round, F.E., 1965. The epipsammon; a relatively unknown freshwater algal association. Br. Phycol. Bull. 2, 456–463. https://doi.org/10.1080/00071616500650071.

Round, F.E., 1971. The growth and succession of algal populations in freshwaters. Mitteilungen—Internationale Vereinigung für Theoretische und Angewandte Limnologie 19, 70–99. https://doi.org/10.1080/05384680.1971.11903924.

Round, F.E., 1981. The Ecology of Algae. Cambridge University Press, Cambridge, p. 651.

Round, F.E., 1993. A review and methods for the use of epilithic diatoms for detecting and monitoring changes in river water quality. HMSO Publications, London, p. 65.

Round, F.E., Crawford, R.M., Mann, D.G., 1990. The diatoms—biology and morphology of the genera. Cambridge University Press, Cambridge, p. 747.

Round, F.E., Happey, C.M., 1965. Persistent, vertical migration rhythms in benthic microflora. Part IV A diurnal rhythm of the epipelic diatom association in non-tidal flowing water. Br. Phycol. Bull. 2, 463–471. https://doi.org/10.1080/00071616500650081.

Rowe, O.F., Sánchez–España, J., Hallberg, K.B., Johnson, D.B., 2007. Microbial communities and geochemical dynamics in an extremely acidic, metal-rich stream at an abandoned sulfide mine (Huelva, Spain) underpinned by two functional primary production systems. Environ. Microbiol. 9, 1761–1771. https://doi.org/10.1111/j.1462-2920.2007.01294.x.

Roy, S., Debnath, M., Ray, S., 2014. Cyanobacterial flora of the geothermal spring at Panifala, West Bengal, India. Phykos 44 (1), 1–8.

Rühland, K.M., Paterson, A.M., Smol, J.P., 2015. Lake diatom responses to warming: reviewing the evidence. J. Paleolimnol. 54, 1–35. https://doi.org/10.1007/s10933-015-9837-3.

Rybak, A.S., Gąbka, M., 2018. The influence of abiotic factors on the bloom-forming alga *Ulva flexuosa* (Ulvaceae, Chlorophyta): possibilities for the control of the green tides in freshwater ecosystems. J. Appl. Phycol. 30, 1405–1416. https://doi.org/10.1007/s10811–017–1301–5.

Saber, A.A., El-Refaey, A.A., Saber, H., Singh, P., Janse van Vuuren, S., Cantonati, M., 2022. Cyanoprokaryotes and algae: classification and habitats. In: El-Sheekh, M., Abomohra, A. (Eds.), Handbook of Algal Biofuels: Aspects of Cultivation, Conversion and Biorefinery. Elsevier, Amsterdam, Netherlands, p. 682.

Sahoo, D., Baweja, P., 2005. General characteristics of algae. In: Sahoo, D., Seckbach, J. (Eds.), The Algae World. In: Cellular Origin, Life in Extreme Habitats and Astrobiology, vol. 26. Springer, Dordrecht, pp. 3–30.

Salonen, K., Rosenberg, M., 2000. Advantages from diel vertical migration can explain the dominance of *Gonyostomum semen* (Raphidophyceae) in a small, steeply-stratified humic lake. J. Plankton Res. 22, 1841–1853. https://doi.org/10.1093/plankt/22.10.1841.

Sambamurty, A.V.S.S., 2005. A Textbook of Algae. I.K. International, New Delhi, p. 336.

Sanders, R.W., 2011. Alternative nutritional strategies in protists: symposium introduction and a review of freshwater protists that combine photosynthesis and heterotrophy. J. Eukaryot. Microbiol. 58, 181–184. https://doi.org/10.1111/j.1550-7408.2011.00543.x.

Sanders, R.W., Porter, K.G., 1988. Phagotrophic phytoflagellates. Adv. Microb. Ecol. 10, 167–192. https://doi.org/10.1007/978-1-4684-5409-3_5.

Schagerl, M., Burian, A., Gruber-Dorninger, M., Oduor, S., Kaggwa, M.N., 2015. Algal communities of Kenyan soda lakes with a special focus on *Arthrospira fusiformis*. Fottea 15 (2), 245–257. https://doi.org/10.5507/fot.2015.012.

Schagerl, M., Kerschbaumer, M., 2009. Autecology and morphology of selected *Vaucheria* species (Xanthophyceae). Aquat. Ecol. 43, 295–303. https://doi.org/10.1007/s10452-007-9163-6.

Schaumburg, J., Schranz, C., Foerster, J., Gutowski, A., Hofmann, G., Meilinger, P., Schneider, S., Schmedtje, U., 2004. Ecological classification of macrophytes and phytobenthos for rivers in Germany according to the water framework directive. Limnologica 34, 283–301. https://doi.org/10.1016/S0075-9511(04)80002-1.

Scheffer, M., Rinaldi, S., Huisman, J., Weissing, F.J., 2003. Why plankton communities have no equilibrium: solutions to the paradox. Hydrobiologia 491, 9–18. https://doi.org/10.1023/A:1024404804748.

Schindler, D.W., 1977. Evolution of phosphorus limitation in lakes. Science 46, 260–262.

Schneider, S.C., Lindstrøm, E.A., 2011. The periphyton index of trophic status PIT: a new eutrophication metric based on non-diatomaceous benthic algae in Nordic rivers. Hydrobiologia 665, 143–155. https://doi.org/10.1126/science.195.4275.260.

Schoeman, F.R., Archibald, R.E.M., 1988. Taxonomic notes on the diatoms (Bacillariophyceae) of the Gross Barmen thermal springs in South West Africa/Namibia. South Afr. J. Bot. 54 (3), 221–256. https://doi.org/10.1016/S0254-6299(16)31322-9.

Schopf, J.W., 2012. The fossil record of cyanobacteria. In: Whitton, B.A. (Ed.), Ecology of Cyanobacteria II: Their Diversity in Space and Time. Springer, Dordrecht, pp. 15–36.

Schwaderer, A.S., Yoshiyama, K., De Tezanos Pinto, P., Swenson, N.G., Klausmeier, C.A., Litchman, E., 2011. Eco-evolutionary differences in light utilization traits and distributions of freshwater phytoplankton. Limnol. Oceanogr. 56, 589–598. https://doi.org/10.4319/lo.2011.56.2.0589.

Seckbach, J. (Ed.), 2007. Algae and Cyanobacteria in Extreme Environments. In: Cellular Origin, Life in Extreme Habitats and Astrobiology. Springer, Dordrecht, p. 786.

Senties, A.G., Espinoza–Avalos, J., Zurita, J.C., 1999. Epizoic algae of nesting sea turtles *Caretta* (L.) and *Chelonia mydas* (L.) from the Mexican Caribbean. Bull. Mar. Sci. 64, 185–188.

Shea, T.B., Sheath, R.G., Chhun, A., Vis, M.L., Chiasson, W.B., Müller, K.M., 2014. Distribution, seasonality and putative origin of the non–native red alga *Bangia atropurpurea* (Bangiales, Rhodophyta) in the Laurentian Great Lakes. J. Great Lake. Res. 40, 27–34. https://doi.org/10.1016/j.jglr.2014.01.004.

Sheath, R.G., Cole, K.M., 1992. Biogeography of stream macroalgae in North America. J. Phycol. 28, 448–460. https://doi.org/10.1111/j.0022-3646.1992.00448.x.

Sheath, R.G., Havas, M., Hellebust, J.A., Hutchinson, T.C., 1982. Effects of long-term natural acidification on the algal communities of tundra ponds at the Smoking Hills, N.W.T., Canada. Can. J. Bot. 60, 58–72. https://doi.org/10.1139/b82-008.

Sheath, R.G., Morrison, M.O., 2004. Epiphytes on *Cladophora glomerata* in the Great Lakes and St. Lawrence seaway with particular reference to the red alga *Chroodactylon ramosum* (= *Asterocytis smargdina*). J. Phycol. 18 (3), 385–391. https://doi.org/10.1111/j.1529-8817.1982.tb03200.x.

Sheath, R.G., Steinman, A.D., 1982. A checklist of freshwater algae of the Northwest Territories, Canada. Can. J. Bot. 60, 1964–1997. https://doi.org/10.1139/b82-245.

Sheath, R.G., Vis, M.L., 2015. Red algae. In: Wehr, J.D., Sheath, R.G., Kociolek, J.P. (Eds.), Freshwater Algae of North America. Academic Press, San Diego, CA, pp. 236–264.

Sheath, R.G., Wehr, J.D., 2015. Introduction to the freshwater algae. In: Wehr, J.D., Sheath, R.G., Kociolek, J.P. (Eds.), Freshwater Algae of North America. Academic Press, San Diego, CA, pp. 1–11.

Shelford, E.J., Suttle, C.A., 2018. Virus-mediated transfer of nitrogen from heterotrophic bacteria to phytoplankton. Biogeosciences 15, 809–819. https://doi.org/10.5194/bg-15-809-2018.

Sherwood, A.R., 2016. Green algae (Chlorophyta and Streptophyta) in rivers. In: Necchi Jr., O. (Ed.), River Algae. Springer International Publishing, Cham, Switzerland, pp. 35–63.

Shi, P., Shen, H., Wang, W., Yang, Q., Xie, P., 2016. Habitat-specific differences in adaptation to light in freshwater diatoms. J. Appl. Phycol. 28, 227–239. https://doi.org/10.1007/s10811-015-0531-7.

Sicko-Goad, L., Stoermer, E.F., 1984. The need for uniform terminology concerning phytoplankton cell size fractions and examples of picoplankton from the Laurentian Great Lakes. J. Great Lake. Res. 10, 90–93. https://doi.org/10.1016/S0380-1330(84)71812-0.

Sieburth, J.Mc.N., Smetacek, V., Lenz, J., 1978. Pelagic ecosystem structure: heterotrophic compartments of the plankton and their relationship to plankton size fractions. Limnol. Oceanogr. 23, 1256–1263. https://doi.org/10.4319/lo.1978.23.6.1256.

Simić, S., Cvijan, M., Ranković, B., 2002. Distribution and ecology of green algae (Chlorophyta) in the watershed of the Timok River (Serbia). Ekologija 37 (1–2), 23–32.

Singh, Y., Gulati, A., Singh, D.P., Khattar, J.I.S., 2018. Cyanobacterial community structure in hot water springs of Indian North-Western Himalayas: a morphological, molecular and ecological approach. Algal Res. 29, 179–192. https://doi.org/10.1016/j.algal.2017.11.023.

Siver, P.A., 2015. Synurophyte algae. In: Wehr, J.D., Sheath, R.G., Kociolek, J.P. (Eds.), Freshwater Algae of North America. Academic Press, San Diego, CA, pp. 607–651.

Siver, P.A., 2019. Potential use of chrysophyte cyst morphometrics as a tool for reconstructing ancient lake environments. Nova Hedwigia, Beiheft 48, 101–112. https://doi.org/10.1127/nova-suppl/2019/115.

Smith, G.M., 1950. The Fresh-water Algae of the United States. McGraw-Hill, New York, NY, p. 719.

Smith, V.H., 1983. Low nitrogen to phosphorus ratios favor dominance by blue-green algae in lake phytoplankton. Science 221, 669–671. https://science.sciencemag.org/content/221/4611/669.

Smol, J.P., 1995. Application of chrysophytes to problems in paleoecology. In: Kristiansen, J. (Ed.), Chrysophyte Algae: Ecology, Phylogeny and Development. Cambridge University Press, Cambridge, pp. 303–329.

Smol, J.P., Stoermer, E.F. (Eds.), 2010. The Diatoms: Applications for the Environmental and Earth Sciences. Cambridge University Press, Cambridge, p. 686.

Sommer, U., Gliwicz, M.Z., 1986. Long range vertical migration of *Volvox* in tropical Lake Cahora Bassa (Mozambique). Limnol. Oceanogr. 31, 650–653. https://doi.org/10.4319/lo.1986.31.3.0650.

Sommer, U., Padisák, J., Reynolds, C.S., Juhász-Nagy, P., 1993. Hutchinson's heritage: the diversity-disturbance relationship in phytoplankton. Hydrobiologia 249, 1–7. https://doi.org/10.1007/BF00008837.

Somogyi, B., Felföldi, T., Tóth, L.G., Bernát, G., Vörös, L., 2020. Photoautotrophic picoplankton—A review on their occurrence, role and diversity in Lake Balaton. Biologia Futura 71, 371–382. https://doi.org/10.1007/s42977-020-00030-8.

Spijkerman, E., 2005. Inorganic carbon acquisition by *Chlamydomonas acidophila* across a pH range. Can. J. Bot. 83, 872–878. https://doi.org/10.1139/b05-073.

Spijkerman, E., 2007. Phosphorus acquisition by *Chlamydomonas acidophila* under autotrophic and osmo-mixotrophic growth conditions. J. Exp. Bot. 58 (15/16), 4195–4202. https://doi.org/10.1093/jxb/erm276.

Stal, L.J., 1995. Physiological ecology of cyanobacteria in microbial mats and other communities. New Phytol. 131, 1–32. https://doi.org/10.1111/j.1469-8137.1995.tb03051.x.

Stancheva, R., Fetscher, A.E., Sheath, R.G., 2012. A novel quantification method for stream-inhabiting, non-diatom benthic algae, and its application in bioassessment. Hydrobiologia 684, 225–239. https://doi.org/10.1007/s10750-011-0986-8.

Stein, J.R., Borden, C.A., 1979. Checklist of freshwater algae of British Columbia. Syesis 12, 3–39.

Steinman, A.D., Lamberti, G.A., Leavitt, P.R., Uzarski, D.G., 2017. Biomass and pigments of benthic algae. In: Hauer, F.M., Lamberti, G.A. (Eds.), Methods in Stream Ecology, vol. 1. Academic Press, San Diego, CA, pp. 223–241.

Stelzer, R.S., Lamberti, G.A., 2001. Effects of N: P ratio and total nutrient concentration on stream periphyton community structure, biomass, and elemental composition. Limnol. Oceanogr. 46, 356–367. https://doi.org/10.4319/lo.2001.46.2.0356.

Sterner, R.W., Elser, J.J., 2002. Ecological Stoichiometry: The biology of elements from molecules to the biosphere. Princeton University Press, Princeton, NJ, p. 439.

Sterner, R.W., Smutka, T.M., McKay, R.M.L., Xiaoming, Q., Brown, E.T., Sherrell, R.M., 2004. Phosphorus and trace metal limitation of algae and bacteria in Lake Superior. Limnol. Oceanogr. 49, 495–507. https://doi.org/10.4319/lo.2004.49.2.0495.

Stevenson, R.J., 1996. An introduction to algal ecology in freshwater benthic habitats. In: Stevenson, R.J., Bothwell, M.L., Lowe, R.L. (Eds.), Algal Ecology: Freshwater Benthic Ecosystems. Academic Press, San Diego, CA, pp. 3–30.

Stevenson, R.J., Bothwell, M.L., Lowe, R.L., 1996. Algal Ecology: Freshwater Benthic Ecosystems. Academic Press, San Diego, CA, p. 753.

Stevenson, R.J., Smol, J.P., 2015. Use of algae in ecological assessments. In: Wehr, J.D., Sheath, R.G., Kociolek, J.P. (Eds.), Freshwater Algae of North America. Academic Press, San Diego, CA, pp. 921–962.

Stewart, W.D.P. (Ed.), 1974. Algal Physiology and Biochemistry. University of California Press, Berkeley, CA, p. 989.

Stewart, A.J., Wetzel, R.G., 1986. Cryptophytes and other microflagellates as couplers in planktonic community dynamics. Arch. Hydrobiol. 106, 1–19.

Stockner, J.G., 1967. Observations of thermophilic algal communities in Mount Rainier and Yellowstone National Parks. Limnol. Oceanogr. 12 (1), 13–17. https://doi.org/10.4319/lo.1967.12.1.0013.

Stoermer, E.F., Sicko-Goad, L., 1977. A new distribution record for *Hymenomonas roseola* Stein (Prymnesiophyceae, Coccolithophoraceae) and *Spiniferomonas Trioralis* Takahashi (Chrysophyceae, Synuraceae) in the Laurentian Great Lakes. Phycologia 16, 355–358. https://doi.org/10.2216/i0031-8884-16-4-355.1.

Stomp, M., Huisman, J., Vörös, L., Pick, F.R., Laamanen, M., Haverkamp, T., Stal, L.J., 2007. Colourful coexistence of red and green picocyanobacteria in lakes and seas. Ecol. Lett. 10 (4), 290–298. https://doi.org/10.1111/j.1461-0248.2007.01026.x.

Stoyneva, M.P., 2003. Survey on green algae of Bulgarian thermal springs. Biologia, Bratislava 58 (4), 563–574.

Sultana, M., Asaeda, T., Manatunge, J., Ablimit, A., 2004. Colonisation and growth of epiphytic algal communities on *Potamogeton perfoliatus* under two different light regimes. N. Z. J. Mar. Freshw. Res. 38, 585–594. https://doi.org/10.1080/00288330.2004.9517264.

Sundbäck, K., Jönsson, B., Nilsson, P., Lindström, I., 1990. Impact of accumulating drifting macroalgae on a shallow-water sediment system: an experimental study. Marine Ecology Progress 58, 261–274. https://doi.org/10.3354/meps058261.

Taipale, S.J., Kainz, M.J., Brett, M.T., 2011. Diet-switching experiments show rapid accumulation and preferential retention of highly unsaturated fatty acids in *Daphnia*. Oikos 120, 1674–1682. https://doi.org/10.1111/j.1600-0706.2011.19415.x.

Takasu, H., Ushio, M., LeClair, J.E., Nakano, S.I., 2015. High contribution of *Synechococcus* to phytoplankton biomass in the aphotic hypolimnion in a deep freshwater lake (Lake Biwa, Japan). Aquat. Microb. Ecol. 75, 69–79. https://doi.org/10.3354/ame01749.

Taxböck, L., Karger, D.N., Kessler, M., Spitale, D., Cantonati, M., 2020. Diatom species richness in Swiss springs increases with habitat complexity and elevation. Water 12, 449. https://doi.org/10.3390/w12020449.

Taylor, B.W., Bothwell, M.L., 2014. The origin of invasive microorganisms matters for science, policy, and management: the case of *Didymosphenia geminata*. Bioscience 64, 531–538. https://doi.org/10.1093/biosci/biu060.

Taylor, J.C., Cocquyt, C., 2016. Diatoms from the Congo and Zambezi basins—methodologies and identification of the genera. ABC Taxa and the Belgian Development Cooperation, Brussels, Belgium, p. 364.

Terrado, R., Pasulka, A.L., Lie, A.A., Orphan, V.J., Heidelberg, K.B., Caron, D.A., 2017. Autotrophic and heterotrophic acquisition of carbon and nitrogen by a mixotrophic chrysophyte established through stable isotope analysis. ISME J. 11, 2022–2034. https://doi.org/10.1038/ismej.2017.68.

Tevena, I., Dzhambazov, B., Koleva, L., Mladenov, R., Schirmer, K., 2005. Toxic potential of five freshwater *Phormidium* species (Cyanoprokaryota). Toxicon 45, 711–725. https://doi.org/10.1016/j.toxicon.2005.01.018.

Thorp, J.H., Delong, M.D., 2002. Dominance of autochthonous autotrophic carbon in food webs of heterotrophic rivers. Oikos 96, 543–550. https://doi.org/10.1034/j.1600-0706.2002.960315.x.

Tiffany, L.H., 1958. Algae, the Grass of Many Waters, second ed. C.S. Thomas, Springfield, IL, p. 199.

Tilman, D., Kilham, S.S., 1976. Phosphate and silicate growth and uptake kinetics of the diatoms *Asterionella formosa* and *Cyclotella meneghiniana* in batch and semicontinuous culture. J. Phycol. 12, 375–383. https://doi.org/10.1111/j.0022-3646.1976.00375.x.

Tilman, D., Kilham, S.S., Kilham, P., 1982. Phytoplankton community ecology: the role of limiting nutrients. Annu. Rev. Ecol. Systemat. 13, 349–372. https://doi.org/10.1146/annurev.es.13.110182.002025.

Tittley, I., Neto, A.I., 2005. The marine algal (seaweed) flora of the Azores: additions and amendments. Bot. Mar. 48 (3), 248–255. https://doi.org/10.1515/BOT.2005.030.

Toplin, J.A., Norris, T.B., Lehr, C.R., McDermott, T.R., Castenholz, R.W., 2008. Biogeographic and phylogenetic diversity of thermoacidophilic cyanidiales in Yellowstone National Park, Japan, and New Zealand. Appl. Environ. Microbiol. 74, 2822–2833. https://doi.org/10.1128/AEM.02741-07.

Toporowska, M., Pawlik-Skowrońska, B., Wojtal, A.Z., 2008. Epiphytic algae on *Stratiotes aloides* L., *Potamogeton lucens* L., *Ceratophyllum demersum* L. and *Chara* spp. in a macrophyte-dominated lake. Oceanol. Hydrobiol. Stud. 37 (2), 51–63. https://doi.org/10.2478/v10009-007-0048-8.

Torres-Ruiz, M., Wehr, J.D., Perrone, A.A., 2007. Trophic relations in a stream food web: importance of fatty acids for macroinvertebrate consumers. J. North Am. Benthol. Soc. 26, 509–522. https://doi.org/10.1899/06-070.1.

Tranvik, L.J., Porter, K.G., Sieburth, J.M., 1989. Occurrence of bacterivory in *Cryptomonas*, a common freshwater phytoplankter. Oecologia 78, 473–476. https://doi.org/10.1007/BF00378736.

Triemer, R.E., Zakryś, B., 2015. Photosynthetic euglenoids. In: Wehr, J.D., Sheath, R.G., Kociolek, J.P. (Eds.), Freshwater Algae of North America. Academic Press, San Diego, CA, pp. 459–483.

Trivedi, P.C., Pandey, S., Bhadauria, S., 2010. Text Book of Microbiology. Aavishkar Publishers, Jaipur, p. 446.

Tromas, N., Taranu, Z.E., Castelli, M., Pimentel, J.S., Pereira, D.A., Marcoz, R., Shapiro, B.J., Giani, A., 2020. The evolution of realized niches within freshwater *Synechococcus*. Environ. Microbiol. 22, 1238–1250. https://doi.org/10.1111/1462-2920.14930.

Tumlison, R., Trauth, S.E., 2006. A novel facultative mutualistic relationship between bufonid tadpoles and flagellated green algae. Herpetol. Conserv. Biol. 1 (1), 51–55.

Turner, M.A., Howell, E.T., Summerby, M., Hesslein, R.H., Findlay, D.L., Jackson, M.B., 1991. Changes in epilithon and epiphyton associated with experimental acidification of a lake to pH 5. Limnol. Oceanogr. 36, 1390–1405. https://doi.org/10.4319/lo.1991.36.7.1390.

Ultsch, G.R., Bradford, D.F., Freda, J., 1999. Physiology: coping with the environment. In: McDiarmid, R.W., Altig, R. (Eds.), Tadpoles: The Biology of Anuran Larvae. University of Chicago Press, Chicago, IL, pp. 189–214.

Urabe, J., Gurung, T.B., Yoshida, T., 1999. Effects of phosphorus supply on phagotrophy by the mixotrophic alga *Uroglena americana* (Chrysophyceae). Aquat. Microb. Ecol. 18, 77–83.

Urabe, J., Gurung, T.B., Yoshida, T., Sekino, T., Nakanishi, M., Maruo, M., Nakayama, E., 2000. Diel changes in phagotrophy by *Cryptomonas* in Lake Biwa. Limnol. Oceanogr. 45, 1558–1563. https://doi.org/10.3354/ame018077.

Vadeboncoeur, Y., Vander Zanden, M.J., Lodge, D.M., 2002. Putting the lake back together: reintegrating benthic pathways into lake food web models: lake ecologists tend to focus their research on pelagic energy pathways, but, from algae to fish, benthic organisms form an integral part of lake food webs. Bioscience 52 (1), 44–54. https://doi.org/10.1641/0006-3568(2002)052[0044:PTLBTR]2.0.CO;2.

Valkanov, A., 1968. Das Neuston. Limnologica 6, 381–408.

Van de Vijver, B., Cocquyt, C., 2009. Four new diatom species from La Calera hot spring in the Peruvian Andes (Colca Canyon). Diatom Res. 2, 209–223. https://doi.org/10.1080/0269249X.2009.9705792.

Van den Hoek, C., Mann, D., Jahns, H.M., 1995. Algae: An Introduction to Phycology. Cambridge University Press, Cambridge, p. 623.

Van den Hoff, J., Bell, E., Whittock, L., 2020. Dimorphism in the Antarctic cryptophyte *Geminigera cryophila* (Cryptophyceae). J. Phycol. 56, 1028–1038. https://doi.org/10.1111/jpy.13004.

Van Ginkel, C.E., Hohls, B.C., Vermaak, E., 2001. A *Ceratium hirundinella* (O.F. Müller) bloom in Hartbeespoort Dam, South Africa. WaterSA 27, 269–276. https://doi.org/10.4314/wsa.v27i2.5000.

Vincent, W.F., 1988. Microbial Ecosystems of Antarctica. Cambridge University Press, Cambridge, p. 304.

Vinebrooke, R.D., Dixit, S.S., Graham, M.D., Gunn, J.M., Chen, Y., Belzile, N., 2002. Whole-lake algal responses to a century of acidic industrial deposition on the Canadian Shield. Can. J. Fish. Aquat. Sci. 59, 483–493. https://doi.org/10.1139/f02-025.

Vis, M.L., Sheath, R.G., 1996. Distribution and systematics of *Batrachospermum* (Batrachospermales, Rhodophyta) in North America. 9. section *Batrachospermum*: description of five new species. Phycologia 35, 124–134. https://doi.org/10.1080/09670269600651371.

Waern, M., 1952. Rocky-shore algae in the Öregund Archipelago. Acta Phytogeogr. Suec. 30, 1–298.

Ward, D.M., Castenholz, R.W., Miller, S.R., 2012. Cyanobacteria in geothermal habitats. In: Whitton, B.A. (Ed.), Ecology of Cyanobacteria II: Their Diversity in Space and Time. Springer, Dordrecht, pp. 39–63.

Ward, D.M., Ferris, M.J., Nold, S.C., Bateson, M.M., 1998. A natural view of microbial biodiversity within hot spring cyanobacterial mat communities. Microbiol. Mol. Biol. Rev. 62, 1353–1370. https://doi.org/10.1128/MMBR.62.4.1353-1370.1998.

Watson, S.B., Satchwlll, T., 2003. Chrysophyte odour production: resource-mediated changes at the cell and population levels. Phycologia 42, 393–405. https://doi.org/10.2216/i0031-8884-42-4-393.1.

Wehr, J.D., 2015. Brown algae. In: Wehr, J.D., Sheath, R.G., Kociolek, J.P. (Eds.), Freshwater Algae of North America. Academic Press, San Diego, CA, pp. 851–871.

Wehr, J.D., 2016. Brown algae (Phaeophyceae) in rivers. In: Necchi Jr., O. (Ed.), River Algae. Springer International Publishing, Cham, Switzerland, pp. 129–151.

Wehr, J.D., Brown, L.M., 1985. Selenium requirement of a bloom-forming planktonic alga from softwater and acidified lakes. Can. J. Fish. Aquat. Sci. 42, 1783–1788. https://doi.org/10.1139/f85-223.

Wehr, J.D., Sheath, R.G., 2015. Habitats of freshwater algae. In: Wehr, J.D., Sheath, R.G., Kociolek, J.P. (Eds.), Freshwater Algae of North America. Academic Press, San Diego, CA, pp. 13–74.

Wehr, J.D., Sheath, R.G., Kociolek, J.P. (Eds.), 2015. Freshwater Algae of North America: Ecology and Classification. Academic Press, San Diego, CA, p. 1050.

Wehr, J.D., Stein, J.R., 1985. Studies on the biogeography and ecology of the freshwater phaeophycean alga *Heribaudiella fluviatilis*. J. Phycol. 21, 81–93. https://doi.org/10.1111/j.0022-3646.1985.00081.x.

Wehr, J.D., Thorp, J.H., 1997. Effects of navigation dams, tributaries, and littoral zones on phytoplankton communities in the Ohio River. Can. J. Fish. Aquat. Sci. 54, 378–395. https://doi.org/10.1139/f96-283.

West, J.A., Kraft, G.T., 1996. *Ectocarpus siliculosus* (Dillwyn) Lyngb. from Hopkins River Falls, Victoria—The first record of a freshwater brown alga in Australia. Muelleria 9, 29–33.

Wetzel, R.G., Ward, A.K., 1992. Primary production. In: Calow, P., Petts, G.E. (Eds.), The Rivers Handbook. I. Hydrological and Ecological Principles. Blackwell Science, Oxford, pp. 354–369.

Whitford, L.A., 1956. The communities of algae in the springs and spring streams of Florida. Ecology 37 (3), 433–442. https://doi.org/10.2307/1930165.

Whitford, L.A., Schumacher, G.J., 1963. Communities of algae in North Carolina streams and their seasonal relations. Hydrobiologia 22, 133–196. https://doi.org/10.1007/BF00039686.

Whittington, J., Sherman, B., Green, D., Oliver, R.L., 2000. Growth of *Ceratium hirundinella* in a subtropical Australian reservoir: the role of vertical migration. J. Plankton Res. 22, 1025–1045. https://doi.org/10.1093/plankt/22.6.1025.

Whitton, B.A., Ellwood, N.T.W., Kawecka, B., 2009. Biology of the freshwater diatom *Didymosphenia*: a review. Hydrobiologia 630, 1–37. https://doi.org/10.1007/s10750-009-9753-5.

Whitton, B.A., Grainger, S.L.J., Hawley, G.R.W., Simon, J.W., 1991. Cell-bound and extracellular phosphatase activities of cyanobacterial isolates. Microb. Ecol. 21, 85–98. https://doi.org/10.1007/BF02539146.

Wilcox, L.W., Wedemayer, G.J., 1991. Phagotrophy in the freshwater, photosynthetic dinoflagellate *Amphidinium cryophilum*. J. Phycol. 27, 600–609. https://doi.org/10.1111/j.0022-3646.1991.00600.x.

Wilkinson, A.N., Zeeb, B., Smol, J.P., 2001. Atlas of chrysophycean cysts, vol. II. Kluwer Academic Publishers, Dordrecht, p. 180.

Willén, E., 1992. Long-term changes in the phytoplankton of large lakes in response to changes in nutrient loading. Nord. J. Bot. 12, 577–587. https://doi.org/10.1111/j.1756-1051.1992.tb01836.x.

Willén, E., 2000. Phytoplankton in water quality assessment—An indicator concept. In: Heinonen, P., Ziglio, G. (Eds.), Hydrological and Limnological Aspects of lake Monitoring. John Wiley and Sons, Chichester, pp. 57–80.

Wilson, S.E., Cumming, B.F., Smol, J.P., 1996. Assessing the reliability of salinity inference models from diatom assemblages: an examination of a 219-lake data set from western North America. Can. J. Fish. Aquat. Sci. 53, 1580–1594. https://doi.org/10.1139/f96-094.

Winder, M., Reuter, J.E., Schladow, S.G., 2009. Lake warming favours small-sized planktonic diatom species. Proc. Biol. Sci. 276, 427–435. https://doi.org/10.1098/rspb.2008.1200.

Wium-Andersen, S., Anthoni, U., Christophersen, C., Houen, G., 1982. Allelopathic effects on phytoplankton by substances isolated from aquatic macrophytes (Charales). Oikos 39, 187–190. https://doi.org/10.2307/3544484.

Woelfl, S., Garcia, P., Duarte, C., 2010. *Chlorella*-bearing ciliates (*Stentor, Ophrydium*) dominate in an oligotrophic, deep North Patagonian lake (Lake Caburgua, Chile). Limnologica 40, 134–139. https://doi.org/10.1016/j.limno.2009.11.008.

Żelazna-Wieczorek, J., 2011. Diatom flora in springs of Lódz Hills (central Poland). In: Witkowski, A. (Ed.), Biodiversity, Taxonomy, and Temporal Changes of Epipsammic Diatom Assemblages in Springs Affected by Human Impact. A.R.G. Gantner Verlag K.G., Ruggell, p. 419.

Żelazna-Wieczorek, J., Lange-Bertalot, H., Olszynski, R.M., Witkowski, A., 2020. Navicula fontana sp. nov., a new freshwater diatom from a limnocrenic spring in central Poland. Phytotaxa 452 (2), 155–164. https://doi.org/10.11646/phytotaxa.452.2.4.

Żelazna-Wieczorek, J., Maninska, M., 2006. Algoflora and vascular flora of a limestone spring in the Warta River valley. Acta Soc. Bot. Pol. 75, 131–143. https://doi.org/10.5586/asbp.2006.016.

Zimba, P.V., Huang, I.S., Gutierrez, D., Shin, W., Bennett, M.S., Triemer, R.E., 2017. Euglenophycin is produced in at least six species of euglenoid algae and six of seven strains of *Euglena sanguinea*. Harmful Algae 63, 79–84. https://doi.org/10.1016/j.hal.2017.01.010.

Zulkifly, S.B., Graham, J.M., Young, E.B., Mayer, R.J., Piotrowski, M.J., Smith, I., Graham, L.E., 2013. The genus *Cladophora* Kützing (Ulvophyceae) as a globally distributed ecological engineer. J. Phycol. 49, 1–17. https://doi.org/10.1111/jpy.12025.

Zwart, G., Kamst-van Agterveld, M.P., Van Der Werff-Staverman, I., Hagen, F., Hoogveld, H.L., Gons, H.J., 2007. Molecular characterization of cyanobacterial diversity in a shallow eutrophic lake. Environ. Microbiol. 7, 365–377. https://doi.org/10.1111/j.1462-2920.2005.00715.x.

CHAPTER

18

Ecology of Algae and Cyanobacteria (Phytoplankton)

Elena Litchman[1,2] *and Paula de Tezanos Pinto*[3]

[1]Kellogg Biological Station and Department of Integrative Biology, Michigan State University, East Lansing, MI, United States [2]Department of Global Ecology, Carnegie Institution for Science, Stanford, CA, United States [3]Instituto de Botánica Darwinion, Consejo Nacional de Investigaciones Científicas y Técnicas, Argentina

OUTLINE

I. Introduction

Phytoplankton, the microscopic eukaryotic algae and Cyanobacteria, living suspended in the water column, are essential primary producers in most aquatic ecosystems, rivers, and lakes. They form the basis of food webs, play important roles in biogeochemical cycling, and significantly affect water quality. Understanding what controls their biomass and community composition is critical for our general understanding of aquatic ecosystems and the ability to predict what these ecosystems will look like in the future. Here we identify major environmental factors that influence the growth and mortality of different groups of phytoplankton. We then outline how biotic interactions, including competition for resources, predation, mutualistic interactions,

Wetzel's Limnology, Fourth Edition
https://doi.org/10.1016/B978-0-12-822701-5.00018-5

and parasitism, modify growth and community structure. Finally, we discuss how changing environmental conditions, including anthropogenic global change, shape present and future phytoplankton communities. We use a trait-based framework to help understand the mechanisms that govern phytoplankton community composition dynamics and how phytoplankton would respond to changing environmental conditions.

Trait-based approaches (approaches based upon heritable characteristics of species) focus on organismal traits and help link an organism's fitness to the environment. Traits can be described as any morphological, physiological, or phenological heritable feature that can be measured at the individual level, from the cell to the entire organism, without reference to the environment or any other level of organization (Garnier et al. 2016). Focusing on traits, rather than species, helps address the mechanisms that structure phytoplankton and other ecological communities.

II. Phytoplankton growth: resources and environmental factors

The composition and dynamics of phytoplankton communities depend upon a multitude of abiotic factors and biotic interactions. Some important factors regulating phytoplankton growth include chemical resources, such as inorganic nutrients and organic compounds; physical variables, such as light and temperature and other abiotic factors including wind, flow, mixing, and pH; and biological factors, such as competition for resources, allelochemical interactions, and mortality due to biotic (predation by grazers and pathogens) and abiotic (flow and sedimentation) causes.

A. Inorganic nutrients

Macronutrients such as nitrogen (Chapter 14) and phosphorus (Chapter 15), and micronutrients (Chapter 16), including iron, molybdenum, and zinc, are key resources for phytoplankton. In addition, certain phytoplankton such as diatoms and chrysophytes also need silica. Many of these nutrients have been described in detail in previous chapters, and hence here we will mostly describe how phytoplankton growth depends on nutrients and highlight relevant physiological traits.

i. Phytoplankton nutrient-dependent growth and uptake

Several physiological traits characterize nutrient uptake and utilization in phytoplankton. The values of those traits, obtained from functions that describe nutrient-dependent growth and uptake, provide information on the ecological strategies of phytoplankton with respect to nutrients. Phytoplankton growth depends on nutrient concentration in a saturating way and is commonly described with the *Monod equation* (a function that describes the dependence of phytoplankton growth on an external resource):

$$\mu = \mu_{\max}\frac{R}{R + K_s} \tag{18-1}$$

where μ is the growth rate (day^{-1}), $\mu_{\max}$ is the maximum growth rate (day^{-1}), R is the external nutrient (resource) concentration (μmol nutrient L^{-1}), and K_s is the *half-saturation constant* (nutrient concentration at which the growth rate is half of the maximum) for growth (μmol nutrient L^{-1}). Note that it is different from the half-saturation constant for uptake (see Eq. 18-2), reflected here by using the capital K.

Phytoplankton nutrient uptake also depends on external nutrient concentration, often described by a saturating function called the *Michaelis–Menten function* (Grover 1991b) (a saturating function of this form, often used to describe enzyme kinetics):

$$V = V_{\max}\frac{R}{R + k_s} \tag{18-2}$$

where V is the uptake rate (μmol nutrient cell^{-1} day^{-1}), $V_{\max}$ is the maximum nutrient uptake rate (μmol nutrient cell^{-1} day^{-1}), k_s is the half-saturation constant for nutrient uptake (μmol nutrient L^{-1}), and R is the external nutrient concentration (μmol nutrient L^{-1}) (Fig. 18-1).

A different model, called the *Droop model* (a function that describes the dependence of phytoplankton growth on intracellular nutrient concentration), assumes that phytoplankton growth rate, μ(Q), depends not on external nutrient availability but on the internal/intracellular nutrient concentration, called the *nutrient quota* Q (intracellular nutrient concentration) (Droop 1973, Grover 1991a, Litchman and Klausmeier 2008):

$$\mu(Q) = \mu_{\infty}\frac{(Q - Q_{\min})}{Q} \tag{18-3}$$

where μ(Q) is the growth rate as a function of internal nutrient concentration, Q (day^{-1}), μ_{∞} is the growth rate at infinite quota (day^{-1}), Q is nutrient quota (intracellular nutrient concentration, μmol nutrient cell^{-1}), and $Q_{\min}$ is the minimum nutrient quota at which growth stops (μmol nutrient cell^{-1}).

The parameters of the equations that describe the uptake and growth for a given nutrient can be viewed as *traits* (heritable characteristics of a species) and are used to characterize and compare growth and nutrient uptake in phytoplankton (μ_{∞}, $Q_{\min}$, $V_{\max}$, and K_s, etc.), and their values determine the performance of species

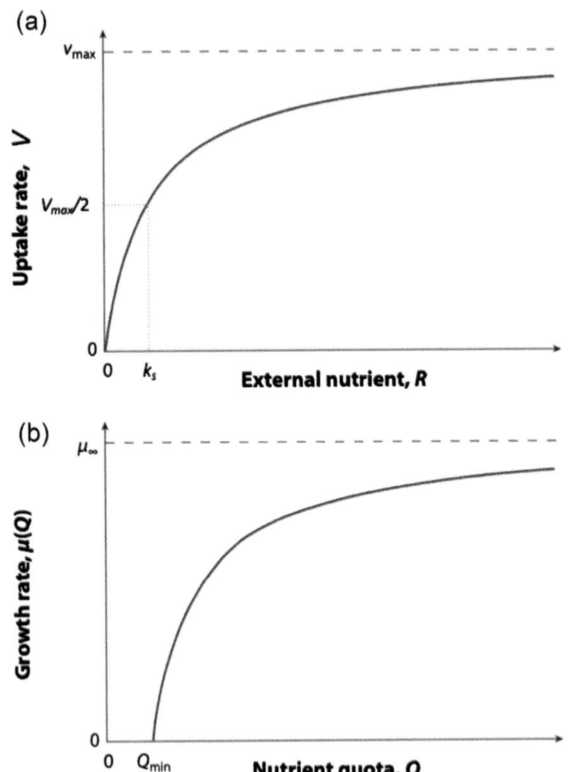

FIGURE 18-1 The dependence of nutrient uptake and growth on nutrient concentration: (a) Michaelis–Menten uptake and (b) Droop's growth function. V is the nutrient uptake rate, V_{max}, is the maximum nutrient uptake rate; k_s, is the half-saturation constant for nutrient uptake, R is the external nutrient concentration; $\mu(Q)$ is the growth rate as a function of internal nutrient concentration Q, μ_∞ is the growth rate of species at an infinite quota; Q_{min} is the minimum nutrient quota at which growth stops. *(From Litchman and Klausmeier [2008].)*

for nutrients (Grover 1997, Litchman and Klausmeier 2008) (Fig. 18-1).

ii. Nutrient trait differences among phytoplankton groups

Nutrient-related traits can vary considerably, depending on light, temperature, and other factors (Edwards et al. 2016, Rhee and Gotham 1981a), yet there are significant differences in those traits across phytoplankton species and taxonomic groups (Edwards et al. 2012). For example, diatoms tend to have low half-saturation constants for phosphorus (P) but high half-saturation constants for nitrogen (N) uptake compared to green algae and Cyanobacteria. Also, green algae grow faster when nutrients are plentiful, compared to diatoms and Cyanobacteria (Edwards et al. 2012).

iii. Nitrogen fixation

Nitrogen (N) fixation is a trait that allows the acquisition of atmospheric nitrogen and its incorporation into other chemical forms (Chapter 14). Among phytoplankton taxonomic groups, the trait of N fixation is found only in Cyanobacteria (Herrero and Flores 2008). Moreover, only a subset of Cyanobacteria taxa is capable of fixing atmospheric N. In freshwater ecosystems the planktonic filamentous forms of the group Nostocales can fix N. This group of Cyanobacteria develops specialized cells, called *heterocytes* (a specialized cell where nitrogen fixation occurs, also known as *heterocysts*), where N fixation takes place (see also Chapter 17). The rate of N fixation per heterocyte is fairly constant (de Tezanos Pinto and Litchman 2010). However, nitrogen fixation is a plastic trait, with higher heterocyte density under N-limited than N-sufficient conditions (de Tezanos Pinto and Litchman 2010).

Nitrogen fixation can provide "new" nitrogen, which can then be released into the environment and consumed by other primary producers and other microbes. The contribution of N fixation is estimated to be more important in high-nutrient (eutrophic) lakes (Howarth et al. 1988), likely because those lakes have a higher abundance of N-fixing Cyanobacteria, compared to low-nutrient lakes.

N-fixing Cyanobacteria will have a competitive advantage in low N situations, as they can complement nitrogen uptake with N fixation. N-fixers, however, seem to have high phosphorus requirements compared to other phytoplankton groups and hence display contrasting N and P competitive abilities (Lenton and Klausmeier 2007). Other trade-offs found relevant to N fixation include a higher demand for light and iron (Agawin et al. 2007, de Tezanos Pinto and Litchman 2010, Ward et al. 2013).

iv. Nutrients and light

Under low irradiance, phytoplankton may require more nutrients to survive (higher minimum quota for nutrients) (Rhee and Gotham 1981b). This is because they use nutrients to synthesize more light-harvesting pigments such as chlorophyll *a* under low light to increase light capture. This suggests a trade-off between light and nutrient competitive abilities. The trade-off should be stronger for N than for P, as pigments are N-rich molecules. Indeed, the experiments of Passarge et al. (2006) found an absence of a trade-off for P and light in phytoplankton, as good phosphorus competitors were also good light competitors.

v. Nutrients and size

Nutrient competitive traits correlate strongly with cell size (Edwards et al. 2012, Marañón et al. 2013). Small-sized phytoplankton have an advantage at low nutrient concentrations due to their high surface-area-to-volume ratio (Reynolds 2006). Larger cells, on the other hand, can use nutrient pulses more efficiently, as maximum nutrient uptake rates increase with cell size

(Edwards et al. 2012, Marañón et al. 2013). Consequently, fluctuating nutrient supply may favor larger phytoplankton (Suttle et al. 1987, Stolte and Riegman 1996, Litchman et al. 2009).

vi. Dissolved inorganic and total nutrient fractions

Nutrients can be found either in dissolved forms (organic or inorganic) or incorporated in biomass (particulate fraction). Total nutrient concentrations are sums of dissolved and particulate fractions. Much debate occurs on which nutrient values limit phytoplankton growth. Reynolds (1998) proposed that dissolved inorganic P limitation is unlikely above 3 $\mu g\ L^{-1}$ soluble reactive phosphorus and that dissolved inorganic N limitation is unlikely above 100 $\mu g\ L^{-1}$ N (the sum of nitrate, nitrite, and ammonia). The concentrations of total nutrients (P and N) limit total biomass of phytoplankton. Thus the maximum plankton biomass that a given ecosystem can attain depends on the total concentration of the limiting nutrient (Chorus and Spijkerman 2021) and, in the absence of other constraints (e.g., light, mortality, etc.), increases with trophic status.

Reynolds emphasized the importance of absolute concentrations yet recognized the value of using N:P ratios to determine which nutrient may be limiting (Chorus and Spijkerman 2021). Trimbee and Prepas (1987) and Downing et al. (2001) found that changes in cyanobacterial dominance might be better explained by the variation in absolute P and N concentrations rather than by the N:P ratios. Smith (1983), however, argued that low nitrogen-to-phosphorus ratios (lower than 29 to 1 by weight) favor dominance by Cyanobacteria and that when this ratio was exceeded, Cyanobacteria tended to be rare. Hence Smith argued that N and P ratios are key drivers of species assembly in phytoplankton communities. Dolman et al. (2012), nevertheless, found that many N-fixing Cyanobacteria were abundant at relatively high N:P ratios. Laboratory and mesocosm experiments showed that changes in N:P ratios can modify phytoplankton community composition (Tilman et al. 1982, de Tezanos Pinto and Litchman 2010). Moreover, large-scale studies used the total nutrient ratios (TN:TP) as an indicator of the N:P supply ratio and also observed that phytoplankton community structure varies with nutrient ratios (Smith 1983). There is still debate about whether the absolute nutrient levels or nutrient ratios better predict phytoplankton composition and biomass. It is noteworthy that different absolute nutrient concentrations can render the same N:P ratio, and hence both the absolute nutrient concentrations and their ratios are necessary to understand how nutrients influence phytoplankton composition and biomass.

vii. Optimal N:P ratios in phytoplankton

The N:P ratios of phytoplankton biomass often converge to 16:1 (by moles). This was first noted by Redfield (1958) for marine phytoplankton, and consequently, it is called the *Redfield ratio* (average particulate N:P ratio of 16:1 observed by Redfield). This ratio is sometimes used to determine if phytoplankton are N or P limited. However, it is not a universal biochemical optimum and varies across species (Geider and Roche 2002, Klausmeier et al. 2004). The N:P ratios are predicted to be lower in fast-growing phytoplankton because high growth rates require more ribosomes that are rich in P (Sterner and Elser 2002, Klausmeier et al. 2004). Consequently, in the environments that select for fast-growing phytoplankton species, such as with high availability of resources, the phytoplankton N:P ratios would be low.

viii. Which nutrient should be controlled?

Strategies to mitigate eutrophication see also Chapters 14 and 15 suggest either the control of one nutrient—in most cases P, less often N (e.g., removal in wastewaters)—or the regulation of both nutrients (e.g., modifying land-use planning and changing agronomic practices), though there is a spirited debate about which approach should be used (reviewed in Chorus and Spijkerman 2021). Decisions should be based not upon a general paradigm of one or both nutrients but rather on a comprehensive and careful site-specific evaluation and definition of targets, as nutrient legacy in water and sediments can vary among water bodies (Chorus and Spijkerman 2021).

B. Organic compounds, vitamins

Vitamins are important organic micronutrients for many algae (Helliwell 2017). Algal species that require vitamins for growth are called *auxotrophic*. In the case of auxotrophic growth the requirements for specific organic compounds are low and these compounds do not contribute significantly to the total cell carbon. Three water-soluble vitamins are identified to be essential for phytoplankton: cobalamin (vitamin B_{12}), thiamine (vitamin B_1), and biotin (vitamin B_7). Vitamin B_{12} and thiamine are required alone or in combination by most of the auxotrophic algae, and B_{12} is needed more often than thiamine. Vitamin B_{12} is required by a substantial number of Cyanobacteria, diatoms, green algae, and dinoflagellates. Biotin is required by a few chrysophytes, dinoflagellates, and euglenoids. Finally, only a few groups show any significant requirements for thiamine.

There is no strict taxonomic correspondence to the known distribution of the requirement for external vitamins (auxotrophy) among different algae. The differences, however, are sufficient to permit some generalizations. Only the xanthophycean algae exhibit no apparent need for these major water-soluble vitamin growth factors, although in several other groups the requirements are rare. Cyanobacteria, Chlorophyceae, Xanthophyceae, and Phaeophyceae have the least number of species requiring vitamins. Most of the species in these groups are strictly autotrophic in metabolism. In contrast, a clear predominance of auxotrophic species occurs in mixotrophic groups, including Chrysophyceae, Dinophyceae, Cryptophyceae, and Euglenophyceae.

C. Light

Light (Chapter 6), as well as nutrients, is an essential resource for phytoplankton. Light significantly influences phytoplankton biomass and community structure, as phytoplankton species have different light requirements (Richardson et al. 1983, Schwaderer et al. 2011). Light capture by phytoplankton decreases light availability (quantity) and may also affect its quality (distribution of different wavelengths).

i. Phytoplankton light traits

Several physiological traits characterize light uptake and utilization in phytoplankton. Their values, obtained from functions that describe light–growth curves, provide information on the ecological strategies with respect to light (Litchman and Klausmeier 2008). Two *growth–irradiance functions* (relationships that describe the dependence of phytoplankton growth on light) are frequently used (Fig. 18-2): Model 1 (Laws and Chalup 1990) is used to characterize phytoplankton growth in response to irradiance, $\mu(I)$, where growth rate becomes saturated at high irradiance; and Model 2 (Eilers and Peeters 1988) is used to characterize $\mu(I)$ if *photoinhibition* (growth rate decline at high irradiance) occurs.

$$\text{Model 1: } \mu(I) = \frac{\mu_{\max} I}{I + \frac{\mu_{\max}}{\alpha}}$$

$$\text{Model 2: } \mu(I) = \frac{\mu_{\max} I}{\frac{\mu_{\max}}{\alpha I_{\text{opt}}^2} I^2 + \left(1 - 2\frac{\mu_{\max}}{\alpha I_{\text{opt}}}\right) I + \frac{\mu_{\max}}{\alpha}}$$

The following light utilization traits, obtained from the abovementioned models, are the parameters of the growth–irradiance curves (Fig. 18-2), where $\mu(I)$ (day^{-1}) is the growth rate as a function of irradiance I, $\mu_{\max}$ (day^{-1}) is the maximum growth rate, α (day^{-1} μmol photons^{-1} m^2 s) is the initial slope of the growth–irradiance curve, and I_{opt} (μmol photons m^{-2} s^{-1}) is the irradiance at which photoinhibition occurs.

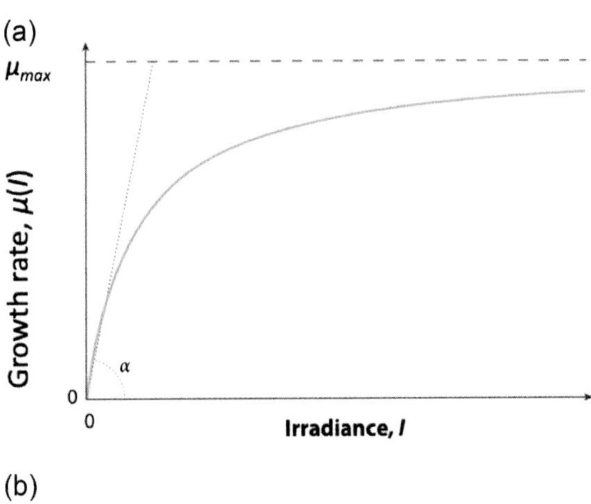

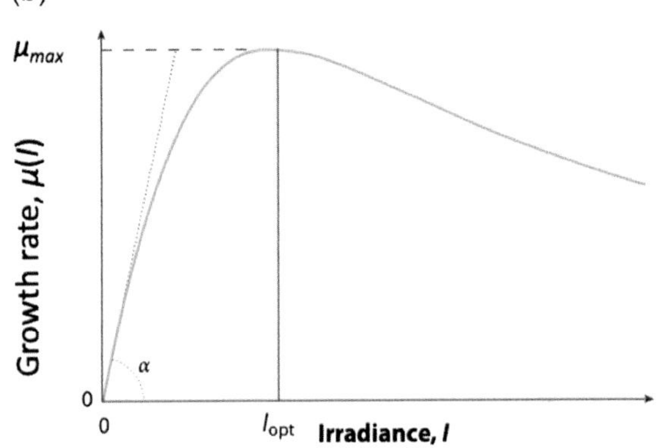

FIGURE 18-2 The dependence of phytoplankton growth rate μ (day^{-1}) on irradiance, I (μmol photons m^{-2} s^{-1}). (a) In the absence of growth inhibition by high irradiance and (b) with growth inhibition by high irradiance (photoinhibition). $\mu_{\max}$ (day^{-1}) is the maximum growth rate, α (day^{-1} μmol photons^{-1} m^2 s) is the initial slope of the growth irradiance curve, I_{opt} (μmol photons m^{-2} s^{-1}) is the irradiance at which photoinhibition occurs. *(From Litchman and Klausmeier [2008].)*

In addition to the light traits derived from the growth–irradiance (described above), other light-related traits considered are maximum photosynthesis (obtained from the photosynthesis–irradiance curves) and the presence of the accessory photosynthetic pigments (other pigments than chlorophyll *a*) that vary among major phytoplankton taxonomic groups and might thus respond in a different way to irradiance.

Different values of light utilization traits can be viewed as adaptations to diverse light environments (Richardson et al. 1983, Langdon 1987). High $\mu_{\max}$ is beneficial under high light conditions (Grover 1991b, Litchman and Klausmeier 2001). Species with high α are adapted to low irradiance and hence have a high efficiency of growth at low light (Langdon 1987). Species adapted to high light tend to have low α. Low I_{opt}

indicates susceptibility to *photoinhibition* at low irradiances (Eilers and Peeters 1988).

Schwaderer et al. (2011) found trade-offs among light traits: such as the negative relationship between I_{opt} and α, which implies that species adapted to efficiently utilizing low light are more susceptible to photoinhibition. This reflects physiological constraints in light utilization traits that preclude the existence of a "superspecies" with respect to light utilization. Similarly, high α may be associated with higher nitrogen demand, presumably to synthesize more light-capturing pigments, such as chlorophyll *a* (Edwards et al. 2015). This can represent a trade-off between light- and nitrogen-competitive abilities.

ii. Differences in light traits across phytoplankton taxonomic groups

Major phytoplankton taxonomic groups differ in their light utilization traits. For example, it is well known that Cyanobacteria are often adapted to low-light environments, whereas green algae are often found in high-light environments (Reynolds 1984).

Indeed, an extensive analysis of light traits (Schwaderer et al. 2011) confirmed that Cyanobacteria are a low-light-adapted group with the highest light utilization efficiency (α) (Fig. 18-3) and most sensitivity to photoinhibition (low I_{opt}). The high α may relate to the environmental conditions that prevailed at the time of cyanobacterial evolution, when solar luminosity was about three-fold lower than at present (Milo 2009). In addition, high α may be due to the structure of cyanobacterial photosynthetic apparatus, as they lack chloroplasts, and therefore light only needs to penetrate the cell wall to reach the thylakoids. These characteristics of the photosynthetic apparatus may also explain the high susceptibility of Cyanobacteria to photoinhibition (low I_{opt}). In addition, Cyanobacteria have other traits that can enhance their light competitive abilities even further, such as *chromatic adaptation* (the ability to adjust the concentration and proportion of their accessory pigments, phycoerythrin and phycocyanin), and the ability to control their position in the water column in stratified conditions, through the synthesis and collapse of gas vesicles (Klemer 1991, Litchman 2003, Huisman et al. 2004). In contrast, green algae show the highest μ_{max} of all phytoplankton groups but low α and high I_{opt}. This combination of traits suggests that green algae are adapted to high-light conditions. Diatoms have relatively low μ_{max} but high α, which makes them adapted to low-light environments. Compared to Cyanobacteria, diatoms have higher I_{opt}, indicating adaptation to conditions with high vertical mixing, where the exposure to both low and high irradiances occurs. Phytoplankton that are *mixotrophic* (able to feed both autotrophically and heterotrophically) display light traits that reflect high irradiance needs and poor light competitive abilities (Schwaderer et al. 2011). Nevertheless, other traits in mixotrophs, such as their ability to move using flagella and the ability to ingest prey, likely compensate for their poor light competitive abilities (Schwaderer et al. 2011).

iii. Light and cell size

Schwaderer et al. (2011) found a trade-off between μ_{max} and cell size, and between α and cell size. This implies that smaller cell sizes grow faster and are more effective at light capture, yet size seems unrelated to photoinhibition. Indeed, the trend that smaller cells might have higher efficiency at using low light was previously proposed (Geider et al. 1986; Fujiki and Taguchi 2002). Likewise, experimental work demonstrated that cell size distribution declines under low light (Claustre and Gostan 1987, de Tezanos Pinto and O'Farrell 2014). Smaller phytoplankton have an advantage under low light conditions because of less pigment self-shading and a shorter path to chloroplasts (Finkel 2001, Malerba et al. 2018).

iv. Behaviors toward light

Phytoplankton species capable of moving in the water column are able to adjust light harvesting through behavioral responses moving either toward or away from the light, when light limited or light inhibited, respectively (Clegg et al. 2003, Klemer 1983, Wallace and Hamilton 1999). Gas vacuoles in Cyanobacteria, and flagella in dinoflagellates, cryptomonads, euglenoids, and chrysophytes, can allow behavioral responses whenever water mixing is low.

v. Light trait plasticity

Phytoplankton can acclimate to changing light intensities, often by regulating pigment concentrations per unit biomass (Geider et al. 1997). In low-light situations phytoplankton frequently display high efficiency of light use but high susceptibility to photoinhibition (Falkowski 1980). In addition, chlorophyll *a* concentrations in cells are higher at low irradiances, which maximizes light use, and the opposite happens at increased irradiances (Falkowski 1980).

vi. Fluctuating light regimes

Phytoplankton face unpredictable light fluctuations because of daily changes in irradiance throughout the day, due to the presence of clouds, and when cells are vertically transported through the mixed layer. Different phytoplankton species respond differently to fluctuations, with diatoms often adapted to fluctuating light

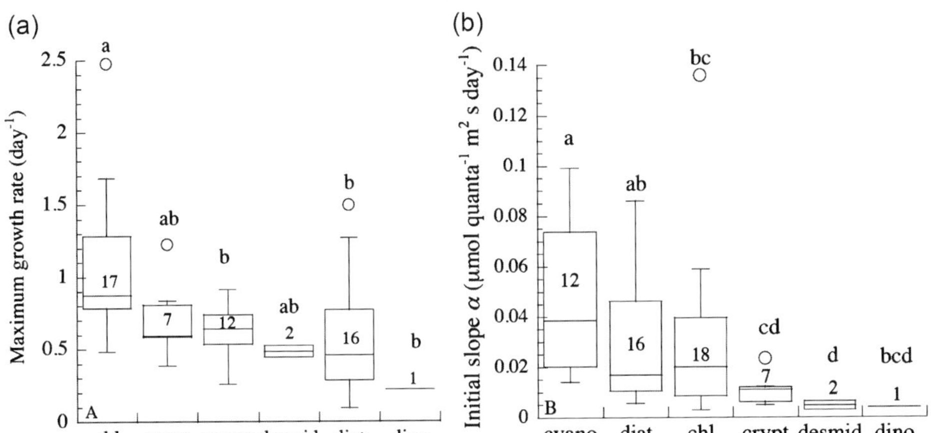

FIGURE 18-3 Light traits: (a) maximum growth rate, μ_{max} (day^{-1}), and (b) initial slope of the growth–irradiance curve, α (day^{-1}µmol photons^{-1} m^2 s), in different phytoplankton groups. Green algae (chl), cryptophytes (crypt), Cyanobacteria (cyano), desmids (desmid), diatoms (diat), and dinoflagellates (dino). Different letters mean traits are significantly different among phytoplankton groups. *(From Schwaderer et al. [2011].)*

(Litchman 1998). Light fluctuations may allow multiple species of phytoplankton to coexist (Litchman and Klausmeier 2001).

D. Temperature

Most physiological and ecological processes, including photosynthesis, respiration, growth, resource acquisition, motility, and sinking, vary with temperature (Eppley 1972, Raven and Geider 1988). Temperature also plays a key role in phytoplankton seasonal succession (Sommer et al. 1986). Temperature dependency will be increasingly important due to global climate change (Strecker et al. 2004).

The relationship between temperature and growth in phytoplankton is a unimodal curve, with a relatively gradual (usually nonlinear) increase up to the optimum temperature, followed by a steep drop above the optimum (Fig. 18-4). Several models are used to describe the relation between growth and temperature (Ahlgren 1987, Briand et al. 2004, Jöhnk et al. 2008). Temperature-related traits include maximum growth rate, μ_{max} (day^{-1}), the optimum temperature for growth or any other physiological or ecological process, T_{opt} (°C), the minimum and maximum temperatures that allow positive growth, CT_{min} (°C) and CT_{max} (°C), respectively, the width of the thermal niche (temperatures where growth is >0), and Q_{10}, the specific difference in growth rate at two temperatures separated by 10°C (e.g., Q_{10} between 10°C and 20°C) (Fig. 18-4).

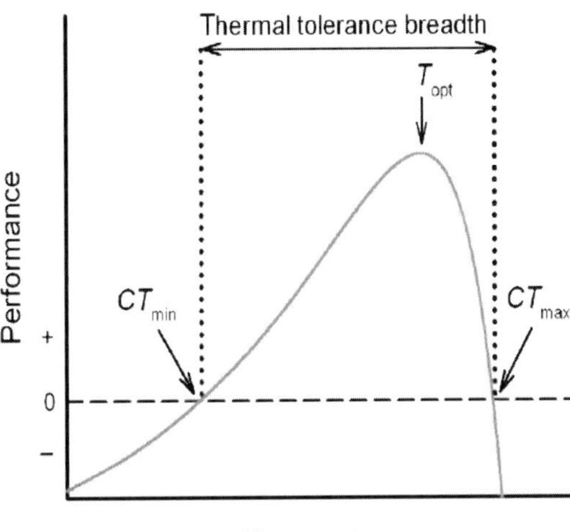

FIGURE 18-4 The dependence of growth rate or other physiological processes on temperature. T_{opt} is the temperature at which the performance is maximal. Thermal tolerance breadth is the width of the thermal niche, and CT_{min} and CT_{max} are the minimum and maximum temperatures, respectively, where the performance is nonnegative. *(From Krenek et al. [2012].)*

i. Temperature traits across phytoplankton groups

Temperature growth optima (T_{opt}) vary among phytoplankton taxonomic groups (Figs. 18-5a and 18-5b). Cyanobacteria have high T_{opt}, compared to other groups, with values ranging from 25°C to 35°C (Robarts and Zohary 1987, Thomas et al. 2016). Likewise, most green algae also have high T_{opt}, within a range similar to Cyanobacteria (Lürling et al. 2013). Diatoms, conversely, have lower T_{opt} (15–25°C) (Thomas et al. 2016). Interestingly, phytoplankton T_{opt} depends on the latitude that the species was isolated from, as values decline with increasing latitudes, suggesting local adaptation (Fig. 18-5b).

ii. Temperature and size

Most studies on the relationship between temperature and cell size in phytoplankton were conducted in

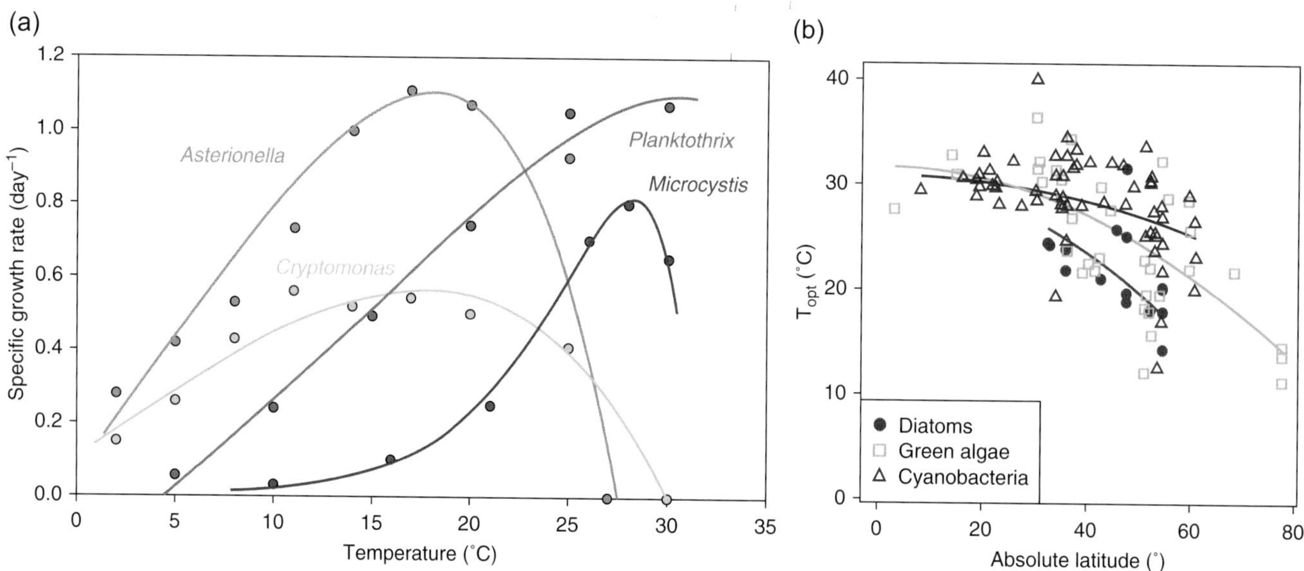

FIGURE 18-5 (a) Thermal performance curves for different phytoplankton species (from Paerl and Huisman 2009) and (b) T_{opt} (°C) for diatoms, green algae, and Cyanobacteria isolated from different latitudes. *(From Thomas et al. [2016].)*

marine environments, and much less is known for freshwater ecosystems (Zohary et al. 2021). As in marine environments, phytoplankton cell or colony size may decrease with increased temperatures (Zohary et al. 2021). However, interacting temperature, grazing, and nutrient concentration may change this relationship and result in larger phytoplankton in warmer conditions (Pomati et al. 2020).

iii. Distribution along temperature gradients

Based upon phytoplankton temperature traits, a potential decline in phytoplankton diversity in warmer scenarios could be expected (Litchman et al. 2010). For example, in a warmer future diatoms and chrysophytes may decline because high temperatures may fall outside their thermal niches. Because Cyanobacteria and green algae have similar temperature optima (Lürling et al. 2013) and similar growth rates (Schwaderer et al. 2011, Lürling et al. 2013), both Cyanobacteria and green algae may become dominant in a warmer future. Cyanobacteria may be more successful than green algae in warmer scenarios, due to other traits, such as their ability to regulate buoyancy and avoid sinking in warmer, more strongly stratified waters, their adaptation to low light environments (such as that created by high algal biomass), or their resistance to grazing due to filamentous and colonial form and toxin production decreasing their edibility.

iv. Cyanobacterial blooms can affect water temperature

Remote sensing studies using satellite images have shown that intense cyanobacterial surface blooms may lead to an increase in the satellite-derived surface water temperatures (Kahru et al. 1993, Ibelings et al. 2003). This in turn may enhance the success of Cyanobacteria in warmer waters, as locally enhanced surface temperatures (in the already warm conditions) would further favor Cyanobacteria over other phytoplankton with lower temperature optima (Hense and Beckmann 2006).

E. Physical and chemical environment

In the previous sections we have discussed how phytoplankton respond to light and temperature. Other physical and chemical factors that can also markedly influence phytoplankton composition and biomass include wind, mixing, flow, inorganic turbidity, pH, and conductivity.

Wind action can stimulate the mixing of the water column and counteract stratification due to temperature difference (Chapters 7 and 8). In shallow lakes wind-induced mixing can reach the bottom of the water body and resuspend sediments with the possible release of nutrients (Chapter 26). The effects of modified wind patterns (increase, decrease) can either act synergistically or antagonistically with increased water temperature: either favoring mixing (e.g., strong, increased winds) or aiding stratification (weak, decreased wind intensity) (Salmaso and Tolotti 2021). Lower wind speeds observed in the last four decades (McVicar et al. 2012) may have promoted water column stability in lakes of the Northern Hemisphere (Woolway et al. 2019). The higher stability of the water column may promote phytoplankton taxa that can regulate their position in the water column through active swimming or buoyancy regulation, such as flagellated phytoplankton (e.g., cryptophytes, chrysophytes, dinoflagellates) and Cyanobacteria, respectively.

Water column mixing affects both light and nutrient availability for phytoplankton. For example, when mixing reaches depths greater than the euphotic zone, the amount of time a phytoplankton remains in the illuminated part of the water column decreases (Reynolds 1997). Stronger thermal stratification of the water column limits the input of nutrients from deeper layers, hence leading to nutrient-poor conditions in surface waters (O'Reilly et al. 2003). Mixing, on the contrary, usually maintains a continuously high level of nutrients.

High flow imposes a high mortality upon phytoplankton. That is, when the flow is higher than the growth rate, phytoplankton biomass would be unable to increase, as losses would exceed growth. Flow is particularly important in rivers (discussed below); high flow would select for species with high growth rates (that would be higher than the mortality imposed by the flow).

Suspended solids can decrease light availability and hence select for low-light-adapted phytoplankton species, such as Cyanobacteria and diatoms. Differences in pH affect the chemical speciation of macronutrients (N, P, C), micronutrients, and trace metals, which affects nutrient bioavailability for phytoplankton.

Anoxic environments (such as those in shallow lakes covered with dense free-floating plants, at the water–sediment interface, or in sites contaminated with organic matter) can indirectly affect phytoplankton through the release of nutrients from sediments. Lack of oxygen can also hinder oxygenic photosynthesis.

The concentration of ions may affect the distribution of phytoplankton, depending upon their tolerances. For example, *Microcystis aeruginosa*, a bloom-forming cyanobacterium, cannot thrive at high conductivities (Martínez de la Escalera et al. 2017).

Humic substances (Chapter 28) alter both the quantity and the quality of light, as they attenuate incoming light and absorb all wavelengths except the red ones. Phytoplankton with phycobiliproteins (phycoerythrin, phycocyanin), capable of absorbing red wavelengths, such as Cyanobacteria, may be favored in humic environments.

F. Factor interactions

i. Nutrients, light, and size interactions

Along the trophic gradient (from oligotrophic to hypereutrophic), there is an opposing trend of nutrient and light availability. For example, when nutrient concentration is low, small-sized phytoplankton cells dominate, as small size and high surface-area-to-volume ratio favor efficient nutrient acquisition (Chisholm 1992) and are often associated with high-light conditions, as observed in oligotrophic systems. Low-light environments also select for small-sized organisms (de Tezanos Pinto and O'Farrell 2014). In extremely light-limited environments with high nutrient availability smaller sizes prevail, hence suggesting that light limitation overrides the effect of high nutrient availability in the phytoplankton size distribution (de Tezanos Pinto and O'Farrell 2014).

ii. Light, temperature, and nutrient interactions

Interactions of light, nutrients, and temperature may affect the physiology and ecology of phytoplankton, changing their relevant traits (Spilling et al. 2015), and cause successional changes (see Seasonal successional patterns). Multistressor studies are essential, as interaction can occur among multiple variables, which can directly or indirectly affect phytoplankton biomass, community composition, diversity, and trait distribution.

III. Phytoplankton in food webs

In the previous sections we have discussed how environmental (abiotic) factors affect phytoplankton. Within food webs, phytoplankton participate in many biotic interactions that affect their growth. For example, they compete for resources among themselves, as well as with other primary producers (macrophytes) and bacteria. They experience mortality by grazers and pathogens (in addition to abiotic factors such as high flow and sedimentation). Phytoplankton may synthesize secondary metabolites that either hinder the growth of other organisms, including other phytoplankton, or enable beneficial interactions with bacteria or other phytoplankton.

A. Competition

In phytoplankton, competition usually involves the consumption of limiting resources, such as light and nutrients, and the dynamics and outcomes of competition influence species abundance and community composition.

Nutrient Competition. Phytoplankton nutrient competition is well documented and has a major effect on community structure. Growth experiments under different nutrient concentrations allow obtaining resource acquisition traits such as μ_{max} (maximum growth rate, day^{-1}) and K_s (the half-saturation constant, μmol nutrient L^{-1}). These traits can be used to calculate

R^*, the minimum concentration of the limiting resource needed by a species to grow at a given mortality or dilution (D, day^{-1}) rate (Fig. 18-6), where:

$$R^* = D\,K_s/(\mu_{max} - D)$$

Species with the lowest R^* for a resource (μmol nutrient L^{-1}) would be able to grow at the lowest resource concentrations and outcompete species that have higher R^* (Tilman et al. 1982). For example, in Fig. 18-6 species A would outcompete species B when competing for one resource, R, because of its lower R^*.

When species compete for two resources, their competitive abilities are characterized by the R^* for each resource. Their R^*s define regions in the two-resource plane where their growth is positive (above their R^*) or negative (below their R^* for either resource). The border between the two regions represents the *zero net growth isocline*, or *ZNGI* (combinations of nutrient concentrations that result in zero net growth) (Fig. 18-7). The ratios of how the two resources are consumed by each species determine the angle of resource consumption vectors. Together, the R^*s, the ZNGIs, and resource consumption vectors determine the outcome of competition between the two species. It could be either exclusion or coexistence and can be determined graphically (Fig. 18-7).

For essential resources such as N and P, if a given species has lower R^*s for both resources, then that species is most competitive. When mortality (e.g., grazing or dilution) increases, the isoclines would move away from the origin, and species would need more resources to reproduce under higher loss rates (R^* increases) (Tilman et al. 1982) (Fig. 18-8).

Species competitive abilities for nutrients (e.g., R^*) depend on various environmental factors, such as temperature or pH (Fig. 18-9). This is in part because physiological traits that determine the competitive ability can change depending on environmental conditions. Therefore the outcome of competition may be different in different conditions, so a species that is a good competitor at low temperatures may not be the best competitor in warmer temperatures.

i. Trade-offs in nutrient competitive abilities

There may be trade-offs in a freshwater phytoplankton's ability to compete for nutrients, where species that are good phosphorus competitors are poor competitors for nitrogen, and vice versa (Edwards et al. 2011). For different taxonomic groups, Tilman et al. (1982) suggested that Cyanobacteria are inferior phosphorus competitors compared to diatoms. Within Cyanobacteria, N-fixers are thought to have contrasting competitive abilities for N and P (Lenton and Klausmeier 2007). Green algae tend to be good N competitors (Rhee and Gotham 1981b). Within diatoms, Tilman (1982) showed a trade-off between silicate and phosphate (Fig. 18-10), and Sommer (1986) between silicate and nitrate competitive abilities.

It is important to note that resource competition theory predicts competitive outcomes at equilibrium, whereas resources in natural habitats fluctuate in space and time. Theories often predict that the number of

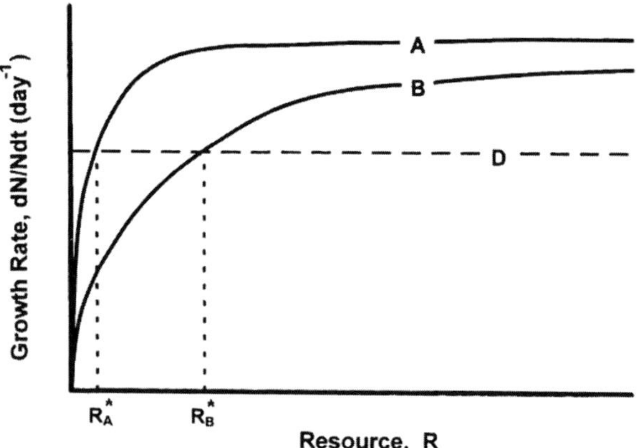

FIGURE 18-6 Monod growth curves for two species of algae (*A* and *B*). The R^* value for each species is the equilibrium resource (nutrient) concentration at specific dilution rates (*D* equals mortality rate at steady state). When the two species, *A* and *B*, compete for a single resource, *R*, the species with the lowest R^* for that resource will win. *(Modified from Kilham and Hecky [1988].)*

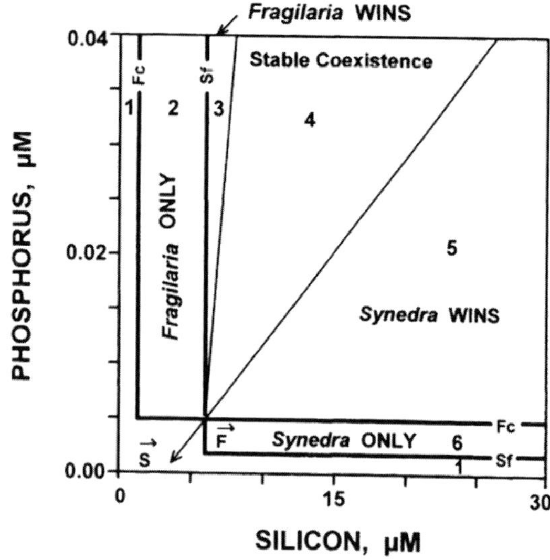

FIGURE 18-7 The predicted outcome of resource (phosphorus and silicon) competition between two freshwater diatoms, *Synedra filiformis* (Sf) and *Fragilaria crotonensis* (Fc). The isoclines of zero net growth (*heavy lines*) for each species cross at an equilibrium point. The consumption vectors (*diagonal lines*) for each species are labeled $\vec{S}$ and $\vec{F}$. Neither species has sufficient nutrients to survive in region 1. In region 2 *Fragilaria* can survive but *Synedra* cannot. In region 3 *Synedra* is outcompeted by *Fragilaria*. Both species stably coexist in region 4. *Fragilaria* is competitively displaced by *Synedra* in region 5. Only *Synedra* can survive in region 6. *(Modified from Kilham and Hecky [1988].)*

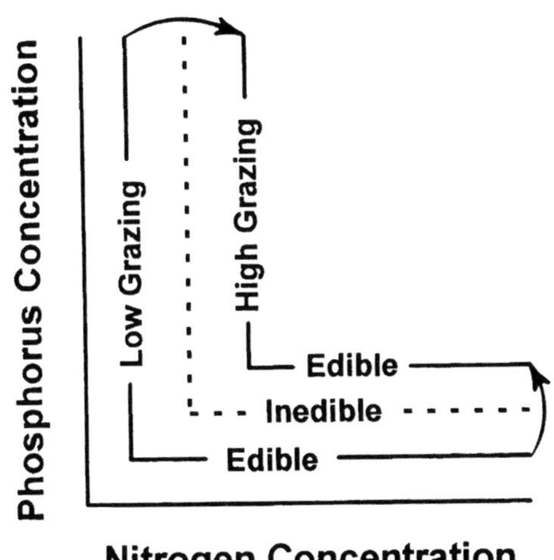

FIGURE 18-8 Hypothetical succession of algal species that results from a shift from low grazing rates to high grazing rates (*arrows*). The isocline of an inedible species (*dashed*) does not vary with grazing pressure, but the isocline of an edible species (*solid lines*) moves away from the axes as grazing pressure increases. The species with its isocline closest to the axes will be the superior competitor and will dominate. *(From Sterner [1989].)*

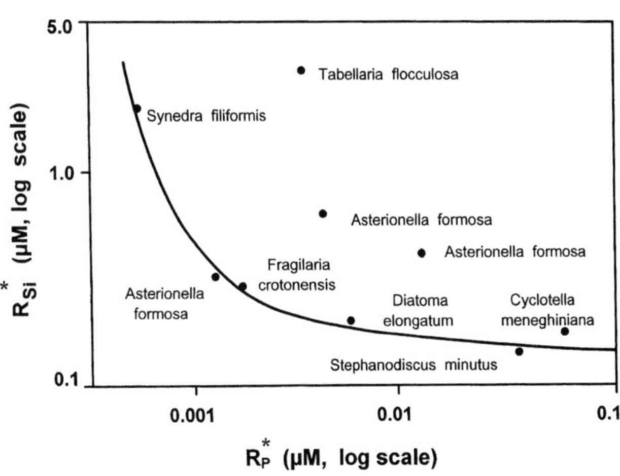

FIGURE 18-10 Differences in nutrient competitive abilities (R^*) (μM nutrient) among freshwater planktonic diatoms at 20°C. R^*_{Si} and R^*_P are the R^* for silica and phosphorus, respectively. *Synedra filiformis*, for example, is the superior competitor for phosphate (lowest R^*_P) but a poor competitor for silicate (higher R^*_{Si}). *(From Kilham and Hecky [1988], based on Tilman et al. [1982].)*

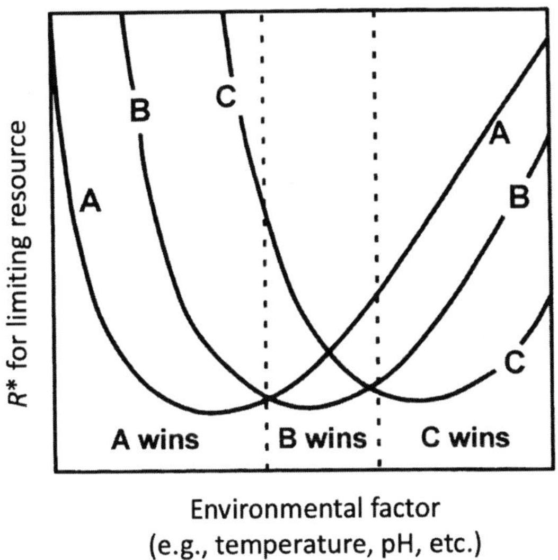

FIGURE 18-9 The dependence of competitive abilities for a resource along an environmental gradient (e.g., temperature, pH, etc.). The species with the lowest requirement (R^*) for a resource at a given level of an environmental factor wins the competition. Note that the identity of the best competitor changes depending on environmental conditions. *(Modified from Tilman et al. [1982].)*

species that can stably coexist at equilibrium cannot exceed the number of limiting resources (Armstrong and McGehee 1980, Tilman 1982). This is known as the *competitive exclusion principle* (Gause 1934). In phytoplankton only a handful of resources can be limiting, yet many more species coexist. This seeming contradiction was highlighted by G. Evelyn Hutchinson, in his famous paper "The paradox of the plankton" (the coexistence of multiple species on few limiting resources) (Hutchinson 1961). Hutchinson himself proposed several explanations for the paradox, including temporal fluctuations in resources or environmental factors. Subsequently, many theoretical and experimental studies showed that many species can coexist on a few limiting resources when resources fluctuate and when species have nonlinear responses to resources (Sommer 1984, Huisman and Weissing 1999, Litchman and Klausmeier 2001).

B. Grazing by zooplankton

Grazing affects both biomass and species composition of phytoplankton (Chapter 20), favoring traits that help avoid predation (reviewed in Lürling 2021). These traits can be morphological, physiological, or behavioral (Litchman and Klausmeier 2008, Pančić and Kiørboe 2018). Phytoplankton morphological antigrazing traits include strategies that provide digestion resistance such as a thickened cell wall that would allow the passage through the grazer's gut (Van Donk et al. 1997) or the presence of mucilage (DeMott et al. 2010) or strategies that provide ingestion resistance, such as large (>30 μm) or small cell size, depending on the grazer. In addition, phenotypic changes such as the development of spines and formation of colonies or cell aggregation in response to grazing pressure or chemicals can occur. Life history traits, such as reproduction, may include high growth rates that would withstand high mortality rates posed by grazing (Sarnelle 1993, 2005). Behavioral traits may include avoidance of predators by timing cyst recruitment (Hansson 1996) and migrating to areas with lower predator pressure (Latta et al. 2009). Developing such plastic responses that provide antigrazer resistance also has costs, such as decreased nutrient uptake due to increased cell sizes, increased sinking rates for colonies (Lürling 2003), and decreased growth rates (Ehrlich et al. 2020), among others. More studies are needed to evaluate the costs and benefits of antigrazing strategies in phytoplankton (Pančić and Kiørboe 2018).

C. Allelopathy

Phytoplankton produce a wide array of *secondary metabolites*, which are chemical compounds that seem nonessential to their core metabolism. Some of these chemicals, often called allelochemicals, may be important in mediating interactions among phytoplankton and with other organisms, such as parasites or grazers. These allelopathic interactions are mostly negative and inhibit the physiological processes of other species. Allelopathic interactions occur when a biochemical compound is released into the environment by a donor species that has an inhibitory effect upon a target species. Legrand et al. (2003) reviewed laboratory studies assessing allelochemicals in phytoplankton and found evidence of *allelopathy* (negative interactions between a donor and a target species) in freshwater phytoplankton for almost all taxonomical groups (dinoflagellates, Cyanobacteria, diatoms, chlorophytes) with target effects on other phytoplankton and bacteria. Allelopathy might enhance phytoplankton species' fitness in their interactions with other autotrophic organisms (Gross 2003) or with bacteria, fungi, and viruses (Legrand et al. 2003). It is, however, challenging to detect allelopathic interactions among planktonic organisms because other interferences (physical, biotic, and chemical) may also happen simultaneously (Inderjit and del Moral 1997). Hence allelopathy is often considered part of competition, as it is not possible to remove allelopathic interactions in competition experiments, whereas competition can be eliminated in allelopathic tests (see below).

Microalgal allelochemicals show much structural variety and include cyclic peptides, alkaloids, organic acids, and long-chain fatty acids (reviewed in Legrand et al. 2003). Allelochemicals can interfere with many processes in the target organisms, often affecting photosynthesis and enzyme activity (Gross 2003) and inhibiting growth (Legrand et al. 2003). Yet, the biochemical characteristics, as well as the metabolic pathways of most allelochemicals, remain uncharacterized (Legrand et al. 2003). While the costs of synthesizing allelochemicals are poorly understood, they likely exist (Legrand et al. 2003).

In aquatic ecosystems most allelopathic exchanges happen between organisms that are in close proximity to each other, either in the benthos or among macrophytes and their attached epiphytes (Gross 2003). Allelopathic interactions in planktonic systems are challenging to address because allelochemicals become diluted in the water column (Legrand et al. 2003). Hence most studies on allelopathy have been run under laboratory conditions. A classic approach to assessing allelopathic exchanges between two phytoplankton species is *cross-culturing* (growing a species in a culture medium rich with cell-free filtrate from another species) (Legrand et al. 2003). Finally, how environmental and biological variables affect the synthesis and excretion of allelochemicals still remains largely uncharacterized. Other secondary metabolites, frequently found in Cyanobacteria, include toxins and odor-producing compounds (see Water quality and harmful algal blooms section).

D. Parasitism and pathogens

Pathogens, together with grazing and sedimentation out of the photic zone, lead to phytoplankton loss and counterbalance growth and productivity. Pathogens that affect phytoplankton include viruses, bacteria, protozoans, and fungi.

Viral phytoplankton pathogens can lyse various types of phytoplankton (Suttle 2000). Viral infections of natural cyanobacterial blooms may contribute to bloom termination (Coulombe and Robinson 1981, Gerphagnon et al. 2015), sometimes removing up to 97% of cyanobacterial production (Tijdens et al. 2008). Indeed, *cyanophages* (viruses that attack Cyanobacteria) are abundant in water bodies and can reach concentrations of about 10^6 viruses mL^{-1} (Suttle 2000).

Bacterial pathogens of phytoplankton affect many species of unicellular and filamentous eukaryotic algae and Cyanobacteria (Shilo 1970, Mayali 2018). Bacteria can affect phytoplankton in several ways: they either enter phytoplankton cells (Caiola and Pellegrini 1984) such as the nonphotosynthetic cyanobacterium *Vampirovibrio chlorellavorus*, an obligate parasite of the green alga *Chlorella* spp. (Soo et al. 2015), engage in cell-to-cell contact (Gumbo and Cloete 2011), or secrete extracellular substances (Choi et al. 2005, Mu et al. 2007) including amino acids, peptides, proteins, or antibiotics that may help lyse the cyanobacterial cells (Gumbo and Cloete 2011).

While many protozoans directly consume phytoplankton, others may parasitize them. Some species of protozoa can attach to phytoplankton cells and remove or digest their contents (Brabrand et al. 1983). Certain herbivorous protozoans were able to clear over 99% chlorophycean algal populations in 1 to 2 weeks (Canter and Lund 1968).

Phytoplankton can also be parasitized by aquatic fungi, which can be external (Chytridiales) or internal (Chytridiomycetes) parasites. There is evidence that chytrid fungi can lead to mass mortalities of phytoplankton, suppressing or delaying its blooms and influencing species assembly and successional changes (Park et al. 2004, Frenken et al. 2017). Chytrid fungal infections occur in most phytoplankton taxonomical groups (reviewed in Park et al. 2004). Parasite prevalence seems to depend on host densities; for example, under low phytoplankton (host) densities, infection prevalence decreases due to lower encounter probabilities (Reynolds 1984). Also, the amount or the composition of extracellular phytoplankton exudates probably affects the susceptibility of phytoplankton to fungal infection. Moreover, environmental factors (light, temperature, and nutrients) affect fungal growth as well (Van Donk and Bruning 1995).

i. Pathogens and ecosystem services

While pathogens are ubiquitous in nature, their effects, either positive or negative, on ecosystem services, for example, drinking water, aquaculture, carbon and nutrient cycling, and aesthetics, are often overlooked (Paseka et al. 2020). From a host perspective, pathogens exert a negative effect, as they reduce host fitness. From a human perspective, pathogens can decrease ecosystem services such as decreased yields that can threaten food security (e.g., aquaculture) and cause species loss, or enhance ecosystem services, for example, through the suppression of harmful algal blooms, hence improving drinking water quality, aquaculture, esthetics, and recreation (Paseka et al. 2020).

E. Mutualism

The *phycosphere* (the space directly adjacent to a phytoplankton cell) is rich in organic compounds that are released by the cell into the aqueous medium and diluted through diffusion and turbulence (Seymour et al. 2017). Symbiotic interactions between phytoplankton and heterotrophic bacteria may occur in the phycosphere, with phytoplankton providing dissolved and particulate carbon and heterotrophic bacteria providing micronutrients via remineralization. Bacteria can also compete with phytoplankton for inorganic nutrients (Seymour et al. 2017). The exchange of metabolites between phytoplankton and bacteria can change the fitness of symbiotic partners, and while it occurs at the microscale, it may have bottom-up effects on the ecosystem (Seymour et al. 2017). The size of the phycosphere is positively correlated with several phytoplankton variables, including cell size, growth, and exudation rate, the diffusivity of the exuded compound, and its background concentration and turbidity (Seymour et al. 2017).

Symbioses in marine environments are rather common and include a symbiosis between diatoms (host) and nitrogen-fixing Cyanobacteria (symbiont), also occurs in freshwater, and between dinoflagellates (host) and several symbionts, including chrysophytes, cryptophytes, and pennate diatoms (Wouters et al. 2009). It may be that, in freshwater ecosystems, symbioses are less common. One such symbiosis is that between the dinoflagellate *Gymnodinium acidotum* (host) and a cryptophyte (symbiont), but both organisms can live in the absence of the other (Wilcox and Wedemayer 1984). Another example of a symbiosis is that between the freshwater dinoflagellate *Durinskia oculata* (host) and a eukaryotic organism derived from a diatom (symbiont) (Kretschmann et al. 2018).

F. Bacterivory and other mixotrophy

Most phytoplankton are obligate *photoautotrophs* (obtain their cell carbon from transformations driven by light energy), as photosynthesis dominates their metabolism. If light is inadequate or unavailable, photoautotrophs may die or become dormant until light becomes available again. However, certain phytoplankton can also be *heterotrophic* (obtain cell carbon from the metabolism of organic substrate). Other phytoplankton organisms can combine phototrophy and heterotrophy and are known as mixotrophic phytoplankton. *Mixotrophy* (the ability to combine phototrophy and heterotrophy) can be important in the survival, competition, and succession of phytoplankton. *Phagotrophic* (able to engulf prey) phytoflagellates can comprise large fractions of phytoplankton communities, with the highest proportions in humic, dystrophic lakes and the smallest proportions in eutrophic lakes (Bird and Kalff 1986, Pålsson and Granéli 2004).

i. Taxonomic groups where mixotrophy occurs and their traits

Mixotrophy is common among several phytoplankton groups including dinoflagellates, cryptophytes, and chrysophytes, with some species being mostly autotrophic while others being mostly heterotrophic (Stoecker 1998). Mixotrophs share common traits: they are unicellular and have flagella; they have been suggested to have poor light competitive abilities (Schwaderer et al. 2011) and low growth rates (Stoecker 1998). Much is still unknown about how mixotrophs regulate phototrophic and phagotrophic modes, both across species and in response to changes in environmental variables such as nutrients, temperature, light, and pH. Moreover, it is unclear if alternating trophic modes (phototrophy or phagotrophy) would alter the costs compared to the simultaneous functioning of both trophic modes (Stoecker 1998). Interestingly, it was observed that when a dinoflagellate grew as a mixotroph, its growth rate was higher than when growing exclusively as a phototroph or a phagotroph (Hansen and Nielsen 1997), suggesting that mixotrophy can increase fitness.

ii. Mixotrophy in food webs and the environment

Mixotrophs are important in many planktonic food webs (Mitra et al. 2016, Selosse et al. 2017). Mixotrophy expands the typical food web interactions so that primary producers can also get nutrients by ingesting bacteria or algae directly (Flynn et al. 2018). Therefore mixotrophic phytoplankton can be primary producers and consumers at the same time. The high relative contribution of mixotrophic phytoplankton to community composition may suggest limited nutrient and/or light availability (Pålsson and Granéli 2004, Ptacnik et al. 2016). Mixotrophy is often associated with oligotrophic habitats, where dissolved nutrients are low and can be supplemented by ingesting prey (Jones 2000, Ptacnik et al. 2016), but mixotrophy occurs in eutrophic waters as well (Flynn et al. 2018).

IV. Temporal variation in phytoplankton communities

A. Seasonal successional patterns

Seasonal patterns and periodicity in phytoplankton biomass are observed in temperate and in polar fresh waters. Phytoplankton succession has been linked with variations in environmental variables, mostly in temperature, light, nutrient concentrations, and loss factors such as predation and parasitism. Because phytoplankton seasonal succession is related to abiotic factors and the stability of the water column, the patterns observed in temperate areas are different from patterns in tropical ecosystems. Each will be discussed separately with the caveat that the processes are similar, but the timing can differ considerably.

While there is much variation in seasonal successional patterns across different lakes and other water bodies, some patterns are reasonably consistent.

i. Dimictic lakes in the temperate zone

For many temperate lakes, the successional patterns were summarized by the *Plankton Ecology Group (PEG)* (Sommer et al. 1986) and are known as the *PEG-model of seasonal succession* (description of typical stages of phytoplankton temporal variation in temperate lakes). Since its original publication in 1986, the model has been updated to include more processes important for species turnover (Sommer et al. 2012). The main patterns of seasonal succession differ somewhat between oligotrophic and eutrophic lakes (Fig. 18-11).

Both types of lakes have a spring bloom of phytoplankton, usually cryptophytes and diatoms, due to seasonally increasing light and high nutrient availability. This spring phytoplankton bloom is followed by a peak in zooplankton fueled by high phytoplankton biomass. The peaks are followed by the so-called *"clear-water" phase* (low abundance of phytoplankton) due to zooplankton grazing. Zooplankton start declining because of the low food (phytoplankton) availability. Low zooplankton abundance allows phytoplankton biomass to increase again, with cryptophytes and large green algae dominating. Species richness and functional diversity also increase, as there are highly edible small species and poorly edible large species. With declining nutrient availability due to summer stratification and the uptake by phytoplankton, algae become nutrient limited. Fast-growing, highly edible species decrease, and larger, less edible colonial Cyanobacteria and dinoflagellates increase. Nitrogen depletion may lead to an increased abundance of nitrogen-fixing Cyanobacteria. Therefore grazing and competition affect phytoplankton community assembly and dynamics. In oligotrophic lakes the "clear water phase" may not be followed by another phytoplankton peak later in the summer, and biomass may remain low.

Later in the season, physical factors become more important. Increasing mixing depth reduces light but increases nutrient availability; consequently, phytoplankton species adapted to being mixed and low light conditions increase in abundance. Diatoms are often the main group that forms fall blooms. The autumnal increase in phytoplankton biomass is followed by a zooplankton peak. Increased mixing and decreasing temperature slow phytoplankton growth, and their biomass declines. Winter may be associated with ice cover in polar and temperate lakes, with reduced mixing and light availability. Phytoplankton biomass is usually low in winter.

In seasonally ice-covered lakes, including large lakes, such as Lake Baikal in Siberia and the Laurentian Great Lakes, there may be under-ice phytoplankton blooms, often consisting of certain diatom species attached to the ice from beneath (Moore et al. 2009, Reavie et al. 2016). The next season, the successional pattern repeats. A decrease in ice cover duration may lead to a decline or a disappearance of under-ice phytoplankton blooms (Wollrab et al. 2021).

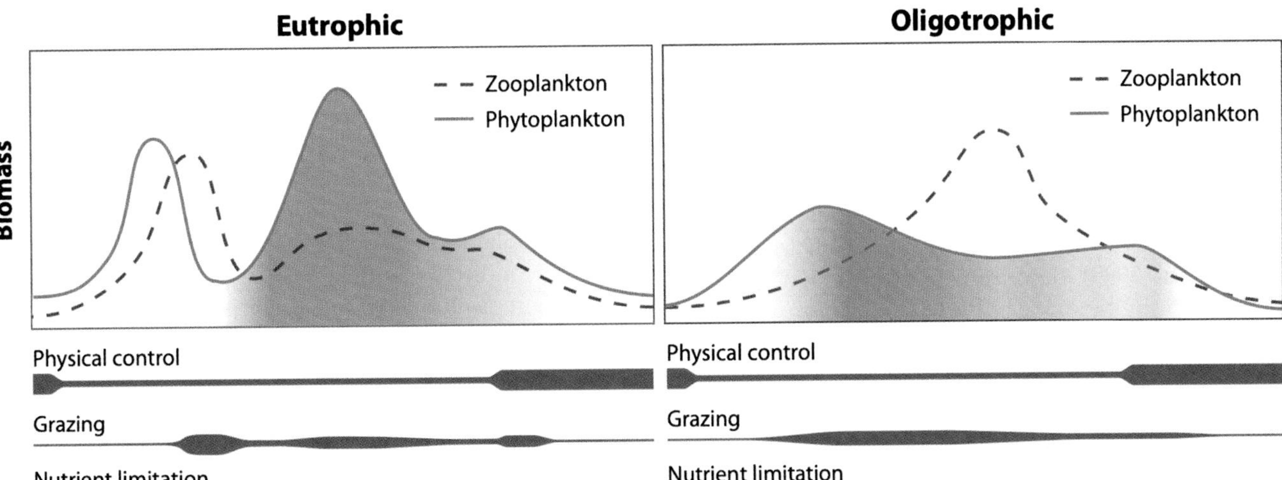

FIGURE 18-11 The typical trajectory of seasonal succession in eutrophic and oligotrophic lakes and major controls of phytoplankton biomass over the season (*dark shading*, phytoplankton inedible for zooplankton; *light shading*, phytoplankton edible for zooplankton). *(From Sommer et al. [2012].)*

Subsequent work showed that, in addition to physical factors, nutrients, and grazing, there are other important drivers of phytoplankton seasonal succession in lakes. The *microbial loop* (carbon and nutrient cycling through microbial components of the pelagic aquatic communities), parasitism, and food quality are among the most important (Sommer et al. 2012). Many phytoplankton species have parasites, and they may delay or decrease phytoplankton blooms, increase species richness by suppressing the dominant competitors, or terminate blooms (Sommer et al. 2012, Frenken et al. 2016). The termination of spring diatom blooms by parasitic chytrid fungi was shown to be accelerated by warming (Frenken et al. 2016). Phytoplankton C:N:P stoichiometry also plays a role in successional patterns by affecting zooplankton growth and reproduction (Sommer et al. 2012). In addition, the concentrations of different fatty acids in phytoplankton also determine phytoplankton nutritional value for zooplankton (Müller-Navarra et al. 2004).

ii. Tropical, subtropical, and warm temperate lakes

Tropical lakes are far less numerous than temperate lakes (see Chapter 4), with some geographic exceptions, such as in Indonesia, which are dominated by lakes of river origin (Lewis 1996). Greater solar irradiance with reduced annual variations results in higher minimum water temperatures. In warm lakes the seasonality in environmental factors and biotic interactions is less pronounced and phytoplankton in winter may not be light limited (Moustaka-Gouni et al. 2014) and, consequently, may have higher winter biomass. The intraseasonal variability in warm lakes may exceed interseasonal variability. Despite some differences, nutrients, light, and grazers control species abundance and dynamics in lakes of different latitudes, from polar to tropical (Lewis 1978, Nan et al. 2020).

iii. Reservoirs

In reservoirs the unpredictable dynamics of inflow and changing flushing processes markedly alter the environmental setting for biotic communities. A reservoir can be thought of as a very dynamic lake where a marked proportion of its volume has the attributes of and functions as a river. Frequently, the riverine part of reservoirs operates like large, turbid rivers where turbulence, high turbidity, and decreased light availability preclude extensive phytoplankton development despite elevated nutrient availability. Nutrient limitations, so characteristic of the successional changes of natural lakes with low-to-moderate productivity, can occur to varying degrees in reservoirs, as utilization and losses of nutrients surpass renewal rates.

Light limitation is a dominant control of productivity in most reservoirs, as it is in many productive natural lakes and rivers. In many cases light limitations in reservoirs result primarily from inorganic clay and silt turbidity. Where light limitation results from high loadings of dissolved organic matter, as in many tropical and subtropical reservoirs, the total and selective spectral light attenuations are quite analogous to those of natural lakes. The succession of phytoplankton communities of the lacustrine portion of reservoirs is generally similar to that found in natural lakes. Disturbance frequencies are much higher in reservoirs than in lakes, with rapid, often irregular, and large changes in water level, flushing rates, turbidity, and mixing.

iv. Other lakes

In permanently ice-covered water bodies in Antarctica the patterns of seasonal succession are much less studied. Antarctica's McMurdo Dry Valley has several such lakes, and Lake Bonney is probably the most well studied of them. The characteristic features include low light availability due to ice cover and high stability of physical and chemical conditions (Priscu et al. 1999). Because of the permanent stratification, the spring bloom is initiated by the increasing solar radiation and not the change in the mixing depth (Priscu et al. 1999). Throughout the year, low light availability selects for mixotrophic phytoplankton and strong stratification benefits motile forms, such as cryptophytes and green flagellates (Spaulding et al. 1994).

B. Long-term changes

i. Alternative stable states

Some shallow lakes may transition between different stable states (Chapter 26), either dominated by phytoplankton or by submerged or floating macrophytes (Scheffer et al. 1993, Scheffer et al. 2003). Such shifts have been documented for both small and large shallow lakes (Scheffer et al. 1993, Scheffer et al. 2003, Janssen et al. 2014). Lakes with macrophyte dominance constitute a clear state, as phytoplankton biomass is low because of competition with macrophytes. Lakes in a clear state can withstand significant increases in nutrient concentrations without changing state, with nutrients absorbed by the vegetation. Further increases in nutrient loading, however, may ultimately result in a shift toward a turbid state with high phytoplankton concentration and low abundance, or lack, of macrophytes. To shift the system back to a clear state, a macrophyte-dominated state, requires a very significant nutrient reduction.

V. Spatial heterogeneity in phytoplankton

A. Vertical distribution of phytoplankton

In aquatic environments light quantity and quality are strongly affected by the aquatic medium. This creates a vertical gradient in light intensity and spectral quality (Kirk 1994) that can affect the vertical distribution of phytoplankton. In addition to the light gradient, there is often an opposite nutrient gradient in the water column, with nutrients supplied by the sediments and decreasing toward the water surface due to uptake by phytoplankton. The opposing light and nutrient gradients can cause vertically heterogeneous phytoplankton distributions. The most common phytoplankton vertical aggregations in lakes are *deep chlorophyll maxima (DCM)* in deep stratified lakes, where the maximum concentration of chlorophyll *a* is at a certain depth in stratified water columns, benthic layers, where algae grow on the bottom of the lake (Fig. 18-12), and surface scums, where algal biomass is high at the lake surface. The aggregations can range from centimeters to meters, depending on the level of vertical mixing (Coon et al. 1987, Lindholm 1992). Vertically heterogeneous patterns in phytoplankton are generally observed under low mixing conditions, either when mixing of the entire water column is hindered by a density gradient or when the kinetic energy (e.g., wind action) is low (Lindholm 1992, Reynolds 1992, Leach et al. 2018).

Nevertheless, poor mixing on its own would not necessarily result in vertically heterogenous phytoplankton distributions; the opposing gradients in light and nutrients play an important role in determining the vertical structure (Davey and Heaney 1989, Klausmeier and Litchman 2001, Leach et al. 2018). Phytoplankton face contrasting gradients of two key resources: light coming from above and nutrients that are often released from the sediments (Klausmeier and Litchman 2001). The competition along the opposing gradients of nutrients and light results in the heterogenous vertical phytoplankton distribution in poorly mixed conditions, where species grow the fastest where they are equally limited by light and by nutrient concentration (Klausmeier and Litchman 2001). Phytoplankton traits, such as nutrient and light competitive abilities, and buoyancy or swimming, thus determine the depth of phytoplankton layers, depending on nutrient and light availability and turbulence (Huisman et al. 2002, Klausmeier and Litchman 2001). With increasing nutrient supply, the DCM occurs closer to the surface and deepens with increased light availability. Under low nutrient conditions, low light attenuation, and shallow water column depths, a benthic layer (algae growing on the bottom) may occur; under intermediate nutrient concentrations in deepwater columns, a DCM is likely; and for high nutrient concentrations, a surface accumulation occurs (Klausmeier and Litchman 2001).

Phytoplankton forming deep DCMs in stratified water bodies are often either filamentous non-N-fixing Cyanobacteria such as *Planktothrix rubescens* and *P. mougeotii* (Reynolds et al. 2002), which have phycobiliproteins as accessory pigments and are low light adapted, or mixotrophic flagellates, such as cryptomonads and chrysophytes (deNoyelles et al. 2016).

B. Horizontal gradients

Variations in physical (flow, currents, turbulence, wind velocity and direction, depth, residence time) and biological variables (competition with macrophytes, grazing) can affect the horizontal distribution of

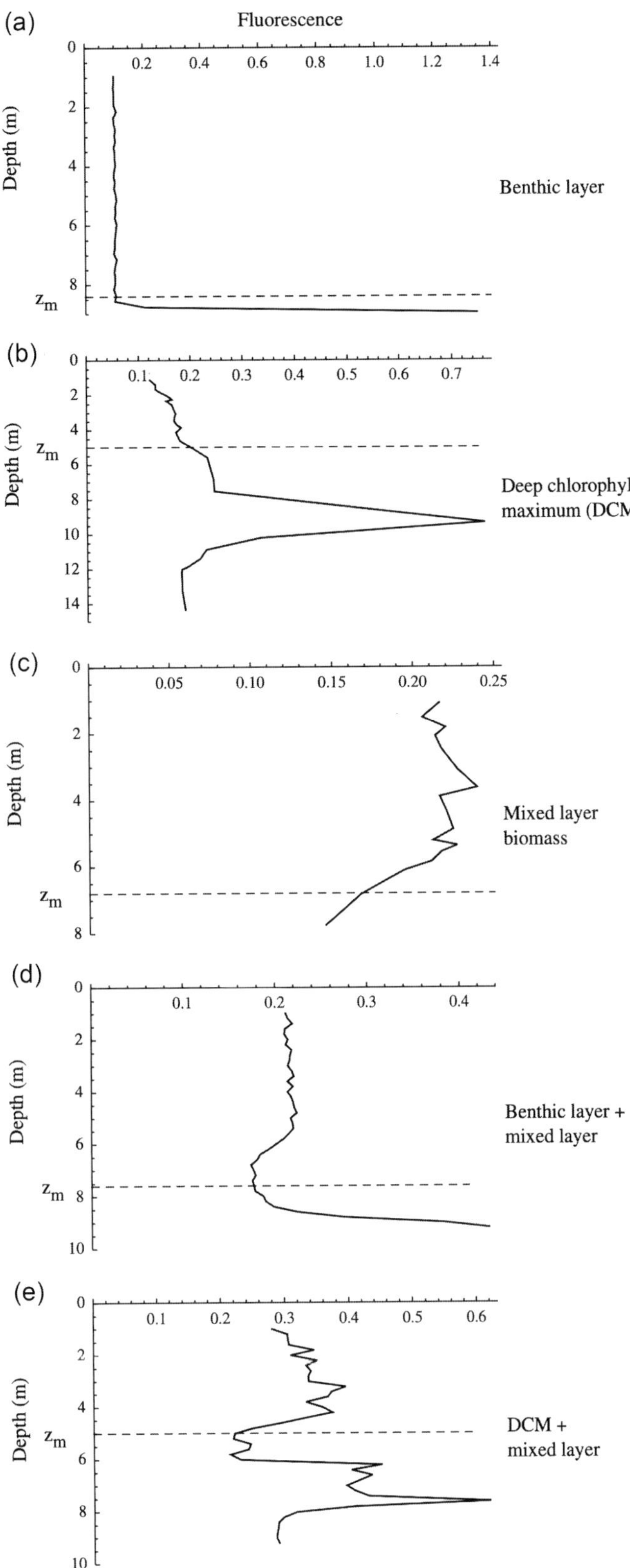

FIGURE 18-12 Examples of phytoplankton vertical distributions observed in stratified lakes in southern Michigan (USA). (a) Benthic layer biomass only; (b) Deep chlorophyll maximum (DCM) biomass only; (c) Mixed layer biomass only; (d) Benthic layer and mixed layer biomass; and (e) DCM and mixed layer biomass. Vertical distributions are chlorophyll *a* fluorescence (relative fluorescence units) profiles and mixed layer depth (*dashed line*), z_m is calculated as the depth where temperature is 1°C cooler than the surface temperature. Profile (a) is from Sherman Lake, Kalamazoo County, Michigan (USA), 12 October 2007. Profile (b) is from Bristol Lake, Barry County, Michigan (USA), 3 October 2007. Profile (c) is from Fine Lake, Barry County, Michigan (USA), 3 October 2007. Profile (d) is from Hogsett Lake, Kalamazoo County, Michigan (USA), 22 October 2007. Profile (e) is from Bassett Lake, Barry County, Michigan (USA), 15 October 2007. *(From Mellard et al. [2011].)*

phytoplankton in freshwater ecosystems. In shallow lakes (Chapter 26) strong spatial differences can be observed depending on the cover of macrophytes, which can result in alternative stable states (Scheffer et al. 1993, Scheffer et al. 2003, de Tezanos Pinto and O'Farrell 2014), with low phytoplankton abundance in areas densely colonized by macrophytes and high(er) phytoplankton biomass in areas without macrophytes (de Tezanos Pinto and O'Farrell 2014).

In rivers phytoplankton composition is mostly shaped by physical variables including flow, depth, suspended solids load, and its impacts on water transparency (Reynolds and Descy 1996). Nutrient availability, however, appears to be a minor controlling factor (Abonyi et al. 2021), as physical variables override the effect of nutrients. Upstream sections are characterized by fast flow, where truly planktonic forms are rare, as the short water residence time hinders phytoplankton population growth and biomass (Abonyi et al. 2021). Downstream, water residence time increases and hence allows the development of truly planktonic forms. In these parts of the river increased nutrient concentration, coupled with turbulent and turbid flow characteristics, generally favors phytoplankton with high growth rate and low light traits (Reynolds 1994, Reynolds and Descy 1996). Therefore, in rivers because of the strong selecting drivers, only few phytoplankton can reach dominance (Reynolds 1994). River phytoplankton in floodplain rivers are also affected by the lake phytoplankton of shallow lakes located along the floodplain (Devercelli and O'Farrell 2013). In rivers planktonic and benthic assemblages of autotrophs are usually intermixed, with benthic organisms being suspended in the plankton and planktonic taxa found in the benthos (Reynolds 1994). The relative amount of benthic species suspended in the water is greater in small-scale streams and in upstream sections of large rivers (Leland 2003).

C. Phytoplankton communities in a landscape

A *metacommunity* is an assemblage of local ecological communities connected by the dispersal of species (Leibold et al. 2004). Understanding how landscape (e.g., patch size, connectivity) influences the characteristics of metacommunities in lakes is still poorly known (Almeida-Gomes et al. 2020). The degree of connectivity among water bodies in the landscape may vary depending upon flooding or nonflooding situations (Devercelli 2009); floods enhance homogeneity in the environmental variables among water bodies in the landscape.

Remote sensing technology is increasingly being used in landscape ecology as it can assess the landscape structure (spatial arrangement of landscape elements), change (spatial and temporal), and function (interaction between the elements in the landscape) (Crowley and Cardille 2020). Satellites can measure several lake properties such as biotic components (algal blooms, phenology of primary producers), water transparency (dissolved organic carbon, turbidity, Secchi disc depth, euphotic depth), hydrology (water level), and temperature (of the epilimnetic layer) (reviewed in Dörnhöfer and Oppelt 2016). In addition, satellites can measure, with high spatial and temporal resolution, phytoplankton chlorophyll *a* and pigments such as phycocyanin (characteristic of Cyanobacteria) and can therefore track phytoplankton biomass at the landscape scale (Kutser 2004, Stumpf et al. 2012, Urquhart et al. 2017). Remote sensing is also very useful for assessing landscape changes (Crowley and Cardille 2020) that can affect the water bodies within their catchments.

VI. Phytoplankton and global environmental change

A. Major stressors/drivers (warming, CO_2, nutrients, land use change, invasive species)

While individual factors, such as nutrients, temperature, CO_2, precipitation, and parasites, affect phytoplankton, these variables are changing simultaneously and might affect phytoplankton interactively. This underscores the difficulty of attributing the observed impacts to a single environmental factor along a complex environmental gradient. The impacts of several stressors might be nonadditive and nonlinear.

Environmental stressors affect different levels of phytoplankton organization (Salmaso and Tolotti 2021). Two key drivers are climate change and eutrophication, which act both at the global and at the regional and local scales (Salmaso and Tolotti, 2021) and may act synergistically. Climate change affects all ecosystems on Earth. The ongoing global warming is resulting in shorter winters and warmer summers over large regions (Dokulil et al. 2006, Woolway et al. 2017).

i. Direct effects of increased temperature

Based upon the optimal temperatures for growth in different taxonomic groups, increasing water temperatures may select for Cyanobacteria or green algae over other taxa (dinoflagellates, cryptomonads, chrysophytes, euglenoids, diatoms) and hence change community composition and possibly reduce diversity (Paerl and Huisman 2008, Paerl and Huisman 2009, Litchman et al. 2010). For individual species, increases in temperature may be critical, depending on their optimal temperature for growth, as growth declines sharply above the optimal growth temperatures (Thomas et al. 2012).

Thus temperature increases above the optimal growth temperature could markedly affect species survival and, consequently, species diversity. Increased temperatures may also lead to smaller phytoplankton sizes and change phytoplankton community size distribution (Zohary et al. 2021).

ii. Indirect effects of increased temperature

Adding to the direct impacts of increased temperatures, warming also increases vertical thermal stratification (Chapter 7). The strength of vertical stratification depends on the density difference of water, and because increased temperatures decrease the density of water (Chapter 3) in the epilimnion (warm water is less dense), they increase the strength and the duration of vertical stratification (earlier summer stratification and later autumnal mixing). Stratified conditions favor motile phytoplankton (e.g., gas-vacuolated Cyanobacteria and flagellated eukaryotes) that can move upward toward light and downward to access higher nutrient levels (Paerl and Huisman 2009) and/or phytoplankton with good nutrient competitive abilities, because under stratified conditions, nutrients can become limiting. Moreover, rising temperatures shorten the duration of ice cover, with earlier ice off in lakes and earlier spring phytoplankton blooms (e.g., Adrian et al. 1999, Wollrab et al. 2021). In addition, shorter duration of ice cover reduces the duration of inverse stratification (when cold, less dense water sits on top, right under the ice). Shorter inverse stratification decreases losses of nonmotile phytoplankton, due to sinking, and results in a larger phytoplankton standing crop at the end of winter that increases rapidly in the spring (Salmaso and Tolotti 2021).

iii. Nutrients

Eutrophication is the main cause of water quality deterioration for the majority of fresh waters (Keiser and Shapiro 2018). As highlighted above, Cyanobacteria are often dominant in eutrophic and hypereutrophic water bodies, particularly when other conditions are favorable (e.g., high temperature, thermal stability, and high residence time) (Salmaso and Tolotti 2021). A high nutrient availability may select for large-sized phytoplankton, though this trend may be counteracted by decreased light availability, which selects for small sizes. Dense surface blooms attenuate incoming irradiance, which could further favor Cyanobacteria, as they are adapted to low light (Schwaderer et al. 2011). Increased nitrogen load can also increase the toxicity of Cyanobacteria (Gobler et al. 2016, Brandenburg et al. 2020).

iv. Increasing CO_2 concentrations

Rising CO_2 levels will have direct and indirect effects on pCO_2. Increased pCO_2 might increase photosynthesis and alter phytoplankton community structure (Hasler et al. 2016). It may also change the pH, potentially making poorly buffered water more acidic, which would influence nutrient availability and may favor certain groups of phytoplankton (Phillips et al. 2015).

v. Land-use change

Changes in land use in the catchment area often significantly affect water bodies and may lead to eutrophication, increased sediment load due to organic and inorganic matter, which decreases oxygen content and light availability, and pollution due to agrochemicals. These changes affect phytoplankton community and trait composition and would select for phytoplankton that are adapted to low light (e.g., Cyanobacteria, diatoms) and high nutrients. Satellites are increasingly being utilized to assess past landscape changes and to predict future landscape changes (Crowley and Cardille 2020).

vi. Changes in hydrology

The declines in water levels, due to drought or due to higher evaporation, as observed, for example, in the Mediterranean region, may result in the disappearance of shallow lakes, both permanent and temporary, and a decline in biodiversity (Barone et al. 2010). The ability to produce cysts or resting stages would allow species persistence in water bodies that face decreased water levels or drought.

B. Water quality and harmful algal blooms

Harmful algal blooms (HABs) consist of toxic phytoplankton that can significantly affect water quality, biodiversity, community structure, and ecosystem functioning. Among freshwater phytoplankton, Cyanobacteria are the main group capable of producing toxic blooms. In marine environments dinoflagellates and diatoms also produce toxins, but in freshwater these groups are generally considered to be nontoxic. Most cyanobacterial HABs negatively affect water quality, fisheries, and other ecosystem functions and may increase in frequency and severity in the future, due to increased nutrient inputs and temperature (Paerl and Huisman 2009).

i. Toxicity

Detailed overviews of the ecology, toxicity, and detrimental effects of Cyanobacteria on health have been documented (e.g., Huisman et al. 2005, Chorus and Welker 2021); hence we will only briefly mention them. All orders of Cyanobacteria can produce cyanotoxins, such as *hepatotoxins* (toxins affecting the liver, e.g., microcystin, cylindrospermopsin), *neurotoxins* (toxins affecting the nervous system, e.g., saxitoxin, anatoxin), *dermatotoxins* (toxins affecting the skin), and others (Bláha et al. 2009). What triggers the production of cyanotoxins is not fully understood, yet it can be affected by several environmental variables, such as light levels and nutrient concentrations (Kardinaal et al. 2007, Chorus and Welker 2021). High concentrations of cyanobacterial cyanotoxins may also correlate with high nitrogen concentrations (Gobler et al. 2016, Brandenburg et al. 2020), probably because cyanotoxin molecules are rich in nitrogen.

ii. Odor

Many bloom-forming planktonic Cyanobacteria produce compounds that can impart odor or taste to water, such as geosmin (filamentous Cyanobacteria, order Nostocales), 2-methylisoborneol (filamentous Cyanobacteria, order Oscillatoriales) (Jüttner and Watson 2007), and ß-cyclocitral; dimethyl disulfide; and dimethyl trisulfide (colonial Cyanobacteria such as *M. aeruginosa*, order Chroococcales) (Watson et al. 2016). The ecological role of odor and taste compounds is still poorly known. Although nontoxic to humans, these compounds severely decrease water quality, increasing the costs of drinking water treatment (Watson et al. 2016). Interestingly, odor and toxicity do not seem correlated (Hrudey et al. 1993).

iii. Cyanobacterial traits and global climate change

Cyanobacterial blooms are likely to increase in the future, in part because several cyanobacterial traits make them well adapted to future conditions (Paerl and Huisman 2009). For example, their high optimal temperatures for growth will ensure faster growth compared to most of the other phytoplankton groups at higher temperatures, their ability to regulate buoyancy would allow optimal position in the water column under stratified situations, the ability for some taxa to fix N could help dominate at low N:P ratios, their high efficiency of light capture (high initial slope of the growth–irradiance curve) would allow growth under increased turbidity due to shading by phytoplankton or inorganic sediments, and their resting stages (akinetes in Nostocales) would enhance population viability and allow persistence of populations under unfavorable environmental conditions.

C. Predicting changes in community composition

While much research has been done to describe phytoplankton community composition under different environmental conditions, we still cannot fully explain and much less predict community composition and dynamics. Predicting how phytoplankton would respond to environmental variation is one of the main challenges for phytoplankton ecologists (Litchman et al. 2010, Kruk et al. 2021).

The way forward is a combination of field observations, laboratory experiments, and mechanistic and statistical models that explicitly relate the presence, growth, or abundance of functional groups with changing environmental variables. Using traits in these predictions helps incorporate the mechanisms that structure phytoplankton communities.

D. Phytoplankton trait evolution

Phytoplankton taxa differ in their functional traits and ecological strategies. Many traits of species also exhibit significant intraspecific variation, and this variation, along with new mutations, provides the basis for selection. Changing environmental conditions exert different selective pressures on phytoplankton that may lead to the evolution of their traits. The selection may act on standing genetic variation or *de novo* mutations. Phytoplankton were shown to adapt to rising temperatures by increasing their optimal temperature for growth and, to a lesser degree, the maximum temperature for growth (Padfield et al. 2016, O'Donnell et al. 2018). Other traits, such as the ability to produce resting stages, may also evolve under changing temperatures (Hinners et al. 2017). Nutrient utilization traits, may also evolve over relatively short time scales (Bernhardt et al. 2020). Evolutionary adaptation to changing conditions may be impaired by various environmental stressors. For example, phytoplankton adaptation to high temperatures may be impeded by nutrient limitation (Aranguren-Gassis et al. 2019).

Considering phytoplankton trait evolution on long time scales may also help establish phylogenetic relationships among phytoplankton taxa and infer missing trait values. Traditionally, phytoplankton phylogenies are built exclusively upon molecular data, but using both sequences and traits in phylogenetic analyses could improve tree resolution (reviewed in Assis 2009). For example, Uyeda et al. (2016) demonstrated that combining genomic and phenotypic datasets in cyanobacterial phylogeny shed light on the link between traits and major events in Earth's history. Deducing unknown trait values of species is based upon the knowledge of the phylogenetic relationships among species and the trait values of similar species (Bruggeman 2011). This approach is useful, as the physiological traits of most species are still unknown.

VII. Summary

1. Introduction

Phytoplankton are major primary producers in lakes, rivers, and many other aquatic ecosystems. Their biomass and community composition affect the rest of the aquatic food webs, biogeochemical cycles, water quality, and ecosystem functioning. Therefore understanding what factors determine phytoplankton growth, biomass, and species composition is essential for our general understanding of aquatic ecosystems. Using phytoplankton eco-physiological traits to identify mechanisms that structure phytoplankton communities is a promising scientific approach. In this chapter we identify traits (morphological, physiological, or phenological heritable features) that help describe the dependence of phytoplankton growth on environmental factors and the relationships with other organisms within aquatic food webs.

2. Phytoplankton growth: Resources and environmental factors

a. Inorganic nutrients (macro- and micronutrients)

Phytoplankton require inorganic nutrients such as nitrogen, phosphorus, iron, and other elements to grow. Phytoplankton growth rate is often a saturating function of dissolved nutrients. The growth also depends on internal nutrient concentrations or cell quotas. The parameters of the growth and uptake functions, such as the maximum rate of growth or uptake and the half-saturation constants, can be viewed as nutrient-related traits. Different phytoplankton groups and individual species vary in their nutrient traits.

b. Organic compounds, vitamins

In addition to inorganic nutrients, many phytoplankton species also require vitamins (defined as auxotrophic) and have the ability to take up organic compounds.

c. Light

Light is an essential resource for phytoplankton. Phytoplankton growth and photosynthetic rates increase and then saturate with increasing light levels; most species get inhibited by high light. There are significant taxonomic differences in light-associated traits. Most Cyanobacteria are adapted to low light as they have high efficiency in light capture (highest initial slopes of the growth–irradiance curve among phytoplankton, α) and lower tolerance of high irradiances (low I_{opt}). Most green algae, on the other hand, appear to be adapted to high light (high I_{opt}) and diatoms appear to be adapted to fluctuating light, through simultaneously being able to grow at low light (high α) and the high tolerance of high irradiances (high I_{opt}).

d. Temperature

Most physiological and ecological processes depend on temperature. Phytoplankton growth rate is often a left-skewed unimodal function of temperature, where the optimum temperature, minimum and maximum temperatures, and the thermal niche width are the traits that characterize the temperature preferences of species. Cryptophytes and diatoms often have lower temperature optima than Cyanobacteria and green algae, and these differences align well with the seasonal succession patterns. Diatoms and cryptophytes often dominate early or late in the season in temperate lakes when temperatures are lower. Green algae and Cyanobacteria tend to dominate in the summer when water temperature is warm.

e. Physical and chemical environment

Besides nutrients, light, and temperature, there are many other physical and chemical factors that influence phytoplankton biomass and taxonomic composition. Wind-induced water column mixing affects phytoplankton access to light and nutrients. Motile species usually benefit from low mixing or stratified conditions because of their ability to regulate their position in the water column, either through active swimming via flagella or passive buoyancy regulation, to access light and nutrients, and gain a competitive advantage. Turbidity and the concentrations of humic substances modify light levels and may be important in determining the outcome of light competition. pH is also an important factor regulating the composition of phytoplankton communities.

f. Factor interactions

Phytoplankton growth and mortality depend on many environmental factors. Those factors often interact to either augment or diminish the effects of individual factors. Many of such interactions are complex and poorly known. They may alter the traits of species complicating the predictions of species distributions based on traits. For example, the optimum temperature for growth may decline under nutrient limitation, so nutrient-limited species may be more sensitive to high temperatures.

3. Phytoplankton in food webs

a. Competition

Phytoplankton compete for inorganic nutrients, such as nitrogen and phosphorus, and light. Species differ in their competitive abilities, which are determined by their eco-physiological traits. A useful measure of competitive abilities for nutrients is the lowest nutrient concentration at which a given species can persist, termed R^*. Species with the lowest R^* wins competition for that nutrient. R^* depends on several eco-physiological traits, such as maximum growth rate, and mortality. Small-celled species tend to have a competitive advantage under low nutrient conditions; consequently, oligotrophic lakes often have small-sized phytoplankton.

b. Predation

Phytoplankton get grazed by various zooplankton that include heterotrophic protists, rotifers, crustaceans (such as copepods and cladocerans), and even some fish. To avoid predation, phytoplankton have evolved many adaptations to avoid being grazed. The traits that increase grazer resistance include increasing cell size, becoming colonial, developing bristles and spikes, or producing toxins.

c. Allelopathy

Some phytoplankton, including many Cyanobacteria, produce secondary metabolites that can inhibit the growth of competitors, such as other species of phytoplankton, or zooplankton grazers. Some of those metabolites are toxic to animals and humans.

d. Parasitism and pathogens

Like most organisms, eukaryotic algae and Cyanobacteria can get infected with parasitic fungi, protozoa, or pathogenic bacteria and viruses. Parasites have the capability to regulate phytoplankton population density, can terminate algal blooms, and modify ecological interactions, such as competition and predation, by differentially affecting different species.

e. Mutualism

Algae often form symbiotic associations with heterotrophic bacteria that live in the layer close to the phytoplankton cell, the phycosphere, or even inside the cells. Many of those associations are mutualistic, where both partners provide benefits to each other. Potentially, these symbiotic interactions can turn competitive or parasitic.

f. Mixotrophy

Many species of eukaryotic phytoplankton rely on different sources of carbon, acquiring it not only through photosynthesis but also through ingesting bacteria or dissolved organic matter. Such taxa are called mixotrophic and were shown to play important ecological and biogeochemical roles in aquatic environments, especially in oligotrophic systems.

4. Temporal variation in phytoplankton communities

a. Seasonal successional patterns

Seasonal changes in environmental factors, such as physical stratification, light, nutrient availability, and temperature, drive successional patterns in phytoplankton. Dimictic lakes in the temperate zone often have a well-documented pattern, with a spring diatom bloom, followed by an increase in zooplankton, that then graze algae down, often leading to a so-called clear water phase. Later in the summer, there is often another increase in phytoplankton, when green algae and Cyanobacteria may become dominant. Weakening of the vertical stratification of the water column may stimulate a smaller fall diatom bloom but the increased mixing and decreasing light and temperature lead to phytoplankton biomass decline. Both phytoplankton and zooplankton biomass and diversity are usually low in winter.

Seasonal succession in tropical or perennially ice-covered lakes is less well studied and general patterns are less well defined.

5. Spatial heterogeneity in phytoplankton

a. Vertical distribution of phytoplankton

Phytoplankton distributions along the vertical dimension of a water body are structured by the opposing gradients of nutrients and light and the degree of vertical mixing. Under thermally stratified conditions, phytoplankton may form heterogeneous distributions, with high biomass at different depths. The depth of their maximum biomass depends on their nutrient and light competitive traits. Good light competitors occur deeper in the water column and good nutrient competitors occur at shallower depths.

b. Horizontal gradients

Phytoplankton are often distributed heterogeneously in the horizontal dimension of the water bodies as well. Spatial variations increase in significance in very small lakes, in very large lakes (e.g., the Laurentian Great Lakes), in proximity to the littoral zone and inlet areas of most lakes, and as the morphometric complexity of lake basins increases.

c. Phytoplankton communities in a landscape

Phytoplankton in most lakes are not isolated from numerous factors at the landscape level. They can disperse across different lakes, thus being affected by the conditions beyond local (lake) conditions. Phytoplankton populations and communities in individual lakes can be viewed as patches partly influenced by dispersal and by the processes at a regional/landscape level (e.g., environmental gradients and land use patterns). Phytoplankton communities were shown to exhibit broad geographic patterns.

6. Phytoplankton and global environmental change

a. Major stressors/drivers

Phytoplankton growth, biomass, and community structure are determined by multiple interacting factors, including the ones that are rapidly changing due to anthropogenic global environmental change (temperature, nutrients). Rising temperatures may stimulate phytoplankton growth and change community composition through a direct enhancement of high temperature-adapted species and groups, such as Cyanobacteria, and indirectly through increasing stratification that may favor certain groups. Increasing nutrient inputs may also lead to higher algal biomass and lead to harmful algal blooms (HABs), especially those linked to Cyanobacteria. The proliferation of HABs in water bodies worldwide decreases water quality and the availability of freshwater resources. Some northern lakes may experience an increase in the dissolved organic matter, input that can change transparency and trigger a cascade of effects, from increased stratification to changes in the food web structure.

b. Phytoplankton trait evolution

Phytoplankton may experience directional selection and their traits may evolve. For example, species can evolve higher temperature optima that would allow them to persist in a warming climate. They may also evolve better competitive abilities. However, the evolutionary responses of species to various selective pressures depend on other factors in a complex way, so predicting the trajectories of phytoplankton trait evolution may be difficult.

Acknowledgments

We thank the editors, Ian Jones and John Smol, for inviting us to contribute to this book and for their comments, which improved this chapter.

References

Abonyi, A., Descy, J.-P., Borics, G., Smeti, E., 2021. From historical backgrounds towards the functional classification of river phytoplankton sensu Colin S. Reynolds: what future merits the approach may hold? Hydrobiologia 848, 131–142.

Adrian, R., Walz, N., Hintze, T., Hoeg, S., Rusche, R., 1999. Effects of ice duration on plankton succession during spring in a shallow polymictic lake. Freshw. Biol. 41, 621–632.

Agawin, N.S.R., Rabouille, S., Veldhuis, M.J.W., Servatius, L., Hol, S., van Overzee, H.M.J., Huisman, J., 2007. Competition and facilitation between unicellular nitrogen-fixing cyanobacteria and non-nitrogen-fixing phytoplankton species. Limnol. Oceanogr. 52, 2233–2248.

Ahlgren, G., 1987. Temperature functions in biology and their application to algal growth constants. Oikos 49, 177–190.

Almeida-Gomes, M., Valente-Neto, F., Pacheco, E.O., Ganci, C.C., Leibold, M.A., Melo, A.S., Provete, D.B., 2020. How does the landscape affect metacommunity structure? A quantitative review for lentic environments. Current Landscape Ecology Reports 5, 68–75.

Aranguren-Gassis, M., Kremer, C.T., Klausmeier, C.A., Litchman, E., 2019. Nitrogen limitation inhibits marine diatom adaptation to high temperatures. Ecol. Lett. 22, 1860–1869.

Armstrong, R.A., McGehee, R., 1980. Competitive exclusion. Am. Nat. 115, 151–169.

Assis, L.C.S., 2009. Coherence, correspondence, and the renaissance of morphology in phylogenetic systematics. Cladistics 25, 528–544.

Barone, R., Castelli, G., Naselli-Flores, L., 2010. Red sky at night cyanobacteria delight: the role of climate in structuring phytoplankton assemblage in a shallow, Mediterranean lake (Biviere di Gela, southeastern Sicily). Hydrobiologia 639, 43–53.

Bernhardt, J.R., Kratina, P., Pereira, A.L., Tamminen, M., Thomas, M.K., Narwani, A., 2020. The evolution of competitive ability for essential resources. Phil. Trans. Biol. Sci. 375, 20190247.

Bird, D.F., Kalff, J., 1986. Bacterial grazing by planktonic lake algae. Science 231, 493–495.

Bláha, L., Babica, P., Maršálek, B., 2009. Toxins produced in cyanobacterial water blooms—toxicity and risks. Interdiscip Toxicol 2, 36–41.

Brabrand, Å., Faafeng, B.A., Källqvist, T., Nilssen, J.P., 1983. Biological control of undesirable cyanobacteria in culturally eutrophic lakes. Oecologia 60, 1–5.

Brandenburg, K., Siebers, L., Keuskamp, J., Jephcott, T.G., Van de Waal, D.B., 2020. Effects of nutrient limitation on the synthesis of N-rich phytoplankton toxins: a meta-analysis. Toxins 12, 221.

Briand, J.F., Leboulanger, C., Humbert, J.F., Bernard, C., Dufour, P., 2004. *Cylindrospermopsis raciborskii* (Cyanobacteria) invasion at mid-latitudes: selection, wide physiological tolerance, or global warming? J. Phycol. 40, 231–238.

Bruggeman, J., 2011. A phylogenetic approach to the estimation of phytoplankton traits. J. Phycol. 47, 52–65.

Caiola, M.G., Pellegrini, S., 1984. Lysis of *Microcystis aeruginosa* (kütz.) by *Bdellovibrio*-like Bacteria. J. Phycol. 20, 471–475.

Canter, H.M., Lund, J.W.G., 1968. The importance of Protozoa in controlling the abundance of planktonic algae in lakes. Proc. Linn. Soc. Lond. 179, 203–219.

Chisholm, S.W., 1992. Phytoplankton size. In: Falkowski, P.G., Woodhead, A.D. (Eds.), Primary Productivity and Biogeochemical Cycles in the Sea. Plenum Press, New York, pp. 213–237.

Choi, H.-J., Kim, B.-H., Kim, J.-D., Han, M.-S., 2005. *Streptomyces neyagawaensis* as a control for the hazardous biomass of *Microcystis aeruginosa* (Cyanobacteria) in eutrophic freshwaters. Biol. Control 33, 335–343.

Chorus, I., Spijkerman, E., 2021. What Colin Reynolds could tell us about nutrient limitation, N:P ratios and eutrophication control. Hydrobiologia 848, 95–111.

Chorus, I., Welker, M. (Eds.), 2021. Toxic Cyanobacteria in Water, second ed. World Health Organization.

Claustre, H., Gostan, J., 1987. Adaptation of biochemical composition and cell size to irradiance in two microalgae: possible ecological implications. Mar. Ecol. Prog. Ser. 40, 167–174.

Clegg, M.R., Maberly, S.C., Jones, R.I., 2003. Behavioural responses of freshwater phytoplanktonic flagellates to a temperature gradient. Eur. J. Phycol. 38, 195–203.

Coon, T.G., Lopez, M., Richerson, P.J., Powell, T.M., Goldman, C.R., 1987. Summer dynamics of the deep chlorophyll maximum in Lake Tahoe. J. Plankton Res. 9, 327–344.

Coulombe, A.M., Robinson, G.G.C., 1981. Collapsing *Aphanizomenon flos-aquae* blooms: possible contributions of photo-oxidation, O_2 toxicity, and cyanophages. Can. J. Bot. 59, 1277–1284.

Crowley, M.A., Cardille, J.A., 2020. Remote sensing's recent and future contributions to landscape ecology. Current Landscape Ecology Reports 5, 45–57.

Davey, M.C., Heaney, S.I., 1989. The control of sub-surface maxima of diatoms in a stratified lake by physical, chemical and biological factors. J. Plankton Res. 11, 1185–1199.

DeMott, W.R., McKinney, E.N., Tessier, A.J., 2010. Ontogeny of digestion in *Daphnia*: implications for the effectiveness of algal defenses. Ecology 91, 540–548.

deNoyelles, F., Smith, V.H., Kastens, J.H., Bennett, L., Lomas, J.M., Knapp, C.W., Bergin, S.P., Dewey, S.L., Chapin, B.R.K., Graham, D.W., 2016. A 21-year record of vertically migrating subepilimnetic populations of *Cryptomonas* spp. Inland Waters 6, 173–184.

de Tezanos Pinto, P., Litchman, E., 2010. The interactive effects of N:P ratios and light on nitrogen-fixer abundance. Oikos 119, 567–575.

de Tezanos Pinto, P., O'Farrell, I., 2014. Regime shifts between free-floating plants and phytoplankton: a review. Hydrobiologia 740, 13–24.

Devercelli, M., 2009. Changes in phytoplankton morpho-functional groups induced by extreme hydroclimatic events in the Middle Parana River (Argentina). Hydrobiologia 639, 5–19.

Devercelli, M., O'Farrell, I., 2013. Factors affecting the structure and maintenance of phytoplankton functional groups in a nutrient rich lowland river. Limnologica 43, 67–78.

Dokulil, M.T., Jagsch, A., George, G.D., Anneville, O., Jankowski, T., Wahl, B., Lenhart, B., Blenckner, T., Teubner, K., 2006. Twenty years of spatially coherent deepwater warming in lakes across Europe related to the North Atlantic Oscillation. Limnol. Oceanogr. 51, 2787–2793.

Dolman, A.M., Rücker, J., Pick, F.R., Fastner, J., Rohrlack, T., Mischke, U., Wiedner, C., 2012. Cyanobacteria and cyanotoxins: the influence of nitrogen versus phosphorus. PLoS One 7, e38757.

Dörnhöfer, K., Oppelt, N., 2016. Remote sensing for lake research and monitoring—recent advances. Ecol. Indicat. 64, 105–122.

Downing, J.A., Watson, S.B., McCauley, E., 2001. Predicting Cyanobacteria dominance in lakes. Can. J. Fish. Aquat. Sci. 58, 1905–1908.

Droop, M.R., 1973. Some thoughts on nutrient limitation in algae. J. Phycol. 9, 264–272.

Edwards, K.F., Klausmeier, C.A., Litchman, E., 2011. Evidence for a three-way tradeoff between nitrogen and phosphorus competitive abilities and cell size in phytoplankton. Ecology 92, 2085–2095.

Edwards, K.F., Thomas, M.K., Klausmeier, C.A., Litchman, E., 2012. Allometric scaling and taxonomic variation in nutrient utilization traits and growth rates of marine and freshwater phytoplankton. Limnol. Oceanogr. 57, 554–566.

Edwards, K.F., Thomas, M.K., Klausmeier, C.A., Litchman, E., 2015. Light and phytoplankton growth: allometry, taxonomic variation, and biogeography. Limnol. Oceanogr. 60, 540–552.

Edwards, K.F., Thomas, M.K., Klausmeier, C.A., Litchman, E., 2016. Phytoplankton growth and the interaction of light and temperature: a synthesis at the species and community level. Limnol. Oceanogr. 61, 1232–1244.

Ehrlich, E., Kath, N.J., Gaedke, U., 2020. The shape of a defense-growth trade-off governs seasonal trait dynamics in natural phytoplankton. ISME J. 14, 1451–1462.

Eilers, P.H.C., Peeters, J.C.H., 1988. A model for the relationship between light intensity and the rate of photosynthesis in phytoplankton. Ecol. Model. 42, 199–215.

Eppley, R.W., 1972. Temperature and phytoplankton growth in the sea. Fish. Bull. Nat. Ocean. Atm. Adm. 70, 1063–1085.

Falkowski, P.G., 1980. Light-shade adaptation in marine phytoplankton. In: Falkowski, P.G. (Ed.), Primary Productivity in the Sea. Plenum Press, New York, pp. 99–119.

Finkel, Z.V., 2001. Light absorption and size scaling of light-limited metabolism in marine diatoms. Limnol. Oceanogr. 46, 86–94.

Flynn, K.J., Mitra, A., Glibert, P.M., Burkholder, J.M., 2018. Mixotrophy in harmful algal blooms: by whom, on whom, when, why, and what next. In: Glibert, P.M., Berdalet, E., Burford, M.A., Pitcher, G.C., Zhou, M. (Eds.), Global Ecology and Oceanography of Harmful Algal Blooms. Springer International Publishing, Cham, pp. 113–132.

Frenken, T., Alacid, E., Berger, S.A., Bourne, E.C., Gerphagnon, M., Grossart, H.P., Gsell, A.S., Ibelings, B.W., Kagami, M., Küpper, F.C., Letcher, P.M., Loyau, A., Miki, T., Nejstgaard, J.C., Rasconi, S., Reñé, A., Rohrlack, T., Rojas-Jimenez, K., Schmeller, D.S., Scholz, B., Seto, K., Sime-Ngando, T., Sukenik, A., Van de Waal, D.B., Van den Wyngaert, S., Van Donk, E., Wolinska, J., Wurzbacher, C., Agha, R., 2017. Integrating chytrid fungal parasites into plankton ecology: research gaps and needs. Environ. Microbiol. 19, 3802–3822.

Frenken, T., Velthuis, M., de Senerpont Domis, L.N., Stephan, S., Aben, R., Kosten, S., Van Donk, E., Van de Waal, D.B., 2016. Warming accelerates termination of a phytoplankton spring bloom by fungal parasites. Global Change Biol. 22, 299–309.

Fujiki, T., Taguchi, S., 2002. Variability in chlorophyll a specific absorption coefficient in marine phytoplankton as a function of cell size and irradiance. J. Plankton Res. 24, 859–874.

Garnier, E., Navas, M.-L., Grigulis, K., 2016. Plant Functional Diversity: Organism Traits, Community Structure, and Ecosystem Properties. Oxford University Press, Oxford, UK.

Gause, G.F., 1934. The Struggle for Existence. Williams and Williams, Baltimore.

Geider, R.J., MacIntyre, H.L., Kana, T.M., 1997. Dynamic model of phytoplankton growth and acclimation: responses of the balanced growth rate and the chlorophyll a: carbon ratio to light, nutrient-limitation and temperature. Mar. Ecol. Prog. Ser. 148, 187–200.

Geider, R.J., Platt, T., Raven, J.A., 1986. Size dependence of growth and photosynthesis in diatoms: a synthesis. Mar. Ecol. Prog. Ser. 30, 93–104.

Geider, R.J., Roche, J.L., 2002. Redfield revisited: variability of C:N:P in marine microalgae and its biochemical basis. Eur. J. Phycol. 37, 1–17.

Gerphagnon, M., Macarthur, D.J., Latour, D., Gachon, C.M.M., Van Ogtrop, F., Gleason, F.H., Sime-Ngando, T., 2015. Microbial players involved in the decline of filamentous and colonial cyanobacterial blooms with a focus on fungal parasitism. Environ. Microbiol. 17, 2573–2587.

Gobler, C.J., Burkholder, J.M., Davis, T.W., Harke, M.J., Johengen, T., Stow, C.A., Van de Waal, D.B., 2016. The dual role of nitrogen supply in controlling the growth and toxicity of cyanobacterial blooms. Harmful Algae 54, 87–97.

Gross, E.M., 2003. Allelopathy of aquatic autotrophs. Crit. Rev. Plant Sci. 22, 313–339.

Grover, J.P., 1991a. Non-steady state dynamics of algal population growth: experiments with two chlorophytes. J. Phycol. 27, 70–79.

Grover, J.P., 1991b. Resource competition in a variable environment: phytoplankton growing according to the variable-internal-stores model. Am. Nat. 138, 811–835.

Grover, J.P., 1997. Resource Competition. Chapman and Hall, London.

Gumbo, J.R., Cloete, T.E., 2011. Light and electron microscope assessment of the lytic activity of *Bacillus* on *Microcystis aeruginosa*. Afr. J. Biotechnol. 10, 8054–8063.

Hansen, P.J., Nielsen, T.G., 1997. Mixotrophic feeding of Fragilidium subglobosum (Dinophyceae) on three species of Ceratium: effects of prey concentration, prey species and light intensity. Mar. Ecol. Prog. Ser. 147, 187–196.

Hansson, L.-a., 1996. Behavioural response in plants: adjustment in algal recruitment induced by herbivores. Proceedings of the Royal Society of London. Series B: Biological Sciences 263, 1241–1244.

Hasler, C.T., Butman, D., Jeffrey, J.D., Suski, C.D., 2016. Freshwater biota and rising pCO^2? Ecol. Lett. 19, 98–108.

Helliwell, K.E., 2017. The roles of B vitamins in phytoplankton nutrition: new perspectives and prospects. New Phytol. 216, 62–68.

Hense, I., Beckmann, A., 2006. Towards a model of cyanobacteria life cycle—effects of growing and resting stages on bloom formation of N2-fixing species. Ecol. Model. 195, 205–218.

Herrero, A., Flores, E., 2008. The cyanobacteria: Molecular biology, Genomics and evolution. Caister Academic Press.

Hinners, J., Kremp, A., Hense, I., 2017. Evolution in temperature-dependent phytoplankton traits revealed from a sediment archive: do reaction norms tell the whole story? Proc. Biol. Sci. 284, 20171888.

Howarth, R.W., Marino, R., Lane, J., Cole, J.J., 1988. Nitrogen-fixation in fresh-water, estuarine, and marine ecosystems. 1. Rates and Importance. Limnol. Oceanogr. 33, 669–687.

Hrudey, S.E., Kenefick, S.L., Best, N., Gillespie, T., Kotak, B.G., Prepas, E.E., Peterson, H.G., 1993. Liver toxins and odour agents in cyanobacterial blooms in Alberta surface water supplies. In: Disinfection Dilemma: Microbiological Control versus Byproducts. Winnipeg, Canada, pp. 383–390.

Huisman, J., Arrayas, M., Ebert, U., Sommeijer, B., 2002. How do sinking phytoplankton species manage to persist? Am. Nat. 159, 245–254.

Huisman, J., Matthijs, H.C.P., Visser, P.M., 2005. Harmful cyanobacteria. Springer, New York.

Huisman, J., Sharples, J., Stroom, J.M., Visser, P.M., Kardinaal, W.E.A., Verspagen, J.M.H., Sommeijer, B., 2004. Changes in turbulent mixing shift competition for light between phytoplankton species. Ecology 85, 2960–2970.

Huisman, J., Weissing, F.J., 1999. Biodiversity of plankton by species oscillations and chaos. Nature 402, 407–410.

Hutchinson, G.E., 1961. The paradox of the plankton. Am. Nat. 95, 137–145.

Ibelings, B.W., Vonk, M., Los, H.F.J., van der Molen, D.T., Mooij, W.M., 2003. Fuzzy modeling of cyanobacterial surface waterblooms: validation with NOAA-AVHRR satellite images. Ecol. Appl. 13, 1456–1472.

Inderjit, del Moral, R., 1997. Is separating resource competition from allelopathy realistic? Bot. Rev. 63, 221–230.

Janssen, A.B.G., Teurlincx, S., An, S., Janse, J.H., Paerl, H.W., Mooij, W.M., 2014. Alternative stable states in large shallow lakes? J. Great Lake. Res. 40, 813–826.

Jöhnk, K.D., Huisman, J., Sharples, J., Sommeijer, B., Visser, P.M., Stroom, J.M., 2008. Summer heatwaves promote blooms of harmful cyanobacteria. Global Change Biol. 14, 495–512.

Jones, R.I., 2000. Mixotrophy in planktonic protists: an overview. Freshw. Biol. 45, 219–226.

Jüttner, F., Watson, S.B., 2007. Biochemical and ecological control of geosmin and 2-methylisoborneol in source waters. Appl. Environ. Microbiol. 73, 4395–4406.

Kahru, M., Leppanen, J.-M., Rud, O., 1993. Cyanobacterial blooms cause heating of the sea surface. Mar. Ecol. Prog. Ser. 101, 1–7.

Kardinaal, W.E.A., Tonk, L., Janse, I., Hol, S., Slot, P., Huisman, J., Visser, P.M., 2007. Competition for light between toxic and nontoxic strains of the harmful cyanobacterium *Microcystis*. Appl. Environ. Microbiol. 73, 2939–2946.

Keiser, D.A., Shapiro, J.S., 2018. Consequences of the clean water act and the demand for water quality. Q. J. Econ. 134, 349–396.

Kilham, P., Hecky, R.E., 1988. Comparative ecology of marine and freshwater phytoplankton. Limnol. Oceanogr. 33, 776–795.

Kirk, J.T.O., 1994. Light and photosynthesis in aquatic ecosystems, second ed. Cambridge University Press, Cambridge.

Klausmeier, C.A., Litchman, E., 2001. Algal games: the vertical distribution of phytoplankton in poorly mixed water columns. Limnol. Oceanogr. 46, 1998–2007.

Klausmeier, C.A., Litchman, E., Daufresne, T., Levin, S.A., 2004. Optimal nitrogen-to-phosphorus stoichiometry of phytoplankton. Nature 429, 171–174.

Klemer, A.R., 1983. Cyanobacterial buoyancy regulation and blooms. Lake Restoration, Protection and Management 234–236.

Klemer, A.R., 1991. Effects on nutritional status on cyanobacterial buoyancy, blooms, and dominance, with special reference to inorganic carbon. Can. J. Bot. 69, 1133–1138.

Krenek, S., Petzoldt, T., Berendonk, T.U., 2012. Coping with temperature at the warm edge—patterns of thermal adaptation in the microbial eukaryote *Paramecium caudatum*. PLoS One 7, e30598.

Kretschmann, J., Žerdoner Čalasan, A., Gottschling, M., 2018. Molecular phylogenetics of dinophytes harboring diatoms as endosymbionts (Kryptoperidiniaceae, Peridiniales), with evolutionary interpretations and a focus on the identity of Durinskia oculata from Prague. Mol. Phylogenet. Evol. 118, 392–402.

Kruk, C., Devercelli, M., Huszar, V.L., 2021. Reynolds functional groups: a trait-based pathway from patterns to predictions. Hydrobiologia 848, 113–129.

Kutser, T., 2004. Quantitative detection of chlorophyll in cyanobacterial blooms by satellite remote sensing. Limnol. Oceanogr. 49, 2179–2189.

Langdon, C., 1987. On the causes of interspecific differences in the growth-irradiance relationship for phytoplankton. Part I. A comparative study of the growth-irradiance relationship of three marine phytoplankton species: *Skeletonema costatum, Olisthodiscus luteus* and *Gonyaulax tamarensis*. J. Plankton Res. 9, 459–482.

Latta, L.C., O'Donnell, R.P., Pfrender, M.E., 2009. Vertical distribution of *Chlamydomonas* changes in response to grazer and predator kairomones. Oikos 118, 853–858.

Laws, E.A., Chalup, M.S., 1990. A microalgal growth model. Limnol. Oceanogr. 35, 597–608.

Leach, T.H., Beisner, B.E., Carey, C.C., Pernica, P., Rose, K.C., Huot, Y., Brentrup, J.A., Domaizon, I., Grossart, H.-P., Ibelings, B.W., Jacquet, S., Kelly, P.T., Rusak, J.A., Stockwell, J.D., Straile, D., Verburg, P., 2018. Patterns and drivers of deep chlorophyll maxima structure in 100 lakes: the relative importance of light and thermal stratification. Limnol. Oceanogr. 63, 628–646.

Legrand, C., Rengefors, K., Fistarol, G.O., Granéli, E., 2003. Allelopathy in phytoplankton—biochemical, ecological and evolutionary aspects. Phycologia 42, 406–419.

Leibold, M.A., Holyoak, M., Mouquet, N., Amarasekare, P., Chase, J.M., Hoopes, M.F., Holt, R.D., Shurin, J.B., Law, R., Tilman, D., Loreau, M., Gonzalez, A., 2004. The metacommunity concept: a framework for multi-scale community ecology. Ecol. Lett. 7, 601–613.

Leland, H.V., 2003. The influence of water depth and flow regime on phytoplankton biomass and community structure in a shallow, lowland river. Hydrobiologia 506, 247–255.

Lenton, T.M., Klausmeier, C.A., 2007. Biotic stoichiometric controls on the deep ocean N:P ratio. Biogeosciences 4, 353–367.

Lewis, W.M., 1978. Dynamics and succession of the phytoplankton in a tropical lake: Lake Lanao, Philippines. J. Ecol. 66, 849–880.

Lewis, W.M., 1996. Tropical lakes: how latitude makes a difference. In: Schiemer, F., Boland, K.T. (Eds.), Perspectives in Tropical Limnology. SPB Academic Publishing, Amsterdam, pp. 43–64.

Lindholm, T., 1992. Ecological role of depth maxima of phytoplankton. Ergeb. Limnol. 35, 33–45.

Litchman, E., 1998. Population and community responses of phytoplankton to fluctuating light. Oecologia 117, 247–257.

Litchman, E., 2003. Competition and coexistence of phytoplankton under fluctuating light: experiments with two cyanobacteria. Aquat. Microb. Ecol. 31, 241–248.

Litchman, E., de Tezanos Pinto, P., Klausmeier, C.A., Thomas, M.K., Yoshiyama, K., 2010. Linking traits to species diversity and community structure in phytoplankton. Hydrobiologia 653, 15–38.

Litchman, E., Klausmeier, C.A., 2001. Competition of phytoplankton under fluctuating light. Am. Nat. 157, 170–187.

Litchman, E., Klausmeier, C.A., 2008. Trait-based community ecology of phytoplankton. Annual Review of Ecology, Evolution, and Systematics 39, 615–639.

Litchman, E., Klausmeier, C.A., Yoshiyama, K., 2009. Contrasting size evolution in marine and freshwater diatoms. Proceedings of the National Academy of the Sciences of the USA 106, 2665–2670.

Lürling, M., 2003. Phenotypic plasticity in the green algae *Desmodesmus* and *Scenedesmus* with special reference to the induction of defensive morphology. Annales de Limnology—International Journal of Limnology 39, 85–101.

Lürling, M., 2021. Grazing resistance in phytoplankton. Hydrobiologia 848, 237–249.

Lürling, M., Eshetu, F., Faasen, E.J., Kosten, S., Huszar, V.L.M., 2013. Comparison of cyanobacterial and green algal growth rates at different temperatures. Freshw. Biol. 58, 552–559.

Malerba, M.E., Palacios, M.M., Palacios Delgado, Y.M., Beardall, J., Marshall, D.J., 2018. Cell size, photosynthesis and the package effect: an artificial selection approach. New Phytol. 219, 449–461.

Marañón, E., Cermeño, P., López-Sandoval, D.C., Rodríguez-Ramos, T., Sobrino, C., Huete-Ortega, M., Blanco, J.M., Rodríguez, J., 2013. Unimodal size scaling of phytoplankton growth and the size dependence of nutrient uptake and use. Ecol. Lett. 16, 371–379.

Martínez de la Escalera, G., Kruk, C., Segura, A.M., Nogueira, L., Alcántara, I., Piccini, C., 2017. Dynamics of toxic genotypes of *Microcystis aeruginosa* complex (MAC) through a wide freshwater to marine environmental gradient. Harmful Algae 62, 73–83.

Mayali, X., 2018. Editorial: metabolic interactions between bacteria and phytoplankton. Front. Microbiol. 9, 727–727.

McVicar, T.R., Roderick, M.L., Donohue, R.J., Li, L.T., Van Niel, T.G., Thomas, A., Grieser, J., Jhajharia, D., Himri, Y., Mahowald, N.M., Mescherskaya, A.V., Kruger, A.C., Rehman, S., Dinpashoh, Y., 2012. Global review and synthesis of trends in observed terrestrial near-surface wind speeds: implications for evaporation. J. Hydrol. 416 (417), 182–205.

Mellard, J.P., Yoshiyama, K., Litchman, E., Klausmeier, C.A., 2011. The vertical distribution of phytoplankton in stratified water columns. J. Theor. Biol. 269, 16–30.

Milo, R., 2009. What governs the reaction center excitation wavelength of photosystems I and II? Photosysnthesis Research 101, 59–67.

Mitra, A., Flynn, K.J., Tillmann, U., Raven, J.A., Caron, D., Stoecker, D.K., Not, F., Hansen, P.J., Hallegraeff, G., Sanders, R., Wilken, S., McManus, G., Johnson, M., Pitta, P., Våge, S., Berge, T., Calbet, A., Thingstad, F., Jeong, H.J., Burkholder, J., Glibert, P.M., Granéli, E., Lundgren, V., 2016. Defining planktonic protist functional groups on mechanisms for energy and nutrient acquisition: incorporation of diverse mixotrophic strategies. Protist 167, 106–120.

Moore, M.V., Hampton, S.E., Izmest'eva, L.R., Silow, E.A., Peshkova, E.V., Pavlov, B.K., 2009. Climate change and the world's "sacred sea"—Lake Baikal, Siberia. Bioscience 59, 405–417.

Moustaka-Gouni, M., Michaloudi, E., Sommer, U., 2014. Modifying the PEG model for Mediterranean lakes—no biological winter and strong fish predation. Freshw. Biol. 59, 1136–1144.

Mu, R.M., Fan, Z.Q., Pei, H.Y., Yuan, X.L., Liu, S.X., Wang, X.R., 2007. Isolation and algae-lysing characteristics of the algicidal bacterium B5. J. Environ. Sci. (China) 19, 1336–1340.

Müller-Navarra, D.C., Brett, M.T., Park, S., Chandra, S., Ballantyne, A.P., Zorita, E., Goldman, C.R., 2004. Unsaturated fatty acid content in seston and tropho-dynamic coupling in lakes. Nature 427, 69–72.

Nan, J., Li, J., Yang, C., Yu, H., 2020. Phytoplankton functional groups succession and their driving factors in a shallow subtropical lake. J. Freshw. Ecol. 35, 409–427.

O'Donnell, D.R., Hamman, C.R., Johnson, E.C., Kremer, C.T., Klausmeier, C.A., Litchman, E., 2018. Rapid thermal adaptation in a marine diatom reveals constraints and trade-offs. Global Change Biol. 24, 4554–4565.

O'Reilly, C.M., Alin, S.R., Plisnier, P.D., Cohen, A.S., McKee, B.A., 2003. Climate change decreases aquatic ecosystem productivity of Lake Tanganyika, Africa. Nature 424, 766–768.

Padfield, D., Yvon-Durocher, G., Buckling, A., Jennings, S., Yvon-Durocher, G., 2016. Rapid evolution of metabolic traits explains thermal adaptation in phytoplankton. Ecol. Lett. 19, 133–142.

Paerl, H.W., Huisman, J., 2008. Climate—blooms like it hot. Science 320, 57–58.

Paerl, H.W., Huisman, J., 2009. Climate change: a catalyst for global expansion of harmful cyanobacterial blooms. Environ. Microbiol. Rep. 1, 27–37.

Pålsson, C., Granéli, W., 2004. Nutrient limitation of autotrophic and mixotrophic phytoplankton in a temperate and tropical humic lake gradient. J. Plankton Res. 26, 1005–1014.

Pančić, M., Kiørboe, T., 2018. Phytoplankton defence mechanisms: traits and trade-offs. Biol. Rev. 93, 1269–1303.

Park, M.G., Yih, W., Coats, D.W., 2004. Parasites and phytoplankton, with special emphasis on dinoflagellate infections. J. Eukaryot. Microbiol. 51, 145–155.

Paseka, R.E., White, L.A., Van de Waal, D.B., Strauss, A.T., González, A.L., Everett, R.A., Peace, A., Seabloom, E.W., Frenken, T., Borer, E.T., 2020. Disease-mediated ecosystem services: pathogens, plants, and people. Trends Ecol. Evol. 35, 731–743.

Passarge, J., Hol, S., Escher, M., Huisman, J., 2006. Competition for nutrients and light: stable coexistence, alternative stable states, or competitive exclusion? Ecol. Monogr. 76, 57–72.

Phillips, J.C., McKinley, G.A., Bennington, V., Bootsma, H.A., Pilcher, D.J., Sterner, R.W., Urban, N.R., 2015. The Potential for CO_2-induced acidification in freshwater: a great lakes case study. Oceanography 28, 136–145.

Pomati, F., Shurin, J.B., Andersen, K.H., Tellenbach, C., Barton, A.D., 2020. Interacting temperature, nutrients and zooplankton grazing control phytoplankton size-abundance relationships in eight Swiss lakes. Front. Microbiol. 10.

Priscu, J.C., Wolf, C.F., Takacs, C.D., Fritsen, C.H., Laybourn-Parry, J., Roberts, E.C., Sattler, B., Lyons, W.B., 1999. Carbon transformations in a perennially ice-covered Antarctic lake. Bioscience 49, 997–1008.

Ptacnik, R., Gomes, A., Royer, S.-J., Berger, S.A., Calbet, A., Nejstgaard, J.C., Gasol, J.M., Isari, S., Moorthi, S.D., Ptacnikova, R., Striebel, M., Sazhin, A.F., Tsagaraki, T.M., Zervoudaki, S., Altoja, K., Dimitriou, P.D., Laas, P., Gazihan, A., Martínez, R.A., Schabhüttl, S., Santi, I., Sousoni, D., Pitta, P., 2016. A light-induced shortcut in the planktonic microbial loop. Sci. Rep. 6, 29286.

Raven, J.A., Geider, R.J., 1988. Temperature and algal growth. New Phytol. 110, 441–461.

Reavie, E.D., Cai, M., Twiss, M.R., Carrick, H.J., Davis, T.W., Johengen, T.H., Gossiaux, D., Smith, D.E., Palladino, D., Burtner, A., Sgro, G.V., 2016. Winter–spring diatom production in Lake Erie is an important driver of summer hypoxia. J. Great Lake. Res. 42, 608–618.

Redfield, A.C., 1958. The biological control of chemical factors in the environment. Am. Sci. 46, 205–221.

Reynolds, C.S., 1984. The ecology of freshwater phytoplankton. Cambridge University Press, Cambridge.

Reynolds, C.S., 1992. Dynamics, selection and composition of phytoplankton in relation to vertical structure in lakes. Ergebnisse Limnology 35, 13–31.
Reynolds, C.S., 1994. The long, the short and the stalled: on the attributes of phytoplankton selected by physical mixing in lakes and rivers. Hydrobiol. (Sofia) 289, 9–21.
Reynolds, C.S., 1997. Vegetation processes in the pelagic: A model for ecosystem theory. Inter-Research, Oldendorf.
Reynolds, C.S., 1998. What factors influence the species composition of phytoplankton in lakes of different trophic status? Hydrobiologia 369, 11–26.
Reynolds, C.S., 2006. The ecology of phytoplankton. Cambridge University Press, Cambridge.
Reynolds, C.S., Descy, J.-P., 1996. The production, biomass and structure of phytoplankton in large rivers. Large Rivers 10, 161–187.
Reynolds, C.S., Huszar, V., Kruk, C., Naselli-Flores, L., Melo, S., 2002. Towards a functional classification of the freshwater phytoplankton. J. Plankton Res. 24, 417–428.
Rhee, G.-Y., Gotham, I.J., 1981a. The effect of environmental factors on phytoplankton growth—temperature and the interactions of temperature with nutrient limitation. Limnol. Oceanogr. 26, 635–648.
Rhee, G.-Y., Gotham, I.J., 1981b. The effect of environmental factors on phytoplankton growth: light and the interactions of light with nitrate limitation. Limnol. Oceanogr. 26, 649–659.
Richardson, K., Beardall, J., Raven, J.A., 1983. Adaptation of unicellular algae to irradiance: an analysis of strategies. New Phytol. 93, 157–191.
Robarts, R.D., Zohary, T., 1987. Temperature effects on photosynthetic capacity, respiration, and growth rates of bloom forming cyanobacteria. N. Z. J. Mar. Freshw. Res. 21, 391–399.
Salmaso, N., Tolotti, M., 2021. Phytoplankton and anthropogenic changes in pelagic environments. Hydrobiologia 848, 251–284.
Sarnelle, O., 1993. Herbivore effects on phytoplankton succession in a eutrophic lake. Ecol. Monogr. 63, 129–149.
Sarnelle, O., 2005. *Daphnia* as keystone predators: effects on phytoplankton diversity and grazing resistance. J. Plankton Res. 27, 1229–1238.
Scheffer, M., Hosper, S.H., Meijer, M.L., Moss, B., Jeppesen, E., 1993. Alternative equilibria in shallow lakes. Trends Ecol. Evol. 8, 275–279.
Scheffer, M., Szabó, S., Gragnani, A., van Nes, E.H., Rinaldi, S., Kautsky, N., Norberg, J., Roijackers, R.M.M., Franken, R.J.M., 2003. Floating plant dominance as a stable state. Proc. Natl. Acad. Sci. USA 100, 4040–4045.
Schwaderer, A.S., Yoshiyama, K., de Tezanos Pinto, P., Swenson, N.G., Klausmeier, C.A., Litchman, E., 2011. Eco-evolutionary patterns in light utilization traits and distributions of freshwater phytoplankton. Limnol. Oceanogr. 56, 589–598.
Selosse, M.A., Charpin, M., Not, F., 2017. Mixotrophy everywhere on land and in water: the grand écart hypothesis. Ecol. Lett. 20, 246–263.
Seymour, J.R., Amin, S.A., Raina, J.-B., Stocker, R., 2017. Zooming in on the phycosphere: the ecological interface for phytoplankton–bacteria relationships. Nature Microbiology 2, 17065.
Shilo, M., 1970. Lysis of blue-green algae by myxobacter. J. Bacteriol. 104, 453–461.
Smith, V.H., 1983. Low nitrogen to phosphorus ratios favor dominance by blue-green algae in lake phytoplankton. Science 221, 669–671.
Sommer, U., 1984. The paradox of the plankton: fluctuations of phosphorus availability maintain diversity of phytoplankton in flow-through cultures. Limnol. Oceanogr. 29, 633–636.
Sommer, U., 1986. Phytoplankton competition along a gradient of dilution rates. Oecologia 68, 503–506.
Sommer, U., Adrian, R., Domis, L.D.S., Elser, J.J., Gaedke, U., Ibelings, B., Jeppesen, E., Lürling, M., Molinero, J.C., Mooij, W.M., van Donk, E., Winder, M., 2012. Beyond the Plankton Ecology Group (PEG) Model: mechanisms driving plankton succession. Annual Review of Ecology, Evolution, and Systematics 43, 429–448.
Sommer, U., Gliwicz, Z.M., Lampert, W., Duncan, A., 1986. The PEG-model of seasonal succession of planktonic events in fresh waters. Archives of Hydrobiology 106, 433–471.
Soo, R.M., Woodcroft, B.J., Parks, D.H., Tyson, G.W., Hugenholtz, P., 2015. Back from the dead; the curious tale of the predatory cyanobacterium Vampirovibrio chlorellavorus. PeerJ 3, e968–e968.
Spaulding, S.A., MCKnight, D.M., Smith, R.L., Dufford, R., 1994. Phytoplankton population dynamics in perennially ice-covered Lake Fryxell, Antarctica. J. Plankton Res. 16, 527–541.
Spilling, K., Ylöstalo, P., Simis, S., Seppälä, J., 2015. Interaction effects of light, temperature and nutrient limitations (N, P and Si) on growth, stoichiometry and photosynthetic parameters of the cold-water diatom *Chaetoceros wighamii*. PLoS One 10, e0126308.
Sterner, R.W., 1989. The role of grazers in phytoplankton succession. In: Sommer, U. (Ed.), Plankton Ecology: Succession in Plankton Communities. Springer, Berlin, pp. 107–170.
Sterner, R.W., Elser, J.J., 2002. Ecological stoichiometry: The biology of elements from molecules to the biosphere. Princeton University Press, Princeton.
Stoecker, D.K., 1998. Conceptual models of mixotrophy in planktonic protists and some ecological and evolutionary implications. Eur. J. Protistol. 34, 281–290.
Stolte, W., Riegman, R., 1996. A model approach for size-selective competition of marine phytoplankton for fluctuating nitrate and ammonium. J. Phycol. 32, 732–740.
Strecker, A.L., Cobb, T.P., Vinebrooke, R.D., 2004. Effects of experimental greenhouse warming on phytoplankton and zooplankton communities in fishless alpine ponds. Limnol. Oceanogr. 49, 1182–1190.
Stumpf, R.P., Wynne, T.T., Baker, D.B., Fahnenstiel, G.L., 2012. Interannual variability of cyanobacterial blooms in Lake Erie. PLoS One 7, e42444.
Suttle, C.A., 2000. Ecological, evolutionary, and geochemical consequences of viral infection of cyanobacteria and eukaryotic algae. In: Hurst, C.J. (Ed.), Viral Ecology. Academic Press, pp. 248–296.
Suttle, C.A., Stockner, J.G., Harrison, P.J., 1987. Effects of nutrient pulses on community structure and cell size of a freshwater phytoplankton assemblage in culture. Can. J. Fish. Aquat. Sci. 44, 1768–1774.
Thomas, M.K., Kremer, C.T., Klausmeier, C.A., Litchman, E., 2012. A global pattern of thermal adaptation in marine phytoplankton. Science 338, 1085–1088.
Thomas, M.K., Kremer, C.T., Litchman, E., 2016. Environment and evolutionary history determine the global biogeography of phytoplankton temperature traits. Global Ecol. Biogeogr. 25, 75–86.
Tijdens, M., Van de Waal, D.B., Slovackova, H., Hoogveld, H.L., Gons, H.J., 2008. Estimates of bacterial and phytoplankton mortality caused by viral lysis and microzooplankton grazing in a shallow eutrophic lake. Freshw. Biol. 53, 1126–1141.
Tilman, D., 1982. Resource competition and community structure. Princeton University Press, Princeton, NJ.
Tilman, D., Kilham, S.S., Kilham, P., 1982. Phytoplankton community ecology: the role of limiting nutrients. Annu. Rev. Ecol. Evol. Syst. 13, 349–372.
Trimbee, A.M., Prepas, E.E., 1987. Evaluation of total phosphorus as a predictor of the relative biomass of blue-green algae with emphasis on Alberta lakes. Can. J. Fish. Aquat. Sci. 44, 1337–1342.
Urquhart, E.A., Schaeffer, B.A., Stumpf, R.P., Loftin, K.A., Werdell, P.J., 2017. A method for examining temporal changes in cyanobacterial

harmful algal bloom spatial extent using satellite remote sensing. Harmful Algae 67, 144–152.

Uyeda, J.C., Harmon, L.J., Blank, C.E., 2016. A comprehensive study of cyanobacterial morphological and ecological evolutionary dynamics through deep geologic time. PLoS One 11, e0162539.

Van Donk, E., Bruning, K., 1995. Effects of fungal parasites on planktonic algae and the role of environmental factors in the fungus-alga relationship. In: Schnepf, E., Wiesner, W., Starr, R.C. (Eds.), Algae, Environment and Human Affairs. Biopress Ltd, Bristol, pp. 223–234.

Van Donk, E., Lürling, M., Hessen, D.O., Lokhorst, G.M., 1997. Altered cell wall morphology in nutrient-deficient phytoplankton and its impact on grazers. Limnol. Oceanogr. 42, 357–364.

Wallace, B.B., Hamilton, D.P., 1999. The effect of variations in irradiance on buoyancy regulation in *Microcystis aeruginosa*. Limnol. Oceanogr. 44, 273–281.

Ward, B.A., Dutkiewicz, S., Moore, C.M., Follows, M.J., 2013. Iron, phosphorus, and nitrogen supply ratios define the biogeography of nitrogen fixation. Limnol. Oceanogr. 58, 2059–2075.

Watson, S.B., Monis, P., Baker, P., Giglio, S., 2016. Biochemistry and genetics of taste- and odor-producing cyanobacteria. Harmful Algae 54, 112–127.

Wilcox, L.W., Wedemayer, G.J., 1984. *Gymnodinium acidotum* Nygaard (Pyrrophyta), a dinoflagellate with an endosymbiotic cryptomonad. J. Phycol. 20, 236–242.

Wollrab, S., Izmest'yeva, L., Hampton, S.E., Silow, E.A., Litchman, E., Klausmeier, C.A., 2021. Climate change–driven regime shifts in a planktonic food web. Am. Nat. 197, 281–295.

Woolway, R.I., Dokulil, M.T., Marszelewski, W., Schmid, M., Bouffard, D., Merchant, C.J., 2017. Warming of Central European lakes and their response to the 1980s climate regime shift. Climatic Change 142, 505–520.

Woolway, R.I., Merchant, C.J., Van Den Hoek, J., Azorin-Molina, C., Nõges, P., Laas, A., Mackay, E.B., Jones, I.D., 2019. Northern Hemisphere atmospheric stilling accelerates lake thermal responses to a warming world. Geophys. Res. Lett. 46, 11983–11992.

Wouters, J., Raven, J.A., Minnhagen, S., Janson, S., 2009. The luggage hypothesis: comparisons of two phototrophic hosts with nitrogen-fixing cyanobacteria and implications for analogous life strategies for kleptoplastids/secondary symbiosis in dinoflagellates. Symbiosis 49, 61–70.

Zohary, T., Flaim, G., Sommer, U., 2021. Temperature and the size of freshwater phytoplankton. Hydrobiologia 848, 143–155.

CHAPTER

19

Zooplankton Communities: Diversity in Time and Space

Stephen J. Thackeray[1] *and Beatrix E. Beisner*[2]

[1]Lake Ecosystems Group, UK Centre for Ecology & Hydrology, Lancaster, UK [2]Department of Biological Sciences, University of Quebec at Montreal, Montreal, Quebec, Canada

OUTLINE

Wetzel's Limnology, Fourth Edition
https://doi.org/10.1016/B978-0-12-822701-5.00019-7

I. Introducing the zooplankton

The zooplankton are defined as the heterotrophic component of the wider plankton community. In freshwater four major groups dominate the zooplankton: (a) *heterotrophic protists* that include ciliates and flagellates; (b) *Rotifera* (*rotifers*) and two subclasses of the Crustacea, (c) the *Cladocera* (*cladocerans*) and (d) the *Copepoda* (*copepods*). In addition, a few ostracods, coelenterates, larval trematode flatworms, gastrotrichs, mites, and the larval stages of certain insects and fish occasionally occur among the true zooplankton, if only for a portion of their life cycles. The phyletic representation and species diversity of zooplankton communities are impoverished in freshwater compared to marine habitats (Lehman, 1988). The great age, depth, and evolutionary continuity of the oceans may be factors in this notable discrepancy, although even ancient lakes support zooplankton communities that are far less diverse than in marine ecosystems.

II. The microzooplankton

The phylogeny of eukaryotic life has undergone a fundamental change in the 21st century (Adl et al., 2019, 2012). As such, we here refer to *heterotrophic protists* rather than the once-widespread term "protozoa". We focus on free-living unicellular, eukaryotic organisms that are *phagotrophic* (capable of feeding by engulfing particles; Finlay and Esteban, 1998a). Though there is variation among taxa, heterotrophic protists may possess a nuclear envelope, eukaryotic ribosomal RNA, and endoplasmic membranes (Fenchel, 1987; Laybourn-Parry, 1992; Patterson, 1992), as well as eukaryotic organelles (mitochondria, chloroplasts, and flagella), histones associated with chromosomal DNA, and an ability to perform *phagocytosis* (the process of engulfing and ingesting particles). Collectively, heterotrophic protists are known to reproduce by binary fission, multiple fission, and sexual reproduction. Growth and generation times are highly variable among free-living protists. Maximum growth rates are found when food availability is not a constraint; under these conditions, populations generally increase directly with temperature within the limnological range and they decrease as body size increases (Taylor and Sanders, 1991). Many protists produce cysts under harsh environmental conditions (e.g., lack of food and desiccation).

Protistan zooplankton are important microbial consumers, influencing organic carbon utilization and nutrient recycling (Finlay and Esteban, 1998a; Chapter 20). Next, we introduce the major groupings within these microzooplankton: heterotrophic flagellates, ciliates, and sarcodines (Fig. 19-1).

A. Flagellates

Flagellates numerically dominate the microzooplankton. Autotrophic and mixotrophic flagellates have already been discussed among the phytoplankton (Chapters 17 and 18), and here we focus only on heterotrophic groups, including heterokont taxa (mainly chrysomonads and bicosoecids), choanoflagellates, kathablepharids, and a number of species, referred to as "orphan lineages" or *incertae sedis*, that cannot be assigned to any of the major groups (Boenigk and Arndt, 2002). The heterotrophic flagellates are differentiated into two general groups on the basis of size: (a) *heterotrophic nanoflagellates* (HNF), below 15 μm; and (b) *large heterotrophic flagellates,* in the range of ≥15–200 μm (Boenigk and Arndt, 2002).

B. Ciliates

Many major ciliate genera are represented in freshwater microzooplankton assemblages. Of the three major groups, the oligotrichs, particularly *Strombidium* and *Halteria,* are found worldwide in lakes across the trophic spectrum (Laybourn-Parry, 1992). The tintinnid ciliates (order Choreotrichida: *Tintinnidium*, *Tintinnopsis*, and *Codonella*) are also widely distributed in temperate to tropical regions. Haptorid ciliates (e.g., *Askenasia* and *Mesodinium*) are similarly distributed broadly and abundantly. Although a few ciliates are *mixotrophic* (capable of photosynthesis and phagotrophy) and supplement nutrition by photosynthesis, most are *holozoic* (ingest organic matter) and feed on bacteria, algae, particulate detritus, and other protists. A few are carnivorous and feed on small metazoans. Ciliates tend to be more significant components of the zooplankton of eutrophic lakes (Beaver and Crisman, 1989).

A number of ciliates are common to the zooplankton, and they can dominate zooplankton numbers, biomass, and grazing in certain situations. Ciliates can move much more rapidly (200–1000 μm s^{-1}) than other protists (0.5–3 μm s^{-1} among those with *pseudopodia*, i.e., temporary arm-like projections; 15–300 μm s^{-1} among those with flagella). Such movement contributes significantly to greater dispersal and higher feeding rates of ciliates.

C. Sarcodines

Sarcodine protozoa (naked and testate amoebae, heliozoans) may occur in the zooplankton of freshwaters but have received less research attention than flagellates

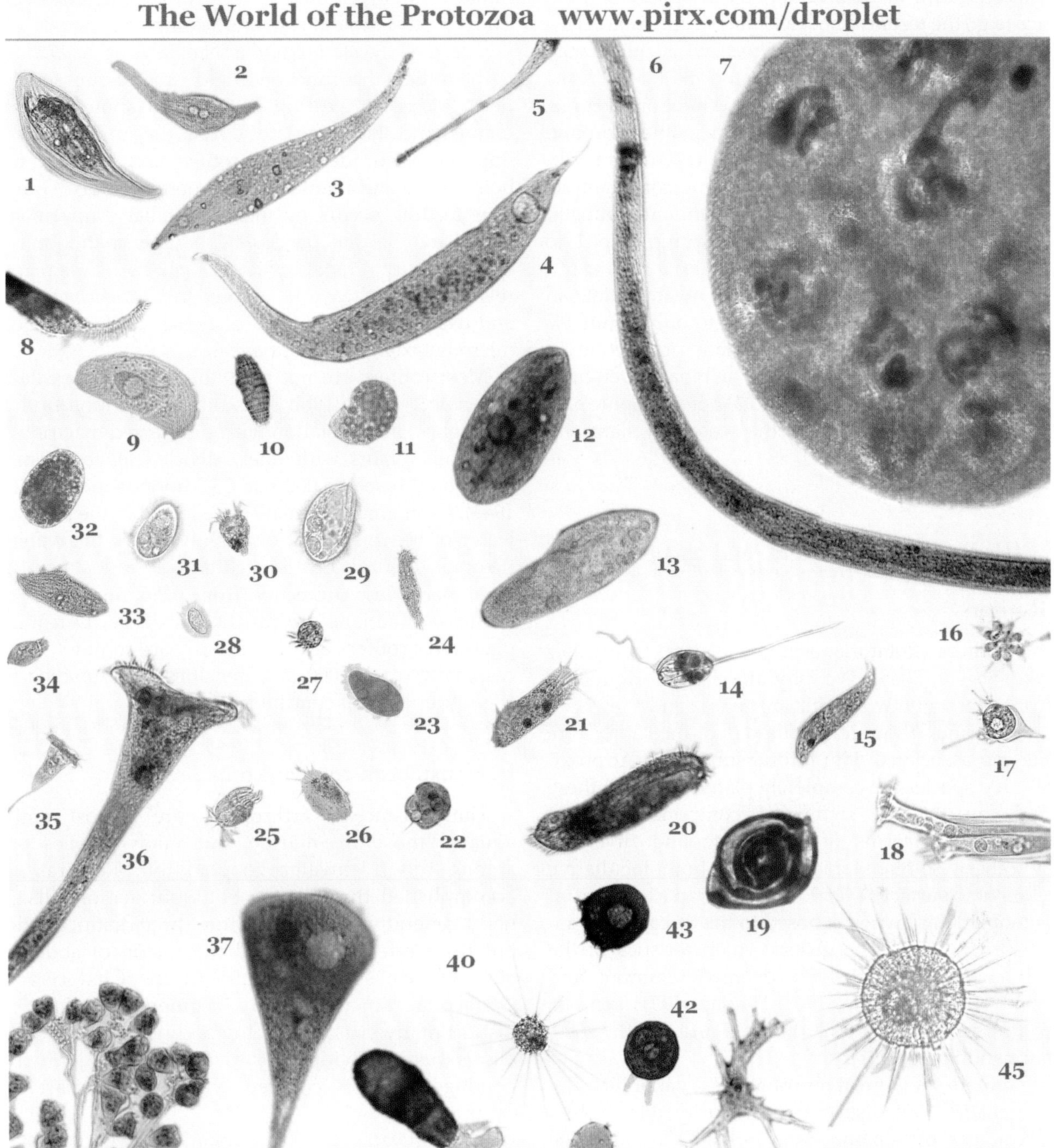

1. Loxophyllum 2. Litonotus 3. Amphileptus 4. Dileptus 5. Lacrymaria 6. Spirostomum 7. Ichtyophthirius 8. Stichotricha 9. Chilodonella 10. Coleps 11. Colpoda 12. Frontonia 13. Paramecium 14. Anisonema 15. Euglena 16. Anthophysis 17. Acineta 18. Thuricola 19. Spiroloculina 20. Urostyla 21. Stylonychia 22. Urocentrum 23. Tetrahymena 24. Amphisiella 25. Diophrys 26. Euplotes 27. Halteria 28. Cyclidium 29. Lembadion 30. Strombidium 31. Pleuronema 32. Nassula 33. Blepharisma 34. Rhabdostyla 35. Vorticella 36, 37. Stentor 38. Carchesium 39. Difflugia 40. Actinophrys 41. Euglypha 42. Acrella 43. Centropyxis 44. Amoeba 45. Actinosphaerium

FIGURE 19-1 A collage of heterotrophic protists (protozoa) showing the diversity of size and form. Copyright © 2003–2008 Piotr Rotkiewicz, https://www.pirx.com/droplet/collage.html.

and ciliates. Some taxa may adopt a truly planktonic existence (e.g., the testate amoeba *Difflugia*, and heliozoans such as *Actinophrys* and *Raphidiophrys*), while others may live attached to sedimenting organic matter or "lake snow" (Finlay and Esteban, 1998a). While, on average, sarcodines might not be as abundant in the plankton as the flagellates and ciliates, seasonal variations in absolute and relative abundance are great; for example, in Lake Constance (central Europe) abundances ranged from below detection limits in the spring (April to mid-June) to summer maxima of up to 6.6 cells mL^{-1} (Zimmermann et al., 1996). Seasonal mean heliozoan production was, however, equivalent to only about 1% of the combined ciliate and flagellate production in this lake. In East Dongting Lake (China) sarcodines comprised between 3 and 35% of total heterotrophic protist abundance, depending on the season of sampling (Wu et al., 2007).

III. Rotifers, Cladocera, and copepods

A. Rotifers

The Rotifera (Rotatoria, estimated >1900 freshwater species [Segers, 2008]) is a phylum clearly originating in freshwater; only two significant genera and a few species are marine. About three-quarters of the rotifers are sessile and associated with littoral substrates. Approximately 100 species are completely planktonic, and these rotifers can form a significant component of the zooplankton, in terms of abundance and diversity. Some otherwise-planktonic species show facultative attachment to surfaces (including the carapace of crustacean zooplankton) when exposed to the threat of predation (Gilbert, 2019). The general characteristics of the group have been extensively detailed (Dumont and Green, 1980; Hutchinson, 1967; Hyman, 1951; Pennak, 1978; Ruttner-Kolisko, 1972; Wallace and Smith, 2009; Wallace and Snell, 1991).

The rotifers show great morphological variability. Rotifers are *eutelic* organisms; upon maturity, they have a fixed number of somatic cells (~1000). In most the body shape tends to be elongated, and regions of the head, trunk, and foot usually are distinguishable (Figs. 19-2 and 19-3). The cuticle is generally thin and flexible, but in some rotifers it is thickened and more rigid and is termed a *lorica*. The anterior end or *corona* of rotifers is ciliated; in some species the periphery is ciliated as well. The movement of the cilia functions both in locomotion, especially among planktonic forms, and in the movement of food particles toward the mouth. The mouth, although variously located, is generally anterior. The digestive system contains a complex muscular pharynx, termed the *mastax*, and a set of jaws or *trophi*, unique to the rotifers that function to seize and disrupt food particles. Along with features of the lorica (spines, ornamentations) and the shape of the foot, the trophi are of key importance in identifying rotifer taxa. Most rotifers, both sessile and planktonic, are nonpredatory. Omnivorous feeding occurs by means of ciliary movement of living and detrital particulate organic matter into the mouth cavity. Predatory taxa, such as *Asplanchna*, are relatively large (e.g., 1–1.5 mm for *Asplanchna sieboldi*) and usually prey upon protists, other rotifers, and other micrometazoa of appropriate size.

Most rotifers are not planktonic but are sessile and associated with littoral substrata. Population numbers are highest in association with submersed macrophytes, especially plants with finely divided leaves; densities commonly reach 25,000 per L (Edmondson, 1946, 1945, 1944). Even greater densities are found in the interstitial water of beach sand at or slightly above the waterline (Pennak, 1940). With reduced sites for attachment and presumably less protection from predation, planktonic rotifer populations are much less dense. Densities of planktonic rotifers of 200 to 300 L^{-1} are common and occasionally reach 1000 L^{-1}; densities rarely exceed 5000 L^{-1} under natural conditions.

B. Crustacean zooplankton

The crustacean arthropods are almost entirely aquatic; most are marine, but many species occur among the freshwater zooplankton. Respiration is accomplished through the body surface or gills. The body generally is separated into three distinct regions, but the tendency is toward the fusion of abdominal and thoracic segments until, in the Cladocera and Ostracoda, apparent body segmentation has been lost. In many crustaceans the body bears paired, usually *biramous* (divided into two branches), jointed appendages, and is covered wholly or in part by a carapace.

In freshwater the truly planktonic Crustacea are dominated almost completely by the Cladocera and Copepoda. Most of the Ostracoda are benthic (cf. Chapter 21), though a few species of *Cypria* and *Heterocypris* are apparently partly planktonic (Fig. 19-4). Only a few insects are planktonic in immature stages; the larvae of *Chaoborus* (Diptera) are a notable example of major importance in the zooplankton.

Several types of Branchiopoda (fairy and clam shrimps, Fig. 19-4) are common inhabitants of shallow

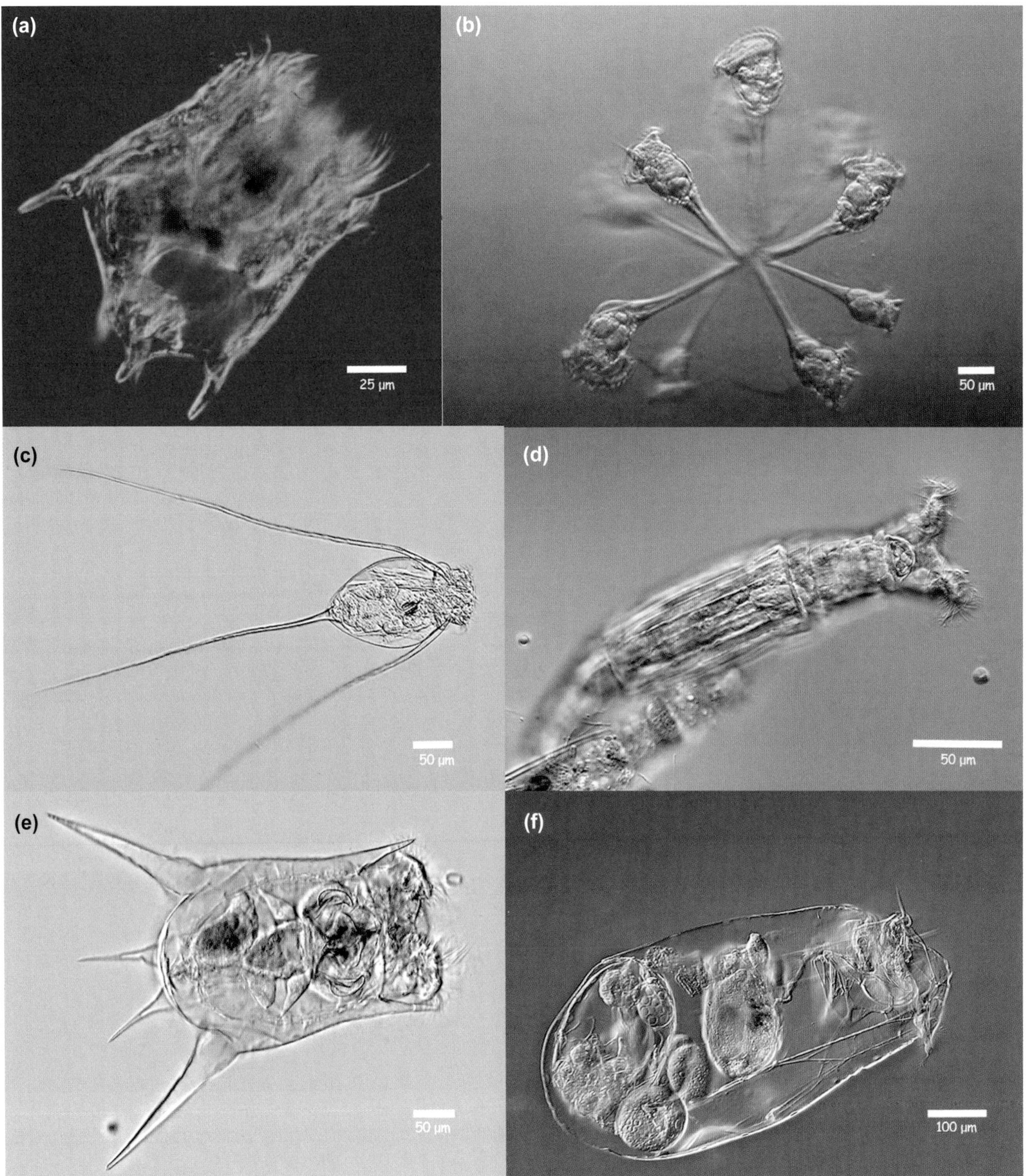

FIGURE 19-2 Examples of rotifer diversity. (a) *Plationus patulus*; (b) *Conochilus unicornis*; (c) *Filinia longiseta longiseta*; (d) *Rotaria rotatoria*; (e) *Brachionus calyciflorus anuraeiformis*; (f) *Asplanchna priodonta*. Image credits: (a) Stephanie Hampton; (b)–(f) Michael Plewka; www.plingfactory.de. Note differences in scale. *(Reproduced from Thackeray (2022).)*

lakes, particularly of temporary, saline inland waters. All members of this group are distinctly segmented and bear many pairs of swimming and respiratory appendages. The tadpole shrimps (Notostraca) are essentially benthic and most often restricted to shallow, temporary lakes of arid regions. The fairy shrimps (Anostraca), lacking a carapace, and the clam shrimps (Conchostraca), compressed laterally with a bivalved

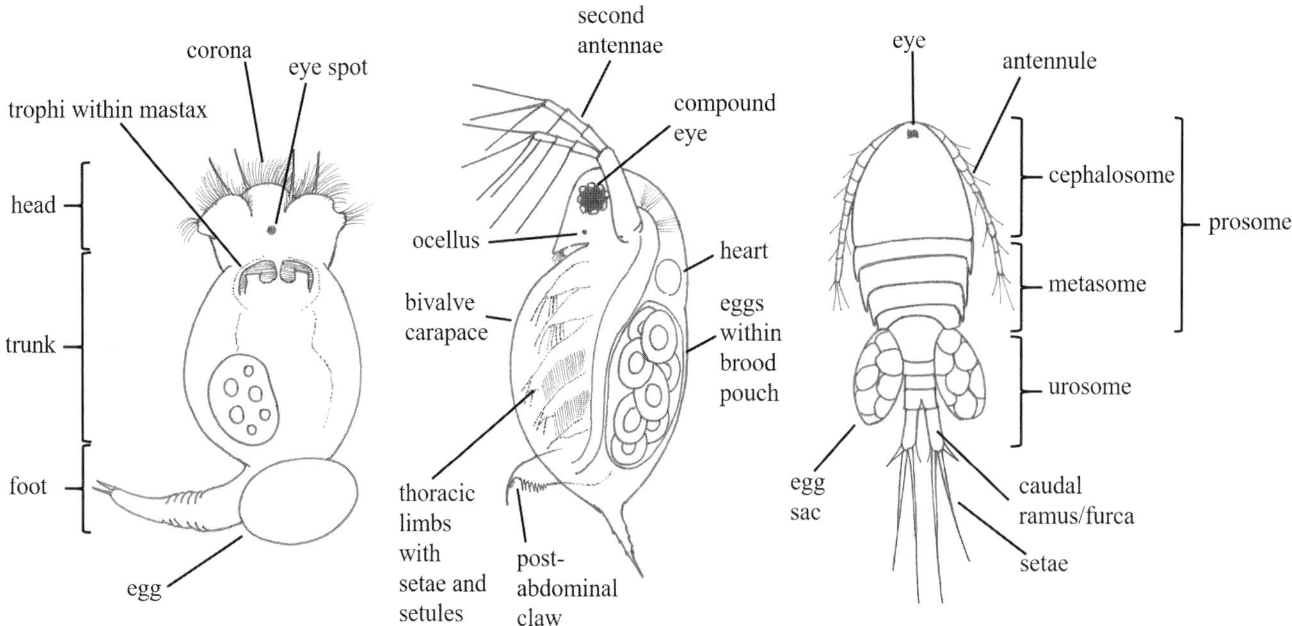

FIGURE 19-3 Line drawings to highlight major anatomical features of rotifers (*Brachionus* sp., *left*), cladocerans (*Daphnia* sp., *middle*), and copepods (*Cyclops* sp., *right*). (*Drawings by S. J. Thackeray.*)

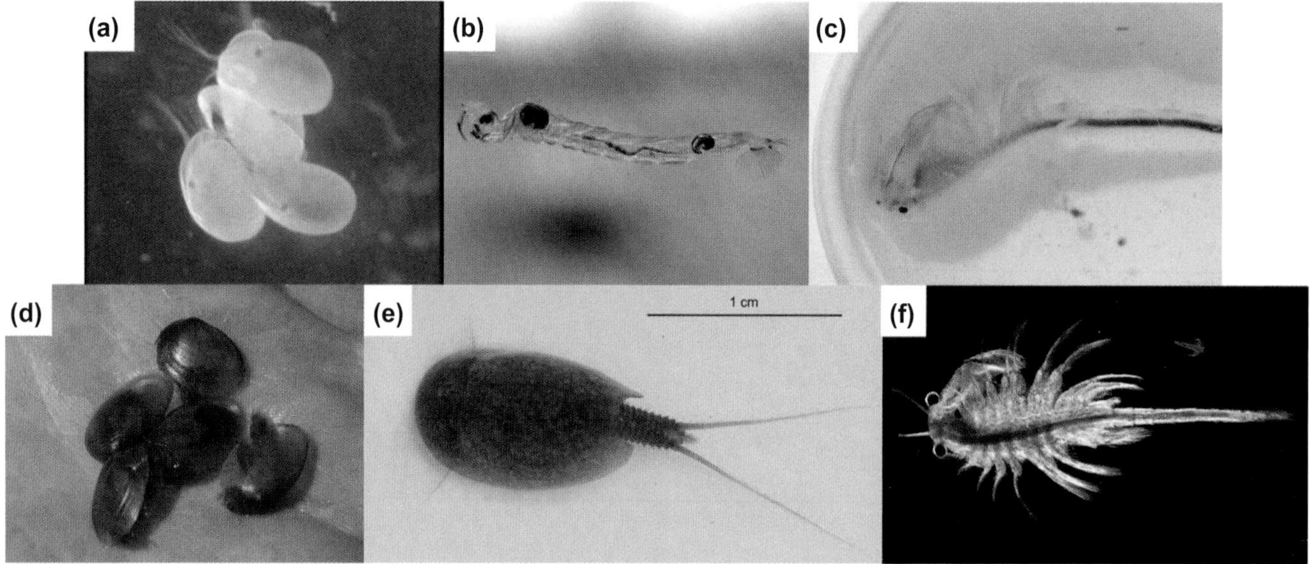

FIGURE 19-4 A selection of aquatic invertebrates. (a) The ostracod *Heterocypris incongruens* (Rossi et al., 2011); (b) *Chaoborus* sp. (photo © Piet Spaans); (c) fairy shrimp (photo © Josh Hull, USFWS); (d) California clam shrimp (*Cyzicus californicus*, photo © Franco Folini); (e) tadpole shrimp (*Lepidurus apus*, photo © Christian Fischer); and (f) brine shrimp (*Artemia monica*, photo © djpmapleferryman).

flexible carapace, are common in the plankton in shallow playa lakes and vernal ponds of semiarid regions. The members of the latter groups are bisexual, although *parthenogenesis* (a form of reproduction where eggs develop into embryos without fertilization, see Section V) is known to occur in the brine shrimp *Artemia* under some conditions. Resistant eggs can be subjected to long periods of desiccation and hatch when wetted again, probably as a result of a reduction in osmotic pressure at the egg surface (Hutchinson, 1967). In semipermanent lakes of more humid regions, hatching and reproductive rates are related strongly to temperature.

i. Cladocera

The Cladocera (class Branchiopoda, estimated >600 species globally [Forró et al., 2008]) includes mainly zooplankton, but there are also benthic species. Nearly all freshwater cladocerans range in size from 0.2 to 3.0 mm, though a few species exceed this, notably the predatory *Leptodora kindtii*. The body is usually covered by a bivalve cuticular carapace; however, this is reduced to covering the brood pouch only in some species (Figs. 19-3 and 19-5). Most have a distinct head (perhaps less obvious for the Holopedidae), with light-sensitive organs, usually a large, *compound eye* (comprising several individual optical units) and a smaller *ocellus* (a simple eyespot). The second *antennae* are large swimming appendages and constitute the primary organs of locomotion. The mouthparts consist of (1) large, chitinized mandibles that grind food particles; (2) a pair of small *maxillules* used to push food between the mandibles; and (3) a median labrum that covers the other mouth parts.

ii. Copepoda

The free-living copepods of this subclass (class Maxillipoda, estimated >2800 freshwater species [Boxshall and Defaye, 2008]) are separable into three distinct groups: the Calanoida, Cyclopoida, and Harpacticoida. Although accurate identification is based largely on morphological details of appendages, several general characteristics delineate the major groups (Williamson, 1991, Figs. 19-3 and 19-6). The body consists of the anterior *prosome* (*cephalothorax*), which is divided into the head region, bearing five pairs of appendages representing antennae and mouth parts, and the thorax, with six pairs of mainly swimming legs. The posterior region of the body, the *urosome*, consists of abdominal segments, the first of which is modified in females as the genital segment, and terminal *caudal rami/furca* (rod-like protrusions) bearing *setae* (hair-like structures).

The harpacticoid copepods are almost exclusively littoral, inhabiting macrovegetation, mosses in particular, and the littoral sediments. Although the cyclopoid copepods are primarily littoral benthic species, those few members that are predominantly planktonic form major components of the copepod zooplankton, especially in small, shallow lakes. The calanoid copepods are almost exclusively planktonic.

IV. Food, feeding, and food selectivity

A. Protist feeding mechanisms

Protists possess a variety of nutritional mechanisms that include autotrophy, mixotrophy, and heterotrophy. As summarized by Capriulo et al. (1991), Fenchel (1991), Gaines and Elbrachter (1987), Nisbet (1984), Posch and Arndt (1996), Radek and Hausmann (1994), Sanders (1991), Sleigh (1989), and Taylor and Sanders (1991) the heterotrophic nutritional modes of free-living protists include:

a. Uptake and assimilation of dissolved organic compounds through the cell plasma membrane (*osmotrophy*) or by pinching off small-sized vesicles (*pinocytosis*)
b. Feeding directly on living or dead particulate organic matter (*phagotrophy*) or by generating water currents to the cell surface where *phagocytosis* occurs, extending a sucking tentacle (some dinoflagellates), *pseudopodial engulfment*, or *endocytosis* (bringing particles into the cell) within the *cytosome* (mouth)
c. Metabolic exchange with *endosymbionts* (organisms living within the protist cell, a type of *mixotrophy*)
d. Combinations of the above mechanisms

i. Flagellates

Heterotrophic flagellates employ a wide variety of feeding mechanisms (Boenigk and Arndt, 2002). These include (a) particle or nutrient uptake by diffusion or interception, active transport across membranes, or endocytosis; (b) mixotrophy, by shifting from autotrophic photosynthesis to heterotrophy in the absence of light or practicing both simultaneously (covered in detail in Chapter 18); (c) the use of one or more flagella to create feeding currents to move food particles to the cytostome; and (d) food capture by filtration (e.g., choanoflagellates).

The range of particle size ingested by heterotrophic nanoflagellates extends from common bacteria to ~30 μm, but there exists feeding selectivity based upon particle size. For example, Chrzanowski and Šimek (1990) found that *Ochromonas danica* and *Bodo*-like heterotrophic flagellates ingested 0.8–1.4 μm^3 bacterial cells with greater frequency than they appeared in the prey population, but that this was not the case for smaller bacterial cells. Capture and ingestion efficiency vary not just with prey size but also with other phenotypic traits such as prey motility and surface biochemical characteristics (Matz et al., 2002; Fig. 19-7).

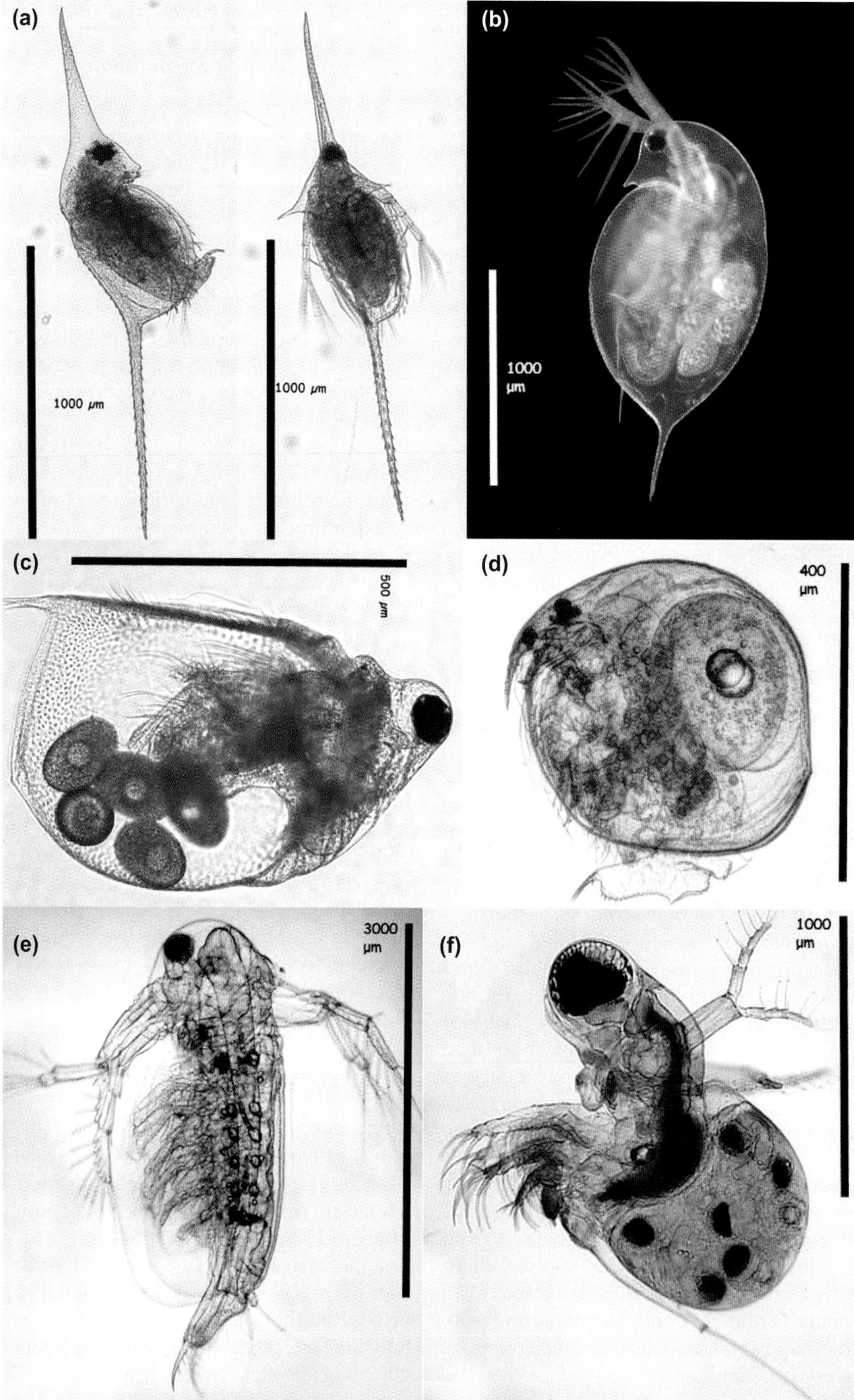

FIGURE 19-5 Examples of cladoceran diversity. (a) *Daphnia lumholtzi*; (b) *Daphnia longispina*; (c) *Scapholeberis* sp.; (d) *Chydorus sphaericus*; (e) *Sida crystallina*; and (f) *Polyphemus pediculus*. Image credits: (a) and (c) K. David Hambright; (b), (d)–(f) Adrian Chalkley. Note differences in scale. *(Reproduced from Thackeray (2022).)*

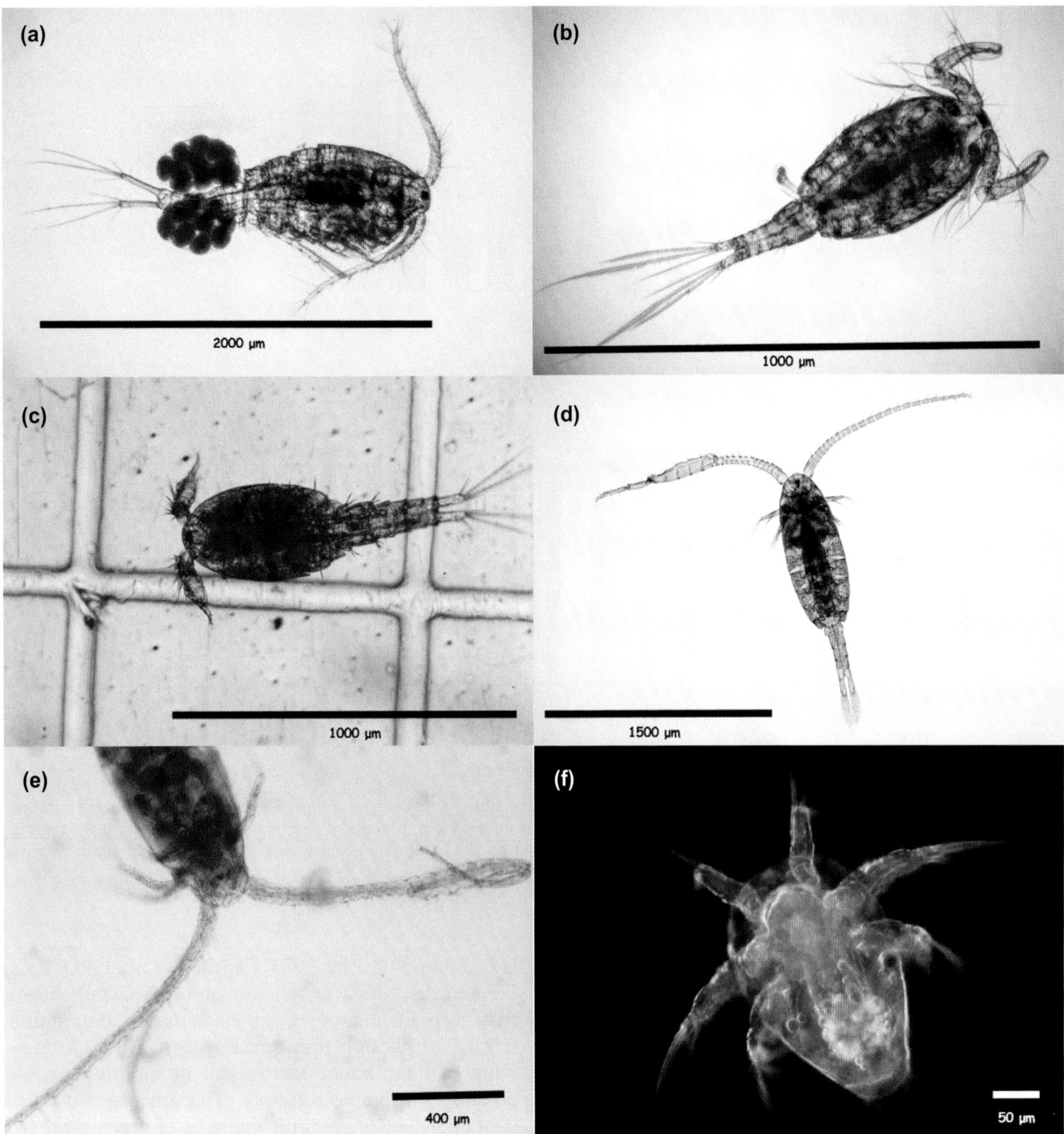

FIGURE 19-6 Examples of copepod diversity. (a) *Cyclops strenuus abyssorum* (female); (b) *Macrocyclops albidus* (male); (c) *Paracyclops fimbriatus*; (d) *Arctodiaptomus wierzejskii* (male); (e) *Eudiaptomus gracilis* (male, close up of antennae); and (f) copepod naupliar (juvenile) stage. Image credits: (a), (b), (d) Kenny Gifford; (c), (e) Stephen Thackeray; (f) Michael Plewka, www.plingfactory.de. Note differences in scale. *(Reproduced from Thackeray (2022).)*

ii. Ciliates

Most free-living ciliates have mouth-like structures and feed on particulate food (Capriulo et al., 1991; Fenchel, 1987; Posch and Arndt, 1996; Skogstad et al., 1987). Organisms ingested include bacteria, coccoid Cyanobacteria, microalgae, diatoms, dinoflagellates, heterotrophic microflagellates, and other ciliates, as well as detrital organic particles in the range of 0.2–20 μm. Many modes

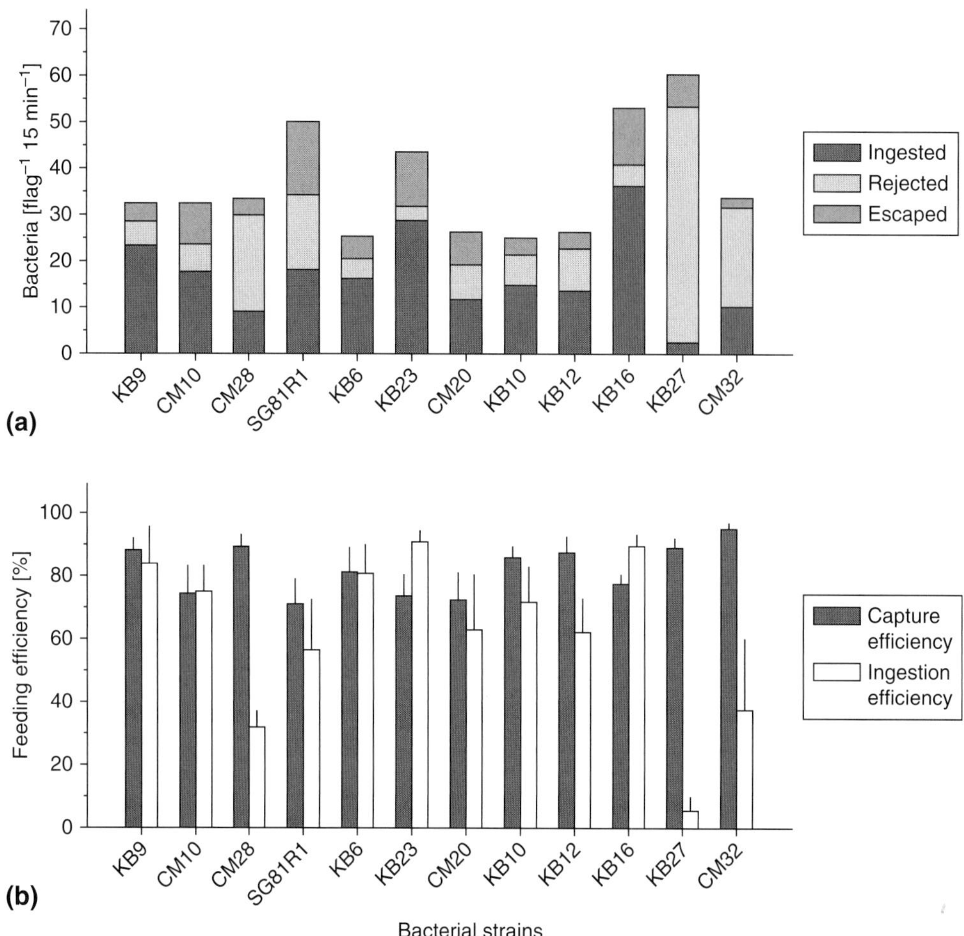

FIGURE 19-7 Flagellate (flag) feeding on bacterial isolates exhibiting different phenotypic properties. (a) Average numbers of escaped, rejected, and ingested bacteria of the total number of flagellate-bacterium contacts and (b) flagellate feeding efficiencies (capture and ingestion efficiency) are given as mean ± standard deviation. The data are based on 12 individual flagellate cells for each bacterial strain within an observation time of 15 min. *(Reproduced from Matz et al. (2002).)*

of food acquisition occur: (a) suspension feeding in which oral membranelles may act as filters; (b) active predatory hunting by random encounter or possibly chemoreception; (c) deposit feeding of particles by phagocytosis; and (d) mixotrophy by algal endosymbionts or with functional chloroplasts (e.g., *Paramecium bursaria* with *Chlorella*, Summerer et al., 2008). The ciliate *Mesodinium rubrum* is completely autotrophic (Lindholm, 1985). Ciliates may feed selectively on the available prey assemblage.

iii. Sarcodines

Testate and naked amoebae utilize pseudopods to engulf bacteria, Cyanobacteria, diatoms, flagellates, other protists, and detrital organic particles as randomly encountered (Canter, 1973; Capriulo et al., 1991). Using the amoeba *Naegleria* sp., Xinyao et al. (2006) showed that even filamentous Cyanobacteria can be engulfed (Fig. 19-8), though prey species vary in their vulnerability to engulfment as a result of morphology, and possibly biochemical factors. The amoeba *Vampyrella* penetrates and ingests the contents of green algal cells (Finlay and Esteban, 1998a). Heliozoans have stiff, radiating pseudopodia, called *axopodia*, which trap any prey with which they collide; a process termed *diffusion feeding*. A few amoeboid protozoa (e.g., *Pelomyxa*) also harbor endosymbiont methanogenic bacteria (van Bruggen et al., 1983). Some small amoebae are raptorial, seizing and ingesting whole prey organisms (Fenchel, 1987).

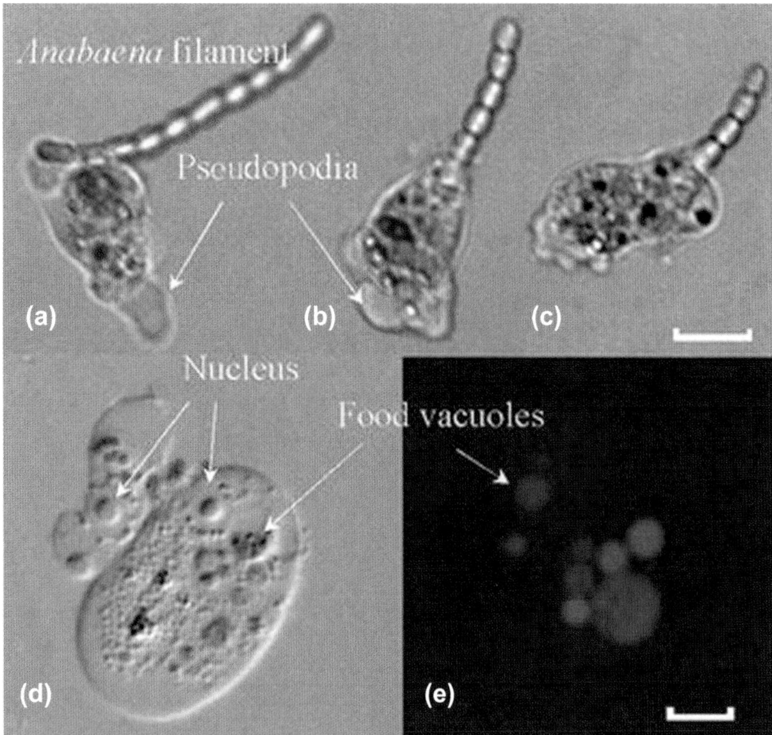

FIGURE 19-8 Process of *Naegleria* sp. grazing on an *Anabaena flos-aquae* (Cyanobacteria) filament. (a) The amoeba aligned itself with and often surrounded the filament while seeking the terminal cell, to which it became attached; (b) the end of the *A. flos-aquae* filament was gradually engulfed by the extension of a cylindrical pseudopodium; (c) six cells of the filament had been ingested. Bar = 10 μm; (d, e) ingested cyanobacterial cells in the food vacuoles of amoebae; (d) differential interference contrast (DIC) picture; and (e) fluorescent picture obtained with a green filter set. Note the correspondence between the dark mass corresponding to the food vacuoles and the region with an altered orange-red fluorescence. Bar = 5 μm. *(Reproduced from Xinyao et al. (2006).)*

B. Rotifer feeding mechanisms

Planktonic rotifers feed largely by sedimenting particles into the mouth orifice by a pulsating action of the coronal cilia (Hutchinson, 1967). Most food particles are small, less than about 12 μm in diameter, although larger cells, up to approximately 50 μm, are sometimes seized, ruptured, and ingested. Rejection behaviors may allow particle ingestion to be regulated; by studying *Brachionus calyciflorus*, Gilbert and Starkweather (1977) could discern three distinct mechanisms by which potential prey could be rejected: (a) bending over the bristle-like cirri of the pseudotrochus to form a screen that prevents particles from entering the funnel-shaped buccal field; (b) rejection of particles in the buccal field before entering the oral canal by ciliary movements; and (c) rejection of particles in the oral canal by the jaws, which push food items back into the buccal field.

Different food types induce different ingestion rates in *Brachionus* (Gilbert and Starkweather, 1978; Starkweather and Gilbert, 1978, 1977). Some rotifer species have been shown to ingest toxin-containing Cyanobacteria and to be able to sustain positive population growth when consuming these prey (Soares et al., 2010). A degree of feeding selectivity has also been demonstrated. Pagano (2008) showed that *B. calyciflorus* may feed preferentially on relatively large prey, switching to smaller prey when larger particles are depleted. The same species has also been shown to discriminate between flagellate and ciliate prey, selecting for *Cryptomonas* more than *Chilomonas*, and *Coleps* more than *Tetrahymena* (Mohr and Adrian, 2002).

While many rotifers suspension-feed, some species are raptorial or draw in cell contents after the cell or body wall of their prey has been punctured. The large predatory rotifer, *Asplanchna*, exemplifies this, feeding on algae, rotifers, and small planktonic crustaceans (Gilbert, 1980).

C. Cladoceran feeding mechanisms

Cladocerans usually have five pairs of legs attached to the ventral part of the thorax. The legs are flattened, bearing numerous hair-like structures termed *setae* and *setules* (Fig. 19-3). Complex movements of these setose legs create a current of water through the valves. This current oxygenates the body surface and forces a stream of food particles anteriorly. The food particles, collected by the setae, enter a ventral food groove between the bases of the legs and are impelled forward toward the mouth, to be mixed there with oral secretions. By comparing the size structure of ambient and ingested food particles with morphological characteristics of cladoceran filter structures (the "mesh size"), some researchers have deduced that the primary food collection mechanism involves simple mechanical filtering. This mechanism would result in the retention of only those food particles larger than the "mesh size" of the thoracic feeding limbs (Brendelberger, 1991; Geller and Müller, 1981; Gophen and Geller, 1984). However, because intersetular spaces on the thoracic limbs and the resulting *Reynolds numbers* (ratios of inertial to viscous forces) are very small ($\approx 10^{-3}$), some pressure difference would be needed to force water through the mesh. For cladocerans, a closed filtering chamber may allow water to be forced through the apparatus with minimal expenditure of energy ($\sim$5% of the total metabolic requirements of the animal) (Brendelberger et al., 1986). Despite these arguments, it has been suggested that food collection is assisted by electrostatic attraction, rather than filtering, and that this mechanism is in turn influenced by the surface charge chemistry of food particles (Gerritsen and Porter, 1982).

The primarily littoral chydorid cladocerans have modified legs that are somewhat prehensile in scraping up larger pieces of detrital material. Feeding by filtration occurs as well. Other cladocerans, such as *Polyphemus*, *Bythotrephes* (Fig. 20-5), and *Leptodora*, are predatory and feed mainly by seizing relatively large particles, such as protozoa, rotifers, and small crustaceans, with their prehensile legs. For *Polyphemus*, there appears to be a role for vision in prey detection (Young and Taylor, 1988).

D. Copepod feeding mechanisms

The mouthparts of harpacticoids (which are not truly planktonic) are adapted for seizing and scraping particles from the sediments and macrovegetation. In the free-living Cyclopoida feeding is raptorial; algal or animal food particles are seized by mouth parts and brought to the mouth. The maxillules hold and pierce the prey and force particles between the mandibles; intermittent oscillating movements macerate some of the food. Some particles are swallowed intact and are differentially digested. Many species of the major cyclopoid genera *Macrocyclops, Acanthocyclops, Cyclops*, and *Mesocyclops* are carnivorous. The food of these carnivores includes microcrustaceans, dipteran larvae, and oligochaetes, many of which are larger than the copepod that preys on them. Cannibalism is known in such species (van den Bosch and Santer, 1993). There are, however, some herbivorous cyclopoids, including many species of *Eucyclops*, some *Acanthocyclops*, and *Microcyclops*, which feed on a variety of algae ranging from unicellular diatoms and other microalgae to long strands of filamentous species. Random encounter appears to be the dominant mode of finding food in both carnivorous and herbivorous species, which search by discontinuous, irregular movements in the water or over the substratum. Herbivorous species apparently employ gustatory chemoreceptor organs, which allow "taste"-based food selection, along with discrimination based upon prey size and motility (Sommer and Sommer, 2006).

While locomotion in many copepods comprises short, jerky swimming movements, the animals being propelled by the rapid movement of most appendages simultaneously, swimming is more continuous in the calanoid copepods. Their gliding movement results from the rotary motion of the antennae and mouth appendages. The movements set up small vortical currents that carry particles to the maxillae. High-speed motion pictures have revealed that calanoid copepods do not strain particles out of the water but instead propel water past the body by flapping four pairs of feeding appendages (Koehl and Strickler, 1981). The second maxillae actively capture "parcels" of that water containing food particles. Particles are then pushed into the mouth by the first maxillae. Selective feeding also exists among the calanoid copepods. For example, dietary differences were found for coexisting *Arctodiaptomus laticeps* and *Eudiaptomus gracilis* (in Windermere, UK; Fryer, 1954). The larger *A. laticeps* fed chiefly on *Melosira italica* (now synonymous with *Aulacoseira italica* Ehrenberg), and the smaller *E. gracilis* consumed mainly minute spherical green algae and particulate detritus. Neither species fed upon the then-abundant diatom *Asterionella formosa*. Competition for food by these two calanoid species was virtually nonexistent. Such small differences in feeding habits, based on morphological and behavioral variations, are common in the calanoids and may be sufficient to separate species into different food niches, even though they occupy the same volume of water.

E. Feeding rates

Several physical and biotic factors affect zooplankton ingestion rates, with at least some organisms possessing a degree of feeding selectivity. These findings have major implications for the effectiveness of food ingestion and differential assimilation by the animals and also for the effects of zooplankton consumption on food populations. At times zooplankton, as a result of direct cropping, can have appreciable effects on phytoplankton populations. The nature of this effect depends upon the structure of the zooplankton assemblage; zooplankton groups (heterotrophic protists, rotifers, cladocerans, copepods) will each consume different subsets of the overall prey assemblage, due to differences in their feeding mechanisms (Sommer and Sommer, 2006).

The *clearance rate* of a zooplankter, measured as a volume per unit time, is defined as the volume of water that is cleared of suspended food particles by the animal in a given time (Wetzel and Likens, 1991). This term does not imply that the volume of water passed over the filtering appendages is known, that all particles of any given type have been removed from the water, or that all particles retained by the filtration apparatus have been consumed. In contrast, *feeding rate*, or *ingestion rate*, is the quantity of food ingested by an animal in a given time (Rigler, 1971; Wetzel and Likens, 1991), measured in terms of the number of cells, volume, dry weight, carbon, nitrogen, or some other relevant aspect of the ingested food.

While microscopic examination of gut contents or fluorescently labeled prey allows confirmation of prey ingestion (Work and Havens, 2003), clearance and ingestion rates can be experimentally determined by measuring food particle removal rates by zooplankton, that is, changes in the concentration of particles, caused by grazing, over time (e.g., Leitão et al., 2018; Sarnelle and Wilson, 2008). Separate rates can be calculated for each prey species, strain, or size class under consideration. However, caution is needed when extrapolating laboratory measurements of clearance and ingestion rates to the natural environment. Laboratory measurements can overestimate *in situ* grazing rates by at least 30–50% because of spatial heterogeneity in natural plankton distributions (Wirick, 1989). The rate of removal of radioactively labeled food particles has also been used to measure zooplankton feeding rates (see review of Rigler, 1971). The latter approach has been extended effectively to natural populations by use of an *in situ* grazing chamber (Haney, 1971) that permits estimates of feeding rates over short periods of time (<10 min) so that ingested food is not excreted during the incubation period of measurement. These methods measure food that is actually taken into the gut after losses, either through active rejection of food or losses of food particles during the maceration process. An alternative *in situ* experimental approach to estimating zooplankton community feeding rates involves the measurement of phytoplankton growth rates in the presence of increasingly concentrated zooplankton communities. The slope of the relationship between phytoplankton growth rate and zooplankton biomass is then taken to estimate the grazing rate (Symons et al., 2012).

Ingestion rates vary as a function of food concentration. The form of this relationship, called the *functional response*, is much debated (e.g., Colina et al., 2016; Jeschke et al., 2004; Sarnelle and Wilson, 2008). For a *Type I functional response*, ingestion rate increases linearly with rising food concentration, up to a threshold *incipient limiting level* (and then remains constant). For a *Type II functional response*, this increase in ingestion rate with food concentration is smoothly curvilinear, and for a *Type III functional response*, it is sigmoidal. Under each functional response type, different assumptions are made about variations in clearance rates and feeding behavior with changing food concentration. McMahon and Rigler (1965) found that, for *Daphnia magna*, the feeding rate is proportional to the concentration of food particles below an incipient limiting level and constant above this threshold food concentration, indicative of a Type I functional response. While Type I responses may indeed be prevalent among filter-feeding zooplankton, feeding behavior is highly variable and all three of the fundamental functional response types have been observed for the freshwater zooplankton, when considering a broad body of literature (Jeschke et al., 2004). It has even been suggested that Type III functional responses may be more common than appreciated to date, given issues of experimental design that might have precluded their detection in many studies (Sarnelle and Wilson, 2008).

Filtering rates have been found to increase with increasing body length (e.g., Burns and Rigler, 1967; Mourelatos and Lacroix, 1990; Peters and Downing, 1984). Brooks and Dodson (1965) emphasized that the filtering rate should be proportional to the square of the body length in filter-feeding Cladocera, though there is in fact variability in the scaling of this relationship (e.g., Burns, 1969a). Predation rates of copepods have also been found to vary with body size (Burns and Gilbert, 1993; Peters and Downing, 1984; Russell et al., 2021).

Filtering rate is also temperature sensitive and, over certain temperature ranges, rates increase with warming (Fig. 19-9). However, temperatures above a given point can result in a decrease in the filtering rate; for example, filtering rates for *D. pulicaria* and *D. ambigua* increased up to 25°C and declined greatly at 30°C (West and Post, 2016). The temperature of maximum filtering rates differs among species; for *D. rosea*, the maximum was found to be about 20°C (Fig. 19-9), for *D. magna* 28°C, and for *D. galeata mendotae* and *D. pulex* 25°C or above (Burns, 1969a; Burns and Rigler, 1967; Geller, 1975; McMahon, 1965). In addition to the effects of temperature, zooplankton feeding can also be affected by physical processes. For example, Serra et al. (2019) demonstrated that *D. magna* clearance rates are altered by turbulence. Up to a critical threshold, clearance rates were found to increase with turbulent dissipation but,

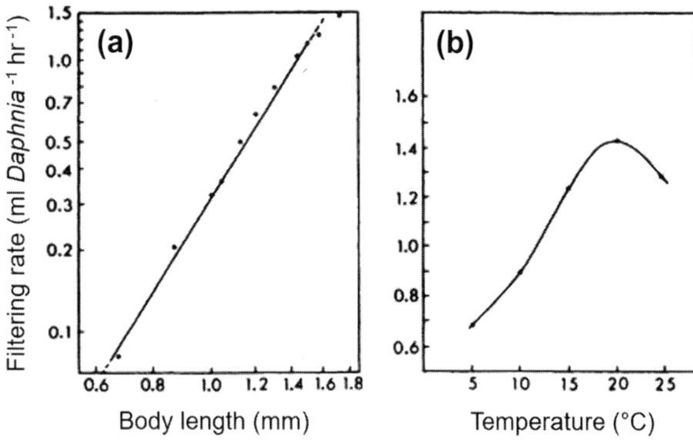

FIGURE 19-9 Relationship between filtering rate and (a) body length at a constant food supply at 20°C, and (b) water temperature (body length 1.65–1.85 mm) of *Daphnia rosea*. *(Modified from Burns and Rigler (1967).)*

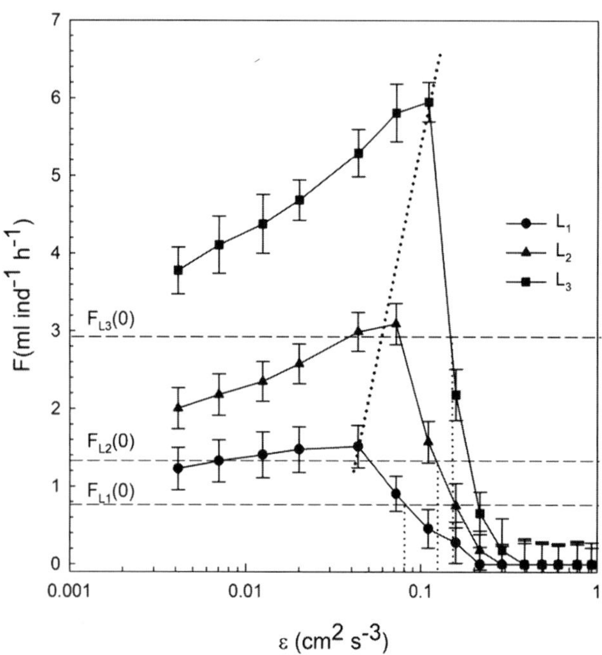

FIGURE 19-10 *D. magna* filtration (F) versus the mean turbulent kinetic energy dissipation rate (ε) for experiments carried out with *D. magna* of three different lengths (L1 = 1.25 mm, L2 = 1.50 mm, and L3 = 1.85 mm). The *horizontal dashed lines* represent the *D. magna* filtration at $\varepsilon = 0$ cm^2 s^{-3} for the three different *D. magna* lengths. The *sloping dashed line* represents the shift to greater ε of the maximum filtration with increasing ε. *(Reproduced from Serra et al. (2019).)*

beyond this, they declined sharply (Fig. 19-10). As a result, turbulence could affect zooplankton population growth and food web interactions.

Since ingestion rates vary among zooplankton species, with body size, food, physical processes, and temperature, *in situ* filtering rates of zooplankton communities show considerable seasonal and spatial variation. Symons et al. (2012), in a study of Canadian subarctic lakes and ponds, estimated that the zooplankton communities of different water bodies could graze between 0 and 13.7% of the total phytoplankton community per day. These community filtering rates were especially high with an increasing biomass of cladocerans, and large *Daphnia* specifically. Seasonal and among-year variations in zooplankton community filtering rates are apparent, for example, with variation in zooplankton community composition in Lake Constance (Tirok and Gaedke, 2006; Fig. 19-11). Shorter-term variation in grazing may also be apparent; filtering rates have been observed to peak during the night, when many cladocerans and copepods migrate to epilimnetic waters (Mourelatos et al., 1989). In natural environments the contribution of each species to the total community clearance rate can be estimated by multiplying individual clearance rates for each species by population density. In Heart Lake (Ontario, Canada) *D. rosea* and *D. galeata* were the most important grazers and together accounted for about 80% of the total annual grazing activity by the larger zooplankton (Haney, 1973). Indeed, cladocerans (especially *Daphnia* species) have often been found to dominate total community grazing (e.g., Crumpton and Wetzel, 1982; Thompson et al., 1982). However, in the absence of substantial populations of large cladocerans, smaller members of the zooplankton (ciliates and rotifers) can make a major contribution to overall grazing rates, especially on small phytoplankton (Lischke et al., 2016; Tirok and Gaedke, 2006; Fig. 19-11).

It should be noted that ingestion does not necessarily result in the complete utilization and assimilation of prey resources by grazers. Algae with durable cell walls (potentially thickened by nutrient limitation and UV stress), gelatinous sheaths, or masses of colonial cells can pass through the zooplankton gut intact and remain viable (Porter, 1976, 1975; Van Donk, 1997; Van Donk et al., 1997). Other species can utilize broken colonies or partially decayed colonies by breaking them into smaller pieces, while intact colonies pass through relatively unaffected. Increases in the growth rates of certain algae after ingestion and passage through grazers may be the result of several interrelated mechanisms. Nutrients may be obtained from the degradation and digestion of other algae; phosphorus, in particular, is probably a major nutrient that is made more available during this process (Porter, 1976).

F. Food quantity and bioenergetics

Several studies have demonstrated the effects of food quantity on zooplankton feeding behavior and bioenergetics. *Daphnia* spp. have been especially well studied and, for these taxa, increases in food concentration can result in numerous such effects, including increased

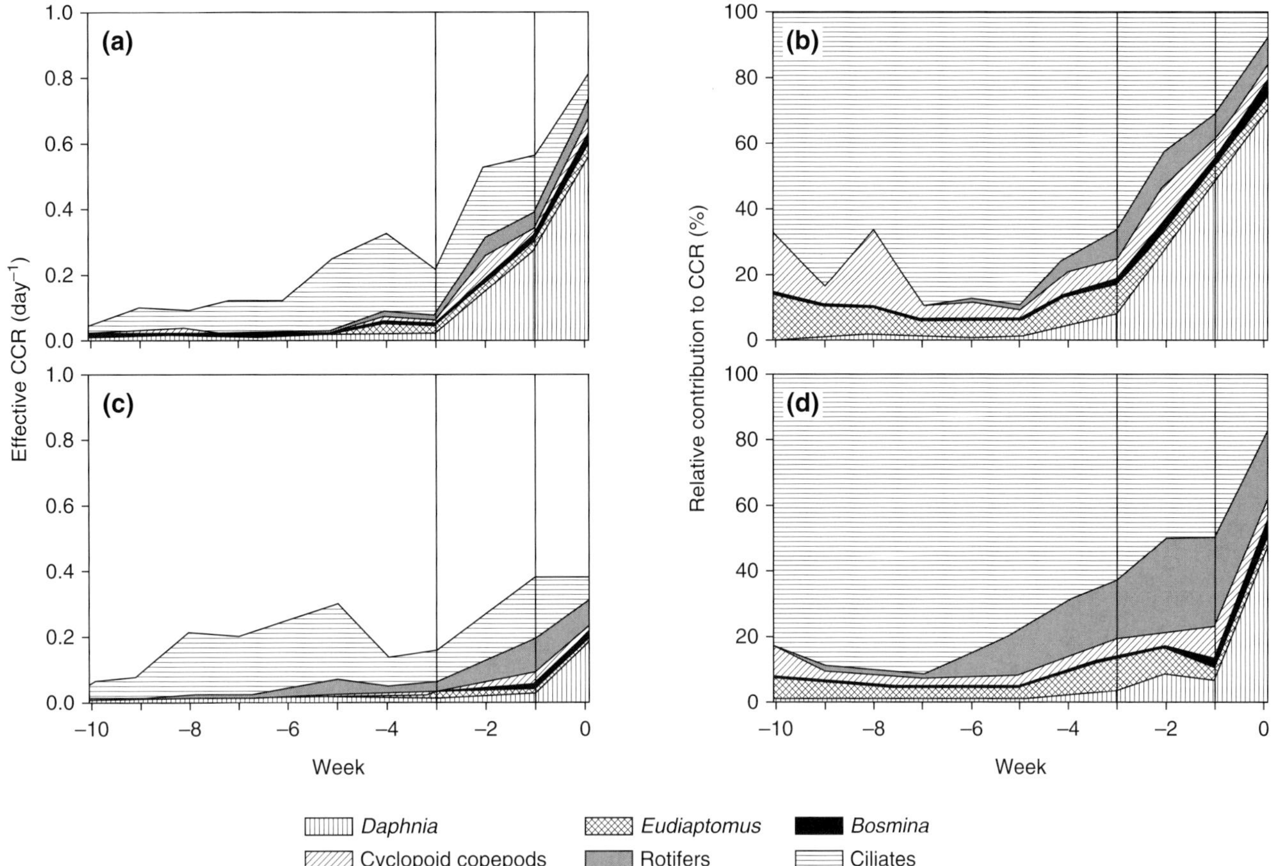

FIGURE 19-11 Calculated mean effective community clearance rate [CCR (day^{-1}), *left*] and relative contribution of the different herbivorous groups to the CCR (*right*) in *Daphnia*-years (a, b) and non-*Daphnia*-years (c, d). Weeks between the vertical lines represent the initiation of the clear-water phase (CWP). In *Daphnia*-years ciliates dominated the CCR until 3 weeks before the CWP and daphnids gained importance during the initiation of the CWP. In non-*Daphnia*-years the CCR was ruled by ciliates with an increasing contribution by rotifers. Daphnids only gained importance in the first week of the CWP (week 0). *(Reproduced from Tirok and Gaedke (2006).)*

assimilation rate (Bohrer and Lampert, 1988), decreased assimilation efficiency (Richman, 1958), reductions in feeding appendage beat rates, and increased rates of food rejection and respiration (Lampert, 1986; Porter et al., 1982). These changes are not necessarily a linear function of food concentration, though, and depend upon whether food concentrations are above or below the incipient limiting level for feeding (see Section IVE); filtering rate decreases and rejection rate increases as food availability increases above this level (Bohrer and Lampert, 1988; Lampert, 1986). Interrelationships among these feeding and bioenergetic parameters allow for the determination of the most energetically costly part of the feeding process: movement of filtering limbs, food rejection via the postabdominal claw, or digestion and biochemical processing of the food (termed "specific dynamic action"). Increased assimilation rates and decreased assimilation efficiencies at higher food concentrations have also been observed for rotifers (Gulati et al., 1987).

Energy acquired from food is allocated to maintaining bodily functions, storage, growth, and reproduction. For some zooplankton, energy allocation among these functions is not fixed and changes in response to environmental conditions (Litchman et al., 2013). For example, studies have investigated the theoretical "rules" that govern these allocations, under conditions of both food sufficiency and starvation. *D. magna* reduce their allocation of energy to growth and reproduction under starvation but, provided the period of starvation is relatively short, they can increase these allocations again when food availability increases and may exhibit "catch-up growth" (Bradley et al., 1991). Energy allocations also vary with predation pressure (Lampert, 1997; see Chapter 20). Energetic trade-offs exist; for example, high investment in current reproduction may come at the cost of future reproduction and survival (Sarma et al., 2002). The optimal energy allocation among competing functions is that which maximizes organism fitness (Litchman et al., 2013).

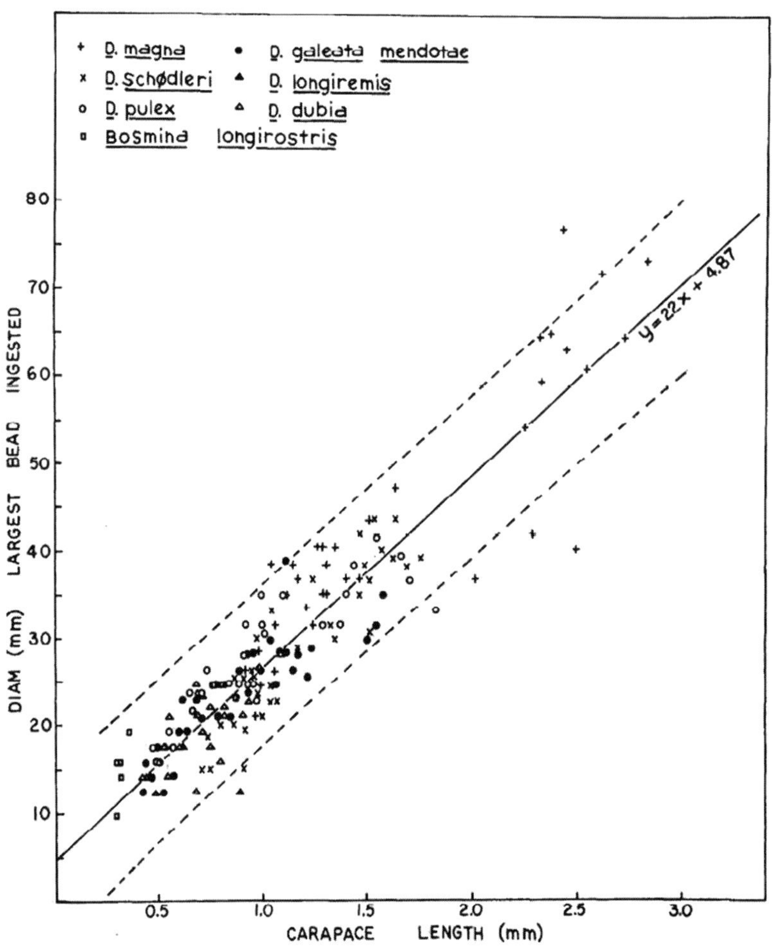

FIGURE 19-12 Relationship between body size and the diameter of the largest particle ingested by seven species of Cladocera. Broken lines equal 95% confidence limits. *(Reproduced from Burns (1968).)*

Given these dependencies, there are individual- and population-level food concentration thresholds for zooplankton. At the level of the individual animal, we can define food thresholds as the amount of food necessary for assimilation to balance metabolic losses, resulting in a zero change in body mass. At the population level, given the dependency of reproduction of food quantity (Section V), thresholds can be defined as the amount of food necessary for reproduction to offset mortality, resulting in zero change in population size. The food concentration dependence of life-history parameters and energy allocation rules have been built into simulation models that capture the dynamics of zooplankton individuals and populations (Rinke and Vijverberg, 2005), which can then be used to project likely population-level consequences of environmental change (Schalau et al., 2008).

In addition to quantity, food quality is of great importance to zooplankton life history, energy allocation, and population dynamics (Zhou et al., 2020). These dependencies are explored in Chapter 20.

G. Feeding selectivity

Zooplankton vary greatly in their degree of feeding selectivity. Though the term "selectivity" implies a degree of active choice among food items, it is important to recognize that variable rates of ingestion upon multiple prey might instead reflect perceptual biases and handling difficulties experienced by zooplankton when confronted by a diverse prey assemblage (Price, 1988).

Some of the apparent selectivity of zooplankton feeding appears to be related to consumer and prey body size. For example, the predation rate of *Cyclops vicinus* is greater for smaller, earlier *instars* (developmental stages) of *Bosmina longirostris* than for larger, later instars (Brandl, 1998). Also, studies of six species of *Daphnia* and the smaller *B. longirostris* using spherical beads within a size range from <1 to 80 μm in diameter (Burns, 1969b, 1968) reveal a positive correlation between increasing grazer body size and increasing size of the largest particle ingested (Fig. 19-12). However, while zooplankton body length may be correlated with the maximal particle size ingested, it has little influence on the ingestion of smaller particles (Bogdan and Gilbert, 1984; Knoechel and Holtby, 1986). Though patterns of feeding behavior and selectivity exist with respect to body size, observations of great variability in filtering rates for phytoplankton species falling within the same size class (Cyr and Curtis, 1999) suggest that much selectivity exists that cannot be predicted by body size or taxonomic group alone.

In addition to size, prey morphology is also important. Some phytoplankton are induced to form colonies, or acquire spines, in the presence of grazing zooplankton (Albini et al., 2019; Hessen and Van Donk, 1993), reducing the efficiency with which they are grazed (Lürling et al., 1997). Large filamentous phytoplankton can also mechanically interfere with the feeding process of cladocerans. Upon exposure to such filaments, *Daphnia* spp. have been shown to reduce the width distance between the margins of the carapace valves and increase the rate with which they reject food through kicks of the postabdominal claw (Gliwicz and Lampert, 1990; Gliwicz and Siedlar, 1980).

Differences in prey swimming behavior may also account for some patterns of feeding selectivity. DeMott and Watson (1991) found that the raptorial cyclopoid copepod *Tropocyclops prasinus* preferred feeding on motile algae, but the suspension-feeding calanoid, *Diaptomus birgei*, did not. By obscuring any chemical gradients that may have led copepods to potential food, the authors also demonstrated that mechanoreception was the dominant mode of prey detection for their study species.

For some zooplankton species, food selection may be based upon taste discrimination between food particles (*chemoreception*). Several copepod species and some smaller-bodied cladocerans and rotifers have been shown to selectively consume plastic spheres coated with phytoplankton exudates, in comparison with unflavored spheres, a tendency that larger cladocerans did not demonstrate (DeMott, 1986; Kerfoot and Kirk, 1991). Evidence also suggests that algae releasing toxic organic compounds are selected against by zooplankton, regardless of the suitability of food size and shape. Certain Cyanobacteria produce and actively release dissolved organic products into the water that reduce thoracic appendage beat rate and filtering rates of Cladocera by 50% or more (Chow-Fraser and Sprules, 1986; Lampert, 1982). The calanoid copepod *Notodiaptomus iheringi*, typical of tropical lowland lakes, has been found to selectively feed upon *Cryptomonas*, when introduced to cocultures containing these prey and the cyanobacterium *Microcystis* (Leitão et al., 2018; Fig. 19-13). It is possible that this selective avoidance of Cyanobacteria even discriminates between strains of the same prey species (Rangel et al., 2020).

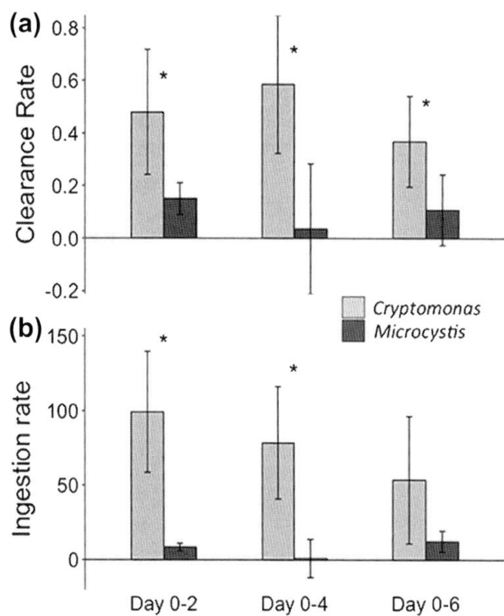

FIGURE 19-13 Mean prey specific rates of (a) clearance (mL copepod^{-1} hour^{-1}) and (b) ingestion (ng copepod^{-1} hour^{-1}) of *Notodiaptomus iheringi* on *Cryptomonas* and *Microcystis*, measured across different time periods (i.e., 0–2, 0–4, 0–6 days). Error bars show the 95% confidence intervals. *Represents a significant difference (i.e., $p < 0.05$) between grazing rates on each prey type for a given time period. *(Reproduced from Leitão et al. (2018).)*

It is important to note that many zooplankton are *omnivorous*, capable of feeding on multiple prey, across trophic levels (Adrian and Schneider-Olt, 1999; Kunzmann et al., 2019; Rao and Kumar, 2002). For example, rotifers (*B. dimidiatus*, *B. plicatilis*) and large ciliates (*Frontonia* sp. and *Condylostoma magnum*) from saline lakes in East Africa consume a range of phytoplankton and small bacterivorous ciliates (Burian et al., 2013). Despite this dietary breadth, feeding rates typically differ among potential prey species. For example, although *E. gracilis* from Lake Constance have been observed to feed upon phytoplankton, heterotrophic flagellates, ciliates, and rotifers, clearance rates were highest for the ciliates (Kunzmann et al., 2019). Similarly, Adrian and Schneider-Olt (1999) found that cladocerans and copepods from Großer Vätersee (Germany) had higher clearance rates for ciliates than for phytoplankton.

V. Reproduction and life histories

A. Protistan growth rates

In the heterotrophic protists both mitotic asexual reproduction (through cell division or budding) and sexual reproduction, involving meiosis, are known. Compared to the metazoan zooplankton, the growth rates of protists are high, and their generation times are short. Nanoflagellates of a size range of ~20–500 μm^{-3} can have generation times of a few hours and reproduce several times per day. Dinoflagellates (size range ~600–50,000 μm^{-3}) rarely have generation times of less than a day, whereas ciliate (size range ~2000–150,000 μm^{-3}) generation times are variable in the range of 0.5 to several times per day. Growth rates are strongly sensitive to environmental conditions. Experimental studies have demonstrated that ciliate growth rates vary with water temperature, peaking at species- and clone-specific optima (Fig. 19-14), though

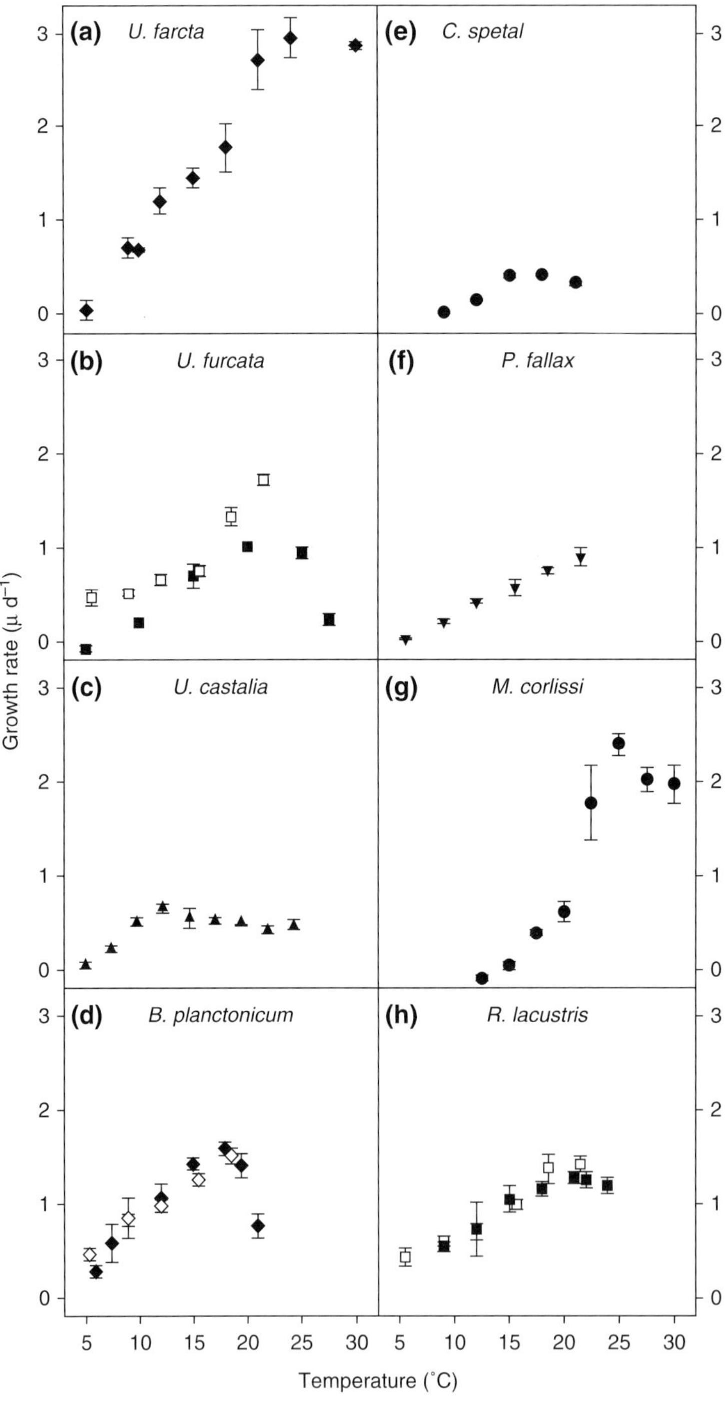

FIGURE 19-14 Population growth rates of prostome (a–e) and oligotrich (f–h) ciliate species. The growth rate μ is the change in population size assuming exponential growth (defined as the difference in log population size over a defined time period). Symbols denote mean values of several replicates, error bars 1 SD. *(Reproduced from Weisse (2006).)*

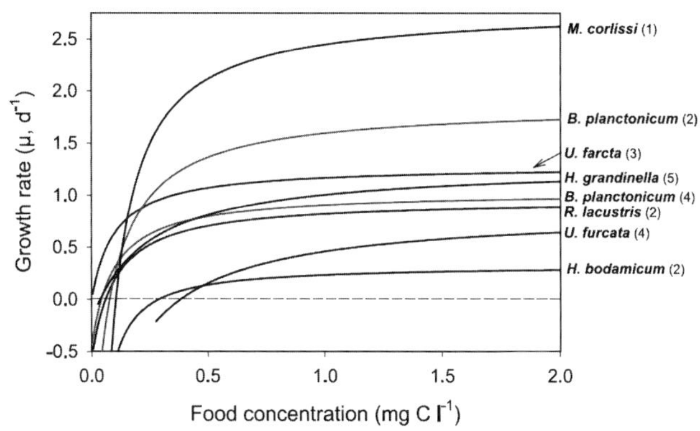

FIGURE 19-15 The relationship between growth rate and food concentration for several freshwater ciliate species. *(Reproduced from Weisse (2006).)*

it appears that many species have wide temperature tolerances (Weisse, 2006). Food concentrations also greatly influence growth rates (Fig. 19-15), with ciliate population growth generally being possible above threshold food concentrations in the range $<0.01-0.1$ mg C L^{-1} (Weisse, 2006). In the natural lake environment protist growth rates will fluctuate greatly over time with changes in temperature, and in the quantity and quality of food resources. Under favorable conditions, heterotrophic protists can achieve population growth rates that are comparable to those of their microbial prey, making it possible for them to impose considerable predation pressure on those prey communities (Finlay and Esteban, 1998a, 2001). Numerous protists have also been observed to form cysts in response to environmental cues (Corliss and Esser, 1974).

B. Rotifers

The life history of typical planktonic rotifers (subclass Monogononta) is termed *cyclic parthenogenesis*. Often many generations pass in which reproduction is parthenogenetic. Under these circumstances, *amictic females*, which are *diploid* (contain paired chromosomes), produce amictic eggs that develop into more amictic females (Fig. 19-16). With an egg-development time of about 1 day under warm, optimal conditions and without the need to encounter males for fertilization, a population of amictic females can develop rapidly in 2–5 days.

However, planktonic rotifers are also capable of sexual reproduction (*mixis*). This strategy is marked by the development of morphologically indistinguishable *mictic* (i.e., sexually reproducing) *females* (Fig. 19-16). The eggs of the mictic females are produced by normal double meiotic division and are thus haploid (containing a single set of chromosomes). If the mictic eggs are not fertilized, they develop rapidly into males. In

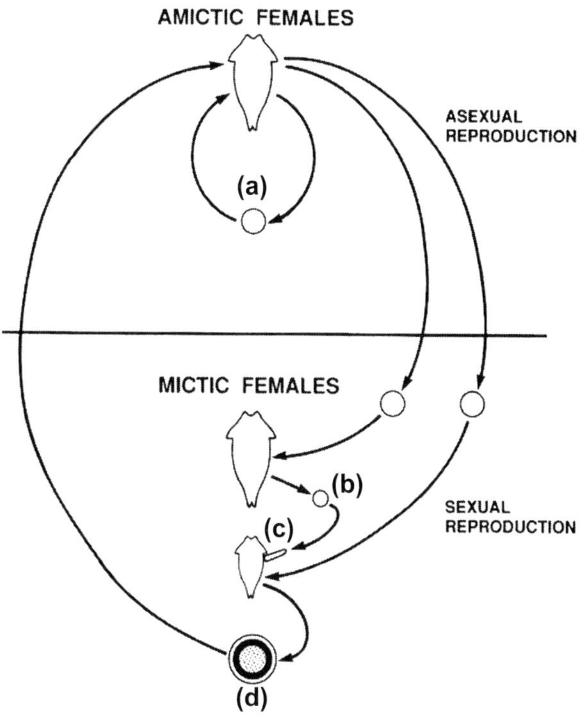

FIGURE 19-16 Typical heterogonic life cycle of monogonont rotifers. Amictic females produce diploid eggs (a) that develop parthenogenetically into females. Mictic females produce haploid eggs (b) that develop parthenogenetically into males (c) or, if fertilized, develop into thick-shelled diapausing embryos called resting eggs (d). *(Reproduced from Gilbert (2003).)*

planktonic species the males are often greatly reduced in size and complexity (i.e., sexual dimorphism; Gilbert, 1993, 2020). The males are short lived and extremely active and are capable of copulation within an hour of hatching. Observations of copulation have shown that insemination occurs when the male pierces the cuticle of the female with its copulatory organ and injects sperm into the body cavity (Gilbert, 1963). If eggs of young mictic females are fertilized, they develop into thick-walled encysted embryos called *resting eggs*. The resting eggs undergo prolonged *diapause* (suspended development) and are highly resistant to adverse environmental conditions. Resting eggs accumulate in the sediments and hatch when their environment changes, often in response to changing temperature, light, and oxygenation (Gilbert, 1995, 2020). The diapause may extend over a period of several weeks or months, though some species can produce resting eggs that hatch over much shorter periods. The resting eggs always result in parthenogenetic amictic females.

The switch from parthenogenesis to sexual reproduction can be triggered by long photoperiods, dietary factors (especially the presence of dietary tocopherol, vitamin E), and crowding (i.e., high population densities), with the latter representing conditions under which the chances of males and females encountering each other would be maximized (Gilbert, 2020, 2003). The propensity of rotifers to switch to sexual reproduction not only varies among species but also among clones and generations and with maternal physiology. Sexual reproduction may peak at specific times of the year or occur throughout the "growing season." Typically, even following appropriate environmental cues, rotifer populations do not switch entirely from asexual to sexual reproduction but instead contain a mix of individuals following each strategy.

In contrast to the life history described for planktonic rotifers, the bdelloids, none of which are planktonic, reproduce primarily by *parthenogenesis*. Clonal reproduction occurs through the mitosis of *oocytes* (immature female sex cells). Though resting eggs are not currently known, bdelloids can enter dormancy upon desiccation, termed *anhydrobiosis* (Ricci, 1998). Despite this life history, bdelloids may still show a high degree of speciation through adaptation to different niches (Birky et al., 2005) and genetic variability through the incorporation of external, nonmetazoan DNA into their genome via horizontal gene transfer (Eyres et al., 2015; Hecox-Lea and Welch, 2018).

Rotifer life history characteristics are strongly sensitive to environmental conditions, such as water temperature and the quality and quantity of available food resources. Indeed, these factors are of major importance in determining seasonal fluctuations in populations, through changes in the balance between reproduction and mortality. In addition to its effects on the rate of egg development, temperature influences the rates of biochemical reactions, feeding, movement, longevity, and fecundity. For example, increased temperatures may increase population growth rate, reduce longevity, and cause an age-specific compression of fecundity (Wenjie et al., 2019; Paraskevopoulou et al., 2020; Yin and Zhao, 2008; Fig. 19-17), as well as affecting the trade-off between egg size and number. Species are variable in their responses to temperature; some are very restricted *stenotherms* (tolerant of a narrow temperature range), while others are *eurythermal* (tolerant of a wide range of temperatures).

A number of studies under both natural and laboratory conditions have shown that food type and quality exert a major influence on the population dynamics of rotifers. For example, peak population size and population growth rate increased with increasing food concentrations (ranging from 0.025×10^6 to 1.6×10^6 *Chlorella* cells mL^{-1}) for *B. angularis, B. calyciflorus, B. patulus, Euchlanis dilatata,* and *Lepadella patella* (Nandini et al., 2007). Food limitation can also result in reduced body and amictic egg size, along with an elongation of the juvenile, prereproductive period (Duncan, 1989). In reality, rotifer life history is likely to be colimited by both food quantity and quality. For example, the population growth, longevity, and fecundity of *B. plicatilis* vary when feeding on different phytoplankton species, representing a range of particle sizes (Yin and Zhao, 2008). *B. calyciflorus sensu stricto* and *B. fernandoi* feeding on cholesterol-supplemented *Synechococcus elongatus* achieve higher population growth than when the cholesterol is absent (Schälicke et al., 2019; Fig. 19-18).

Cross-generational (maternal) effects occur, whereby environmental conditions experienced by the mother during *oogenesis* (egg development) alter the phenotype of their daughters. For example, exposure of mother *B. calyciflorus* to predator (*A. sieboldi*) cues induces protective spine formation in daughter individuals (Gilbert, 1966). It is important to recognize that life history

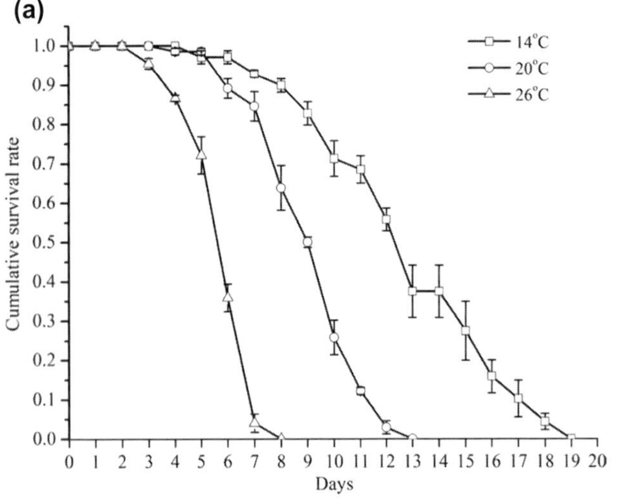

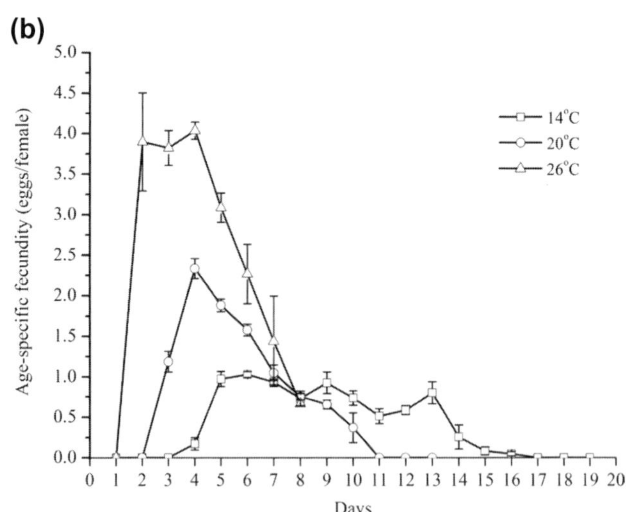

FIGURE 19-17 (a) Effects of temperature on the age-specific survival rate of *E. dilatata*. (b) Effects of temperature on age-specific fecundity of *E. dilatata*. Mean ± SE is shown. *(Reproduced from Wenjie et al. (2019).)*

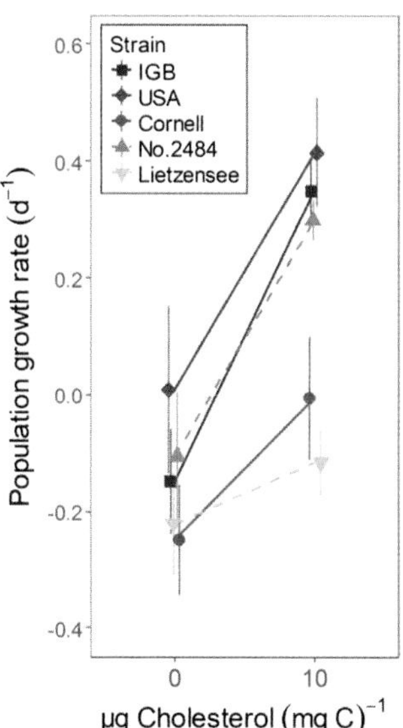

FIGURE 19-18 Population growth rates of five strains of the two rotifer species *Brachionus calyciflorus sensu stricto.* (*red symbols* and *solid lines*) and *Brachionus fernandoi* (*cyan symbols* and *dashed lines*) feeding on sterol-free *Synechococcus elongatus* (3.2 mg C L^{-1}) supplemented with liposomes with cholesterol content (10 μg cholesterol (mg C)$^{-1}$) or without. Values represent means ± standard deviation, number of replicates N = 5. *(Reproduced from Schälicke et al. (2019).)*

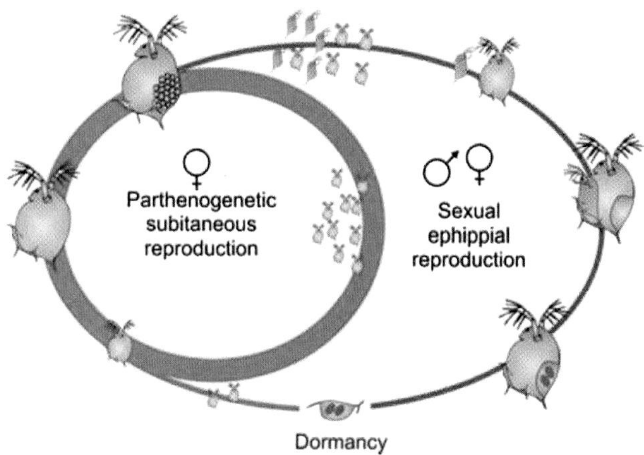

FIGURE 19-19 The cyclically parthenogenetic life cycle of *Daphnia*. During favorable conditions, parthenogenetic reproduction takes place for one to several generations (*green*). Sexual reproduction (*red*) results in the production of long-lived dormant eggs, which can hatch once environmental conditions become favorable again. Some taxa have omitted males from the cycle and produce dormant eggs asexually. *(Reproduced from Decaestecker et al. (2009).)*

responses to environmental conditions vary not only among clearly recognizable species but also among cryptic species, and different genetic strains of the same species (Paraskevopoulou et al., 2020; Schälicke et al., 2019; Yin and Zhao, 2008).

C. Cladocera

Somewhat analogous to that of the rotifers, reproduction in the cladocerans is typically parthenogenetic during a large part of the year (Fig. 19-19). Until interrupted by sexual reproduction, females produce eggs that develop into more parthenogenetic females. The number of eggs produced per clutch varies from 2 or more in the Chydoridae to as many as 40 in the larger Daphniidae. The eggs are deposited into a *brood chamber* or *pouch*, a cavity dorsal to the body bounded by the valves of the carapace. The eggs develop and hatch as a small form of the parent in the brood pouch. As a result, in contrast to the copepods, there are, with the exception of *Leptodora*, no free-living larval forms among the Cladocera. One clutch of eggs is normally released into the brood pouch during each adult instar.

The number of molts is variable. Four preadult instars are common within a range of two (*Bosmina*, *Chydorus*) to about eight (some *Daphnia*). In adult cladocerans the number of molts is more variable and can range from a few to well over 20 instars. Each instar ends with the release of young from the brood pouch, molting, rapid increase in size, and deposition of a new clutch of eggs into the pouch. Therefore an individual can produce several hundred progenies in a lifespan under favorable growing conditions.

Parthenogenesis continues until unfavorable conditions arise, whether they are physical, such as temperature reductions, drying, or short day-length, or biotic, that is, induced through crowding by competition for low food supply or a decrease in the quality of food organisms (Carvalho and Hughes, 1983; Decaestecker et al., 2009; Koch et al., 2009). As the production of parthenogenetic eggs declines, some develop into males, which are smaller and only slightly modified in morphology from females (larger antennules and clasping modifications of the first leg, for copulation). Some females then produce sexual (haploid) eggs, as in the rotifers. Following copulation and fertilization, the carapace surrounding the brood pouch thickens and encloses the egg(s), forming a semielliptical, saddle-shaped *ephippium* (plural *ephippia*, Fig. 19-20). During the subsequent molt, the ephippium is either shed, as in the daphnids, or remains attached, as in the chydorids. If females carrying resting eggs are not fertilized, the resting eggs are resorbed. One species, the Arctic *D. middendorffiana*, is known to be able to produce ephippial eggs parthenogenetically (Zaffagnini and Sabelli, 1972). It is also apparent that the degree of asexuality versus sexuality in the *Daphnia* life cycle varies markedly among species and populations (Decaestecker et al., 2009). Genetic analyses of pelagic *Daphnia* and ephippia have revealed an additional complexity; extensive hybridization can occur among species, as well as

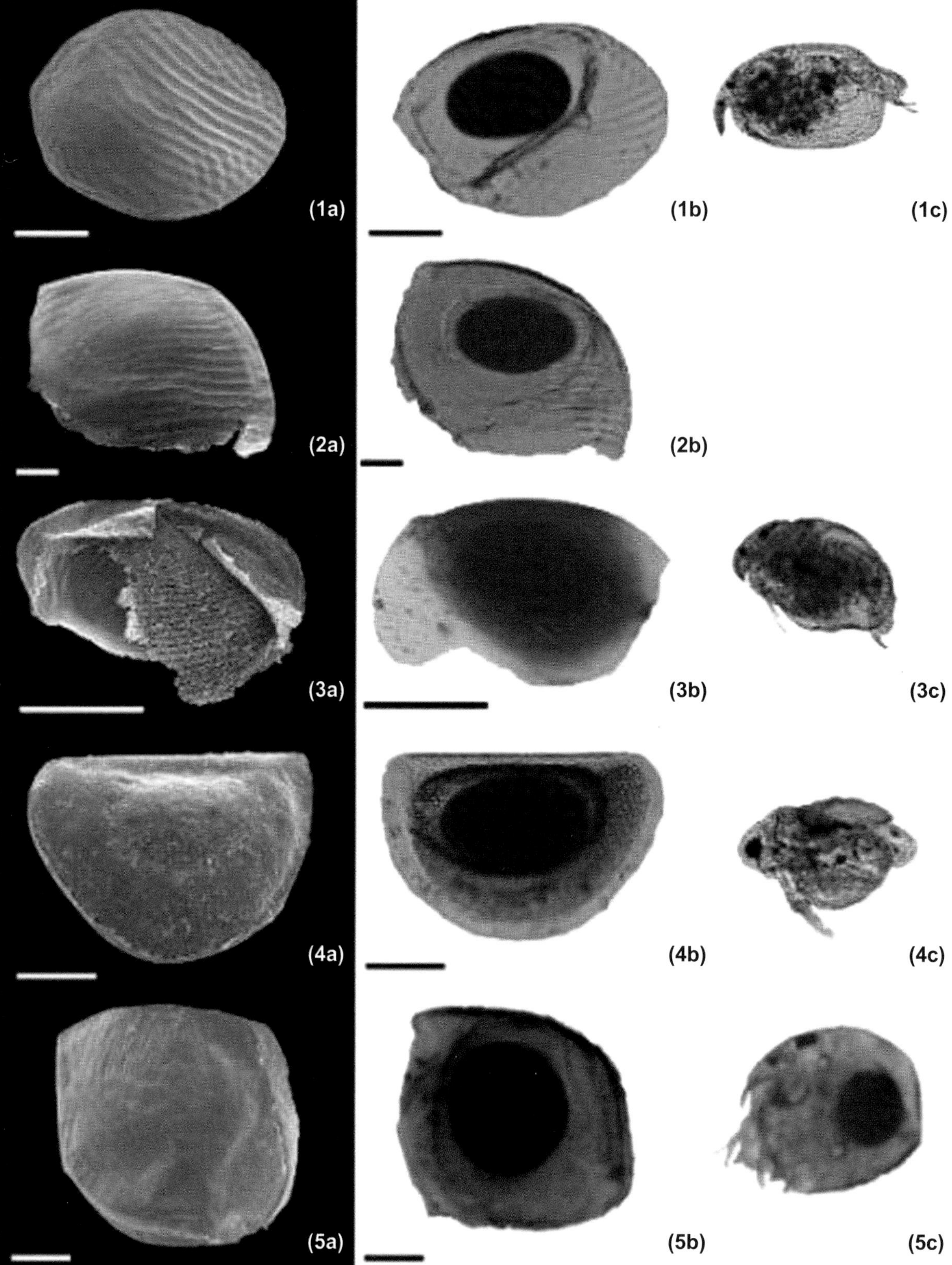

FIGURE 19-20 Micrographs of some Cladocera species and their ephippial ornamentation taken via scanning electron microscope (SEM) (a), light microscope (b), and organism hatched from the egg (c). 1. *Alona quadrangularis*, 2. *Alona rectangula*, 3. *Alona* sp., 4. *Ceriodaphnia pulchella*, 5. *Chydorus sphaericus*. Scale bar 100 μm. **Pictures of organisms extracted from water column. (Modified from Guerrero-Jiménez et al., 2020.)*

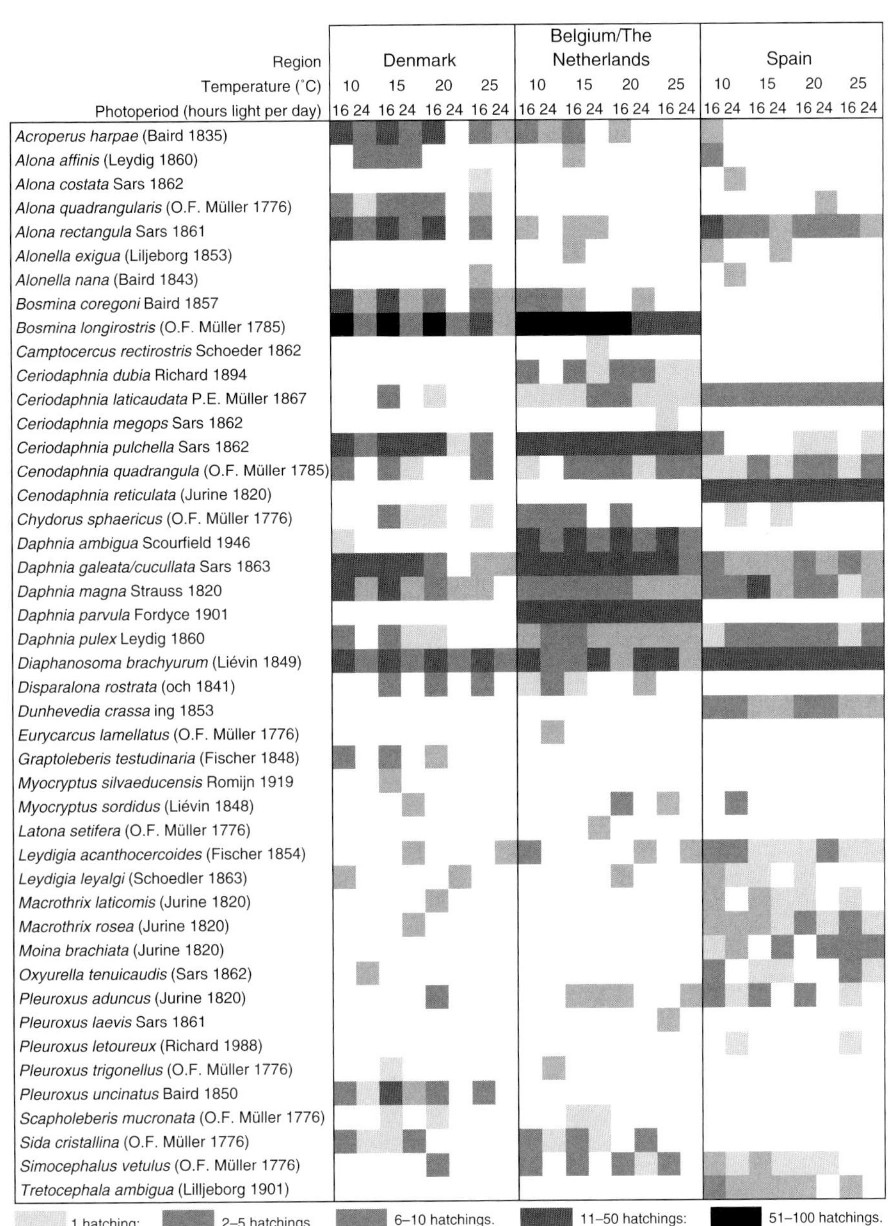

FIGURE 19-21 Categorical number of hatchlings obtained for different temperatures, photoperiods, and regions for several cladoceran species hatched under experimental conditions. *(Reproduced from Vandekerkhove et al. (2005).)*

backcrossing (mating between hybrids and parental species) (Keller and Spaak, 2004).

The ephippia either sink or float and can withstand severe conditions such as freezing or drying. It is not unusual to find large accumulations of ephippia along windward shorelines. It is easy to envision the entanglement and transport of ephippia by birds to other water bodies. Ephippia always hatch into females and in response to specific environmental cues. Vandekerkhove et al. (2005) showed that 45 species of Cladocera gathered from across Europe hatched from ephippia under different combinations of photoperiod and temperature (Fig. 19-21). Cladoceran ephippia may remain viable in sediments for long periods of time, even in excess of 100 years (Cáceres, 1998), providing a means of "temporal escape" from unfavorable conditions.

Water temperature is highly influential for cladoceran life cycles. Longevity, generation time, (post)embryonic development time, age/size at maturity, and time between molts are approximately inversely related to temperature (Brans et al., 2017; Brans and De Meester, 2018; Gillooly, 2000; Pajk et al., 2018; Weetman and Atkinson, 2004). Interestingly, the shift toward faster growth and earlier maturity, at smaller body size, can be viewed as

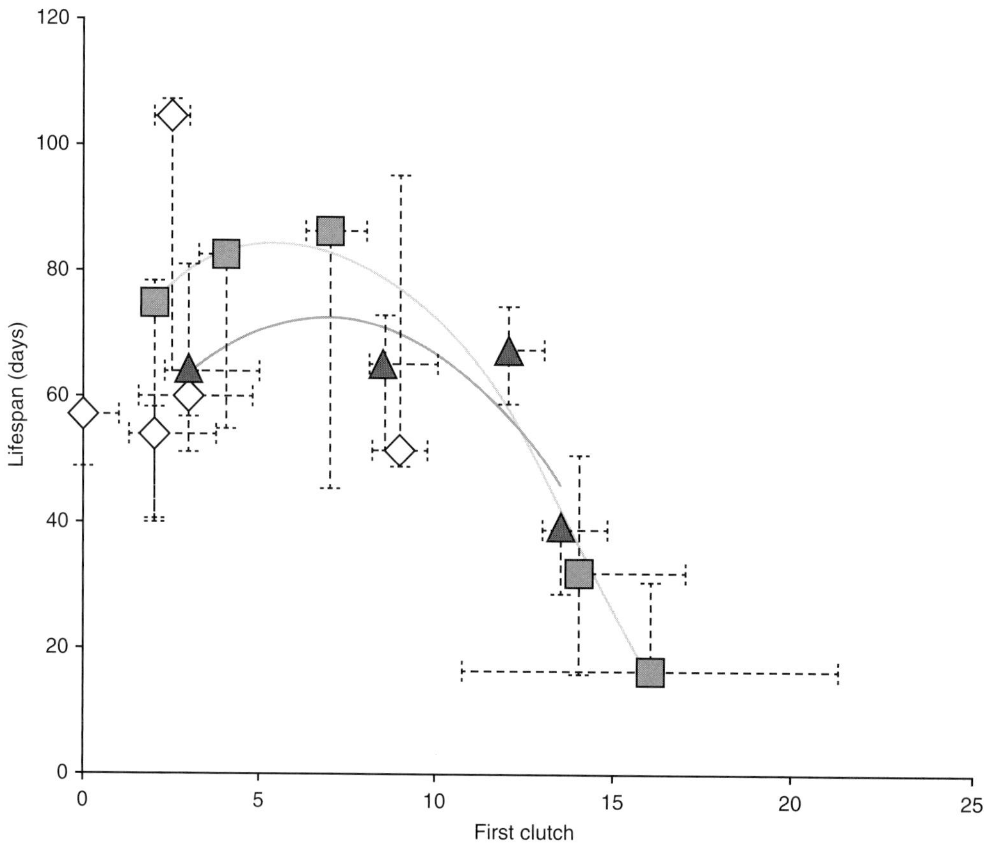

FIGURE 19-22 Median lifespan plotted against median number of offspring in the first clutch, for *Daphnia magna*. Clone D1: *white diamonds*, clone B1: *gray squares*, clone D2: *black triangles*. *(Reproduced from Pietrzak et al. (2010).)*

an increase in the "pace of life"; *common garden experiments* (those that expose genetically distinct populations to the same environmental conditions), involving the exposure of urban and rural *D. magna* populations to different temperatures, have indicated that both plasticity and evolution can underpin such life history variation in the natural environment (Brans and De Meester, 2018). High temperatures can impose physiological stress on cladocerans but there is evidence that, once again, thermal tolerances can be subject to adaptation (Brans et al., 2017). Indeed, Pajk et al. (2018) demonstrate that tropical and subtropical *Diaphanosoma* have higher thermal optima and tolerances than temperate Cladocera, especially *Daphnia*.

Food availability has several effects on life history parameters. Food limitation may result in delayed maturity, reductions in body and clutch size, and population growth (Hakima et al., 2013; Kim and Yan, 2013; Rose et al., 2000). However, under food limitation, cladocerans may "invest" more per offspring, resulting in smaller clutches of individually larger eggs and therefore neonates, which are then more likely to survive food shortage (Guisande and Gliwicz, 1992). Longevity commonly increases with a decrease in food consumption (Kim and Yan, 2013; Pietrzak et al., 2010; Rose et al., 2000), short of starvation, which results in rapid death. Indeed, reduced lifespan under high-food conditions may represent a trade-off with enhanced early reproduction under such conditions (Fig. 19-22). Food quality is also an important determinant of growth and reproduction (Masclaux et al., 2009), a topic covered in more detail in Chapter 20.

Life history responses to environmental conditions vary among clones of any given species (Pajk et al., 2018; Pietrzak et al., 2010). In addition, there is evidence that the life history characteristics of cladocerans are affected not only by the immediate environment but also by the conditions experienced by the previous generation (i.e., maternal effects, Alekseev and Lampert, 2004).

D. Copepods

Reproduction in free-living copepods is similar between taxa, in spite of widely varying species differences in sexual behavior and periodicity of breeding. Some species reproduce throughout the year, others only briefly at specific times of the year. Copepod copulation occurs by a male briefly clasping the female with the transfer of *spermatophores* (packages of sperm) to the ventral side of the female genital segment. Sexual dimorphism in copepods reflects this active role of the male in mate location, capture, and spermatophore transfer. Fertilization may take place immediately or several months after copulation. Fertilized eggs are carried by the female in one (calanoid copepods) or two (cyclopoid copepods) egg sacs. In calanoid copepods eggs are differentiated into *subitaneous* (developing without dormancy) and *resting eggs*, both being fertilized in many species. Resting eggs usually are dropped into the sediments, where they undergo a period of diapause. Clutch size varies greatly in copepods; for example, Maier (1994) reports between 12 and 92 eggs per clutch for European cyclopoid species.

Food availability directly influences the clutch size and other life-history parameters, with shortages leading to smaller clutches, lower egg production rates, smaller body size, and potentially delayed development (Ban, 1994; Hart, 1991; Hart and Bychek, 2011; Tordesillas et al., 2018). In addition to food quantity, copepod life history is also greatly affected by dietary composition, namely, which phytoplankton are consumed, and whether animal prey feature in the diet (Hopp et al., 1997; Nandini and Sarma, 2007; Smyly, 1970). Increases in water temperature can increase the rates of egg and postembryonic development and somatic and population growth and lead to reductions in body size (Ban, 1994; Elmore, 1983; Hart and Bychek, 2011; Herzig, 1983; Liu et al., 2014; Maier, 1989; Smyly, 1974). For example, the time from hatching to reaching the adult stage for *Eodiaptomus japonicus* from Lake Biwa (Japan) dropped from 67.9 to 15.1 days when the temperature increased from 10 to 25°C (Liu et al., 2014). There exist correlations among the life history characteristics of copepods, for example, single and among-species comparisons suggest that egg size, clutch size, lifespan, and extent of sexual size dimorphism have all been found to increase with body size (Maier, 1994).

Copepod eggs hatch into small, free-swimming larvae, termed *nauplii* (Fig. 19-6f), and then develop by molting through a number of subsequent larval stages. The initial nauplius has three pairs of reduced appendages (first and second antennae, mandibles). The successive naupliar stages, six in all, feed, grow, molt, and acquire further appendages. After six naupliar molts, the next molt results in an enlarged and more elongated form, the first *copepodite* instar (morphologically similar to the adult stage but smaller and not yet sexually mature). There are five copepodite stages, during which additional appendages and body segments develop. The sixth and final copepodite stage is the adult.

VI. Trait-based approaches to zooplankton communities

Zooplankton can be classified by their behavioral, morphological, phenological, or physiological characteristics or *functional traits* rather than just their taxon name, based largely on morphological classification. This trait-based approach to understanding zooplankton communities is growing and yielding new insights. Specifically, it allows a comparison of key life history and demographic features, as well as zooplankton community responses to, and effects on, local ecosystems. This is even the case among regions where the list of taxonomic species might differ substantially; it provides a common "currency" and niche-focused approach enabling generalization regarding factors affecting community composition in freshwaters. Functional approaches are especially developed for the crustacean zooplankton, for which a great deal of information on such traits exists from previous studies (Barnett et al., 2007; Hébert et al., 2017, 2016; Hébert and Beisner, 2020; Litchman et al., 2013; Table 19-1). However, functional approaches are

TABLE 19-1 Integrative framework linking traits to levels of organization and ecological parameters of interest. Some traits may be related to specific ecological parameters and thus operate at particular levels of organization ("mono-type" trait); e.g., response traits such as temperature optima or stress tolerance dictating community processes along environmental gradients. Other, more versatile traits may encapsulate information regarding several or all levels of organization (e.g., body size, growth; *)

Level of organization and application	Ecological parameters of interest	Commonly used trait type	Examples of traits for metazooplankton
Ecosystem	Ecosystem functioning (properties and processes)	Effect traits	Body size*, growth*, respiration, excretion, clearance rates
Community	Community structure and assembly processes	Response	Body size*, growth*, temperature optima, stress tolerance (e.g., starvation, hypoxia)
Population	Population dynamics	Demographic	Body size*, growth*, generation time, dispersal
Individual	Individual fitness and performance	Life-history	Body size*, growth*, fecundity, individual performance- or survival-related traits

Adapted from Hébert et al. (2017).

also being developed for other members of the zooplankton. Obertegger et al. (2011) propose that rotifers may be classified into two distinct feeding groups, based upon *trophi* (jaw) morphology and feeding mechanisms: small-particle feeding microphagous taxa and raptorial taxa that capture larger individual prey. The biomass ratio between the groups can then be used as a functional index of community structure. For ciliates, functional approaches are also possible (Weisse, 2017; Zingel and Nõges, 2010), although less well established than for the planktonic crustacea.

VII. Seasonal change and succession in zooplankton communities

A. Heterotrophic protists

Patterns of seasonal succession for heterotrophic protists are complex, characterized by high rates of community turnover and the presence, at any single point in time, of relatively few dominant species, and many more rare ones (Andrushchyshyn et al., 2003; Weisse, 2014). Despite this complexity, annually recurring successional patterns do occur (Zingel and Nõges, 2010). These have been especially well studied in Lake Constance, where total ciliate biomass follows a bimodal seasonal pattern comprising a rapid increase in algivorous taxa with the spring phytoplankton bloom, a decline during the clear water phase, and the subsequent development of a more diverse summer-autumn community (Gaedke and Wickham, 2004; Müller et al., 1991). Heterotrophic nanoflagellates also show strong seasonal variation in abundance, with community turnover occurring from the timing of the spring phytoplankton bloom, through the clear water phase, and into the summer (Cleven and Weisse, 2001; Sonntag et al., 2006). Heterotrophic nanoflagellate abundances have also been observed to increase from the winter mixing period to the summer stratified period (Mukherjee et al., 2017). Recurrent seasonal patterns are not universal. For example, in Lake Tanganyika (East Africa) heterotrophic nanoflagellates and ciliates did not show strong seasonality (De Wever et al., 2007).

Successional patterns arise in response to the complex interplay of changing temperature, food availability, and grazing (Beaver and Crisman, 1989; Sommer et al., 2012). Seasonal variations in ciliate and nanoflagellate abundance and biomass have been found to correlate with fluctuations in suitable food resources, such as bacteria, cryptomonads, and picophytoplankton (James et al., 1995; Mukherjee et al., 2017; Posch et al., 2015; Tirok and Gaedke, 2007; Zingel and Nõges, 2010), implying bottom-up control of their populations. The short generation times of ciliates allow their populations to build rapidly with increases in food availability; population development may lag behind that of their prey by only a week even in cold conditions (Tirok and Gaedke, 2007). Furthermore, based upon data from 20 Florida lakes (USA), Beaver and Crisman (1990) suggested that the seasonality of ciliate abundance and biomass may vary systematically with lake trophic state and thus likely feeding conditions. Predation also shapes protist seasonal succession, contributing to seasonal biomass declines (Tirok and Gaedke, 2007). Importantly, the relative importance of food availability and grazing losses varies over time and among ciliate groups (Gaedke and Wickham, 2004).

Protists have generally not been effectively monitored by "traditional" methods used for phytoplankton and metazoan zooplankton. However, molecular methods are now transforming knowledge of protist seasonal succession, allowing resolution of the community structure of organisms that are difficult to identify microscopically and detection of rare or cryptic species. In a 2-year study of small shallow ponds in France, Simon et al. (2015) compared the dissimilarity of small (5 μm) protist communities collected across different months to reveal clear patterns of seasonality in community structure. Using high-throughput sequencing of ciliate samples from Lake Zurich (Switzerland), Pitsch et al. (2019) showed seasonal changes in the population density of rare and abundant species and successional patterns (Fig. 19-23), while Forster et al. (2021) identified distinct cold- and warm-season protist assemblages.

B. Metazoan zooplankton: estimating birth rates and population growth

For metazoan zooplankton, analyses of demographic parameters produce useful insights into seasonal population change and hence community succession. Such investigations typically quantify egg production (including clutch size and age at first clutch deposition), seasonal growth rates, mortality, the abundance and duration of each instar, and the interval between generations. Instar analyses are easier with copepods than with cladocerans because the naupliar and immature and adult copepodite stages can be distinguished morphologically. Among the cladocerans, and to some extent the rotifers, instars can sometimes be evaluated on the basis of size. To be informative, the time interval between samplings must be shorter than the developmental periods of the organisms being investigated. Assuming stable egg distribution, the number of eggs present in a sample of a zooplankton population at some instant in time represents the increment that would have been added to the population over a period of time equal to the duration of development of the eggs (time from laying to hatching). The population would increase by the number hatched during the time period if no deaths occurred.

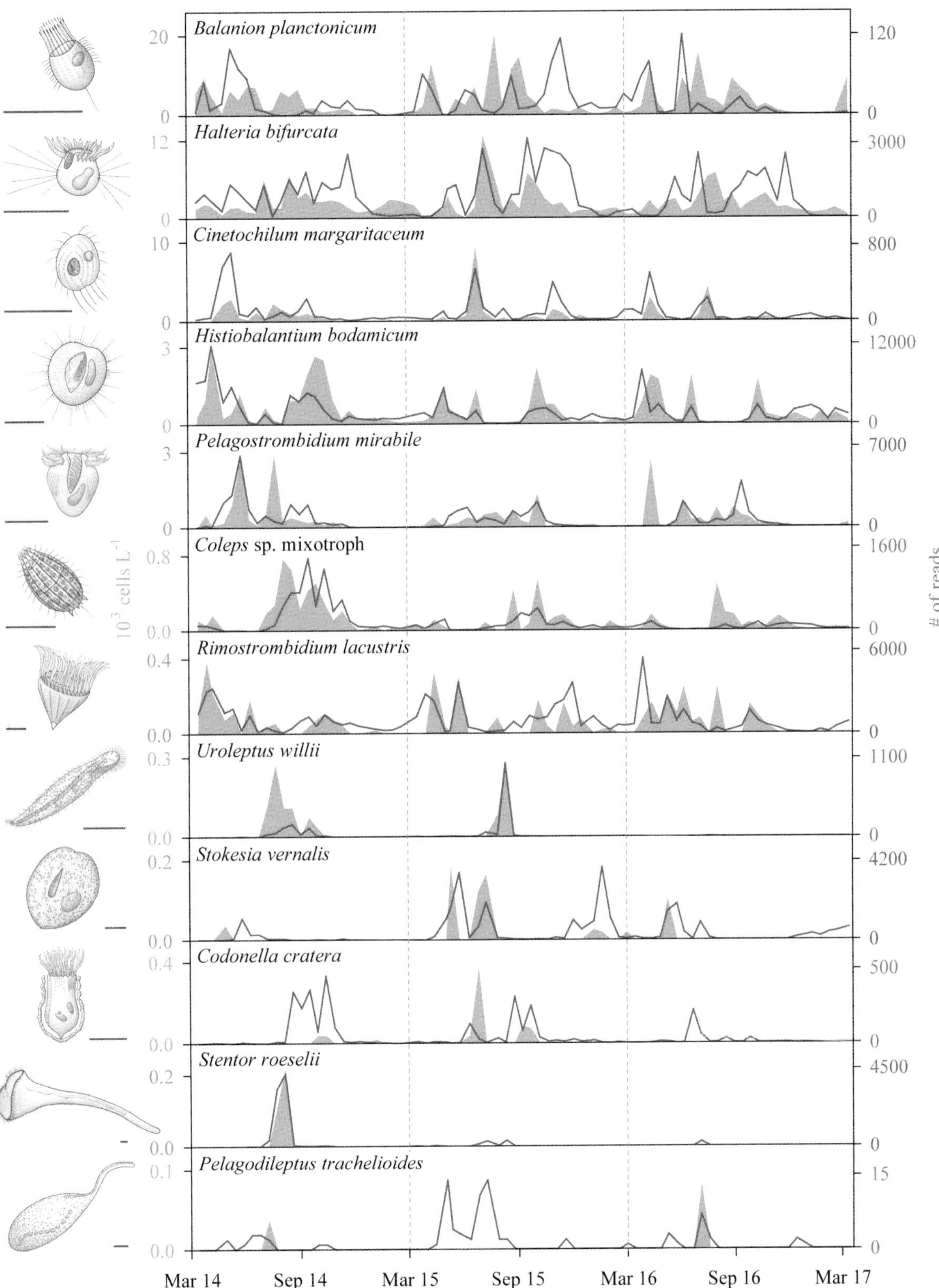

FIGURE 19-23 Seasonal successions of 12 selected ciliate species determined by morphospecies counting of silver-impregnated specimens (areas) and high-throughput sequencing (HTS) of the V9 regions of the 18S rRNA gene (*lines*) over 3 years. The order of species reflects their average abundance based on counting, with the most abundant species on top and the rarest representative on the bottom. Scale bars: 40 μm. *(Reproduced from Pitsch et al. (2019).)*

Birth and death rates may be estimated from reproductive rates and rates of population change. The reproductive rate can be determined indirectly from the ratio of eggs per female (Edmondson, 1974, 1968, 1965, 1960). The ratio of eggs per female (E) observable in a sample at any given moment can be used to calculate a finite population birth rate (B) as eggs per female per day by using:

$$B = \frac{E}{D} \tag{19-1}$$

when D is the duration of development of the eggs, or the mean time an individual spends in the egg stage. This relationship assumes that eggs present at a given time will be added to the population during the next period of time, D. If the age distribution of the eggs is stable, the fraction of eggs hatched during a day is $1/D$. If the population is changing size, the value of B is exact only if $D = 1$; the bias introduced by longer durations can be estimated (Edmondson, 1968).

From the reproductive rate B, the instantaneous growth rate of the population, r, can be calculated based on the conventional exponential growth model, in which positive or negative growth over a short time interval is:

$$N_t = N_0 e^{rt} \text{ and } r = b - d \tag{19-2}$$

in which

N_0 = the size of the initial population at time zero
N_t = the population size at a later time t
r = the growth rate coefficient or intrinsic rate of growth
d = the instantaneous rate of mortality
b = the instantaneous birth rate (calculated from B)

The effective rate of population increase is the difference between the natural logarithms divided by the time increment:

$$r = \frac{\ln N_t - \ln N_0}{t} \tag{19-3}$$

which is, in essence, the difference between natality (b) and mortality (d) over the time interval (t).

Alternatively, birth rate, estimated via the egg-ratio method, gives the daily increment to the population. For a population growing by the amount of E/D in 1 day with no deaths, its growth rate would be:

$$b = \ln(B + 1) \tag{19-4}$$

Here B is estimated at the beginning of the time interval with the assumption that for each female in the population, there will be $(E/D) + 1 = B + 1$ females one day later. Assuming no mortality, the population would grow in 1 day at a rate of:

$$\ln(B + 1) - \ln(1) = \ln(B + 1) \tag{19-5}$$

Thus, by Eq. (19-4), a finite per capita birth rate, B, can be used to estimate the instantaneous rate, b. The calculations assume that B is small; under such circumstances, both B and the E/D are approximately equal to the instantaneous birth rate b.

As pointed out by Edmondson (1968) and rederived by Paloheimo (1974), for moderately large values of r and bD, the finite per capita birth rate and E/D can diverge considerably (Lynch, 1982). Further, the relationships are most applicable to planktonic animals that carry their eggs until hatching, at which time free-swimming progeny are liberated; that is, the eggs are subjected to the same mortality as the adults. Paloheimo (1974) has shown that:

$$E/D = \left(e^{bD} - 1\right)/D \tag{19-6}$$

when D is the development time of the eggs, which is algebraically equivalent to

$$b = \ln[(C_t / N_t) + 1]/D \tag{19-7}$$

when C_t is the total number of eggs counted at time t, which estimates the instantaneous birth rate b from egg counts C_t, where C_t/N_t is the egg ratio. This formula assumes that steady-state conditions prevail, which is not usually the case. Development time is temperature dependent, and in analyses of natural populations this primary factor must be estimated from experimentally determined changes with temperature.

C. Rotifers

Changes in the seasonal distribution of planktonic rotifer populations are complex and generalizations are difficult to make. Even though many rotifers tolerate a wide range of temperatures (i.e., *eurythermal*), species are distinctly seasonal and of two general types: (a) *cold stenotherms* that develop greatest populations in winter and early spring; and (b) species that develop maxima in summer, often with two or more maxima, especially in late summer. As such, patterns of seasonal occurrence, to an extent, reflect thermal tolerances (May, 1983). In Lakes Takvatn and Lombola (Norway) the seasonal succession of rotifers began under the ice in early May with small populations of *Keratella cochlearis* and *Kellicottia longispina*. Following this, the populations of these species grew over the ice-free period to a peak in August/September. During the summer, these species were accompanied by building populations of *Polyarthra*, *Conochilus unicornis*, and *A. priodonta* (Primicerio and Klemetsen, 1999). Total rotifer abundance therefore peaked during the period of warmest water temperatures. Chemical parameters may also influence seasonal rotifer occurrence, for example, oxygen concentrations (Elliott, 1977).

While physicochemical conditions influence rotifer succession, biotic conditions, such as food availability and predation (invertebrates, larval fish), are also of great importance. Seasonal variations in food resource availability influence rotifer fecundity, population growth, and decline; the extent of food limitation varies both over time and among coexisting species such that fluctuating resources can influence which species have competitive dominance over time, and thus community composition (Cordova et al., 2001; González and Frost, 1992; Pauli, 1990). In Lake Michigan (USA) a spring community dominated by the soft-bodied *Synchaeta* was succeeded by a community dominated by spiny taxa (*Kellicottia*, *Keratella*), colonial forms (*Conochilus*), and taxa with evasive swimming responses (*Polyarthra*); this shift correlated with increasing abundances of predatory cyclopoid copepods (Stemberger and Evans, 1984). In the shallow, eutrophic Funada-ike Pond (Japan), considerable turnover occurred in the rotifer community during only a relatively short period (July–September 1990; Fig. 19-24), when physicochemical conditions were stable but food availability and predation pressure varied (Urabe, 1992). Hampton (2005) shows how abiotic and biotic processes combine to influence succession; in Lake Washington (USA) seasonal peaks of *C. hippocrepis* and *C. unicornis* have diverged over a 33-year period due to a temperature-driven lengthening of the plankton growing season, and changes in the seasonal distribution of food resources, as a result of grazing by a cladoceran competitor.

D. Cladocera

Seasonal population variation in the Cladocera varies among species and within a species living in different lake conditions. Some species are perennial and overwinter at low population densities as adults (parthenogenetic females) rather than as resting eggs. Some perennial species exhibit maxima in surface layers only during colder periods in the spring and in the cooler hypolimnetic and metalimnetic strata during summer stratification. The *aestival* (summertime) species that have a distinct diapause in a resting egg stage commonly develop population maxima in the spring and summer when the water is relatively warm. Although one population maximum is typical, a second population peak often occurs in the autumn.

Though cooccurring cladoceran species may show seasonal separation in their abundance peaks, extensive seasonal overlap is possible. An example can be found in the *D. galeata*, *D. hyalina*, and *D. cucullata* in eutrophic Heiligensee (Germany; Adrian and Deneke, 1996). In this lake the relative dominance of the three *Daphnia* species changed during a period of mild winters, from a dominance of the larger *D. galeata* and *hyalina* to the smaller *D. cucullata*. Therefore, while a seasonal *Daphnia* peak remained a dominant feature of the community succession, there was a change in the identity of the species involved.

The timing and magnitude of cladoceran seasonal peaks are affected directly and indirectly by variations in environmental conditions. Mesocosm experiments show that increases in water temperature bring about earlier population maxima (see Section X), while the magnitude of some seasonal abundance peaks (e.g., for *Daphnia*) increases under higher-light conditions (Winder et al., 2012). The latter effect is indirect, occurring via light-driven increases in phytoplankton carrying capacity. Temporal changes in predation pressure may also have effects on patterns of succession (Gliwicz and Pijanowska, 1989; Tessier, 1986; see Chapter 20).

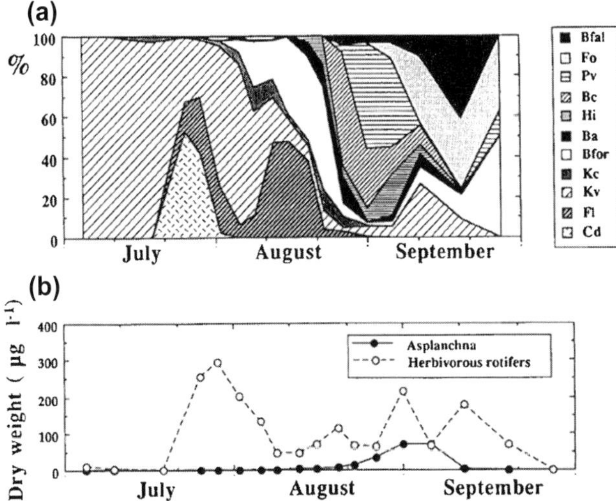

FIGURE 19-24 Temporal changes in the composition and biomass of herbivorous rotifers (a) and biomass of herbivorous rotifers and *A. brightwelli* (b). Samples collected from Funada-ike Pond, Chiba, Japan. *(Reproduced from Urabe (1992).)*

E. Copepods

A number of cyclopoid copepods are known to enter diapause, either at the egg stage or in the copepodite stages, with or without encystment. Consequently, the annual cycle of copepod populations may be interrupted by a diapause that persists from one to several months. In the case of *C. vicinus* in Lake Constance a two-stage dormancy occurs whereby nauplii hatched during the spring phytoplankton bloom develop continuously to the fourth copepodite stage and then enter diapause during the summer period. This diapause terminates late in the subsequent winter. However, at this stage, rates of development through the fifth copepodite stage to adulthood and ultimate reproduction are retarded to a variable degree that increases the chances of the next generation of nauplii hatching during the

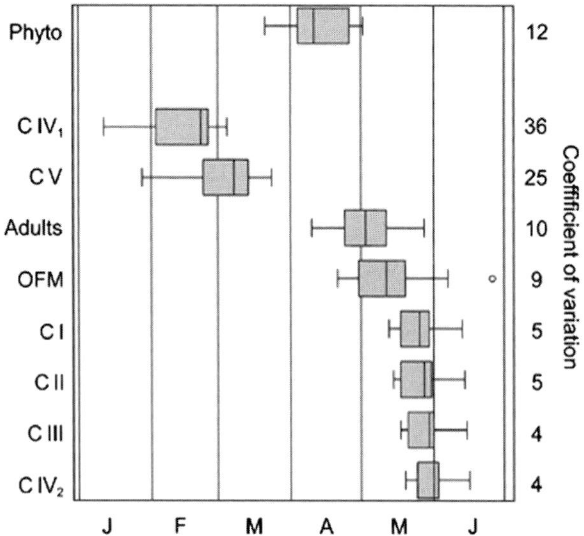

FIGURE 19-25 Timing of the start of the phytoplankton spring bloom and of the maximum abundance of *Cyclops vicinus* stages [copepodite stages (C IV$_1$ and C V), adults, ovigerous females (OFM), and offspring copepodite stages (C I, C II, C III, and C IV$_2$)] across several years in Lake Constance (central Europe), indicated as *box-plots*. Variability of peak timing is shown as coefficients of variation. X-axis indicates months of the year, from January to June. *(Reproduced from Seebens et al. (2009).*

spring phytoplankton bloom, when food resources are abundant (Seebens et al., 2009; Fig. 19-25).

The life cycles of the calanoid copepods are generally somewhat longer than those of the cyclopoids. In the temperate region most species exhibit prolonged reproductive periods with several generations per year that are indistinguishable from one another. Calanoids can show complex responses to changing environmental conditions. In Lake Constance spring abundances of *E. gracilis* declined, while summer abundances increased from 1970 to 1995, which was a period of oligotrophication (Seebens et al., 2007). In copepod populations high fecundity (egg production) does not necessarily translate into high subsequent offspring numbers, due to the effects of variable mortality.

Species typically alternate seasonally in their dominance of the overall copepod community, but these successional patterns can change with environmental conditions. In many central European lakes cyclopoids dominate in spring and calanoids in summer. However, in Lake Geneva (central Europe) in the late 1980s, cyclopoid dominance increased in the summer, at a time when nanophytoplankton food resources also dominated later into the summer (Anneville et al., 2007).

F. Communities

The *Plankton Ecology Group (PEG)* constructed a conceptual model comprising 24 key events that characterize the seasonal succession of phyto- and zooplankton communities (see also Chapter 18). In its original form, most representative of deep temperate lakes (Sommer et al., 1986), the model suggests that zooplankton populations are largely "reset" each winter, increasing in response to the growth of edible phytoplankton in the spring. Smaller zooplankton, with shorter generation times and higher rates of population growth, such as heterotrophic protists and rotifers, increase first, followed by larger cladocerans and copepods (see also Tirok and Gaedke, 2006; Yoshida et al., 2001). It is postulated that the combined grazing of the zooplankton community can drive down concentrations of phytoplankton food resources, ultimately leading to population collapse for the zooplankton themselves, a process that could be accelerated by fish predation in late spring (at least in the case of the larger zooplankton). In the summer a community of smaller-bodied and functionally diverse zooplankton persists, as a result of both size-selective fish predation and the presence of filamentous phytoplankton that interfere with the feeding mechanics of larger species. From autumn through to winter, zooplankton populations decline and enter resting stages, as phytoplankton food resources also decline. The model stresses that the pattern of seasonal zooplankton dynamics depends strongly on lake trophic state.

In an update to the PEG model Sommer et al. (2012) modify this scheme, building upon increasing awareness of the potentially significant "carry-over" effects of overwintering zooplankton on subsequent spring plankton dynamics and on a greater understanding of the ecological role of heterotrophic protists, parasites, food quality, and food web interactions (Chapter 20). A conceptual graphic of the modified PEG model (Fig. 19-26) shows the potentially fundamental alterations to micro- and meta-zooplankton succession that can arise as a result of variation in overwintering populations and fish predation pressure. A global data synthesis has shown that overwintering crustacean zooplankton density is, on average, 25% of summer density, with much variability among lakes and implications for zooplankton dynamics in subsequent seasons (Hampton et al., 2017). Further, changes in climatic conditions can bring about departures from what might be expected under the PEG model (Straile, 2015; see Section X).

The PEG framework is being used to develop an understanding of typical zooplankton seasonality beyond the temperate zone, such as in Arctic and tropical lakes (de Senerpont Domis et al., 2013; Fig. 19-27). Studies of tropical and subtropical lakes show clear evidence of seasonal compositional change; in Feitsui Reservoir (Taiwan) zooplankton community composition showed cyclic within-year variation, and this correlated with

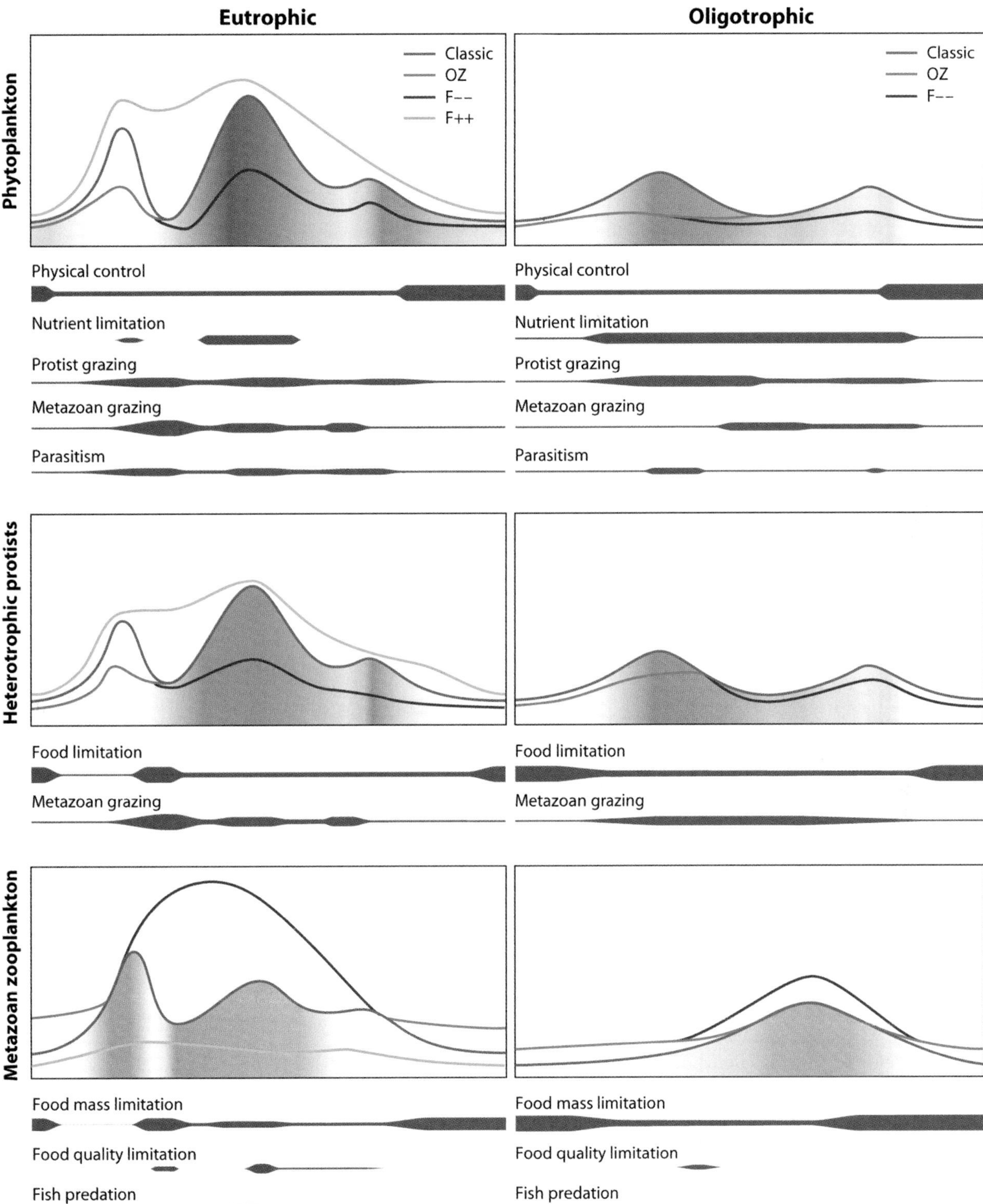

FIGURE 19-26 Seasonal (winter through autumn) biomass patterns of plankton in a eutrophic (*left*) and an oligotrophic (*right*) water body. (*Top*) Phytoplankton, (*middle*) heterotrophic protists, (*bottom*) metazoan plankton. *Gray line*, scenario with moderate fish predation and when overwintering of metazoan plankton is unimportant (the "classic" scenario of the original PEG model); *blue line* (OZ), overwintering zooplankton important; *red line* (F− −), high metazoan density in fishless water bodies; *orange line* (F++), metazoan plankton suppressed by high fish predation. *Shading* indicates the mean vulnerability of phytoplankton and protists against metazoan grazing and of metazoan zooplankton against fish predation in the classic scenario (*light*, low; *dark*, high). The thickness of the horizontal bars indicates the seasonal change in regard to the relative importance of physical factors such as grazing by protists and metazoa, nutrient limitation, fish predation, food mass limitation, and food quality limitation. *(Reproduced from Sommer et al. (2012).)*

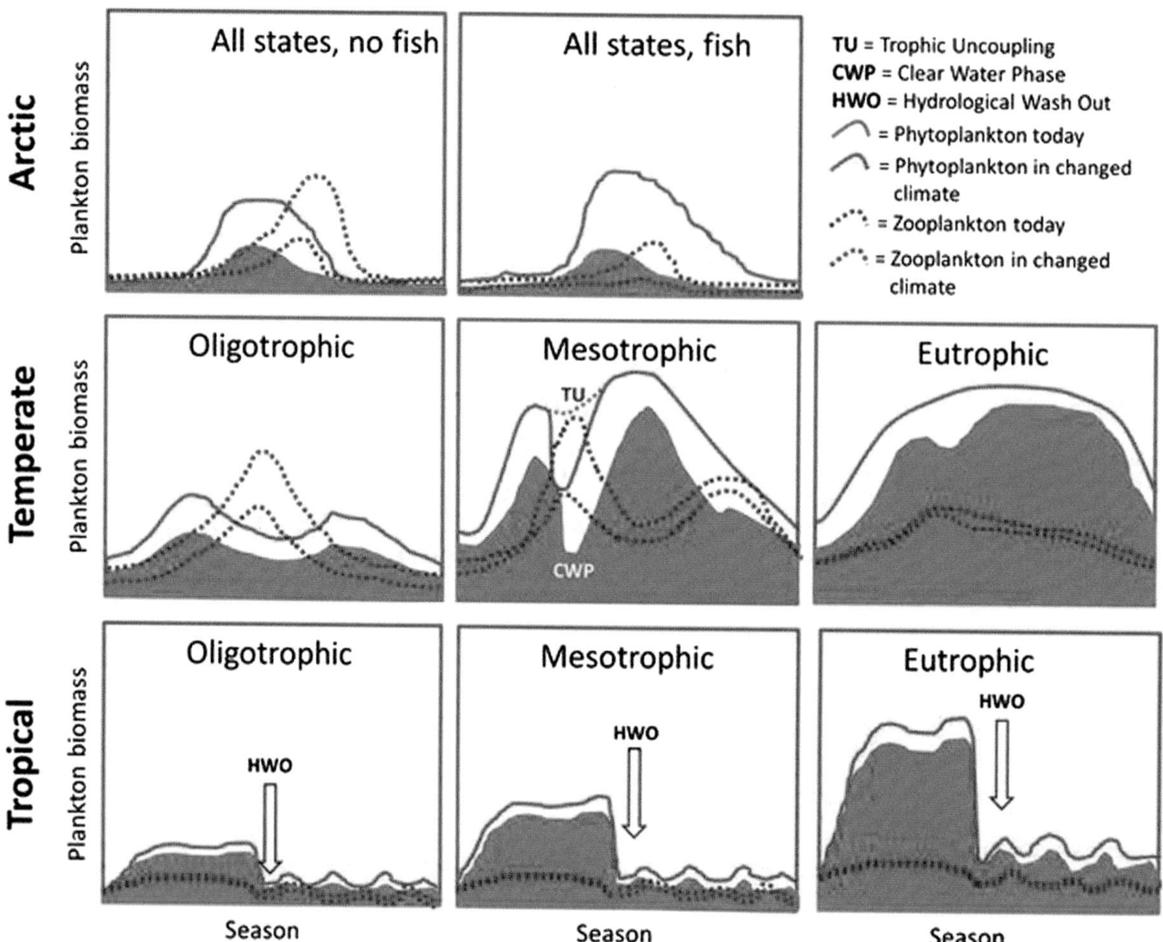

FIGURE 19-27 Conceptual model of the seasonal development of current (*blue area*) and future (*red solid lines*) phytoplankton biomass in polar lakes (*upper panels*), temperate lakes of different trophic states (*middle panels*), and tropical systems (*lower panels*). The generalized expected zooplankton biomass in current and future climate scenarios is indicated by *dark blue* and *brown dotted lines*. *(Reproduced from de Senerpont Domis et al. (2013).)*

changes in physicochemical conditions and food resource composition and availability (Chang et al., 2014). Subtropical and tropical seasonal floodplains and floodplain lakes have distinct zooplankton successional dynamics, linked to variations in water availability. At the Okavango Delta (Botswana), large populations of *Moina micrura*, *Mesocyclops leuckarti*, and *D. laevis* developed during flooding, in response to terrestrial nutrients stimulating the growth of phytoplankton food resources along the flood front (Lindholm and Hessen, 2007). In Ehoma Lake (Nigeria) abundances of cladocerans were higher in the rainy season than in the dry season, and rotifers showed the reverse pattern (Okogwu, 2010).

VIII. Within water body heterogeneity in zooplankton communities

The spatial distribution of zooplankton within freshwaters is generally highly heterogeneous ("patchy"). This heterogeneity is important, not only because it has a bearing on the design of representative sampling programs but also because such patterns can influence ecosystem processes. Based upon a large volume of research, these spatial patterns are known to be hierarchical (or "nested"), with heterogeneity occurring at large (>1 km), coarse (10 m−1 km), fine (1−10 m), and micro (<1 m) scales (Pinel-Alloul, 1995). An important aspect of zooplankton heterogeneity is that the drivers of spatial pattern vary among these scales (the *Multiple Driving Force Hypothesis*; Pinel-Alloul, 1995). While, at large scales, abiotic processes (e.g., water currents, river inflows, basin morphometry, thermal stratification) have a dominant role in driving patchiness, at small scales, biotic processes (e.g., swarming to locate food and mates, or to reduce predation) become more important (Folt and Burns, 1999; Pinel-Alloul, 1995).

Several studies have suggested that water currents influence zooplankton horizontal distribution over large scales, through the passive transport of organisms (Blukacz et al., 2009; George and Winfield, 2000; Thackeray et al., 2004). For example, in Loch Ness (UK) a "conveyor belt" wind-driven circulation pattern results in the downwind accumulation of surface-dwelling *C. strenuus*, and the upwind transport of deeper-dwelling *E. gracilis* by return currents. Over a period of only 72 h, internal wave oscillations triggered variations in the vertical distribution and areal density of crustacean zooplankton in Bautzen Reservoir (Germany; Rinke et al., 2007).

Patches of lower zooplankton abundance can be observed in areas of lakes that have a higher predator abundance, be they vertebrate or invertebrate (George and Winfield, 2000; Lévesque et al., 2010; Masson et al., 2001). However, based upon field data alone, it is not straightforward to determine whether these patterns are driven by predator avoidance behaviors or by the ingestion of zooplankton prey.

Swarming behavior has been observed for several zooplankton species (Colebrook, 1960; Folt and Burns, 1999; George, 1981; Milinski, 1977). For example, free-swimming ostracods (*Heterocypris incongruens*) have been found to swarm around prey during group predation (Rossi et al., 2011). Chemical cues have been implicated in the maintenance of swarms, for example, *Polyphemus* (Wendel and Jüttner, 1997).

There is a temporal element to both patterns in, and drivers of, spatial heterogeneity. For example, the importance of wind-induced currents may vary over time, being a dominant influence only when winds are strong, with biotic drivers assuming more importance under periods of weak wind forcing (Rinke et al., 2009). Even in sheltered lake systems, with limited wind forcing, there can be marked temporal (diel, monthly) changes in zooplankton spatial distribution in both vertical and horizontal planes, as a result of variations in biotic factors such as predation and resource availability (Lévesque et al., 2010). Diel vertical and horizontal migrations also introduce a time-varying element to spatial distribution patterns (see Chapter 20). However, despite the great variability in patterns and drivers of species distribution, persistent within-lake spatial patterns in zooplankton abundance do occur. For example, the declining synchrony of temporal population fluctuations with distance between sampling sites in Lake Constance is indicative of persistent effects of local environmental conditions on zooplankton dynamics even within the same lake (Seebens et al., 2013).

There exist strong vertical gradients in the abundance and composition of heterotrophic protists. In the deep, well-oxygenated, Alpine lake Traunsee (Austria), both ciliates and heterotrophic nanoflagellates were found to be more abundant nearer the surface (Sonntag et al., 2006). Furthermore, the ciliate community structure changed along this depth gradient. Similarly, heterotrophic nanoflagellates were found to be more abundant in the epilimnion than in the hypolimnion of Lake Biwa (Japan), and protist community composition differed between these layers (Mukherjee et al., 2017). Heterotrophic protists can also form subsurface abundance maxima, associated with depth strata that are rich in food resources, such as deep chlorophyll maxima (Comte et al., 2006). Marked vertical heterogeneity is apparent even in shallow pond ecosystems, driven by species responses to physicochemical (especially oxygen) gradients and food resource availability (Finlay and Esteban, 1998b, Fig. 19-28).

Metazoan zooplankton communities also show strong vertical structuring. Studies of Windermere (UK) have shown that crustacean zooplankton community structure varies with depth, coincident with the thermal structure of the water column (Thackeray et al., 2005), and that the constituent species become more vertically aggregated under thermally stratified conditions (Thackeray et al., 2006). Using a whole-lake experimental deepening of the thermocline, Cantin et al. (2011) further demonstrated the link between metazoan zooplankton distribution and water column structure. In their study total zooplankton biomass was more homogeneously distributed with depth under conditions of thermocline deepening. Underlying this "bulk" response were differences among zooplankton size groups, with smaller zooplankton (rotifers, juvenile copepods, small cladocerans) populations tracking the deeper thermocline, and larger zooplankton (large *Daphnia* and copepod species) maintaining their position in the epilimnion despite this change in physical structure. In addition, thermocline deepening affected community composition, favoring rotifers (*Keratella*, *Kellicottia*) and cyclopoid copepods. Relationships between zooplankton distribution and physical water column processes will be the net result of responses to not only physical conditions (temperature, turbulence) but also correlated gradients in other factors (e.g., oxygen concentrations, resource composition and availability). For example, crustacean zooplankton have been shown to avoid lake hypolimnia when oxygen concentrations decline in deepwater (Doubek et al., 2018; Fig. 19-29), and vertical distributions are affected by underwater light gradients, food resource distribution, and population density (George, 1983). Zooplankton vertical distributions likely reflect trade-offs that optimize fitness gain, given vertical gradients in multiple environmental conditions (Lampert, 2005).

IX. Among water body heterogeneity in zooplankton communities

The diversity and community structure of freshwater zooplankton are determined by a complex array of interacting factors. In any specific location the number, identity, and relative dominance of species will be driven by

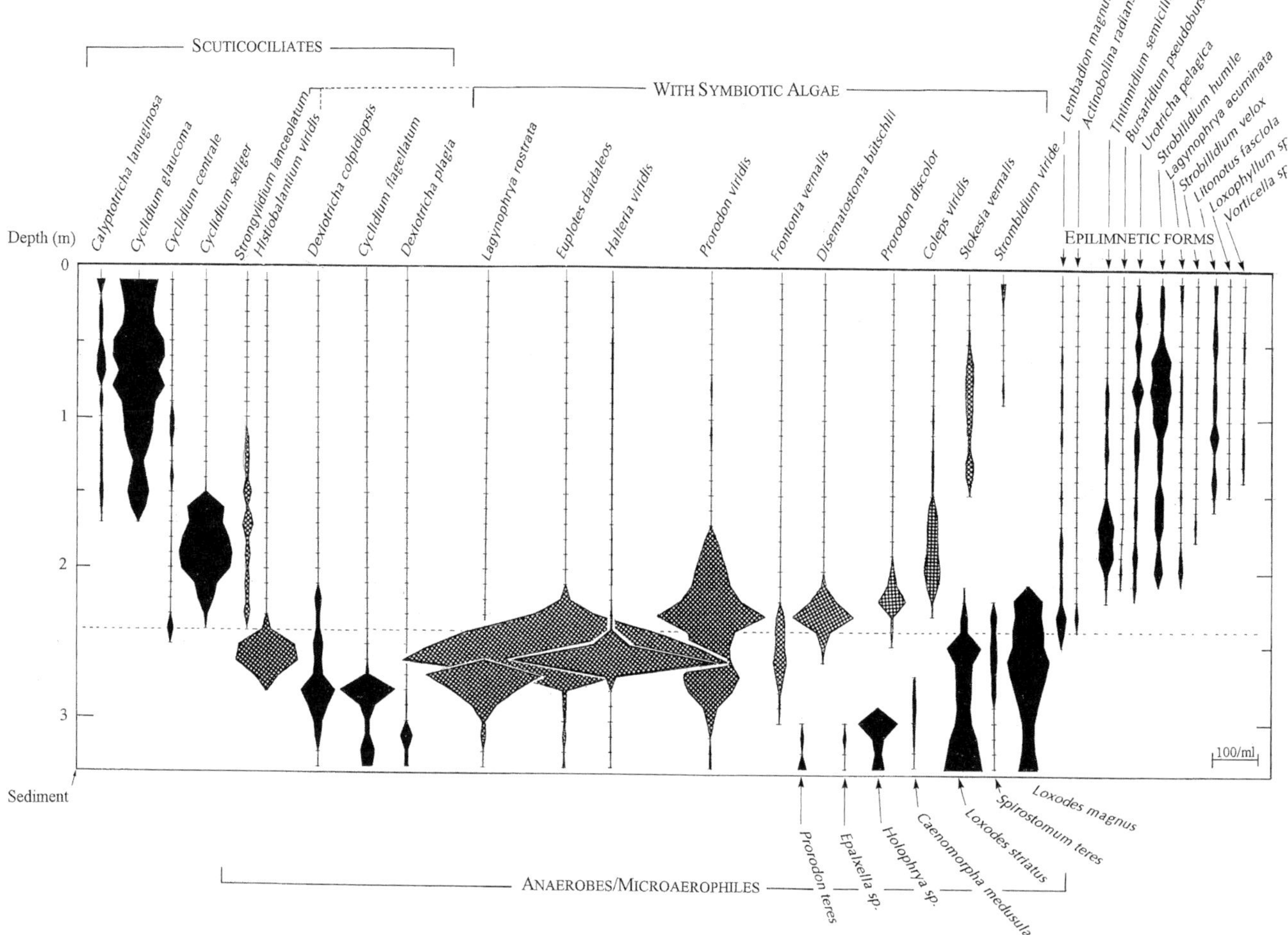

FIGURE 19-28 Vertical distribution of all species of ciliates in the water column of the productive pond, Priest Pot, UK. Hatched kites indicate ciliates with symbiotic algae (i.e., endosymbiotic *Chlorella*, or sequestered chloroplasts of various algae). The *horizontal line* indicates the depth of the oxic-anoxic boundary. The scale bar represents 100 ciliates mL^{-1} for all species apart from *Halteria viridis*, for which it signifies 500 mL^{-1}. *(Reproduced from Finlay and Esteban (1998b).)*

dispersal from other sites and the physicochemical and biological attributes of the habitat in question (Thackeray, 2022). Dispersal is covered in Chapter 20; here we consider the local and regional environmental "filters" that interact with dispersal, to determine which species establish and survive.

A. Physical drivers of diversity

Physical factors play an important role in structuring zooplankton communities. Among these, regional climatic conditions are known to be important, a link that can be inferred from the existence of latitudinal and elevational variations in zooplankton community structure. Examples of such patterns have been found for metazoan zooplankton across Canada (Henriques-Silva et al., 2016) and Norway (Hessen et al., 2006). According to the *species—energy—hypothesis*, higher energy inputs to ecosystems, as a result of higher temperatures and incoming solar radiation, stimulate higher productivity and thus consumer species richness (Lyons and Vinebrooke, 2016; Pinel-Alloul et al., 2013). It should be noted, however, that high levels of potentially damaging UV radiation can limit species richness or select for highly tolerant species (Marinone et al., 2006).

At a more local scale, lake size (surface area) frequently correlates with among-lake variations in species richness, in accordance with the classical theory of island biogeography. Specifically, species richness increases with surface area (Dodson, 1992; Merrix-Jones et al., 2013), possibly since immigration rates and habitat heterogeneity increase for larger lakes, while extinction rates decline. Hydrological regime also influences zooplankton communities. Community composition varies among temporary ponds according to the timing of drying and flooding (Florencio et al., 2020). Furthermore, water residence time influences the composition of communities, since rapid flushing negatively affects

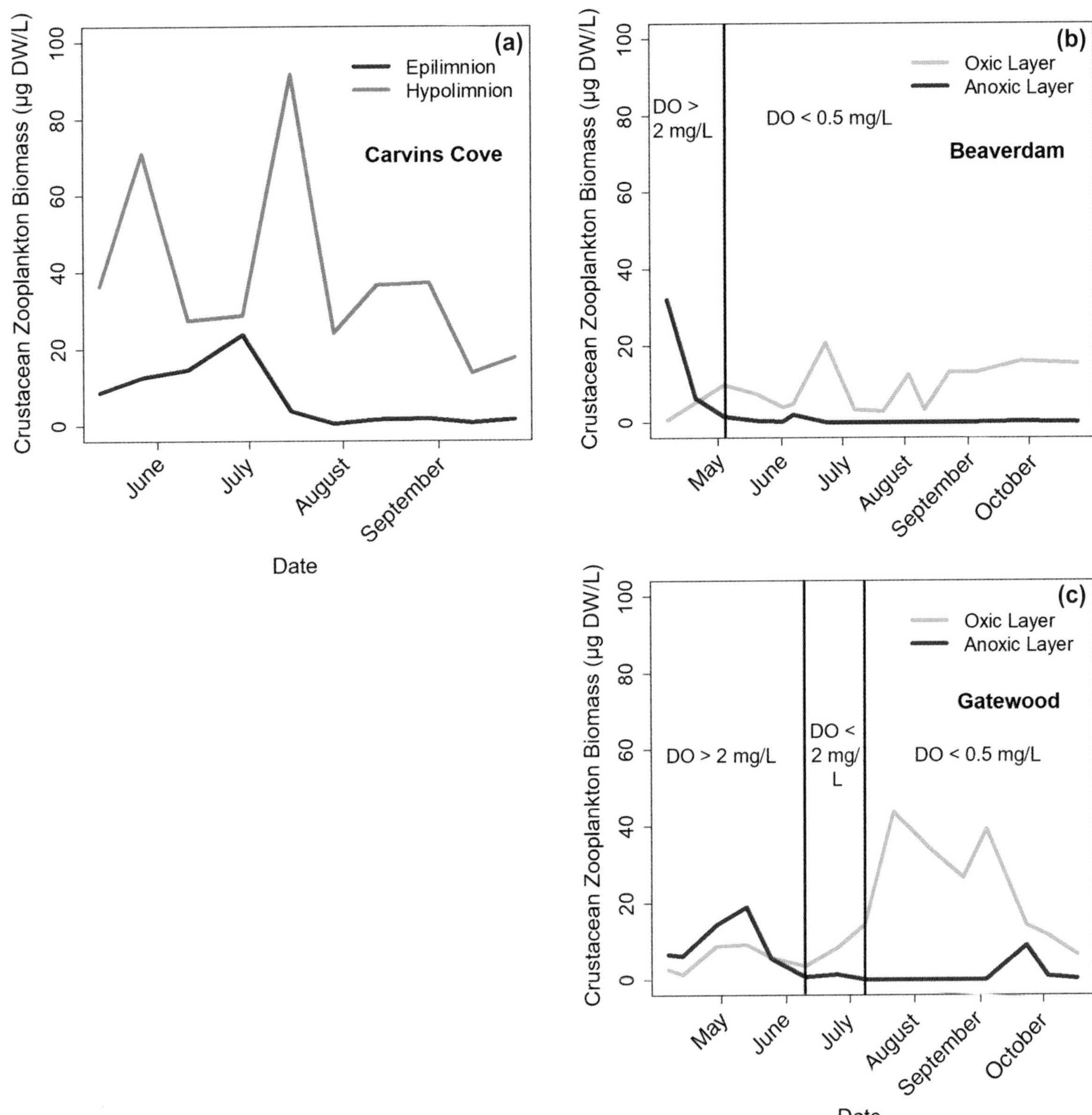

FIGURE 19-29 Biweekly daytime crustacean zooplankton tows of biomass for (a) Carvins Cove (epilimnion + hypolimnion); (b) Beaverdam (oxic and anoxic layers); and (c) Gatewood (oxic, hypoxic, and anoxic layers) reservoirs (USA). The full water column in Carvins Cove remained oxic throughout the monitoring period, and thus the epilimnion + hypolimnion colors are represented by *different shades of blue*. In contrast, the hypolimnion exhibited anoxia in both Beaverdam and Gatewood reservoirs (USA) by the end of the stratified period and is represented by a *red line*. *Vertical black lines* designate sampling days where the hypolimnion either became hypoxic (dissolved oxygen [DO] <2 mg L^{-1}) or anoxic (DO <0.5 mg L^{-1}). The onset of hypoxia occurred simultaneously with anoxia in Beaverdam with our sampling dates, and so hypoxia and anoxia are not differentiated in the Beaverdam panel. *(Reproduced from Doubek et al. (2018).)*

the contribution of longer generation time zooplankton taxa to communities, for example, calanoid copepods versus cladocerans (Doubek et al., 2019), and crustaceans versus rotifers (Obertegger et al., 2007).

B. Chemical drivers of diversity

The chemistry of freshwaters can also influence zooplankton community structure. For example, pH and calcium concentrations, especially, are recognized to have important effects. Metazoan zooplankton richness declines under conditions of extreme pH (Confer et al., 1983; Ivanova and Kazantseva, 2006), and ciliate diversity has been observed to decline with increasing acidity (Beaver and Crisman, 1989). Relatively calcium-rich taxa, such as many larger *Daphnia* spp., may be less abundant in freshwaters with lower calcium concentrations (Jeziorski et al., 2008; Jeziorski and Smol,

2017). Environmental contaminants can influence community composition through differential effects on species of different sensitivities (see also Section X). For example, metal contamination of the Sudbury Lakes (Canada) has influenced metazoan zooplankton composition, negatively affecting sensitive *Daphnia* populations (Valois et al., 2010). Elevated salinity because of climate-driven salinization or road salt contamination (Sinclair and Arnott, 2018) will also shape communities, selecting for the most tolerant species.

C. Biological drivers of diversity

Resource availability is an important driver of community composition. Particularly well studied is the correlation between metazoan species richness and ecosystem productivity, measured in terms of primary production or chlorophyll *a* concentration (Dodson et al., 2000; Hessen et al., 2006). It is hypothesized that higher productivity provides energetic support for a greater variety of species but that, as overall productivity increases, resource quality or habitat conditions deteriorate (e.g., dominance of Cyanobacteria, lower oxygen concentrations, higher predation). The net result is a hump-shaped relationship, with maximal species richness at intermediate productivity. While some studies find evidence of this relationship, others find instead that the relationship is linear. The discrepancy could arise due to sampling uncertainties and the range of productivity that is considered in each study. Furthermore, the nature of the relationship can depend upon whether diversity is expressed in terms of taxonomy or functional traits (Barnett and Beisner, 2007; see Section VI). Among the heterotrophic protists, ciliate richness has been found to increase, and species composition to change, with increasing lake trophic state (Beaver and Crisman, 1989).

Predation also influences zooplankton species composition, which varies among lakes with and without plankton-feeding fish (Knapp et al., 2001; MacLennan et al., 2015). Zooplankton species richness and composition have also been found to vary with fish community structure (Hessen et al., 2006; Jeppesen et al., 2000). The relationships between zooplankton and their predators are covered in more detail in Chapter 20.

X. Zooplankton communities and environmental change

A great many environmental changes affect the abundance, diversity, and dynamics of zooplankton communities. Here, we focus on the specific issues of climate change and environmental contamination.

A. Climate change

There is increasing evidence that climate change is already having a wide range of direct and indirect effects upon freshwater zooplankton (Carter et al., 2017; de Senerpont Domis et al., 2013; Selmeczy et al., 2019; Vadadi-Fülöp et al., 2012). These effects may be manifest through alterations to zooplankton life-history traits, abundance, community composition, seasonality, and distribution. Ultimately, such changes will impact upon food web interactions and ecosystem processes (see Chapter 20).

Shifts in the seasonal timing of biological events (phenological shifts) are one of the most well-recognized ecological impacts of climate change. Such shifts have been demonstrated for freshwater zooplankton, though the extent of change varies among species, even when they inhabit the same lake (Winder and Schindler, 2004). These among-species differences could desynchronize the seasonal dynamics of zooplankton, their food resources, and their predators, with potential food web implications. Mathematical modeling suggests that, dependent upon overwintering strategies (small or large hatches of resting eggs versus active overwintering individuals), *Daphnia* may or may not maintain seasonal synchrony with their phytoplankton food resources (de Senerpont Domis et al., 2007). In Windermere (UK) over 8 decades, the *D. galeata* spring population maximum shifted earlier with rising water temperatures and in response to increasingly early spring phytoplankton blooms, an apparent maintenance of synchrony (Thackeray et al., 2012). In Lake Washington (USA), however, delays to the timing of seasonal *Daphnia* increases during the 1990s resulted in a widening asynchrony with their predators: sockeye salmon fry (Hampton et al., 2006).

Climate change is also associated with long-term changes in zooplankton abundance. Surface water temperatures in the world's deepest lake, Baikal (Russia), increased by 0.2°C per decade over a 60-year period since the mid-1940s. During this time, cladoceran (*Daphnia*, *Bosmina*) populations increased while copepods and rotifers declined (Hampton et al., 2008). Using statistical models that capture both interspecific interactions and the effects of exogenous environmental variables, it was shown that cladoceran populations were being directly affected by the rise in water temperature. Similarly, in Lake Aleknagik, Alaska (USA), summer surface water temperatures increased by 1.4–2.3°C between 1963 and 2009, having positive effects upon cladoceran (*Daphnia*, *Bosmina*, *Holopedium*) populations but negative effects upon calanoid copepods (Carter et al., 2017). Mesocosm experiments have also shown, under more controlled conditions, the effects of warming on zooplankton dynamics (Barneche et al., 2021; Feuchtmayr et al., 2010).

As well as long-term shifts in average conditions, climate change projections also suggest an increased frequency of extreme weather events. Heatwaves can affect zooplankton abundance and seasonal succession, though the effects appear to be variable among sites. Anneville et al. (2010) detected that the 2003 European heatwave altered cladoceran seasonal succession in Lake Geneva but not Lake Annecy (France). The seasonal timing of the heatwave relative to the annual cycle of the lake ecosystem is important. *Bosmina* and cyclopoid copepod populations in Müggelsee (Germany) were affected by a heatwave in 2003 but not by similar events in 2006 and 2007 (Huber et al., 2010). The difference in response was attributed to differences in the seasonal timing of the heatwaves, relative to the specific time periods within which water temperature was most decisive for zooplankton populations.

Using the PEG model as a basis (see Section VII), de Senerpont Domis et al. (2013) have made general predictions of how climate change might be expected to affect the seasonal dynamics of both phyto- and zooplankton in Arctic, temperate, and tropical lakes (Fig. 19-27). In the absence of significant fish populations seasonal peaks in zooplankton populations are expected to grow larger in Arctic lakes, as a result of greater primary productivity driven by lengthening growing seasons. For temperate lakes, the impacts of climate change on zooplankton seasonal development are expected to be strongly dependent on lake trophic state, while in the tropics zooplankton populations are expected to remain low year-round as a result of intense predation pressure from fish.

B. Environmental contaminants

A wide range of environmental contaminants are now known to affect the life history and survival of zooplankton, including agrochemicals such as pesticides, metals from industrial activity, and road salt. High concentrations of pesticides (e.g., Andrade et al., 2021) and metals (e.g., Cu, Cd, Zn, Pb, Hg; Aránguiz-Acuña et al., 2018; Rogalski, 2015) can cause mortality at multiple life stages. However, lower concentrations can also have important nonlethal effects on various species, such as reduced feeding rates, reproduction and growth, and impacts on resting egg production (Aránguiz-Acuña et al., 2018; Pestana et al., 2010; Fig. 19-30). Species differ in their sensitivity to contaminants, and this can lead to changes in community composition. For example, upon exposure to the insecticide esfenvalerate, sensitive *Daphnia* spp. in pond microcosms declined in abundance, while less-sensitive *Simocephalus* spp. did not (Knillmann et al., 2013). Also, differential sensitivity to copper (Cu) toxicity has been suggested as a mechanism promoting the coexistence of *Brachionus* species in some Chilean water bodies (Aránguiz-Acuña et al., 2018). In the natural environment complex mixtures of contaminants can occur, and their combined toxicity may exceed that of each contaminant in isolation. For example, cypermethrin and glyphosate synergistically increase mortality in *Ceriodaphnia dubia* and reduce the abundance of cladocerans and copepods, but not rotifers, in experimental communities (Andrade et al., 2021).

Recently, the potential impacts of plastic pollution have begun to receive much attention. These impacts could be manifest when plastics adhere to the outer body of zooplankton, through internal damage after ingestion, or through toxic effects of chemical leachates or pollutants adsorbed to plastic particles from the surrounding environment. Microplastic particles can be ingested by a wide range of zooplankton taxa, including ciliates, rotifers, and crustaceans (Setälä et al., 2014). Jemec et al. (2016) found, via laboratory experiments, that *D. magna* are capable of ingesting synthetic textile fibers of 300 μm or more in length and that this can lead to increases in mortality. *D. magna* may also be immobilized following ingestion of much smaller 1 μm microplastic particles (Rehse et al., 2016). Nanoparticles (within the size range of 1−100 nm) have also been shown to affect freshwater zooplankton. Exposure to high concentrations of nanoplastic particles can reduce *D. magna* adult and neonate body size, and clutch size, and bring about an increased incidence of neonate malformations (Besseling et al., 2014; Fig. 19-31). The impacts of nanoplastics on this species may be affected by the adsorption of biomolecules, such as proteins, to the plastic particles themselves (Nasser and Lynch, 2016). At present, there is limited understanding of how plastic contaminants might have effects beyond the individual level and on community structure, although experiments by Bosker et al. (2019) suggest declines in *D. magna* population biomass following 21 days of exposure to 1−5 μm plastic particles.

Current research suggests that some zooplankton are capable of evolving a degree of tolerance to

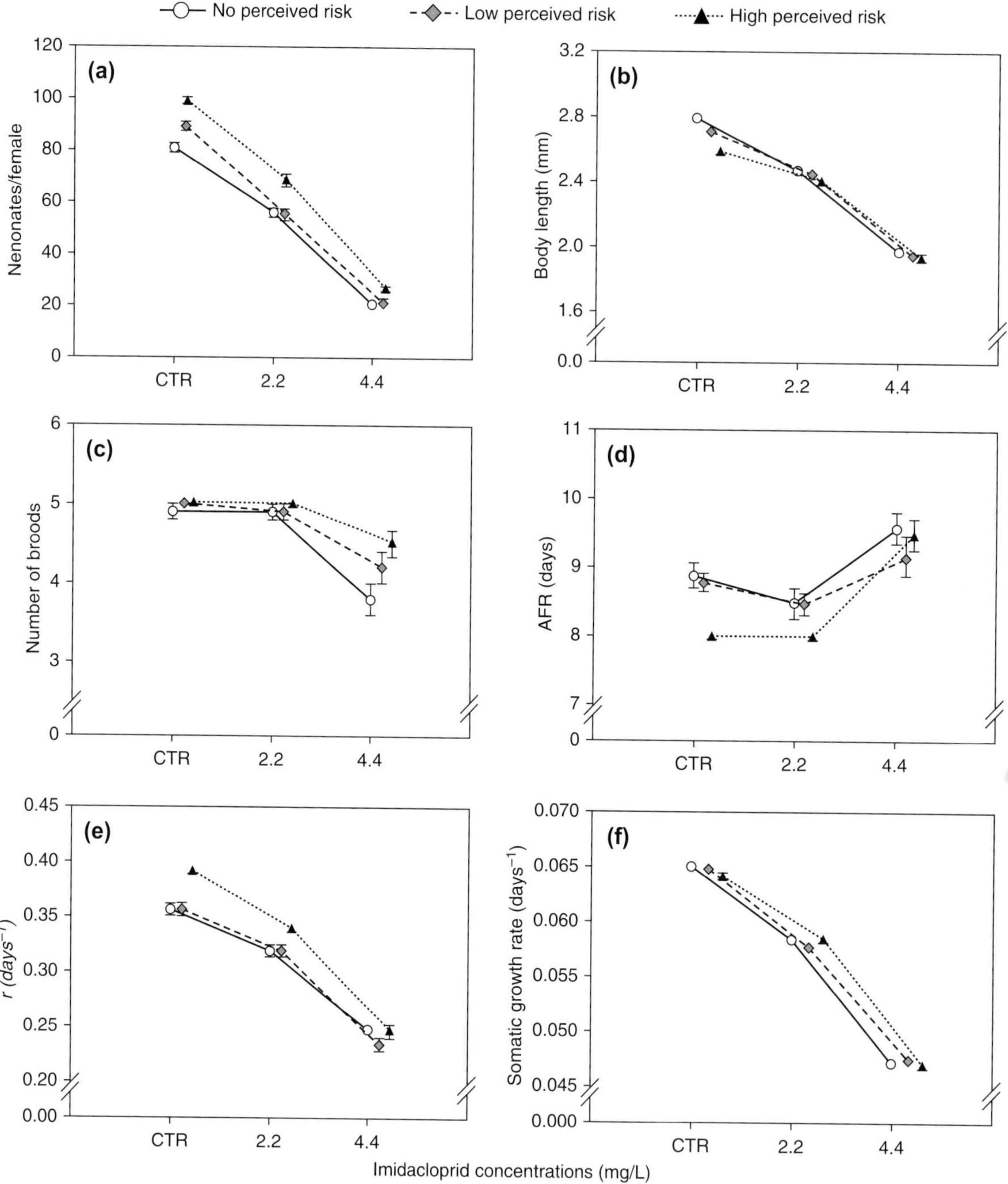

FIGURE 19-30 Effects of imidacloprid and different levels of perceived risk of predation on *D. magna* life-history parameters and somatic growth over 21 days (mean + SE): (a) Total number of neonates produced per female; (b) Size at maturity; (c) Number of broods; (d) Age at first reproduction; (e) Intrinsic rate of natural increase, *r*; and (f) Somatic growth rate. *(Reproduced from Pestana et al. (2010).)*

environmental contaminants, as a result of selection during prior exposure. *Daphnia* populations previously exposed to NaCl contamination suffer lower proportional population declines during subsequent salt contamination events than populations naive to such contamination (Hintz et al., 2019). Similarly, *D. pulex* populations collected from water bodies surrounded by a high percentage coverage of agricultural land appeared to have lower mortality when exposed to the pesticide chlorpyrifos than populations collected from less agricultural areas; an observation consistent with evolved tolerance by local *Daphnia* populations (Bendis and Relyea, 2016).

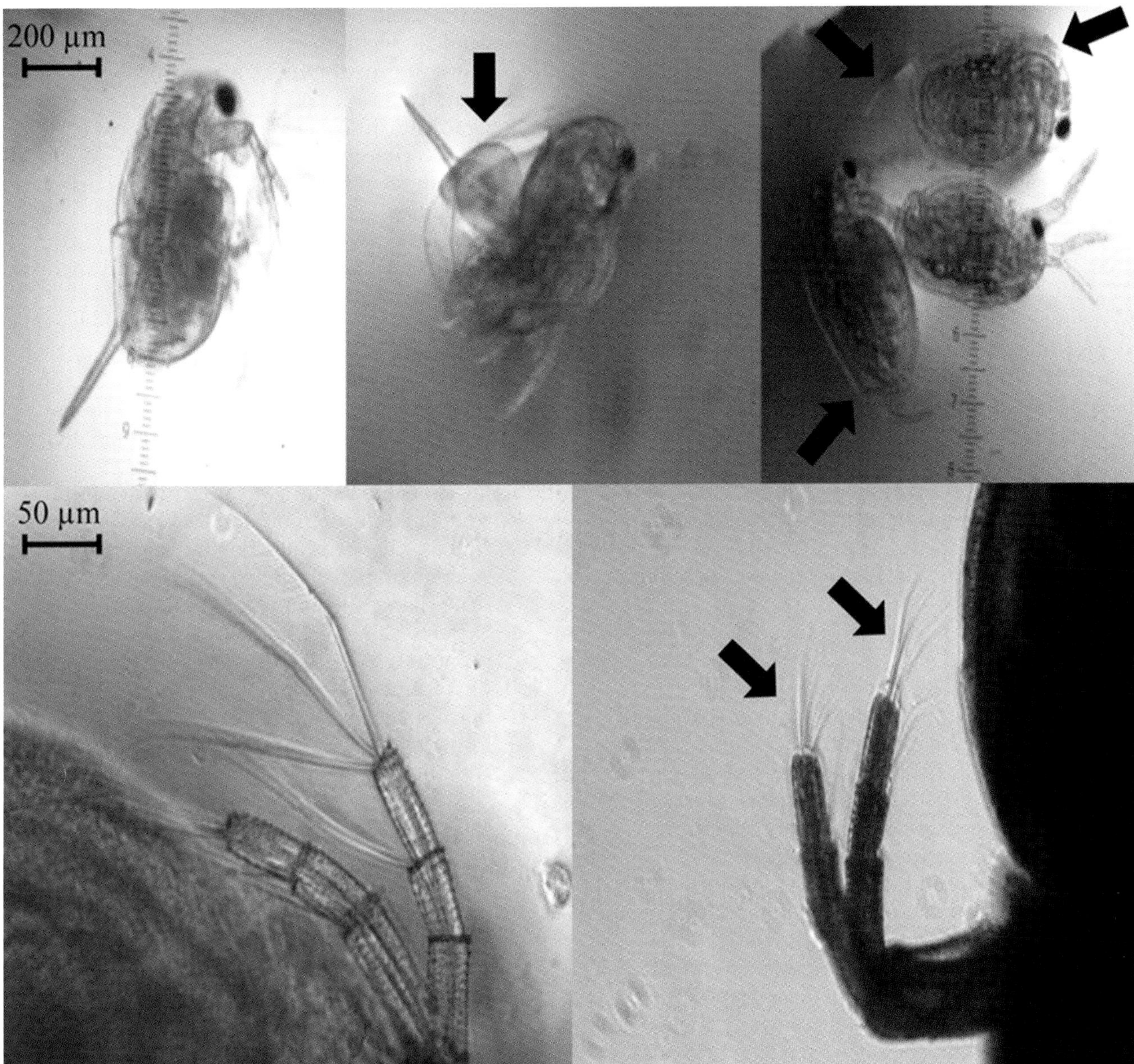

FIGURE 19-31 Malformations in different developmental stages of *Daphnia* neonates. *Top-right*: incompletely developed antenna setae, curved shell spine, and vacuoles around ovary. *Top-middle*: lump in the carapace. *Top-left*: normal developed neonate. *Bottom-right*: short antenna setae. *Bottom-left*: normal developed antenna setae. The *arrows* depict malformed body parts. *(Reproduced from Besseling et al. (2014).)*

XI. Summary

1. The zooplankton are taxonomically diverse and functionally defined as the heterotrophic component of the wider plankton community. This community is dominated by:
   a. Heterotrophic protists
   b. Rotifers
   c. Crustaceans (cladocerans and copepods)
2. Though not animals, many heterotrophic protists may be considered part of the zooplankton, including flagellates, ciliates, and sarcodines (heliozoans, some amoebae).
3. The rotifers are morphologically diverse but frequently have a well-defined head, trunk, and foot. Some species have a thickened cuticle, called a lorica.
4. Cladocerans are usually characterized by having a conspicuous head, light-sensitive organs, swimming antennae, and a bivalve carapace that encloses the body. Eggs are carried within the body, in a brood pouch.
5. In freshwaters most planktonic copepods are members of the Calanoida and Cyclopoida. Their bodies are highly segmented and have conspicuous antennae and swimming and feeding appendages. Eggs are carried externally, in sacs.

6. The zooplankton collectively use several mechanisms to capture their food, including:
   a. Direct interception of small food particles.
   b. Generation of feeding currents, using flagella, cilia, or feeding limbs that direct food toward mouth-like structures.
   c. Filter feeding, possibly assisted by electrostatic attraction of food particles.
   d. Scraping food from surfaces.
   e. Individually seizing larger prey items.
7. Many species are omnivorous but also show a degree of feeding selectivity, based upon body size, prey morphology, swimming behavior, prey toxicity, and "taste."
8. Food consumption can be quantified using clearance rates (the volume of water "processed" by an animal per unit time) and ingestion rates (the quantity of food ingested by an animal per unit time).
9. Ingestion rates for individual zooplankters vary with:
   a. Food concentration (according to a functional response relationship)
   b. Prey type
   c. Body size
   d. Water temperature and turbulence
10. Community ingestion rates vary over space and time and with community composition.
11. In metazoans the energy obtained from food resources is allocated to maintenance, storage, growth, and reproduction. Energy allocations should maximize fitness and vary with environmental conditions.
12. Reproduction may occur through mitotic or meiotic cell division for heterotrophic protists. For the rotifers and cladocerans, reproduction can alternate between asexual and sexual phases (cyclic parthenogenesis), while for freshwater copepods, reproduction is primarily sexual. For all groups, rates of population growth are strongly affected by water temperature and the quantity and quality of available food resources.
13. Zooplankton can be classified by their behavioral, morphological, phenological, or physiological characteristics or functional traits, facilitating generalizable insights into their responses to change and effects on ecosystem processes.
14. Zooplankton community composition and diversity changes over time, in response to variations in the physicochemical environment and biotic factors, such as food resource availability and predation. The dynamics are especially well studied and conceptualized for temperate zone lakes.
15. The spatial distribution of zooplankton within lakes is very heterogeneous, in both the vertical and horizontal dimensions. These patterns can change over time, and the underlying drivers are scale dependent, including:
   a. Physical processes (e.g., water currents, thermal stratification)
   b. Gradients in the chemical environment (e.g., oxygen concentrations)
   c. Patchiness in food resources and the impacts of predators
16. Zooplankton diversity and community structure vary among water bodies, due to among-water body differences in:
   a. The physical environment (e.g., climate, flushing rates, water body morphometry)
   b. The chemical environment (e.g., pH, calcium concentrations, contaminant concentrations)
   c. The biological environment (e.g., productivity, predation)
17. Environmental changes are already having impacts on zooplankton communities:
   a. Climate change is affecting the seasonality of zooplankton communities and driving decadal-scale changes in abundance and composition. Extreme climatological events can also have pronounced effects.
   b. Environmental contaminants such as pesticides, plastics, metals, and road salt have impacts on individual life-history, population size, and community composition.

Acknowledgments

We thank Dania Albini, Stephanie Hampton, and Sally Warring for providing constructive suggestions on the content of this chapter.

References

Adl, S.M., Bass, D., Lane, C.E., Lukeš, J., Schoch, C.L., Smirnov, A., Agatha, S., Berney, C., Brown, M.W., Burki, F., Cárdenas, P., Čepička, I., Chistyakova, L., del Campo, J., Dunthorn, M., Edvardsen, B., Eglit, Y., Guillou, L., Hampl, V., Heiss, A.A., Hoppenrath, M., James, T.Y., Karnkowska, A., Karpov, S., Kim, E., Kolisko, M., Kudryavtsev, A., Lahr, D.J.G., Lara, E., Le Gall, L., Lynn, D.H., Mann, D.G., Massana, R., Mitchell, E.A.D., Morrow, C., Park, J.S., Pawlowski, J.W., Powell, M.J., Richter, D.J., Rueckert, S., Shadwick, L., Shimano, S., Spiegel, F.W., Torruella, G., Youssef, N., Zlatogursky, V., Zhang, Q., 2019. Revisions to the classification, nomenclature, and diversity of eukaryotes. J. Eukaryot. Microbiol. 66, 4–119.

Adl, S.M., Simpson, A.G.B., Lane, C.E., Lukeš, J., Bass, D., Bowser, S.S., Brown, M.W., Burki, F., Dunthorn, M., Hampl, V., Heiss, A., Hoppenrath, M., Lara, E., le Gall, L., Lynn, D.H., McManus, H., Mitchell, E.A.D., Mozley-Stanridge, S.E., Parfrey, L.W., Pawlowski, J., Rueckert, S., Shadwick, L., Schoch, C.L., Smirnov, A., Spiegel, F.W., 2012. The revised classification of eukaryotes. J. Eukaryot. Microbiol. 59, 429–514.

Adrian, R., Deneke, R., 1996. Possible impact of mild winters on zooplankton succession in eutrophic lakes of the Atlantic European area. Freshw. Biol. 36, 757–770.

Adrian, R., Schneider-Olt, B., 1999. Top-down effects of crustacean zooplankton on pelagic microorganisms in a mesotrophic lake. J. Plankton Res. 21, 2175–2190.

Albini, D., Fowler, M.S., Llewellyn, C., Tang, K.W., 2019. Reversible colony formation and the associated costs in *Scenedesmus obliquus*. J. Plankton Res. 41, 419–429.

Alekseev, V., Lampert, W., 2004. Maternal effects of photoperiod and food level on life history characteristics of the cladoceran *Daphnia pulicaria* Forbes. Hydrobiologia 526, 225–230.

Andrade, V.S., Gutierrez, M.F., Reno, U., Popielarz, A., Gervasio, S., Gagneten, A.M., 2021. Synergy between glyphosate and cypermethrin formulations on zooplankton: evidences from a single-specie test and a community mesocosm experiment. Environ. Sci. Pollut. Res. 1–10.

Andrushchyshyn, O., Magnusson, A.K., Williams, D.D., 2003. Ciliate populations in temporary freshwater ponds: seasonal dynamics and influential factors. Freshw. Biol. 48, 548–564.

Anneville, O., Molinero, J.C., Souissi, S., Balvay, G., Gerdeaux, D., 2007. Long-term changes in the copepod community of Lake Geneva. J. Plankton Res. 29, i49–i59.

Anneville, O., Molinero, J.C., Souissi, S., Gerdeaux, D., 2010. Seasonal and interannual variability of cladoceran communities in two peri-alpine lakes: uncoupled response to the 2003 heat wave. J. Plankton Res. 32, 913–925.

Aránguiz-Acuña, A., Pérez-Portilla, P., De la Fuente, A., Fontaneto, D., 2018. Life-history strategies in zooplankton promote coexistence of competitors in extreme environments with high metal content. Sci. Rep. 8, 1–10.

Ban, S., 1994. Effect of temperature and food concentration on post-embryonic development, egg production and adult body size of calanoid copepod *Eurytemora affinis*. J. Plankton Res. 16, 721–735.

Barneche, D.R., Hulatt, C.J., Dossena, M., Padfield, D., Woodward, G., Trimmer, M., Yvon-Durocher, G., 2021. Warming impairs trophic transfer efficiency in a long-term field experiment. Nature 592, 76–79.

Barnett, A., Beisner, B.E., 2007. Zooplankton biodiversity and lake trophic state: explanations invoking resource abundance and distribution. Ecology 88, 1675–1686.

Barnett, A.J., Finlay, K., Beisner, B.E., 2007. Functional diversity of crustacean zooplankton communities: towards a trait-based classification. Freshw. Biol. 52, 796–813.

Beaver, J.R., Crisman, T.L., 1989. The role of ciliated protozoa in pelagic freshwater ecosystems. Microb. Ecol. 17, 111–136.

Beaver, J.R., Crisman, T.L., 1990. Seasonality of planktonic ciliated protozoa in 20 subtropical Florida lakes of varying trophic state. Hydrobiologia 190, 127–135.

Bendis, R.J., Relyea, R.A., 2016. Wetland defense: naturally occurring pesticide resistance in zooplankton populations protects the stability of aquatic communities. Oecologia 181, 487–498.

Besseling, E., Wang, B., Lürling, M., Koelmans, A.A., 2014. Nanoplastic affects growth of *S. obliquus* and reproduction of *D. magna*. Environ. Sci. Technol. 48, 12336–12343.

Birky, C.W., Wolf, C., Maughan, H., Herbertson, L., Henry, E., 2005. Speciation and selection without sex. Hydrobiologia 546, 29–45.

Blukacz, E.A., Shuter, B.J., Sprules, W.G., 2009. Towards understanding the relationship between wind conditions and plankton patchiness. Limnol. Oceanogr. 54, 1530–1540.

Boenigk, J., Arndt, H., 2002. Bacterivory by heterotrophic flagellates: community structure and feeding strategies. Antonie Leeuwenhoek 81, 465–480.

Bogdan, K.G., Gilbert, J.J., 1984. Body size and food size in freshwater zooplankton. Proc. Natl. Acad. Sci. 81, 6427–6431.

Bohrer, R., Lampert, W., 1988. Simultaneous measurement of the effect of food concentration on assimilation and respiration in *Daphnia magna Straus*. Funct. Ecol. 463–471.

Bosker, T., Olthof, G., Vijver, M.G., Baas, J., Barmentlo, S.H., 2019. Significant decline of *Daphnia magna* population biomass due to microplastic exposure. Environ. Pollut. 250, 669–675.

Boxshall, G.A., Defaye, D., 2008. Global diversity of copepods (Crustacea: Copepoda) in freshwater. Hydrobiologia 595, 195–207.

Bradley, M.C., Perrin, N., Calow, P., 1991. Energy allocation in the cladoceran *Daphnia magna* Straus, under starvation and refeeding. Oecologia 86, 414–418.

Brandl, Z., 1998. Feeding strategies of planktonic cyclopoids in lacustrine ecosystems. J. Mar. Syst. 15, 87–95.

Brans, K.I., De Meester, L., 2018. City life on fast lanes: urbanization induces an evolutionary shift towards a faster lifestyle in the water flea *Daphnia*. Funct. Ecol. 32, 2225–2240.

Brans, K.I., Jansen, M., Vanoverbeke, J., Tüzün, N., Stoks, R., De Meester, L., 2017. The heat is on: genetic adaptation to urbanization mediated by thermal tolerance and body size. Glob. Change Biol. 23, 5218–5227.

Brendelberger, H., 1991. Filter mesh size of cladocerans predicts retention efficiency for bacteria. Limnol. Oceanogr. 36, 884–894.

Brendelberger, H., Herbeck, M., Lang, H., Lampert, W., 1986. *Daphnia*'s filters are not solid walls. Arch. Hydrobiol. (Sofia) 107, 197–202.

Brooks, J.L., Dodson, S.I., 1965. Predation, body size, and composition of plankton. Science 150, 28–35.

Burian, A., Schagerl, M., Yasindi, A., 2013. Microzooplankton feeding behaviour: grazing on the microbial and the classical food web of African soda lakes. Hydrobiologia 710, 61–72.

Burns, C.W., 1968. The relationship between body size of filter-feeding Cladocera and the maximum size of particle ingested. Limnol. Oceanogr. 13, 675–678.

Burns, C.W., 1969a. Relation between filtering rate, temperature, and body size in four species of *Daphnia*. Limnol. Oceanogr. 14, 693–700.

Burns, C.W., 1969b. Particle size and sedimentation in the feeding behavior of two species of *Daphnia*. Limnol. Oceanogr. 14, 392–402.

Burns, C.W., Gilbert, J.J., 1993. Predation on ciliates by freshwater calanoid copepods: rates of predation and relative vulnerabilities of prey. Freshw. Biol. 30, 377–393.

Burns, C.W., Rigler, F.H., 1967. Comparison of filtering rates of *Daphnia rosea* in lake water and in suspensions of yeast. Limnol. Oceanogr. 12, 492–502.

Cáceres, C.E., 1998. Interspecific variation in the abundance, production and emergence of *Daphnia* diapausing eggs. Ecology 79, 1699–1710.

Canter, H.M., 1973. A new primitive protozoan devouring centric diatoms in the plankton. Zool. J. Linn. Soc. 52, 63–83.

Cantin, A., Beisner, B.E., Gunn, J.M., Prairie, Y.T., Winter, J.G., 2011. Effects of thermocline deepening on lake plankton communities. Can. J. Fish. Aquat. Sci. 68, 260–276.

Capriulo, G.M., Sherr, E.B., Sherr, B.F., 1991. Trophic behaviour and related community feeding activities of heterotrophic marine protists. In: Reid, P.C., Turley, C.M., Burkill, P.H. (Eds.), Protozoa and Their Role in Marine Processes. Springer Berlin Heidelberg, Berlin, Heidelberg, pp. 219–265.

Carter, J.L., Schindler, D.E., Francis, T.B., 2017. Effects of climate change on zooplankton community interactions in an Alaskan lake. Clim. Change Responses 4, 3.

Carvalho, G.R., Hughes, R.N., 1983. The effect of food availability, female culture-density and photoperiod on ephippia production in *Daphnia magna* Straus (Crustacea: Cladocera). Freshw. Biol. 13, 37–46.

Chang, C.-W., Shiah, F.-K., Wu, J.-T., Miki, T., Hsieh, C., 2014. The role of food availability and phytoplankton community dynamics in the seasonal succession of zooplankton community in a subtropical reservoir. Limnologica 46, 131–138.

Chow-Fraser, P., Sprules, W.G., 1986. Inhibitory effect of *Anabaena* sp. on in situ filtering rate of *Daphnia*. Can. J. Zool. 64, 1831–1834.

Chrzanowski, T.H., Šimek, K., 1990. Prey-size selection by freshwater flagellated protozoa. Limnol. Oceanogr. 35, 1429–1436.

Cleven, E.-J., Weisse, T., 2001. Seasonal succession and taxon-specific bacterial grazing rates of heterotrophic nanoflagellates in Lake Constance. Aquat. Microb. Ecol. 23, 147–161.

Colebrook, J., 1960. Some observations of zooplankton swarms in Windermere. J. Anim. Ecol. 241–242.

Colina, M., Calliari, D., Carballo, C., Kruk, C., 2016. A trait-based approach to summarize zooplankton–phytoplankton interactions in freshwaters. Hydrobiologia 767, 221–233.

Comte, J., Jacquet, S., Viboud, S., Fontvieille, D., Millery, A., Paolini, G., Domaizon, I., 2006. Microbial community structure and dynamics in the largest natural French lake (Lake Bourget). Microb. Ecol. 52, 72–89.

Confer, J.L., Kaaret, T., Likens, G.E., 1983. Zooplankton diversity and biomass in recently acidified lakes. Can. J. Fish. Aquat. Sci. 40, 36–42.

Cordova, S.E., Giffin, J., Kirk, K.L., 2001. Food limitation of planktonic rotifers: field experiments in two mountain ponds. Freshw. Biol. 46, 1519–1527.

Corliss, J.O., Esser, S.C., 1974. Comments on the role of the cyst in the life cycle and survival of free-living protozoa. Trans. Am. Microsc. Soc. 578–593.

Crumpton, W.G., Wetzel, R.G., 1982. Effects of differential growth and mortality in the seasonal succession of phytoplankton populations in Lawrence Lake, Michigan. Ecology 63, 1729–1739.

Cyr, H., Curtis, J.M., 1999. Zooplankton community size structure and taxonomic composition affects size-selective grazing in natural communities. Oecologia 118, 306–315.

de Senerpont Domis, L.N., Elser, J.J., Gsell, A.S., Huszar, V.L., Ibelings, B.W., Jeppesen, E., Kosten, S., Mooij, W.M., Roland, F., Sommer, U., 2013. Plankton dynamics under different climatic conditions in space and time. Freshw. Biol. 58, 463–482.

de Senerpont Domis, L.N., Mooij, W.M., Hülsmann, S., Van Nes, E.H., Scheffer, M., 2007. Can overwintering versus diapausing strategy in *Daphnia* determine match–mismatch events in zooplankton–algae interactions? Oecologia 150, 682–698.

De Wever, A., Muylaert, K., Cocquyt, C., Van Wichelen, J., Plisnier, P.-D., Vyverman, W., 2007. Seasonal and spatial variability in the abundance of auto-and heterotrophic plankton in Lake Tanganyika. Fundam. Appl. Limnol. 170, 49.

Decaestecker, E., De Meester, L., Mergeay, J., 2009. Cyclical parthenogenesis in *Daphnia*: sexual versus asexual reproduction. In: Lost Sex. Springer, pp. 295–316.

DeMott, W.R., 1986. The role of taste in food selection by freshwater zooplankton. Oecologia 69, 334–340.

DeMott, W.R., Watson, M.D., 1991. Remote detection of algae by copepods: responses to algal size, odors and motility. J. Plankton Res. 13, 1203–1222.

Dodson, S., 1992. Predicting crustacean zooplankton species richness. Limnol. Oceanogr. 37, 848–856.

Dodson, S.I., Arnott, S.E., Cottingham, K.L., 2000. The relationship in lake communities between primary productivity and species richness. Ecology 81, 2662–2679.

Doubek, J.P., Campbell, K.L., Doubek, K.M., Hamre, K.D., Lofton, M.E., McClure, R.P., Ward, N.K., Carey, C.C., 2018. The effects of hypolimnetic anoxia on the diel vertical migration of freshwater crustacean zooplankton. Ecosphere 9, e02332.

Doubek, J.P., Carey, C.C., Lavender, M., Winegardner, A.K., Beaulieu, M., Kelly, P.T., Pollard, A.I., Straile, D., Stockwell, J.D., 2019. Calanoid copepod zooplankton density is positively associated with water residence time across the continental United States. PLoS One 14, e0209567.

Dumont, H.J., Green, J., 1980. Rotatoria. In: Proceedings Of The 2nd International Rotifer Symposium Held at Gent, September. Springer Netherlands, pp. 17–21, 1979.

Duncan, A., 1989. Food limitation and body size in the life cycles of planktonic rotifers and cladocerans. Hydrobiologia 186, 11–28.

Edmondson, W.T., 1944. Ecological studies of sessile Rotatoria: Part I. Factors affecting distribution. Ecol. Monogr. 14, 31–66.

Edmondson, W.T., 1945. Ecological studies of sessile Rotatoria, Part II: dynamics of populations and social structures. Ecol. Monogr. 15, 141–172.

Edmondson, W.T., 1946. Factors in the dynamics of rotifer populations. Ecol. Monogr. 16, 357–372.

Edmondson, W.T., 1960. Reproductive rates of rotifers in natural populations. Memorie dell'Istituto Italiano di Idrobiologia 12, 21–77.

Edmondson, W.T., 1965. Reproductive rate of planktonic rotifers as related to food and temperature in nature. Ecol. Monogr. 35, 61–111.

Edmondson, W.T., 1968. A graphical model for evaluating the use of the egg ratio for measuring birth and death rates. Oecologia 1, 1–37.

Edmondson, W.T., 1974. Secondary production. SIL Commun. 20, 229–272, 1953–1996.

Elliott, J.I., 1977. Seasonal changes in the abundance and distribution of planktonic rotifers in Grasmere (English Lake District). Freshw. Biol. 7, 147–166.

Elmore, J.L., 1983. The influence of temperature on egg development times of three species of *Diaptomus* from subtropical Florida. Am. Midl. Nat. 300–308.

Eyres, I., Boschetti, C., Crisp, A., Smith, T.P., Fontaneto, D., Tunnacliffe, A., Barraclough, T.G., 2015. Horizontal gene transfer in bdelloid rotifers is ancient, ongoing and more frequent in species from desiccating habitats. BMC Biol. 13, 1–17.

Fenchel, T., 1987. Ecology of protozoa: The biology of free-living phagotropic protists. Springer-Verlag.

Fenchel, T., 1991. Flagellate design and function. In: The Biology of Free-Living Heterotrophic Flagellates. Clarendon Press, pp. 7–19.

Feuchtmayr, H., Moss, B., Harvey, I., Moran, R., Hatton, K., Connor, L., Atkinson, D., 2010. Differential effects of warming and nutrient loading on the timing and size of the spring zooplankton peak: an experimental approach with hypertrophic freshwater mesocosms. J. Plankton Res. 32, 1715–1725.

Finlay, B.J., Esteban, G.F., 1998a. Freshwater protozoa: biodiversity and ecological function. Biodivers. Conserv. 7, 1163–1186.

Finlay, B.J., Esteban, G.F., 1998b. Planktonic ciliate species diversity as an integral component of ecosystem function in a freshwater pond. Protist 149, 155–165.

Finlay, B.J., Esteban, G.F., 2001. Exploring Leeuwenhoek's legacy: the abundance and diversity of protozoa. Int. Microbiol. 4, 125–133.

Florencio, M., Fernández-Zamudio, R., Lozano, M., Díaz-Paniagua, C., 2020. Interannual variation in filling season affects zooplankton diversity in Mediterranean temporary ponds. Hydrobiologia 847, 1195–1205.

Folt, C.L., Burns, C.W., 1999. Biological drivers of zooplankton patchiness. Trends Ecol. Evol. 14, 300–305.

Forró, L., Korovchinsky, N.M., Kotov, A.A., Petrusek, A., 2008. Global diversity of cladocerans (Cladocera; Crustacea) in freshwater. Hydrobiologia 595, 177–184.

Forster, D., Qu, Z., Pitsch, G., Bruni, E.P., Kammerlander, B., Pröschold, T., Sonntag, B., Posch, T., Stoeck, T., 2021. Lake ecosystem robustness and resilience inferred from a climate-stressed protistan plankton network. Microorganisms 9, 549.

Fryer, G., 1954. Contributions to our knowledge of the biology and systematics of the freshwater Copepoda. Schweiz. Z. Hydrol. 16, 64–77.

Gaedke, U., Wickham, S.A., 2004. Ciliate dynamics in response to changing biotic and abiotic conditions in a large, deep lake (Lake Constance). Aquat. Microb. Ecol. 34, 247–261.

Gaines, G., Elbrachter, M., 1987. Heterotrophic nutrition. In: Taylor, F.J.R. (Ed.), The Biology of Dinoflagellates. Blackwell Scientific, Oxford, pp. 224–268.

Geller, W., 1975. Die Nahrungsaufnahme von *Daphnia pulex* in Abhaengigkeit von der Futterkonzentration, der Temperatur, der Koerpergroesse und dem hungerzustand der Tiere. Arch. Hydrobiol. 48, 47–107.

Geller, W., Müller, H., 1981. The filtration apparatus of Cladocera: filter mesh-sizes and their implications on food selectivity. Oecologia 49, 316–321.

George, D.G., 1981. Zooplankton patchiness. Rep. Freshwat. Biol. Ass. 49, 32–44.

George, D.G., 1983. Interrelations between the vertical distribution of *Daphnia* and chlorophyll *a* in two large limnetic enclosures. J. Plankton Res. 5, 457–475.

George, D.G., Winfield, I.J., 2000. Factors influencing the spatial distribution of zooplankton and fish in Loch Ness, UK. Freshw. Biol. 43, 557–570.

Gerritsen, J., Porter, K.G., 1982. The role of surface chemistry in filter feeding by zooplankton. Science 216, 1225.

Gilbert, J.J., 1963. Contact chemoreception, mating behaviour, and sexual isolation in the rotifer genus *Brachionus*. J. Exp. Biol. 40, 625–641.

Gilbert, J.J., 1966. Rotifer ecology and embryological induction. Science 151, 1234–1237.

Gilbert, J.J., 1980. Feeding in the rotifer *Asplanchna*: behavior, cannibalism, selectivity, prey defenses, and impact on rotifer communities. In: Kerfoot, W.C. (Ed.), Evolution and Ecology of Zooplanktonic Communities, Ed. Univ. Press, New England, Hanover, NH, pp. 158–172.

Gilbert, J.J., 1993. Rotifera. In: Adiyodi, K.G., Adiyodi, R.G. (Eds.), Reproductive Biology of Invertebrates. 5. Sexual Differentiation and Behaviour. Oxford IBH Publishing, New Delhi, pp. 115–136.

Gilbert, J.J., 1995. Rotifera. In: Adiyodi, K.G., Adiyodi, R.G. (Eds.), Reproductive Biology of Invertebrates. 6A. Asexual Propagation and Reproductive Strategies. Oxford IBH Publishing, New Delhi, pp. 231–263.

Gilbert, J.J., 2003. Environmental and endogenous control of sexuality in a rotifer life cycle: developmental and population biology. Evol. Dev. 5, 19–24.

Gilbert, J.J., 2019. Attachment behavior in the rotifer *Brachionus rubens*: induction by *Asplanchna* and effect on sexual reproduction. Hydrobiologia 844, 9–20.

Gilbert, J.J., 2020. Variation in the life cycle of monogonont rotifers: commitment to sex and emergence from diapause. Freshw. Biol. 65, 786–810.

Gilbert, J.J., Starkweather, P.L., 1977. Feeding in the rotifer *Brachionus calyciflorus* I. Regulatory mechanisms. Oecologia 28, 125–131.

Gilbert, J.J., Starkweather, P.L., 1978. Feeding in the rotifer *Brachionus calyciflorus* III. Direct observations on the effects of food type, food density, change in food type, and starvation on the incidence of pseudotrochal screening. SIL Proc. 20, 2382–2388, 1922–2010.

Gillooly, J.F., 2000. Effect of body size and temperature on generation time in zooplankton. J. Plankton Res. 22, 241–251.

Gliwicz, Z.M., Lampert, W., 1990. Food thresholds in *Daphnia* species in the absence and presence of blue-green filaments. Ecology 71, 691–702.

Gliwicz, Z.M., Pijanowska, J., 1989. The role of predation in zooplankton succession. In: Plankton Ecology. Springer, pp. 253–296.

Gliwicz, Z.M., Siedlar, E., 1980. Food size limitation and algae interfering with food collection in *Daphnia*. Arch. Hydrobiol. 88, 155–177.

González, M.J., Frost, T.M., 1992. Food limitation and seasonal population declines of rotifers. Oecologia 89, 560–566.

Gophen, M., Geller, W., 1984. Filter mesh size and food particle uptake by *Daphnia*. Oecologia 64, 408–412.

Guerrero-Jiménez, G., Ramos–Rodríguez, E., Silva-Briano, M., Adabache-Ortiz, A., Conde-Porcuna, J.M., 2020. Analysis of the morphological structure of diapausing propagules as a potential tool for the identification of rotifer and cladoceran species. Hydrobiologia 847, 243–266.

Guisande, C., Gliwicz, Z.M., 1992. Egg size and clutch size in two *Daphnia* species grown at different food levels. J. Plankton Res. 14, 997–1007.

Gulati, R.D., Rooth, J., Ejsmont-Karabin, J., 1987. A laboratory study of feeding and assimilation in *Euchlanis dilatata* lucksiana. Hydrobiologia 147, 289–296.

Hakima, B., Khémissa, C., Boudjéma, S., 2013. Effects of food limitation on the life history of *Simocephalus expinosus* Koch (Cladocera: daphniidae). Egypt. Acad. J. Biol. Sci. B Zool. 5, 25–31.

Hampton, S.E., 2005. Increased niche differentiation between two *Conochilus* species over 33 years of climate change and food web alteration. Limnol. Oceanogr. 50, 421–426.

Hampton, S.E., Galloway, A.W.E., Powers, S.M., Ozersky, T., Woo, K.H., Batt, R.D., Labou, S.G., O'Reilly, C.M., Sharma, S., Lottig, N.R., Stanley, E.H., North, R.L., Stockwell, J.D., Adrian, R., Weyhenmeyer, G.A., Arvola, L., Baulch, H.M., Bertani, I., Bowman Jr., L.L., Carey, C.C., Catalan, J., Colom-Montero, W., Domine, L.M., Felip, M., Granados, I., Gries, C., Grossart, H.-P., Haberman, J., Haldna, M., Hayden, B., Higgins, S.N., Jolley, J.C., Kahilainen, K.K., Kaup, E., Kehoe, M.J., MacIntyre, S., Mackay, A.W., Mariash, H.L., McKay, R.M., Nixdorf, B., Nõges, P., Nõges, T., Palmer, M., Pierson, D.C., Post, D.M., Pruett, M.J., Rautio, M., Read, J.S., Roberts, S.L., Rücker, J., Sadro, S., Silow, E.A., Smith, D.E., Sterner, R.W., Swann, G.E.A., Timofeyev, M.A., Toro, M., Twiss, M.R., Vogt, R.J., Watson, S.B., Whiteford, E.J., Xenopoulos, M.A., 2017. Ecology under lake ice. Ecol. Lett. 20, 98–111.

Hampton, S.E., Izmest'eva, L.R., Moore, M.V., Katz, S.L., Dennis, B., Silow, E.A., 2008. Sixty years of environmental change in the world's largest freshwater lake—Lake Baikal, Siberia. Glob. Change Biol. 14, 1947–1958.

Hampton, S.E., Romare, P., Seiler, D.E., 2006. Environmentally controlled *Daphnia* spring increase with implications for sockeye salmon fry in Lake Washington, USA. J. Plankton Res. 28, 399–406.

Haney, J.F., 1971. An in situ method for the measurement of zooplankton grazing rates. Limnol. Oceanogr. 16, 970–977.

Haney, J.F., 1973. An in situ examination of the grazing activities of natural zooplankton communities. Arch. Hydrobiol. 72, 87–132.

Hart, R.C., 1991. Food and suspended sediment influences on the naupliar and copepodid durations of freshwater copepods: comparative studies on *Tropodiaptomus* and *Metadiaptomus*. J. Plankton Res. 13, 645–660.

Hart, R.C., Bychek, E.A., 2011. Body size in freshwater planktonic crustaceans: an overview of extrinsic determinants and modifying influences of biotic interactions. Hydrobiologia 668, 61–108.

Hébert, M.-P., Beisner, B.E., 2020. Functional trait approaches for the study of metazooplankton ecology. In: Teodosio, M.A., Barbosa, A.B. (Eds.), Zooplankton Ecology. CRC press.

Hébert, M.-P., Beisner, B.E., Maranger, R., 2016. A meta-analysis of zooplankton functional traits influencing ecosystem function. Ecology 97, 1069–1080.

Hébert, M.-P., Beisner, B.E., Maranger, R., 2017. Linking zooplankton communities to ecosystem functioning: toward an effect-trait framework. J. Plankton Res. 39, 3–12.

Hecox-Lea, B.J., Welch, D.B.M., 2018. Evolutionary diversity and novelty of DNA repair genes in asexual Bdelloid rotifers. BMC Evol. Biol. 18, 177.

Henriques-Silva, R., Pinel-Alloul, B., Peres-Neto, P.R., 2016. Climate, history and life-history strategies interact in explaining differential macroecological patterns in freshwater zooplankton. Glob. Ecol. Biogeogr. 25, 1454–1465.

Herzig, A., 1983. The ecological significance of the relationship between temperature and duration of embryonic development in planktonic freshwater copepods. Hydrobiologia 100, 65–91.

Hessen, D.O., Faafeng, B.A., Smith, V.H., Bakkestuen, V., Walseng, B., 2006. Extrinsic and intrinsic controls of zooplankton diversity in lakes. Ecology 87, 433–443.

Hessen, D.O., Van Donk, E., 1993. Morphological changes in *Scenedesmus* induced by substances released from *Daphnia*. Arch. Hydrobiol. 127, 129, 129.

Hintz, W.D., Jones, D.K., Relyea, R.A., 2019. Evolved tolerance to freshwater salinization in zooplankton: life-history trade-offs, cross-tolerance and reducing cascading effects. Philos. Trans. R. Soc. B Biol. Sci. 374, 20180012.

Hopp, U., Maier, G., Bleher, R., 1997. Reproduction and adult longevity of five species of planktonic cyclopoid copepods reared on different diets: a comparative study. Freshw. Biol. 38, 289–300.

Huber, V., Adrian, R., Gerten, D., 2010. A matter of timing: heat wave impact on crustacean zooplankton. Freshw. Biol. 55, 1769–1779.

Hutchinson, G.E., 1967. A treatise on limnology. II. Introduction to lake biology and the limnoplankton. John Wiley & Sons, New York.

Hyman, L.H., 1951. The invertebrates: acanthocephala, aschelminthes, and entoprocta. In: The Pseudocoelomate Bilateria, vol. III. McGraw-Hill Book Co., New York.

Ivanova, M.B., Kazantseva, T.I., 2006. Effect of water pH and total dissolved solids on the species diversity of pelagic zooplankton in lakes: a statistical analysis. Russ. J. Ecol. 37, 264–270.

James, M.R., Burns, C.W., Forsyth, D.J., 1995. Pelagic ciliated protozoa in two monomictic, southern temperate lakes of contrasting trophic state: seasonal distribution and abundance. J. Plankton Res. 17, 1479–1500.

Jemec, A., Horvat, P., Kunej, U., Bele, M., Kržan, A., 2016. Uptake and effects of microplastic textile fibers on freshwater crustacean *Daphnia magna*. Environ. Pollut. 219, 201–209.

Jeppesen, E., Peder Jensen, J., Søndergaard, M., Lauridsen, T., Landkildehus, F., 2000. Trophic structure, species richness and biodiversity in Danish lakes: changes along a phosphorus gradient. Freshw. Biol. 45, 201–218.

Jeschke, J.M., Kopp, M., Tollrian, R., 2004. Consumer-food systems: why type I functional responses are exclusive to filter feeders. Biol. Rev. 79, 337–349.

Jeziorski, A.J., Smol, J.P., 2017. The ecological impacts of lakewater calcium decline on softwater boreal ecosystems. Environ. Rev. 25, 245–253.

Jeziorski, A., Yan, N.D., Paterson, A.M., DeSellas, A.M., Turner, M.A., Jeffries, D.S., Keller, W., Weeber, R.C., McNicol, R.C., Palmer, M.E., McIver, K., Arseneau, K., Ginn, B.K., Cumming, B.F., Smol, J.P., 2008. The widespread threat of calcium decline in fresh waters. Science 322, 1374–1377.

Keller, B., Spaak, P., 2004. Nonrandom sexual reproduction and diapausing egg production in a *Daphnia* hybrid species complex. Limnol. Oceanogr. 49, 1393–1400.

Kerfoot, W.C., Kirk, K.L., 1991. Degree of taste discrimination among suspension-feeding cladocerans and copepods: implications for detritivory and herbivory. Limnol. Oceanogr. 36, 1107–1123.

Kim, N., Yan, N.D., 2013. Food limitation impacts life history of the predatory cladoceran *Bythotrephes longimanus*, an invader to North America. Hydrobiologia 715, 213–224.

Knapp, R.A., Matthews, K.R., Sarnelle, O., 2001. Resistance and resilience of alpine lake fauna to fish introductions. Ecol. Monogr. 71, 401–421.

Knillmann, S., Stampfli, N.C., Noskov, Y.A., Beketov, M.A., Liess, M., 2013. Elevated temperature prolongs long-term effects of a pesticide on *Daphnia* spp. due to altered competition in zooplankton communities. Glob. Change Biol. 19, 1598–1609.

Knoechel, R., Holtby, L.B., 1986. Cladoceran filtering rate: body length relationships for bacterial and large algal particles. Limnol. Oceanogr. 31, 195–199.

Koch, U., von Elert, E., Straile, D., 2009. Food quality triggers the reproductive mode in the cyclical parthenogen *Daphnia* (Cladocera). Oecologia 159, 317–324.

Koehl, M.A.R., Strickler, J.R., 1981. Copepod feeding currents: food capture at low Reynolds number. Limnol. Oceanogr. 26, 1062–1073.

Kunzmann, A.J., Ehret, H., Yohannes, E., Straile, D., Rothhaupt, K.-O., 2019. Calanoid copepod grazing affects plankton size structure and composition in a deep, large lake. J. Plankton Res. 41, 955–966.

Lampert, W., 1982. Further studies on the inhibitory effect of the toxic blue-green *Microcystis aeruginosa* on the filtering rate of zooplankton. Arch. Hydrobiol. 95, 207–220.

Lampert, W., 1986. Response of the respiratory rate of *Daphnia magna* to changing food conditions. Oecologia 70, 495–501.

Lampert, W., 1997. Zooplankton research: the contribution of limnology to general ecological paradigms. Aquat. Ecol. 31, 19–27.

Lampert, W., 2005. Vertical distribution of zooplankton: density dependence and evidence for an ideal free distribution with costs. BMC Biol. 3, 10.

Laybourn-Parry, J., 1992. Protozoan plankton ecology. Chapman and Hall, London.

Lehman, J.T., 1988. Ecological principles affecting community structure and secondary production by zooplankton in marine and freshwater environments. Limnol. Oceanogr. 33, 931–945.

Leitão, E., Ger, K.A., Panosso, R., 2018. Selective grazing by a tropical copepod (*Notodiaptomus iheringi*) facilitates *Microcystis* dominance. Front. Microbiol. 9, 301.

Lévesque, S., Beisner, B.E., Peres-Neto, P.R., 2010. Meso-scale distributions of lake zooplankton reveal spatially and temporally varying trophic cascades. J. Plankton Res. 32, 1369–1384.

Lindholm, T., 1985. *Mesodinium rubrum*—a unique photosynthetic ciliate. Adv. Aquat. Microbiol. 3, 1–48.

Lindholm, M., Hessen, D.O., 2007. Zooplankton succession on seasonal floodplains: surfing on a wave of food. Hydrobiologia 592, 95–104.

Lischke, B., Weithoff, G., Wickham, S.A., Attermeyer, K., Grossart, H.-P., Scharnweber, K., HIlt, S., Gaedke, U., 2016. Large biomass of small feeders: ciliates may dominate herbivory in eutrophic lakes. J. Plankton Res. 38, 2–15.

Litchman, E., Ohman, M.D., Kiørboe, T., 2013. Trait-based approaches to zooplankton communities. J. Plankton Res. 35, 473–484.

Liu, X., Beyrend-Dur, D., Dur, G., Ban, S., 2014. Effects of temperature on life history traits of *Eodiaptomus japonicus* (Copepoda: Calanoida) from Lake Biwa (Japan). Limnology 15, 85–97.

Lürling, M., De Lange, H.J., Van Donk, E., 1997. Changes in food quality of the green alga *Scenedesmus* induced by *Daphnia* infochemicals: biochemical composition and morphology. Freshw. Biol. 38, 619–628.

Lynch, M., 1982. How well does the Edmonson-Paloheimo model approximate instantaneous birth rates? Ecology 63, 12–18.

Lyons, D.A., Vinebrooke, R.D., 2016. Linking zooplankton richness with energy input and insularity along altitudinal and latitudinal gradients. Limnol. Oceanogr. 61, 841–852.

MacLennan, M.M., Dings-Avery, C., Vinebrooke, R.D., 2015. Invasive trout increase the climatic sensitivity of zooplankton communities in naturally fishless lakes. Freshw. Biol. 60, 1502–1513.

Maier, G., 1989. The effect of temperature on the development times of eggs, naupliar and copepodite stages of five species of cyclopoid copepods. Hydrobiologia 184, 79–88.

Maier, G., 1994. Patterns of life history among cyclopoid copepods of central Europe. Freshw. Biol. 31, 77–86.

Marinone, M.C., Marque, S.M., Suárez, D.A., Diéguez, M., del, C., Pérez, P., De Los Ríos, P., Soto, D., Zagarese, H.E., 2006. UV radiation as a potential driving force for zooplankton community structure in Patagonian lakes. Photochem. Photobiol. 82, 962–971.

Masclaux, H., Bec, A., Kainz, M.J., Desvilettes, C., Jouve, L., Bourdier, G., 2009. Combined effects of food quality and temperature on somatic growth and reproduction of two freshwater cladocerans. Limnol. Oceanogr. 54, 1323–1332.

Masson, S., Angeli, N., Guillard, J., Pinel-Alloul, B., 2001. Diel vertical and horizontal distribution of crustacean zooplankton and young of the year fish in a sub-alpine lake: an approach based on high frequency sampling. J. Plankton Res. 23, 1041–1060.

Matz, C., Boenigk, J., Arndt, H., Jürgens, K., 2002. Role of bacterial phenotypic traits in selective feeding of the heterotrophic nanoflagellate *Spumella* sp. Aquat. Microb. Ecol. 27, 137–148.

May, L., 1983. Rotifer occurrence in relation to water temperature in Loch Leven, Scotland. In: Pejler, B., Starkweather, R., Nogrady, T. (Eds.), Biology of Rotifers. Springer Netherlands, Dordrecht, pp. 311–315.

McMahon, J., 1965. Some physical factors influencing the feeding behavior of *Daphnia magna* Straus. Can. J. Zool. 43, 603–611.

McMahon, J., Rigler, F., 1965. Feeding rate of *Daphnia magna* Straus in different foods labeled with radioactive phosphorus. Limnol. Oceanogr. 10, 105–113.

Merrix-Jones, F.L., Thackeray, S.J., Ormerod, S.J., 2013. A global analysis of zooplankton in natural and artificial fresh waters. J. Limnol. 72, e12.

Milinski, M., 1977. Do all members of a swarm suffer the same predation? Z. Tierpsychol. 45, 373–388.

Mohr, S., Adrian, R., 2002. Reproductive success of the rotifer *Brachionus calyciflorus* feeding on ciliates and flagellates of different trophic modes. Freshw. Biol. 47, 1832–1839.

Mourelatos, S., Lacroix, G., 1990. In situ filtering rates of Cladocera: effect of body length, temperature, and food concentration. Limnol. Oceanogr. 35, 1101–1111.

Mourelatos, S., Rougier, C., Pourriot, R., 1989. Diel patterns of zooplankton grazing in a shallow lake. J. Plankton Res. 11, 1021–1035.

Mukherjee, I., Hodoki, Y., Nakano, S., 2017. Seasonal dynamics of heterotrophic and plastidic protists in the water column of Lake Biwa, Japan. Aquat. Microb. Ecol. 80, 123–137.

Müller, H., Schöne, A., Pinto-Coelho, R., Schweizer, A., Weisse, T., 1991. Seasonal succession of ciliates in Lake Constance. Microb. Ecol. 21, 119–138.

Nandini, S., Sarma, S., 2007. Effect of algal and animal diets on life history of the freshwater copepod *Eucyclops serrulatus* (Fischer, 1851). Aquat. Ecol. 41, 75–84.

Nandini, S., Sarma, S., Amador-López, R.J., Bolaños-Muñioz, S., 2007. Population growth and body size in five rotifer species in response to variable food concentration. J. Freshw. Ecol. 22, 1–10.

Nasser, F., Lynch, I., 2016. Secreted protein eco-corona mediates uptake and impacts of polystyrene nanoparticles on *Daphnia magna*. J. Proteomics 137, 45–51.

Nisbet, B., 1984. Nutrition and feeding strategies in protozoa. Croom Helm Ltd, London.

Obertegger, U., Flaim, G., Braioni, M.G., Sommaruga, R., Corradini, F., Borsato, A., 2007. Water residence time as a driving force of zooplankton structure and succession. Aquat. Sci. 69, 575–583.

Obertegger, U., Smith, H.A., Flaim, G., Wallace, R.L., 2011. Using the guild ratio to characterize pelagic rotifer communities. Hydrobiologia 662, 157–162.

Okogwu, O., 2010. Seasonal variations of species composition and abundance of zooplankton in Ehoma Lake, a floodplain lake in Nigeria. Rev. Biol. Trop. 58, 171–182.

Pagano, M., 2008. Feeding of tropical cladocerans (*Moina micrura*, *Diaphanosoma excisum*) and rotifer (*Brachionus calyciflorus*) on natural phytoplankton: effect of phytoplankton size–structure. J. Plankton Res. 30, 401–414.

Pajk, F., Zhang, J., Han, B., Dumont, H.J., 2018. Thermal reaction norms of a subtropical and a tropical species of *Diaphanosoma* (Cladocera) explain their distribution. Limnol. Oceanogr. 63, 1204–1220.

Paloheimo, J.E., 1974. Calculation of instantaneous birth rate. Limnol. Oceanogr. 19, 692–694.

Paraskevopoulou, S., Dennis, A.B., Weithoff, G., Tiedemann, R., 2020. Temperature-dependent life history and transcriptomic responses in heat-tolerant versus heat-sensitive *Brachionus* rotifers. Sci. Rep. 10, 1–15.

Patterson, D.J., 1992. Freeliving freshwater protozoa. CRC Press.

Pauli, H.-R., 1990. Seasonal succession of rotifers in large lakes. In: Tilzer, M.M., Serruya, C. (Eds.), Large Lakes: Ecological Structure and Function. Springer Berlin Heidelberg, Berlin, Heidelberg, pp. 459–474.

Pennak, R.W., 1940. Ecology of the microscopic Metazoa inhabiting the sandy beaches of some Wisconsin lakes. Ecol. Monogr. 10, 537–615.

Pennak, R.W., 1978. Fresh-water invertebrates of the United States. John Wiley & Sons.

Pestana, J.L., Loureiro, S., Baird, D.J., Soares, A.M., 2010. Pesticide exposure and inducible antipredator responses in the zooplankton grazer, *Daphnia magna* Straus. Chemosphere 78, 241–248.

Peters, R.H., Downing, J.A., 1984. Empirical analysis of zooplankton filtering and feeding rates. Limnol. Oceanogr. 29, 763–784.

Pietrzak, B., Grzesiuk, M., Bednarska, A., 2010. Food quantity shapes life history and survival strategies in *Daphnia magna* (Cladocera). Hydrobiologia 643, 51–54.

Pinel-Alloul, P., 1995. Spatial heterogeneity as a multiscale characteristic of zooplankton community. Hydrobiologia 300, 17–42.

Pinel-Alloul, B., André, A., Legendre, P., Cardille, J.A., Patalas, K., Salki, A., 2013. Large-scale geographic patterns of diversity and community structure of pelagic crustacean zooplankton in Canadian lakes. Glob. Ecol. Biogeogr. 22, 784–795.

Pitsch, G., Bruni, E.P., Forster, D., Qu, Z., Sonntag, B., Stoeck, T., Posch, T., 2019. Seasonality of planktonic freshwater ciliates: are analyses based on v9 regions of the 18S rRNA gene correlated with morphospecies counts? Front. Microbiol. 10, 248.

Porter, K.G., 1975. Viable gut passage of gelatinous green algae ingested by *Daphnia*. Int. Ver. Für Theor. Angew. Limnol. Verhandlungen 19, 2840–2850.

Porter, K.G., 1976. Enhancement of algal growth and productivity by grazing zooplankton. Science 192, 1332–1334.

Porter, K.G., Gerritsen, J., Orcutt Jr., J.D., 1982. The effect of food concentration on swimming patterns, feeding behavior, ingestion, assimilation, and respiration by *Daphnia*. Limnol. Oceanogr. 27, 935–949.

Posch, T., Arndt, H., 1996. Uptake of sub-micrometre-and micrometre-sized detrital particles by bacterivorous and omnivorous ciliates. Aquat. Microb. Ecol. 10, 45–53.

Posch, T., Eugster, B., Pomati, F., Pernthaler, J., Pitsch, G., Eckert, E.M., 2015. Network of interactions between ciliates and phytoplankton during spring. Front. Microbiol. 6, 1289.

Price, H.J., 1988. Feeding mechanisms in marine and freshwater zooplankton. Bull. Mar. Sci. 43, 327–343.

Primicerio, R., Klemetsen, A., 1999. Zooplankton seasonal dynamics in the neighbouring Lakes Takvatn and Lombola (Northern Norway). Hydrobiologia 411, 19–29.

Radek, R., Hausmann, K., 1994. Endocytosis, digestion, and defecation in flagellates. Acta Protozool. 33, 127–147.

Rangel, L.M., Silva, L.H., Faassen, E.J., Lürling, M., Ger, K.A., 2020. Copepod prey selection and grazing efficiency mediated by chemical and morphological defensive traits of cyanobacteria. Toxins 12, 465.

Rao, T.R., Kumar, R., 2002. Patterns of prey selectivity in the cyclopoid copepod *Mesocyclops thermocyclopoides*. Aquat. Ecol. 36, 411–424.

Rehse, S., Kloas, W., Zarfl, C., 2016. Short-term exposure with high concentrations of pristine microplastic particles leads to immobilisation of *Daphnia magna*. Chemosphere 153, 91–99.

Ricci, C., 1998. Anhydrobiotic capabilities of bdelloid rotifers. Hydrobiologia 387, 321–326.

Richman, S., 1958. The transformation of energy by *Daphnia pulex*. Ecol. Monogr. 28, 273–291.

Rigler, F.H., 1971. Feeding rates: zooplankton. In: Edmondson, W.T., Winberg, G.G. (Eds.), A Manual on Methods for the Assessment of Secondary Productivity in Fresh Waters. Int. Biol. Program Handbook 17. Blackwell Scientific, Oxford, pp. 228–255.

Rinke, K., Huber, A.M.R., Kempke, S., Eder, M., Wolf, T., Probst, W.N., Rothhaupta, K.-O., 2009. Lake-wide distributions of temperature, phytoplankton, zooplankton, and fish in the pelagic zone of a large lake. Limnol. Oceanogr. 54, 1306–1322.

Rinke, K., Hübner, I., Petzoldt, T., Rolinski, S., König-Rinke, M., Post, J., Lorke, A., Benndorf, J., 2007. How internal waves influence the vertical distribution of zooplankton. Freshw. Biol. 52, 137–144.

Rinke, K., Vijverberg, J., 2005. A model approach to evaluate the effect of temperature and food concentration on individual life-history and population dynamics of *Daphnia*. Ecol. Model. 186, 326–344.

Rogalski, M.A., 2015. Tainted resurrection: metal pollution is linked with reduced hatching and high juvenile mortality in *Daphnia* egg banks. Ecology 96, 1166–1173.

Rose, R.M., Warne, M.S.J., Lim, R.P., 2000. Life history responses of the cladoceran *Ceriodaphnia cf. dubia* to variation in food concentration. Hydrobiologia 427, 59–64.

Rossi, V., Benassi, G., Belletti, F., Menozzi, P., 2011. Colonization, population dynamics, predatory behaviour and cannibalism in *Heterocypris incongruens* (Crustacea: Ostracoda). J. Limnol. 70, 102.

Russell, M.C., Qureshi, A., Wilson, C.G., Cator, L.J., 2021. Size, not temperature, drives cyclopoid copepod predation of invasive mosquito larvae. PLoS One 16, e0246178.

Ruttner-Kolisko, A., 1972. Rotatoria in das zooplankton der binnengewasser. Binnengewasser 26, 146–234.

Sanders, R.W., 1991. Trophic strategies among heterotrophic flagellates. In: Patterson, D.J., Larsen, J. (Eds.), The Biology of Free-Living Heterotrophic Flagellates. Clarendon Press, Oxford, pp. 21–38.

Sarma, S.S.S., Nandini, S., Gulati, R.D., 2002. Cost of reproduction in selected species of zooplankton (rotifers and cladocerans). Hydrobiologia 481, 89–99.

Sarnelle, O., Wilson, A.E., 2008. Type III functional response in *Daphnia*. Ecology 89, 1723–1732.

Schalau, K., Rinke, K., Straile, D., Peeters, F., 2008. Temperature is the key factor explaining interannual variability of *Daphnia* development in spring: a modelling study. Oecologia 157, 531–543.

Schälicke, S., Teubner, J., Martin-Creuzburg, D., Wacker, A., 2019. Fitness response variation within and among consumer species can be co-mediated by food quantity and biochemical quality. Sci. Rep. 9, 16126.

Seebens, H., Einsle, U., Straile, D., 2009. Copepod life cycle adaptations and success in response to phytoplankton spring bloom phenology. Glob. Change Biol. 15, 1394–1404.

Seebens, H., Einsle, U., Straile, D., 2013. Deviations from synchrony: spatio-temporal variability of zooplankton community dynamics in a large lake. J. Plankton Res. 35, 22–32.

Seebens, H., Straile, D., Hoegg, R., Stich, H.-B., Einsle, U., 2007. Population dynamics of a freshwater calanoid copepod: complex responses to changes in trophic status and climate variability. Limnol. Oceanogr. 52, 2364–2372.

Segers, H., 2008. Global diversity of rotifers (Rotifera) in freshwater. Hydrobiologia 595, 49–59.

Selmeczy, G.B., Abonyi, A., Krienitz, L., Kasprzak, P., Casper, P., Telcs, A., Somogyvári, Z., Padisák, J., 2019. Old sins have long shadows: climate change weakens efficiency of trophic coupling of phyto- and zooplankton in a deep oligo-mesotrophic lowland lake (Stechlin, Germany)—a causality analysis. Hydrobiologia 831, 101–117.

Serra, T., Müller, M.F., Colomer, J., 2019. Functional responses of *Daphnia magna* to zero-mean flow turbulence. Sci. Rep. 9, 3844.

Setälä, O., Fleming-Lehtinen, V., Lehtiniemi, M., 2014. Ingestion and transfer of microplastics in the planktonic food web. Environ. Pollut. 185, 77–83.

Simon, M., López-García, P., Deschamps, P., Moreira, D., Restoux, G., Bertolino, P., Jardillier, L., 2015. Marked seasonality and high spatial variability of protist communities in shallow freshwater systems. ISME J. 9, 1941–1953.

Sinclair, J.S., Arnott, S.E., 2018. Local context and connectivity determine the response of zooplankton communities to salt contamination. Freshw. Biol. 63, 1273–1286.

Skogstad, A., Granskog, L., Klaveness, D., 1987. Growth of freshwater ciliates offered planktonic algae as food. J. Plankton Res. 9, 503–512.

Sleigh, M., 1989. Protozoa and other protists. E. Arnold/Hodder & Stoughton, London.

Smyly, W.J.P., 1970. Observations on rate of development, longevity and fecundity of *Acanthocyclops viridis* (Jurine) (Copepoda, Cyclopoida) in relation to type of prey. Crustaceana 18, 21–36.

Smyly, W.J.P., 1974. The effect of temperature on the development time of the eggs of three freshwater cyclopoid copepods from the English Lake District. Crustaceana 27, 278–284.

Soares, M.C.S., Lürling, M., Huszar, V.L.M., 2010. Responses of the rotifer *Brachionus calyciflorus* to two tropical toxic cyanobacteria (*Cylindrospermopsis raciborskii* and *Microcystis aeruginosa*) in pure and mixed diets with green algae. J. Plankton Res. 32, 999–1008.

Sommer, U., Adrian, R., de Senerpont Domis, L., Elser, J.J., Gaedke, U., Ibelings, B., Jeppesen, E., Lürling, M., Molinero, J.C., Mooij, W.M., van Donk, E., Winder, M., 2012. Beyond the Plankton Ecology Group (PEG) model: mechanisms driving plankton succession. Annu. Rev. Ecol. Evol. Syst. 43, 429–448.

Sommer, U., Gliwicz, Z.M., Lampert, W., Duncan, A., 1986. The PEG-model of seasonal succession of planktonic events in fresh waters. Arch. Hydrobiol. 106, 433–471.

Sommer, U., Sommer, F., 2006. Cladocerans versus copepods: the cause of contrasting top-down controls on freshwater and marine phytoplankton. Oecologia 147, 183–194.

Sonntag, B., Posch, T., Klammer, S., Teubner, K., Psenner, R., 2006. Phagotrophic ciliates and flagellates in an oligotrophic, deep, alpine lake: contrasting variability with seasons and depths. Aquat. Microb. Ecol. 43, 193–207.

Starkweather, P.L., Gilbert, J.J., 1977. Feeding in the rotifer *Brachionus calyciflorus*. Oecologia 28, 133–139.

Starkweather, P.L., Gilbert, J.J., 1978. Feeding in the rotifer *Brachionus calyciflorus* IV. Selective feeding on tracer particles as a factor in trophic ecology and in situ technique. SIL Proc 20, 2389–2394, 1922–2010.

Stemberger, R.S., Evans, M.S., 1984. Rotifer seasonal succession and copepod predation in Lake Michigan. J. Gt. Lakes Res. 10, 417–428.

Straile, D., 2015. Zooplankton biomass dynamics in oligotrophic versus eutrophic conditions: a test of the PEG model. Freshw. Biol. 60, 174–183.

Summerer, M., Sonntag, B., Sommaruga, R., 2008. Ciliate-symbiont specificity of freshwater endosymbiotic *Chlorella* (trebouxiophyceae, chlorophyta). J. Phycol. 44, 77–84.

Symons, C.C., Arnott, S.E., Sweetman, J.N., 2012. Grazing rates of crustacean zooplankton communities on intact phytoplankton communities in Canadian Subarctic lakes and ponds. Hydrobiologia 694, 131–141.

Taylor, W.D., Sanders, R.W., 1991. Protozoa. In: Thorp, J.H., Covich, A.P. (Eds.), Ecology and Classification of North American Freshwater Invertebrates. Academic Press, San Diego, pp. 37–93.

Tessier, A.J., 1986. Comparative population regulation of two planktonic Cladocera (*Holopedium gibberum* and *Daphnia catawba*). Ecology 67, 285–302.

Thackeray, S.J., 2022. Zooplankton diversity and variation among lakes. In: Mehner, T., Tockner, K. (Eds.), Encyclopedia of Inland Waters, second ed.vol. 2. Elsevier, Oxford, pp. 52–66. https://doi.org/10.1016/B978-0-12-819166-8.00013-X.

Thackeray, S.J., George, D.G., Jones, R.I., Winfield, I.J., 2004. Quantitative analysis of the importance of wind-induced circulation for the spatial structuring of planktonic populations. Freshw. Biol. 49, 1091–1102.

Thackeray, S.J., George, D.G., Jones, R.I., Winfield, I.J., 2005. Vertical heterogeneity in zooplankton community structure: a variance partitioning approach. Arch. Hydrobiol. 164, 257–275.

Thackeray, S.J., George, D.G., Jones, R.I., Winfield, I.J., 2006. Statistical quantification of the effect of thermal stratification on patterns of dispersion in a freshwater zooplankton community. Aquat. Ecol. 40, 23–32.

Thackeray, S.J., Henrys, P.H., Jones, I.D., Feuchtmayr, H., 2012. Eight decades of phenological change for a freshwater cladoceran: what are the consequences of our definition of seasonal timing? Freshw. Biol. 57, 345–359.

Thompson, J.M., Ferguson, A.J.D., Reynolds, C.S., 1982. Natural filtration rates of zooplankton in a closed system: the derivation of a community grazing index. J. Plankton Res. 4, 545–560.

Tirok, K., Gaedke, U., 2006. Spring weather determines the relative importance of ciliates, rotifers and crustaceans for the initiation of the clear-water phase in a large, deep lake. J. Plankton Res. 28, 361–373.

Tirok, K., Gaedke, U., 2007. Regulation of planktonic ciliate dynamics and functional composition during spring in Lake Constance. Aquat. Microb. Ecol. 49, 87–100.

Tordesillas, D.T., Paredes, P.M.F., Villaruel, K.P.E., Queneri, C.A.A.M., Rico, J.L., Ban, S., Papa, R.D.S., 2018. Effects of food concentration on the reproductive capacity of the invasive freshwater calanoid copepod *Arctodiaptomus dorsalis* (Marsh, 1907) in the Philippines. J. Crustac Biol. 38, 101–106.

Urabe, J., 1992. Midsummer succession of rotifer plankton in a shallow eutrophic pond. J. Plankton Res. 14, 851–866.

Vadadi-Fülöp, C., Sipkay, C., Mészáros, G., Hufnagel, L., 2012. Climate change and freshwater zooplankton: what does it boil down to? Aquat. Ecol. 46, 501–519.

Valois, A., Keller, W., Ramcharan, C., 2010. Abiotic and biotic processes in lakes recovering from acidification: the relative roles of metal toxicity and fish predation as barriers to zooplankton re-establishment. Freshw. Biol. 55, 2585–2597.

van Bruggen, J.J.A., Stumm, C.K., Vogels, G.D., 1983. Symbiosis of methanogenic bacteria and sapropelic protozoa. Arch. Microbiol. 136, 89–95.

van den Bosch, F., Santer, B., 1993. Cannibalism in *Cyclops abyssorum*. Oikos 67, 19–28.

Van Donk, E., 1997. Defenses in phytoplankton against grazing induced by nutrient limitation, UV-B stress and infochemicals. Aquat. Ecol. 31, 53–58.

Van Donk, E., Lürling, M., Hessen, D.O., Lokhorst, G.M., 1997. Altered cell wall morphology in nutrient-deficient phytoplankton and its impact on grazers. Limnol. Oceanogr. 42, 357–364.

Vandekerkhove, J., Declerck, S., Brendonck, L., Conde-Porcuna, J.M., Jeppesen, E., De Meester, L., 2005. Hatching of cladoceran resting eggs: temperature and photoperiod. Freshw. Biol. 50, 96–104.

Wallace, R.L., Smith, H.A., 2009. Rotifera. In: Likens, G.E. (Ed.), Encyclopedia of Inland Waters. Elsevier, Oxford, pp. 689–703.

Wallace, R.L., Snell, T.W., 1991. Rotifera. In: Thorp, J.H., Covich, A.P. (Eds.), Ecology and Classification of North American Freshwater Invertebrates. Academic Press, San Diego, pp. 187–248.

Weetman, D., Atkinson, D., 2004. Evaluation of alternative hypotheses to explain temperature-induced life history shifts in *Daphnia*. J. Plankton Res. 26, 107–116.

Weisse, T., 2006. Freshwater ciliates as ecophysiological model organisms—lessons from *Daphnia*, major achievements, and future perspectives. Arch. Hydrobiol. 167, 371–402.

Weisse, T., 2014. Ciliates and the rare biosphere—community ecology and population dynamics. J. Eukaryot. Microbiol. 61, 419–433.

Weisse, T., 2017. Functional diversity of aquatic ciliates. Eur. J. Protistol. 61, 331–358.

Wendel, T., Jüttner, F., 1997. Excretion of heptadecene-1 into lake water by swarms of *Polyphemus pediculus* (Crustacea). Freshw. Biol. 38, 203–207.

Wenjie, L., Binxia, L., Cuijuan, N., 2019. Effects of temperature on the life history strategy of the rotifer *Euchlanis dilatata*. Zoolog. Sci. 36, 52–57.

West, D.C., Post, D.M., 2016. Impacts of warming revealed by linking resource growth rates with consumer functional responses. J. Anim. Ecol. 85, 671–680.

Wetzel, R.G., Likens, G.E., 1991. Zooplankton feeding. In: Limnological Analyses. Springer, New York, pp. 227–233.

Williamson, C.E., 1991. Copepoda. In: Thorp, J.H., Covich, A.P. (Eds.), Ecology and Classification of North American Freshwater Invertebrates. Academic Press, San Diego, pp. 787–822.

Winder, M., Berger, S.A., Lewandowska, A., Aberle, N., Lengfellner, K., Sommer, U., Diehl, S., 2012. Spring phenological responses of marine and freshwater plankton to changing temperature and light conditions. Mar. Biol. 159, 2491–2501.

Winder, M., Schindler, D.E., 2004. Climate change uncouples trophic interactions in an aquatic ecosystem. Ecology 85, 2100–2106.

Wirick, C.D., 1989. Herbivores and the spatial distributions of the phytoplankton. II. Estimating grazing in planktonic environments. Int. Rev. Gesamten Hydrobiol. Hydrogr. 74, 249–259.

Work, K.A., Havens, K.E., 2003. Zooplankton grazing on bacteria and cyanobacteria in a eutrophic lake. J. Plankton Res. 25, 1301–1306.

Wu, L., Feng, W., Yu, Y., Shen, Y., 2007. Temporal and spatial variations in the species composition, distribution, and abundance of protozoa in East Dongting Lake, China. J. Freshw. Ecol. 22, 655–665.

Xinyao, L., Miao, S., Yonghong, L., Yin, G., Zhongkai, Z., Donghui, W., Weizhong, W., Chencai, A., 2006. Feeding characteristics of an amoeba (Lobosea: *Naegleria*) grazing upon cyanobacteria: food selection, ingestion and digestion progress. Microb. Ecol. 51, 315–325.

Yin, X.W., Zhao, W., 2008. Studies on life history characteristics of *Brachionus plicatilis* O. F. Müller (Rotifera) in relation to temperature, salinity and food algae. Aquat. Ecol. 42, 165–176.

Yoshida, T., Kagami, M., Bahadur Gurung, T., Urabe, J., 2001. Seasonal succession of zooplankton in the north basin of Lake Biwa. Aquat. Ecol. 35, 19–29.

Young, S., Taylor, V.A., 1988. Visually guided chases in *Polyphemus pediculus*. J. Exp. Biol. 137, 387–398.

Zaffagnini, F., Sabelli, B., 1972. Karyologic observations on the maturation of the summer and winter eggs of *Daphnia pulex* and *Daphnia middendorffiana*. Chromosoma 36, 193–203.

Zhou, Q., Lu, N., Gu, L., Sun, Y., Zhang, L., Huang, Y., Chen, Y., Yang, Z., 2020. *Daphnia* enhances relative reproductive allocation in response to toxic *Microcystis*: changes in the performance of parthenogenetic and sexual reproduction. Environ. Pollut. 259, 113890.

Zimmermann, U., Müller, H., Weisse, T., 1996. Seasonal and spatial variability of planktonic heliozoa in Lake Constance. Aquat. Microb. Ecol. 11, 21–29.

Zingel, P., Nõges, T., 2010. Seasonal and annual population dynamics of ciliates in a shallow eutrophic lake. Fundam. Appl. Limnol. 176, 133–143.

C H A P T E R

20

Ecology and Functioning of Zooplankton Communities

Beatrix E. Beisner[1] *and Stephen J. Thackeray*[2]

[1]Department of Biological Sciences, University of Quebec at Montreal, Montreal, QC, Canada [2]Lake Ecosystems Group, UK Centre for Ecology & Hydrology, Lancaster, UK

O U T L I N E

The previous chapter highlighted the structure of freshwater zooplankton communities, including their seasonal and spatial variation. In this chapter we explore in greater detail the ecological functioning of zooplankton communities by considering their productivity, species and food web interactions, ecoevolutionary shifts, and the role of dispersal. We will end with an examination of how modern global changes are expected to modulate zooplankton community functioning.

I. Zooplankton community interactions

A. Predator–prey interactions

Predator–prey interactions in zooplankton communities, from protists to larger metazoans, are highly influenced by the relative size of predator and prey, as well as feeding traits and behavior. While the zooplankton are composed of primarily herbivorous grazers on phytoplankton, they can also correctly be referred to as predators. This is because zooplankton almost always kill or consume their entire prey, as do classical predators like fish higher in the food web. Here we will explore how predator–prey interactions influence the dynamics and stability of zooplankton communities.

One of the most studied interactions in aquatic food webs is that between zooplankton and their phytoplankton prey. The short generation times of plankton mean that such interactions have also provided an ideal testing ground for questions related to more theoretical considerations in ecology such as predator–prey

Wetzel's Limnology, Fourth Edition
https://doi.org/10.1016/B978-0-12-822701-5.00020-3

dynamics and stability. Based on observational data from cross-lake comparisons, greater crustacean community biomass occurs in lakes with more phytoplankton biomass (McCauley and Kalff, 1981) despite the negative interaction predicted by a predator–prey relationship. The cross-lake relationship appears to arise because, generally, greater lake productivity can support more biomass at all trophic levels (Dillon and Rigler, 1974), but this does not discount the effect of zooplankton predation within a single lake or across lakes at the same productivity level, for which we expect the negative detectable effect on phytoplankton. Indeed, experimental work examining these predator–prey dynamics over time shows clearly that crustaceans, rotifers, and other *micrograzers* (<85 μm) can effectively reduce mean phytoplankton biomass or primary production (Carpenter et al., 2001; Elser et al., 1995; Vanni, 1987) or lead to predator–prey cycles (Blasius et al., 2019; McCauley et al., 1999; Yoshida et al., 2003). Plankton predator–prey cycles have been observed in experiments in relation to nutrient enrichment, stage-structure in zooplankton populations, and an absence of inedible phytoplankton prey (McCauley et al., 1999). In nature plankton predator–prey dynamics in ponds and lakes are being constantly disrupted by changing environmental conditions, including the presence of less edible phytoplankton prey and predation by fish and larger invertebrates. Often perturbations occur seasonally, leading to seasonal successional patterns as described in Chapter 19. Underlying this succession, however, is the strong tendency for cyclic predator–prey interactions revealed under controlled conditions in the lab.

Other important predator–prey interactions occur between fish and crustacean zooplankton. In addition to suppressing zooplankton populations directly through predation (Carpenter et al., 2001; Elser et al., 1995; Section IIE), these interactions can alter zooplankton behavior, energetics, and life history that can in turn lead to more complex indirect effects in aquatic food webs (Werner and Peacor, 2003). An example of crustacean life history response to heavy fish predation comes from shifts in egg production because of reduced survival probability of *parthenogenetic females* (individuals producing eggs that hatch immediately) under predation. *Ephippia* (dormant eggs) formation in summer can be a better strategy to ensure long-term survival and fitness than any immediate reproductive gain via egg hatching. Even the presence of chemical compounds, termed *kairomones*, released by fish and invertebrate predators have been shown to induce the production of diapausing ephippial eggs in *Daphnia magna* (Pijanowska and Stolpe, 1996; Slusarczyk, 1995). As an alternative strategy, the age and size at which *Daphnia* spp. reach maturity may decrease upon exposure to fish kairomones (Reede, 1995), increasing the likelihood of surviving to reproduction when visual predators are present.

Fish kairomones can also alter swimming and population aggregation behavior (Jensen et al., 1998) that may reduce vulnerability to predation. *Diel vertical migration* induced by direct fish predation and other predators, or by the presence of their kairomones, can reduce mortality (Gabriel and Thomas, 1988; Gliwicz, 1986; Lampert, 1993; Picapedra et al., 2015) but has implications for crustacean zooplankton individual energy requirements. The energy expenditure in such migrations varies with the swimming and drag characteristics of migrating zooplankton (e.g., Vlymen, 1970; Haury and Weihs, 1976; Enright, 1977; Strickler, 1977; Dawidowicz and Loose, 1992). Among *Daphnia*, a common vertically migrating taxon, the energy expended in swimming is relatively small. Differences in fitness parameters (fecundity, growth) can thus be small, and growth only slightly less among long-distance migrating populations relative to short-distance migrating populations under low food. However, if diel vertical migration occurs between warm epilimnetic and cold hypolimnetic waters, crustacean zooplankton growth can be significantly reduced (by c. 60%) because of the time spent in colder waters avoiding predators (Dawidowicz and Loose, 1992; Loose and Dawidowicz, 1994). Horizontal migration in shallow ecosystems is often also related to visual fish predation, predation which may be impeded by the presence of macrophytes that reduce visibility in the littoral zone (Jeppesen et al., 1998).

Presumably, the smaller size of most rotifers removes much of the pressure of visually oriented predation by fishes. Analyses of the actual cost of ciliary locomotion among rotifers indicated that it is very inefficient and results in well over half of total metabolism (Epp and Lewis, 1984), much greater than the costs of locomotion among crustaceans. The movement of rotifers into the surface waters at midday (e.g., Adeniji, 1978) may, however, reflect an adaptive response to avoid predation by limnetic crustaceans despite the energetic costs.

B. Competitive interactions

There is also evidence of competitive interactions in zooplankton communities that will influence their dynamics and stability. Evidence for *exploitative competition*, occurring when one competitor reduces food resources to levels below the threshold required by its competitor for growth, has become evident in zooplankton communities in several ways. One way to reveal such competitive effects is to test *invasion success*, as done by Shurin (2000). His manipulative experiment demonstrated that ponds with greater zooplankton species richness and densities were more resistant to invasion by novel zooplankton species. Keeping native zooplankton densities low by actively removing individuals in the "Resistance treatment" ponds (Fig. 20-1), he observed greater invasion success by exotic species (3.8 times the number of invader individuals and 16.4 times total biomass of invaders) over the unmanipulated native densities in the "Invasion treatment" ponds. The evidence for a competitive effect of the natives was revealed through the necessity to keep their densities low to preclude their dominance over common resources. Another way to reveal competition among zooplankton is to relax or remove vertebrate predation (usually by fish), either experimentally or naturally. In such cases zooplankton community structure shifts toward larger zooplankton species like *Daphnia*, demonstrating the competitive dominance of large zooplankton over smaller species when the former are not being kept in check by predators (Brooks and Dodson, 1965). Invertebrate predation may accentuate this phenomenon by further reducing small-bodied zooplankton populations that are already disadvantaged by competition with larger zooplankton (Neill, 1984, 1981; Vanni, 1988).

The degree of synchrony in the population dynamics of different crustacean zooplankton species observed in lakes has also provided insight into the importance of competitive interactions under natural conditions that are shifting, often seasonally. Where species alternate in their dominance (*compensatory dynamics*; negatively correlated population sizes), it is thought that exploitative competition is likely an important force. In contrast, where dynamics are *synchronous* (positively correlated population sizes), species respond together to environmental shifts and competition is less obvious. While the evidence for synchrony is strong, and especially at the longer timescales corresponding to seasonality in temperate environments, compensatory dynamics also appear to occur, but mainly at within-season timescales (Keitt, 2008; Vasseur et al., 2014; Vasseur and Gaedke, 2007). For example, in Lake Constance (southern Germany) compensatory dynamics have been observed at 6- to 12-month timescales, suggesting that species are competing, even while the environment is changing (with synchrony observed at >12-month timescales) (Vasseur and Gaedke, 2007). Similarly, in a detailed study of interacting zooplankton population dynamics in Castle Lake (USA), Hoenicke and Goldman (1987) found evidence for exploitative competition among zooplankton, through seasonal changes in densities, relative fecundity, and lipid storage rates, as well as spatial separation of species. At longer, interannual timescales, it appears that the coexistence of *Daphnia* species that are similar in their ecology may occur via the "storage effect," whereby a weaker competitor can continue to persist over the long term (but not be observed each summer) because of certain strong recruitment years from resting (diapausing) eggs in the lake sediments (Cáceres, 1997).

For planktonic rotifers, there is a great deal of variation among species in their threshold food levels (Stemberger and Gilbert, 1985), but these are positively correlated to both rotifer body mass and to population growth rates. Thus the smallest taxa require less food to attain maximum growth rates; they are better adapted to living in food-poor environments than large species that attain maximum growth in food-rich environments. Rotifers also possess the capacity and strategy of producing *diapausing eggs* that can persist through stressful periods such as low-resource conditions (Chapter 19). Diapausing egg production is induced immediately after a brief starvation period at relatively little energetic cost. However, a population producing a high proportion of diapausing eggs has a greatly reduced reproductive potential in the long run because of lower hatching success subsequently (Gilbert and Schreiber, 1995).

Competition can also occur via direct contact between individuals in what is called *interference competition*. The strongest evidence for this type of competition in the zooplankton comes from interactions of rotifers with crustacean zooplankton. While rotifers are suppressed by large *Daphnia* through exploitative competition for shared, limiting food resources, it has also been shown that rotifers cannot become abundant in the presence of large ($\geq$1.2 mm) *Daphnia* and copepods even when shared food resources are abundant (Gilbert, 1988; Lair, 1990; MacIsaac and Gilbert, 1991; Neill, 1984; Stemberger and Gilbert, 1985). Thus, commonly, rotifers appear early in the seasonal succession of zooplankton communities and are succeeded by *Daphnia*. It is typically that only once large *Daphnia* populations have declined again that rotifers can rapidly dominate owing to their short generation times. This exclusion pattern appears to result from a mechanical interference that occurs as smaller rotifers are often swept into the brachial chamber of feeding *Daphnia*, where they are immediately killed, are mortally wounded, or lose attached eggs before being rejected. In some cases the rotifers are even ingested along with phytoplankton prey. Small cladocerans generally do not mechanically interfere with rotifers.

C. Parasitism

There has been an increased focus on the role of parasitism in zooplankton dynamics in recent years. While some zooplankton themselves may adopt a parasitic lifestyle, we will focus here on the interactions of zooplankton with other parasites, as zooplankton

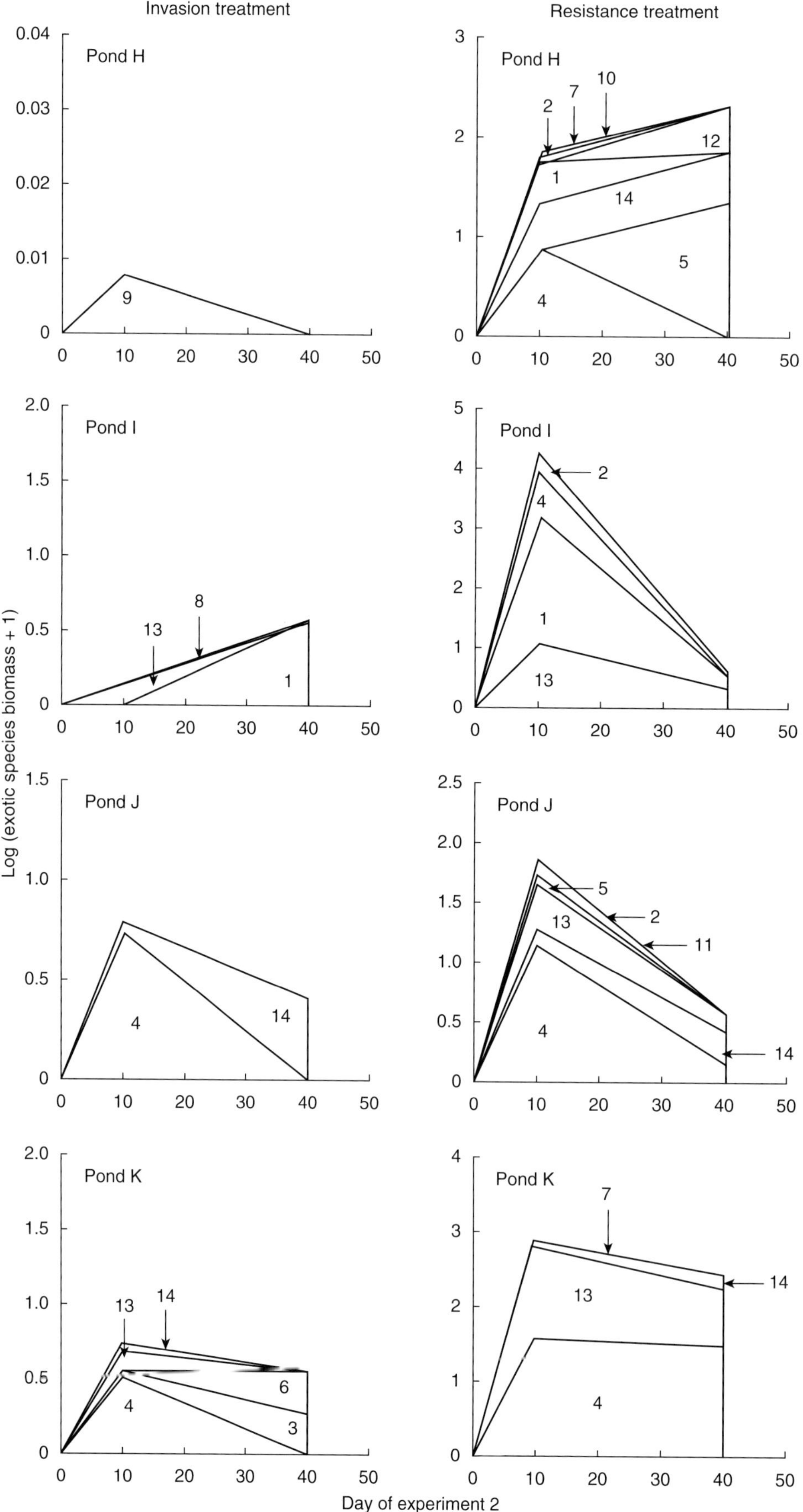

Invasion treatment
Resistance treatment
Pond H
Pond I
Pond J
Pond K
Log (exotic species biomass + 1)
Day of experiment 2

FIGURE 20-1 Results from a study by Shurin (2000) providing evidence for competition between zooplankton. The *left panels* show mean dry biomass (in μg L^{-1}; $\log_{10}$ transformed) of exotic species over time after they were introduced into experimental pond enclosures to create an "Invasion treatment." The *right panels* show the biomass time series of exotics in a "Resistance treatment" in which initial densities of native zooplankton were reduced by filtering the enclosures using a 154-μm mesh net prior to the addition of the exotics. Numbers refer to the following species: (1) *Asplanchna* sp., (2) *Bosmina longirostris*, (3) *Chydorus sphaericus*, (4) *Daphnia dubia*, (5) *Diaphanosoma brachyurum*, (6) *Filinia longiseta*, (7) *Keratella cochlearis*, (8) *Lecane luna*, (9) *Monostyla closterocerca*, (10) *Notholca laurentiae*, (11) *Pompholyx sulcata*, (12) *Sida crystallina*, (13) *Skistodiaptomus oregonensis*, and (14) *Tropocyclops prasinus*. Note that the scales of the y-axes differ among the graphs. *(Source: Fig. 3 in Shurin, J.B., 2000. Dispersal limitation, invasion resistance, and the structure of pond zooplankton communities. Ecology 81, 3074–3086.)*

population dynamics, species diversity, and productivity can be markedly affected by even moderate parasitic infection. For example, fungal parasitism can induce a high percentage of mortality of copepod eggs and female adults (Hoenicke, 1984; Burns, 1985; Garcia et al., 2018). Subsequent depressions of birth rates from such infections can alter population dynamics and function similarly to the way predation does, potentially increasing zooplankton species diversity in unaffected taxa. However, some parasites can also serve as zooplankton prey, especially when they parasitize phytoplankton, and thereby augment zooplankton production (e.g., the *mycoloop* consisting of chytrid fungi parasitizing large, inedible phytoplankton; Kagami et al., 2014).

To date, the *Daphnia*-parasite model system has been the most well studied to determine the effects of parasitism in the zooplankton (Cáceres et al., 2014; Ebert, 2005). This reflects the fact that larger-bodied zooplankton are more likely to be infected, because of higher grazing rates (increasing parasitic spore uptake) and accumulating effects with age (Stirnadel and Ebert, 1997). Common parasites include bacteria (e.g., *Pasteuria ramosa*, *Spirobacillus cienkowskii*), as well as a coccoid (likely a bacterium) that causes white fat cell disease. Fungi (e.g., the endoparasitic *Metschnikowia bicuspidata* and the epibiontic *Amoebidium parasiticum*) appear to be less host specific and usually negatively affect host reproduction and survivorship. Forming the largest group of *Daphnia* parasites are the obligate intracellular and usually tissue-specific Microsporidia (e.g., *Flabelliforma magnivora*). Transmission modes are variable, with some parasites transmitted horizontally between individuals via gut infections but others being vertically transmitted to offspring via ovary infections (Ebert, 2005).

Evidence using "resurrected" (see Section IE) pathogens and *Daphnia* clone hosts from sediments of different ages show that cycles of infection are dampened when the host has had prior exposure to the pathogen (*Pasteuria* bacteria) (Decaestecker et al., 2013). Duncan et al. (2006) similarly demonstrated ecoevolutionary interactions in the infection rates of the *Daphnia* host-parasite system with reduced infection rates after *Daphnia* populations had been exposed to the parasite and presumably evolved some resistance (Fig. 20-2). In other work using resurrected *Daphnia* clones, Auld and Brand (2017) demonstrated that epidemics of *Pasteuria* bacteria were larger under simulated climate change conditions (warming and mixing), leading to losses in host genetic diversity. Recently, studies have begun to explore the dynamics of multipathogen epidemics, demonstrating that the order of infection can also influence epidemic severity (Clay et al., 2020).

The accumulated evidence points to important influences of parasites on the ecological and evolutionary dynamics of *Daphnia* populations, their interaction with other members of lake food webs, as well as processes at the ecosystem level (Cáceres et al., 2014; Decaestecker et al., 2013; Duffy, 2007; Lampert, 2011). Parasitic spores generally settle into the sediments of lakes and cause seasonal outbreaks when zooplankton come into contact with them (Duffy and Hunsberger, 2019; Duncan et al., 2006) and often when fish predation is reduced. Indeed, in fishless ponds epidemics are more common in summer, as is also the case in autumn and winter months in the fish-containing Lake Constance when fish planktivory is reduced (Ebert, 2005). Parasitized *Daphnia* are often less transparent than their healthy counterparts, leading to greater predation rates by planktivorous fish (e.g., Willey et al., 1990). However, epidemics occurring in the autumn may also be related to increasing turbulence in the water column of lakes as stratification degrades: Cáceres et al. (2006) observed autumn fungal epidemics mainly toward the end of summer and into autumn. However, these epidemics only occurred in lakes with steep-sided (V-shaped) basins and not in those in which basin slopes were shallower (U-shaped). This difference was attributed to greater contact of the pelagic zone with sediments in the nearshore area in V-shaped basins relative to lakes with more gently sloping basins.

D. Symbiotic interactions

In general, positive interactions, of which *symbioses* represent the most reciprocal, are the least studied in

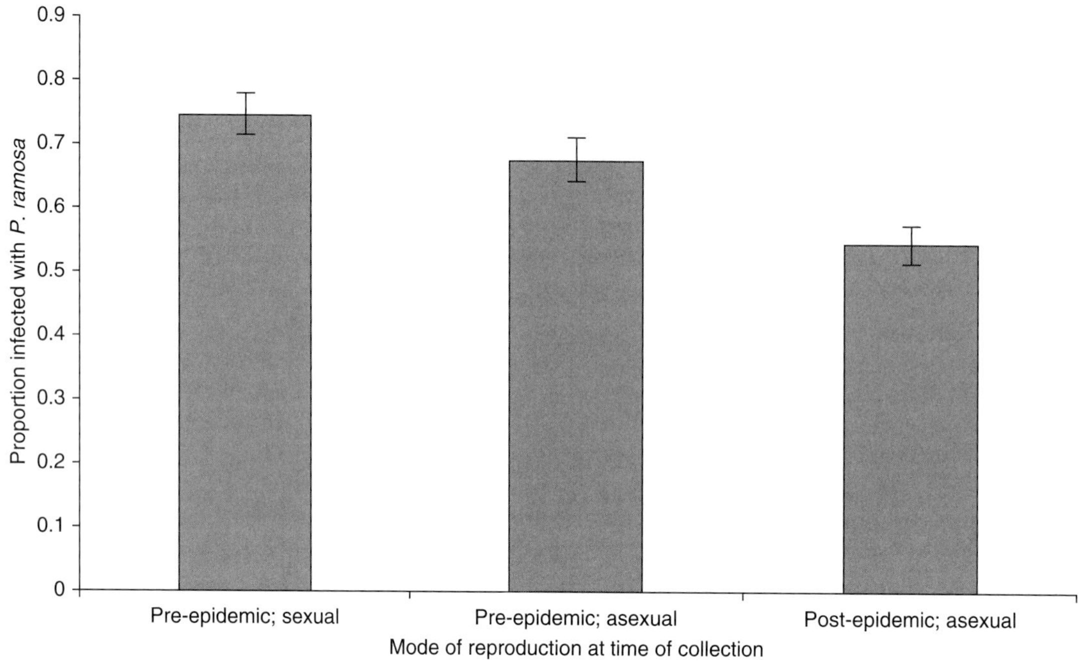

FIGURE 20-2 Comparison of the resistance of *Daphnia* populations to the parasite *Pasteuria ramosa* for hosts collected before and after an epidemic. Variation in mean infection rates (± standard error) were observed in preepidemic individuals based on their mode of reproduction, with the sexually reproducing population having higher infection rates than the asexually reproducing one. *(Source: Fig. 4 in Duncan, A.B., Mitchell, S.E., Little, T.J., 2006. Parasite-mediated selection and the role of sex and diapause in* Daphnia. *J. Evol. Biol. 19, 1183–1189.)*

zooplankton ecology. Most of what we know comes from the microbial food web, wherein long-term mutualistic exchanges (often biochemical) between bacteria and the protist and ciliate zooplankton components have been studied (Dziallas et al., 2012). At these small body sizes, the partners often benefit from a supply of energy or nutrients and may gain protection from predation or environmental threats such as toxins or oxygen radicals. For example, protists often harbor *endosymbionts* (internal symbionts) of bacteria and other protists (also fungi and viruses) that enable them to acquire new metabolic functions (e.g., nitrogen fixation, methanogenesis, photosynthesis) and gain competitive advantages. Ciliates can also harbor both *endo-* and *ectosymbionts* (external symbionts), enabling them to occupy new ecological niches and to be more competitive. In some cases parasitism has led to the "enslavement" of one organism by another; relationships can be hard to define at such small scales and are often complex. Dziallas et al. (2012) provide an overview of the ecological effects on hosts (e.g., nutrient and vitamin supply, protection against parasites, waste degradation, UV protection) and intracellular symbionts (e.g., nutrient supply, increased motility, grazing protection, reduced competition). Ultimately, by enabling some microzooplankton to occupy habitats and niches from which they would be otherwise excluded, symbioses can affect ecosystem processes including sulfur, nitrogen, iron, and matter cycling; photosynthesis; and rates of methanogenesis, among other things.

It is thought that bacteria active as ectosymbionts and endosymbionts (including those within the gut microbiome) benefit from their association with *metazooplankton* (multicellular heterotrophs or metazoans, composed primarily of rotifers and crustaceans) which have been called "microbially diverse and active hotspots" in aquatic environments (Grossart et al., 2009). Bacteria benefit from their association with metazooplankton by reaching new habitats and resources while "hitchhiking" on vertically migrating zooplankton (Grossart et al., 2010), likely also directly utilizing zooplankton waste products. Recent studies demonstrate benefits for the zooplankton as well. In gut microbiota transplants with *Daphnia* it has been shown that tolerance to toxins, such as those produced by Cyanobacteria, can be modulated (Macke et al., 2017). Growth, survival, and fecundity of *Daphnia* is improved when bacteria are present in the environment (Sison-Mangus et al., 2015) or in their food (Callens et al., 2016). Bacterial communities associated with metazooplankton appear to be more conservative in some cladocerans than copepods, and in oligotrophic environments over eutrophic ones (Grossart et al., 2009). Finally, less well studied is the ecology of ciliate epibionts that attach to and live directly on metazooplankton exoskeletons. Epibiont ciliates are widespread and they likely benefit from the swimming of their hosts, which reduces the boundary layer and thereby aids the bacterivorous feeding of the ciliates (Bickel et al., 2012). Future research on the consequences of such symbioses on the long-term population and community dynamics of zooplankton food webs will shed insight into their ecological effects on freshwater ecosystem functioning.

E. Ecoevolutionary shifts

Current evidence is mounting to suggest that evolutionary change can occur in zooplankton populations over periods of years to decades. Such "contemporary" evolution has the potential to alter zooplankton population and community dynamics and results in large part from their relatively rapid population dynamics and morphological plasticity (e.g., *cyclomorphic polymorphism*). Zooplankton communities lend themselves to the experimental study of which ecological shifts can lead to evolutionary ones because the dormant ("diapausing") eggs or cysts can preserve for many years in lake and pond sediments, acting as "time capsules" of adaptations to historical conditions. Maximum viability of dormant eggs has been found to be on the order of 600 to 700 years for *Daphnia* (Frisch et al., 2014), approximately 300 years for some calanoid copepods (Hairston et al., 1995), and 35 to 40 years for rotifers (Marcus et al., 1994). This makes possible a type of study called *resurrection ecology* (Kerfoot et al., 1999), wherein the dormant eggs contained within different layers of sediments (which can be aged using paleolimnological techniques; Chapter 30) can be hatched and the traits (phenotypes or genotypes) of the emerging zooplankton compared. When the ecological history of the lake can also be independently verified (e.g., known stocking of a new fish or exotic species leading to a food web shift; paleolimnological analyses of diatom communities to infer changes in nutrients, salinity, pH, or habitat), then the traits of the hatched individuals can be used in life history comparisons.

There are several excellent examples of resurrected zooplankton egg studies demonstrating the evolution of crustacean zooplankton to various factors including exposure to toxic algae (Hairston et al., 1999a, 1999b), the aforementioned host-parasite interactions (Decaestecker et al., 2013; Duncan et al., 2006) (Section IC), pH change (Derry et al., 2010) eutrophication (Frisch et al., 2014), changes in fish communities (Cousyn et al., 2001), temperature change (Geerts et al., 2015), and multiple stressors (Orsini et al., 2012). For Lake Constance zooplankton communities, Hairston et al. (1999a) were able to show that *Daphnia* genotypes, hatched from eggs obtained from sediments associated with the years prior to the lake's eutrophication (1960s), experienced large reductions in individual juvenile growth rates when experimentally fed a low-quality Cyanobacteria diet over a high-quality diatom one. This was attributed to the fact that these "historic" hatched *Daphnia* had little to no prior exposure to eutrophication-associated Cyanobacteria dominance. On the other hand, *Daphnia* clones hatched from the years amid eutrophication (1978–1980), as well as those just after eutrophication (1992–1997), experienced less reduction in juvenile growth rates when fed Cyanobacteria. Thus they demonstrated that *Daphnia* populations could evolve dietary resistance to low-quality cyanobacterial food in just a few years. Similarly, for a common copepod species (*Leptodiaptomus minutus*), Derry et al. (2010) showed that copepod nauplii resurrected from eggs associated with the period of industrial lake acidification were more tolerant (greater survival) of acidic conditions (pH of 4.7) than individuals resurrected from either prior to industrialization or after recovery of lake pH (to 6.5) when acid-causing emissions had been eliminated.

Evidence for rapid evolution and for the importance of the effects of evolution on ecological dynamics also comes from common-garden and laboratory experiments. Yoshida et al. (2003) demonstrated that rotifer (*Brachionus calyciflorus*) populations feeding on prey consisting of multiple clones of algal protists displayed different predator–prey dynamics (long, fully out-of-phase predator–prey cycles; no extinctions) than those with feeding on a single algal clone (short, quarter period lag predator–prey cycles; rotifer extinction). Thus, where there was a high evolutionary potential, because of the presence of genetic variation through multiple algal clones, they observed an effect of adaptation leading to sustained predator–prey cycles and no extinctions. Establishing the importance of such effects in natural ecosystems, where environmental conditions are also shifting regularly, inducing adaptive responses, remains an ongoing challenge in limnology.

II. Zooplankton food web functioning

The species interactions discussed in Section I form the basis of more complex food webs to which we will now turn our attention.

A. Microbial food web interactions

The main components of the lower zooplankton food web are those implicated in the microbial food web or *microbial loop*. These are predominantly protist zooplankton (heterotrophic flagellates and ciliates) that feed on bacterioplankton (Fig. 20-3). These small

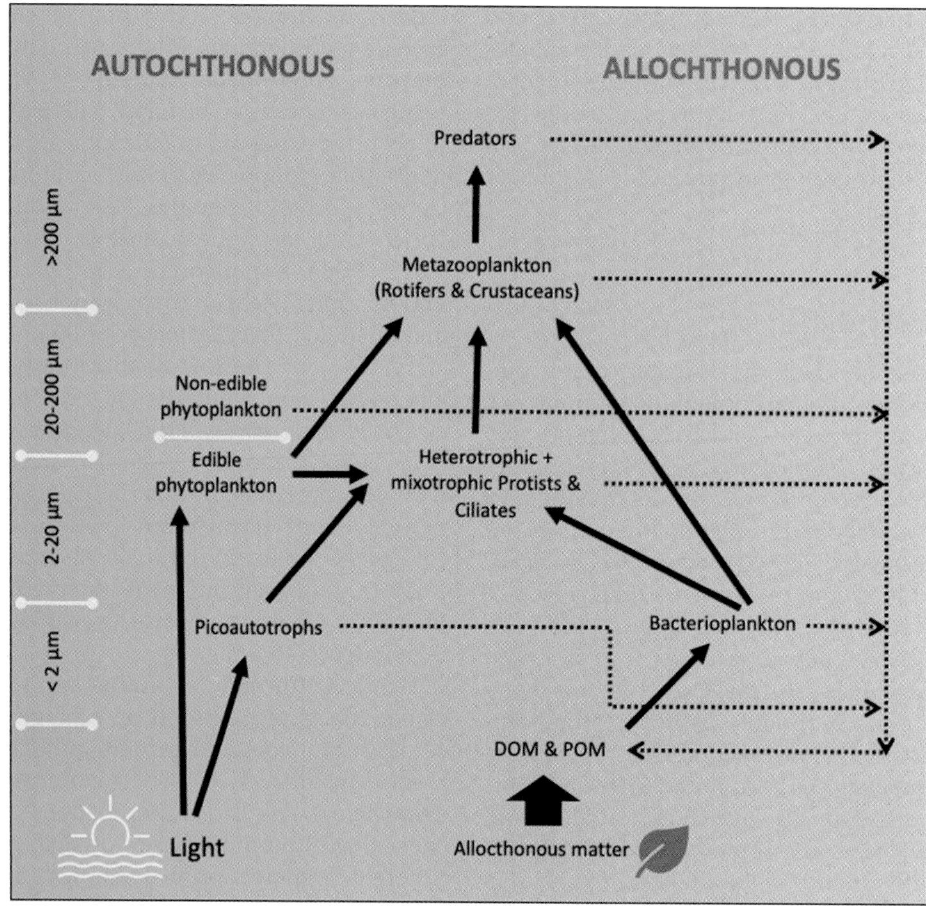

FIGURE 20-3 A simplified schematic of the main pathways of organic matter flow in the pelagic food web linking the microbial food web to the metazooplankton. The *green* (*left*) portion of the figure show pathways with *solid arrows* that depend primarily on photosynthesis (*Green food web*) and the *brown* (*right*) portion highlights those that are driven primarily by allochthonous matter (mostly carbon) input from terrestrial detrital sources (*Brown food web*). *Dotted arrows* show the principal internal sources of dissolved and particulate organic matter (DOM & POM) recycling. The base of the Brown food web is DOM and POM derived from allochthonous terrestrial inputs (*large arrow*) as well as food web internal recycling (*dotted arrows*), while the base of the *Green food web* is sunlight driving photosynthesis. Approximate size of the food web components is according to the log scale indicated at the *left* of the figure. *(Source: Authors' figure compiled and adapted from Jones, 1992 and Rochera and Camacho, 2019.)*

(<5–8 µm) zooplankton consist of nonpigmented, phagotrophic "nano"-flagellates that often dominate heterotrophic planktonic protist communities as well as larger planktonic ciliates (see Chapter 19). Ciliates can be important bacterivores and exhibit high rates of bacterial clearance under both natural and experimental conditions. Some phytoflagellates can also be phagotrophic (mixotrophic protists; Chapter 17) and ingest bacteria at rates comparable to those of the nonpigmented (heterotrophic) flagellates (e.g., Boraas et al., 1988; Sanders and Porter, 1988). Mixotrophic protists show a range of nutritional strategies, however, being more or less phototrophic or phagotrophic (Jones, 2000).

A major function of the microbial food web is to metabolize significant fractions, at least 50% under most conditions, of the *dissolved organic matter* (DOM) and *nonliving particulate organic materials* (POM) (Chapter 28) to gases, soluble nutrients, and living bacterial biomass (Fig. 20-3). This bacterial biomass may be living, dormant, senescing and utilized by other bacteria and viruses, or it may be consumed by bacterivores (primarily flagellate and ciliate protists). Several factors affect rates of bacterivory by protists and ciliates including temperature, light, prey densities, size, and shape. For example, the mixotrophic chrysophyte protist *Ochromonas* sp. grazes increasingly on bacteria (over photosynthesis) as temperatures increase from 13°C to 33°C (Wilken et al., 2013). Size and shape of prey are important, and clearly prey geometry is the first-order determinant of ingestion through mechanical selection by protists. Among both flagellate protists and ciliates, but particularly among the heterotrophic flagellates, there is a tendency to select for larger bacterial cells (Andersson et al., 1986; Chrzanowski and Šimek, 1990;

Šimek and Chrzanowski, 1992), and chemical discrimination among prey is possible (Verity, 1991).

Feeding relationships between ciliates and other protists also occur (not shown explicitly in Fig. 20-3). The nature and quantity of available food have been implicated as major controlling factors in the population dynamics of ciliates. Selectivity of phytoflagellates ingested by ciliates is common (e.g., Takamura and Yasuno, 1983), with food selection largely a function of the size and shape of the protist prey rather than the species. Phagotrophic ciliate ingestion of picophytoplankton often equals that of smaller metazooplankton (copepod nauplii and rotifers). Grazing impacts 20% to 65% of the production of the algal picoautotrophs (<2 μm in size) can occur (e.g., Tirok and Gaedke, 2006).

B. Metazoan food web interactions

The metazoan component of pelagic food webs consists of rotifers and crustacean zooplankton (copepods and cladocerans). As central food web members, metazooplankton link the lower microbial food web to the invertebrate and fish predator trophic levels (Fig. 20-3). Relative to other zooplankton, rotifers generally have high threshold food concentrations in relation to their size (Duncan, 1989), likely restricting their relative importance to benthic particle-rich environments. For this reason, it is only a minority of rotifer species that are implicated in the pelagic zooplankton food web. Predatory rotifers, like *Asplanchna*, are usually large and prey upon protozoa, other rotifers, and other micrometazoa (including cladocerans).

There is evidence for the trophic effects of crustacean zooplankton on rotifers and on each other. The multidecadal time series for zooplankton communities in Lake Constance shows effective population control of both *Daphnia hyalina* and the large predatory rotifer *Asplanchna* by the carnivorous copepod *Cyclops vicinus* (Walz et al., 1987). Furthermore, when predation pressure relaxed in spring as *Cyclops* entered diapause, *Asplanchna* and other rotifers developed rapidly again, before *Daphnia* populations could recover. Other rotifer species were effectively preyed upon by both *Cyclops* and *Asplanchna*. *Cyclops* predation on rotifers has also been revealed in Lake Michigan (USA) zooplankton communities (Stemberger and Evans, 1984). Population dynamics of the predatory cladoceran, *Leptodora*, have been correlated with loss rates and population regulation of cladocerans and copepods across very different lake environments, also showing seasonal shifts between prey types (e.g., Lunte and Luecke, 1990; Branstrator and Lehman, 1991; Hellsten and Stenson, 1995; Levesque et al., 2010).

Invertebrate predation is especially relevant for crustacean and rotifer zooplankton dynamics and diversity in lakes where there are few planktivorous fish. Effects generally occur on zooplankton of a smaller size range than can be utilized effectively by fish. Several studies have shown that invertebrate predation is size selective and can be sufficient to eliminate certain small species, shifting zooplankton communities to larger species (e.g., McQueen, 1969; Dodson, 1970, 1972; Confer, 1971; Confer and Cooley, 1977; Confer and Applegate, 1979; Lynch, 1979; Zaret, 1980; Murtaugh, 1981; Vanni, 1988; Lehman, 1991). For example, rotifer and protist populations are consumed by invertebrate predators, commonly the larval dipteran midge *Chaoborus*, a voracious predator also of crustacean juveniles (e.g., nauplii), and small cladocerans. Detailed studies of *Chaoborus* predation on rotifers in several lakes indicate, however, that the reproductive output of new rotifer individuals may be able to exceed predatory losses (Havens, 1990; Rodusky and Havens, 1996). Other predatory and omnivorous zooplankton, including copepods and some Cladocera genera (*Leptodora*, *Bythotrephes*, *Polyphemus*), are also consumers of protists and rotifers, capable of inducing shifts to larger body sizes in zooplankton communities. On the other hand, when larger invertebrate predators such as *Mysis* sp. (Malacastrocans) have been introduced to lakes (as has been done to support larger fish populations), their predation has led to nearly total suppression of all cladocerans and many copepods (e.g., Murtaugh, 1983; Langeland, 1988), enabling the proliferation of rotifers and a shift to smaller community body sizes.

Invertebrate predation tends to be predominantly tactile because, although the eyes of rotifers, crustaceans, and most insects detect light intensity and movements, they do not form images. *Polyphemus* may be an exception to this rule, using visual cues to chase and leap upon their prey (mostly smaller body-sized Cladocera, including their own young) (Young and Taylor, 1988), although predatory copepods may also use precision attack jumps when capturing prey as do their marine counterparts (Kiørboe et al., 2009). The general relationship of size selectivity and tactile responses in invertebrate predator consumption suggested to led Dodson (1974) to postulate that the shape of prey within the correct size range can influence whether the prey is taken or rejected. Thus *cyclomorphosis* leading to body extensions, such as spines, would have an adaptive advantage for prey subject to invertebrate predation by altering morphology while still permitting growth. Heightened predation by *Leptodora* often coincides with increased spine length in *Bosmina* and is considered an inducible defense (Fig. 20-4) (Sakamoto et al., 2015). Similar responses can be observed to *Chaoborus* larval predation on protists, rotifers, and small cladocerans.

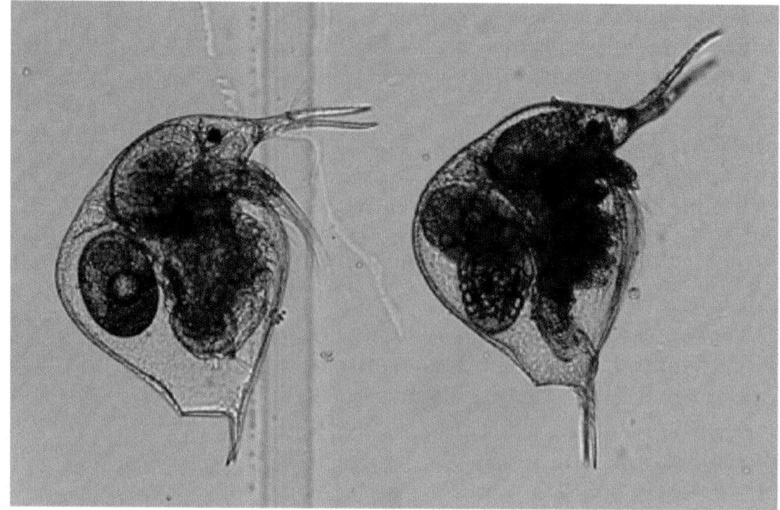

FIGURE 20-4 Undefended (*left*) and induced defensive (*right*) morphologies of adult female *Bosmina fatalis*, in which the latter has longer appendages (antennule and mucrone) and upward antennules. Measurements were body length (*BL*), antennule length (*AL*), and mucrone length (*ML*). *(Source: Fig. 1 in Sakamoto, M., Nagata, T., Ha, J.Y., Kimijima, S., Hanazato, T., Chang, K.W., 2015. Inducible defenses as factor determining trophic pathways in a food web. Hydrobiologia 743, 15–25.)*

In addition to such gross morphological changes, zooplankton can also develop ultrastructural defenses against invertebrate predators, as demonstrated for *D. magna* exposed to *Triops cancriformis* (Rabus et al., 2013).

C. Linking microbial to metazoan food webs

The linking of the microbial to the metazoan zooplankton community is critical to food web functioning as a major fraction of pelagic metabolism and nutrient cycling occurs within the microbial portion. While metazoan zooplankton are mainly consumers of the phytoplankton (autotrophic and mixotrophic protists) in freshwater food webs, they are also capable of ingesting bacteria, both in aggregates and freely in the water (Bern, 1987; Knoechel and Holtby, 1986) (Fig. 20-3). When large cladocerans, such as *Daphnia*, are present in dense populations, nanoflagellate protists can be markedly suppressed. Comparative analyses demonstrate that phytoplankton is the major constituent of the diet of many crustacean zooplankton taxa, with bacterioplankton consumption alone being insufficient to support growth and reproduction (e.g., Pedros-Alio and Brock, 1983; Forsyth and James, 1984; Borsheim and Andersen, 1987; Kankaala, 1988; Sanders et al., 1989; Urabe and Watanabe, 1991; Wylie and Currie, 1991). Clearly, some nutritional component of phytoplankton is needed by the majority of crustacean zooplankton that is critical to growth and reproduction (see Section IID). Thus, where phytoplankton resources are plentiful, cladoceran feeding on bacteria may be in large part incidental to the process of phytoplankton feeding. However, under some conditions, significant (e.g., 25%–70%) portions of bacterial production can be consumed by metazoans, especially rotifers, but also macrozooplankton, especially under eutrophic conditions where bacterial abundance can be high (Hwang and Heath, 1999). Cladoceran and copepod bacterivory can also increase where fish predation is low, permitting dominance of larger-bodied species like *Daphnia*, applying heavy predation pressure not only on protists but also on bacteria (e.g., Jürgens et al., 1994a, 1994b, 1997). Similarly, naupliar stages of copepods can graze heavily on protozoans and bacteria (Merrell and Stoecker, 1998; Roff et al., 1995).

In addition to bacterial uptake of nonliving organic matter (DOM and POM), the numerous pathways present in the microbial food web allow for the recovery of fixed organic carbon (via photosynthesis), often into larger-sized microorganisms that may be more available for consumption by larger organisms (Sherr and Sherr, 1988). Such conversion by microzooplankton of organic matter and energy within the *microbial loop* into sources of higher quality (e.g., fatty acid composition) or more accessible food (e.g., larger particle size) for larger zooplankton has been termed *trophic upgrading* (Klein Breteler et al., 1999). Many direct links exist among algae, bacteria, and other heterotrophic microbes, as discussed previously (Fig. 20-3). The link to metazooplankton occurs when these feed on organisms of the microbial food web, primarily phytoplankton, colorless heterotrophic nanoflagellates, and ciliates larger than 5 μm. In oligotrophic waters that may be dominated by phytoplankton, including *picoautotrophs* too small (c. <5 μm) to be effectively consumed by many zooplankton, the phagotrophic flagellates and ciliates could function as a primary trophic link within the ecosystem.

When larger zooplankton are abundant, mortality rates on ciliates can be high, which would allow an appreciable transfer of carbon and energy from the microbial loop to higher trophic levels. Large differences in rates of ingestion of protists have been observed, and these rates are not necessarily proportional to the size of the prey or predator (Wickham, 1995). Although feeding clearance rates of copepods (mL copepod^{-1} day^{-1}) and other large zooplankton do increase in proportion to body length, under natural conditions many types of prey may be available, potentially reducing ingestion rates. Predation on ciliates, for example, was appreciably less when copepods had alternative, likely preferred, prey species available (Wickham, 1995). Experimental studies both in the laboratory and in the field demonstrated marked increases in daily prey mortality with increases in the size of different species of the cladoceran *Daphnia* (μg per species) (Pace and Vaqué, 1994; Wickham and Gilbert, 1993). Heterotrophic nanoflagellates were suppressed more than ciliates, and growth rates of both the flagellates and especially the ciliates were more severely suppressed by the large *Daphnia*.

Where nutrient levels are low and dissolved organic carbon (DOC) is high (e.g., many oligotrophic northern lakes surrounded by forests or peatlands), the origin of the organic C consumed by some crustacean zooplankton may be linked more to terrestrial than aquatic origins (Carpenter et al., 2005; Karlsson et al., 2003). Thus, instead of relying on a C-resource originating from primarily phytoplankton primary production (*Green food web*; Fig. 20-3), crustaceans are situated in a food web pathway that is based on a C-resource that originates in terrestrial primary production (e.g., tree detritus such as leaves) that makes its way into lakes (*Brown food web*; Fig. 20-3). Isotope marker studies (with stable isotopes: $\delta^{13}C$, $\delta^{15}N$, $\delta^{2}H$ and radioisotope: ^{14}C) have examined the relative contribution to crustacean zooplankton diet composition of terrestrially derived organic matter (*allochthonous*) versus that from lake primary production (*autochthonous*). The way in which allochthonous organic matter gets into crustacean zooplankton was hotly debated for several years with two competing hypotheses: (1) crustaceans directly consume POM from terrestrial detritus, or (2) DOM from terrestrial origins is consumed by bacteria, which in turn are consumed by microzooplankton (protists, ciliates) that are fed upon by crustaceans. A cross-lake study in the northern boreal region of Canada by Berggren et al. (2014) demonstrated that the mechanism depends on the crustacean group: the direct POM pathway appeared to apply best to calanoid copepods, the indirect DOM pathway to cyclopoid copepods, and there was evidence for both direct POM and indirect DOM pathways for cladocerans. The signal for *allochthony* was highest in the Cladocera (31% body C content of terrestrial origin) relative to Copepoda (18% in Cyclopoida and 16% in Calanoida). It should be noted, however, that there is likely to be seasonal variation in the contribution of allochthonous C; Grey et al. (2001) found across mixed zooplankton assemblages that allochthonous contributions to body C content ranged from a low of 4% at the end of summer (September) to a high of 89% in spring (May). In seasonally constrained ecosystems, such as in northern regions, allochthony also appears to play an important role in sustaining zooplankton over the winter months, when lake primary productivity is at its lowest (Rautio et al., 2011). Overall, it appears that when allochthony dominates resource provision in a lake, bacteria can provide an alternative, albeit lower-quality, food source for Cladocera (Kelly et al., 2014). However, the Brown food web leads to lower crustacean zooplankton production because it has high carbon-to-phosphorus (C:P) ratios, indicating that it is low in the limiting nutrient P, while also lacking essential fatty acids (Section IID).

Several studies show that the assimilation efficiency of microbial loop prey by metazooplankton increases with higher energy content of food, as well as temperature (Pechan'-Finenko, 1971; Schindler, 1968). However, the assimilation efficiency of phytoplankton and bacteria by zooplankton is variable, although generally algae are assimilated more efficiently (15%–90%) than bacteria (3%–50%) (Hart and Jarvis, 1993; Monakov and Sorokin, 1961; Ojala et al., 1995; Saunders, 1969; Winberg et al., 1973). Percentage assimilation also varies seasonally with shifts in phytoplankton composition and the availability of alternative prey types. When given mixed diets of green algae and a photosynthetic bacterium, *Ceriodaphnia* assimilated less of the algae than when fed on algae alone (Gophen et al., 1974). Food collection efficiency can also vary in association with shifts in algal composition, most commonly across lake types (oligotrophic to eutrophic). In eutrophic lakes it is common for communities to be dominated by small-bodied zooplankton species. This is because the larger (often filamentous) algae that dominate eutrophic lakes interfere with food collection to a greater extent in larger cladocerans, causing reduced growth and fecundity than in smaller cladoceran species that feed on small particles (Gliwicz, 1980; Gliwicz and Siedlar, 1980). In this way larger cladocerans can experience a reduced efficiency of food collection under eutrophic conditions and may consequently be selected against.

D. Food quality and lipids in zooplankton food webs

While *maintenance (basal) metabolism* requires mainly energy, body growth requires other substances, including essential minerals and compounds, in the diet (Sterner and Hessen, 1994; Sterner and Robinson, 1994). For example, phytoplankton (algae including

Cyanobacteria) with a low growth rate resulting from a low P-availability will contain reduced P and be an inferior food source for cladoceran growth, especially *Daphnia*. However, the C content per unit biovolume will be similar among both low- and high-growth-rate algal food. Thus both are satisfactory as energy sources for maintenance metabolism, but the resulting low P:C ratio in fast-growing phytoplankton will render them of lower quality for body growth that requires essential nutrients like P. Therefore herbivores with high nutrient demands, exemplified by P-rich *Daphnia*, appear frequently to be limited not by food quantity or energy available but by food quality. The role of such ratios, especially C:N:P, in food web functioning is the basis of the study of ecological stoichiometry (e.g., Hessen et al., 2013).

In zooplankton, lipids are dominant energy storage compounds and can constitute a significant portion (to 60%–70%) of their dry mass (Arts, 1999). Lipids in zooplankton and fish are primarily dietary in origin, and consequently the type and amount of lipid contained within zooplankton tissues are often correlated with recent feeding activities and food selectivity. The correlation between food type ingestion and body lipid content can be modified by temperature (Arts et al., 1993) and reproduction (Vanderploeg et al., 1992), resulting in distinct temporal patterns of lipid content in the zooplankton. Storage lipids are composed primarily of nonessential lipids and constitute major energy reserves.

Some *polyunsaturated fatty acids* (PUFAs) and *sterols* are essential for maintaining high growth, survival, and reproductive rates of zooplankton. These *essential fatty acids* (EFAs) include linoleic (18:2ω6), linolenic (18:3ω3), and likely eicosapentaenoic (20:5ω3) acids. The phospholipids, which are abundant in cellular membranes, contain a high percentage of essential PUFA. The absolute quantity of phospholipids in zooplankton tissue is partly under genetic control and relatively constant. Lipid reserves are essential to the survival of all cladoceran life stages when food availability is low (Goulden et al., 1999; Goulden and Henry, 1985; Lampert and Bohrer, 1984).

Food quality in terms of the availability of EFA in the algal food is also critical for the normal growth and survival of *Daphnia* and other crustacean zooplankton (Arts et al., 1992; Holm and Shapiro, 1984). A number of correlational analyses suggest strongly that the nutritional quality of phytoplankton communities and their essential fatty acid content can be an important factor in regulating species and seasonal succession in zooplankton (Ahlgren, 1993; Ahlgren et al., 1990). The lipid content of zooplankton has been found to decrease from spring to early summer and to increase again in late summer and autumn (Arts et al., 1993, 1992; Brett and Müller-Navarra, 1997; Wainman and Lean, 1990), enabling overwinter survival and reproduction (Mariash et al., 2016; Schneider et al., 2017) and thus affecting early spring dynamics. PUFA are critical for maintaining high growth, survival, and reproductive rates of zooplankton. High-food-quality algae are rich in such PUFA (ω3) content (Brett and Müller-Navarra, 1997; DeMott and Müller-Navarra, 1997; Sundbom and Vrede, 1997; Weers and Gulati, 1997). Growth rates of *Daphnia*, for example, can be limited directly by the availability of eicosapentaenoic acid (20:5ω3) within seston used as food (Müller-Navarra, 1995). If fed cryptophytes or green algae with high PUFA content, or if emulsions of PUFAs are added to algal cultures of poor food quality, growth and reproduction rates of zooplankton feeding on these algae can increase markedly.

Finally, toxins can be present in the prey community upon which zooplankton feed, especially those produced by Cyanobacteria. Many metazooplankton can selectively feed on a mixed prey community. Copepods are highly selective feeders and many will thus reject toxic filaments, but rotifers and cladocerans also have some capacity for selective feeding to avoid toxic prey (e.g., Kirk and Gilbert, 1992). While the presence of toxic compounds in the diet of zooplankton may not influence overall zooplankton grazing rates, it does affect grazer survival rates (Wilson et al., 2006), although these may depend on the species or even genotypes of both predator and prey (Lemaire et al., 2012). Generally, a combination of nutritional factors, prey morphology, and to some extent, toxicity are likely involved in determining food quality for most zooplankton.

E. Zooplankton food web interactions with fish predators

Crustacean metazooplankton are consumed primarily by vertebrate fish predators, as well as by invertebrate predators (Section IIB) that are also potential prey for fish. The key component of the interaction between zooplankton and their vertebrate fish predators is prey visibility. Size-biased feeding by fish that select the larger species of a zooplankton community (particularly *Daphnia*) or larger members of a single species has been documented in at least 20 species of fish (reviewed in Gerking, 1994). However, the relationship of size and planktivore predation is not always that simple. Zooplankton prey that are large, but relatively transparent, such as the predatory cladoceran *Leptodora*, are often relatively invisible, reducing predation rates (see, e.g., Costa and Cummins, 1972). Where polymorphic variation in the size of a pigmented eye is present (e.g., *Ceriodaphnia cornuta*), the form with the larger pigmented eye will be heavily consumed by planktivorous fish, despite being otherwise the better competitor (being the dominant conspecific in the absence of fish predation), and despite no variation in body size (Zaret, 1972a, 1972b). Similar results have been observed in the effects of fish predation on *Bosmina* (Hessen, 1985; Zaret and Kerfoot, 1975), with prey selection being related to the large, black-pigmented eye and not body size. Prey movement may also be important: the slower, steady movements of cladocerans render them more vulnerable to fish predation than the jerky, irregular movements of copepods, which can also perform rapid "escape jumps" to evade predators. Although often a response to invertebrate predation, *cyclomorphosis* may also reduce the effectiveness of fish predation when it involves helmet formation (Lampert and Wolf, 1986). The environment can also influence fish predation rates on zooplankton by affecting visibility. Particulate turbidity, especially clay turbidity that is common in reservoirs, reduces visibility and the distance at which predator–prey interactions occur (Abrahams and Kattenfeld, 1997). This constraint will reduce fish predation risk on zooplankton in turbid ecosystems. Similarly, highly stained "brown" waters owing to colored dissolved organic matter (DOM; Chapter 28) may reduce predation pressure on crustacean zooplankton by some fish predators (van Dorst et al., 2018).

When size selection by fish is not in effect and large zooplankters are present, a common observation is that the smaller-sized zooplankton are not generally found to cooccur. Brooks and Dodson (1965; Brooks, 1968) developed the *size-efficiency hypothesis* in an attempt to explain the commonly observed inverse relationship between the abundances of small- and large-bodied herbivorous zooplankton in lakes. According to this hypothesis (Hall et al., 1976):

- **a.** Larger zooplankton filter prey particles (1–15 μm) more efficiently and can also take larger particles than smaller zooplankton. This greater effectiveness of food collection leads to relatively reduced metabolic demands per unit mass, permitting more assimilation to go into egg production by the larger herbivores.
- **b.** Therefore, when the intensity of predation pressure by fish is low, the small planktonic herbivores will be competitively eliminated by large forms (dominance of large cladocerans and calanoid copepods).
- **c.** However, when fish predation is intense, size-dependent predation will eliminate the large forms, allowing the small zooplankton (rotifers, small cladocerans) that escape predation to become the dominants.
- **d.** When predation pressure is moderate, larger zooplankton species are often kept at reduced population sizes that allow the coexistence of smaller competitors.

Generally, vertebrate predation restricts the maximum adult body size of zooplankton, and invertebrate predation often restricts the minimum size. Both of these effects can augment declines in zooplankton productivity caused by food limitations. Minimal food concentrations needed for growth among different species of *Daphnia* decrease as the size of the animals increases (Gliwicz, 1990). As a result, when predation pressure is low, larger cladoceran species should be more successful competitors for food. Large cladoceran species possess greater reproductive rates (intrinsic rates of increase, r) (Goulden et al., 1978). Although having longer time to maturity than smaller species, the larger forms have higher fecundities (number of young produced per female per day, b), which results in higher rates of population increase.

The size-efficiency relationships between fish and zooplankton have major implications for food web structure and have led to research on the potential manipulation of food webs by fish introductions (*biomanipulation*). Fish predation clearly can influence species composition and size structure of zooplankton prey. Morphology, physiology, and behavior of prey can all be affected by size-biased feeding, which can lead to significant evolutionary alterations for predator avoidance. Cyclomorphosis among cladocerans is at least in part a result of predation pressures. Similarly, diel vertical migration and other evasive behavior among zooplankton are partially induced by predation pressures and the presence of predators' kairomones (Lampert, 1993; O'Brien, 1987; Ringelberg, 1991).

Fish predation on zooplankton is unlikely the dominant means of mortality among larger zooplankton size classes (>1.0 mm). Mortality of zooplankton occurs via other factors even under nearly ideal growing conditions (e.g., Boersma et al., 1996), including simple physiological death often after resting egg formation, parasitism, microbial infection, invertebrate predation, and physical degradation (e.g., outflow and death in flowing systems). Thus size-selective predation by fish likely interacts with food limitations and competition to regulate population dynamics within zooplankton communities. For example, population densities of zooplankton of a shallow temperate lake were affected much more by manipulations of food availability than fish predation (Vanni, 1987). Increases in birth rates and declines in mortality rates across cladocerans, some copepods, and rotifers resulted from increased phytoplankton availability. The densities and size of a few cladoceran populations were reduced by planktivorous fish predation, and the size at time of reproduction was smaller. Flexibility in these life history traits allowed the zooplankton community as a whole to withstand intense size-selective predation by planktivorous fish.

Pelagic fish-zooplankton interactions may become more complex in shallow lakes (Chapter 26). Total productivity and particularly fish biomass per unit volume tends to be much higher in shallow lakes than in deep lakes. Similarly, biomass of benthic invertebrates per unit area is greater in shallow lakes, and as a result, fish feeding tends to be less dependent on zooplankton prey (Jeppesen et al., 1997). Fish also tend to shift feeding between zooplankton and benthic animals as changes in prey densities occur seasonally. In pelagic areas of shallow lakes cladocerans tend to be severely impacted by predation as the effectiveness of vertical migration is reduced, unless a strong horizontal migration ensues (Burks et al., 2001). As a result, grazing pressure on pelagic phytoplankton is often reduced. Predation on zooplankton by benthic and planktivorous fish is reduced markedly in shallow lakes rich in submersed macrophyte beds as these littoral areas can thus serve as a refuge for pelagic cladocerans (Burks et al., 2001; Jeppesen et al., 1998), increasing zooplankton grazing of both phytoplankton and bacterioplankton among the submersed macrophyte areas. Much of the nonmacrophyte primary productivity within the littoral areas shifts to the algae and Cyanobacteria that are attached to the surfaces of macrophytes and other substrata. An additional presence of piscivorous fish tends to reduce fish planktivory on zooplankton in the pelagic zones of shallow lakes as pelagic dwelling piscivores induce small planktivorous fish to make greater use of vegetated habitats; increases in the populations of larger cladocerans can thus be observed in pelagic areas (Turner and Mittelbach, 1990).

F. Invasive species

Invasive species are challenging ecological interactions in all types of ecosystems, including in inland waters. In the pelagic food webs within which zooplankton play a critical role, invaders may include organisms at any trophic level, including macrophytes and other littoral species such as gastropods, which can also influence zooplankton community functioning. As the list of invaders making their presence known grows each year, we can only cover a few recent examples of their interaction with zooplankton communities here.

When invasive species influence the structure of the zooplankton habitat, such as invading macrophytes do in shallow lakes, there can be consequences for zooplankton community diversity and functioning. For example, *Eichhornia crassipes* (water hyacinth; Fig. 20-5), native to the Amazon Basin, is an important invader in other tropical regions that can quickly come to cover large areas of water bodies owing to very high reproductive and growth rates. As a free-floating macrophyte, this species can block light for primary production, leading to zooplankton and fish kills in pelagic environments, but also to altered structure of the littoral regions of lakes. Paradoxically, in its native region the presence of this species in ponds and wetlands can increase crustacean zooplankton diversity, both in terms of the number of species present and in terms of the functional traits supported (Stephan et al., 2019): zooplankton that can take advantage of the habitat offered by the macrophyte itself with the capacity to burrow into the plant to feed, and to feed more omnivorously, including on detritus, are favored in its presence. However, in invaded environments where zooplankton do not possess such traits, this species can have more negative effects on native biodiversity (Villamagna and Murphy, 2010).

One of the most studied zooplankton invaders to North American lakes is the European *Bythotrephes longimanus* (also identified as *B. cederstroemi*; Fig. 20-5), commonly known as the spiny water flea. As a predatory cladoceran feeding on other crustacean zooplankton, the *Bythotrephes* invasion has led to reduced zooplankton production and significant changes in several components of the zooplanktonic food web including crustacean size structure, as well as rotifer and protist phytoplankton composition (Fig. 20-6); (Strecker et al., 2011; Strecker and Arnott, 2008). However, some of the compositional shifts (mainly a greater representation of the jelly-clad cladoceran *Holopedium* in place of *Daphnia*) have also been related, in part, to cooccurring long-term declines in calcium (Ca) concentration in softwater lakes associated with recovery from 20th-century industrial acidification (Jeziorski et al., 2008, 2014). Perhaps most interestingly, *Bythotrephes* invasion appears to have altered native crustacean zooplankton diel vertical migration behavior. Unlike most other invertebrate predators that feed via *mechanoreception* (based on touch or pressure changes), *Bythotrephes* appears to be a visual predator, occupying primarily the shallow, well-lit regions of lakes. It is in turn protected in these upper waters from fish predation to some degree by its long spiny tail (Fig. 20-5), especially at larger sizes (Branstrator, 2005). Generally, it appears that the invasion of *Bythotrephes* can induce a strong downward migration of crustacean zooplankton prey during the daylight hours, even where it did not occur previously, but not at night—similar to a fish predation effect (e.g., Lehman and Cáceres, 1993; Pangle and Peacor, 2006). Ultimately, it appears that via this mechanism, *Bythotrephes* invasion results in reduced zooplankton prey population growth rates as the latter must spend more time inhabiting the cooler, unlit,

FIGURE 20-5 Examples of some of the more notorious invasive aquatic species that have had large influences on lake ecosystems, including zooplankton, as discussed in the text. Starting at the *top left corner and moving clockwise*: *Eichhornia crassipes* (Water hyacinth; photo © Hans Hillewaert), *Dreissena polymorpha* (Zebra mussel; photo by Dan Minchin), *Orconectes rusticus* (Rusty crayfish; photo source USFWS—Pacific Region), and *Bythotrephes cederstroemi or longimanus* (Spiny water flea; photo source NOAA Great Lakes Environmental Research Laboratory). *(Source: Photos from Wikimedia.)*

hypolimnetic waters (Pangle et al., 2007). In fact, this nonlethal effect of *Bythotrephes* can be as important as the direct predator effect of this invasive species in some invaded North American lakes, including the Great Lakes.

An increasing number of other invaders, too numerous to discuss in detail here, have had important effects on the plankton communities of lakes, including food web structure and function. Located centrally in North America's geography and as an economic hub for global shipping, the Great Lakes are among the most likely freshwaters to receive invasive species in the world. Combined with the large number of limnologists in this region, aquatic invaders to lakes in this area have received a great deal of scientific scrutiny, including pelagic invaders such as *Bythotrephes*. Other important invaders include benthic organisms such as the zebra mussel (*Dreissena polymorpha*; Fig. 20-5) that consume water column protozoans, influencing community composition and production (Lavrentyev et al., 1995), and the rusty crayfish (*Orconectes rusticus*; Fig. 20-5) that may alter both littoral and pelagic food webs, reducing planktivorous fish populations (Kreps et al., 2016). Globally, fish invaders are in the process of replacing native populations (Olden and Poff, 2004; Villéger et al., 2011), thereby directly and indirectly perturbing freshwater food webs within which zooplankton occupy a central place.

III. Zooplankton productivity

The production rates of specific populations of zooplankton refer to the *net productivity* or the sum of the growth increments of all individuals in the population over a specified time period. This net productivity excludes maintenance losses (respiration and excretion) and includes the growth increments of the organism itself, as well as the biomass produced as gametes and as exuviae during molting. In some invertebrates and some vertebrates the productivity values of specific animal populations are influenced markedly by emigration and immigration (e.g., where dispersal among metacommunities is important). The productivity measurements are complicated when predation causes the removal of a significant portion of the population; this loss to the population can be difficult to quantify accurately.

Traditionally, the manner in which production rates of a specific population are estimated depends

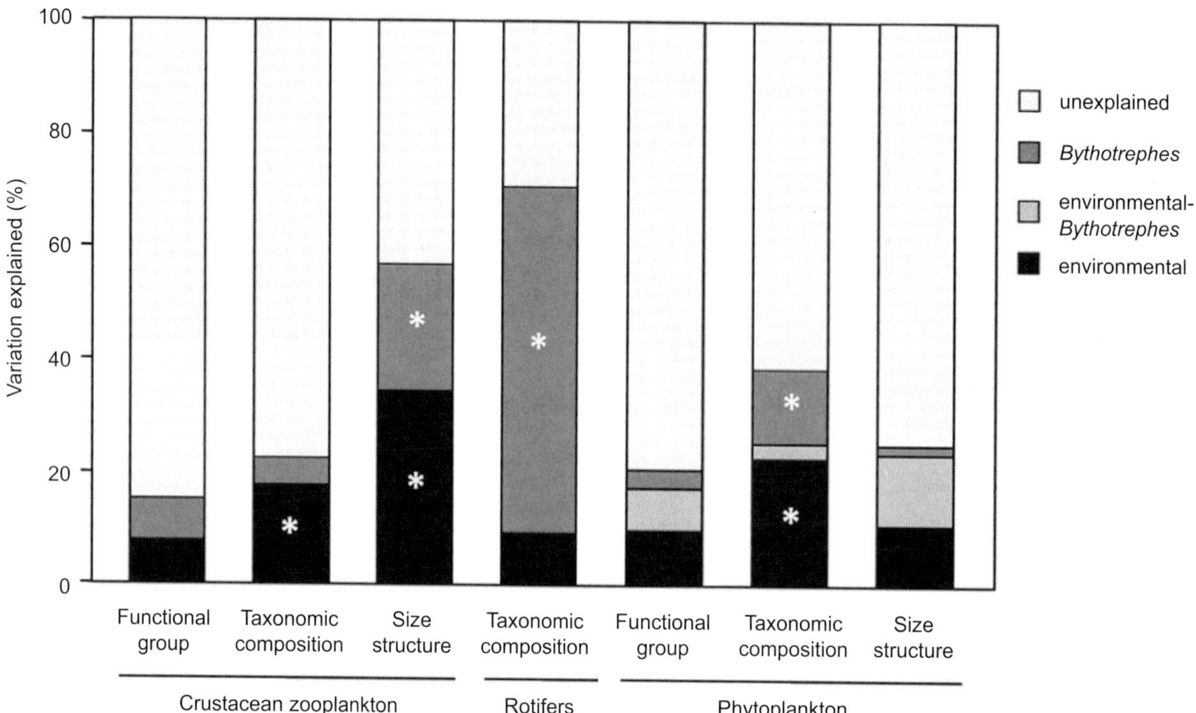

FIGURE 20-6 The relative importance of the invasive (from Europe to North America) predatory cladoceran, *Bythotrephes*, in comparison with the effect of environmental factors varying across lakes and the covariation of these environmental factors with *Bythotrephes*. The differently shaded bars show the proportion of total variation explained by each factor for crustacean zooplankton functional groups, taxonomic composition, and size structure; rotifer taxonomic composition; and phytoplankton functional groups, taxonomic composition, and size structure. The significance of fractions ($P < 0.05$) is indicated by: *. Note that significance testing of the environment–*Bythotrephes* fraction could not be performed. Effects of *Bythotrephes* were mainly associated with significant variation in rotifer and phytoplankton (photosynthetic and mixotrophic protist) community composition followed by crustacean zooplankton size structure, indicating a likely feeding preference on these groups that is limited at larger prey sizes. *(Source: Fig. 1 in Strecker, A.L., Beisner, B.E., Arnott, S.E., Paterson, A.M., Winter, J.G., Johannsson, O.E., Yan, N.D., 2011. Direct and indirect effects of an invasive planktonic predator on pelagic food webs. Limnol. Oceanogr. 56, 179–192.)*

on the particular life cycle, reproductive characteristics, and generation times (discussed at length in Edmondson and Winberg, 1971; Winberg, 1971; Edmondson, 1974; Bottrell et al., 1976; Benke, 1984, 1993; Rigler and Downing, 1984). The productivity of a species with a long life cycle and a short period of reproduction is determined relatively easily if individuals are of the same age or if separate cohorts can be recognized and their biomass determined. In such a situation, after the brief period of reproduction, no recruitment occurs to confound estimates of growth and mortality. When reproduction and recruitment to the population are continuous, as is the case for many zooplankton, the estimation of production is more complicated. Consequently, it is necessary to estimate finite birth and growth rates of individuals and evaluate changes in biomass over the life cycle from birth to death. The reproductive or birth rate can be determined indirectly from the ratio of eggs per female (i.e., *egg-ratio method*) and used to estimate daily increments to the focal population size (see Edmondson, 1960, 1965, 1968, 1974; Paloheimo, 1974).

More recently, new assay methods have been developed to assess crustacean zooplankton community productivity using the production of the molting enzyme *chitobiase* (*N*-acetyl-β-D-glucosaminidase) involved in the production of the new chitinous exoskeleton. Because this enzyme is released into the water when molting occurs, it is possible to measure its turnover rate in the water column. The amount of chitobiase released is related to the size of individuals (Vrba and Macháéek, 1994) and through a series of assays, the concentrations observed in a water body can be used to estimate the rate at which the total biomass of crustaceans is produced (see Sastri et al., 2013).

An index of productivity commonly used is the *P/B coefficient*, the ratio of production (*P*) to biomass (*B*). *B* represents the mass for a population present at one point in time in units of mass per unit area or volume and is therefore a temporary storage of mass or energy. *P* is a flow or flux of mass or energy per area per time. *P/B* is essentially a weighted mean value of biomass growth rates of all individuals in a population with units of inverse time (e.g., year^{-1}) (Benke, 1993). *P/B* ratios can be used as an estimate of the turnover of a population and for comparative purposes of population growth in response to environmental conditions and perturbations. In the case of a population cohort (similar stage animals born in the same time period), the cohort *P/B* ratio is equal to the cohort production divided by the mean cohort biomass (Benke, 1984). The annual *P/B* ratio is the annual production divided by the mean biomass over the entire 12-month period, even if the generation under study was present for less than a year (Wetzel and Likens, 2000).

In populations with a constant age structure and biomass, a rarity in nature, the *P/B* ratio is relatively constant. The ratio varies in most situations when age structure, biomass, and growth are discontinuous; ideally, an average ratio over an interval of time is estimated (e.g., McLaren and Corkett, 1984). Multiplication of the *P/B* ratio by biomass yields an estimation of production during that interval of time. The reciprocal of *P/B*, the *turnover time*, is the average duration of the life of a species under a given set of growth conditions. Biomass turnover times vary from a few hours for protistan zooplankton to a large range of 1.7 days for rotifers to well over 100 days for large zooplankton under poor growth conditions.

Zooplankton productivity varies across aquatic ecosystems of differing primary productivity (Table 20-1). Microzooplankton can represent a significant portion of the total zooplankton productivity, at times exceeding the contributions of the larger zooplankton (Table 20-2). The percentage assimilation of phytoplankton primary productivity by herbivores is nearly always less than 25% (e.g., Table 20-1). Amid the many environmental factors regulating the seasonal production cycle of zooplankton, rates of biomass production are set largely by temperature (Sastri et al., 2013; Shuter and Ing, 1997). Levels of biomass accumulation in zooplankton are determined largely by food resource availability and individual body size. In general, a positive correlation exists between the rates of production of heterogeneous phytoplankton communities and of the heterogeneous micro- and macrozooplankton communities among many different reservoirs and natural lakes (e.g., Canfield and Jones, 1996; Makarewicz and Likens, 1979; Richman et al., 1984; Rublee, 1992; Winberg et al., 1970). The productivity of the filter-feeding zooplankters is distinctly higher than that of the predacious zooplankters (Table 20-1). Variation is great, however, particularly in relation to temperature, food quantity and quality, and other factors. Comparisons over 2 or more years with good sampling show the extent of variation that can occur with the same species in the same lake (Table 20-1; Fig. 20-7). These differences are much greater than can be attributed to sampling and measurement errors.

IV. Zooplankton metacommunity ecology

It has become increasingly clear that dispersal cannot usually be completely ignored when studying the dynamics and functioning of zooplankton in any particular pond or lake. The degree to which dispersal plays a role in local crustacean zooplankton community structure is clearly linked to how connected a particular water body is to other water bodies in a landscape, either through direct waterway connections or via overland distance. In more general ecological language, such water bodies operate as *metacommunities* (Beisner et al., 2006; Cottenie, 2005). Zooplankton metacommunities are sustained by the passive dispersal of different propagules such as resting stages (e.g., ephippia, other dormant eggs, or cysts) by wind, rain, and animal vectors (Cáceres and Soluk, 2002), but also potentially the transfer of live individuals that can occur between stream-connected water bodies where turbulence is minimal (Basu and Pick, 1996; Pace et al., 1992). Molecular markers have aided in the determination of the dispersal potential of zooplankton populations by estimating gene flow between sites (Bohonak and Jenkins, 2003; Zhang et al., 2018). Evidence for the role of dispersal also comes from experiments involving the colonization of mesocosm arrays containing no zooplankton, located at distances from 10 to 200 m from natural ponds, and manipulating the degree to which animal vectors are excluded via netting of different sizes (Cáceres and Soluk, 2002). These experiments show that wind and rain vectors were more important than animal vectors and that there was species-level variation in abilities to disperse or colonize; generally, rotifers colonized earliest and cladocerans later, although a great deal of variation was observed over the 2-year experiment (Fig. 20-8). Finally, the relationship between regional and local richness patterns in crustacean freshwater zooplankton showed a saturating response, indicating that local environmental and species interactions are more important than dispersal. But when expanding the scale of study, the pattern may become more linear, indicating a role for dispersal limitation at regional scales (Shurin et al., 2000).

Many lake and pond crustacean zooplankton function as metacommunities with detectable imprints of both local environmental conditions and spatial factors (distance over which dispersal must occur, often the distance between water bodies) on their compositions (Beisner et al., 2006; Cottenie, 2005; De Bie et al., 2012). It appears that the degree to which dispersal influences community structure corresponds to body size in passive dispersers such as zooplankton (Finlay, 2002), and thus there exists a gradient of effect: protist (autotrophic protists or phytoplankton) and rotifer metacommunities

TABLE 20-1 Examples of productivity of herbivorous and predatory forms of zooplankton. Production of biomass estimated in terms of ash-free dry weight (dw) expressed in g m^{-3} day^{-1} as in this table can be divided by 2 to approximate g C m^{-3} day^{-1} produced (units also sometimes used). Production of biomass estimated in kcal m^{-2} day^{-1} can be roughly converted to g C m^{-2} day^{-1} by assuming that 1 kcal is approximately 0.1 g C.

Type/species	Lake/general productivity	Period of investigation	Production estimates[a] gm $^{-3}$ day $^{-1}$	Production estimates[a] kcal m $^{-2}$ day $^{-1}$	Assimilation % of phyto-plankton production	Average biomass turnover time (days)	Source
Filter feeders							
Cladocera							
Daphnia hyalina	Eglwys Nynydd Reservoir, Wales; eutrophic	Annual, 1970 Annual, 1971	0.57 0.32	0.468	1.7	21.3 15.9	George and Edwards (1974)
	Lake Constance, Germany; mesotrophic	Season (204 days)	0.07–0.1		6.3	29.3	Geller (1985)
	Ardleigh Reservoir, England; eutrophic	Annual, 1981 Annual, 1982	0.023 0.009			16.3 19.5	Mason and Abdul-Hussein (1991)
D. parvula	Severson Lake, MN; eutrophic	Annual	0.010	0.102	0.15	—	Comita (1972)
D. galeata	Lake Constance, Germany; mesotrophic	Season (204 days)	0.07–0.1		4.5	15.7	Geller (1985)
	Lake Esrom, Denmark	Annual		0.329	5.7		Hamburger (1986)
	Lake Esrom, Denmark	Annual	0.0059			10.2	Petersen (1983)
D. galeata mendota	Sanctuary Lake, PA, eutrophic reservoir	May–Nov., 1966 May–Nov., 1967	0.407 0.030	3.026 0.223			Cummins et al. (1969)
	Canyon Ferry Reservoir, MT, eutrophic	April–Sept.	0.114	0.471	4.4	10.0	Wright (1965)
D. schoedleri	Canyon Ferry Reservoir, MT, eutrophic	April–Sept.	0.227	0.943	8.9	6.7	Wright (1965)
D. longispina	Lake Sevan, southern Russia	Annual	0.006			58.9	Meshkova (1952) in Winberg (1971)
D. lumholzi	Lake Samsonvale, Australia	Annual	0.325			7.6	King and Greenwood (1992)
D. rosea	Pond, Japan	Annual	0.044			11.6	Iwakuma et al (1989)
Bosmina longirostris	Severson Lake, MN	Annual	0.007	0.071		—	Comita (1972)
	Ardleigh Reservoir, England; eutrophic	Annual, 1981 Annual, 1982	0.0022 0.0020			18.4 18.6	Mason and Abdul-Hussein (1991)
B. longirostris and *B. coregoni*	Sanctuary Lake, PA	May–Nov., 1966 May–Nov., 1967	0.183 0.067	1.361 0.498			Cummins et al. (1969)
B. meridionalis	Lake Okaro, New Zealand	Annual	0.007			18.7	Forsyth and James (1991)
Ceriodaphnia reticulata	Sanctuary Lake, PA	July–Nov.	0.031	0.154			Cummins et al. (1969)
C. dubia	Lake Okaro, New Zealand	Annual	0.004			6.7	Forsyth and James (1991)
Chydorus sphaericus	Sanctuary Lake, PA	July–Aug., 1966	0.004	0,020			Cummins et al. (1969)
		July–Aug., 1967	0.047	0.233			

Alona affinis	River Thames, England	Annual	0.361			11.7	Robertson (1995)
Leydigia leydigi	Benthic populations in sediments,		0.151			15.7	
Diaparalona rostrata	production per m^2		0.173			16.4	
Cladocera	Naroch Lake, Russia	May–Oct.	0.0026	0.117		13.7	Winberg et al. (1970)
	Myastro Lake, Russia	May–Oct.	0.015	0.403		10.9	
	Batorin Lake, Russia	May–Oct.	0.033	0.484		10.5	
Copepods							
Cyclops strenuus	Buttermere, England; oligotrophic	Annual	0.0004				Smyly (1973)
	Rydal Water, England; eutrophic	Anuual	0.0005				
	Grasmere, England; eutrophic	Annual	0.0006				
	Esthwaite Water, England; eutrophic	Annual	0.0017				
	Lake Sevan, southern Russia	Annual	0.0007			79.3	Meshkova (1952) in Winberg (1971)
C. vicinus	Eglwys Nynydd Reservoir,	Annual, 1970	0.13			22.5	George (1976)
	Wales, eutrophic	Annual, 1971	0.12			19.4	
	Lake Constance, Germany; mesotrophic	Annual	0.003			24.3	Wölfl (1991)
Calamoecia lucasi	Lake Otota, New Zealand; oligotrophic	Annual	0.006			11.3	Green (1976)
Boeckella dilatata	Lake Hayes, New Zealand; eutrophic	Annual	0.032			6.4	Burns (1979)
B. minuta	Lake Samsonvale, Australia	Annual	0.27			12.5	King and Greenwood (1992)
Eudiaptomus graciloides	Naroch Lake, Russia	May–Oct.	0.0010	0.044		24.7	Winberg et al. (1970)
	Myastro Lake, Russia	May–Oct.	0.0065	0.174		20.2	
	Batorin Lake, Russia	May–Oct.	0.0070	0.104		15.4	
	Lake Esrom, Denmark	Seasonal	—	0.10		6.8	Bosselmann (1975)
		Annual		0.178	3.4		Hamburger (1986)
Mesocyclops edax	Severson Lake, MN	Annual	0.0046	0.045		—	Comita (1972)
Diaphanosoma lecuhtenbergianum			0.0067	0.067		—	
Diaptomus siciloides			0.0342	0.341	0.2	—	
D. connexus	Waldsea Lake, Saskatchewan; mesotrophic	Annual	0.29	—		—	Swift and Hammer (1979)
Argyrodiaptomus furcatus	Broa Reservoir, Brazil	Annual	0.021			10.0	Rocha and Mat sumura-Tundisi (1984)

Continued

TABLE 20-1 Examples of productivity of herbivorous and predatory forms of zooplankton. Production of biomass estimated in terms of ash-free dry weight (dw) expressed in g m^{-3} day^{-1} as in this table can be divided by 2 to approximate g C m^{-3} day^{-1} produced (units also sometimes used). Production of biomass estimated in kcal m^{-2} day^{-1} can be roughly converted to g C m^{-2} day^{-1} by assuming that 1 kcal is approximately 0.1 g C.—cont'd

Type/species	Lake/general productivity	Period of investigation	Production estimates[a]		Assimilation % of phyto-plankton production	Average biomass turnover time (days)	Source
			gm^{-3} day^{-1}	kcal m^{-2} day^{-1}			
Calamoecia lucasi	Lake Okaro, New Zealand	Annual	0.004			69.9	Forsyth and James (1991)
Arcthodiaptomus (2 species)	Lake Sevan, southern Russia	Annual	0.0014	—		162	Meshkova (1952) in Winberg (1971)
Lovenula africana	Lake Nakuru, Kenya	Annual	0.055-	0.080	21.0	21.8	Vareschi and Jacobs(1984)
Rotifers							
Keratella quadrata	Severson Lake, MN	Annual	0.0021	0.021		—	Comita (1972)
K. cochlearis			0.0007	0.0074		—	
Filinia longiseta			0.0011	0.0112		—	
Brachionus sp.			0.0075	0.0752		—	
Polyarthra sp			0.0010	0.0103		—	
Rotifers	Naroch Lake, Russia	May–Oct.	0.0027	0.120		1.7	Winberg et al. (1970)
	Myastro Lake, Russia	May–Oct.	0.0024	0.065		3.9	
	Batorin Lake, Russia	May–Oct.	0.0105	0.156		2.6	
Rotifers	Lake Constance, Germany	Annual, 1984 Annual, 1985	0.003 0.0015		3.1	6.7 7.1	Pauli (1991)
Predatory feeders							
Cladocera							
Leptodora kindtii	Sanctuary Lake, PA; eutrophic, shallow	May–Nov., 1966 May–Nov., 1967	0.003 0.013	0.022 0.097		— —	Cummins et al. (1969)
	Lake George, NY; deep oligotrophic	June–Aug.	0.006	—		—	LaRow (1975)
Cladocera	Naroch Lake, Russia	May–Oct.	0.0003	0.013		11.3	Winberg et al. (1972)
	Myastro Lake, Russia		0.0009	0.023		3.9	
	Batorin Lake, Russia		0.0002	0.003		10.8	
Copepods							
Cyclops sp.	Naroch Lake, Russia	May–Oct.	0.0008	0.034		9.7	Winberg et al. (1970)
	Myastro Lake, Russia		0.0023	0.062		19.4	
	Batorin Lake, Russia		0.0094	0.140		14.2	
Rotifers							
Asplancbna priodonta	Naroch Lake, Russia	May–Oct.	0.0014	0.061		2.9	Winberg et al. (1970)
	Myastro Lake, Russia		0.0061	0.163		2.5	
	Batorin Lake, Russia		0.0105	0.156		3.2	

Asplanchna sp.	Severson Lake, MN	Annual	0.0031	0.031		—	Comita (1972)
Synchaeta sp.			0.00009	0.0009		—	
Brachionus dimidiatus	Lake Nakuru, Kenya	Annual	0.109		15.0	2.2	Vareschi and Jacobs (1984)
B. plicatilis			0.288		15.0	1.8	
Insect larvae							
Chaborus punctipennis			0.0001	0.001		—	Comita (1972)
C. flavicans	Pond, Japan	Annual	0.022			15.5	Iwakuma et al. (1989)
Mysids							
Neomysis mercedis	Muriel Lake, British Columbia	Annual (2 yr)	0.073			9.4	Cooper et al. (1992)
	Kennedy Lake, British Columbia	Annual (2 yr)	0.045			9.3	
Mysis relicta	Lake Mjosa, Norway	Annual	0.25			50	Kjellberg et al. (1991)
	Laurentian Great Lakes, USA	Annual	ca. 0.2			30–45	Sell (1982)
Amphipod							
Pontoporeia affinis	Lake Erken, Sweden	Annual	0.118			31	Johnson (1988)

[a]*Conversions estimated using the caloric mean of microconsumers (Cummins and Wuycheck, 1971) when mean depths available.*

TABLE 20-2 Estimates of biomass and production of protists in well-studied lakes[a]. Note that the mg C values can be multiplied by 2 to approximate mg of biomass (ash-free dry weight) for comparison to the data of Table 20-1.

Lake	Trophic status	Seasonal average (mg C m^{-3})	Range of production (mg C m^{-3} day^{-1})	Source
Nanoflagellates (NF)				
Lake Michigan, USA	Oligotrophic	—	0.8–8.4	Carrick et al. (1992)
Heterotrophic NF			2.68	
Phototrophic NF			3.58	
Microflagellates			2.55	
Lake Biwa, Japan	Mesotrophic	—	1.2–8.4	Nagata (1988)
Lake Constance, Germany (5-yr average)	Mesotrophic	2.9–7.4	1.4–3.5	Weisse (1997)
Ciliates				
Lake Constance, Germany	Mesotrophic	—	1.4–2.1	Müller (1989)
Lake Ontario, USA-Canada	Oligotrophic	0.1–35	0–38	Taylor and Johannsson (1991)
Lake Michigan, USA	Oligotrophic	—	7.83	Carrick et al. (1992)

[a]*Multiply the mg C figures by 2 in order to approximate mg of biomass (ash-free dry weight) for comparison to data of Tables 16-17 and 16-20.*

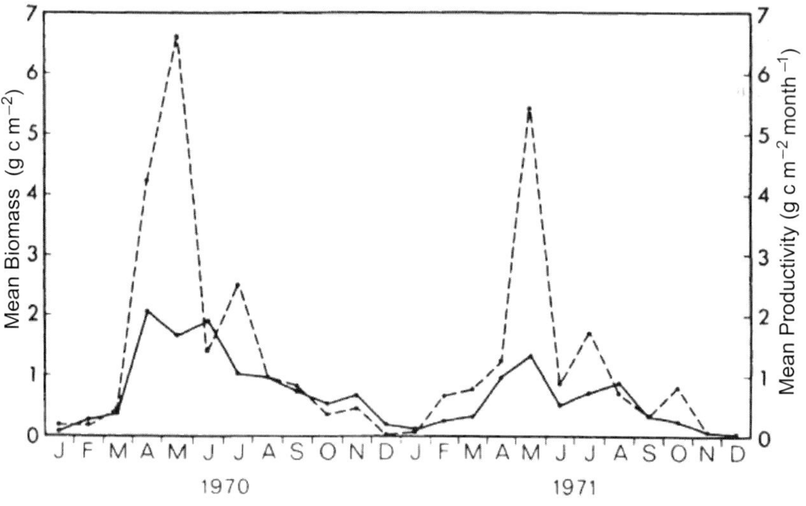

FIGURE 20-7 Seasonal monthly mean biomass (—) and productivity (- - - - -) of *Daphnia hyalina* in g C month^{-2}, calculated from daily measurements, of Eglwys Nydd, Great Britain. (Drawn from data of George and Edwards, 1974). *(Source: Formerly Wetzel Fig. 16-43 from George, D.G., Edwards, R.W., 1974. Population dynamics and production of* Daphnia hyalina *in a eutrophic reservoir. Freshwat. Biol. 4, 445–465.)*

show weak effects of spatial factors (i.e., little dispersal limitation), while larger passive dispersers such as crustacean zooplankton show stronger spatial effects. On the other hand, fish communities generally show even greater spatial effects, as their active dispersal is limited to watercourse connections between lakes and ponds, and because they are behaviorally more responsive to habitat structure than are passively dispersed plankton. Across the range of body sizes including bacteria to fish, spatial distances generally explain more variation in community structure for the largest organisms (fish), while environmental factors explain more variation in the smallest organisms (bacteria). On the other hand, cladoceran zooplankton fall in the middle, operating as metacommunities influenced by both spatial and environmental factors.

There can be variation in metacommunity structure within the crustacean zooplankton as well that goes beyond simple body size considerations. Instead, the life history differences between copepods and cladocerans may influence their community structure and function within lake and pond landscapes. Generally, cladocerans show a greater propensity for metacommunity patterns with dispersal and local environmental selection affecting community composition, while copepods (calanoids) show instead a legacy effect of historical

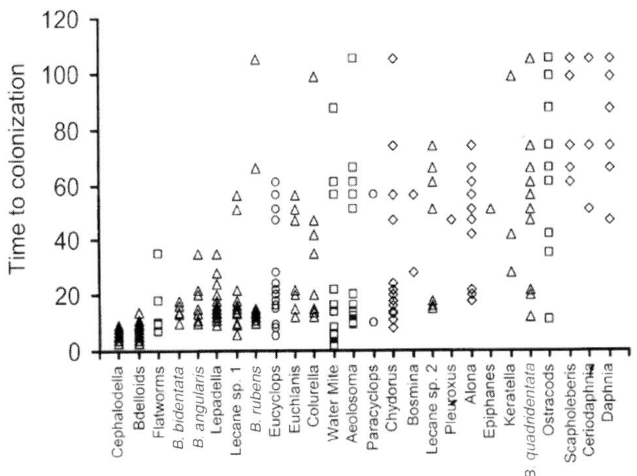

FIGURE 20-8 Time to colonization (weeks) into each novel aquatic habitat (mesocosm) for 26 primarily zooplankton colonist taxa. Taxa are listed in order of average time to colonization. ◇ cladocerans, ○ copepods, △ rotifers, □ all other taxa. *(Source: Fig. 5 in Cáceres, C.E., Soluk, D.A., 2002. Blowing in the wind: a field test of overland dispersal and colonization by aquatic invertebrates. Oecologia 131, 402–408.)*

biogeographical events on their spatial distributions (Leibold et al., 2010). Asexual reproduction and high population growth capacity of cladocerans enable high colonization rates, while copepods are more limited in their colonization capacity because, as obligate sexual reproducers, they must find mates that respond to the same behavioral cues in a new environment.

The dispersal capacity of zooplankton, from protists to crustaceans, has been critical to the reestablishment of communities in lakes where severe perturbations have occurred. A clear example of this is the recovery from the toxic effects of the acidification of lakes (related to industrial emissions of airborne pollutants such as sulfur dioxide and nitrous oxides) in the latter half of the 20th century in north temperate regions. In a study of the recovery of acid-damaged zooplankton communities, Gray et al. (2012) quantified the relative influence of habitat structure (abiotic), dispersal (spatial), and biotic (the presence of invertebrate predators) factors across a lake landscape (Killarney Provincial Park, Ontario, Canada). They observed evidence for mainly dispersal affecting recovery of zooplankton communities, but with influences also of biotic (predation on cladocerans) and habitat (for copepods) factors.

Increasingly, there is also evidence for metacommunity structure in lakes adjacent to large river networks to which these lakes are periodically connected because of strong wet-dry seasonality. In these rivers, particularly common in Brazil and warm temperate China, there is a dynamic connectivity of the mainstem river with adjacent floodplain lakes and streams (Junk, 1989), which is essential to the maintenance of overall biodiversity, with zooplankton assemblages functioning as a metacommunity (e.g., Agostinho et al., 2009; Zhao et al., 2017). Critically, along such rivers, large dams are increasingly destroying these natural metacommunities with negative consequences for large river zooplankton biodiversity (Souza et al., 2019).

V. Global changes and zooplankton community functioning

A. Climate change

Climate change is having important effects on the functioning of zooplankton communities and food webs. It is expected that changes will manifest globally mainly through both the warming and browning (Chapter 28) of freshwaters. Browning appears to result from several processes affecting the mobility of colored DOC from terrestrial to aquatic environments, including extreme precipitation events with climate change, land use alteration, and recovery of soils from acidification (Monteith et al., 2007; Meyer-Jacob et al., 2019).

Increasing temperatures will lead to generally warmer but also less predictable weather conditions, with more extreme precipitation events. Both factors can influence zooplankton interactions with their habitat and with the food web in which they are embedded. In addition to changes in zooplankton seasonal succession (Chapter 19), warmer conditions will generally make plankton predator–prey dynamics less stable, with larger fluctuations and a greater likelihood of extirpations of larger zooplankton like *Daphnia* (Beisner et al., 1996; Wojtal-Frankiewicz, 2012). Such extirpations are also linked to the observation that less edible, colony-forming, and often filamentous Cyanobacteria are favored by warmer conditions (Paerl and Huisman, 2008). Cyanobacterial blooms are especially detrimental to filter-feeding crustacean zooplankton, such as large *Daphnia*, which can become entangled in the filaments in addition to suffering from a shortage of edible food based on prey size (DeMott et al., 2001; Sarnelle et al., 2010). *Daphnia* may also suffer from larger epidemics that reduce genetic diversity under climate change, hampering adaptation to novel conditions (Auld and Brand, 2017). Furthermore, larger zooplankton are also disadvantaged under warmer conditions based purely on energetic reasons (Brown et al., 2004), which should reduce overall community grazing rates (Daufresne et al., 2009; Gillooly and Dodson, 2000). Zooplankton that do persist under warmer conditions will generally grow faster but attain maturity at smaller adult body sizes, and this will lead to further declines in community body size. However, such shifts may allow certain populations to persist where they might not otherwise do so. On the other hand, models and empirical studies (summarized in Jeppesen et al., 2014) also predict increased fish planktivory on zooplankton with warming, which will negatively affect their population sizes and shift their structure to more predator-resistant species. Such predictions assume that planktivorous fish will benefit more from increasing temperatures (DeStasio et al., 1996; Gauthier et al., 2014), although some may actually experience lower winter survival where ice cover is diminishing (Jackson et al., 2007). There is still much to be learned about the role of climate change–induced reduction and changes in ice cover (Sharma et al., 2016) and its influence on zooplankton community functioning, as it is increasingly clear that several zooplankton populations are active through the winter months (e.g., Mariash et al., 2016; Hampton et al., 2017).

More dynamic precipitation patterns, as well as increased terrestrial primary productivity in a warmer world, are expected to lead to widespread *brownification* of lakes (Chapter 28), especially in temperate regions, as increased quantities of colored humic matter enter lakes (de Wit et al., 2016). However, in regions where soils in lake watersheds are recovering from years of acidification associated with industrial activity, browning has been related to the reduction in acidic atmospheric depositions that increases the solubility of terrestrial DOC (Monteith et al., 2007; Meyer-Jacob et al., 2019). Much remains to be determined as to the effects of these shifts that are complicated by being associated with changes in lake thermal stratification, seasonal phenology, hydrology, and warming, all combined with what are often species-specific responses (Adrian et al., 2009). Some have found evidence that browner water may protect zooplankton from visual predators such as fish and some invertebrates (Boeing et al., 2004; van Dorst et al., 2018), as well as from harmful ultraviolet radiation. At the base of the food web, within which zooplankton are embedded, increased DOM may also favor the "Brown food web" (Fig. 20-3), with algal productivity expected to be limited by reduced light penetration, especially in oligotrophic lakes (del Giorgio and Peters, 1994). However, studies also show evidence for increases in both bacterial and phytoplankton productivity, attributable to nutrient inputs that accompany added terrestrial DOM (Jansson et al., 2012; Kelly et al., 2014; Klug, 2002). Thus, in lakes browned in association with climate change, many mixotrophic and heterotrophic protists may benefit from augmented bacterial productivity (Jones, 1992), although these protists may in turn represent a lower quality resource for larger zooplankton (see Section II). However, currently, generalized responses in zooplankton community functioning and stability to climate change are not known as they are subject to several interacting driving factors that warrant much more study.

B. Other global pressures

In addition to climate change and exotic invasive species, freshwaters harboring zooplankton are facing a wide variety of other emerging challenges that are likely to influence zooplankton community functioning through the processes discussed earlier in this chapter (Reid et al., 2018). There is a great deal of active research around these factors of global importance for freshwaters and zooplankton in particular. Agrochemical use, including fertilizers and pesticides (Chapter 19), is increasing, with fertilizers leading to predictable effects on zooplankton communities based on years of study of eutrophication in limnology. Eutrophication effects will be exacerbated by a warming climate, leading to further degradation of zooplankton community functioning. On the other hand, while the effects of pesticides have long utilized certain zooplankton (usually *D. magna* clones) as model organisms, only a few studies to date (e.g., Halstead et al., 2014; Relyea, 2009; Relyea and Hoverman, 2008; Rohr and Crumrine, 2005) have examined single and combined pesticide effects on full pond food webs, including zooplankton communities. While we are far from general conclusions, dependent on dose, time of exposure, and combinations of agrochemicals and heavy metals, generally, many pesticides negatively and directly affect larger body-sized zooplankton but also negatively affect primary production and decomposition at the base of pond food webs (Peters et al., 2013).

Researchers are still attempting to comprehend novel threats to zooplankton community functioning. One of the most critical new challenges for lake zooplankton is from microplastics: highly degradation-resistant polymers (Chapter 19). When consumed, these small particles can potentially accumulate in tissues or the food chain, as can many other contaminants (toxin bioaccumulation) (Jemec et al., 2016). Micropollutants are also perturbing zooplankton community function including heavy metals, pharmaceuticals, and cosmetics. Often such chemicals are also endocrine disruptors that can mimic hormones, including sex-determining ones, with consequences for zooplankton population growth rates (e.g., Peterson et al., 2001; Kashian and Dodson, 2002). Increases in the salinity of freshwater habitats occupied by zooplankton are also being increasingly observed (Chapter 12). Toxicity effects for zooplankton vary among populations and even communities, with sublethal effects on functioning being still far from understood (Sinclair and Arnott, 2018). As with other toxicants, such as mercury, exposure history and food web structure generally play a role in the responses observed by zooplankton. Finally, freshwater diversions and habitat alteration are also an increasing global phenomenon with impacts on zooplankton communities. In particular, a loss of connectivity with potential dispersers is associated with diversions that alter habitat configuration. For example, dams on large rivers that are normally intermittently connected to floodplain lakes can severely impede biodiversity and functioning of zooplankton metacommunities (Section IV).

VI. Summary

1. Predator–prey interactions in zooplankton communities, from protists to larger metazoans, are highly influenced by the relative size of predator and prey, as well as feeding traits and behavior.
   a. Experiments reveal strong predatory effects of zooplankton, but in natural settings transient dynamics are often observed between plankton as conditions shift.
   b. Fish predation can alter zooplankton behavior (e.g., diel vertical migration), energetics, and life history in important ways, even when fish are only detectable through chemical compounds (kairomones).
2. Competitive interactions in zooplankton communities influence dynamics and stability through both exploitative and, in some cases, interference competition.
   a. In natural communities compensatory dynamics (negatively correlated population sizes) suggest that competition is an important factor for populations. On the other hand, synchronous dynamics (positively correlated population dynamics) suggest that environmental shifts are more important than competition.
   b. The same zooplankton community may show these different types of dynamics at different timescales (e.g., within seasons versus interannually).
   c. Competition can help drive seasonal successional patterns.
3. Zooplankton population dynamics, species diversity, and productivity can be markedly affected by even moderate parasitic infection.
   a. Different life stages of zooplankton can be affected by different types of parasites including fungi, trematodes, bacteria, and microsporidia.
   b. Effects on *Daphnia* are the best studied to date and show an evolutionary response, increasing host resistance to parasites as well as relationships with fish predation, water column mixing, and lake bathymetric shape.
4. Most of what we know of positive, symbiotic interactions in the zooplankton is from the microbial food web and involves long-term exchanges (often biochemical) between bacteria and protists and ciliates.
   a. Protists generally harbor endosymbionts (bacteria, other protists, fungi, and viruses), enabling the acquisition of new metabolic functions (e.g., nitrogen fixation, methanogenesis, photosynthesis) and a competitive advantage.
   b. Ciliates can harbor both endo- and ectosymbionts.
   c. Metazooplankton can be considered microbially diverse and active hotspots in aquatic environments, especially beneficial to bacteria ecto- and endosymbionts.
   d. Ecological benefits accrue to both hosts (e.g., nutrient and vitamin supply, protection against parasites, waste degradation, UV protection) and symbionts (e.g., nutrient supply, increased motility, grazing protection, reduced competition).
5. "Contemporary" evolution has the potential to alter zooplankton population and community dynamics and results in large part from their relatively rapid population dynamics and morphological plasticity.
   a. Several studies have taken advantage of preserved diapausing eggs from sediments to examine ecoevolutionary shifts in these communities (resurrection ecology).
   b. Zooplankton communities have been shown to evolve on short timescales to a variety of factors including toxic algae, parasites, acidification, temperature change, eutrophication, fish community shifts, and multiple stressors.
   c. Population dynamics of plankton predators and prey have been shown to be altered by ecoevolutionary shifts.
6. The microbial food web interacts in important ways with the lower zooplankton food web, composed of heterotrophic protists, ciliates, and mixotrophic protists.
   a. These zooplankton influence mostly through predation, the microbes involved in metabolizing dissolved organic matter (DOM) and nonliving particulate organic materials (POM) to gases, soluble nutrients, and living bacterial biomass.
   b. Feeding relationships also occur between ciliates and other protists influencing their population dynamics.
7. The metazoan food web links the lower microbial food web to the invertebrate and fish predator trophic levels.
   a. Rotifers generally play a lesser role overall in biomass and flows than do crustaceans.
   b. Rotifer and crustacean zooplankton can interact via competition and predation to influence population dynamics and seasonal succession.

c. Invertebrate predation is relevant mainly for smaller crustacean and rotifer dynamics and diversity in lakes where there are few planktivorous fish.

8. Linking the microbial to the metazoan food web, feeding on phytoplankton is the major constituent of the diet of crustacean zooplankton, with bacterioplankton consumption alone being inadequate to support growth and reproduction.
   a. Under some conditions, 25% to 50% of bacterial production can be consumed by metazoans, especially under eutrophic conditions, where fish predation is low, or where naupliar stages are abundant.
   b. Trophic upgrading can occur where the microzooplankton (protists, ciliates) essentially package dissolved and particulate organic matter into larger-sized microorganisms that may be more available for consumption by larger organisms.
   c. While many metazooplankton consume carbon (C)-resources originating from phytoplankton primary production (autochthonous matter; Green food chain), crustaceans in many oligotrophic ecosystems rely on C-resources that originate in terrestrial primary production: detritus that makes its way into lakes (allochthonous matter; Brown food chain).
   d. Assimilation efficiency of microbial loop prey by metazooplankton is affected by prey composition and energy content as well as seasonality and temperature.
9. Zooplankton growth is influenced by food quality as well as quantity. Food quality turns on stoichiometric nutrient requirements, lipid content, and type, as well as toxin production.
   a. Lipids represent primary energy storage compounds composing 60% to 70% of zooplankton dry mass.
   b. Essential fatty acids (EFAs) help maintain high growth, survival, and reproduction in crustaceans. A large presence of polyunsaturated fatty acids (PUFAs) ensures a high food quality, but which can vary seasonally.
   c. Toxin-producing prey, especially common in the Cyanobacteria, may be selected against by feeding zooplankton as it can affect survival rates.
10. Fish and invertebrate predators often affect the size structure of the zooplankton communities upon which they feed.
   a. Fish are primarily visual predators, whereas invertebrates are primarily tactile. Factors that render zooplankton more visible (e.g., irregular movements, pigmentation, larger size, high water clarity, lack of diel vertical migration) will increase susceptibility to fish predation.
   b. The size-efficiency hypothesis explains why larger zooplankton that are better competitors for phytoplankton prey are generally absent when fish predation pressure is high. At low fish predation pressure, smaller zooplankton are outcompeted by larger crustaceans. These relationships have been used in the biomanipulation of aquatic food webs.
   c. Fish predation interacts with several other sources of mortality for larger zooplankton including food limitation, physiological death, parasitism, and physical losses.
   d. In shallow lakes additional factors make the fish–zooplankton interactions more spatially complex including the role of macrophytes as refuges and diel horizontal migration. Where piscivores keep planktivorous fish populations in check, larger zooplankton populations and body sizes can occur.
11. Freshwater invasive species can influence zooplankton community functioning directly or indirectly.
   a. Some invasives (e.g., water hyacinth, zebra mussel) influence zooplankton communities indirectly via modifications to the aquatic habitat that reduce phytoplankton primary production. Benthic invaders (e.g., rusty crayfish) can affect fish predators of zooplankton.
   b. Other invasives (e.g., spiny water flea) affect other zooplankton directly through increased epilimnetic predation.
12. Net zooplankton biomass production rates represent the sum of the growth increments of a population (including gamete biomass and exuviae biomass), excluding maintenance losses (respiration and excretion).
   a. Migration and predation can complicate the estimation of productivity in a focal population.
   b. Life cycle, reproductive characteristics, and generation times have traditionally influenced the method used to estimate production rates in a population.
   c. Estimations of chitobiase, a molting enzyme, in zooplankton habitats can provide a simpler real-time community estimate of crustacean biomass production rates.
   d. The ratio of production to biomass (*P/B* coefficient) is a commonly used index that estimates the turnover of zooplankton populations. These coefficients, usually averaged over time, can be compared across lakes, environmental conditions, or perturbations to assess zooplankton population responses.

e. Zooplankton productivity varies with primary productivity, food quality, and temperature, as well as by body size.

13. Dispersal between water bodies interacts with local environmental conditions to influence zooplankton community structure and function, creating metacommunities.
   a. The importance of zooplankton dispersal has been shown both experimentally and using molecular markers, with vectors including wind, rain, and animals.
   b. Dispersal influences on community structure correspond to body size in zooplankton (passive dispersers): protists and rotifers are the best dispersed, while crustaceans are more limited in dispersal capacity.
   c. Dispersal has enabled severely perturbed zooplankton communities (e.g., lake acidification) to recover once environmental conditions were restored.
   d. Metacommunity connections have been shown to also be critical to the maintenance of biodiversity and functioning of large river zooplankton through the loss of periodic connections with floodplain lakes owing to damming.
14. Climate change is likely to influence the functioning of zooplankton communities through both warming and browning of freshwaters.
   a. Warming is likely to increase zooplankton population extirpations through the direct effect of temperature increases on energetics and grazing rate reductions, as well as less stable population dynamics.
   b. Indirect effects of warming on zooplankton can occur via reductions in phytoplankton resource quality (e.g., increasing filamentous Cyanobacteria) and, in some cases, potentially greater fish planktivory.
   c. Lake brownification is occurring through a variety of interacting processes that increase the rate at which colored DOC moves into freshwaters. Ultimate effects on zooplankton community functioning and stability are still unclear, as they will be determined by multiple interacting driving factors including shifts in bacterial versus phytoplankton productivities, thermal stratification, seasonal phenology, hydrology, and warming.
15. A wide variety of other global emerging challenges are likely to influence zooplankton community functioning in the coming years.
   a. Agrochemical use is increasing worldwide, often leading to effects similar to those of eutrophication.
   b. Many pesticides directly negatively affect larger body-sized zooplankton, while others affect zooplankton indirectly by negatively affecting primary production and decomposition rates.
   c. The negative effects of microplastics on zooplankton appear to occur via direct consumption, especially by filter-feeding species.
   d. Other contaminants are making their way into freshwaters with still largely unstudied effects on zooplankton community functioning. These include heavy metals, pharmaceuticals, cosmetics, and road salt.
   e. Habitat configuration alteration, through dam construction primarily, will have impacts on fluvial zooplankton communities as well as their floodplain lakes.

Acknowledgments

We thank Cindy Paquette and Marie-Pier Hébert for their comments on this chapter.

References

Abrahams, M.V., Kattenfeld, M.G., 1997. The role of turbidity as a constraint on predator-prey interactions in aquatic environments. Behav. Ecol. Sociobiol. 40, 169–174.

Adeniji, H.A., 1978. Diurnal vertical distribution of zooplankton during stratification in Kainji Lake, Nigeria. Verh. Internat. Verein. Limnol. 20, 1677–1683.

Adrian, R., O'Reilly, C.M., Zagarese, H., Baines, S.B., Hessen, D.O., Keller, W., Livingstone, D.M., Sommaruga, R., Straile, D., Van Donk, E., Weyhenmeyer, G.A., Winder, M., 2009. Lakes as sentinels of climate change. Limnol. Oceanogr. 54, 2283–2297.

Agostinho, A., Bonecker, C., Gomes, L., 2009. Effects of water quantity on connectivity: the case of the upper Paraná River floodplain. Ecohydrol. Hydrobiol. 9, 99–113.

Ahlgren, G., 1993. Seasonal variation of fatty acid content in natural phytoplankton in two eutrophic lakes. A factor controlling zooplankton species? Verh. Internat. Verein. Limnol. 25, 144–149.

Ahlgren, G., Lundstedt, L., Brett, M., Forsberg, C., 1990. Lipid composition and food quality of some freshwater phytoplankton for cladoceran zooplankters. J. Plankton Res. 12, 809–818.

Andersson, A., Larsson, U., Hagström, Å., 1986. Size-selective grazing by a microflagellate on pelagic bacteria. Mar. Ecol. Prog. Ser. 33, 51–57.

Arts, M.T., 1999. Lipids in freshwater zooplankton: selected ecological and physiological aspects. In: Arts, M.T., Wainman, B.C. (Eds.), Lipids in Freshwater Ecosystems. Springer, New York, pp. 71–90.

Arts, M.T., Evans, M.S., Robarts, R.D., 1992. Seasonal patterns of total and energy reserve lipids of dominant zooplanktonic crustaceans from a hyper-eutrophic lake. Oecologia 90, 560–571.

Arts, M.T., Robarts, R.D., Evans, M.S., 1993. Energy reserve lipids of zooplanktonic crustaceans from an oligotrophic saline lake in relation to food resources and temperature. Can. J. Fish. Aquat. Sci. 50, 2404–2420.

Auld, S.K.J.R., Brand, J., 2017. Simulated climate change, epidemic size, and host evolution across host–parasite populations. Glob. Chang. Biol. 23, 5045–5053.

Basu, B.K., Pick, F.R., 1996. Factors regulating phytoplankton and zooplankton biomass in temperate rivers. Limnol. Oceanogr. 41, 1572–1577.

Beisner, B.E., McCauley, E., Wrona, F.J., 1996. Temperature-mediated dynamics of planktonic food chains: the effect of an invertebrate carnivore. Freshw. Biol. 35, 219–232.

Beisner, B.E., Peres-Neto, P.R., Lindstrom, E., Barnett, A., Longhi, M.L., Lindström, E.S., Barnett, A., Longhi, M.L., 2006. The role of environmental and spatial processes in structuring lake communities from bacteria to fish. Ecology 87, 2985–2991.

Benke, A.C., 1984. Secondary production in aquatic insects. In: Resh, V.H., Rosenberg, D.M. (Eds.), The Ecology of Aquatic Insects. Praeger, New York, pp. 289–322.

Benke, A.C., 1993. Concepts and patterns of invertebrate production in running waters. Verh. Internat. Verein. Limnol. 25, 15–38.

Berggren, M., Ziegler, S.E., St-Gelais, N.F., Beisner, B.E., del Giorgio, P.A., 2014. Contrasting patterns of allochthony among three major groups of crustacean zooplankton in boreal and temperate lakes. Ecology 95, 1947–1959.

Bern, L., 1987. Zooplankton grazing on [methyl-3H]thymidine-labelled natural particle assemblages: determination of filtering rates and food selectivity. Freshw. Biol. 17, 151–159.

Bickel, S.L., Tang, K.W., Grossart, H.P., 2012. Ciliate epibionts associated with crustacean zooplankton in German lakes: distribution, motility, and bacterivory. Front. Microbiol. 3, 1–11.

Blasius, B., Rudolf, L., Weithoff, G., Gaedke, U., Fussmann, G.F., 2019. Long-term cyclic persistence in an experimental predator–prey system. Nature 577, 226–230.

Boeing, W.J., Leech, D.M., Williamson, C.E., Cooke, S., Torres, L., 2004. Damaging UV radiation and invertebrate predation: conflicting selective pressures for zooplankton vertical distribution in the water column of low DOC lakes. Oecologia 138, 603–612.

Boersma, M., van Tongeren, O.F.R., Mooij, W.M., 1996. Seasonal patterns in the mortality of *Daphnia* species in a shallow lake. Can. J. Fish. Aquat. Sci. 53, 18–28.

Bohonak, A.J., Jenkins, D.G., 2003. Ecological and evolutionary significance of dispersal by freshwater invertebrates. Ecol. Lett. 6, 783–796.

Boraas, M.E., Estep, K.W., Johnson, P.W., Sieburth, J.M., 1988. Phagotrophic phototrophs: the ecological significance of mixotrophy. J. Protozool. 35, 249–252.

Borsheim, K.Y., Andersen, S., 1987. Grazing and food size selection by crustacean zooplankton compared to production of bacteria and phytoplankton in a shallow Norwegian mountain lake. J. Plankton Res. 9, 367–379.

Bosselmann, S., 1975. Production of *Eudiaptomus graciloides* in Lake Esrom, 1970. Arch. Hydrobiol. 76, 43–64.

Bottrell, H.H., Duncan, A., Gliwicz, Z.M., Grygierek, E., Herzig, A., Hillbricht-Ilkowska, A., Kurasawa, H., Larsson, P., Weglenska, T., 1976. A review of some problems in zooplankton production studies. Norw. J. Zool. 24, 419–456.

Branstrator, D.K., 2005. Contrasting life histories of the predatory cladocerans *Leptodora kindtii* and *Bythotrephes longimanus*. J. Plankton Res. 27, 569–585.

Branstrator, D.K., Lehman, J.T., 1991. Invertebrate predation in Lake Michigan: regulation of *Bosmina longirostris* by *Leptodora kindtii*. Limnol. Oceanogr. 36, 483–495.

Brett, M., Müller-Navarra, D., 1997. The role of highly unsaturated fatty acids in aquatic foodweb processes. Freshw. Biol. 38, 483–499.

Brooks, L.J., 1968. The effects of prey size selection by lake planktivores. Syst. Biol. 17, 273–291.

Brooks, J.L., Dodson, S.I., 1965. Predation, body size and composition of plankton. Science 150, 28–35.

Brown, J.H., Gillooly, J.F., Allen, A.P., Savage, V.M., West, G.B., 2004. Toward a metabolic theory of ecology. Ecology 85, 1792–1821.

Burks, R.L., Jeppesen, E., Lodge, D.M., 2001. Littoral zone structures as *Daphnia* refugia against fish predators. Limnol. Oceanogr. 46, 230–237.

Burns, C.W., 1979. Population dynamics and production of *Boeckella dilatata* (Copepoda: Calanoida) in Lake Hayes, New Zealand. Arch. Hydrobiol. Suppl. 54, 409–465.

Burns, C.W., 1985. Fungal parasitism in a copepod population: the effects of Aphanomyces on the population dynamics of *Boeckella dilatata* Sars. J. Plankton Res. 7, 201–205.

Cáceres, C.E., 1997. Temporal variation, dormancy, and coexistence: a field test of the storage effect. Proc. Natl. Acad. Sci. U. S. A. 94, 9171–9175.

Cáceres, C.E., Hall, S.R., Duffy, M.A., Tessier, A.J., Helmle, C., MacIntyre, S., 2006. Physical structure of lakes constrains epidemics in *Daphnia* populations. Ecology 87, 1438–1444.

Cáceres, C.E., Soluk, D.A., 2002. Blowing in the wind: a field test of overland dispersal and colonization by aquatic invertebrates. Oecologia 131, 402–408.

Cáceres, C.E., Tessier, A.J., Duffy, M.A., Hall, S.R., 2014. Disease in freshwater zooplankton: what have we learned and where are we going. J. Plankton Res. 36, 326–333.

Callens, M., Macke, E., Muylaert, K., Bossier, P., Lievens, B., Waud, M., Decaestecker, E., 2016. Food availability affects the strength of mutualistic host-microbiota interactions in *Daphnia magna*. ISME J. 10, 911–920.

Canfield, T.J., Jones, J.R., 1996. Zooplankton abundance, biomass, and size-distribution in selected midwestern waterbodies and relation with trophic state. J. Freshw. Ecol. 11, 171–181.

Carpenter, S.R., Cole, J.J., Hodgson, J.R., Kitchell, J.F., Pace, M.L., Bade, D., Cottingham, K.L., Essington, T.E., Houser, J.N., Schindler, D.E., 2001. Trophic cascades, nutrients, and lake productivity: experimental enrichment of lakes with contrasting food webs. Ecol. Monogr. 71, 163–186.

Carpenter, S.R., Cole, J.J., Pace, M.L., Van de Bogert, M., Bade, D.L., Bastviken, D., Gille, C.M., Hodgson, J.R., Kitchell, J.F., Kritzberg, E.S., 2005. Ecosystem subsidies: terrestrial support of aquatic food webs from ^{13}C addition to contrasting lakes. Ecology 86, 2737–2750.

Carrick, H.J., Fahnenstiel, G.L., Taylor, W.D., 1992. Growth and production of planktonic protozoa in Lake Michigan: in situ versus in vitro comparisons and importance to food web dynamics. Limnol. Ocean. 37, 1221–1235.

Chrzanowski, T.H., Šimek, K., 1990. Prey-size selection by freshwater flagellated protozoa. Limnol. Oceanogr. 35, 1429–1436.

Clay, P.A., Duffy, M.A., Rudolf, V.H.W., 2020. Within-host priority effects and epidemic timing determine outbreak severity in co-infected populations. Proceedings. Biol. Sci. 287, 20200046.

Comita, G.W., 1972. The seasonal zooplankton cycles, production and transformations of energy in Severson Lake, Minnesota. Arch. Hydrobiol. 70, 14–66.

Confer, J.L., 1971. Intrazooplankton predation by *Mesocyclops edax* at natural prey densities. Limnol. Oceanogr. 16, 663–666.

Confer, J.L., Applegate, G., 1979. Size-selective predation by zooplankton. Am. Midl. Nat. 102, 378.

Confer, J.L., Cooley, J.M., 1977. Copepod instar survival and predation by zooplankton. J. Fish. Res. Board Canada 34, 703–706.

Cooper, K.L., Hyatt, K.D., Rankin, D.P., 1992. Life history and production of *Neomysis mercedis* in two British Columbia coastal lakes. Hydrobiologia 230, 9–30.

Costa, R.R., Cummins, K.W., 1972. The contribution of *Leptodora* and other zooplankton to the diet of various fish. Am. Midl. Nat. 87, 559.

Cottenie, K., 2005. Integrating environmental and spatial processes in ecological community dynamics. Ecol. Lett. 8, 1175–1182.

Cousyn, C., De Meester, L., Colbourne, J.K., Brendonck, L., Verschuren, D., Volckaert, F., 2001. Rapid, local adaptation of zooplankton behavior to changes in predation pressure in the absence of neutral genetic changes. Proc. Natl. Acad. Sci. USA 98, 6256–6260.

Cummins, K.W., Costa, R.R., Rowe, R.E., Moshiri, G.A., Scanlon, R.M., Zajdel, R.K., 1969. Ecological energetics of a natural population of the predaceous zooplankter *Leptodora kindtii* Focke (Cladocera). Oikos 20, 189–223.

Cummins, K.W., Wuycheck, J.C., 1971. Caloric equivalents for investigations in ecological energetics. Mitt. Internat. Verein. Limnol. 18, 158.

Daufresne, M., Lengfellner, K., Sommer, U., 2009. Global warming benefits the small. Proc. Natl. Acad. Sci. USA 13, 2467–2478.

Dawidowicz, P., Loose, C.J., 1992. Cost of swimming by *Daphnia* during diel vertical migration. Limnol. Oceanogr. 37, 665–669.

De Bie, T., De Meester, L., Brendonck, L., Martens, K., Goddeeris, B., Ercken, D., Hampel, H., Denys, L., Vanhecke, L., Van der Gucht, K., Van Wichelen, J., Vyverman, W., Declerck, S.A.J., 2012. Body size and dispersal mode as key traits determining metacommunity structure of aquatic organisms. Ecol. Lett. 15, 740–747.

de Wit, H.A., Valinia, S., Weyhenmeyer, G.A., Futter, M.N., Kortelainen, P., Austnes, K., Hessen, D.O., Räike, A., Laudon, H., Vuorenmaa, J., 2016. Current browning of surface waters will be further promoted by wetter climate. Environ. Sci. Technol. Lett. 3, 430–435.

Decaestecker, E., De Gersem, H., Michalakis, Y., Raeymaekers, J.A.M., 2013. Damped long-term host-parasite Red Queen coevolutionary dynamics: a reflection of dilution effects? Ecol. Lett. 16, 1455–1462.

del Giorgio, P.A., Peters, R.H., 1994. Patterns in planktonic P:R ratios in lakes: influence of lake trophy and dissolved organic carbon. Limnol. Oceanogr. 39, 772–787.

DeMott, W.R., Gulati, R.D., Van Donk, E., 2001. *Daphnia* food limitation in three hypereutrophic Dutch lakes: evidence for exclusion of large-bodied species by interfering filaments of cyanobacteria. Limnol. Oceanogr. 46, 2054–2060.

DeMott, W., Müller-Navarra, D., 1997. The importance of highly unsaturated fatty acids in zooplankton nutrition: evidence from experiments with *Daphnia*, a cyanobacterium and lipid emulsions. Freshw. Biol. 38, 649–664.

Derry, A.M., Arnott, S.E., Boag, P.T., 2010. Evolutionary shifts in copepod acid tolerance in an acid-recovering lake indicated by resurrected resting eggs. Evol. Ecol. 24, 133–145.

DeStasio, B.T.J., Hill, D.K., Kleinhans, J.M., Nibbelink, N.P., Magnuson, J.J., 1996. Potential effects of global climate change on small north-temperate lakes: physics, fish and plankton. Limnol. Ocean. 41, 1136–1149.

Dillon, P.J., Rigler, F.H., 1974. A test of a simple nutrient budget model predicting the phosphorus concentration in lake water. J. Fish. Res. Board Can. 31, 1771–1778.

Dodson, S.I., 1970. Complementary feeding niches sustained by size-selective predation. Limnol. Oceanogr. 15, 131–137.

Dodson, S.I., 1972. Mortality in a population of *Daphnia rosea*. Ecology 53, 1011–1023.

Dodson, S.I., 1974. Adaptive change in plankton morphology in response to size-selective predation: a new hypothesis of cyclomorphosis. Limnol. Oceanogr. 19, 721–729.

Duffy, M.A., 2007. Selective predation, parasitism, and trophic cascades in a bluegill–*Daphnia*–parasite system. Oecologia 153, 453–460.

Duffy, M.A., Hunsberger, K.K., 2019. Infectivity is influenced by parasite spore age and exposure to freezing: do shallow waters provide *Daphnia* a refuge from some parasites? J. Plankton Res. 41, 12–16.

Duncan, A., 1989. Food limitation and body size in the life cycles of planktonic rotifers and cladocerans. Hydrobiologia 186–187, 11–28.

Duncan, A.B., Mitchell, S.E., Little, T.J., 2006. Parasite-mediated selection and the role of sex and diapause in *Daphnia*. J. Evol. Biol. 19, 1183–1189.

Dziallas, C., Allgaier, M., Monaghan, M.T., Grossart, H.P., 2012. Act together—implications of symbioses in aquatic ciliates. Front. Microbiol. 3, 1–17.

Ebert, D., 2005. Ecology, Epidemiology and Evolution of Parasitism in *Daphnia*. National Library of Medicine (US), National Center for Biotechnology Information, Bethesda, MD.

Edmondson, W.T., 1960. Reproductive rates of rotifers in natural populations. Mem. Ist. Ital. Idrobiol. 12, 21–77.

Edmondson, W.T., 1965. Reproductive rate of planktonic rotifers as related to food and temperature in nature. Ecol. Monogr. 35, 61–111.

Edmondson, W.T., 1968. A graphical model for evaluating the use of the egg ratio for measuring birth and death rates. Oecologia 1, 1–37.

Edmondson, W.T., 1974. Secondary production. Verh. Internat. Verein. Limnol. 20, 229–272.

Edmondson, W.T., Winberg, G.G., 1971. A Manual on Methods for the Assessment of Secondary Productivity in Fresh Waters, second ed. Blackwell Scientific Publications, Oxford.

Elser, J.J., Luecke, C., Brett, B.T., Goldman, C.R., 1995. Effects of food web compensation after manipulation of rainbow trout in an oligotrophic lake. Ecology 76, 52–69.

Enright, J.T., 1977. Problems in estimating copepod velocity. Limnol. Oceanogr. 22, 160–162.

Epp, R.W., Lewis, W.M., 1984. Cost and speed of locomotion for rotifers. Oecologia 61, 289–292.

Finlay, B.J., 2002. Global dispersal of free-living microbial eukaryote species. Science 296, 1061–1063.

Forsyth, D.J., James, M.R., 1984. Zooplankton grazing on lake bacterioplankton and phytoplankton. J. Plankton Res. 6, 803–810.

Forsyth, D.J., James, M.R., 1991. Population dynamics and production of zooplankton in eutrophic Lake Okaro, North Island, New Zealand. Arch. Hydrobiol. 120, 287–314.

Frisch, D., Morton, P.K., Chowdhury, P.R., Culver, B.W., Colbourne, J.K., Weider, L.J., Jeyasingh, P.D., 2014. A millennial-scale chronicle of evolutionary responses to cultural eutrophication in *Daphnia*. Ecol. Lett. 17, 360–368.

Gabriel, W., Thomas, B., 1988. Vertical migration of zooplankton as an evolutionarily stable strategy. Am. Nat. 132, 199–216.

Garcia, R., Jara, F., Steciow, M., Reissig, M., 2018. Oomycete parasites in freshwater copepods of Patagonia: effects on survival and recruitment. Dis. Aquat. Organ. 129, 123–134.

Gauthier, J., Prairie, Y.T., Beisner, B.E., 2014. Thermocline deepening and mixing alter zooplankton phenology, biomass and body size in a whole-lake experiment. Freshw. Biol. 59, 998–1011.

Geerts, A.N., Vanoverbeke, J., Vanschoenwinkel, B., Van Doorslaer, W., Feuchtmayr, H., Atkinson, D., Moss, B., Davidson, T.A., Sayer, C.D., De Meester, L., 2015. Rapid evolution of thermal tolerance in the water flea *Daphnia*. Nat. Clim. Chang. 5, 665–668.

Geller, W., 1985. Production, food utilization and losses of two coexisting, ecologically different *Daphnia* species. Arch. Hydrobiol. Beih. Ergebn. Limnol. 21, 67–79.

George, D.G., 1976. Life cycle and production of *Cyclops vicinus* in a shallow eutrophic reservoir. Oikos 27, 101–110.

George, D.G., Edwards, R.W., 1974. Population dynamics and production of *Daphnia hyalina* in a eutrophic reservoir. Freshwat. Biol. 4, 445–465.

Gerking, S.D., 1994. Feeding Ecology of Fish. Academic Press, San Diego.

Gilbert, J.J., 1988. Suppression of rotifer populations by *Daphnia*: a review of the evidence, the mechanisms, and the effects on zooplankton community structure. Limnol. Oceanogr. 33, 1286–1303.

Gilbert, J.J., Schreiber, D.K., 1995. Induction of diapausing amictic eggs in *Synchaeta pectinata*. Hydrobiologia 313–314, 345–350.

Gillooly, J.F., Dodson, S.I., 2000. Latitudinal patterns in the size distribution and seasonal dynamics of new world, freshwater cladocerans. Limnol. Oceanogr. 45, 22–30.

Gliwicz, Z., 1980. Filtering rates, food size selection, and feeding rates in Cladocerans—another aspect of interspecific competition in filter-feeding zooplankton. In: Kerfoot, W.C. (Ed.), Evolution and Ecology of Zooplankton Communities. University Press of New England, Hanover, NH, pp. 282–291.

Gliwicz, M.Z., 1986. Predation and the evolution of vertical migration in zooplankton. Nature 320, 746–748.

Gliwicz, Z.M., 1990. Food thresholds and body size in cladocerans. Nature 343, 638–640.

Gliwicz, Z.M., Siedlar, E., 1980. Food size limitation and algae interfering with food collection in *Daphnia*. Arch. Hydrobiol. 88, 155–177.

Gophen, M., Cavari, B.Z., Berman, T., 1974. Zooplankton feeding on differentially labelled algae and bacteria. Nature 247, 393–394.

Goulden, C.E., Henry, L., 1985. Lipid energy reserves and their role in Cladocera. In: Meyers, D.G., Strickler, J.R. (Eds.), Trophic Interactions within Aquatic Ecosystems. Westview Press, Boulder, CO, pp. 167–185.

Goulden, C.E., Hornig, L., Wilson, C., 1978. Why do large zooplankton species dominate? Verh. Internat. Verein. Limnol. 20, 2457–2460.

Goulden, C.E., Moeller, R.E., McNair, J.N., Place, A.R., 1999. Lipid dietary dependencies in zooplankton. In: Art, M.T., Wainman, B.C. (Eds.), Lipids in Freshwater Ecosystems. Springer, New York, pp. 91–108.

Gray, D.K., Arnott, S.E., Shead, J.A., Derry, A.M., 2012. The recovery of acid-damaged zooplankton communities in Canadian Lakes: the relative importance of abiotic, biotic and spatial variables. Freshw. Biol. 57, 741–758.

Green, J.D., 1976. Population dynamics and production of the calanoid copepod *Calamoecia lacasi* in a northern New Zealand lake. Arch. Hydrobiol. Suppl. 50, 313–400.

Grey, J., Jones, R.I., Sleep, D., 2001. Seasonal changes in the importance of the source of organic matter to the diet of zooplankton in Loch Ness, as indicated by stable isotope analysis. Limnol. Oceanogr. 46, 505–513.

Grossart, H.P., Dziallas, C., Leunert, F., Tang, K.W., 2010. Bacteria dispersal by hitchhiking on zooplankton. Proc. Natl. Acad. Sci. USA 107, 11959–11964.

Grossart, H.P., Dziallas, C., Tang, K.W., 2009. Bacterial diversity associated with freshwater zooplankton. Environ. Microbiol. Rep. 1, 50–55.

Hairston, N.G.J., Lampert, W., Caceres, C.E., Holtmeier, C., Weider, L., Gaedke, U., Fischer, J., Fox, J., Post, D., 1999a. Rapid evolution revealed by dormant eggs. Nature 401, 446.

Hairston, N.G.J., Perry, L.J., Bohonak, A.J., Fellows, M.Q., Kearns, C.M., Engstrom, D.R., 1999b. Population biology of a failed invasion: paleolimnology of *Daphnia exilis* in upstate New York. Limnol. Oceanogr. 44, 477–486.

Hairston, N.G., Van Brunt, R.A., Kearns, C.M., Engstrom, D.R., 1995. Age and survivorship of diapausing eggs in a sediment egg bank. Ecology 76, 1706–1711.

Hall, D.J., Threlkeld, S.T., Burns, C.W., Crowley, P.H., 1976. The size-efficiency hypothesis and the size structure of zooplankton communities. Annu. Rev. Ecol. Syst. 7, 177–208.

Halstead, N.T., Mcmahon, T.A., Johnson, S.A., Raffel, T.R., Romansic, J.M., Crumrine, P.W., Rohr, J.R., 2014. Community ecology theory predicts the effects of agrochemical mixtures on aquatic biodiversity and ecosystem properties. Ecol. Lett. 17, 932–941.

Hamburger, K., 1986. Energy flow in the populations of *Eudiaptomus graciloides* and *Daphnia galeata* in Lake Esrom. Arch. Hydrobiol. 105, 517–530.

Hampton, S.E., Galloway, A.W.E., Powers, S.M., Ozersky, T., Woo, K.H., Batt, R.D., Labou, S.G., O'Reilly, C.M., Sharma, S., Lottig, N.R., Stanley, E.H., North, R.L., Stockwell, J.D., Adrian, R., Weyhenmeyer, G.A., Arvola, L., Baulch, H.M., Bertani, I., Bowman, L.L., Carey, C.C., Catalan, J., Colom-Montero, W., Domine, L.M., Felip, M., Granados, I., Gries, C., Grossart, H.P., Haberman, J., Haldna, M., Hayden, B., Higgins, S.N., Jolley, J.C., Kahilainen, K.K., Kaup, E., Kehoe, M.J., MacIntyre, S., Mackay, A.W., Mariash, H.L., McKay, R.M., Nixdorf, B., Nõges, P., Noges, T., Palmer, M., Pierson, D.C., Post, D.M., Pruett, M.J., Rautio, M., Read, J.S., Roberts, S.L., Ruecker, J., Sadro, S., Silow, E.A., Smith, D.E., Sterner, R.W., Swann, G.E.A., Timofeyev, M.A., Toro, M., Twiss, M.R., Vogt, R.J., Watson, S.B., Whiteford, E.J., Xenopoulos, M.A., 2017. Ecology under lake ice. Ecol. Lett. 20, 98–111.

Hart, R.C., Jarvis, A.C., 1993. In situ determinations of bacterial selectivity and filtration rates by five cladoceran zooplankters in a hypertrophic subtropical reservoir. J. Plankton Res. 15, 295–315.

Haury, L., Weihs, D., 1976. Energetically efficient swimming behavior of negatively buoyant zooplankton. Limnol. Oceanogr. 21, 797–803.

Havens, K.E., 1990. *Chaoborus* predation and zooplankton community structure in a rotifer-dominated lake. Hydrobiologia 198, 215–226.

Hellsten, M.E., Stenson, J.A.E., 1995. Cyclomorphosis in a population of *Bosmina coregoni*. Hydrobiologia 312, 1–9.

Hessen, D.O., 1985. Selective zooplankton predation by pre-adult roach (*Rutilus rutilus*): the size-selective hypothesis versus the visibility-selective hypothesis. Hydrobiologia 124, 73–79.

Hessen, D.O., Elser, J.J., Sterner, R.W., Urabe, J., 2013. Ecological stoichiometry: an elementary approach using basic principles. Limnol. Oceanogr. 58, 2219–2236.

Hoenicke, R., 1984. The effects of a fungal infection of *Diaptomus novamexicanus* eggs on the zooplankton community structure of Castle Lake, California. Verh. Internat. Verein. Limnol. 22, 573–577.

Hoenicke, R., Goldman, C.R., 1987. Resource dynamics and seasonal changes in competitive interactions among three cladoceran species. J. Plankton Res. 9, 397–417.

Holm, N.P., Shapiro, J., 1984. An examination of lipid reserves and the nutritional status of *Daphnia pulex* fed *Aphanizomenon flos-aquae*. Limnol. Oceanogr. 29, 1137–1140.

Hwang, S.J., Heath, R.T., 1999. Zooplankton bacterivory at coastal and offshore sites of Lake Erie. J. Plankton Res. 21, 699–719.

Iwakuma, T., Shibata, K., Hanazato, T., 1989. Production ecology of phyto- and zooplankton in a eutrophic pond dominated by *Chaoborus flavicans* (Diptera: Chaobolidae). Ecol. Res. 4, 31–53.

Jackson, L.J., Lauridsen, T.L., Søndergaard, M., Jeppesen, E., Søndergaard, M., Jeppesen, E., 2007. A comparison of shallow Danish and Canadian lakes and implications of climate change. Freshw. Biol. 52, 1782–1792.

Jansson, M., Karlsson, J., Jonsson, A., 2012. Carbon dioxide supersaturation promotes primary production in lakes. Ecol. Lett. 15, 527–532.

Jemec, A., Horvat, P., Kunej, U., Bele, M., Kržan, A., 2016. Uptake and effects of microplastic textile fibers on freshwater crustacean *Daphnia magna*. Environ. Pollut. 219, 201–209.

Jensen, K.H., Jakobsen, P.J., Kleiven, O.T., 1998. Fish kairomone regulation of internal swarm structure in *Daphnia pulex* (Cladocera: Crustacea). Hydrobiologia 368, 123–127.

Jeppesen, E., Jensen, J., Sondergaard, M., Lauridsen, T., Pedersen, L.J., Jensen, L., 1997. Top-down control in freshwater lakes: the role of nutrient status, submerged macrophytes and water depth. Hydrobiologia 342/343, 151–164.

Jeppesen, E., Lauridsen, T.L., Kairesalo, T., Perrow, M.R., 1998. Impact of submerged macrophytes on fish-zooplankton interactions in lakes. In: Jeppesen, E., Sondergaard, M., Christoffersen, K. (Eds.), The Structuring Role of Submerged Macrophytes in Lakes. Springer Verlag, New York, pp. 331–338.

Jeppesen, E., Meerhoff, M., Davidson, T.A., Trolle, D., Søndergaard, M., Lauridsen, T.L., Beklioğlu, M., Brucet, S., Volta, P., González-Bergonzoni, I., Nielsen, A., 2014. Climate change impacts on lakes: an integrated ecological perspective based on a multi-faceted approach, with special focus on shallow lakes. J. Limnol. 73, 88–111.

Jeziorski, A., Tanentzap, A.J., Yan, N.D., Paterson, A.M., Palmer, M.E., Korosi, J.B., Rusak, J.A., Arts, M.T., Keller, W.B., Ingram, R., Cairns, A., Smol, J.P., 2014. The jellification of north temperate lakes. Proc. R. Soc. B Biol. Sci. 282, 962–8452.

Jeziorski, A., Yan, N.D., Paterson, A.M., DeSellas, A.M., Turner, M.A., Jeffries, D.S., Keller, B., Weeber, R.C., McNicol, D.K., Palmer, M.E., McIver, K., Arseneau, K., Ginn, B.K., Dumming, B.F., Smol, J.P., 2008. The widespread threat of calcium decline in fresh waters. Science 322, 1374–1377.

Johnson, M.G., 1988. Production by the amphipod *Pontoporeia hoyi* in South Bay, Lake Huron. Can. J. Fish. Aquat. Sci. 45, 617–624.

Jones, R.I., 1992. The influence of humic substances on lacustrine planktonic food chains. Hydrobiologia 229, 73–91.

Jones, R.I., 2000. Mixotrophy in planktonic protists: an overview. Freshw. Biol. 45, 219–226.

Junk, W.J., 1989. The flood pulse concept of large rivers: learning from the tropics. Proc. Int. Large River Symp. Can. Spec. Publ. Fish. Aquat. Sci. 106, 110–127.

Jürgens, K., Arndt, H., Rothhaupt, K.O., 1994. Zooplankton-mediated changes of bacterial community structure. Microb. Ecol. 27, 27–42.

Jürgens, K., Arndt, H., Zimmermann, H., 1997. Impact of metazoan and protozoan grazers on bacterial biomass distribution in microcosm experiments. Aquat. Microb. Ecol. 12, 131–138.

Jürgens, K., Gasol, J.M., Massana, R., Pedrós-Alió, C., 1994. Control of heterotrophic bacteria and protozoans by *Daphnia pulex* in the epilimnion of Lake Cisó. Arch. Hydrobiol. 131, 55–78.

Kagami, M., Miki, T., Takimoto, G., 2014. Mycoloop: Chytrids in aquatic food webs. Front. Microbiol. 5, 166.

Kankaala, P., 1988. The relative importance of algae and bacteria as food for *Daphnia longispina* (Cladocera) in a polyhumic lake. Freshw. Biol. 19, 285–296.

Karlsson, J., Jonsson, A., Meili, M., Jansson, M., 2003. Control of zooplankton dependence on allochthonous organic carbon in humic and clear-water lakes in northern Sweden. Limnol. Oceanogr. 48, 269–276.

Kashian, D.R., Dodson, S.I., 2002. Effects of common-use pesticides on developmental and reproductive processes in *Daphnia*. Toxicol. Ind. Health 18, 225–235.

Keitt, T.H., 2008. Coherent ecological dynamics induced by large-scale disturbance. Nature 454, 331–335.

Kelly, P.T., Solomon, C.T., Weidel, B.C., Jones, S.E., 2014. Terrestrial carbon is a resource, but not a subsidy, for lake zooplankton. Ecology 95, 1236–1242.

Kerfoot, W.C., Robbins, J.A., Weider, L.J., 1999. A new approach to historical reconstruction: combining descriptive and experimental paleolimnology. Limnol. Oceanogr. 44, 1232–1247.

King, C.R., Greenwood, J.G., 1992. The productivity and carbon budget of a natural population of *Daphnia lumholtzi* Sars. Hydrobiologia 231, 197–207.

Kiørboe, T., Andersen, A., Langlois, V.J., Jakobsen, H.H., Bohr, T., 2009. Mechanisms and feasibility of prey capture in ambush-feeding zooplankton. Proc. Natl. Acad. Sci. 106, 12394–12399.

Kirk, K., Gilbert, J.J., 1992. Variation in herbivore response to chemical defenses : zooplankton foraging on toxic cyanobacteria. Ecology 73, 2208–2217.

Kjellberg, G., Hessen, D.O., Nilssen, J.P., 1991. Life history, growth and production of *Mysis relicta* in the large, fiord-type Lake Mjøsa, Norway. Freshw. Biol. 26, 165–173.

Klein Breteler, W.C.M., Schogt, N., Baas, M., Schouten, S., Kraay, G.W., 1999. Trophic upgrading of food quality by protozoans enhancing copepod growth: role of essential lipids. Mar. Biol. 135, 191–198.

Klug, J., 2002. Positive and negative effects of allochthonous dissolved organic matter and inorganic nutrients on phytoplankton growth. Can. J. Fish. Aquat. Sci. 59, 85–95.

Knoechel, R., Holtby, L.B., 1986. Cladoceran filtering rate: body length relationships for bacterial and large algal particles. Limnol. Oceanogr. 31, 195–200.

Kreps, T.A., Larson, E.R., Lodge, D.M., 2016. Do invasive rusty crayfish (*Orconectes rusticus*) decouple littoral and pelagic energy flows in lake food webs? Freshw. Sci. 35, 103–113.

Lair, N., 1990. Effects of invertebrate predation on the seasonal succession of a zooplankton community: a two year study in Lake Aydat, France. Hydrobiologia 198, 1–12.

Lampert, W., 1993. Ultimate causes of diel vertical migration of zooplankton: new evidence for the predator-avoidance hypothesis. Arch. Hydrobiol. Beih. Ergebn. Limnol. 39, 79–88.

Lampert, W., 2011. *Daphnia*: Development of Model Organism in Ecology and Evolution, Excellence in Ecology. International Ecology Institute, Olendorf/Luhu.

Lampert, W., Bohrer, R., 1984. Effect of food availability on the respiratory quotient of *Daphnia magna*. Comp. Biochem. Physiol. Part A Physiol. 78, 221–223.

Lampert, W., Wolf, H.G., 1986. Cyclomorphosis in *Daphnia cucullata*: morphometric and population genetic analyses. J. Plankton Res. 8, 289–303.

Langeland, A., 1988. Decreased zooplankton density in a mountain lake resulting from predation by recently introduced *Mysis relicta*. Verh. Internat. Verein. Limnol. 23, 419–429.

LaRow, E.J., 1975. Secondary productivity of *Leptodora kindtii* in Lake George. N. Y. Am. Midl. Nat. 94, 120–126.

Lavrentyev, P.J., Gardner, W.S., Cavaletto, J.F., Beaver, J.R., 1995. Effects of the zebra mussel (*Dreissena polymorpha* pallas) on protozoa and phytoplankton from Saginaw Bay, Lake Huron. J. Great Lakes Res. 21, 545–557.

Lehman, J.T., 1991. Causes and consequences of cladoceran dynamics in Lake Michigan: implications of species invasion by *Bythotrephes*. J. Great Lakes Res. 17, 437–445.

Lehman, J.T., Cáceres, C.E., 1993. Food-web responses to species invasion by a predatory invertebrate: *Bythotrephes* in Lake Michigan. Limnol. Oceanogr. 38, 879–891.

Leibold, M.A., Economo, E.P., Peres-Neto, P., 2010. Metacommunity phylogenetics: separating the roles of environmental filters and historical biogeography. Ecol. Lett. 13, 1290–1299.

Lemaire, V., Brusciotti, S., van Gremberghe, I., Vyverman, W., Vanoverbeke, J., De Meester, L., 2012. Genotype × genotype interactions between the toxic cyanobacterium *Microcystis* and its grazer, the waterflea *Daphnia*. Evol. Appl. 5, 168–182.

Levesque, S., Beisner, B.E., Peres-Neto, P.R., 2010. Meso-scale distributions of lake zooplankton reveal spatially and temporally varying trophic cascades. J. Plankton Res. 32, 1369–1384.

Loose, C.J., Dawidowicz, P., 1994. Trade-offs in diel vertical migration by zooplankton: the costs of predator avoidance. Ecology 75, 2255.

Lunte, C.C., Luecke, C., 1990. Trophic interactions of *Leptodora* in Lake Mendota. Limnol. Oceanogr. 35, 1091–1100.

Lynch, M., 1979. Predation, competition, and zooplankton community structure: an experimental study. Limnol. Oceanogr. 24, 253–272.

MacIsaac, H.J., Gilbert, J.J., 1991. Discrimination between exploitative and interference competition between cladocera and *Keratella cochlearis*. Ecology 72, 924–937.

Macke, E., Callens, M., De Meester, L., Decaestecker, E., 2017. Host-genotype dependent gut microbiota drives zooplankton tolerance to toxic cyanobacteria. Nat. Commun. 8, 1608.

Makarewicz, J.C., Likens, G.E., 1979. Structure and function of the zooplankton community of Mirror Lake, New Hampshire. Ecol. Monogr. 49, 109–127.

Marcus, N.H., Lutz, R., Burnett, W., Cable, P., 1994. Age, viability, and vertical distribution of zooplankton resting eggs from an anoxic basin: evidence of an egg bank. Limnol. Oceanogr. 39, 154–158.

Mariash, H.L., Cusson, M., Rautio, M., 2016. Fall composition of storage lipids is associated with the overwintering strategy of *Daphnia*. Lipids 52, 83–91.

Mason, C.F., Abdul-Hussein, M., 1991. Population dynamics and production of *Daphnia hyalina* and *Bosmina longirostris* in a shallow, eutrophic reservoir. Freshwat. Biol. 25, 243–260.

McCauley, E., Kalff, J., 1981. Empirical relationships between phytoplankton and zooplankton biomass in lakes. Can. J. Fish. Aquat. Sci. 38, 458–463.

McCauley, E., Nisbet, R.M., Murdoch, W.W., DeRoos, A.M., Gurney, W.S.C., 1999. Large-amplitude cycles of *Daphnia* and its algal prey in enriched environments. Nature 402, 653–656.

McLaren, I.A., Corkett, C.J., 1984. Singular, mass-specific P/B ratios cannot be used to estimate copepod production. Can. J. Fish. Aquat. Sci. 41, 828–830.

McQueen, D.J., 1969. Reduction of zooplankton standing stocks by predaceous *Cyclops bicuspidatus thomasi* in Marion Lake, British Columbia. J. Fish. Res. Board Canada 26, 1605–1618.

Merrell, J.R., Stoecker, D.K., 1998. Differential grazing on protozoan microplankton by developmental stages of the calanoid copepod *Eurytemora affinis* Poppe. J. Plankton Res. 20, 289–304.

Meyer-Jacob, C., Michelutti, N., Paterson, A.M., Cumming, B.F., Keller, W., Smol, J.P., 2019. The browning and re-browning of lakes: divergent lake-water organic carbon trends linked to acid deposition and climate change. Sci. Rep. 9, 1–10.

Monakov, A.B., Sorokin, J.I., 1961. Kolichestvenn'ie dann'ie o pitanii Dafnii. Trudy Inst. Biol. Vodokhranilishch 4, 251–261.

Monteith, D.T., Stoddard, J.L., Evans, C.D., De Wit, H.A., Forsius, M., Høgåsen, T., Wilander, A., Skjelkvåle, B.L., Jeffries, D.S., Vuorenmaa, J., Keller, B., Kopécek, J., Vesely, J., 2007. Dissolved organic carbon trends resulting from changes in atmospheric deposition chemistry. Nature 450, 537–540.

Müller, H., 1989. The relative importance of different ciliate taxa in the pelagic food web of Lake Constance. Microb. Ecol. 18, 261–273.

Müller-Navarra, D., 1995. Evidence that a highly unsaturated fatty acid limits *Daphnia* growth in nature. Arch. Hydrobiol. 132, 297–307.

Murtaugh, P.A., 1981. Size-selective predation on *Daphnia* by *Neomysis mercedis*. Ecology 62, 894–900.

Murtaugh, P.A., 1983. Mysid life history and seasonal variation in predation pressure on zooplankton. Can. J. Fish. Aquat. Sci. 40, 1968–1974.

Nagata, T., 1988. The microflagellate-picoplankton food linkage in the water column of Lake Biwa. Limnol. Oceanogr. 33, 504–517.

Neill, W.E., 1981. Impact of *Chaoborus* predation upon the structure and dynamics of a crustacean zooplankton community. Oecologia 48, 164–177.

Neill, W.E., 1984. Regulation of rotifer densities by crustacean zooplankton in an oligotrophic montane lake in British Columbia. Oecologia 61, 175–181.

O'Brien, W.J., 1987. Planktivory by freshwater fish: thrust and parry in the pelagia. In: Kerfoot, W.C., Sih, A. (Eds.), Predation: Direct and Indirect Impacts on Aquatic Communities. University Press of New England, Hanover, NH, pp. 3–16.

Ojala, A., Kankaala, P., Kairesalo, T., Salonen, K., 1995. Growth of *Daphnia longispina* in a polyhumic lake under various availabilities of algal, bacterial and detrital food. Hydrobiologia 315, 119–134.

Olden, J.D., Poff, N.L., 2004. Ecological processes driving biotic homogenization: testing a mechanistic model using fish faunas. Ecology 85, 1867–1875.

Orsini, L., Spanier, K.I., De Meester, L., 2012. Genomic signature of natural and anthropogenic stress in wild populations of the waterflea *Daphnia magna*: validation in space, time and experimental evolution. Mol. Ecol. 21, 2160–2175.

Pace, M.L., Findlay, S.E.G., Lints, D., 1992. Zooplankton in advective environments—the Hudson River community and a comparative analysis. Can. J. Fish. Aquat. Sci. 49, 1060–1069.

Pace, M.L., Vaqué, D., 1994. The importance of *Daphnia* in determining mortality rates of protozoans and rotifers in lakes. Limnol. Oceanogr. 39, 985–996.

Paerl, H.W., Huisman, J., 2008. Blooms like it hot. Science 320, 57–58.

Paloheimo, J.E., 1974. Calculation of instantaneous birth rate. Limnol. Oceanogr. 19, 692–694.

Pangle, K.L., Peacor, S.D., 2006. Non-lethal effect of the invasive predator *Bythotrephes longimanus* on *Daphnia mendotae*. Freshw. Biol. 51, 1070–1078.

Pangle, K.L., Peacor, S.D., Johannsson, O.E., 2007. Large nonlethal effects of an invasive invertebrate predator on zooplankton population growth rate. Ecology 88, 402–412.

Pauli, H.-R., 1991. Estimates of rotifer productivity in Lake Constance: a comparison of methods. Verh. Internat. Verein. Limnol. 24, 850–853.

Pechan'-Finenko, G.A., 1971. Effectiveness of the assimilation of food by plankton crustaceans (translat. consultants Bureau.). Ekologia 2, 64–72.

Pedros-Alio, C., Brock, T.D., 1983. The impact of zooplankton feeding on the epilimnetic bacteria of a eutrophic lake. Freshw. Biol. 13, 227–239.

Peters, K., Bundschuh, M., Schäfer, R.B., 2013. Review on the effects of toxicants on freshwater ecosystem functions. Environ. Pollut. 180, 324–329.

Peterson, D.L., 1983. Life cycle and reproduction of *Nephelopsis obscura* Verrill (Hirudinea: Erpobdellidae) in permanent ponds of northwestern Minnesota. Freshwat. Invertebr. Biol. 2, 165–172.

Peterson, J.K., Kashian, D.R., Dodson, S.I., 2001. Methoprene and 20-OH-Ecdysone affect male production in *Daphnia pulex*. Environ. Toxicol. Chem. 20, 582.

Picapedra, P.H.S., Lansac-Tôha, F.A., Bialetzki, A., 2015. Diel vertical migration and spatial overlap between fish larvae and zooplankton in two tropical lakes, Brazil. Braz. J. Biol. 75, 352–361.

Pijanowska, J., Stolpe, G., 1996. Summer diapause in *Daphnia* as a reaction to the presence of fish. J. Plankton Res. 18, 1407–1412.

Rabus, M., Söllradl, T., Clausen-Schaumann, H., Laforsch, C., 2013. Uncovering ultrastructural defences in *Daphnia magna*—an interdisciplinary approach to assess the predator-induced fortification of the carapace. PLoS One 8, e67856.

Rautio, M., Mariash, H., Forsström, L., 2011. Seasonal shifts between autocthonous and allocthonous carbon contributions to zooplankton diets in a subarctic lake. Limnol. Oceanogr. 56, 1513–1524.

Reede, T., 1995. Life history shifts in response to different levels of fish kairomones in *Daphnia*. J. Plankton Res. 17, 1661–1667.

Reid, A.J., Carlson, A.K., Creed, I.F., Eliason, E.J., Gell, P.A., Johnson, P.T.J., Kidd, K.A., MacCormack, T.J., Olden, J.D., Ormerod, S.J., Smol, J.P., Taylor, W.W., Tockner, K., Vermaire, J.C., Dudgeon, D., Cooke, S.J., 2018. Emerging threats and persistent conservation challenges for freshwater biodiversity. Biol. Rev. 94, 849–873.

Relyea, R.A., 2009. A cocktail of contaminants: how mixtures of pesticides at low concentrations affect aquatic communities. Oecologia 159, 363–376.

Relyea, R.A., Hoverman, J.T., 2008. Interactive effects of predators and a pesticide on aquatic communities. Oikos 117, 1647–1658.

Richman, S., Bailiff, M.D., Mackey, L.J., Bolgrien, D.W., 1984. Zooplankton standing stock, species composition and size distribution along a trophic gradient in Green Bay, Lake Michigan. Verh. Internat. Verein. Limnol. 22, 475–487.

Rigler, F.H., Downing, J.A., 1984. The calculation of secondary productivity. In: Downing, J.A., Rigler, F.H. (Eds.), A Manual on the Assessment of Secondary Productivity. Blackwell Scientific Publications, Oxford, p. 501.

Ringelberg, J., 1991. A mechanism of predator-mediated induction of diel vertical migration in *Daphnia hyalina*. J. Plankton Res. 13, 83–89.

Robertson, A.L., 1995. Secondary production of a community of benthic Chydoridae (Cladocera: Crustacea) in a large river, UK. Arch. Hydrobiol. 134, 425–440.

Rocha, O., Matsumura-Tundisi, T., 1984. Biomass and production of *Argyrodiaptomus furcatus*, a tropical calanoid copepod in Broa Reservoir, sourthern Brazil. Developments in Hydrobiology 23, 307–311.

Rochera, C., Camacho, A., 2019. Limnology and aquatic microbial ecology of Byers Peninsula: a main freshwater biodiversity hotspot in maritime Antarctica. Diversity 11, 1–20.

Rodusky, A.J., Havens, K.E., 1996. The potential effects of a small *Chaoborus* species (*C. punctipennis*) on the zooplankton of a small eutrophic lake. Arch. Hydrobiol. 138, 11–31.

Roff, J.C., Turner, J.T., Webber, M.K., Hopcroft, R.R., 1995. Bacterivory by tropical copepod nauplii: extent and possible significance. Aquat. Microb. Ecol. 9, 165–175.

Rohr, J.R., Crumrine, P.W., 2005. Effects of an herbicide and an insecticide on pond community structure and processes. Ecol. Appl. 15, 1135–1147.

Rublee, P.A., 1992. Community structure and bottom-up regulation of heterotrophic microplankton in arctic LTER lakes. Hydrobiologia 240, 133–141.

Sakamoto, M., Nagata, T., Ha, J.Y., Kimijima, S., Hanazato, T., Chang, K.W., 2015. Inducible defenses as factor determining trophic pathways in a food web. Hydrobiologia 743, 15–25.

Sanders, R.W., Porter, K.G., 1988. Phagotrophic Phytoflagellates. Adv. Microb. Ecol. 10, 167–192.

Sanders, R.W., Porter, K.G., Bennett, S.J., DeBiase, A.E., 1989. Seasonal patterns of bacterivory by flagellates, ciliates, rotifers, and cladocerans in a freshwater planktonic community. Limnol. Oceanogr. 34, 673–687.

Sarnelle, O., Gustafsson, S., Hansson, L.-A., 2010. Effects of cyanobacteria on fitness components of the herbivore *Daphnia*. J. Plankton Res. 32, 471–477.

Sastri, A.R., Juneau, P., Beisner, B.E., 2013. Evaluation of chitobiase-based estimates of biomass and production rates for developing freshwater crustacean zooplankton communities. J. Plankton Res. 35, 407–420.

Saunders Jr., G.W., 1969. National Research Council. In: Eutrophication: Causes, Consequences, Correctives. The National Academies Press, Washington, D.C.

Schindler, D.W., 1968. Feeding, assimilation and respiration rates of *Daphnia magna* under various environmental conditions and their relation to production estimates. J. Anim. Ecol. 37, 369.

Schneider, T., Grosbois, G., Vincent, W.F., Rautio, M., 2017. Saving for the future: pre-winter uptake of algal lipids supports copepod egg production in spring. Freshw. Biol. 62, 1063–1072.

Schwartz, S.S., Cameron, G.N., 1993. How do parasites cost their hosts? Preliminary answers from trematodes and *Daphnia obtusa*. Limnol. Oceanogr. 38, 602–612.

Sell, D.W., 1982. Size-frequency estimates of secondary production by *Mysis relicta* in Lakes Michigan and Huron. Hydrobiologia 93 (1/2), 69–78.

Sharma, S., Magnuson, J.J., Batt, R.D., Winslow, L.A., Korhonen, J., Aono, Y., 2016. Direct observations of ice seasonality reveal changes in climate over the past 320–570 years. Sci. Rep. 6, 25061.

Sherr, E., Sherr, B., 1988. Role of microbes in pelagic food webs: a revised concept. Limnol. Oceanogr. 33, 1225–1227.

Shurin, J.B., 2000. Dispersal limitation, invasion resistance, and the structure of pond zooplankton communities. Ecology 81, 3074–3086.

Shurin, J.B., Havel, J.E., Leibold, M.A., Pinel-Alloul, B., 2000. Local and regional zooplankton species richness: a scale-independent test for saturation. Ecology 81, 3062–3073.

Shuter, B.J., Ing, K.K., 1997. Factors affecting the production of zooplankton in lakes. Can. J. Fish. Aquat. Sci. 54, 359–377.

Šimek, K., Chrzanowski, T.H., 1992. Direct and indirect evidence of size-selective grazing on pelagic bacteria by freshwater nanoflagellates. Appl. Environ. Microbiol. 58, 3715–3720.

Sinclair, J.S., Arnott, S.E., 2018. Local context and connectivity determine the response of zooplankton communities to salt contamination. Freshw. Biol. 1273–1286.

Sison-Mangus, M.P., Mushegian, A.A., Ebert, D., 2015. Water fleas require microbiota for survival, growth and reproduction. ISME J. 9, 59–67.

Slusarczyk, M., 1995. Predator-induced diapause in *Daphnia*. Ecology 76, 1008–1013.

Smyly, W.J.P., 1973. Bionomics of *Cyclops strenuus abyssorum* sars (Copepoda: Cyclopoida). Oecologia 11, 163–186.

Souza, C.A.D., Vieira, L.C.G., Legendre, P., Carvalho, P.D., Velho, L.F.M., Beisner, B.E., 2019. Damming interacts with the flood pulse to alter zooplankton communities in an Amazonian river. Freshw. Biol. 64, 1040–1053.

Stemberger, R.S., Evans, M.S., 1984. Rotifer seasonal succession and copepod predation in Lake Michigan. J. Great Lakes Res. 10, 417–428.

Stemberger, R.S., Gilbert, J.J., 1985. Body size, food concentration and population growth in planktonic rotifers. Ecology 66, 1151–1159.

Stephan, L.R., Beisner, B.E., Oliviera, S.G.M, Castilho-Noll, M.S.M., 2019. Influence of *Eichhornia crassipes* (Mart) Solms on a tropical microcrustacean community based on taxonomic and functional trait diversity. Water 11, 2423.

Sterner, R.W., Hessen, D.O., 1994. Algal nutrient limitation and the nutrition of aquatic herbivores. Annu. Rev. Ecol. Syst. 25, 1–29.

Sterner, R.W., Robinson, J.L., 1994. Thresholds for growth in *Daphnia magna* with high and low phosphorus diets. Limnol. Oceanogr. 39, 1228–1232.

Stirnadel, H.A., Ebert, D., 1997. Prevalence, host specificity and impact on host fecundity of microparasites and epibionts in three sympatric *Daphnia* species. J. Anim. Ecol. 66, 212.

Strecker, A.L., Arnott, S.E., 2008. Invasive predator, *Bythotrephes*, has varied effects on ecosystem function in freshwater lakes. Ecosystems 11, 490–503.

Strecker, A.L., Beisner, B.E., Arnott, S.E., Paterson, A.M., Winter, J.G., Johannsson, O.E., Yan, N.D., 2011. Direct and indirect effects of an invasive planktonic predator on pelagic food webs. Limnol. Oceanogr. 56, 179–192.

Strickler, J.R., 1977. Observation of swimming performances of planktonic copepods. Limnol. Oceanogr. 22, 165–170.

Sundbom, M., Vrede, T., 1997. Effects of fatty acid and phosphorus content of food on the growth, survival and reproduction of *Daphnia*. Freshw. Biol. 38, 665–674.

Swift, M.C., Hammer, U.T., 1979. Zooplankton population dynamics and *Diaptomus* production in Waldsea Lake, a saline meromictic lake in Saskatchewan. J. Fish. Res. Board Canada 36, 1431–1438.

Takamura, N., Yasuno, M., 1983. Food selection of the ciliated protozoa, *Condylostoma vorticella* (ehrenberg) in Lake Kasumigaura. Japanese J. Limnol. (Rikusuigaku Zasshi) 44, 184–189.

Taylor, W.D., Johannsson, O.E., 1991. A comparison of estimates of productivity and consumption by zooplankton for planktonic ciliates in Lake Ontario. J. Plankton Res. 13, 363–372.

Tirok, K., Gaedke, U., 2006. Spring weather determines the relative importance of ciliates, rotifers and crustaceans for the initiation of the clear-water phase in a large, deep lake. J. Plankton Res. 28, 361–373.

Turner, A.M., Mittelbach, G.G., 1990. Predator avoidance and community structure: interactions among piscivores, planktivores, and plankton. Ecology 71, 2241–2254.

Urabe, J., Watanabe, Y., 1991. Effect of food conditions on the bacterial feeding of *Daphnia galeata*. Hydrobiologia 225, 121–128.

van Dorst, R.M., Gårdmark, A., Svanbäck, R., Beier, U., Weyhenmeyer, G.A., Huss, M., 2018. Warmer and browner waters decrease fish biomass production. Glob. Chang. Biol. 25, 1395–1408.

Vanderploeg, H.A., Gardner, W.S., Parrish, C.C., Liebig, J.R., Cavaletto, J.F., 1992. Lipids and life-cycle strategy of a hypolimnetic copepod in Lake Michigan. Limnol. Oceanogr. 37, 413–424.

Vanni, M.J., 1987. Effects of food availability and fish predation on a zooplankton community. Ecol. Monogr. 57, 61–88.

Vanni, M.J., 1988. Freshwater zooplankton community structure : introduction of invertebrate predators and large herbivores to a small species community. Can. J. Fish. Aquat. Sci. 45, 1758–1770.

Vareschi, E., Jacobs, J., 1984. The ecology of Lake Nakuru (Kenya). Oecologia 61, 83–98.

Vasseur, D.A., Fox, J.W., Gonzalez, A., Adrian, R., Beisner, B.E., Helmus, M.R., Johnson, C., Kratina, P., Kremer, C., de Mazancourt, C., Miller, E., Nelson, W.A., Paterson, M., Rusak, J.A., Shurin, J.B., Steiner, C.F., 2014. Synchronous dynamics of zooplankton competitors prevail in temperate lake ecosystems. Proc. R. Soc. B Biol. Sci. 281, 20140633.

Vasseur, D.A., Gaedke, U., 2007. Spectral analysis unmasks synchronous and compensatory dynamics in plankton communities. Ecology 88, 2058–2071.

Verity, P.G., 1991. Feeding in planktonic protozoans: evidence for non-random acquisition of prey. J. Protozool. 38, 69–76.

Villamagna, A.M., Murphy, B.R., 2010. Ecological and socio-economic impacts of invasive water hyacinth (*Eichhornia crassipes*): a review. Freshw. Biol. 55, 282–298.

Villéger, S., Blanchet, S., Beauchard, O., Oberdorff, T., Brosse, S., 2011. Homogenization patterns of the world's freshwater fish faunas. Proc. Natl. Acad. Sci. U. S. A. 108, 18003–18008.

Vlymen, W.J., 1970. Energy expenditure of swimming copepods. Limnol. Oceanogr. 15, 348–356.

Vrba, J., Macháéek, J., 1994. Release of dissolved extracellular β-N-acetylglucosaminidase during crustacean moulting. Limnol. Oceanogr. 39, 712–716.

Wainman, B.C., Lean, D.R.S., 1990. Seasonal trends in planktonic lipid content and lipid class. Verh. Internat. Verein. Limnol. 24, 416–419.

Walz, N., Elster, H.-J., Mezger, M., 1987. The development of the rotifer community structure in Lake Constance during its eutrophication. Arch. Hydrobiol. Suppl. 74, 452–487.

Waters, T.F., 1977. Secondary production in inland waters. Adv. Ecol. Res. 10, 91–164.

Weers, P.M.M., Gulati, R.D., 1997. Effect of the addition of polyunsaturated fatty acids to the diet on the growth and fecundity of *Daphnia galeata*. Freshw. Biol. 38, 721–729.

Weisse, T., 1997. Growth and production of heterotrophic nanoflagellates in a meso-eutrophic lake. J. Plankton Res. 19, 703–722.

Werner, E.E., Peacor, S.D., 2003. A review of trait-mediated indirect interactions in ecological communities. Ecology 84, 1083–1100.

Wetzel, R.G., Likens, G.E., 2000. Organic matter. In: Wetzel, R.G., Likens, G.E. (Eds.), Limnological Analyses. Springer, New York, pp. 137–146.

Wickham, S.A., 1995. *Cyclops* predation on ciliates: species-specific differences and functional responses. J. Plankton Res. 17, 1633–1646.

Wickham, S.A., Gilbert, J.J., 1993. The comparative importance of competition and predation by *Daphnia* on ciliated protists. Arch. Hydrobiol. 126, 289–313.

Wilken, S., Huisman, J., Naus-Wiezer, S., Van Donk, E., 2013. Mixotrophic organisms become more heterotrophic with rising temperature. Ecol. Lett. 16, 225–233.

Willey, R.L., Cantrell, P.A., Threlkeld, S.T., 1990. Epibiotic euglenoid flagellates increase the susceptibility of some zooplankton to fish predation. Limnol. Oceanogr. 35, 952–959.

Wilson, A.E., Sarnelle, O., Tillmanns, A.R., 2006. Effects of cyanobacterial toxicity and morphology on the population growth of freshwater zooplankton: Meta-analyses of laboratory experiments. Limnol. Oceanogr. 51, 1915–1924.

Winberg, G.G., 1971. Methods for the Estimation of Production of Aquatic Animals. Academic Press, New York.

Winberg, G.G., Alimov, A.F., Boullion, V.V., Ivanova, M.B., Korobtzova, E.V., Kuzmitzkaya, N.K., Nikulina, V.N., Finogenova, N.P., Fursenko, M.V., 1973. Biological productivity of two subarctic lakes. Freshw. Biol. 3, 177–197.

Winberg, G.G., Babitsky, V.A., S. I. Gavrilov, G. V., Bladky, I.S.Z., Kovalevskaya, R.Z., Mikheeva, T.M., P. S. Nevyadomskaya, A.P.O., Petrovich, P.G., Potaenko, J.S., Yakushko, O.F., 1970. Biological productivity of different types of lakes. In: Kajak, Z., Hillbricht-Ilkowska, A. (Eds.), Productivity Problems of Freshwaters. PWN Polish Scientific Publishers, Warsaw, pp. 383–404.

Wojtal-Frankiewicz, A., 2012. The effects of global warming on *Daphnia* spp. population dynamics: a review. Aquat. Ecol. 46, 37–53.

Wölfl, S., 1991. The pelagic copepod species in Lake Constance: abundance, biomass, and secondary productivity. Verh. Internat. Verein. Limnol. 24, 854–857.

Wright, J.C., 1965. The population dynamics and production of *Daphnia* in Canyon Ferry Reservoir, Montana. Limnol. Ocean. 10, 583–590.

Wylie, J.L., Currie, D.J., 1991. The relative importance of bacteria and algae as food sources for crustacean zooplankton. Limnol. Oceanogr. 36, 708–728.

Yoshida, T., Jones, L.E., Ellner, S.P., Fussmann, G.F., Hairston Jr., N.G., 2003. Rapid evolution drives ecological dynamics in a predator-prey system. Nature 424, 303–306.

Young, S., Taylor, V.A., 1988. Visually guided chases in *Polyphemus pediculus*. J. Exp. Biol. 137, 387–398.

Zaret, T.M., 1972a. Predators, invisible prey, and the nature of polymorphism in the Cladocera (Class Crustacea). Limnol. Oceanogr. 17, 171–184.

Zaret, T.M., 1972b. Predator-prey interaction in a tropical lacustrine ecosystem. Ecology 53, 248–257.

Zaret, T.M., 1980. Predation and Freshwater Communities. Yale University Press, New Haven, CT.

Zaret, T.M., Kerfoot, W.C., 1975. Fish predation on *Bosmina longirostris*: body-size selection versus visibility selection. Ecology 56, 232–237.

Zhang, G.K., Chain, F.J.J., Abbott, C.L., Cristescu, M.E., 2018. Metabarcoding using multiplexed markers increases species detection in complex zooplankton communities. Evol. Appl. 11, 1901–1914.

Zhao, K., Song, K., Pan, Y., Wang, L., Da, L., Wang, Q., 2017. Metacommunity structure of zooplankton in river networks: roles of environmental and spatial factors. Ecol. Indic. 73, 96–104.

CHAPTER

21

Benthic Animals

Núria Bonada[1] *and Michael T. Bogan*[2]

[1]Departament de Biologia Evolutiva, Ecologia i Ciències Ambientals, Facultat de Biologia, Institut de Recerca de la Biodiversitat (IRBio), Universitat de Barcelona (UB), Barcelona, Catalonia/Spain [2]School of Natural Resources and the Environment, University of Arizona, Tucson, Arizona, United States

OUTLINE

I. Benthic animal groups

Benthic animals are those present on or near the substrate at the bottom of freshwater ecosystems. Their distribution, abundance, and productivity are determined by several processes, acting from regional to local scales: (1) the historical or biogeographical events that have allowed or prevented a species becoming part of the regional species pool; (2) the dispersal ability of the species to colonize a habitat; (3) the physiological limitations of the species at all stages of the life cycle; (4) the availability of energy resources; and (5) the ability of the species to tolerate competition, predation, and parasitism (e.g., Heino, 2013; Hutchinson, 1959).

Several major methodological problems must be overcome to analyze benthic animal communities effectively (Downing, 1984). First, there is the difficulty of obtaining representative quantitative samples. The distribution of benthic animals within lakes and streams can be extremely heterogeneous, and substrata heterogeneity leads to a patchy, nonrandom distribution that requires extensive replicated sampling. Second, organisms must be separated from the substrate in which they live. Third, the taxonomy of many animal groups is confusing to the nonspecialist; some groups are still very incompletely described. Finally, the emigration and immigration of members of the populations of certain groups, especially among the insects with both aquatic and terrestrial life stages, necessitates more

Wetzel's Limnology, Fourth Edition
https://doi.org/10.1016/B978-0-12-822701-5.00021-5

elaborate sampling methods. Despite these problems, careful and detailed analyses can provide insight into the controlling abiotic and biotic factors of benthic communities.

Benthic organisms are extremely diverse in terms of taxonomic groups, evolutionary histories, and biological and ecological traits (Múrria et al., 2018; Figs. 21-1, 21-2, and 21-4) and are represented by nearly all phyla from protozoans through large macroinvertebrates and vertebrates. In this chapter we describe some of the major benthic protists and animal groups (*meiofauna* and *macroinvertebrates*: the former include invertebrates between 40 μm and 500 μm and the latter those bigger than 500 μm), giving examples of their population and production dynamics in relation to freshwater ecosystems, as well as the factors controlling their distributions. To be practical, we are referring to "benthic animals" but acknowledge that this category includes representatives of the Animal and the Protista kingdoms (i.e., some protozoans). Finally, we provide information on metacommunities of these benthic animals in lentic and lotic ecosystems, and how their biodiversity, taxonomic composition, and functional composition respond to global change.

A. Protozoa

Ciliates and heterotrophic flagellates (and other protozoa, such as ameboid protozoans) are perhaps the least understood groups of benthic organisms that occur in massive numbers on and in surficial sediments (Reiss and Schmid-Araya, 2008) (Fig. 21-2A). Despite the abundant information that we have on their morphology, physiology, genetics, and behavior, scarce information is known concerning their population and community dynamics and productivity. Most protozoans are attached to substrata and occur in the upper 1 cm of sediment.

The diversity of protozoan species, their wide ranges of tolerance to environmental extremes, their varied feeding capabilities (including particulate detritus, bacteria, and other protists), and their large population densities on aerobic and organic-rich sediments point to a significant metabolic role in freshwater ecosystems. Despite their small biomass, their short generation times and rapid turnover suggest that protozoans contribute appreciably to decomposition processes (Dias et al., 2008). In addition, protozoan grazing seems to significantly control biofilm population and composition dynamics (Chapter 25) and alter biofilm morphology (Böhme et al., 2009; Huws et al., 2005). They are particularly important in the metabolism of dissolved and particulate organic matter of sewage-treatment facilities and urban streams (Cairns, 1974; Dias et al., 2008).

Many abiotic factors, such as temperature, flow, turbidity, and pH, affect the distribution and growth of benthic protozoans. Many of them exhibit pronounced summer maxima related to increases in temperature, day length, and organic matter, and microflora serving as food in the surficial sediments (Bark, 1981; Finlay, 1980, 1990). However, dissolved oxygen (Chapter 11) is of paramount importance (Debastiani et al., 2016). Most species are intolerant of low oxygen concentrations and reducing conditions in sediments, but some species tolerate anaerobic conditions. As a result, protozoan species are often segregated vertically in lake sediments along a depth gradient (e.g., Finlay, 1980). In rivers these differential tolerances were used as part of the *Saprobic system*, that is, one of the biomonitoring systems used to evaluate organic pollution in aquatic habitats (Foissner, 1988; Madoni, 2005).

Regarding biotic factors affecting protozoans, most of the evidence for competitive species interactions is indirect under natural conditions (e.g., Cairns and Yongue, 1977). However, predation upon protozoans is common by other protozoan species and by some rotifers, cladocerans, and copepods (Chapters 19 and 20). For example, the experimental introduction of these micrometazoa reduced the abundance of larger protists, mostly ciliates, and certain algae (McCormick and Cairns, 1991). Heterotrophic flagellate abundance increased under these conditions, in part related to declines in ciliates that prey on and compete with the smaller flagellates.

B. Porifera

Most porifers, or *sponges*, are marine, but some occur in freshwater (Fig. 21-2B). Although some species grow to magnificent lobed structures of about half a meter in size (Fig. 21-3), most are small, inconspicuous, and morphologically variable. They live in colonies and grow over submersed substrates, creating a coat of a few millimeters of thickness or forming cushions of erected forms (Manconi and Pronzato, 2008). Growth rates and energy demands generally increase progressively and reach a maximum in midsummer (Melão and Rocha, 1998). In mild climates sponge populations can persist for many years without appreciable change (e.g., Pronzato and Manconi, 1995) until a major disturbance occurs, such as flooding.

Reproduction in freshwater sponges can occur sexually or asexually. The latter includes budding or the production of *gemmules*, i.e., highly resistant resting structures of 100 to 1200 μm in diameter that are made under unfavorable conditions (e.g., low temperatures,

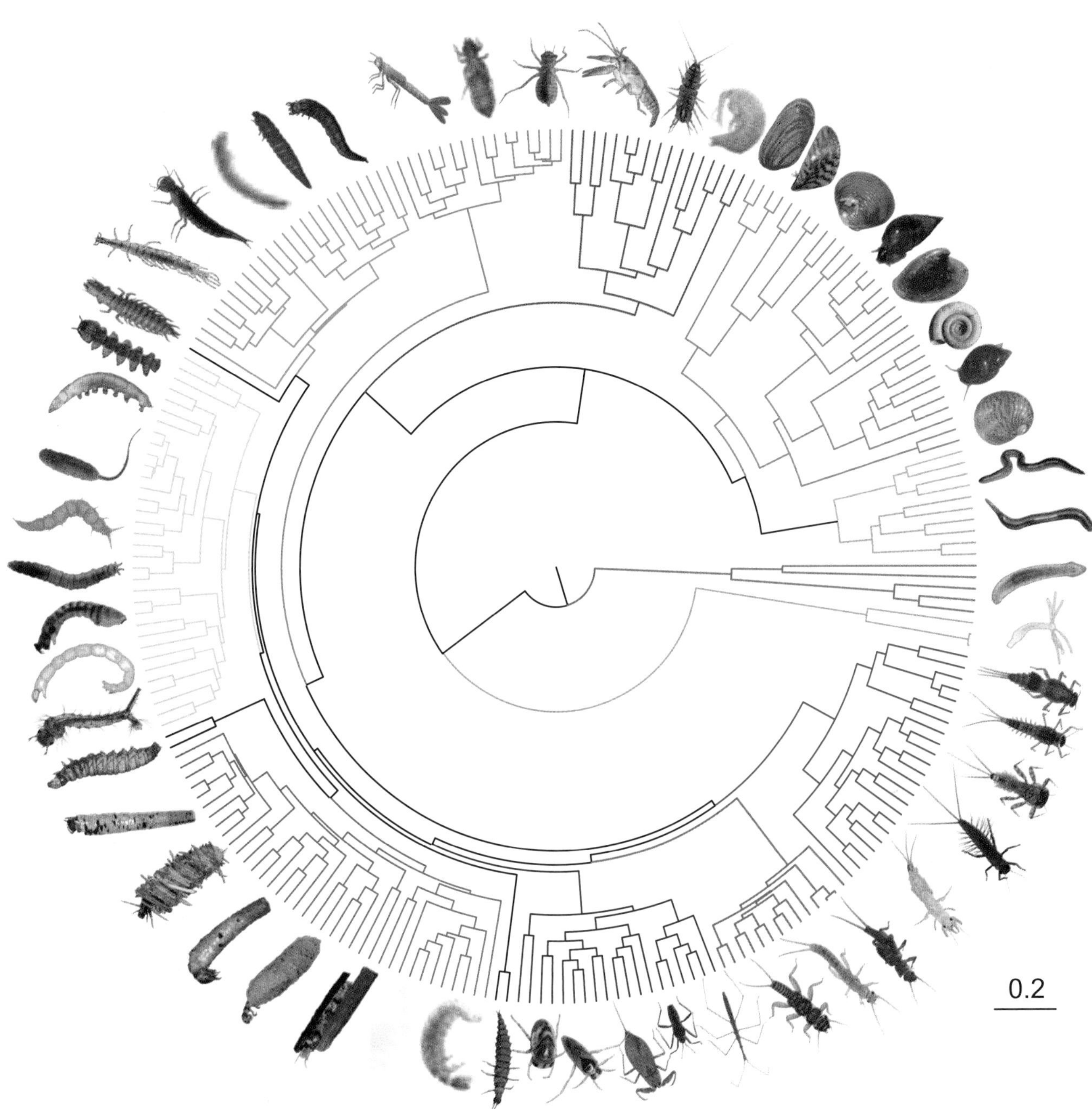

FIGURE 21.1 Phylogenetic tree representing 143 freshwater macroinvertebrate families from Europe generated using three genes (the mitochondrial cytochrome c oxidase subunit I and 16S rRNA genes, and the nuclear 18 rRNA gene). The different colors of the clades correspond to different taxonomic groups. The scale bar of genetic distance indicates the estimated change per nucleotide. *(Modified from Múrria et al., 2018).*

water level) (Gilbert, 1975; Gilbert and Simpson, 1976). Gemmule germination is usually associated with warmer temperatures greater than 15°C in the early spring (Harrison, 1974). Motile larvae move to appropriate substrata, attach, and develop. Some species, particularly those in sublittoral environments with muted seasonality found in ancient lakes, do not form gemmules (Manconi and Pronzato, 2008; Simpson and Gilbert, 1974).

Sponges survive in a wide variety of temperatures, water levels, oxygen concentrations, and freshwater ecosystems and tolerate several types of pollutants (Manconi and Pronzato, 2008). Despite this adaptability, they are restricted largely to waters of moderate silica content (>0.5 mg L^{-1}) (Jewell, 1935). Species distribution is also directly correlated with the calcium content of the water but declines in very hard waters (Strekal and McDiffett, 1974) and to other abiotic factors, such as

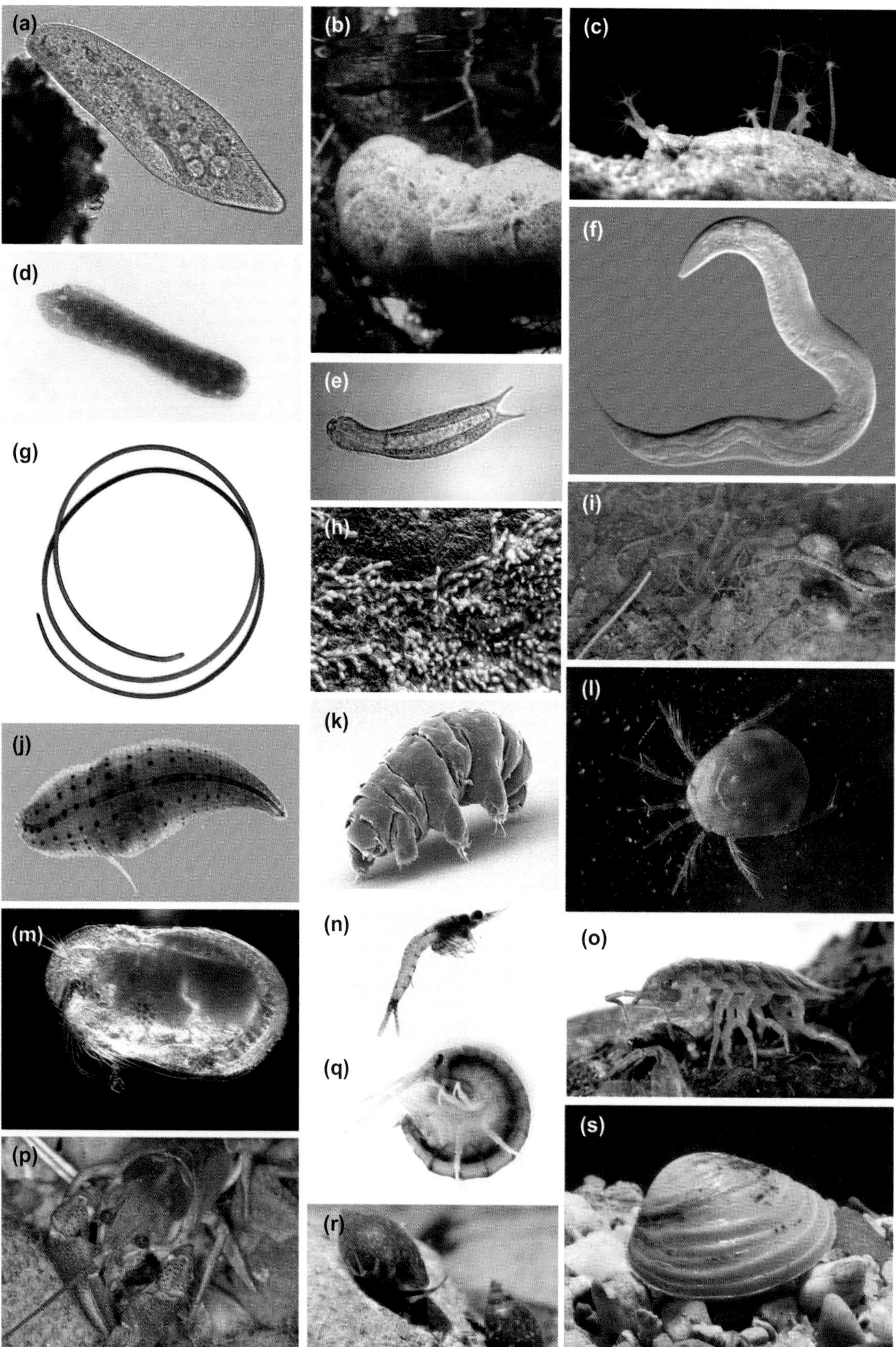

FIGURE 21.2 Examples of the wide variety of benthic animals from freshwater habitats: (a) Protozoa. *(photo credit: Wikipedia.)* (b) Porifera. *(Photo credit: Michael T. Bogan.)* (c) Cnidaria. *(photo credit: Jesús Ortiz, CEN Association.)* (d) Platyhelminthes. *(photo credit: Jesús Ortiz, CEN Association)* (e) Gastrotrichia. *(photo credit: Giuseppe Vago, Wikipedia.)* (f) Nematoda. *(photo credit: Bob Goldstein, Wikipedia.)* (g) Nematomorpha (Gordiaceae). *(photo credit: Jesús Ortiz, CEN Association.)* (h) Bryozoa. *(photo credit: Jesús Ortiz, CEN Association.)* (i) Annelida: Oligochaeta. *(photo credit: Jesús Ortiz, CEN Association.)* (j) Annelida: Hirudinea. *(photo credit: Ulrich Kutschera, Wikipedia.)* (k) Tardigrada. *(photo credit: Wikipedia.)* (l) Acari. *(photo credit: Anatoly Mikhaltsov, Wikipedia.)* (m) Ostracoda. *(photo credit: Anna Syme, Wikipedia.)* (n) Malacostraca: Mysida. *(photo credit: Steven Pothoven, Wikipedia.)* (o) Malacostraca: Isopoda. *(photo credit: André Karwath, Wikipedia.)* (p) Malacostraca: Decapoda. *(photo credit: Jesús Ortiz, CEN Association.)* (q) Malacostraca: Amphipoda. *(photo credit: Jesús Ortiz, CEN Association.)* (r) Mollusca: Gastropoda. *(photo credit: Michael T. Bogan.)* (s) Mollusca: Bivalvia. *(photo credit: Jesús Ortiz.)*

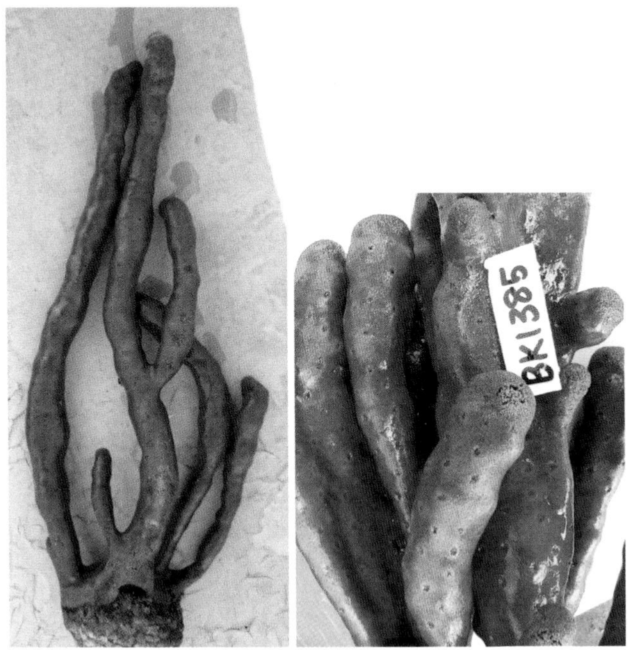

FIGURE 21.3 Examples of the Porifera Lubomirskiidae. This family is endemic to Lake Baikal, southern Siberia (Russia), can reach large sizes (1 m), and occupy a large proportion of the benthos of the lake. It contains 13 species belonging to 4 genera and lives in symbiosis with photosynthetic microorganisms. *(Photo courtesy of Valeria Itskovich).*

water level and temperature (Matteuzzo et al., 2015). These properties, and the production of siliceous spicules, make them good organisms for paleoenvironmental studies (Docio et al., 2021). Major predation upon sponges by other animals is rare. Spicules, and likely organic compounds, can function as deterrents to predators. Several animals, particularly certain immature insects, live within sponges, using them as substrate and to take refuge from predation. Sponges rarely become a major component of benthic communities, except in certain conditions (e.g., 40% of total benthic productivity in the River Thames [UK] [Mann et al., 1972] and in a *Sphagnum* bog pond [Frost, 1978]).

C. Cnidaria

Most cnidarians belong to a predominately marine group that is poorly represented in freshwater (Jankowski et al., 2008). Freshwater representatives primarily correspond to the class Hydrozoa and include fewer than 40 species (Fig. 21-2C). Many cnidarians may alternate between *medusoid* (with tentacles) and *polyp* (without tentacles) stages. The common freshwater *Hydra*, however, only occurs as the tentacled polyp stage, whereas the rare *Craspedacusta* or *Limnocnida* alternate between polyp and medusoid stages. In comparison to *Hydra,* which are solitary polyps, other hydroids, such as *Cordylophora*, form colonies. In one rare case a freshwater cnidarian (*Polypodium hydriforme*, the only species of the class Polypodiozoa) is known to parasitize sturgeon eggs (Jankowski et al., 2008).

Epidermal cells contain *nematocysts* or *stinging cells*, i.e., epidermic capsules that help to capture prey; food particles are then moved to the mouth orifice by the tentacles. Except for the parasitic species, most cnidarians prey upon zooplankton, although polyps can also consume small benthic invertebrates (Dumont, 1994). One species of green *Hydra* (*H. viridissima*) contains symbiotic *Chlorella* algae inside the endodermal epithelial cells. About 10% of the carbon photosynthetically fixed by the algae is released and assimilated by the *Hydra* (Muscatine and Lenhoff, 1963). This photosynthate is clearly of nutritional significance to the *Hydra* (Pardy and White, 1977; Phipps and Pardy, 1982).

Development of cnidarians in lentic ecosystems is rapid in spring, and greatest densities are usually found in early summer and late summer (Cuker and Mozley, 1981; Miller, 1936). Growth rates decrease markedly when populations are crowded (Thorp and Barthalmus, 1975). However, cnidarian densities usually do not become great, and their contribution to total benthic productivity is considered negligible. Most species are found in mesotrophic to eutrophic freshwaters of different types, and some genera are only found in saline lakes (Jankowski et al., 2008). Several species of the genus *Hydra* are known to be sensitive to a large set of pollutants and have been widely used in bioassays (e.g., Quinn et al., 2012).

D. Platyhelminthes

Flatworms are acoelomate benthic animals with only a single opening to the digestive tract (they lack an anus) (Fig. 21-2D). This phylum includes the classes Turbellaria and several others under the monophyletic group Neodermata. Turbellarians are free-living flatworms, whereas Neodermata are exclusively parasitic. Many are small (few millimeters), and the larger taxa (1–5 cm) belong to the order Tricladida and are known as *planarians* (Schockaert et al., 2008).

Turbellarians generally require a firm substratum and possess abundant cilia that assist them in movement over the substrata of shallow lakes and streams. They are mainly predators, and when they encounter small invertebrates or detrital organic matter, the ventral *pharynx* is extruded through the mouth to engulf the material (Young, 1973). Some species contain symbiotic zoochlorellae in their parenchyma and cells of the gastrodermis. Reproduction is either asexual, by budding or fission (Pattee and Persat, 1978), or sexual. A few species are *univoltine* (i.e., one generation per year), but most species are *multivoltine* (i.e., several generations per year), with the number of generations correlated with the availability of habitat and food (Hartmann, 1985).

Most turbellarians, especially the planarians, are negatively phototactic and occur under stones, debris, or macrophytes. A few species are planktonic (Schockaert et al., 2008) or burrow into fine sediments (Vila-Farré and Rink, 2018). Their abundance in lakes is variable within a range from several hundred to 80,000 individuals m^{-2} (Heitkamp, 1982), representing a significant portion of the total benthos productivity. They are rare in temporary waters, but some species have evolved life history strategies to cope with the dry period in pools or springs (Ball et al., 1981). Densities in streams are generally lower but can reach approximately 5000 individuals m^{-2}. Although a number of macroinvertebrates and vertebrates (i.e., plecopterans, trichopterans, newts, and fishes) may consume planarians, they are a minor component of their diet (Reynoldson, 1983), and therefore predation is rarely considered to significantly affect natural populations (Vila-Farré and Rink, 2018). The mucus produced by *rhabdites* (i.e., epidermic structures) assists in locomotion, prevents desiccation, and confers protection from predators. Parasitism by protists, nematodes, and trematodes (one of the parasitic platyhelminths) is high, but mortality effects appear to be low (Armitage and Young, 1990).

The abundance of planarians has been directly correlated with overall ecosystem productivity and the amount of food available, which in turn has been directly related to moderate concentrations of calcium and total dissolved matter (e.g., Reynoldson, 1966, 1983). However, some species are also found in low calcium concentrations (Vila-Farré and Rink, 2018) and even waters that are moderately acidic (pH = 5) or rich in tannins (Grant et al., 2006). Some species are also tolerant to high concentrations of dissolved salts (Reynoldson, 1958) and other chemicals (Stocchino et al., 2017). Additionally, their abundance is inversely related to water velocity, and stream species are more sensitive to high temperatures than lake species (Vila-Farré and Rink, 2018). Interspecific and intraspecific competition is common and depends largely on the distribution and abundance of prey, such as gastropod mollusks, chironomids, plecopteran and ephemeropteran nymphs, amphipods, and oligochaetes, each of which represents the dominant prey of particular species.

E. Gastrotricha

Gastrotrichs represent a very small phylum of diminutive (50–800 μm) pseudocoelomate animals (Fig. 21-2E). They occur in high densities in surficial coarse sediments and biofilm of lakes and streams, and many species move to *hyporheic* sediments (i.e., those between surface and groundwater) during disturbances (Ricci and Balsamo, 2000). Most freshwater representatives belong to the order Chaetonotida (Balsamo et al., 2008). Gastrotrichs commonly attach the posterior end of their spindle-shaped body with adhesive tubes to a substratum and have ventral cilia that assist in locomotion when they move. They feed on bacteria, algae, protozoans, and small detrital particles and are predated by amoebae, cnidarians, and chironomids.

Upon hatching from eggs, young gastrotrichs already contain developing parthenogenetic eggs that can develop rapidly (e.g., 2 days), are quickly laid, and can hatch within a day under favorable conditions (Strayer and Hummon, 1991). This parthenogenesis allows rapid population development, with growth rates of 0.1 to 0.6 per day. Some gastrotrichs live longer and develop into hermaphrodites, and meiotic sexual eggs are produced with genetic recombination. A second type of parthenogenetic egg, a resting egg that is thick shelled and resistant to freezing and desiccation, likely aids in dispersal.

Among the few population analyses, seasonal trends in gastrotrich abundance show differences among species in the same habitats. Generally, abundances are greatest in spring for several species (range 10–50 individuals cm^{-2}) on organic-rich lake sediments; other species reach greater population densities in winter (Kisielewska, 1982; Strayer and Hummon, 1991). Gastrotrich densities can be extremely high, especially in lakes, with a reported range of 10,000 to 2,600,000 individuals m^{-2} (Nesteruk, 1993, 1996). Strayer (1985) estimated the secondary production of gastrotrichs in Mirror Lake, New Hampshire, at c. 100 mg dry matter (DM) m^{-2} $year^{-1}$, which represented <1% of the total production of the zoobenthic community in this lake. They are important organisms in food webs, transferring energy from bacteria and algae to higher trophic levels (Schmid-Araya and Schmid, 2000), but are usually underestimated and overlooked (Ricci and Balsamo, 2000).

Only a few species can tolerate salinity and most of them are sensitive to high flows, but little is known about other environmental factors affecting their distribution (Ricci and Balsamo, 2000). Most likely, their local distribution is influenced by food availability (Swan and Palmer, 2000) and sediment granulometry (Ricci and Balsamo, 2000), but they could also be sensitive to water physicochemistry and pollutants (Hummon and Hummon, 1979).

F. Nematoda

The phylum Nematoda is a significant component of the benthic meiofauna in terms of richness and abundance (Fig. 21-2F). Although many species are parasitic on insects or vertebrates, free-living species are distributed widely in all types of freshwater ecosystems and can tolerate a wide range of environmental conditions (Abebe et al., 2008). The taxonomy of this group of pseudocoelomate animals is difficult and incomplete and has contributed to the relative paucity of understanding of the ecology and productivity of nematodes.

Feeding habits are very diverse among the free-living nematodes (Moens et al., 2004). Some members are strictly detritivores, feeding solely on dead plant or animal particulate matter, or both. Some species are bacterivorous and are prolific on surfaces among organically polluted conditions (Schiemer, 1983). Others are herbivorous and have specialized mouth parts for chewing living or dead plant material or for piercing and sucking cytoplasm from plants (Prejs, 1977a). Mouth parts are more specialized among carnivorous nematodes, enabling them to seize, rasp, and macerate small prey animals, such as other nematodes, protozoans, gastrotrichs, tardigrades, oligochaetes, and small insects.

Nematode populations are usually concentrated at depths of 3 to 4 cm in the sediments (Bretschko, 1973), although some penetrate below 6 cm into the sediments, with considerable vertical separation among species (Traunspurger and Drews, 1996). Most studies have examined community dynamics of nematodes in profundal sediments, where densities can range between 30,000 and greater than 1,000,000 individuals m^{-2} (Pieczyńska, 1959, 1964; Traunspurger, 1996), comprising 40% to 91% of meiofauna (Traunspurger et al., 2006). Despite these high densities, abundance, biomass, production, and species richness in sediments are lower than in marine environments (Eyualem et al., 2004). Nematodes contribute significantly to the decomposition of organic matter, being key players in freshwater food webs between microbes and larger invertebrates and vertebrates (De Ley et al., 2006; Majdi and Traunspurger, 2015).

Environmental factors affecting nematode distribution have not been analyzed in detail, but factors such as grain size, oxygen, temperature, and food availability could be relevant (Traunspurger et al., 2006). Nematodes can tolerate extreme values of acidity, anoxia, or high temperatures where other invertebrates are unable to survive (Traunspurger et al., 2006), offering good opportunities for their use in biomonitoring. Some species also tolerate high episodic nutrient enrichment (Gaudes et al., 2013) or other pollutants, such as metals (Haegerbaeumer et al., 2018), whereas others are very sensitive. The greatest densities of nematodes are found in oligotrophic lakes (Prejs, 1977b, 1977c), and species richness has highest values at intermediate levels of the trophic status of lakes (Traunspurger et al., 2006). The richness values of nematodes in lakes and rivers are similar, but diversity indices are higher in rivers (Traunspurger et al., 2006). Biotic factors, such as predation and competition among nematodes, are reported to affect species distributions in lakes, and Michiels et al. (2004) even found that predation among nematodes increased the number of coexisting species.

G. Nematomorpha (Gordiaceae)

Of the two classes in the phylum Nematomorpha, only Gordiaceae includes freshwater species, and these *gordian worms* are common in many freshwater ecosystems (Fig. 21-2G). As juveniles, they parasitize terrestrial arthropods, but the free-living adult and egg stages are aquatic. Adults occur among littoral vegetation of shallow lakes, wetlands, and streams. They can attain 1 m in length but are usually less than 3 mm in diameter throughout their body length (Poinar, 2008). Larvae transform into cysts that eventually get into immature aquatic insects or other invertebrates and even vertebrates (fish and amphibians). Some species have direct development into adults in the aquatic host, whereas others use it as a *paratenic* host (i.e., an intermediate host where there is no parasite development) until it is consumed by a terrestrial arthropod (Hanelt et al., 2005). Host specificity is not well documented, but not all potential hosts will allow for complete development of gordian worms. When direct development occurs in an aquatic host, sterility or death of the host is induced, and heavy parasitization may affect the population dynamics of host species.

Very little is known about the feeding habits, population dynamics, and general ecology of gordian worms. Except in a few cases, where some species achieve high densities, they tend to occur in low abundance (Poinar, 2008). Understanding their dispersal and distribution is not easy, especially for adults, which often arrive accidently in freshwater ecosystems via terrestrial hosts (Hanelt et al., 2005). Gordian cysts that develop in the paratenic host better indicate the presence of these worms (Hanelt et al., 2001). Finally, gordian worms seem to be very sensitive to pollution (Poinar, 2008).

H. Bryozoa

Colonial bryozoans are small, modular, sessile invertebrates that range in form from small strands of stringy material to large clusters that can weigh several kilograms (Wood, 2015) (Fig. 21-2H). They are often inconspicuous members of the fauna but occasionally form massive colonies in warm lakes, wetlands, and slow-moving portions of larger rivers. The microscopic *zooids* (individual members of the colony) possess ciliated tentacular crowns (named a *lophophore*) that project into the water and create water currents to direct rotifers, small microcrustaceans, and algae (<20 μm) into the mouth (Kaminski, 1984). Some bryozoan colonies also contain algae that are attached to surfaces enclosed by the colonies (Joo et al., 1992). Interestingly, Cyanobacteria can dominate on surfaces enclosed by the colonies, even when they are only minor components of the attached algae on surfaces not colonized by bryozoans. Although there are currently only approximately 125 freshwater bryozoan species known globally (Wood, 2019), the number of undescribed species found recently in the tropics suggests that their true diversity may be much higher (Wood and Okamura, 2017).

Although capable of hermaphroditic sexual reproduction during warm seasons, bryozoan zooids also reproduce asexually by budding new individuals from the mother zooid. Budding permits rapid proliferation on the outer edge of the colonies, which occasionally form amorphous ball-like masses one-quarter of a meter in diameter. These colonies are common in small farm ponds in water less than a meter in depth, and their growth is influenced by fish predation (Dendy, 1963). When protected from fish predation, colonies were branched and more productive; when exposed to fish predation, unbranched, cropped colonies persisted. Many invertebrate animals, from protozoans to large insect larvae (particularly chironomids), live within the colonies of bryozoans. Predators of the bryozoans also consume these coinhabitants.

Under unfavorable environmental conditions, many bryozoans form dormant buds known as *statoblasts.* These statoblasts can remain dormant for years under dry and freezing conditions (Wood, 2015), but their viability may be low (<10%) (Smyth and Reynolds, 1995). The formation of statoblasts assists bryozoans in dispersing through space and time, and their production is approximately proportional to the size of the colony (Karlson, 1992).

Some bryozoan species are associated with macrophyte substrata, whereas others are distributed more ubiquitously (Ricciardi and Reiswig, 1994). Live colonies have been found under ice at water temperatures less than 2°C, although most bryozoans prefer much warmer water (>15°C; Wood, 2015). Growth is often erratic, but luxuriant colonies of *Plumatella* can double in size in as few as 4 days in spring and summer. Predation on actively growing colonies can often be very high, especially by caddisfly larvae and snails.

I. Annelida

Two major groups of the phylum Annelida, or segmented worms, are represented in freshwater ecosystems. The first of these, the Oligochaeta or aquatic *earthworms*, often form a major component of the benthic fauna (Timm and Martin, 2015). The other group, Hirudinea or *leeches*, has predatory habits that influence the population dynamics of other benthic organisms, including oligochaetes (Govedich and Moser, 2015). A third minor group, the Branchiobdellida, consists of leech-like *ectosymbionts* (i.e., the organism lives on the body surface of the host) living only on freshwater crustaceans, primarily on astacid crayfish (Gelder and Williams, 2015). This latter group is of much evolutionary interest but is a minor component of freshwater ecosystems and will not be discussed further here.

i. Oligochaeta

Oligochaetes are segmented, bilaterally symmetrical, hermaphroditic annelids with an anterior ventral mouth and a posterior anus (Fig. 21-2I). Sizes range from less than 1 mm to over 200 cm, but most freshwater forms are less than 5 cm in length. Swimming occurs through rapid rhythmic movements in helical body wave cycles that propel the worm forward (Drewes and Fourtner, 1993). Sexual reproduction predominates in many species under poor environmental conditions, whereas asexual reproduction is more common during favorable conditions (Loden, 1981).

At least a few oligochaete species can be found in almost every freshwater ecosystem, and diversity is often greater in larger lakes, perhaps because of the greater number of different microhabitats. Some species are restricted to relatively oligotrophic waters, whereas others are distributed widely in lakes of greatly differing productivity, from oligotrophic to extremely eutrophic (e.g., Lang and Lang-Dobler, 1980; Särkkä and Aho, 1980).

Nutrition and the availability of food are primary factors influencing the distribution and abundance of oligochaetes. They ingest surficial sediments containing organic matter colonized with bacteria and other microorganisms. Some species actively graze microorganisms growing epiphytically on macrophytic vegetation. Detailed analyses of the gut contents indicate that resource partitioning occurs by differences in worm morphology and by selective feeding on food items, particularly algae (McElhone, 1979). As lakes and streams become organically polluted, it is common to find an abundance of tubificid oligochaetes (Milbrink, 1980).

Many oligochaetes, especially the tubificids, burrow headfirst into the sediments but leave their caudal ends, which contain most of their respiratory appendages, projecting and undulating in the water. The caudal respiratory movements increase in frequency and vigor as the oxygen content of the water decreases. In general, tubificids can adjust their respiration to decreasing oxygen concentration down to a critical level in the range of 10% to 15% saturation. Below this level, feeding and defecation are greatly reduced or cease (Brandt, 1978). Many tubificids can tolerate anaerobic conditions for at least a month or longer but survive only if they are exposed intermittently to some oxygen; they cannot respire anaerobically for any appreciable length of time.

Oligochaetes lack discrete age classes and production is often continuous. The limited data from natural populations suggest that maturation requires as long as a year in some species and up to 4 years in others, but there is much intraspecific variation in development times. In the deep profundal sediments of the Danish eutrophic Lake Esrom, some *Potamothrix hammoniensis* can first reproduce at 3 years of age, but most do not until they are 4 years old (Jónasson and Thorhauge, 1972). The lifespan of this species can reach 5 or more years. Breeding intensifies in late winter and spring as temperatures increase above about 10°C. Most adult oligochaetes die after sexual reproduction.

Oligochaete species frequently segregate vertically within benthic sediments. Naidid oligochaetes are concentrated at the sediment–water interface (Milbrink, 1973). Tubificid oligochaetes are most dense between 2 and 4 cm of sediment depth and occasionally penetrate as deep as 15 cm. The movements of oligochaetes within sediments are important because this activity can disrupt temporarily the oxidized microzone of the sediment–water interface (Chapters 11 and 15), altering rates of chemical exchange between the sediments and overlying water.

ii. Hirudinea

Leeches are commonly encountered in lakes, wetlands, and slower-moving portions of rivers, where they play important trophic roles in aquatic communities (Govedich and Moser, 2015) (Fig. 21-2J). Many leeches are ectoparasites that intermittently consume blood and body fluids of larger vertebrates. Other leeches are predatory on invertebrates, such as oligochaetes, amphipods, snails, and chironomid larvae, and consume their prey entirely (Toman and Dall, 1997).

Most predaceous leeches have an annual or semiannual life cycle, breed once, and then die. Reproduction is usually initiated in the spring or early summer and is influenced by temperature, the density of the populations, and age. Many leeches breed at 1 year of age and die after breeding, although some pass through two or more generations per year (Malecha, 1984). Other species live two years, overwinter, and breed in the second, and die soon thereafter (Elliott, 1973).

After fertilization, eggs are deposited into a cocoon that is usually subsequently laid on or attached to substrata. In the first 3 months after hatching there is heavy mortality of the young due to inadequate feeding as well as predation by other aquatic species (>95% mortality; Young et al., 1995). The swimming activity of leeches in search of food is much greater at night, presumably to avoid visual predators (Angstadt and Moore, 1997) or to follow the diurnal vertical migration of pelagic prey (Blinn and Davies, 1990).

The abundance of leeches is highly variable among different waterbodies, but leech abundance is generally correlated with productivity. More productive habitats may support more diverse macrophytes, with correspondingly greater amounts of invertebrate food sources for the predacious leeches, as well as birds and other vertebrates for the blood-consuming leeches (Sandner and Wilkialis, 1972). Growth and production rates are quite variable in the littoral areas because of the heterogeneity of habitats and associated prey items. In general, however, leech productivity is highest in the upper littoral zone reaches, especially in warmer spring and summer months (Dall, 1987).

J. Tardigrada

The phylum Tardigrada, or *water bears*, are a group of small (100–500 μm) arthropod-like organisms with a bilaterally symmetrical body and four pairs of legs that usually terminate in claws (Fig. 21-2K). Although most tardigrades inhabit wet terrestrial habitats or can live in both terrestrial and aquatic habitats, a small number of species only occur in freshwaters. The total known number of tardigrade species has doubled in recent decades, suggesting much biodiversity remains to be described (Nelson et al., 2015). Tardigrades can play roles as both predators and prey in aquatic food webs (Schmid-Araya and Schmid, 2000) and are known to feed on mosses, algae, bacteria, and detritus (Nelson et al., 2015).

Reproductive strategies of tardigrades vary widely among taxa and populations. Some tardigrades are bisexual, and fertilization usually occurs within the female, but this can vary by population, with some being bisexual and others being unisexual (Kinchin, 1994). Female-only parthenogenic tardigrades also are common in freshwaters; in fact, males have never been observed for one entire clade of 26 species (Nelson et al., 2015). Periodic molting, which requires 5 to 10 days, continues throughout the life of tardigrades.

Very little is known about the population dynamics and productivity of freshwater tardigrades. Some direct correlations have been observed between tardigrade population fluctuations and attached bacteria and algae and leaf litter resources, and inversely with prey nematodes (Kinchin, 1994; Nelson et al., 2015). Arguably, tardigrades are most well-known for their ability to enter quiescence and tolerate extreme environmental conditions (e.g., anoxia, freezing, desiccation). Dormant tardigrades in anhydrobiotic states can even tolerate extreme UV radiation and vacuum conditions associated with outer space. Other strategies for surviving adverse environmental conditions include encystment, when tardigrades metabolize at rates lower than active animals but higher than anhydrobiotic ones, and resting egg stages (Nelson et al., 2015). Little is known about seasonal or interannual fluctuations in tardigrade populations in freshwater or about how population dynamics may be shaped by their numerous strategies for surviving in highly variable environments.

K. Acari

Water mites are a collective group of animals emanating from five groups that are not monophyletic; the Hydrachnida form the largest and most diverse group of freshwater mites (Gledhill, 1985; Smith and Cook, 1991) (Fig. 21-2L). The group is quite diverse, with greater than 6000 species known from lakes, streams, and the interstitial waters beneath benthic zones (Proctor et al., 2015).

Their complex development begins with hexapod larvae that emerge from eggs and become ectoparasites on insects in or close to the water, from which they extract fluids while being passively transported and dispersed. Upon release from the host, water mites commonly enter a resting stage during which development occurs to form an *octopod*, the actively swimming *deutonymph stage* (i.e., a second larval stage). Although water mites are already feeding on prey and growing at this stage, they are not sexually mature. Later, they enter a second resting stage from which, after further development and metamorphosis, they emerge as active, mature adults. After mating and fertilization, eggs are laid in masses in a gelatinous matrix attached to substrata.

Analyses of dominant water mite populations within a complex community of many species in a eutrophic lake in the Netherlands indicated relatively constant population densities of up to 1000 individuals m^{-2} throughout the year, with the highest densities in shallow littoral areas (Davids et al., 1994). Nymphs were abundant from spring through late autumn and declined precipitously in winter. The growth rates of nymphs and adult mites increased with rising temperatures, and the duration of resting stages decreased with increasing temperatures (Butler and Burns, 1995).

The parasitic larvae and predaceous deutonymphs and adults of water mites have direct effects on the size and structure of insect populations in some habitats. Larval water mites may parasitize 20% to 50% of natural populations of aquatic insects (Smith, 1988). This parasitism impairs the vitality, growth, mobility, and fecundity of hosts. Deutonymphs and adults of free-living species of water mites are voracious predators of eggs of insects and fish and larvae of many dipterans and other small insects, ostracods, cladocerans, and copepods (Gledhill, 1985). However, less is known of predation upon water mites. Although some species are found to be consumed by fish, with one new species even being described from specimens found in fish guts (Pešić et al., 2013), other water mites are specifically rejected by fish as distasteful (Proctor and Garga, 2004).

L. Ostracoda

Ostracods are small, bivalved crustaceans usually less than 1 mm in size that are widespread in nearly every nonacidic aquatic habitat (Fig. 21-2M). At least 2000 freshwater species are known globally, and new species are frequently found in undersampled habitats, such as temporary waters, wetlands, and groundwater (Smith et al., 2015). Species distributions are generally influenced by hydroperiod, pH, dissolved oxygen levels, temperature, solute composition, and substrate size (Mesquita-Joanes et al., 2012). Ostracod valves, which superficially resemble clam shells, are held apart when undisturbed. Most ostracods move about on the sediments by beating movements of the antennae and the caudal *ramus* (i.e., caudal appendages). Some species (e.g., *Cypridopsis*) are associated with periphyton on macrophytes and particulate detritus and feed voraciously on attached algae (Mallwitz, 1984). Most ostracods are herbivores and detritivores feeding on bacteria, algae, detritus, and other microorganisms by means of filtration. However, a few ostracod species have been documented consuming the soft tissue, eggs, and juveniles of certain snail species (Smith et al., 2015). The large population numbers of ostracods, commonly in excess of 100,000 m^{-2}, suggest that their role in the metabolism of surficial sediments could be considerable. These high densities also suggest the importance of ostracods to freshwater food webs; they are commonly consumed by waterfowl, fishes, and predaceous aquatic invertebrates (e.g., diving beetles, amphipods; Smith et al., 2015).

Ostracod reproduction is highly variable across species and locations. In some species sexual reproduction is common, but males have not been found for other species, especially in isolated locations such as on Pacific islands (Smith et al., 2015). The time required for egg development is variable, from days to months, and is strongly temperature dependent. The larva hatches as a *nauplius* (i.e., first larval stage of many crustaceans) with a reduced number of appendages and then undergoes a series of growth and molting stages; usually, eight molts are needed to reach the ninth, adult stage, during which morphology becomes more complex and appendages develop. Some species exhibit a single generation per year or may even take multiple years (Smith et al., 2015); others exhibit two or three generations per year (e.g., Mallwitz, 1984).

M. Malacostraca

Of the many species of the malacostracean crustaceans, four groups have distinct freshwater representatives of considerable interest and importance: mysid shrimps, isopods, decapods, and amphipods. Each of these malacostraceans has a definite and fixed number of body segments.

i. Mysida

Mysids, or *opossum shrimps*, are morphologically similar to crayfish but only attain a maximum length of about 3 cm (Fig. 21-2N). Their appendages, however, are elongated, contain abundant setae, and are modified for active swimming. They feed by filtering small zooplankton, phytoplankton, and particulate detritus with their setose appendages.

Gut analyses of *Mysis relicta* showed that adults were voracious predators, feeding on cladocerans and rotifers (Lasenby and Langford, 1973). Several studies have shown that mysids are size selective in their predation of zooplankton (Bowers and Grossnickle, 1978). Large mysids prefer large-sized prey (*Daphnia* and *Epischura*) consistently, whereas small mysids select the smallest zooplankton available. The bivoltine life history of mysids in many lakes results in seasonal variations in population size structure and their impact on zooplankton prey (Murtaugh, 1983). When mysids are introduced into lake ecosystems as a food source for fish, size-selective predation pressure can lead to marked alterations in the zooplankton community (e.g., Northcote, 1991). Selective predation on larger microcrustaceans by mysids can lead to the enhanced development of smaller species because of relaxed competition for food resources. Marked reductions or elimination of some large zooplankton species can thus result in reductions of abundance and growth for some fishes.

Freshwater mysids prefer cold water, and in stratified lakes they are restricted to hypolimnetic strata of less than 15°C. Reproduction occurs only during the colder periods of autumn, winter, and early spring. There are as many as 40 eggs per clutch, and the eggs are kept within the female brood pouch (*marsupium*). After hatching, juveniles are retained within the female for considerable periods of time (up to 3 months), hence the name opossum shrimps. When lake productivity is relatively high, the entire life cycle of mysids takes from 1 to 2 years, but when productivity is low, it may last up to 4 years (Kjellberg et al., 1991).

Members of the two major genera, *Mysis* and *Neomysis*, exhibit distinct diurnal migrations. During the day, the mysids are on the sediments or in the strata immediately overlying the sediments (Herman, 1963). *Mysis relicta* ascends each evening when surface light intensities decrease, but they only migrate into or below the metalimnion, in which temperatures are well below 15°C. Bright moonlight can inhibit the ascent of *Mysis* at night (Bowers and Grossnickle, 1978). Descent occurs when surface light intensities increase again with the following morning. Despite the avoidance of lighted zones, predation on mysids by planktivorous fish can be very high (e.g., Chigbu and Sibley, 1998). Mysid populations can be appreciably decreased during abundant year classes of these fish species.

ii. Isopoda

Isopods, or *sowbugs*, are largely marine or terrestrial. However, isopods occasionally become a significant part of the benthic community of lakes and streams, and approximately 950 species are known globally from freshwaters (Wellborn et al., 2015) (Fig. 21-2O). These small organisms (<2 cm) are flattened dorsoventrally, with seven pairs of well-developed walking legs. Isopods are omnivorous feeders on both plant and animal matter (Willoughby and Marcus, 1979), and for one genus (*Lirceus*), both fungi and bacteria can also be important sources of organic carbon (Findlay et al., 1986). Isopods are generally not active predators on other benthic fauna.

Reproduction is more common during periods of warmer temperatures but otherwise is similar to that of the mysids. The number of eggs per female is fairly high (reaching several hundred), and both eggs and young are retained in the brood pouch for about a month. Relatively little is known about their life cycle and population dynamics; generation time is about 8 to 12 months but quite variable. Isopod distribution in freshwaters is shaped by dissolved mineral concentrations, with many species tolerating high salinity levels and others that are absent when dissolved calcium levels fall too low (<5 mg Ca L^{-1}; Reynoldson, 1961).

iii. Decapoda

Of the decapods, approximately 650 species of freshwater crayfish have been described to date, as well as 800 species and subspecies of freshwater shrimps (Cumberlidge et al., 2015) (Fig. 21-2P). *Crayfish* and *shrimps* are characterized by their approximately cylindrical body, heavily sclerotized translucent shell, and laterally compressed *rostrum* (i.e., a frontal and dorsal projection between the eyes). Their 19 pairs of appendages include well-developed antennae and five pairs of large walking legs, the first three of which are clawed and the first greatly enlarged with a strong pincer claw used for crushing food. Crayfish are omnivorous but often herbivorous on algae and larger aquatic plants; occasionally, they are scavengers (Momot, 1995). Some river species (e.g., *Orconectes*) acquire about two-thirds of their growth and production from allochthonous carbon sources and about a third from benthic invertebrates (Whitledge and Rabeni, 1997). Crayfish can ingest large amounts of herbaceous and detrital materials while searching for and ingesting animal protein, but growth rates are best on a mixture of animal protein with other diets. In contrast, shrimps are primarily grazers and secondarily opportunistic foragers on a variety of detrital plant and animal material (Cumberlidge et al., 2015).

Recently hatched juveniles of crayfish exploit littoral areas in lakes and riffles in streams as areas for food and shelter (Momot, 1984, 1995). The rapid growth of young-of-the-year juveniles compensates somewhat for the high rates of predation by fish. Size-specific predation by fish on crayfish species of different sizes can lead to the eventual elimination of one of the species (DiDonato and Lodge, 1993). Commonly, larger juveniles migrate to deeper water as they develop and then return to littoral and riffle areas as adults. As adults, much of the energy is diverted from growth to reproduction, which can include several cohorts per year.

The lifespan of crayfish is generally fixed, although rapid growth and shorter lifespans occur among crayfish in lower latitudes (Momot, 1984). Crayfish occurring at higher latitudes and colder environments usually live longer and mature later (4–16 years). Development is more rapid in warmer waters, where breeding can occur in the first year but mostly occurs in the second year (Hobbs, 1991, for general details of life histories). In contrast, freshwater shrimps generally only live 1 to 2 years and often incorporate a migratory behavior from brackish to freshwaters during this short life cycle (Cumberlidge et al., 2015). Fish predation can exert a strong influence on many populations of freshwater shrimps (Mace and Rozas, 2018).

The feeding and predatory activities of crayfish can have complex, multitrophic-level effects on the food webs of streams and lakes. For example, dense crayfish can significantly reduce other benthic invertebrates that graze on periphyton, and the reduction of grazing pressure can result in marked increases in periphyton production (Charlebois and Lamberti, 1996). In other cases large attached algae, such as *Cladophora*, can be fed upon directly by the crayfish, which results in altered habitat for benthic invertebrates (Creed, 1994). The grazing activities of crayfish on submersed macrophytes and seedling stages of emergent and floating-leaved macrophytes, either directly or by damaging the plants during predation on macrophyte-associated invertebrate prey, can suppress macrophyte abundance appreciably (e.g., Nyström et al., 1996). At very high densities of crayfish, some species of macrophytes can be eliminated. Abiotic factors, such as water temperature, dissolved oxygen, pH, and calcium concentrations, are important factors in shaping the distributions and abundances of both crayfish and shrimps in freshwaters (Cumberlidge et al., 2015).

iv. Amphipoda

Amphipods, or *scuds*, are represented in freshwaters by at least 1900 species, including an impressive 350 species from Lake Baikal (Russia) alone (Wellborn et al., 2015) (Fig. 21-2Q). Most species are small (5–20 mm), with a laterally compressed, many-segmented body. At the base of their seven pairs of thoracic legs, many have gills that, like the lateral gills on some species, are exposed to currents of water created by the beating of appendages on the abdomen. Amphipods are generally omnivorous substrate feeders that consume bacteria, algae, fungi, and animal and plant remains (e.g., Moore, 1977). Several amphipods possess digestive enzymes that are capable of degrading plant and fungal cell walls (Chamier and Willoughby, 1986). Some amphipods, particularly *Gammarus* spp., can be predacious on living animals (e.g., flatworms, oligochaetes, ostracods) and also cannibalistic (Wellborn et al., 2015).

Fungi colonizing autumn-shed tree leaves are the preferred food of many woodland stream-dwelling amphipods. However, the assimilation of this food varies dramatically, with some taxa incorporating less than 20% of the ingested leaf material, whereas 75% or more of the fungi of decomposing leaves were assimilated in other cases (e.g., Sutcliffe et al., 1981). Assimilation efficiencies vary greatly with the amounts of microbiota attached to and within detrital leaf materials from different plant species. As a result, growth rates of the amphipods feeding on detritus from leaves, such as oaks that are much more resistant to decay, were slower than those feeding on microflora and detritus of more labile leaves, such as from elm (*Ulmus*) or aquatic plants. Amphipods can serve as a dominant macroinvertebrate prey of many fish, both as a seasonal food source and as a year-round staple; some fish feed selectively on larger stages and sizes (MacNeil et al., 1999).

Monoporeia affinis is exceptional among the amphipods in that it migrates extensively into the pelagic zone at night. *Monoporeia* is restricted to cold, relatively oligotrophic waters and migrates only into the upper hypolimnion and metalimnion, in which temperatures are usually less than 15°C (Wells, 1968). Only a portion of the predominantly benthic population migrates into the water column. During daytime, *Monoporeia* is largely on and in the sediments or in the immediately overlying water. Predation can be high at night, especially by the predatory leech *Erpobdella*, which migrates nocturnally in synchrony with *Monoporeia* and feeds vigorously on the amphipod (Blinn and Davies, 1990).

N. Mollusca

Mollusks are unsegmented invertebrates, with a body organized into a muscular foot, a head region, a visceral mass, and a fleshy mantle that secretes a shell of proteinaceous and crystalline calcium carbonate materials. Over 4000 species of *freshwater snails* (Gastropoda) are known (Strong et al., 2008), and approximately 1200 species of *freshwater mussels* and *clams* (Bivalvia) are currently recognized (Bogan, 2008). Both groups face significant conservation challenges, and numerous species with restricted distributions or intolerance for altered habitat conditions have gone extinct in the past century (Cummings and Graf, 2015; Pyron and Brown, 2015).

i. Gastropoda

Among the gastropod snails, shells generally are spirally coiled, whereas those of freshwater limpets are conically shaped (Fig. 21-2R). The gastropod head is distinct, with a pair of contractile tentacles and a ventral mouth with a sclerotized jaw and an internal radula containing numerous teeth. The *radula* (i.e., a structure in the mouth used to scrape food) is extended from the mouth and moves back and forth rapidly, scraping and macerating food particles. Respiration in the snails occurs by gills in many aquatic forms (Prosobranchia) and by pulmonary cavities or "lungs" in the *pulmonate snails*. Cutaneous respiration through the body membranes is common to all freshwater snails (Ghiretti, 1966). Locomotion in snails occurs via muscular movements of the ventral surface of the body (i.e., the "foot").

Most snails feed on algae, detrital particles, and bacteria of the periphyton on submersed substrata. Direct effects of grazing on living macrophytes are probably of minor importance, although a few snails consume submersed macrophytes in sufficient quantities to alter plant growth (Sheldon, 1987). However, snails can have a significant indirect effect on macrophytes by reducing the detrimental effects of epiphytic microbiota that shade and compete for nutrients from the water.

Pulmonate snails are moderately tolerant of desiccation and low levels of dissolved oxygen (McMahon, 1983). Because of the requirements for appreciable calcium for shell generation, most mollusks are found in waters with a dissolved calcium concentration greater than 5 mg L^{-1}, and mortality can be induced when they are moved to waters of lower concentration (Herbst et al., 2008). However, above this concentration, little relationship exists between chemical parameters and gastropod diversity, and substratum, food availability, and disturbance events are known to regulate populations (Lodge et al., 1987).

Predation on snails can markedly affect their abundance. Certain fish adapted to crushing snails, especially sunfish centrarchids, and crayfish are particularly effective in consuming snails that are not able to find refuge (e.g., Brown and DeVries, 1985). Variations in the thickness of snail shells, either within or between species, can selectively alter predation vulnerability; energetic costs to predators increase as shell thickness increases (Stein et al., 1984). Snails can crawl above the water line for several hours to avoid predation (e.g., Alexander and Covich, 1991) or to deposit eggs on moist soil less vulnerable to invertebrate predation.

The life cycle of freshwater snails in temperate regions, particularly the smaller species, tends to be annual (Harman, 1974). One reproductive period may occur in the spring or fall, or two or more reproductive periods may occur throughout the summer, during which the original cohort is replaced or supplemented. Some species overwinter as juveniles or adults and reach maturity the following spring or summer. Larger species tend to have life cycles that extend over 2 to 4 years.

ii. Bivalvia

The bodies of clams and mussels are enclosed in two symmetrical, opposing shell valves (Fig. 21-2S). The body is enclosed by membranous tissue, the *mantle*, which secretes the shell valves. At the posterior end of the clam, the mantle has two openings, known as *siphons*. The lower ciliated siphon draws water into the body cavity, aerating the gills and carrying food particles. Food particles are removed from the water by filtration through the gills and cilia and consist primarily of detritus, microzooplankton, and phytoplankton (Fuller, 1974). The muscular *foot* can be extended from the valves in front of the clam, implanted in the sediment, and then contracted to draw the animal forward. Burrowing activities disturb particles that are then directed by cilia toward the mantle cavity, where they are drawn in. As much as half or more of the total organic carbon assimilated may arise from pedal deposit feeding mechanisms (Burky, 1983).

Bivalves tend to be long-lived and expend appreciable energy for high reproductive rates. Nonpredatory mortality is known to occur but is poorly understood. Predatory losses can be very high by some large invertebrates (e.g., crayfish) and many vertebrates, particularly by birds, molluscivorous fish, and semiaquatic mammals (Fuller, 1974). Two exotic species (Asian clam *Corbicula fluminae* and the zebra mussel *Dreissena polymorpha*) introduced into North America have caused massive destruction of habitat, reduced biodiversity of other mollusks, and appreciable economic damage (Benson and Williams, 2021; Nalepa and Schloesser, 2014).

Freshwater bivalves are nearly all *ovoviviparous* and brood embryos through early development stages in the gill. These embryos hatch first as *trochophore* larvae that develop into *veliger* larvae. In mussels of the Unionacea larvae are called *glochidia* and parasite fish gills. Some species are specific to a particular species of fish. During this parasitic stage, which lasts from about 10 to 30 days, internal development occurs. Once internal development is complete, juveniles detach from their hosts and fall to the sediments. Development ensues if substrata and food conditions are within tolerable limits. Large bivalves live 15 years or longer, whereas the smaller "fingernail clams" (Sphaeriidae) have a longevity of about a year or so. Mortality is exceedingly high in the egg, glochidia, and juvenile stages. The adult populations can be modified greatly by fish predation, certain birds, and a few mammals such as the muskrat (Fuller, 1974).

II. Hexapoda

Hexapoda constitutes the most diverse group of benthic animals, mainly because of the high diversity of aquatic insects that dominate freshwater ecosystems. About 70% of benthic animal species are hexapods (Balian et al., 2008), representing even more than 90% of the total abundances in benthic community surveys (e.g., Trigal et al., 2006). Morphological and molecular data indicate that the Hexapoda is a monophyletic group, closely related to the crustacean class Remipedia (Beutel et al., 2017; Kjer et al., 2006). They are a subphylum of the phylum Arthropoda and include two classes: Entognatha and Insecta. Entognatha includes Collembola, Protura, and Diplura. Only Collembola and Insecta have benthic representatives in freshwaters.

A. Collembola

Collembola, or *springtails*, are primarily small terrestrial hexapods (0.2–8 mm; Deharveng et al., 2008) with varying water dependency levels (Fig. 21-4A). There are no completely aquatic springtails, but a few species survive floating on water surfaces or water films on rocks or are found in vegetation or in soil at the edges of aquatic habitats. Some others can even survive underwater for several days (Thibaud, 1970). The most well-known semiaquatic species is the widespread *Podura aquatica*, which lives on the surface film of water bodies during summer and autumn, feeding on plants, and hibernates in the mud near the shores in winter (Childs, 1915). The remaining water-dependent species are found in snow or subterranean water-saturated habitats (Deharveng et al., 2008). Therefore most springtails in benthic samples are likely caught accidently. Some species can tolerate high-salinity environments.

Morphologically, springtails are characterized by having a ventral tube (named *collophore*), most likely related to fluid exchange but could also be used to adhere to the substrate. Many species also have an abdominal *furca*, a jumping organ, used for locomotion or escape from predation. They have short life cycles and high reproduction rates, with parthenogenesis being common (Beutel et al., 2017). Most species feed on fungal hyphae or leaf litter, but a few are predators of nematodes, rotifers, or other springtails (Cassagnau, 1972). They can become very abundant in soils or in leaf litter (e.g., 670,000 individuals m^{-2} in a snow-free surface of Antarctica or 244,000 individuals m^{-2} in a Scandinavian forest; Petersen and Luxton, 1982), but also on the water surface (e.g., observations of *P. aquatica* estimate c. 60,000 individuals m^{-2}).

B. Insecta

Insects are very abundant and diverse, but most are terrestrial. Of those that are aquatic, nearly all evolved in freshwaters, but others have colonized multiple times from terrestrial environments, such as coleopterans or dipterans (Wiegmann et al., 2011; Múrria et al., 2018) (Fig. 21-4B to K). Some orders are entirely aquatic; others inhabit freshwaters only during certain life stages. Most insects are benthic, living on or burrowing into sediments, or on macrophytic vegetation and plant detritus. The characteristics of insects are well known, and no attempt will be made here to summarize the group differences except to point out salient features of their biological traits.

Odonates (*dragonflies* and *damselflies*), ephemeropterans (*mayflies*), plecopterans (*stoneflies*), and hemipterans (*true bugs*) are orders of winged insects that undergo gradual metamorphosis. In these orders the young are referred to as *nymphs*; their wings develop as external pads, and the organisms increase in size with each molt. The other orders, including dipterans (*flies*), trichopterans (*caddisflies*), megalopterans (*alderflies* and *dobsonflies*), coleopterans (*beetles*), a few species of lepidopterans (*moths* or *aquatic caterpillars*), and neuropterans (*spongeflies*), undergo complete metamorphosis. In this development the wing pads develop internally in early *larval instars* and then evert to the outside in the preadult *instar pupal stage*. In nearly all the major aquatic insect groups, only the immature stages live in the water; the adults and, in some groups, the pupae

FIGURE 21.4 Examples of the wide variety of benthic Hexapoda from freshwater habitats: (a) Collembola. *(photo credit: Christian Fischer, Wikipedia.)* (b) Insecta: Plecoptera. *(photo credit: Michael T. Bogan.)* (c) Insecta: Ephemeroptera. *(photo credit: Michael T. Bogan.)* (d) Insecta: Odonata. *(photo credit: Michael T. Bogan.)* (e) Insecta: Hemiptera. *(photo credit: Michael T. Bogan.)* (f) Insecta: Diptera. *(photo credit: Michael T. Bogan.)* (g) Insecta: Lepidoptera. *(photo credit: Michael T. Bogan.)* (h) Insecta: Megaloptera. *(photo credit: Michael T. Bogan.)* (i) Insecta: Neuroptera. *(photo credit: Matt R. Cover.)* (j) Insecta: Trichoptera. *(photo credit: Jesús Ortiz, CEN Association.)* (k) Insecta: Coleoptera. *(photo credit: Jesús Ortiz, CEN Association.)*

are terrestrial. Only some coleopterans and hemipterans have adapted to the point where both the adults and immature stages can live in water.

i. Life history characteristics

Life histories of aquatic insects are very diverse, and characteristics vary among orders. For example, after fertilization, the eggs of odonates are deposited into water, onto substrata in or near the water, or into submersed parts of macrophytes. Nymphs hatch after about 2 to 5 weeks of egg development. Growth in the nymphal stages is quite variable, especially in relation to temperature and food supply. Within a range of about 6 weeks to 3 years, 9 to about 16 moltings occur. The mature nymphs often leave the water on some emergent substratum to become aerial adults. Of the three primary patterns of life history of odonates, spring emergence is common, and *univoltine* and *semivoltine* (i.e., more than one year to complete the life cycle) species usually emerge in summer.

Ephemeropterans are aquatic as nymphs and terrestrial as adults (Edmunds et al., 1976; Hutchinson, 1993). The adult longevity is very brief (3–4 days), during which no feeding occurs. After mating in flight, fertilized eggs are laid in the water or on submersed objects; the time to hatching varies from a few days to many weeks. Parthenogenesis is widespread, as well as egg diapause. Growth is relatively rapid, from less than 1 mm in length at hatching to about 2 cm in the nymphal stages of many species. From 9 to over 50 instars of molting occur over an average life cycle of about 1 year. A few species live as nymphs for 2 years or longer.

Plecopterans are terrestrial as adults, but in the nymphal stages they are strictly aquatic, and most are restricted to flowing waters of relatively high oxygen concentrations. Fertile eggs, laid over or in the water, require 2 to 3 weeks for hatching in many species and several months among some larger forms. The nymphal instars, from 12 to over 33 moltings, occur in 1 to 3 years. Nymphs tend to be predominately herbivores and detritivores, and a few become carnivorous in later instars.

Hemipterans are essentially terrestrial; some families are semiaquatic and a few families have adapted completely to submersed conditions (e.g., Notonectidae, Corixidae). Most hemipterans overwinter as adults in moist sediments or vegetation (cf. the detailed synthesis of Hutchinson, 1993). Eggs are laid on semiaquatic substrata or in aquatic macrophytes and develop rapidly in 1 to 4 weeks. Nymphs also develop rapidly in 1 to 2 months, commonly with five instar stages, and generally have a 1-year life cycle.

Dipterans are the most diverse group of aquatic insects with complete morphogenesis. Adults are essentially never aquatic, but most of their lives are spent as immature forms in fresh or saline waters. The larval stage, with about three or four growth molts, extends from several weeks to at least 2 years in some species, and many overwinter in the larval stage. Most species have one generation per year, some have two per year, and a few of the species studied have a 2-year life cycle (to 7 years in arctic ponds; Butler, 1982).

Trichopterans generally have a 1-year cycle (Wiggins, 1977). Adults emerge in the warmer periods of the year, often from overlapping cohorts, from May to October. Eggs are dropped or placed on vegetation or laid under water on submersed substrata and develop in about 1 to 3 weeks. Many larvae build beautifully intricate cases from substrate particles of sand, small stones, leaf fragments, or other elements (Mackay and Wiggins, 1979; Wallace and Merritt, 1980). After 5 to 7 larval instars, pupation occurs under water within a cocoon. The pupal stage generally lasts only a few weeks, after which the pupa leaves the cocoon, moves to an aerial substratum, and emerges as an adult. A few species of the closely related Lepidoptera order have aquatic larval stages; most aquatic moth species belong to the crambid group of the family Pyralidae. Many characteristics of the life history of these *aquatic caterpillars* are like those of trichopterans.

Neuropterans and megalopterans are closely related and are often grouped together under Neuroptera. Both groups are primarily terrestrial for all their life cycle. Of the few freshwater species, however, some megalopteran larvae are so large that they may represent a significant portion of the biomass of some benthic communities. Eggs generally are laid on aerial substrates overhanging the water. After a rapid development period (1–2 weeks), the larvae drop into the water and feed actively. The megalopterans have numerous molting instars and most of their life cycle of 1 to 3 years is spent in the larval stage. The pupal stage occurs in soil out of water and lasts up to 1 month. Aquatic neuropteran larvae are relatively rare: Sisyridae are restricted to, and predators on, freshwater sponges, and the few aquatic Osmylidae are predators with amphibious habits.

Such great diversity exists among the aquatic coleopterans that even broad generalization is difficult. Generally, coleopterans adapted for larval and adult existence in water occur among the more phylogenetically primitive groups. The life cycles of many (e.g., Gyrinidae, Haliplidae, Dytiscidae, and Hydrophilidae) are annual, with 3 to 8 larval instars. Eggs laid on or in macrophytes or sediments hatch in 1 to 3 weeks. Larval development is variable (1–8 months), and pupation takes place on some nearby terrestrial substratum. Overwintering generally occurs in the aquatic adult stage.

ii. Morphological and physiological traits

Insects have colonized aquatic environments multiple times (Will and Resh, 2008). Upon their invasion of freshwater environments, aquatic insects encountered the advantage of smaller fluctuations in temperature (daily and annually), but many disadvantages as well, such as living in a completely different physical environment (water vs. air), with decreased oxygen, and greater requirement for osmoregulation. Aquatic insects adapted to these constraints through a multiple set of morphological and physiological traits.

Water density, viscosity, pressure, and surface tension (Chapter 3) are water properties relevant for aquatic insects (Lancaster and Downes, 2013). Aquatic insects can be found in the water column, on the surface of water, or in the benthos. Water column species are not very common but include some species and families of coleopterans, hemipterans, and dipterans. With the exception of the *phantom midge Chaoborus* (which is able to maintain neutral buoyancy in the water column through contracting and expanding the walls of *tracheal sacs* or air sacs; Teraguchi, 1975), aquatic insects must spend energy swimming or rowing at the surface and have modified legs to do so. Body size and the areas in contact with the water surface are key for the animal to be supported by surface tension, and several morphological traits facilitate this (e.g., chemical products that act as water repellent, presence of dense hairy areas) (Lancaster and Downes, 2013). Most aquatic insects, however, are found in the benthos and morphological adaptations become very relevant when considering flowing habitats. Flattened bodies, streamlined shapes, or the presence of morphological structures that reduce dragging (tarsal claws, suckers, hooks, silk-related structures, or fringes of setae that increase substrate contact) are common in aquatic insects from fast-flowing waters. Drag can also be reduced by changing the behavior. Flattened morphologies, however, may reduce drag within the boundary layer but a convex surface may generate lift (Statzner, 2008). In contrast, lentic habitats are dominated by rounded forms and structures that facilitate water movement.

Besides morphological adaptations to move underwater or on water surfaces, aquatic insects also have adaptations to characteristic substrata in freshwater ecosystems. A large diversity exists of substratum types, composition (inorganic to organic), and especially particle sizes, from silt to boulders. Certainly, this range of substratum types has resulted in aquatic insects with adaptations to exploit food resources in each substrate and to use them as refuges from predation. For example, several species of aquatic insects are burrowers and have long and narrow bodies with well-adapted legs and mandibles (Edmunds and McCafferty, 1996).

Aquatic insects either continue to breathe surface air like their terrestrial counterparts or have developed traits that allow them to breathe underwater (Eriksen et al., 1992). Some adult beetles (coleopterans) or some true bugs (hemipterans) make frequent excursions to the surface to acquire atmospheric air directly or carry air stores under water in cavities (*plastrons*) or among pubescent regions and absorb water-dissolved oxygen. Others, such as some hemipterans or dipterans, extend tubular structures to the surface to acquire atmospheric air. Larvae and nymphs of several orders have also developed external gills that exchange gases underwater, independently of the atmospheric air. A few immature dipterans and coleopterans can also penetrate the aerenchymatous tissues of aquatic plants and remove oxygen. Other adaptations include the cutaneous exchange of gases by increased cuticular permeability, but this change requires osmoregulatory mechanisms to retain internal salt concentrations (Ward, 1992). When oxygen becomes very low, aquatic insects have some other physiological adaptations that aim at regulating oxygen consumption independently of the surrounding oxygen concentration (Lancaster and Downes, 2013). However, very few aquatic insects can move to blood-based gas exchange during low oxygen conditions (i.e., hypoxia, Chapter 11). One example is the genus *Chironomus*, which has hemoglobin that allows it to bind oxygen and store it, releasing and transporting it when needed during hypoxic conditions.

Osmoregulation is required in an environment where the external medium can have varied salinity levels. Most aquatic insects are *osmoconformers*, meaning that their inside osmotic concentration matches the outside one. Only species found in hypersaline habitats are *osmoregulators* and have adaptations in their excretory system (e.g., they secrete ions into the rectum) (Lancaster and Downes, 2013). A few aquatic insects have colonized marine habitats and might have similar strategies to avoid high salinity concentrations. Several species of dipterans, hemipterans, lepidopterans, coleopterans, and trichopterans are found in the intertidal zone (and some even in open ocean habitats). One fascinating example is the trichopteran family Chathamiidae, which uses marine organisms as food resources, builds cases from marine red algae, and oviposits its eggs inside of starfish (Winterbourn and Anderson, 1980).

Even though temperature fluctuates less in water than in the terrestrial environment, aquatic insects have adapted to a wide range of temperatures. Nearly all facets of the life history and the distribution of aquatic insects are influenced by temperature: egg development and hatching, growth rates, and voltinism. Temperature is a dominant factor influencing the distribution, diversity, and abundance patterns over elevational gradients and downstream along watercourses.

Entire orders and other major taxa of aquatic insects evolved in cool (e.g., plecopterans) or warm environments (e.g., odonates). Some species also seek microthermal refuges to avoid high or low temperatures. Extremely hot environments, such as thermal springs, are home to dipterans, odonates, hemipterans, and coleopterans (Lancaster and Downes, 2013). The production of heat-shock proteins might allow aquatic insects to survive in these environments (Garbuz et al., 2008). In contrast, cryoprotectants are common in species that live in very cold or frozen water (Danks, 2008). Life cycles can also be synchronized to avoid cold or hot periods depending on the region, and strategies such as dormancy to avoid unfavorable temperatures are common.

Several adaptations have also evolved to cope with the lack of water in temporary freshwater habitats. These include behavioral, life history, morphological, and physiological traits that confer resistance or resilience to drying (Bogan et al., 2017). Some aquatic insects, for example, may develop desiccation-resistant stages as eggs, larvae/nymphs, pupae, or adults through different physiological mechanisms (Strachan et al., 2015) (see section II.B.v): the production of resistant eggs or cysts, anhydrobiosis, diapause, or aestivation/hibernation. These mechanisms are not exclusive to aquatic insects but are also found in many other benthic animals (Watanabe, 2006).

iii. Reproductive traits

Because insects evolved in terrestrial environments, mating and reproducing in freshwaters presents novel challenges (e.g., strong currents in streams, low dissolved oxygen levels in some lakes). Most aquatic insects partly avoid these challenges by having retained the terrestrial adult stage of their ancestors. Once adults emerge from freshwater larval forms, most find mates using many of the same techniques as their fully terrestrial relatives (e.g., synchronous emergence, pheromones, swarming: Lancaster and Downes, 2013). However, plecopterans do not form swarms, and instead, the males of most species will "drum" to create vibrations on terrestrial surfaces and attract females (Tierno de Figueroa et al., 2019).

Strategies diversify considerably after mating occurs and female insects are ready to oviposit. Many aquatic insects searching for water bodies in which to lay eggs use polarized light to detect water from above while in flight (Horváth and Kriska, 2008). Locations for oviposition, and the exact method of egg laying, should maximize the survival of eggs, as well as the early-instar nymphs that hatch from them. Some adult aquatic insects crawl back underwater after mating to carefully cement their eggs on rocks, including many trichopterans and plecopterans, whereas others insert them into submerged vegetation (e.g., some zygopterans and coleopterans: Lancaster and Downes, 2013). Other species are much less deliberate; instead, they broadcast their eggs widely across the surface of the water, or in the air just above the water. The eggs may then float at the surface, be carried to the margins, or settle to the bottom before they hatch (Lancaster and Downes, 2013). Finally, some aquatic insects, including megalopterans and neuropterans, lay their eggs on land or on vegetation that hangs over water bodies (Cover and Bogan, 2015). When these eggs hatch, first-instar larvae will either fall directly into the water or crawl a short distance overland to reach the water body.

Although most aquatic insects have a terrestrial adult stage, many coleopterans and heteropterans have fully aquatic adults, as do a few species of trichopterans, lepidopterans, and other orders. Some strategies used to find mates in terrestrial environments (e.g., pheromones) are less effective in water, so many species have developed acoustic or wave-based methods of signaling for mates. For example, belostomatids do push-ups to create waves that signal for mates (Ohba, 2019), and corixids use their legs and body to make stridulatory noises to "call" to potential mates (Lancaster and Downes, 2013). In fact, the 2-mm-long corixid *Micronecta scholtzi* has been described as one of the loudest insects in the world, relative to its size, when it calls for mates (Sueur et al., 2011).

Some aquatic adult insects lay their eggs on submerged surfaces, whereas others have developed elaborate postzygotic egg care systems (Smith, 1997). Once a male *Abedus herberti* (Belostomatidae) attracts a mate via wave-signaling, they copulate, and then the female attaches fertilized eggs to the back of the male. Due to the large size of the eggs, males must frequently surface and then dive again to ensure that the eggs neither drown nor desiccate (Fig. 21-5). In other belostomatid species, such as *Lethocerus medius*, females lay the eggs on emergent vegetation above the water, and males must tend to these eggs with water droplets to ensure the eggs do not desiccate (Smith and Larsen, 1993). Many additional fascinating case histories of aquatic insect reproduction would surely be revealed with a renewed emphasis on natural history studies.

iv. Trophic mechanisms and food types

The great diversity of food ingested by the aquatic insects and their various feeding mechanisms are often organized within a functional feeding group framework. The three primary classes include *herbivory*, defined as the ingestion of living vascular plant tissue or algae; *carnivory*, the ingestion of living animal tissue; and *detritivory*, the intake of nonliving particulate organic matter and the microorganisms associated with it (Cummins, 1973; Hutchinson, 1993). The organic matter in any of these categories can be ingested either in particulate

FIGURE 21.5 Mating and paternal egg brooding in the giant water bug *Abedus herberti* (Belostomatidae: Hemiptera) in streams from southern Arizona (USA). First (a), the male and female copulate and the female attaches the fertilized egg to the hemielytra of the male. The male must frequently submerge the eggs (b) to ensure that they do not desiccate, but also must surface regularly (c) to ensure that the eggs do not drown.

form (by swallowing, biting, chewing, or scraping) or in dissolved form (by piercing or sucking). Using a combination of dominant food type and feeding mechanism, the following functional feeding groups have been delineated: scrapers (grazers), shredders, collector-gatherers, collector-filterers, plant piercers, and predators (including engulfers and piercers).

Scrapers remove small algae and other microorganisms from benthic surfaces via the use of specialized mouthparts that facilitate "grazing" of the benthic substrate. These feeding activities can help regulate periphyton and biofilm assemblages and create openings for other organisms to inhabit (Cummins et al., 2008). *Shredders* may feed on living plant material (herbivory) and many consume dead leaves and wood (detritivory), such as fallen leaves and twigs (Cummins et al., 2008). The processes of shredding plant material result in smaller particles that frequently go unconsumed by the shredder but can become fine particulate organic matter for other organisms. Beneficiaries of this shredding activity include *collector-gathering* and *collector-filtering* insects, which look for particles to eat in the benthic or pelagic zones or filter directly from the water. A much smaller group of insects, including some trichopteran and coleopteran larvae, are plant piercers, who pierce the cells of algae and plants to obtain food rather than shredding them (Lancaster and Downes, 2013). Finally, *predators* are those insects that consume other living invertebrates, or even vertebrates in the case of hemipterans in the family Belostomatidae (Lytle, 2015). *Engulfer predators* can consume their prey directly by using sharp, specialized mandibles. In contrast, *piercing predators* inject digestive enzymes (and sometimes neurotoxins) into their prey and then slurp up the digested remains via the use of specialized mouthparts (Lancaster and Downes, 2013).

The above functional feeding categories are necessarily broad and general, and individual insects may exhibit one or more modes of feeding depending on food availability, environmental conditions, and life stage (Cummins et al., 2008). Further studies of individual species and ontological and population-level variation in feeding modes and diet preferences would greatly benefit our understanding of the trophic roles aquatic insects play in ecosystems. But even with imperfect knowledge, aquatic insects are responsible for much of the flow of energy, from primary production to the highest trophic levels, in aquatic habitats across the globe.

v. Dispersal mechanisms

Aquatic insects exhibit a wide variety of dispersal mechanisms to persist through environmental changes in dynamic environments as well as to maintain genetic diversity across populations. Most aquatic insects have an aerial adult stage, and this is the stage where a substantial portion of dispersal can occur. Dispersal flights of adult aquatic insects can be as short as a few meters or as long as hundreds to thousands of kilometers, as is the case for some species of odonates (e.g., *Pantala flavescens*: Hobson et al., 2012). Even adult aquatic insects that can only actively disperse short distances (e.g., small species of dipterans) may be carried much longer distances by wind when dispersing aerially (May, 2019). Both short- and long-distance aerial dispersal can be triggered by periods of heavy rainfall, perhaps because this rain provides a cue that new breeding habitats may have filled (Smith and Larsen, 1993; Bogan and Boersma, 2012). Although the study of aerial dispersal was limited for many years, new approaches are allowing for a more robust understanding of the scale and importance of this process. Radio transmitters are now small enough to allow the tracking of individual adults of some species (Wikelski et al., 2006). Another relatively recent approach uses stable isotopes to chemically track adult movement, as has been done for plecopterans and odonates (e.g., Macneale et al., 2005; Hobson et al., 2012; Hallworth et al., 2018).

In addition to dispersal through air, most aquatic insects also disperse passively and actively through the water in their larval stages (and adult stages in some instances). In flowing water insects frequently exhibit drift dispersal, where the current carries them downstream. *Passive drift* occurs when animals are accidently dislodged from benthic surfaces by changing flow dynamics or physical disturbances. In contrast, *active (or behavioral) drift* is when insects choose to leave their location and enter the current. Insects make this choice when food resources in their location are depleted, when novel predators or competitors appear, or when water quality conditions change (Lancaster and Downes, 2013). Although drift primarily occurs in flowing waters, aquatic insects have other methods of dispersing through both lotic and lentic waters. Some species are accomplished swimmers and can move significant distances via undulating motions or legs with morphological adaptations for swimming (Lancaster and Downes, 2013). Other species may move shorter distances via crawling along benthic surfaces, including upstream against the current in lotic systems.

In addition to dispersal through space, some aquatic insects can disperse through time via dormant stages. This adaptation allows populations to persist through extreme environmental conditions without having to search for more stable habitats. For example, some plecopterans and megalopterans have egg or larval diapause stages that allow them to persist through long dry periods (up to 5 years) in intermittent streams (Bogan, 2017; Cover et al., 2015). The harsh desert rock pool environments of the Sahara Desert provide perhaps the most impressive example of dispersal through time. One inhabitant of these pools, the chironomid *Polypedilum vanderplanki*, can survive decades of drying, extreme freezing, and heat and near vacuum-like conditions by entering a dormant stage where their cellular matrix becomes a type of biological glass (Sakurai et al., 2008). Even in more benign environments, aquatic insects commonly enter a short egg or larval diapause to escape unfavorable conditions for one season (e.g., higher water temperatures: Lancaster and Downes, 2013). Thus dispersal through time is an important process for population persistence in numerous species of aquatic insects, even if it is not as commonly studied as dispersal through space.

III. Benthic communities in lakes, wetlands, and ponds

Benthic animals in lakes are found in littoral and profundal habitats. They are exposed to relatively homogeneous conditions of temperature and substratum throughout the year, but significant differences appear with water depth. Factors such as exposure to wind and currents, oxygen, substrate type, food availability and quality, and intra- and interspecific competition vary with depth (Hutchinson, 1993; Jónasson, 1978; Ward, 1992) and determine benthic community composition. For example, oxygen can decrease with depth, especially if lakes become more productive, and substratum heterogeneity is much greater in the littoral than in profundal habitats. The diversity and density of benthic animals thus decrease with depth (Fig. 21-6) (Brinkhurst, 1974; Uutala, 1981) and can even be very low or extirpated in the profundal habitats in highly eutrophic lakes (Jónasson and Lindegaard, 1979). Community composition changes with depth are usually not gradual and distinct groups of taxa often correspond to littoral, sublittoral, and profundal habitats in lakes (Rieradevall et al., 1999; Fig. 21-6).

The profundal habitat is predominantly dominated by oligochaetes and chironomids, regardless of whether the substrata are hard stones or soft sediments (Malmquist et al., 2002; Rieradevall et al., 1999; Wiederholm, 1980). The proportions of oligochaetes and chironomids change with the eutrophic level of the lake; increasing eutrophy results in a higher proportion of oligochaetes in the profundal habitat (Fig. 21-6) (Wiederholm, 1980). In addition to these two groups, the sphaeriid *Pisidium* and several crustaceans (cladocerans, ostracods, or copepods) may also reach high densities in the profundal habitat (Malmquist et al., 2002). In high mountain oligotrophic lakes, species typically found in littoral habitats can also be found in profundal areas, such as the trichopteran *Plectrocnemia laetabilis* (Rieradevall et al., 1999).

The littoral habitat is usually dominated by a wide variety of benthic animals, most of them aquatic insects. Chironomids and oligochaetes still dominate in abundance and diversity in the littoral habitat, but other groups of benthic animals are also present (de Mendoza and Catalan, 2010; Krno et al., 2006; Rieradevall et al., 1999). Chironomids are usually more abundant than oligochaetes, especially in oligotrophic lakes (Boggero and Lencioni, 2006). Noninsect benthic animals are also

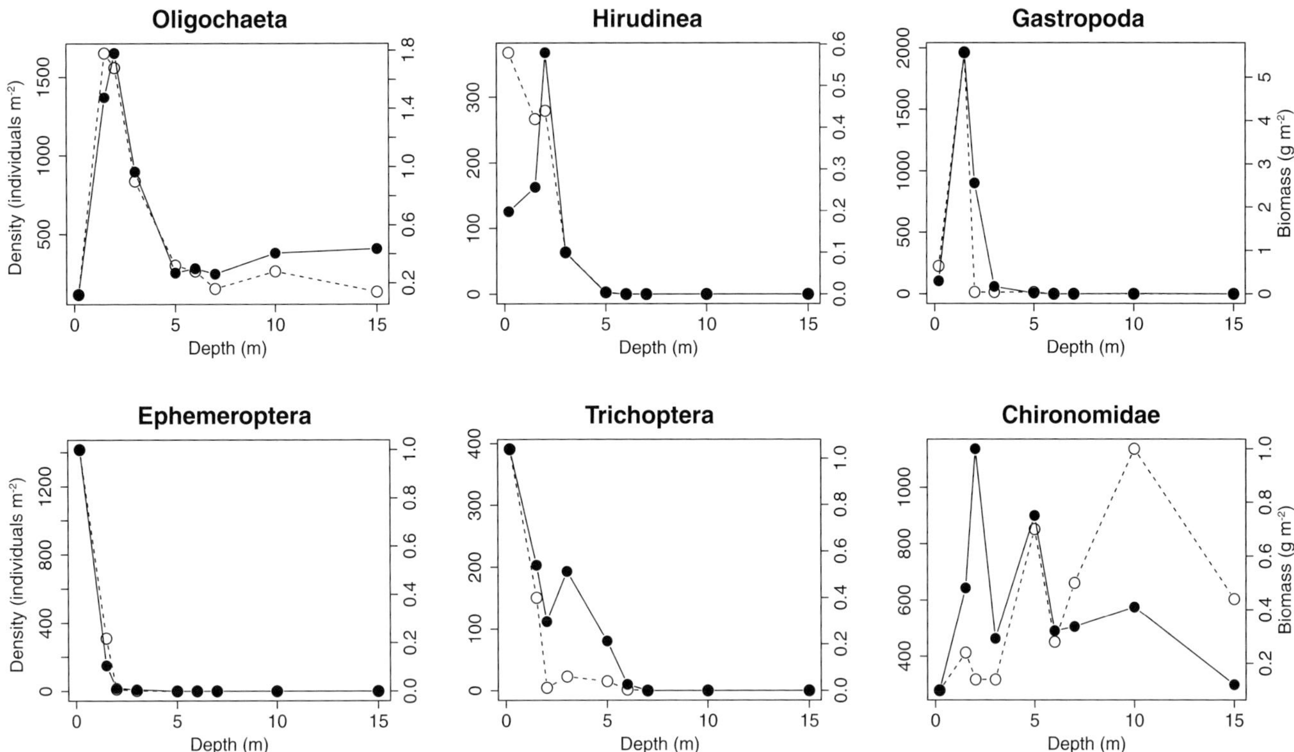

FIGURE 21.6 Average numerical density (individuals m^{-2}), shown by the solid line, and dry weight biomass (g m^{-2}), shown by the dashed line, of the principal groups of benthic animals at different depths of the eutrophic Lake Borrevann in southern Norway. *(Drawn from data in Ökland, 1964)*. Original biomass values, given in wet weight, were converted to an estimated dry mass by ×0.2 (see Winberg, 1971).

common there, such as different species of water mites, leeches, bivalves, gastropods, turbellarians, nematodes, and cnidarians (de Mendoza and Catalan, 2010; Lindegaard, 1992; Fig. 21-6). However, bivalves and gastropods can be absent if lakes are too acidic (Boggero and Lencioni, 2006). As in the profundal habitats, diversity is greater in stony than in fine sediment habitats, and the presence of macrophytes can further increase this diversity (Jeppesen et al., 2001; Rieradevall et al., 1999). Macrophyte coverage, together with other environmental factors, such as temperature or organic matter, partly determine the community composition of benthic animals (Hoffman et al., 1996; Krno et al., 2006; Nyman et al., 2005).

Benthic animals in lakes also respond to external variables, such as those related to altitude. For example, chironomid richness in sub-Arctic lakes peaks in the middle of the altitudinal gradient, where algae also had the maximum diversity, corresponding to a transition between coniferous forests and tundra (Nyman et al., 2005). More specifically, variables such as temperature, sediment organic matter, or pH also contributed to explaining these changes with altitude, indicating that it is not the elevation itself but a set of environmental variables changing with altitude. Similar patterns were also found in mountain lakes in the Pyrenees for the whole benthic communities, with altitude constraining diversity but having fewer effects on abundance, which is most likely related to seasonal- and annual-related variables (de Mendoza and Catalan, 2010). Temperature appears to be an altitudinal-related factor constraining the diversity and composition of many benthic groups in different types of lakes (e.g., de Mendoza and Catalan, 2010; Krno et al., 2006; Nyman et al., 2005). Benthic organic matter is also often negatively correlated to altitude, and lakes with higher amounts of organic matter typically have higher resource availability but sometimes lower levels of oxygen. Other altitude-related factors of interest are the type of substrate (often coarser at higher altitudes), the presence of aquatic vegetation that provides habitat and food (Hoffman et al., 1996), and the presence of fish predators (often lower at higher altitudes) (de Mendoza and Catalan, 2010).

Despite all these altitudinal-related factors that seem to contribute to explaining overall benthic community size and composition, some factors are more influential to certain groups than others (de Mendoza and Catalan, 2010). For example, fish predators are important factors to explain community patterns of coleopterans but may be less relevant for other groups (de Mendoza et al., 2012), although fishless lakes often have higher abundances of some chironomid, trichopterans, and crustaceans (Jeppesen et al., 2017). Interestingly, patterns for taxa widely tolerant of environmental conditions, such as Nematoda, are less dependent on altitude-related factors and may be more dependent on stochastic events (de Mendoza et al., 2016). The relative role of environmental and dispersal-related factors (related to stochastic events) also varies across regions for the same taxonomic group. For example, the distribution of trichopterans in the Pyrenees could be best explained by environmental factors, whereas in boreal lakes dispersal apparently played a more significant role (de Mendoza et al., 2015; Hoffsten, 2004).

Benthic animals in wetlands and ponds are often highly diverse (Biggs et al., 2005; Gopal and Junk, 2000) and include many groups absent from rivers (Batzer et al., 1999). In addition, semiaquatic and terrestrial invertebrates occupying wetland shores are common. Wetlands (Chapter 29) are usually highly dynamic ecosystems with varying hydroperiod (i.e., alternating cycles of wetting and drying conditions, or varying influence of marine water in the case of coastal wetlands) and associated physicochemical parameters (e.g., large fluctuations in oxygen concentration). Wetlands have a mixture of resident and transient species, with diverse biological traits related to dispersal (Boix et al., 2007). Resident species cannot migrate easily among wetlands and are more affected by the local environment, whereas transient species are true colonists that may only be constrained by wetland size and isolation (Hall et al., 2004). Some wetland species can tolerate drying conditions (or recover from them), low-oxygen concentrations, and high salinity values, with a wide range of physiological adaptations such as those described above.

Crustaceans and aquatic insects are probably the most diversified group in wetlands. The former includes representatives of branchiopods, copepods, ostracods, and malacostraceans, whereas the latter mainly includes species of coleopterans, hemipterans, odonates, and dipterans. Other aquatic insects, such as ephemeropterans or trichopterans, are also present but usually in lower abundance and diversity (e.g., Boix et al., 2007). The diversity and relative abundance of these groups depends on abiotic and biotic environmental conditions, geometric constraints, and stochasticity. The hydroperiod, salinity, trophic state, and the presence of aquatic plants or fish are key aspects to explain the local diversity of benthic animals in wetlands and ponds, but size and isolation cannot be ignored (Biggs et al., 2005; Boix et al., 2007; Gascón et al., 2005; Tornero et al., 2018). Temporary wetlands host rich benthic biodiversity, even in non-Mediterranean regions where these habitats are less common (Gleason and Rooney, 2017; Hall et al., 2004). However, benthic animals in most temporary wetlands are usually composed of nested subsets of perennial communities (Baber et al., 2004; Gleason and Rooney, 2017; Ruhí et al., 2013) and, in general, diversity increases with hydroperiod (Baber et al., 2004). However, a few benthic animals are characteristic of wetlands with short hydroperiods, including some anostraceans, notostraceans, and culicids (Gleason and Rooney, 2017; Lillie, 2003). Salinity (Chapter 12) is another influential factor that frequently interacts with the effects of the hydroperiod (Silver et al., 2012; Waterkeyn et al., 2008). Benthic animals in wetlands respond to nutrient concentrations (Silver et al., 2012) but less clear are overall land-use changes (Batzer, 2013). The diversity of benthic animals increases with the increasing abundance of aquatic plants, as they may be used as a food resource, refuge, or habitat (Cañedo-Argüelles and Rieradevall, 2009; Batzer and Wissinger, 1996). As fish abundance increases, benthic animal diversity typically decreases and assemblage composition changes (McLean et al., 2016; Zimmer et al., 2000). Finally, neutral variables, such as wetland size or isolation, also explain benthic diversity patterns in wetlands, as they can be considered islands in the landscape (Hall et al., 2004; Tornero et al., 2018). All the above-mentioned environmental variables usually interact and change across seasons, resulting in seasonal changes of communities (Cañedo-Argüelles and Rieradevall, 2009).

IV. Benthic communities of rivers and streams

Benthic animals have received intensive study in lotic ecosystems. There is reasonable evidence to suggest that many orders and other major taxa of aquatic insects evolved in cool running waters before expanding to warmer running waters and lentic environments (Ward, 1992). Cool flowing waters, where dissolved oxygen concentrations are high and flow minimizes distances for transport to cell membranes, are logical environments to develop respiratory mechanisms and adaptations for extracting dissolved oxygen from the water. Benthic animals in rivers and streams feed largely on algae and other attached microbiota and serve as primary food resources for fish and other vertebrates.

Recent analyses of lotic benthic communities have shifted toward analyses of changes in population and community dynamics and production in response to environmental variables. Environmental variables influencing growth and reproduction include (1) physiological constraints of parameters such as temperature, dissolved oxygen, and osmoregulation; (2) constraints of food acquisition and food quality; (3) biological interactions of predation and intra- and interspecific competition; and (4) physical constraints of changing habitats within stream ecosystems. This same set of factors, as well as mobility (drift, swimming, crawling, and flight), affects the rates of colonization of new or rewatered channels, rivers recovering from pollution, and unstable rivers with fluctuating discharge (Mackay, 1992).

Lotic habitats are generally separable into *riffles* (i.e., areas of fast-flowing water characterized by sediment erosion) and *pools* of depositional habitats (pools, backwaters, and side channels). Flow is a dominant parameter in lotic ecosystems that has led to major morphological and behavioral adaptations of invertebrates in relation to movements, attachment, concealment, feeding, and reproduction (Vogel, 1994). Flow velocity affects benthic habitat structure, dispersal of organisms, their acquisition of essential resources, intra- and interspecific competition, and predation efficacy. If benthic animals live on the surface of the substrata, they usually reside within the roughness layer, where the unique arrangement of sediment particles produces strongly sheared and highly three-dimensional flow patterns (Hart and Finelli, 1999). *Shear stress* (i.e., the force of flowing the water against a solid) is highest in the vertical plane in the wake of boulders, and as a result, turbulence intensity and kinetic energy are greatest in the wake region. Benthic animal communities are significantly richer and more abundant in the wakes than in the front of boulders. Current velocities decline markedly within sediments, and often benthic animals penetrate well into hyporheic interstitial spaces among sand and gravel substrata.

It is common to find the maximum density of lotic benthic animals, particularly oligochaetes, chironomids, amphipods, and microcrustaceans, at a depth of between 5 and 20 cm within porous sediments (Brunke and Gonser, 1999). The depth to which *hyporheic* invertebrates penetrate sediments is likely restricted by the availability of food resources, which generally decrease as the ratio of particulate organic carbon to total fine particles declines. The hyporheic zone can function as a refuge from predation during normal flow conditions and from drying during drought conditions (Stubbington, 2012), although during severe spates of drying, subsurface substrata can be eroded severely with a very high loss (50%–90%) of invertebrates (Palmer et al., 1992). Some invertebrates even inhabit the hyporheic zone under streambeds that are dry at the surface (Dorff and Finn, 2020). Other invertebrates, including several species of plecopterans, spend their entire larval life cycle in the hyporheic zone, many meters laterally from the stream channel, and persist on methane-derived carbon (DelVecchia et al., 2016).

A generalized model of the changes in benthic functional groups and ecological processes along streams from their headwaters to their mouths was established in the *River Continuum Concept* (see also Chapter 5). The concept was originally developed in temperate forested regions where the influence of the terrestrial vegetation is great (Vannote et al., 1980). The River Continuum Concept was later broadened to accommodate regional differences in hydrology, climate, tributaries, location-specific geology and lithology, vegetation, and human-induced factors to encompass broader spatial and temporal scales (e.g., Minshall et al., 1985). More recent attempts to understand lotic ecology and benthic animal distributions at large scales include the *Serial Discontinuity Concept* (Ward and Stanford, 1983), the *Flood Pulse Concept* (Junk et al., 1989), the *Network Dynamics Hypothesis* (Benda et al., 2004), and the *Stream Biome Gradient Concept* (Dodds et al., 2015), and the search for theoretical concepts that fully capture the dynamic nature of lotic community ecology is ongoing (Allen et al., 2020).

Arguably, the concept that most comprehensively conveys the dynamic nature of lotic habitats is the *Natural Flow Regime Concept* (Poff et al., 1997). This concept quantifies the periodicity and intensity of flooding and drying events that shape benthic animal communities in rivers and streams, factors that in turn shape the adaptations that taxa must have to persist in these dynamic and seasonally variable environments (Lytle and Poff, 2004; Tonkin et al., 2017). As our understanding of the myriad intricate linkages between natural flow regimes and benthic animals increases, the potential for restoring degraded or threatened lotic biota increases, even in urban and dammed rivers and streams (Palmer and Ruhí, 2019).

V. Metacommunities of benthic animals

Since the development of the *neutral theory of biodiversity* (Hubbell, 2001), the concept of community ecology has significantly changed our views to explain local species composition. Biodiversity is explained not only by local deterministic environmental factors (abiotic and biotic) but also by regional stochastic factors. Regardless of whether dispersal is a deterministic or a stochastic factor, dispersal processes are nowadays more frequently considered in community ecology studies. Communities are less seen as isolated entities, but rather as ones that interact across the landscape via dispersal, and thus are properly considered *metacommunities*. Despite the criticisms that community ecology received during the 1990s (Lawton, 1999), this paradigm shift has been the basis for the development of modern community ecology theory (Simberloff, 2004). Mirroring population ecology, the concepts of species sorting, ecological drift, speciation, and dispersal processes are now applied to metacommunities (Vellend, 2010; Vellend and Geber, 2005). There are four different models in metacommunity theory (species sorting, mass effects, patch dynamics, and neutral; Leibold et al., 2004) that differ in the relative roles of the environmental (species sorting, including abiotic and biotic filters) and spatial (dispersal) factors. However, these models are not mutually exclusive and do not represent the whole spectrum of metacommunities (Brown et al., 2017). Disentangling the effects of species sorting and dispersal factors, and how they influence fundamental and applied research, is an important challenge in metacommunity ecology (Heino et al., 2015).

The metacommunity concept has also been applied to benthic animal communities, and spatial connectivity differs among freshwater ecosystems. Streams and rivers are hierarchically structured, and networks are dendritic entities, with flow longitudinally connecting metacommunities. In contrast, lentic ecosystems, unless they are connected by streams, have lower connectivity and only rely on organisms actively flying or passively moving from one community to another. The spatial extent is an important factor to understand metacommunity patterns, with dispersal processes being potentially more important in small-scale (mass effects) or large-scale (dispersal limitation) studies, and species sorting prevailing at intermediate scales (Heino et al., 2015). Similarly, the spatial grain considered also determines the influence of species sorting and dispersal factors (Viana and Chase, 2019). Overall, however, species sorting seems to be more important than dispersal processes in freshwater ecosystems (Heino et al., 2015) and includes abiotic and biotic filters (García-Girón et al., 2020). For benthic animals, lakes and wetlands are islands in a matrix of unsuitable habitat. Species sorting and dispersal limitation are common mechanisms that structure lentic metacommunities, but their relative role depends on the dispersal ability of the organisms. The distributions of weak dispersers (e.g., mollusks and crustaceans) are more easily explained by dispersal limitation than strong dispersers (e.g., flying insects) (Heino, 2013). The role of dispersal limitation also increases depending on the occurrence of lentic ecosystems across the landscape, whereas the role of species sorting increases with environmental heterogeneity in the ecosystem. For example, deeper lakes and wetlands have potentially more heterogeneity. *Mass effects* (i.e., a metacommunity model where species massively disperse from optimal to suboptimal habitats), in contrast, are rare phenomena in lentic ecosystems because passive dispersal is much rarer than in lotic systems. There are exceptions, however, including benthic communities in Patagonian wetlands, where winds are strong and have a west-east direction, enhancing the passive dispersal of organisms and determining local composition (Epele et al., 2021).

The local environment and spatial connectivity change along river networks. The dendritic structure of these networks and landscape morphology results in numerous isolated headwaters, with potentially large environmental heterogeneity. Benthic animal metacommunities in these sections are thus best explained by species sorting and dispersal limitation (Altermatt, 2013; Brown and Swan, 2010). Dispersal can be limited because there are no sources of colonizers upstream and flying organisms need to cross several landscape topographic barriers to move across headwaters. In contrast, lowland lotic communities are better connected through the river network because they receive drifting organisms from the whole network upstream, and there are fewer landscape constraints in lower valleys. Mass effects might prevail in these lowland reaches (Brown and Swan, 2010). In comparison to lentic ecosystems, the river network plays a central role in river metacommunities, but with varying effects depending on the dispersal abilities (Tonkin et al., 2018).

Most of the metacommunity studies dealing with benthic animals have focused on analyzing spatial patterns. However, communities also vary with time, with the proportion of species sorting and dispersal processes changing across seasons or years. Temporary rivers and wetlands, for example, are highly dynamic and dispersal events are usually dominant during the rewetting, where animals colonize available habitats (Datry et al., 2016; Sarremejane et al., 2017). In contrast, drying imposes harsh environmental conditions, sorting less adapted species. Long-term metacommunity changes have been also assessed, suggesting that the metacommunities may be relatively stable unless extreme events of flooding or drying occur (Cañedo-Argüelles et al., 2020; Sarremejane et al., 2018). Spatial connectivity decreases during dry years and spatial factors (i.e., through dispersal limitation) become more important, also for strong dispersers, which are usually less affected by spatial factors (Cañedo-Argüelles et al., 2020).

The main limitation of metacommunity studies is the characterization of dispersal processes that reflect spatial effects. They can be divided into endogenous (or intrinsic) and exogeneous (or extrinsic) factors (Bonada et al., 2012). The former considers the different abilities of species to disperse, whereas the latter is related to the landscape, network, or other external constraints that favor or limit dispersal (Tonkin et al., 2018). Strong and weak dispersers might explain mass effects and dispersal limitation metacommunity mechanisms. Similarly, high spatial connectivity might result in mass effects, whereas spatial isolation is related to dispersal limitation. Despite advances to characterize animal dispersal, we still lack quantitative information on the ability of most species to move across the landscape (Bohonak and Jenkins, 2003). Most metacommunity studies, if including endogenous dispersal, classify benthic animals in coarse categories (e.g., aerial/aquatic active/passive dispersal) to obtain a proxy for dispersal potential (Heino et al., 2017). The reality, however, is that there are not always strong differences among dispersal groups, likely because dispersal is not well characterized (Heino et al., 2015). A way to increase the reliability of dispersal estimates would be to use more specific dispersal traits directly or indirectly related to dispersal (e.g., adult wing size; Sarremejane et al., 2020). It would be better to obtain real dispersal values from local experiments or inferred from genetic analyses. Weak dispersers or species in isolated habitats are much more vulnerable to local and regional extinctions than stronger disperses or the same species in highly connected habitats (Cid et al., 2021). However, very few benthic animals have been recognized as being of conservation concern (see www.iucnredlist.org).

Metacommunity dynamics are also important to consider in regard to bioassessments, where current methods are solely based on species sorting processes. However, biomonitoring assessment should also be considered in a metacommunity framework because (1) a species cannot be found in a suitable site if there is dispersal limitation and (2) a species might be found in an unsuitable site, at least on a short-term basis, because of mass effects (Cid et al., 2020, 2021). Furthermore, the location of biomonitoring sites in the network will determine the relative importance of species sorting or dispersal processes (Brown and Swan, 2010). Similarly, the implementation and effectiveness of conservation and restoration measures should be considered in the context of a metacommunity framework (Chase et al., 2020). For example, whether benthic communities are better explained by species sorting or dispersal processes will determine if we should restore local habitats (e.g., to improve water chemistry) or instead work to enhance spatial connectivity (e.g., dam removal).

VI. Benthic animals and global change

Benthic animals are ideal organisms to assess the effects of global change. They are ubiquitous organisms, have intermediate lifespans, respond to general and specific disturbances, and are relatively easy to sample and to identify (Bonada et al., 2006; Rosenberg and Resh, 1993). Despite several limitations that have been identified (e.g., they also respond to seasonal variation), responses of benthic animals to global change have been widely studied. Most benthic animals, and especially macroinvertebrates, have been used to track environmental changes (Birk et al., 2012). Overall, benthic animals respond to a wide range of natural and anthropogenic disturbances, including floods and droughts, wildfires, organic pollution, heavy metals, acidification, salinity, emergent pollutants, habitat fragmentation, or biotic invasions, among others (Raddum and Fjellheim, 2002). All these disturbances result in a general response: richness shows a unimodal response or decrease as disturbance increases, and sensitive species are replaced by tolerant ones (Larsen and Ormerod, 2010). Unless under chronic stress, benthic animals generally recover quickly after disturbance, thanks to their myriad resistance and resilience strategies. This is especially true in regions regularly affected by natural and seasonal disturbances, such as floods and droughts. Many species have adaptations to these natural events and these adaptations confer capabilities to cope with anthropogenic disturbances as well (Soria et al., 2020).

Past global changes also can be understood using benthic animals (e.g., paleolimnology; Chapter 30). Chitin present in many benthic animals is highly resistant to degradation, and exoskeletal remains can be found preserved in sediments thousands or even millions of years old (Stankiewicz et al., 1997). Crustaceans (cladocerans and ostracods) and chironomids are the benthic animals most widely used as proxies for paleoecological reconstruction in lakes (Burge et al., 2017). Other groups, such as trichopterans, sponges, or even rotifers, are also used in lakes where they reach high abundances (Burge et al., 2017; Solem and Birks, 2000). Dormant propagules of crustaceans and rotifers, or body fragments of crustaceans, chironomids, and trichopterans, are accumulated in lake sediments and, if the environmental preferences of the species are known, changes in species composition can be used to infer past climate or environmental conditions of the lake (e.g., temperature, acidity, eutrophication, precipitation; see Chapter 30). Stable isotope analyses are also being used and offer the possibility to reconstruct past food webs (van Hardenbroek et al., 2009).

Most developments regarding benthic animals and global change are related to their use in bioassessments. Many groups are used as bioindicators individually or as part of larger nontaxonomical groups (e.g., macroinvertebrates, meiofauna), and several methods have been developed considering different organizational levels. At the population level, benthic animals are used in bioassays, biomarkers, fluctuating asymmetry, or morphological deformities studies, whereas at the community level, they are used in uni- and multimetric indexes and multivariate methods (Bonada et al., 2006). Functional attributes of benthic animals also have been used in bioassessment methods. Local and regional disturbances, as well as global environmental changes, modify habitat conditions and act as selection pressures that select species with traits more adapted to these new conditions. Methods based on single or multiple biological traits are sometimes considered more reliable than those based on taxonomic composition: they can be applied across large regions because they are not affected by the biogeographical context. Individual traits respond to specific disturbances and thus can discriminate natural and anthropogenic disturbances and directly relate to functional processes in the ecosystem (Bonada et al., 2006; Soria et al., 2020; Statzner et al., 2001). Despite the existence of multiple trait databases, the implementation of trait-based methods in routine biomonitoring is still early in its development.

Bioassessment methods are more developed for lotic communities than lentic ones, likely because the latter have higher heterogeneity and are more difficult to sample (Poikane et al., 2016). Interestingly, most bioassessment methods developed for both ecosystem types are based on the whole macroinvertebrate community, but differences exist when specific groups are considered. Aquatic insects, such as indexes that consider the diversity of ephemeropterans, plecopterans, and trichopterans, have a high reliability to detect disturbance in rivers, whereas noninsect groups, such as oligochaetes, mollusks, or crustaceans, have been more commonly used in lentic ecosystems (Boix et al., 2005; Odountan et al., 2019). Indexes based on chironomids have also been developed for lakes and wetlands (Cañedo-Argüelles et al., 2012; Odountan et al., 2019; Ruse, 2010). Besides macroinvertebrates, methods based on meiofauna also exist, although they are better developed in marine compared to freshwater ecosystems (Balsamo et al., 2012). Because the exclusion of meiofauna interferes with the detection of human impacts (Mbaka et al., 2016), new sampling and identification technologies (e.g., metabarcoding) will result in more integrative and reliable bioassessment methods.

Benthic animals are also used to better understand the effects of future global change and, more specifically, climate change. Macroinvertebrates are again the benthic animals most often used in these predictions, and rivers the most studied ecosystems. Models suggest an overall loss of species richness, shifts in distributional ranges, and an upstream movement of species (Alba-Tercedor et al., 2017; Domisch et al., 2011, 2013; Sáinz-Bariáin et al., 2016). Besides local species extirpations and latitudinal and altitudinal shifts, benthic animals also respond to climate change locally via microevolutionary processes and phenotypic plasticity. For example, life history can be quickly modified in at least some groups by advancing egg hatching or adult emergence, increasing voltinism, or changing size at maturity (Stoks et al., 2014; Van Doorslaer et al., 2010). In addition to these effects of climate change, population sizes can decrease, and this can also influence genetic diversity (Bálint et al., 2011). These trends make headwaters, and species living there, highly vulnerable river sections, which can be further exacerbated by their high habitat isolation (see section V). Benthic animals have differential vulnerability to climate change, and a common approach to assessing species or community vulnerability considers multiple biogeographical, ecological, and biological traits (Conti et al., 2014; Hering et al., 2009; Tierno de Figueroa et al., 2009). For example, cold-stenotherm headwater species with narrow distribution ranges would be much more affected than others and the current habitats of these species will be colonized by downstream species (Domisch et al., 2011, 2013). Benthic communities will thus be reorganized, with consequences also on food webs and ecosystem processes (Woodward et al., 2010).

VII. Summary

1. The distribution, abundance, and productivity of benthic animals are determined by regional and local processes: (a) the historical or biogeographical events that have allowed or prevented a species from becoming part of the regional species pool; (b) the dispersal ability of the species to reach a habitat; (c) the physiological limitations of the species at all stages of the life cycle; (d) the availability of energy resources; and (e) the ability of the species to tolerate competition, predation, and parasitism.
2. The benthic animals (and protozoans) of freshwaters are extremely diverse in terms of species and biological traits. Representatives of nearly every animal phylum occur in, or are associated with, the sediments of lakes and streams.
   a. The most common and diversified groups include freshwater protozoans, porifers, cnidarians, platyhelminths, gastrotrichs, nematodes, nematomorphs, bryozoans, oligochaetes and leeches, tardigrades, acari, mollusks (gastropods and bivalves), and a wide variety of crustaceans (ostracods, malacostraceans) and hexapods.
   b. Besides collembolans, the hexapods include several orders of aquatic insects: odonates, ephemeropterans, plecopterans, hemipterans, dipterans, trichopteran, megalopterans, coleopterans, lepidopterans, and neuropterans.
3. Aquatic insects are typically the most abundant and diverse animal group in the benthos.
   a. They evolved in terrestrial environments but adapted to freshwaters through multiple morphological, physiological, reproductive, and life history traits. Some taxa are fully aquatic, whereas others combine aquatic with terrestrial life stages.
   b. There is also a wide range of feeding strategies in aquatic insects and several functional feeding groups have been recognized: scrapers, shredders,

collector-gatherers, collector-filterers, plant piercers, and predators.

c. Aquatic insects exhibit a wide variety of dispersal mechanisms to persist through environmental changes in dynamic environments as well as to maintain genetic diversity across populations. These mechanisms are usually classified as aquatic active, aquatic passive, aerial active, and aerial passive dispersal. In addition to dispersal through space, some aquatic insects can disperse through time via dormant stages.

4. In lakes benthic animals are found in littoral and profundal habitats. They are exposed to relatively homogeneous conditions of temperature and substratum throughout the year, but significant differences appear with water depth.
   a. The profundal habitat is predominantly dominated by chironomids and oligochaetes, depending on the trophic level of the lake.
   b. The littoral habitat is usually dominated by a wider variety of animals, most of them aquatic insects.
   c. Benthic animals in lakes also respond to external variables, such as temperature, sediment organic matter, pH, substrate, the presence of aquatic vegetation and fish predators, geometric constraints, or stochasticity.
5. In wetlands benthic animals are often highly diverse, with crustaceans and aquatic insects being the most diversified groups. Benthic animals in these habitats have a mixture of resident and transient species and respond to similar environmental variables identified in lakes but also to the changing hydroperiod or salinity.
6. Benthic animals in rivers and streams are strongly affected by flow intensity and variability, directly or indirectly through changes in substrate or other factors. Longitudinal changes of benthic communities along river networks and, at larger scales, are common and resulted in the development of several conceptual frameworks.
7. Benthic animal communities are now considered metacommunities, with compositional patterns explained by local environmental (species sorting) and spatial (dispersal) factors. The relative importance of these factors, however, can change between lotic and lentic ecosystems, longitudinally along river networks, or through time (seasonally and interannually). More recently, the metacommunity concept also is being used in the applied field, to improve biomonitoring approaches and enhance the conservation of threatened animals.
8. Benthic animals are ideal organisms to assess past (e.g., paleolimnology) and current (e.g., biomonitoring) global change effects and to better understand future global change effects. They respond to a wide range of natural and anthropogenic disturbances and community composition change along disturbance patterns. Benthic animals often quickly recover after disturbance, thanks to their myriad resistance and resilience strategies.

Acknowledgments

We are grateful to Cesc Múrria and Valeria Itskovich for providing figures and photos. Special thanks to Kate Boersma, Miguel Cañedo-Argüelles, and Jose María Fernández-Calero for their comments on the earlier version of the chapter, and to John Smol and Ian Jones for inviting us to contribute to this nice and fascinating endeavor. The FEHM (Freshwater Ecology, Hydrology and Management) research group is funded by the "Agència de Gestió d'Ajuts Universitaris i de Recerca" (AGAUR) at the "Generalitat de Catalunya" (2017SGR1643).

References

Abebe, E., Decraemer, W., De Ley, P., 2008. Global diversity of nematodes (Nematoda) in freshwater. Hydrobiologia 595, 67–78.

Alba-Tercedor, J., Sáinz-Bariáin, M., Poquet, J.M., Rodríguez-López, R., 2017. Predicting river macroinvertebrate communities distributional shifts under future global change scenarios in the Spanish Mediterranean area. PLoS One 12, 1–21.

Alexander Jr., J.E., Covich, A.P., 1991. Predation risk and avoidance behavior in two freshwater snails. Biol. Bull. 180, 387–393.

Allen, D.C., Datry, T., Boersma, K.S., Bogan, M.T., Boulton, A.J., Bruno, D., Busch, M.H., Costigan, K.H., Dodds, W.K., Fritz, K.M., Godsey, S.E., 2020. River ecosystem conceptual models and non-perennial rivers: a critical review. Water 7, e1473.

Altermatt, F., 2013. Diversity in riverine metacommunities: a network perspective. Aquat. Ecol. 47, 365–377.

Angstadt, J.D., Moore, W.H., 1997. A circadian rhythm of swimming behavior in a predatory leech of the family Erpobdellidae. Am. Midland Nat. 137, 165–172.

Armitage, M.J., Young, J.O., 1990. A field and laboratory study of the parasites of the triclad *Phagocata vitta* (Dugas). Freshwat. Biol. 24, 101–107.

Baber, M.J., Fleishman, E., Babbitt, K.J., Tarr, T.L., 2004. The relationship between wetland hydroperiod and nestedness patterns in assemblages of larval amphibians and predatory macroinvertebrates. Oikos 107, 16–27.

Balian, E.V., Segers, H., Lévèque, C., Martens, K., 2008. The freshwater animal diversity assessment: an overview of the results. Hydrobiologia 595, 627–637.

Bálint, M., Domisch, S., Engelhardt, C., Haase, P., Lehrian, S., Sauer, J., Theissinger, K., Pauls, S.U., Nowak, C., 2011. Cryptic biodiversity loss linked to global climate change. Nat. Clim. Chang. 1, 313–318.

Ball, I.R., Gourbault, N., Kenk, R., 1981. Planarians (Turbellaria) of temporary waters in eastern North America. Life Sci. Contrib. Roy. Ontario Mus. 127, 27.

Balsamo, M., Semprucci, F., Frontalini, F., Coccioni, R., 2012. Meiofauna as a tool for marine ecosystem biomonitoring. In: Cruzado, A. (Ed.), Marine Ecosystems. InTech, Rijeka, Croatia, pp. 77–104.

Balsamo, M., d'Hont, J.-L., Kisielewski, J., Pierboni, L., 2008. Global diversity of gastrotrichs (Gastrotricha) in fresh waters. Hydrobiologia 595, 85–91.

Bark, A.W., 1981. The temporal and spatial distribution of planktonic and benthic protozoan communities in a small productive lake. Hydrobiologia 85, 239–255.

Batzer, D.P., 2013. The seemingly intractable ecological responses of invertebrates in North American wetlands: a review. Wetlands 33, 1–15.

Batzer, D.P., Rader, R.B., Wissinger, S.A., 1999. Invertebrates in Freshwater Wetlands of North America: Ecology and Management. John Wiley & Sons, Inc., New York, USA.

Batzer, D.P., Wissinger, S.A., 1996. Ecology of insect communities in nontidal wetlands. Annu. Rev. Entomol. 41, 75–100.

Benda, L., Poff, N.L., Miller, D., Dunne, T., Reeves, G., Pess, G., Pollock, M., 2004. The network dynamics hypothesis: how channel networks structure riverine habitats. Bioscience 54, 413–427.

Benson, A.J., Williams, J.D., 2021. Review of the Invasive Asian Clam Corbicula spp. (Bivalvia: Cyrenidae) Distribution in North America, 1924–2019. Geological Survey Scientific Investigations Report 2021–5001, U.S, p. 66.

Beutel, R.G., Yavorskaya, M.I., Mashimo, Y., Fukui, M., Meusemann, K., 2017. The phylogeny of Hexapoda (Arthropoda) and the evolution of megadiversity. Proc. Arthrop. Embryol. Soc. Japan 51, 1–15.

Biggs, J., Williams, P., Whitfield, M., Nicolet, P., Weatherby, A., 2005. 15 years of pond assessment in Britain; results and lessons learned for the work of Pond Conservation. Aquatic Conserv. Mar. Freshw. Ecosyst. 15, 693–714.

Birk, S., Bonne, W., Borja, A., Brucet, S., Courrat, A., Poikane, S., Solimini, A., van de Bund, W., Zampoukas, N., Hering, D., 2012. Three hundred ways to assess Europe's surface waters: an almost complete overview of biological methods to implement the Water Framework Directive. Ecol. Indic. 18, 31–41.

Blinn, D.W., Davies, R.W., 1990. Concomitant diel vertical migration of a predatory leech and its amphipod prey. Freshwat. Biol. 24, 401–407.

Bogan, A.E., 2008. Global diversity of freshwater mussels (Mollusca, Bivalvia) in freshwater. Hydrobiologia 595, 139–147.

Bogan, M.T., 2017. Hurry up and wait: life cycle and distribution of an intermittent stream specialist (*Mesocapnia arizonensis*). Freshw. Sci. 36, 805–815.

Bogan, M.T., Chester, E.T., Datry, T., Murphy, A.L., Robson, B.J., Ruhí, A., Stubbington, R., Whitney, J.E., 2017. Resistance, resilience, and community recovery in intermittent rivers and ephemeral streams. In: Datry, T., Bonada, N., Boulton, A.J. (Eds.), Intermittent Rivers and Ephemeral Streams: Ecology and Management. Elsevier, Inc., Cambridge, USA, pp. 349–376.

Bogan, M.T., Boersma, K.S., 2012. Aerial dispersal of aquatic invertebrates along and away from arid-land streams. Freshw. Sci. 31, 1131–1144.

Boggero, A., Lencioni, V., 2006. Macroinvertebrates assemblages of high altitude lakes, inlets and outlets in the southern Alps. Arch. Hydrobiol. 165, 37–61.

Böhme, A., Risse-Buhl, U., Küsel, K., 2009. Protists with different feeding modes change biofilm morphology. FEMS Microbiol. Ecol. 69, 158–169.

Bohonak, A.J., Jenkins, D.G., 2003. Ecological and evolutionary significance of dispersal by freshwater invertebrates. Ecol. Lett. 6, 783–796.

Boix, D., Gascón, S., Sala, J., Martinoy, M., Gifre, J., Quintana, X.D., 2005. A new index of water quality assessment in mediterranean wetlands based on crustacean and insect assemblages: the case of Catalunya (NE Iberian Peninsula). Aquat. Conserv. 15, 635–651.

Boix, D., Sala, J., Gascón, S., Martinoy, M., Gifre, J., Brucet, S., Badosa, A., López-Flores, R., Quintana, X.D., 2007. Comparative biodiversity of crustaceans and aquatic insects form various water body types in coastal Mediterranean wetlands. Hydrobiologia 584, 347–359.

Bonada, N., Dolédec, S., Statzner, B., 2012. Spatial autocorrelation patterns of stream invertebrates: exogenous and endogenous factors. J. Biogeogr. 39, 56–68.

Bonada, N., Prat, N., Resh, V.H., Statzner, B., 2006. Developments in aquatic insect biomonitoring: a comparative analysis of recent approaches. Annu. Rev. Entomol. 51, 495–523.

Bowers, J.A., Grossnickle, N.E., 1978. The herbivorous habits of *Mysis relicta* in Lake Michigan. Limnol. Oceanogr. 23, 767–776.

Brandt, E., 1978. Anpassungen von *Tubifex tubifex* Müller (Annelida, Oligochaeta) an die Temperatur, den Sauerstoffgehalt und den Ernährungszustand. Arch. Hydrobiol. 84, 302–338.

Bretschko, G., 1973. Benthos production of a high-mountain lake: Nematoda. Verh. Internat. Verein. Limnol. 18, 1421–1428.

Brinkhurst, R.O., 1974. The Benthos of Lakes. Macmillan Press Ltd., London, UK.

Brown, K.M., DeVries, D.R., 1985. Predation and the distribution and abundance of a pulmonate pond snail. Oecologia 66, 93–99.

Brown, B.L., Sokol, E.R., Skelton, J., Tornwall, B., 2017. Making sense of metacommunities: dispelling the mythology of a metacommunity typology. Oecologia 183, 643–652.

Brown, B.L., Swan, C.M., 2010. Dendritic network structure constrains metacommunity properties in riverine ecosystems. J. Anim. Ecol. 79, 571–580.

Brunke, M., Gonser, T., 1999. Hyporheic invertebrates—the clinal nature of interstitial communities structured by hydrological exchange and environmental gradients. J. N. Am. Benthol. Soc. 18, 344–362.

Burge, D.R.L., Edlund, M.B., Frisch, D., 2017. Paleolimnology and resurrection ecology: the future of reconstructing the past. Evol. Appl. 11, 42–59.

Burky, A.J., 1983. Physiological ecology of freshwater bivalves. In: Russell-Hunter, W.D. (Ed.), The Mollusca, vol. 6. Ecology. Academic Press, New York, USA, pp. 281–327.

Butler, M.G., 1982. A 7-year life cycle for two *Chironomus* species in arctic Alaskan tundra ponds (Diptera: Chironomidae). Can. J. Zool. 60, 58–70.

Butler, M.I., Burns, C.W., 1995. Effects of temperature and food level on growth and development of a planktonic water mite. Hydrobiologia 308, 153–165.

Cairns Jr., J., 1974. Protozoans (Protozoa). In: Hart Jr., C.W., Fuller, S.L.H. (Eds.), Pollution Ecology of Freshwater Invertebrates. Academic Press, New York, USA, pp. 1–28.

Cairns Jr., J., Yongue, W.H., 1977. Factors affecting the number of species in freshwater protozoan communities. In: Cairns Jr., J. (Ed.), Aquatic Microbial Communities. Garland, New York, USA, pp. 257–303.

Cañedo-Argüelles, M., Boix, D., Sánchez-Millaruelo, N., Sala, J., Caiola, N., Nebra, A., Rieradevall, M., 2012. A rapid bioassessment tool for the evaluation of the water quality of transitional waters. Estuar. Coast Shelf Sci. 111, 129–138.

Cañedo-Argüelles, M., Rieradevall, M., 2009. Quantification of environment-driven changes in epiphytic macroinvertebrate communities associated to *Phragmites australis*. J. Limnol. 68, 229–241.

Cañedo-Argüelles, M., Gutiérrez-Cánovas, C., Acosta, R., Castro-López, D., Cid, N., Fortuño, P., Munné, A., Múrria, C., Pimentao, A.R., Sarremejane, R., Soria, M., Tarrats, P., Verkaik, I., Prat, N., Bonada, N., 2020. As time goes by: 20 years of changes in the aquatic macroinvertebrate metacommunity of Mediterranean river networks. J. Biogeogr. 47, 1861–1874.

Cassagnau, P., 1972. Un collembole adapté à la predation: *Cephalotoma grandiceps* (Reuter). Nouvelle Rev. Entomol. 2, 5–12.

Chamier, A.-C., Willoughby, L.G., 1986. The role of fungi in the diet of the amphipod *Gammarus pulex* (L.): an enzymatic study. Freshwat. Biol. 16, 197–208.

Charlebois, P.M., Lamberti, G.A., 1996. Invading crayfish in a Michigan stream: direct and indirect effects on periphyton and macroinvertebrates. J. N. Am. Benthol. Soc. 15, 551–563.

Chase, J.M., Jeliazkov, A., Ladouceur, E., Viana, D.S., 2020. Biodiversity conservation through the lens of metacommunity ecology. Ann. N.Y. Acad. Sci. 1469, 86–104.

Chigbu, P., Sibley, T.H., 1998. Predation by longfin smelt (*Spirinchus thaleichthys*) on the mysid *Neomysis mercedis* in Lake Washington. Freshwat. Biol. 40, 295–304.

Childs, G.H., 1915. Some Observations on the Life History of the Water Springtail (*Podura Aquatica*—1758). Retrieved from the University of Minnesota Digital Conservancy. https://hdl.handle.net/11299/177935.

Cid, N., Bonada, N., Heino, J., Cañedo-Argüelles, M., Crabot, J., Sarremejane, R., Soininen, J., Stubbington, R., Datry, T., 2020. A metacommunity approach to improve biological assessments in highly dynamic freshwater ecosystems. Bioscience 70, 427–438.

Cid, N., Erös, T., Heino, J., Singer, G., Jähnig, S.C., Cañedo-Argüelles, M., Bonada, N., Sarremejane, R., Mykrä, H., Sandin, L., Paloniemi, R., Varumo, L., Datry, T., 2021. From meta-system theory to the sustainable management of rivers in the Anthropocene. Front. Ecol. Evol. 20, 49–57.

Conti, L., Schmidt-Kloiber, A., Grenouillet, G., Graf, W., 2014. A trait-based approach to assess the vulnerability of European aquatic insects to climate change. Hydrobiologia 721, 297–315.

Cover, M.R., Seo, J.H., Resh, V.H., 2015. Life history, burrowing behavior, and distribution of *Neohermes filicornis* (Megaloptera: Corydalidae), a long-lived aquatic insect in intermittent streams. West. N. Am. Nat. 75, 474–490.

Cover, M.R., Bogan, M.T., 2015. The minor aquatic insect orders. In: Thorp, J., Rogers, D.C. (Eds.), Ecology and General Biology: Thorp and Covich's freshwater invertebrates. Academic Press, San Diego, USA, pp. 1059–1072.

Creed Jr., R.P., 1994. Direct and indirect effects of crayfish grazing in a stream community. Ecology 75, 2091–2103.

Cuker, B.E., Mozley, S.C., 1981. Summer population fluctuations, feeding, and growth of *Hydra* in an arctic lake. Limnol. Oceanogr. 26, 697–708.

Cumberlidge, N., Hobbs, H.H., Lodge, D.M., 2015. Class Malacostraca, order Decapoda. In: Thorp, J., Rogers, D.C. (Eds.), Ecology and General Biology: Thorp and Covich's Freshwater Invertebrates. Academic Press, San Diego, USA, pp. 797–847.

Cummings, K.S., Graf, D.L., 2015. Class Bivalvia. In: Thorp, J., Rogers, D.C. (Eds.), Ecology and General Biology: Thorp and Covich's Freshwater Invertebrates. Academic Press, San Diego, USA, pp. 423–506.

Cummins, K.W., 1973. Trophic relations of aquatic insects. Annu. Rev. Entomol. 18, 183–206.

Cummins, K.W., Merritt, R.W., Berg, M.B., 2008. Ecology and distribution of aquatic insects. In: Merritt, R.W., Cummins, K.W., Berg, M.B. (Eds.), An Introduction to the Aquatic Insects of North America, fourth ed. Kendall/Hunt Publishing Company, Dubuque, USA, pp. 105–122.

Dall, P.C., 1987. The ecology of the littoral leech fauna (Hirudinea) in Lake Esrom, Denmark. Arch. Hydrobiol. Suppl. 76, 256–313.

Danks, H.V., 2008. Aquatic insect adaptations to winter cold and ice. In: Lancaster, J., Briers, R.A. (Eds.), Aquatic Insects. Challenges to Populations. CAB International, Wallingford, UK, pp. 1–19.

Datry, T., Bonada, N., Heino, J., 2016. Towards understanding the organisation of metacommunities in highly dynamic ecological systems. Oikos 125, 149–159.

Davids, C., Winkel, E.H.T., de Groot, C.J., 1994. Temporal and spatial patterns of water mites in Lake Maarsseveen I. Netherland J. Aquat. Ecol. 28, 11–17.

De Ley, P., Decraemer, W., Abebe, E., 2006. Introduction: Summary of present knowledge and research addressing the ecology and taxonomy of freshwater nematodes. In: Abebe, E., Andrássy, I., Traunspurger, W. (Eds.), Freshwater Nematodes: Ecology and Taxonomy. CAB International, Oxford/Cambridge, UK, pp. 3–30.

de Mendoza, G., Catalan, J., 2010. Lake macroinvertebrates and the altitudinal environmental gradient in the Pyrenees. Hydrobiologia 648, 51–72.

de Mendoza, G., Rico, E., Catalan, J., 2012. Predation by introduced fish constrains the thermal distribution of aquatic Coleoptera in mountain lakes. Freshw. Biol. 57, 803–814.

de Mendoza, G., Traunspurger, W., Palomo, A., Catalan, J., 2016. Nematode distributions as spatial null models for macroinvertebrate species richness across environmental gradients: a case for mountain lakes. Ecol. Evol. 7, 3016–3028.

de Mendoza, G., Ventura, M., Catalan, J., 2015. Environmental factors prevail over dispersal constraints in determining the distribution and assembly of Trichoptera species in mountain lakes. Ecol. Evol. 5, 2518–2532.

Debastiani, C., Meira, B.R., Lansac-Tôha, F.M., Velho, L.F.M., Lansac-Tôha, F.A., 2016. Protozoa ciliates community structure in urban streams and their environmental use as indicators. Braz. J. Biol. 76, 1043–1053.

Deharveng, L., D'Haese, C.A., Bedos, A., 2008. Global diversity of springtails (Collembola; Hexapoda) in freshwater. Hydrobiologia 595, 329–338.

DelVecchia, A.G., Stanford, J.A., Xu, X., 2016. Ancient and methane-derived carbon subsidizes contemporary food webs. Nat. Commun. 7, 1–9.

Dendy, J.S., 1963. Observations on bryozoan ecology in farm ponds. Limnol. Oceanogr. 8, 478–482.

Dias, R.J.P., Wieloch, A.H., D'Agosto, M., 2008. The influence of environmental characteristics on the distribution of ciliates (Protozoa, Ciliophora) in an urban stream of southeast Brazil. Braz. J. Biol. 68, 287–295.

DiDonato, G.T., Lodge, D.M., 1993. Species replacements among *orconectes* crayfishes in Wisconsin lakes: the role of predation by fish. Can. J. Fish. Aquat. Sci. 50, 1484–1488.

Docio, L., Parolin, M., Pinheiro, U., 2021. A contribution to adequate use of freshwater sponges as a proxy in paleoenvironmental studies. Zootaxa 4915 (4), 4915 zootaxa 4.3.

Dodds, W.K., Gido, K., Whiles, M.R., Daniels, M.D., Grudzinski, B.P., 2015. The stream biome gradient concept: factors controlling lotic systems across broad biogeographic scales. Freshw. Sci. 34, 1–19.

Domisch, S., Araújo, M.B., Bonada, N., Pauls, S.U., Jähnig, S.C., Haase, P., 2013. Modelling distribution in European stream macroinvertebrates under future climates. Glob. Chang. Biol. 19, 752–762.

Domisch, S., Jähnig, S.C., Haase, P., 2011. Climate-change winners and losers: stream macroinvertebrates of a submontane region in Central Europe. Freshw. Biol. 56, 2009–2020.

Dorff, N.C., Finn, D.S., 2020. Hyporheic secondary production and life history of a common Ozark stonefly. Hydrobiologia 847, 443–456.

Downing, J.A., 1984. Sampling the benthos of standing waters. In: Downing, J.A., Rigler, F.H. (Eds.), A Manual on Methods for the Assessment of Secondary Productivity in Fresh Waters. IBP Handbook No. 17, second ed. Blackwell Scientific Publications, Oxford, UK, pp. 87–130.

Drewes, C.D., Fourtner, C.R., 1993. Helical swimming in a freshwater oligochaete. Biol. Bull. 185, 1–9.

Dumont, H.J., 1994. The distribution and ecology of the fresh- and brackish-water medusae of the world. In: Dumont, H.J., Green, J., Masundire, H. (Eds.), Studies on the Ecology of Tropical

Zooplankton, Developments in Hydrobiology, 92. Springer, Dordrecht, Netherlands, pp. 1–12.

Edmunds Jr., G.F., Jensen, S.L., Berner, L., 1976. The Mayflies of North and Central America. University of Minnesota Press, Minneapolis, USA.

Edmunds Jr., G.F., McCafferty, W.P., 1996. New field observation on burrowing in Ephemeroptera from around the world. Entomol. News 107, 68–76.

Elliott, J.M., 1973. The life cycle and production of the leech *Erpobdella octoculata* (L.) (Hirudinea: Erpobdellidae) in a Lake District stream. J. Anim. Ecol. 42, 435–448.

Epele, L.B., Dos Santos, D.A., Sarremejane, R., Grech, M.G., Macchi, P.A., Manzo, L.M., Miserendino, M.L., Bonada, N., Cañedo-Argüelles, M., 2021. Blowin' in the wind: wind directionality affects wetland invertebrates metacommunities in Patagonia. Glob. Ecol. Biogeogr. 30, 1191–1203. https://doi.org/10.1111/geb.13294.

Eriksen, C.H., Lamberti, G.A., Resh, V.H., 1992. Aquatic insect respiration. In: Merritt, R.W., Cummins, K.W. (Eds.), An Introduction to the Aquatic Insects of North America. Kendall/Hunt, Dubuque, USA, pp. 29–39.

Eyualem, A., Grizzle, R.E., Hope, D., Thomas, W.K., 2004. Nematode diversity in the Gulf of Maine, USA, and a web-accessible relational database. J. Mar. Biol. Assoc. U.K. 84, 1159–1167.

Findlay, S., Meyer, J.L., Smith, P.J., 1986. Contribution of fungal biomass to the diet of a freshwater isopod (*Lirceus* sp.). Freshwat. Biol. 16, 377–385.

Finlay, B.J., 1980. Temporal and vertical distribution of ciliophoran communities in the benthos of a small eutrophic lock with particular reference to the redox profile. Freshwat. Biol. 10, 15–34.

Finlay, B.J., 1990. Physiological ecology of free-living protozoa. Adv. Microb. Ecol. 11, 1–35.

Foissner, W., 1988. Taxonomic and nomenclatural revision of Sládecek's list of ciliates (Protozoa: Ciliophora) as indicators of water quality. Hydrobiologia 166, 1–64.

Frost, T.M., 1978. Impact of the freshwater sponge *Spongilla lacustris* on a *Sphagnum* bog-pond. Verh. Internat. Verein. Limnol. 20, 2368–2371.

Fuller, S.L.H., 1974. Clams and mussels (Mollusca: Bivalvia). In: Hart Jr., C.W., Fuller, S.L.H. (Eds.), Pollution Ecology of Freshwater Invertebrates. Academic Press, New York, USA, pp. 215–273.

Garbuz, D.G., Zatsepina, O.G., Przhiboro, A.A., Yushenova, I., Guzhova, I.V., Evgen'ev, M.B., 2008. Larvae of related Diptera species from thermally contrasting habitats exhibit continuous up-regulation of heat shock proteins and high thermotolerance. Mol. Ecol. 17, 4763–4777.

García-Girón, J., Heino, J., García-Criado, F., Fernández-Aláez, C., Alahuhta, J., 2020. Biotic interactions hold the key to understanding metacommunity organization. Ecography 43, 1180–1190.

Gascón, S., Boix, D., Sala, J., Quintana, X.D., 2005. Variability of benthic assemblages in relation to the hydrological pattern in mediterranean salt marshes (Empordà wetlands, NE Iberian Peninsula). Arch. Hydrobiol. 163, 163–181.

Gaudes, A., Muñoz, I., Moens, T., 2013. Bottom-up effects on freshwater bacterivorous nematode populations: a microcosm approach. Hydrobiologia 707, 159–172.

Gelder, S.R., Williams, B.W., 2015. Clitellata: Branchiobdellida. In: Thorp, J., Rogers, D.C. (Eds.), Ecology and General Biology: Thorp and Covich's Freshwater Invertebrates, I. Academic Press, London, UK, pp. 551–563.

Ghiretti, F., 1966. Respiration. In: Wilbur, K.M., Yonge, C.M. (Eds.), Physiology of Mollusca, 2. Academic Press, New York, USA, pp. 175–208.

Gilbert, J.J., 1975. Field experiments on gemmulation in the freshwater sponge *Spongilla lacustris*. Trans. Amer. Microsc. Soc. 94, 347–356.

Gilbert, J.J., Simpson, T.L., 1976. Gemmule polymorphism in the freshwater sponge *Spongilla lacustris*. Arch. Hydrobiol. 78, 268–277.

Gleason, J.E., Rooney, R.C., 2017. Pond permanence is a key determinant of aquatic macroinvertebrate community structure in wetlands. Freshw. Biol. 63, 264–277.

Gledhill, T., 1985. Water mites—predators and parasites. Annu. Rep. Freshwat. Biol. Assoc. U.K. 53, 45–59.

Gopal, B., Junk, W.J., 2000. Biodiversity in wetlands: an introduction. In: Gopal, B., Junk, W.J., Davis, J.A. (Eds.), Biodiversity in Wetlands: Assessment, Function and Conservation. Backhuys Publishers, Leiden, The Netherlands, pp. 1–10.

Govedich, F.R., Moser, W.E., 2015. Clitellata: Hirudinida and Acanthobdellida. In: Thorp, J., Rogers, D.C. (Eds.), Ecology and General Biology: Thorp and Covich's Freshwater Invertebrates, vol. I. Academic Press, London, UK, pp. 565–588.

Grant, L.J., Sluys, R., Blair, D., 2006. Biodiversity of Australian freshwater planarians (Platyhelminthes: Tricladida: Paludicola): new species and localities, and a review of paludicolan distribution in Australia. Syst. Biodivers. 4, 435–471.

Haegerbaeumer, A., Höss, S., Heininger, P., Traunspurger, W., 2018. Response of nematode communities to metals and PAHs in freshwater microcosms. Ecotoxicol. Environ. Saf. 148, 244–253.

Hall, D.L., Willig, M.R., Moorhead, D.L., Sites, R.W., Fish, E.B., Mollhagen, T.R., 2004. Aquatic macroinvertebrate diversity of playa wetlands: the role of landscape and island biogeographic characteristics. Wetlands 24, 77–91.

Hallworth, M.T., Marra, P.P., McFarland, K.P., Zahendra, S., Studds, C.E., 2018. Tracking dragons: stable isotopes reveal the annual cycle of a long-distance migratory insect. Biol. Lett. 14, 20180741.

Hanelt, B., Grother, L.E., Janovy Jr., J., 2001. Physid snails as sentinels of freshwater nematomorphs. J. Parasitol. 87, 1049–1053.

Hanelt, B., Thomas, F., Schmidt-Rhaesa, A., 2005. Biology of the phylum Nematomorpha. Adv. Parasitol. 59, 243–305.

Harman, W.N., 1974. Snails (Mollusca: Gastropoda). In: Hart Jr., C.W., Fuller, S.L.H. (Eds.), Pollution Ecology of Freshwater Invertebrates. Academic Press, New York, USA, pp. 275–312.

Harrison, F.W., 1974. Sponges (Porifera: Spongillidae). In: Hart Jr., C.W., Fuller, S.L.H. (Eds.), Pollution Ecology of Freshwater Invertebrates. Academic Press, New York, USA, pp. 29–66.

Hart, D.D., Finelli, C.M., 1999. Physical-biological coupling in streams: the pervasive effects of flow on benthic organisms. Annu. Rev. Ecol. Syst. 30, 363–395.

Hartmann, H.J., 1985. Feeding of *Daphnia pulicaria* and *Diaptomus ashlandi* on mixtures of unicellular and filamentous algae. Verh. Internat. Verein. Limnol. 22, 3178–3183.

Heino, J., 2013. Does dispersal ability affect the relative importance of environmental control and spatial structuring of littoral macroinvertebrate communities? Oecologia 171, 971–980.

Heino, J., Alahuhta, J., Ala-Hulkko, T., Antikainen, H., Bini, L.M., Bonada, N., Datry, T., Erös, T., Hjort, J., Kotavaara, O., Melo, A.S., Soininen, J., 2017. Integrating dispersal proxies in ecological and environmental research in the freshwater realm. Environ. Rev. 25, 334–349.

Heino, J., Melo, A.S., Siqueira, T., Soininen, J., Valanko, S., Bini, L.M., 2015. Metacommunity organisation, spatial extent and dispersal in aquatic systems: patterns, processes and prospects. Freshw. Biol. 60, 845–869.

Heitkamp, U., 1982. Untersuchungen zur Biologie, Ökologie und Systematik limnischer Turbellarien periodischer und perennierender Kleingewässer Sudniedersachsens. Arch. Hydrobiol. Suppl. 64, 65–188.

Herbst, D.B., Bogan, M.T., Lusardi, R.A., 2008. Low specific conductivity limits growth and survival of the New Zealand mud snail from the Upper Owens River, California. West. N. Am. Nat. 68, 324–333.

Hering, D., Schmidt-Kloiber, A., Murphy, J., Lücke, S., Zamora-Munoz, C., López-Rodríguez, M.J., Huber, T., Graf, W., 2009. Potential impact of climate change on aquatic insects: a sensitivity analysis for European caddisflies (Trichoptera) based on distribution patterns and ecological preferences. Aquat. Sci. 71, 3–14.

Herman, S.S., 1963. Vertical migration of the opossum shrimp, *Neomysis americana* Smith. Limnol. Oceanogr. 8, 228–238.

Hobbs, H.H., 1991. Decapoda. In: Thorp, J.H., Covich, A.P. (Eds.), Ecology and Classification of North American Freshwater Invertebrates. Academic Press, San Diego, USA, pp. 823–858.

Hobson, K.A., Anderson, R.C., Soto, D.X., Wassenaar, L.I., 2012. Isotopic evidence that dragonflies (*Pantala flavescens*) migrating through the Maldives come from the northern Indian subcontinent. PLoS One 7 (12), e52594.

Hoffman, R.L., Liss, W.J., Larson, G.L., Deimling, E.K., Lomnicky, G.A., 1996. Distribution of nearshore macroinvertebrates in lakes of the Northern Cascade Mountains, Washington, USA. Arch. Hydrobiol. 136, 363–389.

Hoffsten, P.-O., 2004. Site-occupancy in relation to flight-morphology in caddisflies. Freshw. Biol. 49, 810–817.

Horváth, G.Á., Kriska, G.Y., 2008. Polarization vision in aquatic insects and ecological traps for polarotactic insects. In: Lancaster, J., Briers, R.A. (Eds.), Aquatic Insects: Challenges to Populations. CAB International, Wallingford, UK, pp. 204–229.

Hubbell, S.P., 2001. The Unified Neutral Theory of Biodiversity and Biogeography. Princeton University Press, Princeton, USA.

Hummon, M.R., Hummon, W.D., 1979. Reduction in fitness of the gastrotrich *Lepidodermella squamata* by dilute mine acid water and amelioration of the effect by carbonates. Int. J. Invertebr. Reprod. 1, 297–306.

Hutchinson, G.E., 1959. Homage to Santa Rosalia, or why are there so many kinds of animals? Am. Nat. 93, 145–159.

Hutchinson, G.E., 1993. The Zoobenthos. A Treatise on Limnology, vol. IV. John Wiley & Sons, New York, USA.

Huws, S.A., McBain, A.J., Gilbert, P., 2005. Protozoan grazing and its impact upon population dynamics in biofilm communities. J. Appl. Microbiol. 98, 238–244.

Jankowski, T., Collins, A.G., Campbell, R., 2008. Global diversity of inland water cnidarians. Hydrobiologia 595, 35–40.

Jeppesen, E., Lauridsen, T.L., Christoffersen, K.S., Landkildehus, F., Geertz-Hansen, P., Amsinck, S.L., Sondergaard, M., Davidson, T.A., Rigét, F., 2017. The structuring role of fish in Greenland lakes: an overview based on contemporary and paleoecological studies of 87 lakes from the low and the high Arctic. Hydrobiologia 800, 99–113.

Jeppesen, E., Sondergaard, M., Sondergaard, M., Christoffersen, K. (Eds.), 2001. The Structuring Role of Submerged Macrophytes in Lakes. Springer, New York, USA.

Jewell, M.E., 1935. An ecological study of the freshwater sponges of northern Wisconsin. Ecol. Monogr. 5, 461–504.

Jónasson, P.M., 1978. Zoobenthos of lakes. Verh. Internat. Verein. Limnol. 20, 13–37.

Jónasson, P.M., Lindegaard, C., 1979. Zoobenthos and its contribution to the metabolism of shallow lakes. Arch. Hydrobiol. Beih. Ergebn. Limnol. 13, 162–180.

Jónasson, P.M., Thorhauge, F., 1972. Life cycle of *Potamothrix hammoniensis* (Tubificidae) in the profundal of a eutrophic lake. Oikos 23, 151–158.

Joo, G.-J., Ward, A.K., Ward, G.M., 1992. Ecology of *Pectinatella magnifica* (Bryozoa) in an Alabama oxbow lake: colony growth and association with algae. J. N. Am. Benthol. Soc. 11, 324–333.

Junk, W.J., Bayley, P.B., Sparks, R.E., 1989. The flood pulse concept in river-floodplain systems. Can. J. Fish. Aquat. Sci. 106, 110–127.

Kaminski, M., 1984. Food composition of three bryozoan species (Bryozoa, Phylactolaemata) in a mesotrophic lake. Pol. Arch. Hydrobiol. 31, 45–53.

Karlson, R.H., 1992. Divergent dispersal strategies in the freshwater bryozoan *Plumatella repens*: ramet size effects on statoblast numbers. Oecologia 89, 407–411.

Kinchin, I.M., 1994. The Biology of Tardigrades. Portland Press Ltd., London, UK.

Kisielewska, G., 1982. Gastrotricha of two complexes of peat hags near Siedlce. Fragm. Faunist. 27, 39–57.

Kjellberg, G., Hessen, D.O., Nilssen, J.P., 1991. Life history, growth and production of *Mysis relicta* in the large fiord-type Lake Mjøsa, Norway. Freshwat. Biol. 26, 165–173.

Kjer, K.M., Carle, F.L., Litman, J., Ware, J., 2006. A molecular phylogeny of Hexapoda. Arthropod Syst. Phylo. 64, 35–44.

Krno, I., Sporka, F., Galas, J., Hamerlík, L., Zatovicová, S., Bitusik, P., 2006. Littoral benthic macroinvertebrates of mountain lakes in the Tatra Mountains (Slovakia, Poland). Biologia (Bratisl.) 61 (Suppl. 18), S147–S166.

Lancaster, J., Downes, B.J., 2013. Aquatic Entomology. Oxford University Press, Oxford, UK.

Lang, C., Lang-Dobler, B., 1980. Structure of tubificid and lumbriculid worm communities, and three indices of trophy based upon these communities, as descriptors of eutrophication level of Lake Geneva (Switzerland). In: Brinkhurst, R.O., Cook, D.G. (Eds.), Aquatic Oligochaete Biology. Plenum Press, New York, USA, pp. 457–470.

Larsen, S., Ormerod, S.J., 2010. Combined effects of habitat modification on trait composition and species nestedness in river invertebrates. Biol. Conserv. 143, 2638–2646.

Lasenby, D.C., Langford, R.R., 1973. Feeding and assimilation of *Mysis relicta*. Limnol. Oceanogr. 18, 280–285.

Lawton, J.H., 1999. Are there general laws in ecology? Oikos 84, 77–192.

Leibold, M.A., Holyoak, M., Mouquet, N., Amarasekare, P., Chase, J.M., Hoopes, M.F., Holt, R.D., Shurin, J.B., Law, R., Tilman, D., Loreau, M., Gonzalez, A., 2004. The metacommunity concept: a framework for multi-scale community ecology. Ecol. Lett. 7, 601–613.

Lillie, R.A., 2003. Macroinvertebrate community structure as a predictor of water duration in Wisconsin wetlands. J. Am. Water Resour. Assoc. 39, 389–400.

Lindegaard, C., 1992. Zoobenthos ecology of Thingvallavatn: vertical distribution, abundance, population dynamics and production. Oikos 64, 257–304.

Loden, M.S., 1981. Reproductive ecology of Naididae (Oligochaeta). Hydrobiologia 83, 115–123.

Lodge, D.M., Brown, K.M., Klosiewski, S.P., Stein, R.A., Covich, A.P., Leathers, B.K., Brönmark, C., 1987. Distribution of freshwater snails: spatial scale and the relative importance of physicochemical and biotic factors. Am. Malacol. Bull. 5, 73–84.

Lytle, D.A., 2015. Order Hemiptera. In: Thorp, J., Rogers, D.C. (Eds.), Ecology and General Biology: Thorp and Covich's Freshwater Invertebrates, vol. I. Academic Press, London, UK, pp. 951–963.

Lytle, D.A., Poff, N.L., 2004. Adaptation to natural flow regimes. Trends Ecol. Evol. 19, 94–100.

Mace, M.M., Rozas, L.P., 2018. Fish predation on juvenile penaeid shrimp: examining relative predator impact and size-selective predation. Estuar. Coast 41, 2128–2134.

Mackay, R.J., 1992. Colonization by lotic macroinvertebrates: a review of processes and patterns. Can. J. Fish. Aquat. Sci. 49, 617–628.

Mackay, R.J., Wiggins, G.B., 1979. Ecological diversity in Trichoptera. Ann. Rev. Entomol. 24, 185–208.

Macneale, K.H., Peckarsky, B.L., Likens, G.E., 2005. Stable isotopes identify dispersal patterns of stonefly populations living along stream corridors. Freshw. Biol. 50, 1117–1130.

MacNeil, C., Dick, J.T.A., Elwood, R.W., 1999. The dynamics of predation on *Gammarus* spp. (Crustacea: Amphipoda). Biol. Rev. 74, 375–395.

Madoni, P., 2005. Ciliated protozoan communities and saprobic evaluation of water quality in the hilly zone of some tributaries of the Po River (northern Italy). Hydrobiologia 541, 55–69.

Majdi, N., Traunspurger, W., 2015. Free-living nematodes in the freshwater food web: a review. J. Nematol. 47, 28–44.

Malecha, J., 1984. Cycle biologique de l'hirudinée rhynchobdelle *Piscicola geometra* L. Hydrobiologia 118, 237–243.

Mallwitz, J., 1984. Untersuchungen zur Ökologie litoraler Ostracoden im Schmal- und Lüttauersee (Schleswig-Holstein, BRD). Arch. Hydrobiol. 100, 311–339.

Malmquist, H.J., Ingimarsson, F., Jóhannsdóttir, E.E., Ólafsson, J.S., Gíslason, G.M., 2002. Zoobenthos on the littoral and profundal zones of four Faroese lakes. Ann. Soc. Scient. Faeroensis Suppl. 36, 79–93.

Manconi, R., Pronzato, R., 2008. Global diversity of sponges (Porifera: Spongillina) in freshwater. Hydrobiologia 595, 27–33.

Mann, K.H., Britton, R.H., Kowalczewski, A., Lack, T.J., Mathews, C.P., McDonald, I., 1972. Productivity and energy flow at all trophic levels in the River Thames, England. In: Kajak, Z., Hillbricht-Ilkowska, A. (Eds.), Productivity Problems of Freshwaters. Polish Scientific Publishers, Warszawa-Krakow, Poland, pp. 579–596.

Matteuzzo, M., Volkmer-Ribeiro, C., Varajao, A.F.D.C., Varajao, C.A.C., Alexandre, A., Guadagnin, D.L., Almeida, A.C.S., 2015. Environmental factors related to the production of a complex set of spicules in a tropical freshwater sponge. An. Acad. Bras. Ciênc. 87, 2013–2029.

May, M.L., 2019. Dispersal by aquatic insects. In: Del-Claro, K., Guillermo, R. (Eds.), Aquatic Insects: Behaviour and Ecology. Springer International Publishing, The Netherlands, pp. 35–73.

Mbaka, J.G., M'Erimba, C.M., Karanja, H.T., Mathooko, J.M., Mwaniki, M.W., 2016. Does the exclusion of meiofauna affect the estimation of biotic indices using stream invertebrates? Afr. Zool. 51, 91–97.

McCormick, P.V., Cairns Jr., J., 1991. Effects of micrometazoa on the protistan assemblage of a littoral food web. Freshwat. Biol. 26, 111–119.

McElhone, M.J., 1979. A comparison of the gut contents of two coexisting lake-dwelling Naididae (Oligochaeta), *Nais pseudobtusa* and *Chaetogaster diastrophus*. Freshwat. Biol. 9, 199–204.

McLean, K.I., Mushet, D.M., Renton, D.A., Stockwell, C.A., 2016. Aquatic-macroinvertebrate communities of prairie-pothole wetlands and lakes under a changed climate. Wetlands 36, 1–13.

McMahon, R.F., 1983. Physiological ecology of freshwater pulmonates. In: Russell-Hunter, W.D. (Ed.), The Mollusca, vol. 6. Ecology. Academic Press, Orlando, USA, pp. 359–430.

Melão, M.G.G., Rocha, O., 1998. Growth rates and energy budget of *Metania spinata* (carter 1881) (Porifera, Metaniidae) in Lagoa Dourada, Brazil. Verh. Internat. Verein. Limnol. 26, 2098–2102.

Mesquita-Joanes, F., Smith, A.J., Viehberg, F.A., 2012. The ecology of Ostracoda across levels of biological organisation from individual to ecosystem: a review of recent developments and future potential. Dev. Quat. Sci. 17, 15–35.

Michiels, I.C., Matzak, S., Traunspurger, W., 2004. Maintenance of biodiversity through predation in freshwater nematodes? Nematol. Monogr. Perspect. 2, 723–737.

Milbrink, G., 1973. On the vertical distribution of oligochaetes in lake sediments. Rep. Inst. Freshw. Res. Drottningholm 53, 34–50.

Milbrink, G., 1980. Oligochaete communities in pollution biology: the European situation with special reference to Scandinavia. In: Brinkhurst, R.O., Cook, D.G. (Eds.), Aquatic Oligochaete Biology. Plenum Press, New York, USA, pp. 433–455.

Miller, D.E., 1936. A limnological study of *Pelmatohydra* with special reference to their quantitative seasonal distribution. Trans. Am. Microsc. Soc. 55, 123–193.

Minshall, G.W., Cummins, K.W., Petersen, R.C., Cushing, C.E., Bruns, D.A., Sedell, J.R., Vannote, R.L., 1985. Developments in stream ecosystem theory. Can. J. Fish. Aquat. Sci. 42, 1045–1055.

Moens, T., Yeates, G.W., De Ley, P., 2004. Use of carbon and energy sources by nematodes. In: Cook, R., Hunt, D.J. (Eds.), Nematology Monographs and Perspectives 2. Brill Publishing, Leiden, The Netherlands, pp. 529–545.

Momot, W.T., 1984. Crayfish production: a reflection of community energetics. J. Crustacean Biol. 4, 35–54.

Momot, W.T., 1995. Redefining the role of crayfish in aquatic ecosystems. Rev. Fisheries Sci. 3, 33–63.

Moore, J.W., 1977. Importance of algae in the diet of subarctic populations of *Gammarus lacustris* and *Pontoporeia affnis*. Can. J. Zool. 55, 637–641.

Múrria, C., Dolédec, S., Papadopoulou, A., Volger, A.P., Bonada, N., 2018. Ecological constraints from incumbent clades drive trait evolution across the tree-of-life of freshwater macroinvertebrates. Ecography 41, 1049–1063.

Murtaugh, P.A., 1983. Mysid life history and seasonal variation in predation pressure on zooplankton. Can. J. Fish. Aquat. Sci. 40, 1968–1974.

Muscatine, L., Lenhoff, H.M., 1963. Symbiosis: on the role of algae symbiotic with hydra. Science 142, 956–958.

Nalepa, T.E., Schloesser, D.W., 2014. Quagga and Zebra Mussels: Biology, Impacts, and Control, second ed. CRC Press, Boca Raton, USA.

Nelson, D.R., Guidetti, R., Rebecchi, L., 2015. Phylum Tardigrada. In: Thorp, J., Rogers, D.C. (Eds.), Ecology and General Biology: Thorp and Covich's Freshwater Invertebrates. Academic Press, San Diego, USA, pp. 347–380.

Nesteruk, T., 1993. A comparison of values of freshwater Gastrotricha densities determined by various methods. Acta Hydrobiol. 35, 321–328.

Nesteruk, T., 1996. Density and biomass of Gastrotricha in sediments of different types of standing waters. Hydrobiologia 324, 205–208.

Northcote, T.G., 1991. Success, problems, and control of introduced mysid populations in lakes and reservoirs. Amer. Fish. Soc. Symposium 9, 5–16.

Nyman, M., Korhola, A., Brooks, S.J., 2005. The distribution and diversity of Chironomidae (Insecta: Diptera) in western Finnish Lapland, with special emphasis on shallow lakes. Glob. Ecol. Biogeogr. 14, 137–153.

Nyström, P., Brönmark, C., Granéli, W., 1996. Patterns in benthic food webs: a role for omnivorous crayfish? Freshwat. Biol. 36, 631–646.

Odountan, O.H., Janssens de Bisthoven, L., Abou, Y., Eggermont, H., 2019. Biomonitoring of lakes using macroinvertebrates: recommended indices and metrics for use in West Africa and developing countries. Hydrobiologia 826, 1–23.

Ohba, S.Y., 2019. Ecology of giant water bugs (Hemiptera: Heteroptera: Belostomatidae). Entomol. Sci. 22 (1), 6–20.

Ökland, J., 1964. The eutrophic Lake Borrevann (Norway)—an ecological study on shore and bottom fauna with special reference to gastropods, including a hydrographic survey. Folia Limnol. Scandinavica 13, 337.

Palmer, M.A., Bely, A.E., Berg, K.E., 1992. Response of invertebrates to lotic disturbance: a test of the hyporheic refuge hypothesis. Oecologia 89, 182–194.

Palmer, M., Ruhí, A., 2019. Linkages between flow regime, biota, and ecosystem processes: implications for river restoration. Science 365 (6459).

Pardy, R.L., White, B.N., 1977. Metabolic relationships between green *Hydra* and its symbiotic algae. Biol. Bull. 153, 228–236.

Pattee, E., Persat, H., 1978. Contribution of asexual reproduction to the distribution of triclad flatworms. Verh. Internat. Verein. Limnol. 29, 2372–2377.

Pešić, V., Chatterjee, T., Das, M.K., Bordoloi, S., 2013. A new species of water mite (Acari, Hydrachnidia) from Assam, India, found in the gut contents of the fish *Botia dario* (Botiidae). Zootaxa 3746, 454–462.

Petersen, H., Luxton, M., 1982. A comparative analysis of soil fauna populations and their role in decomposition processes. Oikos 39, 287–388.

Phipps Jr., D.W., Pardy, R.L., 1982. Host enhancement of symbiont photosynthesis in the *Hydra*-algae simbiosis. Biol. Bull. 162, 83–94.

Pieczyńska, E., 1959. Character of the occurrence of free-living Nematoda in various types of periphyton in Lake Tajty. Ekol. Polska (Ser. A) 7, 317–337.

Pieczyńska, E., 1964. Investigations on colonization of new substrates by nematodes (Nematoda) and some other periphyton organisms. Ekol. Polska (Ser. A) 12, 185–234.

Poff, N.L., Allan, J.D., Bain, M.B., Karr, J.R., Prestegaard, K.L., Richter, B.D., Sparks, R.E., Stromberg, J.C., 1997. The natural flow regime. Bioscience 47 (11), 769–784.

Poikane, S., Johnson, R.K., Sandin, L., Schartau, A.K., Solimini, A.G., Urbanic, G., Arbaciauskas, K.S., Aroviita, J., Gabriels, W., Miler, O., Pusch, M.T., Tim, H., Bohmer, J., 2016. Benthic macroinvertebrates in lake ecological assessment: a review of methods, intercalibration and practical recommendations. Sci. Total Environ. 543, 123–134.

Poinar Jr., G., 2008. Global diversity of hairworms (Nematomorpha: Gordiaceae) in freshwater. Hydrobiologia 595, 79–83.

Prejs, K., 1977a. The nematodes of the root region of aquatic macrophytes, with special consideration of nematode groupings penetrating the tissues of roots and rhizomes. Ekol. Polska 25, 5–20.

Prejs, K., 1977b. The species diversity, numbers and biomass of benthic nematodes in central part of lakes with different trophy. Ekol. Polska 25, 31–44.

Prejs, K., 1977c. The littoral and profundal benthic nematodes of lakes with different trophy. Ekol. Polska 25, 21–30.

Proctor, H.C., Garga, N., 2004. Red, distasteful water mites: did fish make them that way? In: Proctor, H.C., Garga, N. (Eds.), Experimental and Applied Acarology. Springer, Dordrecht, The Netherlands, pp. 127–147.

Proctor, H.C., Smith, I.M., Cook, D.R., Smith, B.P., 2015. Subphylum Chelicerata, class Arachnida. In: Thorp, J., Rogers, D.C. (Eds.), Ecology and General Biology: Thorp and Covich's Freshwater Invertebrates. Academic Press, San Diego, USA, pp. 599–660.

Pronzato, R., Manconi, R., 1995. Long-term dynamics of a freshwater sponge population. Freshwat. Biol. 33, 485–495.

Pyron, M., Brown, K.M., 2015. Introduction to Mollusca and the class Gastropoda. In: Thorp, J., Rogers, D.C. (Eds.), Ecology and General Biology: Thorp and Covich's Freshwater Invertebrates. Academic Press, San Diego, USA, pp. 383–421.

Quinn, B., Gagné, F., Blaise, C., 2012. *Hydra*, a model system for environmental studies. Int. J. Dev. Biol. 56, 613–625.

Raddum, G.G., Fjellheim, A., 2002. Species composition of freshwater invertebrates in relation to chemical and physical factors in high mountains in southwestern Norway. Water, Air, & Soil Pollut. 2, 311–328.

Reiss, J., Schmid-Araya, J.M., 2008. Existing in plenty: abundance, biomass and diversity of ciliates and meiofauna in small streams. Freshw. Biol. 53, 652–668.

Reynoldson, T.B., 1958. Observations on the comparative ecology of lake-dwelling triclads in southern Sweden, Finland and northern Britain. Hydrobiologia 12, 129–141.

Reynoldson, T.B., 1961. Observations on the occurrence of *Asellus* (Isopoda, Crustacea) in some lakes of northern Britain. Verh. Internat. Verein. Limnol. 14, 988–994.

Reynoldson, T.B., 1966. The distribution and abundance of lake-dwelling triclads—towards a hypothesis. Adv. Ecol. Res. 3, 1–71.

Reynoldson, T.B., 1983. The population biology of turbellaria with special reference to the freshwater triclads of the British Isles. Adv. Ecol. Res. 13, 235–326.

Ricci, C., Balsamo, M., 2000. The biology and ecology of lotic rotifer and gastrotrichs. Freshw. Biol. 44, 15–28.

Ricciardi, A., Reiswig, H.M., 1994. Taxonomy, distribution, and ecology of the freshwater bryozoans (Ectoprocta) of eastern Canada. Can. J. Zool. 72, 339–359.

Rieradevall, M., Bonada, N., Prat, N., 1999. Substrate and depth preferences of macroinvertebrates along a transect in a Pyrenean high mountain lake (Lake Redó, NE Spain). Limnética 17, 127–134.

Rosenberg, D.M., Resh, V.H. (Eds.), 1993. Freshwater Biomonitoring and Benthic Macroinvertebrates. Chapman & Hall, New York, USA.

Ruhí, A., Boix, D., Gascón, S., Sala, J., Quintana, X.D., 2013. Nestedness and successional trajectories of macroinvertebrate assemblages in man-made wetlands. Oecologia 171, 545–556.

Ruse, L., 2010. Classification of nutrient impact on lakes using the chironomid pupal exuvial technique. Ecol. Indic. 10, 594–601.

Sáinz-Bariáin, M., Zamora-Muñoz, C., Soler, J.J., Bonada, N., Sáinz-Cantero, C.E., Alba-Tercedor, J., 2016. Changes in Mediterranean high mountain Trichoptera communities after a 20-year period. Aquat. Sci. 78, 669–682.

Sakurai, M., Furuki, T., Akao, K.I., Tanaka, D., Nakahara, Y., Kikawada, T., Watanabe, M., Okuda, T., 2008. Vitrification is essential for anhydrobiosis in an African chironomid, *Polypedilum vanderplanki*. Proc. Natl. Acad. Sci. USA 105, 5093–5098.

Sandner, H., Wilkialis, J., 1972. Leech communities (Hirudinea) in the Mazurian and Bialystok regions and the Pomeranian Lake District. Ekol. Polska 20, 345–365.

Särkkä, J., Aho, J., 1980. Distribution of aquatic Oligochaeta in the Finnish lake district. Freshwat. Biol. 10, 197–206.

Sarremejane, R., Cañedo-Argüelles, M., Prat, N., Mykrä, H., Muatka, T., Bonada, N., 2017. Do metacommunities vary through time? Intermittent rivers as model systems. J. Biogeopr. 44, 2752–2763.

Sarremejane, R., Cid, N., Stubbington, R., Datry, T., Alp, M., Cañedo-Argüelles, M., Cordero-Rivera, A., Csabai, Z., Gutiérrez-Cánovas, C., Heino, J., Forcellini, M., Millán, A., Paillex, A., Paril, P., Polásek, M., Tierno de Figueroa, M., Usseglio-Polatera, P., Zamora-Muñoz, C., Bonada, N., 2020. DISPERSE, a trait database to assess the dispersal potential of European aquatic macroinvertebrates. Sci. Data 7, 386.

Sarremejane, R., Mykrä, H., Huttunen, K.-L., Mustonen, K.-R., Marttila, H., Paavola, R., Sippel, K., Veijalainen, N., Muotka, T., 2018. Climate-driven hydrological variability determines inter-annual changes in stream invertebrate community assembly. Oikos 127, 1586–1595.

Schiemer, F., 1983. Comparative aspects of food dependence and energetics of freeliving nematodes. Oikos 41, 32–42.

Schmid-Araya, J.M., Schmid, P.E., 2000. Trophic relationships: integrating meiofauna in to a realistic benthic food web. Freshw. Biol. 44, 149–163.

Schockaert, E.R., Hooge, M., Sluys, R., Schilling, S., Tyler, S., Artois, T., 2008. Global diversity of free living flatworms (Platyhelminthes, "Turbellaria") in freshwater. Hydrobiologia 595, 41–48.

Sheldon, S.P., 1987. The effects of herbivorous snails on submerged macrophyte communities in Minnesota lakes. Ecology 68, 1920–1931.

Silver, C.A., Thompson, J.E., Wong, A.S., Bayley, S.E., 2012. Relationships between wetland macroinvertebrates and waterfowl along an agricultural gradient in the Boreal Transition Zone of western Canada. Northwest. Nat 93, 40–59.

Simberloff, D., 2004. Community ecology: is it time to move on? Am. Nat. 163, 787–799.

Simpson, T.L., Gilbert, J.J., 1974. Gemmulation, gemmule hatching, and sexual reproduction in fresh-water sponges. II. Life cycle events in young, larva-produced sponges of *Spongilla lacustris* and an unidentified species. Trans. Amer. Microsc. Soc. 93, 39–45.

Smith, B.P., 1988. Host-parasite interaction and impact of larval water mites on insects. Annu. Rev. Entomol. 33, 487–507.

Smith, R.L., 1997. Evolution of paternal care in the giant water bugs (Heteroptera: Belostomatidae). In: Choe, J., Crespi, B. (Eds.), The Evolution of Social Behavior in Insects and Arachnids. Cambridge University Press, Cambridge, UK, pp. 116–149.

Smith, I.M., Cook, D.R., 1991. Water mites. In: Thorp, J.H., Covich, A.P. (Eds.), Ecology and Classification of North American Freshwater Invertebrates. Academic Press, San Diego, USA, pp. 523–592.

Smith, A.J., Horne, D.J., Martens, K., Schön, I., 2015. Class Ostracoda. In: Thorp, J., Rogers, D.C. (Eds.), Ecology and General Biology: Thorp and Covich's Freshwater Invertebrates. Academic Press, San Diego, USA, pp. 757–780.

Smith, R.L., Larsen, E., 1993. Egg attendance and brooding by males of the giant water bug *Lethocerus medius* (Guerin) in the field (Heteroptera: Belostomatidae). J. Insect Behav. 6, 93–106.

Smyth, T., Reynolds, J.D., 1995. Survival ability of statoblasts of freshwater Bryozoa found in Renvyle Lough, County Galway. Proc. Royal Irish Acad. 95B, 65–68.

Solem, J.O., Birks, H.H., 2000. Late-glacial and early-holocene Trichoptera (Insecta) from Kråkenes Lake, western Norway. J. Paleolimnol. 23, 49–56.

Soria, M., Gutiérrez-Cánovas, C., Bonada, N., Acosta, R., Rodríguez-Lozano, P., Fortuño, P., Burgazzi, G., Vinyoles, D., Gallart, F., Latron, J., Llorens, P., Prat, N., Cid, N., 2020. Natural disturbances can produce misleading bioassessment results: identifying metrics to detect anthropogenic impacts in intermittent rivers. J. Appl. Ecol. 57, 283–295.

Stankiewicz, B.A., Briggs, D.E.G., Evershed, R.P., Flanney, M.B., Wuttke, M., 1997. Preservation of chitin in 25-million-year-old fossils. Science 276, 1541–1543.

Statzner, B., 2008. How views about flow adaptations of benthic stream invertebrates changed over the last century. Int. Rev. Hydrobiol. 93, 593–605.

Statzner, B., Bis, B., Dolédec, S., Usseglio-Polatera, P., 2001. Perspectives for biomonitoring at large spatial scales: a unified measure for the functional composition of invertebrate communities in European running waters. BAAE 2, 73–85.

Stein, R.A., Goodman, C.G., Marschall, E.A., 1984. Using time and energetic measures of cost in estimating prey value for fish predators. Ecology 65, 702–715.

Strong, E.E., Gargominy, O., Ponder, W.F., Bouchet, P., 2008. Global diversity of gastropods (Gastropoda; Mollusca) in freshwater. Hydrobiologia 595, 149–166.

Stocchino, G.A., Sluys, R., Kawakatsu, M., Sarbu, S.M., Manconi, R., 2017. A new species of freshwater flatworm (Platyhelminthes, Tricladida, Dendrocoelidae) inhabiting a chemoautotrophic groundwater ecosystem in Romania. EJT 342, 1–21.

Stoks, R., Geerts, A.N., De Meester, L., 2014. Evolutionary and plastic responses of freshwater invertebrates to climate change: realized patterns and future potential. Evol. Appl. 7, 42–55.

Strachan, S.R., Chester, E.T., Robson, B.J., 2015. Freshwater invertebrate life history strategies for surviving desiccation. Sci. Rev. 3, 57–75.

Strayer, D., 1985. The benthic micrometazoans of Mirror Lake, New Hampshire. Arch. Hydrobiol. Suppl. 72, 287–426.

Strayer, D.L., Hummon, W.D., 1991. Gastrotricha. In: Thorp, J.H., Covich, A.P. (Eds.), Ecology and Classification of North American Freshwater Invertebrates. Academic Press, San Diego, USA, pp. 173–185.

Strekal, T.A., McDiffett, W.F., 1974. Factors affecting germination, growth, and distribution of the freshwater sponge, *Spongilla fragilis* Leidy (Porifera). Biol. Bull. 146, 267–278.

Strong, E.E., Gargominy, O., Ponder, W.F., Bouchet, P., 2007. Global diversity of gastropods (Gastropoda; Mollusca) in freshwater. Hydrobiologia 595, 149–166.

Stubbington, R., 2012. The hyporheic zone as an invertebrate refuge: a review of variability in space, time, taxa and behaviour. Mar. Freshwater Res. 63, 293–311.

Sueur, J., Mackie, D., Windmill, J.F., 2011. So small, so loud: extremely high sound pressure level from a pygmy aquatic insect (Corixidae, Micronectinae). PLoS One 6, e21089.

Sutcliffe, D.W., Carrick, T.R., Willoughby, L.G., 1981. Effects of diet, body size, age and temperature on growth rates in the amphipod *Gammarus pulex*. Freshwat. Biol. 11, 183–214.

Swan, C.M., Palmer, M.A., 2000. What drives small-scale spatial patterns in lotic meiofauna communities? Freshw. Biol. 44, 109–121.

Teraguchi, S., 1975. Correction of negative buoyancy in the phantom larva, *Chaoborus americanus*. J. Insect Physiol. 21, 1659–1670.

Thibaud, J.M., 1970. Biologie et écologie des collemboles Hypogastruridae édaphiques et cavernicoles. Mem. Mus. Natl. Hist. Nat. Ser. A Zool. 61, 83–201.

Thorp, J.H., Barthalmus, G.T., 1975. Effects of crowding on growth rate and symbiosis in green hydra. Ecology 56, 206–212.

Tierno de Figueroa, J.M., López-Rodríguez, M., Lorenz, A., Graf, W., Schmidt-Kloiber, A., Hering, D., 2009. Vulnerable taxa of European Plecoptera (Insecta) in the context of climate change. Biodivers. Conserv. 19, 1269–1277.

Tierno de Figueroa, J.M., Luzón-Ortega, J.M., López-Rodríguez, M.J., 2019. Drumming for love: mating behavior in stoneflies. In: Del-Claro, K., Guillermo, R. (Eds.), Aquatic Insects: Behaviour and Ecology. Springer International Publishing, The Netherlands, pp. 117–137.

Timm, T., Martin, P.J., 2015. Clitellata: Oligochaeta. In: Thorp, J., Rogers, D.C. (Eds.), Ecology and General Biology: Thorp and Covich's Freshwater Invertebrates. Academic Press, San Diego, USA, pp. 529–549.

Toman, M.J., Dall, P.C., 1997. The diet of *Erpobdella octoculata* (Hirudinea: Erpobdellidae) in two Danish lowland streams. Arch. Hydrobiol. 140, 549–563.

Tonkin, J.D., Altermatt, F., Finn, D.S., Heino, J., Olden, J.D., Pauls, S.U., Lytle, D.A., 2018. The role of dispersal in river network metacommunities: patterns, processes, and pathways. Freshw. Biol. 63, 141–163.

Tonkin, J.D., Bogan, M.T., Bonada, N., Ríos-Touma, B., Lytle, D.A., 2017. Seasonality and predictability shape temporal species diversity. Ecology 98, 1201–1216.

Tornero, I., Boix, D., Bagella, S., Pinto-Cruz, C., Caria, M.C., Belo, A., Lumbreras, A., Sala, J., Compte, J., Gascón, S., 2018. Dispersal mode and spatial extent influence distance-decay patterns in pond metacommunities. PLoS One 13, e0203119.

Traunspurger, W., 1996. Distribution of benthic nematodes in the littoriprofundal and profundal of an oligotrophic lake (Königssee, National Park Berchtesgaden, FRG). Arch. Hydrobiol. 135, 557–575.

Traunspurger, W., Drews, C., 1996. Vertical distribution of benthic nematodes in an oligotrophic lake: seasonality, species and age segregation. Hydrobiologia 331, 33–42.

Traunspurger, W., Michiels, I.C., Abebe, E., 2006. Composition and distribution of free-living freshwater nematodes; global and local perspectives. In: Wilson, M.J., Kakouli-Duarte, T. (Eds.), Nematodes as Environmental Indicators. CAB International, Oxford/Cambridge, UK, pp. 146–171.

Trigal, C., García-Criado, F., Fernández-Aláez, C., 2006. Among-habitat and temporal variability of selected macroinvertebrate metrics in a Mediterranean shallow lake (NW Spain). Hydrobiologia 563, 371–384.

Uutala, A.J., 1981. Composition and secondary production of the chironomid (Diptera) communities in two lakes in the Adirondack mountain region. In: Singer, R. (Ed.), Effect of Acidic Precipitation on Benthos. North American Benthological Society, Hamilton. New York, USA, pp. 139–154.

Van Doorslaer, W., Stoks, R., Swillen, I., Feuchtmayr, H., Atkinson, D., Moss, B., De Meester, L., 2010. Experimental thermal microevolution in community-embedded *Daphnia* populations. Clim. Res. 43, 81–89.

van Hardenbroek, M., Heiri, O., Grey, J., Bodelier, P.L.E., Verbruggen, F., Lotter, A.F., 2009. Fossil chironomid δ13C as a proxy for past methanogenic contribution to benthic food webs in lakes? J. Paleolimnol. 43, 235–245.

Vannote, R.L., Minshall, G.W., Cummins, K.W., Sedell, J.R., Cushing, C.E., 1980. The river continuum concept. Can. J. Fish. Aquat. Sci. 37, 130–137.

Vellend, M., 2010. Conceptual synthesis in community ecology. Q. Rev. Biol. 85, 183–206.

Vellend, M., Geber, M.A., 2005. Connections between species diversity and genetic diversity. Ecol. Lett. 8, 767–781.

Viana, D.S., Chase, J.M., 2019. Spatial scale modulates the inference of metacommunity assembly processes. Ecology 100, e02576.

Vila-Farré, M., Rink, J.C., 2018. The ecology of freshwater planarians. In: Rink, J.C. (Ed.), Planarian Regeneration: Methods and Protocols. Springer, New York, USA, pp. 173–205.

Vogel, S., 1994. Life in Moving Fluids: The Physical Biology of Flow, second ed. Princeton University Press, Princeton, USA.

Wallace, J.B., Anderson, N.H., 1996. Habitat, life history and behavioral adaptations of aquatic insects. In: Merritt, R.W., Cummins, K.W. (Eds.), An Introduction to the Aquatic Insects of North America. Kendall/Hunt Publishing Company, Dubuque, USA, pp. 41–73.

Wallace, J.B., Merritt, R.W., 1980. Filter-feeding ecology of aquatic insects. Ann. Rev. Entomol. 25, 103–132.

Ward, J.V., 1992. Aquatic Insect Ecology: I. Biology and Habitat. John Wiley & Sons, New York, USA.

Ward, J.V., Stanford, J.A., 1983. The serial discontinuity concept of lotic ecosystems. In: Fontained, T.D., Bartell, S.M. (Eds.), Dynamics of Lotic Ecosystems. Ann Arbor Science Publishers, Ann Arbor, USA, pp. 29–42.

Watanabe, M., 2006. Anhydrobiosis in invertebrates. Appl. Entomol. Zool. 41, 15–31.

Waterkeyn, A., Grillas, P., Vanschoenwinkel, B., Brendonck, L., 2008. Invertebrate community patterns in Mediterranean temporary wetlands along hydroperiod and salinity gradients. Freshw. Biol. 53, 1808–1822.

Wellborn, G.A., Witt, J.D., Cothran, R.D., 2015. Class Malacostraca, superorders Peracarida and Syncarida. In: Thorp, J., Rogers, D.C. (Eds.), Ecology and General Biology: Thorp and Covich's Freshwater Invertebrates. Academic Press, San Diego, USA, pp. 781–796.

Wells, L., 1968. Daytime distribution of *Pontoporeia affinis* off bottom in Lake Michigan. Limnol. Oceanogr. 13, 703–705.

Whitledge, G.W., Rabeni, C.F., 1997. Energy sources and ecological role of crayfishes in an Ozark stream: insights from stable isotopes and gut analysis. Can. J. Fish. Aquat. Sci. 54, 2555–2563.

Wiederholm, T., 1980. Use of benthos in lake monitoring. J. Water Poll. Control Fed. 52, 537–547.

Wiegmann, B.M., Trautwein, M.D., Winkler, I.S., Barr, N.B., Kim, J.-W., Lambkin, C., Bertone, M.A., Cassel, B.K., Bayless, K.M., Heimberg, A.M., Wheeler, B.M., Peterson, K.J., Pape, T., Sinclair, B.J., Skevington, J.H., Blagoderov, V., Caravas, J., Kutty, S.N., Schmidt-Ott, U., Kampmeier, G.E., Thompson, F.C., Grimaldi, D.A., Beckenbach, A.T., Courtney, G.W., Friedrich, M., Meier, R., Yeates, D.K., 2011. Episodic radiations in the fly tree of life. Proc. Natl. Acad. Sci. USA 108, 5690–5695.

Wiggins, G.B., 1977. Larvae of the North American Caddisfly Genera. University Toronto Press, Toronto, Canada.

Wikelski, M., Moskowitz, D., Adelman, J.S., Cochran, J., Wilcove, D.S., May, M.L., 2006. Simple rules guide dragonfly migration. Biol. Lett. 2, 325–329.

Will, K.W., Resh, V.H., 2008. Phylogenetic relationships and evolutionary adaptations of aquatic insects. In: Merritt, R.W., Cummins, K.W., Berg, M. (Eds.), An Introduction to the Aquatic Insects of North America, fourth ed. Kendall/Hunt, Dubuque, UAS, pp. 139–156.

Willoughby, L.G., Marcus, J.H., 1979. Feeding and growth of the isopod *Asellus aquaticus* on actinomycetes, considered as model filamentous bacteria. Freshwat. Biol. 9, 441–449.

Winberg, G.G. (Ed.), 1971. Methods for the Estimation of Production of Aquatic Animals. Academic Press, New York, USA.

Winterbourn, M.J., Anderson, N.H., 1980. The life history of *Philanisus plebeius* (Trichoptera: Chathamiidae), a caddisfly whose eggs were found in a starfish. Ecol. Entomol. 5, 293–304.

Wood, T.S., 2015. Phyla Ectoprocta and Entoprocta (bryozoans). In: Thorp, J., Rogers, D.C. (Eds.), Ecology and General Biology: Thorp and Covich's Freshwater Invertebrates. Academic Press, San Diego, USA, pp. 327–345.

Wood, T.S., 2019. Phylum Ectoprocta. In: Thorp, J., Rogers, D.C. (Eds.), Ecology and General Biology: Thorp and Covich's Freshwater Invertebrates. Academic Press, San Diego, USA, pp. 519–529.

Wood, T.S., Okamura, B., 2017. New species, genera, families, and range extensions of freshwater bryozoans in Brazil: the tip of the iceberg? Zootaxa 4306, 383–400.

Woodward, G., Perkins, D.M., Brown, L.E., 2010. Climate change and freshwater ecosystems: impacts across multiple levels of organization. Philos. Trans. R. Soc. Lond. B Biol. Sci. 365, 2093–2106.

Young, J.O., 1973. The prey and predators of *Phaenocora typhlops* (Vejdovsky) (Turbellaria: Neorhabdocoela) living in a small pond. J. Anim. Ecol. 42, 637–643.

Young, J.O., Seaby, R.M.H., Martin, A.J., 1995. Contrasting mortality in young freshwater leeches and triclads. Oecologia 101, 317–323.

Zimmer, K.D., Hanson, M.A., Butler, M.G., 2000. Factors influencing invertebrate communities in prairie wetlands: a multivariate approach. Can. J. Fish. Aquat. Sci. 57, 76–85.

CHAPTER

22

Fish

Erik Jeppesen[1,2,3,4,5], *Pietro Volta*[6] *and Zhigang Mao*[7]

[1]Department of Ecoscience, Aarhus University, Aarhus, Denmark [2]Sino-Danish College, University of Chinese Academy of Sciences, Beijing, P.R. China [3]Limnology Laboratory, Department of Biological Sciences and Centre for Ecosystem Research and Implementation, Middle East Technical University, Ankara, Turkey [4]Institute of Marine Sciences, Middle East Technical University, Erdemli-Mersin, Turkey [5]Institute for Ecological Research and Pollution Control of Plateau Lakes, School of Ecology and Environmental Sciences, Yunnan University, Kunming, China [6]CNR-IRSA Water Research Institute, Verbania Pallanza, Italy [7]State Key Laboratory of Lake Science and Environment, Nanjing Institute of Geography and Limnology, Chinese Academy of Sciences, Nanjing, China

OUTLINE

I. Introduction

Fish are the vertebrates that account for the most species (Mayr, 1963)—15,200 freshwater and ~14,800 marine species have been described (Carrete Vega and Wiens, 2012; Tedesco et al., 2017), compared with 8600 bird species, 6000 reptiles, 4500 mammals, and 2500 amphibians. Furthermore, every year many new fish species are discovered and described, and others are still to be found. Compared with terrestrial tetrapods (i.e., animals having four limbs), fish are at the base of the evolutionary scale of vertebrates, where they occupy the oldest position (c. 400 million years) (Fig. 22-1).

Fish exhibit an astonishing variability in shape and, during the long history of evolution, they have colonized very diverse and remote habitats: from deep ocean depths

Wetzel's Limnology, Fourth Edition
https://doi.org/10.1016/B978-0-12-822701-5.00022-7

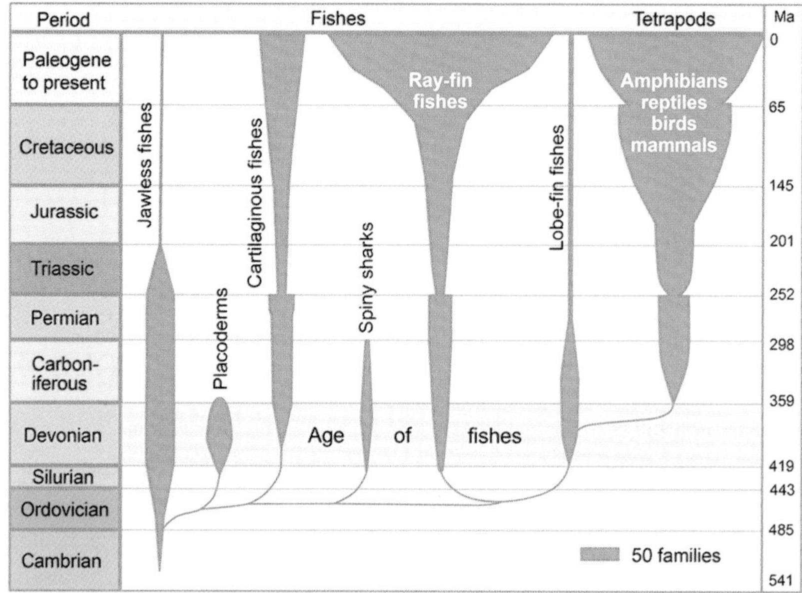

FIGURE 22-1 Evolution of fishes from the Cambrian to the present as a spindle diagram. The widths of the spindles are proportional to the number of families as an approximate estimate of diversity. *(From Benton, 2005.)*

where the pressure reaches thousands of kg cm^{-2} to watercourses and lakes of the Andean Plateau at 5000 m altitude, and from the icy waters of Antarctica to hot springs where the water temperature can reach up to 45°C.

Despite the huge water volume of seas (c. 97% of the total amount of water available on Earth), only 49% of the known fish species live in the sea, whereas 51% live in lakes and rivers that constitute only 0.0093% of the water on Earth. The great fragmentation of freshwater environments, as well as the isolation that occurred during Earth's evolution, led to remarkable species differentiation and endemism.

Being adapted to extreme abiotic and biotic environmental conditions, fish developed greatly different shapes and sizes. The smallest fish known is the cyprinid *Paedocypris progenetica,* with a maximum size of about 1 cm, whereas the largest is the whale shark (*Rhincodon typus*), whose maximum length is about 18 meters.

A. Systematics

Within the Animalia Realm, fish belong to the phylum Chordata and the subphylum Vertebrata, and the fish living nowadays can be divided into the following three classes:

Agnatha: Jawless and cartilaginous eel-like fishes, representing the first vertebrate stock whose line of evolution diverged from that leading to gnathostomes (animals with jaws) at least 500 million years ago. Apart from having no jaws, extant agnathans are characterized by lacking paired appendages. Most of them do have a caudal fin. Fish species included in this group are cyclostomes (e.g., lampreys and hagfish) and the extinct ostracoderms (e.g., Pteraspidomorphi, Thelodonti, Anaspida, and Cephalaspidomorphi).

Chondrichthyes: these fishes have a cartilaginous skeleton, jaws (gnathostomes), cartilaginous branchial arches that are not fully connected to the skull, paired fins that have a stabilizing function while swimming, and two nasal cavities. Chondrichthyes comprises the subclasses of Elasmobranchs (sharks, fin rays, skates, sawfishes) and Holocephalans (rat fishes of the genus Chimaera and the elephant fishes of the genus Callorhinchus).

Osteichthyes: these fishes have a bony skeleton, jaws and branchial arches at least partly ossified but not fully connected to the skull, paired fins, and two nasal cavities. Osteichthyes comprises the subclasses of Sarcopterygii (the basal part of the paired fins has a structure very similar to that of limbs with endoskeletal elements and muscles) and Actinopterygii (paired fins without the aforementioned basal structure). Actinopterygii includes the majority of the freshwater fish families such as Cichlidae, Characidae, Salmonidae, Percidae, Centrarchidae, and Cyprinidae.

B. Morphology and general anatomy of fishes

The features of the liquid medium, that is, density, incompressibility, and viscosity, determine the main morphological characteristics of fish (Gosline, 1971; Lagler et al., 1977).

Water and cellular protoplasm have approximately the same density, and a fish of a certain size has a body mass lower than an organism of the same size on land.

Therefore the skeleton of fish plays a limited role in supporting the body mass; it is simple and light. This is particularly true for cartilaginous fish such as sharks, which have a cartilage density of 1.1 g cm^{-3} and a bone density of approximately 2.0 g cm^{-3}. In bony fish the swim bladder counterbalances the higher density of the skeleton structure that is often needed to support the powerful musculature required for swimming.

Water has low compressibility and high viscosity (Chapter 3). Therefore fish generally have a hydrodynamic shape to reduce, as much as possible, the volume of water moved during swimming and when facing turbulence (Fig. 22-2). As in other vertebrates, fish have a *bilateral symmetry*; that is, the left and right sides of the body mirror each other. Furthermore, the head, the trunk, and the tail form a compact unit with the fins as the only appendices. Schematically, however, the body of a fish can be divided into three regions: the head (from the anterior apex of the muzzle to the posterior limit of the gill arcs), trunk (from the posterior limit of the gill arcs to the anal opening), and the tail (from the anal opening to the end of the caudal fin). Fins are a common structure that almost all fish species possess. All fins have two skeleton components: a supporting endoskeletal structure and an external structure (rays). The unpaired fins, such as the *dorsal fin* (often divided into two portions of which the first is spiny and the other soft), the *caudal fin*, and the *anal fin*, have great importance for the propulsion, the setting of the movement, and the vertical balance of the moving mass. In some fish (for instance, those belonging to the Salmonidae and Characidae families), there is also a small fin called the *adipose fin*, without a ray and apparently without any directional and motor function. The paired fins are the *pectoral* and the *ventral fins* located in the anterior and posterior ends of the fish body, respectively. Both have directional, lifting, and slow-motion functions that help the fish stabilize the body rolling.

The musculature of fish can be divided into three categories: the smooth muscle of the intestine, the heart muscle, and the dominant striated muscles. Fish swim with two types of striated skeletal muscle: *aerobic red muscle* and *anaerobic white muscle*. Aerobic red muscles, which typically maximally represent only about 10% of the total muscle mass, are used for slow to moderate, steady swimming. Anaerobic white muscles form the majority of the myotomal muscle blocks in most fishes, as well as the greatest proportion of body mass (sometimes over 50%), and are used for brief bursts of high-speed swimming.

The fish body is covered by skin that forms a continuous outer layer (transparent over the eyes). The skin is rich in *chromatophores* (staining cells) and cells that secrete mucus, a substance that plays an important protective function. In special pockets that open in the lower layer of the skin (*endoderm*), the scales are formed, that is, small strips of bone tissue and dentine, which cover the body except for the head and fins. Not all species of fish have scales; for example, lampreys, chimeras, and some species belonging to the family Siluridae.

Gills are the primary body organ where the oxygen dissolved in the water enters the bloodstream and from where carbon dioxide is released. Additionally, some fish species such as catfish (*Ameiurus* spp.) and eel (*Anguilla* spp.) can also exchange gas with water through the skin (Hoar et al., 1984a) or though lung-like organs of varying design (e.g., Polypteridae, Osphronemidae).

On each side of the head, the *bony fishes* have a single gill opening that is covered by a bony formation called the *gill operculum*. When the operculum is raised, there is a branchial chamber on each side with four bony arches carrying the breathing elements themselves. The respiratory elements are composed of the four bony branchial arcs, each carrying two orders of *respiratory lamellae*. This structure is connected to the musculature that allows

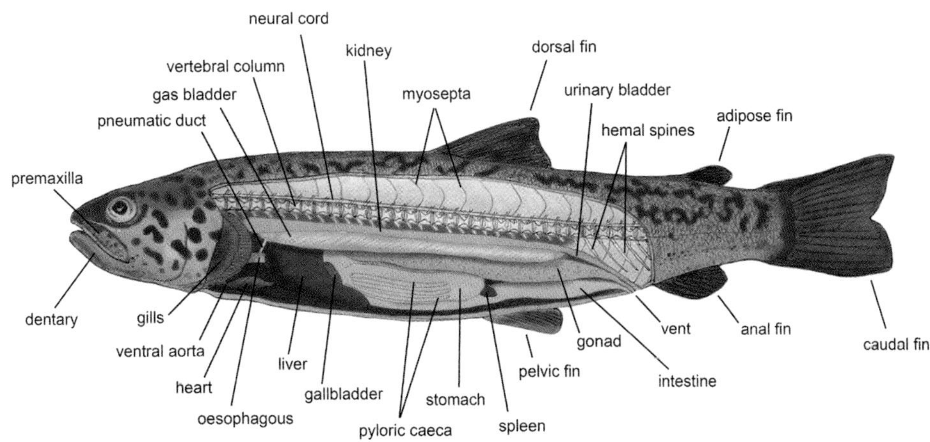

FIGURE 22-2 Generalized overview of the anatomy of a trout (*Salmo marmoratus*). *(Courtesy of Dr. Gianluca Polgar.)*

the movement of the gills themselves, facilitating the water passage and the elimination of residues accumulated within the branchial filaments. The number and total area of filaments are, in general, positively correlated with the oxygen demand of the fish.

The uptake of O_2 and the release of CO_2 are also facilitated by the particular countercurrent system of gills. Indeed, blood in the filament runs against the water flux. This enables the maintenance of an active and constant pressure gradient between the O_2 and CO_2 dissolved in the water and the blood, facilitating the exchange of the two gases between the two environments. Fish breathe molecular O_2 dissolved in the water, the concentration of which decreases with increasing temperature (see Chapter 11). For instance, 1 L of freshwater at 5°C contains about 9 mL of O_2 and about 6 mL at 25°C. Fish use *hemoglobin* to retain O_2 in the blood, where it is up to 25 times more concentrated than in the water. The capacity of hemoglobin to take up and retain the dissolved oxygen or release it depends on the *saturation tension*; saturation tension, in the case of retaining O_2, is the gas pressure at which 95% of the hemoglobin is saturated by oxygen (species-specific characteristic) and, in the case of releasing O_2, it corresponds to 50% of the saturation. Thus the saturation of hemoglobin is 95% at O_2 retention and 50% at O_2 release. The saturation tension is positively correlated with the water temperature; therefore an increase in water temperature leads to an increase in the saturation tension and, consequently, a higher O_2 concentration (the partial pressure in the medium) is required for the O_2 uptake and retention by hemoglobin. The saturation tension is also positively correlated with the CO_2 concentration. Therefore, in those environments where CO_2 reaches high concentrations, there is a risk of fish death because the partial pressure of O_2 in the water is not sufficient to support hemoglobin saturation (because the saturation pressure is higher than normal). In some cases fish could die even if the O_2 concentration in the water is relatively high.

Gills are also the body organ where multiple and intense exchanges of elements and chemical compounds between fish and water occur. Nitrogen compounds, such as ammonia, are excreted through the gills. Such exchanges are very important and complex when fish move between salt and freshwater, and so relatively few species are able to make such movements (i.e., Salmonids) (e.g., Bystriansky et al., 2007; Bystriansky and Schulte, 2011).

Fish growth is a continuous process and the fish length increases during the entire life time of the fish, although the length increments are higher in the first periods of life than once the fish is older. Also, the growth speed depends on water temperature. Fish inhabiting environments with a significant seasonal temperature variation tend to exhibit slow growth in the coldest periods and accelerated growth in the warmer ones. In contrast, fish living in environments with more stable temperatures (such as the tropics) generally show a more continuous growth during the year. Secondary factors, such as reproduction, food availability, and diseases, also influence fish growth. The changes in growth speed, depending on the metabolism rate, can be visually observed in all bony structures of a fish, such as scales, vertebrae, rays, and *otoliths* (otoliths are biomineralized ear stones that contribute to both hearing and vestibular function in fish), as less or more dark patterns, corresponding to more or less condensation of body material. The clearest areas reflect the fastest growth (more diluted bony material) and the dark areas the slowest growth (denser bony material). These records of growth are invaluable for the determination of fish age end reconstruction of body growth trajectories with time.

Weight—length relationships (WLRs, Fig. 22.3) are used for estimating the weight corresponding to a given length, whereas *condition* factors are used for comparing the "condition," "fatness," or "well-being" (Tesch, 1968; Froese, 2006) of fish based on the assumption that heavier fish of a given length are in better condition.

Most fish produce offspring by eggs that are released into the water by females and fecundated by male sperm (Hoar et al., 1984b). However, some fishes are *hermaphroditic* with internal self-fertilization, and some fishes have other rare types of reproduction (Kuwamura et al., 2020); for instance, *Poecilia reticulata* produces juvenile fish (*viviparity*) directly to the water. The fecundity of

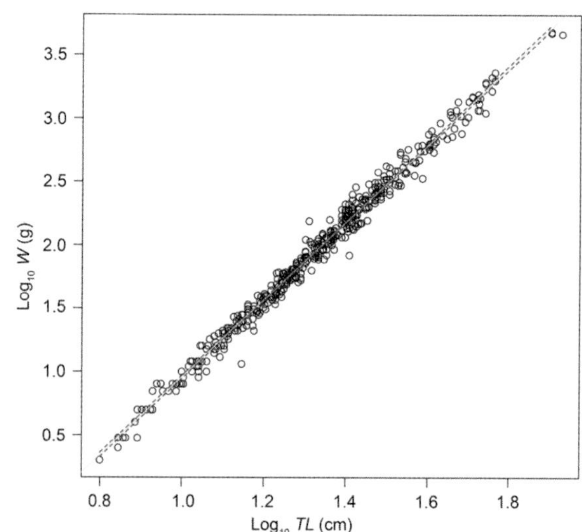

FIGURE 22-3 Linear regression (*grey line*) of the logarithms of body mass (response variable, *W*) on the logarithms of length (explanatory variable, *TL*) of *Salmo marmoratus* (*marble trout*) inhabiting River Toce (North Italy); *dashed lines*: $CI_{95\%}$; adjusted $r^2 = 0.988$, $p < 0.001$, $n = 451$). Equation: $\log_{10}(W) = 3.055 \cdot \log_{10}(TL) - 2.100$ (Polgar et al., 2023).

fish is in general higher than that of terrestrial vertebrates. Seawater fish have the highest absolute fecundity; for instance, the eggs of clupeoids are free floating in the water and those of phytophilic species are released among vegetation, whereas some fish hide their eggs (under sand, gravel, etc.) or protect them (in nests that are actively defended). *Fecundity* can be expressed as an absolute value (number of eggs per fish) or relative value (number of eggs per unit of weight of the fish). The size of the eggs may differ significantly depending on species strategy (*r* and *k*). *K strategists* invest a lot of resources into the survival of a small number of their embryos and larvae and therefore produce a low number of large (yolk-rich) eggs (the eggs of the *Salmo* genus have a diameter of 7–8 mm and the relative fecundity is about 1500–2000 eggs per kg). In contrast, *r strategists* produce a large number of very small eggs (up to 600,000–700,000 per kg of fish biomass).

Most fish are *poikilotherms* and their metabolism and all physiological processes are strongly temperature dependent. The age of sexual maturity of a fish depends on its maximum size that, in general, is negatively correlated with water temperature. In contrast, the speed of fish growth depends positively on temperature, at least up to a certain threshold. In general: (a) small fish species reach sexual maturity before large species; and (b) within the same species, fish in warmer environments (such as shallow lakes or lakes at lower latitudes) tend to grow faster and mature and reproduce earlier than their conspecifics in colder environments. Even fish mortality is highly temperature dependent and, usually, long-living fish species inhabit cold waters and short-living species warm water.

It must be taken into account that, besides intrinsic factors such as species-specific fecundity, natality, and mortality, also extrinsic factors such as food availability, competition with other fish populations, extreme meteorological events, and fishing may contribute to the overall population dynamics of a fish species.

II. Fish communities in natural lakes and streams

Fish of inland waters inhabit an extraordinary variety of habitats spanning from temperate and Arctic to tropical lakes and reservoirs, and from cold water to tropical rivers, desert waters, and caves (Moyle and Cech, 1982).

Biological and physical factors affect differences among and within fish communities (e.g., Schlosser, 1987) at different spatial and temporal scales (Tonn, 1990). On a local and regional scale, physicochemical variables (e.g., Carol et al., 2006; Eckmann, 1995) such as dissolved oxygen as well as biological factors such as predation (Moyle and Vondracek, 1985) and competition (Ross et al., 1985) may interact in shaping fish communities. On a larger spatial scale, however, other factors such as habitat size (Oberdorff et al., 1995; Zhao et al., 2006) or climate (Brucet et al., 2013), together with speciation and dispersal rates (Griffiths, 2006), are major determinants modulating the strength of local-scale factors. Finally, fish communities are susceptible to "disruption" by artificial dispersal in the form of accidental and intentional introductions of nonnative (exotic or invasive) fish species.

Climate, in particular, has been shown to correlate strongly with fish community richness, diversity, density, and size structure. Fish species richness is often higher in warm lakes and streams (González-Bergonzoni et al., 2012; Brucet et al., 2013; Griffiths, 1997; Mandrak, 1995). The same pattern of higher species richness at warmer locations has been found in altitudinal studies worldwide (Amarasinghe and Welcomme, 2002). Investigations suggest that littoral and benthic production promotes fish diversity in lakes (Vander Zanden et al., 2011), which might contribute to the observed higher diversity in the typically plant-associated fish assemblages in warm lakes (Conrow et al., 1990; Delariva et al., 1994; Meerhoff et al., 2007; Teixeira-de Mello et al., 2009).

Secondly, the fish assemblages in many warm lakes, independent of trophic state, are often more functionally diverse (Moss, 2010) and dominated by omnivorous species (Lazzaro, 1997; Teixeira-de Mello et al., 2009) (Fig. 22-4).

Piscivores are species that primarily feed on other fish as adults, *invertivores* feed primarily on benthos or zooplankton, *herbivores* solely on plant or algal material, and *detritivores* instead feed on detritus. *Omnivores* take food more or less equally from at least two trophic levels. Trends with latitude in piscivores, omnivores, and herbivores, but not invertivores, are statistically significant at $p < 0.001$ (regression analysis) (Fig. 22-4 from Moss, 2010).

The biomass and density of fish assemblages tends to increase with increasing ambient temperature, as found in a series of shallow lakes of varying nutrient concentrations along a climate gradient in Europe (Gyllström et al., 2005), and supported by cross-comparison studies of subtropical and temperate lakes (Meerhoff et al., 2007, 2012; Teixeira-de Mello et al., 2009).

A. Fish communities in streams and rivers

The development of the fish fauna in running waters depends on the ability of the fish to cope with watercourse characteristics from upstream to downstream river stretches such as current, turbulence, oxygen concentration, water temperature, and, of course, energy (food) availability (Oberdorff et al., 1995).

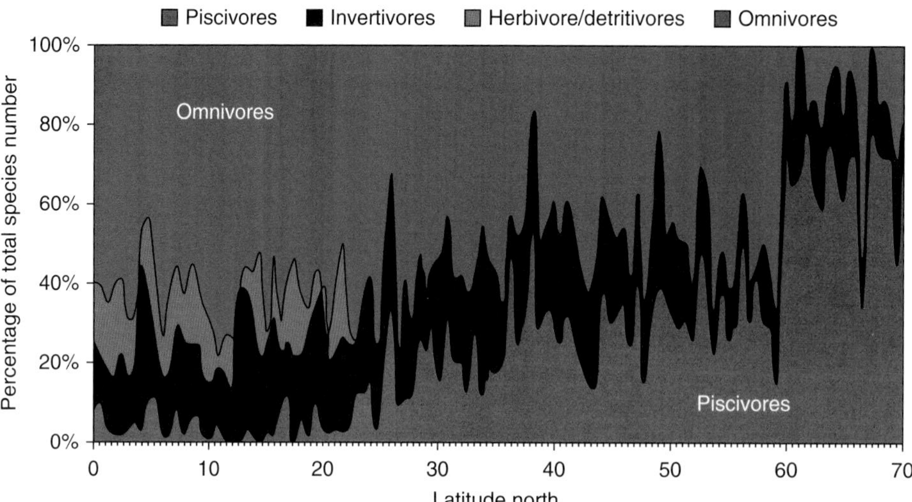

FIGURE 22-4 Changes in the percentage composition of fish communities by trophic category along a latitudinal gradient in South, Central, and North America. Species lists were gleaned from the literature for 120 lake and river sites and food preferences allocated from FishBase and other available literature. The trends were very similar for both rivers and lakes, so data have been combined. *(From Moss, 2010.)*

Along with evidence of changes in fish community composition from upstream to downstream habitats, rivers have been classified into different zones. The best-known river fish zonation is one proposed for European rivers by Huet (1949), who recognized, from upstream to downstream, the *trout zone*, the *grayling zone*, the *barbel zone*, and the *bream zone*, which differ in species richness (increasing from upstream to downstream) and composition (from cold water and sensitive species to warm water and tolerant ones).

Upstream, and often at high altitudes, environments are generally characterized by strong currents, rocky substrates, and high levels of O_2 due to lower water temperatures and/or turbulence induced by the steep slope. Fish generally have a high metabolic rate and a consequently high oxygen demand. Food productivity in upstream environments is relatively poor due to the nutrient-poor conditions. The *productivity/respiration ratio* (P/R) is usually less than 1. Benthic algae such as some diatoms are at the base of the food web together with other food (plant material, insects) derived from the catchment. Fish living in upstream environments are adapted to scraping off diatoms and other attached material clinging to rocks (*Aufwuchs*) (Chapter 25). They often exhibit flattened ventral surfaces and expanded pectoral and pelvic fins to form an effective sucker to remain well anchored to the river bed, where they can feed on the attached epilithon. For instance, stream loaches in Asia and sculpins in North America and in Europe inhabit upstream environments alone or together with salmonids, such as trout and charr, as the latter are able to swim in strong currents and forage on insects.

Changes in stream characteristics, such as a decrease of turbulence due to a less steep stream slope and increased water discharge and food availability, are usually followed by changes in the fish community with increased occurrence of loaches of the genus *Nemacheilus* and catfishes of the family Sisoridae and minnows such as *Labocheilus* spp. Barbels (*Barbus* spp.) are also typical of the lower stretches of this zone in both Europe and in Asia.

Along with changes in morphology (an even lower stream slope, presence of plants, fine substrate) and water physicochemical characteristics (higher water temperature and lower oxygen concentrations, turbidity), the fish community also changes (e.g., Haidvogl et al., 2015). The downstream environment provides opportunities for fish species less dependent upon high concentrations of O_2 and low temperatures, and it permits the colonization of fish species that have deeper and less well-streamlined bodies and a weaker ability to sustain locomotion. Typical fish species of these types of rivers in Europe are those belonging to the family of Cyprinidae, such as carp, tench, chub, and bream. Catfish (bullheads [*Ictalurus* spp.]) and suckers (*Catostomus* spp.) are instead quite typical of lower watercourses in North America. Most riverine fish species are very mobile and can migrate for long distances in order to find suitable conditions for spawning and feeding. For this reason, the disruption of connectivity is a major environmental problem for fish fauna in many rivers.

B. Fish communities in lakes

As for running waters, fish community composition, richness, and diversity in lakes are shaped by different factors acting at different spatial and temporal scales such as lake productivity (energy availability), water temperature, and oxygen, as well as the hydrological connectivity among the lake and its catchment, lake size, volume, and depth (Jeppesen et al., 2000; Mehner et al., 2005, 2007; Volta et al., 2011; Brucet et al., 2013).

In temperate lakes a simple but still effective classification of the fish community includes three typologies: fish communities of deep lakes, shallow lakes, and high altitude (high-latitude) lakes. Fish communities of deep lakes in temperate regions can be approximately separated into two different fish populations: an open-water population, characterized mainly by *planktivorous* fish such as shad (*Alosa* spp.) and whitefish (*Coregonus* spp.) and *predators* such as trout and salmon; and an inshore (littoral and sublittoral) population, constituted mainly by percids, centrarchids, and cyprinids. In deep stratified oligotrophic lakes usually cold and *stenothermal* species, such as salmonids, inhabit the hypolimnion and the metalimnion, whereas warm water and more tolerant species such as percids, esocids (pike [*Esox* spp.]), centrarchids (*Micropterus* spp. and *Lepomis* spp.), and cyprinids are mainly distributed in the epilimnion (Volta et al., 2018; Alexander et al., 2015). Both vertical and horizontal separation among different fish groups is subject to variation depending on season or time of day. For instance, some typical littoral fish, such as pike and pikeperch, can temporarily migrate into open water when searching for food. Also, juveniles of typical littoral fish may migrate into open water in search of zooplankton. Conversely, typical pelagic species, such as coregonids, instead approach the littoral habitats in spring, when lake productivity is still low and zooplankton are less abundant, to prey upon other organisms such as chironomid larvae.

Shallow lakes in temperate regions usually exhibit a fish community that is less rich than in deep lakes. Thus cold stenothermal fish species are usually absent or less abundant because the chemical features of deep and cold water (usually oxygen depleted during summer and autumn) are not suitable for their survival. A typical fish community of a shallow lake in Europe or North America includes percids, esocids, and cyprinids (e.g., Olin et al., 2002; Chu et al., 2003; NatureServe, 2010).

High-altitude or high-latitude lakes are subjected to harsh environmental conditions such as prolonged ice cover, and they are often oligotrophic and, of course, characterized by cold water. For these reasons, they are generally inhabited by a few cold and stenothermal fish species (mainly charr and trout).

The examples given above are a major simplification and mainly describe the conditions and fauna of only the temperate regions of the Northern Hemisphere. There are, however, many typologies of lake classification elaborated by limnologists according to features like temperature, period of water circulation, productivity, and salinity, and each type is capable of supporting a very distinctive fish fauna.

C. Fish ecological guilds and diet

From a functional point of view, fish can be grouped into different *guilds* (e.g., Balon, 1975; Austen et al., 1994) relative to, for instance, habitat preferences, reproductive behavior, thermal preferences, and food preferences. Species are classified as *pelagic* or *benthic* or *rheophilic* depending on their living and feeding habitats. The reproductive guilds include *phytophilic* species, which are those spawning on different parts of living or dead vegetation, and *lithophilic* species, which spawn on clean mineral substrates. Reproductive guilds include also *ariadnophilic* species (i.e., species exhibiting some form of parental care) and *ostracophilic* species (i.e., species spawning in shells).

Fish are classified as being either tolerant or intolerant to any stressor related to lake morphology (habitat), hydrology, or water chemistry (e.g., Karr, 1981) and, according to their thermal preference, as cold water, cool water, or warm water species.

Finally, fish can be classified according to their food preferences. *Herbivores* are those fish with a diet dominated by plants and algae, whereas *carnivores* prey on other animals. More detailed categories include *invertivorous* species, which are those whose adult diet consists of more than 75% insects; *planktivorous* species whose adult diet consists of more than 75% zooplankton and/or phytoplankton; and piscivorous species feeding on fish, at least partly, as adults. If plants and animal material both contribute at least 25% to the diet, the species is considered *omnivorous*. *Benthivorous* species are those whose adult diet contains more than 75% benthic organisms, and *detritivores* are species where detritus is the dominant food source.

In regard to diet, many fish adapt to a wide variety of food sources and often switch from one source to another as environmental and food supply conditions change dietary quality and abundance. The flexibility of the species to switch and to take advantage of the most profitable food source at any particular time, which is termed *trophic adaptability*, makes it very difficult to accurately group fish into specialist or generalist feeding categories. Predatory ontogeny, which implies ontogenetic diet shifts during life (from zooplanktivory to benthivory and, finally, to piscivory), is common in many carnivorous fish such as perch (*Perca fluviatilis*). Specialist fish feed on a restricted diet, whereas generalist fish feed on a broad spectrum of prey species or detrital organic matter. Opportunist fish switch from one food source to another as food populations fluctuate and as their life stages mature. For example, larval fish utilize endogenous sources of nutrients in the yolk sac stages and shift initially to exogenous mouth ingestion of algae as the yolk sac is absorbed, and then to zooplankton, the latter being selected individually, in the later stages of larval feeding. Mortality among larval fish is very high and is, to some extent, attributed to starvation (cf. review of Gerking, 1994). During good reproductive years, however, abundant fish larvae can severely impact zooplankton, particularly larger species such as *Daphnia* (Mills and Forney, 1983). Cannibalism is also quite common in fish, particularly when newly hatched young life stages are seasonally abundant (Smith and Reay, 1991; Pereira et al., 2017).

A number of fish families are predominantly herbivorous and feed on algae and vascular aquatic plants or specialize in reproductive parts (fruits, seeds, and flowers) of higher plants. All herbivorous fish ingest some animal food, which may supplement the largely plant diet. These fish commonly exhibit physical adaptations for the mastication of plant material and partial chemical digestion. *Detritivory* is common among fish (Gerking, 1994). For instance, common carp *Cyprinus carpio* provides an excellent example of *nonmandibular teeth* being used as the primary chewing apparatus. In carp the lower ends of the gill bars have a well-developed musculature that operates two sets of interdigitating teeth so as to grind plants into small pieces before swallowing them. *Pharyngeal teeth* occur in the most fully developed forms of Cyprinidae and Cobitidae, although many other groups also show some degree of abrading or triturating ability with some part of the gill bars. The diet of adult detritivorous fish consists mainly of decomposed organic matter; living organisms associated with the detritus are also consumed, either actively or fortuitously. Early larvae of detritivores are zooplanktivorous and switch to dead organic matter as they mature.

Diversity of feeding behaviour is high among bottom (benthic)—foraging fish. Most species shift among prey types as availabilities change, but certain behavioral adaptations often predominate. Processes include (a) ingesting sediment deposits, (b) scraping periphyton off rock surfaces, (c) grasping invertebrates from surfaces, and (d) crushing mollusks. Extraordinary feeding specializations have evolved among certain fish families, particularly among the nearly 500 endemic cichlid species in East African relict lakes (Greenwood, 1974; Lowe-McConnell, 1987).

Substantial changes in fish faunas occur with increasing productivity, from dominance of salmonids in unproductive lakes to dominance by percids in medium productive lakes and dominance by cyprinids in highly productive lakes (Carpenter and Kitchell, 1993; Leach et al., 1977). Fish biomass increases with productivity (Hanson and Leggett, 1982), whereas the percentage of piscivores decreases in both temperate and subtropical regions (Bachmann et al., 1996; Bays and Crisman, 1983; Hanson and Leggett, 1982; Persson et al., 1988; Quiros, 1990), as also illustrated in Fig. 22-5 from Danish lakes.

In North European lakes a change in the fish community occurs from dominance by perch and pike in mesotrophic lakes to exclusive dominance by cyprinids in eutrophic lakes—first roach (*Rutilus rutilus*) and, at the highest lakewater total phosphorus (TP) concentrations, bream (*Abramis brama*), and such shifts are well documented in a number of studies of North European lakes (Svärdson, 1976; Leach et al., 1977; Persson, 1983; Persson et al., 1988). The superiority of roach in eutrophic lakes is attributed to a high potential growth rate along with a higher predation efficiency on cladocerans (Persson, 1983), as well as an ability to exploit smaller zooplankton prey (Stenson, 1979; Lessmark, 1983). Moreover, cyprinids are omnivorous, whereas large perch are piscivores. Finally, the loss of habitat complexity with the disappearance of submerged macrophytes, and thus increased turbidity (Chapter 26), disfavors percids and, furthermore, augments the intraspecific competition among them (Persson et al., 1988). The enhanced dominance of cyprinids is accompanied by a decline in the average size of both cyprinids and perch (Jeppesen et al., 2000). This decline probably reflects enhanced competition for food, which in turn may be mediated by reduced predation of piscivores on young fish. Thus the fraction of large piscivorous individuals among the perch population declines. In addition, the average weight of pike increases considerably, probably as a result of enhanced cannibalism due to reduced structural complexity (fewer macrophytes) and, consequently, the loss of refuges (Grimm and Backx, 1990). As large pike control small prey fish less efficiently than small pike (Grimm and Backx, 1990), the capability to control young planktivores likely decreases, even though *catch per unit effort* (CPUE) in terms of biomass increases. The changes in the fish community are associated with a major decrease in the zooplankton: phytoplankton ratio (an indicator of grazing efficiency on phytoplankton), an increase in the biomass of phytoplankton (chlorophyll *a* concentrations), reduced Secchi depth, and loss of submerged macrophytes.

III. Size-selective and size-efficiency hypotheses

A. Size selectivity

Numerous studies have demonstrated the importance of planktivorous fish in determining zooplanktonic populations with a distinct shift favoring the survival of species that are smaller in size (Brooks and Dodson, 1965, Gliwicz, 1994; Jeppesen et al., 2004). In other words, for a number of reasons focusing on energetic efficiency, predators selectively consume the largest prey possible within their physical and behavioral capability relative to the abundance of prey.

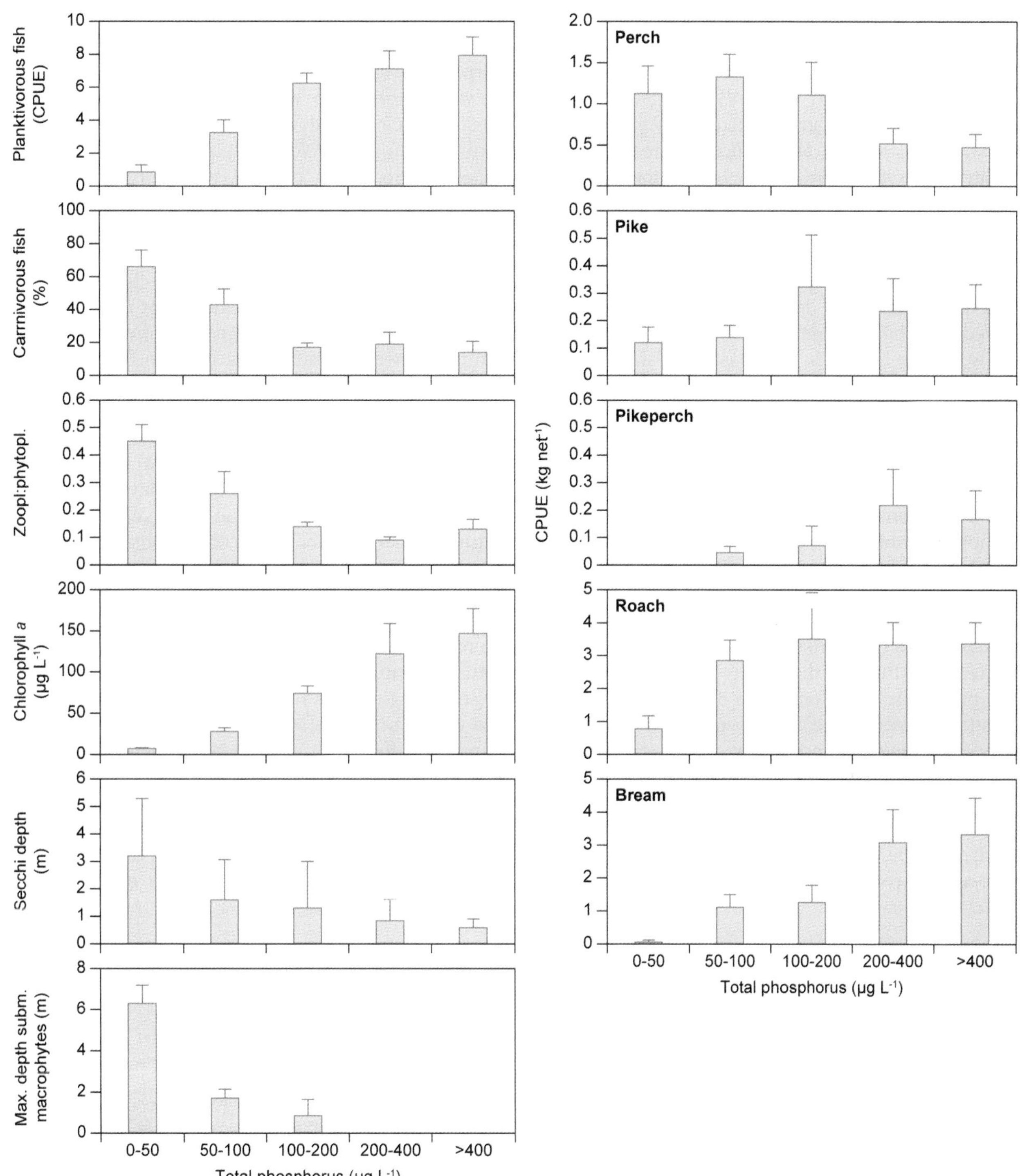

FIGURE 22-5 (*Left*) August biomass of planktibenthivorous fish (measured as catch per net per night [CPUE] in multiple mesh size gill nets, 14 different mesh sizes 6.25–75 mm) versus summer mean lake water total phosphorus (TP) in Danish lakes. Also shown are the percentage of piscivorous fish, summer mean (1 May–1 October) of the zooplankton:phytoplankton biomass ratio, chlorophyll *a*, Secchi depth, and the maximum depth of submerged macrophytes versus the lake water total phosphorus (TP) concentration. Means ± SD of the five TP groups are shown. (*Right*) Biomass (CPUE) of various quantitatively important fish species versus summer mean TP. The upper three species are potentially piscivorous and the last two plankti-benthivorous. *(From Jeppesen et al., 2000.)*

The importance of fish in the regulation of the size and species composition of zooplankters was first demonstrated by Hrbáček and collaborators (Hrbáček, 1958). They found that when large, shallow ponds were stocked heavily with cyprinid fish, the zooplankton community consisted of small cladocerans, such as *Bosmina* and

Ceriodaphnia, and rotifers. Water transparency decreased with the predominant development of small nanoplankton algae. Upon selective removal of the cyprinids, the zooplankton composition suddenly changed to dominance by larger cladocerans, particularly *Daphnia longispina*, whereas rotifer abundance decreased. Also transparency increased as the phytoplankton shifted to smaller densities of larger species.

Another series of simple but definitive studies conducted by Iglesias et al. (2011) involved field observations and fish exclusion experiments in several subtropical Uruguayan lakes. Large-bodied *Daphnia* appeared in a lake where the fish predation pressure was low due to fish kills, whereas small-sized cladocerans were abundant in another lake that exhibited a typical high abundance of fish. Likewise, relatively large cladocerans (e.g., *Daphnia* and *Simocephalus*) appeared in fishless mesocosms of three lakes after only 2 weeks, but their abundance declined again after fish stocking. Dominance of small-sized zooplankton in warm lakes could not be related to high temperature–induced physiological constraints but rather to fish predation.

All planktivorous fish have closely spaced *gill rakers*. Fig. 22-6 shows the closely spaced gill rakers of the first branchial arch of the planktivorous *Alosa pseudoharengus* compared with those of the closely related *A. mediocris*, which primarily feeds on small fish. Studies of the pharyngeal sieves of the planktivorous rainbow trout (*Onchorynchus mykiss*) and yellow perch (*P. flavescens*) in two Michigan lakes demonstrated that trout and perch only removed few zooplankters of a size smaller than 1.3 mm (Galbraith Jr, 1967; Nilsson and Pejler, 1973). It must be emphasized, however, that all freshwater planktivorous fish examined so far actively search for and visually select each plankter that they ingest (Brooks, 1968; Seghers, 1975). Some fish (e.g., bighead carp, *Hypophthalmichthys nobilis*, and bluegills, *L. macrochirus*) gulp water by constantly opening and closing their mouth while swimming (Cremer and Smitherman, 1980; Werner et al., 1981) and in this way ingest aggregations of prey when prey densities are high. However, by the normal respiratory behavior of passing water through the mouth and across the gills, some zooplankters can be collected by gill rakers.

The relationship between size selection of prey and foraging efficiency has been shown for the bluegill sunfish (*L. macrochirus*), a common centrarchid of many temperate waters that selects prey on the basis of size (Werner, 1974; Werner and Hall, 1974; O'Brien et al., 1976; Werner et al., 1981; Mittelbach, 1981). Growth rates increased significantly in direct relation to food size (Hall et al., 1970), and much of this differential growth has been attributed to greater efficiency of foraging with increasing size of prey captured, that is, an improved ratio of expenditure of energy or metabolic cost in obtaining food relative to the return (Pyke et al., 1977). As for bluegill, size selection of prey is related to the optimal allocation of time spent searching for and handling the prey. In other words, the size range of prey that maximizes the energy return per unit of energy expended depends on the costs of searching for and handling times of different prey in the environment. Both the searching ability and prey-handling efficiency increase with increasing fish size (Mittelbach, 1981). The handling time per prey increases exponentially with an increasing ratio of prey size to mouth size (Werner and Mittelbach, 1981). At times of low prey abundance, when the search time is long, prey items of different sizes are eaten as encountered. As

FIGURE 22-6 The alewife, *Alosa pseudoharengus*. *Upper*: a mature specimen, 300 mm in length. *Lower left:* first bronchial arch with closely spaced gill rakers that act as a plankton sieve. *Lower right*: first bronchial arch with widely spaced gill rakers of *A. mediocris*, a species that feeds primarily on small fish. *(From Brooks and Dodson, 1965.)*

prey abundance increases and search time decreases, smaller-sized classes are eaten less frequently or ignored so that the overall return per time increases (see also Brooks, 1968).

Earlier observations of the effects of planktivorous fish on zooplanktonic species composition were confirmed by a study of the changes in the zooplanktonic populations of Crystal Lake, Connecticut (Brooks and Dodson, 1965). Before the introduction of alewife into the lake, zooplankton included the large calanoid copepod *Epischura*, *Daphnia*, and the cyclopoid *Mesocyclops*, as well as numerous smaller *Diaptomus* and *Cyclops* copepods. Some 10 years after the alewife introduction, the larger forms of zooplankters had been replaced and the dominant species included the significantly smaller cladoceran species *Bosmina* and the two cyclopoid copepods,

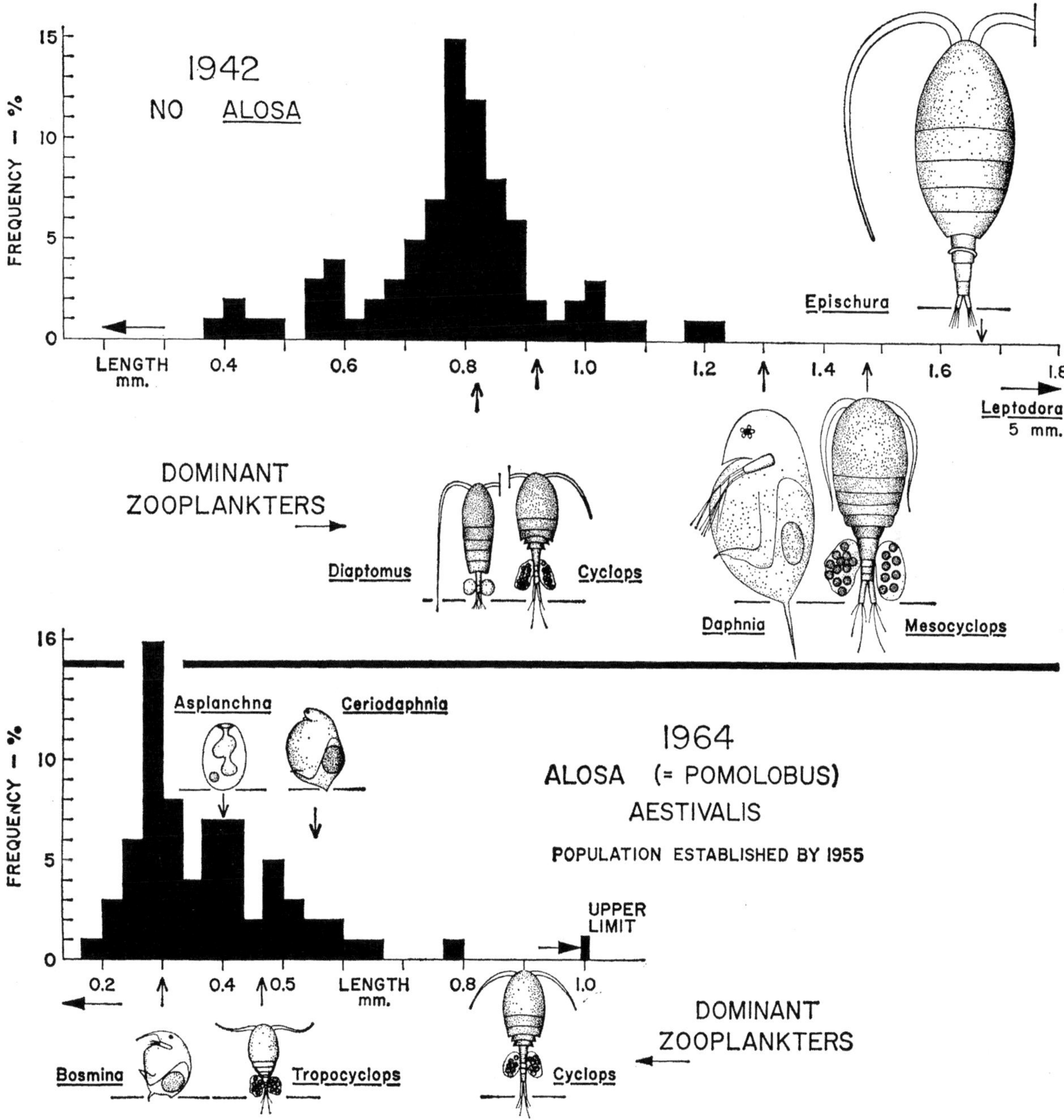

FIGURE 22-7 The composition of the crustacean zooplankton of Crystal Lake, Connecticut, before (1942) and after (1964) a population of *Alosa aestivalis* became well established. Each *square* of the histograms indicates that 1% of the total sample counted was within that size range. Larger zooplankters are not represented because they were relatively rare. The specimens depicted represent the mean size (length from posterior base line to the anterior end) of the smallest mature instar. The arrows indicate the position of the smallest mature instar of each dominant species in relation to the histograms. The predaceous rotifer *Asplanchna* is the only noncrustacean included in this study. *(From Brooks and Dodson, 1965.)*

Tropocyclops and *Cyclops* (Fig. 22-7). The modal length of the numerically dominant forms had shifted from 0.8 to 0.35 mm in the zooplankton assemblages. Larger forms were found only in the littoral zone and near the sediments, areas that are avoided by the pelagic *Alosa*. Such size-feeding relationships have been demonstrated often in both enclosure experiments where fish and zooplankton predation are modified (e.g., Threlkeld, 1988; Lacerot et al., 2013) and in whole-lake ecosystems due to predator introductions or predator alterations (e.g., Elser and Carpenter, 1988; Iglesias et al., 2011; Lemmens et al., 2018).

B. Size-efficiency hypothesis

When size selection by fish is absent and large zooplankters are present, it is common that smaller-sized zooplankton do not cooccur with larger forms. Brooks and Dodson (1965) first proposed the *size-efficiency hypothesis* in an attempt to explain this inverse relationship between the abundances of small- and large-bodied herbivorous zooplankton in freshwater lakes. According to the hypothesis (Hall et al., 1976):

(a) Planktonic herbivorous zooplankton all compete for small particles (1–15 μm) in open waters.
(b) Larger zooplankton filter more efficiently and can also take larger particles.
(c) Therefore, when the intensity of the predation pressure by fish is low, small planktonic herbivores will be competitively eliminated by large forms (dominance by large Cladocera and calanoid copepods).
(d) When fish predation is intense, size-dependent predation will eliminate the larger forms, allowing the smaller zooplankton (rotifers, small Cladocera) that escape predation to become dominant.
(e) When the predation pressure is moderate, larger zooplankton species are often kept at low levels that allow the coexistence of smaller competitors.

The basic assumptions of the size-efficiency hypothesis imply a complex successional pattern in which the optimal body size increases while the range of persisting sizes decreases with declining food concentrations (critically analyzed by Hall et al., 1976; Zaret, 1980). Vertebrate predation restricts the maximum adult body size of zooplankton, and invertebrate predation may restrict the minimum size. Both of these effects can augment a decline in zooplankton productivity caused by food limitation. The minimal food concentrations needed for growth among the different species of *Daphnia* decrease as the size of the animals increases (Gliwicz, 1990). As a result, when the predation pressure is low, larger cladoceran species are expected to be more successful competitors for food. In contrast, the threshold food concentrations of small species of rotifers are lower than among large species, which indicates that smaller rotifers are more energetically efficient per unit of body mass than large species (Stemberger and AuthorAnonymous, 1987).

Fish predation can clearly influence the species composition and the size structure of zooplankton prey. The morphology, physiology, and behavior of prey can all be affected by size-biased feeding, which can lead to significant evolutionary alterations in predator avoidance. The *cyclomorphosis* among cladocerans discussed above is, at least in part, a result of the predation pressure. Similarly, *vertical migration* and other evasive behaviors among zooplankton are partially induced by predation pressure (O'Brien, 1987; Lampert, 1993; Hylander et al., 2009). The migration behavior of many cladocerans upward as darkness occurs and downward as daylight approaches is most common among large, adult zooplankton, particularly egg-carrying females that are most visible to fish. Migration usually occurs when fish are abundant or can also be induced by diurnal variation in UV light. Interestingly, copepod zooplankters, when facing the threat of fish predation, will respond strongly by reducing their pigmentation. Hansson (2004) suggests that there is a trade-off between transparency and pigmentation (red carotenoid pigments), with greater transparency providing crypsis against visual predators such as fish.

The size-efficiency relationships among fish and zooplankton, although not fully understood, have major implications for food web structure and have led to a plethora of research on the potential top-down control of food webs by fish manipulation (*biomanipulation*; see below).

IV. Importance of visibility in predation

Many studies have shown that resuspended particles and high concentrations of humic (dissolved organic carbon, DOC; Chapter 28) substances can severely disturb prey detection by fishes due to their effects on light scattering (Vinyard and O'Brien, 1976; Horppila et al., 2004; Nurminen and Horppila, 2006; Estlander et al., 2010), and climate change is suggested to increase the loading of silt and humic substances to various lakes on the globe (Evans et al., 2005).

The effect of turbidity was demonstrated in a study by Zaret (1972). He found that tropical *Ceriodaphnia cornuta* showed two distinct polymorphic forms of the same body length within the same lake. One form had pointed, hornlike extensions of the exoskeleton on the head, body, and tail regions with a small area of black pigmentation in the compound eye. The other phenotype was unhorned but possessed a large, pigmented eye. Predation by the dominant planktivore fish, silverside (Atherinidae; *Melaniris chagresi*), was more intense on the form with the large, pigmented eye. This form of *C. cornuta* had a superior reproductive potential and a more rapid population growth with greater longevity than that of the horned, small-eyed form. Without fish predation, the large-eyed form can rapidly outcompete the less conspicuous form, but under predation pressure, the form with the small eye, although growing more slowly, can coexist with the large-eyed form because of reduced visibility to the predator.

Similar results were found on the effects of fish predation on *Bosmina* by Zaret and Kerfoot (1975) and Hessen (1985). Prey selection was observed to be related to the large, black-pigmented eye, and body size was of negligible importance. Size-biased feeding by fish that select the larger species of a zooplankton community or larger members of a single species has been documented for at least 20 species of fish (reviewed in Gerking, 1994). Particulate turbidity, especially clay turbidity, which is common to many reservoirs, reduces visibility and the distance at which predator–prey interactions occur (Abrahams and Kattenfeld, 1997). This constraint will reduce the predation risk on zooplankton by fish in turbid aquatic ecosystems.

Reduced light may also have strong effects on the interactions between fishes and their prey, as many fishes are visual foragers and highly dependent on light to detect and consume their prey (De Robertis et al., 2003; Vinyard and O'Brien, 1976). Reduced light intensity may therefore affect the competitive interactions between species because of their differential forage capacities at low light. An illustrative example of the latter is the interaction between European perch and roach, the two most common fish species in European lakes. Perch is a vision-oriented selective predator that depends on good light conditions, whereas the omnivorous roach feeds efficiently at low light intensities (Bohl, 1980), suggesting that turbidity would favor roach (Olin et al., 2010). This potentially has strong implications for the food web structure as adult perch are piscivorous. An illustrative example of this is the investigation by Estlander et al. (2010), who studied a number of Finnish lakes with different levels of transparency (expressed as Secchi depth) and compared the consumption of the preferred food item *Daphnia* relative to their presence in the plankton (*Ivlev index*; Ivlev, 1961). Their results clearly showed that roach was a much better forager on *Daphnia* at low turbidity levels (low Secchi depth), whereas no differences were found in clear water lakes with high Secchi depth (Fig. 22-8).

Turbidity may also affect the risk of fish predation on juvenile fish (Gregory, 1993; the "turbidity as cover" hypothesis). This may favor planktivorous fish as planktivore foraging, as shown above, is less dependent on

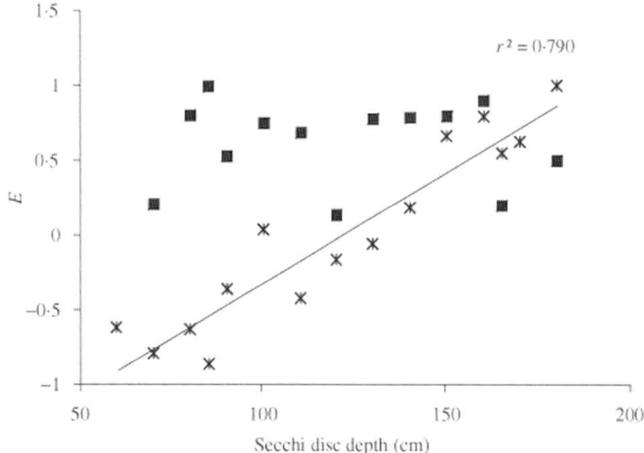

FIGURE 22-8 The relationship between Secchi depth and Ivlev's selectivity index values (E) for *Perca fluviatilis* (*crosses*) and *Rutilus rutilus* (*squares*) for *Daphnia*. If E is 1, foraging is optimal and if approaching −1, daphnids are hardly selected. *(From Estlander et al., 2010.)*

light (De Robertis et al., 2003), and it may potentially also lead to a lower proportion of piscivorous fish in turbid lakes.

V. Trophic cascades (pelagic and benthic food webs)

The term *trophic cascade* originated in studies of marine intertidal organisms (Paine, 1980) and has been widely and ambiguously applied to a number of freshwaters as a foundation for trophic interactions (e.g., Carpenter et al., 1985, Carpenter and Kitchell, 1993). Although formally defined as the propagation of indirect mutualism between nonadjacent levels in a food chain (Schoener, 1993), it is most often applied to well-known pelagic or benthic processes where predation by fish and/or invertebrates alters the zooplankton or benthic invertebrate community structure, and their effectiveness in grazing on phytoplankton and periphyton, respectively. Many studies in enclosures and ponds, as well as whole-lake experiments, have demonstrated significant impacts of planktivorous fish on zooplankton that, in turn, have influenced phytoplankton composition.

Two important points have emerged from analyses of these interactions: (a) the concept of cascading negative trophic interactions is often too simplistic and may fail to operate in natural pelagic ecosystems because of a bevy of compensatory mechanisms that emerge after predatory alterations (e.g., Hansson et al., 1998; Mazumder et al., 1990; McQueen et al., 1989; Ramcharan et al., 1995, 1996; among many others); and (b) if operational for a period, the "cascading" influence may not be sustained for long periods under natural conditions without constant disturbance. Trophic cascades are basically a description of one outcome of indirect interactions out of several possible alternatives (cf. Persson, 1999).

Additional terms and concepts occur in the literature on this subject. *Top-down* and *bottom-up* refer to the regulation of some process or property at different trophic levels, usually among the plankton. Set forth by McQueen et al. (1986, 1989), top-down generally refers to predation processes by invertebrates or fish that influence the zooplankton community structure, which in turn may selectively influence the feeding efficacy on seston. Bottom-up generally refers to resource regulation of growth and production, often beginning with biogeochemical controls of photosynthesis, usually of phytoplankton. By implication, this resource (nutrient or light) limitation of phytoplankton growth can translate to resource (food) limitations for herbivorous zooplankton and consumers of zooplankton. The hypothesis predicts that bottom-up forces are strongest at the bottom of the food chain and that top-down forces are strongest at the top of the food chain.

The concept depicted in Fig. 22-9 now appears to be widely accepted, whether at the community, population, or individual level. Inherent in it is the tacit assumption that the nature of the two impacts is the same, with only the direction, down or up, differing (see Benndorf, 1990; Reynolds, 1994; Drenner and Hambright, 1999; Carpenter et al., 2001; McQueen et al., 2001). For two reasons, however, this assumption can be challenged (Gliwicz, 2002). The first is that top-down and bottom-up forces affect the behavior of an individual animal differently (e.g., a *Daphnia* or a fish), with trade-offs of increased safety versus decreased feeding rates. The second reason is less apparent and has been largely overlooked in the top-down versus bottom-up debate (Gliwicz, 2002). It relates to the fact that rates (e.g., growth rates) are controlled from the bottom up, whereas state variables (e.g., density) at both the population and community levels are controlled from the top down.

Although the bottom-up and top-down impacts are traditionally conceived as compatible with each other, field population density data on two coexisting *Daphnia* species suggest that the nature of the two impacts differs. Rates of change, such as the rate of individual body growth, the rate of reproduction, and each species' population growth rate, are controlled from the bottom up (Fig. 22-10). State variables, such as biomass, individual body size, and population density, are controlled from the top down and are fixed at a specific level regardless of the rate at which they are produced (Fig. 22-10; Gliwicz, 2002).

According to the theory of functional responses, carnivorous and herbivorous predators react to prey density rather than to the rate at which prey are produced or reproduced. The predator's feeding rate (and thus the magnitude of its effect on prey density) should therefore be regarded as a functional response to increasing resource concentration. The disparity between the bottom-up and top-down effects is also apparent in individual decision-making, where a choice must be made between accepting the hazards of hunger and the risks of predation (lost calories versus loss of life). As long as top-down forces are effective, the disparity with bottom-up effects seems evident. In the absence of predation, however, all efforts of an individual become subordinate to the competition for resources.

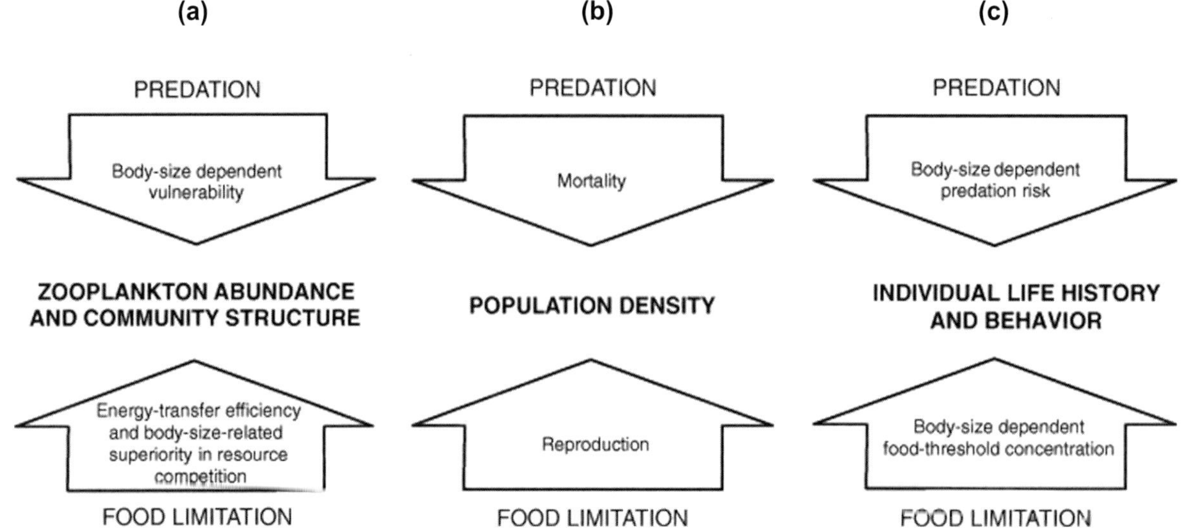

FIGURE 22-9 Diagrammatic representation of the role of *bottom-up* (food limitation) and *top-down* (predation) impacts on zooplankton abundance and community structure (a), population density and age structure (b), and individual behavior and life histories (c). *(From Gliwicz, 2002.)*

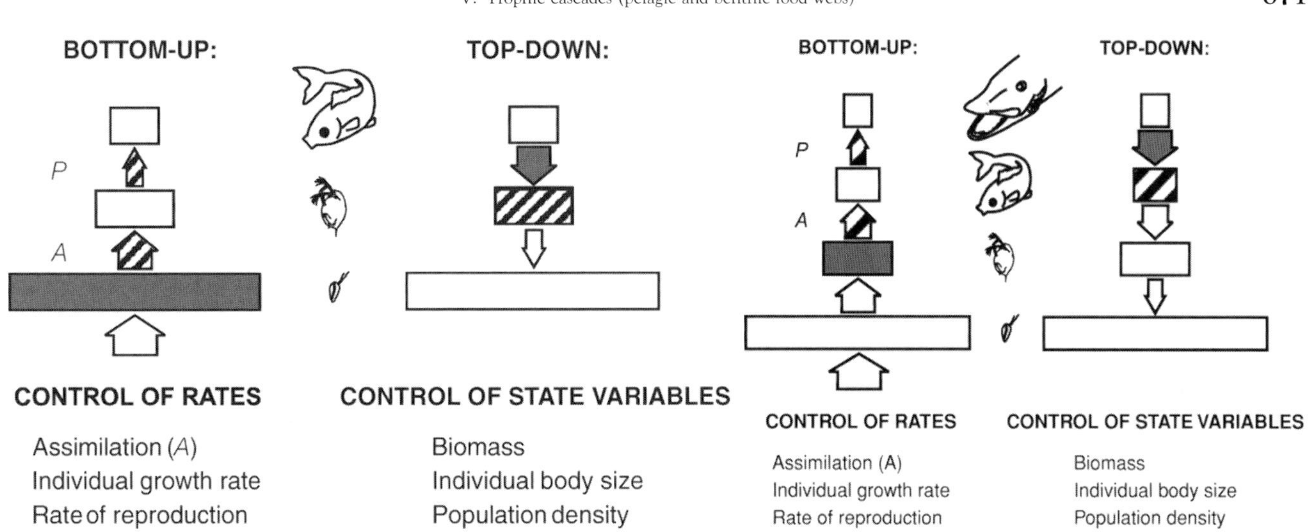

FIGURE 22-10 Diagram of the different nature of *bottom-up* (food limitation) and *top-down* impacts (predation) on zooplankton whose abundance (biomass) is controlled top-down by planktivorous fish and whose process rates (energy/carbon flow) are controlled bottom-up by phytoplankton availability. *Right*: similar diagram for impacts on planktivorous fish whose abundance (biomass) is controlled top-down by piscivores and whose process rates (energy/carbon flow) are controlled bottom-up by the availability of zooplankton prey. A = food assimilation, P = predation. *(Modified from Gliwicz, 2002.)*

Biomass becomes limited from the bottom up as soon as the density of a superior competitor has increased to the carrying capacity of a given habitat (Gliwicz, 2002).

Such a shift in the importance of bottom-up control can be seen in zooplankton in habitats from which fish have been excluded (Gliwicz, 2002).

Top-down control by fish is particularly clear in Arctic and cold alpine freshwaters (O'Brien, 1975; Knapp et al., 2001; Jeppesen et al., 2017). The main reasons are that (1) the lakes are typically transparent, making foraging easy for visually hunting fish under daylight conditions; (2) the potential prey is exposed to predation for a longer time period due to longer generation times; and (3) zooplankton may need and have pigmentation (Rautio and Korhola, 2002) to protect them against UV effects or other oxidative stress, rendering them more visible to fish (Hansson and Hylander, 2009).

A study of 87 lakes in Greenland (Jeppesen et al., 2017) clearly illustrates the top-down control effect of structure (Fig. 22-11). The pelagic zooplankton biomass was, on average, three- to four-fold higher in the fishless lakes and was dominated by large-bodied taxa such as *Daphnia*. The large phyllopod *Branchinecta paludosa* and tadpole shrimp, *Lepidurus arcticus*, were also abundant in some of the lakes without fish, whereas small-bodied crustaceans dominated the lakes with fish. Moreover, fish had a strong impact on the benthic macroinvertebrate communities (Fig. 22-11); the abundance of macroinvertebrates was three-fold higher in the near-shore areas of fishless lakes than in lakes with three-spined sticklebacks (*Gasterosteus aculeatus*). In lakes without fish the abundances of large-bodied *Eurycercus* and *Chironomus* were substantially higher, whereas those of Tanytarsini and *Pisidium* were lower. The increase in the latter two may reflect release from competition from the other, more predation-vulnerable taxa. Branchinecta, Ostracoda and Trichoptera (Limnephilidae), Chironomindae (apart from *Chironomus*), and Tanypodinae were found only in the fishless lakes.

The phytoplankton biomass (expressed as chlorophyll *a*), however, did not differ between lakes with and without fish (Fig. 22-11), likely because nutrient limitation of the phytoplankton was more important than grazing, emphasizing the role of resource control. In areas with higher nutrient levels, however, cascading effects are to be expected. This was shown by experimental mesocosm studies undertaken in sub-Arctic Lake Myvátn, Iceland, where enclosures with fish had higher biomasses of phytoplankton and higher proportions of Cyanobacteria than lakes without fish (Cañedo-Argüelles et al., 2017).

Comparative studies of northern streams with and without fish also reveal strong top-down effects by fish on invertebrates (Meissner and Muotka, 2006). A study in Finland showed that the densities of invertebrate predators were significantly higher in trout-free streams (Meissner and Muotka, 2006). *Baetis* mayflies were less abundant in trout streams, whereas densities of chironomids were positively, although nonsignificantly, related to trout presence (Fig. 22-12). Metaanalysis showed a strong negative impact of trout on invertebrate predators, a negative but variable impact on mobile grazers (mainly mayfly larvae), and a slightly positive impact on chironomid larvae. Being size-selective predators, salmonid fishes have a particularly strong impact on the largest

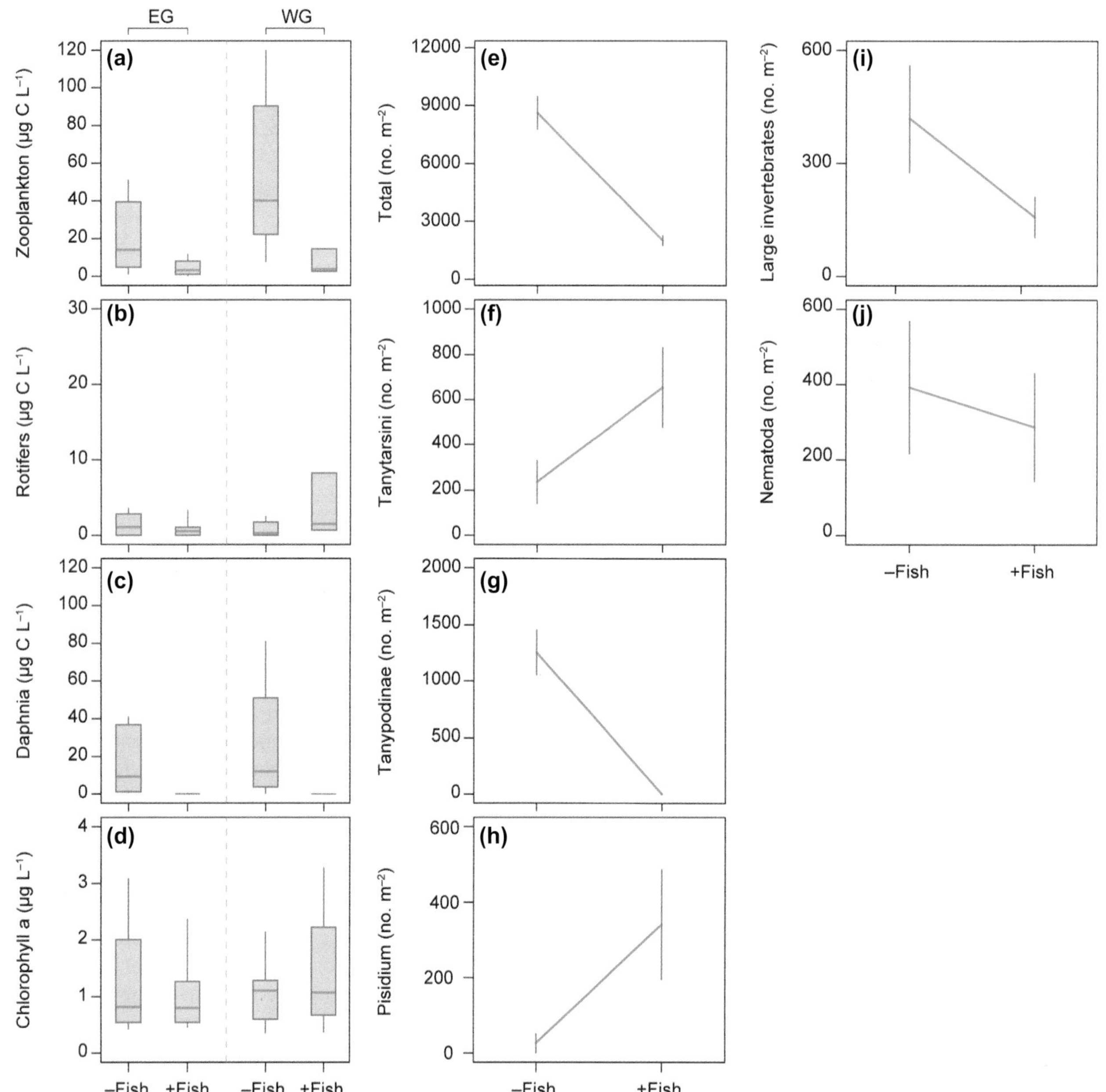

FIGURE 22-11 (a–d): Boxplot (median, 10, 25, 75, and 90% fractiles) of a number of biological and physicochemical variables in lakes without fish and lakes with Arctic charr in North-East Greenland (EG) and West Greenland (WG), respectively. (e–j): abundance (mean ± SD) of various benthic invertebrates in the littoral zone of three lakes with and three without fish (three-spined sticklebacks) in West Greenland. *(Modified from Jeppesen et al., 2017.)*

prey types available, and this effect spans several domains of scale. No cascading effects were found on the periphyton, though that was likely mainly controlled by resources in these nutrient-poor streams.

VI. Specific roles of fish in food webs

Pelagic, littoral, and terrestrial resources can all play a role in supporting fish consumers in lakes and streams. In lakes focus has historically been on pelagic production and processes. However, the benthic and littoral zone fueling of the fish community is also important. Vander Zanden et al. (2011) compiled carbon stable isotope data from 546 fish populations (75 lakes) and calculated the littoral–benthic reliance for each fish species in each lake (Fig. 22-13). The fish littoral–benthic reliance values were averaged by lake to assess the overall benthic reliance of fish species for each lake. Lake-specific *mean benthic–littoral reliance* (BR_L; fish species not weighted according to production or biomass) averaged 57% and was independent of lake morphological

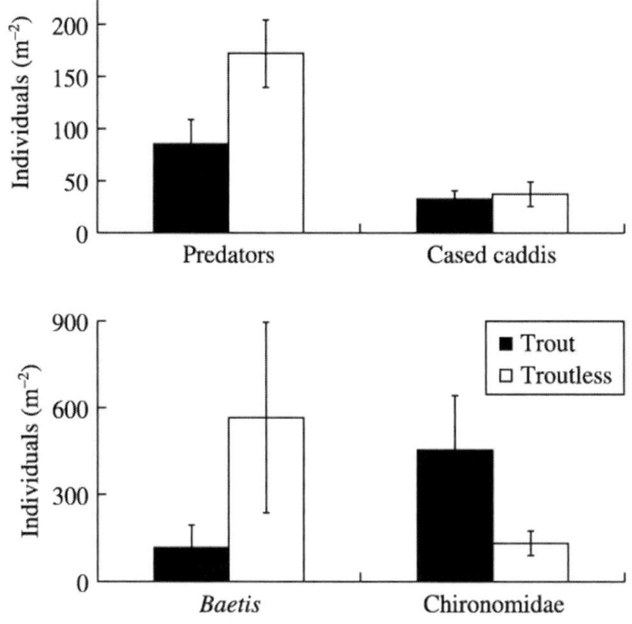

FIGURE 22-12 Benthic densities (mean ± 1 SE) of invertebrate predators, cased caddisflies, baetid mayflies, and chironomid midges in streams with and without brown trout in 24 low-order forest streams in Central Finland. *(From Meissner and Muotka, 2006.)*

and limnological attributes. Benthic algae comprised, on average, 36% of whole-lake primary production. BR_L tended to be high even in large/deep lakes in which benthic algae are a minor contributor to whole-lake primary production. The high littoral–benthic contribution to the individual fish species appears to reflect the high concentration of fish species diversity in the littoral zone. The results are consistent with other works indicating that most fish species inhabit the littoral zone, whereas relatively few exclusively inhabit the pelagic (see Vander Zanden et al., 2011). Vander Zanden et al., (2011) also suggested that it takes less primary production to support a single fish species in the littoral zone than that required to support a species in the pelagic.

Streams, being closely associated with the terrestrial environment, often receive a large amount of allochthonous organic matter. Although earlier studies emphasized the importance of allochthonous inputs to the food web, especially in forested headwater streams (Vannote et al., 1980), recent research using chemical tracers has shown that fish food webs in many systems depend heavily on algal sources, not only for carbon but especially for nitrogen and essential amino and fatty acids (Brett et al., 2017). In tropical and subtropical streams, where light and temperature stimulate algal production, stream and river food webs might be expected to show a stronger dependence on algal carbon (Brett et al., 2017), although the results are ambiguous. Some studies confirm the algal dependence (e.g., Brito et al., 2006; Reis et al., 2020), whereas Neres-Lima et al., (2017) found that the food web was predominately based on allochthonous resources. Terrestrial insects have also been reported to play an important role in the diet of fish (e.g., Rezende and Mazzoni, 2003).

Fish also affect the *food chain length* in lakes and streams. The food chain length is the number of trophic transfers from the base to the top of a food web (Post, 2002), and the chain structure may strongly affect both the lake community structure (Paine, 1980; Pace et al., 1999) and the ecosystem function (Duffy et al., 2005; Schindler et al., 1997). Accordingly, food chain length is considered a central characteristic of ecological communities, and identifying the factors that determine food chain length is a fundamental issue of ecology (Pimm, 1984). The most frequently tested and cited food chain hypotheses are based on energetic considerations, reflecting that energetic efficiencies of trophic interactions are typically low (about 10%). Therefore the number of trophic levels in a given food chain would be limited by energy availability, and longer food chains should therefore occur in more productive habitats (the *productivity hypothesis*; Elton, 1927; Hutchinson, 1959). Schoener (1989) later modified this hypothesis to the *productive-space hypothesis* that predicts that the food chain length increases as a function of *total ecosystem productivity* (the product of ecosystem size and a measure of per unit-size productivity). Tests of these hypotheses in a variety of ecosystems have, however, yielded mixed results (Hoeinghaus et al., 2008), and two studies of temperate lakes (Post et al., 2000; Vander Zanden and Rasmussen, 1999), for example, found no effect of productivity or productive space on the food chain length but instead observed a direct correlation between food chain length and ecosystem size.

Food chain length may also vary with the type of aquatic ecosystem. A global-scale analysis of aquatic food webs found significant differences in food chain length among stream, lake, and marine ecosystems (Vander Zanden and Fetzer, 2007). This difference is well illustrated in the study by Hoeinghaus et al. (2008), who analyzed carbon and nitrogen stable isotope ratios of basal sources and aquatic consumers to estimate the food chain lengths of 10 species-rich aquatic food webs of a large river basin in South America. They examined relationships between food chain length and surrogates of primary production, compared food chain lengths among different aquatic ecosystem types, and identified effects of a common anthropogenic impact (river impoundment) on the aquatic food chain length in the Paran River Basin of South America. In independent analyses they found that the temperature regime, a surrogate of productivity, and ecosystem type significantly affected the food chain length. However, when analyzed together, only ecosystem type explained significant

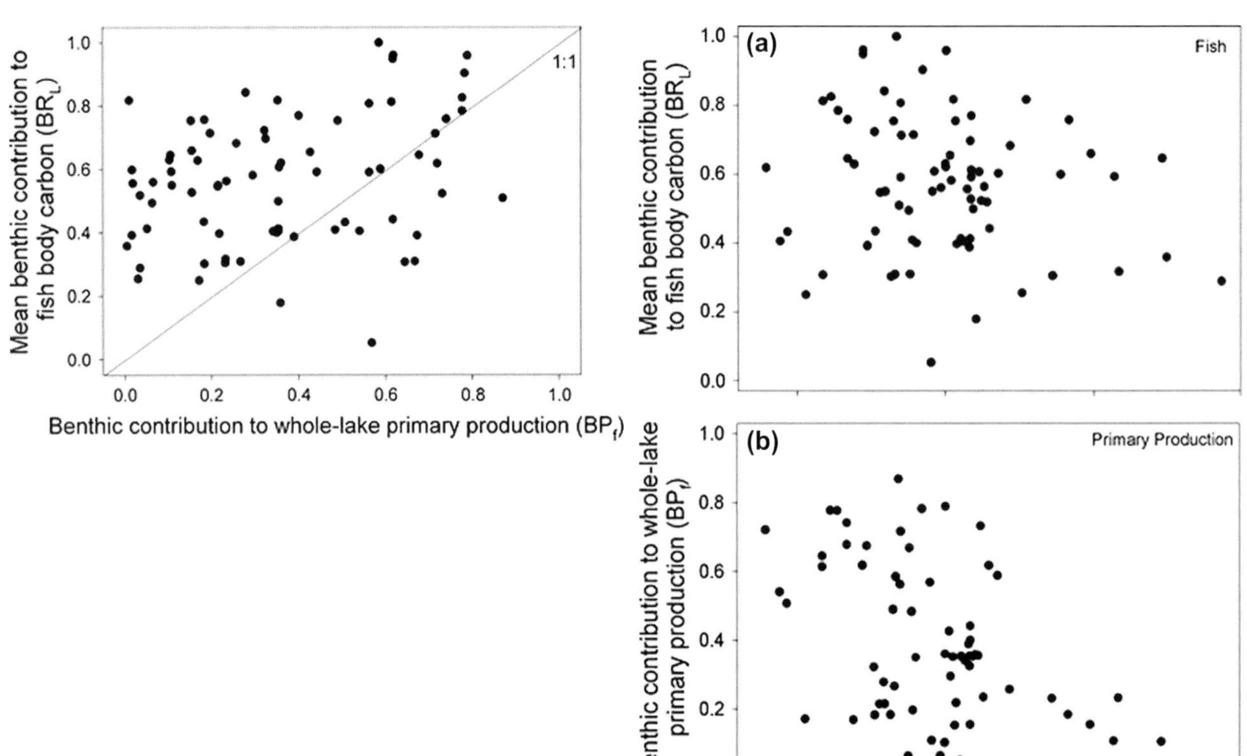

FIGURE 22-13 *Left*: The relationship between benthic algal contribution to whole-lake primary production (BP_f) and the mean littoral–benthic contribution to fish (BR_L) for 75 study lakes. BR_L is the mean of all individual fish species in a lake; species are not weighted according to biomass or production. 1:1 line is shown. *Right*: Relationship between benthic variables and lake mean depth for 75 study lakes. (a): littoral–benthic contribution to fish body carbon (BR_L), (b): benthic algal contribution to whole-lake primary production (BP_f). *(From Vander Zanden et al., 2011.)*

variations in food chain length. The food chain length was shortest upstream with the highest slope, intermediate in low-gradient rivers and rivers below reservoirs, and longest in river impoundments and reservoirs (Fig. 22-14). The proximate mechanism driving this pattern appears to be the body size ratios of primary consumers to apex predators, which differ among trophic pathways (Hoeinghaus et al., 2008).

Hydrogeomorphology was the ultimate mechanism influencing food chain length because it affects the relative importance of the basal carbon sources supporting the higher trophic levels, which through differences in the number of trophic links along the different size-structured pathways appear to drive the observed patterns in food chain length.

VII. Fish production and harvesting

A. Fish production

Fish production is the amount of tissue elaborated per unit of time and area regardless of its fate (Clarke et al., 1946). Production should not be confused with *yield* or *catch*, although these words are sometimes treated as synonyms. In fishery terms yield is that part of the annual production of a population that is harvested by humans or moves from one habitat to another (as in the migration of *Salmo salar* smolts to the sea). Thus yield is dependent upon the level of production and, in a fishery, on the degree of fishing effort (Mann and Penczak, 1986).

Fish productivity has been estimated for a variety of freshwater bodies. Some early models on fish production were constructed using simple morphometric measures like mean depth or lake area (Rounsefell, 1946; Rawson, 1952). Later, the *morphoedaphic index* (MEI), an expression of mean depth and total dissolved solids, was used globally as a fish yield estimator (Ryder, 1965). In the 1970s fish production was frequently linked to phytoplankton (Oglesby, 1977) and benthos (Matuszek, 1978) productivity. Despite all these efforts, probably the most important deficiency of lake fishery production models is that few actually estimate production rates but rather yield or catch (Downing et al., 1990). Thus much literature exists on the yield of fish, but relatively few analyses exist of their growth and mortality over annual periods. Consequently, only a few reasonable estimates of production rates can be made. Growth

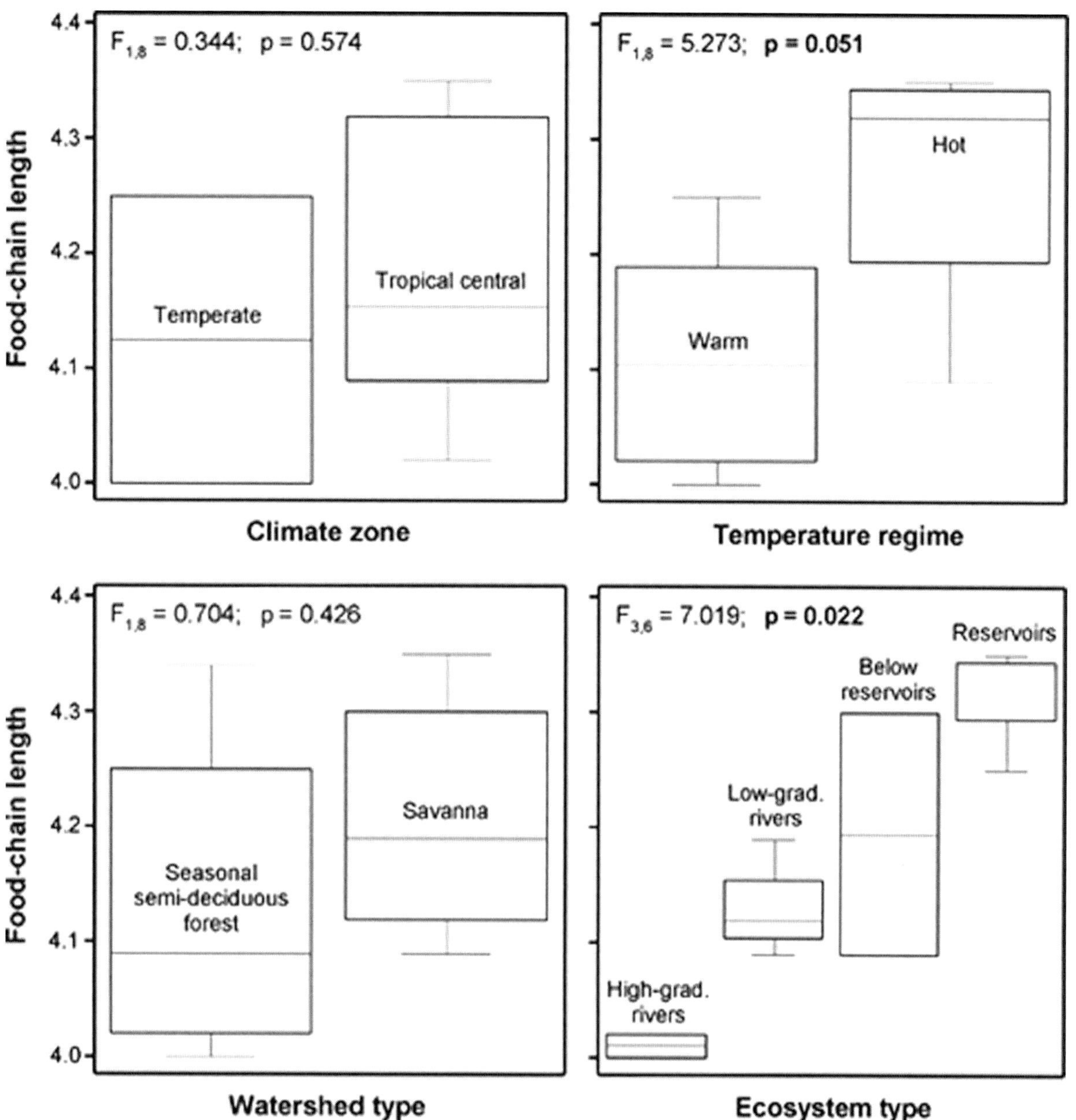

FIGURE 22-14 Box plots depicting independent comparisons of food chain length among different categories of environmental characteristics. Mean (*solid line* within box), quartiles (*box*), and range (*whiskers*) are presented for each category. Significant *p* values are in bold. *(From Hoeinghaus et al., 2008.)*

is highly variable among different species of fish. In general, instantaneous growth is correlated inversely with fish size. At larger sizes, most of the food energy is utilized for maintenance, whereas for young fish, a larger proportion of the energy intake is diverted into growth. Hence the ratio of production to biomass usually decreases with age and increasing size. The reproductive effort of sexually mature members of a fish population can constitute a high proportion of the annual fish production.

Summary estimates of net production rates of various fish groups listed in Table 22-1 demonstrate the general ranges encountered. The production rates are considerably higher in tropical waters than in temperate freshwaters, where growth is restricted to about half of the year. In fertile standing waters the production rates of herbivorous fish in the tropics can reach several hundred g m^{-2} $year^{-1}$ (Chapman, 1978). In standing waters of temperate regions, in which a single species often predominates, the range of rates is from 1 to about 20 g m^{-2} $year^{-1}$. Production rates in streams are usually higher than in standing waters (Table 22-1) and average about 50 g m^{-2} $year^{-1}$ in temperate areas.

TABLE 22-1 Estimates of Rates of Production of Fish from Various Freshwaters[a]

Fish group/ communities	*Annual production (kg ha^{-1} $year^{-1}$)*[b]	*P/B ratios*
Salmonidae (trout, salmon)		
From standing waters	0.21–66	0.62–2.0
From streams	11–300	1.0–5.0
Cottidae (sculpins)	8.0–431	1.2–5.5
Percidae (perch family)	0.91–52	0.35–2.4
Esocidae (pike family)	0.75–14.2	0.32–0.7
Cyprinidae (minnow and carp family)	0.1–915	0.18–1.94
Others	40–625	
Total fish fauna, multispecies		
Temperate zone	90–1980	
Tropical zone	1306–3468	
Planktivores (Russian lakes)	9–24	0.7–0.8
Benthivores (Russian lakes)	13–60	0.4–0.5
Piscivores (Russian lakes)	9–14	0.3–0.4

[a] *After data from numerous sources, particularly Chapman (1967, 1978), Winberg et al. (1970), Waters (1977), and Morgan et al. (1980). Values are expressed as wet weight; approximate conversion values for fishes: 1 g wet weight = 0.2 g dry weight or approximately 1 kcal.*
[b] *kg ha^{-1} $year^{-1}$ = 0.1 g m^{-2} $year^{-1}$.*

Annual *production/biomass (P/B) ratios* for fishes are generally lower than for most invertebrates as a result of their longer lifespans (several years). Relatively high P/B ratios are found in sculpins, for example, which have shorter lifespans (Table 22-1), because the younger, faster-growing stages contribute disproportionately to the population productivity. The annual P/B ratios of a population in an expanding or colonizing stage, such as in a new reservoir or in a stream recovering from flood damage, will be high as growth is high relative to mortality (Waters, 1977). A population that is overcrowded and stunted, as is commonly the case with sunfish populations in temperate lakes, has lower annual P/B ratios, slower growth, and longer lifespans. A very approximate estimate of annual production can be obtained if the annual mean biomass of zooplankton is multiplied by 15 to 20, of benthic animals by 6 to 8, and of fish by 0.5 to 1.0 (1.2 for stream salmonids) (Waters, 1977, Morgan et al., 1980; Benke, 1993).

As production is an integration of population biomass, recruitment, and the instantaneous rates of growth and mortality, it is especially responsive to anthropogenic pressures and environmental variations. From the literature, Downing and Plante (1993) gathered biological production estimates of 100 fish populations in 38 lakes worldwide. Their analyses suggest that fish production is closely correlated with temperature, total phosphorus concentration, chlorophyll *a* concentration, primary production, and pH. Additionally, fish production could be subject to stressors associated with human activities, such as water abstraction, floodplain development, climate change, exotic species introduction, and overexploitation (Carpenter et al., 2011). For instance, recent changes in the distribution and productivity of a number of fish species can be ascribed with high confidence to regional climate and land-use variability (Kao et al., 2020). Climate variations may influence a lake ecosystem directly through changing water temperature and precipitation, and indirectly through changing fish physiology and behavior and inputs of allochthonous-derived nutrients (Magnuson et al., 1997; Brander, 2007). Similarly, land-use changes (e.g., increases in agricultural land-use) could affect lake ecosystems through increasing water withdrawals and nutrient inputs, which could result in decreases in water level and increases in primary productivity (Beeton, 2002; Allan, 2004).

A. Fish harvesting

Fish harvesting is a great threat to the future freshwater fish production. Most inland capture fisheries that rely on natural stock reproduction are now overfished or being fished at their biological limit (Welcomme, 2011). Fishing has become the main source of mortality in many fish stocks and may exceed the natural mortality by more than 400% (Mertz and Myers, 1998). As the harvest intensity increases over a period of years, one classic pattern for a fishery predicts a rise in catch with increasing fishing effort until reaching a relatively high level where the maximum sustainable yield may be harvested over the long term (Allan et al., 2005). Further increases in effort beyond this level are expected to lead ultimately to declining catches and possible collapse. The reality is more complex. Overfishing may not be immediately marked by declines in total catch, even when the target species and long-term sustainability are highly threatened (Allan et al., 2005; Zhou et al., 2014).

As in marine systems, fish harvesting in freshwaters can cause changes in the distribution, demography, and stock structure of individual species and thus fish communities and lake ecosystems. Fishing will reduce the number of larger and older fish and lead to marked declines in the age and size of populations (Penczak and Moliński, 1984; Brander, 2007) (Fig. 22-15). Consequently, and to compensate for increased mortality, fish adjust their life history traits such as reproduction and growth. Data from 37 different commercial fish stocks reveal that, by increasing extrinsic mortality rates, fishing commonly drives the evolution toward earlier sexual maturation at a smaller size and an elevated reproductive effort (Sharpe and Hendry, 2009). Fishing that is selective with respect to size, maturity status, or morphology causes further evolutionary pressures (Heino and Godø, 2002). Such rapid evolution can have important consequences for harvested populations. For example, typical trait changes associated with fishing can slow the demographic recovery. In other words, the recovery of traits may take even longer than the recovery of population size, meaning that the effects of overfishing can continue to depress the harvestable biomass even after demographic sustainability has been achieved (Enberg et al., 2009).

Fish are the main predators in most freshwater systems, and one would expect that removing them might have an impact on lower trophic levels. Cascading effects from piscivorous fish to primary producers have been observed in lakes for decades (McQueen et al., 1989). Loss of apex predators often results in a relaxation of the top-down control of prey populations and stronger top-down control at the next trophic level below. A

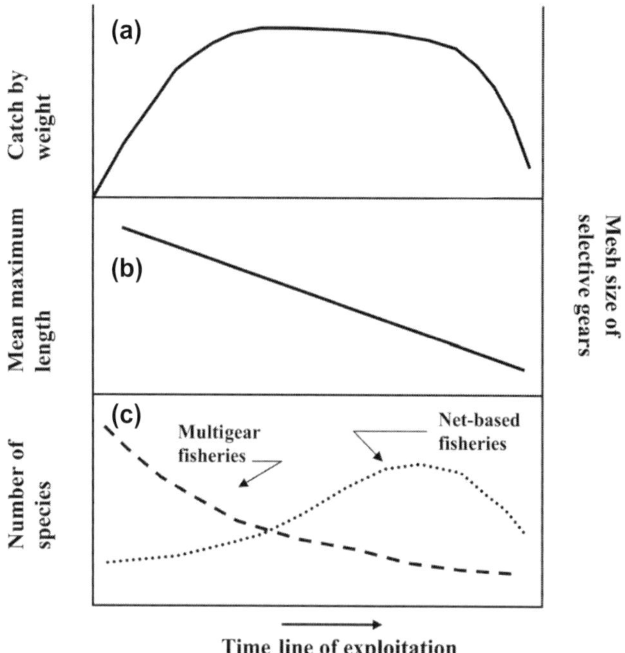

FIGURE 22-15 Characteristics of fishing down, including trends in various parameters, of a multispecies fish assemblage in response to increasing effort: (a) total catch; (b) mean maximum length of assemblage and catch and mesh size of nets; and (c) number of species accessible to net-based and multigear fisheries. *(Adaptation of Allan et al., 2005.)*

striking trophic cascade from fish via zooplankton to phytoplankton has been repeatedly demonstrated in lakes (Hrbacek et al., 1961). A recent analysis of a long-term monitoring data set from Chinese Lake Taihu by Mao et al. (2020) has shown that overfishing of large top-predators in the lake may have created conditions favorable for the development of large populations of small zooplanktivorous fish, notably lake anchovy (*Coilia ectenes taihuensis*) that accounts for nearly 50% of the total annual catch. Such enhanced zooplanktivorous fish predation then negatively affects the resilience of the lake by decreasing the density of large-sized zooplankton and thus the top-down control on phytoplankton.

Freshwater fishes also influence the nutrient dynamics in freshwater ecosystems. One of the most spectacular and best-documented effects of fish on the supply of nutrients is due to the death and decay of great spawning runs of Pacific salmon after entering freshwaters (Allan, 2004). In some regions nutrients derived from decaying salmon appear to be keystone subsidies for vertebrate predators and scavengers, forging an ecologically significant link between aquatic and terrestrial ecosystems (Gende et al., 2002). Therefore a reduction of salmon populations migrating into inland waters because of overfishing and other stressors will have serious implications for a diverse assemblage of aquatic and terrestrial biota, from mink to brown bears, as well as to piscivorous and insectivorous birds (Willson and Halupka, 1995).

Finally, it should be mentioned that overfishing may not always be the sole or even the critical threat. Impacts of human population growth, mechanical habitat destruction, increased water abstraction, introduced species, and climate change on the fisheries might be greater than the effects of overfishing, but the pressures are strongly interrelated. For example, strong interactions occur between the effects of fishing and the effects of climate because fishing reduces the age and size of populations and the biodiversity of lake ecosystems, making both more sensitive to additional stressors such as climate change (Allan, 2004).

B. Recreational fishing

Commercial fisheries have been repeatedly blamed for the global declines in fish stocks. However, the notion that recreational fisheries can also result in declines in fish stocks has been given less attention. Recreational fishing activities are generally considered those where fishing is conducted by individuals for sport and leisure, with a possible secondary motivation of catching fish for personal nutrition support (Cooke et al., 2018; FAO, 1997). The high value of recreational fisheries is commonly recognized in developed countries and, though largely unassessed in developing countries, recreational fisheries appear to be of similar importance there as well (Pitcher and Hollingworth, 2002). Although angling participation rates (e.g., from 1% in southern European countries to more than 40% in Finland) and the proportion of catch retained for consumption (up to 90% in Scandinavia) vary among regions, the yield from recreational fisheries is considerable (Cowx, 1995). Using data from Canada, Cooke and Cowx, (2004) estimated that the potential contribution of recreational fishing around the world may represent 11.5% of the global fish harvest. Moreover, there are many examples in which the recreational harvest rates for individual species exceed those of commercial fisheries (Schroeder and Love, 2002).

Evidence of the negative impacts of recreational fishing harvest in both freshwater and marine systems is mounting rapidly. For example, Post et al., (2000) identified four important inland fisheries in Canada, showing evidence of collapse that could be attributed to recreational fisheries. Most recreational fisheries have no limit on the total effort that a fishery attracts; they are open access. This can result in high local fishing mortality as well as environmental degradation problems. Although many fish captured by anglers are released, there can be substantial postrelease mortality as well as more subtle sublethal effects on growth and fitness (Cooke et al., 2002). Impacts of high and selective fishing mortality (e.g., truncation of the natural age and size structure, depensatory mechanisms, and food web changes) may potentially contribute to the decline of fish stocks and undermine biodiversity and ecological resilience (Lewin et al., 2019). Recreational fisheries are also responsible for an as yet undetermined degree of degradation of fish stocks through fishery enhancement practices or introductions (Cambray, 2003). Environmental degradation from fishing was once attributed primarily to commercial activities, but the recreational sector is now understood to have its fair share of responsibility. Discarded fishing lines or lead-containing fishing tackle can foul birds and other aquatic life or cause environmental contamination (Lewin et al., 2019). The use of live bait organisms that originate from water bodies elsewhere may potentially impact the genetic, species, and ultimately ecosystem diversity because of the release or loss of live (nonnative) bait organisms (Cowx, 2002). Recreational boat traffic and the associated noise pollution, direct hits, fuel spillages, and alterations in the wave climate and water turbidity also contribute to environmental degradation that can affect fisheries loss (Whitfield and Becker, 2014).

To avoid these consequences, anglers' actions must be constrained and coordinated through regulations or collective action. Common management actions used in recreational fisheries, such as setting size-based harvest limits, season closures, or daily limits on what can be taken, may succeed at avoiding recruitment overfishing (Johnston et al., 2010) but do not necessarily solve stakeholder conflicts or optimize angler well-being (Abbott et al., 2018). Relaxing harvest regulations may please many current anglers but at the cost of reduced opportunities for future anglers and possibly commercial fishers. Therefore the challenge for recreational fisheries is to shift away from the poor incentives created by one-size-fits-all harvest regulations and widespread stocking in inland fisheries to policies and regulations that unleash incentives among a vastly more numerous population of highly diverse people (Arlinghaus et al., 2019). Finally, failure to recognize the potential contribution of recreational fishing to fishery declines, environmental degradation, and ecosystem alterations places ecologically and economically important resources at risk. Elevating recreational fisheries to a conservation issue and considering the role of recreational fishing in changing the structure and yield of global fish stocks would promote the development of strategies to increase the sustainability of this activity.

C. Fish stocking

Restoration efforts have been made and the release of hatchery-reared fish is one of the most popular tools to restore fish populations subjected to intensive harvesting or to increase fish production. On almost every continent and for more than a hundred years, the stocking of fishes into natural water systems for commercial and recreational uses has taken place. However, the ecological impacts that such stocking might exert on other organisms in the ecosystem have, in many instances, not been adequately anticipated or considered. The impact of fish stocking on aquatic invertebrates and other taxa only began to be questioned after the advent of a new environmental awareness in the 1960s (Pister, 2001). Since then, a growing body of literature has revealed negative impacts on the biota attributable to introduced fishes, and thus the necessity for wilderness fish stocking is now the subject of widespread debate, especially in view of changing social values and priorities.

Araki and Schmid (2010) summarized 50 years of data accumulated in the scientific literature about the effects of hatchery stocking on wild stock and stock enhancement. There were clear signs of negative effects, including lower survival and reproductive fitness, of stocking fish in the wild and reduced genetic variation in the hatchery populations. However, there were also a few indications of successful stocking where no or little negative effects were found.

The contribution of stocked fish to the population size is difficult to measure because many environmental factors can influence the conditions of natural populations. The population abundance might change with or without fish stocking (Greene, 1951; Wahl et al., 1995). Many fish introductions into freshwater systems had the intention to create recreational or commercial fisheries, whereas a few targeted the conservation of threatened species. As to the latter, broader considerations are needed to achieve the goal, such as the conditions of wild stock and environmental carrying capacity (Ryman, 1991). Although there are examples of successful conservation in which a wild population recovered after a reintroduction program, widespread and long-term fish stocking of lakes and streams has shifted species assemblages and the food web structure in freshwater systems globally. The major undesirable outcomes of this include species endangerment and extinction, lower survival and reproductive fitness, and changes in genetic diversity (Li et al., 1996; Eby et al., 2006; Kostow, 2009).

Fish stocking can generally be divided into the two commonly defined strategies: *new introductions* and *enhancements*. New introduction is the intentional or accidental release of a fish species outside its historically known native range, whereas fish enhancement mostly involves an intentional, and often regular, release of farmed or transferred fish within their past or present native ranges (International Union for Conservation of Nature and Natural Resources, 1987). The establishment of new invasive predators typically leads to one of two outcomes: replacement of native species or an increase in predator species richness.

New exotic (nonnative) fishes often replace native invertebrates and fish through predation and competition. For example, about two-thirds of haplochromine cichlid species in lakes of East Africa have disappeared or are threatened with extinction after the introduction of Nile perch (*Lates niloticus*) (Ligtvoet et al., 1991). Another well-studied case is the stocking of trout into high-altitude, historically fishless lakes throughout Europe and North America (Gliwicz and Rowan, 1984; Donald, 1987). Effects of the introduced trout on the native aquatic fauna have been repeatedly demonstrated (Reimers, 1958; McNaught et al., 1999). Trout introductions typically shift community composition and often lead to the loss of native species, particularly large zooplankton, benthic macroinvertebrates, and amphibians (Fig. 22-16) (Eby et al., 2006). In Alpine lakes trout stocking first eliminated the large zooplankton species that naturally dominate zooplankton assemblages without fish predators, and after this, extirpation of large-bodied crustaceans, rotifers, and small-bodied cyclopoid copepods came to dominate the zooplankton assemblages (Donald, 1987; Donald et al., 2001). Second, by consuming benthic invertebrates and excreting in the pelagic zone, trout transfer nutrients otherwise trapped in the sediment to the water column and thus stimulate pelagic primary production (Schindler et al., 2001). Third, along with their known impact on invertebrates, studies have indicated that the effects of fish stocking can also alter amphibian populations (Drost and Fellers, 1996). In the lakes of the Sierra Nevada the introduction of nonnative trout has led not only to a decline in amphibians but also to a decline in garter snakes. Presence of amphibians is a prerequisite for garter snake persistence in high-elevation portions of the Sierra Nevada and the introduction of trout into an ecosystem can have serious consequences, not just for their prey but also for other predators (Matthews et al., 2002). By contrast, when exotic predators do not replace native species in an ecosystem, they increase the number of top predators. For example, many salmonid species have been introduced into the Great Lakes for recreational fisheries, and five have established populations (Cudmore-Vokey and Crossman, 2000).

New introductions have decreased since the 1960s on a global scale due to increasing environmental awareness and stricter international policies. Fish enhancements have increased since then, and the use of native fishes in stocking programs has become much more likely. However, studies discussing the potential

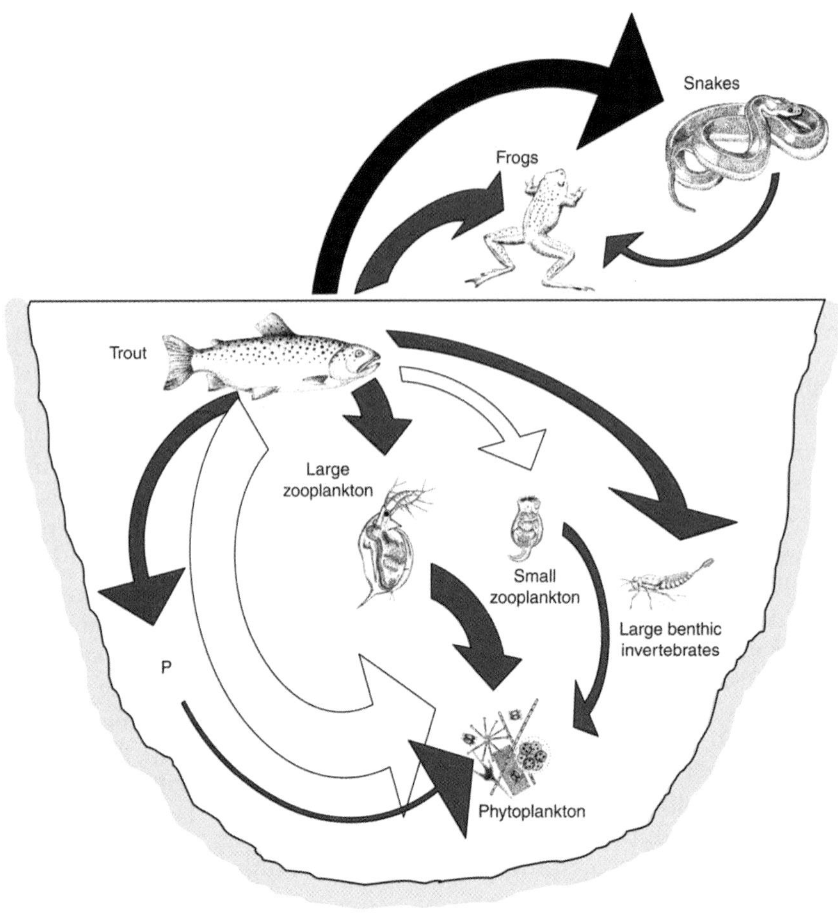

FIGURE 22-16 Effects of trout stocking on the food webs in Sierra Nevada lakes. Stocking fishes into fishless lakes results in a series of effects that cascade through the food web. *Blue lines* denote direct consumption or, in the case of phosphorus, uptake and excretion. *White* and *black lines* indicate positive and negative indirect effects of trout, respectively. *Arrow thickness* gives a coarse indication of the relative strength of the interaction. *(Adaptation of Eby et al., 2006.)*

effects of fish enhancement also raised concerns about topics such as a reduction of wild population abundance via density-dependent mortality (i.e., ecological competition between stocked and wild fish) and genetic introgression from the stocking of genetically maladapted fish in the wild (Araki and Schmid, 2010). For example, Li et al. (1996) found that stocking of walleye in lakes where natural reproduction occurred had no effect on population abundance but decreased the mean weight of fish. Besides, studies finding lower genetic diversity in hatchery populations often emphasized the risk of genetic differentiation and reduction of genetic diversity through the stocking of hatchery fish (Kostow, 2009).

In addition to causing loss of genetic variation within species, stocking can also result in changes in the food web or ecosystem function. Thus new studies increasingly document how fish stocking might increase top-down effects, modify the food web structure, and alter biogeochemical cycles. Stocking of predatory fish has often been used by lake managers in especially Europe and China for the purpose of restoration because of its top-down effects (Mehner et al., 2002; Mao et al., 2020). Thus enhanced abundance of top predators affects the resilience of a lake by decreasing the density of planktivorous fish, thereby increasing the top-down control by zooplankton of phytoplankton. Furthermore, introduced predators can simplify the food web structure by replacing multiple species at the same trophic level or decrease the diversity of lower trophic levels. For example, in lakes of East Africa Nile perch have simplified the food web by replacing hundreds of native consumers (Ligtvoet et al., 1991).

Fish stocking can also produce changes in the biogeochemical cycling within aquatic systems. In Sierra Nevada mountain lakes with trout stocking, increasing phosphorus regeneration by introduced fishes was approximately equal to the atmospheric phosphorus deposition, and thereby substantially increased the phosphorus available for algal production (Schindler et al., 2001). In the meantime, a decline in benthic invertebrate abundances and amphibians resulting from trout introductions also reduced the flux of energy and nutrients to the surrounding terrestrial systems (Cudmore-Vokey and Crossman, 2000).

VIII. Effects of physical modification

A. Dams

The concept of *connectivity* underlies many core questions in ecology because it defines linkages among ecosystem elements in space and time. As awareness of ecological connectivity has grown, the concept has become more widespread in the ecological literature, including aquatic ecology, where it is particularly relevant for rivers (Wiens, 2002). The most important role of connectivity for freshwater fish species lies in its longitudinal dimension. A separation of this longitudinal connectivity results in geographic isolation, which is one of the more pressing factors influencing species distributions (Fahrig and Merriam, 1985). Connectivity interruption has led to declines in the populations of half of the threatened European fish species (Northcote, 1998). It can severely deplete, even extirpate, local fish populations; however, if the connectivity to neighboring populations is maintained, then affected rivers can recover within several years (Howell, 2006).

Fish populations are highly dependent on the characteristics of their aquatic habitat, which supports all their biological functions. Migratory fish require different environments to complete their life cycle and search for food or to avoid adverse conditions (Northcote, 1978). The construction of a dam on a river can block or delay fish migration and thus contribute to a decline and even extinction of migratory species. In 1999 there were 45,000 dams more than 15 m high worldwide, capable of holding about 15% of the total global annual river runoff (World Commission on Dams, 2000). Global modification of water resources by building dams has had a greater effect on riverine fish than any other human activity (Petts, 1984). Dams can alter riverine environments by converting lotic habitats to lentic ones, creating physical barriers, altering natural hydrologic and geomorphic regimes, and disrupting nutrient cycling and sediment transport (Graf, 1999; Stanley and Doyle, 2003). Consequently, dams change the composition and structure of native biotic communities, limit the distribution of species, and block fish migration routes (Benstead et al., 1999; Freeman et al., 2001).

Dams constitute a major threat to freshwater biodiversity worldwide and have led to the documented extirpation and imperilment of many endemic fish species, especially *anadromous* species (those that spend most of their lives in the sea but migrate to freshwater to spawn, such as salmon) (Liermann et al., 2012). Obligate migratory behavior in fishes is a lead trait contributing to their vulnerability to dam obstruction. Dams prevent the migration of fish to feeding or breeding grounds or impede the function of these grounds by changing water depths, flows, and deposition patterns. In addition, competition for spawning sites and food among anadromous fish can increase as dams disconnect, isolate, and reduce the area of habitats (Cambray et al., 1997). For example, in the Shoalhaven River system of Australia, the building of the Tallowa Dam has been identified as the main reason for the extirpation or the depletion of many migrating species (Gehrke et al., 2002).

The concept of obstruction to migration is often associated with high dams. However, even low weirs can constitute a major obstruction to upstream migration. Santucci et al. (2005) investigated the effects of 15 low-head dams on the aquatic biota, habitat, and water quality in a 171 km reach of a midwestern warmwater river in Illinois (USA). The dams were found to impound 55% of the river's surface area within the study reach and influenced the distributions of 30 species of fish by restricting upstream migration. It is interesting to note that some fish species have a special ability to clear obstacles during their upstream movement; thus certain goby species possessing a sucker and enlarged fins are able to cling to the substrate and climb over dams (Mitchell, 1995).

In the first stages of dam development passage through hydropower dam turbines or spillways was not considered to be an important cause of damage to downstream migrating fish. However, a growing body of empirical studies shows that problems associated with downstream migration can also lead to mortality, especially of adult fish (Monten, 1985). Fish passing through hydraulic turbines are subject to various forms of stress (e.g., shocks from moving parts of the turbine or sudden variations in pressure and cavitation) that are likely to cause high mortality (World Commission on Dams, 2000). The mortality rate for migrating fish varies greatly depending on the type of the turbine and the species and size of the fish concerned. For example, mortality in adult eels is generally higher due to their body length. Passage through spillways may be another cause of fish injury or mortality, either directly (e.g., collision injuries) or indirectly (e.g., increased susceptibility of disorientated fish) (Heisey et al., 1996).

Dams can cause shifts in the downstream temperature regime through release of cold water from deep reservoirs. This may have important consequences for river organisms. Decreased temperatures often impair conditions favorable to native species but favor exotics or habitat generalists. For example, cold-water release from high dams of the Colorado River is considered to be the major reason for the decline in native fish abundance (Holden and Stalnaker, 1975). Meanwhile, King et al. (1998) found that the presence of dead and deformed young on their spawn beds in the Olifants River increased after a hypolimnetic water release from Clanwilliam Dam (South Africa), suggesting that changes in their river thermal regime may have had a detrimental effect on the fish larvae. Moreover, hypolimnetic waters could also be anoxic. In some reservoirs millions of fish have died suddenly, probably owing to the upward movement of oxygen-deficient water to near the surface during periods of strong winds or changes in water flow (Fukushima et al., 2017).

To enable fish to pass obstructions, including hydroelectric dams, various designs of fishways or fish ladders have been developed. Where fishways cannot be built, fish may be carried over the barrier by lifts or locks or simply be moved around a dam by trap-and-truck operations (Roscoe and Hinch, 2010). Initial development of effective passage facilities at dams was directed almost entirely at salmonid species. It took decades to develop effective facilities that have had some success in providing passage to salmonids (Williams et al., 2012). Recently, research has begun to focus on developing facilities for upstream nonsalmonid migrants. For example, juvenile paddlefish (*Polyodon spathula*) stocked in the Ohio River were confirmed to have passed through fish locks in both upstream and downstream directions (Barry et al., 2007). Nonetheless, despite advances in fish passage facilities, a large number of fishways still prevent or delay the passage of both target and nontarget species. Assessments of existing adult or juvenile fishways, where they have been made, suggest that migrant fish reject grounds with hydraulic conditions that they determine unsuitable. Even well-designed fish ladders or "nature-like" fish passes for river migrants will fail if incorrectly sited. In Canada and Europe nature-like fishways have been constructed at many obstructions, including even large hydroelectric dams. They look much like a small stream (Fig. 22-17). However, for these passages to work effectively, they

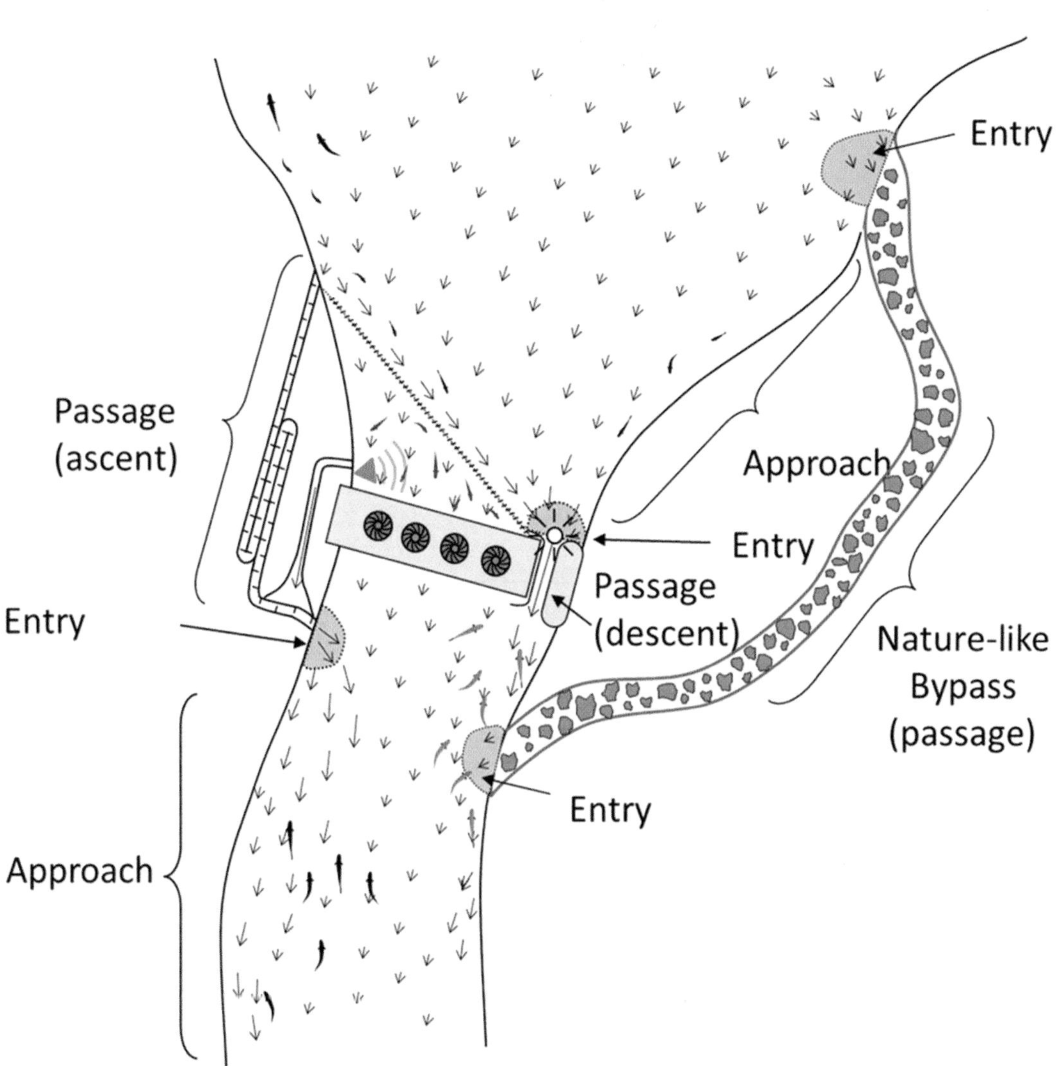

FIGURE 22-17 States of fish passage applied to any obstruction, here illustrated for a powerhouse equipped with separate up- and downstream fishways. *(Adaptation of Castro-Santos and Haro, 2010.)*

need to attract fish to approach, enter, and pass the fishways. Each of these three steps is a discrete task or state, and success or failure to advance through any one of them may occur for a number of reasons, including physical capability and behavioral rejection (Castro-Santos and Haro, 2010; Williams et al., 2012).

Except for fish passages, dam removal is the best option for reconnecting the river as it will eliminate barriers to migration for all types and life stages of fish. Dam removal is gaining credibility as an adaptive management option for dams that have deteriorated physically and are no longer economically practical. More than 1000 dams were removed in the USA by 2014 (O'Connor and Duda, 2015). One of the most widely publicized ecological aspects of dam removal is the elimination of barriers to fish migration. After the removal of four low-head dams in the Baraboo River, Wisconsin (USA), 10 of the 11 occurring species were collected at new sites upstream from the dam, and some of these fish species recolonized rapidly and in large numbers in upstream habitats (Catalano and Bozek, 2007). If dams degrade riverine habitats due to fragmentation, sedimentation, and reduced water quality, dam removal should result in reconnected habitats, restored substrates, and improved water quality. However, after the reorganization of fish communities after dam removals, it may take decades for the restored sites to approximate the community structure of nearby undisturbed sites. For example, Poulos and Chernoff (2017) examined the temporal effects of dam removal on the fish community in the Eightmile River, Connecticut (USA). They found that even 3 years after the removal, the river was still in a recovery stage, failing to approximate the community structure of the reference site.

B. River and lake regulation

Rivers around the globe are increasingly being subjected to various levels of physical alterations and river regulations to provide humans with services such as freshwater, hydropower, flood control, irrigation, and recreation. One of the most important ecological impacts of the physical manipulation of riverine systems is the resulting alterations of the natural flow regime. Natural flow dynamics play a key role in determining the functioning and services of rivers (Bunn and Arthington, 2002). Flow dynamics directly affect the physical and chemical characteristics of the river environment, such as water temperature, habitat heterogeneity, channel geomorphology, and nutrient dynamics (Oliveira et al., 2018). Moreover, they are important variables that may have both individual- and population-level effects on riverine fish, including the success of migration and spawning, the survival of eggs and juveniles, and food production (Murchie et al., 2008).

Regulation of a river can result in a sharp reduction of a migratory population. Any reduction in the river flow during the period of migratory activity can diminish the attractive potential of the river; hence the number of spawners entering the river may decline (World Commission on Dams, 2000). The suppression of the flood regime downstream of an impoundment through flow regulation can cause loss of obligate floodplain spawners by depriving them of their spawning grounds and valuable food supply. For example, Zhong and Power (1996) found that spawning grounds below the dams on the Qiantang and Han Rivers (China) were eliminated with the cumulative effect of diminished peak discharges, stabilized water levels, reduced current velocities, and water temperature. Besides, the degree of variability in the flow regime or specific flow events may play a role in the triggering of reproduction in many riverine fishes. For example, mostly in temperate rivers, rapid but short-lived rises in base flows may initiate spawning (Nesler et al., 1988). If fish reproduction is linked to specific flow events, then disruption of the flow may remove cues for maturation or subsequent spawning.

Another important consequence of river regulation or impoundment is the transformation from lotic to lentic environments. This shift often favors generalist over specialist species and leads to subsequent changes in biotic assemblages (Rahel, 2000). Species adapted to fast-flowing water are especially susceptible to such changes. In slow-flowing or impounded sites aquatic vegetation, periphyton, and macroinvertebrates can be negatively influenced by higher turbidity and sedimentation rates, leading to subsequent reductions of light penetration and changes in substrate composition (Poff et al., 1997). Osmundson et al. (2002) reported their investigations concerning physical habitat and trophic relations in a contiguous portion of the upper Colorado River (USA) inhabited by all life stages of an endangered Colorado pikeminnow (*Ptychocheilus lucius*). They suggested that the availability of prey fish for this native piscivore might, in part, be limited by the reduced standing crops of periphyton and macroinvertebrates resulting from the accumulation of fine sediment in the riverbed, a possible effect of river regulation. Furthermore, some *lithophile* fish species (e.g., Percidae: Etheostomatini) require rocky or gravel habitats, in addition to well-oxygenated flowing water, to spawn and care for their eggs (Simon, 1998). Loss of these habitats due to increased turbidity and sedimentation can render such habitats unsuitable for reproduction even if adults are able to survive.

Many lakes are also being regulated by the increase in hydropower production, flood control, and freshwater

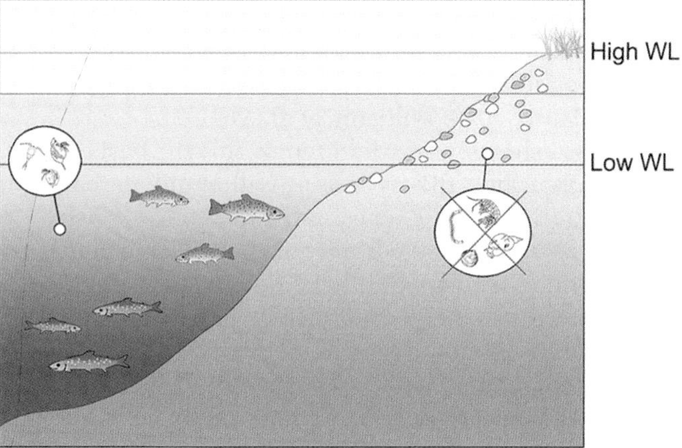

FIGURE 22-18 Schematic illustration of how water-level regulation (WLR) influences lower trophic levels and fish in lakes and reservoirs. The littoral food web compartment is affected by a loss of primary producers and a subsequent change in community composition and density of primary consumers. Sessile taxa become replaced by taxa that can move faster or have physiological adaptations or resting stages that survive desiccation and freezing. The effects of WLR on the pelagic food web compartment are less straightforward because pelagic organisms are less impacted by rising or falling water levels as they can simply "move" with the water level. However, zooplankton communities can be indirectly affected by WLR. For example, nutrient dynamics, water retention time, and other abiotic conditions such as water clarity can cause changes in predator–prey dynamics in the pelagic food web compartment. Many fish species use the littoral zone as feeding, spawning, or nursery grounds, but WLR can make the habitat unavailable when the water levels fall or become unsuitable as a result of macrophyte loss or increased substrate siltation. Due to the reduced littoral resources, the competitive interactions among fish change. Species and individuals that are better at exploiting pelagic or profundal resources gain a competitive edge over littoral specialists. *(Adaptation of Hirsch et al., 2017.)*

supply. The most obvious effect that regulation has on lake ecosystems is a change from natural water-level fluctuations to regulated water levels. These *water-level regulations* (WLRs) often exceed and differ from natural fluctuations in terms of their combined amplitude, rate of change, and frequency. Similar to the alterations of riverine flow dynamics, WLR can also ultimately affect the abundance, growth, niche use, and competitive interactions among fish populations in lakes and reservoirs (Fig. 22-18). Hirsch et al. (2017) summarized several main processes that affect fish when natural water-level fluctuations change into WLR. For example, the degradation or loss of suitable spawning and nursing grounds due to WLR directly affects variations in spawning success and population recruitment. Moreover, the relative changes in the lake's littoral and pelagic food web compartments can have cascading and feedback food web effects. As resources change, competitive and predatory interactions between and among fish species are rearranged.

Water abstraction and transfer, as prevalent activities in river regulation, are rapidly increasing worldwide in order to respond to escalating demands for irrigation, hydropower, and drinking water. Abstraction can alter the hydrological regime of rivers and degrade the habitat quality and may impair the capacity of rivers to support native biota. Because abstraction reduces discharge, its impacts on river ecosystems are somewhat similar to those of drought. Although abstraction can also affect rivers in moist regions where aquatic organisms are adapted to life in permanently flowing waters and thus vulnerable to reduced discharge, its effects on rivers in arid and semiarid ecosystems are stronger (Ledger et al., 2013). For example, rivers in a Mediterranean climate are particularly influenced by water abstraction because they combine a deficit of surface water availability with high human population densities and agricultural development.

The International Union for Conservation of Nature (IUCN) (Smith and Darwall, 2006) has already confirmed abstraction as one of the most important environmental issues in Mediterranean freshwaters and the cause of four of the eight freshwater fish extirpations in this region. Benejam et al. (2010) also investigated six river basins of the Iberian Peninsula and suggested that the effects of water abstraction on fish assemblages are clearly detectable, including reduced population densities, fewer benthic species, and reduced occurrence and abundance of intolerant species. However, water abstraction does not always have "negative" impacts. From a study in the Manuherikia River, New Zealand, Leprieur et al., (2006) suggested that hydrological disturbance associated with human activities benefited a native fish at the expense of an exotic taxon. Low flow conditions associated with water abstraction were considered to be a key factor controlling the spatial distribution of exotic brown trout in the Manuherikia River catchment.

IX. Invasive species and ecosystem consequences

Although all species disperse to some degree, the rate of moving species across the planet has been greatly accelerated by humans (Kolar and Lodge, 2000). This is particularly true for most aquatic species that lack the means for active dispersal into isolated drainages and instead use a variety of transport vectors for passive dispersal.

Invasive alien species (IAS), sometimes referred to as *exotic species*, are species that are introduced, accidently or intentionally, outside of their natural geographic range and that become established and abundant with negative consequences for their new environment. Although several 19th-century naturalists, such as Charles Darwin, mentioned invasive species in their writings, they were only curiosities and not truly perceived as threats to biodiversity. Invasion ecology, that is, the science that studies biological invasions, started in the second half of the 1900s, with Charles S. Elton's (1958) book titled *The Ecology of Invasions by Animals and Plants*. Since then, invasion ecology has grown enormously as a full branch of science (Richardson and Pyšek, 2008).

At present, IAS are recognized as one of the main threats to biodiversity and ecosystem functioning due to their negative impacts at individual, population, community, and ecosystem levels (Gallardo et al., 2020). A reduction in the biological uniqueness (*biotic homogenization*) and extinction of native and endemic species due to competition and predation, as well as habitat modification, genetic changes, and parasite dissemination, are some of the negative effects of the species introduction and invasion. Furthermore, many effects are difficult to predict or occur only in the long term (Vitule et al., 2012) because a time lag of several years can occur from a species introduction to the establishment of a viable population and then to invasiveness. As such, the impacts of alien species introductions are often underestimated (Gherardi et al., 2008; Ricciardi and Kipp, 2008).

Not all nonnative species introduced in a new environment may establish themselves and, among them, not all become invasive and cause biological or economic harm (Colautti and MacIsaac, 2004; Schlaepfer et al., 2011; Williamson, 1996). Nevertheless, current evidence shows that the desirable effects of nonnative species are outweighed by their undesirable effects (Vitule et al., 2012; Simberloff et al., 2013), and a precautionary principle in the risk management of invasive species is usually strongly recommended.

Biological invasion of a water body requires both dispersal and establishment. Dispersal requires a vector and a first inoculation in the new environment. Establishment requires that alien individuals are physiologically adapted to the physical-chemical environment and that they can find adequate food, avoid predators, and reproduce. Only later can they spread and integrate into other nearby sites (Moyle and Marchetti, 2006).

In freshwater ecosystems fishes are frequently introduced as alien species by intentional stocking, involuntary releases, and hitchhiking with other carriers (vectors). For instance, environmental and fishery resource agencies deliberately translocate and stock game fish in lakes worldwide (Kolar and Lodge, 2000; Pelicice et al., 2014). Anglers introduce other species by dumping bait (Litvak and Mandrak, 1993). The aquarium and ornamental fish trades import a stunning variety of fishes and, if released from aquaria, either voluntarily or accidently, are common methods of species invasions (Padilla and Williams, 2004). Also, fish are introduced in order to control unwanted organisms like aquatic vegetation (Chilton and Muoneke, 1992) and mosquitoes (Hildebrand 1919; Bay, 1967). Finally, involuntary fish introduction and spread by water in ships' ballast tanks is now a well-recognized vector for dispersal (Wonham et al., 2000).

The spread of alien fish species in freshwaters experienced a boost in the middle of the 1800s when humans discovered how to transport successfully fecundated eggs in moss and ice for long distances (thousands of kilometers) and time (months) by ships. Salmonids, such as brown trout (*S. trutta*) and Atlantic salmon (*S. salar*), were the first group of fish species to be transported for long distances in the second half of the 1800s (Minard, 2015). The first successful case of long-range transport was in 1864, and the *Yeoman and Australian Acclimatiser Journal* published that, on 24 June 1864, the ship *Norfolk*, which left the London docks on 21 January 1964, arrived in Melbourne on the following 19 April, carrying a load of salmon and brown trout eggs.

Brown trout (*S. trutta*) is an example of an introduced species becoming invasive. It is a cold stenothermal salmonid native to Europe and was introduced for angling purposes in many lakes and rivers of Australia, New Zealand, South America, India, and Africa during the second half of the 1800s to meet the recreational fisheries demand in European colonies. At present, it appears as one of the 100 worst IAS in the Global Invasive Species Database (Lowe et al., 2000).

The impact of alien invasive fish species is multifaceted. One of the most common impacts is the disruption of native species populations, leading to population decline, changes in native species population structure, and in some cases local and global extinction. The mechanisms of interaction are multiple and include direct predation, induced mortality (i.e., via the spread of infectious diseases), competitive exclusion, and introgressive hybridization phenomena between congeneric species. Often, those mechanisms are mediated or boosted by changes in habitat characteristics and/or native fish populations induced by humans, such as by cultural eutrophication or fishing, or by climate changes that, via water warming and changes in the hydrological regime of rivers, tend to favor alien and often more tolerant fish species.

The seminal example of a fish invasion and its effects on biodiversity is that of the Nile perch (*L. niloticus*; Barel et al., 1985). Nile perch is a generalist and large-sized predator fish (up to 1.6 m in length) belonging to the Latidae family (order Perciformes). Nile perch were introduced unofficially into Lake Victoria (Africa) in 1954 and officially in 1962 and 1963. The aim of its introduction was to convert the fish biomass of the lake, consisting of at least 80% endemic small haplochromine cichlids harvested by local populations, to a more valuable and marketable resource such as this large-sized, edible fish (Pringle, 2005). After a period of acclimatization with low densities, lasting approximately 15 to 20 years, the Nile perch population underwent a boom and bust cycle, leading to a population increase. This development was favored by a massive decline in cichlid abundance induced by eutrophication and fishing pressure (Goudswaard et al., 2008; Kolding et al., 2008). Once haplochromine mortality increased up to a certain level, Nile perch were released from this so-called *recruitment depensation* (Walters and Kitchell, 2001). In other words, the decrease of the densities of adult haplochromines controlling young Nile perch by predation increased the recruitment of Nile perch to large-sized stages, and this led to an increase in the predation of adult Nile perch on haplochromines, finally causing a transition from a haplochromines-dominated lake to a Nile perch–dominated lake where the haplochromines no longer control Nile perch recruitment (van de Wolfshaar et al., 2014).

A further mechanism of impact on native, and often also on endemic, species is subtler but no less important, namely the genetic decline induced by hybridization with taxonomically related alien species. This usually happens when humans, through introductions, interrupt the physical separation among species that was maintained for centuries by natural barriers such as mountains and seas. Genetic introgression can increase the likelihood of extinction by reducing the fitness and the ability of a population to adapt to changing environmental conditions (Allendorf and Leary, 2011). For instance, rainbow trout (*Oncorhynchus mykiss*), one of the most important species in freshwater aquaculture and inland fisheries (Crawford and Muir, 2008), has undergone extensive hybridization with the native trout species cutthroat trout (*O. clarki lewisii*), Apache trout (*O. apache*), and Gila trout (*O. gilae*), threatening their persistence in the western United States (Hitt et al., 2003; Weigel et al., 2003; Dowling and Childs, 1992). A similar development has occurred for brown trout, the other salmonids massively stocked for angling purposes worldwide. Besides detrimental effects on the native fish fauna by predation and competition for food and habitats (Townsend, 1996; Lobon-Cervià and Sainz, 2017), brown trout have hybridized with congenerics such as marble trout (*S. marmoratus*), threatening the persistence of the latter in south European rivers flowing into the Adriatic Sea (i.e., Meraner et al., 2007; Sušnik Bajec et al., 2015).

Effects of alien fish species invasions can be identified at different levels of the biological organization: from individuals to populations, communities, and ecosystems. For instance, invaders can alter the behavior of invertebrate species and thereby influence their habitat use and foraging strategies. As reported by (McIntosh and Townsend, 1994), brown trout, when introduced in New Zealand streams, changed the diurnal activity and microhabitat selection of many invertebrate preys, especially those grazing periphyton on rock surfaces on the streambed. Changes in prey behavior have also been identified in lakes where fish have been introduced. A classic example is that of high-altitude fishless lakes where salmonids have been introduced for fishing purposes. Because trout and charr (*Salvelinus* spp.) are visual feeders, the encounter probability with their prey increases as water clarity and light increases. Therefore, to avoid predation, the items of prey change their vertical distribution during the day and move into deeper and darker waters. For instance, Gliwicz and Rowan (1984) reported significant differences in the vertical distribution of the copepod *Cyclops abyssorum tatricus* (Kozminsky) among lakes with and without stocked brook charr (*Salvelinus fontinalis* Mitchill). Where fish were present, the copepod was moving into deeper and darker waters to avoid predation during the day (see also Chapter 20).

In many cases the impact of an invasive fish species is complex and includes not only direct but also cascading effects on food webs, water quality, and the entire ecosystem. Common carp (*C. carpio*) can be used to illustrate the different types of impacts of an IAS in a freshwater environment and how they are often interconnected (Fig. 22-19). The carp is a native fish from the Ponto-Caspian region (Balon, 1995), and common carp is the third most frequently introduced species in freshwaters. It is stocked to enhance fisheries in lakes and ponds and, at present, it is a source of protein for human consumption in many tropical and subtropical areas. Stocking material is relatively cheap and carp are quite resistant to stress and low oxygen concentrations in the water. Furthermore, it is highly appreciated by many recreational fisheries and, for this reason, stocked in many "put and take" (i.e., regularly stocked) lakes for its recreational value. However, despite its

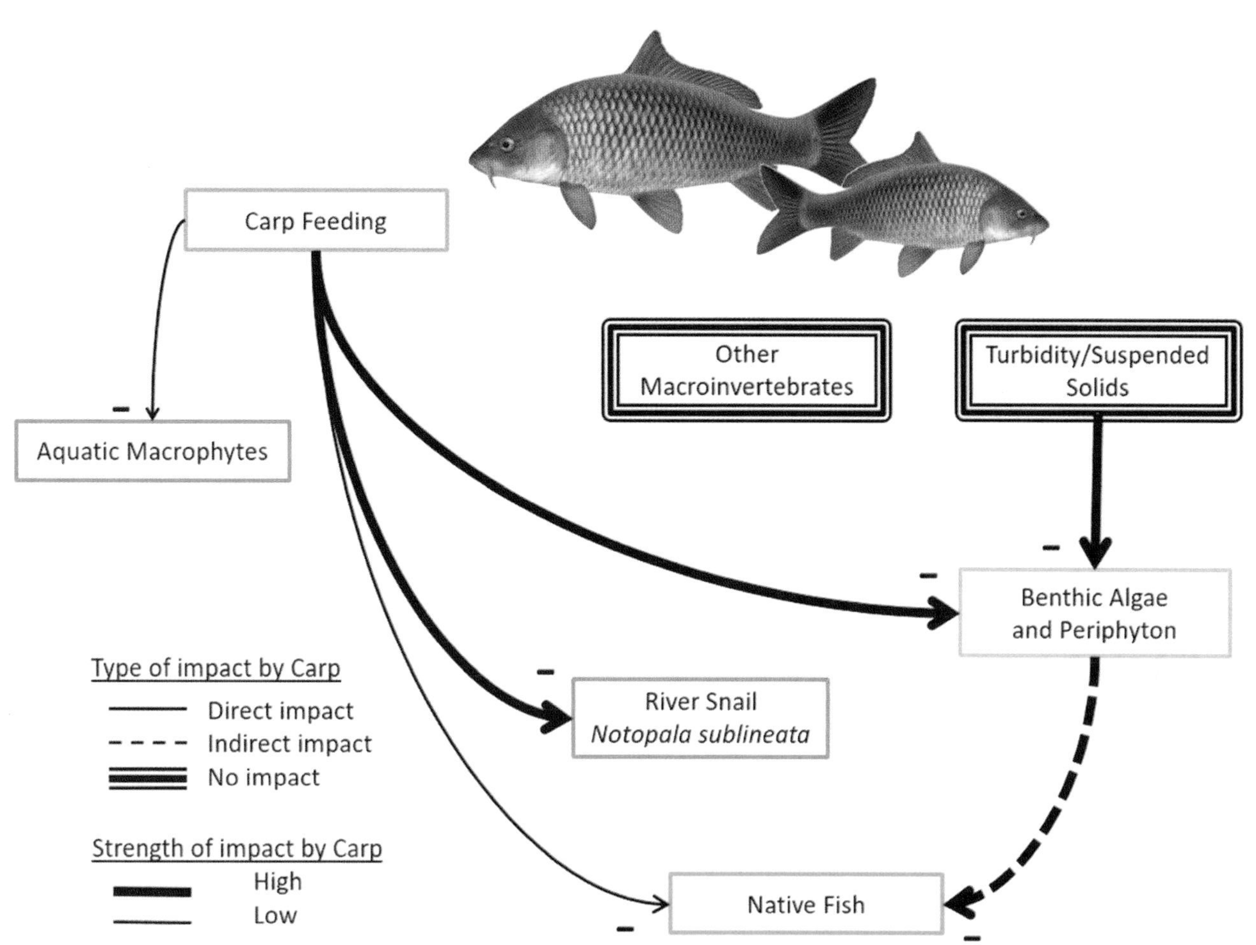

FIGURE 22-19 Refined conceptual model of carp impacts on aquatic ecosystems of dryland river waterholes in Australia. *(From Marshall et al., 2019.)*

importance worldwide, it is considered one of the most pervasive and destructive invasive species (Lowe et al., 2000; Zambrano et al., 2006; Koehn, 2004; Weber and Brown. 2009; Vilizzi et al., 2015; Marshall et al., 2019). Common carp is a zoobenthivorous fish and, especially in shallow aquatic ecosystems, it is considered an "ecosystem engineer" (sensu Jones et al., 1994) because it has strong effects on benthic communities due to its feeding behavior. Common carp negatively affect macrophyte abundance by reducing light availability, increasing siltation rates, ingesting plant matter, and uprooting when feeding (Bajer and Sorensen, 2015; Parkos et al., 2003), and it therefore threatens wetlands that are used by many fish as spawning grounds and nursery habitats (Parkos et al., 2003). Carp also plays an important role in the modification of the nutrient cycling of shallow lakes by increasing nutrients in the water column in two ways: by sediment resuspension and by excretion (Breukelaar et al., 1994; Meijer et al., 1990). Together with the decrease of macrophyte density induced by its feeding behavior, carp introductions can stimulate algal bloom formation and a shift from a clear and macrophyte-dominated to a turbid phytoplankton-dominated state (Chapter 26). Carp, especially when it forms stunted and dense populations, may prey upon zooplankton and, in this way, it releases phytoplankton from zooplankton grazing and therefore reinforces a turbid and phytoplankton-dominated state in shallow lakes (Breukelaar et al., 1994). Additional impacts of the common carp are those on the native species. For instance, Laird and Page (1996) found that common carp competed with ecologically similar species such as carp suckers and buffalo fish, and there is also evidence that common carp prey on the eggs of other fish species (Taylor et al., 1984). Finally, carp may cause nuisance to humans and other animals by stirring up river substrates and reducing aquatic vegetation, rendering waterways unattractive and unsuitable for swimming or undrinkable for livestock (Global Invasive Species Database, 2023).

IAS are increasingly recognized as one of the factors disturbing global biodiversity. Among them, fish play a central role because many fish species have been spread over all inhabited continents with negative effects on native and endemic species and ecosystem functioning. Nevertheless, due to the importance of fish as a source of protein for millions of people, there is an ongoing debate about the need to elucidate and assess both the positive and the negative aspects of fish introduction (e.g., Gozlan, 2008; Vitule et al., 2009) with the aim to reconcile ecology with economics.

X. Restoring lakes by biomanipulation

Biomanipulation, a term introduced by Shapiro et al. (1975), is a type of biological engineering in which manipulations of biota are used to reduce objectionable algal types and biomasses in addition to reductions of nutrient loading. Biomanipulation was originally based on the idea that when the number of planktivorous fish are reduced, the density of large cladoceran zooplankton increases, and their grazing can reduce certain species of planktonic algae and reduce the algal turbidity of the water. The most frequently used biomanipulation method is therefore the removal of planktivorous and benthivorous fish. This method has been extensively used over the past 20 years or so in northern temperate lakes in Europe (Søndergaard et al., 2007). It has been most successful in small shallow lakes but has had variable long-term effectiveness. Removal of a high proportion of the planktivorous and benthivorous fish stock during a 1- to 2-year period has been recommended to avoid regrowth of the original stock and to stimulate the growth of young specimens of fishes that potentially become piscivores when they reach a sufficient size (Hansson et al., 1998; Jeppesen and Sammalkorpi, 2002).

An efficient reduction of the zooplanktivorous fish biomass generally achieves dramatic, short-term cascading effects in eutrophic lakes—a shift occurs to dominance by large zooplankton, phytoplankton biomass (including noxious Cyanobacteria) is reduced, and water transparency improves (Hansson et al., 1998) (Fig. 22-20). Moreover, an increase occurs in benthic feeding and the number of herbivorous waterfowl (Allen et al., 2007; Fox et al., 2019), as does the proportion of piscivorous fish (e.g., perch and pike), in part promoted by the development of submerged macrophytes. Strong cascading effects of fish removal have also been found in ponds, also leading to a decrease in Cyanobacteria (Peretyatko et al., 2012).

In some temperate lakes, however, a reduction of cyanobacterial biomass and improved water clarity after biomanipulation have been achieved without initiating a trophic cascade, which has been attributed to reduced P release from the sediment, not least due to fewer fish foraging at the bottom after the biomanipulation (Horppila et al., 1998).

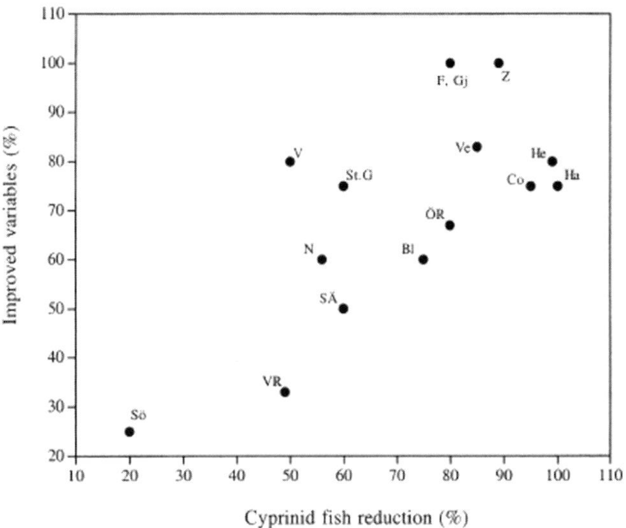

FIGURE 22-20 Proportion of the diagnosis variables (%) that were still improved 5 years after the biomanipulation in relation to the amount of fish taken from the lake. Variables are Secchi depth, chlorophyll *a*, cover of submerged macrophytes, *Daphnia* biomass, total phosphorus, and blue-green algae. Sö, Sövdeborgssjön; VR, Western Basin (Ringsjön); SÄ, Sätoftasjön (Ringsjön); N, Norddiep; Bl, Bleiswijk Zoom; St. G., Lake St. George; V, Væng; ÖR, Eastern Basin (Ringsjön); Ve, Vesijärvi; Co, Cockshoot Broad; Ha, Haugatjern; He, Helgetjern; Z, Zwemlust; Gj, Gjersjøen; F, Finjasjön. *(From Hansson et al., 1998.)*

An alternative or supplementary method to fish removal is stocking of potential piscivores (Benndorf, 1990; Berg et al., 1997; Drenner and Hambright, 1999; Ha et al., 2013), reflecting both a top-down control on planktibenthivorous fish (Carpenter and Kitchell, 1993) and/or by creating a behavioral cascade (Romare and Hansson, 2003). By preying on zooplanktivorous and benthivorous fish, piscivores diminish prey fish abundance, resulting in lower fish-induced resuspension, lower predation on large zooplankton, and lower translocation of nutrients from sediments to water via feeding and excretion. However, stocking of pelagic foraging piscivores, such as zander (*Sander lucioperca* L.), may also affect water clarity more indirectly (behavioral cascade) by forcing prey fish to take refuge in the littoral zone, which releases predation on pelagic zooplankton (Braband and Faafeng, 1994; Romare and Hansson, 2003). However, the success of stocking piscivores to restore lakes has generally been modest unless the species introduced were absent in the lake before (or lost) (Drenner and Hambright, 1999; Jeppesen et al., 2012).

There are only a few long-term studies (>10 years) on the effects of fish manipulation and the results are ambiguous. However, in most studies where strong short-term effects were observed, a gradual return to the turbid state and higher abundance of zooplanktivorous fish occurred after 5 to 10 years. In particular, the trophic cascade effects in the pelagic seem to weaken in the long term due to the return of planktivorous fish; in contrast, the impact on benthic fish apparently lasts longer and may even be permanent. The most comprehensive comparative study of long-term responses so far is from 27 eutrophic Danish lakes (Søndergaard et al., 2008), where the abundance of benthivores, such as bream, was demonstrated to remain lower than before the restoration, creating significantly reduced levels of suspended matter and improved water clarity in the long term. However, the grazing effect, as indicated by the zooplankton:phytoplankton biomass ratio, returned to premanipulation levels and Cyanobacteria returned and became dominant, indicating almost complete nullification of the strong trophic cascade in the pelagic observed 10 to 15 years after the biomanipulation, likely reflecting a return of planktivorous roach. A similar response with long-term effects on bream and mainly short-term effects on roach was observed in biomanipulated Cockshoot Broad, UK (Hoare et al., 2008). Nonetheless, some examples exist of lakes with a persistent clear water state in the long term after an external nutrient loading reduction if the annual mean TP concentrations reach levels of <0.05 mg P L^{-1} for shallow temperate lakes (Jeppesen et al., 2000) and <0.02 mg P L^{-1} for deep temperate lakes (Sas, 1989). However, the internal loading may remain high for decades after the loading reduction (Søndergaard et al., 2013).

Repeated fish removal may be a way to maintain the clear water state if lakes are deemed likely to shift back to the turbid state. Fortunately, the effort needed may be less comprehensive than during the first manipulation for several reasons: (a) large benthivorous fish may suffer from food limitation; (b) small potential piscivores, such as perch, are typically more abundant after the first biomanipulation and tend to shift to piscivory more quickly; (c) presence of submerged macrophytes, seeds, or *turions* (overwintering organs) may facilitate faster development of plants compared with the first manipulation; and (d) the mobile pool of P in the sediment has declined (buried deeper or released via the outlet).

Biomanipulation at warm locations (Mediterranean, subtropical, and tropical lakes) is less likely to be successful for at least four reasons: (a) fast reproduction means high recruitment after fish removal; (b) control by piscivores is generally weak; (c) dominance by omnivores, entailing high predation on large-bodied zooplankton and sediment disturbance; and (d) dominance by small-sized fish (higher predation on zooplankton) (Jeppesen et al., 2012). However, warm lakes also host filter-feeding fish that can consume nuisance algae such as Cyanobacteria, and these have been used as a restoration tool. Typically, various carp and tilapia species are introduced with mixed effects (Arcifa et al., 1986; Jones and Poplawski, 1998; Starling, 1993; Starling et al., 1998). The problem is that they also eat zooplankton and disturb the sediment when foraging and thereby cause nutrient release, and these effects may outperform their filtration of phytoplankton, resulting in phytoplankton enhancement (Attayde and Hansson, 2001; Figueredo and Giani, 2005; Shen et al., 2021). An illustrative example of this was the effect observed of a massive stocking of silver carp and bighead carp conducted to control phytoplankton in Lake Taihu (China) since 2008 (Mao et al., 2020), which has led to an increase in the fish stock with substantial undesired effects on the lake ecosystem (Fig. 22-21). Thus, despite external nutrient loading reduction, phytoplankton biomass and the biomass of nuisance Cyanobacteria have increased, and zooplankton biomass and the size of cladocerans have decreased, as has the zooplankton:phytoplankton biomass ratio, all of which are a sign of a major reduction of grazer control on phytoplankton.

A promising alternative for warm lakes, however, is to combine fish removal with transplantation of submerged macrophytes and capping phosphorus in the sediment (Liu et al., 2018; Zhang et al., 2021). The precipitation of phosphorus and a reduction of internal loading followed by chemical restoration will result in higher water clarity, which, in turn, may reinforce recovery by altering the top-down control of zooplankton by fish through a trophic and behavioral cascade. It is possible that a combination of aluminum treatment or other chemical restoration methods with biomanipulation may have stronger, and perhaps also more long-lasting, effects than when the various methods are applied individually. Synergistic effects may even reduce the cost of restoration as a less extensive chemical and fish manipulation effort will be needed. However, evidence of this is so far limited.

It is important to emphasize that the key measure to restore eutrophicated lakes is reduction of the external nutrient loading. In-lake restoration only serves the purpose of reinforcing recovery, treating symptoms, or improving/maintaining a high environmental quality temporarily until the external loading is significantly reduced.

XI. Climate change impact

Freshwater fish are both directly and indirectly affected by changes in temperature. Being ectotherms, they cannot thermoregulate physiologically but only by moving to areas with appropriate temperatures. Cold-stenothermal species (e.g., Arctic charr, *Salvelinus alpinus*) will shift toward higher latitudes or altitudes, or deeper waters, if possible, but may become locally extinct at the warmest edge of their current distribution

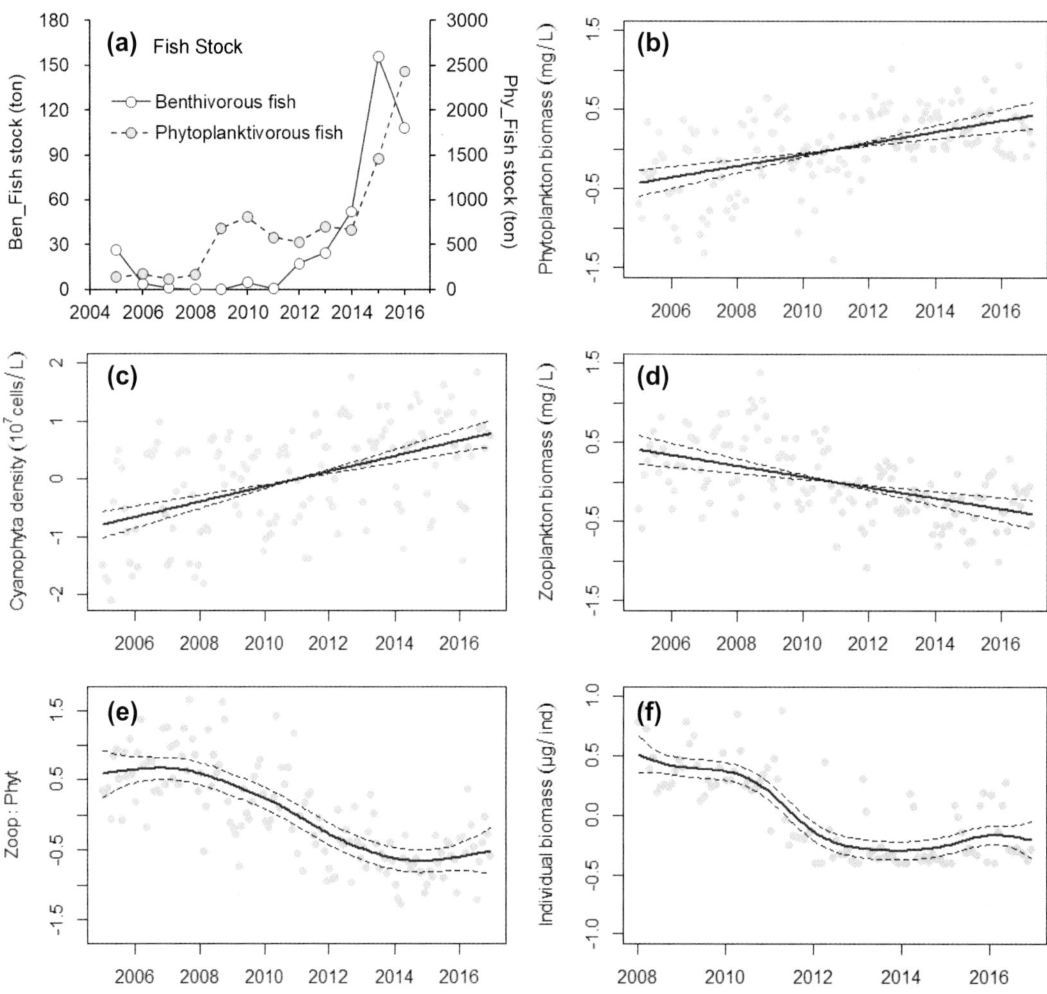

FIGURE 22-21 Time series of fish stock (a), and monthly phytoplankton biomass (b), Cyanobacteria density (c), zooplankton biomass (d), zooplankton-to-phytoplankton biomass ratio (Zoop:Phyt) (e), and mean individual biomass of cladocerans (individual biomass) (f) in Lake Taihu. The *blue solid line* represents the regression line fitted with monthly data using generalized additive mixed models (GAMMs). The *dotted lines* indicate the 95% confidence intervals for the fitted function. Ben_fish and Phyto_fish are benthivorous and phytoplanktivorous fish, respectively. *(Modified from Mao et al., 2020.)*

ranges (Lappalainen and Lehtonen, 1997; Wrona et al., 2006; Graham and Harrod, 2009).

In contrast, eurythermal species exhibiting wide thermal tolerance (e.g., common carp, *C. carpio*) may be able to cope with the new thermal regimes, resulting in a global widening of their habitats (Lappalainen and Lehtonen, 1997). For instance, Souza et al., (2022), by combining standardized catch data of common carp from 378 European lakes with climate data, showed that climate is an important predictor of common carp population viability, which is particularly enhanced under dry conditions and elevated temperatures during spring and the summer months (Fig. 22-22).

By contrast, tropical species typically experience mean temperatures that are close to their physiological optima, and even a small increase in temperature may put them at high risk of extinction (Tewksbury et al., 2008). Many fish species are also adapted to low oxygen concentrations (Holopainen et al., 1997; Soares et al., 2006). When temperature increases, oxygen may drop to critical levels as warm water holds less oxygen and the respiration rates increase (Jeppesen et al., 2020). Global warming may therefore create novel fish assemblages by favoring species that can respond to range shifts in addition to those already locally present within their native range, and this may increase the competition for space and food. As fish species richness is higher in warmer climates (Griffiths, 1997; Amarasinghe and Welcomme, 2002; Zhao et al., 2006; Teixeira-de Mello et al., 2009), richness may increase in present-day cold lakes in a future warmer climate, depending on local conditions, original assemblages and physical barriers to colonization.

Life history traits will also be affected by warmer temperatures. Cross-comparisons of fish populations

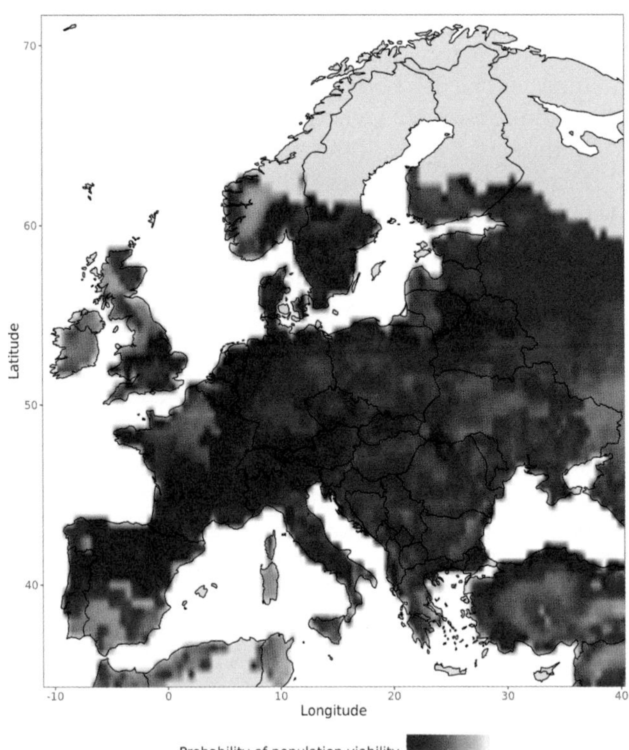

FIGURE 22-22 Projection of the viability of the common carp (*Cyprinus carpio*) population in lakes in Europe. The *grey areas* indicate the regions where the species is currently absent according to the IUCN (International Union for Conservation of Nature). *(From Souza et al., 2022.)*

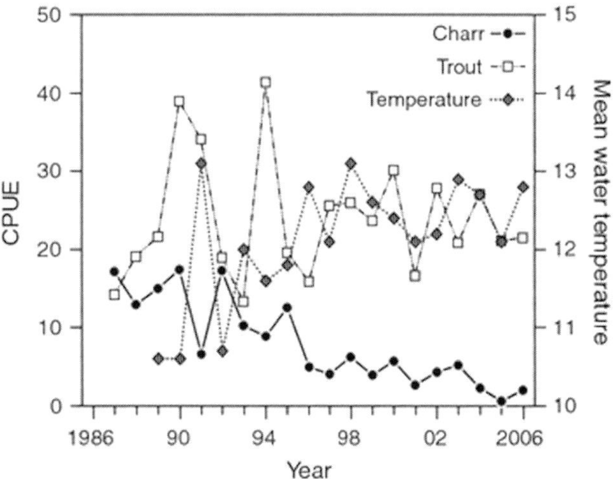

FIGURE 22-23 Catch per unit effort (CPUE, average number of fish caught per net) of Arctic charr and brown trout together with June–September mean temperatures in Lake Elliðavatn. *(Modified by Jeppesen et al., 2012 after Antonsson and Árnason, 2011; IMO, 2011; Malmquist et al., 2009.)*

in similar systems in South America and Europe (Teixeira-de Mello et al., 2009) and within Europe have shown that lower-latitude fish species are often not only individually smaller (Griffiths, 1997; Jeppesen et al., 2010) but also grow faster, mature earlier, have shorter life spans, and allocate less energy to reproduction than species at higher latitudes (Blanck and Lammouroux, 2007). Similar patterns have been found in North America (Mims et al., 2010). Additionally, fish species' vulnerability to climate change generally depends on the most temperature-sensitive life stages. Dahlke et al. (2020) used observational, experimental, and phylogenetic data to assess stage-specific thermal tolerance metrics for 263 freshwater fish species from all climate zones. Their analysis shows that spawning adults and embryos consistently have narrower tolerance ranges than larvae and nonreproductive adults and are most vulnerable to climate warming. Therefore reproductive success is likely at risk under climate change when spawning habitat temperatures exceed the tolerance limit of the most sensitive life stage.

Time series data from a number of European lakes have revealed change attributable to warming in the recent past (Jeppesen et al., 2012). There has been a clear trend toward higher importance of eurythermal species in several of the lakes (Lake Peipsi, Windermere, Lake Geneva, Lake Maggiore, and Lake Constance) as judged from fish harvests and surveys. The cold-stenothermic Arctic charr has been particularly affected by warming, showing a clear decline in the majority of the lakes where its presence is recorded (four out of five, i.e., Lake Elliðavatn, Windermere, Lake Geneva, and Lake Vättern). As shown by Winfield et al. (2010), the thermal problems faced by this species are expectedly more adverse in shallow than in deep lakes due to higher temperatures and lack of a cold hypolimnion refuge. In Iceland charr appears also to suffer from the thermally linked proliferative kidney disease (PKD) (Sterud et al., 2007), as is the case in Lake Elliðavatn (Kristmundsson et al., 2010). This highlights the importance of complex and potential synergetic effects of warming on fish, including both direct effects (e.g., on metabolism and growth) and indirect effects, such as diseases. The population of brown trout, which is a more heat-tolerant species than Arctic charr (Elliott and Elliott, 2010), has remained largely unchanged in Lake Elliðavatn (Fig. 22-23) but has decreased substantially in the warmer Lake Maggiore (Jeppesen et al., 2012).

Although evolutionary responses to global climate change in aquatic ecosystems are poorly documented, a small number of studies indicate that freshwater fishes and other organisms are already exhibiting genetic change (Pauls et al., 2013). For instance, Kovach et al. (2012) used 32 years of genetic data to reveal a genetic change toward earlier migration timing in an adult pink salmon population; average migration time occurs nearly 2 weeks earlier than it did 40 years ago. This rapid microevolution was likely in response to increasing stream temperatures and shifting oceanic conditions. Similarly, they also examined whether genetic diversity (allelic richness) was related to climatic variation in 130 bull trout populations from 24 watersheds across the Columbia River Basin, USA (Kovach et al., 2015). The results strongly suggest that climatic variation influences evolutionary processes in this threatened species and that genetic diversity will likely decrease due to future climate change.

Climate change acts on aquatic ecosystems in concert with other anthropogenic stressors, and together these stressors may have complex, compounded effects on freshwater fishes (Lynch et al., 2016). Some important stressors that are known to interact with climate change are water pollution, stream and river impoundments and flow alterations, altered land use, disease and parasites, and fishing exploitation (Kwak and Freeman, 2010; Staudt et al., 2013). For example, the fish assemblage is not only affected directly by the heating and changes in the thermal stability of the lakes. Numerous recent studies and reviews indicate that warming will exacerbate existing eutrophication problems (Moss et al., 2011), and this will, in a self-amplifying manner, likely further stimulate a shift to small-bodied fish and to the dominance of eurythermal species, which typically tolerate low oxygen levels and high ammonia concentrations (see review in Graham and Harrod, 2009). Therefore we can expect an allied attack by eutrophication and warming in lakes in the future (Moss et al., 2011), and shifts in fish abundance, body size, and composition will be reinforced and stimulated by this process.

XII. Summary

1. Fish are the most species-rich (~30,000) group of vertebrates on Earth, and they exhibit an astonishing variability in size and shape. They have colonized very diverse and remote habitats—from deep ocean depths to high-altitude watercourses and lakes, from polar icy waters to hot springs. About 49% of all known fish species live in the sea, whereas 51% live in lakes and rivers.
2. Fish belong to the phylum Chordata and the subphylum Vertebrata and can be divided into the following three classes: Agnatha, jawless and cartilaginous eel-like fishes; Chondrichthyes, fishes with a cartilaginous skeleton and jaws (gnathostomes), and Osteichthyes, fishes with a bony skeleton. In bony fish the swim bladder counterbalances the higher density of the skeleton structure that is often needed to support the powerful musculature required for swimming.
3. Fish swim with two types of striated skeletal muscles: aerobic red muscle and anaerobic white muscle. The fish body is covered by skin that forms a continuous outer layer. In special pockets that open in the lower layer of the skin (endoderm), scales are formed. Not all species of fish have scales, however.
4. Gills are the primary body organ where the oxygen dissolved in the water enters the bloodstream and from where carbon dioxide is released. Additionally, some fish species can exchange gas with the water through the skin or through lung-like organs of varying design.
5. Fish growth is a continuous process and the fish length increases during the entire lifetime of the fish. Most fish produce offspring by eggs, but some fishes are hermaphroditic with self-fertilization and some have rare types of reproduction. Fish are ectotherms, and their metabolism and all physiological processes are strictly temperature dependent.
6. Fish of inland waters inhabit an extraordinary variety of habitats spanning from Arctic to temperate to tropical lakes and reservoirs, and from cold water to tropical rivers, desert waters, and caves.
   a. Biological, chemical, and physical factors influence differences among and within fish communities in both rivers and lakes, and climate, in particular, has been shown to correlate strongly with fish community richness, diversity, density, and size structure. Fish species richness is often higher in warm lakes and streams. Also, functional diversity is higher in many warm lakes, independent of the trophic state.
   b. Piscivores are species that primarily feed on other fish as adults, invertivores feed primarily on benthos or zooplankton, and herbivores feed solely on plants or algal material. Detritivores are species where detritus is the dominant food source. Omnivores take food more or less equally from at least two trophic levels.
7. Along with evidence of changes in fish community composition from upstream to downstream habitats, rivers are classified into different zones. In temperate lakes a simple but still effective classification of the fish community includes three typologies: fish communities of deep lakes; those of shallow lakes, and those from high-altitude (high-latitude) lakes. From a functional point of view, fish can be grouped into different guilds relative to, for instance, habitat preferences, reproductive behavior, thermal preferences, and food preferences.
8. Planktivorous fish can be important in regulating the abundance and size structure of zooplankton populations. Prey are visually selected, in most cases on an individual basis, although the gill rakers of

certain fish collect some zooplankton as water passes through the mouth and across the gills.

   a. Size selection of prey by fish is governed by the energy return obtained per unit of energy spent at foraging and by the abundance of prey. When prey are abundant, only larger prey are consumed; as prey abundance decreases, smaller prey are taken.
   b. Planktivorous fish select large zooplankters and can eliminate large cladocerans from lakes. Fish predation intensity must be very high to become the main determinant of the zooplankton community but can be sufficient to restructure the zooplankton community on the basis of size. The restructuring can lead to fewer large zooplankton (which typically have high feeding rates) and thus result in reduced grazing of phytoplankton.
   c. When size selection by fish is not in effect, and when large zooplankters are present, smaller-sized zooplankton are generally not found to cooccur with the larger forms. This is likely a result of size-selective predation of smaller zooplankton by invertebrates (copepods, phantom midge larvae, and predaceous Cladocera).

9. Silt, resuspended particles, and high concentrations of humic substances can severely disturb prey detection by fishes due to their effects on the scattering of light. Climate change is suggested to increase the loading of silt and humic substances to various lakes on the globe.
   a. Reduced light may also have strong effects on the interactions between fishes and their prey as many fishes are visual foragers and highly dependent on light to detect and consume their prey. Reduced light intensity may therefore affect the competitive interactions between species because of their differential forage capacity at low light.
   b. Particulate turbidity, especially clay turbidity, reduces the visibility and the distance at which predator–prey interactions occur.
   c. Zooplankton adapt to the reduced risk of predation by turbidity, for instance, by changing their body size and the size and pigmentation of the eyes, the result being a morph with a superior reproductive potential, more rapid population growth, and greater longevity than that of the small-eyed form.
   d. Turbidity affects the risk of fish predation on juvenile fish (the "turbidity as cover" hypothesis), which may favor planktivorous fish as these are less dependent on light, potentially leading to a lower proportion of piscivorous fish.

10. The term "trophic cascade" is most often applied to pelagic or benthic processes where predation by fish and/or invertebrates alters the zooplankton or benthic invertebrate community structure and their effectiveness in grazing on phytoplankton and periphyton, respectively. Additional terms and concepts are "top-down" and "bottom-up" control. Top-down generally refers to predation processes by invertebrates or fish that influence the zooplankton community structure, whereas bottom-up refers to resource regulation (e.g., nutrients) of growth and production, often beginning with biogeochemical controls of photosynthesis.
   a. Top-down control by fish is particularly evident in Arctic and cold alpine freshwaters that are typically transparent, making foraging easy for visually hunting fish. Moreover, the prey is exposed to predation for a longer time period due to longer generation times, and the zooplankton need pigmentation to protect themselves against UV radiation, rendering them more visible to fish.
   b. Comparative studies of northern streams with and without fish also reveal strong top-down effects by fish on invertebrates. Being size-selective predators, salmonid fishes have a particularly strong impact on the largest prey types available.
   c. No cascading effects are found on phytoplankton in lakes and the periphyton in streams in nutrient-poor systems, which are likely being more strongly controlled by resources (i.e., bottom-up factors).

11. Pelagic, littoral, and terrestrial resources can all play a role in supporting fish consumers in lakes and streams.
   a. In lakes focus has historically been on pelagic production and processes. However, the fueling of the fish community from the benthic and littoral zone is also important. A high littoral–benthic contribution to the individual fish species appears to reflect the high fish species diversity in the littoral zone.
   b. Streams, being closely associated with the terrestrial environment, often receive a large amount of allochthonous organic matter. Although earlier studies have emphasized the importance of allochthonous inputs to the food web, especially in forested headwater streams, recent research using chemical tracers has shown that the fish food webs in many systems depend heavily on algal sources, not only for carbon but also especially for nitrogen and essential amino and fatty acids. Terrestrial insects have also been reported to play an important role in the diet of fish.

12. Fish affect the food chain length in lakes and streams. The food chain length is the number of trophic transfers from the base to the top of a food web, and the chain structure may strongly affect both lake community structure and ecosystem function.

a. The most frequently tested and cited food chain hypotheses are based on energetic considerations, reflecting that energetic efficiencies of trophic interactions are typically low (about 10%). Therefore the number of trophic levels in a given food chain will be limited by energy availability.

b. An alternative is the "productive-space hypothesis" predicting that the food chain length increases as a function of total ecosystem productivity (the product of ecosystem size and a measure of per unit-size productivity). Tests of these hypotheses in a variety of ecosystems have, however, yielded mixed results and show that food chain length varies with the type of the aquatic ecosystem. A global-scale analysis of aquatic food webs found significant differences in food chain length among stream, lake, and marine ecosystems and also within systems; for instance, the food chain length was shortest upstream with the highest slope, intermediate in low-gradient rivers and rivers below reservoirs, and longest in river impoundments and reservoirs.

13. Although growth is highly variable among the different fish species, growth rate and size are generally inversely correlated. For young fish, more energy is diverted into growth; for larger fish, more food energy is utilized for maintenance. Fish harvesting in freshwaters, including recreational fishing, can cause changes in the distribution, demography, and stock structure of individual species and thus fish communities and lake ecosystems. The release of hatchery-reared fish is one of the most popular tools for restoring fish populations, but a growing body of evidence has revealed negative impacts on the biota attributable to introduced fishes.

a. Production rates of fish are considerably higher in tropical freshwaters than in temperate waters, where growth is restricted to about half of the year. Annual production-to-mean-biomass ratios are higher among fish with shorter life cycles than among fish species that have slower growth and longer lifespans. Similarly, in cases where young stages contribute disproportionately to population productivity, annual production to biomass (P/B) ratios are greater. In general, annual P/B ratios are highest among planktivorous fishes, lower for benthic feeders, and lowest for carnivores.

b. Most inland fisheries that rely on natural stock reproduction are now overfished or being fished at their biological limit. Fishing reduces the number of larger and older fish and leads to marked declines in the age and size of populations. To compensate for the increased mortality, fish adjust their life history traits and show, for instance, earlier sexual maturation at a smaller size and an elevated reproductive effort. Loss of apex predators or migratory fishes because of overfishing often results in a relaxation of the top-down control of prey populations or a reduction of the nutrient supply into freshwater ecosystems.

c. The global declines of fish populations have been attributed primarily to commercial fisheries, but the recreational fishing sector is also an important factor. Most recreational fisheries have no limit on the total effort that a fishery attracts. This can result in local high and selective fishing mortality, which contributes to the decline of fish stocks and can undermine biodiversity and ecological resilience. Moreover, some serious consequences of recreational activities, such as discarded fishing lines, use of live bait organisms, and recreational boat traffic, are also responsible for environmental degradation.

d. The ecological impacts that fish stocking might exert on other organisms in the ecosystem have not been adequately anticipated or considered. Widespread and long-term fish stocking of lakes and streams has shifted species assemblages and the food web structure in freshwater systems globally. There are clear signs of negative effects, including species endangerment and extirpation, lower survival and reproductive fitness, reduced genetic variation in the hatchery populations, and changes in the food web or ecosystem function.

14. Dams can change the composition and structure of native biotic communities, limit the distribution of species, and block fish migration routes. To avoid this, fishway construction and dam removal are increasingly used in dam management, but their effects are limited. River regulation can alter the natural flow regime and directly affect the physical and chemical characteristics of the river environment. These important variables may have both individual- and population-level effects on riverine fish, including the success of migration and spawning, the survival of eggs and juveniles, and food production.

a. Dams prevent the migration of many endemic fish species, especially anadromous species, to feeding or breeding grounds or impede the function of these grounds by changing water depths, flows, and deposition patterns. Besides, problems associated with downstream migration, such as passing through hydraulic turbines or cold-water release from high dams, can also lead to high mortality of fish.

b. Various designs of fishways or fish ladders have been developed to enable fish to pass dams. Despite advances in fish passage facilities, a large number of fishways still prevent or delay the passage of both target and nontarget species. Except for fish passages, dam removal is the best option for reconnecting the river. However, after the reorganization of fish communities after dam removals, it may take decades for the restored sites to approximate the community structure of nearby undisturbed sites.

c. Reduction of the river flow due to manipulation of riverine systems can result in a sharp reduction of a migratory population. The suppression of the flood regime downstream through flow regulation can also cause loss of obligate floodplain spawners by depriving them of their spawning grounds and valuable food supplies. Another important consequence of river regulation or impoundment is the transformation from lotic to lentic environments. This shift often favors generalist over specialist species adapted to fast-flowing water.

15. IAS are species that are introduced, accidently or intentionally, outside of their natural geographical range and then become established and abundant with negative consequences for their new environment. At present, IAS are recognized as one of the main threats to biodiversity and ecosystem functioning due to their negative impacts at individual, population, community, and ecosystem levels. Not all nonnative species introduced in a new environment may establish themselves, and not all introduced taxa become invasive and cause biological or economic harm. Nevertheless, current evidence shows that the desirable effects of nonnative species are outweighed by their undesirable effects, and a precautionary principle in the risk management of invasive species is usually strongly recommended.

a. Biological invasion of a water body requires both dispersal and establishment. Dispersal requires a vector and a first inoculation in the new environment. Establishment requires that alien individuals are physiologically adapted to the physical-chemical environment and that they can find adequate food, avoid predators, and reproduce. Only later can they spread and integrate into other nearby sites. Fishes are frequently introduced as alien species by intentional stocking, involuntary releases, and hitchhiking with other carriers (vectors).

b. The impact of alien invasive fish species is multifaceted and includes disruption of native species populations by direct predation, induced mortality (i.e., via the spread of infectious diseases), competitive exclusion, and introgressive hybridization phenomena between congeneric species. In many cases the impact of an invasive fish species is complex and includes not only direct but also cascading effects on food webs, water quality, and the entire ecosystem. Brown trout and common carp are well-known examples of introduced species becoming invasive.

16. Biomanipulation is a term covering methods of biological restoration aimed at reducing objectionable algal types and biomass after a reduction of the external nutrient load. The most frequently used biomanipulation method is the removal of planktivorous and benthivorous fish. An alternative or supplementary method to fish removal is the stocking of potentially piscivorous fish.

a. Sufficient removal of zooplanktivorous fish biomass has generally resulted in strong cascading effects, namely dominance by large zooplankton, higher zooplankton grazing, less phytoplankton (and much less noxious Cyanobacteria), and increased transparency, whereas the effects of stocking of piscivores have generally been modest unless the species introduced were absent before stocking.

b. Initial positive effects of biomanipulations may not last long, and so repeated fish removal(s) may be a way to maintain the clear water state if lakes shift back to the turbid state. The effort needed may be less extensive than the initial attempt.

c. Fish removal at warm locations (Mediterranean, subtropical, and tropical lakes) is less likely to benefit zooplankton due to fast reproduction of fish, weak control by piscivores, and dominance by omnivorous and small-sized fish (higher predation on zooplankton), but it may reduce fish-induced resuspension and thereby the nutrient release from the sediment.

d. Warm lakes typically host filter-feeding fish that can consume nuisance algae such as Cyanobacteria, and stocking of various carp and tilapia species has been used as a method of restoration. However, as these fish also eat zooplankton and disturb the sediment when foraging, this approach cannot be recommended. A promising alternative for warm lakes is to combine fish removal with the transplantation of submerged macrophytes and capping phosphorus in the sediment.

It is important, though, to emphasize that the key tool to restoring eutrophicated lakes is the reduction of external nutrient loading. Biomanipulation only serves the purpose of reinforcing recovery, treating symptoms, or improving/maintaining an environmental quality temporarily until the external loading is significantly reduced.

17. Freshwater fish are both directly and indirectly affected by changes in temperature. Being ectotherms, they cannot thermoregulate physiologically but only by moving to areas with appropriate temperatures. Cold-stenothermal species will shift toward higher latitudes or altitudes, or deeper waters, if possible, but may become locally extinct at the warmest edge of their current distribution ranges.

   a. Life history traits will also be affected by warmer temperatures. Lower-latitude fish species are often not only individually smaller but also grow faster, mature earlier, have shorter life spans, and allocate less energy to reproduction than species at higher latitudes.

   b. Fish species' vulnerability to climate change generally depends on the most temperature-sensitive life stages and typically spawning adults and embryos consistently have narrower tolerance ranges than larvae and nonreproductive adults and are most vulnerable to climate warming.

   c. Although evolutionary responses to global climate change in aquatic ecosystems are poorly documented, a small number of studies indicate that freshwater fishes and other organisms are already exhibiting genetic change.

   d. Climate change acts on aquatic ecosystems in concert with other anthropogenic stressors, and together these stressors may have complex, compounded effects on freshwater fishes. Numerous recent studies and reviews indicate that warming will exacerbate existing eutrophication problems, and this will, in a self-amplifying manner, likely further stimulate a shift to small-bodied fish and to dominance of eurythermal species.

Acknowledgments

We wish to thank Martin Søndergaard, Torben Linding Lauridsen, Andreas Severin Berthelsen, Mustafa Korkmaz, Ian Winfield, Massimo Lorenzoni, and Jinley Yu for their valuable comments on the chapter. We further want to thank Anne Mette Poulsen for text editing.

References

Abbott, J.K., Lloyd-Smith, P., Willard, D., Adamowicz, W., 2018. Status-quo management of marine recreational fisheries undermines angler welfare. Proc. Natl. Acad. Sci. USA 115, 8948–8953.

Abrahams, M., Kattenfeld, M., 1997. The role of turbidity as a constraint on predator–prey interactions in aquatic environments. Behav. Ecol. Sociobiol. 40, 169–174.

Alexander, T.J., Vonlanthen, P., Periat, G., Degiorgi, F., Raymond, J.C., Seehausen, O., 2015. Estimating whole-lake fish catch per unit effort. Fish. Res. 172, 287–302.

Allan, J.D., 2004. Landscapes and riverscapes: the influence of land use on stream ecosystems. Annu. Rev. Ecol. Evol. Syst. 35, 257–284.

Allan, J.D., Abell, R., Hogan, Z., Revenga, C., Taylor, B.W., Welcomme, R.L., et al., 2005. Overfishing of inland waters. Bioscience 55 (12), 1041–1051.

Allen, J., Nuechterlein, G., Buitron, D., 2007. Resident nongame waterbird use following biomanipulation of a shallow lake. J. Wildl. Manage. 71, 1158–1162.

Allendorf, F.W., Leary, R.F., 2011. Conservation and distribution of genetic variation in a polytypic species, the cutthroat trout. Conservation Biology 2, 170–184.

Amarasinghe, U.S., Welcomme, R.L., 2002. An analysis of fish species richness in natural lakes. Env. Biol. Fish. 65, 327–339.

Antonsson, Þ., Árnason, F., 2011. Elliðaár 2010. Rannsóknir Á Fiskistofnum Vatnakerfisins. Inst. Freshw. Fish. Report No. VMST/11030 (in Icelandic).

Araki, H., Schmid, C., 2010. Is hatchery stocking a help or harm? Evidence, limitations and future directions in ecological and genetic surveys. Aquaculture 308, S2–S11.

Arcifa, M.S., Northcote, T.G., Froelich, O., 1986. Fish-zooplankton interactions and their effects on water quality of atropical Brazilian reservoir. Hydrobiologia 139, 49–58.

Arlinghaus, R., Abbott, J.K., Fenichel, E.P., Carpenter, S.R., Hunt, L.M., Alós, J., et al., 2019. Opinion: governing the recreational dimension of global fisheries. Proc. Natl. Acad. Sci. USA 116 (12), 5209–5213.

Attayde, J.L., Hansson, L.A., 2001. The relative importance of fish predation and excretion effects on planktonic communities. Limnol. Oceanogr. 46, 1001–1012.

Austen, D.J., Bayley, P.B., Menzel, B.W., 1994. Importance of the guild concept to fisheries research and management. Fisheries 19, 12–20.

Bachmann, R.W., Jones, B.L., Fox, D.D., Hoyer, M.V., Bull, L.A., Canfield Jr., D.E., 1996. Relations between trophic state indicators and fish in Florida (U.S.A.) lakes. Can. J. Fish. Aquat. Sci. 53, 842–855.

Bajer, P.G., Sorensen, P.W., 2015. Effects of common carp on phosphorus concentrations, water clarity, and vegetation density: a whole system experiment in a thermally stratified lake. Hydrobiologia 746, 303–311.

Balon, E.K., 1975. Reproductive guilds of fishes: a proposal and definition. J. Fisheries Res. Board Can. 32, 821–864.

Balon, E.K., 1995. Origin and domestication of the wild carp, *Cyprinus carpio*: from Roman gourmets to the swimming flowers. Aquaculture 129, 3–48.

Barel, C.D.N., Dorit, R., Greenwood, P.H., Fryer, G., Hughes, N., Jackson, P.B.N., et al., 1985. Destruction of fisheries in Africa's lakes. Nature 315, 19–20.

Barry, P.M., Carline, R.F., Argent, D.G., Kimmel, W.G., 2007. Movement and habitat use of stocked juvenile paddlefish in the Ohio River System, Pennsylvania. N. Am. J. Fish. Manage. 27, 1316–1325.

Bay, E.C., 1967. Mosquito control by fish: a present-day appraisal. WHO Chron. 21, 415–423.

Bays, J.S., Crisman, T.L., 1983. Zooplankton and trophic state relationships in Florida lakes. Can. J. Fish. Aquat. Sci. 40, 1813–1819.

Beeton, A.M., 2002. Large freshwater lakes: present state, trends, and future. Env. Cons. 29, 21–38.

Benejam, L., Angermeier, P.L., Munné, A., García-Berthou, E., 2010. Assessing effects of water abstraction on fish assemblages in Mediterranean streams. Freshwat. Biol. 55, 628–642.

Benke, A.C., 1993. Concepts and patterns of invertebrate production in running waters. Verh. Internat. Verein. Limnol. 25, 15–38.

Benndorf, J., 1990. Conditions for effective biomanipulation; conclusions derived from whole-lake experiments in Europe. Hydrobiologia 200/201, 187–203.

Benstead, J.P., March, J.G., Pringle, C.M., Scatena, F.N., 1999. Effects of a low-head dam and water abstraction on migratory tropical stream biota. Ecol. Appl. 9 (2), 656–668.

Benton, M.J., 2005. Vertebrate Paleontology, third ed. Blackwell, Malden, MA, p. 455.

Berg, S., Jeppesen, E., Søndergaard, M., 1997. Pike (*Esox lucius* L.) stocking as a biomanipulation tool 1. Effects on the fish population in Lake Lyng, Denmark. Hydrobiologia 342–343, 311–318.

Blanck, A., Lammouroux, N., 2007. Large-scale intraspecific variation in life-history traits of 44 European freshwater fish. J. Biogeogr. 34, 862–875.

Bohl, E., 1980. Diel pattern of pelagic distribution and feeding in planktivorous fish. Oecologia 44, 368–375.

Braband, A., Faafeng, B., 1994. Habitat shift in roach induced by the introduction of pikeperch. Verh. Internat. Verein. Limnol. 25, 2123.

Brander, K.M., 2007. Global fish production and climate change. Proc. Natl. Acad. Sci. USA 104 (50), 19709–19714.

Brett, M.T., Bunn, S.E., Chandra, S., Galloway, A.W.E., Guo, F., Kainz, M.J., Lau, D.C.P., Moulton, T.P., Power, M.E., Rasmussen, J.B., Taipale, S.J., Thorp, J.H., Wehr, J.D., 2017. How important are terrestrial organic carbon inputs for secondary production in freshwater ecosystems? Freshw. Biol. 62, 833–853.

Breukelaar, A.W., Lammens, E.H.R.R., Klein Breteler, J.G.P., Tátrai, I., 1994. Effects of benthivorous bream (*Abramis brama*) and carp (*Cyprinus carpio*) on sediment resuspension and concentrations of nutrients and chlorophyll a. Freshw. Biol. 32, 113–121.

Brito, E.F., Moulton, T.P., De Souza, M.L., Bunn, S.E., 2006. Stable isotope analysis indicates microalgae as the predominant food source of fauna in a coastal forest stream, southeast Brazil. Austral. Ecol. 31, 623–633.

Brooks, J.L., 1968. The effects of prey size selection by lake planktivores. Syst. Zool. 17 (3), 273–291.

Brooks, J.L., Dodson, S.I., 1965. Predation, body size, and composition of plankton. Science 150, 28–35.

Brucet, S., Pédron, S., Mehner, T., Lauridsen, T.L., Argillier, C., Winfield, I.J., et al., 2013. Fish diversity in European lakes: geographical factors dominate over anthropogenic pressures. Freshw. Biol. 58, 1779–1793.

Bunn, S.E., Arthington, A.H., 2002. Basic principles and consequences of altered hydrological regimes for aquatic biodiversity. Env. Manage. 30, 492–507.

Bystriansky, J.S., Frick, N., Richards, J., Schulte, P., Ballantyne, J., 2007. Wild Arctic char (*Salvelinus alpinus*) upregulate gill Na^+,K^+-ATPase during freshwater migration. Physiol. Biochem. Zool.: Ecol. Evol. Appr. 80 (3), 270–282.

Bystriansky, J.S., Schulte, P.M., 2011. Changes in gill H^+-ATPase and Na^+/K^+-ATPase expression and activity during freshwater acclimation of Atlantic salmon (*Salmo salar*). J. Exp. Biol. 214, 2435–2442.

Cambray, J.A., 2003. Impact on indigenous species biodiversity caused by the globalisation of alien recreational freshwater fisheries. Hydrobiologia 500, 217–230.

Cambray, J.A., King, J.M., Bruwer, C., 1997. Spawning behaviour and early development of the Clanwilliam yellowfish (*Barbus capensis*; Cyprinidae), linked to experimental dam releases in the Olifants River, South Africa. Reg. Riv. Res. Manag. 13, 579–602.

Cañedo-Argüelles, M., Sgarzi, S., Arranz Urgell, I., Quintana, X.D., Ersoy, Z., Landkildehus, F., et al., 2017. Role of predation in biological communities in naturally eutrophic sub-Arctic Lake Mývatn, Iceland. Hydrobiologia 790, 213–223.

Carol, J., Benejam, L., Alcaraz, C., Vila-Gispert, A., Zamora, L., Navarro, E., et al., 2006. The effects of limnological features on fish assemblages of 14 Spanish reservoirs. Ecol. Freshwat. Fish 15, 66–77.

Carpenter, S.R., Cole, J.J., Hodgson, J.R., Kitchell, J.E., Pace, M.L., Bade, D., et al., 2001. Trophic cascades, nutrients, and lake productivity: whole-lake experiments. Ecol. Monogr. 71, 163–186.

Carpenter, S.R., Kitchell, K.L., 1993. The Trophic Cascade in Lakes. Cambridge University Press, Cambridge.

Carpenter, S.R., Kitchell, J.F., Hodgson, J.R., 1985. Cascading trophic interactions and lake productivity. Bioscience 35, 634–639.

Carpenter, S.R., Stanley, E.H., Vander Zanden, M.J., 2011. State of the world's freshwater ecosystems: physical, chemical, and biological changes. Annu. Rev. Environ. Resour. 36, 75–99.

Carrete Vega, G., Wiens, J.J., 2012. Why are there so few fish in the sea? Proc. R. Soc. Lond. Ser. B 279 (1737), 2323–2329.

Castro-Santos, T., Haro, A., 2010. Fish guidance and passage at barriers. In: Domenci, P.B., Kapoor, G. (Eds.), Fish Locomotion: An Eco-ethological Perspective. Science Publishers, Enfield, NH, pp. 62–89.

Catalano, M.J., Bozek, M.A., 2007. Effects of dam removal on fish assemblage structure and spatial distributions in the Baraboo River, Wisconsin. N. Am. J. Fish. Manage. 27, 519–530.

Chapman, D.W., 1967. Production in fish populations. In: Gerking, S.D. (Ed.), The Biological Basis of Freshwater Fish Production. John Wiley & Sons, New York, pp. 3–29.

Chapman, D.W., 1978. Production in fish population. In: Gerking, S.D. (Ed.), Ecology of Freshwater Fish Production. John Wiley & Sons, New York, pp. 5–25.

Chilton III, E.W., Muoneke, M.I., 1992. Biology and management of grass carp (Ctenopharyngodon idella, Cyprinidae) for vegetation control: a North American perspective. Rev. Fish Biol. Fisheries 2, 283–320.

Chu, C., Minns, C.K., Mandrak, N.E., 2003. Comparative regional assessment of factors impacting freshwater fish biodiversity in Canada. Can. J. Fish. Aquat. Sci. 60, 624–634.

Clarke, G.L., Edmondson, W.T., Ricker, W.E., 1946. Dynamics of production in a marine area. Ecol. Monogr. 16 (4), 321–337.

Colautti, R.I., MacIsaac, H.J., 2004. A neutral terminology to define "invasive" species. Diversity and Distributions 10, 135–141.

Conrow, R., Zale, A.V., Gregory, R.W., 1990. Distributions and abundances of early stages of fishes in a Florida lake dominated by aquatic macrophytes. Trans. Am. Fish. Soc. 119, 521–528.

Cooke, S.J., Cowx, I.G., 2004. The role of recreational fishing in global fish crises. Bioscience 54 (9), 857–859.

Cooke, S.J., Schreer, J.F., Dunmall, K.M., Philipp, D.P., 2002. Strategies for quantifying sublethal effects of marine catch-and-release angling– insights from novel freshwater applications. Am. Fish. Soc. Symp. 30, 121–134.

Cooke, S.J., Twardek, W.M., Lennox, R.J., et al., 2018. The nexus of fun and nutrition: recreational fishing is also about food. Fish Fish. 19 (2), 201–224.

Cowx, I.G., 1995. Review of the status and future development of inland fisheries and aquaculture in western Europe. Pages. 25–34. In: O'Grady, K.T. (Ed.), Review of Inland Fisheries and Aquaculture in the EIFAC Area by Subregion and Subsector. Food and Agricultural Organization of the United Nations, Rome.

Cowx, I.G., 2002. Recreational fisheries. In: Hart, P.B.J., Reynolds, J.D. (Eds.), Handbook of Fish Biology and Fisheries, vol. II. Blackwell Science, Oxford (United Kingdom), pp. 367–390.

Crawford, S., Muir, M., 2008. Global introductions of salmon and trout in the genus *Oncorhynchus*: 1870–2007. Rev. Fish Biol. Fisheries 18, 313–344.

Cremer, M.C., Smitherman, R.O., 1980. Food habits and growth of silver and bighead carp in cages and ponds. Aquaculture 20 (1), 57–64.

Cudmore-Vokey, B., Crossman, E.J., 2000. Checklists of the fish fauna of the Laurentian Great Lakes and their connecting channels. Can. Manu. Rep. Fish. Aquat. Sci. 2550, 1–39.

Dahlke, F.T., Wohlrab, S., Butzin, M., Pörtner, H.O., 2020. Thermal bottlenecks in the life cycle define climate vulnerability of fish. Science 369, 65–70.

Delariva, R.L., Agostinho, A.A., Nakatani, K., Baumgartner, G., 1994. Ichthyofauna associated to aquatic macrophytes in the upper Paraná River floodplain. Revta. UNIMAR 16, 41–60.

De Robertis, A., Ryer, C.H., Veloza, A., Brodeur, R.D., 2003. Differential effects of turbidity on prey consumption of piscivorous and planktivorous fish. Can. J. Fish. Aquat. Sci. 60, 1517–1526.

Donald, D.B., 1987. Assessment of the outcome of eight decades of trout stocking in the mountain national parks, Canada. N. Am. J. Fish. Manag. 7, 545–553.

Donald, D.B., Vinebrooke, R.D., Anderson, R.S., Syrgiannis, J., Graham, M.D., 2001. Recovery of zooplankton assemblages in mountain lakes from the effects of introduced sport fish. Can. J. Fish. Aquat. Sci. 58, 1822–1830.

Dowling, T.E., Childs, M.R., 1992. Impact of hybridization on a threatened trout of the southwestern United States. Conserv. Biol. 6, 355–362.

Downing, J.A., Plante, C., 1993. Production of fish populations in lakes. Can. J. Fish. Aquat. Sci. 50, 110–120.

Downing, J.A., Plante, C., Lalonde, S., 1990. Fish production correlated with primary productivity, not the morphoedaphic index. Can. J. Fish. Aquat. Sci. 47, 1929–1936.

Drenner, R.W., Hambright, K.D., 1999. Review: biomanipulation of fish assemblages as a lake restoration technique. Arch. Hydrobiol. 146, 129–165.

Drost, C.A., Fellers, G.M., 1996. Collapse of a regional frog fauna in the Yosemite area of the California Sierra Nevada. Conserv. Biol. 10 (2), 414–425.

Duffy, J.E., Richardson, J.P., France, K., 2005. Ecosystem consequences of diversity depend on food chain length in estuarine vegetation. Ecol. Lett. 8, 301–309.

Eby, L.A., Roach, W.J., Crowder, L.B., Stanford, J.A., 2006. Effects of stocking-up freshwater food webs. Trends Ecol. Evol. 2 (10), 576–584.

Eckmann, R., 1995. Fish richness in lakes of the northeastern lowlands in Germany. Ecol. Freshw. Fish 4, 62–69.

Elliott, J.M., Elliott, J.A., 2010. Temperature requirements of Atlantic salmon *Salmo salar*, brown trout *Salmo trutta* and Arctic charr *Salvelinus alpinus*: predicting the effects of climate change. J. Fish. Biol. 77, 1793–1817.

Elser, J.J., Carpenter, S.R., 1988. Predation-driven dynamics of zooplankton and phytoplankton communities in a whole-lake experiment. Oecologia 76, 148–154.

Elton, C.S., 1927. Animal Ecology. Sidgwick and Jackson.

Elton, C.S., 1958. The Ecology of Invasions by Animals and Plants. Methuen, London.

Enberg, K., Jørgensen, C., Dunlop, E.S., Heino, M., Dieckmann, U., 2009. Implications of fisheries-induced evolution for stock rebuilding and recovery. Evol. Appl. 2, 394–414.

Estlander, S., Nurminen, L., Olin, M., Vinni, M., Immonen, S., Rask, M., et al., 2010. Diet shifts and food selection of perch (*Perca fluviatilis*) and roach (*Rutilus rutilus*) in humic lakes of varying water colour. J. Fish. Biol. 77, 241–256.

Evans, C.D., Monteith, D.T., Cooper, D.M., 2005. Long-term increases in surface water dissolved organic carbon: observations, possible causes and environmental impacts. Envir. Poll. 137, 55–71.

Fahrig, L., Merriam, G., 1985. Habitat patch connectivity and population survival. Ecology 66, 1762–1768.

FAO, 1997. Fisheries Management. FAO Technical Guidelines for Responsible Fisheries. FAO, Rome.

Figueredo, C.C., Giani, A., 2005. Ecological interactions between Nile tilapia (*Oreochromis niloticus*, L.) and the phytoplanktonic community of the Furnas Reservoir (Brazil). Freshwat. Biol. 50, 1391–1403.

Fox, A.D., Balsby, T.J.S., Jørgensen, H.E., Lauridsen, T.L., Jeppesen, E., Søndergaard, M., et al., 2019. Effects of lake restoration on breeding abundance of globally declining common pochard (*Aythya farina* L.). Hydrobiologia 830, 33–44.

Freeman, M.C., Bowen, Z.H., Bovee, K.D., Irwin, E.R., 2001. Flow and habitat effects on juvenile fish abundance in natural and altered flow regimes. Ecol. Appl. 11, 179–190.

Froese, R., 2006. Cube law, condition factor and weight–length relationships: history, meta-analysis and recommendations. J. Appl. Ichthyol. 22, 241–253.

Fukushima, T., Matsushita, B., Subehi, L., Setiawan, F., Wibowo, H., 2017. Will hypolimnetic waters become anoxic in all deep tropical lakes? Sci. Rep. 7, 45320.

Galbraith Jr., M.G., 1967. Size-selective predation on *Daphnia* by rainbow trout and yellow perch. Trans. Am. Fish. Soc. 96, 1–10.

Gallardo, B., Clavero, M., Sànchez, M.I., Vilà, M., 2020. Global ecological impacts of invasive species in aquatic ecosystems. Glob. Chang. Biol. 22, 151–163.

Gehrke, P.C., Gilligan, D.M., Barwick, M., 2002. Changes in fish communities of the Shoalhaven River 20 years after construction of Tallowa Dam, Australia. Riv. Res. Appl. 18, 265–286.

Gende, S.M., Edwards, R.T., Willson, M.F., Wipfli, M.S., 2002. Pacific salmon in aquatic and terrestrial ecosystems. Bioscience 52 (10), 917–928.

Gerking, S.D., 1994. Feeding Ecology of Fish. Academic Press, San Diego.

Gherardi, F., Bertolino, S., Bodon, M., Casellato, S., Cianfanelli, S., Ferraguti, M., Lori, E., Mura, G., Nocita, A., Riccardi, N., Rossetti, G., Rota, E., Scalera, R., Zerunian, S., Tricario, E., 2008. Animal xenodiversity in Italian inland waters: distribution, modes of arrival, and pathways. Biol. Invasions. 10, 435–454. https://doi.org/10.1007/s10530-007-9142-9.

Gilbert, J.J., 1988. Suppression of rotifer populations by Daphnia: a review of the evidence, the mechanisms, and the effects on zooplankton community structure. Limnol. Oceanogr. 33, 1286–1303.

Gliwicz, Z.M., 1990. Food thresholds and body size in cladocerans. Nature 343, 638–640.

Gliwicz, Z.M., 1994. Relative significance of direct and indirect effects of predation by planktivorous fish on zooplankton. Hydrobiologia 272, 201–210.

Gliwicz, Z.M., 2002. On the different nature of top-down and bottom-up effects in pelagic food webs. Freshw. Biol. 47, 2296–2312.

Gliwicz, Z.M., Rowan, M.G., 1984. Survival of *Cyclops abyssorum tatricus* (Copepoda, Crustacea) in alpine lakes stocked with planktivorous fish. Limnol. Oceanogr. 29, 1290–1299.

Global Invasive Species Database. 2023. Species profile: *Cyprinus carpio*. Downloaded from http://www.iucngisd.org/gisd/speciesname/Cyprinus%20carpio on February 26, 2023.

González-Bergonzoni, I., Meerhoff, M., Teixeira-de Mello, F., Baattrup-Pedersen, A., Jeppesen, E., 2012. Meta-analysis shows a consistent and strong latitudinal pattern in fish herbivory across ecosystems. Ecosystems 15, 492–503.

Gosline, W.A., 1971. Functional Morphology and Classification of Teleostean Fishes. University Press of Hawaii, Honolulu, p. 208.

Goudswaard, P.C., Witte, F., Katunzi, E.F.B., 2008. The invasion of an introduced predator, Nile perch (Lates niloticus, L.) in Lake Victoria (East Africa): chronology and causes. Environ. Biol. Fish. 81, 127–139.

Gozlan, R.E., 2008. Introduction of non-native freshwater fish: is it all bad? Fish Fish. 9, 106–115.

Graf, W.L., 1999. Dam nation: a geographic census of American dams and their large-scale hydrologic impacts. Wat. Resour. Res. 35 (4), 1305–1311.

Graham, C.T., Harrod, C., 2009. Implications of climate change for the fishes of the British Isles. J. Fish. Biol. 74, 1143–1205.

Greene, C.W., 1951. Results from stocking brook trout of wild and hatchery strains at Stillwater Pond. Trans. Am. Fish. Soc. 81, 43–52.

Greenwood, P.H., 1974. The cichlid fishes of Lake Victoria, East Africa: the biology and evolution of a species flock. Bull. Br. Mus. (Nat. Hist.) Zool. 6 (Suppl.), 1–134.

Gregory, R.S., 1993. Effect of turbidity on the predator avoidance behaviour of juvenile Chinook salmon (*Oncorhynchus tshawytscha*). Can. J. Fish. Aquat. Sci. 50, 241–246.

Griffiths, D., 1997. Local and regional species richness in North American lacustrine fish. J. Anim. Ecol. 66, 49–56.

Griffiths, A., 2006. Pattern and process in the ecological biogeography of European freshwater fish. J. Anim. Ecol. 75, 734–751.

Grimm, M.P., Backx, J., 1990. The restoration of shallow eutrophic lakes and the role of northern pike, aquatic vegetation and nutrient concentration. Hydrobiologia 200–201, 557–566.

Gyllström, M., Hansson, L.A., Jeppesen, E., Garcıa-Criado, F., Gross, E., Irvine, K., et al., 2005. The role of climate in shaping zooplankton communities of shallow lakes. Limnol. Oceanogr. 50, 2008–2021.

Ha, J.-Y., Saneyoshi, M., Park, H.-D., Toda, H., Kitano, S., Homma, T., Shiina, T., Moriyama, Y., Chang, K.-H., Hanazato, T., 2013. Lake restoration by biomanipulation using piscivore and *Daphnia* stocking; results of the biomanipulation in Japan. Limnology 14, 19–23.

Haidvogl, G., Pont, D., Dolak, H., Hohensinner, S., 2015. Long-term evolution of fish communities in European mountainous rivers: past log driving effects, river management and species introduction (Salzach River, Danube). Aquat. Sciences 77 (3), 395–410.

Hall, D.J., Cooper, W.E., Werner, E.E., 1970. An experimental approach to the production dynamics and structure of freshwater animal communities. Limnol. Oceanogr. 15 (6), 839–928.

Hall, D.J., Threlkeld, S.T., Burns, C.W., Crowley, P.H., 1976. The size-efficiency hypothesis and the size structure of zooplankton communities. Ann. Rev. Ecol. Syst. 7, 177–208.

Hanson, J.M., Leggett, W.C., 1982. Empirical prediction of fish biomass and weight. Can. J. Fish. Aquat. Sci. 39, 257–263.

Hansson, L.-A., 2004. Plasticity in pigmentation induced by conflicting threats from predation and UV radiation. Ecology 85, 1005–1016.

Hansson, L.-A., Annadotter, H., Bergman, E., Hamrin, S.F., Jeppesen, E., Kairesalo, T., et al., 1998. Biomanipulation as an application of food chain theory: constraints, synthesis and recommendations for temperate lakes. Ecosystems 1, 558–574.

Hansson, L.-A., Hylander, S., 2009. Effects of ultraviolet radiation on pigmentation, photoenzymatic repair, behavior, and community ecology of zooplankton. Photochem. Photobiol. Sci. 8, 1266.

Heino, M., Godø, O.R., 2002. Fisheries-induced selection pressures in the context of sustainable fisheries. Bull. Mar. Sci. 70 (2), 639–656.

Heisey, P.G., Mathur, D., Euston, E.T., 1996. Passing fish safely: a closer look at turbine vs. spillway survival. Hydro Rev. 15, 42–50.

Hessen, D.O., 1985. Selective zooplankton predation by pre-adult roach (*Rutilus rutilus*): the size-selective hypothesis versus the visibility-selective hypothesis. Hydrobiologia 124, 73–79.

Hildebrand, S., 1919. Fishes in relation to mosquito control in ponds. Publ. Health Rep. 34 (21), 1113–1128.

Hirsch, P.E., Eloranta, A.P., Amundsen, P.A., Brabrand, Å., Charmasson, J., Helland, I.P., et al., 2017. Effects of water level regulation in alpine hydropower reservoirs: an ecosystem perspective with a special emphasis on fish. Hydrobiologia 794, 287–301.

Hitt, N.P., Frissell, C.A., Muhlfeld, C.C., Allendorf, F.W., 2003. Spread of hybridization between native westslope cutthroat trout, *Oncorhynchus clarki lewisi*, and nonnative rainbow trout, *Oncorhynchus mykiss*. Can. J. Fish. Aquat. Sci. 60, 1440–1451.

Hoar, W.S., Randall, D.J., Brett, J.R. (Eds.), 1984a. Fish Physiology, Vol. 10a. Gills: Anatomy, Gas Transfer and Acid-Base Regulation. Academic Press, New York.

Hoar, W.S., Randall, D.J., Brett, J.R. (Eds.), 1984b. Fish Physiology, Vol. 9a. Reproduction: Behaviour and Fertility Control. Academic Press, New York.

Hoare, D., Phillips, G., Perrow, M., 2008. Review of Biomanipulation Appendix 4 Broads Lake Restoration Strategy.

Hoeinghaus, D.J., Winemiller, K.O., Agostinho, A.A., 2008. Hydrogeomorphology and river impoundment affect food-chain length of diverse neotropical food webs. Oikos 117, 984–995.

Holden, P.B., Stalnaker, C.B., 1975. Distribution and abundance of mainstream fishes of the middle and upper Colorado River Basins, 1967–1973. Trans. Am. Fish. Soc. 104, 217–231.

Holopainen, I.J., Tonn, W.M., Paszkowski, C.A., 1997. Tales of two fish: the dichotomous biology of crucian carp (*Carassius carassius* (L.) in northern Europe. Ann. Zool. Fenn. 34, 1–22.

Horppila, J., Liljendahl-Nurminen, A., Malinen, T., 2004. Effect of clay turbidity and light on the predator–prey interaction between smelts and chaoborids. Can. J. Fish. Aquat. Sci. 61, 1862–1870.

Horppila, J., Peltonen, H., Malinen, T., Luokkanen, E., Kairesalo, T., 1998. Top-down or bottom-up effects by fish—issues of concern in biomanipulation of lakes. Restor. Ecol. 6, 1–10.

Howell, P.J., 2006. Effects of wildfire and subsequent hydrologic events on fish distribution and abundance in tributaries of North Fork John Day River. N. Am. J. Fisheries Manage. 26, 983–994.

Hrbáček, J., 1958. Typologie und Produktivität der teichartigen Gewässer. Verh. Int. Ver. Theor. Ang. Limnol. 13 (1), 394–399.

Hrbacek, J., Dvorakova, M., Korinek, V., Prochazkova, L., 1961. Demonstration of the effect of the fish stock on the species composition of zooplankton and the intensity of metabolism of the whole plankton association. Verh. Internat. Verein. Limnol. 14, 192–195.

Huet, M., 1949. Aperçu des relations entre la pente et les populations piscicoles des eaux courantes. Schweiz. Z. Hydrol. 11, 333–351.

Hutchinson, G.E., 1959. Homage to Santa Rosalia, or why are there so many kinds of animals? Am. Nat. 93, 145–159.

Hylander, S., Larsson, N., Hansson, L.A., 2009. Zooplankton vertical migration and plasticity of pigmentation arising from simultaneous UV and predation threats. Limnol. Oceanogr. 54 (2), 483–491.

Iglesias, C., Mazzeo, N., Meerhoff, M., Lacerot, G., Meerhoff, M., Lacerot, G., et al., 2011. High predation is of key importance for dominance of small-bodied zooplankton in warm shallow lakes: evidence from lakes, fish exclosures and surface sediments. Hydrobiologia 667, 133–147.

IMO, 2011. Icelandic Meteorological Office Database Extraction No, 2011-08-21/01.

International Union for Conservation of Nature and Natural Resources, 1987. IUCN Position Statement on Translocation of Living Organisms: Introductions, Reintroductions and Restocking. IUCN, The World Conservation Union, Species Survival Commission, Gland, Switzerland.

Ivlev, V.S., 1961. Experimental Ecology of the Feeding of Fishes. Yale University Press, New Haven, Connecticut.

Jeppesen, E., Canfield, D.E., Bachmann, R.W., Søndergaard, M., Havens, K.E., Johansson, L.S., et al., 2020. Towards predicting climate change effects on lakes: a comparative study of 1656 shallow lakes from subtropical Florida and temperate Denmark reveals substantial differences in nutrient dynamics, metabolism, trophic structure and top-down control. Inland Waters 10, 197–211.

Jeppesen, E., Jensen, J.P., Søndergaard, M., Fenger-Grøn, M., Bramm, M.E., Sandby, K., et al., 2004. Impact of fish predation on cladoceran body weight distribution and zooplankton grazing in lakes during winter. Freshwat. Biol. 49, 432–447.

Jeppesen, E., Jensen, J.P., Søndergaard, M., Lauridsen, T., Landkildehus, F., 2000. Trophic structure, species richness and biodiversity in Danish lakes: changes along a phosphorus gradient. Freshwat. Biol. 45, 201–213.

Jeppesen, E., Lauridsen, T.L., Christoffersen, K.S., Landkildehus, F., Geertz-Hansen, P., Amsinck, S.L., et al., 2017. The structuring role of fish in Greenland lakes: an overview based on contemporary and paleoecological studies of 87 lakes from the low and the high Arctic. Hydrobiologia 800, 99–113.

Jeppesen, E., Meerhoff, M., Holmgren, K., González-Bergonzoni, I., Teixeira-de Mello, F., Declerck, S.A.J., DeMeester, L., Søndergaard, M., Lauridsen, T.L., Bjerring, R., Conde-Porcuna, J.M., Mazzeo, N., Iglesias, C., Reizenstein, M., Malmquist, H.J., Liu, Z., Balayla, D., Lazzaro, X., 2010. Impacts of climate warming on lake fish community structure and potential effects on ecosystem function. Hydrobiologia 646, 73–90.

Jeppesen, E., Mehner, T., Winfield, I.J., Kangur, K., Sarvala, J., Gerdeaux, D., et al., 2012. Impacts of climate warming on the long-term dynamics of key fish species in 24 European lakes. Hydrobiologia 694, 1–39.

Jeppesen, E., Sammalkorpi, I., 2002. Lakes. In: Perrow, M., Davy, A.J. (Eds.), Handbook of Restoration Ecology, vol. 2. Cambridge University Press, pp. 297–324.

Jeppesen, E., Søndergaard, M., Lauridsen, T.L., Davidson, T.A., Liu, Z.,, Mazzeo, N., et al., 2012. Biomanipulation as a restoration tool to combat eutrophication: recent advances and future challenges. Adv. Ecol. Res. 47, 411–487.

Johnston, F.D., Arlinghaus, R., Dieckmann, U., 2010. Diversity and complexity of angler behaviour drive socially optimal input and output regulations in a bioeconomic recreational-fisheries model. Can. J. Fish. Aquat. Sci. 67, 1507–1531.

Jones, C.G., Lawton, J.H., Shachak, M., 1994. Organisms as ecosystem engineers. Oikos 69, 373–386.

Jones, G.J., Poplawski, W., 1998. Understanding and management of cyanobacterial blooms in subtropical reservoirs of Queensland, Australia. Wat. Sci. Tech. 37, 161–168.

Kao, Y.C., Rogers, M.W., Bunnell, D.B., Cowx, I.G., Qian, S.S., Anneville, O., et al., 2020. Effects of climate and land-use changes on fish catches across lakes at a global scale. Nat. Comm. 11, 2526.

Karr, J.R., 1981. Assessment of biotic integrity using fish communities. Fisheries 6, 21–27.

King, J., Cambray, J.A., Impson, N.D., 1998. Linked effects of dam-released floods and water temperature on spawning of the Clanwilliam yellowfish *Barbus capensis*. Hydrobiologia 384, 4–265.

Knapp, R.A., Matthews, K.R., Sarnelle, O., 2001. Resistance and resilience of alpine lake fauna to fish introductions. Ecol. Monogr. 71, 401–421.

Koehn, J.D., 2004. Carp (*Cyprinus carpio*) as a powerful invader in Australian waterways. Freshw. Biol. 49, 882–894.

Kolar, C.S., Lodge, D.M., 2000. Freshwater nonindigenous species: interactions with other global changes. In: Mooney, H., Hobbs, R.J. (Eds.), Invasive Species in a Changing World. Island Press, Washington DC, pp. 3–30.

Kolding, J., Van Zwieten, P., Mkumbo, O.C., Silsbe, G., Hecky, R., 2008. Are the Lake Victoria fisheries threatened by exploitation or eutrophication? Towards an ecosystem based approach to management. In: Bianchi, G., Skjoldal, H.R. (Eds.), The Ecosystem Approach to Fisheries. FAO, Rome, pp. 309–350.

Kostow, K., 2009. Factors that contribute to the ecological risks of salmon and steelhead hatchery programs and some mitigating strategies. Rev. Fish Biol. Fish. 19, 9–31.

Kovach, R.P., Gharrett, A.J., Tallmon, D.A., 2012. Genetic change for earlier migration timing in a pink salmon population. Proc. R. Soc. B: Biol. Sci. 279 (1743), 3870–3878.

Kovach, R.P., Muhlfeld, C.C., Wade, A.A., Hand, B.K., Whited, D.C., DeHaan, P.W., et al., 2015. Genetic diversity is related to climatic variation and vulnerability in threatened bull trout. Glob. Change Biol. 21, 2510–2524.

Kristmundsson, Á., Antonsson, T., Árnason, F., 2010. First record of proliferative kidney disease in Iceland. Bull. Eu. Ass. Fish Pathol. 30, 35–40.

Kuwamura, T., Sunobe, T., Sakai, Y., Kadota, T., Sawada, K., 2020. Hermaphroditism in fishes: an annotated list of species, phylogeny, and mating system. Ichthyol. Res. 67, 341–360.

Kwak, T.J., Freeman, M.C., 2010. Assessment and management of ecological integrity. In: Hubert, W.A., Quist, M.C. (Eds.), Inland Fisheries Management in North America, third ed. American Fisheries Society, Bethesda, Maryland, pp. 353–394.

Lacerot, G., Kruk, C., Lurling, M., Scheffer, M., 2013. The role of subtropical zooplankton as grazers of phytoplankton under different predation levels. Freshwat. Biol. 58 (3), 494–503.

Lagler, K.F., Bardach, J.E., Miller, R., Passino, D.R.M., 1977. Ichthyology, second ed. Wiley, New York.

Lair, N., 1990. Effects of invertebrate predation on the seasonal succession of a zooplankton community: a two year study in Lake Aydat, France. Hydrobiologia 198, 1–12.

Laird, C.A., Page, L.M., 1996. Non-native fishes inhabiting the streams and lakes of Illinois. Ill Nat. Hist. Surv. Bull. 35 (1), 1–51.

Lampert, W., 1993. Ultimate causes of diel vertical migration of zooplankton: new evidence for the predator-avoidance hypothesis. Arch. Hydrobiol. Beih. Ergebn. Limnol. 39, 79–88.

Lappalainen, J., Lehtonen, H., 1997. Temperature habitats for freshwater fishes in a warming climate. Bor. Environ. Research 2, 69–84.

Lazzaro, X., 1997. Do the trophic cascade hypothesis and classical biomanipulation approaches apply to tropical lakes and reservoirs? Ver. Int. Verein. Limnol. 26, 719–730.

Leach, J.H., Johnson, M.G., Kelso, J.R.M., Harmann, J., Nümann, W., Entz, B., 1977. Responses of percid fishes and their habitats to eutrophication. J. Fish. Res. Board Can. 34, 1964–1971.

Ledger, M.E., Brown, L.E., Edwards, F.K., Milner, A.M., Woodward, G., 2013. Drought alters the structure and functioning of complex food webs. Nat. Clim. Change 3, 223–227.

Lemmens, P., Declerck, S.A.J., Tuytens, K., Vanderstukken, M., De Meester, L., 2018. Bottom-up effects on biomass versus top-down effects on identity: a multiple-lake fish community manipulation experiment. Ecosystems 21, 166–177.

Leprieur, F., Hickey, M.A., Arbuckle, C.J., Closs, G.P., Brosse, S., Townsend, C.R., 2006. Hydrological disturbance benefits a native fish at the expense of an exotic fish. J. Appl. Ecol. 43, 930–939.

Lessmark, O., 1983. Competition between perch (*Perca fluviatilis*) and roach (*Rutilus rutilus*) in south Swedish lakes. Dissertation, University of Lund, Sweden.

Lewin, W.C., Weltersbach, M.S., Ferter, K., et al., 2019. Potential environmental impacts of recreational fishing on marine fish stocks and ecosystems. Rev. Fish. Science & Aquaculture 27, 287–330.

Li, J., Cohen, Y., Schupp, D.H.,, Adelman, I.R., 1996. Effects of walleye stocking on population abundance and fish size. N. Am. J. Fish. Manag. 16 (4), 830–839.

Liermann, C.R., Nilsson, C., Robertson, J., Ng, R.Y., 2012. Implications of dam obstruction for global freshwater fish diversity. Bioscience 62, 539–548.

Ligtvoet, W., Witte, F., Goldschmidt, T., Van Oijen, M.J.P., Wanink, J.H., Goudswaard, P.C., 1991. Species extinction and concomitant ecological changes in Lake Victoria. Neth. J. Zool. 42, 214–232.

Litvak, M.K., Mandrak, N.E., 1993. Ecology of freshwater baitfish use in Canada and the United States. Fisheries 18, 6–13.

Liu, Z., Zhong, P., Zhang, X., Ning, J., Larsen, S.E., Jeppesen, E., 2018. Successful restoration of a tropical shallow eutrophic lake by biomanipulation: strong bottom-up but weak top-down effects recorded. Wat. Res. 146, 88–97.

Lobon-Cervià, J., Sainz, N. (Eds.), 2017. Brown Trout: Biology, Ecology and Management. John Wiley & Sons, Inc, Hoboken, NJ.

Lowe, S., Browne, M., Boudjelas, S., De Poorter, M., 2000. 100 of the World's Worst Invasive Alien Species: A Selection from the Global Invasive Species Database. Published by The Invasive Species Specialist Group (ISSG) a specialist group of the Species Survival Commission (SSC) of the World Conservation Union (IUCN), p. 12. First published as special lift-out in Aliens 12, December 2000. Updated and reprinted version: November 2004.

Lowe-McConnell, R.H., 1987. Ecological Studies in Tropical Fish Communities. Cambridge University Press, Cambridge.

Lynch, A.J., Myers, B.J.E., Chu, C., Eby, L.A., Falke, J.A., Kovach, R.P., et al., 2016. Climate change effects on North American inland fish populations and assemblages. Fisheries 41, 346–361.

Magnuson, J.J., Webster, K.E., Schindler, D.W., Quinn, F.H., Assel, R.A., Bowser, C.J., et al., 1997. Potential effects of climate changes on aquatic systems: Laurentian Great Lakes and Precambrian Shield region. Hydrol. Proc. 11, 825–871.

Malmquist, H.J., Antonsson, þ., Ingvason, H.R., Arnason, F., 2009. Salmonid fish and warming of shallow Lake Elliðavatn in Southwest Iceland. Verhandlungen der Internationale Vereinigung der Limnologie, 30, 1127–1132.

Mandrak, N.E., 1995. Biogeographic patterns of fish species richness in Ontario lakes in relation to historical and environmental factors. Can. J. Fish. Aquat. Sci. 52, 1462–1474.

Mann, R.H.K., Penczak, T., 1986. Fish production in rivers: a review. Pol. Arch. Hydrobiol. 33, 233–247.

Mao, Z.G., Gu, X.H., Cao, Y., Zhang, M., Zeng, Q.F., Chen, H.H., et al., 2020. The role of top-down and bottom-up control for phytoplankton in a subtropical shallow eutrophic lake: evidence based on long-term monitoring and modeling. Ecosystems 23, 1449–1463.

Marshall, J.C., Blessing, J.J., Clifford, S.E., Hodges, K.M., Negus, P.M., Steward, A.L., 2019. Ecological impacts of invasive carp in Australian dryland rivers. Aquat. Conserv. Mar. Freshw. Ecosyst. 29 (11), 1870–1889.

Matthews, K.R., Knapp, R.A., Pope, K.L., 2002. Garter snake distributions in high-elevation aquatic ecosystems: is there a link with declining amphibian populations and nonnative trout populations? J. Herpetol. 36, 16–22.

Matuszek, J.E., 1978. Empirical predictions of fish yields of large North American lakes. Trans. Am. Fish. Soc. 107, 385–394.

May, R.M., 2004. Biological diversity—differences between land and sea. Philos. Trans. R. Soc. B 343, 105–111.

Mayr, E., 1963. Animal Species and Evolution. Belknap Press, Cambridge.

Mazumder, A., Taylor, W.D., McQueen, D.J., Lean, D.R.S., Lafontaine, N.R., 1990. A comparison of lakes and lake enclosures with contrasting abundances of planktivorous fish. J. Plank, Res. 12, 109–124.

McIntosh, A.R., Townsend, C.R., 1994. Inter-population variation in mayfly anti–predator tactics: differential effects. Ecology 75, 2078–2090.

McIntosh, A.R., Townsend, C.R., 1995. Impacts of an introduced predatory fish on mayfly grazing in New Zealand streams. Limnology and Oceanography 40, 1508–1512.

McNaught, A.S., Schindler, D.W., Parker, R.B., Paul, A.J., Anderson, R.S., Donald, D.B., et al., 1999. Restoration of the food web of an alpine lake following fish stocking. Limnol. Oceanogr. 44 (1), 127–136.

McQueen, D.J., Johannes, M.R.S., Post, J.R., Stewart, T.J., Lean, D.R.S., 1989. Bottom-up and top-down impacts of freshwater pelagic community structure. Ecol. Monogr. 59 (3), 289–309.

McQueen, D.J., Post, J.R., Mills, E.L., 1986. Trophic relationships freshwater pelagic ecosystems. Can. J. Fish. Aquat. Sci. 43, 1571–1581.

McQueen, D.J., Ramcharan, C.W., Yan, N.D., Demers, E., Perez-Fuentetaja, A., Dillon, P.J., 2001. The Dorset food web piscivore manipulation project—Part 1: objectives, methods, the physical-chemical setting. Arch. Hydrobiol. Spec. Iss. Adv. Limnol 56, 1–21.

Meerhoff, M., Clemente, J.M., Teixeira de Mello, F., Iglesias, C., Pedersen, A.R., Jeppesen, E., 2007. Can warm climate-related structure of littoral predator assemblies weaken the clear water state in shallow lakes? Glob. Change Biol. 13, 1888–1897.

Meerhoff, M., Teixeira-de Mello, F., Kruk, C., Alonso, C., González-Bergonzoni, I., Pacheco, J.P., et al., 2012. Environmental warming in shallow lakes: a review of effects on community structure as evidenced from space-for-time substitution approaches. Adv. Ecol. Res. 46, 259–350.

Mehner, T., Benndorf, J., Kasprzak, P., Koschel, R., 2002. Biomanipulation of lake ecosystems: successful applications and expanding complexity in the underlying science. Freshwat. Biol. 47, 2453–2465.

Mehner, T., Diekmann, M., Brämick, U., Lemcke, R., 2005. Composition of fish communities in German lakes as related to lake morphology, trophic state, shore structure and human-use intensity. Freshwat. Biol. 50, 70–85.

Mehner, T., Holmgren, K., Lauridsen, T.L., Jeppesen, E., Diekmann, M., 2007. Lake depth and geographical position modify lake fish assemblages of the European "Central Plains" ecoregion. Freshw. Biol. 52, 2285–2297.

Meijer, M.L., de Haan, M.W., Breukelaar, A.W., Buiteveld, H., 1990. Is reduction of the benthivorous fish an important cause of high transparency following biomanipulation in shallow lakes? Hydrobiologia 200/201, 303–315.

Meissner, K., Muotka, T., 2006. The role of trout in stream food webs: integrating evidence from field surveys and experiments. J. Anim. Ecol. 75, 421–433.

Meraner, A., Baric, S., Pelster, B., Dalla Via, J., 2007. Trout (*Salmo trutta*) mitochondrial DNA polymorphism in the centre of the marble trout distribution area. Hydrobiologia 579, 337–349.

Mertz, G., Myers, R.A., 1998. A simplified formulation for fish production. Can. J. Fish. Aquat. Sci. 55 (2), 478–484.

Mills, E.L., Forney, J.L., 1983. Impact on *Daphnia pulex* of predation by young yellow perch in Oneida Lake, New York. Trans. Am. Fish. Soc. 112, 154–161.

Mims, M.C., Olden, J.D., Shattuck, Z.R., Poff, N.L., 2010. Life history trait diversity of native freshwater fishes in North America. Ecol. Freshwat. Fish 19, 390–400.

Minard, P., 2015. Salmonid acclimatisation in Colonial Victoria: improvement, restoration and recreation, 1858–1909. Env. Hist. 21, 177–199.

Minshall, G.W., Cummins, K.W., Sedell, J.R., Cushing, C.E., 1980. The River Continuum Concept. Can. J. Fish. Aquat. Sci. 37, 130–137.

Mitchell, C.P., 1995. Fish passage problems in New Zealand. In: Proceedings of the International Symposium on Fishways, 95. Gifu, Japan, pp. 33–41.

Mittelbach, G.G., 1981. Foraging efficiency and body size: a study of optimal diet and habitat use by bluegills. Ecology 62 (5), 1370–1386.

Monten, E., 1985. Fish and Turbines. Fish Injuries during Passage through Power Station Turbines. Vattenfall, Stockholm, Sweden.

Morgan, N.C., Backiel, T., Bretschko, G., Duncan, A., Hillbricht-Ilkowska, A., Kajak, Z., et al., 1980. Secondary production. In: LeCren, E.D., Lowe-McConnell, R.H. (Eds.), The Functioning of Freshwater Ecosystems. Cambridge Univ. Press, Cambridge, pp. 247–340.

Moss, B., 2010. Climate change, nutrient pollution and the bargain of Dr. Faustus. Freshw. Biol. 55, 175–187.

Moss, B., Kosten, S., Meerhoff, M., Battarbee, R.W., Jeppesen, E., Mazzeo, N., et al., 2011. Allied attack: climate change and nutrient pollution. Inl. Wat. 1, 101–105.

Moyle, P.B., Cech, J.J., 1982. Fishes: An Introduction to Ichthyology. Prentice Hall, Inc., Englewood Cliffs.

Moyle, P.B., Marchetti, M.P., 2006. Predicting invasions success: freshwater fishes of California as a model. Bioscience 56, 515–524.

Moyle, P.B., Vondracek, B., 1985. Persistence and structure of the fish assemblage in a small California stream. Ecology 66, 1–13.

Murchie, K.J., Hair, K.P.E., Pullen, C.E., Redpath, T.D., Stephens, H.R., Cooke, S.J., 2008. Fish response to modified flow regimes in regulated rivers: research methods, effects and opportunities. River. Res. Applic. 24, 197–217.

NatureServe, 2010. Digital Distribution Maps of the Freshwater Fishes in the Conterminous United States.—Version 3.0, Arlington, VA, USA.

Neres-Lima, V., Machado-Silva, F., Baptista, D.F., Oliveira, R.B., Andrade, P.M., Oliveira, A.F., Moulton, T.P., 2017. Allochthonous and autochthonous carbon flows in food webs of tropical foreststreams. Freshw. Biol. 62, 1–12.

Nesler, T.P., Muth, R.T., Wasowicz, A.F., 1988. Evidence for baseline flow spikes as spawning cues for Colorado squawfish in the Yampa River, Colorado. Am. Fish. Soc. Symp. 5, 68–79.

Nikolsky, G.V., 1963. The Ecology of Fishes. Academic Press, London and New York (translated by Russian by L. Birkett).

Nilsson, N.A., Pejler, B., 1973. On the relation between fish, fauna and zooplankton composition in North Swedish lakes. In: Drottningholm, 53. Report of the Institute of Freshwater Research, pp. 51–76.
Northcote, T.G., 1978. Migration strategies and production in freshwater fishes. In: Gerking, S.D. (Ed.), Ecology of Freshwater Fish Production. Blackwell Scientific Publications, Oxford, pp. 326–359.
Northcote, T.G., 1998. In: Jungwirth, M., Schmutz, S., Weiss, S. (Eds.), Migratory Behaviour of Fish and its Significance to Movement through Riverine Fish Passage Facilities. Migration and Fish Bypasses. Fishing News Books, Oxford, pp. 3–18.
Novotna, M., Korinek, V., 1966. Effect of the fish stock on the quantity and species composition of the plankton of two backwaters. Czech. Acad. Publ. Hydrol. Stud. 1, 297–322.
Nurminen, L., Horppila, J., 2006. Efficiency of fish feeding on plant-attached prey—effect of inorganic turbidity and plant-mediated changes in the light environment. Limnol. Oceanogr. 51, 1550–1555.
O'Brien, W.J., 1975. Some aspects of the limnology of the ponds and lakes of the Noatak drainage basin, Alaska. SIL Proceedings 19, 472–479, 1922–2010.
O'Brien, D.P., 1987. Description of escape responses of krill (crustacea: Euphausiacea), with particular reference to swarming behavior and the size and proximity of the predator. J. Crust. Biol. 7 (3), 449–457.
O'Brien, W.J., Slade, N.A., Vinyard, G.L., 1976. Apparent size as the determinant of prey selection by bluegill sunfish (*Lepomis macrochirus*). Ecology 57 (6), 1304–1310.
Oberdorff, T., Guegan, J.F., Hugueny, B., 1995. Global scale patterns of fish species richness in rivers. Ecography 18, 345–352.
O'Connor, J., Duda, J., 2015. 1000 dams down and counting. Science 348, 496–497.
Oglesby, R.T., 1977. Relationship of fish yield to lake phytoplankton standing crop, production and morphoedaphic factors. J. Fish. Res. Board Can. 34, 2271–2279.
Olin, M., Rask, M., Ruuhljärvi, J., Kurkilahti, M., Ala-Opas, P., Ylönen, O., 2002. Fish community structure in mesotrophic and eutrophic lakes of southern Finland: the relative abundances of percids and cyprinids along a trophic gradient. J. Fish. Biol. 60, 593–612.
Olin, M., Vinni, M., Lehtonen, H., Rask, M., Ruuhijärvi, J., Saulamo, K., et al., 2010. Environmental factors regulate the effects of roach *Rutilus rutilus* and pike *Esox lucius* on perch *Perca fluviatilis* populations in small boreal forest lakes. Fish Biol. 76, 1277–1293.
Oliveira, A.G., Baumgartner, M.T., Gomes, L.C., Dias, R.M., Agostinho, A.A., 2018. Long-term effects of flow regulation by dams simplify fish functional diversity. Freshwat. Biol. 63, 293–305.
Osmundson, D.B., Ryel, R.J., Lamarra, V.L., Pitlick, J., 2002. Flow-sediment-biota relations: implications for river regulation effects on native fish abundance. Ecol. Appl. 12 (6), 1719–1739.
Pace, M., Cole, J.J., Carpenter, S.R., Kitchell, J.F., 1999. Trophic cascades revealed in diverse ecosystems. Trends Ecol. Evol. 14, 483–488.
Padilla, D.K., Williams, S.L., 2004. Beyond ballast water: aquarium and ornamental trades as sources of invasive species in aquatic ecosystems. Front. Ecol. Environ. 2, 131–138.
Paine, R.T., 1980. Food webs: linkage, interaction strength and community infrastructure. J. Anim. Ecol. 49, 667–685.
Parkos III, J.J., Santucci Jr., V.J., Wahl, D.H., 2003. Effects of adult common carp (*Cyprinus carpio*) on multiple trophic levels in shallow mesocosms. Can. J. Fish. Aquat. Sci. 60 (2), 182–192.
Pauls, S.U., Nowak, C., Bálint, M., Pfenninger, M., 2013. The impact of global climate change on genetic diversity within populations and species. Molecul. Ecol. 22 (4), 925–946.
Pelicice, F.M., Vitule, J.R.S., Lima, D.P., Orsi, M.L., Agostinho, A.A., 2014. A serious new threat to Brazilian freshwater ecosystems: the naturalization of nonnative fish by decree. Conserv. Lett. 7, 55–60.
Penczak, T., Moliński, M., 1984. Fish production in Oued Sebaou, a seasonal river in North Algeria. J. Fish. Biol. 25, 723–732.
Pereira, L.S., Agostinho, A.A., Winemiller, K.O., 2017. Revisiting cannibalism in fishes. Rev. Fish. Biol. Fisheries 27, 499–513.
Peretyatko, A., Teissier, S., De Backer, S., Triest, L., 2012. Biomanipulation of hypereutrophic ponds: when it works and why it fails. Environ. Monit. Assess. 184, 1517–1531.
Persson, L., 1983. Effects of intra- and interspecific competition on dynamics and size structure of perch *Perca fluviatilis* and a roach *Rutilus rutilus* population. Oikos 41, 126–132.
Persson, L., 1999. Trophic cascades: abiding heterogeneity and the trophic level concept at the end of the road. Oikos 85, 385–397.
Persson, L., Anderson, G., Hamrin, S.F., Johansson, L., 1988. Predation regulation and primary production along the productivity gradient of temperate lake ecosystems. In: Carpenter, S.R. (Ed.), Complex Interactions in Lake Communities. Springer Verlag, New York, pp. 45–65.
Petts, G.E., 1984. Impounded Rivers: Perspectives for Ecological Management. John Wiley & Sons, Chichester.
Pimm, S.L., 1984. The complexity and stability of ecosystems. Nature 307, 321–326.
Pister, E.P., 2001. Wilderness fish stocking: history and perspective. Ecosystems 4, 279–286.
Pitcher, T.J., Hollingworth, C.E., 2002. Fishing for fun: where's the catch? In: Pitcher, T.J., Hollingworth, C.S. (Eds.), Recreational Fisheries: Ecological, Economic, and Social Evaluation. Blackwell Science, Oxford, pp. 1–16.
Poff, N.L.R., Allan, J.D., Bain, M.B., Karr, J.R., Prestegaard, K.L., Richter, B.D., et al., 1997. The natural flow regime. Bioscience 47 (11), 769–784.
Polgar, G., Iaia, M., Sala, P., Khang, T.F., Galafassi, S., Zaupa, S., et al., 2023. Size-age population structure of an endangered and anthropogenically introgressed northern Adriatic population of marble trout (*Salmo marmoratus* Cuv.): insights for its conservation and sustainable exploitation. Peer J. accepted 17 (11), e14991.
Post, D.M., 2002. Using stable isotopes to estimate trophic position: models, methods, and assumptions. Ecology 83, 703–718.
Post, D.M., Pace, M.L., Hairston Jr., N.G., 2000. Ecosystem size determines food-chain length in lakes. Nature 405, 1047–1049.
Poulos, H.M., Chernoff, B., 2017. Effects of dam removal on fish community interactions and stability in the Eightmile River System, Connecticut, USA. Env. Manag. 59, 249–263.
Pringle, R.M., 2005. The origins of the Nile perch in Lake Victoria. Bioscience 55, 780–787.
Pyke, G.H., Pulliam, H.R., Charnov, E.L., 1977. Optimal foraging: a selective review of theory and tests. Quart. Rev. Biol. 52, 137–154.
Quiros, R., 1990. Factors related to variance of residuals in chlorophyll—total phosphorus regressions in lakes and reservoirs of Argentina. Hydrobiologia 200–201, 343–355.
Rahel, F.J., 2000. Homogenization of fish faunas across the United States. Science 288, 854–856.
Ramcharan, C.W., France, R.L., McQueen, D.J., 1996. Multiple effects of planktivorous fish on algae through a pelagic trophic cascade. Can. J. Fish. Aquat. Sci. 53, 2819–2828.
Ramcharan, C.W., McQueen, D.J., Demers, E., Popiel, S.A., Rocchi, A.M., Yan, N.D., et al., 1995. A comparative approach to determining the role of fish predation in structuring limnetic ecosystems. Arch. Hydrobiol. 133, 389–416.
Rautio, M., Korhola, A., 2002. UV-induced pigmentation in subarctic *Daphnia*. Limnol. Oceanogr. 47, 295–299.
Rawson, D.S., 1952. Mean depth and the fish production of large lakes. Ecology 33, 513–521.
Reimers, N., 1958. Conditions of existence, growth, and longevity of brook trout in a small, high-altitude lake of the eastern Sierra Nevada. Cal. Fish Game 44 (4), 319–333.
Reis, A.S., Albrecht, M.P., Bunn, S.E., 2020. Food web pathways for fish communities in small tropical streams. Freshw. Biol. 65, 893–907.

Reynolds, C.S., 1994. The ecological basis for the successful biomanipulation of aquatic communities. Arch. Hydrobiol. 130, 1–33.

Rezende, C.F., Mazzoni, R., 2003. Aspectos da alimentação de Bryconamericus microcephalus (Characiformes, Tetragonopterinae) no córrego Andorinha, Ilha Grande—RJ. Biota Neotropica 3, 1–6.

Ricciardi, A., Kipp, R., 2008. Predicting the number of ecologically harmful exotic species in an aquatic system. Divers. Distrib. 14, 374–380.

Richardson, D.M., Pyšek, P., 2008. Fifty years of invasion ecology—the legacy of Charles Elton. Divers. Distrib. 14, 161–168.

Romare, P., Hansson, L.A., 2003. A behavioral cascade: top predator induced behavioral shifts in planktivorous fish and zooplankton. Limnol. Oceanogr. 48, 1956–1964.

Roscoe, D.W., Hinch, S.G., 2010. Effectiveness monitoring of fish passage facilities: historical trends, geographic patterns and future directions. Fish Fish. 11, 12–33.

Ross, S.T., Matthews, W.J., Echelle, A.A., 1985. Persistence of stream fish assemblages: effects of environmental change. Am. Nat. 126, 24–40.

Rounsefell, G.A., 1946. Fish production in lakes as a guide for estimating production in proposed reservoirs. Copeia 1, 29–40.

Ryder, R.A., 1965. A method for estimating the potential fish production of north-temperate lakes. Trans. Am. Fish. Soc. 94, 214–218.

Ryman, N., 1991. Conservation genetics considerations in fishery management. J. Fish. Biol. 39, 211–224.

Santucci, V.J., Gephard, S.R., Pescitelli, S.M., 2005. Effects of multiple low-head dams on fish, macroinvertebrates, habitat, and water quality in the Fox River, Illinois. N. Am. J. Fish. Manag. 25, 975–992.

Sas, H. (Ed.), 1989. Lake restoration by reduction of nutrient loading. In Expectation, Experiences, Extrapolation. Academia Verlag Richardz GmbH., St. Augustin.

Schindler, D.E., Carpenter, S.R., Cole, J.J., Kitchell, J.F., Pace, M.L., 1997. Influence of food web structure on carbon exchange between lakes and the atmosphere. Science 277, 248–251.

Schindler, D.E., Knapp, R.A., Leavitt, P.R., 2001. Alteration of nutrient cycles and algal production resulting from fish introductions into mountain lakes. Ecosystems 4, 308–321.

Schlaepfer, M.A., Sax, D.F., Olden, J.D., 2011. The potential conservation value of non-native species. Conserv. Biol. 25, 428–437.

Schlosser, I.J., 1987. A conceptual framework for fish communities in small warm water streams. In: Matthews, W.J., Heins, D.C. (Eds.), Community and Evolutionary Ecology of North American Stream Fishes. Univ. of Oklahoma Press, Norman, London, pp. 17–24.

Schoener, T.W., 1989. Food webs from the small to the large. Ecology 70, 1559–1589.

Schroeder, D.M., Love, M.S., 2002. Recreational fishing and marine fish populations in California. Cali. Coop. Ocean. Fish. Invest. Rep. 43, 182–190.

Seghers, B.H., 1975. Role of gill rakers in size-selective predation by lake whitefish, *Coregonus clupeaformis* (Mitchill). Int. Ver. Theor. Ang. Limnol. 19 (3), 2401–2405.

Shapiro, J., Lamarra, V., Lynch, M., 1975. Biomanipulation: an ecosystem approach to lake restoration. In: Brezonik, P.L., Fox, J.L. (Eds.), Proceedings of a Symposium on Water Quality Management through Biological Control. University of Florida, Gainesville, pp. 85–89.

Sharpe, D.M.T., Hendry, A.P., 2009. Life history change in commercially exploited fish stocks: an analysis of trends across studies. Evol. Appl. 2, 260–275.

Shen, R., Gu, X., Chen, H., Mao, Z., Zeng, Q., Jeppesen, E., 2021. Silver carp (*Hypophthalmichthys molitrix*) promote phytoplankton growth by suppressing zooplankton rather than through fish-mediated nutrient recycling: an outdoor mesocosm study. Freshwat. Biol. 66 (6), 1074–1088.

Simberloff, D., Martin, J.L., Genovesi, P., Maris, V., Wardle, D.A., Aronson, J., et al., 2013. Impacts of biological invasions: what's what and the way forward. Trends Ecol. Evol. 28, 58–66.

Simon, T.P., 1998. Assessment of Balon's reproductive guilds with application to Midwestern North American freshwater fishes. In: Simon, T.P. (Ed.), Assessing the Sustainability and Biological Integrity of Water Resources Using Fish Communities. CRC Press, New York, pp. 97–122.

Smith, C., Reay, P., 1991. Cannibalism in teleost fish. Rev. Fish. Biol. Fisheries 1, 41–64.

Smith, K.G., Darwall, W.R.T., 2006. The Status and Distribution of Freshwater Fish Endemic to the Mediterranean Basin. IUCN, Gland, Switzerland.

Soares, M.G.M., Menezes, N.A., Junk, W.J., 2006. Adaptations of fish to oxygen depletion in a central Amazonian floodplain lake. Hydrobiologia 568, 353–367.

Søndergaard, M., Bjerring, R., Jeppesen, E., 2013. Persistent internal phosphorus loading in shallow eutrophic lakes. Hydrobiologia 710, 95–107.

Søndergaard, M., Jeppesen, E., Lauridsen, T.L., Skov, C., Van Nes, E.H., Roijackers, R., et al., 2007. Lake restoration: successes, failures and long-term effects. J. Appl. Ecol. 44, 1095–1105.

Søndergaard, M., Liboriussen, L., Pedersen, A.R., Jeppesen, E., 2008. Lake restoration by fish removal: short and long-term effects in 36 Danish lakes. Ecosystems 11, 1291–1305.

Souza, A.T., Argillier, C., Blabolil, P., Děd, V., Jarić, I., Monteoliva, A.P., Reynaud, N., Ribeiro, F., Ritterbusch, D., Sala, P., Šmejkal, M., Volta, P., Kubečka, J., 2022. Empirical evidence on the effects of climate on the viability of common carp (*Cyprinus carpio*) populations in European lakes. Biological Invasions 24, 1213–1227.

Stanley, E.H., Doyle, M.W., 2003. Trading off: the ecological effects of dam removal. Front. Ecol. Environ. 1, 15–22.

Starling, F.L.R.M., 1993. Control of eutrophication by silver carp (*Hypophthalmichthys molitrix*) in the tropical Paranoá Reservoir (Brasília, Brazil): a mesocosm experiment. Hydrobiologia 257, 143–152.

Starling, F.L.R.M., Beveridge, M., Lazzaro, X., Baird, D., 1998. Silver carp biomass effects on plankton community in Paranoá Reservoir (Brazil) and an assessment of its potential for improving water quality in lacustrine environments. Int. Rev. Hydrobiol. 83, 499–508.

Stemberger, R.S., Gilbert, J.J., 1987. Rotifer threshold food concentrations and the size-efficiency hypothesis. Ecology 68 (1), 181–187.

Stenson, J.A.E., 1979. Predatory-prey relations between fish and invertebrate prey in some forest lakes. Report from the Institute of Freshwater Research. Drottningholm 58, 166–183.

Sterud, E., Forseth, T., Ugedal, O., Poppe, T.T., Jørgensen, A., Bruheim, T., et al., 2007. Severe mortality in wild Atlantic salmon *Salmo salar* due to proliferative kidney disease (PKD) caused by *Tetracapsuloides bryosalmonae* (Myxozoa). Dis. Aquat. Organ. 77, 191–198.

Staudt, A., Leidner, A.K., Howard, J., Brauman, K.A., Dukes, J.S., Hansen, L.J., Paukert, C., Sabo, J., Solórzano, L.A., 2013. The added complications of climate change: understanding and managing biodiversity and ecosystems. Front. Ecol. Environ. 11 (9), 494–501.

Sušnik Bajec, S., Pustovrh, G., Jesenšek, D., Snoj, A., 2015. Population genetic SNP analysis of marble and brown trout in a hybridization zone of the Adriatic watershed in Slovenia. Biol. Conserv. 184, 239–250.

Svärdson, G., 1976. Interspecific population dominance in fish communities of Scandinavian lakes. Report from the Institute of Freshwater Research. Drottningholm 56, 144–171.

Taylor, J.N., Courtenay Jr., W.R., McCann, J.A., 1984. Known impact of exotic fishes in the continental United States. In: Courtenay Jr., W.R., Stauffer, J.R. (Eds.), Distribution, Biology, and Management of Exotic Fish. Johns Hopkins Press, Baltimore, MD, pp. 322–373.

Tedesco, P., Beauchard, O., Bigorne, R., Blanchet, S., Buisson, L., Conti, L., Cornu, J., Dias, M.S., Grenouillet, G., Hugueny, B., Jézéquel, C., Leprieur, F., Brosse, S., Oberdorff, T., 2017. A global database on freshwater fish species occurrence in drainage basins. Sci. Data. 4, 170141.

Teixeira-de Mello, F., Meerhoff, M., Pekcan-Hekim, Z., Jeppesen, E., 2009. Littoral fish community structure and dynamics differ substantially in shallow lakes under contrasting climates. Freshwat. Biol. 54, 1202–1215.

Tesch, F.W., 1968. Age and growth. In: Ricker, W.E. (Ed.), Methods for Assessment of Fish Production in Fresh Waters. Blackwell Scientific Publications, Oxford, pp. 93–123.

Tewksbury, J.J., Huey, R.B., Deutsch, C.A., 2008. Putting the heat on tropical animals. Science 320, 1296–1297.

Threlkeld, S.T., 1988. Planktivory and planktivore biomass effects on zooplankton, phytoplankton, and the trophic cascade. Limnol. Oceanogr. 33 (6), 1362–1375.

Tonn, W.M., 1990. Climate change and fish communities: a conceptual framework. Trans. Am. Fish. Soc. 119, 337–352.

Townsend, C.R., 1996. Invasion biology and ecological impacts of brown trout *Salmo trutta* in New Zealand. Biol. Conserv. 78, 13–22.

Vander Zanden, M.J., Fetzer, W.W., 2007. Global patterns of aquatic food chain length. Oikos 116, 1378–1388.

Vander Zanden, M.J., Rasmussen, J.B., 1999. Primary consumer $\delta^{15}N$ and $\delta^{13}C$ and the trophic position of aquatic consumers. Ecology 80, 1395–1404.

Vander Zanden, M.J., Vadeboncoeur, Y., Chandra, S., 2011. Fish reliance on littoral–benthic resources and the distribution of primary production in lakes. Ecosystems 14 (6), 894–903.

van de Wolfshaar, K.E., HilleRisLambers, R., Goudswaard, K.P.C., Rijnsforp, A.D., Scheffer, M., 2014. Nile perch (*Lates niloticus*, L.) and cichlids (*Haplochromis* spp.) in Lake Victoria: could prey mortality promote invasion of its predator? Theor. Ecol. 7, 253–261.

Vannote, R.L., Minshall, G.W., Cummins, K.W., Sedell, J.R., Cushing, C.E., 1980. The River Continuum Concept. Can. J. Fish. Aquat. Sci. 37, 130–137.

Vilizzi, L., Tarkan, A.S., Copp, G.H., 2015. Experimental evidence from causal criteria analysis for the effects of common carp *Cyprinus carpio* on freshwater ecosystems: a global perspective. Rev. Fish. Sci. Aquacult. 23 (3), 253–290.

Vinyard, G.L., O'Brien, W.J., 1976. Effects of light and turbidity on the reaction distance of bluegill (*Lepomis macrochirus*). J. Fish. Res. Board Can. 33, 2845–2849.

Vitule, J.R.S., Freire, C.A., Simberloff, D., 2009. Introduction of non-native freshwater fish can certainly be bad. Fish Fish. 10, 98–108.

Vitule, J.R.S., Freire, C.A., Vazquez, D.P., Nuñez, M.A., Simberloff, D., 2012. Revisiting the potential conservation value of non-native species. Conserv. Biol. 6, 1152–1155.

Volta, P., Jeppesen, E., Sala, P., Galafassi, S., Foglini, C., Puzzi, C., Winfield, I.J., 2018. Fish assemblages in deep Italian subalpine lakes: history and present status with an emphasis on non-native species. Hydrobiologia 824, 255–270.

Volta, P., Oggioni, A., Bettinetti, R., Jeppesen, E., 2011. Assessing lake typologies and indicator fish species for Italian natural lakes using past fish richness and assemblages. Hydrobiologia 671, 227–240.

Wahl, D.H., Stein, R.A., DeVries, D.R., 1995. An ecological framework for evaluating the success and effects of stocked fishes. Am. Fish. Soc. Symp. 15, 176–189.

Walters, C., Kitchell, J.F., 2001. Cultivation/depensation effects on juvenile survival and recruitment: implications for the theory of fishing. Can. J. Fish. Aquat. Sci. 58, 39–50.

Waters, T.F., 1977. Secondary production in inland waters. Adv. Ecol. Res. 10, 91–164.

Weber, M.J., Brown, M.L., 2009. Effects of common carp on aquatic ecosystems 80 years after "carp as a dominant": ecological insights for fisheries management. Rev. Fisheries Sci. 17 (4), 524–537.

Weigel, D.E., Peterson, J., Spruel, P., 2003. Introgressive hybridization between native cutthroat trout and introduced rainbow trout. Ecol. Appl. 13 (1), 38–50.

Welcomme, R., 2011. Review of the State of the World Fishery Resources: Inland Fisheries. FAO Fisheries and Aquaculture Circular No, Rome, p. 942.

Werner, E.E., 1974. The fish size, prey size, handling time relation in several sunfishes and some implications. J. Fish. Res. Bd. Can. 31 (9), 1531–1536.

Werner, E.E., Hall, D.J., 1974. Optimal foraging and the size selection of prey by the bluegill sunfish (*Lepomis macrochirus*). Ecology 55 (5), 1042–1052.

Werner, E.E., Mittelbach, G.G., 1981. Optimal foraging: field tests of diet choice and habitat switching. Am. Zool. 21 (4), 813–829.

Werner, E.E., Mittelbach, G.G., Hall, D.J., 1981. The role of foraging profitability and experience in habitat use by the bluegill sunfish. Ecology 62, 116–125.

Whitfield, A.K., Becker, A., 2014. Impacts of recreational motorboats on fishes: a review. Mar. Pol. Bull. 83 (1), 24–31.

Wiens, J.A., 2002. Riverine landscapes: taking landscape ecology into the water. Freshw. Biol. 47, 501–515.

Williams, J.G., Armstrong, G., Katopodis, C., Larinier, M., Travade, F., 2012. Thinking like a fish: a key ingredient for development of effective fish passage facilities at river obstructions. Riv. Res. Appl. 28, 407–417.

Williamson, M., 1996. Biological Invasions. Chapman & Hall, London.

Willson, M.F., Halupka, K.C., 1995. Anadromous fish as keystone species in vertebrate communities. Cons. Biol. 9 (3), 489–497.

Winberg, G.G., 1970. Some interim results of Soviet IBP investigations on lakes. In: Kajak, Z., Hillbricht-Ilkowska, A. (Eds.), Productivity Problems of Freshwaters. PWN Polish Scientific Publishers, Warsaw, pp. 363–381.

Winfield, I.J., Hateley, J., Fletcher, J.M., James, J.B., Bean, C.W., Clabburn, P., 2010. Population trends of Arctic charr (*Salvelinus alpinus*) in the U.K.: assessing the evidence for a widespread decline in response to climate change. Hydrobiologia 650, 55–65.

Wonham, M.J., Calrton, J.T., Ruiz, G.M., Smith, L.D., 2000. Fish and ships: relating dispersal frequency to success in biological invasions. Mar. Biol. 136, 1111–1121.

World Commission on Dams (WCD), 2000. Dams and Development: A New Framework for Decisions-Making. Earthscan Publications, London.

Wrona, F.J., Prowse, T.D., Reist, J.D., Hobbie, J.E., Levesque, L.M.J., Vincent, W.F., 2006. Climate change effects on aquatic biota, ecosystem structure and function. Ambio 35, 359–369.

Zambrano, L., Martínez-Meyer, E., Menezes, N., Peterson, A.T., 2006. Invasive potential of common carp (*Cyprinus carpio*) and Nile tilapia (*Oreochromis niloticus*) in American freshwater systems. Can. J. Fish. Aquat. Sci. 63, 1903–1910.

Zaret, T.M., 1972. Predator-prey interaction in a tropical lacustrine ecosystem. Ecology 53, 248–257.

Zaret, T.M., 1980. Predation and Freshwater Communities. Yale University Press, New Haven.

Zaret, T.M., Kerfoot, W.C., 1975. Fish predation on *Bosmina longirostris*: body-size selection versus visibility selection. Ecology 56, 232–237.

Zhang, X., Zhen, W., Jensen, H.S., Reitzel, K., Jeppesen, E., Liu, Z., 2021. Macrophytes (*Vallisneria denseserrulata*) enhance the effect of Phoslock® treatment used to restore shallow freshwater lake ecosystems and water quality. Envir. Poll. 277, 116720.

Zhao, S.Q., Fang, J.Y., Peng, C.H., Tang, Z.Y., Piao, S.L., 2006. Patterns of fish species richness in China's lakes. Glob. Ecol. Biogeogr. 4, 386–394.

Zhong, Y., Power, G., 1996. Environmental impacts of hydroelectric projects on fish resources in China. Reg. Riv. Res. Manag. 12, 81–98.

Zhou, S., Smith, A.D.M., Knudsen, E.E., 2014. Ending overfishing while catching more fish. Fish Fish. 16, 716–722.

CHAPTER

23

Pelagic Bacteria, Archaea, and Viruses

Katherine D. McMahon[1,2] *and Ryan J. Newton*[3]

[1]Department of Civil and Environmental Engineering, University of Wisconsin Madison, Madison, WI, United States [2]Department of Bacteriology, University of Wisconsin Madison, Madison, WI, United States [3]School of Freshwater Sciences, University of Wisconsin Milwaukee, Milwaukee, WI, United States

OUTLINE

I. Overview

Microorganisms underpin virtually all elemental cycling in aquatic ecosystems (Cotner and Biddanda, 2002). In Raymond Lindeman's foundational work *"The Trophic-Dynamic Aspect of Ecology"* (Lindeman, 1942), bacteria and their associated "ooze" are at the center of all lacustrine energy and matter cycling (Fig. 23-1). No one contests the central role of microorganisms in freshwater systems, but surprisingly little is known about the mechanisms governing their diversity and ecology. Most ecosystem scientists treat them as "black boxes" that simply respire and recycle. Aquatic microbial biologists seek to understand constraints on their biodiversity, growth, activity, and evolution, just like any other taxonomic group of organisms studied by "zoologists."

A revolution in our understanding of microbial biodiversity taught us that the prokaryotic organisms embedded in Lindeman's ooze actually included two very distinct domains of life: Bacteria and Archaea (Woese and Fox, 1977). Humans and other eukaryotes

Wetzel's Limnology, Fourth Edition
https://doi.org/10.1016/B978-0-12-822701-5.00023-9

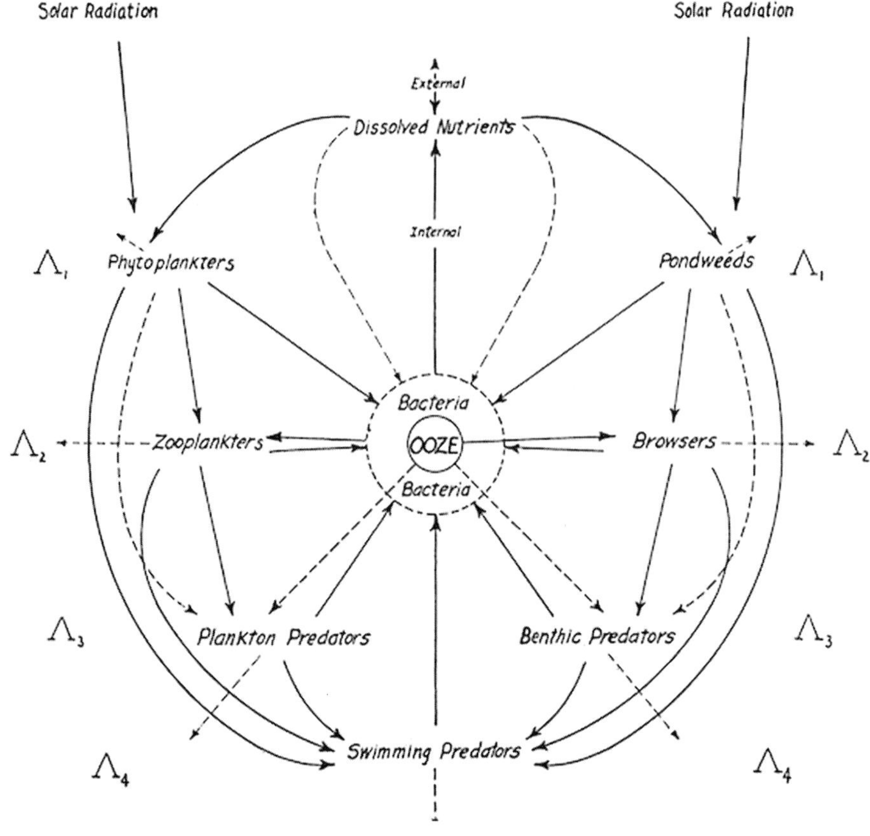

FIGURE 23-1 Microbes are at the center of all biogeochemical processing in ecosystems. Lindeman (1942) referred to them as "ooze" in his landmark publication based on studies of Cedar Bog Lake in Minnesota, USA. *(Reproduced with permission.)*

are more closely related to Archaea than Bacteria are to either of these other two domains. Thus we usually prefer the term "microbes" or the equivalent "microorganisms" to discuss these tiny nucleus-lacking creatures. We recognize that this term is sometimes used as an imprecise definition to include viruses, algae, and even zooplankton. Here we will attempt to be as rigorous as possible by defining "microbes" as Bacteria and Archaea. Unfortunately, this does lead to confusion since "blue-green algae" (or Cyanobacteria) are prokaryotic and bacteria in organization. Since phytoplankton and periphyton (algae and Cyanobacteria, Chapters 17, 18, and 25) are allocated three separate chapters, we will limit this chapter to non-Cyanobacteria.

Microbes are very distinct compared to nearly all known multicellular life forms. They are microscopic and therefore difficult to study. We cannot easily classify them based on their color, number of legs, or body shape. Indeed, microbial morphology is a highly problematic foundation for classification because very closely related taxa may have very different shapes, and distantly related microbes may have the same shape. However, short generation times and large population sizes make microbes compelling model systems for studying basic principles of ecology and evolution that may take decades to millennia to play out in plants and animals. Still, one key constraint on our ability to study microbes is how difficult it can be to grow them in isolation. The famous "great plate count anomaly" introduced by Staley and Konopka (Staley and Konopka, 1985) was inspired by work on Lake Washington (WA, USA) in which they noted that the number of cells recovered on standard microbiological media was 100 times less than the number visible under the microscope. Microbes are finicky creatures and the challenge of bringing them into captivity in the laboratory is widely acknowledged as one of the grand challenges for microbial ecologists.

This chapter focuses on the fundamental characteristics of microbes that define their biodiversity and influence their function in freshwaters. Their central role in the carbon cycle overlaps with Chapter 28 and their inextricable linkages to the nitrogen cycle place them also in Chapter 14. Interactions with other elements are partially covered in Chapters 15 and 16. Sediment microbes are covered mainly in Chapter 27. As noted above, we generally do not discuss Cyanobacteria other than to note their involvement in key ecosystem and biological processes.

II. Basic ecology

A. Abundance, size, and cell composition

Microbes are the most abundant living organisms in freshwater ecosystems (indeed, in all ecosystems). They are also the smallest, as their name implies. Virtually all are invisible to the naked eye. The concentration of cells in lakes and rivers is fairly consistent between 1×10^5 and 5×10^6 cells per mL, though higher values can be found in extremely productive systems in the summer. In general, the numbers and biomass of bacteria increase with increasing productivity and concentrations of inorganic and organic compounds in lakes (Table 23-1). Despite great seasonal variations, bacterial numbers and biomass increase from oligotrophic to eutrophic inland waters. Numbers of bacteria are markedly lower in acidic, dystrophic lakes, which contain high concentrations of humic matter.

The seasonal distribution of bacterial cell numbers is variable from lake-to-lake and from year-to-year within a lake. A few generalizations, however, can be extracted even from cell count data determined with sampling intervals that are far greater than bacterial generation times. Bacterial numbers and biomass are commonly high in the epilimnion, decrease to a minimum in the metalimnion and upper hypolimnion, and increase in the lower hypolimnion, especially if anoxic (Coveney and Wetzel, 1995; Jones, 1978; Pedrós-Alió and Brock, 1982; Simon et al., 1998). Lower numbers are observed under the ice in winter (Bertilsson et al., 2013; Pedrós-Alió and Brock, 1982). A 3-year study of Lake Constance (central Europe) showed little interannual variability but notable within-year variability (Fig. 23-2).

Methods for counting cells and measuring biovolume have many limitations. Simple phase-contrast light microscopy is insufficient because most cells are colorless and clear. Major advances came with *epifluorescence microscopy* (microscopy using a fluorescent light source) coupled with stains that bind to nucleic acids (for example, see Fig. 23-3, and with scanning electron microscopy [Figs. 23-4 to 23-6]). Counts of specific taxa are also possible with fluorescent probes that target specific stretches of DNA that are unique to particular groups of microbes. This technique is known as fluorescent *in situ* hybridization (FISH) (Fig. 23-7). However, many cells are very small and difficult to visualize. Others are attached to particles in multiple layers, making accurate counting impossible. Furthermore, even nucleic acid binding stains might illuminate inactive cells. Thus absolute cell counts do not provide a measure of overall activity because different community members are more or less active (depending on the definition of "activity" being used, of course). Epifluorescence microscopy can be combined with microautoradiography (see Section IIE below) to enumerate cells that incorporate a substrate labeled with a radioisotope (e.g., ^{3}H-thymidine) to estimate the active fraction.

Limnologists frequently use conversion factors to estimate the amount of carbon (C) contained in a natural microbial community. These factors can be either volumetric or per cell. The estimate of 5.6×10^{-13} g of C μm^{-3} is based on microbes cultured in the lab (in pure or mixed culture), microscopy, and analysis of dried biomass elemental composition (Bratbak, 1985). Other published estimates range from 1.2×10^{-13} g of C μm^{-3} (Watson et al., 1977) to 3.5×10^{-13} g of C μm^{-3} (Bjoernsen, 1986) to 10×10^{-13} g of C μm^{-3} (Nagata, 1984). Most of these estimates were derived from marine organisms but are frequently applied to freshwaters. The value is sensitive to the method used for determining biovolume including the fixation protocol used before microscopy. Differences between this widely used conversion factor and the actual volumetric C-content can have significant effects on estimates of production, growth yield, and conversion efficiency. Thus reported values for these important parameters should always be viewed with skepticism. More modern methods based on X-ray microanalysis combined with transmission electron microscopy estimate a per-cell basis of $1.2–3.0 \times 10^{-14}$ g of C per cell (Fukuda et al., 1998) or $2.0–6.0 \times 10^{-14}$ g of C per cell (Fagerbakke et al., 1996). However, C-content is known to increase with cell size, so this conversion is also suboptimal.

The elemental composition of microbial cells in freshwater at any given moment is influenced by many factors. These include ecosystem-level properties such as trophic status but also local factors such as temperature, light, water chemistry, and resource availability. It is well established but worth noting that different taxa have different elemental compositions and respond uniquely to influencing factors (Scott et al., 2012). Many studies assume that microbes are relatively more enriched in N and P as compared to their growth substrates. However, actual field measurements suggest that the often-used *Redfield ratio* (106C:16N:1P; molar ratio) can serve as a first estimate of cell stoichiometry, as shown in a survey of around 120 lakes in the Upper Midwest of the United States (Cotner et al., 2010). That is, bacteria were not particularly enriched in N and P, but rather just slightly more enriched than the general seston pool in these lakes. However, the P content of microbial biomass is quite variable and very sensitive to environmental conditions, especially P availability and growth rate (e.g., as measured using chemostat experiments with isolates) (Godwin et al., 2017; Godwin and Cotner, 2015).

TABLE 23-1 Numbers, Volumes, and Biomass of Bacteria in Inland Waters of Differing Productivity, Generally Measured During the Summer in the Trophogenic Zone

Habitat	Number (10^6 mL^{-1})	Biovolume[a] (μm^3)	Biomass[a] (g m^{-3})	Reference
Oligotrophic				
Schirmacher Lakes, Antarctica	0.018–0.035	0.034	—	Ramaiah, 1995
Lake Fryxell, Antarctica	0.54–4.5	0.13–0.39	—	Konda et al., 1994
Ace, Antarctica	0.13–7.28	—	—	Bell and Laybourn-Perry, 1999
Zelenetskoye, Russia (1970, 1971)	0.175	0.265	0.06	Kuznetsov, 1970
Baikal, Russia	0.20	—	—	Kuznetsov, 1970
Krivoye, Russia (1968, 1969)	0.67	0.43	0.21	Kuznetsov, 1970
Lawrence Lake, Michigan (USA)	2–6	0.09–0.35	0.28	Coveney and Wetzel, 1995
Mirror Lake, New Hampshire (USA)	0.5–7	0.12	—	Ochs et al., 1995
Toolik Lake, Alaska (USA)	0.1–3.1	0.056–0.26	0.01–0.04	Hobbie et al., 1999, 1983
Mesotrophic				
Krasnoye, Russia (1964–1970)	0.70	0.43	0.30	Kuznetsov, 1970
Ladoga, Russia (1977–1989)	0.2–1.2	—	—	Kapustina, 1996
10 Shallow lakes of Florida (mostly mesotrophic) (USA)	1.4–10.5	—	—	Crisman et al., 1984
Lake Biwa, Japan				
North Basin, mesotrophic	1.6–3.4	0.02–0.20	—	Nagata, 1984
South Basin, eutrophic	5.9–8.6	—	—	Nagata, 1984
Eutrophic				
Drivyaty, Russia (1964)	1.84	0.76	1.40	Kuznetsov, 1970
Lake Mendota, USA	0.5–2.0	0.12–0.27	—	Pedrós-Alió and Brock, 1982
Lake Constance, Germany	1–10.0	—	0.016–0.18	Güde et al., 1985
Over 7 years	0.5–10.8	—	—	Simon et al., 1998
Himon-ya Pond, Japan	4.1–19	—	—	Konda, 1984
Lake Valencia, Venezuela	0.1–1.4	—	1.0	Lewis et al., 1986
Reservoirs				
Rybinsk, Russia (1964–1968)	1.70	0.60	1.00	Kuznetsov, 1970
Bratsk, Russia (1965–1972)	0.85	0.90	0.77	Kuznetsov, 1970
Kakhov, Russia (1968)	4.00	0.47	1.90	Kuznetsov, 1970
Dneprodzerzhin, Russia (1968)	3.40	0.65	2.20	Kuznetsov, 1970

TABLE 23-1 Numbers, Volumes, and Biomass of Bacteria in Inland Waters of Differing Productivity, Generally Measured During the Summer in the Trophogenic Zone—cont'd

Habitat	Number (10^6 mL^{-1})	Biovolume[a] (μm^3)	Biomass[a] (g m^{-3})	Reference
Dystrophic				
Chernoe, Russia	1.07	—	—	Kuznetsov, 1970
Lake Botjärn, Sweden	1.2–7.3	0.18–0.20	0.15–0.29	Johansson, 1983
Loch Ness, Scotland	0.23–0.71	—	—	Laybourn-Parry et al., 1994
Rivers				
Kuparuk River, Alaska (USA)	0.3–2.7	0.1–3.1	—	Hobbie et al., 1983
Ogeechee River, Georgia	8–75	0.06–0.18	0.02–1.14	Edwards, 1987
Average values				
Oligotrophic	0.50	0.2–0.4	0.15	Saunders et al., 1980
Mesotrophic	1.00	0.4–1.2	0.70	Saunders et al., 1980
Eutrophic	3.70	0.5–0.9	2.30	Saunders et al., 1980

[a] *Bacterial biovolumes tend to decrease seasonally as water temperatures increase (Chrzanowski et al., 1988) or with intensified bacterivory (Chapter 20). Bacterial cell carbon content can be approximated by a conversion factor of 5.6 × 10^{-13} g of C μm^{-3} (Bratbak, 1985).*

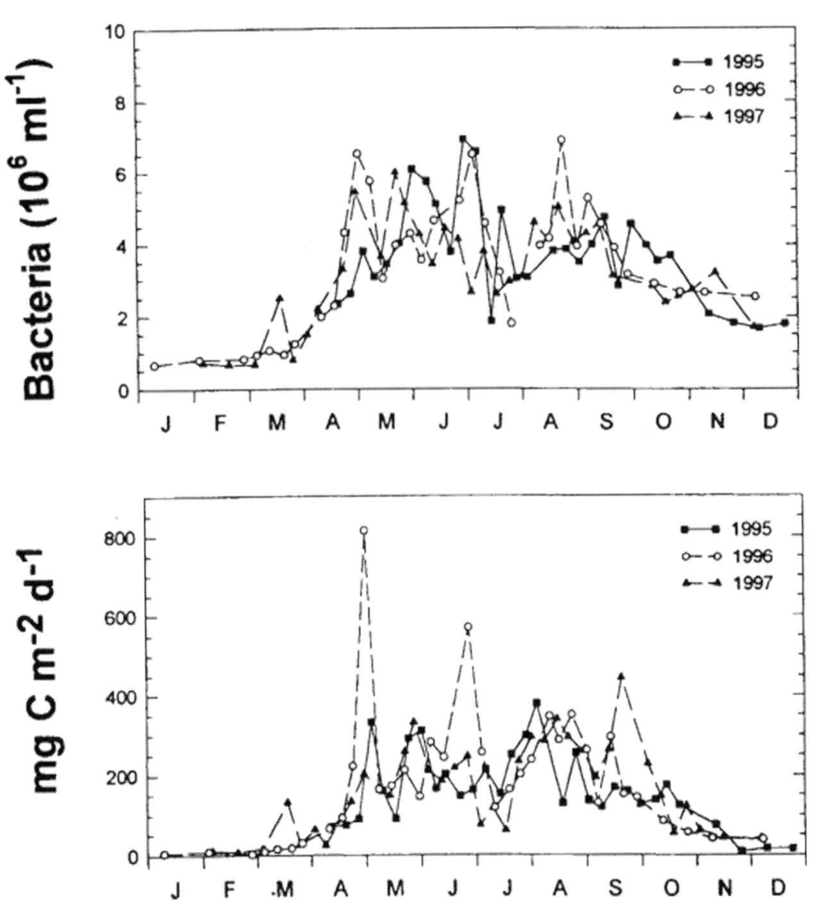

FIGURE 23-2 Bacterial cell counts through time in Lake Constance, Germany.

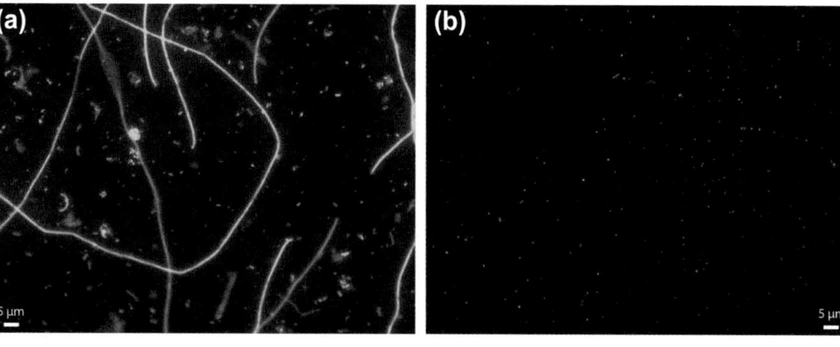

FIGURE 23-3 4′,6-Diamidino-2-phenylindole (DAPI) stained cells in surface water collected from Lake Mendota (WI, USA) and visualized with epifluorescence microscopy when (a) filamentous bacteria are abundant and (b) mainly small bacteria are present. Cells appear a *blue color*. A scale bar for size reference is indicated. Cells were captured on a 0.2 µm filter and were stained using 4′,6-Diamidino-2-phenylindole, which binds to nucleic acids and fluoresces *blue* when viewed under an epifluorescent microscope.

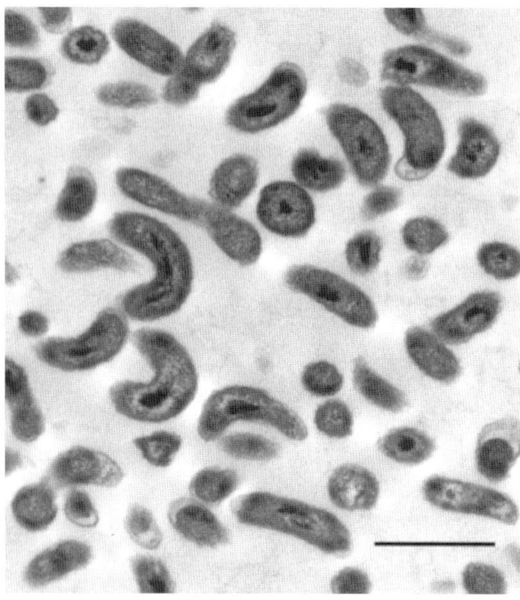

FIGURE 23-4 Transmission electron microscopy of cells from a pure culture of the genus *Limnohabitans* isolated from Lake Mondsee in Austria. The image was obtained at ×20,000 magnification. *(From Hahn et al., 2010.)*

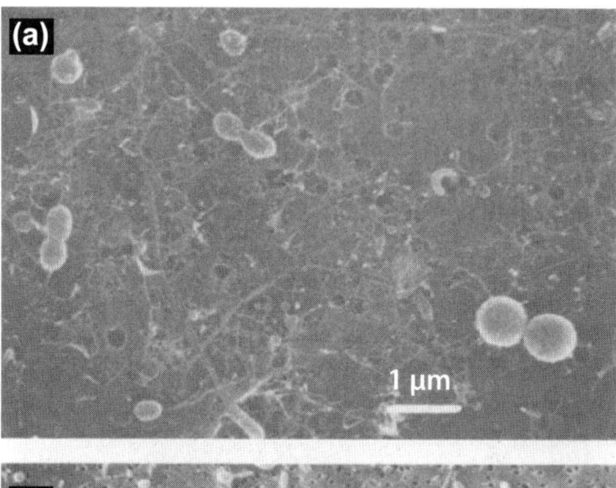

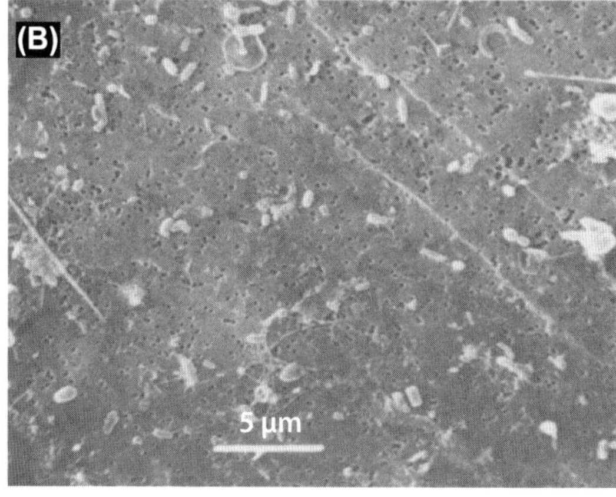

FIGURE 23-5 Scanning electron micrographs (a) & (b) of cells from Lake Mendota in Wisconsin, USA captured on a 0.2-µm filter. *(From Pedrós-Alió and Brock, 1982.)*

B. Classification

Microbiologists classify microbes based on multiple conceptual frameworks. Morphology was used when light microscopes were among the only tools available to study them. Categories include *cocci, rods, vibrio* (curved rods), and *spirals*. However, these distinctions have limited use due to known morphological plasticity by individual microbes. Cultivation in the laboratory can enable classification based on an organism's capacity to metabolize specific substrates (e.g., alcohols, sugars, ammonia). However, this requires the isolation of pure cultures, which is frequently not possible (see below). Fundamental principles of strategies to get energy and build biomass can be used to classify microbes based on "*functional groups*" (Table 23-2). But microbes tend to be rule-breakers and our modern understanding of ecophysiology includes many examples of taxa that were originally classified as, for example, *photolithoautotrophs*, which are in fact *mixotrophic,* meaning that they can sometimes get energy from chemical sources and/or use organic compounds for both carbon and electrons. These distinctions have important implications for carbon and energy flow through freshwater ecosystems.

Indeed, most historical work on freshwater microbes (and microbes in all ecosystems) was conducted using cultivation, microscopy (emphasis on morphological characterizations), bulk process rate measurements,

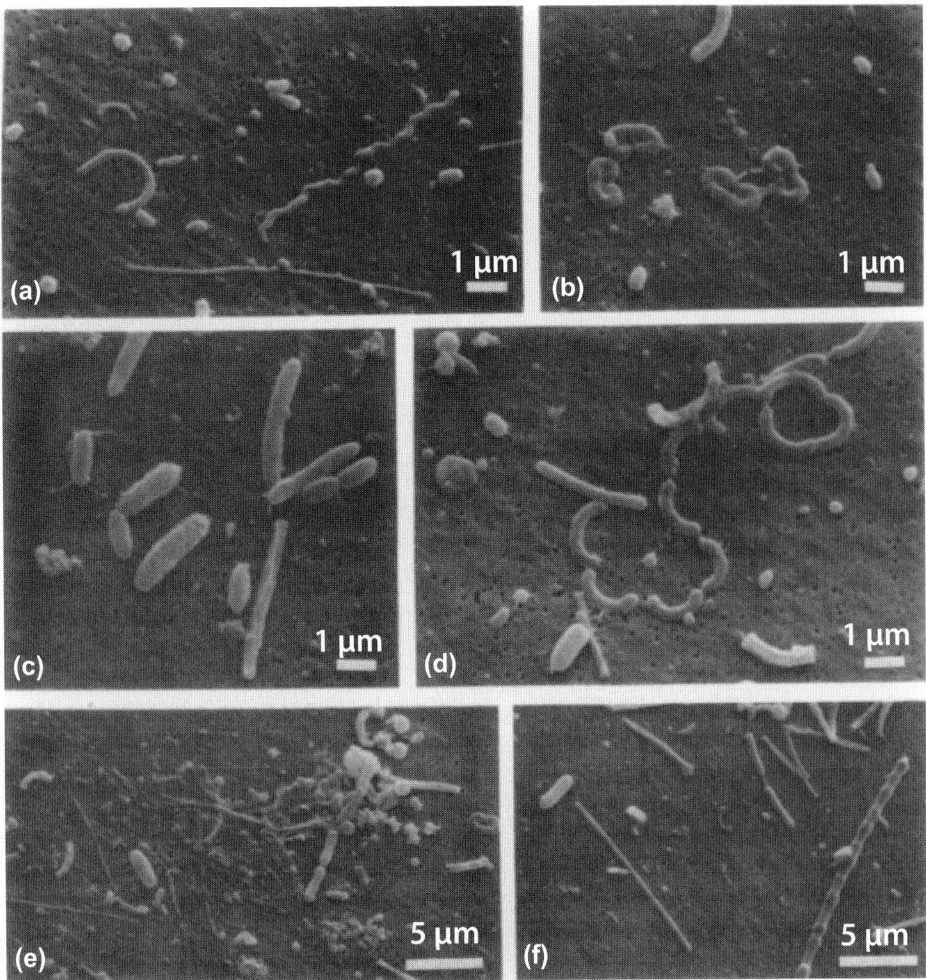

FIGURE 23-6 Scanning electron microscopy of cells from Lake Kinneret in Israel. Images of cells with distinct morphologies are depicted in boxes (a)–(f). *(From Schmaljohann et al., 1987.)*

and isotope tracers. As noted above, none of these provide sufficient insight into the taxonomic composition of microbial communities.

In order to classify microbes, microbial ecologists now most often use a single highly conserved gene, called the *16S ribosomal RNA gene*. This gene can be amplified by *polymerase chain reaction* (PCR) directly out of bulk DNA extracted from environmental samples and then sequenced using a variety of technologies. It is important to understand that the definition of a microbial species is highly controversial, but most often, we use groups of 16S rRNA genes clustered with a minimum of 97% nucleotide sequence identity as a species-like unit. Such groups are referred to as *operational taxonomic units* (OTUs), or *amplicon sequence variants* (ASVs) if they are not clustered.

Microbiology is in a period of transition from classifying based on morphological or physiological traits or 16S rRNA gene sequences to the use of nearly complete genome sequences (Varghese et al., 2015). It is now possible to sequence virtually all DNA extracted from water using the technique called *metagenomics*. In theory, the full complement of genomes present in an environmental sample is called the *metagenome*. The resulting DNA sequences can be computationally pieced together into representative genome sequences that support more rigorous *phylogenomic approaches* to classification (i.e., constructing phylogenetic trees with many genes shared within a taxonomic group). This genome-based work has led to increased scrutiny of how microbes are classified taxonomically. In response, in 2021 official revisions to the microbial taxonomic code designated phylum as an official taxonomic category and standardized naming practices across the taxonomic categories (Oren et al., 2021). The major taxonomic categories are as follows, from most broad to most narrow: Domain → Phylum → Class → Order → Family → Genus → Species. As this chapter is being written, there is an ongoing debate regarding whether to keep long-standing phylum names or adhere to the new conventions, thus altering many names that have appeared in the scientific

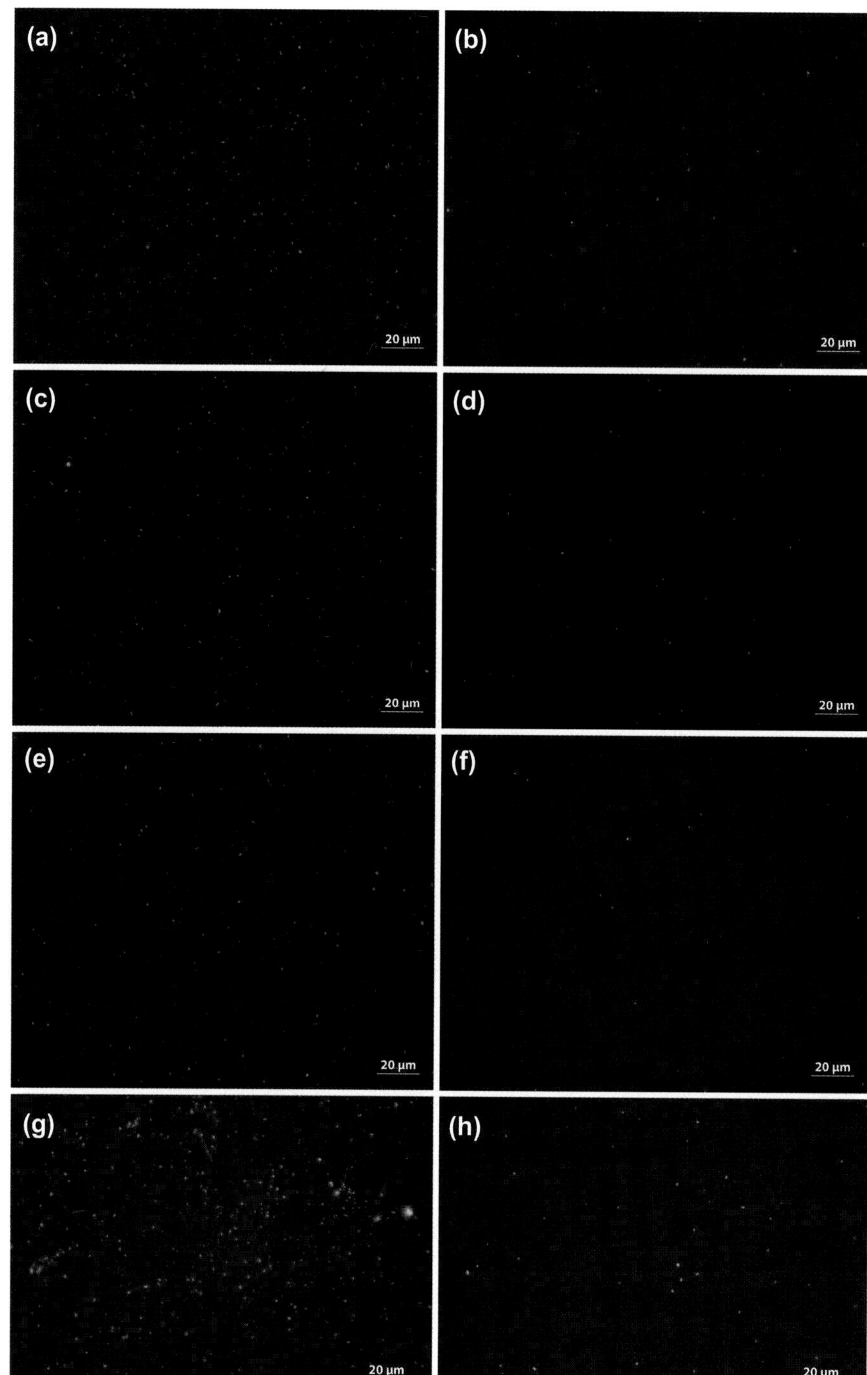

FIGURE 23-7 Fluorescent *in situ* hybridization (FISH) images of major freshwater taxa. *Left-hand panels* are imaged only with DAPI, while *right-hand panels* are in the same field but imaged using the wavelength specific to fluor on the FISH probe. Panels (a) and (b) *Actinobacteria* from the acI-lineage collected from Římov Reservoir in the Czech Republic. Panels (c) and (d) *Alphaproteobacteria* from the LD12 tribe, collected from Lake Mondsee in Austria. Panels (e) and (f) *Betaproteobacteria* from the LD28 tribe, collected from Římov Reservoir. Panels (g) and (h) *Betaproteobacteria* from the RBT group, collected from Lake Zurich in Switzerland. See also Table 23-4 and Newton et al. (2011) to link conventional placeholder names with higher taxonomic classification levels. *(Images generously provided by Dr. Michaela Salcher from the Institute of Hydrobiology, Biology Centre of the Czech Academy of Sciences.)*

TABLE 23-2 Microbes Need Energy, Carbon, and Electrons to Sustain Themselves and Reproduce[a]

Energy	Photo (light)	versus	Chemo (chemical)
Carbon	Auto (inorganic)	versus	Hetero (organic)
Electrons	Litho (inorganic matter)	versus	Organo (organic matter)

[a]*Energy sources refer to how an organism obtains the energy required to synthesize ATP (adenosine triphosphate). Carbon sources refer to the source of carbon atoms used to build biomass. Electron sources refer to the source of reducing power needed to build biomass from comparatively more oxidized compounds.*

TABLE 23-3 Phylum Names for Common Freshwater Microorganisms

Traditional phylum name	*Proposed phylum name*[a]
Actinobacteria	*Actinomycetota*
Bacteroidetes	*Bacteroidota*
Chloroflexi	*Chloroflexota*
Cyanobacteria	*Cyanobacteria*
Firmicutes	*Bacillota*
Nitrospirae	*Nitrospirota*
Nitrospinae	*Nitrospinota*
Planctomycetes	*Planctomycetota*
Proteobacteria	*Pseudomonadota*
Thaumarchaeota	*Thermoproteota*
Verrucomicrobia	*Verrucomicrobiota*

[a]*Proposed names from Oren and Garrity (2021).*

literature for decades. For clarity, we provide in Table 23-3 a list of the old and corresponding new phylum names for those groups discussed in this chapter. We also note that microbiologists adhere to conventions for italicizing names from Phylum to Species, except in the case of recently proposed (but not officially approved) names which use "*Candidatus*" in italics and the genus/species names in regular font (Parker et al., 2019).

C. Community composition and assembly

Most microbial ecologists would argue that freshwater microbial community membership is determined largely by what we would call "bottom-up" factors. These include temperature, light, major ion concentrations, redox status, and pH. Grazing and viral lysis also play a role, of course, but their influence on community assembly and dynamics are less well understood and thought to be less important. Here we focus on the physicochemical drivers and microbe–microbe interactions but return to "top-down" controls later in the chapter.

Our understanding of the "freshwater *microbiome*" has been revolutionized by the development of powerful molecular tools that rely on modern DNA sequencing technologies. The term "microbiome" is most often used to describe a collection of microorganisms (the *microbiota*) and the products of their collective activity (e.g., proteins, lipids, signaling molecules) in a defined habitat (the *biome*), but many definitions persist in the scientific literature (see Berg et al., 2020 for details). Some researchers use the word "microbiome" interchangeably with "metagenome," but most reserve "metagenome" to mean a collection of cooccurring genomes. As described above, the 16S rRNA gene is now the gold standard for classification and sorting microbes into what we consider to be relevant ecological units. Thus most of what we know about species composition in freshwater microbiomes is derived from DNA sequencing, which was not possible until around 1994, and not widely used until around 2000. Research studies published before this time that are based primarily on cultivation for determining microbial composition should be consulted with great caution.

The most intensively studied freshwater microbiomes are those in the upper mixed layer of lakes in temperate regions and our species census from these systems is considered largely complete (Newton et al., 2011). Almost always aerobic with available light, these communities are dominated by a handful of cosmopolitan bacterial families but paradoxically have tremendously high species diversity with a long tail of relatively rare taxa. Archaea are usually rare under these conditions. Based on 16S rRNA genes, the top ten most abundant and prevalent families represent four phyla (Table 23-4). Many of these bacteria persist through much of the year, and most have very small cell sizes with minimal genetic content. Taxa that favor high substrate availability, called *copiotrophs,* have bloom-and-bust cycles and thus are abundant, by definition, relatively briefly in time and are therefore undersampled in the census. Cyanobacteria are an exception and are often the most abundant bacteria particularly in highly productive systems. Their diversity and composition are discussed in Chapters 17 and 18. Many of the taxa considered part of the global freshwater microbiome are found exclusively in freshwater. A few seem to have some salinity tolerance and can be found in estuaries or coastal regions. Very few are found in both marine and freshwater.

The longest and densest continuous record of freshwater community composition is from the epilimnion of Lake Mendota (WI, USA) (Rohwer and McMahon, unpublished). The lake is temperate, eutrophic, and calcareous (hard water with pH ~8.4). The time series consists of 513 samples collected over 19 years, with most time points clustered between April and October. The rank abundance curve of bacterial taxa grouped approximately at the genus level shows a typical uneven distribution, with only around ten taxa comprising 50%

TABLE 23-4 Top Ten Most Abundant and Prevalent Families, Representing the Top Four Most Prevalent Bacterial Phyla Found in Freshwater[a]

Phylum or class	Family (freshwater lineage)	Prevalence
Actinobacteria	*Nanopelagicales (acI)*	100%
Betaproteobacteria	*Comamonadaceae (betI)*	76%
Bacteroidetes	*bacII*	59%
Betaproteobacteria	*Burkholderiaceae (Pnec)*	59%
Actinobacteria	*Acidimicrobiaceae (acIV)*	56%
Verrucomicrobia	*verI*	53%
Bacteroidetes	*bacI*	47%
Bacteroidetes	*bacIII*	47%
Actinobacteria	*acTH1*	44%
Betaproteobacteria	*Methylophilaceae (betIV or LD28)*	41%
Actinobacteria	*Luna1*	41%
Alphaproteobacteria	*Pelagibacteriaceae (LD12)*	38%

[a]Adapted from Newton et al. (2011). Data were compiled from 34 lakes sampled around the world. Prevalence is calculated by dividing the number of lakes that family was found in by the total number of lakes (34). Phylum names used in 2011 are presented in this table, but see Table 23-3 for corresponding newer names.

of the dataset based on relative abundance (Fig. 23-8). This shape is nearly universally found in aquatic bacterial communities. The reader will notice that most of the top ten taxa are members of the families listed in Table 23-4.

Microbial communities are surprisingly dynamic, responding quickly to changing environmental conditions. Their species composition changes on daily time scales, with faster rates of change when water is warm (Eiler et al., 2012; Newton et al., 2006). Seasonal drivers also create repeating community assembly cycles especially in temperate lakes (Kent et al., 2007; Shade et al., 2007). Such dynamics are usually attributed to changes in water temperature, nutrient availability, top-down effects of grazing and viral predation (see below), and potential cascading effects of species–species interactions. Most of our understanding of community dynamics is based on studies of the upper mixed layer (epilimnion), which is especially influenced by meteorological conditions and changing substrate availability driven by primary production. The 19-year Lake Mendota time series described above documents variation in community composition within and across years (Fig. 23-9).

In lakes that stratify thermally the light and redox regimes in the hypolimnion are critical determinants of community composition. No comprehensive synthesis of taxa commonly found in hypolimnia has been conducted, likely because the geochemical structure of the water column creates so much environmental heterogeneity that it is not possible to define a consensus microbiome. Distinct functional groups are expected to be present when oxygen is depleted (Fig. 23-10). The role of microbes in engineering their own habitats is very apparent here, as key activities modify the chemical environment, which in turn selects for specific functional groups. A primary example is the sequential depletion of terminal electron acceptors such as nitrate, ferric iron, manganese, and sulfate during organic matter decomposition (see also Section III). These processes have important implications for the mobilization of nitrogen, phosphorus, and metals in the hypolimnion (Chapters 14–16). A typical depth-profile collected from Lake Mendota, WI, USA, illustrates the different biogeochemically defined layers (Fig. 23-11).

Trophic status and lake water pH are generally considered to be the two most important attributes that shape bacterial community composition (Newton et al., 2011). Cyanobacteria dominate communities in highly eutrophic lakes and are thought to exert strong control over the heterotrophs living alongside them. Lakes with high chromophoric dissolved organic matter (DOM) concentrations (Chapter 28) harbor unique groups, particularly when the pH is relatively low (<6) (Linz et al., 2018; Yannarell and Triplett, 2004). This includes peat-bog systems with open water, which are some of the most numerous freshwater bodies on Earth (Vonk et al., 2015). Light also plays a role, of course. Unique assemblages seem to inhabit hypolimnia when light penetrates to depth and oxygen is not depleted (Paver et al., 2020). A surprising amount of research has been dedicated to studying the effects of water residence time on microbial community composition in lakes. As might be expected, shorter water residence times result in lake communities that more closely resemble rivers (Lindstrom et al., 2005). Other factors such as lake landscape position, which influences the ion composition in the water, are also important (Yannarell and Triplett, 2005). The primary source of organic matter (i.e., autochthonous versus allochthonous) appears to also play a role (Jones et al., 2009; Kritzberg et al., 2006).

As is the case with other limnological parameters, microbial community samples are often collected from a single location within a lake (often the deepest). A few studies have examined horizontal spatial variation. In one study of two very different inland lakes it was found that spatial variation at a scale of 30 m was approximately equivalent to temporal turnover at a scale of 1 day (Jones et al., 2012). Dispersal limitation was also important at scales of greater than 20 m in alpine ponds (Lear et al., 2014). First principles tell us that this should depend also on lake morphometry, fetch, and

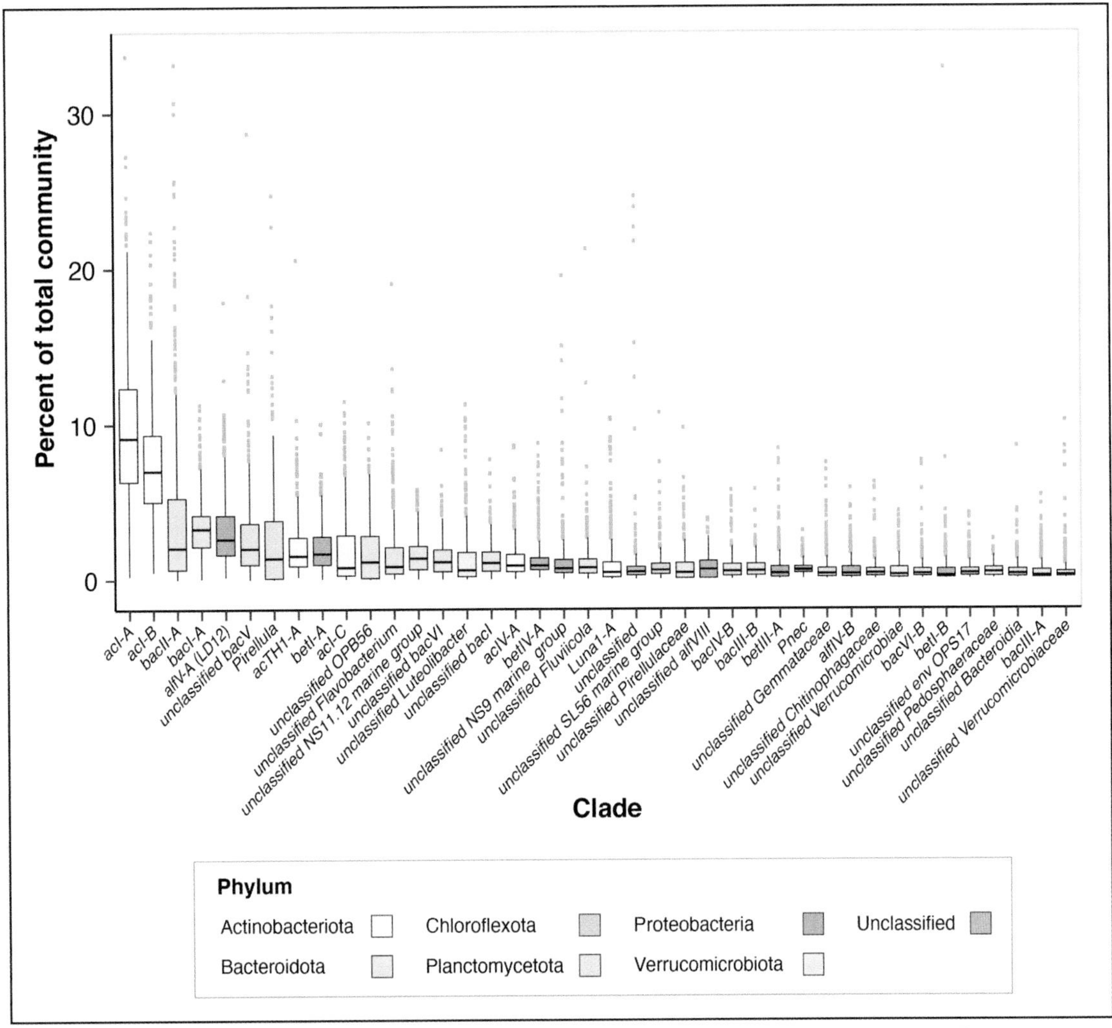

FIGURE 23-8 Rank abundance curve for the top 40 most abundant bacterial taxa detected in Lake Mendota (WI, USA) over 19 years (2000–2019) (Rohwer and McMahon, unpublished). Samples were collected from the upper 12 m (approximately the mixed layer) at regular intervals mainly during the ice-free period. Taxon relative abundances were calculated based on 16S rRNA gene amplicons classified with a custom database (Rohwer et al., 2018). Taxa were grouped at a level that approximates genus ("clade" as defined in Newton et al., 2011). Boxes are colored by bacterial phylum affiliation.

meteorological conditions that may or may not be creating vigorously mixed conditions at the time of sampling. Within the Laurentian Great Lakes, considerable spatial variation can be detected across stations within lakes (Paver et al., 2020). Our understanding of freshwater bacterial biogeography is likely to advance significantly with the study of whole genome sequences instead of the single 16S rRNA gene marker.

Comparatively little is known about microbial communities under ice in frozen lakes or during "shoulder seasons" when ice is forming or melting (Bertilsson et al., 2013). However, efforts focused on specific geochemical cycles (nitrogen and methane) have provided an initial glimpse into linkages between specific taxa and key functions. *Nitrospira* and *Nitrotoga*, two genera involved in nitrite oxidation to nitrate, were more active under ice, as determined using molecular tools (Fournier et al., 2021). Activity measurements using $^{15}N\text{-}NH_4^+$ confirmed that higher ammonia oxidation rates occur under ice and ammonia-oxidizing bacteria (*Nitrosospira*) were only detectable with sensitive molecular techniques during ice cover in an oligotrophic lake (Massé et al., 2019). In contrast, ammonia-oxidizing archaea (e.g., members of the *Thaumarchaeota*) were detectable throughout the year. Members of the enigmatic *Verrucomicrobia* phylum are also more abundant and active in ice-covered lakes, potentially associated with under-ice phytoplankton blooms as determined by the expression of key genes involved in polysaccharide and glycolate degradation (Tran et al., 2018). *Methanotrophs* (organisms that specialize in growth on methane) also seem to be more active (Fournier et al.,

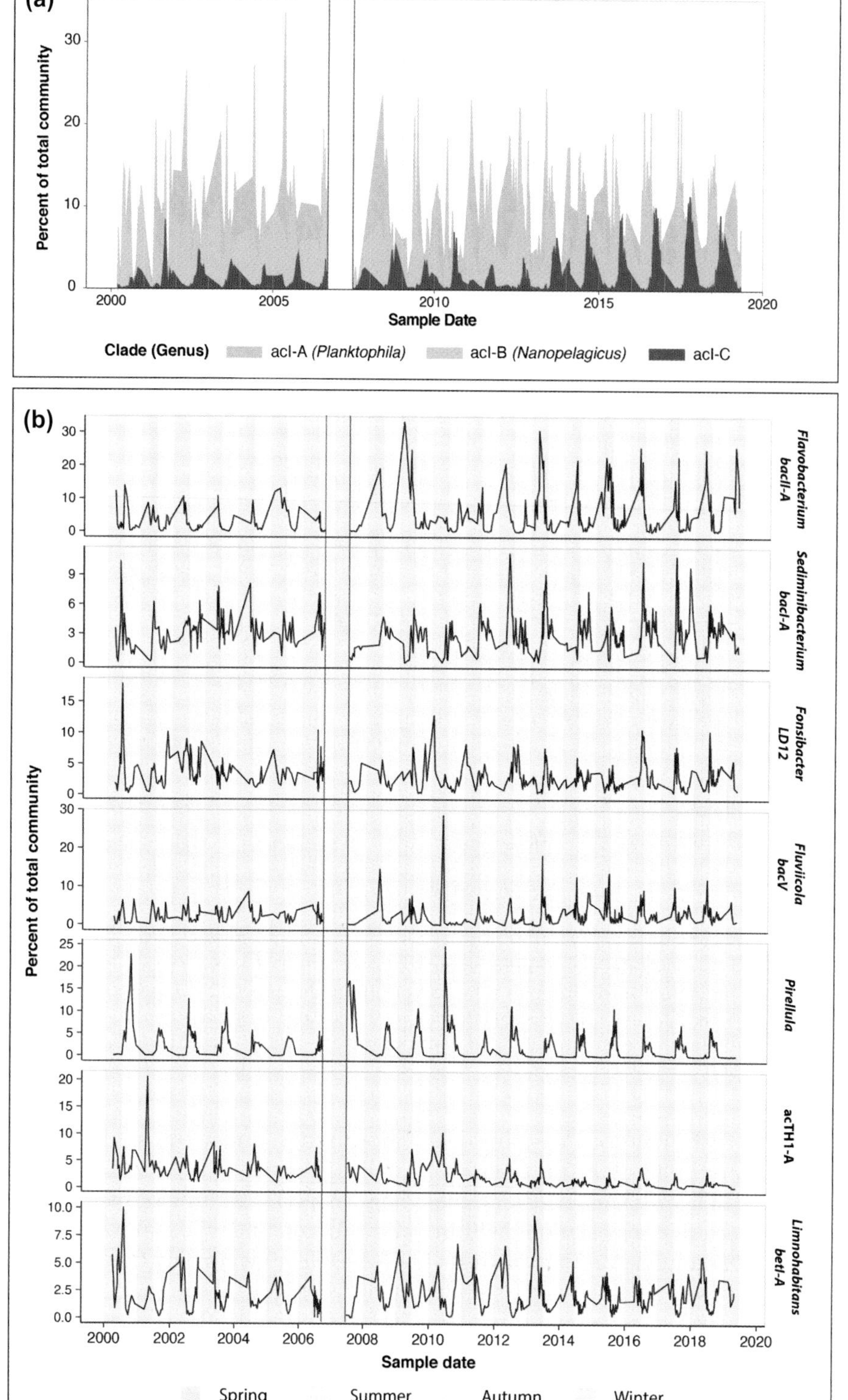

FIGURE 23-9 (a) Time series relative abundance dynamics for 3 of the top 10 most abundant bacterial taxa detected in Lake Mendota (WI, USA) over 19 years (2000–2019) (McMahon, unpublished). The taxa are members of the acI-*Actinobacteria* lineage (*Nanopelagicales*) and represent three distinct genera with contrasting traits. Notably, *Planktophila* (acI-A) and *Nanopelagicus* (acI-B) are nearly always present at high relative abundance, while acI-C (no described genus name) appears only in late summer. Two *vertical lines* indicate a prolonged period (~1 year) of no samples collected. (b) Time series relative abundance dynamics for 7 of the top 10 most abundant bacterial taxa detected in Lake Mendota over 19 years (2000–2019) (Rohwer and McMahon, unpublished). Data were collected across 19 years from the integrated epilimnion of Lake Mendota (total 420 unique time point samples). *Colors* indicate astronomical seasons.

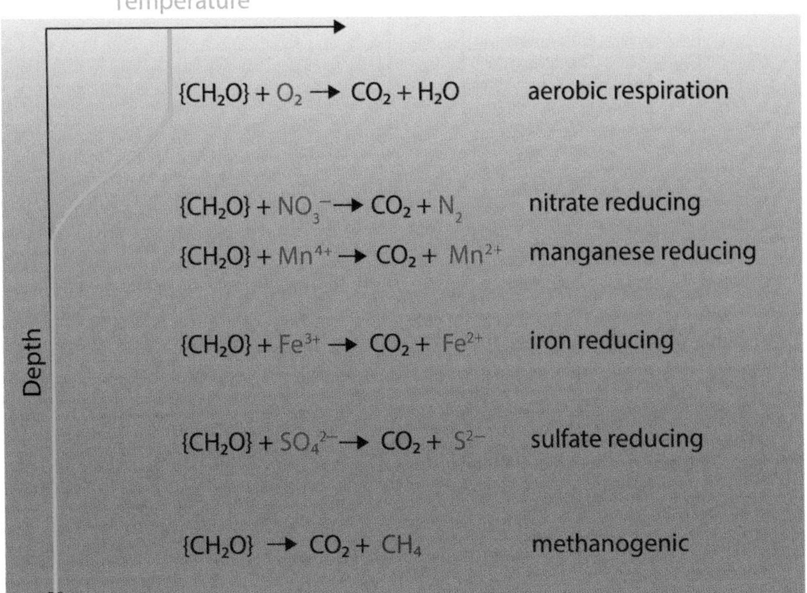

FIGURE 23-10 Idealized biogeochemical cycles in a productive stratified lake during the late summer season with an emphasis on terminal electron acceptor processes. The theoretical carbon source is simplified but represents a large diversity of compounds (sugars, peptides, carboxylic acids, etc.). Terminal electron acceptors (*TEAs*) are colored pink.

2021; Ricão Canelhas et al., 2016) and abundant (Schütte et al., 2016) under ice, suggesting an important role for dampening the methane loading to the atmosphere during spring thaw.

A handful of major rivers have been studied using molecular tools, including the Mississippi (Henson et al., 2018a; Staley et al., 2015) and Columbia (Crump et al., 1999; Fortunato et al., 2012) rivers in the United States, segments of the Amazon River in Brazil (Doherty et al., 2017; Satinsky et al., 2015), the Danube River in Europe (Savio et al., 2015; Winter et al., 2007), and the Yangtze River in China (Wang et al., 2012; Zhang et al., 2020). For the most part, taxa found in flowing waters are similar to those found in lake epilimnia. Biofilms found on rocks and in sediment are distinct and generally considered to comprise a separate "compartment" contributing to biogeochemical cycling in ways that are different from the flowing water (Gautam et al., 2021; Lear et al., 2013).

D. Lifestyles

As with plants and animals, microbes and their ecology can best be understood by considering their lifestyles (Grossart, 2010). An organism's mode of making a living will constrain its range and preferred habitat. Strategies for acquiring needed carbon, energy, and electrons can be used for classification (see Table 23-2) but also define the niches available to each group. For example, an *anoxygenic phototroph* (organisms that photosynthesize without generating oxygen) might use inorganic carbon, light energy, and sulfide to make a living. Sulfide and light are rarely found together in aquatic habitats since light would normally drive oxygenic photosynthesis, inhibiting the biological production of sulfide by sulfate- and sulfur-reducing bacteria. Thus we would expect to find such anoxygenic phototrophs only in a thin plate of water at depths where sufficient light is still available, but sulfide diffuses up from the dark anoxic depths. Knowing about microbial metabolism and corresponding lifestyles goes a long way toward explaining elemental cycling by microbes in freshwaters.

Freshwater microbes, like their marine counterparts, can also be classified based on their general growth strategy. Copiotrophs flourish when resources are abundant. Examples include Cyanobacteria in eutrophic systems, along with heterotrophs responsible for breaking down senescing (i.e., decomposing) blooms such as members of the *Bacteroidetes* (Eiler and Bertilsson, 2004; Schmidt et al., 2016; Xu et al., 2018). These microbes are often found attached to detritus particles that resemble *marine snow* (the shower of organic material continuously falling from the upper to deepwaters of the ocean) and are sources of fresh and highly labile organic matter (Simon et al., 2002). Other particle-associated bacteria that are thought to be copiotrophs include members of the *Verrucomicrobia* and *Planctomycetes* (Allgaier and Grossart, 2006). In contrast, oligotrophs are most successful when resources are scarce. Such bacteria possess physiological and morphological characteristics that maximize their ability to gather nutrients across steep spatial gradients and to use nutrients conservatively. Oligotrophs are generally free living, i.e.,

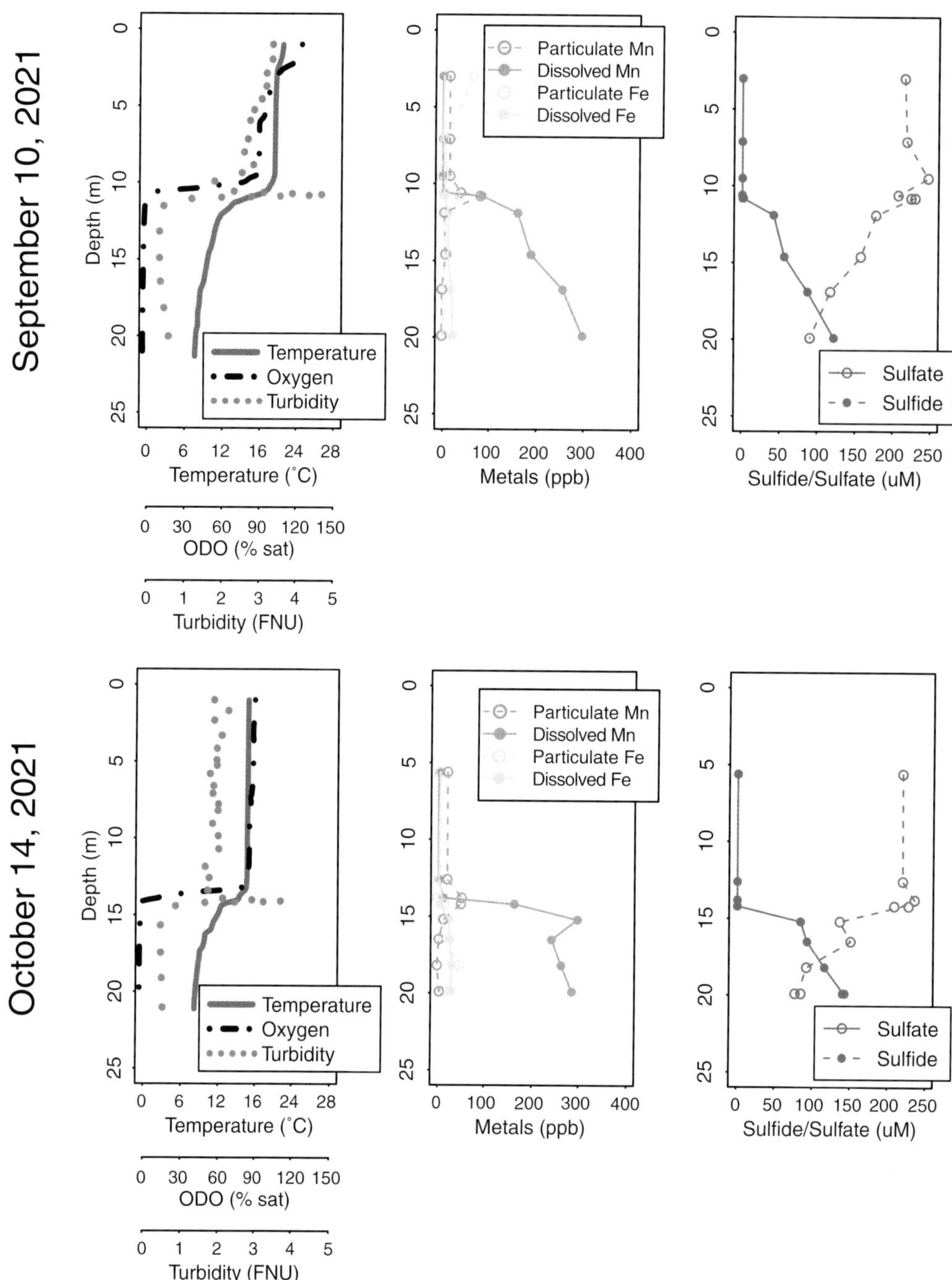

FIGURE 23-11 Geochemical profile with depth, collected from Lake Mendota, WI, USA in 2021 during late thermal stratification. The lake has comparatively high manganese and low iron concentrations. High productivity coupled with high sulfate concentrations leads to significant sulfide accumulation starting in the mid to late summer. Dissolved oxygen (*ODO*) was measured using an optical probe.

not attached to particles. So-called *ultramicrobacteria* such as members of the order *Nanopelagicales* (also known as actinobacterial lineage acI) (Garcia et al., 2013; Neuenschwander et al., 2018) and the genus *Fonsibacter* (also known as LD12) (Henson et al., 2018b) are abundant in nearly all freshwater ecosystems but appear to be oligotrophic taxa based on their cell size and genome content (Ghylin et al., 2014).

Like other nonmicrobial taxa, microbes niche-partition based on resource acquisition strategies. *Chemoorganoheterotrophs* (as defined in Table 23-2) display preferences for different carbon sources such as sugars and carboxylic acids, as discussed in Section II. Radiolabeled tracers can be used to quantify substrate uptake at an aggregate community level but also, when using microscopy, at the individual cell level. When coupled with whole-cell identification methods such as *fluorescence in situ hybridization* (FISH), microbial biologists can link the use of specific substrates to individual taxa. For example, members of the cosmopolitan genus *Fonsibacter* (LD12) specialized in glutamine and glutamate uptake, while acI-*Actinobacteria* seemed to prefer glucose and leucine (Salcher et al., 2013). One particularly relevant aspect of the specialization on various amino acid uptake is that microbes have differential preferences for leucine, an amino acid frequently used for total bacterial productivity measurements. Only 30% of total bacteria in an alpine oligomesotrophic lake were found to incorporate this tracer (Salcher et al., 2010). Thus substrate preferences may have key implications for measurements made to quantify protein synthesis as a proxy for bacteria-driven secondary production and/or overall bacterial activity (Kirchman et al., 1985). Specialization on inorganic energy sources, such as ammonia (nitrifiers), sulfide, or methane (methanotrophs), represents a more restrictive lifestyle. Indeed, the first two are expected to be autotrophs. However, microbes are continually surprising us and each of these textbook-defined guilds has been found to also metabolize other compounds in certain situations, challenging our efforts to link the functions to individual lineages. This is an open and grand challenge in the field of microbial biology. The implications for limnology are major, as we strive to incorporate microbial diversity into our broader understanding of freshwater ecosystems.

III. Microbes and the carbon cycle

A. The microbial loop

Typical food-web diagrams focused on macroscale organisms (all eukaryotes) emphasize the transfer of energy and matter from primary producers up to secondary consumers. Bacteria are typically tacked onto the bottom as "recyclers" or "decomposers." However, in 1983 Farooq Azam et al. proposed a more resolved conceptual model for aquatic ecosystems called the *microbial loop* (Azam et al., 1983), with microbes (Bacteria and Archaea) at the center (Fig. 23-12). Perhaps the most important addition was *dissolved organic matter* (DOM) (Chapter 28), which is produced during the decomposition of nearly all organisms including both

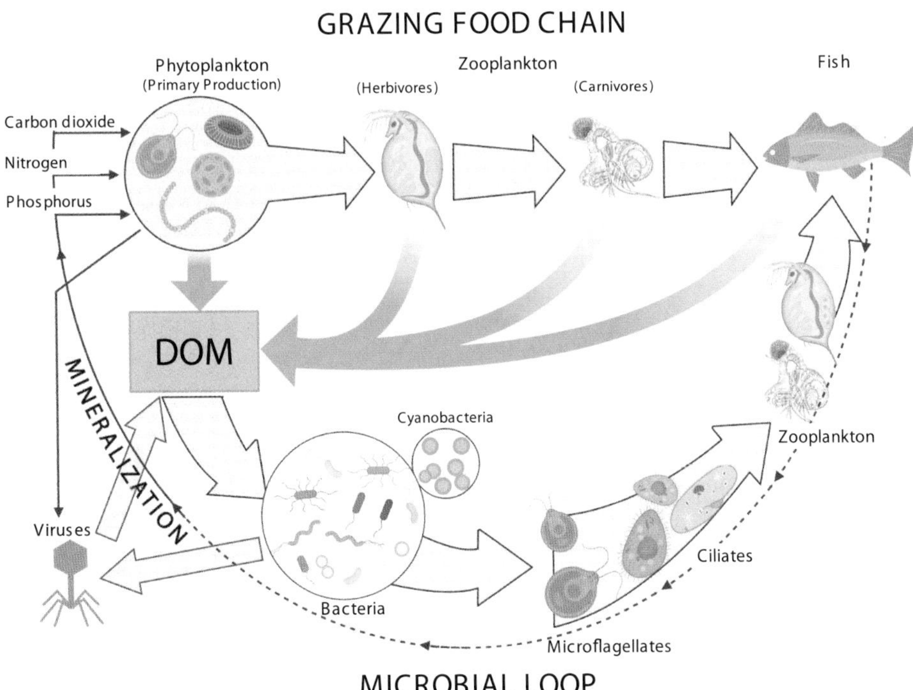

FIGURE 23-12 A cartoon depiction of the microbial loop and its connection to the larger aquatic food web. *Large yellow arrows* indicate the flow of organic carbon and energy to higher trophic levels of the food web. *Large blue arrows* indicate the production of dissolved organic matter (*DOM*) via excretion and other trophic interactions. *Thin dotted arrows* indicate the mineralization of major nutrients contained in organic matter. *Grey arrows* indicate viral mediated release of DOM. *(Adapted from Fuhrman and Caron, 2016 and originally Azam et al., 1983.)*

autochthonous (e.g., algae) and allochthonous (e.g., tree leaves) sources. This DOM is gradually converted into smaller and smaller molecules and/or into microbial biomass. It is ultimately mineralized to dissolved inorganic carbon (DIC, carbonates, and dissolved CO_2; Chapter 13), making it available again for primary production. The key role of viruses in mediating this carbon cycle through cell lysis and subsequent DOC release was also incorporated into the conceptual model and is often referred to as the *viral shunt* (Wilhelm and Suttle, 1999). Phytoplankton also release DOM during photosynthesis and this can be a significant source of energy, carbon, and nitrogen for microbial heterotrophs. Interestingly, the concept of the microbial loop has not been applied as frequently in freshwaters as it has been in marine systems.

Other factors controlling the microbially driven carbon loop are less well studied. For example, parasitic fungi (chytrids) promote the death and lysis of Cyanobacteria, diatoms, and zooplankton (Haraldsson et al., 2018; Klawonn et al., 2021). *Bacterivorous bacteria* (bacteria that eat other bacteria) such as *Bdellovibrio* are both abundant (Ezzedine et al., 2020) and very active, as determined by their overall gene expression levels (Linz et al., 2020).

B. Organic matter decomposition

The central role of microbes in the carbon cycle demands that we recap the dominant factors influencing individual steps in organic matter decomposition with an emphasis on the microbes. *Particulate organic matter* (POM) is a catch-all term for dead plants and animals, fecal pellets, and sometimes aggregates of colloidal material formed through abiotic processes. It is important to note that the distinction between POM and DOM is operational, based on which fractions pass through a 0.45 μm filter or not. In reality, there is a size continuum ranging from multiple molecules stuck together down to small molecules such as acetate.

POM conversion to DOM occurs mainly via the hydrolysis of polymers such as polysaccharides, proteins, lipids, and nucleic acids. Aquatic microbes attach to particles and must secrete enzymes to access the associated monomers or oligomers (e.g., sugars, peptides, amino acids, fatty acids) because the polymers are too large to take up into their cells. In the process of depolymerization (breakdown of larger polymers) some of the smaller organic molecules are released into solution, which makes these molecules accessible to nearby microbes who themselves do not have the capacity to create the extracellular enzymes for depolymerization. As noted above, the microbial community composition on particles is quite distinct from those in bulk water, reflecting different lifestyles of freshwater taxa.

Once organic matter is in the dissolved form, it can undergo both abiotic and biotic transformations. As described in Chapter 28, light drives photooxidation processes that can directly produce inorganic carbon (i.e., DIC) but also break off smaller organic molecules such as carboxylic acids and aldehydes (Cory and Kling, 2018; Kaplan and Cory, 2016). Such molecules are ideal substrates for many aerobic heterotrophs which dominate in photic zones (Bertilsson and Tranvik, 1998, 2000; Tranvik and Bertilsson, 2001). Indeed, it is difficult to measure the rates of photochemical transformation separately from microbial degradation because they occur at the same time. Researchers are actively striving to disentangle these distinct processes in order to better parameterize quantitative carbon cycle models (Cory and Kling, 2018).

POM and DOM are terms created to remind us that these pools are comprised of more than just carbon (Chapter 28). They also include nitrogen, phosphorus, sulfur, and many trace elements such as metals that are essential for microbial biomass construction. An example of DOM that links multiple elemental cycles is *dissolved organic nitrogen* (DON) such as urea and peptides or amino acids (Chapter 14). Urea is found in low concentrations in lakes but can account for up to 50% of total N used by plankton (Solomon et al., 2010). The importance of urease as an enzyme used to liberate ammonia is well recognized among aquatic bacteria. Peptides, amino acids, and chitin monomers (*N*-acetyl-D-glucosamine) appear to be especially important small molecules that are rapidly taken up following secretion by other organisms or cell lysis (Eckert et al., 2013; Goldberg et al., 2015; Salcher et al., 2013). Polyamines are now recognized as a potential DON pool that is rapidly taken up by pelagic bacteria (Krempaska et al., 2018), though the source of these compounds is as yet unknown. Additional discussion of DOM composition and the implications for microbial niche partitioning can be found in section IIE.

It is important to remember that measurements of standing stocks of POM and DOM do not provide a complete picture of the transformations conducted by microbes or the amount of substrate available to them (Moran et al., 2022, 2016). Consider a refrigerator containing beer. At any given moment, we may open the refrigerator and notice which brands and types of beer are in the refrigerator. We may infer that the users of the refrigerator prefer these types of beer. However, a better interpretation is that these types are the *least* preferred because they are still in the refrigerator while all of the favorite types are quickly consumed. Similarly, the most labile and energy-rich compounds are usually favored by microbes and thus may be below any analytical detection limits. Their measured concentrations (or lack thereof) are not good indicators of the elemental and energy flux from DOM to microbes. Fluxes are often estimated through microcosm incubation experiments that involve model carbon compound additions. These compounds can be labeled with radio- or stable-isotopes (e.g., ^{14}C and ^{13}C) to aid in tracking the DOM degradation. Researchers then measure parameters such as CO_2 production, isotope incorporation into biomass, or the amount of isotope remaining in solution. Such experiments still have limitations, including the batch addition of high substrate concentrations versus continuous production via POM breakdown or direct production by phytoplankton. Thus any measured fluxes should still be interpreted with caution. New methods for accurate and relevant measurements of fluxes for use in ecosystem-scale process-based models are sorely needed.

C. Productivity and growth efficiency

The conceptual framework articulated by the microbial loop highlights the role of microbes in rerouting DOM originating from primary producers and all other members of the food web. This has profound implications for ecosystem-scale metabolism and thus can determine whether a system is net heterotrophic versus net autotrophic (see Chapter 28).

A great deal of work has been focused on the conditions under which bacterial production is coupled to phytoplankton primary production. Put another way, when is bacterial growth dependent mainly on fresh phytoplankton exudates or lysis versus some other source such as terrestrially sourced material? Many studies have documented a positive correlation between bacterial production and phytoplankton biomass (Cole et al., 1988). However, these represent broad brush patterns and may underemphasize considerable variation that can be attributed to limitation by specific elements, particularly phosphorus (Coveney and Wetzel, 1992; Simon et al., 1998; Watanabe, 1996) (see also Section III.B). Generally speaking, bacterial production seems to be more frequently limited by phosphorus than nitrogen in freshwaters. This may be because labile DOM in freshwater tends to be rich in nitrogen and microbes seem adept at satisfying both carbon and nitrogen requirements using the same substrates (see also Section V.B). It is important to note that different members of the microbial community may also be in different states of limitation depending on their individual metabolic needs (see also Section IV.B).

Planktonic bacterial production is generally lower during winter than during summer in temperate lakes and reservoirs; this relationship is correlated with low winter temperatures and reduced loading of particulate and dissolved organic substrates from autochthonous (phytoplankton and littoral plants) and allochthonous sources when the soils are frozen. A significant portion of the planktonic bacterial community is thought to be physiologically dormant when conditions are cold and limited organic substrates are available. However, primary production (although much reduced) can and does occur in winter, even under the ice (Hampton et al., 2017). This suggests that bacterioplankton productivity may be patchy under ice, with "hot spots" near where phytoplankton can access enough light to drive measurable carbon fixation. Generalizations are difficult to make as very little is currently known about bacterial communities under the ice (Bertilsson et al., 2013).

Microbes are thought to be predominantly heterotrophs whose role is largely limited to secondary production through the consumption of organic molecules originally synthesized by phytoplankton (Fig. 23-11.). However, many microbes can also fix carbon to generate biomass, thereby living autotrophically. Examples include *photolithoautotrophs* (defined in Table 23-2) such as green sulfur bacteria and purple sulfur bacteria, as well as *chemolithoautotrophs* (defined in Table 23-2) such as nitrifiers and sulfide oxidizers. Thus, unlike phytoplankton, microbial primary productivity can occur under very dim light (Camacho et al., 2001; Di Nezio et al., 2021; Taipale et al., 2011) or even in the dark (Casamayor et al., 2012). Such production is generally much less than that of oxygenic photosynthesis (Parkin and Brock, 1980), though it can account for a substantial amount ($\sim$20%) of carbon sedimentation in some lakes (Parkin and Brock, 1981).

It is important to appreciate that the tools used for measuring productivity have limitations. Most published measurements are based on radiolabel incorporation, with different substrates used to carry the radioisotope. Two widely used substrates are ^{3}H-thymidine and ^{3}H-leucine to infer DNA synthesis and protein synthesis, respectively (Chin-Leo and Evans, 2007). Although they can serve as good proxies for overall bacterial activity, it is well known that some abundant freshwater taxa cannot uptake leucine, for example.

Microbial growth efficiency [bacterial production/(bacterial production + bacterial respiration)] is an important factor in determining whether an aquatic ecosystem is a net source or sink of carbon. It is defined as the fraction of metabolized carbon that is converted into biomass as opposed to being released as carbon dioxide. Many studies have been devoted to calculating this parameter in different kinds of freshwater bodies in an effort to identify drivers such as overall ecosystem primary productivity, pH, temperature, and carbon source (Berggren et al., 2010, 2007; Smith and Prairie, 2004). Growth efficiencies are higher in oligotrophic than eutrophic systems (Biddanda et al., 2001).

The efficiency of bacterial metabolism and growth is highly variable. Although it is theoretically possible for bacteria to convert organic substrates to biomass with high efficiency (>85%) under optimal conditions (Payne and Wiebe, 1978), such high rates are never realized under natural conditions. Growth is rarely in the exponential phase and, with many resource and physical limitations, growth efficiencies are nearly always <50%, and mostly <30% (e.g., Cole and Pace, 1995; Pomeroy and Wiebe, 1988; Schroeder, 1981).

D. Uptake of specific dissolved organic compounds

As much as 80% of DOM in inland waters is composed of organic acids that originate largely from higher aquatic and terrestrial plants. Of these organic acids, some 30–40% are composed of aromatic carbon originating from structural plant tissues (Malcolm, 1990). Concentrations of organic acids are commonly in the range of 4–8 mg-C L^{-1} and can exceed 50 mg-C L^{-1} in wetlands, floodplain waters, and interstitial waters of hydrosoils (Mann and Wetzel, 1995; Wetzel, 1984).

The high concentrations of DOM in inland surface waters result from high photosynthetic productivity associated with lake and river ecosystems, particularly associated with wetland and littoral regions, as well as large loadings of DOM from the decomposition of plant materials within the drainage basin. Similarly, the high photosynthesis and decomposition within the extensive floodplain marginal areas of river ecosystems also serve as source areas of DOM to river runoff water (Wetzel and Ward, 1992). More labile compounds of these heterogeneous organic mixtures are selectively degraded by microbiota as the water is transported along the gradient from land through the littoral to pelagic regions. Residual organic compounds contain high concentrations of the relatively recalcitrant humic substances that originate largely from the partially degraded plant structural tissues (Frimmel and Christman, 1988).

Humic compounds are structurally complex (phenolic linkages; see detailed discussion in Chapter 28) and tend to have long residence times in lakes and streams. In general, their degradation by aquatic microbes proceeds slowly. A portion of the humic materials present in a lake is truly dissolved, but under certain conditions, humic substances can aggregate into colloids or flocs. High-molecular-weight dissolved humic materials readily adsorb to particulate matter (Davis and Gloor, 1981), which may further alter their rates of degradation.

Ultraviolet irradiance (UV) can photolyze portions of proteinaceous and humic macromolecules (Bertilsson and Tranvik, 2000; Geller, 1986; Manny et al., 1971). Detailed chemical analyses of these transformations show that small organic fractions, particularly numerous small fatty acids—such as acetic, formic, citric, pyruvic, malic, and levulinic, among others—were generated by photolysis of the humic substances (Cory and Kling, 2018; Moran and Zepp, 1997; Wetzel et al., 1995). Even before these photolytically generated simple organic substrates were identified, many studies demonstrated that DOM exposed to natural UV radiation exhibited immediate stimulation of and sustained bacterial growth (e.g., Moran and Zepp, 1997; Stewart and Wetzel, 1982; Wetzel et al., 1995).

E. Substrate utilization rates

The relative rates of degradation of the composite pool of hundreds of organic compounds that constitute total DOM by a heterogeneous composite community of hundreds of species of heterotrophic microbes are difficult to quantify in natural waters. In natural systems each microbial population potentially has different abilities to assimilate specific substrates. Measurements of total community production rates of the composite heterotrophic bacterial communities are estimated by rates of incorporation of nucleotides and amino acids into bacterial DNA and proteins, respectively (Chin-Leo and Evans, 2007). Measurements of specific substrate assimilation by bacteria, however, are fraught with methodological problems. The kinetics of organic substrate use by heterogeneous planktonic populations, determined by the uptake rates of radioactivity-labeled organic compounds at substrate concentrations that occur naturally, are instructive in a relative, if not absolute, way.

Parsons and Strickland (1961) employed labeled organic substrates to measure uptake by natural heterotrophic populations in the dark by the relationship

$$v = \frac{cf(S_n + A)}{C_{\mu t}}$$

when

v = rate of uptake (mg C m^{-3} h^{-1})
c = radioactivity of the filtered organisms (cpm, counts per minute)

f = a correction for isotopic discrimination (1.05 or 5% slower uptake of ^{14}C, which has a greater mass than ^{12}C)
S_n = *in situ* concentration (mg L^{-1}) of the organic substrate
A = concentration (mg L^{-1}) of added substrate (labeled and unlabeled)
C = cpm from 1 μ-curie of ^{14}C-labeled substrate on the radioassay instrumentation used
μ = quantity of ^{14}C added to the sample
t = incubation time (h)

The above equation assumes that natural substrate concentrations (S_n) are much less than A. When the uptake of a solute is mediated by a transport system located on or in the cell membrane, the rate of uptake can be described by *Michaelis–Menten kinetics*, when

$$v = \frac{V_{max} \times S}{K_m + S}$$

v = velocity (uptake rate) at a given substrate concentration S
V_{max} = maximum velocity, attained when uptake sites are continually saturated with substrate
K_m = Michaelis constant, which is a measure of the affinity of the uptake system for the substrate. It is equal to the substrate concentration at which the velocity is one-half of the maximum velocity. By definition, $v = V_{max}/2$.

As developed by (Wright and Hobbie, 1966), this nonlinear uptake relation over substrate concentration can be transformed into a linear relationship by the *Lineweaver–Burk equation* to yield

$$\frac{C \times \mu \times t}{c} = \frac{(K_m + S_n)}{V_m} + \frac{A}{V_m}$$

With this equation, data from uptake measurements from algae and bacteria at low substrate concentrations can be plotted as $C\mu t/c$ versus A, giving values for $(K_m + S_n)$ and V_m, where V_m is the maximum measured rate of uptake (Fig. 23-13). The negative intercept on the abscissa is equal to $(K_m + S_n)$, and the reciprocal of the slope is V_m. The ordinate intercept is equivalent to the turnover time (T_t), which is the time required for the complete removal of the natural substrate by the microbes (Hobbie, 1967). The $(K_m + S_n)$ approximates the maximum natural substrate concentration (S_n) if K_m is very small, as is often, but not always, the case. These relationships assume that a constant rate of regeneration of the organic solute is occurring *in situ*, or that steady-state conditions exist. An appreciable portion of ^{14}C organic substrate is respired rapidly by microbes, and corrections for this loss must be made (Hobbie and Crawford, 1969; Wetzel and Likens, 2000).

In contrast to the already discussed nonlinear active uptake velocities at low substrate concentrations, the uptake of organic compounds by natural microbial populations at high substrate concentrations (approximately >0.5 mg L^{-1}) does not exhibit rate limitation kinetics or saturation of uptake sites. Passive uptake velocity continually increases with rising substrate concentrations. The slope of the response line is constant (K_d), derived from diffusion kinetics, and has been used to estimate diffusion uptake by natural populations of algae and bacteria.

Analyses of organic substrate use rates under *in situ* conditions with natural microbial populations have been inconsistent, making it difficult to compare among studies. Additionally, most studies employ only a few simple substrates such as glucose, other sugars and isomers of glucose, acetate, glycolate, and amino acids, which does not closely mimic natural systems. Methodological limitations aside, these studies demonstrate that

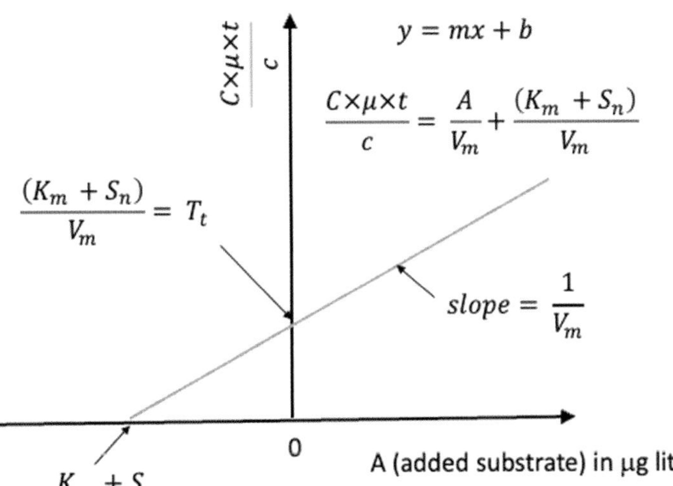

FIGURE 23-13 Graphical analysis of bacterial uptake at low organic substrate concentrations following Michaelis–Menten enzyme kinetics. A plot of $C\mu t/c$ against increasing added substrate concentrations, A, illustrating derivation of (1) maximum natural substrate concentrations, $K_m + S_n$ as μg L^{-1}; (2) maximum velocity, V_m, as μg L^{-1} h^{-1}; and (3) turnover time for substrate regeneration (T_t) in hours. *(After Allen, 1969.)*

dark diffusive algal uptake of simple organic substrates for use in heterotrophic growth at low natural substrate concentrations is almost always <10% of active uptake by bacteria. As stated in another way, the algae are generally ineffective in competing for available organic substrates at substrate concentrations maintained by active bacterial heterotrophic uptake.

Maximum velocities of uptake are quite variable among lakes and different organic substrates but generally occur after spring or following late summer algal maxima. Rates decrease by about an order of magnitude during winter in temperate lakes. Concentrations of substrates such as glucose, acetate, and amino acids remain low throughout the year, for the rate of inputs to the labile dissolved organic pool is balanced roughly by the rate of removal by bacteria. Uptake velocities within a stratified water column tend toward maxima within and below algal maxima in the photic zone, minima in the lower metalimnion—upper hypolimnion. Rates increase again near the sediments.

Use-rates of simple substrates generally increase with greater phytoplankton productivity in an approximately direct relationship (Table 23-5). Great variations are observed in these rates both seasonally and vertically with changes in depth. Uptake rates of simple carbohydrates generally increase with greater microbial densities and biomass and with increasing concentrations of total inorganic nitrogen and total phosphorus (Bowie and Gillespie, 1976; Spencer, 1978). Glucose assimilation rates by planktonic bacteria were found to be nearly an order of magnitude greater in littoral areas where submersed macrophytes were more abundant than in the pelagic areas (Gillespie and Spencer, 1980).

Radioisotope tracing can also be combined with microscopy to visualize the cells that took up the labeled substrate (Brock and Brock, 1966). When combined with probes that cause specifically targeted taxonomic groups to fluoresce, it is possible to link substrate uptake to that taxonomic group. This approach is called *microautoradiography fluorescent in situ hybridization* (MAR-FISH). It has been used in freshwater systems to explore substrate-based niche differentiation (e.g., uptake of glucose versus amino acids; Buck et al., 2009; Salcher et al., 2013) and differences in uptake rates in time and space (Salcher et al., 2010). A primary limitation of the method is the taxonomic resolution afforded by the FISH probes, which target ribosomal RNA. Most probes detect cells that all belong to one family (or possibly genus). It is generally not possible to design probes to differentiate among species or strains within a genus.

IV. Drivers of biogeochemistry

Advancing technologies (e.g., genomics, single-cell biology) have revealed bacterial and archaeal metabolic traits to be complex and the number of organisms capable of specialized energy-generating mechanisms to be far greater than was thought in the decades prior. This section covers the distributions of microorganisms and their interactions with both major nutrients and minor elements found in freshwater.

A. Nitrogen

Bacteria and archaea have high nitrogen demands. Nitrogen (Chapter 14) is one of four major biogenic elements in cells (the others being C, P, and S). Nitrogen is required as it is a primary component of proteins and nucleic acids. Proteins (~55%) and nucleic acids (~25%) make up the majority of a bacterial cell's dry weight (Neidhardt, 1963).

Nitrogen is involved in many important oxidation—reduction reactions because it, unlike most other available elements in freshwater systems, can take on many oxidation states (ranging from −3 to +5). Since nitrogen can exist in a wide range of oxidation states, it plays a key role in energy-generating reactions for many microbes. In freshwater environments as well as other ecosystems (oceans, soils) there is a microbial nitrogen-cycling network that controls a significant fraction of nitrogen transformations (see Chapter 14). In freshwaters this network consists of various bacteria and archaea capable of carrying out at least 14 different nitrogen transformations (Fig. 23-14).

For decades, microbes have been classified according to one of six nitrogen-transforming processes: (1) nitrogen fixation; (2) nitrification; (3) denitrification; (4) anammox; (5) assimilation; and (6) ammonification (see Chapter 14 for further details). However, recent genomic data and follow-up laboratory studies have revealed a large amount of metabolic versatility for many microbes performing nitrogen transformations (Kuypers et al., 2018). For example, many aquatic bacteria can fix dinitrogen gas and perform denitrification simultaneously (Stein and Klotz, 2016). Here we will describe some of the microbes associated with each of the major nitrogen transformations in the freshwater pelagic, but the boundaries between these processes are becoming ever more blurred as new combinations of capabilities are identified within individual microbes.

i. Nitrogen fixation

The capacity for N_2 fixation (known as *diazotrophy*), which is the reduction of nitrogen gas to ammonia (Fig. 23-14; see Chapter 14), is not widespread among freshwater microbes. Given the need for nitrogen by microorganisms, the ability to make inorganic forms of nitrogen from the atmosphere would be beneficial for any microbe. However, nitrogen fixation requires a huge amount of energy to break the triple bond between nitrogen atoms in the N_2 molecule (Kirchman, 2018). These

TABLE 23-5 Comparison of Approximate Rates of Turnover of Organic Substrates by Natural Bacterioplankton in Rivers and in Lakes of Increasing Productivity Based on the Range of Maximum Phytoplanktonic Photosynthetic Rates of Carbon Fixation (P_{max})

Habitat	Substrates	P_{max} (mg C m^{-3} day^{-1})	T_t (h)	Source
Lakes				
Oligotrophic				
Laplandic lakes, Sweden (summer)	Glucose, acetate	1–30	>10,000	Rodhe et al., 1966
Lawrence, Michigan (USA) (annual)	Glucose	1–80	40–300	Wetzel et al., 1972
	Acetate	—	10–120	
Mesotrophic				
Crooked, Indiana (USA) (annual)	Glucose	63–110	80–470	Wetzel, 1968, 1967
	Acetate	—	20–350	
Gravelly Pond, Massachusetts (USA) (summer)	Glycolate	—	60–200	Wright, 1975
Kizaki, Japan	Glucose	—	12–58	Kato and Sakamoto, 1983
	Amino acids	—	3–13	
Kinneret, Israel	Glucose	—		Cavari et al., 1978
0–10 m		—	88	
20–40 m		—	133	
Four Várzea lakes, Brazil (annual)	Glucose	—		Rai, 1979
		Lower	105–>10,000	
Erken, Sweden				
Summer	Glucose	40–130	10–100	Hobbie, 1967
Winter	Glucose	2–20	100–1000	
Lake Mekkojärvi, Finland				
Epilimnion	Leucine	—	48 (9–178)	Münster et al., 1999
	Hydroxybenzoic	—	38 (28–48)	
Hypolimnion	Leucine	—	35 (1–47)	
	Hydroxybenzoic	—	55 (35–68)	
Eutrophic				
Little Crooked, Indiana (USA) (annual)	Glucose	190–205	36–232	Wetzel, 1968, 1967
	Acetate	—	24–190	
Duck, Michigan (USA) (annual)	Glucose	10–320	8–50	Miller, 1972
	Acetate	—	4–40	
Pamlico River (estuary), North Carolina (USA)	Amino acids	—	1.5–26	Hobbie, 1971
Lötsjön, Sweden				
Summer	Glucose	<100	0.4–5	Allen, 1969
Winter	Glucose	<20	20–300	

Continued

TABLE 23-5 Comparison of Approximate Rates of Turnover of Organic Substrates by Natural Bacterioplankton in Rivers and in Lakes of Increasing Productivity Based on the Range of Maximum Phytoplanktonic Photosynthetic Rates of Carbon Fixation (P_{max})—cont'd

Habitat	Substrates	P_{max} (mg C m^{-3} day^{-1})	T_t (h)	Source
Upper Klamath, Oregon (USA) (summer)	Glucose	—	2.4	Wright, 1975
	Acetate	—	2.3	
	Glycine	—	8	
	Glycolate	—	26	
Three Polish lakes, Poland	Palmitic acid	—	7–18	Chróst and Gajewski, 1995
Plußsee, Northern Germany (annual)	Glucose	20–150	6–202	Overbeck, 1975
	Acetate	—	150–290	

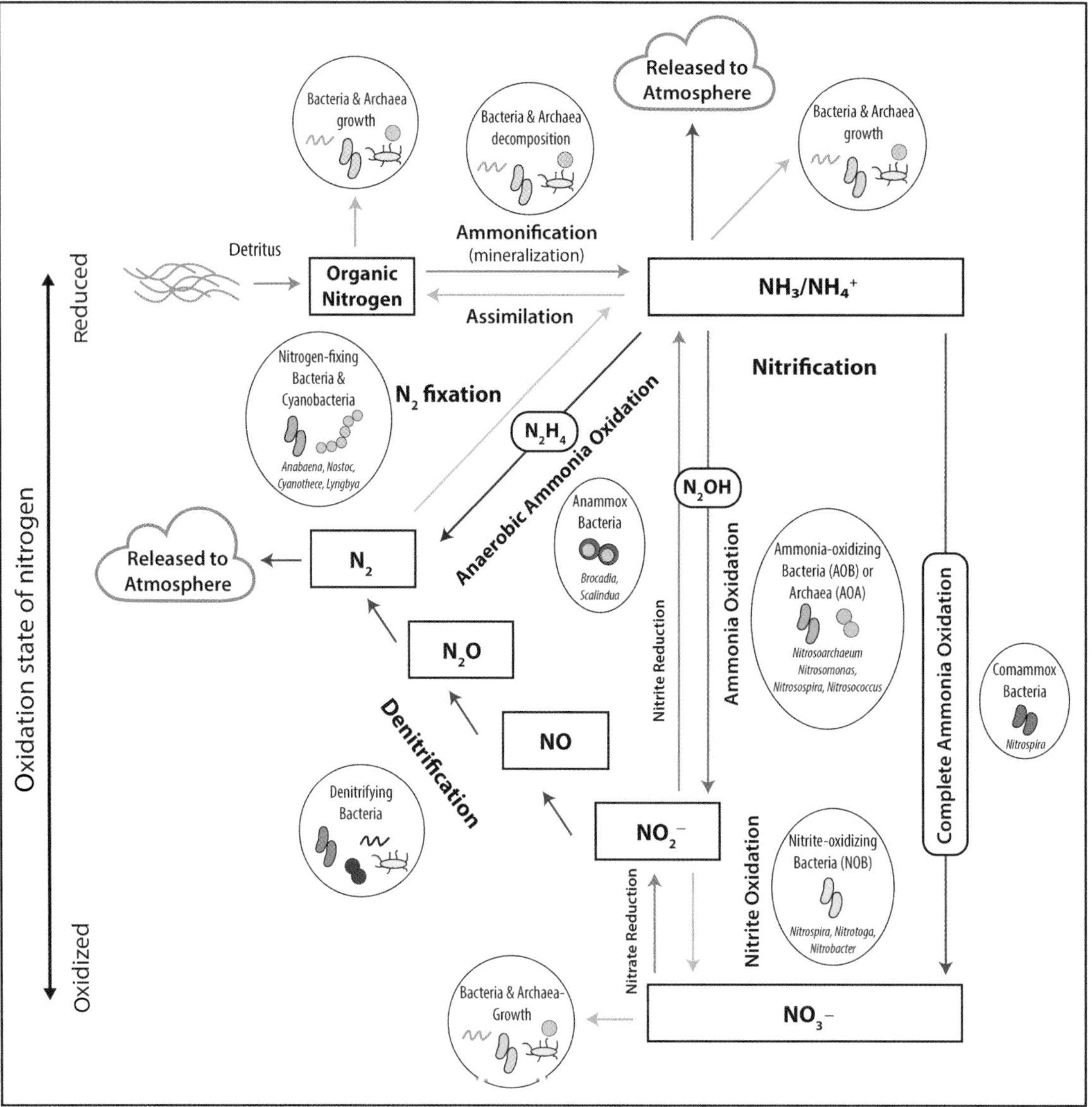

FIGURE 23-14 Microorganism-associated nitrogen transformations. Common freshwater genera are listed for processes where few microbes are known to have that capability. *(Adapted from Kirchman, 2018.)*

energetic costs are thought to have limited the nitrogen-fixing capability to only a few microbes. Nitrogenase is the primary enzyme used by microbes to carry out N_2 fixation (Zehr et al., 2003). Nitrogenase is damaged by oxygen, so freshwater microbes must have specialized strategies to protect its use, such as compartmentalized cells (*heterocysts*) or temporal separation of photosynthetic and nitrogen-fixing activities in the cell. The gene *nifH* is commonly used to identify N_2-fixing bacteria from aquatic samples (Gaby and Buckley, 2012). In the pelagic zone Cyanobacteria dominate among nitrogen-fixing bacteria. Common genera include *Dolichosperumum* (formerly *Anabaena*), *Nostoc*, *Cyanothece*, *Cylindrospermopsis*, and *Lyngbya*.

ii. Nitrification

Nitrification is the process of biotic oxidation of ammonia to nitrate (Fig. 23-14). Nitrification is carried out by various microbes that are collectively called *nitrifiers*. These microorganisms use reduced inorganic forms of nitrogen, such as ammonia, as an energy and electron source and the process of inorganic carbon (CO_2) fixation to meet their cellular carbon demands. Microorganisms that possess this combination of metabolic traits are known as *chemolithoautotrophs*, and this metabolic lifestyle is fairly rare in the diversity of microbial life. Nitrifiers play several important roles in the biochemical cycling of macronutrients across freshwater systems. They mobilize inorganic nitrogen through the conversion of ammonia to nitrate, which then allows for nitrogen removal via denitrification (conversion from nitrate to N_2 gas—see Chapter 14). As fixers of carbon via nonphotosynthetic means, nitrifiers are a major source of carbon fixation in the absence of light (Callieri et al., 2014; Pachiadaki et al., 2017). This process, sometimes referred to as *dark carbon fixation*, can sustain food webs in the deepest parts of freshwater systems where light-derived energy is minimal. Nitrifiers also require a small amount of oxygen for energy generation via the oxidation of ammonia or nitrite, may be growth-inhibited by light, and are relatively slow growing compared to the dominant pelagic heterotrophic microbes (Hatzenpichler, 2012; Kuypers et al., 2018). Although nitrifiers are abundant in most systems, this combination of lifestyle traits limits their distribution in the water column. Most commonly, nitrifiers are found in transition zones where light, readily available organic carbon, and oxygen are present but at low concentrations. This results in nitrifier abundance increasing with depth in most systems (Alfreider et al., 2017; Hugoni et al., 2013; Small et al., 2013; Vissers et al., 2013). Nitrifiers also can be at high abundances in surface sediments where low levels of oxygen are available (Parro et al., 2019).

The understanding of nitrifiers in freshwater systems, especially in the pelagic, is less complete than that of other environments (e.g., wastewater, soil). *Ammonia-oxidizing archaea* (AOA) from the phylum *Thaumarchaeota* dominate the water column of deep oligotrophic lakes, but *ammonia-oxidizing bacteria* (AOB) from the family *Nitrosomonadaceae* tend to dominate shallower and more eutrophic lakes (Auguet et al., 2011; Callieri et al., 2016; Hayden and Beman, 2014; Herber et al., 2020; Hugoni et al., 2013; Mukherjee et al., 2016; Okazaki et al., 2017; Vissers et al., 2013). Freshwater nitrite-oxidizers include only bacteria but are distributed across three bacterial phyla: *Nitrospirae*, *Nitrospinae*, and *Proteobacteria*. The controls on the distribution and niche differentiation among these organisms are poorly understood. Recent evidence from the Laurentian Great Lakes (USA/Canada) suggests that the genus *Nitrospira* prefers deeper, more oligotrophic systems, while the genus *Nitrotoga* prefers shallower, more productive systems (Podowski et al., 2021).

Nitrifiers also span a physiologically and ecologically diverse set of microorganisms. For example, members of the genus *Nitrospira* are known typically as nitrite-oxidizers but recently were shown to be capable of *complete ammonia-oxidation* (known as *comammox*). In the comammox process a single microorganism, rather than a pair of microorganisms, carries out the full oxidation of ammonia to nitrate (Daims et al., 2015; van Kessel et al., 2015). For more than 100 years, it was thought that two separate organisms worked in tandem to complete this process, with ammonia-oxidizers first carrying out the conversion of ammonia to nitrite and then nitrite-oxidizers converting that nitrite to nitrate. The distribution of comammox bacteria in the freshwater pelagic is not well constrained, but it appears these nitrifiers prefer a particle-attached lifestyle, which limits them to the benthos or near-benthos regions. Additional layers of nitrifier complexity were also described recently. Some ammonia-oxidizers and nitrite-oxidizers are now known to be capable of oxidizing nitrogen from alternative sources like urea and cyanate instead of having to rely on free ammonia or nitrite (Boddicker and Mosier, 2018; Koch et al., 2015; Palatinszky et al., 2015), which complicates previous predictions of nitrifier distribution and activities based solely on free ammonia or nitrite concentrations.

iii. Anaerobic ammonia oxidation

Anaerobic ammonia oxidation, also known as *anammox*, is the process of oxidizing ammonia in the absence of oxygen (Fig. 23-14). Unlike aerobic ammonia oxidation, which produces nitrite as an end-product, anammox produces dinitrogen (N_2) gas. Also, nitrite is used instead of oxygen (O_2) as the electron acceptor. Anammox activity is regulated by O_2 and inorganic nitrogen concentrations in the surrounding environment. Anammox is inhibited by O_2 concentrations above 1 μmol L^{-1} (Dalsgaard et al., 2014). As a consequence, anammox bacteria are active primarily in the deep anoxic waters of lakes or in the sediments of lakes and rivers (Crowe et al., 2017). Although anammox activity is ubiquitous in the sediments of lake and river systems (Penton et al., 2006; Yoshinaga et al., 2011; Zhou et al., 2014), the contribution of anammox to N_2 production in the water column is not fully understood but seems to vary spatially and temporally depending on shifting environmental conditions related to available O_2 and nitrogen-containing compounds (Lu et al., 2018). A single study from Lake Tanganyika (eastern Africa) suggests anammox activity in the water column of deep lakes with large anoxic zones could produce significant nitrogen removal via N_2 production (in this case 0.2 Tg per year; Schubert et al., 2006). A metagenomics-based study found genes linked to anammox at similar depths (Tran et al., 2021).

Unlike aerobic ammonia oxidation, anammox is carried out by a very narrow subset of bacteria belonging to the phylum *Planctomycetes* (Oshiki et al., 2016). These bacteria have several unusual features, the most notable of which is the *anammoxosome,* an intracellular structure that contains and separates the anammox reaction from the rest of the cell. The anammoxosome is bounded by a single bilayer membrane containing the lipid ladderane, which is thought to prevent the escape of hydrazine N_2H_4 (Kartal et al., 2012), a highly unstable and toxic intermediate formed during the anammox process and a notable component in rocket fuel. There are only a few bacterial genera known to carry out the anammox process. These include *Brocadia*, *Kuenenia*, *Anammoxoglobus*, *Anammoximicrobium*, *Scalindua*, and *Jettenia* (Kartal et al., 2012; van de Vossenberg et al., 2013). *Brocadia* is thought to be the most common genus in natural freshwater systems (Crowe et al., 2017), but many of the genera seem to be common in freshwater sediments, especially river sediments (Hu et al., 2012; Zheng et al., 2019).

iv. Denitrification and nitrate reduction

Denitrification is an anaerobic respiration process that couples nitrogen reduction to the oxidation of organic compounds (Fig. 23-14). The first step in this multistep process is the *dissimilatory* (energy-generating/non-biomass-producing) reduction of nitrate to nitrite. Nitrite is then reduced to nitric oxide and then nitric oxide to nitrous oxide or dinitrogen (N_2) gas (Kuypers et al., 2018). Oxygen inhibits denitrification. Most organisms capable of one or more steps in the denitrification process are *facultative anaerobes*, meaning they can tolerate both oxic and anoxic environments. Since oxygen, as an electron acceptor, yields more energy than nitrate, oxygenic respiration is preferred by microbes over denitrification. This energy generation disparity means that denitrification does not occur when oxygen is present. Instead, microbes use denitrification only in the anoxic zones of lakes and rivers (Dalsgaard et al., 2014; Kuypers et al., 2018).

Many freshwater microbes are capable of one or more steps in the denitrification pathway. In fact, complete denitrification by a single organism, once thought to be the dominant reduction mechanism for nitrate conversion to N_2, has been found to be the exception. Instead, in anoxic waters it is common to find a network of bacteria and archaea performing various reduction reactions in the overall denitrification process (Graf et al., 2014; Kuypers et al., 2018).

In anoxic waters bacteria are also capable of reducing nitrate to ammonium (Fig. 23-14), a process known as *dissimilatory nitrate reduction to ammonium* (DNRA). DNRA yields about 50% less energy for the cell than denitrification and is a process in direct competition for the resources (nitrate and organic carbon) used in denitrification (Burgin and Hamilton, 2007; Kuypers et al., 2018). So why are bacteria even capable of DNRA? As it turns out, DNRA uses less nitrogen than denitrification to reduce each unit (mole) of carbon, so if the carbon-to-nitrogen ratio is high in the water, then it is advantageous to conserve nitrogen while generating energy via DNRA (Thamdrup, 2012). A wide diversity of bacteria harbor DNRA capabilities (Mohan et al., 2004; Welsh et al., 2014), but the major players in freshwater systems have not been identified. Ultimately the availability of organic carbon and nitrate, including differentiation among microhabitats in the anoxic regions of freshwaters, seems to control whether nitrogen is lost to the atmosphere via denitrification production of N_2 gas or retained as ammonia via DNRA (Palacin-Lizarbe et al., 2019).

v. Ammonium and nitrate assimilation

Ammonium (NH_4^+) is the preferred nitrogen source for heterotrophic bacteria and Cyanobacteria because ammonium has the same oxidation state (−3) as the nitrogen found in many biomolecular cell components, like amino acids (Kirchman, 2018). This preference leads to an extremely fast uptake of ammonium from nearly all freshwater systems. Basically, all microbes are capable of ammonium assimilation via ammonia (NH_3), the unionized form of ammonium. Ammonia readily diffuses across cell membranes where it can be trapped by intracellular enzymes. Many aquatic microbes are also capable of active ammonium transport via specific membrane proteins (Alonso-Sáez et al., 2020; Stewart et al., 2012).

Nitrate uptake and assimilation is also possible for many microbes (Luque-Almagro et al., 2011). Unlike ammonia, nitrate uptake requires transporter proteins to bring the nitrate anion across the cell membrane. Once inside the cell, the nitrate must be reduced to ammonium to be incorporated into cellular biomass (Moreno-Vivián et al., 1999). Energy consumption during this process makes this a less desirable form of nitrogen for many microbes, so nitrate can accumulate in the water column if other forms of nitrogen are readily available.

vi. Ammonification

Ammonification is a process that produces inorganic ammonia from the breakdown of nitrogenous organic matter (Fig. 23-14; Kuypers et al., 2018). For instance, amino acids, DNA, and proteins are nitrogenous containing organic compounds that are released into the environment upon cell death and can be broken down by microbes into various inorganic components like ammonia. These compounds are part of the *dissolved organic nitrogen* (DON) component of freshwaters. Historically, DON was thought mainly to contain compounds that resisted biological degradation (Berman and Bronk, 2003), but it is now known to play an important role in supplying nitrogen to microbes in freshwater. Many pelagic bacteria, including the ubiquitous acI lineage (the genera *Candidatus* Nanopelagicus and *Candidatus* Planktophila), harbor the ability to take up amino acids and polyamine compounds. Differences in these abilities may provide niche differentiation between closely related species that coexist in most surface waters (Ghylin et al., 2014; Neuenschwander et al., 2018). More research is needed to understand the interactions between the DON pool, pelagic bacteria substrate use, and the release of ammonia in freshwater systems.

B. Phosphorus

Phosphorus is an essential element in all living cells (Chapter 15). For freshwater bacteria and archaea, phosphorus is a major component of nucleic acids (DNA and RNA) and membrane lipids, can be stored inside the cell, and is involved in energy transformations (as adenosine triphosphate, ATP) (Sigee, 2005). Phosphorus often limits the growth of microbes in freshwater environments (e.g., Coveney and Wetzel, 1992; Morris and Lewis, 1992; Watanabe, 1996) and thus pelagic microbes have developed several adaptations to survive and grow under low phosphorus regimes. These adaptations include: (1) high-affinity substrate phosphate uptake systems, which allow microbes to scavenge phosphate from the environment when it is at very low concentrations; (2) the ability to store excess phosphorus (luxury uptake), which allows microbes to save phosphorus inside their cell so that it can be used when phosphate is scarce in the environment; (3) cell membrane alterations (i.e., the replacement of phosphorus with other elements) to reduce the amount of phosphorus used in those membranes; and (4) the secretion of enzymes to release phosphorus bound in organic molecules so that it can be transported into the cell (Karl, 2000). In general, because of their size and uptake capabilities, pelagic bacteria are considered to be more N- and P-rich than other plankton. Although this tends to be the case, pelagic bacteria are much more flexible in their cellular phosphorus content than carbon or nitrogen content, which can lead to a very high cellular carbon-to-phosphorus ratio (>250:1; Cotner et al., 2010; Godwin and Cotner, 2015) (Fig. 23-15). This flexibility in cellular stoichiometry suggests pelagic microbes can serve as either nutrient regenerators or consumers depending on whether their cellular carbon to nitrogen/phosphorus content is lower (nutrient regenerator) or higher (nutrient consumer) than that found in the surrounding waters (Cotner et al., 2010).

In phosphorus-limited systems the uptake of phosphate is thought to be dominated by bacteria and archaea because they have higher affinity uptake systems than their eukaryotic competitors (e.g., Cotner and Wetzel, 1992; Currie and Kalff, 1984). Phosphate is often very low in the epilimnion of lakes, but this scarcity is driven by rapid cycling with constant turnover of the microbial biomass rather than stagnant conditions (Pedrós-Alió and Brock, 1982). When excess phosphorus is available, many pelagic microbes have the ability to store it inside the cell as polyphosphate. This phosphorus reservoir can be used later when phosphorus levels are limiting, but it is also used for other cellular functions such as an energy reservoir through the formation of high-energy bonds, metal chelation, buffer against alkaline substances, and buoyancy regulation (Hupfer et al., 2007). Internal phosphorus accumulation appears to be much higher when nitrogen availability is low (N:P < 40:1) (Chrzanowski et al., 1996).

In contrast to nitrogen, there is less differentiation among pelagic microbes in their preference/capabilities to use different phosphorus compounds. Inorganic phosphate is the preferred form of phosphorus for pelagic microbes as it is readily incorporated into cellular biomolecules, but many microbes are also capable of using organic phosphorus-containing compounds such as phosphate esters and nucleotides as a source of phosphate (Valdespino-Castillo et al., 2017). This is accomplished through the use of *phosphatases*, extracellular enzymes that can cleave phosphate from phosphoesters. There are many forms of phosphatases among pelagic microbes, but all act to free phosphate from organic molecules, which can then be taken up readily by the cell (Valdespino-Castillo et al., 2017). A more narrow set of pelagic bacteria are also capable of

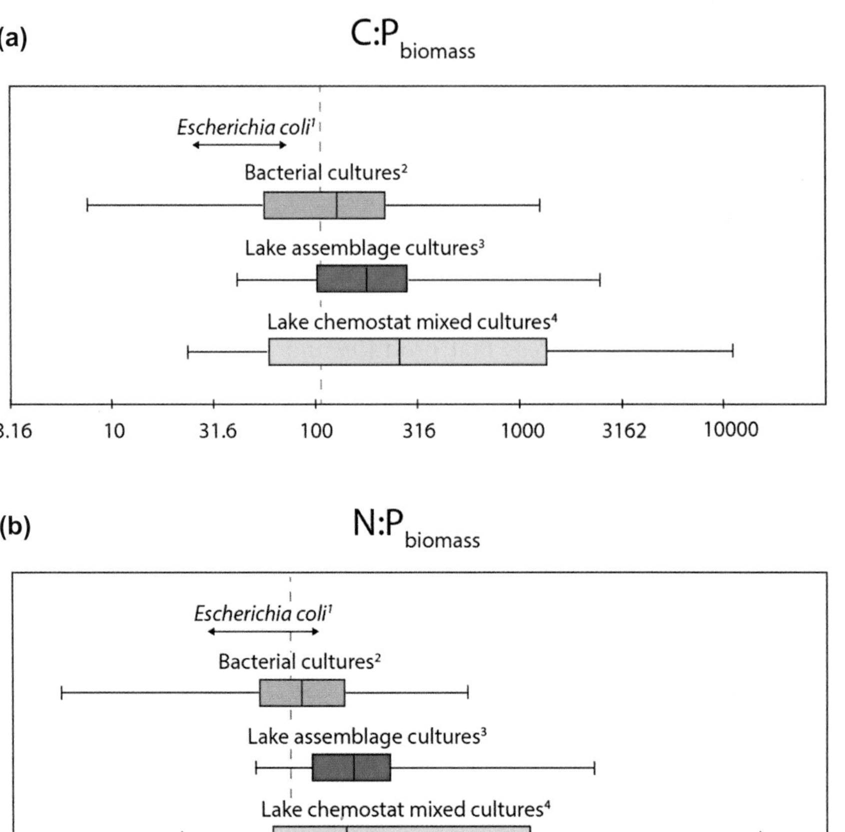

FIGURE 23-15 Ranges of (a) carbon (C) to phosphorus (P), and (b) nitrogen (N) to P ratios in biomass from different bacterial sources, including various studies of: (1) *Escherichia coli* (*E. coli*): isolates grown in culture; (2) Bacterial cultures: individual and assemblage-based cultures from more than a dozen bacterial genera examined in 19 studies as compiled by Godwin and Cotner (2015); (3) Lake assemblage cultures: mixed bacterial assemblage cultures initiated from lake samples as reported in Godwin and Cotner (2014); and (4) Lake chemostat mixed cultures: mixed bacterial assemblage cultures initiated from lake samples and maintained in flow-through chemostats as reported in Godwin and Cotner (2015). The *boxplots* indicate the median (*center line*), the 25% and 75% quantiles (*box edges*), and the minimum/maximum values (whiskers). The *vertical dashed line* indicates the Redfield ratio of C:P = 106 and N:P = 16. *(Adapted from Godwin and Cotner, 2015.)*

releasing phosphate from organic molecules known as phosphonic acids (phosphonates) (Huang et al., 2005). These organophosphorus compounds are also common in cells, particularly in cell membranes. Some bacteria are capable of producing phosphonate lyase enzymes, which can cleave carbon to phosphorus bonds in phosphonates. While nearly all microbes contain phosphatases, the distribution of phosphonate lyase enzymes is less well known and appears to be more restricted (Ilikchyan et al., 2009). Some Cyanobacteria and bacterial heterotrophs are known to have this capability (Kutovaya et al., 2013; Yao et al., 2016), but the extent of this enzyme's distribution among freshwater microbes remains poorly defined.

C. Sulfur

Sulfur (Chapter 16) is one of the four major biogenic elements needed by living organisms (the others being C, N, and P). In cells it is the least abundant of these elements, typically comprising <0.5% of cellular mass (Kirchman, 2018). Sulfur is required as a component of two amino acids, cysteine and methionine, and is present as part of iron−sulfur clusters, which are molecular collections of iron and sulfide found commonly in proteins in all forms of life. The pelagic sulfur cycle is driven by alternating anabolic (using energy to build biomass) and catabolic (breaking down molecules to release energy) microbial activity. In aerobic waters microbes take up inorganic sulfate (SO_4^{2-}) or thiosulfate ($S_2O_3^{2-}$) ions and reduce them to make biomass, such as proteins (Diao et al., 2018). Under anoxic conditions, the decomposition of dead/decaying biomass can result in the conversion of reduced sulfur (−SH-containing compounds) to hydrogen sulfide (H_2S). As this hydrogen sulfide diffuses into the water column, it may be oxidized anaerobically with nitrate or oxidatively to produce a variety of compounds (e.g., sulfate, thiosulfate, elemental sulfur; Diao et al., 2018). In the pelagic zone this cycle is fully realized in eutrophic water bodies, which typically have strong vertical oxygen gradients. In more oligotrophic waters much of the sulfur cycling occurs in the sediments where oxygen is depleted. Complex microbial-mediated sulfur cycles are found also in freshwaters fed primarily by oxygen-depleted groundwater, which can include wetlands, rivers, lakes, and even municipal fountains (Sharrar et al., 2017).

In a similar manner to the phosphorus cycle many microbes are involved in the cycling of sulfur. Most microbes are capable of assimilating inorganic sulfur and incorporating it into biomass (Faou et al., 1990). Protein decomposition in the hypolimnion is carried out by a variety of heterotrophic bacteria and archaea. This is extremely common in the surface sediments, where microbial biomass is several orders of magnitude higher than in the overlying water column.

Under anoxic conditions, sulfate provides a source of oxygen for the oxidation of hydrogen or organic compounds by microbes known as *sulfate-reducing bacteria* (SRB). Common freshwater SRB include members of the genera *Desulfobulbus, Desulfovibrio, Desulfobacterium, Desulfomonile, Desulfuromonas,* and many yet-to-be-named groups (Diao et al., 2018; Kondo et al., 2006; Kubo et al., 2014). SRB are abundant in the water column during anoxic conditions but are at very low levels, if at all, when waters are oxic. The SRB are found consistently near-surface sediments, which is thought to represent a seed bank for their reintroduction into the water column when anoxic conditions are reestablished (Diao et al., 2018). See Table 23-6 for a list of microbe-mediated sulfur reactions and the common freshwater genera involved.

The reduction of sulfate by SRB results in the production of hydrogen sulfide, which diffuses vertically until it reaches the oxycline (Chapter 11). At this oxygen concentration boundary, if light is present, bacteria with specialized metabolisms are capable of oxidizing the H_2S. These bacteria are known as *anaerobic sulfur oxidizers* and they are divided into two primary groups: the *purple sulfur bacteria* (PSB) and the *green sulfur bacteria* (GSB). Both groups oxidize sulfide to sulfur or sulfate in a reaction mediated by light energy. Usually, PSB are found in a layer directly above the GSB (Diao et al., 2018). GSB are adapted to low light levels, like those found in the twilight zone (<1% of surface incident irradiance) of stratified freshwaters (Biebl and Pfennig, 1978; Vila and Abella, 1994).

GSB are anaerobic photoautotrophs that couple anoxic oxidation of sulfide and CO_2 fixation, while PSB are known to tolerate oxygen. GSB contain carotenoids and a variety of bacteriochlorophylls, which dictate niche preferences related to light (Llorens–Marès et al., 2017). GSB cells can appear brown- or green-colored depending on their cellular pigment production. PSB also differ from GSB in that some members can grow in the absence of light, using a chemolithoautotrophic lifestyle, where reduced sulfur compounds are oxidized in low-oxygen conditions (Casamayor et al., 2008) or using a chemoheterotrophic lifestyle. This dual capability allows PSB to bloom at the sediment–water interface, upon the release of sulfide from the sediments (Diao et al., 2018; Peduzzi et al., 2011). *Thiodictyon, Thiocystis,* and *Thiorhodococcus* are common PSB, whereas *Chlorobium* and *Chlorobaculum* are common GSB in freshwater systems (Diao et al., 2018). In lakes with stable chemoclines, such as meromictic lakes, the phototrophic sulfur bacteria can be responsible for significant uptake of inorganic carbon. For example, in Lake Cadagno (Switzerland) phototrophic sulfur bacteria contribute up to 40% of the total inorganic carbon photoassimilation in the lake (Camacho et al., 2001; Parkin and Brock, 1981).

Another group of sulfur-oxidizing bacteria are known as the *colorless sulfur bacteria* (CSB). These bacteria are facultative anaerobes. One group of CSB thrive under *microaerophilic* (low-oxygen) conditions where sulfide is available. These CSB oxidize sulfide or other reduced sulfur compounds (thiosulfate or elemental sulfur) to sulfate but are not dependent on light energy or

TABLE 23-6 Common Freshwater Microorganisms Involved in Sulfur Cycling

Sulfur reaction	Sulfur group	Common freshwater microorganisms	Pelagic habitat
Sulfide oxidation to Elemental Sulfur (H_2S, S^{2-} to S^0)	Colorless Sulfur bacteria (CSB)	*Beggiatoa, Thiothrix, Thiobacillus*	Anoxic waters, sediment surface
Sulfide oxidation to thiosulfate or sulfate (H_2S, S^{2-} to $S_2O_3^{2-}$ or SO_4^{2-})	Purple sulfur bacteria (PSB)	*Thiocystis, Thiodictyon, Thiorhodococcus, Lamprocystis*	Low light, low oxygen conditions, above GSB
	Green sulfur bacteria (GSB)	*Chlorobaculum, Chlorobium*	Low to no light and anoxic, below PSB
	Colorless sulfur bacteria (CSB)	*Arcobacter, Sulfurimonas, Sulfuritalea, Sulfuricurvum*	No light, low oxygen conditions
Thiosulfate or sulfate reduction ($S_2O_3^{2-}$ or SO_4^{2-} to H_2S, S^2)	Sulfate-reducing bacteria (SRB)	*Desulfobacterium, Desulfobulbus, Desulfuromonas, Desulfovibrio, Desulfomonile*	Anoxic waters, sediment surface
Sulfate assimilation	Not applicable	Most microorganisms	Throughout the water column

inhibited by oxygen (Diao et al., 2018). This combination of traits is thought to allow CSB to outcompete other sulfide-oxidizers during periods of reoxygenation, like in water column turnover. The genera *Arcobacter, Sulfurimonas, Sulfuritalea,* and *Sulfurospirillum* are commonly found in freshwater lakes (Biderre-Petit et al., 2011; Diao et al., 2018; Hamilton et al., 2014). The genus *Arcobacter* has many motile members, which allow it to form swarms or mats at the oxic-anoxic interface in some systems (Sievert et al., 2007). The genus *Sulfuritalea* is capable of using nitrate and arsenate to oxidize reduced sulfur (Kojima et al., 2014; Watanabe et al., 2017). The genus *Sulfurospirillum* has members that are capable of both sulfur reduction (Sorokin et al., 2013) and sulfide oxidation (Eisenmann et al., 1995).

A second group of CSB found commonly in freshwaters also oxidizes sulfide but often uses nitrate as the oxidizing agent. These CSB are found typically in or on the surface of sediments and are not particularly common in the freshwater pelagic region. This second CSB group also differs from the first in that the sulfide oxidation reaction produces elemental sulfur, which is deposited inside the cell and stored as sulfur globules (Diao et al., 2018). This sulfur storage makes the cells highly refractory and thus fairly visible (Fig. 23-16). These bacteria also regularly produce filaments and form dense mats of a white or yellow color on sediment surfaces. *Beggiatoa*, *Thiothrix*, and *Thiobacillus* are common genera of freshwater sediment-dwelling CSB (Sharrar et al., 2017). Overall, the CSB contain significant metabolic diversity and more complexity and ecological interactions remain to be discovered.

Collectively, SRB and sulfur-oxidizing bacteria play important roles in regime shifts between oxic and anoxic conditions (Diao et al., 2018). After oxygen is depleted and SRB seeded from the sediments reach high enough concentrations in the water column, it is thought that the SRB stabilize the anoxic conditions by maintaining high concentrations of sulfide. This stability and the ability of some CSB to respire oxygen results in the need for a large oxygen influx (e.g., full water column mixis) to disrupt the microbial community interactions and return the system back to an oxic state.

D. Other elements

Microorganisms in freshwater environments impact the cycling of many elements besides those primarily used to build biomass (i.e., C, N, P, and S). These elements may be referred to as trace elements or micronutrients (Chapter 16). All microorganisms require some of these elements for growth and may use various combinations of these elements to maintain osmotic balance, in active sites in enzymes, or as part of cell membranes. In total, these elements make up only about 1% of the dry weight of a microbial cell (Kirchman, 2018). Additionally, the net charge on most microbial cells is negative, which results in the attraction of microbial cells to many positively charged ions (cations) or compounds. Some microbes can use these ions to generate energy through oxidation−reduction reactions.

In the rest of this section we will cover the microbial impacts on some of the most common trace elements in freshwaters. These elements include iron (Fe), manganese (Mn), magnesium (Mg), nickel (Ni), and zinc (Zn). Other elements used by microbes include copper (Cu), cobalt (Co), molybdenum (Mo), cadmium (Cd), and tungsten (W). In fact, microbes use at least 63 elements that are found in nature (Stolz, 2017). See Table 23-7 for examples of how some of these trace elements are used by microorganisms for cellular components.

Many trace elements show concentration differences between the epilimnion and hypolimnion, with the hypolimnion generally having higher concentrations (Groth, 1971). This concentration difference is due, in part, to uptake by microbes in the epilimnion and release of these elements from decaying matter and detritus in the hypolimnion, but water column

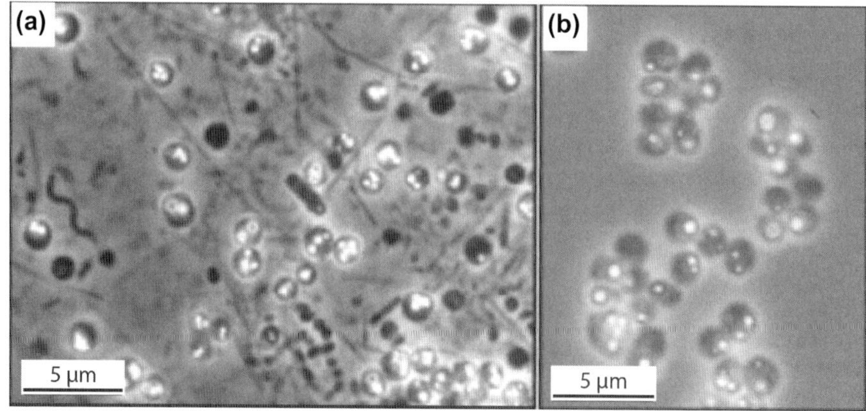

FIGURE 23-16 Phase contrast photomicrograph of (a) the Mahoney Lake (British Columbia, Canada) chemocline microbial community and (b) *Thiohalocapsa* sp. strain ML1 in culture. Intracellular sulfur globules are visible as *white round blobs* in each image. *Scale bars* indicate length in micrometers (μm). *(From Hamilton et al., 2014.)*

TABLE 23-7 Some Common Uses of Trace Elements by Microorganisms

Cell component	Trace element(s)	Biological role	Importance/characteristics
Superoxide dismutase	Fe	Enzyme cofactor	Reduce oxygen stress in cells
Nitrogenase	Fe Mo V	Enzyme used in nitrogen fixation	Moves nitrogen from inert gas to fixed form used in biomass
Nitrite reductase	Fe Cu	Enzyme used in to reduce nitrite	Reduction of nitrite for incorporating ammonia into biomass
Nitrate reductase	Mo	Enzyme used to reduce nitrate	Reduction of nitrate to nitrite for incorporating ammonia into biomass
NADH dehydrogenase	Fe	Used in respiration—electron transfer	For energy generation
Substrate	Fe Mn	Used in oxidation–reduction reactions	For energy generation
Vitamin B_{12}	Co	Involved in DNA and protein production	Nearly all microbes require this molecule, but few can make it
Alkaline phosphatase	Zn	Enzyme used to free phosphate from phosphate esters	Allows cells to scavenge phosphate from more complex molecules

oxidation–reduction potential is the biggest influencing factor. For example, iron is oxidized to Fe^{3+}, a fairly insoluble form in the epilimnion, but exists as Fe^{2+}, a fairly soluble form in the hypolimnion. The oxidation–reduction potential of the water column is important as many of these trace elements play a larger role for microbes when oxygen is depleted, like the conditions found commonly in the hypolimnion of eutrophic lakes or in wetlands.

i. Iron (Fe)

Iron is the fourth most abundant element in the Earth's crust, but most of it is in mineral form and not available to microorganisms. Iron is the most important micronutrient for microbes, as most require it, even though it makes up <0.001% of a cell's biomass (Kirchman, 2018). In most freshwater systems the concentration of dissolved iron in the water column is only a few micromolar (Davison et al., 1982; Nagai et al., 2007), yet these concentrations are typically sufficient to satisfy cellular requirements. Only during more extreme situations, such as during large cyanobacterial blooms or in very large lake systems where much of the pelagic zone has little connection to the terrestrial environment, has iron been noted as a growth-limiting factor (Xu et al., 2013).

Iron is used by microbes in many ways. It is a component of superoxide dismutase (an enzyme used to relieve oxygen stress), nitrogenase (an enzyme involved in nitrogen fixation), nitrite reductase (an enzyme for nitrite reduction), as part of the NADH dehydrogenase complex used in the respiratory electron chain, and as a substrate used by facultative anaerobic bacteria to oxidize organic matter (Sigee, 2005).

In oxygenated surface waters most microbes are capable of scavenging iron from the surrounding environment. Of note are diazotrophic Cyanobacteria, which have a comparably high requirement for iron because it is a component of the nitrogenase enzyme, the enzyme that facilitates nitrogen fixation (Hyenstrand, 2000; Wurtsbaugh and Horne, 1983). In several instances iron limitation has been identified as the growth-limiting factor during cyanobacterial blooms (North et al., 2007; Sterner et al., 2004; Xu et al., 2013) and may be a factor in cyanobacterial dominance over green algae during blooms (Molot et al., 2014; Xu et al., 2013). Bloom-forming Cyanobacteria also produce extracellular *siderophores* (iron-specific chelating agents), which aid in solubilization and uptake of iron (Wilhelm, 1995) and may prevent uptake by other phytoplankton, thus contributing to bloom proliferation (Wilhelm et al., 1996).

In the pelagic zone iron availability for microorganisms often varies seasonally via coupling to the changing oxygen concentrations in the water column. When the water column is oxygenated, iron is largely present in an oxygenated state as soluble Fe^{3+} chelated to DOM (Mortimer, 1941). When oxygen declines, such as in the summer in more eutrophic lakes or in the hyporheic zone of rivers and wetlands, some bacteria reduce iron to oxidize organic matter (Berg et al., 2016). This process does not generate as much energy (a one-electron transfer process) as other oxidizing agents, so typically localized concentrations of manganese ions and nitrate must be depleted before iron oxidation–reduction reactions proceed.

Under oxygen-depleted conditions, the microbes capable of cycling iron are more limited and are used to generate energy through oxidation–reduction reactions rather than uptake for cell components. The microbes known to catalyze these iron transformations are phylogenetically and metabolically diverse. For example, iron may be oxidized *lithotrophically* (use of an inorganic substrate to generate reducing power) by nitrate-reducing microbes such as *Leptothrix*, *Leptospirillum*, or *Siderooxydans* (Berg et al., 2016; Weiss et al., 2007) or by anoxygenic phototrophs using light as an energy source, such as by the genera *Chlorobium* (Heising et al., 1999), *Rhodopseudomonas* (Jiao et al., 2005), and *Rhodobacter* (Berg et al., 2016). These reactions typically occur in the chemoclines of lakes or near sediment–water interfaces in rivers and wetlands. The microorganisms performing iron reduction reactions are also diverse. Some bacteria, like *Geobacter*, *Shewanella*, or *Geothrix*, can use hydrogen (H_2) as a reducing agent (Coates, 1999; Coates et al., 1996; Meshulam-Simon et al., 2007) or others like bacteria in the genus *Rhodoferax* can use organic compounds such as acetate to reduce iron (Finneran et al., 2003).

Rapid iron cycling via coupled iron oxidation and reduction is thought to be common at oxic-anoxic interfaces and has been demonstrated in meromictic lakes, like Lake Cadagno in Switzerland (Berg et al., 2016) and the ferruginous (rust-colored) subbasin of Lake Kivu in Africa (Llirós et al., 2015). Additional measurements across many other types of freshwater systems are needed to confirm if this rapid cycle is widespread, especially in shallower seasonally stratified lakes.

ii. Manganese (Mn)

Manganese plays a central role in microbial enzymes, many of which are related to the central carbon cycle in these cells (Kehres and Maguire, 2003). Although manganese use in enzymes is common, the amount of manganese used is so low that it has little impact on pelagic concentrations. Instead, like iron, manganese can be used by some microbes in oxidation–reduction reactions to generate energy (Nealson and Saffarini, 1994). In fact, manganese cycling by microbes in the pelagic zone closely resembles that of iron and follows changes in the oxidative state of the water. Manganese is insoluble when oxidized (Mn^{3+} and Mn^{4+}) but is soluble in a reduced state (Mn^{2+}). Manganese is reduced at higher redox conditions than iron or sulfur, so it is often reduced to its soluble state in low oxygen conditions before reduced iron or sulfur become available. Manganese oxyhydroxides and oxides often cooccur with iron oxide, resulting in precipitation and the formation of ferromanganese deposits, known as *manganese nodules*. Oneida Lake in New York (USA) is known for its density of manganese nodules in the benthos (Aguilar and Nealson, 1994). Unlike iron, abiotic oxidation of manganese is slow, so microbes are thought to control most manganese cycling (Stein et al., 2001; Tebo et al., 2004).

Many microbes capable of iron oxidation–reduction reactions also are involved in manganese oxidation–reduction reactions. Manganese oxidizers in freshwaters include the genera *Pedomicrobium*, *Hyphomicrobium*, *Leptothrix*, and *Bacillus*, among others (Palermo and Dittrich, 2016; Santelli et al., 2014; Stein et al., 2001). Manganese reducers in freshwater include the genera *Geobacter* (Mahadevan et al., 2011) and *Shewanella* (Hau and Gralnick, 2007), but many bacteria are capable of these transformations in the laboratory. There is not yet a good understanding of which manganese cycling microbes are most prominent overall in the freshwater pelagic zone.

iii. Other elements (Mo, V, Co, Cu, Zn)

Molybdenum (Mo) and vanadium (V) also play a role in the activity of microbes in the freshwater pelagic. Nitrogenase, the key enzyme in nitrogen fixation, is a complex of two proteins, one of which contains either molybdenum or vanadium (Glass et al., 2012; Lee et al., 2010). Molybdenum is present in the most common nitrogenase form and typically is at low concentrations in freshwaters (0.1–13 nmol L^{-1}; Glass et al., 2012) but, aside from a few lakes, has not been shown to have a big impact on nitrogen fixation. Likewise, vanadium quantities appear sufficient to sustain freshwater nitrogen fixation in most systems. However, further work is needed to understand these elemental pools, especially under increasing atmospheric CO_2 conditions, which is altering nitrogen fixation activities (Hungate et al., 2004).

Microbial-driven cycling of other redox-active elements has been noted (e.g., cobalt, copper, and zinc), but much less is known about these elements and how they impact microbes in the freshwater pelagic zone. For instance, copper (Cu) is present in more than a dozen known bacterial enzymes, which include those used for nitrite and nitric oxide reduction and multiple forms of electron transfer (Argüello et al., 2013). Cobalt (Co) is central to the production of vitamin B_{12} (cobalamin), which is required by nearly all organisms for DNA replication and production of certain proteins (Warren et al., 2002). Eukaryotes are not known to make vitamin B_{12}. Production is carried out by a limited number of bacteria and archaea, making competition fierce for this substrate. Primary production may be limited at times by vitamin B_{12} (Cavari and Grossowicz,

1977; Downs et al., 2008), but relatively little is known about the role of cobalt and its potential limitation in freshwaters (Facey et al., 2019). Common metal-reducing bacterial genera, such as *Geobacter* (Mahadevan et al., 2011) and *Shewanella* (Hau and Gralnick, 2007), are capable of the reduction of many trace metals (iron, vanadium, manganese, molybdenum, etc.) but are typically found in freshwater sediments where anoxic conditions persist.

V. Predators and viruses

Freshwater pelagic microbes have growth rates that approximate a doubling of cells every 1–2 days (Simek et al., 2006; Zeder et al., 2009), but seasonally there is relatively little change in their cumulative abundance. This suggests microbial growth is balanced by cell loss (i.e., mortality; Pernthaler, 2005). Like all other known cellular organisms, microbes are subject to top-down forces, such as predation. Predatory organisms one step higher in the food web can eat microbes directly. Also, viruses can infect microbes. Viruses can cause immediate death via cell lysis or a latent (i.e., prolonged) infection that may or may not lead to cell death. Select bacteria can even prey on other bacterial and archaeal cells. These top-down effects represent the primary mechanism for microbial mortality, but the influence of top-down effects on microbial community structure and function is generally less well understood than bottom-up (i.e., nutrient-based) forces. In the following we examine the various groups that prey on or otherwise affect pelagic microbes and how these top-down influences affect biological and ecosystem outcomes across freshwater systems.

A. Grazers

It is now recognized that the grazing activities of protists (see Chapter 20), primarily by *heterotrophic nanoflagellates* (HNFs) and viral lysis, are the two major causes of microbial mortality in freshwaters (Miki and Jacquet, 2008). HNFs are small, flagellated protists that range in size from 3 to 15 mm and are mostly filter feeders that consume particles, including microbes, that are 1–3 μm in size (Gonzalez et al., 1990). *Bacterivory* (i.e., the eating of bacteria and archaea) by these free-living protists is estimated to remove 5–250% of cells per day (Jacquet et al., 2005; Jugnia et al., 2006; Okamura et al., 2012; Sanders et al., 1989) (Tables 23-8 and 23-9).

Competitive interactions among pelagic microbes are altered greatly by the presence of predators (Salcher et al., 2016). Microbial defenses to this predation include size reduction, filament formation, altered cell wall structures, exopolymer production, cell signaling defenses, toxin release, and altered motility patterns (Fig. 23-17). Periods of intense HNF grazing results in a bimodal cell size distribution for microbes, where large

TABLE 23-8 Grazing Rates by Large Zooplankton on Bacterioplankton

Ecosystem	Ingestion rate[a]: bacteria individual^{-1} h^{-1}	Source
Rotifers		
Anuraeopsis fissa	500–920	Ooms-Wilms et al., 1995
Keratella cochlearis	150–230	
Conochilus unicornis	150	
Brachionus angularis	820–910	
Filinia longiseta	5100–7950	
Cladocerans		
Bosmina coregoni	2650	Ooms-Wilms et al., 1995
Bosmina longirostris	3100	
	38–82	Jürgens and Stolpe, 1995
Chydorus spbaericus	5700	Ooms-Wilms et al., 1995
Daphnia cucullata	20,000	
D. magna	475–2780	Jürgens and Stolpe, 1995
D. hyalina	170–900	

[a]*Ingestion rates are expressed as the number of bacteria cleared per individual grazer per hour of grazing time.*

TABLE 23-9 Potential Bacterivory in Lakes Expressed as a Percentage of the Bacterioplankton Production Based on Abundances of Zooplankton and Bacteria and Literature Values of Specific Clearance Rates[a]

Organisms	Clearance rate (μL individual^{-1} h^{-1})[b]	% of bacterial production
Heterotrophic nanoflagellates	0.6–29	3–70
Ciliates	0–0.53	1–19
Rotifers	0.5–10	<7
Cladocera		
Daphnia spp.	12.5–450	1–12
Other cladocerans	10–330	<10
Copepoda	4.6–158	6–12

[a]*Composite from (Chrzanowski and Simek, 1993; Sanders et al., 1989; Simon, 1998; Simon et al., 1998; Tranvik, 1989).*
[b]*Clearance rates are expressed as the volume cleared per individual grazer per hour of grazing time.*

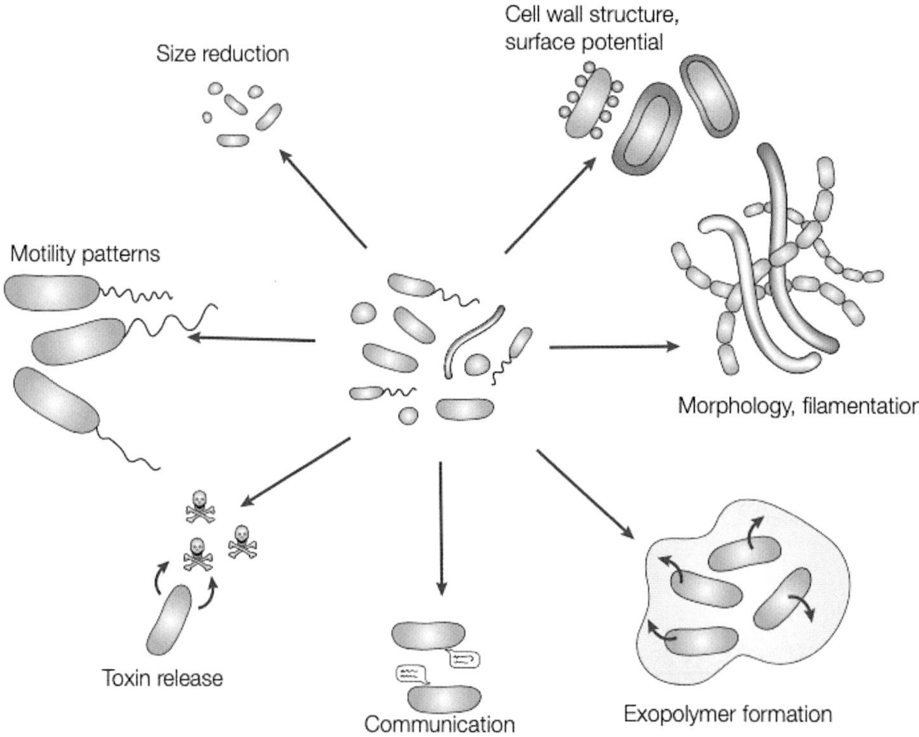

FIGURE 23-17 Depiction of bacterial defense mechanisms to avoid predation. *(From Pernthaler, 2005.)*

filamentous microbes (>10 μm in length) and very small microbes (cell volumes <0.1 μm^3, sometimes called the *ultramicrobacteria*) dominate the microbial biomass (Fig. 23-18; (Jousset, 2012; Pernthaler, 2005). In some cases, during periods of intense grazing, filamentous microbes make up >40% of the total microbial biomass in lake surface waters (Langenheder and Jurgens, 2001; Pernthaler et al., 2004; Schauer and Hahn, 2005). HNF grazing also releases nutrients that are either not used by the grazers or are lost to the environment during the feeding process. This localized liberation of nutrients into the surrounding water is a significant regenerator of nitrogen and phosphorus (Eccleston-Parry and Leadbeater, 1995; Grover and Chrzanowski, 2009; Sherr et al., 1983), which is coveted by nearby microorganisms. As a result, areas of intense grazing are often "hot spots" of localized microbial activity.

There are additional eukaryotic predators that eat microbes, but on average, these predators are thought to play a smaller role in this process. Small (2–3 mm) ciliates are the most prominent grazers of this group, and they seem to have the largest predatory influence in

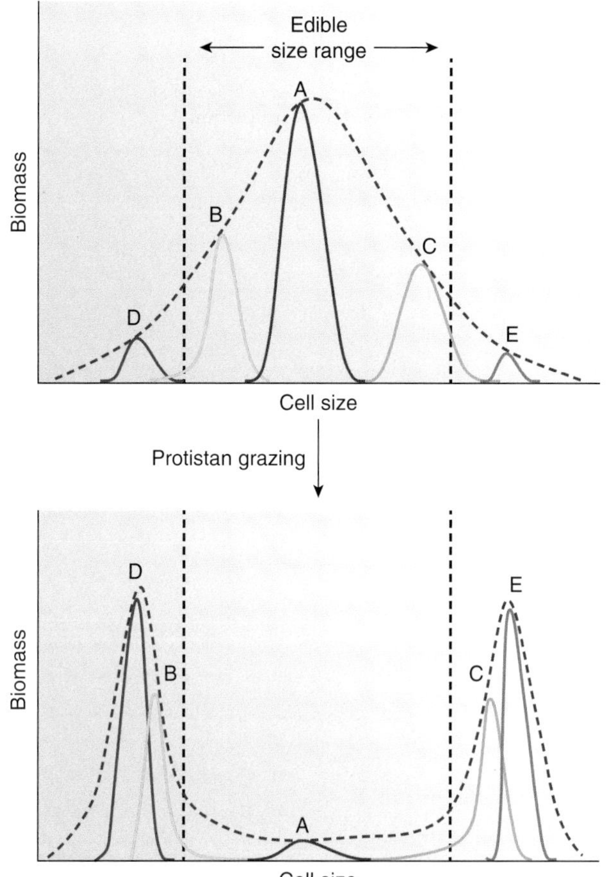

FIGURE 23-18 Representation of the effect of predation on bacterial cell size distribution. Individual bacterial species separated by size distribution are depicted (species A–E). Bacterial cells that are in the size range that can be ingested readily by grazers (species A, B, and C) are disproportionately reduced from the community (species A) unless they can alter their cell size in response to the grazing pressure (species B and C). Species outside of the edible size range are favored during grazing (species D and E). *(From Pernthaler, 2005.)*

very productive systems (see Table 23-7; Nakano et al., 1998; Sherr and Sherr, 2002). The grazing role of even larger protists and zooplankton is not as well understood; some studies have shown that these organisms can consume heterotrophic bacteria and Cyanobacteria (Ger et al., 2014; Güde, 1986; Hwang, 1999; Work, 2003), but generally larger protists and zooplankton are thought to play a minor role in microbial mortality on a yearly scale. Overall, the variability in grazing pressure by different eukaryotic groups is large and is influenced seasonally by the differing chemical composition among lakes, DOM, and other food web interactions (Pernthaler, 2005).

B. Viruses

Viruses themselves are considered microbes by many scientists and are studied in their own right with respect to biodiversity, abundance in freshwater, and population dynamics. Viruses can consist of single- or double-stranded DNA or RNA, and all are surrounded by a protein coat, the *capsid*. Some viruses also have membranes or tails. Viruses cannot replicate (i.e., copy themselves) without a host as they do not possess much of the intracellular machinery and energy-generating mechanisms needed for their own replication (Sime-Ngando, 2014). Likely, all organisms on Earth can be infected by viruses.

Viruses that infect microbes are called *phages*. Phage particles are usually 10–250 nm in diameter and can be visualized both inside cells and as free particles in the water (Sime-Ngando, 2014; Wommack and Colwell, 2000). Virus morphology can be observed using electron microscopy (Fig. 23-19). Estimates of virus particle density are usually obtained with epifluorescence microscopy, where viral particles are stained with a fluorescent dye and then counted using a microscope or flow cytometry. Viral abundance ranges from 10^6 to 10^8 mL^{-1} across most freshwater systems (Auguet et al., 2009; Bergh et al., 1989; Jacquet et al., 2005; Peduzzi and Schiemer, 2004; Thomas et al., 2011; Wommack and Colwell, 2000) and typically exceed microbe concentrations by 10- to 100-fold (Fig. 23-20; Knowles et al., 2016). The full genetic content of viruses in a sample is called the *virome*, which can be accessed via DNA sequencing (see also Section V). Viruses likely represent the greatest reservoir of uncharacterized genetic diversity on Earth (Suttle, 2007, 1994).

Phages represent another primary controller of freshwater microbial biomass and activity (Middelboe et al., 2008; Weinbauer, 2004). Estimates of phage-controlled bacterial mortality rates vary widely with trophic status, with oligotrophic systems exhibiting 1–25% cell mortality per day and highly eutrophic systems having up to 100% mortality rates per day (Heldal and Bratbak, 1991; Mathias et al., 1995; Suttle, 1994; Wommack and Colwell, 2000). On average, it is thought that phages account for 20–40% of daily microbial mortality (Suttle, 2007). Transmission electron microscopy can be used to visualize microbial cells with current phage infections and has revealed that the fraction of infected cells ranges from <1% to 17% (Hennes and Simon, 1995). Phage-induced mortality is believed to be more host targeted than grazing by protists, as phages target cell surface molecules that are often unique to a particular microbial group, but this *host range* (i.e., the number of different hosts that can be infected) varies among phages. Also, several different phages can target the same bacterium (Holmfeldt et al., 2007; Malki et al., 2015). The host range of most phages in freshwater systems is not yet clear, as relatively few have been identified and even fewer have been matched to their hosts (Ghai et al., 2017; Moon et al., 2017).

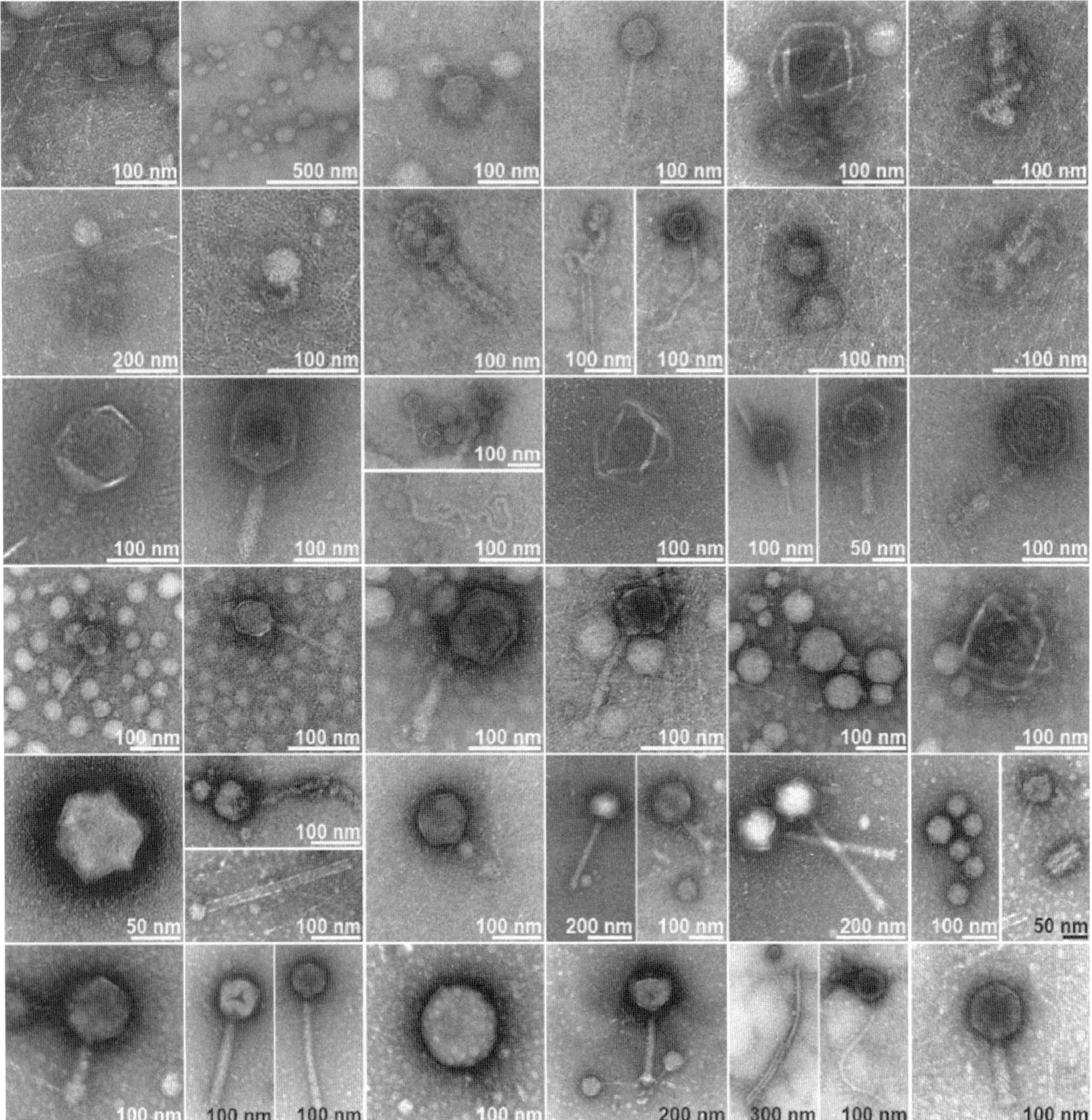

FIGURE 23-19 Transmission electron micrograph of viruses sampled from Feitsui Reservoir in Taiwan. *(From Tseng et al., 2013.)*

In general, viruses can be split into three groups based on life cycle strategy: lytic, lysogenic, and chronic. *Lytic viruses* kill their host via membrane rupture after a large number of viral particles are produced inside the cell. This cell rupture process releases the cell's contents (e.g., nucleic acids, proteins, lipids) into the surrounding water, so in freshwater systems lytic phages play a fundamental role in the cycling of nutrients and organic matter (Pernthaler, 2005; Sime-Ngando, 2014). *Lysogenic viruses* integrate into host DNA and can form long-lived associations with their hosts. This host integration can have both negative and positive consequences. In some cases phages reduce the fitness of their host, whereas in other cases phages carry genes beneficial to their host's competitive abilities and survival (Sime-Ngando, 2014). Less is known about these phages in freshwater systems, but studies have found that lysogenic phages are present in up to 73% of all pelagic microbes (Sime-Ngando and Colombet, 2009). Lower biomass oligotrophic systems appear to have increased numbers of lysogenic phages, presumably as an adaptation to low encounter rates between cells and thus a reduced benefit to fast-acting rupture of the host (Palesse et al., 2014; Stewart and Levin, 1984). *Chronic-cycle*

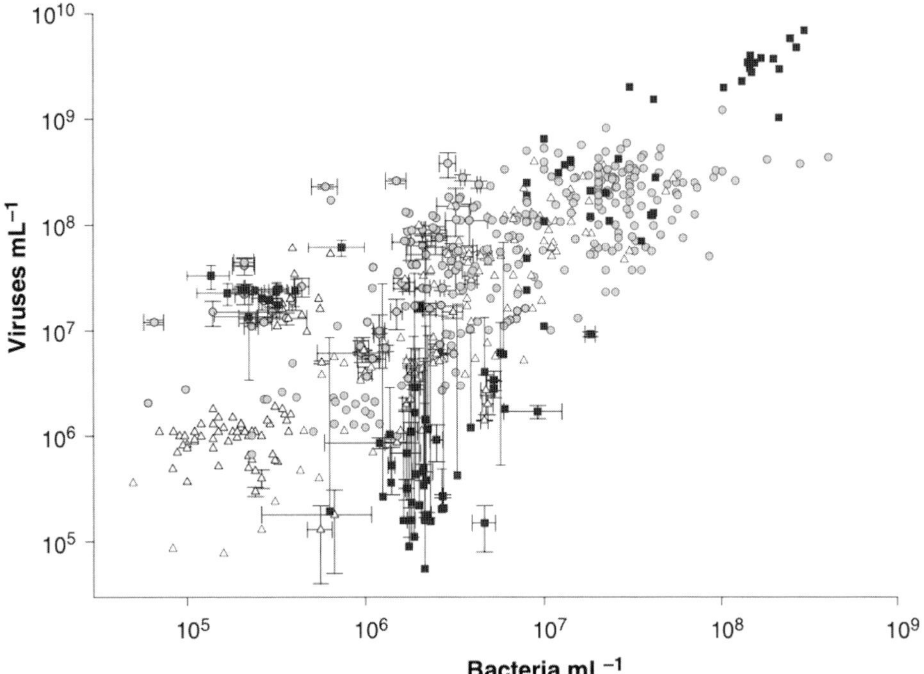

FIGURE 23-20 Relationship between virus abundance and bacterial cell abundance in aquatic systems. *(From Wilhelm and Matteson, 2008.)*

viruses drive their hosts to episodic (i.e., occasional) viral particle production, where viral particles are released via cell budding or extrusion without immediate cell bursting (Sime-Ngando, 2014). Little is known about these phages in aquatic systems.

Phage infection also induces phenotypic changes in the bacterial community as microbes that resist infection are selected and persist, often with corresponding changes to cell surface molecules (Bohannan and Lenski, 2000). The enormity of cell lysis and selection for resistance has resulted in significant coevolution between microbes and their phages (Hampton et al., 2020). Phages also contribute to genomic changes in microbes by acting as a vector of horizontal transfer of DNA (genes) from one microbe to another (Kenzaka et al., 2010; Saye et al., 1987). HNFs can also ingest phages and are thought to graze 1–5% of the daily phage production in some lakes (Bettarel et al., 2005; Gonzalez and Suttle, 1993).

Viruses are difficult entities to classify. There are no universal DNA or gene markers by which to compare all viruses (Breitbart and Rohwer, 2005), so various classification systems have arisen. In freshwater systems extensive viral diversity is being revealed by metagenomic approaches. Freshwater systems appear to contain unique viral communities that are distinct from those found in other ecosystems (Roux et al., 2012), akin to the diversity observed for bacteria and archaea. Common freshwater DNA viruses include many phages, such as the *Myoviridae*, *Microviridae*, *Podoviridae*, and *Siphoviridae*, as well as many that remain unclassified (Coutinho et al., 2020; Green et al., 2015; López-Bueno et al., 2009; Mohiuddin and Schellhorn, 2015; Moon et al., 2020; Roux et al., 2012; Tseng et al., 2013). RNA viruses are also present in freshwaters (Djikeng et al., 2009; Vlok et al., 2019), but these are more likely to infect eukaryotic hosts ranging from single-celled protists to plants, birds, and mammals.

Recently, a new group of viruses was discovered in freshwater amoeba that collectively are known as *giant viruses* (Claverie et al., 2006; La Scola et al., 2003). These viruses have altered the fundamental understanding of what can be considered a virus (Yutin et al., 2014), as they are structurally large (>0.2 μm in diameter) and have DNA-based genomes that are in some cases twice as large as (2.5 Megabases [Mb], where Mb = 1 million bases), and more genetically diverse than, some of the most common freshwater bacteria. These viruses, namely the mimiviruses, pithoviruses, and pandoraviruses, commonly infect amoeba and other phagocytizing protists (Oliveira et al., 2019), many of which are found in freshwater systems. The giant viruses are also prone to their own viral infections by *virophages*. Virophages coinfect protists alongside the giant viruses and then hijack the giant virus replication machinery to replicate themselves (Roux et al., 2017). Both giant viruses and virophages are common in freshwater systems, but little data are available to determine their impact on these ecosystems. In a study by Yau et al. (2011) the activity of virophages was proposed to increase algal bloom frequencies by reducing algal mortality from predators.

C. Other microbial interactions

Some bacteria eat other bacteria or small eukaryotes and are considered predators within the field of microbial ecology. Such bacterial hunting strategies can be categorized into three groups: (1) *epibiotic predation*, where bacteria attach to their prey, puncture the cell, and consume the prey while remaining outside their prey; (2) *endobiotic predation*, where bacteria invade and consume the prey cell from the inside; and (3) *group attack*, where groups of predatory bacteria "hunt in packs" and release enzymes to degrade cell membranes/walls of their prey (Pérez et al., 2016).

Predatory bacteria are common in freshwater systems and appear throughout the pelagic environment (Paix et al., 2019). The genera *Vampirococcus*, *Bdellovibrio*, *Cytophaga*, and *Peredibacter*, as well as what are known as *Bdellovibrio*-like organisms (BALOs), are ubiquitous freshwater predatory bacteria. *Bdellovibrio* and related BALOs are the best studied of predatory bacteria, but little is known about how they impact freshwater ecosystems (Paix et al., 2019). The concentration of predatory bacteria is typically 10^2–10^3 cells per mL (Paix et al., 2019; Williams et al., 2016), which makes them a very small component (1 out of every 1000 or 10,000 cells) of total pelagic bacteria. In experimental settings predatory bacteria are capable of rapid consumption of their prey and large biomass increases (Williams et al., 2016), but it is not clear how often this occurs naturally. Predatory bacteria are thought to control cyanobacterial bloom dynamics in some circumstances (Gumbo et al., 2008; Pal et al., 2020; Rashidan and Bird, 2001), but further work is needed to understand the mechanisms and extent of this control.

The *Melainabacteria* also contain predators. The *Melainabacteria* is a basal lineage of nonphotosynthetic Cyanobacteria that diverged from the photosynthetic Cyanobacteria prior to the evolution of aerobic respiration (Soo et al., 2017). The *Melainabacteria* genus *Vampirovibrio* preys upon the green alga *Chlorella* (Gromov and Mamkaeva, 1972; Soo et al., 2015). *Melainabacteria* are now known to be common in freshwater lakes (Monchamp et al., 2019), but like other bacterial predators, their ecological role remains undefined.

D. Ecology

The removal of predators from microbial assemblages has a significant impact on microbial community composition, tending to favor the growth of a few microbes capable of rapid growth (Salcher et al., 2007; Šimek et al., 2005). When predators are present, the growth of these organisms (sometimes referred to as *microbial weeds*) is suppressed. In viral ecology this concept is known as "killing the winner" (Thingstad, 2000; Winter et al., 2010). The killing-the-winner theory predicts that some microbes have evolved cellular traits that confer a significant competitive advantage for nutrient resources but few defense capabilities, as the mechanisms for rapid nutrient uptake and defense are not fully compatible in a single cell (Pernthaler, 2017). Therefore the rapid growth capabilities of these cells are balanced by rapid cell loss due to predation or viral lysis (Thingstad, 2000).

Predation by protistan grazers is thought to have a more significant impact on total microbial biomass, while phages are considered to have a bigger impact on microbial diversity (Pernthaler, 2005). These differing effects are the product of the predation mechanisms by the two predator groups. Protist grazers are generally capable of feeding on a wide range of microbes, while phages typically have a more limited host range and depend more on cell densities (i.e., high microbial cell densities are needed for phage spread between hosts). The indiscriminate feeding mechanism of protists (i.e., nontargeted) allows them to remove microbial biomass without direct selection for particular taxa, while phages have the biggest impact on the abovementioned fast-growing microbes. Phage lysis of fast-growing microbes is thought to have a disproportionate role in carbon transfer to higher trophic levels (Šimek et al., 2014; Zeder et al., 2009) as the carbon and nutrients in these cells are recycled quickly via rapid growth-to-death processes. Other microbes are thought to specialize on the substrates released from viral lysis and protistan grazing (Eckert et al., 2013; Šimek et al., 2007) and may be considered "microbial vultures."

The balance of the impact of grazing and phage lysis and the development of resistant phenotypes/genotypes is not well understood (Pernthaler, 2017). It is clear that microbial populations can develop rapid resistance to phages (Middelboe et al., 2009), and many pelagic microbes have phenotypic adaptations that resist protistan grazing (Blom and Pernthaler, 2010; Matz and Jürgens, 2005; Šimek and Chrzanowski, 1992; Tarao et al., 2009). In fact, most microbes considered the dominant organisms in the freshwater pelagic zone have significant defense mechanisms (Pernthaler, 2017). These defense-based traits include growth as inedible filamentous cells (Jürgens et al., 1999; Pernthaler et al., 2004; Schauer and Hahn, 2005) and growth as very small cells (the *ultramicrobacteria*; Hahn et al., 2003; Pernthaler et al., 2001; Salcher et al., 2010), with cell wall properties that resist protistan ingestion (Tarao et al., 2009).

There are ongoing debates about the role that viruses play in the carbon and nutrient cycles across freshwater habitats. Certainly, on a daily basis, viruses play an immense role in cell lysis across all freshwater habitats. Cell lysis releases DOM, particulate organic matter (POM), and nutrients into the surrounding waters (Riemann and Middelboe, 2002). This released resource pool may be used directly by heterotrophic microbes, which either enhances temporarily microbial growth and production or is used for energy generation and is respired as CO_2. If microbial respiration or uptake of resources by grazing-resistant microbes is dominant, then viral lysis can decrease the DOM concentration in the lake and thus lower the transfer of energy to higher trophic levels (e.g., fish). This process has been termed the *viral shunt* (Wilhelm and Suttle, 1999). Others have argued that the release of POM could influence the vertical flux of organic matter and therefore reduce the efficiency of carbon burial in sediments (Suttle, 2007). Models in marine systems indicate pelagic viruses increase organic matter recycling, reduce energetic transfer to higher trophic levels, and increase net primary productivity (Weitz et al., 2015).

The interactions between the predatorial grazers and phages, and their prey microorganisms are complex (Miki and Jacquet, 2008) and reflect an ever-changing system where selective pressures act on large organismal population sizes to affect the coevolution of predator and prey. Many have termed this a *biological arms race* (Hampton et al., 2020; Lenski and Levin, 1985; Marston et al., 2012). This "race" is ancient and there is still much to be understood biologically and regarding how these interactions shape pelagic food webs and freshwater ecosystem processes (Pernthaler, 2005).

VI. New discoveries from the era of DNA sequencing

Beginning around 2008, our understanding of freshwater microbial diversity, ecology, and evolution has advanced rapidly due to the development of new tools based on DNA sequencing. So-called "next generation" sequencing has been applied to the more established 16S rRNA gene sequencing to study the taxonomic composition of communities across systems, over time, and in community manipulation experiments. But the most significant advances have been driven by our ability to access the genome sequences of uncultured organisms in freshwater (Grossart et al., 2020). Genomes can be recovered from single cells or from the entire *metagenome* that is defined as the collection of genomes from "all" the organisms in a sample. Not only does genomic information enable more accurate and resolved taxonomic classification, but it is also a basis for predicting an organism's ecophysiology. Gene function predictions can be strung together to predict metabolic pathways involved in energy generation, substrate uptake, terminal electron acceptor reduction, etc. This has profound implications for linking specific taxa to key processes relevant at the ecosystem scale, such as nitrification, POM mineralization, and different kinds of respiration. Here we describe some of the most significant findings derived using this approach, with an emphasis on those most related to fundamental limnology.

A. Diversity and organism life history

Metagenomic-based surveys of microbial diversity in freshwater have mostly confirmed what we knew about community composition based on 16S rRNA gene surveys. Genomes are available for nearly all of the lineages considered to be cosmopolitan (Newton et al., 2011) (Table 23-10).

B. Microbial functions encoded in recovered genomes

Genomes are the blueprint for all of the enzyme-driven transformations that organisms carry out. We can computationally scan a genome and locate putative genes. Computational algorithms yield what are called *annotations* that are the best guess of a gene's function. Functions can then be mapped onto metabolic pathways ranging from central carbon metabolism (e.g., the citric acid cycle) to phosphate uptake to ammonia oxidation. However, these guesses are only as reliable as the quality and completeness of available reference databases linking gene sequences to experimental evidence of function gathered by biochemists. Much remains to be researched regarding basic biochemistry, especially in nonmodel organisms. Still, some predictions are considered to be quite reliable and are the foundation for hypotheses that can be tested in the field and the wet lab. The following text contains examples of recent discoveries facilitated by approaches made possible by high-throughput DNA sequencing.

Rhodopsins: an alternative form for light-driven energy generation?

Rhodopsins are proteins that use light energy to pump protons across the cell membrane. This is important because proton gradients across membranes can be harnessed to generate energy (i.e., ATP). Thus organisms with active rhodopsins may be able to survive or grow using light energy even if they do not have chlorophyll or other pigments and genes traditionally associated with photoautotrophy. Chlorophyll-independent photoautotrophic growth would be a major blindspot in our

TABLE 23-10 Microbial Groups Considered Cosmopolitan That Have Available Genome Sequences

Taxonomic group	Type of genomes available[a]	Citations
Actinobacteria		
Planktophila (acI-A)	SAGs, MAGs, enrichment culture MAGs	Cabello-Yeves et al., 2018; Ghylin et al., 2014; Hamilton et al., 2017; Kang et al., 2017; Linz et al., 2018
Nanopelagicus (acI-B)	SAGs, MAGs, enrichment culture MAGs	Garcia et al., 2013; Ghylin et al., 2014; Hamilton et al., 2017; Linz et al., 2018
Alphaproteobacteria		
Fonsibacter (LD12)	SAGs, MAGs, isolate	Cabello-Yeves et al., 2018; Garcia et al., 2018; Ghylin et al., 2014; Henson et al., 2018b; Linz et al., 2018; Tsementzi et al., 2019; Zaremba-Niedzwiedzka et al., 2013
Betaproteobacteria		
Limnohabitans	SAGs, MAGs, isolate	Kasalický et al., 2018, 2013; Linz et al., 2018; Props and Denef, 2020; Zeng et al., 2012
Methylopumilus (LD28)	SAGs, MAGs, isolate	Cabello-Yeves et al., 2018; Linz et al., 2018; Salcher et al., 2019, 2015
Polynucleobacter	SAGs, MAGs, isolate	Hahn et al., 2016, 2012; Hoetzinger et al., 2017; Linz et al., 2018
Chloroflexi CL-500[b]	MAGs	Cabello-Yeves et al., 2018; Denef et al., 2016; Linz et al., 2018; Mehrshad et al., 2018
Verrucomicrobia	SAGs, MAGs	Cabello-Yeves et al., 2018; He et al., 2017; Linz et al., 2018; Martinez-Garcia et al., 2012; Tran et al., 2018

[a] *MAG, metagenome-assembled genome (i.e., recovered from metagenome assembly); SAG, single-cell-amplified genome (i.e., recovered from a single cell); Isolate, organism is available in pure culture; Enrichment culture, MAG was recovered from an enrichment culture (i.e., less complex than an environmental sample).*

[b] Chloroflexi *were not well represented in the 16S rRNA gene database available to Newton et al. (2011) but available metagenomes suggest they are cosmopolitan.*

understanding of primary production in aquatic ecosystems.

Rhodopsins have been most heavily studied in halophilic Archaea (*bacteriorhodopsin*) and marine bacteria (*proteorhodopsin*) The genomes of many freshwater microbes also encode *proteorhodopsins* (Martinez-Garcia et al., 2012) and members of the highly abundant *Nanopelagicales* order in the *Actinobacteria* (lineage acI) encode a unique *actinorhodopsin* (Sharma et al., 2009, 2008). Some evidence exists that the actinorhodopsin proteins are indeed activated by light, resulting in protons moving from inside the cell to outside (Dwulit-Smith et al., 2018). Gene expression data show that actinorhodopsins are very highly expressed *in situ*

(Hamilton et al., 2017) and that the abundance of their mRNA cycles diurnally (Linz et al., 2020; Wurzbacher et al., 2012). However, it is still unknown whether a proton gradient sufficient to support ATP synthesis is in fact generated in wild freshwater microbes.

The central role of nitrogen-rich carbon sources

As described above, a significant amount of carbon (Chapters 13 and 28) is cycled through dissolved free amino acids (DFAAs) and urea. Isotopic tracers have been widely used to measure DFAA uptake and incorporation into bacterial biomass and to link uptake of specific amino acids to specific taxa (Salcher et al., 2013). Therefore it may not be surprising that freshwater microbial genomes encode many transporters that enable uptake of these compounds (Ghylin et al., 2014; Linz et al., 2018). Among the most cosmopolitan and abundant taxa, *Nanopelagicales* (acI), *Fonsibacter* (LD12), and *Polynucleobacter* encode transporters for amino acids, while the ability to take up oligopeptides is restricted to *Nanopelagicales* (Ghylin et al., 2014).

One surprise found by inspecting genomes was the high abundance of pathways associated with polyamine biosynthesis and uptake/degradation (Linz et al., 2018). Nearly every genome recovered in a survey of eutrophic Lake Mendota (WI, USA) and a humic bog lake had such genes. *Nanopelagicus* (acI-B) members in Lake Mendota were expressing genes predicted to be involved in polyamine uptake (Hamilton et al., 2017). Curiously, *Fonsibacter* (LD12) lack genes for polyamine cycling (Ghylin et al., 2014).

C1 compounds as a carbon source

A surprising number of genomes recovered from freshwater are affiliated with taxa known to metabolize C1 compounds such as methane, methanol, or methylamine. These include members of the *Methylophilaceae* such as *Methylotenera* and *Methylopumilus* (previously known as LD28), which are considerably abundant and prevalent in most lakes (Newton et al., 2011). Genomes from these taxa confirm the presence of genetic machinery to grow on C1 compounds, though some key differences are found between the freshwater representatives and other members of the family (Linz et al., 2018; Salcher et al., 2015). Work with isolates of *Methylopumilus* demonstrated their ability to grow on methanol (Salcher et al., 2015).

Ammonia-oxidizing Archaea

As discussed earlier in this chapter and in Chapter 14, nitrification, the conversion of ammonia to nitrate, plays a key role in the transformation of nitrogen in freshwater systems. For more than a century, it was thought that this process was carried out in two steps by two different groups of bacteria: *ammonia-oxidizing bacteria* (AOB) and *nitrite-oxidizing bacteria* (NOB). Then over the past 25 years, gene-based studies of natural systems changed this perspective dramatically.

The first step in nitrification is the conversion of ammonia to nitrite. It was thought to be carried out by a few bacterial genera (e.g., *Nitrosomonas*, *Nitrosospira*), known collectively as AOB. However, direct recovery of DNA from the world's oceans revealed abundant archaea that seemingly harbored the capability for ammonia oxidation (Delong, 1992; Fuhrman et al., 1992). It is now known that *ammonia-oxidizing archaea* (AOA) dominate aquatic ammonia oxidation at a global scale and are often the dominant ammonia-oxidizers in more oligotrophic freshwater systems (Auguet et al., 2011; Herber et al., 2020; Mukherjee et al., 2016). To date, the controlling factors on the distribution of AOA versus AOB in freshwater systems are not well understood (Salcher et al., 2015). Additionally, AOA have proven metabolically diverse. Traditionally, ammonia-oxidizers were considered highly specialized *chemolithoautotrophs* that oxidized ammonia to generate energy. AOA have proven to be anything but highly specialized. A combination of metagenomic and lab-based studies have revealed that AOA can use urea and cyanate (Palatinszky et al., 2015) as energy sources and that AOA may not even be obligate autotrophs but rather mixotrophs switching between an autotrophic and heterotrophic lifestyle (Qin et al., 2014).

v. Virus diversity

Viromes have been recovered from a number of freshwater ecosystems to date, using metagenomics (Arkhipova et al., 2018; Ghai et al., 2017; Kavagutti et al., 2019; López-Bueno et al., 2009; Moon et al., 2020; Roux et al., 2017, 2012; Skvortsov et al., 2016). However, the study of viral genomes is still very challenging due to our inability to predict gene function and link viral taxa to their hosts without a pure culture. Thus most studies have focused on diversity inventories. Surprisingly, freshwater seems to be a hot spot for giant virus diversity (Schulz et al., 2020).

Viruses in aquatic ecosystems are known to sometimes carry genes that augment or introduce specific functions to their host. One especially intriguing example is the presence of genes encoding two core photosystem proteins found in Cyanobacteria (Ruiz-Perez et al., 2019). The critical gene in methane oxidation, *pmoA*, is found in freshwater phages (Chen et al., 2020). Some freshwater phages encode genes involved in sulfur metabolism (Kieft et al., 2021). The importance of phages for both augmenting microbial metabolism and regulating bacterial numbers has yet to be fully appreciated and understood.

VII. Summary

1. Microbes control nearly all biogeochemical processing in freshwater. Few would contest their fundamental importance, but their size and biodiversity create major barriers to study. Limnologists historically treated microbial communities as an aggregate "black box" functioning as a singular unit, but modern approaches have revealed a formidable level of ecological complexity within that box. Freshwater microbial ecologists are striving to dissect this complexity. As a rule, the field has been limited by available tools and approaches. This has led to major leaps in understanding with each new breakthrough in methodologies available to microbial ecologists.
2. Microbial cell numbers, biomass, and productivity generally increase with increasing photosynthetic productivity of freshwaters.
   - **a.** Cell numbers in pelagic systems are generally consistent within one or two orders of magnitude but vary within this range across seasons, productivity gradients, and with depth.
   - **b.** Free living cells can be very small, resulting in high surface-to-volume ratios that appear to reflect their lifestyle as oligotrophs. Particle-associated taxa are more copiotrophic.
   - **c.** The elemental composition of freshwater microbes is highly plastic but generally conforms to the Redfield ratio as a useful approximation.
3. Accurate and consistent biodiversity inventories have only been possible since the mid-1990s when DNA-based tools were developed. Microbial ecologists use a highly conserved genetic locus called the 16S rRNA gene to identify and classify microbes. The concept of a microbial species is hotly contested in microbiology more broadly, and most microbiologists would argue that current definitions may be lumping taxa too much, leading to a vast undercounting of species richness.
   - **a.** We now know that freshwater microbial communities are incredibly diverse, with hundreds to thousands of species present per milliliter of water. However, the rank abundance curve shows that often only 20–30 species constitute a large majority of each community.
   - **b.** A synoptic assessment of many studies based on this approach reveals a core group of freshwater taxa conserved at approximately the family level. Some of these can be found in virtually every freshwater habitat sampled thus far.
   - **c.** Microbial community assembly and dynamics are thought to be largely driven by bottom-up factors such as resource availability, general water chemistry, and temperature. We do not have a reliable set of "rules" that can be used to predict the structure of microbial communities in freshwater, but some broad patterns are consistent.
4. The role of microbes in the freshwater carbon cycle is important and largely unappreciated by the average limnologist. The conceptual model of the "microbial loop" developed for marine systems is a useful framework for thinking about their roles and should be used more commonly by limnologists.
   - **a.** In many systems microbial growth is fueled by exudates and cell lysis products from phytoplankton but photodegradation of terrestrial carbon is also a major source of labile organic matter supporting their communities.
   - **b.** Microbes drive overall detritus decomposition and element recycling, mineralizing both carbon and essential nutrients such as nitrogen, phosphorus, and sulfur.
   - **c.** Microbial growth efficiency is a parameter frequently calculated by limnologists because it is used to track carbon flow through an ecosystem.
   - **d.** Uptake rates for various substrates of interest, such as sugars, fatty acids, and amino acids, can be calculated using Michaelis–Menten kinetics though the experimental tools for measuring such uptake have serious limitations that should be accounted for.
   - **e.** Radioisotopes can be used for measuring substrate uptake and this can be combined with cell staining tools that help link substrate-use capabilities to specific taxonomic groups.
5. Microbes are also critical for biogeochemical cycles beyond carbon. Their activities can be motivated by the need to generate energy and/or the acquisition of material to build new biomass.
   - **a.** The nitrogen cycle is one of the most complex cycles, after the carbon cycle. Nitrogen fixation is a relatively rare capability among microbes but several diverse taxa outside of the more commonly appreciated Cyanobacteria can fix N_2 in freshwater. Highly specialized taxa are involved in ammonia and nitrite oxidation (nitrification), several of which were completely unknown to microbiologists until the 2000s–2010s. Denitrification capabilities are broadly distributed among many diverse taxa.
   - **b.** In systems that are phosphorus limited, microbes have an advantage due to high-affinity machinery that allows them to compete fiercely with eukaryotic plankton.
   - **c.** Sulfur and various trace elements undergo transformations that are often controlled by specialized taxa in unique compartments of freshwater ecosystems (e.g., anoxic hypolimnia).

6. Microbial communities are also controlled by top-down factors, primarily grazing by protists and lysis by viruses.
   a. Predation by eukaryotes seems to have a larger effect on biomass, while viral lysis influences diversity but can at times equal the impact of predation on biomass.
   b. Grazing has a large influence on the cellular phenotypes of freshwater microorganisms. Most pelagic microbes have significant adaptations that protect against grazing.
   c. Resistance to phages and the ensuing arms race between phages and their hosts is an important dynamic that is yet poorly understood.
7. A revolution in molecular tools has led to a transformation in our understanding of microbial diversity with respect to taxonomy but also functional (metabolic) capabilities. The availability of genomes recovered from uncultured organisms is rapidly accelerating and has already yielded surprises that were invisible without these methods. Highlights include the widespread distribution of rhodopsins that may subsidize chlorophyll-driven production, the importance of C1 compounds in the carbon cycle, the central role of nitrogen-rich DOM, new taxa underpinning the nitrogen cycle, and the incredible diversity of viruses infecting planktonic bacteria, archaea, algae, and protists.

Acknowledgments

We are extraordinarily grateful to Lou LaMartina for assistance with figure generation, Ella Schmidt for friendly review, Dr. Michaela Salcher for sharing FISH images, Dr. Ben Peterson for the lake biogeochemical profile, Dr. Robin Rohwer for unpublished time series data, and Jackie Lemaire for assistance with final proof review. We thank Dr. Shaomei He for recreating the figure illustrating the iconic microbial loop.

References

Aguilar, C., Nealson, K.H., 1994. Manganese reduction in Oneida Lake, New York: estimates of spatial and temporal manganese flux. Can. J. Fish. Aquat. Sci. 51, 185–196.

Alfreider, A., Baumer, A., Bogensperger, T., Posch, T., Salcher, M.M., Summerer, M., 2017. CO_2 assimilation strategies in stratified lakes: diversity and distribution patterns of chemolithoautotrophs: chemoautotrophic CO_2 fixation in lakes. Environ. Microbiol. 19, 2754–2768. https://doi.org/10.1111/1462-2920.13786.

Allen, H.L., 1969. Chemo-organotrophic utilization of dissolved organic compounds by planktic algae and bacteria in a pond. Int. Rev. Ges. Hydrobiol. 54, 1–33.

Allgaier, M., Grossart, H., 2006. Seasonal dynamics and phylogenetic diversity of free-living and particle-associated bacterial communities in four lakes in northeastern Germany. Aquat. Microb. Ecol. 45, 115–128. https://doi.org/10.3354/ame045115.

Alonso-Sáez, L., Morán, X.A.G., González, J.M., 2020. Transcriptional patterns of biogeochemically relevant marker genes by temperate marine bacteria. Front. Microbiol. 11, 465. https://doi.org/10.3389/fmicb.2020.00465.

Argüello, J.M., Raimunda, D., Padilla-Benavides, T., 2013. Mechanisms of copper homeostasis in bacteria. Front. Cell. Infect. Microbiol. 3. https://doi.org/10.3389/fcimb.2013.00073.

Arkhipova, K., Skvortsov, T., Quinn, J.P., McGrath, J.W., Allen, C.C., Dutilh, B.E., McElarney, Y., Kulakov, L.A., 2018. Temporal dynamics of uncultured viruses: a new dimension in viral diversity. ISME J. 12, 199–211. https://doi.org/10.1038/ismej.2017.157.

Auguet, J.C., Montanié, H., Hartmann, H.J., Lebaron, P., Casamayor, E.O., Catala, P., Delmas, D., 2009. Potential effect of freshwater virus on the structure and activity of bacterial communities in the Marennes-Oléron Bay (France). Microb. Ecol. 57, 295–306. https://doi.org/10.1007/s00248-008-9428-1.

Auguet, J.-C., Nomokonova, N., Camarero, L., Casamayor, E.O., 2011. Seasonal changes of freshwater ammonia-oxidizing archaeal assemblages and nitrogen species in oligotrophic alpine lakes. Appl. Environ. Microbiol. 77, 1937–1945. https://doi.org/10.1128/AEM.01213-10.

Azam, F., Fenchel, T., Field, J.G., Gray, J.S., Meyerreil, L.A., Thingstad, F., 1983. The ecological role of water-column microbes in the sea. Mar. Ecol. Prog. Ser. 10, 257–263.

Bell, E.M., Laybourn-Perry, J., 1999. Annual plankton dynamics in an Antarctic saline lake. Freshw. Biol. 41, 507–519.

Berg, J.S., Michellod, D., Pjevac, P., Martinez-Perez, C., Buckner, C.R.T., Hach, P.F., Schubert, C.J., Milucka, J., Kuypers, M.M.M., 2016. Intensive cryptic microbial iron cycling in the low iron water column of the meromictic Lake Cadagno. Environ. Microbiol. 18, 5288–5302. https://doi.org/10.1111/1462-2920.13587.

Berggren, M., Laudon, H., Haei, M., Strom, L., Jansson, M., 2010. Efficient aquatic bacterial metabolism of dissolved low-molecular-weight compounds from terrestrial sources. ISME J. 4, 408–416. https://doi.org/10.1038/ismej.2009.120.

Berggren, M., Laudon, H., Jansson, M., 2007. Landscape regulation of bacterial growth efficiency in boreal freshwaters. Glob. Biogeochem. Cycles 21, n. https://doi.org/10.1029/2006GB002844.

Bergh, Ø., Børsheim, K.Y., Bratbak, G., Heldal, M., 1989. High abundance of viruses found in aquatic environments. Nature 340, 467–468. https://doi.org/10.1038/340467a0.

Berman, T., Bronk, D., 2003. Dissolved organic nitrogen: a dynamic participant in aquatic ecosystems. Aquat. Microb. Ecol. 31, 279–305. https://doi.org/10.3354/ame031279.

Bertilsson, S., Burgin, A., Carey, C.C., Fey, S.B., Grossart, H.P., Grubisic, L.M., Jones, I.D., Kirillin, G., Lennon, J.T., Shade, A., Smyth, R.L., 2013. The under-ice microbiome of seasonally frozen lakes. Limnol. Oceanogr. 58, 1998–2012. https://doi.org/10.4319/lo.2013.58.6.1998.

Bertilsson, S., Tranvik, L.J., 1998. Photochemically produced carboxylic acids as substrates for freshwater bacterioplankton. Limnol. Oceanogr. 43, 885–895.

Bertilsson, S., Tranvik, L.J., 2000. Photochemical transformation of dissolved organic matter in lakes. Limnol. Oceanogr. 45, 753–762. https://doi.org/10.4319/lo.2000.45.4.0753.

Bettarel, Y., Sime-Ngando, T., Bouvy, M., Arfi, R., Amblard, C., 2005. Low consumption of virus-sized particles by heterotrophic nanoflagellates in two lakes of the French Massif Central. Aquat. Microb. Ecol. 39, 205–209. https://doi.org/10.3354/ame039205.

Biddanda, B., Ogdahl, M., Cotner, J., 2001. Dominance of bacterial metabolism in oligotrophic relative to eutrophic waters. Limnol. Oceanogr. 46, 730–739.

Biderre-Petit, C., Boucher, D., Kuever, J., Alberic, P., Jézéquel, D., Chebance, B., Borrel, G., Fonty, G., Peyret, P., 2011. Identification of sulfur-cycle prokaryotes in a low-sulfate lake (Lake Pavin) using aprA and 16S rRNA gene markers. Microb. Ecol. 61, 313–327. https://doi.org/10.1007/s00248-010-9769-4.

Biebl, H., Pfennig, N., 1978. Growth yields of green sulfur bacteria in mixed cultures with sulfur and sulfate reducing bacteria. Arch. Microbiol. 117, 9–16.

Bjoernsen, P.K., 1986. Automatic determination of bacterioplankton biomass by image analysist. Appl. Environ. Microbiol. 51, 119–1204.

Blom, J.F., Pernthaler, J., 2010. Antibiotic effects of three strains of chrysophytes (*Ochromonas, Poterioochromonas*) on freshwater bacterial isolates. FEMS Microbiol. Ecol. 71, 281–290. https://doi.org/10.1111/j.1574-6941.2009.00800.x.

Boddicker, A.M., Mosier, A.C., 2018. Genomic profiling of four cultivated *candidatus* nitrotoga spp. predicts broad metabolic potential and environmental distribution. ISME J. 12, 2864–2882. https://doi.org/10.1038/s41396-018-0240-8.

Bohannan, B.J.M., Lenski, R.E., 2000. Linking genetic change to community evolution: insights from studies of bacteria and bacteriophage. Ecol. Lett. 3, 362–377. https://doi.org/10.1046/j.1461-0248.2000.00161.x.

Bowie, I.S., Gillespie, P.A., 1976. Microbial parameters and trophic status of ten New Zealand lakes. N. Z. J. Mar. Freshw. Res. 20, 343–354.

Bratbak, G., 1985. Bacterial biovolume and biomass estimations. Appl. Environ. Microbiol. 49, 1488–1493. https://doi.org/10.1128/aem.49.6.1488-1493.1985.

Breitbart, M., Rohwer, F., 2005. Here a virus, there a virus, everywhere the same virus? Trends Microbiol. 13, 278–284.

Brock, T.D., Brock, M.L., 1966. Autoradiography as a tool in microbial ecology. Nature 209, 734–736. https://doi.org/10.1038/209734a0.

Buck, U., Grossart, H.-P., Amann, R., Pernthaler, J., 2009. Substrate incorporation patterns of bacterioplankton populations in stratified and mixed waters of a humic lake. Environ. Microbiol. 11, 1854–1865. https://doi.org/10.1111/j.1462-2920.2009.01910.x.

Burgin, A.J., Hamilton, S.K., 2007. Have we overemphasized the role of denitrification in aquatic ecosystems? A review of nitrate removal pathways. Front. Ecol. Environ. 5, 89–96.

Cabello-Yeves, P.J., Zemskaya, T.I., Rosselli, R., Coutinho, F.H., Zakharenko, A.S., Blinov, V.V., Rodriguez-Valera, F., 2018. Genomes of novel microbial lineages assembled from the sub-ice waters of Lake Baikal. Appl. Environ. Microbiol. 84, e02132–17. https://doi.org/10.1128/AEM.02132-17.

Callieri, C., Coci, M., Eckert, E.M., Salcher, M.M., Bertoni, R., 2014. Archaea and bacteria in deep lake hypolimnion: in situ dark inorganic carbon uptake. J. Limnol. 73. https://doi.org/10.4081/jlimnol.2014.937.

Callieri, C., Hernández-Avilés, S., Salcher, M.M., Fontaneto, D., Bertoni, R., 2016. Distribution patterns and environmental correlates of Thaumarchaeota abundance in six deep subalpine lakes. Aquat. Sci. 78, 215–225. https://doi.org/10.1007/s00027-015-0418-3.

Camacho, A., Erez, J., Chicote, A., Florín, M., Squires, M.M., Lehmann, C., Backofen, R., 2001. Microbial microstratification, inorganic carbon photoassimilation and dark carbon fixation at the chemocline of the meromictic Lake Cadagno (Switzerland) and its relevance to the food web. Aquat. Sci. 63, 91–106. https://doi.org/10.1007/PL00001346.

Casamayor, E.O., García-Cantizano, J., Pedrós-Alió, C., 2008. Carbon dioxide fixation in the dark by photosynthetic bacteria in sulfide-rich stratified lakes with oxic-anoxic interfaces. Limnol. Oceanogr. 53, 1193–1203. https://doi.org/10.4319/lo.2008.53.4.1193.

Casamayor, E., Lliros, M., Picazo, A., Barberan, A., Borrego, C., Camacho, A., 2012. Contribution of deep dark fixation processes to overall CO_2 incorporation and large vertical changes of microbial populations in stratified karstic lakes. Aquat. Sci. 74, 61–75. https://doi.org/10.1007/s00027-011-0196-5.

Cavari, B., Grossowicz, N., 1977. Seasonal distribution of vitamin B12 in lake kinneret. Appl. Environ. Microbiol. 34, 120–124. https://doi.org/10.1128/aem.34.2.120-124.1977.

Cavari, B.Z., Phelps, G., Hadas, O., 1978. Glucose concentrations and heterotrophic activity in lake kinneret. Verh Intern. Ver. Limnol 20, 2249–2254.

Chen, L.-X., Méheust, R., Crits-Christoph, A., McMahon, K.D., Nelson, T.C., Slater, G.F., Warren, L.A., Banfield, J.F., 2020. Large freshwater phages with the potential to augment aerobic methane oxidation. Nat. Microbiol. 5, 1504–1515. https://doi.org/10.1038/s41564-020-0779-9.

Chin-Leo, G., Evans, C.T., 2007. Bacterial secondary productivity. In: Hurst, C.J., Crawford, R.L., Garland, J.L., Lipson, D.A., Mills, A.L., Stetzenbach, L.D. (Eds.), Manual of Environmental Microbiology, third ed. ASM Press.

Chróst, R., Gajewski, A., 1995. Microbial utilization of lipids in lake water. FEMS Microbiol. Ecol. 18, 45–50. https://doi.org/10.1016/0168-6496(95)00039-D.

Chrzanowski, T.H., Crotty, R.D., Hubbard, G.J., 1988. Seasonal variation in cell volume of epilimnetic bacteria. Microb. Ecol. 16, 155–163.

Chrzanowski, T., Kyle, M., Elser, J., Sterner, R., 1996. Element ratios and growth dynamics of bacteria in an oligotrophic Canadian shield lake. Aquat. Microb. Ecol. 11, 119–125. https://doi.org/10.3354/ame011119.

Chrzanowski, T.H., Simek, K., 1993. Bacterial-growth and losses due to bacterivory in a mesotrophic lake. J. Plankton Res. 15, 771–785.

Claverie, J.-M., Ogata, H., Audic, S., Abergel, C., Suhre, K., Fournier, P.-E., 2006. Mimivirus and the emerging concept of "giant" virus. Virus Res. 117, 133–144.

Coates, J.D., 1999. *Geothrix fermentans* gen. nov., sp. nov., a novel Fe(III)-reducing bacterium from a hydrocarbon-contaminated aquifer. Int. J. Syst. Bacteriol. 8.

Coates, J.D., Phillips, E.J., Lonergan, D.J., Jenter, H., Lovley, D.R., 1996. Isolation of Geobacter species from diverse sedimentary environments. Appl. Environ. Microbiol. 62, 1531–1536. https://doi.org/10.1128/aem.62.5.1531-1536.1996.

Cole, J., Findlay, S., Pace, M., 1988. Bacterial production in fresh and saltwater ecosystems: a cross-system overview. Mar. Ecol. Prog. Ser. 43, 1–10. https://doi.org/10.3354/meps043001.

Cole, J.J., Pace, M.L., 1995. Bacterial secondary production in oxic and anoxic freshwaters. Limnol. Oceanogr. 40, 1019–1027. https://doi.org/10.4319/lo.1995.40.6.1019.

Cory, R.M., Kling, G.W., 2018. Interactions between sunlight and microorganisms influence dissolved organic matter degradation along the aquatic continuum: interactions between sunlight and microorganisms. Limnol. Oceanogr. Lett. 3, 102–116. https://doi.org/10.1002/lol2.10060.

Cotner, J.B., Biddanda, B.A., 2002. Small players, large role: microbial influence on biogeochemical processes in pelagic aquatic ecosystems. Ecosystems 5, 105–121. https://doi.org/10.1007/S10021-001-0059-3.

Cotner, J., Hall, E., Scott, J., Heldal, M., 2010. Freshwater bacteria are stoichiometrically flexible with a nutrient composition similar to seston. Front. Microbiol. 1, 132. https://doi.org/10.3389/fmicb.2010.00132.

Cotner, J.B., Wetzel, R.G., 1992. Uptake of dissolved inorganic and organic phosphorus compounds by phytoplankton and bacterioplankton. Limnol. Oceanogr. 37, 232–243. https://doi.org/10.4319/lo.1992.37.2.0232.

Coutinho, F.H., Cabello-Yeves, P.J., Gonzalez-Serrano, R., Rosselli, R., López-Pérez, M., Zemskaya, T.I., Zakharenko, A.S., Ivanov, V.G., Rodriguez-Valera, F., 2020. New viral biogeochemical roles revealed through metagenomic analysis of Lake Baikal. Microbiome 8, 163. https://doi.org/10.1186/s40168-020-00936-4.

Coveney, M.E., Wetzel, R.G., 1992. Effects of Nutrients on specific growth-rate of bacterioplankton in oligotrophic lake water cultures. Appl. Environ. Microbiol. 58, 150–156. https://doi.org/10.1128/AEM.58.1.150-156.1992.

Coveney, M.F., Wetzel, R.G., 1995. Biomass, production, and specific growth rate of bacterioplankton and coupling to phytoplankton in an oligotrophic lake. Limnol. Oceanogr. 40, 1187–1200. https://doi.org/10.4319/lo.1995.40.7.1187.

Crisman, T.L., Scheuerman, P., Bienert, R.W., Beaver, J.R., Bays, J.S., 1984. A preliminary characterization of bacterioplankton seasonality in subtropical Florida lakes. Internatioinale Ver. Für Theor. Angew. Limnol. Verhandlungen 22, 620–626.

Crowe, S.A., Treusch, A.H., Forth, M., Li, J., Magen, C., Canfield, D.E., Thamdrup, B., Katsev, S., 2017. Novel anammox bacteria and nitrogen loss from Lake Superior. Sci. Rep. 7, 13757. https://doi.org/10.1038/s41598-017-12270-1.

Crump, B.C., Armbrust, E.V., Baross, J.A., 1999. Phylogenetic analysis of particle-attached and free-living bacterial communities in the Columbia River, its estuary, and the adjacent coastal ocean. Appl. Environ. Microbiol. 65, 3192–3204. https://doi.org/10.1128/AEM.65.7.3192-3204.1999.

Currie, D.J., Kalff, J., 1984. A comparison of the abilities of freshwater algae and bacteria to acquire and retain phosphorus1: algal-bacterial P kinetics. Limnol. Oceanogr. 29, 298–310. https://doi.org/10.4319/lo.1984.29.2.0298.

Daims, H., Lebedeva, E.V., Pjevac, P., Han, P., Herbold, C., Albertsen, M., Jehmlich, N., Palatinszky, M., Vierheilig, J., Bulaev, A., Kirkegaard, R.H., von Bergen, M., Rattei, T., Bendinger, B., Nielsen, P.H., Wagner, M., 2015. Complete nitrification by Nitrospira bacteria. Nature 528, 504–509. https://doi.org/10.1038/nature16461.

Dalsgaard, T., Stewart, F.J., Thamdrup, B., De Brabandere, L., Revsbech, N.P., Ulloa, O., Canfield, D.E., DeLong, E.F., 2014. Oxygen at nanomolar levels reversibly suppresses process rates and gene expression in anammox and denitrification in the oxygen minimum zone off Northern Chile. mBio 5. https://doi.org/10.1128/mBio.01966-14.

Davis, J.A., Gloor, R., 1981. Adsorption of dissolved organics in lake water by aluminum oxide. Effect of molecular weight. Environ. Sci. Technol. 15, 1223–1229. https://doi.org/10.1021/es00092a012.

Davison, W., Woof, C., Rigg, E., 1982. The dynamics of iron and manganese in a seasonally anoxic lake; direct measurement of fluxes using sediment traps: iron and manganese in lakes. Limnol. Oceanogr. 27, 987–1003. https://doi.org/10.4319/lo.1982.27.6.0987.

DeLong, E.F., 1992. Archaea in coastal marine environments. Proc. Natl. Acad. Sci. 89, 5685–5689. https://doi.org/10.1073/pnas.89.12.5685.

Denef, V.J., Mueller, R.S., Chiang, E., Liebig, J.R., Vanderploeg, H.A., 2016. Chloroflexi CL500-11 populations that predominate deep-lake hypolimnion bacterioplankton rely on nitrogen-rich dissolved organic matter metabolism and C_1 compound oxidation. Appl. Environ. Microbiol. 82, 1423–1432. https://doi.org/10.1128/AEM.03014-15.

Di Nezio, F., Beney, C., Roman, S., Danza, F., Buetti-Dinh, A., Tonolla, M., Storelli, N., 2021. Anoxygenic photo- and chemosynthesis of phototrophic sulfur bacteria from an alpine meromictic lake. FEMS Microbiol. Ecol. 97. https://doi.org/10.1093/femsec/fiab010.

Diao, M., Huisman, J., Muyzer, G., 2018. Spatio-temporal dynamics of sulfur bacteria during oxic–anoxic regime shifts in a seasonally stratified lake. FEMS Microbiol. Ecol. 94. https://doi.org/10.1093/femsec/fiy040.

Djikeng, A., Kuzmickas, R., Anderson, N.G., Spiro, D.J., 2009. Metagenomic analysis of RNA viruses in a fresh water lake. PLoS One 4, e7264. https://doi.org/10.1371/journal.pone.0007264.

Doherty, M, Yager, P.L., Moran, M.A., Coles, V.J., Fortunato, C.S., Krusche, A.V., Medeiros, P.M., Payet, J., Richey, J.E., Satinsky, B.M., Sawakuchi, H.O., Ward, N.D., Crump, B.C., 2017. Bacterial Biogeography across the Amazon River-Ocean Continuum. Front Microbiol. 8 (882). https://doi.org/10.3389/fmicb.2017.00882.

Downs, T.M., Schallenberg, M., Burns, C.W., 2008. Responses of lake phytoplankton to micronutrient enrichment: a study in two New Zealand lakes and an analysis of published data. Aquat. Sci. 70, 347–360. https://doi.org/10.1007/s00027-008-8065-6.

Dwulit-Smith, J.R., Hamilton, J.J., Stevenson, D.M., He, S., Oyserman, B.O., Moya-Flores, F., Garcia, S.L., Amador-Noguez, D., McMahon, K.D., Forest, K.T., 2018. acI actinobacteria assemble a functional actinorhodopsin with natively synthesized retinal. Appl. Environ. Microbiol. 84. https://doi.org/10.1128/AEM.01678-18.

Eccleston-Parry, J.D., Leadbeater, B., 1995. Regeneration of phosphorus and nitrogen by four species of heterotrophic nanoflagellates feeding on three nutritional States of a single bacterial strain. Appl. Environ. Microbiol. 61, 1033–1038. https://doi.org/10.1128/aem.61.3.1033-1038.1995.

Eckert, E.M., Baumgartner, M., Huber, I.M., Pernthaler, J., 2013. Grazing resistant freshwater bacteria profit from chitin and cell-wall-derived organic carbon: grazing resistant pelagic scavengers. Environ. Microbiol. 15, 2019–2030. https://doi.org/10.1111/1462-2920.12083.

Edwards, R.T., 1987. Sestonic bacteria as a food source for filtering invertebrates in two southeastern blackwater rivers1: seston in blackwater rivers. Limnol. Oceanogr. 32, 221–234. https://doi.org/10.4319/lo.1987.32.1.0221.

Eiler, A., Bertilsson, S., 2004. Composition of freshwater bacterial communities associated with cyanobacterial blooms in four Swedish lakes. Environ. Microbiol. 6, 1228–1243.

Eiler, A., Heinrich, F., Bertilsson, S., 2012. Coherent dynamics and association networks among lake bacterioplankton taxa. ISME J. 6, 330–342. https://doi.org/10.1038/ismej.2011.113.

Eisenmann, E., Beuerle, J., Sulger, K., Kroneck, P.M.H., Schumacher, W., 1995. Lithotrophic growth of *Sulfurospirillum deleyianum* with sulfide as electron donor coupled to respiratory reduction of nitrate to ammonia. Arch. Microbiol. 164, 180–185.

Ezzedine, J., Jacas, L., Desdevises, Y., Jacquet, S., 2020. Bdellovibrio and like organisms in lake geneva: an unseen elephant in the room? Front. Microbiol. 11. https://doi.org/10.3389/fmicb.2020.00098.

Facey, J.A., Apte, S.C., Mitrovic, S.M., 2019. A review of the effect of trace metals on freshwater cyanobacterial growth and toxin production. Toxins 11, 643. https://doi.org/10.3390/toxins11110643.

Fagerbakke, K., Heldal, M., Norland, S., 1996. Content of carbon, nitrogen, oxygen, sulfur and phosphorus in native aquatic and cultured bacteria. Aquat. Microb. Ecol. 10, 15–27. https://doi.org/10.3354/ame010015.

Faou, A., Rajagopal, B.S., Daniels, L., Fauque, G., 1990. Thiosulfate, polythionates and elemental sulfur assimilation and reduction in the bacterial world. FEMS Microbiol. Lett. 75, 351–381. https://doi.org/10.1111/j.1574-6968.1990.tb04107.x.

Finneran, K.T., Johnsen, C.V., Lovley, D.R., 2003. *Rhodoferax ferrireducens* sp. nov., a psychrotolerant, facultatively anaerobic bacterium that oxidizes acetate with the reduction of Fe(III). Int. J. Syst. Evol. Microbiol. 53, 669–673.

Fortunato, C.S., Herfort, L., Zuber, P., Baptista, A.M., Crump, B.C., 2012. Spatial variability overwhelms seasonal patterns in bacterioplankton communities across a river to ocean gradient. ISME J. 6, 554–563. https://doi.org/10.1038/ismej.2011.135.

Fournier, I.B., Lovejoy, C., Vincent, W.F., 2021. Changes in the community structure of under-ice and open-water microbiomes in urban lakes exposed to road salts. Front. Microbiol. 12, 660719. https://doi.org/10.3389/fmicb.2021.660719.

Frimmel, F.H., Christman, R.F. (Eds.), 1988. Humic Substances and Their Role in the Environment. John Wiley & Sons, Chichester.

Fuhrman, J.A., Caron, D.A., 2016. Heterotrophic Planktonic Microbes: Virus, Bacteria, Archaea, and Protozoa. Manual of Environmental Microbiology, fourth ed. ASM Press, Washington DC, pp. 4.2.2-1–4.2.2-34.

Fuhrman, J.A., McCallum, K., Davis, A., 1992. Novel major archaebacterial group from marine plankton. Nature 356, 148–149. https://doi.org/10.1038/356148a0.

Fukuda, R., Ogawa, H., Nagata, T., Koike, I., 1998. Direct determination of carbon and nitrogen contents of natural bacterial assemblages in marine environments. Appl. Environ. Microbiol. 64, 3352–3358. https://doi.org/10.1128/AEM.64.9.3352-3358.1998.

Gaby, J.C., Buckley, D.H., 2012. A Comprehensive evaluation of PCR primers to amplify the nifH gene of nitrogenase. PLoS One 7, e42149. https://doi.org/10.1371/journal.pone.0042149.

Garcia, S.L., McMahon, K.D., Martinez-Garcia, M., Srivastava, A., Sczyrba, A., Stepanauskas, R., Grossart, H.P., Woyke, T., Warnecke, F., 2013. Metabolic potential of a single cell belonging to one of the most abundant lineages in freshwater bacterioplankton. ISME J. 7, 137–147. https://doi.org/10.1038/ismej.2012.86.

Garcia, S.L., Stevens, S.L.R., Crary, B., Martinez-Garcia, M., Stepanauskas, R., Woyke, T., Tringe, S.G., Andersson, S.G.E., Bertilsson, S., Malmstrom, R.R., McMahon, K.D., 2018. Contrasting patterns of genome-level diversity across distinct co-occurring bacterial populations. ISME J. 12, 742–755. https://doi.org/10.1038/s41396-017-0001-0.

Gautam, A., Lear, G., Lewis, G.D., 2021. Analysis of spatial and temporal variations in bacterial community dynamics within stream biofilms. N. Z. J. Mar. Freshw. Res. 55, 505–523. https://doi.org/10.1080/00288330.2020.1804409.

Geller, A., 1986. Comparison of mechanisms enhancing biodegradability of refractory lake water constituents1: degradation of refractory DOM. Limnol. Oceanogr. 31, 755–764. https://doi.org/10.4319/lo.1986.31.4.0755.

Ger, K.A., Hansson, L.-A., Lürling, M., 2014. Understanding cyanobacteria-zooplankton interactions in a more eutrophic world. Freshw. Biol. 59, 1783–1798. https://doi.org/10.1111/fwb.12393.

Ghai, R., Mehrshad, M., Mizuno, C.M., Rodriguez-Valera, F., 2017. Metagenomic recovery of phage genomes of uncultured freshwater actinobacteria. ISME J. 11, 304–308. https://doi.org/10.1038/ismej.2016.110.

Ghylin, T.W., Garcia, S.L., Moya, F., Oyserman, B.O., Schwientek, P., Forest, K.T., Mutschler, J., Dwulit-Smith, J., Chan, L.-K., Martinez-Garcia, M., Sczyrba, A., Stepanauskas, R., Grossart, H.-P., Woyke, T., Warnecke, F., Malmstrom, R., Bertilsson, S., McMahon, K.D., 2014. Comparative single-cell genomics reveals potential ecological niches for the freshwater acI Actinobacteria lineage. ISME J. 8, 2503–2516. https://doi.org/10.1038/ismej.2014.135.

Gillespie, P.A., Spencer, M.J., 1980. Seasonal variation of heterotrophic potential in Lake Rotoroa. N. Z. J. Mar. Freshw. Res. 14, 15–21.

Glass, J.B., Axler, R.P., Chandra, S., Goldman, C.R., 2012. Molybdenum limitation of microbial nitrogen assimilation in aquatic ecosystems and pure cultures. Front. Microbiol. 3. https://doi.org/10.3389/fmicb.2012.00331.

Godwin, C.M., Cotner, J.B., 2014. Carbon:phosphorus homeostasis of aquatic bacterial assemblages is mediated by shifts in assemblage composition. Aquat. Microb. Ecol. 73, 245–258. https://doi.org/10.3354/ame01719.

Godwin, C.M., Cotner, J.B., 2015. Aquatic heterotrophic bacteria have highly flexible phosphorus content and biomass stoichiometry. ISME J. 9, 2324–2327. https://doi.org/10.1038/ismej.2015.34.

Godwin, C.M., Whitaker, E.A., Cotner, J.B., 2017. Growth rate and resource imbalance interactively control biomass stoichiometry and elemental quotas of aquatic bacteria. Ecology 98, 820–829. https://doi.org/10.1002/ecy.1705.

Goldberg, S.J., Ball, G.I., Allen, B.C., Schladow, S.G., Simpson, A.J., Masoom, H., Soong, R., Graven, H.D., Aluwihare, L.I., 2015. Refractory dissolved organic nitrogen accumulation in high-elevation lakes. Nat. Commun. 6, 6347. https://doi.org/10.1038/ncomms7347.

Gonzalez, J.M., Sherr, E.B., Sherr, B.F., 1990. Size-selective grazing on bacteria by natural assemblages of estuarine flagellates and ciliates. Appl. Environ. Microbiol. 56, 583–589. https://doi.org/10.1128/aem.56.3.583-589.1990.

Gonzalez, J.M., Suttle, C.A., 1993. Grazing by marine nanoflagellates on viruses and virus-sized particles: ingestion and digestion. Mar. Ecol. Prog. Ser. 94, 1–10.

Graf, D.R.H., Jones, C.M., Hallin, S., 2014. Intergenomic comparisons highlight modularity of the denitrification pathway and underpin the importance of community structure for N_2O emissions. PLoS One 9, e114118. https://doi.org/10.1371/journal.pone.0114118.

Green, J., Rahman, F., Saxton, M., Williamson, K., 2015. Metagenomic assessment of viral diversity in Lake Matoaka, a temperate, eutrophic freshwater lake in southeastern Virginia, USA. Aquat. Microb. Ecol. 75, 117–128. https://doi.org/10.3354/ame01752.

Gromov, B., Mamkaeva, K., 1972. Electron microscopic study of parasitism by Bdellovibrio chlorellavorus bacteria on cells of the green alga. Tsitologiia 14, 256–260.

Grossart, H.-P., 2010. Ecological consequences of bacterioplankton lifestyles: changes in concepts are needed: ecological consequences of bacterioplankton lifestyles. Environ. Microbiol. Rep. 2, 706–714. https://doi.org/10.1111/j.1758-2229.2010.00179.x.

Grossart, H.-P., Massana, R., McMahon, K.D., Walsh, D.A., 2020. Linking metagenomics to aquatic microbial ecology and biogeochemical cycles. Limnology and Oceanography S2–S20. https://doi.org/10.1002/lno.11382.

Groth, von P., 1971. Investigations of some trace elements in lakes. Arch. Hydrobiol. 68, 305.

Grover, J.P., Chrzanowski, T.H., 2009. Dynamics and nutritional ecology of a nanoflagellate preying upon bacteria. Microb. Ecol. 58, 231–243. https://doi.org/10.1007/s00248-009-9486-z.

Güde, H., 1986. Loss processes influencing growth of planktonic bacterial populations in Lake Constance. J. Plankton Res. 8, 795–810. https://doi.org/10.1093/plankt/8.4.795.

Güde, H., Haibel, B., Muller, H., 1985. Development of planktonic bacterial populations in a water column of Lake Constance (Bodensee-Öbersee). Arch. Hydrobiol. 105, 59–77.

Gumbo, R.J., Ross, G., Cloete, E.T., 2008. Biological Control of Microcystis Dominated Harmful Algal Blooms Sooas, vol. 9.

Hahn, M.W., 2003. Isolation of strains belonging to the cosmopolitan *Polynucleobacter necessarius* cluster from freshwater habitats located in three climatic zones. Appl. Environ. Microbiol. 69, 5248–5254.

Hahn, M.W., Jezberová, J., Koll, U., Saueressig-Beck, T., Schmidt, J., 2016. Complete ecological isolation and cryptic diversity in Polynucleobacter bacteria not resolved by 16S rRNA gene sequences. ISME J. 10, 1642–1655. https://doi.org/10.1038/ismej.2015.237.

Hahn, M.W., Kasalický, V., Jezbera, J., Brandt, U., Jezberová, J., Šimek, K., 2010. *Limnohabitans curvus* gen. nov., sp. nov., a planktonic bacterium isolated from a freshwater lake. Int. J. Syst. Evol. Microbiol. 60, 1358–1365. https://doi.org/10.1099/ijs.0.013292-0.

Hahn, M.W., Lünsdorf, H., Wu, Q., Schauer, M., Höfle, M.G., Boenigk, J., Stadler, P., 2003. Isolation of novel ultramicrobacteria classified as actinobacteria from five freshwater habitats in Europe and Asia. Appl. Environ. Microbiol. 69, 1442–1451. https://doi.org/10.1128/AEM.69.3.1442-1451.2003.

Hahn, M.W., Scheuerl, T., Jezberova, J., Koll, U., Jezbera, J., Vannini, C., Petroni, G., Wu, Q.L., 2012. The passive yet successful way of planktonic life: genomic and experimental analysis of the ecology of a free-living polynucleobacter population. PLoS One 7, 17.

Hamilton, T.L., Bovee, R.J., Thiel, V., Sattin, S.R., Mohr, W., Schaperdoth, I., Vogl, K., Gilhooly, W.P., Lyons, T.W., Tomsho, L.P., Schuster, S.C., Overmann, J., Bryant, D.A., Pearson, A., Macalady, J.L., 2014. Coupled reductive and oxidative sulfur cycling in the phototrophic plate of a meromictic lake. Geobiology 12, 451–468. https://doi.org/10.1111/gbi.12092.

Hamilton, J.J., Garcia, S.L., Brown, B.S., Oyserman, B.O., Moya-Flores, F., Bertilsson, S., Malmstrom, R.R., Forest, K.T., McMahon, K.D., 2017. Metabolic network analysis and metatranscriptomics reveal auxotrophies and nutrient sources of the cosmopolitan freshwater microbial lineage acI. mSystems 2. https://doi.org/10.1128/mSystems.00091-17.

Hampton, S.E., Galloway, A.W.E., Powers, S.M., Ozersky, T., Woo, K.H., Batt, R.D., Labou, S.G., O'Reilly, C.M., Sharma, S., Lottig, N.R., Stanley, E.H., North, R.L., Stockwell, J.D., Adrian, R., Weyhenmeyer, G.A., Arvola, L., Baulch, H.M., Bertani, I., Bowman, L.L., Carey, C.C., Catalan, J., Colom-Montero, W., Domine, L.M., Felip, M., Granados, I., Gries, C., Grossart, H., Haberman, J., Haldna, M., Hayden, B., Higgins, S.N., Jolley, J.C., Kahilainen, K.K., Kaup, E., Kehoe, M.J., MacIntyre, S., Mackay, A.W., Mariash, H.L., McKay, R.M., Nixdorf, B., Nõges, P., Nõges, T., Palmer, M., Pierson, D.C., Post, D.M., Pruett, M.J., Rautio, M., Read, J.S., Roberts, S.L., Rücker, J., Sadro, S., Silow, E.A., Smith, D.E., Sterner, R.W., Swann, G.E.A., Timofeyev, M.A., Toro, M., Twiss, M.R., Vogt, R.J., Watson, S.B., Whiteford, E.J., Xenopoulos, M.A., 2017. Ecology under lake ice. Ecol. Lett. 20, 98–111. https://doi.org/10.1111/ele.12699.

Hampton, H.G., Watson, B.N.J., Fineran, P.C., 2020. The arms race between bacteria and their phage foes. Nature 577, 327–336. https://doi.org/10.1038/s41586-019-1894-8.

Haraldsson, M., Gerphagnon, M., Bazin, P., Colombet, J., Tecchio, S., Sime-Ngando, T., Niquil, N., 2018. Microbial parasites make cyanobacteria blooms less of a trophic dead end than commonly assumed. ISME J. 12, 1008–1020. https://doi.org/10.1038/s41396-018-0045-9.

Hatzenpichler, R., 2012. Diversity, physiology, and niche differentiation of ammonia-oxidizing Archaea. Appl. Environ. Microbiol. 78, 7501–7510. https://doi.org/10.1128/AEM.01960-12.

Hau, H.H., Gralnick, J.A., 2007. Ecology and biotechnology of the genus *Shewanella*. Annu. Rev. Microbiol. 61, 237–258. https://doi.org/10.1146/annurev.micro.61.080706.093257.

Hayden, C.J., Beman, J.M., 2014. High abundances of potentially active ammonia-oxidizing bacteria and Archaea in Oligotrophic, high-altitude lakes of the Sierra Nevada, California, USA. PLoS One 9, e111560. https://doi.org/10.1371/journal.pone.0111560.

He, S., Stevens, S.L.R., Chan, L.K., Bertilsson, S., Glavina Del Rio, T., Tringe, S.G., Malmstrom, R.R., McMahon, K.D., 2017. Ecophysiology of freshwater Verrucomicrobia inferred from metagenome-assembled genomes. mSphere 2, e00277. https://doi.org/10.1128/mSphere.00277-17, 17.

Heising, S., Richter, L., Ludwig, W., Schink, B., 1999. Chlorobium ferrooxidans sp. nov., a phototrophic green sulfur bacterium that oxidizes ferrous iron in coculture with a "Geospirillum" sp. strain. Arch. Microbiol. 172, 116–124. https://doi.org/10.1007/s002030050748.

Heldal, M., Bratbak, G., 1991. Production and decay of viruses in aquatic environments. Mar. Ecol. Prog. Ser. 72, 205–212. https://doi.org/10.3354/meps072205.

Hennes, K.P., Simon, M., 1995. Significance of bacteriophages for controlling bacterioplankton growth in a mesotrophic lake. Appl. Environ. Microbiol. 61, 333–340. https://doi.org/10.1128/aem.61.1.333-340.1995.

Henson, M.W., Hanssen, J., Spooner, G., Fleming, P., Pukonen, M., Stahr, F., Thrash, J.C., 2018a. Nutrient dynamics and stream order influence microbial community patterns along a 2914 kilometer transect of the Mississippi River. Limnol. Oceanogr. 63, 1837–1855. https://doi.org/10.1002/lno.10811.

Henson, M.W., Lanclos, V.C., Faircloth, B.C., Thrash, J.C., 2018b. Cultivation and genomics of the first freshwater SAR11 (LD12) isolate. ISME J. 12, 1846–1860. https://doi.org/10.1038/s41396-018-0092-2.

Herber, J., Klotz, F., Frommeyer, B., Weis, S., Straile, D., Kolar, A., Sikorski, J., Egert, M., Dannenmann, M., Pester, M., 2020. A single *Thaumarchaeon* drives nitrification in deep oligotrophic Lake Constance. Environ. Microbiol. 22, 212–228. https://doi.org/10.1111/1462-2920.14840.

Hobbie, J.E., 1967. Glucose and acetate in freshwater: concentrations and turnover rates. In: Golterman, H.L., Clymo, R.S. (Eds.), Chemical Environment in the Aquatic Habitat. N. V. Noord-Hollandsche Uitgevers Mattschappij, Amsterdam, pp. 245–251.

Hobbie, J.E., 1971. Heterotrophic bacteria in aquatic ecosystems; some results of studies with organic radioisotopes. In: The Structure and Function of Fresh-Water Microbial Communities. Res. Div. Monogr., vol. 3. Polytechnic Institute, Virginia, pp. 181–194.

Hobbie, J.E., Bahr, M., Rublee, P.A., 1999. Controls on microbial food webs in oligotrophic arctic lakes. Arch. Hydrobiol. 54, 61–76.

Hobbie, J.E., Corliss, T.L., Peterson, B.J., 1983. Seasonal patterns of bacterial abundance in an Arctic Lake. Arct. Alp. Res. 15, 253. https://doi.org/10.2307/1550926.

Hobbie, J.E., Crawford, C.C., 1969. Bacterial uptake of organic substrate: new methods of study and application to eutrophication. Verh Intern. Ver. Limnol 17, 725–730.

Hoetzinger, M., Schmidt, J., Jezberová, J., Koll, U., Hahn, M.W., 2017. Microdiversification of a pelagic polynucleobacter species is mainly driven by acquisition of genomic islands from a partially interspecific gene pool. Appl. Environ. Microbiol. 83, e02266. https://doi.org/10.1128/AEM.02266-16, 16.

Holmfeldt, K., Middelboe, M., Nybroe, O., Riemann, L., 2007. Large variabilities in host strain susceptibility and phage host range govern interactions between lytic marine phages and their *Flavobacterium* hosts. Appl. Environ. Microbiol. 73, 6730–6739. https://doi.org/10.1128/AEM.01399-07.

Hu, B., Shen, L., Zheng, P., Hu, A., Chen, T., Cai, C., Liu, S., Lou, L., 2012. Distribution and diversity of anaerobic ammonium-oxidizing bacteria in the sediments of the qiantang river: anammox in qiantang river sediment. Environ. Microbiol. Rep. 4, 540–547. https://doi.org/10.1111/j.1758-2229.2012.00360.x.

Huang, J., Su, Z., Xu, Y., 2005. The evolution of microbial phosphonate degradative pathways. J. Mol. Evol. 61, 682–690. https://doi.org/10.1007/s00239-004-0349-4.

Hugoni, M., Etien, S., Bourges, A., Lepère, C., Domaizon, I., Mallet, C., Bronner, G., Debroas, D., Mary, I., 2013. Dynamics of ammonia-oxidizing Archaea and Bacteria in contrasted freshwater ecosystems. Res. Microbiol. 164, 360–370. https://doi.org/10.1016/j.resmic.2013.01.004.

Hungate, B.A., Stiling, P.D., Dijkstra, P., Johnson, D.W., Ketterer, M.E., Hymus, G.J., Hinkle, C.R., Drake, B.G., 2004. CO_2 elicits long-term decline in nitrogen fixation. Science 304, 1291. https://doi.org/10.1126/science.1095549, 1291.

Hupfer, M., Gloess, S., Grossart, H., 2007. Polyphosphate-accumulating microorganisms in aquatic sediments. Aquat. Microb. Ecol. 47, 299–311. https://doi.org/10.3354/ame047299.

Hwang, S., 1999. Zooplankton bacterivory at coastal and offshore sites of Lake Erie. J. Plankton Res. 21, 699–719. https://doi.org/10.1093/plankt/21.4.699.

Hyenstrand, P., 2000. Response of pelagic cyanobacteria to iron additions—enclosure experiments from Lake Erken. J. Plankton Res. 22, 1113–1126. https://doi.org/10.1093/plankt/22.6.1113.

Ilikchyan, I.N., McKay, R.M.L., Zehr, J.P., Dyhrman, S.T., Bullerjahn, G.S., 2009. Detection and expression of the phosphonate transporter gene phnD in marine and freshwater picocyanobacteria. Environ. Microbiol. 11, 1314–1324. https://doi.org/10.1111/j.1462-2920.2009.01869.x.

Jacquet, S., Domaizon, I., Personnic, S., Ram, A.S.P., Hedal, M., Duhamel, S., Sime-Ngando, T., 2005. Estimates of protozoan- and viral-mediated mortality of bacterioplankton in lake bourget (france). Freshw. Biol. 50, 627–645. https://doi.org/10.1111/j.1365-2427.2005.01349.x.

Jiao, Y., Kappler, A., Croal, L.R., Newman, D.K., 2005. Isolation and characterization of a genetically tractable photoautotrophic Fe(II)-oxidizing bacterium, *rhodopseudomonas palustris* strain TIE-1. Appl. Environ. Microbiol. 71, 4487–4496. https://doi.org/10.1128/AEM.71.8.4487-4496.2005.

Johansson, J.-Å., 1983. Seasonal development of bacterioplankton in two forest lakes in central Sweden. Hydrobiologia 101, 71–83.

Jones, J.G., 1978. The distribution of some freshwater planktonic bacteria in two stratified eutrophic lakes. Freshw. Biol. 8, 127–140.

Jones, S.E., Cadkin, T.A., Newton, R.J., McMahon, K.D., 2012. Spatial and temporal scales of aquatic bacterial beta diversity. Front. Microbiol. 3. https://doi.org/10.3389/fmicb.2012.00318.

Jones, S.E., Newton, R.J., McMahon, K.D., 2009. Evidence for structuring of bacterial community composition by organic carbon source in temperate lakes. Environ. Microbiol. 11, 2463–2472. https://doi.org/10.1111/j.1462-2920.2009.01977.x.

Jousset, A., 2012. Ecological and evolutive implications of bacterial defences against predators. Environ. Microbiol. 14, 1830–1843. https://doi.org/10.1111/j.1462-2920.2011.02627.x.

Jugnia, L.-B., Sime-Ngando, T., Gilbert, D., 2006. Dynamics and estimates of growth and loss rates of bacterioplankton in a temperate freshwater system: growth and fate of freshwater bacterioplankton. FEMS Microbiol. Ecol. 58, 23–32. https://doi.org/10.1111/j.1574-6941.2006.00145.x.

Jürgens, K., Pernthaler, J., Schalla, S., Amann, R., 1999. Morphological and compositional changes in a planktonic bacterial community in response to enhanced protozoan grazing. Appl. Environ. Microbiol. 65, 1241–1250. https://doi.org/10.1128/AEM.65.3.1241-1250.1999.

Jürgens, K., Stolpe, G., 1995. Seasonal dynamics of crustacean zooplankton, heterotrophic nanoflagellates and bacteria in a shallow, eutrophic lake. Freshw. Biol. 33, 27–38.

Kang, I., Kim, S., Islam, Md.R., Cho, J.-C., 2017. The first complete genome sequences of the acI lineage, the most abundant freshwater Actinobacteria, obtained by whole-genome-amplification of dilution-to-extinction cultures. Sci. Rep. 7, 42252. https://doi.org/10.1038/srep42252.

Kaplan, L.A., Cory, R.M., 2016. Dissolved organic matter in stream ecosystems. In: Stream Ecosystems in a Changing Environment. Elsevier, pp. 241–320. https://doi.org/10.1016/B978-0-12-405890-3.00006-3.

Kapustina, L.L., 1996. Bacterioplankton response to eutrophication in Lake ladoga. Hydrobiologia 322, 17–22.

Karl, D.M., 2000. Phosphorus, the staff of life. Nature 406, 31–32.

Kartal, B., van Niftrik, L., Keltjens, J.T., Op den Camp, H.J.M., Jetten, M.S.M., 2012. Chapter 3—anammox—growth physiology, cell biology, and metabolism. Advances in Microbial Physiology. Academic Press.

Kasalický, V., Jezbera, J., Hahn, M.W., Šimek, K., 2013. The diversity of the limnohabitans genus, an important group of freshwater bacterioplankton, by characterization of 35 isolated strains. PLoS One 8, e58209. https://doi.org/10.1371/journal.pone.0058209.

Kasalický, V., Zeng, Y., P–iwosz, K., Šimek, K., Kratochvilová, H., Koblížek, M., 2018. Aerobic anoxygenic photosynthesis is commonly present within the genus Limnohabitans. Appl. Environ. Microbiol. 84, e02116–e02117. https://doi.org/10.1128/AEM.02116-17.

Kato, K., Sakamoto, M., 1983. The function of the free-living bacterial fraction in the organic-matter metabolism of a mesotrophic lake. Arch. Hydrobiol. 97, 289–302.

Kavagutti, V.S., Andrei, A.-S., Mehrshad, M., Salcher, M.M., Ghai, R., 2019. Phage-centric ecological interactions in aquatic ecosystems revealed through ultra-deep metagenomics. Microbiome 7, 135. https://doi.org/10.1186/s40168-019-0752-0.

Kehres, D.G., Maguire, M.E., 2003. Emerging themes in manganese transport, biochemistry and pathogenesis in bacteria. FEMS Microbiol. Rev. 27, 263–290. https://doi.org/10.1016/S0168-6445(03)00052-4.

Kent, A.D., Yannarell, A.C., Rusak, J.A., Triplett, E.W., McMahon, K.D., 2007. Synchrony in aquatic microbial community dynamics. ISME J. 1, 38–47.

Kenzaka, T., Tani, K., Nasu, M., 2010. High-frequency phage-mediated gene transfer in freshwater environments determined at single-cell level. ISME J. 4, 648–659. https://doi.org/10.1038/ismej.2009.145.

Kieft, K., Zhou, Z., Anderson, R.E., Buchan, A., Campbell, B.J., Hallam, S.J., Hess, M., Sullivan, M.B., Walsh, D.A., Roux, S., Anantharaman, K., 2021. Ecology of inorganic sulfur auxiliary metabolism in widespread bacteriophages. Nat. Commun. 12, 3503. https://doi.org/10.1038/s41467-021-23698-5.

Kirchman, D., 2018. Processes in Microbial Ecology. Oxford University Press.

Kirchman, D., K'Nees, E., Hodson, R., 1985. Leucine incorporation and its potential as a measure of protein synthesis by bacteria in natural aquatic systems. Appl. Environ. Microbiol. 49, 9.

Klawonn, I., Van den Wyngaert, S., Parada, A., Arandia-Gorostidi, N., Whitehouse, M., Grossart, H., Dekas, A., 2021. Characterizing the "Fungal Shunt": parasitic fungi on diatoms affect carbon flow and bacterial communities in aquatic microbial food webs. Proc. Natl. Acad. Sci. U. S. A. 118. https://doi.org/10.1073/pnas.2102225118.

Knowles, B., Silveira, C.B., Bailey, B.A., Barott, K., Cantu, V.A., Cobián-Güemes, A.G., Coutinho, F.H., Dinsdale, E.A., Felts, B., Furby, K.A., George, E.E., Green, K.T., Gregoracci, G.B., Haas, A.F., Haggerty, J.M., Hester, E.R., Hisakawa, N., Kelly, L.W., Lim, Y.W., Little, M., Luque, A., McDole-Somera, T., McNair, K., de Oliveira, L.S., Quistad, S.D., Robinett, N.L., Sala, E., Salamon, P., Sanchez, S.E., Sandin, S., Silva, G.G.Z., Smith, J., Sullivan, C., Thompson, C., Vermeij, M.J.A., Youle, M., Young, C., Zgliczynski, B., Brainard, R., Edwards, R.A., Nulton, J., Thompson, F., Rohwer, F., 2016. Lytic to temperate switching of viral communities. Nature 531, 466–470. https://doi.org/10.1038/nature17193.

Koch, H., Lücker, S., Albertsen, M., Kitzinger, K., Herbold, C., Spieck, E., Nielsen, P.H., Wagner, M., Daims, H., 2015. Expanded metabolic versatility of ubiquitous nitrite-oxidizing bacteria from the genus *Nitrospira*. Proc. Natl. Acad. Sci. 112, 11371–11376. https://doi.org/10.1073/pnas.1506533112.

Kojima, H., Watanabe, T., Iwata, T., Fukui, M., 2014. Identification of major planktonic sulfur oxidizers in stratified freshwater lake. PLoS One 9, e93877. https://doi.org/10.1371/journal.pone.0093877.

Konda, T., 1984. Seasonal variations in four bacterial size fractions from a hypertrophic pond in Tokyo, Japan. Rev. Gesamten Hydrobiol. Hydrogr. 69, 843–858.

Kondo, R., Osawa, K., Mochizuki, L., Fujioka, Y., Butani, J., 2006. Abundance and diversity of sulphate-reducing bacterioplankton in Lake Suigetsu, a meromictic lake in Fukui, Japan. Plankton Benthos Res 1, 165–177. https://doi.org/10.3800/pbr.1.165.

Konda, T., Takii, S., Manabu, F., Kusuoka, Y., Matsumoto, G.I., Torii, T., 1994. Vertical distribution of bacterial population in Lake Fryxell, an antarctic lake. Jpn. J. Limnol. Rikusuigaku Zasshi 55, 185–192.

Krempaska, N., Horňák, K., Pernthaler, J., 2018. Spatiotemporal distribution and microbial assimilation of polyamines in a mesotrophic lake: polyamine variability and bacterial assimilation. Limnol. Oceanogr. 63, 816–832. https://doi.org/10.1002/lno.10672.

Kritzberg, E.S., Cole, J.J., Pace, M.M., Granéli, W., 2006. Bacterial growth on allochthonous carbon in humic and nutrient-enriched lakes: results from whole-lake 13C addition experiments. Ecosystems 9, 489–499. https://doi.org/10.1007/s10021-005-0115-5.

Kubo, K., Kojima, H., Fukui, M., 2014. Vertical distribution of major sulfate-reducing bacteria in a shallow eutrophic meromictic lake. Syst. Appl. Microbiol. 37, 510–519.

Kutovaya, O.A., McKay, R.M.L., Bullerjahn, G.S., 2013. Detection and expression of genes for phosphorus metabolism in picocyanobacteria from the Laurentian Great Lakes. J. Gt. Lakes Res. 39, 612–621. https://doi.org/10.1016/j.jglr.2013.09.009.

Kuypers, M.M.M., Marchant, H.K., Kartal, B., 2018. The microbial nitrogen-cycling network. Nat. Rev. Microbiol. 16, 263–276. https://doi.org/10.1038/nrmicro.2018.9.

Kuznetsov, S.I., 1970. Microflora of lakes and their geochemical activities. Leningr. Izd. Naudaka 440.

La Scola, B., Audic, S., Robert, C., Jungang, L., de Lamballerie, X., Drancourt, M., Birtles, R., Claverie, J.-M., Raoult, D., 2003. A giant virus in Amoebae. Science 299, 2033. https://doi.org/10.1126/science.1081867. 2033.

Langenheder, S., Jurgens, K., 2001. Regulation of bacterial biomass and community structure by metazoan and protozoan predation. Limnol. Oceanogr. 46, 121–134.

Laybourn-Parry, J., Walton, M., Young, J., Jones, R.I., Shine, A., 1994. Protozooplankton and bacterioplankton in a large oligotrophic lake—Loch Ness, Scotland. J. Plankton Res. 16, 1655–1670. https://doi.org/10.1093/plankt/16.12.1655.

Lear, G., Bellamy, J., Case, B.S., Lee, J.E., Buckley, H.L., 2014. Fine-scale spatial patterns in bacterial community composition and function within freshwater ponds. ISME J. 8, 1715–1726. https://doi.org/10.1038/ismej.2014.21.

Lear, G., Washington, V., Neale, M., Case, B., Buckley, H., Lewis, G., 2013. The biogeography of stream bacteria: the biogeography of stream bacteria. Glob. Ecol. Biogeogr. 22, 544–554. https://doi.org/10.1111/geb.12046.

Lee, C.C., Hu, Y., Ribbe, M.W., 2010. Vanadium nitrogenase reduces CO. Science 329, 642. https://doi.org/10.1126/science.1191455, 642.

Lenski, R.E., Levin, B.R., 1985. Constraints on the coevolution of bacteria and virulent phage: a model, some experiments, and predictions for natural communities. Am. Nat. 125, 585–602.

Lewis, W., Frost, T., Morris, D., 1986. Studies of planktonic bacteria in Lake valencia, Venezuela. Arch. Hydrobiol. 106, 289–305.

Lindeman, R.L., 1942. The trophic-dynamic aspect of ecology. Ecology 23, 399–418.

Lindstrom, E.S., Kamst-Van Agterveld, M.P., Zwart, G., 2005. Distribution of typical freshwater bacterial groups is associated with pH, temperature, and lake water retention time. Appl. Environ. Microbiol. 71, 8201–8206.

Linz, A.M., Aylward, F.O., Bertilsson, S., McMahon, K.D., 2020. Time-series metatranscriptomes reveal conserved patterns between phototrophic and heterotrophic microbes in diverse freshwater systems. Limnol. Oceanogr. 65. https://doi.org/10.1002/lno.11306.

Linz, A.M., He, S., Stevens, S.L.R., Anantharaman, K., Rohwer, R.R., Malmstrom, R.R., Bertilsson, S., McMahon, K.D., 2018. Freshwater carbon and nutrient cycles revealed through reconstructed population genomes. PeerJ 6, e6075. https://doi.org/10.7717/peerj.6075.

Llirós, M., García–Armisen, T., Darchambeau, F., Morana, C., Triadó–Margarit, X., Inceoğlu, Ö., Borrego, C.M., Bouillon, S., Servais, P., Borges, A.V., Descy, J., Canfield, D.E., Crowe, S.A., 2015. Pelagic photoferrotrophy and iron cycling in a modern ferruginous basin. Sci. Rep. 5, 13803. https://doi.org/10.1038/srep13803.

Llorens–Marès, T., Liu, Z., Allen, L.Z., Rusch, D.B., Craig, M.T., Dupont, C.L., Bryant, D.A., Casamayor, E.O., 2017. Speciation and ecological success in dimly lit waters: horizontal gene transfer in a green sulfur bacteria bloom unveiled by metagenomic assembly. ISME J. 11, 201–211. https://doi.org/10.1038/ismej.2016.93.

López-Bueno, A., Tamames, J., Velázquez, D., Moya, A., Quesada, A., Alcamí, A., 2009. High diversity of the viral community from an Antarctic Lake. Science 326, 858–861. https://doi.org/10.1126/science.1179287.

Lu, X., Bade, D.L., Leff, L.G., Mou, X., 2018. The relative importance of anammox and denitrification to total N2 production in Lake Erie. J. Gt. Lakes Res. 44, 428–435. https://doi.org/10.1016/j.jglr.2018.03.008.

Luque-Almagro, V.M., Gates, A.J., Moreno-Viván, C., Ferguson, S.J., Richardson, D.J., Roldán, M.D., 2011. Bacterial nitrate assimilation: gene distribution and regulation. Biochem. Soc. Trans. 39, 1838–1843. https://doi.org/10.1042/BST20110688.

Mahadevan, R., Palsson, B.Ø., Lovley, D.R., 2011. In situ to in silico and back: elucidating the physiology and ecology of *Geobacter* spp. using genome-scale modelling. Nat. Rev. Microbiol. 9, 39–50. https://doi.org/10.1038/nrmicro2456.

Malcolm, R.L., 1990. The uniqueness of humic substances in each of soil, stream and marine environments. Anal. Chim. Acta 232, 19–30.

Malki, K., Kula, A., Bruder, K., Sible, E., Hatzopoulos, T., Steidel, S., Watkins, S.C., Putonti, C., 2015. Bacteriophages isolated from Lake Michigan demonstrate broad host-range across several bacterial phyla. Virol. J. 12, 164. https://doi.org/10.1186/s12985-015-0395-0.

Mann, C., Wetzel, R., 1995. Dissolved organic carbon and its utilization in a riverine wetland ecosystem. Biogeochemistry 31, 99–120.

Manny, B.A., Miller, M.C., Wetzel, R.G., 1971. Ultraviolet combustion of dissolved organic nitrogen compounds in lake waters. Limnol. Oceanogr. 16, 71–85. https://doi.org/10.4319/lo.1971.16.1.0071.

Marston, M.F., Pierciey, F.J., Shepard, A., Gearin, G., Qi, J., Yandava, C., Schuster, S.C., Henn, M.R., Martiny, J.B.H., 2012. Rapid diversification of coevolving marine Synechococcus and a virus. Proc. Natl. Acad. Sci. 109, 4544–4549. https://doi.org/10.1073/pnas.1120310109.

Martinez-Garcia, M., Swan, B.K., Poulton, N.J., Gomez, M.L., Masland, D., Sieracki, M.E., Stepanauskas, R., 2012. High-throughput single-cell sequencing identifies photoheterotrophs and chemoautotrophs in freshwater bacterioplankton. ISME J. 6, 113–123. https://doi.org/10.1038/ismej.2011.84.

Massé, S., Botrel, M., Walsh, D.A., Maranger, R., 2019. Annual nitrification dynamics in a seasonally ice-covered lake. PLoS One 14, e0213748. https://doi.org/10.1371/journal.pone.0213748.

Mathias, C.B., Kirschner, A., Velimirov, B., 1995. Seasonal variations of virus abundance and viral control of the bacterial production in a backwater system of the Danube river. Appl. Environ. Microbiol. 61, 3734–3740. https://doi.org/10.1128/aem.61.10.3734-3740.1995.

Matz, C., Jürgens, K., 2005. High motility reduces grazing mortality of planktonic bacteria. Appl. Environ. Microbiol. 71, 921–929. https://doi.org/10.1128/AEM.71.2.921-929.2005.

Mehrshad, M., Salcher, M.M., Okazaki, Y., Nakano, S., Šimek, K., Andrei, A.-S., Ghai, R., 2018. Hidden in plain sight—highly abundant and diverse planktonic freshwater Chloroflexi. Microbiome 6, 176. https://doi.org/10.1186/s40168-018-0563-8.

Meshulam-Simon, G., Behrens, S., Choo, A.D., Spormann, A.M., 2007. Hydrogen metabolism in *Shewanella oneidensis* MR-1. Appl. Environ. Microbiol. 73, 1153–1165. https://doi.org/10.1128/AEM.01588-06.

Middelboe, M., Holmfeldt, K., Riemann, L., Nybroe, O., Haaber, J., 2009. Bacteriophages drive strain diversification in a marine *flavobacterium*: implications for phage resistance and physiological properties. Environ. Microbiol. 11, 1971–1982. https://doi.org/10.1111/j.1462-2920.2009.01920.x.

Middelboe, M., Jacquet, S., Weinbauer, M., 2008. Viruses in freshwater ecosystems: an introduction to the exploration of viruses in new aquatic habitats. Freshw. Biol. 53, 1069–1075. https://doi.org/10.1111/j.1365-2427.2008.02014.x.

Miki, T., Jacquet, S., 2008. Complex interactions in the microbial world: underexplored key links between viruses, bacteria and protozoan grazers in aquatic environments. Aquat. Microb. Ecol. 51, 195–208. https://doi.org/10.3354/ame01190.

Miller, M.C., 1972. The Carbon Cycle in the Epilimnion of Two Michigan Lakes (Ph.D. Dissertation). Michigan State University, East Lansing, MI.

Mohan, S.B., Schmid, M., Jetten, M., Cole, J., 2004. Detection and widespread distribution of the nrfA gene encoding nitrite reduction to ammonia, a short circuit in the biological nitrogen cycle that competes with denitrification. FEMS Microbiol. Ecol. 49, 433–443. https://doi.org/10.1016/j.femsec.2004.04.012.

Mohiuddin, M., Schellhorn, H.E., 2015. Spatial and temporal dynamics of virus occurrence in two freshwater lakes captured through metagenomic analysis. Front. Microbiol. 6. https://doi.org/10.3389/fmicb.2015.00960.

Molot, L.A., Watson, S.B., Creed, I.F., Trick, C.G., McCabe, S.K., Verschoor, M.J., Sorichetti, R.J., Powe, C., Venkiteswaran, J.J., Schiff, S.L., 2014. A novel model for cyanobacteria bloom formation: the critical role of anoxia and ferrous iron. Freshw. Biol. 59, 1323–1340. https://doi.org/10.1111/fwb.12334.

Monchamp, M.-E., Spaak, P., Pomati, F., 2019. Long term diversity and distribution of non-photosynthetic cyanobacteria in Peri-Alpine Lakes. Front. Microbiol. 9, 3344. https://doi.org/10.3389/fmicb.2018.03344.

Moon, K., Kang, I., Kim, S., Kim, S.-J., Cho, J.-C., 2017. Genome characteristics and environmental distribution of the first phage that infects the LD28 clade, a freshwater methylotrophic bacterial group: genome characterization of the first LD28 phage. Environ. Microbiol. 19, 4714–4727. https://doi.org/10.1111/1462-2920.13936.

Moon, K., Kim, S., Kang, I., Cho, J.-C., 2020. Viral metagenomes of lake soyang, the largest freshwater lake in South Korea. Sci. Data 7, 349. https://doi.org/10.1038/s41597-020-00695-9.

Moran, M.A., Kujawinski, E.B., Schroer, W.F., Amin, S.A., Bates, N.R., Bertrand, E.M., Braakman, R., Brown, C.T., Covert, M.W., Doney, S.C., Dyhrman, S.T., Edison, A.S., Eren, A.M., Levine, N.M., Li, L., Ross, A.C., Saito, M.A., Santoro, A.E., Segrè, D., Shade, A., Sullivan, M.B., Vardi, A., 2022. Microbial metabolites in the marine carbon cycle. Nat. Microbiol. 7, 508–523. https://doi.org/10.1038/s41564-022-01090-3.

Moran, M.A., Kujawinski, E.B., Stubbins, A., Fatland, R., Aluwihare, L.I., Buchan, A., Crump, B.C., Dorrestein, P.C., Dyhrman, S.T., Hess, N.J., Howe, B., Longnecker, K., Medeiros, P.M., Niggemann, J., Obernosterer, I., Repeta, D.J., Waldbauer, J.R., 2016. Deciphering ocean carbon in a changing world. Proc. Natl. Acad. Sci. 113, 3143–3151. https://doi.org/10.1073/pnas.1514645113.

Moran, M.A., Zepp, R.G., 1997. Role of photoreactions in the formation of biologically labile compounds from dissolved organic matter. Limnol. Oceanogr. 42, 1307–1316. https://doi.org/10.4319/lo.1997.42.6.1307.

Moreno-Vivián, C., Cabello, P., Martínez-Luque, M., Blasco, R., Castillo, F., 1999. Prokaryotic nitrate reduction: molecular properties and functional distinction among bacterial nitrate reductases. J. Bacteriol. 181, 6573–6584. https://doi.org/10.1128/JB.181.21.6573-6584.1999.

Morris, D.P., Lewis, W.M., 1992. Nutrient limitation of bacterioplankton growth in Lake Dillon, Colorado. Limnol. Oceanogr. 37, 1179–1192. https://doi.org/10.4319/lo.1992.37.6.1179.

Mortimer, C.H., 1941. The exchange of dissolved substances between mud and water in lakes. J. Ecol. 29, 280–329.

Mukherjee, M., Ray, A., Post, A.F., McKay, R.M., Bullerjahn, G.S., 2016. Identification, enumeration and diversity of nitrifying planktonic archaea and bacteria in trophic end members of the laurentian great lakes. J. Gt. Lakes Res. 42, 39–49. https://doi.org/10.1016/j.jglr.2015.11.007.

Münster, U., Heikkinen, E., Likolammi, M., Järvinen, M., Salonen, K., De Haan, H., 1999. Utilisation of polymeric and monomeric aromatic and amino acid carbon in a humic boreal forest lake. Arch Hydrobiol Spec Issues Adv. Limn 54, 105–134.

Nagai, T., Imai, A., Matsushige, K., Yokoi, K., Fukushima, T., 2007. Dissolved iron and its speciation in a shallow eutrophic lake and its inflowing rivers. Water Res. 41, 775–784. https://doi.org/10.1016/j.watres.2006.10.038.

Nagata, T., 1984. Bacterioplankton in Lake Biwa: annual fluctuations of bacterial numbers and their possible relationship with environmental variables. Jpn. J. Limnol. Rikusuigaku Zasshi 45, 126–133. https://doi.org/10.3739/rikusui.45.126.

Nakano, S., Ishii, N., Manage, P., Kawabata, Z., 1998. Trophic roles of heterotrophic nanoflagellates and ciliates among planktonic organisms in a hypereutrophic pond. Aquat. Microb. Ecol. 16, 153–161. https://doi.org/10.3354/ame016153.

Nealson, K.H., Saffarini, D., 1994. Iron and manganese in anaerobic respiration: environmental significance, physiology, and regulation. Annu. Rev. Microbiol. 48, 311–343.

Neidhardt, F.C., 1963. Effects of environment on the composition of bacterial cells. Annu. Rev. Microbiol. 17, 61–86. https://doi.org/10.1146/annurev.mi.17.100163.000425.

Neuenschwander, S.M., Ghai, R., Pernthaler, J., Salcher, M.M., 2018. Microdiversification in genome-streamlined ubiquitous freshwater Actinobacteria. ISME J. 12, 185–198. https://doi.org/10.1038/ismej.2017.156.

Newton, R.J., Jones, S.E., Eiler, A., McMahon, K.D., Bertilsson, S., 2011. A guide to the natural history of freshwater lake bacteria. Microbiol. Mol. Biol. Rev. 75, 36.

Newton, R.J., Kent, A.D., Triplett, E.W., McMahon, K.D., 2006. Microbial community dynamics in a humic lake: differential persistence of common freshwater phylotypes. Environ. Microbiol. 8, 956–970. https://doi.org/10.1111/j.1462-2920.2005.00979.x.

North, R.L., Guildford, S.J., Smith, R.E.H., Havens, S.M., Twiss, M.R., 2007. Evidence for phosphorus, nitrogen, and iron colimitation of phytoplankton communities in Lake Erie. Limnol. Oceanogr. 52, 315–328. https://doi.org/10.4319/lo.2007.52.1.0315.

Ochs, C.A., Cole, J.J., Likens, G.E., 1995. Population dynamics of bacterioplankton in an oligotrophic lake. J. Plankton Res. 17, 365–391. https://doi.org/10.1093/plankt/17.2.365.

Okamura, T., Mori, Y., Nakano, S., Kondo, R., 2012. Abundance and bacterivory of heterotrophic nanoflagellates in the meromictic Lake Suigetsu, Japan. Aquat. Microb. Ecol. 66, 149–158. https://doi.org/10.3354/ame01565.

Okazaki, Y., Fujinaga, S., Tanaka, A., Kohzu, A., Oyagi, H., Nakano, S., 2017. Ubiquity and quantitative significance of bacterioplankton lineages inhabiting the oxygenated hypolimnion of deep freshwater lakes. ISME J. 11, 2279–2293. https://doi.org/10.1038/ismej.2017.89.

Oliveira, G., La Scola, B., Abrahão, J., 2019. Giant virus vs amoeba: fight for supremacy. Virol. J. 16, 126. https://doi.org/10.1186/s12985-019-1244-3.

Ooms-Wilms, A.L., Postema, G., Gulati, R.D., 1995. Evaluation of bacterivory of Rotifera based on measurements of in situ ingestion of fluorescent particles, including some comparisons with Cladocera. J. Plankton Res. 1057–1077.

Oren, A., Arahal, D.R., Rosselló-Móra, R., Sutcliffe, I.C., Moore, E.R.B., 2021. Emendation of rules 5b, 8, 15 and 22 of the international code of nomenclature of prokaryotes to include the rank of phylum. Int. J. Syst. Evol. Microbiol. 71. https://doi.org/10.1099/ijsem.0.004851.

Oren, A., Garrity, G.M., 2021. Valid publication of the names of forty-two phyla of prokaryotes. Int. J. Syst. Evol. Microbiol. 71. https://doi.org/10.1099/ijsem.0.005056.

Oshiki, M., Satoh, H., Okabe, S., 2016. Ecology and physiology of anaerobic ammonium oxidizing bacteria. Environ. Microbiol. 18, 2784–2796. https://doi.org/10.1111/1462-2920.13134.

Overbeck, J., 1975. Distribution pattern of uptake kinetic responses in a stratified eutrophic lake. Verh Intern. Ver. Limnol 19, 2600–2615.

Pachiadaki, M.G., Sintes, E., Bergauer, K., Brown, J.M., Record, N.R., Swan, B.K., Mathyer, M.E., Hallam, S.J., Lopez-Garcia, P., Takaki, Y., Nunoura, T., Woyke, T., Herndl, G.J., Stepanauskas, R., 2017. Major role of nitrite-oxidizing bacteria in dark ocean carbon fixation. Science 358, 1046–1051. https://doi.org/10.1126/science.aan8260.

Paix, B., Ezzedine, J.A., Jacquet, S., 2019. Diversity, dynamics, and distribution of *Bdellovibrio* and like organisms in Perialpine lakes. Appl. Environ. Microbiol. 85. https://doi.org/10.1128/AEM.02494-18.

Pal, M., Yesankar, P.J., Dwivedi, A., Qureshi, A., 2020. Biotic control of harmful algal blooms (HABs): a brief review. J. Environ. Manage. 268, 110687. https://doi.org/10.1016/j.jenvman.2020.110687.

Palacin-Lizarbe, C., Camarero, L., Hallin, S., Jones, C.M., Cáliz, J., Casamayor, E.O., Catalan, J., 2019. The DNRA-denitrification dichotomy differentiates nitrogen transformation pathways in mountain lake benthic habitats. Front. Microbiol. 10, 1229. https://doi.org/10.3389/fmicb.2019.01229.

Palatinszky, M., Herbold, C., Jehmlich, N., Pogoda, M., Han, P., von Bergen, M., Lagkouvardos, I., Karst, S.M., Galushko, A., Koch, H., Berry, D., Daims, H., Wagner, M., 2015. Cyanate as an energy source for nitrifiers. Nature 524, 105–108. https://doi.org/10.1038/nature14856.

Palermo, C., Dittrich, M., 2016. Evidence for the biogenic origin of manganese-enriched layers in Lake Superior sediments: biogenic origin of Mn-enriched sediment layers. Environ. Microbiol. Rep. 8, 179–186. https://doi.org/10.1111/1758-2229.12364.

Palesse, S., Colombet, J., Pradeep Ram, A.S., Sime-Ngando, T., 2014. Linking host prokaryotic physiology to viral lifestyle dynamics in a temperate freshwater lake (lake pavin, france). Microb. Ecol. 68, 740–750. https://doi.org/10.1007/s00248-014-0441-2.

Parker, C.T., Tindall, B.J., Garrity, G.M., 2019. International code of nomenclature of prokaryotes: prokaryotic code (2008 revision). Int. J. Syst. Evol. Microbiol. 69, S1–S111. https://doi.org/10.1099/ijsem.0.000778.

Parkin, T.B., Brock, T.D., 1980. Photosynthetic bacterial production in lakes: the effects of light intensity1: photosynthetic bacterial production. Limnol. Oceanogr. 25, 711–718. https://doi.org/10.4319/lo.1980.25.4.0711.

Parkin, T.B., Brock, T.D., 1981. Photosynthetic bacterial production and carbon mineralization in a meromictic lake. Arch. Hydrobiol. 91, 366–382.

Parro, V., Puente-Sánchez, F., Cabrol, N.A., Gallardo-Carreño, I., Moreno-Paz, M., Blanco, Y., García-Villadangos, M., Tambley, C., Tilot, V.C., Thompson, C., Smith, E., Sobrón, P., Demergasso, C.S., Echeverría-Vega, A., Fernández-Martínez, M.Á., Whyte, L.G., Fairén, A.G., 2019. Microbiology and nitrogen cycle in the benthic sediments of a glacial oligotrophic deep andean lake as analog of ancient martian lake-beds. Front. Microbiol. 10, 929. https://doi.org/10.3389/fmicb.2019.00929.

Parsons, T., Strickland, J., 1961. On the production of particulate organic carbon by heterotrophic processes in sea water. Deep Sea Res. 8, 211–222. https://doi.org/10.1016/0146-6313(61)90022-3.

Paver, S.F., Newton, R.J., Coleman, M.L., 2020. Microbial communities of the laurentian great lakes reflect connectivity and local biogeochemistry. Environ. Microbiol. 22, 433–446. https://doi.org/10.1111/1462-2920.14862.

Payne, W.J., Wiebe, W.J., 1978. Growth yield and efficiency in chemosynthetic microorganisms. Annu. Rev. Microbiol. 32, 155–183. https://doi.org/10.1146/annurev.mi.32.100178.001103.

Pedrós-Alió, C., Brock, T.D., 1982. Assessing biomass and production of bacteria in eutrophic lake Mendota, Wisconsin. Appl. Environ. Microbiol. 44, 203–218. https://doi.org/10.1128/aem.44.1.203-218.1982.

Peduzzi, P., Schiemer, F., 2004. Bacteria and viruses in the water column of tropical freshwater reservoirs. Environ. Microbiol. 6, 707–715. https://doi.org/10.1111/j.1462-2920.2004.00602.x.

Peduzzi, S., Welsh, A., Demarta, A., Decristophoris, P., Peduzzi, R., Hahn, D., Tonolla, M., 2011. *Thiocystis chemoclinalis* sp. nov. and *Thiocystis cadagnonensis* sp. nov., motile purple sulfur bacteria isolated from the chemocline of a meromictic lake. Int. J. Syst. Evol. Microbiol. 61, 1682–1687. https://doi.org/10.1099/ijs.0.010397-0.

Penton, C.R., Devol, A.H., Tiedje, J.M., 2006. Molecular evidence for the broad distribution of anaerobic ammonium-oxidizing bacteria in freshwater and marine sediments. Appl. Environ. Microbiol. 72, 6829–6832. https://doi.org/10.1128/AEM.01254-06.

Pérez, J., Moraleda-Muñoz, A., Marcos-Torres, F.J., Muñoz-Dorado, J., 2016. Bacterial predation: 75 years and counting!: bacterial predation. Environ. Microbiol. 18, 766–779. https://doi.org/10.1111/1462-2920.13171.

Pernthaler, J., 2005. Predation on prokaryotes in the water column and its ecological implications. Nat. Rev. Microbiol. 3, 537–546. https://doi.org/10.1038/nrmicro1180.

Pernthaler, J., 2017. Competition and niche separation of pelagic bacteria in freshwater habitats: niches of freshwater bacterioplankton. Environ. Microbiol. 19, 2133–2150. https://doi.org/10.1111/1462-2920.13742.

Pernthaler, J., Posch, T., Šimek, K., Vrba, J., Pernthaler, A., Glöckner, F.O., Nübel, U., Psenner, R., Amann, R., 2001. Predator-specific enrichment of actinobacteria from a cosmopolitan freshwater clade in mixed continuous culture. Appl. Environ. Microbiol. 67, 2145–2155. https://doi.org/10.1128/AEM.67.5.2145-2155.2001.

Pernthaler, J., Zöllner, E., Warnecke, F., Jürgens, K., 2004. Bloom of filamentous bacteria in a Mesotrophic Lake: identity and potential controlling mechanism. Appl. Environ. Microbiol. 70, 6272–6281. https://doi.org/10.1128/AEM.70.10.6272-6281.2004.

Podowski, J.C., Paver, S.F., Newton, R.J., Coleman, M.L., 2021. Large lakes harbor streamlined free-living nitrifiers. BioRxiv Prepr 50.

Pomeroy, L.R., Wiebe, W.J., 1988. Energetics of microbial food webs. Hydrobiologia 159, 7–18.

Props, R., Denef, V.J., 2020. Temperature and nutrient levels correspond with lineage-specific microdiversification in the ubiquitous and abundant freshwater genus *Limnohabitans*. Appl. Environ. Microbiol. 86, e00140. https://doi.org/10.1128/AEM.00140-20, 20.

Qin, W., Amin, S., Martens-Habbena, W., Walker, C., Urakawa, H., Devol, A., Ingalls, A., Moffett, J., Armbrust, E., Stahl, D., 2014. Marine ammonia-oxidizing archaeal isolates display obligate mixotrophy and wide ecotypic variation. Proc. Natl. Acad. Sci. U. S. A. 111, 12504–12509. https://doi.org/10.1073/pnas.1324115111.

Rai, H., 1979. Microbiology of central Amazon lakes. Amazoniana 6, 583–599.

Ramaiah, N., 1995. Summer abundance and activities of bacteria in the freshwater lakes of Schirmacher Oasis, Antarctica. Polar Biol. 15, 547–553.

Rashidan, K.K., Bird, D.F., 2001. Role of predatory bacteria in the termination of a cyanobacterial bloom. Microb. Ecol. 41, 97–105. https://doi.org/10.1007/s002480000074.

Ricão Canelhas, M., Denfeld, B.A., Weyhenmeyer, G.A., Bastviken, D., Bertilsson, S., 2016. Methane oxidation at the water-ice interface of an ice-covered lake. Limnol. Oceanogr. 61. https://doi.org/10.1002/lno.10288.

Riemann, L., Middelboe, M., 2002. Viral lysis of marine bacterioplankton: implications for organic matter cycling and bacterial clonal composition. Ophelia 56, 57–68. https://doi.org/10.1080/00785236.2002.10409490.

Rodhe, W., Hobbie, J.E., Wright, R.T., 1966. Phototrophy and heterotrophy in high mountain lakes. Verh Intern. Ver. Limnol 16, 302–313.

Rohwer, R.R., Hamilton, J.J., Newton, R.J., McMahon, K.D., 2018. TaxAss: leveraging a custom freshwater database achieves fine-scale taxonomic resolution. mSphere 3, e00327. https://doi.org/10.1128/mSphere.00327-18, 18.

Roux, S., Chan, L.-K., Egan, R., Malmstrom, R.R., McMahon, K.D., Sullivan, M.B., 2017. Ecogenomics of virophages and their giant virus hosts assessed through time series metagenomics. Nat. Commun. 8, 858. https://doi.org/10.1038/s41467-017-01086-2.

Roux, S., Enault, F., Robin, A., Ravet, V., Personnic, S., Theil, S., Colombet, J., Sime-Ngando, T., Debroas, D., 2012. Assessing the diversity and specificity of two freshwater viral communities through metagenomics. PLoS One 7, e33641. https://doi.org/10.1371/journal.pone.0033641.

Ruiz-Perez, C.A., Tsementzi, D., Hatt, J.K., Sullivan, M.B., Konstantinidis, K.T., 2019. Prevalence of viral photosynthesis genes along a freshwater to saltwater transect in Southeast USA. Environ. Microbiol. Rep. 11, 672–689. https://doi.org/10.1111/1758-2229.12780.

Salcher, M.M., Ewert, C., Simek, K., Kasalicky, V., Posch, T., 2016. Interspecific competition and protistan grazing affect the coexistence of freshwater betaproteobacterial strains. FEMS Microbiol. Ecol. 92. https://doi.org/10.1093/femsec/fiv156.

Salcher, M.M., Hofer, J., HorÅÃ¡k, K., Jezbera, J., Sonntag, B., Vrba, J., Å Imek, K., Posch, T., 2007. Modulation of microbial predator–prey dynamics by phosphorus availability: growth patterns and survival strategies of bacterial phylogenetic clades: effects of phosphorus and grazing on bacterial phylogenetic clades. FEMS Microbiol. Ecol. 60, 40–50. https://doi.org/10.1111/j.1574-6941.2006.00274.x.

Salcher, M.M., Neuenschwander, S.M., Posch, T., Pernthaler, J., 2015. The ecology of pelagic freshwater methylotrophs assessed by a high-resolution monitoring and isolation campaign. ISME J. 9, 2442–2453. https://doi.org/10.1038/ismej.2015.55.

Salcher, M.M., Pernthaler, J., Posch, T., 2010. Spatiotemporal distribution and activity patterns of bacteria from three phylogenetic groups in an oligomesotrophic lake. Limnol. Oceanogr. 55, 846–856.

Salcher, M.M., Posch, T., Pernthaler, J., 2013. In situ substrate preferences of abundant bacterioplankton populations in a prealpine freshwater lake. ISME J. 7, 896–907. https://doi.org/10.1038/ismej.2012.162.

Salcher, M.M., Schaefle, D., Kaspar, M., Neuenschwander, S.M., Ghai, R., 2019. Evolution in action: habitat transition from sediment to the pelagial leads to genome streamlining in Methylophilaceae. ISME J. 13, 2764–2777. https://doi.org/10.1038/s41396-019-0471-3.

Sanders, R.W., Porter, K.G., Bennett, S.J., DeBiase, A.E., 1989. Seasonal patterns of bacterivory by flagellates, ciliates, rotifers, and cladocerans in a freshwater planktonic community: lake community bacterivory. Limnol. Oceanogr. 34, 673–687. https://doi.org/10.4319/lo.1989.34.4.0673.

Santelli, C.M., Chaput, D.L., Hansel, C.M., 2014. Microbial communities promoting Mn(II) oxidation in ashumet pond, a historically polluted freshwater pond undergoing remediation. Geomicrobiol. J. 31, 605–616.

Satinsky, B.M., Fortunato, C.S., Doherty, M., Smith, C.B., Sharma, S., Ward, N.D., Krusche, A.V., Yager, P.L., Richey, J.E., Moran, M.A., Crump, B.C., 2015. Metagenomic and metatranscriptomic inventories of the lower Amazon River, May 2011. Microbiome 3 (39). https://doi.org/10.1186/s40168-015-0099-0.

Saunders, G.W., Cummins, K.W., Gak, D.Z., Pieczyńska, E., Straškrabová, V., Wetzel, R.G., 1980. Organic matter and decomposers. In: The Functioning of Freshwater Ecosystems. Cambridge University Press, pp. 341–392.

Savio, D., Sinclair, L., Ijaz, U.Z., Parajka, J., Reischer, G.H., Stadler, P., Blaschke, A.P., Blöschl, G., Mach, R.L., Kirschner, A.K.T., Farnleitner, A.H., Eiler, A., 2015. Bacterial diversity along a 2600 km river continuum. Environ. Microbiol. 17, 4994–5007. https://doi.org/10.1111/1462-2920.12886.

Saye, D.J., Ogunseitan, O., Sayler, G.S., Miller, R.V., 1987. Potential for transduction of plasmids in a natural freshwater environment: effect of plasmid donor concentration and a natural microbial community on transduction in Pseudomonas aeruginosa. Appl. Environ. Microbiol. 53, 987–995. https://doi.org/10.1128/aem.53.5.987-995.1987.

Schauer, M., Hahn, M.W., 2005. Diversity and phylogenetic affiliations of morphologically conspicuous large filamentous bacteria occurring in the pelagic zones of a broad spectrum of freshwater habitats. Appl. Environ. Microbiol. 71, 1931–1940. https://doi.org/10.1128/AEM.71.4.1931-1940.2005.

Schmaljohann, R., Pollingher, U., Berman, T., 1987. Natural populations of bacteria in Lake kinneret—observations with scanning electron and epifluorescence microscopy. Microb. Ecol. 13, 1–12. https://doi.org/10.1007/BF02014959.

Schmidt, M.L., White, J.D., Denef, V.J., 2016. Phylogenetic conservation of freshwater lake habitat preference varies between abundant bacterioplankton phyla: bacterial habitat preferences in freshwater lakes. Environ. Microbiol. 18, 1212–1226. https://doi.org/10.1111/1462-2920.13143.

Schroeder, L.A., 1981. Consumer growth efficiencies: their limits and relationships to ecological energetics. J. Theor. Biol. 93, 805–828.

Schubert, C.J., Durisch-Kaiser, E., Wehrli, B., Thamdrup, B., Lam, P., Kuypers, M.M.M., 2006. Anaerobic ammonium oxidation in a tropical freshwater system (Lake Tanganyika). Environ. Microbiol. 8, 1857–1863. https://doi.org/10.1111/j.1462-2920.2006.01074.x.

Schulz, F., Roux, S., Paez-Espino, D., Jungbluth, S., Walsh, D.A., Denef, V.J., McMahon, K.D., Konstantinidis, K.T., Eloe-Fadrosh, E.A., Kyrpides, N.C., Woyke, T., 2020. Giant virus diversity and host interactions through global metagenomics. Nature 578, 432–436. https://doi.org/10.1038/s41586-020-1957-x.

Schütte, U.M.E., Cadieux, S.B., Hemmerich, C., Pratt, L.M., White, J.R., 2016. Unanticipated geochemical and microbial community structure under seasonal ice cover in a dilute, dimictic Arctic Lake. Front. Microbiol. 7. https://doi.org/10.3389/fmicb.2016.01035.

Scott, J.T., Cotner, J.B., LaPara, T.M., 2012. Variable stoichiometry and homeostatic regulation of bacterial biomass elemental composition. Front. Microbiol. 3. https://doi.org/10.3389/fmicb.2012.00042.

Shade, A., Kent, A.D., Jones, S.E., Newton, R.J., Triplett, E.W., McMahon, K.D., 2007. Interannual dynamics and phenology of bacterial communities in a eutrophic lake. Limnol. Oceanogr. 52, 487–494.

Sharma, A.K., Sommerfeld, K., Bullerjahn, G.S., Matteson, A.R., Wilhelm, S.W., Jezbera, J., Brandt, U., Doolittle, W.F., Hahn, M.W., 2009. Actinorhodopsin genes discovered in diverse freshwater habitats and among cultivated freshwater Actinobacteria. ISME J. 3, 726–737. https://doi.org/10.1038/ismej.2009.13.

Sharma, A.K., Zhaxybayeva, O., Papke, R.T., Doolittle, W.F., 2008. Actinorhodopsins: proteorhodopsin-like gene sequences found predominantly in non-marine environments. Environ. Microbiol. 10, 1039–1056. https://doi.org/10.1111/j.1462-2920.2007.01525.x.

Sharrar, A.M., Flood, B.E., Bailey, J.V., Jones, D.S., Biddanda, B.A., Ruberg, S.A., Marcus, D.N., Dick, G.J., 2017. Novel large sulfur bacteria in the metagenomes of groundwater-fed chemosynthetic microbial mats in the lake huron basin. Front. Microbiol. 8, 791. https://doi.org/10.3389/fmicb.2017.00791.

Sherr, E.B., Sherr, B.F., 2002. Significance of predation by protists in aquatic microbial food webs. Antonie Leeuwenhoek 81, 293–308.

Sherr, B.F., Sherr, E.B., Berman, T., 1983. Grazing, growth, and ammonium excretion rates of a heterotrophic microflagellate fed with four species of bacteria. Appl. Environ. Microbiol. 45, 1196–1201. https://doi.org/10.1128/aem.45.4.1196-1201.1983.

Sievert, S.M., Wieringa, E.B.A., Wirsen, C.O., Taylor, C.D., 2007. Growth and mechanism of filamentous-sulfur formation by *candidatus* arcobacter sulfidicus in opposing oxygen-sulfide gradients. Environ. Microbiol. 9, 271–276. https://doi.org/10.1111/j.1462-2920.2006.01156.x.

Sigee, D., 2005. Freshwater Microbiology: Biodiversity and Dynamic Interactions of Microorganisms in the Aquatic Environment. John Wiley & Sons.

Šimek, K., Chrzanowski, T.H., 1992. Direct and indirect evidence of size-selective grazing on pelagic bacteria by freshwater

Nanoflagellates. Appl. Environ. Microbiol. 58, 3715–3720. https://doi.org/10.1128/aem.58.11.3715-3720.1992.

Simek, K., Hornak, K., Jezbera, J., Masin, M., Nedoma, J., Gasol, J.M., Schauer, M., 2005. Influence of top-down and bottom-up manipulations on the R-BT065 subcluster of beta-proteobacteria, an abundant group in bacterioplankton of a freshwater reservoir. Appl. Environ. Microbiol. 71, 2381–2390.

Simek, K., Hornak, K., Jezbera, J., Nedoma, J., Vrba, J., Straskrabova, V., Macek, M., Dolan, J.R., Hahn, M.W., 2006. Maximum growth rates and possible life strategies of different bacterioplankton groups in relation to phosphorus availability in a freshwater reservoir. Environ. Microbiol. 8, 1613–1624.

Šimek, K., Nedoma, J., Znachor, P., Kasalický, V., Jezbera, J., Hornňák, K., Sed'a, J., 2014. A finely tuned symphony of factors modulates the microbial food web of a freshwater reservoir in spring. Limnol. Oceanogr. 59, 1477–1492. https://doi.org/10.4319/lo.2014.59.5.1477.

Šimek, K., Weinbauer, M.G., Hornák, K., Jezbera, J., Nedoma, J., Dolan, J.R., 2007. Grazer and virus-induced mortality of bacterioplankton accelerates development of *Flectobacillus* populations in a freshwater community. Environ. Microbiol. 9, 789–800. https://doi.org/10.1111/j.1462-2920.2006.01201.x.

Sime-Ngando, T., 2014. Environmental bacteriophages: viruses of microbes in aquatic ecosystems. Front. Microbiol. 5. https://doi.org/10.3389/fmicb.2014.00355.

Sime-Ngando, T., Colombet, J., 2009. Virus et prophages dans les écosystèmes aquatiques. Can. J. Microbiol. 55, 95–109. https://doi.org/10.1139/W08-099.

Simon, M., 1998. Bacterioplankton dynamics in a large mesotrophic lake: II. concentrations and turnover of dissolved amino acids. Fundam. Appl. Limnol. 144, 1–23. https://doi.org/10.1127/archiv-hydrobiol/144/1998/1.

Simon, M., Grossart, H., Schweitzer, B., Ploug, H., 2002. Microbial ecology of organic aggregates in aquatic ecosystems. Aquat. Microb. Ecol. 28, 175–211. https://doi.org/10.3354/ame028175.

Simon, M., Tilzer, M.M., Muller, H., 1998. Simon-1998-grazing-BP.pdf. Arch. Hydrobiol. 143, 385–407.

Skvortsov, T., de Leeuwe, C., Quinn, J.P., McGrath, J.W., Allen, C.C.R., McElarney, Y., Watson, C., Arkhipova, K., Lavigne, R., Kulakov, L.A., 2016. Metagenomic characterisation of the viral community of lough neagh, the largest freshwater lake in Ireland. PLoS One 11, e0150361. https://doi.org/10.1371/journal.pone.0150361.

Small, G.E., Bullerjahn, G.S., Sterner, R.W., Beall, B.F.N., Brovold, S., Finlay, J.C., McKay, R.M.L., Mukherjee, M., 2013. Rates and controls of nitrification in a large oligotrophic lake. Limnol. Oceanogr. 58, 276–286. https://doi.org/10.4319/lo.2013.58.1.0276.

Smith, E.M., Prairie, Y.T., 2004. Bacterial metabolism and growth efficiency in lakes: the importance of phosphorus availability. Limnol. Oceanogr. 49, 137–147.

Solomon, C., Collier, J., Berg, G., Glibert, P., 2010. Role of urea in microbial metabolism in aquatic systems: a biochemical and molecular review. Aquat. Microb. Ecol. 59, 67–88. https://doi.org/10.3354/ame01390.

Sommaruga, R., Psenner, R., 1995. Permanent presence of grazing-resistant bacteria in a hypertrophic lake. Appl. Environ. Microbiol. 61, 3457–3459. https://doi.org/10.1128/aem.61.9.3457-3459.1995.

Soo, R.M., Hemp, J., Parks, D.H., Fischer, W.W., Hugenholtz, P., 2017. On the origins of oxygenic photosynthesis and aerobic respiration in Cyanobacteria. Science 355, 1436–1440. https://doi.org/10.1126/science.aal3794.

Soo, R.M., Woodcroft, B.J., Parks, D.H., Tyson, G.W., Hugenholtz, P., 2015. Back from the dead; the curious tale of the predatory cyanobacterium *vampirovibrio chlorellavorus*. PeerJ 3, e968. https://doi.org/10.7717/peerj.968.

Sorokin, D.Y., Tourova, T.P., Muyzer, G., 2013. Isolation and characterization of two novel alkalitolerant sulfidogens from a thiopaq bioreactor, *Desulfonatronum alkalitolerans* sp. nov., and *Sulfurospirillum alkalitolerans* sp. nov. Extremophiles 17, 535–543. https://doi.org/10.1007/s00792-013-0538-4.

Spencer, M.J., 1978. Microbial activity and biomass relationships in 26 oligotrophic to mesotrophic lakes in South Island, New Zealand. Verh Intern. Ver. Limnol 20, 1175–1181.

Staley, C., Gould, T.J., Wang, P., Phillips, J., Cotner, J.B., Sadowsky, M.J., 2015. Species sorting and seasonal dynamics primarily shape bacterial communities in the Upper Mississippi River. Sci. Total Environ. 505, 435–445. https://doi.org/10.1016/j.scitotenv.2014.10.012.

Staley, J.T., Konopka, A., 1985. Measurement of in situ activities of nonphotosynthetic microorganisms in aquatic and terrestrial habitats. Annu. Rev. Microbiol. 39, 321–346.

Stein, L.Y., Klotz, M.G., 2016. The nitrogen cycle. Curr. Biol. 26, R94–R98. https://doi.org/10.1016/j.cub.2015.12.021.

Stein, L.Y., La Duc, M.T., Grundl, T.J., Nealson, K.H., 2001. Bacterial and archaeal populations associated with freshwater ferromanganous micronodules and sediments. Environ. Microbiol. 3, 10–18. https://doi.org/10.1046/j.1462-2920.2001.00154.x.

Sterner, R.W., Smutka, T.M., McKay, R.M.L., Xiaoming, Q., Brown, E.T., Sherrell, R.M., 2004. Phosphorus and trace metal limitation of algae and bacteria in Lake Superior. Limnol. Oceanogr. 49, 495–507. https://doi.org/10.4319/lo.2004.49.2.0495.

Stewart, F.M., Levin, B.R., 1984. The population biology of bacterial viruses: why be temperate. Theor. Popul. Biol. 26, 93–117.

Stewart, F.J., Ulloa, O., DeLong, E.F., 2012. Microbial metatranscriptomics in a permanent marine oxygen minimum zone: OMZ community gene expression. Environ. Microbiol. 14, 23–40. https://doi.org/10.1111/j.1462-2920.2010.02400.x.

Stewart, a J., Wetzel, R.G., 1982. Influence of dissolved humic materials on carbon assimilation and alkaline-phosphatase activity in natural algal bacterial assemblages. Freshw. Biol. 12, 369–380.

Stolz, J.F., 2017. Gaia and her microbiome. FEMS Microbiol. Ecol. 93, fiw247. https://doi.org/10.1093/femsec/fiw247.

Suttle, C.A., 1994. The significance of viruses to mortality in aquatic microbial communities. Microb. Ecol. 28, 237–243.

Suttle, C.A., 2007. Marine viruses — major players in the global ecosystem. Nat. Rev. Microbiol. 5, 801–812. https://doi.org/10.1038/nrmicro1750.

Taipale, S., Kankaala, P., Hahn, M., Jones, R., Tiirola, M., 2011. Methane-oxidizing and photoautotrophic bacteria are major producers in a humic lake with a large anoxic hypolimnion. Aquat. Microb. Ecol. 64, 81–95. https://doi.org/10.3354/ame01512.

Tarao, M., Jezbera, J., Hahn, M.W., 2009. Involvement of cell surface structures in size-independent grazing resistance of freshwater *Actinobacteria*. Appl. Environ. Microbiol. 75, 4720–4726. https://doi.org/10.1128/AEM.00251-09.

Tebo, B.M., Bargar, J.R., Clement, B.G., Dick, G.J., Murray, K.J., Parker, D., Verity, R., Webb, S.M., 2004. Biogenic manganese oxides: properties and mechanisms of formation. Annu. Rev. Earth Planet Sci. 32, 287–328. https://doi.org/10.1146/annurev.earth.32.101802.120213.

Thamdrup, B., 2012. New pathways and processes in the global nitrogen cycle. Annu. Rev. Ecol. Evol. Syst. 43, 407–428. https://doi.org/10.1146/annurev-ecolsys-102710-145048.

Thingstad, T.F., 2000. Elements of a theory for the mechanisms controlling abundance, diversity, and biogeochemical role of lytic bacterial viruses in aquatic systems. Limnol. Oceanogr. 45, 1320–1328.

Thomas, R., Berdjeb, L., Sime-Ngando, T., Jacquet, S., 2011. Viral abundance, production, decay rates and life strategies (lysogeny versus lysis) in Lake Bourget (France): bacteriophage ecology of Lake Bourget. Environ. Microbiol. 13, 616–630. https://doi.org/10.1111/j.1462-2920.2010.02364.x.

Tran, P.Q., Bachand, S.C., McIntyre, P.B., Kraemer, B.M., Vadeboncoeur, Y., Kimirei, I.A., Tamatamah, R., McMahon, K.D., Anantharaman, K., 2021. Depth-discrete metagenomics reveals the roles of microbes in biogeochemical cycling in the tropical freshwater Lake Tanganyika. ISME J. 15, 1971–1986. https://doi.org/10.1038/s41396-021-00898-x.

Tran, P., Ramachandran, A., Khawasik, O., Beisner, B.E., Rautio, M., Huot, Y., Walsh, D.A., 2018. Microbial life under ice: metagenome diversity and *in situ* activity of Verrucomicrobia in seasonally ice-covered Lakes. Environ. Microbiol. 20, 2568–2584. https://doi.org/10.1111/1462-2920.14283.

Tranvik, L.J., 1989. Bacterioplankton growth, grazing mortality and quantitative relationship to primary production in a humic and a clearwater lake. J. Plankton Res. 11, 985–1000. https://doi.org/10.1093/plankt/11.5.985.

Tranvik, L.J., Bertilsson, S., 2001. Contrasting effects of solar UV radiation on dissolved organic sources for bacterial growth. Ecol. Lett. 4, 458–463.

Tsementzi, D., Rodriguez, L.M., Ruiz-Perez, C.A., Meziti, A., Hatt, J.K., Konstantinidis, K.T., 2019. ecogenomic characterization of widespread, closely-related SAR11 clades of the freshwater genus "candidatus fonsibacter" and proposal of Ca. Fonsibacter lacus sp. nov. Syst. Appl. Microbiol. 42, 495–505.

Tseng, C.-H., Chiang, P.-W., Shiah, F.-K., Chen, Y.-L., Liou, J.-R., Hsu, T.-C., Maheswararajah, S., Saeed, I., Halgamuge, S., Tang, S.-L., 2013. Microbial and viral metagenomes of a subtropical freshwater reservoir subject to climatic disturbances. ISME J. 7, 2374–2386. https://doi.org/10.1038/ismej.2013.118.

Valdespino-Castillo, P.M., Alcántara-Hernández, R.J., Merino-Ibarra, M., Alcocer, J., Macek, M., Moreno-Guillén, O.A., Falcón, L.I., 2017. Phylotype dynamics of bacterial P utilization genes in microbialites and bacterioplankton of a monomictic Endorheic Lake. Microb. Ecol. 73, 296–309. https://doi.org/10.1007/s00248-016-0862-1.

van de Vossenberg, J., Woebken, D., Maalcke, W.J., Wessels, H.J.C.T., Dutilh, B.E., Kartal, B., Janssen-Megens, E.M., Roeselers, G., Yan, J., Speth, D., Gloerich, J., Geerts, W., van der Biezen, E., Pluk, W., Francoijs, K., Russ, L., Lam, P., Malfatti, S.A., Tringe, S.G., Haaijer, S.C.M., Op den Camp, H.J.M., Stunnenberg, H.G., Amann, R., Kuypers, M.M.M., Jetten, M.S.M., 2013. The metagenome of the marine anammox bacterium "*candidatus* scalindua profunda" illustrates the versatility of this globally important nitrogen cycle bacterium. Environ. Microbiol. 15, 1275–1289. https://doi.org/10.1111/j.1462-2920.2012.02774.x.

van Kessel, M.A.H.J., Speth, D.R., Albertsen, M., Nielsen, P.H., Op den Camp, H.J.M., Kartal, B., Jetten, M.S.M., Lücker, S., 2015. Complete nitrification by a single microorganism. Nature 528, 555–559. https://doi.org/10.1038/nature16459.

Varghese, N.J., Mukherjee, S., Ivanova, N., Konstantinidis, K.T., Mavrommatis, K., Kyrpides, N.C., Pati, A., 2015. Microbial species delineation using whole genome sequences. Nucleic Acids Res. 43, 6761–6771. https://doi.org/10.1093/nar/gkv657.

Vila, X., Abella, C.A., 1994. Effects of light quality on the physiology and the ecology of planktonic green sulfur bacteria in lakes. Photosynth. Res. 41, 53–65.

Vissers, E.W., Anselmetti, F.S., Bodelier, P.L.E., Muyzer, G., Schleper, C., Tourna, M., Laanbroek, H.J., 2013. Temporal and Spatial coexistence of Archaeal and bacterial *amoA* genes and gene transcripts in Lake Lucerne. Archaea 2013, 1–11. https://doi.org/10.1155/2013/289478.

Vlok, M., Gibbs, A.J., Suttle, C.A., 2019. Metagenomes of a freshwater charavirus from British Columbia provide a window into ancient lineages of viruses. Viruses 11, 299. https://doi.org/10.3390/v11030299.

Vonk, J.E., Tank, S.E., Bowden, W.B., Laurion, I., Vincent, W.F., Alekseychik, P., Amyot, M., Billet, M.F., Canário, J., Cory, R.M., Deshpande, B.N., Helbig, M., Jammet, M., Karlsson, J., Larouche, J., MacMillan, G., Rautio, M., Walter Anthony, K.M., Wickland, K.P., 2015. Reviews and syntheses: effects of permafrost thaw on Arctic aquatic ecosystems. Biogeosciences 12, 7129–7167. https://doi.org/10.5194/bg-12-7129-2015.

Wang, S., Dong, R.M., Dong, C.Z., Huang, L., Jiang, H., Wei, Y., Feng, L., Liu, D., Yang, G., Zhang, C., Dong, H., 2012. Diversity of microbial plankton across the three gorges dam of the yangtze river, China. Geosci. Front. Times 3, 335–349. https://doi.org/10.1016/j.gsf.2011.11.013.

Warren, M.J., Raux, E., Schubert, H.L., Escalante-Semerena, J.C., 2002. The biosynthesis of adenosylcobalamin (vitamin B12). Nat. Prod. Rep. 19, 390–412. https://doi.org/10.1039/b108967f.

Watanabe, Y., 1996. Limiting factors for bacterioplankton production in mesotrophic and hypereutrophic lakes: estimation by (3H)thymidine incorporation. Jpn. J. Limnol. Rikusuigaku Zasshi 57, 107–117. https://doi.org/10.3739/rikusui.57.107.

Watanabe, T., Miura, A., Iwata, T., Kojima, H., Fukui, M., 2017. Dominance of *Sulfuritalea* species in nitrate-depleted water of a stratified freshwater lake and arsenate respiration ability within the genus: dominance and arsenate respiration of *Sulfuritalea*. Environ. Microbiol. Rep. 9, 522–527. https://doi.org/10.1111/1758-2229.12557.

Watson, S.W., Novitsky, T.J., Quinby, H.L., Valois, F.W., 1977. Determination of bacterial number and biomass in the marine environment. Appl. Environ. Microbiol. 33, 940–946. https://doi.org/10.1128/aem.33.4.940-946.1977.

Weinbauer, M.G., 2004. Ecology of prokaryotic viruses. FEMS Microbiol. Rev. 28, 127–181. https://doi.org/10.1016/j.femsre.2003.08.001.

Weiss, J.V., Rentz, J.A., Plaia, T., Neubauer, S.C., Merrill-Floyd, M., Lilburn, T., Bradburne, C., Megonigal, J.P., Emerson, D., 2007. Characterization of neutrophilic Fe(II)-oxidizing bacteria isolated from the Rhizosphere of wetland plants and description of *ferritrophicum radicicola* gen. nov. sp. nov., and *sideroxydans paludicola* sp. nov. Geomicrobiol. J. 24, 559–570. https://doi.org/10.1080/01490450701670152.

Weitz, J.S., Stock, C.A., Wilhelm, S.W., Bourouiba, L., Coleman, M.L., Buchan, A., Follows, M.J., Fuhrman, J.A., Jover, L.F., Lennon, J.T., Middelboe, M., Sonderegger, D.L., Suttle, C.A., Taylor, B.P., Frede Thingstad, T., Wilson, W.H., Eric Wommack, K., 2015. A multitrophic model to quantify the effects of marine viruses on microbial food webs and ecosystem processes. ISME J. 9, 1352–1364. https://doi.org/10.1038/ismej.2014.220.

Welsh, A., Chee-Sanford, J.C., Connor, L.M., Löffler, F.E., Sanford, R.A., 2014. Refined NrfA phylogeny improves PCR-based *nrfA* gene detection. Appl. Environ. Microbiol. 80, 2110–2119. https://doi.org/10.1128/AEM.03443-13.

Wetzel, R.G., 1967. Dissolved organic compounds and their utilization in two marl lakes. In problems of organic matter determination in freshwater. Hidrol. Kozlony 47, 298–303.

Wetzel, R.G., 1968. Dissolved organic matter and phytoplanktonic productivity in marl lakes. Mitt Intern. Ver. Limnol 14, 261–270.

Wetzel, R.G., 1984. Detrital dissolved and particulate organic carbon functions in aquatic ecosystems. Bull. Mar. Sci. 35, 503–509.

Wetzel, R.G., Hatcher, P.G., Bianchi, T.S., 1995. Natural photolysis by ultraviolet irradiance of recalcitrant dissolved organic matter to simple substrates for rapidbacterial metabolism. Limnol. Oceanogr. 40, 1369–1380. https://doi.org/10.4319/lo.1995.40.8.1369.

Wetzel, R.G., Likens, G.E., 2000. Limnological Analyses, third ed. Springer-Verlag, New York.

Wetzel, R.G., Riche, P.H., Miller, M.C., Allen, H.L., 1972. Metabolism of dissolved and particulate detrital carbon in a temperate hard-water lake. Mem. Ist. Ital. Idrobiol. 29, 185–243.

Wetzel, R.G., Ward, A.K., 1992. Primary production. In: Rivers Handbook. I. Hydrological and Ecological Principles. Blackwell Scientific Publishers, Oxford, pp. 345–369.

Wilhelm, S., 1995. Ecology of iron-limited cyanobacteria: a review of physiological responses and implications for aquatic systems. Aquat. Microb. Ecol. 9, 295–303. https://doi.org/10.3354/ame009295.

Wilhelm, S., Matteson, A., 2008. Freshwater and marine virioplankton: a brief overview of commonalities and differences. Freshw. Biol. 53, 1076–1089. https://doi.org/10.1111/j.1365-2427.2008.01980.x.

Wilhelm, S.W., Maxwell, D.P., Trick, C.G., 1996. Growth, iron requirements, and siderophore production in iron-limited Synechococcus PCC 72. Limnol. Oceanogr. 41, 89–97. https://doi.org/10.4319/lo.1996.41.1.0089.

Wilhelm, S.W., Suttle, C.A., 1999. Viruses and nutrient cycles in the sea—viruses play critical roles in the structure and function of aquatic food webs. Bioscience 49, 8. https://doi.org/10.2307/1313569.

Williams, H.N., Lymperopoulou, D.S., Athar, R., Chauhan, A., Dickerson, T.L., Chen, H., Laws, E., Berhane, T.-K., Flowers, A.R., Bradley, N., Young, S., Blackwood, D., Murray, J., Mustapha, O., Blackwell, C., Tung, Y., Noble, R.T., 2016. Halobacteriovorax, an underestimated predator on bacteria: potential impact relative to viruses on bacterial mortality. ISME J. 10, 491–499. https://doi.org/10.1038/ismej.2015.129.

Winter, C., Bouvier, T., Weinbauer, M.G., Thingstad, T.F., 2010. Trade-offs between competition and defense specialists among unicellular planktonic organisms: the "Killing the Winner" hypothesis revisited. Microbiol. Mol. Biol. Rev. 74, 42–57. https://doi.org/10.1128/MMBR.00034-09.

Winter, C., Hein, T., Kavka, G., Mach, R.L., Farnleitner, A.H., 2007. Longitudinal changes in the bacterial community composition of the Danube River: a whole-river approach. Appl. Environ. Microbiol. 73, 421–431. https://doi.org/10.1128/AEM.01849-06.

Woese, C.R., Fox, G.E., 1977. Phylogenetic structure of the prokaryotic domain: the primary kingdoms. Proc Natl Acad Sci U A 74, 5088–5090.

Wommack, K.E., Colwell, R.R., 2000. Virioplankton: viruses in aquatic ecosystems. Microbiol. Mol. Biol. Rev. 64, 69–114. https://doi.org/10.1128/MMBR.64.1.69-114.2000.

Work, K.A., 2003. Zooplankton grazing on bacteria and cyanobacteria in a eutrophic lake. J. Plankton Res. 25, 1301–1306. https://doi.org/10.1093/plankt/fbg092.

Wright, R.T., 1975. Studies on glycolic acid metabolism by freshwater bacterial: glycolic acid and bacteria. Limnol. Oceanogr. 20, 626–633. https://doi.org/10.4319/lo.1975.20.4.0626.

Wright, R.T., Hobbie, J.E., 1966. Use of glucose and acetate by bacteria and algae in aquatic ecosystems. Ecology 47, 447.

Wurtsbaugh, W.A., Horne, A.J., 1983. Iron in eutrophic Clear Lake, California: its importance for algal nitrogen fixation and growth. Can. J. Fish. Aquat. Sci. 40, 1419–1429.

Wurzbacher, C., Salka, I., Grossart, H.-P., 2012. Environmental actinorhodopsin expression revealed by a new *in situ* filtration and fixation sampler: environmental actinorhodopsin expression unearthed by IFFS. Environ. Microbiol. Rep. 4, 491–497. https://doi.org/10.1111/j.1758-2229.2012.00350.x.

Xu, H., Zhao, D., Huang, R., Cao, X., Zeng, J., Yu, Z., Hooker, K.V., Hambright, K.D., Wu, Q.L., 2018. Contrasting network features between free-living and particle-attached bacterial communities in Taihu Lake. Microb. Ecol. 76, 303–313. https://doi.org/10.1007/s00248-017-1131-7.

Xu, H., Zhu, G., Qin, B., Paerl, H.W., 2013. Growth response of *Microcystis* spp. to iron enrichment in different regions of Lake Taihu, China. Hydrobiologia 700, 187–202. https://doi.org/10.1007/s10750-012-1229-3.

Yannarell, A.C., Triplett, E.W., 2004. Within- and between-lake variability in the composition of bacterioplankton communities: investigations using multiple spatial scales. Appl. Environ. Microbiol. 70, 214–223.

Yannarell, A.C., Triplett, E.W., 2005. Geographic and environmental sources of variation in lake bacterial community composition. Appl. Environ. Microbiol. 71, 227–239. https://doi.org/10.1128/AEM.71.1.227-239.2005.

Yao, M., Elling, F.J., Jones, C., Nomosatryo, S., Long, C.P., Crowe, S.A., Antoniewicz, M.R., Hinrichs, K.-U., Maresca, J.A., 2016. Heterotrophic bacteria from an extremely phosphate-poor lake have conditionally reduced phosphorus demand and utilize diverse sources of phosphorus: phosphorus demand corresponds to environmental supply. Environ. Microbiol. 18, 656–667. https://doi.org/10.1111/1462-2920.13063.

Yau, S., Lauro, F.M., DeMaere, M.Z., Brown, M.V., Thomas, T., Raftery, M.J., Andrews-Pfannkoch, C., Lewis, M., Hoffman, J.M., Gibson, J.A., Cavicchioli, R., 2011. Virophage control of antarctic algal host-virus dynamics. Proc. Natl. Acad. Sci. 108, 6163–6168. https://doi.org/10.1073/pnas.1018221108.

Yoshinaga, I., Amano, T., Yamagishi, T., Okada, K., Ueda, S., Sako, Y., Suwa, Y., 2011. Distribution and diversity of anaerobic ammonium oxidation (anammox) bacteria in the sediment of a eutrophic freshwater lake, lake kitaura, Japan. Microbes Environ 26, 189–197. https://doi.org/10.1264/jsme2.ME10184.

Yutin, N., Wolf, Y.I., Koonin, E.V., 2014. Origin of giant viruses from smaller DNA viruses not from a fourth domain of cellular life. Virology 466–467. https://doi.org/10.1016/j.virol.2014.06.032, 38–52.

Zaremba-Niedzwiedzka, K., Viklund, J., Zhao, W., Ast, J., Sczyrba, A., Woyke, T., McMahon, K., Bertilsson, S., Stepanauskas, R., Andersson, S.G.E., 2013. Single-cell genomics reveal low recombination frequencies in freshwater bacteria of the SAR11 clade. Genome Biol. 14, R130. https://doi.org/10.1186/gb-2013-14-11-r130.

Zeder, M., Peter, S., Shabarova, T., Pernthaler, J., 2009. A small population of planktonic *Flavobacteria* with disproportionally high growth during the spring phytoplankton bloom in a prealpine lake. Environ. Microbiol. 11, 2676–2686. https://doi.org/10.1111/j.1462-2920.2009.01994.x.

Zehr, J.P., Jenkins, B.D., Short, S.M., Steward, G.F., 2003. Nitrogenase gene diversity and microbial community structure: a cross-system comparison. Environ. Microbiol. 5, 539–554. https://doi.org/10.1046/j.1462-2920.2003.00451.x.

Zeng, Y., Kasalický, V., Šimek, K., Koblížek, M., 2012. Genome sequences of two freshwater Betaproteobacterial isolates, Limnohabitans species strains Rim28 and Rim47, indicate their capabilities as both photoautotrophs and ammonia oxidizers. J. Bacteriol. 194, 6302–6303. https://doi.org/10.1128/JB.01481-12.

Zhang, W., Wang, H., Li, Y., Zhu, X., Niu, L., Wang, C., Wang, P., 2020. Bacterial communities along a 4500-meter elevation gradient in the sediment of the Yangtze River: what are the driving factors? Desalination Water Treat. 177, 109–130. https://doi.org/10.5004/dwt.2020.24875.

Zheng, Y., Hou, L., Liu, M., Yin, G., 2019. Dynamics and environmental importance of anaerobic ammonium oxidation (anammox) bacteria in urban river networks. Environ. Pollut. 254, 112998. https://doi.org/10.1016/j.envpol.2019.112998.

Zhou, S., Borjigin, S., Riya, S., Terada, A., Hosomi, M., 2014. The relationship between anammox and denitrification in the sediment of an inland river. Sci. Total Environ. 490, 1029–1036. https://doi.org/10.1016/j.scitotenv.2014.05.096.

CHAPTER

24

Freshwater Plants

Patricia A. Chambers[1] *and Stephen C. Maberly*[2]

[1]Burlington, Ontario, Canada

[2]Lake Ecosystems Group, UK Centre for Ecology & Hydrology, Lancaster, UK

OUTLINE

I. Characteristics of freshwater plants

A. Definition and growth forms

Freshwater plants, often called *aquatic macrophytes*, are challenging to define precisely because of their evolutionary diversity, the spatial and chemical (including salinity) gradients of the habitats in which they live, and the temporal variability of water level. Here we include all freshwater photosynthetic organisms, large enough to see with the naked eye, that actively grow permanently or periodically submerged below, floating on, or growing up through the water surface. Examples include the submerged pondweeds and the floating

https://doi.org/10.1016/B978-0-12-822701-5.00024-0

duckweeds and water lilies, as well as the emergent common reed (*Phragmites australis*), which depends on flooded soils and is integral to the functioning of freshwaters. The definition encompasses *embryophytes* (plants where the embryo is protected within a maternal tissue, largely synonymous with plants that invaded the land) as well as a group of green macroalgae, the charophytes (Charophyceae; also discussed in Chapter 17). The freshwater embryophytes comprise nonvascular plants, namely *hornworts* (Anthocerotophyta), *liverworts* (Marchantiophyta), and *mosses* (Bryophyta), as well as *tracheophytes* with vascular tissues, specifically the pteridophytes (comprising the order Isoetales [*quillworts*] within the Lycopodiopsida and *ferns* Polypodiopsida) and *angiosperms* (Magnoliopsida, flowering plants). The angiosperms comprise monocots (Lilianae) and dicots (a polyphyletic group comprising the nonmonocot angiosperms); The Angiosperm Phylogeny Group, (2016) provides information regarding the position of freshwater species within the angiosperms. We have excluded filamentous algae such as the chlorophyte *Cladophora* or the rhodophyte *Batrachospermum*, as they have very different structures and ecology and are covered in Chapter 17. We also exclude *lichens*, which comprise up to 20,000 species with at least 250 freshwater (largely amphibious) species (1% of total) (Krzewicka et al., 2017), as they are very understudied.

Most freshwater plants derive from land plants (Arber, 1920; Sculthorpe, 1967; Cook, 1999) that in turn evolved from freshwater algae (Wickett et al., 2014). In this respect, freshwater plants are analogous to whales and dolphins: they returned to water after their ancestors evolved to live on land. Some species of freshwater plants have returned back to a terrestrial environment (Ito et al., 2017). For terrestrial plants, adaptations to life on land included (1) managing water content and gas exchange, (2) producing structural strength to counter the low density of air compared to water, and (3) resisting damage by UV radiation that is attenuated in water. These characteristics formed the genetic background of the earliest freshwater plants that then adapted to the very different benefits and constraints of living under water (Graham et al., 2014).

This temporal gradient over evolutionary time of water to land and back to water is related to a spatial gradient between land and water that is blurred by episodic and seasonal changes in water level, including seasonal drying in temporary pools and streams. This also forms a temporal gradient or *hydrosere* with a succession over time from open water to dry land as the open water is filled in with material from the surrounding land and produced internally. The indistinct transition from land to water probably facilitated the invasion of freshwater by land plants. Indeed, Du and Wang, (2014) found evidence for an evolutionary progression from emergent to amphibious to fully submerged leaves in freshwater lineages. The invasion of land plants into water is believed to have occurred independently in at least 100 lineages (Les and Tippery, 2013). Angiosperms from 18 genera can complete their life cycle completely submerged as pollination of flowers can occur under water (Philbrick and Les, 1996), whereas in most genera flowers are produced in air, like their terrestrial ancestors.

The interface between land and water is a gradual cline that produces characteristic zonation patterns (Spence, 1982). Terrestrial and riparian vegetation occur at the top of the gradient, with leaves in the air but roots in soils that are occasionally flooded. These grade into emergent wetland plants with leaves in the air, underwater stems, and roots in flooded soil, and then to freshwater plants where leaves are at the water surface or are completely submerged (Fig. 24-1). The depth range over which this zonation occurs is controlled largely by the penetration of light into water but, in exposed sites, freshwater plants can be excluded from shallow water by intense wave action or unsuitable sediment.

Living at the interface between land and open water, freshwater plants can mediate biogeochemical fluxes between the land, water, and atmosphere. The shallow slopes along many land—water margins are conducive to sedimentation, and fine sand, silt, and organic matter can accumulate on sheltered shores. Submerged soils are often anoxic, altering nutrient fluxes, rates of decomposition, plant metabolism, and community interactions, compared to oxic soils. Emergent or floating species such as *Phragmites australis*, *Equisetum fluviatile*, and *Eichhornia crassipes* are important conduits for methane emission (Bergstrom et al., 2007; Oliveira Junior et al., 2021). Plants also produce physical structure in the littoral that alters hydrodynamic conditions and provides habitats for a wide range of organisms (Section 24.III.D) and controls the balance between production in the littoral and pelagic environments.

Several different schemes exist to classify freshwater plants based upon growth form (see Hutchinson, 1975; Wiegleb, 1991). Growth form is, however, just one of many traits that determine the response of plants to their environment and their influence on ecosystem properties (Kattge et al., 2020). For macrophytes growing on soft sediment, Kautsky (1988) adapted the *C-S-R* (*Competitor-Stress-Ruderal*) model for terrestrial plants of Grime (1979) that is based on the ability of plants to tolerate stress or disturbance by adding a fourth strategy, B, *Biomass storer* that is found when stress is high but disturbance is low. Willby et al. (2000) produced a more comprehensive system for European freshwater species based on 58 attributes including those related to growth form; shoot architecture and flexibility; leaf type, size, and texture; root characteristics; relative

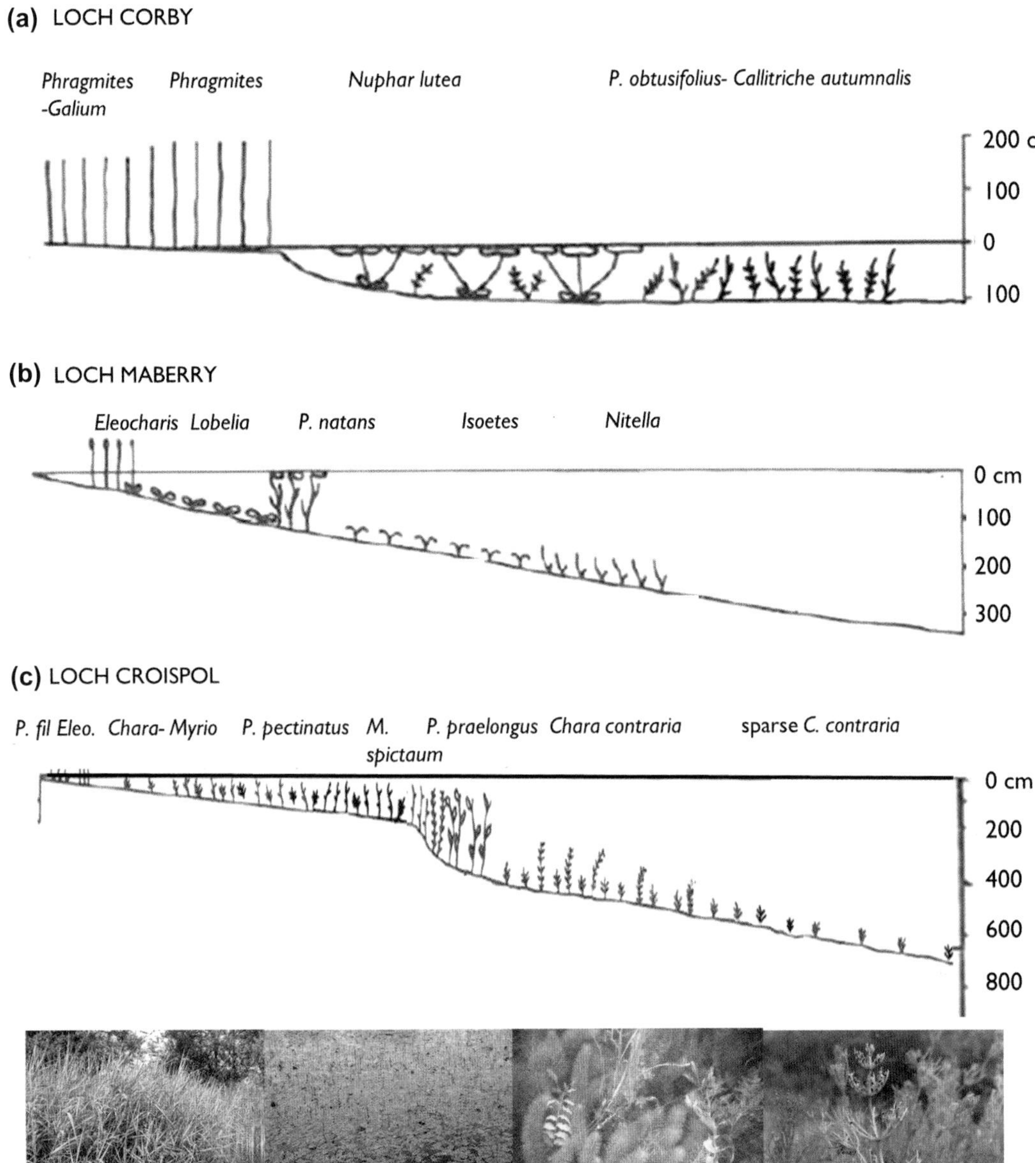

FIGURE 24-1 Examples of changing freshwater plant growth form with depth. Diagrams of Scottish lochs (a) Corby, (b) Maberry, and (c) Croispol are from Spence (1964). *(Photos show [from left to right]: Phragmites australis. [Photo by Klaus van de Weyer, lanaplan GbR.] An extensive bed of Potamogeton natans at Borrans Reservoir, UK. [Photo by the authors.] A community of plants at about 3 m in Loch Borralie, UK. [Photo by the late D.H.N. Spence.] Chara contraria. [Photo by Klaus van de Weyer.])*

below-ground biomass; pollination; and reproductive features.

The diversity of freshwater plant form, life history, and environmental variation means that any classification will be imperfect. For example, a given species may fit into more than one category depending on the environmental conditions. The choice of a classification system will depend, in part, on the purpose and on a trade-off between precision afforded by many categories and generality resulting from few categories. In Table 24-1, we use a simple growth form classification based on plant habit; the usual position of the plants in the water column; and their access to resources in the air, water, and sediment. The different growth forms are illustrated in Fig. 24-2.

i. Emergent plants

Emergent plants grow in swamp forests, fens, and reed swamps (Spence, 1982), as well as along the margins of freshwaters, and can obtain their resources from the sediment, air, and, to a lesser extent, water (Table 24-1). Aerial stems and leaves of emergent

TABLE 24-1 Growth Forms of Freshwater Plants

Name	Description	Resource	Example Species	Photo in Fig. 24-2
Emergent	Rooted with leaves normally above water	A, W, S	*Nelumbo nucifera, Phragmites australis, Typha angustifolia*	a
Amphibious and floating leaved	Rooted with leaves habitually or periodically under, as well as on or above, water	A, W, S	*Nuphar lutea, Trapa natans, Veronica beccabunga*	b, c
Pleustophyte	Floating on the water	A, W	*Eichhornia crassipes, Lemna minor, Salvinia adnata*	d
Submerged—isoetid	Submerged, rooted with leaves in a rosette	W, S	*Isoetes lacustris, Littorella uniflora, Lobelia dortmanna*	e
Submerged—elodeid	Submerged, rooted with leaves on elongated stems	W, S	*Hydrilla verticillata, Potamogeton lucens, Ranunculus fluitans*	f
Submerged—charophyte	Submerged, green macroalgae with rhizoids	W, (S)	*Chara vulgaris, Nitella flexilis, Nitellopsis obtusa*	g
Submerged—haptophyte	Submerged, attached to a substrate but not penetrating it	W	*Fontinalis antipyretica, Hydrostachys polymorpha, Podostemum ceratophyllum*	h
Submerged—free floating	Submerged, free floating within the water column	W	*Aldrovanda vesiculosa, Ceratophyllum demersum, Lemna trisulca*	i

Resource Availability: A, air; W, water; S, sediment.

macrophytes are similar in morphology and physiology to terrestrial plants. Emergent *monocots* (a major group of angiosperms that have seeds with a single seed leaf) such as *Phragmites australis* and *Typha latifolia* produce erect, linear, or lanceolate leaves from an extensive anchoring system of *rhizomes* (underground stems that store carbohydrates and other resources) and roots (Fig. 24-2a). Other species such as the *dicot* (a major polyphyletic group of angiosperms that have seeds with two seed leaves) *Nelumbo nucifera* produce large round or oval leaves above the water surface (Fig. 24-3).

ii. Amphibious and floating-leaved plants

Amphibious and floating-leaved plants can obtain resources from the sediment, water, and air and are the most varied in habit and dependence on water. They range from riparian, marginal, and *hygrophytic* (adapted to wet and moist conditions) plants that experience and tolerate episodic or seasonal changes in water level or live in permanently wet conditions such as adjacent to upland streams or waterfalls. Others, such as *Nymphaea alba* or *Trapa natans*, only have leaves floating on the surface and thus exchange gases with the atmosphere. In contrast, species such as *Nuphar lutea* produce specialized underwater leaves (Fig. 24-2c) as well as floating leaves (Fig. 24-2b). Floating leaves are subject to severe mechanical stresses from wind and waves. Adaptations to these stresses include the tendency to produce leaves that are strong, leathery, and circular or ovoid in shape with an entire margin. Floating leaves have a well-developed dorsoventral organization, in which the *mesophyll* (the middle of the leaf between the upper and lower epidermis) is usually differentiated into an upper photosynthetic palisade mesophyll with closely spaced rectangular cells and a lower spongy mesophyll with extensive lacunae (Arber, 1920; Sculthorpe, 1967; Hutchinson, 1975). Localized masses of spongy tissue aid buoyancy and, in combination with vascular tissues, offer resistance to tearing. Stomata are usually restricted to the upper-leaf epidermis, but a few nonfunctional stomata can be found on the underside of leaves. Amphibious species are often *heterophyllous* with leaves of different shape and structure in air and water (Maberly and Spence, 1989; Li et al., 2019), such as *Nuphar lutea* in

(a)
(b)
(c)
(d)
(e)
(f)
(g)
(h)
(i)

FIGURE 24-2 Growth forms of freshwater plants. (a) Emergent (*Typha angustifolia*); (b) and (c) amphibious/floating (*Nuphar lutea*); (d) pleustophyte (*Lemna minor*); (e) isoetid (*Littorella uniflora*); (f) elodeid (*Elodea nuttallii*); (g) charophyte (*Nitellopsis obtusa*); (h) haptophyte (*Macropodiella garrettii*); (i) free-floating (*Aldrovanda vesiculosa*). *(All photos courtesy Klaus van de Weyer [lanaplan GbR] except [h] courtesy Ehoarn Bidault [Missouri Botanical Garden] and [i] courtesy Lubomir Adamec (Institute of Botany of the Academy of Sciences of the Czech Republic.)*

FIGURE 24-3 A large bed of lotus (*Nelumbo nucifera*) in China. *(Photo by the authors.)*

Fig. 24-2, further supporting the functional significance of leaf shape for life under water.

iii. Pleustophytes (free-floating plants)

Pleustophytes are free-floating at the water surface and can obtain resources from air and water. They include the monocots *Lemna* spp. and *Wolffia* spp. (Fig. 24-2d), which are greatly reduced in size and structural complexity and are usually restricted to small lakes, ponds, and slow-flowing rivers (Laird and Barks, 2018); the aggressive water hyacinth *Eichhornia crassipes*, which is large and can blanket the surface of lakes and ponds; and the ferns (Polypodiophyta) *Salvinia* spp. and *Azolla* spp., which are intermediate in size.

iv. Submerged, isoetids

The isoetid growth form is typified by the lycophyte *Isoetes* as well as dicots such as *Littorella uniflora* (Fig. 24-2e) and *Lobelia dortmanna* and monocots such as *Eleocharis* spp. Isoetid leaves arise from a basal rosette and are usually short, stubby, cylindrical, or awl-like (Fig. 24-4a) with a large cross-sectional area comprising lacunae (Fig. 24-5a) that run down the leaf length (Smolders et al., 2002). They often have a very large proportion of their biomass in roots. In some species, such as *Lobelia dortmanna*, lacunae are continuous from the root to leaf and the sediment is a more important source of material resources, including inorganic carbon for photosynthesis, than the water (see Section 24.II.B). Some isoetids are amphibious and can tolerate exposure to air when water levels fall, further highlighting the impossibility of producing a "water-tight" classification scheme for freshwater plants. Genera such as *Vallisneria* (Fig. 24-4e) also produce leaves in a rosette but are larger than typical isoetids.

v. Submerged, elodeids

Elodeids (Fig. 24-2f) are a species-rich group of largely angiosperms that can also obtain material resources from the sediment and water. They are usually taller than the isoetids and produce long shoots from which leaves arise, and can form dense beds and subsurface canopies. Their leaves have a number of different forms. They can be narrow or linear such as in *Callitriche brutia* (Fig. 24-4b), *Potamogeton obtusifolius* (Fig. 24-4c), or *Ranunculus fluitans* ssp. *penicillatus* (Fig. 24-4d). Some species have larger and broader leaves such as the submerged leaves of *Nuphar lutea* (Figure 24-4g) and some species of *Ottelia*, *Echinodorus*, and *Potamogeton* (Fig. 24-4h). The leaves can have wavey edges as in *Potamogeton crispus* and *Aponogeton crispus* (Fig. 24-4f) and the leaves can be *fenestrated* as in *Aponogeton madagascariensis* (Fig. 24-4i), where the lamina between the veins breaks down, potentially reducing self-shading and minimizing barriers to diffusion. Leaves are often arranged in whorls. Dissected, whorled leaves, as in *Hottonia palustris* (Fig. 24-4j) and *Myriophyllum verticillatum* (Fig. 24-4k), are common in dicots, whereas whorled, entire leaves are more common in monocots such as in *Elodea* spp. or *Hydrilla verticillata* (Figure 24-4l). Frequently, leaves are thin, comprising only an upper and lower epidermis such as in *Elodea* spp. (Fig. 24-5c), or with very limited mesophyll cells with large air spaces such as in some *Ottelia* spp. (Fig. 24-5b).

vi. Submerged, charophytes

Green macroalgae, from the family Charophyceae (see Chapter 17), often comprise a giant internodal cell (up to 20 cm in length) with whorls of smaller cells around it (Fig. 24-6a). Because they have not evolved from land plants, they do not possess roots but do have rhizoids that anchor them to the substrate. This also allows them to obtain at least some nutrients from the sediment (Section 24.II.C).

vii. Submerged, haptophytes

Submerged haptophytes are anchored to a substrate but do not derive nutrients from it and so are mainly

FIGURE 24-4 Leaf forms of submerged freshwater tracheophytes. (a) *Isoetes sinensis*; (b) *Callitriche brutia*; (c) *Potamogeton obtusifolius*; (d) *Ranunculus fluitans* ssp. *penicillatus*; (e) *Vallisneria spiralis*; (f) *Aponogeton crispus*; (g) *Nuphar lutea*; (h) *Potamogeton lucens*; (i) *Aponogeton madagascariensis*; (j) *Hottonia palustris*; (k) *Myriophyllum verticillatum*; (l) *Hydrilla verticillata*. *(Photos courtesy: [a] Qing-Feng Wang, Chinese Academy of Science; [b] and [d] David Fenwick (www.aphotoflora.com); [c] and [g] Klaus van de Weyer (lanaplan GbR); [f], [i], and [j] Tropica Aquarium Plants, www.tropica.com; [l] Hong-Sheng Jiang, Chinese Academy of Sciences; [e], [h], and [k] The authors.)*

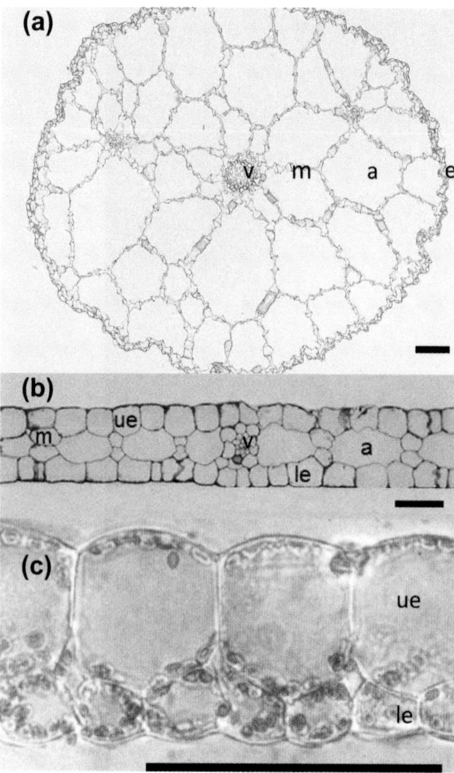

FIGURE 24-5 Cross-section of submerged leaves of: (a) *Littorella uniflora* (photo courtesy Veronika Jandová, Institute of Botany of the Academy of Sciences of the Czech Republic); (b) *Ottelia alismoides* (Han et al., 2020); (c) *Elodea callitrichoides* (Falk and Sitte, 1963). Indicated are: a, air spaces; e, epidermal cells (ue upper epidermis; le, lower epidermis); m, mesophyll cells; and v, vascular bundle. *Scale bar* represents 100 μm.

restricted to the water for their material resources. However, many species become amphibious and can access CO_2 from the air when water levels are low. Primary examples of haptophytes are freshwater mosses and liverworts, whose rhizoids anchor them to a substrate. Another major group of haptophytes is the subtropical and tropical dicot family Podostemaceae (Fig. 24-2h), and the genus *Hydrostachys*, which have fibrous or ribbon-like roots for attachment and grow on rocky substrates in fast-flowing water (Koi et al., 2006).

viii. Submerged, free-floating

A few freshwater plants are free-floating within the water column and are thus reliant solely on water for their material resources. These include *Ceratophyllum* spp. and *Lemna trisulca* that are predominantly found in nutrient-rich sites, and species such as *Aldrovanda vesiculosa* (Fig. 24-2i) and *Utricularia* spp. that are carnivorous and can supplement their nutrient requirements by trapping zooplankton prey (Adamec, 1997).

B. Evolution, phylogeny, and taxonomy

The Charophyta, including green macroalgal genera such as *Chara* and *Nitella* (Fig. 24-6a), are an ancient group that began to colonize land about 500 to 450 million years ago (Domozych et al., 2016). The first embryophytes, the bryophytes comprising hornworts, liverworts, and mosses (Fig. 24-6b - d), are thought to have evolved between about 515 and 470 million years ago (mid-Cambrian to early-Ordovician [Morris et al., 2018]), probably from the Zygnematophyceae, a sister family to the Charophyceae (Wickett et al., 2014). The vegetative stage of the charophytes and bryophytes is haploid, unlike the other freshwater plants that are diploid or polyploid (Lobato-de Magalhães et al., 2021). The first tracheophytes comprising Lycopodiopsida (Fig. 24-6e) and Polypodiopsida (Fig. 24-6f) evolved between about 470 and 420 million years ago (Morris et al., 2018). Angiosperms evolved from other tracheophytes around 209 million years ago (Li et al., 2019). There is still considerable uncertainty over which plants were the first angiosperms (Les, 2015), but a number of freshwater families, such as water lilies and species from the genus *Ceratophyllum*, are ancient and basal in the angiosperm phylogeny (The Angiosperm Phylogeny Group, 2016). In any case, fossil and phylogenetic data show that the return to freshwaters occurred very early in the evolution of the angiosperms and that freshwater plants existed at least in the Lower Cretaceous around 130 million years ago (Gomez et al., 2015).

The taxonomic classification of freshwater plants and algae is constantly being refined and species numbers within different taxonomic levels are frequently revised, so the number of families, genera, and species is not fixed, nor is the demarcation between species and infraspecies in some groups. There is even more uncertainty over the numbers of freshwater taxa because of the difficulty of deciding when a plant is, indeed, a freshwater plant, for the reasons discussed above. Nevertheless, we have provided an overview of the numbers of freshwater plants in the major plant taxa in Table 24-2. Overall, 211 families, 823 genera, and 5006 species of freshwater plants are recognized here (about 4500 species if the charophytes are excluded), representing, respectively, 29%, 15%, and 8% of the global total (Table 24-2). In terms of species the monocots (Fig. 24-6h), followed by the dicots (Fig. 24-6g), are the most numerous. There are also a large number of mosses; however, because they are attached and of small stature, they are particularly difficult to characterize as aquatic or not, as their small stature means that they can be periodically submerged and also frequently inhabit splash

FIGURE 24-6 Phylogenetic groups of freshwater plants. (a) Charophyte (*Nitella translucens*); (b) hornwort (*Phaeoceros carolinianus*); (c) liverwort (*Scapania undulata*); (d) moss (*Fontinalis squamosa*); (e) quillwort (*Isoetes lacustris*); (f) fern (*Pilularia globulifera*); (g) dicot (*Ranunculus fluitans*); (h) monocot (*Ottelia acuminata*). *(All photos courtesy Klaus van de Weyer [lanaplan GbR] except [b] David H. Wagner [Northwest Botanical Institute, USA]; and [h] The authors.)*

TABLE 24-2 Distribution of Freshwater Plant Families, Genera, and Species Among Different Extant Taxonomic Groups, With the Percentage of Freshwater to Total Number of Taxa in Parentheses

Group	Families	Genera	Species	Example Genera	Photo, Fig. 24-6
Charales (charophytes)[1,2]	2 (100%)	6 (100%)	467 (100%)	*Chara, Nitella*	a
Anthocerotophyta (hornworts)[3,4]	4 (80%)	6 (50%)	14 (7%)	*Anthoceros, Phaeoceros*	b
Marchantiophyta (liverworts)[4,5]	45 (54%)	97 (25%)	248 (4%)	*Riccia, Scapania*	c
Bryophyta (mosses)[4,6]	65 (55%)	253 (23%)	766 (6%)	*Fontinalis, Sphagnum*	d
Lycopodiopsida (club mosses)[7,8]	2 (67%)	2 (11%)	131 (10%)	*Isoetes, Lycopodiella*	e
Polypodiopsida (ferns)[7,8]	8 (17%)	12 (4%)	89 (1%)	*Azolla, Pilularia*	f
Angiosperms (total)	85 (20%)	447 (3%)	3291 (1%)		
Dicots[8,9,10]	56 (17%)	240 (2%)	1557 (<1%)	*Myriophyllum, Ranunculus*	g
Lilianae (monocots)[8,9]	29 (38%)	207 (8%)	1734 (2%)	*Ottelia, Potamogeton*	h
TOTAL	**211 (29%)**	**823 (15%)**	**5006 (8%)**		

Notes: 1, Guiry, M.D. & Guiry, G.M. 2021. AlgaeBase. World-wide electronic publication, National University of Ireland, Galway. https://www.algaebase.org; searched on May 10, 2021. 2, The genus *Lamprothamnium*, included here, grows in brackish water. 3, Total numbers were retrieved [May 8, 2021] from the Integrated Taxonomic Information System (ITIS) online database (USGS, 2013). 4, Janice Glime unpublished data; see Glime (2021) for the approach that takes a broad view of freshwater species. 5, Crandall-Stotler et al. (2009). 6, Cox et al. (2010). 7, PPG I (2016). 8, Kevin Murphy pers. comm. and Murphy et al. (2019). 9, Christenhusz and Byng (2016). 10, including basal angiosperms.

zones adjacent to running water. Of the flowering plants, the monocots are the group of plants with the greatest proportion of plants specialized to live in freshwater (about 2% of species) compared to <1% of dicots. The seagrasses are the only terrestrial plants that invaded the oceans, and the very small number of species, approximately 70, all belong to the monocots. There are no freshwater plants within the approximately 1000 species of gymnosperms as we excluded large species such as *Taxodium* spp. in the Florida Everglades.

C. Morphological and anatomical adaptations to life in water

This chapter was written just over a century after the publication of the first modern book on freshwater plants (Arber, 1920), which focused on their structure and morphology, and it is still an important book. Sculthorpe (1967) provides additional detailed information about the anatomy and morphology of freshwater plants. The main morphological and structural adaptations to life in water (following the growth forms in Table 24-1) are outlined below, with more detailed descriptions available in Arber (1920) and Sculthorpe (1967).

Evolutionary pressure has produced a relatively limited array of underwater leaf morphology (Arber, 1920; Sculthorpe, 1967), but the leaves of freshwater plants are among both the smallest and largest of all known leaves and some have the highest *specific leaf area* (SLA, the ratio of leaf area to dry mass) of any species (Diaz et al., 2016). They vary in (projected) area from <1 mm^2 for some pleustophytes such as *Wolffia arrhiza* to leaves of typical area of around 300 mm^2 to the large floating leaves of 44,000 mm^2 of *Nymphaea alba* and, ultimately, to the huge floating leaves of *Victoria amazonica* of nearly 5,000,000 mm^2 (5 m^2). The largest underwater leaves are never as large as the largest floating leaves, presumably because they would be damaged by water currents, but *Ottelia alismoides* has leaves that can reach 13,200 mm^2 in area (Huang et al., 2018) and the underwater leaves of the yellow water lily, *Nuphar lutea*, can have an area of 35,000 mm^2 (Schoelynck et al., 2014).

The morphology of underwater leaves differs from their terrestrial ancestors as well as floating or emergent leaves of heterophyllous amphibious plants (Arber, 1920; Sculthorpe, 1967; Maberly and Gontero, 2018). Stomata evolved in early land plants about 400 million years ago (Edwards et al., 1998) to allow exchange of CO_2 and O_2 while controlling water loss. Most submerged leaves lack functional stomata as the need to restrict water loss is redundant, but the ability to produce them remains in the genome because they are present on the aerial surface of floating leaves and occasionally on both *abaxial* (lower) and *adaxial* (upper) surfaces as in the fern *Marsilea quadrifolia* (Lin and Yang, 1999). A second adaptation that evolved in terrestrial plants to restrict water loss is the production of a translucent layer on the outside of epidermal cells, the *cuticle*. This comprises a range of compounds, including a polymer matrix (*cutin*), polysaccharides, and waxes (Riederer and Schreiber, 2001; Yeats and Rose, 2013), and typically ranges in thickness from 1 to 10 μm. In contrast, submerged freshwater plants, with no need to restrict water loss, but still needing to exchange gases and nutrients over the leaf surface, generally have much thinner cuticles, around 0.1 μm, although species such as *Lobelia dortmanna* that rely on sediment CO_2 (see Section 24.II.B) have much thicker cuticles of about 1 μm (Frost-Christensen et al., 2003). The permeability for water of *Potamogeton lucens* leaves is several thousand times greater than the permeability of terrestrial leaves (reviewed in Riederer and Schreiber, 2001; Schuster et al., 2017), and permeabilities for O_2 and CO_2 in submerged leaves are higher than in aerial or floating leaves of the same species (Frost-Christensen and Floto, 2007).

Some of the differences in leaves acclimated to water or air can be seen in amphibious plants. For example, Klancnik et al. (2014) showed that submerged, floating, and emerged leaves of *Sagittaria sagittifolia* are phenotypically very different despite being genotypically identical (Table 24-3). Submerged leaves have a higher SLA and more *aerenchyma* (lower density tissue with enlarged gas spaces), lack *palisade mesophyll* cells (a layer of photosynthetic cells below the upper epidermis) and stomata, and have a lower pigment content than floating and emergent leaves. The values in Table 24-3 are presented largely on an area basis, which is relevant to areal

TABLE 24-3 Leaves of *Sagittaria sagittifolia* From Three Environments. Values Are Means With Standard Deviation in Parentheses. Anthocyanins, UV-B and UV-A Absorbing Compounds Are Presented as Relative Units

Character	Submerged (s)	Floating (f)	Emerged (e)
SLA ($cm^2\ mg^{-1}$ dry mass; 1-sided area)	0.91 (0.17)^{f,e}	0.40 (0.12)^{s}	0.35 (0.07)^{s}
Tissue density (mg dry mass mm^{-3})	0.066 (0.02)^{f,e}	0.108 (0.03)^{s}	0.116 (0.0)2^{s}
Upper stomata density (mm^{-2})	0^{f,e}	47 (18)^{s,e}	66 (14)^{s,f}
Lower stomata density (mm^{-2})	0^{f,e}	34 (13)^{s,f}	55 (11)^{s,e}
Leaf thickness (μm)	186 (54)^{f,e}	242 (40)^{s}	253 (31)^{s}
Upper cuticle thickness (μm)	0^{f,e}	3.9 (1)^{s}	3.6 (1)^{s}
Upper epidermis thickness (μm)	30.8 (8.2)	28.1 (5)	28.9 (2.7)
Palisade mesophyll (μm)	0^{f,e}	46.3 (11.4)^{s,e}	66.2 (7.4)^{f,s}
Spongy mesophyll (μm)	131 (33)	136 (33)	30 (15)
Lower epidermis thickness (μm)	26.8 (4.2)	30.2 (5.9)	27.7 (5.0)
Lower cuticle thickness (μm)	0^{f,e}	1.9 (0.25)^{s,e}	2.1 (0.21)^{s,f}
Chlorophyll *a* ($\mu g\ cm^{-2}$)	8.5 (1.4)^{f,e}	17.8 (6.0)^{s,e}	34.2 (8.7)^{s,f}
Chlorophyll *b* ($\mu g\ cm^{-2}$)	3.0 (0.3)^{e}	5.5 (0.3)^{e}	19.6 (6.6)^{s,f}
Carotenoids ($\mu g\ cm^{-2}$)	3.0 (0.6)^{f,e}	5.6 (1.5)^{s,e}	9.3 (1.8)^{s,f}
Anthocyanins (units cm^{-2})	0.26 (0.10)^{f,e}	1.44 (0.34)^{s,e}	0.94 (0.22)^{s,f}
UV-B absorbing compounds (units cm^{-2})	2.24 (0.64)^{f,e}	5.94 (2.17)^{s}	8.21 (1.91)^{s}
UV-A absorbing compounds (units cm^{-2})	3.52 (0.87)^{f,e}	11.25 (4.04)^{s,e}	15.69 (3.29)^{s,f}

Superscript letters indicate leaf types that are statistically significantly different ($P \leq 0.05$). Data from (Klancnik et al., 2014). Pigment contents were converted from mg dm^{-2} to μg cm^{-2} by multiplying by 10, and anthocyanins and absorbing compounds converted from units dm^{-2} to units cm^{-2} by dividing by 100.

fluxes, but could be converted to a volume basis which is relevant to volumetric content using the SLA.

Leaves of some riparian plants that are periodically flooded have a hydrophobic external layer that traps an air film around them when submerged. This increases the rate of CO_2 and O_2 exchange, promoting rates of photosynthesis during the day and supporting respiration in the dark (Colmer and Pedersen, 2008; Verboven et al., 2014; Voesenek and Bailey-Serres, 2015). Short-term acclimation of terrestrial plants to flooding, likely linked to reduction of cuticle thickness, also increases the rate of gas transfer between water and the plant (Mommer et al., 2004). Unlike most terrestrial plants, the leaves of submerged freshwater plants have chloroplasts in their epidermal cells and many comprise just two epidermal layers without mesophyll cells except around leaf veins. Laminar submerged leaves are consequently thinner than terrestrial leaves (median thickness 130 versus 240 μm (Enriquez et al., 1996) and so generally have a high (one-sided) SLA with a median of 0.6 $cm^2\ g^{-1}$ compared to 0.2 to 0.05 $cm^2\ g^{-1}$ for terrestrial leaves (Poorter, 2009; Pierce et al., 2012), although with some exceptions such as in freshwater bryophytes (Chmara et al., 2019). When water loss is not a limiting factor, thin leaves with a high surface area to volume ratio help to maximize uptake of light energy, inorganic carbon, and nutrients and also match the amount of photosynthetic material below the surface of the leaf to rates of supply (Black et al., 1981) (see Section 24.II.B).

Leaf form is regulated by environmental conditions. Some plants, such as the isoetid *Littorella uniflora*, produce leaves that are similar in morphology (*homophyllous*) when grown in air or water (Hostrup and Wiegleb, 1991). Nevertheless, Robe and Griffiths, (1998) showed that leaves of *Littorella uniflora* in air were nearly twice as long and thinner than leaves in water and had stomata but a reduced lacunal volume. This regulation is most obvious in *heterophyllous* amphibious species that produce a different leaf form in water compared to air (Maberly and Spence, 1989; Li et al., 2019). Leaves that photosynthesize in air are similar to terrestrial leaves with stomata and substomatal cavities that are lacking in aquatic leaves. Several different environmental factors control the production of aquatic versus aerial leaves including temperature, day length, concentration of CO_2, and water potential (Maberly and Spence, 1989). In *Hippuris vulgaris* the leaf form is controlled by a phytochrome system triggered by the ratio of red to far-red light. Aerial leaves are produced under light with a low red:far-red ratio even when growing in shallow water (Bodkin et al., 1980). Some, or all, of these environmental cues are mediated by hormones. Gibberellic acid promotes the formation of submerged leaves, whereas abscisic acid promotes the formation of aerial leaves. Ethylene promotes shoot elongation and is used as a signaling hormone to respond to flooding (Jackson, 1985).

The leaves and stems of almost all freshwater plants contain large volumes of *lacunae* (air spaces) (Arber, 1920). This produces buoyancy, allowing leaves and stems to float toward the higher light levels near the water surface or, in the case of floating leaves, remain at the water surface. Lacunae also affect the exchange of gases between the plant, water, and sediment (see Section 24.II.B) and scatter light, improving light absorptance (Vogelman et al., 1996). Hartman and Brown (1967) recorded that the percent lacunal volume varied between about 5% in shoots of *Ceratophyllum demersum* and 36% in *Myriophyllum exalbescens* and varied between 27% and 86% of the volume of stems in four other elodeids (Schuette et al., 1994). In the isoetids, *Littorella uniflora* and *Lobelia dortmanna*, the lacunal volumes of leaves were 49% and 56%, respectively (Søndergaard, 1979). The morphology and anatomy of underwater leaves are a response to the opportunities and constraints of their environment and, in particular, to the acquisition of resources.

II. Resource acquisition and physiological responses to environmental conditions

A. Light

Aquatic systems, even the clearest oceans, are essentially shade environments. Light is lost by reflection at the water surface and is then absorbed and scattered within the water column by water molecules, colored dissolved organic matter, phytoplankton, and nonliving particles (Chapter 6; Kirk, 2010). Epiphytes (Chapters 17 and 25) can reduce the amount of light reaching the surface of a plant leaf even further (Sand-Jensen, 1977, 1990). Freshwater plants themselves also attenuate light: dense leaves of amphibious plants or pleustophytes at the water surface can almost completely shade the water column, whereas attenuation by submerged vegetation varies with plant density and architecture in a species-specific way (Su et al., 2019). In contrast, *sunflecks* (bright short-lived flashes of light) caused by wave-focusing can produce subsurface flashes of light up to five times the value at the water surface, each for a few milliseconds (Schubert et al., 2014).

i. Photosynthesis and responses to light levels

Freshwater plants belong to the "green lineage" of photoautotrophs, where chlorophyll *a* is the main light-harvesting pigment, uniquely present at the light reaction centers, whereas chlorophyll *b* is also involved in light absorption. The presence of chloroplasts in the epidermal cells of many submerged plants (Sculthorpe,

1967) and their generally thin leaves (Enriquez, 2005) maximize light absorption. The process of converting light into chemical energy and reducing power is common to all plants (Falkowski and Raven, 2007). Pigment–protein complexes are organized into two linked, light-processing photosystems in the *thylakoids* (membranous sacs) within chloroplasts that split water, releasing oxygen, and produce reducing power as NADPH (nicotinamide adenine dinucleotide phosphate) and chemical energy in the form of ATP (adenosine triphosphate) (Falkowski and Raven, 2007).

A survey of photosynthesis responses to light quantity in aquatic plants (including marine macroalgae) and the consequences of growth in dense communities was undertaken by Binzer et al. (2006). Individual plants had an average *light compensation point* (where net photosynthesis is zero) and I_k (the onset of light limitation) of 22 and 151 μmol photon m^{-2} s^{-1} *photosynthetically active radiation (PAR)*, respectively. For a given species, I_k increases with CO_2 concentration: for example, in the moss *Fontinalis antipyretica* it is fivefold higher when CO_2 is saturating (85 μmol photon m^{-2} s^{-1}) than when CO_2 is at air-equilibrium (17 μmol photon m^{-2} s^{-1}) (Maberly, 1985a). In contrast to individual plants, freshwater plant communities had higher average values of light compensation point and I_k of 119 and 455 μmol photon m^{-2} s^{-1}, respectively, as a consequence of self-shading within the canopy (Fig. 24-7). Moreover, although individual plants were light-saturated on average at 337 μmol photon m^{-2} s^{-1}, communities did not quite reach light saturation even though the highest light values used in the experiments matched the maximum amount that could be received: about 2000 μmol photon m^{-2} s^{-1}.

Leaves of submerged plants acclimate to light levels physiologically and morphologically, as do land plants. In two early seminal papers Spence and Chrystal (1970a, 1970b) showed that *Potamogeton polygonifolius*, found at depths up to 57 cm in Scottish lochs (Spence, 1964), had physiological and morphological characteristics that were very different from *Potamogeton obtusifolius* that grew down to 310 cm. When grown in full sunlight, leaves of *Potamogeton polygonifolius* had lower rates of photosynthesis at low light, higher rates of respiration, and a higher light compensation point than *Potamogeton obtusifolius* (Fig. 24-8). The leaves also had a higher chlorophyll content and a lower specific leaf area and were much thicker. When grown at low light (about 6% of full, unshaded sunlight), shade leaves of *Potamogeton polygonifolius* became more like sun leaves of *Potamogeton obtusifolius*, whereas shade leaves of *Potamogeton obtusifolius* became even more adapted to low light.

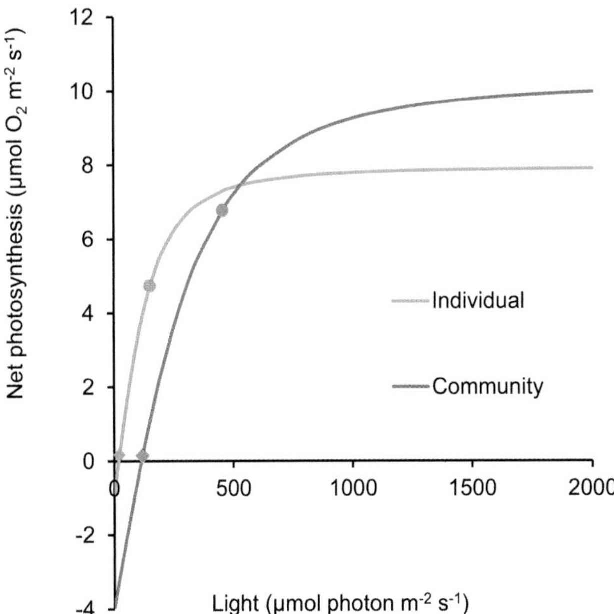

FIGURE 24-7 Net photosynthesis per unit ground area as a function of photon irradiance (PAR). "Individual" refers to leaves or shoots with minimal self-shading, whereas "community" refers to constructed multispecies mixes with a ratio of leaf area to ground area from 2 to 16. The *lower diamonds* on the "individual" and "community" curves identify light compensation points, whereas the *upper circles* identify I_k. Measurements were made between 14°C and 20°C. *(Data were calculated from values in Table 2 of Binzer et al., 2006.)*

In addition to phenotypic variation of leaves to light, shoots become elongated at low light, bringing the growing tip of a shoot closer to the surface and to higher light levels. Thus *Potamogeton obtusifolius*, growing at 3.5 m in Esthwaite Water in the English Lake District, close to its depth limit, was 3.8 times taller, had internodes that were 3 times longer and had a shoot length per unit dry weight that was 5.7 times greater than plants growing at 0.5 m (Maberly, 1993). Model estimates indicated that shoot elongation increased productivity 2.5-fold for plants rooted at 3.5 m but only 1.1-fold for plants rooted at 0.5 m. The morphological changes in shoot architecture were estimated to contribute 36% to increased apical production at 3.5 m, with the remainder resulting from shade acclimation. Plants at 3.5 m did not reproduce sexually, and a similar pattern was found in *Vallisneria spinulosa*, where plants at depth traded off sexual reproduction for asexual reproduction (Li et al., 2018). Movement of chloroplasts within a leaf cell provides a shorter-term (minutes) regulation mechanism to control the amount of light absorbed (Zurzycki, 1955). At low light, chloroplasts are positioned on the surface (*periclinal*) cell wall, parallel to incoming radiation, to maximize the amount of light absorbed. When exposed to high light, chloroplasts move to the

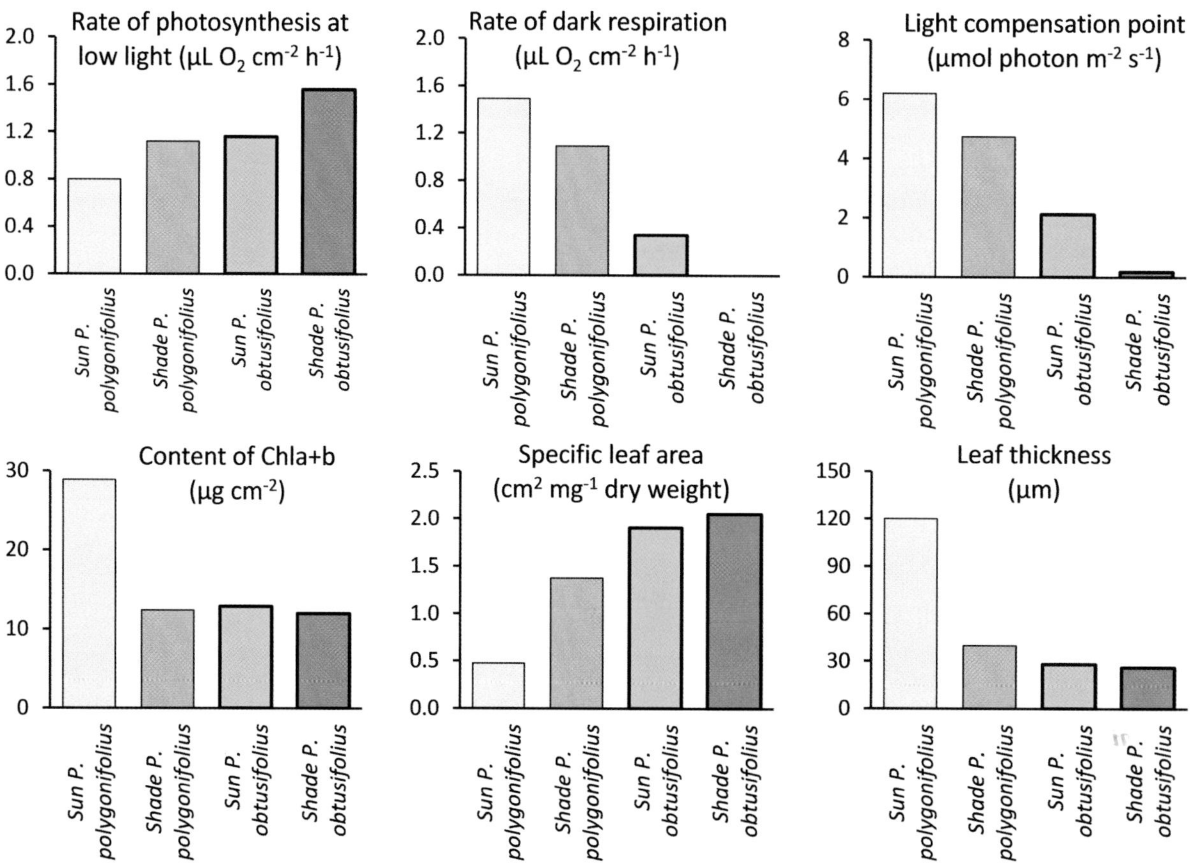

FIGURE 24-8 Comparison of physiological and morphological features of shallow water *Potamogeton polygonifolius* and deepwater *Potamogeton obtusifolius* plants grown in full sun (gold) and shade (6% light, blue) (Spence and Chrystal, 1970b). Light values reported in cal cm^{-2} h^{-1} (380–750 nm) were converted to PAR (400–700 nm) using the published light spectral distribution in Spence and Chrystal (1970a). Low light was 0.216 cal cm^{-2} h^{-1} that equates to a value for PAR of about 10 μmol photon m^{-2} s^{-1}.

perpendicular (*anticlinal*) cell walls, where they experience lower light levels. This is regarded as a way to maximize light absorption when light is low and to minimize it when light is high in order to minimize photoinhibition (Sinclair and Williams, 2001).

ii. Photoinhibition

A striking aspect of the data in Fig. 24-7 is that the rate of photosynthesis does not decline at high light either for individual plants or for communities. High light levels can cause *photoinhibition* by reducing the efficiency of photosynthesis caused by protective thermal dissipation of absorbed photons (dynamic photoinhibition) but also by damaging the proteins in the core of photosystem II faster than they can be repaired. Although in Fig. 24-7, there is no evidence of photodamage at high light, dynamic photoprotective inhibition is likely to be occurring. Factors such as carbon limitation or low temperature can increase photoinhibition at high light by reducing the ability of a leaf to process light energy. Thus, when *Hydrilla verticillata* was exposed to full sunlight for 15 min, light-saturated photosynthesis was reduced by about 50% at a dissolved inorganic carbon concentration of 0.6 mmol L^{-1} but was not inhibited at 2.0 mmol L^{-1} (White et al., 1996). When three species were exposed to light levels up to 14 times those to which they were acclimated, both photoprotection and photodamage occurred (Hussner et al., 2010). In some species there was evidence of photoinhibition even when the light levels were increased 1.4-fold. Laboratory experiments can have the disadvantage that the spectral distribution of light does not match natural conditions; however, for six species of freshwater plants in a small pond, midday photoinhibition occurred at about 1100 μmol photon m^{-2} s^{-1} but was absent when the plants were shaded so that they received about a third of full, unshaded sunlight (Jiang et al., 2018).

iii. UV radiation

Photons of ultraviolet radiation, UV-B (280–315 nm) and UV-A (315–400 nm), are highly energetic, especially UV-B, and may damage DNA and components of the photosystems. Although water rapidly absorbs UV radiation, especially when it contains colored dissolved organic carbon (Scully et al., 1995), leaves at or near the surface may receive high UV doses. Moreover, in clear nonhumic lakes the 1% depth limit for light at 305 nm can be as deep as 27 m (Morris et al., 1995). De Bakker et al. (2005) showed that when *Chara aspera* was exposed to UV-B radiation, growth was reduced and damage to DNA increased but there was no change in the amount of photoprotective UV-B absorbing compounds.

Species that grow in shallow water are more tolerant of UV than those growing in deeper water (Rae et al., 2001; Hanelt et al., 2006). The ability to tolerate UV radiation is linked, in part, to the presence of UV-screening compounds. Nocchi et al. (2020) showed that increased UV-B radiation increased the photoprotective mechanisms via increased flavonoid content and antioxidant activity of floating leaves of *Nymphoides humboldtiana*, which also slightly stimulated their photosynthetic potential. Floating and emergent leaves of *Sagittaria sagittifolia* have a higher content of anthocyanins and UV-absorbing compounds than submerged leaves of the same species (Klancnik et al., 2014; Table 24-3).

B. Inorganic carbon

i. *Constraints on inorganic carbon supply*

Carbon is a macronutrient, typically contributing 35% to 50% of the dry weight of freshwater plants (Demars and Edwards, 2008; Pierce et al., 2012). Inorganic carbon is present in water as carbon dioxide (CO_2) (including carbonic acid, H_2CO_3), bicarbonate (HCO_3^-), and carbonate (CO_3^{2-}), all of which are interconnected by equilibria whose position is controlled largely by pH (Chapter 13). Underwater photosynthesis is obviously not limited by water but instead often by light (see Section 24.II.B) and inorganic carbon. At air equilibrium, the concentration of CO_2 in freshwater is approximately equivalent to that in air and declines with increasing temperature (Maberly and Gontero, 2017). Furthermore, the rate of diffusion of CO_2 through the boundary layer surrounding an underwater leaf is about 10,000 times lower than in air for the same concentration gradient. Large leaves tend to have relatively thick boundary layers around them and this, in conjunction with low rates of diffusion, can impose strong transport limitations of gases and ions between the leaf and the bulk water (Black et al., 1981). Although water flow past a leaf reduces the thickness of the boundary layer, high flow velocities can cause rates of gross photosynthesis to decline and dark respiration to increase, possibly as a result of physical stress (Madsen et al., 1993). The atmospheric CO_2 concentration is close to saturating photosynthesis in terrestrial plants, but as a consequence of transport limitation, half-saturating concentrations of CO_2 in freshwater plants are at least seven times greater than air equilibrium (Maberly and Madsen, 1998). In some situations the concentration of CO_2 in water can exceed air equilibrium many-fold as a result of input from the catchment (Sand-Jensen and Staehr, 2012) or by decomposition of allochthonous or autochthonous organic matter within the river or lake. When productivity is high, however, CO_2 can be removed by photosynthesis much more rapidly than it can be resupplied. In these situations, CO_2 concentrations can fall to about 0.02% of air equilibrium (Maberly and Gontero, 2017). These very low CO_2 concentrations are many times lower than the concentration that half-saturates the primary carbon-fixing enzyme RuBisCO (ribulose-1,5-bisphosphate carboxylase-oxygenase), which, especially when concentrations of oxygen are high, will fix O_2 rather than CO_2, leading to photorespiration and numbers below the CO_2 compensation concentration.

ii. *Strategies to minimize inorganic carbon limitation*

Freshwater macrophytes have different strategies to overcome potential inorganic carbon limitations (Klavsen et al., 2011). These include "avoidance" or simply growing in sites with high CO_2 such as in groundwater-fed rivers (Maberly et al., 2015) or just above the sediment surface where local concentrations of CO_2 can be high (Maberly, 1985b); "exploitation," namely anatomical or morphological features that can tap into areas of high CO_2 availability; and "amelioration," specifically physiological or biochemical processes that minimize inorganic carbon limitation. A widespread exploitation strategy is the possession of floating leaves, part of the evolutionary trajectory into water, that allow access to atmospheric CO_2. There is good evidence that this boosts plant productivity. For example, it promotes flowering in *Nuphar lutea* and *Callitriche hamulata* (= *C. intermedia*; Grainger, 1947), increases soluble carbohydrate reserves in *Hippuris vulgaris* (Janauer and Englmaier, 1986), and stimulates photosynthesis in *Stratiotes aloides* (Prins and Deguia, 1986). A less frequent exploitation strategy involves tapping into the high CO_2 present in most sediments produced by decomposition processes. This was first shown experimentally for the isoetid *Lobelia dortmanna* (Wium-Andersen, 1971) and subsequently for *Littorella uniflora* (Søndergaard and Sand-Jensen, 1979) and is now considered widespread among most isoetids (Winkel and Borum, 2009). It relies on the diffusion of high concentrations of CO_2 from the sediment through lacunae that are continuous from roots to leaves. Diffusion over long distances, even in air, is a slow process and the small stature of isoetids allows this mechanism to be effective. Accordingly, the higher the concentration of CO_2 in the sediment, the longer the leaf length of *Littorella uniflora* that can be supported (Bagger and Madsen, 2004). In contrast, CO_2 from the sediment is a minor contributor to inorganic carbon uptake by tall elodeids (Loczy et al., 1983). However, there is evidence for uptake of CO_2 from the sediment in *Sparganium angustifolium,* which is tall but has an isoetid growth form (Lucassen et al., 2009).

There are three types of amelioration strategies in freshwater plants based on CO_2 concentrating mechanisms (CCMs) that increase the concentration of CO_2 around RuBisCO: use of HCO_3^-, C_4 photosynthesis, and *crassulacean acid metabolism* (CAM). Of these, HCO_3^- use is the most widespread, found in about 44% of the 133 freshwater species currently examined (Iversen et al., 2019). It is not ubiquitous because there are costs associated with the extra energy needed to operate the CCMs (Raven et al., 2014) and because leaves that use HCO_3^- have a lower affinity for CO_2, possibly caused by a lower permeability to prevent CO_2 from leaking out once it has been pumped in as HCO_3^- (Maberly and Madsen, 1998; Madsen and Maberly, 2003). Soft waters with low concentrations of HCO_3^- also have fewer species able to use HCO_3^- than harder water sites (Iversen et al., 2019; Vestergaard and Sand-Jensen, 2000). Consequently, there is a broad biogeographical pattern of HCO_3^- use linked to catchment geology, although waters with naturally high CO_2 concentrations, such as in many streams, have a lower frequency of species that use HCO_3^-. For example, the River Sorgue in France, which is groundwater fed, has high concentrations of CO_2 close to its source and, despite a very high bicarbonate concentration, is dominated by plants restricted to CO_2 such as *Berula erecta* (Maberly et al 2015, Fig. 24-9).

FIGURE 24-9 The River Sorgue in southern France is dominated by large beds of *Berula erecta* near its source. *(Photo by the authors.)*

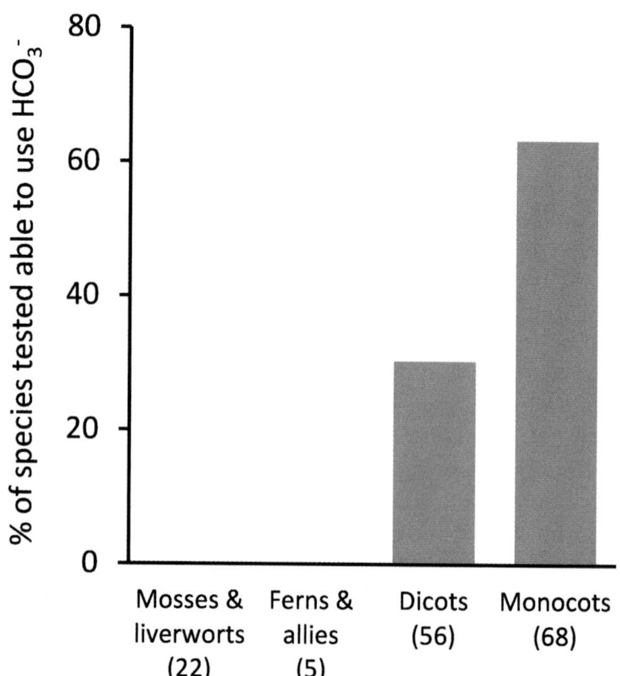

FIGURE 24-10 Percentage of tested species (numbers in parentheses) that are able to use HCO_3. *(Data from Bain and Proctor, 1980; Iversen et al., 2019; Spence and Maberly, 1985.)*

There is a phylogenetic pattern in the frequency of HCO_3^- use (Spence and Maberly, 1985). Thus there are no substantiated reports of HCO_3^- use in mosses, liverworts, lycopods, or pteridophytes and it is more frequent in monocots than in dicots (Fig. 24-10). However, of 65 genera tested, 23 (35%) have species with and without the ability to use HCO_3^-, suggesting that there is a strong ecological component to this characteristic.

Unlike in microalgae, the mechanism, or mechanisms, of HCO_3^- use is poorly understood for freshwater plants. Some broad-leaved species have polar leaves where protons (H^+) are actively excreted at the abaxial surface of the leaf. This converts HCO_3^- to CO_2, some of which then diffuses into the leaf. At the adaxial surface, there is a net excretion of hydroxyl ions, raising the pH and triggering the formation of "marl" or calcium carbonate (Prins et al., 1980). Calcium carbonate precipitation generates protons that also generate CO_2 (McConnaughey, 1991). Charophytes that use HCO_3^- have a similar system with bands of low pH and high pH around their cylindrical cells (Lucas and Smith, 1973). In the monocot *Ottelia alismoides* the use of HCO_3^- relies on an anion exchange protein, SLC4, that appears to be involved in transporting HCO_3^- across the plasma membrane and a periplasmic carbonic anhydrase that facilitates efficient co-diffusion of HCO_3^- through the boundary layer (Huang et al., 2020).

About 7500 terrestrial plants have *C_4 photosynthesis*. It occurs predominantly in places with a high temperature or where CO_2-limitation occurs, triggered by stomatal closure, in hot, dry, or saline conditions (Sage et al., 2018). In the first step, the enzyme PEPC (phosphoenolpyruvate carboxylase) adds HCO_3^- to a C_3 compound, forming a C_4 compound. The C_4 compound is transported to the location of the primary carboxylation enzyme, RuBisCO, where it is *decarboxylated* (broken down, releasing CO_2), producing an elevated concentration of CO_2 around RuBisCO that can then be fixed, and a C_3 molecule that completes the cycle. To prevent a futile cycle, where carboxylation is cancelled out by decarboxylation, these reactions usually occur in separate cells but, recently, terrestrial plants with C_4 photosynthesis in a single cell have been discovered, with the two processes occurring in different parts of one cell (Edwards et al., 2004). C_4 photosynthesis is known to occur in a handful of freshwater plants, all from the monocot family Hydrocharitaceae, specifically *Hydrilla verticillata*, *Egeria densa*, and *Ottelia alismoides* (Casati et al., 2000; Bowes et al., 2002; Bowes, 2011; Zhang et al., 2014). In the first two species C_4 photosynthesis is facultative, triggered by low CO_2, high light, and high temperature, but in mature leaves of *Ottelia alismoides* it appears to be constitutive (Zhang et al., 2014). Probably all freshwater plants with C_4 photosynthesis operate a single cell system, although this is an ongoing area of research (Han et al., 2020).

The enzyme PEPC is also involved in CAM, a well-known syndrome typically found in terrestrial succulent plants from extremely arid areas such as deserts. Here, the separation between carboxylation and decarboxylation occurs temporally rather than spatially. Stomata open at night and the CO_2 is converted into a C_4 compound by PEPC that is stored in the vacuole as malic acid. During the day, the stomata are closed to reduce water loss and the malate is decarboxylated, generating CO_2 internally that is then fixed by RuBisCO. Because this pathway conserves water, it was extremely surprising when the submerged plant, *Isoetes howellii,* was shown to perform underwater CAM (Keeley, 1981). Underwater CAM is probably ubiquitous or at least widespread in submerged species of *Isoetes* and is also found in freshwater *Crassula* such as *C. helmsii.* Currently, about 9% of freshwater plants are known to have CAM (Maberly and Gontero, 2018), often facultatively depending on growth conditions, such as in *Ottelia alismoides,* where CAM is downregulated at high CO_2. *Ottelia alismoides* therefore unusually has three different CCMs, because it can also use HCO_3^- (Huang et al., 2020), raising intriguing questions about how they are regulated and interact. Underwater CAM is not linked to water conservation but to carbon conservation: respiratory CO_2 can be trapped and respiratory CO_2 from the community can be taken up at night (Klavsen and Maberly, 2009). As a result of research on freshwater plants, the purpose of CAM in terrestrial species has broadened to highlight its role as a carbon-conserving, as well as a water-conserving, feature.

C. Nitrogen and phosphorus

Rooted submersed angiosperms are unique in the plant world in that they live in an environment where they can acquire nutrients from both water and bottom sediments. Early researchers debated whether the roots of freshwater angiosperms took up nutrients, especially the macronutrients phosphorus (P) and nitrogen (N), from the sediments or were simply organs of attachment, particularly in the case of submerged plants. An overwhelming accumulation of evidence now demonstrates that roots are essential for the acquisition of macronutrients for freshwater angiosperms, including those that are submersed. Rooted submerged angiosperms, though able to rely on aerenchyma to provide buoyancy and thereby stay erect, have well-structured (albeit reduced) vascular tissue consisting of xylem, phloem, and an endodermis with Casparian strips, important in the regulated flow of ions and water (Seago, 2020). Rooted submersed angiosperms therefore have the capacity to employ *root pressure* (osmotic pressure in the cells of the roots that drives water upward) to drive solute transport from roots to leaves and at rates faster than possible by passive diffusion (reviewed by Stocking, 1956). There is now a general consensus that roots supply the majority of P and N for rooted submersed angiosperms (Barko and Smart, 1980; Carignan and Kalff, 1980; Chambers et al., 1989). However, when P and N concentrations in the water are high, the water can also provide much of the required nutrients. This was identified by Carignan (1982) using an empirical model that predicted that leaf uptake was important when the ratio of sediment to water nutrient availability was low. Subsequent controlled experiments showed that several rooted submerged angiosperm species were able to satisfy their P and N demands by leaf uptake when growing in nutrient-rich waters. For example, a reciprocal transplant experiment showed that the rooted submerged plant *Lagarosiphon major* grew larger and had a greater P content when grown in eutrophic (versus oligotrophic) sediments in nutrient-poor Lake Taupo, New Zealand (Rattray et al. 1991; Fig. 24-11). However, sediment type had no consistent effect on plant growth in nutrient-rich Lake Rotorua (New Zealand): nutrients in the water offset insufficient nutrients in the oligotrophic sediments. It should be noted, however, that the main pathway for uptake of minerals such as calcium, magnesium, sodium, potassium, and sulfate is through the foliage (Barko et al., 1991).

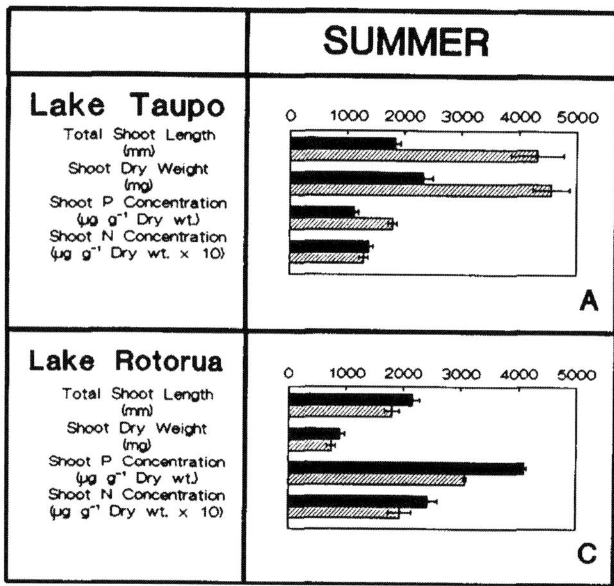

FIGURE 24-11 Shoot characteristics (length, dry weight, P content, and N content) for the rooted submerged plant *Lagarosiphon major* grown in oligotrophic and eutrophic sediments in two lakes of differing trophic status (Lake Taupo, oligotrophic; Lake Rotorua, eutrophic) (Rattray et al., 1991).

Macronutrient absorption and translocation by rooted freshwater plants is driven by root pressure. In the case of emergent vegetation this force is likely further enhanced by *transpiration*, whereby evaporative losses of water from emergent leaves accelerate the flow of dissolved solutes from the roots to upper tissues. Floating angiosperms are, by definition, reliant upon nutrients in the water. Thus the free-floating plants (pleustophytes) that form mats, or floating swamps of tropical regions and large river deltas, employ their roots as well as foliage floating on the water surface to acquire nutrients from the water underlying the mat. These nutrients are tightly recycled among the living foliage, dead and decaying organic matter, and the water (which, in turn, exchanges nutrients with the sediment especially under conditions of anoxia). Freshwater plants without roots or rhizomes (i.e., macroalgae, hornworts, liverworts, and mosses) obtain nutrients by absorption via foliar or rhizoidal structures. For example, the rhizoid-bearing macroalga *Chara* is able to use phosphorus from water (Kufel and Kufel, 2002) or sediments (Wüstenberg et al., 2011), with transport through the plant facilitated by cytoplasmic streaming and polar transport (Raven, 2013).

Considerable information exists on the inorganic chemical composition of freshwater plants, particularly for the macronutrients nitrogen and phosphorus (see review by Hutchinson 1975). This work was originally driven, in large part, by in-lake recovery efforts (i.e., to reduce in-lake nutrients by harvesting plant biomass) and the desire to use freshwater plants as indicators of eutrophication. Early studies by Gerloff and Krombholz (1966), whereby *Vallisneria americana*, *Heteranthera dubia*, *Elodea nuttallii*, and *Zannichellia palustris* were grown under controlled laboratory conditions, showed that the minimum tissue nutrient content associated with maximum yield were about 0.13% dry weight of P and about 1.3% dry weight of N. A more recent analysis by Demars and Edwards (2007) of data from Gerloff (1975) for *Myriophyllum spicatum*, *Egeria nuttallii*, *Ceratophyllum demersum*, and *Lemna minor* gave similar critical nutrient concentrations: $0.10 \pm 0.03\%$ dry mass of P (mean $\pm$ SD) and $1.14 \pm 0.39\%$ dry mass of N for 95% maximum yield, and also $0.16 \pm 0.05\%$ dry mass of P and $1.82 \pm 0.62\%$ dry mass of N for 95% maximum growth. Although several studies reported correlations between plant tissue nutrients (particularly P) and ambient concentrations (Gossett and Norris, 1971; Robach et al., 1995; Carr and Chambers, 1998; Shilla et al., 2006), others have found no relationship (Canfield Jr. and Hoyer, 1988). A major factor determining variability in tissue nutrient concentrations is species identity (Boyd, 1978; Hayati and Proctor, 1990; Thiebaut and Muller, 2003; Waughman, 1980). Demars and Edwards (2008) reported that 378 plant samples collected from 65 sites in Scotland spanning a range of freshwater habitats (lotic to lentic, and large to small) showed an approximate 10-fold range in plant tissue N and P content, with the largest share of variability explained by plant growth form (emergents versus submerged/floating forms versus bryophytes) likely related to taxonomic (mostly species) effects (Fig. 24-12). Physical and chemical factors also influence plant nutrient uptake and tissue nutrient concentrations, including hydrology (Anderson and Mitsch, 2005), seasonality (Gerloff and Krombholz, 1966; Nichols and Keeney, 1976), water nutrient quantity (Gerloff and Krombholz, 1966; Madsen and Cedergreen, 2002), substrate nutrient availability (Angelstein et al., 2009), and light (Cedergreen and Madsen, 2004).

Given that many freshwater plants acquire nutrients from the water and sediments and then incorporate these into their tissues, the question arises as to what extent freshwater plants are a net source or sink of nutrients to their ecosystem. During periods of net growth, the release of phosphorus from healthy shoots is relatively insignificant. For example, a four-month mass-balance study of P entering and exiting a channel with a large undisturbed plant bed showed that the vegetation was not a net source of P to the water column but, if anything, was a slight net sink (Rooney and Kalff, 2003). During the senescence of freshwater plants, a marked translocation of nutrients occurs from the dying tissues to roots and overwintering organs (e.g., Garver et al., 1988). Many perennial emergent plants have

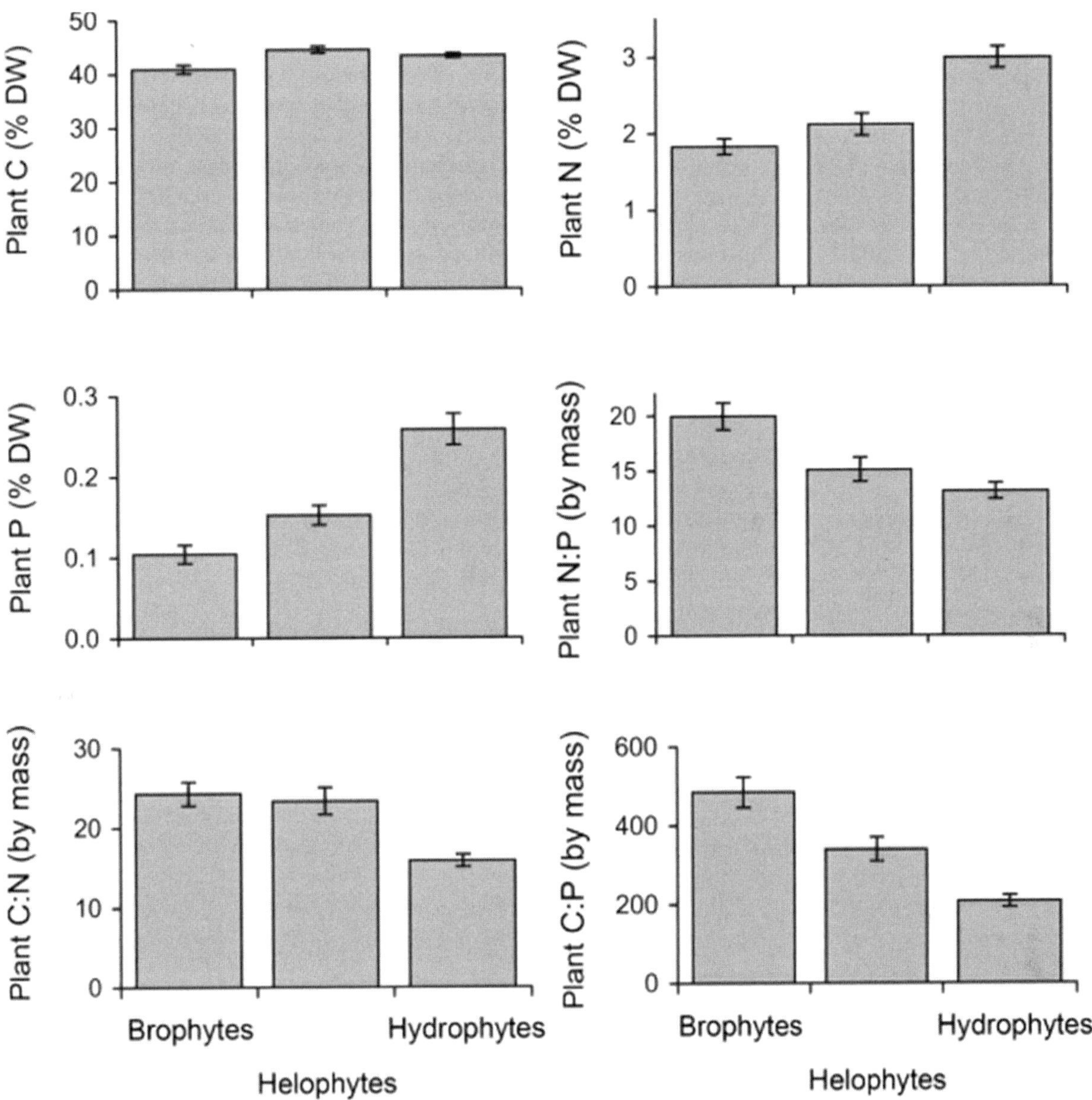

FIGURE 24-12 Tissue carbon (C), nitrogen (N), and phosphorus (P) concentrations (along with C:N:P ratios) for 378 plant samples, grouped according to plant life form (Demars and Edwards, 2008). Results are expressed as mean ± 95% confidence interval.

substantial underground overwintering organs and often 25% to 50% of the above-ground standing stock of P and N is returned to the rhizomes in late summer and fall (Granéli and Solander, 1988). Despite such translocation, release of P to the surrounding water during decay and decomposition can be substantial. For example, large, decaying stands of the submersed plant *Myriophyllum spicatum* in a reservoir in Indiana, USA, contributed about 2% of the annual nitrogen and perhaps 18% of the potential phosphorus loading in comparison to allochthonous loadings (Landers, 1982). Similarly, decomposition of *M. spicatum* was estimated to account for about half of the observed flux of total dissolved P from the littoral to the pelagic zone of Lake Wingra, USA (Carpenter, 1980). Release rates are governed by the population growth dynamics of the plants, their mortality during and after the growth period, and the conditions prevailing at the time of death and decomposition. Release rates are also influenced by the type of vegetation: decomposition and release of nutrients and organic compounds are much faster from submersed and floating-leaved plants than from emergent plants. Some of the release is from immediate autolysis of cells during senescence in addition to simple physical leaching. Depending upon when nutrients and organic compounds enter the ambient water, and how their concentrations are selectively modified and reduced by the metabolic sieve of the submersed plant–epiphyte complex, the inputs can have a marked influence on surface water concentrations and phytoplanktonic growth (e.g., Murray and Hodson, 1985).

D. Anoxia

At equilibrium, the oxygen concentration in water (Chapter 11) is about 30 times lower than in air and, when combined with the 10,000-fold lower rate of diffusion, the supply rate of oxygen is about 300,000 times lower in water than in air (Verberk et al., 2011). During lake stratification, water in the lower hypolimnion can become depleted in oxygen or even anoxic in productive lakes, but low light means that plant shoots and leaves generally do not grow in these regions. However, oxygen depletion can occur elsewhere when the demand for oxygen is high, the rate of supply is restricted, or both. For example, oxygen depletion may occur (1) in dense plant beds where the rate of community respiration exceeds the rate of mixing between the bed and open water (Caraco et al., 2006), (2) where mixing with the atmosphere and photosynthesis is prevented, such as in winter under snow and ice (Agbeti and Smol, 1995), or (3) where a layer of floating-leaved plants (that produce oxygen that is primarily released to the atmosphere) reduce air—water exchange and underwater light and hence restrict photosynthesis underwater (Caraco et al., 2006; Kato et al., 2016). Such reductions in oxygen concentrations can have major detrimental effects on the plants, and other organisms within the water body.

An extreme but widespread case of high demand and low supply of oxygen occurs in lake and river sediments, especially those where organic carbon is rapidly decomposing. Many freshwater sediments are anoxic, at least at depth (Sand-Jensen et al., 1982; Peter et al., 2016). This can limit the survival of species not adapted to these conditions (Moller and Sand-Jensen, 2011) by preventing aerobic respiration of roots and rhizomes, which is more efficient than anaerobic respiration, and by inhibiting mycorrhiza that promote nutrient acquisition by roots. Sediment anoxia also allows the formation of toxic chemical species such as Mn^{2+}, Fe^{2+}, or S^{2-} as a result of redox shifts and the accumulation of toxic products from respiration, such as ethanol (Crawford, 1992). Crawford and Braendle (1996) describe some of the biochemical strategies conferring tolerance to oxygen deprivation. In addition, the extensive lacunae in the leaves, roots, and rhizomes of freshwater plants produce a path for the transport of oxygen to the sediment (Sand-Jensen et al., 1982; Soana and Bartoli, 2013). Thus, in the isoetid *Lobelia dortmanna*, the permeability of oxygen across the roots is two to three times higher than across the leaves and this, combined with the continuous lacunal path from leaves to roots, supports root respiration and permits oxygen to diffuse into the sediment surrounding the roots (Sand-Jensen and Prahl, 1982). The system also permits CO_2 in the sediment to diffuse to the leaf where it can be fixed, and this might be an equally important, or greater, benefit. Other species, such as rice (*Oryza* spp.), restrict radial oxygen loss from roots to the sediment to maintain root oxygen concentrations (Ogorek et al., 2021). In submerged species, oxygen from photosynthesis can move to the roots by diffusion or mass-flow down a concentration or pressure gradient, respectively (Schuette et al., 1994). In amphibious plants with emergent or floating leaves, atmospheric oxygen can be transported to the sediment through the lacunal system. This occurs by *pressurized ventilation* driven by a temperature difference between the environment and a leaf rather than a concentration gradient (Dacey, 1981, 1980; Grosse et al., 1991). For example, in *Nuphar lutea* air is drawn into young leaves, flows through the rhizome, and exits via the old leaves at flow rates of up to 5 L h^{-1}.

III. Growth and distribution

A. Life history

Like all plants, the life history of freshwater plants centers on survival, propagation, and population expansion. However, compared with all other plant groups, aquatic plants exhibit a greater diversity of reproduction systems, with strategies including both vegetative (asexual) and sexual mechanisms (e.g., Li, 2014; Eckert et al., 2016). Vegetative reproduction is the most common form of reproduction and dispersal for freshwater plants. It includes strategies common to terrestrial plants such as production of tubers, stolons, corms, rhizomes, bulbs, and bulbils, as well as methods unique to freshwater plants such as autofragmentation and production of overwintering buds (Fig. 24-13). Vegetative reproduction is an especially effective form of propagation in aquatic environments as it reduces the likelihood of desiccation of propagules and aids in successful dispersal.

Many freshwater macrophytes successfully spread through breakage of shoot fragments caused by external forces such as wave action or animal feeding (a phenomenon known as *allofragmentation*) and by self-initiated abscission (known as *autofragmentation*). *Myriophyllum spicatum* is well-known for its ability to *autofragment*, a process that begins in late summer with formation of adventitious roots on the upper 15—20 cm of stem apices (Kimbel, 1982). These fragments (Fig. 24-13a) contain high concentrations of carbohydrates and, provided they remain in water, are likely to settle and grow into full plants. This method of dispersal undoubtedly

FIGURE 24-13 Photographs of reproduction structures: (a) *Myriophyllum aquaticum* autofragment, (b) *Potamogeton crispus* turion, (c) *Chara* sp. bulbil, (d) *Potamogeton lucens* rhizome, (e) *Sagittaria cuneata* tubers, (f) *S. cuneata* flowers and seeds, and (g) *Ranunculus* flowers. *(Photos a–d courtesy Yang Liu [Chinese Academy of Sciences]; photos e and f courtesy Katherine Standen [University of Saskatchewan]; photo g by the authors.)*

contributed to the rapid spread of *Myriophyllum spicatum* and its development into a nuisance plant upon introduction to North America in the 1940s (Les and Mehrhoff, 1999) and also to the spread of *Elodea canadensis* upon introduction into Europe in the 19th century (Hutchinson, 1975). Colonization by fragmentation appears to depend on the ability of a species to survive without roots and on which part of the plant is abscised. Species with highly dissected leaves and rhizoids instead of roots may be better at acquiring nutrients from the water, whereas shoots with apical tips exhibit better growth and colonization (Riis et al., 2009; Vári, 2013).

Some perennial freshwater plants produce vegetative buds called *turions* that function as overwintering organs and propagules of dispersal. Turions have stem apices reduced to extremely shortened internodes and modified leaves, often to the point of resembling scales (Sculthorpe, 1967) (Fig. 24-13b). They are formed in at least 14 genera of freshwater vascular plants from 9 plant families, mainly in submerged (or amphibious) and free-floating species (Adamec, 2018). Turions are typically produced in response to unfavorable ecological conditions, usually at the end of summer, and sink to the bottom sediment, where they remain until they germinate in spring. An exception to this seasonal pattern is *Potamogeton crispus*, whose turions are produced in early summer, remain dormant over the warm summer period, and then sprout in autumn (Sastroutomo, 1981). Turions are rich in starch as well as free sugars, proteins, lipids, and minerals, and these stores, along with the high photosynthetic capacity of the modified leaves, allow rapid sprouting and growth when conditions are favorable.

In addition to asexual reproduction by unique strategies such as autofragmentation and turion production, many freshwater angiosperms produce the same vegetative propagules as their terrestrial counterparts, namely *corms, rhizomes, tubers*, or *stolons*. Plants with clonal growth can, for example, propagate by sending out horizontal side shoots (i.e., aboveground stolons or underground rhizomes) from which sprout new shoots. With increasing length, the stolons or rhizomes autofragment, separating newer shoots from the parent plant. Reproduction by rhizomes is found in almost all the perennial graminoids as well as many other freshwater angiosperms. Clonal reproduction involving shoot bases or root systems is observed in several species such as the bulbs produced by *Crinum americanum* and corms produced by *Hypoxis* spp. (Grace, 1993).

Although occurring less frequently than vegetative reproduction, all freshwater plants are capable of sexual reproduction. The haploid gametophyte is the conspicuous green stage of the Charophyta and Bryophyta (mosses, liverworts, and hornworts) and it produces eggs and sperm that fuse to form a diploid zygote. In the Charophyta germination of the zygote starts with meiosis, producing a haploid protonema that grows in the haploid plant. In the Bryophyta the diploid zygote grows into a stalked structure bearing a capsule that produces haploid spores that germinate to produce the gametophyte generation.

In the pteridophytes (ferns and allies) and angiosperms the dominant and more conspicuous generation is the diploid sporophyte. The latter produces spores that develop into gametophytes, which, in turn, produce gametes (sperm and eggs) that fuse to form a zygote that grows into the sporophyte. Although most ferns produce identical spores, some freshwater pteridophytes (e.g., *Isoetes*) are *heterosporous*, meaning that they produce both male spores (*microspores*) that develop into male gametophytes with antheridia (male sex organs) and female spores (*megaspores*) that develop into female gametophytes with archegonia (female sex organs). In the angiosperms haploid microspores and megaspores are produced. The microspores develop into a male gametophyte comprising pollen grains with a tube cell and two nonmotile sperm cells. The megaspores develop into a female gametophyte, comprising seven cells including a large central cell with two polar nuclei and an egg cell with one nucleus. During a process unique to angiosperms, known as *double fertilization*, the nucleus of one sperm cell fuses with the nucleus of the haploid egg cell to produce a *diploid zygote*, and the nucleus of the other sperm cell fuses with the two polar nuclei of the large central cell to produce a *triploid endosperm cell*. Both the zygote and the endosperm cell divide by mitosis, producing a diploid embryo (the new immature sporophyte) and triploid endosperm (a food reserve for the embryo). Once this embryonic stage (the seed) is reached, growth is temporarily halted until conditions favor germination.

The reasons for limited sexual reproduction in freshwater plants is a topic of debate with possible explanations including challenges to successful underwater pollination, unsuitable ecological conditions for seed development, and greater likelihood of hybridization resulting in infertile offspring (see discussion in Li, 2014). The return to water by angiosperms resulted in some adaptations to this new environment to increase the likelihood of successful sexual reproduction, including flower modification, new pollination mechanisms such as hydrophily, and water-adapted pollen (Arber, 1920; Sculthorpe, 1967; Philbrick, 1988; Zhang et al., 2010). For example, pollination in the genus *Potamogeton* can occur by *anemophily* (wind pollination), as is the case for many terrestrial plants, but also by *epihydrophily* (pollen grains are transported on the water surface by water currents to floating flower heads) and *hydroautogamy* (self-pollination achieved by the movement of bubble-born pollen from anther to stigma within an open flower). Nevertheless, freshwater environments remain a challenging setting for successful sexual reproduction. Asexual reproduction serves to increase survival under conditions that constrain reproduction. The greater capacity of monocots to reproduce clonally might explain why they make up a high proportion of freshwater angiosperms (Duarte et al., 1994; Grace, 1993). Li (2014) suggested that asexual reproduction in freshwater plants works on a shorter ecological time scale to ensure population maintenance, whereas sexual reproduction works at an evolutionary time scale and is a luxury investment to ensure population restoration from extreme events. An example is the change in vegetation in Poyang Lake Natural Reserve, China, after catastrophic flooding: before flooding, the community was dominated by *Potamogeton malaianus*, whereas *Vallisneria* species predominated during the year after flooding (Cui et al., 2000). This switch was attributed not to the flooding but to the nearly half year's drought after the flood that killed the rhizome system of *Potamogeton malaianus* but not the winter buds of *Vallisneria*, which were found up to 35 cm deep (Xiong and Li, 2002). By about 3 years after the flood, *Potamogeton malaianus* had begun to recover as the result of germination from seed banks (Li et al., 2011). Interestingly, hybrid offspring of Eurasian watermilfoil (*Myriophyllum spicatum*, invasive to North America) and northern watermilfoil (*Myriophyllum sibiricum*, native to North America) can reproduce asexually via fragmentation but also reproduce sexually (LaRue et al., 2013; Thum et al., 2020). Sexual reproduction by hybrids may allow for creation of new genotypes that are more invasive and differ in their response to herbicides.

B. Productivity and seasonal growth

Under natural conditions, the life cycle of freshwater plant populations and the growth and productivity of communities exhibit temporal patterns attuned to regular, often seasonal, patterns in environmental conditions. This is especially evident in temperate regions where freshwater plants initiate growth in spring or early summer once water temperatures reach about 10–12°C and die back when temperatures fall below 3°C (Sculthorpe, 1967). Between these two extremes, seasonal changes in water temperature and day length determine the onset of emergence, timing of peak biomass, the development of perennating structures such as overwintering organs, and, ultimately, the onset of senescence and death (Hutchinson, 1975; Barko et al., 1986; Rooney and Kalff, 2000).

The productivity of freshwater plants is commonly evaluated by measuring changes in biomass (Westlake, 1965). In temperate regions the biomass of submerged and free-floating plants, as well as some floating-leaved species, is often negligible at the onset of the growing season (e.g., seeds or minimal perennating organs, such as turions or rhizomes) compared to at the end of the growing season. For these populations, biomass typically increases in a sigmoid fashion during the growing season (Fig. 24-14 top panel). In this idealized example, gross productivity reaches a plateau and later declines in older tissues; net productivity decreases and becomes negative because respiration continues to increase. Maximum biomass is reached when the current daily net productivity becomes zero and is equal to the maximum cumulative net production (the sum of all daily net productivity values) (Westlake, 1965).

For plants with negligible initial biomass (i.e., true annual plants) and where losses other than respiration are negligible, the seasonal maximum biomass is equal to the maximum annual net production (Fig. 24-14 lower panel, curve A) (Westlake, 1965). However, losses of the current season's production can be appreciable (Fig. 24-14 lower panel, curve E) and production can only be evaluated by detailed analyses of changes in the population demographics and biomass. Compared with annual freshwater plants, emergent plants and most floating-leaved plants (such as waterlilies) have large perennating rooting structures from which vegetative growth emerges in spring. If the perennating organ and new vegetation persist throughout the growing season, maximum annual net production is equal to the seasonal maximum biomass minus the biomass of the perennating structure in spring (Fig. 24-14 lower panel, curve C). However, more commonly, a variable proportion of the initial biomass is lost due to grazing or senescence and decay (curve D), often simultaneous with the emergence of new shoots and decay of others (curve B). Because of this constant growth, senescence, dieback,

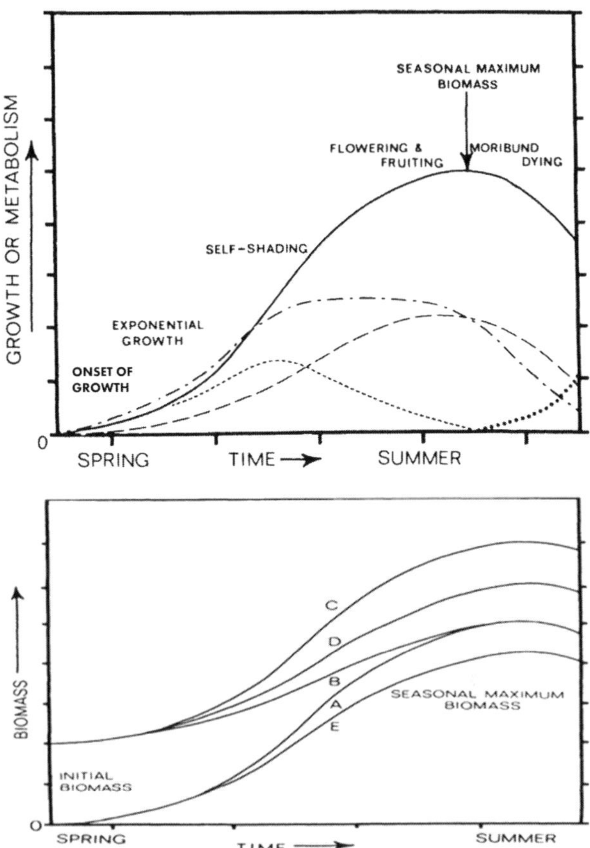

FIGURE 24-14 Growth patterns of freshwater plants. (*Top panel*) Generalized growth and metabolism curves for an annual freshwater plant, showing daily change in ——— total biomass; – · – · – · gross productivity; - - - - - - - net productivity; ············ respiration rate; ············ death losses. (*Lower panel*) Types of growth curves among freshwater plants, showing plant with annual regrowth (i.e., a true annual) (A), plant with obscured annual regrowth (B), plant with spring biomass persisting until seasonal maximum (C), plant with only part of spring biomass persisting until seasonal maximum (D), plant with regrowth and losses from current year's biomass before seasonal maximum (E). *(After Westlake, 1965.)*

and replacement of cohorts of a species population, much of the production does not survive to be measured as terminal maximum biomass. Very large amounts (50%–90%) of annual productivity will be missed by analyzing only maximum biomass. For example, water lilies have numerous cohorts each year, with an average turnover rate of aboveground foliage of four to eight times a year (e.g., Klok and van der Velde, 2017). Leaf longevity averages about 30 days and increases to about 50 days during cooler seasons. True aboveground production rates require summation of the cohort production rates and can result in annual values nearly an order of magnitude higher than would be obtained from areal instantaneous biomass estimates during the growing season. Estimates of the net primary productivity of freshwater plants in comparison to productivity of other ecosystems are shown in Table 24-4.

TABLE 24-4 Estimates of Average Net Primary Productivity for Different Ecosystems in Temperate and Tropical Regions (mg dry mass ha^{-1} $year^{-1}$). Aquatic Systems Are Shown in Bold, Terrestrial in *italics*. (Data are from Westlake (1963), Table 4, which assessed productivity in a uniform way across ecosystems.)

Ecosystem	Temperate	Tropical
Oligotrophic lake phytoplankton	**2**	
Eutrophic lake phytoplankton	**6**	
Freshwater submerged plants	**6**	**17**
Marine submerged plants	**29**	**35**
Reed swamp	**45**	**75**
Agriculture-annual plants	*22*	*30*
Agriculture-perennial plants	*30*	*75*
Terrestrial herbs	*20*	
Forest	*12[a]–28[b]*	*50[c]*

[a]*Deciduous.*
[b]*Coniferous.*
[c]*Rain forest.*

Unlike in temperate zones, seasonal variation in freshwater plant productivity is not always evident in tropical regions because annual changes in temperature and day length are minimal. However, there are cyclical hydrological phenomena in tropical and subtropical regions that occur on an annual, or even longer, frequency. In these regions (which include northern Australia, India, central Africa, central America, and northern and central South America), heavy rainfall during the wet cycle produces a pulse of surface water that fills channels, ponds, and lakes and floods wetlands. During the dry cycle, however, evaporation rates exceed rainfall and water levels can fall such that only the lowest elevations in wetlands and ponds retain surface water. Several of the world's most important wetlands (Chapter 29) occur in tropical wet-dry climates, for example, Pantanal (Brazil), Everglades National Park (USA), Okavango (Botswana), and Kakadu National Park (Australia). Changes in water levels associated with wet-dry cycles lead to periodicity in productivity of freshwater plants in tropical regions. For example, Pettit et al. (2012) observed that freshwater plants in floodplain waterholes in the wet-dry tropics of northern Australia tracked water quantity and quality such that biomass was highest early in the dry cycle and declined as the dry period progressed due to falling water levels and increased turbidity, with any remaining plants flushed out by the first flows of the wet cycle. However, even freshwater plant populations in more permanent tropical freshwaters respond to cyclical changes in water level, although the responses are often species specific and relate to the size of the flood pulse (Martins et al., 2013). For example, in the floodplains of the Paraná and in the Amazon River (Brazil), the biomass of certain freshwater plants increases with water level (*Polygonum stelligerum* in the Paraná, and *Oryza perennis* and *Paspalum repens* in the Amazon), whereas others peak during low water levels (*Eichhornia azurea* in the Paraná, and *Hymenachne amplexicaulis*, several other grass species, and Cyperaceae in the Amazon) (Thomaz et al., 2008). Thus freshwater plants in tropical and subtropical regions respond to periodicity in their environment, although the signals may be less predictable.

On a global scale, water availability is the primary determinant of productivity of terrestrial plants (Hsu et al., 2012). Emergent plants, floating-leaved plants, and pleustophytes experience little or no problems with water availability and this, combined with access to atmospheric CO_2, a readily available source of inorganic carbon, and high light levels, has resulted in their being one of the most productive communities in the world (Westlake, 1963). Temperate and tropical reed swamps may produce about 45 and 75 mg dry mass ha^{-1} y^{-1}, respectively (Westlake, 1963). These values are between four and eight times greater than corresponding values for submerged plants (Table 24-4) and are equivalent to, or exceed, values for terrestrial systems.

C. Physical and chemical controls of diversity and abundance

Although geoclimatic features influence freshwater plant diversity at large spatial scales, a combination of physical, chemical, and biological factors usually determines species composition, diversity, and abundance on a local or regional scale (Fig. 24-15). The physical environment (habitat stability, water temperature, and light availability) is typically the primary determinant for freshwater plant colonization and growth. If the minima are met (i.e., sufficient light for net photosynthesis; wave, wind, and current speeds that do not cause breakage and uprooting), water chemistry then exerts a major influence on plant abundance and composition, both under natural conditions (e.g., freshwater versus saline water; softwater versus hardwater) and when natural waters are modified because of human activity (e.g., acidification and eutrophication). Biological factors (competition, herbivory, allelopathy) can affect freshwater plants in all types of waterbodies; however, their importance in determining community abundance and composition is typically less than that of water chemistry.

Habitat stability is governed by wind and wave action, water movement, the nature of the substrate,

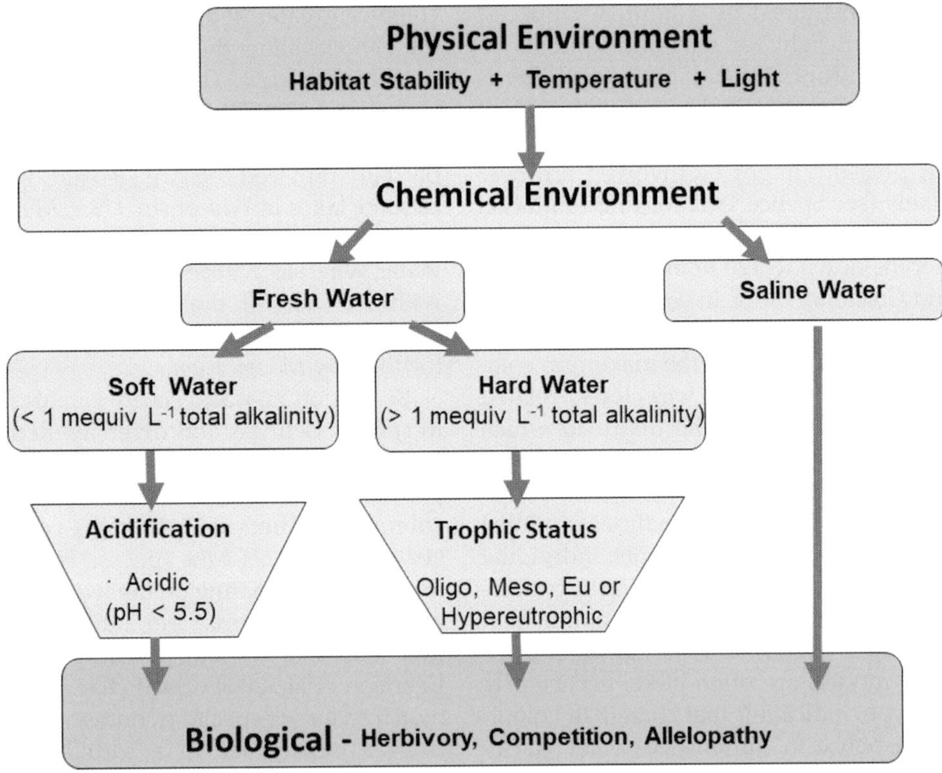

FIGURE 24-15 Conceptual model of the filters (physical, chemical, and biological) that determine the abundance and composition of freshwater plant communities.

and, in some cases, the timing and frequency of flooding or drought. Wind and wave action have long been known to influence the distribution, composition, and abundance of emergent, floating, and submersed plants along the shores of waterbodies (e.g., Hutchinson, 1975; Spence, 1982; Chambers, 1987a). Wind and wave action may affect freshwater plants directly through breakage and uprooting as well as propagule transport and, indirectly, through erosion, deposition, and sorting of bottom sediment. For example, the nearshore depth of submersed plants in four lakes in Quebec, Canada, was found to be strongly correlated with the depth of surface wave mixing (as determined by wind velocity and fetch) (Chambers, 1987a). Points and shallows where wind and wave energy are highest therefore tend to have few (if any) freshwater plants, whereas bays or coves may develop an extensive littoral plant community because of protection from waves and currents. The effects of wind/wave disturbance on freshwater plants may also be modified by lake trophic status: plants may undergo biomechanical changes to adapt to growth in shallow, more exposed conditions when deeper waters become light limited (Zhu et al., 2018; see also Section 24.II.A).

In the case of streams and rivers, current velocity is a major determinant of plant abundance and composition. Slow-flowing waters (<0.2 m s^{-1} base velocity during the growing season) often have high plant biomass dominated by a diverse angiosperm community (with little or no bryophytes or periphyton present), fast-flowing waters ($0.6-0.9$ m s^{-1}) have low biomass composed of bryophytes and a few angiosperm species, whereas the fastest streams (>0.9 m s^{-1}) are dominated by haptophytes such as bryophytes or species from the Podostemaceae that tightly adhere to the boulder substrate (Madsen et al., 2001). Water movement also determines the nature of the bottom substrate and, in turn, freshwater plant abundance and composition. For example, fast-flowing streams have riverbeds of rock, cobble, or boulder, and these are so hard that plant roots are unable to penetrate. In contrast, some soft flocculent sediments are unstable and will not allow plants to attach successfully.

Emergent plants are among the most productive natural plant communities in the biosphere (Westlake, 1963; Table 24-4). Their high rates of production have been attributed to, among many factors, access to (and ability to use) full sunlight with minimal photoinhibition at high irradiance levels and an abundant supply of water (see Section 24.III.B). In contrast, freshwater plants that grow below the water surface for all or part of their life cycle are often constrained by a limited supply of carbon and insufficient light for maximum photosynthesis (Section 24.II.A; Kirk, 2010). Light availability is the main factor restricting the maximum depth a plant can colonize (Middelboe and Markager, 1997). Earlier suggestions that pressure might be involved are now considered unlikely (see Spence 1982 for first refutation of the role of pressure). In very clear lakes, such as Lake Tahoe (USA), depths down to 120 m are colonized by a moss (Frantz and Cordone, 1967). In contrast, the depth limit can be less than 1 m in very turbid lakes and rivers. Søndergaard et al. (2013) found that the maximum colonization depth of submerged plants was closely related ($R^2 = 0.58$) to Secchi depth (i.e., the depth to which ~10% of surface light penetrates), based on analysis of data from 757 lakes and 919 lake-years (i.e., several lakes sampled over multiple years). Other analyses of global data, regional datasets, or even data from individual lakes over time have likewise revealed strong correlations ($0.49 < R^2 < 0.77$) between maximum colonization depth and Secchi depth (Fig. 24-16). The slopes of these various regression models are often alike, particularly for a given plant type, indicating that change in colonization depth in response to diminished water clarity (largely caused by eutrophication) is similar among lakes. Thus the average rooting depth limit of freshwater plants as a percent of surface light is 2.2% for bryophytes, 5% for charophytes and elodeid plants, 12.9% for *Isoetes*, and 16.3% for other isoetid angiosperms (Middelboe and Markager, 1997). Some of these differences in depth limits are likely to be caused by the respiratory burden imposed by nonphotosynthetic tissues such as roots (Sand-Jensen and Madsen, 1991), which are replaced by a smaller biomass of rhizoids in bryophytes and charophytes. Growth form is also important in determining light limits as elodeids can colonize greater depths than isoetids, probably because their shoots allow them to grow up into shallower water with higher light. This contrast was analyzed by Titus and Adams (1979), who observed that the nonnative, canopy-forming elodeid, *Myriophyllum spicatum*, had partially replaced the large isoetid, *Vallisneria americana*, in some lakes in Wisconsin, USA. *Myriophyllum spicatum* can produce 60% of its biomass in the top 30 cm of the water, whereas *V. americana* produces 60% of its biomass within 30 cm of the sediment and thus, despite its greater ability to tolerate low light, was outcompeted for light by *M. spicatum*.

Reduction in water clarity can also result in changes in species richness and diversity. Reduced water transparency, usually driven by nutrient-enhanced phytoplankton growth, is associated with a shift from submerged forms with biomass concentrated near the sediments (e.g., *Chara* spp. or isoetids) to those with leaves near or floating on the water surface (e.g., elodeid or amphibious forms) and, ultimately, emergent vegetation (e.g., *Scirpus* and *Typha* spp.) (Chambers, 1987b; Egertson et al., 2004). Such changes are often accompanied by a loss of species richness and can trigger a switch from a clear-water to a turbid-water state. A clear example of this phenomenon is shown by the 100-year record of changes in freshwater plants in Lake Fure, Denmark (Sand-Jensen et al., 2017). Here, the mesotrophic conditions of 1911 (Secchi depth 5–6 m) were associated with a rich plant community (36 species) dominated by large angiosperms and charophytes along with small angiosperms, moss, and macroalgal species (Fig. 24-17). In 1970 external P inputs peaked at 30 times the initial 1900 values and summer Secchi depths were <3 m. Species richness declined to 21 in 1951 and

FIGURE 24-16 Detransformed regression lines for bryophyte, charophyte, angiosperm, and freshwater plants collectively in relation to Secchi depth. *Regression equations were obtained from: [A] Middelboe and Markager (1997) (note that angiosperm line is for caulescent forms); [B] Søndergaard et al. (2013); [C] May and Carvalho, (2010a); [D] Canfield et al. (1985); [E] Chambers and Kaiff, (1985); [F] Azzella et al. (2014).*

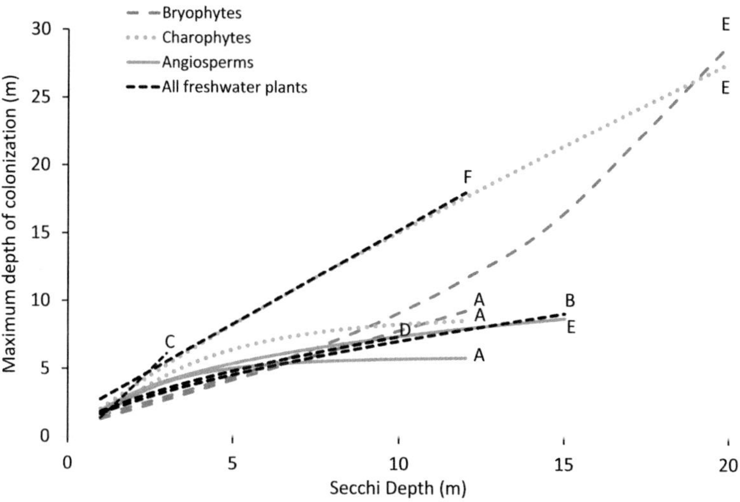

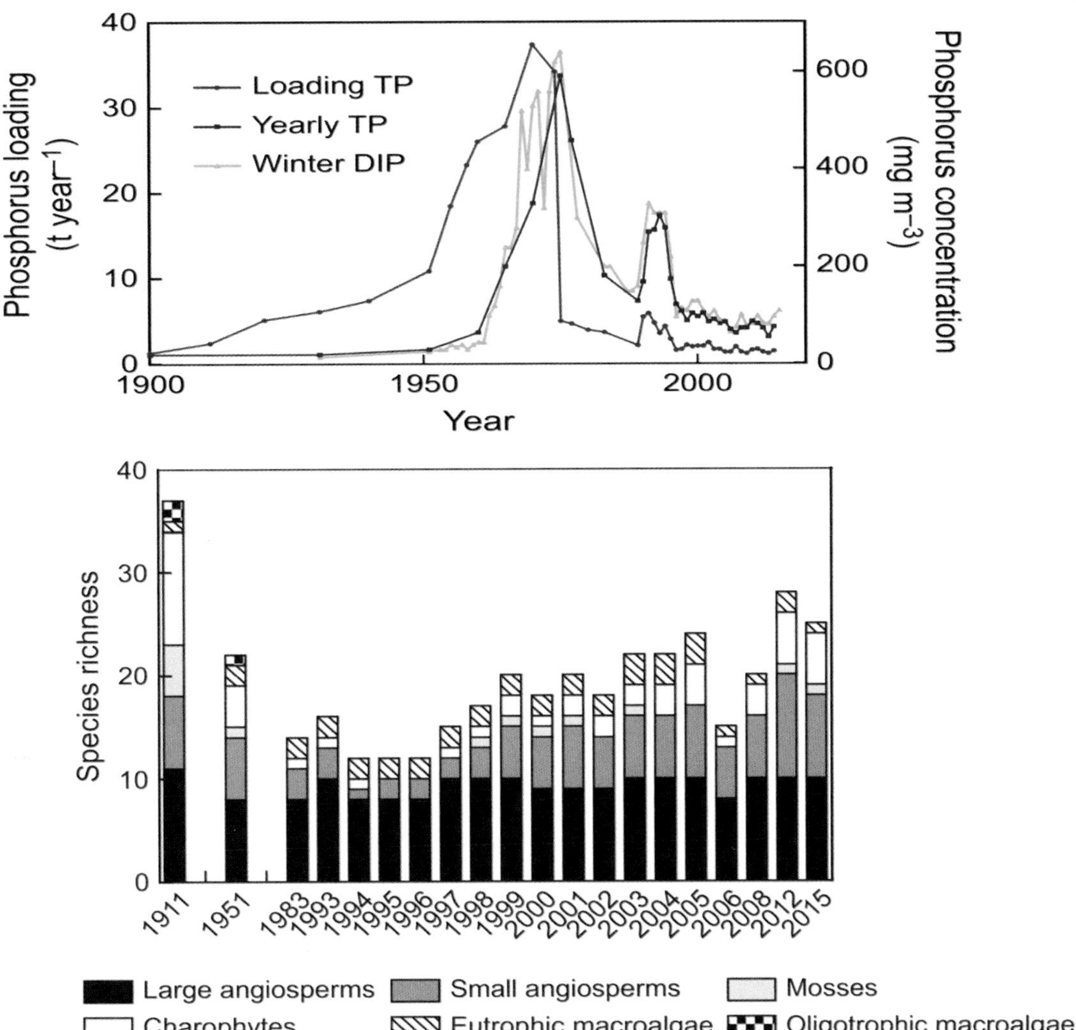

FIGURE 24-17 Changes in water quality and freshwater vegetation in Lake Fure, Denmark, over a 100-year period (from Sand-Jensen et al., 2017). *Top panel*: annual external TP load, mean annual TP, and maximum winter concentration of dissolved inorganic P from 1900 to 2014. *Bottom panel*: species richness of various types of freshwater plants in 19 study years between 1911 and 2015.

reached a minimum of 12 in 1994 to 1996 (about 25 years after the marked reduction in external P loading in 1970) due to loss of all moss, charophyte, and short elodeid angiosperm species. By 2005, diversion and treatment of sewage had reduced P loading to almost 1900 levels, resulting in an increase in summer Secchi depth to almost 4 m and a recovery of several angiosperm and charophyte species (though only one moss species) to a total of 28 species in 2012.

Although physical factors (habitat stability, temperature, and light) are primary determinants of freshwater plant survival, the chemical environment (salinity, alkalinity and pH, nutrients) is an important factor influencing species diversity and abundance (Fig. 24-15). Most limnological investigations have focused on freshwaters rather than saline lakes (see Chapter 12), even though the latter are widespread and found on all continents, including Antarctica, though most common in climatic regions where evaporation exceeds precipitation. Vascular plant species richness decreases with increasing salinity such that in hypersaline lakes the only plants remaining are often submerged *Ruppia* species and several emergent species (Hammer and Heseltine, 1988).

Alkalinity is a major determinant of the types of plants found in freshwaters (Section 24.II.B). Sites with very low alkalinity (<0.2 mequiv L^{-1}) are dominated by species-poor communities mainly comprising isoetids (e.g., *Littorella uniflora, Isoetes lacustris, Lobelia dortmanna*). Isoetids are adapted to conditions of limited carbon (and nutrient) availability, with their extensive root biomass and the unique ability to draw upon dissolved CO_2 in the sediments for photosynthesis (see Section 24.II.B). In contrast, sites with high alkalinity (>1

mequiv L^{-1}) tend to be dominated by vascular plants of the elodeid growth form, many of which can utilize bicarbonate as a carbon source for photosynthesis (albeit with variation in bicarbonate affinity among species). Sites with intermediate alkalinity tend to contain a mix of isoetids and elodeids (Vestergaard and Sand-Jensen, 2000; Iversen et al., 2019).

Most low alkalinity freshwaters are unproductive because the inherently low concentrations of nutrients are insufficient to support substantial plant growth. In contrast, waters with moderate to high alkalinity show a range in trophic status (i.e., nutrient richness), related to both underlying geology (which determines alkalinity and, in turn, influences rock weathering) and human-caused nutrient additions resulting from land-use activities (agriculture, forestry, mining, urban/suburban development) and wastewater discharge (sewage, industrial, and agricultural wastewater). Freshwater algae show a clear increase in abundance in response to added phosphorus (Chapter 15); however, the role of nutrients (nitrogen and phosphorus) in governing the abundance and composition of large freshwater plant communities has long been debated. Although rooted freshwater plants can take up nutrients directly from the water when concentrations are high, the general consensus is that rooted plants obtain most of their nutrients from the sediments (see Section 24.II.C for details). This ability to use sediment nutrients confounds relationships between freshwater plant abundance and composition and trophic status as sediment nutrient concentrations may show no relationship to surface water concentrations and hence trophic status (e.g., Clarke and Wharton, 2001). In addition, other factors such as hydrology and shading may obscure relationships between freshwater plant abundance and composition and trophy. Nevertheless, a few studies have reported increases in freshwater plant abundance due at least in part to increasing nutrients (Chambers and Prepas, 1994; Carr and Chambers, 1998; Mebane et al., 2014) and many others have identified changes in community composition linked to surface water nutrient concentrations and trophic status (Carbiener et al., 1990; Dodkins et al., 2005).

Acknowledgment of the role of nutrients in structuring freshwater plant communities has led to the development of trophic indicators based on freshwater plant composition (Holmes et al., 1999; Schneider and Melzer, 2003; Haury et al., 2006; Willby et al., 2009). The indices vary in their computation, but all are calculated based on species composition at a site and nutrient affinity values (indicative of a species' optimal nutrient conditions) assigned *a priori* to each species. A comparison of two taxonomic indices for freshwater plants, the *Intercalibration Common Metric for lake macrophytes* (ICMLM, based on empirically derived phosphorus optima) and the *Ellenberg Index* (EI, based on expert opinion of nitrogen optima), showed that both metrics were correlated with surface-water TP concentrations based on data from 1474 lake-years from 11 European countries, although the EI proved insensitive to TP > 250 μg L^{-1} (Kolada et al., 2014). The response of both metrics to TP was modified by alkalinity (with ICMLM performing better in lakes with >0.2 mequiv L^{-1} alkalinity, whereas EI was better in lakes with alkalinity <1.0 mequiv L^{-1}), illustrating the need to consider alkalinity as well as nutrients when assessing taxonomic changes in relation to water quality and human activity.

D. Plants as ecosystem engineers

Freshwater plants are integral to the structure and function of the ecosystems in which they grow (Carpenter and Lodge, 1986; van Donk and van de Bund, 2002; Burks et al., 2006). They interact directly and indirectly with organisms in the ecosystem in many ways, including producing structural complexity and altering the habitat in the water and sediment; providing a refuge for other organisms; acting as an attachment surface; supplying energy to the food web; and competing for resources (Fig. 24-18). Different plant species also compete with one another.

Freshwater plants alter the habitat, for example, by reducing flow within dense plant beds in lakes but especially in flowing systems (Gurnell, 2014). In rivers they reduce hydraulic loss of planktonic organisms and in lakes and rivers can cause particles to settle, adding nutrients to the sediment (Sand-Jensen, 1998) and reducing turbidity (Madsen et al., 2001). Dense plant beds can also severely limit the penetration of light in the water column, with species-specific differences determined by both biomass and plant height (Su et al., 2019). Species with floating leaves, or emergent foliage, can have a particularly major effect on the underwater light climate, which can in turn lead to reduced oxygen concentrations or anoxia in the water with dramatic effects on biogeochemistry and biodiversity (Rose and Crumpton, 1996; Caraco et al., 2006) (see Section 24.II.D). Anoxia at the sediment surface and the high pH generated by photosynthesis during the day can lead to the release of phosphate from the sediment to the water (Chapter 15; Mortimer, 1941; Søndergaard, 1988), with consequences for the biota. The structural complexity that plants create has been linked to increased diversity and density of species such as invertebrates (Thomaz et al., 2008). Freshwater plants, along with the concentration of total phosphorus, appear to be linked to taxon richness of many components of shallow lakes (Declerck et al., 2005). Freshwater plants also alter the physical and chemical composition of the sediment, with

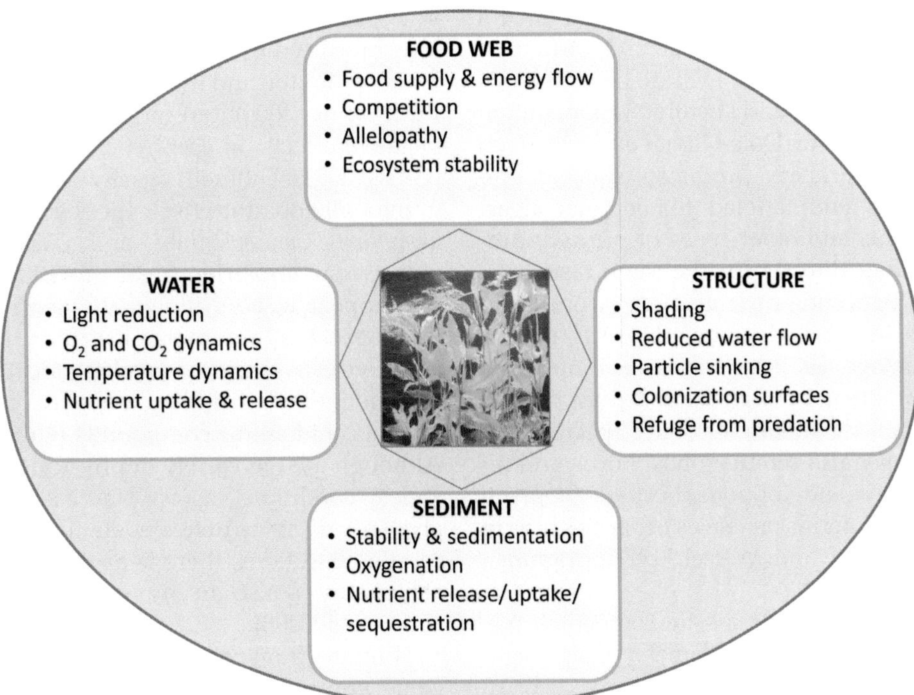

FIGURE 24-18 Effects of freshwater plants on their environment via reciprocal interactions between ecosystem structure, properties of water and sediment, and the food web.

consequences for biological processes and activity of the organisms that live there. For example, roots of the isoetid *Lobelia dortmanna* supply oxygen to the sediment (Sand-Jensen et al., 1982), which can allow aerobic nitrifiers to oxidize ammonium to nitrite and nitrate, in turn allowing anaerobic denitrifiers to denitrify nitrate to nitrogen gas (Petersen and Jensen, 1997), resulting in a loss of nitrogen from the system.

Freshwater plants provide a refuge for zooplankton, reducing their losses by predation (Stansfield et al., 1997; Moss et al., 1998) in a species-specific way (Winfield, 1986). They also provide a refuge for other organisms such as juvenile fish (Persson and Eklov, 1995). This alters trophic interactions among other organisms in the pelagic food web and is one factor responsible for shifting lakes between turbid and clear-water stable states (Chapter 26).

Living freshwater plants are a direct source of energy and elements to many organisms, including mollusks, macrocrustaceans, insects, fish, amphibians, birds, and mammals including humans (Lodge, 1991; Bornette and Puijalon, 2011; Bakker et al., 2016). Freshwater plants in streams and lakes are affected by herbivory at least to a similar extent as amphibious and terrestrial species (Jacobsen and Sand-Jensen, 1992; Sand-Jensen and Jacobsen, 2002; Wood et al., 2017). A comprehensive meta-analysis of the effects of herbivores on plant abundance demonstrated that herbivores reduced (though occasionally increased) plant abundance (Wood et al., 2017). The magnitude of the effect varied with the herbivore taxa involved, and whether or not they were native, and increased with herbivore density and decreased with herbivore species richness. High grazer density can eliminate plants completely. The top-down effect of herbivores on freshwater plants is an overlooked factor that controls freshwater plants, perhaps because of the earlier, erroneous, view that regulation by herbivores was unimportant in freshwater ecosystems.

Living submerged plants can release dissolved organic carbon (DOC; Chapter 28) that can be metabolized by the pelagic microbial community (Wetzel, 1969; Wetzel and Søndergaard, 1998), providing an additional input of resources to the freshwater food web, and also act as *infochemicals* (see below). In addition to consumption of living material, leaf litter from emergent, floating, and submerged species is an important resource that supports macroinvertebrate detritivores and the microbial community (Gessner, 2000; Chimney and Pietro, 2006). Decomposing leaves have been shown to be preferentially colonized by bacteria or fungi depending on the plant species (Mille-Lindblom et al., 2006). Different bacterioplankton are also associated with different species of plant (Zeng et al., 2012). These findings suggest decomposer preferences may relate, at least in part, to allelopathy (Hempel et al., 2009). This provides further evidence for the structuring role of plants in freshwaters with likely consequences for biogeochemical cycles.

Many organisms attach to freshwater plants and benefit from the altered environment and structural complexity of a plant bed. These include eggs of some species of fish, insect larvae, and "benthic" zooplankton from the family Chydoridae (Dole-Olivier et al., 2000), as well as bacteria and primary producers such as epiphytes and attached and tangled filamentous algae. The freshwater plants and other types of primary producers are competing directly for the same resources: light, inorganic carbon, and nutrients. At low concentrations of nutrients in the water, rooted plants have a competitive advantage as they can access nutrients from the sediment, attenuate light, and reduce mixing rates, thereby enhancing sinking of dense species such as diatoms. They may also remove phosphorus and nitrogen by direct uptake and, for phosphorus, by coprecipitation with marl that forms on leaves of species able to use bicarbonate (Otsuki and Wetzel, 1972; Brammer, 1979; Hamilton et al., 2009).

As nutrient concentrations increase, the competitive balance shifts away from plants toward filamentous algae, epiphytes, and phytoplankton, and this can cause a switch to a turbid-water state (Chapter 26). In shallow systems increased growth of epiphytes may be the first step in the reduction of freshwater plant growth (Phillips et al., 1978), followed by increased phytoplankton density because the shading effect of phytoplankton is low in shallow water. In deeper, more exposed lakes, such as Loch Leven in Scotland, phytoplankton might be the primary as well as the ultimate cause of plant decline (Jupp and Spence, 1977). Phytoplankton and epiphytes can outcompete plants for light but also for inorganic carbon (Sand-Jensen and Sondergaard, 1981; Jones et al., 2000; Maberly and Gontero, 2017). Seasonality also plays a role in determining the outcome of the interaction between plants and algae: evergreen plants can lock up nutrients in the water, denying them from phytoplankton, whereas plants that grow seasonally may only have a short window of time to exert their dominance (Sayer et al., 2010). Plants can affect the taxonomic and functional composition of phytoplankton (Barrow et al., 2019) and epiphyton (Wolters et al., 2019), emphasizing the reciprocal interactions between the different photoautotrophs in freshwaters. However, other trophic interactions, such as changes to the fish community, or effects of pollutants (see Section 24.IV.C), may also be important (Phillips et al., 2016).

Water is a good solvent (Chapter 10) and many organic compounds are dissolved in freshwater, including signaling compounds, *infochemicals*, that are released by freshwater plants and affect other organisms. Specifically, many freshwater plants release *allelochemicals* (i.e., chemicals released by one species that have an effect on other species; van Donk and van de Bund, 2002; Gross, 2003). Negative effects of allelochemicals produced by freshwater plants have been reported for zooplankton and insect larvae (Gopal and Goel, 1993; Burks et al., 2000), but most attention has been paid to negative effects on epiphytic algae and phytoplankton (Gross, 2003). Different species of plant appear to differ in their allelopathic effect: species from the genera *Myriophyllum*, *Ceratophyllum*, and *Chara* are particularly effective at impairing algal growth. The algal targets also appear to be differentially sensitive: diatoms and Cyanobacteria are more sensitive than green algae. Moreover, a range of allelopathic compounds have been identified including phenolic compounds, polyphenols, and sulfur compounds (Hilt and Gross, 2008). Although a large variety of physical, chemical, and biological conditions, as well as associated interactions, give an advantage to freshwater plants over algae under certain conditions, there is accumulating evidence that allelopathy is also an important ecological factor (Hilt and Gross, 2008).

Human transport of propagules and anthropogenic environmental change is increasing the number of non-native species in freshwaters and these introductions are forecast to have increasingly important effects on the biodiversity of streams and especially lakes (Sala, 2000; Reid et al., 2019) (see Section 24.V.A). Freshwater plants are particularly successful at colonizing areas outside their native range (Hussner, 2012) because of their ability to propagate vegetatively and because of human movement associated with aquaculture or the aquarium trade. Thus there are well-documented invasions of species such as *Elodea canadensis* from North America to Europe and elsewhere, *Myriophyllum spicatum* from Europe to North America, and *Salvinia adnata* from South America to Africa (see also Section 24.V.A). In China over 150 nonnative aquatic species have been identified, representing around 10% of the native flora (Wang et al., 2016). Upon introduction, these non-natives typically become invasive, especially in the first few decades after their appearance, outcompeting the native flora and hence altering the trophic interactions within freshwaters (Hussner et al., 2021). Over decades, many non-native freshwater plants often become less competitive, perhaps because of increased herbivory (Creed, 1998).

E. Macroecology

Freshwater plants are found worldwide, spanning a latitudinal range of about 80°N to 80°S, including all eight biogeographic areas of the world and seven continents (Fig. 24-19; Chambers et al., 2008) and all thermal

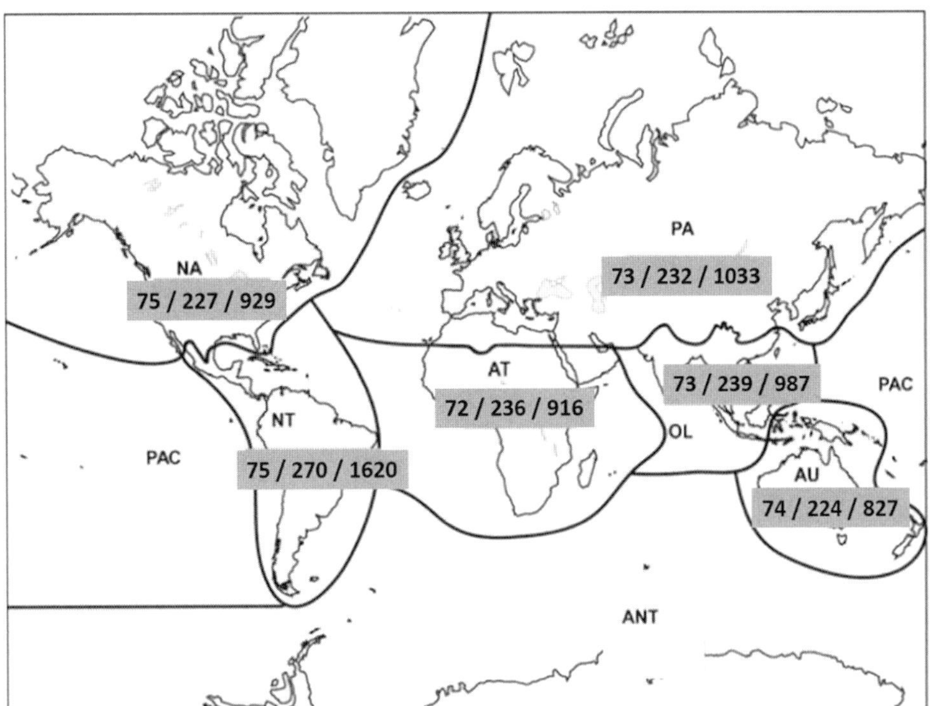

FIGURE 24-19 Diversity of vascular freshwater plants: number of families/genera/species by biogeographic region. PA: Palearctic; NA: Nearctic; NT: Neotropical; AT: Afrotropical; OL: Indo-Malay; AU: Australasian. *(Map from Chambers et al., 2008 with updated data from K.J. Murphy [pers. comm].)*

regimes (Maberly et al., 2020). Freshwater plants have long been classed into floristic assemblages based on species richness and distribution. Groupings based on families recognized those that are cosmopolitan, or almost worldwide, in distribution (e.g., Cyperaceae, Juncaceae, Poaceae) compared with families that are predominately north-temperate (e.g., Potamogetonaceae, Sparganiaceae, Haloragaceae, Elatinaceae, and Hippuridaceae), or pan-tropical (e.g., Podostemaceae, Hydrocharitaceae, Limnocharitaceae, Mayacaceae, Pontederiaceae, and Aponogetonaceae) (Crow, 1993). It should be noted, however, that although families classed as pan-tropical or north-temperate show much higher species richness in these climatic regions, they may still include species that occur outside their climatic region. For example, the family Podostemaceae (riverweeds) is largely found in the Neotropics, Afrotropics, and Indo-Malay regions but has one to seven species in each of the three other bioregions.

Several factors have been suggested to explain the broad distributional ranges of many freshwater plant families and genera including long-distance dispersal by migratory birds, human activity, and continental drift (Arber, 1920; Sculthorpe, 1967; Hutchinson, 1975; Cook, 1985). Les et al. (2003) examined the role of dispersal versus continental drift in the global distribution of freshwater plants. Using molecular estimates of divergence time for 71 freshwater angiosperm species from phylogenetically related freshwater taxa that have discontinuous intercontinental distributions, Les et al. (2003) found that divergence times were far too recent (<30 mya) to implicate continental drift as a determinant of almost all these discontinuous distributions. Even the very ancient *Ceratophyllum demersum* had divergence times of <2.5 mya when specimens from North America, Asia, and Australia were compared, indicating recent dispersal rather than a paleodistribution. Similarly, Cook (1983) considered that about 75% of 61 freshwater plant species and subspecies endemic to Europe and portions of North Africa bordering the Mediterranean evolved after the Ice Age: only about 25% were relicts left by extinction. Long-distance dispersal by birds, human, and other animal activity (both active, through introduction of useful crop plants, and inadvertent) remain viable explanations, with several recent studies suggesting that dispersal resulting from ingestion and gut passage of seeds by waterfowl is a more significant pathway than previously considered (Figuerola and Green, 2002; Lovas-Kiss et al., 2019). The successful long-distance dispersal of freshwater plants has been facilitated by their broad ecological tolerances and plastic responses, enhanced survivorship because of clonal growth, and abundance of easily dislodged propagules (Santamaría, 2002; Les et al., 2003).

Floristic groupings at the family level do not capture the more constrained geographic coverage exhibited by genera and species within a family. For example, a family that contains many species is more likely to be "cosmopolitan," as evidenced by the fact that 12 plant families with 100 or more freshwater plant species are found in every 1 of the 6 biogeographic regions encompassing most of the world's freshwater area (i.e., excluding the Oceania and Antarctic bioregions). Murphy et al. (2019) examined the distribution of freshwater plants at the species level. Of the 3457 species of freshwater ferns and angiosperms analyzed from the six biogeographic regions representing most of the world's freshwater area, only 1.2% (i.e., 42 species) had broad ranges, meaning that they occurred in >50% of the area of the six bioregions (Murphy et al. 2019). In contrast, 78% of all species (i.e., 2697 species) had ranges of <10% of the area of the six bioregions. Thus, in contrast to the wide-ranging distribution of many freshwater plant families and genera, comparatively few species are widely distributed: most freshwater plant species occupy narrow geographical ranges.

It is worth noting that 33% of genera (151 of 457) containing freshwater vascular plants (ignoring any terrestrial species in such genera) are endemic to a bioregion and that 19% of all freshwater vascular species (658 of 3457) are confined to a single $10 \times 10°$ grid cell (i.e., $<1 \times 10^5$ km^2) within a bioregion (Murphy et al., 2019). Many of these endemic genera and species are found within five families: Podostemaceae, Araceae, Isoetaceae, Cyperaceae, and Eriocaulaceae (with 168, 65, 55, 46, and 35 endemic species, respectively). Endemism appears to be favored by warm conditions and is most pronounced in the Neotropical bioregion and poorest (at fivefold less) in the Palearctic. On a smaller geographic scale, endemism is still high in some tropical and subtropical regions but also in some temperate systems: 119 endemic species were recorded by Cook (2004) in South Africa; 100 endemic species were recorded in a region including South Brazil, Uruguay, Paraguay, and North Argentina (Irgang and Gastal Júnior, 2003); 61 endemic species and subspecies were reported for Europe and the portions of North African countries that border the Mediterranean (Cook, 1983); and 38 endemic freshwater plant species were recorded for New Zealand (Coffey and Clayton, 1988). Surprisingly, large ancient lakes such as Baikal and Biwa are poor in endemic freshwater plants: no endemic freshwater plant species have been reported in Lake Baikal, Russia (Kozhova and Izmestéva, 1988), and Lake Biwa, Japan, has only two endemics (Nakajima, 1994).

The distinct distributional patterns of many freshwater plant species have given rise to studies that examined associations between species richness or diversity and environmental clines. An extensive review of broad-scale patterns related to freshwater plant species composition, functional traits, and phylogenetic relationships showed, for example, that species richness of freshwater vascular plants does not decrease linearly from the equator toward the poles but follows a unimodal pattern (Fig. 24-20; Alahuhta et al. 2021). Species richness peaks in tropical to subtropical latitudes (between 10° and 30° absolute), is slightly lower near the equator, and declines steadily from the subtropics toward the poles. In addition, spatial variation in species composition among communities (i.e., beta diversity) was found to be driven by species turnover (i.e., one species replacing another with no change in richness) rather than *nestedness* (i.e., a decrease in species richness in a community that is a subset of the species composition of rich communities within the region). On a global scale, there is evidence that regional patterns in freshwater plant diversity are correlated with geographical gradients such as climate, water availability, and elevation. However, local explanatory variables (e.g., water quality, habitat availability, and hydromorphology) are often important determinants of freshwater plant diversity, particularly at a smaller geographic scale.

IV. Consequences of environmental change

A. Nutrient enrichment and ammonium toxicity

Eutrophication is a natural process that has been greatly accelerated by the discharge of anthropogenic sources of nutrients (P and N) into freshwaters, a phenomenon known as "cultural eutrophication" (Chapters 14 and 15). This surge in nutrient concentrations results in a rapid increase in primary productivity of autotrophs due to alleviation of nutrient limitation and that, in turn, leads to increased water turbidity and decomposition of organic matter, depletion of dissolved oxygen, and changes in abundance and diversity of heterotrophic organisms (see Chapters 19–22). Research on the effects of eutrophication on autotrophs has focused on the phytoplankton and periphyton (Chapters 18 and 25), which have shown more consistent changes in abundance, and often diversity, in response to increased nutrient concentrations. The responses of freshwater plants to eutrophication, however, are not as predictable. In some small shallow lakes there has been a loss of plant cover and decline in species richness and diversity in response to cultural eutrophication, with responses ranging from gradual changes in vegetation composition (e.g., from isoetid to elodeid vegetation; Arts, 2002; Willby et al., 2009), to a shortening of the seasonal duration of plant cover (Sayer et al., 2010), and finally to an abrupt shift from a clear-water state (with abundant and diverse plants) to a turbid-water state (with low transparency and fewer plants, Fig. 24-21; Scheffer,

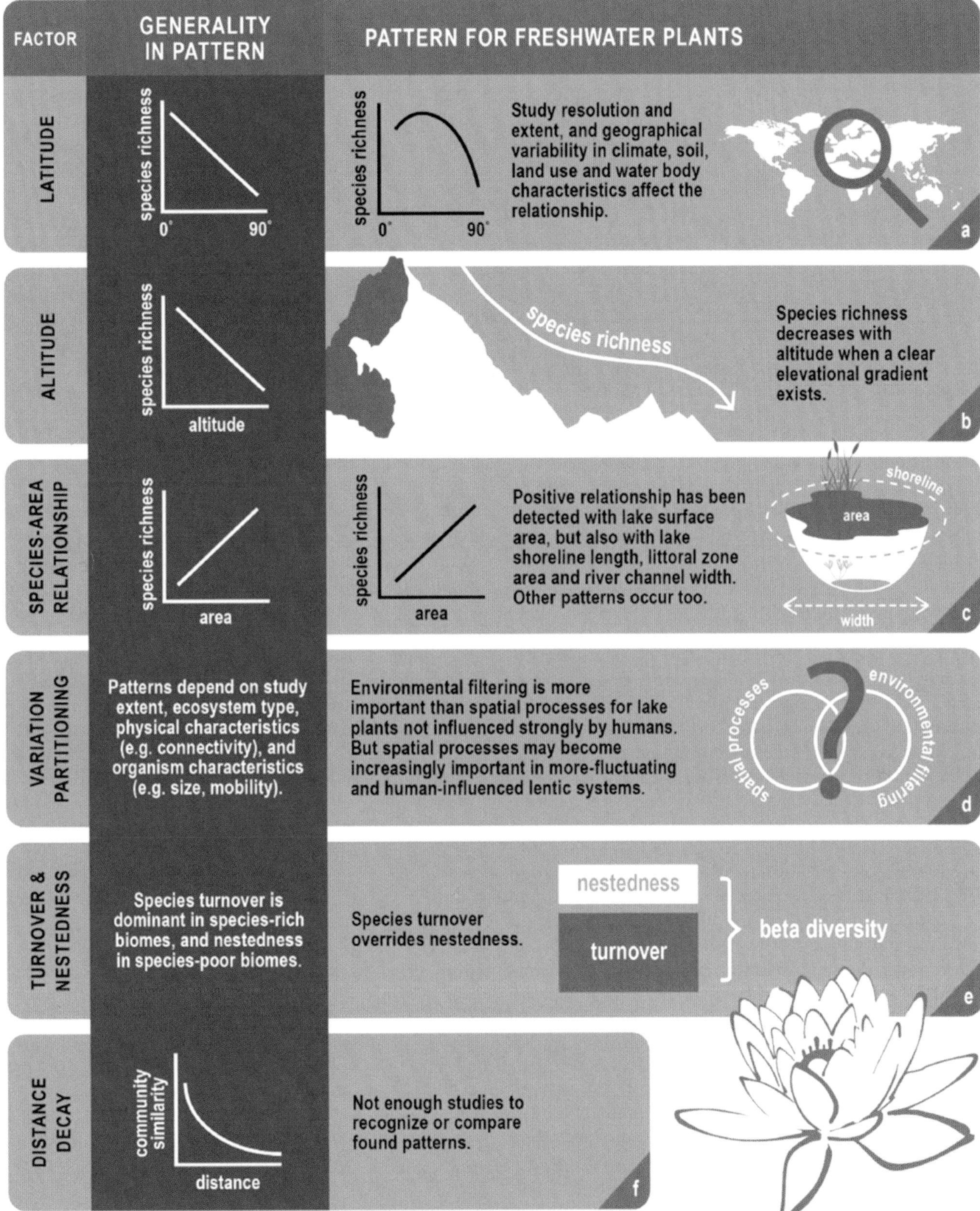

FIGURE 24-20 Examples of large-scale patterns in species richness and other community metrics for freshwater vascular plants (Alahuhta et al., 2021).

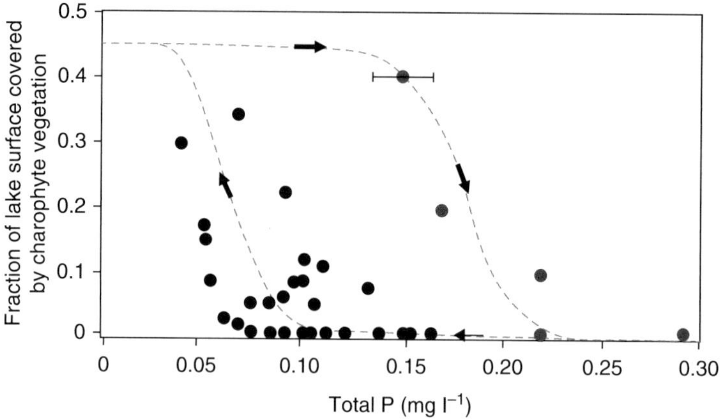

FIGURE 24-21 Response of charophyte macroalgal vegetation in shallow Lake Veluwe, the Netherlands, in response to changes in nutrient loading (Scheffer et al., 2001, originally from Meijer, (2000)). The *red dots* identify the years between the late 1960s and early 1970s when P loading increased and charophyte cover disappeared. The *black dots* show the years when nutrient loading was reduced, eventually resulting in a recovery of the charophytes in the 1990s.

1998), as described more fully in Chapter 26. In contrast to shallow lakes, nutrient addition to rivers has been found to increase biomass of submerged plants, often to the extent that flow is impeded and channel conveyance is reduced as a result of increased plant standing crop (Fig. 24.22) (Carr and Chambers, 1998; Carr et al., 2003; O'Hare et al., 2010). Although excessive inputs of P and N are the underlying cause of cultural eutrophication, the competitive advantages among the autotrophs and, specifically, among the various taxa and growth forms of freshwater plants also relate to physicochemical factors such as water chemistry, water movement, shading, changes in the abundance of grazers, and stochastic weather events (Brönmark and Weisner, 1992; Vestergaard and Sand-Jensen, 2000; Bayley et al., 2007).

Although cultural eutrophication is the major cause of a decline in submersed plants worldwide, ammonium (NH_4^+) toxicity may directly, or in combination with other factors, play a role in the deterioration of freshwater plant populations at extremely polluted sites (Smolders et al., 1996; Clarke and Baldwin, 2002; Cao et al., 2009). Laboratory and field studies have shown that high concentrations of ammonium in water inhibit freshwater plants as a result of disruption of carbon and nitrogen metabolism in order to satisfy greater carbon demand for ammonium detoxification (e.g., Cao et al., 2009). The effects of ammonium depend on environmental conditions (e.g., light availability, temperature) and specific metabolic adaptations of different freshwater plant species. Certain species, such as *Myriophyllum spicatum*, can tolerate high ammonium concentrations by activating a stress response pathway, which is catalyzed by the enzyme glutamate dehydrogenase and detoxifies ammonium at high concentrations (Xian et al., 2020). Few studies have demonstrated ammonium toxicity as the cause of changes in taxonomic composition or abundance in freshwater plant communities, in part because of the interaction with other potential causative factors (e.g., low light). However, results from a survey of lakes in the Yangtze River basin, China, showed that *Vallisneria natans* occurred only at sites where ammonia concentrations were <0.56 mg L^{-1} N (i.e., at 9% of all 609 sites, distributed across 14 of 31 lakes) (Cao et al., 2007). These findings, combined with data from a mesocosm experiment, which showed that plants failed to propagate under high ammonium concentrations (0.81 mg L^{-1} N) because soluble carbohydrate in the rhizomes was depleted to fuel detoxification, suggest that ammonium toxicity played a key role in the loss of *Vallisneria natans* from lakes in the Yangtze basin. Recognition of the fact that certain freshwater plant species can tolerate and assimilate high ammonium concentrations (e.g., *Myriophyllum aquaticum* and *Myriophyllum spicatum*; Gao et al., 2019; Xian et al., 2020) has led to consideration of ammonium-tolerant species as phytoremediators of nitrogen-polluted waters, whereby such plants would be cultivated and then harvested to remove N from the water body.

B. Acidification

Alkalinity and lakewater pH are key variables that control the distribution of freshwater plants (Iversen, 1929; Vestergaard and Sand-Jensen, 2000; Iversen et al., 2019). Some species are typical of naturally soft waters with low alkalinity and typically low pH values. These occur in catchments with low buffering capacity because of their local geology, exacerbated by leaching of organic acids from peat and/or oxidation of iron sulfide to sulfuric acid in volcanic regions. The same oxidation process can also arise from acid-mine drainage associated with lignite or metal mines. Atmospheric deposition, largely of oxides of nitrogen and sulfur (NOx and SOx), but also HCl and ammonium that generates protons when denitrified, is often referred to as "acid rain" or more correctly "acidic precipitation." It can cause acidification over large spatial scales (Stoddard et al., 1999), whereas the nitrogen component can also cause nutrient enrichment (Bergstrom and Jansson, 2006). The greatest ecological effects of acid rain have been seen in poorly buffered soft waters in Northern Europe and North America. In sensitive sites with high deposition rates of acid, as a result of industrialization, burning of fossil fuels, and emissions from agriculture, reductions in pH were detected as early as the mid-19th century (Battarbee, 1990). In affected waters low pH and associated increase in labile aluminum has led to species loss and altered species composition of fish, macroinvertebrate, algal, and freshwater plant communities and has consequently altered the freshwater food web (Christensen et al., 2006; RoTAP, 2012).

There have been winners and losers among the freshwater plants in response to acidification (Farmer 1990). Elodeid species, such as *Myriophyllum alterniflorum*, that grow in a wide range of alkalinities, have been lost from the lowest alkalinity sites of their range (Heitto, 1990). Although isoetids are slightly more tolerant of acid conditions (Arts et al., 1990), species such as *Isoetes lacustris*, *Littorella uniflora*, and *Lobelia dortmanna* have also declined in the Netherlands since the 1950s (Roelofs, 1983). In contrast, the freshwater moss *Sphagnum* (mainly *Sphagnum cuspidatum* and *Sphagnum denticulatum*) and the tall cyperid *Juncus bulbosus* (submerged "fluitans" form) have frequently been observed to appear and increase after acidification (Farmer, 1990; Arts, 2002) and may be partly responsible for the loss of shorter isoetids. *Sphagnum* has an upper pH tolerance of around pH 5.8 (Arts, 2002), so acidification below this threshold is one explanation for the appearance of this genus. Both *Sphagnum* and *Juncus bulbosus* are reliant on free CO_2 in the water as an inorganic carbon source, whereas isoetids have a supplementary source of CO_2 in the sediment and some can perform CAM (Section 24.II.B). So increased CO_2 concentrations in the water may be an additional factor allowing elodeids to invade acidified sites if they can tolerate the low pH (Spierenburg et al., 2009). In sediments that contain some carbonate, acid release of CO_2 is also a likely reason for the profuse growth of *Sphagnum* and *Juncus bulbosus* (Svedang, 1992, 1990).

At some sites, liming the catchment or the lake directly has been undertaken to reverse the detrimental effects of low pH (mainly to reduce the toxic effects of labile aluminum) (Henrikson and Brodin, 1995). Liming has led to the reestablishment of species such as *Fontinalis* spp., *Myriophyllum alterniflorum*, and *Potamogeton* spp. that were extirpated when the lake had become acidified (Brandrud, 2002) and promoted the growth of *Sparganium angustifolium* (Lucassen et al., 2009). It also temporarily stimulated species such as *Juncus bulbosus*, presumably by increasing concentrations of CO_2 in the water.

International legislation, such as the Convention on Long-Range Transboundary Air Pollution (CLRTAP), enacted in 1979 and extended in subsequent protocols, has led to dramatic reductions in sulfur emissions (RoTAP, 2012). In Europe sulfur emissions have decreased by about 80% since the peak in the early 1970s, with smaller reductions in emissions of oxidized nitrogen of about 40% (RoTAP, 2012). Similarly, emissions of sulfur dioxide in Canada and the United States decreased by 69% and 88%, respectively, between 1990 and 2017 (Government of Canada, 2002). This has led to natural chemical recovery and increases in pH and alkalinity at many sites. Although biological responses typically lag chemical ones, there is evidence for slow natural recovery of freshwater plants at some sites. For example, surveys by the UK Acid Waters Monitoring Network (now the Upland Waters Monitoring Network) have found the reappearance of species such as *Callitriche hamulata*, *Chara virgata*, and *Myriophyllum alterniflorum* at five of eight lakes where alkalinity had increased and the mosses *Fontinalis antipyretica*, *Hyocomium armoricum*, and *Hygrohypnum luridum* at four streams (Monteith et al., 2005). At a global scale, anthropogenic activities are causing alkalinity to increase in freshwaters as a result of a large range of processes including climate warming, agriculture, concrete weathering, and reduced acid deposition (Raymond and Hamilton, 2018). The local consequences of this trend will depend on location but may alter the freshwater plant communities further.

C. Pollution

In addition to natural inputs of metals from the catchment, anthropogenic activities such as agricultural and industrial development, among others, have generated a cocktail of substances that enter freshwaters including hydrocarbons, metals, microplastics, nanoparticles, pesticides including herbicides, pharmaceuticals, and surfactants (Ceschin et al., 2021). Pollutants enter water by loss from the catchment, atmospheric deposition in the case of mercury and, in particular, discharges from industry and wastewater treatment works. Many of these materials have direct effects on plants, but others affect them indirectly by altering the food web.

Many metals are essential micronutrients but they can be toxic at high concentrations (Guilizzoni, 1991). Sometimes concentrations of heavy metals or trace metals are naturally high, such as in areas draining catchments where pH is low and metal solubility is high (Pedrozo et al., 2001) or in catchments with rocks with high metal content such as around Lake Ohrid in the Balkans (Minguez et al., 2021). Native populations of plants, such as *Myriophyllum spicatum* in Lake Ohrid, appear to tolerate high concentrations of nickel, chromium, and cobalt. Tolerance might result from genotypic variability because genotypes of this species are differentially tolerant to another metal, copper (Roubeau Dumont et al., 2020). Anthropogenic heavy metal pollution is widespread but its precise biological effects are hard to evaluate because of the complex effect of other environmental conditions on plant toxicity (Guilizzoni, 1991). Similarly, *nanoparticles* (engineered particles 1–100 nm in diameter) containing heavy metals, including silver, cerium, titanium, or zinc, are increasingly used in numerous applications such as medicine, cosmetics, electronics, and manufacturing (Klaine et al., 2008; Lead et al., 2018). Laboratory studies have shown uptake and harmful effects of silver nanoparticles on *Elodea canadensis* (Van Koetsem et al., 2016) and *Spirodela polyrhiza* (Jiang et al., 2017). Although cerium (Ce) nanoparticles entered *Elodea canadensis*, there was no detectable effect under the experimental conditions used (Van Koetsem et al., 2016). There was evidence for inhibition of photosynthesis and induction of oxidative stress by silver nanoparticles in *Spirodela polyrhiza*. However, for a given external concentration, silver (Ag) ions had a greater negative effect than silver nanoparticles. This was linked to ions entering the plant more effectively than particles (Jiang et al., 2017). The concentrations of nanoparticles used in these mechanistic studies probably currently exceed those in most natural situations, apart perhaps from those close to point sources.

Pesticides that do not directly target freshwater plants can nonetheless have indirect, detrimental effects. For example, tributyltin (TBT) was used as an antifouling paint on boats from the 1960s until it was banned globally in 2003. In many of the Norfolk Broads, UK, a wetland area comprising numerous interconnected shallow eutrophic lakes, a shift occurred from clear-water plant-dominated states to turbid-water phytoplankton states (see Chapter 26) in the 1960s to 1970s (Mason and Bryant, 1975). High TBT concentrations have been associated with plant loss via a reduction in mollusks that graze on epiphytes and zooplankton that graze on phytoplankton (Sayer et al., 2010).

Some herbicides are deliberately added to freshwaters to control plant growth (Section 24.V.A), but they can also enter accidently as spray or leaching from adjacent land. It is likely, however, that in most situations herbicides probably do not have a large, persistent effect on freshwater plants (Solomon et al., 1996; Cedergreen and Streibig, 2005). However, the antifouling paints used in the Broads, as discussed above, frequently contained herbicides (Diuron and Irgarol 1051), in addition to TBT, to control algae, and these have also been shown to reduce the growth of a number of plants directly, particularly *Chara vulgaris* (Lambert et al., 2006). Herbicides might therefore have an effect on some freshwater plants when concentrations are particularly high or in specific environments.

In a second example of an indirect effect of pesticides on freshwater plants, a textile mill discharged moth-proofing chemicals into Loch Leven, Scotland. Dieldrin was used from the 1950s until 1964, and chlorphenylid and then flucofenuron from 1964 until discharge ceased in 1988 (D'Arcy et al., 2006). The increase in algal populations over the same time period probably resulted from increased phosphorus input from the mill, and elsewhere, and a reduction in grazing pressure presumed to result from loss of *Daphnia* as a result of the pesticide input. Meanwhile, the biodiversity and density of plants declined dramatically (Jupp and Spence, 1977; (May and Carvalho, 2010b)). Reductions in water transparency caused the maximum colonizable plant depth to decline from 4.9 m in 1905 to 1.5 m in the late 1960s (May and Carvalho, 2010b). Although nutrient enrichment stimulated algal growth, the loss of efficient zooplankton grazers of phytoplankton is also thought to have helped switch the loch from a clear water to a turbid state. After reestablishment of zooplankton once the pesticides had been banned, and a reduction of the phosphorus load, plant diversity and abundance increased and their maximum depth limit is now close to the historic value (Dudley et al., 2012).

Pharmaceuticals and their degradation projects are detectable in most freshwaters, but it is likely that the concentrations would not normally have a major effect on freshwater plants (Cleuvers, 2003; Carvalho et al., 2014). The huge amount of *microplastics* (<5 mm diameter) in the environment is another pollutant of growing recent concern (Petersen and Hubbart, 2021). There are very few studies on the effects of microplastics on freshwater plants, but there is evidence for damage in the carnivorous plants *Utricularia aurea* (Zhou et al., 2020). In this species *polyvinyl chloride particles* (PVC, 10–100 μm), but not the larger *polyethylene particles* (PE, >100 μm), could enter the bladder valves used to catch prey. PVC particles at 50 mg L^{-1} had a negative effect on growth and photosynthesis, but only when bladders were present, suggesting that this is the main route of entry. Studies on more species and concentrations are required to determine the ecological consequence of microplastics on freshwater plants.

There are documented indirect effects of pollutants on freshwater plants, but currently, the magnitude of the effect of pollutants in controlling the biodiversity and abundance of freshwater plants is unclear. This assessment is complicated by variation in the composition and genotypes of species present, interaction between environmental conditions and toxicity, and, perhaps most of all, the largely unknown effect of multiple pollutants interacting with other multiple stressors on plant fitness (Allen et al., 2021). Freshwater plants are able to take up or degrade many pollutants and this is exploited in remediation and the uptake capacity has been used in biomonitoring (Caldwell et al., 2011; Sakakibara et al., 2011; Costa et al., 2018). In their reappraisal of the dynamics of freshwater plants in shallow lakes, Phillips et al. (2016) proposed that, in addition to other factors, pollutants may play a role in the reduction or loss of freshwater plants at some sites. However, further studies are required to elaborate on the circumstances and mechanisms of the processes involved.

D. Climate

The world's climate has shown profound alterations over the past ~100 years. As stated by the International Panel on Climate Change (IPCC, 2021), "The scale of recent changes across the climate system as a whole—and the present state of many aspects of the climate system—are unprecedented over many centuries to many thousands of years." Emissions of greenhouse gases from human activities are primarily responsible for these alterations in climate, causing global warming of about 1.09°C between 2011 and 2020 compared to 1850 and 1900 and a predicted 1.5°C or more of warming during the 21st century unless deep reductions in greenhouse gas emissions are enacted (IPCC, 2021). Concomitant with this are alterations in seasonality (i.e., shortening of winter and an early spring in boreal regions); changes in precipitation patterns (frequency, intensity, and quantity); increased number of hot, sunny days; and unpredictability in the frequency and intensity of extreme events (storms, heatwaves, droughts, floods, and wildfires).

Freshwater plants are directly affected by changes to climate and the hydrologic cycle, especially increases in temperature. Water temperature controls numerous biological and biogeochemical states, structures, and processes including rates of reactions. Increases in average water temperature worldwide in response to climate change (O'Reilly et al., 2015) and projected increases in the intensity and duration of heatwaves (Woolway et al., 2021) have a number of implications for freshwater plants, including making niches more available for nonnative species (Sections 24.III.E and 24.V.D). The speed of future temperature change may exceed natural rates of plant dispersal, preventing required thermal niches from being maintained (Woolway and Maberly, 2020). Temperature also alters the interaction between other factors controlling rates of photosynthesis. For example, the optimum temperature for net photosynthesis when light and inorganic carbon is not, or is weakly, limiting can be over 30°C (Fig. 24-23). However, if one or both of these other factors are limiting, then temperature optima can be dramatically reduced and elevated temperature can lead to reduced productivity or altered metabolic balances (Yvon-Durocher et al., 2010).

Attempts have been made to evaluate the direct and indirect effects of climate change on freshwater plants by examining temporal (interannual and long-term) plant and climate records, undertaking large spatial scale comparisons, and running bioclimatic models. These analyses point to changes in the abundance, composition, and distribution of freshwater plant communities in response to climate change, particularly in cold regions. For example, bioclimatic modeling has indicated that emergent plants will colonize new catchments in the northernmost parts of Finland and increase in cover across all Finnish catchments in response to climate change, in particular the increase in growing degree-days (Fig. 24-24; Alahuhta et al., 2011). In the case of submerged plants, elevated water temperature combined with a longer growing season are likely the most influential direct effects of climate change. *In situ* growth experiments with the elodeid *Callitriche hamulata* indicate that, under a future warmer climate in the Arctic, this plant will expand both its colonization depth and coverage (Lauridsen et al., 2019). In lakes with moderate nutrient availability the positive effect of elevated temperatures may, however, be buffered by shading caused by greater phytoplankton or emergent plant abundance or increased concentration of dissolved organic carbon (Monteith et al., 2007). For example, analysis of data from 83 lakes along a climate gradient in South America indicated that submerged plants tolerate shading better in warmer than in cooler climates, whereas warm lakes may have greater release of sediment-bound P and thus greater phytoplankton abundance than a cooler lake with similar external nutrient loading (Kosten et al., 2011). Hence predicting the competitive outcome between submerged plants and phytoplankton in response to climate change is not straightforward. Even controlled experiments with submerged plant species grown under various climate change scenarios gave results that were species specific. For example, when the canopy-forming *Myriophyllum spicatum* and the bottom-dwelling charophyte *Chara tomentosa* were exposed to higher mean temperatures, *M. spicatum* grew rapidly, whereas *C. tomentosa* was negatively affected, likely because of shading caused by increased phytoplankton (Li et al., 2016). Moreover, when these high temperatures were applied as pronounced fluctuations (to simulate heatwaves), flowering of *M. spicatum* was greatly reduced. Differences among species in their direct and indirect responses to climate change will clearly influence competitive outcomes, and these complex interactions challenge predictions of climate change responses.

FIGURE 24.22 Streams with freshwater plants blocking channel flow. Southern Manitoba, Canada (a) and (b), and southern England (c) and (d) with submerged (*left*) or emergent (*right*) plants. *(Photos of Manitoba streams courtesy [a] Christopher Tyrrell (Milwaukee Public Museum) and [b] Katherine Standen (University of Saskatchewan). Photos of English streams courtesy Peter Scarlett [UK Centre for Ecology & Hydrology].)*

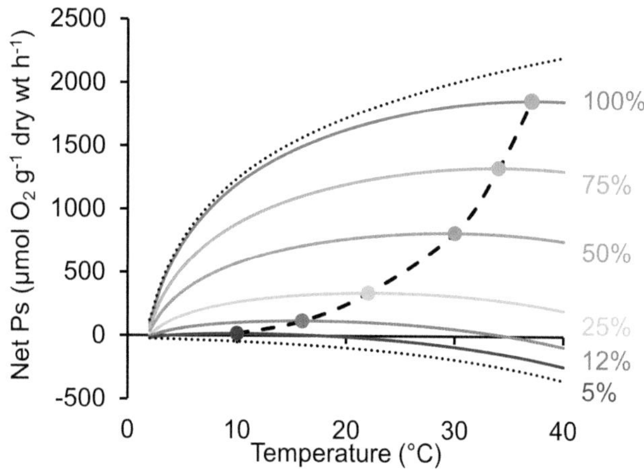

FIGURE 24-23 Response of maximum rates of net photosynthesis (Ps) to temperature at different percent values of maximum gross photosynthesis. The *upper and lower dotted lines* show maximum gross photosynthesis and respiration, respectively. The *ascending dashed line and circles* show the optimum temperature. *(Data for Fontinalis antipyretica [Maberly, 1985a].)*

E. Brownification

The concentration of *dissolved organic carbon* (DOC) (Chapter 28) shows considerable natural variation among freshwaters globally, exceeding 3000-fold, as a consequence of varying climate and catchment characteristics (Sobek et al., 2007). Anthropogenic changes to the climate and land-use, and regional reductions in atmospheric acid deposition, have also caused DOC concentration to increase in many freshwaters. The increase in the colored forms of DOC has been termed *brownification* (Graneli, 2012). This can cause a range of direct and indirect effects on freshwaters including increased strength of thermal stratification, provision of nutrients, and increased light attenuation. For freshwater plants, the reduction in underwater light availability is likely to have the greatest effect. This is illustrated by a large influx of colored DOC to a small Danish lake, Grane Langsø, caused by a storm (Riis and Sand-Jensen, 1998). This influx caused light attenuation to almost double and led to a dramatic reduction in the maximum depth of isoetids, for example, from 5 m for *Isoetes lacustris* in 1955 to 3 m in 1994 and from 3 m for *Lobelia dortmanna* to 1.5 m over the same period. Recent experiments with the amphibious species *Berula erecta* suggested that increased DOC concentrations reduced growth but also interacted with CO_2 concentration and the likely complex interactions with other environmental changes were highlighted (Reitsema et al., 2020). Increased input of organic acids will lower pH and alkalinity, and the decline in charophyte abundance and species richness in some Polish lakes was attributed, at least in part, to acidification caused by increased acid load from the catchment (Bociag et al., 2011). More work is needed to elucidate the potential beneficial (nutrient input and increased concentration of CO_2 from DOC degradation) and detrimental effects (light attenuation and direct harm) of brownification on freshwater plants.

V. Management

A. Control

Although freshwater plants play an important role in the functioning of freshwater ecosystems, their excessive growth may threaten ecosystems and thereby negatively impact the freshwater environment, local economies, and human health. In lowland streams and rivers, freshwater plants are often the most important factor determining flow rates in the channels that they occupy (Watson, 1987). Emergent and submerged vegetation can decrease water speed, raise water levels, and increase the potential for flooding (Fig. 24.22). To maintain high flows and reduce the likelihood of flooding, freshwater plant biomass is often removed by means of cutting or dredging (Fig. 24-25). Although this practice

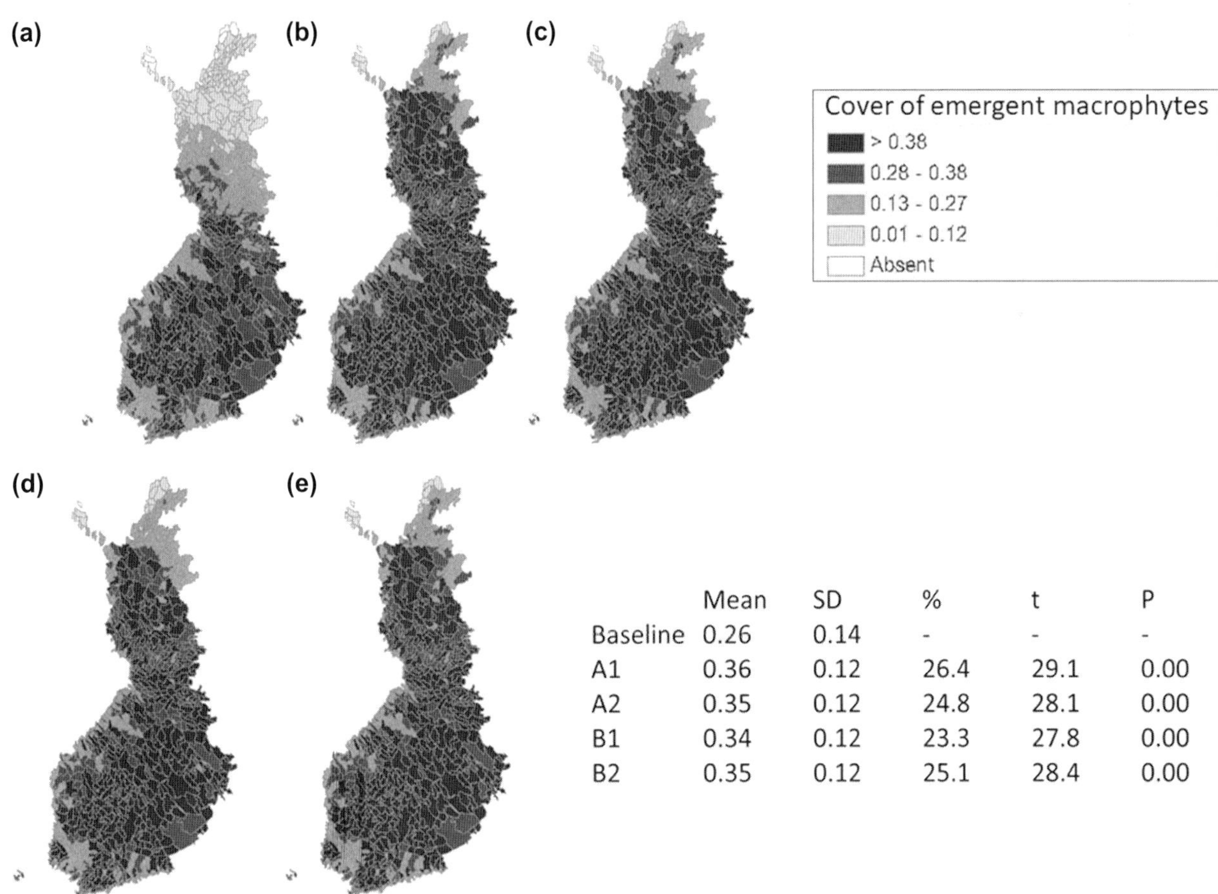

	Mean	SD	%	t	P
Baseline	0.26	0.14	-	-	-
A1	0.36	0.12	26.4	29.1	0.00
A2	0.35	0.12	24.8	28.1	0.00
B1	0.34	0.12	23.3	27.8	0.00
B2	0.35	0.12	25.1	28.4	0.00

FIGURE 24-24 Percent cover of emergent freshwater plants in 848 catchments in Finland in response to climate change (Alahuhta et al., 2011). Panel (a) shows baseline conditions, whereas panels (b), (c), (d), and (e) show expanded plant coverage as predicted in response to four climate scenarios (A1, A2, B1, and B2, respectively). A1 has the highest future greenhouse gas emissions, A2 and B2 scenarios have intermediate values, whereas B1 has the lowest emission levels. Cover data also presented in the table-insert as mean and standard deviation (SD) for all catchments, as well as mean relative change (%) between baseline and the four scenarios. Also given is the statistical significance of the relative changes (paired-sample *t*-test and *P* values).

FIGURE 24-25 Harvesting freshwater plants by cutting and mowing (Buffalo Pound Lake, Saskatchewan, Canada, August 1991). *(photo by the authors.)*

can solve the immediate problem of water conveyance, repeated high-intensity weed cutting may cause suspension and transport of suspended sediment and impairment of ecological quality including changes in freshwater plant diversity, as the cutting causes greater impairment to species with apical meristems versus those with basal growth (Rasmussen et al., 2021). Similar findings of higher turbidity and loss of submerged plant biomass have also been reported in canals with high boat traffic (Willby and Eaton, 1996).

Perhaps because many freshwater plant species exhibit high productivity, broad ecological tolerances, and easily dispersed propagules, several are now considered the worst invasive weeds in the world (Pieterse and Murphy, 1993). Hussner et al. (2017) undertook an intensive review of management and control methods for introduced *invasive alien aquatic plants* (IAAPs). They identified more than 20 species of IAAPs in Europe, New Zealand, and the United States that threaten ecosystems due to excessive growth and cause both ecological and economic impairments. Two of the world's worst freshwater pests are the freshwater fern *Salvinia adnata* and the water hyacinth *Eichhornia crassipes*, both pleustophytes. Originating from South America, these aggressive, competitive species have become serious problems outside their native ranges by covering the surface of lakes and slow-moving rivers in the southern United States; Australia; Southeast Asia; the Pacific; and south, central, and eastern Africa. The water hyacinth, for example, may grow one meter above the water surface and can shade the water from light, leading to deoxygenation and ecosystem degradation (Section 24.III.D; Fig. 24-26). Some of these species form huge floating islands, for example, "the sudd" on the White Nile in South Sudan.

Another serious IAAP is the submerged plant *Hydrilla verticillata*, arguably the most problematic invasive freshwater plant in North America. Native to central and south Asia, it was introduced to Florida in the

FIGURE 24-26 Water hyacinth *Eichhornia crassipes*, growing in a pond in Andhra Pradesh, India *(top, courtesy Ghosh Bobba)*, and in a fish pond in Thailand *(bottom, photo by the authors.)*

1950s via the aquarium trade and is now well established in the southern United States and also found in certain locales in the west coast states of California and Washington. *Hydrilla* forms dense submerged mats of vegetation that may reach the surface, interfering with recreation and destroying fish and wildlife habitat.

Actions aimed at controlling IAAPs can be divided into three categories: physical (i.e., harvesting of plants by cutting/mowing, uprooting, or water-level drawdown); biological (i.e., introduction of a biocontrol agent); and chemical (i.e., herbicide application). Deciding which option is best for a particular situation is complicated: the options differ in their scale of application, cost, and effectiveness, both overall and in relation to the plant species in question, the extent of infestation, and the type of freshwater system (Fig. 24-27; Hussner et al., 2017). The decision to implement an IAAP control program is not one to be taken lightly. For example, actions to control IAAPs in the United States were once estimated to cost $100 million (USD) per year (Pimentel et al., 2000), with the largest share of this money directed at controlling *Hydrilla*. In Florida the management of *Hydrilla* involves the use of aquatic herbicides, biological agents, mechanical harvesting, and physical habitat manipulation. As potential biological controls, four insect and one fish species have been released and although none of the insects have been able to control *Hydrilla* populations adequately, the fish (Chinese grass carp, *Ctenopharyngodon idella*) can be effective. The grass carp is, however, a voracious consumer of most submersed and emergent freshwater plants and, if allowed to grow and reproduce unchecked, would cause deleterious changes to freshwater ecosystems. Therefore only sterile "triploid" grass carp

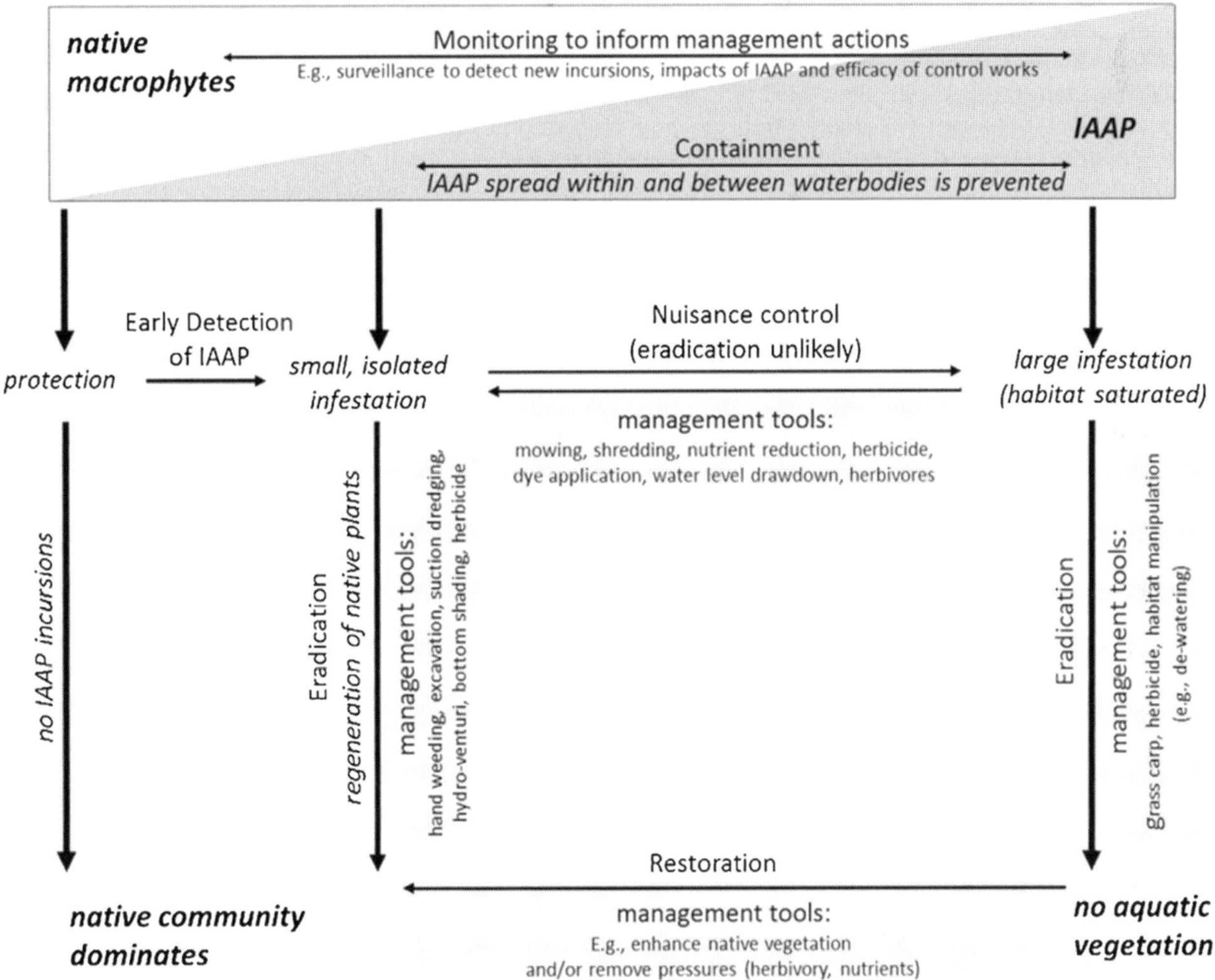

FIGURE 24-27 Decision tool for informing management actions to control introduced invasive alien aquatic plants (IAAPs) (Hussner et al. 2017). The *upper box* identifies the state of the ecosystem, ranging from dominance by native freshwater plants (*white*) to dominance by IAAPs (*gray*). The proportion of native freshwater plants to IAAPs determines the management actions and associated tools (*light grey boxes*) to achieve the desired outcome.

are permitted for freshwater weed control. In contrast to the situation in Florida, *Hydrilla* first appeared in California in 1976. Awareness of the Florida experience, combined with survey and detection, early treatment, public information and awareness, and monitoring (O'Connell, 1997), have enabled managers to restrict *Hydrilla* populations to a few sites in California and these populations are being eradicated. This comparison illustrates the effectiveness of prevention as the first control strategy and, failing this, early detection and rapid response to control IAAP populations and minimize costs.

B. Cultivation

In contrast to the threat posed by invasive freshwater plants, several freshwater plant species are cultivated for human use. Rice (*Oryza* spp.) is the world's most important staple food crop. Rice production is about 500 million tons per year, with China, India, and Bangladesh being the top three producers (World Rice Production 2020/2021; accessed April 29, 2021). Rice is the most important food crop in the developing world and the staple food of more than half of the world's population. Global rice demand is estimated to increase to 555 million tons in 2035. There is concern, however, that current rice production practices may not meet future demands while keeping prices affordable, due to limited possibilities for arable area expansion, fewer water resources for expanding planted area, and competition from both urbanization and other high-value agricultural products. One initiative to tackle this problem is to introduce C_4 biochemistry into C_3 rice in an attempt to boost productivity and nitrogen-use efficiency (von Caemmerer et al., 2012).

Many species of freshwater plants are also cultivated, and commercially traded, for use in aquaria and ponds. The aquarium and pond plant trade is a multimillion-dollar industry, with plants usually exported from tropical and subtropical regions (Padilla and Williams, 2004). Although some plants are harvested from the wild, many are cultivated in large outdoor nurseries in regions with warm climates. There is, however, a move to tissue culture or hydroponic cultivation of freshwater plants, techniques that are more labor intensive but yield healthier plants. Regardless of the cultivation method, cross-border transport of freshwater plants is a serious ecological concern because many IAAPs originate from discarded or escaped aquarium material.

While rice is probably the most widely used freshwater plant by humankind, many other species are used locally or regionally, for example, in pulp production (e.g., *Phragmites australis*); as thatch for house roofs, mats, etc. (e.g., *Cyperus* spp. and *Phragmites australis*), in medicine (e.g., *Alternanthera philoxeroides* and *Sagittaria rhombifolia*); and for aesthetic value (e.g., *Nymphaea* spp., *Hydrocleys* spp., and *Victoria amazonica*). Various freshwater plants, in particular *Eichhornia crassipes, Salvinia adnata, Pistia stratiotes*, and members of the family Lemnaceae, have been tested for use in phytoremediation of domestic, agricultural, and industrial effluent by reducing freshwater contaminants through bioaccumulation in their tissues (e.g., Mustafa and Hayder, 2021). Not surprisingly, several of the species that have high bioaccumulation rates of contaminants are also considered invasive weeds as their fast growth rate and resilience enables rapid uptake of contaminants. Although *phytoremediation* (the use of plants to sequester or degrade contaminants) is a promising technique to treat polluted water, consideration must be given to appropriate disposal of contaminant-rich plant tissue and the prevention of spread of freshwater plants used in phytoremediation.

C. Restoration

Restoration of freshwater vegetation in lakes and rivers has typically focused on reversing the effects of eutrophication, although other stressors (e.g., acidification, habitat modification) can also be responsible for plant loss. In many shallow lakes throughout the world, sustained inputs of excessive nutrients (especially P) have led to shifts from clear water to turbid conditions and loss of submerged plants (Chapter 26). Efforts have been devoted to shallow lake restoration in many countries, but recovery patterns have varied widely. Actions to restore turbid eutrophic lakes can be divided into two types: external measures that aim to reduce external P (and sometimes N) inputs, and internal measures that include *biomanipulation* (primarily removal of herbivorous fish) and physicochemical methods (such as application of chemicals to bind P and dredging of nutrient-rich sediments). After a comprehensive assessment of data from 49 north temperate shallow lakes, Hilt et al. (2018) concluded that reductions in nutrient loading led to an intermediate recovery state characterized by a few freshwater plant species (mainly *Potamogeton* spp.) that can survive in shallow, wave-swept areas and can complete their life cycle by early summer, before being outcompeted by periphyton and phytoplankton (Fig. 24-28). It was only decades after the external nutrient load had been reduced, or when external actions are combined with internal measures, that stable clear-water conditions and a diverse flora were achieved. In contrast, the application of internal lake restoration measures often results in clear-water conditions during the following spring and summer; however, turbid conditions often return within a few years. Variability among lakes in rate of recovery to stable clear-water conditions, and the composition of the

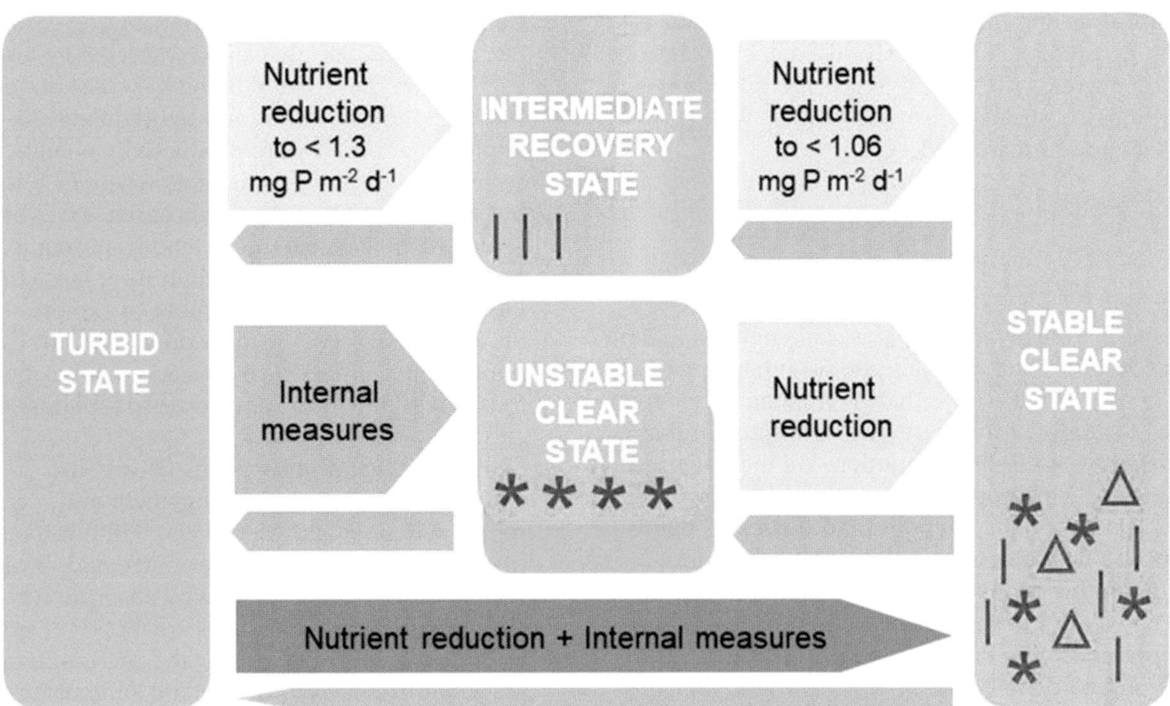

FIGURE 24-28 Response patterns of turbid shallow lakes in north temperate regions to restoration measures. *(From Hilt et al., 2018.)* (*Upper row*) Reduction of P loading is expected to lead, first, to an intermediate recovery state, which permits growth of short-lived plants in spring when clear-water conditions prevail. If P loading is sufficiently reduced or internal measures are also applied, then a stable clear condition and a diverse freshwater plant community will prevail. (*Middle row*) Application of internal measures results in unstable clear-water conditions that will return to turbid state unless P loading is reduced. (*Lower row*) A combination of external and internal restoration measurements can lead directly to clear-water conditions and a diverse freshwater flora.

freshwater plant community, depends not only on the specific restoration measure(s) applied but also stochastic influences (such as winter fish kills, fish predation, or introduction of invasive species) and the availability of propagules and remnant populations.

In addition to actions aimed at reversing eutrophication, restoration of rivers and acidified lakes has also resulted in recovery of freshwater vegetation. A meta-analysis of 40 river restoration projects that considered freshwater plants showed that both biomass and diversity benefited from river restoration, mainly because of measures directed at river widening/rebraiding and restoration of meanders (Kail et al., 2015). Both these measures reduce current velocity and hence create conditions that favor freshwater plant colonization and growth. Restoration of acidified lakes by reducing acidifying emissions (especially SOx, NOx, and NHx) and/or additions of lime has also had moderate success, with a loss of some acidophilous species such as *Sphagnum* and a reestablishment of softwater taxa (Section 24.IV.B; Brouwer et al., 2002). However, the success of acidification restoration measures is strongly affected by hydrologic processes (e.g., lake water residence time) as well as alkalinity-generating mechanisms in the lake (e.g., reduction processes in and cation-exchange with bottom sediments) and catchment (e.g., weathering of soil).

D. Conservation

Although large gaps still exist in our knowledge of freshwater plant abundance and distribution, a number of species are recognized as endangered or vulnerable. As of 2020, the International Union for Conservation of Nature's Red List of Threatened Species (IUCN, 2020) had assessed 3585 plant species that grow in freshwaters (lakes, pools, and flowing water, as well as temporary or permanent wetlands). Of these, five species (0.1%) were identified as extinct or extinct in the wild, 149 species (4.2%) were considered critically endangered, another 250 species (7.0%) were identified as endangered, and 213 species were considered vulnerable (5.9%). These imperilments were largely caused by habitat loss due to urban/residential and agricultural expansion, as well as climate change and severe weather events (particularly droughts). Many of the threats to freshwaters (eutrophication, acidification, pollutants, climate change) lead to reduced plant diversity and will, as a result, threaten the faunal diversity of freshwater ecosystems, favor the establishment and expansion of exotic species at the expense of native species, and challenge our abilities to sustain freshwaters that will support ecosystem and human services.

VI. Summary

1. The term "freshwater plant" is challenging to define because it encompasses all photosynthetic plants, visible to the naked eye, that grow along a cline from land to complete submergence, a distinction that is blurred by water-level variation. We include green macroalgae (charophytes), hornworts, liverworts, mosses, lycopods, polypods, and flowering plants that rely on water to complete their life cycle. Different life forms vary in their access to resources in air, water, and sediment. These include emergent, amphibious, free-floating, and different forms of submerged plants.
2. All freshwater plants, apart from the charophytes, have evolved from terrestrial ancestors. The charophytes are an ancient algal group, the first mosses evolved around 500 million years ago, and flowering plants probably evolved earlier than 120 million years ago. Taxonomic diversity is hard to quantify precisely, but there are approximately 5000 species of freshwater plants spread over 823 genera and 211 families. Flowering plants comprise 66% of all freshwater plant species, and the monocots are a particularly important group.
3. Leaves of freshwater plants include the smallest and largest of all plant leaves. Leaf form has converged to a smaller number of forms than in terrestrial plants. Stomata, which evolved in their terrestrial ancestors, are not usually produced underwater and cuticle thickness is reduced. Laminar leaves are often reduced to an upper and lower epidermis, both with chloroplasts, in response to reduced availability of light and inorganic carbon. Many leaves contain relatively large volumes of air spaces that produce buoyancy and maximize external acquisition, and internal recycling of inorganic carbon.
4. Most freshwaters are shade environments and light availability usually determines plant depth limits. Plants acclimate to light availability via altered physiology, leaf anatomy, and shoot morphology. Photoinhibition and photoprotection can occur at high light especially when plants are limited by other factors such as inorganic carbon. Plants at or near the surface respond to high UV radiation by producing UV-screening compounds.
5. Carbon is a macronutrient, contributing 35%–50% of dry weight. Supply can be restricted by low rates of diffusion that require high CO_2 concentrations to saturate photosynthesis and by frequent CO_2 depletion at sites where productivity is high. Plants respond by growing in habitats where CO_2 is high, by exploiting CO_2 in the air via aerial leaves or in the sediment via lacunae that extend from root to shoot, and also by active CO_2 concentrating mechanisms. Of these mechanisms, the ability to exploit bicarbonate reserves is most prevalent, but C_4 photosynthesis and crassulacean acid metabolism (both well-known in some terrestrial plants) also exist.
6. Nitrogen (N) and phosphorus (P) are also macronutrients, essential for growth and reproduction. Freshwater plants without roots (macroalgae, hornworts, liverworts, and mosses) obtain most of their nutrients by absorption by foliar structures. Roots or rhizomes are, however, essential for nutrient uptake by freshwater pteridophytes and angiosperms, even those that are submersed. Because most freshwater plants obtain P from the bottom sediments, plants can be a source of nutrients to overlying water upon decay.
7. Dense stands of plants can produce anoxia in the water by rapid respiratory uptake of oxygen, exacerbated by rapid light attenuation and reduced exchange of oxygen with the open water or the atmosphere. Many freshwater sediments are anoxic at depth, but the lacunal system can produce oxic zones around the roots via diffusion from the leaves or via ventilation from oxygen in the atmosphere.
8. Like all plants, the life history of freshwater plants centers on survival, propagation, and population expansion. Freshwater plants have, however, a greater variety of vegetative (asexual) reproductive strategies than terrestrial plants: shoot fragmentation and turions in addition to tubers, corms, stolons, and rhizomes. These structures also aid in dispersal and survival during unfavorable conditions.
9. Freshwater, especially emergent, plants are among the most productive natural plant communities in the biosphere. In temperate regions freshwater plants initiate growth once water temperatures reach about 10–12°C, attain peak biomass in late summer, and die back when temperatures fall below 3°C. For plants with negligible biomass in spring (i.e., only rhizomes, turions, or seeds), peak biomass in late summer equals cumulative productivity. In tropical regions temporal variation in freshwater plant communities is linked to wet-dry cycles.
10. Collectively, freshwater plants are found throughout the world. Although several families have broad

regional distributions (pan-tropic or pan-temperate), individual species typically have narrow geographical ranges. Species richness peaks in tropical to subtropical latitudes, is slightly lower near the equator, and declines steadily from the subtropics to the poles.

11. Globally, geographical gradients in climate, water availability, and elevation influence freshwater plant diversity. However, at a regional or local scale, physical, chemical, and biological factors usually determine diversity and abundance.

12. Physical conditions (habitat stability, water temperature, and light availability) determine the suitability of a habitat for plant colonization. If physical conditions are amenable, water chemistry (fresh versus saline; soft versus hard; oligo versus eutrophic) then exerts a major influence on plant abundance and species composition.

13. Freshwater plants are integral to the structure and function of littoral zones and shallow lakes. They produce structural complexity, provide refuges and surfaces for colonization, alter physical and chemical conditions in the water and sediment, and supply energy and nutrients to the food web. They also compete with phytoplankton and epiphytes for resources and interact via allelopathy. Freshwater plants are responsible for generating ecosystem stability and producing clear, rather than turbid, water conditions.

14. Excessive anthropogenic additions of P and N have resulted in a decline in cover and diversity of submerged plants in small lakes due to increased water turbidity caused by surging algal growth. High concentrations of ammonium (NH_4^+) can also be toxic to submersed plants and may contribute to loss of vulnerable species from nitrogen-polluted waters.

15. Some freshwaters are naturally acidic, but anthropogenic atmospheric deposition of oxides of nitrogen and sulfur in the last two centuries has reduced the pH and alkalinity in sensitive sites, causing a loss of many species but an increase in others. Reduced deposition in the last half a century, or management by liming, has started to allow biological recovery at some sites.

16. Agricultural, industrial, and other anthropogenic activities have generated a cocktail of pollutants in freshwaters. These include metals, microplastics, nanoparticles, pesticides including herbicides, pharmaceuticals, and surfactants. Current assessment of exposure to a single pollutant suggests largely local direct effects, but the effect of mixtures is little studied and could be greater. There is evidence for the negative indirect effects of pesticides on freshwater plants mediated by changes to the food web that reduce grazing pressure on competing algae.

17. Freshwater plants are directly affected by a changing climate, for example, by increases in temperature, changes in precipitation regime, and changes to the hydrological cycle. Analysis of long-term records, comparisons among contemporary biogeographic regions, and bioclimatic model predictions point to changes in the abundance, composition, and distribution of freshwater plant communities in response to climate change, particularly in cold climate regions.

18. Because of their high productivity, broad ecological tolerances, and easily dispersed propagules, several freshwater plant species are now considered the world's worst invasive weeds. Control of freshwater invasive weeds is costly and may involve harvesting, introduction of biological control agents, use of herbicides, and habitat manipulation.

19. In contrast to the threat posed by invasive freshwater weeds, several freshwater plants are cultivated for human use. Rice is the most important cultivated freshwater plant and a food staple for more than half of the world's population.

20. Restoration of freshwater vegetation in lakes and rivers has focused on reversing eutrophication. This requires applying external measures to reduce nutrient inputs, internal measures such as manipulating the food web or reducing internal P loading, or a combination of both. Although these efforts can be successful, recovery can take years to achieve.

21. Of 3585 freshwater and wetland plant species assessed, 17.2% have been identified as extinct, vulnerable, or endangered, largely because of habitat loss due to urban/residential and agricultural expansion and climate change and severe weather events (particularly droughts).

22. Many of the anthropogenic threats to freshwaters (eutrophication, acidification, pollutants, climate change, and others) will lead to reduced plant diversity. In turn, this will threaten the faunal diversity of freshwater ecosystems, favor the establishment and expansion of exotic species at the expense of native species, and challenge our abilities to sustain freshwaters that support ecosystem and human services.

Acknowledgments

This chapter is dedicated to the late Professor D.H.N. Spence, our PhD supervisor at the University of St Andrews, Scotland, in the early 1980s. David made a major contribution to the study of the ecology and physiology of freshwater plants. His knowledge and enthusiasm ignited our persisting passion for this topic. We also thank Janice Glime, Elisabeth Gross, Kevin Murphy, and Ole Pedersen for their valuable help and comments. We appreciate the many colleagues who provided photographs, particularly Klaus van de Weyer.

References

Adamec, L., 1997. Mineral nutrition of carnivorous plants: a review. Bot. Rev. 63, 273–299. https://doi.org/10.1007/bf02857953.

Adamec, L., 2018. Ecophysiological characteristics of turions of aquatic plants: a review. Aquat. Bot. 148, 64–77. https://doi.org/10.1016/j.aquabot.2018.04.011.

Agbeti, M., Smol, J., 1995. Winter limnology—a comparison of physical, chemical and biological characteristics in 2 temperate lakes during ice cover. Hydrobiologia 304, 221–234. https://doi.org/10.1007/BF02329316.

Alahuhta, J., Heino, J., Luoto, M., 2011. Climate change and the future distributions of aquatic macrophytes across boreal catchments. J. Biogeogr. 38, 383–393. https://doi.org/10.1111/j.1365-2699.2010.02412.x.

Alahuhta, J., Lindholm, M., Baastrup-Spohr, L., García-Girón, J., Toivanen, M., Heino, J., Murphy, K., 2021. Macroecology of macrophytes in the freshwater realm: patterns, mechanisms and implications. Aquat. Bot. 168, 103325. https://doi.org/10.1016/j.aquabot.2020.103325.

Allen, J., Gross, E.M., Courcoul, C., Bouletreau, S., Compin, A., Elger, A., Ferriol, J., Hilt, S., Jassey, V.E.J., Laviale, M., Polst, B.H., Schmitt-Jansen, M., Stibor, H., Vijayaraj, V., Leflaive, J., 2021. Disentangling the direct and indirect effects of agricultural runoff on freshwater ecosystems subject to global warming: a microcosm study. Water Res. 190, 116713. https://doi.org/10.1016/j.watres.2020.116713.

Anderson, C.J., Mitsch, W.J., 2005. Effect of pulsing on macrophyte productivity and nutrient uptake: a wetland mesocosm experiment. Am. Midl. Nat. 154, 305–319. https://doi.org/10.1674/0003-0031(2005)154[0305:EOPOMP]2.0.CO;2.

Angelstein, S., Wolfram, C., Rahn, K., Kiwel, U., Frimel, S., Merbach, I., Schubert, H., 2009. The influence of different sediment nutrient contents on growth and competition of *Elodea nuttallii* and *Myriophyllum spicatum* in nutrient-poor waters. Fundam. Appl. Limnol 175, 49–57.

Arber, A., 1920. Water Plants: A Study of Aquatic Angiosperms. University Press, Cambridge.

Arts, G.H.P., 2002. Deterioration of Atlantic soft water macrophyte communities by acidification, eutrophication and alkalinisation. Aquat. Bot. 73, 373–393. https://doi.org/10.1016/S0304-3770(02)00031-1.

Arts, G.H.P., Roelofs, J.G.M., Delyon, M.J.H., 1990. Differential tolerances among soft-water macrophyte species to acidification. Can. J. Bot.-Rev. Can. Bot 68, 2127–2134. https://doi.org/10.1139/b90-278.

Azzella, M.M., Bolpagni, R., Oggioni, A., 2014. A preliminary evaluation of lake morphometric traits influence on the maximum growing depth of macrophytes. J. Limnol. 73. https://doi.org/10.4081/jlimnol.2014.932.

Bagger, J., Madsen, T.V., 2004. Morphological acclimation of aquatic *Littorella uniflora* to sediment CO_2 concentration and wave exposure. Funct. Ecol. 18, 946–951. https://doi.org/10.1111/j.0269-8463.2004.00919.x.

Bain, J.T., Proctor, M.C.F., 1980. The requirement of aquatic bryophytes for free CO_2 as an inorganic carbon source, some experimental evidence. New Phytol. 86, 393–400. https://doi.org/10.1111/j.1469-8137.1980.tb01680.x.

Bakker, E.S., Wood, K.A., Pagès, J.F., Veen, G.F.(C.), Christianen, M.J.A., Santamaría, L., Nolet, B.A., Hilt, S., 2016. Herbivory on freshwater and marine macrophytes: a review and perspective. Aquat. Bot. 135, 18–36. https://doi.org/10.1016/j.aquabot.2016.04.008.

Barko, J.W., Adams, M.S., Clesceri, N.L., 1986. Environmental factors and their consideration in the management of submersed aquatic vegetation: a review. J. Aquat. Plant Manag. 24, 1–10.

Barko, J.W., Gunnison, D., Carpenter, S.R., 1991. Sediment interactions with submersed macrophyte growth and community dynamics. Aquat. Bot., Ecology of Submersed Aquatic Macrophytes 41, 41–65. https://doi.org/10.1016/0304-3770(91)90038-7.

Barko, J.W., Smart, R.M., 1980. Mobilization of sediment phosphorus by submersed freshwater macrophytes. Freshw. Biol. 10, 229–238. https://doi.org/10.1111/j.1365-2427.1980.tb01198.x.

Barrow, J.L., Beisner, B.E., Giles, R., Giani, A., Domaizon, I., Gregory-Eaves, I., 2019. Macrophytes moderate the taxonomic and functional composition of phytoplankton assemblages during a nutrient loading experiment. Freshw. Biol. 64, 1369–1381. https://doi.org/10.1111/fwb.13311.

Battarbee, R.W., 1990. The causes of lake acidification, with special reference to the role of acid deposition. Philos. Trans. R. Soc. B-Biol. Sci. 327, 339–347. https://doi.org/10.1098/rstb.1990.0071.

Bayley, S.E., Creed, I.F., Sass, G.Z., Wong, A.S., 2007. Frequent regime shifts in trophic states in shallow lakes on the Boreal Plain: alternative "unstable" states? Limnol. Oceanogr. 52, 2002–2012. https://doi.org/10.4319/lo.2007.52.5.2002.

Bergstrom, A.K., Jansson, M., 2006. Atmospheric nitrogen deposition has caused nitrogen enrichment and eutrophication of lakes in the Northern Hemisphere. Glob. Change Biol. 12, 635–643. https://doi.org/10.1111/j.1365-2486.2006.01129.x.

Bergstrom, I., Makela, S., Kankaala, P., Kortelainen, P., 2007. Methane efflux from littoral vegetation stands of southern boreal lakes: an upscaled regional estimate. Atmos. Environ. 41, 339–351. https://doi.org/10.1016/j.atmosenv.2006.08.014.

Binzer, T., Sand-Jensen, K., Middelboe, A.-L., 2006. Community photosynthesis of aquatic macrophytes. Limnol. Oceanogr. 51, 2722–2733.

Black, M.A., Maberly, S.C., Spence, D.H.N., 1981. Resistance to carbon dioxide fixation in four submerged freshwater macrophytes. New Phytol. 89, 557–568. https://doi.org/10.1111/j.1469-8137.1981.tb02335.x.

Bociag, K., Rekowska, E., Banaś, K., 2011. The disappearance of stonewort populations in *Lobelia* lakes of the Kashubian Lakeland (NW Poland). Oceanol. Hydrobiol. Stud. 40, 30–36. https://doi.org/10.2478/s13545-011-0014-7.

Bodkin, P.C., Spence, D.H.N., Weeks, D.C., 1980. Photoreversible control of heterophylly in *Hippuris vulgaris* L. New Phytol. 84, 533–542. https://doi.org/10.1111/j.1469-8137.1980.tb04560.x.

Bornette, G., Puijalon, S., 2011. Response of aquatic plants to abiotic factors: a review. Aquat. Sci. 73, 1–14. https://doi.org/10.1007/s00027-010-0162-7.

Bowes, G., 2011. Single-cell C_4 photosynthesis in aquatic plants. In: Raghavendra, A.S., Sage, R.F. (Eds.), C_4 Photosynthesis and Related CO_2 Concentrating Mechanisms, Advances in Photosynthesis and Respiration. Springer, pp. 63–80.

Bowes, G., Rao, S.K., Estavillo, G.M., Reiskind, J.B., 2002. C_4 mechanisms in aquatic angiosperms: comparisons with terrestrial C_4 systems. Funct. Plant Biol. 29, 379–392. https://doi.org/10.1071/pp01219.

Boyd, C.E., 1978. Chemical composition of wetland plants. In: Good, R.E., Whigham, D.F., Simpson, R.L. (Eds.), Freshwater Wetlands: Ecological Processes and Management Potential. Academic Press, New York, pp. 155–167.

Brammer, E.S., 1979. Exclusion of phytoplankton in the proximity of dominant water-soldier (*Stratiotes aloides*). Freshw. Biol. 9, 233–249. https://doi.org/10.1111/j.1365-2427.1979.tb01506.x.

Brandrud, T.E., 2002. Effects of liming on aquatic macrophytes, with emphasis on Scandinavia. Aquat. Bot. 73, 395–404. https://doi.org/10.1016/S0304-3770(02)00032-3.

Brönmark, C., Weisner, S.E.B., 1992. Indirect effects of fish community structure on submerged vegetation in shallow, eutrophic lakes: an alternative mechanism. In: Ilmavirta, V., Jones, R.I. (Eds.), The Dynamics and Use of Lacustrine Ecosystems. Developments in Hydrobiology. Springer Netherlands, Dordrecht, pp. 293–301. https://doi.org/10.1007/978-94-011-2745-5_30.

Brouwer, E., Bobbink, R., Roelofs, J.G.M., 2002. Restoration of aquatic macrophyte vegetation in acidified and eutrophied softwater lakes: an overview. Aquat. Bot., Acidification and restoration of soft water lakes and their vegetation 73, 405–431. https://doi.org/10.1016/S0304-3770(02)00033-5.

Burks, R.L., Jeppesen, E., Lodge, D.M., 2000. Macrophyte and fish chemicals suppress *Daphnia* growth and alter life-history traits. Oikos 88, 139–147. https://doi.org/10.1034/j.1600-0706.2000.880116.x.

Burks, R.L., Mulderij, G., Gross, E., Jones, I., Jacobsen, L., Jeppesen, E., Van Donk, E., 2006. Center Stage: the crucial role of macrophytes in regulating trophic interactions in shallow lake wetlands. In: Bobbink, R., Beltman, B., Verhoeven, J.T.A., Whigham, D.F. (Eds.), Wetlands: Functioning, Biodiversity Conservation, and Restoration, Ecological Studies. Springer Berlin Heidelberg, Berlin, Heidelberg, pp. 37–59. https://doi.org/10.1007/978-3-540-33189-6_3.

Caldwell, E.F., Duff, M.C., Ferguson, C.E., Coughlin, D.P., 2011. Plants as bio-monitors for Cs-137, Pu-238, Pu-239,240 and K-40 at the savannah river site. J. Environ. Monit. 13 (1410). https://doi.org/10.1039/c0em00610f.

Canfield Jr., D.E., Hoyer, M.V., 1988. Influence of nutrient enrichment and light availability on the abundance of aquatic macrophytes in Florida streams. Can. J. Fish. Aquat. Sci. 45, 1467–1472. https://doi.org/10.1139/f88-171.

Canfield, D.E.J., Langeland, K.A., Linda, S.B., Haller, W.T., 1985. Relations between water transparency and maximum depth of macrophyte colonization in lakes. J. Aquat. Plant Manage 25–28.

Cao, T., Xie, P., Ni, L., Wu, A., Zhang, M., Wu, S., Smolders, A.J.P., Cao, T., Xie, P., Ni, L., Wu, A., Zhang, M., Wu, S., Smolders, A.J.P., 2007. The role of NH+4 toxicity in the decline of the submersed macrophyte *Vallisneria natans* in lakes of the Yangtze River basin, China. Mar. Freshw. Res. 58, 581–587. https://doi.org/10.1071/MF06090.

Cao, T., Xie, P., Ni, L., Zhang, M., Xu, J., 2009. Carbon and nitrogen metabolism of an eutrophication tolerative macrophyte, *Potamogeton crispus*, under NH4+ stress and low light availability. Environ. Exp. Bot. 66, 74–78. https://doi.org/10.1016/j.envexpbot.2008.10.004.

Caraco, N., Cole, J., Findlay, S., Wigand, C., 2006. Vascular plants as engineers of oxygen in aquatic systems. Bioscience 56, 219. https://doi.org/10.1641/0006-3568(2006)056[0219:VPAEOO]2.0.CO;2.

Carbiener, R., Trémolières, M., Mercier, J.L., Ortscheit, A., 1990. Aquatic macrophyte communities as bioindicators of eutrophication in calcareous oligosaprobe stream waters (Upper Rhine plain, Alsace). Vegetatio 86, 71–88. https://doi.org/10.1007/BF00045135.

Carignan, R., 1982. An empirical model to estimate the relative importance of roots in phosphorus uptake by aquatic macrophytes. Can. J. Fish. Aquat. Sci. 39, 243–247. https://doi.org/10.1139/f82-034.

Carignan, R., Kalff, J., 1980. Phosphorus sources for aquatic weeds: water or sediments? Science 207, 987–989. https://doi.org/10.1126/science.207.4434.987.

Carpenter, S.R., 1980. Enrichment of Lake Wingra, Wisconsin, by submersed macrophyte decay. Ecology 61, 1145–1155. https://doi.org/10.2307/1936834.

Carpenter, S.R., Lodge, D.M., 1986. Effects of submersed macrophytes on ecosystem processes. Aquat. Bot. 26, 341–370. https://doi.org/10.1016/0304-3770(86)90031-8.

Carr, G.M., Bod, S.A.E., Duthie, H.C., Taylor, W.D., 2003. Macrophyte biomass and water quality in Ontario rivers. J. North Am. Benthol. Soc. 22, 182–193. https://doi.org/10.2307/1467991.

Carr, G.M., Chambers, P.A., 1998. Macrophyte growth and sediment phosphorus and nitrogen in a Canadian prairie river. Freshw. Biol. 39, 525–536. https://doi.org/10.1046/j.1365-2427.1998.00300.x.

Carvalho, P.N., Basto, M.C.P., Almeida, C.M.R., Brix, H., 2014. A review of plant-pharmaceutical interactions: from uptake and effects in crop plants to phytoremediation in constructed wetlands. Environ. Sci. Pollut. Res. 21, 11729–11763. https://doi.org/10.1007/s11356-014-2550-3.

Casati, P., Lara, M.V., Andreo, C.S., 2000. Induction of a C_4-like mechanism of CO_2 fixation in *Egeria densa*, a submersed aquatic species. Plant Physiol 123, 1611–1621. https://doi.org/10.1104/pp.123.4.1611.

Cedergreen, N., Madsen, T.V., 2004. Light regulation of root and leaf NO^{3-} uptake and reduction in the floating macrophyte *Lemna minor*. New Phytol. 161, 449–457. https://doi.org/10.1046/j.1469-8137.2003.00936.x.

Cedergreen, N., Streibig, J.C., 2005. The toxicity of herbicides to non-target aquatic plants and algae: assessment of predictive factors and hazard. Pest Manag. Sci. 61, 1152–1160. https://doi.org/10.1002/ps.1117.

Ceschin, S., Bellini, A., Scalici, M., 2021. Aquatic plants and ecotoxicological assessment in freshwater ecosystems: a review. Environ. Sci. Pollut. Res. 28, 4975–4988. https://doi.org/10.1007/s11356-020-11496-3.

Chambers, P.A., 1987a. Nearshore occurrence of submersed aquatic macrophytes in relation to wave action. Can. J. Fish. Aquat. Sci. 44, 1666–1669. https://doi.org/10.1139/f87-204.

Chambers, P.A., 1987b. Light and nutrients in the control of aquatic plant community structure. II. In situ observations. J. Ecol. 75, 621–628. https://doi.org/10.2307/2260194.

Chambers, P.A., Kaiff, J., 1985. Depth distribution and biomass of submersed aquatic macrophyte communities in relation to Secchi depth. Can. J. Fish. Aquat. Sci. 42 (4), 701–709. https://doi.org/10.1139/f85-090.

Chambers, P.A., Lacoul, P., Murphy, K.J., Thomaz, S.M., 2008. Global diversity of aquatic macrophytes in freshwater. In: Balian, E.V., Lévêque, C., Segers, H., Martens, K. (Eds.), Freshwater Animal Diversity Assessment, Developments in Hydrobiology. Springer Netherlands, Dordrecht, pp. 9–26. https://doi.org/10.1007/978-1-4020-8259-7_2.

Chambers, P.A., Lacoul, P., Murphy, K.J., Thomaz, S.M., 2008. Global diversity of aquatic macrophytes in freshwater. Hydrobiologia 595, 9–26. https://doi.org/10.1007/s10750-007-9154-6.

Chambers, P.A., Prepas, E.E., 1994. Nutrient dynamics in riverbeds: the impact of sewage effluent and aquatic macrophytes. Water Res. 28, 453–464. https://doi.org/10.1016/0043-1354(94)90283-6.

Chambers, P.A., Prepas, E.E., Bothwell, M.L., Hamilton, H.R., 1989. Roots versus shoots in nutrient uptake by aquatic macrophytes in flowing waters. Can. J. Fish. Aquat. Sci. 46, 435–439. https://doi.org/10.1139/f89-058.

Chimney, M.J., Pietro, K.C., 2006. Decomposition of macrophyte litter in a subtropical constructed wetland in south Florida (USA). Ecol. Eng. 27, 301–321. https://doi.org/10.1016/j.ecoleng.2006.05.016.

Chmara, R., Szmeja, J., Robionek, A., 2019. Leaf traits of macrophytes in lakes: interspecific, plant group and community patterns. Limnologica 77, 125691. https://doi.org/10.1016/j.limno.2019.125691.

Christenhusz, M.J.M., Byng, J.W., 2016. The number of known plants species in the world and its annual increase. Phytotaxa 261, 201. https://doi.org/10.11646/phytotaxa.261.3.1.

Christensen, M.R., Graham, M.D., Vinebrooke, R.D., Findlay, D.L., Paterson, M.J., Turner, M.A., 2006. Multiple anthropogenic stressors cause ecological surprises in boreal lakes. Glob. Change Biol. 12, 2316–2322. https://doi.org/10.1111/j.1365-2486.2006.01257.x.

Clarke, E., Baldwin, A.H., 2002. Responses of wetland plants to ammonia and water level. Ecol. Eng. 18, 257–264. https://doi.org/10.1016/S0925-8574(01)00080-5.

Clarke, S.J., Wharton, G., 2001. Using Macrophytes for the Environmental Assessment of Rivers: The Role of Sediment Nutrients (R&D Technical Report E1-S01/TR). Environment Agency.

Cleuvers, M., 2003. Aquatic ecotoxicity of pharmaceuticals including the assessment of combination effects. Toxicol. Lett. 142, 185–194. https://doi.org/10.1016/S0378-4274(03)00068-7.

Coffey, B.T., Clayton, J.S., 1988. New Zealand Waterplants: A Guide to Plants Found in New Zealand Freshwaters. N. Z. Waterplants Guide Plants Found N. Z. Freshw.

Colmer, T.D., Pedersen, O., 2008. Underwater photosynthesis and respiration in leaves of submerged wetland plants: gas films improve CO_2 and O_2 exchange. New Phytol. 177, 918–926. https://doi.org/10.1111/j.1469-8137.2007.02318.x.

Cook, C.D.K., 2004. Aquatic and Wetland Plants of Southern Africa. Backhuys Publishers, Leiden, The Netherlands.

Cook, C.D.K., 1983. Aquatic plants endemic to Europe and the Mediterranean. Bot. Jahrbucher Syst. Pflanzengesch. Pflanzengeogr 103, 539–582.

Cook, C.D.K., 1985. Range extensions of aquatic vascular plant species. J. Aquat. Plant Manag. 23, 1–6.

Cook, C.D.K., 1999. The number and kinds of embryo-bearing plants which have become aquatic: a survey. Perspect. Plant Ecol. Evol. Syst 2, 79–102. https://doi.org/10.1078/1433-8319-00066.

Costa, M.B., Tavares, F.V., Martinez, C.B., Colares, I.G., Martins, C., de, M.G., 2018. Accumulation and effects of copper on aquatic macrophytes *Potamogeton pectinatus* L.: potential application to environmental monitoring and phytoremediation. Ecotoxicol. Environ. Saf. 155, 117–124. https://doi.org/10.1016/j.ecoenv.2018.01.062.

Cox, C., Goffinet, B., Wickett, N., Boles, S., Shaw, A., 2010. Moss diversity: a molecular phylogenetic analysis of genera. Phytotaxa 9, 175–195.

Crandall-Stotler, B., Stotler, R.E., Long, D.G., 2009. Phylogeny and classification of the marchantiophyta. Edinb. J. Bot. 66, 155–198. https://doi.org/10.1017/S0960428609005393.

Crawford, R.M.M., 1992. Oxygen availability as an ecological limit to plant distribution. Adv. Ecol. Res. 23, 93–185. https://doi.org/10.1016/s0065-2504(08)60147-6.

Crawford, R.M.M., Braendle, R., 1996. Oxygen deprivation stress in a changing environment. J. Exp. Bot. 47, 145–159. https://doi.org/10.1093/jxb/47.2.145.

Creed, R.P., 1998. A biogeographic perspective on Eurasian watermilfoil declines: additional evidence for the role of herbivorous weevils in promoting declines? J. Aquat. Plant Manag. 36, 16–22.

Crow, G.E., 1993. Species diversity in aquatic angiosperms: latitudinal patterns. Aquat. Bot. 44, 229–258. https://doi.org/10.1016/0304-3770(93)90072-5.

Cui, X.H., Zhong, Y., Li, W., Chen, J.K., 2000. The effect of catastrophic flood on biomass and density of three dominant aquatic plant species in the Poyang Lake. Acta Hydrobiol. Sin 24, 322–325.

Dacey, J.W.H., 1980. Internal winds in water lilies: an adaptation for life in anaerobic sediments. Science 210, 1017–1019. https://doi.org/10.1126/science.210.4473.1017.

Dacey, J.W.H., 1981. Pressurized ventilation in the yellow waterlily. Ecology 62, 1137–1147. https://doi.org/10.2307/1937277.

D'Arcy, B.J., May, L., Long, J., Fozzard, I.R., Greig, S., Brachet, A., 2006. The restoration of Loch Leven, Scotland, UK. Water Sci. Technol. 53, 183–191. https://doi.org/10.2166/wst.2006.311.

De Bakker, N.V.J., Van Bodegom, P.M., Van De Poll, W.H., Boelen, P., Nat, E., Rozema, J., Aerts, R., 2005. Is UV-B radiation affecting charophycean algae in shallow freshwater systems? New Phytol. 166, 957–966. https://doi.org/10.1111/j.1469-8137.2005.01377.x.

Declerck, S., Vandekerkhove, J., Johansson, L., Muylaert, K., Conde-Porcuna, J.M., Van der Gucht, K., Pérez-Martínez, C., Lauridsen, T., Schwenk, K., Zwart, G., Rommens, W., López-Ramos, J., Jeppesen, E., Vyverman, W., Brendonck, L., De Meester, L., 2005. Multi-group biodiversity in shallow lakes along gradients of phosphorus and water plant cover. Ecology 86, 1905–1915. https://doi.org/10.1890/04-0373.

Demars, B.O.L., Edwards, A.C., 2007. Tissue nutrient concentrations in freshwater aquatic macrophytes: high inter-taxon differences and low phenotypic response to nutrient supply. Freshw. Biol. 52, 2073–2086. https://doi.org/10.1111/j.1365-2427.2007.01817.x.

Demars, B.O.L., Edwards, A.C., 2008. Tissue nutrient concentrations in aquatic macrophytes: comparison across biophysical zones, surface water habitats and plant life forms. Chem. Ecol. 24, 413–422. https://doi.org/10.1080/02757540802534533.

Diaz, S., Kattge, J., Cornelissen, J.H.C., Wright, I.J., Lavorel, S., Dray, S., Reu, B., Kleyer, M., Wirth, C., Prentice, I.C., Garnier, E., Boenisch, G., Westoby, M., Poorter, H., Reich, P.B., Moles, A.T., Dickie, J., Gillison, A.N., Zanne, A.E., Chave, J., Wright, S.J., Sheremet'ev, S.N., Jactel, H., Baraloto, C., Cerabolini, B., Pierce, S., Shipley, B., Kirkup, D., Casanoves, F., Joswig, J.S., Guenther, A., Falczuk, V., Rueger, N., Mahecha, M.D., Gorne, L.D., 2016. The global spectrum of plant form and function. Nature 529, 167–171. https://doi.org/10.1038/nature16489.

Dodkins, I., Rippey, B., Hale, P., 2005. An application of canonical correspondence analysis for developing ecological quality assessment metrics for river macrophytes. Freshw. Biol. 50, 891–904. https://doi.org/10.1111/j.1365-2427.2005.01360.x.

Dole-Olivier, M.-J., Galassi, D.M.P., Marmonier, P., Creuzé Des Châtelliers, M., 2000. The biology and ecology of lotic microcrustaceans. Freshw. Biol. 44, 63–91. https://doi.org/10.1046/j.1365-2427.2000.00590.x.

Domozych, D.S., Popper, Z.A., Sorensen, I., 2016. Charophytes: evolutionary giants and emerging model organisms. Front. Plant Sci. 7, 1470. https://doi.org/10.3389/fpls.2016.01470.

Du, Z.-Y., Wang, Q.-F., 2014. Correlations of life form, pollination mode and sexual system in aquatic angiosperms. PLoS One 9, e115653. https://doi.org/10.1371/journal.pone.0115653.

Duarte, C.M., Planas, D., Penuelas, J., 1994. Macrophytes, taking control of an ancestral home. In: Margalef, R. (Ed.), Limnology Now: A Paradigm of Planetary Problems. Elsevier Sci., New York, pp. 59–79.

Dudley, B., Gunn, I.D.M., Carvalho, L., Proctor, I., O'Hare, M.T., Murphy, K.J., Milligan, A., 2012. Changes in aquatic macrophyte communities in Loch Leven: evidence of recovery from eutrophication? Hydrobiologia 681, 49–57. https://doi.org/10.1007/s10750-011-0924-9.

Eckert, C.G., Dorken, M.E., Barrett, S.C.H., 2016. Ecological and evolutionary consequences of sexual and clonal reproduction in aquatic plants. Aquat. Bot. 135, 46–61. https://doi.org/10.1016/j.aquabot.2016.03.006.

Edwards, G.E., Franceschi, V.R., Voznesenskaya, E.V., 2004. Single-cell C_4 photosynthesis versus the dual-cell (Kranz) paradigm. Annu. Rev. Plant Biol. 55, 173–196. https://doi.org/10.1146/annurev.arplant.55.031903.141725.

Edwards, D., Kerp, H., Hass, H., 1998. Stomata in early land plants: an anatomical and ecophysiological approach. J. Exp. Bot. 49, 255–278. https://doi.org/10.1093/jexbot/49.suppl_1.255.

Egertson, C.J., Kopaska, J.A., Downing, J.A., 2004. A century of change in macrophyte abundance and composition in response to agricultural eutrophication. Hydrobiologia 524, 145–156. https://doi.org/10.1023/B: HYDR.0000036129.40386.ce.

Enriquez, S., 2005. Light absorption efficiency and the package effect in the leaves of the seagrass *Thalassia testudinum*. Mar. Ecol. Prog. Ser. 289, 141–150. https://doi.org/10.3354/meps289141.

Enriquez, S., Duarte, C.M., Sand-Jensen, K., Nielsen, S.L., 1996. Broad-scale comparison of photosynthetic rates across phototrophic organisms. Oecologia 108, 197–206.

Falk, H., Sitte, P., 1963. Zellfeinbau bei Plasmolyse: I. Der Feinbau der Elodea-Blattzellen. Protoplasma 57, 290–303. https://doi.org/10.1007/BF01252061.

Falkowski, P.G., Raven, J.A., 2007. Aquatic Photosynthesis, second ed. Princeton University Press, Princeton.

Farmer, A., 1990. The effects of lake acidification on aquatic macrophytes—a review. Environ. Pollut 65, 219–240. https://doi.org/10.1016/0269-7491(90)90085-Q.

Figuerola, J., Green, A.J., 2002. Dispersal of aquatic organisms by waterbirds: a review of past research and priorities for future studies. Freshw. Biol. 47, 483–494. https://doi.org/10.1046/j.1365-2427.2002.00829.x.

Frantz, T.C., Cordone, A.J., 1967. Observations on deepwater plants in Lake Tahoe, California and Nevada. Ecology 48, 709–714. https://doi.org/10.2307/1933727.

Frost-Christensen, H., Floto, F., 2007. Resistance to CO_2 diffusion in cuticular membranes of amphibious plants and the implication for CO_2 acquisition. Plant Cell Environ. 30, 12–18. https://doi.org/10.1111/j.1365-3040.2006.01599.x.

Frost-Christensen, H., Jogensen, L.B., Floto, F., 2003. Species specificity of resistance to oxygen diffusion in thin cuticular membranes from amphibious plants. Plant Cell Environ. 26, 561–569. https://doi.org/10.1046/j.1365-3040.2003.00986.x.

Gao, J., Ren, P., Zhou, Q., Zhang, J., 2019. Comparative studies of the response of sensitive and tolerant submerged macrophytes to high ammonium concentration stress. Aquat. Toxicol 211, 57–65. https://doi.org/10.1016/j.aquatox.2019.03.020.

Garver, E.G., Dubbe, D.R., Pratt, D.C., 1988. Seasonal patterns in accumulation and partitioning of biomass and macronutrients in *Typha* spp. Aquat. Bot. 32, 115–127. https://doi.org/10.1016/0304-3770(88)90092-7.

Gerloff, G.C., 1975. Nutritional Ecology of Nuisance Aquatic Plants (No. EPA 660-3-75-027). University of Wisconsin, Madison.

Gerloff, G.C., Krombholz, P.H., 1966. Tissue analysis as a measure of nutrient availability for the growth of angiosperm aquatic plants. Limnol. Oceanogr. 11, 529–537. https://doi.org/10.4319/lo.1966.11.4.0529.

Gessner, M.O., 2000. Breakdown and nutrient dynamics of submerged *Phragmites* shoots in the littoral zone of a temperate hardwater lake. Aquat. Bot. 66, 9–20. https://doi.org/10.1016/S0304-3770(99)00022-4.

Gomez, B., Daviero-Gomez, V., Coiffard, C., Martin-Closas, C., Dilcher, D.L., 2015. *Montsechia*, an ancient aquatic angiosperm. Proc Natl Acad Sci U A 112, 10985–10988. https://doi.org/10.1073/pnas.1509241112.

Gopal, B., Goel, U., 1993. Competition and allelopathy in aquatic plant communities. Bot. Rev. 59, 155–210. https://doi.org/10.1007/BF02856599.

Gossett, D.R., Norris, W.E., 1971. Relationship between nutrient availability and content of nitrogen and phosphorus in tissues of the aquatic macrophyte, *Eichornia crassipes* (Mart.) Solms. Hydrobiologia 38, 15–28. https://doi.org/10.1007/BF00036789.

Government of Canada, P.S. and P.C, 2002. Canada–United States Air Quality Agreement : Progress Report. : En85–1e-PDF—Government of Canada Publications—Canada.Ca [WWW Document] (accessed 5.19.21). http://publications.gc.ca/site/eng/9.506241/publication.html.

Grace, J.B., 1993. The adaptive significance of clonal reproduction in angiosperms: an aquatic perspective. Aquat. Bot. 44, 159–180. https://doi.org/10.1016/0304-3770(93)90070-D.

Graham, L., Lewis, L.A., Taylor, W., Wellman, C., Cook, M., 2014. Early terrestrialization: transition from algal to bryophyte grade. In: Hanson, D.T., Rice, S.K. (Eds.), Photosynthesis in Bryophytes and Early Land Plants, Advances in Photosynthesis and Respiration. Springer Science +Business Media. Dordercht, The Netherlands, pp. 9–28. https://doi.org/10.1007/978-94-007-6988-5_2.

Grainger, J., 1947. Nutrition and flowering of water plants. J. Ecol. 35, 49–64.

Graneli, W., 2012. Brownification of lakes. In: Bengtsson, L., Herschy, R.W., Fairbridge, R.W. (Eds.), Encyclopedia of Lakes and Reservoirs. Springer Netherlands, Dordrecht, pp. 117–119. https://doi.org/10.1007/978-1-4020-4410-6_256.

Granéli, W., Solander, D., 1988. Influence of aquatic macrophytes on phosphorus cycling in lakes. Hydrobiologia 170, 245–266. https://doi.org/10.1007/BF00024908.

Grime, J.P., 1979. Plant Strategies and Vegetation Processes. Wiley, Chichester, New York.

Gross, E.M., 2003. Allelopathy of aquatic autotrophs. Crit. Rev. Plant Sci. 22, 313–339. https://doi.org/10.1080/713610859.

Grosse, W., Buchel, H.B., Tiebel, H., 1991. Pressurized ventilation in wetland plants. Aquat. Bot. 39, 89–98. https://doi.org/10.1016/0304-3770(91)90024-y.

Guilizzoni, P., 1991. The role of heavy metals and toxic amterials in the physiological ecology of submersed macrophytes. Aquat. Bot 41, 87–109. https://doi.org/10.1016/0304-3770(91)90040-C.

Gurnell, A., 2014. Plants as river system engineers. Earth Surf. Process. Landf. 39, 4–25. https://doi.org/10.1002/esp.3397.

Hamilton, S.K., Bruesewitz, D.A., Horst, G.P., Weed, D.B., Sarnelle, O., 2009. Biogenic calcite–phosphorus precipitation as a negative feedback to lake eutrophication. Can. J. Fish. Aquat. Sci. 66, 343–350. https://doi.org/10.1139/F09-003.

Hammer, U.T., Heseltine, J.M., 1988. Aquatic macrophytes in saline lakes of the Canadian prairies. In: Melack, J.M. (Ed.), Saline Lakes, Developments in Hydrobiology. Springer Netherlands, Dordrecht, pp. 101–116. https://doi.org/10.1007/978-94-009-3095-7_6.

Han, S., Maberly, S.C., Gontero, B., Xing, Z., Li, W., Jiang, H., Huang, W., 2020. Structural basis for C_4 photosynthesis without Kranz anatomy in leaves of the submerged freshwater plant *Ottelia alismoides*. Ann. Bot 125, 869–879. https://doi.org/10.1093/aob/mcaa005.

Hanelt, D., Hawes, I., Rae, R., 2006. Reduction of UV-B radiation causes an enhancement of photoinhibition in high light stressed aquatic plants from New Zealand lakes. J. Photochem. Photobiol. B Biol. 84, 89–102. https://doi.org/10.1016/j.jphotobiol.2006.01.013.

Hartman, R.T., Brown, D.L., 1967. Changes in internal atmosphere of submersed vascular hydrophytes in relation to photosynthesis. Ecology 48, 252–258. https://doi.org/10.2307/1933107.

Haury, J., Peltre, M.-C., Trémolières, M., Barbe, J., Thiébaut, G., Bernez, I., Daniel, H., Chatenet, P., Haan-Archipof, G., Muller, S., Dutartre, A., Laplace-Treyture, C., Cazaubon, A., Lambert-Servien, E., 2006. A new method to assess water trophy and organic pollution — the Macrophyte Biological Index for Rivers (IBMR): its application to different types of river and pollution. Macrophytes Aquat. Ecosyst. Biol. Manag. Proc. 11th Int. Symp. Aquat. Weeds Eur. Weed Res. Soc. Hydrobiologia 153–158. https://doi.org/10.1007/978-1-4020-5390-0_22.

Hayati, A.A., Proctor, M.C.F., 1990. Plant distribution in relation to mineral nutrient availability and uptake on a wet-heath site in south-west England. J. Ecol. 78, 134–151.

Heitto, L., 1990. A Macrophyte Survey in Finnish Forest Lakes Sensitive to Acidification, International Association of Theoretical and Applied Limnology—Proceedings, vol. 24. E Schweizerbart'sche Verlagsbuchhandlung, Stuttgart. Pt 1.

Hempel, M., Grossart, H.P., Gross, E.M., 2009. Community composition of bacterial biofilms on two submerged macrophytes and an artificial substrate in a pre-alpine lake. Aquat. Microb. Ecol. 58, 79–94. https://doi.org/10.3354/ame01353.

Henrikson, L., Brodin, Y.-W., 1995. Liming of Surface Waters in Sweden — a Synthesis. In: Henrikson, L., Brodin, Y.W. (Eds.), Liming of Acidified Surface Waters. Springer Berlin Heidelberg, Berlin, Heidelberg, pp. 1–44. https://doi.org/10.1007/978-3-642-79309-7_1.

Hilt, S., Alirangues Nuñez, M.M., Bakker, E.S., Blindow, I., Davidson, T.A., Gillefalk, M., Hansson, L.-A., Jans, J.H., Janssen, A.B.G., Jeppesen, E., Kabus, T., Kelly, A., Köhler, J., Lauridsen, T.L., Mooij, W.M., Noordhuis, R., Phillips, G., Rücker, J., Schuster, H.-H., Søndergaard, M., Teurlincx, S., van de Weyer, K., van Donk, E., Waterstraat, A., Willby, N., Sayer, C.D., 2018. Response of submerged macrophyte communities to external and internal restoration measures in North Temperate Shallow Lakes. Front. Plant Sci. 9. https://doi.org/10.3389/fpls.2018.00194.

Hilt, S., Gross, E.M., 2008. Can allelopathically active submerged macrophytes stabilise clear-water states in shallow lakes? Basic Appl. Ecol. 9, 422–432. https://doi.org/10.1016/j.baae.2007.04.003.

Holmes, N.T.H., Newman, J.R., Chadd, S., Rouen, K.J., Saint, L., Dawson, F.H., 1999. Mean Trophic Rank: A User's Manual. Environment Agency, UK.

Hostrup, O., Wiegleb, G., 1991. Anatomy of leaves of submerged and emergent forms of *Littorella uniflora* (L.) Ascherson. Aquat. Bot., Physiological Ecology of Aquatic Macrophytes 39, 195–209. https://doi.org/10.1016/0304-3770(91)90032-Z.

Hsu, J.S., Powell, J., Adler, P.B., 2012. Sensitivity of mean annual primary production to precipitation. Glob. Change Biol. 18, 2246–2255. https://doi.org/10.1111/j.1365-2486.2012.02687.x.

Huang, W., Han, S., Jiang, H., Gu, S., Li, W., Gontero, B., Maberly, S.C., 2020. External α-carbonic anhydrase and solute carrier 4 are required for bicarbonate uptake in a freshwater angiosperm. J. Exp. Bot. 71, 6004–6014. https://doi.org/10.1093/jxb/eraa351.

Huang, W.M., Shao, H., Zhou, S.N., Zhou, Q., Fu, W.L., Zhang, T., Jiang, H.S., Li, W., Gontero, B., Maberly, S.C., 2018. Different CO_2 acclimation strategies in juvenile and mature leaves of *Ottelia alismoides*. Photosynth. Res. 138, 219–232. https://doi.org/10.1007/s11120-018-0568-y.

Hussner, A., 2012. Alien aquatic plant species in European countries. Weed Res. 52, 297–306. https://doi.org/10.1111/j.1365-3180.2012.00926.x.

Hussner, A., Heidbüchel, P., Coetzee, J., Gross, E.M., 2021. From introduction to nuisance growth: a review of traits of alien aquatic plants which contribute to their invasiveness. Hydrobiologia 848, 2119–2151. https://doi.org/10.1007/s10750-020-04463-z.

Hussner, A., Hoelken, H.P., Jahns, P., 2010. Low light acclimated submerged freshwater plants show a pronounced sensitivity to increasing irradiances. Aquat. Bot. 93, 17–24. https://doi.org/10.1016/j.aquabot.2010.02.003.

Hussner, A., Stiers, I., Verhofstad, M.J.J.M., Bakker, E.S., Grutters, B.M.C., Haury, J., van Valkenburg, J.L.C.H., Brundu, G., Newman, J., Clayton, J.S., Anderson, L.W.J., Hofstra, D., 2017. Management and control methods of invasive alien freshwater aquatic plants: a review. Aquat. Bot. 136, 112–137. https://doi.org/10.1016/j.aquabot.2016.08.002.

Hutchinson, G.E., 1975. A treatise on limnology. In: Limnological Botany, vol. 3. Wiley, New York.

IPCC, 2021. Climate Change 2021: The Physical Science Basis. Contribution of Working Group I to the Sixth Assessment Report of the Intergovernmental Panel on Climate Change. Cambridge University Press.

Irgang, B.E., Gastal Júnior, C.V.S., 2003. Problemas taxonômicos e distribuição geográfica de macrófitas aquáticas do sul do Brasil. In: Thomaz, S.M., Bini, L. (Eds.), Ecologia e Manejo de Macrófitas. Eduem, Maringá, pp. 163–169.

Ito, Y., Tanaka, N., Barfod, A.S., Kaul, R.B., Muasya, A.M., Garcia-Murillo, P., De Vere, N., Duyfjes, B.E.E., Albach, D.C., 2017. From terrestrial to aquatic habitats and back again: molecular insights into the evolution and phylogeny of *Callitriche* (Plantaginaceae). Bot. J. Linn. Soc. 184, 46–58. https://doi.org/10.1093/botlinnean/box012.

IUCN, 2020. The IUCN Red List of Threatened Species. Version 2020-3.

Iversen, J., 1929. Studien uber die pH-Verhaltnisse danischer Gewasser und ihren Einfluss auf die Hydrophyten-Vegetation. Bot. Tidskr 40, 277–326.

Iversen, L.L., Winkel, A., Baastrup-Spohr, L., Hinke, A.B., Alahuhta, J., Baattrup-Pedersen, A., Birk, S., Brodersen, P., Chambers, P.A., Ecke, F., Feldmann, T., Gebler, D., Heino, J., Jespersen, T.S., Moe, S.J., Riis, T., Sass, L., Vestergaard, O., Maberly, S.C., Sand-Jensen, K., Pedersen, O., 2019. Catchment properties and the photosynthetic trait composition of freshwater plant communities. Science 366, 878–881. https://doi.org/10.1126/science.aay5945.

Jackson, M., 1985. Ethylene and responses of plants to soil waterlogging and submergence. Annu. Rev. Plant Physiol. Plant Mol. Biol. 36, 145–174. https://doi.org/10.1146/annurev.pp.36.060185.001045.

Jacobsen, D., Sand-Jensen, K., 1992. Herbivory of invertebrates on submerged macrophytes from Danish fresh-waters. Freshw. Biol. 28, 301–308. https://doi.org/10.1111/j.1365-2427.1992.tb00588.x.

Janauer, G.A., Englmaier, P., 1986. The effects of emersion on soluble carbohydrate accumulations in *Hippuris vulgaris* L. Aquat. Bot. 24, 241–248. https://doi.org/10.1016/0304-3770(86)90060-4.

Jiang, H.S., Yin, L.Y., Ren, N.N., Zhao, S.T., Li, Z., Zhi, Y., Shao, H., Li, W., Gontero, B., 2017. Silver nanoparticles induced reactive oxygen species via photosynthetic energy transport imbalance in an aquatic plant. Nanotoxicology 11, 157–167. https://doi.org/10.1080/17435390.2017.1278802.

Jiang, H.S., Zhang, Y., Yin, L., Li, W., Jin, Q., Fu, W., Zhang, T., Huang, W., 2018. Diurnal changes in photosynthesis by six submerged macrophytes measured using fluorescence. Aquat. Bot. 149, 33–39. https://doi.org/10.1016/j.aquabot.2018.05.003.

Jones, J.I., Eaton, J.W., Hardwick, K., 2000. The influence of periphyton on boundary layer conditions: a pH microelectrode investigation. Aquat. Bot. 67, 191–206.

Jupp, B.P., Spence, D.H.N., 1977. Limitations on macrophytes in a eutrophic lake, Loch Leven. 1. Effects of phytoplankton. J. Ecol. 65, 175–186. https://doi.org/10.2307/2259072.

Kail, J., Brabec, K., Poppe, M., Januschke, K., 2015. The effect of river restoration on fish, macroinvertebrates and aquatic macrophytes: a meta-analysis. Ecol. Indic 58, 311–321. https://doi.org/10.1016/j.ecolind.2015.06.011.

Kato, Y., Nishihiro, J., Yoshida, T., 2016. Floating-leaved macrophyte (*Trapa japonica*) drastically changes seasonal dynamics of a temperate lake ecosystem. Ecol. Res. 31, 695–707. https://doi.org/10.1007/s11284-016-1378-3.

Kattge, J., Boenisch, G., Diaz, S., Lavorel, S., Prentice, I.C., Leadley, P., Tautenhahn, S., Werner, G.D.A., Aakala, T., Abedi, M., Acosta, A.T.R., Adamidis, G.C., Adamson, K., Aiba, M., Albert, C.H., Alcantara, J.M., Alcazar, C.C., Aleixo, I., Ali, H., Amiaud, B., Ammer, C., Amoroso, M.M., Anand, M., Anderson, C., Anten, N., Antos, J., Apgaua, D.M.G., Ashman, T.-L., Asmara, D.H., Asner, G.P., Aspinwall, M., Atkin, O., Aubin, I., Baastrup-Spohr, L., Bahalkeh, K., Bahn, M., Baker, T., Baker, W.J., Bakker, J.P., Baldocchi, D., Baltzer, J., Banerjee, A., Baranger, A., Barlow, J., Barneche, D.R., Baruch, Z., Bastianelli, D., Battles, J., Bauerle, W., Bauters, M., Bazzato, E., Beckmann, M., Beeckman, H., Beierkuhnlein, C., Bekker, R., Belfry, G., Belluau, M., Beloiu, M., Benavides, R., Benomar, L., Berdugo-Lattke, M.L., Berenguer, E., Bergamin, R., Bergmann, J., Carlucci, M.B., Berner, L., Bernhardt-Roemermann, M., Bigler, C., Bjorkman, A.D., Blackman, C., Blanco, C., Blonder, B., Blumenthal, D., Bocanegra-Gonzalez, K.T., Boeckx, P., Bohlman, S., Boehning-Gaese, K., Boisvert-Marsh, L., Bond, W., Bond-Lamberty, B., Boom, A., Boonman, C.C.F., Bordin, K., Boughton, E.H., Boukili, V., Bowman, D.M.J.S., Bravo, S., Brendel, M.R., Broadley, M.R.,

Brown, K.A., Bruelheide, H., Brumnich, F., Bruun, H.H., Bruy, D., Buchanan, S.W., Bucher, S.F., Buchmann, N., Buitenwerf, R., Bunker, D.E., Buerger, J., Burrascano, S., Burslem, D.F.R.P., Butterfield, B.J., Byun, C., Marques, M., Scalon, M.C., Caccianiga, M., Cadotte, M., Cailleret, M., Camac, J., Julio Camarero, J., Campany, C., Campetella, G., Campos, J.A., Cano-Arboleda, L., Canullo, R., Carbognani, M., Carvalho, F., Casanoves, F., Castagneyrol, B., Catford, J.A., Cavender-Bares, J., Cerabolini, B.E.L., Cervellini, M., Chacon-Madrigal, E., Chapin, K., Chapin, F.S., Chelli, S., Chen, S.-C., Chen, A., Cherubini, P., Chianucci, F., Choat, B., Chung, K.-S., Chytry, M., Ciccarelli, D., Coll, L., Collins, C.G., Conti, L., Coomes, D., Cornelissen, J.H.C., Cornwell, W.K., Corona, P., Coyea, M., Craine, J., Craven, D., Cromsigt, J.P.G.M., Csecserits, A., Cufar, K., Cuntz, M., da Silva, A.C., Dahlin, K.M., Dainese, M., Dalke, I., Dalle Fratte, M., Dang-Le, A.T., Danihelka, J., Dannoura, M., Dawson, S., de Beer, A.J., De Frutos, A., De Long, J.R., Dechant, B., Delagrange, S., Delpierre, N., Derroire, G., Dias, A.S., Diaz-Toribio, M.H., Dimitrakopoulos, P.G., Dobrowolski, M., Doktor, D., Drevojan, P., Dong, N., Dransfield, J., Dressler, S., Duarte, L., Ducouret, E., Dullinger, S., Durka, W., Duursma, R., Dymova, O., E-Vojtko, A., Eckstein, R.L., Ejtehadi, H., Elser, J., Emilio, T., Engemann, K., Erfanian, M.B., Erfmeier, A., Esquivel-Muelbert, A., Esser, G., Estiarte, M., Domingues, T.F., Fagan, W.F., Fagundez, J., Falster, D.S., Fan, Y., Fang, J., Farris, E., Fazlioglu, F., Feng, Y., Fernandez-Mendez, F., Ferrara, C., Ferreira, J., Fidelis, A., Finegan, B., Firn, J., Flowers, T.J., Flynn, D.F.B., Fontana, V., Forey, E., Forgiarini, C., Francois, L., Frangipani, M., Frank, D., Frenette-Dussault, C., Freschet, G.T., Fry, E.L., Fyllas, N.M., Mazzochini, G.G., Gachet, S., Gallagher, R., Ganade, G., Ganga, F., Garcia-Palacios, P., Gargaglione, V., Garnier, E., Luis Garrido, J., Luis de Gasper, A., Gea-Izquierdo, G., Gibson, D., Gillison, A.N., Giroldo, A., Glasenhardt, M.-C., Gleason, S., Gliesch, M., Goldberg, E., Goeldel, B., Gonzalez-Akre, E., Gonzalez-Andujar, J.L., Gonzalez-Melo, A., Gonzalez-Robles, A., Graae, B.J., Granda, E., Graves, S., Green, W.A., Gregor, T., Gross, N., Guerin, G.R., Guenther, A., Gutierrez, A.G., Haddock, L., Haines, A., Hall, J., Hambuckers, A., Han, W., Harrison, S.P., Hattingh, W., Hawes, J.E., He, T., He, P., Heberling, J.M., Helm, A., Hempel, S., Hentschel, J., Herault, B., Heres, A.-M., Herz, K., Heuertz, M., Hickler, T., Hietz, P., Higuchi, P., Hipp, A.L., Hirons, A., Hock, M., Hogan, J.A., Holl, K., Honnay, O., Hornstein, D., Hou, E., Hough-Snee, N., Hovstad, K.A., Ichie, T., Igic, B., Illa, E., Isaac, M., Ishihara, M., Ivanov, L., Ivanova, L., Iversen, C.M., Izquierdo, J., Jackson, R.B., Jackson, B., Jactel, H., Jagodzinski, A.M., Jandt, U., Jansen, S., Jenkins, T., Jentsch, A., Jespersen, J.R.P., Jiang, G.-F., Johansen, J.L., Johnson, D., Jokela, E.J., Joly, C.A., Jordan, G.J., Joseph, G.S., Junaedi, D., Junker, R.R., Justes, E., Kabzems, R., Kane, J., Kaplan, Z., Kattenborn, T., Kavelenova, L., Kearsley, E., Kempel, A., Kenzo, T., Kerkhoff, A., Khalil, M.I., Kinlock, N.L., Kissling, W.D., Kitajima, K., Kitzberger, T., Kjoller, R., Klein, T., Kleyer, M., Klimesova, J., Klipel, J., Kloeppel, B., Klotz, S., Knops, J.M.H., Kohyama, T., Koike, F., Kollmann, J., Komac, B., Komatsu, K., Koenig, C., Kraft, N.J.B., Kramer, K., Kreft, H., Kuehn, I., Kumarathunge, D., Kuppler, J., Kurokawa, H., Kurosawa, Y., Kuyah, S., Laclau, J.-P., Lafleur, B., Lallai, E., Lamb, E., Lamprecht, A., Larkin, D.J., Laughlin, D., Le Bagousse-Pinguet, Y., le Maire, G., le Roux, P.C., le Roux, E., Lee, T., Lens, F., Lewis, S.L., Lhotsky, B., Li, Y., Li, X., Lichstein, J.W., Liebergesell, M., Lim, J.Y., Lin, Y.-S., Linares, J.C., Liu, C., Liu, D., Liu, U., Livingstone, S., Llusia, J., Lohbeck, M., Lopez-Garcia, A., Lopez-Gonzalez, G., Lososova, Z., Louault, F., Lukacs, B.A., Lukes, P., Luo, Y., Lussu, M., Ma, S., Pereira, C.M.R., Mack, M., Maire, V., Makela, A., Makinen, H., Mendes Malhado, A.C., Mallik, A., Manning, P., Manzoni, S., Marchetti, Z., Marchino, L., Marcilio-Silva, V., Marcon, E., Marignani, M., Markesteijn, L., Martin, A., Martinez-Garza, C., Martinez-Vilalta, J., Maskova, T., Mason, K., Mason, N., Massad, T.J., Masse, J., Mayrose, I., McCarthy, J., McCormack, M.L., McCulloh, K., McFadden, I.R., McGill, B.J., McPartland, M.Y., Medeiros, J.S., Medlyn, B., Meerts, P., Mehrabi, Z., Meir, P., Melo, F.P.L., Mencuccini, M., Meredieu, C., Messier, J., Meszaros, I., Metsaranta, J., Michaletz, S.T., Michelaki, C., Migalina, S., Milla, R., Miller, J.E.D., Minden, V., Ming, R., Mokany, K., Moles, A.T., Molnar, A., Molofsky, J., Molz, M., Montgomery, R.A., Monty, A., Moravcova, L., Moreno-Martinez, A., Moretti, M., Mori, A.S., Mori, S., Morris, D., Morrison, J., Mucina, L., Mueller, S., Muir, C.D., Mueller, S.C., Munoz, F., Myers-Smith, I.H., Myster, R.W., Nagano, M., Naidu, S., Narayanan, A., Natesan, B., Negoita, L., Nelson, A.S., Neuschulz, E.L., Ni, J., Niedrist, G., Nieto, J., Niinemets, U., Nolan, R., Nottebrock, H., Nouvellon, Y., Novakovskiy, A., Nystuen, K.O., O'Grady, A., O'Hara, K., O'Reilly-Nugent, A., Oakley, S., Oberhuber, W., Ohtsuka, T., Oliveira, R., Ollerer, K., Olson, M.E., Onipchenko, V., Onoda, Y., Onstein, R.E., Ordonez, J.C., Osada, N., Ostonen, I., Ottaviani, G., Otto, S., Overbeck, G.E., Ozinga, W.A., Pahl, A.T., Paine, C.E.T., Pakeman, R.J., Papageorgiou, A.C., Parfionova, E., Paertel, M., Patacca, M., Paula, S., Paule, J., Pauli, H., Pausas, J.G., Peco, B., Penuelas, J., Perea, A., Luis Peri, P., Petisco-Souza, A.C., Petraglia, A., Petritan, A.M., Phillips, O.L., Pierce, S., Pillar, V.D., Pisek, J., Pomogaybin, A., Poorter, H., Portsmuth, A., Poschlod, P., Potvin, C., Pounds, D., Powell, A.S., Power, S.A., Prinzing, A., Puglielli, G., Pysek, P., Raevel, V., Rammig, A., Ransijn, J., Ray, C.A., Reich, P.B., Reichstein, M., Reid, D.E.B., Rejou-Mechain, M., Resco de Dios, V., Ribeiro, S., Richardson, S., Riibak, K., Rillig, M.C., Riviera, F., Robert, E.M.R., Roberts, S., Robroek, B., Roddy, A., Rodrigues, A.V., Rogers, A., Rollinson, E., Rolo, V., Roemermann, C., Ronzhina, D., Roscher, C., Rosell, J.A., Rosenfield, M.F., Rossi, C., Roy, D.B., Royer-Tardif, S., Rueger, N., Ruiz-Peinado, R., Rumpf, S.B., Rusch, G.M., Ryo, M., Sack, L., Saldana, A., Salgado-Negret, B., Salguero-Gomez, R., Santa-Regina, I., Carolina Santacruz-Garcia, A., Santos, J., Sardans, J., Schamp, B., Scherer-Lorenzen, M., Schleuning, M., Schmid, B., Schmidt, M., Schmitt, S., Schneider, J.V., Schowanek, S.D., Schrader, J., Schrodt, F., Schuldt, B., Schurr, F., Selaya Garvizu, G., Semchenko, M., Seymour, C., Sfair, J.C., Sharpe, J.M., Sheppard, C.S., Sheremetiev, S., Shiodera, S., Shipley, B., Shovon, T.A., Siebenkaes, A., Carlos, S., Silva, V., Silva, M., Sitzia, T., Sjoman, H., Slot, M., Smith, N.G., Sodhi, D., Soltis, P., Soltis, D., Somers, B., Sonnier, G., Sorensen, M.V., Sosinski, E.E., Soudzilovskaia, N.A., Souza, A.F., Spasojevic, M., Sperandii, M.G., Stan, A.B., Stegen, J., Steinbauer, K., Stephan, J.G., Sterck, F., Stojanovic, D.B., Strydom, T., Laura Suarez, M., Svenning, J.-C., Svitkova, I., Svitok, M., Svoboda, M., Swaine, E., Swenson, N., Tabarelli, M., Takagi, K., Tappeiner, U., Tarifa, R., Tauugourdeau, S., Tavsanoglu, C., te Beest, M., Tedersoo, L., Thiffault, N., Thom, D., Thomas, E., Thompson, K., Thornton, P.E., Thuiller, W., Tichy, L., Tissue, D., Tjoelker, M.G., Tng, D.Y.P., Tobias, J., Torok, P., Tarin, T., Torres-Ruiz, J.M., Tothmeresz, B., Treurnicht, M., Trivellone, V., Trolliet, F., Trotsiuk, V., Tsakalos, J.L., Tsiripidis, I., Tysklind, N., Umehara, T., Usoltsev, V., Vadeboncoeur, M., Vaezi, J., Valladares, F., Vamosi, J., van Bodegom, P.M., van Breugel, M., Van Cleemput, E., van de Weg, M., van der Merwe, S., van der Plas, F., van der Sande, M.T., van Kleunen, M., Van Meerbeek, K., Vanderwel, M., Vanselow, K.A., Varhammar, A., Varone, L., Vasquez Valderrama, M.Y., Vassilev, K., Vellend, M., Veneklaas, E.J., Verbeeck, H., Verheyen, K., Vibrans, A., Vieira, I., Villacis, J., Violle, C., Vivek, P., Wagner, K., Waldram, M., Waldron, A., Walker, A.P., Waller, M., Walther, G., Wang, H., Wang, F., Wang, W., Watkins, H., Watkins, J., Weber, U., Weedon, J.T., Wei, L., Weigelt, P., Weiher, E., Wells, A.W., Wellstein, C., Wenk, E., Westoby, M., Westwood, A., White, P.J., Whitten, M., Williams, M., Winkler, D.E., Winter, K., Womack, C., Wright, I.J., Wright, S.J., Wright, J., Pinho, B.X., Ximenes, F., Yamada, T., Yamaji, K., Yanai, R., Yankov, N., Yguel, B., Zanini, K.J., Zanne, A.E., Zeleny, D., Zhao, Y.-P., Zheng, Jingming, Zheng, Ji, Zieminska, K., Zirbel, C.R., Zizka, G., Zo-Bi, I.C., Zotz, G., Wirth, C., 2020. TRY plant trait database—enhanced coverage and open access. Glob. Change Biol. 26, 119–188. https://doi.org/10.1111/gcb.14904.

Kautsky, L., 1988. Life strategies of aquatic soft bottom macrophytes. Oikos 53, 126–135. https://doi.org/10.2307/3565672.

Keeley, J.E., 1981. *Isoetes howelli*—a submerged aquatic CAM plant. Am. J. Bot. 68, 420–424. https://doi.org/10.2307/2442779.

Kimbel, J.C., 1982. Factors influencing potential intralake colonization by *Myriophyllum spicatum* L. Aquat. Bot. 14, 295–307. https://doi.org/10.1016/0304-3770(82)90104-8.

Kirk, J.T.O., 2010. Light and Photosynthesis in Aquatic Ecosystems, third ed. Cambridge University Press, Cambridge. https://doi.org/10.1017/CBO9781139168212.

Klaine, S.J., Alvarez, P.J.J., Batley, G.E., Fernandes, T.F., Handy, R.D., Lyon, D.Y., Mahendra, S., McLaughlin, M.J., Lead, J.R., 2008. Nanomaterials in the environment: behavior, fate, bioavailability, and effects. Environ. Toxicol. Chem. 27 (1825). https://doi.org/10.1897/08-090.1.

Klancnik, K., Pancic, M., Gaberscik, A., 2014. Leaf optical properties in amphibious plant species are affected by multiple leaf traits. Hydrobiologia 737, 121–130. https://doi.org/10.1007/s10750-013-1646-y.

Klavsen, S.K., Maberly, S.C., 2009. Crassulacean acid metabolism contributes significantly to the in situ carbon budget in a population of the invasive aquatic macrophyte *Crassula helmsii*. Freshw. Biol. 54, 105–118. https://doi.org/10.1111/j.1365-2427.2008.02095.x.

Klavsen, S.K., Madsen, T.V., Maberly, S.C., 2011. Crassulacean acid metabolism in the context of other carbon-concentrating mechanisms in freshwater plants: a review. Photosynth. Res. 109, 269–279. https://doi.org/10.1007/s11120-011-9630-8.

Klok, P.F., van der Velde, G., 2017. Plant traits and environment: floating leaf blade production and turnover of waterlilies. PeerJ 5, e3212. https://doi.org/10.7717/peerj.3212.

Koi, S., Fujinami, R., Kubo, N., Tsukamo, I., Inagawa, R., Imaichi, R., Kato, M., 2006. Comparative anatomy of root meristem and root cap in some species of Podostemaceae and the evolution of root dorsiventrality. Am. J. Bot. 93, 682–692. https://doi.org/10.3732/ajb.93.5.682.

Kolada, A., Willby, N., Dudley, B., Nõges, P., Søndergaard, M., Hellsten, S., Mjelde, M., Penning, E., van Geest, G., Bertrin, V., Ecke, F., Mäemets, H., Karus, K., 2014. The applicability of macrophyte compositional metrics for assessing eutrophication in European lakes. Ecol. Indic 45, 407–415. https://doi.org/10.1016/j.ecolind.2014.04.049.

Kosten, S., Jeppesen, E., Huszar, V.L.M., Mazzeo, N., Nes, E.H.V., Peeters, E.T.H.M., Scheffer, M., 2011. Ambiguous climate impacts on competition between submerged macrophytes and phytoplankton in shallow lakes. Freshw. Biol. 56, 1540–1553. https://doi.org/10.1111/j.1365-2427.2011.02593.x.

Kozhova, O.M., Izmestéva, L.R., 1988. Lake Baikal: Evolution and Biodiversity. Backhuys Publishers, Leiden, The Netherlands.

Krzewicka, B., Smykla, J., Galas, J., Sliwa, L., 2017. Freshwater lichens and habitat zonation of mountain streams. Limnologica 63, 1–10. https://doi.org/10.1016/j.limno.2016.12.002.

Kufel, L., Kufel, I., 2002. Chara beds acting as nutrient sinks in shallow lakes—a review. Aquat. Bot. Ecology of Charophytes 72, 249–260. https://doi.org/10.1016/S0304-3770(01)00204-2.

Laird, R.A., Barks, P.M., 2018. Skimming the surface: duckweed as a model system in ecology and evolution. Am. J. Bot. 105, 1962–1966. https://doi.org/10.1002/ajb2.1194.

Lambert, S.J., Thomas, K.V., Davy, A.J., 2006. Assessment of the risk posed by the antifouling booster biocides Irgarol 1051 and diuron to freshwater macrophytes. Chemosphere 63, 734–743. https://doi.org/10.1016/j.chemosphere.2005.08.023.

Landers, D.H., 1982. Effects of naturally senescing aquatic macrophytes on nutrient chemistry and chlorophyll *a* of surrounding waters. Limnol. Oceanogr. 27, 428–439. https://doi.org/10.4319/lo.1982.27.3.0428.

LaRue, E.A., Grimm, D., Thum, R.A., 2013. Laboratory crosses and genetic analysis of natural populations demonstrate sexual viability of invasive hybrid watermilfoils (*Myriophyllum spicatum* × *M. sibiricum*). Aquat. Bot. 109, 49–53. https://doi.org/10.1016/j.aquabot.2013.04.004.

Lauridsen, T.L., Mønster, T., Raundrup, K., Nymand, J., Olesen, B., 2019. Macrophyte performance in a low arctic lake: effects of temperature, light and nutrients on growth and depth distribution. Aquat. Sci. 82 (18). https://doi.org/10.1007/s00027-019-0692-6.

Lead, J.R., Batley, G.E., Alvarez, P.J.J., Croteau, M.-N., Handy, R.D., McLaughlin, M.J., Judy, J.D., Schirmer, K., 2018. Nanomaterials in the environment: behavior, fate, bioavailability, and effects—an updated review: nanomaterials in the environment. Environ. Toxicol. Chem. 37, 2029–2063. https://doi.org/10.1002/etc.4147.

Les, D.H., 2015. Water from the rock: ancient aquatic angiosperms flow from the fossil record. Proc. Natl. Acad. Sci. U. S. A 112, 10825–10826. https://doi.org/10.1073/pnas.1514280112.

Les, D.H., Crawford, D.J., Kimball, R.T., Moody, M.L., Landolt, E., 2003. Biogeography of discontinuously distributed hydrophytes: a molecular appraisal of intercontinental disjunctions. Int. J. Plant Sci. 164, 917–932. https://doi.org/10.1086/378650.

Les, D.H., Mehrhoff, L.J., 1999. Introduction of Nonindigenous Aquatic Vascular Plants in Southern New England: A Historical Perspective 20.

Les, D.H., Tippery, N., 2013. In time and with water... the systematics of alismatid monocotyledons. In: Wilkin, P., Mayo, S.J. (Eds.), Early Events in Monocot Evolution. Cambridge University Press, Cambridge, pp. 118–164.

Li, G., Hu, S., Hou, H., Kimura, S., 2019. Heterophylly: phenotypic plasticity of leaf shape in aquatic and amphibious plants. Plants 8, 420. https://doi.org/10.3390/plants8100420.

Li, L., Lan, Z., Chen, J., Song, Z., 2018. Allocation to clonal and sexual reproduction and its plasticity in *Vallisneria spinulosa* along a water-depth gradient. Ecosphere 9, e02070. https://doi.org/10.1002/ecs2.2070.

Li, S.C., Liu, W.Z., Liu, H., Li, W., 2011. Soil seed bank and ecological implications for the lakeshore marsh of Bang Lake, a part of Poyang Lake. Plant Sci. J 29, 164–170.

Li, W., 2014. Environmental opportunities and constraints in the reproduction and dispersal of aquatic plants. Aquat. Bot., Special Issue: In Honour of George Bowes: Linking Terrestrial and Aquatic Botany 118, 62–70. https://doi.org/10.1016/j.aquabot.2014.07.008.

Li, Z., He, L., Zhang, H., Cordero, P., Ekvall, M., Hollander, J., Hansson, L.-A., 2016. Climate warming and heat waves affect reproductive strategies and interactions between submerged macrophytes. Glob. Change Biol. 23. https://doi.org/10.1111/gcb.13405.

Lin, B.-L., Yang, W.-J., 1999. Blue light and abscisic acid independently induce heterophyllous switch in *Marsilea quadrifolia*. Plant Physiol 119, 429–434.

Lobato-de Magalhães, T., Murphy, K., Efremov, A., Chepinoga, V., Davidson, T.A., Molina-Navarro, E., 2021. Ploidy state of aquatic macrophytes: global distribution and drivers. Aquat. Bot. 173, 103417. https://doi.org/10.1016/j.aquabot.2021.103417.

Loczy, S., Carignan, R., Planas, D., 1983. The role of roots in carbon uptake by the submersed macrophytes *Myriophyllum spicatum*, *Vallisneria americana*, and *Heteranthera dubia*. Hydrobiologia 98, 3–7. https://doi.org/10.1007/bf00019244.

Lodge, D.M., 1991. Herbivory on freshwater macrophytes. Aquat. Bot. 41, 195–224. https://doi.org/10.1016/0304-3770(91)90044-6.

Lovas-Kiss, Á., Sánchez, M.I., Wilkinson, D.M., Coughlan, N.E., Alves, J.A., Green, A.J., 2019. Shorebirds as important vectors for plant dispersal in Europe. Ecography 42, 956–967. https://doi.org/10.1111/ecog.04065.

Lucas, W.J., Smith, F.A., 1973. Formation of alkaline and acid regions at surface of *Chara corallina* cells. J. Exp. Bot. 24, 1–14. https://doi.org/10.1093/jxb/24.1.1.

Lucassen, E.C.H.E.T., Spierenburg, P., Fraaije, R., Smolders, A., Roelofs, J., 2009. Alkalinity generation and sediment CO_2 uptake influence establishment of *Sparganium angustifolium* in softwater lakes. Freshw. Biol. 54, 2300–2314. https://doi.org/10.1111/j.1365-2427.2009.02264.x.

Maberly, S., Spence, D., 1989. Photosynthesis and photorespiration in freshwater organisms—amphibious plants. Aquat. Bot. 34, 267–286. https://doi.org/10.1016/0304-3770(89)90059-4.

Maberly, S.C., 1993. Morphological and photosynthetic characteristics of *Potamogeton obtusifolius* from different depths. J. Aquat. Plant Manag. 31, 34–39.

Maberly, S.C., 1985a. Photosynthesis by *Fontinalis antpyretica* 1. Interaction between photon irradiance, concentration of carbon dioxide and temperature. New Phytol. 100, 127–140. https://doi.org/10.1111/j.1469-8137.1985.tb02765.x.

Maberly, S.C., 1985b. Photosynthesis by *Fontinalis antipyretica* 2. Assessment of environmental factors limiting photosynthesis and production. New Phytol. 100, 141–155. https://doi.org/10.1111/j.1469-8137.1985.tb02766.x.

Maberly, S.C., Berthelot, S.A., Stott, A.W., Gontero, B., 2015. Adaptation by macrophytes to inorganic carbon down a river with naturally variable concentrations of CO_2. J. Plant Physiol. 172, 120–127. https://doi.org/10.1016/j.jplph.2014.07.025.

Maberly, S.C., Gontero, B., 2017. Ecological imperatives for aquatic CO_2-concentrating mechanisms. J. Exp. Bot. 68, 3797–3814. https://doi.org/10.1093/jxb/erx201.

Maberly, S.C., Gontero, B., 2018. Trade-offs and synergies in the structural and functional characteristics of leaves photosynthesizing in aquatic environments. Leaf Platf. Perform. Photosynth. Adv. Photosynth. Respir. 44, Springer International Publishing AG, part of Springer Nature 2018. In: Adams III, W.W., Terashima, I. (Eds.), pp. 308–334. https://doi.org/10.1007/978-3-319-93594-2_11.

Maberly, S.C., Madsen, T.V., 1998. Affinity for CO_2 in relation to the ability of freshwater macrophytes to use $HCO^-{}_3$. Funct. Ecol. 12, 99–106. https://doi.org/10.1046/j.1365-2435.1998.00172.x.

Maberly, S.C., O'Donnell, R.A., Woolway, R.I., Cutler, M.E.J., Gong, M., Jones, I.D., Merchant, C.J., Miller, C.A., Politi, E., Scott, E.M., Thackeray, S.J., Tyler, A.N., 2020. Global lake thermal regions shift under climate change. Nat. Commun. 11, 1232. https://doi.org/10.1038/s41467-020-15108-z.

Madsen, T.V., Cedergreen, N., 2002. Sources of nutrients to rooted submerged macrophytes growing in a nutrient-rich stream. Freshw. Biol. 47, 283–291. https://doi.org/10.1046/j.1365-2427.2002.00802.x.

Madsen, J.D., Chambers, P.A., James, W.F., Koch, E.W., Westlake, D.F., 2001. The interaction between water movement, sediment dynamics and submersed macrophytes. Hydrobiologia 444, 71–84. https://doi.org/10.1023/A:1017520800568.

Madsen, T., Enevoldsen, H., Jorgensen, T., 1993. Effects of water velocity on photosynthesis and dark respiration in submerged stream macrophytes. Plant Cell Environ. 16, 317–322. https://doi.org/10.1111/j.1365-3040.1993.tb00875.x.

Madsen, T.V., Maberly, S.C., 2003. High internal resistance to CO_2 uptake by submerged macrophytes that use HCO_3^-: measurements in air, nitrogen and helium. Photosynth. Res. 77, 183–190. https://doi.org/10.1023/a:1025813515956.

Martins, S.V., Milne, J., Thomaz, S.M., McWaters, S., Mormul, R.P., Kennedy, M., Murphy, K., 2013. Human and natural drivers of changing macrophyte community dynamics over 12 years in a neotropical riverine floodplain system. Aquat. Conserv. Mar. Freshw. Ecosyst. 23, 678–697. https://doi.org/10.1002/aqc.2368.

Mason, C.F., Bryant, R.J., 1975. Changes in the ecology of the Norfolk broads. Freshw. Biol. 5, 257–270. https://doi.org/10.1111/j.1365-2427.1975.tb00139.x.

May, L., Carvalho, L., 2010. Maximum growing depth of macrophytes in Loch Leven, Scotland, United Kingdom, in relation to historical changes in estimated phosphorus loading. Hydrobiologia 646, 123–131. https://doi.org/10.1007/s10750-010-0176-0.

May, L., Spears, B.M., 2012. Managing ecosystem services at Loch Leven, Scotland, UK: actions, impacts and unintended consequences. Hydrobiologia 681, 117–130. https://doi.org/10.1007/s10750-011-0931-x.

McConnaughey, T., 1991. Calcification in *Chara corallina*: CO_2 hydroxylation generates protons for bicarbonate assimilation. Limnol. Oceanogr. 36, 619–628. https://doi.org/10.4319/lo.1991.36.4.0619.

Mebane, C.A., Simon, N.S., Maret, T.R., 2014. Linking nutrient enrichment and streamflow to macrophytes in agricultural streams. Hydrobiologia 722, 143–158. https://doi.org/10.1007/s10750-013-1693-4.

Meijer, M.L., 2000. Biomanipulation in the Netherlands—15 Years of Experience. Wageningen Univ., Wageningen, The Netherlands, p. 208 pp..

Middelboe, A., Markager, S., 1997. Depth limits and minimum light requirements of freshwater macrophytes. Freshw. Biol. 37, 553–568. https://doi.org/10.1046/j.1365-2427.1997.00183.x.

Mille-Lindblom, C., Fischer, H., Tranvik, L.J., 2006. Litter-associated bacteria and fungi—a comparison of biomass and communities across lakes and plant species. Freshw. Biol. 51, 730–741. https://doi.org/10.1111/j.1365-2427.2006.01532.x.

Minguez, L., Gross, E.M., Vignati, D.A.L., Romero Freire, A., Camizuli, E., Gimbert, F., Caillet, C., Pain-Devin, S., Devin, S., Guérold, F., Giambérini, L., 2021. Profiling metal contamination from ultramafic sediments to biota along the Albanian shoreline of Lake Ohrid (Albania/Macedonia). J. Environ. Manage 291, 112726. https://doi.org/10.1016/j.jenvman.2021.112726.

Moller, C.L., Sand-Jensen, K., 2011. High sensitivity of *Lobelia dortmanna* to sediment oxygen depletion following organic enrichment. New Phytol. 190, 320–331. https://doi.org/10.1111/j.1469-8137.2010.03584.x.

Mommer, L., Pedersen, O., Visser, E.J.W., 2004. Acclimation of a terrestrial plant to submergence facilitates gas exchange under water. Plant Cell Environ. 27, 1281–1287. https://doi.org/10.1111/j.1365-3040.2004.01235.x.

Monteith, D.T., Hildrew, A.G., Flower, R.J., Raven, P.J., Beaumont, W.R.B., Collen, P., Kreiser, A.M., Shilland, E.M., Winterbottom, J.H., 2005. Biological responses to the chemical recovery of acidified fresh waters in the UK. Environ. Pollut 137, 83–101. https://doi.org/10.1016/j.envpol.2004.12.026.

Monteith, D.T., Stoddard, J.L., Evans, C.D., de Wit, H.A., Forsius, M., Hogasen, T., Wilander, A., Skjelkvale, B.L., Jeffries, D.S., Vuorenmaa, J., Keller, B., Kopacek, J., Vesely, J., 2007. Dissolved organic carbon trends resulting from changes in atmospheric deposition chemistry. Nature 450, 537. https://doi.org/10.1038/nature06316. U9.

Morris, J.L., Puttick, M.N., Clark, J.W., Edwards, D., Kenrick, P., Pressel, S., Wellman, C.H., Yang, Z., Schneider, H., Donoghue, P.C.J., 2018. The timescale of early land plant evolution. Proc. Natl. Acad. Sci. U. S. A. 115, E2274–E2283. https://doi.org/10.1073/pnas.1719588115.

Morris, D.P., Zagarese, H., Williamson, C.E., Balseiro, E.G., Hargreaves, B.R., Modenutti, B., Moeller, R., Queimalinos, C., 1995. The attenuation of solar UV radiation in lakes and the role of dissolved organic carbon. Limnol. Oceanogr. 40, 1381–1391.

Mortimer, C., 1941. The exchange of dissolved substances between mud and water in lakes I and II. J. Ecol. 29, 280–329.

Moss, B., Ryszard, K., Measey, G.J., 1998. The effects of nymphaeid (*Nuphar lutea*) density and predation by perch (*Perca fluviatilis*) on the zooplankton communities in a shallow lake: effects of lilies and perch on zooplankton. Freshw. Biol. 39, 689–697. https://doi.org/10.1046/j.1365-2427.1998.00322.x.

Murphy, K., Efremov, A., Davidson, T.A., Molina-Navarro, E., Fidanza, K., Crivelari Betiol, T.C., Chambers, P., Tapia Grimaldo, J., Varandas Martins, S., Springuel, I., Kennedy, M., Mormul, R.P., Dibble, E., Hofstra, D., Lukács, B.A., Gebler, D.,

Baastrup-Spohr, L., Urrutia-Estrada, J., 2019. World distribution, diversity and endemism of aquatic macrophytes. Aquat. Bot. 158, 103127. https://doi.org/10.1016/j.aquabot.2019.06.006.

Murray, R.E., Hodson, R.E., 1985. Annual cycle of bacterial secondary production in five aquatic habitats of the Okefenokee Swamp ecosystem. Appl. Environ. Microbiol. 49, 650–655.

Mustafa, H.M., Hayder, G., 2021. Recent studies on applications of aquatic weed plants in phytoremediation of wastewater: a review article. Ain Shams Eng. J. 12, 355–365. https://doi.org/10.1016/j.asej.2020.05.009.

Nakajima, T., 1994. Lake Biwa. Ergeb. Limnol. 44, 43–54.

Nichols, D.S., Keeney, D.R., 1976. Nitrogen nutrition of *Myriophyllum spicatum*: uptake and translocation of 15N by shoots and roots. Freshw. Biol. 6, 145–154. https://doi.org/10.1111/j.1365-2427.1976.tb01598.x.

Nocchi, N., Duarte, H.M., Pereira, R.C., Konno, T.U.P., Soares, A.R., 2020. Effects of UV-B radiation on secondary metabolite production, antioxidant activity, photosynthesis and herbivory interactions in *Nymphoides humboldtiana* (Menyanthaceae). J. Photochem. Photobiol., B 212, 112021. https://doi.org/10.1016/j.jphotobiol.2020.112021.

O'Connell, R.A., 1997. The state of California's noxious weed eradication programs. presented at the California exotic pest plant council 1997. Symposium 10.

Ogorek, L.L.P., Pellegrini, E., Pedersen, O., 2021. Novel functions of the root barrier to radial oxygen loss—radial diffusion resistance to H_2 and water vapour. New Phytol. 231 (4), 1365–1376. https://doi.org/10.1111/nph.17474.

O'Hare, M.T., Clarke, R.T., Bowes, M.J., Cailes, C., Henville, P., Bissett, N., McGahey, C., Neal, M., 2010. Eutrophication impacts on a river macrophyte. Aquat. Bot. 92, 173–178. https://doi.org/10.1016/j.aquabot.2009.11.001.

Oliveira Junior, E.S., van Bergen, T.J.H.M., Nauta, J., Budiša, A., Aben, R.C.H., Weideveld, S.T.J., de Souza, C.A., Muniz, C.C., Roelofs, J., Lamers, L.P.M., Kosten, S., 2021. Water hyacinth's effect on greenhouse gas fluxes: a field study in a wide variety of tropical water bodies. Ecosystems 24, 988–1004. https://doi.org/10.1007/s10021-020-00564-x.

O'Reilly, C.M., Sharma, S., Gray, D.K., Hampton, S.E., Read, J.S., Rowley, R.J., Schneider, P., Lenters, J.D., McIntyre, P.B., Kraemer, B.M., Weyhenmeyer, G.A., Straile, D., Dong, B., Adrian, R., Allan, M.G., Anneville, O., Arvola, L., Austin, J., Bailey, J.L., Baron, J.S., Brookes, J.D., de Eyto, E., Dokulil, M.T., Hamilton, D.P., Havens, K., Hetherington, A.L., Higgins, S.N., Hook, S., Izmest'eva, L.R., Joehnk, K.D., Kangur, K., Kasprzak, P., Kumagai, M., Kuusisto, E., Leshkevich, G., Livingstone, D.M., MacIntyre, S., May, L., Melack, J.M., Mueller-Navarra, D.C., Naumenko, M., Noges, P., Noges, T., North, R.P., Plisnier, P.D., Rigosi, A., Rimmer, A., Rogora, M., Rudstam, L.G., Rusak, J.A., Salmaso, N., Samal, N.R., Schindler, D.E., Schladow, S.G., Schmid, M., Schmidt, S.R., Silow, E., Soylu, M.E., Teubner, K., Verburg, P., Voutilainen, A., Watkinson, A., Williamson, C.E., Zhang, G.Q., 2015. Rapid and highly variable warming of lake surface waters around the globe. Geophys. Res. Lett. 42, 10773–10781. https://doi.org/10.1002/2015gl066235.

Otsuki, A., Wetzel, R.G., 1972. Coprecipitation of phosphate with carbonates in a marl lake. Limnol. Oceanogr. 17, 763–767. https://doi.org/10.4319/lo.1972.17.5.0763.

Padilla, D., Williams, S., 2004. Beyond ballast water: aquarium and ornamental trades as sources of invasive species in aquatic ecosystems. Front. Ecol. Environ. 2, 131–138. https://doi.org/10.1890/1540-9295(2004)002[0131:BBWAAO]2.0.CO;2.

Pedrozo, F., Kelly, L., Diaz, M., Temporetti, P., Baffico, G., Kringel, R., Friese, K., Mages, M., Geller, W., Woelfl, S., 2001. First results on the water chemistry, algae and trophic status of an Andean acidic lake system of volcanic origin in Patagonia (Lake Caviahue). Hydrobiologia 452, 129–137. https://doi.org/10.1023/a:1011984212798.

Persson, L., Eklov, P., 1995. Prey refuges affecting interactions between piscivorous perch and juvenile perch and roach. Ecology 76, 70–81. https://doi.org/10.2307/1940632.

Peter, S., Isidorova, A., Sobek, S., 2016. Enhanced carbon loss from anoxic lake sediment through diffusion of dissolved organic carbon: DOC flux from ANOXIC lake sediments. J. Geophys. Res. Biogeosciences 121, 1959–1977. https://doi.org/10.1002/2016JG003425.

Petersen, F., Hubbart, J.A., 2021. The occurrence and transport of microplastics: the state of the science. Sci. Total Environ. 758, 143936. https://doi.org/10.1016/j.scitotenv.2020.143936.

Petersen, N.R., Jensen, K., 1997. Nitrification and denitrification in the rhizosphere of the aquatic macrophyte *Lobelia dortmanna* L. Limnol. Oceanogr. 42, 529–537. https://doi.org/10.4319/lo.1997.42.3.0529.

Pettit, N.E., Jardine, T.D., Hamilton, S.K., Sinnamon, V., Valdez, D., Davies, P.M., Douglas, M.M., Bunn, S.E., et al., 2012. Seasonal changes in water quality and macrophytes and the impact of cattle on tropical floodplain waterholes. Mar. Freshw. Res. 63, 788–800. https://doi.org/10.1071/MF12114.

Philbrick, C.T., 1988. Evolution of underwater outcrossing from aerial pollination systems: a hypothesis. Ann. Mo. Bot. Gard 75, 836–841. https://doi.org/10.2307/2399371.

Philbrick, C.T., Les, D.H., 1996. Evolution of aquatic angiosperm reproductive systems. Bioscience 46, 813–826. https://doi.org/10.2307/1312967.

Phillips, G.L., Eminson, D., Moss, B., 1978. A mechanism to account for macrophyte decline in progressively eutrophicated freshwaters. Aquat. Bot. 4, 103–126. https://doi.org/10.1016/0304-3770(78)90012-8.

Phillips, G., Willby, N., Moss, B., 2016. Submerged macrophyte decline in shallow lakes: what have we learnt in the last forty years? Aquat. Bot. 135, 37–45. https://doi.org/10.1016/j.aquabot.2016.04.004.

Pierce, S., Brusa, G., Sartori, M., Cerabolini, B.E.L., 2012. Combined use of leaf size and economics traits allows direct comparison of hydrophyte and terrestrial herbaceous adaptive strategies. Ann. Bot 109, 1047–1053. https://doi.org/10.1093/aob/mcs021.

Pieterse, A.H., Murphy, K.J., 1993. Aquatic Weeds, second ed. Oxford University Press, Oxford.

Pimentel, D., Lach, L., Zuniga, R., Morrison, D., 2000. Environmental and economic costs of nonindigenous species in the United States. Bioscience 50, 53–65. https://doi.org/10.1641/0006-3568(2000)050[0053:EAECON]2.3.CO;2.

Poorter, L., 2009. Leaf traits show different relationships with shade tolerance in moist versus dry tropical forests. New Phytol. 181, 890–900. https://doi.org/10.1111/j.1469-8137.2008.02715.x.

PPG, I., 2016. A community-derived classification for extant lycophytes and ferns: PPG I. J. Syst. Evol. 54, 563–603. https://doi.org/10.1111/jse.12229.

Prins, H.B.A., Deguia, M.B., 1986. Carbon source of the water soldier, *Stratiotes aloides* L. Aquat. Bot. 26, 225–234. https://doi.org/10.1016/0304-3770(86)90023-9.

Prins, H.B.A., Snel, J.F.H., Helder, R.J., Zanstra, P.E., 1980. Photosynthetic HCO_3^- utilization and OH^- excretion in aquatic angiosperms: light induced pH changes at the leaf surface. Plant Physiol 66, 818–822. https://doi.org/10.1104/pp.66.5.818.

Rae, R., Rae, R., Hanelt, D., Hanelt, D., Hawes, I., Hawes, I., 2001. Sensitivity of freshwater macrophytes to UV radiation: relationship to depth zonation in an oligotrophic New Zealand lake. Mar. Freshw. Res. 52 (1023). https://doi.org/10.1071/MF01016.

Rasmussen, J.J., Kallestrup, H., Thiemer, K., Alnøe, A.B., Henriksen, L.D., Larsen, S.E., Baattrup-Pedersen, A., 2021. Effects of different weed cutting methods on physical and hydromorphological conditions in lowland streams. Knowl. Manag. Aquat. Ecosyst. 10. https://doi.org/10.1051/kmae/2021009.

Rattray, M.R., Howard-Williams, C., Brown, J.M.A., 1991. Sediment and water as sources of nitrogen and phosphorus for submerged rooted aquatic macrophytes. Aquat. Bot. 40, 225–237. https://doi.org/10.1016/0304-3770(91)90060-I.

Raven, J.A., 2013. Polar auxin transport in relation to long-distance transport of nutrients in the Charales. J. Exp. Bot. 64, 1–9. https://doi.org/10.1093/jxb/ers358.

Raven, J.A., Beardall, J., Giordano, M., 2014. Energy costs of carbon dioxide concentrating mechanisms in aquatic organisms. Photosynth. Res. 121, 111–124. https://doi.org/10.1007/s11120-013-9962-7.

Raymond, P.A., Hamilton, S.K., 2018. Anthropogenic influences on riverine fluxes of dissolved inorganic carbon to the oceans. Limnol. Oceanogr. Lett. 3, 143–155. https://doi.org/10.1002/lol2.10069.

Reid, A.J., Carlson, A.K., Creed, I.F., Eliason, E.J., Gell, P.A., Johnson, P.T.J., Kidd, K.A., MacCormack, T.J., Olden, J.D., Ormerod, S.J., Smol, J.P., Taylor, W.W., Tockner, K., Vermaire, J.C., Dudgeon, D., Cooke, S.J., 2019. Emerging threats and persistent conservation challenges for freshwater biodiversity. Biol. Rev. 94, 849–873. https://doi.org/10.1111/brv.12480.

Riederer, M., Schreiber, L., 2001. Protecting against water loss: analysis of the barrier properties of plant cuticles. J. Exp. Bot. 52, 2023–2032. https://doi.org/10.1093/jexbot/52.363.2023.

Reitsema, R.E., Wolters, J.-W., Preiner, S., Meire, P., Hein, T., De Boeck, G., Blust, R., Schoelynck, J., 2020. Response of submerged macrophyte growth, morphology, chlorophyll content and nutrient stoichiometry to increased flow velocity and elevated CO_2 and dissolved organic carbon concentrations. Front. Environ. Sci. 8. https://doi.org/10.3389/fenvs.2020.527801.

Riis, T., Madsen, T.V., Sennels, R.S.H., 2009. Regeneration, colonisation and growth rates of allofragments in four common stream plants. Aquat. Bot. 90, 209–212. https://doi.org/10.1016/j.aquabot.2008.08.005.

Riis, T., Sand-Jensen, K., 1998. Development of vegetation and environmental conditions in an oligotrophic Danish lake over 40 years. Freshw. Biol. 40, 123–134. https://doi.org/10.1046/j.1365-2427.1998.00338.x.

Robach, F., Hajnsek, I., Eglin, I., Trémolières, M., 1995. Phosphorus sources for aquatic macrophytes in running waters: water or sediment? Acta Bot. Gallica 142, 719–731. https://doi.org/10.1080/12538078.1995.10515296.

Robe, W.E., Griffiths, H., 1998. Adaptations for an amphibious life: changes in leaf morphology, growth rate, carbon and nitrogen investment, and reproduction during adjustment to emersion by the freshwater macrophyte *Littorella uniflora*. New Phytol. 140, 9–23. https://doi.org/10.1046/j.1469-8137.1998.00257.x.

Roelofs, J., 1983. Impact of acidification and eutrophication on macrophyte communities in soft waters in The Netherlands 1. field observations. Aquat. Bot. 17, 139–155. https://doi.org/10.1016/0304-3770(83)90110-9.

Rooney, N., Kalff, J., 2000. Inter-annual variation in submerged macrophyte community biomass and distribution: the influence of temperature and lake morphometry. Aquat. Bot. 68, 321–335. https://doi.org/10.1016/S0304-3770(00)00126-1.

Rooney, N., Kalff, J., 2003. Submerged macrophyte-bed effects on water-column phosphorus, chlorophyll a, and bacterial production. Ecosystems 6, 797–807. https://doi.org/10.1007/s10021-003-0184-2.

Rose, C., Crumpton, W.G., 1996. Effects of emergent macrophytes on dissolved oxygen dynamics in a prairie pothole wetland. Wetlands 16, 495–502. https://doi.org/10.1007/BF03161339.

RoTAP, 2012. Review of Transboundary Air Pollution (RoTAP): Acidification, Eutrophication, Ground Level Ozone and Heavy Metals in the UK. Centre for Ecology & Hydrology on Behalf of Defra and the Devolved Administrations, Place of Publication Not Identified.

Roubeau Dumont, E., Larue, C., Michel, H.C., Gryta, H., Liné, C., Baqué, D., Maria Gross, E., Elger, A., 2020. Genotypes of the aquatic plant *Myriophyllum spicatum* with different growth strategies show contrasting sensitivities to copper contamination. Chemosphere 245, 125552. https://doi.org/10.1016/j.chemosphere.2019.125552.

Sage, R.F., Monson, R.K., Ehleringer, J.R., Adachi, S., Pearcy, R.W., 2018. Some like it hot: the physiological ecology of C_4 plant evolution. Oecologia 187, 941–966. https://doi.org/10.1007/s00442-018-4191-6.

Sakakibara, M., Ohmori, Y., Ha, N.T.H., Sano, S., Sera, K., 2011. Phytoremediation of heavy metal-contaminated water and sediment by *Eleocharis acicularis*. CLEAN—Soil Air Water 39, 735–741. https://doi.org/10.1002/clen.201000488.

Sala, O.E., 2000. Global biodiversity scenarios for the year 2100. Science 287, 1770–1774. https://doi.org/10.1126/science.287.5459.1770.

Sand-Jensen, K., 1977. Effect of epiphytes on eelgrass photosynthesis. Aquat. Bot. 3, 55–63. https://doi.org/10.1016/0304-3770(77)90004-3.

Sand-Jensen, K., 1990. Epiphyte shading: its role in resulting depth distribution of submerged aquatic macrophytes. Folia Geobot. Phytotaxon 25, 315–320. https://doi.org/10.1007/BF02913033.

Sand-Jensen, K., 1998. Influence of submerged macrophytes on sediment composition and near-bed flow in lowland streams. Freshw. Biol. 39, 663–679.

Sand-Jensen, K., Bruun, H.H., Baastrup-Spohr, L., 2017. Decade-long time delays in nutrient and plant species dynamics during eutrophication and re-oligotrophication of Lake Fure 1900–2015. J. Ecol. 105, 690–700. https://doi.org/10.1111/1365-2745.12715.

Sand-Jensen, K., Jacobsen, D., 2002. Herbivory and growth in terrestrial and aquatic populations of amphibious stream plants. Freshw. Biol. 47, 1475–1487. https://doi.org/10.1046/j.1365-2427.2002.00890.x.

Sand-Jensen, K., Madsen, T.V., 1991. Minimum light requirements of submerged fresh-water macrophytes in laboratory growth experiments. J. Ecol. 79, 749–764. https://doi.org/10.2307/2260665.

Sand-Jensen, K., Prahl, C., 1982. Oxygen exchange with the lacunae and across leaves and roots of the submerged vascular macrophyte, *Lobelia dortmanna* L. New Phytol. 91, 103–120. https://doi.org/10.1111/j.1469-8137.1982.tb03296.x.

Sand-Jensen, K., Prahl, C., Stokholm, H., 1982. Oxygen release from roots of submerged aquatic macrophytes. Oikos 38, 349–354. https://doi.org/10.2307/3544675.

Sand-Jensen, K., Sondergaard, M., 1981. Phytoplankton and epiphyte development and their shading effect on submerged macrophytes in lakes of different nutrient status. Int. Rev. Gesamten Hydrobiol. 66, 529–552. https://doi.org/10.1002/iroh.19810660406.

Sand-Jensen, K., Staehr, P.A., 2012. CO2 dynamics along Danish lowland streams: water-air gradients, piston velocities and evasion rates. Biogeochemistry 111, 615–628. https://doi.org/10.1007/s10533-011-9696-6.

Santamaría, L., 2002. Why are most aquatic plants widely distributed? Dispersal, clonal growth and small-scale heterogeneity in a stressful environment. Acta Oecol. 23, 137–154. https://doi.org/10.1016/S1146-609X(02)01146-3.

Sastroutomo, S.S., 1981. Turion formation, dormancy and germination of curly pondweed, *Potamogeton crispus* L. Aquat. Bot. 10, 161–173. https://doi.org/10.1016/0304-3770(81)90018-8.

Sayer, C.D., Davidson, T.A., Jones, J.I., 2010. Seasonal dynamics of macrophytes and phytoplankton in shallow lakes: a eutrophication-driven pathway from plants to plankton? Freshw. Biol. 55, 500–513. https://doi.org/10.1111/j.1365-2427.2009.02365.x.

Scheffer, M., 1998. Ecology of Shallow Lakes. Kluwer Academic Publishers.

Scheffer, M., Carpenter, S., Foley, J.A., Folke, C., Walker, B., 2001. Catastrophic shifts in ecosystems. Nature 413, 591–596. https://doi.org/10.1038/35098000.

Schneider, S., Melzer, A., 2003. The trophic index of macrophytes (TIM)—a new tool for indicating the trophic state of running waters. Int. Rev. Hydrobiol. 88, 49–67. https://doi.org/10.1002/iroh.200390005.

Schoelynck, J., Bal, K., Verschoren, V., Penning, E., Struyf, E., Bouma, T., Meire, D., Meire, P., Temmerman, S., 2014. Different morphology of

Nuphar lutea in two contrasting aquatic environments and its effect on ecosystem engineering. Earth Surf. Process. Landf. 39, 2100–2108. https://doi.org/10.1002/esp.3607.

Schubert, H., Sagert, S., Forster, R.M., 2014. Evaluation of the different levels of variability in the underwater light field of a shallow estuary. Helgol. Mar. Res. 55, 12–22. https://doi.org/10.1007/s101520000064.

Schuette, J.L., Klug, M.J., Klomparens, K.L., 1994. Influence of stem lacunar structure on gas transport: relation to the oxygen transport potential of submersed vascular plants. Plant Cell Environ. 17, 355–365. https://doi.org/10.1111/j.1365-3040.1994.tb00304.x.

Schuster, A.-C., Burghardt, M., Riederer, M., 2017. The ecophysiology of leaf cuticular transpiration: are cuticular water permeabilities adapted to ecological conditions? J. Exp. Bot. 68, 5271–5279. https://doi.org/10.1093/jxb/erx321.

Scully, N.M., Lean, D.R.S., McQueen, D.J., Cooper, W.J., 1995. Photochemical formation of hydrogen peroxide in lakes: effects of dissolved organic carbon and ultraviolet radiation. Can. J. Fish. Aquat. Sci. 52, 2675–2681. https://doi.org/10.1139/f95-856.

Sculthorpe, C.D., 1967. The Biology of Aquatic Vascular Plants. Edward Arnold Publishers, London.

Seago, J.L., 2020. Revisiting the occurrence and evidence of endodermis in angiosperm shoots. Flora 273, 151709. https://doi.org/10.1016/j.flora.2020.151709.

Shilla, D.A., Asaeda, T., Kian, S., Lalith, R., Manatunge, J., 2006. Phosphorus concentration in sediment, water and tissues of three submerged macrophytes of Myall Lake, Australia. Wetl. Ecol. Manag. 14, 549–558. https://doi.org/10.1007/s11273-006-9007-5.

Sinclair, J., Williams, T., 2001. Photosynthetic energy storage efficiency, oxygen evolution and chloroplast movement. Photosynth. Res. 70, 197–205. https://doi.org/10.1023/A:1017998517335.

Smolders, A.J.P., Lucassen, E.C.H.E.T., Roelofs, J.G.M., 2002. The isoetid environment: biogeochemistry and threats. Aquat. Bot. 73, 325–350. https://doi.org/10.1016/S0304-3770(02)00029-3.

Smolders, A.J.P., Roelofs, J.G.M., Den Hartog, C., 1996. Possible causes for the decline of the water soldier (*Stratiotes aloides* L.) in The Netherlands. Arch. Hydrobiol. 327–342. https://doi.org/10.1127/archiv-hydrobiol/136/1996/327.

Soana, E., Bartoli, M., 2013. Seasonal variation of radial oxygen loss in *Vallisneria spiralis* L.: an adaptive response to sediment redox? Aquat. Bot. 104, 228–232. https://doi.org/10.1016/j.aquabot.2012.07.007.

Sobek, S., Tranvik, L.J., Prairie, Y.T., Kortelainen, P., Cole, J.J., 2007. Patterns and regulation of dissolved organic carbon: an analysis of 7,500 widely distributed lakes. Limnol. Oceanogr. 52, 1208–1219. https://doi.org/10.4319/lo.2007.52.3.1208.

Solomon, K.R., Baker, D.B., Richards, R.P., Dixon, K.R., Klaine, S.J., La Point, T.W., Kendall, R.J., Weisskopf, C.P., Giddings, J.M., Giesy, J.P., Hall, L.W., Williams, W.M., 1996. Ecological risk assessment of atrazine in North American surface waters. Environ. Toxicol. Chem. 15, 31–76. https://doi.org/10.1002/etc.5620150105.

Søndergaard, M., 1979. Light and dark respiration and the effect of the lacunal system on refixation of CO_2 in submerged aquatic plants. Aquat. Bot. 6, 269–283. https://doi.org/10.1016/0304-3770(79)90065-2.

Søndergaard, M., 1988. Seasonal variations in the loosely sorbed phosphorus fraction of the sediment of a shallow and hypereutrophic lake. Environ. Geol. Water Sci. 11, 115–121. https://doi.org/10.1007/BF02587770.

Søndergaard, M., Phillips, G., Hellsten, S., Kolada, A., Ecke, F., Mäemets, H., Mjelde, M., Azzella, M.M., Oggioni, A., 2013. Maximum growing depth of submerged macrophytes in European lakes. Hydrobiologia 704, 165–177. https://doi.org/10.1007/s10750-012-1389-1.

Søndergaard, M., Sand-Jensen, K., 1979. Carbon uptake by leaves and roots of *Littorella uniflora* (L.) Aschers. Aquat. Bot. 6, 1–12. https://doi.org/10.1016/0304-3770(79)90047-0.

Spence, D.H.N., 1964. The macrophytic vegetation of lochs, swamps and associated fens. In: The Vegetation of Scotland. Oliver & Boyd, Edinburgh, pp. 306–425.

Spence, D., 1982. The zonation of plants in freshwater lakes. Adv. Ecol. Res. 12, 37–125. https://doi.org/10.1016/S0065-2504(08)60077-X.

Spence, D.H.N., Chrystal, J., 1970a. Photosynthesis and zonation of freshwater macrophytes. 1. Depth distribution and shade tolerance. New Phytol. 69, 205–215. https://doi.org/10.1111/j.1469-8137.1970.tb04064.x.

Spence, D.H.N., Chrystal, J., 1970b. Photosynthesis and zonation of fresh-water macrophytes. 2. Adaptability of species of deep and shallow water. New Phytol. 69, 217. https://doi.org/10.1111/j.1469-8137.1970.tb04065.x, 217.

Spence, D.H.N., Maberly, S.C., 1985. Occurrence and ecological importance of HCO_3^- use among aquatic higher plants. In: Lucas, W.J., Berry, J.A. (Eds.), Inorganic Carbon Uptake by Aquatic Photosynthetic Organisms. American Society of Plant Physiologists, Rockville, Maryland, pp. 125–143.

Spierenburg, P., Lucassen, E.C.H.E.T., Lotter, A.F., Roelofs, J.G.M., 2009. Could rising aquatic carbon dioxide concentrations favour the invasion of elodeids in isoetid-dominated softwater lakes? Freshw. Biol. 54, 1819–1831. https://doi.org/10.1111/j.1365-2427.2009.02229.x.

Stansfield, J.H., Perrow, M.R., Tench, L.D., Jowitt, A.J.D., Taylor, A.A.L., 1997. Submerged macrophytes as refuges for grazing Cladocera against fish predation: observations on seasonal changes in relation to macrophyte cover and predation pressure. Hydrobiologia 342–343, 229–240. https://doi.org/10.1007/978-94-011-5648-6_25.

Stocking, C.R., 1956. Guttation and bleeding. In: Adriani, M.J., Aslyng, H.C., Burström, H., Geiger, R., Gessner, F., Härtel, O., Huber, B., Hülsbruch, M., Kalle, K., Kern, H., Killian, C., Kisser, J.G., Kramer, P.J., Lemée, G., Levitt, J., Meyer, B.S., Mothes, K., Pisek, A., Ruttner, F., Stålfelt, M.G., Stiles, W., Stocker, O., Stocking, C.R., Straka, H., Thornthwaite, W.C., Troll, C., Ullrich, H., Veihmeyer, F.J. (Eds.), Pflanze und Wasser / Water Relations of Plants, Handbuch der Pflanzenphysiologie / Encyclopedia of Plant Physiology. Springer, Berlin, Heidelberg, pp. 489–502.

Stoddard, J.L., Jeffries, D.S., Lukewille, A., Clair, T.A., Dillon, P.J., Driscoll, C.T., Forsius, M., Johannessen, M., Kahl, J.S., Kellogg, J.H., Kemp, A., Mannio, J., Monteith, D.T., Murdoch, P.S., Patrick, S., Rebsdorf, A., Skjelkvale, B.L., Stainton, M.P., Traaen, T., van Dam, H., Webster, K.E., Wieting, J., Wilander, A., 1999. Regional trends in aquatic recovery from acidification in North America and Europe. Nature 401, 575–578. https://doi.org/10.1038/44114.

Su, H., Chen, J., Wu, Y., Chen, J., Guo, X., Yan, Z., Tian, D., Fang, J., Xie, P., 2019. Morphological traits of submerged macrophytes reveal specific positive feedbacks to water clarity in freshwater ecosystems. Sci. Total Environ. 684, 578–586. https://doi.org/10.1016/j.scitotenv.2019.05.267.

Svedang, M., 1990. The growth dynamics of *Juncus bulbosus* L.—a strategy to avoid competition. Aquat. Bot. 37, 123–138.

Svedang, M., 1992. Carbon-dioxide as a factor regulating the growth dynamics of *Juncus bulbosus*. Aquat. Bot. 42, 231–240.

The Angiosperm Phylogeny Group, 2016. An update of the Angiosperm Phylogeny Group classification for the orders and families of flowering plants: APG IV. Bot. J. Linn. Soc. 181, 1–20. https://doi.org/10.1111/boj.12385.

Thiebaut, G., Muller, S., 2003. Linking phosphorus pools of water, sediment and macrophytes in running waters. Annales De Limnologie - International Journal of Limnology 39 (4), 307–316. https://doi.org/10.1051/limn/2003025.

Thomaz, S.M., Dibble, E.D., Evangelista, L.R., Higuti, J., Bini, L.M., 2008. Influence of aquatic macrophyte habitat complexity on invertebrate abundance and richness in tropical lagoons. Freshw. Biol. 53 (2), 358–367. https://doi.org/10.1111/j.1365-2427.2007.01898.x.

Thomaz, S.M., Esteves, F.A., Murphy, K.J., dos Santos, A.M., Caliman, A., Guariento, R.D., 2008. Aquatic Macrophytes in the Tropics: Ecology of Populations and Communities, Impacts of Invasions and Use by Man. Encyclopedia of Life System Support. UNESCO.

Thum, R.A., Chorak, G.M., Newman, R.M., Eltawely, J.A., Latimore, J., Elgin, E., Parks, S., 2020. Genetic diversity and differentiation in populations of invasive Eurasian (*Myriophyllum spicatum*) and hybrid (*Myriophyllum spicatum* × *Myriophyllum sibiricum*) watermilfoil. Invasive Plant Sci. Manag. 13, 59–67. https://doi.org/10.1017/inp.2020.12.

Titus, J., Adams, M., 1979. Coexistence and the comparative light relations of the submersed macrophytes *Myriophyllum spicatum* L. and *Vallisneria americana* Michx. Oecologia 40, 273–286. https://doi.org/10.1007/BF00345324.

USGS, L.B.T., 2013. Integrated taxonomic information system (ITIS). https://doi.org/10.5066/F7KH0KBK.

van Donk, E., van de Bund, W.J., 2002. Impact of submerged macrophytes including charophytes on phyto- and zooplankton communities: allelopathy versus other mechanisms. Aquat. Bot. 72, 261–274. https://doi.org/10.1016/s0304-3770(01)00205-4.

Van Koetsem, F., Xiao, Y., Luo, Z., Du Laing, G., 2016. Impact of water composition on association of Ag and CeO_2 nanoparticles with aquatic macrophyte *Elodea canadensis*. Environ. Sci. Pollut. Res. 23, 5277–5287. https://doi.org/10.1007/s11356-015-5708-8.

Vári, Á., 2013. Colonisation by fragments in six common aquatic macrophyte species. Fundam. Appl. Limnol. Arch. Für Hydrobiol. 183, 15–26. https://doi.org/10.1127/1863-9135/2013/0328.

Verberk, W.C.E.P., Bilton, D.T., Calosi, P., Spicer, J.I., 2011. Oxygen supply in aquatic ectotherms: partial pressure and solubility together explain biodiversity and size patterns. Ecology 92, 1565–1572.

Verboven, P., Pedersen, O., Ho, Q.T., Nicolai, B.M., Colmer, T.D., 2014. The mechanism of improved aeration due to gas films on leaves of submerged rice. Plant Cell Environ. 37, 2433–2452. https://doi.org/10.1111/pce.12300.

Vestergaard, O., Sand-Jensen, K., 2000. Alkalinity and trophic state regulate aquatic plant distribution in Danish lakes. Aquat. Bot. 67, 85–107. https://doi.org/10.1016/s0304-3770(00)00086-3.

Voesenek, L.A.C.J., Bailey-Serres, J., 2015. Flood adaptive traits and processes: an overview. New Phytol. 206, 57–73. https://doi.org/10.1111/nph.13209.

Vogelman, T.C., Nishio, J.N., Smith, W.K., 1996. Leaves and light capture: light propagation and gradients of carbon fixation within leaves. Trends Plant Sci. 1, 65–70. https://doi.org/10.1016/S1360-1385(96)80031-8.

von Caemmerer, S., Quick, W.P., Furbank, R.T., 2012. The development of C_4 rice: current progress and future challenges. Science 336, 1671–1672. https://doi.org/10.1126/science.1220177.

Wang, H., Wang, Q., Bowler, P., Xiong, W., 2016. Invasive aquatic plants in China. Aquat. Invasions 11, 1–9. https://doi.org/10.3391/ai.2016.11.1.01.

Watson, D., 1987. Hydraulic effects of aquatic weeds in U.K. rivers. Regul. Rivers Res. Manag. 1, 211–227. https://doi.org/10.1002/rrr.3450010303.

Waughman, G.J., 1980. Chemical aspects of the ecology of some south German peatlands. J. Ecol. 68, 1025–1046.

Westlake, D.F., 1963. Comparisons of plant productivity. Biol. Rev. 38, 385–425. https://doi.org/10.1111/j.1469-185X.1963.tb00788.x.

Westlake, D.F., 1965. Some basic data for investigations of the productivity of aquatic macrophytes. Mem Ist Ital Idrobiol 18 (Suppl. 1), 229–248.

Wetzel, R.G., 1969. Factors influencing photosynthesis and excretion of dissolved organic matter by aquatic macrophytes in hard-water lakes: with 8 figures and 10 tables in the text. SIL Proc. 1922–2010 (17), 72–85. https://doi.org/10.1080/03680770.1968.11895828.

Wetzel, R.G., Søndergaard, M., 1998. Role of submerged macrophytes for the microbial community and dynamics of dissolved organic carbon in aquatic ecosystems. In: Jeppesen, E., Søndergaard, M., Søndergaard, M., Christoffersen, K. (Eds.), The Structuring Role of Submerged Macrophytes in Lakes, Ecological Studies. Springer New York, New York, NY, pp. 133–148. https://doi.org/10.1007/978-1-4612-0695-8_7.

White, A., Reiskind, J.B., Bowes, G., 1996. Dissolved inorganic carbon influences the photosynthetic responses of *Hydrilla* to photoinhibitory conditions. Aquat. Bot. 53, 3–13. https://doi.org/10.1016/0304-3770(95)01008-4.

Wickett, N.J., Mirarab, S., Nguyen, N., Warnow, T., Carpenter, E., Matasci, N., Ayyampalayam, S., Barker, M.S., Burleigh, J.G., Gitzendanner, M.A., Ruhfel, B.R., Wafula, E., Der, J.P., Graham, S.W., Mathews, S., Melkonian, M., Soltis, D.E., Soltis, P.S., Miles, N.W., Rothfels, C.J., Pokorny, L., Shaw, A.J., DeGironimo, L., Stevenson, D.W., Surek, B., Villarreal, J.C., Roure, B., Philippe, H., dePamphilis, C.W., Chen, T., Deyholos, M.K., Baucom, R.S., Kutchan, T.M., Augustin, M.M., Wang, J., Zhang, Y., Tian, Z., Yan, Z., Wu, X., Sun, X., Wong, G.K.-S., Leebens-Mack, J., 2014. Phylotranscriptomic analysis of the origin and early diversification of land plants. Proc. Natl. Acad. Sci. U. S. A 111, E4859–E4868. https://doi.org/10.1073/pnas.1323926111.

Wiegleb, G., 1991. Die Lebens- und Wuchsformen der makrophytischen Wasserpflanzen und deren Beziehungen zu Ökologie, Verbreitung und Vergesellschaftung der Arten. Tuexenia 11, 135–148.

Willby, N.J., Abernethy, V.J., Demars, B.O.L., 2000. Attribute-based classification of European hydrophytes and its relationship to habitat utilization. Freshw. Biol. 43, 43–74. https://doi.org/10.1046/j.1365-2427.2000.00523.x.

Willby, N.J., Eaton, J.W., 1996. Backwater habitats and their role in nature conservation on navigable waterways. Hydrobiologia 340, 333–338. https://doi.org/10.1007/BF00012777.

Willby, N., Pitt, J.A., Phillips, G., 2009. The Ecological Classification of UK Lakes Using Aquatic Macrophytes.

Winfield, I.J., 1986. The influence of simulated aquatic macrophytes on the zooplankton consumption rate of juvenile roach, *Rutilus rutilus*, rudd, *Scardinius erythrophthalmus*, and perch, *Perca fluviatilis*. J. Fish. Biol. 29, 37–48. https://doi.org/10.1111/j.1095-8649.1986.tb04997.x.

Winkel, A., Borum, J., 2009. Use of sediment CO_2 by submersed rooted plants. Ann. Bot 103, 1015–1023. https://doi.org/10.1093/aob/mcp036.

Wium-Andersen, S., 1971. Photosynthetic uptake of free CO_2 by roots of *Lobelia dortmanna*. Physiol. Plant 25, 245–248. https://doi.org/10.1111/j.1399-3054.1971.tb01436.x.

Wolters, J.-W., Reitsema, R.E., Verdonschot, R.C.M., Schoelynck, J., Verdonschot, P.F.M., Meire, P., 2019. Macrophyte-specific effects on epiphyton quality and quantity and resulting effects on grazing macroinvertebrates. Freshw. Biol. 64, 1131–1142. https://doi.org/10.1111/fwb.13290.

Wood, K.A., O'Hare, M.T., McDonald, C., Searle, K.R., Daunt, F., Stillman, R.A., 2017. Herbivore regulation of plant abundance in aquatic ecosystems. Biol. Rev. 92, 1128–1141. https://doi.org/10.1111/brv.12272.

Woolway, R.I., Jennings, E., Shatwell, T., Golub, M., Pierson, D.C., Maberly, S.C., 2021. Lake heatwaves under climate change. Nature 589. https://doi.org/10.1038/s41586-020-03119-1, 402-+.

Woolway, R.I., Maberly, S.C., 2020. Climate velocity in inland standing waters. Nat. Clim. Change 10, 1124–1129. https://doi.org/10.1038/s41558-020-0889-7.

Wüstenberg, A., Pörs, Y., Ehwald, R., 2011. Culturing of stoneworts and submersed angiosperms with phosphate uptake exclusively from an artificial sediment. Freshw. Biol. 56, 1531–1539. https://doi.org/10.1111/j.1365-2427.2011.02591.x.

Xian, L., Zhang, Y., Cao, Y., Wan, T., Gong, Y., Dai, C., Ochieng, W.A., Nasimiyu, A.T., Li, W., Liu, F., 2020. Glutamate dehydrogenase plays an important role in ammonium detoxification by submerged macrophytes. Sci. Total Environ. 722, 137859. https://doi.org/10.1016/j.scitotenv.2020.137859.

Xiong, B.H., Li, W., 2002. Winter buds of *Vallisneria* in Banghu and Zhonghuchi, two lakes in Poyang Lake nature sanctuary. Acta Hydrobiol. Sin. 26, 19–24.

Yeats, T.H., Rose, J.K.C., 2013. The formation and function of plant cuticles. Plant Physiol 163, 5–20. https://doi.org/10.1104/pp.113.222737.

Yvon-Durocher, G., Jones, J.I., Trimmer, M., Woodward, G., Montoya, J.M., 2010. Warming alters the metabolic balance of ecosystems. Philos. Trans. R. Soc. B-Biol. Sci. 365, 2117–2126. https://doi.org/10.1098/rstb.2010.0038.

Zeng, J., Bian, Y., Xing, P., Wu, Q.L., 2012. Macrophyte species drive the variation of bacterioplankton community composition in a shallow freshwater lake. Appl. Environ. Microbiol. 78, 177–184. https://doi.org/10.1128/AEM.05117-11.

Zhang, X., Gituru, R.W., Yang, C., Guo, Y., 2010. Exposure to water increased pollen longevity of pondweed (*Potamogeton* spp.) indicates different mechanisms ensuring pollination success of angiosperms in aquatic habitat. Evol. Ecol. 24, 939–953. https://doi.org/10.1007/s10682-010-9351-z.

Zhang, Y., Yin, L., Jiang, H.-S., Li, W., Gontero, B., Maberly, S.C., 2014. Biochemical and biophysical CO_2 concentrating mechanisms in two species of freshwater macrophyte within the genus *Ottelia* (Hydrocharitaceae). Photosynth. Res. 121, 285–297. https://doi.org/10.1007/s11120-013-9950-y.

Zhou, J., Cao, Y., Liu, X., Jiang, H., Li, W., 2020. Bladder entrance of microplastic likely induces toxic effects in carnivorous macrophyte *Utricularia aurea* Lour. Environ. Sci. Pollut. Res. 27, 32124–32131. https://doi.org/10.1007/s11356-020-09529-y.

Zhu, G., Yuan, C., Di, G., Zhang, M., Ni, L., Cao, T., Fang, R., Wu, G., 2018. Morphological and biomechanical response to eutrophication and hydrodynamic stresses. Sci. Total Environ. 622–623, 421–435. https://doi.org/10.1016/j.scitotenv.2017.11.322.

Zurzycki, J., 1955. Chloroplast arrangement as a factor in photosynthesis. Acta Soc. Bot. Pol. 24, 27–63.

CHAPTER

25

Benthic Algae and Cyanobacteria of the Littoral Zone

Yvonne Vadeboncoeur[1] *and Rex Lowe*[2]

[1]Department of Biological Sciences, Wright State University, Dayton, Ohio United States

[2]Department of Biological Sciences, Bowling Green State University, Bowling Green, Ohio, United States; Center for Limnology, University of Wisconsin, Madison, Wisconsin United States

OUTLINE

The bottoms of aquatic ecosystems are structurally complex habitats comprised of inorganic sediments (ranging in size from microscopic clay particles to expanses of exposed bedrock), dead organic material (from fine detritus to entire trees), and living organisms (from minute mollusk shells to dense macrophyte forests). These varied surfaces are living space for *microbial biofilms*—structured aggregations of diverse microorganisms that exchange metabolites within a collective, extracellular matrix (ECM) of slime. The biofilms on any submerged surface that is exposed to light will include, and often be dominated by, photosynthetic organisms that we gather collectively under the imperfect linguistic canopy of "benthic algae." The most abundant benthic algae are Cyanobacteria, diatoms, and green algae (Chapters 17 and 18). *Periphyton* and *aufwuchs* are more inclusive terms for the organic and living components, both heterotrophic and autotrophic, of aquatic biofilms. This chapter focuses on benthic littoral algae in lakes.

I. Littoral benthic algae and Cyanobacteria

A. Littoral habitats

Algal communities reflect the complexity of the littoral zones in which they occur, and littoral zone structure is determined by regional geology, terrestrial-aquatic linkages, and lake size (Lowe, 1996; Cantonati and Lowe, 2014; Fig. 25-1). The chemistry and structure of illuminated submerged surfaces, the ultimate

Wetzel's Limnology, Fourth Edition
https://doi.org/10.1016/B978-0-12-822701-5.00025-2

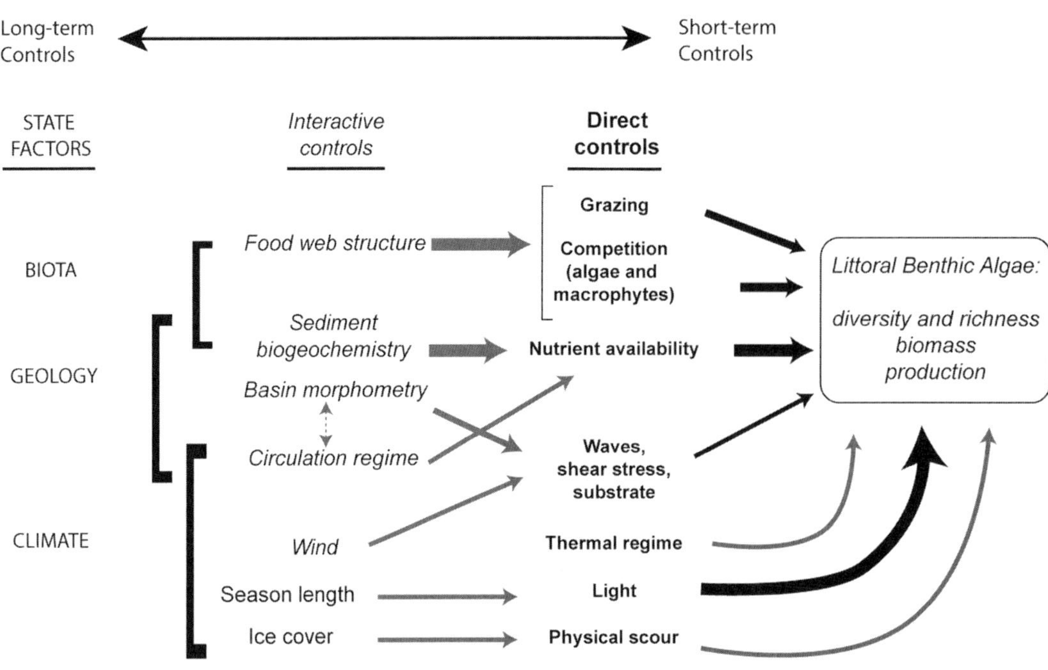

FIGURE 25-1 The taxonomic composition, biomass, and productivity of littoral benthic algae are affected by long- and short-term controls. The biota, geology, and climate of a region determine the algal species pool available and the organisms that will compete with and prey upon the algae. Regional geology and history affect the size and shape of the lake and the dominant geologic parent material. Lake morphometry, combined with terrestrial vegetation and climate, determines substratum composition, nutrient dynamics, water clarity, and maximum wave height. Climate affects season length and circulation regime, which in turn affect the temporal variation in benthic algal dynamics. *(Modified from Vadeboncoeur et al., 2021.)*

requirement for the occurrence of benthic algae in aquatic ecosystems, strongly determines which algae are most likely to persist in that space. The availability of surfaces varies by region and their distribution within lakes changes with depth. For example, rocks and dead trees may dominate in the upper littoral zones of Canadian Shield lakes, while flocculent organic detritus and living bog vegetation may be the most abundant surfaces for algal growth at the edge of prairie pothole lakes or subarctic lakes.

Benthic algae are categorized based on the surfaces that they grow upon (see also Chapter 17). *Epipelic algae*, or *epipelon*, grow on soft, unconsolidated, organic sediments that dominate the bottoms of most lakes (Figs. 25-2b and 25-3a–c), while *epipsammic algae* (i.e., *epipsammon*) grow on inorganic sand. Unconsolidated, often nutrient-rich, sediments are not physically stable. Motile epipelic algae rapidly respond to sediment deposition or suspension by moving back up to the sediment–water interface (SWI). Conversely, cobbles, boulders, and bedrock provide a relatively physically stable, but nutritionally inert, growth surface for *epilithic algae*, or *epilithon* (Fig. 25-2a and Figs. 25-3d, e). Algae growing on living organisms, including submerged and floating vegetation (*epiphytic algae* or *epiphyton*; Figs. 25-3b, c, f) and animal shells and cases (*epizootic algae* or *epizoan*; Figs. 25-3d, f), take advantage of the nutrients that living organisms excrete or leak (Burkholder, 1996; Ings et al., 2010). Dead terrestrial vegetation, especially trees (*epixylic algae* or *epixylon*; Fig. 25-2b) and leaves, are not particularly nutrient rich but support a variety of microbial and fungal decomposers in addition to algae (Tank and Webster, 1998; Vadeboncoeur and Lodge, 2000). Floating clouds of unattached, loosely aggregated algae and cyanobacteria are called *metaphyton* (Fig. 25-2c) and occur in littoral areas of many lakes, ponds, and wetlands and the floodplain areas of rivers.

While regional geology, terrestrial vegetation, and lake biota are long-term controls on the potential substrata for benthic algae (Fig. 25-1), lake size and morphology determine the relative abundance of erosional versus deposition habitat and their distribution with respect to depth. Although we restrict our discussion to lakes, rivers are a mosaic of shallow and deep, erosional and depositional habitats. Wetlands are primarily shallow depositional habitats with abundant aquatic vegetation that provides surfaces for periphyton growth.

i. Erosional shores

The highest light availability occurs at lake edges, which are energetically dynamic habitats subject to waves, currents, and water-level fluctuations. The size and frequency of wind waves scale directly with fetch

FIGURE 25-2 Benthic algal assemblages reflect the physical structure of lake littoral zones. (a) Periphytic algae in the high-energy littoral zone of Lake Huron (Canada/USA; surface area [SA] = 59,656 km^2) form low biomass, thin skins on the rocks. (b) The soft organic sediments and dead trees in the littoral zone of Sparkling Lake (Wisconsin, USA; SA = 0.64 km^2) support thick, diatom-dominated biofilms, and macrophytes. (c) In a small (SA = 150 m^2) pond high nutrients and still water promote the development of metaphyton. *Mougeotia*, a green alga, begins growth attached to the bottom but rises to the surface each day when high photosynthesis rates trap oxygen. (d) Thin cyanobacterial biofilms lie atop the soft sediments in the deep epilimnion (23 m) of Lake Huron. *(Photos: (a) Leon Katona; (b) Shawn Devlin; (c) Yvonne Vadeboncoeur; (d) Biddanda et al. (2012).)*

(Chapter 8). Breaking waves sort lakebed sediments, carrying away fine clays and detritus and leaving behind denser, larger particles such as sand, cobbles, and boulders. The proportion of the shoreline exposed to wind waves, and the size and depth of influence of those waves, scales directly with lake size. Thus the world's largest lakes (e.g., the Laurentian and African Great Lakes, Lake Baikal) have large expanses of sand, cobble, and bedrock in upper littoral habitats (Fig. 25-2a). In contrast, small, protected lakes in forested landscapes may have none of these substrates (Fig. 25-2b). Many algae, especially green algae and Cyanobacteria with strong basal attachments, as well as adnate diatoms, exploit the physically challenging erosional nearshore habitats where light is plentiful.

ii. Depositional shores

Inlets, bays, and entire shorelines of very small, forest lakes are protected from wind and large waves. In relatively flat, lowland landscapes, fine sediments and detritus accumulate throughout the basin, including shallow nearshore habitats (e.g., Fig. 25-2b). Depositional upper littoral habitats are composed of varying mixtures of sand, mud, and organic detritus that provide a nutrient-rich surface for algal growth. Long-lived emergent reeds and cattails along shorelines are a relatively perennial, vertically oriented organic surface for algal growth. Macrophytes also continuously generate detritus, much of which accumulates in the sediments forming the base of a detrital food web (James et al., 2000; Mehner et al., 2016). Historically, submerged trees were a common nearshore substrate in forested landscapes, but humans now often intentionally remove dead trees from lakes (Christensen et al., 1996).

iii. Middle- and deep-littoral habitats

Below the influence of breaking waves, light increasingly limits benthic algal productivity (Vadeboncoeur et al., 2014) and strongly determines assemblage composition (Round, 1964, 1972; Stevenson and Stoermer, 1981; Kingston et al., 1983; Stevenson et al., 1985). With distance from the shore, terrestrial inputs (e.g., trees and leaves, emergent vegetation) have less direct influence on the composition of the lake bottom. Rather, deposited fine sediments and detritus dominate the lake bed, with occasional expanses of bedrock or boulders in steep areas and in areas exposed to currents. The macrophyte community transitions to submerged taxa that provide an excellent, if ephemeral, substrate for benthic algae (Burkholder, 1996). Submerged macrophyte (Chapter 24) beds are a common

FIGURE 25-3 Benthic algae grow on any submerged illuminated surface. (a) Epipelic diatoms impart a golden hue to organic sediments and grow in rounded spires in a shallow river. Oxygen bubbles from photosynthesis become trapped in the biofilm. Fungi and algae colonize a submerged leaf. (b) The motile cyanobacterium, *Oscillatoria*, growing epiphytically on filamentous green algae and extending into an elongated spire. (c) The cyanobacterium *Nostoc* has several morphologies, including the globular clumps shown here. A dying macrophyte is coated with diatoms. (d) A larval caddisfly crawls across a turf of epilithic green algae growing on boulders. (e) Fronds of epilithic green algae in the splash zone of Lake Erie (USA/Canada). (f) Diatoms growing on sand, epiphytically on the macrophyte *Isoetes*, and epizootically on the shell of a mussel *Pyganodon grandis*. *(Photos: (a)–(d), (f) Leon Katona; (e) Rex Lowe.)*

feature of small and shallow lakes (Chapter 26) and occur in protected, low-slope regions in large lakes.

iv. Littoral-profundal habitats

The littoral-profundal zone is a deep transitional zone that often spans the metalimnion and the upper hypolimnion in stratified lakes (Chapter 7). This benthic habitat is rarely studied, although Kingston et al. (1983) described a species-rich diatom community below the thermocline in Grand Traverse Bay, Lake Michigan, USA. However, our observations and rare reports from the literature (Fig. 25-2d; Biddanda et al., 2012) indicate that the sediments of this extremely low light, depositional habitat is often covered by a thin (<2 mm) tissue of motile filamentous Cyanobacteria (*Microcoleus, Phormidium, Oscillatoria, Lyngbya*) interlaced with highly motile, diatoms. Mosses, Charophyceae, green algae, and motile Cyanobacteria/diatoms can be abundant in the low-disturbance, low-temperature, low-light, littoral profundal habitat.

B. Taxonomic composition of attached algal assemblages

Benthic algal assemblages in lakes are dominated by three different divisions (phyla) of algae including green algae (Chlorophyta), diatoms (Bacillariophyta), and blue-green algae or Cyanobacteria (Cyanophyta). A few taxa of red algae (Rhodophyta) and brown algae (Phaeophyta) occur in freshwaters but rarely dominate algal biomass (Chapter 17). Interstitial spaces in beach sands and unconsolidated sediments provide a habitat for several genera of motile flagellated algae. This habitat has not been extensively studied but contains species of *Euglena* and cryptomonad genera.

i. Green algae

Green algae (Chlorophyta) are evolutionarily related to terrestrial plants, with chlorophylls *a* and *b* as dominant pigments. Similar to plants, green algae store accumulated photosynthetic products as starch. The cell wall is constructed of cellulose and may have an additional layer of pectin. Although green algae are palatable to primary consumers, many of the filamentous genera are large and unwieldy and thus difficult for grazers to ingest. However, taxa such as *Cladophora* serve as important hosts to epiphytic diatoms, a preferred food for grazers. Large growths of filamentous green algae greatly increase the functional surface area of the littoral zone, increasing the habitat available for smaller, epiphytic algae and invertebrates.

Green algae in the high-energy littoral zones normally have strong mechanisms of attachment to the substrate. *Ulothrix* (Fig. 25-4a) are among the first epilithic algae to appear in the spring. Filaments have a basal holdfast cell and each filament elongates by cell division, resulting in filaments several centimeters long. *Ulothrix* reproduces episodically by cloning. Each cell in the filament except the holdfast can produce 16 flagellated swimming spores (zoospores) that can attach to substrata in the littoral zone, allowing *Ulothrix* to spread quickly around the margin of a lake, producing a green ring of filaments similar to a "bathtub ring." *Cladophora glomerata* (Figs. 25-3e and 25-4b) is another common filamentous alga in high-energy wave zones of lakes. This species clones episodically, with each cell producing many swimming spores that can colonize adjacent shoreline. *Cladophora glomerata* is a very common component of the littoral algae in the Laurentian Great Lakes that provide a habitat for epiphytic diatoms (Fig. 25-4b).

Stigeoclonium (Fig. 25-4c), *Chaetophora*, and *Coleochaete* (Figs. 25-4d–f) are morphologically adapted to withstand physical disturbance and grazing pressure in the littoral zone and can become locally abundant where scraping grazers are abundant. *Stigeoclonium* are *heterotrichous*, producing two types of filaments: prostrate filaments adhere very tightly to rocks, and upright filaments emerge from the prostrate filaments. Upright filaments are readily consumed by grazers but the prostrate filaments resist grazing. *Coleochaete* also has a prostrate thallus with erect filaments, but the erect filaments have a cellulose collar at their base that protects them from grazing to some extent (Fig. 25-4f). *Chaetophora,* a third genus with two filament types, is often covered in a firm mucilaginous sheath that is resistant to grazers.

Free, unattached filamentous green algae can become abundant in protected habitats without intense wave action. Green algae in the family Zygnemataceae, including *Spirogyra* (Fig. 25-4g), *Mougeotia* (Fig. 25-4h), and *Zygnema* (Fig. 25-4i), are often very abundant in quiescent habitats. Species in these genera all have an external layer of pectin over the cellulose cell wall and, unlike *Cladophora*, feel very slippery. The pectin on the cell wall discourages colonization by epiphytic algae; thus the filaments retain a verdant green color. These metaphytic taxa trap oxygen bubbles in the mass of filaments, which may cause the algal mats to become buoyant (Fig. 25-2c). These proliferations are often called "pond silk" or "frog spit."

Areas of reduced wave energy often support diverse assemblages of nonfilamentous green algae, particularly *Chlorococcales*. These algae can become suspended in the water column during periods of wave action but fall on littoral sediments during calm periods. There are many species among several genera in this group of algae including *Ankistrodesmus, Actinastrum, Botryococcus, Desmodesmus, Scenedesmus, Coelastrum, Crucigenia, Dictyosphaerium, Kirchneriella, Nephrocytium, Oocystis, Pediastrum, Quadrigula*, and *Tetraedron* (Figs. 25-5a–n).

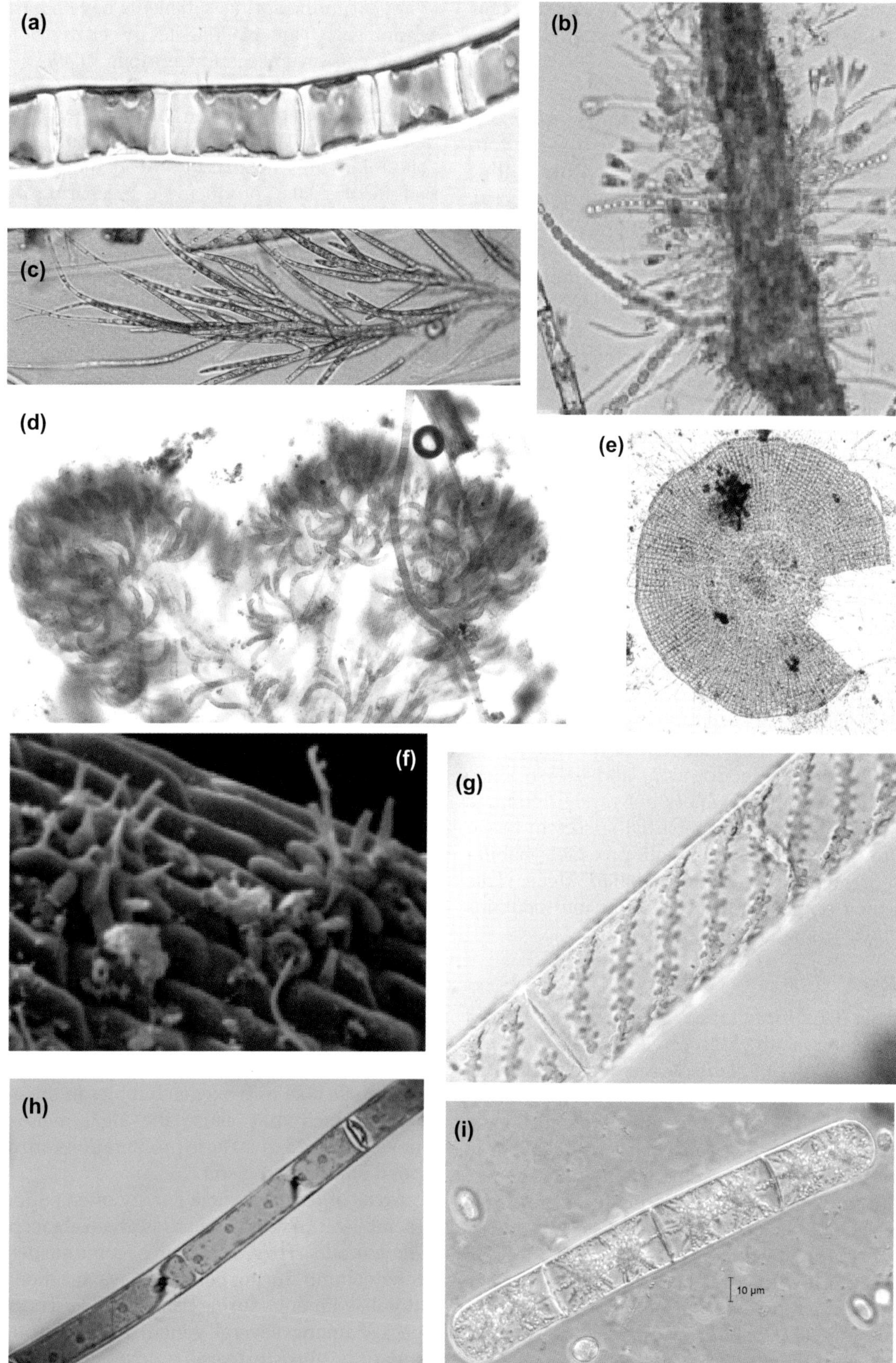

FIGURE 25-4 Benthic filamentous green algae (Chlorophyta) common in the littoral zone: (a) *Ulothrix* sp.; (b) *Cladophora glomerata* covered with epiphytes, primarily diatoms; (c) *Stigeoclonium* sp.; (d) *Chaetophora* sp.; (e) *Coleochaete scutata* filaments forming a disc. This morphology often occurs as epiphytes on vascular plants; (f) Scanning electron micrograph of *Coleochaete* sp. illustrating the prostrate portion of the filament epiphytic on an aquatic vascular plant; (g) *Spirogyra* sp.; (h) *Mougeotia* sp.; (i) *Zygnema* sp. *(Photos by Rex Lowe.)*

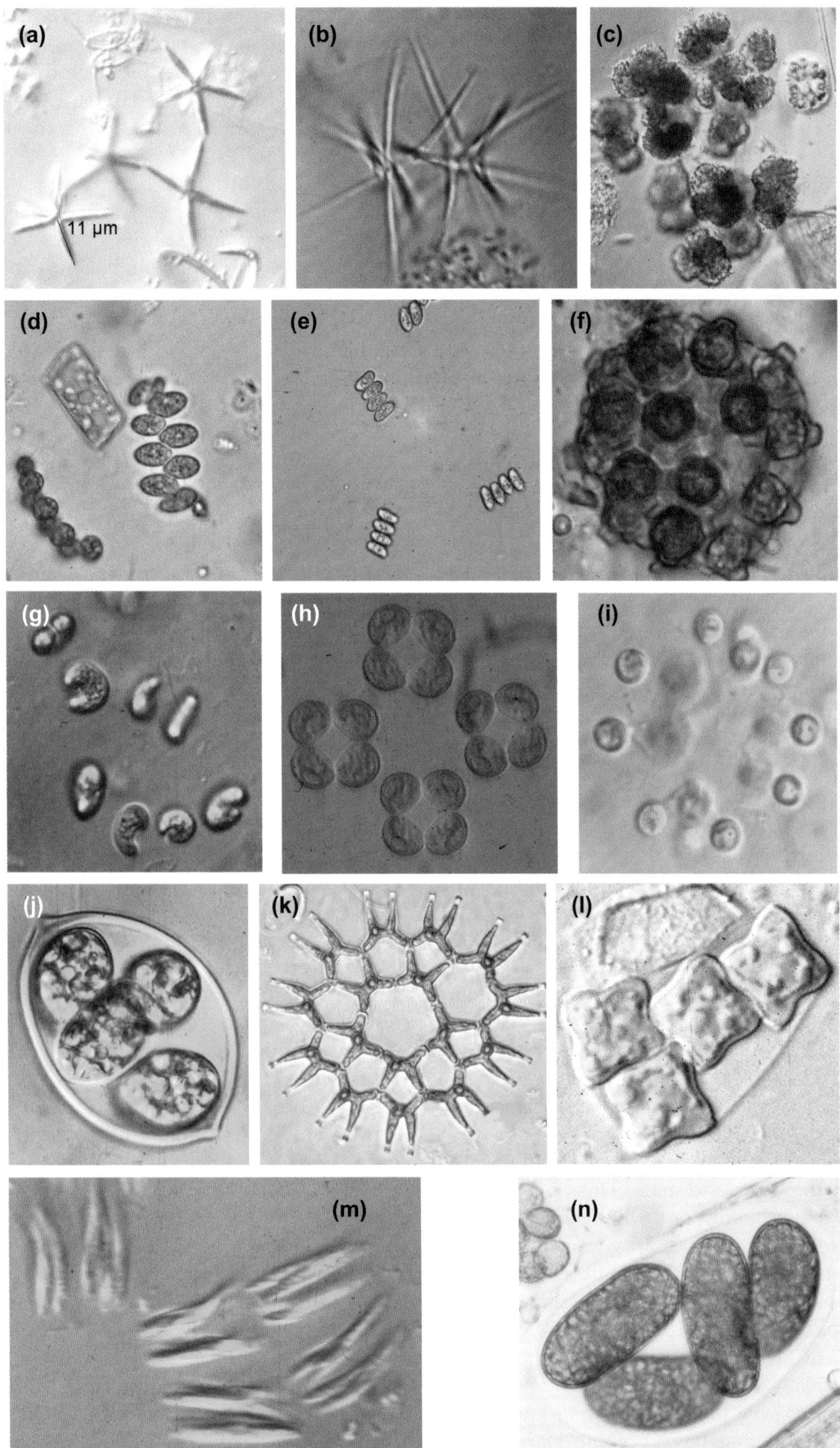

FIGURE 25-5 Colonial green algae of the littoral zone. (a) *Actinastrum* sp.; (b) *Ankistrodesmus* sp.; (c) *Botryococcus* sp.; (d) *Scenedesmus ovalternus*; (e) *Desmodesmus* sp.; (f) *Coelastrum* sp.; (g) *Kirchneriella* sp.; (h) *Crucigenia* sp.; (i) *Dictyosphaerium* sp.; (j) *Oocystis* sp.; (k) *Pediastrum* sp.; (l) *Tetraedron minimum*; (m) *Quadrigula* sp.; (n) *Nephrocytium* sp. *(Photos by Rex Lowe.)*

Additionally, desmids such as *Closterium, Cosmarium, Euastrum,* and *Staurastrum* are common in quiet and low pH littoral habitats, especially areas with organic sediments and in macrophyte beds.

ii. Diatoms

Diatoms (Bacillariophyta; Figs. 25-6 and 25-7) are a ubiquitous component of benthic littoral algal assemblages and often impart a golden hue to sediments (Fig. 25-3a). Diatoms have a rigid silica cell wall, called a *frustule*, composed of two overlapping halves called *valves*. Many benthic diatoms have a slit, called a *raphe*, on one or both valves by which the diatom can attach to or move over surfaces. The dominant photosynthetic pigments in diatoms are chlorophylls *a* and *c*, but these are usually obscured by carotenoid pigments, especially fucoxanthin, that give diatoms a golden-brown color. One of the end products of photosynthesis in diatoms is oil. Diatom oil is rich in omega-3 essential fatty acids, which are critical to the development of many consumers, and has also been targeted by industries developing sources of renewable energy.

Lake littoral zones can hold hundreds of different species of diatoms displaying a variety of ecological preferences. Some diatom species are sessile, prostrate, and tightly attached (Figs. 26-a, c, and d). Other species are loosely attached either as single cells, on the ends of long stalks, or linked together in long filaments. Many species are highly motile with morphologies that allow them to exploit specific habitats. There may also be species that are periodically suspended in the water column. The most common tightly attached prostrate genera of diatoms are *Cocconeis* (Fig. 25-6a) and *Achnanthidium* (Fig. 25-6b). Both of these genera have a raphe on only one valve of the cell. *Cocconeis pediculus* has a flexed mitten-like cell (Fig. 25-6c), and it epiphytizes filamentous green algae like *Cladophora* (Fig. 25-6d). *Cocconeis pediculus* is often dominant where *Cladophora* is abundant, and its tight prostrate form makes it difficult to access by grazers. *Achnanthidium* can be prostrate like *Cocconeis* but sometimes it attaches by a short stalk.

The production of a stalk allows some diatom species to rise above benthic turfs and mats. This increases access to light in a dense biofilm or turf. However, stalked diatoms are more easily consumed by grazers and detached by disturbance than are tightly attached benthic diatoms. The most common genera of stalked diatoms include species of *Gomphonema* (Fig. 25-6e), *Gomphoneis, Cymbella* (Fig. 25-6f), *Rhoicosphenia* (Fig. 25-6g), *Achnanthidium*, and *Didymosphenia* (Fig. 25-6h). Under certain conditions, *Didymosphenia* produces copious quantities of mucilaginous stalks that have become alarming in many lakes and rivers of the world (Fig. 25-6i; Bothwell et al., 2014). *Encyonema* is similar in morphology to *Cymbella*, but *Encyonema* cells do not have an apical pore field, the structure necessary to produce stalks. Instead cells of *Encyonema* cooperatively produce a hollow mucilaginous tube that the cells glide through (Fig. 25-6n), elevating them from the surface of the substrate. The tube may offer them protection from micrograzers, but this is speculative.

Motile benthic diatoms occupy a wide variety of microhabitats. *Navicula* (Fig. 25-6j) and *Nitzschia* (Fig. 25-6k) are common genera, and species display a wide variety of morphologies. In epipelic habitats highly motile diatom genera like *Gyrosigma* can respond to light, moving up to the surface of the sediment after burial (Figs. 25-6o, p, and 25-7b; Greenwood et al., 1999). Species in the genus *Epithemia* are unique because they contain endosymbiotic nitrogen-fixing cyanobacteria (Figs. 25-6l, m). *Epithemia* frequently dominate habitats where the ambient N:P is low, and *Epithemia* can be an important source of organic nitrogen for grazers (Furey et al., 2012). Epipsammic diatoms are abundant in lakes with sandy littoral zones. Experiments show that epipsammic diatoms can liberate silicon from sand grains, which is important when dissolved silicon is low (Carrick and Lowe, 2007). Sand grain mineralogy partly determines the attached diatom flora (Krejci and Lowe, 1986).

iii. Cyanobacteria

Cyanobacteria, or blue-green algae, were the first photosynthetic organisms on Earth and are prokaryotic bacteria bearing chlorophyll *a*. In addition to chlorophyll, Cyanobacteria also have the water-soluble pigment, the phycobilin phycocyanin, imparting their typical blue-green color. A diverse group of Cyanobacteria are found in the littoral zone. Some Cyanobacterial taxa can produce toxins (Wood et al., 2020) and others fix atmospheric nitrogen (Chapters 14, 17, and 18). Nitrogen-fixing Cyanobacteria often have a specialized cell called the *heterocyte* (Figs. 25-8a, c). The heterocyte is a thick-walled cell providing an anaerobic environment where the nitrogen-fixing enzyme nitrogenase can function. The benthic prostrate genus *Dichothrix* (Fig. 25-8b) is common on hard substrates in high-energy wave zones, especially in the Great Lakes. *Dichothrix* and a similar genus, *Calothrix* (Fig. 25-8c), are both nitrogen fixers with a terminal heterocyte on each filament. *Dichothrix* is prostrate and tightly attached, while *Calothrix* is more loosely attached with erect filaments. Two additional nitrogen-fixing benthic cyanobacteria are common in lakes where N:P is relatively low, *Anabaena* and *Nostoc*. *Anabaena* (Fig. 25-8a) forms solitary or loosely aggregated filamentous arrangements, while *Nostoc* (Fig. 25-3c) filaments are contained in a tough mucilaginous sheath. *Nostoc* sheathed colonies are often visible to the naked eye ranging in size from a few millimeters to several centimeters wide. *Nostoc pruniforme* can produce very large colonies called "mare's eggs."

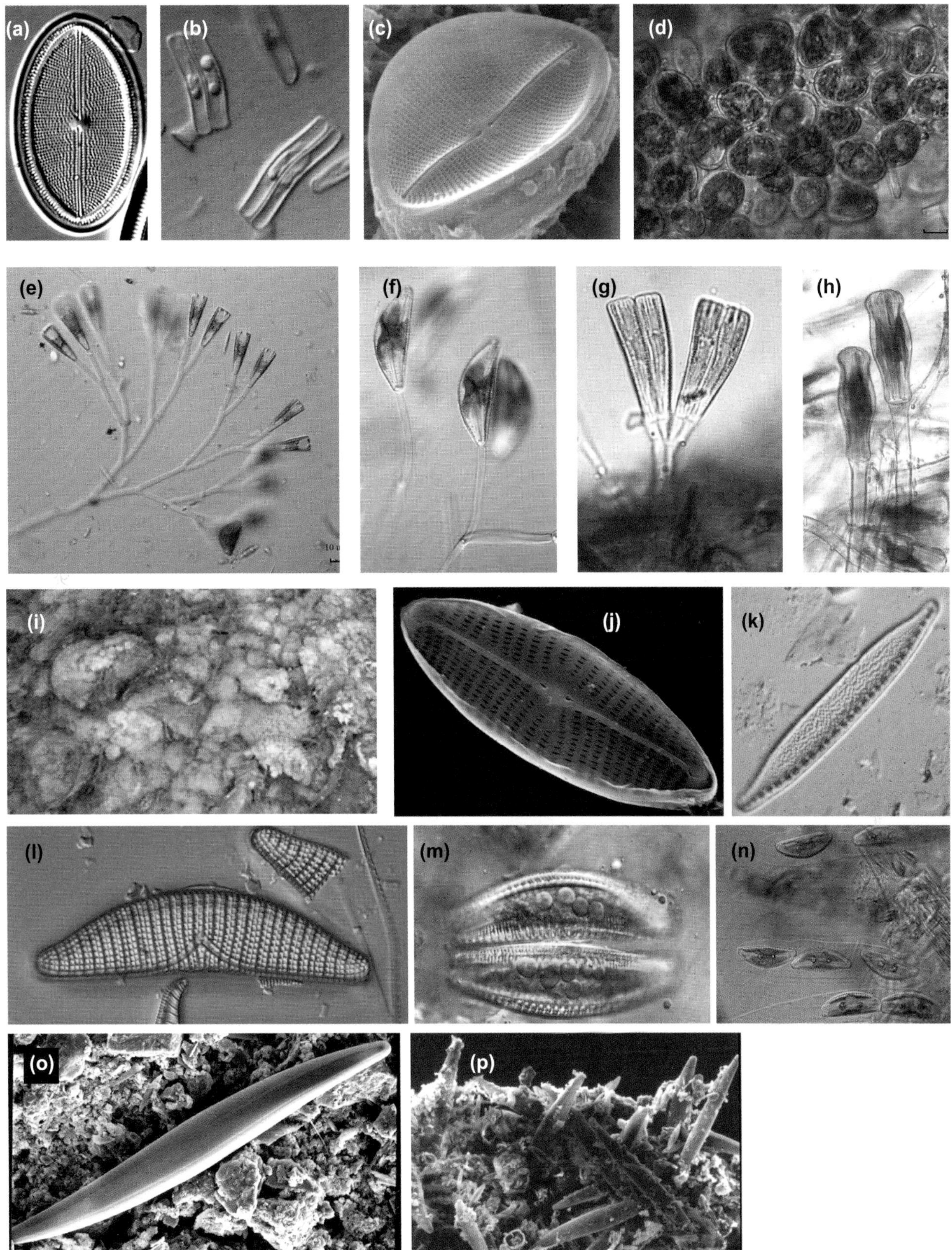

FIGURE 25-6 Benthic diatoms common in littoral habitats: (a) *Cocconeis placentula*; (b) *Achnanthidium minutissimum*; (c) scanning electron micrograph (SEM) of *Cocconeis pediculus*; (d) *Cocconeis pediculus epiphytic on Cladophora*; (e) *Gomphonema* sp. on mucilaginous stalks; (f) *Cymbella* sp. on stalks; (g) stalked *Rhoicosphenia* sp.; (h) *Didymosphenia* sp. on stalks; (i) massive growth of *Didymosphenia*; (j) SEM of *Navicula* sp.; (k) *Nitzschia* sp.; (l) *Epithemia* sp.; (m) *Epithemia* displaying internal endosymbiotic Cyanobacteria; (n) *Encyonema* in gelatinous tubes; (o) *Gyrosigma* sp. SEM; (p) SEM of *Gyrosigma* emerging from sediment. *(Photos by Rex Lowe.)*

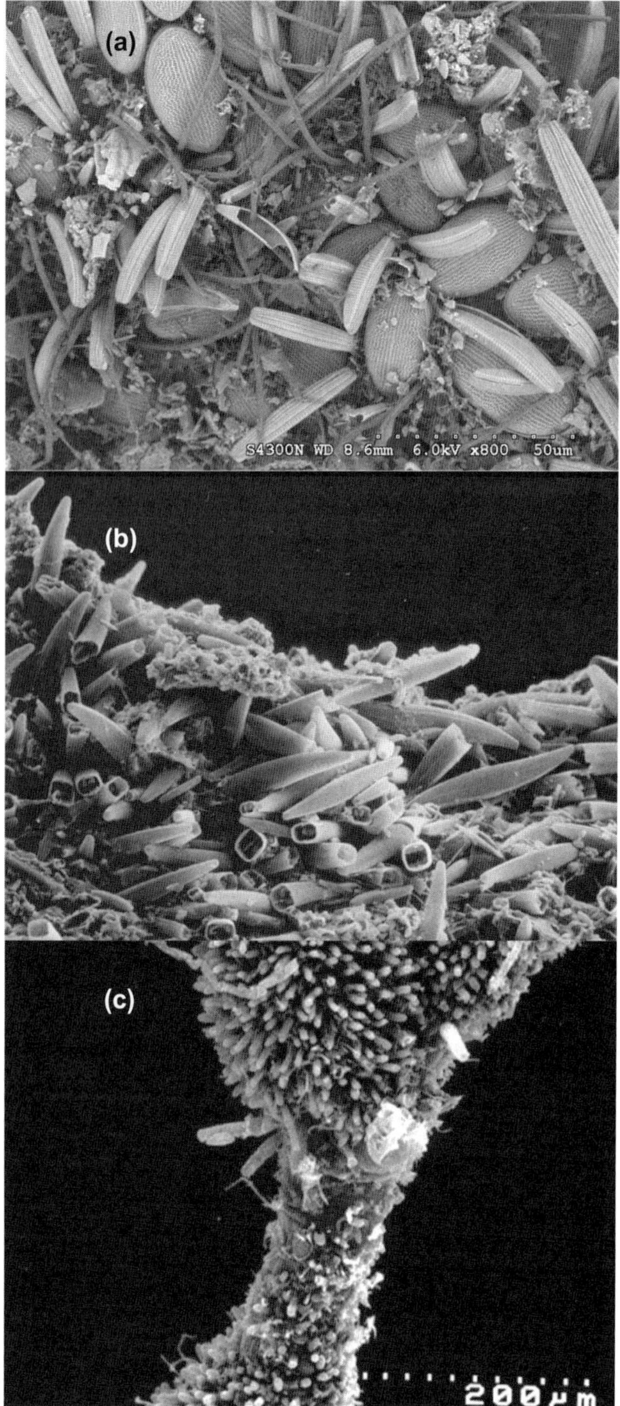

FIGURE 25-7 Scanning electron micrographs (SEMs) illustrating the structure of algal biofilms. (a) A diatom assemblage growing epiphytically on aquatic moss in the Great Smoky Mountains (USA). *Cocconeis* and *Eunotia* and filamentous Cyanobacteria dominate. (b) A freeze-fracture sample of epipelic diatoms growing on sediments in the Little Miami River, Ohio, USA. (c) The brown algae *Pleurocladia* growing covering a filament of *Cladophora* in Lake Michigan, USA. *(Photos by Rex Lowe.)*

Several benthic cyanobacterial genera lack heterocytes. Among them, *Homeothrix* is tightly attached and forms clusters of small tapering filaments, while *Microseira wollei* is an invasive species in North America that forms thick, extensive mats on muddy lake bottoms (Figs. 25-8d, e). Storms frequently dislodge these proliferations and wave action rolls sections of the mats into small spherical floating colonies. Interstitial habitats often support nonmotile cyanobacteria such as *Chroococcus* (Fig. 25-8f) and *Merismopedia* (Fig. 25-8g).

iv. Red algae

Although primarily restricted to marine habitats, red algae (Rhodophyta) do occur in freshwater ecosystems. *Batrachospermum* (Fig. 25-9a), a complex filamentous alga, is most common in areas of low light and low temperature. During the summer in northern Michigan (USA), *Batrachospermum* is common on hard substrates near the top of the thermocline. It also occurs deep in acidic bogs on hard substrates. *Bangia atropurpurea* (Fig. 25-9b) is an invasive red alga that was first reported from the Laurentian Great Lakes in 1970 (Kishler and Taft, 1970). It probably invaded from coastal areas of the North Atlantic. In the Laurentian Great Lakes *Bangia atropurpurea* grows on rocks above the splash zone, where filaments are periodically wetted by waves (Sheath and Cole, 1980). This periodic drying can increase the concentration of dissolved solids around the alga, perhaps mimicking salt water habitats. A third red algal and nonnative genus, *Chroodactylon* (Fig. 25-9c), epiphytizes *Cladophora* in the Laurentian Great Lakes (Sheath and Morison, 1982).

v. Brown algae

Brown algae (Phaeophyta) are almost entirely marine, occurring as large leathery seaweeds. Nonetheless, some brown algae occur in the littoral zone of freshwater lakes. *Sphacelaria* has been infrequently reported. *Pleurocladia lacustris* (Fig. 25-9d) has recently been reported in the Laurentian Great Lakes associated with *Cladophora* following the invasion of Dreissenid mussels (Fig. 25-7c).

C. Structure and development of benthic algal assemblages

Benthic algal development on bare substrates follows an often-predictable pattern over time (Azim and Asaeda, 2005). New surfaces (e.g., a freshly scoured rock, a newly fallen tree, or an emerging macrophyte leaf) are first colonized by a thin biofilm of bacteria and Archaea that excrete extracellular slime, possibly facilitating benthic algal colonization. The first algal

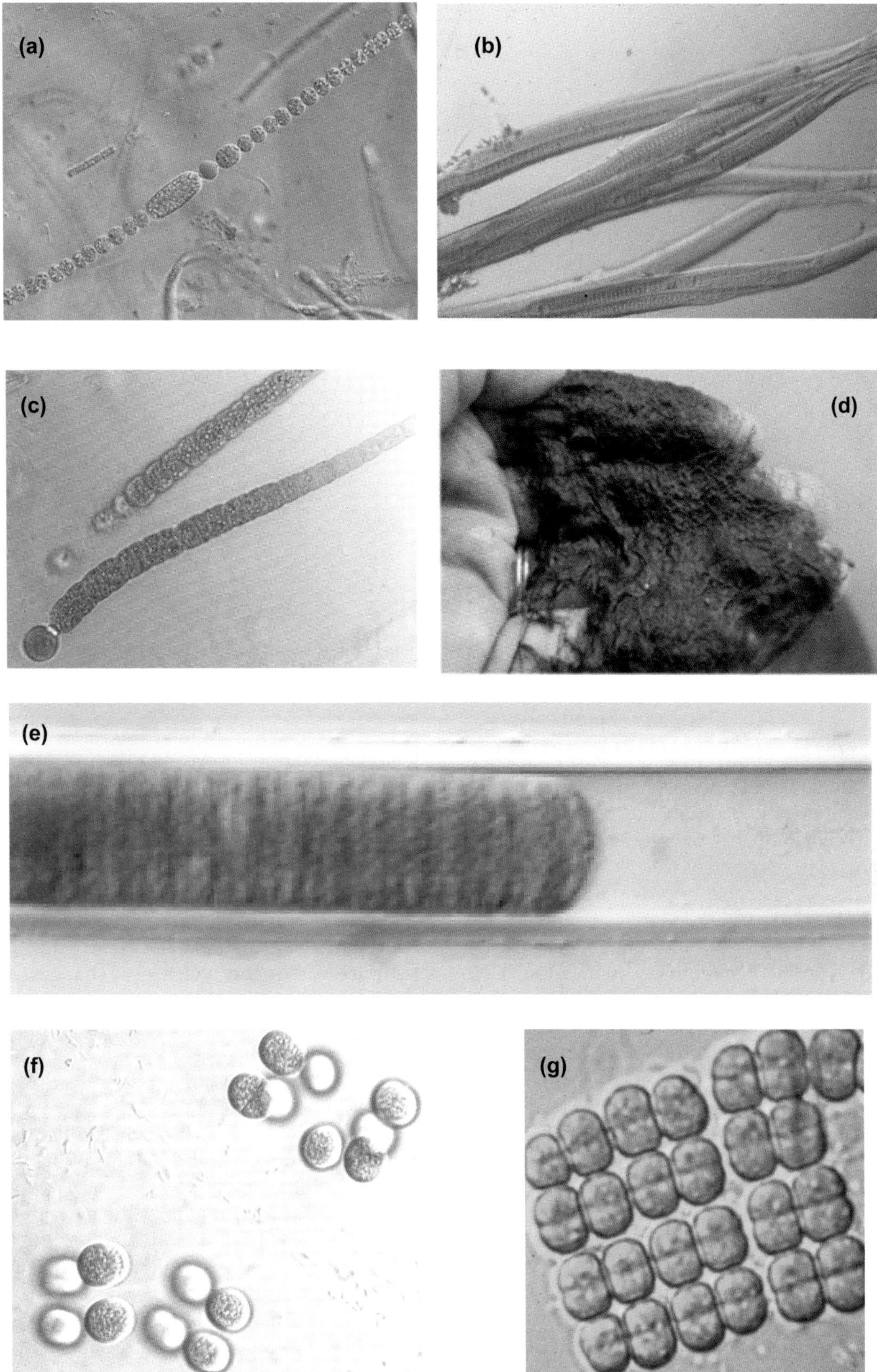

FIGURE 25-8 Littoral-benthic Cyanobacteria: (a) *Anabaena* sp.; (b) *Dichothrix* sp.; (c) *Calothrix* sp.; (d) Macroscopic mat of *Microseira wollei*; (e) *Microseira wollei*; (f) *Chroococcus* sp.; (g) *Merismopedia* sp. *(Photos by Rex Lowe.)*

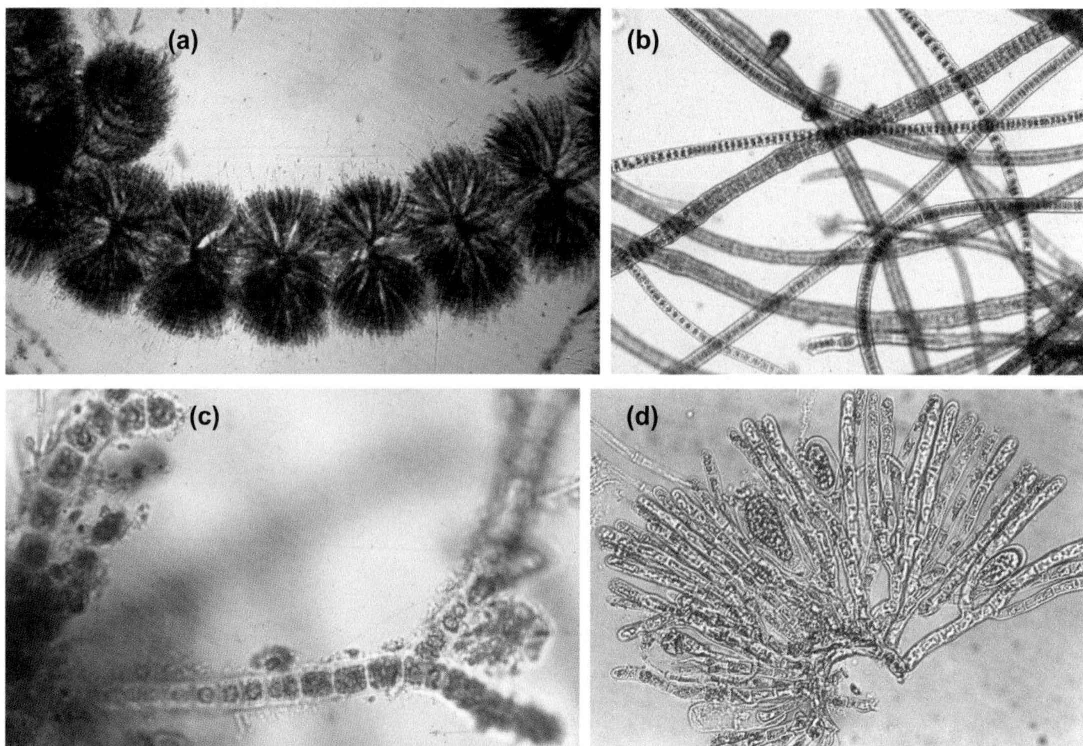

FIGURE 25-9 Red algae (a–c) and a brown alga (d) that are sometimes found in the littoral zone. (a) *Batrachospermum* sp.; (b) *Bangia* sp.; (c) *Chroodactylon* sp.; (d) *Pleurocladia lacustris*. *(Photos by Rex Lowe.)*

colonizers are often low-growing, *adnate* (closely appressed to the substrate) diatoms such as *Achnanthidium minutissimum* that thrive in the relatively unobstructed light environment. Subsequently, stalked or rosette-forming diatoms, such as *Gomphonema* and *Synedra*, and attached Cyanobacteria provide vertical structure (Hoagland et al., 1982). Motile diatoms move throughout the developing biofilm and persist during all stages of development. Eventually, filamentous Cyanobacteria, filamentous green algae, and filamentous diatoms create more vertical, three-dimensional structures. These filamentous algae provide new surfaces that light-loving, small, attached diatoms and Cyanobacteria exploit. A progressive competition for light has begun, and this vertical development continues with new algae extending further and further into the water. Chemical and light gradients develop within the biofilm, with cells at the base of the mat exposed to anoxia and low light. Ultimately, stressful conditions experienced by algae attached directly to hard substrates may lead to sloughing from the substrate, opening up the possibility for the succession cycle to begin again.

The plethora of studies on the colonization of new hard substrata likely reflects the overwhelming influence of riverine studies on our understanding of periphyton. In river experiments, researchers often deploy uncolonized artificial substrata to try to standardize benthic algal assemblage composition (Cattaneo and Amireault, 1992). Whereas stream beds are regularly turned over by scouring floods, the availability of completely uncolonized mineral substrata is probably relatively rare in lakes. Even cobbles and sand grains tumbled about in the wave-washed upper littoral will maintain a thin skin of bacteria and adnate diatoms (Fig. 25-10a). The majority of euphotic zone surfaces in lakes are below the influence of surface waves and water-level changes, and biofilms may be more perennial than in streams, with mats that grow or shrink with changing light, temperature, and grazing pressure.

i. Biofilms

The term *biofilm* refers to a consortium of microorganisms, both autotrophic and heterotrophic, attached to a surface. Biofilms are united by an *ECM* that gives the biofilm cohesion and mediates the exchange of metabolites (Wetzel, 1993). Unconsolidated sediments and rocks in lakes are covered by biofilms, but the nature of that biofilm changes across gradients of light and disturbance. In high-disturbance or extremely low-light habitats, these biofilms may form slimy *skins*, less than a millimeter thick, that do not extend beyond individual substrate particles (e.g., a sand grain or cobble; Fig. 25-10a). In sandy habitats diatoms may form thin golden *crusts*, loosely binding sand grains (Forehead et al.,

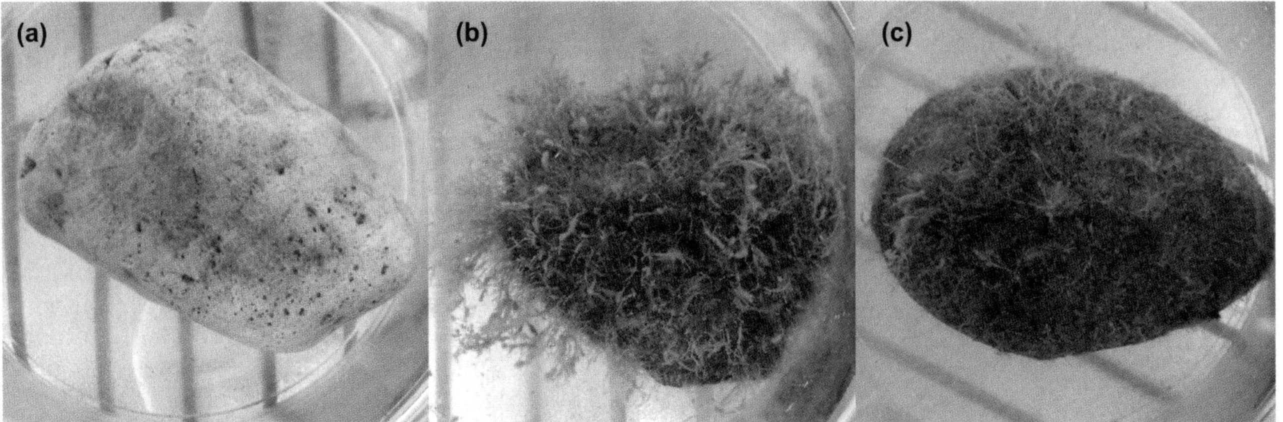

FIGURE 25-10 Wave action and light often determine the benthic algal biomass in the littoral zones of large lakes. Epilithic algae collected in midsummer at one site in the erosional zone of eutrophic western Lake Erie (USA/Canada). (a) At 0.5 m depth, thin skins of Cyanobacteria and diatoms coat the cobble. Wave disturbance prevents high biomass accrual. (b) A dense turf of the green alga, *Cladophora glomerata*, develops at 1.5 m, where wave disturbance is less and light is relatively abundant. (c) At 4 m, light is limiting and the cobbles accumulate less biomass. *(Photos by Leon Katona.)*

2013; Hofmann et al., 2020). In high-light habitats with organic sediments, benthic algae and Cyanobacteria can contribute to a laminated or stratified assemblage of taxa that extends a few centimeters into the sediments. Laminated biofilms have distinct taxonomic layers owing to the rapid changes in light and chemistry with distance from the biofilm surface (Hawes et al., 2014). The ECM exuded by bacteria and algae (Scott et al., 2014) in these relatively thick biofilms can stabilize unconsolidated sediments.

Under very low light conditions (e.g., around the transition zone between the metalimnion and the hypolimnion), stratified biofilms may transition to thin (~1 mm), autotroph-dominated sheets of motile Cyanobacteria and *raphid* diatoms (a *raphe* is a slit in the frustule that allows diatoms to move across surfaces) that surf along the top of watery, unconsolidated sediments (Fig. 25-2d; Biddanda et al., 2012; Snider et al., 2017). These motile assemblages appear to lack the characteristic ECM of biofilms and are unattached to the sediments, but like laminated biofilms, the cells are organized along a horizontal plane and oriented along the lake bottom or migrate through the sediments on a diurnal cycle in search of nutrients.

We have no established names to distinguish among the different forms of biofilm, and we know little of the functional consequences of these biogenic structures for sediment stability and nutrient flow across the SWI. Nor do we know how the physical-chemical characteristics of the sediments determine the taxonomic compositions of the communities that they support.

ii. Turfs

Although all periphyton assemblages begin as a biofilm, competition for light eventually fosters the establishment of upright forms with a basal attachment that projects above the biofilm surface. As upright attached green algae, such as *Cladophora* and *Oedogonium*, become established, the biofilm transitions into a more turf-like structure. *Turfs* can be less than 1 cm tall (Figs. 25-3d and 25-10c) or extend tens of centimeters into the overlying water (Fig. 25-3e). Extracellular matrices are less important for turfs than biofilms, and the turf interface with the water column is a highly irregular surface dictated by the vertical growth of many macroalgae (Figs. 25-3e and 25-10b, c).

iii. Metaphyton

Metaphyton is an inclusive term for clouds of unattached benthic filamentous algae in the water column. Unlike phytoplankton, metaphyton are not dispersed and free-floating in a mixed water column. Rather, metaphyton are macroscopic aggregations of filamentous algae that can originate either from a rapidly growing mass of a single taxon that is not attached to a substrate, or from detached turfs of algae. Many taxa that form metaphyton do not have a holdfast and tend toward a *reticulate* (netlike) growth form. The cyanobacterium *Microseira wollei* and the green algae *Enteromorpha*, *Spirogyra*, and *Mougeotia* often form reticulate colonies loosely associated with the lake bottom (Hudon et al., 2014; Hawes and Smith, 1994). As biomass accrues and benthic assemblages become increasingly dominated by vertically oriented organisms, turfs become increasingly subject to the lift and drag of waves and currents. Rapid photosynthesis in the aggregation traps oxygen bubbles causing the algae to detach and form a floating surface scum (Fig. 25-2c). Clouds of metaphyton can transiently appear in summer in any quiescent littoral zone, but persistent metaphyton assemblages are characteristic of prairie pot hole lakes, wetlands, and acidic lakes (Goldsborough and Robinson, 1996).

D. Abiotic growth-regulating resources

Organisms require different resources to live, and the relative availability of different resources determines the productivity and taxonomic composition of benthic algal assemblages. Primary producers require light and carbon dioxide that fuel photosynthesis. Algae also require nutrients such as nitrogen (N, Chapter 14), phosphorus (P, Chapter 15), silicon (Si), and micronutrients (Chapter 16) to build enzymes, proteins, cell membranes, and cell walls. Ultimately, algae need space in which to grow and enough time to complete their life cycles and reproduce. The availability of these resources depends upon the physical-chemical characteristics of the lake but also upon where in the lake algae are growing. The relative importance of space, light, and nutrients in determining the algal composition and productivity rates is dynamic and context dependent.

i. Space

Benthic organisms are limited by the quantity of surfaces available for colonization and growth. In any illuminated habitat, lake bottom roughness, or *rugosity*, determines surface area for periphyton growth. Flat, horizontal lake bottoms, such as homogenous organic sediments in a bay, provide much less surface area for growth than a rocky, steep slope (Loeb et al., 1983). The seasonal emergence of macrophytes and macroalgae such as *Cladophora* increases the habitat space by orders of magnitude for smaller epiphytic diatoms and Cyanobacteria. Although bottom rugosity and macrophyte density can increase benthic surface area, substratum composition also determines the amount of surface area across which incoming solar radiation is distributed. At a given incident light intensity, flat and horizontal surfaces will receive more energy per square meter of substrate than structurally complex rugose benthic habitats at the same depth (Loeb et al., 1983). Furthermore, no matter how much surface area is available, periphytic algae develop by successive generations growing over foundations of their predecessors. Eventually, competition for space during algal growth and succession rapidly becomes a competition for that all-important resource, light.

ii. Light

Within lakes, light and depth are often the best predictor of benthic algal biomass, productivity, and taxonomic composition (Round, 1964; Vadeboncoeur et al., 2008, 2014; Hofmann et al., 2020). Variation in light intensity is predictable on temporal (day of year, time of day) and spatial (latitude, depth) scales. The intensity of light hitting the lake surface (I_0) is also dependent on current weather. The rapidity with which light is attenuated (see also Chapter 6) in the water column is quantified by the downwelling light attenuation coefficient k_d, which determines the light intensity at depth z, I_z (where $I_z = I_0e^{-kdz}$). Among-lake variation in water clarity, and hence variation in habitat availability for benthic photoautotrophs, is determined by substances suspended in the water column including colored dissolved organic carbon (DOC) (Chapter 28), phytoplankton, and suspended particulate minerals (e.g., glacial till).

Once light reaches benthic algae, it is rapidly attenuated within the biofilm itself, with k_d in the biofilm being up to 1000 times higher than in the water column (Fig. 25-11; Krause-Jensen and Sand-Jensen, 1998; Dodds et al., 1999). Thus, as biomass accumulates, the addition of cells on the surface of the biofilm has a relatively small influence on total photosynthesis (Krause-Jensen and Sand-Jensen, 1998; Higgins et al., 2008a). In biofilms, area-specific photosynthesis asymptotes at a relatively low algal biomass because rapidly growing cells at the surface of the biofilm shade underlying cells, reducing their photosynthetic performance (Dodds et al., 1999).

Light intensity and wavelength distribution affect taxonomic composition of algae because different taxa differ in their capacity to use specific wavelengths. Ultraviolet radiation in shallow, clear water can damage algal cells (Vinebrooke and Leavitt, 1999). Green algae appear to need higher light intensities (Donahue et al., 2003). However, because many filamentous green algae are an ideal substrate for diatoms, they are often epiphytized by other algae that limit light penetration to the host.

iii. Nutrient sources for periphyton

Algal taxa differ in their requirements for the macronutrients N, P, and Si, and spatiotemporal variation in the relative availability of these nutrients leads to variation in periphyton community structure. The addition of nutrients often alters competitive interactions among species, and biomass accumulation is accompanied by taxonomic shifts (Donahue et al., 2003). In this section we discuss nutrient sources and nutrient demand across broad habitat and taxonomic categories. Benthic algae can acquire nutrients from the water column, the groundwater, and the substratum on which they grow. Some Cyanobacteria (Figs. 25-8a–c), as well as diatoms that host endosymbiotic Cyanobacteria (Fig. 25-6m), can fix dinitrogen gas (N_2). As biofilms accumulate biomass, mineralization of nutrients within the algal mat

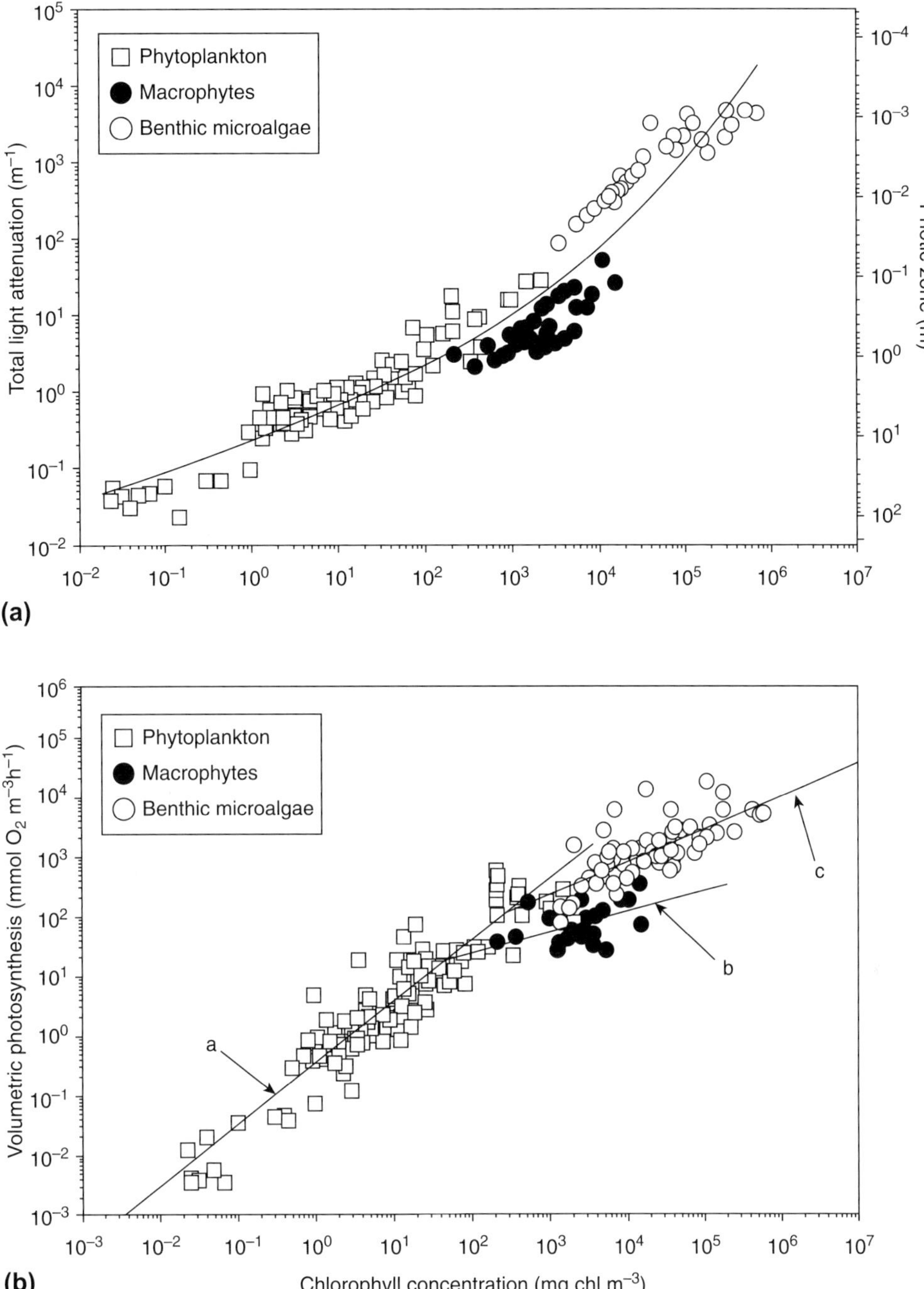

FIGURE 25-11 (a) A comparison of light attenuation coefficients and photic zone depths in natural assemblages of phytoplankton, macrophytes, and benthic algal biofilms. Cells in biofilms and benthic algal turfs are more concentrated than cells in phytoplankton assemblages or macrophyte stands. Thus volumetric chlorophyll concentrations and rates of light attenuation are much higher in periphyton compared with the other two functional groups of autotrophs. (b) Volumetric photosynthesis (mmol O_2 m^{-3} h^{-1}) asymptotes at high chlorophyll concentrations typical of benthic microalgae owing to rapid light attenuation in the periphyton mat. *(Modified from Krause-Jensen and Sand-Jensen, 1998.)*

(recycling) becomes an increasingly important nutrient source (Mulholland et al., 1991; Wetzel, 1996; Borchardt, 1996; Burkholder, 1996).

The rocky littoral zones of clear oligotrophic lakes often support a large complement of nitrogen-fixing Cyanobacteria in the Rivulareaceae, especially *Calothrix* and *Tolypothrix* (Reuter et al., 1983; Vinebrook and Leavitt, 1999; Higgins et al., 2003a; Diehl et al., 2018). Ample light is essential because nitrogen fixation is energetically costly, and high photosynthetic rates fuel nitrogen fixation. Cyanobacteria with basal heterocytes (e.g., *Calothrix*) and diatoms in the Rhopalodiaceae with endosymbiotic N-fixing Cyanobacteria are common in the nearshore zones of oligotrophic, clear lakes

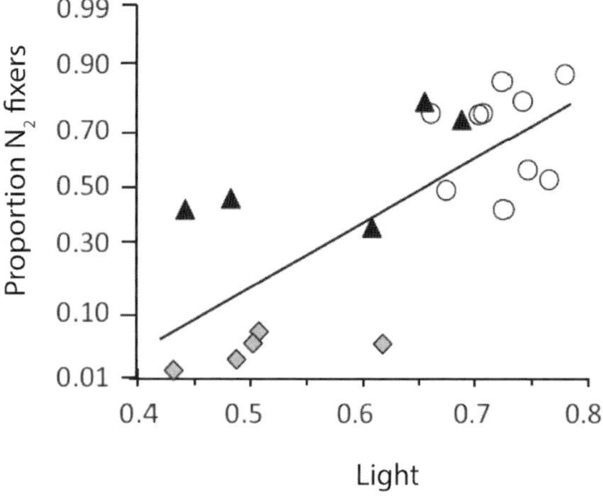

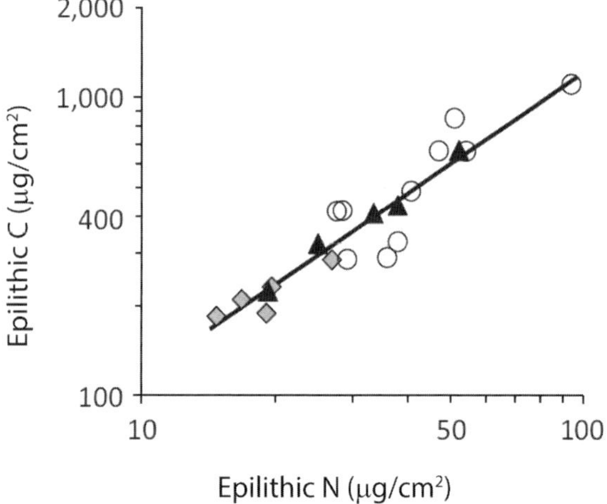

FIGURE 25-12 Nitrogen-fixing Cyanobacteria (N_2-fixers) are a common constituent of epilithic algal assemblages in nutrient-poor lakes. (Top) In boreal and subarctic lakes the proportion of epilithic biomass represented by N_2-fixers increases as a function of the proportion of incoming light available at 0.35 m depth. Nitrogen fixation is an energy-demanding process, and high light drives high rates of photosynthesis. (Bottom) Epilithic algal biomass was limited by light, which controlled the rate of N accumulation in epilithic algal biomass. Each data point represents a lake. *Symbols* denote latitudinal regions (*diamonds*: 60°N, *triangles*: 64°N, *circles*: 68°N). Proportions are shown on an arcsine-square root transformed scale. *(From Diehl et al., 2018.)*

such as Lake Tahoe (USA), Lake Tanganyika (East Africa), and Lake Malawi (East Africa) (Reuter et al., 1983; Higgins et al., 2003a). Periphytic assemblages in high-elevation and boreal lakes are also dominated by N-fixing Cyanobacteria (Vinebrooke and Leavitt, 1999; Fleming and Prufert-Bebout, 2010; Diehl et al., 2018). The common occurrence and even dominance of N-fixing benthic algae in clear, oligotrophic lakes indicates that N-fixation is critical to nutrient dynamics under high-light, low-nutrient conditions (Fig. 25-12); these littoral assemblages rich in N-fixers may have been the norm in oligotrophic waters before anthropogenic nitrogen became abundant and pervasive (Vitousek et al., 1997). Benthic N-fixation is rarely included in whole-lake N budgets, but in Lake Malawi, one of the world's largest lakes by volume, benthic N-fixation may account for 35% of the N inputs into the epilimnion (Higgins et al., 2003b).

The water column surrounding benthic algae is a source of nutrients for algae growing on the outermost surface of biofilms, for algal turfs that have dendritic or foliose growth forms, and for metaphyton. The rate of nutrient uptake from the water column depends, in part, on the thickness of the boundary layer across which the nutrients must diffuse (Borchardt, 1996; Burkholder, 1996) and the surface area of the algal assemblage relative to its volume (Steinman et al., 1992). Biofilms, especially thick, stratified biofilms, can exploit nutrients in the water column only through diffusion-dependent exchange across an interface that, relative to turfs, is relatively homogenous and impervious (Wetzel, 1993). Thus only cells at the outermost surface of the biofilm directly take up water column nutrients. Conversely, the irregular surface of algal turfs that are dominated by upright and filamentous algae provides much more direct access to dissolved substances in the water column. Water movement disrupts or erodes boundary layers around cells protruding from the surface and increases the rate of nutrient delivery to benthic surfaces (Stevenson, 1996).

Although most monitoring programs focus on the water column as the most relevant nutrient source for algae, many lakes are fed by groundwater, and groundwater nutrients can fuel benthic algal growth. Attached algae growing around groundwater springs in a softwater lake in northern Wisconsin (USA) had taxonomic compositions associated with high P availability and had higher biomass than assemblages at low groundwater discharge sites (Hagerthey and Kerfoot, 1998). Groundwater pollution has transformed littoral Lake Baikal (Russia) by causing massive outbreaks of both native and nonnative filamentous algae (Timoshkin et al., 2018). Groundwater nutrient sources need to be better integrated into monitoring programs that aim to manage nutrient pollution and periphyton growth.

Attached algae colonize plants and animals, and periphyton can acquire nutrients directly from living hosts such as macrophytes, macroalgae, and mollusks. Rooted macrophytes translocate sediment nutrients to their stems and foliage, and N and P leaking from these structures can be taken up by epiphytes (Burkholder, 1996). Radioactive P tracers showed that filamentous green algae and filamentous Cyanobacteria could acquire up to 40% of their P from host macrophytes, while prostrate and adnate diatoms could acquire as much as 75% of their P from sediment nutrients translocated by macrophytes (Moeller et al., 1988). ^{15}N additions to sediments demonstrated that macrophytes were a trivial source of N for epiphytes (Song et al., 2017), but these experiments were conducted at relatively high water-column nutrient concentrations (>1 mg N L^{-1}) and may not be applicable to nutrient-poor lakes. Excessive epiphyte growth often contributes to the senescence and eventual death of macrophytes. The rapid decomposition of macrophytes may release a pulse of nutrients to associated epiphytes, but any benefit is offset by the disintegration of the substrate on which the algae grow and their subsequent transport to the lake bottom (Wetzel, 1996).

The hard shells and carapaces of mollusks and crustaceans and the retreats and cases of caddisflies (Trichoptera) are excellent surfaces for microalgal growth (Lukens et al., 2017). Invasive mussels (*Dreissena* spp.) in the Laurentian Great Lakes provide both a novel substrate and nutrients for attached filamentous algae (Francoeur et al., 2017). Detailed budgets using stable isotope analysis indicate that psychomyiid caddisflies increase the abundance of their periphyton food that grows on their retreats because the caddisflies' excretory products fertilize the algae (Ings et al., 2010). Animal excretion is a critical nutrient source for periphyton even if the animals consume, rather than host, the algae (Section 25.IIC).

Epipelic algae flourish in the spatially and temporally dynamic, high-nutrient microzone of the *SWI*. Nutrient concentrations in unconsolidated sediments often exceed that in the water column by several orders of magnitude (Carey and Rydin, 2011), and epipelic and epipsammic algae exploit pore-water nutrients. Even in an oxygen-rich epilimnion, sediment oxygen concentrations decline precipitously within 1 mm of the SWI affecting the mobility and solubility of P and oxidation form of N (NH_4, NO_3, NO_2). At night, heterotrophic metabolic activity is not offset by photosynthetic oxygen production by epipelic algae, and extreme gradients at the SWI drive a vertical flux of P from the sediments to the water column (Carlton and Wetzel, 1988; Wood et al., 2015; Benelli et al., 2018). Dissolved inorganic carbon concentrations are also higher in the sediment than in the water column owing to microbial respiration and the influence of groundwater, and this provides a dissolved C source for epipelic algal photosynthesis (Vadeboncoeur and Lodge, 1998). Experimental evidence at multiple spatial scales demonstrates that epipelic algae rely substantially on the sediment pore-water for nutrients (Carlton and Wetzel, 1988), and epipelic algae are rarely positively influenced by additions of nutrients to the overlying water (Vadeboncoeur et al., 2001; Vadeboncoeur and Lodge, 2000; Fig. 25-13).

iv. Linking nutrient criteria to benthic algae

Limnology is dominated by paradigms developed for plankton, and most monitoring programs attempt to link periphyton biomass to water-column nutrients.

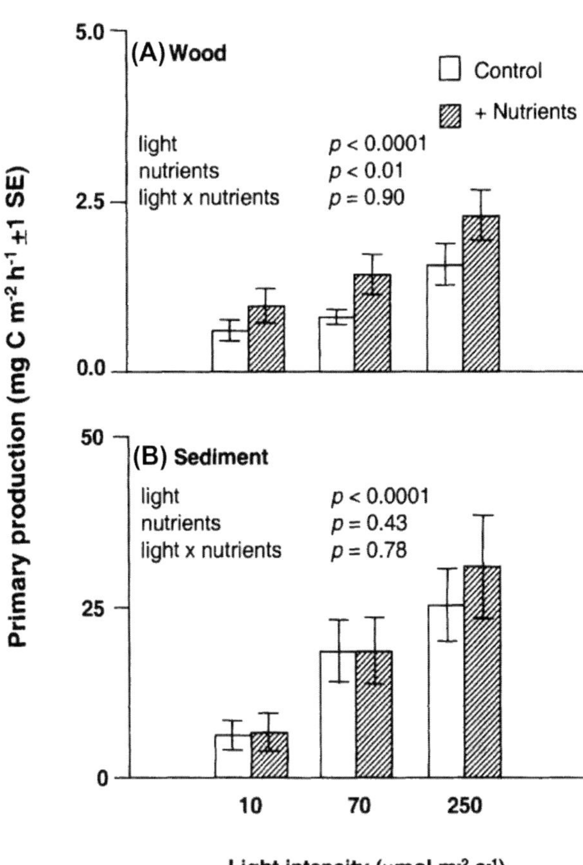

FIGURE 25-13 Periphyton growing on natural substrata, wood, and organic sediments were exposed to three light intensities and two different water column nutrient concentrations in a 25-day laboratory mesocosm experiment. Nutrient treatments were fertilized daily with nitrogen and phosphorus. (a) Productivity of algae on wood responded to increasing light and nutrients because wood does not provide nutrients for algae. (b) Primary productivity on sediments increased with increased light intensity but did not respond to water column nutrient additions. Productivity on nutrient-rich sediments was 10 times the productivity on wood. *(From Vadeboncoeur and Lodge, 2000.)*

Applying these nutrient paradigms to periphyton is inappropriate because: (1) the useful autocorrelation between phytoplankton biomass and total nutrients does not have a corollary with benthic algae because the source of nutrients (water column) is in a different habitat than the variable of interest (benthic algae); (2) metrics of total nutrients in the water column include indices of the biomass of a competitor of benthic algae (phytoplankton); (3) the water column is not the only source of nutrients for periphyton; and (4) chlorophyll *a* is an imperfect index of periphyton biomass (Steinman et al., 2017).

Efforts at relating periphyton biomass to typical water quality parameters are confounded by the diverse sources of nutrients for benthic algae. The water column is the primary source of nutrients for metaphyton and turfs that are dominated by filamentous green alga.

Nevertheless, there is rarely a tight positive relationship between nutrients and periphyton biomass. Metaphyton and other filamentous benthic algae can efficiently remove inorganic nutrients from the water, causing high algal biomass to be associated with low water column dissolved nutrients (Dodds, 2003; Vadeboncoeur et al., 2021). The water column is often the least important source of nutrients for periphyton growing on the soft, relatively nutrient-rich sediments that dominate littoral substrata, decoupling benthic algal biomass and productivity from water column nutrients (DeNicola et al., 2003; Vadeboncoeur et al., 2001, 2003; Karlsson et al., 2009). Furthermore, eutrophic lakes with high water-column nutrient concentrations have high concentrations of phytoplankton that shade epipelic algae. Most chlorophyll—nutrient relationships for periphyton are derived from rivers, but they are informative with respect to lakes. In rivers epilithic chlorophyll—nutrient regressions are weakly positive but highly variable (Fig. 25-14; Dodds and Whiles, 2010, Fig. 18-11; Miltner, 2010).

Benthic chlorophyll-nutrient relationships are also problematic because chlorophyll *a* is a poor metric of periphyton biomass across the gradient of light intensities associated with depth in the littoral zone. Cells growing under low light conditions synthesize more chlorophyll to efficiently harvest the available light, leading to highly variable chlorophyll-to-biomass ratios in periphyton across a depth gradient. Furthermore, biofilms accrue vertically, and as cells at the base of the mat die, their chlorophyll may degrade to pheophytin. Degraded chlorophyll can be analytically detected and accounted for (i.e., with an acidification step in fluorometric or spectrophotometric extractive analyses), but under cold or anoxic conditions in sediments, the chlorophyll may not degrade, thus inflating estimates of living algal biomass. Finally, living phytoplankton cells settle to the deepest areas of stratified lakes and contribute nonbenthic chlorophyll to sediment chlorophyll estimates (Carrick, 2004). For these reasons, benthic chlorophyll, and its relationship to water-column nutrient concentrations, needs to be interpreted with caution.

v. Assessing nutrient limitation of benthic algae

Despite the difficulty of constraining nutrient—periphyton regressions at the ecosystem scale (Fig. 25-14), scientists and managers wish to know how nutrients affect periphyton and need to be able to assess when excessive periphyton growth is likely to constitute a nuisance or health threat to humans (DeNicola and Kelly, 2014; Poikane et al., 2016; Wood et al., 2020). Two tools that are widely used to assess nutrient effects on benthic algae are using algal taxa as bioindicators and experimentally manipulating nutrient availability to periphyton using *nutrient-diffusing substrata (NDS)*.

There is a long history of relating the taxonomic composition of benthic diatom assemblages to nutrient criteria (Lowe and Pan, 1996). Whether algae are collected from natural substrata (Passy, 2008), cultured on artificial substrata (McCormick et al., 1996), or assessed using paleolimnological approaches (Donahue et al., 2003; Pla-Rablés and Catalan, 2018), diatoms are robust indicators of resource availability and long-term water quality (Fig. 25-15). However, a high level of expertise is required to analyze diatom taxonomic composition, and it is difficult to communicate the meaning of the results in an accessible manner to stakeholders and the public (DeNicola and Kelly, 2014; Poikane et al., 2016). Furthermore, the public is often concerned about accumulations of algal biomass or potentially toxic benthic algal blooms (Hudon et al., 2014; Wood et al., 2020; Vadeboncoeur et al., 2021), and diatom taxonomic indicators do not provide

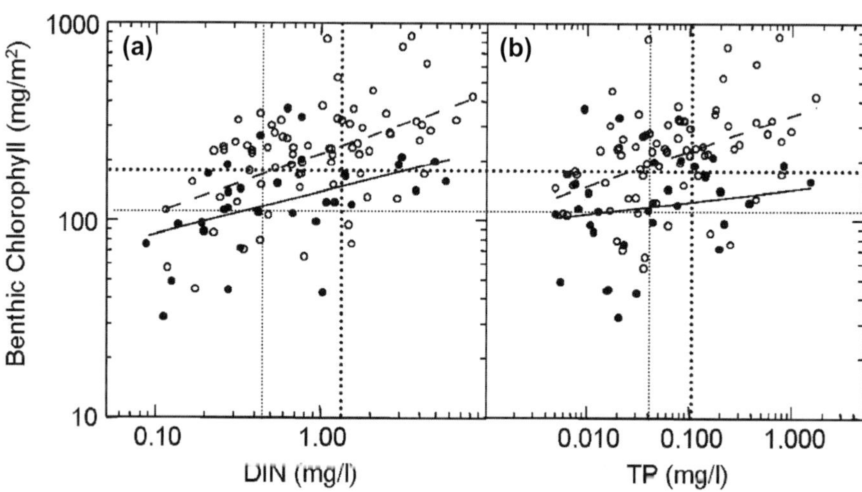

FIGURE 25-14 Summer epilithic chlorophyll concentrations in Ohio rivers (USA) with respect to (a) dissolved inorganic nitrogen (*DIN*) and (b) total phosphorus (*TP*) concentrations. *Open circles* have a canopy gap of less than 40° (less light) and *closed circles* have a canopy gap of >40° (i.e., more light). Similar data are not available for periphyton in lakes, but it is notable that nutrient—chlorophyll relationships are highly variable because of the strong influence of factors such as disturbance regime and grazers. *(From Miltner, 2010.)*

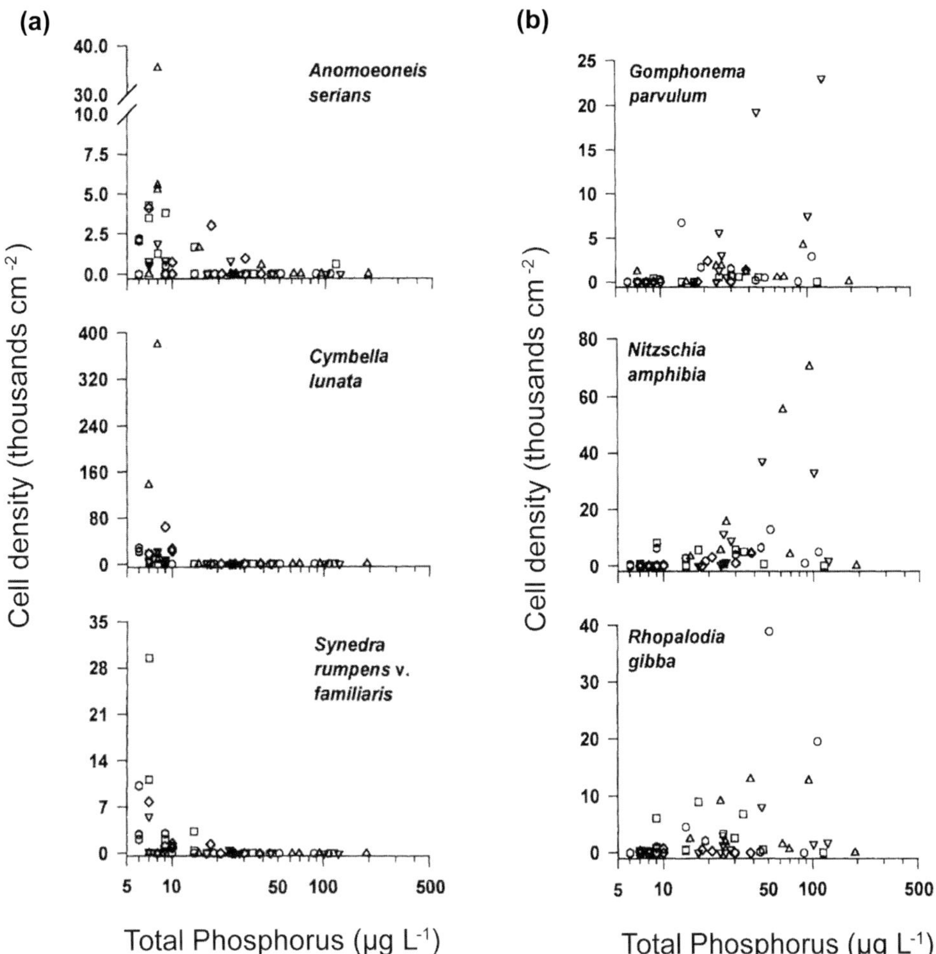

FIGURE 25-15 Benthic diatom taxonomic composition grown on artificial substrata across a phosphorus (P) gradient in the Florida Everglades (USA). (a) Some diatoms were most abundant at sites with low P concentrations and decreased as nutrient availability increased. (b) Other taxa attained higher densities as P concentrations increased. *(From McCormick et al., 1996.)*

information on the green algae and Cyanobacteria that contribute to nuisance and toxic blooms. Thus monitoring programs that rely only on diatoms need to expand to integrate other indicators of ecosystem function (DeNicola and Kelly, 2014; Poikane et al., 2016).

Nutrient limitation of the growth (productivity) of entire algal assemblages can be assessed with NDS. NDS provide an experimental assessment of which nutrient (e.g., N, P), or a combination of nutrients (e.g., N + P), limits algal biomass (Fairchild et al., 1985). Individual or combinations of nutrients are added to an agar matrix, and the nutrient cocktail is allowed to solidify in small plastic pots or centrifuge tubes, which are then capped with a glass fiber filter or a glass frit (Tank et al., 2006). The experimental pots are affixed to a metal bar and deployed in the aquatic environment for several weeks (Fig. 25-16a). After the *in situ* incubation period, the glass frits or filters are retrieved and analyzed for photosynthetic productivity, species composition, pigment concentration, and/or ash-free dry mass (AFDM). For stream periphyton, experimental deployments of NDS suggest colimitation by N and P is the most common condition, but it is also common to have no nutrient limitation (Dodds and Whiles, 2010). NDS data from Lake Baikal (Russia) and the Laurentian Great Lakes (Canada/USA) indicate colimitation by N and P, or N limitation (Cooper et al., 2016; Ozersky et al., 2018), while in Lake Tanganyika periphyton appeared to be limited by P, or colimited by N and P (McIntyre et al., 2006). In most experiments changes in biomass in response to nutrient amendments are accompanied by taxonomic changes because providing nutrients alters the outcome of competition among algae. *Pulse amplitude modulated (PAM) fluorometry* can provide information on photosynthetic responses to NDS experiments and increase the speed and sensitivity with which nutrient responses are detected (Whorley and Francoeur, 2013).

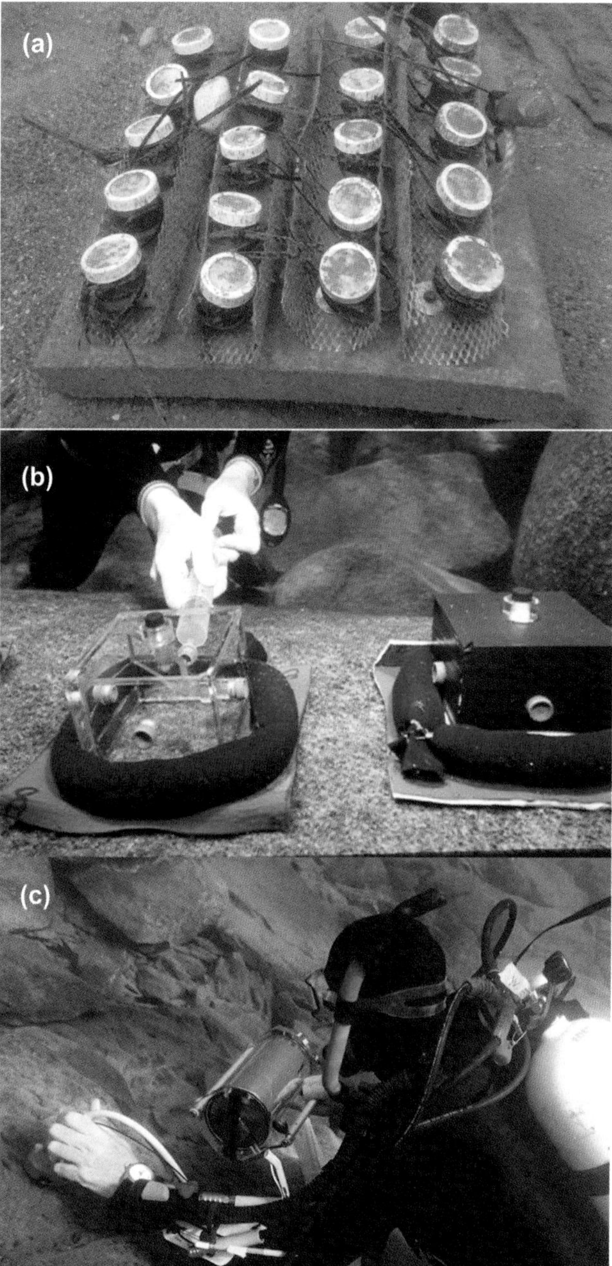

FIGURE 25-16 *In situ* experiments and incubations are often employed to monitor benthic algal dynamics. (a) Nutrient-diffusing substrata (*NDS*) are deployed in Lake Superior (Canada/USA) to assess periphyton responses to different nutrients. *(Photo Ted Ozersky.)* (b) Light and dark incubation chambers allow researchers to measure photosynthesis using bulk oxygen exchange methods. *(Photo Brant Allen.)* (c) The underwater pulse amplitude modulated (*PAM*) fluorometer monitors photosynthesis-irradiance responses of benthic algae. *(Photo Elliot Gaines.)*

E. Growth-regulating mortality and losses

The accrual of algal biomass on surfaces in lakes is limited by self-shading, disturbance, and removal by grazers. These sources of mortality and loss affect the biomass, productivity, and taxonomic composition of periphyton. Light is an overwhelming driver of variation in algal biomass and productivity in lake littoral zones. Algae that grow on the surface of biofilms or grow epiphytically on larger algal cells affect light transmission (both quantity and quality) to the algal cells below them, causing senescence of cells at the base of the mat.

i. Disturbance

Disturbance regimes determine the accumulation of algal biomass in space and time by agitating and sorting substrata. Variation in algal taxonomic composition tracks these changes in substrate, and temporal variation in wave intensity or winter ice scour will affect the amount of biomass that develops in shallow habitats. Waves constitute a disturbance when they physically move the substratum, burying or shearing the algae from the surface. Wave action transports organic detritus away from the lake edge leaving behind rocks or unstable sand. Waves suspend and transport sand particles in the direction of dominant wave motion or current. Adnate algae attached tightly to sand grains experience repeated intervals of exposure to the surface and burial, limiting biomass accumulation (Forehead and Thompson, 2010). Waves that have sufficient energy to move and tumble cobbles will prevent the luxuriant growth of periphyton in the high-light, nearshore environment, even when nutrients are sufficient. For example, along the shallow, rocky shoreline of nutrient-rich Lake Erie (Canada/USA), algal biomass was lowest at depths <0.5 m and was highest at intermediate depths where light was still sufficient but disturbance did not scour algae from the cobbles (Fig. 25-10).

High-energy shorelines composed of boulders and bedrock sometimes support the prolific growth of green algae and Cyanobacteria with strong basal attachments. Prolonged periods of low wave action, warm temperatures, and increased irradiance during summer months can foster rapid algal growth and biomass accumulation (Fig. 25-17a). However, when wind and waves return, these assemblages are often sheared from the rocks and either transported downslope or deposited on beaches (Fig. 25-17b; Higgins et al., 2008b). This beach fouling is unsightly and can harbor bacteria that are harmful to humans (Hudon et al., 2014; Timoshkin et al., 2016).

Winter ice accumulation is greatest at the lake edge, and the ice permeates the surface of the littoral sediments, sheering algae from rocks and cobbles. In very large lakes ice along the shoreline is moved by wind and waves, causing an annual scouring of accumulated algae in rocky habitats. The effects of climate-driven reductions in ice cover (Sharma et al., 2019) have been explored mainly by paleolimnologists using diatoms to track long-term changes in climate (Chapter 30), and

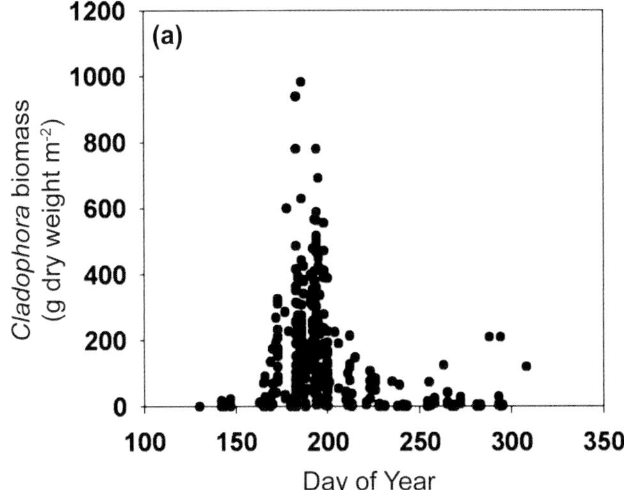

FIGURE 25-17 Benthic algal biomass undergoes seasonal cycles of growth and senescence. (a) In the eastern basin of Lake Erie (USA/Canada) biomass of the filamentous green alga, *Cladophora glomerata*, peaks in midsummer with as much as 1 kg dry mass m^{-2} (samples collected at <3 m from 1976 to 2006). (b) Nuisance filamentous algal blooms (*FABs*) can cause beach fouling when the algal turfs dislodge and wash ashore. *(From Higgins et al., 2008b.)*

especially in polar regions (e.g., Smol, 1988; Douglas and Smol, 2010). However, the diminishment of this annual cleaning and resetting of littoral periphyton communities may be altering the annual phenology of periphyton succession and allowing nuisance filamentous algae to develop earlier in the year (Vadeboncoeur et al., 2021).

ii. Grazing

Many animals consume benthic algae (Section IIIC). Grazers reduce algal biomass, alter the composition of periphyton, and interrupt algal succession (Steinman et al., 1987; Steinman, 1996). Grazing is a significant source of mortality for periphyton. Rather than algae accumulating on surfaces, new algal growth is often consumed by grazing animals and shunted up the food web to higher trophic levels. Most periphyton grazing studies are conducted in riverine habitats rather than in lake littoral zones. However, across all field experiments, grazers may reduce benthic algal biomass by an average of 55% (Hillebrand, 2009). Grazers also reduce spatial variation in algal biomass (Hillebrand, 2008), converting algal production into grazer biomass rather than allowing it to accumulate as algal biomass (Hill et al., 2010; Marks et al., 2000).

Mobile grazers at high densities track benthic primary productivity creating relatively uniform algal biomass across extensive habitat and productivity gradients. Reduced spatial variation in algal biomass is the aggregate effect of individual grazers increasing feeding frequency and duration when they encounter patches with high algal standing crops (Hart, 1981; Kohler, 1984). Densities of grazing cichlid fish in Lake Tanganyika decline markedly with depth (Munubi et al., 2018) because algal productivity is highest in the shallowest water. Grazing cichlids aggregate in shallow waters where light and algal productivity are highest. However, algal biomass varied little over the 8 m depth gradient because grazing kept algal biomass uniformly low (Munubi et al., 2018).

Grazers reduce the rate of biomass accumulation, and this effect is strongest at low algal biomass when algal growth rates are often very high (Abe et al., 2007; Figs. 25-18 and 25-19). Grazers strongly affect the taxonomic composition of periphyton by selecting motile and loosely attached diatoms, which are smaller and more easily manipulated and consumed than filamentous Cyanobacteria and green algae. Nevertheless, diatoms persist at low levels in the presence of grazers (Abe et al., 2007). Typically, when grazers are abundant, firmly attached filamentous Cyanobacteria (e.g., *Calothrix*, *Homoeothrix*) or appressed filamentous green algae (*Stigeoclonium*) become more prevalent (Kupferberg, 1997; Abe et al., 2007; Rober et al., 2011). When grazers are excluded, or in areas of low grazer densities, filamentous green algae often dominate algal biomass (Harrison and Hildrew, 2001).

Many studies that test for simultaneous effects of nutrients and grazers conclude that grazers have a more profound effect on algal biomass than does the addition of nutrients (Fig. 25-20; Hill et al., 1992, Hillebrand, 2002; McIntyre et al., 2006; Rober et al., 2011). As biomass accumulates, the influence of nutrient delivery rate to the biofilm diminishes owing to increasing light limitation. Grazers seek out and consume high algal biomass, and biomass removal by grazers ameliorates light limitation. When grazers are experimentally excluded, biomass rapidly increases, and area-specific primary productivity sometimes increases, but biomass-specific primary productivity usually decreases owing to light limitation within the algal mat (Steinman et al., 1992).

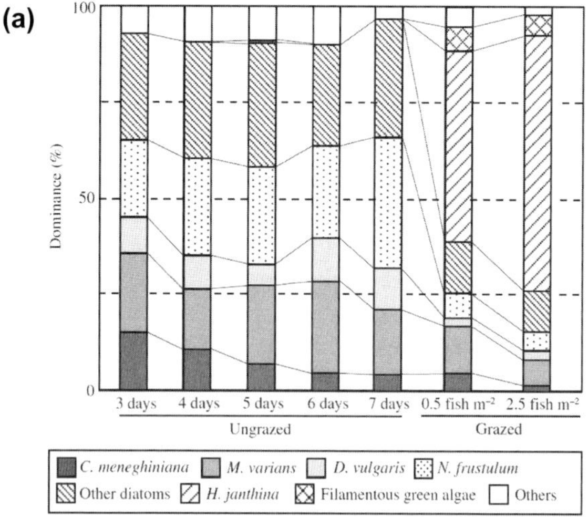

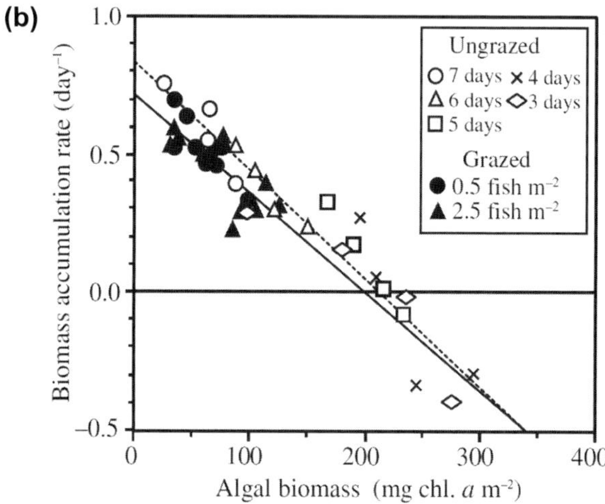

FIGURE 25-18 (a) Grazing fish transform periphyton assemblages in Japanese streams from dominance by diatoms to dominance by the cyanobacterium *Homeothrix janthina.* (b) As biomass accumulates on surfaces, growth rate of the algae slows. Grazing fish promote high growth rate of benthic algae by maintaining the biofilms at low biomass. *(From Abe et al., 2007.)*

F. Seasonal dynamics

Seasonal variation in temperature and light affect algal growth rates (DeNicola, 1996), and the phenology of grazers influences biomass accumulation. Annual and seasonal cycles of epipelic algae are complex (e.g., Round, 1964, 1972; Ilmavirta et al., 1977; Sheath and Hellebust, 1978; Romagoux, 1979). Epipelic diatom growth increases in early spring, and productivity peaks in midsummer (Gruendling, 1971; Brothers et al., 2013). This pattern is similar to that of planktonic species except that diatoms are often dominant components of benthic assemblages throughout the year.

The seasonal dynamics of epiphytic algae on submerged macrophytes are complex because the surface area available for colonization and growth is continually changing as the macrophytes grow, senesce, and enter detrital phases. Computations of the population changes of the total epiphytic algal community are rare and depend upon the supporting macrophyte communities (Kowalczewski, 1975; Cattaneo and Kalff, 1980; Müller, 1996). For example, changes in the biomass of attached algae associated with a zone of the emergent bulrush *Scirpus acutus* were more than an order of magnitude lower and out of phase with those of algae attached to substrates in a shallow zone dominated by submersed plants (*Najas flexilis* and *Chara*) (Allen, 1971).

II. Metabolic interactions in littoral communities

Benthic algae interact directly or indirectly with other littoral organisms, from microbes to fish. Benthic algae exchange metabolites and metabolic gases with heterotrophic microbes, influencing nutrient and carbon flux within the biofilm. Periphytic algae exploit the same resources as phytoplankton and macrophytes, setting the stage for competition between different functional groups of primary producers in the lakes. Finally, many animals live on or near surfaces, and their metabolic wastes fertilize periphyton.

A. Interactions between algae and heterotrophic bacteria in periphyton

The establishment, growth, senescence, and death of the microorganisms that make up periphyton are driven by metabolic processes that convert resources from the environment into living cells. Resources and waste products are rapidly exchanged between the autotrophic and heterotrophic organisms living in close proximity to periphyton. At the simplest level, waste products from photosynthesis create an oxygen-rich environment that supports microbes using aerobic metabolic pathways, while bacterial respiration produces CO_2 used by algae in photosynthesis. In addition, algae provide a labile source of DOC (Chapter 28) that fuels microbial metabolism, while bacteria mineralize the nitrogen and phosphorus necessary for algal growth. Both algae and bacteria contribute to the ECM that strongly mediates the movement of material within biofilms (Scott et al., 2014). The contributions of heterotrophic relative to autotrophic processes in biofilm metabolism depend upon the architecture and chemistry of the substrate on which the biofilm develops and upon light availability.

The metabolic waste products of heterotrophic denizens of periphyton are valuable resources for their autotrophic counterparts and vice versa. Free oxygen is a toxic waste product of photosynthesis that the algae must export from the chloroplast. However, it is a

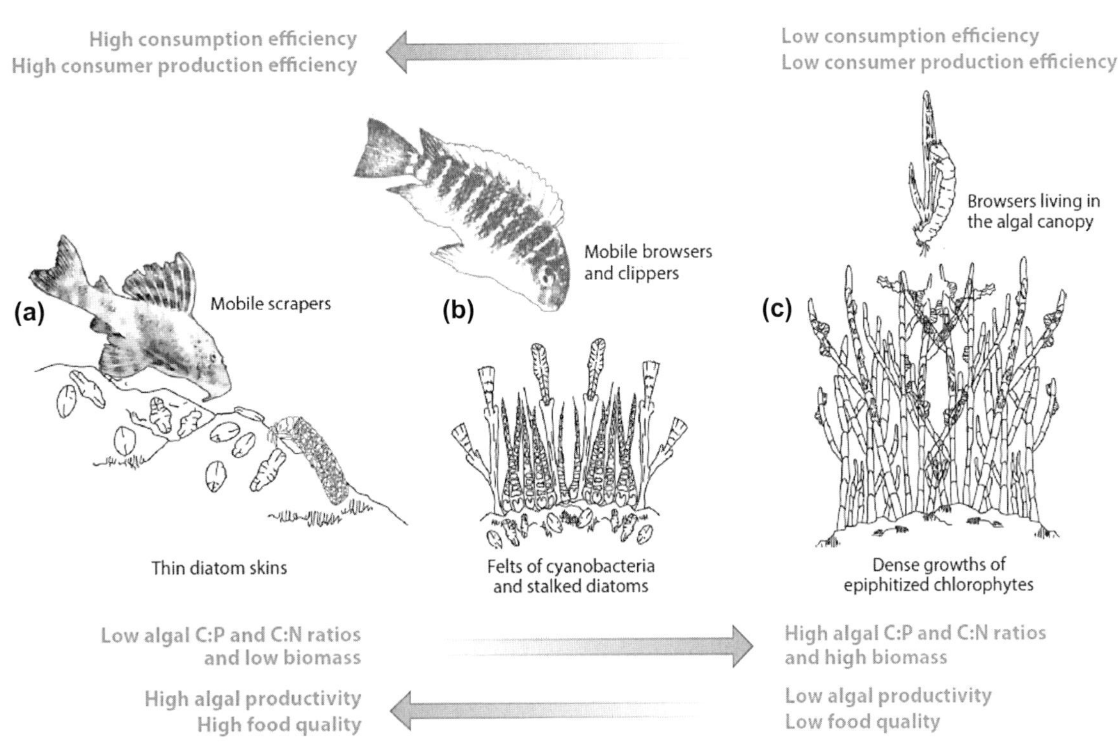

FIGURE 25-19 Grazers reduce benthic algal biomass, affect taxonomic composition, increase biomass-specific productivity, and tend to increase algal food quality (reduce carbon to nutrient ratios) of algae. (a) Mobile grazers with scraping mouth parts can maintain barely perceptible, thin skins of rapidly growing, nutritious diatoms. Algal quality is high due to high concentrations of nitrogen and phosphorus relative to carbon (low C:N). (b) Cyanobacteria in the family Rivulariaceae and stalked diatoms persist in the presence of grazers that tear and comb loose algae from low-growing biofilms on rocks. (c) When grazing pressure is relaxed, green algae such as *Cladophora* form dense turfs that provide habitat for small grazers. *(From Vadeboncoeur and Power, 2017.)*

necessary electron acceptor used in aerobic respiration by algae and heterotrophs. Algal cells leak photosynthate that heterotrophic microbes use for respiration and growth. Upon death, cell contents are lysed and degraded by bacteria. When bacteria or fungi in the periphyton break down dead organisms (including algae), they mineralize the nitrogen and phosphorus from dead cells. The newly liberated nutrients fuel algal growth. In turn, bacteria also produce CO_2 and certain organic micronutrients (e.g., vitamin B_{12}) (Wetzel, 1969) that can serve as growth factors for algae and other primary producers. Bacterial-algal interactions are a rich area for new research and molecular and tracer techniques can illuminate these fascinating interactions (Gubelit and Grossart, 2020).

B. Interactions among benthic algae and other primary producers

Unlike the immediate and intimate interactions with heterotrophic constituents of biofilms, benthic algal interactions with other functional groups of primary producers in the littoral zone are often indirect. Phytoplankton, macrophytes, and benthic algae all require light for photosynthesis, but access to light differs among these groups. Furthermore, the ability of different primary producers to use light for growth depends upon the availability of nutrients in the sediments and water column. In an ecological version of the children's game of "rock, paper, scissors," each functional group has its context-specific advantages for acquiring light and nutrients. Phytoplankton in the water column can shade periphyton and macrophytes but cannot directly access sediment nutrients; macrophytes can access sediment nutrients and grow vertically toward the light but can be overgrown by periphyton; periphyton have access to sediment nutrients but are shaded by phytoplankton and macrophytes. These competitive interactions for light and nutrients ultimately determine the *autotrophic structure* of lakes, or the distribution of primary production between benthic and planktonic habitats.

i. Interactions with macrophytes

Macrophytes, which include aquatic vascular plants and macroscopic algae with plant-like morphologies (Charales), provide ephemeral structure for periphyton. Submerged, emergent, and floating-leaved macrophytes

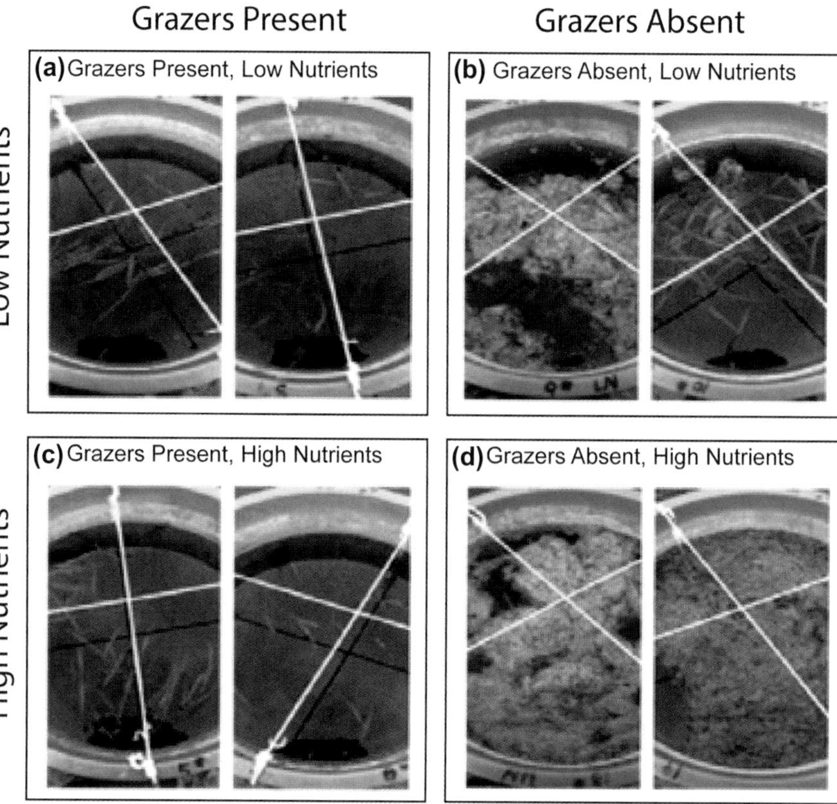

FIGURE 25-20 The grazing snail (*Radix swinhoei)* strongly determined the response of epiphyton and metaphyton to nutrient addition in a 30-day mesocosm experiment with the macrophyte *Vallisneria denseserrulata*. Snails reduced the biomass of epiphytes on *Vallisneria* in both the low-nutrient (a) and high-nutrient (c) treatments. Filamentous green algae (*Spirogyra* spp. and *Oedogonium* spp.) only accumulated in mesocosms without snails (b and d) and were more abundant in the high-nutrient (d) treatment. *(From Yang et al., 2020.)*

have different structures (Chapter 24) and vary in their effect on light availability to periphyton. Submerged macrophytes increase habitat for periphyton. For instance, the macrophyte *Scirpus subterminalis* provided ~24 m^2 of surface for growth of epiphyton per square meter of lake bottom (Burkholder and Wetzel, 1989). When epiphytic biofilms develop on macrophytes, the algae exploit the upward growth of the macrophyte, which provides access to more light than if the periphyton were growing on a nonliving, inert substratum on the bottom of the lake. The macrophytes leak metabolites that can be a source of N and P for the algae (Burkholder, 1996). Dense macrophyte beds further facilitate benthic algal growth by reducing turbulence and absorbing wave energy, providing a habitat in which metaphyton can flourish (Fig. 25-20; Yang et al., 2020).

Interactions between macrophytes and benthic algae for light are context dependent. Epiphytic algae reduce macrophyte growth rate because the epiphytes intercept light before it reaches the plant leaf and reduce CO_2 flux to the plant surface (Fig. 25-21; Jones et al., 2000). The development of epiphytic biofilms on macrophyte leaves can accelerate the senescence and death of the leaves. These negative effects of epiphytes on aquatic plants are substantially offset by grazers, especially snails (Fig. 25-20), which can reduce epiphytic biofilms to thin skins (Jones and Sayer, 2003). Macrophytes shade epipelic algae growing on sediments, but reduction in light availability may be offset by macrophytes stabilizing the sediments.

ii. Interactions with phytoplankton

Phytoplankton are dispersed throughout the water column and can intercept downwelling radiation before either benthic algae or macrophytes have the opportunity to do so. However, nutrient sources for phytoplankton are largely limited to nutrients dissolved in the water column, and the ability of phytoplankton to intercept light before submerged macrophytes or periphyton is of little consequence when there are few nutrients to support phytoplankton growth. When water-column nutrients are low, epipelic algae and macrophytes can fully exploit sediment nutrients and the light reaching the lake bottom. High phytoplankton biomass reduces light to periphyton, causing a decline in whole-lake periphyton productivity as lakes become

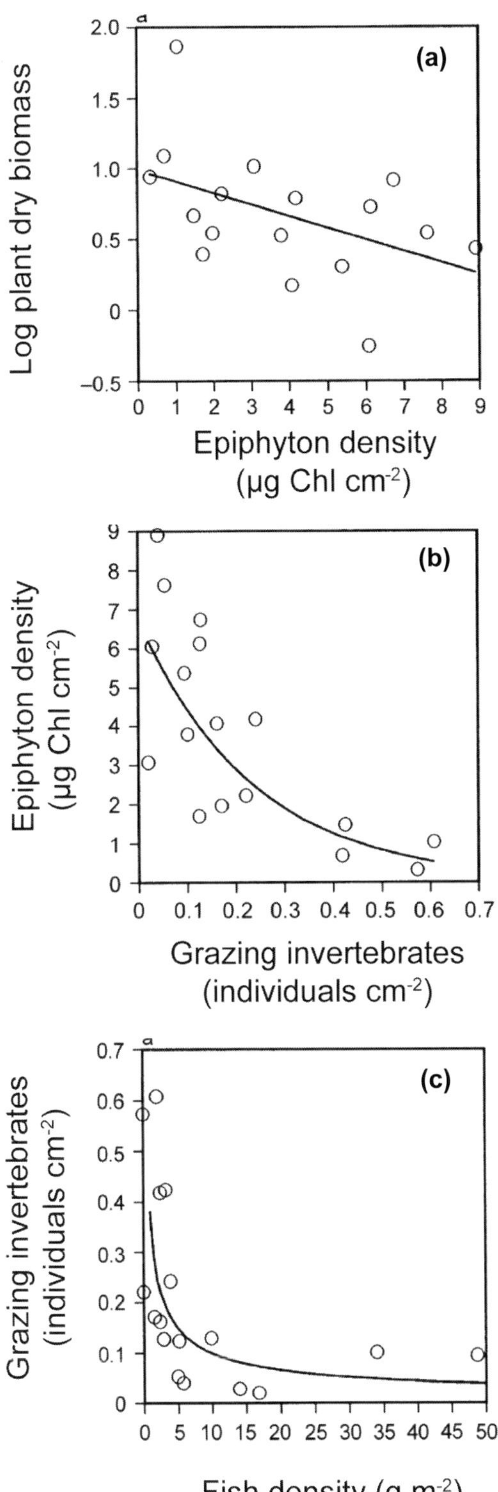

FIGURE 25-21 Direct and indirect effects in littoral food chains. (a) Epiphytic algae have a weak negative effect on macrophyte density in a lake's littoral zone. (b) Grazing invertebrates may benefit macrophytes by increasing light availability through reduction in epiphyton growth. (c) Consumption by fish reduces the density of grazing invertebrates. *(From Jones and Sayer, 2003.)*

more eutrophic (Vadeboncoeur et al., 2001, 2003; Vasconcelos et al., 2016).

Most algal taxa are either benthic or planktonic, but many can exploit both habitats (Hansson, 1996; Carey et al., 2014). Many flagellated planktonic algae (e.g., dinoflagellates, cryptomonads, and chrysophytes) have sediment resting stages or cysts that reinoculate the water column when conditions are favorable for growth. In addition, viable phytoplankton may settle to the sediments, perhaps to access nutrients, and then be recruited back to the plankton (Hansson, 1996). Similarly, recruitment from the sediment of Cyanobacteria (e.g., *Gloeotrichia*, *Anabaena*, and *Aphanizomenon*) can substantially augment planktonic populations (Carey et al., 2014). Viable planktonic diatoms are common in the benthos of large lakes, and their contributions to spring and autumn diatom blooms associated with mixing events need to be quantified (Carrick, 2004).

C. Metabolic interactions with animals

Animals that live in the littoral zones of lakes often associate closely with surfaces because littoral structures provide hiding places for mobile vertebrates and invertebrates, and most fish species consume zoobenthic invertebrates or algae from the littoral zone (Vadeboncoeur et al., 2011; Vander Zanden et al., 2011). Grazing invertebrates and fishes live in or on benthic surfaces in order to scrape or sweep periphyton from those surfaces. When grazing animals digest the algae that they consume, they necessarily excrete N and P in their urine and feces.

Excretion by grazers or other animals can create hotspots of nutrient cycling that lead to spatial variation in algal productivity. Spatiotemporal variation in the distribution of mobile grazers can create areas of high resource availability (Capps and Flecker, 2013a). Soluble reactive phosphorus and ammonium concentrations were 66% and 41% higher in areas of a tropical river where introduced loricariid catfish aggregated during the day, compared with adjacent areas without catfish aggregations (Capps and Flecker, 2013a). Sessile grazers and retreat builders directly affect benthic algae through selective feeding but also by excreting nutrients (Furey et al., 2012). For example, the NH_4 excreted by the caddisfly, *Tinodes waeneri*, was an important N source for the algae growing on the galleries of the larvae (Ings et al., 2010). The stoichiometric demand of grazers alters the N:P ratios of algae. Thus nutrient recycling by the P-rich armored catfish, *Ancistrus triradiatus*, increased periphyton N:P ratios more than tadpoles, which have a lower P demand (Knoll et al., 2009).

Animals are a constant source of small amounts of nutrients, and individual excretion rates need to be multiplied by population densities to quantify grazer contributions to ecosystem nutrient budgets (Vanni and McIntyre, 2016). Grazers that reach high densities can be surprisingly important to nutrient recycling. In the African Rift Valley lakes >45% of periphyton nutrient demand can be met by excretion and defecation by fishes (Andre et al., 2003; McIntyre et al., 2007). Lake Myvatn (which translates to "midge-water") in Iceland is named for the chironomid midges that live in its sediments; emerging adults form immense clouds that fill the sky. In a single growing season chironomid midges in the sediments can recycle the N and P equal to the annual external loading of these nutrients (Herren et al., 2017). This nutrient cycling leads to a strong positive relationship between chironomid densities and benthic algal primary productivity but not algal biomass (Herren et al., 2017). Species invasions and extinctions may be altering nutrient availability to benthic algae (McIntyre et al., 2007; Capps and Flecker, 2013b).

III. Functional roles of periphyton in lakes

Periphytic algae are primary producers that contribute to ecosystem function by fixing CO_2 into simple sugars, respiring fixed carbon; taking up N, P, and Si to produce macromolecules (proteins, phospholipids) for growth and metabolism; and providing food for consumers. Although there is a long history of including littoral dynamics in assessments of lake ecosystem function (Lindeman, 1942), over time, studies began to focus almost exclusively on dynamics in the limnetic, or open-water, portions of lakes (Vadeboncoeur et al., 2002). The role of littoral zones was minimized by the simple expedient of not including them in calculations (Vander Zanden and Vadeboncoeur, 2020), but littoral zones are critical habitats for invertebrates and fish, and benthic algae form the base of littoral food chains. Littoral and open-water habitats are linked through nutrient cycling and food webs, even in very large lakes (Schindler and Schuerell, 2002; Corman et al., 2010; Vander Zanden et al., 2011). Benthic algae contribute to lake metabolism, affect nutrient flux to the water column through oxygen production at the *SWI*, and provide a high-quality, concentrated basal resource in lake food webs.

A. Quantifying littoral-benthic algae

The spatial complexity of littoral zones (Figs. 25-2 and 25-3) creates a fascinating template upon which to characterize the diverse roles of benthic algae in lakes. This complexity can be daunting when quantifying benthic algal biomass and productivity. However, with care, and an appreciation for the strong effects of light and substrate, benthic algae can be quantified at the ecosystem scale and integrated into models of lake ecosystem function.

i. Biomass measurements

Periphyton biomass dynamics in space and time provides insights into benthic algal contributions to ecosystem structure. Biomass estimates through pigment analysis (especially chlorophyll *a*), *AFDM*, and direct cell counts are often employed, but each method has its strengths and weaknesses (Steinman et al., 2017). Pigment analysis using *high-performance liquid chromatography (HPLC)*, combined with pigment: carbon estimates for different algal taxa, provides comprehensive estimates of taxonomic biomass and assemblage composition (Vinebrooke and Leavitt, 1999). Despite the rich information provided by HPLC, this approach is uncommon due to complexity and cost. Rather, periphyton biomass is typically estimated using chlorophyll *a*. All algae contain chlorophyll *a* and chlorophyll can provide an index of algal biomass when used with care (Steinman et al., 2017).

Chlorophyll *a* concentrations are less than 1 mg m^{-2} early in biofilm development or on non-nutrient-diffusing substrates such as wood and rocks. In dense biofilms such as those that develop on nutrient-rich sediments or on macrophytes, chlorophyll *a* concentrations typically exceed 100 mg m^{-2} (Hawes and Smith, 1994; Higgins et al., 2008b; Vadeboncoeur et al., 2014). *Dry mass (DM)*, corrected for inorganic sediment and inorganic cellular constituents (AFDM), is often used in conjunction with chlorophyll *a* to characterize periphyton biomass on mineral substrata. DM measurements include organic matter from bacteria and micrograzers associated with periphyton but are a robust and useful index of filamentous algae (Higgins et al., 2008b). Over an annual cycle, DM of filamentous green algae can range from nearly 0 g DM per square meter to over a kg DM per square meter (Fig. 25-17).

Counting algal cells with microscopy is time consuming and requires expertise, but it provides more granular ecological information than chlorophyll *a* or AFDM alone. Quantitative samples of algae are counted and converted to biomass using standardized equations for different shapes (Hillebrand et al., 1999). Algal taxonomic composition may affect the role of algae in biogeochemical cycling and food webs, and molecular approaches are broadening ecologists' ability to use algae as indicators of water quality and ecosystem state.

ii. Primary productivity

Measurements of algal biomass, whether derived from pigment analysis, AFDM, or taxonomic composition, provide a snapshot of the net balance of the algal growth and loss processes but are of limited value in assessing ecosystem function. Grazers can quickly shunt algal carbon into lake food webs, leaving behind thin, inconspicuous biofilms that are nevertheless critical to food webs (Vadeboncoeur and Power, 2017; Olofsson et al., 2021). Primary productivity is a better index than biomass of the potential role of benthic algae in lake metabolism and food webs. To measure phytoplankton productivity, limnologists often inoculate water samples with a ^{14}C tracer and incubate the water in light and dark bottles (Chapter 9). Although ^{14}C was adapted for use with periphyton (Revsbech et al., 1981), the concentrated nature of periphyton biofilms or turfs makes the use of the bulk oxygen exchange method very tractable (Fig. 25-16b; Higgins et al., 2008b; Godwin et al., 2014; Vadeboncoeur et al., 2014).

Using bottles and chambers and either ^{14}C or oxygen exchange, productivity can be estimated for phytoplankton, periphyton, and macrophytes at different depths and extrapolated to the entire lake (Schindler et al., 1973; Welch and Kalff, 1974; Brothers et al., 2013). Late in the 20th century, whole-lake gas exchange experiments and whole-lake tracer additions (Hesslein et al., 1980) provided direct measurements of production and respiration at the ecosystem level, but whole-lake oxygen exchange data cannot distinguish different primary producer functional groups, and it often underestimated littoral zone production (Fig. 25-22; Van de Bogert et al., 2007; Brothers et al., 2013). The increased availability of oxygen probes (sondes) that can be deployed for entire seasons has normalized the use of open-water gas exchange methods, but the common practice of deploying single probes on buoys in the deepest, central part of the lake exacerbates the problem of minimizing littoral zone production (Brothers et al., 2017; Brothers and Vadeboncoeur, 2021). Chamber approaches are still critical for measuring periphyton contributions to whole-lake primary production, but the deployment of multiple sondes in the nearshore littoral

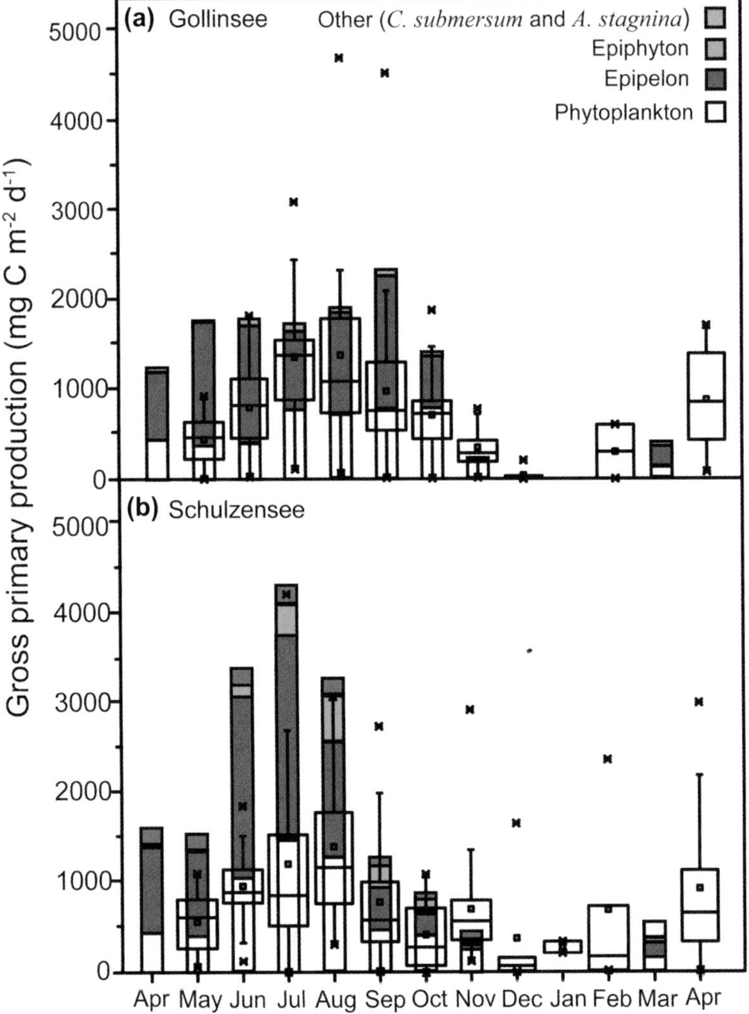

FIGURE 25-22 Gross primary production of phytoplankton, epipelic periphyton, epiphyton, and macrophytes (submerged *Ceratophyllum submersum*) in eutrophic, phytoplankton-dominated Gollinsee, and eutrophic, macrophyte-rich Schulzensee in Germany. *Columns* are based primarily upon direct fluorescence data and production models; *box plots* are whole-lake primary production measured with oxygen probes deployed at the center of the lake. Phytoplankton and periphyton in both lakes underwent seasonal cycles of primary productivity associated with light and temperature. (a) In phytoplankton-dominated Gollinsee epipelic algal production was less seasonally variable than phytoplankton, owing to depressed productivity during the summer when phytoplankton production biomass was high. (b) High light penetration and consequently high epipelic primary production resulted in macrophyte-dominated Schulzensee having higher primary production than Gollinsee. *(From Brothers et al., 2013.)*

zone can provide insights into littoral-benthic productivity (Van de Bogert et al., 2007).

Areal rates of benthic algal productivity are strongly determined by *photon flux density (PFD)* (the rate at which photons hit a surface) at the SWI (DeNicola et al., 2003; Vadeboncoeur et al., 2003, 2008; Karlsson et al., 2009; Godwin et al., 2014). There is a curvilinear relationship between PFD and photosynthesis that can be mathematically described by a *photosynthesis–irradiance (PE)* curve (see also Chapter 18). If PE relationships are determined across a depth gradient and combined with midday estimates of maximum photosynthesis rates, total photic zone benthic algal productivity can be assessed in much the same way that chamber estimates at multiple depths are used to estimate phytoplankton productivity at the ecosystem scale (Fee, 1969; Devlin et al., 2016). Creating PE curves for benthic algae using chamber methods can be prohibitively time consuming, but submersible *PAM fluorometers* can rapidly assess the relative responses of benthic algae to increasing light *in situ* (Fig. 25-16c; Hawes et al., 2014; Devlin et al., 2016). These approaches, combined with modeling, have helped us understand how productivity changes over time and space and how much benthic algae contribute to whole lake metabolism (Vadeboncoeur et al., 2008; Jäger and Diehl, 2014).

iii. Spatial and temporal variation in biomass and productivity

Season, substratum, disturbance, and grazers regulate within-lake variation in benthic algal biomass and productivity (Hillebrand, 2008, 2009; Figs. 25-10, 25-17 to 25-22). Estimates of whole-lake primary production that rely on extrapolations from chamber measurements are most sensitive to variation in primary productivity associated with substratum (Vadeboncoeur et al., 2001, 2006) and with accurately parameterizing variation in PE curves with depth (Vadeboncoeur et al., 2008; Genkai-Kato et al., 2012; Devlin et al., 2016). Within a lake, productivity on non-NDS such as wood and cobbles is low relative to productivity on nutrient-rich sediments (Vadeboncoeur et al., 2001, 2003; Fig. 25-13). With few exceptions (e.g., Liboriussen and Jeppesen, 2003; Brothers et al., 2013), estimates of annual primary production are based on data from the summer months. Algal productivity increases with increasing temperature, but seasonal changes in light often coincide with temperature driving a midsummer maximum in benthic algal, especially epipelic and epiphytic, productivity (Fig. 25-22). However, benthic algal productivity in eutrophic lakes peaks earlier in the year, and sometimes during winter, because high phytoplankton biomass during the summer months depresses epipelic productivity (Liboriussen and Jeppesen, 2003; Ogdahl et al., 2010; Brothers et al., 2013).

B. Benthic algal contributions to lake metabolism

Lake metabolism refers to the total rate of inorganic carbon fixation by primary producers (*gross primary production, GPP*) and respiration of fixed carbon by all autotrophic (R_a) and heterotrophic (R_h) organisms (see also Chapter 28). Lake metabolism is usually quantified on an annual or daily time scale, and metabolic fluxes are largely controlled by primary producers and microbes (Dodds and Cole, 2007). Functional groups of photoautotrophic organisms in lakes include phytoplankton, benthic algae, and macrophytes. The relative contribution of these three functional groups to lake metabolism, specifically GPP, is termed *autotrophic structure* (Higgins et al., 2014) and depends on latitude, lake morphometry, water clarity, and the distribution of nutrients. *Net primary production* (NPP = GPP-R_a) is the amount of carbon accumulated in autotrophic biomass in a year. NPP is an estimate of autotrophic carbon available to higher trophic levels, but NPP is difficult to measure in aquatic ecosystems because respiration by autotrophs and heterotrophs cannot be differentiated. The difference between GPP and ecosystem R ($R_a + R_h$) is termed *net ecosystem production (NEP)* and provides an index of the relative magnitude of autotrophic versus heterotrophic metabolic pathways in lakes. NEP of periphyton can be negative, even when periphyton GPP is quite high, because benthic algae grow so closely associated with heterotrophic microbes.

It is a mistake to assume that a low or negative NEP is indicative of low benthic algal productivity or a lack of importance of algae in the food web (Brothers and Vadeboncoeur, 2021). Attached algae grow rapidly and are avidly consumed by grazing arthropods and mollusks, as well as a plethora of infaunal grazers, whose respiration cannot be differentiated from that of the algae or associated bacteria. A negative NEP in the benthos simply means that bacteria and metazoans are respiring sources of carbon, such as terrestrial leaves, terrestrial DOC, or settled phytoplankton. This use of allochthonous carbon is arguably consequential for organismal energy budgets, especially those of bacteria, but it is no evidence that autochthonous algal production is unimportant to lake ecosystem function (Brothers and Vadeboncoeur, 2021).

i. Periphyton contributions to autotrophic structure

In spite of their often-cryptic appearance, the concentrated growth form of biofilms can yield surprisingly high rates of productivity, meaning that periphytic contributions to autotrophic structure are easily, and often, overlooked. In Lake Ontario (Canada/USA) benthic primary production per square meter of lake surface in littoral zone biofilms exceeded area-specific phytoplankton production to a depth of 12 m (Malkin et al., 2010). Thus it took a column of phytoplankton 12 m thick to equal the primary production rates achieved by a biofilm growing on sediments at the bottom of that water column.

Robert Wetzel's pioneering and detailed work on Lawrence Lake in Michigan (USA) demonstrated that littoral zones could dominate the autotrophic structure of lakes. Macrophytes in Lawrence Lake undergo seasonal cycles of abundance such that their contributions to autotrophic structure exceed that of phytoplankton (phytoplankton = 430 kg C $ha^{-1}y^{-1}$, versus 540 kg C $ha^{-1}y^{-1}$ for macrophytes) (Wetzel et al., 1972; Burkholder and Wetzel, 1989). However, the seasonal macrophyte growth in Lawrence Lake provided substrate for epiphytic algae that contributed an additional 2296 kg C ha^{-1} y^{-1} to whole-lake primary production. Macrophytes are particularly abundant and consequential in shallow lakes (Chapters 24 and 26), where they both fix carbon and provide surface area for periphyton growth. Epipelic algae dominate benthic algal contributions to primary production in many lakes (Vadeboncoeur et al., 2001, 2003; Liboriussen and Jeppesen, 2003; Ask et al., 2009; Brothers et al., 2013; Fig. 25-22) because soft sediments are usually the most extensive substratum in littoral zones. Furthermore, sediments have high rates of area-specific primary productivity relative to more nutrient-poor substrata (Fig. 25-13).

Low-nutrient, clear lakes can have high rates of littoral primary production because high light penetration leads to high light at the SWI, but whether periphyton dominates autotrophic structure in clear lakes depends on lake morphometry (the distribution of sediment with respect to depth) and lake volume. *Depth ratio* (mean depth/maximum depth) is an approximate, but informative, index of the distribution of sediments with respect to light (Carpenter, 1983), which strongly affects relative contributions of periphyton to whole-lake primary production (Vadeboncoeur et al., 2008). Periphyton tends to make larger contributions to autotrophic structure in lakes with low depth ratio (<0.5) because, for a given lake volume, a larger proportion of the lake bottom is exposed to light than in lakes with a high depth ratio.

In addition to the depth ratio, lake size (volume) affects autotrophic structure. Lake volume, on average, increases with maximum depth, and as volume increases, surface area:volume ratios typically decrease. Thus, in deep tectonic lakes, periphyton contribute less than 1% of whole lake primary production simply due to the overwhelming abundance of open water relative to benthic habitat (Vadeboncoeur et al., 2008; O'Reilly, 2006), but periphyton may make substantial contributions to autotrophic structure in the nearshore region of large glacially formed lakes such as the Laurentian Great Lakes (Brothers et al., 2016). In small, clear lakes, benthic primary production may account for close to 100% of primary production (Fig. 25-23; Vadeboncoeur et al., 2003). Most lakes in the world are small and relatively shallow (mean depth <5 m), which maximizes the potential for benthic contributions to autotrophic structure.

ii. Littoral productivity and eutrophication

Eutrophication is a process by which excess surface-water N and P stimulate the production of phytoplankton. In addition to contributing to autotrophic structure, phytoplankton strongly affect the rate of light absorption in the water column. Reductions in water

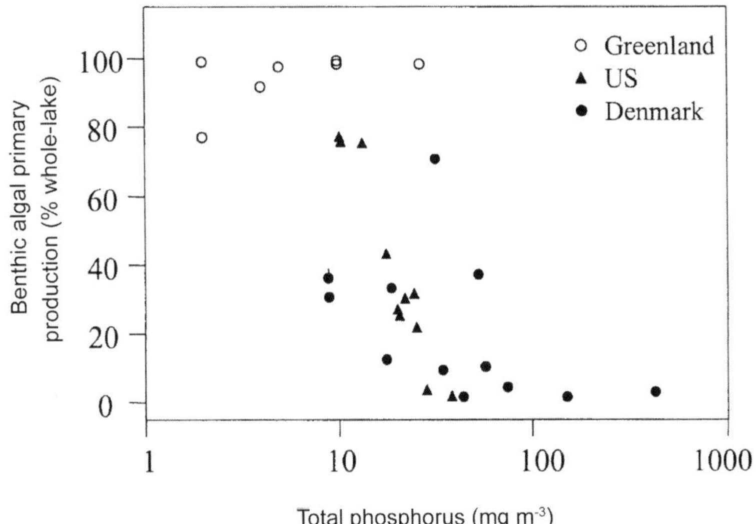

FIGURE 25-23 Benthic algal production can dominate autotrophic structure in shallow oligotrophic Arctic lakes, but in shallow eutrophic Danish lakes phytoplankton are responsible for the majority of whole-lake primary production. In stratified lakes in the US both phytoplankton and periphyton contributed to whole-lake primary production. As total phosphorus increases, phytoplankton greatly reduce light penetration, constricting the littoral zone and limiting epipelic primary production. *(From Vadeboncoeur et al., 2003.)*

clarity associated with high phytoplankton biomass negatively affect periphyton. In low-gradient landscapes most lakes are small, fed predominantly by surface water (rivers), and have short water retention times and high rates of exchange between sediment nutrient pools and the water column. These small and shallow lakes are often afflicted by eutrophication from agriculture and sewage. Polluted surface water promotes phytoplankton growth in shallow lakes. Although periphyton can make up >80% of whole-lake primary production in shallow, oligotrophic Arctic and boreal lakes (Vadeboncoeur et al., 2001, 2003; Ask et al., 2009), the autotrophic structure of small lakes in anthropogenic landscapes is overwhelmingly dominated by phytoplankton because high phytoplankton concentrations intercept light before it can reach benthic algae (Fig. 25-23; DeNicola et al., 2003; Vadeboncoeur et al., 2003), and this reduces epipelic algal production. Unlike epipelic algae, periphyton growing on wood, macrophytes, or rocks may have a positive response to increased water column nutrients at low levels of nutrient enrichment (Fig. 25-13; Vadeboncoeur and Lodge, 2000; Liboriussen and Jeppesen, 2006). Littoral *filamentous algal blooms (FABs)* (Fig. 25-17) are occurring in large, clear lakes, and this has been interpreted as an early indicator of eutrophication through groundwater pollution (Kann and Falter, 1989; Perillon et al., 2017; Timoshkin et al., 2018). However, based on water quality metrics, FABs lakes are usually oligotrophic (Vadeboncoeur et al., 2021).

Macrophytes can mitigate the effect of surface water nutrients on phytoplankton. Shallow lakes with macrophytes typically have clearer water (less phytoplankton biomass) than comparable lakes without macrophytes (Chapter 26). Macrophytes can also enhance filamentous algal biomass because macrophytes reduce water movement in the littoral zone and provide structural support for metaphyton. The negative effects of macrophytes on phytoplankton combined with the facilitation of periphyton growth in the presence of macrophytes lead to alternative stable states in shallow lakes in which either littoral primary producers (macrophytes and periphyton) or phytoplankton dominate autotrophic structure in these lakes (Chapter 26). In the absence of macrophytes epipelic algae may regulate transitions between alternative stable states (Genkai Kato et al., 2012; Jäger and Diehl, 2014), but in very eutrophic lakes phytoplankton shade both macrophytes and periphyton, and autotrophic structure is overwhelmingly dominated by phytoplankton.

iii. Benthic algal productivity and dissolved organic carbon

Colored DOC (Chapter 28) is one of the strongest determinants of light penetration in lakes (Williamson et al., 1999; Godwin et al., 2014). Whether DOC more negatively affects phytoplankton or periphyton productivity depends upon lake morphometry. High DOC has a stronger negative effect on periphyton compared to phytoplankton in steep-sided lakes with high depth ratios (Vadeboncoeur et al., 2008) because DOC will vastly constrict the littoral zone in steep-sided lakes. However, littoral zone structure can offset the effects of morphometry. The sediments of steep-sided, brown-colored, bog lakes in Finland are not well lit and do not support epipelic photosynthesis (Vesterinen et al., 2016). Nevertheless, surrounding bog vegetation and emergent and submerged macrophytes provided substrate for periphyton. Periphyton contributions to whole-lake primary production varied between 50% and 100% in lakes in which the littoral zone constituted between 15% and 25% of lake surface area (Vesterinen et al., 2016).

When epipelic productivity dominates autotrophic structure, the negative effects of DOC can cause whole-lake primary production in lakes to be limited by light, not nutrients (Karlsson et al., 2009). Most current conceptual models of lake productivity assume that water-column nutrients control total primary production. However, in small and moderate-sized oligotrophic lakes with organic, nutrient-rich sediments, epipelic productivity often dominates whole-lake primary production (Vadeboncoeur et al., 2001, 2003; Ask et al., 2009). DOC concentrations, not water-column nutrients, can determine whole-lake primary production in oligotrophic lakes (Karlsson et al., 2009). Fish production in these lakes was directly related to the amount of light hitting the lake bottom because fish depended upon the consumption of zoobenthos that fed on benthic algae (Karlsson et al., 2009; Fig. 25-24).

The negative effect of DOC loading and "brownification" on benthic algae can indirectly benefit phytoplankton by increasing nutrient flux to the water column (Kazanjian et al., 2021). Benthic algae regulate P flux across the SWI by producing oxygen through photosynthesis (Carlton and Wetzel, 1988; Wood et al., 2015). When epipelic algal productivity is suppressed by DOC, increased anoxia at the SWI moves P from the sediments to the water column (Genkai Kato et al., 2012; Brothers et al., 2014). Internal loading of P to the water column causes increases in phytoplankton that further shade out benthic algae and may promote eutrophication (Brothers et al., 2014; Vasconcelos et al., 2016, 2018).

C. Benthic algal contributions to food webs

Benthic algae are the primary food for many zoobenthic invertebrates and algivorous fish (Hecky and Hesslein, 1995). Predatory fishes feed on these benthic

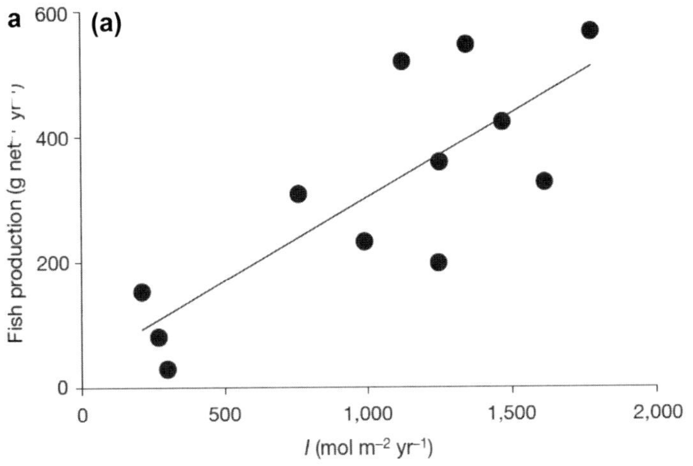

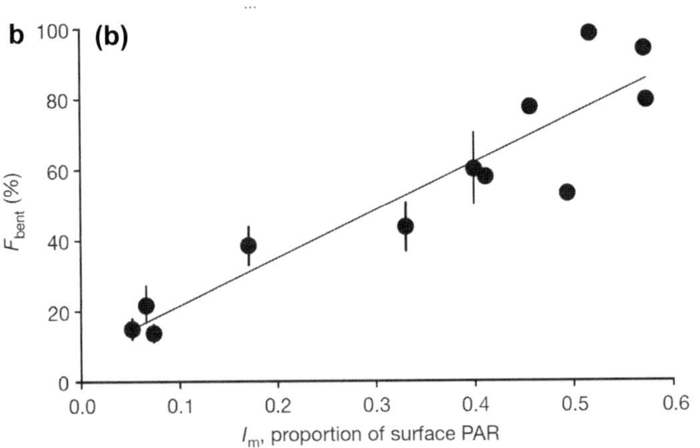

FIGURE 25-24 (a) Fish production in small boreal lakes was directly correlated with the average irradiance in the lake during the ice-free season, I, which controlled whole-lake primary production. Fish production was estimated based on the mass and growth rates of fish collected in standardized gill net sets. (b) Benthic algal contributions to fish body carbon (based on stable isotopes of carbon) are a function of the mean illumination of the lake (I_m). As light availability in the littoral zone increases, fish reliance on carbon fixed by benthic algae increases. The data are from relatively small boreal lakes in which I_m is controlled by concentrations of colored DOC. *(From Karlsson et al., 2009.)*

primary consumers, entraining carbon fixed by benthic algae into lake food webs (Vander Zanden and Vadeboncoeur, 2002; Vander Zanden et al., 2011). Furthermore, most fish species feed on either littoral food chains or linked littoral and open-water food chains (Vander Zanden and Vadeboncoeur, 2002; Sierszen et al., 2006).

i. Primary consumer diversity

Compared with phytoplankton, benthic algae grow in dense aggregations, concentrating biomass in relatively small volumes (Krause-Jensen and Sand-Jensen, 1998; Malkin et al., 2010). The physically compact growth forms of biofilms and turfs mean that benthic algae can be consumed directly by organisms ranging in size from a single cell to fish tens of centimeters long and weighing over 1 kg. Single-celled protozoans, rotifers, nematodes, and early instars of insects live within periphyton assemblages, consuming individual algal cells and grazing on the ECM (e.g., Furey et al., 2012). Many insect larvae, such as caddisflies and mayflies, graze by sweeping loosely attached benthic algae into their mouths. Gastropod mollusks and larval insects with scraping mouthparts effectively reduce biofilms to just a few cell layers thick (Fig. 25-19). Although grazing zooplankton generally rely on phytoplankton, zooplankton in clear Arctic lakes with low phytoplankton biomass are supported primarily by benthic algae (Rautio and Vincent, 2006). Vertebrate grazers include larval anurans (frog tadpoles) (Knoll et al., 2009; Whiles et al., 2013) and algivorous fish (Hori et al., 1993; Vadeboncoeur et al., 2011). In contrast to the large size range of periphyton grazers, zooplankton that graze directly on phytoplankton are typically much smaller than 5 mm long. The high diversity and broad size range of species that feed directly on benthic algae means that there are many pathways by which carbon fixed by benthic algae can flow into lake food webs (Hecky and Hesslein, 1995).

ii. Food quality

One reason for the diversity of grazers on benthic algae is that, like phytoplankton (Brett et al., 2009), benthic algae are digestible and highly nutritious. Thus organisms that feed on benthic algae have high production efficiency (growth/consumption) (Vadeboncoeur and Power, 2017; Trochine et al., 2020). Diatoms and Cyanobacteria have fewer indigestible structural compounds than green algae, which contain cellulose. But even green algae lack the difficult-to-digest lignin associated with terrestrial detritus derived from leaves or wood. Cyanobacteria are protein rich and diatoms have high concentrations of energy-dense oils including the essential fatty acid *eicosapentaenoic acid (EPA)*. EPA is crucial for the neurological development of many organisms and must be obtained from the diet (Torres-Ruiz et al., 2007; Guo et al., 2016a). The high food quality of benthic microalgae combined with the concentrated growth means that periphyton is readily and efficiently converted into consumer biomass, especially compared to terrestrial and aquatic plant detritus (James et al., 2000; Guo et al., 2016b; Mehner et al., 2016; Trochine et al., 2020).

Trophic efficiency is the proportion of biomass in a given trophic level that is consumed by higher trophic levels. Between 15% and 90% of net annual benthic microalgal primary production is consumed by grazers, which is among the highest consumption efficiencies of any food chain based on living (as opposed to detrital) primary producers (Cebrian, 1999).

iii. Littoral-pelagic food web linkages

Benthic primary consumers accumulate the essential fatty acids synthesized by diatoms and construct animal bodies based on carbon fixed by benthic algae. Secondary and tertiary consumers, especially fish, consume organisms that eat benthic algae. Stable isotopes of C, N, and H; fatty acids; and compound-specific stable isotopes provide natural tracers that allow researchers to quantify the importance of benthic algae to higher trophic levels. Beginning with Hecky and Hesslein's (1995) classic comparative study of fish diets in lakes using stable isotopes, evidence has accumulated that benthic algae are a dominant carbon pathway in lakes regardless of how much or how little they contribute to whole-lake primary production (Vander Zanden et al., 2011; Vander Zanden and Vadeboncoeur, 2020; Figs. 25-24 and 25-25).

Grazers that feed on benthic algae are consumed by fish (Vander Zanden and Vadeboncoeur, 2002). Even in the world's largest lakes, the majority of fish live and feed in littoral zones either exclusively or in addition to open water and profundal habitats (Vadeboncoeur et al., 2011; Sierszen et al., 2006, 2014). Most fish species feed on zoobenthos during some life stage, and benthic algae are the primary resource for, and support the production of, a wide variety of zoobenthos (Devlin et al., 2013; Herren et al., 2017; Trochine et al., 2020). Stable isotope analysis of fish from lakes around the world, including the largest lakes, demonstrate the widespread direct (e.g., algivores in tropical lakes) and indirect reliance of fish on carbon fixed by benthic algae (Figs. 25-24 and 25-25; Hecky and Hesslein, 1995; Fry et al. 1999; Vander Zanden and Vadeboncoeur, 2002, 2020; Karlsson et al., 2009; Vander Zanden et al., 2011).

When benthic algae are incorporated into models of lake food webs, we see that food chains based on phytoplankton alone provide the trophic basis of production for only a few fish species (Vander Zanden and Vadeboncoeur, 2002), though in large lakes these few species can be highly productive and support large open water commercial fisheries. In contrast, periphyton contribute to the production of most fish species in lakes, but not necessarily most fish production; the fish species that live in the littoral zone have lower population densities than the open water pelagic species supported by phytoplankton-based food chains (Vander Zanden et al., 2011). Food web linkages between littoral and pelagic habitats stabilize trophic cascades in lakes and buffer long-lived consumers from spatiotemporal variation in food availability (Vadeboncoeur et al., 2005; Stewart et al., 2017). Conservation of littoral zones is critical to aquatic conservation because lake food webs are highly integrated.

IV. Littoral benthic algae in a changing world

Humans have increased the amount of limiting nutrients in lakes and altered the Earth's climate. Littoral zones are experiencing changes in nutrient loading, wind, rainfall patterns, stratification regimes, average and maximum temperatures, and the duration of the growing season (e.g., Strayer and Findlay 2010). Some of these changes are altering (often increasing) the flux of DOC to lakes (Schindler et al., 1996), and increases in DOC will likely have a net negative effect on benthic algae and promote eutrophication (Brothers et al., 2014; Godwin et al., 2014; Vasconcelos et al., 2016, 2018; Diehl et al., 2018). Increases in DOC are expected to alter ecosystem function in many boreal and subarctic lakes, attenuating the influence of littoral zone processes.

Climate change alters stratification patterns, temperature regimes, and winter ice cover in lakes (Sharma et al., 2019). Extended growing seasons and reduced ice cover may increase the likelihood of benthic algal blooms. Increased CO_2 in the water may also promote the growth of filamentous green algae (Anderson and Anderson, 2006). In an era of global warming more research is needed on the direct influence of temperature on benthic algae (Trochine et al., 2014; Mahdy et al., 2015; Hao et al., 2018; Oleksy et al., 2020). Temperature and nutrients interact to affect the phenology of attached algal communities in shallow lakes with macrophytes (Trochine et al., 2014), but an increase in ambient temperature of 4.5°C did not strongly affect periphyton biomass on macrophytes and artificial substrata (Hao

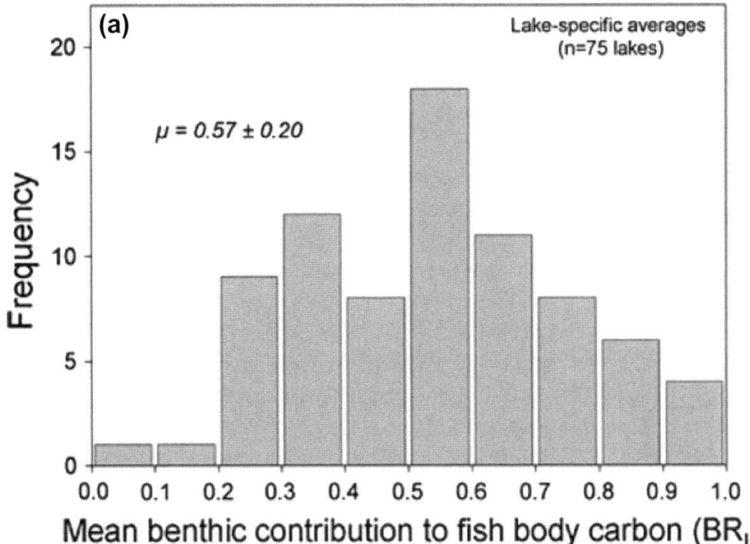

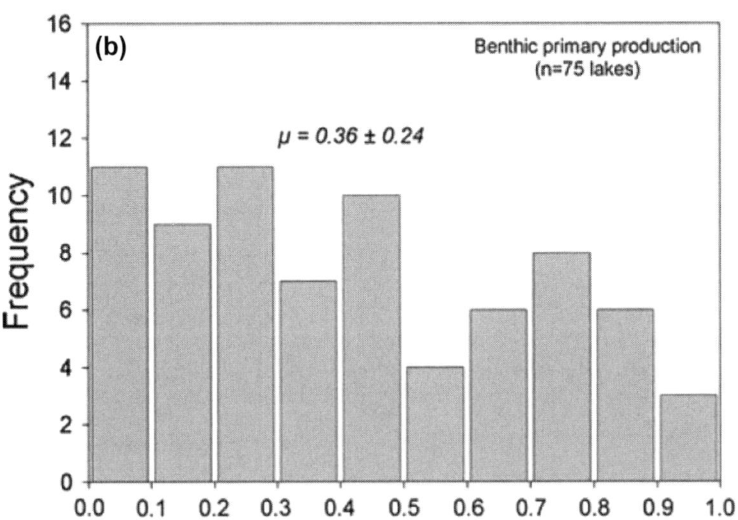

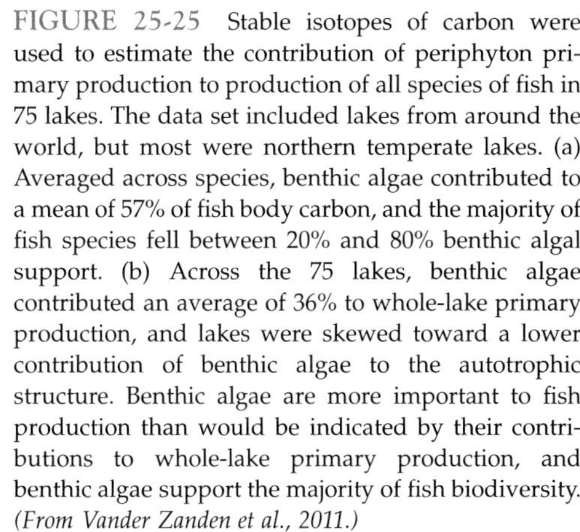
FIGURE 25-25 Stable isotopes of carbon were used to estimate the contribution of periphyton primary production to production of all species of fish in 75 lakes. The data set included lakes from around the world, but most were northern temperate lakes. (a) Averaged across species, benthic algae contributed to a mean of 57% of fish body carbon, and the majority of fish species fell between 20% and 80% benthic algal support. (b) Across the 75 lakes, benthic algae contributed an average of 36% to whole-lake primary production, and lakes were skewed toward a lower contribution of benthic algae to the autotrophic structure. Benthic algae are more important to fish production than would be indicated by their contributions to whole-lake primary production, and benthic algae support the majority of fish biodiversity. *(From Vander Zanden et al., 2011.)*

et al., 2018). Based on paleolimnological evidence, increases in temperature and nutrients seem to promote the growth of filamentous algae in remote mountain lakes (Oleksy et al., 2020). In Icelandic thermal streams whose mean summer temperatures varied from 2°C to 30°C, periphyton biomass increased with ambient stream temperature (Fig. 25-26). Benthic secondary production of invertebrates responded strongly and positively to this increase in resource supply, but consumption did not keep pace with the increase in algal biomass production associated with increased temperature (Junker et al., 2020). The interactive effects of temperature, nutrients, and grazing on benthic algae in lakes is a rich area for future research (Kazanjian et al., 2018).

In the past decade littoral FABs have been increasingly observed in clear oligotrophic lakes (Wood et al., 2020; Vadeboncoeur et al., 2021). Often, these lakes are quite remote and have minimal obvious signs of human impact (Rosenberger et al., 2008; Oleksy et al., 2020). Sometimes the source of the problem is identifiable. Groundwater pollution has transformed portions of the littoral zone of Lake Baikal (Russia), promoting the explosive growth of filamentous green alga *Spirogyra* (Timoshkin et al., 2016, 2018; Volkova et al., 2018). The littoral zones of the lower Laurentian Great Lakes are carpeted by dense growths of the green alga *Cladophora* because invasive Dreissenid mussels have translocated nutrients from the water column to the benthos and have increased water clarity, both of which favor FABs

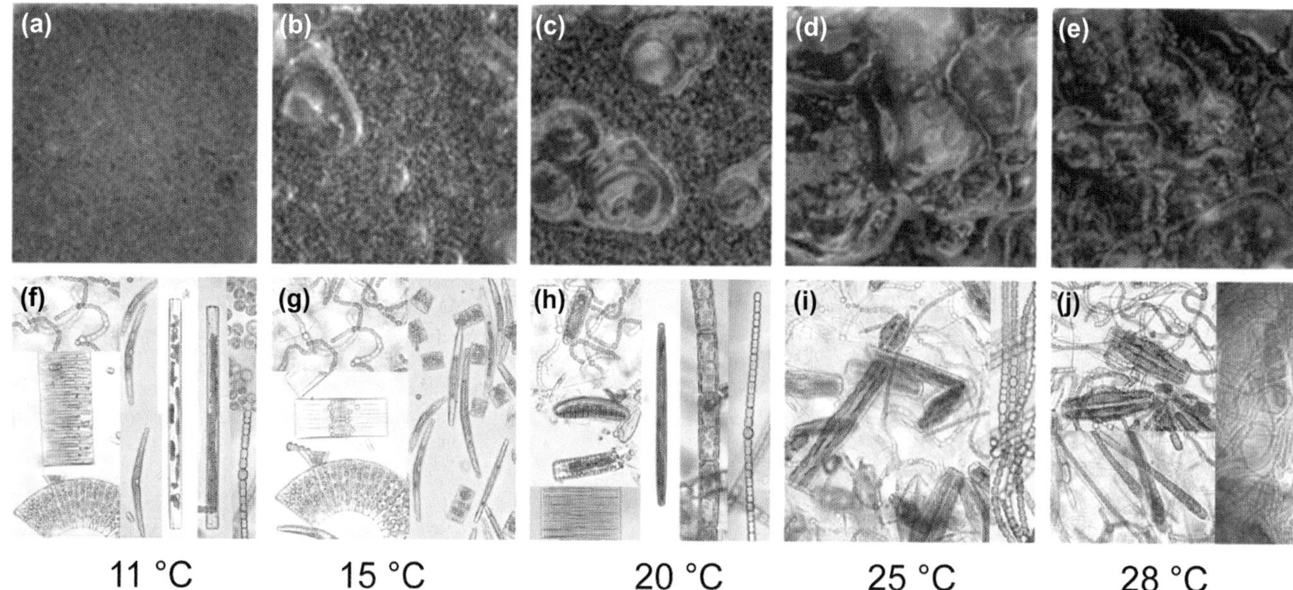

FIGURE 25-26 Temperature directly controls the growth rate of benthic algae. In the Hengill region of Iceland periphyton growing on tiles accumulate biomass much faster at warmer temperatures in experimental stream channels with a range of temperatures. (a)–(e) Increasing biomass with increasing temperature. (f)–(j) As temperatures warm, taxonomic composition shifts, and Cyanobacteria contribute more to total biomass. *(Photo credit: Tiles (a–e), James M. Hood; Light microscope images (f–i), Paula C. Furey.)*

(Hecky et al., 2004; Fig. 25-17). However, the occurrence of FABs in remote mountain lakes is more puzzling and may be caused by increased atmospheric nutrient deposition, a change in stratification and mixing regimes, or a reduction in grazing intensity (Rosenberger et al., 2008; Yamamuro et al., 2019; Oleksy et al., 2020; Vadeboncoeur et al., 2021). Responding to FABs and to climate change is hampered by the historical lack of research in littoral zones (Schindler and Scheurell, 2002; Vander Zanden and Vadeboncoeur, 2020), even though these nearshore habitats are where humans interact most intensively with lakes.

V. Summary

1. Benthic algae and Cyanobacteria grow on illuminated, submerged substrata in lake littoral zones. Littoral zones have diverse microhabitats including areas of sand, rock, organic sediments, macrophytes, and submerged wood. The distribution and abundance of these substrates depend on the terrestrial landscape and water movement.
2. Benthic algae are categorized by the types of substrata upon which they grow including epilithon (rocks), epipelon (organic sediments), epipsammon (sand), epiphyton (aquatic plants and macroalgae), epixylon (wood), and epizoan (animals).
3. Over 80% of all algal species are associated with benthic habitats.
    - **a.** Green algae are a common component of benthic algal communities. Filamentous green algae are common in high-light, high-energy habitats, while single-celled species are more common in quiescent habitats.
    - **b.** Diatoms occur and often dominate on all substrata throughout the littoral zones. Species can be sessile and tightly appressed to substrates, motile, stalked, or filamentous.
    - **c.** Cyanobacteria are common in all habitats and tolerate a wide range of light and temperature. Some Cyanobacteria taxa fix nitrogen, and others produce toxins.
    - **d.** Red and brown algae occur in some freshwater habitats but are uncommon.
4. Biofilms are layered aggregations of bacteria and algae that secrete an ECM that both sequesters ions and isolates the microorganisms from the water column. Turfs are more three-dimensionally complex aggregations that are dominated by upright and filamentous algae. Metaphyton are unattached clouds of benthic algae that form amorphous clouds above the sediments.
5. Resources regulating growth include both external resources in the overlying medium, such as light and nutrients in the surrounding water, and internal resources, such as nutrients that are obtained from the substrata to which the periphyton communities are attached.
    - **a.** Space affects the productivity of the benthic algal assemblages. Submerged macrophytes and

filamentous algae greatly increase space for epiphytic algae.

b. Light availability influences periphyton production seasonally and across depth gradients. Light is rapidly attenuated within biofilms, and as biofilms thicken, area-specific productivity asymptotes. Light limitation can cause algal senescence at the base of the mat and subsequent sloughing. Light limitation by epiphytes can kill the leaves of a macrophyte host.

c. Benthic algae can acquire nutrients from sediment pore-water, groundwater, macrophytes, animals, or the water column. Some benthic Cyanobacteria and diatoms fix nitrogen.

d. Increases in nutrients, particularly phosphorus and nitrogen, in the overlying water commonly result in shifts in species composition and increases in growth and biomass of attached algae on chemically inert substrata (e.g., rocks and wood).

e. Water movement reduces boundary layer thickness between the periphyton and the overlying water, increasing nutrient exchange with the overlying water.

f. Nutrient limitation of periphyton is usually determined experimentally with NDS. Water quality metrics used for phytoplankton are poorly correlated with benthic algal biomass.

6. Mortality losses for attached algae are dominated by disturbance and grazing.
   a. Physical disruptions of substrata by waves and currents and by changes in water level can reduce benthic algal biomass in erosional zones of lakes.
   b. Grazing of periphyton by animals (insect larvae, gastropod mollusks, crayfish, and fishes) strongly reduces benthic algal biomass but tends to have a smaller effect on area-specific productivity. Diatoms are favored by grazers but are able to persist in the face of heavy grazing. In the absence of grazing filamentous green algae often gain ascendancy.
7. In temperate regions benthic algal biomass and productivity change with seasonal changes in incident light and temperature.
   a. Epipelic diatoms frequently exhibit population maxima in the spring and summer.
   b. Seasonal population dynamics of epiphytic algae are more variable and change as the surface area available for colonization of the supporting macrophytes changes seasonally.
8. Algae and bacteria in biofilms exchange metabolites and nutrients. Bacterial production is increased by algal exudates of organic substrates and oxygen. Benthic algae use CO_2, nutrients, and vitamins from heterotrophic microbes.
9. Benthic algae compete with macrophytes and phytoplankton for light and nutrients. Benthic algae and macrophytes have access to sediment nutrients, which are often the largest pool of nutrients in the lake. Benthic algae can acquire nutrients from macrophytes and can overgrow macrophytes, affecting their CO_2 and light acquisition. When water column nutrients are high, phytoplankton can shade benthic algae and macrophytes.
10. Algae growing on animal shells, carapaces, and retreats exploit the nutrients excreted by animals. Animal excretion can change the stoichiometry of benthic algal assemblages. Aggregations of animals can create nutrient hotspots for benthic algae, and animal excretion can supply a substantial proportion of benthic algal demand for nutrients.
11. It is necessary to account for the physical heterogeneity of littoral zones in order to quantify benthic algal contributions to ecosystem function.
    a. Estimates of benthic algal biomass using pigment analysis, taxonomic composition, or AFDM can provide information on the role of benthic algae in a lake.
    b. Estimates of benthic algal productivity allow researchers to assess littoral contributions to lake metabolism and food webs.
    c. Molecular techniques will help us understand the algal and microbial diversity supported by complex littoral habitats.
12. Benthic algae can dominate whole-lake primary production (autotrophic structure), especially in small, clear lakes. Eutrophication diminishes benthic contributions to autotrophic structure. Oxygen probes deployed in the middle of lakes underestimate benthic contributions to lake metabolism.
13. Benthic algae are eaten by a wide size range of primary consumers, and they contribute to the growth of the majority of fish and invertebrate diversity in lakes. Benthic algal contributions to lake food webs are usually greater than would be predicted based on benthic algal contributions to whole-lake primary production.
14. Climate warming and increased mobilization of nutrients will tend to increase the growth and biomass accumulation of benthic algae. In subarctic and boreal regions increased loading of DOC will tend to reduce the role of epipelic algae in lake ecosystem function. Benthic FABs are increasingly common in clear lakes worldwide.

References

Abe, S., Uchida, K., Nagumo, T., Tanaka J., J., 2007. Alterations in the biomass-specific productivity of periphyton assemblages mediated by fish grazing. Freshw. Biol. 52, 1486–1493.

Allen, H.L., 1971. Primary productivity, chemo-organotrophy, and nutritional interactions of epiphytic algae and bacteria on macrophytes in the littoral of a lake. Ecol. Monogr. 41, 97–127.

Anderson, T., Anderson, F.O., 2006. Effects of CO_2 concentration on growth of filamentous algae and *Littorella uniflora* in a Danish softwater lake. Aquat. Bot. 84, 267–271.

Andre, E.R., Hecky, R.E., Duthie, H.C., 2003. Nitrogen and phosphorus regeneration by cichlids in the littoral zone of Lake Malawi, Africa. J. Great Lakes Res. 29 (Suppl. 2), 190–201.

Ask, J., Karlsson, J., Persson, L., Ask, P., Byström, P., Jansson, M., 2009. Whole-lake estimates of carbon flux through algae and bacteria in benthic and pelagic habitats of clear-water lakes. Ecology 90, 1923–1932.

Azim, M.E., Asaeda, T., 2005. Periphyton structure, diversity, and colonization. In: Azim, M.E., Verdegem, M.C.J., van Dam, A.A., Beveridge, M.C.M. (Eds.), Periphyton: Ecology, Exploitation, and Management. CABI Publishing, Wallingford, UK, pp. 15–33.

Benelli, S., Bartoli, M., Zilius, M., Vybernaite-Lubiene, I., Ruginis, T., Petkuviene, J., Fano, E.A., 2018. Microphytobenthos and chironomid larvae attenuate nutrient recycling in shallow-water sediments. Freshw. Biol. 63, 187–201.

Biddanda, B.A., Nold, S.C., Dick, G.J., Kendall, S.T., Vail, J.H., Ruberg, S.A., Green, C.M., 2012. Rock, water, microbes: underwater sinkholes in Lake Huron are habitats for ancient microbial life. Nature Education Knowledge 3, 13.

Borchardt, M.A., 1996. Nutrients. In: Stevenson, R.J., Bothwell, M.L., Lowe, R.L. (Eds.), Algal Ecology: Freshwater Benthic Ecosystems. Academic Press, San Diego, pp. 183–227.

Bothwell, M.L., Taylor, B.W., Kilroy, C., 2014. The Didymo story: the role of low dissolved phosphorus in the formation of *Didymosphenia geminata* blooms. Diatom Res. https://doi.org/10.1080/0269249X.889041.

Brett, M.T., Kainz, M.J., Taipale, S.J., Seshan, H., 2009. Phytoplankton, not allochthonous carbon, sustains herbivorous zooplankton production. Proc. Natl. Acad. Sci. USA 106, 21197–21201.

Brothers, S.M., Hilt, S., Meyer, S., Köhler, J., 2013. Plant community structure determines primary productivity in shallow, eutrophic lakes. Freshw. Biol. 58, 2264–2276. https://doi.org/10.1111/fwb.12207.

Brothers, S., Kazanjian, G., Köhler, J., Scharfenberger, U., Hilt, S., 2017. Convective mixing and high littoral production established systematic errors in the diel oxygen curves of a shallow, eutrophic lake. Limnol Oceanogr. Methods 15, 429–435. https://doi.org/10.1002/lom3.10169.

Brothers, S., Köhler, J., Attermeyer, K., Grossart, H.P., Mehner, T., Meyer, N., Scharnweber, K., Hilt, S., 2014. A feedback loop links brownification and anoxia in a temperate, shallow lake. Limnol. Oceanogr. 59, 1388–1398.

Brothers, S., Vadeboncoeur, Y., 2021. Shoring up the foundations of production to respiration ratios in lakes. Limnol. Oceanogr. 66, 2762–2778.

Brothers, S., Vadeboncoeur, Y., Sibly, P., 2016. Benthic algae compensate for phytoplankton losses in large aquatic systems. Glob. Change Biol. 22, 3865–3873.

Burkholder, J.M., 1996. Interactions of benthic algae with their substrata. In: Stevenson, R.J., Bothwell, M.L., Lowe, R.L. (Eds.), Algal Ecology: Freshwater Benthic Ecosystems. Academic Press, San Diego, pp. 253–297.

Burkholder, J.M., Wetzel, R.G., 1989. Microbial colonization on natural and artificial macrophytes in a phosphorus limited, hardwater lake. J. Phycol. 25, 55–65.

Cantonati, M., Lowe, R.L., 2014. Lake benthic algae: toward an understanding of their ecology. Freshw. Sci. 33, 475–486.

Capps, K.A., Flecker, A.S., 2013a. Invasive fishes generate biogeochemical hotspots in a nutrient-limited system. PLoS One 8 (1), e54093. https://doi.org/10.1371/journal.pone.0054093.

Capps, K.A., Flecker, A.S., 2013b. Invasive aquarium fish transform ecosystem nutrient dynamics. Proc. R. Soc. B 280, 20131520. https://doi.org/10.1098/rspb.2013.1520.

Carey, C., Rydin, E., 2011. Lake trophic status can be determined by the depth distribution of sediment phosphorus. Limnol. Oceanogr. 56, 2051–2063.

Carey, C.C., Weathers, K.C., Ewing, H.A., Greer, M.L., Cottingham, K.L., 2014. Spatial and temporal variability in recruitment of the cyanobacterium *Gloeotrichia echinulate* in an oligotrophic lake. Freshw. Sci. 33, 577–592.

Carlton, R.G., Wetzel, R.G., 1988. Phosphorus flux from lake sediments: effect of epipelic algal oxygen production. Limnol. Oceanogr. 33, 562–570.

Carpenter, S.R., 1983. Lake geometry: implications for production and sediment accretion rates. J. Theor. Biol. 105, 273–286.

Carrick, H.J., 2004. Algal distribution patterns in Lake Erie: implications for oxygen balances in the eastern basin. J. Great Lakes Res. 30, 133–147.

Carrick, H.J., Lowe, R.L., 2007. Nutrient limitation of benthic algae in Lake Michigan: the role of silica. J. Phycol. 43, 228–234.

Cattaneo, A., Amireault, M.C., 1992. How artificial are artificial substrata for periphyton? J. N. Am. Benthol. Soc. 11, 244–256.

Cattaneo, A., Kalff, J., 1980. The relative contribution of aquatic macrophytes and their epiphytes to the production of macrophyte beds. Limnol. Oceanogr. 25, 280–289.

Cebrian, J., 1999. Patterns in the fate of production in plant communities. Am. Nat. 154, 449–468.

Christensen, D.L., Herwig, B.R., Schindler, D.E., Carpenter, S.R., 1996. Impacts of lakeshore residential development on coarse woody debris in north temperate lakes. Ecol. Appli. 6, 1143–1149.

Cooper, M.J., Costello, G.M., Francoeur, S.N., Lamberti, G.A., 2016. Nitrogen limitation of algal biofilms in coastal wetlands of Lakes Michigan and Huron. Freshw. Sci. 35, 25–40.

Corman, J.R., McIntyre, P.B., Kuboja, B., Mbemba, W., Fink, D., Wheeler, C.W., Gans, C., Michel, E., Flecker, A.S., 2010. Upwelling couples chemical and biological dynamics across the littoral and pelagic zones of Lake Tanganyika, East Africa. Limnol. Oceanogr. 55, 214–224.

DeNicola, D.M., 1996. Periphyton responses to temperature at different ecological levels. In: Stevenson, R.J., Bothwell, M.L., Lowe, R.L. (Eds.), Algal Ecology: Freshwater Benthic Ecosystems. Academic Press, San Diego, pp. 149–181.

DeNicola, D.M., de Eyto Wemaere, A., Irvine, K., 2003. Production and respiration of epilithic algal communities in Irish lakes of different trophic status. Arch. Hydrobiol. 157, 67–87.

DeNicola, D.M., Kelly, M., 2014. Role of periphyton in ecological assessment of lakes. Freshw. Sci. 33, 619–638.

Devlin, S.P., Vander Zanden, M.J., Vadeboncoeur, Y., 2013. Depth-specific variation in carbon isotopes demonstrates resource partitioning among the littoral zoobenthos. Freshw. Biol. 11, 2389–2400.

Devlin, S.P., Vander Zanden, M.J., Vadeboncoeur, Y., 2016. Littoral-benthic primary production estimates: sensitivity to simplifications with respect to periphyton productivity and basin morphometry. Limnol Oceanogr. Methods. https://doi.org/10.1002/lom3.10080.

Diehl, S., Thomsson, G., Kahlert, M., Guo, J., Karlsson, J., Liess, A., 2018. Inverse relationship of epilithic algae and pelagic phosphorus in unproductive lakes: roles of N_2 fixers and light. Freshw. Biol. 63, 662–675.

Dodds, W.K., 2003. The role of periphyton in phosphorus retention in shallow freshwater systems. J. Phycol. 39, 840–849.

Dodds, W.K., Biggs, B.J.F., Lowe, R.L., 1999. Photosynthesis-irradiance patterns in benthic microalgae: variations as a function of assemblage thickness and community structure. J. Phycol. 35, 42–53.

Dodds, W.K., Cole, J.J., 2007. Expanding the concept of trophic state in aquatic ecosystems: it's not just the autotrophs. Aquat. Sci. 69, 427–439. https://doi.org/10.1007/s00027-007-0922-1.

Dodds, W.K., Whiles, M.R., 2010. Freshwater Ecology, second ed. Elsevier, San Diego.

Donahue, W.F., Turner, M.A., Findlay, D.L., Leavitt, P.R., 2003. The role of solar radiation in structuring the shallow benthic communities of boreal forest lakes. Limnol. Oceanogr. 48, 31–47.

Douglas, M.S.V., Smol, J.P., 2010. Freshwater diatoms as indicators of environmental change in the High Arctic. In: Smol, J.P., Stoermer, E.F. (Eds.), The Diatoms: Applications for the Environmental and Earth Sciences, second ed. Cambridge University Press, Cambridge, pp. 249–266.

Fairchild, W., Lowe, R.L., Richardson, W.B., 1985. Algal periphyton growth on nutrient-diffusing substrates: an *in situ* bioassay. Ecology 66, 465–472.

Fee, E.J., 1969. A numerical model for the estimation of photosynthetic production, integrated over time and depth, in natural waters. Limnol. Oceanogr. 14, 906–911.

Fleming, E.D., Prufert-Bebout, L., 2010. Characterization of cyanobacterial communities from high-elevation lakes in the Bolivian Andes. J. Geophys. Res. 115, G00D07. https://doi.org/10.1029/2008JG000817.

Forehead, H.I., Thomson, P.A., 2010. Microbial communities of subtidal shallow sandy sediments change with depth and wave disturbance, but nutrient exchanges remain similar. Mar. Ecol. Prog. Ser. 414, 11–26.

Forehead, H., Thomson, P., Kendrick, G.A., 2013. Shifts in composition of microbial communities of subtidal sandy sediments maximise retention of nutrients. FEMS Microbiol. Ecol. 83, 279–298.

Francoeur, S.N., Winslow, K.A.P., Miller, D., Peacor, S.D., 2017. Mussel-derived stimulation of benthic filamentous algae: the importance of nutrients and spatial scale. J. Great Lakes Res. 43, 69–79.

Fry, B., Mumford, P.L., Tam, F., Fox, D.D., Warren, G.L., Havens, K.E., Steinman, A.D., 1999. Trophic position and individual feeding histories of fish from Lake Okeechobee, Florida. Can. J. Fish. Aquat. Sci. 56, 590–600.

Furey, P.C., Lowe, R.L., Power, M.E., Campbell-Craven, A.M., 2012. Midges, Cladophora and epiphytes: shifting interactions through succession. Freshw. Sci. 31, 93–107.

Genkai-Kato, M., Vadeboncoeur, Y., Liboriussen, L., Jeppesen, E., 2012. Benthic-planktonic coupling, regime shifts, and whole-lake primary production in shallow lakes. Ecology 93, 619–631. https://doi.org/10.1016/0921-4526(94)91066-9.

Godwin, S.C., Jones, S.E., Weidel, B.C., Solomon, C.T., 2014. Dissolved organic carbon concentration controls benthic primary production: results from in situ chambers in north-temperate lakes. Limnol. Oceanogr. 59, 2112–2120.

Goldsborough, L.G., Robinson, G.G.C., 1996. Patterns in wetlands. In: Stevenson, R.J., Bothwell, M.L., Lowe, R.L. (Eds.), Algal Ecology: Freshwater Benthic Ecosystems. Academic Press, San Diego, pp. 77–117.

Gruendling, G.K., 1971. Ecology of the epipelic algal communities in marion lake, British Columbia. J. Phycol. 7, 239–249.

Gubelit, Y.I., Grossart, H.P., 2020. New methods, new concepts: what can be applied to freshwater periphyton? Front. Microbiol. 11, 1275.

Guo, F., Kainz, M.J., Sheldon, F., Bunn, S.E., 2016a. The importance of high quality algal food sources in stream food webs—current status and future perspectives. Freshw. Biol. 61, 815–831.

Guo, F., Kainz, M.J., Valdez, D., Sheldon, F., Bunn, S.E., 2016b. The effect of light and nutrients on algal food quality and their consequent effect on grazer growth in subtropical streams. Freshw. Sci. 35, 1202–1212.

Hagerthey, S.E., Kerfoot, W.C., 1998. Groundwater flow influences the biomass and nutrient ratios of epibenthic algae in a north temperate seepage lake. Limnol. Oceanogr. 43, 1127–1242.

Hansson, L.A., 1996. Algal recruitment from lake sediments in relation to grazing, sinking, and dominance patterns in the phytoplankton community. Limnol. Oceanogr. 41, 1312–1323.

Hao, B., Fabrin Roejkjaer, A., Wu, H., Cao, Y., Jeppesen, E., Li, W., 2018. Responses of primary producers in shallow lakes to elevated temperature: a mesocosm experiment during the growing season of *Potamogeton crispus*. Aquat. Sci. 80, 34. https://doi.org/10.1007/s00027-018-0585-0.

Harrison, S.S., Hildrew, A.G., 2001. Epilithic communities and habitat heterogeneity in a lake littoral. J. Anim. Ecol. 70, 692–707.

Hart, D.D., 1981. Foraging and resource patchiness: field experiments with a grazing stream insect. Oikos 37, 46–52.

Hawes, I., Giles, H., Doran, P.T., 2014. Estimating photosynthetic activity in microbial mats in an ice-covered Antarctic lake using automated oxygen microelectrode profiling and variable chlorophyll fluorescence. Limnol. Oceanogr. 59, 674–688.

Hawes, I., Smith, R., 1994. Seasonal dynamics of epilithic periphyton in oligotrophic Lake Taupo, New Zealand. N. Z. J. Mar. Freshwater Res. 28, 1–12.

Hecky, R.E., Hesslein, R.H., 1995. Contributions of benthic algae to lake food webs as revealed by stable isotope analysis. J. N. Am. Benthol. Soc. 14, 631–653.

Hecky, R.E., Smith, R.E., Barton, D.R., Guildford, S.J., Taylor, W.D., Charlton, M.N., Howell, T., 2004. The nearshore phosphorus shunt: a consequence of ecosystem engineering by dreissenids in the Laurentian Great Lakes. Can. J. Fish. Aquat. Sci. 61, 1285–1293.

Herren, C.M., Webert, K.C., Drake, M.D., Vander Zanden, M.J., Einarsson, A., Ives, A.R., Gratton, C., 2017. Positive feedback between chironomids and algae creates net mutualism between benthic primary consumers and producers. Ecology 98, 447–455.

Hesslein, R.H., Broecker, W.S., Quay, P.D., Schindler, D.W., 1980. Whole-lake radiocarbon experiment in an oligotrophic lake at the Experimental Lakes Area, northwestern Ontario. Can. J. Fish. Aquat. Sci. 37, 454–463.

Higgins, S.N., Althouse, B., Devlin, S.P., Vadeboncoeur, Y., Vander Zanden, M.J., 2014. Potential for large-bodied zooplankton and dreissenids to alter the productivity and autotrophic structure of lakes. Ecology 95, 2257–2267.

Higgins, S.N., Kling, H.J., Hecky, R.E., Taylor, W.D., Bootsma, H.A., 2003a. The community composition, distribution, and nutrient status of epilithic periphyton at five rocky littoral zone sites in Lake Malawi, Africa. J. Great Lakes Res. 29 (Suppl. 2), 181–189. https://doi.org/10.1016/S0380-1330(03)70547-4.

Higgins, S.N., Hecky, R.E., Taylor, W.D., 2003b. Epilithic nitrogen fixation in the rocky littoral zones of Lake Malawi, Africa. Limnol. Oceanogr. 46, 976–982.

Higgins, S.N., Hecky, R.E., Guildford, S.J., 2008a. The collapse of benthic macroalgal blooms in response to self-shading. Freshw. Biol. 53, 2557–2572.

Higgins, S.N., Malkin, S.Y., Howell, E.T., Guildford, S.J., Campbell, L., Hiriart-Baer, V., Hecky, R.E., 2008b. An ecological review of *Cladophora glomerata* (chlorophyta) in the Laurentian Great Lakes. J. Phycol. 44, 839–854.

Hill, W.R., Boston, H.L., Steinman, A.D., 1992. Grazers and nutrients simultaneously limit lotic primary productivity. Can. J. Fish. Aquat. Sci. 49 (3), 504–512.

Hill, W.R., Smith, J.G., Stewart, A.J., 2010. Light, nutrients and herbivore growth in oligotrophic streams. Ecology 91, 518–527.

Hillebrand, H., 2002. Top-down versus bottom-up control of autotrophic biomass—a meta-analysis on experiments with periphyton. J. N. Am. Benthol. Soc. 21, 349–369.

Hillebrand, H., 2008. Grazing regulates the spatial variability of periphyton biomass. Ecology 89, 165–173.

Hillebrand, H., 2009. Meta-analysis of grazer control of periphyton biomass across aquatic ecosystems. J. Phycol. 45, 798–806.

Hillebrand, H., Dürselen, C.-D., Kirschtel, D., Pollingher, U., Zohary, T., 1999. Biovolume calculation for pelagic and benthic microalgae. J. Phycol. 35, 403–424.

Hoagland, K.D., Roemer, S.C., Rosowski, J.R., 1982. Colonization and community structure of two periphyton assemblages, with emphasis on the diatoms (Bacillariophyceae). Am. J. Bot. 69, 188–213.

Hofmann, A.M., Geist, J., Nowotny, L., U. Raeder, J., 2020. Depth-distribution of lake benthic diatom assemblages in relation to light availability and substrate: implications for paleolimnological studies. J. Paleolimnol. 64, 315–334.

Hori, M., Gashagaza, M.M., Nshombo, M., Kawanabe, H., 1993. Littoral fish communities in Lake Tanganyika: irreplaceable diversity supported by intricate interactions among species. Conserv. Biol. 7, 657–666.

Hudon, C., De Sève, M., Cattaneo, A., 2014. Increasing occurrence of the benthic filamentous cyanobacterium *Lyngbya wollei*: a symptom of freshwater ecosystem degradation. Freshw. Sci. 33, 606–618.

Ilmavirta, V., Jones, R.I., Kairesalo, T., 1977. The structure and photosynthetic activity of pelagial and littoral plankton communities in Lake Pääjärvi, southern Finland. Ann. Bot. Fennici 14, 7–16.

Ings, N.L., Hildrew, A.G., Grey, J., 2010. Gardening by the psychomyiid caddisfly *Tinodes waeneri*: evidence from stable isotopes. Oecologia 163, 127–139.

Jäger, C.G., Diehl, S., 2014. Resource competition across habitat boundaries: asymmetric interactions between benthic and pelagic producers. Ecol. Monogr. 84, 287–302. https://doi.org/10.1890/13-0613.1.

James, M.R., Hawes, I., Weatherhead, M., Carmen Stanger, C., Gibbs, M., 2000. Carbon flow in the littoral food web of an oligotrophic lake. Hydrobiologia 441, 93–106.

Jones, J.I., Eaton, J.W., Hardwick, K., 2000. The influence of periphyton on boundary layer pH conditions: a microelectrode investigation. Aquat. Bot. 67, 191–206.

Jones, J.I., Sayer, C.D., 2003. Does the fish-invertebrate-periphyton cascade precipitate plant loss in shallow lakes? Ecology 84, 2155–2167.

Junker, J.R., Cross, W.F., Benstead, J.P., Huryn, A.D., Hood, J.M., Nelson, D., Gíslason, G.M., Ólafsson, J.S., 2020. Resource supply governs the apparent temperature dependence of animal production in stream ecosystems. Ecol. Lett. https://doi.org/10.1111/ele.13608.

Kann, J., Falter, C.M., 1989. Periphyton as indicators of enrichment in Lake Pend Oreille, Idaho. Lake Reserv. Manag 5, 39–48.

Karlsson, J., Byström, P., Ask, J., Ask, P., Persson, L., Jansson, M., 2009. Light limitation of nutrient-poor lake ecosystems. Nature 460, 506–509. https://doi.org/10.1038/nature08179.

Kazanjian, G., Brothers, S., Köhler, J., Hilt, S., 2021. Incomplete recovery of a shallow lake from a natural browning event. Freshw. Bio. 66, 1089–1100.

Kazanjian, G., Velthuis, M., Aben, R., Stephan, S., Peeters, E.T., Frenken, T., Touwen, J., Xue, F., Kosten, S., Van de Waal, D.B., de Senerpont Domis, L.N., Van Donk, E., Hilt, S., 2018. Impacts of warming on top-down and bottom-up controls of periphyton production. Sci. Rep. 8, 9901.

Kingston, J.C., Lowe, R.L., Stoermer, E.F., Ladewski, T.B., 1983. Spatial and temporal distribution of benthic diatoms in northern Lake Michigan. Ecology 64, 1566–1580.

Kishler, J., Taft, C.E., 1970. *Bangia atropurpurea (Roth) A.* In western Lake Erie. Ohio J. Sci. 70, 56–57.

Knoll, L.B., McIntyre, P.B., Vanni, M.J., Flecker, A.S., 2009. Feedbacks of consumer nutrient recycling on producer biomass and stoichiometry: separating direct and indirect effects. Oikos 118, 1732–1742.

Kohler, S.L., 1984. Search mechanism of a stream grazer in patchy environments: the role of food abundance. Oecologia 62, 209–218.

Kowalczewski, A., 1975. Algal primary production in the zone of submerged vegetation of a eutrophic lake. Verh. Internat. Verein. Limnol. 19, 1305–1308.

Krause-Jensen, D., Sand-Jensen, K., 1998. Light attenuation and photosynthesis of aquatic plant communities. Limnol. Oceanogr. 43, 396–407.

Krejci, M.E., Lowe, R.L., 1986. Spatial and temporal variation of epipsammic diatoms in a spring-fed brook. J. Phycol. 23, 585–590.

Kupferberg, S.J., 1997. Facilitation of periphyton production by tadpole grazing: functional differences between species. Freshw. Biol. 37, 427–439.

Liboriussen, L., Jeppesen, E., 2003. Temporal dynamics in epipelic, pelagic and epiphytic algal production in a clear and a turbid shallow lake. Freshw. Biol. 48, 418–431.

Liboriussen, L., Jeppesen, E., 2006. Structure, biomass, production and depth distribution of periphyton on artificial substratum in shallow lakes with contrasting nutrient concentrations. Freshw. Biol. 51, 95–109.

Lindeman, R.L., 1942. The trophic-dynamic aspect of ecology. Ecology 23, 399–418.

Loeb, S.L., Reuter, J.E., Goldman, C.R., 1983. Littoral zone production of oligotrophic lakes. The contributions of phytoplankton and periphyton. In: Wetzel, R.G. (Ed.), Periphyton of Freshwater Ecosystems. W. Junk Publ., The Hague, pp. 161–167.

Lowe, R.L., 1996. Periphyton patterns in lakes. In: Stevenson, R.J., Bothwell, M.L., Lowe, R.L. (Eds.), Algal Ecology: Freshwater Benthic Ecosystems. Academic Press, San Diego, pp. 57–76.

Lowe, R.L., Pan, Y., 1996. Benthic algal communities as biological monitors. In: Stevenson, R.J., Bothwell, M.L., Lowe, R.L. (Eds.), Algal Ecology: Freshwater Benthic Ecosystems. Academic Press, San Diego, pp. 705–739.

Lukens, N.R., Kraemer, B.M., Constant, V., Hamann, E.J., Michel, E., Socci, A.M., Vadeboncoeur, Y., McIntyre, P.B., 2017. Animals and their epibiota as net autotrophs: size scaling of epibiotic metabolism on snail shells. Freshw. Sci. 36, 307–315. https://doi.org/10.1086/691438.

Mahdy, A., Hilt, S., Filiz, N., Beklioglu, M., Hejzlar, J., Ozkundakci, D., Papastergiadou, E., Scharfenberger, U., Sorf, M., Stefandis, K., Tuvikene, L., Zingel, P., Sondergaard, M., Jeppesen, E., Adrian, R., 2015. Effects of water temperature on summer periphyton biomass in shallow lakes: a pan-European mesocosm experiment. Aquat. Sci. 77, 499–510.

Malkin, S., Bokaniov, S.A., Smith, R.E., Guildford, S.J., Hecky, R.E., 2010. *In situ* measurements confirm the seasonal dominance of benthic algae over phytoplankton in nearshore primary production of a large lake. Freshw. Biol. 55, 2468–2483.

Marks, J.C., Power, M.E., Parker, M.S., 2000. Flood disturbance, algal productivity, and interannual variation in food chain length. Oikos 90, 20–27.

McCormick, P.V., Rawlik, P.S., Lurding, K., Smith, E.P., Sklar, F.H., 1996. Periphyton-water quality relationships along a nutrient gradient in the northern Florida Everglades. J. N. Am. Benthol. Soc. 15, 433–449.

McIntyre, P.B., Jones, L.E., Flecker, A.S., Vanni, M.J., 2007. Fish extinctions alter nutrient recycling in tropical freshwaters. Proc. Natl. Acad. Sci. U.S.A. 104, 4461–4466.

McIntyre, P.B., Michel, E., Olsgard, M., 2006. Top-down and bottom-up controls on periphyton biomass and productivity in Lake Tanganyika. Limnol. Oceanogr. 51, 1514–1523.

Mehner, T., Attermeyer, K., Brauns, M., others, 2016. Weak response of animal allochthony and production to enhanced supply of terrestrial leaf litter in nutrient-rich lakes. Ecosystems 19, 311–325. https://doi.org/10.1007/s10021-015-9933-2.

Miltner, R.J., 2010. A method and rationale for deriving nutrient criteria for small rivers and streams in Ohio. Environ. Manage. 45, 842–855.

Moeller, R.E., Burkholder, J.M., Wetzel, R.G., 1988. Significance of sedimentary phosphorus to a rooted submerged macrophyte (*Najas flexilis*) and its algal epiphytes. Aquat. Bot. 32, 261–281.

Mulholland, P.J., Steinman, A.D., Palumbo, A.V., Elwood, J.W., Kirschtel, D.B., 1991. Role of nutrient cycling and herbivory in regulating periphyton communities in laboratory streams. Ecology 72, 966–982.

Müller, U., 1996. Production rates of epiphytic algae in a eutrophic lake. Hydrobiologia 330, 37–45.

Munubi, R.N., McIntyre, P.B., Vadeboncoeur, Y., 2018. Do grazers respond to or control food quality? Cross-scale analysis of algivorous fish in littoral Lake Tanganyika. Oecologia. https://doi.org/10.1007/s00442-018-4240-1.

Ogdahl, M., Lougheed, V.L., Stevenson, R.J., Steinman, A.D., 2010. Influences of multi-scale habitat on metabolism in a coastal Great Lakes watershed. Ecosystems 11, 222–238.

Oleksy, I.A., Baron, J.S., Leavitt, P.R., Spaulding, S.A., 2020. Nutrients and warming interact to force mountain lakes into unprecedented ecological states. Proc. R. Soc. B 287, 20200304.

Olofsson, M., Power, M.E., Stahl, D.A., Vadeboncoeur, Y., Brett, M.T., 2021. Cryptic constituents: the paradox of high flux–low concentration components of aquatic ecosystems. Water 13, 2301. https://doi.org/10.3390/w13162301.

O'Reilly, C.M., 2006. Seasonal dynamics of periphyton in a large tropical lake. Hydrobiologia 533, 293–301.

Ozersky, T., Volkova, E.A., Bondarenko, N.A., Timoshkin, O.A., Malnik, V.V., M Domysheva, V., Hampton, S.E., 2018. Nutrient limitation of benthic algae in Lake Baikal, Russia. Freshw. Sci. 37, 472–482.

Passy, S.I., 2008. Continental diatom biodiversity in stream benthos declines as more nutrients become limiting. Proc. Natl. Acad. Sci. U.S.A. 105, 9663–9667.

Perillon, C., Pöschke, F., Lewandowski, J., Hupfer, M., Hilt, S., 2017. Stimulation of epiphyton growth by lacustrine groundwater discharge to an oligo-mesotrophic hardwater lake. Freshw. Sci. 36, 555–570.

Pla-Rabés, S., Catalan, J., 2018. Diatom species variation between lake habitats: implications for interpretation of paleolimnological records. J. Paleolimnol. 60, 169–187.

Poikane, S., Kelly, M., Cantonati, M., 2016. Benthic algal assessment of ecological status in European lakes and rivers: challenges and opportunities. Sci. Total Environ. 568, 603–613.

Rautio, M., Vincent, W.F., 2006. Benthic and pelagic food resources for zooplankton in shallow high-latitude lakes and ponds. Freshw. Biol. 51, 1038–1052.

Reuter, J.E., Loeb, S.L., Goldman, C.R., 1983. Nitrogen fixation in periphyton of oligotrophic Lake Tahoe. In: Wetzel, R.G. (Ed.), Periphyton of Freshwater Ecosystems. Springer, Netherlands, pp. 101–109.

Revsbech, N.P., Jørgensen, B.B., Brix, O., 1981. Primary production of microalgae in sediments measured by oxygen microprofile, $H^{14}CO_3^-$ fixation, and oxygen exchange methods. Limnol. Oceanogr. 26, 717–730.

Rober, A.R., Wyatt, K.H., Stevenson, R.J., 2011. Regulation of algal structure and function by nutrients and grazing in a boreal wetland. J. N. Am. Benthol. Soc. 30, 787–796.

Romagoux, J.-C., 1979. Caracteristiques du microphytobenthos d'un lac volcanique meromitique (Lac Pavin, France). I. Biomasse chlorophyllienne et déterminisme du cycle annuel. Int. Rev. ges. Hydrobiol. 64, 303–343.

Rosenberger, E.E., Hampton, S.E., Fradkin, S.C., Kennedy, B.P., 2008. Effects of shoreline development on the nearshore environment in large deep oligotrophic lakes. Freshw. Biol. 53, 1673–1691.

Round, F.E., 1964. The ecology of benthic algae. In: Jackson, D.F. (Ed.), Algae and Man. Plenum Press, New York, pp. 138–184.

Round, F.E., 1972. Patterns of seasonal succession of freshwater epipelic algae. Brit. Phycol. J. 7, 213–220.

Schindler, D.W., Bayley, S.E., Parker, B.R., Beaty, K.G., Cruikshank, D.R., Fee, E.J., Schindler, E.U., Stainton, M.P., 1996. The effects of climatic warming on the properties of boreal lakes and streams at the Experimental Lakes Area, northwestern Ontario. Limnol. Oceanogr. 41, 1004–1017.

Schindler, D.W., Frost, V.E., Schmidt, R.V., 1973. Production of epilithiphyton in two lakes of the experimental lakes area, northwestern Ontario. J. Fish. Res. Board Can. 30, 1511–1524.

Schindler, D.E., Scheuerell, M.D., 2002. Habitat coupling in lake ecosystems. Oikos 98, 177–189.

Scott, C.E., Jackson, D.A., Zimmerman, A.P., 2014. Environmental and algal community influences on benthic algal extracellular material in Lake Opeongo, Ontario. Freshw. Sci. 33, 568–576.

Sharma, S., Blagrave, K., Magnuson, J.J., O'Reilly, C.M., Oliver, S., Batt, R.D., Magee, M.R., Straile, D., Weyhenmeyer, G.A., Winslow, L., Woolway, R.I., 2019. Widespread loss of lake ice around the Northern Hemisphere in a warming world. Nat. Clim. Change 9, 227–231.

Sheath, R.G., Cole, K.M., 1980. Distribution and salinity adaptations of *Bangia atropurpurea* (rhodophyta), a putative migrant into the laurentian great lakes. J. Phycol. 16, 412–420.

Sheath, R.G., Hellebust, J.A., 1978. Comparison of algae in the euplankton, tychoplankton, and periphyton of a tundra pond. Can. J. Bot. 56, 1472–1483.

Sheath, R.G., Morison, M.O., 1982. Epiphytes on Cladophora glomerata in the great lakes and St. Lawrence seaway with particular reference to the red alga *Chroodactylon ramosum* (=*Asterocytis smargdina*). J. Phycol. 18, 385–391.

Sierszen, M.E., Hrabik, T.R., Stockwell, J.D., Cotter, A.M., Hoffman, J.C., Yule, D.L., 2014. Depth gradients in food-web processes linking habitats in large lakes: Lake Superior as an exemplar ecosystem. Freshw. Biol. 99, 2122–2136.

Sierszen, M.E., Peterson, G.S., Scharold, J.V., 2006. Depth specific patterns in benthic-planktonic food web relationships in Lake Superior. Can. J. Fish. Aquat. Sci. 63, 1496–1503.

Smol, J.P., 1988. Paleoclimate proxy data from freshwater arctic diatoms. Verh. Int. Ver. Limnol. 23, 837–844.

Snider, M.J., Biddanda, B.A., Lindback, M., Grim, S.L., Dick, G.J., 2017. Versatile photophysiology of compositionally similar cyanobacterial mat communities inhabiting submerged sinkholes of Lake Huron. Aquat. Microb. Ecol. 79, 63–78.

Song, Y., Wang, J., Gao, Y., Qin, B., 2017. Nitrogen incorporation by epiphytic algae via *Vallisneria natans* using ^{15}N tracing in sediment with increasing nutrient availability. Aquat. Microb. Ecol. 80, 93–99.

Steinman, A.D., 1996. Effects of grazers on freshwater benthic algae. In: Stevenson, R.J., Bothwell, M.L., Lowe, R.L. (Eds.), Algal Ecology: Freshwater Benthic Ecosystems. Academic Press, San Diego, pp. 341–373.

Steinman, A.D., Lamberti, G.A., Leavitt, P., Uzarski, D.G., 2017. Biomass and pigments of benthic algae. In: Hauer, R., Lamberti, G. (Eds.), Methods in Stream Ecology, , third ed.vol. 1. Elsevier Press, Burlington, pp. 223–241.

Steinman, A.D., Mulholland, P.J., Hill, W.R., 1992. Functional responses associated with growth form in stream algae. J. N. Am. Benthol. Soc. 11, 229–243.

Steinman, A.D., McIntire, C.D., Lowry, R.R., 1987. Effect of herbivore type and density on chemical composition of algal assemblages in laboratory streams. J. N. Am. Benthol. Soc. 6, 189–197.

Stevenson, R.J., 1996. The stimulation and drag of current. In: Stevenson, R.J., Bothwell, M.L., Lowe, R.L. (Eds.), Algal Ecology: Freshwater Benthic Ecosystems. Academic Press, San Diego, pp. 321–340.

Stevenson, R.J., Singer, R., Roberts, D.A., Boylen, C.W., 1985. Patterns of benthic algal abundance with depth, trophic status, and acidity in poorly buffered New Hampshire lakes. Can. J. Fish. Aquat. Sci. 42, 1501–1512.

Stevenson, R.J., Stoermer, E.F., 1981. Quantitative differences between benthic algal communities along a depth gradient in Lake Michigan. J. Phycol. 17, 29–36.

Stewart, S.D., Hamilton, D.P., Baisden, W.T., Dedua, M., Verburg, P., Duggan, I.C., Hicks, B.J., Graham, B.S., 2017. Variable littoral-pelagic coupling as a food-web response to seasonal changes in pelagic primary production. Freshw. Biol. 62, 2008–2025.

Strayer, D.L., Findlay, S.E.G., 2010. Ecology of freshwater shore zones. Aquat. Sci. 72, 127–163.

Tank, J.L., Bernot, M.J., Rosi-Marshall, E.J., 2006. Chapter 10. Nitrogen limitation and uptake. In: Hauer, F.R., Lamberti, G.A. (Eds.), Methods in Stream Ecology, second ed. Academic Press, Burlington.

Tank, J.L., Webster, J.R., 1998. Interaction of substrate and nutrient availability on wood biofilm processes in streams. Ecology 79, 2168–2179.

Timoshkin, O.A., Moore, M.V., Kulikova, N.N., Tomberg, I.V., Malnik, V.V., Shimaraev, M.N., Troitskaya, E.S., Shirokaya, A.A., Sinyukovich, V.N., Zaitseva, E.P., Domysheva, V.M., 2018. Groundwater contamination by sewage causes benthic algal outbreaks in the littoral zone of Lake Baikal (East Siberia). J. Great Lakes Res. 44, 230–244.

Timoshkin, O.A., Samsonov, D.P., Yamamuro, M., Moore, M.V., Belykh, O.I., Malnik, V.V., Sakirko, M.V., Shirokaya, A.A., Bondarenko, N.A., Domysheva, V.M., Fedorova, G.A., 2016. Rapid ecological change in the coastal zone of Lake Baikal (East Siberia): is the site of the world's greatest freshwater biodiversity in danger? J. Great Lakes Res. 42, 487–497.

Torres-Ruiz, M., Wehr, J.D., Perrone, A.A., 2007. Trophic relations in a stream food web: importance of fatty acids for macroinvertebrate consumers. J. N. Am. Benthol. Soc. 26, 509–522.

Trochine, C., Guerrieri, M.E., Liboriussen, L., Lauridsen, T.L., Jeppesen, E., 2014. Effects of nutrient loading, temperature regime and grazing pressure on nutrient limitation of periphyton in experimental ponds. Freshw. Biol. 59, 905–917.

Trochine, C., Diaz Villanueva, V., Brett, M.T., 2020. The ultimate peanut butter on crackers for Hyalella: diatoms on macrophytes rather than bacteria and fungi on conditioned terrestrial leaf litter. Freshw. Biol. 66, 599–614.

Vadeboncoeur, Y., Devlin, S.P., McIntyre, P.B., Vander Zanden, M.J., 2014. Is there light after depth? Distribution of periphyton chlorophyll and productivity in lake littoral zones. Freshw. Sci. 33, 524–536.

Vadeboncoeur, Y., Jeppesen, E., Vander Zanden, M.J., Schierup, H.-H., Christoffersen, K., Lodge, D.M., 2003. From Greenland to green lakes: cultural eutrophication and the loss of benthic energy pathways in lakes. Limnol. Oceanogr. 48, 1408–1418.

Vadeboncoeur, Y., Kalff, J., Christoffersen, K., Jeppesen, E., 2006. Substratum as a driver of variation in periphyton chlorophyll and productivity in lakes. J. N. Am. Benthol. Soc. 25, 379–392.

Vadeboncoeur, Y., Lodge, D.M., 1998. Dissolved inorganic carbon sources for epipelic algae: sensitivity of primary production estimates to spatial and temporal distribution of ^{14}C. Limnol. Oceanogr. 43, 1222–1226.

Vadeboncoeur, Y., Lodge, D.M., 2000. Periphyton production on wood and sediments: substratum-specific response to laboratory and whole-lake nutrient manipulations. J. N. Am. Benthol. Soc. 19, 68–81.

Vadeboncoeur, Y., Lodge, D.M., Carpenter, S.R., 2001. Whole-lake fertilization effects on the distribution of primary production between benthic and pelagic habitats. Ecology 82, 1065–1077.

Vadeboncoeur, Y., McCann, K.S., Vander Zanden, M.J., Rasmussen, J.B., 2005. Effects of multi-chain omnivory on the strength of trophic control. Ecosystems 8, 692, 683.

Vadeboncoeur, Y., McIntyre, P.B., Vander Zanden, M.J., 2011. Borders of biodiversity: life at the edge of the world's large lakes. Bioscience 61, 526–537.

Vadeboncoeur, Y., Moore, M.V., Stewart, S.D., Chandra, S., Atkins, K.S., Baron, J.S., Bouma-Gregson, K., Brothers, S., Francoeur, S.N., Genzoli, L., Higgins, S.N., Hilt, S., Katona, L.R., Kelly, D., Oleksy, I.A., Ozersky, T., Power, M.E., Roberts, D., Smits, A.P., Tromboni, F., Vander Zanden, M.J., Volkova, E.A., Waters, A.S., Wood, S.A., Yamamuro, M., 2021. Blue waters, green bottoms: benthic Filamentous Algal Blooms (FABs) are an emerging threat to clear lakes worldwide. Bioscience 71, 1011–1027.

Vadeboncoeur, Y., Peterson, G., Vander Zanden, M.J., Kalff, J., 2008. Benthic algal production across lake-size gradients: interactions among morphometry, nutrients and light. Ecology 89, 2542–2552.

Vadeboncoeur, Y., Power, M.E., 2017. Attached algae: the cryptic base of inverted trophic pyramids in fresh waters. Ann. Rev. Ecol. Evol. Syst. 48, 258–279.

Vadeboncoeur, Y., Vander Zanden, M.J., Lodge, D.M., 2002. Putting the lake back together: reintegrating benthic pathways into lake food web models. Bioscience 52, 44–55.

Van de Bogert, M.C., Carpenter, S.R., Cole, J.J., Pace, M.L., 2007. Assessing pelagic and benthic metabolism using free water measurements. Limnol Oceanogr. Methods 5, 145–155.

Vander Zanden, M.J., Vadeboncoeur, Y., 2002. Fish as integrators of benthic and pelagic food chains in lakes. Ecology 83, 2152–2161.

Vander Zanden, M.J., Vadeboncoeur, Y., 2020. Putting the lake back together 20 years later: what in the benthos have we learned about habitat linkages in lakes? Inland Waters 10, 305–321.

Vander Zanden, M.J., Vadeboncoeur, Y., Chandra, S., 2011. Fish reliance on littoral-benthic resources and the distribution of primary production in lakes. Ecosystems 14, 894–903.

Vanni, M.J., McIntyre, P.B., 2016. Predicting nutrient excretion of aquatic animals with metabolic ecology and ecological stoichiometry: a global synthesis. Ecology 97, 3460–3471.

Vasconcelos, F.R., Diehl, S., Rodríguez, P., Hedström, P., Karlsson, J., Byström, P., 2016. Asymmetrical competition between primary producers in a warmer and browner world. Ecology 97, 2580–2592.

Vasconcelos, F.R., Diehl, S., Rodríguez, P., Karlsson, J., Byström, P., 2018. Effects of terrestrial organic matter on aquatic primary production as mediated by pelagic-benthic resource fluxes. Ecosystems 21, 1255–1268.

Vesterinen, J., Devlin, S.P., Syväranta, J., Jones, R.I., 2016. Accounting for littoral primary production by periphyton shifts a highly humic boreal lake towards net autotrophy. Freshw. Biol. 61, 265–276.

Vinebrooke, R.D., Leavitt, P.R., 1999. Phytobenthos and phytoplankton as potential indicators of climate change in mountain lakes and ponds: a HPLC-based pigment approach. J. N. Am. Benthol. Soc. 18, 15–33.

Vitousek, P.M., Mooney, H.A., Lubchenco, J., Melillo, J.M., 1997. Human domination of Earth's ecosystems. Science 277, 494–499.

Volkova, E.A., Bondarenko, N.A., Timoshkin, O.A., 2018. Morphotaxonomy, distribution and abundance of *Spirogyra* (Zygnematophyceae, Charophyta) in Lake Baikal, East Siberia. Phycologia 57, 298–308.

Welch, H.E., Kalff, J., 1974. Benthic photosynthesis and respiration in Char Lake. J. Fish. Res. Board Can. 31, 609–620.

Wetzel, R.G., 1969. Factors influencing photosynthesis and excretion of dissolved organic matter by aquatic macrophytes in hard-water lakes. Verh. Internat. Verein. Limnol. 17, 72–85.

Wetzel, R.G., 1993. Microcommunities and microgradients: linking nutrient regeneration, microbial mutualism, and high sustained aquatic primary production. Netherlands J. Aquat. Ecol. 27, 3–9.

Wetzel, R.G., 1996. Benthic algae and nutrient cycling in lentic freshwater ecosystems. In: Stevenson, R.J., Bothwell, M.L., Lowe, R.L. (Eds.), Algal Ecology: Freshwater Benthic Ecosystems. Academic Press, San Diego, pp. 641–667.

Wetzel, R.G., 2001. Limnology: Lake and River Ecosystems, third ed. Academic Press, San Diego.

Wetzel, R.G., Rich, P.H., Miller, M.C., Allen, H.L., 1972. Metabolism of dissolved and particulate detrital carbon in a temperate hard-water lake. Mem. Ist. Ital. Idrobiol. 29 (Suppl. l.), 185–243.

Whiles, M.R., Hall, R.O., Dodds, W.K., Verburg, P., Huryn, A.D., Pringle, C.M., Lips, K.R., Kilham, S.S., Colón-Gaud, C., Rugenski, A.T., Peterson, S., Connelly, S., 2013. Disease-driven amphibian declines alter ecosystem processes in a tropical stream. Ecosystems 16, 146–157.

Whorley, S.B., Francoeur, S.N., 2013. Active fluorometry improves nutrient-diffusing substrata bioassay. Freshw. Sci. 32, 108–115.

Williamson, C.E., Morris, D.P., Pace, M.L., Olson, O.G., 1999. Dissolved organic carbon and nutrients as regulators of lake ecosystems: Resurrection of a more integrated paradigm. Limnol. Oceanogr. 44, 795–803.

Wood, S.A., Depree, C., Brown, L., McAlister, T., Hawes, I., 2015. Entrapped sediments as a source of phosphorus in epilithic cyanobacterial proliferations in low nutrient rivers. PLoS One 10 (10), e0141063.

Wood, S.A., Kelly, L.T., Bouma-Gregson, K., Humbert, J.-F., Laughinghouse IV, H.D., Lazorchak, J., McAllister, T.G., McQueen, A., Pokrzywinski, K., Puddick, J., Quiblier, C., Reitz, L.A., Ryan, K.G., Vadeboncoeur, Y., Zastepa, A., Davis, T.W., 2020. Toxic benthic freshwater cyanobacterial proliferations: challenges and solutions for enhancing knowledge and improving monitoring and mitigation. Freshw. Biol. 65, 1824–1842.

Yamamuro, M., Komuro, T., Kamiya, H., Kato, T., Hasegawa, H., Kameda, Y., 2019. Neonicotinoids disrupt aquatic food webs and decrease fishery yields. Science 366, 620–623.

Yang, L., He, H., Guan, B., Yua, J., Yaoa, Z., Zhen, W., Yina, C., Wang, Q., Jeppesen, E., Liu, Z., 2020. Mesocosm experiment reveals a strong positive effect of snail presence on macrophyte growth, resulting from control of epiphyton and nuisance filamentous algae: implications for shallow lake management. Sci. Total Environ. 705, 135958.

CHAPTER

26

Shallow Lakes and Ponds

Mariana Meerhoff[1,2] *and Meryem Beklioğlu*[3]

[1]Department of Ecology and Environmental Management, Centro Universitario Regional del Este, Universidad de la República, Maldonado, Uruguay [2]Department of Ecoscience, Aarhus University, Aarhus, Denmark
[3]Limnology Laboratory, Department of Biological Sciences and Centre for Ecosystem Research and Implementation (EKOSAM), Middle East Technical University, Ankara, Turkey

OUTLINE

I. Origins and distribution

Shallow lakes and ponds normally occur in high abundance in lowland areas of very gentle relief. These are temporary or permanent, natural or artificial, inland ecosystems that are shallow enough to potentially allow light penetration to the the bottom sediment adequate to support phytosynthesis of higher plants and benthic algae and whose water column is not stably stratified for significant periods (i.e., are polymictic). Turbidity from abiotic or biotic sources may prevent light from reaching the sediments, but lakes or ponds are sufficiently shallow for this potential condition to occur (usually ≤3 m).

The water column of a shallow lake can be fully mixed by wind-induced turbulence, whereas temperature-induced mixing is more common in ponds (Brönmark and Hansson, 2005). Ponds, however, may have more complex mixing patterns, ranging from well mixed to rarely mixed, depending on their surface area. Surface area is thus both a pragmatic and functional way of distinguishing between shallow lakes and ponds, ponds being those water bodies covering a size range between 1 m^2 and a few ha, most typically below 5 ha, and with less than 30% coverage of emergent vegetation (Richardson et al., 2022). As will be shown later in the chapter, system size makes a difference in many aspects, whereas, in others, ponds and lakes are similar.

As discussed in Chapter 4, these shallow basins can originate naturally from processes that have caused modest depressions in the landscape. Certainly, some processes were major geological disturbances, such as glacial retreat or tectonic plate movements. Other depressions were formed in relatively flat regions from altered river courses or wind deflation. Small depressions in impermeable or semipermeable terrain can be filled with water throughout the year or during certain times of the year. The precipitation and

Wetzel's Limnology, Fourth Edition
https://doi.org/10.1016/B978-0-12-822701-5.00026-4

evapotranspiration patterns of a given region, with spring snow melting and rain-filled depressions, may affect the duration of water in ponds from being permanent to temporary or semipermanent shallow waters. Hundreds of thousands of small, often temporary, ponds and lakes occur in the floodplains of major river ecosystems. These floodplain impoundments are especially common adjacent to Arctic and tropical rivers, where large areas of the land are flooded seasonally for several months. Tectonic-formed, ice-formed, and fluvially formed lakes and ponds are indeed not the only ways of lake and pond formation but are the most common ways. In cold regions, such as vast areas of the Arctic, shallow thermokarst or thaw ponds are also common in permafrost regions. Such ponds are highly dynamic systems, especially with recent climatic warming (Smith et al., 2005). In addition, humans have created millions of shallow lakes and ponds, either fortuitously or intentionally, as we modify the landscape through construction and agriculture. In many developed and some developing countries, most of the current ponds are the legacy of several anthropogenic activities (e.g., by mining) in the last two or three centuries (Oertli, 2018).

Aquatic ecosystems are archetypal society-biosphere interfaces: civilizations not only have settled near surface waters but have shaped them according to their needs (Naiman et al., 1995). On a local level, aquatic ecosystems provide resources and supporting services, such as absorbing, cycling, and transporting nutrients and waste (Grizzetti et al., 2016). Both natural and human-made shallow waters may be especially critical for providing resources and services for industries, irrigated agriculture and aquaculture (mostly of fishes), water storage for human and animal consumption, and recreational or aesthetic purposes, both in rural landscapes and in urban areas. Land cover and land use are thus fundamental drivers of change in shallow lakes and ponds. The enormous changes in land use that have occurred since the second half of the 20th century have led to both the creation and the destruction or deterioration of a large proportion of shallow lakes and ponds worldwide. Creation of ponds as a nature-based solution to the observed loss of biodiversity, loss of nature for recreation, flood control, or to trap nutrients within agricultural landscapes, among other processes in both rural and urban landscapes, has been promoted in several places (e.g., central and northern Europe; Rannap et al., 2009). At the same time, climate warming and/or water extraction are generating many shallow ecosystems by glacier melting and ice retreat, permafrost thaw, and the contraction of large rivers and their floodplains.

Most natural and artificial lakes are small and relatively shallow (Messager et al., 2016). Despite being neglected in many textbooks and, more importantly, in conservation policies (Grasel et al., 2018), lakes and ponds are the most abundant freshwaters in the world (Downing et al., 2006). Current estimations of the number of lakes and ponds reach over 117 million when including lakes smaller than 1 ha (Verpoorter et al., 2014) (see Chapter 4). However, the accuracy of the method used for this estimate implied that the lower bound for the analysis was ponds of 0.2 ha, and therefore the global importance of small shallow water bodies is still highly likely underrepresented. Shallow lakes and ponds have a disproportionate role in aquatic biodiversity (Oertli et al., 2009) and ecosystem services (Hilt et al., 2017). Shallow lakes and ponds are also the most common aquatic ecosystems in cities and urban areas, where they provide opportunities for recreation and culture, and at the same time they can be hotspots for biodiversity (Oertli and Parris, 2019) (Fig. 26-1).

II. Characteristics

The characteristics of a particular shallow lake or a pond are the result of physical, chemical, and biological processes acting and interacting at different spatial and temporal scales, on the landscape, the catchment, and the water body itself. The climate regime, biogeographic processes, and the historic and current land use at the regional or catchment levels determine the abiotic framework within which the biological communities live and interact (Brönmark and Hansson, 2005). Large-scale abiotic factors for a given location on the globe or region include solar radiation, temperature, precipitation patterns, and wind-induced turbulence. Other abiotic factors result from geophysical characteristics and biogeographical and other historical and current processes occurring in the catchment, including land cover and land- and water-use changes. Landscape features may affect lake communities in two different ways. First, geophysical characteristics have direct implications as they largely determine the bedrock, soils, water residence time, carbon and nutrient concentrations, humic content, acidity, and indirectly, the concentration and dynamics of oxygen levels. Local biodiversity and the functioning of a given lake or pond are also largely influenced by the potential connectivity with other freshwaters, including dispersal possibilities. Other abiotic and biotic characteristics that affect the biodiversity and overall functioning of shallow lakes and ponds occur at the water body level, as will be discussed in Section IV.

Lowland areas tend to accumulate terrestrial organic matter and nutrients, and appreciable amounts of these substances can be transported to shallow lakes and ponds. In areas of relatively small volume and a large

FIGURE 26-1 Different shallow lakes and ponds. From *left* to *right*, and from *top* to *bottom*: first column: pondscape of glacier-melt ponds (Ilulissat region, Greenland), water lilies in a warm temperate shallow lake (Lake Poyrazlar, Kocaeli, Turkey), cyanobacterial bloom in a subtropical shallow lake (Lake Sauce, Maldonado, Uruguay); second column: cold temperate shallow lake under a clear water regime (Silkeborg, Denmark), Mediterranean shallow lake (Lake Gölhisar, Burdur, Turkey); third column: large Mediterranean shallow lake (Lake Beysehir, Konya, Turkey), naturally eutrophic pond in Kruger National Park (South Africa); fourth column: artificial pond used for irrigation (Ayas, Turkey), artificial pond used for cattle (Rocha, Uruguay), fish farm pond (Mato Grosso, Brazil). Photo credits, ordered as above: first column: Mariana Meerhoff, Meryem Beklioğlu, Guillermo Goyenola; second column: M. Meerhoff, İdil Çakıroğlı; third column: M. Beklioğlu, M. Meerhoff; fourth column: Feride Avcı, Franco Teixeira-de Mello, Victor Duque.

surface-to-volume ratio, the loading of nutrients per unit volume and per unit of area can be high. The importance of loading of nutrients to primary productivity of lakes, particularly phosphorus and nitrogen, has been discussed in detail in several previous chapters (Chapters 18, 24, and 25). Not only are the nutrient loadings proportionally higher in shallow systems than in deep ones, but the losses of nutrients to potential depositories, such as the sediments or outflow, are lower and the rates of nutrient recycling faster in shallow lakes and ponds. Shallow water bodies can store organic matter and nutrients in their sediments, but the internal loading (particularly of phosphorus, P; Chapter 15) can easily release P under anoxic conditions (Søndergaard et al., 2003, 2013), which becomes available to the benthic and pelagic biota through thermal mixing, wind resuspension, and/or bioturbation by benthic fishes and macroinvertebrates. Ponds and shallow lakes thus tend to have strong connectivity with the atmosphere, surrounding land, and benthos (Fig. 26-2). As a result of this connectivity, productivity (e.g., bacterial, primary, and secondary) in ponds and shallow lakes is often greater than in large and/or deep lakes.

Thermal stratification (Chapter 7) in deep lakes is important for losses of nutrients, particularly P, to the sediments. Stratification has strong effects on the living conditions for the biota, as not only temperature but also oxygen concentration differs with water depth (Chapter 11). Shallow lakes and ponds tend to have a low percentage, if any, of total water volume in a thermally distinct hypolimnion. Water commonly mixes for long periods in shallow lakes and ponds, and the intense interaction between the water column and the sediments is highly relevant to the functioning of these ecosystems (Vadeboncoeur et al., 2002). However, thermal stratification may occur frequently but typically does not last for an appreciable length of time (e.g., weeks). Thermal stratification patterns depend on the surface area, with ponds being subject to more frequent stratification than larger, shallow lakes. Stratification also depends on the existence of barriers against wind action (such as the closed canopy of a forest, or adjacent topographic structures). The occurrence of stratification, even if short-term, affects oxygen profiles and can promote the anoxic release of phosphorus and methane from the sediments. Despite very small ponds (of <0.1 ha) currently representing less than 9% of lakes and ponds by area, they account for more than 15% of global CO_2 emissions and 40% of diffusive CH_4 emissions (Holgerson and Raymond, 2016). The high emissions from those very small ponds likely result from the often high phytoplankton production and organic matter—rich sediments, their shallow depths, high sediment and edge to water volume ratios, and stratification

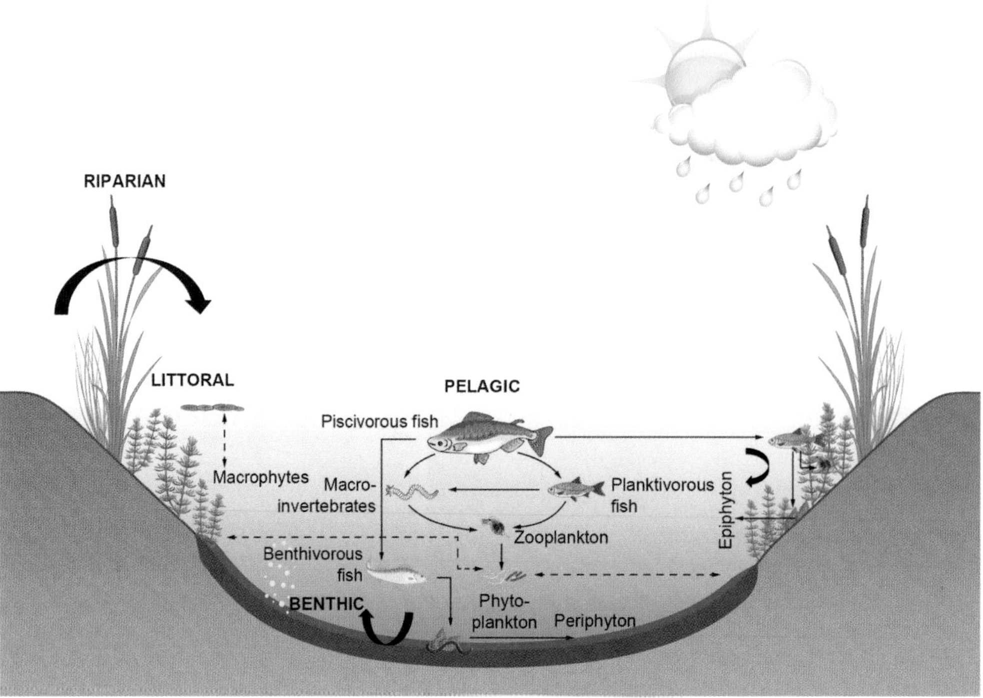

FIGURE 26-2 The different habitats of a shallow lake, including the nearshore littoral zone, the open-water pelagic zone, and the benthic zone with the sediments, and some of the most important processes for lake structure and functioning. Feeding interactions between representative organisms from the main communities of the "classical food web" and competitive interactions between major primary producers are shown with *solid* and *broken arrows*, respectively. *Curved arrows* highlight habitat coupling and major fluxes of nutrients, matter, and information among lake zones and through the riparian zone with the surrounding terrestrial environment, whereas *bubbles* represent a potential efflux of greenhouse gases. Subsidies from the lake to the land and to neighboring water bodies may be locally important. *(Modified with permission from Fig. 3 in Kosten and Meerhoff, 2014,* ELS Encyclopedia of Life Science. *John Wiley & Sons. Drawing by M. Meerhoff and T. Christensen.)*

and mixing dynamics (Holgerson and Raymond, 2016). With climate warming, and particularly with heat waves, thermal stratification is enhanced, with stronger effects in the smaller and wind-protected systems.

Nutrients and light availability are the key factors that regulate both photosynthetic productivity and the diversity and growth of organisms that depend on that productivity. Nutrient supply does not influence phytoplankton productivity in shallow lakes in the same manner discussed for deep lakes. This is due to the importance of littoral, benthic, and pelagic habitat coupling in shallow lakes (Liboriussen and Jeppesen, 2003; Rautio and Vincent, 2006; Schindler and Scheuerell, 2002) and to competitive interactions between phytoplankton, benthic periphyton, and different types of plants and their attached microbiota, among other processes Chapter 24 (Fig. 26-2). Those other processes include defense adaptations; behavioral and top-down control involving fish, zooplankton, and macroinvertebrates; and their grazing pressure on phytoplankton, periphyton, and plants.

In shallow lakes, the littoral zone (understood as the zone where aquatic plants and benthic periphyton can develop) can and often does extend over the entire lake or pond basin. Littoral vegetation, which has much higher rates of organic matter production than phytoplankton per unit area, often dominates the production of organic matter within shallow lakes and ponds. Basin heterogeneity can lead to large variability in the distribution and productivity of higher vegetation and of the abundant microbiota associated with the surface of these larger plants and their particulate detritus. The type of macrophyte colonization is also highly variable. The common depth distribution and zonation of plants in large lakes, as discussed earlier (Chapter 24), can apply similarly in shallow lakes and ponds but usually on a greatly contracted depth gradient, simply because the light is often rapidly attenuated with depth. When light availability is high, primary production is possible both in the water column and the benthos, either by benthic periphyton or higher aquatic plant vegetation. The submersed plant community can then be dominated by short-growing species (e.g., charophytes), but when the underwater light climate is relatively less favorable, plant communities may shift to tall-growing submersed plant species (e.g., the pondweed *Potamogeton* spp.). Under eutrophic conditions, free-floating plants can also occur particularly along shallow, wind-protected zones, but they can even cover entire lake surfaces if environmental conditions are favorable (Sculthorpe, 1967). The dominant primary producer in a given lake or pond largely determines its trophic web structure, ecosystem state (Scheffer et al., 1993), and processes (e.g., Jeppesen et al., 2016), with consequences for ecosystem goods and services (Hilt et al., 2017). As will be discussed later, submersed plants are a key component of shallow lakes and ponds and markedly influence the structure and biological dynamics (Burks et al., 2006; Carpenter and Lodge, 1986; Jeppesen et al., 1998).

III. Alternative states theory revisited

The theoretical background for the idea that ecosystems could have *alternative stable states* (or *alternative regimes*, as later preferred) appeared long before the empirical data supported it (Holling, 1973). Alternative regimes are contrasting, stable regimes of a given ecosystem over certain environmental conditions. The study of shallow lakes had a great impulse after many lines of empirical and theoretical knowledge, built over many years by different research groups (Jeppesen et al., 1998; Moss, 1990; Scheffer, 1998; Scheffer et al., 1993), collated their observations to synthesize the alternative equilibria or stable states hypothesis for temperate shallow lakes (Scheffer et al., 1993). The whole concept of alternative regimes in shallow lakes is strongly based not only on the theory of alternative equilibria but also on the trophic cascades and competition theories. This idea has been criticized arguing that it is not generalizable to all shallow lakes, its empirical evidence is scarce and geographically limited, and it lacks predictability (Capon et al., 2015; Hillebrand et al., 2020). Empirical data analysis has also indicated that alternative regimes may be observed over short-time periods but that, in the long term, gradual changes prevail (Davidson et al., 2023). The overall stability of the alternative regimes and consequently the usefulness of the idea for lake management has also been questioned.

A. Theoretical framework

Several ecosystem types may present different configurations or regimes, with contrasting community structures and functioning patterns. Nongradual or nonlinear shifts from one regime to another have often been observed in several ecosystems. In this framework, shifts in ecosystem states are often illustrated as marbles residing and moving on stability landscapes with different cups or basins of stability, two in its simplest version (Fig. 26-3). The actual state of the system is represented by the current position of the marble, and the basins or cups represent domains of attraction for the marble once it is near. The marble can move in the landscape via two mechanisms: either the marble is pushed in an unchanged landscape, or the topology of the landscape itself is modified, making the marble move (Beisner et al., 2003). The first mechanism represents a change in the state of the ecosystem in a relatively constant environment (such as stochastic changes in densities or biomass), and the second represents a change in key external conditions or environmental drivers (Beisner et al., 2003; van Nes et al., 2016).

Many natural and anthropogenic conditions that are relevant for ecosystems often change gradually over time. Ecosystems frequently respond in a linear way to such changes. In other cases there might be no apparent response until the condition approaches a certain *threshold* where the ecosystem state can respond strongly. In these scenarios only one equilibrium exists for each condition. However, the occurrence of *alternative equilibria* implies that two different regimes may occur over the same range of the condition (i.e., two cups in the landscape). A sudden drastic switch (also often called *critical transition* or *catastrophic shift* when *the ecosystem is pushed beyond its capacity to recover*) from a regime to an alternative state can happen after apparently little change in the external condition (Fig. 26-3).

One of the theoretical implications is that, to induce a switch back to the initial state, it is not enough to restore the condition to the level before the shift. The condition must be reversed to a lower threshold before a shift back might be possible. This pattern is known as *hysteresis*, and those different threshold values are known as *bifurcation points* (Folke et al., 2004; Scheffer et al., 2001) (Fig. 26-3). A *tipping point* is a specific type of bifurcation point; it refers to a critical threshold where the ecosystem state changes abruptly and often irreversibly.

Within the range of conditions where alternatives are possible, a perturbation must occur to produce the shift. Stochastic events, such as pest outbreaks, mortality events caused by fire, droughts or severe winters, extreme climatic events, or some anthropogenic activities, can affect the ecosystem state directly (i.e., the marble) with sufficient amplitude so as to promote a regime shift. Stochastic events can sometimes also promote fluctuations in environmental conditions (i.e., the landscape), increasing the likelihood of a regime shift.

The likelihood of a shift depends not only on the level of perturbation but also on the capacity of the ecosystem to withstand the perturbation without experiencing a major change (*resistance*) or to recover the same functionality once a change has been experienced (*resilience*). Resilience is one of those concepts that most people would have at least an intuitive idea about its meaning

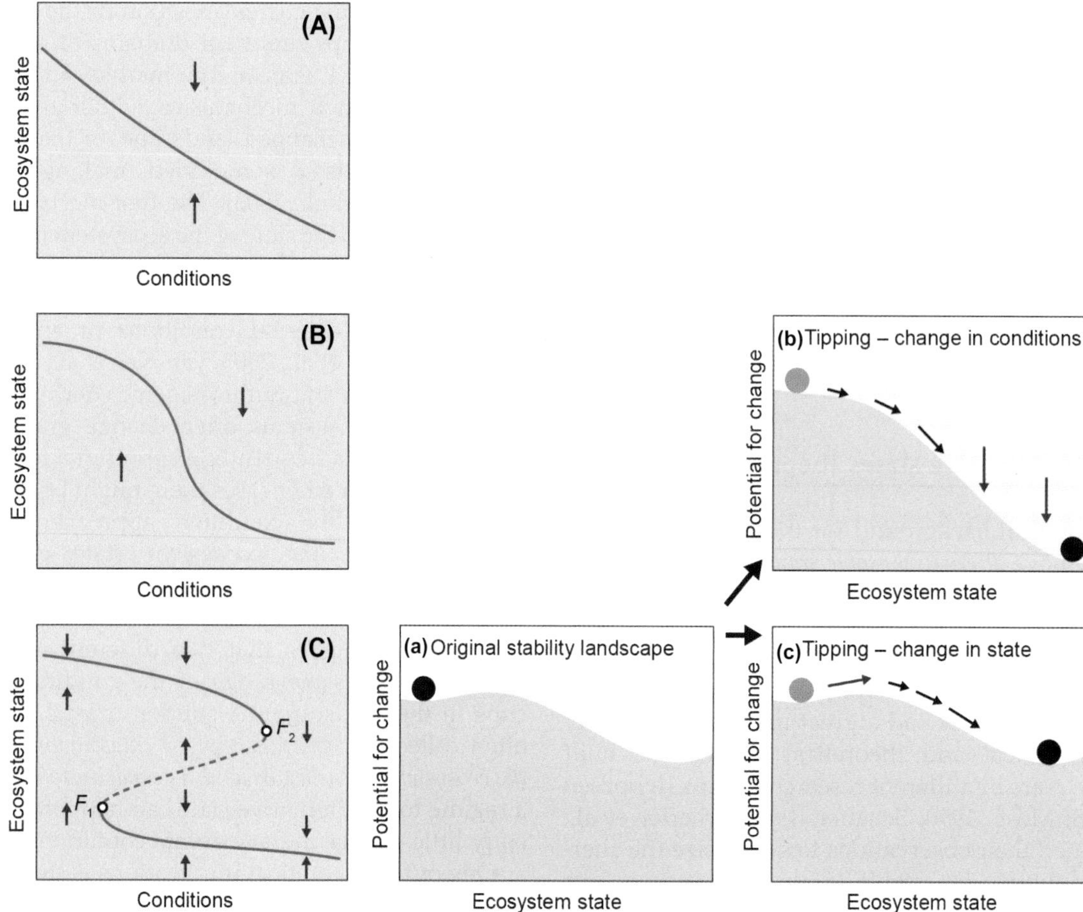

FIGURE 26-3 Different theoretical ways in which ecosystem states can respond to external changes. *Left-hand side*: as external conditions change, only one equilibrium exists for each condition in (A) (smooth response) and (B) (abrupt response after a threshold), whereas in (C), three equilibria can occur over a range of the condition. *Arrows* indicate the direction in which the ecosystem moves when it is not in equilibrium (*dashed line*), and F1 and F2 are bifurcation points where a slight incremental change in conditions may induce a catastrophic shift. *Right-hand side*: the original stability landscape and current ecosystem state *a* may shift after changes in external conditions make the current state unstable *b* or, alternatively, after the state of the ecosystem is pushed by a large enough perturbation *c*. *Red arrows* show how the system changes; *black arrows* show the accelerated change of the system. *(Published with permission from* Scheffer et al., 2001, Nature [left] *and from* van Nes et al., 2016, TREE.*)*

but whose rigorous definition is hard to agree upon. See, for example, Donohue et al. (2016), Folke (2016), Hillebrand et al. (2018), and Hodgson et al. (2015). Resilience is used here as a broad concept, often addressing resistance or multiple dimensions of ecosystem stability. Once established, the contrasting regime can be maintained by a series of *feedbacks*, which are often called stabilizing (or negative) if they minimize or prevent a change, and self-reinforcing (or positive) when they magnify a change (van Nes et al., 2016). The stability of each of the potential alternative regimes relies on the occurrence of such stabilizing mechanisms that make the ecosystem more or less resilient (Holling, 1973) or resistant (Hillebrand et al., 2018).

Besides changes in richness and the overall state of the ecosystem, the functional diversity associated with key components of trophic webs may change with a regime shift, further promoting changes in ecosystem functioning. Such regime shifts can thus often have negative consequences for communities and ecosystems, and therefore there is a need for determining the critical values, for instance, of nutrient and pollutant loads, that trigger a shift. Despite considerable research efforts, *early-warning signals* of an approaching shift are hard to obtain (Scheffer, 2020; Scheffer et al., 2009; Wang et al., 2012). It is notoriously difficult to determine and even more difficult to predict the thresholds for a given ecosystem before the occurrence of the shifts (Groffman et al., 2006; Hillebrand et al., 2020). For both theoretical and management purposes, it is essential to understand the basic functioning of ecosystems, and the major disturbances, both stochastic events and slowly changing drivers (such as land use, nutrient loads, and climate warming, among others), that influence the key response variables.

B. Contrasting regimes in shallow lakes

i. Community shifts along a nutrient gradient

Eutrophication is among the most pervasive processes affecting freshwaters worldwide, with nutrient loading being the key environmental condition that drives regime shifts in shallow lakes and ponds. In the extremes of the nutrient gradient, contrasting regimes are typically expected to occur with a high probability. Under oligotrophic conditions, clear water is expected because primary production is largely restricted to the benthic zone and dependent on the nutrients stored in the sediments. The benthic primary production, mostly by periphyton (Chapter 25), often contributes a large fraction of the total autotrophic productivity of oligotrophic systems, such as Arctic ponds and lakes, with increasing dominance toward the poles (Vadeboncoeur et al., 2003). The coupling processes between the water column and the benthos provide carbon sources other than phytoplankton that support the secondary production of zooplankton in Arctic shallow lakes and ponds. Benthic processes can then be relatively more important than planktonic processes for the functioning of these ecosystems (Rautio et al., 2011). Submersed vegetation (Chapter 24) utilize nutrients from the sediments, with charophyte macroalgae being common in calcareous waters, whereas isoetid submersed plants are common in more acidic waters. Such plants and macroalgae are typically small, growing close to the sediments, and are vulnerable to shading. Shading by phytoplankton communities is generally modest at these nutrient levels. At least in temperate zones, submersed macrophyte (or benthic micro- and macroalgae) dominance is often therefore the most likely condition in nutrient-poor shallow lakes.

Under higher phosphorus and nitrogen levels due either to naturally richer and more reactive soils in the catchment or to increased loading (promoted by urbanization, industries, and intensification of agricultural land use), lake communities change (Jeppesen et al., 2000). More nutrients may also enter the aquatic system due to the local loss of natural barriers such as wetlands and riparian vegetation. In temperate regions the freshwater plant community is gradually replaced with larger and taller submersed plants, as well as increasing densities of water lilies (Nymphaeaceae) if lake physical characteristics allow. The larger submersed plants are much more productive and still derive most of their nutrients from the sediments. Biodiversity, in general, is unexpectedly high in the presence of submersed plants (Declerck et al., 2005). Although the phytoplankton communities expand and proliferate under conditions of increasing nutrients, they can potentially coexist with certain highly competitive, eutrophic submersed plants that grow rapidly and concentrate photosynthetic tissues near the surface (provided that underwater light climate is good enough for plants early in the growing season). Each of these two communities can become the dominant primary producer in shallow lakes under mesotrophic and eutrophic conditions, as will be elaborated on later.

When nutrient loading increases further, the trophic web structure of shallow lakes can change dramatically (Jeppesen et al., 2000). As clearly documented in Northern Hemisphere temperate lakes, fish communities increase in total biomass but become dominated by smaller species and individuals, with a large share of planktivorous and benthivorous species, with low numbers or even the absence of predatory fishes (Chapter 22). Due to a combination of bottom-up and top-down processes, the zooplankton community then typically changes toward dominance of less efficient grazers, with a reduction in mean body size and the loss or significant reduction of large-bodied organisms such as *Daphnia* and other cladocerans (Chapter 20). Macroinvertebrates are also heavily preyed upon by fishes. Submersed plants become scarce and may eventually disappear (Balls et al., 1989), likely due to competition with phytoplankton and shading by epiphytic periphyton (Phillips et al., 1978) that are no longer controlled from the top down by grazing macroinvertebrates (Jones and Sayer, 2003). The pelagic zone thus becomes the largest contributor to the ecosystem's primary production, with phytoplankton typically dominating under high and very high nutrient conditions (Liboriussen and Jeppesen, 2003; Vadeboncoeur et al., 2003) (Chapter 25). Phytoplankton dominance is thus the most likely regime under nutrient-rich conditions.

ii. Regime shifts under mesotrophic and eutrophic conditions

Besides identifying the major role of nutrient loading, the alternative states theory highlighted the importance of some biological communities as significant internal drivers of change and the role of biotic interactions such as trophic cascades and competition as stabilizers of a particular regime or state (Scheffer et al., 1993). In summary, the idea states that, over a wide range of nutrient (commonly phosphorus) concentrations, alternative equilibria may occur, although with different likelihoods depending on the history of the lake, its trophic web structure, and the climate regime. Submersed plant-dominated and phytoplankton-dominated states were the two alternative regimes originally suggested for temperate shallow lakes (Jeppesen et al., 1998; Scheffer, 1998; Scheffer et al., 1993).

The *submersed plant-dominated regime* is characterized by high water clarity (Canfield et al., 1984) and is associated with high biodiversity in temperate (Declerck et al., 2005), subtropical (Kruk et al., 2009), and tropical zones (Thomaz et al., 2008). The submersed-plant regime is also associated with potentially more ecosystem services being provided (Hilt et al., 2017; Janssen et al., 2021), including effects on the carbon cycle. Rooted submersed plants potentially decrease methane (CH_4) emissions by creating oxidizing conditions in the sediment and water column (Davidson et al., 2015). Emergent parts of rooted plants may, however, also directly transport CH_4 from the sediments to the atmosphere. Empirical evidence overall suggests that abundant submersed plants may decrease CO_2 efflux (Jeppesen et al., 2016) and tend to decrease CH_4 emissions, compared to the turbid state.

Under certain conditions, other regimes can also occur with self-stabilizing properties. In eutrophic and particularly hypertrophic waters, free-floating plants can also become dominant,replacing submersed plants (Scheffer et al., 2003) and phytoplankton (de Tezanos Pinto and O'Farrell, 2014) (Fig. 26-4). These plants uptake nutrients solely from the water, and therefore their nutrient concentration requirements are high. A potential expansion of free-floating plants also requires that air temperatures are high enough to avoid winter kills due to freezing and that water bodies are not too sensitive to wind actions, particularly at their initial stages of growth (Sculthorpe, 1967). The dominance of such plants is therefore more likely to occur in tropical and subtropical regions (O'Farrell et al., 2003) and in ditches, ponds, and relatively small or protected shallow lakes (Meerhoff and Jeppesen, 2009). However, free-floating plants can expand as a thick layer covering hundreds or thousands of hectares under favorable conditions (e.g., invasion by the water fern *Salvinia molesta* D. Mitch in Lake Kariba, Africa, or by the water hyacinth *Pontederia crassipes* [Mart.] Solms in Lake Victoria and in many tropical locations outside its native South America), promoting anoxic and acidic conditions underwater with often mass mortality of fishes and macroinvertebrates. Under a *free-floating plant—dominated regime*, the food web structure is simplified, and the strength of trophic interactions weakens, leading to a shift from a top-down to a bottom-up control of the food web (Moi et al., 2021). With intermediate covers and fluctuating environments, however, free-floating plants can either decrease (Meerhoff et al., 2003) or locally promote (Agostinho et al., 2007) biodiversity. A dominance of free-floating plants can also have strong effects on carbon emissions, although net effects are likely to be determined by local conditions. In tropical floodplain lakes, areas dominated by free-floating plants can act as net CO_2 sinks due to carbon fixation by the plants, whereas open waters generally emit CO_2 (Peixoto et al., 2016). Due to the reduced exchange between the atmosphere and the water column, a large proportion of the CH_4 produced in the system may be oxidized below the plants, although also methanogenesis can be enhanced due to high organic matter production, particularly under anoxic conditions (Kosten et al., 2016).

At the higher end of the nutrient gradient, Cyanobacteria may also become dominant and promote self-stabilizing conditions. Changes in meteorological conditions leading to changes in thermal stratification patterns, as well as indirect changes in potential grazing pressure by zooplankton communities, together with an efficient use of resources and high light efficiency by several species, are among the mechanisms that can facilitate the dominance of Cyanobacteria within the *phytoplankton-dominated state* (Dokulil and Teubner, 2000). Cyanobacterial dominance is particularly enhanced by increasing water temperatures (Kosten et al., 2012). The increasing frequency, duration, and intensity of blooms likely select for better-adapted zooplankton that coexist with, rather than control, Cyanobacteria (Ger et al., 2014). In particular, species of the Oscillatoriaceae group are shade tolerant, which explains their potential dominance when the water has become sufficiently turbid (Scheffer, 1998). Many species within this group can regulate their position in the water column, shading other phytoplankton groups and submersed plants. Phytoplankton blooms are associated with a net uptake of CO_2 (Jeppesen et al., 2016) but with a high emission of N_2O (Wang et al., 2006). The dominance of Cyanobacteria, in particular, seems associated with high emissions of greenhouse gases (GHGs) in general, and especially of CH_4 during the decomposition of decaying blooms (Yan et al., 2017).

The dominance of each primary producer under nutrient-rich conditions depends on the outcome of the asymmetric competition among taxa for light and nutrients, which is affected by both external and internal processes (Fig. 26-4). Free-floating plants are superior competitors for light than both submersed plants and phytoplankton, whereas phytoplankton are superior competitors to submersed plants. Submersed plants are superior competitors for nutrients than phytoplankton and free-floating plants, given their access to nutrients from both water and sediments, whereas the outcome of the competition for nutrients between phytoplankton and free-floating plants likely depends on the functional characteristics of the dominant species in each group.

Each of the above regimes is, to a certain extent, stabilized by a series of physical, chemical, and biological processes and interactions, as will be described later.

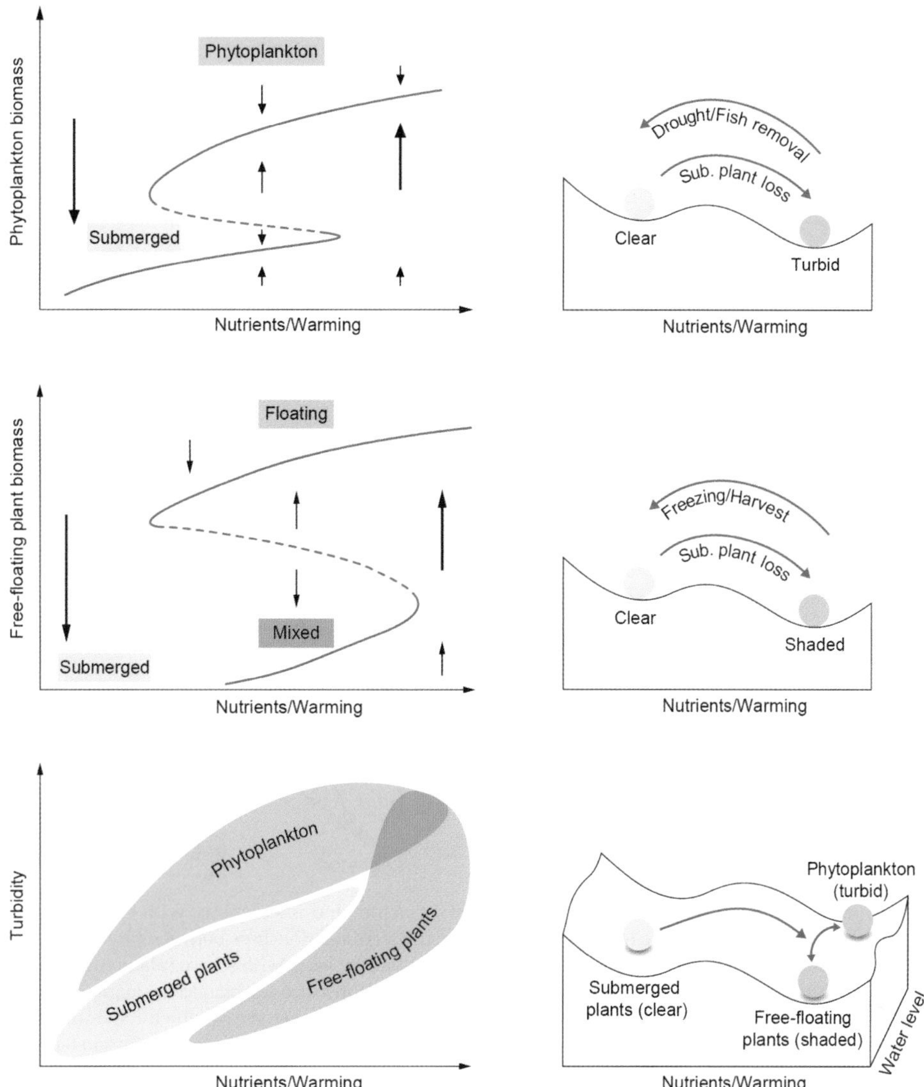

FIGURE 26-4 Alternative regimes in shallow lakes. *Left-hand side*: effect of nutrient loading on the equilibrium biomass of phytoplankton (*top*) and free-floating plants (*middle*) with respect to the biomass of submersed plants. The *arrows* indicate the direction of change if the system is out of equilibrium (i.e., the dashed equilibrium is unstable). Shifts to an alternative regime occur as vertical transitions in the scheme. *Lower panel*: probability of occurrence and dominance of the three alternative communities along gradients of nutrients and warming and of turbidity. *Right-hand side*: regime shift transitions and some key drivers between two or three likely states over a stability landscape. (Left-hand side panels: *published with permission from Fig. 3 in Meerhoff and Jeppesen, 2009,* Encyclopedia of Inland Waters; *redrawn after Scheffer et al., 1993,* Trends Ecol. Evol. *8, 275–279; Scheffer et al., 2003, PNAS 100: 4040–4045, and after Meerhoff and Mazzeo, 2004,* Ecosistemas 2004/2.)

Particularly in the alternative equilibria between phytoplankton and submersed plants, both regimes are self-reinforced by feedbacks involving water clarity (Scheffer et al., 1993). Clear water promotes the development of submersed plants, creating a positive feedback that further creates conditions for clear water. With variations, water clearing effects of submersed plants have been observed in shallow lakes under different climate regimes (Kosten et al., 2009) (Fig. 26-5). In contrast, water turbidity prevents the development of submersed plants, indirectly favoring phytoplankton growth, which further increases water turbidity even further preventing the development of plants. Free-floating plants promote strong shading conditions for both submersed plants and phytoplankton. In this case light limitation overrides the high nutrient availability (and the low grazing pressure by the depauperate zooplankton) that could promote phytoplankton growth (de Tezanos Pinto and O'Farrell, 2014). When free-floating plant dominance is naturally or artificially interrupted, however, phytoplankton biomass often thrives, and although submersed plants could potentially develop, this does not generally happen (de Tezanos Pinto and O'Farrell, 2014; Meerhoff and Jeppesen, 2009). The outcome of

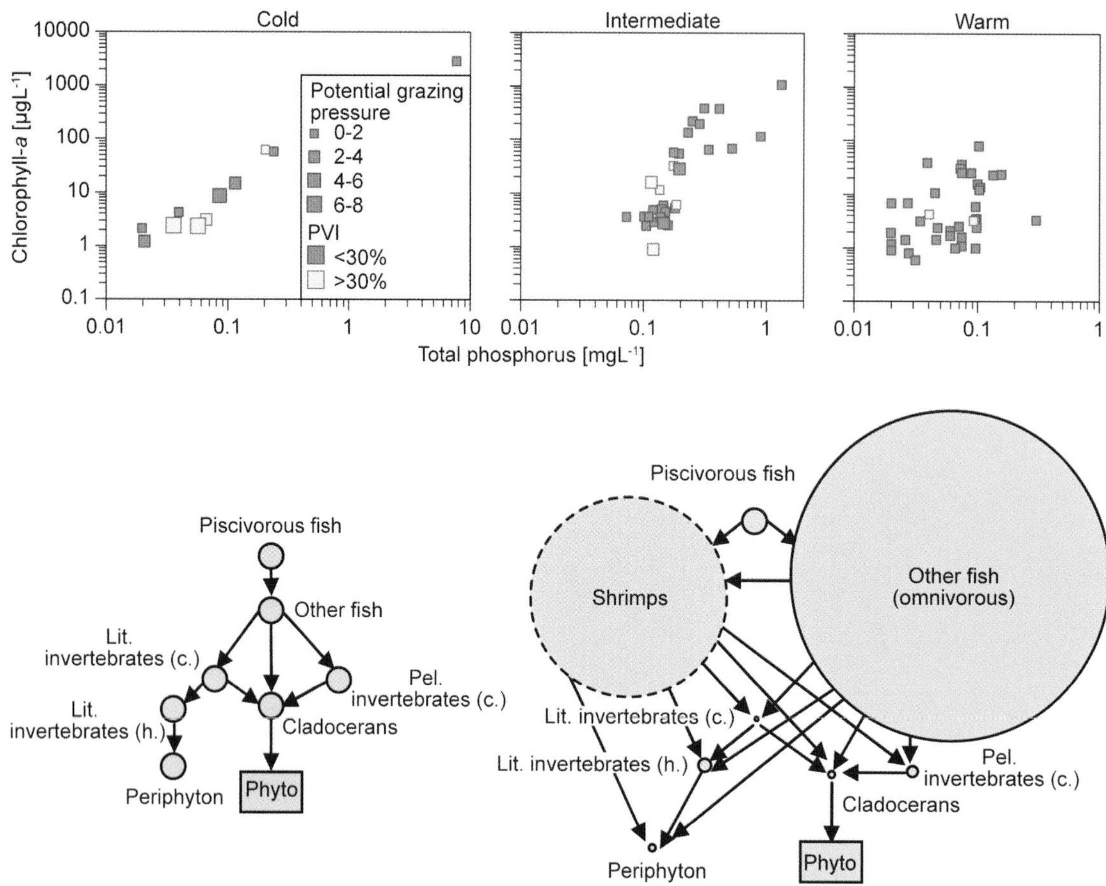

FIGURE 26-5 Role of climate on aquatic plants effects and on trophic structure and interactions, with consequences for water clarity and feedback mechanisms that may stabilize shallow lake regimes. *Upper section*: different relationships between phytoplankton biomass (chlorophyll *a* concentration) and total phosphorus concentration (on logarithmic scales) in shallow lakes with scarce (plant volume inhabited [PVI] <30%) or abundant (PVI >30%) plant growth and with various degrees of potential grazing pressure (estimated as zooplankton:phytoplankton biomass ratio), in warm, intermediate, and cold climate regions of South America. The nutrient level above which the dominance of submersed plants becomes rare is lower in warm regions than in colder regions. *Lower section*: trophic web structure and interactions in temperate and warm shallow lakes with comparable phytoplankton (phyto) biomass and limnological characteristics. The densities in the subtropics are augmented or decreased in relationship to those in the temperate lakes. The consumers were classified as intermediate herbivores (h.) such as cladocerans, and other invertebrates of littoral (Lit.) and pelagic (Pel.) zones, intermediate carnivores (c.), intermediate omnivores, and top carnivores (i.e., piscivorous fish). Except for fish, the same taxa shared the same trophic classification in both climate zones. Shrimp relative density is dotted due to the typical absence of shrimp in temperate lakes. *(Upper figure modified and published with permission from Fig. 6 in Kosten et al., 2009,* Ecosystems, *12, 1117–1129.* Lower *figure published with permission from Fig. 2 in Meerhoff et al., 2007a,* Global Change Biol. *13, 1888–1897.)*

the pair-wise competitive interaction between submersed plants, phytoplankton, and free-floating plants may theoretically also be affected by climate, either directly (Kosten et al., 2011) or via climate-related food web structure (Meerhoff et al., 2007a) or water-level dynamics (Coops et al., 2003), as will be described below. Functional characteristics of the species involved, and the initial biomasses of each group, may also affect those competitive interactions between the three potentially dominant types of primary producers.

iii. Drivers of regime shifts

The likelihood of a shift from a submersed plant dominated to a free-floating plant–dominated or a phytoplankton-dominated regime increases with increasing nutrient concentrations, with total phosphorus typically considered the main driver but total nitrogen also being significant (González-Sagrario et al., 2005). However, besides nutrient loading, some other external factors may increase or decrease the resilience of a given regime and, respectively, decrease or increase the likelihood of a shift. The predominant climate regime, and expectedly also climate change, will increase or decrease the likelihood of certain regimes (Fig. 26-4). Climate effects interact with nutrient loading: for example, the nutrient level above which the dominance of submersed plants becomes rare differs among climate regions, with lower nutrient levels potentially triggering shifts in warmer regions (Beklioğlu et al., 2007; Kosten et al., 2009) (Fig. 26-5).

Water-level fluctuations are also major drivers of regime shifts. In areas prone to high evaporation and irregularities in precipitation (or in systems subject to intense water extractions), shallow lakes and ponds can experience high water-level fluctuations and frequent regime transitions (Beklioğlu et al., 2017; Coops et al., 2003). Regardless of water turbidity, low water-level periods may allow submersed plants to develop (Ersoy et al., 2020). A marked reduction in water-level may, however, expose plants to air either in summer or in winter leading to their collapse. In arid and semiarid regions decreases in water-level are associated with an increase in cyanobacterial dominance (Brasil et al., 2016). Similar ecosystem changes may occur in floodplain lakes with marked hydrological changes. For instance, in a tropical shallow floodplain lake in the Pantanal (Brazil), submersed macrophytes decreased with falling water-levels, likely due to the negative conditions generated by the low oxygen saturation and high ammonium concentrations. In this kind of floodplain lake the flood pulse regime also determines the temporal dynamics of benthic fish migration, with high biomasses of benthic fish promoting a turbid regime by sediment resuspension during the low-water periods (Mormul et al., 2012). Strong reductions in water-level can also facilitate a shift from a free-floating plant dominance to a phytoplankton (and particularly Cyanobacteria) dominance (O'Farrell et al., 2011), most likely due to the high sensitivity of free-floating plants to the consequent higher water conductivity. In contrast, the submersed plants may expand during high water-level periods, likely due to the arrival of new propagules brought by the flooding river water (Loverde-Oliveira et al., 2009).

Besides large water-level fluctuations, single events that may damage submersed plants and thus reduce the stability of this regime include violent storms or increases in salinity. The introduction or arrival of large benthic or herbivorous fish (such as grass carp and common carp), waterfowl, or mammals such as otters and muskrats, or outbreaks of herbivorous insects, can also destroy or remove large numbers of plants. People can also influence submersed vegetation either directly by mechanical harvest or application of herbicides purposefully for control or accidently via agricultural runoff, or by damage from boat motors (Moss et al., 1996). Turbid conditions have been purposely generated throughout history by stocking plant-eating fish as a way to enhance fish production and fisheries, particularly in South east Asia (Jia et al., 2013). In many places around the world such shifts have been driven by biological invasions; for example, by crayfish and herbivorous carp (or by invasive plants in the reverse shifts), especially in the tropics and other warm locations (Schooler et al., 2011; Thomaz et al., 2015).

Indirect effects that weaken the submersed plant-dominated state via trophic cascades may occur by changes in the top of the food web (i.e., piscivorous fish) with consequent increases in the biomass of planktivorous and/or benthivorous fishes followed by decreases in biomass of zooplankton and macroinvertebrates (Burks et al., 2006). Excessive removal of large-bodied, piscivorous fish and selective winter or summer kills of piscivorous fishes may thus positively impact phytoplankton or epiphytic periphyton abundance sufficiently as to weaken the dominance of plants and induce a shift to dominance by phytoplankton. The introduction of benthivorous fish (such as common carp) can also lead to sharp increases in water turbidity due to sediment resuspension after intense prey search activity of the fishes (Zambrano et al., 2001).

Importantly, there is not one single critical nutrient level for maintaining clear water, but rather such levels are likely specific and context dependent. Different lakes may tolerate quite different nutrient levels before becoming turbid (e.g., Kong et al., 2017). After nutrients, lake depth and water-level fluctuations play a major role in determining the probability of a lake being dominated by plants or by phytoplankton. Low water depths often permit a better light regime near the sediments and therefore more easily permit the development of submersed plants in ponds and shallow lakes than in deeper lakes (Coops et al., 2003). Therefore the theoretical nutrient concentration (and water turbidity) threshold for a regime shift is higher in shallow lakes than in deeper lakes.

Lake size also affects the probability of a particular regime, with smaller systems being more prone to be dominated by plants and to have high biodiversity, despite potentially higher water nutrient concentrations than similar larger lakes (Søndergaard et al., 2005). Many large, shallow lakes often lack plants, regardless of their nutrient concentrations, likely because the effect of the winds on sediment destabilization and resuspension is particularly strong in such systems (Crisci et al., 2017). These often-turbid lakes generally present low productivity (as phytoplankton are also light limited) and low biodiversity. Small lakes are in general less sensitive to wind action, and therefore sediment resuspension and water turbidity are expectedly lower than in similar large lakes. A weaker nutrient control on phytoplankton biomass occurs in ponds and small shallow lakes compared with larger lakes, supporting the view that major shifts in the functional coupling of grazers and phytoplankton occur along a lake-size gradient (Tessier and Woodruff, 2002). This is related to fish being frequently absent from small lakes and ponds as a consequence of frequent anoxic events or, depending on the regional climate, also due to alternations between drying out and freezing solid, leading to fish kills. As stated above, the absence or scarcity of planktivorous and benthivorous fishes contributes to more stabilized sediments and releases grazer zooplankton and macroinvertebrates from predation, indirectly leading to better growth conditions for plants (Scheffer et al., 2006).

Geomorphic conditions, such as surrounding hills or trees that minimize wave action and currents that would resuspend sediments and increase abiotic turbidity, can help maintain submersed plant dominance. Other factors relevant to potential regime shifts include lake spatial heterogeneity. For instance, a nonuniform lake depth profile tends to reduce the chance of large-scale shifts between alternative regimes (Scheffer and van Nes, 2007). Because submersed plants are less affected by water turbidity at shallower sites, shifts in local states will not happen simultaneously within a spatially heterogeneous lake.

Whatever the cause, once the shift has occurred, restoring prechange conditions (such as actions to decrease nutrient concentrations in the water) is typically insufficient to reinstate the previous regime. It is also important to highlight that the processes described above help understand the likelihood of regime shifts to happen and the new regimes to be maintained under certain environmental conditions, but nature often does not follow our theoretical expectations. Stochastic factors may also occur, leading to unexpected shifts between regimes. Moreover, many ecosystems, such as lakes in extreme climate regimes where there are strong temperature fluctuations and where winter or summer fish kills or summer desiccation are common, are often intrinsically far from any equilibrium state and are subjected to high disturbance regimes as a rule rather than as an exception (e.g., Loverde-Oliveira et al., 2009), (Rautio et al., 2011; Scheffer and van Nes, 2007).

iv. Feedback mechanisms of the submersed plant–dominated regime

Submersed plant dominance can be maintained by several buffer and feedback mechanisms that directly or indirectly promote the clear water state (Canfield et al., 1984; Jeppesen et al., 1998; Scheffer et al., 1993). Their relative importance varies with the species involved and with the regional and local characteristics of lakes and ponds.

1. Stabilization of sediments and water column. Submersed plants stabilize sediments with their root structure, decreasing wind-induced turbidity and thus helping to maintain an adequate underwater light regime for photosynthesis (Barko and James, 1998; Horppila and Nurminen, 2003). This is in turn favorable for the growth of more submersed plants. Besides, the reduction in wave action due to mechanical resistance offered by the plant beds facilitates the sedimentation of phytoplankton cells and detritus or abiotic material. Other plant types, such as emergent plants in the littoral zone, may also reduce wind-induced resuspension (Horppila and Nurminen, 2005), indirectly favoring submersed plants.
2. Competition for nutrients. Sequestering of phosphorus and nitrogen by submersed plants above nutritional requirements reduces availability to phytoplankton and periphyton (Stephen et al., 1998). Because of the uptake of combined nitrogen by submersed plants and especially epiphytic periphyton communities, nitrogen availability can become particularly restricted to phytoplankton. Within submersed plant beds, anoxic areas occur near the sediments, which leads to appreciable bacterial denitrification and losses of nitrogen to the atmosphere (Veraart et al., 2011).
3. Indirect enhancement of grazing on phytoplankton. Submersed plants can often provide daytime refuges from fish predation for large pelagic zooplankton, such as the keystone species *Daphnia* and other cladocerans, that effectively reduce phytoplankton biomass (Schriver et al., 1995; Timms and Moss, 1984). This refuge effect of submersed plants has been described mostly in temperate and cold shallow lakes and explains the diel horizontal migration pattern of large cladocerans there (e.g., Burks et al., 2002; Lauridsen and Lodge, 1996). The *pelagic trophic cascade* (that is, the effects on the biomass of lower trophic levels provoked by the feeding activity of a higher trophic level), from piscivorous fish to phytoplankton, is a key process that may either stabilize or weaken the plant-dominated state. However, the refuge effect of submersed plants for zooplankton becomes weaker or even disappears in subtropical and Mediterranean shallow lakes (e.g., Meerhoff et al., 2007b; Tavsanoğlu et al., 2012). In warm, shallow lakes the abundance of different guilds of fishes among plants is substantially larger than in otherwise similar temperate lakes (Teixeira-de Mello et al., 2009), which, together with the presence of predatory macroinvertebrates (González-Sagrario and Balseiro, 2010), has been stated as the most likely explanation for the lack, or weak refuge, of submersed plants for grazer zooplankton in warm climates (Meerhoff et al., 2007b).
4. Indirect decrease of epiphytic periphyton. Submersed plants have a large surface area that provides a habitat for the prolific development of epiphytic periphyton. Although periphyton development can suppress light availability to the plants (Phillips et al., 1978, 2016), plant-associated macroinvertebrates (e.g., some insect larvae and snails) can be effective in

grazing periphyton and reducing shading to the supporting plants. More so than by nutrients, grazer macroinvertebrates are controlled from the top down by benthivorous and molluscivorous fishes (Brönmark and Weisner, 1992), which are in turn partly controlled by piscivorous fishes. This *littoral or benthic trophic cascade* is one of the key processes that may either stabilize or weaken the plant-dominated state (Jones and Sayer, 2003), being often more important than the pelagic trophic cascade (Mormul et al., 2018).

5. Direct inhibition of phytoplankton through *allelopathy* (see also Chapter 24). Although much less studied than other mechanisms, organic compounds that potentially suppress the growth of phytoplankton and periphyton can be released from several submersed macrophyte species (van Donk and van de Bund, 2002), many of which typically occur in temperate shallow systems (Hilt and Gross, 2008) and in the subtropics (Vanderstukken et al., 2011). Suppression of phytoplankton growth by these compounds of plants has been demonstrated under laboratory conditions and in mesocosm experiments, but such interactions are more difficult to show in the field (Gross et al., 2007). This can be partly explained by the modification of these compounds after they are released in the water, by interaction with natural organic compounds and bacteria. Phytoplankton species exhibit differential sensitivity against allelochemicals released by submersed plants, whereas in general, epiphytic species are apparently less sensitive despite their closer physical contact with the plants and their importance for macrophyte growth due to shading (Hilt and Gross, 2008).
6. Occasional poor conditions for many fishes, leading to major decreases in total fish biomass, may promote, through trophic cascading effects, a stronger grazing pressure of zooplankton and macroinvertebrates on phytoplankton and periphyton competitors, respectively. The intense metabolism of the submersed plants and epiphytic microbiota, as well as the large deposits of decomposing organic detritus, frequently cause marked diurnal fluctuations of dissolved oxygen, pH, and other parameters in the littoral waters (Barko and James, 1998). Some of these conditions can suppress some fish and minimize their predation on zooplankton and macroinvertebrate grazers, leading to cascading effects ultimately favorable to plants. As discussed above, the smaller the area of the lake, the higher the chances of fish kills happening and of plants developing (Scheffer et al., 2006).
7. The overall higher biodiversity that occurs in the plant-dominated state acts as a positive feedback and buffer mechanism, by offering propagules, seeds, and inoculum of different species and functional groups (e.g., of plants, macroinvertebrates, and zooplankton) that may generate viable populations once conditions are favorable. The "biological memory" of the system, mainly stored in the sediments in the form of seeds, eggs, and resting stages, can then act as insurance increasing the resilience of the lake against external perturbations.

v. Feedback mechanisms of the phytoplankton-dominated regime

In shallow lakes and ponds the phytoplankton-dominated conditions where submersed plants are suppressed or eliminated can also be maintained by several feedback mechanisms:

1. Mixing of the water column and wind-induced sediment resuspension. Due to the lack of plants producing different levels of physical structure, the water column experiences common mixing with resuspension of surficial sediments in the open habitat. As a result, abiogenic and biogenic turbidity increases with a marked reduction of light penetration, limiting the chances for submersed plants to grow and persist. The loose, flocculent sediment is also a poor substratum for submersed macrophyte colonization.
2. High internal nutrient loading. Nutrients and particularly phosphorus (P) accumulate in the sediments and can be released back into the water column via different mechanisms (Søndergaard et al., 2003). The frequently intermittent but sometimes permanently anoxic conditions in the sediments, which can take place at night and/or during warm periods, may promote the chemical release of stored P, which is easily available for phytoplankton (Chapter 15). The loose sediments and associated nutrients are easily disturbed by wind-induced resuspension or by bioturbation by benthivorous fishes (Dantas et al., 2019). The high internal nutrient loading of shallow lakes and ponds is one of the main reasons why determining the nutrient thresholds for regime shifts (typically considering lake water P concentrations) is extremely difficult.
3. Competition for light and nutrients. The development of prolific phytoplankton communities reduces light but also nutrients such as inorganic carbon available for submersed plants (Søndergaard and Moss, 1997). This condition is particularly effective for the development of small algal and cyanobacterial species with a high capacity for light absorption and low light compensation points.
4. Indirect pelagic trophic cascading effects (Carpenter et al., 1985). Many piscivorous fishes are poorly

adapted to turbid conditions, lacking macrophyte habitats and plant-associated food items. Gradually, piscivorous fish reproduction and recruitment declines due to increased water turbidity and limited access to food, and to some extent also due to fluctuations in oxygen concentration. Also, the high phytoplankton biomass settling down to the sediment increases the fresh organic matter or detritus, thus indirectly favoring benthic detritivorous macroinvertebrates. These benthic subsidies and coupling of grazer and detritivore food chains indirectly enhance benthic fish biomass (Attayde and Ripa, 2008). A large proportion of benthivores and small zooplanktivores then often dominates the fish communities as turbidity increases, leading via top-down effects to a reduced zooplankton biomass (often shifting size structure toward smaller body–sized groups) with a reduced grazing pressure on phytoplankton.

5. At very high phosphorus and combined nitrogen concentrations in the water, green algae can effectively compete with nitrogen-fixing Cyanobacteria, depending on lake characteristics. Once the cyanobacterial populations are established, however, several species and functional groups generate a series of conditions that facilitate their self-perpetuation (e.g., water turbidity, low-oxygen conditions that promote sediment-P release, stability of the water column, and liberation of toxins and allelopathic substances, among others; Dokulil and Teubner, 2000).
6. Growth early in the season. In temperate lakes early spring growth of phytoplankton may continue if fish predation reduces zooplankton grazing pressure and nutrients are easily available. High initial biomasses of phytoplankton facilitate their competition against later-growing submersed plants.

vi. Feedback mechanisms of the free-floating plants–dominated regime

Free-floating plants comprise many species of vastly different sizes, from tiny lemnids to large-bodied water hyacinth (Chapter 24). Physical effects of different floating plants may be similar, but the effects on the chemical environment, and particularly on nutrient availability, vary due to their different nutritional requirements. Research on the potential buffer mechanisms and the drivers of regime shifts is substantially less advanced than that on the previously described regimes (de Tezanos Pinto and O'Farrell, 2014). Due to the higher sensitivity of free-floating plants to external factors, the stability of these feedbacks seems weaker than those of submersed plants and phytoplankton.

1. Competition for light and nutrients. Several free-floating species, particularly large plants (such as *Pistia stratiotes* L. and *Pontederia crassipes*), can luxuriously uptake nutrients and store them in their biomass. Under dense cover of free-floating plants, most of the incoming light can be attenuated (de Tezanos Pinto et al., 2007). Both phytoplankton and submersed plant photosynthesis is thus hindered below the dense floating mats, despite potentially high nutrient concentrations. Climate warming is expected to further promote the growth of free-floating plants under eutrophic conditions (Netten et al., 2010).
2. Favorable physical and chemical conditions for the maintenance of free-floating plants. The water column under dense floating mats can become anoxic and pH may decrease markedly compared to plant-free habitats. Anoxic conditions promote the release of phosphorus from the sediments, increasing its availability for the floating plants. The temperature below the mats of free-floating plants is reduced, despite the fact that the dense cover of floating plants can promote higher surface temperatures, thus creating a favorable environment for their growth.

IV. Biodiversity

What biodiversity really is, what determines biodiversity at different spatial and temporal scales, and whether biodiversity patterns occur and can be detected are interesting and important questions for all ecosystems. However, shallow lakes and ponds are particularly good model ecosystems to explore these issues (De Meester et al., 2005). *Biodiversity* can be broadly defined as the biological variety and variability of life and the ecosystems where biota occurs and could be considered at different biological levels of organization, ranging from genes to ecosystems (Brönmark and Hansson, 2005). Freshwaters harbor a unique and diverse set of species and functional groups. About 15% of all current animal species live in different freshwater systems. Approximately 70,000 freshwater species from 570 families and 16 phyla have been described (Strayer, 2013), and more remain to be described and/or reclassified considering the rapid advances in taxonomic identification including molecular techniques and environmental DNA approaches.

Based on species–area relationships for true islands and due to their small size, ponds and small shallow lakes have traditionally been considered as providing insignificant biodiversity to the regional species pool compared to larger ecosystems such as large lakes and rivers (Céréghino et al., 2008). Shallow lakes and ponds have, however, lent support for the *habitat amount hypothesis*, which predicts that, in equal-sized sample sites (ponds and lakes), species richness should increase with the amount of habitats in that landscape and not with the mere area of each sample site (Fahrig, 2013). It is thus increasingly clear that shallow lakes and ponds contribute disproportionately more to biodiversity at a regional level than any other freshwater ecosystem (e.g., ponds support c. 70% of the freshwater species pool in European landscapes) (Williams et al., 2004). Particularly under harsh climatic conditions, such as in the Arctic and Antarctic, small lakes and ponds represent local hotspots of biodiversity and production (Smol, 2016), which are often sustained by microbial benthic mats as systems often lack higher aquatic vegetation and fish (Izaguirre et al., 2020; Rautio et al., 2011).

The high contribution to local and regional diversity of ponds and small shallow lakes is also because they often strongly differ in species composition among each other (Oertli et al., 2002) as well as to the occurrence of species that are specific to these water bodies. The compositional dissimilarity among lakes and ponds, called *beta* (β) *diversity*, is determined by abiotic and biotic filtering mechanisms as well as dispersal, as opposed to *alpha* (α) *diversity*, which accounts for the species richness in each water body. Therefore, although many individual ponds may contain relatively few species (α-diversity), these ecosystems represent an enormous variety of abiotic and biotic conditions allowing for an overall diversity increase in ecological communities (β diversity), which, in turn, results in a greater contribution to landscape-level biodiversity (*gamma* [γ] *diversity*) than those of larger but homogeneous sites (Hassall, 2014; Williams et al., 2004).

The relatively high beta diversity, particularly of ponds but also of shallow lakes, is no doubt related to their high variability in ecosystem characteristics or local abiotic and biotic filters (e.g., as shown for aquatic plants; Alahuhta et al., 2017). Such local filters include the temperature regime, pH, salinity, nutrient availability, hydroperiod, water-level fluctuations, habitat stability and complexity, trophic structure, and biotic interactions, among others (Fig. 26-6). However, multiple layers of processes operate at different spatiotemporal scales, determining colonization, speciation, or extinction in a given water body at a given time. Current

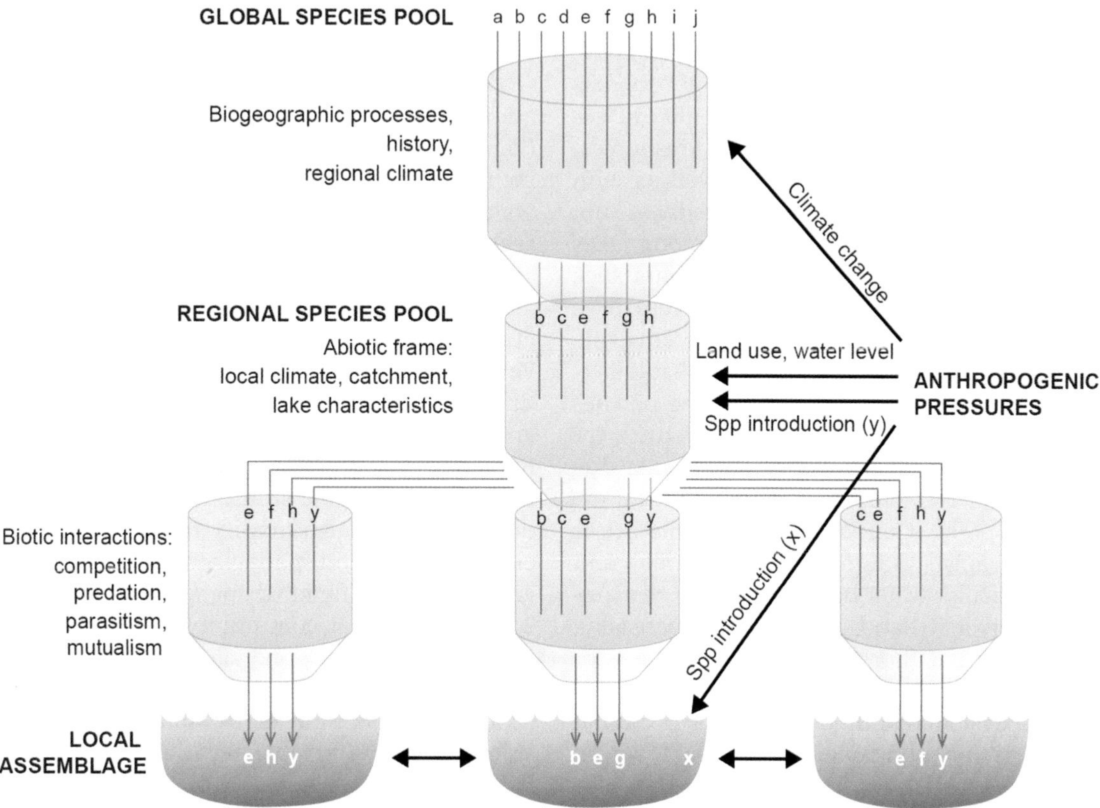

FIGURE 26-6 Conceptual model showing how "filters" operating at different spatial scales determine species sorting and community structure of a given shallow lake or pond, and how different anthropogenic activities add new filters or modify natural filters. The importance of a landscape perspective and the potential connectivity and dispersal with other lakes and ponds for alpha, beta, and gamma biodiversity is highlighted. *(Modified with kind permission from Fig. 1 in Kosten and Meerhoff, 2014,* ELS Encyclopedia of Life Science. *John Wiley and Sons. Originally based on Brönmark and Hansson, 2005.)*

local biodiversity is also determined by large-scale filters such as large biogeographic and historical scales (e.g., glaciation, continental drifts) and regional processes, such as dispersal barriers and climatic conditions (e.g., precipitation and temperature patterns) (Fig. 26-6).

In the last few decades it has been more fully recognized that population persistence, community dynamics, and biodiversity of a local ecosystem are deeply affected by its position in the landscape and the potential connectivity with other systems on a landscape scale (Leibold et al., 2004). Current understanding has evolved into a *process-based metacommunity framework* to unite the local and regional scales of *ecological community dynamics theory* (Thompson et al., 2020). The metacommunity framework has therefore become a powerful tool to disentangle the relative importance of local environmental factors, including abiotic (environmental heterogeneity) and biotic (e.g., competition, predation, mutualism) variables, and landscape attributes (spatial filter affecting dispersal of individuals between habitats) in shaping community structure and biodiversity in shallow lakes and ponds.

Lakes and ponds in a shared landscape therefore constitute a network (often referred to as a *pondscape*), with many species having metapopulations across the landscape (Hill et al., 2017; Oertli and Parris, 2019; Pereira et al., 2011). The occurrence, spatial arrangement, and potential connectivity with other water bodies mediate the dispersal of organisms and the flux of energy, materials, and information (e.g., genetic resources) between and within lakes and ponds, as well as with other ecosystem types. Effective connectedness not only is a function of geographic distances among lakes and ponds but also depends on the regional abundance of species, which may, in turn, depend on the occurrence of various water bodies in the landscape (Horváth et al., 2019). Landscape connectivity in fragmented habitats is usually associated with dispersal ability among patches that facilitates or impedes movements and thus enhances local persistence in source–sink systems, due to higher chances of recolonization following local extinctions (Hanski, 1999). Connectivity may thus act as an insurance against external perturbations and stressors on a particular water body, increasing or maintaining local biodiversity, and consequently increasing resistance and resilience (Loreau et al., 2003). A loss or decrease in connectivity can affect biodiversity and therefore the ecosystem stability to external perturbations and stressors, including climate change. However, high connectedness can, under some circumstances, also negatively affect the biodiversity of shallow lakes and ponds (Scheffer et al., 2012). This could happen through increased dispersal among sites resulting in reduced compositional dissimilarity (reduced beta diversity) and particularly in cases of anthropogenic disturbances in the landscape (e.g., point source pollution, the introduction of exotic species, fire, eutrophication, and toxic Cyanobacteria inoculum), that end up affecting all connected systems simultaneously (Fig. 26-6).

The body sizes and dispersal modes of organisms are important functional traits that drive metacommunity structure. Organisms ranging in size from bacteria to phytoplankton to macroinvertebrates can be efficient dispersers either through active (e.g., flight in adult freshwater insects) or passive dispersal mechanisms via transport by animal vectors or water and wind, often involving desiccation-resistant stages (Bilton et al., 2001; Padial et al., 2014). Larger organisms, such as plants and fish, have fewer effective dispersers in aquatic environments. For instance, the role of dispersal for crustacean zooplankton community structure seems linked to the connectedness of a particular water body (Chapter 20). In sets of interconnected lakes, crustacean zooplankton and fish can be more constrained by dispersal-based processes than bacteria and phytoplankton (Beisner et al., 2006). In contrast, in well-connected small geographic areas local environmental factors associated with alternative regimes may still promote differences in local zooplankton community structure despite the high dispersal rates (Cottenie et al., 2003).

Dispersal therefore interacts with local filtering mechanisms to determine the structure of local communities (Fig. 26-6). Local abiotic factors relevant to biodiversity include the hydrological dynamics and particularly water-level fluctuations. These are the result of the regional climate (balance between precipitation and evapotranspiration patterns) and local geology, and also of anthropogenic uses and management (Coops et al., 2003). Natural water-level fluctuations typically vary among climatic regions, with substantially different patterns in polar, temperate, Mediterranean, and tropical and subtropical lakes, leading to different functioning of shallow lakes and ponds including the likelihood of regime shifts. The hydrological dynamics of a region can also modify the relative importance of environmental and spatial factors for different biological groups with different dispersal potentials (Dias et al., 2016). From a metacommunity perspective, floods facilitate dispersal, thus enhancing the role of the local environment in shaping local community structure, whereas droughts enhance the relative importance of spatial variables. Floods reduce spatial variability among connected lakes and ponds, whereas droughts allow for systems to follow different temporal trajectories due to local stochastic variation (Thomaz et al., 2007). Effects of climate change on water-level fluctuations are also expected to differ substantially among regions (Coops et al., 2003) and this may have strong implications for shallow lakes and ponds, especially temporary ponds (see below).

Other local factors in shallow lakes and ponds, such as productivity (often associated with the likelihood of the clear or turbid water regimes) and some key biotic interactions, can largely explain community structure, including taxonomic, functional, and size diversity. Among biotic interactions, the presence of predators (mostly fish) and plants have a strong species sorting impact, as has been described in previous sections. In shallow mesoeutrophic lakes with large cover and volume occupied by submersed plants, the clear water regime is associated with higher total taxonomic richness, particularly in temperate regions (Declerck et al., 2005) but also in other climate regions (Kruk et al., 2009; Thomaz and Cunha, 2010). Aquatic vegetation plays a key role in local biodiversity by increasing habitat complexity and heterogeneity and by modulating predator–prey interactions promoting coexistence (Kovalenko et al., 2012). However, in warm lakes (e.g., subtropical and Mediterranean shallow lakes), the net positive effect of submersed plants on biodiversity may be comparably weaker than in similar temperate lakes due to frequent high fish predation pressure that affects the biomass and richness of several groups of organisms (Brucet et al., 2010; Meerhoff et al., 2007a).

The effects of in-lake processes on biodiversity may thus outweigh the effects of lake surface area and potential connectivity, which are considered fundamental by island biogeography theory (Scheffer et al., 2006). Below a certain surface area, isolated shallow lakes and ponds often have a well-developed plant (or benthic) community but host low densities and low richness of fish, or even lack them. Because of the low fish biomass and consequent low predation pressure, together with the abundant submersed vegetation and associated high spatial heterogeneity, a relatively high richness of zooplankton, plants, macroinvertebrates, amphibians, and birds often occurs in ponds and small lakes (Oertli et al., 2002; Scheffer et al., 2006; Søndergaard et al., 2005) (Fig. 26-7). Therefore a network of small shallow lakes and ponds (e.g., pondscape) contributes disproportionately more to biodiversity at a regional level than few larger freshwater ecosystems.

In shallow lakes and ponds, besides bacteria, the aquatic groups hosting the higher number of species are crustaceans, rotifers, insects, and oligochaetes, together with submersed plants, microalgae, and Cyanobacteria, either as free living or attached (Brönmark and Hansson, 2005). Among the many taxa that depend on ponds, perhaps the best known are the charismatic groups, particularly the amphibians, waterfowl, dragonflies, and aquatic plants, which are important and visible contributors to pond biodiversity. Semiaquatic mammals, such as muskrats, capybaras, and otters, can be frequent inhabitants of shallow lakes, ponds, and wetlands in different locations. However, snails, microcrustaceans, aquatic mites, springtails, chironomids, and a high diversity of Hemiptera and Coleoptera are also relevant for pond biodiversity.

V. Variations in the theme: temporary and urban ponds

Shallow lakes and ponds can be classified according to their hydrological dynamics into permanent and temporary. Characteristics of permanent natural ponds mostly follow the ecological dynamics predominating in the littoral zone of shallow lakes, as has been discussed in the previous sections. The distinctive aspects of the ecology of different types of ponds, namely temporary and urban ponds, are the focus of this section.

Temporary ponds are a naturally widespread habitat occurring, often in abundance, in all biogeographical regions. In the cold, dry climate conditions of the circumpolar Arctic, temporary ponds represent a large proportion of the enormous number of shallow lakes and ponds that occur there. In Europe temporary ponds are highly common, from the boreal snow-melt pools of northern Scandinavia to the seasonally inundated coastal dune pools of southern Spain and other Mediterranean areas. They are also quite common in warm dry and semidry climates (such as Australia) and in lowlands near the large coastal lagoons along the Atlantic coast and the floodplains of large rivers in South America and Asia.

The key environmental factor driving the structure of ecological communities in temporary ponds is *hydroperiod*, which is the length of time that ponds have water. Some temporary ponds only dry out for a few weeks in hot summers, whereas in others the dry period may last for several years. The timing of the wet and dry period also differs depending on the predominant climatic, geological, and land- and water-use features of a given region. A typical temporary pond has no inflow or outflow and is dependent on precipitation and surface water run-off. Other ponds are connected to nearby permanent streams or lakes. Basins may fill up in spring and dry out and stay dry throughout the year until the next spring (*temporary vernal ponds*). These systems are extensively widespread on a global scale and are especially common in regions with Mediterranean-type climates (found in five regions of the world). On the other hand, other temporary ponds have water throughout the year and only dry out during summer (called *temporary autumnal ponds*). In shallow ponds from cold environments the water column and part of

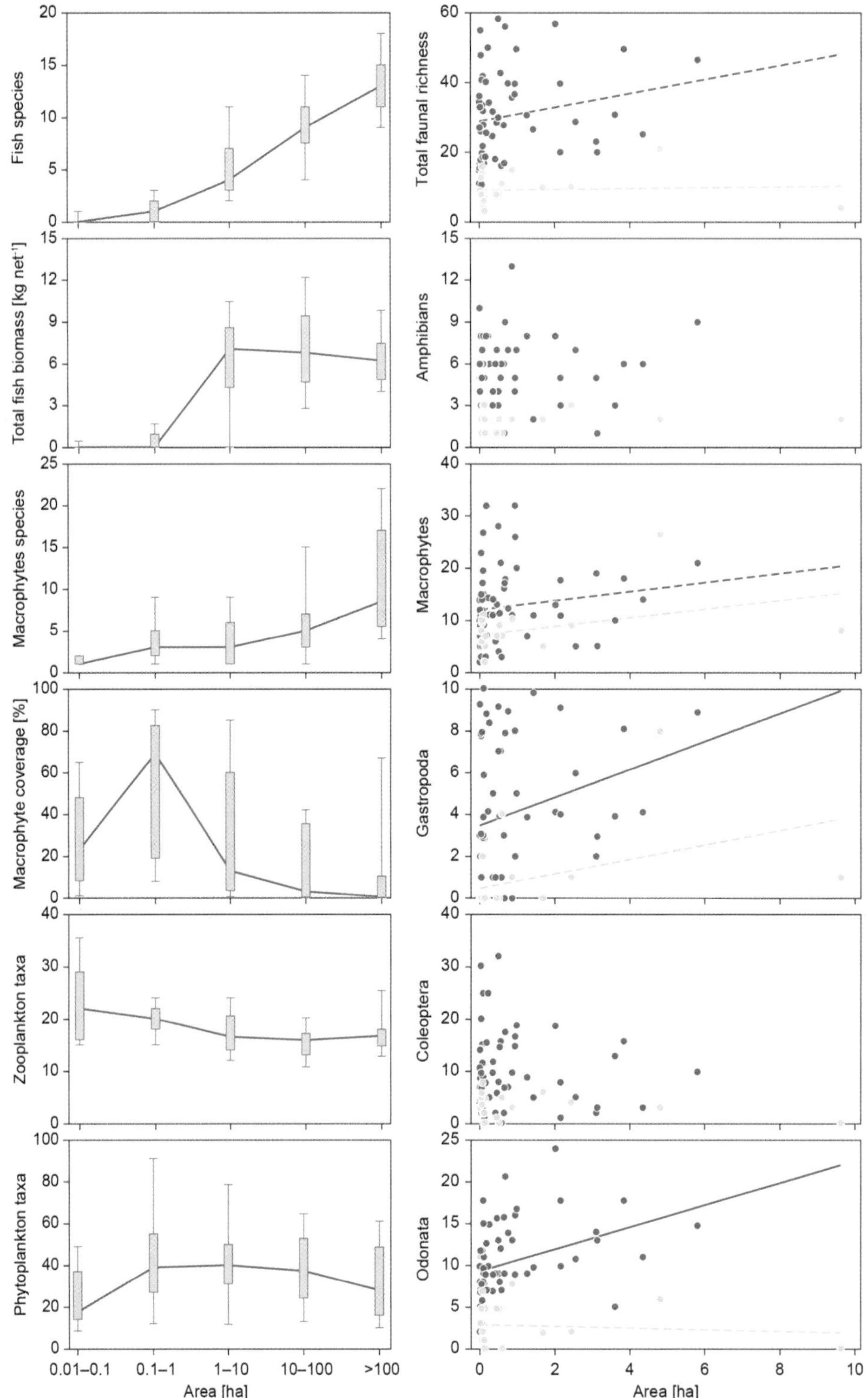

FIGURE 26-7 Taxonomic richness in shallow lakes and ponds. *Left-hand side*: importance of surface area for taxonomic richness of plants, fish, zooplankton, and phytoplankton and for submersed plant cover and fish biomass, in temperate shallow lakes and ponds ($n = 796$ lakes and ponds from Denmark). Each box represents the 25th and 75th percentiles, the horizontal line the mean value, and the *top* and *bottom* of the *thin line* the 90th and 10th percentiles, respectively. *Right-hand side*: focus on the relationship between area and taxonomic richness in temperate ponds (Switzerland), showing total fauna (five groups), amphibians, aquatic vegetation, and macroinvertebrates Gastropoda, Coleoptera, and Odonata. The *dark symbols* and *lines* represent lower-altitude ponds ($n = 60$), and the *light symbols* and *lines* higher-altitude ponds ($n = 20$ ponds, altitude >1400 m.a.s.l.). Significant linear relationships are indicated by *solid lines* and significant relationships after log transformation by *broken lines*, whereas absence of lines indicates nonsignificant relationships even with log-log transformation. *(Redrawn with kind permission from Figs. 5 and 7 in Søndergaard et al., 2005,* Arch. Hydrobiol. *162, 143–165* (left) *and from Fig. 3 in Oertli et al., 2002,* Biol. *Conserv. 104[1], 59–70.)*

the sediments freeze solid during winter (Hobbie, 1980). These ponds may contain water at all times of the year and thus would be considered permanent water bodies. Because biological activity essentially ceases during the winter period, these waters have been termed *aestival ponds* (Welch, 1952). Organisms of aestival ponds are adapted to tolerate interruption of development by freezing until subsequent thawing periods. The physiological rigors of freezing without desiccation are less severe than being exposed to a drying temporary pond where both freezing and desiccation can occur (Daborn and Clifford, 1974).

The marginal vegetation growing adjacent to, or within, the ponds influences physical, chemical, and biological features, from thermal stratification patterns to loading of nutrients and organic matter. Often, the predominant organic loading occurs as dissolved organic matter leached from the surrounding vegetation and soils. Especially during dry periods, terrestrial vegetation encroaches inward of the ponds. Particulate organic matter is predominantly allochthonous via leaf fall from trees or wetlands and earlier pond vegetation. These rich organic nutrient sources foster the development of biological communities of shredders or detritivores together with aerobic bacteria when the pond is filled up again. In addition, the annual disturbance imposed by the drying period allows for accelerated aerobic degradation and oxidation of organic matter, as well as for the release of more nutrients and GHGs (Boon et al., 1997; Obrador et al., 2018) than would occur in permanently inundated ponds.

As a result, upon reflooding, nutrient availability is high with increased initial productivity. When exposed to appreciable light, this nutrient-rich environment provides ideal circumstances for plant and benthic and planktonic algal growth, which may dominate the metabolism in ponds, as was emphasized in the treatment of the ecology and metabolism of littoral and floodplain environments (Chapters 24 and 25) and the sediments (Chapter 27). Furthermore, the land around a pond often provides habitats for the terrestrial stages of several species (e.g., amphibians, many insects).

Abiotic conditions in temporary ponds can be unpredictable and harsh, especially during the dry period. Temporary ponds in especially dry regions can face large water-level fluctuations, which might be coupled with strong physical and chemical changes. Linked to the loss of water, there are decreases in habitat volume and increases in insolation with subsequent links to water temperature, dissolved oxygen level, primary productivity, pH, and water chemistry (Williams, 1996). The temperature may have diel fluctuations together with dissolved oxygen concentrations and with large fluctuations in water chemistry. Water losses by evaporation can also result in increasing salinity. With increasing salinity, the potential of submersed plants for maintaining clear water is reduced, and often conditions are suboptimal for several aquatic organisms (Brucet et al., 2009; Brucet et al., 2012). In arid regions, in particular, concentrations of salts can reach extraordinary levels before drying and further constrain communities leaving only a few tolerant species (Chapter 12; Williams, 1996).

Organisms living in such irregular, harsh habitats must have many structural, behavioral, physiological, and evolutionary adaptations to survive or avoid drought and drought-induced changes in physical and chemical environments. In permanent shallow lakes and ponds predators are important in regulating prey populations, whereas in temporary ponds predation pressure is often low owing to the usual absence of fish (excluding some African and South American killifish species that bury themselves deep into the wet mud to stay in a dormant phase or leave their eggs for future rewetting conditions) or macroinvertebrate predators, which may arrive late in the water cycle. Thus organisms adapted to survive the dry periods have an advantageous habitat with rich food sources and low predation pressure, being able to build up high population densities. Under these conditions, a requisite for all organisms, and particularly for larger animals, is a rapid rate of development during the wet phase. Development and locating a mate must be completed in a few weeks, often accelerated by rapid changes in physical (e.g., increasing temperatures and ultraviolet intensities), chemical (increasing salinity and dissolved organic matter, reduced dissolved oxygen concentrations), and biological (increasing susceptibility to predation as the fish hatch, such as in tropical and subtropical temporary ponds) conditions in the receding ponds. Additionally, species of temporary ponds have a marked seasonality in their life cycles that is coupled with the probability of the inundation cycle of the ponds. Artificial elimination of the flooding-drying periodicity can alter or eliminate needed cues for oviposition, embryonic development, and hatching, among dominant taxa.

There is a diverse range of species with many adaptations, especially centered around their dispersal ability. Several animal groups are permanent residents and capable only of passive dispersal, which aestivate and overwinter in the sediments of dry ponds, either as stages resistant to desiccation (certain *Turbellaria* [cysts and egg cocoons], Bryozoa [statoblasts], Anostraca, Cladocera [ephippia], Copepoda [eggs], and Ostracoda), resistant eggs (Oligochaeta, Hirudinoidea, Decapoda, and mollusks), or in seed banks. Others are capable of some dispersal (certain Ephemeroptera, Coleoptera, Trichoptera, and Diptera), oviposit on the water in spring, and then aestivate and overwinter in the dry basin in various stages of the life cycle. On the other hand, some animals (certain Ephemeroptera, Odonata, Hemiptera, Coleoptera, Diptera, and amphibians) that have well-developed powers of dispersal leave the disappearing pond and pass the dry phases in permanent waters. Some of these animals subsequently return to oviposit in the temporary pond in the following spring. Temporary ponds (e.g., Mediterranean temporary ponds) are often considered hotspots of biodiversity, with the occurrence of endemic or rare species at regional scales due to their inherent intra- and interannual fluctuations in limnological features, promoting heterogeneity in the structure and dynamics (Parra et al., 2021).

Shallow lakes and ponds can also be classified as urban based or non–urban based on whether they are found in urban settings. Urban shallow lakes and ponds are largely constructed by people (Oertli, 2018), often in low-elevation or wet areas. These artificial ecosystems often provide a particularly wide range of ecosystem services to society, including water supply for several purposes, hydrological regulation, nutrient retention, fish production, wildlife protection, recreation, education, religious celebration, and research. They often present different environmental characteristics than nonurban (seminatural/agricultural) ponds. Urban lakes and ponds commonly have concrete margins, a synthetic base, reduced vegetation cover, and lower connectivity to other water bodies and are subject to run-off from residential and industrial developments, greatly increasing the nutrient loading and the concentration of contaminants. In addition to this, urban water bodies are frequently smaller than natural lakes, despite them showing a large variation in size (>10 ha for some park lakes).

Urban ponds appear to be on the rise through intensified urbanization (Hassall, 2014; Oertli and Parris, 2019). These small water bodies collectively harbor a significant proportion of global biodiversity and could be priority ecosystems for conservation (Oertli, 2018; Oertli et al., 2005; Williams et al., 2004), considering urbanization or built-up areas are expected to almost triple in surface area and surpass the rural areas by mid this century (Seto et al., 2012).

The same prejudice as with shallow lakes and ponds, in general, has lingered for urban ponds, which were considered to host significantly lower biodiversity than nonurban ponds, based on the anticipation from terrestrial and lotic habitats' homogenized or reduced taxonomic richness in urban areas. Despite the considerable anthropogenic pressures on urban ponds, recent studies have demonstrated that ponds located within an urban setting can provide important habitats for a wide range of taxa, including macroinvertebrates (Hill et al., 2015), waterfowl, and amphibians (Hamer and McDonnell, 2008) and that, contrary to predictions, gamma diversity is similar to nonurban ponds (Hill et al., 2016). In urban ponds Oligochaeta and Chironomidae species are common, consistent with processes of historical disturbance and subsequent recolonization by disturbance-tolerant taxa. Urban lakes and ponds can host a considerable portion of the national species pool for plants and some macroinvertebrate taxa (e.g., Hirudinea, Gastropoda, and Tricladida), but more sensitive taxa (e.g., Plecoptera and Ephemeroptera) tend to be excluded. In contrast, urban ecosystems can also favor the establishment of invasive species (Shochat et al., 2010).

VI. Restoration

This section aims at describing the current state of knowledge on, and the range of available techniques for, the restoration of shallow lakes, especially under eutrophic and hypertrophic conditions (Fig. 26-8). Lake ecosystems around the world are suffering the consequences of intense anthropogenic pressure, particularly land-use changes leading to eutrophication and hydrological alterations. Excessive loadings of nutrients from human activities in the catchments (e.g., intensive agriculture, urbanization, and industrialization), together with the loss of connectivity and of natural buffer mechanisms such as riparian vegetation, have pushed many shallow lake ecosystems into a eutrophic state with low biodiversity and massive primary production, most frequently a phytoplankton-dominated or free-floating plant–dominated regime. Besides the loss of biodiversity and of water quality and recreation potential, eutrophied lakes and ponds have a higher carbon footprint than less impacted systems. As described in Sections II and III, recovery from such conditions is extremely unlikely without external perturbations (Suding et al., 2004). As with other ecosystems, there is a growing need for effective restoration approaches to achieve desired ecological states.

A slightly different approach to restoration often applies to rural and urban ponds (Oertli and Parris, 2019). In rural areas *ghost ponds* can often be identified and restored by digging under agricultural, currently filled, soils and allowing for biota recolonization from the seed bank still present in the former sediment (Alderton et al., 2017). The aim of this restoration is to promote a recovery of biodiversity in intensively used landscapes. In urban areas, besides enhancing water quality by a reduction of loading of nutrients, salts, or pollutants, the suggested framework to help improve the quality of urban habitats includes having larger pond sizes like nonurban ones, improving pond margins with indentation, generating slopes of different angles and with large drawdown zones, allowing existing hydroperiods, and promoting the presence of aquatic

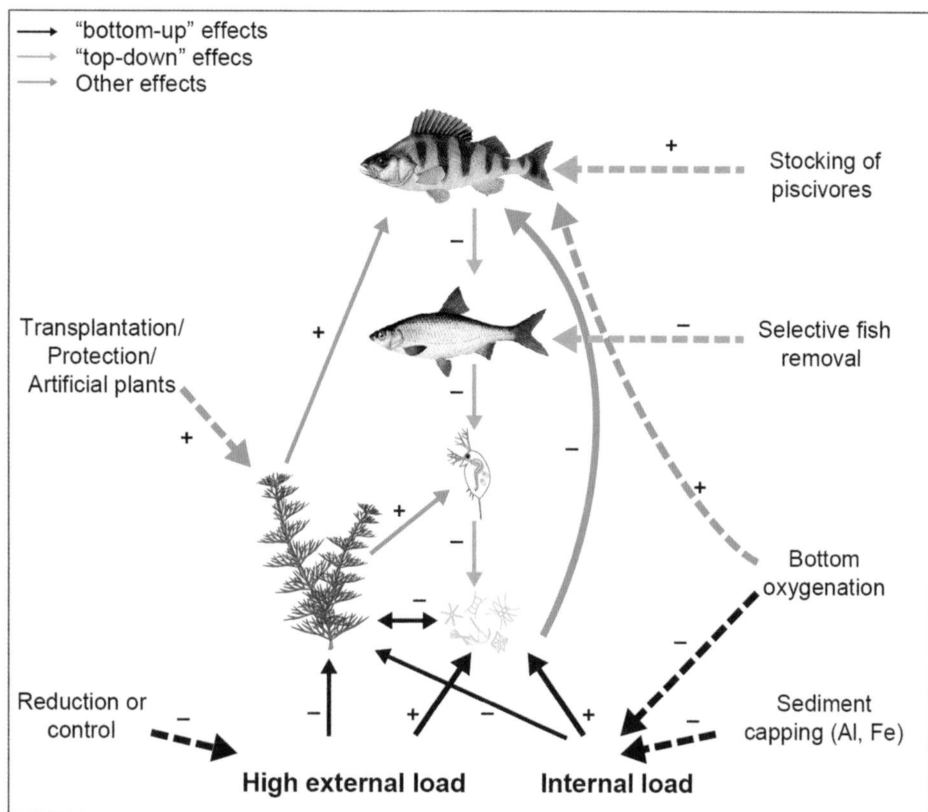

FIGURE 26-8 Conceptual model on the interactions between nutrients and the main components of a classic lake food web as affected by biomanipulation to enhance trophic cascade effects and other restoration measures. The nature of the interactions is highlighted, whether bottom-up (via nutrients, *black*), top-down (via consumption, *red*), or others (e.g., physical or chemical mechanisms; *green*). The simultaneous application of both biomanipulation and a treatment to reduce internal nutrient loading may generate synergistic effects. *(With kind permission from Diagram 1 in Jeppesen et al., 2012,* Adv. Ecol. Res. *47, 411–488.)*

plants (emergent, submersed, or floating) while eliminating nonnative or invasive species.

A. External measures

Priority measures for restoration are those addressed at reducing *external nutrient loading* to lakes (Cooke et al., 2016). Such approaches include phosphorus (P) stripping and occasionally nitrogen (N) removal at urban sewage works and sewage diversion, tertiary treatments of industrial sewage waters, and the use of phosphate-free detergents. These nutrient sources are called *point sources* as they are easily identified and could be relatively easily addressed if enough resources are devoted.

The agricultural system is, however, responsible for most of the nutrient loading reaching water bodies in a spatially diffused way (Moss, 2008), called *diffuse sources*. Some of the measures that need to be taken to reduce nutrient loading in a significant manner include better (and more humane) management of animals including fodder characteristics, manure storage, and treatment of effluents, strict and accurate fertilization plans according to crop needs, soil characteristics and local current and future climates, and green cover of agricultural fields in winter, where appropriate. In addition, nutrient retention and N loss in lake catchments can be enhanced by reestablishing wetlands and a natural riparian zone (Vidon et al., 2010), stabilizing river banks to reduce erosion, and allowing natural hydrological dynamics (Jeppesen et al., 2011). In several locations major reductions in nutrient loadings have been achieved via urban and sewage nutrient load reductions, as well as by modifying land use, agricultural, and forestry practices (Søndergaard et al., 2007).

Understanding lake biogeochemical cycles is a prerequisite for planning restoration strategies. Climate change predictions for a given locality and expectations for nutrient mineralization and runoff should be considered to avoid hindering recovery (Ockenden et al., 2017). However, significant reductions in external nutrient loading may still not guarantee a successful recovery, due to chemical or biological within-lake resilience mechanisms (Section III).

B. In-lake measures

Although not addressing the ultimate cause of the problem (i.e., nutrient sources), lake internal measures designed to either keep nutrients locked in sediments or remove them from the lake ecosystem, or to facilitate regime shifts via manipulation of trophic interactions, are often more feasible and less costly. In-lake restoration measures act as perturbations that may temporarily shift the current regime by removing some of the buffer mechanisms of the turbid-water state. Such restoration measures therefore reinforce recovery, treat symptoms, or improve/maintain a high environmental quality, until the external loading can be significantly reduced and with it, the likelihood of returning shallow lakes to the desired set of conditions can increase.

Interestingly, within the set of in-lake measures, the popularity of *biomanipulation* (i.e., purposely managing biotic communities in an aquatic system) has declined since the 1990s, whereas inactivation of P internal loading has shown a clear upsurge in applications over the last decade (Jilbert et al., 2020). Dual methods combining biological and physicochemical techniques to make restoration more efficient and long-lasting are promising, although climate-related particularities should be considered to secure success (Araújo et al., 2016). It is important to emphasize, though, that the key measure to restore eutrophied lakes is treating the cause of the problem, that is, the reduction in external nutrient loading.

i. Biomanipulation of fish communities

Biological resilience of the eutrophic turbid regime largely involves fish-related mechanisms, as the typically high biomass of plankti-benthivorous fishes can delay the return of submersed plants via physical mechanisms and trophic cascading effects (Jeppesen et al., 2012). Based on food web theory (Carpenter et al., 2001; Shapiro et al., 1975), biomanipulation aims at improving water clarity through a major reduction in plankti-benthivorous fish biomass, leading to a higher biomass of large-bodied zooplankton and greater grazing on phytoplankton. A reduction in benthivorous fish biomass may also facilitate a transition to clear water by mechanical and trophic effects, as mentioned in earlier sections. In northern temperate lakes biomanipulation can yield long-lasting success in recovery when the in-lake total phosphorus (TP) concentration is moderate, and fish removal is sufficiently extensive (e.g., >75% of typically existing biomass in European temperate shallow lakes) over a short period of time (e.g., 1–2 years) (Jeppesen and Sammalkorpi, 2002). Often, fish removal must be repeated to maintain clear-water conditions. Although the success seems weaker, an alternative or supplementary method to fish removal is stocking of potential piscivores (Benndorf, 1995; Hansson et al., 1998) with the intention to also promote linear trophic cascades that end up with lower phytoplankton biomass (Carpenter and Kitchell, 1996).

It is not clear whether the fish manipulation approach used in northern cold-temperate shallow lakes can be successfully applied in warm lakes (i.e., tropical, subtropical, and Mediterranean) (Jeppesen et al., 2012), as several key differences in the fish community may represent an obstacle. Such differences typically include a high fish taxonomic and functional richness, a frequent high density and biomass, omnivory as the predominant feeding mode with potentially weaker trophic cascading effects, and frequent reproduction during the year leading to a small-size structure of the community in warm lakes (González-Bergonzoni et al., 2012; Teixeira-de Mello et al., 2009). All these characteristics promote a much higher predation pressure on large zooplankton and macroinvertebrate grazers in warm lakes compared to similar cold or temperate lakes (Jeppesen et al., 2012, and see references quoted there). Exceptions to the limited usefulness of the classic biomanipulation in tropical and subtropical shallow lakes may occur if the water remains turbid owing to enhanced bottom-up control and sediment resuspension by benthivorous fish feeding (e.g., carp). In such cases extensive removals of benthivorous fish biomass can promote a transition to a clear-water state even in warm lakes (Beklioğlu et al., 2017; Liu et al., 2018).

Fish biomanipulation, complemented with bottom oxygenation as a dual treatment, has been tried with success in deep lakes promoting reduced internal loading and additional improvements to piscivorous fish biomass (Jeppesen et al., 2012). The same approach could be useful to reduce P internal loading and enhance biological responses under anoxic sediment conditions in hypertrophic shallow lakes (Fig. 26-8).

ii. Biomanipulation of submersed plants

Given the fundamental structuring role of submersed plants in shallow lakes and ponds, their reestablishment and protection are considered key for long-term recovery (Moss, 1990). Although reduced external nutrient loading and favorable light conditions can be achieved through biomanipulation or other means, submersed plants often do not readily return to former levels of biomass and composition. This may occur because of herbivory from fish or waterfowl in the initial stages of growth or lack of enough propagule or seed banks in the lake. Furthermore, to reestablish naturally occurring submersed plants requires long-term, stable clear-water conditions. In such situations submersed plant transplantation and protection as a type of biomanipulation have been tried with some success in northern cold lakes for several decades (Moss et al., 1996). This is a time- and resource-demanding approach but may be regarded as the only feasible restoration option when the required reductions in nutrient loading cannot be addressed (Qiu et al., 2001). Together with other measures, submersed plant transplantation as a restoration measure is increasingly used in tropical and subtropical regions (e.g., in China, Yu et al., 2016), despite the fact that the overall stabilizing buffer role of plants seems weaker in warmer climates (see Section III).

iii. Inactivation of in-lake P legacy

The chemical resilience of the eutrophic, turbid-water state is largely due to P release from the sediment pool accumulated during high loading periods (Søndergaard et al., 2003). The legacy of P from the terrestrial soils to the freshwater sediment continuum following the control of external nutrient sources in the catchment can delay lake recovery for years or decades (Sharpley et al., 2013). Depending on the loading history and local conditions, in northern European shallow lakes this internal P loading may typically persist for 10 to 15 years after the external loading reduction before new equilibrium conditions are established (Jeppesen et al., 2005). The delay might be even longer in warmer regions owing to additional constraints related to prolonged hydraulic retention time (Beklioğlu et al., 2017).

The sediments of eutrophic lakes build up a high pool of P that may result in high internal loading of P and, due to the lack of natural P-binding capacity, increases the mobility of the surface-sediment P pool (Søndergaard et al., 2003). Artificial enhancement of the sediment binding capacity for P, often termed *inactivation*, has been used as an in-lake restoration measure since the late 1960s. This group of methods usually employs minerals that occur naturally in freshwater sedimentary environments. Aluminum (Al), iron (Fe), and calcium (Ca) are a few elements that are used to inactivate sediment P through binding (Chapter 15), each having specific requirements and environmental conditions to be effective.

Since the early 2000s, other measures employed to trap P involve the addition of P-binding metals to clay material that is then introduced to lakes and allowed to precipitate and mix with the sediments. Both lanthanum (La) (e.g., Phoslock) and Al, together with clay materials, and special combinations of metals/materials are being used. For example, "flock and lock" (Al and lanthanum [La]-modified clay) is used to improve sediment P-binding and coagulation/precipitation of phytoplankton blooms with the clay material. These methods have achieved various degrees of reduction of sediment P release in the short term (Lürling et al., 2016); however, they need to be tested to determine long-term effectiveness and the cost-benefit relationship from environmental and economic points of view.

iv. Cyanobacterial harmful algal blooms (CyanoHABs) management

The most distinctive phenomenon associated with lake eutrophication is cyanobacterial blooms (often called *CyanoHABs*). There is growing evidence that the spatial and temporal incidence of harmful algal blooms is increasing, posing potential risks to ecosystem structure, function, biodiversity, and human health. A paradigm of strict P limitation has guided lake management strategies over previous decades; however, current understanding implies that in many large and especially shallow lake ecosystems, blooms are sensitive to N inputs (but see Chapters 14 and 15) (e.g., summer-fall blooms of nondiazotrophic toxic *Microcystis* complex in Lake Taihu, China) (Paerl and Huisman, 2008; Paerl et al., 2020). Especially for Cyanobacteria-dominated systems, dual-nutrient management is thus essential when planning future restoration programs (Paerl et al., 2020).

Besides a stringent reduction in N and P loading, specific measures to target Cyanobacteria include hydrological measures such as altering freshwater flow and flushing, dredging, chemical applications, and introduction of selective grazers (Paerl and Barnard, 2020). Lake restoration practices solely based on grazer control by large-bodied *Daphnia* through biomanipulation can be effective in controlling several Cyanobacteria strains but do not seem to be sufficient to control the overgrowth of all cyanobacterial diversity (Urrutia-Cordero et al., 2016). Additionally, climate change is further driving CyanoHAB proliferation (Paerl and Huisman, 2008) through several mechanisms acting at different spatial and time scales (Jeppesen et al., 2014). Controlling harmful cyanobacterial blooms in the face of climate change–induced warming or droughts should thus inevitably include more stringent nutrient reductions.

VII. Climate change, land-use change, and the biodiversity crisis

Shallow and small freshwater systems are among the most threatened ecosystems in the world. Their small volume and large surface:volume ratio make these ecosystems highly vulnerable to hydrological constraints, land uses in the catchment, anthropic direct uses, and climate change and variability. Small lakes and ponds are particularly vulnerable because they are often neglected in conservation policies and monitoring programs.

Although eutrophication, mostly due to land-use changes, is likely still the major disturbance of these ecosystems worldwide (Moss, 2008; Teffera et al., 2017), other major disturbances include drainage, the disconnection from other freshwater systems and uncoupling of habitats (for instance, by removing littoral vegetation or by hydrological modifications in floodplain landscapes), nonnative species introductions, acidification, brownification, excessive water removal, and salinization (Brönmark and Hansson, 2005). New threats are represented by a rapid increase of new entities entering into natural ecosystems, such as micro- and nanoplastics, and emerging contaminants such as pharmaceuticals, personal care products, drugs of abuse, and pesticides, as found in many shallow lakes, even in apparently pristine or seminatural areas (Griffero et al., 2019). The ecological consequences of several of these impacts have been studied for many years in different places around the world and using a myriad of approaches, from controlled microcosm or mesocosm experiments to whole-lake manipulations and from paleolimnological to modeling studies (Jeppesen et al., 2014). However, most studies have addressed the effects of each disturbance in isolation or in simple combinations, thus rendering it difficult to understand potential interactions among many simultaneous impacts, often occurring at different times or spatial scales. Ecosystems rarely recover from a perturbation before they are affected by another one. Multiple stressors on these ecosystems are the rule rather than the exception, yet multistress effects have not been adequately assessed, obscuring management efforts (Nõges et al., 2016).

Climate change can affect all the anthropogenic impacts described above and others, with cumulative and often synergistic effects potentially leading to unpredictable emergent properties. Globally, temperature and precipitation patterns are predicted to change markedly because of climate change. Different components of climate change will impact different aspects of biodiversity and levels of biological organization, from mutation rates inside individual organisms to frequency of catastrophes inside biomes (Bellard et al., 2012). Changes in range distribution of many species toward high latitudes or high altitudes, changes in phenology leading to potential mismatches between environment and biota or between interacting species, and changes in mean body size at population and community levels are among the most widely described biotic responses to climate warming (Parmesan and Yohe, 2003). However, although climate change is a global phenomenon, its impacts are highly asymmetric, with different expectations in different regions.

Due to climate change, many shallow lakes and ponds are already shrinking in size and depth or even completely drying out, as found in the Arctic (Smol and Douglas, 2007). Some areas are already experiencing increases in mean precipitation and changes in seasonal precipitation patterns (Marengo et al., 2009). The regions with a cold or hot semiarid climate and the Mediterranean climate zone are expected to be strongly affected by climate change. A 25% to 30% decrease in precipitation coupled with an increase in evaporation is expected by the end of the 21st century in the Mediterranean region, to be accompanied by an even stronger reduction in surface runoff of up to 30% to 40%. This will lead to major disruptions in the water budget of lakes and ponds and the hydroperiod of temporary ponds and can trigger soil and lake salinization (Jeppesen et al., 2020), as well as eutrophication (Coppens et al., 2016; Menezes et al., 2019). The proportion of land mass that is affected by drought is expected to double in this century (IPPC, 2014).

Climate change has already affected the duration and extent of the hydroperiod of many existing temporary ponds, whereas some previously permanent water bodies may become temporary (Yılmaz et al., 2021). Shorter hydroperiods and consequent greater water-level fluctuations in ponds and shallow lakes can result in shortened growing seasons, thus imposing rigid time constraints for reproduction and population growth and increasing the risk of extinction of species and habitats loss (Pinceel et al., 2018).

The above projections do not consider the concurrent increases in water abstraction for crop irrigation and animal farming, expected as a consequence of the increasing human population and global ongoing diet shift toward a higher proportion of animal products such as meat and milk. Global warming together with more intense irrigation of crops will lead to increased salinization of soils and aquatic ecosystems in the dry climate zones, even to the extent that entire lakes and rivers may dry out temporarily or permanently. In Mediterranean and semiarid tropical waters, for instance, such drying events may result in the loss of numerous *endemic species* (i.e., species that are present only in a small geographical area) that evolved during periods where other parts of the world were temporarily ice covered due to glaciation (Yılmaz et al., 2021).

Aiming to feed the rising human populations with animal protein has led to a skyrocketing increase in inland aquaculture production, for which earthen fish ponds remain the most common type. In 2018inland fish ponds and their productions contributed to more than 60% of the world's farmed fish production with an annual growth rate of 5% to 11% (FAO, 2020). Although fish ponds may contribute to high regional (gamma) diversity for several taxonomic groups (Wezel et al., 2014), in most cases fish farming can represent a major threat to freshwater quality and biodiversity due to the use of many nonnative fish species and

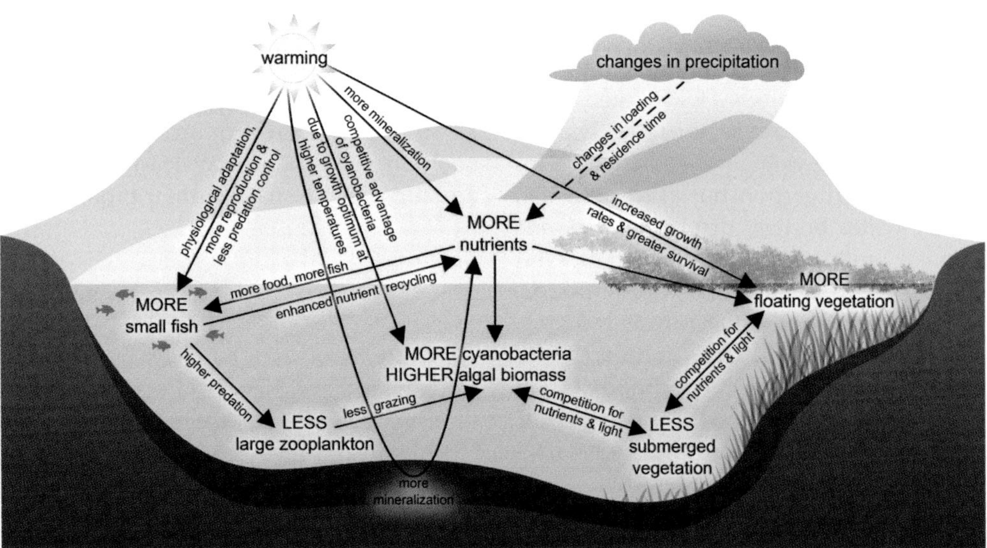

FIGURE 26-9 Schematic view of the main feedback effects of climate change on eutrophication and its symptoms. *Dashed line* indicates that changes in precipitation regimes may either lead to more or less nutrient and organic carbon loading, depending on local and regional circumstances. Warming directly intensifies water stratification and nutrient mineralization increasing nutrient concentrations and directly or indirectly promotes several biological changes leading to a higher likelihood of free-floating plant and phytoplankton dominance. *(With kind permission from Fig. 1 in Moss et al., 2011,* Inland Waters *1, 101–105. Drawing by Alan R. Joyner.)*

unsustainable commercial practices, which conflicts with all targets of the Convention on Biological Diversity (Lima Junior et al., 2018).

One of the most important expectations for ponds and shallow lakes is the additive or even synergistic effect of climate change, and of warming in particular, with eutrophication, both through enhancement of nutrient availability and of several eutrophication symptoms (Moss et al., 2011) (Fig. 26-9). Climate change can directly increase nutrient loading to ponds and lakes by affecting hydrological patterns (due to modified precipitation and evapotranspiration patterns) and runoff of nutrients and organic matter (Ockenden et al., 2017), by enhancing fire frequency with consequent large loads of organic carbon and ashes, and by enhancing nutrient cycling through higher organic matter mineralization rates with warming (Jeppesen et al., 2009). Warming-induced physical changes, such as an increase in lake surface temperature, enhanced evaporation, stronger stratification patterns, and lower solubility of oxygen, favor the development of phytoplankton and cyanobacterial blooms. Biotic communities and their interactions can also be directly and indirectly affected by climate warming. A reduction in population and community mean body size is expected for fish and zooplankton through different direct and indirect mechanisms, overall leading to lower grazing pressure on phytoplankton and a weakening of submersed plants and several of their feedback mechanisms (Meerhoff et al., 2012). The effects of warming may differ, however, under different contexts such as the availability of nutrients, and higher temperatures might favor piscivorous fish at low phosphorus loading but suppress them at high phosphorus loading. Warming may thus either enhance or reduce the strength of trophic cascades, with potentially major implications for water clarity and ecosystem state of shallow lakes and ponds (Nõges et al., 2018). Through various processes operating from the watershed to the lake itself, climate warming is overall expected to increase the likelihood of undesired shifts and thus reduce the theoretical nutrient threshold for the shifts to occur (Jeppesen et al., 2014). In particular, a future increase in frequency and magnitude of cyanobacterial blooms, and likely also in toxicity, is expected (Kosten et al., 2012; Mantzouki et al., 2018; Paerl and Huisman, 2008) (Fig. 26-9). The responses of individual lakes and ponds to these external disturbances are affected by their current climatic context and associated hydrological dynamics and trophic web structure. Shallow lakes and ponds located in warm climate regions (subtropical, tropical, Mediterranean) seem much more sensitive than cold lakes to some external impacts, such as warming, water-level modifications, and nutrient loading increases (Meerhoff and Jeppesen, 2009).

Eutrophic conditions, in turn, most likely favor the release of GHGs from different lake compartments, thus generating positive feedbacks between eutrophication and climate change, particularly through the release of CH_4 but also of CO_2 (Davidson et al., 2018; Li et al., 2021; Yan et al., 2017). Drying and rewetting periods with exposed sediments, which likely will become more frequent and long-lasting due to climate change, enhance the conditions for significant releases of GHG from shallow lakes and ponds (Holgerson and Raymond, 2016; Obrador et al., 2018). Fish ponds in particular appear to be major avenues of GHG (Kosten et al., 2020). This expanding aquaculture sector relies heavily on the application of nutrient-rich aquafeeds, which increase nutrient loading and carbon burial in both fish ponds and adjacent water bodies, driving fish ponds into becoming major anthropogenic sources of CH_4 and N_2O emissions (Yuan et al., 2019). Fish ponds, together with other freshwater aquaculture systems, release more than 80% of the total CH_4 emitted due to anoxic conditions under nutrient-rich conditions (Yuan et al., 2019). This unprecedented growth of fish ponds, leading to water quality deterioration and high GHG emissions, requires urgent consideration and management.

The unique threats to critical ecosystem services and the ecology of freshwater systems (including connectivity across scales and the high levels of endemism) make them a distinct ecological realm whose explicit recognition has important consequences for applied conservation and management policies. Freshwater biodiversity is disproportionately threatened and underprioritized relative to the marine and terrestrial biota. The 2020 Living Planet Index survey highlighted that more than 3740 monitored freshwater populations, representing 944 species of mammals, birds, amphibians, reptiles, and fishes, have declined by an average of 84% (range: 77%–89%), between 1970 and 2016. This is equivalent to 4% per year. Shallow lakes and ponds are particularly underrepresented in biodiversity research and conservation programs. This imposes serious challenges to local conservation, management, and restoration programs and stresses the need to generate local knowledge, besides closing the huge gap between conservation science and policy decisions (Azevedo-Santos et al., 2017).

Most of the current policies are not strong enough to protect small lakes and ponds (Biggs et al., 2017; van Rees et al., 2021). Concerted research and policy actions are needed at a global scale to safeguard freshwater life and its associated ecosystem services, requiring a coherent and far-reaching framework. Van Rees et al. (2021) generated several key recommendations for the conservation of global freshwater biodiversity that would contribute to halting and reversing the rapid global decline of freshwater biodiversity if widely followed (Fig. 26-10). To date, however, there exists no such specific guidance for addressing the freshwater biodiversity crisis, while actions to halt this crisis have been inadequate (IPBES, 2019). However, the 2022 United Nations Biodiversity Conference (COP15) of the Parties to the UN Convention on Biological Diversity (CBD) held in Montreal, Canada, has marked a hopeful agreement, which includes protecting 30% of land and oceans by 2030 ("30 by 30") and 22 other targets intended to reduce biodiversity loss. A similar decision, though with a lower percentage (20%), has been taken by the European Commision. If implemented well, such an agreement may lead to a better protection of aquatic ecosystems and their biodiversity.

VIII. Summary

1. Millions of shallow lakes and ponds occur in lowland areas and in floodplain regions adjacent to river ecosystems, representing the most abundant freshwater ecosystems worldwide. Artificial shallow lakes and ponds are also very common, constructed as a result of different anthropogenic activities throughout history.
2. These shallow waters are seldom thermally stratified for long periods, and thus there is a tight connection between the water column and the sediments, as well as between the littoral zone and the open-water region. Light penetration to much of the sediment surface can, at least theoretically, support the growth of submersed plants or benthic algae over much of the basin. Loadings of nutrients are proportionally higher, losses are lower, and rates of nutrient recycling are faster in shallow lakes and ponds than in deep ones.
3. Complex competitive interactions for light and nutrients among different plant types, attached microbiota, and phytoplankton occur in shallow lakes and ponds. The outcome of such competition largely determines the environmental state, water quality, biodiversity, and carbon fluxes, among other ecosystem processes and services.
4. Contrasting regimes may occur with self-stabilizing mechanisms and different context-dependent likelihoods. The likelihood of a shift from clear to turbid waters increases with increasing nutrient loading and with climate warming, but it is also dependent on several local lake characteristics.
   a. Submersed-plant dominance prevails when nutrient concentrations of the water are low or moderate, if not prevented by external processes.
   b. As nitrogen and phosphorus concentrations increase, regime transitions may occur over a wide range of nutrient concentrations. Submersed-plant dominance can be maintained by several physical, chemical, and biological feedback mechanisms that at least in the short-term, suppress the development of profuse phytoplankton communities. A shift to phytoplankton-dominated conditions, where submersed plants are suppressed, can result from

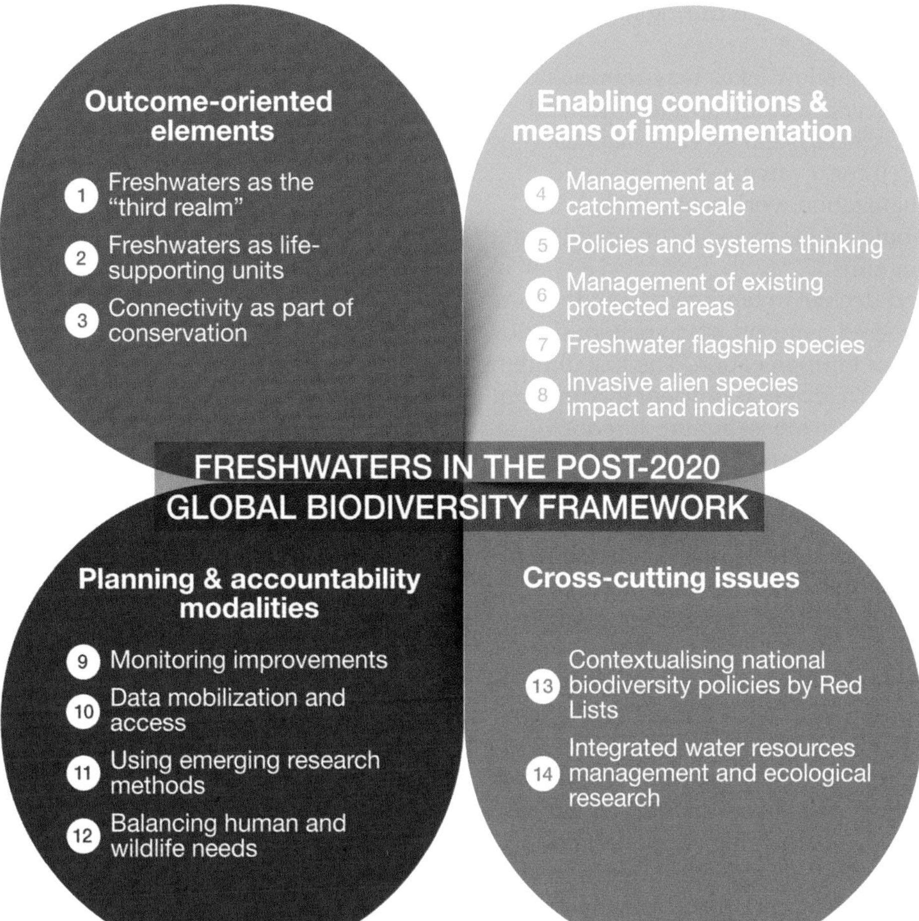

FIGURE 26-10 Recommendations for a new global biodiversity framework based on the European experience. These 14 special recommendations for protecting freshwater biodiversity are grouped around the four clusters of the Global Biodiversity Framework (GBF) for the post-2020 period, following the end of the Convention on Biological Diversity (CBD; 2011–2020). *(With kind permission from Fig. 1 in van Rees et al., 2020,* Conservation Letters *14, 1.)*

different physical or biological perturbations largely involving plants and fishes.

c. The phytoplankton-dominated condition, the most likely regime under high nutrient conditions, is also maintained by a series of internal feedbacks. In the long term, however, nutrient levels largely drive phytoplankton biomass. Cyanobacterial dominance can represent a particular self-stabilizing regime.

d. Free-floating plants can represent an alternative regime under high to extremely high nutrient concentrations, particularly in warm climates and/or small water bodies, as these plants are very sensitive to low temperatures and winds.

5. Shallow lakes and ponds host a disproportionate taxonomic and functional richness considering their size and volume owing to a tight habitat coupling and large variability in physical and chemical conditions in each of the water bodies. Collectively, they represent hotspots of biodiversity at the landscape level, even those water bodies that are naturally temporary.

6. Temporary ponds are shallow waters that are flooded periodically, largely from precipitation and immediate terrestrial runoff, and are dry for periods of the year.

   a. Sources of organic matter are largely allochthonous, and benthic microbial metabolism dominates in these ecosystems.

   b. Losses of water are largely by evaporation and result in marked, rapid changes in ionic composition, nutrient availability, and desiccation rate.

   c. The organisms adapted to living in these irregular, transient habitats have many structural, behavioral, and physiological adaptations for surviving (e.g., aestivation) or avoiding (e.g., through dispersal) drought and desiccation. Very rapid rates of development during the wet phase are essential.

7. Urban ponds and shallow lakes may represent hotspots of biodiversity in towns and cities and provide other key ecosystem services such as recreation and cultural value. Given the rapid increase in urbanization worldwide, these ecosystems should be the focus of conservation and restoration programs.
8. The widespread decline in biodiversity and water quality has triggered many efforts to restore shallow lakes mostly from eutrophication. The key ultimate measure is to reduce external nutrient loading, but several in-lake measures exist aiming at reducing the internal P loading or at manipulating key biological communities, such as fish and submersed plants, to promote favorable albeit likely unstable conditions for plants and water clarity. Higher biodiversity and lower carbon emissions may be a side effect of successful restoration. Pond restoration involves other sets of measures.
9. Many anthropogenic impacts simultaneously affect shallow lakes and ponds, with unpredicted effects due to their potential addition or synergy. Eutrophication remains the most pervasive impact on a global scale, but old perturbations continue while new threats are emerging. The effects of climate change are increasingly clear at different spatial scales and biological organization levels.
10. From the poles to the tropics, many shallow lakes and ponds have already become saltier or disappeared due to climate warming and water extraction in arid and semiarid areas. Many more are expected to disappear or change dramatically in the coming decades. Eutrophication and climate change are expected to interact positively, leading to a loss of biodiversity, enhancement of cyanobacterial blooms, and a higher carbon footprint than at present.
11. Freshwater biodiversity is disproportionately threatened and underprioritized relative to marine and terrestrial biota. Populations of several taxonomic and functional groups of freshwater organisms have declined by more than 80% since 1975, with an estimated average annual rate of 4%. Such figures highlight that an unfortunate mass extinction is prone to occur unless a globally coordinated effort takes immediate action.

Acknowledgments

We dedicate this chapter to the memory of our dear Prof. Brian Moss, an always inspiring researcher and communicator of the enormous ecological and cultural value of shallow lakes and ponds. We would like to acknowledge the constructive comments and the friendly reviews by Sandra Brucet, Sarian Kosten, and Roger P. Mormul. We also deeply thank the graphical assistance of Tinna Christensen and Juana Jacobsen.

References

Agostinho, A.A., Thomaz, S.M., Gomes, L.C., Baltar, S.L., 2007. Influence of the macrophyte *Eichhornia azurea* on fish assemblage of the Upper Paraná River floodplain (Brazil). Aquat. Ecol. 41 (4), 611–619.

Alahuhta, J., Kosten, S., Akasaka, M., Auderset, D., Azzella, M.M., Bolpagni, R., Bove, C.P., Chambers, P.A., Chappuis, E., Clayton, J., 2017. Global variation in the beta diversity of lake macrophytes is driven by environmental heterogeneity rather than latitude. J. Biogeogr. 44 (8), 1758–1769.

Alderton, E., Sayer, C.D., Davies, R., Lambert, S.J., Axmacher, J.C., 2017. Buried alive: aquatic plants survive in "ghost ponds" under agricultural fields. Biol. Conserv. 212, 105–110.

Araújo, F., Becker, V., Attayde, J.L., 2016. Shallow lake restoration and water quality management by the combined effects of polyaluminium chloride addition and benthivorous fish removal: a field mesocosm experiment. Hydrobiologia 778 (1), 243–252.

Attayde, J.L., Ripa, J., 2008. The coupling between grazing and detritus food chains and the strength of trophic cascades across a gradient of nutrient enrichment. Ecosystems 11 (6), 980–990.

Azevedo-Santos, V.M., Fearnside, P.M., Oliveira, C.S., Padial, A.A., Pelicice, F.M., Lima, D.P., Simberloff, D., Lovejoy, T.E., Magalhaes, A.L., Orsi, M.L., 2017. Removing the abyss between conservation science and policy decisions in Brazil. Biodivers. Conserv. 26 (7), 1745–1752.

Balls, H., Moss, B., Irvine, K., 1989. The loss of submerged plants with eutrophication. I. Experimental design, water chemistry, aquatic plant and phytoplankton biomass in experiments carried out in ponds in the Norfolk Broad. Freshw. Biol. 22, 71–87.

Barko, J.W., James, W.F., 1998. Effects of submerged aquatic macrophytes on nutrient dynamics, sedimentation, and resuspension. In: Jeppesen, E., Søndergaard, M.S., Søndergaard, M., Christoffersen, K. (Eds.), The Structuring Role of Submerged Macrophytes in Lakes, 131. Springer, pp. 197–217.

Beisner, B., Peres-Neto, P., Lindstro, E., Barnett, B., Longhi, M., 2006. The role of environmental and spatial processes in structuring lake communities from bacteria to fish. Ecology 87 (12), 2985–2991.

Beisner, B.E., Haydon, D.T., Cuddington, K., 2003. Alternative stable states in ecology. Front. Ecol. Environ. 1 (7), 376–382.

Beklioğlu, M., Bucak, T., Coppens, J., Bezirci, G., Tavsanoğlu, Ü.N., Çakıroğlu, A.İ., Levi, E.E., Erdoğan, S., Filiz, N., Özkan, K., 2017. Restoration of eutrophic lakes with fluctuating water levels: a 20-year monitoring study of two inter-connected lakes. Water 9 (2), 127.

Beklioğlu, M., Romo, S., Kagalou, I., Quintana, X., Bécares, E., 2007. State of the art in the functioning of shallow Mediterranean lakes: workshop conclusions. Hydrobiologia 584 (1), 317–326.

Bellard, C., Bertelsmeier, C., Leadley, P., Thuiller, W., Courchamp, F., 2012. Impacts of climate change on the future of biodiversity. Ecol. Lett. 15 (4), 365–377.

Benndorf, J., 1995. Possibilities and limits for controlling eutrophication by biomanipulation. Int. Rev. Gesamten Hydrobiol. 80, 519–534.

Biggs, J., Von Fumetti, S., Kelly-Quinn, M., 2017. The importance of small waterbodies for biodiversity and ecosystem services: implications for policy makers. Hydrobiologia 793 (1), 3–39.

Bilton, D.T., Freeland, J.R., Okamura, B., 2001. Dispersal in freshwater invertebrates. Annu. Rev. Ecol. Systemat. 32 (1), 159–181.

Boon, P.I., Mitchell, A., Lee, K., 1997. Effects of wetting and drying on methane emissions from ephemeral floodplain wetlands in south-eastern Australia. Hydrobiologia 357 (1), 73–87.

Brasil, J., Attayde, J.L., Vasconcelos, F.R., Dantas, D.D., Huszar, V.L., 2016. Drought-induced water-level reduction favors cyanobacteria blooms in tropical shallow lakes. Hydrobiologia 770 (1), 145–164.

Brönmark, C., Hansson, L.A., 2005. The Biology of Lakes and Ponds, second ed. Oxford University Press.

Brönmark, C., Weisner, S.E.B., 1992. Indirect effects of fish community structure on submerged vegetation in shallow, eutrophic lakes—an alternative mechanism. Hydrobiologia 243, 293–301.

Brucet, S., Boix, D., Gascón, S., Sala, J., Quintana, X.D., Badosa, A., Søndergaard, M., Lauridsen, T.L., Jeppesen, E., 2009. Species richness of crustacean zooplankton and trophic structure of brackish lagoons in contrasting climate zones: north temperate Denmark and Mediterranean Catalonia (Spain). Ecography 32 (4), 692–702.

Brucet, S., Boix, D., Nathansen, L.W., Quintana, X.D., Jensen, E., Balayla, D., Meerhoff, M., Jeppesen, E., 2012. Effects of temperature, salinity and fish in structuring the macroinvertebrate community in shallow lakes: implications for effects of climate change. PLoS One 7 (2), e30877.

Brucet, S., Boix, D., Quintana, X.D., Jensen, E., Nathansen, L.W., Trochine, C., Meerhoff, M., Gascón, S., Jeppesen, E., 2010. Factors influencing zooplankton size structure at contrasting temperatures in coastal shallow lakes: implications for effects of climate change. Limnol. Oceanogr. 55 (4), 1697–1711.

Burks, R.L., Lodge, D.M., Jeppesen, E., Lauridsen, T.L., 2002. Diel horizontal migration of zooplankton: costs and benefits of inhabiting littoral zones. Freshw. Biol. 47, 343–365.

Burks, R.L., Mulderij, G., Gross, E., Jones, J.I., Jacobsen, L., Jeppesen, E., Van Donk, E., 2006. Center stage: the crucial role of macrophytes in regulating trophic interactions in shallow lake wetlands. In: Bobbink, R., Beltman, B., Verhoeven, J.T.A., Whigham, D.F. (Eds.), Wetlands: Functioning, Biodiversity Conservation, and Restoration, vol. 191. Springer-Verlag, pp. 37–59.

Canfield, D.E.j., Shireman, J.V., Colle, D.E., Haller, W.T., Watkins, C.E.I., Maceina, M.J., 1984. Prediction of chlorophyll a concentrations in Florida lakes: importance of aquatic macrophytes. Can. J. Fish. Aquat. Sci. 41, 497–501.

Capon, S.J., Lynch, A.J.J., Bond, N., Chessman, B.C., Davis, J., Davidson, N., Finlayson, M., Gell, P.A., Hohnberg, D., Humphrey, C., Kingsford, R.T., 2015. Regime shifts, thresholds and multiple stable states in freshwater ecosystems: a critical appraisal of the evidence. Sci. Total Environ. 534, 122–130.

Carpenter, S., Cole, J.J., Hodgson, J.R., Kitchell, J.E., Pace, M.L., Bade, D., Cottingham, K.L., Essington, T.E., Houser, J.N., Schindler, D.E., 2001. Trophic cascades, nutrients, and lake productivity: whole-lake experiments. Ecol. Monogr. 71 (2), 163–186.

Carpenter, S., Kitchell, J.F. (Eds.), 1996. The Trophic Cascade in Lakes. Cambridge University Press.

Carpenter, S.R., Kitchell, J.F., Hodgson, J.R., 1985. Cascading trophic interactions and lake productivity. Bioscience 35, 635–639.

Carpenter, S.R., Lodge, D.M., 1986. Effects of submersed macrophytes on ecosystem processes. Aquat. Bot. 26, 341–370.

Céréghino, R., Biggs, J., Oertli, B., Declerck, S., 2008. The ecology of European ponds: defining the characteristics of a neglected freshwater habitat. Hydrobiologia 597, 1–6.

Cooke, G.D., Welch, E.B., Peterson, S., Nichols, S.A., 2016. Restoration and Management of Lakes and Reservoirs. CRC Press.

Coops, H., Beklioglu, M., Crisman, T.L., 2003. The role of water-level fluctuations in shallow lake ecosystems—workshop conclusions. Hydrobiologia 506 (1–3), 23–27.

Coppens, J., Özen, A., Tavsanoğlu, Ü.N., Erdoğan, S., Levi, E.E., Yozgatlıgil, C., Jeppesen, E., Beklioğlu, M., 2016. Impact of alternating wet and dry periods on long-term seasonal phosphorus and nitrogen budgets of two shallow Mediterranean lakes. Sci. Total Environ. 563, 456–467.

Cottenie, K., Michels, E., Nuytten, N., De Meester, L., 2003. Zooplankton metacommunity structure: regional vs. local processes in highly interconnected ponds. Ecology 84 (4), 991–1000.

Crisci, C., Terra, R., Pacheco, J.P., Ghattas, B., Bidegain, M., Goyenola, G., Lagomarsino, J.J., Méndez, G., Mazzeo, N., 2017. Multi-model approach to predict phytoplankton biomass and composition dynamics in a eutrophic shallow lake governed by extreme meteorological events. Ecol. Model. 360, 80–93.

Daborn, G.R., Clifford, H.F., 1974. Physical and chemical features of an aestival pond in western Canada. Hydrobiologia 44 (1), 43–59.

Dantas, D.D., Rubim, P.L., de Oliveira, F.A., da Costa, M.R., de Moura, C.G., Teixeira, L.H., Attayde, J.L., 2019. Effects of benthivorous and planktivorous fish on phosphorus cycling, phytoplankton biomass and water transparency of a tropical shallow lake. Hydrobiologia 829 (1), 31–41.

Davidson, T.A., Audet, J., Jeppesen, E., Landkildehus, F., Lauridsen, T.L., Søndergaard, M., Syväranta, J., 2018. Synergy between nutrients and warming enhances methane ebullition from experimental lakes. Nat. Clim. Change 8 (2), 156–160.

Davidson, T.A., Audet, J., Svenning, J.C., Lauridsen, T.L., Søndergaard, M., Landkildehus, F., Larsen, S.E., Jeppesen, E., 2015. Eutrophication effects on greenhouse gas fluxes from shallow-lake mesocosms override those of climate warming. Global Change Biol. 21 (12), 4449–4463.

Davidson, T.A., Sayer, C.D., Jeppesen, E., Søndergaard, M., Lauridsen, T.L., Johansson, L.S., Baker, A., Graeber, D., 2023. Bimodality and alternative equilibria do not help explain long-term patterns in shallow lake chlorophyll-a. Nat. Comm. 14, 398.

De Meester, L., Declerck, S., Stoks, R., Louette, G., Van de Meutter, F., De Bie, T., Michels, E., Brendonck, L., 2005. Ponds and pools as model systems in conservation biology, ecology and evolutionary biology. Aquat. Conserv. Mar. Freshw. Ecosyst. 15 (6), 715–725.

de Tezanos Pinto, P., Allende, L., O'Farrell, I., 2007. Influence of free-floating plants on the structure of a natural phytoplankton assemblage: an experimental approach. J. Plankton Res. 29 (1), 47–56.

de Tezanos Pinto, P., O'Farrell, I., 2014. Regime shifts between free-floating plants and phytoplankton: a review. Hydrobiologia 740 (1), 13–24.

Declerck, S., Vandekerkhove, J., Johansson, L., Muylaert, K., Conde-Porcuna, J.M., Van Der Gucht, K., Pérez-Martínez, C., Lauridsen, T.L., Schwenk, K., Zwart, G., Rommens, W., López-Ramos, J., Jeppesen, E., Vyverman, W., Brendonck, L., De Meester, L., 2005. Multi-group biodiversity in shallow lakes along gradients of phosphorus and water plant cover. Ecology 86 (7), 1905–1915.

Dias, J.D., Simões, N.R., Meerhoff, M., Lansac-Tôha, F.A., Velho, L.F.M., Bonecker, C.C., 2016. Hydrological dynamics drives zooplankton metacommunity structure in a neotropical floodplain. Hydrobiologia 781 (1), 109–125.

Dokulil, M.T., Teubner, K., 2000. Cyanobacterial dominance in lakes. Hydrobiologia 438, 1–12.

Donohue, I., Hillebrand, H., Montoya, J.M., Petchey, O.L., Pimm, S.L., Fowler, M.S., Healy, K., Jackson, A.L., Lurgi, M., McClean, D., 2016. Navigating the complexity of ecological stability. Ecol. Lett. 19 (9), 1172–1185.

Downing, J.A., Prairie, Y., Cole, J., Duarte, C., Tranvik, L., Striegl, R.G., McDowell, W., Kortelainen, P., Caraco, N., Melack, J., 2006. The global abundance and size distribution of lakes, ponds, and impoundments. Limnol. Oceanogr. 51 (5), 2388–2397.

Ersoy, Z., Scharfenberger, U., Baho, D.L., Buçak, T., Feldmann, T., Hejzlar, J., Levi, E.E., Mahdy, A., Nõges, T., Papastergiadou, E., Stefanidis, K., Šorf, M., Søndergaard, M., Trigal, C., Jeppesen, E., Beklioğlu, M., 2020. Impact of nutrients and water level changes on submerged macrophytes along a temperature gradient: a pan-

European mesocosm experiment. Global Change Biol. 26 (12), 6831–6851.

Fahrig, L., 2013. Rethinking patch size and isolation effects: the habitat amount hypothesis. J. Biogeogr. 40 (9), 1649–1663.

FAO, 2020. World fisheries and aquaculture, Sustainability in action.

Folke, C., 2016. Resilience (republished). Ecol. Soc. 21 (4).

Folke, C., Carpenter, S., Walker, B., Scheffer, M., Elmqvist, T., Gunderson, L., Holling, C.S., 2004. Regime shifts, resilience, and biodiversity in ecosystem management. Annu. Rev. Ecol. Evol. Syst. 35, 557–581.

Ger, K.A., Hansson, L.A., Lürling, M., 2014. Understanding cyanobacteria-zooplankton interactions in a more eutrophic world. Freshw. Biol. 59 (9), 1783–1798.

González-Bergonzoni, I., Meerhoff, M., Davidson, T.A., Teixeira-de Mello, F., Baattrup-Pedersen, A., Jeppesen, E., 2012. Meta-analysis shows a consistent and strong latitudinal pattern in fish omnivory across ecosystems. Ecosystems 15 (3), 492–503.

González-Sagrario, M.A., Balseiro, E., 2010. The role of macroinvertebrates and fish in regulating the provision by macrophytes of refugia for zooplankton in a warm temperate shallow lake. Freshw. Biol. 55 (10), 2153–2166.

González-Sagrario, M.A., Jeppesen, E., Goma, J., Søndergaard, M., Jensen, J.P., Lauridsen, T.L., Lankildehus, F., 2005. Does high nitrogen loading prevent clear-water conditions in shallow lakes at moderately high phosphorus concentrations? Freshw. Biol. 50, 27–41.

Grasel, D., Mormul, R.P., Bozelli, R.L., Thomaz, S.M., Jarenkow, J.A., 2018. Brazil's native vegetation protection law threatens to collapse pond functions. Pers. Ecol. Conserv. 16 (4), 234–237.

Griffero, L., Alcántara-Durán, J., Alonso, C., Rodríguez-Gallego, L., Moreno-González, D., García-Reyes, J.F., Molina-Díaz, A., Pérez-Parada, A., 2019. Basin-scale monitoring and risk assessment of emerging contaminants in South American Atlantic coastal lagoons. Sci. Total Environ. 697, 134058.

Grizzetti, B., Lanzanova, D., Liquete, C., Reynaud, A., Cardoso, A., 2016. Assessing water ecosystem services for water resource management. Environ. Sci. Pol. 61, 194–203.

Groffman, P.M., Baron, J.S., Blett, T., Gold, A.J., Goodman, I., Gunderson, L.H., Levinson, B.M., Palmer, M.A., Paerl, H.W., Peterson, G.D., 2006. Ecological thresholds: the key to successful environmental management or an important concept with no practical application? Ecosystems 9 (1), 1–13.

Gross, E.M., Hilt, S., Lombardo, P., Mulderij, G., 2007. Searching for allelopathic effects of submerged macrophytes on phytoplankton-state of the art and open questions. Hydrobiologia 584, 77–88.

Hamer, A.J., McDonnell, M.J., 2008. Amphibian ecology and conservation in the urbanising world: a review. Biol. Conserv. 141 (10), 2432–2449.

Hanski, I., 1999. Metapopulation Ecology. Oxford University Press.

Hansson, L.-A., Annadotter, H., Bergman, E., Hamrin, S.F., Jeppesen, E., Kairesalo, T., Luokkanen, E., Nilsson, P.-Å., Søndergaard, M., Strand, J., 1998. Biomanipulation as an application of food-chain theory: constraints, synthesis, and recommendations for temperate lakes. Ecosystems 1 (6), 558–574.

Hassall, C., 2014. The ecology and biodiversity of urban ponds. Wiley Interdisciplinary Reviews: Water 1 (2), 187–206.

Hill, M., Mathers, K., Wood, P., 2015. The aquatic macroinvertebrate biodiversity of urban ponds in a medium-sized European town (Loughborough, UK). Hydrobiologia 760 (1), 225–238.

Hill, M.J., Biggs, J., Thornhill, I., Briers, R.A., Gledhill, D.G., White, J.C., Wood, P.J., Hassall, C., 2017. Urban ponds as an aquatic biodiversity resource in modified landscapes. Global Change Biol. 23 (3), 986–999.

Hill, M.J., Ryves, D., White, J.C., Wood, P., 2016. Macroinvertebrate diversity in urban and rural ponds: implications for freshwater biodiversity conservation. Biol. Conserv. 201, 50–59.

Hillebrand, H., Donohue, I., Harpole, W.S., Hodapp, D., Kucera, M., Lewandowska, A.M., Merder, J., Montoya, J.M., Freund, J.A., 2020. Thresholds for ecological responses to global change do not emerge from empirical data. Nature Ecol. Evol. 4 (11), 1502–1509.

Hillebrand, H., Langenheder, S., Lebret, K., Lindström, E., Östman, Ö., Striebel, M., 2018. Decomposing multiple dimensions of stability in global change experiments. Ecol. Lett. 21 (1), 21–30.

Hilt, S., Brothers, S., Jeppesen, E., Veraart, A.J., Kosten, S., 2017. Translating regime shifts in shallow lakes into changes in ecosystem functions and services. Bioscience 67 (10), 928–936.

Hilt, S., Gross, E.M., 2008. Can allelopathically active submerged macrophytes stabilise clear-water states in shallow lakes? Basic Appl. Ecol. 9 (4), 422–432.

Hobbie, J.E., 1980. Limnology of tundra ponds, Barrow, Alaska. Dowden, Hutchinson & Ross.

Hodgson, D., McDonald, J.L., Hosken, D.J., 2015. What do you mean, "resilient"? Trends Ecol. Evol. 30 (9), 503–506.

Holgerson, M.A., Raymond, P.A., 2016. Large contribution to inland water CO_2 and CH_4 emissions from very small ponds. Nat. Geosci. 9 (3), 222–226.

Holling, C.S., 1973. Resilience and stability of ecological systems. Annu. Rev. Ecol. Systemat. 4, 1–23.

Horppila, J., Nurminen, L., 2003. Effects of submerged macrophytes on sediment resuspension and internal phosphorus loading in Lake Hiidenvesi (southern Finland). Water Res. 37 (18), 4468–4474.

Horppila, J., Nurminen, L., 2005. Effects of different macrophyte growth forms on sediment and P resuspension in a shallow lake. Hydrobiologia 545, 167–175.

Horváth, Z., Ptacnik, R., Vad, C.F., Chase, J.M., 2019. Habitat loss over six decades accelerates regional and local biodiversity loss via changing landscape connectance. Ecol. Lett. 22 (6), 1019–1027.

IPBES, 2019. Global assessment report on biodiversity and ecosystem services of the intergovernmental science-policy platform on biodiversity and ecosystem services. E. S. Brondizio, J. Settele, S. Díaz, H. T. Ngo (editors). IPBES secretariat, Bonn.

IPCC (Intergovernmental Panel on Climate Change), 2014. Summary for policymakers. climate change 2014: impacts, adaptation, and vulnerability. Contribution of working group II to the 5th assessment report of the Intergovernmental Panel on Climate Change. Cambridge University Press, Cambridge (UK), pp. 1–32.

Izaguirre, I., Allende, L., Romina Schiaffino, M., 2020. Phytoplankton in Antarctic lakes: biodiversity and main ecological features. Hydrobiologia 1–31.

Janssen, A.B., Hilt, S., Kosten, S., de Klein, J.J., Paerl, H.W., Van de Waal, D.B., 2021. Shifting states, shifting services: linking regime shifts to changes in ecosystem services of shallow lakes. Freshw. Biol. 66 (1), 1–12.

Jeppesen, E., Beklioğlu, M., Özkan, K., Akyürek, Z., 2020. Salinization increase due to climate change will have substantial negative effects on inland waters: a call for multifaceted research at the local and global scale. Innovation 1 (2), 100030.

Jeppesen, E., Jensen, J.P., Søndergaard, M., Lauridsen, T.L., 2005. Response of fish and plankton to nutrient loading reduction in eight shallow Danish lakes with special emphasis on seasonal dynamics. Freshw. Biol. 50 (10), 1616–1627.

Jeppesen, E., Kronvang, B., Meerhoff, M., Søndergaard, M., Hansen, K.M., Andersen, H.E., Lauridsen, T.L., Liboriussen, L., Beklioğlu, M., Özen, A., Olesen, J.E., 2009. Climate change effects on runoff, catchment phosphorus loading and lake ecological state, and potential adaptations. J. Environ. Qual. 38 (5), 1930–1941.

Jeppesen, E., Kronvang, B., Olesen, J., Audet, J., Søndergaard, M., Hoffmann, C., Andersen, H., Lauridsen, T.L., Liboriussen, L., Larsen, S., Beklioğlu, M., Meerhoff, M., Özen, A., Özkan, K., 2011. Climate change effects on nitrogen loading from cultivated catchments in Europe: implications for nitrogen retention, ecological state of lakes and adaptation. Hydrobiologia 663 (1), 1–21.

Jeppesen, E., Meerhoff, M., Davidson, T.A., Trolle, D., SondergaarD, M., Lauridsen, T.L., Beklioğlu, M., Brucet Balmaña, S., Volta, P., González-Bergonzoni, I., Nielsen, A., 2014. Climate change impacts on lakes: an integrated ecological perspective based on a multi-faceted approach, with special focus on shallow lakes. J. Limnol. 73, 84–107.

Jeppesen, E., Jensen, J.P., Soendergaard, M., Lauridsen, T., Landkildehus, F., 2000. Trophic structure, species richness and biodiversity in Danish lakes: changes along a phosphorus gradient. Freshw. Biol. 45 (2), 201–218.

Jeppesen, E., Sammalkorpi, I., 2002. 13 lakes. Handbook of Ecological Restoration 2, 297.

Jeppesen, E., Søndergaard, M., Lauridsen, T.L., Davidson, T.A., Liu, Z., Mazzeo, N., Trochine, C., Özkan, K., Jensen, H.S., Trolle, D., Landkildehus, F., Starling, F., Larsen, S., Lazzaro, X., Meerhoff, M., 2012. Biomanipulation as a restoration tool to combat eutrophication: recent advances and future challenges. Adv. Ecol. Res. 47, 411–488.

Jeppesen, E., Søndergaard, M., Søndergaard, M., Christoffersen, K., 1998. The structuring role of submerged macrophytes in lakes. Springer Verlag.

Jeppesen, E., Trolle, D., Davidson, T.A., Bjerring, R., Søndergaard, M., Johansson, L.S., Lauridsen, T.L., Nielsen, A., Larsen, S.E., Meerhoff, M., 2016. Major changes in CO_2 efflux when shallow lakes shift from a turbid to a clear water state. Hydrobiologia 778 (1), 33–44.

Jia, P., Zhang, W., Liu, Q., 2013. Lake fisheries in China: challenges and opportunities. Fish. Res. 140, 66–72.

Jilbert, T., Couture, R.-M., Huser, B.J., Salonen, K., 2020. Preface: restoration of eutrophic lakes: current practices and future challenges. Hydrobiologia 847 (21), 4343–4357.

Jones, J.I., Sayer, C.D., 2003. Does the fish-invertebrate-periphyton cascade precipitate plant loss in shallow lakes? Ecology 84, 2155–2167.

Kong, X., He, Q., Yang, B., He, W., Xu, F., Janssen, A.B., Kuiper, J.J., van Gerven, L.P., Qin, N., Jiang, Y., Liu, W., Yan, C., Bai, Z., Zhang, M., Kong, F., Janse, J.H., Mooij, W.M., 2017. Hydrological regulation drives regime shifts: evidence from paleolimnology and ecosystem modeling of a large shallow Chinese lake. Global Change Biol. 23 (2), 737–754.

Kosten, S., Almeida, R.M., Barbosa, I., Mendonça, R., Muzitano, I.S., Oliveira-Junior, E.S., Vroom, R.J., Wang, H.J., Barros, N., 2020. Better assessments of greenhouse gas emissions from global fish ponds needed to adequately evaluate aquaculture footprint. Sci. Total Environ. 748, 141247.

Kosten, S., Huszar, V.L., Bécares, E., Costa, L.S., van Donk, E., Hansson, L.A., Jeppesen, E., Kruk, C., Lacerot, G., Mazzeo, N., De Meester, L., Moss, B., Lürling, M., Noges, T., Romo, S., Scheffer, M., 2012. Warmer climates boost cyanobacterial dominance in shallow lakes. Global Change Biol. 18 (1), 118–126.

Kosten, S., Jeppesen, E., Huszar, V.L., Mazzeo, N., van Nes, E.H., Peeters, E.T., Scheffer, M., 2011. Ambiguous climate impacts on competition between submerged macrophytes and phytoplankton in shallow lakes. Freshw. Biol. 56 (8), 1540–1553.

Kosten, S., Lacerot, G., Jeppesen, E., da Motta Marques, D., van Nes, E.H., Mazzeo, N., Scheffer, M., 2009. Effects of submerged vegetation on water clarity across climates. Ecosystems 12 (7), 1117–1129.

Kosten, S., Meerhoff, M., 2014. Lake communities. In: ELS Encyclopedia of Life Science. John Wiley & Sons.

Kosten, S., Piñeiro, M., de Goede, E., de Klein, J., Lamers, L.P., Ettwig, K., 2016. Fate of methane in aquatic systems dominated by free-floating plants. Water Res. 104, 200–207.

Kovalenko, K.E., Thomaz, S.M., Warfe, D.M., 2012. Habitat complexity: approaches and future directions. Hydrobiologia 685 (1), 1–17.

Kruk, C., Rodríguez-Gallego, L., Meerhoff, M., Quintans, F., Lacerot, G., Mazzeo, N., Scasso, F., Paggi, J.C., Peeters, E.T., Marten, S., 2009. Determinants of biodiversity in subtropical shallow lakes (Atlantic coast, Uruguay). Freshw. Biol. 54 (12), 2628–2641.

Lauridsen, T.L., Lodge, D.M., 1996. Avoidance of *Daphnia magna* by fish and macrophytes: chemical cues and predator-mediated use of macrophyte habitat. Limnol. Oceanogr. 41, 794–798.

Leibold, M.A., Holyoak, M., Mouquet, N., Amarasekare, P., Chase, J.M., Hoopes, M.F., Holt, R.D., Shurin, J.B., Law, R., Tilman, D., 2004. The metacommunity concept: a framework for multi-scale community ecology. Ecol. Lett. 7 (7), 601–613.

Li, Y., Shang, J., Zhang, C., Zhang, W., Niu, L., Wang, L., Zhang, H., 2021. The role of freshwater eutrophication in greenhouse gas emissions: a review. Sci. Total Environ. 768, 144582.

Liboriussen, L., Jeppesen, E., 2003. Temporal dynamics in epipelic, pelagic and epiphytic algal production in a clear and a turbid shallow lake. Freshw. Biol. 48 (3), 418–431.

Lima Junior, D.P., Magalhães, A.L.B., Pelicice, F.M., Vitule, J.R.S., Azevedo-Santos, V.M., Orsi, M.L., Simberloff, D., Agostinho, A.A., 2018. Aquaculture expansion in Brazilian freshwaters against the Aichi Biodiversity Targets. Ambio. 47 (4), 427–440.

Liu, Z., Hu, J., Zhong, P., Zhang, X., Ning, J., Larsen, S.E., Chen, D., Gao, Y., He, H., Jeppesen, E., 2018. Successful restoration of a tropical shallow eutrophic lake: strong bottom-up but weak top-down effects recorded. Water Res. 146, 88–97.

Loreau, M., Mouquet, N., Gonzalez, A., 2003. Biodiversity as spatial insurance in heterogeneous landscapes. Proc. Natl. Acad. Sci. USA 100 (22), 12765–12770.

Loverde-Oliveira, S.M., Huszar, V.L.M., Mazzeo, N., Scheffer, M., 2009. Hydrology-driven regime shifts in a shallow tropical lake. Ecosystems 12 (5), 807–819.

Lürling, M., Mackay, E., Reitzel, K., Spears, B.M., 2016. Editorial-a critical perspective on geo-engineering for eutrophication management in lakes. Water Res. 97, 1–10.

Mantzouki, E., Lürling, M., Fastner, J., de Senerpont Domis, L., Wilk-Woźniak, E., Koreivienė, J., Seelen, L., Teurlincx, S., Verstijnen, Y., Krztoń, W., 2018. Temperature effects explain continental scale distribution of cyanobacterial toxins. Toxins 10 (4), 156.

Marengo, J.A., Jones, R., Alves, L.M., Valverde, M.C., 2009. Future change of temperature and precipitation extremes in South America as derived from the PRECIS regional climate modeling system. Int. J. Climatol. 29 (15), 2241–2255.

Meerhoff, M., Clemente, J.M., Teixeira-de Mello, F., Iglesias, C., Pedersen, A.R., Jeppesen, E., 2007a. Can warm climate-related structure of littoral predator assemblies weaken the clear water state in shallow lakes? Global Change Biol. 13, 1888–1897.

Meerhoff, M., Iglesias, C., Teixeira-de Mello, F., Clemente, J.M., Jensen, E., Lauridsen, T.L., Jeppesen, E., 2007b. Effects of habitat complexity on community structure and predator avoidance behaviour of littoral zooplankton in temperate versus subtropical shallow lakes. Freshw. Biol. 52 (6), 1009–1021.

Meerhoff, M., Jeppesen, E., 2009. Shallow lakes and ponds. In: Likens, G. (Ed.), Encyclopedia of Inland Waters, 2. Elsevier, pp. 645–655.

Meerhoff, M., Mazzeo, N., Moss, B., Rodríguez-Gallego, L., 2003. The structuring role of free-floating versus submerged plants in a subtropical shallow lake. Aquat. Ecol. 37 (4), 377–391.

Meerhoff, M., Teixeira-de Mello, F., Kruk, C., Alonso, C., Gonzalez-Bergonzoni, I., Pacheco, J.P., Lacerot, G., Arim, M., Beklioğlu, M., Brucet, S., Goyenola, G., Iglesias, C., Mazzeo, N., Kosten, S., Jeppesen, E., 2012. Environmental warming in shallow lakes: a review of potential changes in community structure as evidenced from space-for-time substitution approaches. Adv. Ecol. Res. 46, 259–349.

Menezes, R.F., Attayde, J.L., Kosten, S., Lacerot, G., Coimbra e Souza, L.C., Costa, L.S., da, S.L., Sternberg, L., dos Santos, A.C., de Medeiros Rodrigues, M., Jeppesen, E., 2019. Differences in food webs and trophic states of Brazilian tropical humid and semi-arid shallow lakes: implications of climate change. Hydrobiologia 829 (1), 95–111.

Messager, M.L., Lehner, B., Grill, G., Nedeva, I., Schmitt, O., 2016. Estimating the volume and age of water stored in global lakes using a geo-statistical approach. Nat. Commun. 7, 13603.

Moi, D.A., Alves, D.C., Antiqueira, P.A.P., Thomaz, S.M., Teixeira-de Mello, F., Bonecker, C.C., Rodrigues, L.C., Garcia-Rios, R., Mormul, R.P., 2021. Ecosystem shift from submerged to floating plants simplifying the food web in a tropical shallow lake. Ecosystems 24 (3), 628–639.

Mormul, R.P., Ahlgren, J., Brönmark, C., 2018. Snails have stronger indirect positive effects on submerged macrophyte growth attributes than zooplankton. Hydrobiologia 807 (1), 165–173.

Mormul, R.P., Thomaz, S.M., Agostinho, A.A., Bonecker, C.C., Mazzeo, N., 2012. Migratory benthic fishes may induce regime shifts in a tropical floodplain pond. Freshw. Biol. 57 (8), 1592–1602.

Moss, B., 1990. Engineering and biological approaches to the restoration from eutrophication of shallow lakes in which aquatic plant communities are important components. Hydrobiologia 200/201, 367–377.

Moss, B., 2008. Water pollution by agriculture. Phil. Trans. Biol. Sci. 363 (1491), 659–666.

Moss, B., Kosten, S., Meerhoff, M., Battarbee, R.W., Jeppesen, E., Mazzeo, N., Havens, K., Lacerot, G., Liu, Z., De Meester, L., Scheffer, M., 2011. Allied attack: climate change and eutrophication. Inland Waters 1 (2), 101–105.

Moss, B., Madgwick, J., Phillips, G.L., 1996. A guide to the restoration of nutrient-enriched shallow lakes. Broads Authority and Environment Agency (CE).

Naiman, R.J., Magnuson, J.J., McKnight, D.M., Stanford, J.A., Karr, J., 1995. Freshwater ecosystems and their management: a National Initiative. Science 270, 584–585.

Netten, J.J., Arts, G.H., Gylstra, R., van Nes, E.H., Scheffer, M., Roijackers, R.M., 2010. Effect of temperature and nutrients on the competition between free-floating *Salvinia natans* and submerged *Elodea nuttallii* in mesocosms. Fund. Appl. Limnol. 177 (2), 125.

Nõges, P., Argillier, C., Borja, Á., Garmendia, J.M., Hanganu, J., Kodeš, V., Pletterbauer, F., Sagouis, A., Birk, S., 2016. Quantified biotic and abiotic responses to multiple stress in freshwater, marine and ground waters. Sci. Total Environ. 540, 43–52.

Nõges, T., Anneville, O., Guillard, J., Haberman, J., Järvalt, A., Manca, M., Morabito, G., Rogora, M., Thackeray, S.J., Volta, P., Winfield, I., Nõges, P., 2018. Fisheries impacts on lake ecosystem structure in the context of a changing climate and trophic state. J. Limnol. 77 (1).

O'Farrell, I., Sinistro, R., Izaguirre, I., Unrein, F., 2003. Do steady state assemblages occur in shallow lentic environments from wetlands? Hydrobiologia 502 (1–3), 197–209.

O'Farrell, I., Izaguirre, I., Chaparro, G., Unrein, F., Sinistro, R., Pizarro, H., Rodríguez, P., de Tezanos Pinto, P., Lombardo, R., Tell, G., 2011. Water level as the main driver of the alternation between a free-floating plant and a phytoplankton dominated state: a long-term study in a floodplain lake. Aquat. Sci. 73 (2), 275–287.

Obrador, B., von Schiller, D., Marcé, R., Gómez-Gener, L., Koschorreck, M., Borrego, C., Catalán, N., 2018. Dry habitats sustain high CO_2 emissions from temporary ponds across seasons. Sci. Rep. 8 (1), 1–12.

Ockenden, M.C., Hollaway, M.J., Beven, K.J., Collins, A., Evans, R., Falloon, P., Forber, K.J., Hiscock, K., Kahana, R., Macleod, C., 2017. Major agricultural changes required to mitigate phosphorus losses under climate change. Nat. Commun. 8 (1), 1–9.

Oertli, B., 2018. Freshwater biodiversity conservation: the role of artificial ponds in the 21st century. Aquat. Conserv. Mar. Freshw. Ecosyst. 28 (2), 264–269.

Oertli, B., Biggs, J., Cereghino, R., Grillas, P., Joly, P., Lachavanne, J.B., 2005. Conservation and monitoring of pond biodiversity: introduction. Aquat. Conserv. Mar. Freshw. Ecosyst. 15, 535–540.

Oertli, B., Céréghino, R., Hull, A., Miracle, R., 2009. Pond conservation: from science to practice. In: Pond Conservation in Europe. Springer, pp. 157–165.

Oertli, B., Joye, D.A., Castella, E., Juge, R., Cambin, D., Lachavanne, J.-B., 2002. Does size matter? The relationship between pond area and biodiversity. Biol. Conserv. 104 (1), 59–70.

Oertli, B., Parris, K.M., 2019. Toward management of urban ponds for freshwater biodiversity. Ecosphere 10 (7), e02810.

Padial, A.A., Ceschin, F., Declerck, S.A., De Meester, L., Bonecker, C.C., Lansac-Tôha, F.A., Rodrigues, L., Rodrigues, L.C., Train, S., Velho, L.F., 2014. Dispersal ability determines the role of environmental, spatial and temporal drivers of metacommunity structure. PLoS One 9 (10), e111227.

Paerl, H., Huisman, J., 2008. Blooms like it hot. Science 320, 57–58.

Paerl, H.W., Barnard, M.A., 2020. Mitigating the global expansion of harmful cyanobacterial blooms: moving targets in a human-and climatically-altered world. Harmful Algae 96, 101845.

Paerl, H.W., Havens, K.E., Xu, H., Zhu, G., McCarthy, M.J., Newell, S.E., Scott, J.T., Hall, N.S., Otten, T.G., Qin, B., 2020. Mitigating eutrophication and toxic cyanobacterial blooms in large lakes: the evolution of a dual nutrient (N and P) reduction paradigm. Hydrobiologia 847 (21), 4359–4375.

Parmesan, C., Yohe, G., 2003. A globally coherent fingerprint of climate change impacts across natural systems. Nature 421, 37–42.

Parra, G., Guerrero, F., Armengol, J., Brendonck, L., Brucet, S., Finlayson, M., Gomes-Barbosa, L., Grillas, P., Jeppesen, E., Ortega, F., Vega, R., Zohary, T., 2021. The future of temporary wetlands in drylands under the global change. Inland Waters 11 (4), 445–456.

Peixoto, R., Marotta, H., Bastviken, D., Enrich-Prast, A., 2016. Floating aquatic macrophytes can substantially offset open water CO_2 emissions from tropical floodplain lake ecosystems. Ecosystems 19 (4), 724–736.

Pereira, M., Segurado, P., Neves, N., 2011. Using spatial network structure in landscape management and planning: a case study with pond turtles. Landsc. Urban Plann. 100 (1–2), 67–76.

Phillips, G., Eminson, D., Moss, B., 1978. A mechanism to account for macrophyte decline in progressively eutrophic freshwaters. Aquat. Bot. 4, 103–126.

Phillips, G., Willby, N., Moss, B., 2016. Submerged macrophyte decline in shallow lakes: what have we learnt in the last forty years? Aquat. Bot. 135, 37–45.

Pinceel, T., Buschke, F., Weckx, M., Brendonck, L., Vanschoenwinkel, B., 2018. Climate change jeopardizes the persistence of freshwater zooplankton by reducing both habitat suitability and demographic resilience. BMC Ecol. 18 (1), 1–9.

Qiu, D., Wu, Z., Liu, B., Deng, J., Fu, G., He, F., 2001. The restoration of aquatic macrophytes for improving water quality in a hypertrophic shallow lake in Hubei Province, China. Ecol. Eng. 18 (2), 147–156.

Rannap, R., Lohmus, A., Briggs, L., 2009. Restoring ponds for amphibians: a success story. In: Pond Conservation in Europe. Springer, pp. 243–251.

Rautio, M., Dufresne, F., Laurion, I., Bonilla, S., Vincent, W.F., Christoffersen, K.S., 2011. Shallow freshwater ecosystems of the circumpolar Arctic. Ecoscience 18 (3), 204–222.

Rautio, M., Vincent, F., 2006. Benthic and pelagic food resources for zooplankton in shallow high-latitude lakes and ponds. Freshw. Biol. 51, 1038–1052.

Richardson, D.C., Holgerson, M.A., Farragher, M.J., Hoffman, K.K., King, K., Alfonso, M.B., Andersen, M.R., Cheruveil, K.S., Coleman, K.A., Farruggia, M.J., Fernandez, R.L., 2022. A functional definition to distinguish ponds from lakes and wetlands. Sci. Rep. 12 (1), 1–13.

Scheffer, M., 1998. Ecology of Shallow Lakes. Chapman & Hall.

Scheffer, M., 2020. In: Critical Transitions in Nature and Society. Princeton University Press.

Scheffer, M., Bascompte, J., Brock, W.A., Brovkin, V., Carpenter, S.R., Dakos, V., Held, H., Van Nes, E.H., Rietkerk, M., Sugihara, G., 2009. Early-warning signals for critical transitions. Nature 461 (7260), 53–59.

Scheffer, M., Carpenter, S., Foley, J.A., Folke, C., Walker, B., 2001. Catastrophic shifts in ecosystems. Nature 413, 591–596.

Scheffer, M., Carpenter, S.R., Lenton, T.M., Bascompte, J., Brock, W., Dakos, V., Van de Koppel, J., Van de Leemput, I.A., Levin, S.A., Van Nes, E.H., 2012. Anticipating critical transitions. Science 338 (6105), 344–348.

Scheffer, M., Hosper, H.S., Meijer, M.L., Moss, B., Jeppesen, E., 1993. Alternative equilibria in shallow lakes. Trends Ecol. Evol. 8 (8), 275–279.

Scheffer, M., Szabó, S., Gragnani, A., van Nes, E., Rinaldi, S., Kautsky, N., Norberg, J., Roijackers, R., Franken, R., 2003. Floating plant dominance as a stable state. Proc. Natl. Acad. Sci. USA 100, 4040–4045.

Scheffer, M., Van Geest, G., Zimmer, K., Jeppesen, E., Søndergaard, M., Butler, M., Hanson, M., Declerck, S., De Meester, L., 2006. Small habitat size and isolation can promote species richness: second-order effects on biodiversity in shallow lakes and ponds. Oikos 112 (1), 227–231.

Scheffer, M., van Nes, E.H., 2007. Shallow lakes theory revisited: various alternative regimes driven by climate, nutrients, depth and lake size. Hydrobiologia 584, 455–466.

Schindler, D.E., Scheuerell, M.D., 2002. Habitat coupling in lake ecosystems. Oikos 98 (2), 177–189.

Schooler, S.S., Salau, B., Julien, M.H., Ives, A.R., 2011. Alternative stable states explain unpredictable biological control of *Salvinia molesta* in Kakadu. Nature 470 (7332), 86–89.

Schriver, P., Bøgestrand, J., Jeppesen, E., Søndergaard, M., 1995. Impact of submerged macrophytes on fish-zooplankton-phytoplankton interactions: large-scale enclosure experiments in a shallow eutrophic lake. Freshw. Biol. 33, 255–270.

Sculthorpe, C.D., 1967. The Biology of Aquatic Vascular Plants. Edward Arnold.

Seto, K.C., Güneralp, B., Hutyra, L.R., 2012. Global forecasts of urban expansion to 2030 and direct impacts on biodiversity and carbon pools. Proc. Natl. Acad. Sci. USA 109 (40), 16083–16088.

Shapiro, J., Lammara, V., Lynch, M., 1975. Biomanipulation: an ecosytem approach to lake restoration. In: Brezonik, P.L., Fox, J.L. (Eds.), Water Quality Management through Biological Control. University of Florida, pp. 85–96.

Sharpley, A., Jarvie, H.P., Buda, A., May, L., Spears, B., Kleinman, P., 2013. Phosphorus legacy: overcoming the effects of past management practices to mitigate future water quality impairment. J. Environ. Qual. 42 (5), 1308–1326.

Shochat, E., Lerman, S.B., Anderies, J.M., Warren, P.S., Faeth, S.H., Nilon, C.H., 2010. Invasion, competition, and biodiversity loss in urban ecosystems. Bioscience 60 (3), 199–208.

Smith, L.C., Sheng, Y., MacDonald, G.M., Hinzman, L.D., 2005. Disappearing Arctic lakes. Science 308 (5727), 1429, 1429.

Smol, J.P., 2016. Arctic and sub-Arctic shallow lakes in a multiple stressor world: a paleoecological perspective. Hydrobiologia 778, 253–272.

Smol, J.P., Douglas, M.S., 2007. Crossing the final ecological threshold in high Arctic ponds. Proc. Natl. Acad. Sci. USA 104 (30), 12395–12397.

Søndergaard, M., Bjerring, R., Jeppesen, E., 2013. Persistent internal phosphorus loading during summer in shallow eutrophic lakes. Hydrobiologia 710 (1), 95–107.

Søndergaard, M., Jensen, J.P., Jeppesen, E., 2003. Role of sediment and internal loading of phosphorus in shallow lakes. Hydrobiologia 506 (1–3), 135–145.

Søndergaard, M., Jeppesen, E., Jensen, J.P., 2005. Pond or lake: does it make any difference? Arch. Hydrobiol. 162 (2), 143–165.

Søndergaard, M., Jeppesen, E., Lauridsen, T., Skov, C., Van Nes, E., Roijackers, R., Lammens, E., Portielje, R., 2007. Lake restoration: successes, failures and long-term effects. J. Appl. Ecol. 44 (6), 1095–1105.

Søndergaard, M., Moss, B., 1997. Impact of submerged macrophytes on phytoplankton in shallow freshwater lakes. In: Jeppesen, E., Søndergaard, M., Søndergaard, M., Christoffersen, K. (Eds.), The Structuring Role of Submerged Macrophytes in Lakes. Springer Verlag, pp. 115–132.

Stephen, D., Moss, B., Phillips, G., 1998. The relative importance of top-down and bottom-up control of phytoplankton in a shallow macrophyte-dominated lake. Freshw. Biol. 39, 699–713.

Strayer, D.L., 2013. Endangered freshwater invertebrates. In: Levin, S.A. (Ed.), Encyclopedia of Biodiversity. Academic Press, San Diego, California.

Suding, K.N., Gross, K.L., Houseman, G.R., 2004. Alternative states and positive feedbacks in restoration ecology. Trends Ecol. Evol. 19 (1), 46–53.

Tavsanoğlu, N.Ü., Çakiroğlu, I.A., Erdoğan, S., Meerhoff, M., Jeppesen, E., Beklioğlu, M., 2012. Sediments, not plants, offer the preferred refuge for *Daphnia* against fish predation in Mediterranean shallow lakes: an experimental demonstration. Freshw. Biol. 57 (4), 795–802.

Teffera, F.E., Lemmens, P., Deriemaecker, A., Brendonck, L., Dondeyne, S., Deckers, J., Bauer, H., Gamo, F.W., De Meester, L., 2017. A call to action: strong long-term limnological changes in the two largest Ethiopian Rift Valley lakes, Abaya and Chamo. Inland Waters 7 (2), 129–137.

Teixeira-de Mello, F., Meerhoff, M., Pekcan-Hekim, Z., Jeppesen, E., 2009. Substantial differences in littoral fish community structure and dynamics in subtropical and temperate shallow lakes. Freshw. Biol. 54 (6), 1202–1215.

Tessier, A.J., Woodruff, P., 2002. Cryptic trophic cascade along a gradient of lake size. Ecology 83 (5), 1263–1270.

Thomaz, S.M., Bini, L.M., Bozelli, R.L., 2007. Floods increase similarity among aquatic habitats in river-floodplain systems. Hydrobiologia 579 (1), 1–13.

Thomaz, S.M., Cunha, E.R.d., 2010. The role of macrophytes in habitat structuring in aquatic ecosystems: methods of measurement, causes and consequences on animal assemblages' composition and biodiversity. Acta Limnol. Bras. 22 (2), 218–236.

Thomaz, S.M., Dibble, E.D., Evangelista, L.R., Higuti, J., Bini, L.M., 2008. Influence of aquatic macrophyte habitat complexity on invertebrate abundance and richness in tropical lagoons. Freshw. Biol. 53 (2), 358–367.

Thomaz, S.M., Mormul, R.P., Michelan, T.S., 2015. Propagule pressure, invasibility of freshwater ecosystems by macrophytes and their ecological impacts: a review of tropical freshwater ecosystems. Hydrobiologia 746 (1), 39–59.

Thompson, P.L., Guzman, L.M., De Meester, L., Horváth, Z., Ptacnik, R., Vanschoenwinkel, B., Viana, D.S., Chase, J.M., 2020. A process-based metacommunity framework linking local and regional scale community ecology. Ecol. Lett. 23 (9), 1314–1329.

Timms, R.M., Moss, B., 1984. Prevention of growth of potentially dense phytoplankton populations by zooplankton grazing in the presence of zooplanktivorous fish, in a shallow wetland ecosystem. Limnol. Oceanogr. 29 (3), 472–486.

Urrutia-Cordero, P., Ekvall, M.K., Hansson, L.A., 2016. Controlling harmful cyanobacteria: taxa-specific responses of cyanobacteria to grazing by large-bodied daphnia in a biomanipulation scenario. PLoS One 11 (4), e0153032.

Vadeboncoeur, Y., Jeppesen, E., Vander Zanden, M.J., Schierup, H.H., Christoffersen, K., Lodge, D.M., 2003. From Greenland to green lakes: cultural eutrophication and the loss of benthic pathways in lakes. Limnol. Oceanogr. 48 (4), 1408–1418.

Vadeboncoeur, Y., Vander Zanden, M.J., Lodge, D.M., 2002. Putting the lake back together: reintegrating benthic pathways into lake food web models. Bioscience 52 (1), 44–54.

van Donk, E., van de Bund, W.J., 2002. Impact of submerged macrophytes including charophytes on phyto- and zooplankton communities: allelopathy versus other mechanisms. Aquat. Bot. 72 (3–4), 261–274.

van Nes, E.H., Arani, B.M., Staal, A., van der Bolt, B., Flores, B.M., Bathiany, S., Scheffer, M., 2016. What do you mean, "tipping point"? Trends Ecol. Evol. 31 (12), 902–904.

van Rees, C.B., Waylen, K.A., Schmidt-Kloiber, A., Thackeray, S.J., Kalinkat, G., Martens, K., Domisch, S., Lillebø, A.I., Hermoso, V., Grossart, H.P., 2021. Safeguarding freshwater life beyond 2020: recommendations for the new global biodiversity framework from the European experience. Conserv. Lett. 14 (1), e12771.

Vanderstukken, M., Mazzeo, N., Van Colen, W., Declerck, S.A., Muylaert, K., 2011. Biological control of phytoplankton by the subtropical submerged macrophytes *Egeria densa* and *Potamogeton illinoensis*: a mesocosm study. Freshw. Biol. 56 (9), 1837–1849.

Veraart, A.J., de Bruijne, W.J., de Klein, J.J., Peeters, E.T., Scheffer, M., 2011. Effects of aquatic vegetation type on denitrification. Biogeochemistry 104 (1), 267–274.

Verpoorter, C., Kutser, T., Seekell, D.A., Tranvik, L.J., 2014. A global inventory of lakes based on high-resolution satellite imagery. Geophys. Res. Lett. 41, 6396–6402.

Vidon, P., Allan, C., Burns, D., Duval, T.P., Gurwick, N., Inamdar, S., Lowrance, R., Okay, J., Scott, D., Sebestyen, S., 2010. Hot spots and hot moments in riparian zones: potential for improved water quality management. J. Am. Water Resour. Assoc. 46 (2), 278–298.

Wang, H., Wang, W., Yin, C., Wang, Y., Lu, J., 2006. Littoral zones as the "hotspots" of nitrous oxide (N_2O) emission in a hyper-eutrophic lake in China. Atmos. Environ. 40 (28), 5522–5527.

Wang, R., Dearing, J.A., Langdon, P.G., Zhang, E., Yang, X., Dakos, V., Scheffer, M., 2012. Flickering gives early warning signals of a critical transition to a eutrophic lake state. Nature 492 (7429), 419–422.

Welch, P.S., 1952. Limnology, 2nd ed. McGraw-Hill Book Co.

Wezel, A., Oertli, B., Rosset, V., Arthaud, F., Leroy, B., Smith, R., Angélibert, S., Bornette, G., Vallod, D., Robin, J., 2014. Biodiversity patterns of nutrient-rich fish ponds and implications for conservation. Limnology 15 (3), 213–223.

Williams, D.D., 1996. Environmental constraints in temporary fresh waters and their consequences for the insect fauna. J. North Am. Benthol. Soc. 15 (4), 634–650.

Williams, P., Whitfield, M., Biggs, J., Bray, S., Fox, G., Nicolet, P., Sear, D., 2004. Comparative biodiversity of rivers, streams, ditches and ponds in an agricultural landscape in Southern England. Biol. Conserv. 115 (2), 329–341.

Yan, X., Xu, X., Wang, M., Wang, G., Wu, S., Li, Z., Sun, H., Shi, A., Yang, Y., 2017. Climate warming and cyanobacteria blooms: looks at their relationships from a new perspective. Water Res. 125, 449–457.

Yılmaz, G., Çolak, M.A., Özgencil, İ.K., Metin, M., Korkmaz, M., Ertuğrul, S., Soyluer, M., Bucak, T., Tavsanoğlu, Ü.N., Özkan, K., Akyürek, Z., Beklioğlu, M., Jeppesen, E., 2021. Decadal changes in size, salinity and biota in lakes in Konya Closed Basin, Turkey, subjected to increasing water abstraction for agriculture and climate change. Inland Waters 11 (4), 538–555.

Yu, J., Liu, Z., Li, K., Chen, F., Guan, B., Hu, Y., Zhong, P., Tang, Y., Zhao, X., He, H., Zeng, H., Jeppesen, E., 2016. Restoration of shallow lakes in subtropical and tropical China: response of nutrients and water clarity to biomanipulation by fish removal and submerged plant transplantation. Water 8 (10), 438.

Yuan, J., Xiang, J., Liu, D., Kang, H., He, T., Kim, S., Lin, Y., Freeman, C., Ding, W., 2019. Rapid growth in greenhouse gas emissions from the adoption of industrial-scale aquaculture. Nat. Clim. Change 9 (4), 318–322.

Zambrano, L., Scheffer, M., Martinez-Ramos, M., 2001. Catastrophic response of lakes to benthivorous fish introduction. Oikos 94 (2), 344–350.

CHAPTER

27

Sediments and Microbiomes

Warwick F. Vincent[1], *Michio Kumagai*[2] *and Raoul-Marie Couture*[3]

[1]Department of Biology and Centre for Northern Studies (CEN), Laval University, Quebec City, Quebec, Canada

[2]Research Center for Lake Biwa and Environmental Innovation, Ritsumeikan University, Kusatsu, Shiga, Japan

[3]Department of Chemistry and Centre for Northern Studies (CEN), Laval University, Quebec City, Quebec, Canada

OUTLINE

The term *sediments* in the limnological literature refers to inorganic and organic particles in suspension (Fig. 27-1), as well as to their accumulation as deposits at the bottom of lakes and rivers (Fig. 27-2). Sediments play many roles in aquatic ecosystems, including as habitats for bottom-dwelling animals, as substrates for water plants and associated biota, and, in suspension, as controls on underwater light availability, solar heating, and water color. They are especially important in biogeochemical cycles as sinks for carbon storage, sites of gas and nutrient exchange, and habitats for complex communities of diverse microbes, or *microbiomes*, defined as the complete assemblages of microbes in specific habitats, increasingly revealed in terms of structure and function by molecular methods (Grossart et al., 2020).

Sediment microbiomes transform carbon, nitrogen, sulfur, iron, and other elements across redox states, drive elemental fluxes between sediments and the surrounding or overlying water, and affect the net exchange of oxygen, carbon dioxide, methane, and other gases with the atmosphere. *Metagenomic methods* (nucleic acid sequence analysis of the entire microbial community) and other *omics approaches* (total analysis of specific groups of biomolecules) are producing new insights into the diversity, location, and controls on these

Wetzel's Limnology, Fourth Edition
https://doi.org/10.1016/B978-0-12-822701-5.00027-6

biogeochemical processes in lake and river sediments. Environmental change, including climate warming, has wide-ranging impacts on sediment microbiomes, which in turn affect water quality, ecosystem services, and aquatic habitats.

I. Sediment characterization

A. Sources

The ensemble of particles in suspension, both living and dead, is referred to as *seston*, *suspended particulate matter* (SPM), or *total suspended solids or sediments* (TSS), usually with an arbitrary cut-off of minimum particle size that is determined by the filter used to collect the material. For example, commonly used glass fiber filters of grade GF/F have a nominal pore size of 0.7 μm, while those of grade GF/C have a nominal pore size of 1.2 μm. The total mass of particles is expressed in terms of milligrams of dry weight per liter, or it may be assessed as *water turbidity* using a *nephelometer*, an optical instrument that measures light scattering, with the results expressed as *nephelometric turbidity units* (NTUs) calibrated against known standards.

FIGURE 27-1 "Sediments" refers to particles in suspension, such as here in the turbid downstream waters of the Saint Lawrence River, Canada, and to their accumulation as bottom deposits. *(Photo credit: Warwick Vincent.)*

Particles arrive in lakes and rivers from a great variety of sources (Fig. 27-2; see also Chapter 30), including sources outside the aquatic ecosystem (*allochthonous*), for example, soil particles and terrestrial vegetation, as well as sources within the ecosystem (*autochthonous*), such as phytoplankton and aquatic plants. One of the simplest ways of classifying particles is in terms of size. For river systems, a doubling scale is used, from the finest silt and clay particles to large rocks (Table 27-1). Stream ecologists separate organic particles into *coarse particulate organic matter* (CPOM; trapped by a 1 mm sieve) and *fine particulate organic matter* (FPOM; passing through a 1 mm sieve), and often there is an increase in the ratio of FPOM to CPOM with river order, with larger particles broken down by benthic invertebrates and microbial activity as they move downstream.

Particle density (mass per unit particle volume) varies greatly among particle types and is of major importance

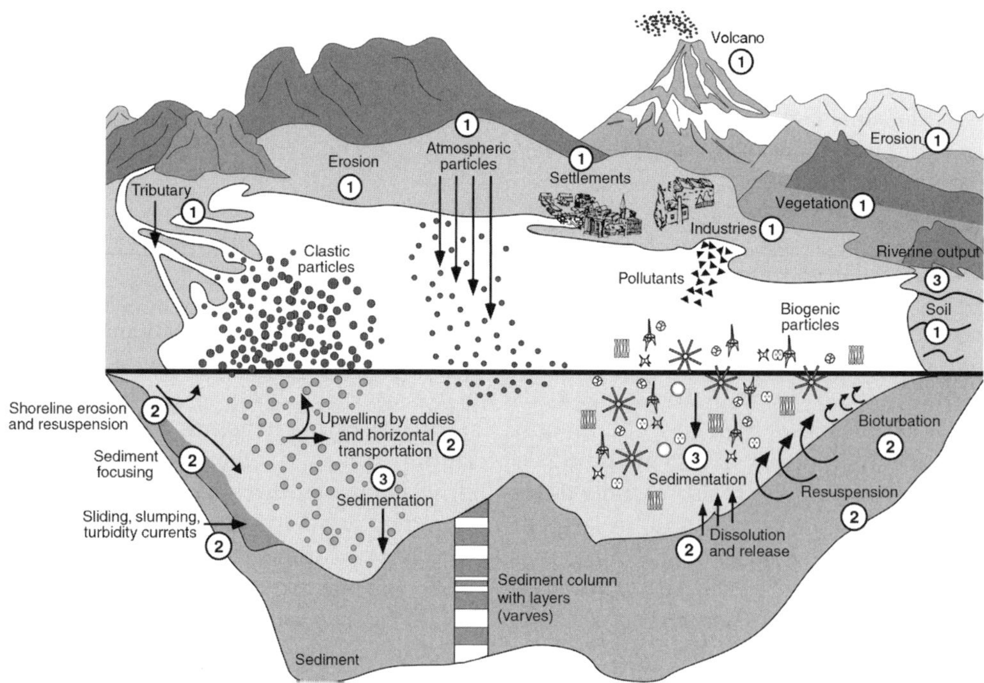

FIGURE 27-2 The diverse sources and fate of lake sediments. (1) Sources of suspended sediments; (2) transport and transformation processes; and (3) transfer from the water column to the bottom sediments and outflow. *Clastic particles* are fragments of weathered rock. *(From Bloesch, 2009.)*

TABLE 27-1 Riverbed Substrates Classified in Terms of Size. In the Wentworth Grain Size Scale Shown Here, Each Size Class Differs by a Factor of Two (Wentworth, 1922). The Additional Category of Clays is Defined According to ISO Soil Standards as Particles with a Diameter of Less Than 0.002 mm (<2 μm)

Large materials	*Size range (mm)*	*Small materials*	*Size range (μm)*
Cobble		**Sand**	
Large	128–256	Very coarse	1000–2000
Small	64–128	Coarse	500–1000
Gravel		Medium	250–500
Very coarse	32–64	Fine	125–250
Coarse	16–32	Very fine	62.5–125
Medium	8–16	**Silt**	<62.5
Fine	4–8		
Very fine	2–4		

because, along with particle size, it dictates the rate at which particles sink and thereby accumulate on the bottom. Specifically, the sinking velocity (V) of small particles affected by drag in water is described by *Stokes' Law*, valid for particles <100 μm (Kumagai, 1988):

$$V = (2\,g\,\Delta\rho\,r^2) / (9\,\nu\varphi_p)$$

where $\Delta\rho$ is the difference in density between the particle and the surrounding water, r is the radius of the particle, g is acceleration due to gravity, ν is kinematic water viscosity (viscosity divided by water density), and φ_p is the coefficient of form resistance (ratio of the sinking velocity of a sphere of the same volume and density to that of the particle; Davey and Walsby, 1985). For biological particles, sinking rates range from millimeters (mm) per day for picocyanobacteria (or even less with cell surface features that increase φ_p; del Mar Aguilo-Ferretjans et al., 2021), to meters per day for long cyanobacterial filaments (Davey and Walsby, 1985) and greater than 10 m h^{-1} for diatom aggregates (Ploug et al., 2010), while mineral particles have faster sinking rates as a result of their higher density. The sinking rate of large microbial aggregates (*marine and freshwater snow*; Section IVA) in the sea ranges from 1 to more than 40 m h^{-1} but cannot be estimated from Stokes' Law because the particles are porous, highly variable in composition and nonspherical, and their *Reynolds number* (gravitational effects relative to viscous forces) is high.

Sediment accumulation rates in lakes vary as a function of their algal and macrophyte biomass stocks, and thus trophic status, and also depend on the extent of catchment erosion and sediment transport by their inflows. The rate of accumulation therefore varies greatly among lakes: for example, around 0.2 mm per year in deep, oligotrophic Lake Baikal (Russia) (Klump et al., 2020), 3–7 mm per year in eutrophic Lake Mendota in Wisconsin (USA) (Walsh et al., 2019), and up to 12 mm per year in shallow lakes of the Yangtze River basin (China) due to eutrophication and intensive land use (Xu et al., 2017). There are also large internal differences in sedimentation rates over different time scales, from seasonal and annual to throughout the geological history of the lake. Sediment redistribution and horizontal transport results in *sediment focusing* (see Chapter 30), whereby greater rates of accumulation occur at the deepest points of lakes (Likens and Davis, 1975), to an extent that is influenced by the shape of the lake basin (Blais and Kalff, 1995).

B. Chemical characterization

The chemical properties of aquatic sediments and their associated environments can be studied at a range of temporal and spatial scales, each with their associated sampling methods (Fig. 27-3). *In situ* moorings provide detailed information about near-sediment conditions; for example, moored oxygen loggers have revealed fluctuations at timescales from minutes to seasonal and interannual (Staehr et al., 2010), including variable periods of exposure of the sediments to bottom water anoxia (e.g., Bégin et al., 2021; Deshpande et al., 2015; Rabaey et al., 2021). Water column profilers provide *fine-structure* (cm) resolution of near-bottom gradients, while microelectrodes allow the analysis of dissolved oxygen, nitrate, and other solutes in the boundary layer and surficial sediments over millimeter scales or less (e.g., Sweerts and de Beer, 1989; Couture et al., 2016; Zhu et al., 2020b; see also Section III).

The pore water chemistry of sediments is a sensitive indicator of biogeochemical processes taking place down the profile. The volume of pore water varies among and within sediments and is determined by *sediment porosity, ϕ*, defined as:

$$\varphi = Vv/Vt$$

where Vv is the volume of pore space (void or interstitial space) and Vt is the total volume of sediment. In freshwater bottom sediments ϕ ranges from less than 0.5 in compact sands to more than 0.95 in loose, organic surface sediments. Sediment pore waters can be sampled by coring followed by centrifugation of subsamples (e.g., Pogodaeva et al., 2017; Anschutz and Charbonnier, 2021) or by dialysis plates (*peepers*; Fig. 27-3c) left in place for days or weeks to equilibrate (e.g., Couture et al., 2016; Clayer et al., 2020).

Bulk properties of the sediments are measured on samples obtained by sediment collectors moored in the water column for days to weeks, and on bottom samples collected by sediment coring instruments (Chapter 30),

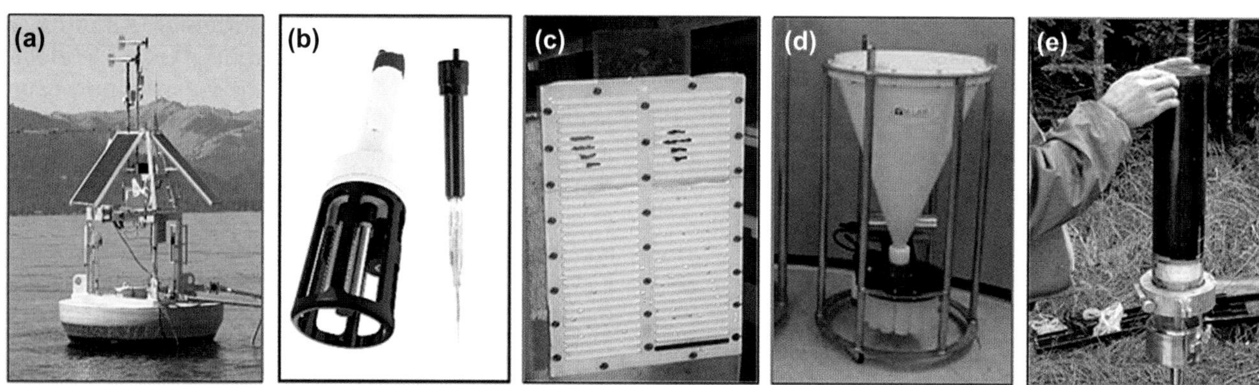

FIGURE 27-3 Methods to study the chemical properties of sediments and their overlying waters at different time and space scales. (a) Mooring for continuous measurements by sensors located at different depths in the lake; (b) fine structure profiler for water column observations, and microelectrode for oxygen measurements across the sediment–water interface; (c) dialysis system that is inserted into the sediments to sample pore waters at different depths; (d) sediment trap for mooring in the lake and collecting sinking particles; and (e) a lake sediment core being subsampled according to depth and age. *(Photo credits: [a] NASA, Lake Tahoe; [b] RBR Ltd, Ottawa, Canada, and Unisense Ltd.; [c] André Tessier; [d] McLane Research Laboratories Inc.; and [e] André Tessier.)*

with the resultant cores then subsampled according to depth and age of deposition. These sediment samples can be characterized in terms of their elemental composition (including stable isotope signals), elemental ratios, or the presence of specific compounds and particles, including contaminants such as metals, organic pollutants, and microplastics. Information can then be obtained at very high resolution (mm or less) by core scanners that measure hyperspectral reflectance (e.g.,- Ghanbari et al., 2020), micro–X-ray fluorescence (e.g., Cuven et al., 2010), and computed tomography (e.g., Emmanouilidis et al., 2020).

The pore waters of sediments typically show large gradients down through the sediment profile, consistent with variations in microbiome composition and function with depth (Section VC). For example, biogeochemical analysis and modeling of sediment water in oligotrophic Lake Tantaré in southern Quebec, Canada, revealed three layers (Fig. 27-4; Clayer et al., 2020). Oxygen penetrated through the upper layer, to a depth of around 5 mm, aided by *bioirrigation* (see below; Couture et al., 2016). Within this layer, methane declined toward the surface while the $\delta^{13}C-CH_4$ values increased toward the surface, both features consistent with the importance of *methanotrophy* (see below) in aerobic surface sediments (Bastviken et al., 2008). Sulfate dropped rapidly, indicating the role of sulfate-reducing bacteria at greater depth, while dissolved iron increased, implying the

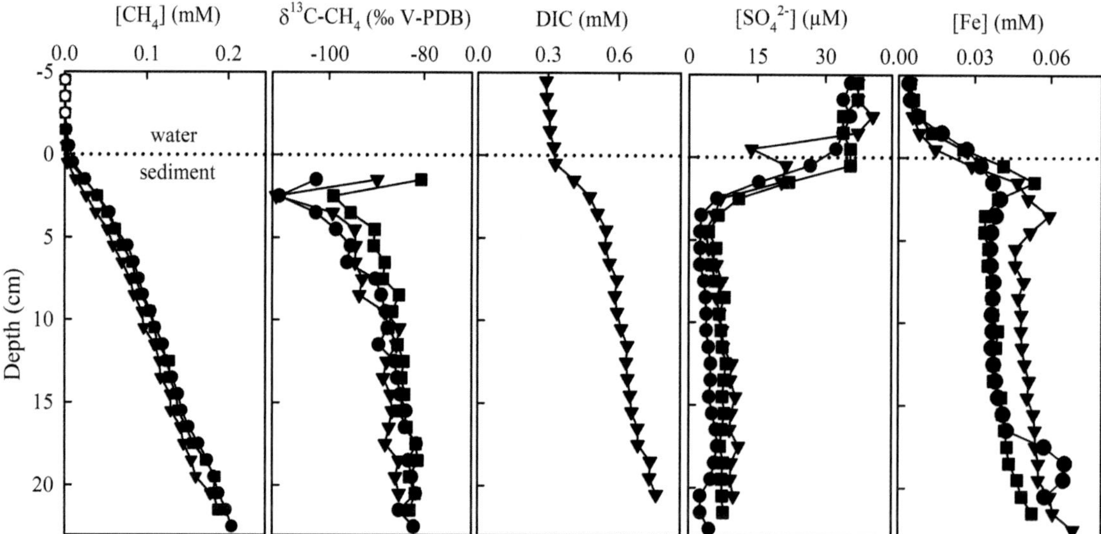

FIGURE 27-4 Replicate pore water profiles in the sediments of an oligotrophic lake in southern Quebec, Canada. The depth scale is in centimeters, the *dotted line* represents the sediment–water interface, and the *shaded area* is the middle depth zone (see text). *(Adapted from Clayer et al., 2020.)*

increasing importance of solid Fe(III) reduction at depth. Like methane, dissolved inorganic carbon (DIC) increased with depth dues to CO_2-producing *fermentation processes* (microbial metabolism of organic carbon in the absence of oxygen) in the deeper anaerobic layers.

The bulk properties of lake sediments reflect their autochthonous and allochthonous origins, and elemental compositions and ratios, such as the *C:N ratio* (total carbon to total nitrogen content), are highly variable among lakes and also vary with sediment depth (see below). In general, molar C:N ratios for algae are in the range of 4 to 10, while terrestrial vascular plants have ratios of 20 or higher (Meyers and Teranes, 2002; see also Chapter 28). For example, in a set of eight Polish mountain lakes the sediments ranged from fine sands with 0.5% organic carbon to *gyttja* (olive-black, gel-like ooze) with up to 50% organic carbon. The low C:N ratios in these sediments showed that in-lake primary production was likely the primary source of organic matter for lakes located above the treeline, while terrestrial plant fragments were the main organic matter constituents in dystrophic forest lakes (Gąsiorowski and Sienkiewicz, 2013). In lake surveys the analysis of C:N and *C:P ratios* (total carbon to total phosphorus content) of sedimenting particles (seston) show large variations, with overlapping C:N ratios among trophic states but a tendency toward lower C:P ratios with increasing levels of nutrient enrichment (Table 27-2).

Silica is present in sediments as a component of mineral particles but also as *biogenic silica* associated with siliceous microfossils, notably diatoms (Chapter 30). These algae take up dissolved silicate (Chapters 16 and 17) to produce their silica cell walls (frustules), and biogenic silica analyses show that inland waters and their sediments are major retention sites for silicon via diatom sedimentation, thereby substantially reducing the dissolved silica fluxes to the ocean (Harrison et al., 2012). In hardwater lakes carbonates are present not only as particles washed in from the catchment but also through biogenic processes. Carbonates may be precipitated during algal photosynthesis (visible in many lakes as chalky surface waters or *lake whitings*, Chapter 13) and as crusts over certain *charophytes* (a group of freshwater algal macrophytes; Chapters 17 and 24), thereby contributing to inorganic carbon and coprecipitated phosphorus storage in the sediments (Otsuki and Wetzel, 1972; Kufel et al., 2020).

TABLE 27-2 Carbon to Nitrogen (C:N) and Carbon to Phosphorus (C:P) Ratios in Lake Particles (Seston)

Trophic status	*TP range*	*C:N*	*C:P*	*n*
Oligotrophic	<10	18 (2.4–61)	152.6 (4–1426)	338
Mesotrophic	10–25	15 (1.2–58)	65 (121–386)	293
Eutrophic	25–50	11 (0.9–38)	43.8 (4.5–267.7)	331
Hypertrophic	>50	8.5 (2.4–61)	22.8 (0.5–88)	268

The values (average, minimum, and maximum) are for relative molar abundance by trophic status in 1230 samples from the US EPA 2007 Lake Survey. *N*: Number of lakes in each category, separated by total phosphorus concentration (TP, in μg L^{-1}). *Source: Modified from Maranger et al. (2018).*

Metal concentrations in sediments also vary widely, including over time associated with mining, agricultural, urban, and restoration activities; for example, the Laurentian Great Lakes (Aliff et al., 2020). At highly contaminated lake and river sites, sediment concentrations can exceed 1 μg of mercury, 21 μg of cadmium, 670 μg of copper, 9000 μg of zinc, and 71,000 μg of lead per gram of sediment (Luoma and Rainbow, 2008). Metal analysis by *X-ray fluorescence* (micro-XRF) scans of sediment cores can provide resolution at the submillimeter level to detect short-term fluctuations as well as long-term trends in environmental conditions (e.g., Antoniades et al., 2011). Sulfur occurs as sulfate and organic sulfur compounds in upper sediments, with reduced sulfide forms increasingly important at depth (Couture et al., 2016; Fakhraee et al., 2017).

Among other contaminants, *microplastics* have become a notable component of lake and river sediments throughout the world. These are synthetic polymers in the form of beads, fibers, foams, films, and fragments and are typically sampled and counted in the broad size range 1 μm to 5 mm. In freshwater environments the most common particles are fibers, and the primary sources are waste treatment plants and laundry water (Li et al., 2020). Large rivers are major conduits of microplastics, with an estimated 25% of the global fluxes to the world ocean carried by 14 major rivers (Lebreton et al., 2017), and these particles are incorporated into river sediments in large concentrations. For example, some of the highest concentrations of microplastics have been recorded in the Saint Lawrence River downstream of the city of Montreal (range 65–7562 plastic particles per kg dry weight of sediment; Crew et al., 2020). Microplastic particles can be ingested by sediment-dwelling invertebrates (e.g., Fueser et al., 2020), with unknown effects. These particles eventually break down into *nanoplastics*, which, because of their small size (1–1000 nm), have different properties to *microplastics*, including large reactive surface areas per unit mass (Gigault et al., 2021). Nanoplastics can remain in suspension as *colloids* (dispersed insoluble particles) but are also incorporated into larger microbial aggregates that eventually sink to the sediments. They are likely to be an increasingly major fraction of the total plastic pollution that enters the aquatic food web, again with unknown consequences.

C. Diagenesis

Diagenesis refers to the physical and biogeochemical changes that take place in the bottom sediments once the sinking particles have been deposited. Sediments become more compact with time, typically showing a decrease in porosity with depth and an associated increase in the ratio of sediment particle mass to wet sediment volume (*apparent* or *dry-bulk density*). This is due to the loss of water that is pushed out by gravitational effects as the sediments accumulate and settle, combined with the loss of organic matter by decomposition (e.g., Gälman et al., 2008; Maier et al., 2013).

The biogeochemical changes are primarily mediated by the sediment microbiome and occur most rapidly in the surface strata where oxygen and labile organic matter are both in greatest abundance. Burrowing animals accelerate these transformation processes by mixing the surficial sediments (*bioturbation*) and drawing oxygen deeper into the sediment profile while exposing reduced substances to oxidants (*bioirrigation*). Tunnel and tube-forming chironomid larvae are especially well known for this effect and are viewed as efficient *ecosystem engineers* (organisms that change the properties of their environment) because of their aeration effects on sediment biogeochemical processes (Fig. 27-5).

At a given depth in the sediment column, changes in concentration of the solid (e.g., total particulate carbon) and dissolved (e.g., dissolved inorganic carbon) constituents of aquatic sediments through time can be described by the diagenetic equation (Berner, 1980), which can be expanded and rearranged as (Boudreau, 1999):

$$\sum R = \frac{\partial}{\partial z}\left(\varphi(D_S + D_B)\frac{\partial[\mathrm{C}]}{\partial z}\right) + \varphi\alpha\left([\mathrm{C}]_{\mathrm{irrigation}} - [\mathrm{C}]\right)$$

where the left-hand term is the net rate of change ("net reaction rate") of the modeled constituent. In the right-hand terms, z is sediment depth, ϕ is sediment porosity, D_S is molecular diffusion through the sediments corrected for *tortuosity* (the ratio of actual diffusion path length to the straight-line distance between start and end points), D_B is the rate of bioturbation, [C] is the concentration of the modeled constituent, α is a bioirrigation coefficient, and $[\mathrm{C}]_{\mathrm{irrigation}}$ is the concentration of the constituent in the overlying waters that is drawn into the sediments by bioirrigation. In practice, both D_B and α are difficult to estimate accurately because these effects vary among benthic animal species, locations in the lake, and seasons, and they are therefore usually approximated. This equation assumes that advection is negligible, but in some sediments there may be bubbling, water flows, and venting (e.g., at sites in Lake Biwa, Japan; Kumagai et al., 2021), which would disrupt and potentially eliminate the vertical gradient in solute concentrations.

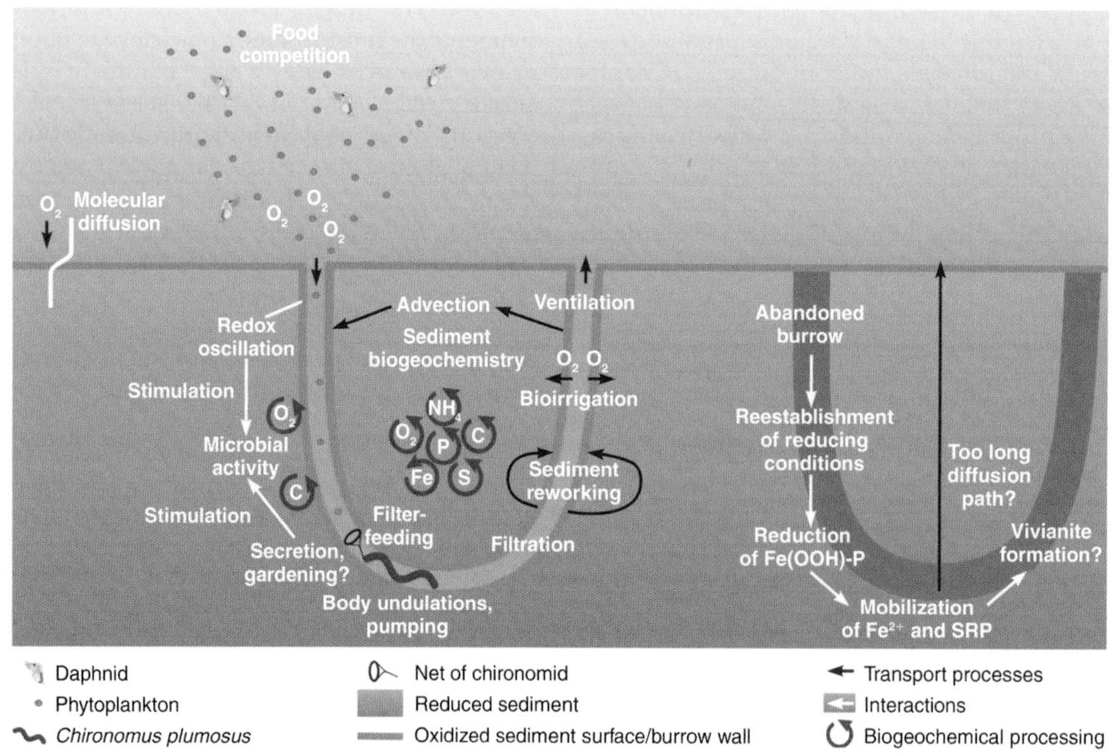

FIGURE 27-5 The biogeochemical effects of chironomid larvae that burrow into lake sediments and draw water into their tubes. *(From Hölker et al., 2015.)*

The biogeochemical effects of diagenesis result in large changes in the chemical properties of sediments with sediment depth, both in pore water chemistry (as illustrated in Fig. 27-4) and in bulk elemental composition. The organic content typically decreases with depth, while the C:N ratio may increase with depth associated with faster bacterial mineralization of nitrogen than carbon. In a northern Swedish lake, for example, 20% of the carbon and 30% of the nitrogen were removed in the first 5 years after deposition, giving rise to the pattern of increasing C:N with depth (Gälman et al., 2008).

D. Effluxes

There is a close reciprocal interaction between benthic sediments and the overlying water column. Particles arrive by sedimentation, with or without horizontal advection by density currents, seiche-induced movements, and, in the case of rivers, bed-load flows. These deposits can be partially eroded, with some of the particles resuspended and redistributed by currents and mixing (Section II). Additionally, substances are exchanged between the sediments and overlying water by *diffusion* across the benthic boundary layer (Section III), or in the case of some gases such as methane and nitrogen, by the production of bubbles that then float up toward the lake or river surface (*ebullition*), where the gases are released into the atmosphere (Section V). Connections between the atmosphere and the sediments also occur in the opposite direction, by dissolution of oxygen and other gases from the overlying air into the water, and their transport by mixing to the sediments, where they may be consumed by biogeochemical processes.

The total solute concentration is often higher in the bottom waters of lakes as a result of sediment biogeochemistry and the release of dissolved materials from the sediment surface into the boundary layer (compounded by density flows). The decomposition of organic materials (e.g., carbohydrates; CH_2O) in surface sediments results in the production of CO_2 that in turn reacts with carbonate minerals in the sediments to release bicarbonate and calcium ions:

$$CH_2O_{(s)} + O_{2(aq)} \rightarrow CO_{2(aq)} + H_2O_{(l)} \leftrightarrow H^+_{(aq)} + HCO^-_{3(aq)}$$

$$CO_{2(aq)} + H_2O_{(l)} + CaCO_{3(s)} \leftrightarrow Ca^{2+}_{(aq)} + 2HCO^-_{3(aq)}$$

where (s) refers to solid phase, (l) to liquid, and (aq) to dissolved aqueous phase. This effect is observed in the bottom tens of centimeters of the water column adjacent to the sediments by a drop in pH and increase in ion concentrations, and thus conductivity, and can be detected by water column profilers (Fig. 27-3b). The increased concentrations of dissolved calcium and bicarbonate result in greater water density (Mortimer and Mackereth, 1958), and this solute-induced effect caused by sediment respiration may contribute to density flows over the lakebed (MacIntyre et al., 2018; Cortés and MacIntyre, 2020).

Benthic geochemical and microbial processes can also result in other changes in the chemical composition of the water immediately above the sediments. These include depleted oxygen due to sediment respiration and nitrate release by nitrification in aerobic sediments (e.g., Vincent and Downes, 1981), although this flux may be small relative to planktonic nitrification in deep lakes such as Lake Tahoe (Beutel and Horne, 2018). Similarly, reduced substances such as methane, hydrogen, sulfide, iron (Fe(II)), ammonium (NH_4^+), and manganese (Mn(III)) can accumulate above anaerobic sediments. Laboratory experiments under alternating *redox potentials* (Chapter 10) have shown the capacity of wet soil mixtures to sequester certain metals such as chromium under both oxic and anoxic conditions but to release other metals such as manganese and iron under reducing conditions (Couture et al., 2015). Alternation of reducing and oxidizing conditions may be especially important in controlling the biogeochemistry of environments subject to hydrological fluctuations, for example, wetlands with changing water levels (Peiffer et al., 2021). Organic carbon is also released from lake and river sediments, both under anaerobic and aerobic conditions (Lau and del Giorgio, 2020).

The release of inorganic phosphate from anoxic sediments (*internal loading*; Chapter 15) can accelerate or worsen eutrophication and has therefore been a subject of particular interest to limnologists and environmental managers. This process is sustained chiefly by sedimentary processes, and its intensity is a function of both historical inputs to the sediment and modern biogeochemical processes, driven by microbial decomposition of organic carbon. The forms of phosphorus in aquatic sediments and the mechanisms controlling its release have received much attention from geochemists (Katsev et al., 2006; Orihel et al., 2017; Markelov et al., 2019). Charged dissolved ions such as those of P, Fe, metals, and metalloids have a strong binding affinity for the charged mineral surfaces of Fe and Mn (oxy)hydroxides. As a result, the reductive dissolution of those minerals is often a trigger for measurable return fluxes of phosphate (Fig. 27-6). Similarly, return fluxes of reduced Fe, transported by Fe-binding ligands, can sustain the iron demand by Cyanobacteria (Du et al., 2019). Metalloids and trace elements, such as molybdenum, a cofactor of nitrogen-fixing enzymes, are also diffused from sediments to the water column (Chappaz et al., 2008), with potential effects on lake productivity. Silicate, a key nutrient for diatom growth, can likewise be supplied to the water column via diffusion from the sediment (Conley et al., 1988).

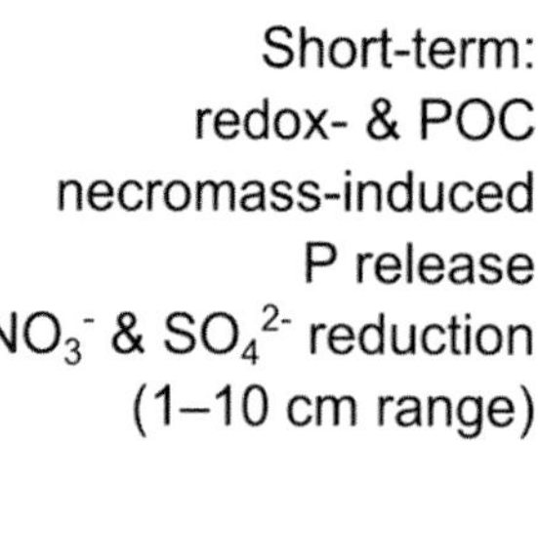

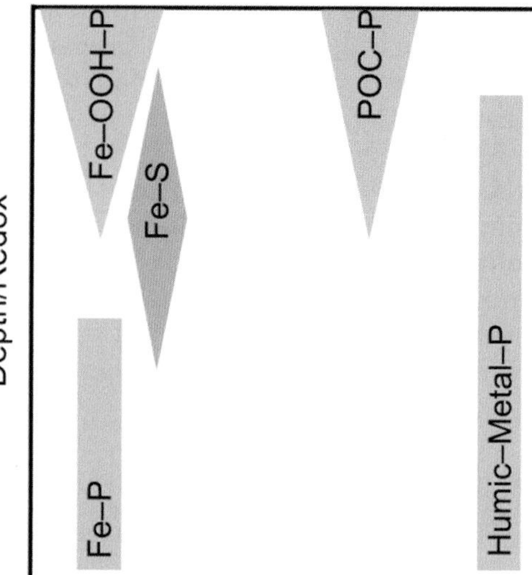

FIGURE 27-6 Key forms of phosphorus and their distribution in lake sediments. Fe–OOH–P: redox-sensitive iron oxyhydroxides with adsorbed P, such as ferrihydrite and goethite; POC–P: particulate organic carbon with P, including *necromass* (detrital organic material). Humic–P: humic–metal–phosphate complexes. Fe–P: iron phosphate minerals such as vivianite. Not shown: calcium phosphate minerals, in hard-water lakes. *(From Orihel et al., 2017.)*

E. Storage

Carbon and other elements enter and are lost from bottom sediments via the processes described above, but there is a net accumulation of materials, at rates that vary greatly among lakes and within a given lake, over the course of its geological evolution in the landscape. Sediment storage is favored by the low temperatures of the bottom waters of most lakes, along with the absence of oxygen in most of the sediment profile, which further slows the microbial breakdown of organic matter. In the deepwaters of lakes, accumulation is additionally favored by the absence of resuspension caused by direct wind-induced mixing (see Section II).

Lake sediments are characterized by higher organic carbon accumulation rates than in marine systems, in part due to the higher primary production and burial rates in lake ecosystems (Meyers and Ishiwatari, 1993), along with higher inputs of less labile, catchment-derived organic matter, and lower oxygen availability (Sobek et al., 2009). In eutrophic lakes, primary production dominates the flux of organic carbon, and their shallow water columns mean that a higher proportion of the primary production reaches the sediments. In boreal lakes rich in dissolved organic carbon of terrestrial origin, flocculation gives rise to particles that have little biological reactivity, and this allochthonous matter can account for as much as 87% of the accumulated carbon in bottom sediments (Guillemette et al., 2017). Carbon storage rates in inland waters appear to be increasing, in part as a result of human activities, specifically soil erosion, river dams, and eutrophication, but also through broader landscape mechanisms that are still poorly defined (Heathcote et al., 2015).

The total carbon and nutrient storage in lake sediments is substantial. Global estimates suggest that 0.15 peta grams (Pg) of carbon are buried in inland waters each year, which is of a similar magnitude to carbon burial rates in the oceans despite their vastly greater area. Some 40% of the inland water burial is calculated to be stored in reservoirs, with higher burial rates predicted in warm and dry regions (Mendonça et al., 2017). Burial efficiency, defined as the ratio of organic carbon burial to organic carbon delivery by primary production and allochthonous sources, varies greatly among lakes. In Lake Baikal this ratio is 0.2 or less, with around 50% of the carbon recycling process taking place within the sediments and largely via conversion to CO_2 (Klump et al., 2020). By contrast, the burial efficiency in lakes with high inputs of less labile, allochthonous carbon and reduced oxygen exposure time (estimated as the mean oxygen penetration depth divided by the sedimentation rate) can be very high, for example, up to 0.9 in Lake Wohlen, Switzerland, a reservoir receiving high riverine inputs of organic matter (Sobek et al., 2009).

Eutrophic lake sediments receive large inputs of carbon due to in-lake primary production (for example, farm ponds, compounded by inputs from erosion; Downing et al., 2008), and although much of this organic material may be rapidly broken down by microbial processes, the remaining carbon for long-term storage may still be considerable. In a comparative study of lake sediments in the Te Arawa Lake District of the central North Island of New Zealand, (Santoso et al., 2017) applied the diagenetic equation to analyze organic carbon content as a function of depth:

$$G_z = G_m e^{-\alpha z} + G_{inf}$$

where G_z is percent organic carbon at depth z (cm) in the sediments, G_m is the labile carbon in the initially deposited sediments, and G_{inf} is the nonlabile residual carbon. The decay parameter α (cm^{-1}) can be defined as k_m/w (Crill and Martens, 1987), where k_m is the first-order rate constant (year^{-1}) and w is the rate of sediment accumulation in cm year^{-1}. Consistent with this model, G showed an exponential decline with depth in each of the sediments. The most eutrophic lake, Lake Okaro, had the lowest carbon burial rate, indicative of the rapid decomposition of a substantial fraction of the large annual sediment fluxes, yet also had the largest accumulated carbon stocks.

II. Resuspension and redeposition of sediments

A. Lakes

There is a continuous exchange of particles between the lakebed and sediments in suspension. This is especially pronounced in the shallow littoral zone of lakes where wave-induced motions are sufficient to resuspend the smaller particles (Fig. 27-7). Currents and mixing transport these finer particles offshore where they have a higher probability of sinking to depths beyond the limits of resuspension. This results in a gradient of sediment conditions from inshore to offshore, with an offshore limit for fine sediment resuspension that has been referred to as the *mud deposition boundary* (Rasmussen and Rowan, 1997). The removal of finer sediments from the littoral zone by this process is called *sediment winnowing*, and the net offshore transport contributes toward greater sediment accumulation in the deepest part of the lake (*sediment focusing*; see also Chapter 30).

Sediment resuspension is greatest in shallow lakes with long fetches, and it can be highly seasonal. In shallow dimictic lakes, full water column circulation in spring and fall may be accompanied by the resuspension of fine sediments such as biogenic carbonates, giving rise to periods of high water column turbidity (Fig. 27-8). Mixing-induced resuspension of sediments can cause rapid changes in the optical and chemical properties of lakewaters, including a large reduction in light availability for benthic photosynthesis (e.g., Bégin et al., 2021) and peaks in total phosphorus concentrations. TSS values may also increase due to seiche movements (Chapter 8) that cause sediment resuspension and turbidity in the benthic boundary layer and due to internal waves that break against the sloping inshore sediments of the lake basin.

B. Rivers

Rivers have much higher suspended sediment concentrations than most lakes because of the inorganic and organic particles washed in from their catchments, the continuous input of particles from riverbank erosion, and the resuspension of particles from the riverbed by the flowing, turbulent waters (Fig. 27-1).

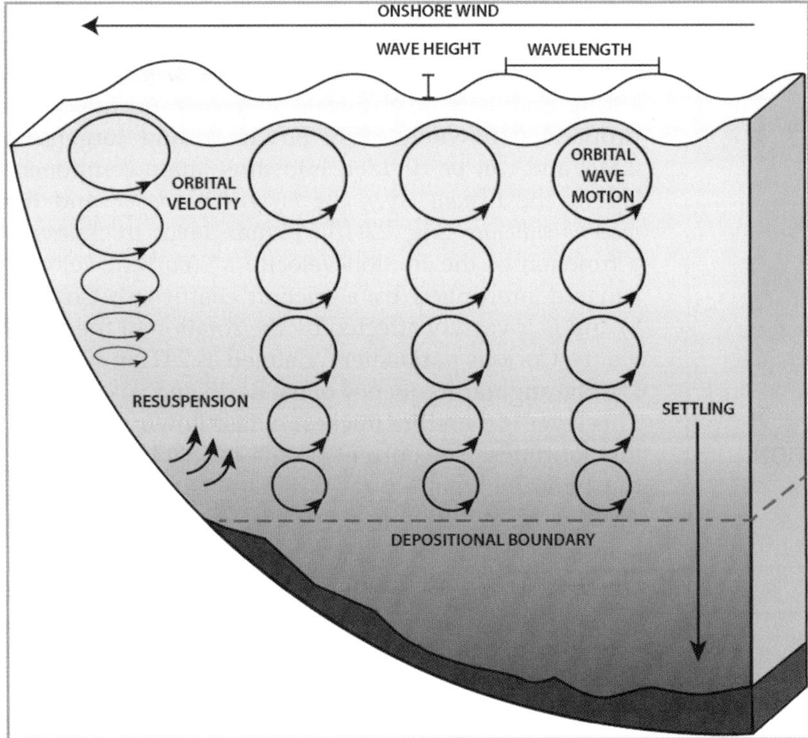

FIGURE 27-7 Resuspension of sediments by wave action is especially pronounced in the inshore zone of the littoral zone, preventing the accumulation of fine sediments. These orbital motions are flattened ellipsoids inshore, while offshore, they are circular and do not extend to the bottom. *(Based on information in Constantin and Villari, 2008; Laenen and LeTourneau, 1996; Rasmussen and Rowan, 1997.)*

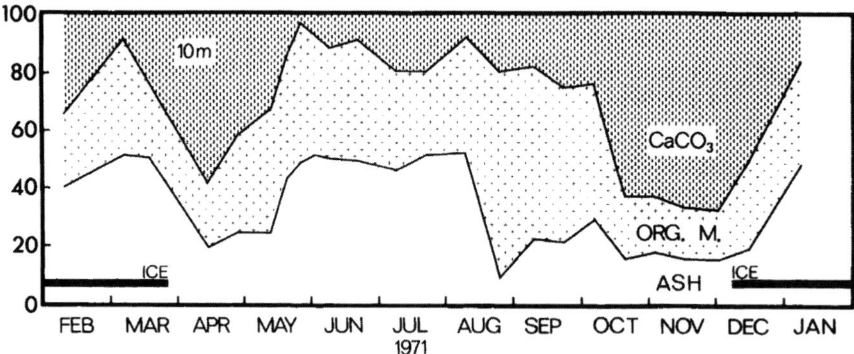

FIGURE 27-8 Resuspension of fine carbonate particles during spring and fall mixing in Lawrence Lake, Michigan (USA). The *vertical axis* is the percent contribution to the total dry weight mass of particles collected in a sediment trap installed at 10 m depth for 2-week intervals (maximum depth is 12.5 m). *(From White and Wetzel, unpublished.)*

Sediments are a key determinant of river ecosystem structure and function (Wohl et al., 2015), and they can be greatly modified by human activities. The installation of dams can radically deplete river sediment loads by allowing the decantation of particles, while altered land-use patterns and climate change can result in higher rates of sediment delivery to river ecosystems (Section VI).

Sediments are mostly transported by the water columns of rivers (the *suspended load*), but a portion of the total sediment load occurs near the bottom, where sediment particles are resuspended and tumble downstream over the riverbed. This *bed load* increases at higher water flows as a result of their greater power for erosion as well as for sediment resuspension (Fig. 27-9). At river flows below the *fall velocity*, sediment particles sink out and are deposited on the riverbed, while at river flows above the fall velocity, particles remain in suspension. At flows above the *erosion velocity* (shown as a band in Fig. 27-9), particles are eroded from river sediments and added to the suspended sediment load. Both the fall velocity and erosion velocity are nonlinear functions of sediment size.

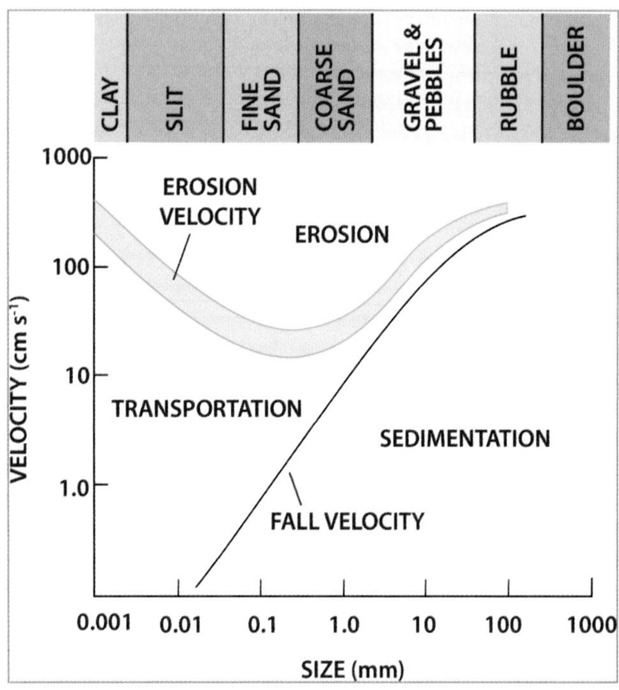

FIGURE 27-9 Transport, erosion, and sedimentation of particles as a function of river velocity and particle size. *(From Allan and Castillo, 2007.)*

III. Benthic boundary layer

A. Importance and structure

The *benthic boundary layer* is the layer of water directly above the sediment surface in which the transport of dissolved constituents becomes progressively controlled by molecular diffusion rather than advection and turbulent mixing. It is often characterized by strong gradients of physical, chemical, and biological properties (Bowden, 1978; Boudreau and Jorgensen, 2001) and can be divided into three main component layers: the *Ekman layer*, the *logarithmic layer*, and the *viscous sublayer* (Fig. 27-10). Ekman layer thickness is a function of the friction velocity u^* (current velocity squared multiplied by a friction coefficient; Chapter 8) and is inversely affected by the rotation of the Earth via the Coriolis parameter f, defined as $2\,\Omega \sin \varphi$, where Ω is the angular frequency of the Earth and φ is latitude. This layer is therefore thickest in fast-flowing waters at low latitudes. The current profile in this layer follows the typical Ekman spiral structure with depth, with the overall mass transport (*Ekman transport*) in the benthic boundary layer to the left side in the Northern Hemisphere due to dominance of Kelvin (basin scale) waves, and to the right side in the Southern Hemisphere. This transport process plays an important role in carrying materials horizontally in the Ekman layer and in the dissipation of momentum from the outer flow (e.g., Duck and Foster, 2001). The eddy viscosity coefficient in the Ekman layer is of order 10^{-4} to 10^{-3}

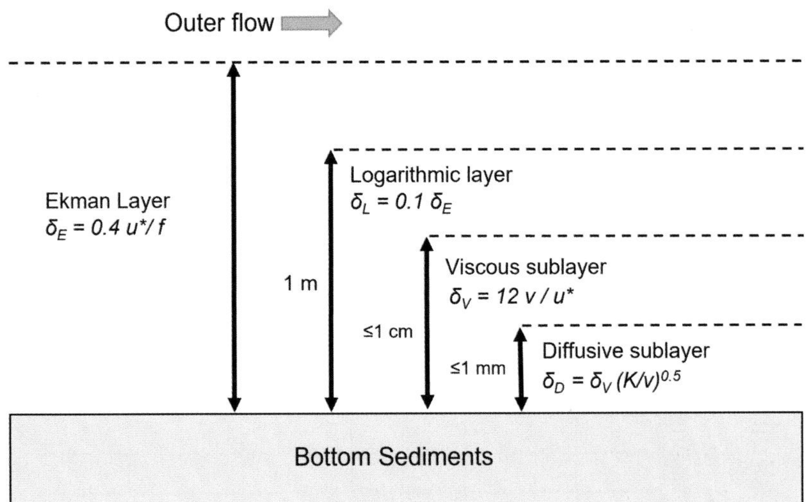

FIGURE 27-10 Schematic representation of the benthic boundary layer showing the approximately log-scale subdivisions. In Lake Biwa the Ekman layer may be up to 10 m thick. *(Based on information in Boudreau and Jorgensen, 2001; Bowen et al., 2020.)*

$m^2\ s^{-1}$, which gives a thickness of the Ekman layer in Lake Biwa (latitude 35.3°N), Japan, of between 1 and 10 m.

The logarithmic layer is a transition region of turbulent motion, with a thickness of around 10% of that of the Ekman layer. In the viscous sublayer vertical transport is by molecular rather than turbulent diffusion, and the horizontal flow (U) at each depth (z) can be derived as:

$$U(z) = \frac{{u_*}^2 z}{\nu_M}$$

where ν_M is the kinematic viscosity coefficient. By dimensional analysis, the thickness of this layer (δ_V) can be solved as:

$$\delta_V = A\frac{\nu_M}{u_*}$$

where *A is* a constant derived from experimentation and is often around 12 (Soulsby, 1983). This gives thicknesses of the viscous sublayer in lakes in the range of several millimeters to several centimeters, and given the slow exchange by molecular diffusion to the logarithmic layer above, this layer can have chemical concentrations that differ markedly from the Ekman layer. Finally, the diffusive sublayer in which molecular diffusion controls the fluxes between the sediments and the overlying water may be a fraction of a millimeter in thickness (δ_D), determined by the ratio between molecular diffusivity (K) and viscosity (ν).

The thickness of the benthic boundary changes with time. At short timescales, oscillating motions above the lakebed caused by seiches (Chapter 8) can cause periodic fluctuations. At longer timescales, seasonal changes in water column stability give rise to large variations in Ekman layer thickness (δ_E). Weatherly and Martin, (1978) derived the following relationship to describe this seasonal effect in the ocean:

$$\delta_E = \frac{1.3u_*}{f\left[1+\left(\frac{N^2}{f^2}\right)\right]^{\frac{1}{4}}}$$

Thus the thickness of the boundary layer decreases as stratification increases in summer (buoyancy frequency, N, becomes greater) and increases as stratification decreases in spring and fall (N becomes smaller).

B. Case study at Lake Biwa, Japan

The use of microelectrodes (Jørgensen and Revsbech, 1985) has provided an increasingly refined view of the benthic boundary in lakes and its role in oxygen dynamics. A set of observations from Lake Biwa illustrates the nature and temporal variations of this layer (Fig. 27-11). Dissolved oxygen (Chapter 11) was measured by glass microsensors with 40 to 60 μm outside tip diameters (Unisense OX-50) from a submersible, automated observation platform that was lowered to just above the sediments at 90 m depth in the lake. The oxygen and temperature sensors were lowered by a stepping motor at 0.1 mm intervals from 10 mm above to 10 mm below the water–sediment interface over a total profiling time of around 1 h.

The profiles taken three times of the year in Lake Biwa revealed that the sediments of Lake Biwa were at or near anoxia on all dates of measurement, despite oxygenated waters (from 2 to >10 mg $O_2\ L^{-1}$) within 1 cm of the sediments (Fig. 27-11). The viscous sublayer extended 1–3 mm above the sediments and was characterized by sharp dissolved oxygen gradients that

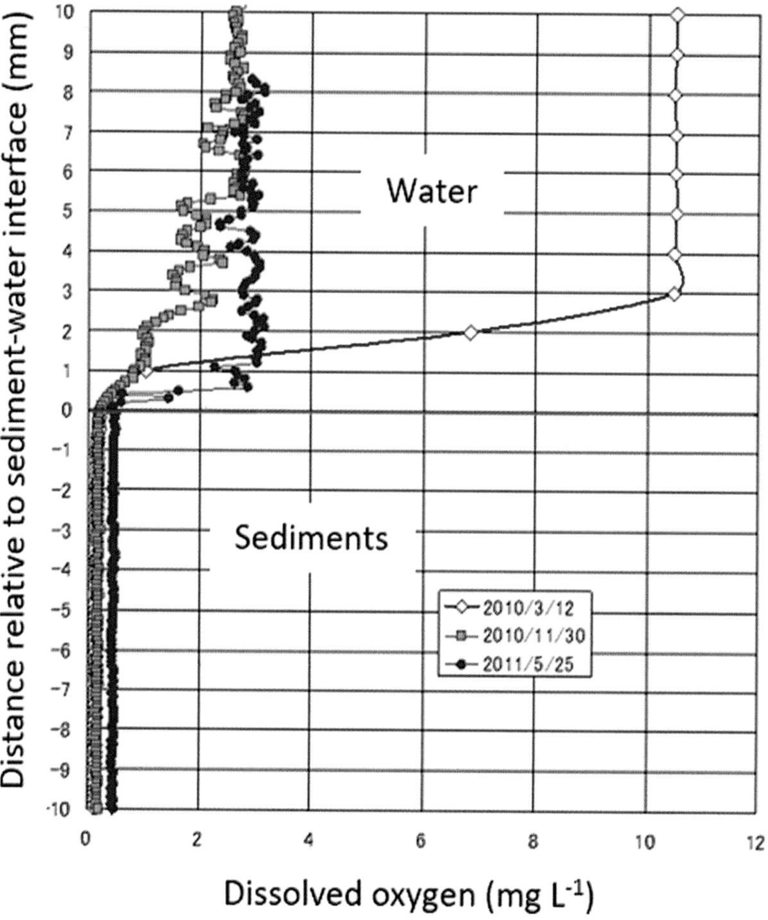

FIGURE 27-11 Dissolved oxygen profiles measured in the vicinity of the water–sediment interface at 90 m depth in Lake Biwa. *(From M. Kumagai, unpublished data.)*

could be used to calculate a diffusive flux into the sediments. The molecular diffusivity of dissolved oxygen in water is of the order of 10^{-9} m^2 s^{-1}, and for the average dissolved oxygen gradient between water and sediments of 1 mg O_2 L^{-1} mm^{-1}, as shown in Fig. 27-11, the dissolved oxygen flux through the interface between water and sediments would be around 10^{-6} g m^{-2} s^{-1}, which would be inadequate to meet the sediment oxygen demand.

IV. Sediment microbiomes

Microbes live in the aquatic environment in four ways: inside other organisms, as parasites or symbionts; as free-living planktonic organisms; in multispecies aggregates; and as biofilms that coat solid substrata. The first two lifestyles certainly occur in sediments, for example, viruses inside bacterial cells, symbiotic bacteria within ciliates, and free-swimming nanoflagellates in the water-filled interstices between sediment particles. However, it is aggregates and biofilms that have attracted special attention from limnologists interested in microbial biodiversity and the functioning of sediment microbiomes. In both cases the microbiome is composed of diverse species within a matrix of organic polymers.

Microbial aggregates occur in suspension and are transported via turbulent mixing and currents, as well as by sinking to the lakebed or riverbed. *Biofilms* range from the coating around mineral and organic particles in bottom sediments to thick phototrophic films and mats that occur over stable bottom substrates exposed to sunlight. Microbial biofilms also control biogeochemical transformations in the *hyporheic zone* of rivers, and in the organic-rich detritus at the surface of littoral and river sediments. Genomic approaches to the analysis of each of these microbiome types have revealed their complex community structure and multiple biological pathways of mass and energy flows (Grossart et al., 2020; see Chapter 23). These approaches also offer the opportunity to identify key intermediates that may be present in low concentrations, even below the limits of analytical detection, yet may play an important role as biogeochemical hubs that have fast turnover rates (Vigneron et al., 2021).

A. Microbial aggregates in suspension

Although microbial cells in the water columns of lakes and seas are mostly solitary and free-living cells (colonies or filaments for some taxa), they also occur in multispecies aggregates. In some environments, such as turbid rivers and the freshwater–saltwater transition zone of estuaries, the microbiomes associated with these suspended sediments may comprise a major fraction of total water column biomass and biological activity. The larger aggregates in the sea have been referred to as *marine snow* (Alldredge and Silver, 1988), and, by analogy, similarly large freshwater aggregates have been called *lake snow* (Grossart and Simon, 1993) and *river snow* (Böckelmann et al., 2000). These and the more common smaller particles (typically collected on a 3 μm pore-size filter) may be consumed by filter feeders or may ultimately sink and be consolidated within bottom sediments. The aggregates may include, in addition to microbial cells, mineral particles and various organic constituents including *transparent extracellular polymers* (TEPs). The latter are abundant gel particles, largely composed of polysaccharides, that can be visualized under the microscope by specific staining protocols. Microplastics also provide physical habitats for an attached microbiome (the *plastisphere*; Zettler et al., 2013) in the aquatic environment, and nanoplastics (Section IC) are an increasingly important component of the ensemble of particles in suspension (Gigault et al., 2021). Suspended aggregates have been considered a separate lifestyle for microbial functioning because the microenvironmental and other ecological conditions differ substantially from those experienced by free-living planktonic cells and those in biofilms attached to surfaces (Cai, 2020).

Microbial aggregates have much higher cell concentrations than free-living populations in the surrounding water. For example, in Lake Constance, Germany, suspended flocs and particles had cell concentrations that were about 100 times higher than the free-living cells in the lakewater (Grossart and Simon, 1993). The suspended sediment communities often contain many of the same species found in the free-living community, suggesting exchange between the two. However, particle communities may be enriched or depleted in certain taxa. For example, in the Shenzhen River-Bay system in south China, 16S rRNA gene amplicon sequencing of size-fractionated samples showed that the particles >3 μm in size had lower proportions of Betaproteobacteria than those in the free-living communities, but higher relative abundances of anaerobic archaeal taxa, including Bathyarchaeota, Methanobacteria, and Methanomicrobia. Modeling indicated that both deterministic and stochastic processes influenced the assembly of bacterial and archaeal communities, but that stochastic effects were more pronounced for bacteria (Wang et al., 2020). Bacterial aggregates in rivers, lakes, and the coastal ocean often have a markedly higher taxonomic richness than the free-living community (Bižic-Ionescu et al., 2014; Blais et al., 2022; Savio et al., 2015; Schmidt et al., 2020), which has been attributed to the greater diversity of microhabitats within such particles.

Biogeochemical modeling of sinking particles in marine environments has shown how different particle layers may have *microzones* that favor different microbes and processes (Bianchi et al., 2018). Oxygen, nitrate, and sulfate are sequentially used as electron accepters, with a central core of anoxic conditions in the middle of the particle caused by microbial oxygen consumption and the slow molecular diffusion of oxygen across the external boundary layer and outer particle layers. Bianchi et al., (2018) argue that the extent of anoxia in the ocean is more widespread than appreciated because of these inner particle microenvironments; this results in much higher rates of denitrification and other reductive processes compared with estimates based simply on the volume of fully anoxic water columns.

Studies on soil particles have shown the importance of redox gradients (Masue-Slowey et al., 2013) and similar zonation might be expected in lake and river particles, particularly in the larger size fractions. For example, ^{15}N-isotopic experiments on suspended sediments from the Yangtze and Yellow Rivers (China) showed that denitrification could take place even when the particles were surrounded by oxic waters, indicating microzones inside the particles that provide suitable redox conditions (Fig. 27-12). The resultant denitrifying activity would increase estimates of nitrogen losses via denitrification by 25% to 120% relative to estimates based on biogeochemical measurements of the riverbed sediments (Xia et al., 2017).

The *estuarine transition zone*, where rivers discharge into the sea, is an environment that provides an

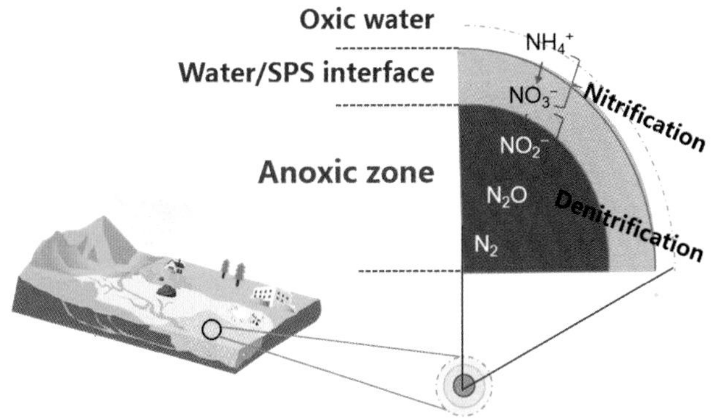

FIGURE 27-12 Microzones associated with suspended sediment (SPS) particles and the spatial partitioning of nitrification and denitrification. *(From Xia et al., 2017.)*

especially particle-rich habitat for microbial aggregates. Particles can be concentrated in this region by estuarine recirculation processes (Frenette et al., 1995) as well as by flocculation of organic-coated particles due to salinity effects (Mosley and Liss, 2020). Observations show that much of the microbial biomass and biological activity is associated with the particle fraction and that the species composition differs from the planktonic microbiome in the surrounding waters (e.g., Crump et al., 1999; Garneau et al., 2009; Herfort et al., 2017).

B. Photosynthetic biofilms and microbial mats

In many lakes and streams, the bottom sediments are coated with a cohesive layer of microbes that are embedded in a *polymeric gel* (Fig. 27-13). This mucilaginous structure provides a variety of microhabitats, and biofilm microbiomes are composed of diverse species in terms of function as well as taxonomy. Where light is available for photosynthesis, these communities are often dominated by Cyanobacteria along with eukaryotic phototrophs such as diatoms and green algae (collectively referred to as *periphyton*; Chapter 25). They range in thickness from membranous biofilms that are less than a few hundred micrometers thick to well-developed mats that are several millimeters or even centimeters in thickness, with multiple layers that differ in microbial composition and function.

The gel materials encasing the microbial populations are referred to as *extracellular polymeric substances* (EPS) or as an *extracellular matrix* (ECM) and are mostly composed of organic molecules exuded by the microbes. The dominant components in this matrix are long polysaccharide chains that act as the architectural framework for the mat (*scaffolding*), but there are numerous other constituents that are exuded by living cells or released by broken and lysed cells, including free DNA (*environmental DNA*), lipids, and proteins (Table 27-3). Given the importance of this matrix for the survival and growth of the microbes, and the many emergent properties of the resultant complex such as habitat structure and diversity, adhesion to surfaces, resource capture by sorption, enzyme retention, social interactions, and toxin resistance (Flemming et al., 2016), there is considerable interest in the exact biochemical composition of this *matrixome*, the ensemble of all molecules in the extracellular matrix (Karygianni et al., 2020).

Freeze-fracture microscopy of microbial mats shows that the ECM is an open structure with a large volume of water-filled space (de los Ríos et al., 2004), as indicated in Fig. 27-13. The interstitial water sampled from microbial mats may contain orders of magnitude higher concentrations of inorganic and organic solutes relative to bulk concentrations in the overlying freshwater (Villeneuve et al., 2001; Wood et al., 2015), indicating that these benthic microbes live in microhabitats that differ greatly in their chemical properties from the lake or river macroenvironment.

Some of the most luxuriant mats and films occur in extreme environments such as polar lakes and rivers (Vincent, 2000; Ramoneda et al., 2021) and hot spring environments (Ward et al., 1998; Prieto-Barajas et al., 2018),

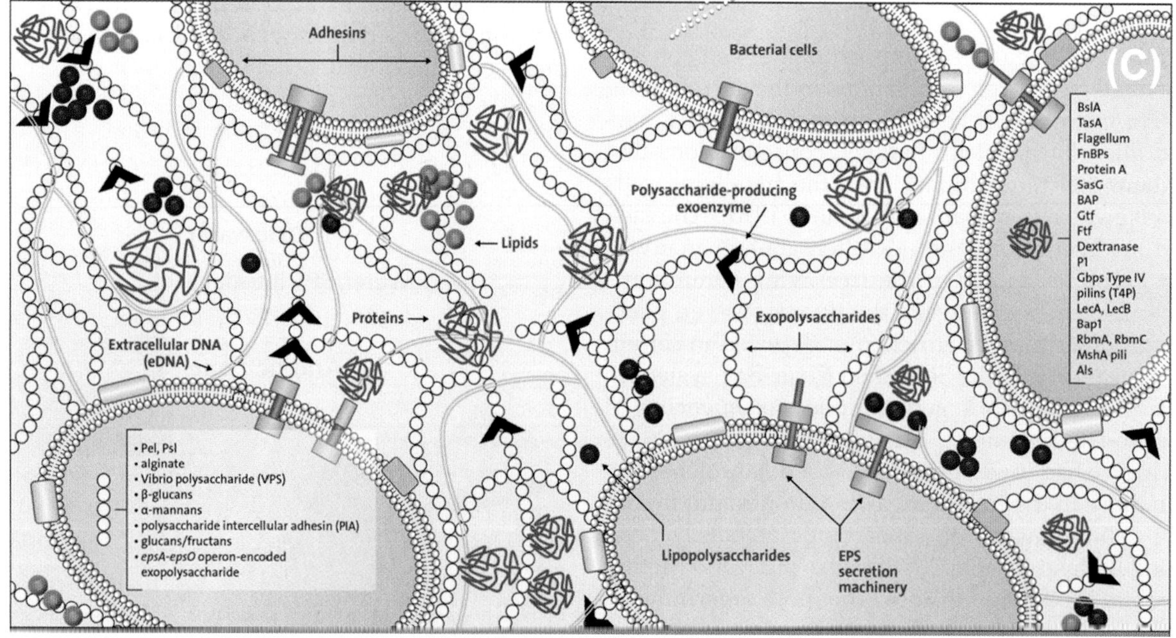

FIGURE 27-13 The complex biochemical composition of the extracellular matrix of a model bacterial biofilm. *(From Karygianni et al., 2020.)*

TABLE 27-3 Examples of Polymeric Substances and Biogenic Minerals Found in Microbial Biofilms and Mats

Compound	*Location*	*Function*
Polysaccharides		
Exopolysaccharides	Extracellular	Adhesion, scaffolding, stability, sorption
Glucans/fructans	Extracellular	Adhesion, scaffolding, stability, cell–cell binding
Mannans	Extracellular/ cellular	Mannan–glucan complexes, scaffolding, protection
Alginate	Extracellular	Adhesion, scaffolding, water/nutrient retention, protection
Proteins		
Surface-layer proteins	Extracellular	Surface hydrophobicity, protection
Biofilm-proteins	Extracellular	Adhesion, cell–cell binding, scaffolding
Flagella	Cell-associated	Motility, attachment, sensing
Lectins	Extracellular/ cellular	Adhesion, cell–cell binding, stability, cytotoxins
Enzymes	Extracellular	Hydrolysis, nutrient, and carbon mobilization
Nucleic acids		
eDNA	Extracellular	Scaffolding, adhesion, gene transfer, biochemical interactions
Lipids		
Glycerolipids	Cell walls, extracellular	Carbon source
Teichoic acid	Extracellular/ cellular	Adhesion, cohesion, protection
Lipopolysaccharides		
LPSs (endotoxins)	Cellular/ extracellular	Adhesion, colonization
Biogenic minerals		
Silica	Cellular/detritus	Cell walls (diatoms), physical substrates
Carbonates	Extracellular	Inorganic carbon chemistry, pH buffering, rigidity
Iron oxides	Extracellular	Redox chemistry, pigmentation

Source: From Karygianni et al. (2020) and other sources.

where grazing and other losses are muted, allowing large microbial biomasses to accumulate (Fig. 27-14). Metagenomic analysis of these communities has shown that they contain diverse taxa with a variety of mechanisms for capturing energy and carbon.

In a comparison of two aerobic, shallow-water lake communities in northern Canada, both microbial mats were up to several millimeters in thickness and dominated in terms of biomass by filamentous Cyanobacteria (Vigneron et al., 2018a). However, the metagenomes revealed numerous other taxa, especially Bacteroidetes, Planctomycetes, and Alphaproteobacteria, along with Acidobacteria; Chloroflexi; Beta-, Delta-, and Gammaproteobacteria; and Verrucomicrobia (Fig. 27-15). Most of the Deltaproteobacteria identified in the mats were related to bacterivorous nonsulfate-reducing lineages of Bdellovibrionales, NB1j, and Myxococcales. Eukaryotes accounted for 4.4% ± 0.6% of the rRNA gene reads and included diatoms, green algae, Cryptomycota (potential parasites), heterotrophic Pezizomycotina (fungi in the phylum Ascomycetes), ciliates and Cercozoa (heterotrophic protists), and metazoa (Nematoda, Rotifera, and Platyhelminthes). Viral sequences accounted for 0.017% ± 0.001% of total reads in the metagenomes, and more than 80% of these were tailed bacteriophages (Myoviridae, Podoviridae, and Siphoviridae). Based on closest affiliations, the main hosts of these phages would be Proteobacteria and Cyanobacteria, consistent with their predominance in the microbial mats.

In these high-latitude phototrophic mats, genes coding for cyanobacterial pigments and photosynthetic carbon fixation were well represented, as expected (Vigneron et al., 2018a). However, there was an unexpected degree of solar energy utilization via other pathways, with genes involved in bacteriochlorophyll synthesis, indicating the additional importance of aerobic anoxygenic phototrophy, and genes coding for microbial rhodopsins (including heliorhodopsin), distributed through several phyla. Organisms able to use light for energy-related processes accounted for up to 85% of the total microbial community, with 15% to 30% attributable to Cyanobacteria and 55% to 70% attributable to other bacteria. There was also genomic evidence of multiple pathways for inorganic carbon uptake and for the degradation of diverse organic carbon substrates. The picture that emerges is one of microbial richness and complexity, with energy and carbon flows sustained by solar energy as well as by intense recycling processes within the biofilm matrix (Fig. 27-15).

One of the most striking examples of mat microbiomes is found in ice-covered Lake Untersee in Antarctica, where Cyanobacteria form benthic mounds up to 70 cm high (Fig. 27-14). In terms of their morphology and internal layering, these microbial structures closely resemble the Precambrian *stromatolites* (layered rocks produced by microbial activity) that

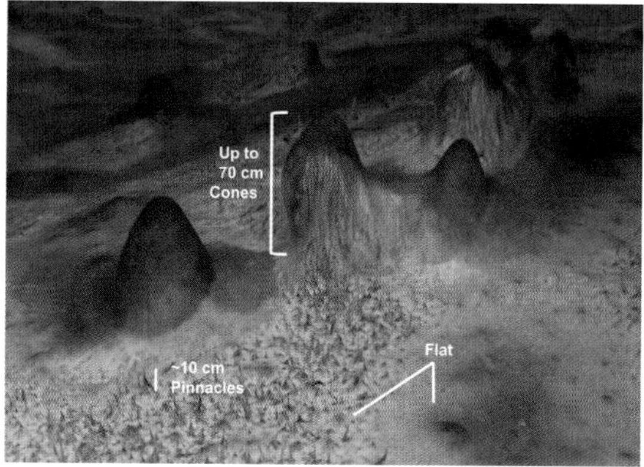

FIGURE 27-14 Cyanobacterial microbiomes in polar aquatic environments. *Left*: Microbial biofilm in a stream at Cape Discovery, Ellesmere Island, Canadian High Arctic. *(Photo credit: Warwick Vincent, from Vincent et al., 2011.)* *Right*: Benthic microbial mats on the bottom of ice-capped Lake Untersee, Antarctica. *(Photo credit: Dale Andersen, from Greco et al., 2020.)*

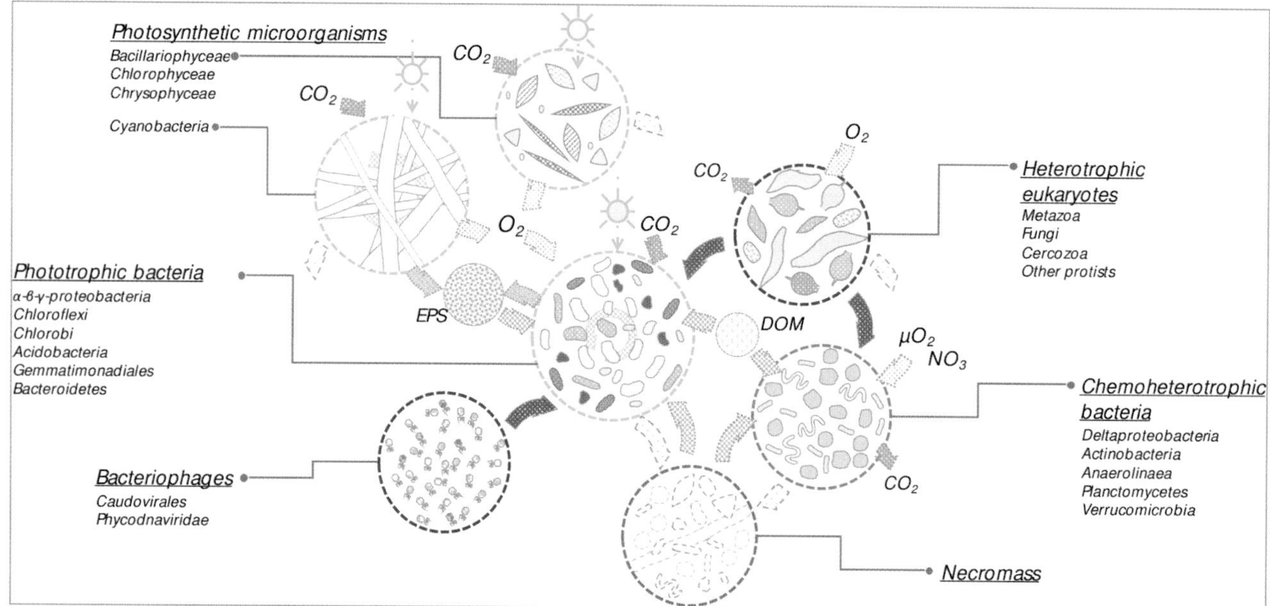

FIGURE 27-15 The complex microbiome of high-latitude microbial mats, with diverse microbial groups and multiple pathways of energy and carbon flow. *(From Vigneron et al., 2018a.)*

were widely distributed in shallow seas throughout much of the early history of life on Earth (Andersen et al., 2011). Analysis of ribosomal genes in these large microbial structures (Greco et al., 2020) showed that most prevalent bacterial phyla were Cyanobacteria (49.7% of the reads), Proteobacteria (23.6%), Verrucomicrobia (8.8%), Planctomycetes (4.9%), and Actinobacteria (4.2%). The 18S rRNA gene analyses indicated that the eukaryote assemblages were dominated by Ciliophora (31.7% of all reads), Chlorophyta (19.7%), Fungi (19.6%), and Cercozoa (15.2%). Unlike most cyanobacterial mats and films, including elsewhere in Antarctica, diatoms were conspicuously absent from the benthic mat communities of Lake Untersee, which has been attributed to its unusually alkaline pH, in the range of 9.8 to 10.6. Yet despite these extreme chemical conditions along with 3 m of perennial ice cover and several months each year of winter darkness, these thick benthic mats harbor large microbial populations of diverse species and functions.

Nitrogen-fixing Cyanobacteria are commonly found in autotrophic biofilms in lakes and rivers, including the taxa *Nostoc*, *Anabaena*, *Calothrix*, *Rivularia*, and *Scytonema*. Species of *Nostoc*, including the cosmopolitan morphospecies *N. commune*, form surface crusts and films in rivers and wetlands and are also common as membranous sheets and mucilaginous balls that lie loosely over the bottom of lakes and ponds. These range in size from a few hundred micrometers in diameter to larger, grape-like spheres. In some Greenland lakes spherical red-colored colonies of *Nostoc* up to 5 cm in diameter accumulate over the sediments and are referred to locally as "sea tomatoes" (Trout-Haney et al., 2021).

Comparison of *Nostoc* genomes has shown that there is a large genetic variation, with certain features common to all species including desiccation tolerance and a high production of diverse secondary metabolites (Jungblut et al., 2021). This study also drew attention to the presence and diversity of the viral defense system CRISPR-cas in *Nostoc*, and virus-host interactions may be a control on *Nostoc* cell populations and nutrient release. Many nonphotosynthetic bacteria occur in association with *Nostoc*. For example, the microbiome of a *Nostoc microbialite* (microbially produced mineral aggregates or structures; see Theisen et al., 2015) in Laguna Larga, a freshwater lagoon in southern Chile, contained more than 50 bacterial taxa including the sulfate reducers *Desulfomicrobium* and *Sulfospirillum* (Graham et al., 2014). Large spherical *Nostoc* colonies sampled in a lake and stream on the Andean plateau were found to have an internal microbiome of reduced species diversity, mostly Alpha- and Betaproteobacteria, relative to that on the outside of the colony, suggesting that the internal gel environment exerts a stronger selective pressure for species that would thrive in the higher nutrient conditions than the outside oligotrophic waters (Aguilar et al., 2019).

Biofilms dominated by the filamentous, nonnitrogen-fixing Cyanobacteria genus *Microcoleus* (including some taxa previously classified as *Phormidium*) are common on streambeds and riverbeds throughout the world. This phototroph and its associated microbiome of heterotrophs and other phototrophs can form mats up to several millimeters in thickness, even in waters that have low phosphorus concentrations, and some species are toxic (McAllister et al., 2016). A detailed genomic analysis of *Microcoleus* biofilms from the rock surfaces of a New Zealand river showed that they were dominated by Cyanobacteria and diatoms (also evident from the microscopic analysis, Fig. 27-16), with numerous associated bacterial taxa (Tee et al., 2020).

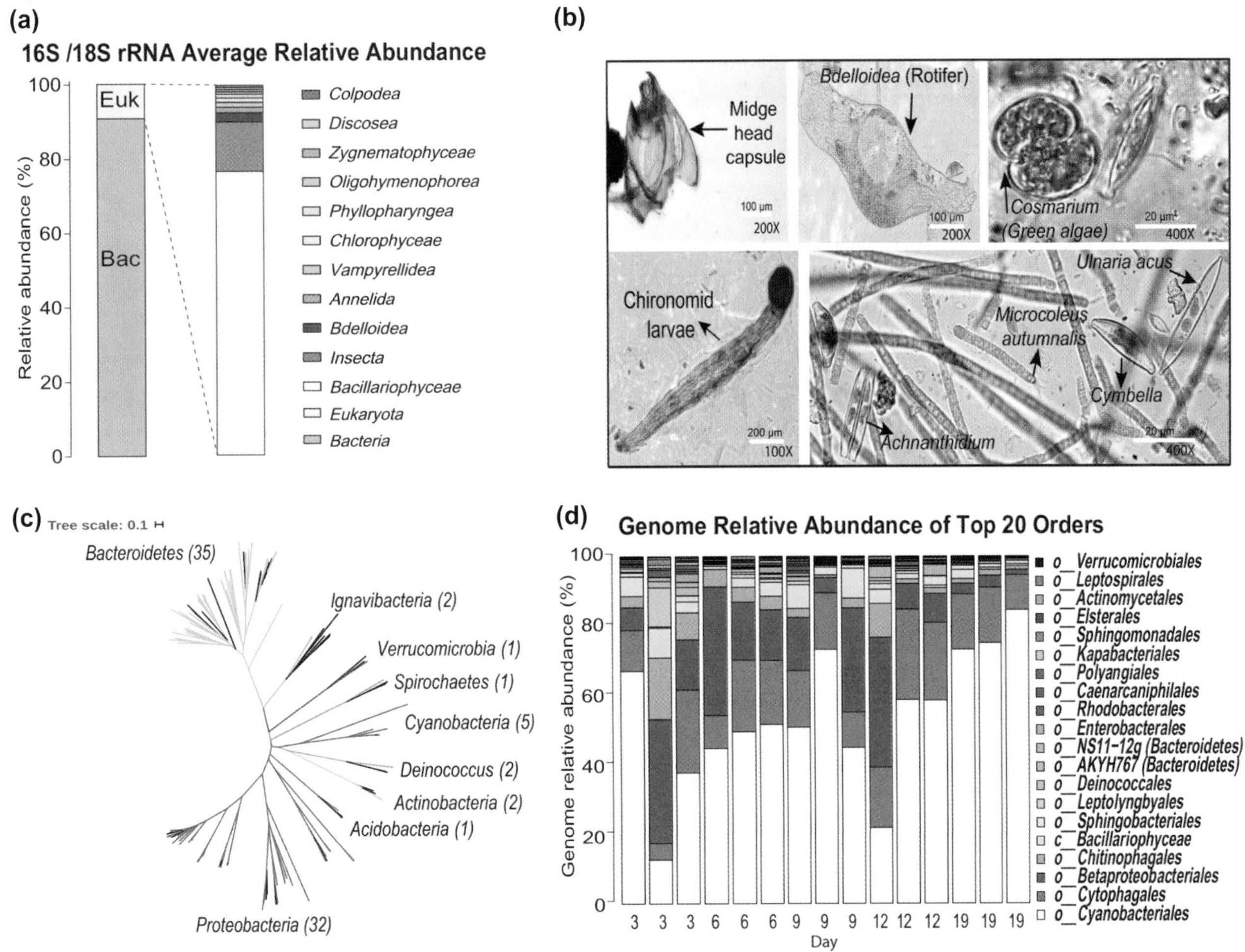

FIGURE 27-16 The *Microcoleus* microbiome. This microbial community, with its associated metazoans, produces mats and films over the riverbed. These analyses are for samples from the Wai-iti River, New Zealand. (a) Proportion of bacteria (Bac), eukaryotes (Euk), and the main eukaryote groups based on rRNA gene sequencing; (b) microscopy images of the community; (c) the phylogenetic distribution of 81 metagenome-assembled genomes (MAGs); and (d) changes through time in taxonomic relative abundance. *(From Tee et al., 2020.)*

The diversity of the latter was indicated by the analysis of 91 *metagenome-assembled genomes* (MAGs; whole genomes derived from the analysis of environmental DNA samples), which spanned eight different phyla in addition to Cyanobacteria (Fig. 27-16). These analyses showed that the heterotrophic bacteria had various functions for the breakdown of organic materials associated with the biofilms, including genes for cellulases, chitinases, hydrolases, proteases, and ribonucleases.

The *Microcoleus* populations had diverse mechanisms for nitrogen and phosphorus uptake and storage, allowing them to proliferate and outcompete other taxa (Tee et al., 2020). These mechanisms included nitrate and urea transporters; enzymes for the production and use of the nitrogen storage compound *cyanophycin*; transporters for inorganic phosphorus, organic phosphates, and *phosphonates* (organic phosphorus compounds containing the group $-CH_2-PO_3H_2$); mineralization enzymes; and enzymes for the production and use of polyphosphate storage molecules. The results suggested that the *Microcoleus* populations met at least part of their phosphorus requirements by the breakdown of organic phosphorus compounds within the biofilm and were therefore not entirely dependent on dissolved inorganic phosphorus from the overlying river water. Studies on another riverbed microbial mat community also dominated by filamentous Cyanobacteria found that dissolved reactive phosphorus levels inside the mat matrix were 320 times higher than in the river water. This was ascribed to sediment particle trapping by the mat and mobilization of phosphorus from the particles during the strong diurnal fluctuations in oxygen and pH within the mat caused by photosynthesis and respiration (Wood et al., 2015).

Thick biofilms can show large vertical differences in their microhabitat characteristics, species composition, and functioning (Fig. 27-17). Typically, there is a transition from bright-light communities in the surface layer, often rich in carotenoids and other photoprotective pigments, to dim light–adapted populations including Cyanobacteria with high cellular concentrations of light-capturing phycobiliproteins. These communities may be underlain in turn by anaerobic species including photosynthetic sulfur bacteria, denitrifiers, and sulfate reducers (Fig. 27-17). In some microbial mats, *methanogens* (see below) may also be present and seasonally active (Mohit et al., 2017). Metagenomic analysis of microbial mats in a highly stratified lake in Antarctica has shown how the diversity of genes changes vertically within the mat, but with large differences in the pattern

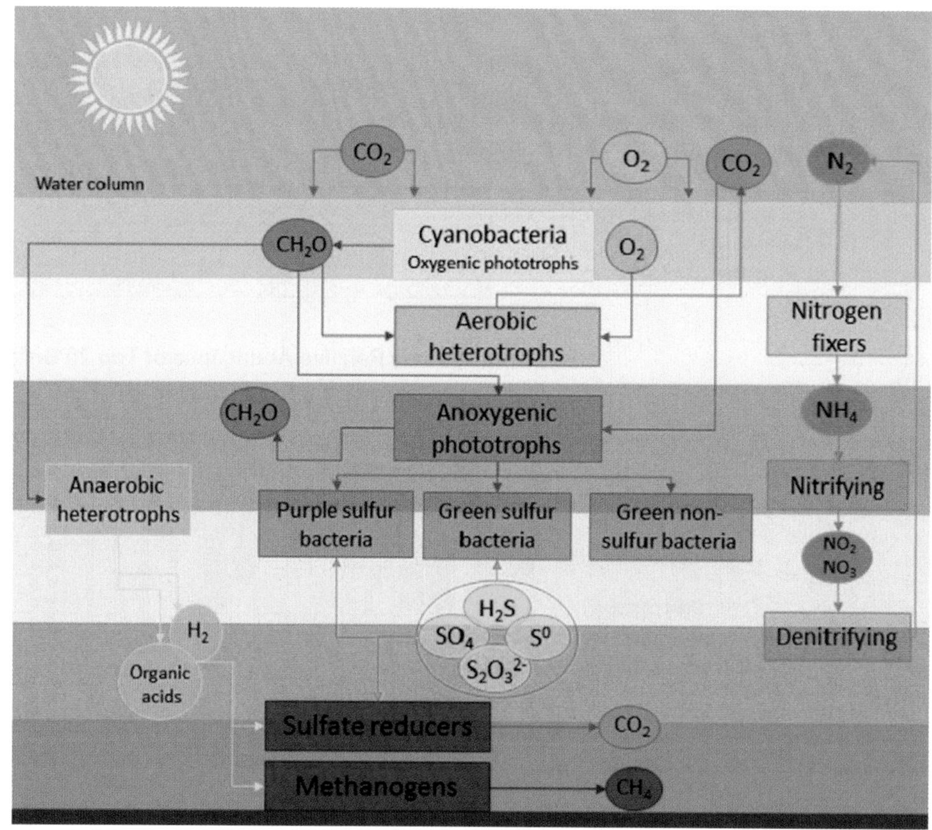

FIGURE 27-17 Potential layering of microbial mats in species composition and function, with aerobic surface layers and anaerobic underlayers. *(From Prieto-Barajas et al., 2018.)*

of change with depth and oxygen conditions in the lake (Dillon et al., 2020). Certain components of the microbiome may be motile and be able to swim or glide up and down the mat profile to optimize their use of light and nutrient resources while minimizing their exposure to damaging bright light and UV radiation. Mat phototrophs that exhibit this gliding behavior include filamentous Cyanobacteria (e.g., Quesada and Vincent, 1997) and pennate diatoms (e.g., Johnson et al., 1997).

Microelectrode studies have revealed the strong vertical gradients in the environmental properties of mats and films, which provide opportunities for diverse microbes to exploit different niches. These studies have also shown that there can be large changes through time in biofilm properties and gradients. Some of the most pronounced examples are in the microbial communities of geothermal springs, where high temperatures allow rapid physiological responses to changes in the ambient light regime. In Mushroom Spring, Yellowstone National Park, the 4-mm-thick mat microbiome has a low species diversity, likely restricted by the continuously hot temperatures (58°C), as well as by geochemical effects, that is composed of the unicellular cyanobacterium *Synechococcus*, several photoheterotrophic or mixotrophic *Chloroflexi*, photoheterotrophic acidobacteria (*Chloracidobacterium*), *Chlorobi*, and aerobic heterotrophic taxa (Klatt et al., 2011). Over the 24-h cycle, the oxygen profile in the mat shifted from near-surface aerobic conditions associated with photosynthetic oxygen production during the day to respiration and complete anoxia during the night (Revsbech et al., 2016). This was accompanied by a remarkable 3-unit shift in pH units in the mat, from pH 9 associated with intense photosynthesis during the day, dropping to pH 6 or below during the nighttime period when CO_2 production by respiration dominated (Fig. 27-18).

C. The hyporheic zone of rivers

The term *hyporheos* was first coined by Traian Orghidan in 1955 from the Greek roots *hypo-* (under) and *-rheos* (flow) as an "effort to convince skeptical minds that the near-stream alluvium encloses a rich and unique ecosystem" (Käser, 2010). The application of molecular and biogeochemical techniques has confirmed the distinct biological richness and importance of the hyporheic microbiome.

The *hyporheic zone*, sometimes referred to as a *transient storage zone* (TSZ; Cardenas, 2015), refers to the uppermost stratum of a riverbed in which there is mixing of river water and groundwater (Fig. 27-19). This zone of slowly flowing water is absent from streams flowing over rock or compacted clays, but it typically extends to a depth of several centimeters or more, depending on the porosity of the substrate, and often also extends out laterally. In hydrological terms *hyporheic flow* is defined as the transport of river water through sediments in flow paths that return to the surface, with a 10% contribution of river water considered the lower bound of the hyporheic zone. This mixture of flowing waters with different geochemical and microbiological properties means that the hyporheic zone is a microbial habitat of high biological diversity and diverse biogeochemical functions, including oxidation of organic carbon, mineralization of nutrients, denitrification, metal oxide precipitation, and bioremediation, with the potential to substantially modify river water chemistry (Boano et al., 2014). Oxygen concentrations are often below saturation (Käser, 2010), reflecting these intense microbial activities as well as the prolonged *water residence time* (how long the water remains in the hyporheic zone before returning to the stream channel).

As for lake sediments (Section I), the analysis of sediment pore water in the hyporheos may show strong

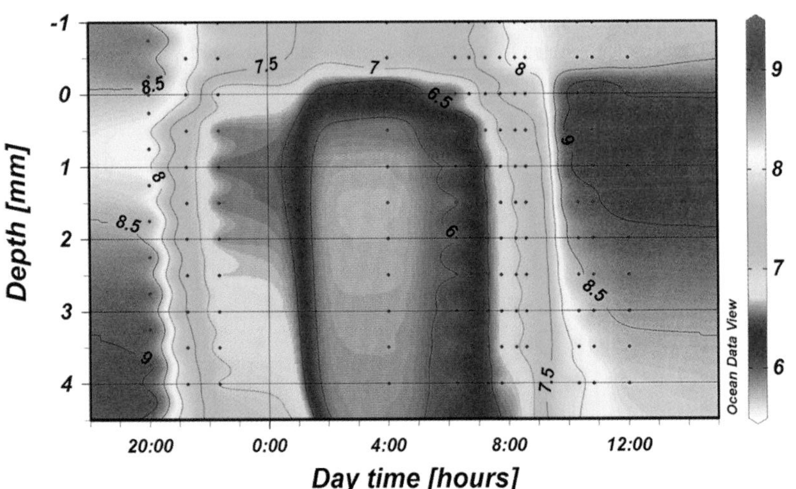

FIGURE 27-18 Variations in pH over the 24-hour diel cycle in a hot spring microbial mat in Yellowstone National Park. *(From Revsbech et al., 2016.)*

gradients with depth, depending on the residence time and extent of depletion of oxygen and other oxidants.

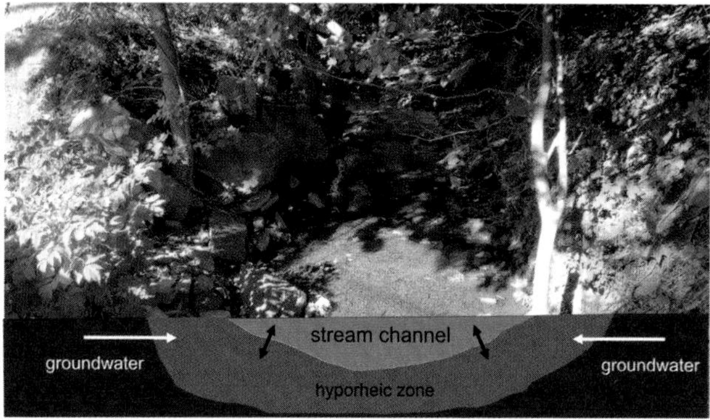

FIGURE 27-19 The hyporheic zone, which receives groundwater and exchanges water vertically and laterally with the stream channel. The hyporheic microbiome plays a major role in stream biogeochemistry. *(Photo credit: Warwick Vincent; based on a conceptual design by Karen Jackson, Clemson University.)*

For example, in the sandy bed of the Nasseys River in France downstream of an industrial waste discharge point, the pore water profiles showed evidence of low redox potentials at depth, with depletion of sulfate and nitrate; increases in dissolved iron, manganese, and inorganic phosphorus; and an intermediate zone of nitrite accumulation implying intense nitrification at the interface between ammonium supply from decomposition processes below and oxygen brought in by the hyporheic flow from above (Fig. 27-20). Residence time is a key variable controlling the effect of microbial and geochemical reactions on water chemistry, and Gooseff, (2010) suggests that this aspect could be incorporated into spatial definitions of the hyporheos, in recognition of the continuum of flow paths and associated residence times.

Battin et al., (2016) have described the microbiomes coating streambeds and associated hyporheic environments as a *microbial skin* that interacts intensely with the water flowing downstream via surface and

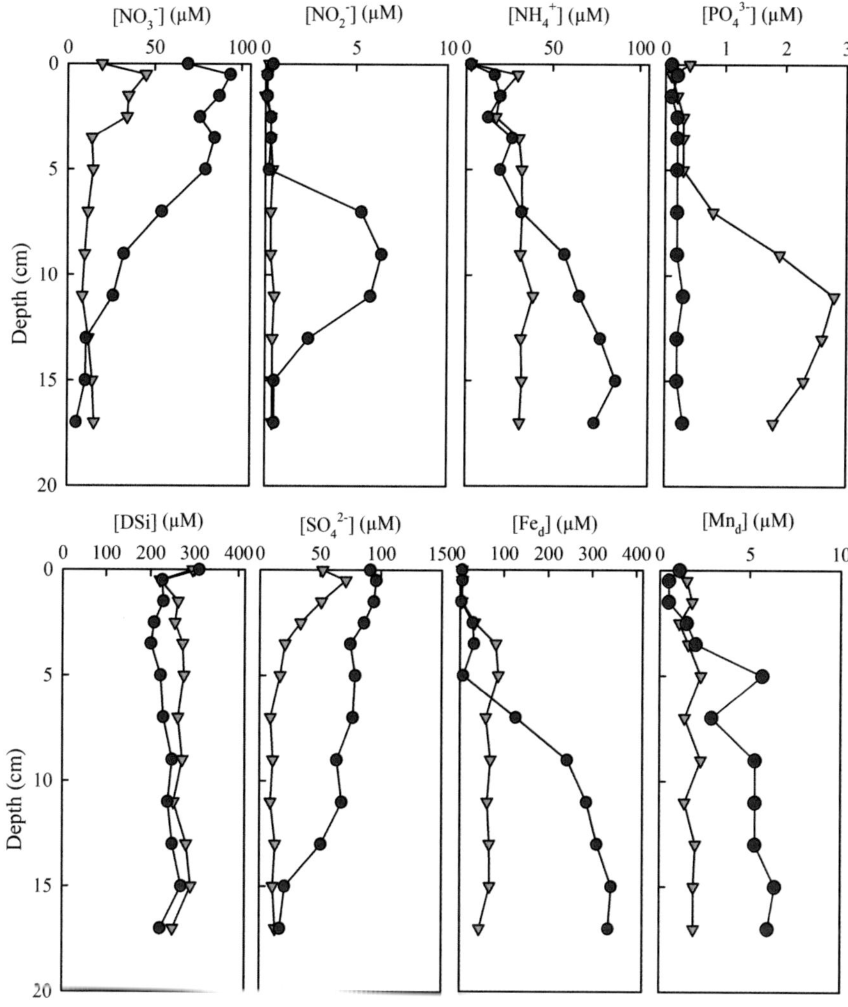

FIGURE 27-20 Solute concentrations in the interstitial waters of a riverbed above (orange triangles) and below an industrial discharge (blue circles). *(From Anschutz and Charbonnier, 2021.)*

subsurface pathways. Their calculations draw attention to the vast collective surface area and microbial cell abundance in these sediment habitats at a global level: potentially up to 5 m^2 of sediment surface area available for microbial colonization for each square meter of catchment land surface area. Assuming up to 10^9 microbial cells per square centimeter of sediment surface, this would translate to 50 trillion cells in the streambed to receive and process matter from each square meter of catchment land surface area. The bacterial communities are embedded within protective biofilm matrices, typically dominated by Bacteriodetes and Proteobacteria. Two classes of Bacteriodetes, Flavobacteria and Sphingobacteria, appear to be important in stream biofilms and include taxa that degrade biopolymers such as cellulose and chitin. There may be pockets or strata with different microhabitat properties and microbiome compositions.

Rivers affected by organic pollutants are likely to have a much higher proportion of anaerobic taxa and microaerophilic species that depend on reduced compounds. For example, a metagenomic study of sediments in the Ganges and Yamuna rivers, in India, showed that the ammonium oxidizer *Nitrosopumilus* accounted for around 60% of the archaeal reads and occurred along with methanogens, sulfate reducers, and iron-reducing bacteria (Samson et al., 2019). Analysis of genomic diversity and metabolic restrictions or bottlenecks in the hyporheos offers new opportunities to connect genes to the ecosystem and thereby establish a better understanding of river biogeochemistry (Nelson et al., 2020).

Given its role in carbon, nutrient, and contaminant processing, there is increasing attention directed toward the hyporheic zone in stream restoration plans. Studies in an urban floodplain restoration in the city of Seattle, USA, showed no significant change in the microbiome structure of suspended and particle-associated microbial communities between stream reaches with and without hyporheic restoration, but dissolved organic carbon and microbial metabolism were higher, along with higher hyporheic invertebrate density and taxonomic richness (Morley et al., 2021). This benthic region of the river deserves special attention for the ecosystem service role (Chapter 2) that it can play in reducing human impacts (Lewandowski et al., 2019), as well as in the biogeochemical structure, connectivity, and functioning of hydrological networks from upland streams to the sea (Battin et al., 2008).

C. Lake sediment microbiomes

Bacterial profiles in lakes typically follow the distribution of organic carbon availability (Fig. 27-21). The highest densities occur at the surface of the lakebed, with cell concentrations rising from 10^5 to 10^7 cells mL^{-1} in the overlying water column to up to 10^9 cells mL^{-1} (or g^{-1}) in the sediments. Within the sediments, there is typically an exponential decline with depth, reflecting the loss of oxygen for aerobic processes, and the increasing loss of labile organic carbon substrates with increasing duration of diagenesis down the sediment profile. As for the microbial aggregates and phototrophic biofilms, genomic analysis is revealing a complex microbiome with numerous taxa, and sharp gradients in functional as well as taxonomic composition that control the patterns of interstitial solute concentrations (Section IB). The microbial communities coat the sediment particles and so, at the microscopic level, may be considered thin biofilms over particle surfaces in which there are strong interactions between the cells, the particles, and the aqueous and gel microenvironments in the interstitial spaces between the sediment particles. These microenvironments are likely to be much more heterogeneous than analysis of solute concentrations suggests, and they are conducive to a wide diversity of potential niches and microbial interactions, from *syntrophy* (biogeochemical interdependence) and *mutualism* (coexistence with positive interactions between species) to *competitive inhibition* and *parasitism*.

A comprehensive study of microbial community composition in the sediments of oligomesotrophic Lake Stechlin in northern Germany indicated three distinct zones that differed in terms of microbiome structure (Wurzbacher et al., 2017). The amplicon

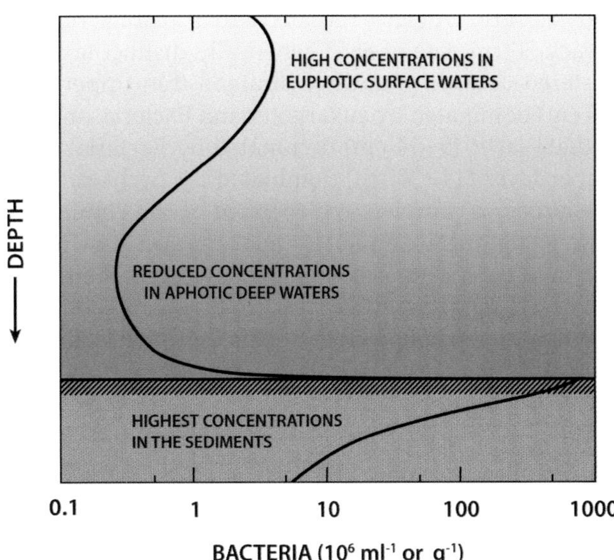

FIGURE 27-21 General distribution of bacterial cell concentrations in the water and sediments of a lake, as measured by fluorescent staining and counts. The depth scale is in meters for the water column and centimeters for the sediments.

analysis of rRNA genes revealed a total average richness in each 30-cm core of 8545 genetically distinct taxa that clustered according to three strata: a thin upper layer (<5 cm) dominated by eukaryotes and Bacteria, an intermediate layer (5–14 cm) dominated by Bacteria, and a deeper layer (14–30 cm) dominated by Archaea. There was almost a complete replacement of the community between the surface and bottom layers, and these microbial gradients were accompanied by large changes in other sediment properties with depth. Dissolved refractory organic carbon decreased exponentially with sediment depth, as did total DNA concentration, with an inferred half-life of 22 years. For the 30-cm cores, Bacteria plus Archaea cell concentrations averaged $1.8 \pm 0.5 \times 10^9$ cells mL^{-1} of wet sediment, with highest concentrations in the surface layer where heterotrophic production rates were also highest (up to 282 μg C mL^{-1} day^{-1}) declining to near-zero rates below 10 cm. Oxygen penetrated to a mean depth of 4.6 ± 1.4 mm, with depletion of nitrate and nitrite at the sediment surface, reduction of sulfate to a minimum plateau below 5 cm, and an increase in soluble reactive phosphorus, ammonium, and CH_4 with depth. As might be expected, the taxonomic diversity for these microbial communities was highest in the redox-stratified surface 5 cm layer and then declined with depth.

The abundance of microbial cells with sediment depth can be estimated by direct cell counts, for example, by *fluorescence microscopy*, or by *quantitative polymerase chain reaction* (qPCR) analysis that targets specific gene sequences and thereby allows phylogenetic groups to be differentiated and quantified. For example, in a comparative analysis of Swiss lake sediments across a trophic gradient, from oligotrophic to eutrophic waters, there was a consistent pattern of rapidly decreasing bacterial abundance with depth in the sediments (Fig. 27-22; Han et al., 2020), while Archaea increased with depth. This resulted in an order of magnitude decrease in the ratio of Bacteria to Archaea, thus increasing the relative importance of the archaeal community with depth in the sediment. This effect is analogous to organic-rich lakes that freeze over and lose their oxygen in winter, with an associated 15-fold increase in the abundance of Archaea, dominated by methanogens (Vigneron et al., 2019).

In all of the Swiss lake sediments the phylogenetic richness and diversity of Bacteria was greater than for Archaea. There was an overall correlation between bacterial gene copies and total organic carbon, and also between archaeal gene copies and total organic carbon. Consistent with the trend of decreasing concentrations of labile carbon down the sediment profile, in all lakes there was a decrease in short-chain *n*-alkanes, fatty acids, and algal steroids with depth in the sediment. In contrast, the contribution of more refractory materials from higher plants, the ratio of long-chain alkanes, and lignin phenols either increased or showed no clear pattern with depth.

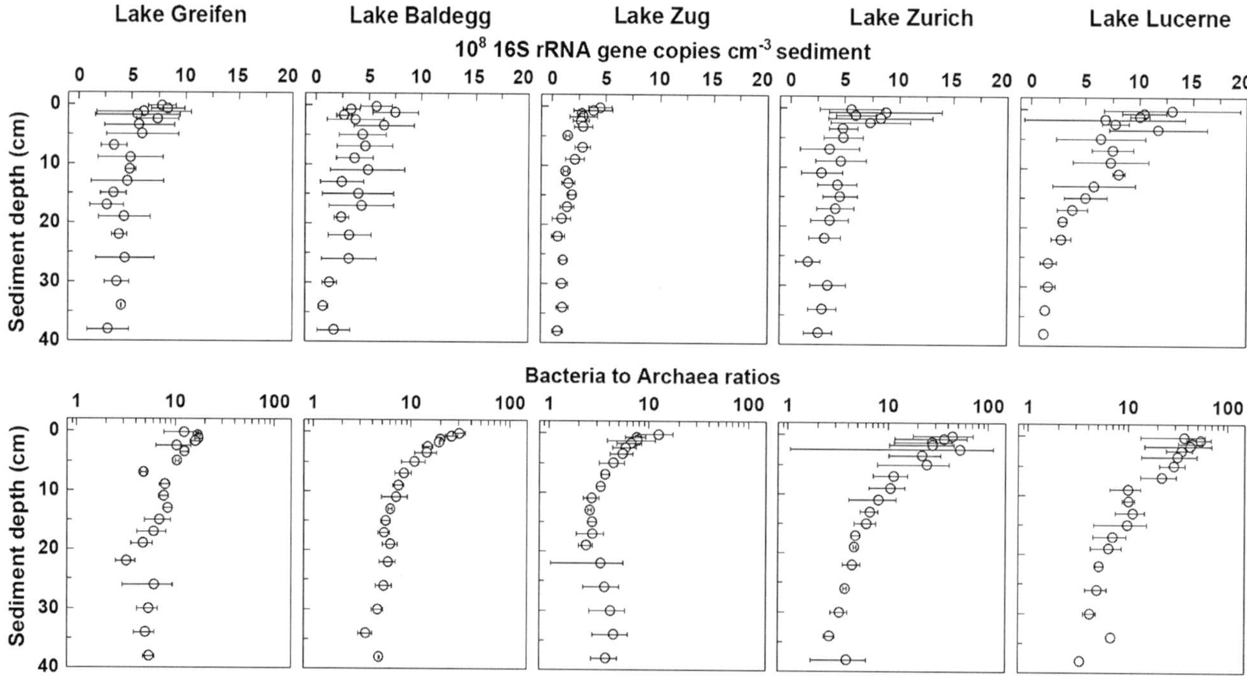

FIGURE 27-22 The distribution of Bacteria (upper panels) and the ratio of Bacteria to Archaea in the top 40 cm of sediments in five Swiss lakes covering a range of trophic states. The values were obtained by qPCR of 16S rRNA genes. *(From Han et al., 2020.)*

Despite the differences in trophic status, there were strong similarities at the class and phylum levels in microbial composition and depth trends in the sediments of the five Swiss lakes. Specifically, there was a decline in Beta- and Gammaproteobacteria, Bacteroidetes, Verrucomicrobia, Cyanobacteria, and Acidobacteria, accompanied by an increase in Deltaproteobacteria, Chloroflexi, Firmicutes, Planctomycetes, Actinobacteria, Armatimonadetes, Acetothermia, and Aminicenantes with sediment depth at almost all sites. For the Archaea, there were overall decreases in Pacearchaeota and Woesearchaeota with depth, and increases in Thermoplasmata, Altiarchaeales, Diapherotrites, and Lokiarchaeota. There was evidence of differences among trophic states, for example, with higher abundance of nitrifying bacteria in an oligotrophic lake, and differences at the order level for Bacteria between oligotrophic and eutrophic sediments (Han et al., 2020).

Deep drilling of marine sediments has revealed the existence of potentially active microbial communities to sediment depths in excess of 2 km in the subseafloor, the so-called *deep biosphere*. There have been few equivalent studies in deep inland water sediments, but the evidence to date suggests that viable microbial communities can occur to great depth. In Lake Van, a large, deep soda lake in Turkey (alkalinity 155 meq L^{-1}, pH 9.81, salinity 21.4‰), *varved* subfloor sediments (annual bands caused by mineral precipitation and primary production) showed a gradual decline but continuous persistence of bacteria throughout the sediment profile, from around 10^6 cells per milliliter in the surficial sediments to 10^3 to 10^4 cells per milliliter at 100-m depth, and large differences among microlayers (Kallmeyer et al., 2015).

There can be large spatial differences in microbial population size and community structure within lakes, and especially between the littoral and pelagic zones. The inshore region is characterized by vascular plants growing in the water (*aquatic macrophytes*; Chapter 24), as well as shoreline and stream inputs of vascular plant debris from terrestrial sources. These carbon sources differ from the sinking phytoplankton cells that accumulate in offshore sediments and include complex structural polymers such as *lignin* along with lower-molecular-weight organic molecules. Aquatic hyphomycetes are well known for their ability to decompose terrestrial plant leaves (Suberkropp and Klug, 1980) and thereby provide a food source for macroinvertebrates (Bärlocher, 1985; Marks, 2019), while ascomycetes can slowly degrade woody debris (Jones et al., 2014; Sridhar, 2020). Both fungal groups play important roles in lake sediments as well as streambeds (Grossart et al., 2019). There is evidence that many of the fungal taxa that decompose dead leaves in streams may enter the water with the leaves, while bacterial colonization of the leaves largely takes place within the water (Hayer et al., 2022).

High methane efflux rates are often observed in the littoral zone, which in part reflects the abundant carbon sources. In a littoral mesocosm experiment with artificial sediments amended with leaf litter (Yakimovich et al., 2020), moderate levels of enrichment by terrestrial organic carbon resulted in higher methane pore water concentrations. This effect was greater in sunlight-exposed incubations in an oligotrophic lake, potentially due to photochemical processes that are known to break down complex organic materials into more biodegradable products (e.g., Laurion and Mladenov, 2013; Bowen et al., 2020). Molecular analyses of the littoral mesocosm sediments showed that five orders of methanogens were represented, with greater than 90% dominance by two taxa in the Methanobacteriales, carbon dioxide–reducing methanogens that use hydrogen or formate as the reducing agent (see the following section).

V. Microbiome processes

A. Gas bubbles and fluxes

One of the most obvious indicators of the intense microbiological activity that takes place at the bottom of lakes and rivers is the production of gases. Some of these gases are conspicuous by their smell, notably hydrogen sulfide derived from sulfate reduction in anoxic sediments (e.g., in urban canals; Jantharadej et al., 2021). Most noticeable, however, is the process of *ebullition*, the production of gas bubbles in the sediment that then rise up through the water column and release their contents to the atmosphere.

The bubbles arriving at the lake or river surface are usually enriched in methane (see also Chapter 28), which reflects the combined effect of high rates of biological methane production (*methanogenesis*; Section VD) in the sediments and the low solubility of this gas in water (1.44 mmol L^{-1} at 20°C and 1 atm; Kaye and Laby, 1986). Microbial fermentation processes in the sediments may also produce large quantities of carbon dioxide; however, this gas is highly soluble (38.41 mmol L^{-1}) relative to methane. It is therefore less prone to bubble formation or diffusion from water into the methane bubbles, and carbon dioxide emissions from lakes are typically dominated by diffusive fluxes. For example, in a study of 32 reservoirs ebullition accounted on average for only 0.63% of total CO_2 emissions, but more than 50% of the total CH_4 emissions in 75% of the reservoirs (Beaulieu et al., 2020). Similarly, nitrous oxide is fairly soluble (23.71 mmol L^{-1}), and transport of this powerful greenhouse gas by ebullition to the atmosphere tends to be small relative to that by diffusion.

In four temperate latitude streams passing through farmland, nitrous oxide ebullition was less than 0.1% of diffusive losses of this gas, while ebullition contributed 20% to 67% of the methane emissions (Baulch et al., 2011).

Gas bubbles can often be observed associated with actively photosynthesizing benthic plants and biofilms and are likely due to oxygen production. These bubbles may create sufficient buoyancy to cause detachment and flotation of some phototrophic communities (Parker et al., 1982; Howard et al., 2018). Oxygen is poorly soluble (1.34 mmol L^{-1}), and methane bubbles passing up through the water column can scavenge dissolved oxygen and cause oxygen losses to the atmosphere, even when the oxygen concentrations in the lakewater are below air-equilibrium values (Koschorreck et al. 2017). Methane bubbles can also scavenge other gases including nitrogen (produced by denitrification as well as entering from the atmosphere), argon, mercury, hydrogen, and carbon monoxide (e.g., Poissant et al., 2007). For two shallow eutrophic reservoirs, the mean composition of gas bubbles was 63 ± 17% N_2 + Ar; 19 ± 23% CH_4; 17 ± 10% O_2; and 0.3 ± 0.5% CO_2, reflecting the combination of sediment microbial activity, solubility, and scavenging effects (Koschorreck et al. 2017). Conversely, in less methane-rich waters some of the methane in bubbles can diffuse out into the surrounding lakewater during their passage through the water column (McGinnis et al., 2006).

Ebullition is often the dominant pathway of methane release to the atmosphere from lakes and rivers, with the highest effluxes measured in rivers (especially high over organic permafrost), river impoundments (reservoirs), and aquaculture ponds (Table 27-4). Yang et al., (2020) note that China has 25,700 km^2 of aquaculture ponds (23% of the global total), and the cumulative CH_4 production from these waters is therefore considerable. Ebullition can also be high in water bodies on degrading permafrost (thermokarst lakes and ponds), with ebullition hotspots apparent from bubbles trapped in the ice over winter (Fig. 27-23; Anthony Walter et al. 2021).

Within any water body, there are large spatial as well as temporal variations in ebullition rates (DelSontro et al., 2015, 2018; Wik et al., 2016, 2018), and it is difficult to obtain accurate lake-wide estimates of methane fluxes. In general, bubble fluxes are higher in ponds and in the shallow inshore waters of lakes where organic matter concentrations in the sediments are higher, temperatures are warmer, and the hydrostatic pressure is less (see Section IV). The proportion of nitrogen in sampled bubbles can be used as an index of diffusive versus ebullitive CH_4 fluxes and as a way to identify ebullition hotspots; faster methanogenesis rates will cause greater scavenging depletion of nitrogen in the sediments (*nitrogen stripping*) by the methane bubbles, and the resultant bubbles will be depleted in nitrogen relative to methane. Langenegger et al., (2019) have shown that combining this information with other key observations such as depth of sediment-bubble formation can lead to solution of the complete mass balance of CH_4 and flux pathways.

TABLE 27-4 Examples of Ebullition Flux Rates for Methane from Different Aquatic Environments and Percent Contribution to the Total Diffuse plus Ebullition Methane Flux. The Observations are Ordered from Highest Maximum Rates to Lowest

Site	*CH_4 ebullition*		*Source*
	mmol m^{-2} day^{-1}	*% total*	
Tibet permafrost rivers[a]	12, 734	>95	Zhang et al., (2020)
USA reservoirs[b]	1.4, 233	>50	Beaulieu et al., (2020)
Boreal rivers[c]	4.4, 151	72	Campeau et al., (2014)
River impoundments	46–120	–	Wilkinson et al., (2015)
Tibet peatland lake	54 ± 29	74	Zhu et al., (2016)
Aquaculture ponds, China[d]	32.6, 12.0	>90	Yang et al., (2020)
Urban lakes, China	11 ± 30	80	Wang et al., (2021)
Pond sediments[e]	10.1, 7.1	95, 90	Aben et al., (2017)
Amazon floodplain	1.6–9.8	20–53	Bartlett et al., (1990)
Boreal waters, Quebec[f]	1.2, 4.6	56	DelSontro et al., (2016)
Permafrost lakes[g]	0.7, 4.5	50, 59	Sepulveda-Jauregui et al., (2015)
Tundra ponds[h]	2.4	76	Prėskienis et al., (2021)
Kettle lake, Switzerland	0.8–2.1	10–16	Langenegger et al., (2019)
Subarctic lakes, Sweden[i]	1.3	66	Wik et al., (2018)
Tropical reservoir[j]	1.1	99	DelSontro et al., (2011)
Peatland thermokarst lakes	<0.1–0.8	<10	Matveev et al., (2016)
Lake Taihu, China[k]	0.17 ± 0.05	64	Xiao et al., (2017)

[a]*Mean and maximum for streams and rivers across organic-rich permafrost on the Tibet Plateau;*
[b]*median and maximum for 536 reservoirs;*
[c]*mean and maximum;*
[d]*annual means for two consecutive years;*
[e]*peak rates for indoor mesocosms with and without +4°C warming, respectively;*
[f]*means for 3 lakes (inshore) and 10 shallow ponds, respectively, in the boreal zone;*
[g]*averages for 32 non-Yedoma and 8 Yedoma permafrost lakes, respectively;*
[h]*average for 30 ponds;*
[i]*average flux for three lakes during the ice-free season;*
[j]*from annual mass balance calculations for a 2500 km^2 subbasin;*
[k]*annual average for the whole lake.*

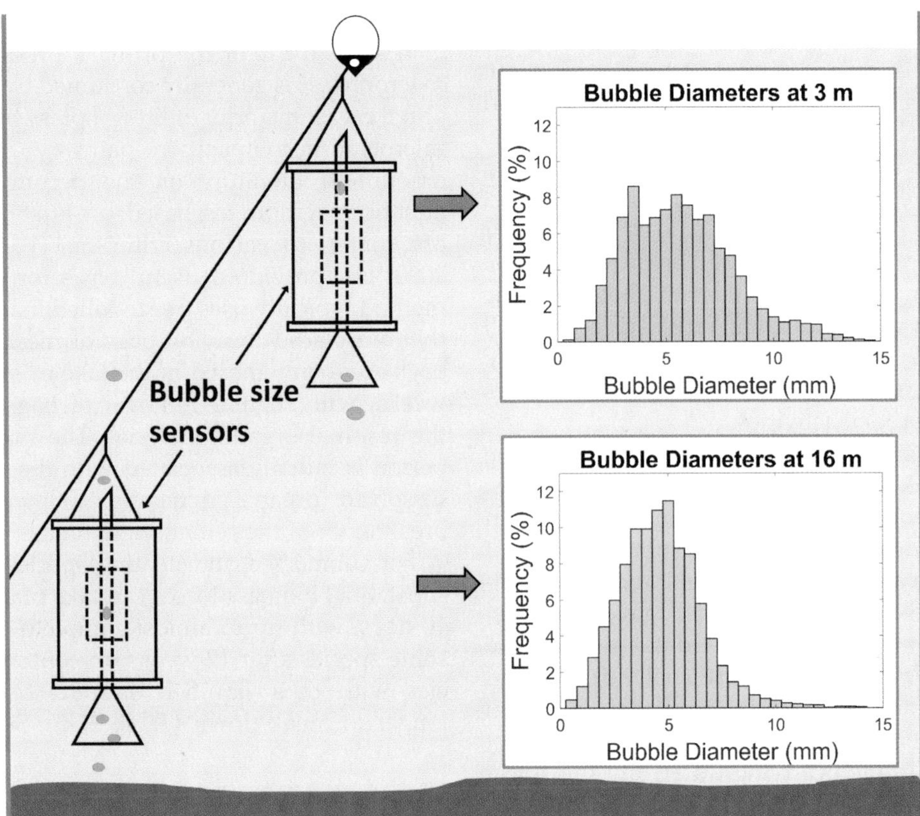

FIGURE 27-23 Methane bubbles. Above: Bubbles in lake ice, and holes in the ice caused by intense bubbling "hotspots" Walter Anthony et al., 2021. Below: Use of submerged, in situ cameras to determine the size distribution of bubbles released from the sediments of a eutrophic lake (Delwiche and Hemond, 2017).

In general, larger bubbles contain more methane and less nitrogen and indicate regions of the highest rates of sediment methanogenesis. These biogeochemical hotspots can be detected by bubble trains captured in lake ice, and at the most intense sites (for example, over *taliks*, unfrozen deep sediments in thermokarst lakes), the ebullition may even inhibit freeze-up (Fig. 27-23; Walter Anthony et al., 2021). Automated *in situ* cameras have been used to measure the size of bubble trains released from lake sediments, and the resultant image analyses show that bubble size is highly variable. In a set of automated measurements at eight sites in Upper Mystic Lake, a eutrophic water body near Boston, USA, the bubbles varied in size from less than 1 to more than 10 mm, with averages in the range of 4.5 to 8.1 mm (Fig. 27-23). There were large variations among sites, indicating the importance of local sediment conditions, and in general, the highest total ebullition fluxes correlated with average bubble size (Delwiche and Hemond, 2017). Bubble fluxes have also been shown to increase as an Arrhenius-type function of sediment temperatures, with high *Q_{10} values* (factor of increase in rates induced by warming of 10°C) indicating strong thermal control of methane production rates (Wik et al., 2014).

B. Oxygen conditions and fluxes

The sediment microbiome exerts a strong effect on oxygen levels in the benthic boundary layer by oxygen production in photosynthetic biofilms (Section IVB) and by microbial respiration due to the decomposition of organic matter in surface sediments. Phototrophic biofilms on river and stream beds can give rise to marked diurnal fluctuations in oxygen saturation, for example, up to 110% air-equilibrium in the subalpine Ybbs River (Austria) falling to 95% at night (Segatto et al., 2020). Phototrophic biofilms in lakes can be supersaturated with oxygen during the day, but there is a strong downward flux of oxygen to subsurface heterotrophic strata, and oxygen concentrations throughout the biofilms fall below saturation at night (Carlton and Wetzel, 1987). Littoral primary production by such biofilms may be sufficient to drive even humic lakes toward net autotrophy (Vesterinen et al., 2016), but the net flux of oxygen to the overlying water may be minimal or negative because of respiration in and below the biofilm. In many stream and lake sediments with photosynthetic communities, oxygen production rates are often well below the rates of CO_2 efflux, indicating the importance of allochthonous carbon and alternate electron acceptors in supporting microbial activities.

Microbial respiration in lake sediments is often the main oxygen sink in hypolimnia and shallow lakes, and it therefore requires special attention in lake oxygen models. For example, an oxygen consumption rate of 250 mg O_2 m^{-2} day^{-1} at the lower boundary of oxic sediments and zero-order loss kinetics gave the best modeling fit to microelectrode data across the sediment–water interface of two artificially oxygenated water bodies (Man et al., 2020). Similarly, microbial respiration in river sediments can exert a strong effect on oxygen levels in the overlying water, and this effect is especially pronounced in response to organic effluents. In the Ganges River (Ganga in Hindustani), which passes through India and Bangladesh and is the third largest river by discharge in the world, severe pollution has resulted in extreme sediment oxygen demand rates of 2000 to 16,800 mg O_2 m^{-2} day^{-1}; at some locations, this draws down oxygen at the sediment–water interface to anoxic and hypoxic (<2 mg O_2 L^{-1}) conditions and results in high rates of sediment phosphorus release (Jaiswal and Pandey, 2019).

In the littoral zone of lakes macrophytes can have a range of effects on sediment oxygen dynamics. Tall beds of aquatic plants restrict wind-induced mixing and currents and can lead to thermal stratification and sediment anoxia within dense macrophyte stands (Vilas et al., 2017). Conversely, during periods of active photosynthesis, there may be a net flux of oxygen via the leaves of the plants through the roots into the sediments, which stimulates aerobic processes and gives rise to a distinct microbial community in the *rhizosphere*, the microhabitat near the roots. For example, in Lake Taihu (China) the rhizosphere microbiome of two emergent macrophyte species had less representation of anaerobic taxa than in the bulk sediments, and a higher diversity of taxa that were organized into more interconnected cooccurrence networks (Huang et al., 2020a). Finally, the organic litter from decomposing macrophytes in lakes may accelerate sediment respiration and methanogenesis. This may be especially the case in ice-covered lakes during winter when there is a seasonal die-off of plant populations (see below).

Less than 5% of the primary production of aquatic macrophytes is generally consumed by grazing animals, and the vast majority of littoral plant biomass enters the microbial decomposition pathway. There is a large amount of literature on the decomposition rates of aquatic macrophytes based on *litter bag assays*. These are subject to various criticisms (Wetzel and Likens, 2000) but provide a useful basis for comparison. The method typically uses freeze-killed macrophyte samples that are placed in nylon bags of mesh size 2 mm. The bags are then tethered in the lake or stream for days to weeks, with regular removal of bags and analysis of the residual organic material. The mass loss over this period is initially associated with the rapid leaching of dissolved organic materials, followed by microbial breakdown of the remaining biomass.

For submersed and floating species of macrophytes, substantial biomass losses can take place within the first 10 days, with up to almost complete decomposition of some species after 12 weeks. In contrast, emergent species with their lignified support tissues take much longer to break down (Table 27-5). Similarly, in rivers

TABLE 27-5 Examples of Aquatic Macrophyte Decomposition Rates Measured as the Percent Decrease in Organic Matter Dry Weight

Plant type	*Loss of dry weight (%)*	
	After 7–10 days	*After 12 weeks*
Submersed		
Elodea nuttallii	5–30	25–95
Myriophyllum heterophyllum	28–40	88
Floating-leaved		
Lemnaceae	20	90
Nymphaea odorata	15	45
Emergent		
Juncus effusus	5	30
Typha latifolia	5	25

and streams the woody materials that play such an important role in structuring the ecosystem (Wohl et al., 2019) break down very slowly, and their contribution to sediment respiration rates and oxygen fluxes is likely to be negligible relative to that fueled by dissolved and particulate organic carbon from more labile allochthonous materials and from autochthonous production.

Macrophyte decomposition in sediments may have implications for the seasonal oxygen dynamics of the overlying lakewater. In winter experiments with two emergent macrophytes species (*Typha orientalis* and *Phragmites australis*) in mesocosms containing littoral water and sediments from Hengshui Lake, North China, substantial decomposition occurred over winter (15% loss of mass after 10 days and >60% loss after 100 days) and dissolved oxygen dropped to near-zero beneath the ice cover (Wei et al., 2020). In a series of high-frequency dissolved oxygen measurements in eight shallow lakes in Minnesota, USA, the clear lakes in winter had less oxygen, a higher frequency of anoxic periods, and higher oxygen depletion rates; the latter were positively correlated with the peak summer macrophyte biomass, suggesting that winter decomposition of senescent macrophytes in the sediments can exert a major control on oxygen conditions (Rabaey et al., 2021).

C. Degradation processes and the redox ladder

In well-lit environments, such as ponds, shallow streams, and the littoral zone of lakes, photosynthesis drives the production of organic matter, including via phototrophic biofilms (Section IVB). Biogeochemical dynamics in the sediment below any surface photosynthetic layer are driven, directly or indirectly, by the microbial degradation of organic carbon. There is thus a transfer of energy from reduced carbon compounds to the sediment microbiome.

The proportion of organic carbon in the sediments from different origins and composition will determine its *reactivity* (tendency to decompose), which changes throughout the season and from year to year. On average, this reactivity decreases by more than 10-fold for each 10-fold increase in the age of the carbon reaching the sediment (Arndt et al., 2013; Katsev and Crowe, 2015). A decrease in reactivity is observed during sediment burial as described in Section IC or, for example, when organic carbon ages during its transport through a hydrological network (Battin et al., 2008; Mostovaya et al., 2016; Vachon et al., 2017).

Once on the sediment floor, the degradation of organic carbon continues via enzymatic pathways with different microorganisms, oxidants, and intermediate compounds. Microbial metabolism involves energy harvesting and can only proceed if there is a favorable energetic drive. During photosynthesis, for instance, this drive is provided by solar radiation. During chemotrophy, the drive is provided by a thermodynamic disequilibrium between two redox-active chemical species.

The distance from disequilibrium can be expressed as the *Gibbs energy of reaction* (ΔG_r), that is, the energy that can be used to reach equilibrium (Amend and LaRowe, 2019). When a disequilibrium arises in the presence of an electron source such as organic carbon and an electron acceptor such as oxygen, ΔG_r is calculated from the Gibbs energies of half-reactions; these describe the complete mineralization (oxidation) of organic compounds coupled to the reduction of terminal electron acceptors (LaRowe and Van Cappellen, 2011). In practice, it is difficult to estimate ΔG_r values for the full range of possible combinations between carbon molecules and terminal electron acceptors in the environment. Nevertheless, quantitative results from field measurements, laboratory experiments, and numerical models are providing such estimates.

In lakes and rivers molecular oxygen is generally the most thermodynamically favorable oxidant. Once it is consumed by accepting electrons from reduced organic carbon, the ongoing oxidation of that carbon proceeds via the next most thermodynamically favorable pathway. In the sediment column there is a vertical hierarchy of electron-accepting processes known as the *thermodynamic* or *redox ladder* (Fig. 27-24), a concept first formulated by Champ et al., (1979) and Froelich et al., (1979). In their view organic carbon oxidation is coupled to the utilization of terminal electron acceptors stepwise down the ladder in the order of O_2, NO_3^-, Mn(IV), Fe(III), and SO_4^{2-}, followed by methanogenesis and/or fermentation (see also Chapters 10 and 16). Depending on the degradation pathway, organic carbon is directly oxidized to CO_2, partly oxidized to intermediate compounds, or reduced to CH_4. Collectively, the suite of processes contributing to organic carbon degradation during its burial in the sediment is referred to as "early diagenesis" (Berner, 1980).

The concept of a thermodynamic ladder during early diagenesis has been fundamental to understanding the distribution of microbial activities in anoxic water columns and sediments. This understanding has evolved over time, from the early view that assumed a fixed sequence of electron-accepting processes described above to one that acknowledges a continuously reorganizing network of processes. The latter is because the diverse carbon sources in sediments give different energy yields depending on both the carbon source and the electron acceptor, leading to a wide array of combinations between electron donors and acceptors (Bethke et al., 2011; LaRowe et al., 2012). These combinations

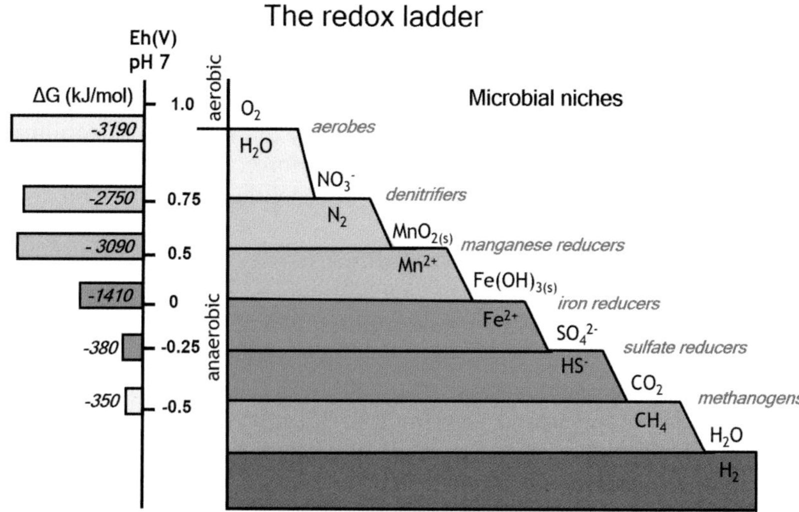

FIGURE 27-24 The thermodynamic ladder of redox couples in aquatic sediments. Microbial reactions of decreasing Gibbs Free Energy change (ΔG) occur with increasing depth in the sediments, and the most energetic electron accepters are sequentially depleted down the profile. *(Adapted from a diagram by Karen Vaughan, 2019 [by permission].)*

modulate the quantity of available energy that microbes can take advantage of to drive their metabolism. Energy yield, and thus favorability of one redox couple over another, is further dependent on the ambient conditions of temperature, pH, and ionic strength. For example, at equal concentrations of Fe(III) and SO_4^{2-} in a volume of sediment and using acetate as a carbon source, a shift in pH from 4 to 7 will shift thermodynamic favorability away from Fe(III) reduction and toward SO_4^{2-} reduction because of the high proton consumption needs for Fe(III) reduction (Paper et al., 2021).

Dissolved terminal electron acceptors such as O_2, NO_3^-, and SO_4^{2-} are supplied via diffusion from the bottom water to the sediment column (Fig. 27-10) and are taken up by microbes directly from the aqueous phase. In contrast, solid phases such as Mn and Fe oxides are supplied via sinking particles. To use these minerals as electron acceptors, microorganisms must dissolve the minerals and transfer electrons from distant surfaces to the cell. This is achieved via a variety of extracellular substrates, such as humic acids, reduced organic molecules, and electrically conductive biomaterials (Shi et al., 2016). Electron pathways through the cells, from electron donors to often insoluble acceptors, are mediated by a variety of cellular mechanisms and extracellular organic compounds, chiefly *quinones* (Fig. 27-25; Lau and del Giorgio, 2020) and *flavins* (classes of cyclic organic molecules; Markelova et al., 2018). Quinones and flavins may transport electrons, effectively diffusing them in the pore water and coupling electron sources and sinks over short distances.

Organic ligands (organic molecules that form stable complexes with metals) are an important control on oxide dissolution processes and ambient redox conditions (Lotfi-Kalahroodi et al., 2021). Microbial exudates released by bacteria for mineral dissolution affect the changes in redox potential observed in natural waters, overriding the theoretical redox potential values expected when specific redox couples are at equilibrium. The electron transfer between exudates and an ORP electrode explains the otherwise puzzlingly low Eh values measured in the presence of abundant oxidants such as oxygen and nitrate (Markelova et al., 2018).

D. Methanogenesis

Large quantities of organic matter are degraded under anaerobic conditions through biological methane production in the sediments of lakes, ponds, wetlands, rivers, and streams, as well as in waste-treatment facilities such as oxidation ponds, anaerobic lagoons, and septic tanks. *Methanogenesis* is a complex multistep process that involves six unusual coenzymes and the expression of around 200 genes to produce these coenzymes and associated proteins. This complexity and number of genes may be the reason that this process is limited to specific Archaea, with little evidence for horizontal gene transfer (Lyu et al., 2018). Two primary groups of methanogens can be distinguished by their pathways of methane production: *hydrogenotrophic methanogens*, which oxidize H_2, formate, or alcohols and reduce CO_2 to CH_4; and *acetoclastic methanogens*, which split acetate to produce CH_4 and CO_2. A third group of methanogens uses methylated compounds including methanol, trimethylamine, and dimethyl sulfate.

All three pathways use the enzyme methyl-coenzyme M reductase in their final step to produce methane, and the *mcrA* gene coding for a subunit of this enzyme is used as a molecular marker to detect methanogens. This unusual nickel-containing enzyme is extremely

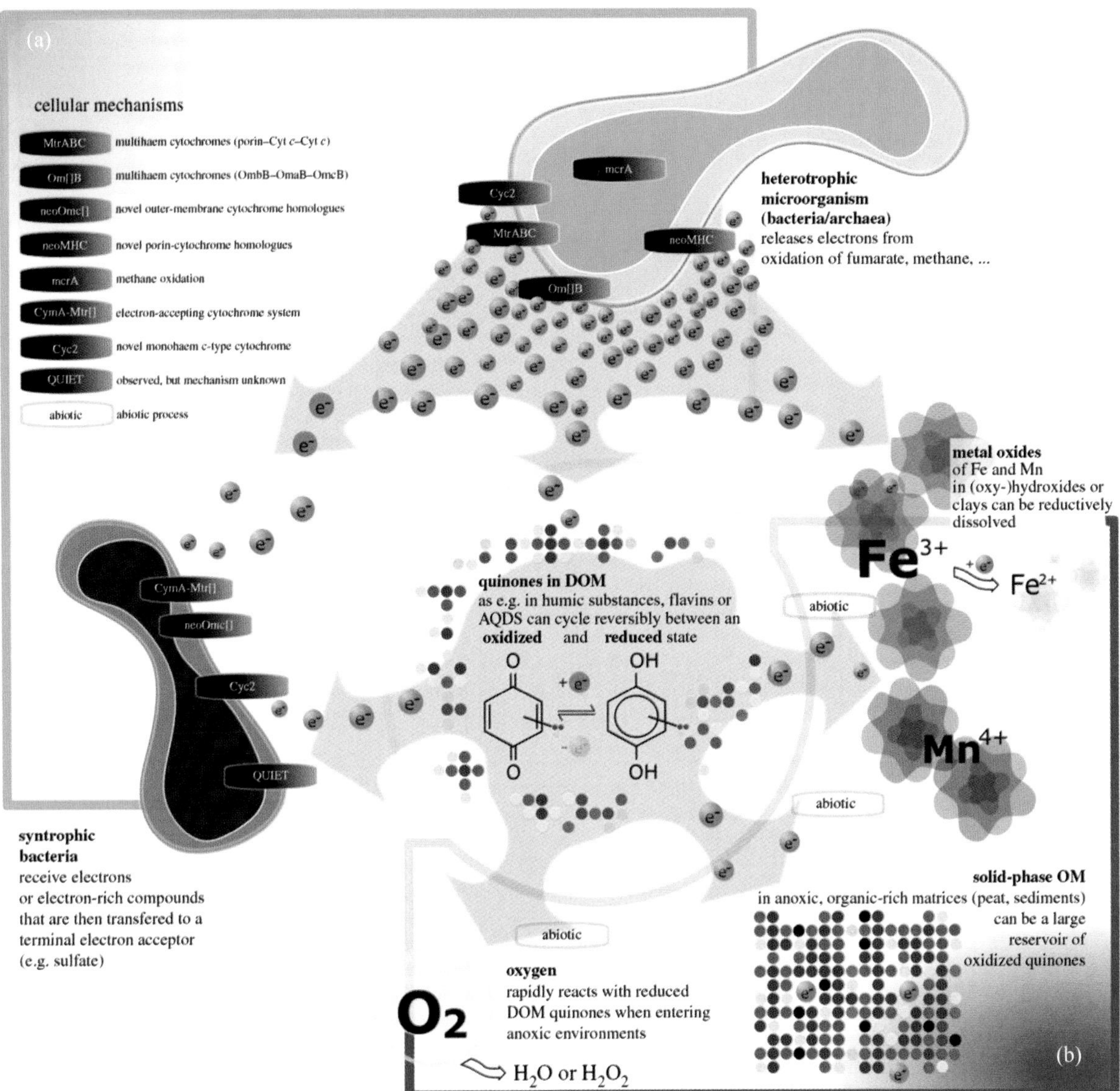

FIGURE 27-25 Postulated electron flux routes in anaerobic metabolism mediated by dissolved organic matter (DOM). *(From Lau and Del Giorgio, 2020.)*

sensitive to oxygen, and methanogenesis is most intense in anoxic habitats (Fig. 27-26). Methanogens are typically located below the surface of aerated sediments or throughout the sediment column beneath anoxic hypolimnia, where planktonic methanogens also occur (e.g., in thermokarst lakes and ponds; Crevecoeur et al., 2016). Stable isotope analysis of carbon can be used to assess the relative importance of the different pathways for methanogenesis. In a study on Canadian north-temperate lakes, for example, this analysis indicated that the strong ebullition fluxes of methane from inshore, organic sediments were primarily derived from the acetoclastic pathway, while the hydrogenotrophic pathway played a greater role in deeper, offshore sediments (Thottathil and Prairie, 2021).

Early studies showed that in eutrophic lakes a large percentage of total sediment decomposition is via methanogenesis (Table 27-6). Only a fraction of this methane, however, enters the atmosphere because of oxidation processes in the sediments and water column (Section VE). The transfer of methane to the atmosphere takes place via three mechanisms (Fig. 27-27; see also Chapter 28): (1) sediment production of methane bubbles that rise to the lake surface (the *ebullition flux*; Section VA); (2) diffusion into the water column and then across the air–water interface (the *diffusive flux*); and (3) diffusion into the roots of plants, transport through their stems, and release from their leaves into the surrounding air. The plants involved in this third process include emergent aquatic macrophytes (Sebacher et al., 1985;

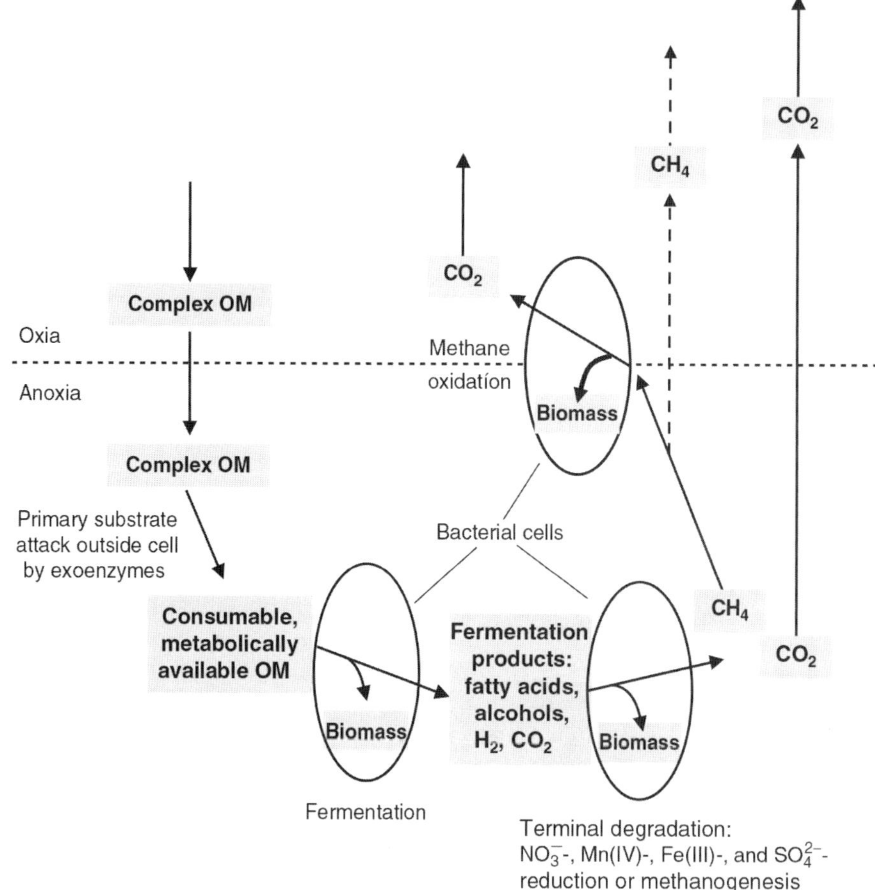

FIGURE 27-26 Production and consumption of methane in sediments. Bacterial fermentation of organic matter (OM) produces degradation products that are then used by Archaea in methanogenesis. Much of this methane may be oxidized to carbon dioxide by methanotrophs at and near the sediment surface, or in the water column. *(From Bastviken, 2009.)*

TABLE 27-6 Early Studies Indicating the Large Fraction of Particulate Organic Carbon Decomposed to Methane in Eutrophic Lake Sediments (from Canada and the USA), and the Much Lower Fraction Transferred to the Atmosphere as Methane

Lake	*Period*	*% Conversion*	*% Loss*
Frains Lake, MI	Summer	59	28
Lake 227, ON	Annual	55	<10
Lake Mendota, WI	Summer	54	c. 5
Wintergreen Lake, MI	Summer	39	17
Third Sister Lake, MI	Summer	36	6
Lake Washington, WA	Feb., Oct.	20	c. 2

% Conversion: The % of input particulate organic carbon converted to methane in lake sediments. % Loss: The % of input particulate organic carbon lost from the lake as methane.

Bergström et al., 2007) and trees in flooded forests (Pangala et al., 2017).

Ebullition and diffusion may be accelerated by wave action that disturbs inshore sediments in the littoral zone (Hofmann et al., 2010), and fluxes may be especially high when the methane accumulated in anoxic bottom waters over summer is brought to the surface by fall mixing (Kankaala et al., 2006) or when methane trapped beneath and within the ice in winter is released by ice-out and mixing (Matveev et al., 2019; Matthews et al., 2020). Large amounts of methane can be produced in hydroelectric reservoirs (see below), with transfer into the atmosphere occurring not only across the reservoir air–water interface but also when the water flows through turbines, spillways, and the downstream river (Guérin et al., 2006).

Methanogenesis and ebullition are highly variable among lakes and among sites within lakes. In general, methane production is strongly responsive to temperature. Studies at Lake Mendota (USA), for example, showed that methanogenesis was related to *in situ* temperature and could be greatly stimulated by warming lake sediments to temperatures well above ambient (Zeikus and Winfrey, 1976). More recent studies, however, have drawn attention to the complex effects of temperature on methane dynamics. Experimental warming of deep sediment samples from Lake Stechlin (Germany) and Lake Geneva (France/Switzerland) indicated that

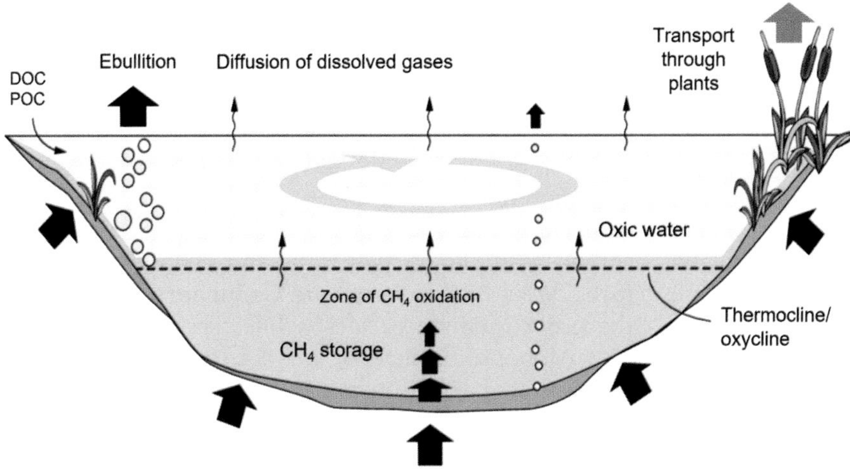

FIGURE 27-27 Three mechanisms of methane efflux from lakes to the atmosphere. *(Modified from Bastviken et al., 2004 and Wik et al., 2016.)*

at higher temperatures, methane oxidation could balance and even exceed methane production (Fuchs et al., 2016), while in a shallow pond experiment this balance was not observed and methane emissions increased with warming (Zhu et al., 2020a). Analysis of methane emissions from water bodies of different sizes in southern Quebec (Canada) showed that CH_4 fluxes from ponds increased with increasing sediment temperature; however, this relationship was not observed in the lakes and was modulated by trophic status (DelSontro et al., 2016).

Depth is a critical variable for methane ebullition (Bastviken et al., 2004). In shallow waters the total dissolved gas pressure can exceed the hydrostatic plus atmospheric pressure, allowing bubbles to form (see Langenegger et al., 2019). Additionally, the greater availability of organic material for decomposition in littoral sediments is conducive to methanogenesis. The microbial conversion efficiency of macrophyte biomass to methane under anoxic conditions varies greatly among plant species and correlates with the nutrient status (C/N ratio) in the plant tissues (Grasset et al., 2019). Consistent with the importance of depth as well as nutrient enrichment, the methane efflux rates in 32 reservoirs distributed across different landscapes in the USA were related to reservoir morphology and watershed agricultural land use (Beaulieu et al., 2020).

Unexpectedly high methane concentrations have been observed in the aerobic offshore waters of lakes where methanogen activities seem unlikely, and these puzzling results have generated considerable discussion. In part, this effect may be associated with methane released from inshore sediments and then advected offshore by currents (DelSontro et al., 2018). Additionally, the capacity for oxic methane production has been confirmed in many aerobic microorganisms including Cyanobacteria (Bižić et al., 2020) and may

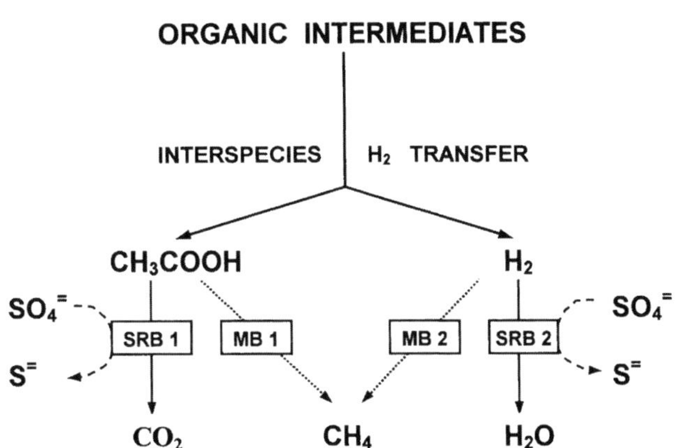

FIGURE 27-28 Competition for acetate and hydrogen by two groups of sulfate-reducing bacteria (SRB 1 and 2) and methanogens (MB 1 and 2).

be especially important in deep lakes (Günthel et al., 2021).

Given the greater thermodynamic advantage of sulfate reduction over methanogenesis (Fig. 27-24), methane production may be suppressed in anaerobic high-sulfate waters, with sulfate reducers outcompeting methanogens for the utilization of acetate and hydrogen (Fig. 27-28). Dissimilatory sulfate reduction can account for large fluxes of organic carbon; for example, up to 50% of anaerobic carbon mineralization in wetlands, with a dampening effect on methane effluxes to the atmosphere given this competitive interaction with methanogens (Pester et al., 2012).

E. Methane oxidation

Methane concentrations in lakes and rivers and the net efflux to the atmosphere are strongly regulated by the activity of methane oxidizers, specifically *methanotrophs*; this latter term is restricted to microbes that can grow on methane as their sole carbon and energy source rather than cooxidizing methane along with other reduced substrates (e.g., some ammonia–oxidizing bacteria). Methanotrophs typically occur in oxygenated habitats but close to anaerobic environments where methane is being produced. In aerobic lake and river sediments they can account for >25% of the total prokaryotic cell population and can greatly reduce and even largely suppress methane evasion fluxes (Table 27-6).

Many if not most methanotrophs still cannot be cultivated in the laboratory, and metagenomic analysis of environmental DNA, which obviates the need for cultivation, has provided important insights into their diversity and function (Smith and Wrighton, 2019). For example, the ubiquitous filamentous bacterium *Crenothrix* was well known for more than a century but eluded cultivation; it was only through genomic analysis of environmental samples, in combination with other chemical and imaging techniques, that it was confirmed to be a methanotroph.

To date, three bacterial phyla are known to contain methanotrophs, and all representatives contain the enzyme complex methane monooxygenase that oxidizes a C–H bond in methane with oxygen. The gene coding for the particulate form of this enzyme (*pmoA*) is often used as a marker to identify methanotrophs in sediments and other environments. In addition, certain Archaea (clades within the Euryarchaeota related to the orders Methanosarcinales and Methanomicrobiales) oxidize methane in the absence of molecular oxygen by reversing the pathway that methanogens use to produce methane. These *anaerobic methanotrophs* (ANME) use the methanogenesis enzyme methyl-coenzyme reductase, and the same associated gene (*mcrA*) is used to analyze these Archaea. ANMEs occur in lake sediments (Deutzmann et al., 2014) and make a variable contribution to methane oxidation among lakes.

Much of the work to date on lake sediment methane oxidation has focused on two groups of abundant Proteobacteria that differ taxonomically as well as in their carbon assimilation pathways and cellular ultrastructure. *Type 1 methanotrophs* are Gammaproteobacteria in the order Methylococcales, while *type 2 methanotrophs* are Alphaproteobacteria in the order Rhizobiales. Studies at Lake Constance (Germany/Switzerland/Austria; Rahalkar et al., 2009) have shown differences in population size and depth distribution between inshore and offshore sediments (Fig. 27-29). In the deep offshore zone of this lake, 90% of the methane produced by the sediments is oxidized by aerobic methanotrophs, while in sediments of the more productive littoral zone, much of the methane is directly lost to the atmosphere through ebullition. These spatial variations are also accompanied by temporal changes. For example, microbiological observations of the sediment–water interface at Lake Bourget, France, have shown that methanotroph abundance and species composition vary seasonally as a function of oxygen conditions (Lyautey et al., 2021). A third group of methane-oxidizing bacteria, *verrumicrobial methanotrophs*, occur in acid environments and display a high degree of metabolic versatility, with abilities to use methane, hydrogen gas, carbon dioxide, nitrogen gas, and possibly hydrogen sulfide (Schmitz et al., 2021).

Molecular studies on methanotrophs in the Amazonian floodplain have revealed diverse taxa that include

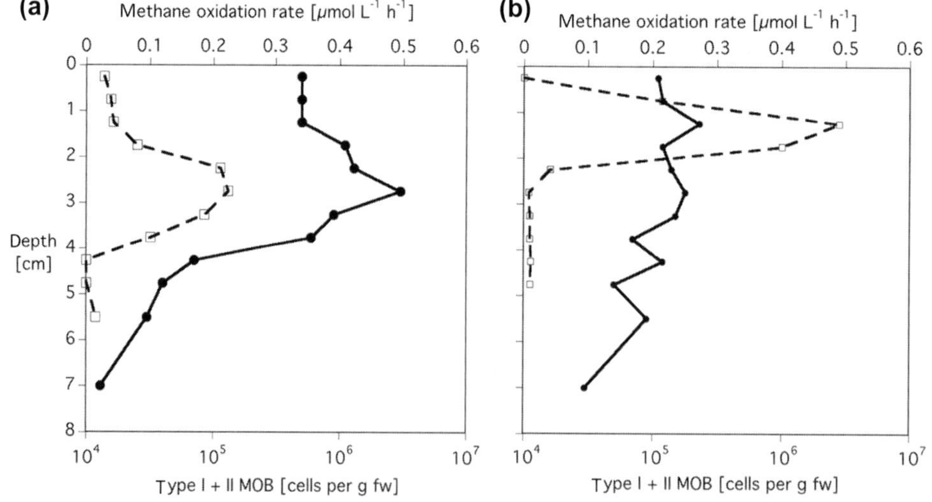

FIGURE 27-29 Distribution of methanotrophs and methane oxidation activity in the littoral (*left*) and profundal sediments of Lake Constance. *Dashed line*: cell abundance of methane-oxidizing bacteria (MOB) per gram fresh-weight (fw) of sediment; *solid line*: methane oxidation rates. *(From Rahalkar et al., 2009.)*

representatives from Type 1, Type 2, and verrumicrobial groups, as well as anaerobic methanotrophs (Gontijo et al., 2021). The taxa vary in their distribution and seasonality, indicating diverse metabolic functionality in this seasonally flooded aquatic-terrestrial environment. This vast region is known to be a globally significant source of methane to the atmosphere, and methanotrophs likely play a key role in modulating these emissions.

F. Other gas dynamics

In addition to oxygen, carbon dioxide, and methane, other gases are produced and consumed in lake and river sediments. In the nitrogen cycle this includes dinitrogen (N_2), the end product of denitrification, and the powerful greenhouse gas nitrous oxide (N_2O). The latter is an intermediate in nitrification and denitrification, and its fluxes from the sediments may be modulated by nitrate availability (Saarenheimo et al., 2015). Molecular hydrogen (H_2) is formed in the sediments under anaerobic conditions by various pathways including the fermentative degradation of carbohydrates, cellulose, and hemicelluloses to fatty acids, especially acetic acid (Akhlaghi and Najafpour-Darzi, 2020), but is rapidly consumed because of its high reactivity, including via methanogenesis. Carbon monoxide (CO) may be produced by fermentative reactions and is consumed by aerobic and anaerobic bacteria, but in general, it is a minor end product and is more substantially produced in surface waters by the photochemical breakdown of humic materials. For example, in a reach of the Saint Lawrence River (Canada) where the sediments were enriched by industrial and municipal wastes, the methane and carbon dioxide ebullition fluxes were 3.73 and 0.19 mg m^{-2} h^{-1}, respectively, while the carbon monoxide flux was only 1.23 μg m^{-2} h^{-1} and the hydrogen flux was 0.0012 μg m^{-2} h^{-1} (Poissant et al., 2007). Hydrogen sulfide (H_2S) is produced in the sulfur cycle by sulfate reduction under anaerobic conditions and is consumed in phototrophic biofilms by photosynthetic sulfur bacteria, as well as in lake and river sediments by S-oxidizers such as the common filamentous bacterium *Beggiatoa* (Holmer and Storkholm, 2001). As for other sedimentary microbial processes, genomic analyses of the sulfur cycle are revealing a remarkable diversity of microbiome taxa and *syntrophic* (cross-feeding) assemblages (Vigneron et al., 2018b).

Molecular oxygen is a master variable for all gas dynamics in lake and river sediments by affecting the balance between aerobic (e.g., aerobic decomposition, methanotrophy, nitrification, sulfur oxidation, iron oxidation) and anaerobic processes (e.g., fermentation, methanogenesis, sulfate reduction, denitrification). Microbial activities over the redox ladder (Section VC) depend on the supply of both oxidants and reductants, and changes in oxygen availability can uncouple these supplies. For example, in Lake Taihu (China), sediment regions of high algal bloom accumulation have reduced rates of denitrification; this has been attributed to oxygen depletion by microbial decomposition in these organic-enriched sediments and therefore a reduced nitrification supply of nitrate to denitrifiers (Zhu et al., 2020b). Thus sediment oxygen levels are strongly regulated by microbial respiration, and in turn, they modulate the gas exchanges with overlying lakewater.

VI. Implications of environmental change

Human activities have wide-ranging effects on the suspended sediments of lake and river ecosystems and on biogeochemical processes that take place in the bottom sediments of these environments. The limnological literature on these subjects is increasingly vast, and this section provides an overview of sediment impacts caused by three major types of human-induced environmental change: land use, dams, and climate change.

A. Land-use impacts

Intensive land use through agriculture and urbanization can lead to large shifts in sediment dynamics, with implications for lake and river ecosystem services. In part, this is through the increased delivery of allochthonous particles by greater overland flow, accelerated soil erosion in catchments, and less trapping of particles by terrestrial vegetation. For example, high sedimentation rates of the shallow lakes of the basins of the Yangtze and Yellow Rivers (China) have been ascribed to defective land-use practices (Xu et al., 2017). Changes in runoff patterns due to human activities on land can also influence the stability of streambed sediments. An example is in the Chesapeake Bay area, where in precolonial times there was a low input of sediment and a balance between sediment input and depositional processes (Noe et al., 2020). That balance was disrupted by large-scale deforestation, which led to loss of soils from disturbed lands, as well as increased flows, sediment transport, and downstream "legacy deposits" of sediment. With urbanization, impervious surfaces have reduced the infiltration of rainwater and overland flows have further increased, causing increased sediment loss and erosion of riverbanks and of the legacy deposits.

Autochthonous production by aquatic plants and algae is a second important source of particles to lake sediments that is affected by land-use practices. Increased nutrient runoff, especially of phosphorus and nitrogen, can lead to the luxuriant growth of submerged macrophyte beds and phytoplankton blooms. Given the ease of microbial decomposition of this within-lake biomass, the resultant eutrophication may be accompanied by increased rates of bacterial respiration that draw down oxygen levels, resulting in sediment anoxia and nutrient release via the mechanisms described above. This vicious circle of increased internal nutrient loading causing ever greater degrees of eutrophication is a positive feedback process that is most effectively arrested by attention to land-use practices.

Urbanization also leads to increased transfer of pollutants to aquatic ecosystems, and many of these materials accumulate in sediments. Microplastics and nanoplastics are among the most conspicuous pollutants at present and are best viewed as sentinel indicators of the contaminating cocktails that are entering lake and river ecosystems from urban sources. For example, a study of a hydrological system in China showed that microplastic concentrations in river sediments throughout the network were positively correlated with human population density and the percentage of built-up land use (Huang et al., 2020b). A survey of microplastic pollution throughout China has drawn attention to their wide distribution throughout inland water sediments, from alpine lakes on the Tibetan Plateau to lowland lakes and rivers, and the pressing need to build on and strengthen existing mitigation measures (Fu et al., 2020). Such mitigation is likely to have large collateral benefits in controlling the many associated urban pollutants and contaminants that affect sediment and water quality in the aquatic environment.

B. Impacts of dams

Humankind has a long history of damming waterways to create artificial lakes and ponds and to control water flow for agriculture and other uses, but in the 20th century large-scale projects for hydroelectricity and navigation became symbols of progress and brought large economic benefits along with major expansion of impounded lakes and associated environmental impacts (Vincent, 2018). Dam building continues today at an accelerated pace, especially in developing countries, and with the added important argument that hydroelectricity offers a green alternative to fossil fuel–based power generation.

Dams have a broad range of effects on river ecosystem functioning, and some of the most striking consequences are on sediment dynamics. Impounded lakes act as decantation basins, and therefore the sediment transport downstream of a dam installation is markedly less than in the wild river state. In a modeling study of 47 of the world's major river deltas, Dunn et al., (2019) estimated that the majority would receive much less sediment delivery in the future, on average 38% less by the end of this century. This reduction will be mostly as a result of new dams and land-use changes, which in combination with rising sea levels is putting delta ecosystems and settlements at increasing risk of catastrophic flooding and loss. The impacts to date are already considerable. A remote sensing analysis of suspended sediment fluxes by rivers to the world ocean from 1984 to the present showed a 49% decline in the Northern Hemisphere, attributed to sediment trapping by dams, while riverine sediment fluxes in the Southern Hemisphere increased by 41%, attributed to intensive land-use impacts (Dethier et al., 2022).

Dammed reservoirs act as "in-stream" reactors that increase the hydraulic residence time of river water and result in the differential loss of nutrients to sediments (Maavara et al., 2020). Dissolved silica may be removed more rapidly than nitrogen and phosphorus as a result of diatom growth, sinking, and permanent storage in reservoir sediments, resulting in lower N:Si ratios in waters delivered to the coastal ocean. This effect was most strikingly observed in the Danube River, where damming caused a >60% reduction in dissolved silica and large changes in N:Si ratios. This shift in the stoichiometry of nutrient supply is thought to have caused a shift from diatoms to flagellates and coccolithophores in the receiving coastal waters of the Black Sea (Huang et al., 2020b).

Enhanced sedimentation in dammed reservoirs also results in the accumulation of organic carbon, compounded in the early life of a reservoir by flooded soils and terrestrial vegetation. This makes reservoir sediments ideal sites for intense microbial decomposition, anoxia, sulfate reduction, and methanogenesis (Fig. 27-30; Table 27-4). These rates are likely to vary greatly over the course of the reservoir cycle (Maavara et al., 2020).

An additional biogeochemical issue related to dam building is the mobilization of mercury contained in flooded soils and vegetation. Anaerobic microbial processes in the newly flooded sediments convert the mercury from inorganic forms to methylmercury (MeHg), a neurotoxin that can be bioaccumulated in aquatic food webs. Various strategies have been considered to reduce this production of MeHg (Mailman et al., 2006). Detailed studies in the La Grande Hydroelectric complex in Quebec, Canada, showed that total mercury (THg) increased in all fish species after impoundment (Bilodeau et al., 2017), peaking at levels 2 to 8 times higher than in natural lakes of the area within 4 to 11 years (nonpiscivorous fish) and 9 to 14 years (piscivorous fish). The return to natural levels took up to three decades, in the absence of additional flooding. Bioaccumulation of mercury has been widely observed in the predatory fish of South American hydroelectric reservoirs, with MeHg levels often exceeding public health

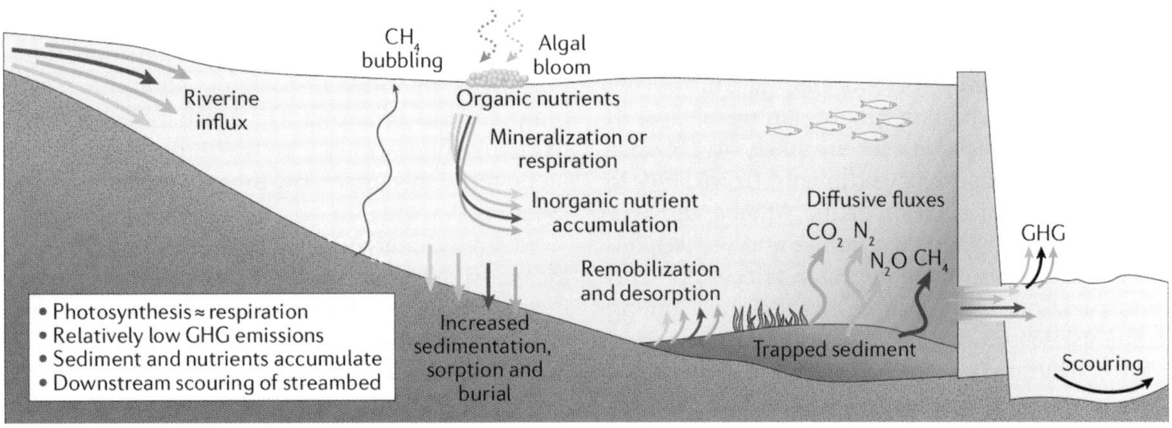

FIGURE 27-30 The key role of sedimentation and sediments for biogeochemical processes in reservoirs. In the earlier stages of the reservoir life cycle, sediment processes are dominated by the soil and vegetation of flooded lands. In the later stages of the cycle, large accumulations of organic matter accelerate anaerobic processes and greenhouse gas fluxes increase. *(From Maavara et al., 2020.)*

guidelines and hypolimnetic withdrawal resulting in downstream transport of mercury-containing particles from the deeper, anoxic waters immediately behind the dam (Pestana et al., 2019).

Mobilization of mercury is also an issue for small hydropower systems based on run-of-the-river (RoR) dams (Fig. 27-31). Microbiome analysis of ponded sediments behind RoR dams shows that the genetic capacity for MeHg production is associated with sediment methanogens, sulfate reducers, and fermenters. The RoR pondage favors the accumulation of sediments, organic matter, and mercury, and these effects are compounded by forest fires and logging (Millera Ferriz et al., 2021).

C. Climate change impacts

Climate change is exerting ever-increasing pressure on water quality, along with direct nutrient inputs from agriculture, land-use change, and excessive water use (Mack et al., 2019). Ongoing climate warming will have effects on lake and river sediments in three ways: by affecting the mobilization and transport of allochthonous sediment particles; by direct warming effects on lake and river sediments; and by amplifying eutrophication effects on lake sediments through increased algal sedimentation fluxes and changes in lake and river hydrodynamics.

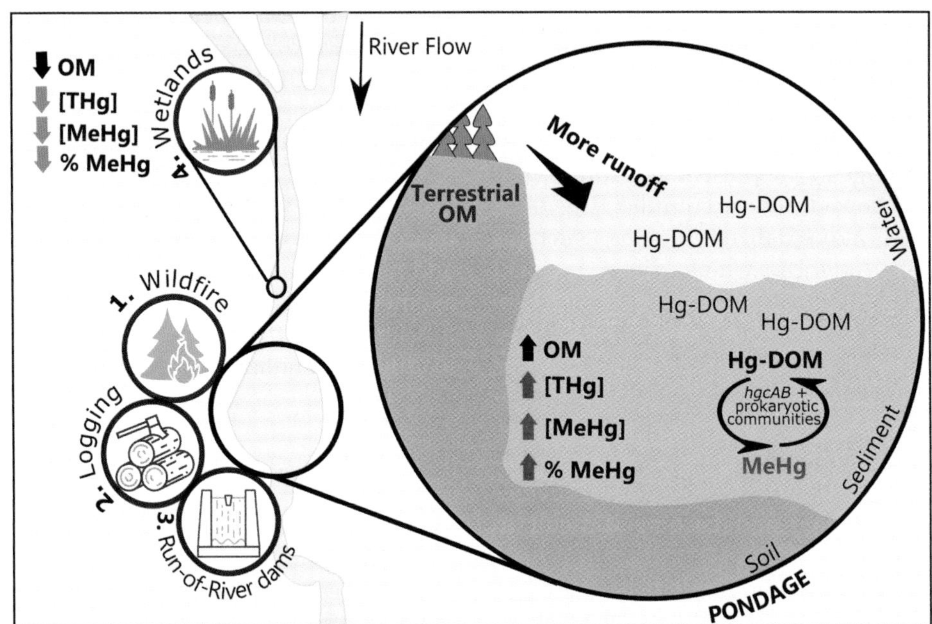

FIGURE 27-31 Sediment accumulation and microbial transformation of its mercury content behind run-of-the-river dams. OM: organic matter. DOM: dissolved organic matter. THg: total mercury. MeHg: methylmercury. %MeHg: MeHg as % of THg. *(From Millera Ferriz et al., 2021.)*

Climate projections show that annual precipitation patterns will change in many parts of the world, with an increasing frequency of extreme storm events (IPCC, 2018) and, in high northern latitudes, a shift from snow to rain. These effects are likely to increase the erosion and transport of sediments from land to rivers and ultimately to lakes. For the Athabasca River catchment in northern Canada, warmer and wetter conditions are projected to have complex effects on sediment mobilization, with large variations among subcatchments according to land use (Shrestha and Wang, 2018). The greatest increases in sediment transport in a warming climate were projected for agricultural lands, indicating that the effects of climate change need to be placed in the context of other environmental changes taking place. Similarly, in the modeling analysis of large river deltas, dam and land-use effects on reducing sediment loads greatly exceeded the increases associated with climate change (Dunn et al., 2019).

The risk of fires is increasing with climate warming, and these have a major influence on sediment mobilization. The removal of stabilizing vegetation results in accelerated erosion, and the rapid transport of particles by unimpeded runoff. Additionally, fires cause the release of metals from soil organic matter and vegetation, which may then be transported into receiving waters (Abraham et al., 2017). This includes mercury, although the exact transformations, transport pathways, and fate of this element in burned forested catchments remain uncertain (Sever, 2021). The increased frequency of forest fires with climate warming may result in amplified effects: a study of a montane lake in the Yukon boreal zone, Canada, showed that a second fire in a burned catchment resulted in greater metal loading and sedimentation rates than in the previous fire (Pelletier et al., 2022).

Lake management measures must increasingly account for climate change in the present and future, and local actions need to consider the accumulated pools of legacy nutrients. In industrialized regions, such as Europe and North America, where direct nutrient loading has been declining due to environmental legislation, lakes and their watersheds have accumulated large pools of legacy nutrients due to anthropogenic impacts (Van Meter et al., 2016), and these may be partially mobilized with climate warming. In industrializing areas, such as parts of China (Tong et al., 2020), urgent measures are needed to mitigate external loading before large internal loads emerge and become difficult to reverse.

Warmer temperatures, compounded in northern lakes by shorter periods of lake ice (e.g., Guo et al., 2020), are likely to result in warmer inshore sediment and riverbed temperatures, which will accelerate all of the biogeochemical reactions described above. This is likely to increase oxygen consumption rates, compounded by lower oxygen solubility at higher temperatures (Chapter 11). Methanogenesis is known to be highly responsive to warming; however, these effects could be offset by increases in methanotrophy (Section IV).

Several climate-related processes are likely to favor increased algal production in inland waters, thereby increasing the delivery of autochthonous organic particles to lake and river sediments, and favoring more intense decomposition, oxygen depletion, and methanogenesis (Beaulieu et al., 2019). These effects include longer and more stable thermal stratification, increased nutrient inputs to lakes through greater runoff and erosion, heatwaves that promote algal blooms (Bartosiewicz et al., 2021), and increased inputs of dissolved organic matter ("browning," Chapter 28) that absorbs solar radiation and reinforces thermal stratification (Pilla and Couture, 2021; see Chapter 6). The interactive effects of warming and transparency loss are likely to override the effects of atmospheric warming alone, resulting in increased carbon storage in lake sediments because of colder bottom temperatures, but also increased oxygen drawdown and methanogenesis as a result of prolonged stratification (Bartosiewicz et al., 2019).

The precarious oxygen balance at the lake sediment—water interface, as illustrated by the measurements in the benthic boundary layer of Lake Biwa (Section IIIB), may be especially vulnerable to changes in natural aeration processes as well as increased organic inputs. For example, models for Lake Tahoe (USA) show that climate warming could lengthen the number of years between deep mixing events and ultimately drive the sediments toward anoxia and phosphorus release (Beutel and Horne, 2018).

These observations and modeling results all show that inland water sediments of the future will be subject to greater environmental pressure in terms of particle inputs, organic loading, and oxygen depletion, with the tendency toward increased anoxia, methanogenesis, and nutrient release. In warmer climates land-use management will be more important than ever, requiring a much greater attention to particle, organic, and nutrient loading to maintain water quality.

VII. Summary

1. The term "sediments" refers to particles in suspension or seston, and to the deposits of these particles on the lakebed or riverbed. Benthic sediments play key roles in aquatic ecosystems by providing habitats for plants and animals, locations for intense microbial activity in biogeochemical

cycles, sites of long-term storage of carbon and nutrients, and use as controls on oxygen and other elements in the overlying water.

2. Sediments are composed of organic matter in various states of degradation: mineral particles washed in from catchments, atmospheric deposits, pollutants such as microplastics and nanoplastics, and biogenic mineral particles such as silicates in diatom frustules and carbonates precipitated by photosynthetic activities.
3. Annual sediment accumulation rates in lakes vary according to trophic status and sediment transport by inflows and range from tenths of a millimeter in oligotrophic water bodies to more than 10 mm per year in shallow eutrophic lakes influenced by land disturbance. Sediment porosity varies greatly among sites, and it decreases with sediment depth as the particles settle and microbial decomposition processes reduce the organic content.
4. Organic carbon produced by phytoplankton is more labile than the structural tissue of macrophytes, and partially degraded material from the littoral zone is distributed throughout the lake. C:N ratios in sediments are affected by the relative importance of algal versus catchment inputs, as well as by the input from decomposing aquatic macrophytes. Seston C:P ratios tend to decrease with increasing degrees of lake enrichment.
5. There is a continuous exchange between bottom deposits and sediments in suspension, with seasonal variations in magnitude and direction (deposition or resuspension).
   a. In the shallow littoral zone of lakes, wave-induced motions are often sufficient to resuspend smaller particles. Advection offshore results in sediment winnowing (loss of fine particles from inshore sites) and contributes to sediment focusing (greater sediment accumulation at deeper offshore sites).
   b. In rivers, sediments are mostly transported in the water column (suspended load) but a portion of the total sediment load occurs near the bottom, where sediment particles are resuspended and move downstream over the riverbed. This bed load increases at higher water flows, with increased erosion and particle resuspension.
6. The interstitial waters of sediments typically show large solute gradients down through the sediment profile, consistent with variations in microbial processes with depth and the biogeochemical transformation of sediments during diagenesis. These changes occur most rapidly in the surface strata where both oxygen supply and labile organic matter concentrations are highest. Burrowing animals such as chironomid larvae accelerate these transformation processes by mixing and aerating the surficial sediments.
7. The chemical composition of the water immediately above the sediments differs from the rest of the water column as a result of benthic chemical and microbial processes. Nitrate may accumulate over aerobic sediments, while depleted oxygen levels along with higher concentrations of dissolved inorganic phosphorus and reduced substances such as methane, ammonium, hydrogen sulfide, iron (Fe(II)), and manganese (Mn(III)) occur over anaerobic sediments.
8. Strong chemical gradients can occur across the benthic boundary layer, defined as the layer adjacent to the lake, river, or ocean bed where the flow is affected by processes occurring at the sediment—water interface. This is divided into three main sublayers that differ in their physical and chemical properties: the Ekman layer, the logarithmic layer, and the viscous sublayer. The latter includes a thin diffusive sublayer immediately over the sediments.
9. Four types of microbiomes, defined as the full assemblage of microbes in a habitat (Bacteria, Archaea, microbial eukaryotes including algae and fungi, and viruses), can be identified in association with sediments, and their taxonomic and functional complexity is increasingly revealed by the application of molecular methods.
   a. Multispecies aggregates in suspension. Microbial diversity in these assemblages is often higher than in the free-living plankton community, and microzones inside the particles can provide suitable redox conditions for anaerobic processes such as denitrification, even in oxygenated waters.
   b. Photosynthetic biofilms and microbial mats. These are complex species assemblages embedded in an extracellular matrix of polysaccharides, proteins, and other compounds (the matrixome), with phototrophs such as Cyanobacteria and diatoms in combination with numerous heterotrophic taxa. Metagenomic analyses have revealed multiple pathways of energy acquisition and carbon flow in these communities, which experience a chemical microenvironment that differs greatly from the overlying lake and river water.
   c. Hyporheic microbiomes. The hyporheic zone refers to the subsurface stratum of a riverbed in which there is a mixing of river water and groundwater. This mixture of flowing waters with different geochemical and microbiological properties results in a microbial habitat of high biological diversity and diverse biogeochemical

functions. Aquatic hyphomycetes play an important role on and below the streambed in decomposing leaves and wood particles, and as a food source for invertebrates.

d. Lake sediment microbiomes. The highest bacterial population densities and production rates occur in surficial sediments, with orders of magnitude greater cell concentrations than in the overlying lakewater and an exponential decline with depth down the sediment profile. Littoral sediments are enriched in macrophyte and terrestrial detritus, with higher bacterial decomposition rates and active fungal specialists.

10. Sediment biogeochemical processes produce gas bubbles that rise to the lake or river surface (ebullition). These contain methane to an extent that reflects the intensity of sediment methanogenesis, which is stimulated by warming. The highest rates occur at sites with organic-rich sediments, including reservoirs, permafrost thaw waters, littoral zones, and shallow ponds.

11. The sediment microbiome exerts a strong effect on oxygen concentrations in the benthic boundary layer, which even in deep lakes such as Lake Biwa, Japan, can drop to hypoxic or anoxic levels. Winter decomposition of senescent macrophytes in the sediments may severely deplete oxygen levels.

12. The sequence of biogeochemical reactions during sediment diagenesis can be expressed in terms of the thermodynamic ladder of redox couples. The most energetic electron accepters are sequentially depleted down the profile, and microbial reactions of decreasing Gibbs Free Energy change occur with increasing depth in the sediments, for example, sulfate reduction before methanogenesis. However, the ladder sequence is modified by environmental variables such as pH and organic substrates. In anaerobic environments quinones in the dissolved organic carbon pool act as important electron shuttles, coupling electron sources and sinks over short distances.

13. Large quantities of organic matter are degraded in sediments under anaerobic conditions through biological methane production by Archaea. In eutrophic lake sediments this process of methanogenesis may consume more than 60% of the total carbon inputs. Methane evasion to the atmosphere occurs via three mechanisms: ebullition, diffusion, and transport through plants, with the latter especially important in wetlands. Emission fluxes may be accelerated by wave action that disturbs littoral sediments and can be especially high when the methane accumulated in anoxic bottom waters over summer is mixed to the surface in fall, or when methane produced beneath the ice in winter is released by ice-out and mixing.

14. Methane concentrations and net effluxes to the atmosphere are regulated by methane-oxidizers (methanotrophs). In aerobic lake and river sediments, methanotrophs can account for more than 25% of the total prokaryotic cell population and can greatly reduce and even largely suppress methane evasion. High rates of methane oxidation as well as methane evasion can occur during lake overturn in fall. Methane can also accumulate under ice cover and support methane oxidation processes, which contribute toward winter oxygen loss.

15. Other biologically active gases in lake and river sediments include nitrous oxide, molecular hydrogen, carbon monoxide, and hydrogen sulfide, which are kept at low net efflux rates because of their rapid consumption by the sediment microbiome. Molecular oxygen is a master variable for all gas dynamics in lake and river sediments and controls the balance between aerobic and anaerobic processes.

16. Environmental change through human activities has the potential to greatly alter inland water ecosystems, via multiple effects on sediment processes:

a. Land-use impacts, resulting in increased delivery of particles by greater overland flow, accelerated soil erosion, and less trapping by vegetation. A conspicuous effect of urbanization is the transport of microplastics and nanoplastics into waterways throughout the world, and their accumulation in bottom sediments can be viewed as a sentinel indicator of urban pollution.

b. Dams, which act as decantation basins and as sequential batch reactors that prolong water residence times, trap sediments and change nutrient ratios, with implications for downstream coastal ecosystems. The limnological effects of dams also include increased rates of methanogenesis from the trapped sediments, and the microbial production and release of the neurotoxin methylmercury, which can then bioaccumulate through the food web.

c. Climate impacts, through increased input of particles from catchments, direct warming of sediments, and promotion of sediment anoxia, nutrient release, and eutrophication. The increased frequency of wildfires will also increase the transport of particles and metals, including mercury. In a warmer climate land-use management will be more important than ever, requiring greater attention to sediment particle loading to maintain water quality.

Acknowledgments

We acknowledge support for our limnological research from the Natural Sciences and Engineering Research Council of Canada (NSERC); the Canada First Research Excellence Fund (CFREF) Sentinel North; the Networks of Centres of Excellence ArcticNet; the Canada Research Chair program; and Grants-in-Aid for Scientific Research (20244079) and JST SICORP (JPMJCR1804), Japan. We thank Dermot Antoniades, Marie-Amélie Blais, Tonya DelSontro, David Hamilton, Anne Jungblut, Isabelle Laurion, and Alex Matveev for helpful review comments; Amanda Toperoff for figure graphics; and Ian Jones and John Smol for their editorial guidance and support.

References

Aben, R.C., Barros, N., Van Donk, E., Frenken, T., Hilt, S., Kazanjian, G., et al., 2017. Cross continental increase in methane ebullition under climate change. Nat. Commun. 8, 1682.

Abraham, J., Dowling, K., Florentine, S., 2017. Risk of post-fire metal mobilization into surface water resources: a review. Sci. Total Environ. 599, 1740–1755.

Aguilar, P., Dorador, C., Vila, I., Sommaruga, R., 2019. Bacterial communities associated with spherical *Nostoc* macrocolonies. Front. Microbiol. 10, 483.

Akhlaghi, N., Najafpour-Darzi, G., 2020. A comprehensive review on biological hydrogen production. Int. J. Hydrogen Energy 45, 22492–22512.

Aliff, M.N., Reavie, E.D., Post, S.P., Zanko, L.M., 2020. Anthropocene geochemistry of metals in sediment cores from the Laurentian Great Lakes. PeerJ 8, e9034.

Allan, J.D., Castillo, M.M., 2007. Stream Ecology: Structure and Function of Running Waters, second ed. Springer Science & Business Media, Dordrecht, The Netherlands.

Alldredge, A.L., Silver, M.W., 1988. Characteristics, dynamics and significance of marine snow. Prog. Oceanogr. 20, 41–82.

Amend, J.P., LaRowe, D.E., 2019. Minireview: demystifying microbial reaction energetics. Environ. Microbiol. 21, 3539–3547.

Andersen, D.T., Sumner, D.Y., Hawes, I., Webster-Brown, J., McKay, C.P., 2011. Discovery of large conical stromatolites in Lake Untersee, Antarctica. Geobiology 9, 280–293.

Anschutz, P., Charbonnier, C., 2021. Sampling pore water at a centimeter resolution in sandy permeable sediments of lakes, streams, and coastal zones. Limnol Oceanogr. Methods 19, 96–114.

Antoniades, D., Francus, P., Pienitz, R., St-Onge, G., Vincent, W.F., 2011. Holocene dynamics of the Arctic's largest ice shelf. Proc. Natl. Acad. Sci. USA 108, 18899–18904.

Arndt, S., Jørgensen, B.B., LaRowe, D.E., Middelburg, J., Pancost, R., Regnier, P., 2013. Quantifying the degradation of organic matter in marine sediments: a review and synthesis. Earth Sci. Rev. 123, 53–86.

Bärlocher, F., 1985. The role of fungi in the nutrition of stream invertebrates. Bot. J. Linn. Soc. 91, 83–94.

Bartlett, K.B., Crill, P.M., Bonassi, J.A., Richey, J.E., Harriss, R.C., 1990. Methane flux from the Amazon River floodplain: emissions during rising water. J. Geophys. Res. Atmos. 95, 16773–16788.

Bartosiewicz, M., Maranger, R., Przytulska, A., Laurion, I., 2021. Effects of phytoplankton blooms on fluxes and emissions of greenhouse gases in a eutrophic lake. Water Res. 196, 116985.

Bartosiewicz, M., Przytulska, A., Lapierre, J.F., Laurion, I., Lehmann, M.F., Maranger, R., 2019. Hot tops, cold bottoms: synergistic climate warming and shielding effects increase carbon burial in lakes. Limnol. Oceanogr. 4, 132–144.

Bastviken, D., 2009. Methane. In: Likens, G.E. (Ed.), Encyclopedia of Inland Waters, vol. 2. Elsevier, Oxford, pp. 783–805.

Bastviken, D., Cole, J.J, Pace, M.L., Tranvik, L., 2004. Methane emissions from lakes: dependence of lake characteristics, two regional assessments, and a global estimate. Global Biogeochem. Cycles 18, GB4009.

Bastviken, D., Cole, J.J., Pace, M.L., Van de Bogert, M.C., 2008. Fates of methane from different lake habitats: connecting whole-lake budgets and CH_4 emissions. J. Geophys. Res.: Biogeosciences 113, G2.

Battin, T.J., Besemer, K., Bengtsson, M.M., Romani, A.M., Packmann, A.I., 2016. The ecology and biogeochemistry of stream biofilms. Nat. Rev. Microbiol. 14, 251.

Battin, T.J., Kaplan, L.A., Findlay, S., Hopkinson, C.S., Marti, E., Packman, A.I., et al., 2008. Biophysical controls on organic carbon fluxes in fluvial networks. Nat. Geosci. 1, 95–100.

Baulch, H.M., Dillon, P.J., Maranger, R., Schiff, S.L., 2011. Diffusive and ebullitive transport of methane and nitrous oxide from streams: are bubble-mediated fluxes important? J. Geophys. Res. 116, G04028.

Beaulieu, J.J., DelSontro, T., Downing, J.A., 2019. Eutrophication will increase methane emissions from lakes and impoundments during the 21st century. Nat. Commun. 10, 1375.

Beaulieu, J.J., Waldo, S., Balz, D.A., Barnett, W., Hall, A., Platz, M.C., White, K.M., 2020. Methane and carbon dioxide emissions from reservoirs: controls and upscaling. J. Geophys. Res.: Biogeosciences 125, e2019JG005474.

Bégin, P.N., Tanabe, Y., Kumagai, M., Culley, A.I., Paquette, M., Sarrazin, D., Uchida, M., Vincent, W.F., 2021. Extreme warming and regime shift toward amplified variability in a far northern lake. Limnol. Oceanogr. 66, S17–S29.

Bergström, I., Mäkelä, S., Kankaala, P., Kortelainen, P., 2007. Methane efflux from littoral vegetation stands of southern boreal lakes: an upscaled regional estimate. Atmos. Environ. 41, 339–351.

Berner, R.A., 1980. Early Diagenesis: A Theoretical Approach. Princeton University Press, New York.

Bethke, C.M., Sanford, R.A., Kirk, M.F., Jin, Q.S., Flynn, T.M., 2011. The thermodynamic ladder in geomicrobiology. Am. J. Sci. 311, 183–210.

Beutel, M.W., Horne, A.J., 2018. Nutrient fluxes from profundal sediment of ultra-oligotrophic Lake Tahoe, California/Nevada: implications for water quality and management in a changing climate. Water Resour. Res. 54, 1549–1559.

Bianchi, D., Weber, T.S., Kiko, R., Deutsch, C., 2018. Global niche of marine anaerobic metabolisms expanded by particle microenvironments. Nat. Geosci. 11, 263–268.

Bilodeau, F., Therrien, J., Schetagne, R., 2017. Intensity and duration of effects of impoundment on mercury levels in fishes of hydroelectric reservoirs in northern Québec (Canada). Inland Waters 7, 493–503.

Bižić, M., Klintzsch, T., Ionescu, D., Hindiyeh, M.Y., Günthel, M., Muro-Pastor, A.M., et al., 2020. Aquatic and terrestrial cyanobacteria produce methane. Sci. Adv. 6, eaax5343.

Bižic-Ionescu, M., Zeder, M., Ionescu, D., Orlić, S., Fuchs, B.M., Grossart, H.-P., Amann, R., 2014. Comparison of bacterial communities on limnic versus coastal marine particles reveals profound differences in colonization. Environ. Microbiol. 17, 3500–3514.

Blais, J.M., Kalff, J., 1995. The influence of lake morphometry on sediment focusing. Limnol. Oceanogr. 40, 582–588.

Blais, M.-A., Matveev, A., Lovejoy, C., Vincent, W.F., 2022. Size-fractionated microbiome structure in subarctic rivers and a coastal plume across DOC and salinity gradients. Front. Microbiol. 12, 760282.

Bloesch, J., 2009. Sediments of aquatic ecosystems. In: Likens, G.E. (Ed.), Encyclopedia of Inland Waters, Vol. 1. Elsevier, Oxford, pp. 479–490.

Boano, F., Harvey, J.W., Marion, A., Packman, A.I., Revelli, R., Ridolfi, L., Wörman, A., 2014. Hyporheic flow and transport processes: mechanisms, models, and biogeochemical implications. Rev. Geophys. 52, 603–679.

Böckelmann, U., Manz, W., Neu, T.R., Szewzyk, U., 2000. Characterization of the microbial community of lotic organic aggregates ("river snow") in the Elbe River of Germany by cultivation and molecular methods. FEMS Microbiol. Ecol. 33, 157–170.

Boudreau, B.P., 1999. Metals and models: diagenic modelling in freshwater lacustrine sediments. J. Paleolimnol. 22, 227–251.

Boudreau, B.P., Jorgensen, B.B. (Eds.), 2001. The Benthic Boundary Layer: Transport Processes and Biogeochemistry. Oxford University Press, New York.

Bowden, K.F., 1978. Physical problems of the benthic boundary layer. Geophys. Surv. 3, 255–296.

Bowen, J.C., Kaplan, L.A., Cory, R.M., 2020. Photodegradation disproportionately impacts biodegradation of semi-labile DOM in streams. Limnol. Oceanogr. 65, 13–26.

Cai, Y.M., 2020. Non-surface attached bacterial aggregates: a ubiquitous third lifestyle. Front. Microbiol. 11, 557035.

Campeau, A., Lapierre, J.F., Vachon, D., del Giorgio, P.A., 2014. Regional contribution of CO_2 and CH_4 fluxes from the fluvial network in a lowland boreal landscape of Québec. Global Biogeochem. Cycles 28, 57–69.

Cardenas, M.B., 2015. Hyporheic zone hydrologic science: a historical account of its emergence and a prospectus. Water Resour. Res. 51, 3601–3616.

Carlton, R.G., Wetzel, R.G., 1987. Distributions and fates of oxygen in periphyton communities. Can. J. Bot. 65, 1031–1037.

Champ, D.R., Gulens, J., Jackson, R.E., 1979. Oxidation–reduction sequences in ground water flow systems. Can. J. Earth Sci. 16, 12–23.

Chappaz, A., Gobeil, C., Tessier, A., 2008. Geochemical and anthropogenic enrichments of Mo in sediments from perennially oxic and seasonally anoxic lakes in Eastern Canada. Geochem. Cosmochim. Acta 72, 170–184.

Clayer, F., Gélinas, Y., Tessier, A., Gobeil, C., 2020. Mineralization of organic matter in boreal lake sediments: rates, pathways, and nature of the fermenting substrates. Biogeosciences 17, 4571–4589.

Conley, D.J., Quigley, M.A., Schelske, C.L., 1988. Silica and phosphorus flux from sediments: importance of internal recycling in Lake Michigan. Can. J. Fish. Aquat. Sci. 45, 1030–1035.

Constantin, A., Villari, G., 2008. Particle trajectories in linear water waves. J. Math. Fluid Mech. 10, 1–18.

Cortés, A., MacIntyre, S., 2020. Mixing processes in small arctic lakes during spring. Limnol. Oceanogr. 65, 260–288.

Couture, R.-M., Charlet, L., Markelova, E., Madé, B.T., Parsons, C.T., 2015. On–off mobilization of contaminants in soils during redox oscillations. Environ. Sci. Technol. 49, 3015–3023.

Couture, R.M., Fischer, R., Van Cappellen, P., Gobeil, C., 2016. Non-steady state diagenesis of organic and inorganic sulfur in lake sediments. Geochem. Cosmochim. Acta 194, 15–33.

Crevecoeur, S., Vincent, W.F., Lovejoy, C., 2016. Environmental selection of planktonic methanogens in permafrost thaw ponds. Sci. Rep. 6, 31312.

Crew, A., Gregory-Eaves, I., Ricciardi, A., 2020. Distribution, abundance, and diversity of microplastics in the upper St. Lawrence River. Environ. Pollut. 260, 113994.

Crill, P.M., Martens, C.S., 1987. Biogeochemical cycling in an organic-rich coastal marine basin. 6. Temporal and spatial variations in sulfate reduction rates. Geochem. Cosmochim. Acta 51, 1175–1186.

Crump, B.C., Armbrust, E.V., Baross, J.A., 1999. Phylogenetic analysis of particle-attached and free-living bacterial communities in the Columbia River, its estuary, and the adjacent coastal ocean. Appl. Environ. Microbiol. 65, 3192–3204.

Cuven, S., Francus, P., Lamoureux, S.F., 2010. Estimation of grain size variability with micro X-ray fluorescence in laminated lacustrine sediments, Cape Bounty, Canadian High Arctic. J. Paleolimnol. 44, 803–817.

Davey, M.C., Walsby, A.E., 1985. The form resistance of sinking algal chains. Br. Phycol. J. 20, 243–248.

de Los Ríos, A., Ascaso, C., Wierzchos, J., Fernández-Valiente, E., Quesada, A., 2004. Microstructural characterization of cyanobacterial mats from the McMurdo Ice Shelf, Antarctica. Appl. Environ. Microbiol. 70, 569–580.

del Mar Aguilo-Ferretjans, M., Bosch, R., Puxty, R.J., Latva, M., Zadjelovic, V., Chhun, A., et al., 2021. Pili allow dominant marine cyanobacteria to avoid sinking and evade predation. Nat. Commun. 12, 1857.

DelSontro, T., Boutet, L., St-Pierre, A., del Giorgio, P.A., Prairie, Y.T., 2016. Methane ebullition and diffusion from northern ponds and lakes regulated by the interaction between temperature and system productivity. Limnol. Oceanogr. 61, S62–S77.

DelSontro, T., del Giorgio, P.A., Prairie, Y.T., 2018. No longer a paradox: the interaction between physical transport and biological processes explains the spatial distribution of surface water methane within and across lakes. Ecosystems 21, 1073–1087.

DelSontro, T., Kunz, M.J., Kempter, T., Wüest, A., Wehrli, B., Senn, D.B., 2011. Spatial heterogeneity of methane ebullition in a large tropical reservoir. Environ. Sci. Technol. 45, 9866–9873.

DelSontro, T., McGinnis, D.F., Wehrli, B., Ostrovsky, I., 2015. Size does matter: importance of large bubbles and small-scale hot spots for methane transport. Environ. Sci. Technol. 49, 1268–1276.

Delwiche, K.B., Hemond, H.F., 2017. Methane bubble size distributions, flux, and dissolution in a freshwater lake. Environ. Sci. Technol. 51, 13733–13739.

Deshpande, B.N., MacIntyre, S., Matveev, A., Vincent, W.F., 2015. Oxygen dynamics in permafrost thaw lakes: anaerobic bioreactors in the Canadian subarctic. Limnol. Oceanogr. 60, 1656–1670.

Dethier, E.N., Renshaw, C.E., Magilligan, F.J., 2022. Rapid changes to global river suspended sediment flux by humans. Science 376, 1447–1452.

Deutzmann, J.S., Stief, P., Brandes, J., Schink, B., 2014. Anaerobic methane oxidation in a deep lake. Pro. Natl. Acad. Sci. U. S. A 111, 18273–18278.

Dillon, M.L., Hawes, I., Jungblut, A.D., Mackey, T.J., Eisen, J.A., Doran, P.T., Sumner, D.Y., 2020. Environmental control on the distribution of metabolic strategies of benthic microbial mats in Lake Fryxell, Antarctica. PLoS One 15, e0231053.

Downing, J.A., Cole, J.J., Middelburg, J.J., Striegl, R.G., Duarte, C.M., Kortelainen, P., et al., 2008. Sediment organic carbon burial in agriculturally eutrophic impoundments over the last century. Global Biogeochem. Cycles 22, GB1018.

Du, X.L., Creed, I.F., Sorichetti, R.J., Trick, C.G., 2019. Cyanobacteria biomass in shallow eutrophic lakes is linked to the presence of iron-binding ligands. Can. J. Fish. Aquat. Sci. 76, 1728–1739.

Duck, P.W., Foster, M.R., 2001. Spin-up of homogeneous and stratified fluid. Annu. Rev. Fluid Mech. 33, 231–263.

Dunn, F.E., Darby, S.E., Nicholls, R.J., Cohen, S., Zarfl, C., Fekete, B.M., 2019. Projections of declining fluvial sediment delivery to major deltas worldwide in response to climate change and anthropogenic stress. Environ. Res. Lett. 14, 084034.

Emmanouilidis, A., Messaris, G., Ntzanis, E., Zampakis, P., Prevedouros, I., Bassukas, D.A., Avramidis, P., 2020. CT scanning, X-ray fluorescence: non-destructive techniques for the identification of sedimentary facies and structures. Rev. Micropaleontol. 67, 100410.

Fakhraee, M., Li, J.Y., Katsev, S., 2017. Significant role of organic sulfur in supporting sedimentary sulfate reduction in low-sulfate environments. Geochimica et Cosmochica. Acta 213, 502–515.

Flemming, H.C., Wingender, J., Szewzyk, U., Steinberg, P., Rice, S.A., Kjelleberg, S., 2016. Biofilms: an emergent form of bacterial life. Nat. Rev. Microbiol. 14, 563.

Frenette, J.J., Vincent, W.F., Dodson, J.J., Lovejoy, C., 1995. Size-dependent variations in phytoplankton and protozoan community structure across the St. Lawrence River transition region. Mar. Ecol. Prog. Ser. 120, 99–110.

Froelich, P., Klinkhammer, G.P., Bender, M.L., Luedtke, N.A., Heath, G.R., Cullen, D., et al., 1979. Early oxidation of organic matter in pelagic sediments of the eastern equatorial Atlantic: suboxic diagenesis. Geochem. Cosmochim. Acta 43, 1075–1090.

Fu, D., Chen, C.M., Qi, H., Fan, Z., Wang, Z., Peng, L., Li, B., 2020. Occurrences and distribution of microplastic pollution and the control measures in China. Mar. Pollut. Bull. 153, 110963.

Fuchs, A., Lyautey, E., Montuelle, B., Casper, P., 2016. Effects of increasing temperatures on methane concentrations and methanogenesis during experimental incubation of sediments from oligotrophic and mesotrophic lakes. J. Geophys. Res.: Biogeosciences 121, 1394–1406.

Fueser, H., Mueller, M.T., Traunspurger, W., 2020. Ingestion of microplastics by meiobenthic communities in small-scale microcosm experiments. Sci. Total Environ. 746, 141276.

Gälman, V., Rydberg, J., Sjöstedt-de Luna, S., Bindler, R., Renberg, I., 2008. Carbon and nitrogen loss rates during aging of lake sediment: changes over 27 years studied in varved lake sediment. Limnol. Oceanogr. 53, 1076–1082.

Garneau, M.È., Vincent, W.F., Terrado, R., Lovejoy, C., 2009. Importance of particle-associated bacterial heterotrophy in a coastal Arctic ecosystem. J. Mar. Syst. 75, 185–197.

Gąsiorowski, M., Sienkiewicz, E., 2013. The sources of carbon and nitrogen in mountain lakes and the role of human activity in their modification determined by tracking stable isotope composition. Water, Air, Soil Pollut. 224, 1498.

Ghanbari, H., Jacques, O., Adaïmé, M.É., Gregory-Eaves, I., Antoniades, D., 2020. Remote sensing of lake sediment core particle size using hyperspectral image analysis. Rem. Sens. 12, 3850.

Gigault, J., El Hadri, H., Nguyen, B., Grassl, B., Rowenczyk, L., Tufenkji, N., et al., 2021. Nanoplastics are neither microplastics nor engineered nanoparticles. Nat. Nanotechnol. 16, 501–507.

Gontijo, J.B., Paula, F.S., Venturini, A.M., Yoshiura, C.A., Borges, C.D., Moura, J.M.S., et al., 2021. Not just a methane source: amazonian floodplain sediments harbour a high diversity of methanotrophs with different metabolic capabilities. Mol. Ecol. 30, 2560–2572.

Gooseff, M.N., 2010. Defining hyporheic zones—advancing our conceptual and operational definitions of where stream water and groundwater meet. Geography. Compass 4, 945–955.

Graham, L.E., Knack, J.J., Piotrowski, M.J., Wilcox, L.W., Cook, M.E., Wellman, C.H., et al., 2014. Lacustrine *Nostoc* (Nostocales) and associated microbiome generate a new type of modern clotted microbialite. J. Phycol. 50, 280–291.

Grasset, C., Abril, G., Mendonça, R., Roland, F., Sobek, S., 2019. The transformation of macrophyte-derived organic matter to methane relates to plant water and nutrient contents. Limnol. Oceanogr. 64, 1737–1749.

Greco, C., Andersen, D.T., Hawes, I., Bowles, A., Yallop, M.L., Barker, G., Jungblut, A.D., 2020. Microbial diversity of pinnacle and conical microbial mats in the perennially ice-covered Lake Untersee, East Antarctica. Front. Microbiol. 11, 3173.

Grossart, H.P., Massana, R., McMahon, K.D., Walsh, D.A., 2020. Linking metagenomics to aquatic microbial ecology and biogeochemical cycles. Limnol. Oceanogr. 65, S2–S20.

Grossart, H.P., Simon, M., 1993. Limnetic macroscopic organic aggregates (lake snow): occurrence, characteristics, and microbial dynamics in Lake Constance. Limnol. Oceanogr. 38, 532–546.

Grossart, H.P., Van den Wyngaert, S., Kagami, M., Wurzbacher, C., Cunliffe, M., Rojas-Jimenez, K., 2019. Fungi in aquatic ecosystems. Nat. Rev. Microbiol. 17, 339–354.

Guillemette, F., von Wachenfeldt, E., Kothawala, D.N., Bastviken, D., Tranvik, L.J., 2017. Preferential sequestration of terrestrial organic matter in boreal lake sediments. J. Geophys. Res.: Biogeosciences 122, 863–874.

Guérin, F., Abril, G., Richard, S., Burban, B., Reynouard, C., Seyler, P., Delmas, R., 2006. Methane and carbon dioxide emissions from tropical reservoirs: significance of downstream rivers. Geophys. Res. Lett. 33, L21407.

Günthel, M., Donis, D., Kirillin, G., Ionescu, D., Bižić, M., McGinnis, D.F., et al., 2021. Reply to "Oxic methanogenesis is only a minor source of lake-wide diffusive CH_4 emissions from lakes". Nat. Commun. 12, 1205.

Guo, M., Zhuang, Q., Tan, Z., Shurpali, N., Juutinen, S., Kortelainen, P., Martikainen, P.J., 2020. Rising methane emissions from boreal lakes due to increasing ice-free days. Environ. Res. Lett. 15, 064008.

Han, X., Schubert, C.J., Fiskal, A., Dubois, N., Lever, M.A., 2020. Eutrophication as a driver of microbial community structure in lake sediments. Environ. Microbiol. 22, 3446–3462.

Harrison, J.A., Frings, P.J., Beusen, A.H., Conley, D.J., McCrackin, M.L., 2012. Global importance, patterns, and controls of dissolved silica retention in lakes and reservoirs. Global Biogeochem. Cycles 26, GB2037.

Hayer, M., Wymore, A.S., Hungate, B.A., Schwartz, E., Koch, B.J., Marks, J.C., 2022. Microbes on decomposing litter in streams: entering on the leaf or colonizing in the water? ISME J. 16, 717–725.

Heathcote, A.J., Anderson, N.J., Prairie, Y.T., Engstrom, D.R., del Giorgio, P.A., 2015. Large increases in carbon burial in northern lakes during the Anthropocene. Nat. Commun. 6, 1694.

Herfort, L., Crump, B.C., Fortunato, C.S., McCue, L.A., Campbell, V., Simon, H.M., et al., 2017. Factors affecting the bacterial community composition and heterotrophic production of Columbia River estuarine turbidity maxima. Microbiol. Open 6, e00522.

Hofmann, H., Federwisch, L., Peeters, F., 2010. Wave-induced release of methane: littoral zones as source of methane in lakes. Limnol. Oceanogr. 55, 1990–2000.

Hölker, F., Vanni, M.J., Kuiper, J.J., Meile, C., Grossart, H.P., Stief, P., et al., 2015. Tube-dwelling invertebrates: tiny ecosystem engineers have large effects in lake ecosystems. Ecol. Monogr. 85, 333–351.

Holmer, M., Storkholm, P., 2001. Sulphate reduction and sulphur cycling in lake sediments: a review. Freshw. Biol. 46, 431–451.

Howard, E.M., Forbrich, I., Giblin, A.E., Lott III, D.E., Cahill, K.L., Stanley, R.H., 2018. Using noble gases to compare parameterizations of air-water gas exchange and to constrain oxygen losses by ebullition in a shallow aquatic environment. J. Geophys. Res.: Biogeosciences 123, 2711–2726.

Huang, Y., Tian, M., Jin, F., Chen, M., Liu, Z., He, S., et al., 2020a. Coupled effects of urbanization level and dam on microplastics in surface waters in a coastal watershed of Southeast China. Mar. Pollut. Bull. 154, 111089.

Huang, R., Zeng, J., Zhao, D., Cook, K.V., Hambright, K.D., Yu, Z., 2020b. Sediment microbiomes associated with the rhizosphere of emergent macrophytes in a shallow, subtropical lake. Limnol. Oceanogr. 65, S38–S48.

Humborg, C., Ittekkot, V., Cociasu, A., Bodungen, B.v., 1997. Effect of Danube River dam on Black Sea biogeochemistry and ecosystem structure. Nature 386, 385–388.

IPCC, 2018. Global Warming of 1.5°C. Intergovernmental Panel on Climate Change Special Report. World Meteorological Organization, Geneva, Switzerland. https://www.ipcc.ch/sr15/.

Jaiswal, D., Pandey, J., 2019. Anthropogenically enhanced sediment oxygen demand creates mosaic of oxygen deficient zones in the Ganga River: implications for river health. Ecotoxicol. Environ. Saf. 171, 709–720.

Jantharadej, K., Limpiyakorn, T., Kongprajug, A., Mongkolsuk, S., Sirikanchana, K., Suwannasilp, B.B., 2021. Microbial community compositions and sulfate-reducing bacterial profiles in malodorous urban canal sediments. Arch. Microbiol. 203, 1981–1993.

Johnson, R.E., Tuchman, N.C., Peterson, C.G., 1997. Changes in the vertical microdistribution of diatoms within a developing periphyton mat. J. North Am. Benthol. Soc. 16, 503–519.

Jones, E.G., Hyde, K.D., Pang, K.L. (Eds.), 2014. Freshwater Fungi and Fungal-like Organisms. Walter de Gruyter GmbH & Co KG, Berlin.

Jørgensen, B.B., Revsbech, N.P., 1985. Diffusive boundary layers and the oxygen uptake of sediments and detritus 1. Limnol. Oceanogr. 30, 111–122.

Jungblut, A.D., Raymond, F., Dion, M.B., Moineau, S., Mohit, V., Lovejoy, C., et al., 2021. Genomic diversity and CRISPR-cas systems in the cyanobacterium *Nostoc* in the High Arctic. Environ. Microbiol. 23, 2955–2968.

Kallmeyer, J., Grewe, S., Glombitza, C., Kitte, J.A., 2015. Microbial abundance in lacustrine sediments: a case study from Lake Van, Turkey. Int. J. Earth Sci. 104, 1667–1677.

Kankaala, P., Huotari, J., Peltomaa, E., Saloranta, T., Ojala, A., 2006. Methanotrophic activity in relation to methane efflux and total heterotrophic bacterial production in a stratified, humic, boreal lake. Limnol. Oceanogr. 51, 1195–1204.

Karygianni, L., Ren, Z., Koo, H., Thurnheer, T., 2020. Biofilm matrixome: extracellular components in structured microbial communities. Trends Microbiol. 28, 668–681.

Käser, D.H., 2010. A new habitat of subsurface waters: the hyporheic biotope, by Traian Orghidan (1959). Fundam. Appl. Limnol 176, 291–302.

Katsev, S., Crowe, S.A., 2015. Organic carbon burial efficiencies in sediments: the power law of mineralization revisited. Geology 43, 607–610.

Katsev, S., Tsandev, I., L'Heureux, I., Rancourt, D.G., 2006. Factors controlling long-term phosphorus efflux from lake sediments: exploratory reactive-transport modeling. Chem. Geol. 234, 127–147.

Kaye, G.W.C., Laby, T.H., 1986. Tables of Physical and Chemical Constants, Fifteenth ed. Longman, New York.

Klatt, C.G., Wood, J.M., Rusch, D.B., Bateson, M.M., Hamamura, N., Heidelberg, J.F., et al., 2011. Community ecology of hot spring cyanobacterial mats: predominant populations and their functional potential. ISME J. 5, 1262–1278.

Klump, J.V., Edgington, D.N., Granina, L., Remsen III, C.C., 2020. Estimates of the remineralization and burial of organic carbon in Lake Baikal sediments. J. Great Lake. Res. 46, 102–114.

Koschorreck, M., Hentschel, I., Boehrer, B., 2017. Oxygen ebullition from lakes. Geophys. Res. Lett. 44, 9372–9378.

Kufel, L., Strzałek, M., Biardzka, E., Becher, M., 2020. Carbon and nutrients transfer from primary producers to lake sediments—a stoichiometric approach. Limnologica 83, 125794.

Kumagai, M., 1988. Predictive model for resuspension and deposition of bottom sediment in a lake. Jpn. J. Limnol. 49, 185–200.

Kumagai, M., Robarts, R.D., Aota, Y., 2021. Increasing benthic vent formation: a threat to Japan's ancient lake. Sci. Rep. 11, 4175.

Laenen, A., LeTourneau, A.P., 1996. Upper Klamath Basin nutrient-loading study—estimate of wind-induced resuspension of bed sediment during periods of low lake elevation, 95–414 U.S. Geol. Surv. Open. File. Rep 12. Portland, Oregon.

Langenegger, T., Vachon, D., Donis, D., McGinnis, D.F., 2019. What the bubble knows: lake methane dynamics revealed by sediment gas bubble composition. Limnol. Oceanogr. 64, 1526–1544.

LaRowe, D.E., Dale, A.W., Amend, J.P., Van Cappellen, P., 2012. Thermodynamic limitations on microbially catalyzed reaction rates. Geochem. Cosmochim. Acta 90, 96–109.

LaRowe, D.E., Van Cappellen, P., 2011. Degradation of natural organic matter: a thermodynamic analysis. Geochem. Cosmochim. Acta 75, 2030–2042.

Lau, M.P., Del Giorgio, P., 2020. Reactivity, fate and functional roles of dissolved organic matter in anoxic inland waters. Biol. Lett. 16, 20190694.

Laurion, I., Mladenov, N., 2013. Dissolved organic matter photolysis in Canadian Arctic thaw ponds. Environ. Res. Lett. 8, 035026.

Lebreton, L.C.M., van der Zwet, J., Damsteeg, J.W., Slat, B., Andrady, A., Reisser, J., 2017. River plastic emissions to the world's oceans. Nat. Commun. 8, 15611.

Lewandowski, J., Arnon, S., Banks, E., Batelaan, O., Betterle, A., Broecker, T., et al., 2019. Is the hyporheic zone relevant beyond the scientific community? Water 11, 2230.

Li, C., Busquets, R., Campos, L.C., 2020. Assessment of microplastics in freshwater systems: a review. Sci. Total Environ. 707, 135578.

Likens, G.E., Davis, M.B., 1975. Post-glacial history of Mirror Lake and its watershed in New Hampshire, USA: an initial report. Int. Vereinigung für theoretische angewandte Limnologie Verh. 19, 982–993.

Lotfi-Kalahroodi, E., Pierson-Wickmann, A.C., Rouxel, O., Marsac, R., Bouhnik-Le Coz, M., Hanna, K., Davranche, M., 2021. More than redox, biological organic ligands control iron isotope fractionation in the riparian wetland. Sci. Rep. 11, 1933.

Luoma, S.N., Rainbow, P.S., 2008. Metal Contamination in Aquatic Environments: Science and Lateral Management. Cambridge University Press, New York.

Lyautey, E., Billard, E., Tissot, N., Jacquet, S., Domaizon, I., 2021. Seasonal dynamics of abundance, structure, and diversity of methanogens and methanotrophs in lake sediments. Microb. Ecol. 82, 559–571.

Lyu, Z., Shao, N., Akinyemi, T., Whitman, W.B., 2018. Methanogenesis. Curr. Biol. 28, R727–R732.

Maavara, T., Chen, Q., Van Meter, K., Brown, L.E., Zhang, J., Ni, J., Zarfl, C., 2020. River dam impacts on biogeochemical cycling. Nat. Rev. Earth Environ. 1, 103–116.

MacIntyre, S., Cortés, A., Sadro, S., 2018. Sediment respiration drives circulation and production of CO_2 in ice-covered Alaskan arctic lakes. Limnol. Oceanography. Letters 3, 302–310.

Mack, L., Andersen, H.E., Beklioğlu, M., Bucak, T., Couture, R.-M., Cremona, F., et al., 2019. The future depends on what we do today—projecting Europe's surface water quality into three different future scenarios. Sci. Total Environ. 668, 470–484.

Maier, D.B., Rydberg, J., Bigler, C., Renberg, I., 2013. Compaction of recent varved lake sediments. GFF 135, 231–236.

Mailman, M., Stepnuk, L., Cicek, N., Bodaly, R.D., 2006. Strategies to lower methyl mercury concentrations in hydroelectric reservoirs and lakes: a review. Sci. Total Environ. 368, 224–235.

Man, X., Bierlein, K.A., Lei, C., Bryant, L.D., Wüest, A., Little, J.C., 2020. Improved modeling of sediment oxygen kinetics and fluxes in lakes and reservoirs. Environ. Sci. Technol. 54, 2658–2666.

Maranger, E., Jones, S.E., Cotner, J.B., 2018. Stoichiometry of carbon, nitrogen and phosphorus through the freshwater pipe. Limnol. Oceanography. Letters 3, 89–101.

Markelov, I., Couture, R.M., Fischer, R., Haande, S., Van Cappellen, P., 2019. Coupling water column and sediment biogeochemical dynamics: modeling internal phosphorus loading, climate change responses, and mitigation measures in Lake Vansjø, Norway. J. Geophys. Res.: Biogeosciences 124, 3847–3866.

Markelova, E., Parsons, C.T., Couture, R.M., Smeaton, C.M., Madé, B., Charlet, L., Van Cappellen, P., 2018. Deconstructing the redox cascade: what role do microbial exudates (flavins) play? Environ. Chem. 14, 515–524.

Marks, J.C., 2019. Revisiting the fates of dead leaves that fall into streams. Ann. Rev. Ecol. Evol. Syst. 50, 547–568.

Masue-Slowey, Y., Ying, S.C., Kocar, B.D., Pallud, C.E., Fendorf, S., 2013. Dependence of arsenic fate and transport on biogeochemical heterogeneity arising from the physical structure of soils and sediments. J. Environ. Qual. 42, 1119–1129.

Matthews, E., Johnson, M.S., Genovese, V., Du, J., Bastviken, D., 2020. Methane emission from high latitude lakes: methane-centric lake classification and satellite-driven annual cycle of emissions. Sci. Rep. 10, 12465.

Matveev, A., Laurion, I., Deshpande, B.N., Bhiry, N., Vincent, W.F., 2016. High methane emissions from thermokarst lakes in subarctic peatlands. Limnol. Oceanogr. 61, S150–S164.

Matveev, A., Laurion, I., Vincent, W.F., 2019. Winter accumulation of methane and its variable timing of release from thermokarst lakes on subarctic peatlands. J. Geophysical Res. Biogeosciences 124, 3521–3535.

McAllister, T.G., Wood, S.A., Hawes, I., 2016. The rise of toxic benthic *Phormidium* proliferations: a review of their taxonomy, distribution, toxin content and factors regulating prevalence and increased severity. Harmful Algae 55, 282–294.

McGinnis, D.F., Greinert, J., Artemov, Y., Beaubien, S.E., Wüest, A., 2006. Fate of rising methane bubbles in stratified waters: how much methane reaches the atmosphere? J. Geophys. Res.: Oceans 111, C09007.

Mendonça, R., Müller, R.A., Clow, D., Verpoorter, C., Raymond, P., Tranvik, L.J., Sobek, S., 2017. Organic carbon burial in global lakes and reservoirs. Nat. Commun. 8, 1694.

Meyers, P.A., Ishiwatari, R., 1993. Lacustrine organic geochemistry—an overview of indicators of organic matter sources and diagenesis in lake sediments. Org. Geochem. 20, 867–900.

Meyers, P.A., Teranes, J.L., 2002. Sediment organic matter. In: Last, W.M., Smol, J.P. (Eds.), Tracking Environmental Change Using Lake Sediments. Springer, Dordrecht, pp. 239–269.

Millera Ferriz, L., Ponton, D.E., Storck, V., Leclerc, M., Bilodeau, F., Walsh, D.A., Amyot, M., 2021. Role of organic matter and microbial communities in mercury retention and methylation in sediments near run-of-river hydroelectric dams. Sci. Total Environ. 774, 145686.

Mohit, V., Culley, A., Lovejoy, C., Bouchard, F., Vincent, W.F., 2017. Hidden biofilms in a far northern lake and implications for the changing Arctic. npj Biofilms and Microbiomes 3, 17.

Morley, S.A., Rhodes, L.D., Baxter, A.E., Goetz, G.W., Wells, A.H., Lynch, K.D., 2021. Invertebrate and microbial response to hyporheic restoration of an urban stream. Water 13, 481.

Mortimer, C.H., Mackereth, F.J.H., 1958. Convection and its consequences in ice-covered lakes. Int. Vereinigung für theoretische angewandte Limnologie Verh. 13, 923–932.

Mosley, L.M., Liss, P.S., 2020. Particle aggregation, pH changes and metal behaviour during estuarine mixing: review and integration. Mar. Freshw. Res. 71, 300–310.

Mostovaya, A., Koehler, B., Guillemette, F., Brunberg, A.K., Tranvik, L.J., 2016. Effects of compositional changes on reactivity continuum and decomposition kinetics of lake dissolved organic matter. J. Geophys. Res. Biogeosciences 121, 1733–1746.

Nelson, W.C., Graham, E.B., Crump, A.R., Fansler, S.J., Arntzen, E.V., Kennedy, D.W., Stegen, J.C., 2020. Distinct temporal diversity profiles for nitrogen cycling genes in a hyporheic microbiome. PLoS One 15, e0228165.

Noe, G.B., Cashman, M.J., Skalak, K., Gellis, A., Hopkins, K.G., Moyer, D., et al., 2020. Sediment dynamics and implications for management: State of the science from long-term research in the Chesapeake Bay watershed. WIRES Water, 7, e1454.

Orihel, D.M., Baulch, H.M., Casson, N.J., North, R.L., Parsons, C.T., Seckar, D.C., Venkiteswaran, J.J., 2017. Internal phosphorus loading in Canadian fresh waters: a critical review and data analysis. Can. J. Fish. Aquat. Sci. 74, 2005–2029.

Otsuki, A., Wetzel, R.G., 1972. Coprecipitation of phosphate with carbonates in a marl lake. Limnol. Oceanogr. 17, 763–767.

Pangala, S.R., Enrich-Prast, A., Basso, L.S., Peixoto, R.B., Bastviken, D., Hornibrook, E.R.C., et al., 2017. Large emissions from floodplain trees close the Amazon methane budget. Nature 552, 230–234.

Paper, J.M., Flynn, T.M., Boyanov, M.I., Kemner, K.M., Haller, B.R., Crank, K., et al., 2021. Influences of pH and substrate supply on the ratio of iron to sulfate reduction. Geobiology 19, 405–420.

Parker, B.C., Simmons Jr., G.M., Wharton Jr., R.A., Seaburg, K.G., Love, F.G., 1982. Removal of organic and inorganic matter from Antarctic lakes by aerial escape of bluegreen algal mats. J. Phycol. 18, 72–78.

Peiffer, S., Kappler, A., Haderlein, S.B., Schmidt, C., Byrne, J.M., Kleindienst, S., Vogt, C., Richnow, H.H., Obst, M., Angenent, L.T., Bryce, C., 2021. A biogeochemical–hydrological framework for the role of redox-active compounds in aquatic systems. Nat. Geosci. 14, 264–272.

Pelletier, N., Chételat, J., Sinon, S., Vermaire, J.C., 2022. Wildfires trigger multi-decadal increases in sedimentation rate and metal loading to subarctic montane lakes. Sci. Total Environ. 824, 153738.

Pestana, I.A., Azevedo, L.S., Bastos, W.R., de Souza, C.M.M., 2019. The impact of hydroelectric dams on mercury dynamics in South America: a review. Chemosphere 219, 546–556.

Pester, M., Knorr, K.H., Friedrich, M.W., Wagner, M., Loy, A., 2012. Sulfate-reducing microorganisms in wetlands—fameless actors in carbon cycling and climate change. Front. Microbiol. 3, 72.

Pilla, R.M., Couture, R.M., 2021. Attenuation of photosynthetically active radiation and ultraviolet radiation in response to changing dissolved organic carbon in browning lakes: modeling and parametrization. Limnol. Oceanogr. 66, 2278–2289.

Ploug, H., Terbrüggen, A., Kaufmann, A., Wolf-Gladrow, D., Passow, U., 2010. A novel method to measure particle sinking velocity in vitro, and its comparison to three other in vitro methods. Limnol Oceanogr. Methods 8, 386–393.

Pogodaeva, T.V., Lopatina, I.N., Khlystov, O.M., Egorov, A.V., Zemskaya, T.I., 2017. Background composition of pore waters in Lake Baikal bottom sediments. J. Great Lake. Res. 43, 1030–1043.

Poissant, L., Constant, P., Pilote, M., Canario, J., O'Driscoll, N., Ridal, J., Lean, D., 2007. The ebullition of hydrogen, carbon monoxide, methane, carbon dioxide and total gaseous mercury from the Cornwall Area of Concern. Sci. Total Environ. 381, 256–262.

Prėskienis, V., Laurion, I., Bouchard, F., Douglas, P.M.J., Billett, M.F., Fortier, D., Xu, X., 2021. Seasonal patterns in greenhouse gas emissions from lakes and ponds in a High Arctic polygonal landscape. Limnol. Oceanogr. 66, S117–S141.

Prieto-Barajas, C.M., Valencia-Cantero, E., Santoyo, G., 2018. Microbial mat ecosystems: structure types, functional diversity, and biotechnological application. Electron. J. Biotechnol. 31, 48–56.

Quesada, A., Vincent, W.F., 1997. Strategies of adaptation by Antarctic cyanobacteria to ultraviolet radiation. Eur. J. Phycol. 32, 335–342.

Rabaey, J.S., Domine, L.M., Zimmer, K.D., Cotner, J.B., 2021. Winter oxygen regimes in clear and turbid shallow lakes. J. Geophys. Res. Biogeosciences 126 e2020JG006065.

Rahalkar, M., Deutzmann, J., Schink, B., Bussmann, I., 2009. Abundance and activity of methanotrophic bacteria in littoral and profundal sediments of Lake Constance (Germany). Appl. Environ. Microbiol. 75, 119–126.

Ramoneda, J., Hawes, I., Pascual-García, A.,J., Mackey, T.,Y., Sumner, D., Jungblut, A., 2021. Importance of environmental factors over habitat connectivity in shaping bacterial communities in microbial mats and bacterioplankton in an Antarctic freshwater system. FEMS Microbiol. Ecol. 97, fiab044.

Rasmussen, J.B., Rowan, D.J., 1997. Wave velocity thresholds for fine sediment accumulation in lakes, and their effect on zoobenthic biomass and composition. J. North Am. Benthol. Soc. 16, 449–465.

Revsbech, N.P., Trampe, E., Lichtenberg, M., Ward, D.M., Kühl, M., 2016. In situ hydrogen dynamics in a hot spring microbial mat during a diel cycle. Appl. Environ. Microbiol. 82, 4209–4217.

Saarenheimo, J., Rissanen, A.J., Arvola, L., Nykänen, H., Lehmann, M.F., Tiirola, M., 2015. Genetic and environmental controls on nitrous oxide accumulation in lakes. PLoS One 10, e0121201.

Samson, R., Shah, M., Yadav, R., Sarode, P., Rajput, V., Dastager, S.G., et al., 2019. Metagenomic insights to understand transient influence of Yamuna River on taxonomic and functional aspects of bacterial and archaeal communities of River Ganges. Sci. Total Environ. 674, 288–299.

Santoso, A.B., Hamilton, D.P., Hendy, C.H., Schipper, L.A., 2017. Carbon dioxide emissions and sediment organic carbon burials across a gradient of trophic state in eleven New Zealand lakes. Hydrobiologia 795, 341–354.

Savio, D., Sinclair, L., Ijaz, U.Z., Parajka, J., Reischer, G.H., Stadler, P., et al., 2015. Bacterial diversity along a 2600 km river continuum. Environ. Microbiol. 17, 4994–5007.

Schmidt, M.L., Biddanda, B.A., Weinke, A.D., Chiang, E., Januska, F., Props, R., Denef, V.J., 2020. Microhabitats are associated with diversity–productivity relationships in freshwater bacterial communities. FEMS Microbiol. Ecol. 96, fiaa029.

Schmitz, R.A., Peeters, S.H., Versantvoort, W., Picone, N., Pol, A., Jetten, M.S., Op den Camp, H.J., 2021. Verrucomicrobial methanotrophs: ecophysiology of metabolically versatile acidophiles. FEMS Microbiol. Rev. 45, fuab00745.

Sebacher, D.I., Harriss, R.C., Bartlett, K.B., 1985. Methane emissions to the atmosphere through aquatic plants. J. Environ. Qual. 14, 40–46.

Segatto, P.L., Battin, T.J., Bertuzzo, E., 2020. Modeling the coupled dynamics of stream metabolism and microbial biomass. Limnol. Oceanogr. 65, 1573–1593.

Sepulveda-Jauregui, A., Walter Anthony, K.M., Martinez-Cruz, K., Greene, S., Thalasso, F., 2015. Methane and carbon dioxide emissions from 40 lakes along a north–south latitudinal transect in Alaska. Biogeosciences 12, 3197–3223.

Sever, M., 2021. Inner Workings: big wildfires mobilize mercury. What are the risks to surface water? Proc. Natl. Acad. Sci. USA 118, e2110558118.

Shi, L., Dong, H., Reguera, G., Beyenal, H., Lu, A., Liu, J., Yu, H.Q., Fredrickson, J.K., 2016. Extracellular electron transfer mechanisms between microorganisms and minerals. Nat. Rev. Microbiol. 14, 651–662.

Shrestha, N.K., Wang, J., 2018. Predicting sediment yield and transport dynamics of a cold climate region watershed in changing climate. Sci. Total Environ. 625, 1030–1045.

Smith, G.J., Wrighton, K.C., 2019. Metagenomic approaches unearth methanotroph phylogenetic and metabolic diversity. Curr. Issues Mol. Biol. 33, 57–84.

Sobek, S., Durisch-Kaiser, E., Zurbrügg, R., Wongfun, N., Wessels, M., Pasche, N., Wehrli, B., 2009. Organic carbon burial efficiency in lake sediments controlled by oxygen exposure time and sediment source. Limnol. Oceanogr. 54, 2243–2254.

Soulsby, R.L., 1983. The Bottom Boundary Layer of Shelf Seas. Elsevier Oceanography Series, pp. 189–266, 35.

Sridhar, K.R., 2020. Dimensions, diversity and ecology of aquatic mycobiome. Kavaka 54, 10–23.

Staehr, P.A., Bade, D., Van de Bogert, M.C., Koch, G.R., Williamson, C., Hanson, P., et al., 2010. Lake metabolism and the diel oxygen technique: state of the science. Limnol Oceanogr. Methods 8, 628–644.

Suberkropp, K., Klug, M.J., 1980. The maceration of deciduous leaf litter by aquatic hyphomycetes. Can. J. Bot. 58, 1025–1031.

Sweerts, J.P.R., de Beer, D., 1989. Microelectrode measurements of nitrate gradients in the littoral and profundal sediments of a meso-eutrophic lake (Lake Vechten, The Netherlands). Appl. Environ. Microbiol. 55, 754–757.

Tee, H.S., Waite, D., Payne, L., Middleditch, M., Wood, S., Handley, K.M., 2020. Tools for successful proliferation: diverse strategies of nutrient acquisition by a benthic cyanobacterium. ISME J. 14, 2164–2178.

Theisen, C.H., Sumner, D.Y., Mackey, T.J., Lim, D.S.S., Brady, A.L., Slater, G.F., 2015. Carbonate fabrics in the modern microbialites of Pavilion Lake: two suites of microfabrics that reflect variation in microbial community morphology, growth habit, and lithification. Geobiology 13, 357–372.

Thottathil, S.D., Prairie, Y.T., 2021. Coupling of stable carbon isotopic signature of methane and ebullitive fluxes in northern temperate lakes. Sci. Total Environ. 777, 146117.

Tong, Y., Wang, M., Peñuelas, J., Liu, X., Paerl, H.W., Elser, J.J., et al., 2020. Improvement in municipal wastewater treatment alters lake nitrogen to phosphorus ratios in populated regions. Proc. Natl. Acad. Sci. USA 117, 11566–11572.

Trout-Haney, J.V., Ritger, A.L., Cottingham, K.L., 2021. Benthic cyanobacteria of the genus *Nostoc* are a source of microcystins in Greenlandic lakes and ponds. Freshw. Biol. 66, 266–277.

Vachon, D., Prairie, Y.T., Guillemette, F., del Giorgio, P.A., 2017. Modeling allochthonous dissolved organic carbon mineralization under variable hydrologic regimes in boreal lakes. Ecosystems 20, 781–795.

Van Meter, K.J., Basu, N.B., Veenstra, J.J., Burras, C.L., 2016. The nitrogen legacy: emerging evidence of nitrogen accumulation in anthropogenic landscapes. Environ. Res. Lett. 11, 035014.

Vesterinen, J., Devlin, S.P., Syväranta, J., Jones, R.I., 2016. Accounting for littoral primary production by periphyton shifts a highly humic boreal lake towards net autotrophy. Freshw. Biol. 61, 265–276.

Vigneron, A., Cruaud, P., Alsop, E., de Rezende, J.R., Head, I.M., Tsesmetzis, N., 2018b. Beyond the tip of the iceberg; a new view of the diversity of sulfite-and sulfate-reducing microorganisms. ISME J. 12, 2096–2099.

Vigneron, A., Cruaud, P., Culley, A.I., Couture, R.M., Lovejoy, C., Vincent, W.F., 2021. Genomic evidence for sulfur intermediates as new biogeochemical hubs in a model aquatic microbial ecosystem. Microbiome 9, 46.

Vigneron, A., Cruaud, P., Mohit, V., Martineau, M.J., Culley, A.I., Lovejoy, C., Vincent, W.F., 2018a. Multiple strategies for light-harvesting, photoprotection, and carbon flow in high latitude microbial mats. Front. Microbiol. 9, 2881.

Vigneron, A., Lovejoy, C., Cruaud, P., Kalenitchenko, D., Culley, A., Vincent, W.F., 2019. Contrasting winter versus summer microbial communities and metabolic functions in a permafrost thaw lake. Front. Microbiol. 10, 1656.

Vilas, M.P., Marti, C.L., Adams, M.P., Oldham, C.E., Hipsey, M.R., 2017. Invasive macrophytes control the spatial and temporal patterns of temperature and dissolved oxygen in a shallow lake: a proposed feedback mechanism of macrophyte loss. Front. Plant Sci. 8, 2097.

Villeneuve, V., Vincent, W.F., Komarek, J., 2001. Community structure and microhabitat characteristics of cyanobacterial mats in an extreme High Arctic environment: Ward Hunt Lake. Nova Hedwigia Beih. 123, 199–224.

Vincent, W.F., 2000. Cyanobacterial dominance in the polar regions. In: Whitton, B.A., Potts, M. (Eds.), The Ecology of Cyanobacteria. Kluwer Academic Press, The Netherlands, pp. 321–340.

Vincent, W.F., 2018. Lakes—A Very Short Introduction. Oxford University Press, U.K.

Vincent, W.F., Downes, M.T., 1981. Nitrate accumulation in aerobic hypolimnia: relative importance of benthic and planktonic nitrifiers in an oligotrophic lake. Appl. Environ. Microbiol. 42, 565–573.

Vincent, W.F., Fortier, D., Lévesque, E., Boulanger-Lapointe, N., Tremblay, B., Sarrazin, D., et al., 2011. Extreme ecosystems and geosystems in the Canadian High Arctic: Ward Hunt Island and vicinity. Ecoscience 18, 236–261.

Walsh, J.R., Corman, J.R., Munoz, S.E., 2019. Coupled long-term limnological data and sedimentary records reveal new control on water quality in a eutrophic lake. Limnol. Oceanogr. 64, S34–S48.

Walter Anthony, K.M., Lindgren, P., Hanke, P., Engram, M., Anthony, P., Daanen, R.P., et al., 2021. Decadal-scale hotspot methane ebullition within lakes following abrupt permafrost thaw. Environ. Res. Lett. 16, 035010.

Wang, Y., Pan, J., Yang, J., Zhou, Z., Pan, Y., Li, M., 2020. Patterns and processes of free-living and particle-associated bacterioplankton and archaeaplankton communities in a subtropical river-bay system in South China. Limnol. Oceanogr. 65, S161–S179.

Wang, G., Xia, X., Liu, S., Zhang, L., Zhang, S., Wang, J., Xi, N., Zhang, Q., 2021. Intense methane ebullition from urban inland waters and its significant contribution to greenhouse gas emissions. Water Res. 189, 116654.

Ward, D.M., Ferris, M.J., Nold, S.C., Bateson, M.M., 1998. A natural view of microbial biodiversity within hot spring cyanobacterial mat communities. Microbiol. Mol. Biol. Rev. 62, 1353–1370.

Weatherly, G.L., Martin, P.J., 1978. On the structure and dynamics of the oceanic bottom boundary layer. J. Phys. Oceanogr. 8, 557–570.

Wei, Y., Zhang, M., Cui, L., Pan, X., Liu, W., Li, W., Lei, Y., 2020. Winter decomposition of emergent macrophytes affects water quality under ice in a temperate shallow lake. Water 12, 2640.

Wentworth, C.K., 1922. A scale of grade and class terms for clastic sediments. J. Geol. 30, 377–392.

Wetzel, R.G., Likens, G.E., 2000. Limnological Analyses, Third ed. Springer, New York.

Wik, M., 2016. Emission of Methane from Northern Lakes and Ponds. Doctoral Thesis. Stockholm University, Department of Geological Sciences, Stockholm, Sweden, p. 42.

Wik, M., Johnson, J.E., Crill, P.M., DeStasio, J.P., Erickson, L., Halloran, M.J., et al., 2018. Sediment characteristics and methane ebullition in three subarctic lakes. J. Geophys. Res.: Biogeosci. 123, 2399–2411.

Wik, M., Thornton, B.F., Bastviken, D., MacIntyre, S., Varner, R.K., Crill, P.M., 2014. Energy input is primary controller of methane bubbling in subarctic lakes. Geophys. Res. Lett. 41, 555–560.

Wik, M., Varner, R.K., Walter Anthony, K., MacIntyre, S., Bastviken, D., 2016. Climate-sensitive northern lakes and ponds are critical components of methane release. Nat. Geosci. 9, 99–105.

Wilkinson, J., Maeck, A., Alshboul, Z., Lorke, A., 2015. Continuous seasonal river ebullition measurements linked to sediment methane formation. Environ. Sci. Technol. 49, 13121–13129.

Wohl, E., Bledsoe, B.P., Jacobson, R.B., Poff, N.L., Rathburn, S.L., Walters, D.M., Wilcox, A.C., 2015. The natural sediment regime in rivers: broadening the foundation for ecosystem management. Bioscience 65, 358–371.

Wohl, E., Kramer, N., Ruiz-Villanueva, V., Scott, D.N., Comiti, F., Gurnell, A.M., et al., 2019. The natural wood regime in rivers. Bioscience 69, 259–273.

Wood, S.A., Depree, C., Brown, L., McAllister, T., Hawes, I., 2015. Entrapped sediments as a source of phosphorus in epilithic cyanobacterial proliferations in low nutrient rivers. PLoS One 10, e0141063.

Wurzbacher, C., Fuchs, A., Attermeyer, K., Frindte, K., Grossart, H.P., Hupfer, M., et al., 2017. Shifts among Eukaryota, Bacteria, and Archaea define the vertical organization of a lake sediment. Microbiome 5, 41.

Xia, X., Liu, T., Yang, Z., Michalski, G., Liu, S., Jia, Z., Zhang, S., 2017. Enhanced nitrogen loss from rivers through coupled nitrification-denitrification caused by suspended sediment. Sci. Total Environ. 579, 47–59.

Xiao, Q.T., Zhang, M., Hu, Z.H., Gao, Y.Q., Hu, C., Liu, C., et al., 2017. Spatial variations of methane emission in a large shallow eutrophic lake in subtropical climate. J. Geophysical. Res. Biogeosciences 122, 1597–1614.

Xu, M., Dong, X., Yang, X., Chen, X., Zhang, Q., Liu, Q., et al., 2017. Recent sedimentation rates of shallow lakes in the middle and lower reaches of the Yangtze River: patterns, controlling factors and implications for lake management. Water 9, 617.

Yakimovich, K.M., Orland, C., Emilson, E.J., Tanentzap, A.J., Basiliko, N., Mykytczuk, N.C., 2020. Lake characteristics influence how methanogens in littoral sediments respond to terrestrial litter inputs. ISME J. 14, 2153–2163.

Yang, P., Zhang, Y., Yang, H., Guo, Q., Lai, D.Y., Zhao, G., et al., 2020. Ebullition was a major pathway of methane emissions from the aquaculture ponds in southeast China. Water Res. 184, 116176.

Zeikus, J.G., Winfrey, M.R., 1976. Temperature limitation of methanogenesis in aquatic sediments. Appl. Environ. Microbiol. 31, 99–107.

Zettler, E.R., Mincer, T.J., Amaral-Zettler, L.A., 2013. Life in the "plastisphere": microbial communities on plastic marine debris. Environ. Sci. Technol. 47, 7137–7146.

Zhang, L., Xia, X., Liu, S., Zhang, S., Li, S., Wang, J., et al., 2020. Significant methane ebullition from alpine permafrost rivers on the East Qinghai–Tibet Plateau. Nat. Geosci. 13, 349–354.

Zhu, D., Wu, Y., Chen, H., He, Y., Wu, N., 2016. Intense methane ebullition from open water area of a shallow peatland lake on the eastern Tibetan Plateau. Sci. Total Environ. 542, 57–64.

Zhu, Y., Purdy, K.J., Eyice, Ö., Shen, L., Harpenslager, S.F., Yvon-Durocher, G., et al., 2020a. Disproportionate increase in freshwater methane emissions induced by experimental warming. Nat. Clim. Change 10, 685–690.

Zhu, L., Shi, W., Van Dam, B., Kong, L., Yu, J., Qin, B., 2020b. Algal accumulation decreases sediment nitrogen removal by uncoupling nitrification-denitrification in shallow eutrophic lakes. Environ. Sci. Technol. 54, 6194–6201.

C H A P T E R

28

Organic Carbon Cycling and Ecosystem Metabolism

Erin R. Hotchkiss[1] *and Tonya DelSontro*[2]

[1]Department of Biological Sciences, Virginia Polytechnic Institute and State University, Blacksburg, Virginia, United States [2]Department of Earth and Environmental Sciences, University of Waterloo, Waterloo, Ontario, Canada

OUTLINE

I. Overview

Freshwater carbon cycling reflects organic matter (OM) and other energy input dynamics, governs the flow of energy through food webs and to downstream ecosystems, and is sustained by external and nonliving organic matter sources in most freshwater ecosystems. Research on the trophic structure of aquatic ecosystems was historically dominated by evaluations of energy fixation by primary producers of pelagic communities and the transfer efficiencies of algal–cyanobacterial energy sources to higher trophic levels, especially for lake food webs. Early descriptions of energy flow in ecosystems emphasized how energy transfers and losses across trophic levels limit the trophic complexity of communities (Lindeman, 1942; Hutchinson, 1959; Wetzel, 1995). These relationships between pelagic trophic structure and energy fluxes, however, solely focused on *particulate organic carbon* (POC) and the ingestion and utilization of *particulate organic matter* (POM). Early work on feeding–POM relationships quantified variations in sizes of ingested POM, morphological aspects of ingestion (e.g., filtration and gape), avoidance of ingestion (e.g., transparency/visibility, interference by cellular or body projections), and the capabilities of organisms to move within the pelagic zone in relation to refuges or escape from predators. Many organisms ingest variable amounts of particulate *detritus (dead OM)*. Particulate detritus is typically a much larger portion of total POM compared to living POM of

Wetzel's Limnology, Fourth Edition
https://doi.org/10.1016/B978-0-12-822701-5.00028-8

primary producers (e.g., Saunders, 1972). Thus quantitative measures of consumption, assimilation, and residual egestion of detrital POM and its associated microbes are critical for aquatic ecosystem budgets and food web energetics.

From the inception of evolving ecological constructs, early research on trophic dynamics, particularly in lakes, emphasized integration and interdependence among biotic components internally, but these dynamics were entirely predation based (see Chapter 20). As the metabolism of community components was analyzed with increasing accuracy, however, estimates of *organic carbon* (OC) flux rates and pathways identified a number of complexities and inconsistencies that could not be explained within conventional food-web paradigms. Early carbon budgets demonstrated the importance of considering the fluxes and reactivity of *dissolved organic carbon* (DOC) from terrestrial, wetland, and littoral production. For example, several lake carbon budgets demonstrated that most DOC was less reactive and derived from structural tissues of terrestrial and higher aquatic plants, even though higher aquatic plants colonized only small portions of the benthic area. Furthermore, most heterotrophic respiration of OM occurs in sediments, which contributes to the net evasion of *carbon dioxide* (CO_2) to the atmosphere. We note here that as perspectives on OM reactivity, persistence, and biological versus physical availability continue to evolve (e.g., Marín-Spiotta et al., 2014; Lehmann and Kleber, 2015), updates to this edition of Wetzel's *Limnology* include removing most references to "recalcitrant" or "labile" OM. Instead, we focus on relative OM ***reactivity***, how reactivity may differ among OM sources and composition, and what that means for ecosystem metabolism and food web energetics.

Up to >90% of the total OM produced within ecosystems or imported to ecosystems from external sources is *metabolized* (i.e., consumed and used for growth, maintenance, and reproduction) but is never consumed by metazoans. Recognition of this reality has been agonizingly slow, or, alternatively, it is acknowledged but ignored by many who focus on predation despite the dominant role of microbial processes in governing ecosystem OC fluxes and metabolism. Predation is often not the prevailing cause of the mortality of organisms. In addition, and also of great ecosystem significance, assimilation efficiencies of ingested food are modest at best under natural conditions (usually <<50%), and much of the ingested OM is released or egested as both soluble and particulate detrital OM. Effective management of aquatic ecosystems is difficult and certainly imprecise if most of the metabolism, energetic fluxes, and control mechanisms of that microbial metabolism are poorly understood and separated from higher trophic levels.

The evaluation of OC budgets and ecosystem metabolism, particularly among low- and medium-order streams, has long demonstrated that running waters are heterotrophic (i.e., ecosystem respiration exceeds primary production and is subsidized by external OC inputs; Duarte and Prairie, 2005; Battin et al., 2008; Hall et al., 2016). These evaluations, more recently supported by stable isotope tracer studies and an increasing number of whole-ecosystem metabolism estimates, assisted in the erosion of the long-held belief that all lakes were autotrophic (i.e., photosynthesis exceeds respiration at the ecosystem scale). We now have substantial evidence that most ecosystems are sustained by a combination of both externally and internally produced OM sources (e.g., del Giorgio et al., 1997; Duarte and Prairie, 2005; Cole et al., 2011). Nonetheless, despite OC budgets that indicate up to 99% of OM fluxes within aquatic ecosystems are detrital based, the predation-based paradigms still often prevail as the primary constructs of ecosystem operations, whereas in reality they are the minority. Essentially, all inland water ecosystems are microbially based heterotrophic ecosystems in which respiration of OM within lakes and streams exceeds—usually greatly exceeds—internal autotrophic production.

Over the last several decades, a general consensus has emerged concerning the functions of dissolved OM (DOM) and POM in aquatic ecosystems, which can be highlighted by two important changes in thought. First, sealing a lake or river off from its drainage basin would drastically alter metabolism and biogeochemical cycling within it. Although lakes and rivers are referred to as ecosystems, the influences of their drainage basin, particularly imported detrital OM, intimately link aquatic ecosystems with their surrounding terrestrial and aquatic environments. Furthermore, a lake and associated inlet rivers themselves are only individual components of a larger landscape unit, the catchment, which must include the entire drainage basin of a lake or river. Inland waters are typically treated as integrated ecosystems with the increasing appreciation that the metabolism of lakes and rivers depends to a major extent on metabolism in the adjacent drainage basin. Second, commonality of function is found amid the plethora of individual habitats and biotic diversity (Wetzel, 1995; 2003); this is especially evident in the recognition that processes are functionally similar in freshwater and marine ecosystems. Differences between these ecosystems are apparent because of differences in the sources, histories, and reactivity of OM, which affect the rates of OM cycling, not the biogeochemical pathways themselves.

The study of OC cycling and ecosystem metabolism in inland waters entered a new era in the 21st century. Over the last few decades, sensors used to estimate rates of photosynthesis and respiration (i.e., ecosystem metabolism) in aquatic ecosystems have become less costly and more reliable, while computational power and process-based models have drastically improved. As technological and computational capacities allow us to estimate daily and annual rates of carbon cycling and fluxes using long-term (i.e., multiyear) time series of high-frequency data across multiple ecosystems, the characterization of annual "metabolic regimes" (*sensu* Bernhardt et al. 2018) has expanded across different ecosystem types, biomes, and geographic regions. As continued studies of aquatic carbon cycling confirm that most ecosystems are heterotrophic and detritus plays an outsized role in supporting food webs, the increasing focus on the role of inland waters in the global carbon cycle generated a new conceptual model of freshwater ecosystems as "active players" (not "passive pipes") in transporting, burying, and emitting carbon (Cole et al., 2007; Battin et al., 2009). We now know that most inland waters are net sources of CO_2 and methane (CH_4) to the atmosphere (Bastviken et al., 2011; Raymond et al., 2013; Stanley et al., 2016), and we also have a better understanding of the degree to which metabolism within inland waters generates some portion of those emissions (e.g., Cole and Caraco, 2001; Battin et al., 2008; Hotchkiss et al., 2015). Furthermore, many limnologists now recognize the need to integrate larger-scale metaecosystem and global datasets with studies of isolated ecosystems (e.g., Tranvik et al., 2018; Webb et al., 2019). These ongoing developments in our perspectives of aquatic ecosystem carbon cycling support the notion that inland waters are: (1) open and heterotrophic with respect to allochthonous inputs and metabolism; and (2) net emitters of greenhouse gases to the atmosphere.

II. Organic matter composition

The OM of soils and inland waters consists of a heterogeneous mixture of plant, microbial, and animal products, either alive or as fragments in various stages of decomposition. OM found in aquatic ecosystems is either *autochthonous*, meaning produced within the ecosystem, or *allochthonous*, meaning of an external, often terrestrial, source from the catchment (see Table 28-1 for relevant terminology and Section III for more details on sources of OM). In inland waters, OM can be present as DOM and POM (Fig. 28-1), which is separated via filtration using pore sizes ranging from 0.2 to 0.7 μm (Thurman, 1985; Dittmar and Stubbins, 2014). Living OM constitutes a minute fraction of total OM. The majority of OM is detrital (i.e., not living) and in dissolved form, although DOM is operationally defined based on the filtration and can include colloids, viruses, and small bacteria (Fig. 28-2). DOM is thus a complex mixture of allochthonous and autochthonous material that includes carbon (C) as well as varying quantities of other elements, including oxygen (O), hydrogen (H), nitrogen (N), sulfur (S), and phosphorus (P) (Fig. 28-1). Even though OM is composed of elements other than carbon, the terms *OM* and *OC*, as well as their dissolved (DOM and DOC) and particulate (POM and POC) fractions (Fig. 28-1), are often used interchangeably in the literature, and our use of OC versus OM here will depend on how other studies measured and reported carbon fluxes.

As total OM in aquatic ecosystems increases, the percentage in the dissolved fraction (DOM) increases disproportionately to that of the particulate fraction (POM). DOM consists of approximately 50% carbon (Birge and Juday 1934), but the exact fraction varies depending on OM source and history. DOM of allochthonous origin contains a lower percentage of N relative to C (C:N ~50:1), while that produced autochthonously by algae and other primary producers has a higher initial N content (C:N ~12:1). Therefore the C:N of OM increases as the proportion of allochthonous DOM increases. Much of allochthonous DOC is dominated by fulvic and humic acids (described below) that originate from the structural tissues of higher plants. While we cannot cover the rich history of OM composition research in its entirety here, further reading of notable early work on OM in various environments includes Breger (1963), Kononova (1966), Swain (1963), and Vallentyne (1957). For additional resources on OM and DOM in inland waters, select reviews of the topic include Creed et al. (2015), Dittmar and Stubbins (2014), Fellman et al. (2010), Findlay and Sinsabaugh (2003), and Thurman (1985).

A. Dissolved organic matter (DOM)

Historically, DOM was separated into two fractions—*humic substances* (*HS*) and *nonhumic substances* (NHS)—based on solubility in acid, where the resulting material has general properties associated with each fraction's reactivity (Table 28-1) (Aiken et al., 1985; Thurman, 1985). Some of the scientific community, however, has moved away from this terminology and the investigation of extracted components of DOM, and toward investigating DOM composition and reactivity in its bulk form, thanks to new analytical techniques and

TABLE 28-1 Dissolved Organic Matter (DOM) Terminology

Term	What it usually means
Allochthonous, allochthonous-like, terrestrial	Refers to DOM that is derived externally from outside the aquatic ecosystem (e.g., from soils or humified plant material)
Autochthonous, autochthonous-like	DOM that is derived within the aquatic ecosystem produced by microbes, algae, or aquatic macrophytes
Aromatic, aromaticity	Property of DOM containing many ring-shaped planar structures that can be measured using a proxy that correlates with proportion of aromatic carbon
Aliphatic	Compounds within the DOM pool that are not aromatic
Lignin, lignin-like	Phenolic compound found in terrestrial plants. Some aquatic organisms may contain lignin-like compounds.
Humic, humic-like	Chemical class, based on extraction technique or solubility
Fulvic, fulvic-like	Chemical class, based on extraction technique or solubility
Microbial, microbial-like, microbially derived, protein-like, tryptophan-like, tyrosine-like	Usually refers to DOM that is derived from *in situ* processes and whose composition is largely simple, small, and of proteinaceous compounds
Labile[a,b]	Easily degraded or easily broken apart through abiotic (e.g., light) and/or biotic transformations
Bioavailable, biolabile	Components of the DOM which are degradable or useable by microbial organisms
Recalcitrant[b]	Not easily degraded, persistent
Refractory[b]	Not degradable
Reactive[a], reactivity	Used to describe abiotic (e.g., light) and chemical effects of DOM (e.g., DOM acting as a proton and metal ion buffer), as well as biological transformations of DOM (e.g., bioreactive)
Chromophoric DOM (CDOM)	Refers to components of the DOM that absorb light or parts of the solar spectrum
Transparent/bleached	Highly photoprocessed

[a]*Sometimes used interchangeably.*
[b]*Terms not readily used any longer; 'Reactivity' is instead preferred.*
Updated from Xenopoulos et al. (2021)

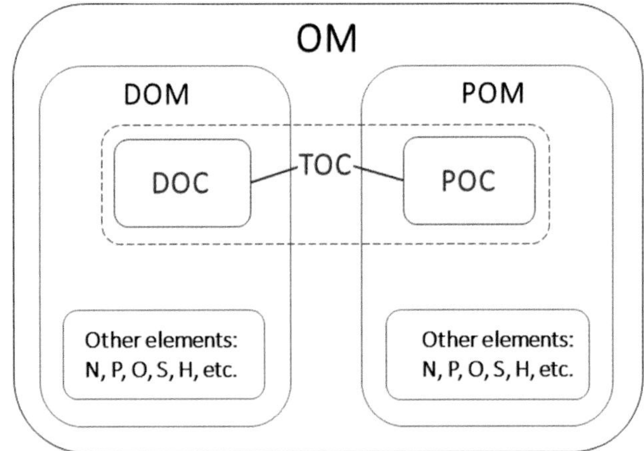

FIGURE 28-1 Forms of organic matter found in aquatic ecosystems: total organic matter (OM), which can be dissolved (DOM) or particulate organic matter (POM); total organic carbon (TOC), which can be dissolved (DOC) or particulate organic carbon (POC). Both DOM and POM can contain dissolved or particulate forms of other elements, such as nitrogen (N), phosphorus (P), oxygen (O), sulfur (S), and hydrogen (H). *(Adapted from Pagano et al. (2014) by D. Kothawala.)*

tools. The evolution of optical methods and mass spectrometry has revealed substantially more information regarding the structure and cycling of DOM in recent decades (Dittmar and Stubbins, 2014; McCallister et al., 2018). We provide a brief overview below of how new knowledge is shifting perspectives and terminology related to OM composition.

i. Changing OM paradigms and humic-like substances

The characterization and use of HS originated in the soil sciences, with the term *humus* used to describe degradation products of organic material in soils and HS used to define the fractions of humus that can be isolated firstly by extraction in an alkali solution (0.1 mol L^{-1} NaOH) (Swift, 1996). HS extracted from soils and sediments were then separated into three categories: (1) *humic acids* (*HA*), which precipitate in water upon acidification (pH 2) but are soluble at higher pH; (2) *fulvic acids* (*FA*), which are soluble at any pH; and (3) *humin*,

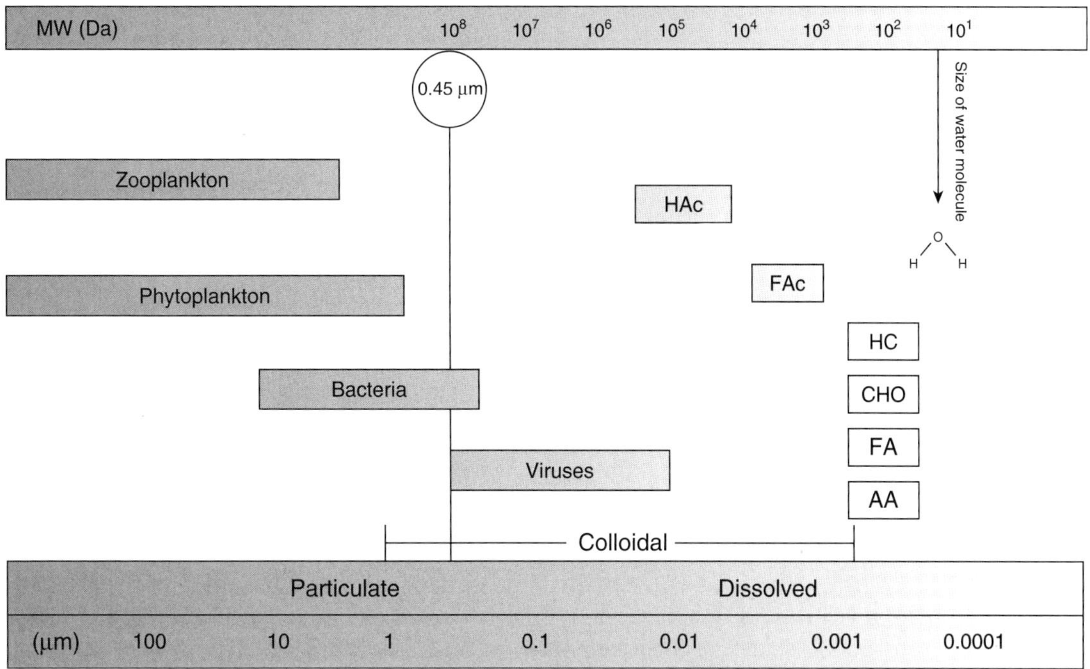

FIGURE 28-2 Molecular weight (MW) in daltons (Da) of organic matter (OM) found in aquatic ecosystems, including humic acids (HAc), fulvic acids (Fac; FA elsewhere in the text of this chapter), hydrocarbons (HC), carbohydrates (CHO), fatty acids (FA), and amino acids (AA). Size (μm) of various OM components, including organisms, with the colloidal fraction range and a typical filter size (0.45 μm) used to distinguish between dissolved and particulate OM shown. *(From Thurman (1985) via Artifon et al. (2019).)*

which is insoluble in water and in dilute solutions at any pH. The International Humic Substances Society (IHSS) updated the definition of FA to include substances that are soluble in alkali as well as acid solutions but also adsorb to a nonionic resin (IHSS, 2007). We now know soil humin contains recognizable classes of molecules, so it is no longer considered an HS but is still within the definition of humus (Hayes and Swift, 2020). NHS includes recognizable debris fragments with carbohydrates, proteins, peptides, amino acids, lipids, lignins, fats, waxes, resins, pigments, other organic substances such as hydrocarbons, and trace compounds (Thurman, 1985). The instantaneous concentrations of individual components of NHS are often low because of their higher reactivity (Table 28-1) and rapid turnover rate.

The HS components in inland waters, namely HA and FA, are operationally defined and well integrated into past OM research (e.g., McKnight and Aiken, 1998; Aiken et al., 1992; Swift, 1996). Reverse osmosis is sometimes used to concentrate the OM prior to extraction (Serkiz and Perdue, 1990). HA and FA are in dissolved form, yet within a colloidal size range, which allows them to provide the characteristic yellow-brown color of some freshwaters (Fig. 28-3; Thurman, 1985). FA is the most soluble and prevalent of the two in freshwater ecosystems (40–60% of DOM; McKnight and Aiken, 1998).

Many now suggest that the HS fractionation scheme is arbitrary, as each fraction is a heterogeneous mixture of amorphous and degraded organic debris, whose specific compositions vary with source material and state of degradation. Researchers agree that a few properties of HS are common across ecosystems. Most agree that HS are dark colored (yellow to black), which is reflected in the various colors that aquatic waters can display (Fig. 28-3), and moderate to high in molecular weight, ranging from hundreds to hundreds of thousands of daltons (Fig. 28-2; Piccolo, 2002). The complex chemistry of HS and presence in higher quantities than NHS has led researchers to believe that HS are less biologically reactive (sometimes referred to as "recalcitrant"; Table 28-1). However, lakes rich in HS also contain more reactive, bioavailable DOM (Tranvik 1988). Thus a discussion regarding the terminology and methods traditionally used to distinguish HS and NHS, and whether HS even exist, continues (e.g., Hayes and Swift, 2020; Kleber and Lehmann, 2019; Lehmann and Kleber, 2015; Olk et al., 2019a, 2019b). Since HS can include bioavailable compounds, it may be that extraction methods created artifacts because of the extreme pH conditions used or

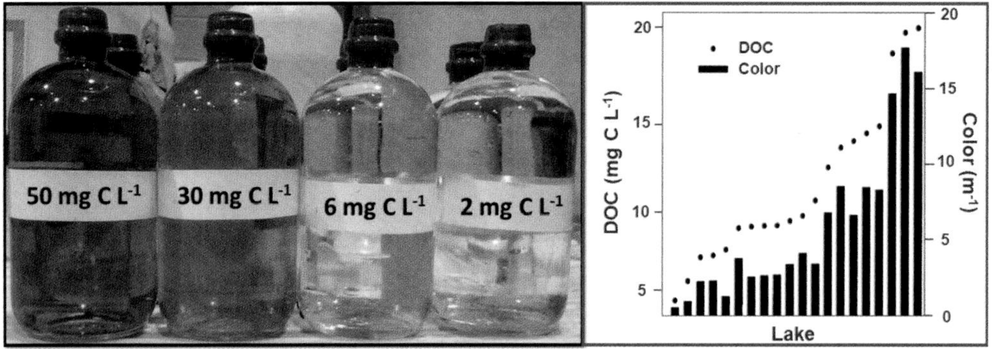

FIGURE 28-3 *Left*: Samples from tributaries of the Moise River in the Cote-Nord region of Quebec, Canada, with a dissolved organic carbon (*DOC*) concentration range of 2–50 mg C L^{-1}. (Photo credit: R. Hutchins, 16 July 2013.) *Right*: Average DOC (mg C L^{-1}) and color (m^{-1}; based on a wavelength-specific absorption coefficient at 440 nm) in 20 lakes in northern Michigan, USA (ranked according to DOC concentration). *(Adapted from Pace and Cole (2002).)*

that extractions did not completely separate HS from NHS (Lehmann and Kleber, 2015). Moreover, some suggest that the very existence of HS should be questioned because no single compositional structure has been discovered for HS. HS fractions are indeed operationally defined and not considered to be homogeneous substances, but HS can be considered members of a family of molecules with similar properties (Hayes and Swift, 2020; Kelleher and Simpson, 2006). Ultimately, limnologists and biogeochemists have moved toward investigating the composition, structure, and reactivity of bulk DOM rather than extracted HS, using new and evolving methods that we highlight below (Kothawala et al., 2021; McCallister et al., 2018).

For further reading on this topic, comprehensive HS reviews include Aiken et al. (1985), Frimmel and Christman (1988), Hayes et al. (1989), Stevenson (1982), Tan (2003), and the special issue associated with Weber et al., (2018). Historical work on aquatic HS can be found in Gjessing (1976), Hessen and Tranvik (1998), Keskitalo and Eloranta (1999), Perdue and Gjessing (1990), and Suffet and MacCarthy (1988).

ii. DOM characterization

DOM is a heterogeneous mixture of detritus that reflects the sources and cycling of OM within and across ecosystem boundaries. Multiple approaches have been developed to analyze various properties of DOM (see reviews by Minor et al., 2014 and McCallister et al., 2018). The following three methods are frequently used to analyze aquatic DOM, and each provides different types of information, ranging from bulk chemical properties such as color or fluorescence to structural properties that include functional groups, molecular mass to charge ratios (m/z), elemental composition (i.e., C:H:O and potentially N, S, and P), and assigned molecular formulas based on m/z. From a limnological perspective, information about the composition of DOM contributes most to our understanding of freshwater ecosystems when linked with the inputs, reactivity, and ecosystem-scale cycling of OM. Integrating results from different methods can provide more detailed insights into not only the composition of DOM but also its reactivity and fate.

Optical analyses

The optical properties of DOM provide information about bulk chemical properties, such as reactivity and source, but optical analysis methods cannot provide direct elemental or structural information. The colored (or chromophoric) fraction of DOM (*colored dissolved organic carbon* or *CDOM*) absorbs light (ultraviolet and visible) and the slope of the absorption spectrum can provide information about DOM source and subsequent processing (Helms et al., 2008). The *specific UV absorbance* (SUVA) at a given wavelength (often 254 nm) normalized for DOC concentration can be used to approximate DOM aromaticity (Table 28-1; Weishaar et al., 2003). Upon exposure to light, a portion of DOM will become excited and emit light at a longer wavelength. Multiple excitation and emission wavelengths can be analyzed to produce an *excitation–emission matrix* (EEM). The *fluorescent DOM* (*FDOM*) fraction will exhibit fluorophore peaks associated with chemical structures that may also be related to source and reactivity (McKnight et al., 2001; Murphy et al., 2014). Using a multivariate approach like *parallel factor analysis* (PARAFAC) with EEMs helps classify different fluorophore peaks according to 2 (Stedmon and Bro, 2008), 6 (Stubbins et al., 2014), or even 13 components (Cory and McKnight, 2005), spanning humic-,

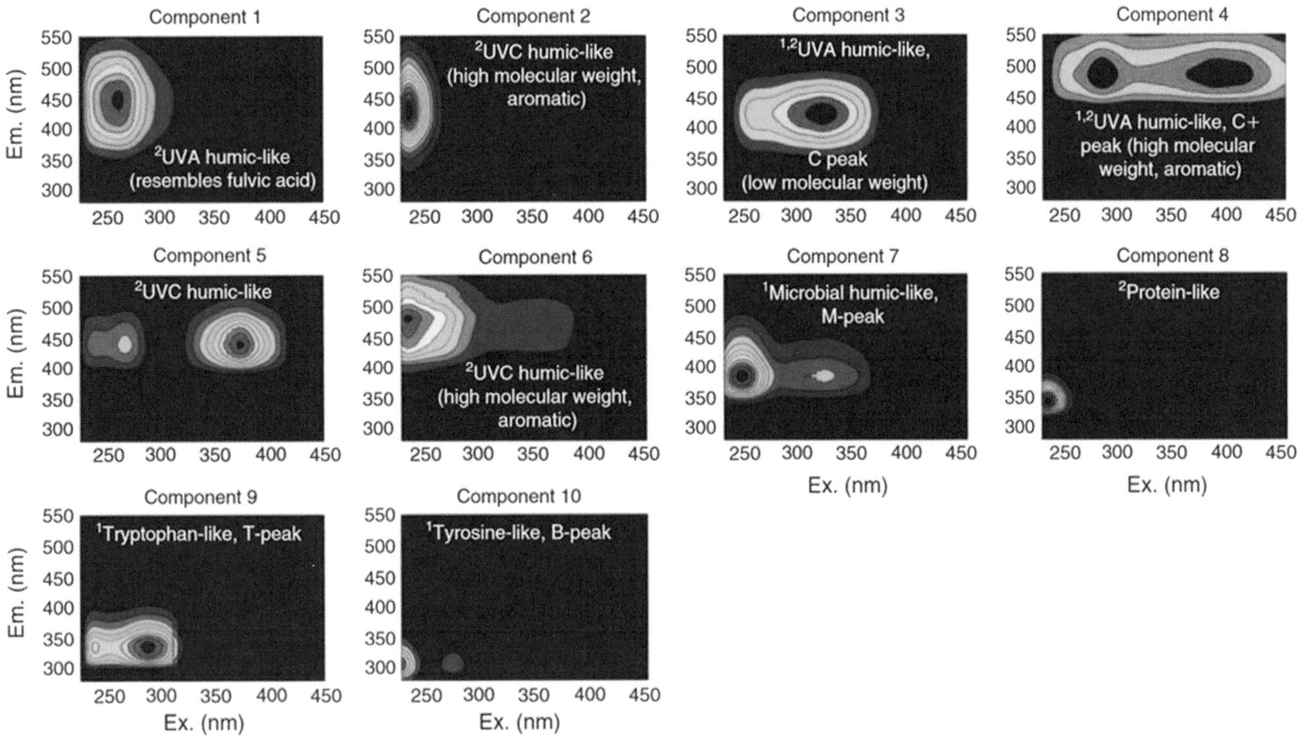

FIGURE 28-4 Examples of excitation–emission (EX-EM) matrices (*EEMs*) of parallel factor analysis (*PARAFAC*) components. Here, *components 1–7* are humic-like (i.e., less reactive, recalcitrant) and *components 8–10* are protein-like (i.e., more reactive, labile). *(From Cuss et al. (2019) and references therein.)*

fulvic-, protein-, and quinone-like fractions, among others (Fig. 28-4; Table 28-1; Cuss et al., 2019).

Relating DOM fluorophore peaks with environmental drivers of peak variability within or across ecosystems provides insights into DOM source, composition, reactivity, or fate. For example, conducting PARAFAC on DOM samples from 560 lakes in Sweden, Kothawala et al. (2014) found that lakes with longer water residence times and more upstream water bodies in their catchment had more protein-like DOM than humic-like or terrestrially derived DOM, with protein-like DOM likely sourced from algal or heterotrophic microbial processes. The PARAFAC analysis thus suggested that DOM composition controlled its reactivity, with the aromatic humic-like fractions being more reactive and the protein-like components being more persistent or resistant to degradation (Fig. 28-5). Often, CDOM and FDOM analyses are combined with other methods, such as those discussed next or incubations of water samples, to provide more structural, molecular, or reactivity information about the DOM. In a study of the five largest rivers in the Republic of Korea (South Korea), Shin et al. (2016) used PARAFAC and biological assay incubations to investigate the role of monsoons on DOM composition and reactivity, respectively. Aromatic terrestrial-like components increased in all rivers following summer monsoons, and the overall reactivity of river DOM decreased, suggesting monsoon-related DOM inputs from the catchment are less reactive than DOM inputs during other times of the year.

Nuclear magnetic resonance (NMR) and high-performance liquid chromatography (HPLC)

One of the most detailed current approaches for analyzing DOM composition is *nuclear magnetic resonance* (NMR) spectroscopy. NMR uses the resonance frequency and magnetic properties of atomic nuclei (^{1}H, ^{13}C, ^{15}N, ^{31}P) to gather information about the presence of functional groups and the cycling of the element being analyzed (i.e., H, C, N, or P; Fig. 28-6) (Abdulla et al., 2010; Mopper et al., 2007). This is the only method of the three highlighted here that can directly resolve DOM structure or fractions thereof, but it cannot resolve structural information for individual molecules. Combining NMR with *high-performance liquid chromatography* (HPLC) or other techniques for

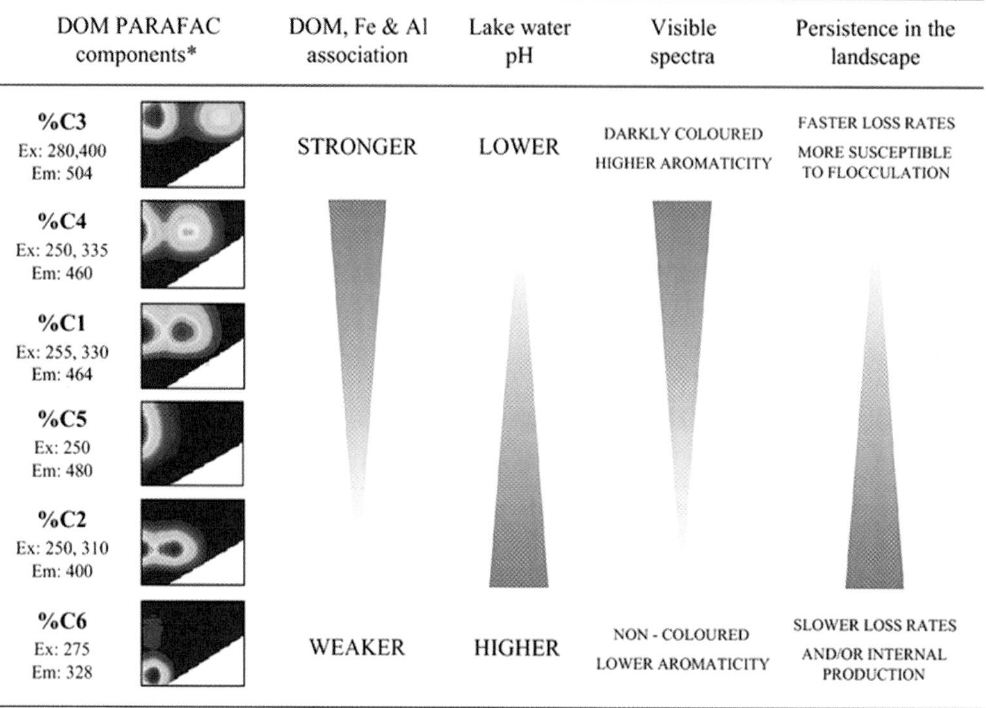

FIGURE 28-5 Parallel factor analysis (*PARAFAC*) of dissolved organic matter (*DOM*) from 560 lakes across Sweden displayed according to emission wavelength (longer to shorter) and PARAFAC DOM component (C1–C6) relative to the importance of each component to different physicochemical properties, including metal association, pH, color, aromaticity, and persistence in the landscape. *(From Kothawala et al. (2014).)*

fractional separation of DOM provides a more detailed analysis of the structural components of DOM (Simpson et al., 2011). NMR analysis reveals the relative percentage of aromatic, aliphatic (Table 28-1), and carbohydrate compound classes, as well as specific functional groups such as carboxyl, hydroxyl, or carbonyl (Mopper et al., 2007). Mopper et al. (2007) and Simpson et al. (2004) provide more detailed reviews of NMR.

While used more often in marine than freshwater DOM analyses to date, NMR has revealed interesting aspects of OM structure in relation to freshwater biogeochemistry. For example, logging increased the amount of C in carboxylic groups in lakes in Quebec (Canada), which enabled greater DOM–mercury interactions (O'Driscoll et al., 2006). Carboxyl-rich alicyclic molecules, which are assumed to be less reactive (Hertkorn et al., 2006), dominated DOM from three lakes in Minnesota (USA) and appeared to be selectively preserved with increasing water residence time (Cao et al., 2018). High-molecular-weight isolates of freshwater DOM can be indistinguishable from that of marine DOM: a large fraction of freshwater DOM resembles algal-derived polysaccharides of marine DOM, suggesting autochthonous freshwater DOM may persist over longer time periods in marine environments (Repeta et al., 2002).

Fourier-transform ion cyclotron resonance mass spectrometry (FT-ICR-MS)

Fourier-transform ion cyclotron resonance mass spectrometry (FT-ICR-MS) provides information about DOM at the molecular level. FT-ICR-MS used with a soft ionization technique like electrospray ionization can provide a molecular mass within 0.1 mDa (milli-daltons) to tens of thousands of intact molecules. A molecular formula including C, H, O, N, P, and S atoms can be assigned to each individual molecule based on their mass (Dittmar and Stubbins, 2014; Koch et al., 2007). FT-ICR-MS analyses have further supported the notion that DOM is a heterogeneous mixture, with thousands of different compounds identified thus far. Also interesting is that most DOM molecules identified are of low molecular weight, ~500 daltons or less, and thus within the size range capable of being taken up by microbes (Dittmar and Stubbins, 2014; Stubbins et al., 2014). Importantly, FT-ICR-MS is restricted to a specific range of masses and definitive structural formulas cannot be determined. Moreover, thousands to millions of structural isomers may exist in a single DOM sample. Basic structural components, however, can be gleaned from the density of elemental bonds or elemental ratios, like the *aromaticity index* based on the density of C–C bonds present in a molecule (Koch and Dittmar, 2006). The

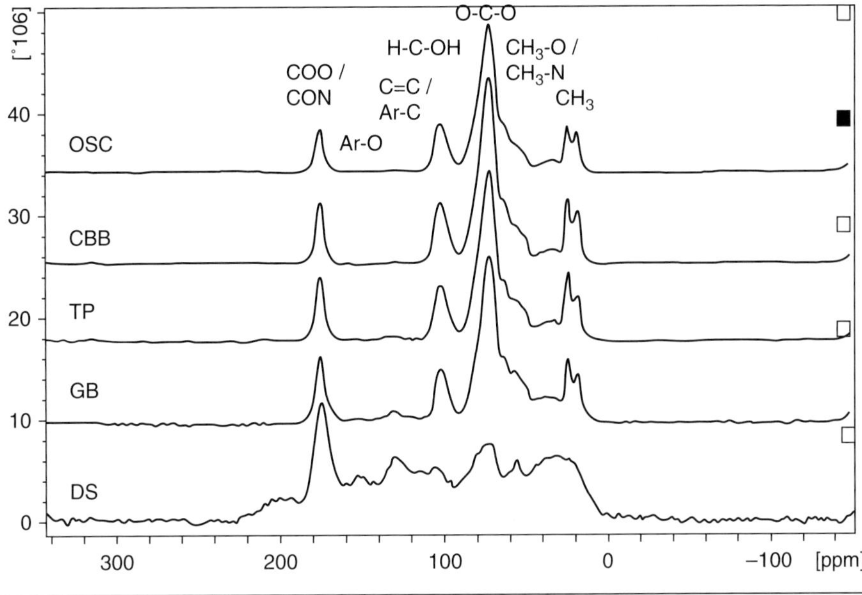

Site	CH_3 (0–45) ppm	CH_3–O (45–60) ppm	Total aliphatic	HCOH (60–90) ppm	O–C–O (90–120) ppm	Total carbohydrate	C=C/Ar–C (120–140) ppm	Ar–C (140–160) ppm	Total aromatic	COO/CON (160–190) ppm	C=O (190–220) ppm
DS	20	7	27	18	13	31	11	8	19	18	5
GB	19	10	29	44	11	55	3	2	5	10	1
TP	17	9	26	47	13	60	2	1	3	10	1
CBB	16	8	24	52	14	66	1	0	1	9	0
OSC	15	7	22	54	14	68	1	0	1	8	1

FIGURE 28-6 Example of cross-polarization/magic angle spinning (*CP/MAS*) solid-state ^{13}C nuclear magnetic resonance (*NMR*) spectra of high-molecular-weight dissolved organic matter (*DOM*) from five study sites (OSC, CBB, TP, GB, DS) in Abdulla et al. (2010). The graph indicates the chemical functional groups found in DOM at each field site. Relative carbon (C) percentage in each DOM functional group and compound class (total aliphatic, carbohydrate, and aromatic) as determined from the spectra are given in the table.

detailed elemental information provided by FT-ICR-MS, including ratios of H:C and O:C in *Van Krevelen plots* (i.e., 2- or 3-D plots of the atomic ratios of elements in the compounds of interest), can be used to map distributions and differences in structural groups like aromatics, aliphatics, and phenolics (Fig. 28-7a; Hutchins et al., 2017; Kellerman et al., 2018) or compound classes like proteins, lipids, and carbohydrates (Fig. 28-7b; Smith et al., 2018). Mopper et al. (2007) and Dittmar and Stubbins (2014) are two useful resources to learn more about FT-ICR-MS analyses and applications.

Optical methods are increasingly used alongside FT-ICR-MS in aquatic DOM studies to provide source and fate context to the molecular information provided by mass spectrometry. For example, optical spectroscopy and FT-ICR-MS were used to analyze samples of macrophytes, algae, sediments, and DOM in two eutrophic lakes in China, Lakes Taihu and Dianchi, to identify sources of lake DOM (Liu et al., 2020). Autochthonous DOM was more aliphatic and less oxidized than allochthonous DOM, and algae contained more lipids, while the macrophytes contained more lignin and tannins. Interestingly, the dominant source of lake DOM differed between the two eutrophic lakes that experience frequent harmful algal blooms, with Lake Taihu being dominated by DOM with allochthonous characteristics and Lake Dianchi by algal-like DOM (Liu et al., 2020).

Discrepancies in linking OM composition, reactivity, and persistence among studies and study sites support the notion that DOM is not universal in composition across or within ecosystems. A combined PARAFAC and FT-ICR-MS analysis of samples from 109 Swedish lakes was used to investigate the reactivity of DOM along environmental gradients (Fig. 28-8; Kellerman et al., 2015). More reactive DOM appeared to be allochthonous and colored and contain more oxidized and aromatic compounds, whereas more persistent DOM appeared to be produced *in situ*, corresponded to PARAFAC component C6 (Fig. 28-8), and included more reduced, aliphatic, and N-enriched compounds (Kellerman et al., 2015). PARAFAC, FT-ICR-MS, and size exclusion chromatography also revealed distinct compositional changes in DOM along a soil-stream-river continuum

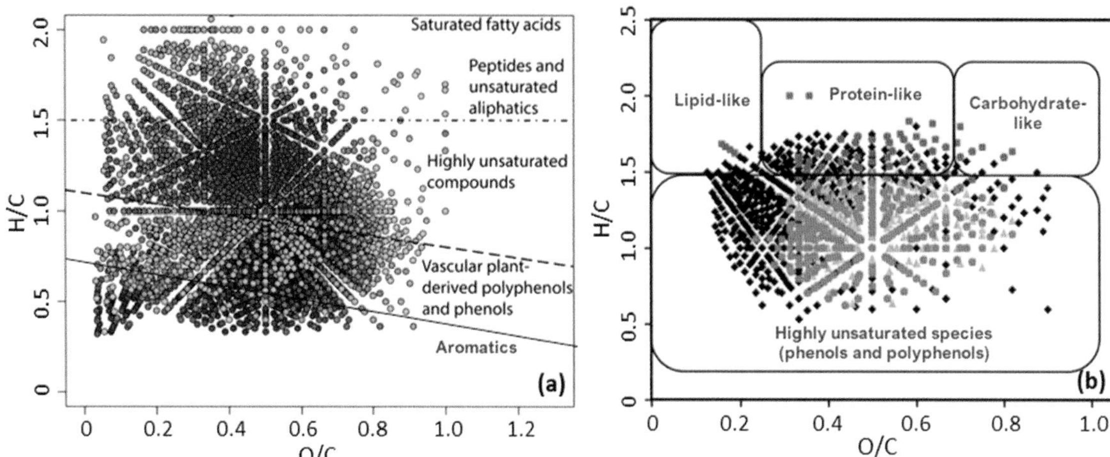

FIGURE 28-7 Examples of Van Krevelen plots of Fourier-transform ion cyclotron resonance mass spectrometry (FT-ICR-MS) data showing (a) structural groups (adapted from Hutchins et al., 2017) and (b) compound classes (from Smith et al., 2018). Colors in (a) represent correlation analysis results but, effectively, *red* is associated with soil water dissolved organic matter (*DOM*) and *blue* with stream and river DOM. Colors in (b) represent different molecular compositions. The axes represent ratios of hydrogen to carbon atoms (H/C) and oxygen to carbon (O/C) atoms in the molecular formulas found, which help delineate compound groups among the samples.

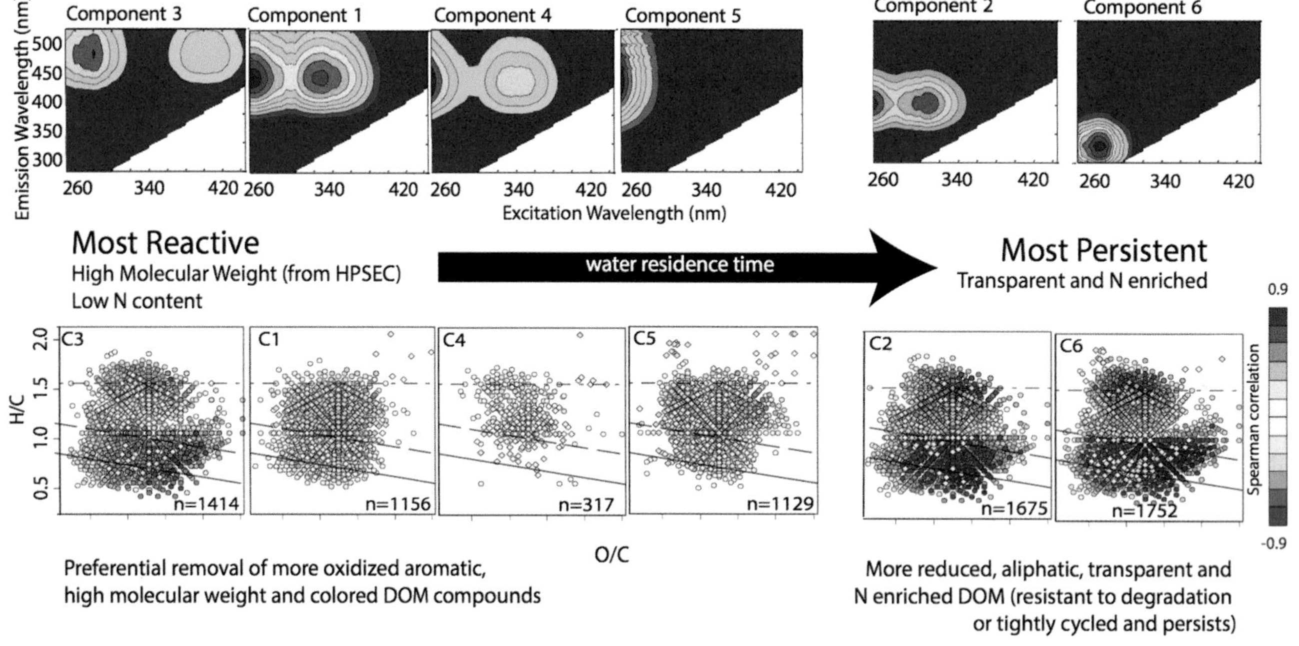

FIGURE 28-8 Results from the optical parallel factor analysis (*PARAFAC*) approach (*top*), which yields underlying fluorescence components (labeled as C or Component 1–6), provides reactivity information about dissolved organic matter (*DOM*) from 109 Swedish lakes. These are compared to molecular Fourier-transform ion cyclotron resonance mass spectrometry (FT-ICR-MS) results (*bottom*) of those lakes and show how compositional changes in DOM are linked to changes in reactivity along a water residence time gradient. Reactivity and molecular weight (determined by high-performance size-exclusion chromatography—HPSEC) decrease from *left* to *right*, while nitrogen (N) content increases. Axes in lower row of figures represent ratios of hydrogen to carbon atoms (H/C) and oxygen to carbon (O/C) atoms in the molecular formulas found, which help delineate compound groups amongst the samples. *(Adapted from Kellerman et al. (2015) by D. Kothawala.)*

in Quebec, Canada (Hutchins et al., 2017). In those samples, aliphatic and protein compounds corresponding to the C6 PARAFAC component were removed or transformed at the soil–stream interface rather than being persistent, as found in Kellerman et al. (2015). Much of the water we analyze has already lost its most reactive OM components and the composition of the remaining OM plays a major role in its reactivity and fate.

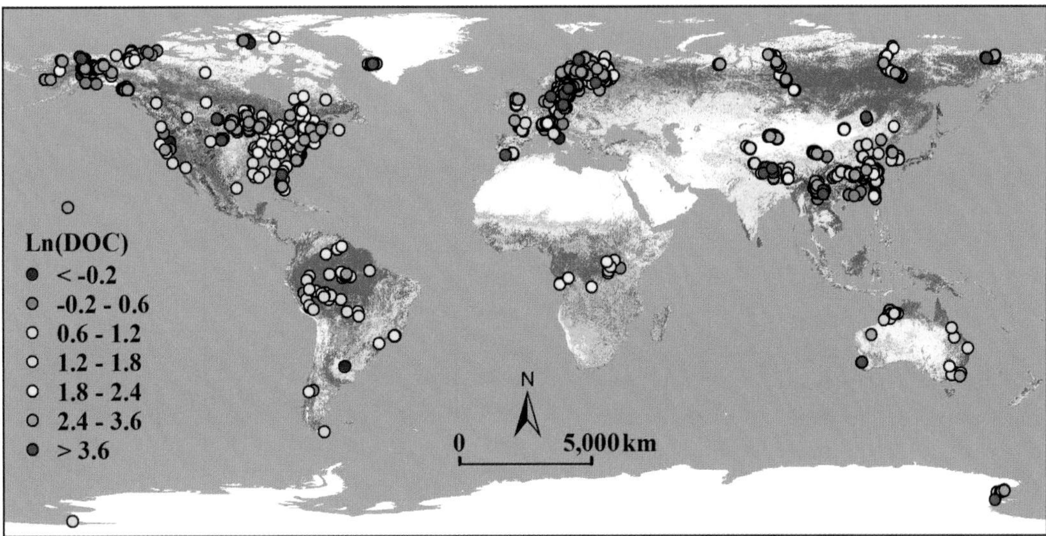

FIGURE 28-9 Data compiled by Zhou et al. (2018) of dissolved organic carbon (DOC) concentrations from >3000 lakes and rivers. Concentrations shown in natural log-transformed DOC [mg L^{-1}], which is equivalent to the following untransformed ranges in mg L^{-1}: <0.8, 0.8–1.8, 1.8–3.3, 3.3–6, 6–11, 11–36.6, >36.6.

iii. Dissolved organic carbon (DOC) concentrations

While the characterization of DOM is useful for understanding sources and persistence of OM, estimates of DOC and POC concentrations are necessary when constructing ecosystem carbon budgets and comparing ambient carbon pools with biophysical transformations. To separate DOC from POC in bulk samples, water is filtered in the same manner as for DOM. While DOC concentrations vary across ecosystems, they tend to stay within a range spanning two to three orders of magnitude. One of the first multi-ecosystem surveys of aquatic OM pools found DOC ranging from 1 to 30 mg C L^{-1} in 529 lakes of Wisconsin, USA (Birge and Juday, 1934). Higher DOC concentrations were found in smaller, more productive ecosystems or those with inputs from marshes or peat. Since then, while tens of thousands of more measurements of DOC have been made and methods have improved, remarkably similar DOC values have been found in Wisconsin lakes in more recent studies (e.g., 0.8–32.1 mg C L^{-1} in Hanson et al., 2007; 1.6–29 mg C L^{-1} in Xenopoulos et al., 2003). Global-scale meta-analyses of lake DOC concentrations show larger ranges in DOC concentrations, with DOC values in >7000 lakes on 6 continents ranging from 0.1 to 332 mg C L^{-1}, but with a mean of 7.58 mg C L^{-1} (Sobek et al., 2007). Direct DOC measurements of >3000 lakes and rivers across all continents found a range of 0.1–100 mg C L^{-1} (Fig. 28-9; Zhou et al., 2018), although DOC concentrations up to 300 mg C L^{-1} are possible in arid or evaporative climates (Mulholland, 2003). Rivers and streams tend to have lower DOC concentrations (0–50 mg C L^{-1}; Massicotte et al., 2017; Mulholland, 2003). A compilation of >12,000 unique DOC concentrations across the aquatic continuum of inland waters to the ocean suggests DOC is highest in wetlands and lakes and decreases through rivers and estuaries out to the ocean (Fig. 28-10; Massicotte et al., 2017).

B. Particulate organic matter (POM)

POM is operationally defined as the undissolved OM consisting of particles larger than the pore size of filter used to isolate DOM (typically 0.45 μm pore size, but ranges from 0.22 to 0.7 μm) or particles caught in suspended sediment traps. POM consists of fragments of undecomposed biomass that can be rapidly degraded in the water column (Grossart and Simon, 1998), sourced from within the ecosystem or from the catchment, and assessed using stable and radiocarbon isotopes or C:N ratios (e.g., McClelland et al., 2016). POM gets deposited at the bottom of static and more slowly flowing water, descending at a rate related to the size and density of the particles. In lakes, POM that settled in the bottom can be resuspended back into the water column, particularly near boundaries where turbulence occurs (e.g., Ostrovsky and Yacobi, 2010). In flowing waters, POM can experience a series of deposition and resuspension events (e.g., Newbold et al., 2005) and provides an

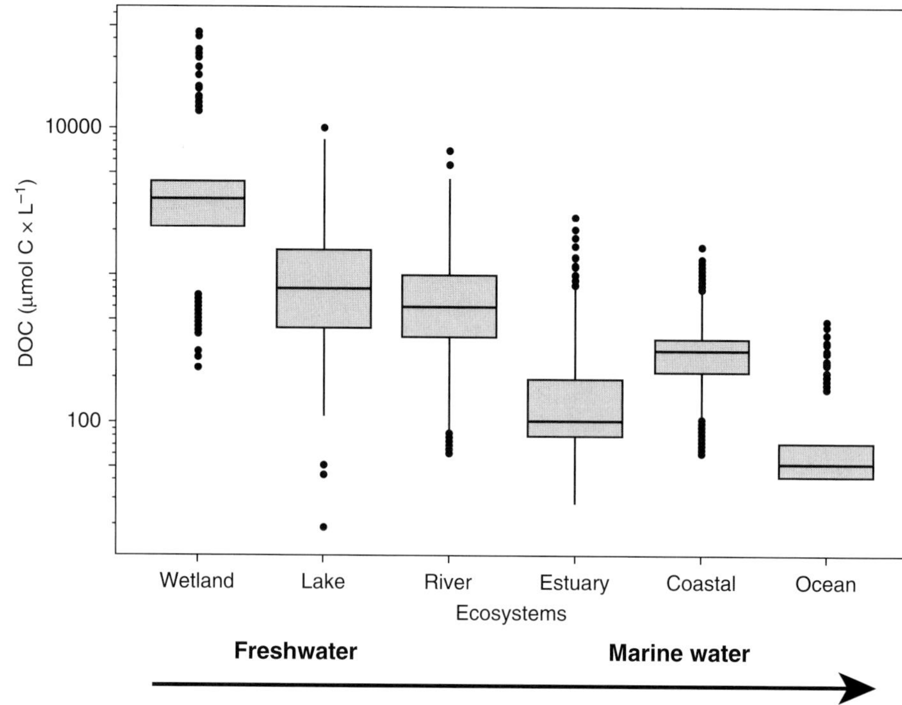

FIGURE 28-10 *Boxplots* of >12,000 unique dissolved organic carbon (*DOC*) concentrations across the aquatic continuum from wetlands to the ocean from Massicotte et al. (2017). Labeled tick marks of 100 and 10,000 μmol C L^{-1} are equivalent to 1.2 and 120 mg C L^{-1}.

important substrate for microbial assemblages and metabolism while suspended in aggregates and transported in larger rivers (Battin et al., 2008; Richardson et al., 2013).

i. Particulate organic carbon (POC) concentrations

POM can be analyzed in much the same way as DOM for composition and structure and, for the most part, will be composed of similar elements and functional groups. Like DOM, POM consists of a significant amount of POC, which can be calculated according to the mass lost upon combustion at 550°C (the "ash-free dry mass"), whereby the OC is oxidized to CO_2 and lost from the sample. POC concentrations and contributions to total organic carbon (TOC) vary across ecosystems and over time according to environmental conditions (e.g., discharge; Wang et al., 2012) but are typically less than DOC (Tranvik et al., 2009). Across 30 different aquatic ecosystems in Sweden (streams, rivers, and lakes), POC was often <5 mg C L^{-1} and contributed ~9% to TOC (Attermeyer et al., 2018). Across >1100 gauge stations in the United States, riverine POC concentrations ranged from <1.3 to >15.1 mg C L^{-1} but 71.5% of all values were relatively low (<1.3 mg C L^{-1}; Fig. 28-11; Yang et al., 2016). Remote sensing may be a useful method for monitoring the spatiotemporal dynamics of POC concentrations in large ecosystems, such as in Lake Taihu (China), where POC ranged from 0.7 to 15 mg C L^{-1} (Jiang et al., 2015), although uncertainties may be high when estimating aquatic carbon species using only satellite data.

C. Formation of POC and interactions with DOC

Different sources of POC will be discussed in Section III, but DOC, particularly in the colloidal size range, can aggregate by several processes to form POC. *Flocculation*, a generic term covering all aggregation processes, refers to the formation of a particle with a diameter larger than ~1 μm, on which gravity dominates over colloidal interactions (Gregory, 1989). Flocculation occurs when particles collide and can adhere, for example, by van der Waals forces or hydrophobic reactions. The smallest colloids (<100 nm) coagulate into aggregates in the size range of 0.1–10 μm and particles >10 μm rapidly sediment. Colloid constituents can include biological debris (e.g., algae), iron oxyhydroxides, calcium carbonate ($CaCO_3$), clays, amorphous silica, polysaccharides, and soil-derived HS (Buffle and Leppard, 1995). Several physical processes can influence colloidal collision and adhesion, such as Brownian motion, shear flows, differential settling based on size, aggregation at the air–water interface due to rising bubbles, and scavenging of smaller colloids by filtration through rapidly sinking macroaggregates (i.e., "lake snow"; O'Melia and Tiller, 1993).

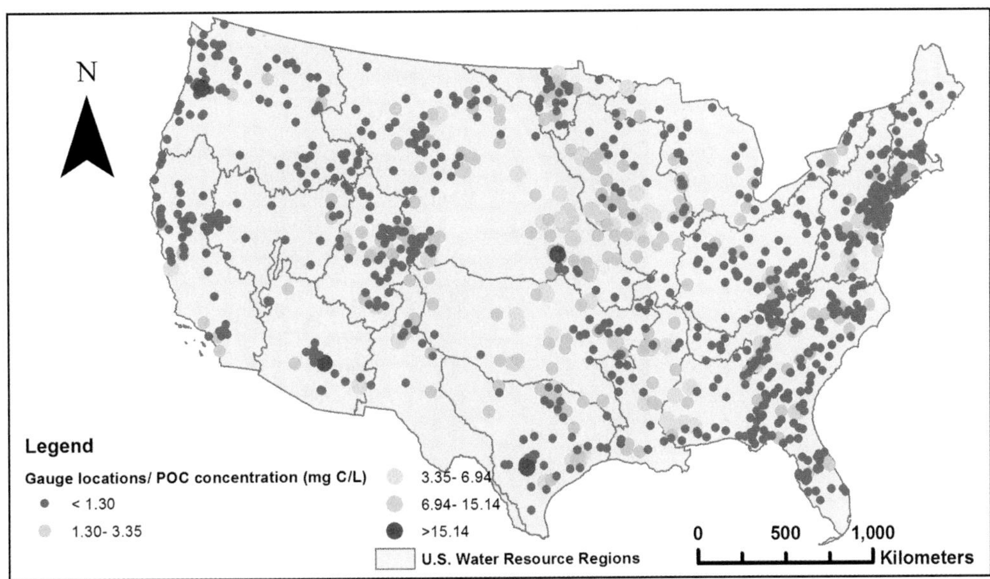

FIGURE 28-11 Particulate organic carbon (*POC*) concentrations (mg C L^{-1}) at >1100 gauge stations along rivers in the United States. *(From Yang et al. (2016).)*

Both biotic and abiotic mechanisms influence flocculation. The bacterial degradation of bioavailable DOC may result in flocculation settling of POC (Von Wachenfeldt et al., 2009). If associated with inorganic particulate matter, DOC will also sediment (e.g., when adsorbed to clay or $CaCO_3$ particles; Groeneveld et al., 2020), which may be a relevant pathway for POC formation, particularly in ecosystems receiving higher DOC inputs (Von Wachenfeldt and Tranvik, 2008) or with shorter water residence times (Evans et al., 2017). POC can also be formed abiotically by photochemical formation from DOC. For example, up to 22% of DOC loss in boreal aquatic ecosystems was attributed to light-mediated DOC flocculation forming POC (Von Wachenfeldt et al., 2008). Although the exact mechanisms or large-scale impact on carbon budgets for photochemical flocculation are not well constrained (Porcal et al., 2013), recent evidence suggests it may contribute significantly to carbon budgets in high OM boreal ecosystems (Attermeyer et al., 2018).

No matter the formation process, POC sedimentation is an important means of transferring OC from source to sediments, thus sequestering some of the settled carbon for potentially long periods of time (e.g., Ferland et al., 2014; Sobek et al., 2009; role of sediments in ecosystem carbon budgets discussed in Section IV). POC and DOC detrital fractions are merely arbitrary divisions within a smooth continuum of large particles to small molecules (Rich and Wetzel, 1978) and the transformation between the two fractions can occur in either direction. For example, POC can be transformed into DOC via the dissolution of POC in sediments (Peter et al., 2017) or POC degradation in the water column (Attermeyer et al., 2018). Although the composition of detritus modifies the way in which it is used and its effects on the environment, fluxes of detrital OC carry energy from its point of origin to its place of transformation or burial.

III. Organic matter (OM) sources

The biogeochemical cycles and food web energetics of freshwater ecosystems are sustained by a combination of internally fixed and externally derived OM (autochthonous and allochthonous sources, respectively). Past research identifying the relative importance of autochthonous and allochthonous OM to carbon cycling within inland waters often assumed autochthonous OM fixed by freshwater primary producers during photosynthesis was more reactive and biologically available than allochthonous OM of terrestrial origin. More recent work, including research supported by emerging techniques used to characterize OM composition (Section II), suggests that both autochthonous and allochthonous OM sources include compounds readily metabolized by freshwater organisms. How accessible OM is to microbes (e.g., buried in sediments or dissolved in water), OM nutrient content (i.e., the C:N:P content of OM relative to the availability of limiting nutrients in the surrounding environment), and the residence time of water in ecosystems (e.g., Catalán et al., 2016) may be more important drivers of OM "availability" than OM source alone (Marín-Spiotta et al., 2014; Lehmann and Kleber, 2015; Fig. 28-12). Indeed, just as previous editions of this textbook addressed changing terminology

FIGURE 28-12 An updated perspective of how scientists consider how the persistence and reactivity of organic matter (*OM*) in soils and aquatic ecosystems is influenced by physical and biological characteristics of ecosystems (i.e., "pore spaces," "water," and "biota") in addition to the matrix of OM itself. *(From Marín-Spiotta et al. (2014).)*

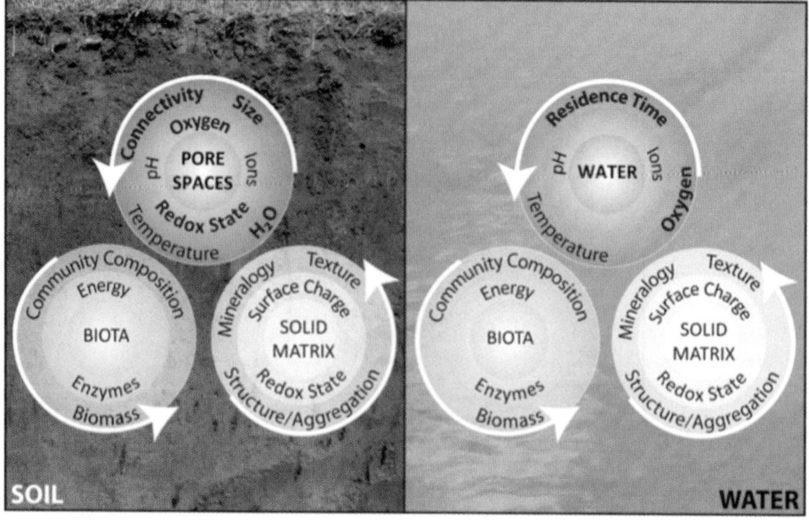

related to OM degradation by microbes (e.g., "The word *recalcitrance* connotes contumacious reduction in capabilities for microbial degradation, usually related to the chemical structure of the organic matter and is used in preference to *refractory*. Although interpretations vary, the term *refractory* often implies chemical resistance and nondegradability, which is not true in these natural ecosystems."; Wetzel, 2001), our updates to this text use terminology based on our emerging understanding of OM sources and their role in carbon cycling. We thus refer to different OM sources as more or less *reactive* based on what we know about their biophysical *reactivity* and/or *persistence* in freshwater ecosystems.

The sources of OM transported to and cycled in freshwater ecosystems are diverse. The major inputs to DOC pools are: (1) photosynthetic inputs through exudation or lysis of cells; (2) allochthonous DOC inputs, composed largely of substances originating from terrestrial, wetland, and littoral higher plant tissues that can be more resistant to rapid bacterial degradation; (3) bacterial degradation and chemosynthesis of OM with subsequent release of DOC; and (4) DOC from excretions of secondary producers, zooplankton, and larger animals that ingest living or detrital OM, which are often quantitatively negligible but still play an important role in ecosystem energetics.

Production of DOC and POC is a result of autotrophic synthesis and partial heterotrophic metabolism. Measurements of DOC and POC concentrations in freshwater ecosystems are biased toward less reactive compounds, that is, compounds that are relatively stable chemically, of low solubility, more slowly degraded by microbes, and thus more likely to persist in freshwater ecosystems long enough to be sampled and analyzed compared to more reactive (i.e., more labile or biologically available) organic compounds. Indeed, more reactive components cycle rapidly at low equilibrium concentrations but can represent major carbon pathways and energy fluxes in freshwater ecosystems. The degree to which more reactive compounds are autochthonous or allochthonous in origin is still not well known: how much and why some autochthonous and allochthonous OM persists in transport instead of being metabolized remains an active area of research. Finally, the role of detritus (dead OM of autochthonous or allochthonous origin) is a critical but still often overlooked component of freshwater food webs and biogeochemical cycles, and discussed at the end of this section in recognition of the unique importance of detritus in freshwater ecosystem carbon cycling.

A. Autochthonous organic matter sources

The structure and productivity of inland waters is addressed more comprehensively in Chapter 9 of this text, but here we briefly review key knowledge related to autochthonous OM sources. Autochthonous OM sources sustain many important biogeochemical and food web processes within ecosystems that are supplemented by allochthonous inputs (Minshall, 1978; Thorp and Delong, 2002; Pace et al., 2004) and include: (1) benthic photosynthetic POM and DOM sources from active exudation, decomposition, and lysis of littoral macrophytes and attached algae and Cyanobacteria as well as biofilm assemblages on other substrates throughout the benthic zone; and (2) primary producers of the pelagic zone, primarily algal and cyanobacterial phytoplankton.

There are few direct estimates of autochthonous OM inputs in freshwater ecosystems. We can roughly estimate net primary production from whole-ecosystem

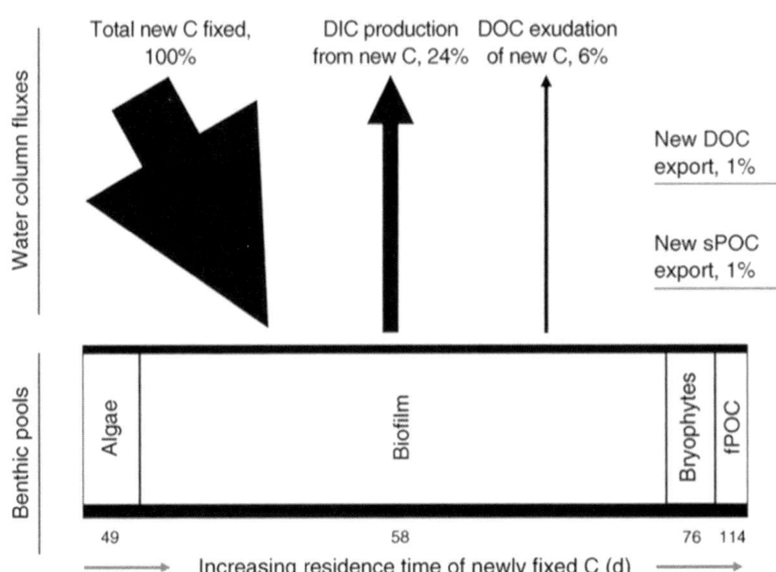

FIGURE 28-13 The sources, fluxes, and turnover time of autochthonous organic matter (*OM*) measured using a pulse-chase ^{13}C-dissolved inorganic carbon experiment in French Creek, Wyoming, USA. The different sizes of arrows and pools are scaled relative to the total new C fixed (water column fluxes) or total benthic OM pool (consisting of algae, biofilm, bryophytes, and fine particulate organic carbon [*OC*]), respectively. *(From Hotchkiss and Hall (2015).)*

metabolism (gross primary production minus respiration by the autotrophs themselves; Section V), but autotrophic respiration is still poorly constrained at the ecosystem scale (but see Solomon et al., 2013; Hall and Beaulieu, 2013) and the biomass of actively photosynthesizing primary producers is difficult to measure directly. Stable isotope tracer and OM leaching experiments have offered some insights into the magnitude of DOM exudation by autotrophs (~15% of primary production: Cole et al., 1982; Baines and Pace, 1991; Lyon and Ziegler, 2009; Hotchkiss and Hall, 2015) and the turnover times of autochthonously produced OM in ecosystems (e.g., 49–76 days in a productive mountain stream, with turnover times depending on autochthonous OM source; Hotchkiss and Hall, 2015; Fig. 28-13). Autochthonous DOC exudation may yield higher ecosystem respiration during the day than at night due to rapid uptake and respiration of more reactive autochthonous OC (Tobias et al., 2007; Hotchkiss and Hall, 2014). A large portion of autochthonous OM inputs also support secondary production of higher trophic levels in freshwater food webs, even in ecosystems with substantial detrital support of food web energetics (e.g., Thorp and Delong, 2002; Finlay et al., 2002; Carpenter et al., 2005; Solomon et al., 2008).

Nutrient enrichment alters the relative contributions to ecosystem production by phytoplankton, submerged and emergent littoral and wetland macrophytes, benthic primary producers, and eulittoral algae (Chapters 17–18, 24, 25). The shift in dominance with respect to the productivity of these major groups results in differential contributions to the DOC pool (Fig. 28–14). If the drainage basin is not manipulated extensively by human activities (see Section V), allochthonous inputs of DOC to an ecosystem are defined by characteristics of its drainage basin and are relatively constant on a long-term annual basis; the relative contribution of allochthonous contributions to freshwater ecosystems consequently decreases as autochthonous sources increase.

The increasing number of annual and multiannual datasets of aquatic ecosystem metabolism offer insight into how the timing of autochthonous OM inputs may differ among ecosystems and throughout the year. Some aquatic ecosystems have peaks of autochthonous inputs during spring and autumn seasons, before and after terrestrial leaf out, and when terrestrial productivity is low and therefore intercepting fewer nutrients before they can be exported to inland waters (Roberts et al., 2007; Savoy et al., 2019; Myrstener et al., 2021). Many ecosystems have highest autochthonous production during summer, in synchrony with environmental drivers of peak terrestrial productivity (Solomon et al., 2013; Savoy et al., 2019). Finally, some ecosystems, especially those in regions of the globe that experience less seasonality, may have "aseasonal" and relatively constant autochthonous OM production year-round (Savoy et al., 2019).

B. Allochthonous organic matter sources

The amounts and chemical composition of DOM and POM inputs to an ecosystem depend on ecosystem size and position in the landscape. Allochthonous inputs also change seasonally with the volume of flow, the growth and decay cycles of the terrestrial and wetland vegetation through which runoff flows, and other

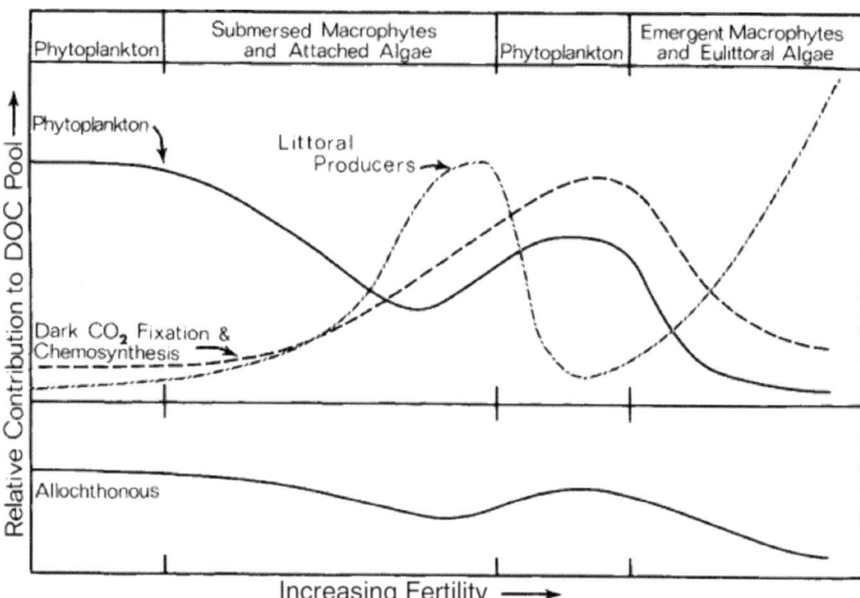

FIGURE 28-14 Generalized relative contributions of phytoplankton, littoral producers, dark and chemosynthetic CO_2 fixation, and allochthonous sources of dissolved organic carbon (*DOC*) to lakes of increasing fertility (i.e., nutrient status) in the progression of dominating pelagic and littoral primary producers.

factors, especially climatic variations. Small inland waters typically receive more allochthonous inputs because they have a higher surface-area-to-volume ratio than large lakes, reservoirs, and rivers; a high proportion of the surface area and edges of smaller ecosystems consists of littoral zones and adjacent wetland, riparian, and floodplain areas with productive aquatic and terrestrial vegetation tolerant of wet soils.

Allochthonous and littoral sources of dissolved and particulate detrital carbon form major inputs to freshwater ecosystems, and these sources of organic compounds and energy markedly influence the metabolism of organisms. In streams, DOC is decomposed by both the benthic microflora and the planktonic bacteria (Cummins et al., 1972; Wetzel and Manny, 1972; Dahm, 1981), with especially high rates of allochthonous carbon cycling in benthic zones. In the relatively static waters of lakes gravity acts as an important selective agent by which sedimentation displaces a major portion of carbon and metabolism to the sediments (Vallentyne, 1962; Wetzel et al., 1972). To adequately assess the importance of the complex carbon cycle that is central to both the structure and function of inland waters, one must know the productivity of all components of the ecosystem and have an understanding of the origins and metabolism of OC from all living and detrital sources.

Terrestrial plants contribute much of the allochthonous OM inputs to aquatic ecosystems. The OC of plant structural tissue residues is transformed by microbial use and degradation both on land at the sites of growth and while OM is being transported by runoff, subsurface, and surface water. Much of the OM produced in the terrestrial portions of a drainage basin will remain on land and decompose there (e.g., Fisher and Likens, 1973). Export of non- or partially-decomposed OM is mostly as DOM in surface runoff and subsurface flow paths, the concentrations of which are variable over space and time once they enter freshwater ecosystems (e.g., McDowell and Likens, 1988; Mulholland et al., 1990; Zarnetske et al., 2018; Gómez-Gener et al., 2021). DOM inputs from land to inland waters result from direct leaching from living vegetation or from soluble compounds carried in runoff from dead plant material in various stages of decomposition (Fig. 28-15). Seasonally, the concentration of DOC in soil leachates can be inversely correlated with temperature. Additionally, warm temperate and tropical communities are more productive than those of temperate regions, but oxidation of OM from forests of any region is rapid and reduces DOC concentrations in leachates. POM can fall directly into water from overhanging tree canopies, be transported by runoff water, particularly from floodplains, or be windblown into freshwater ecosystems. Foliage from trees and ground vegetation can provide significant inputs of OM, both as POM and as leached DOM from dead POM (Marks, 2019).

The vertical structure of soils influences the concentration, age, and mobilization of OM from land to inland waters. Highest DOC concentrations often occur in upper soil horizons (O/A in Fig. 28-16a), as a result of inputs from throughfall of the vegetation canopy and decomposition of surface POM (Fig. 28-16a), and tend to be greater in forest soils among coniferous vegetation than among hardwood trees. Concentrations of DOC decline steeply in lower soil horizons (B and C in

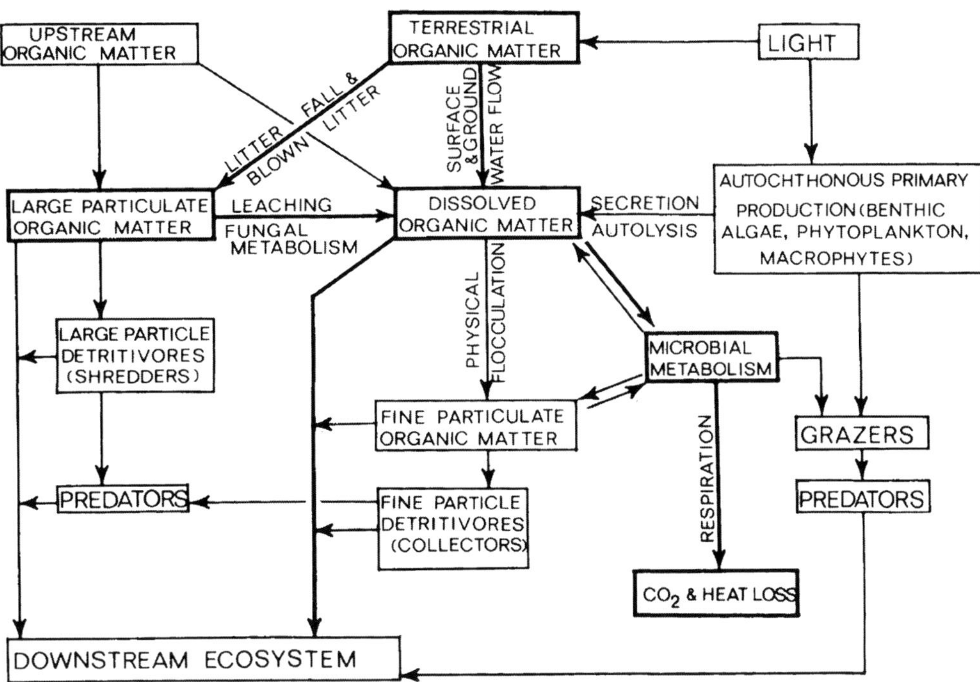

FIGURE 28-15 Simplified compartment model of the structure of an idealized stream ecosystem. *Heavier lines* indicate dominant transport and metabolic pathways of organic matter. *(Composite of modified figures after Fisher and Likens (1973) and Cummins et al. (1972).)*

Fig. 28-16a), primarily because of adsorption and coprecipitation in mineral soils with iron and aluminum (McDowell and Wood, 1984; David and Vance, 1991; Kaiser and Zech, 1998). Where soils do not have an iron-rich clay horizon, large quantities of DOC, particularly hydrophobic DOC, can be released and exported to inland waters. Organic acids of root exudates from terrestrial vegetation can also alter hydrophobic arrangements, which can mobilize and release organic compounds adsorbed to the soil. As DOC moves laterally and reaches floodplain and riparian soils, further changes occur (Fig. 28-16b). The DOC moving through soil matrices is subject to continual microbial degradation and may become progressively less reactive if not replenished by additional OM inputs along the flow path. For example, up to half of the DOC leached from soils is readily biodegradable by soil or freshwater microbes (Baker et al., 2000). Despite microbial carbon metabolism in soil and groundwater flow paths, DOC concentrations can be higher in groundwater than in surface water in landscapes with high OM content (Ford and Naiman, 1989).

Expanding hydrologic connections across landscapes during high flow increase the export of DOC to recipient inland waters. During low-flow periods, DOC concentrations in soils increase as long as lower hydrologic connectivity between soils and inland waters persists. High water percolation through soils during rapid snowmelt and precipitation events allows for subsequent rapid declines in DOC concentrations as soil pore spaces and subsurface flow paths are flushed (Lewis and Saunders, 1989; Hornberger et al., 1994). Maximum DOC concentrations typically appear in receiving inland waters soon thereafter and rapidly decline as the rivers return to base flows. As the area of floodplains increases in lowland river corridors, stream DOC concentrations often increase during high flows as DOC-rich water originating in wetlands is transported to the channel (e.g., Dalva and Moore, 1991). In large rivers, concentrations and total export of DOC are coupled to high-flow periods when floodplains are inundated, often in winter. In tropical rivers, DOC concentrations can be more variable due to reduced seasonality of temperature, variable precipitation and flooding periods, and the duration of floodplain inundation. Although the initial flush of DOC from the land or wetlands can be high following snowmelt or a large precipitation event, sustained DOC concentrations in runoff are often reduced from the dilution from low-DOC water inputs.

More recent work has characterized how DOC concentrations respond to changes in precipitation or stream flow, including analyses of multiyear *concentration–discharge*, or *C-Q*, dynamics. One study of >1000 rivers in the United States estimated that 80%

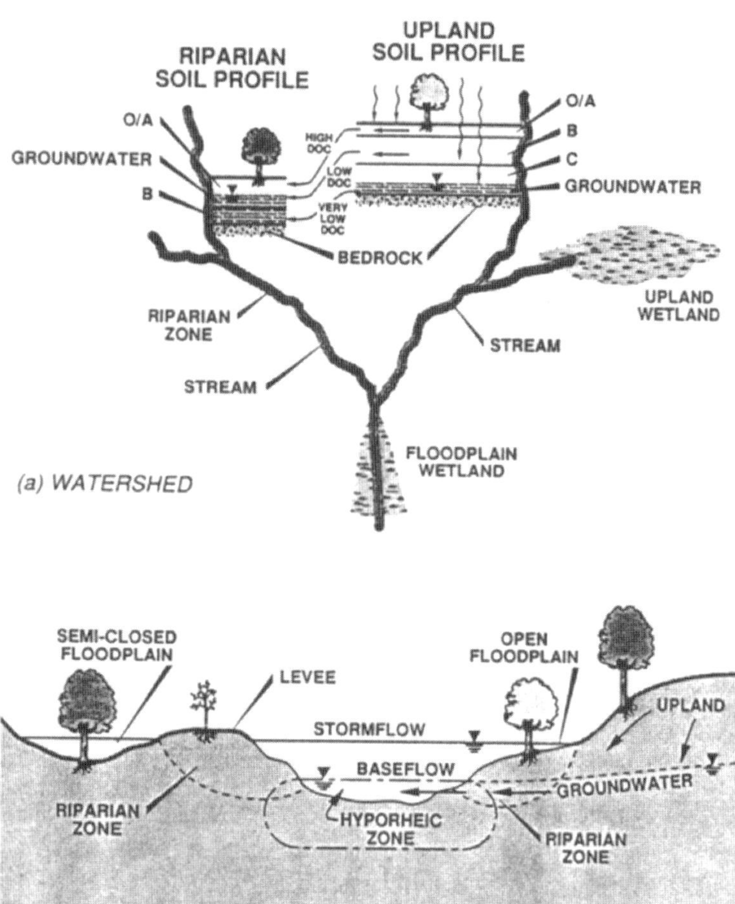

FIGURE 28-16 Potential dissolved organic carbon (*DOC*) sources and hydrologic linkages with the stream/river at (a) the drainage basin (watershed) and (b) floodplain/channel spatial scales. *(Extracted from Mulholland et al. (1990).)*

of all DOC-Q dynamics were transport limited, meaning DOC fluxes in those rivers increased as flows increased (Zarnetske et al., 2018). A forested boreal stream in Sweden had similar DOC-Q dynamics as most transport-limited streams and rivers in the United States, but higher-flow dynamics differed between snowmelt and snow-free periods of the year: DOC inputs increased with higher flows during snow-free periods, but higher flow periods during snowmelt had relatively unchanging DOC (Gómez-Gener et al., 2021). Furthermore, nearby streams draining a wetland-dominated catchment and a lake had mostly stable (i.e., chemostatic) DOC with variable discharge (Gómez-Gener et al., 2021), a result that reinforces the importance of capturing DOC-Q patterns across different landscape types, ecosystem interfaces, and hydrologic periods to better understand the range of different DOC input dynamics across the globe. Large precipitation events also increase allochthonous DOC loading into lakes. For example, around 50% of annual terrestrially-derived DOC inputs to temperate lakes in the northern United States arrived during extreme precipitation events and played a measurable role in lake metabolism, especially in lakes with short water residence times (Zwart et al., 2017).

In summary, much of the DOM in freshwater ecosystems originates from lignin and cellulose and related structural precursor compounds of higher plants. These substances are abundantly produced at the landscape scale; that is, the entire drainage basin and all OM produced photosynthetically within it. The productivity of terrestrial vegetation and aquatic plants associated with the land–water interface is usually several orders of magnitude greater than that of algae. Organic substances from higher plant tissues are abundant, chemically complex, and typically have slower biological degradation rates compared to less structurally complex OM sources (Thurman, 1985). During oxidative and anaerobic degradation, OM compounds are modified by microbial activities in detrital masses, including standing dead tissues that can remain in an oxidative environment for months or years. Many of the DOM compounds released from the partial decomposition of plant tissues and from associated microbial degradation

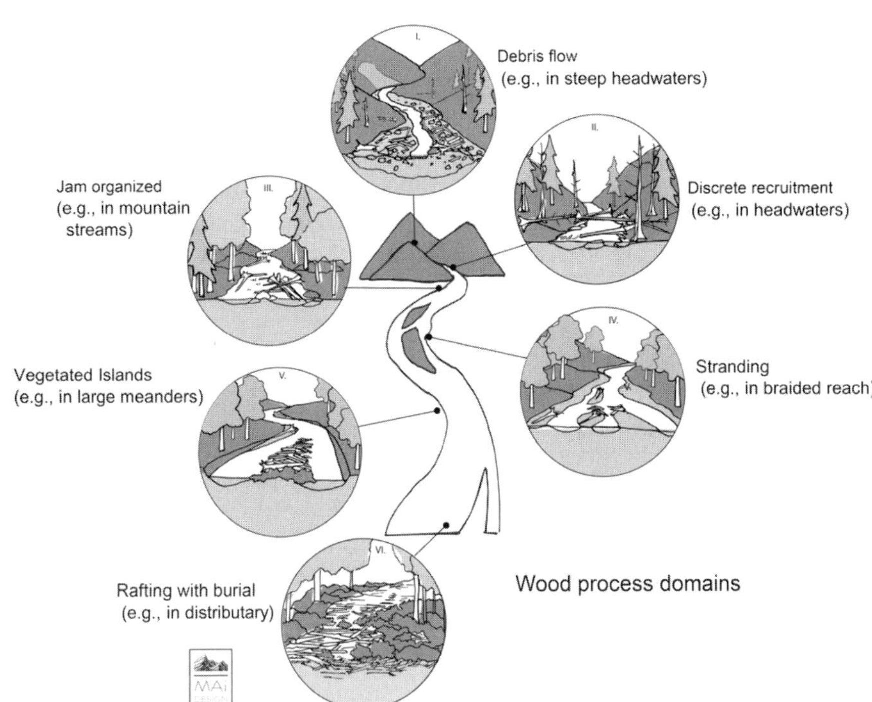

FIGURE 28-17 Proposed subset of wood process domains within a river network, where each domain has specific wood regime characteristics related to wood delivery, physical arrangement, and residence time in streams and rivers, as described in Wohl et al. (2019).

products are leached and partially degraded *en route* toward recipient freshwater ecosystems. Microbial OM modification continues during partial decomposition in terrestrial soils and hydrosoils of wetlands. Once within land–water interface zones, water containing DOM moves, often diffusely, through dense aggregations of living emergent and submersed aquatic plants and massive amounts of particulate, largely plant derived, detritus (extensively reviewed in Wetzel, 1990). The enormous surface areas of these habitats support large aggregations of rapidly growing microbial communities. During transport through the microbial metabolic sieves associated with wetland and littoral areas, appreciable further selective degradation of more reactive DOM constituents occurs before final movement into the receiving lake body or river channel.

Allochthonous POM in freshwater ecosystems include leaves, soils, coarse woody debris, pollen, feces, dead terrestrial organisms, and other materials deposited by terrestrial organisms when they are using waterways. Movement of OM, alive or dead, across ecosystem boundaries is considered a *metaecosystem flux* (Gounand et al., 2018). Cross-ecosystem OM fluxes sustain biogeochemical cycles and food webs within and downstream of the ecosystems that first receive them (Vannote et al., 1980; Marks, 2019). Differences in leaf properties (e.g., physical structure, chemistry) are linked to their decomposition rates as well as distinct pathways of energy flow and ultimate fates (Marks, 2019). Beyond leaf litter, there are many examples of how metaecosystem OM transfers alter recipient freshwater ecosystem food webs. The types of animals on the landscape can alter the nature of *cross-ecosystem subsidies*: for example, feces of domestic livestock and native mammals provide distinct OM subsidies with different C:N content to freshwater ecosystems, which may influence rates of in-stream metabolism (Masese et al., 2020). The experimental exclusion of terrestrial arthropods that would otherwise fall into waterways altered fish-feeding behavior and primary production within a Japanese stream (Nakano et al., 1999). Furthermore, wildebeest carcasses in the Mara River of Kenya contribute to rapid and multiyear biogeochemical cycling in rivers after drowning during migratory river crossings (Subalusky et al., 2017). Seasonal pollen inputs can also contribute a substantial amount of OM that is incorporated into consumer biomass in some landscapes (Masclaux et al., 2013). Finally, the input, transport, slow decomposition, and storage of trees and smaller pieces of woody debris are critical to the physical structure, biogeochemical cycling, and ecology of rivers (Wohl et al., 2019; Fig. 28-17).

Leaf OM falling into inland waters, particularly in forest-canopied small streams (first to fourth order), is a major input of OM (cf. Chapters 9 and 27; Vannote et al., 1980). Mean DOC concentrations in many small streams are directly related to the mass of leaf litter (Mulholland, 1997; Meyer et al., 1998). In-stream generation of DOC from leaf litter trapped and stored with stream sediments contributes about 20% of the daily DOC exported from many forested headwater streams. In the temperate zone this source of DOC is seasonal—greatest during autumn and winter and least during spring and summer. The remainder of DOC emanates largely from the floodplain vegetation and its degradation as well as from terrestrial sources.

Leaf litterfall, pollen, and other POC inputs can be significant OM sources in small lakes or ponds in heavily forested areas, but this input is usually small in comparison to other external carbon sources. Domestic and industrial POM pollution by organic wastes is more often transported to lakes and reservoirs via stream and river inflows rather than via groundwater sources. Early types of organic pollution, partial degradation *en route*, and effects on stream properties were comprehensively discussed by Hynes (1960, 1970). The degradation of OM in ecosystems that receive a high proportion of allochthonous OM should be more variable and require a greater diversity of metabolic pathways than microbial degradation in ecosystems that receive OM predominantly from autotrophic sources. It should be stressed, however, that many lakes derive much of their OC and nutrients from terrestrial and wetland sources of the surrounding drainage basin. The magnitude of allochthonous inputs is difficult to measure in full, but measurement approaches using stable isotopes (e.g., Pace et al., 2004; Carpenter et al., 2005; Cole et al., 2011; Karlsson et al., 2012) and whole-stream metabolism (Section IV) have confirmed the importance of allochthonous OM inputs for food webs and carbon cycling in lakes as well as streams and rivers.

C. The role of detritus

Detritus is all dead OC, distinguishable from living organic and inorganic carbon. *Detritus* consists of OC lost by nonconsumptive means from any trophic level (includes egestion, excretion, secretion, and so forth) or inputs from external sources that enter and cycle in the ecosystem (i.e., allochthonous OC). Detritus and associated terms presented here are defined to ensure terminology is consistent with the ecosystem concept and to avoid perpetuating the oversight of nonpredatory (i.e., not direct consumption) pathways in many ecosystem budgets. At the scale of an ecosystem, there is no energetic difference between DOC lost from a phytoplankter or feces or other exudates lost from an animal. Reactivity and uptake rates may differ, but functionally the OC and energy are the same, regardless of whether it is in dissolved or particulate form. The bacterial component is not combined with detritus because this interferes with the general applicability of the term to situations in which detritus is not simply ingested. Such cases include the use of detrital energy in the regeneration of nutrients such as CO_2, N, and P; algal heterotrophy; losses by adsorption; flocculation; precipitation with $CaCO_3$; chelation of elements; and so forth. Detritus is dead, and therefore living egested material (bacteria) is not included as detrital OM.

The *detritus food web pathways* are any route by which chemical energy contained within detrital OC becomes available to biota. Detrital food web pathways must include the cycling of detrital OC, both dissolved and particulate, to the biota by direct heterotrophy of DOC or absorption and ingestion. By definition, the "detritus food web pathway" emphasizes the actual trophic linkage between the nonliving detritus and living organisms, and recognizes the metabolic activities of bacteria attached to detrital substrates as a trophic transfer. The special case of a detrital food web pathway in which detrital energy is subsequently transferred by noncarbon substrates in an anaerobic environment is the *detrital electron flux*. Such energy may reenter the biota (via *chemosynthesis*, the process by which some microbes get energy from chemical reactions, often in the absence of sunlight) or mediate chemical or physical phenomena, or both, such as increasing the availability of inorganic nutrients in an anaerobic hypolimnion or in the sediments. In this case the term *flux* specifies the flow of electrons, not carbon, to alternate electron acceptors in the absence of molecular oxygen.

Detritus, as a component of the environment, can also affect facets of the chemical and physical environment without clear, defined energetic transformations of the detritus itself or associations with oxidation–reduction reactions. There are also several indirect effects by which detrital OC can influence and regulate the total energy and carbon flux of an aquatic ecosystem, including adsorption onto the surfaces of, and coprecipitation of, dissolved organic compounds with inorganic particulate matter (e.g., clays, $CaCO_3$) as well as complexing of inorganic nutrients by dissolved organic substances (e.g., humic-like substances).

To capture the suite of different compounds that contribute to ecosystem energetics, it is necessary to include living and detrital autochthonous and allochthonous OM sources. A large portion of the particulate and dissolved detrital OM that enters inland waters from allochthonous sources has undergone microbial stripping of more reactive compounds, meaning much of detrital OM is in the form of substances that are relatively resistant to further microbial degradation. Consequently, most organisms in freshwater ecosystems rely on a mixture of OM sources to support their energetic requirements for cellular maintenance, growth, and reproduction.

Early attempts to integrate producer, consumer, and decomposer components of ecosystems coupled the plant and animal components qualitatively as complex food webs, in which the feeding or trophic relationships among organisms were interconnected. Although these analyses provide descriptors of biotic components, the food-production and food-consumption processes are very dynamic and constantly change with differing environmental conditions and during the life histories of the organisms. So many organisms are involved in ecosystem functions that it is exceedingly difficult to quantify the roles of even the major actors of ecological interactions (e.g., Hutchinson, 1959); however, quantifying carbon fluxes within communities of freshwater ecosystems allows tracking OC cycling within the ecosystem before it is respired (primarily via decomposition) as CO_2. Much functional information on the operation of freshwater ecosystems has emerged from quantitative analyses of rates of carbon cycling and the identification of parameters that regulate these rates.

In summary, the trophic dynamic structure of aquatic ecosystems depends operationally on a dynamic detrital structure (Wetzel et al., 1972; Rich and Wetzel, 1978; Wetzel, 1995). From the standpoint of carbon fluxes, most energy and OC in ecosystems is dead, from a mixture of OM sources, and undergoing microbial degradation, the rates of which are variable and serially decrease with losses in OM reactivity. In lakes, reservoirs, and rivers, detrital OM is the main supportive metabolic base of carbon and energy. Since most lakes of the world are small, much of the autochthonous production of detrital OM originates from benthic littoral and wetland vegetation, which augments the loadings of DOM from terrestrial origins. Similarly, detrital heterotrophic metabolism dominates in streams, where the major OM sources are usually allochthonously derived terrestrial plant material.

Common to all aquatic ecosystems is the dominance of detrital metabolism, which gives ecosystems a fundamental metabolic stability. The trophic structure above the producer-decomposer level, with all of its complexities of population fluctuations, metabolism, and behavior, has a relatively small impact on the total carbon flux of the ecosystem. While plant-derived OM is one of the most commonly studied examples of detrital subsidies to aquatic ecosystems, dead organisms provide pulses of nutrient-rich OM (lower C:N than plant OM) that are critical to ecosystem processes (Bilby et al., 1996; Subalusky et al., 2017; Benbow et al., 2020). The slower, relatively consistent degradation of dissolved and particulate detritus by microorganisms underlies the more sporadic autochthonous metabolism that responds rapidly to, and depends to a greater extent on, environmental fluctuations (Wetzel, 1995). Autochthonous primary production is often small and variable, but in combination with allochthonous inputs, autotrophically produced detritus drives ecosystem dynamics. The functional operation of lentic and lotic ecosystems converges at this point of similarity in detrital metabolism, independent of whether the detrital OM was originally alive on land or in water.

IV. Organic matter (OM) cycling

A. Whole-ecosystem metabolism

Ecosystem metabolism is the fixation and breakdown of OC structures through photosynthesis and respiration, respectively. The processes of photosynthesis and respiration are fundamental to the storage and use of energy, from the cellular to ecosystem scale, and offer insights into the sources and amounts of resources available for higher trophic levels in freshwater food webs (e.g., Marcarelli et al., 2011). Tracing the rates and balance of metabolic processes in ecosystems allows us to obtain estimates of OM reactivity, sources, and fate that we might otherwise miss from water chemistry analyses alone, as much of the most reactive OM will be taken up by microbes and respired before it can be sampled and analyzed. This section reviews key processes included in whole-ecosystem metabolism (i.e., metabolic processes at the whole-ecosystem scale, not water column bioassays or benthic incubations), similarities and differences among ecosystem types, and our emerging understanding of how much carbon is cycled in freshwater ecosystems.

The fate of carbon in ecosystems can be traced from *gross primary production* (GPP), the total fixation of organic compounds via photosynthesis, through respiratory losses, burial, and export (Lovett et al., 2006; Fig. 28-18). All living things require energy, which, at the cellular level, is obtained through the breakdown of OC structures and use of energy released for cellular maintenance and growth. Both primary producers and heterotrophs respire. Together, all respiration by biota within the boundaries of an ecosystem is *ecosystem respiration* ($ER = R_a + R_h$, where R_a is *autotrophic respiration* and R_h is *heterotrophic respiration*).

The OC remaining after respiration by autotrophs themselves is *net primary production* ($NPP = GPP - R_a$). We rarely have measurements of R_a that allow for whole-ecosystem calculations of NPP; NPP estimates are likely most feasible (but nontrivial) in ecosystems where the community of primary producers is dominated by macrophytes and bryophytes, the biomass of which can be quantified for seasonal and annual NPP estimates (e.g., Fisher and Carpenter, 1976). After losses

FIGURE 28-18 Cycling and fates of autochthonous and allochthonous carbon in ecosystems. *GPP* is gross primary production; *NPP* is net primary production; R_a, R_h, and R_e are autotrophic, heterotrophic, and ecosystem respiration (referred to as ER in this text), respectively; and *NEP* is net ecosystem production. *(From the original figure legend in Lovett et al. (2006): "The shaded area contains the components of the NEP of the system.")*

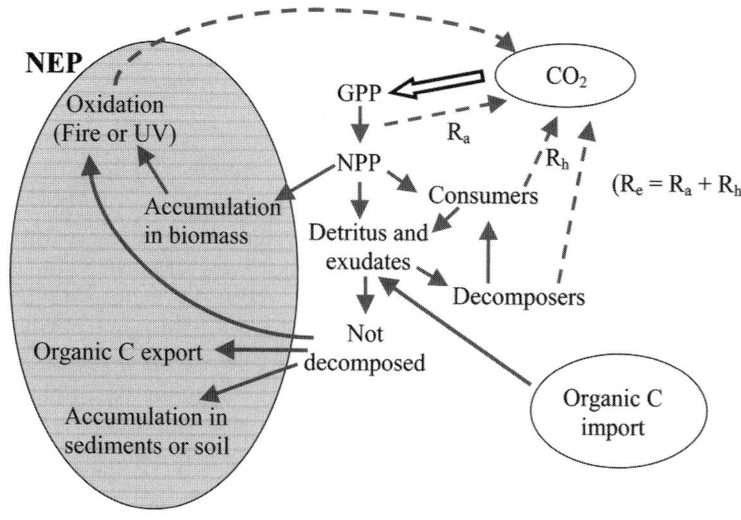

from R_a, potential fates of NPP include: direct consumption; assimilation into biomass; entering the detrital pool via DOC exudation or death of primary producers, consumers, and decomposers; respiration to CO_2; oxidation by abiotic processes (e.g., photo-oxidation); burial; and export from the ecosystem (Lovett et al., 2006; Fig 28-18).

Estimates of ER allow us to calculate *net ecosystem production* (NEP = GPP − ER), which provides quantitative evidence for the balance between the fixation and respiratory breakdown of OC (Woodwell and Whittaker, 1968; Lovett et al., 2006; Hall and Hotchkiss, 2017). If NEP is positive, the ecosystem is *autotrophic*, GPP exceeds ER (or GPP/ER > 1), and the ecosystem is accumulating and storing autochthonous OM. If NEP is negative, the ecosystem is *heterotrophic*, ER exceeds GPP (or GPP/ER < 1), and the ecosystem is subsidized by allochthonous OM from terrestrial and upstream aquatic ecosystems (Odum, 1956; Hall and Hotchkiss, 2017). Autochthonous OM fixed earlier in the year may subsidize ER at a later date and is sometimes miscategorized as "external OM" because it was not recently produced, but annual estimates of metabolism in most ecosystems should resolve this issue and provide appropriate metrics to quantify how much externally sourced OM is sustaining ER. One early expression of NEP (Woodwell and Whittaker, 1968) was modified by Woodwell et al. (1973) to account for the import (NP_{alloch}) and export (NP_{export}) of net production from external sources, where NEP = $(GPP + NP_{alloch}) - (R_a + R_h + NP_{export})$. Both formulations of NEP can be leveraged to interpret fates of OM, either as living biota or dead organic storage, after ER. The latter definition of NEP is less commonly used, as it specifically subtracts organic exports, thus limiting the application of the term to only that material that remains inside ecosystem boundaries.

Odum (1956) proposed how one might use diel (24 h) changes in dissolved oxygen to estimate GPP and ER in flowing waters. Much of Odum's research took place in productive freshwater springs and marine communities, allowing for observations of large diel swings in dissolved O_2 that inspired what we now describe as the "open water," "open channel," or "whole-ecosystem" method of estimating metabolism. Odum's foundational 1956 paper set the stage for the core metabolism modeling approach still used today, despite increases in model complexity and advances in how we estimate GPP and ER using maximum likelihood or Bayesian parameter estimation techniques in lakes (e.g., Van de Bogert et al., 2007; Winslow et al., 2016;) and streams/rivers (e.g., Holtgrieve et al., 2010; Grace et al., 2015; Hall and Hotchkiss, 2017; Appling et al., 2018). In addition to presenting the core equation still used for metabolism calculations today (the areal rate of dissolved O_2 change = GPP − ER +/− air−water gas exchange of O_2), Odum (1956) also highlighted how GPP and ER can be used to map energy flow and fate in ecosystems; predicted how wastewater and other OM inputs will alter ecosystem metabolism downstream; and proposed how we might classify different ecosystems as autotrophic and heterotrophic based on the magnitude and balance of GPP and ER (Fig. 28-19).

In freshwater ecosystems heterotrophy is the rule, not the exception. ER exceeds GPP in most inland waters, and the increasing number of annual metabolism estimates only provides stronger support for the importance of allochthonous OM sources in subsidizing ER. While standing waters (lakes, ponds, reservoirs) are, on average, more metabolically balanced than running waters (streams, rivers), most are still supported by some fraction of allochthonous OM (Hoellein et al., 2013; Solomon et al., 2013; Fig. 28-20). Smaller streams

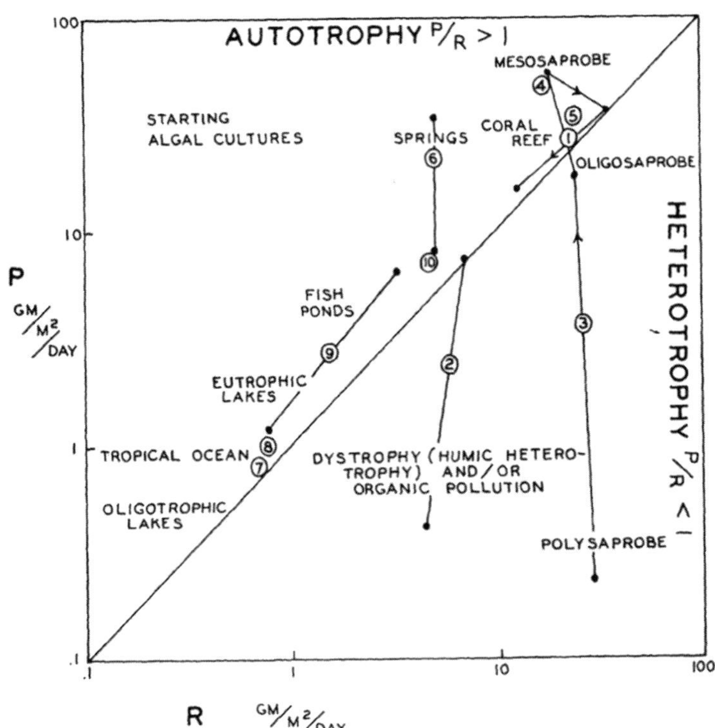

FIGURE 28-19 The original diagram from Odum (1956) describing the functional classifications of aquatic communities based on the magnitude and balance of gross primary production (GPP in this chapter and "P" in the figure) and ecosystem respiration (ER in this chapter and "R" in the figure), where heterotrophic ecosystems have ER > GPP and GPP/ER < 1 while autotrophic ecosystems have GPP > ER and GPP/ER > 1. Ecosystems that are in metabolic balance (GPP = ER) fall along the *solid line*. The terms meso-, oligo-, and polysaprobe refer to communities living in high-oxygen low-OM, moderate-oxygen and moderate-OM, and low-oxygen high-OM environments, respectively. Specific examples are described in more detail in Odum (1956) and references therein.

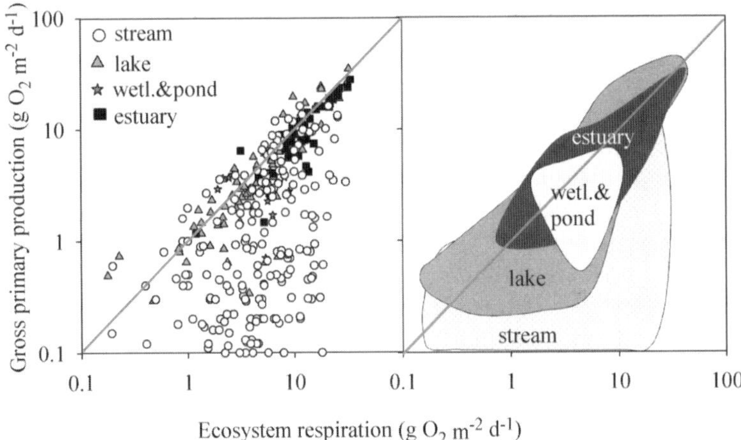

FIGURE 28-20 Data compiled by Hoellein et al. (2013) for a synthesis of ecosystem metabolism estimates from streams, lakes, wetlands, ponds, and estuaries. The *solid line* in each plot is where gross primary production and ecosystem respiration are equal.

are typically more heterotrophic than larger rivers, likely due to a combination of higher allochthonous inputs and lower light in streams than rivers (Vannote et al., 1980; Webster, 2007; Hall et al., 2016). Annual and multiyear whole-ecosystem metabolism datasets are increasingly providing new information needed to categorize the timing of OM and light inputs, flow changes, and metabolism (e.g., Uehlinger, 2006; Solomon et al., 2013; Bernhardt et al., 2022), and associated insights into the variability and drivers of ER within and among freshwater ecosystems (e.g., Roberts et al., 2007; Solomon et al., 2013; Demars, 2019). Water residence time has been proposed as a master variable to predict differences in total OC metabolism within and among ecosystems (Catalán et al., 2016; Raymond et al., 2016), as the opportunities for microbes to encounter and metabolize OC will be higher the longer water remains in an ecosystem (Battin et al., 2008; Catalán et al., 2016; Raymond et al., 2016) even if instantaneous metabolic rates are lower. Differences in stream, river, lake, and reservoir metabolism, through the lens of water residence time gradients among ecosystems, are summarized in

TABLE 28-2 Ecosystem Metabolism Characteristics Across Small Streams to Large Lakes, Framed Around Differences in Water Residence Time (WRT)

Low WRT (e.g., small streams) ←	→ High WRT (e.g., large lakes)
Metabolic behavior is dominated by organic carbon (OC) loads	Metabolic behavior is dominated by trophic state
Events are a dominant time-scale feature	Integration of longer time periods is the dominant time-scale feature (long lags)
Allochthonous OC pool is relatively young	Allochthonous OC pool is relatively old
Metabolism of allochthonous OC ≥ 0.1 day^{-1}	Metabolism of allochthonous OC is ~ 0.001 day^{-1}
Particulate OC can be a dominant constituent of the metabolized OC pool, because of high energy	Particulate OC is rarely a dominant constituent of the metabolized OC pool, because of low energy
Large gradient in biologically reactive solute concentrations (e.g., carbon, nutrients) at landscape scales	Smaller gradient in biologically reactive solute concentrations
Closely coupled with surrounding landscape	Loosely coupled with surrounding landscape
Allochthonous OC is a significant component of metabolic budgets	Autochthonous OC is a significant component of metabolic budgets in ecosystems with decadal and longer WRT

Updated from Hotchkiss et al. (2018)

Table 28-2 (Hotchkiss et al., 2018). Light, temperature, and nutrients, in addition to CO_2 and OC substrate availability and redox state, also govern rates of metabolism in inland waters (e.g., Bernot et al., 2010; Solomon et al., 2013; Hoellein et al., 2013). Below we highlight additional examples of DOC and POC inputs, metabolism, and fate in inland waters and how these measurements inform carbon budgets at the ecosystem scale. Some of the ways human activities alter OC metabolism and fate are highlighted in Section V.

i. Dissolved organic carbon and metabolism

Inputs of dissolved organic substrates to inland waters are approximately equal to their rates of microbial degradation. Extracellular release of DOC by living and senescing phytoplanktonic algae and littoral algae and macrophytes supports higher epilimnetic DOC concentrations in many ecosystems (Section III-A) and can contribute to diel changes in algal-derived DOC and bacterial metabolism at the whole-ecosystem scale (Kaplan and Bott, 1982; 1989). Decomposition of more reactive organic compounds often is very rapid (e.g., <48 h), and their dynamics would not be delineated by the sampling frequency (e.g., weekly) employed for more generalized measurements of DOC pools. Indeed, bacterial biomass responses to changes in DOC inputs occur at timescales ranging from hours to weeks (Bott et al., 1984; Kaplan and Bott, 1989). Phytoplankton productivity and allochthonous sources from the drainage basin are the primary sources of DOC in oligotrophic waters. In larger bodies of water, phytoplankton photosynthesis can be a major contributor to the DOC pool. In smaller lakes and streams littoral and benthic primary production can be more important sources of DOC than pelagic GPP. Humic-like substances, including less reactive fulvic acids, are generated by algae and contribute to the diverse compounds that make up DOM (McKnight et al., 1991; 1994). Algal-derived OM compounds are potentially important to metabolism in littoral and benthic areas, where attached and sessile algal productivity is often several orders of magnitude greater than that of phytoplankton (e.g., Wetzel, 1990; 1996). Fresh DOM leached from leaf litter tends to be reactive and some portion degrades quickly once dissolved and more physically accessible to microbes (as reviewed in Mineau et al., 2016). Fresh soil leachates may have a similar reactivity in freshwater ecosystems as the algal leachates generated from adjacent rivers (Hotchkiss et al., 2014), but many comparisons of DOM uptake find higher overall reactivity of algal than terrestrial DOC sources (e.g., Guillemette et al., 2013).

A large portion, usually >90%, of the OM imported from allochthonous and littoral/wetland sources to aquatic ecosystems is in dissolved or colloidal form. Although some DOM may aggregate and shift to a particulate form that can sediment out of the water, most imported DOM is dispersed within the water. That dispersion of DOM is important owing to its retention in zones of higher metabolism or modification by physical processes (e.g., photolysis; Cory and Kling, 2018), which may not be the case with the POM subject to sedimentation. An appreciable input of DOC enters the upper strata of lakes and reservoirs during the period of summer stratification, and these inputs, from both autochthonous and allochthonous sources, consistently fluctuate to a greater extent in the upper strata, especially near the surface, than do concentrations in the hypolimnion. In reservoirs, allochthonous inputs of DOC can exceed by several times the amounts of POC and DOC produced autochthonously (e.g., Romanenko, 1967). In some ecosystems, point-source inputs with different DOC concentrations or reactivities, such as a river inflow to a reservoir or a wetland-influenced stream tributary to a river, cause longitudinal spatiotemporal variation in DOC inputs and metabolism within a single ecosystem or fluvial network.

High rates of microbial DOM degradation occur within freshwater ecosystems despite the lower

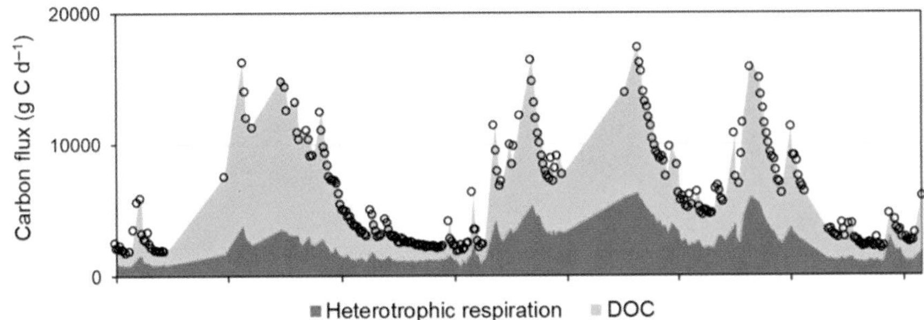

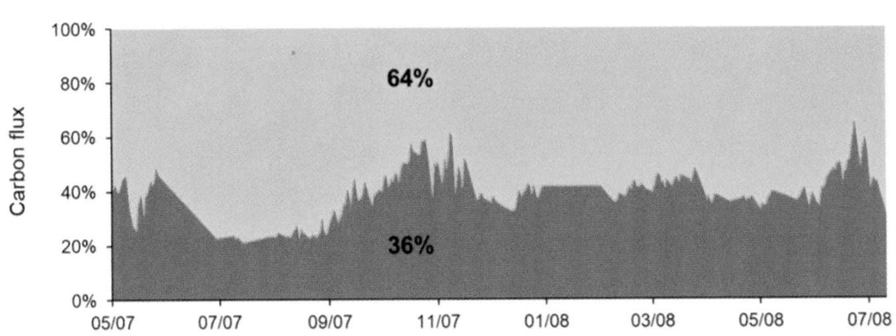

FIGURE 28-21 The magnitude and proportions of heterotrophic respiration (R_h) and dissolved organic carbon (*DOC*) fluxes in Cain Birn, a first-order stream in Scotland. The x-axis is time (month/year), with a range of May 2007 to July 2008. *Open circles* represent days with respiration estimates. *(From Demars (2019).)*

biological reactivity of much of the OM pool. The bulk DOC in lakes and streams is often dominated by apparently recalcitrant humic-like substances, which historically led to the belief that the bulk of DOC in freshwater ecosystems was resistant to microbial degradation. However, we now know that bacterial production depends to a great extent upon humic-like materials from allochthonous sources and is stimulated by high-flow events that deliver humic-like organic influxes into lakes and streams (e.g., Geller, 1985; Meyer et al., 1987; Gremm and Kaplan, 1998; Volk et al., 1997; Bergström and Jansson, 2000). An experiment using stable isotope labeled DOC to track OC fluxes and fate found that a small portion (~9%) of in-stream DOC was very reactive and traveled <300 m downstream before it was taken up by microbes, while a larger portion of semireactive DOC traveled an average of 4.5 km downstream before microbial uptake (Kaplan et al., 2007). Because of the lower reactivity of many dissolved organic compounds, DOM can reside within ecosystems with longer water residence times for months to years before it is transformed or exported (Kellerman et al., 2015; Catalán et al., 2016; Kellerman et al., 2018). Water column DOC loss rates are relatively slow, but they are consistently in the range of 0.5–2% per day under different environmental conditions (Wetzel, 2001; Catalán et al., 2016; Mineau et al., 2016). In fact, most estimates of DOC reactivity and freshwater metabolism are likely biased low because they come from bioassays and ambient measurements of concentration, not whole-ecosystem measurements of *in situ* process rates (e.g., Mineau et al., 2016; Plont et al., 2022). For example, simulations using ecosystem-scaled DOC uptake rates applied to a river network estimated that 27–45% of DOC inputs are removed via metabolism while in transport downstream, which is a much higher removal rate than water column bioassay DOC uptake estimates predict (Mineau et al., 2016).

Large seasonal and daily variations in hydrodynamics, particularly in streams and rivers, influence water residence time, OM transport, benthic scouring, and ecosystem metabolism (e.g., Uehlinger, 2006; Demars, 2019; Bernhardt et al., 2018). Daily and seasonal variations in water levels are smaller in most lakes and reservoirs than in streams and rivers, and DOM metabolism may therefore also be less variable. Pulses of DOC may be associated with an increase in ER in freshwaters draining catchments that release reactive DOM during high flow periods (e.g., Demars, 2019; Fig. 28-21). One notable example, Demars (2019), leveraged a multiyear dataset of stream carbon fluxes and metabolism to estimate the fate of DOC and characterize DOC–ER relationships. Allochthonous DOC and ER both increased with discharge in two Scottish streams, and 23% of annual DOC inputs were respired during only one hour of transport through a small stream (Fig. 28-21; Demars, 2019). In the Amazon, heavy precipitation storm events transport more reactive DOM, whereas DOM sourced from wetlands and groundwater flow paths tends to be hydrophobic and less reactive after selective biological utilization of the more labile components (McClain and Richey, 1996). Increases in terrestrially derived DOC also support higher ER in lakes (e.g., Zwart et al., 2016), but interactions between

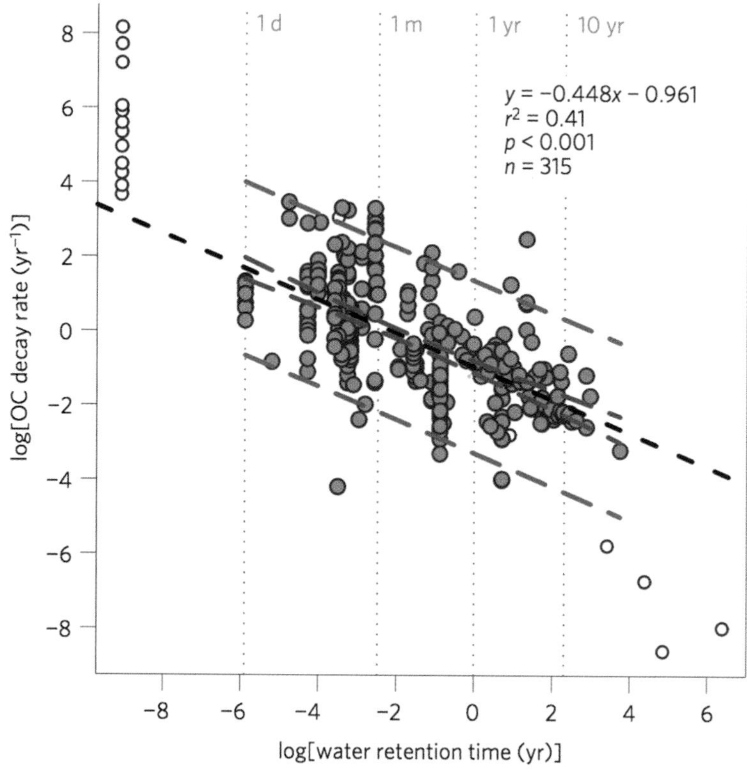

FIGURE 28-22 The relationship between log-transformed water retention time (*WRT*) and organic carbon (*OC*) decay rates in aquatic ecosystems. *Open symbols* are ocean (high WRT) and artificial substrates (low WRT), which were excluded from the regression analyses. *Blue dashed lines* are the 95% confidence interval from a linear regression; *red dashed lines* are the 95% prediction interval. *(From Catalán et al. (2016).)*

higher DOC, increased ER, and potential decreases in GPP due to light–DOC interactions may limit GPP and change the metabolic balance in ecosystems with higher DOC concentrations (Ask et al., 2009). Water residence time, which is variable over time and longer in most lakes and reservoirs than in streams and rivers, is frequently identified as a critical driver of OC metabolism across inland waters (Catalán et al., 2016; Hotchkiss et al., 2018; Table 28-2, Fig. 28-22).

Dissolved macromolecules in freshwater ecosystems are often of considerable age (decades to centuries) but are mixed with variable and rapidly changing inputs of younger OM substances. Mixtures of different DOC sources and reactivities may have nonadditive effects on whole-ecosystem DOC metabolism through what is referred to as a *priming effect*, where the input of more reactive DOM sometimes primes additional metabolism of less reactive DOM compared to ambient rates of uptake (Guenet et al., 2010; Hotchkiss et al., 2014). For example, Guenet et al. (2014) found a 12% increase in the consumption of soil-derived DOM by freshwater microbes when bioreactive glucose was added to experimental microcosms. Not all studies find evidence for priming effects in all of their experimental tests of mixed DOC sources (e.g., Bengtsson et al., 2014; Hotchkiss et al., 2014, Kelso et al., 2020). Consequently, the relative importance of mixed DOC inputs and potential priming effects on whole-ecosystem freshwater carbon cycling is still unresolved (e.g., Kelso et al., 2020).

ii. Particulate organic carbon and metabolism

Generalizations on the cycling of POC by planktonic communities are difficult to make because of the paucity of data collected under natural conditions (e.g., not collected from incubations or experimental mesocosms). POC concentrations in pelagic zones are usually considerably larger (double or greater) in eutrophic lakes than in less productive waters. Algal carbon commonly increases nonlinearly with increasing ecosystem productivity, and algal cell size is often greater in more eutrophic lakes. Estimates of the replacement of algal cell carbon in pelagic lake POC can be made from measurements of the biomass of algal carbon and NPP (Miller, 1972). Compensating for respiratory carbon losses, the daily net accumulation of POC can be estimated from NPP. The total suspended epilimnetic POC of Lawrence Lake (USA) had an average replacement time (turnover) of 40.7 days (range, 8.1–544 days). In the POC pool a mean of 83 mg C m^{-2} (range, 30–241 mg C m^{-2}) was algal cell carbon that was replaced by primary productivity in 1.1 days (range, 0.30–2.55 days) during the ice-free seasons. Algal carbon had an annual mean replacement time of 3.6 days. Replacement times of algal carbon by NPP are usually larger (slower

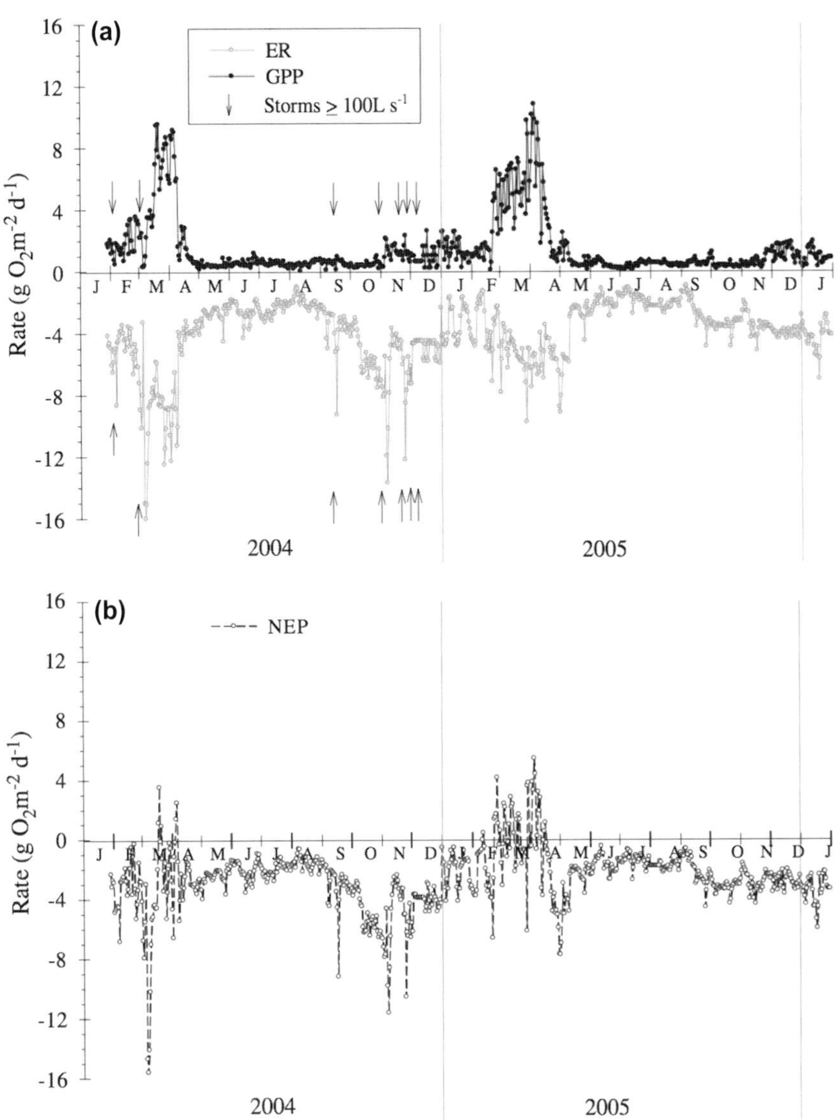

FIGURE 28-23 (a) Daily gross primary production (*GPP*) and ecosystem respiration (*ER*); and (b) net ecosystem production (*NEP*), with *arrows* noting events of storm-induced discharge >100 L s^{-1} within the 2-year time series, in Walker Branch, Tennessee, USA. *(From Roberts et al. (2007).)*

turnover times) in eutrophic than in oligotrophic waters, but the increases in replacement times are not proportional to increases in the carbon content of algae. Hence, in less productive waters, algae can photosynthesize a greater carbon flux per cell.

POM inputs support both pelagic and benthic respiration in freshwater ecosystems. Most allochthonous POM inputs (e.g., wood, leaves, organisms, fine particulates) have a longer residence time within streams and rivers compared to DOM but are still more likely to be transported downstream than decomposed near the location they entered (Webster et al., 1999). Finer OM particles appear to be especially important for whole-stream metabolism. For example, POC was 15 times more reactive than DOC in assay incubations from boreal lakes, streams, and rivers, and had an average half-life of 17 days (Attermeyer et al., 2018). Pelagic respiration increased up to two-fold in stream water incubations with sustained POC suspension (Richardson et al., 2013). Incubation-based POC metabolism estimates found an average in-stream POC turnover time of 10 days (Richardson et al., 2013). An analysis of the seasonality and storm-driven responses of stream metabolism in a forested stream found higher rates of ER in autumn after leaf fall as well as after high flow events (Roberts et al., 2007; Fig. 28-23). Not all POM inputs will change carbon cycling directly: allochthonous POM inputs to streams draining thawing permafrost release N and P, which can enhance carbon metabolism if microbes are nutrient limited, but appear to contain lower reactivity POC and may thus alter ecosystem carbon budgets through both nutrient-stimulation of metabolism and increased POC burial in depositional environments (Shakil et al., 2020).

Sources of POC shift in their relative contributions to the total POC and OC pools depending on ecosystem type and productivity. GPP by phytoplanktonic and littoral autotrophs is a major contributor to lake, pond, and reservoir POC, while allochthonous POC inputs are relatively small in contrast to DOC, with some exceptions (e.g.,small lakes or ponds in forested areas, reservoirs that are small in volume in relation to inputs and flowthrough). The relative importance of littoral POC production often increases in the transition from nutrient-limited conditions of oligotrophic lakes to light limitations imposed by biogenic turbidity and thus higher POC inputs by emergent macrophytes and associated littoral microflora in more eutrophic ecosystems. In oligotrophic to moderately productive freshwater ecosystems with longer water residence times, the observed depth-time distributions of pelagic POC follow the productivity and biomass distribution of phytoplankton rather closely during stratified periods. When Lawrence Lake, for example, is stratified, the POC maximum lags behind highs in productivity from several days to about 2 weeks, especially when sedimentation of plankton is slowed in thermal density gradients of the metalimnion. During ice cover, GPP decreases, inputs of allochthonous POC are lower, and concentrations of POC thus decline. POC often increases during periods of circulation or flow disturbance when sediments are resuspended into the water. In less productive lakes hypolimnetic POC generally does not increase unless the lower hypolimnion becomes anaerobic. In hypereutrophic lakes that receive large inputs of planktonic and littoral POC, the hypolimnia are rapidly rendered anoxic and bacterial productivity contributes to marked increases in POC.

iii. Anaerobic metabolism in inland waters

In many ecosystems only low intensities of light reach the sediments; this fact, combined with the continual sedimentation of predominantly dead POC, ensures that benthic metabolism is primarily heterotrophic and detrital (Rich and Wetzel, 1978). Limited oxygen penetration into sediments, combined with rapid oxygen uptake when it is present, means that much benthic metabolism is anaerobic (cf. Chapter 27), even when hypolimnetic oxygen depletion does not occur (Sobek et al. 2017). Hutchinson (1941) termed this metabolism *pelometabolism* and described its importance relative to *hydrometabolism* that occurs in the free water of a lake. CO_2 produced by anaerobic detrital pelometabolism represents the escape of the oxidized product of an oxidation–reduction reaction. The continued production of CO_2 in the absence of oxygen indicates that the respiratory quotient (CO_2 release:O_2 uptake) of benthic metabolism is >1 (see further discussion at the end of this section) and that alternate electron acceptors are being reduced instead of molecular oxygen (e.g., Müller et al., 2012). Thus alternative terminal electron acceptors are receiving energy that was originally captured during photosynthesis and transferred to sediments. For the most part, recent productivity supports more benthic metabolism than older (i.e., >10 years) productivity (e.g., Matzinger et al., 2010).

The accumulation of electrons in aquatic sediments must be interpreted as functional with respect to the ecosystem (Wetzel et al., 1972; Rich and Wetzel, 1978). The *detrital electron flux* into and ultimately out of the sediments originates from the same photosynthetic energy source as does a predator's food. Presumably, evolution, or at least integration, of the ecosystem has incorporated this energy flow into its overall function to the same extent as the more recognized flow through predator–prey systems. Reduced ions, radicals, and molecules are generally more soluble than their oxidized form; as a result, the reducing conditions favor an increase in the net rate of diffusion out of sediments. Upon entry of these products into the oxidized layers of sediments, further oxidation may proceed either chemically or biologically. If the original detrital reduction involved a nutrient (sulfate, nitrate, or ferric phosphate) or a fermentation product, the detrital energy has altered both nutrient regeneration and translocation. If the subsequent hydrometabolic oxidation is biological (i.e., chemosynthesis), the detrital energy has been reintroduced into biota. To further understand the consequences of benthic detrital electron fluxes, some basic considerations of the possible alternatives of benthic metabolism are worth evaluating. Material enters the sediments by sedimentation because it is particulate—that is, it is relatively insoluble. If the material remains insoluble, it can enter the permanent sediments of the lake and reduce the lifespan of the lake by increasing sediment accumulation.

To escape the sediments, materials must become more soluble, gaseous, or both (Chapter 27). Presumably, cellulose and similar materials constitute much of the organic material that enters the sediments. The intermediary metabolic products of such compounds are increasingly soluble and result in CO_2 that is both gaseous and soluble. Very reduced carbon compounds like CH_4 are also gaseous and volatile, soluble, or both. In one reservoir in Virginia (USA), experimental manipulations of dissolved oxygen induced clear differences in terminal electron acceptor pathways compared to a reference reservoir that experienced ambient oxygen declines during summer stratification, which altered the accumulation of CO_2 relative to CH_4 within each reservoir (McClure et al., 2020; Fig. 28-24). The loss of DOC from sediments is not well understood, but organic acids, acetic acid in particular, are common products and POC can dissolve into DOC (Peter et al., 2017). In

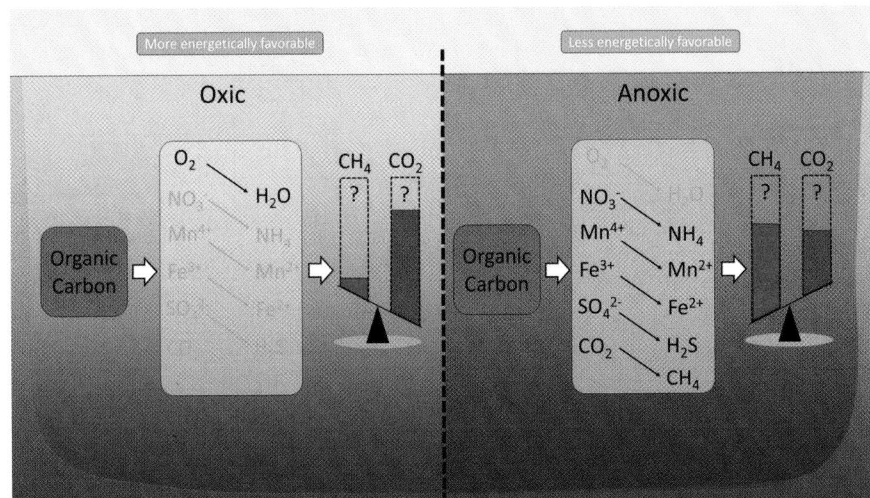

FIGURE 28-24 Illustration of organic carbon (*OC*) metabolism pathways in oxic and anoxic conditions, with alternative terminal electron acceptor (*TEA*) metabolism pathways and potential outcomes for CO_2 versus CH_4 emissions to the atmosphere. *(From McClure et al. (2020).)*

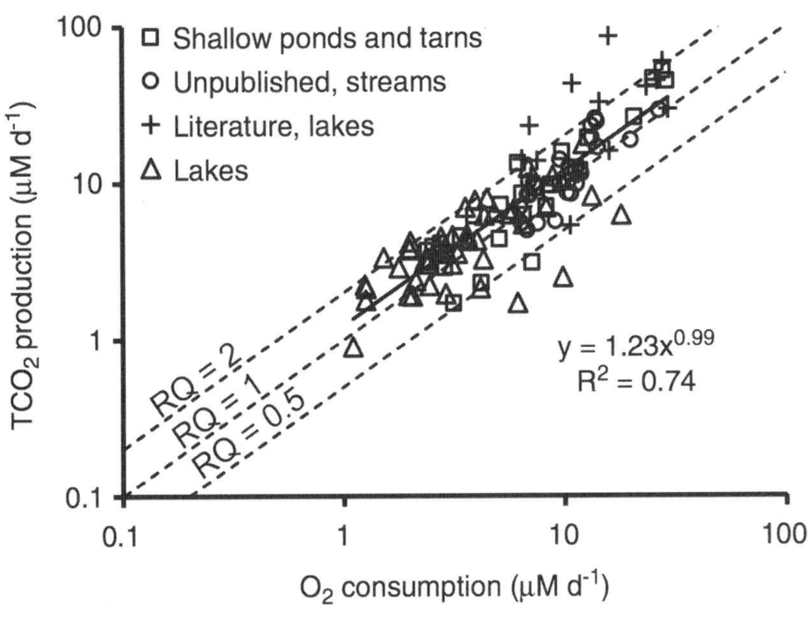

FIGURE 28-25 Relationship between total CO_2 (TCO_2) released and O_2 consumed (i.e., the *RQ* or respiratory quotient) during incubations of water from 110 different freshwater ecosystems. *(From Berggren et al. (2011).)*

terms of CO_2 and CH_4 production, a carbohydrate (e.g., cellulose) requires one molecule of oxygen for each carbon atom to be oxidized to CO_2 and water, while CH_4 requires two carbon atoms to be released for each molecule of oxygen. The removal of oxygen via the production of CO_2 reduces the mass of the molecule appreciably because oxygen is the heaviest atom in cellulose; thus there is a net mass loss of sediment.

The reduction–oxidation (redox) potential in sediments brought about by anaerobic metabolism represents a measure of electron activity that has been dissociated from biochemical (enzymatic) "constraints" or reaction specificity (Rich and Wetzel, 1978). Consequently, these electrons may diffuse out of the sediments in association with a number of inorganic and organic compounds. The exhaustion of a particular class of electron acceptors causes the benthic redox potential to become more negative relative to molecular oxygen; this continues until new acceptors precipitate back into the sediments following hydrometabolic oxidation as highly insoluble compounds (e.g., FeS or CuS). On the other hand, the carbon and oxygen in the form of CO_2 are the relatively massive products of the oxidized component of anaerobic redox reactions (Fig. 28-24). Thus the net diffusion of reduced compounds out of sediments is controlled by redox potential, and the net diffusion controls the export of mass—that is, carbon and oxygen—from sediments. In this manner, the lifespan of the lake is affected directly by the reduction of mass through losses of large quantities of gaseous

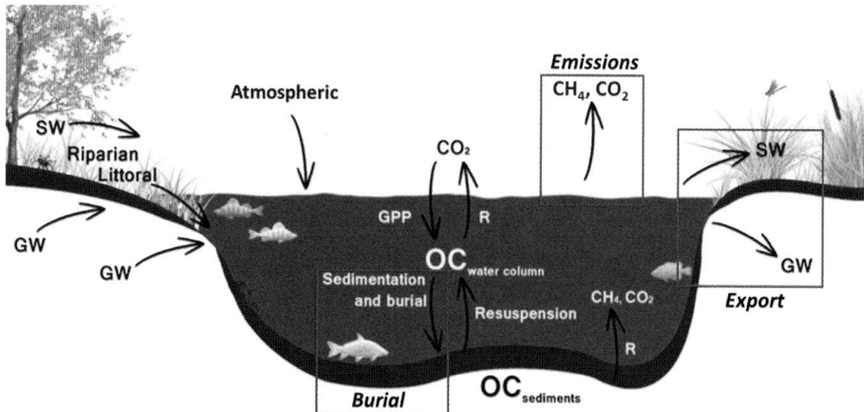

FIGURE 28-26 Conceptual diagram of dominant organic carbon (*OC*) fluxes (*black arrows*) in aquatic ecosystems. Three fates of C in *red boxes* are: emissions, storage/burial in sediments, and export to groundwater (*GW*) and surface water (*SW*). *GPP* is gross primary production; *R* is OC mineralization (i.e., respiration). Note this diagram does not include inorganic carbon cycling such as reactions of CO_2 with carbonates (see Chapter 13). *(Adapted from Hanson et al. (2015).)*

carbon and oxygen. The rate of the detrital electron flux is fundamental to the eutrophication rates of ecosystems as bioavailable nutrients are also released from sediments (Smolders et al., 2006).

The relative importance of detrital electron flux in benthic metabolism is reflected in the changes in ratios of CO_2 evolution to consumption of molecular oxygen. The oxygen uptake by benthic sediments (also referred to as *sediment oxygen demand*) typically exceeds the concomitant evolution of CO_2. Oxidations of proteins and fats result in *respiratory quotients* (RQ = CO_2 released/O_2 consumed) of less than unity, as has been demonstrated often by caloric combustions of benthic organisms. An RQ value of 0.85, generally accepted as an average value (Ohle, 1952; Hutchinson, 1957), was used in estimates of lake productivity by hypolimnetic CO_2 accumulation (Chapter 13). Anaerobic fermentative metabolism by bacteria produces excess CO_2 and volatile organic compounds, such as CH_4, that diffuse out of sediments. Sediments have an *oxygen debt* that is indicated by low and negative redox potentials and nonbiological chemical uptake of molecular oxygen by reduced substrates that were formed under anaerobic conditions (Rich, 1975). This suggests that microbial metabolism is continuing, while oxidative reactions using molecular oxygen as the terminal electron acceptor are impeded by limited O_2 diffusion (Wetzel et al., 1972). RQ at the benthic community level, which includes anaerobic bacteria that use electron acceptors other than molecular O_2, would be higher than the RQ of organisms undergoing aerobic metabolism. Mean RQ estimated from stream and lake water column incubations was 1.2 and varied with the composition of organic compounds (Berggren et al., 2011; Fig. 28-25). Additionally, *in situ* RQ values likely vary seasonally because of mixing during turnover and changes in redox gradients in response to varying rates of oxygen diffusion. In Dunham Pond, Connecticut (USA), Rich (1975) found that hypolimnion RQ varied inversely with oxygen availability from <1 after spring turnover and oxygen renewal to nearly 3 under anoxic conditions during the latter portion of summer stratification. Similar results have been found in a number of other lakes (e.g., Rich and Devol, 1978; Rich, 1980; Mattson and Likens, 1993). Under anaerobic conditions, high RQs are the result of the oxidation of OC to CO_2 and CH_4 during the reduction of alternate electron acceptors, which then appear as oxidizable substrates. Ultimately, settling OM quantity and reactivity, dissolved oxygen levels, and the flux of reduced substances are all major controls on the redox potential driving the biogeochemical dynamics and benthic metabolism in sediments and hypolimnia.

B. Fluxes, fates, and whole-ecosystem budgets

The whole-ecosystem carbon budget of a single aquatic ecosystem or several connected systems includes various carbon fluxes and fates in addition to metabolic processing. In general, carbon is transported into or produced within an ecosystem (Section III), where it can be metabolically transformed (Section IV-A), and either the original or transformed carbon is (1) stored by burial within sediments; (2) emitted from the water surface; or (3) transported/exported downstream (Fig. 28-26). The balance of these three fates, which can vary significantly across ecosystems (e.g., Tranvik et al., 2009), is important when establishing the carbon footprint of inland waters globally but also when determining whether an individual ecosystem is a carbon source or sink. Values for all three fates are not always measured concomitantly but are necessary to provide complete whole-ecosystem OC budgets (examples below) as well as integrative assessments of all carbon fates (Vachon et al., 2021). Moreover, there is often a mismatch between the spatial and temporal scales of OC fate estimates; for example, if emissions are reported on daily timescales in multiple locations but burial and export are reported annually from single locations.

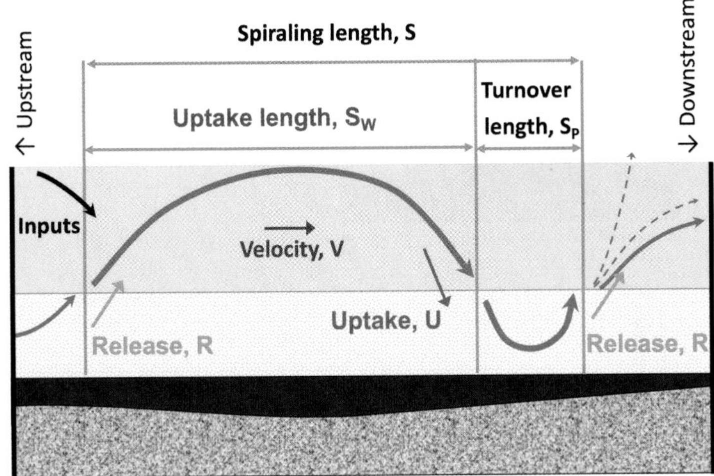

FIGURE 28-27 Illustration of different spiraling metrics used to estimate the biological demand and spiraling lengths of organic carbon (OC) in streams and rivers, from upstream inputs and sediment release to the location of uptake and respiration, and eventual downstream transport or emission of carbon gases generated from OC metabolism. *(Figure adapted by E.R. Hotchkiss from an example provided by R.A. Sponseller.)*

Ideally, sampling discrepancies should be resolved or at least acknowledged when estimating carbon fates in any ecosystem.

Fluxes and Fates

The fate of carbon in freshwater ecosystems depends on its form. Gaseous forms like CO_2 and CH_4 can be emitted, POC can be buried, and any of these forms, in addition to DOC, can be transported. The three fates of carbon are closely linked, and the fate allocation balance can change when carbon is transformed multiple times between various forms within an ecosystem or when environmental parameters favor one fate over another. The fate of carbon is relevant at multiple scales—within an individual ecosystem, across several ecosystems regionally or continentally, and globally. Global estimates are most relevant for the role inland waters play in our climate system. Local (i.e., ecosystem) and regional scale studies provide detailed insight into the dynamics of each fate, as well as the intrinsic and/or extrinsic factors that influence carbon cycling dynamics, and are useful for understanding how carbon fates will be altered by environmental changes (Vachon et al., 2021).

The decomposition of photosynthetically fixed OC by microbes of littoral and benthic sediments, which probably dominates in most inland waters of the world since a vast majority of lakes and streams are small (Chapter 4), relegates the role of animals as "decomposers" to a lesser category, and the burden of heterotrophic metabolism is shifted to the microbes. In large lakes and rivers in which the OM inputs from littoral and external sources are proportionately low, or in hypereutrophic lakes and large rivers in which phytoplanktonic densities reduce light to the point of excluding much of the littoral flora, animals may assume a somewhat greater importance in the degradation of POC. The bulk of OC is dissolved; the fate of DOM that is decomposed and respired is almost completely microbial driven, with some photolysis, and happens throughout ecosystems.

The simultaneous downstream transport and metabolism of OC provides an opportunity to quantify the fate of OC after it enters waterways. One way to quantify OC transport/transformation dynamics is through *OC spiraling* calculations (Newbold et al., 1982; Hall et al., 2016; Plont et al., 2020), where the average length OC travels downstream before being taken up and respired (i.e., spiraled) is governed by both ecosystem metabolism and hydrologic transport (Fig. 28-27). OC spiraling lengths are shorter when R_h exceeds OC transport, and longer when hydrologic transport dominates relative to R_h. While this approach was proposed 40 years ago (Newbold, 1982) and has great promise for elucidating the role of in-stream processes on C dynamics, few studies have included OC spiraling estimates in their analyses of stream OC or DOC dynamics (but see: Kaplan et al. 2007; Griffiths et al., 2012; Hall et al., 2016; Demars, 2019; Plont et al., 2020). One recent synthesis of OC metabolism and transport in North American streams estimated that, on average, OC travels 21 m to 37 km before being metabolized, with shorter spiraling lengths in streams draining landscapes with more pristine land cover than streams draining urban and agricultural landscapes (Plont et al., 2020). Published spiraling lengths for rivers across the United States range from 38 to 1190 km (Hall et al., 2016). Even fewer published studies have estimated spiraling lengths over multiday or multiseason time series (but see: Griffiths et al., 2012; Demars, 2019) or in streams beyond North America (e.g., Lisboa et al., 2016). The increasing frequency of metabolism

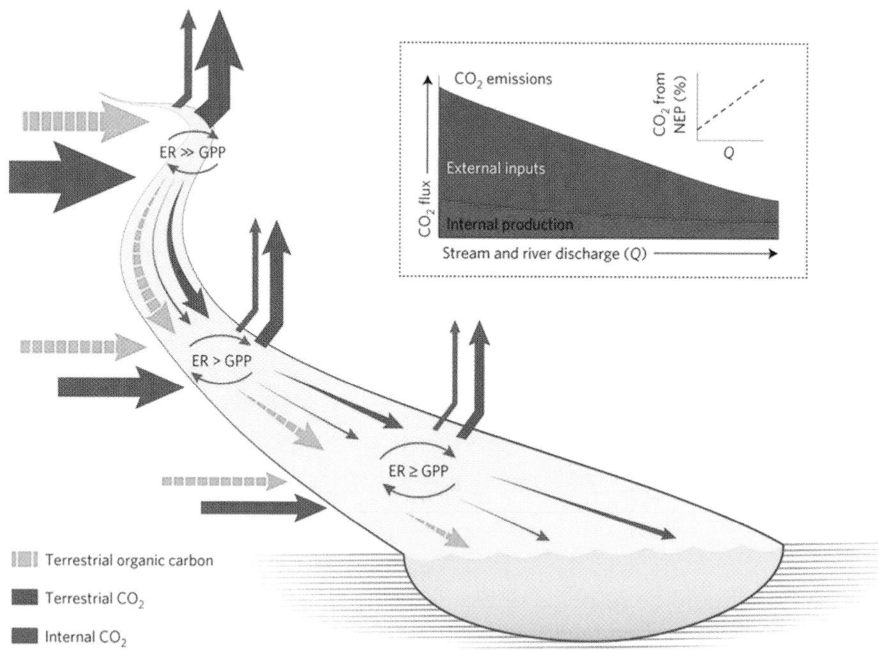

FIGURE 28-28 Predicted changes in the sources and processes controlling carbon transport and emissions along a continuum of headwater streams to large rivers. While small streams have higher in-stream metabolism production of CO_2, they also have larger CO_2 inputs derived from soil respiration, and external sources of CO_2 dominate emissions. Larger rivers are less heterotrophic and thus generate less *in situ* CO_2 from metabolism, but also have fewer external CO_2 inputs, and thus should emit a larger % of CO_2 derived from in-stream metabolism. ER and GPP are ecosystem respiration and gross primary production, respectively. *(From Hotchkiss et al. (2015).)*

measurements made across multiple seasons and continents will provide future opportunities to characterize OC transport, metabolism, and fate at broader spatiotemporal scales if OC fluxes are measured concomitantly. OC spiraling measurements have been limited to streams and rivers, but OC spiraling calculations for ponds, lakes, and reservoirs may yield exciting possibilities for broader comparisons of OC reactivity, transport, and spiraling within and among networks of different freshwater ecosystem types.

i. Carbon emissions

Inland waters can both emit and take up CO_2, depending on season, connections with the landscape, and the net balance between GPP and ER at the ecosystem scale. A portion of aquatic CO_2 is produced *in situ*, but a large fraction of freshwater carbon emissions also come from respiration elsewhere in the catchment (e.g., Jones and Mulholland, 1998; Cole and Caraco, 2001; Weyhenmeyer et al., 2015; Bogard and del Giorgio, 2016). The transition from GPP and ER during the day to only ER at night can also generate higher CO_2 emissions during the nighttime than during the day as CO_2 uptake by GPP pauses and CO_2 production by ER continues (e.g., Rocher-Ros et al. 2020). CO_2 fluxes from inland waters are driven by hydrodynamics and level of CO_2 saturation, and stream and river emissions can be an order of magnitude higher than emissions from lakes (Raymond et al., 2013). At least 30% of stream and river CO_2 emissions are derived from in-stream metabolism, with a higher proportion of external CO_2 respired elsewhere in the catchment contributing to CO_2 emissions from smaller streams compared to larger rivers (Hotchkiss et al., 2015; Fig. 28-28). Some of the world's largest rivers, such as the Amazon (Sawakuchi et al., 2017), Congo (Borges et al., 2019), and Mississippi (Crawford et al., 2016), are significant carbon emitters because of their large surface area and also because they flow through fringing wetlands and developed landscapes that supply additional CO_2 and OM, some of which is emitted or respired and emitted while in transport.

CH_4 can also be sourced from the catchment as well as internally produced (see Chapter 27). Diffusive CH_4 emissions from lakes and reservoirs are typically <50 mg C m^{-2} day^{-1} but can be orders of magnitude higher (DelSontro et al., 2018; Sanches et al., 2019). Net CH_4 uptake in lakes and reservoirs has not been observed and zero fluxes of CH_4 likely happen but are not reported. On average, stream and river diffusive CH_4 fluxes are on the order of 10^2 mg C m^{-2} day^{-1} and some ecosystems have negative diffusive CH_4 fluxes (Stanley et al., 2016; Bretz et al. 2021). CH_4 is also emitted via *ebullition* (i.e., bubbling) and plants

FIGURE 28-29 CH_4 bubbles frozen in a thermokarst lake near Cherskii, Russia. *(Photo credit: Katey Walter-Anthony, October 2003.)*

(Chapter 27). Ebullitive CH_4 fluxes can be highly variable within an individual ecosystem, even spanning four orders of magnitude up to 10^4 mg C m^{-2} day^{-1} at some locations (Linkhorst et al., 2021). Such high rates are typically confined to hot spot regions, however, and the average ebullitive CH_4 flux extrapolated across the entire surface area of a water body will usually range from 10 to 10^3 mg C m^{-2} day^{-1} (DelSontro et al., 2018; Sanches et al., 2019). Temperature (Yvon-Durocher et al., 2014; Aben et al., 2017) and productivity (Deemer et al., 2016; DelSontro et al., 2018) are important drivers of CH_4 diffusion and ebullition from lakes and reservoirs. CH_4 ebullition in fluvial ecosystems is measured less often and does not appear to be as prevalent as in lentic ecosystems, but it does occur with an average of around 10 mg C m^{-2} day^{-1} (Stanley et al., 2016; Robison et al., 2021). Plant-mediated CH_4 emissions in lakes average from 10 to 10^2 mg m^{-2} day^{-1} (Milberg et al., 2017, Zhang et al., 2019), but data supporting plant-mediated flux estimates are still limited.

Both CO_2 and CH_4 can be stored in the hypolimnion (Bastviken et al., 2004) or beneath ice (Denfeld et al., 2018) and then released via diffusion following turnover or ice melt, respectively. Release of CH_4 from the melting of bubbles frozen in ice is also important in northern lake-rich regions (Fig. 28-29; Walter et al., 2007). Gas emissions from the drying and rewetting of littoral sediments due to droughts or management actions are an understudied but likely significant aspect of aquatic carbon emissions (Kosten et al., 2018; Marcé et al., 2019; Keller et al., 2020). Finally, reservoirs exhibit special emission behaviors that may be spatially disconnected from within-reservoir sampling locations, such as degassing at hydroelectric turbines (Roehm and Tremblay, 2006) and in downstream rivers (Guérin et al., 2006).

ii. Carbon burial and storage

Carbon, specifically POC, can be stored and buried in the bottom sediments of lakes, reservoirs, and rivers if resuspension, remineralization, or dissolution to DOC does not occur (also see Chapter 27). Buried carbon is the largest carbon pool in lakes and is related to the age of the water body (Prairie and Cole, 2009). For example, the sediments of Lake Tanganyika in East Africa, one of the oldest lakes on Earth, are estimated to hold as much carbon as the entire terrestrial biosphere (Alin and Johnson 2007). There are, however, two timescales relevant to burial: (1) long-term sequestration of carbon that considers the entire sediment column and operates over hundreds to thousands of years; and (2) short-term sequestration within shallower sediments that operates on an annual basis over tens of years. Here we will discuss short-term burial (from here on referred to as "burial"), which is assessed in the first tens of centimeters of sediments and is the relevant timescale for assessing ongoing cycling of OC in aquatic ecosystems.

Carbon burial is the result of OC sedimentation and burial efficiency, the latter of which is calculated as the ratio of buried carbon to deposited carbon. Burial efficiency can range from 3% to 93% (Sobek et al., 2009), with the most efficient ecosystems being those with high sedimentation rates that transfer OC quickly to anoxic and often cold sediments that are not conducive to substantial remineralization. Extremely high OC sedimentation rates could, however, lead to high rates of methanogenesis and ebullition that reduces the carbon burial efficiency of an ecosystem (Sobek et al., 2012). Oxygen exposure time is a driver of high carbon burial efficiency (Sobek et al., 2009) and more work is needed to understand the impact of temperature on burial as it should reduce OC mass accumulation (Gudasz et al., 2010), but some studies found that burial positively correlated with temperature (Clow et al., 2015; Mendonça et al., 2017).

Burial rates are typically reported on an annual basis and vary over four orders of magnitude, with rates in reservoirs generally higher than in lakes (Fig. 28-30; Mendonça et al., 2017). Reservoirs are exceptional carbon burial sites (Downing et al., 2008; Isidorova et al., 2019), as are small ponds (Taylor et al., 2019). Work on carbon burial in rivers and streams is less common because the remobilization of sediments in flowing channels tends to inhibit long-term burial, but river floodplains store large amounts of carbon (Sutfin et al., 2016; Wohl et al., 2012).

iii. Carbon export

Carbon remaining in an ecosystem (after burial or emission) may be exported, with DOC (and DIC) carbon

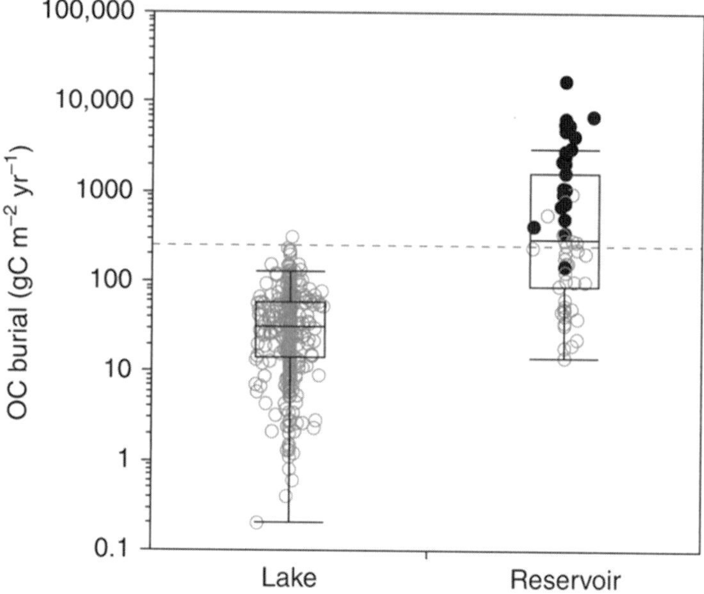

FIGURE 28-30 Organic carbon (*OC*) burial rates in lakes versus reservoirs. *Filled circles* are small agricultural ponds and the *dashed line* is the average of all systems. *(From Mendonça et al. (2017).)*

species most readily available for transport. Carbon export by rivers and streams has dominated carbon transport studies because of their shorter water residence times and global significance. Fluvial waters transport a fraction of the carbon fixed in the biosphere to the oceans, where it has the potential to be stored as part of the oceanic carbon sink (Mulholland, 2003). The spatiotemporal variability of carbon export from river networks is predominantly controlled by hydrology and carbon substrate availability. Global riverine OC export is correlated with runoff and productivity, with rates ranging from <1 g C m^{-2} year^{-1} in arid regions to up to 10 g C m^{-2} year^{-1} in wetter, more productive regions (Mulholland, 2003 and references therein). Mean DOC export from networks of smaller rivers and streams can be higher (up to 25 g C m^{-2} year^{-1}; Mulholland, 2003; Wallin et al., 2015), especially in carbon-rich headwater catchments. Export rates are often reported per annum, thereby averaging over within-year variability associated with seasonal and event-based changes in hydrology. Without high-frequency monitoring, large pulses of carbon export during high runoff events that disproportionately contribute to annual carbon transport and metabolism in larger rivers may be missed (Raymond and Saiers, 2010; Raymond et al., 2016). POC can also be exported and may respond differently to sudden hydrologic changes than DOC. For example, in Amazon headwaters >90% of POC was exported during storm events compared to only 32% of DOC (Johnson et al., 2006).

Lakes and reservoirs can also export significant amounts of carbon despite having longer water residence times than streams and rivers. One study of ~80,000 lakes in 21 catchments across Sweden reported catchment export rates to the sea of <5 g C m^{-2} year^{-1} (Algesten et al., 2004), but studies of individual lakes found much higher carbon export rates—from 13 to >500 g C m^{-2} year^{-1} in Scandinavia (Einola et al., 2011; Sobek et al., 2006) and from 5 to 183 g C m^{-2} year^{-1} in Minnesota, USA (Stets et al., 2010). The highest values in these studies came from lakes high in allochthonous DOC with surface water outflows and were either headwaters, small relative to their catchment, or in proximity to a wetland, thus having higher carbon inputs. Lakes with longer water residence times may favor remineralization or another fate over transport. For example, a modeling study found that 60% of allochthonous DOC is exported from northern USA lakes with residence times less than 1 year, while 60% of DOC is remineralized in ecosystems with residence times greater than 6 years (Hanson et al., 2011). CO_2 and CH_4 can also be transported out of a standing water body, which is of particular concern for dammed reservoirs (Guérin et al., 2006).

iv. Global flux estimates

The amount and forms of carbon moving through inland waters on an annual and global basis are important components of the global carbon cycle (Cole et al., 2007; Tranvik et al., 2009; Tranvik et al., 2018). Inland waters emit ~2 Pg C year^{-1} (Raymond et al., 2013), with ~0.5 Pg C year^{-1} as CO_2 coming from lakes, ponds, and impoundments globally (Holgerson and Raymond, 2016; DelSontro et al., 2018). CH_4 emissions from all inland waters may add an additional 0.15 Pg C $year^{-1}$ to the atmosphere (Rosentreter et al., 2021). Global estimates for carbon burial in lakes and reservoirs range from 0.06 to 0.25 Pg C $year^{-1}$ (Mendonça et al., 2017; Anderson et al., 2020) but could be significantly more if floodplains are included. Global export of total carbon from inland waters is estimated to be ~0.95 Pg C $year^{-1}$, about half of which is OC (Regnier et al., 2013). Based on these estimates, carbon emissions are the dominant flux of the three fates globally. Aquatic carbon emissions have notable consequences for our climate system, especially when the global warming potential of CH_4 is taken into account (85 times more potent as a greenhouse gas than CO_2 over a 20-year timeframe; Myhre et al., 2013), as emissions offset some of the terrestrial and ocean carbon sinks (Bastviken et al., 2011; Butman et al., 2016; DelSontro et al., 2018). Global estimates of carbon fates continuously evolve with new data and approaches to upscaling, most recently due to a rapidly growing number of gas concentration and emission measurements across inland waters that required multiple upwards revisions of the carbon inputs needed to sustain inland water burial, emission, and export fluxes (Drake et al., 2018).

v. Whole-ecosystem budgets

Quantitative analyses of the major OC transformations and fates in aquatic ecosystems (i.e., budgets) include the following general characteristics: (1) the dominant pool of OC is DOC; (2) three major sources of POC occur—allochthonous POC, and two distinct major zones of autochthonous POC from littoral and benthic versus pelagic zones; (3) allochthonous inputs from the drainage basin and exports from a lake or river occur largely as DOC and represent a major source of C for metabolism; (4) detrital metabolism occurs principally in benthic zones (Chapter 27), where the majority of POC is decomposed to CO_2 and CH_4, and in the pelagic zones during sedimentation; and (5) OC has three major fates in inland waters: burial in sediments, metabolism and emission as CO_2 or CH_4 at the air–water interface, or export to downstream ecosystems.

Comprehensive whole-ecosystem OC budgets are relatively rare because they require extensive measurements of spatially variable components. At the minimum, an OC budget should include: (1) surface water, groundwater, and external (e.g., leaf litter) inputs of allochthonous OC; (2) sediment burial, emission, and export via major outflows; and (3) autochthonous OC inputs and all OC losses due to GPP and ER, respectively (Fig. 28-26; Hanson et al., 2011). Often, only the components needed to address a specific research question are measured, but to identify the parameters and anthropogenic disturbances regulating OC metabolism at an ecosystem level, we must assess all carbon fluxes among all ecosystem components and throughout the year. Here we present a few examples of carbon budgets to illustrate various ways to construct them and how they can contribute to our understanding of ecosystem and metaecosystem OM dynamics.

One of the most comprehensive carbon budgets and also one of the first to highlight the role of detrital carbon cycling relative to living carbon is that of the hardwater Lawrence Lake (maximum depth of 12.6 m, surface area ~0.05 km^2) in Michigan, USA (Wetzel et al., 1972). Autochthonous OC inputs and losses were measured in detail, including estimates of primary production and excretion of phytoplankton, macrophytes, epiphytic and epipelic algae, as well as respiration, in the three major lake habitats—littoral, pelagic, and benthic. The budget also included allochthonous DOC and POC inputs, sedimentation/burial, $CaCO_3$ precipitation, and DOC and POC outflows (Table 28-3; Fig. 28-31). The single largest flow of OC was POC from the littoral zone, which was transported to sediments in deeper parts of the lake, where it was supplemented by pelagic POC. Microbial decomposition in benthic sediments converted a significant fraction of OC to CO_2, which was released to the photic zone and presumably the atmosphere, although atmospheric emissions were not directly measured. A much smaller fraction of benthic OC was resuspended and used by the pelagic community, and a similarly small amount was permanently buried. Predatory and nonpredatory impacts of zooplankton were also assessed but found to be negligible (Crumpton and Wetzel, 1982). While zooplanktonic grazing can alter algal populations at certain times of the year, the analyses in Lawrence Lake demonstrated the importance of algal losses by sedimentation, which on an annual basis were much greater than losses by grazing. Overall, the carbon budget of Lawrence Lake was nearly balanced with only ~10 g C m^{-2} $year^{-1}$ more inputs than outputs, and would likely be more balanced if atmospheric CO_2 and CH_4 emissions had been directly measured. Note that while calcite precipitation was accounted for in this budget, DIC inputs were not but can be important in the total carbon balance of ecosystems (e.g., Finlay et al., 2010; Stets et al., 2009).

Mass balance approaches are commonly used for carbon budget calculations (e.g., Stets et al., 2010). The study of Lake Gäddtjärn (maximum depth, 11 m; surface area, 0.07 km^2) in central Sweden is an example of developing a carbon budget using mass balance and NEP estimates, although the metabolism estimates were only possible in this DOC-rich lake during peak summer when biological signals in the O_2 data were less swamped out by physical processes (Chmiel et al., 2016). Additional modeling and water incubations were employed to support missing NEP estimates. The remaining budget components were IC and OC inflows

TABLE 28-3 Total Annual Budget of Carbon Fluxes for Lawrence Lake, Michigan, USA[a]

Components	$g\ C\ m^{-2}\ year^{-1}$	Percentage
Inputs		
Autochthonous		
Phytoplankton	43.4	19.1
Submersed macrophytes	87.9	38.8
Epiphytic algae	37.9	16.8
Epipelic algae	2.0	0.9
Algal secretion and autolysis	14.7	6.5
Littoral plant secretion	5.5	2.5
Heterotrophy	2.8	1.2
Dark CO_2 fixation	7.1	3.1
Allochthonous		
Stream and groundwater DOC	21.0	9.3
Stream POC	4.1	1.8
Shoreline litter	0.01	0.0
Total Inputs	226.4	100.0
Outputs		
Respiration		
Benthic respiration	117.5	54.6
Bacterial respiration of DOC	20.6	9.6
Bacterial respiration of POC	8.6	4.0
Algal respiration	13.0	6.1
Permanent sedimentation	14.8	6.9
Coprecipitation of DOC with $CaCO_3$	2.0	1.0
Outflow		
Dissolved	35.8	16.5
Particulate	2.8	1.3
Total Outputs	215.1	100.0

[a]*From data in Wetzel et al. (1972); Otsuki and Wetzel (1974); Burkholder and Wetzel (1989).*

and outflows, measurements of CO_2 and CH_4 emissions, laboratory estimates of photolysis, and detailed sediment analyses for benthic metabolism and burial. This study focused on identifying the role of sediments in a whole-lake carbon budget and concluded that OC remineralization and burial in sediments were minor components of the annual basin-wide carbon budget, representing only 4% of inputs and 1% of outputs, respectively (Table 28-4; Chmiel et al., 2016). In this small lake, stream inflow and export dominated the C budget (60% and 68%, respectively), but groundwater inputs and gas emissions also constituted large fractions of the budget (25% of input and 31% of output, respectively). While they did not provide the autochthonous detail that Wetzel et al. (1972) did for Lawrence Lake, Chmiel et al. (2016) managed to balance the carbon budget of Lake Gäddtjärn within 5% using a budget approach like that outlined by Hanson et al. (2015). This seasonally and spatially integrated budget revealed that sediments may not play an important role in every lake and that groundwater inputs and gas emissions should not be neglected in freshwater ecosystem carbon budgets.

In a data review and simulation of OM inputs, cycling, and export at the scale of the Little Tennessee River network, Webster (2007) used OM measurements, ecosystem metabolism estimates, and whole-stream budgets from small streams and larger rivers to identify longitudinal changes in inputs and fates of OC in the forested mountains of North Carolina, USA. As with many budgets, these simulations do not include all carbon forms and fates: they did not include DOC (which is low but varies with flow in streams in this region; Meyer and Tate, 1983) or inorganic carbon, including CO_2. Historically, inorganic carbon was often not included in stream carbon budgets because much of the research in this subfield of limnology focused on how detrital and algal OM controlled stream and river ecosystem processes. However, results from a recent study of small streams in the upper Little Tennessee River network can be used to predict that CO_2 transport and losses are dominated by CO_2 derived from external sources (e.g., soil and root respiration), not in-stream metabolism (Bretz et al., 2021). At the scale of the entire Little Tennessee River network, POM inputs consisted of 19% leaves and 81% GPP, despite GPP being very low in the headwater streams (Webster, 2007; Fig. 28-32). Fates of OM inputs included 51% exported (of which 45% was autochthonous and 6% was allochthonous), 28% respired by autotrophs, and 21% respired by heterotrophs (including R_h of *coarse benthic OM* [CBOM], *fine benthic OM* [FBOM], and *seston OM* in transport). Moving from headwaters to the Little Tennessee River, the dominant OM inputs shift from leaves to GPP (Fig. 28-33a), and major outputs transition from allochthonous OM transport and R_h of CBOM to autochthonous OM transport and R_a of autochthonous OM (Fig. 28-33b). While these stream-to-river patterns largely confirm those proposed by Vannote et al. (1980) in their formation of the *River Continuum Concept* (see Chapter 5), this network-scale budget synthesis offers unique insights into different OM inputs and fates in smaller streams as well as larger rivers downstream within the same catchment (Webster, 2007).

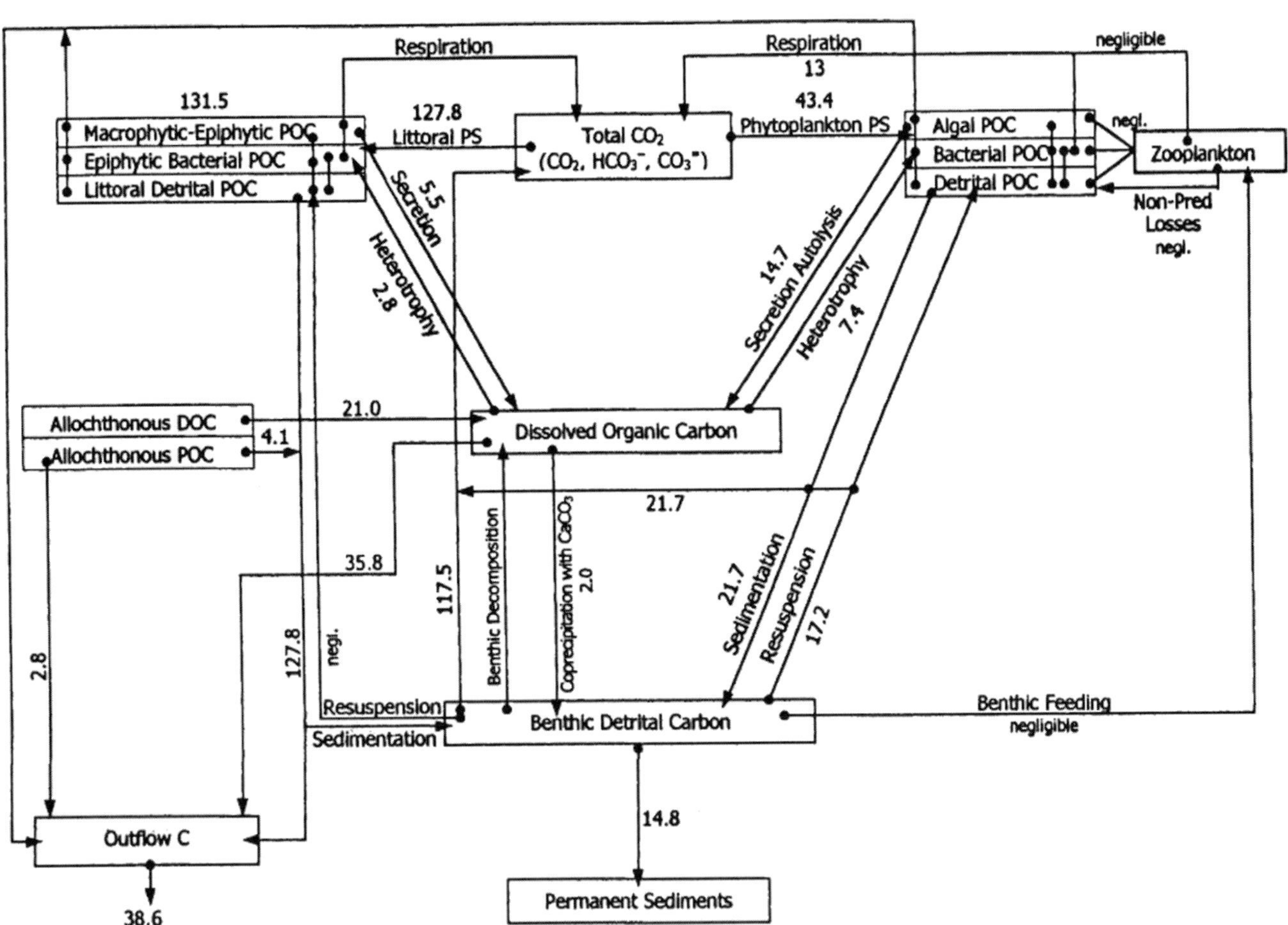

FIGURE 28-31 Detrital structure and flux of organic carbon (g C m^{-2} $year^{-1}$) of Lawrence Lake, southwestern Michigan (USA). DOC, dissolved organic carbon; *POC*, particulate organic carbon; *PS*, photosynthesis. *(After Wetzel et al. (1972).)*

TABLE 28-4 Annual Carbon Budget of Lake Gäddtjärn, Sweden

Components	t C $year^{-1}$	g C m^{-2} $year^{-1}$	Percentage
Inputs			
Stream inflow	13.5	193	60
Groundwater inflow	5.5	79	25
Net ecosystem production	2	29	9
Sediment remineralization	1	14	4
Photomineralization	0.4	6	2
Atmospheric deposition	Negligible	Negligible	—
Total Inputs	22.4	321	100
Outputs			
Gas emission	6.5	93	31
Burial	0.3	4	1
Stream export	14.5	207	68
Total Outputs	21.3	304	100

From Chmiel et al. (2016)

There are multiple drivers of carbon dynamics in aquatic ecosystems, such as morphometry of lakes (i.e., depth and size), macrophyte coverage, water chemistry, productivity, inflow and outflow discharge, catchment-to-area ratio, surrounding topography and biomass, and climate. Dominant drivers are ecosystem specific and can vary over time. While our knowledge of aquatic carbon dynamics has increased substantially in the past several decades, future work is needed to test the relative importance of allochthonous versus autochthonous OC for metabolism and food webs; characterize source-specific, time-varying, and ecosystem-scale OC mineralization rates and fates; and, of course, predict the current and future effects of environmental changes on OC cycling (Hanson et al., 2015). Modeling efforts can provide a framework for constructing carbon budgets, but more intensive sampling efforts and monitoring will continue to be crucial for developing comprehensive budgets necessary to make long-term predictions of freshwater carbon cycling and responses to anthropogenic changes.

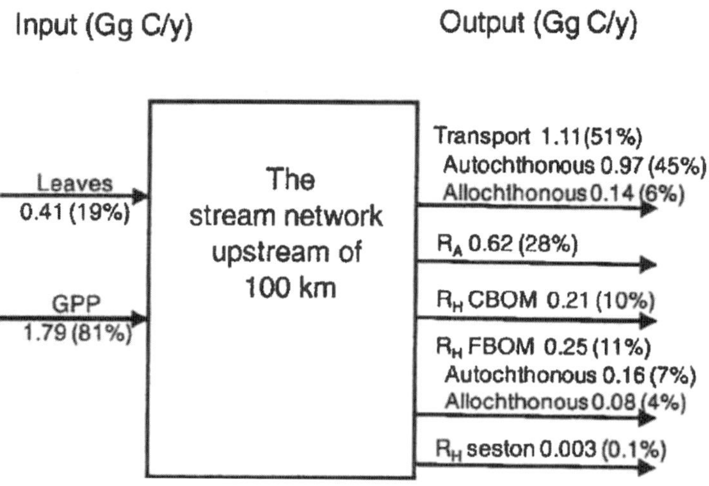

FIGURE 28-32 Particulate organic matter (*OM*) budget for the upper Little Tennessee River network, USA. *GPP* is gross primary production, R_A is autotrophic respiration, R_H is heterotrophic respiration, *CBOM* is coarse benthic OM, and *FBOM* is fine benthic OM. % in parentheses are based on total inputs or outputs. *(From Webster (2007).)*

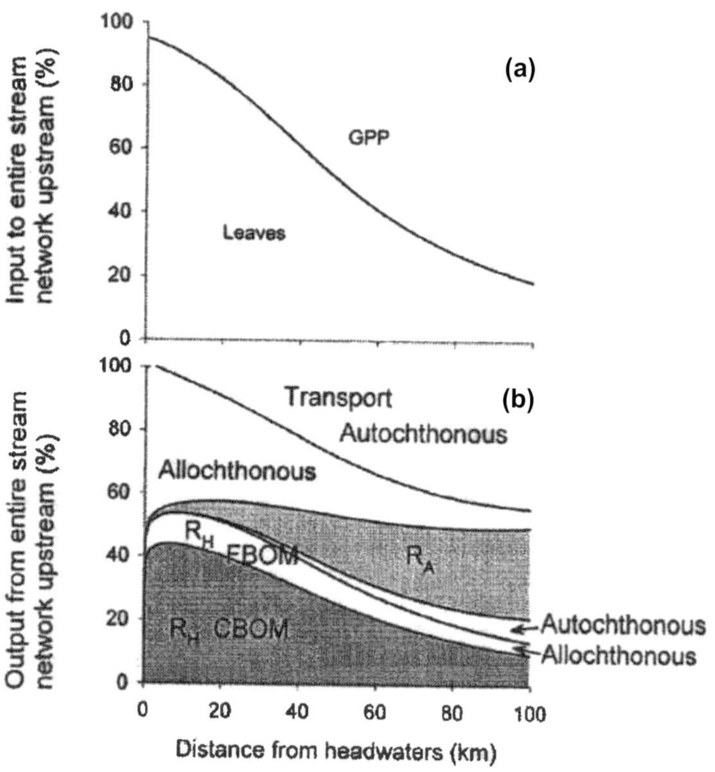

FIGURE 28-33 Change in the particulate organic matter inputs and outputs (i.e., fates) along the first 100 km of the Little Tennessee River Network, USA. (a) The percent of inputs as GPP (gross primary production) or leaves changes with distance from headwaters. (b) The output and fate of upstream network inputs also changes with distance from headwaters, and includes autochthonous and allochthonous transport, R_A (autotrophic respiration), R_H (heterotrophic respiration) of allochthonous and autochthonous CBOM (coarse benthic organic matter) and FBOM (fine benthic OM). *(From Webster (2007).)*

V. Anthropogenic changes to organic matter (OM) dynamics

The primary thesis of this chapter is that DOM regulates biogeochemical cycling and detrital OM provides a fundamental thermodynamic stability to freshwater ecosystems. Alterations of the environment by human activities change OM cycling and fluxes in aquatic ecosystems. While some of the environmental concerns highlighted in the previous edition of this text have improved or worsened (e.g., acid rain and ozone depletion or climate change, respectively) and new challenges have been identified (e.g., emerging contaminants), a suite of different human activities continues to alter the sources, fluxes, and metabolism of OM in freshwater ecosystems. We acknowledge that we cannot cover all environmental changes and biogeochemical consequences of human activities within the scope of this

TABLE 28-5 Interactions Between Global Trajectories in Environmental Changes (Warming, Land-Use Change, Flow Regulation) and Drivers of Ecosystem Metabolism (Light, Thermal, and Flow Regimes).

		Global trajectories		
		Global warming	**Land-use change**	**Flow regulation**
Drivers of Ecosystem Metabolism	**Light Regime**	• Longer ice free period • Earlier leaf out • Later leaf fall	• Increased sediment load • Reduced riparian shading	• Reduced sediment load • Altered water depth • Altered riparian vegetation
	Thermal Regime	• Higher mean annual temperature	• Reduced riparian shading • Channel incision enhances bank shading	• Hypolimnetic release alters natural thermal regimes • Retention in shallow basins raises temperatures
	Flow Regime	• Increasing drought severity • Alter timing of floods • Earlier, reduced snowmelt	• Stormwaters and tile drainage increase flood frequency and magnitude of peak flows • Withdrawals reduce baseflow • Leaks enhance baseflow	• Hydropeaking increases frequency and magnitude of flood peaks • Dams alter timing of seasonal floods and change baseflows

From Bernhardt et al. (2018)

chapter section, but we use this space to highlight several major anthropogenic drivers of current and future changing OM sources, fluxes, and cycling.

Anthropogenic environmental changes alter carbon inputs and cycling in freshwater ecosystems. Greenhouse gas concentrations in the atmosphere continue to rise, largely due to anthropogenic combustion of fossil fuels, and therefore our atmosphere and biosphere are warming (IPCC, 2021). Climate change increases temperatures and precipitation extremes (IPCC, 2021) while shifting the magnitude and timing of terrestrial productivity (Ballantyne et al., 2012; Piao et al., 2019), thus altering many biological and physical processes that contribute to the carbon biogeochemistry of inland waters. Elevated levels of CO_2 in the atmosphere can increase rates of carbon uptake by terrestrial photosynthesis (Ballantyne et al., 2012). Higher terrestrial photosynthesis from CO_2 fertilization may lead to freshwater primary and secondary production being limited by nutrient availability, as C:N ratios of plant material and structural tissues, particularly lignin content, increase (Wetzel, 2003). DOM leached into surface waters from plants grown in elevated-CO_2 environments may also contain larger amounts of photoreactive dissolved humic-like substances (Wetzel, 2003). How changes in the OM composition of allochthonous inputs to freshwater ecosystems will alter food web energetics and biogeochemical processes remains an active area of research. Other human activities disrupt hydrologic flow paths through damming and other flow modifications (Pekel et al., 2016). Given our current understanding of the environmental drivers of ecosystem metabolism (e.g., light, temperature, and flow regimes) and the trajectories of different environmental changes (e.g., global warming, land-use change, and flow regulation), we can predict how anthropogenic changes have and will continue to alter freshwater ecosystem metabolism (Table 28-5; Bernhardt et al., 2018).

Freshwater ecosystems are simultaneously recovering from some anthropogenic stressors while other stressors continue or intensify, including: anthropogenic climate change, ozone loss and associated UV damage, hydrologic modifications, nutrient pollution, and acid precipitation and deposition. Anthropogenic emissions, precipitation, and deposition of sulfur and nitrogen compounds lowered the pH of many inland waters, decreased DOM inputs from land, and increased water transparency (Likens and Bormann 1974; Driscoll and Van Dreason, 1993; Gjessing et al., 1998). Previous periods of reduced ozone in the stratosphere, in part associated with releases of chlorofluorohydrocarbon (CFC) compounds used in refrigeration and propellant devices, resulted in large increases in the amounts of UV-B and, to a lesser extent, UV-A reaching the surface of the Earth (Kerr and McElroy, 1993; Madronich, 1994), which likely accelerated the photolytic degradation of macromolecules of DOM to CO_2. Successful policy changes decreased acid precipitation and deposition (Stoddard et al., 1999) and reduced the emission of ozone-depleting chemicals (Chipperfield et al., 2017), which has altered ion exchange in soils, increased soil OM exports to aquatic ecosystems, and changed light–OM dynamics within inland waters. While we do not detail the progress and ongoing challenges of these environmental changes that are now "in recovery" (or at least not worsening), we mention them here to provide historical context for ever-changing environmental stressors and to note potential interactions between the legacies of higher UV radiation and acidification with continued climate change intensification (e.g., Anesio and Granéli, 2003; Williamson et al., 2019). Ongoing and emerging environmental changes that may alter the timing and magnitude of OM fluxes and cycling in freshwaters include shifts in land cover; increased temperatures; hydrologic alterations; changing DOM concentrations and composition; altered nutrient and sediment loads; and increasing contaminant loads throughout waterways.

A. Changes in land cover

Soil OC is one of the largest pools of carbon globally, surpassing vegetation and atmospheric pools (Scharlemann et al., 2014). Thus past and ongoing changes in land cover from more pristine forests, grasslands, and wetlands to agricultural, residential, or industrial environments have altered the transport of soil OC and other OC sources to and through aquatic ecosystems (Regnier et al., 2013). Not only can fluxes of OM into aquatic ecosystems be modified by land-use change (e.g., Stanley et al., 2012), but so can OM composition (e.g., Wilson and Xenopoulos, 2009; Roebuck et al., 2020; Xenopoulos et al., 2021) and OC age (e.g., Butman et al., 2015), likely driven by changing hydrologic flow paths and OM sources (Barnes et al., 2018). For example, changes to a landscape through deforestation can alter nutrient loads (e.g., Mattsson et al., 2009), light availability (e.g., Masese et al., 2017), and temperature (e.g., Burrows et al., 2014), with varied consequences for OC cycling in aquatic ecosystems (Burrows et al., 2014; O'Driscoll et al., 2016; Masese et al., 2017).

Enhanced nutrient and sediment loading from agricultural landscapes and the trapping of OC in reservoirs has led to a globally relevant amount of OC sequestration in freshwater sediments over the last century (Anderson et al., 2020). Long-term burial of carbon may offset increases in carbon emissions resulting from the degradation of allochthonous carbon mobilized by land-use change, although global values for all these components are still not well constrained (Butman et al., 2018). Ultimately, variability in OM, nutrients, and environmental conditions resulting from land cover changes can impact rates of ecosystem metabolism and the metabolic balance of ecosystems (e.g., Bernot et al., 2010; Finlay, 2011; Fuß et al., 2017; Masese et al., 2017). In a comparison of metabolism in small streams across the United States and Puerto Rico during summer low flow conditions, streams draining more developed or agriculturalized catchments had higher GPP than streams draining reference conditions (e.g., forests, grasslands), likely due to increases in light and nutrients (Bernot et al., 2010; Finlay, 2011). These patterns are similar in tropical streams of Costa Rica and Kenya, where streams draining agricultural land cover had higher GPP (due to higher light, nutrients, and/or temperature) and ER (mostly due to higher GPP) than streams draining less modified landscapes (Ortega-Pieck et al., 2017; Masese et al., 2017). Across multiple geographic regions, both GPP and ER are often higher in streams draining human-dominated landscapes compared to reference streams, while land cover–associated differences in GPP and ER are less apparent in larger rivers (Finlay, 2011).

B. Warming temperatures

Many biological and physical processes scale with temperature, including leaching of DOM from parent material, GPP, ER, and organism growth rates. Anthropogenic climate change is increasing water temperatures (Kaushal et al., 2010) and inland waters are increasingly vulnerable to losing ice cover during part or all of historically ice-covered seasons (Sharma et al., 2019; Yang et al., 2020). Warmer and ice-free waterways retain lower concentrations of dissolved gases, including oxygen needed to sustain aerobic respiration (e.g., Jane et al., 2021). CO_2 may become an increasingly significant limiting factor for GPP (e.g., Hamdan et al., 2018) as shorter periods of ice cover reduce CO_2 accumulation under ice through spring (e.g., Denfeld et al., 2018) and springtime GPP may thus decrease without high under-ice CO_2 concentrations. Many mesocosm- or ecosystem-scale warming experiments have identified warming-induced increases in GPP (Demars et al., 2011; Hood et al., 2017; Yvon-Durocher et al., 2017; Fig. 28-34) and/or ER (Demars et al., 2011; Yvon-Durocher et al., 2017; Fig. 28-34). Studies comparing stream metabolism across a gradient of ambient temperatures predict there may be differential metabolic responses to warming depending on whether sites had lower or higher prewarming temperatures and metabolism (Song et al., 2018). ER responses to temperature may be stronger than GPP–temperature relationships, thus making some ecosystems more heterotrophic (i.e., more negative NEP) in response to long-term warming (Demars et al., 2011). Detrital pool carbon cycling may also increase with warming: a global assessment of leaf litter breakdown in streams and rivers predicted a 5–21% increase in mean decomposition rates with an increase in temperatures of 1–4°C (Follstad Shah et al., 2017) and increasing mineralization of boreal lake sediments in response to warming could decrease annual OC burial by 4–27% (Gudasz et al., 2010). Functional redundancy, species-specific responses, or habitat-specific responses to warming may generate a neutral effect of warming on ecosystem-scale carbon cycling and food web energy fluxes (e.g., Rodríguez et al., 2016; Padfield et al., 2017; Nelson et al., 2020). Relative increases in allochthonous versus autochthonous OC sources may also yield contrasting ER–temperature relationships

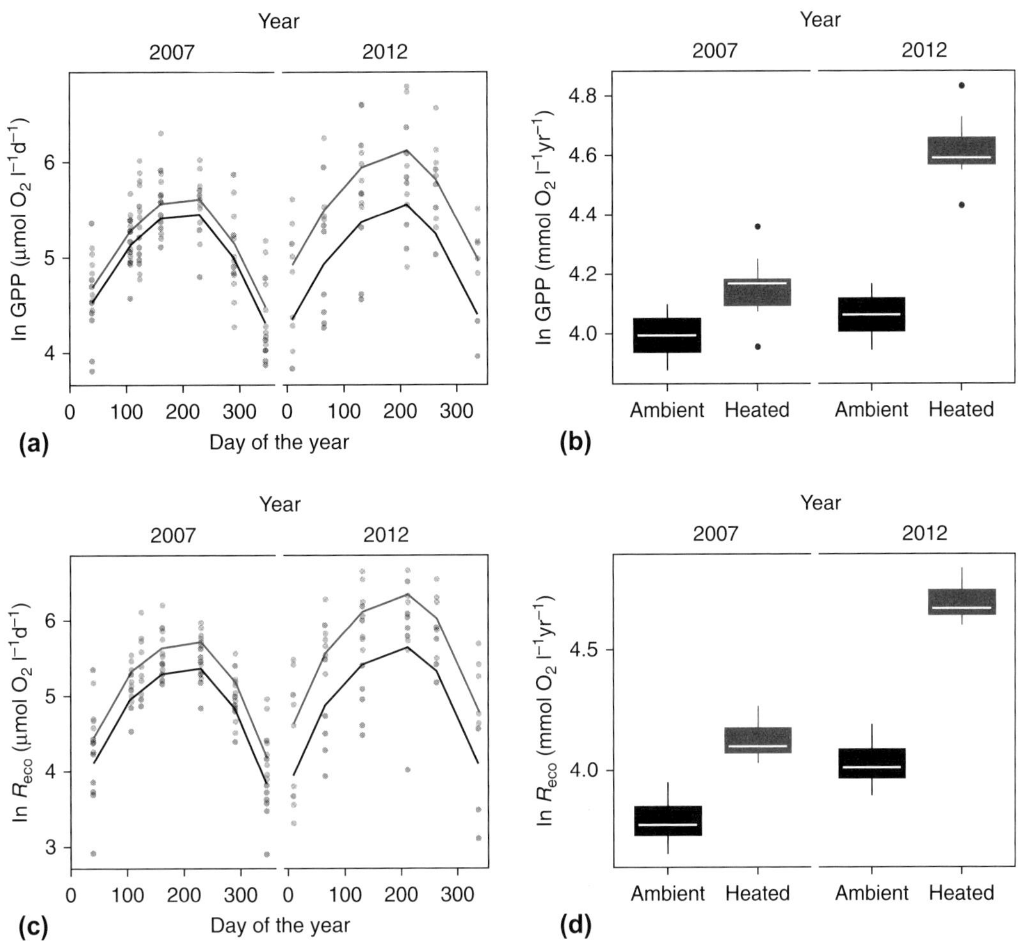

FIGURE 28-34 Differences in pond gross primary production (*GPP*) and ecosystem respiration (R_{eco} in figure; ER in this chapter) in response to experimental warming from 2007 to 2012. Warmed treatments are in *red*; ambient treatments are in *black*. *Solid lines* in panels (a) and (c) are from the best fitting generalized additive mixed models. *(From Yvon-Durocher et al. (2017).)*

(Jane and Rose, 2018). Clearly, earlier work testing biotic responses to changing temperatures in more controlled laboratory environments (often with single species or simplified communities) has been critical for making predictions about the ecosystem-scale consequences of warming, but long-term warming responses at the ecosystem scale may yield unexpected results due contrasting species responses to changes in water temperatures.

C. Hydrologic alterations

The presence and dynamics of inland waters across the globe have been significantly altered by direct anthropogenic flow modifications as well as climate change. Damming, diversion, burial, and extraction have modified global surface water distributions, creating new types of water bodies in some locations and eliminating others (Pekel et al., 2016). Glacial melt due to climate warming has led to more surface water and higher discharge in some regions (Pekel et al., 2016; Do et al., 2017); intermittency of surface water inundation and flow is intensifying due to climate and direct flow modifications (Messager et al., 2021; Zipper et al., 2021); and rivers across the globe have increasing flow variability and fragmentation that are mismatched with the needs of endemic biota and potential rates of ecosystem metabolism (Palmer and Ruhi, 2019). Climate change has amplified temperature and precipitation extremes, increasing the intensity and/or duration of droughts and large precipitation events across the globe (IPCC, 2021).

Because hydrologic connections and water residence time drive OM inputs, metabolism, and transport in aquatic ecosystems (Catalán et al., 2016; Zarnetske et al., 2018; Asmala et al., 2019), hydrologic changes alter the sources, fluxes, and cycling of OC in transport through restored rivers (e.g., Palmer and Ruhi, 2019; Arroita et al., 2019) and intermittently connected ecosystems (e.g., Bretz et al., 2021). Human-induced hydrologic changes during water storage and release in dammed fluvial networks can also impact DOM quality and thus cycling (Ulseth and Hall, 2015; Bianchi et al., 2017; He et al., 2020). Higher flow events may increase allochthonous OM supply along fluvial networks (Casas-Ruiz et al., 2020) and, after periods of drought, can lead to higher rates of primary production that induce anoxia and fish kills in lakes (e.g., Kragh et al., 2020). Changing hydrologic connections between aquatic ecosystems sustains a high diversity of DOM compounds (Kellerman et al., 2014). Thus, in regions where hydrologic connectivity and OM inputs are changing, such as in northern latitudes (Johnston et al., 2020; Beel et al., 2021), we expect to see ongoing and future changes in the sources, composition, and cycling of OM (Creed et al., 2018). Finally, hydrologic alterations, whether due to droughts or direct flow management, can enhance carbon emissions when dried aquatic sediments are exposed to higher oxygen concentrations that support increased respiration and higher emissions upon rewetting (e.g., Kosten et al., 2018; Marcé et al., 2019).

D. Changing nutrient and sediment inputs

Nutrient and sediment loading to freshwaters increases as a result of various land use and climatic changes, and can significantly impact drivers and processes involved in OC cycling. Eutrophication has been an issue in freshwaters since the mid-1800s, initially due to inputs from urbanized landscapes (Jenny et al., 2016) and later because of excess P in detergents (Schindler, 2006), as well as P and N loading from fertilizers, wastewater, and other human activities (Schindler, 2006; Paerl et al., 2016; Jenny et al., 2016). Excess nutrients stimulate productivity that may induce algal blooms, cause shifts in food web dynamics, and generate low-oxygen conditions through decomposition of detrital OM, all of which have implications for OC cycling and fate in inland waters (Beaulieu et al., 2019; Wurtsbaugh et al., 2019). Human-mediated inputs of nutrients to inland waters can increase the relative autochthonous contributions to food webs through increases in aquatic GPP (Hadwen and Bunn, 2004; Kritzberg et al., 2006). Multiyear increases in nutrient inputs to forested streams, where detrital OM inputs from the terrestrial landscape are critical resources for consumer secondary production (Wallace et al., 1997; Wallace et al., 2015), increase rates of in-stream breakdown of detrital POC (Rosemond et al., 2015). Nutrient-associated changes in metabolism and carbon fluxes are expected to increase with continued land cover change and flow modifications (e.g., Bernot et al., 2010; Finlay, 2011; Bernhardt et al., 2018), but the degree to which enhanced or suppressed metabolic rates will interact with other stressors or induce low-oxygen periods and anaerobic OC metabolism (e.g., Genzoli and Hall, 2016; Jenny et al., 2016; Blaszczak et al., 2019) is still an active area of research.

Anthropogenic activities mobilize billions of metric tons of sediment annually through streams and rivers, half of which is retained behind dams (Syvitski et al., 2005), and accelerate sediment accumulation in lakes and reservoirs (Dearing and Jones, 2003; Baud et al., 2021). Enhanced sedimentation in lakes and reservoirs can translate into more long-term OC burial (Kastowski et al., 2011; Dietz et al., 2015) but also in more carbon metabolism and emissions in the short term (Sobek et al., 2012). Carbon metabolism can also be disrupted by increased sediment loading because of changes to light, oxygen, temperature, pollutants adsorbed to incoming sediment, and physical scouring or burial of benthic communities (Donohue and Garcia Molinos 2009; Bernhardt et al., 2018; Blaszczak et al., 2019).

E. Changing OM concentrations and composition

Human activities have increased export of OC from soils to inland waters (Larsen et al., 2011; Regnier et al., 2013; Wohl et al., 2017); altered the composition of OM inputs to freshwaters from human-modified landscapes (Stanley et al., 2012; Xenopoulos et al., 2021; Fig. 28-35) and thawing permafrost (e.g., Mann et al., 2015); and changed locations of OC storage within fluvial networks as river−riparian−floodplain disconnections decrease OC storage and reservoir construction increases OC burial (Wohl et al., 2017). Clear-cutting of forests within the drainage basins of small streams reduces DOC inputs and transport (Meyer and Tate, 1983; Grieve, 1990). Interactions between warming temperatures, land use change, alterations to the trophic structure within ecosystems, shifts in terrestrial production, and changing soil chemistry and hydrology are projected to increase and alter the composition of DOM inputs to inland waters, with potentially significant consequences for lake food web energetics in response to changing OM composition (Creed et al., 2018; Fig. 28-36). In contrast, management actions that decrease OM inputs to waters can decrease the magnitude and variability of ER as those ecosystems recover from past OM pollution (Bernhardt et al., 2018; Arroita et al., 2019). The likely net effect of anthropogenic changes is an increase in carbon export from inland waters to the ocean when floodplain and riparian modifications are included in continental carbon budgets (Wohl et al., 2017).

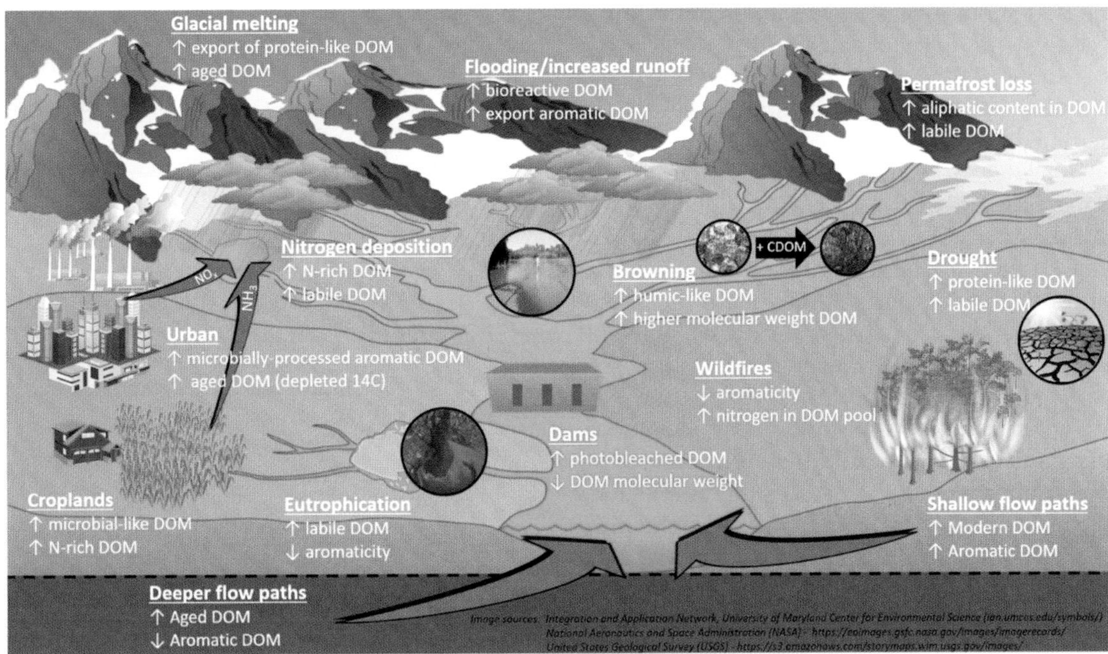

FIGURE 28-35 Anthropogenic changes to organic carbon (*OC*) in freshwater ecosystems. *(From Xenopoulos et al. (2021).)*

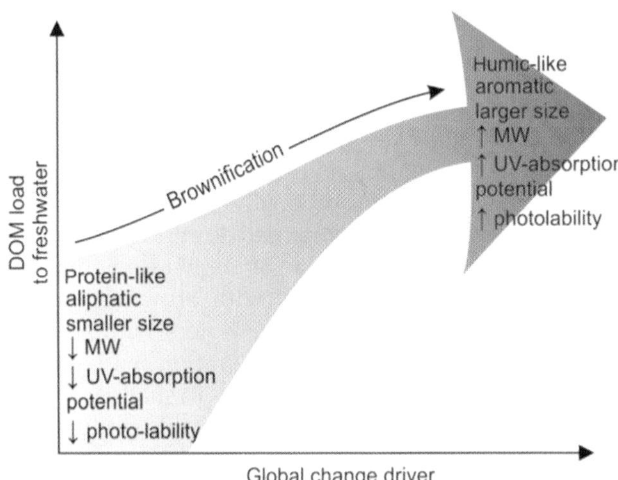

FIGURE 28-36 Predicted effects of anthropogenic global changes on dissolved organic matter (*DOM*) inputs (load) and properties in northern lakes. Global change stressors are hypothesized to increase allochthonous OM inputs as well as change the composition of that OM, which may increase in size and molecular weight (*MW*), include more humic-like and aromatic structures, and be more susceptible to photodegradation. *(From Creed et al. (2018).)*

F. Emerging contaminants and other pollutants

The chemicals and materials humans use to control pests, clean surfaces, treat medical ailments, enhance personal appearances, pack materials, improve farm yields, and conduct a range of other industrial and agricultural operations enter freshwater ecosystems through multiple pathways. Contaminants of known and emerging concern (e.g., pharmaceuticals and personal care products, nanoparticles, plastics, and pesticides) can persist or accumulate in freshwater ecosystems and food webs (Reid et al., 2019). Material contaminants in freshwater (e.g., plastics) serve as physical habitats for biofilms (Arias-Andres et al., 2019), and biofilm communities on anthropogenic substrates appear to differ from those on natural substrates (e.g., Hoellein et al., 2017; Kettner et al., 2017; Rummel et al., 2017). Research on contaminants in inland waters has largely focused on their distribution and concentration or the lethality of specific compounds in controlled laboratory settings, but some recent work has employed laboratory and *in situ* studies to assess ecological consequences. Primary producer biomass and photosynthesis typically decrease in response to pharmaceutical or pesticide exposure (e.g., Rosi-Marshall et al., 2013; Shaw et al., 2015; Gallagher and Reisinger, 2020; Rumschlag et al., 2020), but caffeine and antihistamines may stimulate biofilm photosynthesis (Shaw et al., 2015). Contaminant exposure often suppresses biofilm and plankton respiration (e.g., Rosi-Marshall et al., 2013; Shaw et al., 2015; Rumschlag et al., 2020) and OM breakdown (Hunter et al., 2021), but many experiments have found no cumulative effects of contaminants on OM processing or respiration (e.g., Shaw et al., 2015; Gallagher and Reisinger, 2020) or even increased CO_2 production (Jonsson et al., 2015). Microbes in urban areas subjected to chronic exposure seem to be resilient to the effects of contaminant exposure (unaltered primary production and respiration or no net change in ecosystem function), likely due to the development of more resistant microbial communities in those ecosystems (Rosi et al., 2018; Jepsen et al., 2019; McClean and Hunter, 2020). However, even rates of CO_2 and CH_4 production by microbial communities in naïve sediments did not differ significantly from control treatments after exposure to single or multiple antibiotics (Gray and Bernhardt, 2022). Many inland waters in urban and agricultural landscapes are also getting saltier (see Chapter 12) due to winter deicing, fertilizer application, and other human activities on the landscape and within sewersheds (Kaushal et al., 2018a), with still largely untested impacts on ecosystem-scale processes like metabolism (Berger et al., 2019). As novel chemicals continue to enter our waterways and the stoichiometry of elements shifts in response to different human activities, we must improve how we assess the interactions between environmentally relevant levels of chemical exposure and biological effects (e.g., Diamond and Burton, 2021). Future work integrating isolated contaminant effects with the biological consequences of exposure to contaminant mixtures will improve our understanding of how freshwater communities and carbon cycling processes are changing in response to chemical stressors (Bernhardt et al., 2017; Peace et al., 2021).

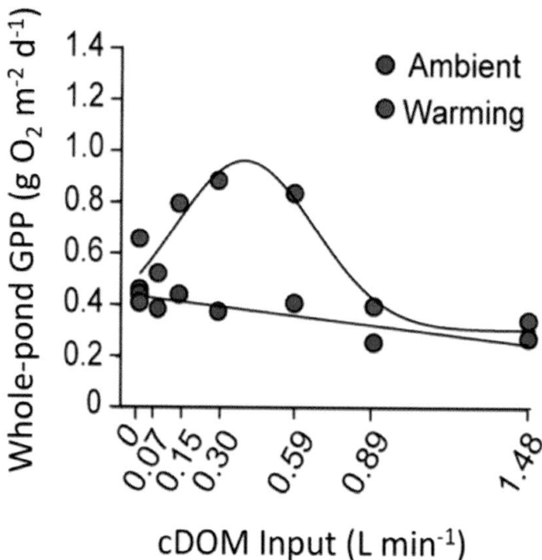

FIGURE 28-37 Whole-pond gross primary production (*GPP*) in both ambient (*blue circles*) and heated (*red circles*) enclosures along a gradient of colored dissolved organic matter (*cDOM*) inputs. *(From Hamdan et al. (2021).)*

G. Multiple anthropogenic stressors

The stressors influencing OC cycling in freshwater ecosystems are numerous and interacting. Flow modifications change the distribution and connectivity of streams, lakes, rivers, wetlands, and reservoirs across landscapes. Anthropogenic climate change is altering temperatures, precipitation, snowpack, terrestrial productivity, and, consequently, terrestrial OM inputs to freshwaters. Furthermore, human activities modify land use, sediment dynamics, the stoichiometry of carbon and nutrients available for biological processes, the introduction and loss of species, and the persistence of anthropogenic chemicals and other pollutants in freshwater ecosystems.

By testing the consequences of more than one environmental change stressor at a time or finding mismatches between model predictions and data, recent work has highlighted how tracking multiple environmental changes is required to fully understand how carbon cycling is changing in inland waters. For example, warming is occurring at the same time as other environmental changes that alter the light and nutrient environment of inland waters: increased OM inputs (i.e., brownification) and/or nutrient loading (i.e., eutrophication). Recent increases in littoral algal blooms cannot be explained by models of lake eutrophication alone (Vadeboncoeur et al., 2021), highlighting the need for ongoing research on how multiple environmental changes alter ecosystem carbon cycling. GPP in experimentally warmed ponds only increased when ponds received intermediate increases in allochthonous DOM (Hamdan et al., 2021; Fig. 28-37), suggesting GPP may peak in warming ecosystems when moderate increases in DOM provide limiting nutrients for metabolism but DOM inputs are not high enough to block primary producer access to light. Additionally, inland waters may experience more stable thermal stratification in the future due to a combination of higher temperatures and lower water transparency with brownification or eutrophication. Warming surface waters with stronger stratification and lower light penetration may actually promote the cooling of bottom waters in shallow ecosystems (Bartosiewicz et al., 2019). More stable stratification can also lengthen periods of anoxia, which may increase carbon burial while producing and emitting more CH_4 (Bartosiewicz et al., 2019).

As anthropogenic actions continue to generate different combinations of "chemical cocktails" in inland waters (*sensu* Kaushal et al., 2018b), characterizing the interactions between the transport and cycling of OC and other chemicals is a current research need in limnology. Some chemical mixtures may minimize or enhance the emergent effects of two or more combined stressors. For example, increasing biological activity in response to high-nutrient and high-OM wastewater effluent may offset some effects of emerging contaminants on river biofilm metabolism (Aristi et al., 2015), whereas flow intermittency could make primary producers more sensitive to contaminant exposure while decreasing bacterial sensitivity to contaminants (Corcoll et al., 2015). How the changing composition of "chemical cocktails" may interact with other environmental stressors and alter carbon cycling in freshwater ecosystems represents a critical knowledge gap in our current understanding of freshwater ecosystem function.

VI. Summary

1. Organic matter (OM) in aquatic ecosystems is a heterogeneous mixture of living and decomposing plant, microbial, and animal products derived from within the system (autochthonous) or from the surrounding catchment (allochthonous).
2. OM is separated into dissolved (DOM) and particulate (POM) fractions via filtration, as is organic carbon (DOC and POC). Most OM in aquatic ecosystems is DOM. OM consists of approximately 50% carbon with varying quantities of O, H, N, S, and P.
   a. Historically, aquatic DOM was separated into humic and nonhumic fractions based on their solubility in acid. Many limnologists still include "humic-like" DOM as a compositional category, but the ongoing development of new methods improves our ability to characterize DOM composition beyond bulk concentrations and laboratory extractions.
   b. No universal DOM characterization method exists, but recent developments have improved our ability to characterize elements of composition, structure, and reactivity in bulk DOM samples using: (1) optical analyses of colored and fluorescent DOM; (2) nuclear magnetic resonance (NMR) spectroscopy; and (3) Fourier-transform ion cyclotron resonance mass spectrometry (FT-ICR-MS).
3. In lakes, DOC concentrations are typically in the 0.1–100 mg C L^{-1} range but values up to 300 mg C L^{-1} are possible. Most rivers and streams have lower DOC concentrations (e.g., 0–50 mg C L^{-1}). Higher DOC typically causes yellow to brown coloring of freshwaters.
4. POM consists of living and dead biomass fragments that accumulate in bottom sediments and are easily mobilized under turbulent conditions.
   a. A large fraction of POM is POC, with concentrations typically below 10 mg C L^{-1}.
   b. DOC can be generated from POC during OM breakdown. POC can also be formed from DOC via flocculation, interactions with inorganic matter, and bacterial degradation.
5. The biogeochemical cycles and food web energetics of freshwater ecosystems are sustained by a combination of internally fixed and externally derived OM (autochthonous and allochthonous sources, respectively). Most organisms in freshwater ecosystems rely on a mixture of allochthonous and autochthonous OM sources to support their energetic requirements.
6. Major inputs to the DOC pool are: (1) photosynthetic inputs through exudation and lysis; (2) allochthonous DOC, composed largely of substances originating from terrestrial and wetland/littoral higher plant tissues that can be more resistant to rapid bacterial degradation; and (3) bacterial degradation and chemosynthesis of OM with subsequent release of DOC; much smaller amounts, usually quantitatively negligible but qualitatively of potential importance, include (4) DOC from excretions of secondary producers, zooplankton, and higher animals, that have ingested living or detrital OM.
7. Allochthonous DOM inputs typically have low nitrogen relative to carbon content (organic C:N approximately 50:1), resulting from a large proportion of structural compounds from terrestrial and higher aquatic plants and from the selective decomposition of more reactive compounds as DOM is transported by water flow. DOM derived from autochthonous production within water bodies has a higher initial nitrogen content, with a C:N of about 12:1.
8. Autochthonous OM sources sustain biogeochemical and food web processes within ecosystems that are supplemented by allochthonous inputs and include: (1) benthic photosynthetic POM and DOM sources from active exudation, decomposition, and lysis of littoral macrophytes and attached algae and Cyanobacteria as well as biofilm assemblages on other substrates throughout the benthic zone; and (2) primary producers of the pelagic zone, primarily algal and cyanobacterial phytoplankton.
   a. Inputs of allochthonous OM are mostly as DOM in surface runoff and subsurface flow paths, the concentrations of which are variable over space and time once they enter freshwater ecosystems. The vertical structure of soils and terrestrial-aquatic interface dynamics influences the concentrations, age, and mobilization of OM from land to inland waters. Expanding hydrologic connections across landscapes during higher flows typically increase the export of DOC to recipient inland waters.
   b. Allochthonous POM in freshwater ecosystems include leaves, soils, coarse woody debris, pollen, feces, and other materials deposited by terrestrial organisms when they are using waterways, as well as terrestrial organisms themselves. Allochthonous POM can fall directly into water

(e.g., overhanging tree canopies), be transported by water, or be windblown into ecosystems.

9. Detritus is all dead organic carbon (OC), distinguishable from living OC and inorganic carbon. Detritus consists of OC lost by nonconsumptive means from any trophic level (includes egestion, excretion, secretion, and so forth) or inputs from sources external to the ecosystem that enter and cycle in the ecosystem (i.e., allochthonous OC).
   a. Detritus food web pathways are any route by which chemical energy contained within detrital OC becomes available to biota. From the standpoint of carbon fluxes, most OC in ecosystems is dead, from a mixture of OM sources, and is undergoing microbial degradation, at rates that are variable and serially decrease with losses in OM reactivity.
   b. Common to all aquatic ecosystems is the dominance of detrital metabolism, which gives ecosystems a fundamental metabolic stability.

10. Ecosystem metabolism is the fixation and breakdown of OC structures through photosynthesis and respiration, respectively. Photosynthesis and respiration are fundamental to the storage and use of energy, from the cellular to ecosystem scale, and offer insights into the sources and amounts of resources available for higher trophic levels in freshwater food webs.
   a. Tracing the rates and balance of metabolic processes in ecosystems allows us to obtain estimates of carbon reactivity, sources, and fate that we might otherwise miss from water chemistry analyses alone, as much of the most reactive OM will be taken up by microbes and respired before we can sample it from ambient water.
   b. The fate of carbon in ecosystems can be traced from gross primary production (GPP), the total fixation of organic compounds via photosynthesis, through respiratory losses. Together, all respiration by biota is ecosystem respiration ($ER = R_a + R_h$; R_a = autotrophic respiration and R_h = heterotrophic respiration). The OC remaining after R_a is net primary production ($NPP = GPP - R_a$).
   c. Net ecosystem production ($NEP = GPP - ER$) reflects the difference between the fixation and respiratory breakdown of OC. If NEP is positive, the ecosystem is autotrophic, GPP exceeds ER ($GPP/ER > 1$), and the ecosystem is accumulating and storing autochthonous OM. If NEP is negative, the ecosystem is heterotrophic, ER exceeds GPP ($GPP/ER < 1$), and the ecosystem is subsidized by allochthonous OM inputs from terrestrial and upstream aquatic ecosystems.
   d. In freshwater ecosystems, heterotrophy is the rule, not the exception. ER exceeds GPP in most inland waters. The increasing number of annual metabolism estimates only provides stronger support for the importance of allochthonous OM sources in subsidizing ER.
   e. Light, temperature, and nutrients, in addition to CO_2 and OC substrate availability and redox state, govern rates of metabolism in inland waters.
   f. Annual and multiannual datasets of aquatic ecosystem metabolism offer insight into how the timing and magnitude of autochthonous OM inputs differs among ecosystems and throughout the year: summer peaks, spring/autumn productivity peaks before/after leaf out and leaf fall, or aseasonal and relatively constant autochthonous OM inputs year-round.

11. Water residence time has been proposed as a master variable to predict differences in total OC metabolism within and among ecosystems, as the opportunities for microbes to encounter and metabolize OC will be higher the longer water remains in an ecosystem.

12. Inputs of dissolved organic substrates to inland waters are approximately equal to their rates of microbial degradation. High rates of microbial DOM degradation occur within freshwater ecosystems despite lower biological reactivity of much of the OM pool.
   a. Our current understanding of DOC metabolism in inland waters depends somewhat on the methods we use. Water column DOC loss rates seem to be relatively slow but consistently in the range of 0.5–2% per day under different environmental conditions.
   b. Simulations using ecosystem-scaled DOC uptake rates (i.e., benthic and pelagic) estimated that 27–45% of DOC inputs are removed via whole-ecosystem metabolism while in transport through a river network, which is a much higher removal rate than water column bioassay DOC uptake estimates predict.

13. Seasonal and daily variations in hydrodynamics, particularly in streams and rivers, influence water residence time, OM transport, benthic scouring, and ecosystem metabolism.
   a. Pulses of DOC may be associated with higher ER in freshwaters draining catchments that release reactive DOM during higher flow periods.
   b. Interactions between DOM, increased ER, and potential decreases in GPP due to light–DOM

interactions may limit GPP and change the metabolic balance in ecosystems with higher DOM.

14. The balance between OC metabolism and transport, referred to as OC spiraling by those working in streams and rivers, provides a way to quantify the average distance OC travels downstream before being taken up and respired by microbes. Published estimates of OC spiraling lengths in streams and rivers range from 21 m to >1000 km.

15. Sources of POC shift in their relative contributions to the total POC and OC pools depending on ecosystem type and productivity. POC concentrations in pelagic zones usually are considerably larger (double or greater) in eutrophic lakes than in less productive waters. GPP by phytoplanktonic and littoral autotrophs is a major contributor to lake, pond, and reservoir POC, while allochthonous POC inputs are usually relatively small in contrast to DOC.

16. Most allochthonous POM inputs (e.g., wood, leaves, and fine particulates) have a longer residence time than DOM but are still more likely to be transported downstream than decomposed at their input site. Rates of POC decomposition in recipient waters can vary with the organic composition of the material but are usually slower (e.g., leaves in weeks; woody material in years) than those of DOC (days to weeks).

17. The trophic dynamic structure of freshwater ecosystems depends on detrital resources. Detrital metabolism is slow and more evenly sustained by a large OM reserve.

18. Relatively small portions of allochthonous and autochthonous OM reach higher trophic levels; most is degraded by microbial and abiotic (e.g., photo) processes.

19. Most benthic metabolism is heterotrophic, microbial, and anaerobic, even when hypolimnetic oxygen depletion does not occur.

   a. Detrital POC from allochthonous, littoral, and pelagic sources is often metabolized under anoxic conditions in benthic sediments.

   b. Increases in autochthonous inputs in relation to allochthonous inputs result in more decomposition in the sediments than in the water column.

   c. The continued production of CO_2 by anaerobic benthic metabolism, and release of this oxidized product to the overlying water, results in benthic and hypolimnetic respiratory quotients (CO_2 released/O_2 uptake) >1.

20. The reduction of alternate electron acceptors (nitrate, iron and manganese oxides, sulfate, and various organic substrates) instead of molecular oxygen results in an accumulation of electrons in the sediments and the receipt of energy originally transferred during photosynthesis. This flow, termed detrital electron flux, into and out of the sediments derives its energy from photosynthesis, just as a predatory animal does from its food, and is a significant proportion of the total energy inputs and transfer processes because most OM becomes detrital in an ecosystem without ever passing through an animal.

21. Carbon in aquatic ecosystems is subject to different fates after or in place of metabolic transformation—burial in sediments, emission to the atmosphere, or export.

   a. Fate also depends on form—gaseous forms are emitted or exported, dissolved forms are most readily transported, and particulate forms are buried or exported.

   b. The balance of carbon inputs and losses determines whether the ecosystem is a carbon source or sink.

22. Carbon emission is in the form of CO_2 or CH_4. CO_2 is mostly emitted via diffusion, while CH_4 is emitted via diffusion, ebullition (i.e., bubbling), and plant-mediated fluxes.

   a. Emission rates range over several orders of magnitude, with source availability, hydrodynamics, morphology, oxygen, and temperature being important drivers.

   b. Essentially all freshwater ecosystems are heterotrophic net CO_2 producers, with evasion of CO_2 to the atmosphere in excess of autochthonous production and related CO_2 fixation.

   c. The contribution of *in situ* respiration versus external CO_2 sources to emissions appears to vary with ecosystem size, metabolic balance, and different connections with adjacent ecosystems.

23. Sedimentation rate and oxygen exposure time are drivers of carbon burial efficiency. Lakes and reservoirs bury carbon more efficiently than fluvial ecosystems.

24. DOC, POC, CO_2, and CH_4 transport fluxes depend on hydrology and substrate availability.

25. Globally, carbon emission estimates dominate over burial and transport, but estimates for all carbon fates are continuously being updated with new information from diel, multiseason, multiannual, and more geographically expansive carbon flux measurements.

26. Whole-ecosystem carbon budgets are rare because of the considerable amount of data needed to fully characterize all carbon fluxes and fates but should include: (1) allochthonous OC inputs; (2) autochthonous OC inputs; and (3) outputs.

   a. Many carbon budgets focus on OM inputs, metabolism, food web fluxes, and export but often exclude inorganic carbon fluxes and emissions.

b. Different examples of whole-ecosystem carbon budgets exist with various levels of detail. For example, the Lawrence Lake (USA) budget included multiple levels of autochthonous OC inputs, while the Lake Gäddtjärn (Sweden) example used high-resolution data to estimate metabolism and focused on carbon outputs like emission and burial.

c. A synthesis of OM flux estimates within the Little Tennessee River (USA) network provides a unique opportunity to track changes in OM sources and fates across a network of small mountain streams to the main river.

d. More intensive sampling and monitoring of all carbon forms and fates is required to construct models to predict changes in carbon budgets and cycling due to anthropogenic pressures.

27. Alterations of the environment by human activities change OM cycling and fluxes in aquatic ecosystems.

28. Land-use change impacts aquatic OM cycling directly by modifying OM loading, composition, and burial.

a. Indirect impacts of land-use change include changes in nutrient and sediment loading, light, hydrology, and temperature, all of which influence ecosystem metabolism.

b. Across multiple geographic regions, both GPP and ER are often higher in ecosystems draining human-dominated landscapes compared to reference vegetation, while land cover–associated differences in GPP and ER can be less apparent in larger ecosystems.

29. Many biological and physical processes scale with temperature, including leaching of DOM from parent material, metabolism, and organismal production.

a. While some research has identified warming-induced increases in GPP and/or ER, there may be a longer-term trend where functional redundancy or different species-specific or habitat-specific responses to warming generate a neutral effect of warming on ecosystem-scale carbon cycling and food web energy fluxes.

b. ER responses to temperature may be stronger than GPP–temperature relationships, thus making some ecosystems more heterotrophic with warming.

30. Hydrologic alterations from anthropogenic flow modifications (e.g., damming and extraction) and climate change (e.g., glacial melt, droughts, higher intensity precipitation events, increased flow intermittency) alter OM quantity, reactivity, and fate.

31. Excess nutrients can lead to eutrophication, which results in higher productivity and changes to food webs. Increased nutrient inputs to ecosystems accelerate decomposition rates of detrital OM. Eutrophication can also govern changes in redox potential and enhance carbon emissions from freshwater ecosystems.

32. Higher sediment loads disrupt metabolism by changing light, oxygen, and temperature regimes; mobilizing pollutants; and enhancing carbon burial and emissions.

33. Human activities change the export of OM from soils to inland waters and alter the composition of OM inputs to freshwaters from human-modified landscapes and thawing permafrost.

a. Past concerns about declines in OM inputs to freshwaters due to acid deposition have transitioned to ongoing research testing the consequences of increasing DOM (i.e., brownification) of inland waters.

b. Changes in OM inputs will alter the locations of OC storage, metabolism, and emissions within fluvial networks (e.g., river–riparian–floodplain disconnections decrease OC storage and reservoir construction increases OC burial).

34. The chemicals and materials humans use enter freshwater ecosystems through multiple pathways.

a. Microbial communities in ecosystems subjected to chronic contaminant exposure may be relatively resilient to the effects of contaminant pollution.

b. Recent research has found evidence for both stimulation and suppression of GPP and ER with exposure to pharmaceuticals and pesticides, supporting the need for continued research on the metabolic responses to single contaminants and contaminant mixtures at environmentally relevant concentrations.

35. The stressors influencing OC cycling in freshwater ecosystems are numerous and interacting.

a. By testing the consequences of more than one environmental change stressor at a time or finding mismatches between model predictions and data, recent work has highlighted how tracking multiple environmental changes is required to fully understand how carbon cycling is changing in inland waters.

b. How changing mixtures of pollutants interact with other environmental stressors to alter carbon cycling in ecosystems represents a critical knowledge gap in our understanding of carbon cycling in inland waters.

Acknowledgments

We thank the following colleagues for providing feedback on our updates to this chapter: Kristen Bretz, Austin Gray, Dolly Kothawala, Sebastian Sobek, Lars Tranvik, and Dominic Vachon. This chapter update was partially supported by an award from the National Science Foundation to ERH (NSF DEB 1754237). Finally, we thank Robert Wetzel for sharing his extensive knowledge of inland waters through his research publications and earlier editions of this text.

References

Abdulla, H.A.N., Minor, E.C., Dias, R.F., Hatcher, P.G., 2010. Changes in the compound classes of dissolved organic matter along an estuarine transect: a study using FTIR and ^{13}C NMR. Geochim. Cosmochim. Acta 74, 3815–3838.

Aben, R.C.H., Barros, N., Van Donk, E., Frenken, T., Hilt, S., Kazanjian, G., Lamers, L.P.M., Peeters, E.T.H.M., Roelofs, J.G.M., De Senerpont Domis, L.N., Stephan, S., Velthuis, M., Van De Waal, D.B., Wik, M., Thornton, B.F., Wilkinson, J., Delsontro, T., Kosten, S., 2017. Cross continental increase in methane ebullition under climate change. Nat. Commun. 8, 1–8.

Aiken, G.R., McKnight, D.M., Thorn, K.A., Thurman, E.M., 1992. Isolation of hydrophilic organic acids from water using nonionic macroporous resins. Org. Geochem. 18, 567–573.

Aiken, G., McKnight, D., Wershaw, R., MacCarthy, P., 1985. Humic Substances in Soil, Sediment, and Water-Geochemistry, Isolation, and Characterization. John Wiley & Sons, Inc, New York, NY, USA.

Algesten, G., Sobek, S., Bergström, A.K., Ågren, A., Tranvik, L.J., Jansson, M., 2004. Role of lakes for organic carbon cycling in the boreal zone. Glob. Chang. Biol. 10, 141–147.

Alin, S.R., Johnson, T.C., 2007. Carbon cycling in large lakes of the world: a synthesis of production, burial, and lake-atmosphere exchange estimates. Global Biogeochem. Cycles 21, 1–12.

Anderson, N.J., Heathcote, A.J., Engstrom, D.R., 2020. Anthropogenic alteration of nutrient supply increases the global freshwater carbon sink. Sci. Adv. 6 eaaw2145.

Anesio, A.M., Granéli, W., 2003. Increased photoreactivity of DOC by acidification: implications for the carbon cycle in humic lakes. Limnol. Oceanogr. 48, 735–744.

Appling, A.P., Hall, R.O., Yackulic, C.B., Arroita, M., 2018. Overcoming equifinality: leveraging long time series for stream metabolism estimation. J. Geophys. Res. Biogeosciences 123, 624–645.

Arias-Andres, M., Rojas-Jimenez, K., Grossart, H.P., 2019. Collateral effects of microplastic pollution on aquatic microorganisms: an ecological perspective. Trends Anal. Chem. 112, 234–240.

Aristi, I., von Schiller, D., Arroita, M., Barceló, D., Ponsatí, L., García-Galán, M.J., Sabater, S., Elosegi, A., Acuña, V., 2015. Mixed effects of effluents from a wastewater treatment plant on river ecosystem metabolism: subsidy or stress? Freshw. Biol. 60, 1398–1410.

Arroita, M., Elosedi, A., Hall, R.O., 2019. Twenty years of daily metabolism show riverine recovery following sewage abatement. Limnol. Oceanogr. 64, 577–592.

Artifon, V., Zanardi-Lamardo, E., Fillmann, G., 2019. Aquatic organic matter: classification and interaction with organic microcontaminants. Sci. Total Environ. 649, 1620–1635.

Ask, J., Karlsson, J., Persson, L., Ask, P., Byström, P., Jansson, M., 2009. Terrestrial organic matter and light penetration: effects on bacterial and primary production in lakes. Limnol. Oceanogr. 54, 2034–2040.

Asmala, E., Carstensen, J., Räike, A., 2019. Multiple anthropogenic drivers behind upward trends in organic carbon concentrations in boreal rivers. Environ. Res. Lett. 14, 124018.

Attermeyer, K., Catalán, N., Einarsdottir, K., Freixa, A., Groeneveld, M., Hawkes, J.A., Bergquist, J., Tranvik, L.J., 2018. Organic carbon processing during transport through boreal inland waters: particles as important sites. J. Geophys. Res. Biogeosciences 123, 2412–2428.

Baines, S.B., Pace, M.L., 1991. The production of dissolved organic matter by phytoplankton and its importance to bacteria: patterns across marine and freshwater systems. Limnol. Oceanogr. 36, 1078–1090.

Baker, M.A., Valett, H.M., Dahm, C.N., 2000. Organic carbon supply and metabolism in a shallow groundwater ecosystem. Ecology 81, 3133–3148.

Ballantyne, A.P., Alden, C.B., Miller, J.B., Tans, P.P., White, J.W.C., 2012. Increase in observed net carbon dioxide uptake by land and oceans during the past 50 years. Nature 488, 70–72.

Barnes, R.T., Butman, D.E., Wilson, H.F., Ramond, P.A., 2018. Riverine export of aged carbon driven by flow path depth and residence time. Environ. Sci. Technol. 52, 1028–1035.

Bartosiewicz, M., Przytulska, A., Lapierre, J.F., Laurion, I., Lehmann, M.F., Maranger, R., 2019. Hot tops, cold bottoms: synergistic climate warming and shielding effects increase carbon burial in lakes. Limnol. Oceanogr. Lett. 4, 132–144.

Bastviken, D., Cole, J., Pace, M., Tranvik, L., 2004. Methane emissions from lakes: dependence of lake characteristics, two regional assessments, and a global estimate. Global Biogeochem. Cycles 18, GB4009.

Bastviken, D., Tranvik, L.J., Downing, J.A., Crill, P.M., Enrich-Prast, A., 2011. Freshwater methane emissions offset the continental carbon sink. Science 331, 50.

Battin, T.J., Kaplan, L.A., Findlay, S., Hopkinson, C.S., Marti, E., Packman, A.I., Newbold, J.D., Sabater, F., 2008. Biophysical controls on organic carbon fluxes in fluvial networks. Nat. Geosci. 1, 95–100.

Battin, T.J., Luyssaert, S., Kaplan, L.A., Aufdenkampe, A.K., Richter, A., Tranvik, L.J., 2009. The boundless carbon cycle. Nat. Geosci. 2, 598–600.

Baud, A., Jenny, J.-P., Francus, P., Gregory-Eaves, I., 2021. Global acceleration of lake sediment accumulation rates associated with recent human population growth and land-use changes. J. Paleolimnol. 66, 453–467.

Beaulieu, J.J., DelSontro, T., Downing, J.A., 2019. Eutrophication will increase methane emissions from lakes and impoundments during the 21st century. Nat. Commun. 10, 1–5.

Beel, C.R., Heslop, J.K., Orwin, J.F., Pope, M.A., Schevers, A.J., Hung, J.K.Y., Lafrenière, M.J., Lamoureux, S.F., 2021. Emerging dominance of summer rainfall driving high arctic terrestrial-aquatic connectivity. Nat. Commun. 12, 1–9.

Benbow, M.E., Receveur, J.P., Lamberti, G.A., 2020. Death and decomposition in aquatic ecosystems. Front. Ecol. Evol. 8, 17.

Bengtsson, M.M., Wegner, K., Burns, N.R., Herberg, E.R., Wanek, W., Kaplan, L.A., Battin, T.J., 2014. No evidence of aquatic priming effects hyporheic zone microcosms. Sci. Rep. 4, 5187.

Berger, E., Frör, O., Schäfer, R.B., 2019. Salinity impacts on river ecosystem processes: a critical mini-review. Philos. Trans. R. Soc. B Biol. Sci. 374, 20180010.

Berggren, M., Lapierre, J.-F., del Giorgio, P.A., 2011. Magnitude and regulation of bacterioplankton respiratory quotient across freshwater environmental gradients. ISME J. 6, 984–993.

Bergström, A.-K., Jansson, M., 2000. Bacterioplankton production in humic Lake Örträsket in relation to input of bacterial cells and input of allochthonous organic carbon. Microb. Ecol. 39, 101–115.

Bernhardt, E.S., Heffernan, J.B., Grimm, N.B., Stanley, E.H., Harvey, J.W., Arroita, M., Appling, A.P., Cohen, M.J., McDowell, W.H., Hall, R.O., Read, J.S., Roberts, B.J., Stets, E.G.,

Yackulic, C.B., 2018. The metabolic regimes of flowing waters. Limnol. Oceanogr. 63, S99–S118.

Bernhardt, E.S., Rosi, E.J., Gessner, M.O., 2017. Synthetic chemicals as agents of global change. Front. Ecol. Environ. 15, 84–90.

Bernhardt, E.S., Savoy, P., Vlah, M.J., Appling, A.P., Koenig, L.E., Hall, R.O., Arroita, M., Blaszczak, J.R., Carter, A.M., Cohen, M., Harvey, J.W., Heffernan, J.B., Helton, A.M., Hosen, J.D., Kirk, L., McDowell, W.H., Stanley, E.H., Yackulic, C.B., Grimm, N.B., 2022. Light and flow regimes regulate the metabolism of rivers. Proc. Nat. Acad. Sci. 119, e2121976119.

Bernot, M.J., Sobota, D.J., Hall, R.O., Mulholland, P.J., Dodds, W.K., Webster, J.R., Tank, J.L., Ashkenas, L.R., Cooper, L.W., Dahm, C.N., Gregory, S.V., Grimm, N.B., Hamilton, S.K., Johnson, S.L., Mcdowell, W.H., Meyer, J.L., Peterson, B., Poole, G.C., Valett, H.M., Arango, C., Beaulieu, J.J., Burgin, A.J., Crenshaw, C., Helton, A.M., Johnson, L., Merriam, J., Niederlehner, B.R., O'brien, J.M., Potter, J.D., Sheibley, R.W., Thomas, S.M., Wilson, K., 2010. Inter-regional comparison of land-use effects on stream metabolism. Freshw. Biol. 55, 1874–1890.

Bianchi, T.S., Butman, D., Raymond, P.A., Ward, N.D., Kates, R.J.S., Flessa, K.W., Zamora, H., Arellano, A.R., Ramirez, J., Rodriguez, E., 2017. The experimental flow to the Colorado river delta: effects on carbon mobilization in a dry watercourse. J. Geophys. Res. Biogeosciences 122, 607–627.

Bilby, R.E., Fransen, B.R., Bisson, P.A., 1996. Incorporation of nitrogen and carbon from spawning coho salmon into the trophic system of small streams: evidence from stable isotopes. Can. J. Fish. Aquat. Sci. 53, 164–173.

Birge, E.A., Juday, C., 1934. Particulate and dissolved organic matter in inland lakes. Ecol. Monogr. 4, 440–474.

Blaszczak, J.R., Delesantro, J.M., Urban, D.L., Doyle, M.W., Bernhardt, E.S., 2019. Scoured or suffocated: urban stream ecosystems oscillate between hydrologic and dissolved oxygen extremes. Limnol. Oceanogr. 64, 877–894.

Bogard, M.J., del Giorgio, P.A., 2016. The role of metabolism in modulating CO_2 fluxes in boreal lakes. Global Biogeochem. Cycles 30, 1509–1525.

Borges, A.V., Darchambeau, F., Lambert, T., Morana, C., Allen, G.H., Tambwe, E., Toengaho Sembaito, A., Mambo, T., Wabakhangazi, J.N., Descy, J.P., Teodoru, C.R., Bouillon, S., 2019. Variations in dissolved greenhouse gases (CO_2, CH_4, N_2O) in the Congo River network overwhelmingly driven by fluvial-wetland connectivity. Biogeosciences 16, 3801–3834.

Bott, T.L., Kaplan, L.A., Kuserk, F.T., 1984. Benthic bacterial biomass supported by streamwater dissolved organic matter. Microb. Ecol. 10, 335–344.

Breger, I., 1963. Organic Geochemistry. Macmillan Company, New York, NY, USA.

Bretz, K.A., Jackson, A.R., Rahman, S., Monroe, J.M., Hotchkiss, E.R., 2021. Integrating ecosystem patch contributions to stream corridor carbon dioxide and methane fluxes. J. Geophys. Res. Biogeosciences 126, e2021JG006313.

Buffle, J., Leppard, G.G., 1995. Characterization of aquatic colloids and macromolecules. 1. Structure and behavior of colloidal material. Environ. Sci. Technol. 29, 2169–2175.

Burkholder, J.M., Wetzel, R.G., 1989. Epiphytic microalgae on natural substrata in a hardwater lake: seasonal dynamics of community structure, biomass and ATP content. Arch. für Hydrobiol. Suppl. Monogr. Beitrage 83, 1–56.

Burrows, R.M., Magierowski, R.H., Fellman, J.B., Clapcott, J.E., Munks, S.A., Roberts, S., Davies, P.E., Barmuta, L.A., 2014. Variation in stream organic matter processing among years and benthic habitats in response to forest clearfelling. For. Ecol. Manag. 327, 136–147.

Butman, D., Stackpoole, S., Stets, E., McDonald, C.P., Clow, D.W., Striegl, R.G., 2016. Aquatic carbon cycling in the conterminous United States and implications for terrestrial carbon accounting. Proc. Natl. Acad. Sci. U. S. A. 113, 58–63.

Butman, D., Striegl, R., Stackpoole, S., del Giorgio, P., Prairie, Y., Pilcher, D., Raymond, P., Paz Pellat, F., Alcocer, J., 2018. Chapter 14: inland waters. In: Cavallaro, N., Shrestha, G., Birdsey, R., Mayes, M.A., Najjar, R.G., Reed, S.C., Romero-Lankao, P., Zhu, Z. (Eds.), Second State of the Carbon Cycle Report (SOCCR2): A Sustained Assessment Report. U.S. Global Change Research Program, Washington, DC, USA, pp. 568–595.

Butman, D., Wilson, H.F., Barnes, R.T., Xenopoulos, M.A., Raymond, P.A., 2015. Increased mobilization of aged carbon to rivers by human disturbance. Nat. Geosci. 8, 112–116.

Cao, X., Aiken, G.R., Butler, K.D., Mao, J., Schmidt-Rohr, K., 2018. Comparison of the chemical composition of dissolved organic matter in three lakes in Minnesota. Environ. Sci. Technol. 52, 1747–1755.

Carpenter, S.R., Cole, J.J., Pace, M.L., Van de Bogert, M., Bade, D.L., Bastviken, D., Gille, C.M., Hodgson, J.R., Kitchell, J.F., Kritzberg, E.S., 2005. Ecosystem subsidies: terrestrial support of aquatic food webs from ^{13}C addition to contrasting lakes. Ecology 86, 2737–2750.

Casas-Ruiz, J.P., Spencer, R.G., Guillemette, F., von Schiller, D., Obrador, B., Podgorski, D.C., Kellerman, A.M., Hartmann, J., Gómez-Gener, L., Sabater, S., Marcé, R., 2020. Delineating the continuum of dissolved organic matter in temperate river networks. Global Biogeochem. Cycles 34, e2019GB006495.

Catalán, N., Marcé, R., Kothawala, D.N., Tranvik, L.J., 2016. Organic carbon decomposition rates controlled by water retention time across inland waters. Nature Geosci 9, 501–504.

Chipperfield, M.P., Bekki, S., Dhomse, S., Harris, N.R.P., Hassler, B., Hossaini, R., Steinbrecht, W., Thiéblemont, R., Weber, M., 2017. Detecting recovery of the stratospheric ozone layer. Nature 549, 211–218.

Chmiel, H.E., Kokic, J., Denfeld, B.A., Einarsdóttir, K., Wallin, M.B., Koehler, B., Isidorova, A., Bastviken, D., Ferland, M.-È., Sobek, S., 2016. The role of sediments in the carbon budget of a small boreal lake. Limnol. Oceanogr. 61, 1814–1825.

Clow, D.W., Stackpoole, S.M., Verdin, K.L., Butman, D.E., Zhu, Z., Krabbenhoft, D.P., Striegl, R.G., 2015. Organic carbon burial in lakes and reservoirs of the conterminous United States. Environ. Sci. Technol. 49, 7614–7622.

Cole, J.J., Caraco, N.F., 2001. Carbon in catchments: connecting terrestrial carbon losses with aquatic metabolism. Mar. Freshw. Res. 52, 101–110.

Cole, J.J., Carpenter, S.R., Kitchell, J., Pace, M.L., Solomon, C.T., Weidel, B., 2011. Strong evidence for terrestrial support of zooplankton in small lakes based on stable isotopes of carbon, nitrogen, and hydrogen. Proc. Natl. Acad. Sci. U.S.A. 108, 1975–1980.

Cole, J.J., Likens, G.E., Strayer, D.L., 1982. Photosynthetically produced dissolved organic carbon: an important carbon source for planktonic bacteria. Limnol. Oceanogr. 27, 1080–1090.

Cole, J.J., Prairie, Y.T., Caraco, N.F., McDowell, W.H., Tranvik, L.J., Striegl, R.G., Duarte, C.M., Kortelainen, P., Downing, J.A., Middelburg, J.J., Melack, J., 2007. Plumbing the global carbon cycle: integrating inland waters into the terrestrial carbon budget. Ecosystems 10, 172–185.

Cory, R.M., Kling, G.W., 2018. Interactions between sunlight and microorganisms influence dissolved organic matter degradation along the aquatic continuum. Limnol. Oceanogr. Lett. 3, 102–116.

Cory, R.M., McKnight, D.M., 2005. Fluorescence spectroscopy reveals ubiquitous presence of oxidized and reduced quinones in dissolved organic matter. Environ. Sci. Technol. 39, 8142–8149.

Corcoll, N., Casellas, M., Huerta, B., Gausch, H., Acuña, V., Rodríguez-Mozaz, S., Serra-Compte, A., Barceló, D., Sabater, S., 2015. Effects of flow intermittency and pharmaceutical exposure on the structure and metabolism of stream biofilms. Sci. Total Environ. 503–504, 159–170.

Crawford, J.T., Loken, L.C., Stanley, E.H., Stets, E.G., Dornblaser, M.M., Striegl, R.G., 2016. Basin scale controls on CO_2 and CH_4 emissions from the Upper Mississippi River. Geophys. Res. Lett. 43, 1973–1979.

Creed, I.F., Bergström, A.-K., Trick, C.G., Grimm, N.B., Hessen, D.O., Karlsson, J., Kidd, K.A., Kritzberg, E., McKnight, D.M., Freeman, E.C., Senar, O.E., Andersson, A., Ask, J., Berggren, M., Cherif, M., Giesler, R., Hotchkiss, E.R., Kortelainen, P., Palta, M., Vrede, T., Wehenmeyer, G., 2018. Global change-driven effects on dissolved organic matter composition: implications for food webs of northern lakes. Global Change Biol. 24, 3692–3714.

Creed, I.F., McKnight, D.M., Pellerin, B.A., Green, M.B., Bergamaschi, B.A., Aiken, G.R., Burns, D.A., Findlay, S.E., Shanley, J.B., Striegl, R.G., Aulenbach, B.T., 2015. The river as a chemostat: fresh perspectives on dissolved organic matter flowing down the river continuum. Can. J. Fish. Aquat. Sci. 72, 1272–1285.

Crumpton, W.G., Wetzel, R.G., 1982. Effects of differential growth and mortality in the seasonal succession of phytoplankton populations in Lawrence Lake, Michigan. Ecology 63, 1729–1739.

Cummins, K.W., Klug, M.J., Wetzel, R.G., Petersen, R.C., Suberkropp, K.F., Manny, B.A., Wuycheck, J.C., Howard, F.O., 1972. Organic enrichment with leaf leachate in experimental lotic ecosystems. Bioscience 22, 719–722.

Cuss, C.W., Donner, M.W., Noernberg, T., Pelletier, R., Shotyk, W., 2019. EEM-PARAFAC-SOM for assessing variation in the quality of dissolved organic matter: simultaneous detection of differences by source and season. Environ. Chem. 16, 360.

Dahm, C.N., 1981. Pathways and mechanisms for removal of dissolved organic carbon from leaf leachate in streams. Can. J. Fish. Aquat. Sci. 38, 68–76.

Dalva, M., Moore, T.R., 1991. Sources and sinks of dissolved organic carbon in a forested swamp catchment. Biogeochemistry 15, 1–19.

David, M.B., Vance, G.F., 1991. Chemical character and origin of organic acids in streams and seepage lakes of central Maine. Biogeochemistry 12, 17–41.

Dearing, J.A., Jones, R.T., 2003. Coupling temporal and spatial dimensions of global sediment flux through lake and marine sediment records. Glob. Planet. Change 39, 147–168.

Deemer, B.R., Harrison, J.A., Li, S., Beaulieu, J.J., DelSontro, T., Barros, N., Bezzera-Neto, J.F., Powers, S.M., dos Santos, M.A., Vonk, J.A., 2016. Greenhouse gas emissions from reservoir water surfaces: a new global synthesis. Bioscience 66, 949–964.

del Giorgio, P.A., Cole, J.J., Cimbleris, A., 1997. Respiration rates in bacteria exceed phytoplankton production in unproductive aquatic systems. Nature 385, 148–151.

DelSontro, T., Beaulieu, J.J., Downing, J.A., 2018. Greenhouse gas emissions from lakes and impoundments: upscaling in the face of global change. Limnol. Oceanogr. Lett. 3, 64–75.

Demars, B.O.L., 2019. Hydrological pulses and burning of dissolved organic carbon by stream respiration. Limnol. Oceanogr. 64, 406–421.

Demars, B.O.L., Russell Manson, J., Ólafsson, J.S., Gíslason, G.M., Gudmundsdóttir, R., Woodward, G., Reiss, J., Pichler, D.E., Rasmussen, J.J., Friberg, N., 2011. Temperature and the metabolic balance of streams. Freshw. Biol. 56, 1106–1121.

Denfeld, B.A., Baulch, H.M., del Giorgio, P.A., Hampton, S.E., Karlsson, J., 2018. A synthesis of carbon dioxide and methane dynamics during the ice-covered period of northern lakes. Limnol. Oceanogr. Lett. 3, 117–131.

Diamond, J., Burton, G.A., 2021. Moving beyond the term "contaminants of emerging concern". Environ. Toxicol. Chem. 40, 1527–1529.

Dietz, R.D., Engstrom, D.R., Anderson, N.J., 2015. Patterns and drivers of change in organic carbon burial across a diverse landscape: insights from 116 Minnesota lakes. Global Biogeochem. Cycles 29, 708–727.

Dittmar, T., Stubbins, A., 2014. 12.6—dissolved organic matter in aquatic systems. In: Holland, H.D., Turekian, K.K. (Eds.), Treatise on Geochemistry. Elsevier, Oxford, pp. 125–156.

Do, H.X., Westra, S., Leonard, M., 2017. A global-scale investigation of trends in annual maximum streamflow. J. Hydrol. 552, 28–43.

Donohue, I., Garcia Molinos, J., 2009. Impacts of increased sediment loads on the ecology of lakes. Biol. Rev. 84, 517–531.

Downing, J.A., Cole, J.J., Middelburg, J.J., Striegl, R.G., Duarte, C.M., Kortelainen, P., Prairie, Y.T., Laube, K.A., 2008. Sediment organic carbon burial in agriculturally eutrophic impoundments over the last century. Global Biogeochem. Cycles 22, GB1018.

Drake, T.W., Raymond, P.A., Spencer, R.G.M., 2018. Terrestrial carbon inputs to inland waters: a current synthesis of estimates and uncertainty. Limnol. Oceanogr. Lett. 3, 132–142.

Driscoll, C.T., Van Dreason, R., 1993. Seasonal and long-term temporal patterns in the chemistry of Adirondack lakes. Water Air Soil Poll 67, 319–344.

Duarte, C.M., Prairie, Y.T., 2005. Prevalence of heterotrophy and atmospheric CO_2 emissions from aquatic ecosystems. Ecosystems 8, 862–870.

Einola, E., Rantakari, M., Kankaala, P., Kortelainen, P., Ojala, A., Pajunen, H., Mäkelä, S., Arvola, L., 2011. Carbon pools and fluxes in a chain of five boreal lakes: a dry and wet year comparison. J. Geophys. Res. Biogeosciences 116, 3009.

Evans, C.D., Futter, M.N., Moldan, F., Valinia, S., Frogbrook, Z., Kothawala, D.N., 2017. Variability in organic carbon reactivity across lake residence time and trophic gradients. Nat. Geosci. 10, 832–835.

Fellman, J.B., Hood, E., Spencer, R.G.M., 2010. Fluorescence spectroscopy opens new windows into dissolved organic matter dynamics in freshwater ecosystems: a review. Limnol. Oceanogr. 55, 2452–2462.

Ferland, M.-E., Prairie, Y.T., Teodoru, C., del Giorgio, P.A., 2014. Linking organic carbon sedimentation, burial efficiency, and long-term accumulation in boreal lakes. J. Geophys. Res. Biogeosci. 119, 836–847.

Findlay, S.E.G., Sinsabaugh, R.L. (Eds.), 2003. Aquatic Ecosystems: Interactivity of Dissolved Organic Matter. Academic Press, San Diego, CA.

Finlay, J.C., 2011. Stream size and human influences on ecosystem production in river networks. Ecosphere 2, art87.

Finlay, J.C., Khandwala, S., Power, M.E., 2002. Spatial scales of carbon flow in a river food web. Ecology 83, 1845–1859.

Finlay, K., Leavitt, P.R., Patoine, A., Patoine, A., Wissel, B., 2010. Magnitudes and controls of organic and inorganic carbon flux through a chain of hard-water lakes on the northern great plains. Limnol. Oceanogr. 55, 1551–1564.

Fisher, S.G., Carpenter, S.R., 1976. Ecosystem and macrophyte primary production of the fort river, massachusetts. Hydrobiologia 49, 175–187.

Fisher, S.G., Likens, G.E., 1973. Energy flow in bear brook, new hampshire: an integrative approach to stream ecosystem metabolism. Ecol. Monogr. 43, 421–439.

Follstad Shah, J.J., Kominoski, J.S., Ardón, M., Dodds, W.K., Gessner, M.O., Griffiths, M.O., Griffiths, N.A., Hawkins, C.P., Johnson, S.L., Lecerf, A., Leroy, C.J., Manning, D.W., Rosemond, A.D., Sinsabuagh, R.L., Swan, C.M., Webster, J.R., Zeglin, L.H., 2017. Global synthesis of the temperature sensitivity of leaf litter breakdown in streams and rivers. Global Change Biol. 23, 3064–3075.

Ford, T.E., Naiman, R.J., 1989. Groundwater-surface water relationships in boreal forest watersheds: dissolved organic carbon and inorganic nutrient dynamics. Can. J. Fish. Aquat. Sci. 46, 41–49.

Frimmel, F., Christman, R., 1988. Humic Substances and Their Role in the Environment. John Wiley & Sons, Inc, Chichester, UK.

Fuß, T., Behounek, B., Ulseth, A.J., Singer, G.A., 2017. Land use controls stream ecosystem metabolism by shifting dissolved organic matter and nutrient regimes. Freshw. Biol. 62, 582–599.

Gallagher, M.T., Reisinger, A.J., 2020. Effects of ciprofloxacin on metabolic activity and algal biomass of urban stream biofilms. Sci. Total Environ. 706, 135728.

Geller, A., 1985. Degradation and formation of refractory DOM by bacteria during simultaneous growth on labile substrates and persistent lake water constituents. Schweiz. Z. Hydrol. 47, 27–44.

Genzoli, L., Hall, R.O., 2016. Shifts in Klamath river metabolism following a reservoir cyanobacterial bloom. Freshw. Sci. 35, 795–809.

Gjessing, E.T., 1976. Physical and Chemical Characteristics of Aquatic Humus. Ann Arbor Science Publs., Ann Arbor, MI.

Gjessing, E.T., Riise, G., Lydersen, E., 1998. Acid rain and natural organic matter (NOM). Acta Hydrochim. Hydrobiol. 26, 131–136.

Gómez-Gener, L., Hotchkiss, E.R., Laudon, H., Sponseller, R.A., 2021. Integrating discharge-concentrations dynamics across carbon forms in a boreal landscape. Water Resour. Res. 57, e2020WR028806.

Gounand, I., Little, C.J., Harvey, E., Altermatt, F., 2018. Cross-ecosystem carbon flows connecting ecosystems worldwide. Nat. Commun. 9, 4825.

Grace, M.R., Giling, D.P., Hladyz, S., Caron, V., Thompson, R.M., Mac Nally, R., 2015. Fast processing of diel oxygen curves: estimating stream metabolism with BASE (Bayesian single-station Estimation). Limnol Oceanogr. Methods 13, 103–114.

Gray, A.D., Bernhardt, E., 2022. Are nitrogen and carbon cycle processes impacted by common stream antibiotics? A comparative assessment of single vs. mixture exposures. PLoS One 17 (1), e0261714.

Gregory, J., 1989. Fundamentals of flocculation. Crit. Rev. Environ. Control 19, 185–230.

Gremm, T.J., Kaplan, L.A., 1998. Dissolved carbohydrate concentration, composition, and bioavailability to microbial heterotrophs in stream water. Acta Hydrochim. Hydrobiol. 26, 167–171.

Grieve, I.C., 1990. Seasonal, hydrological, and land management factors controlling dissolved organic carbon concentrations in the loch fleet catchments, southwest Scotland. Hydrol. Processes 4, 231–239.

Griffiths, N.A., Tank, J.L., Royer, T.V., Warner, T.J., Frauendorf, T.C., Rosi-Marshall, E.J., Whiles, M.R., 2012. Temporal variation in organic carbon spiraling in Midwestern agricultural streams. Biogeochemistry 108, 149–169.

Groeneveld, M., Catalán, N., Attermeyer, K., Hawkes, J., Einarsdóttir, K., Kothawala, D., Bergquist, J., Tranvik, L., 2020. Selective adsorption of terrestrial dissolved organic matter to inorganic surfaces along a boreal inland water continuum. J. Geophys. Res. Biogeosciences 125, e2019JG005236.

Grossart, H., Simon, M., 1998. Significance of limnetic organic aggregates (lake snow) for the sinking flux of particulate organic matter in a large lake. Aquat. Microb. Ecol. 15, 115–125.

Gudasz, C., Bastviken, D., Steger, K., Premke, K., Sobek, S., Tranvik, L.J., 2010. Temperature-controlled organic carbon mineralization in lake sediments. Nature 466, 478–481.

Guenet, B., Danger, M., Abbadie, L., Lacroix, G., 2010. Priming effect: bridging the gap between terrestrial and aquatic ecology. Ecology 91, 2850–2861.

Guenet, B., Danger, M., Harrault, L., Allard, B., Jauset-Alcala, M., Bardoux, G., Benest, D., Abbadie, L., Lacroiz, G., 2014. Fast mineralization of land-born C in inland waters: first experimental evidences of aquatic priming effect. Hydrobiologia 721, 35–44.

Guérin, F., Abril, C., Richard, S., Burban, B., Reynouard, C., Seyler, P., Delmas, R., 2006. Methane and carbon dioxide emissions from tropical reservoirs: significance of downstream rivers. Geophys. Res. Lett. 33, L21407.

Guillemette, F., McCallister, S.L., del Giorgio, P.A., 2013. Differentiating the degradation dynamics of algal and terrestrial carbon within complex natural dissolved organic carbon in temperate lakes. J. Geophys. Res. Biogeosciences 118, 963–973.

Hadwen, W.L., Bunn, S.E., 2004. Tourists increase the contribution of autochthonous carbon to littoral zone food webs in oligotrophic dune lakes. Mar. Freshw. Res. 55, 701–708.

Hall, R.O., Beaulieu, J.J., 2013. Estimating autotrophic respiration in streams using daily metabolism data. Freshw. Sci. 32, 507–516.

Hall, R.O., Hotchkiss, E.R., 2017. Stream metabolism. In: Hauer, F.R., Lamberti, G.A. (Eds.), Methods in Stream Ecology: Volume 2: Ecosystem Function, third ed. Academic Press, pp. 219–233.

Hall, R.O., Tank, J.L., Baker, M.A., Rosi-Marshall, E.J., Hotchkiss, E.R., 2016. Metabolism, gas exchange, and carbon spiraling in rivers. Ecosystems 19, 73–86.

Hamdan, M., Byström, P., Hotchkiss, E.R., Al-Haidarey, M.J., Ask, J., Karlsson, J., 2018. Carbon dioxide stimulates lake primary production. Sci. Rep. 8, 10878.

Hamdan, M., Byström, P., Hotchkiss, E.R., Al-Haidarey, M.J., Karlsson, J., 2021. An experimental test of climate change effects in northern lakes: increasing allochthonous organic matter and warming alters autumn primary production. Freshw. Biol. 66, 815–825.

Hanson, P.C., Carpenter, S.R., Cardille, J.A., Coe, M.T., Winslow, L.A., 2007. Small lakes dominate a random sample of regional lake characteristics. Freshw. Biol. 52, 814–822.

Hanson, P.C., Hamilton, D.P., Stanley, E.H., Preston, N., Langman, O.C., Kara, E.L., 2011. Fate of allochthonous dissolved organic carbon in lakes: a quantitative approach. PLoS One 6, e21884.

Hanson, P.C., Pace, M.L., Carpenter, S.R., Cole, J.J., Stanley, E.H., 2015. Integrating landscape carbon cycling: research needs for resolving organic carbon budgets of lakes. Ecosystems 18, 363–375.

Hayes, M., MacCarthy, P., Malcolm, R., Swift, R., 1989. Humic Substances II: In Search of Structure. John Wiley & Sons, Inc, Chichester, UK.

Hayes, M.H., Swift, R.S., 2020. Vindication of humic substances as a key component of organic matter in soil and water. Adv. Agron. 163, 1–37.

He, D., Wang, K., Pang, Y., He, C., Li, P., Li, Y., Xiao, S., Shi, Q., Sun, Y., 2020. Hydrological management constraints on the chemistry of dissolved organic matter in the three gorges reservoir. Water Res. 187, 116413.

Helms, J.R., Stubbins, A., Ritchie, J.D., Minor, E.C., Kieber, D.J., Mopper, K., 2008. Absorption spectral slopes and slope ratios as indicators of molecular weight, source, and photobleaching of chromophoric dissolved organic matter. Limnol. Oceanogr. 53, 955–969.

Hertkorn, N., Benner, R., Frommberger, M., Schmitt-Kopplin, P., Witt, M., Kaiser, K., Kettrup, A., Hedges, J.I., 2006. Characterization of a major refractory component of marine dissolved organic matter. Geochim. Cosmochim. Acta. 70, 2990–3010.

Hessen, D.O., Tranvik, L.J. (Eds.), 1998. Aquatic Humic Substances, Ecological Studies. Springer Berlin Heidelberg, Berlin, Heidelberg.

Hoellein, T.J., Bruesewitz, D.A., Richardson, D.C., 2013. Revisiting Odum (1956): a synthesis of aquatic ecosystem metabolism. Limnol. Oceanogr. 58, 2089–2100.

Hoellein, T.J., McCormick, A.R., Hittie, J., London, M.G., Scott, J.W., Kelly, J.J., 2017. Longitudinal patterns of microplastic concentration and bacterial assemblages in surface and benthic habitats of an urban river. Freshw. Sci. 36, 491–507.

Holgerson, M.A., Raymond, P.A., 2016. Large contribution to inland water CO_2 and CH_4 emissions from very small ponds. Nat. Geosci. 9, 222–226.

Holtgrieve, G.W., Schindler, D.E., Branch, T.A., A'mar, Z.T., 2010. Simultaneous quantification of aquatic ecosystem metabolism and reaeration using a Bayesian statistical model of oxygen dynamics. Limnol. Oceanogr. 55, 1047–1063.

Hood, J.M., Benstead, J.P., Cross, W.F., Huryn, A.D., Johnson, P.W., Gíslason, G.M., Junker, J.R., Nelson, D., Óladsson, J.S., Tran, C., 2017. Increased resource use efficiency amplifies positive response of aquatic primary production to experimental warming. Global Change Biol. 24, 1069–1084.

Hornberger, G.M., Bencala, K.E., McKnight, D.M., 1994. Hydrological controls on dissolved organic carbon during snowmelt in the snake river near Montezuma, Colorado. Biogeochemistry 25, 147–165.

Hotchkiss, E.R., Hall, R.O., 2014. High rates of daytime respiration in three streams: use of $\delta^{18}O$-O_2 and O_2 to model diel ecosystem metabolism. Limnol. Oceanogr. 59, 798–810.

Hotchkiss, E.R., Hall, R.O., 2015. Whole-stream ^{13}C tracer addition reveals distinct fates of newly fixed carbon. Ecology 96, 403–416.

Hotchkiss, E.R., Hall, R.O., Baker, M.A., Rosi-Marshall, E.J., Tank, J.L., 2014. Modeling priming effects on microbial consumption of dissolved organic carbon in rivers. J. Geophys. Res. Biogeosciences 119, 982–995.

Hotchkiss, E.R., Hall, R.O., Sponseller, R.A., Butman, D., Klaminder, J., Laudon, H., Rosvall, M., Karlsson, J., 2015. Sources of and processes controlling CO_2 emissions change with the size of streams and rivers. Nat. Geosci. 89, 696–699.

Hotchkiss, E.R., Sadro, S., Hanson, P.C., 2018. Toward a more integrative perspective on carbon metabolism across lentic and lotic inland waters. Limnol. Oceanogr. Lett. 3, 57–63.

Hunter, W.R., Williamson, A., Sarneel, J.M., 2021. Using the tea bag index to determine how two human pharmaceuticals affect litter decomposition by aquatic microorganisms. Ecotoxicology 30, 1272–1278.

Hutchins, R.H.S., Aukes, P., Schiff, S.L., Dittmar, T., Prairie, Y.T., del Giorgio, P.A., 2017. The optical, chemical, and molecular dissolved organic matter succession along a Boreal soil-stream-river continuum. J. Geophys. Res. Biogeosciences 122, 2892–2908.

Hutchinson, G.E., 1941. Limnological studies in connecticut. IV. The mechanisms of intermediary metabolism in stratified lakes. Ecol. Monogr. 11, 21–60.

Hutchinson, G.E., 1957. A Treatise on Limnology. I. Geography, Physics, and Chemistry. John Wiley & Sons, New York, p. 1015.

Hutchinson, G.E., 1959. Homage to Santa Rosalia, or why are there so many kinds of animals? Am. Nat. 93, 145–159.

Hynes, H.B.N., 1960. The Biology of Polluted Waters. Liverpool University Press, Liverpool, p. 202.

Hynes, H.B.N., 1970. The Ecology of Running Waters. University of Toronto Press, Toronto, p. 555.

IHSS, 2007. What Are Humic Substances? [WWW Document]. Int. Humic Subst. Soc. URL (accessed 1.4.21). http://humic-substances.org/what-are-humic-substances-2/.

IPCC, 2021. Climate change 2021: the physical science basis. In: Masson-Delmotte, V., Zhai, P., Pirani, A., Connors, S.L., Péan, C., Berger, S., Caud, N., Chen, Y., Goldfarb, L., Gomis, M.I., Huang, M., Leitzell, K., Lonnoy, E., Matthews, J.B.R., Maycock, T.K., Waterfield, T., Yelekçi, O., Yu, R., Zhou, B. (Eds.), Contribution of Working Group I to the Sixth Assessment Report of the Intergovernmental Panel on Climate Change. Cambridge University Press, Cambridge and New York, USA, p. 2391.

Isidorova, A., Mendonça, R., Sobek, S., 2019. Reduced mineralization of terrestrial OC in anoxic sediment suggests enhanced burial efficiency in reservoirs compared to other depositional environments. J. Geophys. Res. Biogeosciences 124, 678–688.

Jane, S.F., Hansen, G.J.A., Kraemer, B.M., Leavitt, P.R., Mincer, J.L., North, R.L., Pilla, R.M., Stetler, J.T., Williamson, C.E., Woolway, R.I., Arvola, L., Chandra, S., DeGasperi, C.L., Diemer, L., Dunalska, J., Erina, O., Flaim, G., Grossart, H.-P., Hambright, K.D., Hein, C., Hejzlar, J., Janus, L.L., Jenny, J.-P., Jones, J.R., Knoll, L.B., Leoni, B., Mackay, E., Matsuzaki, S.-I.S., McBride, C., Müller-Navarra, D.C., Paterson, A.M., Pierson, D., Rogora, M., Rusak, J.A., Sadro, S., Saulnier-Talbot, E., Schmid, M., Sommaruga, R., Thiery, W., Verburg, P., Wathers, K.C., Weyhenmeyer, G.A., Yokota, K., Rose, K.C., 2021. Widespread deoxygenation of temperate lakes. Nature 594, 66–70.

Jane, S.F., Rose, K.C., 2018. Carbon quality regulates the temperature dependence of aquatic ecosystem respiration. Freshw. Biol. 63, 1407–1419.

Jenny, J.-P., Francus, P., Normandeau, A., Lapointe, F., Perga, M.-E., Ojala, A., Schimmelmann, A., Zolitschka, B., 2016. Global spread of hypoxia in freshwater ecosystems during the last three centuries is caused by rising local human pressure. Global Change Biol. 22, 1481–1489.

Jepsen, R., He, K., Blaney, L., Swan, C., 2019. Effects of antimicrobial exposure on detrital biofilm metabolism in urban and rural stream environments. Sci. Total Environ. 666, 1151–1160.

Jiang, G., Ma, R., Loiselle, S.A., Duan, H., Su, W., Cai, W., Huang, C., Yang, J., Yu, W., 2015. Remote sensing of particulate organic carbon dynamics in a eutrophic lake (Taihu Lake, China). Sci. Total Environ. 532, 245–254.

Johnson, M.S., Lehmann, J., Selva, E.C., Abdo, M., Riha, S., Couto, E.G., 2006. Organic carbon fluxes within and streamwater exports from headwater catchments in the southern Amazon. Hydrol. Processes 20, 2599–2614.

Johnston, S.E., Striegl, R.G., Bogard, M.J., Dornblaser, M.M., Butman, D.E., Kellerman, A.M., Wickland, K.P., Podgorski, D.C., Spencer, R.G., 2020. Hydrologic connectivity determines dissolved organic matter biogeochemistry in northern high-latitude lakes. Limnol. Oceanogr. 65, 1764–1780.

Jones, J.B., Mulholland, P.J., 1998. Carbon dioxide variation in a hardwood forest stream: an integrative measure of whole catchment soil respiration. Ecosystems 1, 183–196.

Jonsson, M., Ershammar, E., Fick, J., Brodin, T., Klaminder, J., 2015. Effects of an antihistamine on carbon and nutrient recycling in streams. Sci. Total Environ. 538, 240–245.

Kaiser, K., Zech, W., 1998. Rates of dissolved organic matter release and sorption in forest soils. Soil Sci. 163, 714–725.

Kaplan, L.A., Bott, T.L., 1982. Diel fluctuations of DOC generated by algae in a piedmont stream. Limnol. Oceanogr. 27, 1091–1100.

Kaplan, L.A., Bott, T.L., 1989. Diel fluctuations in bacterial activity on streambed substrata during vernal algal blooms: effects of temperature, water chemistry, and habitat. Limnol. Oceanogr. 34, 718–733.

Kaplan, L.A., Wiegner, T.N., Newbold, J.D., Ostrom, P.H., Gandhi, H., 2007. Untangling the complex issue of dissolved organic carbon uptake: a stable isotope approach. Freshw. Biol. 5, 855–864.

Karlsson, J., Berggren, M., Ask, J., Byström, P., Jonsson, A., Laudon, H., Jansson, M., 2012. Terrestrial organic matter support of lake food webs: evidence from lake metabolism and stable hydrogen isotopes of consumers. Limnol. Oceanogr. 57, 1042–1048.

Kastowski, M., Hinderer, M., Vecsei, A., 2011. Long-term carbon burial in European lakes: analysis and estimate. Global Biogeochem. Cycles 25, GB3019.

Kaushal, S.S., Gold, A.J., Bernal, S., Newcomer Johnson, T.A., Addy, K., Burgin, A., Burns, D.A., Coble, A.A., Hood, E., Lu, Y., Mayer, P., Minor, E.C., Schrother, A.W., Vidon, P., Wilson, H., Xenopoulos, M.A., Doody, T., Galella, J.G., Goodling, P., Haviland, K., Haq, S., Wessel, B., Wood, K.L., Jaworski, N., Belt, K.T., 2018b. Watershed 'chemical cocktails': forming novel elemental combinations in Anthropocene fresh waters. Biogeochemistry 141, 281–305.

Kaushal, S.S., Likens, G.E., Jaworski, N.A., Pace, M.L., Sides, A.M., Seekell, D., Belt, K.T., Secor, D.H., Wingate, R.L., 2010. Rising stream and river temperatures in the United States. Front. Ecol. Environ. 8, 461–466.

Kaushal, S.S., Likens, G.E., Pace, M.L., Utz, R.M., Haq, S., Gorman, J., Grese, M., 2018a. Freshwater salinization syndrome on a continental scale. Proc. Natl. Acad. Sci. U.S.A. 115, E574–E583.

Kelleher, B.P., Simpson, A.J., 2006. Humic substances in soils: are they really chemically distinct? Environ. Sci. Technol. 40, 4605–4611.

Keller, P.S., Catalán, N., von Schiller, D., Grossart, H.P., Koschorreck, M., Obrador, B., Frassl, M.A., Karakaya, N., Barros, N., Howitt, J.A., Mendoza-Lera, C., Pastor, A., Flaim, G., Aben, R., Riis, T., Arce, M.I., Onandia, G., Paranaíba, J.R., Linkhorst, A., del Campo, R., Amado, A.M., Cauvy-Fraunié, S., Brothers, S., Condon, J., Mendonça, R.F., Reverey, F., Rõõm, E.I., Datry, T., Roland, F., Laas, A., Obertegger, U., Park, J.H., Wang, H., Kosten, S., Gómez, R., Feijoó, C., Elosegi, A., Sánchez-Montoya, M.M., Finlayson, C.M., Melita, M., Oliveira Junior, E.S., Muniz, C.C., Gómez-Gener, L., Leigh, C., Zhang, Q., Marcé, R., 2020. Global CO_2 emissions from dry inland waters share common drivers across ecosystems. Nat. Commun. 11, 1–8.

Kellerman, A.M., Dittmar, T., Kothawala, D.N., Tranvik, L.J., 2014. Chemodiversity of dissolved organic matter in lakes driven by climate and hydrology. Nat. Commun. 5, 1–8.

Kellerman, A.M., Guillemette, F., Podgorski, D.C., Aiken, G.R., Butler, K.D., Spencer, R.G.M., 2018. Unifying concepts linking dissolved organic matter composition to persistence in aquatic ecosystems. Environ. Sci. Technol. 52, 2538–2548.

Kellerman, A.M., Kothawala, D.N., Dittmar, T., Tranvik, L.J., 2015. Persistence of dissolved organic matter in lakes related to its molecular characteristics. Nat. Geosci. 8, 454–457.

Kelso, J.E., Rosi, E.J., Baker, M.A., 2020. Towards more realistic estimates of DOM decay in streams: Incubation methods, light, and non-additive effects. Freshw. Sci. 39, 559–575.

Kerr, J.B., McElroy, C.T., 1993. Evidence for large upward trends of ultraviolet-B radiation linked to ozone depletion. Science 262, 1032–1034.

Keskitalo, J., Eloranta, P., 1999. Limnology of Humic Waters. Backhuys Publishers, Leiden, NL.

Kettner, M.T., Rojas-Jimenez, K., Oberbeckmann, S., Labrenz, M., Grossart, H.-P., 2017. Microplastics alter composition of fungal communities in aquatic ecosystems. Environ. Microbiol. 19, 4447–4459.

Kleber, M., Lehmann, J., 2019. Humic substances extracted by alkali are invalid proxies for the dynamics and functions of organic matter in terrestrial and aquatic ecosystems. J. Environ. Qual. 48, 207–216.

Koch, B.P., Dittmar, T., 2006. From mass to structure: an aromaticity index for high-resolution mass data of natural organic matter. Rapid Commun. Mass Spectrom. 20, 926–932.

Koch, B.P., Dittmar, T., Witt, M., Kattner, G., 2007. Fundamentals of molecular formula assignment to ultrahigh resolution mass data of natural organic matter. Anal. Chem. 79, 1758–1763.

Kononova, M., 1966. Soil Organic Matter: Its Nature, its Role in Soil Formation and in Soil Fertility. Pergamon Press, Oxford.

Kosten, S., van den Berg, S., Mendonça, R., Paranaíba, J.R., Roland, F., Sobek, S., Van Den Hoek, J., Barros, N., 2018. Extreme drought boosts CO_2 and CH_4 emissions from reservoir drawdown areas. Inland Waters 8, 329–340.

Kothawala, D.N., Kellerman, A.M., Catalán, N., Tranvik, L.J., 2021. Organic matter degradation across ecosystem boundaries: the need for a unified conceptualization. Trends Ecol. Evol. 36, 113–122.

Kothawala, D.N., Stedmon, C.A., Müller, R.A., Weyhenmeyer, G.A., Köhler, S.J., Tranvik, L.J., 2014. Controls of dissolved organic matter quality: evidence from a large-scale boreal lake survey. Global Change Biol. 20, 1101–1114.

Kragh, T., Martinsen, K.T., Kristensen, E., Sand-Jensen, K., 2020. From drought to flood: sudden carbon inflow causes whole-lake anoxia and massive fish kill in a large shallow lake. Sci. Total Environ. 739, 140072.

Kritzberg, E.S., Cole, J.J., Pace, M.M., Granéli, W., 2006. Bacterial growth on allochthonous carbon in humic and nutrient-enriched lakes: results from whole-lake ^{13}C addition experiments. Ecosystems 9, 489–499.

Larsen, S., Andersen, T., Hessen, D.O., 2011. Climate change predicted to cause severe increase of organic carbon in lakes. Global Change Biol. 17, 1186–1192.

Lehmann, J., Kleber, M., 2015. The contentious nature of soil organic matter. Nature 528, 60–68.

Lewis, W.M., Saunders, J.F., 1989. Concentration and transport of dissolved and suspended substances in the Orinoco River. Biogeochemistry 7, 203–240.

Likens, G.E., Bormann, F.H., 1974. Acid rain: a serious regional environmental problem. Science 184, 1176–1179.

Lindeman, R.L., 1942. The trophic-dynamic aspect of ecology. Ecology 23, 399–418.

Linkhorst, A., Paranaíba, J.R., Mendonça, R., Rudberg, D., DelSontro, T., Barros, N., Sobek, S., 2021. Spatially resolved measurements in tropical reservoirs reveal elevated methane ebullition at river inflows and at high productivity. Global Biogeochem. Cycles 35, e2020GB006717.

Lisboa, L.K., Thomas, S., Moulton, T.P., 2016. Reviewing carbon spiraling approach to understand organic matter movement and transformation in lotic ecosystems. Acta. Limnol. Bras. 28, e14.

Liu, S., He, Z., Tang, Z., Liu, L., Hou, J., Li, T., Zhang, Y., Shi, Q., Giesy, J.P., Wu, F., 2020. Linking the molecular composition of autochthonous dissolved organic matter to source identification for freshwater lake ecosystems by combination of optical spectroscopy and FT-ICR-MS analysis. Sci. Total Environ. 703, 134764.

Lovett, G.M., Cole, J.J., Pace, M.L., 2006. Is net ecosystem production equal to ecosystem carbon accumulation? Ecosystems 9, 152–155.

Lyon, D.R., Ziegler, S.E., 2009. Carbon cycling within epilithic biofilm communities across a nutrient gradient of headwater streams. Limnol. Oceanogr. 54, 439–449.

Madronich, S., 1994. Increases in biologically damaging UV-B radiation due to stratospheric ozone reductions: a brief review. Arch. Hydrobiol. Beih. Ergebn. Limnol. 43, 17–30.

Mann, P.J., Eglinton, T., McIntyre, C., Zimov, N., Davydova, A., Vonk, J.E., Holmes, R.M., Spencer, R.G.M., 2015. Utilization of ancient permafrost carbon in headwaters of Arctic fluvial networks. Nat. Commun. 6, 7856.

Marcarelli, A.M., Baxter, C.V., Mineau, M.M., Hall, R.O., 2011. Quantity and quality: unifying food web and ecosystem perspectives on the role of resource subsidies in freshwaters. Ecology 92, 1215–1225.

Marcé, R., Obrador, B., Gómez-Gener, L., Catalán, N., Koschorreck, M., Arce, M.I., Singer, G., von Schiller, D., 2019. Emissions from dry inland waters are a blind spot in the global carbon cycle. Earth Sci. Rev. 188, 240–248.

Marín-Spiotta, E., Gruley, K.E., Crawford, J., Atkinson, E.E., Miesel, J.R., Greene, S., Cardona-Correa, C., Spencer, R.G.M., 2014. Paradigm shifts in soil organic matter research affect interpretations of aquatic carbon cycling: transcending disciplinary and ecosystem boundaries. Biogeochemistry 117, 279–297.

Marks, J.C., 2019. Revisiting the fate of dead leaves that fall into streams. Annu. Rev. Ecol. Evol. Syst. 50, 547–568.

Masclaux, H., Perga, M.-E., Kagami, M., Desvilettes, C., Bourdier, G., Bec, A., 2013. How pollen organic matter enters freshwater food webs. Limnol. Oceanogr. 58, 1185–1195.

Masese, F.O., Kiplagat, M.J., Gonzalez-Quijano, C.R., Subaluskly, A.L., Dutton, C.L., Post, D.M., Singer, G.A., 2020. Hippopotamus are distinct from domestic livestock in their resource subsidies to and effects on aquatic ecosystems. Proc. Royal R. Soc. B Biol. Sci. 287, 20193000.

Masese, F.O., Salcedo-Borda, J.S., Gettel, G.M., Irvine, K., McClain, M.E., 2017. Influence of catchment land use and seasonality on dissolved organic matter composition and ecosystem metabolism in headwater streams of a Kenyan river. Biogeochemistry 132, 1–22.

Massicotte, P., Asmala, E., Stedmon, C., Markager, S., 2017. Global distribution of dissolved organic matter along the aquatic continuum: across rivers, lakes and oceans. Sci. Total Environ. 609, 180–191.

Mattson, M.D., Likens, G.E., 1993. Redox reactions of organic matter decomposition in a soft water lake. Biogeochemistry 19, 149–172.

Mattsson, T., Kortelainen, P., Laubel, A., Evans, D., Pujo-Pay, M., Räike, A., Conan, P., 2009. Export of dissolved organic matter in relation to land use along a European climatic gradient. Sci. Total Environ. 407, 1967–1976.

Matzinger, A., Müller, B., Niederhauser, P., Schmid, M., Wüest, A., 2010. Hypolimnetic oxygen consumption by sediment-based reduced substances in former eutrophic lakes. Limnol. Oceanogr. 55, 2073–2084.

McCallister, S.L., Ishikawa, N.F., Kothawala, D.N., 2018. Biogeochemical tools for characterizing organic carbon in inland aquatic ecosystems. Limnol. Oceanogr. Lett. 3, 444–457.

McClain, M.E., Richey, J.E., 1996. Regional-scale linkages of terrestrial and lotic ecosystems in the Amazon basin: a conceptual model for organic matter. Large Rivers 10 (1–4), 111–125.

McClean, P., Hunter, W.R., 2020. 17α-ethynylestradiol (EE2) limits the impact of ibuprofen upon respiration by streambed biofilms in a sub-urban stream. Environ. Sci. Pollut. Res. 27, 37149–37154.

McClelland, J., Holmes, R., Peterson, B., Raymond, P., Striegl, R., Zhulidov, A., Zimov, S., Zimov, N., Tank, S., Spencer, R., Staples, R., Gurtovaya, T., Griffin, C., 2016. Particulate organic carbon and nitrogen export from major Arctic rivers. Glob. Biogeochem. Cycles 30, 629–643.

McClure, R.P., Schreiber, M.E., Lofton, M.E., Chen, S., Krueger, K.M., Carey, C.C., 2020. Ecosystem-scale oxygen manipulations alter terminal electron acceptor pathways in a eutrophic reservoir. Ecosystems 24, 1281–1298.

McDowell, W.H., Likens, G.E., 1988. Origin, composition, and flux of dissolved organic carbon in the Hubbard Brook Valley. Ecol. Monogr. 58, 177–195.

McDowell, W.H., Wood, T., 1984. Podzolization: soil processes control dissolved organic carbon concentrations in stream water. Soil Sci. 137, 23–32.

McKnight, D.M., Aiken, G.R., 1998. Sources and age of aquatic humus. In: Hessen, D.O., Tranvik, L.J. (Eds.), Aquatic Humic Substances: Ecology and Biogeochemistry. Springer-Verlag, New York, pp. 9–39.

McKnight, D.M., Aiken, G.R., Smith, R.L., 1991. Aquatic fulvic acids in microbially based ecosystems: results from two Antarctic desert lakes. Limnol. Oceanogr. 36, 998–1006.

McKnight, D.M., Andrews, E.D., Smith, R.L., Dufford, R., 1994. Aquatic fulvic acids in algal-rich Antarctic ponds. Limnol. Oceanogr. 39, 1972–1979.

McKnight, D.M., Boyer, E.W., Westerhoff, P.K., Doran, P.T., Kulbe, T., Andersen, D.T., 2001. Spectrofluorometric characterization of dissolved organic matter for indication of precursor organic material and aromaticity. Limnol. Oceanogr. 46, 38–48.

Mendonça, R., Müller, R.A., Clow, D., Verpoorter, C., Raymond, P., Tranvik, L.J., Sobek, S., 2017. Organic carbon burial in global lakes and reservoirs. Nat. Comm. 8, 1–7.

Messager, M.L., Lehner, B., Cockburn, C., Lamouroux, N., Pella, H., Tockner, K., Trautmann, T., Watt, C., Datry, T., 2021. Global prevalence of non-perennial rivers and streams. Nature 594, 391–397.

Meyer, J.L., Edwards, R.T., Risley, R., 1987. Bacterial growth on dissolved organic carbon from a blackwater river. Microb. Ecol. 13, 13–29.

Meyer, J.L., Tate, C.M., 1983. The effects of watershed disturbance on dissolved organic carbon dynamics of a stream. Ecology 64, 33–44.

Meyer, J.L., Wallace, J.B., Eggert, S.L., 1998. Leaf litter as a source of dissolved organic carbon in streams. Ecosystems 1, 240–249.

Milberg, P., Törnqvist, L., Westerberg, L.M., Bastviken, D., 2017. Temporal variations in methane emissions from emergent aquatic macrophytes in two boreonemoral lakes. AoB Plants 9, plx029.

Miller, M.C., 1972. The Carbon Cycle in the Epilimnion of Two Michigan Lakes. Ph.D. Diss. Michigan State University, p. 214.

Mineau, M.M., Wollheim, W.M., Buffam, I., Findlay, S.E.G., Hall, R.O., Hotchkiss, E.R., Koenig, L.E., McDowell, W.H., Parr, T.B., 2016. Dissolved organic carbon uptake in streams: a review and assessment of reach-scale measurements. J. Geophys. Res. Biogeosciences 121, 2019–2029.

Minor, E.C., Swenson, M.M., Mattson, B.M., Oyler, A.R., 2014. Structural characterization of dissolved organic matter: a review of current techniques for isolation and analysis. Environ. Sci. Process. Impacts 16, 2064–2079.

Minshall, G.W., 1978. Autotrophy in stream ecosystems. Bioscience 28, 767–771.

Mopper, K., Stubbins, A., Ritchie, J.D., Bialk, H.M., Hatcher, P.G., 2007. Advanced instrumental approaches for characterization of marine dissolved organic matter: extraction techniques, mass spectrometry, and nuclear magnetic resonance spectroscopy. Chem. Rev. 107, 419–442.

Mulholland, P.J., 1997. Dissolved organic matter concentrations and flux in streams. J. N. Am. Benthol. Soc. 16, 131–140.

Mulholland, P.J., 2003. Large-scale patterns in dissolved organic carbon concentration, flux, and sources. In: Aquatic Ecosystems. Academic Press, Cambridge, pp. 139–159.

Mulholland, P.J., Dahm, C.N., David, M.B., Di Toro, D.M., Fisher, T.R., Hemond, H.F., Kögel-Knabner, I., Meybeck, M.H., Meyer, J.L., Sedell, J.R., 1990. What are the temporal and spatial variations of organic acids at the ecosystem level. In: Perdue, E.M., Gjessing, E.T. (Eds.), Organic Acids in Aquatic Ecosystems. John Wiley & Sons, Chichester, pp. 315–329.

Müller, B., Bryant, L.D., Matzinger, A., Wüest, A., 2012. Hypolimnetic oxygen depletion in eutrophic lakes. Environ. Sci. Technol. 46, 9964–9971.

Murphy, K.R., Stedmon, C.A., Wenig, P., Bro, R., 2014. OpenFluor—an online spectral library of auto-fluorescence by organic compounds in the environment. Anal. Methods 6, 658–661.

Myhre, G., Shindell, D., Bréon, F.-M., Collins, W., Fuglestvedt, J., Huang, J., Koch, D., Lamarque, J.-F., Lee, D., Mendoza, B., Nakajima, T., Robock, A., Stephens, G., Takemura, T., Zhang, H., 2013. Anthropogenic and natural radiative forcing. In: Stocker, T., Qin, D., Plattner, G.-K., Tignor, M., Allen, S.K., Boschung, J., Nauels, A., Xia, Y., Bex, V., Midgley, P. (Eds.), Climate Change: The Physical Science Basis. Contribution of Working Group I to the Fifth Assessment Report of the Intergovernmental Panel on Climate Change. Cambridge University Press, Cambridge, UK and New York, NY, USA, pp. 659–740.

Myrstener, M., Gómez-Gener, L., Rocher-Ros, G., Giesler, R., Sponseller, R.A., 2021. Nutrients influence seasonal metabolic patterns and total productivity of Arctic streams. Limnol. Oceanogr. 66, 5182–5196.

Nakano, S., Miyasaka, H., Kuhara, N., 1999. Terrestrial-aquatic linkages: riparian arthropod inputs alter trophic cascades in a stream food web. Ecology 80, 2435–2441.

Nelson, D., Benstead, J.P., Huryn, A.D., Cross, W.F., Hood, J.M., Johnson, P.W., Junker, J.R., Gíslason, G.M., Ólafsson, J.S., 2020. Thermal niche diversity and trophic redundancy drive neutral effects of warming on energy flux through a stream food web. Ecology 1010, e02952.

Newbold, J.D., Mulholland, P.J., Elwood, J., O'Neill, R.V., 1982. Organic carbon spiralling in stream ecosystems. Oikos 38, 266–272.

Newbold, J.D., Thomas, S.A., Minshall, G.W., Cushing, C.E., Georgian, T., 2005. Deposition, benthic residence, and resuspension of fine organic particles in a mountain stream. Limnol. Oceanogr. 50, 1571–1580.

O'Driscoll, C.O., O'Conner, M., Asam, Z.-u.-Z., de Eyto, E., Brown, L.E., Xiao, L., 2016. Forest clearfelling effects of dissolved oxygen and metabolism in peatland streams. J. Environ. Manage. 166, 250–259.

O'Driscoll, N.J., Siciliano, S.D., Peak, D., Carignan, R., Lean, D.R.S., 2006. The influence of forestry activity on the structure of dissolved organic matter in lakes: implications for mercury photoreactions. Sci. Total Environ. 366, 880–893.

O'Melia, C., Tiller, C., 1993. Physicochemical aggregation and deposition in aquatic environments. In: Bubble, J., van Leeuwen, H. (Eds.), Environmental Particles, vol. 2. Lewis Publishers, Ann Arbor, MI, USA, pp. 353–385.

Odum, H.T., 1956. Primary production in flowing waters. Limnol. Oceanogr. 1, 102–117.

Ohle, W., 1952. Die hypolimnische Kohlendioxyd-Akkumulation als produktionsbiologischer Indikator. Arch. Hydrobiol. 46, 153–285.

Olk, D.C., Bloom, P.R., De Nobili, M., Chen, Y., McKnight, D.M., Wells, M.J.M., Weber, J., 2019a. Using humic fractions to understand natural organic matter processes in soil and water: selected studies and applications. J. Environ. Qual. 48, 1633–1643.

Olk, D.C., Bloom, P.R., Perdue, E.M., McKnight, D.M., Chen, Y., Farenhorst, A., Senesi, N., Chin, Y.-P., Schmitt-Kopplin, P., Hertkorn, N., Harir, M., 2019b. Environmental and agricultural relevance of humic fractions extracted by alkali from soils and natural waters. J. Environ. Qual. 48, 217–232.

Ortega-Pieck, A., Fremier, A.K., Orr, C.H., 2017. Agricultural influences on the magnitude of stream metabolism in humid tropical headwater streams. Hydrobiologia 799, 49–64.

Ostrovsky, I., Yacobi, Y.Z., 2010. Sedimentation flux in a large subtropical lake: spatiotemporal variations and relation to primary productivity. Limnol. Oceanogr. 55, 1918–1931.

Otsuki, A., Wetzel, R.G., 1974. Calcium and total alkalinity budgets and calcium carbonate precipitation of a small hard-water lake. Arch. Hydrobiol. 73, 14–30.

Pace, M.L., Cole, J.J., 2002. Synchronous variation of dissolved organic carbon and color in lakes. Limnol. Oceanogr. 47, 333–342.

Pace, M.L., Cole, J.J., Carpenter, S.R., Kitchell, J.F., Hodgson, J.R., Van de Bogert, M.C., Bade, D.L., Kritzberg, E.S., Bastviken, D., 2004. Whole-lake carbon-13 additions reveal terrestrial support of aquatic food webs. Nature 427, 240–243.

Padfield, D., Lowe, C., Buckling, A., Ffrench-Constant, R., Team, S.R., Jennings, S., Shelley, F., Ólafsson, J., Yvon-Durocher, G., 2017. Metabolic compensations constrains the temperature dependence of gross primary production. Ecology Lett. 20, 1250–1260.

Paerl, H.W., Scott, J.T., McCarthy, M.J., Newell, S.E., Gardner, W.S., Havens, K.E., Hoffman, D.K., Wilhem, S.W., Wurtsbaugh, W.A., 2016. It takes two to tango: when and where dual nutrient (N & P) reductions are needed to protect lakes and downstream ecosystems. Environ. Sci. Technol. 50, 10805–10813.

Pagano, T., Bida, M., Kenny, J., 2014. Trends in levels of allochthonous dissolved organic carbon in natural water: a review of potential mechanisms under a changing climate. Water 6, 2862–2897.

Palmer, M., Ruhi, A., 2019. Linkages between flow regime, biota, and ecosystem processes: implications for river restoration. Science 365, 1264.

Peace, A., Frost, P.C., Wagner, N.D., others, 2021. Stoichiometric ecotoxicology for a multisubstance world. Bioscience 71, 132–147.

Pekel, J.F., Cottam, A., Gorelick, N., Belward, A.S., 2016. High-resolution mapping of global surface water and its long-term changes. Nature 540, 418–422.

Perdue, E., Gjessing, E., 1990. Organic Acids in Aquatic Ecosystems. John Wiley & Sons, Inc, New York, NY, USA.

Peter, S., Agstam, O., Sobek, S., 2017. Widespread release of dissolved organic carbon from anoxic boreal lake sediments. Inland Waters 7, 151–163.

Piao, S., Liu, Q., Chen, A., Janssens, I.A., Fu, Y., Dai, J., Liu, L., Lian, X., Shen, M., Zhu, X., 2019. Plant phenology and global climate change: current progresses and challenges. Global Change Biol. 25, 1922–1940.

Piccolo, A., 2002. The supramolecular structure of humic substances: a novel understanding of humus chemistry and implications in soil science. Adv. Agron. 75, 57–134.

Plont, S., O'Donnell, B.M., Gallagher, M.T., Hotchkiss, E.R., 2020. Linking carbon and nitrogen spiraling in streams. Freshw. Sci. 39, 126–136.

Plont, S., Riney, J., Hotchkiss, E.R., 2022. Integrating perspectives on dissolved organic carbon removal and whole-stream metabolism. J Geophys. Res. Biogeosciences 127, e2021JG006610.

Porcal, P., Dillon, P.J., Molot, L.A., 2013. Photochemical production and decomposition of particulate organic carbon in a freshwater stream. Aquat. Sci. 75, 469–482.

Prairie, Y.T., Cole, J.J., 2009. Carbon, unifying currency. In: Likens, G.E. (Ed.), Encyclopedia of Inland Waters. Elsevier, Oxford, pp. 743–746.

Raymond, P.A., Hartmann, J., Lauerwald, R., Sobek, S., McDonald, C., Hoover, M., Butman, D., Striegl, R., Mayorga, E., Humborg, C., Kortelainen, P., Dürr, H., Meybeck, M., Ciais, P., Guth, P., 2013. Global carbon dioxide emissions from inland waters. Nature 503, 355–359.

Raymond, P.A., Saiers, J.E., 2010. Event controlled DOC export from forested watersheds. Biogeochemistry 100, 197–209.

Raymond, P.A., Saiers, J.E., Sobczak, W.V., 2016. Hydrological and biogeochemical controls on watershed dissolved organic matter transport: pulse-shunt concept. Ecology 97, 5–16.

Regnier, P., Friedlingstein, P., Ciais, P., Mackenzie, F.T., Gruber, N., Janssens, I.A., Laruelle, G.G., Lauerwald, R., Luyssaert, S., Andersson, A.J., Arndt, S., Arnosti, C., Borges, A.V., Dale, A.W., Gallego-Sala, A., Goddéris, Y., Goossens, N., Hartmann, J., Heinze, C., Ilyina, T., Joos, F., LaRowe, D.E., Leifeld, J., Meysman, F.J.R., Munhoven, G., Raymond, P.A., Spahni, R., Suntharalingam, P., Thullner, M., 2013. Anthropogenic perturbation of the carbon fluxes from land to ocean. Nat. Geosci. 6, 597–607.

Reid, A.J., Carlson, A.K., Creed, I.F., Eliason, E.J., Gell, P.A., Johnson, P.T., Kidd, K.A., MacCormack, T.J., Olden, J.D., Ormerod, S.J., Smol, J.P., Taylor, W.W., Tockner, K., Vermaire, J.C., Dudgeon, D., Cooke, S.J., 2019. Emerging threats and persistent conservation challenges for freshwater biodiversity. Biol. Rev. 94, 849–873.

Repeta, D.J., Quan, T.M., Aluwihare, L.I., Accardi, A.M., 2002. Chemical characterization of high molecular weight dissolved organic matter in fresh and marine waters. Geochim. Cosmochim. Acta. 66, 955–962.

Rich, P.H., 1975. Benthic metabolism of a soft-water lake. Verh. Internat. Verein. Limnol. 19, 1023–1028.

Rich, P.H., 1980. Hypolimnetic metabolism in three Cape Cod lakes. Am. Midland Nat. 1 (104), 102–109.

Rich, P.H., Devol, A.H., 1978. Analysis of five North American lake ecosystems. VII. Sediment processing. Verh. Internat. Verein. Limnol. 20, 598–604.

Rich, P.H., Wetzel, R.G., 1978. Detritus in lake ecosystems. Amer. Nat. 112, 57–71.

Richardson, D.C., Newbold, J.D., Aufdenkampe, A.K., Taylor, P.G., Kaplan, L.A., 2013. Measuring heterotrophic respiration rates of suspended particulate organic carbon from stream ecosystems. Limnol Oceanogr. Methods 11, 247–261.

Roberts, B.J., Mulholland, P.J., Hill, W.R., 2007. Multiple scales of temporal variability in ecosystem metabolism rates: results from 2 years of continuous monitoring in a forested headwater stream. Ecosystems 10, 588–606.

Robison, A.L., Wollheim, W.M., Turek, B., Bova, C., Snay, C., Varner, R.K., 2021. Spatial and temporal heterogeneity of methane ebullition in lowland headwater streams and the impact on sampling design. Limnol. Oceanogr. 66, 4063–4076.

Rocher-Ros, G., Sponseller, R.A., Bergström, A.-K., Myrstener, M., Giesler, R., 2020. Stream metabolism controls diel patterns and evasion of CO_2 in Arctic streams. Global Change Biol. 26, 1400–1413.

Rodríguez, P., Byström, P., Geibrink, E., Hedström, P., Rivera Vasconcelos, F., Karlsson, J., 2016. Do warming and humic river runoff alter the metabolic balance of lake ecosystems? Aquat. Sci. 78, 717–725.

Roebuck, J.A., Seidel, M., Dittmar, T., Jaffé, R., 2020. Controls of land use and the river continuum concept on dissolved organic matter composition in an anthropogenically disturbed subtropical watershed. Environ. Sci. Tech. 54, 195–206.

Roehm, C., Tremblay, A., 2006. Role of turbines in the carbon dioxide emissions from two boreal reservoirs, Québec, Canada. J. Geophys. Res. 111, D24101.

Roley, S.S., Tank, J.L., Griffiths, N.A., Hall, R.O., Davis, R.T., 2014. The influence of floodplain restoration on whole-stream metabolism in an agricultural stream: insights from a 5-year continuous data set. Freshw. Sci. 33, 1043–1059.

Romanenko, V.I., 1967. Sootnoshenie mezhdu fotosintezom fitoplanktona i destruktsiei organicheskogo veshchestva v vodokhranilishchakh. Trudy Inst. Biol. Vnutrennikh Vod 15, 61–74.

Rosemond, A.D., Benstead, J.P., Bumpers, P.M., Gulis, V., Kominoski, J.S., Manning, D.W.P., Suberkropp, K., Wallace, J.B., 2015. Experimental nutrient additions accelerate terrestrial carbon loss from stream ecosystems. Science 347, 1142–1145.

Rosentreter, J.A., Borges, A.V., Deemer, B.R., Holgerson, M.A., Liu, S., Song, C., Melack, J., Raymond, P.A., Duarte, C.M., Allen, G.H., Olefeldt, D., Poulter, B., Battin, T.I., Eyre, B.D., 2021. Half of global methane emissions come from highly variable aquatic ecosystem sources. Nat. Geosci. 14, 225–230.

Rosi, E.J., Bechtold, H.A., Snow, D., Rojas, M., Reisinger, A.J., Kelly, J.J., 2018. Urban stream microbial communities show resistance to pharmaceutical exposure. Ecosphere 9, e02041.

Rosi-Marshall, E.J., Kincaid, D.W., Bechtold, H.A., Royer, T.V., Rojas, M., Kelly, J.J., 2013. Pharmaceuticals suppress algal growth and microbial respiration and alter bacterial communities in stream biofilms. Ecol. Appl. 23, 583–593.

Rummel, C.D., Jahnke, A., Gorokhova, E., Kühnel, D., Schmitt-Jansen, M., 2017. Impacts of biofilm formation on the fate and potential effects of microplastic in the aquatic environment. Environ. Sci. Technol. Lett. 4, 258–267.

Rumschlag, S.L., Mahon, M.B., Hoverman, J.T., Raffel, T.R., Carrick, H.J., Hudson, P.J., Rohr, J.R., 2020. Consistent effects of pesticides on community structure and ecosystem function in freshwater systems. Nat. Commun. 11, 1–9.

Sanches, L.F., Guenet, B., Marinho, C.C., Barros, N., de Assis Esteves, F., 2019. Global regulation of methane emission from natural lakes. Sci. Rep. 9, 255.

Saunders, G.W., 1972. Potential heterotrophy in a natural population of *Oscillatoria agardhii* var. *Isothrix* Skuja. Limnol. Oceanogr. 17, 704–711.

Savoy, P., Appling, A.P., Heffernan, J.B., Stets, E.G., Read, J.S., Harvey, J.W., Bernhardt, E.S., 2019. Metabolic rhythms in flowing waters: an approach for classifying river productivity regimes. Limnol. Oceanogr. 64, 1835–1851.

Sawakuchi, H.O., Neu, V., Ward, N.D., Barros, M. de L.C., Valerio, A.M., Gagne-Maynard, W., Cunha, A.C., Less, D.F.S., Diniz, J.E.M., Brito, D.C., Krusche, A.V., Richey, J.E., 2017. Carbon dioxide emissions along the lower Amazon River. Front. Mar. Sci. 4, 76.

Scharlemann, J.P., Tanner, E.V., Hiederer, R., Kapos, V., 2014. Global soil carbon: understanding and managing the largest terrestrial carbon pool. Carbon Manag. 5, 81–91.

Schindler, D.W., 2006. Recent advances in the understanding and management of eutrophication. Limnol. Oceanogr. 51, 356–363.

Serkiz, S.M., Perdue, E.M., 1990. Isolation of dissolved organic matter from the Suwannee River using reverse osmosis. Water Res. 24, 911–916.

Shakil, S., Tank, S.E., Kokelj, S.V., Vonk, J.E., Zolkos, S., 2020. Particulate dominance of organic carbon mobilization from thaw slumps on the Peel Plateau, NT: quantification and implications for stream systems and permafrost carbon release. Env. Res. Lett. 15, 114019.

Sharma, S., Blagrave, K., Magnuson, J.J., O'Reilly, C.M., Oliver, S., Batt, R.D., Magee, M.R., Straile, D., Weyhenmeyer, G.A., Winslow, L., Woolway, R.I., 2019. Widespread loss of lake ice around the Northern Hemisphere in a warming world. Nat. Clim. Change 9, 227–231.

Shaw, L., Phung, C., Grace, M., 2015. Pharmaceuticals and personal care products alter growth and function in lentic biofilms. Environ. Chem. 12, 301–306.

Shin, Y., Lee, E.J., Jeon, Y.J., Hur, J., Oh, N.H., 2016. Hydrological changes of DOM composition and biodegradability of rivers in temperate monsoon climates. J. Hydrol. 540, 538–548.

Simpson, A.J., Mcnally, D.J., Simpson, M.J., 2011. NMR spectroscopy in environmental research: from molecular interactions to global processes. Prog. Nucl. Magn. Reson. Spectrosc. 58, 97–175.

Simpson, A.J., Tseng, L.-H., Simpson, M.J., Spraul, M., Ulrich, B., Kingery, W.L., Kelleher, B.P., Hayes, M.H.B., 2004. The application of LC-NMR and LC-SPE-NMR to compositional studies of natural organic matter. Analyst 129, 1216–1222.

Smith, H.J., Tigges, M., D'Andrilli, J., Parker, A., Bothner, B., Foreman, C.M., 2018. Dynamic processing of DOM: insight from exometabolomics, fluorescence spectroscopy, and mass spectrometry. Limnol. Oceanogr. Lett. 3, 225–235.

Smolders, A.J.P., Lamers, L.P.M., Lucassen, E.C.H.E.T., Van der Velde, G.J.G.M., Roelofs, J.G.M., 2006. Internal eutrophication: how it works and what to do about it—a review. Chem. Ecol. 22, 93–111.

Sobek, S., DelSontro, T., Wongfun, N., Wehrli, B., 2012. Extreme organic carbon burial fuels intense methane bubbling in a temperate reservoir. Geophys. Res. Lett. 39, L01401.

Sobek, S., Durisch-Kaiser, E., Zurbrugg, R., Wongfun, N., Wessels, M., Pasche, N., Wehrli, B., 2009. Organic carbon burial efficiency in lake sediments controlled by oxygen exposure time and sediment source. Limnol. Oceanogr. 54, 2243–2254.

Sobek, S., Gudasz, C., Koehler, B., Tranvik, L.J., Bastviken, D., Morales-Pineda, M., 2017. Temperature dependence of apparent respiratory quotients and oxygen penetration depth in contrasting lake sediments. J Geophys. Res. Biogeosciences 122, 3076–3087.

Sobek, S., Söderbäck, B., Karlsson, S., Andersson, E., Brunberg, A.K., 2006. A carbon budget of a small humic lake: an example of the importance of lakes for organic matter cycling in boreal catchments. Ambio 35, 469–475.

Sobek, S., Tranvik, L.J., Prairie, Y.T., Cole, J.J., 2007. Patterns and regulation of dissolved organic carbon: an analysis of 7,500 widely distributed lakes. Limnol. Oceanogr. 52, 1208–1219.

Solomon, C.T., Bruesewitz, D.A., Richardson, D.C., Rose, K.C., Van de Bogert, M.C., Hanson, P.C., Kratz, T.K., Larget, B., Adrian, R., Leroux Babin, B., Chiu, C.-Y., Hamilton, D.P., Gaiser, E.E., Hendricks, S., Istvánovics, V., Laas, A., O'Donnell, D.M., Pace, M.L., Ryder, E., Staehr, P.A., Torgersen, T., Vanni, M.J., Weathers, K.C., Zhu, G., 2013. Ecosystem respiration: drivers of daily variability and background respiration in lakes around the globe. Limnol. Oceanogr. 58, 849–866.

Solomon, C., Carpenter, S., Cole, J., Pace, M., 2008. Support of benthic invertebrates by detrital resources and current autochthonous primary production: results from a whole-lake ^{13}C addition. Freshw. Biol. 53, 42–54.

Song, C., Dodds, W.K., Rüegg, J., Argerich, A., Baker, C.L., Bowden, W.B., Douglas, M.M., Farrell, K.J., Flinn, M.B., Garcia, E.A., Helton, A.M., Harms, T.K., Jia, S., Jones, J.B., Koenig, L.E., Kominoski, J.S., McDowell, W.H., McMaster, D., Parker, S.P., Rosemond, A.D., Ruffing, C.M., Sheehan, K.R., Trentman, M.T., Whiles, M.R., Wollheim, W.M., Ballantyne, F., 2018. Continental-scale decrease in net primary productivity in streams due to climate warming. Nat. Geosci. 11, 415–420.

Stanley, E.H., Casson, N.J., Christel, S.T., Crawford, J.T., Loken, L.C., Oliver, S.K., 2016. The ecology of methane in streams and rivers: patterns, controls, and global significance. Ecol. Monogr. 86, 146–171.

Stanley, E.H., Powers, S.M., Lottig, N.R., Buffam, I., Crawford, J.T., 2012. Contemporary changes in dissolved organic carbon (DOC) in human-dominated rivers: is there a role for DOC management? Freshw. Biol. 57, 26–42.

Stedmon, C.A., Bro, R., 2008. Characterizing dissolved organic matter fluorescence with parallel factor analysis: a tutorial. Limnol Oceanogr. Methods 6, 572–579.

Stets, E.G., Striegl, R.G., Aiken, G.R., 2010. Dissolved organic carbon export and internal cycling in small, headwater lakes. Global Biogeochem. Cycles 24, GB4008.

Stets, E.G., Striegl, R.G., Aiken, G.R., Rosenberry, D.O., Winter, T.C., 2009. Hydrologic support of carbon dioxide flux revealed by whole-lake carbon budgets. J. Geophys. Res. Biogeosciences 114, 1–14.

Stevenson, F., 1982. Humus Chemistry: Genesis, Composition, Reactions. John Wiley & Sons, Inc, New York, NY, USA.

Stoddard, J.L., Jeffries, D.S., Lükewille, A., Clair, T.A., Dillon, P.J., Driscoll, C.T., Forsius, M., Johannessen, M., Kahl, J.S., Kellogg, J.H., Kemp, A., Mannio, J., Monteith, D.T., Murdoch, P.S., Patrick, S., Rebsdorf, A., Skjelkvåle, B.L., Stainton, M.P., Traaen, T., van Dam, H., Webster, K.E., Wleting, J., Wilander, A., 1999. Regional trends in aquatic recovery from acidification in North America and Europe. Nature 401, 575–578.

Stubbins, A., Lapierre, J.F., Berggren, M., Prairie, Y.T., Dittmar, T., del Giorgio, P.A., 2014. What's in an EEM? Molecular signatures associated with dissolved organic fluorescence in boreal Canada. Environ. Sci. Technol. 48, 10598–10606.

Subalusky, A.L., Dutton, C.L., Rosi, E.J., Post, D.M., 2017. Annual mass drownings of the Serengeti wildebeest migration influence nutrient cycling and storage in the Mara River. Proc. Natl. Acad. Sci. U.S.A. 114, 7647–7652.

Suffet, I., MacCarthy, P., 1988. Aquatic humic substances. Advances in Chemistry. American Chemical Society, Washington, DC.

Sutfin, N.A., Wohl, E.E., Dwire, K.A., 2016. Banking carbon: a review of organic carbon storage and physical factors influencing retention in floodplains and riparian ecosystems. Earth Surf. Process. Landforms 41, 38–60.

Swain, F., 1963. Geochemistry of humus. In: Breger, I. (Ed.), Organic Geochemistry. Macmillan Company, New York, NY, USA, pp. 87–147.

Swift, R.S., 1996. Organic matter characterization. In: Sparks, D.L. (Ed.), Methods of Soil Analysis: Part 3 Chemical Methods. Madison, WI, pp. 1011–1069.

Syvitski, J.P., Vörösmarty, C.J., Kettner, A.J., Green, P., 2005. Impact of humans on the flux of terrestrial sediment to the global coastal ocean. Science 308, 376–380.

Tan, K.H., 2003. Humic Matter in Soil and the Environment: Principles and Controversies. CRC Press, New York.

Taylor, S., Gilbert, P.J., Cooke, D.A., Deary, M.E., Jeffries, M.J., 2019. High carbon burial rates by small ponds in the landscape. Front. Ecol. Environ. 17, 25–31.

Thorp, J.H., Delong, M.D., 2002. Dominance of autochthonous autotrophic carbon in food webs of heterotrophic rivers. Oikos 96, 543–550.

Thurman, E.M., 1985. Organic Geochemistry of Natural Waters, vol. 2. Springer Science and Business Media LLC, Hingham, MA.

Tobias, C.R., Böhlke, J.K., Harvey, J.W., 2007. The oxygen-18 isotope approach for measuring aquatic metabolism in high productivity waters. Limnol. Oceanogr. 52, 1439–1453.

Tranvik, L.J., 1988. Availability of dissolved organic carbon for planktonic bacteria in oligotrophic lakes of differing humic content. Microb. Ecol. 16, 311–322.

Tranvik, L.J., Cole, J.J., Prairie, Y.T., 2018. The study of carbon in inland waters-from isolated ecosystems to players in the global carbon cycle. Limnol. Oceanogr. Lett. 3, 41–48.

Tranvik, L.J., Downing, J.A., Cotner, J.B., Loiselle, S.A., Striegl, R.G., Ballatore, T.J., Dillon, P., Finlay, K., Fortino, K., Knoll, L.B., Kortelainen, P.L., Kutser, T., Larsen, S., Laurion, I., Leech, D.M., McCallister, S.L., McKnight, D.M., Melack, J.M., Overholt, E., Porter, J.A., Prairie, Y., Renwick, W.H., Roland, F., Sherman, B.S., Schindler, D.W., Sobek, S., Tremblay, A., Vanni, M.J., Verschoor, A.M., Wachenfeldt, E., Von, Weyhenmeyer, G.A., 2009. Lakes and reservoirs as regulators of carbon cycling and climate. Limnol. Oceanogr. 54, 2298–2314.

Uehlinger, U., 2006. Annual cycle and inter-annual variability of gross primary production and ecosystem respiration in a floodprone river during a 15-year period. Freshw. Biol. 51, 938–950.

Ulseth, A.J., Hall, R.O., 2015. Dam tailwaters compound the effects of reservoirs on the longitudinal transport of organic carbon in an arid river. Biogeosciences 12, 4345–4359.

Vachon, D., Sponseller, R.A., Karlsson, J., 2021. Integrating carbon emission, accumulation and transport in inland waters to understand their role in the global carbon cycle. Global Change Biol. 27, 719–727.

Vadeboncoeur, Y., Moore, M.V., Stewart, S.D., Chandra, S., Atkins, K.S., Baron, J.S., Bouma-Gregson, K., Brothers, S., Francoeur, S.N., Genzoli, L., Higgins, S.N., Hilt, S., Katona, L.R., Kelly, D., Oleksy, I.A., Ozersky, T., Power, M.E., Roberts, D., Smits, A.P., Timoshkin, O., Tromboni, F., Vander Zanden, M.J., Volkova, E.A., Waters, S., Wood, S.A., Yamamuro, M., 2021. Blue waters, green bottoms: benthic filamentous algal blooms are an emerging threat to clear lakes worldwide. Bioscience 71, 1011–1027.

Vallentyne, J.R., 1957. The molecular nature of organic matter in lakes and oceans, with lesser reference to sewage and terrestrial soils. J. Fish. Res. Board Canada 14, 33–82.

Vallentyne, J.R., 1962. Solubility and the decomposition of organic matter in nature. Arch. Hydrobiol. 58, 423–434.

Van de Bogert, M.C., Carpenter, S.R., Cole, J.J., Pace, M.L., 2007. Assessing pelagic and benthic metabolism using free water measurements. Limnol Oceanogr. Methods 5, 145–155.

Vannote, R.L., Minshall, G.W., Cummins, K.W., Sedell, J.R., Cushing, C.E., 1980. The river continuum concept. Can. J. Fish. Aquat. Sci. 37, 130–137.

Volk, C.J., Volk, C.B., Kaplan, L.A., 1997. Chemical composition of biodegradable dissolved organic matter in streamwater. Limnol. Oceanogr. 42, 39–44.

Von Wachenfeldt, E., Bastviken, D., Tranvik, L.J., 2009. Microbially induced flocculation of allochthonous dissolved organic carbon in lakes. Limnol. Oceanogr. 54, 1811–1818.

Von Wachenfeldt, E., Sobek, S., Bastviken, D., Tranvik, L.J., 2008. Linking allochthonous dissolved organic matter and boreal lake sediment carbon sequestration: the role of light-mediated flocculation. Limnol. Oceanogr. 53, 2416–2426.

Von Wachenfeldt, E., Tranvik, L.J., 2008. Sedimentation in boreal lakes—the role of flocculation of allochthonous dissolved organic matter in the water column. Ecosystems 11, 803–814.

Wallace, J.B., Eggert, S.L., Meyer, J.L., Webster, J.R., 1997. Multiple trophic levels of a forest stream linked to terrestrial litter inputs. Science 277, 102–104.

Wallace, J.B., Egger, S.L., Meyer, J.L., Webster, J.R., 2015. Stream invertebrate productivity linked to forest subsidies: 37 stream-years of reference and experimental data. Ecology 96, 1213–1228.

Wallin, M.B., Weyhenmeyer, G.A., Bastviken, D., Chmiel, H.E., Peter, S., Sobek, S., Klemedtsson, L., 2015. Temporal control on concentration, character, and export of dissolved organic carbon in two hemiboreal headwater streams draining contrasting catchments. J. Geophys. Res. Biogeosciences 120, 832–846.

Walter, K.M., Smith, L.C., Chapin, F.S., 2007. Methane bubbling from northern lakes: present and future contributions to the global methane budget. Philos. Trans. A. Math. Phys. Eng. Sci. 365, 1657–1676.

Wang, X., Ma, H., Li, R., Song, Z., Wu, J., 2012. Seasonal fluxes and source variation of organic carbon transported by two major Chinese Rivers: the Yellow River and Changjiang (Yangtze) River. Global Biogeochem. Cycles 26, GB2025.

Webb, J.R., Santos, I.R., Maher, D.T., Finlay, K., 2019. The importance of aquatic carbon fluxes in net ecosystem carbon budgets: a catchment-scale review. Ecosystems 22, 508–527.

Weber, J., Chen, Y., Jamroz, E., Miano, T., 2018. Preface: humic substances in the environment. J. Soils Sediments 18, 2665–2667.

Webster, J.R., 2007. Spiraling down the river continuum: stream ecology and the U-shaped curve. J. N. Am. Benthol. Soc. 26, 375–389.

Webster, J.R., Benfield, E.F., Ehrman, T.P., Schaeffer, M.A., Tank, J.L., Hutchens, J.J., D'Angelo, D.J., 1999. What happens to allochthonous material that falls into streams? A synthesis of new and published information from Coweeta. Freshw. Biol. 41, 687–705.

Weishaar, J.L., Aiken, G.R., Bergamaschi, B.A., Fram, M.S., Fujii, R., Mopper, K., 2003. Evaluation of specific ultraviolet absorbance as an indicator of the chemical composition and reactivity of dissolved organic carbon. Environ. Sci. Technol. 37, 4702–4708.

Wetzel, R.G., 1990. Land-water interfaces: metabolic and limnological regulators. Verh. Internat. Verein. Limnol. 24, 6–24.

Wetzel, R.G., 1995. Death, detritus, and energy flow in aquatic ecosystems. Freshw. Biol. 33, 83–89.

Wetzel, R.G., 1996. Benthic algae and nutrient cycling in lentic freshwater ecosystems. In: Stevenson, R.J., Bothwell, M.L., Lowe, R.L. (Eds.), Algal Ecology: Freshwater Benthic Ecosystems. Academic Press, San Diego, pp. 641–667.

Wetzel, R.G., 2001. Limnology: Lake and River Ecosystems. Academic Press, San Diego, p. 1006.

Wetzel, R.G., 2003. Dissolved organic carbon: detrital energetics, metabolic regulators, and drivers of ecosystem stability of aquatic ecosystems. In: Findlay, S., Sinsabaugh, R. (Eds.), Dissolved Organic Matter in Aquatic Ecosystems. Academic Press, San Diego.

Wetzel, R.G., Manny, B.A., 1972. Decomposition of dissolved organic carbon and nitrogen compounds from leaves in an experimental hard-water stream. Limnol. Oceanogr. 17, 927–931.

Wetzel, R.G., Rich, P.H., Miller, M.C., Allen, H.L., 1972. Metabolism of dissolved and particulate detrital carbon in a temperate hard-water lake. Mem. Ist. Ital. Idrobio. 29, 185–243.

Weyhenmeyer, G.A., Kosten, S., Wallin, M.B., Tranvik, L.J., Jeppesen, E., Roland, F., 2015. Significant fraction of CO_2 emissions from boreal lakes derived from hydrologic inorganic carbon inputs. Nat. Geosci. 8, 933–939.

Williamson, C.E., Neale, P.J., Hylander, S., Rose, K.C., Figueroa, F.L., Robinson, S.A., Häder, D.-P., Wängberg, S.-Å., Worrest, R.C., 2019. The interactive effects of stratospheric ozone depletion, UV radiation, and climate change on aquatic ecosystems. Photochem. Photobiol. Sci. 18, 717–746.

Wilson, H.F., Xenopoulos, M.A., 2009. Effects of agricultural land use on the composition of fluvial dissolved organic matter. Nat. Geosci. 2, 37–41.

Winslow, L.A., Zwart, J.A., Batt, R.D., Dugan, H.A., Woolway, R.I., Corman, J.R., Hanson, P.C., Read, J.S., 2016. LakeMetabolizer: an R package for estimating lake metabolism from free-water oxygen using diverse statistical methods. Inland Waters 6, 622–636.

Wohl, E., Dwire, K., Sutfin, N., Polvi, L., Bazan, R., 2012. Mechanisms of carbon storage in mountainous headwater rivers. Nat. Commun. 3, 1263.

Wohl, E., Hall, R.O., Lininger, K.B., Sutfin, N.A., Walters, D.M., 2017. Carbon dynamics of river corridors and the effects of human alterations. Ecol. Monogr. 87, 379–409.

Wohl, E., Kramer, N., Ruiz-Villanueva, V., Scott, D.N., Comiti, F., Gurnell, A.M., Piegay, H., Lininger, K.B., Jaeger, K.L., Walters, D.M., Fausch, K.D., 2019. The natural wood regime in rivers. Bioscience 69, 259–273.

Woodwell, G.M., Rich, P.H., Hall, C.A.S., 1973. The carbon cycle of estuaries. In: Woodwell, G.M., Pecan, E.V. (Eds.), Carbon and the Biosphere. Proc. 24th Brookhaven Symposium in Biology. U.S. Atomic Energy Commission. Symp. Ser. CONF-720510, Brookhaven, NY, pp. 221–240.

Woodwell, G.M., Whittaker, R.H., 1968. Primary production in terrestrial ecosystems. Am. Zool. 8, 19–30.

Wurtsbaugh, W.A., Paerl, H.W., Dodds, W.K., 2019. Nutrients, eutrophication and harmful algal blooms along the freshwater to marine continuum. Wiley Interdisciplinary Reviews: Water 6, e1373.

Xenopoulos, M.A., Barnes, R.T., Boodoo, K.S., Butman, D., Catalán, N., D'Amario, S.C., Fasching, C., Kothawala, D.N., Pisani, O., Solomon, C.T., Spencer, R.G.M., Williams, C.J., Wilson, H.F., 2021. How humans alter dissolved organic matter composition in freshwater: relevance for the Earth's biogeochemistry. Biogeochemistry 154, 323–348.

Xenopoulos, M.A., Lodge, D.M., Frentress, J., Kreps, T.A., Bridgham, S.D., Grossman, E., Jackson, C.J., 2003. Regional comparisons of watershed determinants of dissolved organic carbon in temperate lakes from the Upper Great Lakes region and selected regions globally. Limnol. Oceanogr. 48, 2321–2334.

Yang, X., Pavelsky, T.M., Allen, G.H., 2020. The past and future of global river ice. Nature 577, 69–73.

Yang, Q., Zhang, X., Xu, X., Asrar, G.R., Smith, R.A., Shih, J.S., Duan, S., 2016. Spatial patterns and environmental controls of particulate organic carbon in surface waters in the conterminous United States. Sci. Total Environ. 554–555, 266–275.

Yvon-Durocher, G., Allen, A.P., Bastviken, D., Conrad, R., Gudasz, C., St-Pierre, A., Thanh-Duc, N., Del Giorgio, P.A., 2014. Methane fluxes show consistent temperature dependence across microbial to ecosystem scales. Nature 507, 488–491.

Yvon-Durocher, G., Hulatt, C.J., Woodward, G., Trimmer, M., 2017. Long-term warming amplifies shifts in the carbon cycle of experimental ponds. Nat. Clim. Change 7, 209–213.

Zarnetske, J.P., Bouda, M., Abbott, B.W., Saiers, J., Raymond, P.A., 2018. Generality of hydrologic transport limitation of watershed organic carbon flux across ecoregions of the United States. Geophys. Res. Lett. 45, 11702–11711.

Zhang, M., Xiao, Q., Zhang, Z., Gao, Y., Zhao, J., Pu, Y., Wang, W., Xiao, W., Liu, S., Lee, X., 2019. Methane flux dynamics in a submerged aquatic vegetation zone in a subtropical lake. Sci. Total Environ. 672, 400–409.

Zhou, Y., Davidson, T.A., Yao, X., Zhang, Y., Jeppesen, E., de Souza, J.G., Wu, H., Shi, K., Qin, B., 2018. How autochthonous dissolved organic matter responds to eutrophication and climate warming: evidence from a cross-continental data analysis and experiments. Earth-Science Rev. 185, 928–937.

Zipper, S.C., Hammond, J.C., Shanafield, M., Zimmer, M., Datry, T., Jones, C.N., Kaiser, K.E., Godsey, S.E., Burrows, R.M., Blaszczak, J.R., Busch, M.H., Price, A.N., Boersma, K.S., Ward, A.S., Costigan, K., Allen, G.H., Krabbenhoft, C.A., Dodds, W.K., Mims, M.C., Olden, J.D., Kampf, S.K., Burgin, A.J., Allen, D.C., 2021. Pervasive changes in stream intermittency across the United States. Environ. Res. Lett. 16, 084033.

Zwart, J.A., Craig, N., Kelly, P.T., Sebestyen, S.D., Solomon, C.T., Weidel, B.C., Jones, S.E., 2016. Metabolic and physiochemical responses to a whole-lake experimental increase in dissolved organic carbon in a north-temperate lake. Limnol. Oceanogr. 61, 723–734.

Zwart, J.A., Sebestyen, S.D., Solomon, C.T., Jones, S.E., 2017. The influence of hydrologic residence time on lake carbon cycling dynamics following extreme precipitation events. Ecosystems 20, 1000–1014.

CHAPTER

29

Wetlands

Shuqing An[1,2], Shenglai Yin[2,3], Jos T.A. Verhoeven[4] and Nasreen Jeelani[1]

[1]School of Life Science and Institute of Wetland Ecology, Nanjing University, Nanjing, China [2]Nanjing University Ecological Research Institute of Changshu, Suzhou, China [3]College of Life Science, Nanjing Normal University, Nanjing, China [4]Ecology and Biodiversity, Department of Biology, Utrecht University, Utrecht, the Netherlands

OUTLINE

I. Wetlands and their global distribution

A. Wetland definitions

Scientists, policymakers, and lawyers have attempted to define wetlands from different professional angles. Scientists are more likely to focus on the processes, functions, and ecological characteristics, whereas policymakers and lawyers are prone to focus on the services, regulations, and preventions for modifying or exploiting wetland resources.

Scientists generally accept a comprehensive definition that emphasizes ecological characteristics such as vegetation and environmental condition (Cowardin, 1979):

> *"Wetlands are lands transitional between terrestrial and aquatic systems where the water table is usually at or near the surface or the land is covered by shallow water. Wetlands must have one or more of the following three attributes: (1) at least periodically, the land supports predominantly hydrophytes; (2) the substrate is predominantly undrained hydric soil; and (3) the substrate is saturated with water or covered by shallow water at some time during the growing season of each year."*

The Convention on Wetlands of International Importance especially as Waterfowl Habitat, commonly

Wetzel's Limnology, Fourth Edition
https://doi.org/10.1016/B978-0-12-822701-5.00029-X

known as the Ramsar Convention, also provided a definition that has been broadly accepted by international conservationists and policymakers (Keddy, 2010):

> *"Areas of marsh, fen, peatland, or water, whether natural or artificial, permanent or temporary, with static water or flowing, fresh, brackish, or salt including areas of marine water, the depth of which at low tide does not exceed six meters."*

Thus wetlands can be seen as the collection of ecosystems located in the transition zones between land and open water. They commonly possess the characteristics of both aquatic and terrestrial ecosystems, especially with three principal components, which are the hydrological conditions of permanent or periodical inundation, unique physicochemical environment, and wetland biota (Fig. 29-1).

Hydrological processes determine the physicochemical conditions in a wetland, creating unique and fluctuating redox potentials in the sediment. Both hydrological processes and physicochemical conditions also determine the composition of local biota because flora and fauna must tolerate these conditions. The biota can also influence hydrological processes and the physicochemical environment. For example, vegetation can change the water flow and inundation by modifying wetland topography. Similarly, plants can release oxygen through their roots in submerged soils, thereby altering the redox conditions in the sediment and altering biogeochemical cycles.

B. Wetland classification

Wetlands can be classified into several large groups according to different criteria, such as inland and coastal wetlands, or freshwater and brackish wetlands, or natural and human-made wetlands. These general classifications can be further subdivided with one or more factors such as location, water salinity, vegetation type, or hydrologic characteristics. According to the Ramsar inventory, the main wetlands types can be classified as *riverine*, *lake*, *marshes*, *peatlands*, *coastal*, and *human-made wetlands* (Fig. 29-2; Gardner et al., 2018).

i. Riverine wetlands

Riverine wetlands are located in channels that are either naturally or artificially created and periodically or continuously contain flowing water. Riverine wetlands recharge lakes and carry fertile sediments that enrich floodplains and marshes. Water velocities are important variables influencing the physical characteristics and biotic structures in riverine wetlands. With the fast velocity and shallow water typical for lower-order streams (see Chapter 4), rocks are often the main substrates and dissolved oxygen is near saturation (see Chapter 11). In contrast, riverine wetlands in high-order river stretches (see Chapter 4) are characterized by slow water velocity and floodplains with frequent inundations with sediments of sand and/or mud. The flora and fauna of these communities are adapted to life in these different substrates.

Floodplains, which are one of the most important types of riverine wetlands, are usually adjacent to river channels, are at least occasionally flooded, and are often characterized by abundant vegetation such as trees, shrubs, moss, and emergent aquatic plants. Floodplains can improve water quality, recharge groundwater reserves, and reduce the severity and frequency of floods. Since flooding can enhance the breakdown of organic matter,

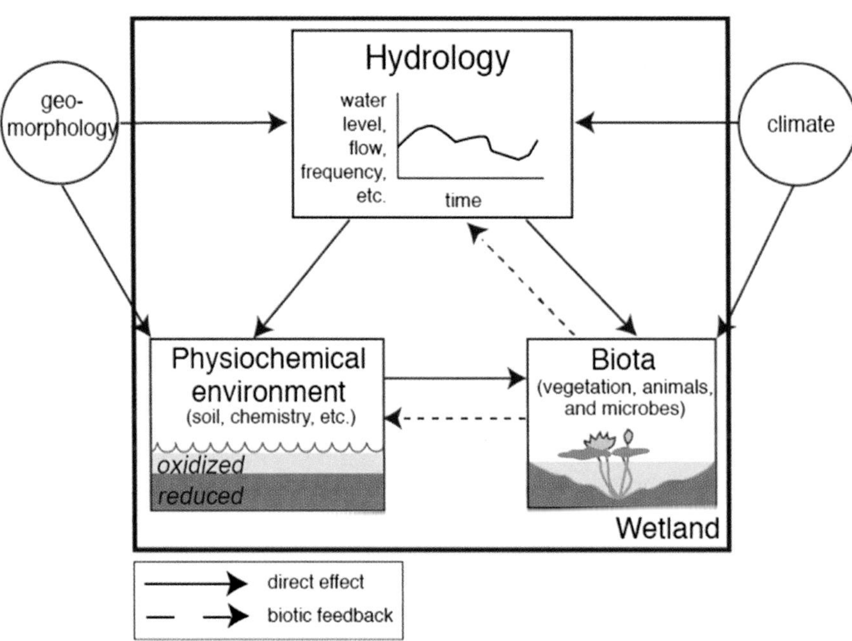

FIGURE 29-1 The three principal components of wetlands (Mitsch and Gosselink, 2015).

FIGURE 29-2 Main wetland types. (a) Riverine wetland in Karnataka, India; (b) Lake wetland in Gifu Prefecture, Japan; (c) Freshwater marsh in Washington, USA; (d) Peatland in Scotland, UK; (e) Coastal wetland in Massachusetts, USA; (f) Human-made wetland in Hebei, China. ((a)–(e) are from Wikipedia and (f) is provided by Shuqing An.)

floodplain soils are typically fertile, and so floodplains are often farmed in many parts of the world (Janse et al., 2008).

ii. Lake wetlands

Lake wetlands are usually located in areas where water depths are shallow and gradually become deeper to contain standing water or slowly flowing water. Lake wetlands are important sources of drinking water, food, and irrigation for agriculture, and they can regulate river flow, trap sediments and nutrients, and regulate local climate. Generally, two alternative states can be found among shallow lake wetlands, macrophyte-dominated lakes and phytoplankton-dominated lakes (see Chapter 26). The switching between these two states largely depends on the loading of nutrients such as nitrogen and phosphorus. When a macrophyte-dominated lake receives nutrients in quantities that exceed a critical value, eutrophication happens, and the lake may switch from a clearwater state with submerged macrophytes to a turbid phytoplankton-dominated lake. This results in many management challenges, as returning to the clearwater state often requires a long period of lower critical nutrient loading (Janse et al., 2008).

iii. Marshes

Marshes are typically depressional wetlands with mineral soils where groundwater discharge or surface water inflow leads to temporary or permanently wet conditions. This inundation may result in anoxic conditions and slow decomposition, thereby accumulating a considerable amount of organic matter. Marshes are often vegetated by herbaceous or shrubby plants that are adapted to the saturated soil condition.

iv. Peatlands

Peatlands (or sometimes referred to as *mires*) occur mostly in colder northern areas and in tropical areas with very high rainfall. They cover approximately 2–4% of the world's land area (Xu et al., 2018) and usually are vegetated by *Sphagnum* mosses, sedges, and shrubs. Peatlands contain the greatest terrestrial carbon pool on Earth, since their waterlogged condition slows down considerably the decomposition of dead plants and mosses, and thereby accumulate a thick layer of organic peat. It has been estimated that the global peatland soils contain approximately 6×10^{11} to 7×10^{11} tons of carbon (Turetsky et al., 2015), which exceeds that of global vegetation ($\sim 5.6 \times 10^{11}$ tons) and approaches the total amount of carbon in the atmosphere ($\sim 8.5 \times 10^{11}$ tons). This carbon storage indicates the importance of peatlands in regulating the global carbon cycle (Dise, 2009).

Peatlands and *mires* are collective terms that can be further classified into *bogs* (or *moss* in British definition) or *fens*, differentiated according to water and nutrient sources, as well as by rather subtle floristic variations. "Mire" is equivalent to the Swedish *myr* and the German *moor*. All mires are in areas where the groundwater table is permanently at or near the surface of the ground. Mires are commonly separated into *minerotrophic mires* (or fens), in which water and nutrients are supplied from groundwater, surface sources, and atmospheric precipitation. The water of fens tends to be base rich ($pH > 5.5$) and commonly supports herbaceous vegetation. In *ombrotrophic mires* (bogs) water-table recharge and nutrient inputs originate entirely from atmospheric precipitation. Water is of very low salinity and tends to be acidic ($pH < 5.5$). A *carr* refers to a mire or portion of a mire that is dominated by woody shrubs or trees among the herbaceous vegetation on a relatively stable peat mass (e.g., a fen carr could be dominated by trees, such as species of *Alnus*, *Fraxinus*, and *Betula*, with a dense understory of fen herbaceous plants).

v. Coastal wetlands

Coastal wetlands, such as mangroves and *salt marshes*, are crucial habitats for many waterbird species to rest and feed during seasonal migration, but they can also provide commercial opportunities to humans. *Mangroves* only occur in the intertidal zone in tropical and subtropical climates. Like peatlands, mangroves store large amounts of carbon due to their high productivity and low decomposition rates. Mangrove species are well recognized by their tangled roots, which can protect shorelines from coastal erosion and storms, and can create suitable habitats for various aquatic species such as oysters, mussels, mudskippers, and lemon sharks. *Coastal salt marshes* are located in the intertidal zones in nontropical climates, playing important roles in protecting the upland area by reducing the effects of storms and flooding.

vi. Human-made wetlands

Apart from the natural wetlands described above, *human-made wetlands* are another important category. The surface area of human-made wetlands recently accounted for at least 12% of the total global wetlands extent (Gardner et al., 2018) and continues to increase. These wetlands are designed and created by integrating and optimizing water flows (subsurface and surface flow), and vegetation and microbiome composition, to fulfill certain types of human needs. In general, the human-made wetlands can be classified into three major categories: *constructed wetlands*, *food-producing wetlands*, and *landscape or buffer wetlands*. Constructed wetlands are used primarily to purify industrial wastewater, domestic wastewater, and/or surface runoff from agricultural land. The food-producing wetlands are developed for agricultural or aquacultural use and would include rice paddies, fishponds, and certain reservoirs. Meanwhile, landscape wetlands are constructed to provide a broader range of services, including water quality improvement, biodiversity conservation, and human recreation (An and Wang, 2014).

C. Global distribution and extent

Wetlands can be found all over the world, except for Antarctica, but most are located in boreal and tropical regions (Fig. 29-3). Asia and North America have the largest areas of wetlands, which constitute 31.8% and 27.1% of the global wetlands area, respectively, followed by Latin America and the Caribbean (15.8%), Europe (12.5%), Africa (9.9%), and Oceania (2.9%). The extent of global wetlands was estimated to be more than 12.1 million km^2, of which 54% of the area was permanently inundated and 46% seasonally inundated. Marshes and peatlands contribute 53% of the total inland wetlands, together with 33% lake wetlands and 5% riverine wetlands, whereas another 9% are forested wetlands. For the coastal and marine wetlands, salt marshes and mangroves account for 34% and 8% of the total area; other types include 17% coral reefs, 11% seagrass beds, and 2% coastal deltas (Davidson et al., 2018).

D. Biogeochemical cycles

Biogeochemical cycles in wetlands are fundamentally affected by hydrological conditions, such as fluctuating water levels. The top layer of the wetland shows periodic changes in water saturation, which can result in alterations of oxic and anoxic conditions. This creates dynamic redox conditions with consequences for the biogeochemical cycles of nitrogen, iron, sulfur, carbon, and phosphorus, as well as other elements (see Chapters 10 and 14–16).

Water enters wetlands from three main sources: precipitation, surface water flow, and groundwater flow. Direct rainfall is an important source and for peat bogs, it is the only source of water, and therefore these systems are very nutrient poor and acidic. Surface water usually originates from drainage of rain toward streams or melting snow and glaciers and is driven by gravity toward the wetlands. Groundwater can discharge to a wetland surface, but wetlands can also recharge the groundwater through downward flows.

Fluctuation of the water table determines the biogeochemical cycles in wetlands by altering the oxygen availability in the sediment. When the water table rises, the sediment becomes subject to anaerobic conditions, whereas when the water table falls, the sediment experiences aerobic conditions. The decline of oxygen availability reduces the rates of decomposition and mineralization of organic matter, which releases nutrients such as nitrogen and phosphorus for carbon fixation (Verhoeven, 2009).

Although permanent or periodical inundation slows down the decomposition rate, anaerobic bacteria will continue to mineralize the easily degradable organic matter by using a suite of electron acceptors to replace

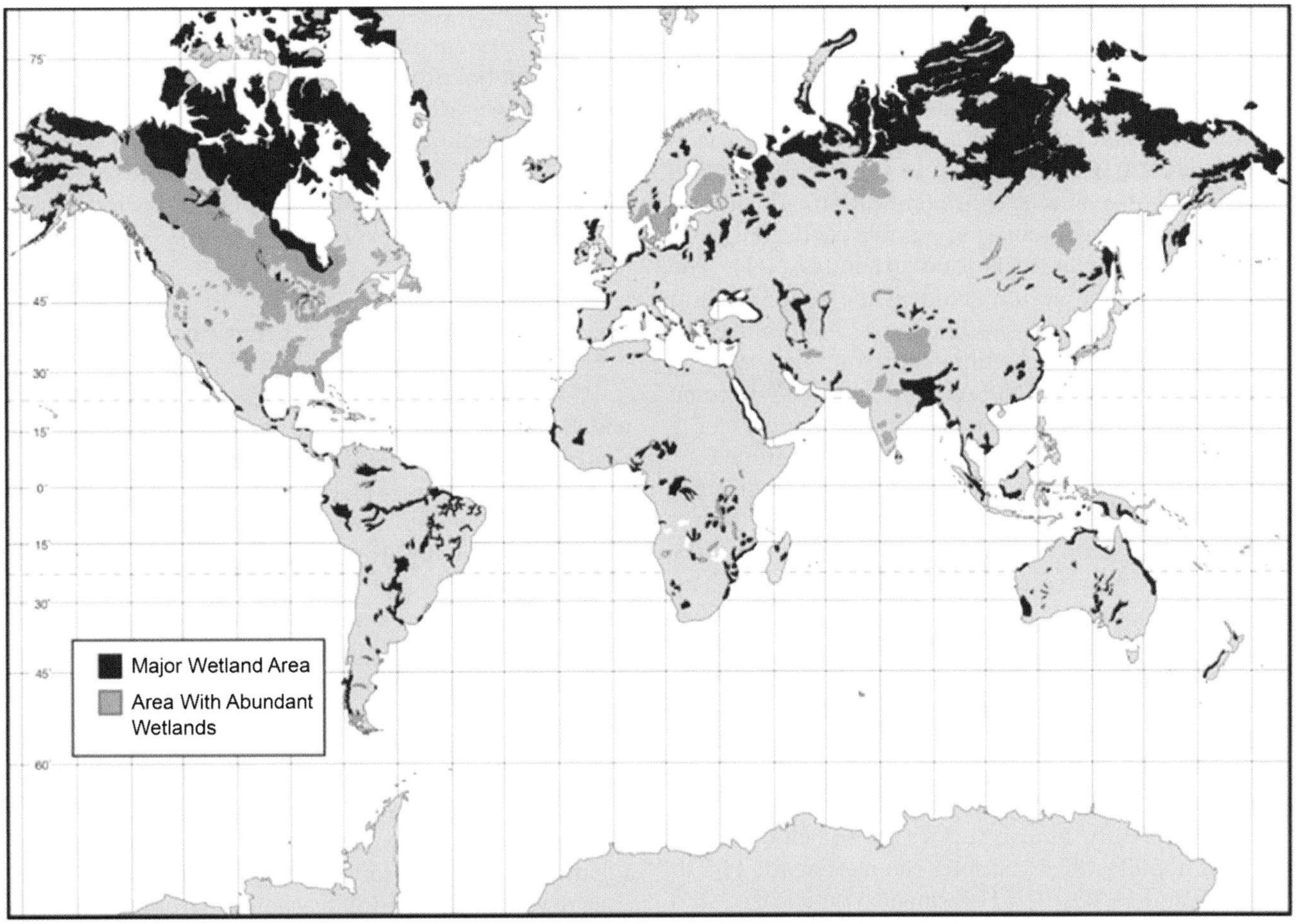

Major Wetland Area
Area With Abundant Wetlands

FIGURE 29-3 General distribution of global wetlands (Mitsch and Gosselink, 2015).

oxygen. The various electron acceptors are, in theory, used successively while the redox potential decreases (Fig. 29-4).

When the inundation occurs, oxygen is quickly consumed in the sediment, and fermentation of organic matter breaks down the high-molecular-weight carbohydrates into low-molecular-weight organic complexes, while the redox potential gradually drops. As soon as the potential has dropped to 250 mV, nitrate becomes the main electron acceptor. Organic matter is oxidized,

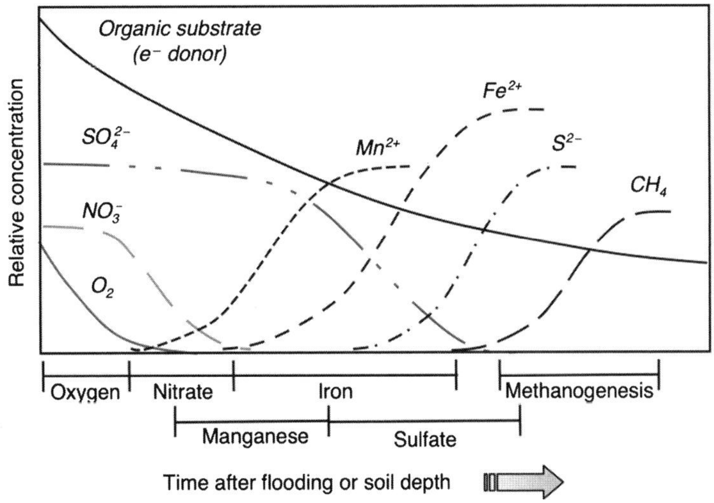

FIGURE 29-4 Relative sequence of electron acceptors as a function of time after flooding a soil (Reddy and DeLaune, 2008).

while nitrate is reduced to N_2O and N_2. This process is called denitrification (Chapter 14) and often contributes to water quality improvement.

Nitrogen in wetlands becomes available by nitrogen fixation by certain blue-green algae and N-fixing bacteria (Chapter 14). Decomposition of organic matter also makes nitrogen available. Ammonification, which happens under both anaerobic and aerobic conditions, consumes the organic nitrogen and generates NH_4^+. The NH_4^+ can either be accumulated in the sediment and absorbed by plants or be oxidized to NO_3^- by nitrifying bacteria (Mitsch and Gosselink, 2015).

After the depletion of nitrate, Mn^{4+} and Fe^{3+} become the main electron acceptors at approximately +225 to −110 mV. The Mn^{4+} is reduced slightly before Fe^{3+} and generates Mn^{2+}. The reduction of Fe^{3+} has substantial consequences for the availability of phosphorus, because it can bind with phosphates more strongly than Fe^{2+}, so this reduction decreases the phosphate sorption and increases its availability (Chapter 15).

The depletion of Fe^{3+} decreases the redox potential to −100 to −200 mV, when sulfate becomes the main electron acceptor. Sulfate can be delivered to wetlands either by atmospheric deposition, surface runoff, or organic litter from the surrounding plants. Some anaerobic bacteria can reduce the sulfate to gaseous H_2S. When the H_2S diffuses to the upper layer, chemoautotrophic and photosynthetic bacteria can oxidize the H_2S to sulfur (S) and H_2SO_4. The H_2S can also be released into the atmosphere as well, causing the characteristic wetland smell of "rotten eggs."

Anaerobic respiration can take place in inundated wetland sediments, where the mineralization to CO_2 and methanogenesis are the main processes (Yin et al., 2015). While this mineralization happens when the redox potential is relatively high (250 mV), methanogenesis only happens under anaerobic conditions when the redox potential drops to <−200 mV, and the other electron acceptors are depleted. Methanogens use CO_2 and methyl compounds to generate CH_4, which can then diffuse into the upper layers of the wetland soil, where it can be oxidized by methanotrophic bacteria or be released into the atmosphere (Fig. 29-5).

II. Wetland functions

Wetlands offer many valuable ecosystem services, which can be summarized into four general categories: *regulating services*, *provisioning services*, *cultural services*, and *supporting services* (Table 29-1).

Wetlands regulate a series of essential ecological processes and life-supporting functions. For example, wetlands affect climate by sequestering atmospheric carbon as peat layers and can affect microclimate by evaporation and heat absorption. Moreover, wetlands are widely recognized as furthering water purification, as they can decrease the nitrogen and phosphorus concentrations in domestic and agricultural wastewater by either denitrification, absorption to build biomass, or storage in the sediment. In fact, the building of *human-made wetlands* (i.e., constructed wetlands) is becoming a common approach to remove pollutants such as agriculture runoff and landfill leachates (An and Verhoeven, 2019). Regulating services also include flow regulation, flood mitigation, coastal protection, and wastewater purification (Gardner et al., 2018).

Wetlands provide material benefits for humans in many other ways. Wetland primary production rates are among the highest in the world, especially in

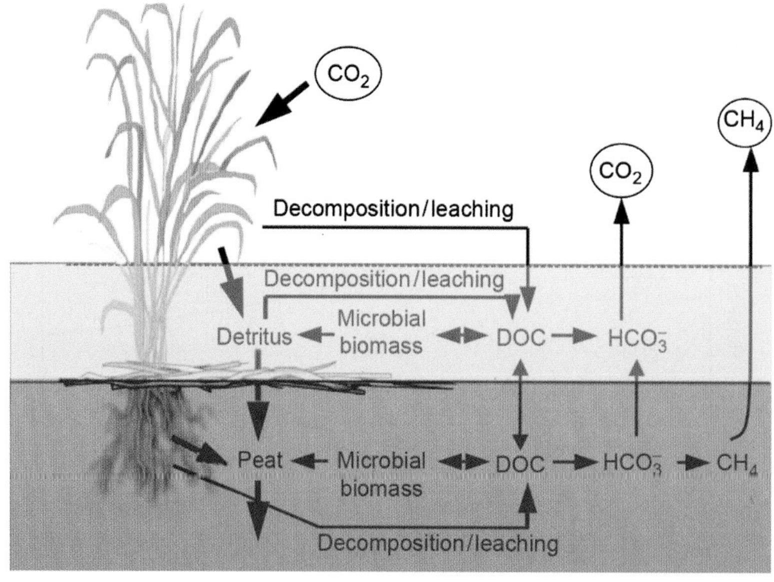

FIGURE 29-5 Schematic diagram of the carbon biogeochemical cycle in wetlands (Reddy and DeLaune, 2008).

TABLE 29-1 Some of the Major Ecosystem Services From Wetlands[a]

	Inland wetlands				Coastal/marine wetlands		Human-made wetlands		
Wetland types/services	**Riverine**	**Lake**	**Peatland**	**Marsh**	**Saltmarsh**	**Mangrove**	**Reservoir**	**Rice paddy**	**Aqua ponds**
Regulating services									
Climate	L	H	H	H	H	H	M	L	na
Hydrological	H	H	M	M	M	H	H	M	na
Pollution control	H	M	M	H	H	H	L	L	na
Erosion protection	M	M	M	M	M	H	L	M	na
Natural hazards	M	H	M	H	H	H	L	L	na
Provisioning services									
Food	H	H	H	H	H	H	M	H	H
Freshwater	H	H	L	M	L	na	M	na	na
Fiber and fuel	M	M	H	H	L	H	L	na	L
Biochemical products	L	?	?	L	L	L	?	na	?
Genetic materials	L	L	?	?	L	L	L	L	L
Cultural services									
Spiritual and inspirational	M	H	M	M	?	L	M	L	na
Recreational	H	H	L	M	?	?	H	L	na
Aesthetic	M	M	L	M	M	M	H	M	na
Educational	H	H	M	M	L	L	H	L	L
Supporting services									
Biodiversity	H	H	H	H	M	M	M	M	L
Soil formation	H	L	H	H	M	M	L	M	na
Nutrient cycling	H	L	H	H	M	M	L	M	L
Pollination	L	L	L	L	L	M	L	L	na

[a]*From Gardner et al. (2018).*
H = high values; *L* = low values; *M* = medium values; *na* = not applicable; and ? = not known.

swamps and marshes, with comparable rates to tropical forests (Gardner et al., 2018). For example, annual fishery harvests from inland wetlands increased from 2 million tons in 1950 to over 11.6 million tons in 2012 (Gardner et al., 2018). In fact, the aquaculture in the wetlands is the fastest growing food production sector; the contribution to global fish production rose from 45.7% in 2008 to 62.5% in 2018 (FAO, 2020). Wetlands also provide crops and areas for hunting and other types of food gathering.

Wetlands also support a wide spectrum of cultural services by providing areas for recreation and ecotourism. For example, in China a total of 14 Ramsar sites have initiated ecotourism and ecoeducation activities. In addition to protecting the exceptional biodiversity of these wetlands, they have attracted approximately 14.5 million visitors and brought 2.7 billion RMB yuan of income to local communities in 2019 alone (www.forestry.gov.cn). Wetlands also have important cultural and spiritual significance, such as the sacred lakes in the Himalayan region and terraces of rice paddies in Asia.

Wetlands are critical for maintaining biodiversity by providing habitat for a wide variety of animals. Importantly, wetlands provide the major habitat for most of the world's waterbirds, as feeding and breeding grounds, as well as key habitat for migratory species. Furthermore, wetlands provide habitats for approximately 8226 species of amphibians (McDonough, 2014), especially for species that depend on rivers and streams. Clearly, wetlands are crucial habitats for sustaining biodiversity (Gardner et al., 2018).

III. Wetland destruction

Despite their clear importance, natural wetlands are experiencing extensive declines around the world, with approximately 35% of wetland surface area lost between 1970 and 2015 (Gardner et al., 2018). The main drivers for this decline include drainage for conversion to other types of land use, excavation, deforestation, and biological invasions, as discussed further below. The rate of decline is three times the rate of the global loss of forest cover (Gardner et al., 2018).

A. Drainage and conversion

Wetlands provide approximately 42,000 km^3 of renewable water resources every year, of which nearly 93% has been extracted for human use, most of which (70%) is for agriculture (FAO, 2011). The intensive reclamation of wetlands for human and agricultural use can significantly lower the regional water table and drain the wetland. Approximately half of the world's coastal wetlands have been converted to agriculture, aquaculture, urban, or industrial land. Unfortunately, this trend in wetland destruction continues to grow with the increase in human population size and intensification of agricultural land use.

B. Excavation of peatlands

Global forested and tropical peatlands have been reduced by 25–28% during 2007–2015 (Gardner et al., 2018), due mainly to drainage, flooding, and increased fire events, but most importantly by peat excavation. A striking example is the Zoigê peatland on the Qinghai–Tibet Plateau, where 19 Gt of carbon are stored in the peat layers (Yang et al., 2017). Local people, however, have begun to excavate the peat for fuel use as the growing population has high fuel demands for factories, schools, and apartment dwellings. In Zoigê county the peatland area shrank by 30% from 1974 to 2007. In Hongyuan county, next to Zoigê, about 9500 tons of peat are consumed for heating and cooking every year. Given the slow rate of peat regeneration, these losses cannot be easily recovered (Yang et al., 2017).

C. Mangrove deforestation

Mangrove forests, which are located in tropical and subtropical regions, support high biodiversity and store large amounts of carbon (Donato et al., 2011). However, at least 30% of the global mangrove area has been deforested for the aquaculture industry over the past 30 years, most of which has occurred in Southeast Asian countries. For example, in Indonesia and the Philippines, where the most extensive conversion has occurred, aquaculture has accounted for 48.6% and 36.7% of their total mangrove loss from 2000 to 2012 (Richards and Friess, 2016). In addition, rice farming and oil palm plantations are two other dominant drivers for mangrove deforestation, which have caused up to 22% and 16% of mangrove loss, respectively, in Southeast Asia from 2000 to 2012 (Richards and Friess, 2016).

D. Biological invasions

Biological invasions are a global problem and a major threat to all ecosystems. Although the mechanisms of biological invasion are not always fully understood, the list of invasive species is growing.

Wetlands provide many examples of issues related to invasive species. For example, the list of invasive species impacting wetlands in China includes plants (e.g., *Eichhornia crassipes*, *Alternanthera philoxeroides*, and *Spartina alterniflora*), mammals (e.g., *Myocastor coypus*, *Ondatra zibethicus*), fishes (e.g., *Oreochromis mossambicus*), insects (e.g., *Lissorhoptrus oryzophilus*), and other invertebrates (e.g., *Procambarus clarkii*) and amphibians (e.g., American bullfrog, *Rana catesbeiana*).

The invasive species may be transported to new habitats by natural drivers, such as wind, flood, or currents, or be introduced unintentionally attached to visitors or boats, or can be intentionally introduced as target species for restoration or management purposes. For example, the water hyacinth *Eichhornia crassipes* was introduced into China in the last century as an ornamental plant, but its rapid growth and expansion caused many negative consequences for wetlands, causing massive die-off of the local flora and fauna. This invasion has also negatively impacted agriculture, fish production, and navigation. A second example is the smooth cordgrass *Spartina alterniflora*, which was introduced to the coastline of China for erosion control, soil amelioration, and dike protection in 1979 (Yang et al., 2016; Xia et al., 2020). It has now invaded almost the entire eastern coastline, colonized 249.1 km^2 in seven Ramsar sites, and has drastically changed the structure and composition of the local coastal wetlands.

IV. Climate change

A. Response and feedback

Climate change has also been identified as one of the major threats to wetlands. Rising temperatures, as well as changes in precipitation intensity and frequency, along with changes in evaporation rates, affect hydrological processes and biogeochemical cycles, as well as carbon exchange and community composition in wetlands. These changes can lead to wetland degradation, with concomitant losses in ecosystem function and services. For example, climate change might promote phosphorus release from wetland sediment by 30% in the coming three decades (Ockenden et al., 2017), which would further impact water purification.

Climate change can also shift wetlands from carbon sinks to sources. This is especially true in permafrost regions where the frozen ground impedes bacterial activity and keeps the carbon stored in the frozen soil. Rising temperatures, however, can lead to permafrost thaw, which increases water and oxygen availability in the soil, and accelerates microbial activities, thereby releasing large amounts of CO_2 and CH_4 into the atmosphere. This consequence has caused a great deal of concern because of the positive feedback between temperature rises and greenhouse gas release.

There are, however, still many uncertainties regarding the consequences of climate change and wetlands. For example, climate change can cause vegetation succession from a *Sphagnum*-dominated wetland to vascular plants, if higher temperature coincides with more precipitation (Desta et al., 2012). Therefore, in this case, the peatland may maintain its ecosystem service as a carbon sink by increasing overall primary production and carbon storage. Another example of the uncertainties of climate change effects is the extent of change in the main water sources (i.e., the Yellow River, the Yangtze River, and the Lancang River) of the Qinghai–Tibet Plateau. From the early 1990s to 2004, the wetland area declined by 0.8%, whereas in the following 8 years, the trend reversed itself with a 2% increase in area. Researchers concluded that this reversal was linked to a climate transition from warm-dry to warm-wet conditions (Tong et al., 2014).

B. Climate change mitigation and wetlands

Although wetlands may release large amounts of carbon with warming temperatures, improving wetland conservation and management can play an important role in conserving or improving a wetland's capacity for carbon storage and in mitigating climate change. For example, there is growing evidence that the conservation of *blue carbon*, that is, carbon captured by coastal wetlands especially mangroves, salt marshes and seagrasses, and *teal carbon*, that is, carbon captured by inland freshwater wetlands, can contribute to global carbon management (Nahlik and Fennessy, 2016). Taking advantage of blue carbon to mitigate climate change is particularly interesting to policymakers and conservationists because it has almost no negative consequences and provides additional benefits such as enhanced fishery resources (Zinke, 2020). Thus the Food and Agriculture Organization of the United Nations and Wetlands International have provided guidance for the conservation and management of wetlands (Joosten et al., 2012).

V. Protection and restoration

Considering the importance of wetlands, coupled with the increasing threats that these ecosystems are exposed to, a number of international and national efforts have been undertaken to promote the conservation, protection, and management of wetlands. We summarize some of these efforts below.

A. The Ramsar Convention

To better conserve and wisely use wetlands, the Convention on Wetlands of International Importance Especially as Waterfowl Habitat, also known as the Ramsar Convention, was adopted and signed in 1971. Contracting parties must take serious actions to conserve, protect, and promote the wise use of wetlands in their countries. This Convention requires parties (1) to formulate planning so as to better conserve and wisely use wetlands, (2) to establish nature reserves in wetlands and promote training programs for wetland research and management, (3) to consult with other contracting parties to implement the Convention, and (4) to comply with the commitments of the Convention (Ramsar, 2016).

The Ramsar Convention has become the largest and most influential international treaty for wetland conservation. By the end of 2020, 171 contracting parties had joined the Convention, and 2418 wetlands, covering more than 2.5 million km^2, had been included in the List of Wetlands of International Importance, known as Ramsar sites.

B. National and regional policies

Since over half of the wetlands had already been drained for agriculture and urban development in the United States by the 1980s, the government started to formulate a series of laws to reverse this alarming rate of wetlands loss. The best-known and perhaps most influential policy is the *No Net Loss of Wetlands*, asking stakeholders who are benefiting from wetland exploitation to build or replace impacted wetlands with a replacement wetland of the same size with similar functions and values. In other words, wetlands impacted by human activities must be compensated for by the creation or restoration of a new wetland. *Wetlands Mitigation Banking* is another policy that is widely accepted by conservationists, policymakers, and stakeholders to further ensure wetland conservation. Under this arrangement, wetlands are created, restored, or preserved in exchange as a condition for development approval. These have led to a boom in wetland restoration projects and achieved some positive outcomes, including an increase in some wetland types in the United States (Levrel et al., 2017; Maron et al., 2018).

The Chinese government began considering wetland conservation as a national priority in the 1990s in an attempt to reverse the rapid wetland decline and to better conserve ecosystem functions and services. For example, the *National Wetland Conservation Plan* (2002–2030) was issued in 2003 and aimed to establish 713 wetlands reserves, protect more than 90% of natural wetlands by 2030, and restore 14,000 km^2 of natural wetlands. Moreover, the *Ecological Compensation Scheme* is the primary mechanism to fund wetland conservation in China, by evaluating the value of ecosystem functions and services, to balance the benefits and economic costs of wetlands conservation among stakeholders, governments, and local communities. In addition, since the policy of *Ecological Civilization Construction* was adopted as a national strategy in 2012, coastal wetlands have been considerably expanded from 2012 to 2018, although the total area of coastal wetlands still decreased by 27.9% compared to the area in 1984 (Wang et al., 2021).

In Europe wetland ecosystems were among the first nature conservation subjects that were considered in European policy. This started with the *Communication on the Wise Use and Conservation of Wetlands* in 1994, which identified the obstacles that negatively affected wetlands. More importantly, it also provided a strategic plan to tackle these obstacles. In 2000 this policy was replaced by the *Water Framework Directive*, which became the most substantial piece of EU water legislation, aimed at protecting and restoring various wetland ecosystems, among other water resources, to ensure a safe drinking water supply and other important services.

VI. Examples of restoration and protection projects

A. Changshu, China

Changshu is located in the southeast of China, bordering the Yangtze River to the north and Taihu Lake to the south. It has an area of 1276 km^2 and a population density of 1183 people km^{-2}. The landscape of Changshu is a typical composite wetland ecosystem. These wetlands, however, were previously converted into agricultural lands such as rice paddies as well as fishing ponds, and this drastic change resulted in wetland degradation, biodiversity loss, and even local climate change. Meanwhile, rapid industrialization and urbanization made the water unsafe.

To confront this environmental deterioration, ecological restoration projects started in the 1980s, firstly by converting paddy fields and farmlands back to wetlands. In the next few decades lakes began to be restored as well. Today, Changshu sets a positive example of the value of wetland restoration by integrating ecological conservation, food production, ecotourism, and other livelihoods (Fig. 29-6).

A key aspect of the Changshu wetlands restoration strategy included a comprehensive multilevel wetland conservation system, which included national wetland parks, provincial wetland parks, wetland conservation zones, and wetland communities. This proved to be an effective way for densely populated, economically developed, and high-land-use regions to achieve both wetland conservation and urban development, with safeguards to protect and restore wetlands. Changshu was accredited as 1 of the 18 Wetland Cities across the globe during Ramsar COP13 in 2018.

B. Room for the River, the Netherlands

Another example of a large initiative for the restoration of the area and functional quality of wetlands is the program *Room for the River*, launched by the government of the Netherlands in 2006 (Nienhuis, 2008). This ambitious plan to restore the connections between river channels and floodplain areas had been discussed for a long time. However, it became urgent when, in 1995, the discharge of the rivers Rhine and Meuse became so high that dikes were at the point of collapsing, which would have been a major disaster. It became necessary to evacuate approximately 250,000 people and their husbandry within 24 h in the Land van Maas en Waal. There was wide recognition that the two lower basins of the rivers Rhine (locally called Waal) and Meuse (Maas) had been too drastically engineered to facilitate navigation. The river channel had been straightened and cut off from its floodplains by elevated banks and, further from the channel, by artificial high levees, made of sand and clay.

The above system ensured sufficient water depth in the channel and left the remainder of the floodplain, between the banks and the high levees, dry during summer, so farmers used it for cattle grazing and crop growing. However, the continued confinement of the river channel into narrower courses, coupled with the more extreme discharge peaks resulting from enhanced runoff from hardened surfaces in cities and drained lands in the catchments, and increasingly extreme weather events, created a vulnerable and dangerous situation. The Physical Planning Core Decision "Room for the River" was ratified in the Dutch parliament in 2006.

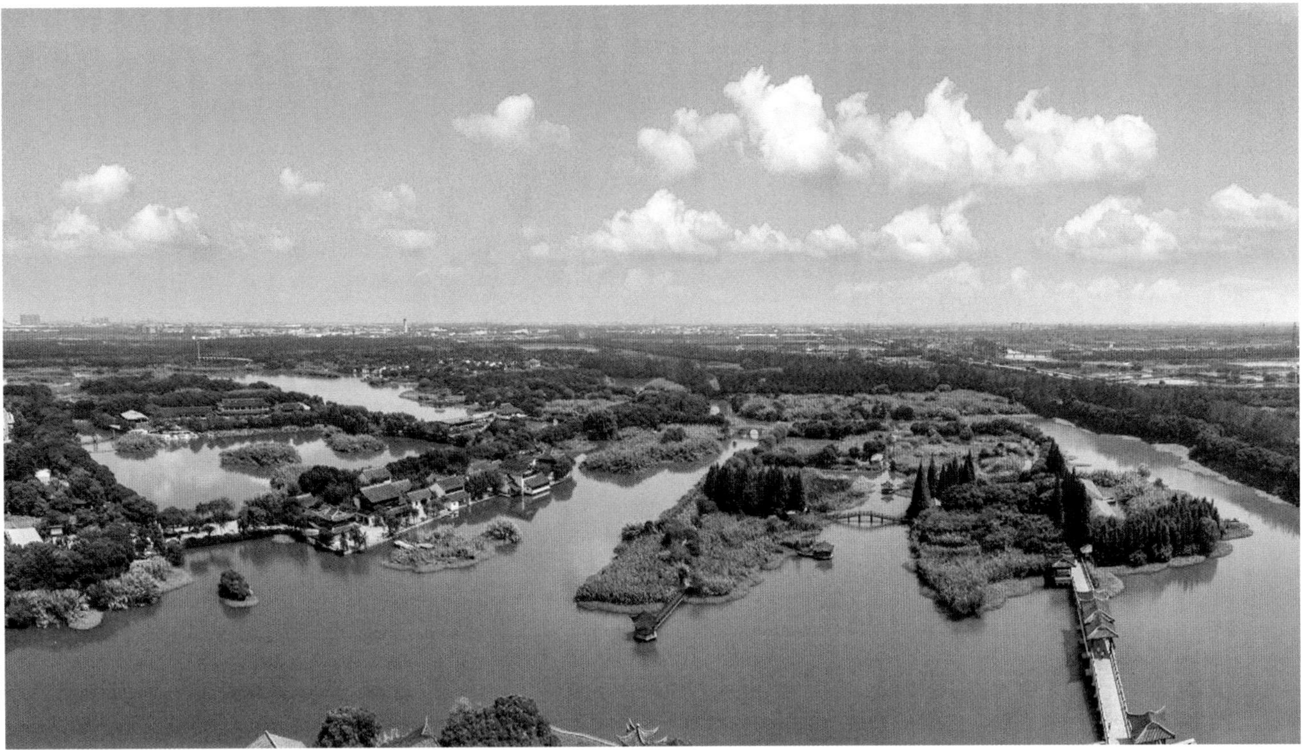

FIGURE 29-6 Shajiabang Wetland Park in Changshu, China (From Changshu Wetlands Station).

The program encompassed 37 projects along the rivers Rhine, Meuse, and IJssel, in which the embankments and the active floodplains were excavated to a lower level. In addition, the levees were repositioned more inland at several locations, thus increasing the area of regularly flooded floodplains and the frequency and duration of flooding. This also strongly increased the capacity for effective river discharge and now protects more than 4 million people from flood risks. The program was completed in 2019 (Fig. 29-7).

The Room for the River program has also resulted in a drastic increase in the biodiversity of the floodplain areas. Many characteristic species that were recently almost extirpated have returned. In addition, the natural functioning of the river-floodplain connections has been restored and has contributed to landscape attractiveness and environmental health.

C. Everglades, Florida, USA

Everglades National Park is an important type of wetland situated in a transitional bioclimatic zone between subtropical and tropical regions. There are nine distinct habitats in the Park, including pine Rocklands, coastal lowlands, and marine waters. The Park is best known for its mangroves, sawgrass prairies, and freshwater sloughs that draw water from Lake Okeechobee.

Lake Okeechobee, a large, shallow lake located in central/southern Florida, is also called the "Everglades' liquid heart." Over the past few decades, the lake and its watershed have been subjected to significant increases in nutrients, especially phosphorus and nitrogen, from agricultural and urban activities. Excess phosphorus enhanced the growth of vegetation and caused eutrophication, posing a substantial threat to the ecosystem and the associated biodiversity (Havens and James, 2005).

Several projects have been implemented to reduce substantially the phosphorus load into the lake. As part of a larger restoration plan, *stormwater treatment areas* (STA) were established over large parcels of land; these are constructed wetlands that remove and assimilate nutrients through plant growth and convert the biomass into soil (Schade-Poole and Möller, 2016; Khare et al., 2019). Five STAs are functional south of Lake Okeechobee that are currently removing excess nutrients from agricultural runoff water and, in some cases, runoff from urban tributaries, before discharging it into the Everglades and other natural areas. Approximately 275 km^2 of land south of Lake Okeechobee have been converted from agricultural land to STAs, yielding 255 km^2 of effective treatment wetlands.

Hydrologic conditions in the Everglades are characterized by an annual cycle of distinctly wet and dry seasons. In fact, water managers and ecologists measure

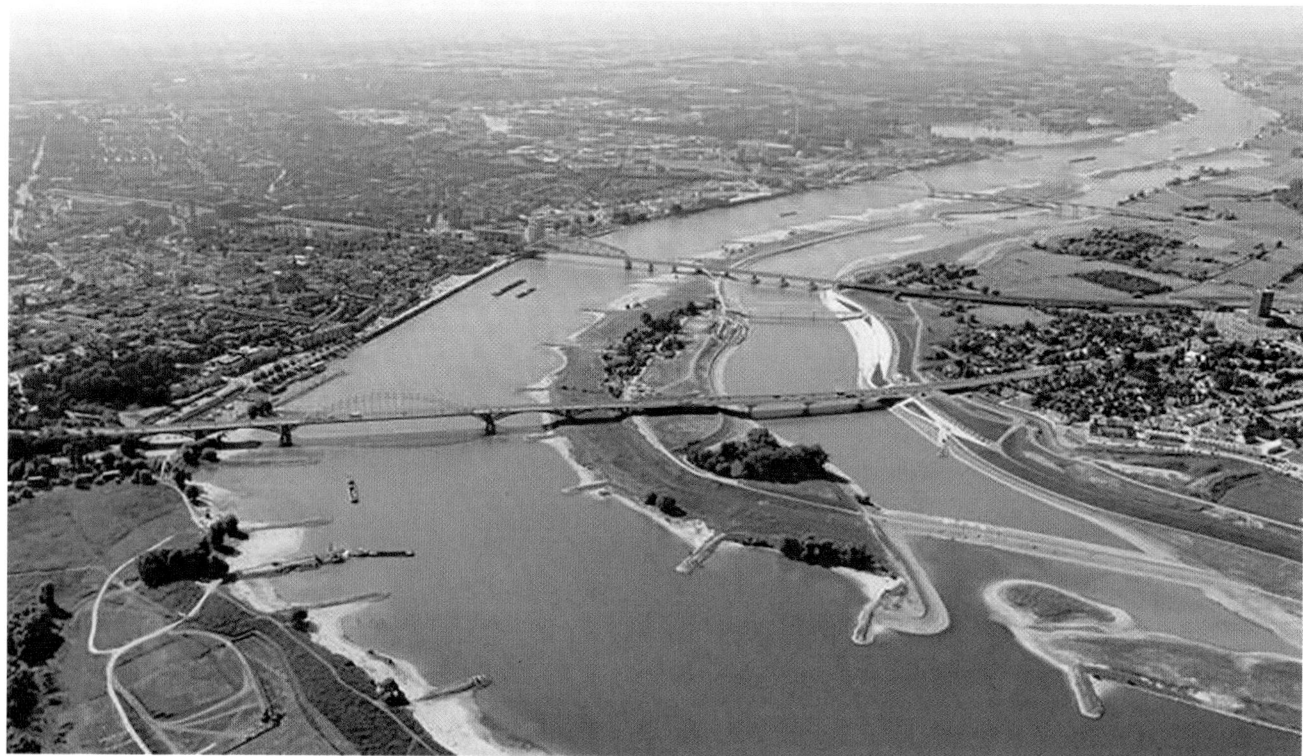

FIGURE 29-7 A prime example of a Room for the River project is the "Spiegelwaal near Lent," Nijmegen, the Netherlands, completed in 2018. (Licenced under the Creative Commons Attribution-Share Alike 4.0 International license).

time using a *water year* (WY) synchronized with the annual hydrologic cycle instead of the calendar year. The WY begins on May 1 of the preceding calendar year and ends on April 30. In WY 2019 and WY 2020 the Everglades STAs treated an average of 1.4 km^3 of water and retained 143 tons of phosphorus per year, which equated to an 80% phosphorus reduction, and produced an average outflow phosphorus concentration of 24 $\mu g\ L^{-1}$ (Mitsch et al., 2015; Zhao and Piccone, 2020).

The State of Florida and the US Environmental Protection Agency reached a consensus plan on new restoration strategies for improving water quality in the Everglades. Under these strategies, the South Florida Water Management District is implementing a technical plan to complete several projects that includes more than 26.3 km^2 of new STAs and 0.14 km^3 of additional water storage through the construction of *flow equalization basins* (FEBs). A FEB is a constructed storage used to capture and store peak stormwater runoff. Water managers can move water from FEBs into STAs at a steady rate that enhances the operation of phosphorus treatment. In addition, a robust Science Plan, including a framework to develop and coordinate scientific research related to phosphorus reduction in the Everglades' STAs, is now ongoing (South Florida Water Management District, 2017).

In addition, a novel patented method of *floating aquatic vegetative tilling* (FAVT) wetland treatment facility has been established in a 2.15 km^2 habitat to treat local agricultural runoff from the Hendry Hilliard Water Control District, the East Caloosahatchee River, and Lake Okeechobee. FAVT systems utilize an innovative approach to enhance nitrogen and phosphorus removal from the surface water; the process is cost-effective and utilizes floating aquatic plants (Fig. 29-8, Water & Soil Solutions LCC, 2015). Many species used as FAV are known to rapidly assimilate nitrogen and phosphorus, but their high nutrient uptake rate can only be sustained if the plants are maintained at an optimal density. Instead of harvesting the plants periodically, the total biomass is rapidly incorporated directly into the soil by way of tilling during the dry seasons. FAVT systems therefore operate similarly to a conventional treatment wetland by storing phosphorus in the soil, but they accomplish phosphorus removal more efficiently and at a significantly faster rate. The Fisheating Creek facility, located in the Lake Okeechobee watershed, is comprised of 1 km^2 of floating aquatic vegetation and submerged aquatic

FIGURE 29-8 Floating aquatic vegetation, water hyacinth *Eichhornia crassipes* and water lettuce *Pistia stratiotes*, in the Everglades (Photo by Tom DeBusk).

vegetation communities. From July 2019 to June 2020, a total of 7.3 km^3 was treated, removing 1.5 tons of total P and 5.0 tons of total N (Florida Department of Environmental Protection, 2014).

VII. Summary

1. Wetlands can be regarded as a collection of ecosystems that are transition zones between land and water.
   a. They can be found on all continents, except for Antarctica. The estimated extent of global wetlands is about 12.1 million km^2, more than 50% of which are located in Asia and North America.
   b. Their functioning is governed by specific hydrological conditions, physicochemical environments with typical redox cycles, and characteristic wetland biota.
2. Natural wetlands can generally be divided into riverine, lake, marsh, coastal wetlands, and peatlands. Each type has its own communities adapted to the local biogeochemical conditions.
   a. The riverine and lake wetlands are important resources for food, agriculture irrigation, and drinking water supply.
   b. Marshes are wetlands situated on mineral soil, such as saltmarshes and floodplains.
   c. Peatlands have organic soils and accumulate considerable amounts of carbon, thus playing an important role in regulating the global carbon cycle.
   d. Coastal wetlands, and in particular mangroves, can also store carbon, protect the shoreline from erosion and storms, and provide significant commercial value.
3. Human-made wetlands are usually constructed to purify wastewater, supply food, and provide other ecosystem services, such as building materials, recreation, and ecotourism.
4. Hydrological conditions (e.g., inundation) fundamentally determine biogeochemical cycles in wetlands by creating gradients of oxygen availability in space and time.
   a. High and fluctuating water levels create gradients in space and time in the availability of oxygen.
   b. When aerobic decay processes deplete the oxygen in the sediment, anaerobic bacteria take over and use a series of electron acceptors to replace oxygen for mineralization. These electron acceptors, which are nitrate, Mn^{4+}, Fe^{3+}, sulfate, and carbon dioxide, are being used successively with a gradual decrease of the redox potential. Part of the end products of the anaerobic processes are gaseous, for example, nitrogen gas, nitrous oxide, hydrogen sulfide, methane, and carbon dioxide.
5. Due to human impacts, wetlands are declining around the world, with an approximately 35% decrease in surface area between 1970 and 2015. The main drivers for this drastic loss include (1) drainage and conversion for agricultural use; (2) peat excavation for fuel use; (3) mangrove deforestation for aquaculture, rice paddies, and palm plantations; and (4) biological invasions.
6. Recent climate change is another major threat to wetlands. A changing climate will fundamentally

alter the biogeochemical cycle in wetlands; however, many uncertainties remain.

a. On the one hand, an increase in wetland primary production is likely with higher temperatures if this coincides with more precipitation. On the other hand, droughts may lead to a total desiccation of affected wetlands.

b. Wetland conservation, restoration, and protection can become effective approaches to mitigate climate change, by protecting and enhancing the wetland's function as a carbon store.

7. To counter the global decline in wetlands, international and national efforts have been undertaken.

a. The Ramsar Convention is the largest and most influential international treaty for wetland conservation.

b. Other national and regional policies, such as the No Net Loss in the United States, the National Wetland Conservation Plan and the policy of Ecological Civilization Construction in China, and the Water Framework Directive in the EU, have all played positive roles in wetland conservation and restoration.

c. Three examples of restoration projects include the Wetland City (i.e., accredited by Ramsar), Changshu, in China, which demonstrated a comprehensive multilevel wetland conservation system; the Room for the River project in the Netherlands, which included an ambitious plan to restore the connections between river channels and floodplain areas and to increase biodiversity; and projects in the Everglades in the United States, which represent restoration strategies to improve water quality while conserving natural wetland ecosystems.

Acknowledgments

We sincerely thank Ms. Yaqin Yao for her help in proofreading and coordinating. We also thank Dr. Lu Xia, Dr. Yajun Qiao, and Dr. Wen Yang for their contribution to collecting writing materials, and Ms. Shujie Yin for processing images.

References

An, S., Verhoeven, J.T.A., 2019. Wetlands: Ecosystem Services, Restoration and Wise Use. Springer, New York. https://doi.org/10.1007/978-3-030-14861-4.

An, S., Wang, L., 2014. Wetland Restoration: Shanghai Dalian Lake Project. Springer, New York. https://doi.org/10.1007/978-3-642-54230-5.

Cowardin, L.M., 1979. Classification of Wetlands and Deepwater Habitats of the United States. Fish andWildlife Service. US Department of the Interior. Available from http://www.npwrc.usgs.gov/resource/wetlands/classwet/index.htm.

Davidson, N.C., Fluet-Chouinard, E., Finlayson, C.M., 2018. Global extent and distribution of wetlands: Trends and issues. Mar. Freshw. Res. https://doi.org/10.1071/MF17019.

Desta, H., Lemma, B., Fetene, A., 2012. Aspects of climate change and its associated impacts on wetland ecosystem functions: a review. J. Am. Sci. 8 (10), 582–596.

Dise, N.B., 2009. Peatland response to global change. Science 326 (5954), 810–811. https://doi.org/10.1126/science.1174268.

Donato, D.C., Kauffman, J.B., Murdiyarso, D., Kurnianto, S., Stidham, M., Kanninen, M., 2011. Mangroves among the most carbon-rich forests in the tropics. Nat. Geosci. 4 (5), 293–297. https://doi.org/10.1038/ngeo1123.

FAO, 2011. The State of the World's Land and Water Resources for Food and Agriculture: Managing Systems at Risk. Earthscan, London, UK. Available from https://www.fao.org/3/i1688e/i1688e.pdf.

FAO, 2020. The State of World Fisheries and Aquaculture. Rome, Italy. Available from https://www.fao.org/documents/card/en/c/ca9229en.

Florida Department of Environmental Protection, 2014. Basin Management Action Plan for the Implementation of Total Maximum Daily Loads for Total Phosphorus by the Florida Department of Environmental Protection in Lake Okeechobee. Tallahassee, FL. Available from https://floridadep.gov.

Gardner, C., Finlayson, C., Davidson, N., Fennessy, S., Coates, D., van Damm, A., et al., 2018. GlobalWetland Outlook: State of theWorld's Wetlands and Their Services to People. Available from https://www.global-wetland-outlook.ramsar.org.

Havens, K.E., James, R.T., 2005. The phosphorus mass balance of Lake Okeechobee, Florida: Implications for eutrophication management. Lake. Reserv. Manag. 21 (2), 139–148. https://doi.org/10.1080/07438140509354423.

Janse, J.H., Janse, J.H., de Senerpont Domis, L.N., Scheffer, M., Lijklema, L., van Liere, L., et al., 2008. Critical phosphorus loading of different types of shallow lakes and the consequences for management estimated with the ecosystem model PCLake. Limnologica 38 (3–4), 203–219. https://doi.org/10.1016/j.limno.2008.06.001.

Joosten, H., Tapio-Biström, M.L., Tol, S., 2012. Peatlands: Guidance for Climate Change Mitigation through Conservation, Rehabilitation and Sustainable Use. Food and Agriculture Organization of the United Nations, Rome. Available from http://www.fao.org/documents.

Keddy, P.A., 2010. Wetland Ecology: Principles and Conservation. Cambridge University Press, Cambridge, UK. https://doi.org/10.1017/CBO9780511778179.

Khare, Y., Naja, G.M., Stainback, G.A., Martinez, C.J., Paudel, R., van Lent and, T., 2019. A phased assessment of restoration alternatives to achieve phosphorus water quality targets for Lake Okeechobee, Florida, USA. Water 11 (2), 327. https://doi.org/10.3390/w11020327.

Levrel, H., Scemama, P., Vaissière, A.C., 2017. Should we be wary of mitigation banking? Evidence regarding the risks associated with this wetland offset arrangement in Florida. Ecol. Econ. 135, 136–149. https://doi.org/10.1016/j.ecolecon.2016.12.

Maron, M., Brownlie, S., Bull, J.W., Evans, M.C., von Hase, A., Quétier, F., et al., 2018. The many meanings of no net loss in environmental policy. Nat. Sustain. 1 (1), 19–27. https://doi.org/10.1038/s41893-017-0007-7.

McDonough, K., 2014. Amphibian species of the world: an online reference (version 6). Ref. Rev. 28 (6), 32. https://doi.org/10.1108/RR-05-2014-0125

Mitsch, W.J., Gosselink, J.G., 2015. Wetlands, fifth ed. Wiley Online Library, Hobroken, NJ, USA.

Mitsch, W.J., Zhang, L., Marois, D., Song, K., 2015. Protecting the Florida Everglades wetlands with wetlands: can stormwater phosphorus be reduced to oligotrophic conditions? Ecol. Eng. 80, 8–19. https://doi.org/10.1016/j.ecoleng.2014.10.006.

Nahlik, A.M., Fennessy, M.S., 2016. Carbon storage in US wetlands. Nat. Commun. 7, 13835. https://doi.org/10.1038/ncomms13835.

Nienhuis, P.H., 2008. Environmental History of the Rhine-Meuse Delta: An Ecological Story on Evolving Human-Environmental Relations Coping with Climate Change and Sea-Level Rise. Springer Science & Business Media. https://doi.org/10.1007/978-1-4020-8213-9.

Ockenden, M.C., Hollaway, M.J., Beven, K.J., Collins, A.L., Evans, R., Falloon, P.D., et al., 2017. Major agricultural changes required to mitigate phosphorus losses under climate change. Nat. Commun. 8 (1), 1–9. https://doi.org/10.1038/s41467-017-00232-0.

Ramsar, 2016. An Introduction to the Ramsar Convention on Wetlands, seventh ed. Ramsar Convention Secretariat, Gland, Switzerland. Available from https://www.ramsar.org/sites/default/files/documents/library.

Reddy, K.R., DeLaune, R.D., 2008. Biogeochemistry of Wetlands: Science and Applications. CRC Press, Boca Raton, FL. https://doi.org/10.1201/9780203491454.

Richards, D.R., Friess, D.A., 2016. Rates and drivers of mangrove deforestation in Southeast Asia, 2000–2012. Proc. Natl. Acad. Sci. 113 (2), 344–349. https://doi.org/10.1073/pnas.1510272113.

Schade-Poole, K., Möller, G., 2016. Impact and mitigation of nutrient pollution and overland water flow change on the Florida Everglades, USA. Sustainability 8 (9), 940. https://doi.org/10.3390/su8090940.

South Florida Water Management District, 2017. Quick Facts on Restoration Strategies for Clean Water for the Everglades. Available from. https://www.sfwmd.gov/sites/default/files/documents.

Tong, L., Xu, X., Fu, Y., Li, S., 2014. Wetland changes and their responses to climate change in the "Three-River Headwaters" region of China since the 1990s. Energies 7 (4), 2515–2534. https://doi.org/10.3390/en7042515.

Turetsky, M.R., Benscoter, B., Page, S., Rein, G., van der Werf, G.R., Watts, A., 2015. Global vulnerability of peatlands to fire and carbon loss. Nat. Geosci. 8 (1), 11–14. https://doi.org/10.1038/ngeo2325.

Verhoeven, J.T.A., 2009. Wetland biogeochemical cycles and their interactions. The Wetlands Handbook. Wiley. https://doi.org/10.1002/9781444315813. Ch. 12.

Wang, X., Xiao, X., Xu, X., Zou, Z., Chen, B., Qin, Y., et al., 2021. Rebound in China's coastal wetlands following conservation and restoration. Nat. Sustain. https://doi.org/10.1038/s41893-021-00793-5.

Water & Soil Solutions LCC, 2015. Implementation of floating aquatic vegetative tilling technology in the Caloosahatchee River Watershed: Task 20 deliverable final report. Available from https://www.fdacs.gov/content/download/76296/file/21121_FinalReport_Site_1.pdf.

Xia, L., Yang, W., Geng, Q., Jeelani, N., An, S., 2020. Research on *Spartina alterniflora* using molecular biological techniques: an overview. Mar. Freshw. Res. 71 (12), 1564–1571. https://doi.org/10.1071/MF19255, 2015.

Xu, J., Morris, P.J., Liu, J., Holden, J., 2018. PEATMAP: Refining estimates of global peatland distribution based on a meta-analysis. Catena 160, 134–140. https://doi.org/10.1016/j.catena.2017.09.010.

Yang, G., Peng, C., Chen, H., Dong, F., Wu, N., Yang, Y., et al., 2017. Qinghai–Tibetan Plateau peatland sustainable utilization under anthropogenic disturbances and climate change. Ecosyst. Health. Sustain. 3 (3), e01263. https://doi.org/10.1002/ehs2.1263.

Yang, W., Jeelani, N., Leng, X., Cheng, X., An, S., 2016. *Spartina alterniflora* invasion alters soil microbial community composition and microbial respiration following invasion chronosequence in a coastal wetland of China. Sci. Rep. 6 (1), 1–13. https://doi.org/10.1038/srep26880.

Yin, S., An, S., Deng, Q., Zhang, J., Ji, H., Cheng, X., 2015. Spartina alterniflora invasions impact CH_4 and N_2O fluxes from a salt marsh in eastern China. Ecol. Eng. 81, 192–199. https://doi.org/10.1016/j.ecoleng.2015.04.044.

Zhao, H., Piccone, T., 2020. Large scale constructed wetlands for phosphorus removal, an effective nonpoint source pollution treatment technology. Ecol. Eng. 145, 105711. https://doi.org/10.1016/j.ecoleng.2019.105711.

Zinke, L., 2020. The colours of carbon. Nat. Rev. Earth Environ. 1 (3), 141. https://doi.org/10.1038/s43017-020-0037-y.

CHAPTER

30

Paleolimnology: Approaches and Applications

Irene Gregory-Eaves[1] *and John P. Smol*[2]

[1]Department of Biology, McGill University, Montreal, Quebec, Canada [2]Paleoecological Environmental Assessment and Research Lab (PEARL), Department of Biology, Queen's University, Kingston, Ontario, Canada

OUTLINE

In the distant, past most alterations of aquatic ecosystems have been brought about by natural climatic dynamics that were driven largely by changes in the shape of the Earth's orbit around the sun or solar emission variability. More recently, changes have been greatly accelerated by human activities, which have affected drainage basins (watersheds), atmospheric inputs, water budgets, and nutrient loading; all these factors can result in changes in the structure and functioning of inland waters. One of the greatest challenges faced by limnologists (and environmental scientists in general) is the lack of direct long-term monitoring data. In fact, for many environmental issues, scientists are often only alerted to issues once a problem has developed (e.g., fish dying or algal blooms). Even when data are available, typically the monitoring window covers only a very short period (e.g., 3 years or less; Fig. 30-1) and rarely provides sufficient data to assess what conditions were like before problems developed (i.e., background or reference conditions) or information on the trajectory of long-term ecosystem change (Smol, 2019).

Sediments provide a natural record that can yield insights long into the past, even when no monitoring occurred. Lakes function as natural traps for sediment, integrating materials from both outside and within the lake. Once materials entering or produced in the lake settle to its bottom, energy levels and hydrodynamics are usually insufficient to transport materials out of the basin (Chapter 8). As a result, a stratigraphic depth-time profile develops in the accumulating sediment layers, leaving records of climatic and human-induced changes in the environment in the mud of lakes, ponds, reservoirs, and rivers (Cohen, 2003; Smol, 2008). Most of the work to date on sediment cores has been conducted in lakes (as opposed to rivers or reservoirs), and thus this chapter will reflect this bias.

Paleolimnology is the multidisciplinary science that uses the physical, chemical, and biological information preserved in sedimentary profiles to reconstruct environmental and ecological histories. These natural archives may potentially record both short-term (e.g., seasonal) and long-term (e.g., decadal, millennial) changes. The sedimentary record is neither perfect nor

Wetzel's Limnology, Fourth Edition
https://doi.org/10.1016/B978-0-12-822701-5.00030-6

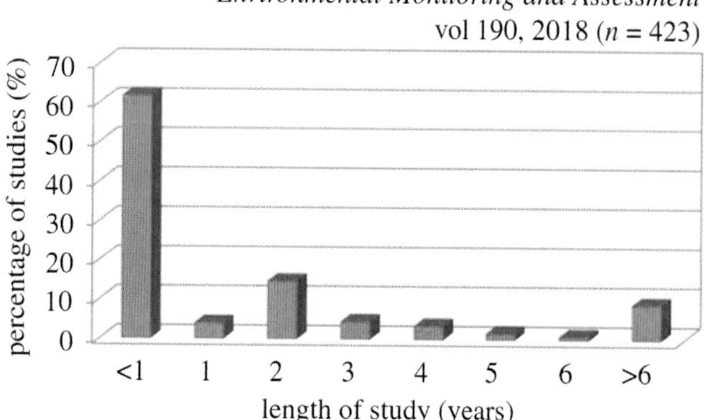

FIGURE 30-1 The length of time devoted to aquatic studies published in 2018 in the journal *Environmental Monitoring and Assessment*. This analysis provides evidence that most environmental studies collect data over a period of less than 1 year, and highlights that often we lack long-term datasets. *(Figure from Smol, 2019.)*

complete, and so paleolimnologists also study processes that can alter the record (e.g., *diagenesis*, which refers to the transformation of the sediment postdeposition through physical, chemical, or biological processes). The ultimate goal is to gain insights into past ecosystem conditions, shifts in aquatic composition, and changes in ecosystem services. In addition, insights can be gained into the possible future trajectory of the ecosystem (Deevey, 1984; Levy, 2017; Reavie, 2019). The methodologies used by paleolimnologists are reasonably well established and summarized in Last and Smol (2001a, 2001b) and Smol et al. (2001a, 2001b), while the main numerical approaches are reviewed in Birks et al. (2012).

A lake is located at the receiving end of a drainage basin and the air above, and thus lake sediments are comprised of materials from within the lake or its watershed and airshed (Fig. 30-2). Materials that are transported into the lake from the surrounding drainage basin or that enter from the air (e.g., leaf fall or dust) are considered *allochthonous*. In contrast, materials that are generated within the lake (e.g., dead organic matter, chemical precipitates such as carbonates, siliceous diatom frustules) and often settle to the lake's bottom

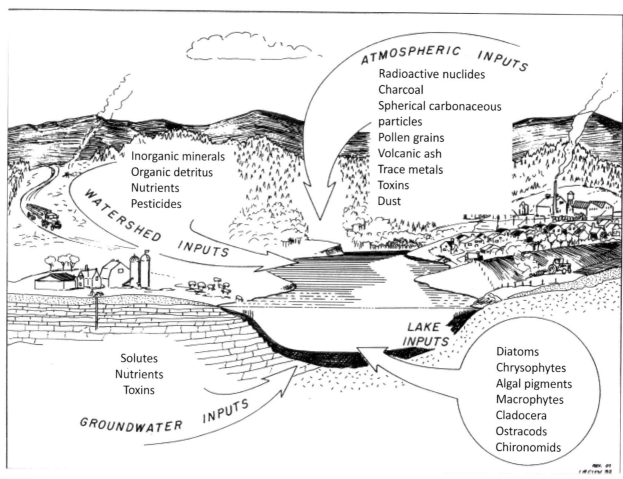

FIGURE 30-2 The complex pathways and sources of material that can ultimately be incorporated into lake sediments. Inputs arising from outside of the lake are known as allochthonous sources (i.e., atmospheric, watershed, and groundwater inputs). Inputs to sediments arising from within the lake are known as autochthonous sources. *(Modified figure from Smol, 2008.)*

are considered *autochthonous*. Each of the components of lake sediments, both allochthonous and autochthonous materials, can provide information on the environmental history of a lake and the surrounding region. Insights gained from studying allochthonous indicators (e.g., such as the field of *palynology*, which studies pollen grains and spores) can be very useful in terms of understanding how the lake may have been affected by climatic and/or land-use changes in the watershed. For example, land-use changes have been inferred by studying variation in *coprophilous ascospores* (spores associated with fecal material that indicate how herbivore densities have changed) or by examining the composition of terrestrial plant pollen assemblages found in the lake sediments (Belle et al., 2016). To disentangle climatic forcing from land-use changes, one needs to compare sites within a region that were known to have little versus more substantial land-use changes based on analyses of historical records, current land-use patterns, or other long-term datasets. In contrast, many autochthonous indicators are useful for understanding how the lake has responded to changes in the watershed or climate, but the most complete perspective is gained by studying a complement of allochthonous and autochthonous indicators.

I. Stratigraphy and geochemistry

A. Sedimentary record

Many lakes were formed as a result of glacial activities, but other key processes that gave rise to lakes include tectonic movements, volcanic activity, and coastal or river dynamics (see Chapter 4). Most lakes of earlier glacial periods have been obliterated by succeeding glaciations. It is only because of the most recent retreat of the last major glaciation phase that so many lakes currently exist; thus the longest sediment records one can hope to retrieve from most lakes in previously glaciated regions are between 15,000 and 20,000 years old. However, even longer records can be obtained in ancient lakes of the rift valleys of Africa and Asia (Livingstone, 1975), or in some special cases where past glacial advances did not destroy lake sediment stratigraphy (Axford et al., 2009). For example, sediment records spanning 12,000,000 years have been examined from Lake Baikal, Russia (Kashiwaya et al., 2001).

The composition of lake sediments is influenced to a great extent by the geomorphology, climate, and geology of the lake and its drainage basin, as well as biological processes therein. For example, in arid regions where there is no direct drainage to the ocean (i.e., endorheic regions), one commonly finds inland saline lakes (termed *athalassic* lakes; see Chapter 12); these sediments may be characterized by rich concentrations of precipitated salts. Saltier waters are also habitats for a distinct composition of algal and invertebrate communities, called *halophiles*, and so past salinity changes could be traced in the sediment record by tracking the changes in the morphological remains of the algae and invertebrates or with biogeochemical proxies.

Water movements within lakes play a role in distributing particulates according to their size and density. *Sediment focusing* is one such process, where particles are transported to the central depression(s) of a lake basin (Chapter 27), and arises from three major water movements (Hilton, 1985): (1) peripheral wave action that generates turbulence and resuspends fine sediments in littoral areas that are subsequently redeposited in deeper areas; (2) intermittent, random redistribution of sediments of the entire basin when the lake is not stratified, particularly during autumnal overturn; and (3) movement of sedimentary material parallel to a sloping bed (sliding) or movement initiated by a rotational failure of the sediments (*slumping*), resulting in material being transported by flows from shallow regions to deeper waters. Little sediment will accumulate on slopes >14%, and thus lake morphometry dictates the dominant process of sediment redistribution (Håkanson, 1981). The overall result is a general sediment gradient of coarser particles nearshore and a greater proportion of finer particles under deeper waters. However, various mechanisms can disrupt this gradient. For example, deposits of sediments in littoral regions or river deltas can accumulate until they become gravitationally unstable and then slump in landslide fashion or move as *turbidity flows* (i.e., *turbidites*) to deeper portions of the basin where they come to overlie younger sediments. Often these slumps can be identified by studying the lithology of the sediment record by looking for massive, unsorted deposits (Osleger et al., 2009). In addition to the physical structure of lake sediments, chemical properties (e.g., heavy metals, carbon and organic pollutants), biochemicals (e.g., pigments, lipids, sterols, and DNA), and morphological remnants of specific organisms (e.g., diatoms, cladocerans, and chironomids) are preserved in sediment archives, which are discussed in more detail below.

A key factor in using paleolimnology to reconstruct past environmental histories is to extract a relatively undisturbed sediment core, such that the sediment on top is the youngest and sediment below is the oldest. Consequently, areas with slow water flow and regular sediment accumulation, away from steep slopes and riverine inputs, are chosen to retrieve sediment cores. For lakes, this is typically the deepest part of the lake, where water movement is slowest and sediment redistribution is minimal. Hence a critical early step in any paleolimnological study is retrieving a sediment core.

The tools used for sediment core retrieval depend on the desired temporal extent of the study and the physical lake setting (Glew et al., 2001). However, the top 30–50 cm of sediment of most lakes can be retrieved using a relatively simple gravity corer (Glew et al., 2001). To collect cores that span millennial sequences or to retrieve records from deep lakes (>50 m), one requires a more specialized operation (Fig. 30-3). More information on field techniques is summarized in Last and Smol (2001a, 2001b).

B. Dating of sediments

An important prerequisite for most paleolimnological studies is to establish a reliable depth-time profile. Several methods have been used to determine the age of sedimented materials in lake deposits (Table 30-1). The dating notation varies as a function of the time scale and methods used. Although traditionally the terms Anno Domini (AD) and Before Christ (BC) referred to years within the past several thousands of years, we now typically use the terms Common Era (CE) to refer to the number of years since AD 1. When applying radiocarbon methods, Before the Present (BP) refers to the number of years before the reference year of 1950 CE (0 BP).

For most records spanning the postglacial period, an important dating technique for aging materials is based on measuring their radioactive carbon content (^{14}C). Radiocarbon is formed in the upper atmosphere by the reaction of nitrogen (^{14}N) with neutrons (n): $^{14}N + n \rightarrow {}^{14}C + {}^{1}H$, where ^{14}C has 8 neutrons and 6 protons. Radioactive carbon decays back to ^{14}N with the emission of a β^- particle and has a half-life of ~5700 years (Hajdas et al., 2021). Generally, older, deeper sediments are depleted in radiocarbon relative to younger sediments. Radiocarbon techniques are not typically used for dating sediments over the past few hundred years because of the plateau in atmospheric ^{14}C activity between 1650 and 1950 CE (Trumbore et al., 2016), due in part to the burning of old carbon (fossil fuels) and release of this carbon into the atmosphere (i.e., known as the *Suess effect*). In addition, increased solar variability that modulated the flux of neutrons from cosmic rays is believed to have further influenced the ratio of $^{14}C/^{12}C$ in the atmosphere between 1650 and 1950 CE (Paterne et al., 2020). Between 1945 and 1963, there were numerous aboveground nuclear weapons tests that greatly enriched the atmosphere with ^{14}C (Hajdas et al., 2021), leading to a spike in atmospheric ^{14}C activity that has since been declining. With appropriate calibrations, ^{14}C dating can be applied to help delineate dates of very recent stratigraphical sequences (e.g., Turney et al., 2018). However, more commonly, alternate geochronological methods are applied to sediments spanning the last few hundred years (see below).

The concentration of ^{14}C in the atmosphere is a very small part of Earth's atmospheric CO_2 reservoir (~1 in

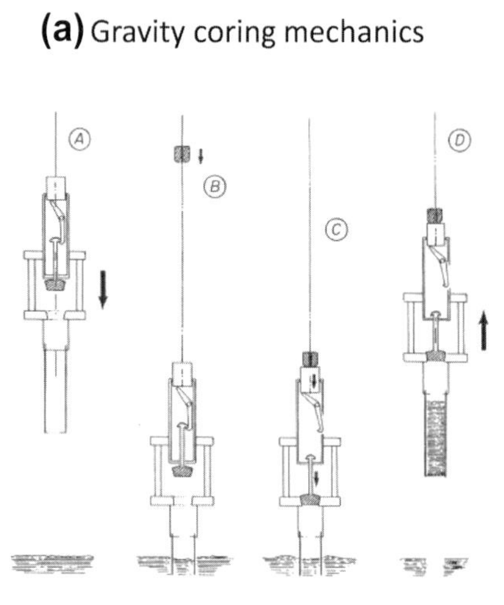

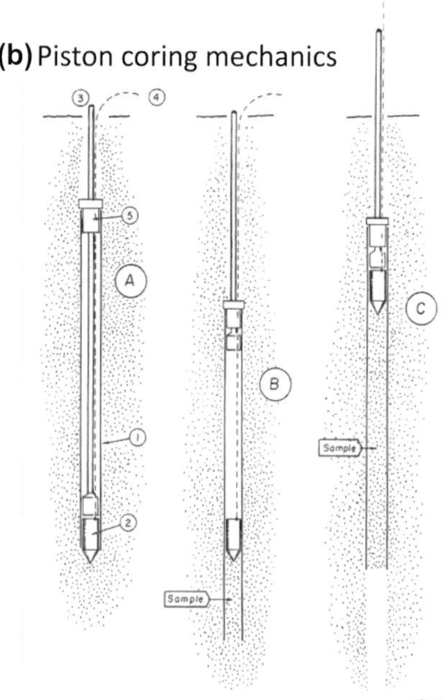

FIGURE 30-3 The mechanics of the two most widely applied lake sediment coring approaches. (a) Gravity coring is designed to retrieve the recent history of sediment deposition (usually <50 cm cores) and obtain an undisturbed sediment–water interface; and (b) piston coring is designed to obtain longer sediment records (often several meters;). Often these approaches are paired with one another. *(Source: [a] Smol, 2008* and *[b] Glew et al., 2001.)*

TABLE 30-1 Geochronological Methods That Can Be Applied to Sediment Archives

1. Radiometric techniques		
Marker	**Half-life (years)**	**Typical application (in terms of time window) or requirements**
^{14}C	5700 ± 30	~300–55,000 years BP
^{210}Pb	22.3	~1850 CE–present
^{137}Cs	30.2	~1963 CE coinciding with peak in nuclear bomb testing (and 1986 CE over parts of Europe linked to Chernobyl accident)
2. Volcanic ash layers (tephras)	–	Requires previous identification of dated ash layer
3. Pollen horizons	–	Requires previous identification of dated horizon (usually over last 10,000 years BP)
4. Spherical charcoal particles	–	Requires previous dating of documented changes (usually over the last ~150 years)

BP = Before Present; CE = Common Era.

10^{12} parts). For ^{14}C dating in its simplest form, it is assumed that ^{14}C produced in the atmosphere is in rapid equilibrium with the terrestrial CO_2 reservoirs and that the balance among the carbon reservoirs has been constant over time. During photosynthesis, $^{14}CO_2$ is photosynthetically incorporated into organic matter in proportion to its availability. When organisms die and remains are buried in sediments, the ^{14}C contained in the organic matter continues to decay back to ^{14}N with the emission of a β^- particle. The half-life of ^{14}C disintegration (i.e., ~5700 years) is constant, and when the residual specific activity of ^{14}C in the carbon of organic matter is accurately radioassayed, it provides an estimate of the age of the sample. The decay rate of ^{14}C should permit age determinations back to a limit of about 75,000 years, but technical detection limits of assaying for negative beta radiation of ^{14}C above background radiation usually limit ^{14}C-age determinations to 40,000–55,000 years (Hajdas et al., 2021). With large samples and where an enrichment process is applied, ^{14}C ages can be extended (Björck and Wohlfarth, 2001).

As stated above, a key assumption of ^{14}C dating is that the production of ^{14}C has been nearly constant over time, but in fact we know there has been variability in the supply of neutrons produced by cosmic radiation and thus corrections need to be applied. The *cosmic-ray flux* in the upper atmosphere is influenced by (1) the intensity of the Earth's magnetic field and (2) short-term changes in solar wind magnetic properties (Stuiver and Quay, 1980). Several correction factors have been developed to account for these sources of variation including dendrochronological techniques assessing ^{14}C levels in the wood of annual tree rings. Calibrations of ^{14}C dating measurements of older samples have been made by careful comparisons with other dating techniques ($^{230}Th/^{234}U$ ratios, thermoluminescence, magnetic dating). There are now several calibration curves to choose from, depending on the hemisphere one's study site is situated in and the ecosystem of study (marine versus inland waters; Reimer et al., 2020).

A second correction may also be applied to sediment records retrieved from hardwater lakes that are set in watersheds with appreciable quantities of geologically old, ^{14}C-deficient carbonates (e.g., limestone basins). In these sites the inorganic carbon fixed during photosynthesis is often not in equilibrium with atmospheric ^{14}C activity and may also include old carbon from the drainage basin. Carbonates washed directly into the lakes from the watershed can also contribute to the old carbon pool. The old ^{14}C-deficient carbon dilutes the contemporary carbon originating from atmospheric sources and thus dated material from such lakes can give spuriously old dates. Corrections are only approximate, and often the radiocarbon-dating values from these lakes must be compared with independent chronological measures (MacDonald et al., 1991; Björck and Wohlfarth, 2001). However, dating macrofossils or charcoal from terrestrial plant materials (if available in the sediment record) can circumvent this issue because these plants relied on atmospheric CO_2 when living (Douglas et al., 2016).

Dating of more recent sediment archives typically involves analyses of lead-210 (^{210}Pb), but other dating tools may also be used (including identifying known volcanic ash layers (*tephras*) or pollen horizons). ^{210}Pb is part of the uranium decay series, where over thousands of years, uranium-238 decays into radium-226 (^{226}Ra) in soils and then over approximately another 1000 years decays to radon-222. Radon-222 is a gas and escapes to the atmosphere, where it decays to ^{210}Pb within days. ^{210}Pb then enters a lake via precipitation or dry deposition and is eventually incorporated into the sediments. The portion of ^{210}Pb that enters the lake via the atmospheric pathway is termed *unsupported ^{210}Pb* since it was produced from radium located outside the sediments. However, there is a small and relatively constant portion of ^{210}Pb that arises from *in situ* decay of ^{226}Ra within the sediments (termed the *supported ^{210}Pb* fraction). The unsupported ^{210}Pb fraction will decrease with depth in the sediments because of radioactive decay, and provided that both the input of ^{210}Pb to a lake and the residence time of the lake are constant, then a dating profile can be generated (Appleby, 2001). Since the radioactive half-life of ^{210}Pb (22.3 years) is relatively short, this dating technique is limited to estimating ages over the past ~150 years. Several age modeling techniques can be used to estimate sediment ages from the ^{210}Pb supported and unsupported activities, which make different assumptions about ^{210}Pb fluxes as well as sediment sources and redistribution (Appleby, 2001).

Cesium-137 (^{137}Cs) was first released with the deployment of atomic bombs in 1945 but only increased to noticeable concentrations since 1954 because of aboveground atomic bomb testing. Concentrations of ^{137}Cs peaked in 1963, at the heights of the so-called "Cold War" between the United States and the former Soviet Union, but then declined rapidly following a treaty banning atmospheric nuclear testing. Tracking changes in ^{137}Cs has been used as a secondary marker for dating recently deposited sediments and assumes that during rain events, ^{137}Cs becomes attached to particles, which are quickly transported from the drainage basin to lake deposits (<1 year). Additionally, ^{137}Cs falling directly upon the lake surface is adsorbed onto suspended particulate matter and sedimented. However, the ^{137}Cs dating approach is not without its challenges. First, ^{137}Cs can be somewhat mobile in organic-rich sediments, and there is evidence of molecular diffusion and readsorption of ^{137}Cs in sediment profiles (Davis et al., 1984; Foster et al., 2006; Klaminder et al., 2012). Second, the nuclear accident at Chernobyl in 1986 created a new maximum in ^{137}Cs in some locations that may obscure the 1963 peak (Mitchell et al., 1983; Appleby et al., 1993; Callaway et al., 1996). However, in some lakes from areas like northern Europe, two distinct ^{137}Cs peaks are commonly observed, and thus the more recent one can be attributed to the Chernobyl nuclear accident peak (particularly when independently corroborated with ^{210}Pb dating).

In both the ^{210}Pb and ^{137}Cs dating techniques for recently deposited sediments, it is assumed that the disturbance of the sediment stratigraphy is small. In some cases disturbance by water movements and biological redistribution of sediments by benthic macroinvertebrates (Chapter 21), termed *bioturbation*, can be sufficient to obscure dating chronology, particularly in shallow lakes (e.g., Bennion et al., 2010). However, in most cases careful interpretation of recent dating analyses, combined with other paleolimnological data, provides much insight into the reconstruction of past lake events (e.g., Bouchard et al., 2017).

C. Physical properties of lake sediments

A wide suite of physical properties of lake sediments can provide useful information regarding the integrity of a sediment core (e.g., evidence of a slump event), the processes that led to the deposition of sediments, and potential sources of sediments. Sediment density, which can be measured through percent water quantification or through *computed tomography* (CT scan) or *X-ray scanning*, provides information about the structure of the core archive and can provide a researcher additional confidence that the temporal integrity of the core has remained intact. *Magnetic susceptibility*—a measure of the sediment's relative attraction to a magnet—is useful in evaluating probable sources of sedimented materials in lakes (Sandgren and Snowball, 2001). Inorganic material washed into the lake from the drainage basin tends to have a higher magnetic susceptibility signature and thus this measure is often used to infer erosion. From magnetic susceptibility time series, interpreted together with the chronology of sedimentation rates and other proxy indicators, a researcher can draw conclusions about watershed deforestation, afforestation, and/or agricultural disturbances. However, one also needs to consider potential remobilization of redox-sensitive metals (e.g., iron and manganese) and postsediment deposition, as these metals have relatively high magnetic susceptibility signatures (Liu et al., 2012). Finally, magnetic susceptibility has been used to align cores collected from the same basin, but this work should ideally be ground-truthed with additional proxies.

Some lake deposits exhibit distinct laminations in which temporal differences in the composition and quantities of suspended matter laid down create fine alternations of light- and dark-colored layers. In some cases the laminations are not visible to the naked eye but can be discerned using X-ray or other imaging analyses. When a pair of laminae (a dark and a light couplet) represent a period of 1 year, the lamination is called a *varve*. Varves are generally restricted to sediments of lakes that have not been physically disturbed after deposition and experience seasonal variation in sediment sources. For example, varves may appear in glacial environments, where relatively coarse sediments are deposited in recipient lakes with the inflow of meltwater in the spring and summer (Lamoureaux, 2001). Finer sediments are deposited over these during the winter when the lake is covered by ice. Similar bimodal deposition can be found in temperate lakes, particularly in meromictic lakes in which the sediments are not disturbed by water circulation (e.g., Zolitschka et al., 2015). In these cases records of seasonal differences can be found in the deposition of diatoms, tree pollen, clay, iron oxides, and/or calcium carbonate (Simola, 1977). Varves provide an independent, high-resolution means of dating sediments and varve presence and thickness may be related to former climate conditions and/or trophic conditions (Zolitschka et al., 2015; Jenny et al., 2016).

Many recent lake sediment profiles contain charcoal, fly ash, and various particles from industrial processes. Wood burning, either from forest fires or from agricultural, domestic, or industrial sources, has been studied for many years through the analysis of *charcoal particles* in the sediment record (Whitlock and Larsen, 2001; Mustaphi and Pisaric, 2014). Atmospherically derived particles originating from fossil fuel combustion, nonferrous metal smelting, or iron and steel manufacturing often include a magnetic fraction that can be characterized and approximately quantified by their magnetic susceptibility (Oldfield, 1991; Dearing, 1999). Alternatively, small particles (i.e., ~30 μm) originating from incomplete coal and oil combustion, known as fly-ash *spheroidal carbonaceous particles* (SCPs), can be identified and counted under the microscope; such particles are regularly preserved in lake sediments and do not originate from natural processes. SCPs can be abundant in recent sediments when the coring site is near combustion-based pollution sources. Such soot particles provide a sedimentary record that can reflect the history and intensity of coal and oil combustion (Renberg and Wik, 1984; Rose, 2001). SCP deposition increased in the middle of the 19th century and has risen progressively since, with some reduced loading coincident with emission controls in recent times across North America and Europe (Rose, 2015). Quantification of SCP particles in regions of well-documented industrial histories has been applied as a complementary sediment core dating technique.

D. Inorganic chemistry

Chemical constituents of the sediments have been used to determine changes in limnological processes, atmospheric deposition, drainage basins, and climate (Engstrom and Wright, 1984; Boyle, 2001). Interpretations of chemical analyses from dated sediment cores often assume that, at least on a long-term basis, the stratigraphic changes observed reflect changes in inputs to the lake and/or metabolic transformations that have occurred within the lake. For example, the relative intensity of soil erosion is generally reflected in the concentrations of elements primarily associated with *clastic* minerals, which are fragments of rock transported from elsewhere (e.g., titanium [Ti] and aluminum [Al]; Boës et al., 2011). Another commonly measured element in sediment cores is phosphorus, which has been measured in the form of sorbed components of amorphous iron oxides, in discrete mineral phases, or as organically bound phosphorus. However, the inorganic phosphorus–iron complex is largely responsible for the exchange of dissolved phosphorus in many sediments, and the retention of phosphorus in lake sediments is affected by redox conditions, sediment mixing, and pH (Chapter 15). Thus it can be difficult to interpret total phosphorus profiles generated from direct sediment measurements as robust archives of past conditions (Ginn et al., 2012), but certain P fractions (namely HCl–P and organic-P fractions) are more stable in sediments (Ostrofsky, 2012). Many researchers instead infer past water column phosphorus dynamics from the application of modern calibration models (discussed below) to *subfossil* (i.e., remains of organisms that have not undergone the complete fossilization process) profile data.

Studying the dynamics of the inorganic chemical composition of sediments, particularly biologically nonessential elements (e.g., Ti, Al), allows one to develop estimates of erosion and allochthonous inputs. For example, the application of *high-resolution X-ray fluorescence scanning* to a core from Lago Petén Itzà, Guatemala (Obrist-Farner and Rice, 2019), close to the Mayan settlement of Nixtun-Ch'ich', shows a clear increase in magnetic susceptibility as well as numerous elements associated with erosion (e.g., titanium [Ti], iron [Fe], and silica [Si]; Fig. 30-4). Using the periods defined by archeological analyses of Nixtun-Ch'ich', it is evident that the greatest changes over the course of the entire sediment record occurred during the Mayan occupation, which suggests that enhanced allochthonous inputs were associated with their settlement and agricultural

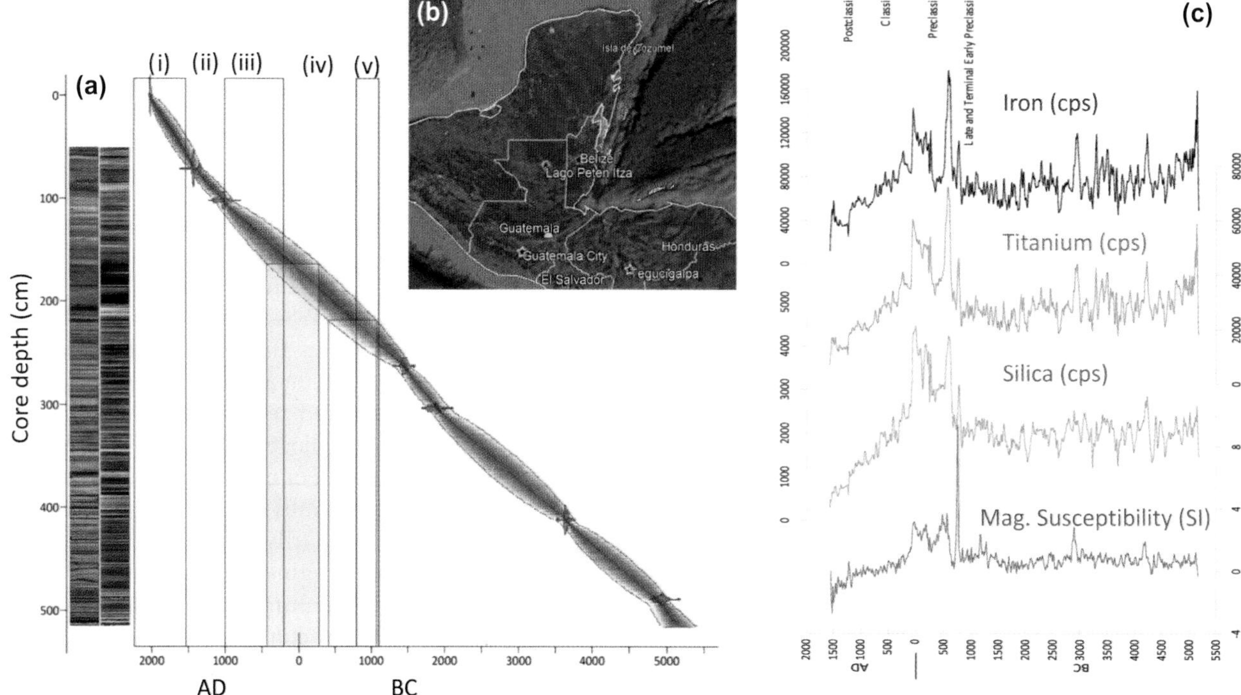

FIGURE 30-4 (a) The sediment core image, density scan, and chronology based on six radiocarbon dates from Lago Petén Itzà (plotted alongside the cultural eras defined as: (*i*) Modern; (*ii*) Postclassic; (*iii*) Classic; (*iv*) Preclassic; and (*v*) Late and Terminal Early Preclassic; (b) map showing the study site in northern Guatemala; and (c) the magnetic susceptibility (SI units, which is a relative measure) and geochemical profiles measured in counts per second (cps) on an X-ray fluorescence core scanner. Note that the surface portion of the sediment record is not presented. *(Figure modified from one presented in Obrist-Farner and Rice, 2019.)*

activities, or fires and/or droughts (Obrist-Farner and Rice, 2019).

The chemical composition of sediments can also be used to track contaminants released from a variety of anthropogenic sources, such as from mining or other industries (Blais et al., 2015). For example, a short sediment core from Lake Orta, Italy, recorded a history of multiple stressors over the past century (Fig. 30-5). Initial contamination of Lake Orta was associated with a rayon factory that discharged wastes with elevated concentrations of copper and ammonium sulfate (Manca and Comoli, 1995). Once in the lake, the ammonium sulfate was biologically oxidized and produced a large acidification effect. New electroplating industries on the shores of Lake Orta also contributed chromium, copper, zinc, and nickel. Measures were put in place to minimize the discharge of pollutants in the 1980s. In addition, Lake Orta was treated with lime to increase the pH and this led to a reduction in heavy metal concentrations. The sediment record from Lake Orta captures many of the historical dynamics described, with the water and sediment Cu time series showing similar dynamics (Tolotti et al., 2018). Differences between these records may be in part because the sediment records capture a lake-wide and temporally integrated signature, whereas the water column data is based on a pelagic sampling effort (and likely conducted only during the summer). The changes in Lake Orta's chemical conditions had a profound impact on its biology, much of which was also captured in sediment archives (discussed below).

Paleolimnologists may also choose to conduct broader regional studies of many lakes and thus trade off detailed temporal sampling for a "snap-shot" approach. Paleolimnologists often use the term "*top-bottom analyses*" whereby proxy data from the surface sediments (i.e., the "top" sample) represent recent conditions and are compared to those from a sediment slice from predisturbance times (i.e., the "bottom" sample). This top-bottom approach has been effectively used for regional assessments, beginning with the early paleolimnological work on acid rain (Cumming et al., 1992). Likewise, Camarero et al. (2009) conducted a study of 275 alpine and Arctic lakes across Europe and measured trace elements in surface (contemporary) sediments and sediments taken at a depth of 15–17 cm, which represented preindustrial conditions in their study lakes (Fig. 30-6). The authors then compared the enrichment in heavy metals between surface and preindustrial samples, weighting each interval against the Ti concentration in that interval to account

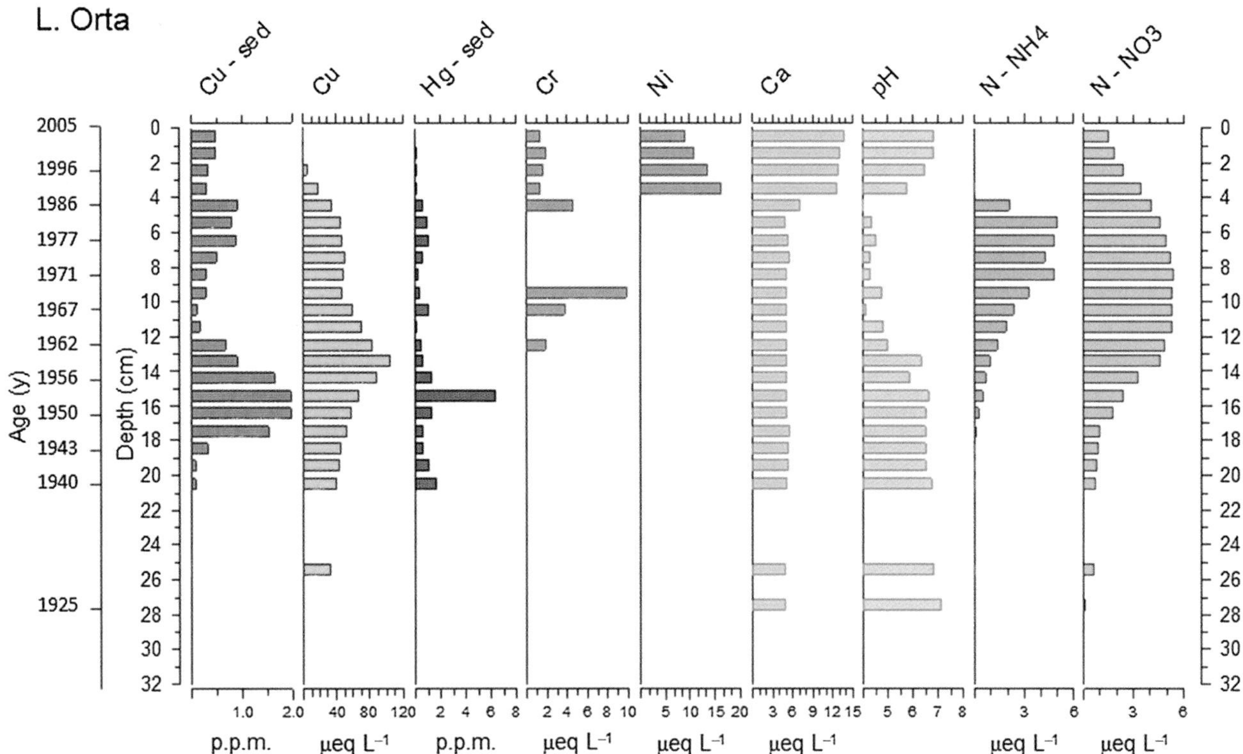

FIGURE 30-5 Time series spanning the last century of numerous elements measured in the water column (measured as μeq L^{-1}) and in the sediment core (as ppm) from Lake Orta, Italy. This lake is a site with one of the longest records of paired data of both heavy metal concentrations measured in the water column and corresponding sediment intervals. *(Figure from Tolotti et al., 2018.)*

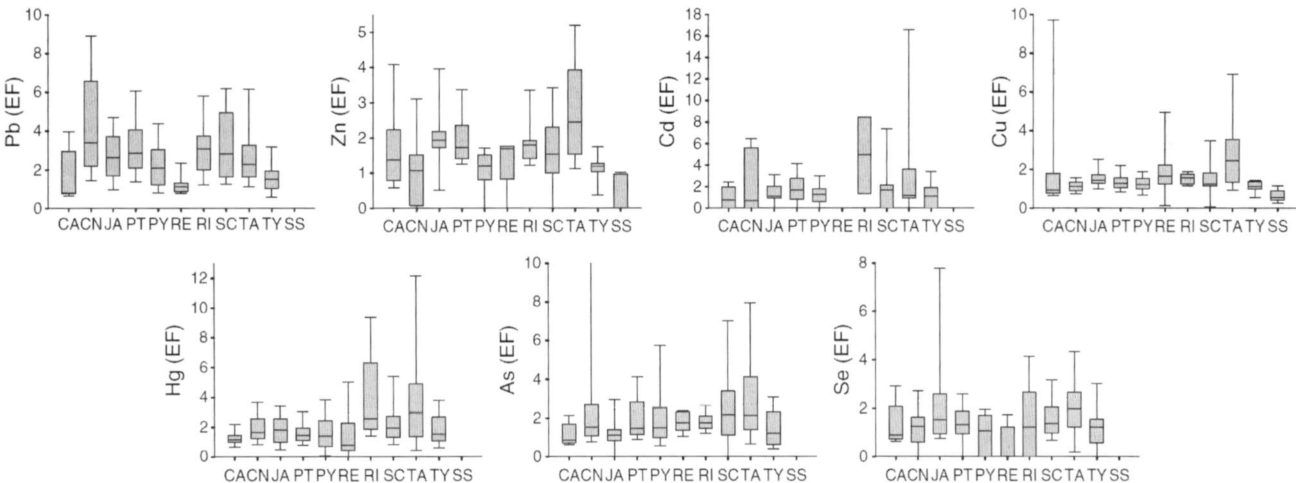

FIGURE 30-6 Enrichment factors (EFs), calculated as $EF = ([Metal]_{top}/[Ti]_{top})/([Metal]_{bottom}/[Ti]_{bottom})$, for a suite of heavy metals measured in surface (*top*) sediments and preindustrial (*bottom*) sediments across 11 lake regions in Europe. On average, Pb has the greatest enrichment factors across all lakes, followed by Hg and As. *(Figure from Camarero et al., 2009.)* The codes along the x-axis refer to different lake districts: CA, Central Alps; CN, Central Norway; JA, Julian Alps; PT, Piemonte-Ticino; PY, Pyrenees; RE, Retezat; RI, Rila Mountains; SC, Scotland; TA, Tatras; TY, Tyrol; SS, Greenland.

for natural geological weathering that might differ through time and among sites. Given that these were all remote sites where heavy metal contamination would be driven by airborne deposition, it is interesting that many sites have concentrations of elements that are generally associated with moderately contaminated sites. Ideally, top-bottom work is paired with the study of several continuous records from the region to generate a portrait of the continuous temporal dynamics of the region.

E. Organic constituents

There is a broad range of organic compounds in lake sediments (Meyers and Teranes, 2001; Blais et al., 2015). For many years, scientists were just quantifying the concentration of bulk organic matter or organic carbon, as well as inorganic carbon, using relatively simple analyses (i.e., through weight loss after ignition analyses; Heiri et al., 2001). Stable isotope analyses of lake sediments have also been widely applied for several decades, and C, N, O, H, and Si stable isotopes are among the most extensively studied. For example, $\delta^{15}N$ values are enriched in top predators (e.g., carnivores, piscivores) and their waste products (Peterson and Fry, 1987). As such, $\delta^{15}N$ values have been used in sediment records to trace past human or animal population dynamics, and often ground-truthed with historical data or measurements from other paleolimnological indicators (e.g., Finney et al., 2000; Bunting et al., 2007; Duda et al., 2020). Recently, there has been a move toward analyzing the stable isotopic signature locked within subfossils or within specific compounds so that one can gain greater insight into the ecological nature of shifts in stable isotopes over time (van Hardenbroek et al., 2019).

There has also been a move toward measuring distinct spectra of bulk sediment using *visible–near-infrared spectroscopy* to infer past lakewater total and dissolved organic carbon (DOC) concentrations (Meyer-Jacob et al., 2017, 2019, 2020). Understanding whether the browning of waters (Chapter 28) observed in many northern lakes over the past few decades is a recent phenomenon or part of a rebrowning cycle following ecosystem acidification has been studied using this approach. In particular, analyses of sediment cores collected from the Sudbury region of Canada have shown that, with the intensive 20th-century acidification of the region, inferred DOC concentrations declined in lakes set in poorly buffered catchments (Meyer-Jacob et al., 2019, 2020). With legislation brought in to control emissions of SO_x and NO_x, many lakes returned to a more neutral pH and the inferred DOC concentrations have also rebounded. In contrast, lakes set in a nearby region with lower acid deposition rates showed more subtle declines in lakewater DOC, but these values have rebounded to higher-than-preindustrial values, which may be due to recent climate change (Meyer-Jacob et al., 2019).

Another major advancement in paleolimnology is applying compound specific analyses that are being discovered through the application of an *omics approach*. Specifically, an omics approach refers to a holistic study of biological systems that blends chemistry, multivariate analyses, and *a priori* knowledge of the ecosystems (Bell and Blais, 2019). The most common groups of organic compounds analyzed in lake sediments typically include lipids (sterols, stanols, *n*-alkanes, and fatty acids), lignins, polyaromatic hydrocarbons (PAHs) and other toxic pollutants (such as insecticides and PCBs), pigments, and DNA. Plastic particles can also be added to this list, which themselves are organic molecules (i.e., carbon-based molecules) and generally designed to be resistant to decay; plastic particles are increasingly studied in recognition of their growing distribution across landscapes and through time. The detection and quantification of these different organic compounds can provide insights into a wide variety of environmental issues, as well as provide information related to the transformation of organic matter (see Table 30-2 for examples). We will reserve discussion of pigments and DNA for the section on biological proxies, as they are commonly used to identify the presence of particular taxa.

Sterols and their breakdown products, the *stanols*, represent a group of paleolimnological markers that are of emerging interest because they can be reflective of very targeted groups (e.g., *coprostanol*, a sterol that is quite abundant in human feces; Bell and Blais, 2020). Sterols originate as a key component of cell membranes in microbes, plants, and animals. One of the earlier paleolimnological studies to analyze the coprostanol marker in a sediment archive was from a remote island in Sweden (D'Anjou et al., 2016). The Holocene record retrieved from Lilandsvatnet Lake was based on a study of numerous biomarkers, including several fecal biomarkers (e.g., 5β-stigmastanol) as well as *PAHs* (which provide information on the region's fire history) and a ratio of *n-alkanes* that provide information on the relative amount of forest versus grassland cover. The collection of biomarkers from Lilandsvatnet Lake demonstrated a clear transition from a relatively wild landscape to a human-occupied ecosystem around 2250 calendar years before present, at which point all fecal and pyrolytic biomarkers increased and the dominant vegetation cover shifted from forests to grasslands (Fig. 30-7). Subsequent variation in fecal biomarkers was moderately correlated to a temperature reconstruction based on tree-ring analyses (*dendrochronology*), suggesting that at least part of the dynamics in the human population at this site was related to climate change. However, the authors were careful to point out that the temporal variation in fecal biomarker concentrations was not a direct indicator of human population size but rather should be interpreted as a first-order approximation.

The examination of many organic compounds in sediments must include an evaluation of whether the organic compound originated from within or outside of the lake basin, and whether the organic compounds underwent degradation during and/or after sedimentation. The first question may be addressed by conducting

TABLE 30-2 Groups of Some Organic Compounds and Stable Isotopes Commonly Used in Paleolimnology, With Example Applications

Groups	Subgroups	Example of the indicator's application in paleolimnology
Lipids	Long chain *n*-alkanes	*n*-Alkanes with 25–33 C molecules; relative contributions from terrestrial leafy vegetation to grasses expressed as the relative abundance of $([C_{25}] + [C_{27}] + [C_{29}])/([C_{29}] + [C_{31}])$; where the number after the C refers to the Carbon chain length
	Fatty acids (FAs)	Polyunsaturated FA 20:5ω3 and 22:6ω3; unique to certain phytoplankton
	Sterols and stanols	Cannabinol; unique to hemp
Lignin-phenols		Cinnamyl/vanillyl ratio; indicator of woody to nonwoody plants
Polyaromatic hydrocarbons (PAHs)		PAH concentration and composition are complementary fire markers to charcoal and can also be used to track fossil fuel emissions
Stable isotopes	$\delta^{13}C$	Bulk sediment $\delta^{13}C$ measurements; used to infer past changes in algal productivity or C sources
	$\delta^{15}N$	Bulk sediment $\delta^{15}N$ measurements; used to infer past changes in animal (e.g., anadromous salmon, seabirds) or wastewater inputs
	$\delta^{18}O$	Cellulose $\delta^{18}O$ measurements; used to infer hydrological changes

regional calibration studies where lake and watershed features are well known. To address the issue of preservation, there is a growing body of literature that summarizes the relative stability of different compounds as well as what conditions favor their stability (e.g., Berke, 2018; Derrien et al., 2020). Multiproxy studies can also shed light on how the lake ecosystem and its preservation environment may have changed over time, thus constraining the interpretation of the organic marker of interest.

With plastics, we have the opposite issue in that they are very resistant to decay and thus have experienced the undesirable outcome of being transported to water bodies and integrated into food webs (D'Avignon et al., 2021). Crawford and Quinn (2016) defined *microplastic particles* as being small synthetic polymers that are less than 5 mm along their longest dimension but greater than 1 μm (particles smaller than 1 μm are termed *nanoplastics*). Given the exponential growth in plastic use since ~1950 CE (Geyer et al., 2017) and their resistance to decay, we are now finding microplastics in most habitats around the world, including lakes and rivers. For example, surface sediments from the St. Lawrence River (Canada/USA) had densities of microplastics that were on the same order of magnitude as those reported for other aquatic habitats situated in other heavily urbanized regions (Crew et al., 2020). There are only a handful of microplastic studies that have been conducted in dated lake sediment cores (reviewed in Bancone et al., 2020), but we anticipate that this area will receive substantial attention in the years to come.

II. Biological indicators

A. Morphological remains

Paleolimnology has a long history that began with analyses of morphological remains of aquatic and terrestrial organisms. Preservation of organisms in sediments is typically incomplete, and the extent of preservation depends both upon the type of organism and the environmental conditions that prevailed at the time of sedimentation. Similar to the focusing effects of sediment particles and organic matter discussed earlier,

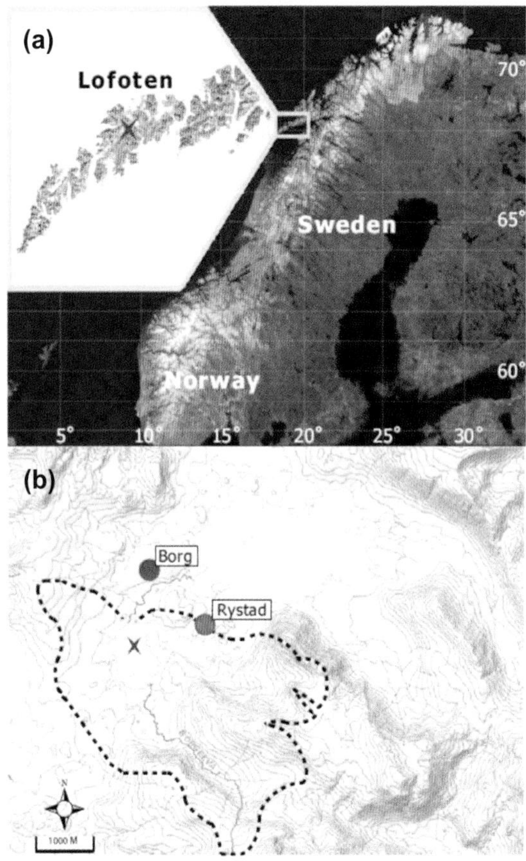

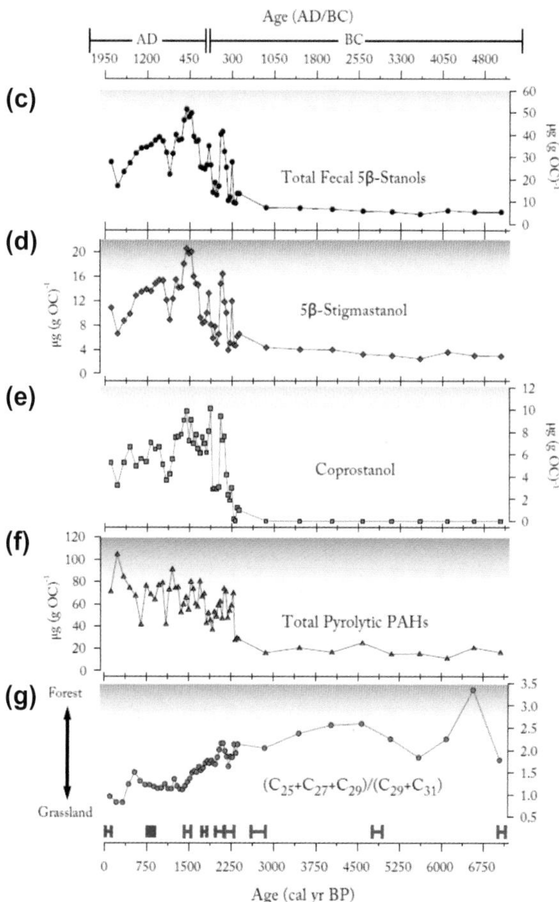

FIGURE 30-7 Holocene sediment record from Sweden showing the trajectory of numerous biomarkers and providing evidence of transition from a relatively natural landscape to a human-occupied ecosystem around 2250 calendar years before present (cal yr BP). Panel (a) shows the location of the study region; (b) shows the watershed and coring location, as well as two archeological sites; (c)–(g) show time series of the following biomarkers: total fecal 5β-stanols, 5β-stigmastanol, coprostanol, total pyrolytic PAHs, and an *n*-alkanes index from leaf waxes (C_{25} + C_{27} + C_{29})/(C_{29} + C_{31}). *(Figure from D'Anjou et al., 2016.)* The radiocarbon dates are shown at the bottom of panel (g) (i.e., time span reflects two standard deviation error estimates), where dated macrofossils are highlighted in *blue* and tephras in *red*.

remains of inshore organisms are moved offshore by wind-generated water movements and currents. The amount of such movement and displacement varies widely with different drainage basin and lake basin characteristics (e.g., basin slope, maximum depth) as well as the size of morphological subfossil. Pollen and spores of terrestrial plants are frequently abundant, as are frustules of diatoms and chrysophyte cysts (Frey, 1974, 1988; Cohen, 2003; Smol, 2008). Although remnants of some other groups of algae can occur (e.g., *Pediastrum* cell nets, *Chara* oocysts, cyanobacterial akinetes), many groups are not reliably preserved as identifiable subfossils, and for those groups, biogeochemical and DNA approaches are typically used.

Cladocerans and chironomid midges are the animals that generate the most abundant and diversified zoological remains. Many subfossils can be identified down to species, particularly among the morphologically distinctive diatoms, desmids, Cladocera, ostracods, and several insect groups. When only one stage in the life cycle is preserved (e.g., resting eggs of rotifers), linking these remains to the species of origin is more difficult and in some cases simply cannot be done (although genetic methods could advance this work in the future; see the following DNA section). Comprehensive reviews on biological proxies used in paleolimnology include Cohen (2003), Seppa and Bennett (2003), Smol (2008), Domaizon et al., (2017), and chapters in Smol et al. (2001a, 2001b).

i. Pollen and spores

Plant pollen and spores are produced in great abundance, but only a few of those produced fulfill their reproductive function. As pollen types are dispersed, they are well mixed by atmospheric turbulence broadly in proportion to the quantity and composition of the parent vegetation in and surrounding lakes, although different plants produce different amounts of pollen grains. Most pollen grains have a heavy protective layer (*exine*) and are resistant to decay. The taxonomy of pollen is based largely on the pore shapes and numbers as well as the structure and sculpture of exine. Generally, the taxonomy of pollen is relatively well known for vegetation of many parts of the world. The study of pollen grains and spores, collectively referred to as the field of *palynology*, has many applications in the paleoenvironmental sciences, but most notably in the reconstruction of terrestrial vegetation, from which past climatic and other environmental information can be gleaned (e.g., impacts of past agriculture and effects of plant pathogens). Since pollen is relatively abundant in sediments, small aliquots of sediments are sufficient for analysis. As a result, analyses can be performed at very close intervals within the stratigraphy of sedimentary deposits. The pollen assemblage provides an index of vegetation in and surrounding the lake and can reflect changes that have occurred in the vegetation through time. Pollen analyses have also been used to study macrophyte dynamics, but such pollen are often not very abundant and require careful calibration (Zhao et al., 2006). Many detailed works are available on the principles of pollen analysis, including pollen identification and interpretation of pollen data (e.g., Fægri and Iversen, 1991; Moore et al., 1991; Bennett and Willis, 2001; Seppä, 2013). A good introduction to palynology remains the classic book by Birks and Birks (1980).

Changes in pollen and spores in a sediment profile provide a vegetation history from which presumptive evidence for climatic and direct watershed vegetational changes can be obtained. These vegetational changes in the landscape and catchment area provide insights into temperature and rainfall changes, soil development, and changes in the drainage basin caused by human activities, as well as other drivers. For example, in Eastern North America, the arrival of European settlers is often identified in lake sediment records by the increase in ragweed (*Ambrosia* spp.) pollen as well as other weedy species. Rapid changes in vegetational succession, such as those caused by catastrophic pathogen outbreaks, are also readily tracked in the pollen record of lake sediments (Waller, 2013). Finally, widespread congruence between accelerated lake sedimentation rates and reduced abundance of tree pollen was identified by synthesizing pollen diagrams from 632 lakes worldwide; these results highlight that land clearance dating back thousands of years had a significant effect on landscape erosion (Jenny et al., 2019).

ii. Plant macrofossils

Plant *macrofossils* are subfossils from terrestrial and aquatic plants that are large enough to be visible to the naked eye and are frequently preserved in lake sediments, especially in nearshore environments. Common examples include seeds, megaspores, and fragments of leaves, rhizomes, flowers, and woody tissue (see reviews by Birks and Birks, 1980; Hannon and Gaillard, 1997; Gaillard and Birks, 2013). Seeds and other reproductive parts are often identifiable to the species level. Such macrofossils can supplement the pollen record, as many pollen grains cannot be identified to the species level, and some pollen types are poorly preserved. In addition, many aquatic macrophytes typically produce such low quantities of pollen that they are not routinely represented in pollen counts. Importantly, as pollen grains can often be transported long distances by wind or other vectors, pollen analyses generally provide a regional record of vegetation change. Given that macrofossils are larger and less easily transported, they represent much more local (i.e., watershed level) changes in vegetation. However, the limited transport of plant macrofossils can result in a heterogeneous distribution of macrofossils within a lake. Moreover, because the quantity of plant macrofossils per volume of sediment is usually very low, one must examine large quantities of sediment to acquire sufficient numbers. In general, the most powerful interpretations are made when both groups of indicators are used simultaneously (e.g., Birks and Birks, 2001).

Quantitative analyses of the stratigraphy of plant macrofossils assist in the reconstruction of past lake level, changes in primary productivity, and other variables. Some of the strongest lines of evidence for lake-level change are the advance and retreat of littoral vegetation deduced from analyses of macrophyte fossils and coarse organic matter; such inferences are often based on known physiological and ecological characteristics of the plants (e.g., Birks, 2001; Birks and Birks, 1980; Hannon and Gaillard, 1997; Gaillard and Birks, 2013). Terrestrial macrofossils can also provide insights into plant succession that have occurred in and surrounding lakes (MacDonald et al., 2000), as well as human activities (e.g., wood chips from a sawmill; Engstrom et al., 1985).

iii. Algae and other primary producers

Several types of algae are well preserved as subfossils in lake sediments, but by far the most common are the diatoms (Bacillariophyceae, Chapter 17), which have siliceous cell walls (called *frustules*, which are composed of two *valves*; Smol and Stoermer, 2010). Diatoms are useful paleoecological indicators because their remains commonly occur in abundance and are often well preserved, and most can be identified to the species level or lower (e.g., subspecies) from cell wall characteristics (Fig. 30-8). Additionally, given that diatoms are present in numerous aquatic habitats, and that the physiological and ecological characteristics of many taxa are reasonably well known, reconstructions of probable past lake

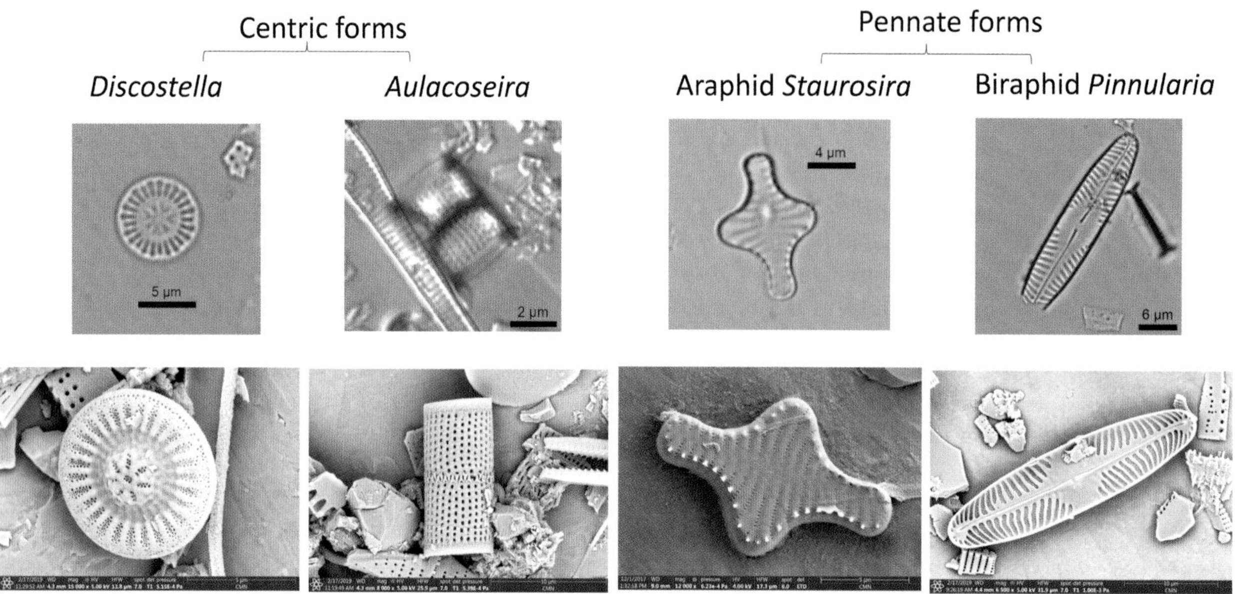

FIGURE 30-8 Light micrographs (*top*) and scanning electron micrographs (*bottom*) of diatoms showing a diversity of forms and taxonomic groups. *(Images from Katherine Griffiths.)*

conditions can be made from sedimentary diatom assemblages. The ecological optima and tolerances of many taxa have been further refined by paleolimnologists using surface sediment calibration or training sets, as described below.

Diatoms have been used extensively to assess natural as well as anthropogenic environmental changes. Many diatom species respond quickly to changes in habitat characteristics, particularly their chemical environment, so preserved assemblages record much ecological information (e.g., Smol and Stoermer, 2010). Shifts in preserved assemblages can therefore be used to infer long-term ecological changes resulting from eutrophication, acidification, climatic change, and many other environmental issues (Smol and Stoermer, 2010). Sedimentary diatoms have also been used extensively to evaluate changes in climate-related variables (reviewed in Rühland et al., 2015). Likewise, identifying littoral (epiphytic or benthic) versus planktonic-dwelling species in sediment core layers can yield further insight into past conditions of lake morphometry, water-level fluctuations, and shifts between littoral versus planktonic dominance of primary productivity (e.g., Wolin and Stone, 2010).

1. Surface sediment calibration sets and interpretation

Much of what we know about the environmental optima and tolerances of diatoms (as well as other paleolimnological indicators) has been derived from the use of *surface sediment calibration* or *training sets*. The overall approach is fairly straightforward and involves establishing a relationship between the sediment diatom assemblages found in the surface sediments of a set of lakes and the environmental conditions in those lakes over the recent past (i.e., when the sediment would have accumulated). Ideally, the lakes would span a large range for the environmental variable(s) of interest. The established relationship can then be used as a reference, allowing a paleolimnologist to say that if diatom assemblage X is found in a sediment core, it is likely that the environmental condition when the sediment accumulated was Y. For example, if a paleolimnologist wished to relate diatom assemblages to past lakewater pH, they would first choose a set of calibration lakes (typically $n > 50$ lakes). Environmental data (e.g., pH and other key variables) for all calibration lakes would be measured during field sampling or compiled from existing measurements. The paleolimnologist would then collect the most recent sediments (the surface 0.5 cm or 1 cm depth, representing the last few years of sediment accumulation), typically with a surface sediment corer (Glew et al., 2001). The analyst would then need to identify and count the diatom taxa in these surface sediment assemblages. Data analyses would then relate the environmental data matrix to the diatom assemblage matrix using a variety of statistical techniques (Birks et al., 2012). The statistical relationship where one uses the subfossil assemblage to infer the environmental condition is called a *transfer function*. The application of a pH transfer function to diatom stratigraphies from Adirondack lakes of New York (USA) indicated that many lake core records generated in the 1980s and 1990s have shown considerable acidification since the 1850s (Kingston and Birks, 1990; Cumming et al., 1994; Smol et al., 1998). Many other reconstructions of historical lake pH exist based on observations of changes from alkaliphilous to acidophilous sedimentary diatom remains (reviewed in Battarbee et al., 2010). Paleolimnology played a key role in showing that most cases of lake acidification had occurred primarily because of atmospheric deposition of acid pollution as rain, snow, and particulate precipitation (Smol, 2008). *Weighted-averaging models* have also been developed to infer lakewater total phosphorus from diatom assemblages preserved in sediments (e.g., reviewed in Hall and Smol, 2010), as well as other important variables (e.g., lakewater salinity, Fritz et al., 2010).

Other siliceous algal subfossils include those produced by chrysophyte algae (Chapter 17), of which many produce scales and all of whom produce resting stages in the form of endogenously formed siliceous cysts (referred to as *statospores* in the earlier literature, but now referred to as *stomatocysts* or simply cysts; Duff et al., 1995; Wilkinson et al., 2001). The cysts are spherical or oval and often the walls are ornamented; these and other features (such as collar structure) allow for fairly specific taxonomic identification (Cronberg, 1986; Smol, 1988, 1995). In addition, some chrysophyte taxa, such as those in the commonly encountered genera *Mallomonas* and *Synura*, are also characterized by silica scales, which, like diatom valves, are well preserved in lake sediments and are taxonomically diagnostic (Smol, 1995). Scales and cysts are common in the sediments of many lakes and have been used in a variety of paleolimnological applications, such as reconstructing lakewater pH, specific conductivity, and chemical constituents (Siver, 1993; Duff et al., 1995, 1997; Wilkinson et al., 2001; Zeeb and Smol, 2001). Chrysophytes tend to be associated with relatively oligotrophic lakes, so declines in their abundance are associated with an oligotrophication, and as a result, a ratio of subfossil diatom frustules to chrysophycean cysts has been suggested as a potentially useful index of trophic status in temperate lakes (Smol, 1985), although changes in other lake conditions can alter this ratio.

As with all paleolimnological indicators, one must consider *diagenesis* (e.g., differential rates of dissolution of the diatom frustules or chrysophyte scales and cysts during and after sedimentation) (Ryves et al., 2006, 2009). However, in the vast majority of examples preservation is excellent and paleolimnological assessments are highly reproducible and verifiable.

Other primary producers are also preserved as subfossils but are less commonly used in paleolimnology. Certain green algae (e.g., *Pediastrum*; Turner et al., 2016), cysts of dinoflagellates (e.g., McCarthy and Krueger, 2013), and *akinetes* and *heterocysts* of Cyanobacteria (e.g., van Geel et al., 1994), as well as other *nonpollen palynomorphs* (e.g., van Geel et al., 2002), are preserved in lake sediments. However, less is known of the physiology and ecology of these algae than is known for many diatoms and chrysophytes, and as a result, interpretation is often less rigorous than it is with changes in diatom assemblages, and the stratigraphic analysis of these algae is often most useful in multiproxy studies, when changes in numerous parameters preserved in the sedimentary archive are studied.

iv. Animal remains

Nearly all groups of animals leave at least some identifiable morphological remains in lake sediments. The most abundant animal subfossils are those of Cladocera and nonbiting midges (Chironomidae), which produce remains made of *chitin* that preserves well in lake sediments (Fig. 30-9). Under favorable circumstances, shells of ostracods, shells or cases of testaceous rhizopods, spicules of sponges, egg cocoons of neorhabdocoele Turbellaria, resting eggs of rotifers, bryozoan statoblasts, copepod spermatophores, and oribatid mites can also be found (Frey, 1964; Smol et al., 2001b). Similar to diatoms and other bioindicators, most analyses of animal remains rely on the interpretations of changes in the relative frequencies (%) of individual species or larger taxonomic groups. The remains of larger organisms (e.g., fish scales and bones) are generally less abundant (Davidson et al., 2003) and require larger volumes of sediments for the recovery of significant numbers. Therefore cores must be analyzed at coarser intervals than for subfossils.

1. Nonbiting midges (chironomids) and related insect indicators

The larvae of Chironomidae are often well represented in lake sediments by their chitinized head capsules, which can often be identified to genera or even lower. Because most larval chironomids inhabit the profundal regions of a lake, they provide important environmental information on deepwater habitats, such as changes in deepwater oxygen levels (Brodersen and Quinlan, 2006). For example, some species of chironomids (e.g., *Chironomus*) are commonly distributed in the profundal zone of eutrophic lakes, which often experience low-dissolved-oxygen concentrations at the end of the stratified period. Others (e.g., some species of *Tanytarsus*) are more common in lakes with higher dissolved oxygen in the hypolimnion at the end of the stratified period. Therefore, if a lake becomes more eutrophic with decreasing oxygen content in the hypolimnion, one might expect, for example, a shift in the profundal midge assemblage from the dominance of taxa that require high amounts of oxygen to those that tolerate decreasing oxygen concentrations. Transfer functions based on the oxygen optima of chironomid taxa have been developed whereby past deepwater oxygen levels can be estimated based on the assemblage composition

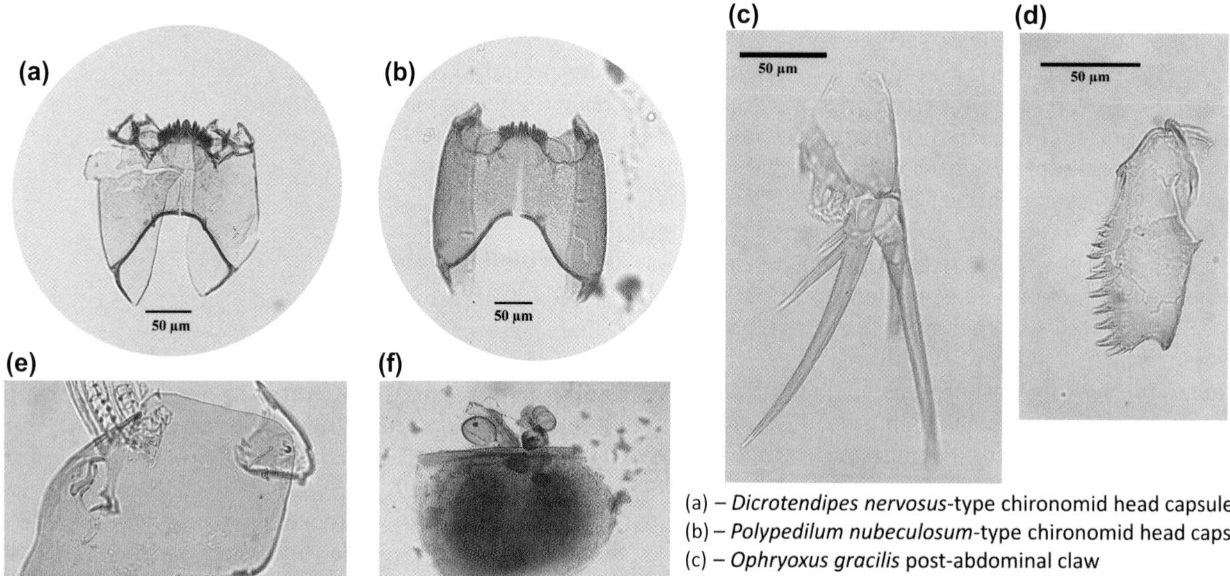

FIGURE 30-9 Digital micrographs of larval chironomid head capsules (a) and (b) and subfossil cladocerans remains (c)–(f) commonly found in lake sediment samples. *(Images from Brigitte Simmatis.)*

of head capsules (e.g., Quinlan and Smol, 2002). In addition to deepwater oxygen reconstructions, chironomid assemblage composition is known to also be closely linked to climate-related variables such as temperature. Beginning with the first chironomid-based transfer function developed over 30 years ago (Walker et al., 1991), this paleoclimatic inference tool has been applied globally.

Other chitinous insect subfossils have been used in paleolimnological assessments, although they are far less common than chironomids. For example, the chitinized mandibles of *Chaoborus* larvae can be identified to the species level. Some taxa, like *Chaoborus americanus*, rarely coexist with fish, so an increased abundance of this taxon's mandibles in sediments likely indicates the loss of fish populations (Sweetman and Smol, 2006). Another group of insects used in paleolimnology are black fly (Simuliidae), whose larval subfossils have been used to track past hydrological changes, such as those associated with river flow (Currie and Walker, 1992). Other related insect indicators and applications have been reviewed by Walker (2001).

2. Cladocerans

Exoskeletons and *ephippia* (i.e., *resting eggs*) of cladoceran zooplankton preserve reasonably well in lake sediments and thus have been studied in the paleolimnological record in great detail (Korhola and Rautio, 2001). In terms of exoskeleton remains the pelagic cladocerans tend to be numerically dominant in sediments collected from the profundal zone of lakes, but it is not uncommon to detect a rich diversity of littoral taxa as well. Considerable evidence exists showing that littoral faunal remains are redistributed by currents within lake basins, where they become integrated with remains of planktonic Cladocera (Kerfoot, 1981; Frey, 1988). The populations recorded in profundal sediments therefore represent a reasonable integration of community dynamics over habitats and seasons. Overall, changes in preserved cladoceran assemblages reflect shifts in food quantity and quality, habitat, competition, or predation (Chen et al., 2011; Korosi et al., 2013; Griffiths et al., 2019).

Numerous insights about past conditions have been gained from analyzing cladoceran remains from sediment records (e.g., Hofmann, 1987; Alric et al., 2013; Nevalainen and Rautio, 2014). For example, changes in species diversity of littoral chydorid communities, as pioneered by Goulden (1969b), have been used to assess the responses of the communities to disturbances. Changes in the morphology of subfossil cladocerans have been associated with changes in invertebrate predators (Chen et al., 2011; Korosi et al., 2013). Likewise, greater relative abundances of *Daphnia* compared to *Bosmina* in sediments may occur when zooplanktivorous fish are absent or when fish are present in low densities when macrophytes are co-occurring (Davidson et al., 2010). However, in softwater lakes decreases in the relative abundance of larger daphnids may also reflect low calcium concentrations, as these conditions have been shown experimentally to be an important driver of calcium-rich *Daphnia* species abundances (Jeziorski et al., 2008). More recently, paleolimnologists have been making direct measurements of cladoceran subfossil pigmentation to infer past underwater UV conditions (Nevalainen and Rautio, 2014).

In addition to analyses of carapace remains, there is a substantial body of work focusing on the ephippia of cladocerans to understand changes in their morphology, abundance, chemistry, or population structure. For example, using a large suite of lakes from Denmark, Greenland, and New Zealand, Jeppesen et al. (2002) demonstrated that *Daphnia* ephippia size was related to the organism's adult body size; these factors were in turn strongly related to planktivorous fish abundance. Size-fish predation relationships were further improved by including total phosphorus concentrations as another predictor. Jeppesen et al. (2003) later showed that the proportion of *Daphnia* to bosminid ephippia was significantly and inversely related to fish density and positively related to total phosphorus. Direct measurements of the chemistry of ephippia have also provided insight into food sources, trophic position, and ecotoxicological exposures (Wyn et al., 2007; Perga, 2011). Finally, there has been a substantial body of work aimed at understanding how the population structure of cladocerans has changed over time. In some cases the relative abundances of sexual versus asexual morphs have been used to gain insight into how different groups responded to past stressors (Sarmaja-Korjonen, 2003, 2004), whereas in other studies the "fossil" populations have been hatched and used in laboratory studies to define their responses to key stressors (Burge et al., 2018).

3. Ostracods

Ostracods (also spelled ostracodes), which are characterized by calcareous carapaces (formed by two valves or shells), are often well preserved in sediments of hardwater lakes. As reviewed by Holmes (2001), many taxa have specific environmental optima and tolerances that can be linked to changes in habitat, pH, salinity, hydrology, and climate-related variables. Interestingly, past limnological conditions are reflected in the isotopic and geochemical composition of the ostracod shell (Schwalb, 2003). The study of ostracods as past indicators of ecological change has been undertaken in many parts of the world (Horne et al., 2012).

4. Other zoological proxy data

Other zoological groups are also used, although much less commonly than those reviewed thus far. Examples include protozoa, rotifer eggs, sponge spicules, bryozoans statoblasts, oribatid mites, and mollusks, as well as other invertebrate remains (see chapters in Smol et al., 2001a, 2001b). It is typically not practical to include all groups in paleolimnological assessments, and so most studies focus on proxy data that will be most relevant to the research questions posed; however, it is clearly understood that additional proxy data always strengthen paleoenvironmental interpretations (Birks and Birks, 2006).

B. Pigments

Pigments are cosmopolitan in their distribution among primary producers (phytoplankton, pigmented bacteria, benthic algae, macrophytes, and land plants). Fossil pigments of biota have been investigated extensively as organic constituents of sediments (Brown, 1969; Leavitt and Hodgson, 2001). Upon senescence and death, photosynthetic pigments undergo molecular transformations in which ions (e.g., the central magnesium atom of chlorophylls) and side groups (long-chain terpene alcohol, phytol) are lost progressively during physical or biological degradation. There is a tendency for the stability of degradation products to increase, that is, decrease in solubility and increase in relative resistance to further microbial and physical decomposition. Chlorophyll *a*, for example, degrades to pheophytin *a* with the loss of magnesium; pheophytin *a* then degrades to pheophorbide *a* with the loss of the phytyl group. The alternative sequence in which the phytyl group is lost first leads to the formation of the intermediate product, chlorophyllide *a*.

There is a general positive correlation between long-term growth and biomass of photosynthetic microbes and their associated pigments. If pigments are reasonably preserved by the time they are buried in sediments, the stratigraphic distribution of pigments offers the possibility of identifying relative changes in the abundance of different primary producers over time and, potentially, changes in environmental conditions of the water and sediments at the time of their deposition.

The number of pigments recorded in sediments is large. Nondegraded chlorophylls are generally less abundant than pheophytins, chlorophyllides, pheophorbides, and bacteriochlorophyll degradation products, which have all been identified and quantified. Additionally, a wide array of *carotenoids* is preserved, some of which are highly specific to groups of organisms such as Cyanobacteria or certain families of organisms (Leavitt and Hodgson, 2001). Certain algal and bacterial carotenoids, as well as bacteriochlorophyllous degradation products of purple photosynthetic and green sulfur bacteria (Chapter 23), have been recorded in sediment profiles and associated with events that led to eutrophication and meromixis (e.g., Hodgson et al., 1998). Myxoxanthin, myxoxanthophyll, oscillaxanthin, and other carotenoids have been used to infer long-term dynamics in Cyanobacteria (Züllig, 1989; Taranu et al., 2015).

A wide number of analyses on different types of lakes have reported correlations between sedimentary pigments and changes in trophic state. For example, Bunting et al. (2007) conducted comparative analyses of water column phytoplankton and water chemistry time series spanning 30 years with paired analyses of sedimentary pigment and isotopes from Loch Neagh, Northern Ireland. This site experienced eutrophication due to agricultural runoff, which was confirmed by identifying strong correlations between nutrient loading in the watershed (particularly nitrogen loading) and water column chlorophyll *a* or cell counts of dominant Cyanobacteria as well as sedimentary N isotopic signatures and diagnostic pigment time series (Fig. 30-10). More recently, a synthesis of paleolimnological data from across the north temperate to subarctic zones demonstrated that eutrophication of a large population of lakes is leading to the expansion of bloom-forming Cyanobacteria (measured as myxoxanthophyll; Taranu et al., 2015), an observation that has been corroborated with evidence from remote sensing of lakes over the past 30 years (Ho et al., 2019). Cyanobacterial pigments and their toxins have also been extracted from sediments dating back thousands of years. For example, in a Guatemalan lake that served as an important ecosystem for the Maya, the sediment archive shows evidence of harmful algal blooms during ancient times (Waters et al., 2021).

Over recent years, there has been a considerable expansion of analytical methods available to measure cyanobacterial and algal pigments. The new methods are typically faster, are nondestructive, and facilitate high-resolution studies. In particular, *visible reflectance spectroscopy* (VRS) has been extensively used over the past decade to measure total chlorophyll *a* (including its isomers and main breakdown products; Michelutti and Smol, 2016). Calibration work has shown that measurements of pigments from freeze-dried mud with a VRS instrument between 650 and 700 nm are strongly correlated with concentrations measured from sediment extracts that are measured with a *high-performance liquid chromatography* (HPLC) instrument (previously the main methodology for detailed pigment studies; Wolfe et al., 2006, Michelutti et al., 2010; Rydberg et al., 2020). A comparable calibration exercise was carried out for cyanobacterial pigments, and a proof-of-concept method is described in Favot et al. (2020). In a similar vein *hyperspectral core scanners* have been applied to track past changes in algal "green" pigments (i.e., the sum of chlorophyll *a* and pheophytin *a*) and shown to be strongly correlated to these same pigments measured via HPLC (Schneider et al., 2018). The advantages of hyperspectral core scanners are that they can provide data at very high spatial resolution (40–200 μm) and in a rapid and nondestructive analysis (Zander et al., 2022)

Nonetheless, the interpretation of fossil pigments as a measure of qualitative and quantitative changes of former microbial populations must be done critically, as with all proxy data. Many complex abiotic and biotic interactions alter the diagenesis of pigments before and after arriving at the sediments (Sanger, 1988; Cuddington and Leavitt, 1999). For example, chlorophyllous pigments decompose readily to pheopigments in senescent leaves and soils (Hoyt, 1966). Most of the particulate contributions of chlorophyll and carotenoid derivatives in woodland soils, swamps, ponds, and lakes are autochthonous (Sanger, 1988). Differential degradation of pigments during and after sedimentation is a more serious problem that is difficult to resolve. Typically, paleolimnologists examine the relative changes within a single pigment or focus their analyses on a restricted set of the more stable pigments (e.g., Moorehouse et al., 2014).

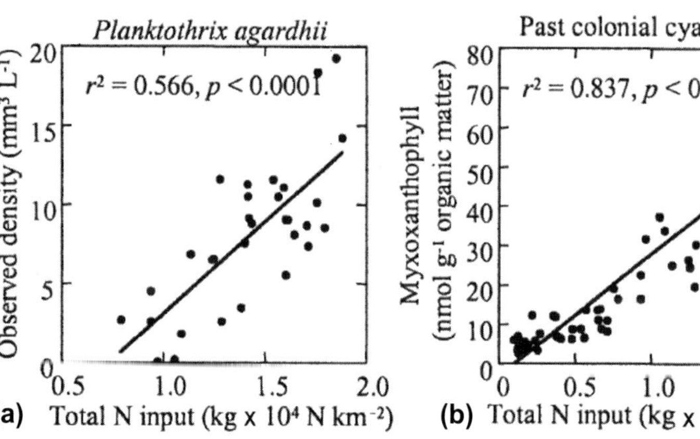

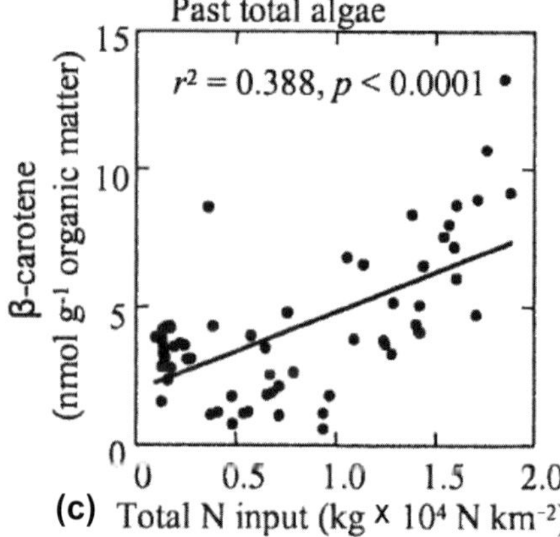

FIGURE 30-10 Comparison of Cyanobacteria and total phytoplankton production metrics from the water column (a), as well as from sedimentary pigment concentrations (b) and (c), with loads of historical watershed nitrogen shown on the x-axis. *(Figure from Bunting et al., 2007.)*

Overall, numerous studies have shown that pigment records can reliably be used to infer large changes in past lake primary production. Nonetheless, these records need to be carefully interpreted in the context of complementary data that can be used to evaluate other potential factors including changes in chemical oxidation, light climate, and herbivory. In general, meromictic lakes provide ideal preservation conditions, followed by lakes with fast sedimentation rates (Sanger et al., 1988).

C. DNA

A relatively new and fast-growing area of research in paleoecology is the study of DNA preserved in lake sediments (*sedimentary DNA*) and other natural archives. One of the initial studies demonstrating the potential of this field was the detection of woolly mammoth DNA extracted from a permafrost stratigraphical sequence, which allowed researchers to conclude that the population persisted for several thousands of years longer than previously thought based on macrofossil research (Haile et al., 2009; Fig. 30-11). DNA can be preserved in sediments for thousands of years and genetic analyses of extracts from stratigraphical sequences can be used to trace long-term changes in a large range of organisms (Capo et al., 2021). In some cases sedimentary DNA studies can open new avenues of paleolimnological study, as some taxa do not leave morphological subfossils or chemical markers (beyond DNA). Critical to any form of sedimentary DNA analyses are precautions

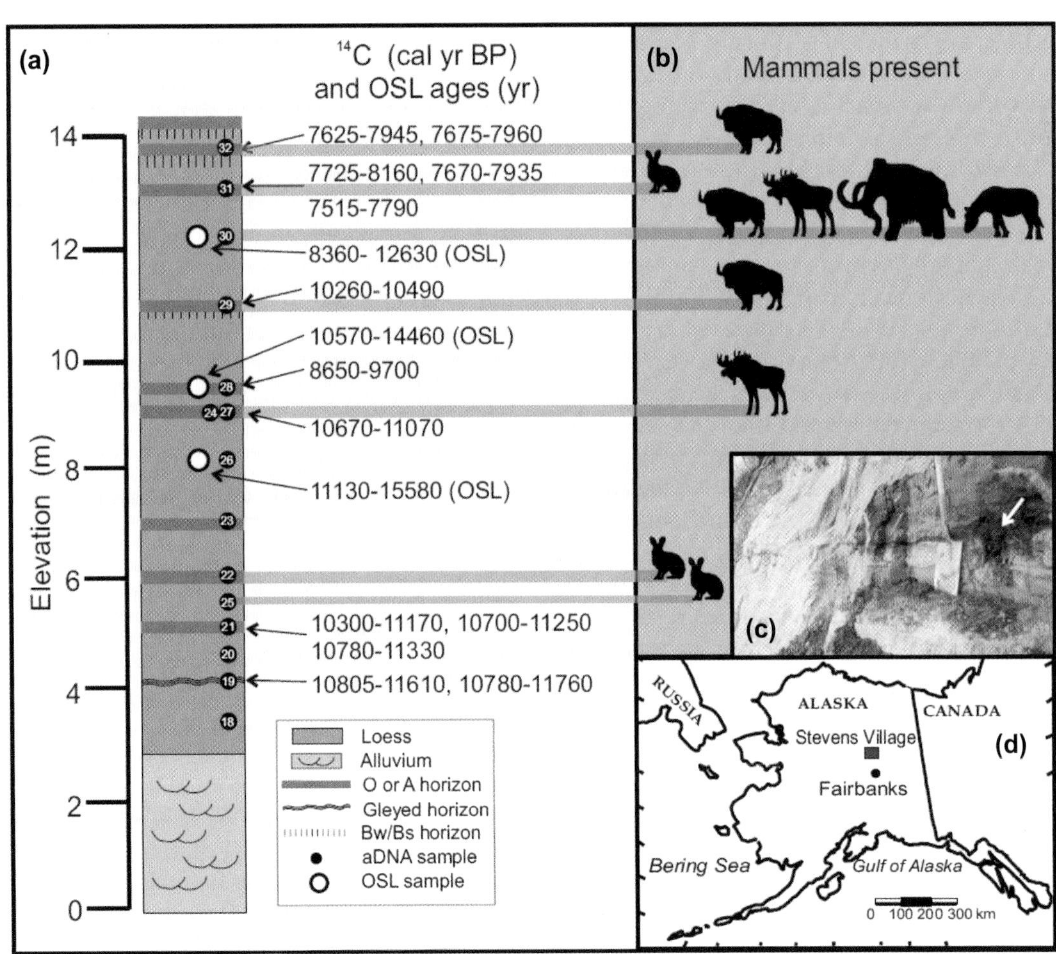

FIGURE 30-11 Analysis of ancient DNA (aDNA) preserved in a permafrost sequence shows the changes in mammal assemblages at a site in Alaska (USA) over 11,500 calendar years BP. (a) Dated permafrost sequence spanning 14 m, where the highest elevation samples are the most recent. Stratigraphy shows changes in the matrix composition from alluvium (i.e., river deposit of clay, silt, and sand) to loess (i.e., wind deposit, often yellow-gray in color), and soil horizons (i.e., O or A = organic to less organic soils, gleyed = soils saturated with water; Bw/Bs = subsoil horizons). Open circles refer to the position of optically stimulated luminescence dates (OSL; a form of dating that allows investigators to estimate the time since mineral grains were last exposed to sunlight). Black circles indicate the position of aDNA samples, many of which are associated with a radiocarbon date (age range reported based on two standard deviation units); (b) highlights the presence of different mammals based on the sequencing of aDNA; (c) photograph of the permafrost section under study, with the white arrow indicating the presence of a buried shrub root; (d) map of Alaska, USA (latitude 51–72° north), showing the position of the permafrost section under study from Stevens Village. *(Figure modified from original presented in Haile et al., 2009.)*

that should be followed to ensure that no modern DNA contaminates the older sediment sample, as detailed in Domaizon et al. (2017).

To date, a wide range of sedimentary DNA studies have been completed, including the study of population or community dynamics of taxa ranging from microbiota to large mammals. Like any relatively new field of study, there is a considerable amount of calibration work required to understand the strengths and limitations of the approach. Some of the calibration work has compared classical paleolimnological indicators such as subfossil diatom or pollen assemblages with data from sedimentary DNA extracts. For example, Stoof-Leichsenring et al. (2012) studied a sediment core spanning the past ~200 years and showed broad similarities between analyses of subfossil diatoms and DNA, with significant congruence detected in species richness trends across analytical platforms. Nonetheless, there were still taxa unique to each method of analysis, suggesting that these approaches are complementary.

A common technique for surveying the taxonomic composition of sedimentary DNA is to select a marker gene and then amplify this region using *polymerase chain reaction* (PCR); *marker genes* (also known as *barcodes*) are regions of DNA that are flanked by highly conserved sequences (i.e., present in most taxa), but within the flanking regions are variable DNA sequences that reflect a wide diversity of taxa within a taxonomic group of interest. The DNA amplification generates exponential copies of DNA for sequencing, and then the sequences are sorted taxonomically through a bioinformatic pipeline; this overall approach is known as *metabarcoding*. A recent metabarcoding study by Monchamp et al. (2017) demonstrated that the genetic signature of the cyanobacterial community preserved in lake sediment archives was consistent with the dynamics recorded in the long-term phytoplankton counts from the water column, providing confidence that sediment records faithfully preserved biodiversity dynamics. Interestingly, though, Monchamp et al. (2017) detected more taxa in the sediment record than monitoring data, which may be in part due to the integrative nature of sediments (which preserve taxa from beyond the pelagic zone, such as littoral and profundal sites) and/or the presence of *cryptic taxa* (taxa that can be discerned genetically but cannot be distinguished morphologically). Overall, there are numerous analytical decisions to make in paleogenetic studies, but a recent review of the literature by a large group of experts can help guide some of this decision making (Capo et al., 2021).

Further explorations of DNA preserved in lake sediments are being gained by applying *metagenomics* (also known as *shotgun sequencing*), whereby all DNA extracted from a sample is sequenced and analyzed, without targeting a particular gene region. This technique can be advantageous over metabarcoding as there is less bias introduced in amplifying only particular sequences present in the sample. An early application of metagenomics to lake sediment records was by Pedersen et al. (2016), who examined cores that extended back into the Late Pleistocene from sites that were within an ice-free corridor in northern North America (an unusual habitat, 12,000–15,000 years before present). This multiproxy study provided complementary evidence of the succession of flora and fauna (including mammoth and several fish species) from their study sites and concluded that their study region was unlikely to have acted as a key human migration route during the Late Pleistocene. More recently, work by Garner et al. (2020) put forth a proof-of-concept approach to pair the study of the water column with lake sediments using metagenomics, such that investigators could potentially distinguish: (1) microbes that lived solely in the water column; (2) those that lived in the water column and were then preserved in lake sediments; (3) and finally, those only found in lake sediment (and thus more likely to be living in sediments). As there are several global efforts afoot to develop more complete genetic reference libraries reflecting the diversity of life, along with new analyses of DNA preservation and improved laboratory protocols, the study of DNA in lake sediments has the potential to grow substantially in the future.

III. Case studies

Paleolimnological studies have evolved over the last century such that most current studies adopt a multiple indicator approach. There are many excellent case studies to choose from in the field of paleolimnology, but herein we highlight two multi-indicator studies that reflect different aspects of global change (i.e., climate change and biological invasions).

The first case study is a synthesis of paleolimnological records from across the Arctic where 26 authors contributed analyses of subfossil diatom, chrysophyte, cladoceran, and chironomid assemblages (Smol et al., 2005). In total, Smol et al. (2005) examined 55 sediment cores that dated back to at least 1850 CE, with most records collected at sites without any evidence of local human impact. The authors recorded a substantial temporal turnover in their assemblages (measured as the length of the first axis from a *detrended canonical correspondence analysis* constrained to time) but also showed a trend of enhanced turnover at higher latitudes (Fig. 30-12). Following consistent paleolimnological approaches, the

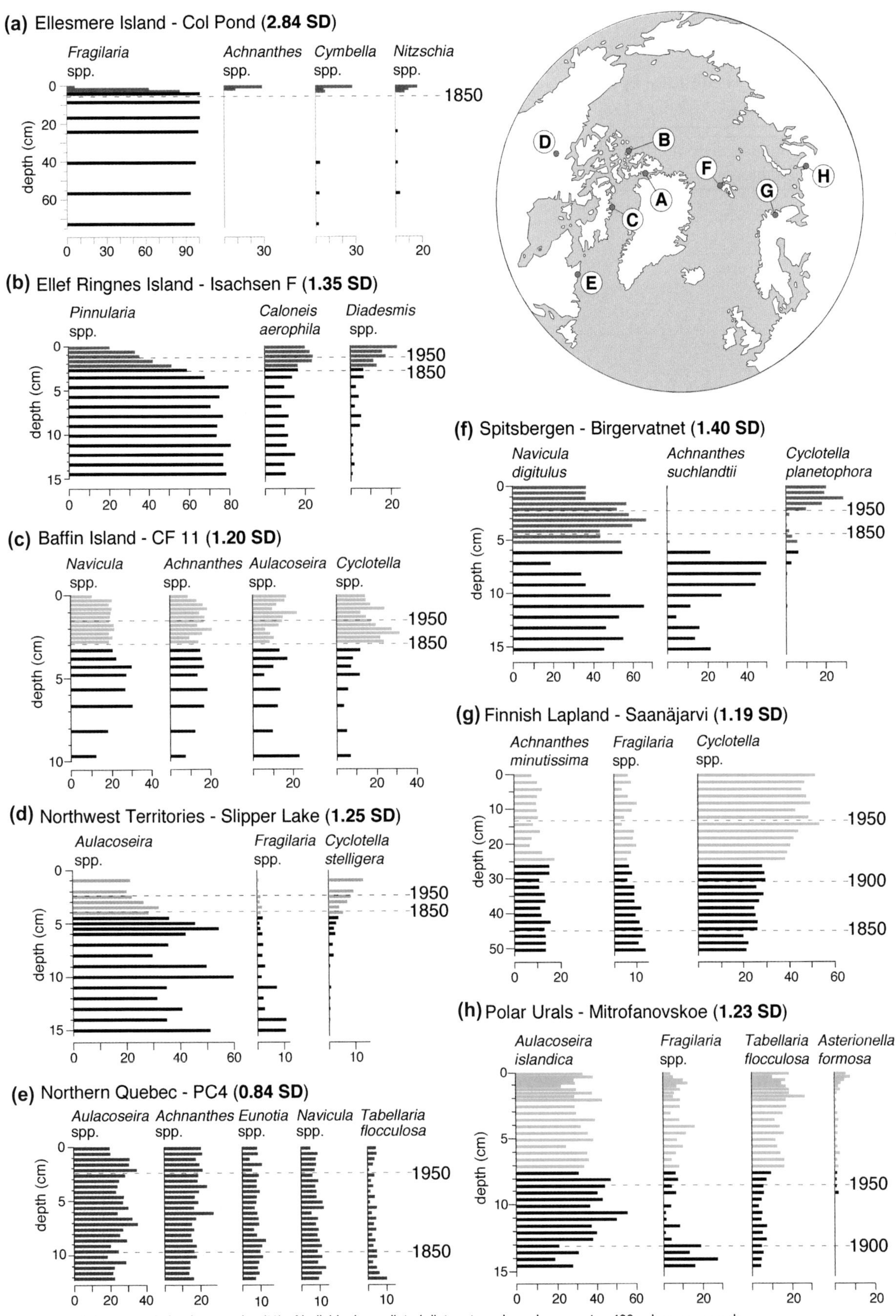

FIGURE 30-12 Diatom analyses of sediment cores collected from locations across the Arctic (a)–(h); x-axis shows relative abundances of dominant taxa). Profiles with red highlights show the greatest change in temporal turnover (measured as beta-diversity in standard deviation [SD] units) over the past ~150 years. Sites with more modest changes are shown in orange and then green. Relatively complacent records are shown in blue and were concentrated in northern Quebec, where climate change has only been observed in the last few decades, mostly after the cores were collected. Subsequent coring of lakes from regions that were slow to warm (such as the Hudson Bay Lowlands), but have subsequently warmed, record similar diatom changes in the uppermost sediments (see text for details). Overall, there is a trend of greater turnover in the most northerly latitudes (based on the beta-diversity metric calculated for each profile and presented as a function of latitude in panel [i]), which matches the accelerated change in temperature measured since 1948 CE. *(From Smol et al., 2005.)*

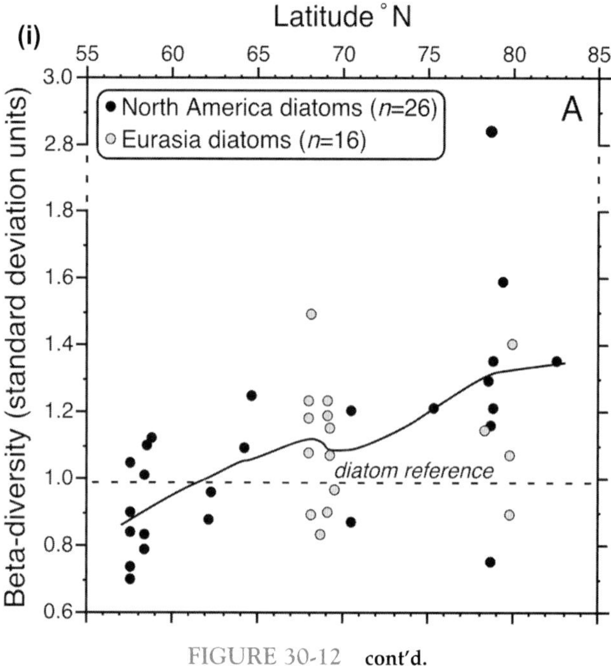

FIGURE 30-12 **cont'd.**

authors considered numerous potential drivers for the observed changes (e.g., persistent organic pollutants, land-use change, atmospheric nutrient deposition), but climate change was the most parsimonious explanation based on the known ecology of the taxa involved and the heightened magnitude and timing of changes in the more northerly latitudes. It is clear from the paleolimnological time series that substantial changes in aquatic communities have occurred since preindustrial times that can be clearly linked to climate-related variables, such as declining lake ice cover.

Focusing on a different set of indicators on the opposite side of the planet is a paleolimnological study that traced the invasion history and associated ecological impacts from an introduced rabbit population (Ficetola et al., 2018). On the study island (the main island of Kerguelen Archipelago, near Antarctica), historical records documented that rabbits were introduced in 1864 and modern ecological studies have associated rabbit grazing with changes in the local vegetation. However, the full set of dynamics between the date of introduction and the current day were poorly understood. To address this knowledge gap, Ficetola et al. (2018) detected the rabbit invasion from a continuous sediment record through the analysis of rabbit DNA and coprophilous fungal spores (i.e., *Sporormiella*, which have been used as a proxy to track the density of herbivores). Interestingly, no rabbit DNA was detected before 1941 CE (as many as eight replicates per interval were examined), and the authors found that rabbit DNA detection was significantly correlated with the influx of *Sporormiella* spores (Fig. 30-13; Ficetola et al., 2018). Given that rabbits are voracious herbivores, it is not surprising that the invasion had broad ecological effects including altering the dominance of plant taxa (based on plant metabarcoding) and accelerating erosion, which was measured as enhanced sedimentation rates (which peaked around the time when the rabbits were first broadly detectable across replicate samples). Alternative drivers were also considered, and indeed climate was found to be of secondary importance. In particular, a variance partitioning analysis demonstrated that both rabbit occurrence and climatic variables were significant predictors of past plant assemblages, but more variation was explained by the rabbit dynamics. Overall, paleolimnological studies have generated critical long-term data, which are key in identifying the impacts of global change drivers on lakes and their watersheds.

IV. Summary

1. Paleolimnology allows the reconstruction of past lake communities and biogeochemistry as well as interactions between the lake and its watershed and/or airshed. While most studies are completed on lakes, where sediment accumulation is typically less disturbed, paleolimnological approaches can be adapted to the study of rivers and other inland waters.
   a. Disturbance to a lake ecosystem is often linked to changes in the drainage basin. Materials within sediments include both proxy indicators derived from the drainage basin (erosional inputs, pollen, and other organic matter from vegetation), the atmosphere (metals, nutrients, organic compounds, and particles), and from within the lake (biological matter and chemical precipitates).
   b. Changes in lake biota can be correlated with changes in external inputs and geochemical proxy dynamics.
   c. Analyses of lake sediments, particularly when coupled with contemporary data, allow investigators to draw inferences about responses of the lake communities to natural and human-induced environmental changes (e.g., eutrophication, acidification, and climate change).
2. As with all natural archives, the sediment record is incomplete. Variable inputs from the drainage basin, redistribution of sediments within the lake basin, and differential preservation always require a critical interpretation of the record. Nonetheless, a large amount of reliable environmental information is recorded in sedimentary profiles.

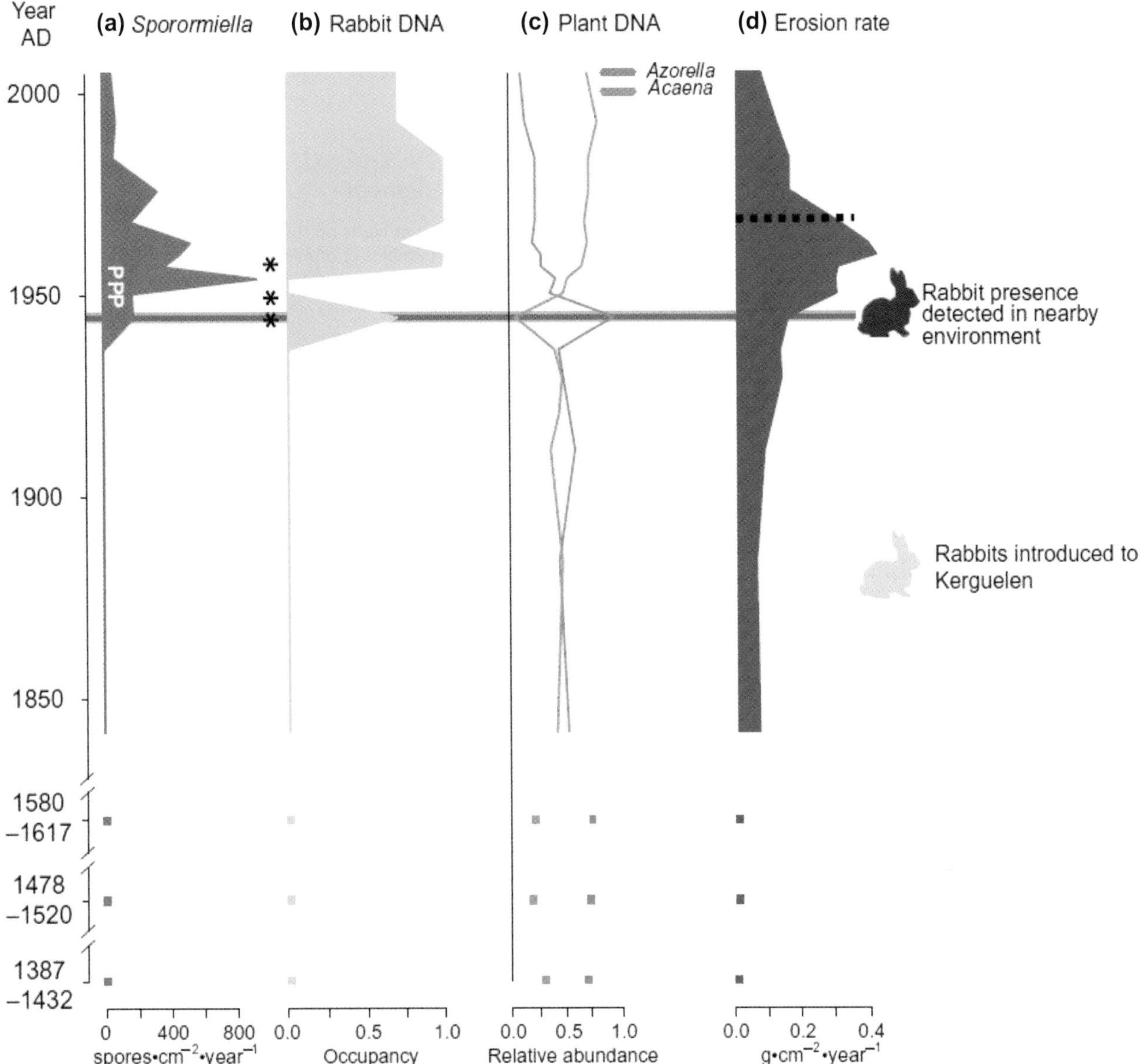

FIGURE 30-13 A multiproxy paleolimnological study from a sub-Antarctic environment (Kerguelen Islands) where an invasive rabbit was introduced. The rabbit presence was detected in the core archive through both direct (sedimentary DNA) and indirect analyses (i.e., fungal dung spores, *Sporormiella*, and Podospora [shown as PPP on the *Sporormiella* graph]; * indicates the presence of multicellular ascospores of *Sporormiella*). There is a clear association between the increased occupancy of rabbits and changes in watershed plant structure and erosion rates. Panels (a)–(d) show the flux of *Sporormiella*, occupancy estimate of rabbit DNA, relative abundance of plant DNA, and erosion rates, respectively. The *dashed line* on panel (d) shows the abrupt change determined from statistical analyses. *(Figure from Ficetola et al., 2018.)*

3. Evaluations of past events in the historical development of a lake require an accurate sediment chronology. Several methods have been used to determine the age of sediment records.
   a. The ^{14}C (radiocarbon) dating technique permits age estimates over the past ~40,000 years or somewhat longer. Numerous variables in the synthesis of ^{14}C and its incorporation into organic matter require careful time corrections to provide accurate dates.
   b. The age of recent sediments (up to ~150 years) can be estimated by the content of short-lived radioisotopes (e.g., ^{210}Pb).
   c. Some lake deposits are laid down in distinct laminae. When annual laminations (couplets) are distinguishable, they are called varves. Under ideal depositional circumstances, as is found in certain meromictic lakes and in lakes where annually pulsed loading occurs during thermal

stratification, varves permit direct age determinations.

4. Lake sediments contain numerous organic compounds that were produced either autochthonously or came from allochthonous sources.
   a. Some compounds (e.g., lipid derivatives) are specifically characteristic of terrestrial plants, algae, or bacteria. Analyses of the proportions of these compounds in sediments provide insight into the differences in the proportion of allochthonous versus autochthonous sources of the organic matter found in sediments.
   b. Certain pigments and their degradation products are well preserved in sediments, and their composition and remnant concentrations have been used to interpret past changes in trophic state as well as provide insights into the dynamics of particular algal and bacterial groups.
5. Many biological organisms are preserved in lake sediments as morphological remains or as geochemical signatures (including DNA).
   a. Plant pollen and spores are commonly deposited into lakes in proportion to changes in the composition and quantity of parent plants in or surrounding lakes. Most pollen grains preserve well once they have settled into sediments, where they can be identified, sometimes to the species level.
   b. Plant macrofossils (e.g., seeds, megaspores, and tissue fragments) are frequently recorded in sediments but are less abundant than microfossils. Compared to pollen grains (which typically represent regional signals), plant macrofossils are often more representative of the local conditions. Plant macrofossils assist in the reconstruction of past changes in plant succession, water levels, water chemistry, and regional climate from known physiological and ecological characteristics of the plants.
   c. Siliceous frustules of diatoms are the most common algal microfossils preserved in lake sediments. Diatoms in sediment profiles have been studied extensively because they generally preserve well and can be identified to species, and their ecological characteristics are typically well studied.
   d. The most widely studied and abundant animal remains in lake sediments are from the Cladocera and Chironomidae insect larvae. Many of these subfossils can be identified to the species level from their chitinous exoskeleton (cladoceran) and head-capsule (chironomid) remnants.
6. The study of DNA extracted from lake sediments is a rapidly developing field of paleolimnology and is expanding the range of taxa and types of questions that can be investigated. Careful calibration and new analytical methods will strengthen this area in the years to come.

Acknowledgments

Many helpful comments on this chapter were provided by Dr. Peter Douglas and members of our research labs. IGE and JPS acknowledge support from the Canada Research Chair program, which supports their research activities.

References

Alric, B., Jenny, J.-P., Berthon, V., Arnaud, F., Pignol, C., Reyss, J.-L., et al., 2013. Local forcings affect lake zooplankton vulnerability and response to climate warming. Ecology 94 (12), 2767–2780.

Appleby, P.G., Richardson, N., Smith, J.T., 1993. The use of radionuclide records from Chernobyl and weapons test fallout for assessing the reliability of ^{210}Pb in dating very recent sediments. Verhandlungen des Internationalen Verein Limnologie 25, 266–269.

Appleby, P.G., 2001. Chronostratigraphic techniques in recent sediments. In: Last, W.M., Smol, J.P. (Eds.), Tracking Environmental Change Using Lake Sediments. Volume 1: Basin Analysis, Coring, and Chronological Techniques. Kluwer Academic Publishers, Dordrecht, The Netherlands, pp. 171–203.

Axford, Y., Briner, J.P., Cooke, C.A., Francis, D.R., Michelutti, N., Miller, G.H., et al., 2009. Recent changes in a remote Arctic lake are unique within the past 200,000 years. Proc. Natl. Acad. Sci. USA 106, 18443–18446.

Bancone, C.E.P., Turner, S.D., Ivar do Sul, J.A., Rose, N.L., 2020. Paleoecology of microplastic contamination. Front. Environ. Sci. 8, 574008.

Battarbee, R.W., Charles, D.F., Bigler, C., Cumming, B.F., Renberg, I., 2010. Diatoms as indicators of surface-water acidity. In: Smol, J.P., Stoermer, E.F. (Eds.), The Diatoms: Applications for the Environmental and Earth Sciences, second ed. Cambridge University Press, Cambridge, United Kingdom, p. 686.

Belle, S., Verneaux, V., Millet, L., Etienne, D., Lami, A., Musazzi, S., et al., 2016. Climate and human land-use as a driver of Lake Narlay (Eastern France, Jura Mountains) evolution over the last 1200 years: implication for methane cycle. J. Paleolimnol. 55, 83–96.

Bell, M.A., Blais, J.M., 2019. -Omics" workflow for paleolimnological and geological archives: a review. Sci. Total Environ. 672, 438–455.

Bell, M.A., Blais, J.M., 2020. Paleolimnology in support of archeology: a review of past investigations and a proposed framework for future study design. J. Paleolimnol. 65, 1–32.

Bennett, K.D., Willis, K.J., 2001. Pollen. In: Smol, J.P., Birks, H.J.B., Last, W.M. (Eds.), Tracking Environmental Change Using Lake Sediments. Volume 3: Terrestrial, Algal, and Siliceous Indicators. Kluwer Academic Publishers, Dordrecht, pp. 5–32.

Bennion, H., Sayer, C.D., Tibby, J., Carrick, H.J., 2010. Diatoms as indicators of environmental change in shallow lakes. In: Smol, J.P., Stoermer, E.F. (Eds.), The Diatoms: Applications for the Environmental and Earth Sciences, second ed. Cambridge University Press, Cambridge, United Kingdom, pp. 152–173.

Berke, M.A., 2018. Reconstructing terrestrial paleoenvironments using sedimentary organic biomarkers. In: Croft, D.A., Su, D., Simpson, S.W. (Eds.), Methods in Paleoecology: Reconstructing Cenozoic Terrestrial Environments and Ecological Communities. Vertebrate Paleobiology and Paleoanthropology, pp. 121–149.

Birks, H.J.B., Birks, H.H., 1980. Quaternary paleoecology. Edward Arnold, London, United Kingdom, p. 289.

Birks, H.H., 2001. Plant macrofossils. In: Smol, J.P., Birks, H.J.B., Last, W.M. (Eds.), Tracking Environmental Change Using Lake Sediments. Volume 3: Terrestrial, Algal, and Siliceous Indicators. Kluwer Academic Publishers, Dordrecht, pp. 49–74.

Birks, H.H., Birks, H.J.B., 2001. Future uses of pollen analysis must include plant macrofossil analysis. J. Biogeogr. 27, 31–35.

Birks, H.H., Birks, H.J.B., 2006. Multi-proxy studies in palaeolimnology. Veg. Hist. Archaeobotany 15, 235–251.

Birks, H.J.B., Lotter, A.F., Juggins, S., Smol, J.P. (Eds.), 2012. Tracking environmental change using lake sediments. In: Data Handling and Numerical Techniques, vol. 5. Springer, Dordrecht, p. 745.

Björck, S., Wohlfarth, B., 2001. ^{14}C chronostratigraphic techniques in paleolimnology. In: Last, W.M., Smol, J.P. (Eds.), Tracking Environmental Change Using Lake Sediments. Volume 1: Basin Analysis, Coring, and Chronological Techniques. Kluwer Academic Publishers, Dordrecht, pp. 205–245.

Blais, J.M., Rosen, M., Smol (Eds.), J.P., 2015. Environmental contaminants: Using natural archives to track sources and long-term trends of pollution. Springer, Dordrecht, p. 509.

Boës, X., Rydberg, J., Martinez-Cortizas, A., Bindler, R., Renberg, I., 2011. Evaluation of conservative lithogenic elements (Ti, Zr, Al, and Rb) to study anthropogenic element enrichments in lake sediments. J. Paleolimnol. 46, 75–87.

Bouchard, F., MacDonald, L.A., Turner, K.W., Thienpont, J.R., Medeiros, A.S., Biskaborn, B.K., et al., 2017. Paleolimnology of thermokarst lakes: a window into permafrost landscape evolution. Arctic Science 3, 91–117.

Boyle, J.F., 2001. Inorganic geochemical methods in paleolimnology. In: Last, W.M., Smol, J.P. (Eds.), Tracking Environmental Change Using Lake Sediments. Volume 2: Physical and Geochemical Methods. Kluwer Academic Publishers, Dordrecht, pp. 83–141.

Brodersen, K.P., Quinlan, R., 2006. Midges as palaeoindicators of lake productivity, eutrophication and hypolimnetic oxygen. Quat. Sci. Rev. 25, 1995–2012.

Brown, G.W., 1969. Predicting temperatures of small streams. Water Resource Research 5, 68–75.

Bunting, L., Leavitt, P.R., Gibson, C.E., McGee, E.J., Hall, V.A., 2007. Degradation of water quality in Lough Neagh, Northern Ireland, by diffuse nitrogen flux from a phosphorus-rich catchment. Limnol. Oceanogr. 52 (1), 354–369.

Burge, D.R.L., Edlund, M.B., Frisch, D., 2018. Paleolimnology and resurrection ecology: the future of reconstructing the past. Evolutionary Applications 11, 42–59.

Callaway, J.C., DeLaune, R.D., Patrick Jr., W.H., 1996. Chernobyl ^{137}Cs used to determine sediment accretion rates at selected northern European coastal wetlands. Limnol. Oceanogr. 41, 444–450.

Camarero, L., Botev, I., Muri, G., Psenner, R., Rose, N., Stuchlik, E., 2009. Trace elements in alpine and arctic lake sediments as a record of diffuse atmospheric contamination across Europe. Freshw. Biol. 54, 2518–2532.

Capo, E., Giguet-Covex, C., Rouillard, A., Nota, K., Heintzman, P.D., Vuillemin, A., et al., 2021. Lake sedimentary DNA research on past terrestrial and aquatic biodiversity: overview and recommendations. Quaternary 4 (1), 6.

Chen, G., Selbie, D.T., Finney, B.P., Schindler, D.E., Bunting, L., Leavitt, P.R., Gregory-Eaves, I., 2011. Long-term zooplankton responses to nutrient and consumer subsidies arising from migratory sockeye salmon *Oncorhynchus nerka*. Oikos 120, 1317–1326.

Cohen, A.S., 2003. Paleolimnology: The history and evolution of lake systems. Oxford University Press, New York, USA, p. 500.

Crawford, C.B., Quinn, B., 2016. Microplastic pollutants, first ed. Elsevier Limited, Amsterdam, p. 316.

Crew, A., Gregory-Eaves, I., Ricciardi, A., 2020. Where does all the plastic go? Distribution, abundance and diversity of microplastic in the Upper St. Lawrence River. Environmental Pollution 260, 113994.

Cronberg, G., 1986. Chrysophycean cysts and scales in lake sediments: a review. In: Kristiansen, J., Andersen, R.A. (Eds.), Chrysophytes: Aspects and Problems. Cambridge University Press, Cambridge, pp. 281–315.

Cuddington, K., Leavitt, P.R., 1999. An individual-based model of pigment flux in lakes: implications for organic biogeochemistry and paleoecology. Can. J. Fish. Aquat. Sci. 56, 1964–1977.

Cumming, B.F., Smol, J.P., Kingston, J.C., Charles, D.F., Birks, H.J.B., Camburn, K.E., et al., 1992. How much acidification has occurred in Adirondack region (New York, USA) lakes since pre-industrial time? Can. J. Fish. Aquat. Sci. 49, 128–141.

Cumming, B.F., Davey, K., Smol, J.P., Birks, H.J., 1994. When did Adirondack Mountain lakes begin to acidify and are they still acidifying? Canadian Journal of Fisheries and Aquatic Science 51, 1550–1568.

Currie, D.C., Walker, I.R., 1992. Recognition and paleohydrologic significance of fossil black fly larvae with a key to Nearctic genera (Diptera: Simuliidae). J. Paleolimnol. 7, 37–54.

D'Anjou, R.M., Bradley, R.S., Balascio, N.L., Finkelstein, D.B., 2016. Climate impacts on human settlement and agricultural activities in northern Norway revealed through sediment biogeochemistry. Proc. Natl. Acad. Sci. USA 109, 20332–20337.

Davidson, T.A., Sayer, C.D., Perrow, M.R., Tomlinson, M.L., 2003. Representation of fish communities by scale sub-fossils in shallow lakes: implications for inferring percid-cyprinid shifts. J. Paleolimnol. 30, 441–449.

Davidson, T.A., Sayer, C.D., Perrow, M., Bramm, M., Jeppesen, E., 2010. The simultaneous inference of zooplanktivorous fish and macrophyte density from sub-fossil cladoceran assemblages: a multivariate regression tree approach. Freshw. Biol. 55, 546–564.

D'Avignon, G., Gregory-Eaves, I., Ricciardi, A., 2021. Microplastics in lakes and rivers: an issue of emerging significance to limnology. Environ. Rev. https://doi.org/10.1139/er-2021-0048.

Davis, R.B., Hess, C.T., Norton, S.A., Hanson, D.W., Hoagland, K.D., Anderson, D.S., 1984. ^{137}Cs and ^{210}Pb dating of sediments from soft-water lakes in New England (U.S.A.) and Scandinavia, a failure of ^{137}Cs dating. Chem. Geol. 44, 151–185.

Dearing, J.A., 1999. Holocene environmental change from magnetic proxies in lake sediments. In: Maher, B.A., Thompson, R. (Eds.), Quaternary Climates, Environments and Magnetism. Cambridge University Press, Cambridge, pp. 231–278.

Deevey Jr., E.S., 1984. Stress, strain, and stability of lacustrine ecosystems. In: Haworth, E.Y., Lund, J.W.G. (Eds.), Lake Sediments and Environmental History. University of Minnesota Press, Minneapolis, pp. 203–229.

Derrien, M., Choi, H., Jard, E., Shin, K.-H., Hur, J., 2020. Do early diagenetic processes affect the applicability of commonly used organic matter source tracking tools? An assessment through controlled degradation end-member mixing experiments. Water Res. 173, 115588.

Domaizon, I., Winegardner, A., Capo, E., Gauthier, J., Gregory-Eaves, I., 2017. DNA-based methods in paleolimnology: New opportunities for investigating long-term dynamics of lacustrine Biodiversity. J. Paleolimnol. 58, 1–21.

Douglas, P.M., Brenner, M., Curtis, J.H., 2016. Methods and future directions for paleoclimatology in the Maya Lowlands. Global Planet. Change 138, 3–24.

Duda, M.P., Allen-Mahé, S., Barbraud, C., Blais, J.M., Boudreau, A., Bryant, R., et al., 2020. Linking 19th century European settlement to the disruption of a seabird's natural population dynamics. Proc. Natl. Acad. Sci. USA 117 (51), 32484–32492.

Duff, K., Zeeb, B., Smol, J.P., 1995. Atlas of chrysophycean cysts. Kluwer Academic Publishers, Dordrecht, p. 189.

Duff, K.E., Zeeb, B.A., Smol, J.P., 1997. Chrysophyte cyst biogeographical and ecological distributions: A synthesis. J. Biogeogr. 24, 791–812.

Engstrom, D.R., Wright Jr., H.E., 1984. Chemical stratigraphy of lake sediments as a record of environmental change. In: Haworth, E.Y., Lund, J.W.G. (Eds.), Lake Sediments and Environmental History. University of Minnesota Press, Minneapolis, pp. 11–67.

Engstrom, D.R., Swain, E.B., Kingston, J.C., 1985. A palaeolimnological record of human disturbance from Harvey's Lake, Vermont: Geochemistry, pigments and diatoms. Freshw. Biol. 15, 261–288.

Fægri, K., Iversen, J., 1991. Textbook of pollen analysis. John Wiley and Sons, Chichester, United Kingdom, p. 328.

Favot, E.J., Hadley, K.R., Paterson, A.M., Michelutti, N., Watson, S.B., Zastepa, A., et al., 2020. Using visible near-infrared reflectance spectroscopy (VNIRS) of lake sediments to estimate historical changes in cyanobacterial production: Potential and challenges. J. Paleolimnol. 64, 335–345.

Ficetola, G.F., Poulenard, J., Sabatier, P., Messager, E., Gielly, L., LeLoup, A., et al., 2018. DNA from lake sediments reveals long-term ecosystem changes after a biological invasion. Sci. Adv. 4 eaar4292.

Finney, B.P., Gregory-Eaves, I., Sweetman, J., Douglas, M.S.V., Smol, J.P., 2000. Impacts of climatic change and fishing on Pacific salmon abundance over the past three hundred years. Science 290, 795–799.

Foster, I.D.L., Mighall, T.M., Proffitt, H., Walling, D.E., Owens, P.N., 2006. Post-depositional ^{137}Cs mobility in the sediments of three shallow coastal lagoons, SW England. J. Paleolimnol. 35, 881–895.

Frey, D.G., 1964. Remains of animals in Quaternary lake and bog sediments and their interpretation. Archiv fur Hydrobiologie Beihefte Ergebnisse der Limnologie 2, 114.

Frey, D.G., 1974. Paleolimnology. Mitteilungen: Internationale Vereinigung für theoretische und angewandte Limnologie, vol. 20, pp. 95–123.

Frey, D.G., 1988. Littoral and offshore communities of diatoms, cladocerans and dipterous larvae, and their interpretation in paleolimnology. J. Paleolimnol. 1, 179–191.

Fritz, S.C., Cumming, B.F., Gasse, F., Laird, K.R., 2010. In: Smol, J.P., Stoermer, E.F. (Eds.), The Diatoms: Applications for the Environmental and Earth Sciences, second ed. Cambridge University Press, Cambridge, pp. 186–208.

Gaillard, M.-J., Birks, H., 2013. Paleolimnological applications. In: Elias, S.A., Mock, C. (Eds.), Encyclopedia of Quaternary Science, second ed., vol. 2. Elsevier, Amsterdam, Netherlands, pp. 657–673. Paleobotany.

Garner, R., Gregory-Eaves, I., Walsh, D., 2020. Sediment metagenomes as time capsules of lake microbiomes. mSphere 5 (6), e00512–e00520.

Geyer, R., Jambeck, J.R., Law, K.L., 2017. Production, use, and fate of all plastics ever made. Sci. Adv. 3, e1700782.

Ginn, B.K., Rühland, K.M., Young, J.D., Hawryshyn, J., Quinlan, R., Dillon, P.J., Smol, J.P., 2012. The perils of using sedimentary phosphorus concentrations for inferring long-term changes in lake nutrient levels: Comments on Hiriart-Baer et al., 2011. J. Great Lake. Res. 38, 825–829.

Glew, J.R., Smol, J.P., Last, W.M., 2001. Sediment core collection and extrusion. In: Last, W.M., Smol, J.P. (Eds.), Tracking Environmental Change Using Lake Sediments. Volume 1: Basin Analysis, Coring, and Chronological Techniques. Kluwer Academic Publishers, Dordrecht, pp. 73–105.

Goulden, C.E., 1969b. Temporal changes in diversity. In: Woodwell, G.M., Smith, H.H. (Eds.), Diversity and Stability in Ecological Systems. Brookhaven Symposia in Biology, vol. 22, pp. 96–102.

Griffiths, K., Winegardner, A., Beisner, B., Gregory-Eaves, I., 2019. Cladoceran assemblage changes across the Eastern United States as recorded in the sediments from the National Lakes Assessment, USA. Ecol. Indicat. 96, 368–382.

Hajdas, I., Ascough, P., Garnett, M.H., Fallon, S.J., Pearson, C.L., Quarta, G., Spalding, K.L., Yamaguchi, H., Yoneda, M., 2021. Radiocarbon dating. Nature Reviews 1, 62.

Haile, J., Froese, D.G., MacPhee, R.D.E., Roberts, R.G., Arnold, L.J., Reyes, A.V., et al., 2009. Ancient DNA reveals late survival of mammoth and horse in interior Alaska. Proc. Natl. Acad. Sci. USA 106, 22352–22357.

Håkanson, L., 1981. A manual of lake morphometry. Springer-Verlag, New York, p. 78.

Hall, R.I., Smol, J.P., 2010. Diatoms as indicators of lake eutrophication. In: Smol, J.P., Stoermer, E.F. (Eds.), The Diatoms: Applications for the Environmental and Earth Sciences, second ed. Cambridge University Press, Cambridge, pp. 122–151.

Hannon, G.E., Gaillard, M.-J., 1997. The plant-macrofossil record of past lake-level changes. J. Paleolimnol. 18, 15–28.

Heiri, O., Lotter, A.F., Lemcke, G., 2001. Loss on ignition as a method for estimating organic and carbonate content in sediments: Reproducibility and comparability of results. J. Paleolimnol. 25, 101–110.

Hilton, J., 1985. A conceptual framework for predicting the occurrence of sediment focusing and sediment redistribution in small lakes. Limnol. Oceanogr. 30, 1131–1143.

Ho, J.C., Michalak, A.M., Pahlevan, N., 2019. Widespread global increase in intense lake phytoplankton blooms since the 1980s. Nature 574, 667–670.

Hodgson, D.A., Wright, S.W., Tyler, P.A., Davies, N., 1998. Analysis of fossil pigments from algae and bacteria in meromictic Lake Fidler, Tasmania, and its application to lake management. J. Paleolimnol. 19, 1–22.

Hofmann, W., 1987. Cladocera in space and time: Analysis of lake sediments. Hydrobiologia 145, 315–321.

Holmes, J.A., 2001. Ostracoda. In: Smol, J.P., Birks, H.J.B., Last, W.M. (Eds.), Tracking Environmental Change Using Lake Sediments. Developments in Paleoenvironmental Research, vol. 4. Springer, Dordrecht, pp. 125–151.

Horne, D.J., Holmes, J.A., Rodriguez-Lazaro, J., Viehberg, F.A., 2012. Ostracoda as proxies for Quaternary climate change, vol. 17. Elsevier Science & Technology, Burlington, p. 379.

Hoyt, P.B., 1966. Chlorophyll-type compounds in soil. II. Their decomposition. Plant Soil 25, 313–328.

Jenny, J.P., Normandeau, A., Francus, P., Taranu, Z.E., Gregory-Eaves, I., Lapointe, F., et al., 2016. Urbanization was the leading cause for the historical spread of hypoxia across European lakes, not intensified agriculture. Proc. Natl. Acad. Sci. USA 113, 12655–12660.

Jenny, J.-P., Koirala, S., Gregory-Eaves, I., Francus, P., Niemann, C., Ahrens, B., et al., 2019. Human and climate global-scale imprint on sediment transfer during the Holocene. Proc. Natl. Acad. Sci. USA 116, 22972–22976.

Jeppesen, E., Jensen, J.P., Amsinck, S., Landkildehus, F., Lauridsen, T., Mitchell, S.F., 2002. Reconstructing the historical changes in *Daphnia* mean size and planktivorous fish abundance in lakes from the size of *Daphnia* ephippia in the sediment. J. Paleolimnol. 27, 133–143.

Jeppesen, E., Jensen, J.P., Jensen, C., Faafeng, B., Hessen, D.O., Søndergaard, M., et al., 2003. The impact of nutrient state and lake depth on top-down control in the pelagic zone of lakes: A study of 466 lakes from the temperate zone to the Arctic. Ecosystems 6, 313–325.

Jeziorski, A., Yan, N.D., Paterson, A.M., DeSellas, A.M., Turner, M.A., Jeffries, D.S., et al., 2008. The widespread threat of calcium decline in fresh waters. Science 322, 1374–1377.

Kashiwaya, K., Ochia, S., Sakai, H., Kawai, T., 2001. Orbit-related long-term climate cycles revealed in a 12-Myr continental record from Lake Baikal. Nature 410, 71–74.

Kerfoot, W.C., 1981. Long-term replacement cycles in cladoceran communities: A history of predation. Ecology 62, 216–233.

Kingston, J.C., Birks, H.J.B., 1990. Dissolved organic carbon reconstructions from diatom assemblages in PIRLA project lakes, North America. Phil. Trans. Biol. Sci. 327, 279–288.

Klaminder, J., Appleby, P., Crook, P., Renberg, I., 2012. Post-deposition diffusion of ^{137}Cs in lake sediment: Implications for radiocaesium dating. Sedimentology 59, 2259–2267.

Korhola, A., Rautio, M., 2001. Cladocera and other branchiopod crustaceans. In: Smol, J.P., Birks, H.J.B., Last, W.M. (Eds.), Tracking Environmental Change Using Lake Sediments. Volume 4: Zoological Indicators. Kluwer Academic Publishers, Dordrecht, pp. 5–41.

Korosi, J.B., Kurek, J., Smol, J.P., 2013. A review on utilizing *Bosmina* size structure archived in lake sediments to infer historic shifts in predation regimes. J. Plankton Res. 35, 444–460.

Lamoureux, S., 2001. Varve chronology techniques. In: Last, W.M., Smol, J.P. (Eds.), Tracking Environmental Change Using Lake Sediments. Volume 1: Basin Analysis, Coring, and Chronological Techniques. Kluwer Academic Publishers, Dordrecht, pp. 247–260.

Last, W.M., Smol, J.P. (Eds.), 2001a. Tracking environmental change using lake sediments. In: Basin Analysis, Coring, and Chronological Techniques, vol. 1. Kluwer Academic Publishers, Dordrecht, p. 548.

Last, W.M., Smol, J.P. (Eds.), 2001b. Tracking environmental change using lake sediments. In: Physical and Geochemical Methods, vol. 2. Kluwer Academic Publishers, Dordrecht, p. 504.

Leavitt, P.L., Hodgson, D.A., 2001. Sedimentary pigments. In: Smol, J.P., Birks, H.J.B., Last, W.M. (Eds.), Tracking Environmental Change Using Lake Sediments. Volume 3: Terrestrial, Algal, and Siliceous Indicators. Kluwer Academic Publishers, Dordrecht, pp. 295–325.

Levy, S., 2017. Paleoecology—Looking to the past to inform the future. Bioscience 67 (9), 791–798.

Liu, Q., Roberts, A.P., Larrasoana, J.C., Banerjee, S.K., Guyodo, Y., Tauxe, L., Oldfield, F., 2012. Environmental magnetism: Principles and applications. Rev. Geophys. 50, RG4002.

Livingstone, D.A., 1975. Late Quaternary climatic change in Africa. Annual Review of Ecology, Evolution, and Systematics 6, 249–280.

MacDonald, G.M., Beukens, R.P., Kieser, W.E., 1991. Radiocarbon dating of limnic sediments: A comparative analysis and discussion. Ecology 73, 1150–1155.

MacDonald, G.M., Velichko, A.A., Kremenetski, C.V., Borisova, O.K., Goleva, A.A., Andreev, A.A., et al., 2000. Holocene treeline history and climate change across Northern Eurasia. Quaternary Research 53, 302–311.

Manca, P., Comoli, P., 1995. Temporal variations of fossil Cladocera in the sediments of Lake Orta (N. Italy) over the last 400 years. J. Paleolimnol. 14, 113–122.

McCarthy, F.M.G., Krueger, A.M., 2013. Freshwater dinoflagellates in palaeolimnological studies: *Peridinium* cysts as proxies of cultural eutrophication in the SE Great Lakes region of Ontario, Canada. In: Lewis, J.M., Marret, F., Bradley, L. (Eds.), Biological and Geological Perspectives of Dinoflagellates. The Micropalaeontological Society, Special Publications. Geological Society, London, pp. 133–139.

Meyer-Jacob, C., Michelutti, N., Paterson, A.M., Monteith, D., Yang, H., Weckström, J., Smol, J.P., Bindler, R., 2017. Inferring past trends in lake-water organic carbon concentrations in northern lakes using sediment spectroscopy. Environmental Science & Technology 55, 13248–13255.

Meyer-Jacob, C., Michelutti, N., Paterson, A.M., Cumming, B.F., Keller, W., Smol, J.P., 2019. The browning and re-browning of lakes: Divergent lake-water organic carbon trends linked to acid deposition and climate change. Sci. Rep. 9, 16676.

Meyer-Jacob, C., Labaj, A.L., Paterson, A.M., Edwards, B.A., Keller, W., Cumming, B.F., Smol, J.P., 2020. Re-browning of Sudbury (Ontario, Canada) lakes now approaches pre-acid deposition lake-water dissolved organic carbon levels. Sci. Total Environ. 725, 138347.

Meyers, P.A., Teranes, J.L., 2001. Sediment organic matter. In: Last, W.M., Smol, J.P. (Eds.), Tracking Environmental Change Using Lake Sediments. Volume 2: Physical and Geochemical Methods. Kluwer Academic Publishers, Dordrecht, pp. 239–269.

Michelutti, N., Blais, J.M., Cumming, B.F., Paterson, A.M., Rühland, K., Wolfe, A.P., Smol, J.P., 2010. Do spectrally inferred determinations of chlorophyll *a* reflect trends in lake trophic status? J. Paleolimnol. 43, 205–217.

Michelutti, N., Smol, J.P., 2016. Visible spectroscopy reliably tracks trends in paleo-production. J. Paleolimnol. 56, 253–265.

Mitchell, J.K., Mostaghimi, S., Freeny, D.S., McHenry, J.R., 1983. Sediment deposition estimation from cesium-137 measurements. Water Resource Bulletin 19, 549–555.

Monchamp, M.-E., Wasler, J.-C., Pomati, D., Spaak, P., 2016. Sedimentary DNA reveals cyanobacterial community diversity over 200 years in two perialpine lakes. Applications in Environmental Microbiology 82, 6472–6482.

Moore, P.D., Webb, J.A., Collinson, M.E., 1991. Pollen analysis. Blackwell Scientific Publications, Oxford, p. 216.

Moorehouse, H., McGowan, S., Jones, M.D., Barker, P., Leavitt, P.R., Grayshaw, S.A., Haworth, E.Y., 2014. Contrasting effects of nutrients and climate on algal communities in two lakes in the Windermere catchment since the late 19th century. Freshw. Biol. 59, 2605–2620.

Mustaphi, C.J.C., Pisaric, M.F.J., 2014. A classification for macroscopic charcoal morphologies found in Holocene lacustrine sediments. Prog. Phys. Geogr. 38 (6), 734–754.

Nevalainen, L., Rautio, M., 2014. Spectral absorbance of benthic cladoceran carapaces as a new method for inferring past UV exposure of aquatic biota. Quat. Sci. Rev. 84, 109–115.

Obrist-Farner, J., Rice, P.M., 2019. Nixtun-Ch'ich' and its environmental impact: Sedimentological and archaeological correlates in a core from Lake Petén Itzá in the southern Maya lowlands, Guatemala. J. Archaeol. Sci.: Reports 26, 101868.

Oldfield, F., 1991. Environmental magnetism—a personal perspective. Quat. Sci. Rev. 10, 73–85.

Osleger, D.A., Heyvaeart, A.C., Stoner, J.S., Verosub, K.L., 2009. Lacustrine turbidites as indicators of Holocene storminess and climate: Lake Tahoe, California and Nevada. J. Paleolimnol. 42 (1), 103–122.

Ostrofsky, M.L., 2012. Differential post-depositional mobility of phosphorus species in lake sediments. J. Paleolimnol. 48, 559–569.

Paterne, M., Michel, É., Hatté, C., Dutay, J.-C., 2021. Carbon-14. In: Ramstein, G., Landais, A., Bouttes, N., Sepulchre, P., Govin, A. (Eds.), Paleoclimatology. Frontiers in Earth Sciences. Springer, Cham, pp. 51–71.

Pedersen, M., Ruter, A., Schweger, C., Friebe, H., Staff, R.A., Kjeldsen, K.K., et al., 2016. Postglacial viability and colonization in North America's ice-free corridor. Nature 537, 45–49.

Perga, M.-E., 2011. Taphonomic and early diagenetic effects on the C and N stable isotope composition of cladoceran remains: Implications for paleoecological studies. J. Paleolimnol. 46, 203–213.

Peterson, B.J., Fry, B., 1987. Stable isotopes in ecosystem studies. Annu. Rev. Ecol. Systemat. 18, 293–320.

Quinlan, R., Smol, J.P., 2002. Regional assessment of long-term hypolimnetic oxygen changes in Ontario (Canada) shield lakes using subfossil chironomids. J. Paleolimnol. 27, 249–260.

Reavie, E.D., 2019. Paleolimnology supports aquatic management by providing early warnings of stressor impacts. Lake Reservoir Manag. 36 (3), 210–217.

Reimer, P.J., Austin, W.E.N., Bard, E., Bayliss, A., Blackwell, P.G., Ramsey, C.B., et al., 2020. The IntCal20 Northern Hemisphere radiocarbon age calibration curve (0–55 cal kBP). Radiocarbon 62 (4), 725–757.

Renberg, I., Wik, M., 1984. Dating recent lake sediments by soot particle counting. Verhandlungen des Internationalen Verein Limnologie 22, 712–718.

Rose, N.L., 2001. Fly-ash particles. In: Last, W.M., Smol, J.P. (Eds.), Tracking Environmental Change Using Lake Sediments. vol. 2: Physical and Geochemical Methods. Kluwer Academic Publishers, Dordrecht, pp. 319–349.

Rose, N.L., 2015. Spheroidal carbonaceous fly ash particles provide a globally synchronous stratigraphic marker for the Anthropocene. Environmental Science & Technology 49, 4155–4162.

Rühland, K.M., Paterson, A.M., Smol, J.P., 2015. Diatom assemblage responses to warming: Reviewing the evidence. J. Paleolimnol. 54, 1–35.

Rydberg, J., Cooke, C.A., Tolu, J., Wolfe, A.P., Vinebrooke, R.D., 2020. An assessment of chlorophyll preservation in lake sediments using multiple analytical techniques applied to the annually laminated lake sediments of Nylandssjön. J. Paleolimnol. 64, 379–388.

Ryves, D.B., Battarbee, R.W., Juggins, S., Fritz, S.C., Anderson, N.J., 2006. Physical and chemical predictors of diatom dissolution in freshwater and saline lake sediments in North America and West Greenland. Limnol. Oceanogr. 51, 1355–1368.

Ryves, D.B., Battarbee, R.W., Fritz, S.C., 2009. The dilemma of disappearing diatoms: Incorporating diatom dissolution data into palaeoenvironmental modelling and reconstruction. Quat. Sci. Rev. 28, 120–136.

Sanger, J.R., 1988. Fossil pigments in paleoecology and paleolimnology. Palaeogeography, Palaeoclimatology, Palaeoecology 62 (1), 343–359.

Sandgren, P., Snowball, I., 2001. Application of mineral magnetic techniques to paleolimnology. In: Last, W.M., Smol, J.P. (Eds.), Tracking Environmental Change Using Lake Sediments. Volume 2: Physical and Geochemical Methods. Kluwer Academic Publishers, Dordrecht, The Netherlands, pp. 217–235.

Sarmaja-Korjonen, K., 2003. Chydorid ephippia as indicators of environmental change—biostratigraphical evidence from two lakes in southern Finland. Holocene 13, 691–700.

Sarmaja-Korjonen, K., 2004. Chydorid ephippia as indicators of past environmental changes—a new method. Hydrobiologia 526, 129–136.

Schneider, T., Rimer, D., Butz, C., Grosjean, M., 2018. A high-resolution pigment and productivity record from the varved Ponte Tresa basin (Lake Lugano, Switzerland) since 1919: Insight from an approach that combines hyperspectral imaging and high-performance liquid chromatography. J. Paleolimnol. 60, 381–398.

Schwalb, A., 2003. Lacustrine ostracodes as stable isotope recorders of late-glacial and Holocene environmental dynamics and climate. J. Paleolimnol. 29, 265–351.

Seppä, H., Bennett, K.L., 2003. Quaternary pollen analysis: Recent progress in palaeoecology and palaeoclimatology. Prog. Phys. Geogr. 27, 548–579.

Seppä, H., 2013. Pollen analysis, principles. In: Elias, S.A., Mock, C. (Eds.), Encyclopedia of Quaternary Science, second ed., vol. 3. Elsevier, Amsterdam, Netherlands, pp. 794–804.

Simola, H., 1977. Diatom succession in the formation of annually laminated sediment in Lovojärvi, a small eutrophicated lake. Annals Botanica Fennici 18, 160–168.

Siver, P.A., 1993. Inferring the specific conductivity of lake water with scaled chrysophytes. Limnol. Oceanogr. 38, 1480–1492.

Smol, J.P., 1985. The ratio of diatom frustules to chrysophycean statospores: A useful paleolimnological index. Hydrobiologia 123, 199–208.

Smol, J.P., 1988. Chrysophycean microfossils in paleolimnological studies. Palaeogeogr. Palaeoclimatol. Palaeoecol. 62, 287–297.

Smol, J.P., 1995. Application of chrysophytes to problems in paleoecology. In: Sandgren, C., Smol, J.P., Kristiansen, J. (Eds.), Chrysophyte Algae: Ecology, Phylogeny and Development. Cambridge University Press, Cambridge, pp. 303–329.

Smol, J.P., 2008. Pollution of lakes and rivers: A paleoenvironmental perspective, second ed. Blackwell Publishing, Oxford, UK, p. 389.

Smol, J.P., 2019. Under the radar: Long-term perspectives on ecological changes in lakes. Proc. Biol. Sci. 286, 20190834.

Smol, J.P., Cumming, B.F., Dixit, A.S., Dixit, S.S., 1998. Tracking recovery patterns in acidified lakes: A paleolimnological perspective. Restor. Ecol. 6, 318–326.

Smol, J.P., Birks, H.J.B., Last, W.M. (Eds.), 2001a. Tracking environmental change using lake sediments. In: Terrestrial, Algal, and Siliceous Indicators, vol. 3. Kluwer Academic Publishers, Dordrecht, p. 371.

Smol, J.P., Birks, H.J.B., Last, W.M. (Eds.), 2001b. Tracking environmental change using lake sediments. In: Zoological Indicators, vol. 4. Kluwer Academic Publishers, Dordrecht, p. 217.

Smol, J.P., Wolfe, A.P., Birks, H.J.B., Douglas, M.S.V., Jones, V.J., Korhola, A., et al., 2005. Climate-driven regime shifts in the biological communities of arctic lakes. Proc. Natl. Acad. Sci. USA 102, 4397–4402.

The Diatoms. In: Smol, J.P., Stoermer, E.F. (Eds.), 2010. Applications for the Environmental and Earth Sciences, second ed. Cambridge University Press, Cambridge, p. 667.

Stoof-Leichsenring, K.R., Epp, L.S., Trauth, M.H., Tiedemann, R., 2012. Hidden diversity in diatoms of Kenyan Lake Naivasha: A genetic approach detects temporal variation. Mol. Ecol. 21, 1918–1930.

Stuiver, M., Quay, P.D., 1980. Changes in atmospheric carbon-14 attributed to a variable sun. Science 207, 11–19.

Sweetman, J.N., Smol, J.P., 2006. Reconstructing past shifts in fish populations using subfossil *Chaoborus* (Diptera: Chaoboridae) remains. Quat. Sci. Rev. 25, 2013–2023.

Taranu, Z.E., Gregory-Eaves, I., Leavitt, P., Bunting, L., Buchaca, T., Catalan, J., et al., 2015. Acceleration of cyanobacterial dominance in north temperate-subarctic lakes during the Anthropocene. Ecol. Lett. 18, 375–384.

Trumbore, S.E., Sierra, C.A., Hicks Pries, C.E., 2016. Radiocarbon nomenclature, theory, models, and interpretation: Measuring age, determining cycling rates, and tracing source pools. In: Schuur, E.A.G., Druffel, E.R.M., Trumbore, S.E. (Eds.), Radiocarbon and Climate Change. Springer International Publishing, Switzerland, pp. 45–82.

Tolotti, M., Dubois, N., Milan, M., Perga, M.-E., Straile, D., Lami, A., 2018. Large and deep perialpine lakes: A paleolimnological perspective for the advance of ecosystem science. Hydrobiologia 824, 291–321.

Turner, F., Zhu, L.P., Lü, X.M., Peng, P., Ma, Q.F., Wang, J.B., et al., 2016. *Pediastrum* sensu lato (Chlorophyceae) assemblages from surface sediments of lakes and ponds on the Tibetan Plateau. Hydrobiologia 771, 101–118.

Turney, C.S.M., Palmer, J., Maslin, M.A., Hogg, A., Fogwill, C.J., Southon, J., Fenwick, P., Helle, G., Wilmshurst, J.M., McGlone, M., Ramsey, C.B., Thomas, Z., Lipson, M., Beaven, B., Jones, R.T., Andrews, O., Hua, Q., 2018. Global peak in atmospheric radiocarbon provides a potential definition for the onset of the Anthropocene epoch in 1965. Sci. Rep. 8, 3293.

van Geel, B., Mur, R., Ralska-Jasiewiczowa, M., Goslar, M., 1994. Fossil akinetes of *Aphanizomenon* and *Anabaena* as indicators for medieval phosphate-eutrophication of Lake Gosciaz (Central Poland). Rev. Palaeobot. Palynol. 83, 97–105.

van Geel, B., 2002. Non-pollen palynomorphs. In: Smol, J.P., Birks, H.J.B., Last, W.M. (Eds.), Tracking Environmental Change Using Lake Sediments. Developments in Paleoenvironmental Research, vol. 3. Springer, Dordrecht, pp. 99–119.

van Hardenbroek, M., Heiri, O., Leng, M.J., 2019. Stable isotopes in biological and chemical fossils from lake sediments: Developing and calibrating palaeoenvironmental proxies. Quat. Sci. Rev. 218, 157–159.

Walker, I.R., 2001. Midges: Chironomidae and related Diptera. In: Smol, J.P., Birks, H.J.B., Last, W.M. (Eds.), Tracking Environmental Change Using Lake Sediments. Volume 4: Zoological Indicators. Kluwer Academic Publishers, Dordrecht, pp. 43–66.

Walker, I.R., Smol, J.P., Engstrom, D.R., Birks, H.J.B., 1991. An assessment of Chironomidae as quantitative indicators of past climatic change. Can. J. Fish. Aquat. Sci. 48, 975–987.

Waller, M., 2013. Drought, disease, defoliation and death: Forest pathogens as agents of past vegetation change. J. Quat. Sci. 28, 336–342.

Waters, M.N., Brenner, M., Curtis, J.H., Romero-Oliva, C.S., Dix, M., Cano, M., 2021. Harmful algal blooms and cyanotoxins in Lake Amatitlán, Guatemala, coincided with ancient Maya occupation in the watershed. Proc. Natl. Acad. Sci. USA 118 e2109919118.

Whitlock, C., Larsen, C.P.S., 2001. Charcoal as a fire proxy. In: Smol, J.P., Birks, H.J.B., Last, W.M. (Eds.), Tracking Environmental Change Using Lake Sediments. Volume 3: Terrestrial, Algal, and Siliceous Indicators. Kluwer Academic Publishers, Dordrecht, pp. 75–97.

Wilkinson, A.N., Zeeb, B., Smol, J.P., 2001. Atlas of chrysophycean cysts, vol. 2. Kluwer Academic Publishers, Dordrecht, p. 180.

Wolfe, A.P., Vinebrooke, R.D., Michelutti, N., Rivard, B., Das, B., 2006. Experimental calibration of lake-sediment spectral reflectance to chlorophyll *a* concentrations: Methodology and paleolimnological validation. J. Paleolimnol. 36, 91–100.

Wolin, J.A., Stone, J.R., 2010. Diatoms as indicators of water-level change in freshwater lakes. In: Smol, J.P., Stoermer, E.F. (Eds.), The Diatoms: Applications for the Environmental and Earth Sciences, second ed. Cambridge University Press, Cambridge, UK, p. 667.

Wyn, B., Sweetman, J.N., Leavitt, P.R., Donald, D.B., 2007. Historical metal concentrations in lacustrine food webs revealed using fossil ephippia from daphnia. Ecol. Appl. 17, 754–764.

Zander, P.D., Wienhues, G., Grosjean, M., 2022. Hyperspectral imaging for *in situ* biogeochemical analysis of lake sediment cores: Review of recent developments. Journal of Imaging 8, 58.

Zeeb, B.A., Smol, J.P., 2001. Chrysophyte scales and cysts. In: Smol, J.P., Birks, H.J.B., Last, W.M. (Eds.), Tracking Environmental Change Using Lake Sediments. Volume 3: Terrestrial, Algal, and Siliceous Indicators. Kluwer Academic Publishers, Dordrecht, pp. 203–223.

Zhao, Y., Sayer, C.D., Birks, H.H., Hughes, M., Peglar, S.M., 2006. Spatial representation of aquatic vegetation by macrofossils and pollen in a small and shallow lake. J. Paleolimnol. 35, 335–350.

Zolitschka, B., Francus, P., Ojala, A.E.K., Schimmelmann, A., 2015. Varves in lake sediments—a review. Quat. Sci. Rev. 117, 1–41.

Züllig, H., 1989. Role of carotenoids in lake sediments for reconstructing trophic history during the late Quaternary. J. Paleolimnol. 2, 23–40.

CHAPTER

31

Inland Waters: The Future of Limnology is Interdisciplinary, Collaborative, Inclusive, and Global

Sapna Sharma[1], *Stephanie E. Hampton*[2] *and Ismael Kimirei*[3]

[1]Biology Department, York University, Toronto, Ontario, Canada [2]Biosphere Sciences and Engineering, Carnegie Institution for Science, Pasadena, CA, United States [3]Tanzania Fisheries Research Institute, Dar es Salaam, Tanzania

OUTLINE

I. Access to clean freshwater is a human right

Access to clean freshwater is essential for human survival. We rely on water for hydration, food production, sanitation, and transportation (Chapter 2). Yet, freshwater in lakes, rivers, and groundwater accounts for less than 1% of the world's water supply. Over 2 billion people worldwide do not have access to safe drinking water in their homes, contributing to 3.5 million deaths annually (United Nations Children's Fund and World Health Organization, 2019). The United Nations has declared that access to clean freshwater is a basic human right. The Sustainable Development Goal Target 6.1 has called for "equitable access to safe and affordable drinking water by 2030" (United Nations, 2015).

The ability to easily access clean freshwater would be transformational to enhancing the quality of life for billions of people worldwide by increasing longevity, improving living standards, decreasing health risks, and promoting social and environmental justice. Currently, water scarcity is faced by millions of people because of drought, population growth, and environmental degradation. For example, residents of some of the world's most populated cities and their surrounding areas, including Mexico City, Sao Paulo, and Cape Town, have lost access to running water in their homes (BBC, 2018). Women and children in sub-Saharan Africa spend hours each day collecting water, although it presents a safety risk and precludes further economic opportunities. In addition, many communities around the world

Wetzel's Limnology, Fourth Edition
https://doi.org/10.1016/B978-0-12-822701-5.00031-8

only have access to *unpotable water* (water that is unsafe to drink). For example, the prevalence of water borne diseases in Africa is a testament to poor water quality (Nkoko et al., 2011). In 2014, residents of Flint, Michigan (USA) were informed that the public water supply was contaminated with lead, beginning a multi-year water crisis that eroded public trust in drinking water safety. Many indigenous communities in Canada, a country with about 9 million lakes (Verpoorter et al., 2014), continue to have boiling water advisories. Clearly, even in highly economically developed nations, marginalized communities do not always have access to clean freshwater.

Although numbers vary (see Chapter 4), Verpoorter et al. (2014) estimated there to be 118 million lakes on Earth. However, freshwater is distributed unequally, and primarily in northern regions of the planet (Verpoorter et al., 2014; Chapter 4). Climate change, land-use changes, population growth, and increased water scarcity all contribute to the degradation of water quality. Furthermore, racial, socioeconomic, and cultural inequalities are readily apparent in patterns of access to clean freshwater worldwide. Global environmental degradation further increases the risk of water scarcity, deterioration in water resources, and declines in biodiversity. For example, fishing in lakes is important across many cultures while also providing critical protein to surrounding communities. Sterner et al. (2020) estimated that 1.35 billion tonnes of fish are harvested from just the 25 largest lakes in the world. For example, the local Buryat, Evenk, and Tuva peoples in Siberia rely on hunting and fishing species found only in Lake Baikal, including the Baikal omul (*Coregonus migratorius*), a whitefish, and the nerpa (*Pusa sibirica*), the only freshwater seal. The fish caught in the lakes of East Africa provide a key source of protein for tens of millions of people (Bootsma and Hecky, 1993; Mölsä et al., 1999). In this chapter we hope to bring global, diverse, inclusive, and multidisciplinary perspectives to highlight key research avenues to safeguard our invaluable freshwater resources.

A. Need for better integration between disciplines

No universal set of management actions will optimize ecosystem services in every lake, stream, or river. Specific challenges vary widely across lakes. For example, just as there are many routes through which a lake can eutrophy, there are also many routes by which mercury mobilizes and biomagnifies (Chen et al., 2018). Further, a water body's climatic, geomorphic, and socioeconomic context can radically affect the efficacy of any specific remedy (e.g., reoligotrophication), due to the action of multiple interacting stressors (Birk et al., 2020; Jeppesen et al., 2005). Yet a new approach to limnology is beginning to transform our ability to address global water challenges, as it is increasingly engaged worldwide: integrated teamwork across freshwater disciplines that have, for many decades, grown strong in parallel and now are beginning to work together.

In 1995 a group of prominent aquatic scientists recognized the fragmentation of the freshwater sciences as a formidable challenge to addressing key fundamental and applied questions that underpin sustainability of freshwater resources (Naiman et al., 1995). In the past several decades, efforts to bridge historical disciplinary divides have soundly demonstrated the value of further broadening limnology. For example, historically, limnologists may have circumscribed their studies of water quality criteria that directly affect human health, to include primarily those that involve alterations in "natural" cycles, such as nutrients and carbon. More recent attention to "less natural" yet ubiquitous pollutants such as pharmaceuticals (e.g., alprazolam [Xanax] and fluoxetine [Prozac]) and personal care products (Meyer et al., 2019; Robson et al., 2020; Rosi-Marshall and Royer, 2012), petroleum products associated with stormwater, or other anthropogenic "chemicals" (Fig. 31-1; Bernhardt et al., 2017) highlights the vital need for limnology to embrace topics once considered to be the domain of ecotoxicology. Further, the history of limnology has shown us that, although it may have been uncommon for limnologists to become involved in solving the problems they identify, their dedication to these interdisciplinary and cross-sector engagements can have an outsized impact on water quality and human well-being (e.g., Edmondson, 1996; Kaushal et al., 2010; Schindler, 2006). For environmental solutions regarding issues such as water treatment, restoration, or conservation actions, limnologists should step forward alongside civil engineers, environmental engineers, hydrologists, ecotoxicologists, and fisheries biologists, as well as concerned citizens, to offer help to decision makers. Modern limnological research and training increasingly seeks to bridge historical gaps between disciplines and societal sectors and to engage limnologists in finding solutions.

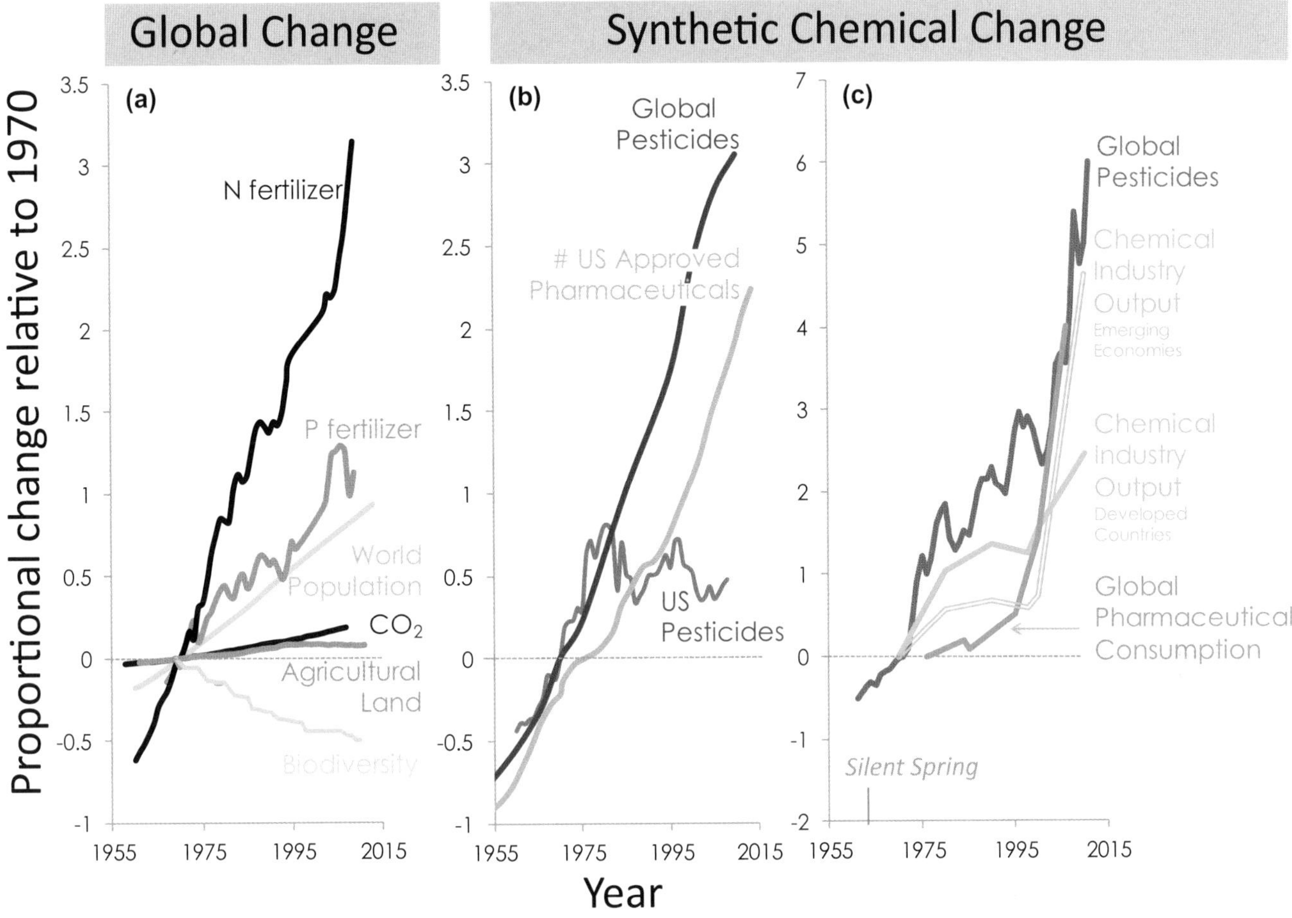

FIGURE 31-1 Proportional change of drivers relative to the 1970s of (a) global environmental change; (b) diversity of pesticides and pharmaceuticals in the United States; and (c) global trade value of synthetic chemicals, including pesticides and pharmaceuticals. *Silent Spring* refers to the publication of an influential environmental science book (Carson, 1962). *(From Bernhardt et al., 2017.)*

II. Integrated approaches to water quality management and science

A. Threats to water quality from local to global sources

As described in various chapters of this book, water quality in lakes and rivers is sensitive to anthropogenic threats, in part because of the proximity and importance of freshwater to human populations (Fig. 31-2). These threats can be categorized as biological (e.g., species invasions, pathogens), physical (e.g., temperature, precipitation, solar radiation changes, hydrological changes such as flow, habitat alteration), chemical (e.g., nutrient enrichment, organic contamination, pharmaceutical inputs), and human extraction of resources (e.g., mining, overfishing) (Fig. 31-3; Jenny et al., 2020). These threats may act independently or in concert to impact ecosystems by changing rates of primary production, promoting or inhibiting growth, altering reproductive abilities and fecundity, changing behavior, affecting interactions with other species, or contributing to mortality (Jenny et al., 2020; Vörösmarty et al., 2010). Below, we expand on a few of the threats affecting our freshwaters today in many regions of the world.

i. Climate change

Climate change poses one of the largest threats to our environment and, as has been documented within many of the chapters in this book, has wide-ranging impacts on our lakes and rivers (Whitehead et al., 2009; Woolway and Maberly, 2020). In lakes climate change contributes to the loss of ice cover (Magnuson et al., 2000; Sharma et al., 2019), increases in surface water temperatures (O'Reilly et al., 2015; Schneider and Hook, 2010), alterations in lake mixing regimes (Woolway and Merchant, 2019), and acceleration in lake evaporation (Wang et al.,

FIGURE 31-2 Human use of lakes can vary widely, depending on accessibility and regional needs. (a) Remote in Siberia, ancient Lake Baikal holds 20% of the world's surface freshwater, situated in a largely roadless watershed where most of its shoreline is accessible only by boat (Stephanie Hampton). (b) The African Rift Lakes support the largest lentic freshwater fishery in the world and provide a key protein source for tens of millions of East Africans (Ishmael Kimirei). (c) Lake Titicaca shares its catchment between Bolivia and Peru, and the ecosystem services it provides are threatened by an increasing human population, expanding agriculture, water-level manipulation, and mining. *CC-BY-SA Diego Delso.* (d) Altogether, the Laurentian Great Lakes hold 20% of the world's surface freshwater and provide a great variety of ecosystem services including fisheries and water for agriculture and industry as well as drinking water for millions of people (CC-BY-NC-ND Laurent Gass).

2018). Climate change has also contributed to ice loss in rivers (Yang et al., 2020) and warming water temperatures (Kaushal et al., 2010), in addition to alterations in amount and timing of stream flow (Arnell and Gosling, 2013) through increased flashiness (the speed at which the flow of a stream increases or decreases following precipitation), floods, and droughts (Whitehead et al., 2009).

The changes in the physical aspects of water bodies can have wide-reaching impacts on the ecology of our lakes and rivers, particularly because the velocity of climate change is higher in standing inland waters than in terrestrial or marine systems (Woolway and Maberly, 2020). For example, increases in water temperatures can influence the distribution of freshwater fishes (Chu et al., 2005; Comte and Olden, 2017), with range expansion for nonnative warm water fishes (Sharma et al., 2007), range contractions for cool water and cold-water fishes (Hansen et al., 2017; Wenger et al., 2011), and ultimately decreased fish productivity (O'Reilly et al., 2003; Verburg et al., 2003). In addition, earlier ice-off in the spring and onset of stratification can contribute to increased cyanobacterial blooms, even in remote, oligotrophic lakes (Favot et al., 2019) and later into the fall (Winter et al., 2011).

ii. Nutrient enrichment

Nutrient inputs, particularly phosphorus and nitrogen, can lead to well-understood enrichment of plant and other organic productivity in freshwaters (Chapters 14 and 15). Excess nutrient inputs can lead to *cultural eutrophication*, which often results in enhanced rates of decomposition, decreased water transparency, and decreased hypolimnetic oxygen concentrations (*hypoxia*) and in certain conditions can greatly reduce or eliminate suitable habitat for many species of plants and animals (Carpenter, 2005). We will further discuss eutrophication within Section IV. More recently, increased inputs of dissolved organic matter are currently a concern as lakes experience brownification and decreased water clarity (Chapters 6 and 28)

iii. Pollutants

Where human activity contributes organic waste and nutrients, other pollutants are sure to be found. For a

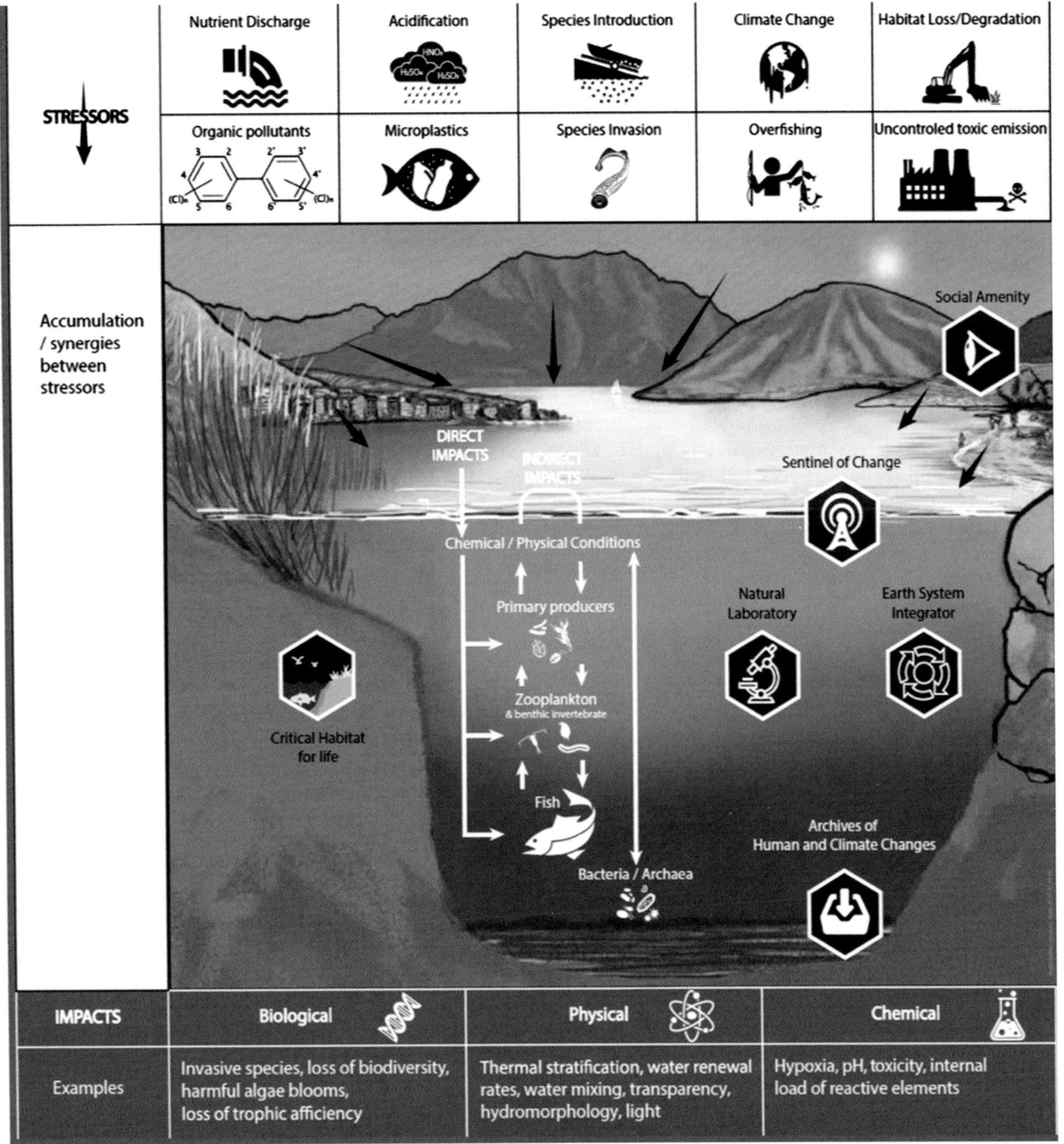

FIGURE 31-3 Environmental stressors directly and indirectly impacting lakes. *(From Jenny et al., 2020.)*

relatively small number of such pollutants, such as the *macropollutants* that occur at concentrations of mg L^{-1} (Schwarzenbach et al., 2006), the source, behavior, and potential treatment approaches are relatively well understood, yet the magnitude of the problem may be unknown. For example, salt is a growing problem for inland waters in industrialized cold temperate regions of the world where it is used to combat ice on roads (Dugan et al., 2017; Chapter 12), and studies of these systems now join a diverse literature on salinization that has accompanied land-use change and unsustainable extraction of water resources (Scanlon et al., 2007). As the extent of the problems of macropollutants is revealed, the challenges are to predict ecosystem responses, to optimize treatment, management, and policies (Schwarzenbach et al., 2006).

It is far more difficult to assess the effects of the thousands of synthetic and naturally trace element *micropollutants* and *microplastics* (typically $\mu g\ L^{-1}$ or $mg\ L^{-1}$) associated with human activity and now globally ubiquitous in freshwater (Bernhardt et al., 2017; Schwarzenbach et al., 2006). With thousands of pesticides, pharmaceuticals, and industrial and consumer products approved for use in a regulatory environment that varies across geopolitical boundaries, the scope of the research problem is daunting. Toxicological effects customarily have been assessed for one chemical on one species, focusing on mortality, an approach that does not readily enable extrapolation to community- or ecosystem-level effects, sublethal effects, or considerations for multiple stressors. Schwarzenbach et al. (2006) suggested that the most parsimonious approach to the problem is to focus on understanding generalizable phenomena across micropollutants and environments. We will better understand and predict the behavior of micropollutants if we consider these problems in terms of key system- and compound-specific generalizable properties and reactivities such as adsorption to solid phases, partitioning between solid and aqueous phases, or the formation of complexes in solution, as well as abiotic and biotic transformations.

iv. Conversion of land use to agriculture and urban landscapes

Water quality of rivers and lakes is dependent on the nature of the surrounding environment. Rivers and lakes that are surrounded by intact forests or environments, for example, generally tend to have better water quality and a thriving biodiversity. Conversely, aquatic ecosystems surrounded by poorly managed or excessively converted and exploited environments have poor quality waters. Such global patterns highlight the connections of land and water—both above and below ground—that argue for interdisciplinary water quality management at the level of watershed and beyond. Rapid urbanization can increase the extent of impermeable surfaces and increase soil erosion, such that runoff over land occurs more quickly and without the benefit of microbial processes in the soil that can reduce pollutants before they reach ground and surface waters. Similarly, conversion to agriculture and subsequent poor management of these lands frequently results in enhanced soil erosion and runoff that degrades water quality with increased turbidity, organic matter, nutrient fertilizer, and other chemicals used to maximize agricultural production. Both these scenarios illustrate the central role of healthy soils in buffering environmental quality from degradation caused by human land uses. A focus on healthy soil can put into perspective most land-use choices, as healthy soil filters water and promotes productive aboveground growth of forests and crops alike (Lal, 2015), highlighting the need to manage watersheds rather than just water within its visible boundaries. Because of the need to feed the ever-growing human population, pressure for agriculture is likely to continue growing and already contributes disproportionately to water quality problems around the globe, including in Finland (Ekholm and Mitikka, 2006), China (Qin et al., 2007; Wang et al., 2019), East and Southern Africa (Kimirei et al., 2017, 2020), and around the world (Moslenko et al., 2020).

v. Threats to biodiversity

Species invasion and introduction—both intentional and accidental—have caused irreversible changes to the biodiversity of aquatic ecosystems around the globe, effects that may be particularly dramatic because establishments of exotic species are more likely to succeed in systems already disturbed in other ways by humans (Dudgeon et al., 2006; van Rees et al., 2021). Humans have facilitated invasions directly by moving "hitchhikers" among water bodies on small watercraft, and at large scales by global trade such as those associated with ballast water and the aquarium trade. Species range expansions have been enabled by anthropogenic climate change and habitat disturbance. Intentional introductions of exotic species may have different objectives, chief among which is to either occupy an unoccupied niche or as biological control of certain species. Such introductions and their effects are well illustrated in African freshwater systems, many of which harbor extraordinary biodiversity and endemism. For example, the Lake Tanganyika sardine, *Limnothrissa miodon*, was introduced into Lake Kariba and Lake Kivu to utilize the pelagic space that was unoccupied at the time (Kolding et al., 2019). For Lake Victoria, the introduction of Nile perch (*Lates niloticus*) was intended to convert smaller and bony fishes into a larger and economically valuable species (Sayer et al., 2018). Although these objectives may have been met and results have been viewed as positive to many local fishers and the riparian countries' economies, the ecological states of lakes subjected to the introductions were profoundly disrupted, with substantial losses in native biodiversity (e.g., Witte et al., 2000). Similarly, the introduction and establishment of Nile perch and tilapia (*Oreochromis niloticus*) in Lake Victoria changed species diversity and dominance of the lake's fisheries and simplified its food web (Kolding et al., 2008; Witte et al., 2000). Before the invasion, Lake Victoria had a multispecies fishery with over 100 species in the catches, which were dominated by haplochromine cichlids (Kishe-Machumu et al., 2015; Pringle, 2005; Witte et al., 2000). The introduced species outcompeted the native taxa for food and habitat, thus narrowing their niche breadth and leading to possible extinction of some native species (Marshall et al., 2018;

Witte et al., 2000). The fishery ultimately became dominated by these two species and the cyprinid *Rastrineobola argentea* (Kayanda et al., 2009; Ogutu-Ohwayo, 1990).

vi. Overfishing

Overfishing simply means the act of catching or removing too many fish in an aquatic system such that the breeding population becomes too depleted to recover. Inland fisheries—mainly concentrated in tropical and subtropical regions (Funge-Smith and Bennett, 2019)—contribute significantly to human well-being by ensuring food security and poverty alleviation (Funge-Smith and Bennett, 2019). However, like in many ocean fisheries (Worm et al., 2009; Ye et al., 2013), inland fisheries are depleted due to overfishing, habitat loss, and/or pollution (Hilborn et al., 2003), a scenario that jeopardizes the livelihood of many, especially, impoverished communities (Bennett et al., 2021; O'Meara et al., 2021; Weyl et al., 2010). Cases of overfishing of inland waters are global (Mulimbwa et al., 2019; Ngor et al., 2018; Post et al., 2002; Sarvala et al., 2020) and may be quite dramatic; for example, overfishing is suspected to endanger over 76% of all native fish species in Lake Victoria (Sayer et al., 2018). Overfishing is caused by many factors, including poor management of fishing capacity/effort, high demand for fish caused by high population growth, and change in fishing technology—especially in poorly managed fisheries where control over the introduction of new technology is very limited or nonexistent. Although in the industrialized world, fishing in inland waters is often recreational, fishing in most of the global south is a lifeline and the sustenance of many communities (Cooke et al., 2016b). In Europe, North America, and some African countries, access to fishing is controlled (i.e., *right-based*—a fisheries management regime in which access to the fishery is controlled by use rights. Use rights may include not only the right to fish but may also specify any or all of the following: [1] how the fishing may be conducted [e.g., the vessel and gear]; [2] where people may fish; [3] when people may fish; and [4] how much fish people may catch). In contrast, fisheries management in many African countries allows fishing by anyone wishing to do so unrestricted (i.e., *open access*—a condition of a fishery in which anyone who wishes to fish may do so). Open access can easily result in a "tragedy of the commons" problem where personal gains outweigh communal benefits.

B. Interacting multiple environmental stressors

A key knowledge gap remains in our understanding of the interactions between climate change and other environmental stressors, which may be amplified or weakened in the presence of climate change that may lead to ecological surprises (Christensen et al., 2006). Endangered species of Pacific salmon typify the problem. Many species and populations are locally extirpated or at historically low abundances in human-dominated watersheds. In the Pacific Northwest of North America, decades of research have focused on four main threats to salmon populations—the four Hs: habitat degradation in streams and rivers, hydropower dams affecting migration and altering habitat, harvest by commercial and recreational fisheries, and hatcheries producing juveniles that may outcompete or introduce disease to, rather than augment, native salmon populations. All of these threats interact in some way with warmer temperatures, changing snowpack and precipitation, altered ocean conditions, and seasonality changes associated with climate change in the Pacific Northwest. In the past decade another lethal threat has become clear—the slurry of chemicals associated with stormwater runoff, the water that runs off of impermeable surfaces and into streams. Chemical contaminants associated with both motor vehicles and roadways together number in the thousands (Du et al., 2017; Peter et al., 2018), and exposure to stormwater can induce "urban spawner mortality syndrome" in coho salmon within hours, wherein erratic swimming, gasping, and disequilibrium is followed by death (Scholz et al., 2011). A single storm event can account for 60% to 100% of prespawning mortality of salmon in urban streams (Scholz et al., 2011). The specific chemical agents inducing the syndrome remained unclear for a decade, but research demonstrated that filtration of the stormwater through bioretention media comprising sand and compost removes pollutants to a degree that the syndrome does not appear in coho salmon, and aquatic invertebrates are protected as well (McIntyre et al., 2015; Spromberg et al., 2016). Finally, a compound derived from tires was identified as the most deadly component of stormwater (Tian et al., 2021). This case illustrates (1) the need to understand multiple stressors; (2) the ability to take effective conservation action (i.e., roadside bioretention stormwater interception) even before the causative agents are entirely understood; and (3) the benefits of multiple disciplines converging on research and solutions.

III. Restoration of aquatic ecosystems

In the previous section we outlined some of the ways aquatic ecosystems can be degraded and some methods of prevention. Here we focus on approaches to restoration. The most common type of degradation is contamination of the water and biota by inorganic and organic pollutants. In the case of toxic substances, remedial measures are difficult or impractical once the substances are dispersed (Wetzel, 1992). Indeed, dispersion and dilution are unfortunately relied upon as a common corrective, remedial measure. Once the loading of the pollutant to the aquatic ecosystem has ceased or been reduced appreciably, natural water renewal rates are relied upon to dilute the contaminants to acceptable levels. Lakes are more likely to function as sinks for pollutants than are rivers, such that pollutants tend to accumulate and potentially become more problematic over time. This phenomenon is magnified when the soils in the watershed are also permeated with *legacy pollutants* (e.g., Jarvie et al., 2013), such as phosphorus.

Restoration and remediation actions have largely been focused on correcting eutrophication associated with excess nutrients, and pollution with toxic materials. Nutrient loadings, particularly of phosphorus and nitrogen, have resulted in nutrient enrichments and excessive growth of algae and macrophytic plants. This accelerated eutrophication process has led to a spectrum of remedial techniques to mitigate the problems and return the lake or reservoir to some state of improved, lower productivity. Another type of degradation includes the introduction of toxic materials such as heavy metals, chlorinated hydrocarbons, and radioactive materials. The sources of all of these pollutants are often diffuse (non-point), which makes their control very difficult.

When one attempts to restore a disturbed aquatic ecosystem to some former state or condition, the question arises as to what the objectives of such recovery are given the challenges of restoring ecosystems to their previous state. The objective may be the complex task of restoring specific habitats for important native species or, more commonly, restoring water quality to acceptable levels.

A. Lake Management and Restoration

Lake management and restoration have focused on problems particularly associated with excessive nutrient loading and poor land management, and a great deal of research exists on the topic (Cooke et al., 2016a). Only a brief summary of management and restoration is presented here, although we acknowledge that freshwater systems are susceptible to multiple environmental stressors.

i. Eutrophication

Eutrophication is a process leading to increased biological productivity from the excessive addition of dissolved and particulate inorganic and organic materials to lakes and reservoirs (Cooke et al., 2016a). The cause of the enhanced productivity in lakes and reservoirs is most frequently enhanced phosphorus availability (Chapter 15), although other elements, particularly nitrogen (e.g., Dodds et al., 1989; Maberly et al., 2020; Chapter 14) can become the dominant limiting nutrient for phytoplanktonic growth in certain waters when phosphorus supplies are sufficient. Lake restoration efforts have been directed largely toward reducing the loading of phosphorus, and more recently, to a lesser extent, nitrogen, to the surface waters by advanced wastewater treatment, diversion, land management, or reducing the phosphorus load in wastewater by restricting the phosphorus content of detergents (e.g., Abell et al., 2010; Edmondson, 1996; Jeppesen et al., 2005). If releases of sediment phosphorus stores become significant, in part stimulated by low redox potentials that have resulted from organic matter loading, production, and deposition (Chapter 15), attempts are often made to control the availability or recycling of nutrients by physical or chemical methods within the lake or reservoir.

Prevention of nutrient pollution and eutrophication is clearly the most prudent long-term solution. Often, reductions in nutrient loadings require evaluation of *diffuse* (e.g., agricultural land drainage) and *point sources* (e.g., sewer pipes) within the drainage basin and a systematic multifaceted program of reduction and control (Cooke et al., 2016a). Once the loadings of nutrients and water are known, a *nutrient budget* or a *process-based model* can be used to evaluate the dynamics of annual inputs and outputs of substances. The budget permits the estimation of changes in algal biomass that could result from increases or decreases of nutrient loadings, water residence time, or increased mean depth as may occur from sediment removal. Restoration of lakes from excessive productivity from eutrophication is typically accomplished by (1) nutrient loading reduction and (2) control or removal of plant biomass, as discussed below.

i. Nutrient Control

Seven common methods are used, singly or in combination, to reduce nutrient availability or suitable habitat for photosynthesis (Cooke et al., 2016a; Klapper, 1999).

1. *Nutrient Removal by Advanced Treatment and Land Management.* Regulation and reduction in external loadings of nutrients to eutrophic lakes is the best of long-term corrective measures, without which treatments within the lakes will have minimal lasting results. Advanced wastewater treatment is only practical where sewage is collected in a wastewater system where phosphorus in the water can then be reduced to concentrations that will not appreciably alter lake productivity when the treated water is returned to the stream, reservoir, or lake. Agricultural land-management practices, including manure storage and changes in tillage, and green infrastructure approaches in urban settings, can also reduce nutrient loading.
2. *Nutrient Diversion.* Occasionally diversion of major external nutrient loadings has been adequate to restore a eutrophic lake—Lake Washington in Seattle

(USA) is a well-documented example (Edmondson, 1996; Schindler, 2006)—although results are not always straightforward (see Jeppesen et al., 2005). In situations where the primary source or sources of nutrient loadings are defined, the wastewater or stormwater may be diverted without appreciable alteration of lake or reservoir hydrology. Diversion is most likely to be successful in rapidly flushed lakes (low retention time) and where internal nutrient loading from sediments is small.

3. *Hypolimnetic Withdrawal.* Nutrient-enriched hypolimnetic water can be removed by large-scale siphoning, pumping, or deepwater discharge at the dam. Residence time of hypolimnetic water is reduced and the oxygen content of water overlying sediments may be increased with a concomitant decrease in internal loading of phosphorus from the sediments (Chapter 15). If designed and regulated carefully, it is possible to have hypolimnetic removal without thermal destratification, but this may contribute to reduced water quality downstream.
4. *Dilution and Flushing.* If a major source of nutrient-poor water is available, addition of such water to an enriched lake or reservoir may be adequate to dilute nutrient levels to suppress algal productivity. If sufficiently large quantities can be added, the water alone can be adequate to flush out algal cells and maintain lower productivity by dilution. Lower algal biomass can be maintained even with nutrient-rich water if flow is adequate to wash out the algae faster than algal growth rates. Both methods require the availability of a suitable water supply, particularly during seasons of high productivity. These methods are sometimes used in coordinated management with induced lake circulation, chemical precipitation, and biological manipulations.
5. *Phosphorus Precipitation and Inactivation.* Recovery from eutrophication (*oligotrophication*), in which essential nutrients are reduced to growth-limiting concentrations, can be a slow process if lake water renewal times are long and if nutrient releases from sediments and entrainments of nutrient-rich hypolimnetic water into the epilimnion are large. Phosphorus is adsorbed to and/or coprecipitated with clay and carbonate particles (Chapter 15). A similar process can be induced by adding aluminum compounds to the water and the sediments. Aluminum sulfate (*alum*) or sodium aluminate, or both, form a precipitate of aluminum phosphate or a colloid of aluminum hydroxide, depending upon the alkalinity and pH conditions. In both cases phosphorus is bound and scavenged to the sediments without aluminum toxicity if circumneutral pH is maintained. Further additions of these aluminum compounds to the sediments inactivate and slow phosphorus migration into the overlying water.
6. *Sediment Oxidation.* High bacterial respiration of organic matter in sediments results in anaerobic conditions in nearly all sediments. These reducing conditions lead to reduced iron and the release of associated phosphate into interstitial sediments and overlying waters (Chapter 15). Nitrate can serve as an alternative electron acceptor for oxygen, delaying the reduction of iron and release of phosphate. Thus artificial addition of nitrate to stimulate denitrification and oxidation of organic matter, ferric chloride to remove hydrogen sulfide and precipitate phosphorus, and lime (calcium carbonate) to raise the pH of the hypolimnetic water and/or of sediments can result in the inactivation of phosphorus release from sediments. However, nitrogen also is a significant pollutant, limiting its uses to rather restricted circumstances.
7. *Hypolimnetic Oxygenation.* Maintaining an oxygenated environment in the bottom waters of lakes and reservoirs reduces nutrient release and mobilization of heavy metals that compromise water quality. However, oxygenation can create unintended ecological consequences by altering diverse physical and chemical dynamics, and the design of new aeration systems is an active area of research and development. In recent decades novel hypolimnetic oxygenation systems, including direct aeration (Singleton et al., 2007), Speece Cone contact chambers (McGinnis and Little, 1998), and side-stream supersaturation systems (Gerling et al., 2014), have been deployed in lakes and reservoirs around the world (Preece et al., 2019). These novel oxygenation systems have exhibited varying success in adding oxygen to bottom waters while maintaining thermal stratification. When successful, hypolimnetic oxygenation systems have decreased nutrient loads, reduced the mobilization of iron and manganese from the sediments, and reduced the biomass and frequency of Cyanobacteria blooms in lakes dominated by internal loading. However, hypolimnetic oxygenation systems still require thorough evaluation of their sustainability, energy usage, and unknown effects on ecosystem processes (Carey et al., 2018; Gerling et al., 2014).

ii. Control of Macrophyte Biomass

Excessive growth of aquatic macrophytes (Chapter 24), particularly of certain exotic nuisance species such as the water hyacinth (*Eichhornia crassipes*), hydrilla (*Hydrilla verticillata*), and the Eurasian watermilfoil (*Myriophyllum spicatum*), can curtail or eliminate the use of lakes, reservoirs, and river ecosystems by humans. Entire disciplines have emerged that are directed to the control and management of aquatic macrophytes. We summarize five common approaches to control and manage excessive development of aquatic macrophytes (see Cooke et al., 2016a).

1. *Drawdown of Water Level.* The drawdown of water level is a multipurpose restoration and management technique for ponds and reservoirs to control certain aquatic plants and to modify habitats for the management of fish populations. Exposure of aquatic macrophytes to dry or freezing conditions for adequate (i.e., species-specific and variable) periods of time can kill the plants. Negative effects can include the losses of, or damage to, important adjacent wetlands and losses of important benthic communities, as well as releasing the greenhouse gas methane.
2. *Mechanical Removal.* Aquatic macrophyte communities that have developed to nuisance densities have been controlled via numerous manual and mechanical methods. Mechanized devices effectively drag, dredge, and till the sediments by mowing and tillage machinery or by suction and diver-operated dredging equipment. Harvested plants are preferably removed from the water, thereby simultaneously removing appreciable amounts of nutrients.
3. *Plant and Sediment Shading.* Sediments and submersed macrophytes can be covered with opaque sheeting or dense screen materials that are impenetrable by plants growing from root systems. These methods are limited to small areas of intensive use where macrophytes are not desired.
4. *Biological Controls.* Introduction of phytophagous insects and fish or plant pathogens such as fungi and viruses have been used as species-specific control agents to reduce the success and biomass of targeted macrophytes. The objective is to establish an equilibrium between the control organism and its target plant at an acceptable level of plant biomass (reviewed in Cooke et al., 2016a).
5. *Chemical Controls.* A number of general and moderately selective herbicides have been effectively applied to the control of aquatic macrophytes, particularly where invasive macrophytes have altered aquatic habitats or other ecosystem services. Such compounds are often used in concert with other control methods.

Although removing macrophytes may alleviate some of the issues with cultural eutrophication, there is always the danger of simultaneously shifting a shallow lake system from a clear-water, macrophyte-dominated state to a phytoplankton-dominated state with the potential for cyanobacterial blooms, as discussed in Chapter 26.

For all of these actions that focus on the water body itself, it is worth emphasizing that where pollution pressure is expected to continue, whether from legacy pollution in the watershed or continued poor land-use management, diverse efforts to reduce watershed inputs are vital. Riparian buffers, wetland construction, and landowner education and engagement can help to reduce pollutant loading over time and reduce dependency on the above services becoming a long-term, repeated drain on management resources.

B. River management and restoration

A fundamental aspect of both river management and restoration is to protect or restore the riparian floodplain areas of streams and rivers. In many cases streams and rivers have been channelized by straightening and deepening the channel, particularly in lowland agricultural areas to increase drainage. Accompanying the loss of stream length is a loss of riffles and pools, a loss of riparian floodplains and wetlands, and a loss of riparian vegetation. These changes lessen opportunities for energy to dissipate along the stream or river, thus increasing the mechanical energy per unit weight of groundwater (*hydraulic head*). Increased flows reduce habitat diversity, particularly in the sediments and adjacent wetland, with a catastrophic reduction in biodiversity (e.g., Petersen et al., 1992). Loss of riparian floodplains and wetlands decreases water tables, increases rates of water runoff, enhances nutrient losses from adjacent land, and increases scouring and sediment yield to the channel. Water retention capacities are greatly reduced, and as a result, during high precipitation events, flooding is very common and much more severe than would be the case if portions of the riparian and floodplain environments were retained. It is estimated that drainage basins comprising 5% to 10% wetlands are capable of providing a 50% reduction in peak flood period compared to those drainage basins that have none.

The relative stability of flowing water ecosystems and communities living within them is indicated by their resistance to disturbance as well as their rate of recovery from disturbance (Webster et al., 1983; Yount and Niemi, 1990). After a disturbance, such as a flooding event or alteration of channel stability, lotic communities and ecosystems can recover relatively rapidly because: (1) life histories of communities are commonly adapted for rapid recolonization and reestablishment of disturbed areas; (2) unaffected internal refugia as well as unaffected upstream and downstream populations can serve as seed inocula for reestablishment; (3) flushing characteristics assist in dilution and replacement of polluted regions; and (4) streams and rivers have frequent natural disturbances and are adapted to human-induced perturbations.

Major alterations, such as channelization, that severely alter sediment and flow characteristics will result in much longer recovery times.

Several management and restoration measures for streams and rivers are commonly used (see Boon et al., 2000; Brierley and Fryirs, 2013; Eiseltová and Biggs, 1995; Gore, 1985; Palmer et al. 2005; Petersen et al., 1990, 1992), some of which are summarized here.

1. *Buffer Strips.* Plant and microbial communities on a *buffer strip* of land, about 10 to 50 m wide, along each side of the river channel function to stabilize banks and as a metabolic filter for nutrients and other substances potentially released from land. These narrow strips of vegetated riparian land can retain and reduce nutrient loadings to the channel by 65% to 100%. Hence buffer zones and riparian strips function as barriers to the eutrophication of streams, aid in channel stability, and have considerable ecological influence by the provision of *habitat patch variability.* Variations in the natural revegetation of the buffer strips are advantageous to enhance biodiversity and habitat differences. Restoration methods are also used to encourage the development of wetlands at juncture points of small tributaries or drainage ditches to the main channel. Further, buffer strips may provide channel shading to reduce water temperatures and mitigate climate change effects.
2. *Reduction of Channel Side–Slope.* Reduction of the slope gradient at the edge of the channel, particularly when dredging is periodically required, will encourage channel meandering and floodplain revegetation and lead to a more natural riparian system.
3. *Channel Migration and Floodplain Development.* Reversion of the channel to natural migrations and meandering morphology increases channel length, alters path morphology, and greatly increases habitat diversity. The erosional and depositional process will seek a meander frequency of about five to seven channel widths, dissipates flow energy, and allows for greatest habitat variability. Because of the greater time that water spends in contact with sediment surfaces, nutrient retention and spiraling properties are optimized.

IV. Broadening our global perspectives and voices

A. Integrating new technologies

Continual advances in remote sensing technology, including satellite, optical, and microwave remote sensing, in addition to unoccupied aerial vehicles such as drones, offer exciting opportunities to further understand freshwaters around the world, including in remote regions that previously were relatively unexplored. Despite improvements in the resolution and capacity of *remote sensing* technology, it can still be difficult to assess small lakes and rivers, mountainous areas, and cloudy regions, not to mention the limnological and ecological characteristics below the water's surface. Integrating data from disparate research disciplines and participants, including *in situ* and remote sensing observations from both scientists and community scientists, process-based modeling, and experiments, are necessary to understand underlying mechanisms driving broad-scale limnological patterns at larger limnological landscapes beyond political boundaries (i.e., Sharma et al., 2020; Thackeray and Hampton, 2020; Woolway et al., 2020).

Lake ice is an example of an opportune study for scientists to integrate their knowledge from disparate disciplines (Fig. 31-4). For example, lake ice has been recorded by Arctic Indigenous communities and community scientists for decades to centuries in hundreds of lakes around the Northern Hemisphere, dating as far back as 1443 when Shinto priests began recording the date of ice formation on Lake Suwa (Arakawa, 1954; Benson et al., 2000; Knopp et al., 2020; Magnuson et al., 2000; Sharma et al., 2016). Further, the proliferation and advances in remote sensing in the past few decades have allowed further observations of ice in large, remote lakes previously lacking *in situ* studies and thereby broadly expanding the spatial scale at which ice can be observed in lakes (i.e., Cai et al., 2017; Latifovic and Pouliot, 2007; Zhang and Pavelsky, 2019). In conjunction, process-based models and winter experiments have delved into further understanding the mechanistic drivers of lake ice dynamics (i.e., Hrycik and Stockwell, 2021; Kirillin et al., 2012; Leppäranta, 2010). Paleolimnological perspectives can extend these records even farther back in time (Chapter 30). Integrating these disparate disciplines provides a unique opportunity to: (1) validate, calibrate, and extend existing ice records; (2) develop global mechanistically driven statistically based lake ice models; (3) move the research field beyond simply studying ice phenology (timing of ice formation and ice melt); (4) expand the capacity and our abilities to address critical knowledge gaps on lake ice loss; and (5) acquire a global understanding of freshwater ice dynamics over vast spatial and temporal scales (Sharma et al., 2020).

Such integration of technologies has the potential not only to correct limitations in any single technology but also to inspire new types of questions and to generate new insight through simultaneous study of the small and large scales. Our ability to make observations at larger scales has improved in recent decades, particularly through remote sensing, at the same time that our

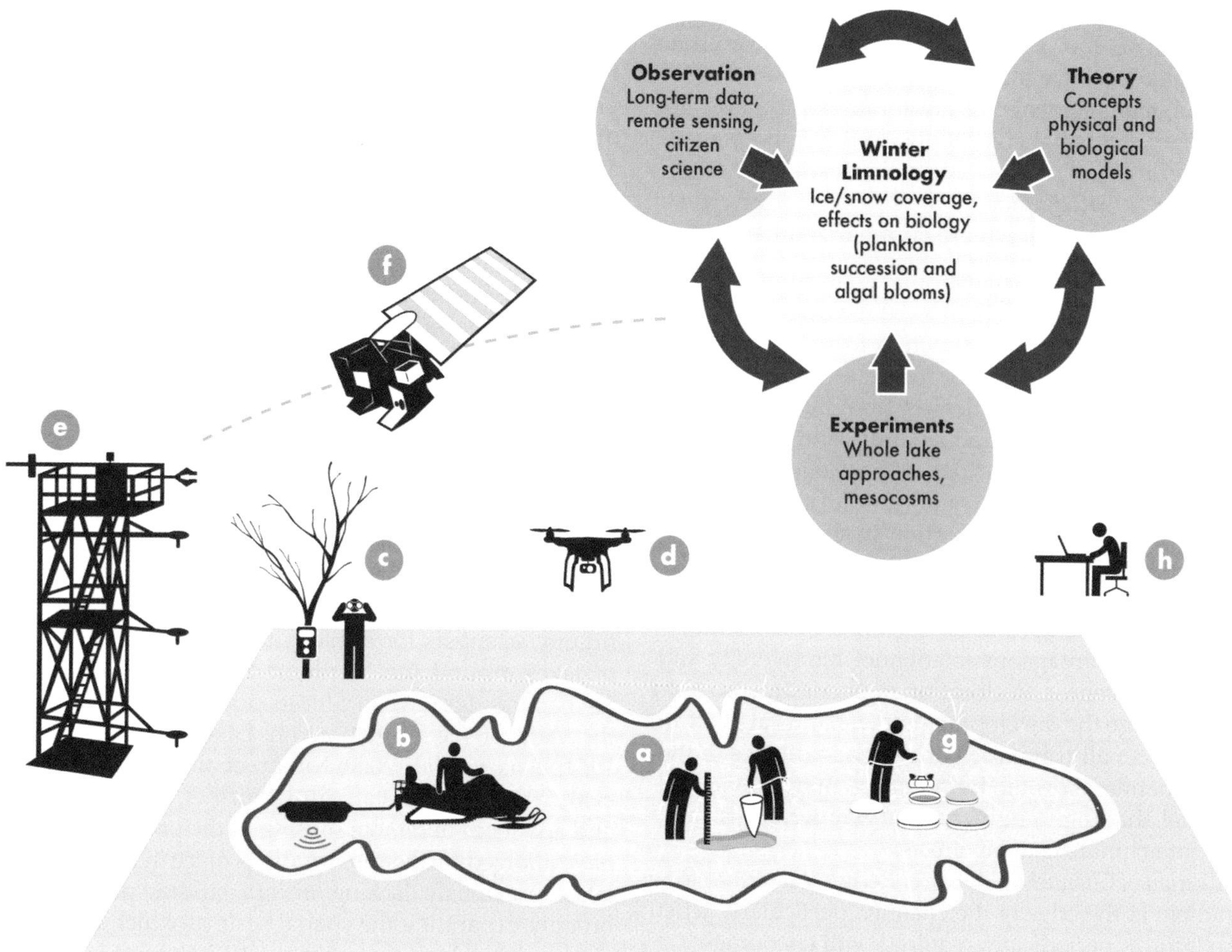

FIGURE 31-4 Methodological tools to study lake ice dynamics, including (*a*) ice cores, (*b*) ground-penetrating radar, (*c*) shoreline observations, (*d*) drones, (*e*) high-frequency climate sensors, (*f*) spaceborne remote sensing instruments, (*g*) experiments, and (*h*) process-based mechanistic models. *(From Sharma et al., 2020.)*

ability to study fine-scale patterns has improved as well. Rapidly advancing approaches to *environmental DNA* hold the potential to revolutionize biodiversity studies, and metabolomics can provide detailed characterizations of functional diversity. Together the comparison or merging of multiple techniques may enable larger scale study of epigenetic effects, rapid evolution, or more complex microbiome and food web interactions. Such complementarities among technologies can be realized in a variety of distinct manners. In some cases data gathered through different approaches may be used in parallel as separate characterizations of the same focal phenomenon, essentially viewing the same problem through different lenses (*comparative complementarity*; Fig. 31-5; Thackeray and Hampton, 2020), or where the different technologies enhance the capabilities of another (*translational complementarity*; Fig. 31-5; Thackeray and Hampton, 2020). Causal mechanisms may be suggested with the strategic deployment of multiple technologies (*causal complementarity*; Fig. 31-5; Thackeray and Hampton, 2020), and the importance of the ecological and spatiotemporal context within which focal biological responses are interpreted can be better understood (*contextual complementarity*; Fig. 31-5; Thackeray and Hampton, 2020). Research approaches each have their own operational scales and resolution of time and space, such that there is potential for research integration to enable the flow of knowledge among these scales (*scaling complementarity*; Fig. 31-5; Thackeray and Hampton, 2020). Understanding genotype–phenotype–environment interactions that affect large-scale spatial patterns and temporal dynamics in Cyanobacteria toxin production, for example, has implications for predicting disruption to water safety in a warming world (Thackeray and Hampton, 2020).

Although the proliferation and speed of advances in such a variety of technologies can be intimidating, no single researcher needs to master all or even most of them. As limnologists increasingly embrace collaboration that enables knowledge integration, we need

(a) Comparative complementarity

Method 1

S_1 S_2

Method 2

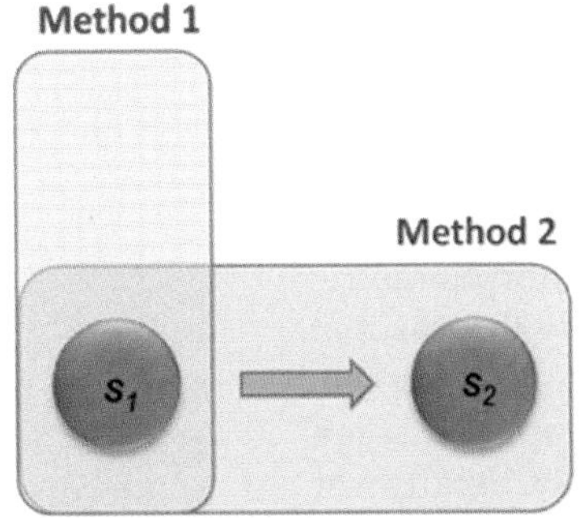

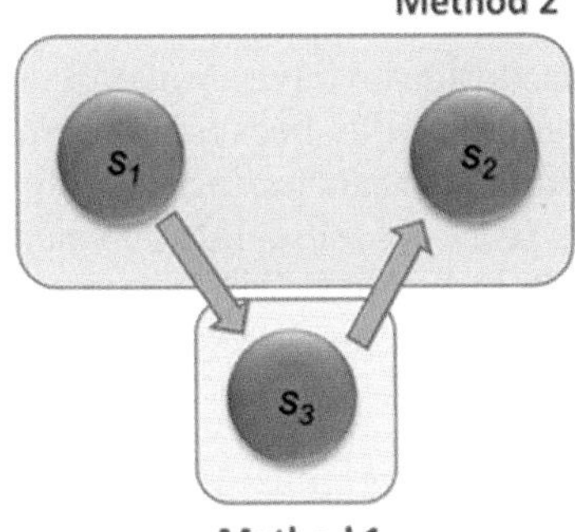

(d) Contextual complementarity

Method 1

Method 2

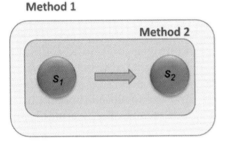

S_1 S_2

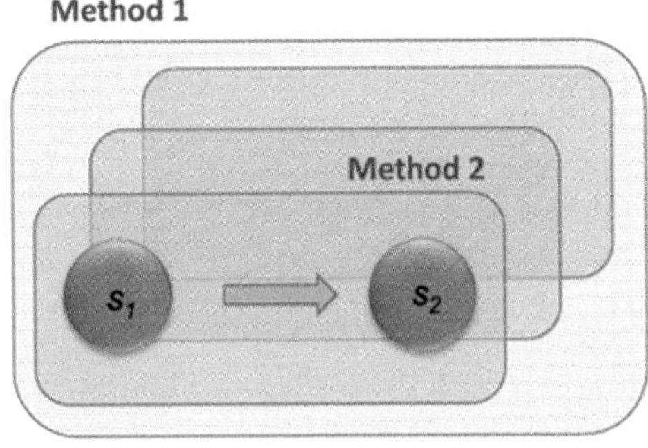

FIGURE 31-5 Research approaches can complement each other in multiple distinct ways when deployed together in a study (see text for details). The schematic considers the case of studying the relationship between two variables: a global change driver (e.g., water temperature, S1) and an ecological response (S2). In the case of causal complementarity an intermediate variable (S3) causally links S1 and S2. Rectangles indicate the operational domain of each approach, that is, which variables can be observed using each method. Integration of different approaches may enable (a) observation of the same phenomenon using more than one approach, increasing confidence in the observed relationship; (b) the use of data from one approach to extend the capabilities of another; (c) the combination of approaches in order to fill empirical gaps in causal inference; (d) understanding the spatiotemporal context for observations; and (e) the application of empirical models from one location, based upon one method, to make predictions at other locations. *(From Thackeray and Hampton, 2020.)*

primarily to resist developing silos within subdisciplines or centered on promising new technologies. Limnologists have long been justifiably proud of being inherently interdisciplinary scientists, whatever our specialties may be, expected to understand the basics of biology, physics, geosciences, and chemistry. As our subdisciplines broaden and deepen, active attention to improving our knowledge and our conversancy across the subdisciplines of limnology can propel science forward through appropriate collaborations that expand rather than contract around specific approaches (Hampton et al., 2015).

B. Limnology is global

From the earliest studies of Lake Geneva (Lac Leman) in central Europe, where François-Alphonse Forel (1841–1912) christened our discipline "limnology" at the turn of the 20th century, to the continuing evolution of limnology across the European continent, and its later development across the plains of the Midwestern United States and the lakes of the Canadian Shield, limnology traces its roots to Europe and North America where the center of gravity has largely remained (Egerton, 2014). Although the history is illustrious, the limitations of this perspective are increasingly recognized. North temperate lakes have implicitly dominated the development of our empirical and theoretical studies, which have been developed through the lens of a very restricted set of cultures and epistemologies. The science of limnology has grown throughout the world in recent decades, and, as outlined in previous chapters, we have learned that the paradigm based on north temperate lakes does not necessarily apply across different regions and continents, such as the lakes of Australia, Africa, and Asia. As we have opportunities to take comparative, global approaches to limnological studies, the differences that emerge across regions are striking. For example, a 2015 synthesis (O'Reilly et al., 2015) of *in situ* and satellite-derived summer lake water temperatures from 235 lakes highlighted that lakes in different regions of the world can have similar trends in summer lake water temperatures because of similarities in the drivers affecting lakes, including not only common climate variables (e.g., air temperature, solar radiation, and cloud cover) but also local lake-specific geomorphometric characteristics (e.g., lake depth). Lakes with seasonal ice cover warm twice as fast as the global average, whereas some lakes of the Tibetan Plateau and some Arctic lakes have been cooling since 1989 because of glacial meltwater inputs as air temperatures in these regions warm and glaciers melt (O'Reilly et al., 2015). Essentially working across the globe allows limnologists to observe much longer environmental gradients and improves our abilities to capture nonlinearities and interactions among drivers and lake responses that are not present in a single region. Further, scientists in north temperate zones, where water is comparatively abundant and not typically a source of subsistence protein, simply may have asked very different questions or focused on different attributes than would be the case elsewhere.

The inclusion of diverse voices from scientists worldwide is essential to sparking scientific creativity, cross-pollinating ideas to manage our aquatic systems, and ultimately increasing the resilience of limnology as a discipline. Scientists from around the world have a lot to learn from each other, as well as from indigenous knowledge-keepers, as we move beyond our narrow local or regional perspectives. Advances in grassroots limnological networks, such as the Global Lake Ecological Observatory Network (GLEON; Weathers et al., 2013), remote sensing technologies, and increased accessibility of open-access data, work toward a global, more accessible field. Remote sensing is revolutionizing the field of limnology by capturing data from lakes and rivers worldwide and openly sharing data and ideas through collaborations. The movement toward open-access data is essential to democratizing data and making the aquatic sciences more inclusive and accessible. In the United States, for example, federal agencies have taken a relatively aggressive stance on open data, through programs such as the Environmental Protection Agency's National Aquatic Resource Surveys and the National Science Foundation's National Ecological Observatory Network, as well as its policies on data sharing for its grantees, enabling teams of aquatic researchers to provide leadership in continental and global studies. Many other regions are taking similar actions. As open data practices become the norm rather than the exception, our science and our stewardship can increasingly flourish through global teamwork.

Water is a scarce resource, and we have a global responsibility to safeguard this resource as access to clean freshwater is a human right. The inclusion of global and diverse voices into limnology is essential to protect our freshwater resources around the world.

V. Summary

1 Access to clean freshwater is a basic human right and essential for human survival.
   a. Interdisciplinary collaborations are crucial to address global water challenges.
2 Freshwater systems are sensitive to multiple environmental stressors:
   a. Climate change reduces ice cover, warms water temperatures, accelerates evaporation rates, and alters species distributions and fish productivity.
   b. Nutrient enrichment through increased inputs of phosphorus and nitrogen can lead to cultural eutrophication and degraded water quality.
   c. Pollutants may disrupt ecosystem functioning.
   d. Conversion of land use to agriculture and urban landscapes degrades water quality around the world.
   e. Introduction of invasive species and overfishing are a threat to the biodiversity of native species.
   f. Interaction of multiple environmental stressors may lead to ecological surprises.
3 Eutrophication from nutrient enrichments causes excessive growth of algae and macrophytic plants. Common remediation efforts include the reduction of nutrients and macrophyte removal within the contaminated waters.
4 Buffer strips, reduction of channel slide-slopes, channel migration, and floodplain development are common management and restoration techniques for rivers and streams.
5 Continual advances in remote sensing technology, interdisciplinary collaboration, and diverse, inclusive teams are key components to advancing the field of limnology globally.

Acknowledgments

We thank Alessandro Filazzola, Ryan McClure, Michael Meyer, Luke Moslenko, and Stephen Thackeray for providing revisions on the early drafts of the chapter.

References

Abell, J.M., Özkundakci, D., Hamilton, D.P., 2010. Nitrogen and phosphorus limitation of phytoplankton growth in New Zealand lakes: implications for eutrophication control. Ecosystems 13, 966–977.

Arakawa, H., 1954. Fujiwhara on five centuries of freezing dates of Lake Suwa in the Central Japan. Arch. Meteorol. Geophys. Bioklimatol. B 6, 152–166.

Arnell, N.W., Gosling, S.N., 2013. The impacts of climate change on river flow regimes at the global scale. J. Hydrol. 486, 351–364.

BBC, 2018. The 11 Cities Most Likely to Run Out of Drinking Water, like Cape Town. https://www.bbc.com/news/world-42982959. (Accessed 11 February 2018).

Bennett, A., Basurto, X., Virdin, J., et al., 2021. Recognize fish as food in policy discourse and development funding. Ambio 50, 1–9.

Benson, B.J., Magnuson, J.J., Sharma, S., 2000. Global Lake and River Ice Phenology Database. NSIDC National Snow Ice Data Center, Boulder, Colorado, USA. https://doi.org/10.7265/N5W66HP8. Updated 2020, Version 1.

Bernhardt, E.S., Rosi, E.J., Gessner, M.O., 2017. Synthetic chemicals as agents of global change. Front. Ecol. Environ. 15, 84–90.

Birk, S., Chapman, D., Carvalho, L., et al., 2020. Impacts of multiple stressors on freshwater biota across spatial scales and ecosystems. Nat. Ecol. Evol. 4, 1060–1068.

Boon, P.J., Davies, B., Petts, G., 2000. Global Perspectives on River Conservation: Science, Policy and Practice. John Wiley & Sons.

Bootsma, H.A., Hecky, R.E., 1993. Conservation of the African great lakes: a limnological perspective. Conserv. Biol. 7, 644–656.

Brierley, G.J., Fryirs, K.A., 2013. Geomorphology and River Management: Applications of the River Styles Framework. John Wiley & Sons.

Cai, Y., Ke, C.-Q., Duan, Z., 2017. Monitoring ice variations in Qinghai Lake from 1979 to 2016 using passive microwave remote sensing data. Sci. Total Environ. 607, 120–131.

Carey, C.C., Doubek, J.P., McClure, R.P., Hanson, P.C., 2018. Oxygen dynamics control the burial of organic carbon in a eutrophic reservoir. Limnol. Oceanogr. Lett. 3, 293–301.

Carpenter, S.R., 2005. Eutrophication of aquatic ecosystems: bistability and soil phosphorus. Proc. Natl. Acad. Sci. 102, 10002–10005.

Carson, R., 1962. Silent Spring. Houghton Mifflin.

Chen, M.M., Lopez, L., Bhavsar, S.P., Sharma, S., 2018. What's hot about mercury? Examining the influence of climate on mercury levels in Ontario top predator fishes. Environ. Res. 162, 63–73.

Christensen, M.R., Graham, M.D., Vinebrooke, R.D., Findlay, D.L., Paterson, M.J., Turner, M.A., 2006. Multiple anthropogenic stressors cause ecological surprises in boreal lakes. Glob. Chang. Biol. 12, 2316–2322.

Chu, C., Mandrak, N.E., Minns, C.K., 2005. Potential impacts of climate change on the distributions of several common and rare freshwater fishes in Canada. Divers. Distrib. 11, 299–310.

Comte, L., Olden, J.D., 2017. Climatic vulnerability of the world's freshwater and marine fishes. Nat. Clim. Chang. 7, 718–722.

Cooke, G.D., Welch, E.B., Peterson, S., Nichols, S.A., 2016a. Restoration and Management of Lakes and Reservoirs. CRC Press.

Cooke, S.J., Arlinghaus, R., Johnson, B.M., Cowx, I.G., 2016b. Recreational fisheries in inland waters. Freshw. Fish. Ecol. 449–465.

Dodds, W.K., Johnson, K.R., Priscu, J.C., 1989. Simultaneous nitrogen and phosphorus deficiency in natural phytoplankton assemblages: theory, empirical evidence, and implications for lake management. Lake Reserv. Manag. 5, 21–26.

Du, B., Lofton, J.M., Peter, K.T., Gipe, A.D., James, C.A., McIntyre, J.K., et al., 2017. Development of suspect and non-target screening methods for detection of organic contaminants in highway runoff and fish tissue with high-resolution time-of-flight mass spectrometry. Environ. Sci. Process. Impacts 19, 1185.

Dudgeon, D., Arthington, A.H., Gessner, M.O., et al., 2006. Freshwater biodiversity: importance, threats, status and conservation challenges. Biol. Rev. 81, 163–182.

Dugan, H.A., Bartlett, S.L., Burke, S.M., Doubek, J.P., Krivak-Tetley, F.E., Skaff, N.K., et al., 2017. Salting our freshwater lakes. Proc. Natl. Acad. Sci. 114, 4453–4458.

Edmondson, W.T., 1996. Uses of Ecology: Lake Washington and Beyond. University of Washington Press.

Egerton, F.N., 2014. History of ecological sciences, part 50: formalizing limnology, 1870s to 1920s. Bull. Ecol. Soc. Am. 95, 131–153.

Eiseltová, M., Biggs, J., 1995. Restoration of Stream Ecosystems: An Integrated Catchment Approach. International Waterfowl and Wetlands Research Bureau Gloucester.

Ekholm, P., Mitikka, S., 2006. Agricultural lakes in Finland: current water quality and trends. Environ. Monit. Assess. 116, 111–135.

Favot, E.J., Rühland, K.M., DeSellas, A.M., Ingram, R., Paterson, A.M., Smol, J.P., 2019. Climate variability promotes unprecedented cyanobacterial blooms in a remote, oligotrophic Ontario lake: evidence from paleolimnology. J. Paleolimnol. 62, 31–52.

Funge-Smith, S., Bennett, A., 2019. A fresh look at inland fisheries and their role in food security and livelihoods. Fish Fish. 20, 1176–1195.

Gerling, A.B., Browne, R.G., Gantzer, P.A., Mobley, M.H., Little, J.C., Carey, C.C., 2014. First report of the successful operation of a side stream supersaturation hypolimnetic oxygenation system in a eutrophic, shallow reservoir. Water Res. 67, 129–143.

Gore, J.A., 1985. Restoration of Rivers and Streams. Butterworth, United States.

Hampton, S.E., Anderson, S.S., Bagby, S.E., et al., 2015. The Tao of open science for ecology. Ecosphere 6, 1–13.

Hansen, G.J.A., Read, J.S., Hansen, J.F., Winslow, L.A., 2017. Projected shifts in fish species dominance in Wisconsin lakes under climate change. Glob. Chang. Biol. 23, 1463–1476.

Hilborn, R., Branch, T.A., Ernst, B., Magnusson, A., Minte-Vera, C.V., Scheuerell, M.D., Valero, J.L., 2003. State of the world's fisheries. Annu. Rev. Environ. Resour. 28, 359–399.

Hrycik, A.R., Stockwell, J.D., 2021. Under-ice mesocosms reveal the primacy of light but the importance of zooplankton in winter phytoplankton dynamics. Limnol. Oceanogr. 66, 481–495.

Jarvie, H.P., Sharpley, A.N., Spears, B., Buda, A.R., May, L., Kleinman, P.J.A., 2013. Water quality remediation faces unprecedented challenges from "legacy phosphorus". Environ. Sci. Technol. 47 (16), 8997–8998.

Jenny, J.-P., Anneville, O., Arnaud, F., Baulaz, Y., Bouffard, D., Domaizon, I., et al., 2020. Scientists' warning to humanity: rapid degradation of the world's large lakes. J. Great Lakes Res. 46, 686–702.

Jeppesen, E., Søndergaard, M., Jensen, J.P., Havens, K.E., Anneville, O., Carvalho, L., et al., 2005. Lake responses to reduced nutrient loading—an analysis of contemporary long-term data from 35 case studies. Freshw. Biol. 50, 1747–1771.

Kaushal, S.S., Likens, G.E., Jaworski, N.A., Pace, M.L., Sides, A.M., Seekell, D., et al., 2010. Rising stream and river temperatures in the United States. Front. Ecol. Environ. 8, 461–466.

Kayanda, R., Taabu, A.M., Tumwebaze, R., Muhoozi, L., Jembe, T., Mlaponi, E., et al., 2009. Status of the major commercial fish stocks and proposed species-specific management plans for Lake Victoria. African J. Trop. Hydrobiol. Fish. 21, 15–21.

Kimirei, I.A., Semba, M., Mwakosya, C., Mgaya, Y.D., Mahongo, S.B., 2017. Environmental changes in the Tanzanian part of Lake Victoria. In: Mgaya, Y.D., Mahongo, S.B. (Eds.), Lake Victoria Fisheries Resources. Springer, pp. 37–59.

Kimirei, I.A., Mubaya, C.P., Ndebele-Murisa, M., Kaaya, L., Mangadze, T., Mwedzi, T., Kushata, J.N.T., 2020. Trends in ecological changes: implications for East and southern Africa. In: Ndebele-Murisa, M., Kimirei, I.A., Mubaya, C.P. (Eds.), Ecological Changes in the Zambezi River Basin. CODESRIA, p. 49.

Kirillin, G., Leppäranta, M., Terzhevik, A., Granin, N., Bernhardt, J., et al., 2012. Physics of seasonally ice-covered lakes: a review. Aquat. Sci. 74, 659–682.

Kishe-Machumu, M.A., Voogd, T., Wanink, J.H., Witte, F., 2015. Can differential resurgence of haplochromine trophic groups in Lake

Victoria be explained by selective Nile perch, *Lates niloticus* (L.) predation? Environ. Biol. Fishes 98, 1255–1263.

Klapper, H., 1999. Control of Eutrophication in Inland Waters. Ellis Horwood, p. 337.

Knopp, J.A., Levenstein, B., Watson, A., Ivanova, I., Lento, J., 2020. Systematic review of documented Indigenous Knowledge of freshwater biodiversity in the circumpolar. Freshw. Biol. https://doi.org/10.1111/fwb.13570.

Kolding, J., van Zwieten, P., Marttin, F., Funge-Smith, S., Poulain, F., 2019. Freshwater Small Pelagic Fish and Fisheries in the Major African great lakes and Reservoirs in Relation to Food Security and Nutrition. FAO Fisheries and Aquaculture Technical Paper No. 642. Licence: CC BY-NC-SA 3.0 IGO. FAO, Rome, p. 124.

Kolding, J., Zwieten, P. van, Mkumbo, O., Silsbe, G., Hecky, R., 2008. Are the Lake Victoria fisheries threatened by exploitation or eutrophication? Towards an ecosystem-based approach to management. In: Bianchi, G., Skjoldal, H.R. (Eds.), The Ecosystem Approach to Fisheries. Food and Agriculture Organization of the United Nations, p. 309.

Lal, R., 2015. Restoring soil quality to mitigate soil degradation. Sustainability 7, 5875–5895.

Latifovic, R., Pouliot, D., 2007. Analysis of climate change impacts on lake ice phenology in Canada using the historical satellite data record. Remote Sens. Environ. 106, 492–507.

Leppäranta, M., 2010. Modelling the formation and decay of lake ice. In: George, G. (Ed.), The Impact of Climate Change on European Lakes. Springer, pp. 63–83.

Maberly, S.C., Pitt, J.-A., Davies, P.S., Carvalho, L., 2020. Nitrogen and phosphorus limitation and the management of small productive lakes. Inland Waters 10, 159–172.

Magnuson, J.J., Robertson, D.M., Benson, B.J., Wynne, R.H., Livingstone, D.M., Arai, T., et al., 2000. Historical trends in lake and river ice cover in the Northern Hemisphere. Science 289, 1743–1746.

Marshall, J., Davison, A.J., Kopf, R.K., Boutier, M., Stevenson, P., Vanderplasschen, A., 2018. Biocontrol of invasive carp: risks abound. Science 359, 877.

McGinnis, D.F., Little, J.C., 1998. Bubble dynamics and oxygen transfer in a speece cone. Water Sci. Technol. 37, 285–292.

McIntyre, J.K., Davis, J.W., Hinman, C., Macneale, K.H., Anulacion, B.F., Scholz, N.L., Stark, J.D., 2015. Soil bioretention protects juvenile salmon and their prey from the toxic impacts of urban stormwater runoff. Chemosphere 132, 213–219.

Meyer, M.F., Powers, S.M., Hampton, S.E., 2019. An evidence synthesis of pharmaceuticals and personal care products (PPCPs) in the environment: imbalances among compounds, sewage treatment techniques, and ecosystem types. Environ. Sci. Technol. 53, 12961–12973.

Mölsä, H., Reynolds, J.E., Coenen, E.J., Lindqvist, O.V., 1999. Fisheries research towards resource management on Lake Tanganyika. Hydrobiologia 407, 1–24.

Moslenko, L., Blagrave, K., Filazzola, A., Shuvo, A., Sharma, S., 2020. Identifying the influence of land cover and human population on chlorophyll a concentrations using a pseudo-watershed analytical framework. Water 12, 3215.

Mulimbwa, N., Sarvala, J., Micha, J.-C., 2019. The larval fishery on *Limnothrissa miodon* in the Congolese waters of Lake Tanganyika: impact on exploitable biomass and the value of the fishery. Fish. Manag. Ecol. 26, 444–450.

Naiman, R.J., Magnuson, J.J., Stanford, J.A., McKnight, D.M., 1995. The Freshwater Imperative: A Research Agenda. Island Press.

Ngor, P.B., McCann, K.S., Grenouillet, G., So, N., McMeans, B.C., Fraser, E., Lek, S., 2018. Evidence of indiscriminate fishing effects in one of the world's largest inland fisheries. Sci. Rep. 8, 1–12.

Nkoko, D.B., Giraudoux, P., Plisnier, P.-D., Tinda, A.M., Piarroux, M., Sudre, B., et al., 2011. Dynamics of cholera outbreaks in great lakes region of Africa, 1978–2008. Emerg. Infect. Dis. 17, 2026.

Ogutu-Ohwayo, R., 1990. The decline of the native fishes of lakes Victoria and Kyoga (East Africa) and the impact of introduced species, especially the Nile perch, *Lates niloticus*, and the Nile tilapia, *Oreochromis niloticus*. Environ. Biol. Fishes 27, 81–96.

O'Meara, L., Cohen, P.J., Simmance, F., Marinda, P., Nagoli, J., Teoh, S.J., Funge-Smith, S., Mills, D.J., Thilsted, S.H., Byrd, K.A., 2021. Inland fisheries critical for the diet quality of young children in sub-Saharan Africa. Glob. Food Sec. 28, 100483.

O'Reilly, C.M., Alin, S.R., Plisnier, P.-D., Cohen, A.S., McKee, B.A., 2003. Climate change decreases aquatic ecosystem productivity of Lake Tanganyika, Africa. Nature 424, 766–768.

O'Reilly, C.M., Sharma, S., Gray, D.K., Hampton, S.E., Read, J.S., Rowley, R.J., Schneider, P., Lenters, J.D., McIntyre, P.B., Kraemer, B.M., Weyhenmeyer, G.A., Straile, D., Dong, B., Adrian, R., Allan, M.G., Anneville, O., Arvola, L., Austin, J., Bailey, J.L., Baron, J.S., Brookes, J.D., de Eyto, E., Dokulil, M.T., Hamilton, D.P., Havens, K., Hetherington, A.L., Higgins, S.N., Hook, S., Izmest'eva, L.R., Joehnk, K.D., Kangur, K., Kasprzak, P., Kumagai, M., Kuusisto, E., Leshkevich, G., Livingstone, D.M., MacIntyre, S., May, L., Melack, J.M., Mueller-Navarra, D.C., Naumenko, M., Noges, P., Noges, T., North, R.P., Plisnier, P.-D., Rigosi, A., Rimmer, A., Rogora, M., Rudstam, L.G., Rusak, J.A., Salmaso, N., Samal, N.R., Schindler, D.E., Schladow, S.G., Schmid, M., Schmidt, S.R., Silow, E., Soylu, M.E., Teubner, K., Verburg, P., Voutilainen, A., Watkinson, A., Williamson, C.E., Zhang, G., 2015. Rapid and highly variable warming of lake surface waters around the globe. Geophys. Res. Lett. 42 (10), 773–781.

Palmer, M.A., Bernhardt, E.S., Allan, J.D., Lake, P.S., Brooks, G.A.S., Carr, J., et al., 2005. Standards for ecologically successful river restoration. J. Appl. Ecol. 42, 208–217.

Peter, K.T., Tian, Z.Y., Wu, C., Lin, P., White, S., Du, B.W., et al., 2018. Using high-resolution mass spectrometry to identify organic contaminants linked to urban stormwater mortality syndrome in coho salmon. Environ. Sci. Technol. 52 (18), 10317.

Petersen Jr., R.C., Petersen, L.B.M., Lacoursière, J.O., et al., 1990. Restoration of lowland streams: the building block model. Vatten 46, 244–249.

Petersen, R.C., Petersen, L.B.M., Lacoursiere, J., Boon, P.J., Calow, P., Petts, G.E., 1992. A building-block model for stream restoration. In: Boon, P.J., Calow, P., Petts, G.E. (Eds.), River Conservation and Management. John Wiley & Sons Ltd., pp. 293–309

Post, J.R., Sullivan, M., Cox, S., Lester, N.P., Walters, C.J., Parkinson, E.A., Paul, A.J., Jackson, L., Shuter, B.J., 2002. Canada's recreational fisheries: the invisible collapse? Fisheries 27, 6–17.

Preece, E.P., Moore, B.C., Skinner, M.M., Child, A., Dent, S., 2019. A review of the biological and chemical effects of hypolimnetic oxygenation. Lake and Reserv. Manag. 35, 229–246.

Pringle, R.M., 2005. The origins of the nile perch in Lake Victoria. Bioscience 55, 780–787.

Qin, B., Xu, P., Wu, Q., Luo, L., Zhang, Y., 2007. Environmental issues of Lake Taihu, China. In: Qin, B., Liu, Z., Havens, K. (Eds.), Eutrophication of Shallow Lakes with Special Reference to Lake Taihu. Springer, China, pp. 3–14.

Robson, S.V., Rosi, E.J., Richmond, E.K., Grace, M.R., 2020. Environmental concentration of pharmaceuticals alter metabolism, denitrification, and diatom assemblages in artificial streams. Freshw. Sci. 39, 256–267.

Rosi-Marshall, E.J., Royer, T.V., 2012. Pharmaceutical compounds and ecosystem function: an emerging research challenge for aquatic ecologists. Ecosystems 15, 867–880.

Sarvala, J., Helminen, H., Ventelä, A.-M., 2020. Overfishing of a small planktivorous freshwater fish, vendace (*Coregonus albula*), in the boreal lake Pyhäjärvi (SW Finland), and the recovery of the population. Fish. Res. 230, 105664.

Sayer, C.A., Máiz-Tomé, L., Darwall, W.R.T., 2018. Freshwater Biodiversity in the Lake Victoria Basin: Guidance for Species Conservation, Site protection, Climate Resilience and Sustainable Livelihoods. International Union for Conservation of Nature.

Scanlon, B.R., Jolly, I., Sophocleou, M., Zhang, L., 2007. Global impacts of conversions from natural to agricultural ecosystems on water resources: Quantity versus quality. Water Resour. Res 43. https://doi.org/10.1029/2006WR005486.

Schindler, D.W., 2006. Recent advances in the understanding and management of eutrophication. Limnol. Oceanogr. 51, 356–363.

Schneider, P., Hook, S.J., 2010. Space observations of inland water bodies show rapid surface warming since 1985. Geophys. Res. Lett. 37.

Scholz, N.L., Myers, M.S., McCarthy, S.G., et al., 2011. Recurrent die-offs of adult coho salmon returning to spawn in Puget Sound lowland urban streams. PLoS One 6, e28013.

Sharma, S., Jackson, D.A., Minns, C.K., Shuter, B.J., 2007. Will northern fish populations be in hot water because of climate change? Glob. Chang. Biol. 13, 2052–2064.

Sharma, S., Magnuson, J.J., Batt, R.D., Winslow, L.A., Korhonen, J., Aono, Y., 2016. Direct observations of ice seasonality reveal changes in climate over the past 320–570 years. Sci. Rep. 6, 25061.

Schwarzenbach, R.P., Escher, B.I., Fenner, K., Hofstetter, T.B., Johnson, C.A., Von Gunten, U., Wehrli, B., 2006. The challenge of micropollutants in aquatic systems. Science 313, 1072–1077.

Sharma, S., Blagrave, K., Magnuson, J.J., O'Reilly, C.M., Oliver, S., Batt, R.D., Magee, M.R., Straile, D., Weyhenmeyer, G.A., Winslow, L., Woolway, R.I., 2019. Widespread loss of lake ice around the Northern Hemisphere in a warming world. Nat. Clim. Chang. 9, 227–231.

Sharma, S., Meyer, M.F., Culpepper, J., Yang, X., Hampton, S., Berger, S.A., Brousil, M.R., Fradkin, S.C., Higgins, S.N., Jankowski, K.J., Kirillin, G., Smits, A.P., Whitaker, E.C., Yousef, F., Zhang, S., 2020. Integrating perspectives to understand lake ice dynamics in a changing world. J. Geophys. Res. Biogeosci. 125, (8) e2020JG005799.

Singleton, V.L., Gantzer, P., Little, J.C., 2007. Linear bubble plume model for hypolimnetic oxygenation: full-scale validation and sensitivity analysis. Water Resour. Res. 43. https://doi.org/10.1029/2005WR004836.

Spromberg, J.A., Baldwin, D.H., Damm, S., McIntyre, J.K., Huff, M., Sloan, C.A., Anulacion, B.F., Davis, J.W., Scholz, N.L., 2016. Widespread coho salmon spawner mortality in western U.S. urban watersheds: lethal stormwater impacts are reversed by soil bioinfiltration. J. Appl. Ecol. 53, 498.

Sterner, R.W., Keeler, B., Polasky, S., Poudel, R., Rhude, K., Rogers, M., 2020. Ecosystem services of Earth's largest freshwater lakes. Ecosyst. Serv. 41, 101046.

Thackeray, S.J., Hampton, S.E., 2020. The case for research integration, from genomics to remote sensing, to understand biodiversity change and functional dynamics in the world's lakes. Glob. Chang. Biol. 26, 3230–3240.

Tian, Z., Zhao, H., Peter, K.T., et al., 2021. A ubiquitous tire rubber—derived chemical induces acute mortality in coho salmon. Science 371, 185–189.

United Nations, 2015. Transforming Our World: The 2030 Agenda for Sustainable Development. https://sdgs.un.org/2030agenda.

United Nations Children's Fund and World Health Organization, 2019. Progress on Household Drinking Water, Sanitation and Hygiene 2000–2017. Special Focus on Inequalities. United Nations Children's Fund (UNICEF) and World Health Organization.

van Rees, C.B., Waylen, K.A., Schmidt-Kloiber, A., et al., 2021. Safeguarding freshwater life beyond 2020: recommendations for the new global biodiversity framework from the European experience. Conserv. Lett. 14, e12771.

Verburg, P., Hecky, R.E., Kling, H., 2003. Ecological consequences of a century of warming in Lake Tanganyika. Science 301, 505–507.

Verpoorter, C., Kutser, T., Seekell, D.A., Tranvik, L.J., 2014. A global inventory of lakes based on high-resolution satellite imagery. Geophys. Res. Lett. 41, 6396–6402.

Vörösmarty, C.J., McIntyre, P.B., Gessner, M.O., et al., 2010. Global threats to human water security and river biodiversity. Nature 467, 555–561.

Wang, W., Lee, X., Xiao, W., Liu, S., Schultz, N., Wang, Y., Zhang, M., Zhao, L., 2018. Global lake evaporation accelerated by changes in surface energy allocation in a warmer climate. Nat. Geosci. 11, 410–414.

Wang, M., Strokal, M., Burek, P., Kroeze, C., Ma, L., Janssen, A.B.G., 2019. Excess nutrient loads to Lake Taihu: opportunities for nutrient reduction. Sci. Total Environ. 664, 865–873.

Weathers, K.C., Hanson, P.C., Arzberger, P., Brentrup, J., Brookes, J., Carey, C.C., Gaiser, E., Hamilton, D.P., Hong, G.S., Ibelings, B., Istvanovics, V., Jennings, E., Kim, B., Kratz, T., Lin, F.P., Muraoka, K., O'Reilly, C., Rose, K.C., Ryder, E., Zhu, G., 2013. The Global Lake Ecological Observatory Network (GLEON): the evolution of grassroots network science. Limnol. Oceanogr. 22, 71–73.

Webster, J.R., Gurtz, M.E., Hains, J.J., Meyer, J.L., Swank, W.T., Waide, J.B., Wallace, J.B., 1983. Stability of stream ecosystems. In: Barnes, J.R., Minshall, G.W. (Eds.), Stream Ecology. Springer, pp. 355–395.

Wenger, S.J., Isaak, D.J., Luce, C.H., et al., 2011. Flow regime, temperature, and biotic interactions drive differential declines of trout species under climate change. Proc. Natl. Acad. Sci. 108, 14175–14180.

Wetzel, R.G., 1992. Clean water: a fading resource. Hydrobiologia 243, 21–30.

Weyl, O.L.F., Ribbink, A.J., Tweddle, D., 2010. Lake Malawi: fishes, fisheries, biodiversity, health and habitat. Aquat. Ecosyst. Health Manag. 13, 241–254.

Whitehead, P.G., Wilby, R.L., Battarbee, R.W., Kernan, M., Wade, A.J., 2009. A review of the potential impacts of climate change on surface water quality. Hydrol. Sci. J. 54, 101–123.

Winter, J.G., DeSellas, A.M., Fletcher, R., Heinstsch, L., Morley, A., Nakamoto, L., Utsumi, K., 2011. Algal blooms in Ontario, Canada: increases in reports since 1994. Lake Reserv Manag 27, 107–114.

Witte, F., Msuku, B.S., Wanink, J.H., Seehausen, O., Katunzi, E.F.B., Goudswaard, P.C., Goldschmidt, T., 2000. Recovery of cichlid species in Lake Victoria: an examination of factors leading to differential extinction. Rev. Fish Biol. Fish. 10, 233–241.

Woolway, R.I., Maberly, S.C., 2020. Climate velocity in inland standing waters. Nat. Clim. Chang. 10, 1124–1129.

Woolway, R.I., Merchant, C.J., 2019. Worldwide alteration of lake mixing regimes in response to climate change. Nat. Geosci. 12, 271–276.

Woolway, R.I., Kraemer, B.M., Lenters, J.D., Merchant, C.J., O'Reilly, C.M., Sharma, S., 2020. Global lake responses to climate change. Nat. Rev. Earth Environ. 388–403.

Worm, B., Hilborn, R., Baum, J.K., et al., 2009. Rebuilding global fisheries. Science 325, 578–585.

Yang, X., Pavelsky, T.M., Allen, G.H., 2020. The past and future of global river ice. Nature 577, 69–73.

Ye, Y., Cochrane, K., Bianchi, G., Willmann, R., Majkowski, J., Tandstad, M., Carocci, F., 2013. Rebuilding global fisheries: the World Summit Goal, costs and benefits. Fish Fish. 14, 174–185.

Yount, J.D., Niemi, G.J., 1990. Recovery of lotic communities and ecosystems from disturbance—a narrative review of case studies. Environ. Manage. 14, 547–569.

Zhang, S., Pavelsky, T.M., 2019. Remote sensing of lake ice phenology across a range of lakes sizes, ME, USA. Remote Sens 11, 1718.

Index

Note: Page numbers followed by *f* indicate figures and *t* indicate tables.

D

E

F

G

H

M

U

V